# Magnetism and Magnetic Materials—1975

(21st Annual Conference—Philadelphia)

**AIP Conference Proceedings**
Series Editor: Hugh C. Wolfe
**Number 29**

# Magnetism and Magnetic Materials—1975

(21st Annual Conference—Philadelphia)

Editors

**J. J. Becker**
General Electric

**G. H. Lander**
Argonne National Laboratory

**J. J. Rhyne**
National Bureau of Standards

**American Institute of Physics**
New York 1976

## CONFERENCE ORGANIZATION

### General Chairman

R. L. White

### Steering Committee

R. L. White, Chairman
R. E. Watson, Secretary
D. I. Gordon, Treasurer
E. W. Pugh, Chairman-Elect
J. M. Lommel
H. C. Wolfe (AIP)
C. D. Graham

### Program Committee

R. M. White) Co-Chairmen
K. Lee
J. J. Becker
H. N. Bertram
M. Blume
W. J. L. Buyers
Gilbert Chin
A. S. Edelstein
L. Falicov
R. B. Frankel
F. Holtzberg
R. M. Josephs
J. S. Kouvel
A. J. Kurtzig
G. H. Lander
R. M. MacFarlane
D. Mills
R. Orbach
J. J. Rhyne
P. M. Richards
J. C. Slonczewski
G. Vella-Coleiro
J. H. Wernick

### Local Committee

B. F. Stein, Chairman
R. L. Coren
T. Egami
G. Finke
P. Flanders
C. D. Graham
T. Lubensky
T. Mihalisin
D. B. Oat
L. S. Onyshkevych

### Publications Committee

J. J. Becker; G. H. Lander; J. J. Rhyne

### Publicity Chairman

I. S. Jacobs

### Advisory Committee

#### Chairman

J. M. Lommel

#### Term Expires 1975

W. A. Baker
W. D. Doyle
D. I. Gordon
E. M. Gyorgy
V. Jaccarino
S. Kern
C. J. Kriessman
J. M. Lommel
R. L. White
W. P. Wolf

#### Term Expires 1976

S. H. Charap
P. L. Donoho
F. B. Hagedorn
I. S. Jacobs
A. V. Pohm
E. W. Pugh
M. P. Sarachik
B. F. Stein
E. J. Torok
P. E. Werner

#### Term Expires 1976

J. L. Archer
G. Bate
C. D. Graham
K. Lee
C. E. Patton
J. J. Rhyne
P. W. Schumate
R. E. Watson
M. K. Wilkinson
R. Wolfe

### Sponsoring Society Representatives

J. M. Lommel (IEEE)
H. C. Wolfe (AIP)

### Co-operating Society Representatives

J. W. Shilling (Met.Soc.AIME)
D. H. Jones (ASTM)
J. O. Dimmock (ONR)

### IUPAP Representative

G. T. Rado



L.C. Catalog Card No. 76-10931
ISBN 0-88318-128-2
ERDA CONF–751209

TWENTY-FIRST ANNUAL CONFERENCE

on

MAGNETISM AND MAGNETIC MATERIALS

December 9-12, 1975
Philadelphia, Pennsylvania

Sponsored by

The American Institute of Physics
The Magnetics Society
of the
Institute of Electrical and Electronics Engineers

In Co-operation with

The Office of Naval Research
The American Physical Society
The Metallurgical Society
of the
American Institute of Mining, Metallurgical, and Petroleum Engineers
The American Society for Testing and Materials

The Conference is especially grateful to

THE OFFICE OF NAVAL RESEARCH

for its support of the expenses of foreign and interdisciplinary speakers
under Contract N00014-76-G0006

Contributions to the Conference from the following firms
are gratefully acknowledged:

Ampex Corporation
Applied Magnetics Corporation
The Arnold Engineering Company
Ceramic Magnetics, Inc.
Digital Equipment Corporation
E. I. du Pont de Nemours and Company, Inc.
Electronic Memories & Magnetics Corporation
Eriez Magnetics, Inc.
Ford Motor Company
General Electric Company
General Magnetic Company
G.T.E. Automatic Electric Company
International Business Machines Corporation
International Telephone & Telegraph Corporation
Magnetic Metals Company
Magnetics Division of Spang Industries, Inc.
National Micronetics, Inc.
Nippon Steel Corporation
The Permanent Magnet Company, Inc.
Pfizer Inc.
Raytheon Company
RCA Laboratories
Siemens Corporation
Texas Instruments, Inc.
Trans-Tech, Inc.
Sperry Univac, Inc.
United States Steel Corporation
Walker Scientific, Inc.
Westinghouse Electric Corporation
Xerox Corporation

All papers in this volume, and in previous Proceedings of the Conference on Magnetism and Magnetic Materials published in this series, have been reviewed for technical content. The selection of referees, review guidelines, and all other editorial procedures are in accordance with standards prescribed by the American Institute of Physics.

TABLE OF CONTENTS

Section 5 BUBBLE TRANSLATION

Translational Velocities and Ballistic Overshoot of Bubbles in Garnet Films ..... 58
A. P. Malozemoff, J. C. Slonczewski and J. C. DeLuca
Overshoot in Magnetic Bubble Translation ..... 64
G. P. Vella-Coleiro
Bubble Domain Translational Motion in the Presence of an In-Plane Field ..... 65
R. M. Josephs and B. F. Stein, W. R. Bekebrede
The Speed Difference Between an S=0 and an S=1 Bubble in the Presence of an In-Plane Magnetic Field ..... 67
T. Hsu, B. R. Brown and M. D. Montgomery
Wall State Stability During Translational Motion ..... 69
B. R. Brown
Bubble Response to Harmonic Gradient Drive ..... 71
O. Voegeli, C. A. Jones and J. A. Brown
Domain Wall States in Annealed Low Q Garnet Films ..... 72
J. L. Su, E. B. Moore, T. L. Hsu, and B. A. Calhoun
Temperature Stability of Bubble Domain Wall States ..... 74
T. Obokata, K. Yamaguchi and K. Asama
Dynamic Properties of Magnetic Bubbles in Amorphous GdCoCu Films ..... 76
Robert I. Potter, V. J. Minkiewicz, Kenneth Lee, and P. A. Albert

Section 6 BUBBLE DOMAIN PHYSICS

Stability and Dynamics of Microwave Generated Ring Domains ..... 78
H. Dötsch
The Spin Wave Spectrum of Very Hard Magnetic Bubbles ..... 84
A. A. Thiele
Bubble Lattice and Spider-Web-Like Domain Pattern in MnBi Platelet ..... 84
Tetsuzo Kusuda, Shigeo Honda and Tadayoshi Ideshita
Initial Rapid Domain Wall Motion in Magnetic Bubble Garnet Materials ..... 85
G. J. Zimmer, L. Gal, F. B. Humphrey
High-Speed Wall Motion in Bubble Films With In-Plane Anisotropy ..... 87
Ernst Schlömann
Wall Structure of Cylindrically Symmetric Magnetic Domains ..... 89
E. Della Torre, C. Hegedüs and G. Kádár

Section 7 BUBBLE MATERIALS

High Speed Bubble Garnets Based on Large Gyromagnetic Ratios (High g) ..... 91
R. C. LeCraw, S. L. Blank, G. P. Vella-Coleiro and R. D. Pierce
Study of Defects in Reduced LPE Bubble Garnet Films ..... 95
R. C. LeCraw, E. M. Gyorgy and R. Wolfe
Magnetic Properties of Rare Earth (Gd, Dy, Ho, Er)-Cobalt Amorphous Films ..... 97
H. Jouve, J. P. Rebouillat, R. Meyer
The Static and Dynamic Properties of Magnetic Bubble Domains in Sm, Ca, Ge-Substituted Yttrium Iron Garnet Films ..... 99
P. F. Tumelty, R. Singh, and M. A. Gilleo
Germanium-Substituted Bubble Domain Garnet Films For Use Over an Extended Temperature Range and at a High Data Rate ..... 101
D. M. Heinz, R. G. Warren and M. T. Elliott
Reproducible Growth and Bubble Properties of Rare-Earth-Substituted YCaGeIG Films ..... 103
T. Obokata, H. Tominaga, T. Mori and H. Inoue
$(LuSm)_3Fe_{5-x}Ga_xO_{12}$ Garnet Films for Small Bubble Diameters ..... 105
J. T. Carlo, D. C. Bullock, R. E. Johnson and S. G. Parker
Magnetic Properties of Sputter Deposited GdCoCu Amorphous Bubble Films ..... 107
V. J. Minkiewicz, P. A. Albert, Robert I. Potter and C. R. Guarnieri
Lattice Mismatch Between Pb-Substituted EuIG LPE Films and NdGG Substrates ..... 109
T. S. Plaskett, E. Klokholm and D. C. Cronemeyer

Section 8 ANISOTROPY AND MAGNETOSTRICTION

Variations of Magnetic Anisotropy Within Epitaxial Films of $Y_{2.85}La_{0.15}Fe_{3.75}Ga_{1.25}O_{12}$ Obtained From Spin Wave Resonance ..... 111
B. Hoekstra and J. M. Robertson, G. Bartels
The Temperature Dependence of the Uniaxial Anisotropy of $Gd_{1-x-y}Co_xMo_y$ Amorphous Alloy Films on Glass Substrates ..... 113
P. Chaudhari and D. C. Cronemeyer
Growth-Induced Anisotropy in Yttrium Iron Garnet Films Grown by Liquid Phase Epitaxy ..... 115
M. T. Elliott and H. L. Glass
Evidence for Non-Cubic Magnetostriction in Bubble Garnets ..... 116
M. H. Yang and M. W. Muller

The Structure of Amorphous Gd-Co Alloy Films....188
C. N. J. Wagner, N. Heiman, T. C. Huang, A. Onton
W. Parrish
The Effect of Co on the Anisotropy of Amorphous Gd-Fe Films....190
R. C. Taylor
Room Temperature Composition Dependence of the Anisotropy Energy of $Ho_xTb_{1-x}Fe_2$ Single Crystals....191
C. M. Williams, N. C. Koon, and J. B. Milstein
Rhombohedral Distortion in Highly Magnetostrictive Laves Phase Compounds....192
A. E. Clark and J. R. Cullen, O. D. McMasters, and
E. R. Callen
Mössbaür and Magnetization Study of $Tb(Fe_{1-x}Al_x)_2$....194
R. S. Preston, S. P. Taneja
S. M. Drensky, A. E. Dwight, C. W. Kimball and L. R. Sill
Magnetic Properties of $Gd_{1-x}Fe_x$ Films....196
C. Vittoria, P. Lubitz and J. Schelleng

Section 12 TRANSITION METAL GLASSES

Amorphous Alloys as Soft Magnetic Materials II....198
E. M. Gyorgy, H. J. Leamy, R. C. Sherwood, & H. S. Chen
Domain Observations in an Amorphous Iron-Nickel Alloy....204
Joseph J. Becker
Magnetic Core Loss and Internal Stress in Metallic Glasses....206
R. C. O'Handley
Magnetostriction as a Function of Glass Formers in Amorphous Alloys of Fe-P-B-Al....208
H. A. Brooks and H. S. Chen
Kinetics of Reorientation of Magnetically Induced Anisotropy in Amorphous $Ni_{40}Fe_{40}P_{14}B_6$....209
F. E. Luborsky
The Effect of Heat Treatment on the Curie Temperature of a Metallic Glass....211
H. J. Leamy, E. M. Gyorgy, R. C. Sherwood, T. Wakiyama, & H. S. Chen
Magnetic Properties of Amorphous $Fe_{40}Ni_{40}P_{14}B_6$....214
C. L. Chien, R. Hasegawa
Magnetization, Anisotropy and Coercivity of a Glassy Metallic Alloy....216
R. Hasegawa
Annealing Effects in Amorphous Magnetic Alloys....218
C. D. Graham, Jr. T. Egami, Robert S. Williams and Y. Takei
Temperature Dependence of "Magnetic Anisotropy" in Amorphous Alloys....220
T. Egami and P. J. Flanders
Domain Structure During Magnetization of an Annealed and Elastically Strained Amorphous Ni-Fe-P-B Alloy....222
J. W. Shilling

Section 13 SPIN GLASS ALLOYS

Recent Developments in the Theory of Spin Glasses....224
D. Sherrington
Onset of Magnetic Order in V-Fe and Cu-Ni....228
Helmut Claus
$\mu^+$ Spin Precession Experiments in Spin Glasses....229
A. T. Fiory
Mössbauer and Magnetization Measurements in FeCr Solid Solutions....232
B. D. Dunlap and A. T. Aldred, R. J. Nemanich and C. W. Kimball
Theory of Systems With Random Bond Ising Interactions in the Paramagnetic Phase....232
R. A. Tahir-Kheli and R. J. Elliott
A Calorimetic Investigation of the Spin Glass: CuMn....233
L. E. Wenger and P. H. Keesom
Magnetic Ordering of $Au_{0.92}Fe_{0.08}$: An Ultrasonic Investigation....235
G. F. Hawkins, T. J. Moran, and R. L. Thomas
Mictomagnetism in Fe-Al Alloys....236
R. D. Shull, H. Okamoto and Paul A. Beck
Low Temperature Magnetic Properties of the Spin Glass System PtMn....237
E. F. Wassermann and J. L. Tholence
The Hydrogenation of PdCr Alloys....239
J. A. Mydosh
Mössbauer Measurements in Spin-Glass Alloy $(Pd_{0.5}Ag_{0.5})_{0.99}Fe_{0.01}$....241
S. De Benedetti, J. A. Rayne and A. Zangwill, R. A. Levy

Section 14 SPIN EXCITATION IN DISORDERED SYSTEMS

Excitations of Substitutionally Disordered Antiferromagnets ........ 243
R. A. Cowley
The Coherent Potential Approximation for the Diluted Anisotropic Antiferromagnet ........ 247
William K. Holcomb
The Observation of Structure in the Neutron Scattering from Antiferromagnetic $Mn_{0.32}Zn_{0.68}F_2$ ........ 248
E. C. Svensson, W. J. L. Buyers and T. M. Holden, D. A. Jones
Calculations of Spin Waves in a Randomly Diluted Two-Dimensional Antiferromagnet ........ 250
R. Alben and M. F. Thorpe
Resonant Perturbation of Spin Waves in $KMn_{0.99}Mg_{0.01}F_3$ AND $KMn_{0.96}Zn_{0.04}F_3$ ........ 252
R. H. March, E. C. Svensson and T. M. Holden, R. Stedman
D. A. Jones
Spin Waves in $KNi_{0.75}Mn_{0.25}F_3$ ........ 254
G. J. Coombs and R. A. Cowley, D. A. Jones, G. Parisot
and D. Tocchetti
Magnetic Excitations in Antiferromagnetic $Ni_{1-x}Co_x0$ ........ 255
V. Wagner, D. Tocchetti and B. Hennion
Crystal Field in Liquid Cerium and Praseodymium ........ 257
A. H. Millhouse and A. Furrer

Section 15 SPIN EXCITATIONS

Ground- and Excited-State Spin Waves in $PrAl_2$ ........ 259
H.-G. Purwins, W.J.L. Buyers, T. M. Holden and E. C. Svensson
Theory of Singlet-Doublet Excitations in Praseodymium ........ 261
P. Bak
A Study of the Frequency Dependent Susceptibility of the Compound, $CeAl_3$, by Inelastic Neutron Scattering ........ 263
T. Brun and S. Sinha, A. S. Edelstein and R. Majewski
H. R. Child
Structure, Moment, and Magnetic Excitations of the Transparent Ferromagnet $Rb_2CrCl_4$ ........ 263
M. T. Hutchings, P. Day, M. Fair and A. K. Gregson
Magnetic Excitations in Holmium Phosphide ........ 264
A. Furrer, E. Kaldis
Spin-Wave and Critical Scattering of Neutrons From Cobalt Disulfide ........ 266
M. Iizumi and J. W. Lynn, A. Ohsawa, H. Ito
Investigation of $FeBO_3$ By Brillouin Scattering from Thermal Magnons ........ 268
W. Jäntz, J. R. Sandercock and W. Wettling
Virtual Spin Waves in Itinerant Ferromagnets Above $T_c$ ........ 270
M. A. Klenin, J. A. Hertz
Magnon Modes of a Gyrotropic Magnetic Bar ........ 272
T. M. Sharon and A. A. Maradudin
Magnetic Scattering of Neutrons by a Relativistic Atom in the $\ell^N$ Configuration ........ 274
C. Stassis and H. W. Deckman

Section 16 SPIN DENSITY

Electron Spin Polarization in Ferromagnetic Metals ........ 276
R. Meservey, P. M. Tedrow and D. Paraskevopoulos
Photoelectron-Spin Polarization,-Spin Fluctuation and - Paramagnon Interactions and Energy Spectra in Photoemission from Itinerant, Narrow d-Band Magnetic Metals: Inclusion of Ten-Fold, Spin-Orbital, d-Electron Degeneracy ........ 281
Edward Siegel
A Localized Moment Model for Paramagnetism in $Ni_3Al$ ........ 282
A. Parthasarathi and Paul A. Beck
Magnetic Susceptibility of Ni-Aℓ Alloys at High Temperatures ........ 284
Sigurds Arajs and J. R. Kelly
Magnetic Moment Distribution in $Ni_3Al$ ........ 285
G. P. Felcher, J. S. Kouvel, A. E. Miller
Detection of Local Moments in the 3d Series by Electron Energy-Loss Spectra of $3p \rightarrow 3d$ Transitions ........ 286
Mary Beth Stearns and Lee A. Feldkamp
Thermal Expansion, Volume Magnetostriction and Shift of the Curie Temperature With Pressure of $Ni_3Al$ and Ni-Pt Alloys ........ 288
J. J. M. Franse, T. F. M. Kortekaas and N. Buis
Hyperfine Fields at Interstitial Postive Muon Sites in Ferromagnetic Materials ........ 290
P. Jena
Moment Disturbances in Ni-Cu Alloys ........ 292
R. A. Medina and J. W. Cable

Section 17 KONDO PROBLEMS REVISITED



Section 18 NARROW BAND MAGNETISM



Section 19 DILUTE ALLOYS I



Section 20 DILUTE ALLOYS II

Section 24 PHASE TRANSITIONS



Section 25 MULTICRITICAL POINTS; CRITICAL DYNAMICS

Section 26 CRITICAL PHENOMENA

First-Order Transitions, Symmetry, and the $\epsilon$ -Expansion ..... 474
D. Mukamel, S. Krinsky, and P. Bak
Thermodynamic Behavior Near Valence Instabilities ..... 479
R. D. Parks and J. M. Lawrence
Critical Scattering Scaling Functions and the Measurement of $\eta$ ..... 483
Craig A. Tracy
Spin Dynamics and Critical Fluctuations of a Site-Random, Two-Dimensional Antiferromagnet: $Rb_2Mn_{0.5}Ni_{0.5}F_4$ 487
J. Als-Nielsen, R. J. Birgeneau, G. Shirane

Section 27 CRITICAL PHENOMENA, RANDOM AND UNIFORM SYSTEMS

Randomly Dilute Two Dimensional Ising Models ..... 488
R. Fisch and A. B. Harris
Critical Effects of Regularly Spaced Point Defects ..... 490
M. E. Fisher and Helen Au-Yang
Ising Ferromagnet With Quenched Impurities ..... 490
E. Stoll, T. Schneider
Critical Exponents Associated With Isolated Impurities in Ferromagnets ..... 491
T. K. Bergstresser
Effect of Impurities on Hyperfine Critical Exponents ..... 493
R. M. Suter and C. Hohenemser
Critical Behavior of the Band Gap in Ferromagnetic Semiconductors ..... 495
J. S. Helman and I. Balberg and S. Alexander
New Mössbauer Study of the Curie Point in Iron ..... 497
M. A. Kobeissi and C. Hohenemser
Absolute Measurement of the Critical Scattering Cross Section in Cobalt ..... 499
C. J. Glinka and V. J. Minkiewicz and L. Passell
Critical Behavior of Dilute Magnets With Site-Randomness ..... 501
Tatuo Kawasaki
Series Studies of Critical Exponents in Continuous Dimensions ..... 502
J. P. Van Dyke and William J. Camp

Section 28 LINEAR CHAIN SYSTEMS

Spin-Peierls Transitions in Heisenberg Antiferromagnetic Linear Chain Systems ..... 504
J. W. Bray, H. R. Hart, Jr., L. V. Interrante, I. S. Jacobs, J. S. Kasper
G. D. Watkins, and S. H. Wee, and J. C. Bonner
Magnetic Ordering in $NiZrF_6.6H_2O$ ..... 505
M. Karnezos, D. Meier and S. A. Friedberg
Ultrasonic Attenuation in Linear Antiferromagnet $CsNiCl_3$ ..... 505
D. P. Almond and J. A. Rayne
Poly(Metal Phosphinates): Antiferromagnetism in Disordered Linear Polymers ..... 506
J. C. Scott, T. S. Wei, A. F. Garito, and A. J. Heeger,
H. D. Gillman and Piero Nannelli
Metamagnetism in Single Crystal $\gamma$-Co(Pyridine)$_2$Cl$_2$ ..... 510
S. Foner and R. B. Frankel, W. M. Reiff and H. Wong, G. J. Long
Theory of Two Spin Infrared Absorption in One Dimensional Heisenberg Antiferromagnets ..... 511
M. Drawid and J. W. Halley
Quantitative Studies of Magnetic Cooling on a Magnetic System Which Obeys the Third Law ..... 512
J. D. Johnson, J. C. Bonner

Section 29 TRANSPORT AND ELECTRONIC PROPERTIES

The Extraordinary Hall Effect: Intuitive Theory and Experimental Verification ..... 514
J.-N. Chazalviel and I. Solomon
Hall Effect and Transport in 3d Ferromagnetic Metals ..... 520
R. V. Coleman
Anisotropic Magnetoresistance in Ferromagnetic 3d Ternary Alloys ..... 526
T. R. McGuire, R. D. Hempstead and S. Krongelb
Magneto-Resistivity of Beta-Cerium ..... 527
P. Burgardt, S. Legvold, J. H. Queen and K. A. Gschneidner, Jr.
Pressure Dependence of the Electronic Structure of Nickel ..... 529
J. R. Anderson and Peter Heimann, J. E. Schirber, D. R. Stone
Elastic Anisotropy in the TSDW Phase of Chromium ..... 530
M. O. Steinitz, D. J. Stanley and E. Fawcett
Electronic Structure and Magnetic Properties of Scandium ..... 531
Shashikala G. Das
Magnetic and Electric Properties of MnSb ..... 532
Tu Chen, W. Stutius, and J. W. Allen, G. R. Stewart,
Crystal Electric Field Effects on the Transport Properties of Intermetallic Rare Earth Systems ..... 534
Y. H. Wong and B. Lüthi

Section 30 SURFACE PROPERTIES OF MAGNETIC MATERIALS

Magnetic Characterization of Semi-Amorphous Nickel Catalysts and Their Methanation Activity ..... 536
L. N. Mulay, R. C. Everson, O. P. Mahajan, and P. L. Walker, Jr.
Temperature Studies of the Hyperfine Magnetic Field in Thin Iron Films ..... 538
Robert J. Semper, C. L. Chien and J. C. Walker

Section 37 RESONANCES IN FILMS AND MAGNETIC MATERIALS

## PREFACE

This volume comprises the permanent record of the 21st Annual Conference on Magnetism and Magnetic Materials, held in Philadelphia December 9-12, 1975. In it approximately 350 speakers, contributing and invited, have set down their principal argument and conclusions. To keep the Proceedings down to one volume, even one very large volume, each author has been severely constrained in the length of the article he may publish. Despite the somewhat cryptic nature of the papers forced by the length constraint, the Conference Proceedings constitutes a uniquely useful encapsulation of information on what was being done, and who was doing it, in magnetism and magnetic materials as of the Fall of 1975.

The Proceedings are useful in many ways. Perhaps the most common mode of usage is by one specialist to read exactly what his fellow specialists are doing in an area of mutual expertise. For such a user, the short communication is not cryptic at all, but packed with information, implications and insights. For each of us there are perhaps half a dozen to a dozen papers in the Proceedings that we can be truly said to read with understanding, and it is for these articles that we first scan the Proceedings when it comes.

A second and somewhat more leisurely use of the Proceedings is to obtain an overall perspective on research in magnetism - where it is now and where it is going. Comparing the present volume with one from the late fifties brings out striking evidence of the shifts of interest as knowledge develops and, probably more important, new technologic utilizations appear. Comparing the present volume with last year's or that of two years ago, clear trends are evident, even on so short a time scale. Though bubbles are still a very important topic, they do not dominate the Conference as they did two years ago. Conversely, other topics have grown in representation; almost every morning and every afternoon this year there were sessions on amorphous magnetism, and in only slightly lesser abundance, sessions on critical phenomena. Thus it is with a mature but active field; no longer explosive growth but a continuing evolution of emphasis and activity.

A third valuable kind of information provided by these Proceedings is the "who" part; information on who the bellwether scientists are in each area of activity. A scientist is frequently confronted by the need or desire to know what the current state of knowledge is on some phenomenon with which he has little first-hand acquaintance. The most difficult and time-consuming step in the acquisition of the required information is, in my experience at least, identifying the "first key reference", that reference which provides access to the names and publications of current leading research workers in the appropriate field. In the field of magnetism, my favorite route to the "first key reference" is to cut the latest Proceedings of the Conference on Magnetism and Magnetic Materials. In this day of computerized "key word" indices this approach is perhaps Neanderthalic, but I still recommend it. Instead of hundreds of references, of variable quality, relevance and accessibility, I have at my finger tips, through the wisdom of the Program Committee and of the referees recruited by the Publication Chairmen, the golden residue I really need.

For the prompt appearance of these Proceedings, and for the existence of the Conference of which these are the Proceedings, we are indebted to many people. Most obviously we are indebted to the Publications Committee Co-Chairmen, the editors of this volume, Mssrs. J. J. Becker, G. H. Lander and J. J. Rhyne, who devoted countless hours before, during, and after the Conference shepherding authors, referees, and publishers alike. We are indebted to the American Institute of Physics, and to Hugh Wolfe in particular, for the crucial role they play in assembling and producing the Proceedings. We owe gratitude to the Program Committee under Chairmen K. Lee and R. M. White, for the task of shaping the program through inviting key speakers and the less pleasant task of winnowing out contributed papers judged to be below standards. On their heads rests the quality of the program preserved in these Proceedings. Not evident in the Proceedings, but enormously evident at the Conference itself, were the Herculean labors of the Local Committee, chaired by B. F. Stein. I know of no Conference which has gone more smoothly than this one, a tribute to unseen effort by a dedicated group. To the Steering Committee, and to the Advisory Committee, who provide every year management and advice to this Conference, we are also in debt.

The Proceedings of this Conference propagate in time and space the scientific content of a meeting at which only a fraction of the total interested parties could be present. There is evidence that this fraction is a decreasing fraction, not only at our Conference but in many scientific fields. At the 1975 $M^3$ Conference we had a near-record high in papers delivered and a near-record low in attendance. With such a trend upon us, the Proceedings acquire increased importance as the vehicle by which advances in magnetism and magnetic materials are communicated. I am honored to write the preface of these Proceedings, which do that job so admirably.

Robert L. White
General Chairman
Stanford University
Stanford, CA 94305

COMMENTS ON UNITS IN MAGNETISM

L. H. Bennett, C. H. Page, and L. J. Swartzendruber
National Bureau of Standards
Gaithersburg, Maryland 20234

ABSTRACT

Suggestions are given on how to express magnetic quantities in SI units.

INTRODUCTION

Perusing the 1974 M[3] Conference Proceedings indicates that, at the present time, Systeme Internationale (SI) units are avoided by most leading scientists and engineers in the field of magnetism. Throughout the Proceedings, almost universal preference is displayed for the cgs electromagnetic system (or for the Gaussian system, which gives an equivalent description of magnetic quantities). However, usage of SI units in the field of magnetism will undoubtedly increase with time. One barrier to increased usage is the present lack of standardized and agreed upon relationships between magnetic quantities within the SI. In this paper we will tentatively propose notation and definitions for those relationships most frequently used by experimentalists, with the hope that this will help stimulate the magnetism community to make their views known on preferred definitions.

SOME CONSIDERATIONS ON THE TWO SYSTEMS

One major property of the Gaussian (and the cgs emu) system, considered an advantage by some and a disadvantage by others, is that B and H have the same numerical value in empty space. Changing to the SI, where not only do B and H have different units in empty space, but also different numerical magnitudes, puts one somewhat in the position of Casimir's[1] mythical tangenometrists who decided that, "The volumetric displacement of empty space - although equal to unity - had the dimension Archimedes per Euclid".

The SI is a "rationalized" system, whereas the Gaussian is unrationalized. Thus, when magnetic susceptibilities are converted between the two systems a factor of $4\pi$ is involved. Further factors of 10 are involved depending on whether volume, mass, or molar susceptibility is in question. This gives considerable latitude for errors and ambiguities in data compilations, handbooks, and treatises which attempt to convert existing numerical values to SI units, and numerous examples of such errors can be found. For example, in the recent treatise on magnetic materials by Heck[2], who endeavors to use SI units as much as possible, a table of paramagnetic susceptibilities apparently gives the rationalized mass susceptibility for Pt in $cm^3/g$, the unrationalized mass susceptibility for $\gamma$-Fe in $cm^3/g$, and the rationalized volume susceptibility for Li (dimensionless). Since these differences in units are not listed in the table, an unsuspecting user could easily be misled. As most commonly used with SI, the relation between B, H, and M is defined as $B=\mu_0(H+M)$, $\chi=M/H$. Some authors[3] exhibit the $\mu_0$ associated with the SI explicitly by replacing, H by $B/\mu_0$, giving $\chi=\mu_0 M/B$. This is, of course, approximately correct for the small susceptibilities found in most diamagnetic and paramagnetic materials, but could be misapplied to superparamagnetic or ferromagnetic materials.

Table 1

Symbols and names for magnetic quantities in SI and cgs, Gaussian (or cgs emu).

| Symbol | Name: cgs emu | Name: SI |
|---|---|---|
| B | flux density magnetic induction | flux density (magnetic induction) |
| H | magnetic field strength | magnetic field strength |
| M | magnetization | magnetization |
| J | --- | magnetic polarization |
| $\chi$ | volume susceptibility | --- |
| $\kappa$ | --- | rationalized volume susceptibility |
| $\chi_\rho$ | mass susceptibility | --- |
| $\kappa_\rho$ | --- | rationalized mass susceptibility |
| $\chi_{mole}$ | molar susceptibility | --- |
| $\kappa_{mole}$ | --- | rationalized molar susceptibility |
| m | magnetic moment | magnetic moment |
| $\mu_B$ | Bohr magneton | Bohr magneton |

RECOMMENDATIONS

In order to ease conversion from Gaussian (and cgs emu) to SI units, the names, definitions, and symbols for magnetic quantities should be standardized. This requires agreeement within the magnetism community. Our current recommendations are summarized in the Tables.

Table 1 lists recommended symbols and names for magnetic quantities in SI and cgs emu. When using SI units to express susceptibility, we believe it would be useful to label it 'rationalized' and give it the symbol $\kappa$, reserving $\chi$ for the non-rationalized cgs emu system. What we have labeled the "volume susceptibility" in Table 1 is often referred to simply as just "susceptibility". The

TABLE 2

Corresponding equations in SI and cgs Gaussian (or cgs emu). In this table, F refers to force, W refers to the energy of a magnetic dipole in a field, w refers to the volume energy density. Other symbols are defined in Table 1.

| Gaussian (or cgs emu) | SI | |
|---|---|---|
| $B = H + 4\pi M$ | $B = \mu_0\ (H + M)$ | (1) |
| | $B = \mu_0\ H + J$ | (2) |
| $\chi = M/H$ | $\kappa = M/H$ | (3) |
| $F = \chi VH \dfrac{\partial H}{\partial x}$ | $F = \mu_0 \kappa VH\ \partial H/\partial x$ | (4) |
| $W = -mB\cos\theta$ | $W = -mB\cos\theta$ | (5) |
| $w = \dfrac{BH}{8\pi}$ | $w = \frac{1}{2}BH$ | (6) |

Table 3. Conversion from Gaussian to S.I. Units

| Multiply the Number for | | by | To Obtain the Number for | |
|---|---|---|---|---|
| Gaussian Quantity | Unit | | SI Quantity | Unit |
| flux density, B | G | $10^{-4}$ | flux density, B | T(≡Wb/m$^2$≡Vs/m$^2$) |
| magnetic field strength, H | Oe | $10^3/4\pi$ | magnetic field strength, H | A/m |
| volume susceptibility, χ | emu/cm$^3$(dimensionless) | $4\pi$ | rationalized volume susceptibility, κ | dimensionless |
| mass susceptibility, $\chi_\rho$ | emu/g (≡cm$^3$/g) | $4\pi\cdot10^{-3}$ | rationalized mass susceptibility, $\kappa_\rho$ | m$^3$/kg |
| molar susceptibility,* $\chi_{mole}$ | emu/mol (≡cm$^3$/mol) | $4\pi\cdot10^{-6}$ | rationalized molar susceptibility, $\kappa_{mole}$ | m$^3$/mol |
| magnetization, M | G or Oe | $10^3$ | magnetization, M | A/m |
| | | $4\pi\cdot10^{-4}$ | magnetic polarization, J | T |
| magnetization, 4πM | G or Oe | $10^3/4\pi$ | magnetization, M | A/m |
| | | $10^{-4}$ | magnetic polarization, J | T |
| magnetization, M | $\mu_B$/atom or $\mu_B$/form. unit, etc.** | 1 | magnetization, M | $\mu_B$/atom or $\mu_B$/form. unit, etc.** |
| magnetic moment of a dipole, m | erg/G | $10^{-3}$ | magnetic moment of a dipole, m | J/T (≡Am$^2$) |
| demagnetizing factor, N | dimensionless | $1/4\pi$ | rationalized demagnetizing factor, N | dimensionless |

* Also called atomic susceptibility. Molar susceptibility is preferred since atomic susceptibility has also been used to refer to the susceptibility per atom.

** "Natural" units, independent of unit system. However, the numerical value of the Bohr magneton does depend on the unit system.

introduction of the symbol J (where $J = \mu_0 M$) in the SI is useful due to the controversy[4] over whether one should define $B=\mu_0(H+M)$ or $B=\mu_0 H+M$. Further, the symbol J and the associated name 'magnetic polarization', are in current use[5].

Table 2 compares several of the more important equations in the field of magnetism. Eqs. (1) and (2) define the recommended usage of the symbols M and J in SI, as mentioned above. In both Gaussian and SI units, the volume susceptibility, defined by Eq. (3), is dimensionless and is the ratio of M to H, (both with magnitudes which will change by a factor of 4π upon rationalization). Eq. (4) gives the force on a material placed in a magnetic field gradient. (This equation involves certain assumptions and is most useful for small samples with small susceptibilities.) Eq. (5) gives the energy of a (point) magnetic moment in a magnetic field, and Eq. (6) gives the volume energy density associated with a magnetostatic field.

Table 3 gives numerical factors for converting between the two unit systems. The conversions for flux density, B, and susceptibility, χ and κ, are independent of the conventions adopted, i.e. whether B=H+M, $B=\mu_0 H+M$, etc. Other conversions will depend on these conventions. One problem for those not thoroughly familar with current magnetic unit usage is that 'emu' is not really a unit but rather a flag to describe the unit system being used. Often, though not always, a dimensional anaylsis on susceptibility units may be performed if 'emu' is replaced by cm$^3$. Another problem which undoubtedly gives further difficulty to the uninitiated is the variety of units used for the same quantity in the Gaussian system. For example, in the 1974 M$^3$ conference proceedings we find the following units used for "magnetization": G, Oe, emu/g, $\mu_B$/atom, B.M./FORMULA UNIT, $\mu_B$/impurity, G cm$^3$/g, emu/cm$^3$, and emu: and for "susceptibility" we find the following variety of units: emu/g, emu/cm$^3$, emu/mole, emu/g kOe, emu/gm-At. V, and emu/Oe mole.

To convert an equation given in the Gaussian system to the corresponding equations in the SI, Table 4 can often be useful. For example, in the Gaussian system the magnetization can be considered as the magnetic moment per unit volume,

TABLE 4

Substitutional Symbols for Equations

To convert an equation in Gaussian units to a corresponding equation in SI, replace the symbols in the column labled Gaussian by the combination of symbols in the column labeled SI. Symbols representing quantities with units involving only volume, force, energy, and length transform directly.

| Gaussian Quality | Gaussian symbol | SI symbol |
|---|---|---|
| flux density | B | $\sqrt{4\pi/\mu_0}\ B$ |
| magnetic field | H | $\sqrt{4\pi\mu_0}\ H$ |
| magnetization | M | $\sqrt{\mu_0/4\pi}\ M$, or $\sqrt{1/4\pi\mu_0}\ J$ |
| volume susceptibility | χ | $(1/4\pi)\kappa$ |
| magnetic moment | m | $\sqrt{\mu_0/4\pi}\ m$ |

Table 5

Important Fundamental Constants

| Quantity | Gaussian | SI |
|---|---|---|
| $\mu_o$, permeability of free space | 1 (dimensionless) | $4\pi \times 10^{-7}$ H/m $\left(\equiv \frac{Tm}{A} \equiv \frac{Vs}{Am}\right)$ |
| $\mu_B$, Bohr magneton | $9.274078(36) \times 10^{-21} \frac{erg}{G}$ | $9.274078(36) \times 10^{-24} \frac{J}{T} \left(\equiv Am^2\right)$ |
| $\mu_N$, Nuclear magneton | $5.050824(20) \times 10^{-24}$ erg/G | $5.050824(20) \times 10^{-27}$ J/T |

$$M = \frac{m}{V} \quad (1)$$

where M is the magnetization in G, m is an appropriate magnetic moment in erg/G, and V is an appropriate volume in $cm^3$. Using the substitutions of Table 4 we have

$$\sqrt{\frac{\mu_o}{4\pi}}\, M = \frac{\sqrt{\mu_o/4\pi}\, m}{V} \quad (2)$$

which reduces to

$$M = \frac{m}{V} \quad (3)$$

Thus the magnetization in our suggested SI system can also be considered as the magnetic moment per unit volume, with magnetization in A/m, dipole moment in J/T, and volume in $m^3$. Table 5 gives the numerical value of three important fundamental magnetic constants in the two unit systems, and Table 6 compares demagnetizing coefficients, N, for several familar shapes, where the defining equation for N for both systems is

$$H = H_o - NM \quad (4)$$

with H the magnetic field strength within the magnetized body and $H_o$ the applied magnetic field strength.

## DISCUSSION

There are currently several systems of electromagnetic equations that may be used with SI units[4,6]. In order to apply SI units in the field of magnetism with a minimum of confusion, agreement and uniformity in symbols and definitions would be extremely helpful. Here we have suggested such a set of symbols and definitions which covers most of the quantities of current interest to those who publish in the $M^3$ proceedings. We would emphasize that this set is possibly not the one most desirable to a majority of magneticians. It was selected as one which appeared to us to be most in conformity with current international usage. An example of an alternative system would be the SI analog of a rationalized 'Gaussian' system. In such a system B, H, and M would be given the relation B=H+M, and H and M would also have units of 'tesla'. This would overcome the problem, troublesome to some, of giving B and H different numerical values in a vacuum. Another possibility, favored by Coleman[7], is the "SI electric" in which one defines $B=H+\mu_o M$ as the general relationship between B, H and M. In this system the unit for B and H is tesla and the unit for M is $Am^{-1}$, again giving B and H the same numerical value in empty space. However, both of these systems have the advantage (or disadvantage) found in the Gaussian system that B and H have the same numerical value in empty space.

Many of the details listed in the Tables given here depend on the particular SI relationship adopted for magnetic quantities. However, whichever relationships are adopted, the conversions for magnetic induction and susceptibility listed in Table 3 will remain valid, and the use of the proper unit and of the term 'rationalized' whenever susceptibility values are given in SI units would do much to reduce the possibility for errors and misinterpretation.

TABLE 6

Demagnetizing Coefficients, N, for homogeneous isotropic bodies of various shapes.

| Shape | N Gaussian (unrationalized) | N SI (rationalized) |
|---|---|---|
| ∥ to axis of long needle | 0 | 0 |
| ⊥ to axis of long needle | $2\pi$ | 1/2 |
| sphere | $4\pi/3$ | 1/3 |
| ⊥ to plane of a thin disc | $4\pi$ | 1 |

## REFERENCES


1. H. G. B. Casimir, Helv. Physica Acta 41, 741 (1968).
2. C. Heck, "Magnetic Materials and Their Applications", (Crane, Russak and Company, Inc., New York, 1974), p. 16.
3. For example, C. Kittel, "Introduction to Solid State Physics", Fourth Edition, (J. Wiley and Sons, Inc., New York, 1971), p. 499.
4. L. J. Giacoletto, IEEE Trans. on Magnetics 4, 1134 (1974).
5. For example, H. Zijlstra, Philips Tech. Rev. 34, 193 (1974).
6. See, for example, C. H. Page, Amer. J. of Phys. 38, 421 (1970); H. V. Stopes-Roe, Nature 222, 500 (1969); 224, 579 (1969); R. Green, Geophys. Prosp. 16, 1 (1968).
7. J. E. Coleman, Amer. J. Phys. 41, 221 (1973).

# NMR IN SUPERFLUID $^3He$ *

Alexander L. Fetter
Institute of Theoretical Physics, Department of Physics, Stanford University, Stanford, CA 94305

ABSTRACT

Evidence is presented to support the assumption of spin-triplet p-wave Cooper pairing in the A and B phases of $^3He$, and the particular form of the order parameter in each phase follows from available experimental data. Leggett's description of the spin dynamics shows that the macroscopic condensate qualitatively alters the NMR properties, for the spin density experiences an additional coherent torque that depends on the orientation of the orbital angular momentum of the Cooper pairs. This torque produces many new phenomena, including a temperature-dependent shift in the transverse CW resonance signal away from the Larmor frequency, resonant absorption in longitudinal CW excitation, and ringing behavior following a sudden perturbation in the applied magnetic field.

In 1972, experiments at Cornell[1] discovered peculiar features along the melting curve of $^3He$ when it was cooled below 2.7 mk. Since that time, the phenomenon has been studied in several laboratories.[2] As a result of these investigations and intense theoretical efforts,[3] liquid $^3He$ is now thought to enter a superfluid state, mediated by Cooper pairs as in metallic superconductors. The present article will review the arguments to support this conclusion, with particular emphasis on the magnetic properties of superfluid $^3He$. Since the He atoms remain in the electronic spin-singlet ground state, any magnetism necessarily comes from the nuclear spin 1/2 and corresponding nuclear magnetic moment; as a corollary, no conduction currents can occur, and effects like Meissner flux expulsion are wholly absent. Thus detection of NMR in $^3He$ is rather straightforward, in contrast to the situation in superconducting metals.

## BASIC EXPERIMENTAL FACTS

It is helpful to review the phase diagram of $^3He$, whose critical temperature is about 3.2 K. The liquid phase extends from this temperature down to absolute zero, because of the large zero-point motion, with the solid forming only at elevated pressure. When examined in detail, the liquid-solid phase boundary has a clear minimum at $\approx 0.3$ K and $\approx 30$ atm (see Fig. 1), which is readily understood from the Clausius-Clapeyron relation $dp/dT = (s_\ell - s_s)/(v_\ell - v_s)$ for the slope of the melting curve. Thermodynamic stability here requires $v_\ell > v_s$; on the other hand, $s_\ell$ varies approximately linearly with T, like an ideal Fermi gas, but $s_s$ remains essentially constant at the classical value $k_B \ln 2$ per nuclear spin. Thus dp/dT becomes negative at low temperatures, as seen in Fig. 1. This phenomenon (the Pomeranchuk effect) gives rise to compressional cooling if T initially lies below the minimum, and the original Cornell experiments all used such Pomeranchuk cells to reach the millidegree region. At very low temperatures, $^3He$ exhibits two distinct new phases, conventionally called A and B (see Fig. 2). Although the Pomeranchuk cooling is restricted to the melting curve, the remaining region has been explored with adiabatic demagnetization of CMN[4] and nuclear (copper) spins.[5] Careful experiments[2] have shown the following behavior: (1) The line $T_c$ represents a second-order transition, with a jump in specific heat like that in metallic superconductors. In contrast,

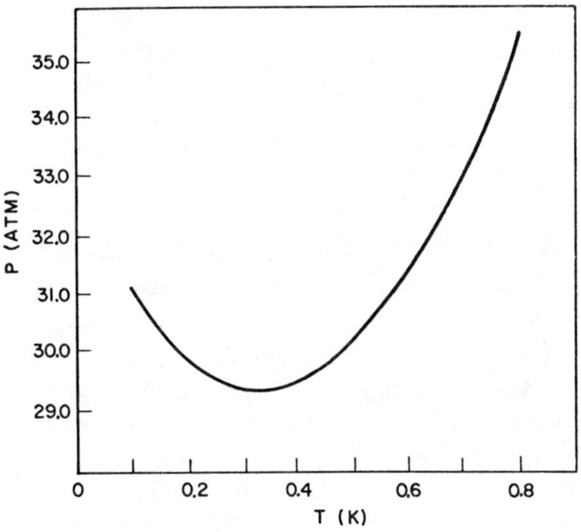


Fig. 1. Melting curve of $^3He$.

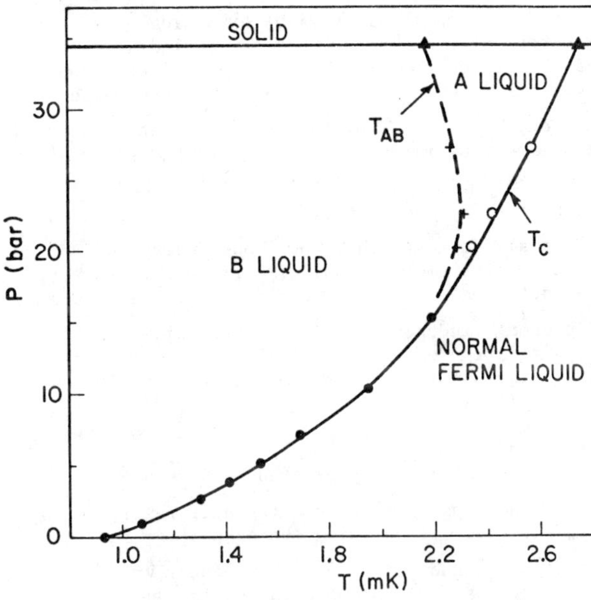


Fig. 2. Phase diagram of liquid $^3He$ at low temperature in a magnetic field of 320 gauss. For zero field, the line $T_{AB}$ intersects the line $T_c$ at a "polycritical point" ($P \approx 21$ bar and $T \approx 2.4$ mk) (after Ref. 5; see also Fig. 1 of Ref. 2).

the line $T_{AB}$ is a first-order transition. (2) The viscosity drops on entering the new phases, analogous to the situation in $^4He$. (3) New sound modes can propagate in the A and B phases, even though the large viscosity strongly attenuates ordinary sound in the normal state. These waves are similar to fourth sound in superfluid $^4He$, and they provide rather direct proof of superfluidity in $^3He$. (4) The static magnetic susceptibility $\chi_A$ in the A phase is essentially unchanged from the normal value $\chi_n$, whereas $\chi_B$ falls toward a smaller but nonzero value with decreasing temperature.

## COOPER PAIRING

These various phenomena can be successfully explained with a generalized theory analogous to the BCS description of superconducting electrons. First, we assume the existence of an energy gap Δ and an order parameter that characterizes the Cooper pairs. Moreover, the low transition temperature ($T \approx 2.7$ mK) implies that the gap $\Delta \approx k_B T_c$ is smaller than in metals by ≈ 1000. On the other hand, the usual BCS relation for the coherence length

$$\xi_0 = O(\hbar v_F / k_B T_c) \tag{1}$$

suggests that $\xi_0$ is ≈ 150 Å, only somewhat smaller than in metals because $v_F$ now refers to the massive atoms. Since $\xi_0$ far exceeds a typical interatomic separation, a mean-field (Ginzburg-Landau) theory should describe the phase transition in $^3$He. For comparison, note that the analogous length in superfluid $^4$He is thought to be only a few Ångstroms.

To study the form of the order parameter in $^3$He, we first recall the situation in metals, where the Cooper pairs are spin singlets (antiparallel pairing) in orbital s states. As T falls below $T_c$, the relative fraction of condensed pairs (measured, for example, in the strength of the flux expulsion) rises and approaches unity at T = 0. Since the binding energy is finite at low temperatures (≈ Δ per pair), a weak magnetic field with $\mu_B H \ll \Delta$ cannot reorient those spins bound in pairs, and the static magnetic susceptibility thus falls dramatically in the superconducting state. As noted previously, this behavior differs greatly from that of $^3$He-A ($\chi_A \approx \chi_n$), which first of all suggests that the Cooper pairs in $^3$He must be spin triplets (parallel pairing). Moreover, we may note that the $m_s = 0$ component of the triplet has an energy independent of the external magnetic field and is therefore magnetically inert. Consequently, the pairing in the A phase must involve only the $m_s = +1$ or $-1$ components, because $\chi_A$ remains close to $\chi_n$. Such a coherent superposition of ↑↑ and ↓↓ pairs was originally suggested by Anderson and Morel.[6] The B phase behaves quite differently, with $0 < \chi_B < \chi_n$, consistent with the assumption that all three magnetic substates contribute equally; this possibility was first proposed by Balian and Werthamer.[7]

As a result, we are led to conclude that the pairing in $^3$He occurs in spin-triplet states with S = 1. To incorporate this feature, Balian and Werthamer[7] generalized the usual BCS gap to a 2×2 matrix $\Delta_{\alpha\beta}$ in spin space, and the triplet nature implies that the matrix is symmetric. It must therefore be a linear combination of the three symmetric Pauli matrices 1, $\sigma_1$, and $\sigma_3$, with three complex coefficients. More conveniently, it is usually parametrized as

$$\Delta_{\alpha\beta} = i(\underset{\sim}{\Delta}\cdot\underset{\sim}{\sigma}\sigma_2)_{\alpha\beta} \tag{2}$$

where $\underset{\sim}{\Delta}$ is a complex vector in spin space that is somewhat analogous to the polarization vector of a photon. In most cases of interest (unitary states, see below), $\underset{\sim}{\Delta}$ lies perpendicular to the total spin of the pair, with $\underset{\sim}{\Delta}\cdot\underset{\sim}{S} = 0$.

In addition to the symmetric spin dependence, the orbital properties of the Cooper pairs must conform to the Pauli principle, which demands antisymmetry under exchange of all coordinates of the pair. To make this notion precise, recall that Δ is proportional to the amplitude for finding a pair with wave vectors $\pm\underset{\sim}{k}$ in a spin triplet

$$\Delta_{\alpha\beta}(\underset{\sim}{k}) \propto \langle a_{\underset{\sim}{k}\alpha} a_{-\underset{\sim}{k}\beta} \rangle \ . \tag{3}$$

Since the relevant region of momentum space lies close to the Fermi surface, $\Delta_{\alpha\beta}(\underset{\sim}{k})$ in fact depends only on the direction $\hat{k}$, and the Pauli principle then allows only spherical harmonics with odd L. In the simplest case of L = 1, it is not difficult to see that $\underset{\sim}{\Delta}(\hat{k})$ may be rewritten as a <u>linear</u> function of $\hat{k}$ with a complex 3×3 matrix as coefficient. The necessity for this complicated structure follows from the presence of two unit angular momenta $\underset{\sim}{L}$ and $\underset{\sim}{S}$, each with three magnetic substates.

In general, the theory becomes quite intricate because the triplet pairing renders physical quantities spin dependent. Fortunately, the observed gap function usually has a special property known as unitarity, in which case the actual energy gap at a point $\hat{k}$ on the Fermi surface becomes independent of spin, given by $\underset{\sim}{\Delta}(\hat{k})^*\cdot\underset{\sim}{\Delta}(\hat{k}) \equiv |\underset{\sim}{\Delta}(\hat{k})|^2$. In the A phase, $\underset{\sim}{\Delta}(\hat{k})$ is thought to have a factored (ABM) structure $\underset{\sim}{\Delta}(\hat{k}) = \underset{\sim}{\Delta}[(\hat{m}+i\hat{n})\cdot\hat{k}]$ with $\underset{\sim}{\Delta}$ a constant vector lying in the xy plane perpendicular to the quantization axis.[6,8] Here $\hat{m}$ and $\hat{n}$ are orthogonal unit vectors with $\hat{\ell} = \hat{m}\times\hat{n}$ lying along the orbital angular momentum of the Cooper pairs. A simple calculation shows that the squared energy gap has an angular dependence $|\underset{\sim}{\Delta}(\hat{k})|^2 = \Delta^2[1-(\hat{k}\cdot\hat{\ell})^2]$; it vanishes along the directions $\pm\hat{\ell}$, and these nodes would in principle affect low-temperature properties like the specific heat. In practice, however, the A phase does not extend to sufficiently low temperatures to observe such features. In contrast, the B phase is thought to have a very different order parameter of the form $\underset{\sim}{\Delta}(\hat{k}) = \Delta R\hat{k}$, with R a real 3×3 orthogonal matrix representing an arbitrary rotation.[7] The corresponding energy gap is isotropic, so that the thermodynamic properties of the B phase should resemble those in a BCS superconductor.

## STATIC MAGNETIC PHENOMENA

The Cooper pairs in $^3$He involve two unit angular momenta $\underset{\sim}{S}$ and $\underset{\sim}{L}$, whose orientations depend on various perturbations.[3] There are three distinct classes:

1. External perturbations that act on $\underset{\sim}{S}$ (for example, magnetic fields).
2. External perturbations that act on $\underset{\sim}{L}$ (for example, flows, walls, electric fields).
3. Nuclear dipole-dipole interactions that couple $\underset{\sim}{L}$ and $\underset{\sim}{S}$.

If all these perturbations vanished, then the total Hamiltonian would be invariant under separate rotations of the spin and orbital variables. This is the approximation we have considered so far, because the quantization axes for the orbital motion (the $\hat{k}$ variables) and the spin could be chosen independently. In practice, however, experiments occur in an external environment that defines preferred spatial directions associated with perturbations of the types 1 and 2. For concreteness, consider an external magnetic field $\underset{\sim}{H}$. In this case a classical magnetization density $\underset{\sim}{M}$ would precess about the instantaneous field $\underset{\sim}{H}(t)$, as is familiar from NMR. This behavior may be derived from a phenomenological energy density[9]

$$E_0 = M^2/2\chi - \underset{\sim}{M}\cdot\underset{\sim}{H} \ , \tag{4}$$

where χ is the susceptibility. In a static field $\underset{\sim}{H}_0$, the magnetization density adjusts itself to minimize the energy $E_0$, yielding the expected equilibrium relation $\underset{\sim}{M} = \chi\underset{\sim}{H}_0$. For more general time-dependent fields, the usual commutation relations for angular momentum immediately give

$$d\underset{\sim}{M}/dt = (i\hbar)^{-1}[\underset{\sim}{M},E_0] = \gamma \underset{\sim}{M}\times\underset{\sim}{H} \;, \tag{5}$$

where $\gamma$ is the gyromagnetic ratio. This equation exhibits the precession of $\underset{\sim}{M}$ about the total external field; with a weak rf field $\underset{\sim}{H}'$ perpendicular to $\underset{\sim}{H}_0$, it accounts for the resonant absorption of energy at the Larmor frequency $\gamma H_0$.

The above treatment indicates that the applied $\underset{\sim}{H}$ field provides the quantization axis for the total spin of the Cooper pairs and hence the magnetization $\underset{\sim}{M}$. Analogously, other perturbations can be shown to orient the orbital angular momentum of the pair, but the two vectors still remain unrelated at this level of approximation. In this regard, Leggett[9] observed that the nuclear dipole-dipole interaction energy plays a crucial role, for it couples the directions of $\underset{\sim}{S}$ and $\underset{\sim}{L}$. Leggett's description is elegant and simple, and its basic ideas are easily explained. Consider first the interaction energy between two fixed parallel magnetic dipoles with total dipole moment $\gamma\hbar\underset{\sim}{S}$ (see Fig. 3); configuration a is known to have lower energy than configuration b. If the dipoles execute slow orbital rotation keeping their individual axes fixed, configuration c ($\underset{\sim}{L} \perp \underset{\sim}{S}$) evidently has lower mean energy than configuration d ($\underset{\sim}{L} \parallel \underset{\sim}{S}$). Thus the dipole energy $E_D$ depends on the relative orientation of $\underset{\sim}{L}$ and $\underset{\sim}{S}$, favoring the orientation with $\underset{\sim}{L}$ perpendicular to $\underset{\sim}{S}$. This description applies directly to the $^3$He atoms in a Cooper pair. Moreover, the quantum coherence of the macroscopic condensate locks all the Cooper pairs together with a common $\underset{\sim}{S}$ and a common $\underset{\sim}{L}$. As a result, the total dipole energy exceeds that for a single pair by a factor comparable with the number of pairs $\approx N(\Delta/k_BT_c)^2$, which increases linearly with $T_c - T$ in the vicinity of $T_c$. This is the basic observation; it means that the dipole energy acquires a macroscopic value and must be included in computing the dynamical motion of the spin density and magnetization. As mentioned above, such behavior originates in the macroscopic quantum condensate; it thus provides another example of London's long-range order familiar in superfluid $^4$He and in superconductors.

Leggett made these qualitative ideas precise by evaluating the dipole energy in a condensed state with spin-triplet p-wave pairing; he found[9]

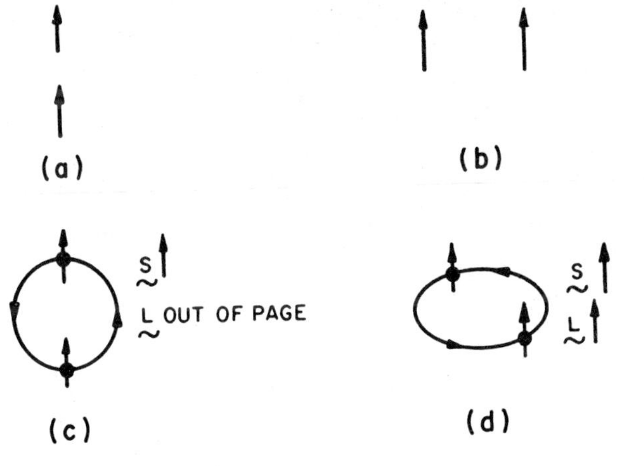


Fig. 3. Various configurations of two parallel magnetic dipoles.

$$E_D = 5g_D(4\pi)^{-1} \int d\Omega_k (3\hat{k}_\mu \hat{k}_\nu - \delta_{\mu\nu})\Delta_\mu(\hat{k})^* \Delta_\nu(\hat{k}) \;, \tag{6}$$

where the integral is over the Fermi surface and $g_D$ is a definite constant. As expected, this energy couples the orbital variables $\hat{k}$ with the total spin $\underset{\sim}{S}$ of the pair through the vector $\underset{\sim}{\Delta}$. Equation (6) is easily evaluated for any particular state; for example, the A phase yields

$$(E_D)_A = -2g_D(\hat{\ell}\cdot\underset{\sim}{\Delta})^2 + \text{const} \;, \tag{7}$$

explicitly demonstrating that $E_D$ is minimum when the orbital angular momentum of the Cooper pair ($\hat{\ell}$) lies parallel to $\underset{\sim}{\Delta}$ and hence perpendicular to the total spin $\underset{\sim}{S}$ of the pair (recall that $\underset{\sim}{\Delta}\cdot\underset{\sim}{S} = 0$ for both A and B phases). More generally, $E_D$ introduces a dependence on the relative orientation of $\underset{\sim}{L}$ and $\underset{\sim}{S}$ that can significantly affect the dynamics of the spin density and resulting magnetization.

To incorporate these dipole contributions, Leggett merely augments the previous phenomenological energy $E_0$ by the dipole energy $E_D$, which depends on the functional form of the order parameter $\underset{\sim}{\Delta}(\hat{k})$ through an integral over the Fermi surface

$$E = M^2/2\chi - \underset{\sim}{M}\cdot\underset{\sim}{H} + E_D\{\underset{\sim}{\Delta}(\hat{k})\} \;. \tag{8}$$

Here the last term contains all the specific properties of superfluid $^3$He, including the orientational effects of external perturbations or walls on the orbital angular momentum of the Cooper pairs; as expected, $E_D$ vanishes in the normal state because $\underset{\sim}{\Delta} = 0$, and it increases nearly linearly for $T_c - T \ll T_c$. Clearly, Eq. (8) is more complicated than that for a normal fluid. Consider first the simplest example of a static field $\underset{\sim}{H}_0$. Since the magnetization must adjust to minimize E and not just $E_0$, it need not lie along $\underset{\sim}{H}_0$ but instead can be tipped by dipole torques that depend on the specific state in question and on the external perturbations. If the system is then disturbed by a small transverse rf field $\underset{\sim}{H}' \perp \underset{\sim}{H}_0$, the same dipole torques can shift the resonant absorption of energy away from the Larmor frequency, leading to a rich variety of NMR effects.

## DYNAMIC MAGNETIC PHENOMENA

The total energy E serves as an effective Hamiltonian for the motion of the spin density $\underset{\sim}{M}/\gamma$ and the order parameter $\underset{\sim}{\Delta}(\hat{k})$. In this way, Leggett derived the coupled equations

$$d\underset{\sim}{M}/dt = \gamma\underset{\sim}{M}\times\underset{\sim}{H} + \gamma\underset{\sim}{R}_D\{\underset{\sim}{\Delta}(\hat{k})\} \;, \tag{9a}$$

$$d\underset{\sim}{\Delta}(\hat{k})/dt = \gamma\underset{\sim}{\Delta}(\hat{k})\times(\underset{\sim}{H} - \underset{\sim}{M}/\chi) \;, \tag{9b}$$

where $\underset{\sim}{R}_D$ is the additional dipole torque. It depends on the instantaneous configuration of $\underset{\sim}{\Delta}(\hat{k})$, which itself precesses about the instantaneous vector $\underset{\sim}{H} - \underset{\sim}{M}/\chi$. Evidently, the coupled nonlinear motion can become extremely complicated, and only a few simple examples will be mentioned here.

1. Bulk CW resonance with no external perturbation on $\underset{\sim}{L}$. In this case, a static magnetic field $\underset{\sim}{H}_0$ induces a net magnetization $\underset{\sim}{M}_0 = \chi\underset{\sim}{H}_0$, and the Cooper pairs initially orient themselves to minimize the total energy E, ensuring that the dipole torque vanishes in equilibrium. If a weak rf field is then applied, the orbital angular momentum of the Cooper pairs will remain fixed, owing to their large size and moment of inertia ($\approx m_3\xi_0^2$ with $\xi_0 \approx 150$ Å). On the

other hand, $\underset{\sim}{\Delta}$ and $\underset{\sim}{S}$ will change, keeping $\underset{\sim}{\Delta}\cdot\underset{\sim}{S} = 0$, and the resulting linearized equations describe the weak-field response. A somewhat lengthy analysis yields the following predictions.

a. "Transverse resonance" ($\underset{\sim}{H}' \perp \underset{\sim}{H}_0$). This is the usual NMR configuration, and a normal system would invariably resonate at the Larmor frequency $\gamma H_0$. In the superfluid, however, the dipole energy produces additional torques, and the resonance is shifted from $\gamma H_0$. Using the expressions for the A and B phases, Leggett obtained the following values for the transverse resonance frequency in the A and B phases, respectively,[9]

$$(\omega_t^{\,2})_A = (\gamma H_0)^2 + \Omega_A^{\,2}(T);\ (\omega_t^{\,2})_B = (\gamma H_0)^2. \qquad (10)$$

Note that the B phase turns out to resonate at the usual Larmor frequency, whereas the A phase has a temperature-dependent (but field-independent) shift given by

$$\Omega_A^{\,2}(T) = 4\gamma^2 g_D \Delta^2(T)/\chi\ . \qquad (11)$$

Since $\Delta^2 \propto T_c - T$ for T near $T_c$, the deviation in $\omega_t^{\,2}$ should increase linearly as the system enters the superfluid phase. This effect was first observed in 1972;[10] it provided the original impetus for Leggett's calculations.

b. "Longitudinal resonance" ($\underset{\sim}{H}' \parallel \underset{\sim}{H}_0$). In a normal system, longitudinal excitation would not lead to resonant absorption, so that its detection in liquid $^3$He served as strong evidence for the superfluidity. A straightforward calculation leads to the following expressions for the longitudinal resonance frequency in the two phases (assuming $\Delta_A^{\,2} = \Delta_B^{\,2}$):

$$(\omega_\ell)_A = \Omega_A(T);\quad (\omega_\ell)_B \approx (5\chi_A/2\chi_B)^{1/2}\Omega_A(T) \qquad (12)$$

where $\Omega_A$ is again given in Eq. (11) and $\chi$ denotes the static susceptibility in the appropriate phases. These relations (10) and (12) constitute rather precise tests of the assumed structure of the A and B phases. Figures 4 and 5 show examples of their verification.[11,12]

2. Ringing following a sudden perturbation. A different class of experiments monitors the time-dependent magnetization $\underset{\sim}{M}(t)$ following a sudden change in the configuration. This technique was developed by Wheatley and collaborators,[13,14] who stepped the d.c. magnetic field from $H_0\hat{z}$ to $H\hat{z}$. In a normal system, the magnetization would merely relax to the new value $\chi H\hat{z}$ with the characteristic time $T_1$. Superfluid $^3$He behaves very differently, however, exhibiting time-dependent oscillations whose frequency and magnitude depend on the size of the step $\Delta H$. The existence of such ringing can be understood directly from Leggett's equations (9a) and (9b). Immediately after the step is applied, the vector $\underset{\sim}{H} - \underset{\sim}{M}/\chi$ no longer vanishes, and $\underset{\sim}{\Delta}(\hat{k})$ therefore starts to rotate about the z axis. The resulting oscillatory torque $\underset{\sim}{R}_D$ then induces a time-dependent component of $M_z$. If $\gamma\Delta H$ is much less than the characteristic dipole frequency $\Omega_A(T)$, then the signal rings at the same frequency as the longitudinal CW resonance; these and other predictions[15-17] have been verified in considerable detail.[13,14] More recently, Osheroff and Corruccini[18] have used pulsed fields to tip $\underset{\sim}{M}$ away from its original direction along $\underset{\sim}{H}_0$. The resulting finite value of $\underset{\sim}{H}_0 - \underset{\sim}{M}/\chi$ again causes $\underset{\sim}{\Delta}(\hat{k})$ to rotate, and the induced oscillatory component of M rings at a frequency of order $\Omega_A(T)$.

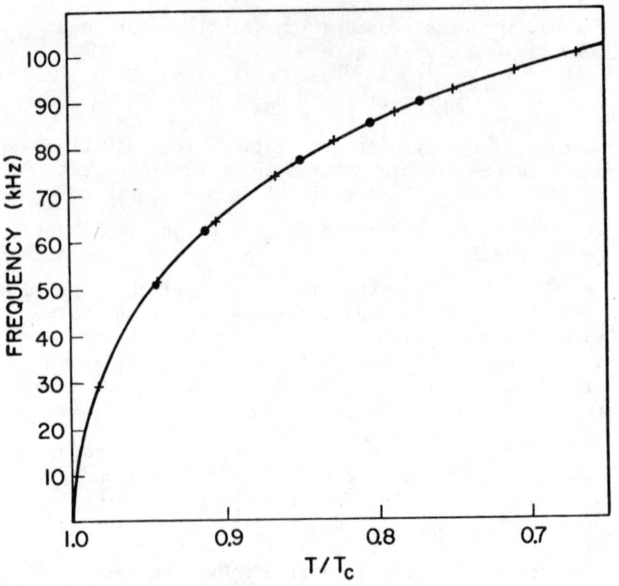


Fig. 4. Comparison of transverse and longitudinal resonance frequencies in $^3$He-A. Plus signs denote $(2\pi)^{-1}[(\omega_t^{\,2})_A - (\gamma H_0)^2]^{1/2}$ and dots denote $(\omega_\ell)_A/2\pi$ (after Ref. 11).

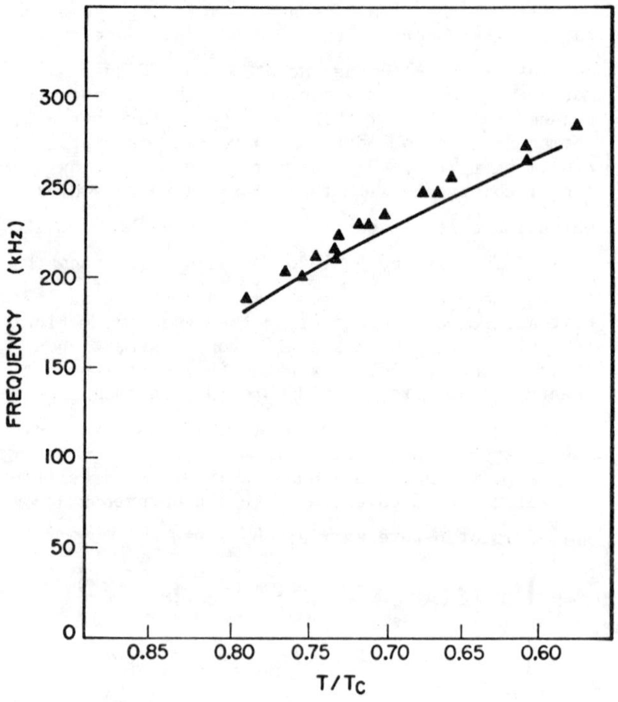


Fig. 5. Comparison of longitudinal resonance frequencies in $^3$He-A and $^3$He-B. Experimental points denote $(\omega_\ell)_B/2\pi$ and solid line represents $(5\chi_A/2\chi_B)^{1/2}(\omega_\ell)_A/2\pi$ (after Ref. 12).

For large deviations from equilibrium, nonlinearities become significant, and the detailed agreement between theory[19] and experiment[18] strongly supports Leggett's theory.

At present, most NMR experiments have used bulk uniform fluid, where the Cooper pairs initially orient their spin and orbital angular momentum to eliminate the dipole torques. In the presence of walls,[20,21] hydrodynamic flow,[22] or electric[23] fields, however, the total energy contains additional terms that depend on the various perturbations. The subsequent minimization of the total free energy can rotate $\underset{\sim}{L}$ from its original direction, and the net dipole torque need not vanish, even in equilibrium. The dynamic phenomena correspondingly become even more varied. For example, both the A and B phases should display longitudinal- and transverse-resonance shifts that depend on the orientation and strength of the perturbations, because of the dipole coupling between $\underset{\sim}{S}$ and $\underset{\sim}{L}$. Crowds of interesting experiments come to mind,[12,24,25] and we may expect many new observations. If past experience serves as a reliable guide, they will continue to confirm Leggett's theory of spin dynamics in superfluid $^3$He.

## REFERENCES

*Research sponsored in part by NSF grant MPS 75-08516.

1. D. D. Osheroff, R. C. Richardson, and D. M. Lee, Phys. Rev. Lett. 28, 885-888 (1972).
2. J. C. Wheatley, Rev. Mod. Phys. 47, 415-470 (1975) reviews the experimental situation in great detail.
3. A. J. Leggett, Rev. Mod. Phys. 47, 331-414 (1975) reviews the theoretical situation in great detail.
4. T. J. Greytak, R. T. Johnson, D. N. Paulson, and J. C. Wheatley, Phys. Rev. Lett. 31, 452-455 (1973).
5. A. I. Ahonen, M. T. Haikala, M. Krusius, and O. V. Lounasmaa, Phys. Rev. Lett. 33, 628-631 (1974).
6. P. W. Anderson and P. Morel, Phys. Rev. 123, 1911-1934 (1961).
7. R. Balian and N. R. Werthamer, Phys. Rev. 131, 1553-1564 (1963).
8. P. W. Anderson and W. F. Brinkman, Phys. Rev. Lett. 30, 1108-1111 (1973); W. F. Brinkman and P. W. Anderson, Phys. Rev. A8, 2732-2734 (1973).
9. A. J. Leggett, Phys. Rev. Lett. 29, 1227-1230 (1972); 31, 352-355 (1973); Ann. Phys. (N.Y.) 85, 11-55 (1974).
10. D. D. Osheroff, W. J. Gully, R. C. Richardson, and D. M. Lee, Phys. Rev. Lett. 29, 920-923 (1972).
11. D. D. Osheroff and W. F. Brinkman, Phys. Rev. Lett. 32, 584-587 (1974).
12. D. D. Osheroff, Phys. Rev. Lett. 33, 1009-1012 (1974).
13. R. A. Webb, R. L. Kleinberg, and J. C. Wheatley, Phys. Rev. Lett. 33, 145-148 (1974); Phys. Lett. 48A, 421-422 (1974).
14. R. A. Webb, R. E. Sager, and J. C. Wheatley, Phys. Rev. Lett. 35, 615-617 (1975); 35, 1010-1013 (1975).
15. K. Maki and T. Tsuneto, Prog. Theor. Phys. 52, 773-776 (1974).
16. W. F. Brinkman, Phys. Lett. 49A, 411-412 (1974).
17. K. Maki and C. R. Hu, J. Low Temp. Phys. 18, 377-385 (1975); 19, 259-268 (1975).
18. D. D. Osheroff and L. R. Corruccini, Phys. Lett. 51A, 447-448 (1975).
19. W. F. Brinkman and H. Smith, Phys. Lett. 51A, 449-450 (1975); 53A, 43-44 (1975).
20. V. Ambegaokar, P. G. deGennes, and D. Rainer, Phys. Rev. A9, 2676-2685 (1974).
21. W. F. Brinkman, H. Smith, D. D. Osheroff, and E. I. Blount, Phys. Rev. Lett. 33, 624-627 (1974).
22. P. G. deGennes and D. Rainer, Phys. Lett. 46A, 429-430 (1974).
23. J. M. Delrieu, J. de Physique Lett. 35, L189-L192 (1974); 36, L22 (1975).
24. S. Takagi, J. Phys. C8, 1507-1515 (1975).
25. A. L. Fetter, Phys. Lett. 54A, 63-65 (1975).

## SUPERSTRONG MAGNETIC FIELDS AND NEUTRON STARS

Malvin Ruderman
Physics Department, Columbia University, New York, N.Y. 10027

### ABSTRACT

The observed spin down rates of pulsars (rapidly spinning isolated neutron stars) imply that their stellar surface magnetic fields exceed $10^{12}$ G. In such superstrong fields atoms shrink in each dimension by over an order of magnitude and the binding energy of atoms into molecular chains can exceed that of all the electrons in an isolated atom. The stellar surface layer forms a dense ($10^4 - 10^5$ g cm$^{-3}$) tightly bound (10 kev/atom) anisotropic conductor which gives the star a sharp edge even at temperatures of $10^7$ °K.

Strong magnetospheric electric fields are insufficient to pull ions from such a surface. Instead they generate electron-positron discharges in the $10^{12}$ G field of the near magnetosphere. These particles are probably the source of current flow in the magnetosphere and of many of the observed properties of pulsar radiation.

References and details are contained in the Proceedings of the Conference on the Role of Magnetic Fields in Physics and Astrophysics, Copenhagen, 1974,[1] V. Canuto, ed.

[1] Annals of the New York Academy of Sciences 257, 127 (1975)

# MAGNETIC COUPLING BETWEEN LIQUID $He^3$ AND ELECTRON SPINS IN SOLIDS

D. L. Mills†
Department of Physics, University of California, Irvine, Ca. 92664
M. T. Beal-Monod*
Physique des Solides, University Paris-Sud, 91405 Orsay, France

ABSTRACT

We review briefly the experimental and theoretical literature on the contribution from spin-spin interactions to the thermal boundary conductance (Kapitza conductance) between liquid $He^3$ and solids. This contribution to the boundary conductance has its origin in the interaction of the $He^3$ nuclear magnetic moment with electron spins in solids. We summarize some new predictions of the behavior of the magnetic boundary conductance at temperatures below the onset of superfluid order in $He^3$. Also, we present initial results of our study of the longitudinal relaxation of liquid $He^3$ nuclear spins in dipolar contact with electron spins in a paramagnet.

If a substance is placed in contact with a magnetic cooling salt at very low temperatures, the rate at which it may be cooled is controlled by the thermal boundary resistance (Kaptiza resistance).[1] If the dominant contribution to the boundary conductance comes from coupling to the phonons in the cooling salt, then for either a liquid or a solid in contact with it, the Kapitza resistance increases dramatically (like $T^{-3}$) as the temperature is lowered. As a consequence, one's ability to carry out experiments at very low temperature can be limited by the very long equilibration times required to cool the substance under study.

A number of years ago, Wheatley discovered[2] that when liquid $He^3$ is placed in contact with paramagnetic CMN, as the temperature is lowered $R_K$ reached a maximum, then decreased to become proportional to T well below the maximum. This remarkable observation shows that as the temperature is lowered, the thermal contact between $He^3$ and paramagnetic CMN progressively improves, in sharp contrast to the behavior described above. Furthermore, more recent work shows that at fixed temperature, the boundary resistance is decreased even more by application of a small magnetic field.[3]

In his original paper, Wheatley proposed that this anomalous behavior has its origin in the magnetic coupling of the $He^3$ nuclear spin to the paramagnetic electron spins in the substrate. However, no detailed theoretical studies of this mechanism appeared until the work of Leggett and Vuorio.[4] These authors explored a description of the energy transfer between $He^3$ spins coupled to a paramagnet via a short ranged contact interaction. They showed that such a mechanism indeed explains both the temperature and magnetic field dependence observed for liquid $He^3$ coupled to CMN. Furthermore, they find that when the contact interaction is adjusted in strength to mimic dipolar coupling between nuclear spins and the electron spins in the CMN, the magnitude of the calculated boundary resistance is in good qualitative accord with the data.[5]

The basic process envisioned by Leggett and Vuorio is one in which the $He^3$ nuclear spin transfers its Zeeman energy to the electron spins in the solid, with the kinetic energy of the $He^3$ atoms an energy reservoir that absorbs the difference between the electrinic and nuclear Zeeman energy. This energy transfer is most efficient when the characteristic energy of the electron spins in the solid ($g\mu_B H$ for paramagnetic spins in an external field) matches the characteristic energy ($k_B T$, the average excitation energy $\epsilon_k - \epsilon_F$ of a $He^3$ atom) of the kinetic energy reservoir. From this description, the energy transfer between a paramagnet and $He^3$ is optimal when $g\mu_B H \sim k_B T$.

This picture is valid for a free Fermi gas in magnetic contact with the substrate. A more general description is to consider the spin fluctuations in the (exhanced enhanced) $He^3$ in contact with thermal fluctuations of the spin system in the substrate.

Our own interest in this area was stimulated by new measurements from the Wheatley group.[6] In a study of the thermal boundary resistance between liquid $He^3$ and (impure) platinum, these authors found $R_K \sim T^{-2}$, a result distinctly different from the value appropriate to coupling to phonons, and also different from the result appropriate to liquid $He^3$ in contact with paramagnetic CMN. It was suggested initially that this new behavior has its origin in the coupling between the $He^3$ spins, and the spins of the conduction electrons in the metal.

Our first investigation explored this possibility[7] within the framework of a formalism that treated fully the dipolar interaction between the $He^3$ and the electron spins in the solid, without replacing it by an effective contact interaction, and we also required the wave function of the $He^3$ particles to vanish at the solid/liquid interface. We examined the effect of magnetic coupling to the conduction electrons and also reanalyzed the theory for the case of coupling to paramagnetic local moments, to explore the sensitivity of the result of Leggett and Vuorio to the assumptions used by them. In the latter case, we recovered results quite identical to those of Leggett and Vuorio, save for the numerical prefactor; the dipolar interaction indeed behaves like an effective contact interaction with a range the order of $\lambda_F$, the Fermi wavelength of the $He^3$ spins.

While this is true for the case of $He^3$ quasi-particles coupled to paramagnetic local moments, the long-ranged character of the dipole interaction plays a critical role in the description of magnetic coupling to the conduction electrons. In the language of spin fluctuation theory, long wavelength spin fluctuations in the Fermi fluid set up macroscopic dipolar fields, and the long wavelength fluctuations on each side of the interface couple strongly as a consequence. The resulting contribution to the Kapitza conductance varies with temperature like $(T^3 \ln T)^{-1}$ rather than $T^{-3}$ expected for a contact interaction. Nonetheless, quantitative estimates based on the resulting formula show that coupling between $He^3$ nuclear spins and conduction electron spins is too inefficient by many orders of magnitude to explain the data on Pt.

We find, however, that magnetic impurities in the Pt matrix can indeed produce a $T^{-2}$

variation of the Kapitza resistance, for temperatures below the ordering temperature of the impurities; at low impurity concentrations, we presume the impurities will order into a spin glass configuration.[8] Thus, we suggest that the data presented in reference (6) may be explained by magnetic coupling between the $He^3$ nuclei, and an ordered array of magnetic impurities.

More generally, for $He^3$ nuclei in contact with local moments present as impurities in a metal, one expects[8] that above the ordering temperature $T_o$ of the impurities, the Leggett-Vuorio behavior $R_K \sim T$ obtains. As the temperature is lowered toward $T_o$, $R_K$ decreases to pass through a minimum for T near $T_o$, and if the spins order into a spin glass configuration one has $R_K \sim T^{-2}$ for $T \ll T_o$. The precise behavior of $R_K$ below $T_o$ depends on the nature of the order present below $T_o$. For example, for a ferromagnetically aligned but spatially disordered array of spins (as in Pd Fe for concentrations greater than ~100 ppm), one expects $R_K \sim T^{-5/2}$ for $T \ll T_o$ while $R_K \sim T^{-3/2}$ for coupling to a ferromagnetically aligned array of spins arranged on a regular crystal lattice.[8]

The suggestion that the $T^{-2}$ variation of $R_K$ observed by Bishop et al.[6] has its origin in coupling to local moments ordered in a spin glass configuration appears consistent with the susceptibility data; the susceptibility has a maximum somewhat above the temperature range employed in the measurements.[6]

Avenel et al.[9] have also reported Kapitza resistance data for Au Gd and Pd Fe samples. This data may also be interpreted within the framework outlined above. It is important to realize that in the temperature regime explored in this work, the phonon contribution to the Kapitza conductance is comparable to that from spin-spin coupling. Thus before one can analyze the temperature variation of the magnetic contribution to $R_K$, the phonon contribution must be subtracted from the data.[8]

Thus, it appears that the data in reference (6) and reference (9) may be interpreted in a simple unified manner. The magnetic contribution to the thermal boundary resistance has its origin in the coupling of the $He^3$ nuclei to local moments in the solid. One can realize a variety of temperature variations in $R_K$ when this coupling provides the dominant fraction of the observed boundary conductance. The observed temperature variation depends on whether one is above, below or near the ordering temperature, and the nature of the order present.[10] We should emphasize that the data available presently is extremely sketchy. Before the predictions of the theory can be checked out properly, one needs a systematic study of the Kapitza conductance between liquid $He^3$ and several samples of a given alloy system, with controlled amounts of impurity. It is extremely helpful to measure the temperature variation of $\chi$ through the temperature range where $R_K$ is explored, as Bishop et al.[6] have done. We hope more data will become available in the near future.

We conclude this review with a discussion of a most interesting theoretical paper by Guyer.[11] Guyer has explored the effect of the long ranged character of the dipole-dipole interaction on $R_K$ by an approximate method quite different than the approach used by Leggett and Vuorio, and subsequently by us. In addition, Guyer predicted that for paramagnetic solid $He^3$ in contact with paramagnetic local moments (CMN), $R_K$ should vary as $T^{-2}$. This prediction has been verified experimentally by Reinstein and Zimmerman.[12] Here the magnetic coupling seems extremely efficient. The energy transfer rate between the two systems is dominated by the magnetic coupling below ~ 60 mK, while in the case of $He^3$ against magnetic solids, the magnetic coupling does not begin to dominate until ~ 30 mK.

In the above remarks, we have provided a brief review of the recent literature on the role played by magnetic coupling between liquid (or solid) $He^3$ and electron spins in solids on lowering the thermal boundary resistance between the two systems at very low temperatures. In the remainder of this paper, we summarize some of our more recent work. In the present note, we present our principal conclusions; the details will be presented in full elsewhere.

The theoretical and experimental studies cited above confine their attention to normal $He^3$ in contact with a magnetic substrate. It is important to know the effect of the superfluid transition in the $He^3$ on the observed boundary resistance. In collaboration with Maki, we have explored this question theoretically, within the framework of a simple model.[13] This study confines its attention to the case of superfluid $He^3$ in contact with paramagnetic electron spins.

To carry out a complete calculation of this effect, it is necessary to know the detailed behavior of the superfluid order parameter in the near vicinity of the surface. At this time, very little may be said about the behavior of the order parameter, or the nature of spin fluctuations near the boundary between superfluid $He^3$ and a solid. In our work, we have presumed that order parameter remains uniform right up to the surface. The dynamical spin susceptibility of the superfluid has been approximated by the expressions appropriate to the bulk fluid. These functions may be found in the paper by Maki and Ebisawa.[14] For the case of the anisotropic A phase, we have oriented the vector $\vec{d}$ normal to the surface, as simple physical considerations suggest must be the case. (The A phase is an equal admixture of spin triplet pairs with azimuthal quantum number for total spin m=+1, and pairs with m=-1. The vector $\vec{d}$ is parallel to the quantization axis used for the spins.)

Before we present the results of our study, we pause for a brief comment. The results below are computed within the framework of the assumptions outlined in the preceding paragraph. These assumptions may prove oversimplified, when experimental studies of the boundary resistance between superfluid $He^3$ and CMN become available. We note that deviations from our predictions in principle contain information about the behavior of the superfluid order parameter very near the surface; the study of the magnetic contributions to $R_K$ probe magnetic properties of both the substrate and the fluid very near (within a few Fermi wavelengths typically) of the surface. This is a feature of this phenomenon that has yet to be exploited experimentally, and we believe it is a most promising direction to pursue. One should shift part of the focus in this area away from the (very important) technological question of improving thermal contact between materials at very low temperatures over to the physics one can learn from the phenomenon. The literature on the theory of magnetism very near surfaces is by now a rich one, but very little data of any kind is available. We return to this point later.

Our study of the magnetic contribution to the thermal boundary conductance has been carried out for the case where the $He^3$ is coupled to the substrate via an isotropic contact interaction, and also by the full dipolar interaction. The results are as follows, where $R_K^N$ is the expression for the boundary conduction of the normal fluid, extrapolated to below the superfluid ordering temperature:

(i) The A Phase of $He^3$; Isotropic Contact Interaction

$$\frac{R_K^N}{R_K^A} = \begin{cases} 1+\frac{80}{9}\left(1-\frac{T}{T_c}\right), & T \text{ near } T_c \\ \frac{1}{9}\left(\frac{\pi T}{\Delta}\right)^2\left(1+\frac{1}{3}\ln\left(4\sqrt{6}\,\frac{\Delta}{T}\right)\right), & T\to 0 \end{cases} \tag{1}$$

(ii) The B Phase of $He^3$; Contact Interaction

$$\frac{R_K^N}{R_K^B} = \begin{cases} 1+\frac{32}{3}\left(1-\frac{T}{T_c}\right), & T \text{ near } T_c \\ 2f(\Delta)\left(1+\frac{\Delta}{6T}\,(1-f(\Delta))\ln\left(\frac{8\Delta}{T}\right)\right), & T\to 0 \end{cases} \tag{2}$$

where $f(x) = [e^x + 1]^{-1}$ is the Fermi function, and $\Delta$ is the average of the gap over the Fermi surface of the $He^3$.

(i) The A Phase of $He^3$; Dipolar Coupling

$$\frac{R_K^N}{R_K^B} = \begin{cases} 1+\frac{20}{3}\left(1-\frac{T}{T_c}\right), & T \text{ near } T_c \\ \frac{1}{9}\left(\frac{\pi T}{\Delta}\right)^2\left(1+\frac{1}{2}\ln\left(4\sqrt{6}\,\frac{\Delta}{T}\right)\right), & T\to 0 \end{cases} \tag{3}$$

(ii) The B Phase of $He^3$; Dipolar Coupling

$$\frac{R_K^N}{R_K^B} = \begin{cases} 1+\frac{32}{3}\left(1-\frac{T}{T_c}\right), & T \text{ near } T_c \\ 2f(\Delta)\left(1+\frac{\Delta}{6T}\,(1-f(\Delta)\ln\left(\frac{8\Delta}{T}\right)\right), & T\to 0\,. \end{cases} \tag{4}$$

We presume the B phase is a reliazation of the Balian-Werthamer state in these calculations.

The results displayed above show that as the temperature is lowered below the superfluid transition temperature, the thermal boundary resistance should drop below the value appropriate to the normal state, thus improving the thermal contact between the salt and the superfluid $He^3$. However, at lower temperatures, $R_K^N/R_K^S$ increases dramatically; the superfluid pairing supresses spin fluctuations by virtue of the gap opened up in the excitation spectrum, and the energy transfer efficiency is reduced as a consequence. This effect is much more dramatic in the B phase ($R_K^N/R_K^{(B)} \sim \exp[\Delta/k_BT]$) than in the A phase ($R_K^N/R_K^A \to T^{-2}\ln(\Delta/T)$). The anisotropic nature of the spin fluctuations in the superfluid A phase has the consequence that $R_K^N/R_K^A$ is sensitive to the difference between dipolar and the contact interaction. In the isotropic B phase, we find the ratio insensitive to the difference. We hope that data in the behavior of the thermal boundary resistance below the onset of superfluidity in $He^3$ will become available soon.

We wish to conclude this paper by returning again to the question of what one might learn from the study of magnetic interactions between liquid $He^3$ and magnetic substrates. As remarked above, under most circumstances (these are exceptions that have been discussed[7,8]), even if the $He^3$ interacts with substrate spins via the dipolar interaction, the effective range of the dipolar interaction is the order of the Fermi wavelength $\lambda_F$ of the $He^3$. The $He^3$ atom thus samples primarily the outermost atomic layer of spins, and conversely the spins in the solid "see" the few Angstroms of the fluid closest to the surface. Thus, the magnetic coupling may be employed in principle either to probe the liquid $He^3$ near the surface (as discussed earlier), or to probe the magnetic properties of the substrate very near the surface. There are a variety of questions one may pose in this regard. We have seen that $R_K$ below the superfluid transition in the $He^3$ may provide information about the behavior of the superfluid order parameter very near the surface. In these concluding remarks, we focus on the information on the information one may obtain about the spins in the surface of the solid. As remarked earlier, considerable theoretical attention has been devoted to this area, but very little data exists.

In an earlier paper,[7] we suggested that one may investigate questions such as this through the study of the longitudinal relaxation time ($T_1$) of the $He^3$ spins in contact with a magnetic substrate. The magnetic coupling responsible for the decreased boundary resistance at very low temperatures should also enhance the longitudinal relaxation rate of $He^3$ nuclear spins excited by an r.f. field. It is already well known that in the normal state of liquid $He^3$, $T_1$ is usually dominated by the contribution from wall relaxation. However, this phenomenon has not been studied systematically except for one recent work cited below. The study of $T_1$ offers several advantages over the study of $R_K$ as a probe of spin behavior near surfaces. The Kapitza boundary resistance contains contributions from phonons in addition to the magnetic contribution. At all but the very lowest temperatures ($T \leqslant 30$ mK) in all cases studied so far, phonons make the dominant contribution to the boundary resistance. Thus, the Kapitza resistance is a useful probe of spin behavior in surfaces only in a restricted range of temperatures. In contrast, $T_1$ is sensitive only to the magnetic coupling, and should serve as a probe of the magnetic coupling even at temperatures well above 30mK.

Following our suggestion, Saito[15] has studied nuclear resonance in $He^3$ against the salt $Ca(NH_3)_4\,SO_4\cdot H_2O$ over a temperature range that spans a phase transition in the salt. He observes a spectacular increase in signal intensity near the ordering temperature of the substrate. From this work, one can see that the study of NMR in liquid $He^3$ placed in contact with a magnetic substrate can indeed prove a powerful new probe of the behavior of spins near solid surfaces, although a detailed interpretation of the data must await the study of simpler salts and the development of a more complete theoretical structure.

As a first step in this direction, we have developed a simple theory of the contribution to $T_1$ from wall relaxation of spins in a paramagnetic substance. We conclude with a sketch of this work. While the investigation is not fully complete, a number of general conclusions are apparent.

Consider (normal) $He^3$ placed between two plane parallel walls, one at $x = -L/2$ and one at $x = + L/2$. Suppose the nuclear resonance is partially saturated by a spatially uniform r f pulse at time $t = 0$. Then for subsequent times, we suppose the longitudinal magnetization develops in time via a diffusion equation

$$\frac{\partial M}{\partial t} = D\frac{\partial^2 M}{\partial x^2} - \frac{1}{T_1^{(o)}} M \qquad (5)$$

subject to boundary conditions

$$D\frac{\partial M}{\partial x} \pm pM = 0 \quad , \quad x = \pm \tfrac{1}{2}L \ . \qquad (6)$$

Here D is the spin diffusion constant, $T_1^{(o)}$ the bulk longitudinal relaxation time, and p a "spin pinning" parameter we discuss in more detail below.

The magnetization $M(x,t)$ for $t > 0$ is given by

$$M(x,t) = M_o \exp\left[-t\left(\frac{1}{T_1^{(o)}} + \gamma\right)\right]\cos\left[\left(\frac{\gamma}{D}\right)^{\frac{1}{2}} x\right] \qquad (7)$$

where $M_o$ depends on the strength of the exciting pulse, and $\gamma$ is found from the boundary conditions. One has two limits:

(i) The limit of strong spin pinning $p >> D/L$.

Here one finds $(\gamma/D)^{\frac{1}{2}} = \pi/L$ and the observed relaxation time is

$$\frac{1}{T_1} = \frac{1}{T_1^{(o)}} + \frac{\pi^2 D}{L^2} \qquad (8)$$

independent of p. In this limit, spin flip scattering at the walls pins the magnetization to zero at the surface, with the consequence that $T_1$ is independent of p.

(ii) The limit of weak spin pinning $p << D/L$

Here we have $\gamma = 2p/L$, so the observed relaxation rate is

$$\frac{1}{T_1} = \frac{1}{T_1^{(o)}} + \frac{2p}{L} \ . \qquad (9)$$

The above considerations are straightforward, and one requires information about p in order to explore the effect of the magnetic coupling on $T_1$. In a recent investigation, we have constructed a microscopic theoretical description of p, for the model we used in our re-analysis of the Leggett-Vuorio mechanism for $He^3$ against a paramagnetic wall.[7] There are a number of striking results that we find in this work.

Before we summarize the results, we comment on the method we used to obtain them. We began by constructing an expression for spin flip scattering of a $He^3$ quasi-particle from a wall it sees as an inpenetrable barrier. A formalism developed elsewhere proves useful for this purpose.[15] Then from this we may deduce an expression for the time rate of change of the magnetization produced by the spin-flip scattering.

Our principal results are as follows:

(1) Suppose the spin Hamiltonian of the substrate is left invariant by rotation of the electron spins relative to the crystal axes. Then in this case, for dipolar coupling, p depends on the angle $\theta$ between the Zeeman field and the normal to the surface. The functional dependence of $p(\theta)$ on $\theta$ is

$$p(\theta) = p(0)[1 + \alpha_2(H)\sin^2\theta + \alpha_4(H)\sin^4\theta] \qquad (10)$$

where $\alpha_2(H) \approx 0$ for $\mu_e H << k_B T$, while $\alpha_4(H)$ remains finite as $H \to 0$, with $\mu_e$ the magnetic moment of the electron spins in the wall. For isotropic spin-spin coupling, with the same assumptions, we find p independent of $\theta$.

Thus, the study of the variation of the wall induced contribution to $T_1$ in a suitable configuration should enable one to see quite directly whether the $He^3$-electron spin coupling is truly dipolar in nature. While many theories presume the coupling is dipolar in nature, there is at present no hard experimental evidence to support this, save for the rough agreement between the calculated and measured values of the magnetic contributions to $R_K$.

(2) At $\theta = 0$, for a paramagnet, we find[16] (for dipolar coupling)

$$p(\theta) = 4\,\frac{n_s \mu_e^2 \mu_N^2 m k_F}{k_B T\,\hbar^2}\,\lambda \int_{-\infty}^{\infty} d\Omega\,\Omega\, n(\Omega)[1 + n(\Omega)] \qquad (11)$$
$$\times\ [A_\perp(\Omega) + A_\parallel(\Omega)] \ ,$$

where $n_s$ is the number of surface spins/unit area, $\mu_e$ and $\mu_N$ the electron and $He^3$ nuclear moments, $A_\perp(\Omega)$ and $A_\parallel(\Omega)$ the spectral densities of fluctuations for a single spin parallel and perpendicular to the Zeeman field, $\lambda$ a dimensionless constant of order unity, m and $k_F$ the $He^3$ mass and Fermi wave vector respectively. The spectral densities are defined as in earlier work.[7,8]

If we consider low fields where $A_\perp \cong A_\parallel$, and suppose the structure in $A(\Omega)$ is confined to frequencies low compared to $k_B T$ (valid for a paramagnet well above its ordering temperature), we find the estimate

$$p(o) = 8\,\frac{n_s \mu_e^2 \mu_N^2\, m k_F \lambda}{\hbar^4}\, k_B T \int_{-\infty}^{+\infty} \frac{d\Omega}{\Omega} A(\Omega) \qquad (12a)$$

$$= \frac{8}{3}\,\frac{n_s \mu_e^2 \mu_N^2\, m k_F \lambda}{\hbar^3}\; S(S+1) \qquad (12b)$$

The last step follows from application of a sum rule on the spectral density.

When the result of Eq. (12b) is combined with Eq. (9), we expect a temperature independent supplement to the relaxation rate from spin flip scattering from the wall. It is of considerable interest to extend this calculation to the near vicinity of the ordering temperature. We are exploring this question at the moment.

## REFERENCES


(1) See p. 425 of J. Wilks, Liquid and Solid Helium, (Clarendon Press, Oxford 1967).

(2) J. C. Wheatley, Phys. Rev. 165, 304 (1968).

(3) J. H. Bishop et al., J. Low Temp. Phys. 10, 379 (1973).

(4) A. J. Leggett and M. Vuorio, J. Low Temp. Phys. 3, 359 (1970).

(5) However, as we have pointed out, Leggett and Vuorio appear to have overestimated the effect of exchange enhancement on the calculated boundary conductance. A revised estimate (still approximate) is presented in Section III of reference (8).

(6) J. H. Bishop et al., Proc. of the Thirteenth Int'l. Conf. on Low Temp. Phys., Boulder, Colorado, 1972.

(7) D. L. Mills and M. T. Beal-Monod, Phys. Rev. A10, 343 (1974).

(8) D. L. Mills and M. T. Beal-Monod, Phys. Rev. A10, 2473 (1974).

(9) O. Avenel et al., Phys. Rev. Letters 31, 76 (1973).

(10) In addition, in the absence of impurity-impurity interactions, the temperature variation of $R_K$ depends on the ratio $T/T_K$ or $T/T_{LSF}$, where $T_K$ and $T_{LSF}$ are the Kondo and local spin fluctuation temperatures of the isolated impurity. For a dilute alloy, we expect $R_K \sim T$ for $T >> T_K$, $T_{LSF}$ (Leggett-Vuorio regime), and for $T << T_K$ or $T_{LSF}$, $R_K \sim T^{-3}$.

(11) R. A. Guyer, J. Low Temp. Phys. 10, 157 (1972).

(12) L. E. Reinstein and G. O. Zimmerman, Phys. Rev. Letters 34, 458 (1975).

(13) K. Maki, M. T. Beal-Monod and D. L. Mills, preprint entitled "Magnet Kapitza Conductance Between Liquid $He^3$ in Its Superfluid Phases and Magnetic Substances" (to be published).

(14) K. Maki and H. Ebisawa, J. Low Temp. Phys. 15, 213 (1974).

(15) E. Evans and D. L. Mills, Phys. Rev. B5, 4126 (1972).

(16) This result has been derived assuming the $He^3$ to be a gas of non-interacting quasi-particles, i.e. as in all the preceding work, we do not take explicit account of exchange enhancement effects. We have previously estimated crudely (reference 8) that one should enhance the magnetic contribution to the Kapitza boundary conductance by a factor of $\approx 25$ when exchange enhancement effects are included. The enhancement factor in $p(\theta)$ should be close in value to that which appears in $R_K^{-1}$.


# THE PHYSICAL BASIS OF GEOMAGNETISM AND GEOMAGNETIC REVERSALS


Eugene H. Levy*
Department of Planetary Sciences
Lunar and Planetary Laboratory
University of Arizona
Tucson, Arizona 85721



## ABSTRACT

Geomagnetism is the most familiar example of astrophysical magnetic field production. Our current understanding of the principles of magnetic field generation is briefly reviewed and the major outstanding problems are outlined, with particular emphasis on the Earth. The geomagnetic reversal phenomenon is an important feature of the field's behavior. Several possibilities for the origin of geomagnetic reversals are discussed.


## INTRODUCTION

In an early physical science experiment, in the 16th century, William Gilbert had a piece of lodestone machined into the shape of a sphere. Upon exploring the magnetic field produced by the spherical magnet, through the use of a small magnetized needle, he noticed a strong similarity between the magnetic action of his so called "terella" on the needle and the action of Earth's magnetism on mariners' compasses and dipping circles. As a result of this similarity, Gilbert concluded that Earth's field was produced by an internal dipole.

The systematic analysis of the geomagnetic field which was begun by Gilbert some 400 years ago has continued, with increasing sophistication, to the present day. As a result, we have a fair picture of the detailed behavior of the field during the past few hundred years. For our present purposes that picture can be summarized briefly[1]. The geomagnetic field originates within the earth (apart from small contributions from magnetospheric currents, driven largely by the solar wind, which need not concern us here). The intensity of the field at Earth's surface ranges between about one half gauss and one gauss. The field is largely dipolar; the axis of the dipole is aligned roughly with the rotation axis of Earth, deviating about ten degrees at the present time. The dipole is approximately geocentric. The intensity of the dipole part of the field has evidently decreased about five per cent in the last hundred years[2]. Superposed on the smooth dipole field are about a dozen irregular magnetic features. The intensity of this irregular part of the field is of the order of twenty per cent of the dipole field. The features are of roughly continental size, that is up to ten thousand kilometers across. The magnetic features are not stationary in time. Individual features grow and decay at rates which range up to a maximum of one or two thousandths of a gauss per year, but which are typically less than that. This data suggests that individual features appear and disappear at random intervals of about a thousand years.

The magnetic field lines are largely constrained to move with the core fluid by virtue of the fluid's high conductivity. Thus it is generally thought that the evolving magnetic features are associated with the random appearance of convection eddies in the liquid iron core of the earth. The inferred lifetime of the magnetic features suggests that the velocity of an overturning cell of core fluid is between $10^{-1}$ and $10^{-2}$cm $sec^{-1}$. In addition to the random growth and decay of the individual features, the entire pattern of magnetic irregularities drifts to the west at a rate of about 0.2° longitude per year. It is commonly thought that the drift is a consequence of relative motion between the earth's mantle and the outer layers of the core and that the core is in a state of nonuniform rotation. With this interpretation the total nonuniform rotation rate of the core is about a tenth of a centimeter per second. Bullard[3] suggested that the nonuniform rotation arises from conservation of angular momentum of the fluid in the convecting core, so that the interior of the core would rotate faster than the exterior. This is consistent with the observations.

In recent years techniques have been developed to extract information about the state of the geomagnetic field in distant, prehistoric times from the paleomagnetic record locked in old rocks[1,4]. Through the use of these techniques, a coherent picture of the gross long term behavior of the field has emerged. Paleomagnetic measurements suggest that, through much of geological history, the field has remained predominantly dipolar and that the irregular component has been similar to what it is at present. The dipole axis is, on the average, aligned with the earth's rotation axis. Deviations from exact alignment seem to be distributed randomly. So far as can be discerned the strength of the field has, on the average, been the same as it is today. However, the intensity appears to fluctuate widely about its mean value, the fluctuations having an amplitude of perhaps 50% or more and time scales of a few thousands to a few tens of thousands of years.

Perhaps the most surprising aspect of the field's behavior that has become known through the paleomagnetic studies is the geomagnetic reversal. From time to time the geomagnetic field abruptly and spontaneously changes sign. The reversal events appear to be distributed randomly in time[5] with intervals between episodes ranging from about $10^5$ to $10^7$ years. The duration of the actual reversal event is relatively short. The behavior of the field is shown schematically in Figure 1. The intensity declines smoothly to a small but non-zero value and then returns to its original average. The time for this to happen seems to be about 20,000 years. The direction of the field changes much more suddenly as is illustrated in Figure 1b. A most reasonable interpretation of the reversal data suggests that the dipole part of the field vanishes during a reversal and subsequently reappears with opposite sign. The irregular component probably remains present throughout.

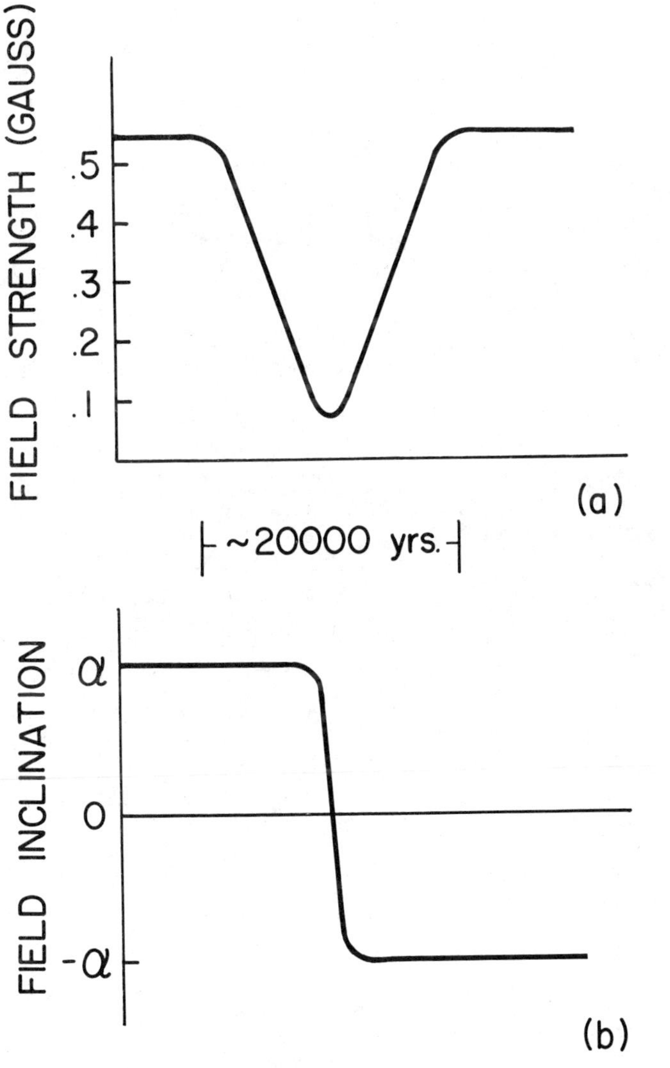


Figure 1. A schematic description of a geomagnetic reversal event. (a) The local field intensity falls rapidly to a small fraction of its usual value. (b) Very rapidly, the local field vector changes direction by 180°.

## GENERATION OF THE GEOMAGNETIC FIELD

Elsasser[6], after a careful study, pointed out that a hydromagnetic dynamo in the earth's core appeared to offer the only tenable explanation for the origin of the geomagnetic field. In a hydromagnetic dynamo, the movement of an electrically conducting fluid in the presence of a magnetic field generates the currents which produce the field. Thus the field is generated through a self-consistent, boot-strap process. The field generation process is energized by the forces which produce the fluid motions. The effect of a hydromagnetic dynamo is to amplify and maintain a magnetic field, but it does not account for the initial production of the field. In general, one must assume the prior existence of an, arbitrarily weak, seed field to get the dynamo started. Large scale, weak magnetic fields pervade the universe, evidently resulting from early cosmological processes and from the effects of galaxy formation. Additionally, on smaller scales, there are several thermoelectric and electrochemical effects which can produce the weak, initial fields in individual objects.[7]

Within a large-scale, electrically conducting fluid, Maxwell's equations reduce to the familiar hydromagnetic equation

$$\frac{\partial \vec{B}}{\partial t} = \vec{\nabla}\times(\vec{V}\times\vec{B}) + \eta\nabla^2\vec{B}; \qquad (1)$$

$\eta \equiv c^2/(4\pi\sigma)$ is known as the magnetic diffusivity, and $\sigma$ is the electrical conductivity of the fluid. The first term on the right hand side of equation 1 describes the motion of the magnetic field lines as they are carried about by the moving fluid. The second term describes diffusion of the field lines through the fluid.

The effect of a specified fluid flow on a magnetic field is given by equation 1. The first demonstration of a regenerative fluid flow was given by Parker in 1955[8]. Parker pointed out that the Coriolis force, due to the earth's rotation, caused the convective cells in the core to be cyclonic in much the same way that large atmospheric convection cells are cyclonic. He suggested that the convergence of the fluid convection caused the flow to have, on the average, right-handed helicity in the northern hemisphere of the core and left-handed helicity in the south. We will briefly review the physical arguments to show the regeneration of a magnetic field by cyclonic convection and the nonuniform rotation discussed above. This simple example will serve as a point of departure for our further discussion. The composite fluid flow is shown in Figure 2. The radius of Earth's core is approximately half that of the entire globe. The dipole field varies as $r^{-3}$, so that in the core the dipole field is about 5 gauss. The effect of the nonuniform rotation on the dipole field is illustrated in Figure 3. The magnetic field lines, constrained to move with the fluid, are stretched around to form strong bands of toroidal magnetic field. If the nonuniform rotation velocity is about 0.1cm $sec^{-1}$, as suggested by the drifting magnetic irregularities at the surface and if $\eta \simeq 10^5 cm^2 sec^{-1}$, equation 1 suggests that the toroidal field is several hundred gauss. (The toroidal field vanishes outside of the conducting core and thus cannot be observed at the surface.) Thus if there is substantial nonuniform rotation in the core, the field there is predominantly toroidal. Letting the nonuniform rotation velocity be $V_\phi$, and denoting the toroidal field by $B_\phi$ and the remaining poloidal field by $\vec{B}_p = (B_r, B_\theta, 0)$, straightforward manipulation of equation 1 reveals that the generation of $B_\phi$ is given by

$$\left(\frac{\partial}{\partial t} - \eta\nabla^2\right)B_\phi = \vec{B}_p\cdot\vec{\gamma}\ , \qquad (2)$$

where

$$\vec{\gamma} \equiv \left(\frac{\partial V_\phi}{\partial r} - \frac{V_\phi}{r}\right)\vec{1}_r + \frac{1}{r}\left(\frac{\partial V_\phi}{\partial \theta} - V_\phi \cot\theta\right)\vec{1}_\theta\ . \qquad (3)$$

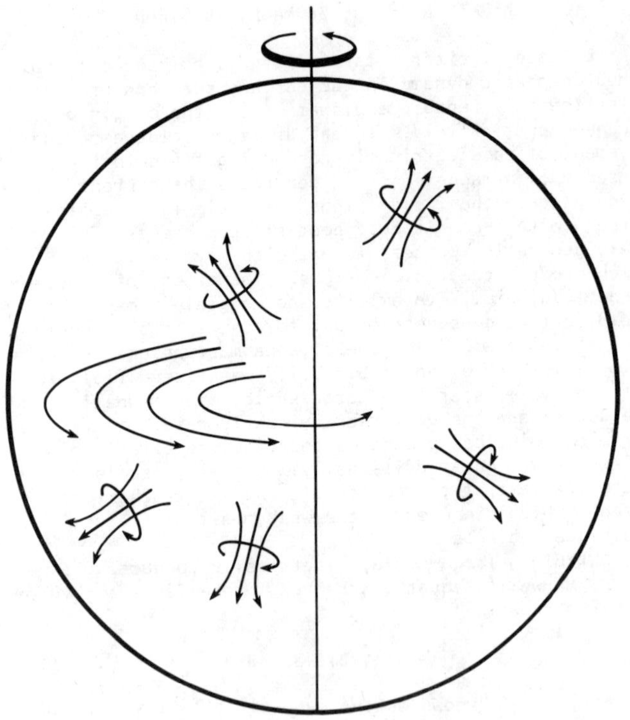

Figure 2. A simple model for the fluid motion in Earth's core consists of nonuniform rotation with angular velocity increasing inward. In addition, the convection cells are cyclonic because of the Coriolis force. The net helicity of the convection is thought to be positive in the northern hemisphere and negative in the south.

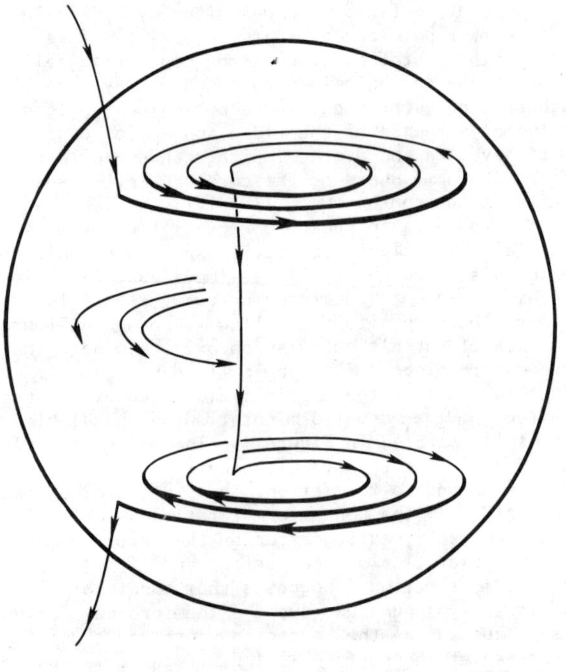

Figure 3. The nonuniform rotation stretches the dipolar field lines, $B_p$, around to generate strong bands of toroidal magnetic field $B_\phi$.

Now consider the interaction of the cyclonic convection with the strong toroidal magnetic field. The distortion and twisting of a line of toroidal field by a rising convective cell in the northern hemisphere is depicted in Figure 4. As can be seen directly from this illustration, the curl of the field in the resultant loop corresponds to an electrical current flowing along and in the direction opposite to the original toroidal field. The effect of a multitude of such

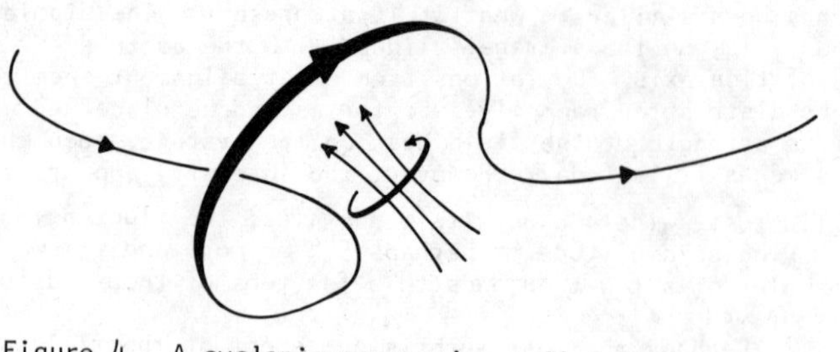

Figure 4. A cyclonic convective cell is shown distorting a toroidal field line into a meridional loop of poloidal field.

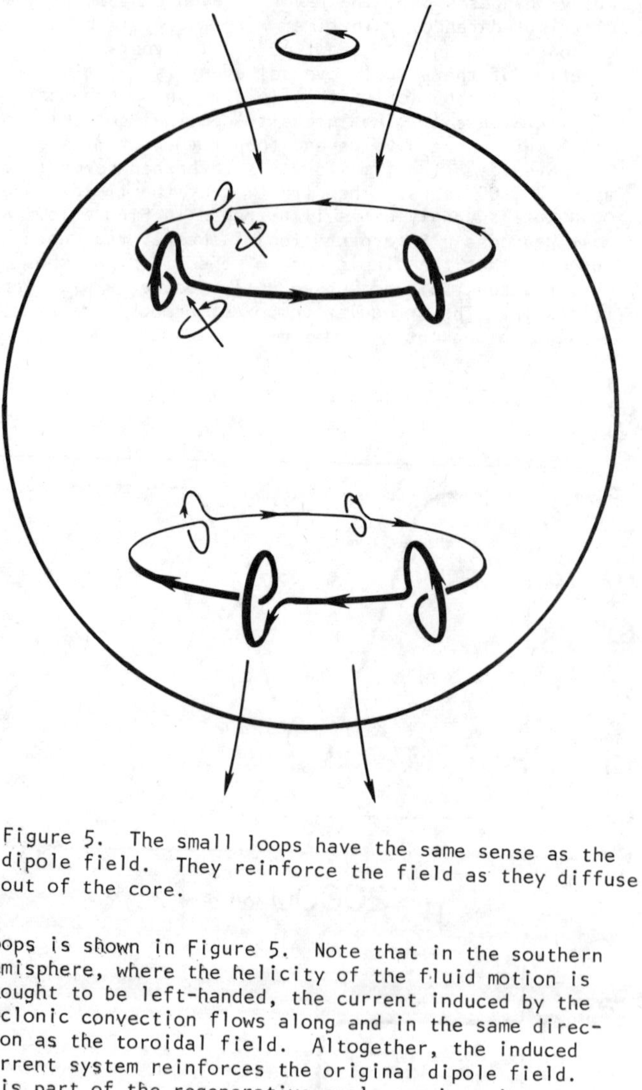

Figure 5. The small loops have the same sense as the dipole field. They reinforce the field as they diffuse out of the core.

loops is shown in Figure 5. Note that in the southern hemisphere, where the helicity of the fluid motion is thought to be left-handed, the current induced by the cyclonic convection flows along and in the same direction as the toroidal field. Altogether, the induced current system reinforces the original dipole field. This part of the regenerative cycle can be written conveniently in terms of a vector potential $\vec{A}_\phi = A_\phi \vec{1}_\phi$. where

$$\vec{B}_p = \vec{\nabla}\times\vec{A}_\phi \ . \qquad (4)$$

The resulting expression is[8,9,10,11]

$$\left(\frac{\partial}{\partial t} - \eta\nabla^2\right) A_\phi = \Gamma B_\phi \ , \qquad (5)$$

where $\Gamma$ is a suitable average of the fluid velocity in the cyclonic convective motions. Equations 2, 4 and 5 describe the entire regenerative cycle that we have discussed. While this particular type of fluid motion is not unique in its capacity to regenerate a magnetic field, the other known regenerative flows are basically similar to it. It is worth pointing out here that the overall sign of the magnetic field is not relevant to the working of the regeneration process. This follows immediately from the fact that the set of dynamo equations 2, 4 and 5 are homogeneous in the field quantities; it can also be seen from careful scrutiny of the illustrations. Indeed equation 1 shows that any hydromagnetic dynamo does not depend on the sign of the field. So far we have only considered kinematical dynamo problems, that is, generation of a field by a specified fluid flow. However, even when dynamical effects are included, the sign of the field has no effect, since the magnetic field enters the dynamical equations only through the Lorentz force $\vec{j}\times\vec{B}/c = 4\pi\vec{\nabla}\times\vec{B}\times\vec{B}/c$. Thus, in a hydromagnetic dynamo, the sign of the field is not intrinsic to the generation process, it is a result of the imposed initial conditions or subsequent accident.

Consider the dynamo equations for the case of a stationary field, where $\partial/\partial t = 0$. In dimensional form they become:

$$\frac{\eta B_\phi}{R^2} = \gamma B_p; \quad \frac{\eta A_\phi}{R^2} = \Gamma B_\phi; \quad \frac{A_\phi}{R} = B_p , \qquad (6)$$

where R is the radius of the dynamo region. In order for this set of three equations to be consistent, it is necessary that

$$N \equiv \frac{\Gamma\gamma R^3}{\eta^2} = 1 . \qquad (7)$$

The dimensionless quantity N is commonly referred to as the dynamo number; it parametrizes the solutions of the dynamo equations. Solutions of the dynamo equations reveal many dynamo states corresponding to different values of N. There is however a lower bound on N, below which no magnetic field generation can occur. This is a general consequence[12] of the hydromagnetic equation 1. The simplest sequence of dynamo states is that set corresponding to the stationary magnetic fields. These stationary fields occur at discrete values of N. In general, states corresponding to increasing values of the dynamo number have increasing contributions from higher order multipole components of the field[13]. In addition to the stationary states, there is also a class of oscillating magnetic fields[14,15,16,17].

## DYNAMICAL EFFECTS

Until now, most of the progress that has been made in our understanding of magnetic field generation in natural objects has been through the simplified kinematical approach used above, in which a fluid velocity is specified. Some advances have occurred in our knowledge of the more complete dynamical questions. The dynamical problem divides quite naturally into a number of separate questions. Although, rigorously, none of the questions can be considered independently of the others, it is convenient at our present stage of understanding to regard them as independent. The most obvious question is of course the ultimate origin of the fluid motions which generate magnetic fields. There are two concerns here. One is the energy source for the motion. Despite the clear evidence that the earth's core moves (the motion is reflected in the time varying magnetic irregularities at the surface), and despite the existence of a number of proposals, we have no satisfactory picture of the origin of the core motions at the present time. Secondly, given the existence of an energy source to drive convection, we have been unable to adequately delimit the conditions under which the convective motions acquire the character to generate a magnetic field. Although it is not rigorously proved, it seems that cyclonic motions, similar to those discussed in the previous section, are required for the generation of magnetic fields. This suggests that magnetic fields may be produced only in rotating objects. This idea is supported by both observations and existing calculations. Several linearized studies of magnetic field production in rotating objects have been carried out[10,18,19] which formally demonstrate the role of rotation in the generation process.

Another major dynamical question is that of the equilibrium intensity of naturally generated magnetic fields. For many years it has been argued[6,20,21] that the final state of naturally generated magnetic fields consists of an equipartition between the field energy density and the energy density of the fluid. There is, in fact, no reason for such an equipartition to occur in the large scale fields of astrophysical objects. The equilibrium of the field and fluid is governed by the dynamo equations together with the equation of motion of the fluid. Thus one expects the final state balance to result from equilibration between the forces which drive the fluid motion and the forces which result from the stress of the magnetic field. In the core of the earth, for example, equipartition of the fluid kinetic energy density with just the energy density of the 5 gauss poloidal field in the core would require a fluid velocity of about 0.4cm $sec^{-1}$, an order of magnitude faster than the velocities thought to occur there. On the other hand, the Coriolis force, which is responsible for the cyclonic nature of the fluid flow, is easily adequate to balance the stress of the geomagnetic field[22,23]. Again several linearized studies have been carried out on the equilibrium balance of dynamo fields[24,25] but the nonlinear problem has so far remained intractable.

The magnetic stresses, in addition to enforcing an equilibration in final states of dynamo magnetic fields, often have surprising dynamical effects. For example, it has been shown that a large fraction of stationary dynamo magnetic field states are made dynamically unstable by the action of the magnetic field stress[23].

## REMARKS

We have closely confined the preceding discussion to the geomagnetic field. However the same physical ideas carry over to magnetic field generation in a great many astrophysical objects. Since there is not the space here to discuss these objects, we refer the interested reader to some of the literature[26,27,28,29,30].

## GEOMAGNETIC REVERSALS

In principle there are two general ways by which the geomagnetic field can change its sign. The conceptually simplest reversal mechanism involves temporary interruption of the normal regeneration process. This could occur if the velocity of the core fluid falls to a value too low to maintain the regenerative process[31] or alternatively if the flow temporarily falls into a nonregenerative configuration[32,33]. Presumably after the fluctuation, when the fluid flow lapses back to its normal regenerative state, a new field is generated with its sign corresponding to whatever random, stray fields are present at that time. All such passive geomagnetic reversal schemes fail for a simple reason. Consider the extreme case when the fluid velocity falls to zero. Equation 1 shows that the magnetic field diffuses away with a characteristic e-folding time $\tau \approx R^2\eta^{-1}$. In Earth's core $\tau \sim 20{,}000$ years. Since the stray astrophysical fields near Earth are of the order of $10^{-4}$ gauss, it would take some ten e-folding times for even the 5 gauss poloidal part of the core field to fall to a value low enough to be overwhelmed. However, as we noted in the Introduction, geomagnetic reversals take place much more rapidly than that. An individual reversal event is substantially completed in about 20,000 years.

Another possibility, of course, is that the geomagnetic dynamo operates in one of the purely oscillatory modes which we mentioned earlier. The solar magnetic field seems to be an example of such an oscillatory dynamo field. However the oscillatory modes can be easily set aside. The oscillatory states have oscillation times that are typically of the order of or shorter than the resistive decay time $\tau$. In the earth's core this is about 20,000 years. By contrast

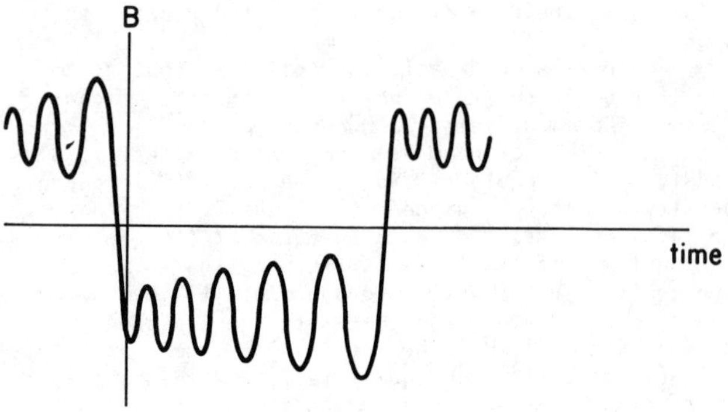


Figure 6. Geomagnetic reversals may be caused by growing dynamical oscillations of the field about its nonzero mean. When the amplitude of the oscillation grows as large as the mean, the field reverses.

the lengths of polarity intervals of the geomagnetic field are typically in the range $10^5$ to $10^6$ years. Furthermore the geomagnetic field does not have the aspect of regular oscillations. The polarity intervals have randomly distributed durations and in the time between reversal events the field is, on the average, steady.

The remaining tenable possibilities for the geomagnetic reversal mechanism are of two kinds: deterministic and stochastic. By deterministic we mean that the fluid-field system in the earth's core is in such a state as to inevitably change sign from time to time. This possibility arose originally through studies of the behavior of a pair of coupled unipolar inductors[34]. It was found that, under some conditions, the magnetic field behaved in the peculiar manner shown in Figure 6. The field oscillated overstably about a nonzero mean until the amplitude of the oscillations exceeded the mean intensity whereupon the field changed sign and the sequence repeated. Even this simple system is sufficiently nonlinear that the lengths of the mean polarity intervals may in fact be distributed randomly[35]. Such behavior has not yet been fully demonstrated for homogeneous dynamos like that in the earth's core; but a recent large scale numerical study of a simplified model has revealed an instance of overstable oscillations[36]. While the numerical convergence of these calculations has not been demonstrated, there is no reason to exclude the possibility of such behavior for the geomagnetic field. Such an overstable evolution of the field should eventually be discernible in the paleomagnetic data if it occurs; so far as we are aware the available evidence is not yet decisive.

The secular variations in the surface geomagnetic field suggest that the core convection is to some extent random. It is clear then that the geomagnetic reversal may be the result of the random character of the fluid motions. Suppose, for example, that the dynamo occasionally suffers temporary excursions into an oscillatory state[14,15] as a result of a temporary change in the convection velocity. Such a phenomenon is an obvious candidate for producing geomagnetic reversals. Temporary oscillations could also be brought about by a fluctuation in the geometrical distribution of the convection cells. If oscillatory excursions are the cause of reversals then one might expect to also see, in the paleomagnetic record, interludes during which the field oscillates for more than a half cycle. To our knowledge such events have neither been searched for nor found. A crude, oversimplified estimate[37] suggests that one might expect 5% as many full oscillations as half oscillations.

Recall from the second section that in order for the dipole field to be reinforced by the regeneration process the toroidal field must be of the correct sign (see Figure 7). There are stationary solutions to the dynamo equations which have toroidal field of both signs in each hemisphere[13]. For these states the net production of dipole field is a balance between a large

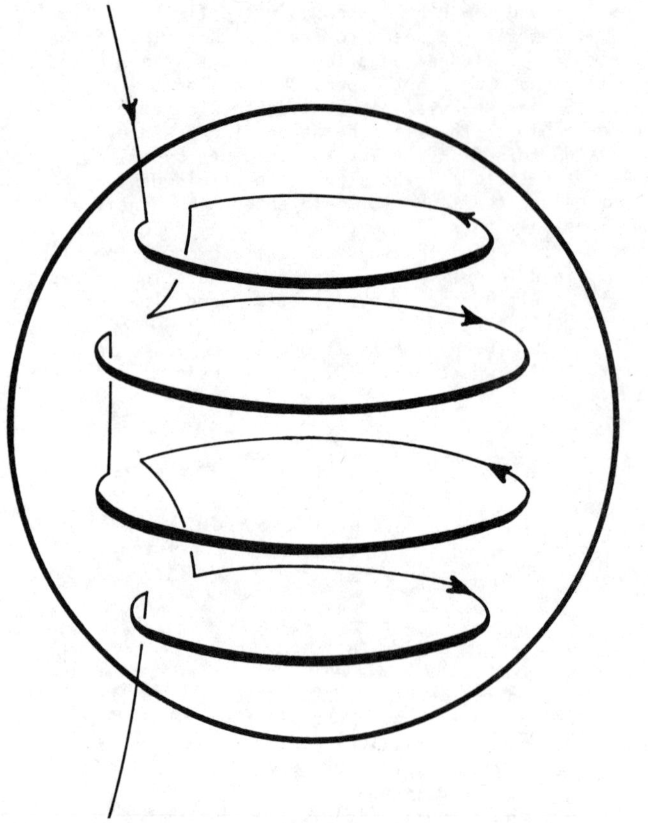

Figure 7. Some stationary magnetic field states have normal toroidal field at high latitudes, from which the dipole field is reinforced, and reverse toroidal field at low latitudes from which the field is degraded. The stationary state of the field results from a balance in favor of reinforcement.

amount of regenerative dipole flux produced from the normal toroidal field and a somewhat smaller amount of degenerative dipole flux produced from the reversed toroidal field. The reversed toroidal field typically predominates at low latitudes while, the high latitude region of the core contains normal, regenerative toroidal field. A temporary fluctuation of only thirty or forty per cent in the distribution of convection cells, so that an excess falls at low latitudes, is sufficient to change the sign of the dipole field[13]. Subsequent relaxation back to the usual convective state will then maintain the reversed field until the next fluctuation again changes the polarity. The most easily generated stationary dynamo fields do not have the broad regions of reverse toroidal flux which facilitate easy reversal of the field by fluctuations in the distribution of convection cells. Although these states can also have their polarity changed by such fluctuations, larger fluctuations are generally required[38,39].

Before concluding we should make a few remarks about the origin of fluctuations, and perhaps longer term changes, in the distribution of core convection. First by assuming a model for the core motion consisting of randomly appearing, statistically independent, convective eddies, one can estimate the appearance of fluctuations in a simple way. On this basis one expects very large fluctuations in the distribution of convection cells from time to time, with an average separation which is about right to account for the reversals[38]. On the other hand nonlinear interaction of the convection cells may render the assumption of statistical independence invalid and very large fluctuations may not occur. Even with much of the statistical independence removed, fluctuations large enough to reverse the stationary states having broad regions of reverse toroidal flux are likely to occur[13].

It has recently been realized that the body of Earth is far more dynamic than had previously been thought. In particular, the mantle, which had once been assumed to be stationary, is now known to convect. It is also thought that there may exist transient channels through which heat is convected rapidly from the core to the crust. Thus the core-mantle boundary is inhomogeneous and evolving in time. Since transport of heat through this boundary is likely to be a significant factor controlling core convection, the way is opened for variations in the convective state to be brought on by changes in the boundary conditions. These possibilities are currently in the realm of speculation.

## SUMMARY

Studies of magnetic field generation by specified fluid flows support the idea that the geomagnetic field is generated by a hydromagnetic dynamo in the molten iron core. The major outstanding problems are dynamical and concern the origin and character of the fluid flow and the effects of the magnetic stresses on the equilibrium state of the field and fluid. The geomagnetic reversal phenomenon can be understood as a result of random fluctuations in the fluid velocity or as a result of a dynamically overstable equilibrium which drives the field into reversal from time to time. The unanswered question here is whether the reversals are accidents resulting from large fluctuations in the random fluid motion or whether they are certain events which are an inevitable result of the field's equilibrium.

## REFERENCES

*Work supported in part by the National Aeronautics and Space Administration under Grant No. NSG 7168.

[1]A more detailed discussion of the history of the geomagnetic field is given by R. R. Doell, J. Appl. Phys., 40, 945(1969).
[2]R. R. Doell and A. Cox, Advan. Geophys., 8, 221 (1961).
[3]E. C. Bullard, Proc. Roy. Soc. London, 197A, 433 (1949).
[4]See the review by S. K. Banerjee, J. Appl. Phys., 41, 966 (1970).
[5]A. Cox, J. Geophys. Res., 73, 3247 (1968); Science, 163, 237 (1969); J. Geophys. Res., 75, 7501 (1970).
[6]W. M. Elsasser, See: Rev. Mod. Phys., 22, 1 (1950) for references.
[7]W. M. Elsasser, Phys. Rev., 55, 489 (1939); D. R. Inglis, Rev. Mod. Phys., 27, 212 (1955).
[8]E. N. Parker, Astrophys. J., 122, 293 (1955).
[9]E. N. Parker, Astrophys. J., 162, 665 (1970).
[10]M. Steenbeck, F. Krause, and K.-H. Rädler, Z. Naturforsch, 21a, 369 (1966).
[11]S. I. Braginskii, Soviet Physics JETP, 20, 726 (1965).
[12]G. E. Backus, Ann. Phys., 4, 372 (1958).
[13]E. H. Levy, Astrophys. J., 175, 573 (1972).
[14]S. I. Braginskii, Geomag. i Aeron., 4, 732 (1964).
[15]E. N. Parker, Astrophys. J., 164, 491 (1971).
[16]P. H. Roberts, Phil. Trans. Roy. Soc. London, 272A, 663 (1972).
[17]W. Deinzer, H.-U. v. Kusserow and M. Stix, Astron. and Astrophys., 36, 69 (1974).
[18]H. K. Moffatt, J. Fluid Mech., 44, 705 (1970).
[19]D. Gubbins, S. I. Appl. Math., 53, 137 (1974).
[20]G. K. Batchelor, Proc. Roy. Soc. London, 201A, 405, (1950).
[21]S. Chandrasekhar, Proc. Roy. Soc. London, 204A, 435, (1950).
[22]E. C. Bullard and H. Gellman, Phil. Trans. Roy. Soc. London, 247A, 213 (1954).
[23]E. H. Levy, Astrophys. J., 187, 361 (1974).
[24]H. K. Moffatt, J. Fluid Mech., 53, 385 (1972).
[25]F. H. Busse, Geophys. J. R.A.S., 42, 437 (1975).
[26]E. N. Parker, Astrophys. J., 160, 383 (1970); Ibid., 163, 255 (1971).
[27]S. I. Vainshtein and A. A. Ruzmaikin, Sov. Astr. - A. J., 15, 714 (1972).
[28]E. H. Levy and W. K. Rose, Nature, 250, 40 (1974)
[29]E. H. Levy and W. K. Rose, Astrophys. J., 193, 419, (1974).
[30]E. H. Levy, Ann. Rev. Earth Planet. Sci., 4, (1976), in press.
[31]H. Takeuchi, J. Phys. Earth, 4, 11 (1956).
[32]T. Nagata, J. Geomag. Geoelectr., 21, 701 (1968).
[33]F. E. M. Lilley, Nature, 227, 1336 (1970).
[34]T. Rikitake, Proc. Cambridge Phil. Soc., 54, 89 (1958).
[35]T. Rikitake, Phys. Earth Planet. Interiors, 6, 340, (1972).
[36]D. Gubbins, Geophys. J. R.A.S., 42, 295 (1975).
[37]E. H. Levy, unpublished.
[38]E. N. Parker, Astrophys. J., 158, 815 (1969).
[39]E. H. Levy, Astrophys. J., 171, 621 (1972); Ibid., 171, 635 (1972).

## GROWTH-INDUCED ANISOTROPY

**H. Callen**
**University of Pennsylvania, Philadelphia, Pennsylvania 19174**

ABSTRACT

The occurrence of a non-cubic anisotropy, related to the orientation of the growth face during crystal formation, is vital to garnet bubble memories; the effect is also present in amorphous GdCo bubble films. In addition, the effect has great fundamental interest as the only trace in the bulk material of the dynamics of the crystal growth process. The detailed mechanisms of the growth-induced anisotropy will be described, with experimental evidence for each of the mechanisms in particular systems. All of the mechanisms are subject to the same symmetry considerations which, in turn, determine the symmetry properties of the induced anisotropy. Hence, symmetry does not distinguish among the mechanisms. Simple parametrization of the symmetry theory gives rise to the "two-parameter model". Experimental data increasingly deviates from this simple representation. The dynamics of the growth process on the crystal face will be discussed in some detail, with particular emphasis on the sources of the deviations from the two-parameter model.

## TUTORIAL ON MAGNETIC RECORDING THEORY

J.C. Mallinson
Ampex Corporation, Redwood City, Ca. 94063

ABSTRACT

In this talk, the more fundamental issues confronting magnetic recording theory will be reviewed in a fashion which should be of interest to both recording specialists and other magneticians. The absence of a usable, microscopic, vector model of the recording medium magnetization is the principal theoretical problem in both recording and other magnetic studies. M-H curve fitting or, better, the Preisach function approach[1] yields only limited information on the recorded signal and none on the medium noise. In contrast, the theory of recording head fringing fields appears to be nearly complete; recent results on narrow pole-tip heads[2] and on pole-tip partial saturation effects[3] will be reviewed. After a brief discussion of theory of the anhysteretic[4] and thermo-remanent processes, some results of large scale modeling[5] of the write process will be presented. A salient finding of these experiments is a rotating vector magnetization pattern which may not be understood fully without a vector M-H prescription; the rotating magnetization patterns give rise to curious symmetries[6] of the medium flux. Whereas the minimum additive noise due to the particle statistics in the medium may be deduced from first principles[7], multiplicative (or modulation) noises due to physical aggregations or magnetic collective behavior of particles presently defy simple prediction. It is possible to conclude, however, that provided such aggregations are small in comparison with the trackwidth, the signal-to-noise ratio (SNR) will decrease by only 3 dB per halving of the trackwidth but by 6 dB per halving of the wavelength. More information per unit area can, therefore, be recorded by using narrow tracks and low linear densities than with wide tracks and high linear density. With current $\gamma$-$Fe_2O_3$ media, the SNR8 limited areal densities are somewhat above $10^7$ bits/sq. in.[8], a factor of ten higher than in current practice.

1. Schwantke, G., Audio Eng. Soc. Jour., 9, 37-47 (1961).
2. Lindholm, D.A., IEEE Trans. MAG-11 (Sept. 1975).
3. Goron, A.I., Telecommunications and Radio Eng. 6, 28/29, 123-25 (1974).
4. Bertram, H.N., Journal De Physique, 32 (Supplement), C684-5 (Feb.-Mar. 1971).
5. Tjaden, D.L.A., Philips Tech. Rev. 25, 319-329 (1963-64).
6. Mallinson, J.C., IEEE Trans. MAG-9, 4, 678-82 (1973).
7. Mallinson, J.C., IEEE Trans. MAG-5, 3, 182-86 (1969).
8. Mallinson, J.C., IEEE Trans. MAG-10, 2, 368-73 (1974).

# THE APPLICATION OF JOSEPHSON JUNCTIONS TO COMPUTER STORAGE AND LOGIC ELEMENTS AND TO MAGNETIC MEASUREMENTS

John Clarke
Department of Physics, University of California, and
Materials and Molecular Research Division,
Lawrence Berkeley Laboratory, Berkeley, California 94720

ABSTRACT

The dc and ac Josephson effects and the behavior of the current-voltage characteristics of Josephson junctions are briefly reviewed. The Josephson junction cryotron is described. Two cryotrons have been incorporated into a NDRO memory cell with a switching time of 80ps, a power-delay product of $10^{-16}$J, and an area of $900\mu m^2$. DRO memory cells involving the storage of a single flux quantum have estimated switching speeds of 50ps and a dissipation in continuous operation of 40nW. Latching logic elements with a switching time of 200ps and a power-delay product of 5fJ have been built. Non-latching devices and the flux shuttle are described. The use of dc and rf SQUIDs in conjunction with flux transformers for magnetic measurements is discussed. Presently achieved sensitivities are: magnetic field, $10^{-15}THz^{-1/2}$, magnetic field gradient, $10^{-13}Tm^{-1}Hz^{-1/2}$; magnetic susceptibility, $2x10^{-11}$ e.m.u. for $1cm^3$ at $10^{-1}$T; voltage, $10^{-15}VHz^{-1/2}$. The highest energy resolution reported for a SQUID-transformer combination is $8x\ 10^{-30}JHz^{-1}$ in a frequency range from $2x10^{-2}$Hz to 1kHz. At lower frequencies the noise power spectrum is 1/f. The performance of the computer and measuring devices is compared with that of conventional (non-superconducting) devices.

## INTRODUCTION

In this article I shall briefly review the use of Josephson junctions as computer elements and as magnetometers for the measurement of magnetic fields, magnetic field gradients, magnetic susceptibilities, and voltages. These two areas of application are at very different stages of development. A variety of ingenious and elegant Josephson devices have been operated as computer storage and logic elements, and their very high speed and very low power dissipation have been clearly demonstrated. The potential of these devices in future computers seems to me to be extremely promising, but of course (to my knowledge) no superconducting computer has yet been built. Thus this field is still in a state of rapid development and innovation. I have tried to list most of the important devices that have appeared in the literature and to give a reasonably complete set of references. On the other hand, Josephson junction magnetometers are well-established instruments, and are commercially available. The literature on this subject is vast, and I have confined the references to a representative selection that should enable the interested reader to find his way into the field.

## JOSEPHSON JUNCTIONS

The classic Josephson tunnel junction[1,2,3] consists of two superconductors separated by an insulating barrier typically 20Å to 30Å thick through which both single ("normal") and paired ("superconducting") electrons can tunnel. Junctions are fabricated by evaporating or sputtering a strip of superconductor onto an insulating substrate, oxidizing the strip, and depositing a second film of superconductor[ Fig. 1(a)]. In most practical applications it is desirable for the transition temperature, $T_c$, of the superconductor to be a factor of two or more higher than the boiling temperature of liquid helium, 4.2K. At 4.2K the energy gap, $\Delta$, of the superconductor is then close to its zero temperature value (about 1.4mV in Pb), and is relatively independent of temperature. A typical current-voltage (I-V) characteristic is shown in Fig. 1(b) for $T \lesssim T_c/2$. As the current I through the junction is increased from zero, no voltage is developed for $I < I_m$, the critical current. In this zero voltage regime, the current is carried by superfluid electron pairs that tunnel through the barrier. At $I = I_m$ the voltage jumps discontinuously to a value of approximately $2\Delta/e$. As I is increased and then reduced again, the voltage does not return to zero at $I = I_m$, but remains at approximately $2\Delta/e$ until a transition occurs at a relatively small value of current.[4-6] The amount of hysteresis is determined by the value of the parameter $\beta_c = 2\pi I_m R^2C/\phi_0$(R is the normal state resistance of the junction, C is the junction capacitance, and $\phi_0 = h/2e \approx 2x10^{-15}$Wb is the flux quantum). If the junction is resistively shunted so that $\beta_c \lesssim 1$, the hysteresis vanishes, and the I-V characteristic of Fig. 1(c) is obtained.

For $T \lesssim T_c/2$, $I_m = \alpha\pi\Delta/2eR$[7], where $\alpha$ is usually between 0.5 and 1.0. Critical current densities as high as $30kAcm^{-2}$ have been reported.[8] If a magnetic field H is applied parallel to one of the strips in Fig. 1(a), the critical current is modulated according to the relation[3] $I_m(H) = I_m(0)|\sin(\pi\phi/\phi_0)/(\pi\phi/\phi_0)|$, provided the width of the strip w is small compared with the Josephson penetration depth[1,2] $\lambda_j = [\phi_0/2\pi\mu_0(2\lambda + d)j_m]^{1/2}$. Here, $\lambda$ is the penetration depth of the superconductor (~500Å), d is the barrier thickness, $j_m$ is the critical current density, and $\phi = Hw(2\lambda + d)$. In the limit $\lambda_j \ll w$, the tunneling supercurrent no longer flows uniformly through the junction but is confined to a region of width $\sim\lambda_j$ near the edges of the junction. The junction is said to be "self field limited." In the same way, an external magnetic field is screened by supercurrents flowing in the Josephson penetration depth. As we shall see later, the modulation effects of the magnetic field are considerably modified.

At non-zero voltages, the supercurrent oscillates at a frequency[1] $f = V/\phi_0$. This oscillation persists out to voltages of at least $2\Delta/e \sim 10^{12}$ Hz in lead or niobium at 4.2K. One of the manifestations of the ac Josephson effect is the presence of constant voltage current steps[9] on the I-V characteristic. These steps occur when the frequency of

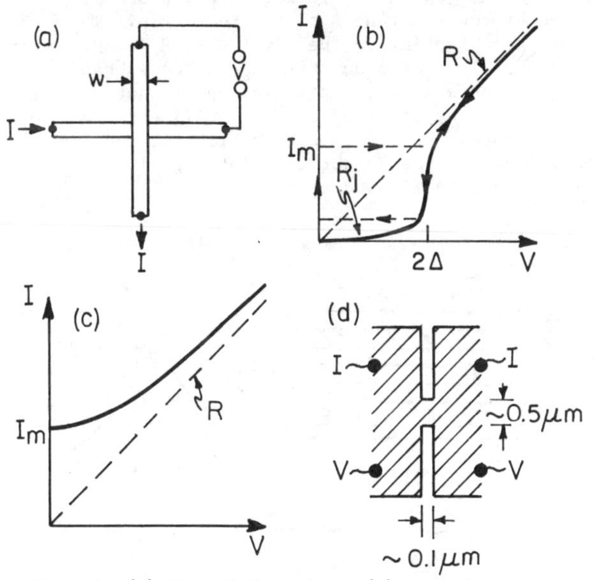


Fig. 1 (a) Tunnel junction; (b) I-V characteristic of tunnel junction with identical superconductors; (c) I-V characteristic of shunted tunnel junction; (d) Anderson-Dayem bridge.

the Josephson current matches the frequency of one at the electromagnetic modes of the cavity formed by the junction. The lower limit for frequencies that can propagate is the plasma frequency[1] $\omega_j = (2\pi j_m d/\phi_0 \varepsilon)^{1/2}$ ($\varepsilon$ is the barrier dielectric constant).

The materials most commonly used for practical devices are a lead alloy ($T_c$~7K to 8K) and niobium ($T_c \approx 9.5$K). Sophisticated techniques for fabricating Pb-PbOx-Pb junctions have been developed in which the oxide barrier is grown by an rf discharge in oxygen.[10-13] Nb-NbOx-Pb[14] and Nb-NbOx-Nb[15] junctions have also been made, while Van Duzer and co-workers have successfully fabricated Pb-Te-Pb[16] and Pb-Si-Pb[17] junctions. Photoresist techniques[18] have been used to produce evaporated lead strips with widths of about 1μm. The stability of these junctions during prolonged storage at room temperature and thermal cycling between room and helium temperatures is reported to be excellent. All of the devices for computer applications have utilized tunnel junctions. Magnetometers currently use tunnel junctions, point contacts, or bridges. The point contact[19] consists of a sharp niobium point pressed against a niobium block. In the Anderson-Dayem bridge[20] [Fig. 1(d)], a bridge of submicron dimensions is fabricated in a film of superconductor. In one version[21], the bridge is "weakened" by a normal metal underlay or overlay that lowers its transition temperature.

COMPUTER APPLICATIONS

The possibility of using a Josephson junction as a cryotron for high speed low dissipation storage and logic computer elements was first suggested by Matisoo.[22] Subsequently Anacker[23] designed a random access memory module based on these devices with non-destructive readout. Anderson[24] and McCumber[25] proposed the use of junctions for the control and movement of single flux quanta. After describing experiments to determine the switching speed of a Josephson junction I will discuss the principle of the cryotron, and its use as a memory cell and logic element. I will then briefly describe devices based on single flux quanta.

Matisoo[22,26] measured the risetime of transitions from the zero voltage state to the single-particle tunneling state [Fig. 1(b)] using 125μm square and 250μm square junctions, and found an upper bound of 800ps. Subsequently Zappe and Grebe[27] measured risetimes of 60ps using 100x125μm junctions. Jutzi et al.[28] estimated a risetime of less than 24ps in a 8.5x20μm junction. This time was probably limited by the reactances of the test chip. The theoretical risetime in the limit $2\Delta \ll I_m R_j e$, approximately[28] $C\Delta/eI_m$, was about 6ps ($R_j$ is the subgap tunneling resistance). It would appear that switching times on the order of a few ps will eventually be obtainable. However, as we shall see, the junction switching time is not the limiting time in logic devices fabricated so far.

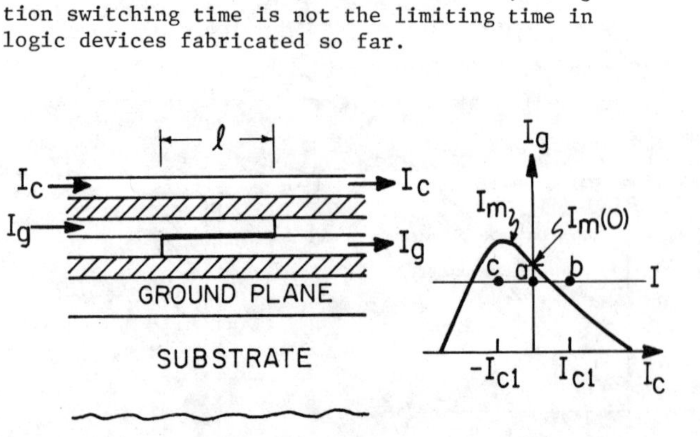


Fig. 2 (a) In-line junction cryotron; (b) $I_m$ vs. $I_c$ for in-line cryotron.

A junction may be used as an in-line cryotron[22,23] [Fig. 2(a)], in which a given control current $I_c$ switches a larger gate current $I_g$. A superconducting ground plane is deposited on a substrate, followed by an insulating layer. The in-line junction is fabricated on the insulator, and coated with a second insulating film. Finally a control line is evaporated over the junction. The length of the junction ℓ (the overlap of the two superconducting films) is larger than $\lambda_j$, so that the junction is self-field limited. The variation of junction critical current $I_m$ with control current in the upper film is then asymmetrical as shown in Fig. 2(b). The self-field generated by the gate current significantly depresses the critical current of the junction. If the control and gate currents flow in opposite directions there is a value of control current for which the field in the junction is zero, and for which the critical current is a maximum. The curve shown in Fig. 2(b) represents a "skewed" version of the central peak of the diffraction pattern for the non-self-field limited junction, and in practice higher order maxima are also observed. The shape of the $I_m$-$I_c$ curve has been investigated in great detail.[29]

If the values of $I_g$ and $I_c$ are chosen to be inside the curve in Fig. 2(b), the junction will be in the zero voltage state; if they are outside, the the junction will be in the finite voltage state. Suppose that the junction is initially in the zero voltage state ("a") with a gate current I and $I_c = 0$. If a control current $I_{c1}$ is now applied, the critical current of the junction will fall below I, and the junction will switch to a finite voltage ("b"). If instead a control current $-I_{c1}$ is applied, the junction remains in the zero voltage state ("c").

Two cryotrons can be used to make a memory cell cell[23] (Fig. 3) that has nondestructive readout and that can be incorporated into a random access memory. The two cryotrons or write gates (WG), which ideally have identical characteristics, are mounted symmetrically on a superconducting loop to which a word line is connected. A bit line overlays the two junctions. A "1" or a "0" is stored respectively as a clockwise or anticlockwise persistent circulating supercurrent . Suppose that initially there are no circulating or applied currents, and that one wishes to write a "0". A current pulse $I_w$ is applied to the word line so that, because of symmetry, a current $I_w/2$ flows through each junction. $I_w$ is below the critical current $I_m(0)$ of either junction. Simultaneously a bit current pulse $I_b$ is applied to the bit line in the left-to-right direction. The magnetic fields generated by $I_w/2$ and $I_b$ cancel at WG1 but add at WG0 so that its critical current is reduced

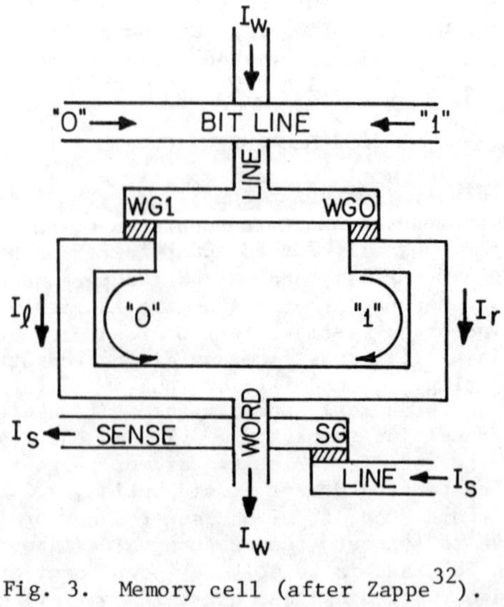


Fig. 3. Memory cell (after Zappe[32]).

to below $I_w/2$. WGO thus makes a transition into the resistive state. The current that was carried by WGO is transferred to the left hand arm of the cell until the current flowing through WGO falls below the value required for the junction to return to the zero voltage state. The cell becomes superconducting again with $I_r \approx 0$ and $I_\ell \approx I_w$. When the bit and word currents are removed, the flux trapped in the cell, approximately $(L/2)I_w$, is conserved. Thus a circulating supercurrent of about $I_w/2$ is established in an anticlockwise direction to represent a stored "0".

Suppose a subsequent attempt is made to write a "0". The word current establishes currents $I_\ell \approx I_w$ and $I_r \approx 0$. The field generated by the application of $I_b$ is insufficient to drive either junction out of the zero voltage state, and on removing the word and bit currents, the "0" remains. On the other hand, suppose that one wishes to write a "1". A bit current in the right-to-left direction causes WG1 to make a transition into the resistive state, while WGO remains in the zero voltage state. The word current is thus transferred into the right arm of the cell. Upon removal of the bit and word pulses, a clockwise supercurrent of approximately $I_w/2$ remains, representing a "1". It is important to realize that no transitions occur unless the word and bit currents are applied simultaneously. Thus a single cell may be addressed from a large array of memory cells with common word and bit lines.

Circulating currents in the memory cell are control currents for the sense gate (SG). A current pulse $I_s$ is applied to the sense line. In the absence of a word current pulse, a circulating current ($\sim I_w/2$) of either polarity generates insufficient field for SG to switch to the resistive state. However suppose that a word current pulse is applied simultaneously with $I_s$. If the cell contains a "0", $I_r \approx 0$ and the SG remains in the zero voltage state. On the other hand, if the cell contains a "1", $I_r \approx I_w$, the critical current of SG is exceeded, and the resultant voltage is detected. The sense operation does not affect the current stored in the cell, so that readout is non-destructive. It is evident that the sensing operation can be readily performed in an array.

To make a working memory cell, there are a number of practical problems that must be overcome. For example, self-induced steps[9] can disrupt the switching process. It is important that the LCR circuit formed by one junction and the superconducting loop during the switching process not be underdamped.[6,30,31] Zappe[6] has shown that critical damping occurs when $2R_j = (L/C)^{\frac{1}{2}}$, and has discussed the choice of the various circuit parameters and their tolerances. Zappe[32] has operated cells with junctions of size 110x85μm with $C \approx 220$pF, $R_j \approx 0.25\Omega$, and $L \approx 37$pH, and achieved a writing time of 600ps with a repetition rate of 1GHz. The energy dissipated during each write cycle was less than $2\times10^{-15}$J. Broom et al.[8] have operated a considerably smaller cell with junctions of size 4x5μm and an overall cell size of about 900μm$^2$. They achieved a writing time of less than 80ps and an energy dissipation of less than $10^{-16}$J per writing cycle.

Logic circuits based on the cryotron have been demonstrated by Henkels[33] and Herrell.[34] A circuit for AND, OR, INVERT, and CARRY functions is illustrated in Fig. 4(a). The tunneling gate is connected to two superconducting strip lines of characteristic impedance $Z_o$ terminated by a matching resistor $R = 2Z_o$. Thus when the junction is switched to a voltage $2\Delta$, a current $2\Delta/eR$ is established in the strip line, and can be read by a sense junction (not shown). The gate junction has three control lines A, B, and C. In operation, the junction is biased with a current I in the zero voltage state at "a" in Fig. 4(b). The application of a current i in the same direction as I to any one of the control lines A, B, or C causes the junction to switch to a non-zero voltage state ("b"). This is the OR function. A current i opposed to I in one or two of the control lines will not switch the gate ("c" or "d"), but in all three lines will switch the junction to the resistive state ("e"): This is the AND operation.

Henkels[33] and Herrell[34] have reported a logic delay of less than 200ps and a power-delay product of about 5fJ, representing a power dissipation of about 20μW. More sophisticated circuits have also been operated. Herrell has reported a one-bit adder[35] with a 500ps delay time, and four-bit multiplier[36] with a multiplication time of 27ns, and Yao and Herrell[37] have built an eight-bit shift register.

These logic devices are latching: The gate junction cannot be returned to the zero voltage state by removing or reversing the control current, but only by momentarily switching off the junction bias current. This may be a serious disadvantage. However, Chan et al.[38] have proposed a logic gate that automatically unlatches after each operation. Baechtold et al.[39,40] have reported a non-latching circuit involving two junction gates that has a switching time of 60ps and a dissipation of about 17μW. Zappe[41] has operated a three-junction logic device that can be operated in both latching and non-latching modes. With hysteretic junctions, the device operates in a latching mode in essentially the same way as the single-junction logic element. However, by resistively shunting the junctions, Zappe[41] eliminated the hysteresis and obtained non-latching operation with a logic delay of 235ps and a power dissipation of <40nW in continuous operation.

The idea[24,25] of using a single flux quantum $\phi_o$ in memory and logic elements was first exploited by Fulton and co-workers in the flux shuttle.[42-44] Two basic configurations are possible. In one, a self-field limited junction ($\ell \gg \lambda_j$) supports one (or more) $\phi_o$ [Fig. 5(a)] by means of circulating supercurrents.[1,2,45,46] In the other, a flux quantum is stored in a superconducting ring containing two junctions.[42,43,47,48]

The flux shuttle of Fulton and Dunkleberger[44] [Fig. 5(b)] consists of a film with crossarms evaporated onto a wider oxidized film to form a long tunnel junction. Suppose a current is passed between leads 4 and 5 (for example). The current generates a field in the junction that attracts

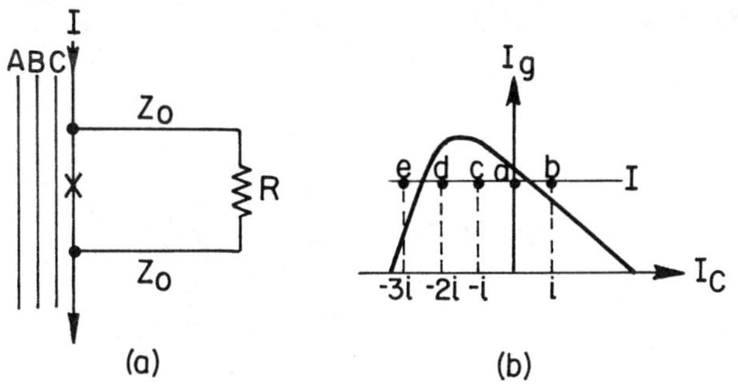


Fig. 4 (a) Logic circuit; (b) $I_g$ vs. $I_c$ for gate (after Herrell[34]).

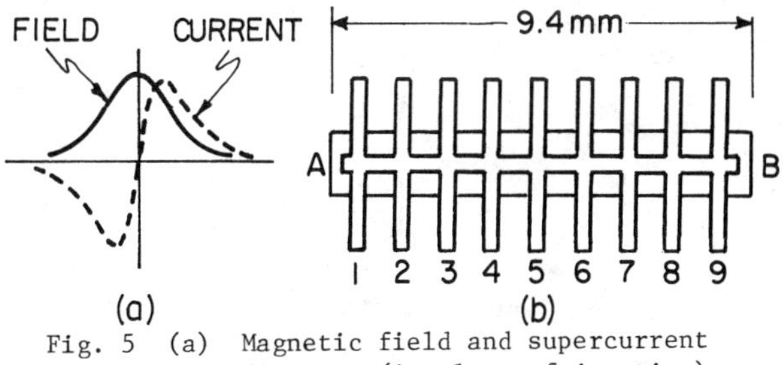


Fig. 5 (a) Magnetic field and supercurrent profiles vs. distance (in plane of junction) across vortex; (b) flux shuttle (after Fulton and Dunkleberger[44]).

(repels) flux quanta of the same(opposite) field sense, thus providing a potential well(hill). A shift register was operated in the following mode. Pairs of leads attached to 6-7, 5-6, 4-5, and 3-4 were each fed a sinusoidal current with relative phases 0°, 140°, 270°, and 360°. A flux quantum that originated in the well 6-7 when that well was deepest was then transferred successively to wells 5-6, 4-5, and 3-4 in one clock cycle. The flux quantum was written when the well at 6-7 was deepest by applying a short pulse to leads 7-9, the current in lead 7 opposing the clock current. The pulse introduced several flux quanta into the wells 6-9. The pulse amplitude was chosen so that when the pulse ended a single $\phi_0$ remained in well 6-7. This quantum was then successively shifted along the register to well 3-4. As the clock advanced further, the well at 4-5 became a potential hill, while the well depth at 3-4 became zero. The flux quantum was thus expelled towards A. A bias current I was applied to leads 1-A whose magnitude was somewhat below the critical current of the junction formed by the overlap of crossarm 1 and the lower strip. The arrival of the flux quantum lowered the critical current below I, and a detectable voltage appeared across the junction.

The flux shuttle has been tested only at 300-700Hz. However, it is believed that operation at frequencies approaching the Josephson plasma frequency ($10^{10}$Hz to $10^{11}$Hz) should be possible. The energy dissipation per shift is of order $I_c\phi_0 \sim 10^{-18}$J for $I_c \sim$ 1mA.

Gueret[48,49] has operated a single self-field limited junction as a memory cell in which a "0" and a "1" are represented by the absence and presence of one flux quantum respectively. A ring containing two junctions has been used as a single flux quantum memory cell.[47,48] Zappe[47] obtained an estimated switching time of about 50ps. The single quantum cells have destructive readout.

The performances of a number of Josephson memory and logic elements are summarized in Table I. For comparison, approximate values for several semiconductor devices are included in the table. The Josephson devices offer the combined advantages of high speed and low dissipation. For example, a buffer memory with a density of $10^4$ NDRO Josephson cells[8] per $cm^2$ would have a dissipation of only 10mW $cm^{-2}$ even if all cells were in continuous operation; this heat load can be easily handled by liquid helium.

## MAGNETIC MEASUREMENTS

The basic device for the measurement of magnetic fields is the SQUID (Superconducting QUantum Interference Device). There are two versions: The dc SQUID and the rf SQUID. The dc SQUID was the first to be developed, but subsequently the rf SQUID became the more popular, and is now available commercially from several companies. There are several review articles that describe both types of SQUID in detail.[50-53]

| Device | Switching speed (ps) | Power x delay (J) |
|---|---|---|
| NDRO memory[8] | 80 | $10^{-16}$ |
| Single $\phi_o$ memory[47] | 50 | $10^{-18}$ |
| Latching logic[34] | 200 | $5\times10^{-15}$ |
| N-L complementary[39] | 60 | $10^{-15}$ |
| N-L 3-jn.[41] | 235 | $10^{-17}$ |
| Flux shuttle[43](expected) | 10-100 | $\sim 10^{-18}$ |
| TTL | $10^4$ | $10^{-10}$ |
| EFL | 3,000 | $3\times10^{-12}$ |
| n-MOS | $10^5$ | $10^{-11}$ |
| IIL | $10^4$ | $10^{-13}$ |

Table I Approximate switching times and power-delay products for Josephson and semiconductor devices (semiconductor values from ref. 34 and T. van Duzer, private communication. N-L = non-latching).

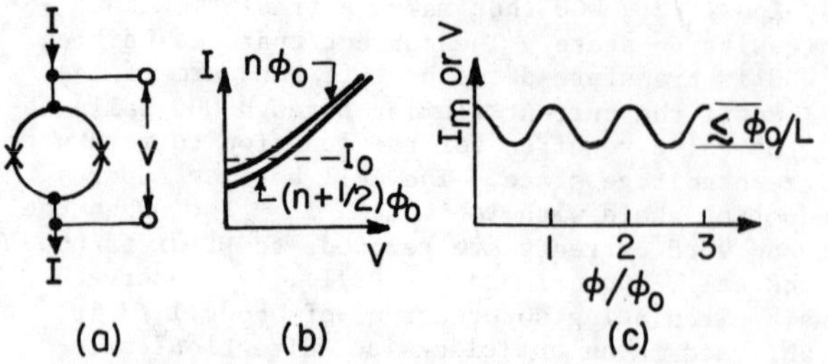


Fig. 6 (a) Dc SQUID; (b) modulation of I-V characteristic by magnetic flux; (c) modulation of $I_m$ or V by magnetic flux.

The dc SQUID[54] consists of a superconducting ring on which are mounted two junctions [Fig. 6(a)] with non-hysteretic I-V characteristics. A dc current $I_0$ biases the SQUID at a non-zero voltage [Fig. 6(b)]. As the magnetic flux threading the ring is slowly changed, the critical current of the two junctions $I_m$ oscillates with a period $\phi_0$ [Fig. 6(c)], being a maximum for $\phi = n\phi_0$ and a minimum for $\phi = (n + ½)\phi_o$ (n is an integer). The voltage across the SQUID at a constant current bias is correspondingly periodic in the applied flux. The most sensitive dc SQUID is that of Clarke, Goubau, and Ketchen.[55] Their SQUID has a cylindrical geometry and involves two shunted Nb-NbOx-Pb junctions evaporated on a 3mm o.d. quartz tube. Typical parameters are: SQUID inductance ~1nH; parallel resistance of shunts 0.4Ω; $I_m \approx 5\mu$A (both junctions); $\beta_c \approx 0.5$; voltage modulation depth, $V[(n + ½)\phi_o] - V[n\phi_o]$, ~0.5μV.

In practice, a flux change of much less than $\phi_0$ can be detected by using the SQUID as a flux null detector in a feedback circuit. A 100kHz flux modulation is superimposed on the quasistatic flux $\phi$ in the SQUID. The amplitude of the 100kHz voltage across the SQUID is then periodic in $\phi$. This voltage is amplified by an LC resonant circuit with a superconducting coil, further amplified by conventional (room temperature) electronics, and lock-in detected. The output of the lock-in is also periodic in the applied flux. This output is fed back as a current into a coil inside the SQUID to produce a flux that cancels any flux applied to the SQUID. The feedback current is proportional to the flux change applied to the SQUID. The dynamic range of the closed-loop system is $\sim 10^6\phi_0$, and the maximum slewing rate is $\sim 10^5\phi_0 sec^{-1}$. Noise and drift measurements were made with the SQUID in a superconducting shield that screened out external trum is white from $2\times10^{-2}$Hz to about 2kHz, with an rms value of typically $3.5\times10^{-5}\phi_0 Hz^{-1}$, corresponding to a magnetic field resolution of about $10^{-14}THz^{-1/2}$. This noise limit is set by the Johnson noise in the junction shunts. Below $2\times10^{-2}$Hz, the noise power spectrum is approximately 1/f. The long term drift is typically $2\times10^{-5}\phi_0 h^{-1}$.

The rf SQUID[56,57] [Fig. 7(a)] consists of a single Josephson junction on a superconducting ring. The ring is coupled to the inductor of a tank circuit whose resonant frequency is typically 30MHz.

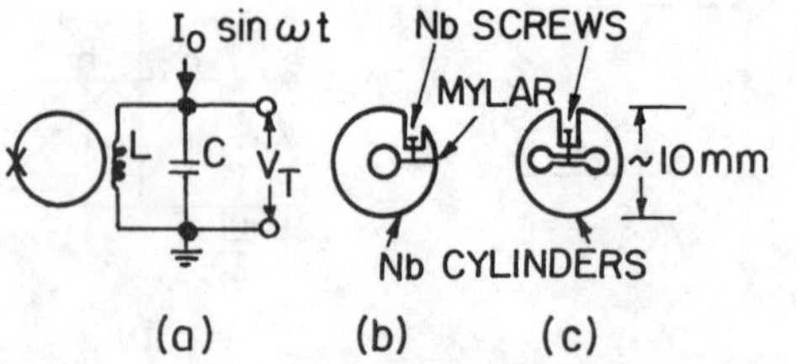


Fig. 7 (a) Rf SQUID; (b) and (c) point contact rf SQUIDs.

An ac current at the resonant frequency is applied to the tank circuit, and adjusted to the appropriate amplitude. The amplitude of the 30MHz voltage across the tank circuit is periodic in the flux $\phi$ applied to the SQUID, the period again being $\phi_0$. The 30MHz signal is amplified and demodulated to produce a quasistatic voltage that is periodic in the flux. A second, lower frequency flux (typically 100kHz) modulates the flux in the SQUID which is then used in a feedback circuit in much the same way as the dc SQUID. Zimmerman and co-workers,[56] have pioneered the use of point contacts in a variety of rf SQUIDs machined from niobium cylinders. The first version [Fig. 7(b)] consisted of a niobium tube with a slit down one side connected by an adjustable niobium screw.[56] A later version [Fig. 7(c)] contained two holes and a single point contact.[51,56] Zimmerman[58] has also operated a "fractional turn" rf SQUID, while more recently point contact rf SQUIDS with a toroidal geometry have become commercially available.[59,60] An alternative rf SQUID design was evolved by Mercereau and co-workers.[57] A superconducting film is evaporated around a quartz tube, and a single bridge is fabricated in the film. This type of SQUID is also commercially available.[60,61]

The noise in rf SQUIDs has been extensively investigated.[62] In practice, rf SQUIDs with an inductance $\lesssim 10^{-9}$H and an rf frequency of 30MHz have a resolution of about $10^{-4}\phi_0\text{Hz}^{-\frac{1}{2}}$ in the white noise region. The intrinsic noise is proportional to (rf frequency)$^{-\frac{1}{2}}$, and a lower intrinsic noise has been observed in rf SQUIDS operated at higher frequencies. Gaerttner (unpublished) has achieved a resolution of $7\text{x}10^{-6}\phi_0\text{Hz}^{-\frac{1}{2}}$ in a SQUID of unspecified inductance with an rf frequency of 440MHz.

To make magnetic measurements, it is often convenient to use a superconducting flux transformer in conjunction with a dc or rf SQUID that is itself shielded from external magnetic fields [Fig. 8(a)]. A magnetic field applied to the pick-up loop generates a persistent supercurrent that in turn produces a flux in the flux-locked SQUID. The diameter of the pick-up loop is typically 50mm, while the secondary coil coupled to the SQUID usually has several turns so that its inductance is equal to that of the pick-up loop. With proper design,[58,63] the flux transformer can enhance the sensitivity of a SQUID by an order of magnitude,[58] although in practice the noise limitation is set by fluctuations in the earth's magnetic field. The flux transformer may also be used as a gradiometer[64] [Fig. 8(b)]. The two pick-up coils are carefully balanced so that a uniform magnetic field induces no supercurrent. A gradient $\partial H_z/\partial z$ induces a flux in the SQUID. A resolution approaching $10^{-13}\text{Tm}^{-1}\text{Hz}^{-\frac{1}{2}}$ is possible. Transformers measuring off-diagonal gradients ($\partial H_x/\partial y$ etc.) and second-derivatives ($\partial^2 H_z/\partial z^2$) have also been operated.[65,66] The magnetic field resolution of SQUIDs is appreciably higher than that obtainable using other instruments: The fluxgate and proton precession magnetometers have sensitivities of about $10^{-10}\text{T Hz}^{-\frac{1}{2}}$, while the pumped cesium vapor magnetometer has a sensitivity of about $10^{-11}\text{T Hz}^{-\frac{1}{2}}$.[67]

The magnetic susceptibility of a sample may be measured by placing it in the pick-up loop of a flux transformer, or in one of the loops of a gradiometer, and applying a magnetic field to it. In order to obtain a sufficiently stable magnetic field, it is necessary to use a superconducting magnet in a persistent field mode and to stabilize the field with a superconducting shield.[68] Systems in which the sample temperature can be varied from 4.2K to 300K or higher have been operated.[59,69] A susceptibility resolution of $10^{-10}$e.m.u. for a 1cm$^3$ sample in a field of $10^{-2}$T was achieved by Cukauskas et al.,[68] and of $2\text{x}10^{-11}$ e.m.u. for a 1 cm$^3$ sample in a field of $10^{-1}$ T has been achieved commercially.[60] The limit on these measurements is set by vibration and temperature drifts rather than by SQUID noise. By way of comparison, Foner[70] has reported a resolution of $10^{-13}$ e.m.u.g$^{-1}$ at 1T using a vibrating sample magetometer, and a resolution of $10^{-11}$ e.m.u.g$^{-1}$ at 10T has been achieved commercially using a Faraday balance.[71] The resolution of the SQUID system could probably be much improved if the filling factor of the sample in the pick-up coil were increased, and/or if the vibration of the sample could be reduced so that a larger magnetic field could be used.

The SQUID can be used as a voltmeter[51,72-74] by connecting a voltage source in series with a known resistor and a superconducting coil coupled to the SQUID. At 4.2K, the voltage resolution can be limited by Johnson noise in the resistor for values of resistance less than a few ohms. For example, for a resistance of $10^{-8}\Omega$, the resolution is $\sim 10^{-15}\text{VHz}^{-\frac{1}{2}}$.

To conclude, I shall comment on the figure of merit[63] used to compare SQUIDs. The sensitivity is often quoted in flux $\text{Hz}^{-\frac{1}{2}}$, but this is insufficient information to properly characterize the SQUID. Consider a coil (the secondary of a flux transformer) of inductance $L_c$ coupled to the SQUID with mutual inductance $M_c$. If the SQUID mean square flux resolution is $\langle\phi_N{}^2\rangle$(per $\text{Hz}^{\frac{1}{2}}$), the current resolution in the coil is $\langle I_N{}^2\rangle = \phi_N{}^2/M_c{}^2$, corresponding to an energy resolution $L_c\langle I_N{}^2\rangle/2 = \langle\phi_N{}^2\rangle/(2M_c{}^2/L_c)$. This quantity is an appropriate figure of merit for all SQUID applications[63] in the limit of very low frequency. The smallest energy resolution reported is that of Clarke et al.[55], $8\text{x}10^{-30}\text{JHz}^{-1}$.

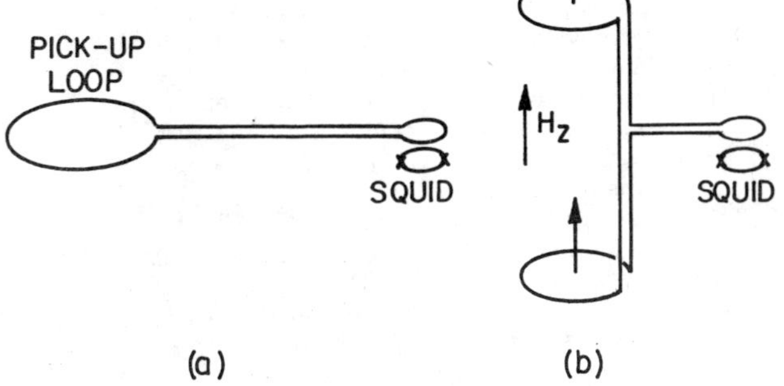


Fig. 8 (a) Flux transformer; (b) gradiometer.

ACKNOWLEDGEMENTS

I am indebted to Drs. W. Anacker, R. F. Broom, T. A. Fulton, P. Guéret, W. H. Henkels, Th. O. Mohr, T. van Duzer, P. Wolfe, and H. H. Zappe for informative discussions on computer devices. This work was support by the USERDA.

REFERENCES

1. B. D. Josephson, Phys. Lett. 1, 251-253 (1962); Rev. Mod. Phys. 36, 216-220 (1964); Adv. Phys. 14, 419-451 (1965).
2. P. W. Anderson, Lectures on the Many Body Problems, (ed. E. Caianello, Academic Press, New York, 1964) Vol. 2, 113-135.
3. P. W. Anderson and J. M. Rowell, Phys. Rev. Lett. 10, 230-232 (1963).
4. W. C. Stewart, Appl. Phys. Lett. 12, 277-280 (1968); D. E. McCumber, J. Appl. Phys. 39, 3113-3118 (1968).
5. P. K. Hansma, G. I. Rochlin, and J. N. Sweet, Phys. Rev. B4, 3003-3014 (1971).
6. H. H. Zappe, J. Appl. Phys. 44, 1371-1377 (1973).
7. V. Ambegaokar and A. Baratoff, Phys. Rev. Lett. 10, 486-489 (1963).
8. R. F. Broom, W. Jutzi, and Th. O. Mohr, IEEE Trans. Magn. MAG-11, 755-758 (1975).
9. M. D. Fiske, Rev. Mod. Phys. 36, 221-222 (1964).
10. W. Schroen, J. Appl. Phys. 39, 2671-2678 (1968).

11. J. H. Greiner, J. Appl. Phys. 42, 5151-5155 (1972); ibid. 45, 32-37 (1974).
12. J. H. Greiner, S. Basavaiah, and I. Ames, J. Vac. Sci. and Tech., 11, 81 (1974).
13. S. Basavaiah, J. M. Eldridge, and J. Matisoo, J. Appl. Phys. 45, 457-464 (1974).
14. J. E. Nordman, J. Appl. Phys. 40, 2111-2115 (1969); J. E. Nordman and W. H. Keller, Phys. Lett. 36A, 52-53 (1971); L. O. Mullen and D. B. Sullivan, J. Appl. Phys. 40, 2115-2117 (1969); R. Graeffe and T. Wiik, J. Appl. Phys. 42, 2146-2147 (1971); K. Schwidtal, J. Appl. Phys. 43, 202-208 (1972); P. K. Hansma, J. Appl. Phys. 45, 1472-1473 (1974); S. Owen and J. E. Nordman, IEEE Trans Magn. MAG-11, 774-777 (1975).
15. R. B. Laibowitz and J. J. Cuomo, J. Appl. Phys. 41, 2748 -2750 (1970); R. B. Laibowitz and A. F. Mayadas, Appl. Phys. Lett. 20, 254-256 (1970); G. A. Hawkins and John Clarke, Bull. Am. Phys. Soc. 20, 332 (1975), and to be published in J. Appl. Phys. (1976); R. F. Broom, R. Jaggi, R. B. Laibowitz, Th. O. Mohr, and W. Walter, Proc. LT14, Helsinki, Finland, 172-175 (1975).
16. J. Seto and T. Van Duzer, Appl. Phys. Lett. 19, 488-491 (1971).
17. C. L. Huang and T. Van Duzer, Appl. Phys. Lett. 25, 753-756 (1974); IEEE Trans. Magn. MAG-11, 766-769 (1975).
18. K. E. Drangeid, R. F. Broom, W. Jutzi, Th. O. Mohr, A. Moser and G. Sasso, Intl. Solid-State Circuits Conference Digest, 68-69 (1971); K. Grebe, I. Ames and A. Ginzberg, J. Vac. Sci. Technol., 11, 458 (1974).
19. J. E. Zimmerman and A. H. Silver, Phys. Rev. 141, 367-375 (1966); R. A. Buhrman, S. F. Strait, and W. W. Webb, J. Appl. Phys. 42, 4527-4528 (1971); J. E. Zimmerman, Proc. Applied Superconductivity Conference, Annapolis, Maryland, 544-561 (1972).
20. P. W. Anderson and A. H. Dayem, Phys. Rev. Lett. 13, 195-197 (1964).
21. H. A. Notarys and J. E. Mercereau, Proc. Intl. Conf. Science Superconductivity, Stanford, California, 424-431 (1969).
22. J. Matisoo, Proc IEEE 55, 172-180 (1967).
23. W. Anacker, IEEE Trans. Magn. MAG-5, 968-971 (1969).
24. P. W. Anderson, Physics Today, 23, 23-29 (November 1970).
25. D. E. McCumber, J. Appl. Phys. 39, 2503-2508 (1968).
26. J. Matisoo, Appl. Phys. Lett. 9, 167-168 (1966); Proc. IEEE 55, 2052-2053 (1967).
27. H. H. Zappe and K. R. Grebe, IBM J. Res. Dev. 15, 405-407 (1971); J. Appl. Phys. 44, 865-874 (1973).
28. W. Jutzi, Th. O. Mohr, M. Gasser, and H. P. Geschwind, Electron. Lett. 8, 589-590 (1972).
29. C. S. Owen and D. J. Scalapino, Phys. Rev. 164, 538-544 (1967); J. Matisoo, J. Appl. Phys. 40, 1813-1820 (1969); K. Schwidtal, Phys. Rev. B2, 2526-2532 (1970); S. Basavaiah and R. F. Broom, IEEE Trans. Magn. MAG-11, 759-762 (1975).
30. T. A. Fulton, Appl. Phys. Lett. 19, 311-313 (1971).
31. T. A. Fulton and R. C. Dynes, Solid State Comm. 9, 1069-1073 (1971).
32. H. H. Zappe, IEEE J. Solid-State Circuits SC-10, 12-19 (1975).
33. W. H. Henkels, IEEE Trans. Magn. MAG-10, 860-863 (1974).
34. D. J. Herrell, IEEE J. Solid State Circuits, SC-9, 277-282 (1974).
35. D. J. Herrell, IEEE Trans. Magn. MAG-10, 864-867 (1974).
36. D. J. Herrell, "An Experimental Multiplier Circuit based on Superconducting Josephson Devices", to be published.
37. Y. L. Yao and D. J. Herrell, Tech. Digest of Int. Electron Devices Mtg., Washington, D. C., Dec. 9-11, 1974.
38. H. W. Chan, W. Y. Lum, and T. Van Duzer, IEEE Trans. On Magn. MAG-11, 770-773 (1975).
39. W. Baechtold, Th. Forster, W. Heuberger, and Th. O. Mohr, Electron. Lett. 11, No. 10 (1975).
40. W. Baechtold, Proc. 1975 Solid State Circuits Conference.
41. H. H. Zappe, Appl. Phys. Lett. 27, 432-434 (1975).
42. T. A. Fulton, Applied Superconductivity Conf., Anapolis, Maryland, 1972, 575-580.
43. T. A. Fulton, R. C. Dynes, and P. W. Anderson, Proc. IEEE 61, 28-35 (1973).
44. T. A. Fulton and L. N. Dunkleberger, Appl. Phys. Lett. 22, 232-233 (1973).
45. P. W. Anderson, Prog. in Low Temp. Phys. (ed. C. J. Gorter, North Holland, Amsterdam, 1967) 5, 1.
46. T. A. Fulton and R. C. Dynes, Solid State Comm. 12, 57-61 (1973).
47. H. H. Zappe, Appl. Phys. Lett. 25, 424-426 (1974).
48. P. Guéret, Appl. Phys. Lett. 25, 426-428 (1974).
49. P. Guéret, IEEE Trans. Magn. MAG-11, 751-754 (1975).
50. W. W. Webb, IEEE Trans. Magn. MAG-8, 51-60(1972).
51. R. P. Giffard, R. A. Webb, and J. C. Wheatley, J. Low Temp. Phys. 6, 533-610(1972).
52. J. Clarke, Proc. IEEE 61, 8-19 (1973).
53. J. Clarke, Science, 184, 1235-1242 (1974).
54. R. C. Jaklevic, J. Lambe, A. H. Silver, and J. E. Mercereau, Phys. Rev. Lett. 12, 159-160(1964).
55. J. Clarke, W. M. Goubau, and M. B. Ketchen, IEEE Trans. Magn. MAG-11, 724-727 (1975); Appl. Phys. Lett. 27, 155-156 (1975); Proc. LT14, Helsinki, Finland, 214-217 (1975).
56. J. E. Zimmerman, P. Thiene, and J. T. Harding, J. Appl. Phys., 41, 1572-1580 (1970).
57. J. E. Mercereau, Rev. Phys. Appl., 5, 13-20(1970); M. Nisonoff, Rev. Phys. Appl., 5, 21-24(1970);
58. J. E. Zimmerman, J. Appl. Phys., 42, 4483-4488 (1971).
59. S.H.E. Corp.
60. S.C.T. Inc.
61. Develco Inc.
62. For a comprehensive article and a list of references, see L. D. Jackel and R. A. Buhrman, J. Low Temp. Physics, 19, 201-246 (1975).
63. J. H. Claassen, J. Appl. Phys. 46, 2268-2275 (1975).
64. J. E. Zimmerman and N. V. Frederick, Appl. Phys. Lett. 19, 16-19 (1971).
65. J. E. Opfer, IEEE Trans. Magn. MAG-9, 536 (1974).
66. D. Brenner, S. J. Williamson, and L. Kaufman, Proc. LT14, Helsinki, Finland, 4, 266-269 (1975).
67. Values are order of magnitude only. Sources for sensitivity of non-superconducting instruments for use in field applications are: P. Hood in Mining and Groundwater Geophysics (ed. L. W. Manley, Dept. of Energy, Mines, and Resources, Ottawa, Canada, 1969), 3-31; and several articles in Rev. Phys. Appl. 5, No. 1 (1970).
68. E. J. Cukauskas, D. A. Vincent, and B. S. Deaver, Jr., Rev. Sci. Instrum. 45, 1-6 (1974).
69. H. E. Hoenig, R. W. Wang, G. R. Rossman, and J. E. Mercereau, Proc. Appl. Superconductivity Conf., Annapolis, 581-587 (1972).
70. For recent improvements and references, see S. Foner, Rev. Sci. Instrum., 46, 1425-1426 (1975).
71. Oxford Instruments.
72. J. Clarke, Phil. Mag. 13, 115-127 (1966).
73. J. E. Lukens, R.J. Warburton, and W.W. Webb, J. Appl. Phys. 42, 27-30 (1972).
74. A. Davidson, R. S. Newbower, and M. R. Beasley, Rev. Sci. Instrum. 45, 838-846 (1975).

# ON THE OPERATION OF A PICKAX REPLICATE/TRANSFER GATE*

S. K. Singh, W. C. Hubbell and D. C. Bullock,
Texas Instruments Incorporated,
P. O. Box 5936, M. S. 145, Dallas, Texas 75222

## ABSTRACT

The design of a pickax replicate/transfer gate has been reported.[1] Here we report on the operation of this element with respect to variations in the current pulse parameters (amplitude A, width $t_w$, and position $t_p$), inplane drive field amplitude and bias field. This element was operated at 100 KHz on $(YSm)_3(FeGa)_5O_{12}$ garnet with a nominal bubble diameter of 5 μm and was configured in the major loop to detector path on a major/minor loop chip. The following values of pulse parameters provide replicate and transfer bias field margins equal to that of loop propagation (16 Oe) while maintaining acceptable tolerances: Replicate: A = 75 mA, $t_w$ = 2.0 μsec and $t_p$ = 5.4 μsec (zero μsec corresponds to the inplane field parallel to the long axis of the pickax element); Transfer: A = 75 mA, $t_w$ = 4.0 μsec and $t_p$ = 3.8 μsec. With these pulse parameters, reducing the drive field amplitude from 40 to 25 Oe reduces the bias field margin for replicate and transfer by 3 and 2 Oe, respectively.

## INTRODUCTION

Bubble memory chips using major/minor loop organization and long stretcher detectors require both replication and annihilation operations. The performance of chips where these functions are accomplished by a single replicate/annihilate element consisting of Permalloy chevrons and a hairpin conductor loop has been described.[2] This element has two disadvantages. It uses "stretch" and "cut" pulses requiring a bipolar current driver, and the stretch pulse can cause nucleation of bubbles in low $H_k$ materials. The latter can severely limit the temperature range of device operation. Both disadvantages can be circumvented by eliminating the stretch pulse. An element that accomplishes this was described by Bonyhard, et al.[1] It uses a pickax-shaped Permalloy element to stretch the bubbles. This element is placed in the major loop of the chip and interfaces with the chevron track leading into the detector. The annihilate function is achieved by transferring bubbles from the major loop into the detector and the guardrail.

Here we describe the operation of this element with respect to variations in the current pulse parameters, the inplane drive field amplitude and bias field. The current pulse parameters are amplitude A, width $t_w$, and pulse position within the rotating field cycle $t_p$ (Figure 1). The results presented here are typical of a number of chips of this nature.

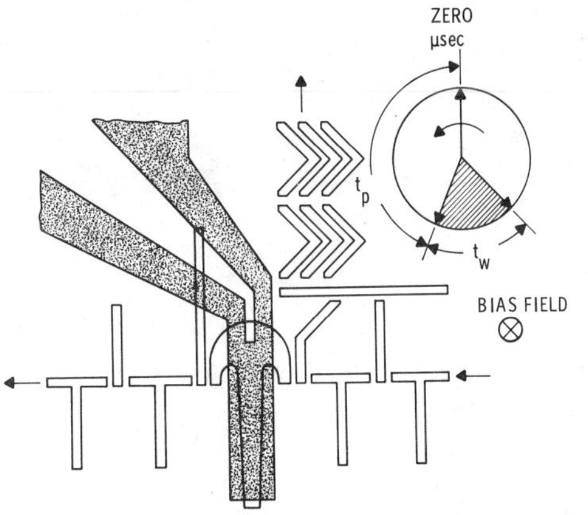


Fig. 1. Pickax Replicate/Transfer Gate, Nominal Bar Width is 2.5 μm.

The pickax element is shown in Figure 1. Bubbles propagating from right to left are stretched as they move along the upper curved part of the element. A current pulse is applied to the overlying conductor loop to cut the domain for the replicate operation. For transfer, a pulse is applied at an earlier time to block the motion of the strip in the horizontal direction and favor its motion in the vertical direction, thus routing it along the chevron track that leads into the detector. Test circuits were fabricated on $(YSm)_3(FeGa)_5O_{12}$ garnet supporting 5 μm bubbles. Permalloy was sputtered (4,000 to 5,000 Å thick) over a 1 μm thick silicon oxide spacer and had a coercivity of < 2 Oe. All testing was done at 100 KHz inplane drive field frequency and the chip temperature was maintained at (30 ± 1)°C. The bias field is directed such that bubbles are attracted to negative poles.

## EXPERIMENTAL RESULTS

Figure 2 shows the minimum pulse amplitude, A min, required for replicate operation as a function of pulse width ($t_w$) for $t_p$ = 5.3 μsec and a 40 Oe inplane drive field amplitude. For pulses wider than 0.2 μsec the minimum pulse amplitude required is constant at 50 mA. Decreasing the pulse width increases the pulse amplitude. Thus, pulses wider than 0.2 μsec should be used. Figure 3 shows the permissible values of $t_p$ and $t_w$ which allow replication for two different bias fields (125 Oe and 130 Oe). Any combination of ($t_p$, $t_w$) within the curves is acceptable. At a bias field of 125 Oe, the pulse position is unaffected by the pulse amplitude for the range 0.2 μsec $\leq t_w \leq$ 3.5 μsec. Raising the bias field reduces the allowable range of the pulse position. This is a result of an incomplete stretching of bubble domains as they traverse the pickax element. This figure shows that $t_w \geq$ 1.5 μsec is required to achieve the greatest tolerance on pulse position for both bias fields. Figure 4 shows $t_p$ as a function of $H_b$ for two values of pulse amplitude (60 mA and 150 mA). The acceptable range of replicate pulse positions decreases as the bias field increases. Furthermore, variation in the pulse amplitude has only a small effect on this dependence. Optimal pulse parameters for replicate operation are A = 75 mA, $t_w$ = 2.0 μsec, and $t_p$ = 5.4 μsec. With these pulse parameters, the bias field margin for replicate operation was found to equal the major loop propagation margin (16 Oe).

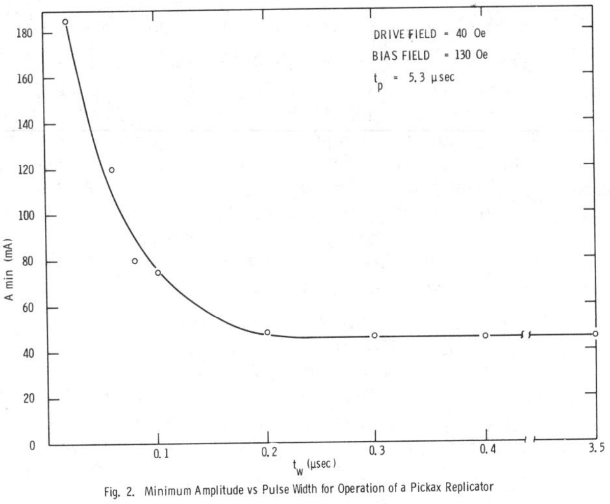


Fig. 2. Minimum Amplitude vs Pulse Width for Operation of a Pickax Replicator

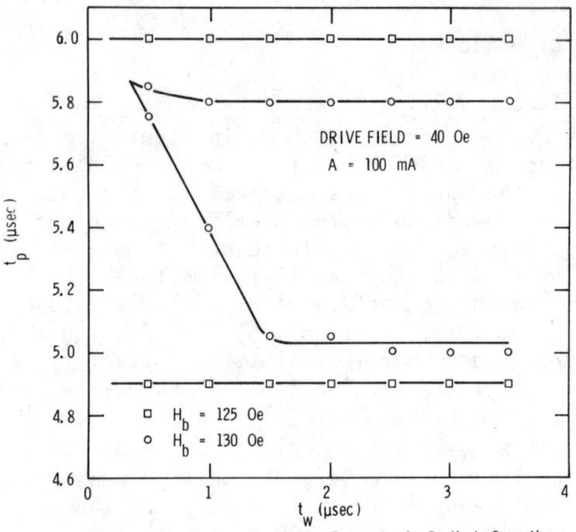


Fig. 3. $t_p$ vs $t_w$, $H_b$ as a Parameter for Replicate Operation

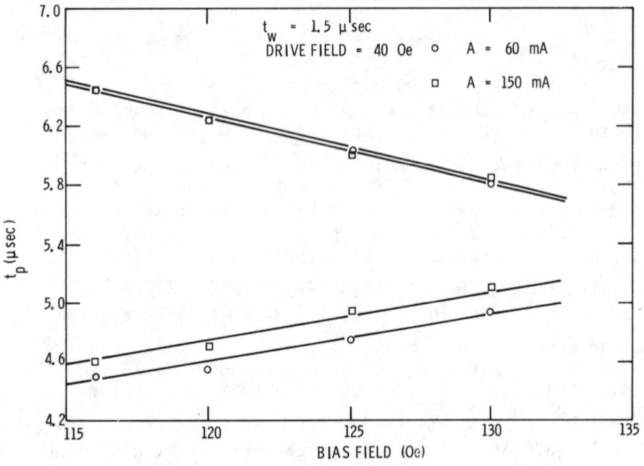


Fig. 4. $t_p$ vs Bias Field, A as a Parameter for Replicate Operation

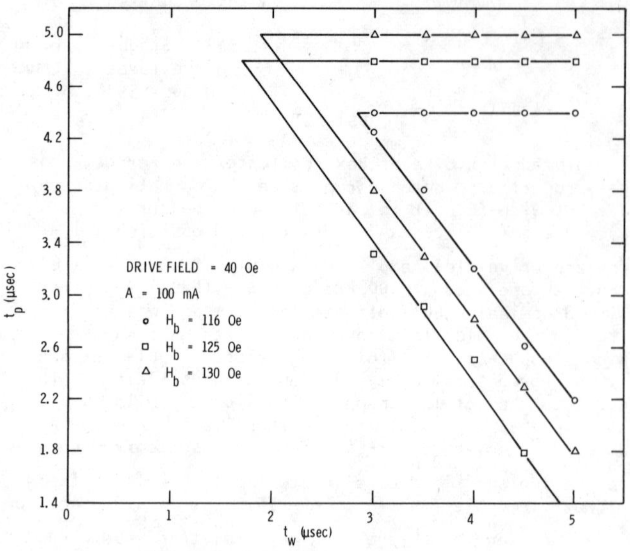


Fig. 5. $t_p$ vs $t_w$, Bias Field as a Parameter for Transfer Operation

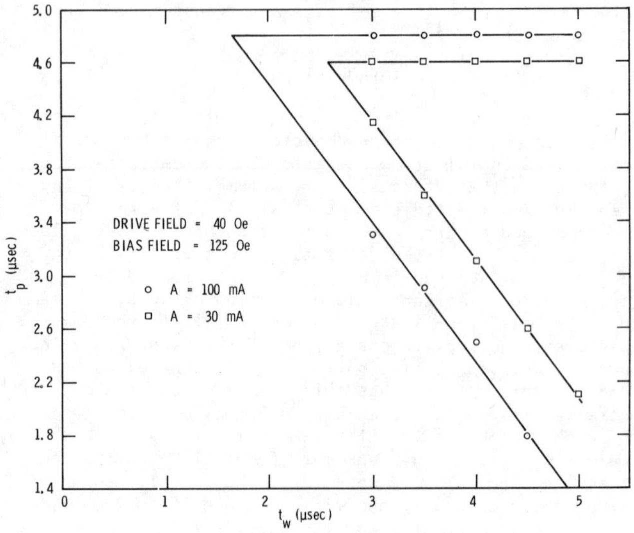


Fig. 6. $t_p$ vs $t_w$, A as a Parameter for Transfer Operation

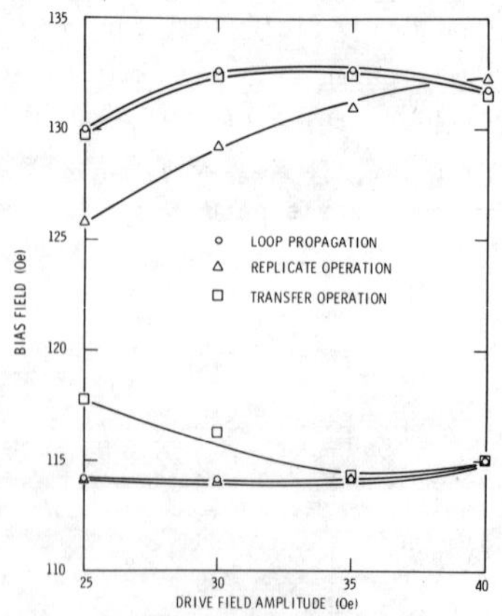


Fig. 7. Bias Field Margin vs Drive Field Amplitude for Replicate and Transfer Operations

The transfer operation requires pulse amplitudes and widths greater than 15 mA and 2.5 μsec, respectively. Figure 5 shows how the range of acceptable values of $t_p$ and $t_w$ varies with bias field. Increasing the pulse width increases the range of $t_p$ for transfer operation. At a given bias field, there is a latest time ($t_{p\ell}$) in the cycle before which the pulse must begin (e.g., 4.4 μsec for a bias field of 116 Oe) for the transfer operation. This time delay $t_{p\ell}$ is independent of the pulse width. The higher the bias field, the higher is the value of $t_{p\ell}$. For a given pulse width, increasing the bias field reduces the range of $t_p$. Qualitatively, these results can be understood as follows. The transfer pulse blocks the right to left motion of a bubble (Figure 1) and creates a potential well below the conductor. This keeps the bubble trapped below the conductor until the magnetic poles due to the chevrons become attractive to the bubble (after a 90° rotation of the field). The bubble then propagates on the chevron track. In order for the transfer operation to succeed, the pulse must be applied before the bubble passes below the conductor. The latest time ($t_{p\ell}$) before which a pulse must be applied, therefore, is determined only by the time at which the leading edge of the bubble arrives at the conductor and; therefore it is independent of the pulse width. In addition, lowering the bias field strips the bubble causing the leading edge of the bubble to reach the conductor earlier in time, thus requiring $t_{p\ell}$ to be smaller for lower bias fields. Since the drive field must rotate by 90° (2.5 μsec) before the chevron track becomes attractive to a bubble at the top of the pickax element, the transfer function requires pulse widths greater than 2.5 μsec. Applying a pulse earlier in time does not affect the motion of a bubble which is propagating towards the peak of the pickax element. Therefore, for a given bias field, one can widen the $t_p$ range by selecting larger pulse widths.

Figure 6 shows the allowable values of $t_p$ and $t_w$ for transfer for two pulse amplitudes. Increasing the pulse amplitude from 30 to 100 mA increases the range for $t_p$. Optimal pulse parameters selected for transfer operation are $t_p$ = 3.8 μsec, $t_w$ = 4 μsec, A = 75 mA (equals A for replicate pulse for driver simplicity). With these pulse parameters, the bias margin for transfer equals the major loop propagation margin (16 Oe).

With the pulses set at their optimal values, replicate and transfer bias margins were measured for drive field amplitudes between 25 Oe and 40 Oe. These data are shown in Figure 7. Major loop propagation margin is also given for comparison. The lower margin for replicate and the upper margin for transfer track the corresponding loop propagation margins. At 30 Oe drive field, the combined effects of the replicate and transfer operations produce an overall 4 to 5 Oe reduction in operating bias field range.

The above test results establish that the pickax element has adequate bias field margins and control pulse tolerances. Since it eliminates the stretch pulse, it is preferred over the chevron hairpin replicate/annihilate element.

ACKNOWLEDGMENTS

The authors are pleased to thank G. G. Sumner for providing the garnet material and M. S. Shaikh for processing the samples. The technical help of H. F. Collins, D. J. Marshall and J. W. Taylor is acknowledged. We are indebted to F. G. West for his valuable expert criticism and helpful discussions during this work.

REFERENCES

*Work partially supported by U. S. Air Force Avionics Laboratory

1. Bonyhard, et al., AIP Conference Proceedings No. 18, 100 (1973)
2. Bonyhard, et al., IEEE Transactions on Magnetics, MAG-9, 433 (1973)

## ON SPONTANEOUS NUCLEATION IN FIELD ACCESSED BUBBLE DEVICES

M.H. Kryder, C.H. Bajorek, and R.J. Kobliska
IBM Thomas J. Watson Research Center, PO Box 218, Yorktown Heights, New York 10598

ABSTRACT

The dependence of the in-plane drive field at which bubble domains spontaneously nucleate in field accessed bubble devices has been investigated as a function of $H_k-4\pi M_s$ and of spacer thickness between the bubble film and permalloy propagation elements. The experiments were carried out on amorphous GdCoMo bubble films with T-bar and Y-bar structures. For a given structure and spacer thickness the nucleation field increases linearly with $H_k-4\pi M$. Larger spacer thicknesses also lead to increased nucleation fields. A model based on the Stoner-Wohlfarth astroid is compared to these data and found to be useful in explaining the qualitative trends, but to be in poor quantitative agreement. It is concluded that since the drive field required in a device is proportional to $4\pi M_s$, $Q-1 = (H_k-4\pi M_s)/4\pi M_s$ must be greater than some minimum value for a given device structure and spacer thickness to permit reliable device operation. This paper has been submitted for publication in the IEEE Transactions on Magnetics.

## DISTRIBUTION OF NORMAL COMPONENT OF MAGNETIZATION IN ELEMENTS OF T-TYPE PERMALLOY CIRCUITS

Prof. G.S. Krinchik, U.N. Shamatov, E.E. Chepurova
Faculty of Physics, Moscow State University, Moscow, 117234 (USSR)

For the first time the distribution of normal (perpendicular to the elements plane) component of magnetization M in T and I bars of efficient bubble domain propagation circuits at various orientations and values of external magnetic field H has been studied by means of magnetooptic micromagnetometer method, where previously measurements of horizontal component of magnetization M have been obtained.*

Parts have been discovered, where M emerges from the plane of bars, which are magnetostatic wells for the bubble. While H is oriented along I, two parts with M have been registered with dimensions of about 1/8 that of I. When H is oriented perpendicularly to a short section of T, one magnetostatic well is observed in its centre, and when H is oriented at an angle of 45° to the short section of T, three magnetostatic wells appear. The distribution of M in the short section of T in case of longitudinal magnetization is similar to that of M in I. It has been found that M in the whole range of operate fields does not exceed 0, $IM_s$.

*Krinchik etc., Proceedings of XX MMM Conf., San Francisco

# MAGNETORESISTIVE PROPERTIES OF A CHEVRON STRETCHER DETECTOR

W. C. Hubbell, S. K. Singh and F. G. West, Texas Instruments Incorporated
P. O. Box 5936, M. S. 145, Dallas, Texas 75222

## ABSTRACT

To better understand how bubbles affect a resistance change in a Permalloy chevron stretcher detector, a more detailed knowledge of the magnetoresistive properties of such elements in the absence of bubbles is required. Results of systematic resistance measurements on a single element as a function of magnitude, H, and direction of an inplane dc magnetic field are presented for 10 Oe < H < 10,000 Oe.

## INTRODUCTION

There is a considerable interest in the use of thick film chevron expander magnetoresistive elements for the detection of magnetic bubbles. Such detectors are composed of a long column of electrically interconnected Permalloy chevron propagation elements. Because of the resulting complex geometry and the attendant possibilities for complex domain structures in the Permalloy, the magnetoresistive properties in the presence of a rotating inplane magnetic field, H, are not well understood. Furthermore, the detailed manner in which a bubble modifies this magnetoresistive behavior when it is being sensed is even more obscure. To better understand this behavior, the magnetoresistance has been studied for quasistatic inplane fields by slowly rotating a sample in the field of an electromagnet. This procedure has permitted a systematic characterization of the magnetoresistive waveform, i.e., the dependence of the element resistance on the orientation of the inplane field, over a large range of field values, 10 Oe < H < 10,000 Oe.

## EXPERIMENT

The magnetoresistive element studied is depicted in Figure 1(a). It consists of 320 interconnected chevrons and is capable of propagating and detecting 5 μm diameter bubbles which have been expanded along the length of the element. The nominal dimensions are shown in Figure 1(b). Elements of this type were patterned in zero magnetostriction Permalloy (81 Ni - 19 Fe) on glass substrates. The Permalloy was sputter deposited at elevated temperatures to a thickness of ∿ 4,500 A before being ion milled into the desired geometry. This process is identical to that usually used in bubble devices. Typical resistance values for these elements are approximately 1,000 ohms.

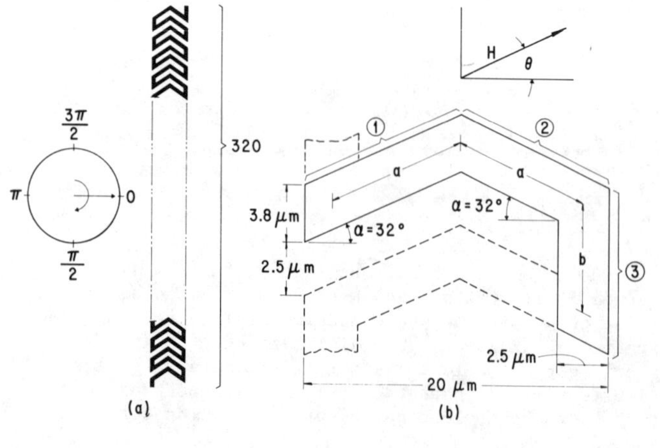


Fig. 1. Permalloy Chevron Expander Detector
(a) Column of Interconnected Chevrons,
(b) Detail of One Chevron-interconnect Element of Detector.

The sample was mounted in a rotation apparatus placed within the gap of a large electromagnet with the magnetoresistive element forming one arm of a Wheatstone bridge. The periodic bridge imbalance associated with the rotation angle, Θ (defined in Figure 1b), of the element relative to the field was monitored by a recorder. Plots of resistance change versus Θ were made for values of H in the range 10 Oe < H < 10,000 Oe starting with the lower values first. At these lower field values, H < 20 Oe, measurements were made after the prior application of a dc field at various orientations relative to the detector element.

## RESULTS

The measured waveforms of the magnetoresistance of a typical element are shown in Figures 2 and 3 for selected values of the inplane field, H. These waveforms show substantial variations in both shape and amplitude. In the range 5,000 Oe < H < 10,000 Oe, the resistance has a cos2Θ dependence (Figure 2a) with maxima occurring when H is perpendicular to the long axis of the element, i.e., Θ = 0, π, etc. The peak-to-peak variation of resistance, 2B, is 8 Ω for a 1,000 Ω element. Figures 2 (b), (c) and (d) illustrate the progressive amplitude reduction and distortion of the cos2Θ waveform for successively lower values of H. For H = 250 Oe, 2B is reduced to 3.8 Ω. Although severe distortion is occurring in this range, the maxima still occur at Θ = 0, π, etc. and the 2Θ periodicity remains.

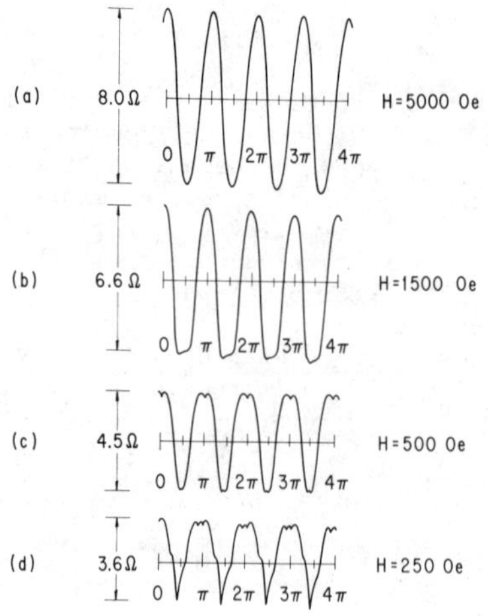


Fig. 2. Magnetoresistive Waveforms Representative of High Inplane Fields.

The characteristic behavior in the range 50 Oe < H < 100 Oe is exhibited in Figure 3(a). There are two important departures from the high field case; (1) the waveform now has a large 4Θ component with maxima now occurring at Θ = 0, π/2, π, 3π/2, etc. and (2) the amplitude of the resistance variation is substantially reduced. In this range the peak-to-peak variation of the resistance exhibits its lowest value, 2.1 Ω.

The field values 20 Oe < H < 50 Oe encompass the range of rotating drive fields typically used with 5 μm bubble memory devices. In this range the magnetoresistance is again periodic in 2Θ. This behavior is sometimes referred to as a 2ω dependence, i.e., if the frequency of a rotating drive field is ω, the frequency of the magnetoresistive waveform is 2ω. The waveform shown in Figure 3(b) for quasistatic rotation of the element relative to the field is essentially identical to the magnetoresistance waveform observed in a 100 kHz rotating field. In this range of fields the resistance maxima occur at π/2 and 3π/2 representing a phase shift

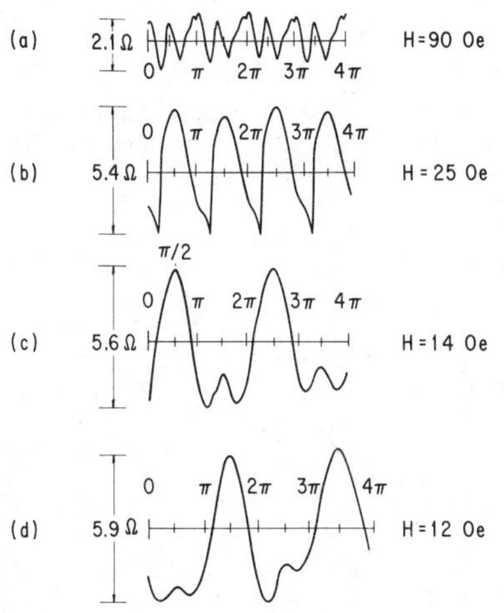


Fig. 3. Magnetoresistive Waveforms Representative of Low Inplane Fields.

of π/2 from the high field behavior. Also, the peak-to peak variation in resistance has increased from the region with 4Θ periodicity.

If H is reduced below about 20 Oe the waveform undergoes a rather abrupt transition to one where the principal component is Θ dependent. In this case the maxima can occur at either π/2 or 3π/2 depending on the magnetic history. For instance, when a field of only a few oersteds is applied along the long axis of the element ($\Theta = \pi/2$) and then removed prior to the rotating sample measurement, the subsequent waveform in low field (H = 12 to 14 Oe) has a maximum at $\Theta = \pi/2$ (Figure 3c). On the other hand, when a prior field of reverse polarity is applied, the resulting low field waveform has a maximum at $\Theta = 3\pi/2$ (Figure 3d). A typical low field waveform has a peak-to-peak resistance variation of 5.8 Ω.

## DISCUSSION

The high field behavior of the magnetoresistance (5,000 Oe < H < 10,000 Oe) is consistent with calculations based on the assumption that H is sufficient to saturate the Permalloy. The angular dependence of the resistance for the chevron expander detector can be calculated from the following approximate expression

$$R(\Theta) = N\left(\frac{\Delta R}{\Delta \ell}\right)\left\{\left(1 + \frac{1}{2}\frac{\Delta\rho}{\rho}\right)(2a+b) + \frac{1}{2}\frac{\Delta\rho}{\rho}(2a\cos 2\alpha - b)\cos 2\Theta\right\}, \quad (1)$$

where a, b and α are defined in Figure 1(b), $\left(\frac{\Delta R}{\Delta \ell}\right)$ is the resistance per unit length and N is the total number of chevrons in the detector. $\rho = \rho_\perp$ is the resistivity which results when the field is normal to the current density. The term $\Delta\rho$ is defined by $\Delta\rho = \rho_{||} - \rho_\perp$ where $\rho_{||}$ is the resistivity when the field is parallel to the current density. Equation (1) is of the form $R(\Theta) = A + B\cos 2\Theta$ where the coefficient B of the $\cos 2\Theta$ term corresponds to one-half of the peak-to-peak variation of the resistance in Figure 2(a). The algebraic sign of this coefficient determines the position of the maxima and minima. Explicitly this coefficient is

$$B = \frac{1}{2}N\left(\frac{\Delta R}{\Delta \ell}\right)\left(\frac{\Delta\rho}{\rho}\right)(2a\cos 2\alpha - b). \quad (2)$$

In order to compare $B\cos 2\Theta$ with the waveform in Figure 2(a) the following values were used in eq (2):

$\left(\frac{\Delta R}{\Delta \ell}\right) = 1.23 \times 10^3$ Ω/cm, measured

$\left(\frac{\Delta\rho}{\rho}\right) = 3\%$, from Ref 2

N = 320
a = 10.3 μm
b = 6.4 μm
α = 32°

(N, a, b and α: defined by sample geometry)

The resulting value of B is ∿ 1.6 Ω. Although eq (1) is approximate, it gives the proper order of magnitude for B and correctly predicts the positions of the maxima (Θ = 0, π, 2π, etc.). An exact derivation for R(Θ) requires integrating over the finite width of the Permalloy elements. Furthermore, the term $2a\cos 2\alpha - b$ is extremely sensitive to the experimental values of a, b, and α when $2a\cos 2\alpha \sim b$. Considering these limitations, the agreement between the calculated and experimental values for B seems to be reasonable.

For values of H < 5,000 Oe domains are presumably present since the sample is not saturated. The saturation demagnetizing field for a long Permalloy strip of width 2.5 μm and thickness 0.45 μm in the transverse direction is ∿ 1800 Oe. The complex geometry and edge effects probably increase the saturation field to the higher value indicated.

When domains are present, the directions of magnetization relative to the current density depend on the nature of the domain formation. This apparently reduces the degree to which the magnetization follows the direction of H and thereby results in a reduction in the amplitude and distortion of the magnetoresistance waveform. This trend is seen in the waveforms (Figure 2) for successively lower fields in the range 100 Oe < H < 5,000 Oe.

The rate at which the peak-to-peak variation of the resistance decreases for successively lower values of H would suggest that for fields less than ∿ 40 Oe the resistance variation should be negligible. However, this is not the case as is seen in Figure 3 where the waveforms become larger for even lower values of H.

The fact that the resistance change is not negligible in low fields, as well as the Θ to 2Θ transition can be explained qualitatively by a remnant domain structure which depends on the magnetic history in the manner observed experimentally. The proposed domain structure embodies magnetization fanning[3] and is presented in more detail in another paper.[4]

## ACKNOWLEDGMENTS

The authors would like to thank J. W. Taylor for fabricating the test samples.

## REFERENCES

1. A. H. Bobeck, I. Danylchuk, F. C. Rossol and W. Strauss, IEEE Trans. Mag., Vol. MAG-9, 474 - 480 (Sept. 1973).
2. David A. Thompson, Lubomyr T. Romankiw and A. F. Mayadas, IEEE Trans. Mag., MAG-11, 1039 (1975).
3. I. S. Jacobs and C. P. Bean, Phys. Rev. 100, 1060 (1955).
4. F. G. West, W. C. Hubbell and S. K. Singh, "ω - 2ω Transition in a Chevron Stretcher Detector," paper submitted to this conference.

# ω - 2ω TRANSITION IN A CHEVRON STRETCHER DETECTOR

F. G. West, W. C. Hubbell and S. K. Singh, Texas Instruments Incorporated
P. O. Box 5936, M. S. 145, Dallas, Texas 75222

## ABSTRACT

The ω - 2ω transition[1] in the magnetoresistance waveform at 100 KHz was investigated for a long Permalloy chevron stretcher detector in the absence of bubbles propagating through the structure. When the Permalloy is demagnetized by decreasing the rotating inplane field, H, in the absence of any dc fields, the waveform of the resistance variation is 2ω for $0 < H < 40$ Oe, where ω is the angular frequency of rotation. When the Permalloy is ac demagnetized in the presence of a small dc field applied along the long axis of the detector element, the resulting waveform has an ω dependence for $0 < H \lesssim 20$ Oe, becoming 2ω for $20 < H < 40$ Oe. In the ω regime, the phase angles of the rotating inplane field at which resistance maxima occur depend on the direction in which the dc field was applied during demagnetization. A domain model based on magnetization fanning is proposed to explain these results.

## INTRODUCTION

Knowledge of the magnetoresistive properties of Permalloy stretcher detectors is required to understand how bubble fields affect these elements. Observations of the resistance R(Θ) as a function of the direction, Θ, of a slowly rotating inplane magnetic field, H, are reported in a separate paper[2] in which it was shown that a $\Theta \rightarrow 2\Theta$ transition occurs for fields in the range of interest for field access memory devices. To study these effects in greater detail and for 100 KHz rotating fields, experiments were conducted to examine R(Θ), or more precisely the R(ωt) waveforms, after ac demagnetization in the presence of inplane dc magnetic fields, $H_p$. The resulting observations along with a qualitative explanation in terms of a fanned magnetization structure[3] are reported.

## EXPERIMENT

The geometry of the Permalloy detector element chosen for this study is shown in Figure 1 of Reference 2. The element consists of a column of 320 interconnected Permalloy chevrons (apex angle = 116°) fabricated on a glass substrate by the same deposition and lithographic process used to make bubble memory devices on garnet films. The coercivity of the Permalloy is less than 2 Oe. The device is mounted on a ceramic substrate for insertion into a pair of orthogonal coils which produce an inplane rotating magnetic field, H, of frequency 100 KHz and amplitude 0 to 40 Oe. This drive coil pair is placed inside a pair of Helmholtz coils which produce a dc inplane field between 0 and 30 Oe. Provision is made for passing a dc current through the detector element, the output of which is monitored directly on an oscilloscope.

The experiments consist of demagnetizing the Permalloy in dc inplane fields of various magnitudes and directions, reducing the dc inplane field to zero, and then recording the magnetoresistive waveform for the sample for different amplitudes of a 100 KHz rotating field. Demagnetization was accomplished by first applying a dc inplane field of desired magnitude, $H_p$, and orientation, ϕ, (Figure 3(c)) and then slowly reducing the magnitude of the 100 KHz rotating inplane field from 40 to 0 Oe. Care was taken to ensure that the sample was aligned so that the inplane component of the earth's field is also at an angle ϕ. $H_p$ denotes the sum of the inplane component of the earth's field and the dc field produced by the dc Helmholtz coils. Observations were also made for sinusoidally oscillating inplane fields oriented separately parallel ($\Theta = \pi/2$) and perpendicular ($\Theta = 0$) to the long axis of the detector element. In Figure 3(c) Θ is the instantaneous orientation of the rotating field vector, or the fixed orientation of the sinusoidal field.

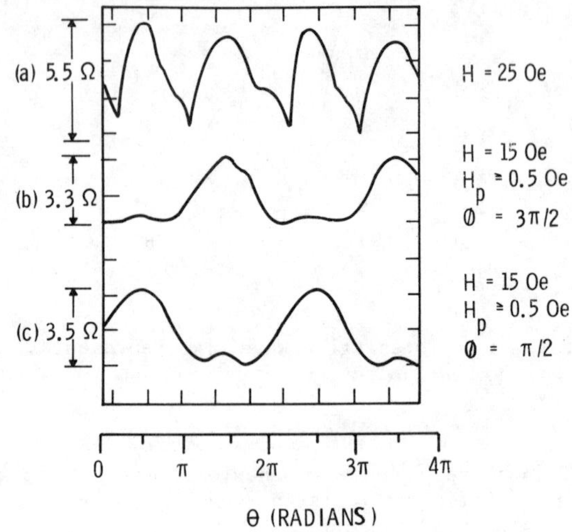


Fig. 1. Magnetoresistance Waveforms at 100 KHz For A Chevron Stretcher Detector.

## RESULTS

For rotating fields of angular frequency ω: (1) When the Permalloy is demagnetized with $H_p \sim 0$, the principal component of the R(ωt) waveform is periodic in 2ωt over the range $0 \leq H \leq 40$ Oe; Figure 1(a). (2) For $H \lesssim 20$ Oe, and $H_p$ equal to the earth's field, or greater, the principal component is periodic in ωt with the maxima of R(ωt) occurring at $\omega t = \pi/2$ or $\omega t = 3\pi/2$, accordingly as the Permalloy is demagnetized with $H_p$ applied along $\phi = \pi/2$ or $\phi = 3\pi/2$, respectively; Figures 1(b) and 1(c). (3) For $H \gtrsim 20$ Oe, the principal component is periodic in 2ωt, irrespective of the previous magnetic history. Thus when the Permalloy is demagnetized in the presence of a weak field $H_p$, a transition from a strong ω component to a strong 2ω component occurs at $H \sim 20$ Oe.

For uniaxial oscillating fields of angular frequency ω: (1) When the oscillating field is along the $\Theta = \pi/2, 3\pi/2$ axis, the behavior is essentially identical to the behavior for all the cases described for a rotating field. (2) When the oscillating field is along $\Theta = 0$, the magnetoresistive waveform is 2ω periodic for $0 < H < 40$ Oe, with the resistance minima occurring when the instantaneous value of the oscillating field is zero.

## DISCUSSION

Consider a Permalloy element carrying a current i. The resistance change of the element depends on rotation of the magnetization relative to the direction of current flow so that the resistance is maximum when the magnetization is fully aligned parallel or antiparallel to the current direction. Consider the fanned magnetization structure[3] shown in Figure 2(a), where there is a net component of magnetization parallel to the direction of current flow. The application of a magnetic field H as shown in Figure 2(b) reduces the degree of fanning and thereby increases the component of magnetization parallel to the current axis thus increasing the resistance. Conversely, the resistance decreases when the fanning angle increases by applying a field in the opposite direction.

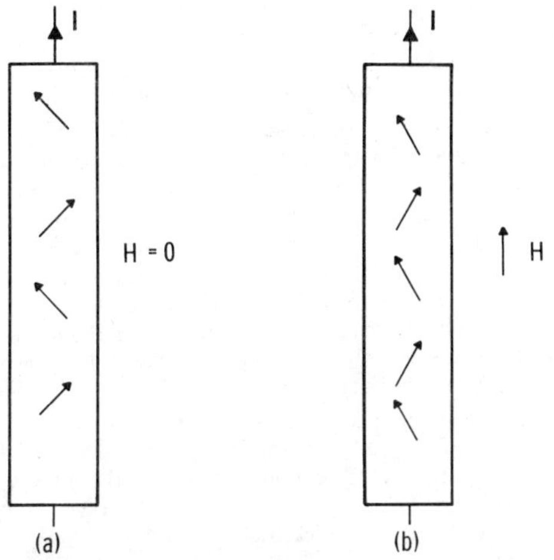


Fig. 2. Schematic Representation of Magnetization Fanning in a Thin Permalloy Strip.

Remanent domain states with magnetization fanning of the types shown schematically in Figures 3(a) and 3(b) are consistent with the experimentally observed magnetoresistive waveforms. These remanent states can be established by demagnetizing the Permalloy in the presence of a dc field oriented parallel to the long axis of the detector; i.e., along $\pi/2$ or $3\pi/2$. Once such states are established, moderate inplane rotating fields will cause the net magnetization along the direction of current flow in the various segments of the element to increase or decrease depending on the instantaneous field orientation. For the state shown in Figure 3(a), a rotating field which is insufficient to switch the magnetization into the opposite remanent state (Figure 3(b)) causes the element resistance to attain a minimum value when the field is oriented along $\pi/2$ and a maximum along $3\pi/2$. Thus, as the field rotates, the resistance is periodic with the same frequency as the rotating field. The same argument applies to the case of the other remanent state depicted in Figure 3(b), except that the maxima now occur at $\pi/2$ and the minima at $3\pi/2$. This model is consistent with the results in Figure 1 where the $\omega$ dependent waveforms exhibit a phase shift of $\pi$ radians when the Permalloy is demagnetized in a dc field first applied along the $3\pi/2$ direction and then along the $\pi/2$ direction.

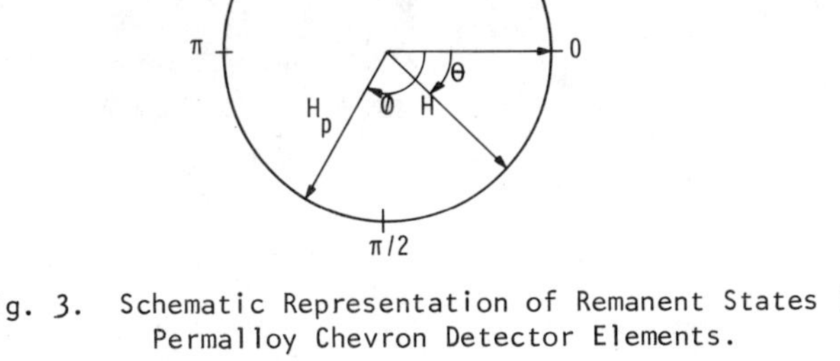

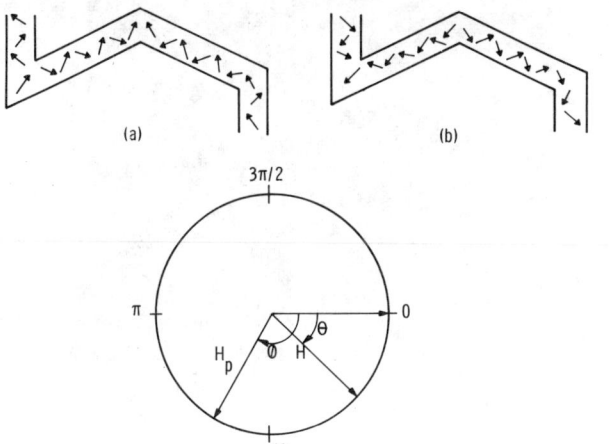

Fig. 3. Schematic Representation of Remanent States in Permalloy Chevron Detector Elements.

If the magnitude of the rotating inplane field is large enough, it will switch the fanned structure from one state, Figure 3(a) or 3(b), to the other. Thus when $0 < \Theta < \pi$, beginning for example with the initial state depicted in 3(b), the resistance will initially go through a maximum at $\Theta = \pi/2$. However, when $\Theta$ is in the range $\pi < \Theta < 2\pi$, the fanned structure will switch to that depicted in 3(a) and the resistance will experience another maximum at $\Theta = 3\pi/2$. For each revolution, the resistance experiences two maxima and the frequency of the waveform is twice that of the rotating field, i.e., $2\omega$. Thus, as the magnitude of the drive field is increased, the magnetoresistive waveform due to a rotating magnetic field of frequency $\omega$ changes from $\omega$ to $2\omega$.

This model for the magnetization process is verified by the essentially identical behavior observed with uniaxial oscillating fields, where it is evident that only a component of field along the $\pi/2$, $3\pi/2$ axis produces the $\omega \rightarrow 2\omega$ transition. Only for this condition can the remanent magnetization associated with the states shown in Figure 3 change by a domain rotation (fanning) process. Such a process is required to explain the smooth variation in resistance which is characteristic of the waveforms.

When demagnetization is carried out with $H_p \simeq 0$, the demagnetized state should be a mixture of the two states shown in Figure 3. In this case, due to the competing effects of the two states, only a small (or zero) amplitude waveform should be observed until the uniaxial oscillating field exceeds that required for the $\omega \rightarrow 2\omega$ transition. This effect was observed. It was found that even the field due to the detector current influenced the amplitude. Therefore, to unequivocally determine the effect of small $H_p$, the current was turned off during the demagnetization cycle.

## CONCLUSIONS

The $\omega \rightarrow 2\omega$ transition observed in the magnetoresistance waveform for a chevron stretcher detector element is explained by the existence of an initial remanent state which is established by an inplane dc magnetic field. When the Permalloy is demagnetized by decreasing the rotating inplane field, a dc field no larger than the earth's field is sufficient to establish this state. The observed waveform with $\omega$ periodicity is consistent with a fanning magnetization process. The $2\omega$ periodic waveform is consistent with a combination of fanning and switching between alternate states. Both processes also agree with previously observed Permalloy behavior in the presence of magnetization dispersion[4] (i.e., ripple). The large demagnetizing fields associated with narrow Permalloy device elements strongly favor a magnetization fanning process once remanent states like those in Figure 3 are established. The random distribution in anisotropy directions on a microscopic scale (to which dispersion is attributed) assures that such states will form under the appropriate conditions.

## REFERENCES

1. Andrew H. Bobeck, Irnej Danylchuk, Frederick C. Rossol and Walter Strauss, IEEE Trans. MAG-9, 474 (1973).
2. W. C. Hubbell, S. K. Singh and F. G. West, "Magnetoresistive Properties of a Chevron Stretcher Detector", paper submitted to this conference.
3. I. S. Jacobs and C. P. Bean, Phys. Rev. 100 1060 (1955).
4. F. G. West, Proceedings of the (Leuven) Symposium on Electric and Magnetic Properties of Thin Metallic Layers, Brussels, 1961, p. 243.

# PERMALLOY FILM FOR BUBBLE DETECTION AND PROPAGATION

S. Sakai, S. Matsuyama and M. Segawa
Fujitsu Laboratories Ltd., Kawasaki, Japan

## ABSTRACT

Permalloy films for bubble circuits were investigated. Films were prepared by the vacuum evaporation of Ni-Fe alloys containing 70% to 90% Ni onto heated glass substrates (150°C to 340°C). The films were then etched to form 10-μm wide strips. From observations of the Bitter patterns, it was shown that an anisotropy was induced in the negative magnetostrictive films. They showed high magnetoresistive sensitivity. Some films had a stripe domain structure and exhibited higher magnetoresistance, but the coercive force was too large for bubble propagation. For a practical bubble circuit, 84% Ni-Fe films evaporated onto 250°C substrates were selected by optimizing the magnetoresistive sensitivity and the coercivity. Using this material, a practical bubble circuit was fabricated for 3-μm bubbles with the serpentine detector of 375 Ω, and an output sensitivity of 1 mV/mA was obtained at a driving field of 40 Oe.

## INTROCUCTION

One reason that Permalloy film has been used for magnetic bubble circuits is its low coercive force.[1] The introduction of the thick film serpentine detector proposed by Bobeck[2] requires that the Permalloy have a high magnetoresistance (MR) in addition to the low coercivity. Komenou et al. reported that a Permalloy film prepared with a low deposition rate (3 Å/sec) and a high substrate temperature (320°C), have a high MR sensitivity for a bubble detector.[3] Various Fe-Ni alloy compositions prepared using different film deposition conditions were investigated and the most favorable bubble circuit material has been determined by optimizing the MR sensitivity and the coercivity.

## EXPERIMENT

Specimens were prepared by the E-beam evaporation of 70 to 90% Ni-Fe alloy with a deposition rate of 3 Å/sec onto Pyrex glass substrates at various temperatures. The Ni content in the film is reduced by about 2% from the source ingot. The evaporated films were chemically etched to form serpentine detectors and 10-μm wide by 150-μm long test strips. The change in resistance was measured in the longitudinal direction of the strip by applying magnetic fields in the longitudinal ($\Delta R_{/\!/}/R$) and transverse ($\Delta R_{\perp}/R$) directions.

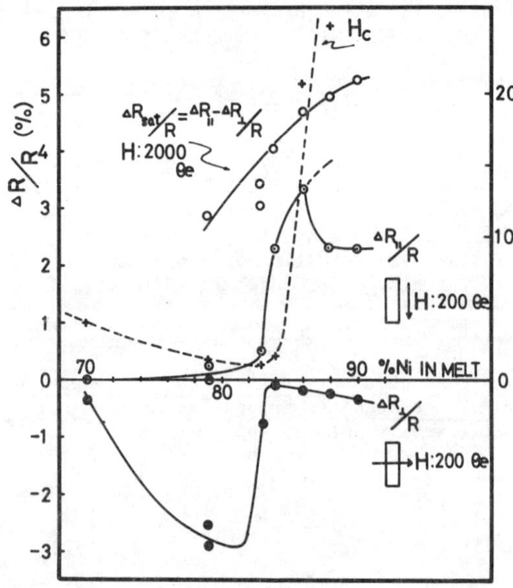

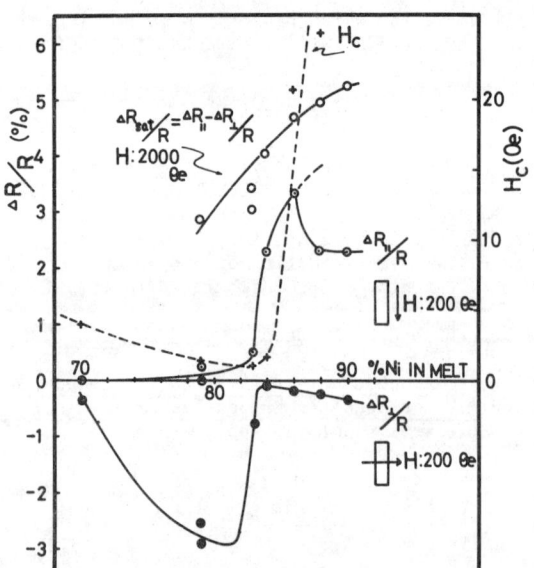

Fig. 1. Ni concentration dependence of $\Delta R/R$ and Hc of the strip. $\Delta R_{/\!/}/R$ and $\Delta R_{\perp}/R$ are measured under the applied field of 200 Oe in the longitudinal and the transverse direction, respectively. R is measured after removal of the applied field.

The results of measurements of $\Delta R_{/\!/}/R$ and $\Delta R_{\perp}/R$ of the strips are plotted in Fig. 1, where R is measured in the residual magnetization state, and the applied field is 200 Oe, which is approximately the same as the $4\pi M_s$ of the bubble material.

The saturation value $\Delta R_{sat}/R = (\Delta R_{/\!/} - \Delta R_{\perp})/R$ of specimens was measured by applying 2000 Oe. The measured value agrees with the value for bulk Ni-Fe alloys.[4] It indicates the film specimens do not have any special structural deviations from bulk materials.

The dependences of $\Delta R/R$ and $H_c$ upon substrate temperature is illustrated in Fig. 2.

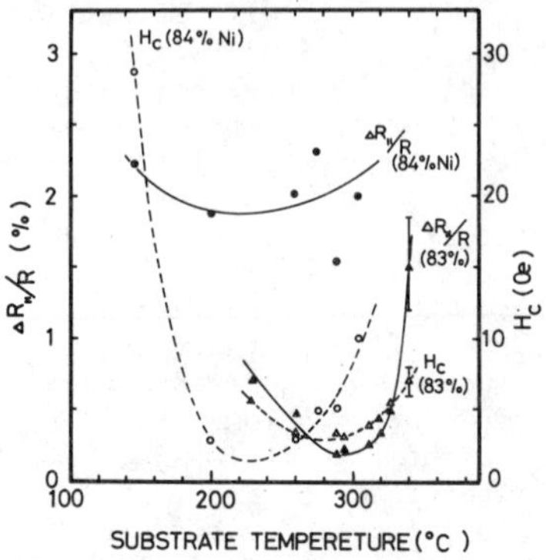


Fig. 2. Dependence of $\Delta R_{/\!/}/R$ and $H_c$ on substrate temperature.

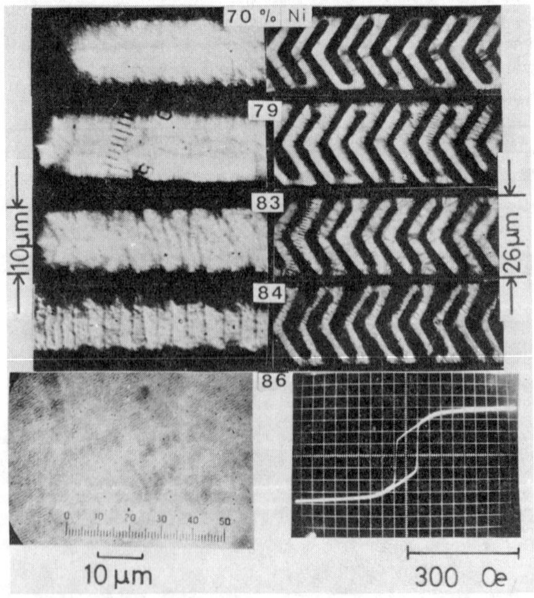

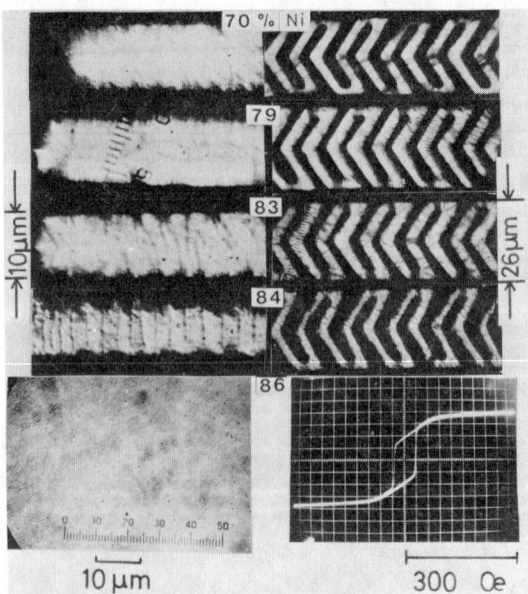

Fig. 3. Bitter patterns of 70 to 86% Ni-Fe strips after a magnetic field of 120 Oe has been applied along longitudinal axis of strips and removed. In serpentines, the same domain structures are observed.

## DISCUSSION

Fig. 1 shows a remarkable change of $\Delta R/R$ vs composition at the zero magnetostrictive composition (81% Ni, 19% Fe). The photographs in Fig. 3 show Bitter patterns of these strips after a magnetic field of 120 Oe has been applied and removed. They indicate that films with Ni contents greater than 81% (negative magnetostrictive film) have a weak uniaxial magnetic anisotropy with its easy axis across the strip. The same domain structure exists in a serpentine detector, as shown in Fig. 3. This anisotropy is induced by the interaction between the negative magnetostriction and the anisotropic reduction of residual tensile stress caused by the removal of the film surrounding the strip

Strips with Ni contents less than 81% exhibit a longitudinal anisotropy. The change in the direction of the anisotropy corresponds to the change of the sign of the magnetostrictive constant. The large $\Delta R_{\parallel}/R$ in the Ni-rich films and the $\Delta R_{\perp}/R$ in the Ni-poor films are explained by the transverse and longitudinal anisotropies, respectively.

In 86% Ni films, stripe domains were observed as shown in Fig. 3.[5] These domains indicate the existence of a perpendicular anisotropy which could be caused by a large negative magnetostrictive constant. In films with a Ni content greater than 86%, $\Delta R_{\parallel}/R$ is reduced, as shown in Fig. 1, because the applied field of 200 Oe is not sufficient to saturate films with stripe domains. The domains were also found in low substrate temperature samples of 83% Ni.[6]

As shown in Fig. 2, $\Delta R_{\parallel}/R$ of the film deposited onto higher temperature substrates (above 320°C) have larger values. In these films crystal grains grow large and nodular surfaces are observed, as shown in Fig. 4. It is considered to concern the perpendicular magnetization component caused by perpendicularly grown columnar grains.[7]

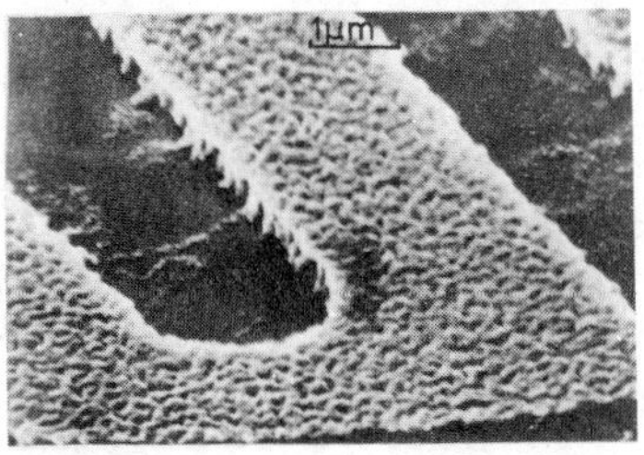


Fig. 4. Scanning electron micrograph of the film deposited from 83% Ni ingot onto a substrate at 325°C.

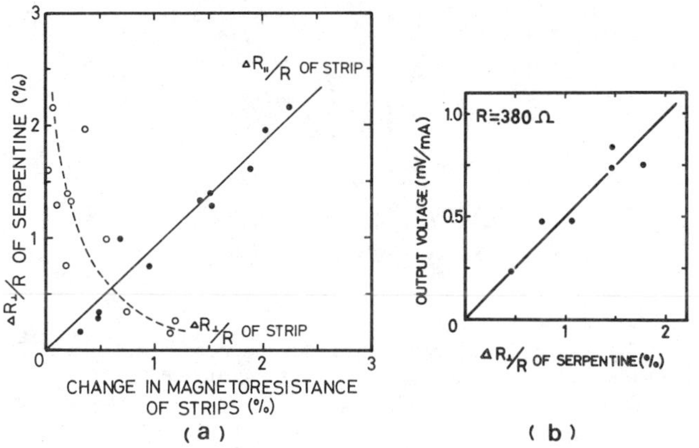


Fig. 5. The relation between the output voltage of the serpentine detector and magnetoresistance change of the strip in various films.

## PERMALLOY FOR BUBBLE CIRCUIT

Fig. 5 (a) illustrates the relation between the MR change of serpentine detectors and the $\Delta R_{\parallel}/R$ and $\Delta R_{\perp}/R$ of strips for various films. A field of 200 Oe is applied across the serpentine elements. Since the output voltage is proportional to the $\Delta R_{\perp}/R$ of various serpentine samples as shown in Fig. 5 (b), the films which have large $\Delta R_{\parallel}/R$ in strips, i.e. negative magnetostrictive films, are concluded to be suitable for serpentine detectors.

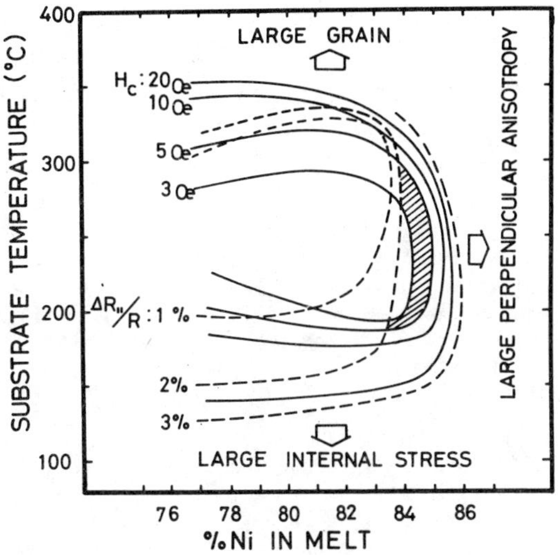


Fig. 6. Plot of magnetoresistive sensitivity and coercive force in the Ni concentration and substrate temperature plane.

The films which have stripe domain structures exhibit high MR changes. This is attributable to the rotation of the perpendicular component of the magnetization. However, in these films the coercive force is so high that they are not suitable for practical bubble circuits.

The results of the above discussion are plotted in Fig. 6 where contour lines of $H_c$ and $\Delta R_{\parallel}/R$ are illustrated. The hatched region indicates $\Delta R_{\parallel}/R \gtrsim 2\%$ and $5\ \text{Oe} \gtrsim H_c \gtrsim 3\ \text{Oe}$, which is suitable for use in practical bubble circuits. Considering the bubble stability at the bit position during the start/stop operation, we have estimated the most suitable $H_c$ is about 4 Oe.

## CONCLUSION

In conclusion, Permalloy with 84% Ni evaporated onto substrate at about 250°C at a deposition rate of 3 Å/sec is most favorable. Using this condition, a practical bubble circuit was fabricated including a serpentine detector for 3-μm bubbles, and an output sensitivity of 1 mV/mA through 375Ω was obtained at a drive field of 40 Oe.[8]

The authors are deeply indebted to Prof. S. Uchiyama, Nagoya Univ., for many valuable discussions.

## REFERENCES

1. J. P. Reekstin, et al., J. Vac. Sci. Tech., 10, 847, (1973).
2. A. H. Bobeck et al., LEEE Trans. Magn., MAG-9, 474, (1973).
3. K. Komenou, H. Nakajima and K. Asama, AIP Conf. Proc., 24, 554, (1974).
4. R. M. Bozorth, "Ferromagnetism", 8th, Van Nostrand, New Jersey, U.S.A., (1964), 758.
5. N. Saito, H. Fujiwara and Y. Sugita, J. Phys. Soc. Japan, 19, 1116, (1964).
6. M. M. Hanson, et al., Appl. Phys. Letters, 9, 99, (1966).
7. T. Iwata, R. J. Prosen and B. E. Gran, J. Appl. Phys., 37, 1285, (1966).
8. S. Igarashi, et al., "DESIGN OF A 3 μm BUBBLE 80-kbit MEMORY CHIP", paper submitted to this conference.

# DEVICE CHARACTERIZATION OF A COMPLETE SET OF PASSIVE BUBBLE LOGIC FUNCTIONAL ELEMENTS*

C. T. M. Chang, Texas Instruments Incorporated,
P. O. Box 5936, M. S. 145, Dallas, Texas 75222

ABSTRACT

This paper reports the operating characteristics of a set of passive logic elements. The elements were fabricated on $(YSm)_3(GaFe)_5O_{12}$ garnet using identical chip fabrication parameters. The set consists of generator, annihilator, bubble splitter, AND/OR gate, and bubble crossover. The experimental results show that all elements except the crossover operate with an overlapping bias range of 8 to 10 Oe in a 40 Oe drive field at 100 kHz.

## INTRODUCTION

In a recent work, Lee and Chang[1] discussed magnetic bubble logic using elements reported in several previous publications. While the abilities of these elements to perform logic operations, in principle, were described; the important question of physical realizability of some of these elements was not discussed and no operating characteristics of the elements were given. Although the bias field versus drive field margins of some of the logic devices were previously reported, the diversity of the material and processing parameters associated with them has prevented ascertaining whether or not they can be made to operate together to perform magnetic bubble logic. This paper reports on the operating margins of a set of logic elements fabricated on the same garnet material using identical chip fabrication parameters. The set consists of generator, annihilator,[2] bubble splitter,[3] AND/OR gate[4] and bubble crossover. All elements in the set are passive and require no conductor control. The set is "complete" in that collectively they are capable of carrying out any logic operation.[5] The bubble inverter was excluded from the set because (i) no satisfactory approach to realize it was found and, (ii) its function can be achieved by generating and routing the bubble signal and its complement together and interchanging them whenever inverting operations are required.

The experimental devices were fabricated on $(YSm)_3(FeGa)_5O_{12}$ garnet slices approximately 5 μm thick which support 5 μm diameter bubbles. The $SiO_2$ spacer and Permalloy film are 10,000 Å and 4,500 Å thick, respectively. The width of the Permalloy elements and the gaps between elements in the propagating circuits are nominally 2.5 μm and 1.25 μm, respectively.

All the parameters are those successfully employed for fabricating memory devices in our laboratory. Characterization was done at both 1 Hz and 100 kHz. Since the results are similar for both frequencies, only data for 100 kHz are presented here, with the exception of the crossover element which operated only quasi-statically.

## EXPERIMENTAL RESULTS

AND/OR Gate. Bonyhard, et al, reported an AND/OR gate test circuit which operated with 6 Oe bias field range[4] at 100 kHz. A similar device designed for 5 μm bubbles was fabricated and is shown in the photomicrograph of Figure 1. The gate only includes those elements in the vicinity of a, b, c and d. The other elements constitute bubble return paths only. Input bubbles come from paths Aa and Bb and are labeled as A and B input bubbles, respectively. When a single bubble is located at either a or b, it passes through c and gives rise to the A + B output from the gate area. However, the simultaneous appearance of bubbles at a and b causes the two to compete for position c one quarter of a drive cycle later. The competition results in the A bubble occupying c and B bubble being repelled to point d. Thus, one obtains bubbles at the A + B and A·B outputs.

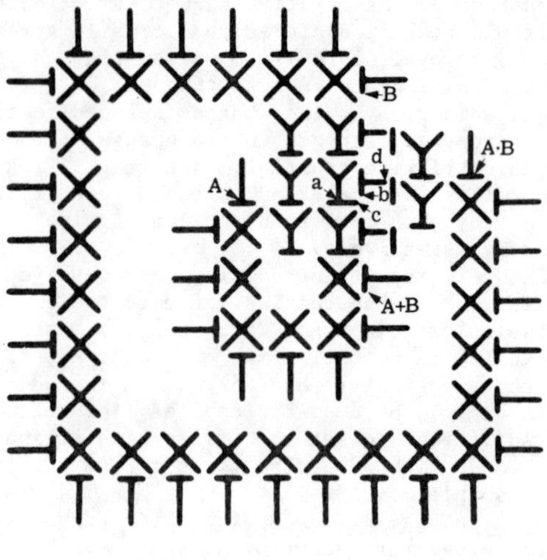


Fig. 1. Photomicrograph of the AND/OR Gate.

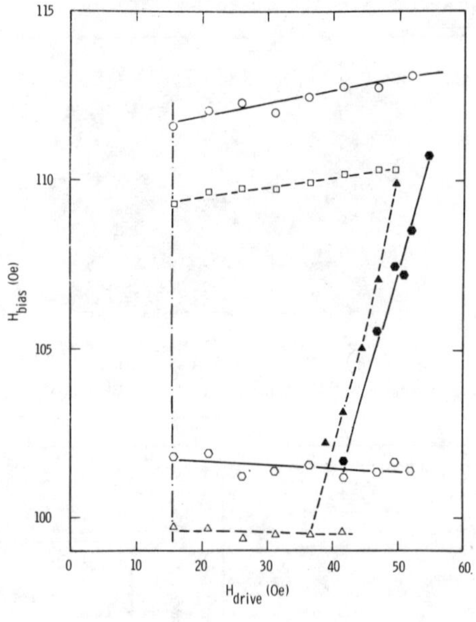


Fig. 2. Bias Field Versus Drive Field Margin of the AND/OR Gate Shown in Figure 1.

Characterization of this device confirmed the failure modes for the AND/OR gate reported by Bonyhard, et al,[4] and it was found that both the upper and lower margins of the complete circuit at 100 kHz depend on how the outer return loop is populated. In Figure 2, the open circles and the open hexagons are the measured upper and lower margins, respectively, for a sparsely populated outer loop, i.e., bubbles no closer than every other bit position. The open squares and the open triangles are the corresponding data points for the fully populated outer loop. Failures at low drive field are due to shifts in bubble position. Although no failure was observed at the higher drive field when sparsely populated, failure was observed (> 42 Oe near stripout and 48 Oe near collapse) when fully populated. In addition, for start-stop operation, an upper limit on the allowable drive field was observed. This is due to the disappearance of bubbles near d when the field is stopped such that it is pointing to the top of Figure 1.

These measurements are shown by the solid triangles and hexagons for the fully and sparsely populated outer loops, respectively.

Generator and Annihilator. In many logic operations, it is often necessary to generate a continuous string of bubbles and to annihilate bubbles. We have fabricated and tested several generator and annihilator designs. Examples of these are shown in the inset of Figure 3, where a generator and an annihilator,[2] marked by G and A, respectively, are joined by T-bars. The operating margins of this circuit are presented in Figure 3, where the upper margin was limited by the collapse of the seed bubble on the generator disc, and the lower margin was due to the escape of bubbles from the annihilator.

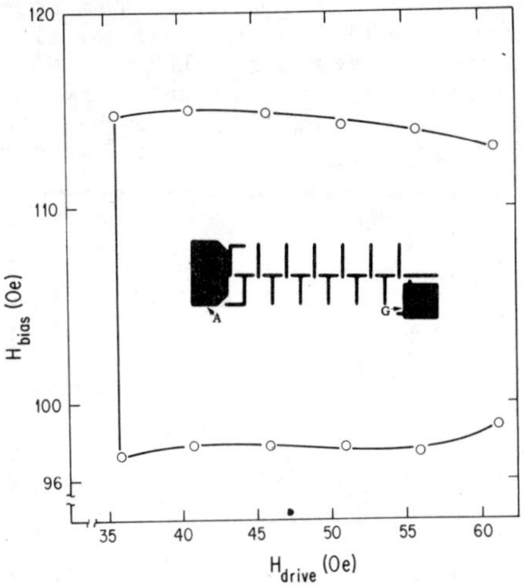


Fig. 3. Bias Field Versus Drive Field Margin of the Generator-Annihilator Circuit Shown in the Inset.

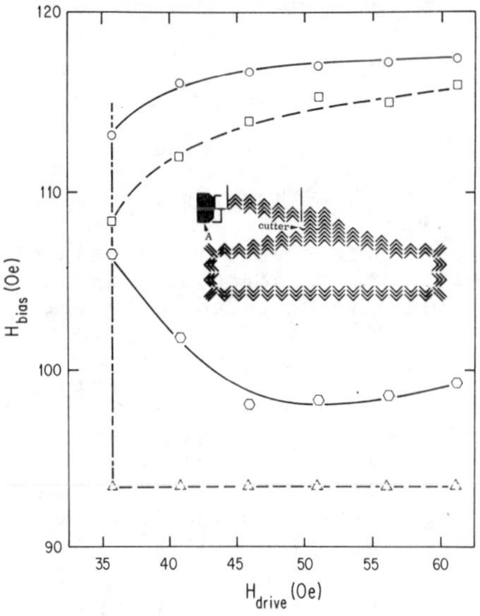


Fig. 4. Bias Field Versus Drive Field Margin of the Bubble Splitter Circuit Shown in the Inset.

The Bubble Splitter. A bubble splitter is used to copy a data stream so that the fanout to separate data paths can be accomplished. A photomicrograph of an element similar to that reported by Nelson[3] is shown in the inset of Figure 4. Unlike the Nelson design,[3] the present splitter uses 90° chevrons. The operating margins are also shown in Figure 4. A shift in bias margins similar to that reported by Nelson[3] was seen when comparing fully populated (squares and triangles) versus sparsely populated loops (circles and hexagons). The upper margins result from bubble collapse at the cutter. The lower margin for the sparsely populated circuit (hexagons) is due to bubble stripping. Stripout of bubbles in the fully populated circuit was not observed. The lower margin for the fully populated circuit (triangles) was limited by the interference of stray bubbles from the adjacent test circuits on the chip. Failure at low drive field is again due to shifts in bubble position.

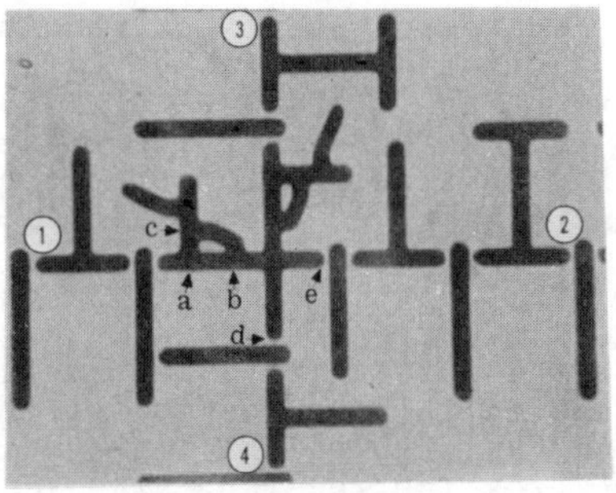


Fig. 5. The Bubble Crossover Element.

Crossover. A crossover element enables bubbles on intersecting data paths to continue traveling along their respective paths. We have designed and fabricated an experimental crossover which is shown in Figure 5. Bubbles from tracks 1 - 2 and 3 - 4 pass through the crossover region at different times within one drive field cycle. This results in spatial separation of bubbles from the two tracks and minimizes the bubble interaction at the crossover. Crossover operation was observed at approximately 16 Oe quasistatic drive field over a narrow bias range at 111.5 Oe. At a higher drive field, bubbles from point 1 go to output point 4. At lower drive fields, bubbles become trapped in the vicinity of a-b-c. At higher bias fields, the bubbles collapse at positions in between points a and b. For lower bias fields, bubbles were found to strip out across points d and e resulting in incorrect crossover operations. The operation was also found to be sensitive to the data patterns in the immediate vicinity of the crossover element.

## CONCLUSION

The results presented above show that the AND/OR gate, the bubble splitter and the crossover were all sensitive to the data patterns. This limits the overlapping bias range for all elements presented here, except the crossover, to 8 Oe in 40 Oe drive field for 100 kHz operation. However, this overlapping bias range can be increased to more than 10 Oe by requiring that no more than every other position in any bubble path be populated. Although the crossover element has not been reliably implemented, this experience suggests that bubble logic can be successfully implemented.

## ACKNOWLEDGEMENTS

The author would like to thank G. G. Sumner for providing the garnet material, and M. S. Shaikh and J. W. Taylor for fabricating the Permalloy circuits.

## REFERENCES

*Work partially supported by U. S. Air Force Avionics Laboratory.

1. S. Y. Lée and H. Chang, IEEE Trans. Magnetics, MAG-10, 1059 (1975).
2. W. F. Druyvesteyn et al, Inst. Phys. Conf. Ser. No. 25, 1975, p. 53.
3. T. J. Nelson, AIP Conf. Proc. 24, 624 (1975).
4. P. I. Bonyhard, et al, IEEE Trans. Magnetics, MAG-9, p. 708 (1973).
5. F. G. West, et al, Air Force Avionics Lab. Tech. Rept. # AFAL-TR-74-196, Sept. 1974, p. 23.

# CORROSION AND MAGNETORESISTANCE OF PERMALLOY-RHODIUM ALLOYS

James C. Suits
IBM Research Laboratory, San Jose, CA 95193
D. W. Rice
Materials Science, IBM General Products Division, San Jose, CA 95193

## ABSTRACT

Bulk samples of permalloy-rhodium alloys have been fabricated according to the formula $(Ni_{.78}Fe_{.22})_{1-x}Rh_x$ where x is 0, 0.1, 0.2, and 0.3. It was found that the resistance of permalloy to atmospheric corrosion is significantly improved by the addition of rhodium. By contrast, a bulk sample of permalloy with 30 at.% ruthenium showed no improvement in corrosion resistance. The room temperature magnetic moment decreases with increasing rhodium content up to x=0.2 as expected for simple dilution assuming zero moment for the rhodium. The addition of 10 at.% rhodium to permalloy increases the resistivity by a factor of three and reduces the magnetoresistance anisotropy by nearly an order of magnitude. The resistivity and magnetoresistance changes are attributed to increased electron scattering caused by a rhodium 4d band.

## INTRODUCTION

Permalloy is useful in many applications which require high magnetic permeability. Some of these applications also require corrosion resistance. For example, thin film heads for inductive magnetic recording must fly very close to the recording surface and must not corrode with time. Other examples are the use of permalloy for magnetoresistive sensors and for bubble device personalization.

One potential technique for improving the resistance of magnetic materials to atmospheric corrosion is the addition of noble metals. Preferrential oxidation of the magnetic component can present a more noble surface to the atmosphere. Also, solution depolarization in the adsorbed aqueous phase can be enhanced by the noble metal additive. One difficulty with this method may be that the magnetic properties and/or the magnetoresistive properties deteriorate severely before sufficient noble metal is incorporated for corrosion resistance. To investigate this approach bulk samples of permalloy have been fabricated with Rh or Ru added.

## EXPERIMENTAL

Samples of permalloy with Rh added were made by arc melting according to the following formula:

$$(Ni_{.78}Fe_{.22})_{1-x}Rh_x$$

where x=0, 0.1, 0.2, and 0.3. A single sample of permalloy with 30 at.% ruthenium was also made at the same time. After arc melting, the ingots were cut in half and the surface of one half was rough polished, and the other half was machined into bars for electrical measurements. Also, some powder was filed from each ingot for vibrating sample magnetometer measurements. The powder, bars, and half ingots were annealed at 900°C for one hour, furnace cooled to 400°C, then annealed at 600°C for one hour, and air cooled.(1). The surfaces of the half ingots which were previously rough polished were then given a fine polish using alumina powder and water, rinsed with alcohol and dried in compressed air. These air exposed surfaces were analyzed by Auger electron spectroscopy.

The composition of the atmosphere for the corrosion test was designed to accelerate corrosion processes encountered in an urban business office environment. The composition, in addition to purified air, is: $SO_2$ -300 ppb (parts per billion), $NO_2$ -480 ppb, $O_3$ -170 ppb, $H_2S$ -15 ppb, $Cl_2$ -3 ppb, and 70% relative humidity at a temperature of 23°C. The exposure was for 24 hours.

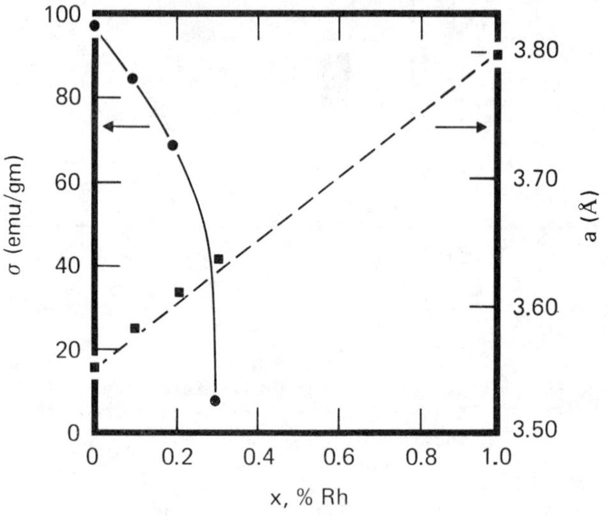


**Figure 1.** Magnetization and lattice constant.

## RESULTS

X-ray diffractometer analysis of the surfaces of the polished half ingots showed the samples with Rh added to be polycrystalline, single phase, fcc solid solutions. The lattice constant increases approximately according to Vegard's law (see Fig. 1). The sample with x=0.2 showed a nearly random orientation of polycrystals, while the other samples showed a strong (100) orientation.

In addition to Rh, substitution with Ru was attempted. The sample with 30 at.% Ru was shown by x-rays diffraction to consist of two phases - a fcc phase with a=3.582 A and a hexagonal phase. The fact that Rh but not Ru can be substituted into the Ni-Fe lattice is suggested by the Ni-Rh and Ni-Ru binary phase systems(2) which show complete solid solubility in the former case and very limited solubility in the latter.

The magnetization σ, measured at room temperature in a field of 18 kOe using a vibrating sample magnetometer, is shown in Fig. 1. The rapid decrease of σ between 20 and 30 at.% Rh is probably due to the lowering of $T_c$ to near room temperature for x=0.3. This is also indicated by the lack of saturation of the M-H loop for x=0.3. Table I shows 4πM calculated from the measured σ and the x-ray density. Also shown, is the

TABLE I

PROPERTIES OF $(Ni_{.78}Fe_{.22})_{1-x}(^{Rh}_{Ru})_x$

| Element added | x=0 – | x=0.1 Rh | x=0.2 Rh | x=0.3 Rh | x=0.3 Ru |
|---|---|---|---|---|---|
| Crystallographic | | | | | |
| Lattice type | fcc | fcc | fcc | fcc | fcc+hcp |
| Lattice constant (Å) | 3.552 | 3.582 | 3.614 | 3.640 | – |
| Calculated density | 8.604 | 9.037 | 9.430 | 9.848 | – |
| Magnetic | | | | | |
| σ(20C,20kOe)emu/gm | 97.2 | 81.9 | 68.5 | 7.1* | – |
| $4\pi M$ (Gauss) | 10,500 | 9,300 | 8,110 | – | – |
| $\bar{\mu}(\mu_B)$ | 1.01 | 0.92 | 0.82 | – | – |
| $H_c$ (Oe) | 0.5 | 1.0 | 2.7 | 1.5-3.0 | – |
| Corrosion | | | | | |
| Dark field reflectivity | | | | | |
| before (mv) | 76 | 152 | 212 | 192 | – |
| after (mv) | 3,400 | 168 | 328 | 164 | |
| Visible corrosion | severe | slight | slight | none | severe |
| Electrical | | | | | |
| Resistivity (μΩ-cm) | 17.8 | 59.8 | 120 | 85.7 | – |
| $\frac{(\rho_{\parallel}-\rho_{\perp})}{\rho_o}$ (H=18kOe) | 3.1% | 0.4% | 0.02% | 0.00% | – |

*not saturated

average moment per atom $\bar{\mu}$. The reduction in $\bar{\mu}$ is that expected for simple dilution by a non-magnetic solute, i.e. incorporation of 10 at.% Rh reduces the average moment by about 10%.

The coercive force $H_c$ is given in Table I. $H_c$ remains low with Rh dilution.

The results of the atmospheric corrosion test are given in Table I. The dark field reflectivity is the intensity of reflected light measured under dark field conditions on a microscope. It is a measure of the light scattered diffusely from the surface. It was found that the scattered light from the pure permalloy was increased by nearly two orders of magnitude by the corrosion product on the surface. Light field examination by microscope shows the entire surface to have become rough and non-metallic looking. The other samples show little or no corrosion by scattered light (Table I). Light field microscope examination showed corrosion product on about 10% of the sample surface for the x=0.1 and 0.2 samples, and no visible corrosion for x=0.3. The extent to which the slight corrosion of the x=0.1 and 0.2 samples is a function of sample inhomogeniety was not investigated, although for the x=0.1 sample the only visible corrosion was associated with grain boundaries. It is remarkable, and not anticipated, that only 10 at.% Rh can have such a large effect on the atmospheric corrosion of permalloy. Auger surface analysis of the air formed films show that Rh is enhanced in the surface region compared to the other cations (Fe,Ni). This is most likely due to the preferrential oxidation of the magnetic components and may be the reason for the improved atmospheric stability of the alloy.

The sample of permalloy with 30 at.% ruthenium showed severe corrosion. This is undoubtedly due to the lack of solid solubility of Ru in $Ni_{.78}Fe_{.22}$.

The resistivity and magnetoresistance anisotropy $\Delta\rho/\rho$ (the difference in resistivity measured parallel and perpendicular to H divided by the zero field resistivity) are given in the Table. The field dependence of $\Delta\rho/\rho$ is shown in Fig. 2. We have seen that the addition of 10 at.% Rh to permalloy only changes the magnetization by about 10%; however, it is seen that much larger changes take place in the electrical properties.

These changes may be attributed to the addition of a 4d band by the Rh which overlaps the Ni-Fe 3d density of states near the Fermi level (see Fig. 3). This new band is not expected to be exchange split. Its hybridization with the 3d band is doubtful in view of the evidence that in such alloys the bands remain relatively independent (see Ni-Cu for example[3]). This increase in the density of d states at the Fermi level contributes to the increase in resistivity with Rh addition, since in addition to 4s-3d-4s scattering, one also has (4s,5s)-4d-(4s,5s) scattering.

Smit[4] has shown that the resistivity of a transition metal alloy may be represented by a four resistor network (see lower part of Fig. 3). The s-s scattering is small (therefore a small s-s resistor) for both spin directions due to the low

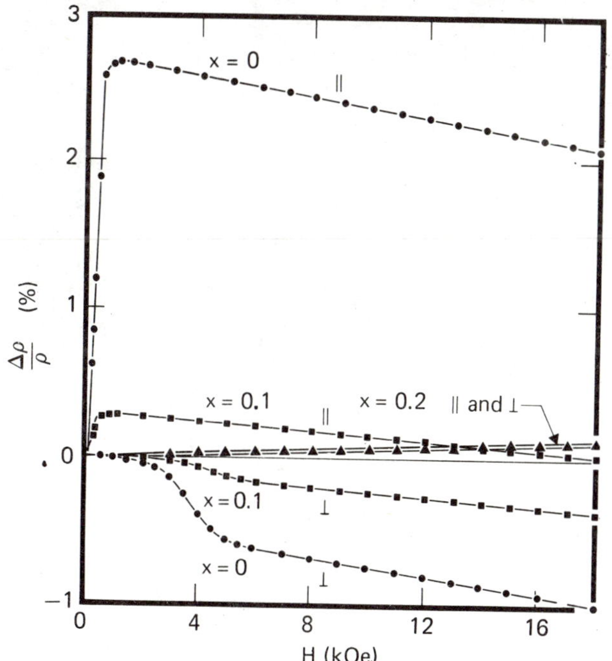


Figure 2. Magnetoresistance versus field.

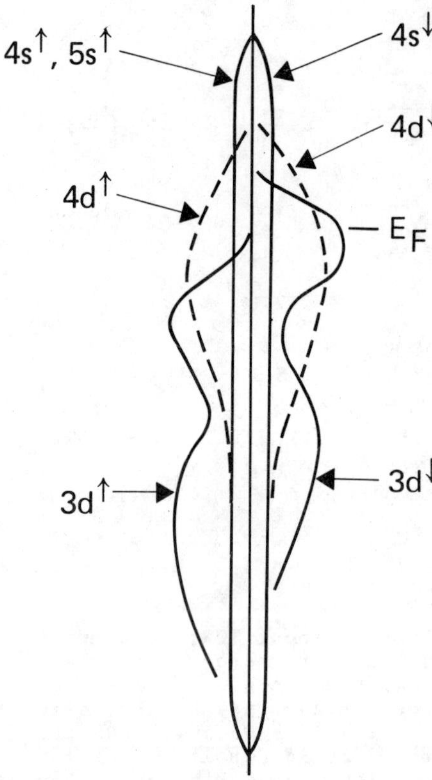


**Figure 3a.** Schematic band structure.

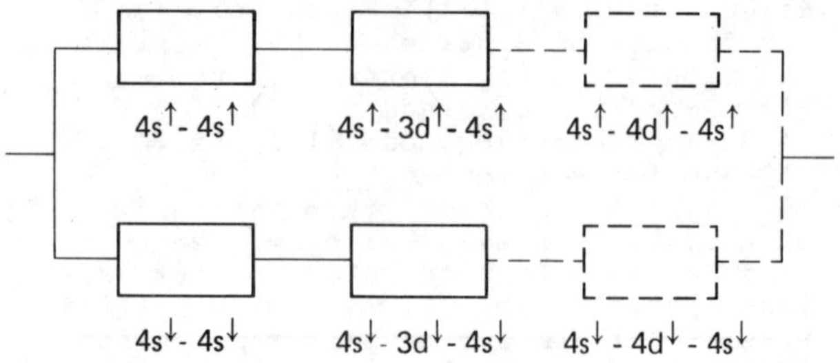


**Figure 3b.** Equivalent network.

density of 4s states. The s-d-s scattering for 3d states is small for spin up electrons because the 3d↑ band is nearly full and large for spin down electrons; therefore, for pure permalloy the conduction is large via the upper branch of the network. Adding Rh adds two additional resistors representing scattering into the 4d band and both resistors are large since this band is not nearly full. The large increase in $\rho$ and therefore a decrease in $\Delta\rho/\rho$ may be understood on this basis. This model thus predicts that $\Delta\rho/\rho$ will generally decrease for substitutions of 4d and 5d elements.

## CONCLUSIONS

Bulk samples of permalloy with Rh added show improved resistance to atmospheric corrosion. This improvement is obtained with relatively small changes in magnetic moment and coercive force. This result may have application in areas where corrosion resistant high permeability materials are important, such as in thin film recording heads. On the other hand, Rh addition causes a strong decrease in magnetoresistance anisotropy which suggests that it may not be useful in magnetoresistive applications.

## ACKNOWLEDGEMENTS

The authors wish to thank G. Guthmiller, R. Tremoureux, and C. Breen for technical assistance.

## REFERENCES

1. R. M. Bozorth, Ferromagnetism (D. Van Nostrand Co., New York 1956).
2. F. A. Shunk, Constitution of Binary Alloys, Second Supplement (McGraw-Hill Book Co., New York 1969) p. 551.
3. S. Hufner, G. K. Wertheim, R. L. Cohen, and J. H. Wernick, Phys. Rev. Letters, 28, 488 (1972).
4. J. Smit, Physica XVI, 612 (1951).

# DOMAIN WALL OBSERVATION OF PERMALLOY OVERLAY BARS BY INTERFERENCE CONTRAST TECHNIQUE

Issa Khaiyer, T.H. O'Dell

Department of Electrical Engineering, Imperial College
London S.W.7, England

## ABSTRACT

Direct observation of the domain wall structure in real-size Permalloy overlay bars on the glass substrate has been made. This has been achieved by the application of the Interference Contrast Technique after Nomarski. The experimental measurements of the lateral displacement of the $180^\circ$ domain wall due to the application of an in-plane field have been taken and compared to the theoretical results obtained by the domain wall model. The overall experimental observation and results obtained confirm this type of device modelling.

## INTRODUCTION

Since the development of the field access devices the theoretical study of the magnetostatic field and magnetization in thin film Permalloy overlay bars has been approached by different types of device modelling which is briefly summarized in [2].

Even though each one of these models explains some of the experimental results, their application in field analysis and device optimization could be seriously limited since they ignore the domain wall pattern inside the real size overlay bars, a physical phenomenon in their structure which has not been known clearly up to this time. Recently a device modelling has been proposed based upon the Bitter pattern observation of the domain wall pattern of the larger overlay bars[3]. The comparatively accurate results obtained by this model have been an encouraging motive to search into different techniques with which the observation of the domain wall pattern in these small elements would be possible. The conventional techniques such as normal or dark field illumination failed to produce any desirable results, but the application of Interference Contrast Technique after Nomarski[1] produced excellent results and clear images of the ferromagnetic fluid lines formed on the domain walls of the Permalloy overlay bars.

## EXPERIMENT

The device employed is a standard colour interference contrast equipment (e.g. Reichert, Wein) and the relevant theory may be found in reference [1]. A critical layer of diluted commercial ferromagnetic fluid[4] was produced between the Permalloy overlay bars (on a glass substrate) and a cover glass. With Interference Contrast Technique (ICT) any unevenness in level of the order of milimicron resolution can be easily observed. The image is visualized with a pronounced three dimensional effect against the background. The thickness of the ferrofluid lines formed on the domain walls are in the range of 1. micron depending upon the density of the ferrofluid and can easily be detected by this technique, because the refractive index of the fluid depends upon the particle concentration [5].

The photographs taken from the domain walls of a simple I and T pattern bar and a bubble generator are shown in Fig. 1. The $90^\circ$ domain walls near the bars ends where a closure domain usually exists, do not form in these bars due to the circular shape of the edges. The magnetic moments rotate gradually parallel to the bar edges due to the self demagnetization effect causing an overall magnetization reversal from one section to the opposite one, Fig. 2.

Dimensions in microns:
I bar : 48 x 8 x 0.5
T bar Top : 30 x 8 x 0.5
Dia. = $48^{\mu m}$

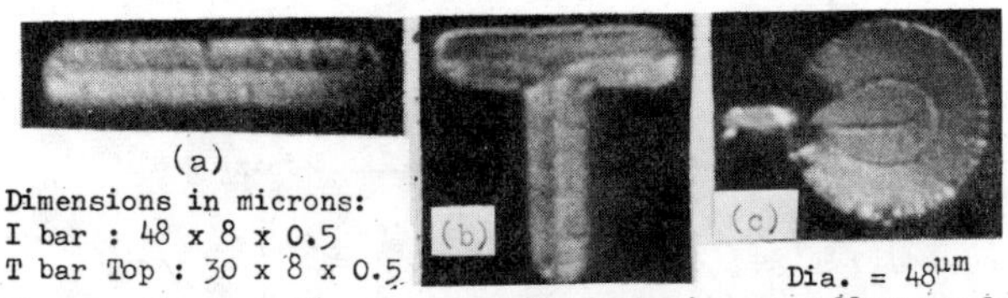

Fig. 1. Domain wall pattern of the real size permalloy overlay bars on glass substrate. (a) Simple I bar, (b) T bar, (c) Bubble generator. The scales of the photographs are the same in (a) and (b), but smaller in (c).

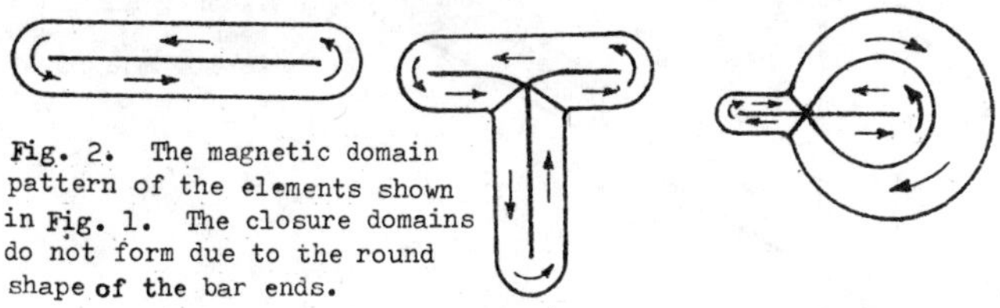

Fig. 2. The magnetic domain pattern of the elements shown in Fig. 1. The closure domains do not form due to the round shape of the bar ends.

The thickness of the bars on different samples varies between 0.5 - 1.0 microns.

When a uniform in-plane field is applied parallel to the bar the $180^\circ$ domain walls move laterally bowing slightly in the middle. The photographs in Fig. 3 show this. The slight curvature in $180^\circ$ domain wall

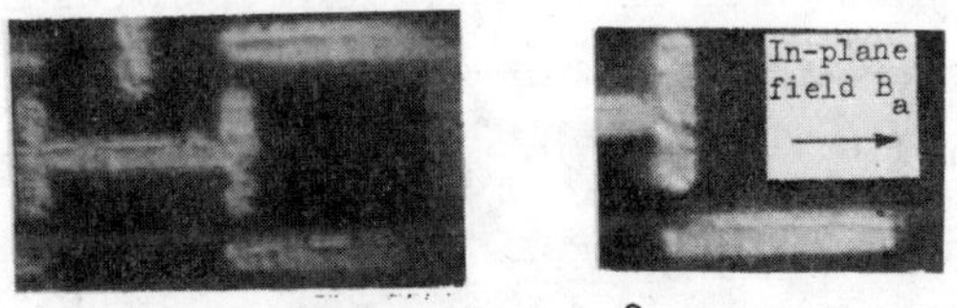

Fig. 3. The lateral movement of the $180^\circ$ domain wall due to the in-plane field $B_a$. There is a slight curvature in $180^\circ$ domain wall. Application of a bias field sharpens the domain walls since the walls are of Bloch type

is not taken into account by the domain wall model proposed previously for a simple I bar[3]. This is because that bar (21 x 4.5 x 0.46 $\mu m^3$) is magnetized by 24% under 17.0e in-plane field and its $180^\circ$ domain wall at this stage, can be approximated as a straight line. However, for a very accurate analysis the curvature in $180^\circ$ domain wall should be taken into account.

The lateral displacement of the two different points on the $180^\circ$ domain wall is measured as a function of $B_a$ the applied in-plane field parallel to the bar and the results are plotted in Fig. 4. It may be noted that the lateral movement of the point 1 is less and slower than that of the point 2 and the same applies for the net magnetization of the bar at these points.

A similar behaviour is reported on the net magentization of the bar at points 1 and 2 in Ref. [6], but the reason for it is incorrectly attributed to the influence of the additional fields from the ends of the neighbouring T elements.

The information obtained by ICT about the domain wall movement of a T bar due to an in-plane field is rather interesting. When $B_a$ is parallel to the top of the T bar the $180^\circ$ domain wall moves laterally as expected. When $B_a$ is parallel to the stem, the $180^\circ$ domain wall at the stem moves also laterally but the point A moves on a semi circle like the point B as shown in Fig. 5.

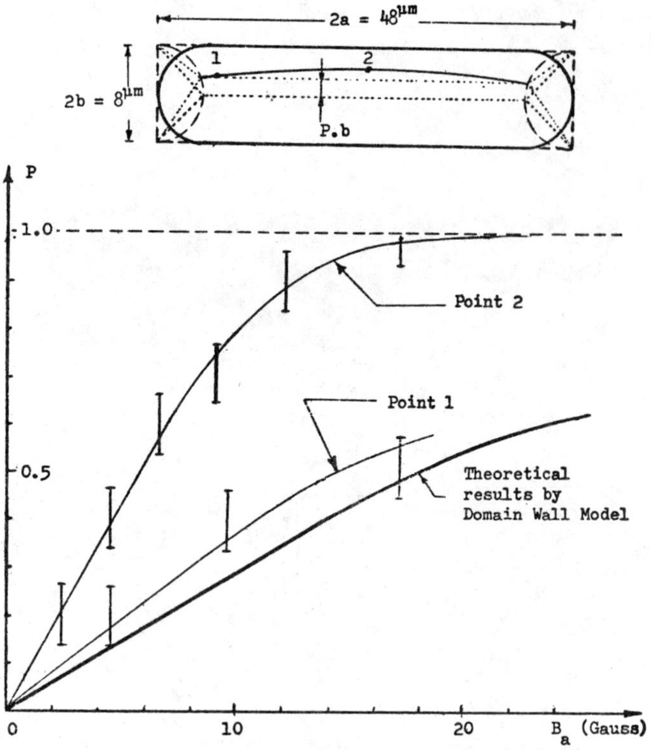


Fig. 4. Experimental measurements of the lateral displacement for two different points on the 180° domain wall of a simple I bar (48 x 8 x 0.5 μm³). This graph shows that the different values of the net magnetization at points 1 and 2, measured by magneto-optical Kerr effect[6] are simply related to the domain wall behaviour of the bar due to an applied parallel in-plane field and not the influence of the additional fields from the ends of the neighouring T elements.

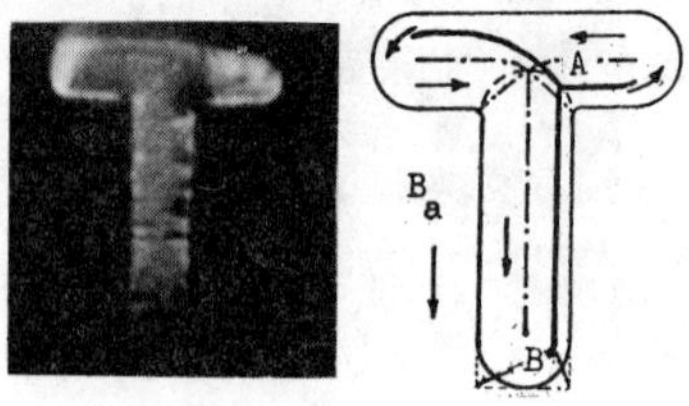


Fig. 5. Domain wall movement of a T bar due to an in-plane field $B_a$ = 7 Gauss applied to its stem. The photograph is taken from a T bar 30μm long at the top, 35μm long at the stem and 8μm in width.

This occurs because the 180° domain wall AB resists stretching since it aquires more energy.
As a result, the position of the 180 domain wall in the top of the T becomes disturbed resulting in a consequent disturbance in the magnetization of this part. One side of the 180° domain wall moves up to allow for an increase in the magnetization of the section that follows the expanded section of the stem, in order to keep the variation in the magnetization minimum (divM = 0) while the other side of the 180° domain wall moves down to follow the contracted section of the stem. But since there is no in-plane field applied parallel to the top of the T bar, the net magnetization in this direction remains zero. If a net magnetization is measured along the top of the T bar, the magnetization profile will be similar qualitatively to that obtained by the magneto-optical Kerr effect measurement[6], Fig. 6. As it may be seen this is due to the position of the domain wall in the T bar which is based on a minimum energy principle and not the interaction of the elements as described in [6].

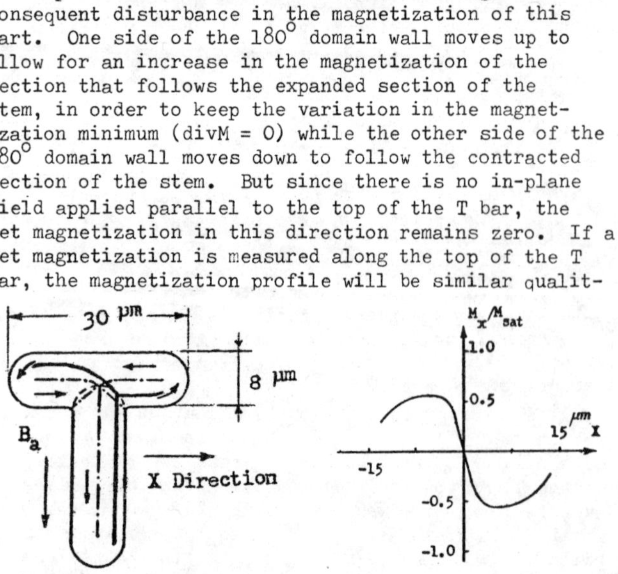

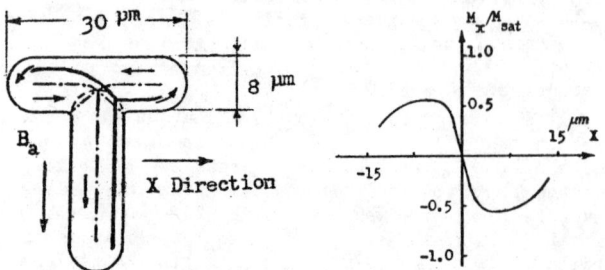

Fig. 6. The net magnetization distribution along the top of a T bar when an in-plane field $B_a$ is applied parallel to its stem.

Acknowledgement

The authors would like to thank Dr. M.E. Jones from Post Office Research Laboratory, London, England for the permalloy samples on glass substrate.

References

1 - G. Nomarski, A.R. Weil , "Application a la métallographie des méthodes interferentielles à deux ondes polariseées", Revue de Metallurgie, Ll1 (1955), p. 121, 134.

2 - Issa Khaiyer, "Device optimization and theoretical explanation of the bubble domain propagation in bubble devices", ICEE Conf. Oct. 27-30, 1975, paper B6-3, Pahlavi University, Shiraz, Iran.

3 - Issa Khaiyer, "Models for I and T pattern permalloy overlay bars based upon Bitter pattern observation of domain wall movements", Intermag Conf. 1975, paper 21-5, to be published in March 1976 issue of IEEE Trans. on Magnetics.

4 - Water base ferrofluid - type A01 - 200 of Ferrofluid catalogue, Ferrofluidics Corporation.

5 - L.P. Landau, E.M. Lifshitz, "Electrodynamics of Continuous media", Pergamon Press Ltd., 1960 p. 299-302.

6 - G.S. Krinchik, E.E. Chopurova and U.N. Shamatov, V.K. Raev and A.K. Andreev, "Magnetization Reversal study in permalloy bubble-propagation circuits with Magneto-Optical Equipment of Micron Resolution", AIP Conf. proc. 24, San Fransisco 1974.

## ANALYTICAL MODEL OF A BUBBLE SWITCH

M. S. Cohen, W. J. Hsieh, and G. S. Almasi
IBM T. J. Watson Research Center
Yorktown Heights, New York 10598

ABSTRACT

A model is described for a bubble-domain switch constructed from a Y-bar, in which the current through a conductor crossing the base leg of the Y determines whether a bubble passes from one arm to either (a) the base or (b) the opposing arm of the Y. Switching experiments were conducted with a rotating field, as well as with a static dc in-plane field for ease in checking the theory. As expected, the current required for switching is found to increase linearly with in-plane field amplitude.

The analytical model is based on a theory[1] of effective bubble bias field originating in the interaction of the bubble with the Permalloy of the Y in the presence of the in-plane field; this theory is extended to take into account the effect of the current in the switch conductor. Excellent agreement is found with the static experiments, e.g., at 40 Oe the theory predicts and the experiments verify a switch current of 20 ma for 7-8 μm diameter bubbles. The model is extended to geometrical variations of the original design; in particular, the heuristically derived scaling law[2] for decreasing bubble diameters was investigated more rigorously and found to be valid.

1. G. S. Almasi and Y. S. Lin to be published.
2. M. S. Cohen and H. Chang, Proc. IEEE, 63, 1196 (1975).

# EFFECTS OF INTERFACIAL MIXING ON THE PROPERTIES OF RF SPUTTERED NiFe FILMS

P. S. Ho, K. Y. Ahn and J. E. Lewis
IBM T. J. Watson Research Center, Yorktown Heights, New York 10598

## ABSTRACT

Auger Electron Spectroscopy in conjunction with ion sputtering technique has been used to measure the composition profile in Au/NiFe and NiFe/$SiO_2$ films. Interfacial mixing was found to exist as a result of biased sputtering and its extent was affected by bias voltage and substrate materials. Changing the bias voltage from 0 to 60 volts in sputtering NiFe films of 600Å thickness on $SiO_2$ can reduce the resistivity by 40% and $H_c$ by about 80% but increase the magneto-resistive effect from 1% to 2.5%. Also as a result of mixing at the Au/NiFe interface, the magnetic moment has been observed to be diluted by about 80% in a 500Å NiFe film deposited on a glass substrate at 0 to 60 volts. The degree of dilution was significantly reduced if deposition was made on a $SiO_2$ substrate or at higher voltages. An intermediate layer of 100Å $SiO_2$ was found to be effective in eliminating the mixing between Au and NiFe.

## INTRODUCTION

In using RF sputtering process for film fabrication we have observed that the magnetic properties of NiFe films can be quite strongly affected by the sputtering conditions. It is known[1,2] that during RF sputtering, a certain amount of resputtering occurs at the film surface. Resputtering is expected to have the most effect on the film composition at interfaces between layers of dissimilar materials. The extent of this effect depends upon the bias voltage between the film and the target and the substrate material. For controlling film properties, the effect may be quite important since the interfacial inhomogeneity can extend through a major portion of the thin film. In this paper, we will report some preliminary results from a study of the effect of sputtering conditions on the magnetic and electrical properties of permalloy (80Ni/20Fe) films. Auger Electron Spectroscopy (AES) was used in conjunction with ion sputtering to measure the surface composition and interfacial profiles.

## EXPERIMENTAL PROCEDURES

The RF sputtering system used for film preparation was a commercial Randex 2400 ATS system equipped with three 8 in. diameter targets and a 450 ℓ/s turbomolecular pump. It was usually operated at a power level of 250 watts with an Ar atmosphere of 20 micron pressure. During deposition, the substrate holder was water-cooled but the actual temperature on the substrate surface may have reached 200°C as inferred from other measurements. Typical deposition rates were 120Å/min for NiFe, 100Å/min for $SiO_2$ and 330Å/min for Au.

All Auger measurements were made by using a Thin Film Analyzer manufactured by Physical Electronics Industries, Ltd. The electron beam was operated at 5 keV and on occasion defocused to reduce the intensity to avoid charge build-up in the insulator substrate. A Faraday cup was used to align the electron beam to the center portion of the sputter crater, an important step in obtaining accurate composition profiles for thin-layered films. The sputter gun was operated at 1 keV under an Ar atmosphere of $5\times10^{-5}$ torr with an intensity of about $5\times10^{13}$ ions/cm$^2$/sec. This combination of sputtering conditions was found to be a good compromise in limiting effects of sputter damage on profile measurement while providing a suitable sputtering rate.[3] For measuring the composition profile, a Multiplexor unit[4] was used to monitor simultaneously the peak-to-peak heights of the following Auger lines: O(510 eV), Fe(598 eV), Ni(848 eV), Si(1619 eV) and Au(2024 eV).

## RESULTS AND DISCUSSION

Auger and magnetic measurements were made on two types of magnetic structures; Au on NiFe and NiFe on $SiO_2$. For the first set, the permalloy films were all prepared by RF sputtering at a bias voltage of -60 volts on substrates of Corning 0211 glass a or thermally oxidized Si wafers. The magnetic properties of this set of films are summarized in Part A of Table I. These films did not exhibit noticeable magnetic anisotropy, so it was difficult to measure the anisotropy field $H_k$. In Table I, we list instead the switching field $H_c'$ along the hard magnetic axis.

Among the magnetic properties, the magnetic thickness of the permalloy film shows the most noticeable variation with the bias voltage for Au deposition on 0211 glass substrate. With this substrate, the magnetic thickness was about four times higher at -120 volts than at 0 and -60 volts. Such a strong

Table I. Results of Magnetic and Auger Measurements

| | A. Au/NiFe Films | | | | | | B. NiFe/Sputtered $SiO_2$ Films | | | | | |
|---|---|---|---|---|---|---|---|---|---|---|---|---|
| Substrate Materials | $SiO_2$/Si | | | 0211 glass | | | Si | | | 0211 glass | | |
| Bias for top layer deposition (volts) | 0 | -60 | -120 | 0 | -60 | -120 | 0 | -60 | -120 | 0 | -60 | -120 |
| Interfacial Width* (A) | 170 ±30 | 185 ±15 | 210 ±15 | 810 ±80 | 900 ±100 | 310 ±80 | 175 ±15 | 120 ±15 | 105 ±15 | 380 ±80 | 345 ±50 | 480 ±100 |
| Magnetic Thickness of NiFe (A) | 440 | 450 | 380 | 100 | 90 | 390 | 600 | 680 | 560 | 600 | 680 | 560 |
| $H_c'$(Oe) | 1.4 | 1.8 | 1.8 | 1.4 | 1.4 | 1.7 | 5.2 | 0.8 | 1.2 | 5.5 | 0.8 | 1.0 |
| $H_k$**(Oe) | 1.6 | 1.8 | | 1.6 | 1.8 | | 4 | 3.8 | 4 | 4 | 3.8 | 4 |
| $\rho_o$(μΩ-cm) | | | | | | | 55 | 32 | 32.5 | 55 | 32 | 32.5 |
| $\frac{\Delta\rho}{\rho}$ (%) | | | | | | | 1.1 | 2.5 | 2.6 | 1.1 | 2.5 | 2.6 |

*Refer to the Au/NiFe interface in A and the NiFe/$SiO_2$ interface in B. **$H_c'$ for Au/NiFe films.

variation was not observed for films on Si substrates. Since the magnetic thickness was measured from the hysteresis loop, it is directly related to the total effective magnetic moment of the film. With a nonmagnetic Au top layer, the total magnetic moment of the permalloy film can be reduced if mixing occurs between Au and NiFe. This was confirmed by Auger sputtering measurement on the composition profile. For comparison, we plot in Fig. 1 typical profiles measured in samples deposited simultaneously on thermally oxidized Si and Corning 0211 glass substrates. The Au and Ni profiles clearly indicate extensive mixing of the Au and NiFe layers deposited on the glass substrate. The extent of the mixing depends on the bias voltage, as indicated by data in Table I showing the decrease of the interfacial width as a function of the bias voltage. Even though interfacial mixing is not extensive on the thermal $SiO_2$ substrate, it still reduces the total magnetic moment of the NiFe film since the effective magnetic thickness was observed to be consistently less than the actual film thickness by about 100Å.

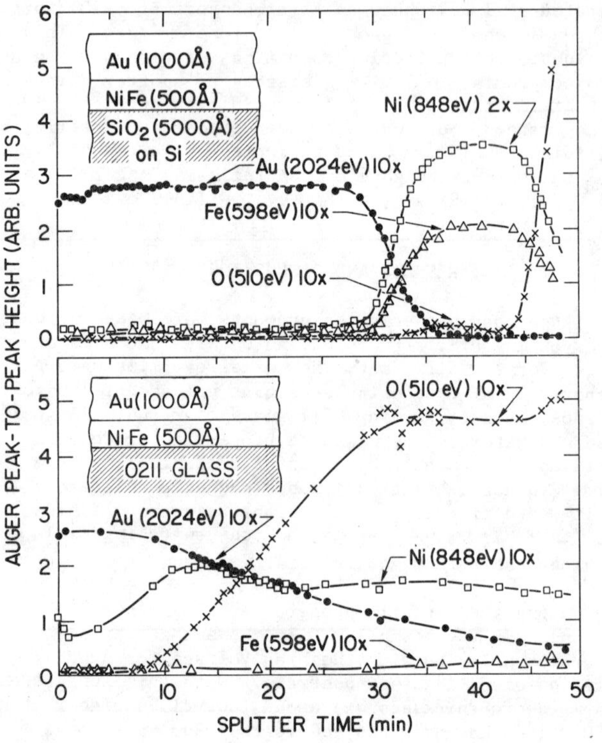


Fig. 1 Composition profiles of Au/NiFe films observed by Auger sputtering technique. Bias voltages for sputtering: Au, -60 volts, NiFe, -60 volts.

The exact cause for the mixing is not clear at present, although resputtering near the film surface is expected to be important. Of particular interest is the effect of oxygen on the resputtering process and its dependence on the bias voltage. (Note the early detection of the oxygen peak in the glass-substrate film.) The substrate surface temperature which would be higher on the glass substrate due to poorer heat conduction can also promote interfacial mixing. We can rule out the possibility that the extent of the inhomogeneity observed on glass substrates is due to sputtering damage introduced during Auger measurement since the interfacial width observed on $SiO_2$ substrates which should be equally affected is much smaller. In fact, we have measured the depth resolution by sputtering evaporated films on $SiO_2$ substrates and found an interfacial width of 80Å at a depth of 300Å and 160Å at a depth of 4000Å. These widths are smaller than that observed on sputtered films. In addition, the change in the magnetic thickness is not caused by variation in the NiFe film since composition calibration by electron microprobe showed all samples having the same composition with an uncertainty of ± 3%.

The second set of films had the structure of about 600Å NiFe on 1000Å sputtered $SiO_2$ on two types of substrates, Si and Corning 0211 glass. All the $SiO_2$ layers were RF sputtered at a bias of -60 volts. For this set of samples, because of the insulating $SiO_2$ underlayer, we obtained magneto-resistance data in addition to other magnetic measurements. The results are summarized in Part B of Table I and typical profiles measured on films with NiFe sputtered at 0 volts are plotted in Fig. 2. The data indicates essentially no difference in magnetic and magneto-resistive properties for films deposited on Si and glass substrates at a given bias voltage. It appears that the effect of the substrate material can be effectively masked by a sputtered $SiO_2$ layer of about 1000Å in thickness. On the other hand, the bias voltage still seems to be important in controlling the magnetic as well as the magneto-resistive properties judging from the data at zero bias and high bias voltages (-60 and -120 volts). Auger profiling revealed rather similar profiles for films on different substrates, although the interfacial width is still higher for the glass substrate. Among various factors considered, we found that this may be related to the oxygen content in the NiFe film. As seen from Fig. 2, the oxygen signal is relatively uniform in the film and its level was found to decrease with the bias voltage. Unfortunately, it was difficult to calibrate the absolute concentration of oxygen in the NiFe film due to its presence in the $SiO_2$ substrate. However, in judging from the height of the Auger peak observed in films on Si substrate, the oxygen content in the zero-bias film was at least twice as high as the films with bias voltage. Since oxygen is known to cause an increase in $H_c$[5] and $\rho_o$,[6] the observed voltage dependence of the oxygen level is consistent with the trend of changes in the film magnetic properties.

We have checked the idea that whether the interfacial mixing between Au and NiFe can be reduced by in-

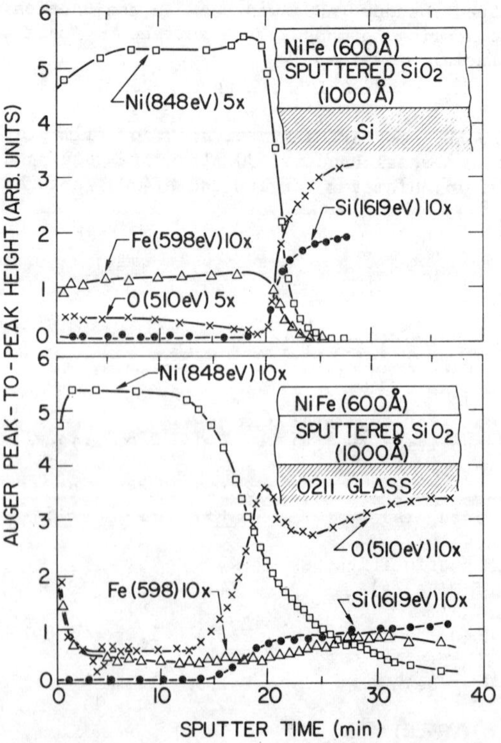


Fig. 2 Composition profiles of $NiFe/SiO_2$ films. Bias voltages for sputtering NiFe, 0 volts, $SiO_2$, -60 volts.

corporating an intermediate barrier. We found that a layer of $SiO_2$ as thin as 100Å can be very effective in reducing the mixing while maintaining the magnetic properties of NiFe unchanged.

In conclusion, the composition inhomogeneity observed at interfaces of layered films can affect the magnetic and electrical properties and its extent depends on the bias voltage, substrate material and film structure. In using the RF sputtering method for device fabrication, it is important to control such composition inhomogeneity in order to prepare magnetic films with required properties.

REFERENCES


1. J. L. Vossen, J. Vac. Sci. Tech. 8, 812-830
2. G. K. Wehner and G. S. Anderson, Ch. 3, Handbook of Thin Film Tech. Ed. by Maissel and Glang, McGraw Hill, N. Y. 1970.
3. P. S. Ho and J. E. Lewis, to be published.
4. P. W. Palmberg, J. Vac. Sci. Technol. 9, 160 (1972).
5. M. S. Cohen, M.I.T. Lincoln Lab., Quarterly Report Solid State Research, 67, July 15, 1961.
6. S. Krongelb, J. of Electronic Materials 2, 227 (1973) and unpublished work of NiFe films prepared in uhv by K. Y. Ahn.


# TEMPERATURE DEPENDENCE OF THE NUCLEATE GENERATOR USED IN BUBBLE DEVICES


I. Danylchuk and A. H. Bobeck
Bell Laboratories, Murray Hill, N. J. 07974


ABSTRACT


The effects of temperature on the operation of the nucleate generator have been extensively studied. The experimental test procedure is outlined and the test results on the generator current as a function of temperature are discussed. A temperature profile in the nucleate region due to an instantaneous heat source is derived both experimentally and theoretically. The calculated thermal time constant was verified by an experiment in which the generator was preheated by a non-nucleating current pulse and then driven by a nucleation pulse in which the delay and amplitude were varied to generate a time-temperature profile.


INTRODUCTION

The purpose of this study is to determine the effect of temperature on the operation of the nucleate generator used to input data in bubble circuits. Test procedures used in determining the minimum nucleate generator current, $I_G$, for the epitaxial garnet $Y_{2.75}Sm_{.25}Ga_{1.3}Fe_{3.7}O_{12}$ are described. These data are then used to infer transient heating effects when generators are operated at various duty cycles. The nucleate generator[1] is shown in Fig. 1. It is from a 28 μm period bubble circuit and was processed with 1) a 4000Å Al/Cu conductor, 2) an 8000Å $SiO_2$ spacing layer, and 3) 4000Å permalloy chevron elements. The geometry of the generator has been so designed that nucleation takes place on the upper chevron element within the current loop. The polarity, amplitude and width of the current pulse are chosen so that the field directly beneath the conductor loop is sufficient ($\sim H_k$) to nucleate a stable bubble domain.

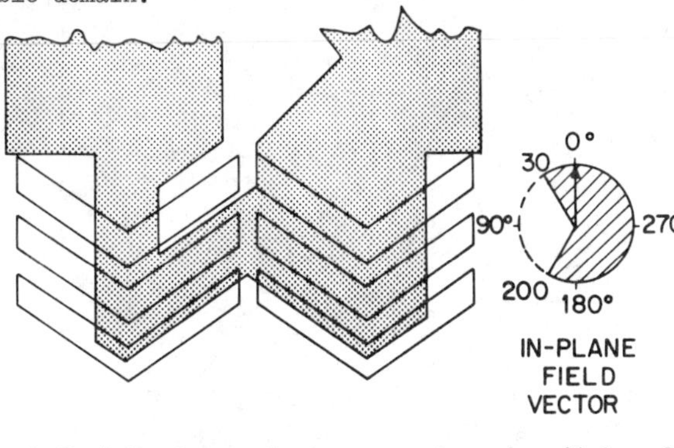


Fig. 1-Sketch of a nucleate generator circuit based on chevron propagate elements. Also, in-plane field vector showing generator pulse timing.

Domain nucleation becomes progressively easier to achieve as the ambient temperature increases. The conditions for magnetization instability in the Stoner-Wohlfarth[2] sense are given by

$$(H_z - H_{bias})^{2/3} + (H_{xy})^{2/3} = H_k^{2/3} \quad (1)$$

where $H_k$ is the uniaxial anisotropy field and $H_z$ and $H_{xy}$ are normal and in-plane field components produced in the bubble epitaxial layer by the rotating in-plane field and applied control currents both directly and via permalloy elements. The uniaxial anisotropy field vs temperature is plotted in Fig. 2.[3] One does expect the minimum generate current to temperature track $H_k$ and this is generally so.

The permalloy chevrons in the generator of Fig. 1 complicate an exact understanding of its operation but are, of course, necessary to propagate nucleated domains into the bubble circuit proper. At elevated temperatures the local fields produced by the permalloy elements themselves are sufficient to cause nucleation, i.e., the generator current falls to zero and reliable bubble propagation no longer exists. Our studies have shown that the onset of spontaneous nucleation due to permalloy features separated from the epi surface by 8000Å will take place when $H_k$ has dropped to 600 Oe. This limits operation to an upper temperature of approximately 60°C for the garnet used in these tests.

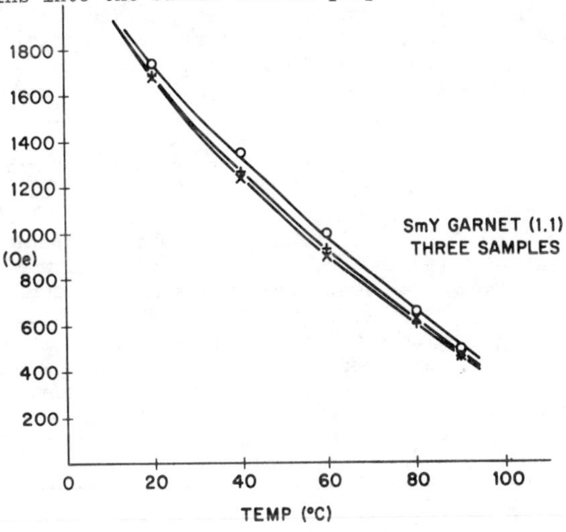


Fig. 2-Plot of $H_k$ as a function of temperature T for three representative SmY garnet samples.

EXPERIMENTAL TEST PROCEDURE AND RESULTS

The sample to be tested is mounted on a sapphire holder and positioned in the test fixture. The electrical contacts to the generator circuit are made with

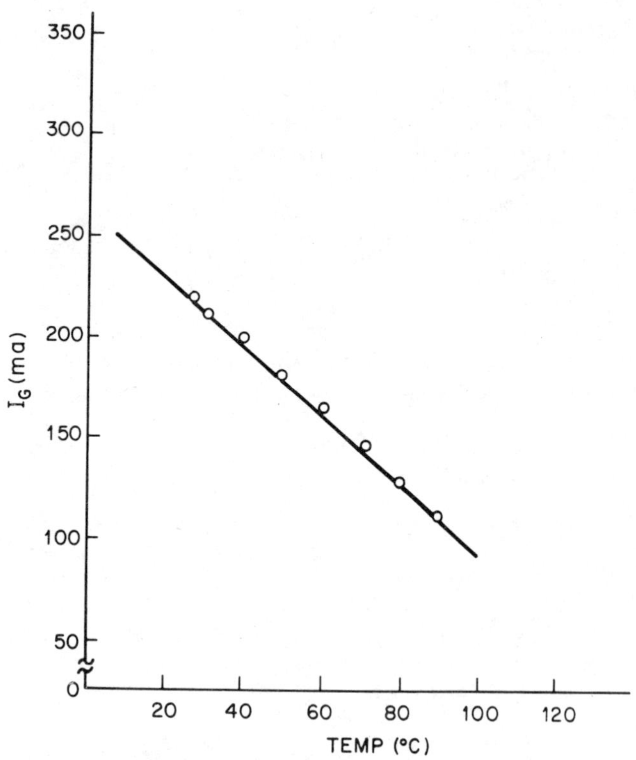


Fig. 3-Plot of minimum nucleate generator current, $I_G$, as a function of temperature, T, for a typical bubble chip.

a pair of adjustable probes (Electroglas). An iron-constantine thermocouple is attached to the body of the test fixture in the immediate vicinity of the sapphire plate, thus temperature is sensed within 2-3°C of the true sample temperature.

In the present test station, using a liquid heat exchanger--LO TEMPROL 154 of Precision Scientific--a temperature excursion from -10°C to 100°C can be readily achieved. However, frosting of the chip prevented accurate and reliable data collection much below 9°C.

The procedure of data taking is as follows: First, all domains are annihilated, i.e., the chip is cleared of all strips and bubbles by applying a 200 Oe bias field; then the bias field is lowered to the operating level (90-100 Oe) and the nucleate pulse amplitude is slowly raised to a level where only a single bubble (or a small strip) appears in the generator loop which encloses a chevron element. An in-plane field is not used. With this procedure, very reproducible values of the minimum nucleate generator current are obtained.

Figure 3 shows a plot of minimum generator current $I_{G\ MIN}$ as a function of the sample temperature T. The relation is linear with a negative slope. When many samples are tested there is a large dispersion in both $I_{G\ MIN}$ and the slope. The range of the slope extends from -0.6 to -2.0 mA/°C, and the value of $I_{G\ MIN}$ at room temperature falls between 110 mA and 210 mA.

GENERATOR CURRENT AS A FUNCTION OF REPETITION RATE

In the normal operation of a device a long string of "one's", or bubbles, can be generated. At the higher operating frequencies, e.g., 1 MHz, the pulsed currents in the generator conductor cause a cumulative ohmic heating and the temperature rise affects the minimum nucleate current. Figure 4 shows a plot of $I_{G\ MIN}$ as a function of repetition rate of the generate pulses. The decrease in the amplitude of the $I_{G\ MIN}$ indicates that the garnet area enclosed by the generator loop has been subjected to an appreciable temperature rise. The points corresponding to various temperature levels have been transferred onto the curve in Fig. 4 from Fig. 3 ($I_{G\ MIN}$ vs T) showing, for example, that the temperature rise for a 3 MHz rep. rate is approximately 60°C. At this point the minimum nucleate current value has decreased from 220 mA to 100 mA.

Note that the temperature rise at 1 MHz operation is approximately 30°C.

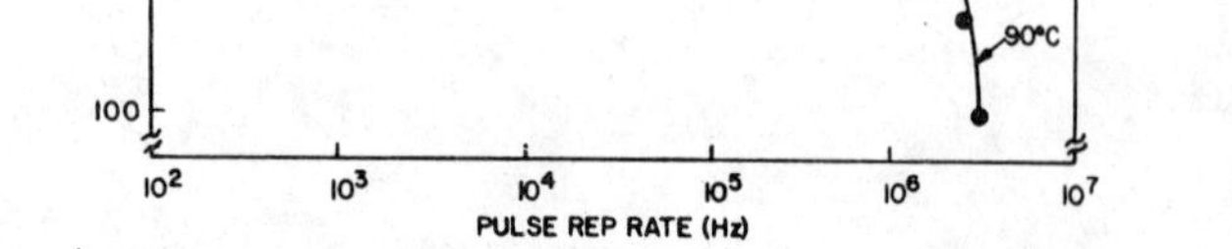

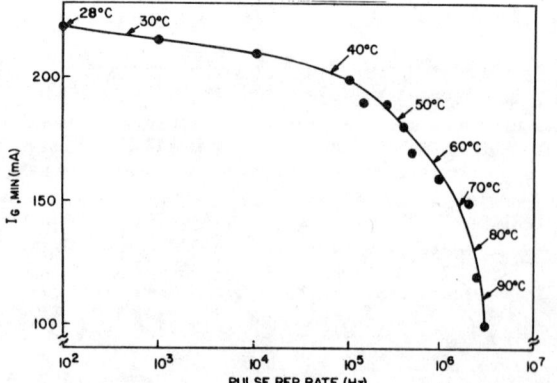

Fig. 4-Plot of $I_G$ minimum as a function of the generator pulse repetition rate. The temperature indicated on the curve are inferred from the data of Fig. 3.

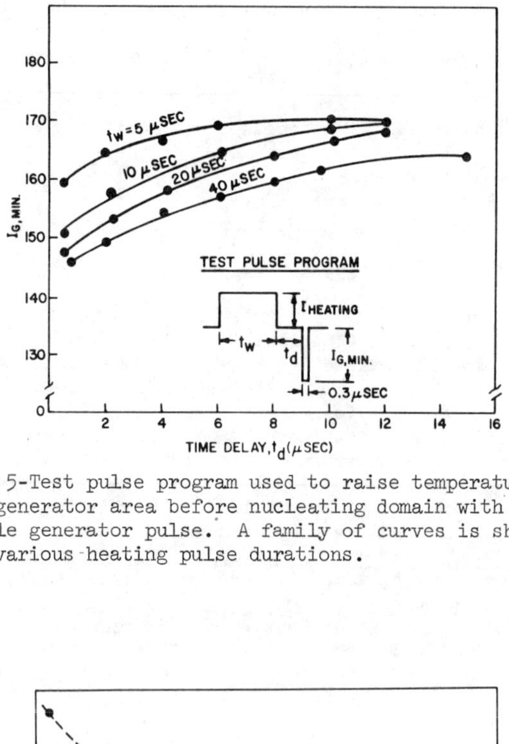


Fig. 5-Test pulse program used to raise temperature of the generator area before nucleating domain with a single generator pulse. A family of curves is shown for various heating pulse durations.

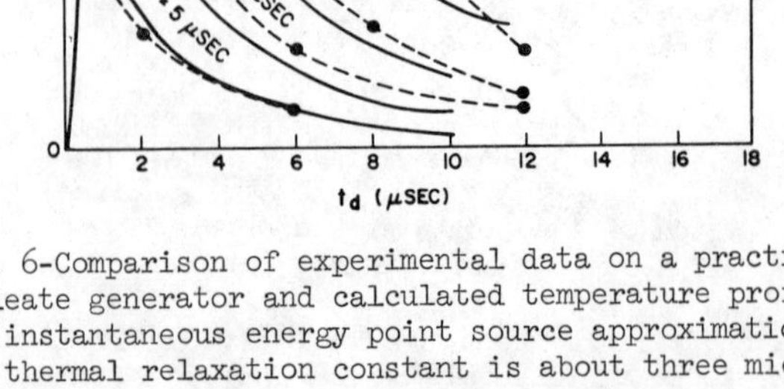

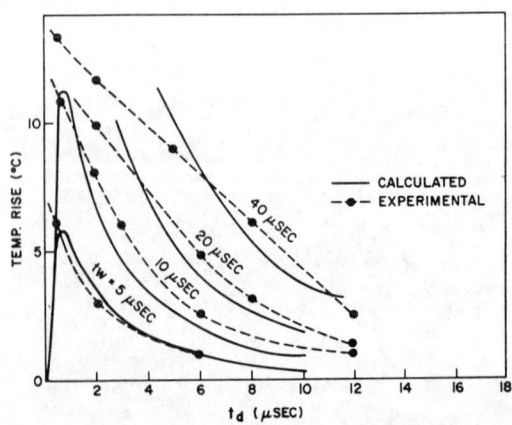

Fig. 6-Comparison of experimental data on a practical nucleate generator and calculated temperature profiles for instantaneous energy point source approximation. The thermal relaxation constant is about three microseconds.

TEMPERATURE PROFILE IN THE NUCLEATE REGION DUE TO AN INSTANTANEOUS APPROXIMATE POINT SOURCE OF HEAT

To gain an insight into the thermal transient response in the vicinity of the nucleate generator the following experiment was performed. First a "preheat" pulse of fixed amplitude and for a set of pulse widths $t_w$ was applied to inject an ohmic heating energy burst into the generator region. The amplitude of this preheat pulse was fixed at 0.14 amperes and had a polarity opposite to that used to nucleate domains. The nucleate pulse $I_{G\ MIN}$ was then applied at various delay times, $t_d$, following the first pulse. The family of curves so obtained are plotted in Fig. 5. The values of $I_{G\ MIN}$ of Fig. 5 can be translated into equivalent temperatures using Fig. 3 and the results are given in Fig. 6. The curves of Fig. 6 show the rate with which $I_{G\ MIN}$ increases to its original value at ambient temperature (≈28°C) thus giving the thermal relaxation time for the garnet sample tested.

A theoretical family of curves is also plotted in Fig. 6. For these calculations we have used the method of Carslow and Jaeger[4] for solving the differential equation of heat conduction assuming an instantaneous point source of heat Q. The transient temperature profile is calculated for the surface of the semisphere seen in Fig. 7. The value of Q is used as a parameter

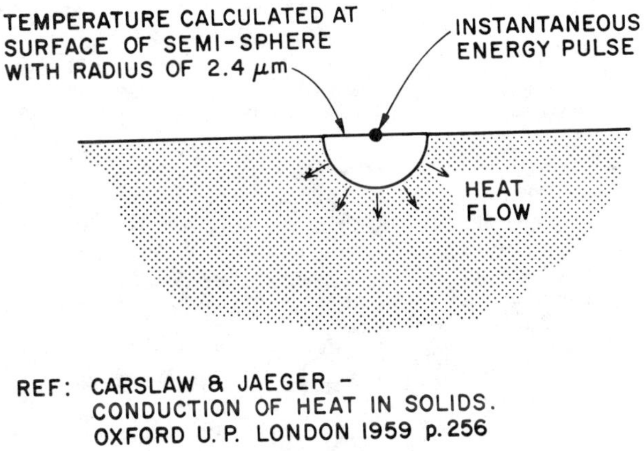


Fig. 7-Physical model used to calculate the temperature profile at the surface of a semisphere. It is assumed that all of the input energy escapes via conduction into the semi-infinite solid.

in generating a family of curves, shown in solid lines, Fig. 6.

The expression for temperature as a function of time is

$$T = \frac{Q}{8(\pi k t)^{3/2}} e^{-\frac{r^2}{4kt}}$$

where t = time in sec., r = radius of the semisphere in cm, k = thermal diffusivity of the garnet in $cm^2/sec$ (k=0.025) and q = point source strength in °C-$cm^3$ ($4\pi r^3 T_0$). After several trial calculations we find that $r = 2.4 \times 10^{-4}$ cm results in a fairly good fit of the theoretical to the experimental curves. For large Q one does not expect to get good agreement between theory and experiment simply because $t_w$ must be many thermal time constants in width and the mathematical model is less applicable. It is not possible to reduce $t_w$ by increasing the amplitude of the preheat pulse since this will cause domain nucleation.

CONCLUSION

Our results indicate that as we turn from 100 kHz operation toward higher data rates careful attention will have to be given to local transient temperature effects in such control functions as generation, replication, annihilation, etc. The nucleate generator described in this article has a 30°C temperature rise when operated continuously, i.e., generating all "one's" at 1 MHz. In a practical bubble device one must also generate a single isolated "one" reliably. For that value of $I_G$ when operated continuously the temperature rise is 45°C. Furthermore we show that the analysis of heat conduction in garnets based on an instantaneous point source of heat agrees with results obtained experimentally using a pulse program to measure the effect of local heating on domain nucleation.

REFERENCES

1. T. J. Nelson, Y. S. Chen and J. E. Geusic, IEEE Trans. Magnetics MAG-9, 289-293 (1973).
2. E. C. Stoner and E. P. Wohlfarth, Phil. Trans. Roy. Soc. A240, 74-77 (1948).
3. Unpublished data of R. J. Peirce.
4. H. S. Carslaw and J. C. Jaeger, Conduction of Heat in Solids, Oxford Univ. Press, London (1959), p. 256.
5. N. Goldberg, IEEE Trans. Magnetics MAG-3, 605 (1967).

# A SIMPLIFIED FOURIER SERIES METHOD FOR THE CALCULATION OF MAGNETOSTATIC INTERACTIONS IN BUBBLE CIRCUITS*

D. B. Dove, J. K. Watson, E. Huijer and H. Ma
Electrical Engineering, University of Florida, Gainesville, Florida 32611

ABSTRACT

A procedure is being developed for analyzing bubble circuits, based on periodic arrays of I-bars as building blocks for more complex elements. This paper describes a method for solving the basic I-bar magnetostatic problem, based on the simultaneous solution of Poisson's equation and an equation of magnetic equilibrium. The solution uses a Fourier series approach which lends itself to spatially nonuniform applied fields, as caused by bubbles. The mathematical formulation of the bar geometry is in 3-dimensions, with magnetization modelled along one axis, the major axis dominating. A numerical simplification results from analytically averaging the demagnetizing fields across the transverse axis, an 80-harmonic calculation of an I-bar array requires 5 seconds of computer time. Results compare well with others for I-bars in uniform fields and for the potential well of a bubble on an I-bar. In a new result with implications for analysis of logic gates, energy minima are found for two bubbles on opposite ends of an I-bar.

## INTRODUCTION

Several major approaches have been reported for the analysis of the forces acting on a bubble in field access permalloy bar propagation or logic circuits. Numerical techniques have been employed by Copeland[1] and by George and Archer[2] with considerable success in treating both isolated bar and bubble-bar interactions. George and Hughes have extended the two-dimensional numerical method to a variety of propagation elements.[3] Recently Almasi and collaborators[4] have introduced a reluctance model in an attempt to obtain a more flexible approach to circuit design, utilizing the results of the more exact numerical method where needed. A third method utilizing a Fourier series expansion of the magnetization reported by Lin,[5] offers analytical advantages but requires considerable computing capability when applied with full three dimensional generality.

The present method achieves a substantial simplification of the Fourier series by using a one-dimensional approximation of the magnetization, based upon two considerations. (a) For a given applied field, the induced response of the bar is much larger for fields along the bar length than for fields applied across the bar, and (b) the external fields that arise from the induced distributions of magnetization are attenuated with distance relatively much more rapidly for the case of distributions across the width.

## ANALYTICAL METHODS

This mathematical formulation is based on the Fourier series solution of the magnetic Poisson's equation for an elementary component of a permalloy propagation element. Previous workers[5,6] have shown that the Fourier series approach provides a means of solving the magnetostatic problem directly without iteration, expressing the harmonics of the demagnetizing field in terms of the harmonics of magnetization.

Referring to Fig. 1, the bar of dimensions $\ell \times w \times t$ is periodic with periods $\lambda_x$, $\lambda_y$. The 3-dimensional geometry is described in the z-direction by a Fourier transform, with results imbedded within the below potential functions (1), (3); in the y-direction by a cosine series with harmonics $a_j$; and in the x-direction by another cosine array indicated in (2) by the T on the right-hand side. The array T also includes local smoothing multipliers. For a field applied in the x-direction, the magnetization is assumed to have only an x-component array of harmonics $M_x$ in (2), represented in (1) and (3) by the saturation $M_s$ as suitably multiplied by the sum of sine and cosine harmonics $m_i'$, $m_i''$, respectively. With these approximations, the following solution of Poisson's equation has been derived for the magnetic potential in terms of an arbitrary distribution of magnetization along the bar, valid for $|z| < t/2$.

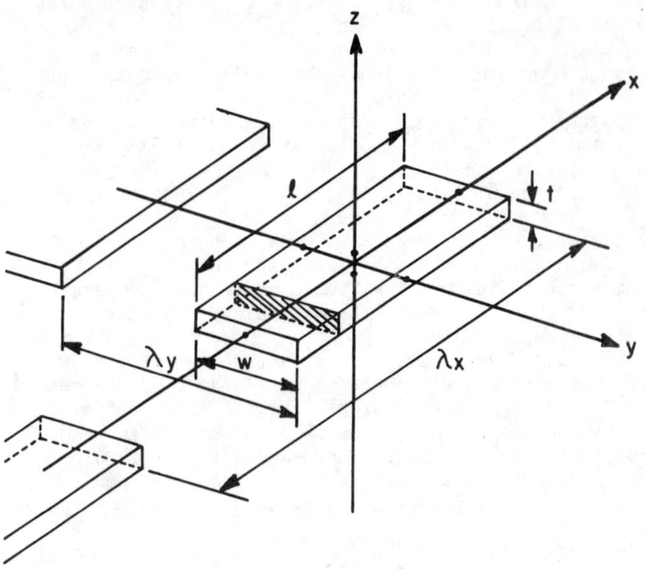


Fig. 1 Periodical array of I-bars employed in SFS method. Magnetization and fields are averaged and assumed constant over a cross-section.

$$V(x,y,z) = \sum_{i=1} \sum_{j=0} K_{ij} [1 - \exp(-\pi t \sqrt{\frac{i^2}{\lambda_x^2} + \frac{j^2}{\lambda_y^2}}) \cdot \cosh(2\pi z \sqrt{\frac{i^2}{\lambda_x^2} + \frac{j^2}{\lambda_y^2}})]$$

$$K_{ij} = \frac{2iM_s a_j}{\lambda_x (\frac{i^2}{\lambda_x^2} + \frac{j^2}{\lambda_y^2})} \cdot [-m_i'' \cos(\frac{2\pi i x}{\lambda_x}) + m_i' \sin(\frac{2\pi i x}{\lambda_x})] \cos(\frac{2\pi j y}{\lambda_y}) \qquad (1)$$

From this equation the average demagnetizing field $H_{dx}$ acting at distance x along the bar is readily obtained by integration of $-\partial V/\partial x$ over the bar cross-section, e.g. the shaded part of Fig. 1. The equation for local equilibrium is then taken as

$$M_x(x) = \chi T(x) (H_{ax}(x) + H_{dx}(x)) \qquad (2)$$

where $\chi$ is magnetic susceptibility and T(x) is unity for $-\ell/2 \leq x \leq \ell/2$ and is zero elsewhere. By expressing the applied field $H_{ax}$ as a Fourier series this equation reduces to a set of linear simultaneous equations that are readily solved numerically for the coefficients of magnetization. It can be noted that material susceptibility is retained explicitly, that the demagnetizing field problem is solved automatically, and that the response of the bar is readily calculated for nonuniform applied field, as due to a bubble.

After the coefficients of magnetization are found, a similar use of $-\partial V/\partial x$ from (1) will give explicitly the field acting on a neighboring propagation element. It follows that T and I-bar interactions can be analyzed. This method therefore provides a means of systematically analyzing propagation and other circuits in an interacting block approach. Similarly, use of $-\partial V/\partial z$ from (3) gives explicitly the z-direction field which can act on a bubble, valid for $|z| > t/2$.

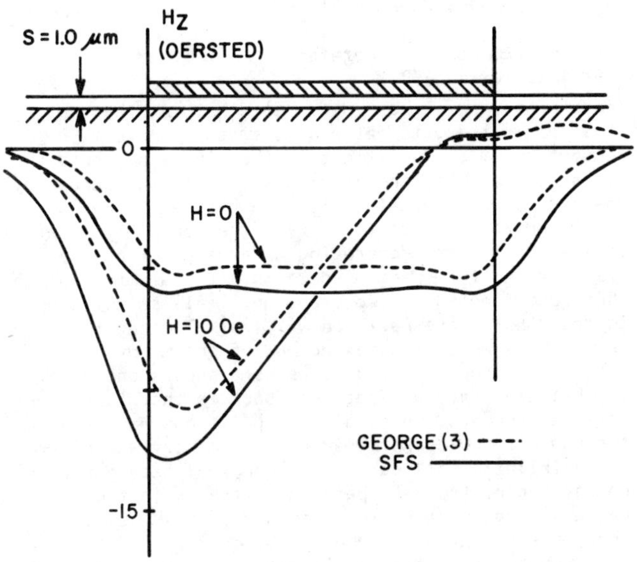


Fig. 2 Comparison of potential energy, expressed in equivalent field magnitude as a function of bubble position under a 3 x 15 x 0.4μm I-bar. $4\pi M_g$ = 200 gauss, R = 3μm, h = 3μm; $4\pi M_p$ = $10^4$ gauss, X = 400 chosen to be of the order of $M_s/H_c$.

$$V(x,y,z) = \sum_{i=1} \sum_{j=0} K_{ij} \exp\left(-2\pi|z|\sqrt{\frac{i^2}{\lambda_x^2}+\frac{j^2}{\lambda_y^2}}\right) \cdot \sinh\left(\pi t\sqrt{\frac{i^2}{\lambda_x^2}+\frac{j^2}{\lambda_y^2}}\right) \quad (3)$$

This method is largely analytical but requires the use of a computer to obtain the required series coefficients. The required computer processing is fast and inexpensive; representative computer times (IBM 370/165) are indicated below.

## RESULTS

We have calculated the response to uniform applied field of various geometry I-bars for comparison with other workers. Each calculation required about 5 seconds. Agreement is within a few percent of calculations by Lin[5] and by George and Archer[2] and of measurements by Krinchik et al[7], but a discrepancy of a factor of 4 was noted with the pioneering calculation by Copeland[1] for fields parallel to the bar long dimension. A more complete comparison of calculated and measured responses is available elsewhere[8].

Figure 2 shows the calculated interaction energy of a rigid 6μm diameter bubble as a function of bubble location 1μm beneath an I-bar of dimensions 3 x 15 x 0.4μm. This geometry was chosen to permit direct comparison with the recent 2-dimensional calculation of George and Hughes[3]. The computer time to calculate the energy for 27 bubble locations, for two values of applied field, was 40 seconds.

In a preliminary step toward analysis of bubble logic gates, Figure 3 shows an example of two bubbles interacting through an I-bar, for bars of different length. For sufficiently long bars, bubbles can remain at both ends of the bar in shallow energy minima. If a field is applied along the bar, one bubble would be repelled from the bar. For a bar as short as 5μm, however, the bubble-bubble repulsion exceeds the bubble-bar attraction so that the bubbles repel each other away from the bar, even in the absence of an external field.

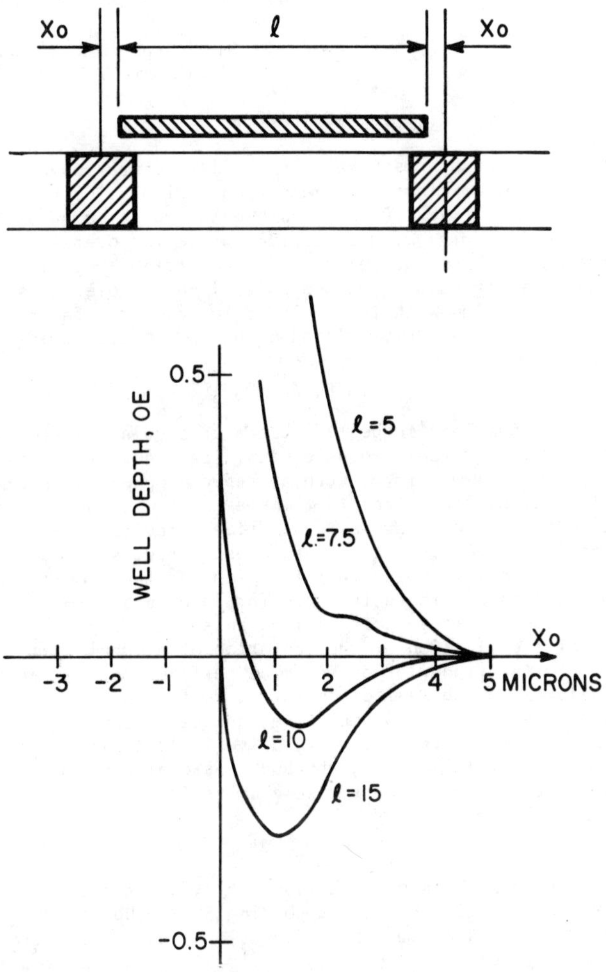


Fig. 3
Potential energy for two bubbles acting on opposite ends of an I-bar 2 x ℓ x 0.4μm. $4\pi M_g$ = 150g, R = 1.5μm, h = 8μm, s = 0.466μm, $\chi$ = 400.[9]

The authors appreciate permission by P. K. George to use his unpublished results in Figure 2.

## REFERENCES

*Supported by National Science Foundation Grant GK-40491.

1. J. A. Copeland, J. Appl. Phys., 43, 1905 (April 1972).
2. P. K. George, J. L. Archer, AIP Conf. Proc., 18, 116 (1974).
3. P. K. George, A. J. Hughes, "Bubble Domain Field Access Device Modelling, Part 2," (Feb 7, 1975), Rockwell Int. Report X74-1094/501.
4. G. S. Almasi et al, AIP Conf. Proc., 24, 635 (1975).
5. Y. S. Lin, IEEE Trans., MAG-8, 375 (1972).
6. D. B. Dove, BSTJ XLVI, 1527 (1967).
7. G. S. Krinchik et al., AIP Conf. Proc., 24, 649 (1975).
8. D. B. Dove, J. K. Watson, H. R. Ma, E. Huijer (submitted to Appl. Phys. Lett.)

# OPERATION OF A BUBBLE MEMORY CHIP WITH A TRIANGULAR DRIVE FIELD*

S. K. Singh and W. C. Hubbell, Texas Instruments Incorporated,
P. O. Box 5936, M. S. 145, Dallas, Texas 75222


## ABSTRACT

This paper describes the complete operation of a 20 kbit major/minor loop memory chip with a 100 KHz triangular drive. Data are presented which compare the triangular drive with sinusoidal drive for operational characteristics such as bias field margin, longevity, bubble signal amplitude and control pulse tolerances. Chip operation with triangular drive compares favorably with sinusoidal drive although the power requirement is higher.


## INTRODUCTION

It has been suggested that a triangular waveform drive coil current can have significant advantages over a sinusoidal waveform, such as reduced power consumption, simplified stop/start implementation, and elimination of the need for resonant drive circuits.[1] A triangular drive, by avoiding resonant circuits, should provide better phase stability with varying temperature as well as eliminate the need for large precision capacitors. These potential advantages have prompted careful evaluation of the operation of a bubble memory chip with triangular drive. The performance parameters evaluated are bias field margins, control pulse tolerances, bubble signal amplitude, and data longevity. The effects of variations in drive field amplitude on the bias field margins, the bubble signal amplitude, and data longevity were also examined.

## EXPERIMENT

A 20 kbit major/minor loop chip, processed on $(YSm)_3(FeGa)_5O_{12}$ garnet supporting 5 μm bubbles, was used for this evaluation. Operational characteristics for chips of this type have previously been reported[2,3] for sinusoidal drive fields. The chip has two levels of metallization: an Al-Cu layer delineated to form a nucleate generator, $-sign transfer gates and a hairpin-chevron replicate-annihilate gate, and a Permalloy layer which forms the T-bar propagation elements and the chevron expander detector. The performance characteristics of this chip are typical of chips from many slices.

The drive coil provided an inplane field uniform to within 10% with a vertical component less than 2% of the inplane field over the chip area. The dc bias coil provided a field uniformity of 99.8% over the chip area. The triangular driver used power switching transistors to impress a square wave voltage waveform across the coil, thus producing a triangular current in the coil. The performance of the bubble memory chip was examined for two different orientations of the drive field components. In one case, denoted by a diamond Lissajous' figure, the drive field components are alternately parallel and perpendicular to the T-bar propagation elements. In the other case, denoted by a square Lissajous' figure, the drive field components oscillated at an angle of 45° relative to the T-bar elements. These Lissajous' figures are the locus of the two orthogonal components of the inplane drive field (In the case of sinusoidal drive the Lissajous' figure is simply a circle). The two different Lissajous' figures were obtained by keeping the same triangular current waveforms in the drive coils but changing the chip orientation within the drive coils by 45°. The chip was operated in a continuous mode to eliminate stop/start effects. All data was obtained with a 100 KHz bubble propagation rate and at a chip temperature of (30 ± 1)°C.

For the purpose of direct comparison, data was also taken on the same chip in the same drive coil set for a sinusoidal drive. This provided a basis for an unequivocal comparison of the two drive schemes.

The bias margins for simple minor loop propagation as well as for complete read/write cycle operation were measured for worst case data patterns. At the upper bias margin the data pattern consisted of about 90% occupation of bit positions in the minor loop data storage area. For the lower bias margin the occupancy was reduced to about 10% with the bubbles being spaced widely apart. The write part of the read/write operation consisted of first generating a sequence of bubbles, transferring this page of data into the minor loops and then repeating these two steps until all selected pages in the minor loops had been written. To read the pages, the bubbles were transferred out of the minor loops after which replicated bubbles were subsequently detected while the data was returned back to the minor loops. This was repeated until all bit positions were read. At the end of the read/write cycle the memory was cleared by annihilating all bubbles at the replicate port. A comparison of the data pattern written with the data pattern read revealed the combined errors from all of these steps. Minor loop propagation margin data and full read/write cycle margin data was taken for the two triangular drive modes, i.e., diamond and square Lissajous' figures, as well as for sinusoidal drive. In the case of the triangular drive, the amplitude of the drive field is defined as the minimum distance from the origin to a point on the Lissajous' figure.

## RESULTS

The bias field margins versus drive field amplitude for minor loop propagation and read/write operation (which includes generate, transfer, replicate, detect, and annihilate functions) are shown in Figures 1a and b for sinusoidal and triangular drive, respectively. When the minimum field produced by the triangular drive equals the radius of the circular locus, i.e., the square locus circumscribes the circular locus, the triangular drive yields a wider bias field operating range at low drive fields than the sinusoidal drive. This is a result of the higher instantaneous value of the drive field amplitude achieved with the triangular drive for this condition. Differences in the bias field margins for the two square Lissajous' figures obtained with a triangular drive (Figure 1b) are small. At high drive fields, the bias field margins for minor loop propagation and read/write operation with triangular drive are approximately equal to those with sinusoidal drive.

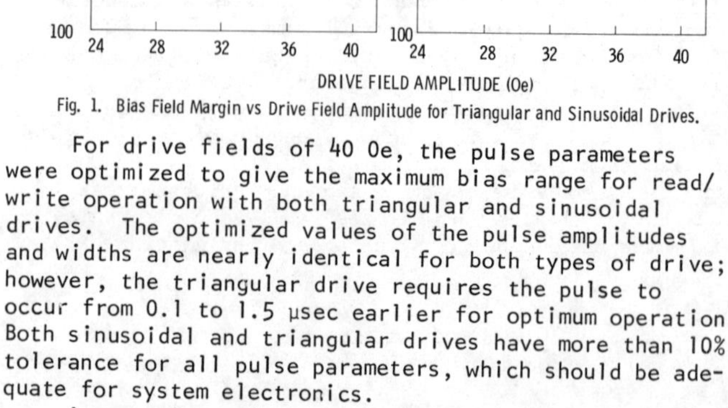

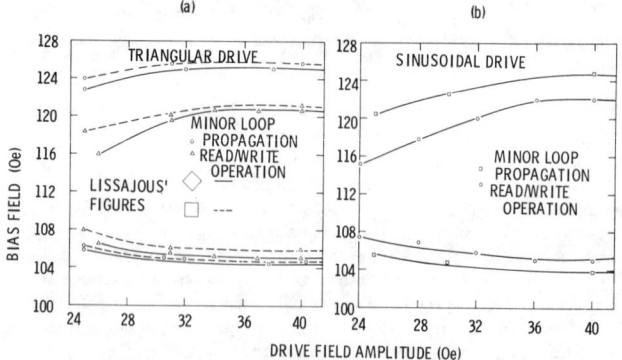

Fig. 1. Bias Field Margin vs Drive Field Amplitude for Triangular and Sinusoidal Drives.

For drive fields of 40 Oe, the pulse parameters were optimized to give the maximum bias range for read/write operation with both triangular and sinusoidal drives. The optimized values of the pulse amplitudes and widths are nearly identical for both types of drive; however, the triangular drive requires the pulse to occur from 0.1 to 1.5 μsec earlier for optimum operation. Both sinusoidal and triangular drives have more than 10% tolerance for all pulse parameters, which should be adequate for system electronics.

A comparison of the bias margins obtained with a sinusoidal drive of 40 Oe was made for two other values of triangular drive field. When the locus of the circle and square enclose equal areas (i.e., for a circular drive field of 40 Oe and a triangular drive field of 35 Oe), the margins are still found to be comparable (16.9 Oe). This condition corresponds to equal power dissipation in the drive coils. On the other hand when the circular locus circumscribes the square with the corners of the square touching the circle (i.e., for a circular drive field of 40 Oe and a triangular drive of 28 Oe), the triangular drive margin is reduced by about 30% (11.8 Oe). Thus, there seems to be no advantage in coil power dissipation with the triangular drive.

Longevity data for memory cycling for two different drive field amplitudes is shown in Figures 2a and b. The bias margin degradation is approximately equal for both drives. For a triangular drive with an amplitude of 40 Oe, the bias field range degradation is 0.25 Oe/decade. These data include operation of all memory functions, and the data were taken for a memory operating in a typical read/write mode.

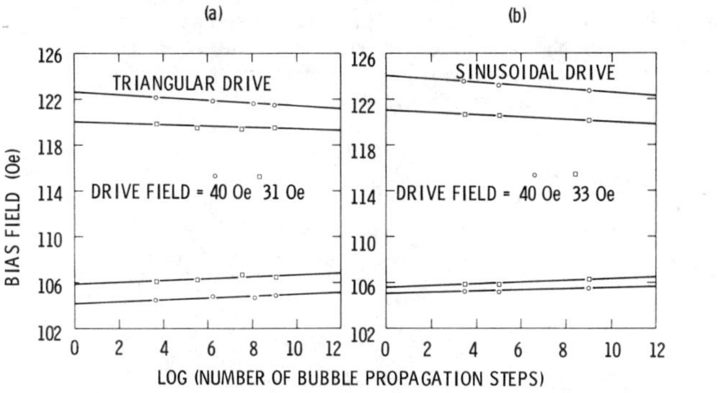


Fig. 2. Longevity Data for Memory Cycling with Triangular and Sinusoidal Drives.

A plot of bubble signal amplitude versus bias field is shown in Figure 3 for drive fields of 32 and 40 Oe. In the case of the triangular drive the bubble signal is superimposed on a $d\phi/dt$ pickup that is a square wave of approximately 1 mV peak-to-peak. The bubble signatures are similar for both sinusoidal and triangular drives.

## CONCLUSIONS

In summary, a triangular drive does not degrade bubble signal amplitudes, the longevity characteristics,

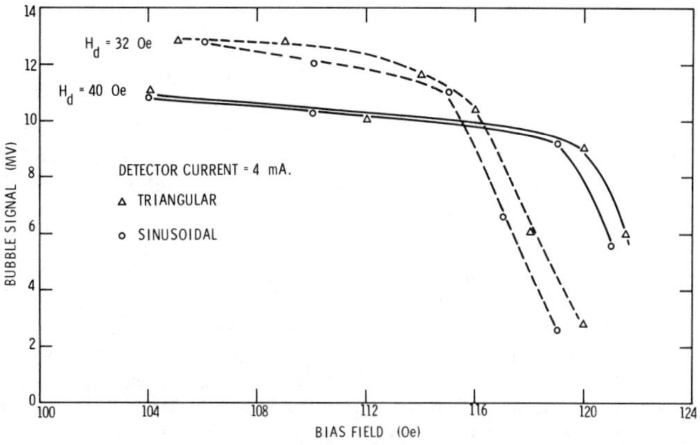


Fig. 3. BUBBLE SIGNAL VS BIAS FIELD FOR TRIANGULAR AND SINUSOIDAL WAVEFORMS

or the control pulse tolerances. To preserve the bias field margin for memory cycling, the amount of power a triangular drive will dissipate in the drive coils will be the same as, or up to 33% more than, a sinusoidal drive. This increase in coil power must, therefore, be traded off against the potential systems advantages.

## ACKNOWLEDGMENTS

The authors are pleased to thank H. F. Collins for designing and building the electronic test systems and the coil drivers. We are indebted to F. G. West for his valuable expert criticism and helpful discussions during this work.

## REFERENCES

*Work partially supported by U. S. Air Force Avionics Laboratory.

1. Yamagishi, et al., IEEE Trans. on Magnetics, MAG-11, 16 (Jan. 1975).
2. Bonyhard, et al., IEEE Trans. on Magnetics, MAG-9, No. 3, 433 (1973).
3. S. K. Singh, "Characterization of a 20 K-bit Major/Minor Loop Memory Chip", presented at the International Conference on Magnetic Bubbles, San Jose, Calif., Dec. 1974.
4. P. W. Shumate, P. C. Michaelis and R. J. Pierce, AIP Conference Proc. No. 18, Part 1, page 140 (1973).

# COMPACT REPLICATE/TRANSFER SWITCH FOR FIELD ACCESS BUBBLE DEVICES

I. S. Gergis, L. R. Tocci and J. L. Archer
Electronics Research Division, Rockwell International, Anaheim, CA 92803

## ABSTRACT

A compact transfer/replicate switch has been developed which is also capable of bubble annihilation. The switch utilizes a new 180 degree multi-chevron corner adapted to a conventional T-I propagation pattern (Figure 1) and permits two bit spacing of the minor loops on the access track resulting in minimum access time for any given data block. A two level version of the switch has been employed in a 64K bit block access chip design.[1] The manipulation of the domains is accomplished in a fashion similar to an all chevron, one level switch previously introduced by other workers.[2] The propagation margin of the minor loop at -25°, 25°, and 70°C is 16, 17, and 13 Oe respectively, for a drive field of 45 Oe at 100 kHz. The annihilation function overlaps the propagation margin completely over this temperature range. At -25°, 25°, and 70°C replicate-out has a margin of 13, 15, and 10 Oe; replicate-in has 12, 13.5, and 10 Oe; and transfer-out has 8, 14, and 9 Oe. All the switching functions have at least 40 degree margins.

1 - J.E. Ypma, I.S. Gergis, and J.L. Archer, paper 3A-7, this conference.
2 - T.J. Nelson, AIP Conference Proc. No. 18, p.95 (1973).

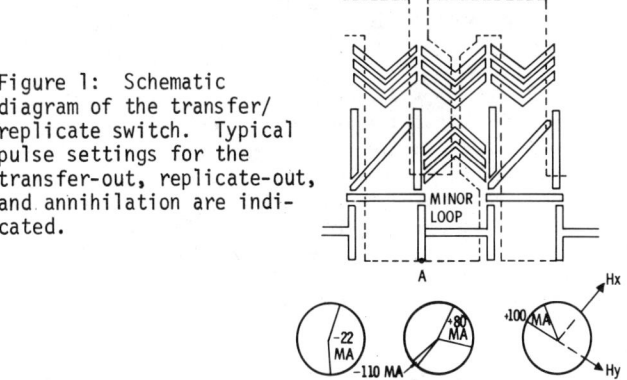

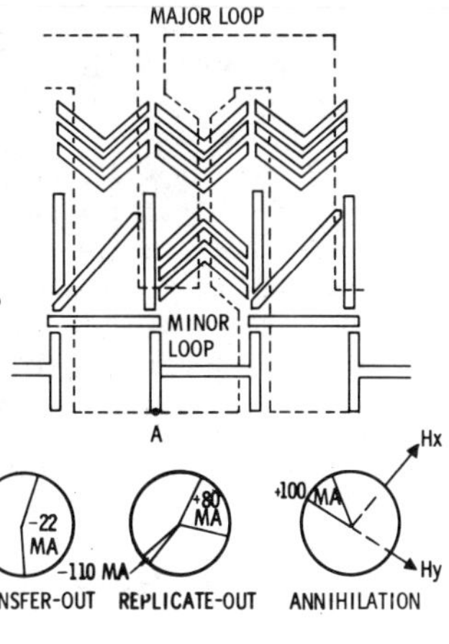


Figure 1: Schematic diagram of the transfer/replicate switch. Typical pulse settings for the transfer-out, replicate-out, and annihilation are indicated.

# DESIGN OF A 3 μm BUBBLE 80-kbit MEMORY CHIP

S. IGARASHI, K. IGARASHI, A. HIRANO, S. ORIHARA,
and K. YAMAGISHI
FUJITSU LABORATORIES LTD., Kawasaki, Japan

## ABSTRACT

A 3 μm bubble memory chip has been developed based on the established 6 μm bubble technology [1]. As the saturation magnetization of 3 μm bubbles is nearly the same as that of 6 μm bubbles, propagation patterns are scaled down to 1/2 the size of those for 6 μm bubbles. However, the propagation of 3 μm bubbles in the H-I pattern minor-loops results in a remarkable increase in the minimum drive field, which is caused by anomalous bubble propagation. From the experimental study of the potential well depth under the H pattern, it is concluded that the anomalous propagation is caused by the magnetizing effect of bubbles on the patterns. This explanation can be justified by the fact that no anomalous propagation occurs when the H pattern is cut into two T patterns at its center.

A 80-kbit major-minor memory chip was designed using broken H patterns. An overall operating margin of ~8 Oe for a drive field of 40 Oe at 500 kHz was measured on a 282-bit major-minor loop chip with the same design.

## INTRODUCTION

Minimizing the bubble size makes it possible to increase the bit density and operating speed at the expense of increased saturation magnetization which causes a stronger magnetizing effect on the permalloy propagation patterns and results in requiring a higher drive field [2][3]. However, reducing the bubble size to the 3 μm level can be achieved by appropriate selection of LPE garnet film parameters without introducing this annoying problem [4]. We investigated the propagation patterns for 3 μm bubbles whose saturation magnetization $4\pi Ms$ is nearly the same as that of 6 μm bubbles [1]. The propagation patterns for 3 μm bubbles were scaled down to 1/2 the size of those for 6 μm bubbles based on the consideration that the nearly same value of $4\pi Ms$ and the same shape anisotropy of the patterns would not cause a difference between the propagations of 6 μm and 3 μm bubbles.

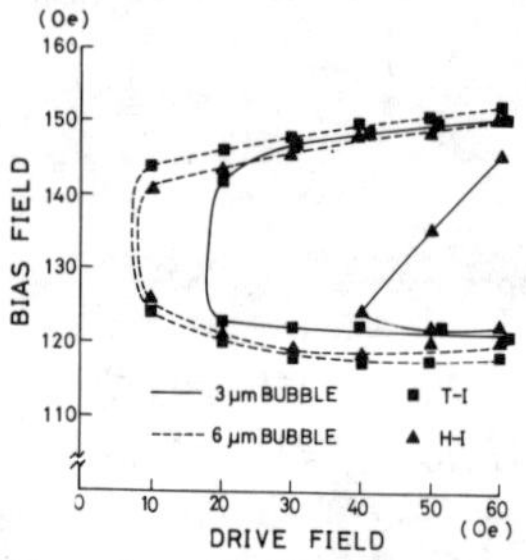


| THICK. (μm) | Hcol (Oe) | 4πMs (Gauss) | PATTERN(μm) a | b | c | g | p | THICK(Å) |
|---|---|---|---|---|---|---|---|---|
| 2.95 | 135 | 269 | 2 | 10 | 18 | 1 | 14 | 2000 |
| 6.20 | 141 | 254 | 4 | 20 | 36 | 2 | 28 | 4000 |

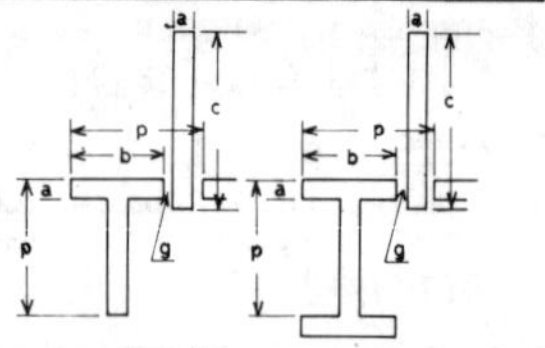

Fig. 1. Difference between the operating margins of 6 μm and 3 μm bubbles in the patterns with the same shape anisotropy.

As shown in Fig. 1, however, the quasi-static propagation of 3 μm bubbles in the H-I pattern minor-loops resulted in a remarkable increase in the minimum drive field, which is caused by the anomalous bubble propagation; the bubble cannot propagate from the I pattern to the H pattern when the H pattern is already occupied by another bubble and transfers to an adjacent minor-loop via the I pattern. This phenomenon is scarcely observed in the propagation of 6 μm bubbles.

## EXPERIMENTS ON MINOR-LOOP PROPAGATION

Two types of LPE garnet films were prepared for the experiments to investigate the anomalous propagation and to optimize the design of minor-loop propagation patterns. One is $(YEuYbCa)_3(FeGe)_5O_{12}$ for 3 μm bubbles and the other is $(YEuYb)_3(FeAl)_5O_{12}$ for 6 μm bubbles, and their characteristics are listed in the tables in Figs. 1 and 2. Hard bubbles were suppressed by ion implantation. T-I, H-I, broken H-I, and chevron pattern closed loops were fabricated on the same chip, where the broken H-I pattern is a H-I pattern with a slit at the center of the H pattern. Patterns are varied in period from 14 to 20 μm every 2 μm for 3 μm bubbles and from 28 to 40 μm every 4 μm for 6 μm bubbles. Pattern width are fixed at 2 μm for 3 μm bubbles and 4 μm for 6 μm bubbles. Pattern gap are fixed at 1 μm for 3 μm bubbles and 2 μm for 6 μm bubbles. The thickness of $SiO_2$ spacer between the garnet film and the permalloy patterns are 6000Å for 3 μm bubbles and 1.2 μm for 6 μm bubbles.

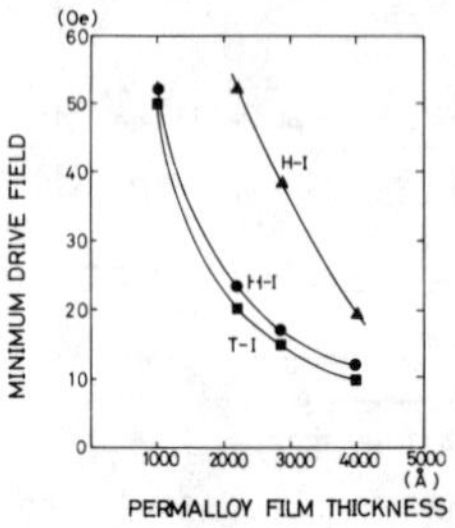


| | PERMALLOY THICK(Å) | Hc(Oe) | GARNET THICK | Hcol | 4πMs |
|---|---|---|---|---|---|
| A | 1050 | 2.2 | 2.91 | 137 | 269 |
| B | 2000 | 1.8 | 2.95 | 135 | 269 |
| C | 2850 | 1.9 | 3.22 | 137 | 266 |
| D | 4000 | 1.7 | 3.09 | 139 | 266 |

Fig. 2. Dependence of the minimum drive field of the T-I, H-I, broken H-I patterns on the permalloy film thickness.

Fig. 2 shows the dependence of the minimum drive field of quasi-static propagation on the permalloy film thickness used in the T-I, H-I, and broken H-I patterns. The minimum drive field is defined as the drive field required to produce operating margin window of ±5% of the collapse field of the free bubbles. It shows that the minimum drive field increases with a decrease in the permalloy film thickness and it is most noticeable in the H-I patterns due to the anomalous propagation mentioned above.

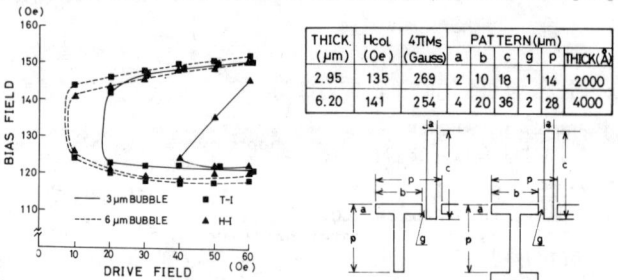

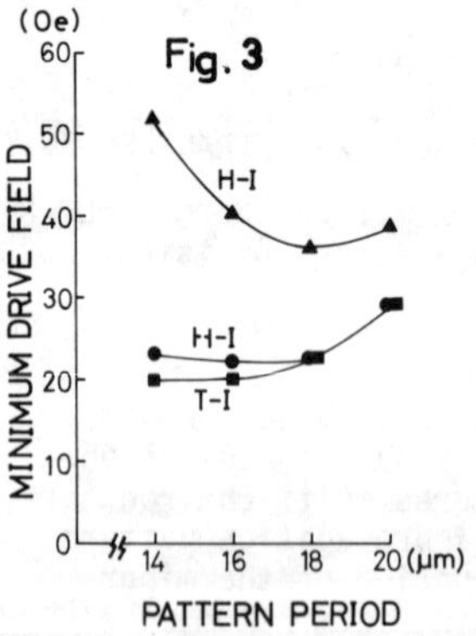


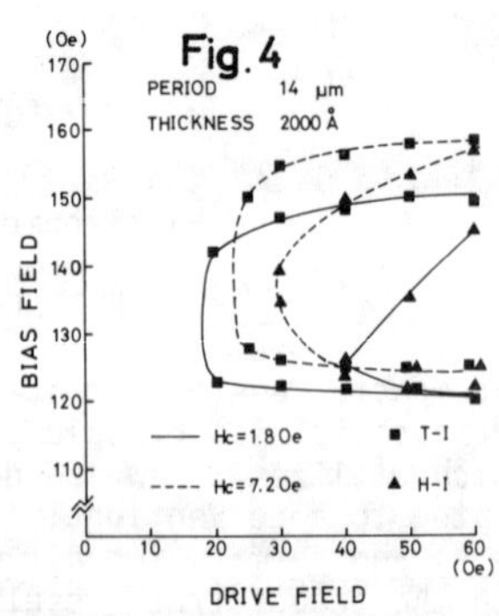


Fig. 3. Dependence of the minimum drive field of the T-I, H-I, and broken H-I patterns on their period.
Fig. 4. Difference of the margin curves when the coercivity of the permalloy film is varied.

Fig. 3 shows the dependence of the minimum drive field of quasi-static propagation on the period of the T-I, H-I, and broken H-I patterns. The minimum drive field of the H-I pattern appears to be improved with an increase in its pattern period, but it is not enough when compared to those for other two patterns.

Fig. 4 shows the difference of the margin curves of the T-I and H-I patterns when the coercivity of the permalloy film is varied. Note that the propagation in the H-I pattern using a permalloy film with slightly high coercivity is significantly improved.

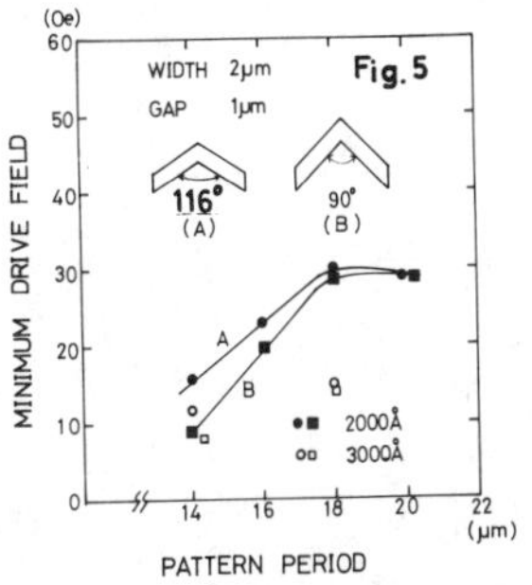


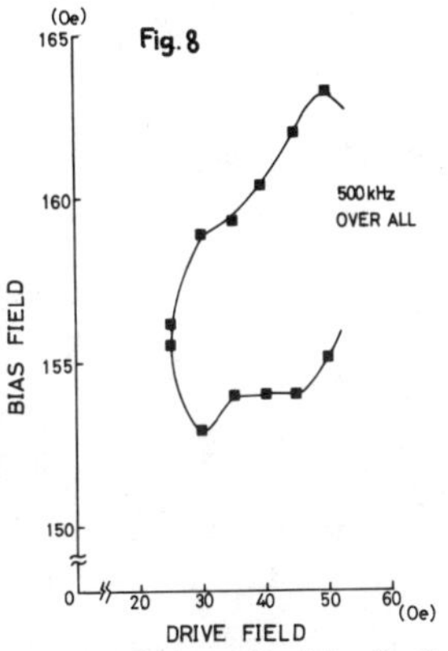


Fig. 5. Dependence of the minimum drive field of the chevron patterns on their period and shape.
Fig. 8. Operating margin of a 282-bit major-minor memory chip with the same design as a 80-kbit memory chip at 500 kHz.

Fig. 5 shows the dependence of the minimum drive field of the chevron patterns on their pattern period and shape. Anomalous propagation similar to that in the H-I patterns is also observed in the chevron patterns; when all patterns are occupied by bubbles, bubbles cannot be propagated or sometimes are repelled out of the propagation path at a lower drive field.

DISCUSSION ON ANOMALOUS BUBBLE PROPAGATION

We compared the potential well depth under the isolated H and T patterns by measuring the difference between the collapse fields of the bubbles trapped under the pattern and the free bubbles. The procedure is as follows; (1) produce bubbles around the patterns to be measured with large rotating field of 60 Oe at 100 Hz, (2) trap bubbles under the patterns, slowly reducing the rotating field to zero, (3) apply a given in-plane field parallel to the pattern arms, and (4) measure the difference between the collapse fields of the trapped and the free bubbles.

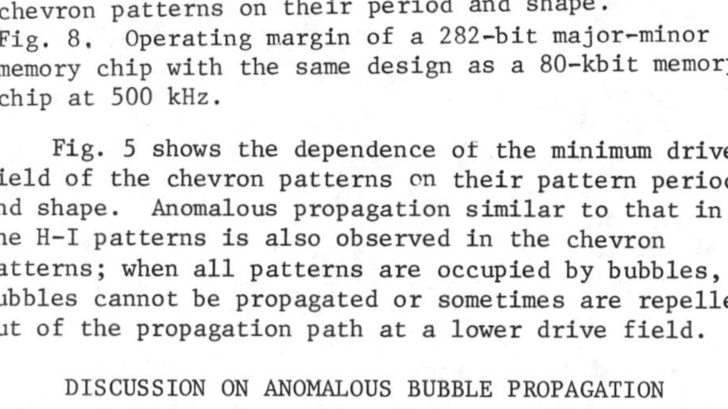

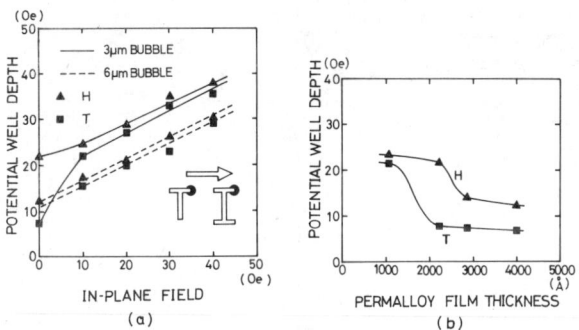

Fig. 6 (a) Dependence of the potential well depth of the H and T patterns on applied in-plane field. (b) Dependence of the potential well depth of the H and T patterns measured in the absence of in-plane field on the permalloy film thickness.

Fig. 6 (a) shows the dependence of the potential well depth of the H and T patterns corresponding to 3 µm and 6 µm bubbles on the applied in-plane field. In the patterns for 6 µm bubbles, there is no difference in the potential well depth between the two pattern shapes, while in the patterns for 3 µm bubles there is a distinct difference below the in-plane field of 10 Oe. This may explain the difference between the operating margins of 6 µm and 3 µm bubbles in terms of the pattern shape as shown in Fig. 1.

Fig. 6 (b) shows the dependence of the potential well depth of the H and T patterns on the permalloy film thickness measured in the absence of an in-plane field. The H pattern is much more magnetized by the bubble stray field than the T pattern, and this

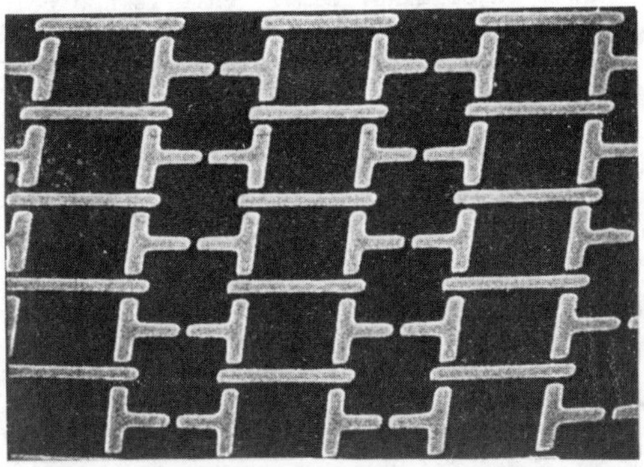

Fig. 7. SEM photograph of the broken H-I pattern minor-loops employed in a 80-kbit major-minor memory chip.

tendency becomes more noticeable with the decrease in the permalloy film thickness. With a film thickness of 1000Å, however, both H and T patterns are magnetized nearly to saturation. Comparing Fig. 6 to Fig. 2, the potential well depth and the minimum drive field exhibit a similar incremental dependence on the decrease in the permalloy film thickness. This implies that, in thinner patterns, they are magnetized by the bubble stray field, which requires further externally applied field to overcome the magnetizing effect and to propagate bubbles, and the minimum drive field is increased. The anomalous bubble propagation is also explained by the magnetizing effect of the bubbles on the patterns. That is, when one of two propagation paths of the H pattern is occupied by a bubble, an increased amount of magnetization works at the opposite side of the propagation path through the pattern so as to degrade the attractive force of the other bubble.

Summerizing these experimental results, our conclusions are as follows.

(1) Simply scaling down is not always possible in designing propagation patterns even when the $4\pi Ms$ of the bubble film is kept constant.

(2) The Magnetizing effect of the bubble on the pattern depends on the size, shape, and coercivity of the pattern as well as on the pattern thickness and the saturation magnetization of the bubble material.

DESIGN OF A 80-kbit MAJOR-MINOR MEMORY CHIP

The anomalous propagation in H-I patterns is improved by increasing the pattern thickness, but bubbles are stripped out easily at the portions where many multi-directional magnetic poles may concentrate such as at the transfer gates, and the 90° - and 180° -turns. The permalloy film thickness of 3000∼3500Å is considered to be the most appropriate from the viewpoint of the total margin. Anomalous propagation is avoid almost completely by cutting the H pattern into two T patterns, that is a broken H pattern.

Based on the results obtained with the experiments on the basic propagation of 3 µm bubbles mentioned above, a 80-kbit major-minor memory chip was designed. The broken H pattern is employed in the minor-loop and the period, bar width and pattern gap are 14 µm, 2 µm, 1 µm, respectively as shown in the SEM photograph, Fig. 7. The transfer gates and the splitter are scaled down from the design for 6 µm bubbles and the H pattern in the transfer gate is also cut into two at its center. Chevron patterns in the major-loop and stretcher-detecter are 116° chevrons and 90° rectangular chevrons respectively. The number of minor-loops is 142 and they consist of 128 storage loops, 4 loops for address information, 2 timing loops, and 8 redundant loops. The chip size is 4.2 x 5.2 mm including bonding pads.

The operating margin of this design was investigated with small capacity chips of the same design. Fig. 8 shows the operating margin of a 282-bit major-minor chip at 500 kHz. This margin window is comparable to that of 6 µm bubbles chips operated at 100 kHz.

A 80-kbit major-minor memory chip is now being fabricated and the good operating margin of ~8 Oe for a 40 Oe drive field at 500 kHz is expected.

REFERENCES


1. S. Orihara, et al., IEEE Trans Magn., MAG-11, (1975)
2. P. K. George, et al., IEEE Trans Magn., MAG-10, 821, (1974)
3. M. H. Kryder, et al., IEEE Trans Magn., MAG-10, 825, (1974)
4. T. Obokata, et al., submitted to the 1975 3M Conf.


## DATA UNRELIABILITY IN BUBBLE MEMORY DEVICES CAUSED BY SPONTANEOUS ANNIHILATION


T. T. Chen, L. R. Tocci and J. L. Archer
Electronics Research Division, Rockwell International, Anaheim, California 92803


ABSTRACT


The unreliability of the bubble devices caused by spontaneous bubble annihilation in a large capacity chip is different from that measured in a small test loop because of the direct interaction between neighboring bubbles. This direct interaction can be separated from the indirect interaction by measuring the individual bit decay rates for different word patterns. Measurement on a 16 micron 20 K bit chip shows that the bubble annihilation rate caused by direct interaction can be several orders higher than that caused by indirect bubble interaction through the remanence effect in the permalloy overlay. Because of this strong pattern dependence, the inverse of the initial slope of the error accumulation (mean step to failure, MSTF) should be used as the comparison parameter to evaluate the long term reliability effects in large capacity bubble memory devices.


## REVIEW OF BUBBLE DEVICE MODELING


G.S. Almasi
Thomas J. Watson Research Center, P.O. Box 218, Yorktown Heights, New York 10598


ABSTRACT


This paper reviews the status of quantitative models for field-access bubble domain devices. At present, quasistatic single-bubble margins can be predicted with reasonable accuracy. The essential effects which such a model must include are discussed first, and then used as a basis for comparing the approaches published so far. Inclusion of bubble-bubble interactions and long-term statistical effects is discussed briefly. The paper concludes with some comments on device optimization and scaling.

# 64K FAST ACCESS CHIP DESIGN

J. E. Ypma, I. S. Gergis and J. L. Archer
Electronics Research Division, Autonetics Group
Rockwell International
3370 Miraloma Ave., Anaheim, CA 92803

## ABSTRACT

A 65,664 bit, 4 μm, minor loop chip of 128 loops of 513 bits is described. The goals of fast read and write access, and high average data rates are shown to lead to a decision to use a replicate switch of compact design to minimize loop length and data houskeeping. Operation of the device is described, followed by an example of a read and write of a single block of data and examples of read/write of multiple blocks. Use of modulo 128 and modulo 151 block address assignments illustrate means of obtaining either gapless and gapped record delivery.

FIGURE 1: 65664 BIT CHIP

## INTRODUCTION

The bubble domain memory chip shown in Figure 1 is one which was designed to meet several goals. These goals were: To create a 65K bit chip, of approximately square dimensions, which exhibited good read and write access time, a potential for a high sustained data rate, minimum housekeeping, and finally, maximum flexibility to the user system in choice of block size.

To meet these goals several decisions were made:

1) Minor loops with replication to the main track were used rather than transfer switches.

a) Data return housekeeping is eliminated giving the user more flexibility in choosing block spacing.

b) Data removal from main storage during readout is avoided, minimizing the possible loss of a data block through system error or power transient.

2) A replicate switch[1] which fits in a single period was used to minimize access time by limiting the loop length to one traversal up and down the device (513 bits).

3) A dual generator and detector was used to:

a) Allow alternate bit operation in the main track, improving margins.

b) Maintain data flow at field rate.

c) Allow a single replication to create two streams of alternate data which fold into a biphase output channel.

4) A mutually prime number ratio between minor loop size and minor loop quantity was used to allow continuous delivery of data in any block multiple from one 128-bit block through the entire 65664-bit capacity of 513 blocks.

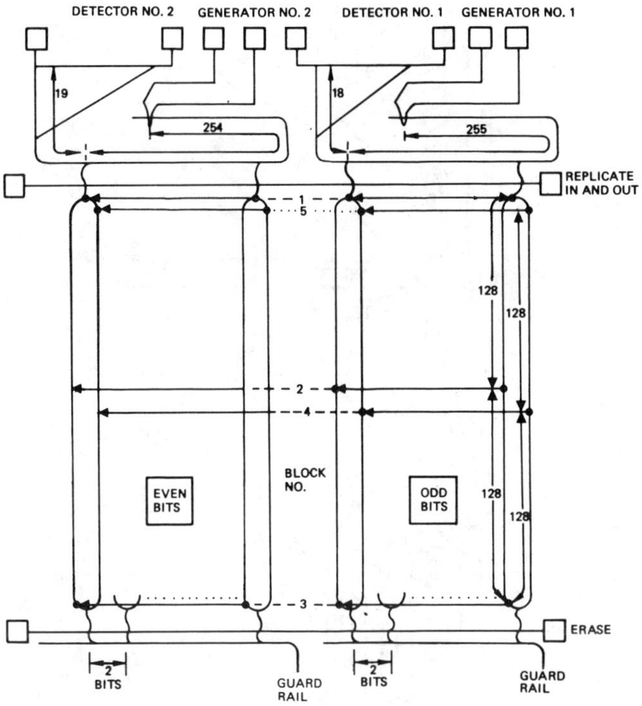


FIGURE 2: 64K CHIP ORGANIZATION

5) A separate erase path was installed to allow continuous erase-write and to minimize write address time

6) The generator path was lengthened to align with the erase coincidence. This allows an early acceptance of input data, thus minimizing system transfer time.

## CHIP DESCRIPTION

The diagram shown in Figure 2 corresponds directly with the photo in Figure 1. Each record is stored in two halves, with the even bits written in the left half field, and the odds in the right half as illustrated by the arrows lying across the minor loop. A one-bit difference in the generator paths aligns the even and odd streams opposite the replicator at the same step, allowing a single replicate pulse to enter all data. During read, the unequal paths to the detectors re-skew the data so that it is once again ordered as a simple serial stream.

## SINGLE RECORD READ-WRITE

The sequences required to write or read a block of data are as follows: During write, the first action is that of erasing data from the space to be rewritten. Thus write access time is the time required to bring the proper record to the erase elements. The long generator path allows system data transfer to follow immediately. Data replication to the storage field occurs later, independent of further system interaction. During read, access time is the sum of time required to bring the read record to the front switches plus the 18-bit travel to the detector.

When performing either a single read or write, the ratio of data bits transferred to total cycle time taken is low. An average read access of 275 steps preceeds 128 bits of data. A ratio of $128 \div (275 + 128) = 0.310$ indicates a low net data rate. A need for better ordering of data to improve average speed is needed.

MODULO 128 OPERATION TO CONSECUTIVE ADDRESSES

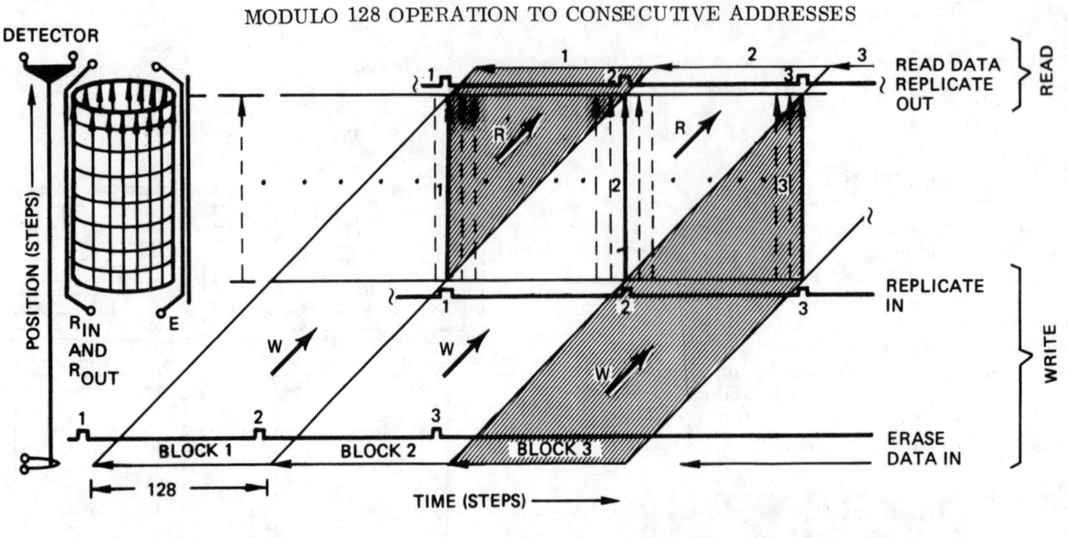


FIGURE 3: MODULO 128 OPERATION TO CONSECUTIVE ADDRESSES

MODULO 151 OPERATION TO CONSECUTIVE ADDRESSES

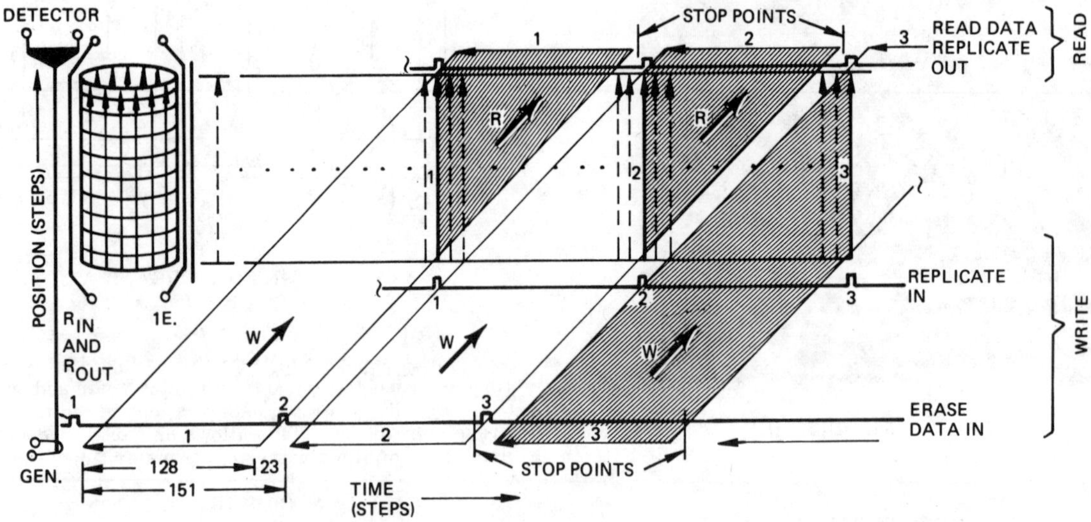


FIGURE 4: MODULO 151 OPERATION TO CONSECUTIVE ADDRESSES

MODULO 128 DATA ORGANIZATION

One means of improving net data rate is to increase the size of the block transferred with each access. This can be accomplished by assigning consecutive record addresses on the minor loop in sites that are 128 bits apart[2]. Now, as each 128 bit record is transferred, the next is immediately available. The lack of a common factor in the numbers 128 and 513 ensures that the whole sequence of 513 numbered records interleave as if the data were in a simple continuous loop. Figure 3 diagrams this operation, illustrating both the read and write modes. The ordinate represents position on the main track, the abscissa represents time, and the "sheaf of arrows" represent data in the minor loops which cycles without progression. In this mode, a group of four 128 bit records, could be interpreted as a single block of 512 bits. So used, the data delivered per access are increased fourfold. The ratio of data bits delivered to total steps is improved to $512 \div (275 + 512) = 0.650$. Larger block sizes yield ratios which approach 1.0.

MODULO 151 DATA ORGANIZATION

An alternative approach to record assignment is the use of a modulo which is greater than the 128-bit record length and still prime to 513. Figure 4 illustrates modulo 151 operation. Here data blocks are spaced, by the 23 step excess count. While some sacrifice of overall speed is made, two benefits are achieved over modulo 128. First, the choice of an excess which is larger than the detector delay (19 bits), places the read-replicate pulses between records. This eliminates it as a source of detector signal interference. Secondly, a stopping point between readout of consecutive records exists where the previous record is

| Specifications | |
|---|---|
| Capacity: | 65664 bits (128 loops of 513 bits) |
| Speed: | 300 kHz |
| Size: | 0.250" x 0.250" |
| Storage Area Density: | $2.0 \cdot 10^6$ bits/in.$^2$ |
| Bubble Size: | 4 μm |
| Net Chip Density | $1.0 \cdot 10^6$ bits/in.$^2$ |
| Read Access: | Avg 275 Steps (918 μs) Max 531 Steps (1770 μs) |
| Write Access: | Avg 256 Steps (853 μs) Max 514 Steps (1713 μs) |
| Temperature Range: | -25 to +75°C |
| Major Elements: | Generator/Replicate-Transfer/Detector |
| Number of Leads: | 12 |

TABLE 1

fully read while the succeeding is not yet loaded into the main track. This stop capability allows convenient speed buffering to the system thru intermittent operation, without imposing start-stop criteria on the main track or detector. Incremental 128-bit blocks can be transferred as a subset of a larger total block in keeping with the system speed requirements. Compared to modulo 128, data ratio for a 512-bit block drops from 0.650 to 512 ÷ (275 + 3 X 23 + 512) = 0.598.

## CONCLUSIONS

The chip described represents a good compromise between bubble element characteristics and system requirements. In particular, the use of a close-spaced replicator as a primary coupling element leaves a number of options open to the system designer which are not evident in a major-minor system. Modulo 128 and modulo 151 are only two examples of how this chip can be used. The reader is invited to explore modulo 179. Here the erase track can be ignored with the detector side replicator performing all functions: replicate in, replicate out and erase.

## REFERENCES


1) Gergis, I.S., L.R. Tocci, J. L. Archer, "A Compact Transfer/Replicate Switch for Field Access Bubble Devices," MMM Proceedings, 1975.

2) Ypma, John Edward, AFIPS NCC Proceedings, Volume 44, pp. 523-528, 1975.


# FIELD-ACCESSED BUBBLE LATTICE DEVICES


K. B. Mehta and O. Voegeli
IBM General Products Division, San Jose, California 95193


## ABSTRACT


A field-accessed scheme for bubble lattice devices is discussed, which uses deposited permalloy patterns driven by a rotating in-plane field. The method provides distributed driving forces on the lattice and does not impair the packing density advantages of the lattice concept. Operational characteristics of different designs are compared.


## INTRODUCTION

Bubble lattice devices have been discussed[1] in which, for data accessing, the storage lattice is translated by means of deposited current conductors spaced several bubble columns apart.[2] Lattice translation relies on an elastic distortion of the lattice and, consequently, local bubble velocities vary significantly.

This paper describes an alternative translation method: field accessing. The motivation for this approach is to simplify the device structure while providing more uniformly distributed driving forces on the lattice. To be viable, however, the propagation pattern should not impose lithographic resolution requirements which impair the density advantage of the bubble lattice scheme. While this constraint excludes utilization of conventional (T and I bars, etc.) patterns, it can be met in bubble lattice devices because bit positions need not be defined by the propagation pattern. They are retained by bubble interactions in the lattice.

Resulting translation patterns consist of permalloy bars which are sequentially magnetized by a rotating in-plane field. Two mutually (roughly) perpendicular sets of parallel bars are employed. Three such pairs are depicted in Fig. 1 in terms of their major axes locations relative to lattice coordinates. In each case, the sequentially attractive bar tips (e.g., 1-2-3-4) are successively displaced along the translation direction by 1/4 lattice spacing $a_0$. The same is true for the four corresponding repulsive tips which are located between bubble positions. Bubble attraction, as well as repulsion, is thus involved in the translation mechanism with the sign of translation dependent on the sense of field rotation.

The permalloy bars comprising a translation pattern are spaced by some integer lattice constants and arranged in a regular array. Since there exists a choice in tip location and bar spacing, various arrangements can be designed. Different configurations utilizing the

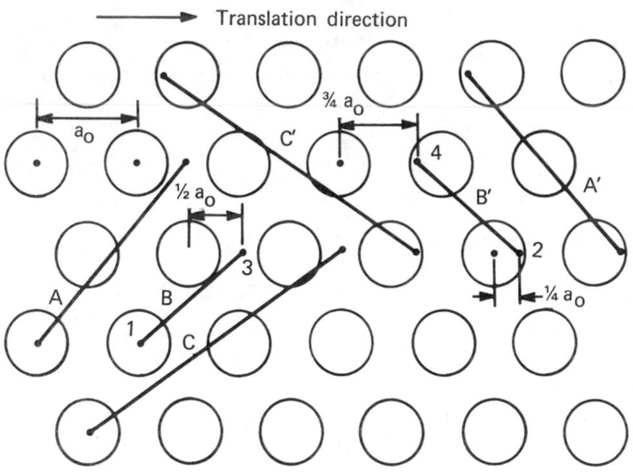


Fig. 1 Major axes of propagation elements relative to bubble lattice. Pairs A-A', B-B', C-C' represent different choices for lattice translation patterns.

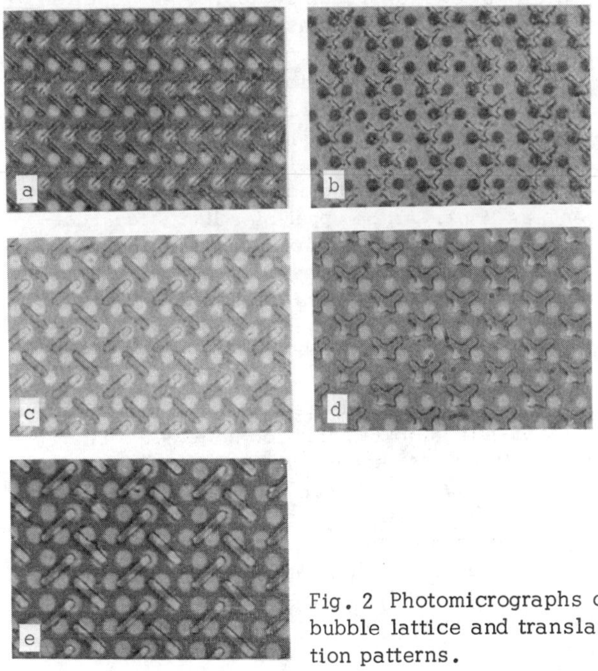


Fig. 2 Photomicrographs of bubble lattice and translation patterns.

bar pair B-B' have been investigated. B-B' was chosen for conceptual simplicity and because it allows for the largest fraction of driven bubbles.

## EXPERIMENTAL RESULTS

Figure 2 shows the five translation patterns evaluated experimentally. They differ with respect to width, spacing and relative positions of the propagation bars (B) and (B') as follows:

| Pattern | Width B & B' | Spacing longitudinal | Spacing transverse | Drive | Position |
|---|---|---|---|---|---|
| (a) | $.2a_0$ | $a_0$ | $2a_0$ | 1/2 | separate |
| (b) | $.2a_0$ | $2a_0$ | $a_0$ | 1/2 | cross |
| (c) | $.4a_0$ | $2a_0$ | $2a_0$ | 1/4 | separate |
| (d) | $.4a_0$ | $3a_0$ | $a_0$ | 1/3 | cross |
| (e) | $.4a_0$ | $3a_0$ | $2a_0$ | 1/6 | separate |

All patterns have a unit dimension, $a_0 = 12.5\ \mu m$. "Drive" refers to the fraction of bubbles attracted to bar tips at each 1/4 cycle of field rotation. "Position" indicates whether the bars B and B' "cross" or are located between "separate" rows of bubbles.

To evaluate device performance as a function of pattern geometry, as well as permalloy and spacer thickness, patterns (a) through (d) were fabricated on EuYIG wafers with the permalloy thickness varying between 500 and 2000 Å, and the $SiO_2$ spacer varying in thickness between 2 and 6 $\mu m$. Each translation pattern consists of a rectangular ($32\ a_0 \times 64\ a_0$) array with no provisions for lattice isolation or buffer region control. For device evaluation, first an "amorphous" bubble array is generated through application of a large (800 Oe) a-c in-plane field with the bubble density controlled by the d-c bias field. A regular bubble lattice is then established using a small bias field modulation while applying a nominal (10 Oe) in-plane field to magnetize one set of propagation bars. Lattice translation is observed within the central region of the pattern at frequencies up to 100 Hz and direction reversal after 1 to 10 translation steps. While the devices do not include controlled buffer regions at both ends of the lattice, complying stripe or bubble domain configurations do form within a few translation cycles. Their function is aided by retaining a small bias field modulation.

All devices produce variable translation when the lattice constant matches the periodicity of the translation pattern. This occurs over a range of bias fields where interactions between the lattice and the translation pattern retain the lattice period constant with bias field. Stripe-out of the bubble domains takes place at the lower bias limit. At the upper bias limit, local lattice perturbations occur instead of coherent translation, because the smaller bubble diameter to lattice constant ratio provides insufficient lattice rigidity. As the upper bias limit is lower with thin separation layers, a minimum separation of 2 $\mu m$ is desirable in combination with a 2000 Å thick permalloy pattern.

While the functional characteristics of the four patterns differ with regard to translation uniformity and magnitude of transverse excursions, these differences are presumably dependent on bar length rather than pattern geometry. Considering lithographic resolution requirements, the separate bar arrangement is favored, as best translation characteristics are obtained with the wide bar pattern (c), while for the cross arrangement, best characteristics are observed with the narrow bar pattern (d).

Based on the above evaluations, an optimized device (e), having separate bar arrangement, was fabricated with the following improvements: A wafer was chosen having a 5.6% longer strip width so that, for better rigidity, the lattice constant matches the unit dimension of the translation pattern at low values of bias field. To minimize transverse lattice expansions, the bar dimensions were adjusted to provide collinear lattice positions along the translation axis at successive 90° drive field rotations. The associated increase of bar length by 5 $\mu m$ also necessitated a one-unit dimension wider bar spacing along the translation axis to attain adequate bar separation. Permalloy thickness is 2000 Å, spacer thickness is 4 $\mu m$.

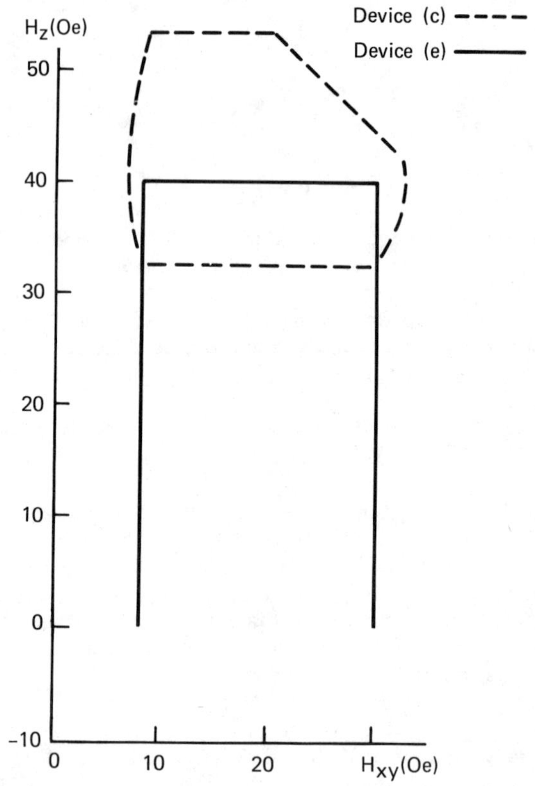


Fig. 3 Quasistatic bias ($H_z$) and drive field ($H_{xy}$) margins measured with bias modulating field of 9 Oe for device (c) and 3 Oe for device (e).

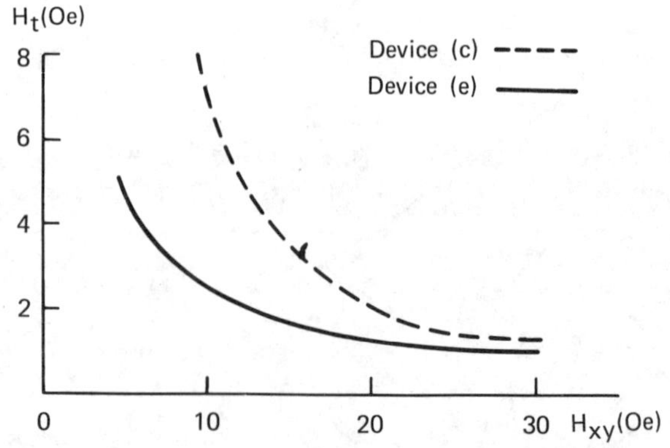


Fig. 4 Bias modulation ($H_t$) and drive field ($H_{xy}$) thresholds for lattice translation.

Figure 3 shows comparative bias and drive field margins, with improvement signified by increased bias margins and the reduction in required bias field modulation. Drive margins are limited by failure to propagate and stripe out at large drive fields. Threshold values for reliable lattice translation as a function of drive field amplitude and bias field modulation are shown in Fig. 4. Measurements of comparative bubble trajectories are shown in Fig. 5 for four adjacent bubble positions with no noticeable lattice distortions and transverse excursions in the optimized device.

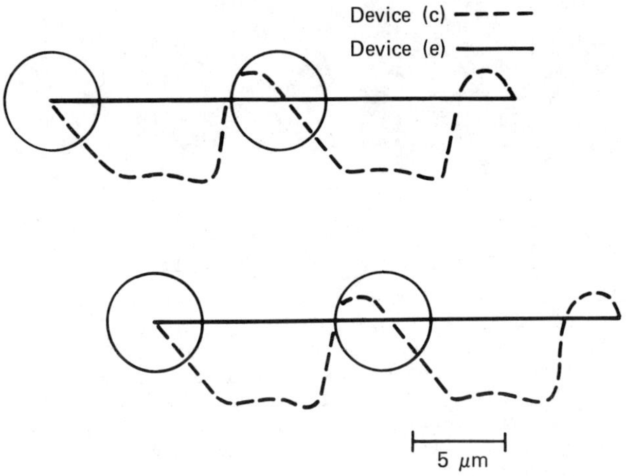


Fig. 5 Trajectories of four adjacent bubbles for one rotation of drive field.

CONCLUSIONS

Field-accessed bubble lattice translation has been demonstrated with translation patterns whose smallest dimensions equal .4 lattice constant. Compared to the conductor width of .43 lattice constant in the current access scheme,[3] pattern fabrication need not impair the density aspect of the lattice concept. The field-access approach could, conceivably, enhance wall state coding stability and operating speed because of more coherent lattice translation, reduced on-chip power dissipation, and may lead to a simple device configuration.

ACKNOWLEDGEMENT

We wish to thank R. F. Beeck, J. C. Hill, T. O. Montelbano and S. F. Vogel for device fabrications; J. S. Eggenberger, D. D. Rienstra and P. F. Swartzle for instrumentation support, and A. M. Desouches for measurement assistance.

REFERENCES

1. O. Voegeli, B. A. Calhoun and L. L. Rosier, AIP Conf. Proc. 24, 617 (1974).
2. L. L. Rosier, et al., AIP Conf. Proc. 24, 620 (1974).
3. J. S. Eggenberger, AIP Conf. Proc. 24, 622 (1974).

# FABRICATION OF 2-4 μm MAGNETIC BUBBLE MEMORIES WITH ELECTRON BEAM NEGATIVE RESIST LITHOGRAPHY*

J. N. Sweet, D. H. Schroeder, J. K. Maurin, J. R. Adams
Sandia Laboratories, Albuquerque, New Mexico 87115

## ABSTRACT

Electron lithographic techniques with co-polymer (Co-P) negative resist and ion milling have been used to fabricate small prototype 2 μm and 4 μm, single level bubble test circuits. The 4 μm circuits were used to evaluate two different generator and two different detector designs while the 2 μm circuits were used to study small bubble propagation characteristics. The minimum interelement spacing which could be routinely produced was about 0.8 μm which is just adequate for reliable 2 μm bubble propagation.

## INTRODUCTION

To fabricate field access bubble memory devices with bubble sizes $< 4$ μm, it is generally acknowledged that electron lithographic techniques are required since pattern linewidths are $< 2.5$ μm and interelement spaces $< 1.3$ μm. If large numbers of wafers are to be processed directly in an electron beam system, short writing times are a necessity and hence an electron resist with a sensitivity $\sim 10^{-6}$ C/cm$^2$ or better is desirable. In addition, writing speed may be minimized if a negative electron resist is used as an etch mask since the beam need then only be addressed to the ≈ 25% of the chip area covered by permalloy. Once the pattern is defined in resist, it may be delineated by ion milling, a technique commonly used in bubble circuit fabrication because of its inherent high etching accuracy and high yield characteristics.

We have fabricated a number of 4 μm and 2 μm 1-level bubble memory test circuits by electron beam lithography and evaluated their performance under a variety of operating conditions.

## ELECTRON LITHOGRAPHY

The pattern definition was performed with an ETEC computer-controlled SEM microfabrication system which could expose a pattern, point by point, in a 1 mm$^2$ area on an 8K x 8K grid, corresponding to a resolution ~ 0.1 μm. The electron resist used in all of our experiments was a co-polymer of glycidyl methacrylate and ethyl acrylate first reported by Thompson and Feit[1] and later discussed extensively by Thompson.[2] The resist had a sensitivity $\sim 5 \times 10^{-7}$ C/cm$^2$ for 70% developed thickness and a contrast ~ 1.1. All exposures were made at 10 kV with a beam current of $1.7 \times 10^{-11}$ amp and an exposure time of 1.2 μs/point. This beam current, exposure time, and sensitivity correspond to use of a beam with an effective diameter ≈ 700 Å.

Two major problems encountered in the use of Co-P resist for definition of bubble circuits are line broadening and the high ion-milling rate of the resist. Line broadening is believed to occur because of lateral electron scattering in the resist layer[3] which limits the minimum linewidth and interelement spacing that can be routinely achieved. Electron beam exposure of nearby elements separated by a gap can result in exposure of resist in the gap region if the interelement spacing is reduced sufficiently. In Fig. 1a a micrograph of a permalloy T-bar pattern is shown which is typical of the routine resolution capability of this resist. The circuit dimensions are: period $\lambda \approx 13$ μm; linewidth L ≈ 1.4 μm; and gap S ≈ 1.0 μm. These dimensions are marginally suitable for propagation of 2 μm bubbles in that S is about 50% too large. The gap size is critical, and we find in practice that gap enlargement severely degrades the propagation margins.

The second major problem is that of retaining sufficient permalloy in the ion-milled pattern for reliable bubble propagation. The resist is spun on and after prebaking the thickness is ≈ 6000 Å. After re-

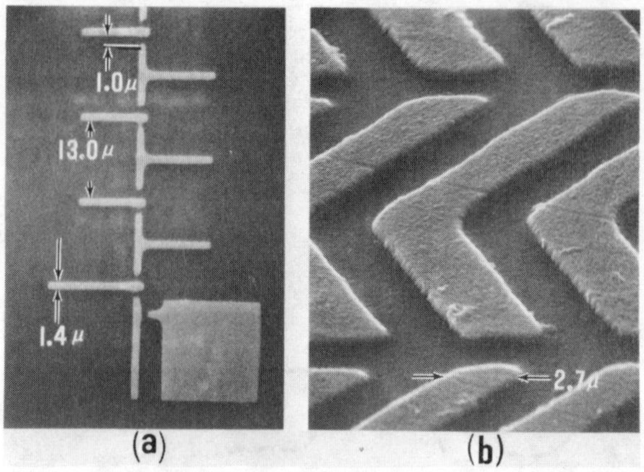


Fig. 1(a) T-bar circuit for 2 μm bubble propagation
(b) Chevron structure for 4 μm bubbles

sist exposure at the critical dose level and development of the pattern, approximately 3000 Å of resist remains. A variety of postbaking treatments and chemical crosslinking processes have been used in an attempt to make the resist harder and decrease its ion milling rate. With the best of these, the milling rate for Co-P was ≈ 200 Å/min in a Veeco Microetch system operated at 300 V, $8 \times 10^{-5}$ Torr(Ar), and a current density ≈ 0.6 mA/cm$^2$. Since the rf sputtered permalloy films milled at about 100 Å/min, it was possible to retain ≈ 1500 Å of permalloy after milling. In most of our processing a 500 Å Cr layer was deposited on the permalloy prior to lithography. This layer was chemically etched prior to ion milling of the permalloy pattern and it thus served as an additional etch mask. With the Cr layer, about 1800-2000 Å of permalloy could be retained in the final pattern. Fig. 1b shows an SEM micrograph of typical 4 μm bubble chevron propagation elements.

## TEST CIRCUIT FABRICATION

Several shift registers of 20 to 100 bit capacity have been designed and fabricated using all permalloy, single-level techniques. These shift registers offer generation, propagation, storage, and detection as required for a buffer memory. Since the observed propagation bias margins are much wider than either the generator or detector margins, most of our design effort has been directed toward increasing the latter margins.

These shift registers have been operated at quasi-static speeds and at speeds greater than 10 kHz. No attempt was made to suppress hard bubble creation although collapse field measurements indicated their presence. Thus no attempt was made to achieve high-speed (100 kHz) operation.

## PERFORMANCE OF E-LITH BUBBLE CIRCUITS

The best 4 μm circuits operated to date have been fabricated on $Y_{2.35}Eu_{0.65}Ga_{1.2}Fe_{3.8}O_{12}$ garnet with a zero bias strip width of 4.1 μm and a normal bubble collapse field of 68 Oe. Fig. 2 shows a typical 100 bit 4 μm bubble circuit and a margin curve for a similar 60 bit circuit. The circuit period is 16 μm, the permalloy linewidth and thickness are 2.5 and 0.2 μm respectively, while the interelement gaps are about 1.0 μm. The permalloy is separated from the garnet by 0.3 μm of sputtered $SiO_2$. The high bias failure mode is failure of the domains to expand in the detector region while at low bias the failure mode is strip out at either the generator or in the propagation track.

Two types of 103 chevron guardrail detectors were

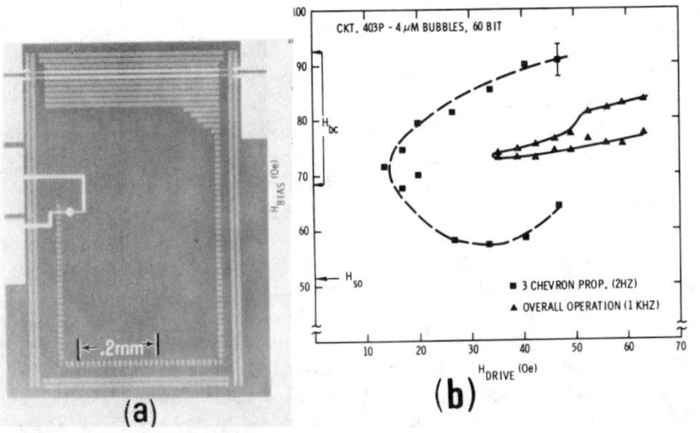


Fig. 2(a) Typical 4 μm 100 bit test circuit with disk generator and guardrail detector
(b) 4 μm circuit margin curve

tested, a conventional "end-connected" design and a center connected, or herringbone design. These were similar to those shown in Figs. 3c and 3a respectively in the article of Bobeck, et.al.[4]

The conventional detectors provided a 40% greater signal than the herringbone detector. In contrast, Bobeck[4] found, for 6 μm circuits, a much greater signal strength difference. However, the conventional detector also broke up the expanded domain as it left the first conductor strip and entered the second while the herringbone detector did not exhibit this characteristic. This domain breakup did simplify the detector electronics as each bubble produced an output signal independent of adjacent bubbles. However, since the total operating margin of the herringbone detector

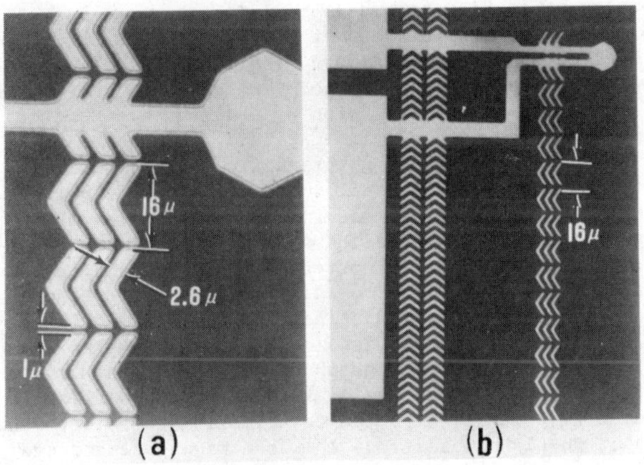


Fig. 3 4 μm bubble circuit generators (a) Dot generator (b) Loop generator

was greater, it was chosen for later designs.

The dot generator[5] shown in Fig. 3a was compared to the loop generator of Fig. 3b. In general, the dot generator will operate over a wider margin than the loop generator and, in addition, requires only a single polarity stretching pulse of 50 ma for 2 μs while the loop generator requires a dual polarity pulse of approximately the same amplitude and duration to both stretch the seed bubble into the track and, 180° later, to cut this stretched domain so that additional bubbles are not introduced into the track. The loop generator also tends to spontaneously insert bubbles in the track at low bias, thus the dot generator appears to be a superior all permalloy generator design.

All of the 2 μm circuits have been fabricated on $Y_{2.0}Eu_{0.5}Tm_{0.5}Ga_{0.5}Fe_{4.5}O_{12}$ material. The circuit reported on here was on a wafer with a zero bias stripwidth of 1.9 μm, bubble collapse of 213 Oe and $4\pi M$ of 413 G. The circuit is similar to that shown in Fig. 2a, the period is 8 μm, with the permalloy elements linewidth, separation and thickness equal to 1.2, 0.7, and 0.2 μm respectively while the permalloy is separated from the garnet by 0.4 μm of sputtered $SiO_2$.

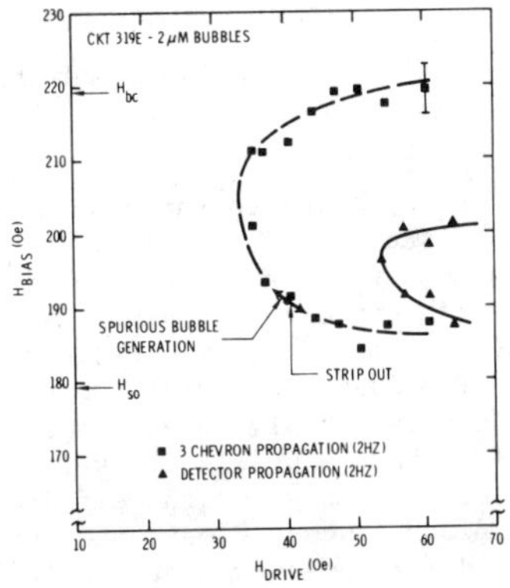


Fig. 4 2 μm bubble track and detector propagation margins for a circuit with a 10 μm period

The performance of the 2 μm devices to date has been limited to quasi-static propagation; no reliable generation or detection has been obtained. The margin curves shown in Fig. 4 indicate relatively good propagation in the storage loop while expansion in the detector region occurs only over a narrow bias window. The circuit from which these data were obtained incorporated both the loop generator and the conventional detector, which, in the 4 μm circuit, were both inferior circuit elements. Generation could not be controlled either by stripping out the bubbles at high bias or by blocking bubbles at low bias due to burnout of the generator with currents of about 10 ma corresponding to a current density of $\approx 5 \times 10^6$ A/cm$^2$.

Detector operation could not be established without an active generator; however, at low bias fields and high drive fields, the bubbles will fully strip out, thus the detectors should function within very narrow bias limits.

## SUMMARY

The fabrication and design techniques described here are adequate for 4 μm bubble shift registers and were found to be satisfactory for propagation of 2 μm bubbles. However, the overall 4 μm circuit operation margins were small (~ 8%) and required relatively high drive fields (~ 60 Oe) while the 2 μm circuits were not operational due to the lack of generation capability.

The fabrication of fully operational 2 μm circuits using the methods described here will require higher resist resolution and either a significant decrease in the resist milling rate or an alternate milling procedure in order to define 0.3 to 0.4 μm thick permalloy propagation elements.

*This work supported by U. S. Energy Research and Development Administration.

## REFERENCES


1. L. F. Thompson, E. D. Feit, "Organic Coatings and Plastics Preprint," Chicago ACS Meeting, April 1973.
2. L. F. Thompson, Solid State Tech., 17 #7, 27 (1974), 17 #8, 41 (1974).
3. R. D. Heidenrich, L. F. Thompson, E. D. Feit, C. M. Melliar-Smith, J. Appl. Phys. 44 4039 (1973).
4. A. H. Bobeck, I. Danylchuk, F. C. Rossol, W. Strauss, IEEE Trans. Magn., MAG-9 474 (1973).
5. P. K. George, IEEE Trans. Magn., MAG-10 1136 (1974).

# TRANSLATIONAL VELOCITIES AND BALLISTIC OVERSHOOT OF BUBBLES IN GARNET FILMS

A. P. Malozemoff, J. C. Slonczewski and J. C. DeLuca
IBM Research Center, Yorktown Heights, New York 10598

ABSTRACT

We report on high speed photography studies of bubble propagation by pulsed gradient fields. We have examined garnet films of different compositions, as-grown and ion-implanted, with different thicknesses and degrees of crystallographic alignment. In all cases we find that for moderate gradient drives, the bubble moves at an average velocity which is roughly independent of the drive strength and exhibits a ballistic overshoot after the end of the pulse. A theory for the effects is based on the notion that Bloch lines nucleate and store the kinetic energy needed for the overshoot. Calculations of the velocity-momentum relation for a moving bubble indicate that the average velocity should be higher than in the plane wall or radial case and close to the Bloch line instability velocity $V_{po}$ for a plane wall.

## I. INTRODUCTION

In a recent paper[1] we have reported that ballistic overshoot can occur in the field gradient propagation of magnetic bubbles. In this experiment[2] a pulse of gradient field of known duration T is applied to a bubble, resulting in a net displacement $X_\infty$. In early work the average bubble velocity was taken to be $X_\infty/T$ on the assumption that the bubble moved only while the gradient field was on. In fact, however, our high speed photography measurements showed that the transient bubble motion can persist for a considerable time after the pulse terminates and that therefore the actual velocity can be considerably different from $X_\infty/T$. Furthermore, we reported that the true translational velocity in one garnet film was saturated with drive field. We call the motion after the end of the pulse "ballistic overshoot" because the bubble appears to coast in the absence of any external driving force. Recently, Patterson et al.[3] have also seen overshoot in a gradient propagation experiment.

In Section II of this paper we present further experimental data showing that these effects are quite a general characteristic of garnet bubble films. In particular, we report on different YIG compositions, on ion-implanted versus as-grown films, on films of similar composition but varying thickness, and on films with varying degrees of crystallographic misorientation. We compare the translational saturation velocity to radial saturation velocity measurements. Furthermore we describe a new effect associated with crystallographic misorientation, namely the dependence of the ballistic overshoot on the direction of motion in the sample.

In another recent paper[4] we have introduced a simple model for the ballistic overshoot effect, based on the notion of the wind-up of Bloch lines while the pulse is on and their unwinding when the pulse turns off. In Section III of this paper we present a more detailed model. We introduce the concept of the total linear momentum of a bubble and derive the equation of motion for the time rate of change of the momentum. We use the Bloch line approximation to derive the dependence of bubble velocity on the momentum, and we present numerical calculations of the Bloch line structure. Finally we estimate the critical and saturation velocities, and in Section IV we compare these theoretical predictions to experiment.

The significance of our experimental results and theoretical interpretation can be seen in the light of previous work on domain wall inertia and velocity saturation in garnet bubble materials. A one-dimensional theory of domain wall inertia was first introduced by Döring.[5] Similar one-dimensional models are consistent with experiments on bubble materials, under the conditions of large thicknesses (bulk measurements) or high in-plane fields.[6,7,8] But much larger masses were observed in thin films at low or zero in-plane field[7,8,9] in straight wall oscillation experiments. Schlömann[10] and Hubert[11] formulated a theory accounting for this large "small-signal" mass, but Cape, Hall and Lehman[12] first proposed theoretically that a giant mass effect could also occur in bubble motion because of the wind-up of Bloch lines. Our experiments have now shown that ballistic effects much larger than predicted from Döring's theory do indeed occur in bubble motion, although the detailed mechanism is different from that of Cape et al.[12]

Another important aspect of domain wall dynamics in bubble materials was the observation of a saturation velocity in a number of experiments including dynamic bubble collapse,[13] straight wall motion[14] and stripe and bubble expansion.[15,16,17] One of us earlier proposed a mechanism for the saturation velocity involving the nucleation and propagation of Bloch lines.[18] However Vella-Coleiro et al.[19] reported bubble propagation velocities monotonically increasing with drive field to values far above the saturation velocity seen in dynamic bubble collapse. These results indicated that "dynamic conversion," that is, Bloch line nucleation and propagation, usually did not occur in bubble translation. Scatter in the data suggested that random defects permitted occasional dynamic conversion.[20] However, these conclusions stemmed from the simple assumption $V=X_\infty/T$ in the data analysis. Our experimental results now show that velocity saturation occurs in bubble translation and raise the possibility that the high velocities reported previously were an artifact of the bubble overshoot. Our theory shows that these results may be explained by a mechanism similar to that proposed in connection with plane wall or radial bubble motion.

## II. EXPERIMENTAL PROCEDURES AND RESULTS

### A. Material Parameters

Magnetic films were grown epitaxially on [111]-oriented substrates. All the films in our study support bubbles whose diameters d=2r fall in the range between 3 and 8 μm. Material parameters for a series of films are given in Table I and were determined as follows: The thickness h was determined from optical interference. The spontaneous magnetization $M_s$ and length parameter $\ell$ were measured by the Fowlis-Copeland technique.[21] Uniaxial anisotropy $K_u$ and gyromagnetic ratio $\gamma$ were determined from ferromagnetic resonance. The Gilbert damping parameter $\alpha$ was determined in an approximate manner from the slope of the apparent translational velocity vs. drive curve, corrected for coercivity as discussed elsewhere.[4] The coercive field $H_c$ was measured from an optical hysteresis loop measurement and is 0.25±0.1 Oe for all samples. Many of the samples had a tilt $\Delta_c$ in the crystallographic [111] axis away from the surface normal, which was measured by x-ray diffraction. Such a tilt can be expected to give rise to an orthorhombic or in-plane anisotropy $K_p$ whose magnitude and easy axis were obtained from homogeneous nucleation measurements.[22,23] A prediction for the azimuthal direction $\phi_p$(theor.) of the in-plane easy axis was obtained by combining the measurement of the crystallographic misorientation with a phenomenological theory of growth anisotropy.[22]

### B. Bubble Propagation Measurements

For the pulsed field gradient propagation experiment, an S=0 bubble,[24] that is, one which propagates normal to the gradient pulse lines and presumably contains two Bloch lines, was selected. A priming gradient pulse whose strength was held close to the coercive limit, so as not to disturb the initial state of the bubble, was used to position the bubble at its starting position at the center of the pulse lines. Pulses of varying strengths were used and the positions of the bubble as a function of time were recorded by means of high-speed photography in the same manner as previously reported.[1] All bubble translation data were obtained using proper bias compensation[1] to eliminate gross distortions of the bubble due to a decreasing bias field as the bubble moved down the field gradient.[25] The gradient field rise and fall times were generally 15 nsec but no significant effect on the bubble motion was observed when the rise and fall were varied from 15 to 120 nsec in sample 1 of the Table, using a 2 Oe, 0.5 μsec

TABLE: Material Parameters of Garnet Films (symbols defined below**)

| Sample Number | 1 | 2 | 3 | 4 | 5 | 6 | 7 |
|---|---|---|---|---|---|---|---|
| Nominal Composition | $Eu_{0.65}$ | $Eu_{0.65}$ | $Eu_{0.65}Tb_{0.04}$ | $Gd_{0.9}Tm_{1.1}$ | $Gd_{0.9}Tm_{1.1}$ | $Sm_{0.4}$ | $Sm_{0.1}Ca_{0.98}$ |
| (YIG) | $Ga_{1.2}$ | $Ga_{1.15}$ | $Ga_{1.15}$ | $Ga_{0.75}$ | $Ga_{0.75}$ | $Ga_{1.2}$ | $Ge_{0.98}$ |
| h(μm) | 4.3 | 4.9 | 5.2 | 3.8 | 5.4 | 5.1 | 5.8 |
| 4πM(G) | 160 | 183 | 179 | 157 | 163 | 167 | 139 |
| ℓ(μm) | 0.77 | 0.40 | 0.56 | 0.84 | 0.81 | 0.80 | 0.92 |
| $K_u$(ergs/cm$^3$) | 9030 | 6400 | 10200 | 8800 | 8950 | 14000 | 4880 |
| $\gamma$x$10^7$Oe$^{-1}$sec$^{-1}$ | 1.21 | 1.38 | 1.27 | 1.27 | 1.27 | 1.89 | 1.81 |
| $K_p$(ergs/cm$^3$±150) | 130 | 380 | 180 | <400 | 490 | <350 | * |
| $\alpha_{max}$ | 0.05 | 0.025 | 0.08 | 0.045 | 0.05 | 0.2 | 0.1 |
| $\Delta_c$(±.05°) | 0.15 | 0.6 | 0.2 | 0.4 | 0.75 | 0.4 | 0.45 |
| $\phi_p$(exptl,±10°) | * | 87 | 150 | * | 124 | 64 | * |
| $\phi_p$(theor,± 5°) | * | 89 | 165 | 147 | 140 | 70 | 150 |
| $V_t$(±50 cm/sec) | 800 | 900 | 800 | 1100 | 1000 | 850 | 1760 |
| $V_{r-}$(±100 cm/sec) | 925 | 1175+ | 880 | 1400 | 1150 | 1300 | 2100 |
| $V_{r+}$(±100 cm/sec) | 925 | 1550+ | 1220 | 1400 | 1425 | 1800 | 2100 |
| $\mu_h$(cm/sec Oe) | 6.5 | 11.8 | 6 | 9.5 | 8.5 | 14 | 8 |
| $V_{oo}$(cm/sec) | 360 | 280 | 210 | 480 | 360 | 310 | 810 |
| $V_{po}$(cm/sec) | 1200 | 950 | 710 | 1630 | 1220 | 1060 | 2780 |
| $\mu_1$(cm/sec Oe) | 17 | 15 | 34 | 14 | 23 | 80 | 80 |
| $\mu_2$(cm/sec Oe) | 3 | 1.5 | 3 | 2.5 | 3 | 11 | 13 |

*not measurable, +no clean saturation in this sample

**h=film thickness, M=spontaneous magnetization, $\ell=(AK_u)^{1/2}/\pi M^2$=characteristic length $K_u$=uniaxial anisotropy, $\gamma$=gyromagnetic ratio, $K_p$=in-plane component of anisotropy, $\alpha_{max}$=bound on Gilbert damping parameter, $\Delta_c$=tilt angle of [111]-axis, $\phi_p$=azimuthal orientation of in-plane easy axis, $V_t$=translational saturation velocity, $V_{r-/+}$=radial expansion velocity for (minor/major) axis, $\mu_h$=high drive mobility in radial expansion, $V_{(o/p)o}$=theoretical radial (saturation/peak) velocity for an isolated plane wall, $\mu_1$=mobility of a wall with a single horizontal Bloch line (Eq. 4.1), $\mu_2$=mobility of a wall undergoing free precession.

compensated pulse.

Results on a series of 7 films of different compositions are shown in Table I and Figs. 1-8. Data on film 2 of Table I have been shown in earlier publications.[1,4] These films include YIG doped with EuGa, EuTbGa, GdTmGa, SmGa and SmCaGe. Figures 1-5 show the position $x_T$ of the bubble at the end of the drive pulse and its final position $x_\infty$, as a function of drive strength. The data in Figs. 1-5 do not include very low drives, where, particularly for high mobility films, there is a sudden drop off of $x_T$ and $x_\infty$ as the drive is reduced towards the coercive threshold. For all samples, a saturation in $x_T$ appeared at sufficiently high drive, even in the high damping SmGa sample as shown in Figs. 1-5. The saturation velocity $V_t$ was obtained from $x_T/T$ and is given in the Table. The EuGa, EuTbGa, GdTmGa and SmGa compositions were ion-implanted at dosages sufficient to cause hard bubble suppression and show roughly the same saturation velocity as the as-grown films, as shown for sample 1 in Figs. 1 and 2. Further evidence for the saturation velocity is shown in Fig. 6, where the position of the bubble is plotted as a function of time for sample 1. The velocities during and after the pulse appear somewhat different from each other but are roughly constant as a function of drive field. The data indicate that during the ballistic overshoot phase the bubble coasts at a constant velocity until it rather abruptly stops. Bubble position as a function of time during the pulse has also been measured (e.g. see Fig. 4 of ref. 1 which shows the equivalent results, namely, the bubble position at the end of a pulse of varying length); the results are linear within experimental accuracy.

Fig. 7 shows the transient bubble shapes at the end of the drive pulse, as a function of pulse strength. The bubbles tend to become elliptical with their major axis perpendicular to the direction of motion. As shown in the figure, the ellipticity increases monotonically with drive strength. The ellipticity depends strongly on the bias compensation, while the position $x_T$ depends only weakly on the compensation. Earlier Vella-Coleiro et al.[25] reported bubble strip-out in the absence of compensation and no distortion with compensation. However, we find that some degree of ellipticity can persist even with "correct" compensation,[1] as shown in Fig. 7. The figure also shows the area of the bubble at the end of the pulse as a function of pulse strength. The area is sensitive to the bias compensation, but even at correct compensation the area increases a small amount in most samples.

We have also investigated the size of the overshoot as a function of the direction of motion in samples 1, 2, 5 and 6 of the Table. In sample 1, which is crystallographically well-aligned, we found no noticeable effect, but in the other samples which are more misaligned, we found a variation in $x_\infty$ (but not in $x_T$) with a $2\phi$ periodicity as shown in Fig. 8 for sample 5. The maxima were found to lie within experimental error along the easy axes of in-plane anisotropy, and are given in the Table as $\phi_p$ (exptl.).

Two series of EuGaYIG films were prepared to examine the thickness dependence of the saturation velocity. The first series included four films grown under identical conditions but for varying times to give the different thicknesses. The second series consisted of

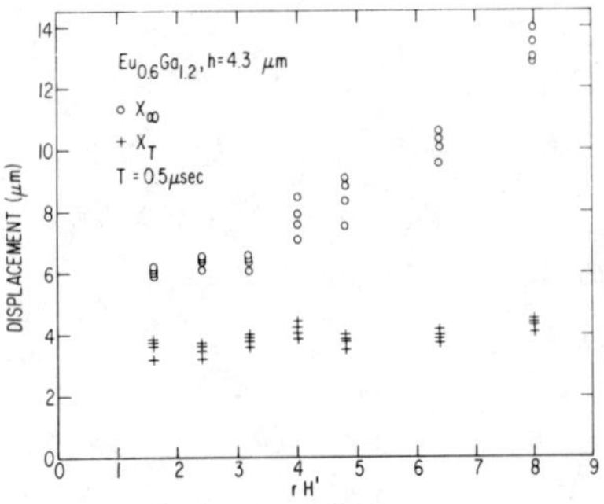


Fig. 1. Bubble displacement as a function of gradient drive in Sample 1 of the Table. $x_T$ is the displacement at the end of the pulse of length T, and $x_\infty$ is the final displacement.

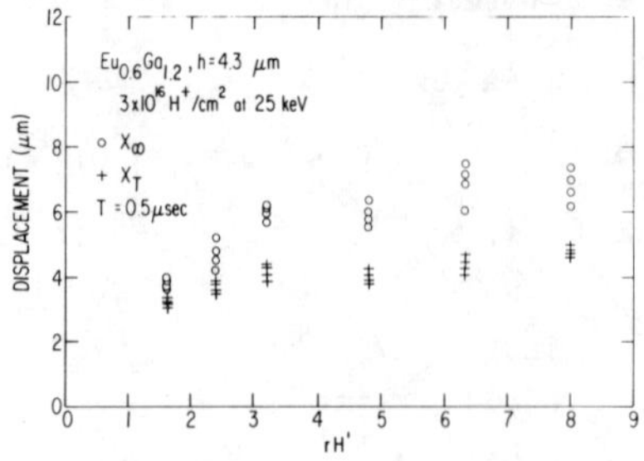


Fig. 2. Same as Fig. 1, sample 1, ion-implanted with protons.

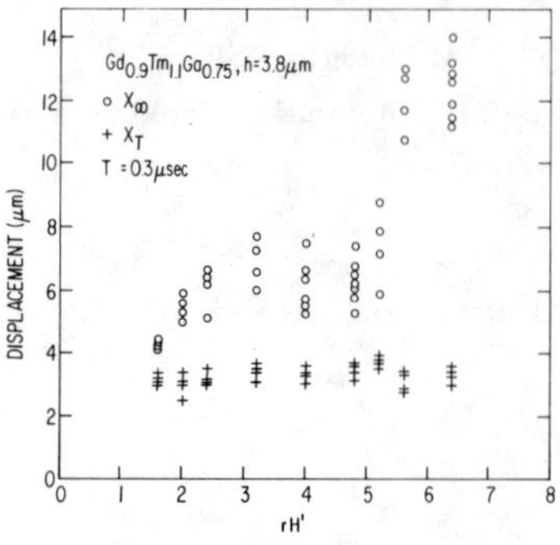


Fig. 3. Same as Fig. 1, sample 4.

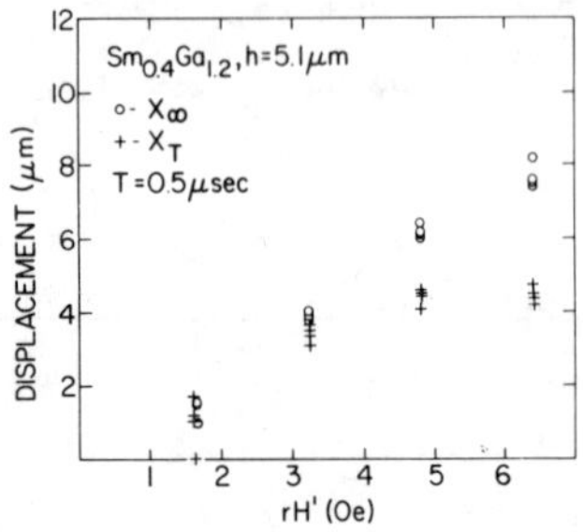


Fig. 4. Same as Fig. 1, sample 5.

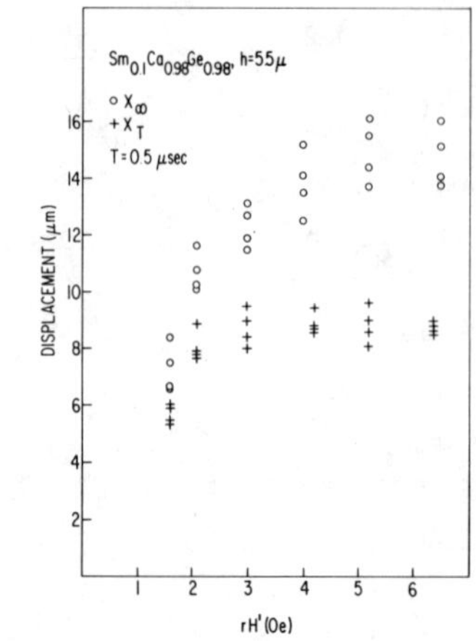


Fig. 5. Same as Fig. 1, sample 7.

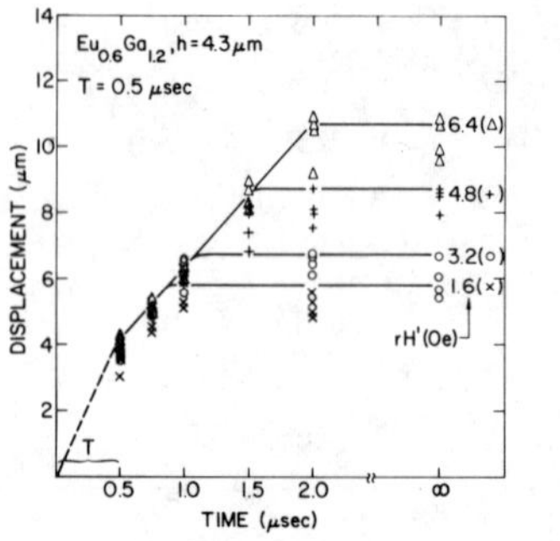


Fig. 6. Bubble displacement as a function of time for four different gradient drives in sample 1.

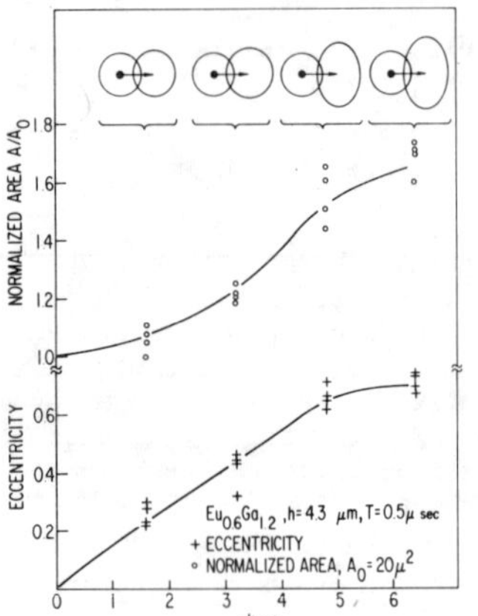


Fig. 7. Bubble eccentricity $[1-d_{min}^2/d_{max}^2]^{1/2}$ and normalized area $d_{max}d_{min}/d_o^2$ at the end of the gradient drive pulse, as a function of drive. The sketches are taken from actual double exposure photographs giving the initial and transient positions of the bubbles.

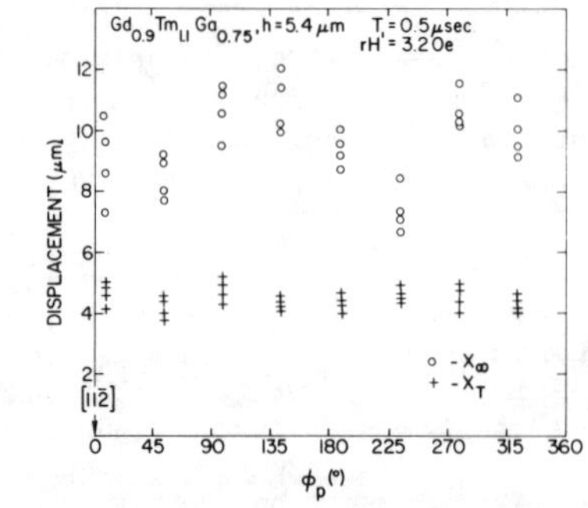


Fig. 8. Bubble displacement $x_T$ and $x_\infty$, as in Fig. 1, as a function of direction of propagation in sample 5.

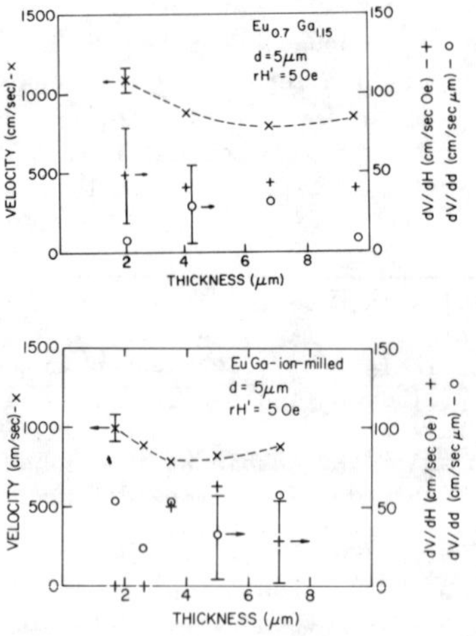


Fig. 9. Saturation velocity and its dependence on bubble size and drive, in two series of films of constant composition but varying thickness. dV/dd is increase in velocity per decrease in diameter at constant drive. dV/dH is the mobility in the saturation region.

a single film, diced and then ion-milled to five different thicknesses using 500 eV, 0.6 ma/cm$^2$ argon bombardment. These films were gallium-backed to insure good heat sinking during the milling. The first series was grown on substrates crystallographically aligned to 0.5°, the second on a substrate aligned to 0.1°.

The measurements were performed by measuring $X_T$ at different diameters and drives, and the results were interpolated to determine a velocity at constant diameter and drive, given in Fig. 9. Both series show similar results in that the velocity tends to rise for very thin films. Furthermore the velocity tends to drop as the diameter increases, and the slope is plotted in Fig. 9. It is of order 50 cm/sec per micron of change in bubble size. The high drive mobility is also plotted and is always less than 60 cm/sec Oe in these series. Considering experimental error, the dependence of the saturation velocity on bubble size and drive is very weak. These results underlie the procedure in Table I of listing saturation velocities without specifying bubble size or drive strength.

C. Bubble Expansion Measurements

We have also measured the saturation velocity in bubble expansion,[16,17] to compare with the translation results. An S=0 bubble was positioned at the center of a pulse coil. A current pulse several tenths of a microsecond long and with 20 nsec rise times was applied to the coil to create a negative bias pulse. As has been shown in previous work,[16,17] the expanding bubbles can distort in complicated ways and the velocity can vary during the motion if the field is too weak or the pulse is too long. In our work we have used large enough drives (25-200 Oe) and short enough pulses to prevent such distortions and velocity variations. The change in bubble radius $\delta r$ after a given time T was measured by high speed photography, and the velocity was taken to be $\delta r/T$.

The small slope found in the velocity with bias field, which is a high-drive mobility $\mu_h$, and the velocity back extrapolated to zero drive $V_r$, are recorded in the Table. Furthermore in samples with larger crystallographic tilt, the expansion was elliptical, with different velocities along the major and minor axes. The asymmetric effects on $V_r$ are labeled in the Table by subscripts + for major axis and - for minor axis. From earlier work[17] we expect that the minor axis should correspond to the easy axis of in-plane anisotropy, and the direction of this axis is listed in the Table as $\phi_p$(exptl.). The result coincides within experimental

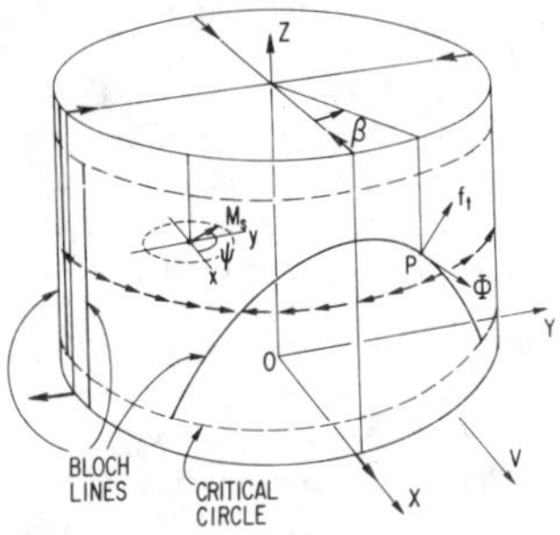


Fig. 10. Structure of bubble domain with a curved Bloch line, moving at velocity V.

error with $\phi_p$ obtained from overshoot (see IIIB) and from growth anisotropy theory.

## III. THEORY OF BUBBLE TRANSLATION

### A. Bubble Momentum and its Equation of Motion

Our analysis of bubble translation is analogous to a previous treatment of radial motion.[18] We begin with the principle that the dynamical variable $p_n = 2M_s\psi/\gamma$ is the effective canonical momentum per unit area of a domain wall.[26] Here $\psi$ is the azimuthal angle describing the orientation of the magnetic moment within the domain wall in the plane of the film, measured from a direction fixed in the medium, as shown in Fig. 10. By convention $p_n$ is directed from the $M_z = M_s$ side of the wall to the $M_z = -M_s$ side, so that in the case of a cylinder domain with $M_z = -M_s$, it is directed radially inward. In the language of Hamilton's equations of motion, $p_n$ is canonically conjugate to the normal wall displacement at any given point on a wall surface of general shape.

Let us now consider the total linear momentum $P_x$ conjugate to the displacement x of the bubble center. We expect $P_x$ to be given by projecting the normal momentum $p_n$ onto the x axis and integrating over all of the wall surface. If the position variable z and the angle $\beta$ shown in Fig. 10 describe the position of a point on the cylinder, then we have

$$P_x = 2\pi M_s \gamma^{-1} rh\bar{\psi}_x \qquad (3.1)$$

where, for convenience, we have defined the variable

$$\bar{\psi}_x = -(\pi h)^{-1} \int_0^h dz \int_0^{2\pi} d\beta\ \psi(\beta,z)\cos\beta + \bar{\psi}_{xo}. \qquad (3.2)$$

$\bar{\psi}_x$ may be interpreted as twice the mean projected precession angle of the wall moment. The physically meaningless constant $\bar{\psi}_{xo}$ may be arbitrarily assigned a convenient value. Equation (3.2) is valid for the case S=0, which is under consideration, and Figure 10 indicates the pair of vertical Bloch lines whose presence is required in this case. [For other state numbers S≠0, a more general definition is needed because then $\psi(\beta) \neq \psi(\beta+2\pi)$].

According to a principle of mechanics, one has the relation

$$\frac{dP_x}{dt} = F_x \qquad (3.3)$$

where $F_x$ is the total effective force on the domain. A general proof of this relation for domains will be given elsewhere. For the state S=0, the force can be written[27]

$$F_x = 2\pi rhM_s[H - H_{cb}\,\mathrm{sgn}\dot{x} - \mu^{-1}\dot{x}] \qquad (3.4)$$

Here $H = -rdH_z/dx$ measures the gradient; $H_{cb}$ is the bubble coercivity, related to the plane wall coercivity $H_c$ by $H_{cb} = 4H_c/\pi$; $\dot{x} = dx/dt$ is the instantaneous velocity, sgn $\dot{x}$ is ±1 depending on the sign of $\dot{x}$; and $\mu$ is the linear mobility $\gamma\Delta\alpha^{-1}$, where $\Delta$ is the Bloch wall width parameter $\sqrt{A/K_u} \approx \pi\ell M_s^2/K_u$. In (3.4) we have ignored extra damping due to Bloch lines.[28,29]

Combining Eqs. (3.1), (3.3) and (3.4) we find the important relation[1]

$$\dot{\bar{\psi}}_x = \gamma(H - H_{cb}\,\mathrm{sgn}\dot{x} - \mu^{-1}\dot{x}) \qquad (3.5)$$

according to which the mean rate of precession is proportional to the sum of effective fields arising from the applied gradient, coercivity and viscous drag. Now, if the forces on the bubble are sufficiently small it is reasonable to assume that its motion approaches a quasi-steady state. In other words, at any time t a bubble with $\bar{\psi}_x = \bar{\psi}_x(t)$ has the same velocity $V(\bar{\psi}_x)$ that it would have in a state of free and steady motion at the same $\bar{\psi}_x$. Thus a knowledge of the relation $V(\bar{\psi}_x)$ allows one to combine the equation

$$\dot{x} = V(\bar{\psi}_x) \qquad (3.6)$$

with Eq. (3.5) to obtain by numerical integration the transient bubble motion, including the motion at drives for which uniform motion is not stable.

### B. Calculation of Bloch Line Structure

Previously the overshoot effect in bubble propagation was interpreted in terms of the ad hoc assumption that $V(\dot{\bar{\psi}}_x=0)=0$ and $V(\dot{\bar{\psi}}_x>0)=V_s$ where $V_s$ is a constant "saturation velocity."[4] This assumption, together with Eqs. (3.5) and (3.6) explained many features of the experiments. Here we attempt to deduce the function $V(\bar{\psi}_x)$ from, essentially, the Landau-Lifschitz equation. In principle the procedure is to assume a value for V, solve equations of motion for $\vec{M}(\vec{x})$ or $\psi(\beta,z)$, then evaluate the integral (3.2) to obtain $\bar{\psi}_x(V)$.

This task is lightened by approximating the function $\psi(\beta,z)$ with the Bloch line model, according to which $\psi$ at all points takes on one of its local static equilibrium values except along the Bloch lines which serve as "phase" boundaries for different orientations.[18] This model has been used to treat radial bubble motion[18,30] and its validity has been tested by comparison with more general calculations.[11,30] For bubble translation, some of the Bloch lines are curved as first described qualitatively by Hagedorn.[4,20,29]

To calculate the shape of a moving Bloch curve, consider any point P on the curve moving with velocity $\vec{V}$. This motion implies the precession of electron spins in the neighborhood of the line. To provide the torques responsible for this precession a force per unit length[28,29]

$$\vec{f} = 2M_s\vec{\Phi}\times\vec{V}/\gamma \qquad (3.7)$$

must be provided by some combination of exchange stiffness, anisotropy and magnetic dipole energy, if dissipation is neglected. Here $\vec{\Phi}$ is a vector tangent to the curve whose magnitude is

$$\Phi = |\psi_1 - \psi_2| = 2\phi\ , \qquad (3.8)$$

where $\psi_1$ and $\psi_2$ are the static values of $\psi$ within the surface on either side of the Bloch line, near P. The schematic plot of $\psi$ versus z, for constant $\beta$, shown in Fig. 11, illustrates this. If $\psi$ is measured from OX in Fig. 10, then

$$\psi_{1,2} = \pm\phi + \beta,\quad \phi = \cos^{-1} H_r(z)/8M_s \qquad (3.9)$$

where ± $\phi$ is the local static deflection of the wall mo-

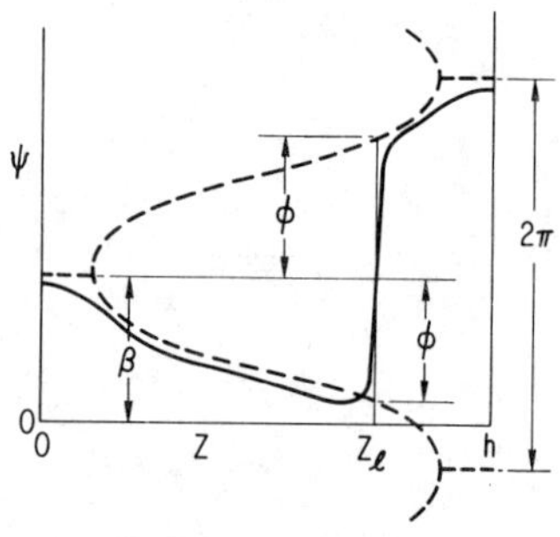


Fig. 11. Wall-moment angle ψ versus position along film normal z, with a Bloch line at $z=z_\ell$. (Schematic)

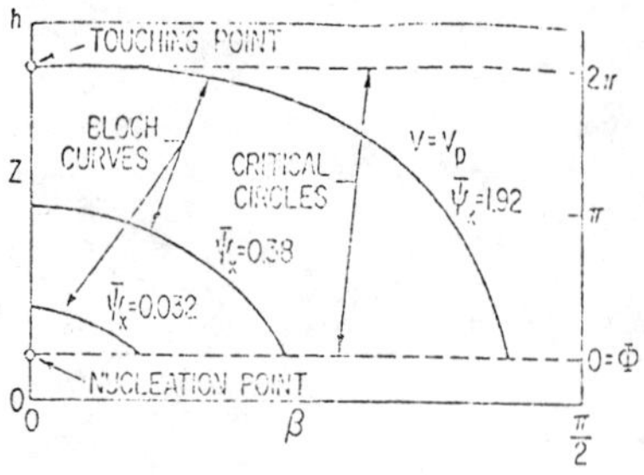


Fig. 12. Flattened quarter of bubble-domain wall showing critical circles and computed Bloch curves for the case h=r.

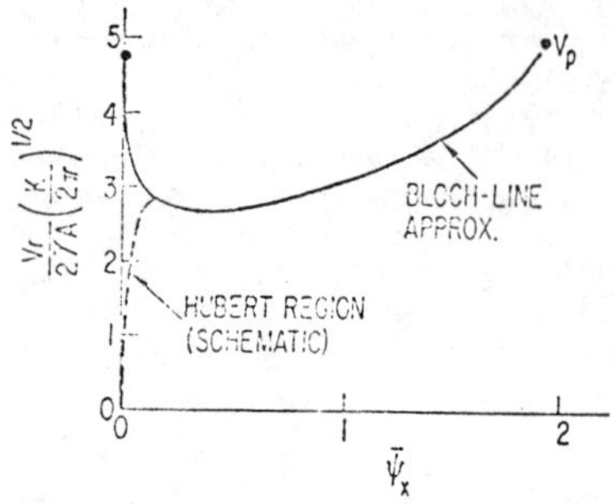


Fig. 13. Dimensionless velocity versus dimensionless momentum $\bar{\psi}_x$ for the case h=r and stray field of Eq. (3.14).

ment away from the wall tangent by the radial component of stray field $H_r(z)$ at the Bloch line position $z=z_\ell$.[18,20] The direction of $\vec{\Phi}$ is given by that of $\vec{\nabla}\theta\times\vec{\nabla}\psi$, where the polar angle $\theta=\cos^{-1}M_z/M_s$ varies through the wall to give the conventional Bloch wall profile.[29]

The dynamical equilibrium of a Bloch line is understood by reference to Fig. 10. The strong physical constraint requiring the Bloch line to remain within the wall provides the wall-normal component of $\vec{f}$ and need not be considered further. The component $f_t$, tangential to the wall surface S and normal to the Bloch line, derives from the energy of the Bloch curve configuration. It includes one term due to the Bloch line curvature κ within the tangent plane plus one due to the dependence of Bloch line energy density u on z owing to the dependence of stray field on z.[18] It is written

$$f_t = u\kappa - (du/dz)\cos\alpha \, , \qquad (3.10)$$

where α is the angle between $f_t$ and the z axis. The energy is[18]

$$u = 8AM_s(2\pi/K_u)^{1/2}(\sin\phi - \phi\cos\phi) \qquad (3.11)$$

under the condition $|H_r|\leq 8M_s$. According to Eq. (3.7), $f_t$ must balance the dynamic reaction caused by the wall-normal component of domain velocity $V_n=V\cos\beta$.

We let $ds=(dz^2+r^2d\beta^2)^{1/2}$ be an increment of Bloch line length. Equations (3.7) and (3.10) combine to give the differential equations

$$u(d^2z/ds^2) + wr(d\beta/ds)=0$$
$$u(d^2\beta/ds^2) - w(dz/ds)=0 \qquad (3.12)$$
$$w = 4M_s\gamma^{-1}V\phi\cos\beta - r(d\beta/ds)(du/dz) \qquad (3.13)$$

whose solution β(s), z(s) describes the shape of the Bloch line on the +x or front side of the bubble. The Bloch line is bounded by the critical circle satisfying $H_r(z)=8M_s$, where φ=0 and u=0. The symmetrically disposed Bloch line on the -x side is similarly bounded by the critical circle $H_r(z)= -8M_s$.

For ease of computation we use the model stray field

$$H_r = 4M_s \ln[z/(h-z)] \, , \qquad (3.14)$$

which is actually correct for a plane wall. In the case of a radially collapsing bubble, the critical velocity has been shown[18] to increase by only 18% from r=∞ to r=h because of the changes in the stray field. This indicates that within the range r>h, use of (3.14) for the translational case will not greatly obscure the effects of Bloch line curvature or inhomogeneity in the normal velocity component. The curves in Fig. 12 show the portions of a family of Bloch lines, computed for 3 values of V, lying in one quadrant $(0\leq\beta\leq\frac{\pi}{2})$ of the cylinder surface.

C. Velocity-Momentum Relation for a Translating Bubble

In the Bloch line approximation, Eq. (3.2) becomes

$$\bar{\psi}_x = -(2/\pi h)\iint \phi(z)\cos\beta dz d\beta \qquad (3.15)$$

where φ is given by Eq. (3.9). The integration in this instance is carried over the wall surface enclosed by each of the two curved Bloch lines and the respective critical circles. The choice of $\bar{\psi}_{xo}$ in Eq. (3.2) is such that $\bar{\psi}_x=0$ when the only Bloch lines are the initial pair of vertical ones located at $\beta= -\frac{\pi}{2}$, where they are restrained by the gyrotropic force when $\dot{x}>0$. Using the Bloch lines computed above for the stray field (3.14) and the special case h=r, we find the velocity-momentum relation in the range $0<\bar{\psi}_x<\bar{\psi}_{xp}=2.44$, as shown in Fig. 13.

The result V>0 at $\bar{\psi}_x=0$ is unphysical. It is caused by a breakdown of the Bloch line approximation in the neighborhood of the bounding critical circle, where the theoretical Bloch line thickness tends to infinity.[18,20,31] Since the Bloch line nucleates at a critical circle (Fig. 12), the greatest error of the Bloch line approximation is at small $\bar{\psi}_x$. Hubert's more general solution[11] of the related problem of a moving plane wall indicates that V rises steeply from zero, as indicated schematically in the figure.

The shape of $V(\bar{\psi}_x)$ resembles the corresponding curve for radial motion,[18] with the principal difference that the minimum lies higher, as would be expected from the added velocity contribution of the Bloch line curvature. The maximum velocity, occuring at $\bar{\psi}_x=\bar{\psi}_{xp}$, is $V_p=24.9\gamma A/hK^{1/2}$. This velocity is only slightly greater, by 4%, than the maximum velocity $V_{po}$ for a flat wall,[18] because the Bloch line curvature is very small at the critical or "touching point," where the Bloch line touches the second critical circle satisfying φ=π (see Figs. 11 and 12). Extension of the calculations to the range r<h will require a numerical representation of $H_r(z)$ in place of Eq. (3.14) and thus increase the computational complexity. Judging from the radial case,[18] we expect the maximum velocity to increase as the diameter d decreases at constant h, but the average velocity (taken over $0\leq\bar{\psi}_x\leq 2.44$) will be considerably less sensitive to d.

Upon combining Eq. (3.5) with the $\dot{x}=V(\bar{\psi}_x)$ relation of Fig. 13, we expect gradient propagation behavior similar to that predicted for radial or plane wall motion.[18,28] If

$$H < \mu^{-1}V_p + H_{cb} \qquad (3.16)$$

then the bubble moves steadily with the conventional velocity

$$V = \mu(H-H_{cb}), \; V<V_p \, . \qquad (3.17)$$

If H exceeds the critical value (3.16), then $\bar{\psi}_x(t)$ increases beyond $\bar{\psi}_{xp}$ and one can expect some sort of cyclic Bloch line generation process. However, much as in the case of the plane wall, the details of what happens when $\bar{\psi}_x$ exceeds $\bar{\psi}_{xp}$ are not clear. There are several possibilities:

1) On one hand, the Bloch line model suggests an instability[18] associated with the nucleation of a new Bloch line at the "touching point." Such a process, if continued, would ultimately lead to nested networks of Bloch lines of the type suggested by Hagedorn.[20] In this case we expect the velocity to remain in the same range as shown in Fig. 13, as long as the number of Bloch lines is small. We estimate the average velocity by taking an average over the first cycle from $\bar{\psi}_x=0$ to $\bar{\psi}_{xp}$; this gives $V_{ave,1}=16.5\gamma A/hK^{1/2}$, for the case h=r. Furthermore, such nested Bloch lines should unwind

after the end of the drive pulse, thus causing a ballistic overshoot, according to Eq. (3.5), of roughly

$$(x_\infty - x_T)|_1 = V_{ave,1}T(\mu H - \mu H_{cb} - V_{ave,1})/(\mu H_{cb} + V_{ave,1}) \quad (3.18)$$

In this case the final position of the bubble should lie close to a straight line as a function of drive:[4]

$$x_\infty = \mu HT[V_{ave,1}/(V_{ave,1} + \mu H_{cb})] \quad (3.19)$$

2) On the other hand, Hubert's plane-wall calculation[11] suggests that Bloch lines can "punch-through" to the film surface, in accordance with another suggestion by Hagedorn.[20] In this case the Bloch line segments are likely to straighten up under the influence of Bloch line tension into a pair of vertical unwinding Bloch lines. The drive field causes the wall moments in these Bloch lines to precess in the same sense. Since the senses of twist differ they move apart to the flanks of the bubble (where $\beta=\pm\pi/2$), and $\bar\psi_x$ increases from $\bar\psi_{xp}$ to 4. However, during this time V=0, for both terms in Eq. (3.10) vanish for straight vertical Bloch lines. Beyond $\bar\psi_x=4$, V will again increase as new Bloch lines become nucleated. Thus in this case we expect rather abrupt velocity changes with a period of order $4\gamma^{-1}H^{-1}$, for large H. The velocity, averaged over a full cycle (from $\bar\psi_x=0$ to 4) is $V_{ave,2}=10.1\gamma A/hK^{1/2}$, for the case h=r. In this process the overshoot is reduced because the vertical Bloch lines are metastable on the sides of the bubble.[4] The maximum overshoot that can be obtained occurs when the pulse terminates just before a pair of Bloch lines has reached the touching points, for only these Bloch lines will retract under the influence of their line tension. The overshoot, according to (3.5), is roughly

$$(x_\infty - x_T)|_2 = \mu\gamma^{-1}\,\delta\bar\psi_x[1 + (\mu H_{cb}/V_{ave,1})]^{-1} \quad (3.20)$$

where in this case the maximum decrease $\delta\bar\psi_x$ in $\bar\psi_x$ as the Bloch lines retract is just $\bar\psi_{xp}=2.44$.

Still other alternatives, too complex to pursue in this paper, are afforded by the possibility that one of the curved Bloch lines would nucleate a new Bloch line or would punch through, while the area enclosed by the other would shrink. In this case the velocity would remain closer to $V_{ave,1}$, while the maximum overshoot would however be $(x_\infty - x_T)|_2$. The situation may be further complicated by in-plane anisotropy, by the presence of surface layers which may impede or induce punch-through, and by the transient bubble ellipticity.

## IV. COMPARISON OF THEORY WITH EXPERIMENT

As discussed in IIIC, the theory suggests an abrupt break to non-linear behavior at the critical field given in (3.16). Our experimental results show no evidence for such a peak, which should appear as a peak in $x_T$ as a function of H. Rather the $x_T$ values rise smoothly to their saturated value. One possibility is that the peak is masked by fluctuations of static coercivity in the low damping samples where the critical field is low, but even in the higher damping EuTbGa and SmGa samples there is no peak. Other possible explanations are defects, or surface layers which modify the stray field, or a breakdown in the Bloch line approximation at the critical velocity because of the steep gradient of $H_r$. Thiele[31] has predicted that flexural vibrations of a wall may couple strongly to motion at velocities of order $V_{po}$, and this effect may also smear out the peak. Such wall modes might be invoked to explain the velocity saturation as well, but they apparently do not account for ballistic overshoot; so we do not consider this model any further.

Considering previous attempts to explain non-linear effects in bubble dynamics, the observed saturation velocities of our samples are in good agreement with our predicted average velocity $V_{ave,1}=0.69V_{po}$, as can be seen from the Table, where $V_{po}$ is listed for each sample. The poorer agreement with $V_{ave,2}=0.42V_{po}$ appears to rule out the "punch-through" possibility (2) discussed in Section IIIC, as does our failure to observe periodic velocity fluctuations. On the other hand the final positions as a function of drive do not form a straight line, except in the SmGa sample (see Figs. 1-5). Thus the nesting possibility (1) of IIIC must also be ruled out. Clearly the real situation is a more complex one in which a certain amount of punch-through occurs, but not enough to prevent sizable overshoots. It is interesting that in samples 1, 3 and 4 there is a plateau in $x_\infty$ at low drives (see Fig. 3). Solving for $\delta\bar\psi_x$ in Eq. (3.20), we find that the overshoot in the plateau region of all three samples corresponds to a reduction in $\bar\psi_x$ of close to 4. While we cannot easily explain this result, we note that these three samples have the best crystallographic alignment (see Table). In the other samples $x_\infty$ as a function of drive shows no plateau. Now, these data were taken in a direction of maximum overshoot, which was found to correspond roughly to the in-plane easy axis (see Fig. 8). The increased overshoot can be understood qualitatively from the fact that the in-plane anisotropy can provide another force on the Bloch lines, in addition to the stray field and line tension forces of Eq. (3.10). It is easy to see that the in-plane anisotropy tends to push Bloch lines towards the nearest point of intersection of the bubble circumference with the diameter lying in the easy axis direction. As the Bloch lines move in this way, $\bar\psi_x$ is decreased, and, according to Eq. (3.21), the overshoot is increased.

Our interpretation is further complicated by the fact that the bubbles become elliptical and expand in area during their motion, as shown in Fig. 7. However in one way this effect provides further evidence for our theoretical interpretation, for the distortion can be understood qualitatively as the effect of the wall-normal component of gyrotropic force (3.7) of the nucleated Bloch lines.

Finally we comment briefly on the radial expansion results of the Table. The radial saturation velocities, averaged over the major and minor axes ($V_r=\frac{1}{2}V_{r+}+\frac{1}{2}V_{r-}$) are all larger than - but within a factor of two of - the translation saturation velocities $V_t$. This makes it doubtful that the radial saturation velocities are due to the simple mechanism in which a single Bloch line is propagated at a time,[18] for the line tension force should make the translational velocities larger. Indeed the predicted radial saturation velocity $V_{oo}$, given in the Table, is far too low in each case. Furthermore, the high drive mobilities $\mu_h$ may be compared to formulas previously derived for the effect of a single Bloch line propagating at a time[28]

$$\mu_1 = \gamma\Delta[\alpha + (\pi^2\Delta Q^{1/2}/2h\alpha)]^{-1}\ , \quad (4.1)$$

and for the case of free precession[26]

$$\mu_2 = \gamma\Delta(\alpha + \alpha^{-1})^{-1}\ . \quad (4.2)$$

In all cases except SmCaGe, the observed mobilities are bracketed by the two calculated mobilities $\mu_1$ and $\mu_2$ given in the Table. This result also suggests that the actual dynamic process is more complex than the single Bloch line process. That the mechanisms of the radial and translational saturation velocities reported in the Table are different is, of course, not surprising considering the order of magnitude difference in drive in the two experiments.

The present paper, together with previous studies of dynamic collapse in garnet films,[13-20,32] supports a consistent picture of mean velocity, both radial and translational, for long pulses of less than 10 oersted amplitude. As the drive increases beyond the coercive force, the velocity is initially linear as in the classical theory up to a peak velocity $V_p$ corresponding to instability of moving wall structure. Beyond this peak the velocity saturates at a constant value less than $V_{po}$ for $\alpha=0$, or increases linearly [as in Eq. (4.1)] for $\alpha>0$. Numerical values of radial and translational velocity differ by a factor of less than 2 in this context. There exist, however, reports of very high velocities at very short pulse widths and high amplitudes[33,34]. The quasi-steady-state [Eq. (3.6)] and Bloch-line [Eq. (3.7)] concepts relied upon in our interpretation become invalid for sufficiently short times and great amplitudes so that the full explanation of these effects must include other considerations.

The authors thank E. Giess and R. Atkin for preparing garnet films, R. Anderson and T. L. Hsu for preparing the ion-milled thickness series, M. Rozmus for the radial expansion measurements, W. Haag for the homogeneous nucleation measurements, D. C. Cronemeyer for the FMR measurements, L. Buszko and B. E. Argyle for the coercivity measurements, and A. Segmuller, J. Angilello and J. Karasinski for x-ray diffraction measurements.

REFERENCES

1. A. P. Malozemoff and J. C. DeLuca, Appl. Phys. Lett. 26, 719 (1975).
2. G. P. Vella-Coleiro and W. J. Tabor, Appl. Phys. Lett. 21, 7 (1972); J. A. Cape, J. Appl. Phys. 43, 3551 (1972).
3. R. W. Patterson, A. J. Braginski and F. B. Humphrey, IEEE Trans. Magn. MAG-11, 1094 (1975).
4. A. P. Malozemoff and J. C. Slonczewski, IEEE Trans. Magn. MAG-11, 1091 (1975).
5. W. Döring, Z. Naturforschung 3a, 373 (1948).
6. G. P. Vella-Coleiro, D. H. Smith and L. G. Van Uitert, J. Appl. Phys. 43, 2428 (1972).
7. F. H. deLeeuw and J. M. Robertson, J. Appl. Phys. 46, 3182 (1975).
8. B. E. Argyle and A. P. Malozemoff, AIP Conf. Proc. 10, 344 (1973).
9. R. W. Shaw, J. W. Moody and R. M. Sandfort, J. Appl. Phys. 45, 2672 (1974).
10. E. Schlömann, AIP Conf. Proc. 18, 183 (1974).
11. A. Hubert, J. Appl. Phys. 46, 2276 (1975).
12. J. A. Cape, W. F. Hall and G. W. Lehman, J. Appl. Phys. 45, 3572 (1972).
13. B. E. Argyle, J. C. Slonczewski and A. F. Mayadas, AIP Conf. Proc. 5, 175 (1972).
14. H. Callen, R. M. Josephs, J. A. Seitchik and B. F. Stein, Appl. Phys. Lett. 21, 366 (1972); F. H. deLeeuw, J. Appl. Phys. 45, 3106 (1974).
15. T. M. Morris and A. P. Malozemoff, AIP Conf. Proc. 18, 242 (1974).
16. G. J. Zimmer, T. M. Morris and F. B. Humphrey, IEEE Trans. Magn. MAG-10, 651 (1974).
17. A. P. Malozemoff and K. Papworth, J. Phys. D. 8, 1149 (1975).
18. J. C. Slonczewski, J. Appl. Phys. 44, 1759 (1973).
19. G. P. Vella-Coleiro, F. B. Hagedorn, Y. S. Chen and S. L. Blank, Appl. Phys. Lett. 22, 324 (1973).
20. F. B. Hagedorn, J. Appl. Phys. 45, 3129 (1974).
21. D. C. Fowlis and J. A. Copeland, AIP Conf. Proc. 10, 458 (1972).
22. A. P. Malozemoff and J. C. DeLuca, J. Appl. Phys. 45, 4586 (1974).
23. A. Hubert, A. P. Malozemoff and J. C. DeLuca, J. Appl. Phys. 45, 3562 (1974).
24. J. C. Slonczewski, A. P. Malozemoff and O. Voegeli, AIP Conf. Proc. 10, 458 (1973).
25. G. P. Vella-Coleiro, F. B. Hagedorn and S. L. Blank, Appl. Phys. Lett. 26, 69 (1975).
26. J. C. Slonczewski, Int. J. Magnetism 2, 85 (1972).
27. A. A. Thiele, A. H. Bobeck, E. Della Torre and U. F. Gianola, Bell Syst. Tech. J. 50, 711 (1971).
28. J. C. Slonczewski, J. Appl. Phys. 45, 2705 (1974).
29. A. A. Thiele, J. Appl. Phys. 45, 377 (1974).
30. E. Schlömann, AIP Conf. Proc. 10, 478 (1973); J. Appl. Phys. 44, 1850 (1973).
31. A. A. Thiele, Phys. Rev.B 7, 391 (1973).
32. A. P. Malozemoff, J. Appl. Phys. 44, 5080 (1973).
33. G. P. Vella-Coleiro, AIP Conf. Proc. 24, 595 (1975).
34. G. J. Zimmer, L. Gál and F. B. Humphrey, these Proceedings.

# OVERSHOOT IN MAGNETIC BUBBLE TRANSLATION

G. P. Vella-Coleiro
Bell Laboratories, Murray Hill, N.J. 07974

## ABSTRACT

We have investigated the occurrence of overshoot[1] in the translational motion of magnetic bubble domains using high speed photography. The apparatus utilizes a low power pulsed argon laser as a light source and an image intensifier to expose high speed film in a single shot. The optical power density incident on the sample is sufficiently low that there occurs no observable perturbation of the bubble motion by the light pulse. Thus we are able to obtain data in the form of triple exposures where the first exposure records the initial position of the bubble, the second exposure records an intermediate position, for example at the end of a field gradient pulse, and the third exposure records the final bubble position. In a LuGdAl garnet film having a thickness of 9.4 μm and a bubble diameter of 6.1 μm, and using field gradient pulses 500 nsec in duration, we obtained results similar to those of Ref. 1, i.e., at high drive fields the bubble overshoots by several μm and the displacement during the pulse is only weakly dependent on drive field. However, when 100 nsec wide pulses were used, overshoot was absent in a large fraction of the observations, and no velocity saturation appeared. Average velocities during the pulse up to 5000 cm/sec were observed in this case. Measurements made as a function of pulse width showed that the bubble velocity is time-dependent, the average velocity during the pulse decreasing from 4200 cm/sec at 100 nsec pulse width to 1700 cm/sec at 500 nsec pulse width at a fixed drive field of 9.3 Oe across the bubble diameter.

We have also found a correlation between the amount of overshoot and the bubble displacement during the pulse using a pulse width of 500 nsec and a drive field of 6.3 Oe. Smaller than average displacements during the pulse were accompanied by larger than average overshoots, and vice versa. These results indicate that the number of Bloch lines generated is not constant but varies substantially from pulse to pulse.

Bubble motion in a YEu garnet film with large g-value ($g \geqslant 30$) was also investigated. Because of the large gyromagnetic ratio, the Slonczewski critical velocity in this film has the relatively high value of 30,000 cm/sec,[2] which is much larger than the highest bubble translational velocity we have achieved of 6000 cm/sec. Thus no overshoot is to be expected and we have not observed any.

A fuller account of the results described above will be published elsewhere.

REFERENCES

1. A. P. Malozemoff and J. C. DeLuca, Appl. Phys. Lett. 26, 719 (1975).
2. G. P. Vella-Coleiro et al., Appl. Phys. Lett. 26, 722 (1975).

# BUBBLE DOMAIN TRANSLATIONAL MOTION IN THE PRESENCE OF AN IN-PLANE FIELD

R. M. Josephs and B. F. Stein
Sperry Univac, P. O. Box 500, Blue Bell, Pa. 19422
W. R. Bekebrede
Sperry Research Center, Sudbury, Mass. 01776

ABSTRACT

The dependence of collapse field, velocity, and trajectory angle on in-plane field has been measured on a set of $(YLaEu)_3(FeGa)_5O_{12}$ films ion implanted at energies ranging from 25 to 200 Kev. At zero in-plane field, the S number was always ~1. At increasing values of in-plane field, S was found to increase, then undergo a sharp downward transition, and eventually decrease to zero. The transition occurred at a value of in-plane field for which the collapse field was a minimum. These results are interpreted in terms of a bubble containing two Bloch points with good agreement between the model properties and the data.

It was first reported by Bullock[1] that bubbles in ion implanted garnet films did not propagate along the field gradient until a sufficiently large in-plane field was applied. Under certain conditions he also observed the simultaneous presence of bubbles having several different deflection angles. More recently Hasegawa[2] reported transport data on an unimplanted film showing that the observed distribution in S numbers depended on the value of the in-plane field. These data were interpreted in terms of the Bloch point theory of Slonczewski.[3] Hsu[4,5] has measured the relative stability of S=0 and S=1 bubbles in implanted samples and has used the Bloch point picture to explain the results. Smith et al[6] have observed a substantial increase in velocity in implanted films at a value of in-plane field for which the S=0 conversion occurred. In the present work, the dependence of S on in-plane field has been measured for a set of films implanted at different energies. The data are compared with the predictions of the Bloch point theory and offer additional insight into the nature of the domain wall structure in implanted films.

Data were taken on four films cut from LPE wafers[7] of $Y_{2.38}La_{.09}Eu_{.53}Fe_{3.9}Ga_{1.1}O_{12}$ for which the room temperature magnetic parameters were $h=6.0\mu$, $4\pi M_s=117$G, collapse field $H_0=60.2$ Oe prior to ion implantation, $H_k=850$ Oe, $\gamma=0.76\gamma_e$ (FMR), easy axis tilt $<0.9°$, and $\mu\sim1,000$cm/sec-Oe in the direction of the gradient for S=1 bubbles in zero in-plane field. The four films were implanted at a constant dosage of $1.5 \times 10^{14}$ $Ne^{+}_{20}/cm^2$ at energies of 25, 50, 100, and 200 Kev respectively. Following implantation, static and dynamic measurements of the bubble properties were made in a polarizing microscope modified to fit within the 6" diameter pole pieces of the electromagnet which provided the in-plane field $H_y$. The microscope had several degrees of freedom, allowing the easy axis of a sample to be oriented perpendicular to the field of the magnet to within 0.1°. The transport apparatus has been described previously.[8] The relative orientations of $H_y$, the bias field $H_1$, and the bubble trajectory are shown in Fig. 1. In all this work, $H_y$ was within 2° of a [110] direction.

In the transport measurements the motion was not restricted to the midpoint of the circuit. The pulse amplitude was chosen to give an appreciable distance (30$\mu$) between the collapse and run-out positions and the trajectory measurements were bounded by these limits. The amplitude was such that the drive field was 1.5, 2.5, and 4 Oe at the beginning, midpoint, and end of the trajectory. The pulse width was fixed at 0.41$\mu$s. Single pulses were used to march the bubble across the circuit with the bubble co-ordinates being noted after each pulse. The trajectory was taken as the best straight line through these co-ordinates (excluding the occasionally observed erratic behavior on the first pulse following a trajectory reversal). Since $H_0$ depended on $H_y$, $H_1$ was adjusted, at each value of $H_y$, such that $(H_0 - H_1)$ was always 6.8 Oe. All the data were taken on bubbles generated by cutting worms at $H_y$=0. The procedure was to measure the trajectory at $H_y$=0, then to reduce the pulse amplitude to a value just greater than the minimum value for motion and bring the bubble back to the starting position. $H_y$ was then increased by the desired amount and the trajectory measured. For a given bubble this procedure was repeated until the motion was parallel to the gradient.

The static and dynamic results are illustrated in Fig. 2 where the curves are drawn roughly through the data points. The dependence of $H_0$ on $H_y$ is noticeably different from the linear dependence observed in unimplanted samples. It is seen that $H_0$

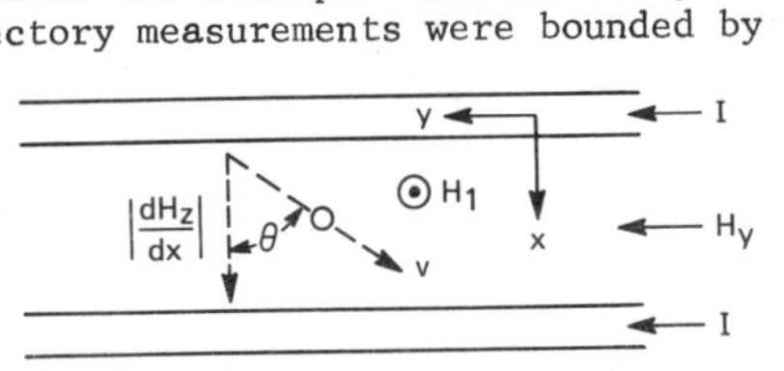


Fig. 1 Experimental configuration.

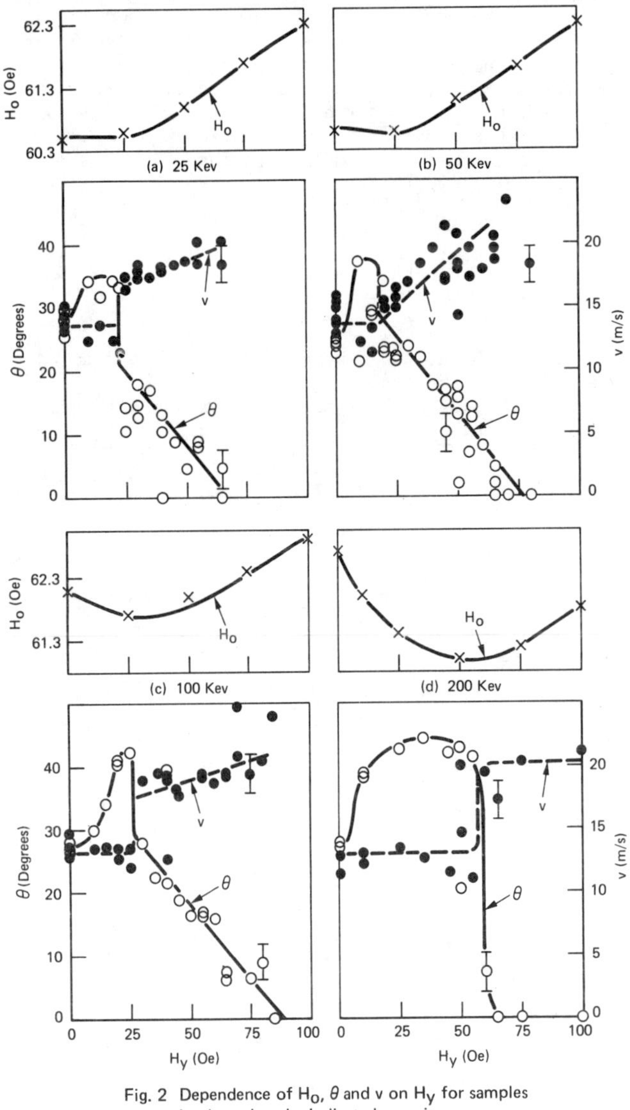


Fig. 2 Dependence of $H_0$, θ and v on $H_y$ for samples implanted at the indicated energies.

at $H_y=0$ increased with increasing implantation energy and that the depth of the minimum in the $H_o$ versus $H_y$ curves also increased. In general, the trajectorie were straight lines. The only departure from this behavior was that occasionally, in the vicinity of the transition, the path might consist of two distinct linear segments. For all the samples, the trajectory angle $\theta$ was $26^o \pm 3^o$ at $H_y=0$ and increased with increasing $H_y$. For the 25 and 50 Kev implants, there was a modest increase in $\theta$ followed by a transition and a monotonic decrease to $0^o$ as $H_y$ increased. For the 100 Kev implant, the increase in $\theta$ was more pronounced, but again there was a transition followed by a gradual decrease to $0^o$. In the case of the deepest implant (200 Kev) $\theta$ increased to and remained at the largest angle observed in these samples before making an abrupt transition to $0^o$. In each of the samples, the transition in $\theta$ occurred at a value of $H_y$ for which $H_o$ was a minimum. Fig. 2 also shows the $H_y$ dependence of the total velocity v of the bubbles as they crossed the center-line of the drive circuit, i.e., at a drive field of 2.5 Oe. For all the samples, v was roughly independent ($\sim$13 m/s) of $H_y$ for all $H_y$ below the transition. For those samples which exhibited a gradual decrease in $\theta$ following the transition, v increased monotonically achieving a value of $\sim$20 m/s at $\theta=0^o$. In the 200 Kev implanted sample, in which the $\theta$ value fell abruptly to zero, v correspondingly increased abruptly to 20 m/s.

The dependence of $H_o$ on $H_y$ reflected the influence of $H_y$ on the closure domain in the implanted layer.[9] As $H_y$ increased, this domain diminished in size, thus reducing its keepering effect on the bubble and thereby decreasing $H_o$. After the closure domain was annihilated, the linear dependence of $H_o$ on $H_y$, characteristic of unimplanted samples, was again observed (Fig. 2). Since the transitions in the dynamic behavior were observed at the same value of $H_y$ for which $H_o$ was a minimum, it is felt that the abrupt changes in $\theta$ and v were associated with the disappearance of the closure domain.

The S number[10] was calculated from Eq. (1)

$$S = \frac{\gamma d \Delta H_z \sin\theta}{8v} \quad (1)$$

The values of v, $\theta$, and d were taken from the data shown in Fig. 2 and from the measured values of the static diameter at each value of $H_y$ (in the case of elliptical bubbles observed at the higher values of

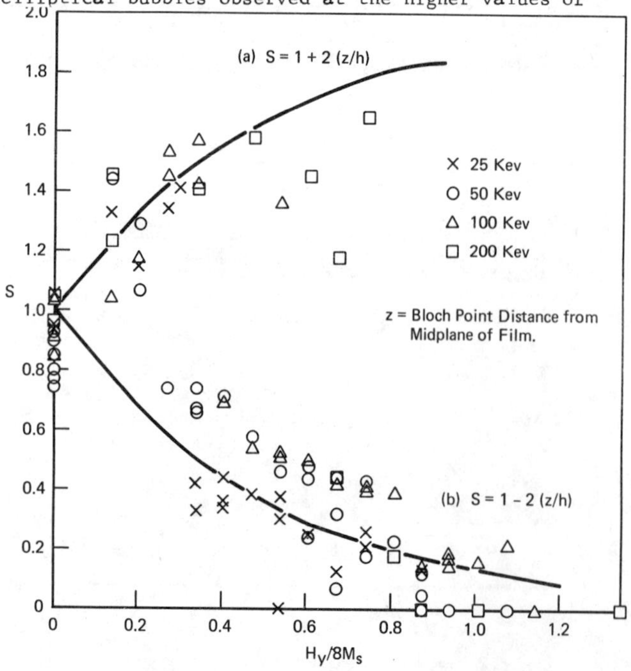


Fig. 3 Dependence of S on $H_y/8M_s$ for the samples described in Figure 2. The solid curves are the theoretical predictions of the 2 BP model.

$H_y$, the diameter was taken as the geometric mean of the major and minor dimensions). The dependence of S on $H_y$ for all the samples is shown in Fig. 3. Because S was determined from several measured parameters, the uncertainty in this quantity can be as large as 25%. All the bubbles generated in $H_y=0$ were S=1 bubbles. The unexpected feature is that S increased substantially before decreasing to 0. This behavior can not be explained if the S=1 bubbles had a simple Bloch wall structure. However, it is consistent with a more complicated wall structure including two Bloch points (BPs). The equilibrium positions of these postulated Bloch points were determined by a balance between $H_y$ and the radial component of the bubble stray[11]field. Curves (a) and (b) in Fig. 3 show the theoretical dependences of S on $H_y$ for two possible cases (shown in Fig. 4) of a bubble containing 2 BPs; these results were calculated from Fig. 8 of Ref. 11 (with d/h=1). More specifically, the model illustrated in Fig. 4 consists of a pair of Bloch lines each having a BP at the midplane (at $H_y=0$) with a closure domain at the top. In Fig. 4 (a), the spins at the side of the wall are assumed to be parallel to the magnetization in the closure domain, although the specific interaction favoring this orientation is not yet clear. As $H_y$ increases, the BPs move apart and an S=2 section develops in the region between the BPs. The resultant S number of the bubble is the average of the S numbers of the 3 sections weighted by the fraction of the film thickness they respectively occupy. When a value of $H_y$ is reached such that the closure domain disappears, there is a transition to the structure shown in Fig. 4 (b) where the spins at the side of the wall are now parallel to $H_y$. In this case, the configuration in the region between the BPs is an S=0 structure which expands, at the expense of the S=1 sections, as $H_y$ increases.

In conclusion, the structure of the S=1 bubble has been shown to be more complicated than expected. To account for the dependence of S on $H_y$, it was necessary to introduce a domain wall structure involving two Bloch points.

The authors wish to thank H. Callen and W. D. Doyle for many helpful discussions.

1. D.C. Bullock, AIP Conf. Proc. 18, 232 (1973).
2. R. Hasegawa, AIP Conf. Proc. 24, 515 (1975).
3. J.C. Slonczewski, AIP Conf. Proc, 24, 613 (1975).
4. T. Hsu, AIP Conf. Proc. 24, 624 (1975).
5. T. Hsu, presented at 1975 Intermag, London.
6. A.B. Smith, M. Kestigian, and W.R. Bekebrede, Mat. Res. Bull. 10, 303 (1975).
7. B.F. Stein and M. Kestigian, to be published in J. Crystal Growth.
8. R.M. Josephs, Appl. Phys. Lett. 25, 244 (1974).
9. R. Wolfe and J.C. North, Appl. Phys. Lett. 25, 122 (1974).
10. J.C. Slonczewski, A.P. Malozemoff, and O. Voegeli, AIP Conf. Proc. 10, 458 (1972). This formula was derived for a constant diameter and applies only at the center-line. Applied elsewhere, it breaks down and predicts a curved trajectory in disagreement with our observations.
11. J.C. Slonczewski, J. Appl. Phys. 44, 1759 (1973).

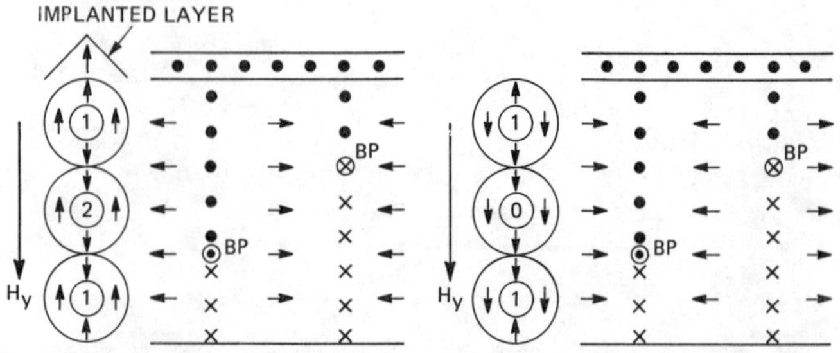


Fig. 4 Model of the bubble domain wall having 2 BPs where (a) is the structure resulting in increasing S with increasing $H_y$ and (b) is the structure having decreasing S with increasing $H_y$. The S numbers are shown at the center of each bubble section. The dots and crosses inside the BP circles indicates the direction of $H_y$ at that point.

# THE SPEED DIFFERENCE BETWEEN AN S=0 AND AN S=1 BUBBLE IN THE PRESENCE OF AN IN-PLANE MAGNETIC FIELD

T. Hsu, B. R. Brown and M. D. Montgomery
IBM Research Laboratory, San Jose, California 95193

## ABSTRACT

Using the bubble translation technique, we have measured the speed as a function of drive field of an S=0 and an S=1 bubble in the presence of an in-plane magnetic field. The speed of both states increases with the application of an in-plane field, but the increase for an S=1 bubble is much less than the increase for an S=0 bubble. For a given in-plane field, the ratio of the speed for an S=0 bubble to that of an S=1 bubble increases with increasing gradient field. For a given drive field, this ratio increases linearly with increasing in-plane field. This effect can be observed in either as-grown or ion-implanted bubble films. Although the deflection angle of an S=0 and an S=½ bubble are different, they have essentially the same speed.

## INTRODUCTION

An S=1 bubble can be differentiated from an S=0 bubble by the difference in deflection angle (θ) observed during translation in a gradient drive field. In the absence of an in-plane magnetic field ($H_{ip}$), both S=0 and S=1 states exhibit essentially the same speed for a given drive field. We use the word speed throughout the text to represent the magnitude of the total velocity of a bubble. Recently, it has been reported[1,2] that the speed of an S=0 bubble increases substantially with the application of a large in-plane field. We have measured the speed of an S=0 bubble and an S=1 bubble in the range of in-plane field where both states are stable and have found that the increase in speed of an S=1 bubble due to the presence of an in-plane magnetic field is considerably less than that observed for an S=0 bubble. The experimental techniques used in the measurements will be described in the next section and will be followed with the results obtained and a discussion of the results. We have made measurements on two garnet compositions, $(YEu)_3(GaFe)_5O_{12}$ and $(YSmCa)_3(GeFe)_5O_{12}$. Both as-grown and ion-implanted samples were used in the measurements.

## EXPERIMENTAL TECHNIQUES

We have used two techniques to measure the bubble speed ($|V|$) as a function of the gradient field across a bubble, ($\Delta H_z$), and the in-plane field, $H_{ip}$. The first technique used is the bubble transport method first reported by Vella-Coleiro and Tabor.[3] In principle, it is possible to transport the bubble under conditions of constant $\Delta H_z$ by means of a bias field compensation loop; however, in practice, overshoot effects[4] limit seriously the degree to which precise compensation can be achieved. Nevertheless, this method is still probably the most widely used method to measure the bubble speed as a function of drive field, $\Delta H_z$. It has been our experience that reproducible measurements require that extreme care be taken to prepare the bubble for measurement. The starting position of a bubble must be exactly midway between the drive conductors. The bubble was positioned for measurement by using short, low amplitude pulses to approach the midway point. Using this technique, it was possible to avoid the deflection angle "offset" effect often observed upon reversal of the bubble motion. For a given $\Delta H_z$, the θ of a bubble was first determined by taking multiple-exposure photographs of the trajectory. The duration of the drive pulse required to translate the bubble a distance of one bubble diameter was determined using a video micrometer to accurately measure the displacement of the bubble.

The second method of measurement which we call the "rocking" technique also uses a conductor overlay; the principle is the same as the first technique except that instead of using one single translation step, the bubble is rocked back and forth by repetitively alternating the polarity of the drive current. A conductor loop carrying a dc current is used to generate a potential well which centers the rocking bubble. The conductor overlay used in our measurements is shown in Fig. 1. The measurement procedure involved first selecting a minimum pulse amplitude to start the bubble rocking. The pulse width was then adjusted to make the two rocking bubble images tangent to each other which in effect means that the bubble is translated a distance of exactly one bubble diameter. Thus, a measurement of the pulse width required to produce a displacement of one bubble diameter enables one to determine the speed at that particular value of pulse amplitude which in turn can be related to the applied gradient field. This procedure was repeated with increasing values of pulse amplitude. In this manner, a $|V|$ versus $\Delta H_z$ curve could be obtained in a relatively short time. A variation of this procedure was to apply a series of small current pulses to the conductor loop superimposed on the dc current before each gradient field pulse to move the bubble back to the center of the loop so that the starting position of the bubble was always midway between the two parallel conductors used to produce the gradient field. As described by Brown,[5] this method has been found to be very useful in investigating the stability of bubble states. During these measurements the deflection angle was readily visible. If needed, more angular resolution could be obtained by increasing the pulse width and thereby the displacement.

In general, the collapse field of a bubble is a function of the applied in-plane field, $H_{ip}$, and if there is a (111) axis tilt, the direction of $H_{ip}$. The (111) axis tilt of all samples used in these measurements was less than 0.3°. The collapse and runout fields of a bubble were measured as a function of $H_{ip}$ and the mean value of the collapse and runout fields was adopted as the bias field for making all measurements. A further complication arises from the fact that a circular bubble becomes elliptical in the presence of an in-plane field. For the purpose of later calculations, the major and minor axis of a bubble were measured for each value of $H_{ip}$ used.

## RESULTS AND DISCUSSIONS

Fig. 1 shows a photo micrograph of an S=0 and an

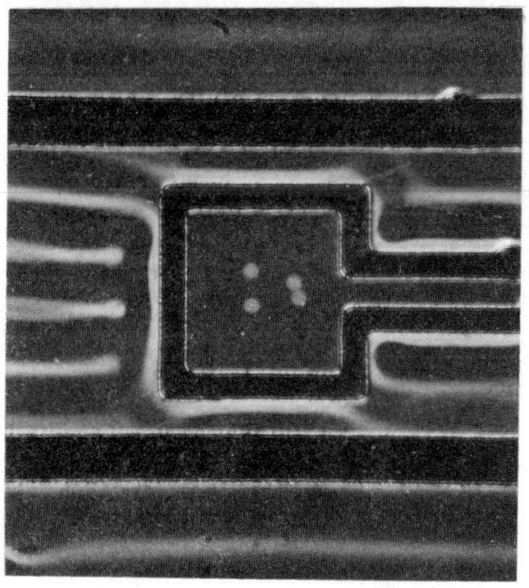

Fig. 1 Picture of an actual conductor overlay used in the rocking technique. To the left is the trajectory of an S=0 bubble. To the right, an S=1. Film composition: $(YSmCa)_3(GeFe)_5O_{12}$.

$S=1$ bubble in a $(YSmCa)_3(GeFe)_5O_{12}$ film rocked by a series of current pulses in the presence of a 70 Oe $H_{ip}$. As can be seen from the stripe patterns outside the centering loop, the $H_{ip}$ was applied perpendicular to the direction of the gradient field. Each rocking pulse was 400 nsec long and produced a gradient field across the bubble of 3.5 Oe. The difference in displacement or bubble speed between the two bubble states is clearly shown in Fig. 1.

Using the bubble transport technique, the $|V|$ versus the $\Delta H_z$ relation was measured for both an $S=0$ and an $S=1$ bubble at three different values of $H_{ip}$. The results are shown in Fig. 2. This was taken using bias field compensation during translation. In measurements where the bias field was not compensated, we found a 15% to 20% increase in the measured speed at high $\Delta H_z$. It can be noted from Fig. 2 that the speed of both bubble states

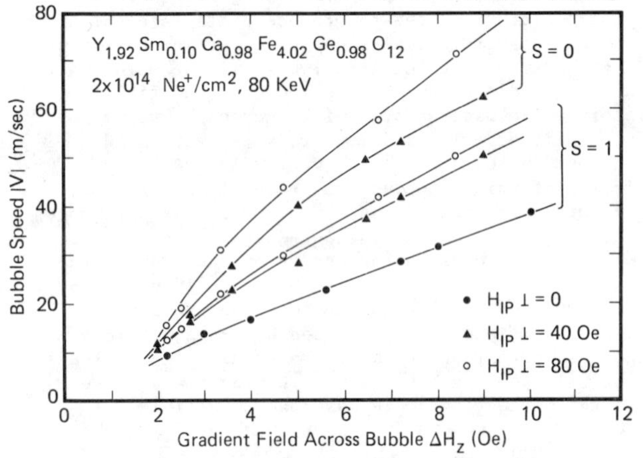


Fig. 2 Bubble speed as a function of gradient drive field for three different values of in-plane magnetic field.

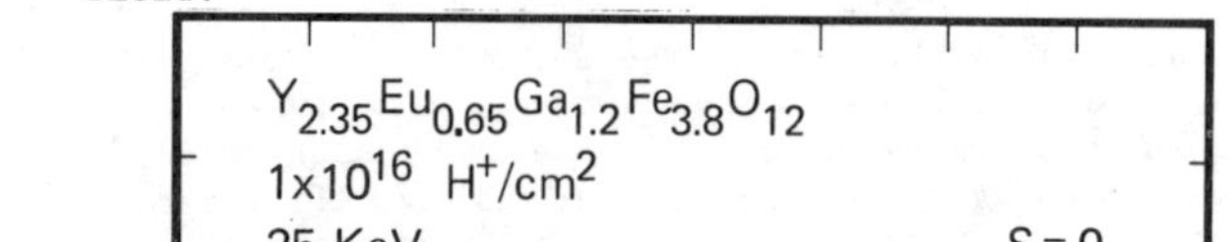

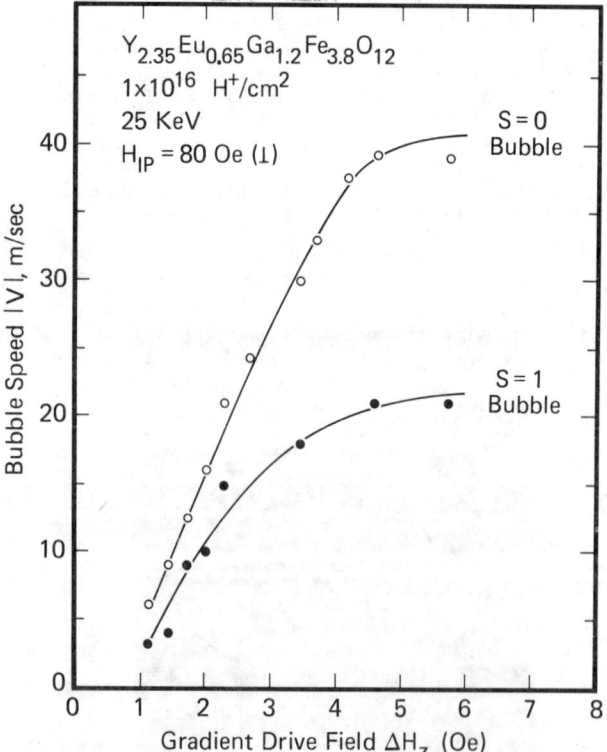

Fig. 3 Bubble speed as a function of gradient drive field for an $S=0$ and an $S=1$ bubble in the presence of an in-plane field of 80 Oe applied perpendicular to the direction of the gradient field.

increased substantially with the application of an $H_{ip}$, but the increase was less for an $S=1$ bubble than for an $S=0$ bubble. We first observed this effect in ion-implanted samples; however, similar results have also been obtained in as-grown samples. It was thus concluded that the effect must be related to the structure of the domain wall rather than the coupling between the domain wall and the capping layer. The effect has also been observed in other garnet compositions. Fig. 3 shows the speed difference in a $(YEu)_3(GaFe)_5O_{12}$ sample in the presence of an $H_{ip}$ of 80 Oe.

In the above measurements, $H_{ip}$ was applied perpendicular to the direction of $\Delta H_z$. Bullock[1] reported that the enhancement of the speed of an $S=0$ bubble by an in-plane field was greater when the in-plane field was perpendicular to the applied field gradient as compared with the parallel orientation. Our results are consistent with Bullock's; however, for the range of in-plane fields used in our measurements, the orientation dependence of the in-plane field was small. By contrast, we found that in the case of an $S=1$ bubble, the speed enhancement is greater when the in-plane field is parallel to the applied gradient field, but since the differences were small (10%-20%), we are not prepared to attach significance to this.

The transport technique is a useful method to measure the velocity of a bubble at a given $H_{ip}$ for several values of $\Delta H_z$. However, to study the dynamic response of a bubble as a function of $\Delta H_z$ and the direction and amplitude of $H_{ip}$, the "rocking" technique has been found to be a much more powerful tool. Since the scatter of both speed and deflection angle is automatically averaged out, it is very easy to obtain the dynamic response as a function of driving conditions. Under the drive conditions where a bubble state was stable, the $|V|$ versus $\Delta H_z$ relation measured by the rocking technique was found to be almost identical to that measured by the transport technique. Dynamic conversion effects[7] can be easily observed in the rocking experiment as a bubble jumps from one state to another and causes abrupt changes in both the $|V|$ and $\theta$ of a bubble. This usually occurs when the drive field exceeds a certain threshold value. In materials that show velocity saturation at high $\Delta H_z$, the $\theta$ of an $S=1$ bubble was observed to decrease with increasing drive. This is expected from the theoretical prediction of the deflection angle as given by the formula[6]

$$\sin\theta = \frac{8\ S \times V}{\gamma d \Delta H_z}$$

Using the rocking experiment, we observed another effect which may prove to be very important in understanding domain wall dynamics. An $S=1$ bubble in an ion-implanted sample was first rocked back and forth at $H_{ip}=0$ and then $H_{ip}$ was slowly increased. When $H_{ip}$ reached a value between 7 and 10 Oe, we observed a sudden 20% to 30% drop in speed. The speed of the "new" $S=1$ bubble was found to slowly increase with increasing $H_{ip}$. This effect was reversible at about the same magnitude of $H_{ip}$. Our speculation is that the "new" $S=1$ bubble contains a pair of unwinding Bloch lines (BL's)[8] situated on opposite sides of the bubble in the direction of $H_{ip}$. The magnetostatic energy of the domain wall is presumably reduced by the introduction of a pair of unwinding lines; however, we are somewhat surprised by the large drop in speed observed. It is our belief that under conditions where $H_{ip}$ is greater than 10 Oe all $S=1$ bubbles studied by the above measurements had this kind of wall structure.

By using either the transport method or the rocking technique, we were able to measure the speed of the two states as a function of $H_{ip}$. We found that, for a given drive field, the ratio of the speed of an $S=0$ bubble to that of an $S=1$ bubble increased linearly with increasing in-plane field.

An $S=\frac{1}{2}$ state is a term that we use to represent those bubbles that have two BL's having a Bloch point[9] (BP) on one of them. This state is frequently observed in ion-implanted samples with an $H_{ip}$ bias. In the presence of an $H_{ip}$, the exact S number is not equal to $\frac{1}{2}$ and the $\theta$ decreases with increasing $H_{ip}$. In the samples used in these measurements, the $\theta$ of an $S=\frac{1}{2}$ bubble was found to vary between 7° and 15° depending on the magnitude of $H_{ip}$. However, in terms of bubble speed, the S=0 and S=½ bubbles are indistinguishable.

## SUMMARY

We have measured the speed of an S=1 and an S=0 bubble in the presence of an $H_{ip}$ by using two different techniques. The results show that the velocity enhancement produced by an in-plane field is much greater for an S=0 bubble than for an S=1 bubble. The observed effect may possibly be useful in bubble lattice devices[10] for the discrimination of bubble states. The speed ratio of the two bubble states has been found to be directly proportional to $H_{ip}$. The use of the rocking technique has led to some interesting observations. It is our belief that in the presence of a modest $H_{ip}$, (>10 Oe), and S=1 bubble actually contains a pair of unwinding BL's. As yet, we have no satisfactory explanation to account for the observed difference in bubble speed.

## ACKNOWLEDGEMENTS

The authors wish to thank B. A. Calhoun for many useful discussions and L. L. Rosier for the critical reading of the manuscript.

## REFERENCES

1. D. C. Bullock, AIP Conf. Proc. 18, 232 (1973).
2. F. H. deLeeuw and J. M. Robertson, AIP Conf. Proc. 24, 601 (1974).
3. G. P. Vella-Coleiro and W. J. Tabor, Appl. Phys. Lett. 21, 7 (1972).
4. A. P. Malozemoff and J. C. DeLuca, Appl. Phys. Lett. 26, 719 (1975).
5. B. R. Brown, Paper 5A-5, to be published in the 1975 3M Proceedings.
6. G. P. Vella-Coleiro, AIP Conf. Proc. 18, 217 (1973).
7. J. C. Slonczewski, J. Appl. Phys. 44, 1759 (1973).
8. O. Voegeli and B. A. Calhoun, IEEE Trans. on Mag. MAG-9, 617 (1973).
9. J. C. Slonczewski, AIP Conf. Proc. 24, 613 (1974).
10. O. Voegeli, B. A. Calhoun, L. L. Rosier, and J. C. Slonczewski, AIP Conf. Proc. 24, 617 (1974).

# WALL STATE STABILITY DURING TRANSLATIONAL MOTION

B. R. Brown
IBM Research Laboratory, San Jose, California 95193

## ABSTRACT

An overlay technique is used to measure S-state stability of bubbles in YEuGaFeO films. It is found that for comparable drive amplitudes, translational motion is more severe than radial motion. Additionally, the direction of in-plane field and easy axis tilt are significant. The need for ion-implantation, while having little effect on stable drive amplitude, is reaffirmed as necessary for state control. Reduction of film thickness increases stability, but is offset by increased coercivity.

## INTRODUCTION

Regions of wall state stability have been mapped as a function of field gradient translational drive and in-plane field amplitude for 5 μm YEuGaFeO films at room temperature. A comparison is made to radial motion stability investigated earlier by Hsu,[1] then results of experiments to relate stability to in-plane field direction, ion-implantation, and film thickness are presented.

The impetus for this study stems from the coding scheme of the bubble lattice file[2] which uses wall magnetization revolution number, S, to carry information in each bubble of a fully populated array. Slonczewski, Malozemoff, and Voegeli show that information so stored may be detected because the sine of the propagation angle is proportional to S.[3] Hsu has demonstrated controlled generation and switching of S=0 and 1 states using rapid wall motion and in-plane fields.[4] Useful storage of information in a lattice also requires the S number to remain unchanged over the distance it may translate from generation to detection.

A change of S-state requires nucleation or annihilation of Bloch lines and/or points. Although the forces controlling domain wall structure are known, an exact mathematical application to state stability is generally intractable. Even if this were not the case, experimental measurement is appropriate because wall states, for low S-numbers, are easily observed; whereas, magnetic parameters required by theory are not easily measured. Nevertheless, approximate results from theory give clues to the significant parameters for stability. For example, Argyle, Slonczewski, and Mayadas[5] predict a critical velocity for onset of Bloch line generation:

$$V_p = 24\gamma A h^{-1} k_u^{-1/2}. \quad (1)$$

Slonczewski[6] gives a relation for Bloch point energy:

$$W_{bp} = 2\pi A^{3/2} k_u^{-1/2} (\ln Q + \text{constants}). \quad (2)$$

For this report, the magnetic parameters A, $k_u$, γ, and Q are maintained fairly constant by restriction to films of composition $Y_{2.35}Eu_{0.65}Ga_{1.2}Fe_{3.8}O_{12}$; however, the effect of thickness, h, is reported in a later section.

## EXPERIMENTAL TECHNIQUE

Experiments were conducted using a polarizing microscope equipped with x, y, and z coils. Fig. 1 shows the overlay conductor configuration and trajectory end points of a bubble rocking back and forth in response to alternate gradient pulses in the parallel conductors. The square loop typically carries 80 mA DC, adding approximately 1 kA/m (4π Oe) to the bias field while creating a fairly soft potential well to confine and isolate the bubble under test from surrounding bubble and stripe domains. In addition, eight pulses of 400 ns duration and 30 mA amplitude are superimposed on the DC before each gradient pulse to tickle the bubble back to the center so that it is not collapsed by the opposite gradient pulse. Drive gradients to 0.8 kA/m per 5 μm bubble diameter can be achieved at a 100 kHz rate. Bias compensation is not used; however, the gradient pulse width is adjusted (usually between 0.2 and 1 μs) to achieve a constant displacement of 3 μm from the center.

The usual experimental routine is to adjust the in-plane field magnitude to find the limit for which the state will survive $5 \times 10^6$ translations at a given drive level. States are determined by viewing the trajectory angle. Typically, the S=1 propagates at 30° to the gradient, the S=0 is parallel; while, depending upon the in-plane field, the S=1/2 varies from about 5° to 15°.

## RADIAL VS TRANSLATIONAL STABILITY

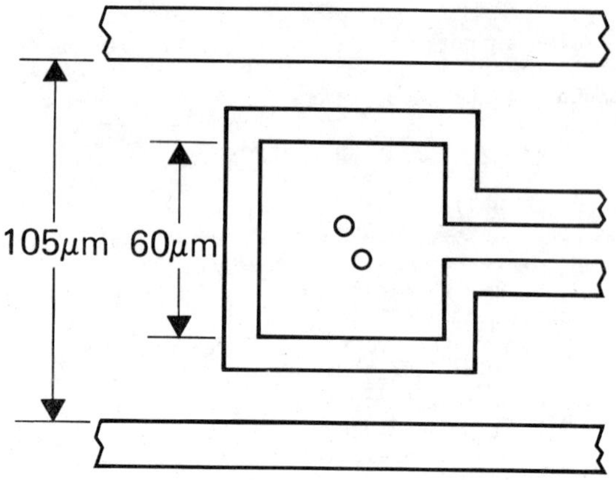


**Figure 1.** Overlay conductor pattern and bubble trajectory.

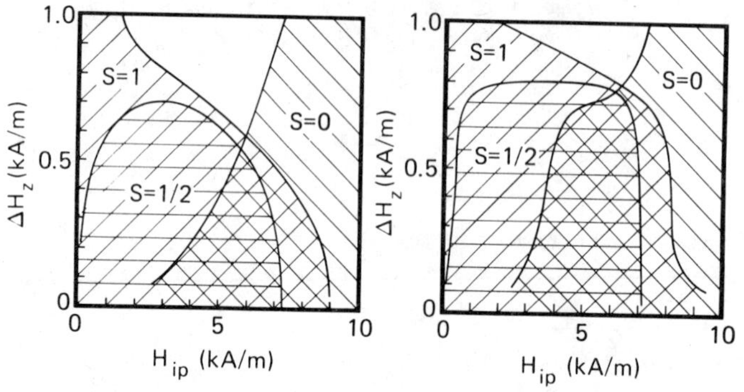


**Figure 2.** Radial stability maps for 10ns z-field rise time (a), and 200ns (b).

For comparison purposes, both radial and translational stability were measured on one YEuGaFeO sample implanted with 2 x $10^{14}$ $O^+/cm^2$ at 100 keV. Figs. 2a and 2b show radial stability maps for negative (expanding) bias field steps of 400 ns duration with rise times of 10 and 200 ns respectively. For the slower rise time, stable regions for all states are almost independent of drive up to 0.7 kA/m. At this amplitude, the bias reaches stripe-out. Otherwise, bias field changes at

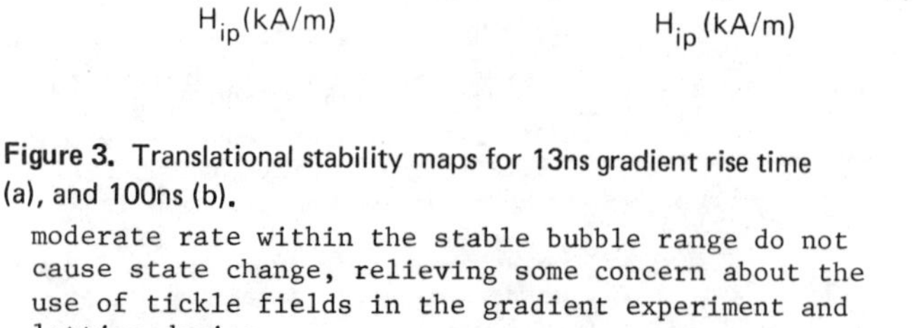

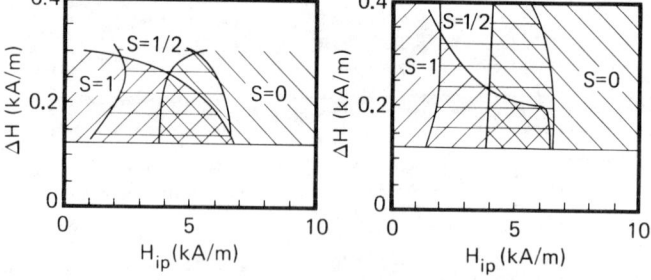

**Figure 3.** Translational stability maps for 13ns gradient rise time (a), and 100ns (b).

moderate rate within the stable bubble range do not cause state change, relieving some concern about the use of tickle fields in the gradient experiment and lattice device.

Figs. 3a and 3b show translational stability maps for the same sample with gradient rise times of 13 and 100 ns. Here we note less difference attributable to rise time, but that the S=1 state is quite unstable under relatively modest gradient drive.

IN-PLANE FIELD DIRECTION

In Figs. 3a and 3b, the in-plane field is perpendicular to the drive gradient. Fig. 4 is identical to

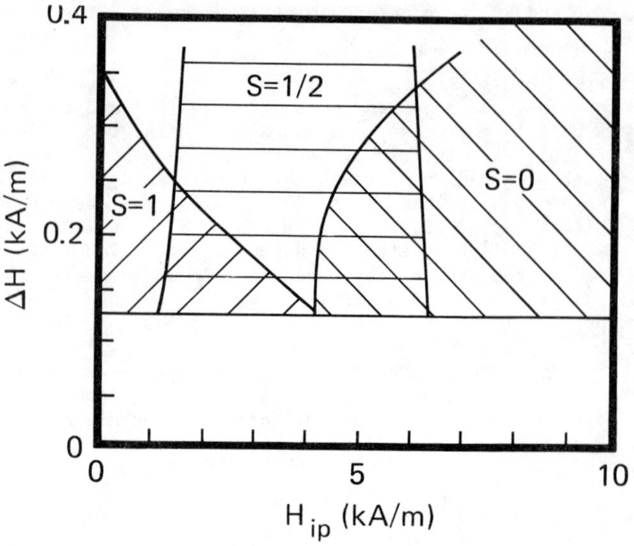


**Figure 4.** Similar to 3a, in-plane field parallel gradient.

3a except that the in-plane field is now parallel to the gradient. Comparison shows a further reduction in the S=1 stability region, although the S=1/2 is somewhat increased while S=0 is nearly unchanged.

Although the two in-plane field directions exhibit different stability, it is perhaps more remarkable that the difference is not greater. The in-plane field tends to force the Bloch lines to opposite ends of a bubble diameter aligned to the in-plane field. A field paralle to the gradient favors Bloch lines at the leading and trailing edge where translational motion causes the most distortion and rapid change of spin direction.

ION-IMPLANTATION

Fig. 5 is the stability map of an as-grown film which also demonstrates an effect of easy axis tilt. In this experiment, the in-plane field and tilt were perpendicular to the gradient; and both positive and negative directions of in-plane field were tested. Two characteristics are observed. The stability peak is offset by the tilt; but more significantly, the state stability is scarcely differentiated by the in-plane field.

This result is consistent with Hsu's model for state control with an exchange coupled layer.[4] His model uses the in-plane domain structure of an ion-implanted layer, exchange coupled with the domain, as a handle for the in-plane field to control the state. The qualitative difference between as-grown and implanted stability indicates that direct interaction of the in-plane field with wall magnetization is much less significant. Furthermore, the scarcity of S=1/2 states observed is in agreement with his model for Bloch points in as-grown films.

Otherwise, the maximum drive for which state stability exists is similar to the implanted film. This indicates that the in-plane layer functions more to control what type of state changes occur rather than at what drive gradient or wall velocity.

FILM THICKNESS

To test stability dependence on film thickness, maps were made on several chips from one LPE wafer which had been ion-milled to different thicknesses. Fig. 6 shows the maximum drive for S=1 stability and the minimum drive required to overcome coercivity as a function of thickness for an in-plane field of 3 kA/m. Stability improves with decreasing thickness but not as fast as the $h^{-1}$ dependence of $V_p$ might predict. We also note that coercivity of these samples increases rapidly with decreasing thickness offsetting the gain.

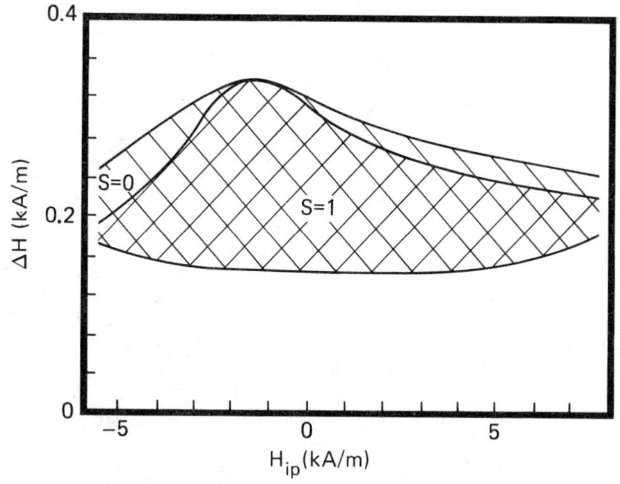


**Figure 5.** Stability map for tilted, as-grown sample.

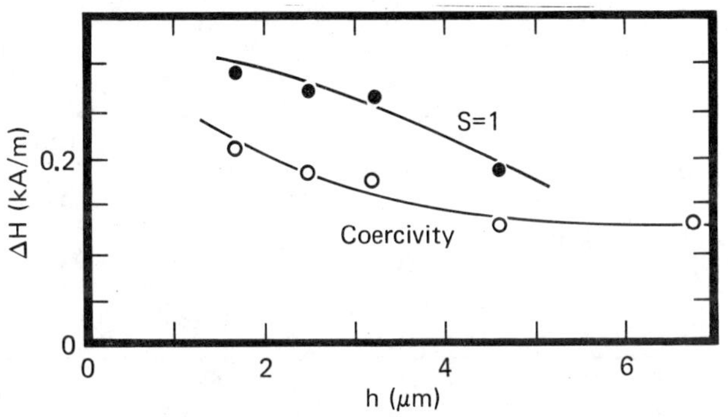


**Figure 6.** S=1 stability and coercivity vs film thickness.

STATISTICAL SIGNIFICANCE OF RESULTS

The stability maps were obtained by observing $5 \times 10^6$ translations, a number chosen as a compromise between patience and infinity. A reasonable question might be to ask to what extent the maps would change if a different number were used?

For a partial answer, the lifetime of an S=1 state was measured for 8 runs at each of 17 combinations of in-plane field between 6.5 and 8.5 kA/m and drive between 0.15 and 0.28 kA/m. Average lifetime ranged from $3 \times 10^2$ to $3 \times 10^7$ empirically fitting the relation; $\exp(144 - 150\Delta H - 14H_{ip})$. Thus, an order of magnitude change in lifetime requires a 0.015 kA/m change in drive or 0.16 kA/m change in in-plane field. Since these quantities are less than the accuracy of the maps, an order of magnitude change in the number of steps observed would not be significant.

The standard deviation of lifetimes for each set of runs was nearly equal to the average. This suggests that the states are subject to exponential decay; i.e., a finite probability of conversion with each step.

CONCLUSIONS

From the experiments, it appears that translational stability is generally more critical than radial motion stability; but both may be conveniently measured using overlay conductors. Translational stability is sensitive to the direction of in-plane field and easy axis tilt. Ion-implantation shows relatively little effect on the maximum stable drive level but is needed for state control via in-plane field. Decreasing the film thickness improves stability somewhat, but at the expense of increased coercivity.

In no case, however, did the YEuGaFeO films tested show a region of overlapping stability between the S=1 and S=1/2 or 0 states at drives greater than 2.5 times the minimum drive required to overcome coercivity. This limit may restrict the operating margins of a practical lattice device significantly. However, the theory for $V_p$ and $W_{bp}$ suggests changes in magnetic parameters A, $k_u$, and $\gamma$, which are likely to improve stability and can be obtained in films of different composition.

ACKNOWLEDGEMENTS

The author is indebted to T. Hsu for his helpful collaboration and efforts to provide samples for the experiments, and to M. D. Montgomery and D. Y. Saiki for their technical assistance.

REFERENCES

1. T. Hsu, paper 1-8, 1975 Intermag Conf., London.
2. O. Voegeli, B. A. Calhoun, and L. L. Rosier, AIP Conf. Proc. 24, 617 (1974).
3. J. C. Slonczewski, A. P. Malozemoff, and O. Voegeli, AIP Conf. Proc. 10, 458 (1972).
4. T. Hsu, AIP Conf. Proc. 24, 624 (1974).
5. B. E. Argyle, J. C. Slonczewski, and A. F. Mayadas, AIP Conf. Proc. 5, 175 (1971).
6. J. C. Slonczewski, AIP Conf. Proc. 24, 613 (1974).

## BUBBLE RESPONSE TO HARMONIC GRADIENT DRIVE

O. Voegeli, C.A. Jones and J.A. Brown
IBM General Products Division, San Jose, California 95193

ABSTRACT

The bubble response to harmonically varying field gradients has been measured in EuYIG samples as a function of drive amplitude and frequency, as well as in-plane field magnitude. The length and direction of resulting domain trajectories depend on the detailed wall state S = n, l, p; where n is the revolution number of the wall magnetization, l is the number of Bloch lines and p is the number of Bloch points in the wall. Longitudinal and transverse velocities for $0 \leq n \leq 1$ states containing Bloch points and/or pairs of unwinding Bloch lines are given. Also presented are transition thresholds between different states versus the drive parameters.

# DOMAIN WALL STATES IN ANNEALED LOW Q GARNET FILMS

J. L. Su, E. B. Moore, T. L. Hsu, and B. A. Calhoun
IBM Corporation, San Jose, California 95193

## ABSTRACT

This paper reports on the lowering of Q by thermal annealing and its effect on hard bubble suppression, control of bubble states, and reduction of coercivity. Garnet films of nominal composition $Y_{2.35}Eu_{0.65}Ga_{1.2}Fe_{3.8}O_{12}$ were annealed in either nitrogen or oxygen atmospheres. Annealing in $N_2$ at 1250°C yields almost identical reduction in Q value of a film as annealing in $O_2$ at 1350°C for the same length of time. Annealing in $O_2$ at 1350°C for 50 min. and 100 min. reduces Q from about 6 to 3 and to 1, respectively. After the first 50 min. annealing, the coercivity is reduced from 0.12 to 0.04 Oe. Hard bubble suppression is obtained when $Q \lesssim 2.1$ in films approximately 4 micron thick and at higher Q values in thicker films. Using bubble translation we have observed that bubbles generated by the pulse modulated bias field are mostly S=1 in low Q films. S=0 bubbles can be reliably generated in these films with the application of an in-plane magnetic field together with pulse modulated bias field.

## INTRODUCTION

Control of bubble states[1] has been demonstrated in garnet films with a capping layer. Smith et al reported that as-grown LPE films with Q<4.5 exhibit no hard bubbles[2]. LeCraw et al reported suppression of hard bubbles in LPE garnet films by inert atmosphere annealing[3]. Annealing behavior and temperature dependence of the uniaxial anisotropy energy $K_u$ in Sm-YIGG and $Y_{2.4}Eu_{0.6}Ga_{1.2}Fe_{3.8}O_{12}$ LPE films were investigated by Hagedorn et al.(4,5)

In bubble lattice devices[6] only two stable bubble states are desired. We have found that, in $Y_{2.35}Eu_{0.65}Fe_{3.8}Ga_{1.2}O_{12}$ films, the desired control of the bubble states can be obtained suitably by adjusting Q. In this paper, we describe the annealing procedures used and the magnetic properties of the annealed films.

## EXPERIMENTAL

The films used in the experiment were obtained in the form of one-inch diameter wafers which were diced into rectangular chips. Film properties are presented in Table I. For each set of annealing conditions virgin chips were used. Annealing was performed in a horizontal tube furnace with the ends of the tube closed loosely with glass wool. The gas inlet end was more tightly packed to provide a positive unidirectional gas flow. The garnet chips were supported on a thin alumina plate contained in a small platinum boat. A platinum wire was used to pull the boat to a cooler section of the tube to effect a quench when the procedure called for it. Temperature was monitored with a thermocouple placed directly under the boat.

In all runs the furnace and contents were brought to temperature from ambient in a uniform manner with the gas flowing. The last 100 degrees was achieved in 5-7 minutes by manipulation of furnace power and the set point. After annealing in nitrogen, the samples were furnace cooled to 1050°C, soaked for two hours in oxygen and then quenched. After annealing in oxygen, the samples were cooled in the furnace. These heat treatments resulted in little change from the as-grown values of 4¶Ms.

The LPE film parameters were measured at room temperature by standard techniques. The uniaxial[8] anisotropy field $H_k$ was determined visually under a polarizing microscope by applying a d.c. in-plane magnetic field necessary to make contrast between adjacent stripe domains disappear. The applied in-plane field was along a <110> direction of the sample. The coercivity $H_c$ was determined from the motion of stripe domains in an a.c. bias field[7]. The bubble states were determined from the angle between the field gradient and direction of bubble translation[9].

## RESULTS

The original film properties of the samples used in the annealing experiment is given in Table 1. The anistropy energy $K_u$ measured at room temperature for samples annealed in oxygen at 1200, 1275, and 1350°C, respectively, falls exponentially with annealing time as shown in Fig. 1. Annealing in nitrogen at temperatures between 1150 and 1250°C had approximately the same effect

Table I As Grown Film Properties

| Film # | h(μm) | $W_s$(μm) | $H_o$(Oe) | $4\P M_s$(G) | Q |
|---|---|---|---|---|---|
| A | 4.21 | 5.51 | 72.6 | 167.7 | 5.8 |
| B | 3.80 | 5.28 | 68.7 | 165.6 | 5.6 |
| C | 3.90 | 5.34 | 67.2 | 160.3 | 5.7 |
| D | 6.20 | 6.55 | 82.8 | 165.7 | 6.5 |
| E | 8.40 | 7.20 | 91.7 | 162.1 | 8.0 |

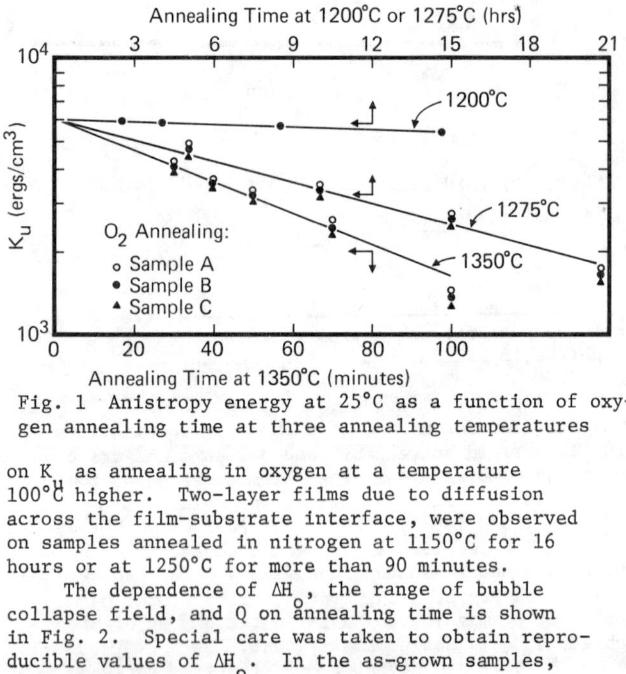


Fig. 1 Anistropy energy at 25°C as a function of oxygen annealing time at three annealing temperatures

on $K_u$ as annealing in oxygen at a temperature 100°C higher. Two-layer films due to diffusion across the film-substrate interface, were observed on samples annealed in nitrogen at 1150°C for 16 hours or at 1250°C for more than 90 minutes.

The dependence of $\Delta H_o$, the range of bubble collapse field, and Q on annealing time is shown in Fig. 2. Special care was taken to obtain reproducible values of $\Delta H_o$. In the as-grown samples, $\Delta H_o$ was 27 Oe and after annealing in nitrogen at 1250°C or in oxygen at 1350°C for 70 min., $\Delta H_o$ decreased to <1.2 Oe. Each point in Fig. 2 represents the average value for samples A, B, and C which have similar as-grown properties. The same change in magnetic properties of these films after annealing was observed, suggesting good reproducibility of the annealing process. The effect of film thickness on hard bubble suppression is indicated in Fig. 3. Samples were annealed in oxygen at 1275°C for various lengths of time to obtain different Q values. For the same Q, thicker films exhibit smaller values of $\Delta H_o$.

Annealing in both oxygen and nitrogen reduced the coercivity from 0.12 Oe in as-grown films to 0.04 Oe. The time required to achieve this reduction for a particular temperature and atmos-

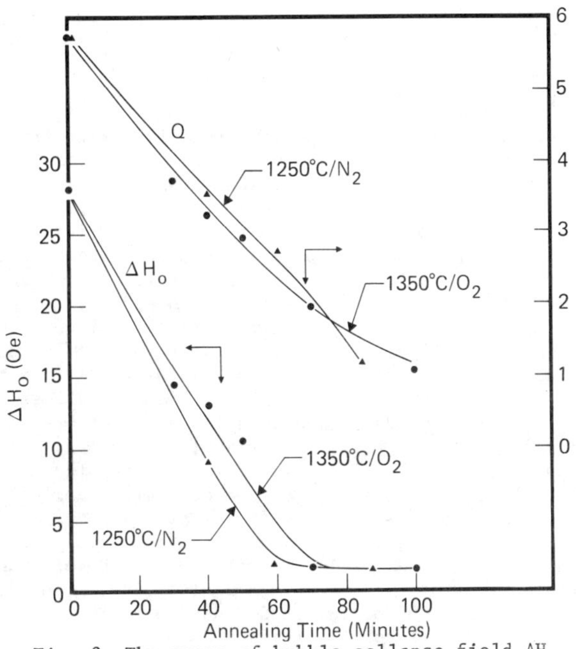


Fig. 2 The range of bubble collapse field $\Delta H_o$ and Q at 25°C vs annealing time

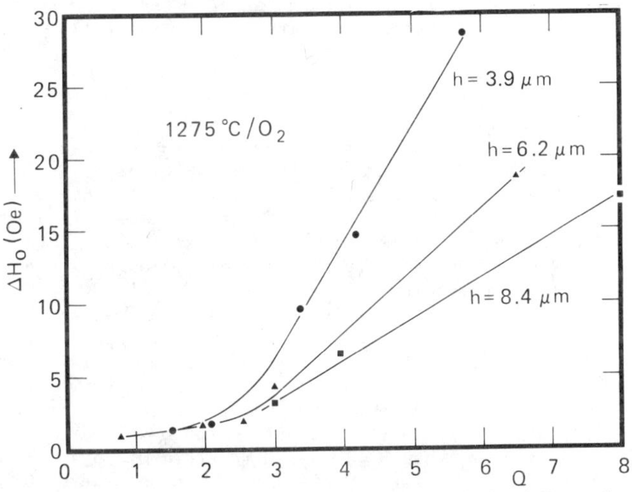


Fig. 3 $\Delta H_o$ as a function of Q for 3.9, 6.2, and 8.4-μm thick films

phere was approximately the same as the time required to reduce Q to 2. Additional annealing produced no further change in $H_c$. The minimum drive field for bubble translation also decreased from 1.2 to 0.54 Oe.

The Q of a chip annealed in nitrogen at 1150°C for four hours was reduced from 6.5 to 2.3. Bubbles generated in this chip by a pulse modulated field (PMBF) with no in-plane field had a $\Delta H_o$ <1.0 Oe. Dynamically, they were all S=1 bubbles and propagated in an angle of 50° from the direction of the field gradient. S=0 bubbles could be reliably generated by applying an in-plane field and rapid wall motion[1]. In this chip, S=0 bubbles, once generated, were stable without any in-plane field bias. In the 1350°C oxygen annealing series, the Q of a chip was reduced to about 1.1 after 100 min. of annealing. PMBF in the absence of an in-plane field generated mostly S=1 bubbles with an angle of 65° and a few S=0 and S=-1 bubbles. PMBF in the presence of 100 Oe in-plane field generated only S=0 bubbles. The S=0 bubbles were stable during translation in the absence of an in-plane field. In the same oxygen

Fig. 4 Two-layer film formed by excessive time and temperature in nitrogen (1150°C for 16 hrs.)

annealing series, the Q of a chip was reduced to 2.0 after 70 min. of annealing. PMBF generated about equal distributions of S=1, S=0, and S=-1 bubbles. S=0 bubbles could not be reliably controlled with an in-plane field of 100 Oe.

## DISCUSSION

The better state control obtained with nitrogen annealing may be due to the formation of a second layer by gallium diffusion from the substrate; however, care must be used to avoid excessive diffusion which produces the deleterious effects shown in Fig. 4.

As suggested in Ref. 1, the hard bubble suppression in low Q films may be due to the ability of the stray field to twist the magnetization near the surface into the plane of the film. This is similar to the situation in ion-implanted films; however, unlike the ion-implanted films, S=0 bubbles in these annealed low Q films are stable in the absence of an in-plane field.

## ACKNOWLEDGEMENTS

We wish to thank G. Galli and H. Turk for supplying us with the LPE films.

## REFERENCES

1. T. L. Hsu, AIP Conf. Proc. 24, 624 (1974).
2. A. B. Smith, M. Kestigian, and W. R. Bekebrede, AIP Conf. Proc. 18, 167 (1973).
3. R. C. LeCraw, E. M. Gyorgy, and R. Wolf, Appl. Phys. Lett. 24, 573 (1974).
4. F. B. Hagedorn, J. Appl. Phys. 45, 3123 (1974)
5. F. B. Hagedorn, S. L. Blank, and R. L. Barns, Appl. Phys. Lett. 22, 209 (1973)
6. O. Voegeli, B. A. Calhoun, L. L. Rosier, and J. C. Slonczewiski, AIP Conf. 24, 617 (1974)
7. P. W. Shumate, IEEE TRans, Magn. MAG-7, 586 (1971)
8. A. J. Kurtzig and F. B. Hagedorn, IEEE Trans. Magn. MAG-7, 473 (1971)
9. G. P. Vella-Coleiro and W. J. Taylor, Appl. Phys. Lett. 21, 7 (1972)

# TEMPERATURE STABILITY OF BUBBLE DOMAIN WALL STATES

T. Obokata, K. Yamaguchi and K. Asama
FUJITSU LABORATORIES LTD., Kawasaki, Japan

## ABSTRACT

It is well known that bubbles with revolution numbers S=1 and S=0 can exist in a capped garnet film. The stability of these two types of bubbles is considered to be a very important problem, especially when they are used for binary coding of bubble lattice devices. This paper describes the temperature stability of these two types of bubbles. The sample used was $(YGdYb)_3(FeGa)_5O_{12}$ with a bubble diameter of 3 μm. An S=1 bubble was stable up to about 100°C, although the deflection angle θ increased linearly from 40° at room temperature to 70° at 100°C. An S=0 bubble was generated by applying an in-plane magnetic field $H_p$=100 Oe and after reducing $H_p$ to a certain value, temperature dependences of the deflection angle θ and the bubble velocity V were measured at a constant drive field $\nabla Hz=3.5\times10^3$ Oe/cm. When $H_p$=40 Oe, θ remained 0° up to 80°C, indicating that the S=0 state was stable up to 80°C. When $H_p$=20 Oe, on the other hand, θ changed discretely from 0° to 27° at about 55°C, thereafter increasing with temperature until it reached 40° at 80°C. The S value of this wall state was calculated to be 0.25. A similar variation was observed when $H_p$=30 Oe. These S values smaller than 1 can be explained by assuming the thermal creation of one Bloch point in one of the two vertical Bloch lines in the domain wall.

## INTRODUCTION

As is generally known, in the absence of an in-plane magnetic field, the S=1 bubbles are the most stable bubbles in a capped garnet film. These bubbles move at an angle to the magnetic field gradient. However, by applying an in-plane magnetic field S=1 bubbles can be controllably converted to S=0 bubbles which propagate along the field gradient. Stabilities of these two types of bubbles are considered to be of great importance, especially so when they are used for binary coding of bubble lattice devices[1]. Hsu[2] has studied the stability of an S=0 bubble as a function of an in-plane magnetic field by applying a z-field pulse to the bubble to produce domain wall motion. The purpose of this paper is to discuss the stability of these two kinds of bubbles against temperature variation.

## EXPERIMENT AND RESULTS

The sample used was $(YGdYb)_3(FeGa)_5O_{12}$ grown by an LPE technique on $Gd_3Ga_5O_{12}$ substrate. This film has a thickness of h=3.8 μm, a characteristic length $\ell$=0.40 μm, a saturation magnetization $4\pi Ms$=200 Gauss and an anisotropy field $H_k$=700 Oe. In order to suppress hard bubbles in this sample, $7\times10^{13}$ $Ne^+/cm^2$ were implanted at 100 keV. A pair of parallel strip lines which were 10 μm wide and separated by a distance of 100 μm produced gradient fields when pulse currents were applied. A bubble 3 μm in diameter was nucleated by a pulse field and subjected to the gradient field. Midway between the two parallel conductors bubbles were moved by about 10 μm with 0.5 μsec pulse currents. A pair of Helmholtz coils were used to apply an in-plane dc magnetic field $H_p$ parallel to the direction of the magnetic field gradient. The gradient of the magnetic field was kept constant at $\nabla Hz=3.5\times10^3$ Oe/cm. In this experiment the stability of bubble domain wall states was studied by moving a bubble with a few steps of pulsed gradient field. The stability for many pulsed steps applied is the subject for a future study.

As is shown in Fig. 1, the deflection angle θ shows a hysteresis curve with variation of $H_p$: an S=1 bubble can be switched to an S=0 bubble by applying an in-plane field $H_p$ of larger than 80 Oe and this S=0 state is maintained even when $H_p$ is reduced from 80 Oe to less than 10 Oe, but an S=0 bubble is converted again to an S=1 bubble at about $H_p$=0 Oe. The temperature stability of the S=1 bubble was measured at $H_p$=0 Oe (point A in Fig. 1). Figure 2 shows temperature dependences of the deflection angle θ and the revolution number S, which was calculated by the well known formula[3],

$$S=r^2\gamma\nabla Hz\sin\theta/2V \qquad (1)$$

where r is the bubble radius and $\gamma$ is the gyromagnetic ratio, which is assumed to be $\gamma=1.64\times10^7$ $sec^{-1}$ $Oe^{-1}$. The S=1 state was maintained up to about 100°C, although the deflection angle increased almost linearly from 40° at room temperature to 70° at 100°C. This increase of θ is due to the increase of bubble velocity V with temperature. As is shown in Fig. 1, S=0 bubbles were generated by applying an in-plane magnetic field of 100 Oe parallel to the direction of the pulse field gradient. Thereafter the in-plane magnetic field was decreased to certain values, B,C and D in Fig. 1, at which we measured temperature dependences of θ and V. Figure 3 shows the variation of θ and S with temperature. When $H_p$=40 Oe, θ remained 0° up to 80°C, indicating that the S=0 state was stable up to 80°C.

When $H_p$=20 Oe, however, θ changed discontinuously at about 55°C from 0° to 27°, then increasing further with temperature until it reached 40° at 80°C. S values of these wall states with θ≠0° were calculated to be 0.25 according to Eq.(1). A similar result was obtained when $H_p$=30 Oe. At this value of $H_p$, θ jumped at about

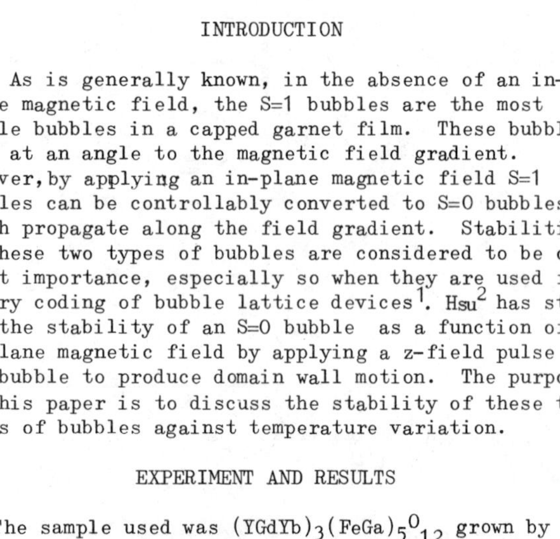


Fig. 1 Hysteresis loop of deflection angle θ vs in-plane magnetic field $H_p$.

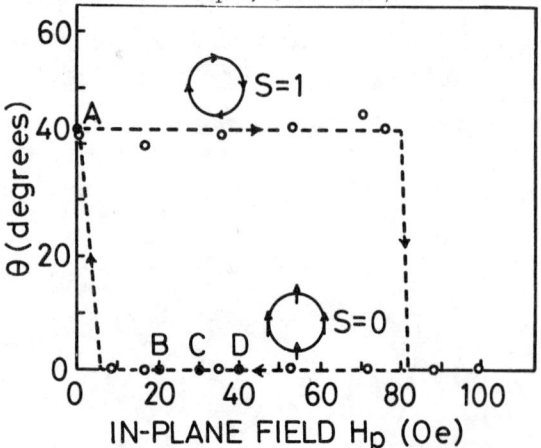

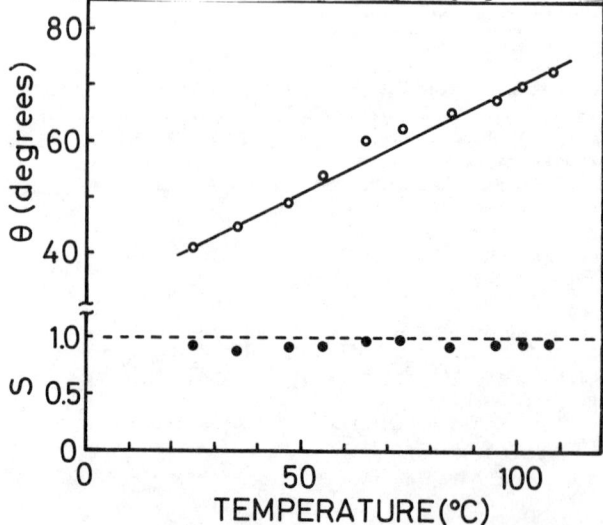

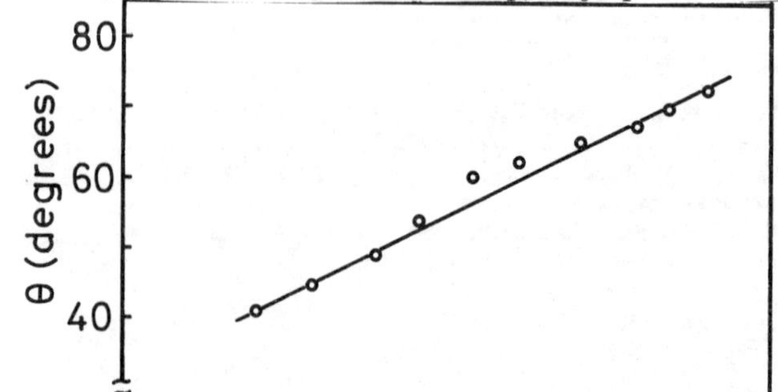


Fig. 2 Temperature dependences of θ and S for S=1 bubble.

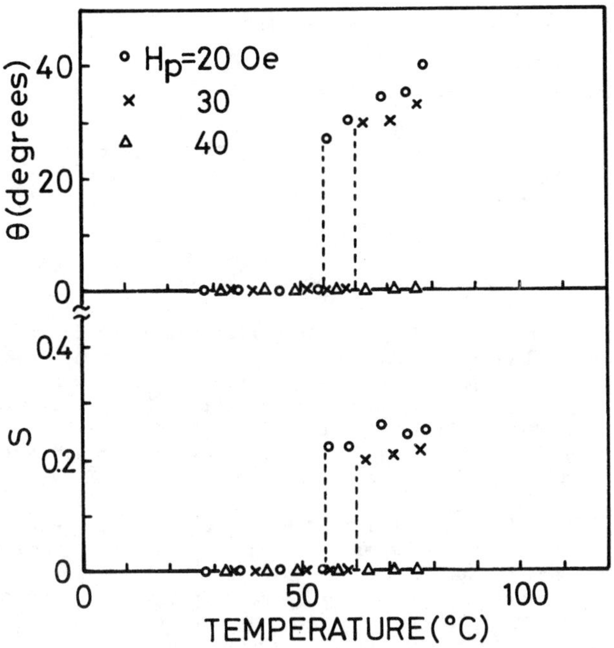


Fig. 3 Temperature dependences of θ and S for S=0 bubble.

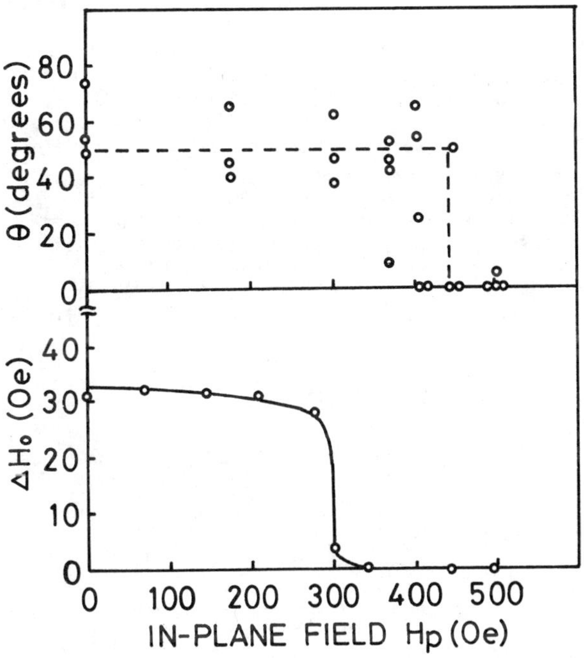


Fig. 4 Fluctuations of deflection angle θ and collapse field Ho as a function of in-plane field applied temporarily. ( ΔHo=Homax - Homin)

65°C abruptly from 0° to 30°, thereafter increasing slightly to about 35° at 77°C. In this case S was calculated to be 0.21.

The stability of the bubble domain wall states in the garent film without a capping layer, namely, the as-grown garnet film, was then studied. While hard bubbles with a large number of Bloch lines can exist in this type of film, the number of Bloch lines can be reduced to one pair by applying a large in-plane field (but less than anisotropy field). Figure 4 shows the change in the measured fluctuations of the deflection angle and the collapse field of bubbles when an in-plane magnetic field is applied prior to application of the pulsed gradient field. As is shown in this figure, hard bubbles can be converted to S=0 bubbles when Hp larger than 450 Oe is applied. The temperature stability of S=0 bubbles thus formed was measured in the same manner as mentioned above. In this case the deflection angle remained 0°, i.e., the S=0 state was

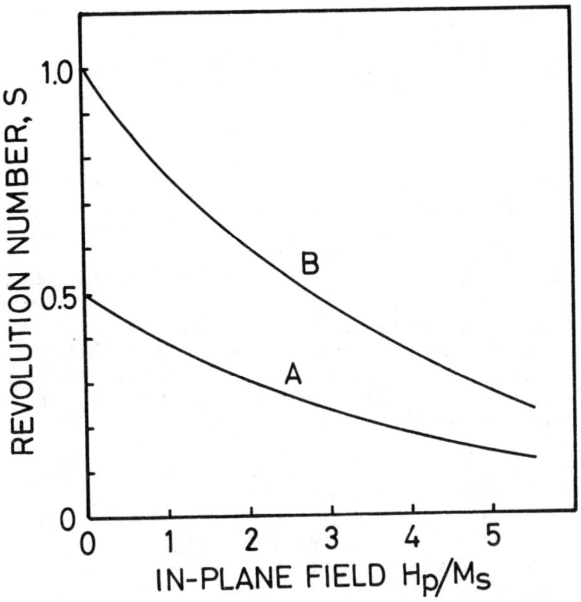


Fig. 5 Effect of in-plane field on S. Line A corresponds to the case with one Bloch point in one of the two Bloch lines of negative revolution. Line B represents the case with one Bloch point in each of the two Bloch lines.

maintained up to above 100°C even when Hp=10 Oe.

## DISCUSSION AND CONCLUSION

In the case of a capped garnet film, the S=0 state was observed to be switched to the S≠0 state at a certain temperature when the applied in-plane field was 20 to 30 Oe. The occurrence of such a state with a value of S smaller than 1 can be explained by assuming that a Bloch point is thermally excited in one of the two vertical Bloch lines in the domain wall. Recently Slonczewski[4] introduced Bloch points in order to explain half-integer values of S appearing at Hp=0 Oe, which were actually observed by Bullock[5] and Hasegawa[6]. Figure 5 shows the effect of the in-plane field on S calculated by Hasegawa. From this figure, the values of S for Hp=20 Oe and 30 Oe are predicted to be 0.36 and 0.3, respectively. These values are qualitatively consistent with the values of 0.25 and 0.21 for Hp=20 Oe and 30 Oe, respectively, calculated for thermally activated wall states as is shown in Fig. 3. According to Slonczewski, it is likely that a Bloch point can be excited at room temperature or above. The excitation of a Bloch point was not observed in the case of the garnet film without any capping layer, indicating that a Bloch point was thermally nucleated at the interface of the bubble film and the capping layer.

From these results it can be concluded that the S=1 state in the capped film of $(YGdYb)_3(FeGa)_5O_{12}$ is stable up to 100°C, but that an in-plane magnetic field of about 40 Oe or more is necessary to maintain the S=0 state stable up to at least 80°C.

## ACKNOWLEDGMENT

The authors wish to thank K. Kashiro for his assistance in measurements and H. Tominaga for critical reading of the manuscript. They also thank M. Nakamura, F. Iwai, and S. Kojima for their encouragement.

## REFERENCES

1. O. Voegeli, B.A. Calhoun, L.L. Rosier and J.C. Slonczewski, AIP Conf. Proc. 24, 617 (1975).
2. T.L. Hsu, AIP Conf. Proc. 24, 624 (1975).
3. J.C. Slonczewski, A.P. Malozemoff and O. Voegeli, AIP Conf. Proc. 10, 458 (1973).
4. J.C. Slonczewski, AIP Conf. Proc. 24, 613 (1975).
5. D.C. Bullock, AIP Conf. Proc. 18, 232 (1974).
6. R. Hasegawa, AIP Conf. Proc. 24, 615 (1975).

# DYNAMIC PROPERTIES OF MAGNETIC BUBBLES IN AMORPHOUS GdCoCu FILMS

Robert I. Potter, V. J. Minkiewicz, Kenneth Lee, and P. A. Albert
IBM Research Laboratory, San Jose, California 95193

ABSTRACT

We have measured bubble velocity v as a function of drive field for nominally 1.2 μm diameter bubbles in amorphous GdCoCu films, and have extracted from these data the dynamic coercive force $H_c$, mobility μ, and information about the domain wall state S. Our results are in the range $H_c$=2 to 5 Oe, v=1000 to 2000 cm/sec at 25 Oe drive field, μ=100 to 300 cm/sec-Oe, and we have reproducibly propagated bubbles at zero and non-zero angles with respect to the field gradient in a way consistent with the existence of S=0 and S=1 wall states.

## I. INTRODUCTION

Amorphous Gadolinium - transition metal alloys are of interest as potential bubble device materials.[1,2] We report here the results of pulsed field gradient experiments on 1.0 to 1.6 μm diameter bubbles in amorphous RF sputter deposited GdCoCu films. The fabrication and composition dependence of the static magnetic properties such as 4πM are described in a companion paper.[3] Of interest here are dynamic properties such as mobility, especially in view of very large values reported[4] for the GdCo and GdCoAu systems, and bubble deflection angle with respect to the applied field gradient, which is related to the detailed structure or state of the domain wall. Considered also is the influence of magnetic annealing on coercivity and wall state.

## II. EXPERIMENTAL PROCEDURE AND RESULTS

A cross section of the apparatus is shown in Fig. 1. The polarizing, reflected light microscope is of inverted design with the objective pointing upward. Magnetic domains in the sample are observed via the polar Kerr effect in the usual way. Integral with the microscope frame is a water cooled, air core, solenoidal magnet (not shown in Fig. 1) that generates the approximately 1 kOe dc bias field required for these materials.

The pulsed field gradient is generated by two parallel conductors as described by Vella-Coleiro and Tabor.[5] The conductors are 2-3 μm thick, separated 50 μm center-to-center and deposited upon a 0.2 mm

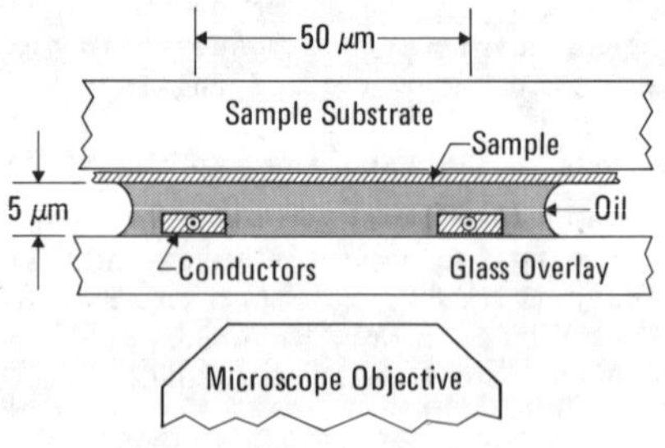


Fig. 1 Cross section of experimental apparatus.

thick glass substrate that is subsequently attached to the microscope stage. The sample is held in proximity to this glass overlay by a thin film of microscope objective immersion oil as shown in Fig. 1. We have found this to be a reliable technique that allows rapid survey of several samples without the bother and risk of depositing insulation and conductors directly onto the electrically conductive sample. The depth of field of the 50X cover glass corrected microscope objective is sufficiently small that the distance between the near sides of the sample and conductors can be conveniently measured via the calibrated focus knob on the microscope. This distance, typically 5-7 μm, is taken as the midplane to midplane distance because the sample and conductors are of comparable thickness. The field gradient at the symmetry plane x=0 located equidistant from each conductor is, in MKS units,

$$\frac{\partial H_z}{\partial x} = \frac{4I}{\pi D^2}\,\frac{1-(2z/D)^2}{[1+(2z/D)^2]^2} \qquad (1)$$

where the x direction is perpendicular to the conductors, D is the separation between conductors, z the separation between sample and conductors, and I the current. The first factor in Eq. (1) is more conveniently expressed as 6.4 I Oe/μm when D=50 μm and the second is 0.888 when z/D=0.1. Possible eddy current damping caused by the metallic sample is not considered.

The conductors are series connected to guarantee equal current in each, which is necessary in order that $H_z$(pulse)=0 at x=0, and comprise the center element of a 50 Ω π-pad. Pulse polarity is reversed by interchanging the remotely located 50 Ω termination resistor and the pulse generator, a Velonex model 380 with unipolar output and 10 A capability. Joule heating of the conductors limits the pulse duration to 150 ns. Pulse rise time measured at the conductors is approximately 12 ns.

Behavior typical of amorphous GdCoCu is shown in Fig. 2, where the coordinates of a bubble are plotted after each pulse of a series of first positive and then negative pulses. A striking feature is the strong dependence of velocity (displacement per pulse) on location with respect to the center line. There exists in this experiment, in addition to the pulsed field gradient, a large pulsed z field that excites a propagation mode in which bubble diameter is not constant.[6] This field increases with distance from the center line to a maximum of about 400 Oe or almost half the dc bias field. Addition of a second pair of

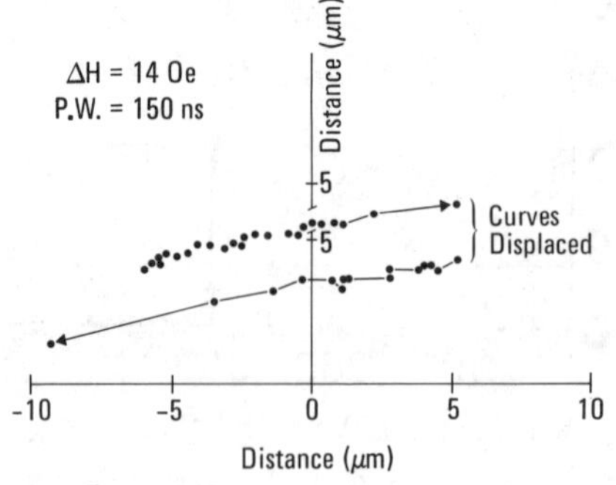


Fig. 2 Bubble coordinates after each of a series of pulses. Vertical axis is parallel to and centered between conductors. Bubble diameter is 1 μm.

ramp-current driven compensation conductors[5] that cancel this field is impractical when dealing with materials such as ours. Consequently, a stripe-out mode of propagation exists on the "downhill" side of the center line that results in apparent velocities up to an order of magnitude larger than seen on the opposite side. Occasionally, due to the relatively high wall motion coercive force, such bubbles are left as short stripes with major axis along the propagation direction. All data reported here were obtained by propagating bubbles over a distance of about 5 μm toward the center line and computing the average displacement per pulse.

A second feature of Fig. 2 is the scatter in bubble location. In part this is experimental; the coordinates are obtained with an estimated precision of ±1/4 μm from photographs. However, we visually observe the displacement per pulse to be far from smooth and regular. There apparently exist material inhomogeneities that pin a bubble for one or two pulses, or laterally deflect it from its average trajectory by as much as a micrometer.

Wall motion coercivity $H_c$ and mobility $\mu$ are related to bubble velocity v and drive field $\Delta H=(\partial H_z/\partial x)d$, where d is the bubble diameter, by the well known result[7]

$$v = \frac{\mu}{2}\left(\Delta H - \frac{8H_c}{\pi}\right), \ \Delta H \geq \frac{8H_c}{\pi} . \qquad (2)$$

A typical velocity versus drive field plot is shown in Fig. 3 and more detailed results for additional samples appear in Table I. Note that the velocity is linear up to $\Delta H$=36 Oe and the mobility is an order of magnitude lower than that reported[4] for similar amorphous films.

As Table I indicates, we have found in agreement with others[2,8] that magnetic annealing lowers the coercivity. However, we do not observe a significant decrease in uniaxial anisotropy constant $K_u$. The decrease in Q is due to an increase in $4\pi M$ at room temperature, which in turn is attributed to a shift in compensation temperature caused by preferential oxidation of Gd during annealing. The magnetic annealing treatment is 2 hr. at 200°C, $10^{-6}$ Torr and 20 kOe applied field followed by cooling with field on. We regard sample D as suggestive but not definitive evidence that the presence of a magnetic field during annealing is beneficial.

Apparent in Fig. 2 is a gyrotropic deflection angle $\phi$=10° with sign[9] corresponding to a positive wall state S; that is, the bias field points out of

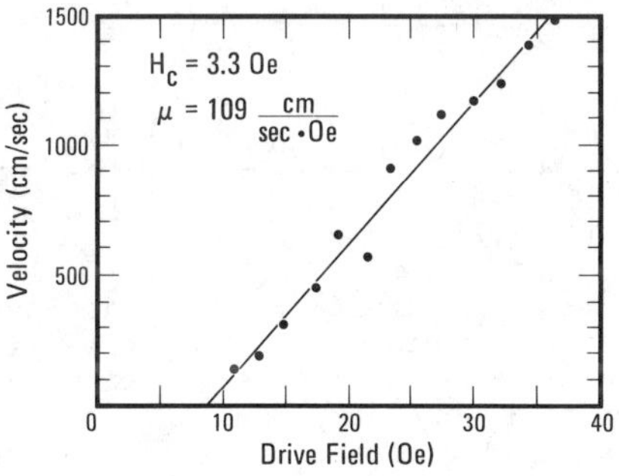


Fig. 3 Velocity versus drive field $\Delta H$ for a representative unannealed sample.

Table I. Dynamic and related properties of four amorphous GdCoCu samples before and after magnetic annealing. Sample D was annealed in zero field.

| Sample | Thickness (μm) | Stripe Width (μm) | $4\pi M$ (gauss) | $K_u$ ($\times 10^6$ erg/cm$^3$) | Q | $H_c$ (dynamic) (Oe) | μ (cm/sec-Oe) |
|---|---|---|---|---|---|---|---|
| A | 2.29 | 1.5 | 1460 | 0.51 | 6.0 | 5 | 200 |
| A (mag. ann.) | | 1.2 | 1750 | 0.42 | 3.4 | 2.1 | 300 |
| B | 1.24 | 1.1 | 1060 | 0.35 | 7.8 | unmeasurable | |
| B (mag. ann.) | | 0.9 | 1390 | 0.30 | 3.9 | 3.8 | 325 |
| C | 1.24 | 1.1 | 930 | 0.30 | 7.3 | unmeasurable | |
| C (mag. ann.) | | 0.9 | 1600 | 0.33 | 2.8 | 2.0 | 92 |
| D | 2.29 | 1.6 | - | - | - | 3.8 | 205 |
| D (thermal ann.) | | - | - | - | - | 5.8 | 170 |

the plane of the figure and the negative field gradient toward the right. The wall state is related to the deflection angle $\phi$ by[9]

$$S = \frac{\gamma d \Delta H}{8v} \sin\phi \qquad (3)$$

where $\gamma$ is the gyromagnetic ratio, d the bubble diameter, $\Delta H$ the drive field, and v the velocity. Assuming $\gamma=1.76\times10^7$ $Oe^{-1}$ $sec^{-1}$ (the free electron value), d=1.0 μm, $\Delta H$=14 Oe, and v=275 cm/sec, we obtain S=1.95, which is sufficiently close to 2 that one might identify this as an S=2 bubble. However, the uncertainties in all of these numbers, especially d, which actually enters the formula quadratically, are such that this identification is tentative.

Additional information on state identification was sought through histograms of number of bubbles with deflection angle $\phi$ versus $\phi$. These data were obtained with the aid of a low light level TV camera and monitor. A minimum of five coordinate pairs, not always corresponding to consecutive pulses, were used for each deflection angle measurement and for these measurements alone the full propagation distance indicated in Fig. 2 was employed. Under these conditions the standard deviation in deflection angle can be shown to be 1.2° if the standard deviation in coordinates is 0.25 μm. Aberrations in the TV system add both systematic and statistical (because of varying locations on the screen where measurements were taken) errors to the above. The overall precision is estimated to be about ±3°.

Shown in Fig. 4 are deflection angle histograms for one sample both before and after magnetic annealing. According to Eq. (3) the deflection angle for the S=1 state in the annealed sample should be in the range 5 to 10°. The factor of three increase in the average deflection angle upon magnetic annealing can be understood on the basis of a measured factor of three increase in the ratio μ/d. The salient feature of these histograms is a broad distribution of deflection angles extending, in the case of the annealed sample, to about 20°. The lack of well-defined peaks coupled with roughly a factor of two uncertainty in the calculated deflection angles precludes positive state identification, if indeed discrete and time independent states even exist in this material. The spread in both histograms is, however, sufficient to cover both the S=0 and S=1 states.

In summary, we have measured the wall mobility and dynamic coercive force for amorphous GdCoCu films by bubble translation experiments and presented what

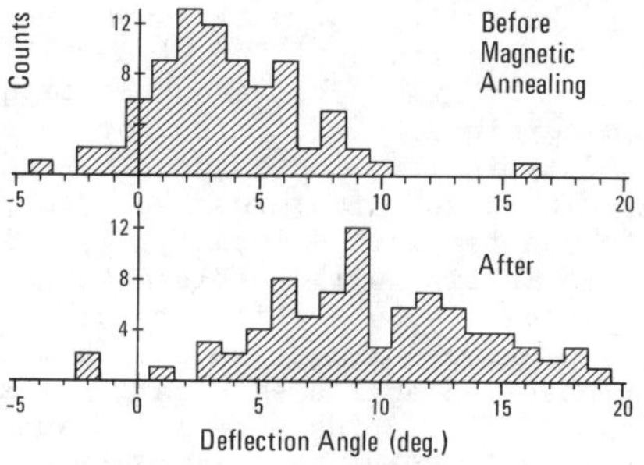


Fig. 4 Number of bubbles observed at deflection angle $\phi$ versus $\phi$ before and after magnetic annealing.

we believe to be the first conclusive data on the existence of gyrotropic deflection in an amorphous material.

ACKNOWLEDGEMENTS

We thank G. Guthmiller for many eyestraining hours of data acquisition, D. Saiki for fabrication of conductor overlays, and H. L. Hu for helpful advice.

REFERENCES

1. P. Chaudhari, J. J. Cuomo, and R. J. Gambino, IBM J. Res. Develop. 17, 66 (1973).
2. R. J. Gambino, P. Chaudhari, and J. J. Cuomo, AIP Conf. Proc. 18, 578 (1973).
3. V. J. Minkiewicz, P. A. Albert, R. I. Potter, and C. R. Guarnieri, this Proceedings.
4. M. H. Kryder and H. L. Hu, AIP Conf. Proc. 18, 213 (1973).
5. G. P. Vella-Coleiro and W. J. Tabor, Appl. Phys. Letters 21, 7 (1972).
6. R. M. Josephs and B. F. Stein, AIP Conf. Proc. 24, 598 (1974).
7. A. A. Thiele, Bell System Tech. J. 50, 725 (1971).
8. R. Hasegawa, R. J. Gambino, J. J. Cuomo, and J. F. Ziegler, J. Appl. Phys. 45, 4036 (1974).
9. A. P. Malozemoff, J. Appl. Phys. 44, 5080 (1973).

# STABILITY AND DYNAMICS OF MICROWAVE GENERATED RING DOMAINS

H. Dötsch
Philips GmbH Forschungslaboratorium Hamburg
2 Hamburg 54, W. Germany

ABSTRACT

In LPE garnet films strong spin precessions can be excited locally by microwaves near the ferrimagnetic resonance frequency. When the precession angle θ exceeds a critical value the spins tip over and create a bubble within which the spins keep on precessing due to the microwave excitation. With further increasing θ a second critical value is reached where the spins again tip over and create a bubble within the first one, i.e. a ring domain. Continuing in this way systems of many concentric rings can be generated which are stable even in the case of vanishing microwave power. However, single ring domains in films of low coercivity can only be stabilized by microwaves. To a first approximation the stabilizing forces on these domains are due to the precession induced change of the dc magnetization inside the ring.

If the microwaves stabilizing a single ring domain are amplitude modulated a corresponding radial motion of the ring is induced. From this motion one can derive the eigenfrequency and velocity of the ring.

## INTRODUCTION

In LPE garnet films which are currently used for magnetic bubble devices the ferrimagnetic resonance can be excited by microwaves. The linewidths are as low as a few Oe if no strong relaxers are present in the film. The resonance frequency is mainly determined by the anisotropy and the bias field and typically lies between 0.2 and 4 GHz. By special microwave structures the spin precession in the film can be excited locally, i.e. in regions comparable in size with bubble domains. In contrast to bulk materials, where the precession angle does not exceed a few degrees, we obtain in the magnetic films precession angles even larger than 90°. In regions where the spins precess at such large angles magnetic bubble and ring domains will be generated if the bias field is adjusted properly. The domains thus created are then subjected to forces due to the spin precession within these domains or their surroundings.

## EXPERIMENTAL SET UP

Fig. 1a) shows schematically a cross section through our experimental set up. The magnetic film is pressed on top of a short circuited microslot line of a few microns in width, which is etched into a metal film. The dashed lines indicate the rf magnetic field of the slot. The in-plane component of this field excites the spin precession in the film. Fig. 1b) shows a top view of the central part. In addition to the slot line we use a coplanar waveguide which is short circuited close to the short circuit of the slot line. It has two purposes: 1) It serves as a monitor to observe the ferrimagnetic resonance transmission. 2) It compresses the rf current of the slot line to the small region between the two short circuits. Therefore we obtain in this region a local maximum of the rf magnetic field. The halfwidth diameter of this excitation region varies between 10 and 20 µm and is thus comparable in size with bubble domains in the films we use in the experiments. The material parameters of these films are listed in Table I.

The domains are observed by the double Faraday effect (fig. 1a): Polarized light falls in nearly perpendicular to the film plane, is reflected at the metal surface, and passes the film a second

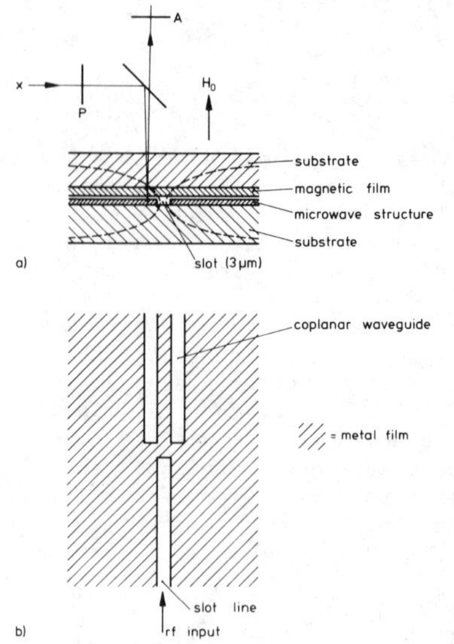


Fig.1. Cross section (a) and top view (b) of the experimental set up.

Table I. Parameters of the films, having the composition $Y_{3-x}\,La_x\,Fe_{5-y}\,Ga_y\,O_{12}$, which have been used in the experiments. (t=thickness, l=characteristic length, $4\pi M_s$=saturation magnetization, $K_1$= cubic and $K_u$=uniaxial anisotropy constant.)

| Sample | x | y | t [µm] | l [µm] | $4\pi M_s$ [G] | $K_1$ [erg/cm³] | $K_u$ [erg/cm³] |
|---|---|---|---|---|---|---|---|
| 1 | 0.15 | 1.4 | 3.6 | 2.1 | 51 | -495 | 1350 |
| 2 | 0.14 | 1.23 | 2.3 | 0.69 | 98 | -820 | 2040 |
| 3 | | 1.2 | 4.7 | 0.5 | 153 | -950 | 2550 |

time, so that the Faraday rotation is doubled. The light is then detected by a TV camera after having passed the analyzer A, set at 45° with respect to the polarizer P. The output signal of the camera can be displayed on an oscilloscope for any selected scanning line. From this signal we can compute the Faraday rotation and thus the precession angle θ in the film. To a first approximation θ is given by

$$\cos\theta = 1 - 2(I(0^\circ) - I(\theta))/(I(0^\circ) - I(180^\circ)) \qquad (1)$$

where $I(\theta)$ is the TV signal induced by the light which passed through those regions of the crystal where the spins precess at an angle θ, while $I(0^\circ)$ and $I(180^\circ)$ correspond to regions where the spins are parallel or antiparallel oriented with respect to the bias field.

For large bias fields the film becomes magnetically saturated and the ferrimagnetic resonance frequency $f_R$ is approximately given by

$$f_R = \gamma(H_o - 4\pi M_s + H_a) \qquad (2)$$

where $\gamma$, $H_o$, and $H_a$ are the gyromagnetic ratio, bias field, and anisotropy field respectively. The resonance frequency decreases with decreasing bias field as shown in fig. 2. At $H_o = H_s$ domains nucleate in the film and the frequency increases again. Within a few MHz the resonance frequency is independent of the orientation of domains with respect to the microwave structure and there occurs no discontinuity of $f_R$ at $H_s$. The resonance transmission, however, depends very critically on the orientation of the domains as shown in fig. 3: If stripe domains are aligned parallel to the microwave lines we observe a high transmission while at perpendicular alignment the transmission is very low, but occurs nearly at the same frequency. In view of this experiment it is obvious that magneto-

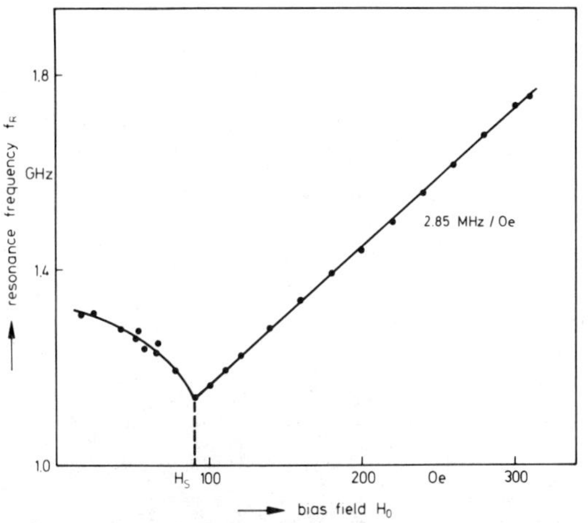


Fig. 2. Measured ferrimagnetic resonance frequency of sample No. 3 versus bias field.

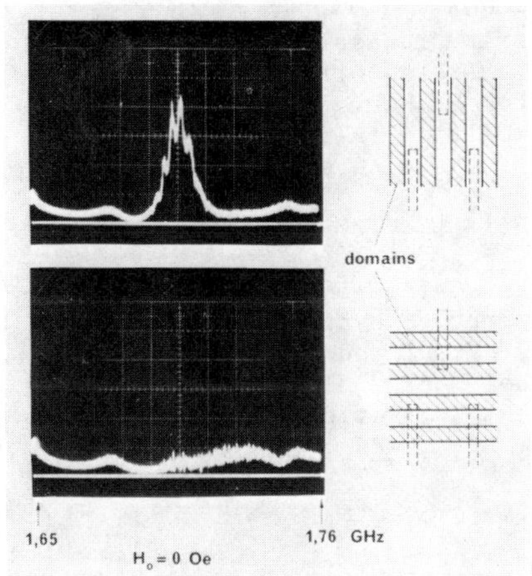


Fig. 3. Resonance transmission in sample No. 1 at two different orientations of parallel stripe domains with respect to the microwave structure.

static spin waves which are responsible for the transmission do not pass domain walls. Furtheremore it is observed that strong spin precessions can also be excited off resonance in a wide frequency band of several 100 MHz bandwidth around the resonance frequency.

## GENERATION OF BUBBLE AND RING DOMAINS

In bulk materials of low linewidth the precession angle $\Theta$ is strongly limited by parametric excitation of short wavelength spin waves; typically $\Theta$ does not exceed a few degrees. In thin films, however, the instability threshold is much higher as shown by Bendik, Kalinikos and Chartorizhskii[1]. Furtheremore the threshold can also be increased by exciting the spin precession locally because in this case the precession relaxes much stronger to magnetostatic spin waves outside the excitation region (two magnon process) than to short wavelength spin waves inside the excitation region (three and four magnon processes). Thus we can obtain very high precession angles in the small excitation region between the two short circuits of the microwave structure (fig. 1b). At high bias fields we observe precession angles up to 150°. However, if the bias field is lowered below the collapse field of bubble domains, such large values of $\Theta$ cannot be reached anymore. As soon as $\Theta$ exceeds a critical value $\Theta_{c1}$ the spins tip over and create a bubble. The measured values of $\Theta_{c1}$ are shown in fig. 4 by the dashed curve. These critical angles are nearly independent of the bias field and equal about 44°.

The spins keep on precessing in the bubble thus created, because the resonance frequency is

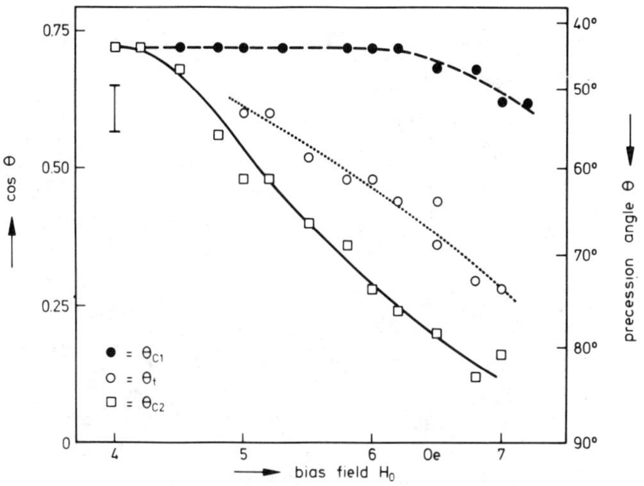


Fig. 4. Measured critical precession angles for bubble and ring generation induced by microwave fields in sample No. 1.

nearly unchanged (fig. 2 and 3), albeit with a larger precession angle $\Theta_t$. The measured values of $\Theta_t$ are shown by the dotted curve in fig. 4. This increase of the precession angle after bubble generation is probably caused by a local build up of energy due to a reflection of magnetostatic spin waves at the domain wall (fig. 3). If at high bias fields the precession angle is further increased, a second critical value $\Theta_{c2}$ is reached where the spins again tip over and create a new bubble within the bubble domain created first, i.e. we obtain a ring domain. The measured values of $\Theta_{c2}$ are shown in fig. 4 by the solid curve. In the low field range, however, $\Theta_{c1}$ and $\Theta_{c2}$ are equal. This behaviour can be explained as follows: At low bias fields the generated bubble has a large diameter which is even larger than that of a static bubble due to the spin precession in its interior[2]. As in this case the excitation region is small compared with such a large bubble we have approximately the same situation as before when a bubble was created in a saturated film.

Therefore at low bias fields the spin precession at the angle $\Theta_{c1}$ is unstable and the spins continue to tip over, creating every time a new bubble concentric with the bubbles created before. Thus we obtain consecutively a bubble, a ring domain, a bubble within a ring (ring bubble), two concentric rings and so on. Every time a bubble is created all ring diameters increase due to the magnetostatic repulsion of the bubble. But with an increasing number of rings the magnetostatic reaction of the ring system on the bubble at the centre increases too, so that finally the centre-bubble becomes small and the bubble generation stops. It can again start if the bias field is further lowered, causing the rings to expand, or by increasing the precession angle. In this way we can obtain a large system of concentric rings; two examples are presented in fig. 5 (slightly different microwave structures have been used in these cases). Such ring systems are stable at low bias fields and even at vanishing microwave power. If such a static ring system is subjected to an increasing bias field all ring diameters decrease and the rings collapse one after the other starting with the innermost ring.

If at high bias fields the bubble diameter becomes much smaller than the excitation region a created bubble will not stay any longer at that position. It is ejected out of the excitation region along a spiral path as shown in [3]. Two examples of this effect are given in fig. 6 where bubbles have been created in between concentric systems of rings; in fig. 6b) bubbles are circulating around the excitation region in the two innermost rings.

The microwave generation of a bubble domain in

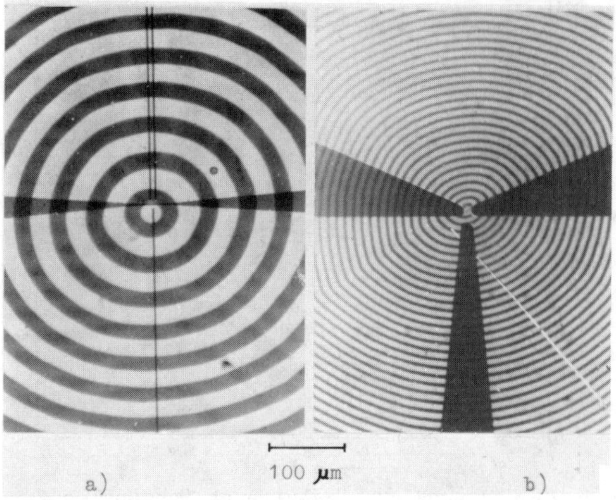


Fig. 5. Systems of concentric ring domains; (a) sample No. 1 and (b) sample No. 2.

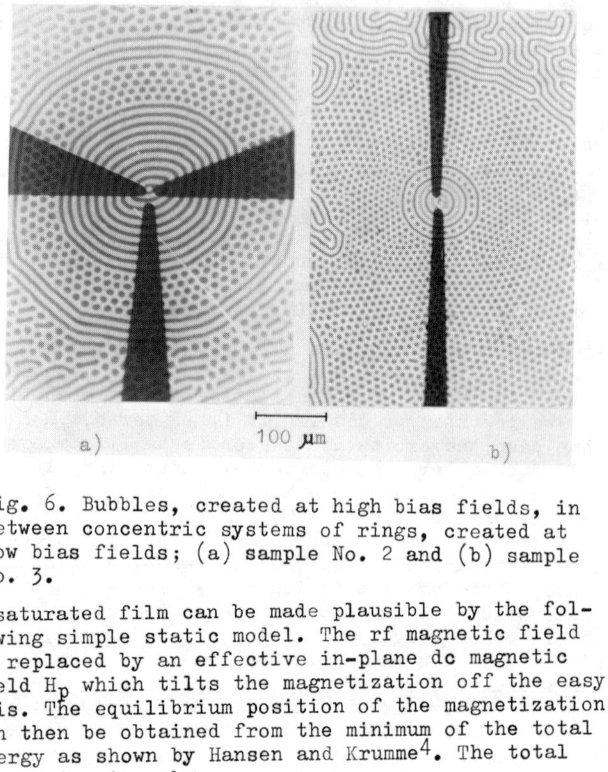


Fig. 6. Bubbles, created at high bias fields, in between concentric systems of rings, created at low bias fields; (a) sample No. 2 and (b) sample No. 3.

a saturated film can be made plausible by the following simple static model. The rf magnetic field is replaced by an effective in-plane dc magnetic field $H_p$ which tilts the magnetization off the easy axis. The equilibrium position of the magnetization can then be obtained from the minimum of the total energy as shown by Hansen and Krumme[4]. The total energy is given by

$$E_{tot} = -\vec{M}(\vec{H}_p + \vec{H}_o) + 2\pi M_s^2 \cos^2\theta + K_u \sin^2\theta + K_1(\alpha_1^2\alpha_2^2 + \alpha_1^2\alpha_3^2 + \alpha_2^2\alpha_3^2) \tag{3}$$

where $\theta$ is the angle between the magnetization vector $\vec{M}$ and the film normal (corresponding to the precession angle) and the $\alpha_i$ are the direction cosines of $\vec{M}$ with respect to the cubic axes.

Applying $H_p$ along the $[11\bar{2}]$ direction in the $(1\bar{1}0)$ plane, which is an easy plane of magnetization, we obtain the results shown in fig. 7. With decreasing in-plane field $\theta$ decreases. At $H_p$=-300 Oe the total energy for $\theta_1 = -22^o$ and $\theta_2 = -151^o$ is the same; thus both states are degenerate and the spins can jump between $\theta_1$ and $\theta_2$. As $H_p$ is large compared with $H_o$ the angles $\theta_1$ and $\theta_2$ do not depend strongly on the bias field.

The measured value $\theta_{c1} = 44^o$ has to be compared with the calculated values $|\theta_1| = 22^o$ and 180 - $|\theta_2| = 29^o$. The experimental value is somewhat larger due to the existence of an energy barrier between the two states characterized by $\theta_1$ and $\theta_2$.

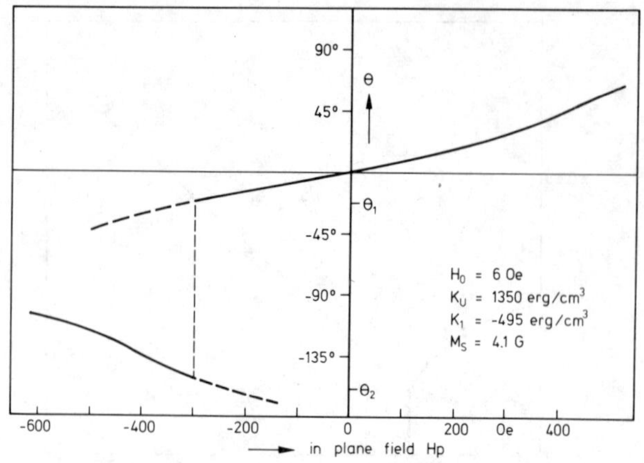


Fig. 7. Calculated equilibrium angle $\theta$ of the magnetization with respect to the $[111]$ direction if an in-plane dc magnetic field $H_p$ is applied along the $[11\bar{2}]$ direction in the plane of a $[111]$ garnet film. (Sample No. 1.)

For spin precessions within bubbles which are not large compared with the excitation region the simple model of equ. (3) can no longer be applied. The magnetostatic energy changes considerably and larger angles are necessary to create a ring domain.

## STABILITY OF SINGLE RING DOMAINS

Single static ring domains have been observed before by Druyvesteyn and de Jonge[5]. For these rings they calculated the total energy $E_{tot}$; extending their calculation to the more general case of n concentric static rings one obtains:

$$E_{tot} = 8\pi^2 M_s^2 t^3 \sum_{j=k}^{2n} \Big[ r_j l/t + (-1)^j (h-1) r_j^2 - 2r_j^3 \left(F(1,1/r_j) - 4/(3\pi)\right) + 4 \sum_{i=j+1}^{2n} (-1)^{i+j} r_j r_i^2 \left(F(r_j/r_i,0) - F(r_j/r_i,1/r_i)\right)\Big] \tag{4}$$

where k determines whether there is a bubble at the centre (k=0) or not (k=1); h is the reduced bias field $h=H_o/4\pi M_s$ and the $r_i$ are the reduced radii of the rings, measured in units of t. The function F is:

$$F(a,b) = \int_0^\infty J_1(ax)\, J_1(x) \exp(-bx)\, dx/x^2 \tag{5}$$

and $J_1$ is the Bessel function of the first order. The minimum of $E_{tot}$ determines the ring radii.

The theoretical dependence of the diameters of a single static ring on the bias field can be obtained from equ. (4) with n=k=1. For the sample No. 1 we obtain the result shown by the dashed line in fig. 8. The stability range is very narrow, about 0.1% of $4\pi M_s$. However, the double headed arrow in fig. 8 indicates the range of the bias field where single static ring domains can be observed experimentally; they are stable due to the coercivity of the material which is neglected in the calculations. If the bias field is varied beyond these static stability limits the rings suddenly collapse or run out.

At higher bias fields microwave stabilized single rings are observed. For the same sample No.1 the measured dependence of the ring diameter on the bias field is shown in fig. 8 by the solid curves; the parameter p is the input power. Along these curves the coercivity is strongly reduced. The curves terminate at high bias fields where the rings collapse and at low bias fields where the rings become elliptically distorted or where bubble domains are generated inside the rings. The stability range of these single ring domains extends far

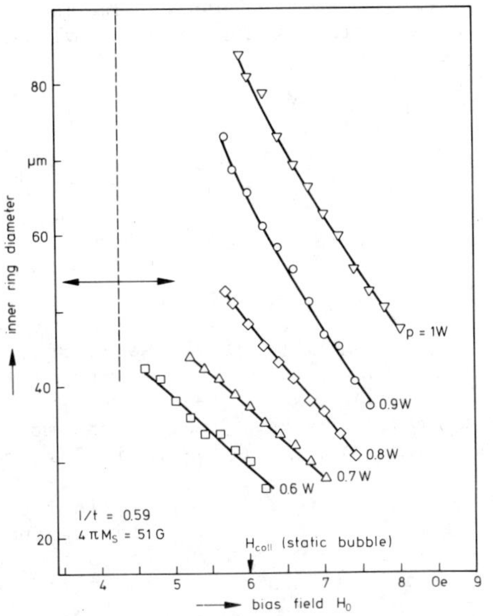


Fig. 8. Measured dependence of the inner ring diameter of microwave stabilized single ring domains on the bias field for various values of the input power p, using sample No. 1. Dashed line and double headed arrow are theory for static single ring domains and extended range where they are observed due to coercivity. (See text.)

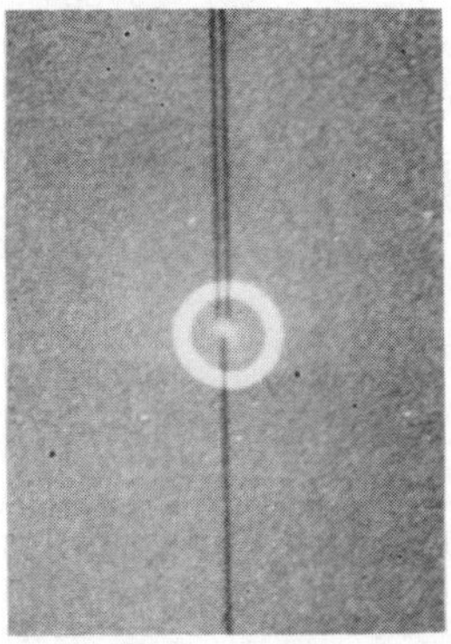


Fig. 9. Photograph of a microwave stabilized single ring domain. (Sample No. 1.)

beyond the collapse field $H_{coll}$ = 6.0 Oe of static bubbles.

Fig. 9 shows a photograph of a microwave stabilized single ring domain. The bright region in the interior clearly points to a strong spin precession inside the ring. Fig. 10 gives the spatial distribution of the precession angle along a diameter of the ring. The maximum precession angle $\Theta_{max}$ at the centre is about 90°; it does not change very much with the bias field.

In order to get a plausible idea of the stabilizing forces of microwaves on single ring domains we use a very simple heuristic model. It is a pure static model and only takes into account the precession induced change of the dc magnetization inside the ring. A hint for that model can be obtained from the static system of a ring bubble. Fig. 11 shows the calculated field dependence of the inner diameter of a static bubble and of a static ring bubble of sample No. 1 where equ. (4) was used with n=k=1 and with n=1, k=0 respectively. The static stability range of the ring bubble is much larger than that of the ring, particularly for the higher bias fields. This higher stability against collapse is due to the magnetostatic repulsion of the bubble at the centre; on the other hand the bubble does not affect very much the stability against run out. In fig. 11 the stability curves end at the corresponding collapse fields. To get a further approximation to the measurements the bubble inside the ring is assumed fixed at its collapse radius. The ring domain around this bubble is then stable

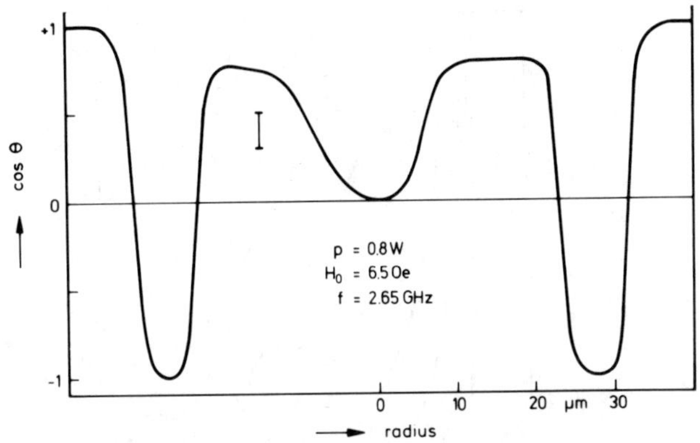


Fig. 10. Measured spatial variation of the precession angle Θ along a diameter of a microwave stabilized ring domain. (Sample No. 1.)

up to high bias fields as shown in fig. 12 by the dashed curve marked 'fixed ring bubble'. (Note the large difference in scaling of the horizontal axes of fig. 11 and 12.) The stability curve of such an artificial ring already comes close to the region where experimentally microwave stabilized ring domains are observed; three experimental curves of fig. 8 have been included in fig. 12 (solid curves).

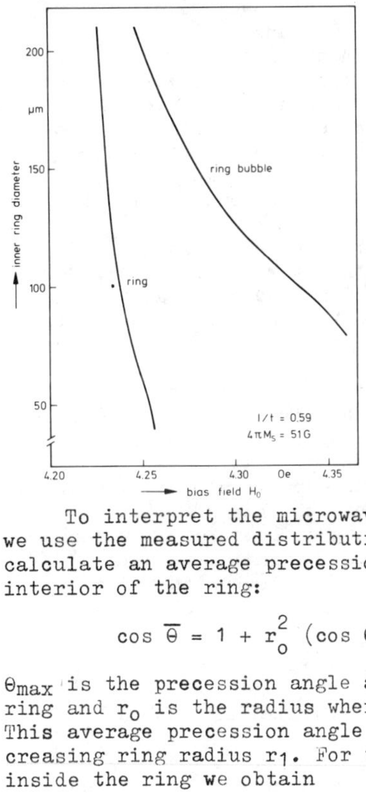


Fig. 11. Calculated dependence of the inner ring diameter on the bias field for a static ring and a static ring bubble. (Sample No. 1.)

To interpret the microwave enforced stability we use the measured distribution of Θ in order to calculate an average precession angle $\overline{\Theta}$ over the interior of the ring:

$$\cos\overline{\Theta} = 1 + r_o^2\,(\cos\Theta_{max} - 1)/r_1^2 \qquad (6)$$

$\Theta_{max}$ is the precession angle at the centre of the ring and $r_o$ is the radius where $\cos\Theta = (\cos\Theta_{max})/2$. This average precession angle decreases with increasing ring radius $r_1$. For the dc magnetization M inside the ring we obtain

$$M = M_s\cos\overline{\Theta} \qquad (7)$$

and for the total energy $E_{tot}$

$$E_{tot} = 8\pi^2M_s^2t^3\Big[\,l(r_1+r_2)/t + (h-1)(r_2^2-qr_1^2) \qquad (8)$$
$$- 2r_1^3q^2(F(1,1/r_1)-4/(3\pi)) - 2r_2^3(F(1,1/r_2)-4/(3\pi))$$
$$- 4r_1r_2^2q(F(r_1/r_2,0) - F(r_1/r_2,1/r_2))\Big]$$

where $q = (1+\cos\overline{\Theta})/2$.

In equ. (8) the dependence of the wall energy on $\overline{\Theta}$ (the first term in the bracket) has been neglected because the actual precession angle at the inner wall is less than $\overline{\Theta}$. Using for $r_o$ and $\Theta_{max}$ the values which can be derived from the measurements for p=0.8 W we obtain the solution shown in fig. 12 by the dashed curve marked $c_1$. However, for

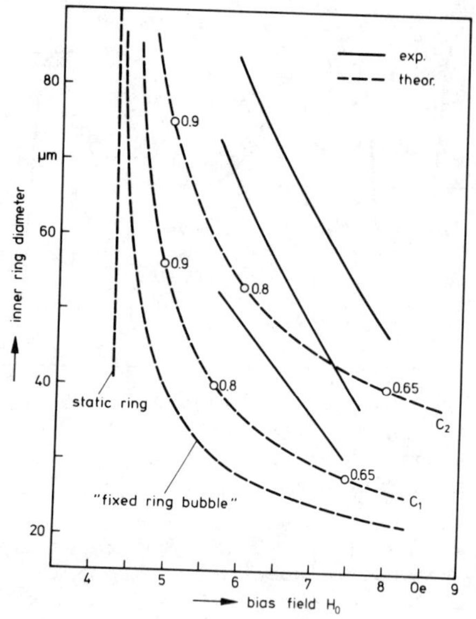


Fig. 12. Calculated dependence of the inner ring diameter of single rings on the bias field for various cases explained in the text. (Sample No.1.)

the other values of the input power we obtain curves which are close to $c_1$ because $r_0$ and $\Theta_{max}$ do not change very much with the input power between p=0.7 and p=1W. On the other hand the measurements show that the precession angle in the vicinity of the inner domain wall increases slightly with increasing input power. Due to the finite resolution of our set up we cannot measure this effect quantitatively yet; taking it into account qualitatively by a larger effective radius $r_0$ in equ. (6), we obtain e.g. the dashed curve marked $c_2$ in fig. 12 where $r_0$ is 33% larger than for the curve $c_1$. Some values of cos $\overline{\Theta}$ are noted at the calculated curves.

In the low field region the dependence of the calculated ring diameter on the bias field is about the same as the measured one (fig. 12); in the high field region, however, the calculated ring diameter decreases too slowly with increasing bias field as compared with the measurements. This behaviour is due to the strong increase of $\overline{\Theta}$ at small ring radii (equ. (6)). Obviously the experimental curves are not as strongly influenced by that effect. It seems also evident that the ring diameter depends much more on the precession in the immediate neighbourhood of the wall than on the precession in the interior of the ring. In this case it is important whether the spin waves are reflected at the wall with a change of phase or not.

## DYNAMIC BEHAVIOUR OF SINGLE RING DOMAINS

A single ring domain has two domain walls; for radial oscillations we therefore have two eigenmodes, one in which the ring radii are in phase and one in which they have opposite phases. The eigenfrequencies of these modes can be estimated by the second derivatives of the total energy in the two directions $r_2 = r_1 + (r_{20}-r_{10})$ and $r_2 = -r_1 + (r_{20}+r_{10})$ where zero denotes the equilibrium values. It turns out that for our crystals the frequency of the second mode is much higher than that of the first one. At the static equilibrium the ratio of the two eigenfrequencies is about 100; if we use equ. (8) of our simple model to calculate this ratio for the microwave stabilized rings we find that it is smaller but still at least 20. Thus if the microwaves stabilizing a single ring domain are sinuoussoidally amplitude modulated at low modulation frequencies the ring will oscillate rigidly, i.e. with approximately constant ring width.

To analyse these oscillations we can use the procedure developed by Kaczér and Tomás[6] to study the oscillations of single bubble domains. Assuming a harmonic oscillation of the ring one obtains for the amplitude $\delta r$:

$$\delta r/\,\delta r_0 = f_0^2\,((f_0^2 - f^2)^2 + 4b^2f^2)^{-1/2} \qquad (9)$$

where $f_0$ is the eigenfrequency, f the modulation frequency, b is the damping term, and $\delta r_0$ is the amplitude at f=0. The damping term b is connected with the straight wall mobility $\mu$ by

$$b = M_s/(\mu\, m) \qquad (10)$$

where m is the Döring-mass of the wall. Fig. 13 shows the experimental curves of $\delta r/\,\delta r_0$ for various bias fields. During all measurements $\delta r_0$ and $r_{10}$, the ring radius at zero modulation amplitude, had been kept constant while the ratio $\delta r_0/r_{10}$ was about 0.2.

It was attempted to fit equ. (9) to the experimental curves by the method of least squares in order to determine $f_0$ and b simultaneously. However, it turned out that the minimum of the root-mean-square depends so weakly on $f_0$ and b that the results are very inaccurate. Using the most accurate measurements at $H_0$ = 29 and 31 Oe we obtain values for the mobility $\mu$ which vary between $2\cdot10^3$ and $10\cdot10^3$ cm/(s Oe). The calculated value of $\mu$, using the relation $\mu = \gamma\,\Delta/\alpha$ where $\Delta$ is the wall width and $\alpha$ the Gilbert damping parameter which was determined from the ferrimagnetic resonance ($\alpha$ = 0.0035), is $\mu = 6.8\cdot10^3$ cm/(s Oe); the corresponding damping term b is 480 kHz. Using this calculated damping term we could fit equ. (9) to the measurements and obtained for the eigenfrequencies the results shown in fig. 14. The decrease of $f_0$ with increasing bias field is expected, because the curvature of the energy surface (equ. (4) and (8)) at the minimum decreases too when we approach the collapse field of the ring. (Note that the collapse field of a static bubble is 28 Oe.)

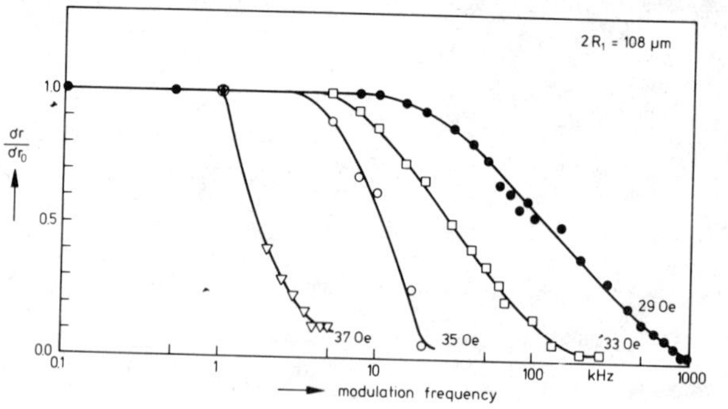


Fig. 13. Measured oscillation amplitude of a single ring versus modulation frequency for some values of the bias field. (Sample No. 2.)

In another experiment we measured the collapse time of single ring domains by switching off the microwaves. The critical switch-off time $T_c$ during

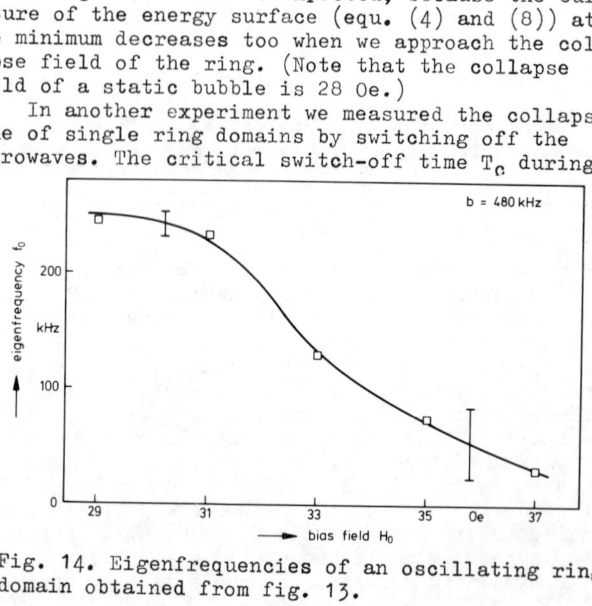


Fig. 14. Eigenfrequencies of an oscillating ring domain obtained from fig. 13.

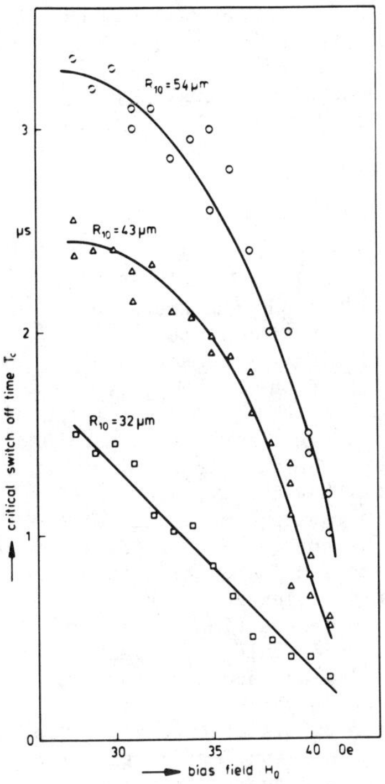


Fig. 15. Critical switch-off time $T_c$ of a single ring domain versus bias field for three different starting radii. (Sample No. 2.)

which the rings reach the collapse radius is shown in fig. 15 for three different starting radii.

From the data of fig. 15 we can calculate a mean velocity $\overline{v}$ of the collapsing ring by

$$\overline{v} = (R_{10} - R_{1coll})/T_c \qquad (11)$$

As $R_{1coll}$ is not known for this dynamic process we use the statically measured collapse radius in equ. (11). In this way we obtain the mean velocities shown in fig. 16. These velocities increase strongly with increasing bias field. From the derivative of equ. (4) with respect to $r_1$ one can calculate an effective drive field for the collapsing rings and thus, using the mobility derived above, a mean velocity. These calculated velocities lie in the same range as those obtained from equ. (11).

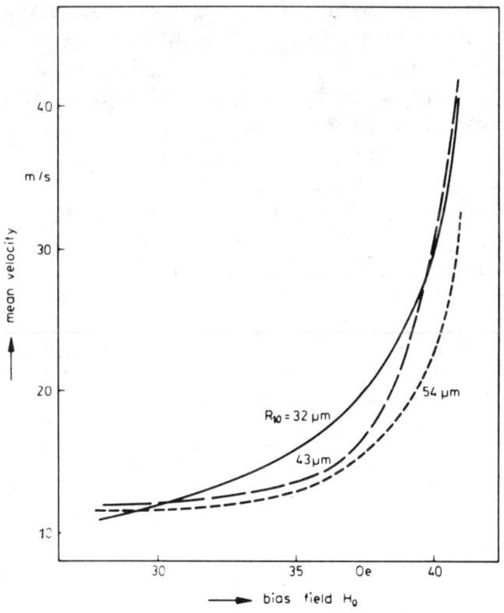


Fig. 16. Mean velocities calculated from the data of fig. 15.

## CONCLUSION

The presented technique provides a new and simple tool for generating and stabilizing ring domains which are otherwise highly unstable. However, there are still large problems in understanding the basic interactions between precessing spins and domain walls. Especially dynamic effects (e.g. spin wave radiation pressure) have not been taken into account yet. The modulation technique may be applied to measure the frequency response of domain walls, or it may be used to check the quality of bubble films with respect to point defects where a large area can be inspected at a time. Possibly the patterns of concentric ring systems may find an application in light diffraction experiments.

## ACKNOWLEDGEMENTS

I thank Dr. H.J.Schmitt, Dr. W.Schilz, and Dr. P.Hansen for many stimulating discussions, W.Klossner for technical assistance, and Dr. W. Tolksdorf, G. and I.Bartels for the preparation of the garnet films.

## REFERENCES

1) O.G.Bendik, B.A.Kalinikos, and D.N.Chartorizhskii, Sov. Phys. Solid St. 16, 1785 (1975).
2) To be published.
3) H.Dötsch and H.J.Schmitt, Appl. Phys. Lett. 24, 442 (1974).
4) P.Hansen and J.P.Krumme, J. Appl. Phys. 44, 2847 (1973).
5) W.F.Druyvesteyn and F.A. de Jonge, Philips Res. Repts. 25, 415 (1975).
6) J.Kaczér and I.Tomás, phys. stat. sol. (a) 10, 619 (1972).

## THE SPIN WAVE SPECTRUM OF VERY HARD MAGNETIC BUBBLES

A.A. Thiele
Bell Laboratories, Murray Hill, N.J. 07974

ABSTRACT

The spin wave spectrum of an infinite planar magnet domain wall containing closely packed Bloch lines (as occurs in a very hard magnetic bubble) has been obtained. Although the wave functions are hypergeometric functions the scattering coefficients connecting the two asymptotic regions contain functions no more complex than hyperbolic tangents. The functions are normalized and a representation of the Dirac delta function in terms of them is given. It is found that a spin wave impinging on the domain wall is partially reflected and partially transmitted. In transmission, the component of the impinging spin wave k vector parallel to the Bloch lines and the k vector magnitude are conserved. The k vector component parallel to the wall but transverse to the Bloch lines is transformed according to $k_y \rightarrow k_y - 4\pi/s$ where s is the spatial period of the Bloch pairs. When these conditions cannot be simultaneously satisfied, the spin waves are totally reflected. The Bloch lines also remove part of spin wave spectrum localized to the wall which may account for the anomalously high mobilities observed in very hard bubbles. An expression for spin wave radiation pressure which in some cases differs from the usually accepted expression is developed. In general it is concluded that the understanding of dynamics experiments is enhanced by consideration of the second order (spin wave) theory in addition to the usual first order theory.

## BUBBLE LATTICE AND SPIDER-WEB-LIKE DOMAIN PATTERN IN MnBi PLATELET

TETSUZO KUSUDA, SHIGEO HONDA and TADAYOSHI IDESHITA
Department of Electronics, Hiroshima University, Hiroshima, Japan.

ABSTRACT

When the single crystal MnBi platelet thicker than 3μm is pricked by a needle under zero field, it is demagnetized dynamically producing the bubble lattice. The bubble lattice formation is suppressed by an external field; a radial stripe pattern or a spider web like pattern depending on the field polarity is formed. The phase velocity of the wall propagation during lattice formation was estimated as about 50 m/s by the pulse field method. The data obtained is not the instantaneous velocity but the average velocity for cxtremely fluctuating motion, and is almost the same order of magnitude as the critical velocity by Walker (150 m/s for tip motion). And this phase velocity is much larger than the Slonczewski's threshold velocity $V_p$ for already nucleated Bloch line. The most plausible explanation for this discrepancy is that the bubble lattice will be formed as the result of the relaxation oscillation which is occurring during the whole period of bubble lattice formation.

Once the reverse domain is nucleated, the elongated strip domain will expand due to the internal field acting on the wall from the saturated domain. Because the internal field at this cylindrical domain is much larger than the Walker instability field, the maximum velocity obtainable for this pure Bloch wall is limited by the Walker velocity. This high velocity should be maintained until the dynamic conversion will take place. Assuming the transit time for high speed mode as 10 ns which is experimentally observed by Vella Coleiro, the wall displacement during this time is 1.5μm,which is very close to one segment length of the hexagonal bubble lattice and this value should be noted for the independence of the film thickness.

As soon as the winding of vertical Bloch line happens, the vertical Bloch line will be left behind in the flank of the strip, with the right-handed and left-handed lines being segregated on opposite sides of the strip. When the film becomes thicker, the rate of generation of Bloch lines becomes stronger and the closely packed bunching of Bloch line should happen at the tip of the moving wall. At the same time, the momentum of the moving bubble increases with proportion to vertical Bloch lines stored at domain tip. Very soon the critical moment for abrupt annihilation of closely packed Bloch line should come. Presumably, this phenomenon is very similar to the phenomenon which is observed in superhard bubble and called as static conversion by Philips research group.

# INITIAL RAPID DOMAIN WALL MOTION IN MAGNETIC BUBBLE GARNET MATERIALS*

G. J. Zimmer, L. Gal
Central Research Institute for Physics, H-1525, Budapest POB 49, Hungary

F. B. Humphrey
California Institute of Technology, Pasadena, California 91125

## ABSTRACT

The initial motion of domain walls in Tm-doped bubble garnet material subjected to a step change in bias field has been investigated on bubbles by transient photography and on walls by optical averaging. It was found that the average velocity of radially expanding bubble walls varied greatly during the first 20 nsec. Velocities in the range from 22 m/sec to 0.5 m/sec were observed with no correlation noted between pulses. On demagnetized stripe walls, the fast mode velocity is a maximum of 18 m/sec for about 20 nsec at 10 Oe changing smoothly to 14 m/sec for about 10 nsec at 70 Oe. Attempts to explain these results using the Walker analysis were unsuccessful. It was found that once the sample converted to the slow mode of wall motion, the fast motion could not be initiated unless the walls were at rest for some time.

## INTRODUCTION

Various attempts have been made to characterize the initial stages of wall motion and the Walker breakdown or dynamic conversion that might be occurring. Vella-Coleiro(1), using the bubble collapse technique, found a characteristic time for dynamic conversion and a peak in the average velocity. Other investigations (2,3,4) have observed an initial fast motion of walls in garnet materials more directly using magneto-optic averaging methods(5). Qualitatively, these observations all gave similar results. The wall moves fast initially then slows or becomes oscillatory. Some overlap between Moody et al(3) and Vella-Coleiro(1) for Q=3 samples, indicate an inconsistency. Vella-Coleiro found velocities very close to the Walker limit ($V_w$) for times like 10 nsec and displacements as large as 1 µm; whereas Moody et al(3) found only $V_w/5$. Since the creditability of the bubble collapse method has been shown(6), a closer look at this initial phase of domain wall motion by direct observation with high-speed photography and by magneto-optical averaging performed on the same sample seemed to be justified.

## EXPERIMENTAL

A matched pair of nominally identical Tm-doped garnet LPE samples, one ion-implanted and the other not, was used. Relevant parameters were given previously(6). A sampling optical microscope(7) was used to take single exposure photographs of the bubble transient shape in four photograph combinations. First, the bubble was photographed in equilibrium with the bias field and then 20 nsec after the 10% point of the leading edge of a 170 nsec bias-field-reducing pulse. Again, it was photographed in equilibrium with the bias field and finally 170 nsec after the application of the same size bias field pulse used for the second photographs. This sequence, then, compares the radial expansion during 20 nsec and during 170 nsec on two different but sequential pulses. For the data in this paper, 25 sequences were taken for a fixed set of parameters and each bubble in the sequence was measured 25 times independently using a semi-automatic measuring microscope. The deviation of the bubble radius for each set was σ = 0.03 micron.

The magneto optic signal averaging experiment was similar to others(2-5). A continuous 20 mw 0.1 mm diameter He-Ne laser beam (unfocused) was directed through the garnet layer. Advantage is taken of the repetitious nature of the sample by placing the detector about 4 m from the garnet so that only the 0'th order (undifracted) beam was incident on the analyzer, greatly increasing the quality of the dark condition. A high gain photomultiplier is used in conjunction with a sampling amplifier with 3 nsec aperture. The magnetic field pulse is a few hundred nanoseconds long with a 7 nsec rise time and a 900 Hz repetition rate. The sample gate runs at twice this frequency so that a sample is taken at a known time during the fast rise pulse and another when the pulse is absent(2). The 900 Hertz square wave out of the sample gate has an amplitude proportional to the light level on the photomultiplier during the sample time. Therefore, in principle, it would be possible to evaluate the change in magnetization from this signal. However, to keep the light level on the photomultiplier as low as possible, a null method was used. In this method the domain structure is adjusted, quasi-statically, to match the dynamic movement. The sample gate output is amplified with a phase-sensitive detector amplifier and used to automatically adjust the amplitude of an additional 900 Hz, low rise time (>100µsec), square wave bias modulation sychronized to alternate with the fast rise time pulse and to be either at full magnitude or zero when the sample gate opens. It has been shown(2) that as long as this quasi-static bias modulation is $< 0.1\ (4\pi M)$, the stripe width will change but the periodicity will be preserved. The dynamic configuration of the domain pattern at a particular instant after the application of the fast rise time step change in bias field is then represented by the magnitude of the quasi-static current. In these experiments, the quasi-static bias change was always $< 0.05\ (4\pi M)$; superimposed on the quasi-static pulse was a free running 100 Kc, 10 Oe, 100 nsec pulse train to reduce the effect of coercive force. The response time of the optical system is 3 nsec.

## RESULTS

Typical results from the transient photographs are shown in Fig. 1. The average velocity of a bubble wall during the first 20 nsec (dots) and during the first 170 nsec (crosses) is shown for 25 independent four frame sequences. Each sequence was handled independently. Each point is obtained by measuring the transient diameter and subtracting it from the starting diameter for that particular expansion. Each of the

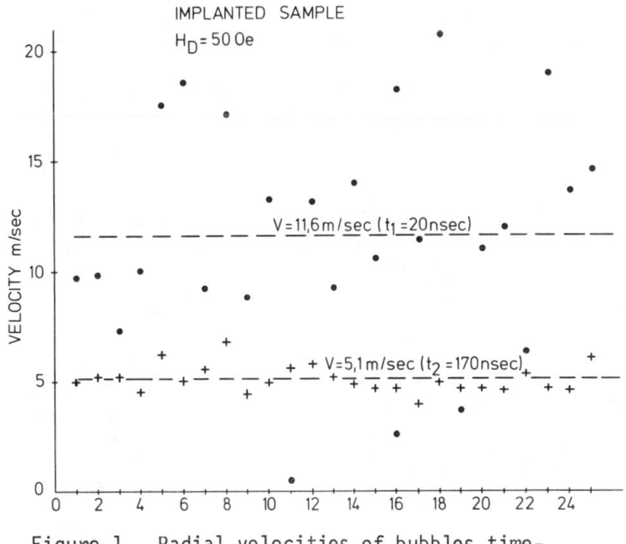


Figure 1. Radial velocities of bubbles time-averaged over the first 20 nsec (dots) and 170 nsec (crosses) of motion found in 25 independent expansions.

four photographs in a sequence was measured 25 times to minimize scatter due to the measuring technique. The deviation for a particular diameter was $\sigma$=0.03μm. In Fig. 1 the large scatter during the first 20 nsec can be clearly seen as well as the lack of correlation between this velocity and the average velocity for 170 nsec taken on the very next pulse. The scatter of the 170 nsec velocities is consistent with the assumption that it is caused by the extremes in initial velocity seen in the 20 nsec points. Also indicated is the 25 point average showing that the velocity measured during the first 20 nsec averages twice that measured during 170 nsec. The scatter in the velocity during the first 20 nsec was found in all the data with extremes of 22 m/sec (Walker velocity 30 m/sec) and 0.5 m/sec recorded.

The averaging experiment showed this same initial fast motion with a transition to a slower mode. A summary of the fast motion for the implanted sample is shown in Fig. 2. Here the initial velocity, calculated from the maximum slope of the observed average displacement at various time curves, is shown as a function of drive. It can be seen that, after a steeply rising early portion, the average velocity reaches a maximum of about 18 m/sec at 10 Oe and gradually decreases to 13.5 m/sec at 70 Oe. The fast period lasts about 20 nsec for 10 Oe drive decreasing to about 10 nsec at 70 Oe drive. The transition from fast to slow mode was found to be the most abrupt at the maximum velocity with the steep portion shorter and the transition softer as the drive increased. The slow mode saturated at 5 m/sec above about 12 Oe in agreement with earlier findings[6]. Data on the non-implanted sample was similar with more scatter.

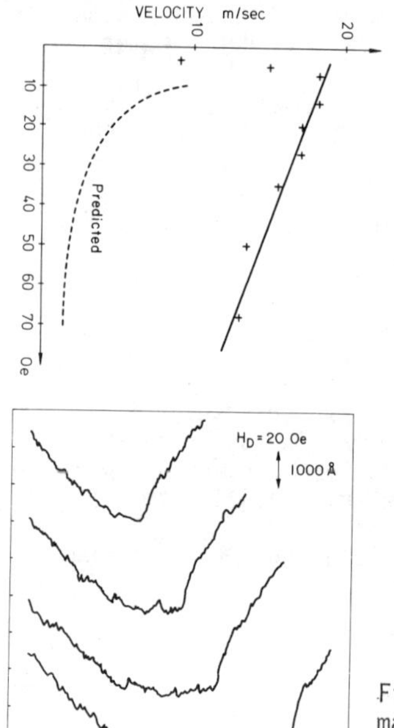


Figure 2. Initial velocity vs drive observed by optical averaging. Dotted line is the average velocity as predicted[9].

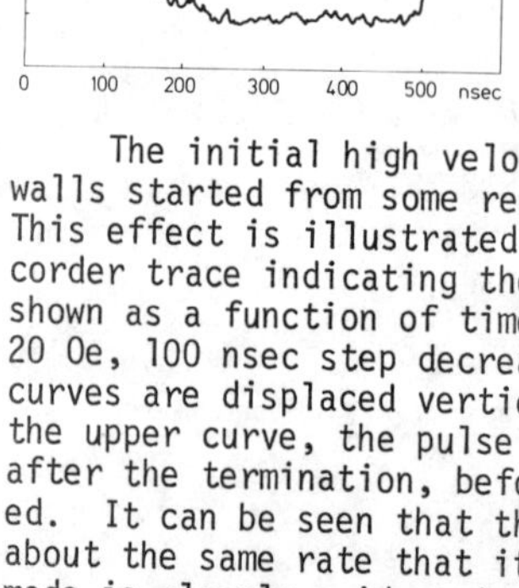


Figure 3. Position of domain walls vs time with different delays between pulses

The initial high velocity was only observed if the walls started from some reasonable equilibrium state. This effect is illustrated in Fig. 3. Here the recorder trace indicating the position of the walls is shown as a function of time after the completion of a 20 Oe, 100 nsec step decrease in bias field. The curves are displaced vertically for comparison. For the upper curve, the pulse field is reapplied 200 nsec after the termination, before equilibrium is established. It can be seen that the bubble expands again at about the same rate that it was shrinking. The fast mode is clearly evident in the lower curve presumably because the walls had time to reach some equilibrium structure so that the fast mode could be excited. The fast rise and the break over to a slower mode that is seen at 500 nsec on the lower curve is typical of the data from which the velocity (slope of fast change portion) was reported in Fig. 2 except that the integration time constant was longer, and the time scale was greatly expanded. It should be noted that the long lifetime of the fast portion seen here is electronic in origin.

## DISCUSSION

Domain walls in bubble material that are subjected to a step change in bias field exhibit an initial fast movement, then convert to a slower movement and stay converted until the motion has stopped for some time. Reversing the direction of wall motion is not enough to dissipate the slow structure. Other results on the slow motion are consistent with a more detailed study[6] and will not be discussed here. For bubbles starting from at rest, there is considerable scatter from pulse to pulse in the expansion during the first 20 nsec. As indicated by the average movement during this time, it is more likely that the conversion time rather than the velocity varies randomly from event to event[1], although the possibility that both vary has not been excluded.

Since the fast mode is related to a structure characteristic of a wall at rest and a one-dimensional model should be much faster than a two-dimensional one, it is tempting to apply the essentially identical results of Schryer and Walker [9] or Bourne and Bartran [10] to this early phase motion. For the samples used here, the critical field is 3 Oe so that both 15 and 50 Oe drive fields should be in the oscillatory region. The wall is expected to reach maximum velocity ($V_W$ = 30 m/sec) in 5 and 1.5 nsec and come to a stop in 10 and 3 nsec, respectively, then move backwards. The velocity averaged over 20 nsec in a single event should not come close to $V_W$ but it does. Also, the ensemble average of the wall positions should show considerable structure as a function of time, especially for the low drives or else the phasing between walls is somehow lost and the velocity averages out. Although the expected structure is well within the space and time resolution of the present experiments, none was found. The structureless initial velocity must then be identified with the average velocity predicted[9,10] and is shown as a dotted line on Fig. 2. As it is evident that neither the drive dependence or the velocity magnitude is close to the experimental data, it is concluded, therefore, that the existing one-dimensional models do not account for the initial fast portion of the domain wall motion.

## REFERENCES AND FOOTNOTES

* Supported in part by the National Science Foundation.
(1) G. P. Vella-Coleiro, AIP Conf. Proc. 24, 595 (1974)
(2) B. E. Argyle and A. P. Malozemoff, AIP Conf. Proc. 10, 344 (1972)
(3) J. W. Moody, R. W. Shaw, R. M. Sandfort and R. L. Stermer, IEEE Trans. MAG-9, 377 (1973)
(4) F. H. de Leeuw, J. Appl. Phys. 45, 3106 (1974)
(5) J. A. Seitchik, W. D. Doyle and G. K. Goldberg, J. Appl. Phys. 42, 1272 (1971)
(6) G. J. Zimmer, L. Gal, K. Vural and F. B. Humphrey, J. Appl. Phys. 46, 4976 (1975).
(7) F. B. Humphrey, IEEE Trans. MAG-11, 1679 (1975).
(8) R. W. Shaw, D. E. Hill, R. M. Sandfort and J. W. Moody, J. Appl. Phys. 44, 2346 (1973)
(9) N. L. Schryer and L. R. Walker, J. Appl. Phys. 45, 5406 (1974)
(10) H. C. Bourne and D. S. Bartran, IEEE Trans. MAG-10 1081 (1974)

# HIGH-SPEED WALL MOTION IN BUBBLE FILMS WITH IN-PLANE ANISOTROPY*

Ernst Schlömann
Raytheon Research Division, Waltham, MA 02154

## ABSTRACT

The effect of an in-plane anisotropy on the static and dynamic wall structure has been calculated for walls that are either parallel ($\psi = 0$) or perpendicular ($\psi = \pi/2$) to the secondary (i.e., in-plane) easy axis. The results are described in terms of a parameter $\eta = \Delta/2\pi M_0^2$ where $\Delta$ is the strength of the in-plane anisotropy and $M_0$ the saturation magnetization. For $\psi = 0$ the critical points of the static wall structure move towards the film surface as $\eta$ increases. For $\psi = \pi/2$ they move towards the mid-plane of the film where they merge for $\eta \to 1$. The critical velocity at which wall motion becomes unstable is strongly dependent on $\eta$. For $\eta < 1$ the lowest velocity threshold is due to Bloch-line nucleation at a critical point. It is given by

$$v_{crit}(\eta) = v_{crit}(0)\,(1 \pm \eta)^{-\frac{1}{2}} \cosh^2(1 \pm \eta)/\cosh^2(1)$$

where $v_{crit}(0)$ is the critical velocity previously calculated by Slonczewski [J. Appl. Phys. 44, 1759 (1973)] and the upper (lower) sign corresponds to $\psi = 0$ ($\psi = \pi/2$). For $\psi = \pi/2$, $\eta > 1$ the velocity threshold is due to punch-through of the already existing Bloch line. For $\psi = 0$ the critical velocity generally increases with increasing $\eta$. For $\psi = \pi/2$ the nucleation threshold at first decreases with increasing $\eta$ and then increases. The punch-through threshold is substantially independent of $\eta$. For $\eta \gg 1$ the critical velocities for a film are substantially the same as for an infinite medium.

## INTRODUCTION

It has previously been suggested that the limiting velocity for domain-wall motion in bubble films is strongly influenced by the presence of an in-plane anisotropy.[1,2] The previous theoretical work is based on a one-dimensional wall model, which is appropriate for walls in an infinite medium, but is apt to be misleading when applied to walls in thin films. The analysis described in the present paper is based upon the two-dimensional wall model which has had considerable

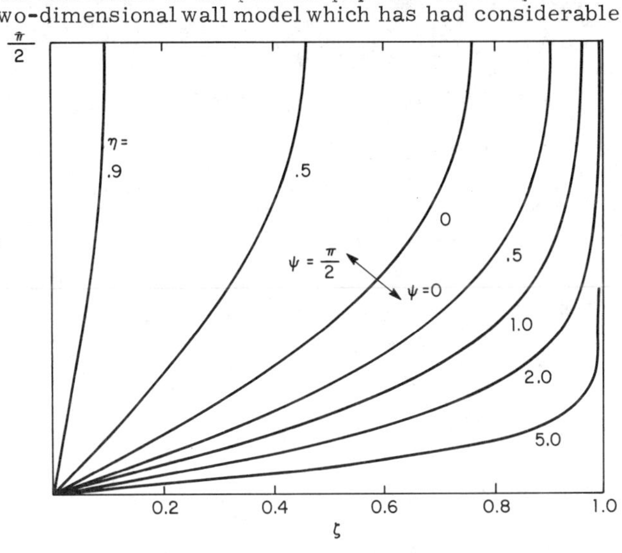


Fig. 1. Wall structure functions $\phi(\zeta)$ for the static case. The curves correspond to different values of the anisotropy parameter $\eta = \Delta/2\pi M_o^2$ and different wall orientations relative to the secondary easy axis. $\zeta = 0$ corresponds to the mid-plane of the film, $\zeta = 1$ to its surface. $\phi(\zeta) = \pi/2$ between the critical point and the surface.

success in accounting for the high-velocity behavior of walls in bubble films without in-plane anisotropy.[3]

In the following it is assumed that the density of magnetic anisotropy energy is described by

$$\mathscr{E}_{anis} = K_u \sin^2\theta + \Delta \sin^2\theta \sin^2\hat{\phi} \tag{1}$$

where $\theta$ is the polar angle of the magnetization vector relative to the film normal and $\hat{\phi}$ the azimuthal angle relative to a "secondary easy axis" which is in the plane of the film. Only plane walls are considered and it is assumed that the wall is oriented at an angle $\psi$ to the secondary easy axis. The azimuthal angle of the magnetization relative to the wall plane is therefore $\phi = \hat{\phi} + \psi$. For simplicity only the cases $\psi = 0$ and $\psi = \pi/2$ are considered.

For a plane wall in an infinite medium having the orthorhombic anisotropy described by (1) the critical velocity is[1,2,4]

$$v_{crit} = \gamma(H_a D)^{\frac{1}{2}} \{[1 + \sigma(1+\eta)]^{\frac{1}{2}} - 1\}\;, \quad \psi = 0 \tag{2}$$

$$= \gamma(H_a D)^{\frac{1}{2}} \left|(1+\sigma)^{\frac{1}{2}} - (1+\sigma\eta)^{\frac{1}{2}}\right|\;, \quad \psi = \pi/2 \tag{3}$$

where $\sigma = 4\pi M_0/H_a$, $\eta = \Delta/2\pi M_0^2$. Here $\gamma$ is the gyromagnetic ratio, $H_a = 2K_u/M_0$ the anisotropy field, $D = 2A/M_0$, $M_0$ is the saturation magnetization, and A the exchange stiffness constant.

## STATIC WALL STRUCTURE

The static wall structure is such that the total magnetic energy is minimal. The relevant contributions to the magnetic energy are the exchange-, anisotropy-, and dipolar energy. The former two are "local" energies in the sense that they are represented by a single, three-dimensional integral over an energy density function. The dipolar energy is in general "non-local", but it can be represented by a quasi-local approximation when the film thickness is large compared to the wall width.[3,5]

For this case the wall energy per unit area E is

$$\tilde{E} = \tilde{a} + \tilde{a}^{-1}\langle 1 + \tilde{c}^{-2}\phi'^2 + \sigma[\sin^2\phi + \eta\sin^2(\phi - \psi) - 2\sin\phi f(\zeta)]\rangle. \tag{4}$$

The tilde (~) indicates a reduced notation[5] in which lengths are expressed in units of $(D/H_a)^{\frac{1}{2}}$, velocity in units of $\gamma(H_a D)^{\frac{1}{2}}$, and energy per unit wall area in units of $M_0(H_a D)^{\frac{1}{2}} = 2(K_u A)^{\frac{1}{2}}$.

It has been assumed that, apart from the scale factor $\tilde{a}$, the x-dependence of the wall structure in the film is the same as that in an infinite medium. The angular brackets denote the average over $\zeta = z/c$ (where 2c is the film thickness) and the prime denotes differentiation with respect to $\zeta$. The energy $\tilde{E}$ must be minimized with respect to the scale factor $\tilde{a}$ and with respect to the function $\phi(\zeta)$.

The function $f(\zeta)$ describes the dependence of the magnetic field due to surface charges (suitably averaged) upon the coordinate perpendicular to the film plane. In the simplest approximation[5] $f(\zeta) = \tanh^{-1}\zeta$.

Figure 1 shows the static wall structure as calculated on the basis of (4) for $\psi = 0$, $\pi/2$ and various values of $\eta$. For the cases considered the derivative term in (4) is of secondary importance and has been neglected. In this approximation the function $\phi(\zeta)$ which describes the static wall structure consists of two branches, which meet at the "critical point" $\zeta_0$, defined by $f(\zeta_0) = 1 \pm \eta$ where the upper sign corresponds to $\psi = 0$, the lower to $\psi = \pi/2$.

For $\psi = 0$, the in-plane anisotropy tends to pull the wall magnetization into the plane of the wall. In this case, the critical points move towards the film surface as $\eta$ is increased. For $\psi = \pi/2$, the in-plane anisotropy tends to pull the wall magnetization perpendicular to the plane of the wall. In this case, the critical points of the static wall structure move towards the mid-plane of the film as $\eta$ is increased. For $\eta = 1$, the two critical points merge at $\zeta = 0$, in other words, the static wall contains a fully developed, but rather wide, Bloch line. As $\eta$ increases beyond 1 the width of the Bloch line decreases and the wall energy increases.

For $\psi = \pi/2$ it is necessary to consider another static wall type in addition to the "twisted wall" just described. This other wall type is an "untwisted wall" in which the wall magnetization is aligned with the secondary easy axis and therefore at right angles to the wall plane. The untwisted wall becomes energetically favored over the twisted wall when $\eta$ is sufficiently large.

## DYNAMIC WALL STRUCTURE AND CRITICAL VELOCITY

For a given wall velocity $\tilde{v}$ the wall structure is generally determined by the variational principle[5]

$$\delta[\tilde{E} - 2\tilde{v} < \phi >] = 0 \qquad (5)$$

where $\tilde{E}$ is the wall energy considered as a functional of the wall-structure function $\phi(\zeta)$. For the present case $\tilde{E}$ is given by (4).

The Euler equation of the variational principle (5) is in general a nonlinear second order differential equation. No rigorous solutions are known, but it can be shown[3,5,6] that for $\eta \ll 1$ functions characterized by a Bloch line are approximate solutions provided that the Bloch-line width is very small compared to the film thickness. The approximation also becomes invalid as the Bloch line approaches one of the critical points. The critical velocity calculated on this basis is represented by the equation given in the abstract.

Figure 2 shows the various threshold velocities for relatively small $\eta$ and $\psi = 0, \pi/2$. It may be seen that the results based on the infinite medium theory differ considerably from those derived for films by taking the magnetic stray field into account. In the calculation of the latter curves it has been assumed that the ratio of film thickness h to Bloch-line width $\delta_\ell$ is 15. The curves are drawn as solid lines in the region where they are considered to be reasonably good approximations. For larger values of $\eta$ and $\psi = 0$ the theoretical threshold velocity calculated for a film approaches that appropriate in an infinite medium [see (2)].

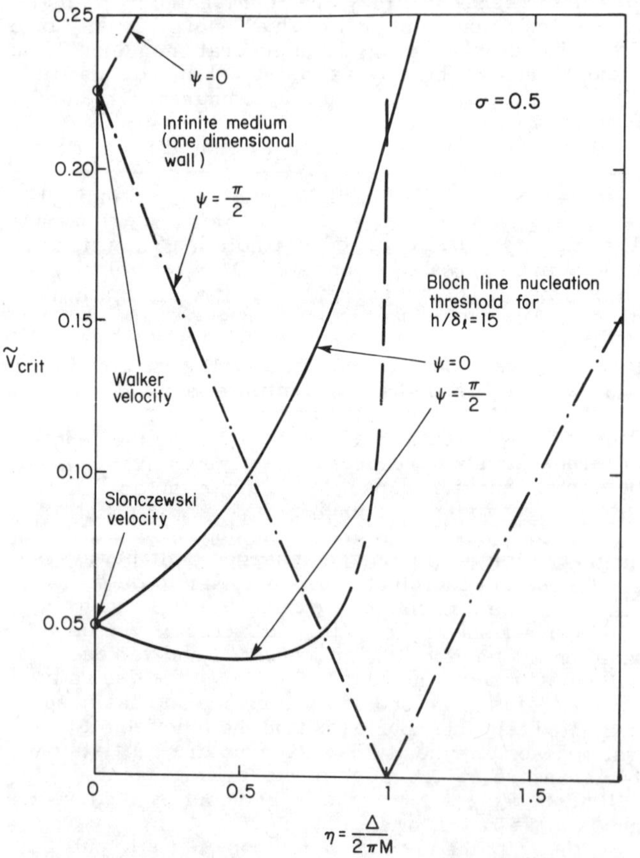


Fig. 2. Threshold velocities for relatively small values of the in-plane anisotropy, $\sigma = .5$, and $h/\delta_\ell = 15$.

## DISCUSSION

The results of the theory described in the present paper indicate that the critical velocity at which wall motion becomes unstable can be more than an order of magnitude larger in films with an in-plane anisotropy than in similar films without such an anisotropy. A partial experimental confirmation of these theoretical results is the observation of very high wall velocities (in excess of 9000 m/sec) in ortho-ferrites.[7,8] In these materials, an anisotropy of the type described by (1) is present because of the orthorhombic crystal structure. Gyorgy and Hagedorn[9] have shown that conventional micromagnetic theory such as used in the present paper is to some extent applicable to the orthoferrites even though these materials have two magnetic sublattices which are canted with respect to each other due to a Dzyaloshinskii interaction. It has been argued by Konishi et al.,[7,10] however, that the wall velocities observed in orthoferrites are too high to be explainable in terms of the in-plane anisotropy. These authors have, therefore, proposed a mechanism that involves the exchange coupling between the canted sublattices.

Another qualitative indication of the effectiveness of in-plane anisotropy is an experiment by Tabor, et al.[11] in which an epitaxial film without in-plane anisotropy was compared with a platelet cut from a bulk sample. The measurements on the platelet indicated that it had a modest in-plane anisotropy, but in other respects, the two samples were quite similar. The critical velocities deduced from bubble collapse experiments were determined to be at least nine times higher in the platelet (with in-plane anisotropy) than in the epi-film. In bubble translation experiments, the difference in the behavior of the two samples was much smaller, however.

A detailed paper on this subject will be published in J. Appl. Phys.

## REFERENCES

* Supported by the Office of Naval Research. A detailed description of this work has been submitted to J. Appl. Phys.

1. F. B. Hagedorn, AIP Conf. Proc. 5, Part 1, 72 (1972).
2. A. A. Thiele, Phys. Rev. B7, 391 (1973).
3. J. C. Slonczewski, J. Appl. Phys. 44, 1759 (1973).
4. In Ref. 1 only the case $\psi = 0$ is considered. In Ref. 2 both of the cases are considered but no explicit formula for the critical velocity is given.
5. E. Schlömann, J. Appl. Phys. 44, 1837 (1973); 44, 1850 (1973); 45, 369 (1974).
6. A. Hubert, J. Appl. Phys. 46, 2276 (1975).
7. S. Konishi, T. Kawamoto and M. Wada, IEEE Transactions, MAG-10, 642 (1974).
8. C. H. Tsang and R. L. White, AIP Conf. Proc. 24, 749 (1975).
9. E. M. Gyorgy and F. B. Hagedorn, J. Appl. Phys. 39, 88 (1968).
10. S. Konishi and T. Miyama, AIP Conf. Proc. 24, 740 (1975).
11. W. J. Tabor, G. P. Vella-Coleiro, F. B. Hagedorn and L. G. Van Uitert, J. Appl. Phys. 45, 3617 (1974).

# WALL STRUCTURE OF CYLINDRICALLY SYMMETRIC MAGNETIC DOMAINS

E. Della Torre,* C. Hegedüs and G. Kádár
Central Research Institute for Physics
Budapest, Hungary

## ABSTRACT

The magnetostatic stability of cylindrical magnetic domains in uniaxial magnetic films, having domain walls whose structure is independent of circumferential position but is otherwise quite general, is investigated numerically using finite difference techniques. This micromagnetic model is capable of analyzing materials with low $q=2K_u/\mu M_s^2$, however, since it permits only cylindrically symmetric solutions it cannot examine strip-bubble transitions. It is never the less capable of calculating the amount of wall bulging, Schlömann-DeBonte surface effects, change in wall energy with curvature, variation in stability as domain wall thickness approaches the domain diameter, and the parameters at bubble collapse.

## INTRODUCTION

A domain theory for cylindrical magnetic domains has been presented by Thiele.[1] He discussed wall bulging and wall width effects qualitatively due to the limitation of his model. He further assumed that wall energy was independent of wall curvature. Schlömann[2], using a simple parametric model, investigated the effect of surface poles on the domain walls and suggested the existence of anomalous walls. DeBonte[3] in a four parameter variational model, and Hubert[4] in a finite difference model elaborated on these results; however, all these analyses were for a plane wall.

Since the corrections introduced by these various effects appeared to be significant in certain limiting cases, it was felt that a detailed micromagnetic analysis was justified. The model presented here is sufficiently general to shed some light on all these questions; however, due to the limitation of memory capacity and running time the model is only two-dimensional. Thus, only bubbles with detail cylindrical symmetry can be analyzed and strip-bubble transitions are beyond the scope of this model. One can analyze garnet bubbles with a continuous Bloch wall but not orthoferrite bubbles.

The model, in terms of the coordinate system of Fig. 1, assumes that the magnetization, $\vec{M}$, and consequently the demagnetization field, $\vec{H}_D$, and other variables are not a function of α. Furthermore, the demagnetizing field will not have an α component. Since the magnitude of the magnetization is constant, the magnetization can be described by the two coordinates: the polar angle θ, and the azimuthal angle ϕ. Thus the fields have only two components and are a function of only two coordinates, though not the same two.

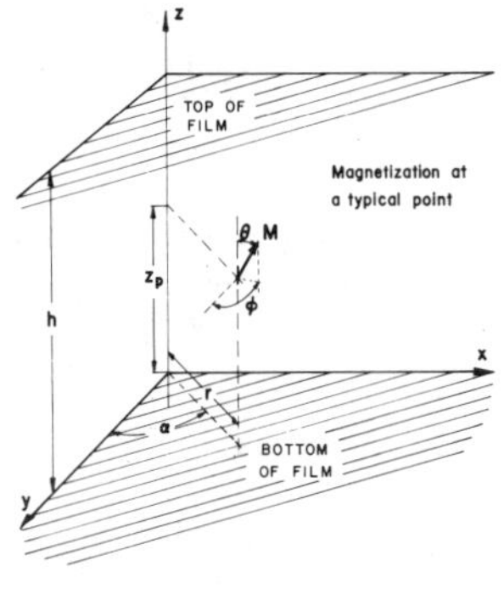


Fig. 1 Coordinate system used.

## THE COMPUTER MODEL

An initial solution is calculated using Thiele's theory[1]. The model assumes that for $r>r_{max}$ the magnetization in the film is given by $\vec{M} - M_s\vec{I}_z$ and for $r<r_{min}$ the magnetization is $\vec{M} = -M_s\vec{I}_z$. The main program chooses $r_{max}$ and $r_{min}$ so that sufficient resolution is obtained in the wall. In this region it is assumed that the magnetization is arbitrary and the program varies it in such a way that the total energy is minimized. There are four contributions to this energy: the exchange energy, the magnetocrystalline anisotropy energy, the demagnetizing energy and the external bias energy. The subroutine, MAGCAL, which performs this minimization is illustrated in Fig. 2.

This subroutine iteratively improves the solution until a termination criterion is met. The resolution of the finite difference calculation and the termination criterion are variable and controlled by the main program.

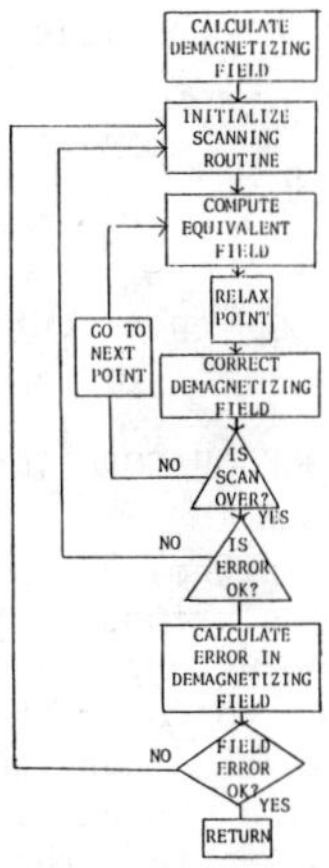


Fig. 2 MAGCAL SUBROUTINE

## DEMAGNETIZING FIELD CALCULATION

The most difficult and time-consuming task is the calculation of the demagnetizing field. Two types of calculations are used in this program, to minimize computation time, which are depicted in Fig. 2 by the boxes labelled "CALCULATE DEMAGNETIZING FIELD" and "CORRECT DEMAGNETIZING FIELD" respectively. The first calculation is used when nothing is known about the demagnetizing field and one has to integrate over the entire film. The second calculation is used during the relaxation process to update the demagnetizing field due to perturbations introduced by varying the magnetization. Since there is a possibility that an excessive error can build up in the second type of calculation, the first type of calculation must be performed periodically to correct the results.

In the first calculation one computes at all points of interest the effect of all the magnetic poles in the system. The poles outside the region of interest are uniform surface poles on the top and bottom planes of the film and their effect can be expressed in terms of elliptic integrals. The calculation of elliptic integrals is performed by the algorithm of Bulirsch.[5] Since the pole distribution in the region of interest, which is equal to the divergence· of M, is arbitrary this contribution must be calculated numerically as a double integral.

The second type of calculation is performed when the demagnetizing field has to be corrected after the magnetization of a point in the lattice has been varied in the relaxation procedure. This field is that of a dipolar ring and can also be expressed in terms of elliptic integrals.

## EQUIVALENT FIELD

The equivalent field, computed by the subroutine depicted by the box labelled "COMPUTE EQUIVALENT FIELD", is the sum of the demagnetizing field, the bias field and the exchange field. The anisotropy energy is the only energy that cannot be represented by an equivalent field since it has more than one minimum energy configuration in general. In the relaxation subroutine "RELAX POINT" the magnetization is chosen so that the energy associated with that point is minimized in the presence of the equivalent field and simple uniaxial anisotropy. Higher order uniaxial anisotropy can be introduced in this program at this point without any further complications.

Since the exchange energy per unit volume is given by:

$$w_{ex} = -A\,\vec{I}_M \cdot \nabla^2 \vec{I}_M \qquad (1)$$

where $\vec{I}_M$ is a unit vector in the direction of the magnetization, then it is possible to define an exchange field by

$$\vec{H}_{ex} = \frac{A}{\mu_o M_s} \nabla^2 \vec{I}_M \qquad (2)$$

so that the exchange energy per unit volume is given by $-\mu_o \vec{M} \cdot \vec{H}_{ex}$.

## CONVERGENCE OF SOLUTION

The iterative algorithm used in the solution generates a sequence of solutions whose limit is presumably the desired solution. The algorithm used adjusts the magnetization to lower the total energy of the system if the magnetic configuration does not satisfy the system equation. Since the energy of the system has a lower bound, in fact the desired solution is the greatest lower bound, this algorithm converges and the limit is the solution if it is unique.

If more than one solution exists then the algorithm will terminate at any of the possible solutions, and the solution might not be the desired one. In the present case for the same bias field there exist up to three possible solutions with the desired symmetry: the film saturated in the direction of the bias field, a bubble and the film saturated in the opposite direction of the bias field. By insisting that for $r = 0$, $\vec{M} = -M_s\vec{I}_z$ and for $r = \infty$, $\vec{M} = M_s\vec{I}_z$, the two saturated solutions are ruled out. Thus, this program will always search for a bubble even if one is not stable. It is assumed Thiele's theory will warn us when the strip-bubble transistion has occured. If the magnetic structure is sufficiently altered Thiele's theory will have to be modified. This model permits the bubble to collapse but not to zero radius. Thus, bubble collapse will be said to have occurred if the average bubble radius decreases discontinuously to an unreasonably small value.

## RESULTS

The program has been tested on several digital computers. In the limiting case of high q and no wall anomalies, the model yields results in agreement with Thiele's theory.

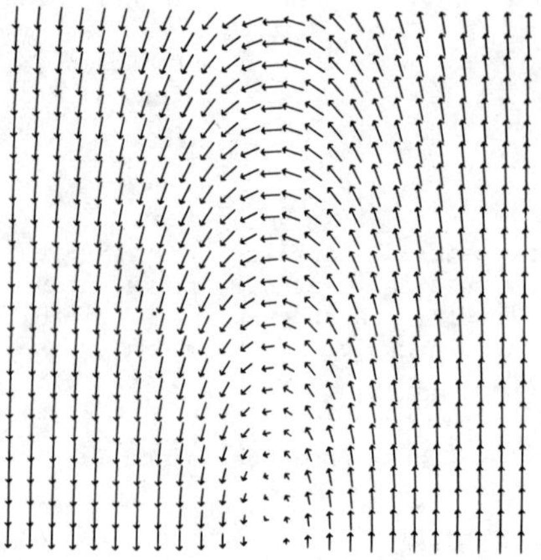

Fig. 3a

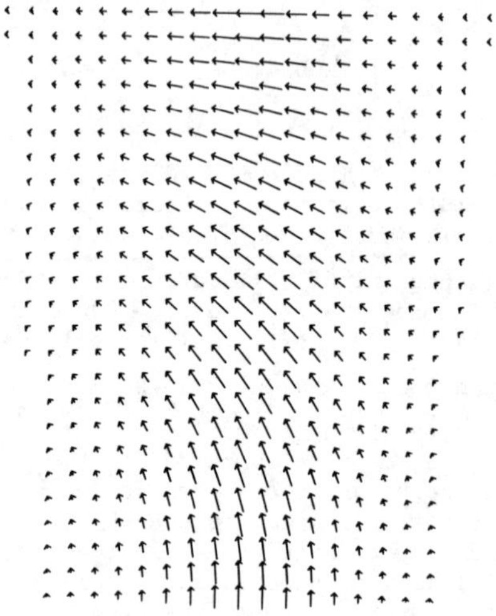

Fig. 3b

Figure 3 illustrates a typical magnetization distribution obtained by this program for a material whose Q is 16.7 when the thickness-material length ratio is 4. At each physical point Fig. 3a and 3b show the projection of the magnetization on the r-z plane and r-α plane respectively. N.D. vertical scale compressed by a factor of 10.

## ACKNOWLEDGEMENTS

The authors would like to acknowledge that this work was supported by the Hungarian Academy of Sciences and the National Research Council of Canada. We should also like to thank L. Gal and G. Zimmer for many helpful discussions.

## REFERENCES

1. A.A. Thiele, Bell Sys. Tech. J. 46, 3287 (1969).
2. E. Schlömann, J. Appl. Phys., 44, 1837, 1850, (1973).
   J. Appl. Phys., 45, 369, (1974).
3. W.J. DeBonte, IEEE Tran. Magn., MAG-11, 3 (1975).
4. A. Hubert, J. Appl. Phys., 46, 2276 (1975).
5. R. Bulirsch, Numerische Math., 13, 305 (1969).

*On leave from McMaster University, Hamilton, Canada.

# HIGH SPEED BUBBLE GARNETS BASED ON LARGE GYROMAGNETIC RATIOS (HIGH g)

R. C. LeCraw, S. L. Blank, G. P. Vella-Coleiro and R. D. Pierce
Bell Laboratories, Murray Hill, New Jersey 07974

ABSTRACT

An approach to overcoming the problem of dynamic conversion in high-mobility bubble garnets is described based on large gyromagnetic ratios (high-g factors). In a film of $Eu_{1.45}Y_{0.45}Ca_{1.1}Fe_{3.9}Si_{0.6}Ge_{0.5}O_{12}$, a g factor greater than 30 has been obtained, which increases the usable domain wall velocity before onset of dynamic conversion by more than an order of magnitude over comparable bubble garnet films with g approximately 2. The temperature dependence of the important bubble parameters has been measured and a simple bias magnet constructed which matches the steeper than usual variation of the bubble collapse field with temperature. Two different methods of hard bubble suppression are described, one involving short oxygen anneals at ∿1050°C, and the other ion implantation.

\- - - - - - -

When useful bubble garnet materials with relatively high mobilities, ∿1000 cm/secOe or greater, became available,[1-3] another limitation on achieving high bubble velocities was observed. Experimentally, erratic propagation of bubble domains was observed during repetitive bubble transport measurements.[4] This was attributed to the conversion of a normal bubble domain into a relatively complex state, similar to a hard bubble, during rapid displacement. A model for this effect, called dynamic conversion, was given by Hagedorn,[5] who extended previous work of Slonczewski[6] and Thiele.[7]

The critical or limiting velocity is given by[6]

$$V_p = 24\ \gamma A/hK_u^{\frac{1}{2}}, \qquad (1)$$

where $\gamma = ge/2mc$ is the gyromagnetic ratio, A is the exchange constant, h is the film thickness, and $K_u$ is the uniaxial anisotropy constant. (In a ferrimagnet the g in $\gamma$ is actually the effective g, $g_{eff}$, but for convenience g will be used here.)

Thus far attempts to maximize $V_p$ have involved keeping h and $K_u$ as small as possible consistent with other bubble requirements, and using as little diamagnetic substitution as possible to achieve the required reduced magnetization, e.g., Ge in preference to Ga.[8,9] This has the effect of keeping the exchange constant A as large as possible. Another attempt was a three-layer film described by Hagedorn,[5] the purpose of the thin middle diamagnetic layer being to suppress undesirable motions of Bloch lines which were believed to lead to the dynamic instability.

Some success in suppressing dynamic conversion has been achieved by keeping the thickness small,[10] but this is a severely limiting boundary condition. The elimination of dynamic conversion of bubbles by using Permalloy-coated garnet films has also been proposed.[11] However, we believe these latter results are not yet conclusive, particularly in light of the discovery of the bubble overshoot effect.[12]

It was finally realized that the gyromagnetic ratio $\gamma = ge/2mc$ in Eq. (1) had not been considered sufficiently for increasing the critical velocity, possibly because of the customary assumption that $g \approx 2$. However, a garnet system involving Eu together with a diamagnetic substitution on tetrahedral sites was shown by LeCraw, Remeika, and Matthews[13] to produce very large values of g where

$$g = (M_{Eu} + M_{Fe})\left(\frac{M_{Eu}}{g_{Eu}} + \frac{M_{Fe}}{g_{Fe}}\right)^{-1}. \qquad (2)$$

Here $M_{Eu}$ and $M_{Fe}$ are the magnetizations of the Eu and total Fe sublattices, respectively, and $g_{Eu}$ and $g_{Fe}$ are the g factors for the Eu and Fe ions, respectively. Because of the J = 0 ground state of Eu, $g_{Eu} >> 2$, and hence g in Eq. (2) becomes very large as $M_{Fe} \to 0$. The total moment does not vanish as $M_{Fe} \to 0$ because of the induced Eu moment, which results from its exchange coupling almost exclusively to tetrahedral Fe ions. These results are shown in Figs. 3-5 of Ref. 13. Only Eu of the rare earth (RE) series is effective in this way. The other possibly usable rare earth iron garnets, those with line widths no greater than Sm, i.e., Sm, Gd, Er, Tm, Yb have $g_{RE} \cong 2$. Thus the denominator in Eq. (2) can become zero (high g) only at very nearly the same point at which the numerator becomes zero (zero total moment).

The expected influence of the mobility μ and g on wall velocity is shown schematically in Fig. 1. This shows that even with large mobilities, high-g factors are necessary to achieve large usable velocities. Experimentally we have found that μ is essentially independent of g, which indicates that if the simple model for domain wall motion is used, α in the Gilbert equation is proportional to g. This observation is consistent with resonance linewidth measurements on films with widely differing g factors, where ΔH is observed to be essentially independent of g. It is also consistent with an expression derived by Pierce and LeCraw[14] for the effective phenomenological damping constant of a multiple sublattice system in which one sublattice contains ions with relatively large damping.

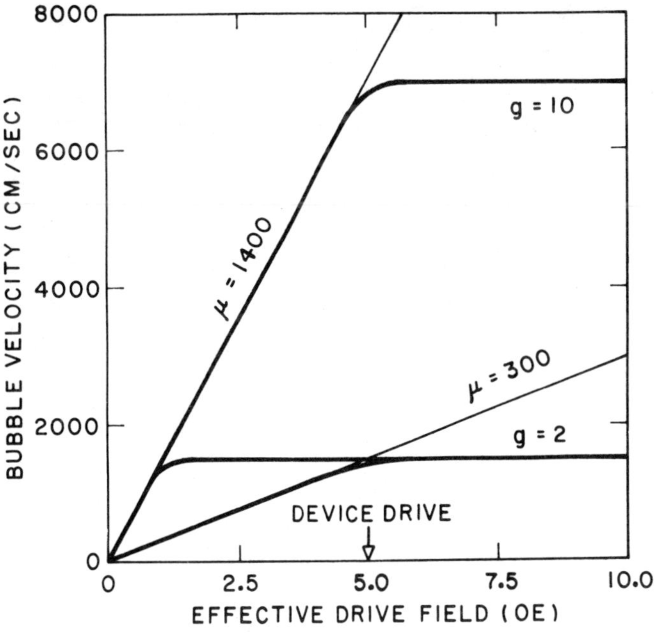


Fig. 1 Schematic representation showing how both mobility and g influence the bubble velocity.

TABLE I

Pertinent Bubble Properties of a Film of $Eu_{1.45}Y_{0.45}Ca_{1.1}Fe_{3.9}Si_{0.6}Ge_{0.5}O_{12}$

| | |
|---|---|
| Thickness | 4.23 μm |
| Demagnetized Strip Width | 5.18 μm |
| Collapse Field | 100.2 Oe |
| Material Length | 0.64 μm |
| Curie Point | 466°K |
| Coercivity | 0.4 Oe |
| $4\pi M$ | 218 G |
| Lattice Constant | 12.385 Å |
| $2K_u/M$ | 1500 Oe |
| Mobility | 1500 cm/secOe |
| g | >30 |

Thus a series of LPE films was grown containing Eu, with Ge-Ca instead of Ga used to reduce $M_{Fe}$ to study the effects of high g factors on bubble velocities.[15] Brief details of the growth conditions are given in Ref. 15, and the details of the phase equilibria[16] and the growth kinetics[9,16] in systems containing divalent-tetravalent ion substitutions are discussed elsewhere. Table I shows the properties measured on a typical high-g film.

It should be noted that $2K_u/M$ and g given in Table I were measured by microwave resonance techniques,[17] the lower limit on g being determined by the microwave frequency of 17.5 GHz. It should also be pointed out that these large g factors are not strongly temperature dependent since when $M_{Fe}$ is compensated to be approximately zero by diamagnetic tetrahedral site substitution, the small remaining $M_{Fe}$ is only slowly varying with T.

Using such a high-g film, the propagation data at 1 and 2 MHz shown in Fig. 2 were obtained with a TX-type circuit having a period of 28.8 μm. The parallel margins indicate that there is no discernible limit in the number of error-free propagation steps. Two

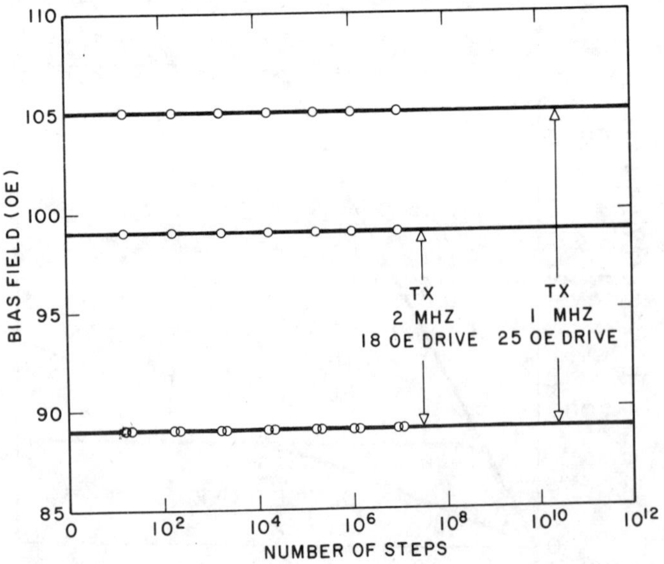


Fig. 2 Bias field margins vs. number of steps propagated at 1 and 2 MHz rates for a film of $Eu_{1.45}Y_{0.45}Ca_{1.1}Fe_{3.9}Si_{0.6}Ge_{0.5}O_{12}$.

MHz was the limit of the operating range of the propagation drive circuitry. At this frequency the available rf drive was 18 Oe compared to 25 Oe at 1 MHz. For the 28.8 μm circuit period, 2 MHz is approaching the mobility-limited operating frequency at this drive. However, from Eq. (1) it can be calculated that dynamic conversion effects for $g > 30$ would not have occurred until above 10 MHz.

A striking confirmation of the effect of g on the critical domain wall velocity has been observed on a 9-μm-thick film of the same composition as that in Table I together with a film similar in all other parameters but with less Ca and a g of 1.07. For the latter film the critical velocity is ≈1000 cm/sec, whereas for the 9-μm film with $g > 30$, $V_p \approx 30{,}000$ cm/sec. Velocities as high as 60,000 cm/sec were observed on the high-g film, which is probably the largest domain wall velocity yet observed in a magnetic garnet.[18]

Thus far we have reviewed briefly what has been published previously on high-g bubble garnets. We will now consider later developments:

Another confirmation of the greatly increased suppression of dynamic conversion by high g factors has been observed recently by G. P. Vella-Coleiro[19] in noting the absence of bubble overshoot in a high-g film during his investigations of bubble motion using very high speed photography. This result together with the 2 MHz bubble propagation rate with flat bias margins out to $10^7$ steps (Fig. 2), and the 60,000 cm/sec domain wall velocity all combine to give evidence which strongly supports the effect of high g factors on dynamic conversion.

## HARD BUBBLE SUPPRESSION

In the absence of dynamic conversion it was expected from theoretical considerations that a material with $g \gtrsim 20$ would not exhibit hard bubbles. Compared to the usual YSmCa-type garnets it was much more difficult to produce hard bubbles by the usual pulsing or rapid demagnetization techniques in the high g garnets. Yet some hard bubbles were produced in all of the as-grown films.

When the high-g samples were annealed in $O_2$ at 1050°C for 0.5 h, however, the hard bubbles were eliminated. This was repeated on several different high-g samples in both $O_2$ and $N_2$. With the sample used for the data in Table I and Fig. 2, the annealing time was 0.75 h in $O_2$ at 1050°C. (Extra time was used to be certain). These short anneals may relieve some highly localized strains acting as pinning points with which the domain walls interact in a complicated way to produce hard bubbles as the domain walls "snap off" these points. However, only a very small decrease in overall coercivity is observed for this short anneal at a temperature which is only slightly above the growth temperature. Etching off 3 μm of the film in 1 μm steps was tried to determine if the surface played a role, but no differences were seen. It should be noted that the above short $O_2$ anneal does not eliminate hard bubbles in the same class of EuYCa-type samples but with less Ca and g = 1.07. Even several hours in $O_2$ at 1050°C did not suffice. Further work needs to be done to understand this striking effect.

The short $O_2$ anneals which effectively eliminate hard bubbles in high-g samples should not be confused with the recently described technique of suppression of hard bubbles in the usual type of bubble garnets by inert atmosphere annealing. There the temperatures are higher and the times are longer. This process is reasonably well understood.[20]

Ion implantation has also been used successfully to suppress hard bubbles in high-g films, although the range of dosage, $1\text{-}3\times10^{14}$ Ne/cm$^2$ at 100 keV is only about one-third as wide as with the usual YSmGa and YSmCa-type films.[21] This is qualitatively consistent with the negative magnetostriction constant, estimated to be $\lambda_{111} \cong -0.3\times10^{-6}$, using the method of R. L. White.[22] This value is almost an order of magnitude lower than $\lambda_{111}$ for the usual bubble garnets not containing Eu. By using Ferrofluid we have confirmed the existance of a thin top layer with planar magnetization, as has been observed in other ion implanted garnets with negative magnetostriction.[23]

## TEMPERATURE DEPENDENCE

In order to effectively utilize the considerable increase in possible bubble device speed with high-g films, it is necessary to know and allow for the temperature dependence of the important bubble parameters. Figures 3 and 4 show the temperature dependence of the collapse field $H_o$, the anisotropy field $H_k$, the magnetization $4\pi M$, the material length $\ell$, the demagnetized strip width w, and the quality factor $Q = H_k/4\pi M$ from 0 to 100°C. These data were obtained on a typical high-g film of the approximate composition given in Table I. The film was 6.6 μm thick. The quantities directly measured were $H_o$, $H_k$, w and the film thickness t, from which the other parameters were calculated using well-known relationships. Values of the wall energy $\sigma$, the uniaxial anisotropy constant $K_u$, and the exchange constant A can also be calculated at each temperature. At 25°C they are, respectively, 0.23 erg/cm$^2$, $1.2\times10^4$ erg/cm$^3$, and $2.6\times10^{-7}$ erg/cm.

The temperature dependence of $H_o$ is particularly important since the permanent magnet which provides the bias field for a bubble device must track this field. Figure 3 shows that $H_o(T)$ is very nearly linear with a temperature coefficient at 50°C of -0.58%/°C. Similar high-g samples grown from different melts show very nearly the same linearity and slope. While $H_o(T)$ is steeper than would ordinarily be desired, another important de-

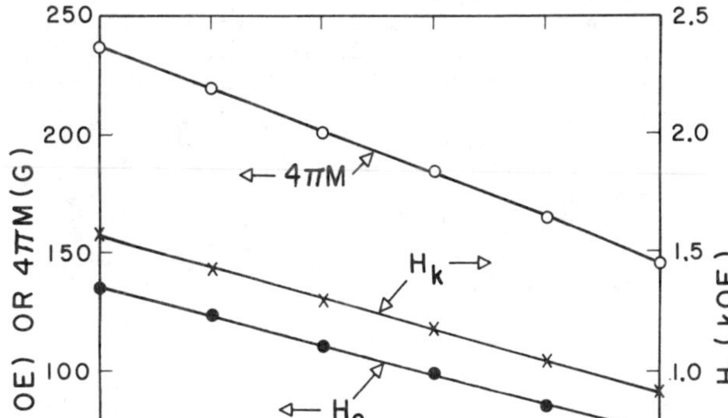


Fig. 3 The collapse field $H_o$, the anisotropy field $H_k$, and the magnetization $4\pi M$ vs. temperature for a high-g film of $Eu_{1.45}Y_{0.45}Ca_{1.1}Fe_{3.9}Si_{0.6}Ge_{0.5}O_{12}$.

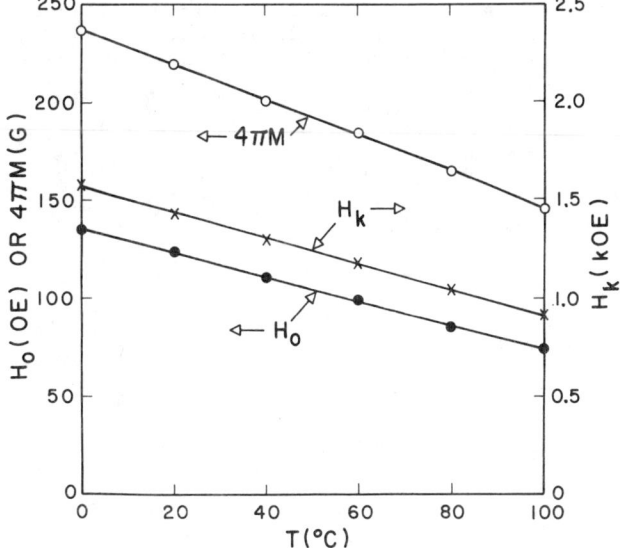

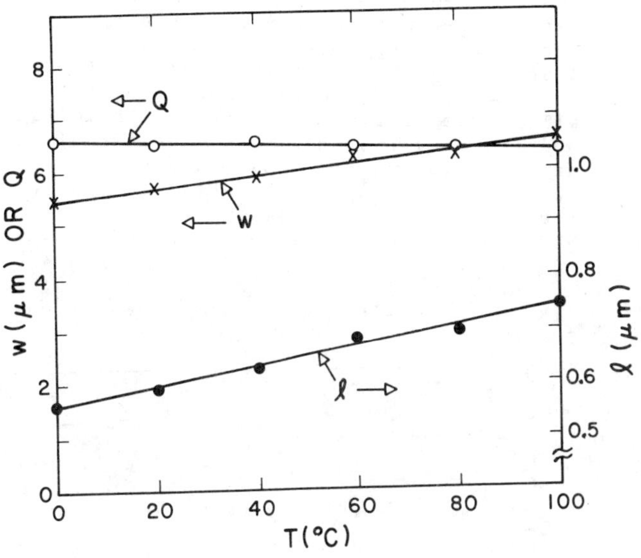


Fig. 4 The material length $\ell$, the demagnetized strip width w, and the quality factor $Q = H_k/4\pi M$ vs. temperature for the same film as in Fig. 3.

vice parameter, Q, remains nearly constant instead of decreasing strongly with T as observed in most bubble materials. This behavior occurs because $4\pi M$ decreases rapidly enough to compensate for the usual rapid decrease of $K_u$ with T.

In a typical YSmCa-type film with a similar Curie point the variation of $H_o$ with temperature is much slower than -0.58%/°C. We believe the difference is because of the following: In the present material with $g \geq 20$, the net magnetization of the iron sublattices is approximately zero and hence the net moment is essentially all from the Eu ions. LeCraw, Remeika and Matthews[13] showed that the exchange field acting on the Eu is almost entirely due to the tetrahedral iron ions. Thus one would expect, based on the temperature independent paramagnetic susceptibility $\chi_{Eu}$ of Eu, an observed moment proportional to $M_{tet}$, which varies much more slowly than -0.58%/°C. Such behavior can be expected at lower temperatures. However, the J = 1 multiplet levels of Eu, which average ∿500°K above the J = 0 ground state, begin to be occupied at room temperature and $\chi_{Eu}$ becomes temperature dependent. Thus the temperature dependence of $H_o$ and $4\pi M$ shown in Fig. 3 are determined by the combined effects of $M_{tet}(T)$ and $\chi_{Eu}(T)$, making them steeper than in the usual YSmCa-type films where the net moment is dominated by the tetrahedral iron.

Several methods of constructing a bias magnet to match the slope of $H_o$ have been considered. One such method uses the following principle. If two permanent magnets with widely different linear temperature coefficients (TC) are combined in opposing fashion, the TC of the resultant field can be made larger than the TC of either magnet. This is possible because the magnets can be adjusted to exactly cancel at some arbitrary temperature. Then at other temperatures, the TC of the net field is determined by the temperature dependence of the difference in the magnitudes of the two component fields. Thus, a magnet system with a desired TC can be constructed by properly selecting the magnitudes of the opposing fields and the cancellation temperature.

Satisfactory operation from 23 to 100°C of a high-g chip has been achieved using a magnet structure of this type.[24] However, a single permanent magnet material with the desired temperature coefficient results in a less cumbersome and physically simpler device. Such a ferrite material has recently been developed at these laboratories by F. J. Schnettler, E. M. Gyorgy and R. D. Pierce and will be reported on separately. A device module using this new ferrite material and a high-g chip has been assembled and tested at 100 kHz, yielding highly satisfactory results. The temperature coefficient measured in the air gap of the ferrite bias magnet structure was -0.6%/°C at 50°C compared to -0.58%/°C for the bubble collapse field $H_0$ in Fig. 3, and like $H_0$, the slope was quite linear from 0 to 100°C.

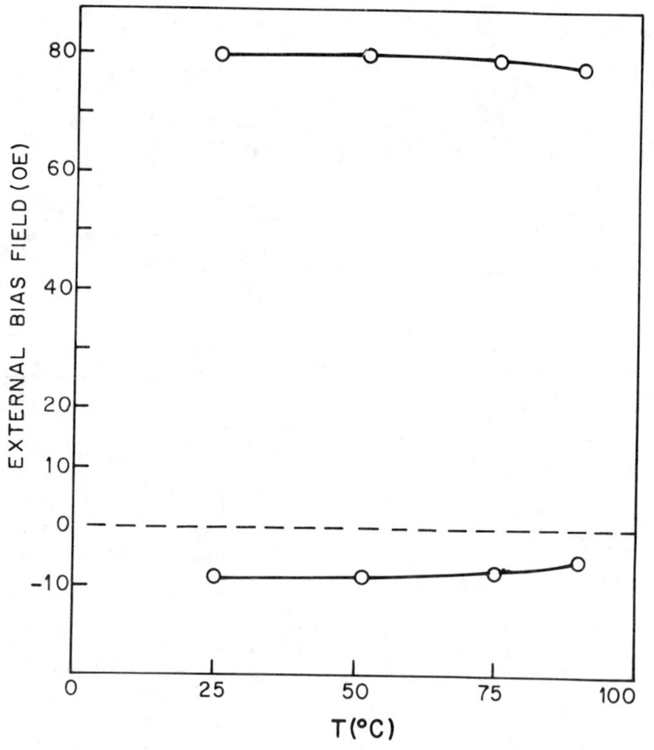


Fig. 5 External bias values for operation at 100 kHz of the module containing the high-g chip. The external margins are much larger than the chip margin of ~12 Oe, because of the shielding effect of the U-shaped metallic magnetic yoke of the ferrite bias magnet structure. The almost flat margins show that the ferrite bias field at the chip is tracking closely the fall-off of the bubble collapse field. (The zero not being in the middle of the margins indicates only that the ferrite bias magnet was set a few Oe too low).

Figure 5 shows that the module will operate over the temperature range with no external field. It also shows the measured values of external field over which the module will operate. The external field margins shown are much larger than the actual margins, ~12 Oe, for a complete circuit on this chip, because of the shielding effect from the U-shaped metallic magnetic yoke which supports the ferrite permanent magnet pieces to form the air gap for the chip. The approximately horizontal external field margins show that the ferrite bias field at the chip is tracking closely the fall-off of the bubble collapse field. The fact that the zero external field point is not in the middle of the range indicates only that the ferrite bias magnet was set a few Oe too low. This can be adjusted easily. These data are all taken for steady state operation. Transient operation of the device will require further study.

## CONCLUSIONS

High g factors arising from Eu in the iron garnets have been shown to suppress dynamic conversion in bubble garnets. The hard bubble problem can be eliminated either by short oxygen anneals at ~1050°C or by ion implantation. A ferrite permanent magnet material is now available which matches the steeper than usual temperature dependence of the bubble collapse field.

There are two principal limitations: For $g \gtrsim 20$, the minimum bubble diameter is ~5 μm. This corresponds to Ca = 1.1, or $M_{Fe} = 0$, in the system $(EuYCa)_3(FeGeSi)_5O_{12}$. Bubble diameters of ~3.5 μm have been achieved with $g \approx 6$, which yields a three times higher dynamic conversion frequency than usual bubble materials. This was done by increasing the Ca > 1.1 which increases $M_s$ by adding to the Eu moment some moment from the net iron lattice, since now the net iron moment is dominated by the octahedral sites instead of the tetrahedral sites as when Ca < 1.1.

The maximum mobility for a high-g garnet with a bubble size of ~5 μm is ≈1600 cm/secOe, this being determined primarily by the Eu damping. The amount of Eu cannot be greatly decreased, for although g can still be high, $M_s$ would decrease which would increase the bubble size. Because of the high dynamic conversion frequencies, however, this system offers the highest presently obtainable bubble device speeds for bubbles in the 3 to 6 μm range.

## ACKNOWLEDGMENTS

We wish to thank A. D. Butherus for the permanent magnet designs and W. Strauss for the measurements of the device margins vs. temperature. We also thank W. B. Venard, W. A. Biolsi and R. J. Peirce for very helpful technical assistance, and R. Wolfe, J. W. Nielsen and F. B. Hagedorn for their continuing useful discussions.

## REFERENCES

1. E. A. Giess, B. A. Calhoun, E. Klokholm, T. R. McGuire and L. L. Rosier, Mat. Res. Bull. 6, 317 (1971).
2. W. A. Bonner, J. E. Geusic, D. H. Smith, F. C. Rossol, L. G. Van Uitert and G. P. Vella-Coleiro, J. Appl. Phys. 43, 3226 (1972).
3. J. W. Nielsen, S. L. Blank, D. H. Smith, G. P. Vella-Coleiro, F. B. Hagedorn, R. L. Barns and W. A. Biolsi, J. Electron. Mat'l. 3, 693 (1974).
4. G. P. Vella-Coleiro, F. B. Hagedorn, Y. S. Chen and S. L. Blank, Appl. Phys. Lett. 22, 324 (1973).
5. F. B. Hagedorn, J. Appl. Phys. 45, 3129 (1974).
6. J. C. Slonczewski, J. Appl. Phys. 44, 1759 (1973).
7. A. A. Thiele, Phys. Rev. B7, 391 (1973).
8. S. Geller, H. J. Williams, G. P. Espinosa and R. C. Sherwood, Bell Syst. Tech. J. 43, 565 (1964).
9. W. A. Bonner, J. E. Geusic, D. H. Smith, L. G. Van Uitert and G. P. Vella-Coleiro, Mat'l. Res. Bull. 8, 1223 (1973).
10. F. B. Hagedorn, S. L. Blank and R. J. Peirce, Appl. Phys. Lett. 26, 206 (1975).
11. R. Suzuki and Yutaka Sugita, Appl. Phys. Lett. 26, 587 (1975).
12. A. P. Malozemoff and J. C. DeLuca, Appl. Phys. Lett. 26, 719 (1975).

13. R. C. LeCraw, J. P. Remeika and H. Matthews, J. Appl. Phys. 36, 901 (1965).
14. R. D. Pierce and R. C. LeCraw (unpublished).
15. R. C. LeCraw, S. L. Blank and G. P. Vella-Coleiro, Appl. Phys. Lett. 26, 402 (1975).
16. S. L. Blank, J. W. Nielsen and W. A. Biolsi, presented at the Annual Meeting of the Electrochemical Society, Dallas, Texas, October 1975 and submitted for publication.
17. R. C. LeCraw and R. D. Pierce, AIP Conf. Proc. 5, 200 (1972).
18. G. P. Vella-Coleiro, S. L. Blank and R. C. LeCraw, Appl. Phys. Lett. 26, 722 (1975).
19. G. P. Vella-Coleiro, paper this conference.
20. R. C. LeCraw, E. M. Gyorgy and R. Wolfe, Appl. Phys. Lett. 24, 573 (1974).
21. R. Wolfe, J. C. North and Y. P. Lai, Appl. Phys. Lett. 22, 683 (1973).
22. R. L. White, IEEE Trans, Mag. MAG-9, 606 (1973).
23. R. Wolfe and J. C. North, Appl. Phys. Lett. 25, 122 (1974).
24. A. D. Butherus and W. Strauss, private communication.

# STUDY OF DEFECTS IN REDUCED LPE BUBBLE GARNET FILMS

R. C. LeCraw, E. M. Gyorgy and R. Wolfe
Bell Laboratories, Murray Hill, New Jersey 07974

ABSTRACT

Aluminum deposited on LPE bubble garnet films, which are then heated for 0.5 hour at 450°C, has been used to study reduction-associated defects in garnets. This treatment leaves the garnet darkened but magnetically unchanged. Subsequent heating at ∿600°C with the Al removed, causes a controllable reduction in magnetization due to Ga-Fe redistribution, accelerated by the defects introduced at 450°C. Defects introduced into the same garnet films by annealing in nitrogen without aluminum at 1250°C have distinctly different characteristics. Thus the existance of at least two different types of reduction-associated defects in garnets has been demonstrated.

\- - - - - - -

It has been shown that when Si is deposited on Ga-containing LPE bubble garnet films which are then heated in the range 600 to 700°C, rapid changes in the magnetization, $M_s$, occur under the Si. This was attributed to oxygen vacancies created at the Si-garnet interface. The oxygen vacancies or some other kind of reduction-associated defects then diffuse through the film. The defects cause local distortions of the lattice which accelerate the interchange of Ga and Fe ions between octahedral and tetrahedral sites, thus changing $M_s$. Highly localized control of $M_s$ was obtained in this way.[1]

We have found that for studying the nature of this effect, Al is much more useful than Si because defects can be produced readily at the Al-garnet interface at 400-450°C rather than 600-700°C. These defects diffuse through the film thickness in times of the order of hours at substantially lower temperatures (∿425°C) than are required (∿600°C) for the Ga-Fe redistribution to occur.

In a typical experiment, a 5000 Å layer of Al was deposited by e-beam evaporation on a 6 μm thick film of $Y_{2.6}Sm_{0.4}Fe_{3.8}Ga_{1.2}O_{12}$. Three samples from this wafer were initially given the same heat treatment, i.e., 0.5 h at 450°C in nitrogen. The Al was then etched off of each sample. Sample 1 was then heated at 600°C in $N_2$, which greatly decreased $M_s$ as shown by the large decrease in the room temperature bubble collapse field, $H_o$, in Fig. 1(a). The sample was then put in $O_2$ for 3 h at 1000°C, and $H_o$ returned very nearly to its original value.

With Sample 2, Fig. 1(b), after the initial Al treatment, 1.6 μm was etched off by hot phosphoric acid. $H_o$ decreased by 9 Oe due to the decrease in film thickness. The sample

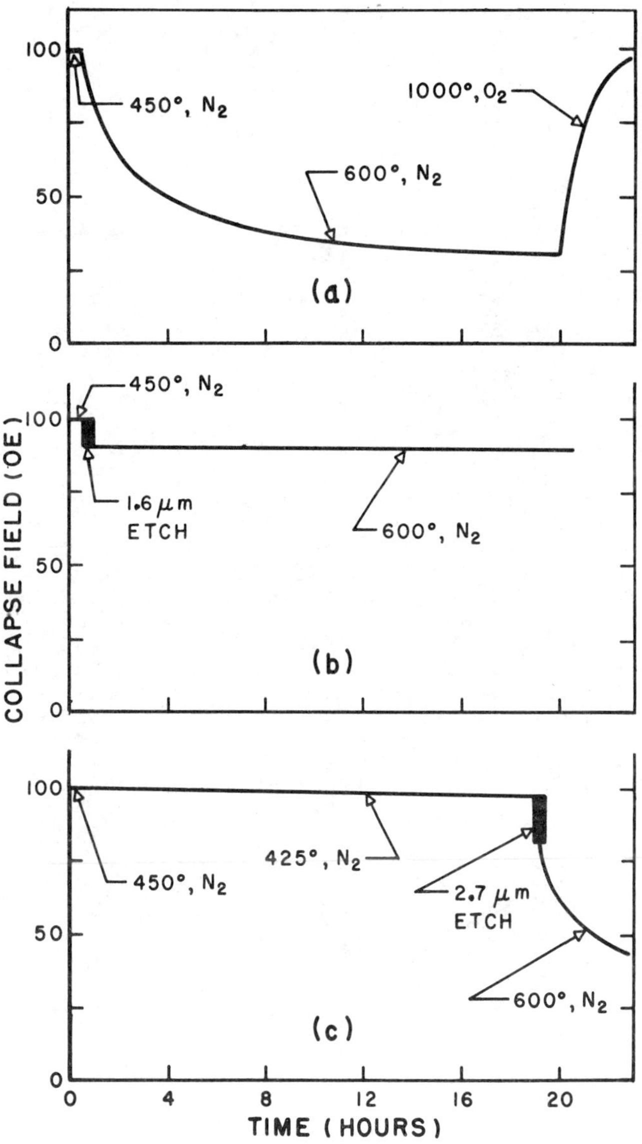


Fig. 1 Bubble collapse field, $H_o$, vs. heating time for three samples from a 6 μm thick wafer of composition $Y_{2.6}Sm_{0.4}Fe_{3.8}Ga_{1.2}O_{12}$. Each of the samples was coated with 5000 Å of Al and heated for 0.5 h in nitrogen. Then the Al was etched off each sample and the heat treatments shown in the figure were continued. The two vertical dark bars in (b) and (c) show the drops in collapse field upon etching the samples. The curves are best fits to the experimental points.

was then heated at 600°C in $N_2$ for 20 h and $H_o$ remained unchanged. This shows that the original defects produced by the Al were confined to the top 1.6 μm of the film.

After the 0.5 h Al treatment at 450°C of Sample 3 Fig. 1(c), the sample was heated at 425°C in $N_2$ for 20 h which caused only 3 Oe decrease in $H_o$. Then 2.7 μm was etched off which decreased $H_o$ by 15 Oe. This would remove all the defects unless they had diffused through most of the thickness. The sample was then heated at 600°C for 4 h in $N_2$ causing a large decrease in $H_o$. The very small decrease in $H_o$ after a long anneal at 425°C together with the large decrease in $H_o$ in 4 h at 600°C after 2.7 μm had been etched off showed that the defects originally produced at the top surface had diffused through the sample thickness in a time and at a temperature which allowed only negligible Ga-Fe redistribution to take place.

Thus using Al, samples having a high defect concentration but an essentially unchanged $M_s$ can be produced. This is a kind of "memory" effect since the sample is different after the initial one-half hour at 450°C, i.e., defects have been introduced, but with very little change in its magnetic properties until it is heated at ∿600°C or higher. This "memory" effect can also be used for localized control of $M_s$.

Experiments have also been performed with a 4000 Å pre-spacer of $SiO_2$ between the Al and the garnet. Even though anisotropy effects due to stress from the Al were observed in the garnet upon heating in the range 300-450°C with this spacer present,[2] no reduction defects were observed in the garnet even upon heating at 600°C with the Al present.

The defects produced at 450°C with Al directly on the garnet visibly darken the garnet film while also greatly enhancing the Ga-Fe redistribution rate. This darkening can be removed upon heating in $O_2$ or air at 600°C for 0.5 h with the Al off. Defects can also be introduced into garnet films by annealing in $N_2$ at 1250°C for several hours. This treatment also darkens the films, but the darkening cannot be removed by subsequent annealing even in $O_2$ at 1250°C for 24 h. Also, after the $N_2$ anneal there is no enhancing of the Ga-Fe redistribution rate at 600°C. (Some increase in the Ga-Fe redistribution rate is observed in the nitrogen treated sample at 800°C). Thus the defects introduced by the two different treatments, Al at 450°C and $N_2$ at 1250°C, are two distinctly different types.

There are three probable types of defects that can be produced in the garnet lattice upon reduction. If the initial crystal is stoichiometric only two types are probable: a vacancy in the oxygen lattice, or a cation interstitial with the oxygen lattice defect-free. Both defects of course require the presence of $Fe^{2+}$ to maintain charge neutrality. Substantial experimental evidence,[3] however, indicates that flux grown iron garnets are not stoichiometric but rather contain $Fe^{4+}$. The presence of $Fe^{4+}$ may be compensated by either some $Pb^{2+}$ or cation vacancies. If cation vacancies are present in the initial material, then the low temperature reduction using Al may produce vacancies in the oxygen lattice without removing the cation vacancies. Thus, this third type of defect has both cation vacancies and vacancies in the oxygen lattice. Intuitively this defect state would appear to lead to the most rapid Ga-Fe redistribution rate at a given temperature. At present the experimental data do not permit us to choose from the above three possibilies.

ACKNOWLEDGMENTS

The authors would like to thank J. W. Nielsen for his suggestion of the third possible defect mechanism. We also thank S. L. Blank and R. D. Pierce for many helpful discussions.

REFERENCES

1. R. C. LeCraw, P. A. Byrnes, Jr., W. A. Johnson, H. J. Levinstein, J. W. Nielsen, R. R. Spiwak and R. Wolfe, IEEE Trans. Mag., MAG-9, 422 (1973).

2. R. Wolfe and W. A. Johnson, 1975 Magnetism and Magnetic Materials Conf., December 9-12, 1975, Philadelphia, Pa.

3. D. L. Wood and J. P. Remeika, J. Appl. Phys. 37, 1232 (1966); 38 1038 (1967).

# MAGNETIC PROPERTIES OF RARE EARTH (Gd, Dy, Ho, Er)-COBALT AMORPHOUS FILMS

H. JOUVE[†], J.P. REBOUILLAT[*], R. MEYER[†]
[†]L.E.T.I., C.E.N.G., 85-X, 38041-GRENOBLE-Cedex, France
[*]Laboratoire de Magnétisme, C.N.R.S., 166-X, 38042-GRENOBLE-Cedex, France

## ABSTRACT

Magnetization data are presented on D.C. sputtered amorphous $RCo_x$ films where R = Gd, Dy, Ho, Er and x = 2.1 - 4.9. The Curie temperatures of the four $RCo_{3.5}$ alloys are found to be very close to each other confirming the absence of charge transfer in the amorphous state. At low temperature a 15 - 20 % reduction in the R contribution to the magnetization is observed, probably due to local anisotropy fluctuations. Stripe domains are obtained at room temperature in Gd, Dy, Ho compounds ; Gd and Ho films exhibit similar anisotropy energies. The existence of perpendicular anisotropy seems to be independent of the R constituent, but is influenced by the bias conditions during deposition.

## INTRODUCTION

Among the factors governing the properties of the rare earth-transition metal amorphous films, the preparation method[1], the constituents[2,3] involved and the composition are of major importance. In this work systematic magnetic measurements are presented on four series of $RCo_x$ alloys prepared by D.C. sputtering where R = Gd, Dy, Ho, Er and x ∿ 2.1 - 4.9. The R elements have been choosen as having different 4f orbital symmetries. The composition range studied gives low room temperature magnetization favorable for the appearence of bubble domains.

## FILM PREPARATION

The films are deposited from melted targets by D.C. sputtering onto water cooled substrates with a deposition rate of 150 Å/min. The targets of nominal composition $RCo_2$ and $RCo_3$ are obtained by H.F. fusion. The film composition is monitored by the variation of the bias voltage from -50 v to -150 v. Atomic absorption is used for composition analysis.

## TEMPERATURE DEPENDENCE OF MAGNETIZATION

The thermal variation of the spontaneous magnetization for samples of composition close to $RCo_{3.5}$ are presented in figure 1. The points refer to the low temperature data obtained by the extraction method in fields up to 65 kOe. For temperatures higher than 300 K (solid and dashed lines) a dynamic force balance was used. The temperature was raised at a rate of 2-6 K/sec in order to reach the Curie temperature $T_c$, before a total alteration of the amorphous phase occurs. The dashed lines represent the region where the recristallisation starts, extrapolation is used above 600 K where the experimental curves are perturbed by the partial formation of cristallized compounds. Nevertheless a net dip is observed in the experimental curve which is assumed to be the $T_c$ of the remaining amorphous part of the sample.
The experimental values of the absolute saturation magnetization M(0) exp., the compensation temperature $T_{comp}$ and $T_c$ are listed in Table I. The M(0)exp results for Gd sample are interpreted assuming an absolute moment for the Co atoms equal to 1.65 $\mu_B$ and 7 $\mu_B$ for Gd atoms (see ref. 4). The same assumption for the Co atomic moment is maintained for the other alloys (R = Dy, Ho, Er) considering the values obtained in amorphous $YCo_2$ and $YCo_5$ ($M(0)_{Co}$ = 1.6 and 1.5 $\mu_B$)[2]. The rare earth atomic moments $M(0)_R$ thus deduced (see Table I) are found to be 15 - 20 % smaller than the ionic moment $g_J J$. This reduction is accounted for by a non colinear spin arrangement : the picture is that of a fully aligned Co spin system and a weakly coupled R system,

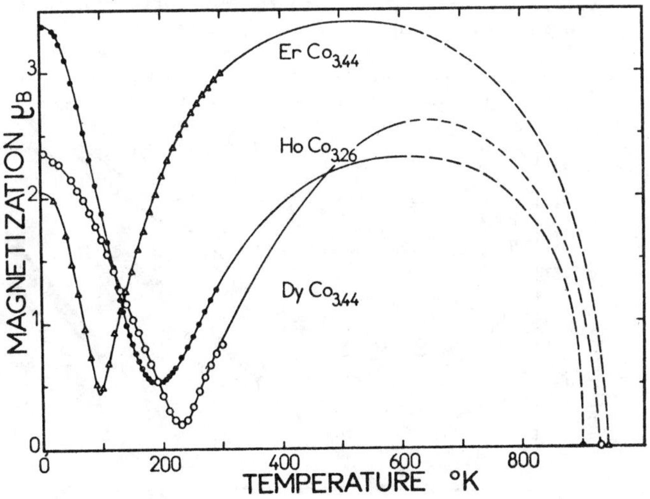


Figure 1 : Magnetization thermal variation for $RCo_x$ samples (for $GdCo_{3.55}$ curve, see ref.4, Fig. 1).

TABLE I

| R | | Gd | Dy | Ho | Er |
|---|---|---|---|---|---|
| x | | 3.55 | 3.44 | 3.26 | 3.44 |
| $M(0)_{exp}$ | ($\mu_B$) | 1.21 | 2.36 | 3.38 | 2.13 |
| $M(0)_{Co}$ | ($\mu_B$/Co atom) | 1.65 | 1.65 | 1.65 | 1.65 |
| $M(0)_R$ | ($\mu_B$/R atom) | 7 | 8.04 | 8.76 | 7.81 |
| $g_J J \mu_B$ | ($\mu_B$) | 7 | 10 | 10 | 9 |
| $T_{comp}$ | (°K) | 230 | 228 | 190 | 94 |
| $T_c$ | (°K) | 880 | 930 | 900 | 940 |
| $H_{Co-Co}$ | ($10^{-6}$ Oe) | 24 | 25 | 24 | 25 |
| $H_{Co-R}$ | ($10^{-6}$ Oe) | 1.6 | 0.94 | 0.60 | 0.47 |
| | | | 0.63 | 0.49 | 0.34 |
| $H_c$ 4.2 K | (Oe) | <10 | 1-2.$10^3$ | 2.$10^2$ | 5.$10^2$ * |
| $H_c$ 300 K | (Oe) | <10 | 2.$10^2$ | <10 | <10 |

* (20 K)

in which R moments are misaligned by the random local anisotropy field originating in the structural disorder[5]. Recent Mössbauer studies on $DyCo_{3.5}$ samples confirm the spatial dispersion of Dy moments[6]. Although this statement is strongly dependent on the accuracy of the composition analysis the difference observed between Gd and the other rare earth on various samples is considered to be significant (a non-reduced R-moment would imply a 5 % error in composition analysis).

The high values of $T_c$ indicate the preponderance of 3d correlations. As a first approximation, the total exchange field $H_{Co-Co}$ (Table I) is thus deduced in the molecular field model $H_{Co-Co} = 3kT_c/M_{Co}$. The similar $T_c$ values indicate that the 3d band filling is not disturbed by the rare earth conduction electrons contrary to the behaviour in the crystalline compounds[7].

The thermal variation of the R moment is assumed to follow the Brillouin function $B_J$, where J is the angular moment. Neglecting the R - R interactions, only the molecular field $H_{Co-R}$ originated by the Co spins on the R moments is taken into account. At $T_{comp}$ the R sublattice magnetization is equal to that of Co, for which the $M(0)_{Co}$ saturation value is taken as due to the low $T_{comp}/T_c$ ratio (0.1 - 0.26). Two $H_{Co-R}$ values are calculated in table I, with the assumption that the R thermal law is $M(0)_R\ B_J$ for the upper ones and $g_J J \mu_B\ B_J$ for the lower ones. This last situation reflects the possibility of R moment alignement with increasing temperature.

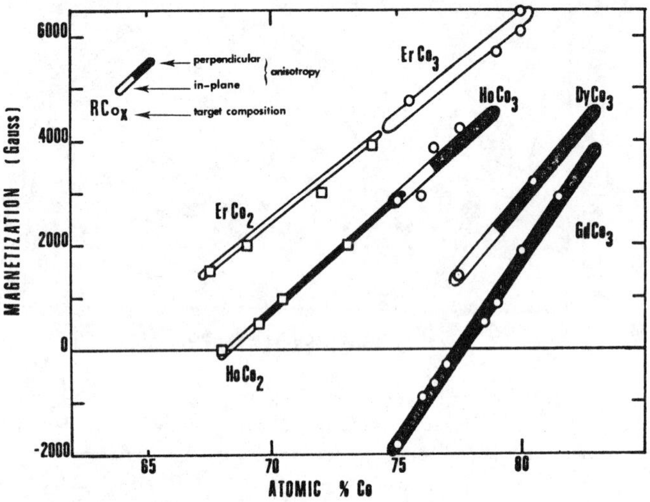


Figure 2 : Room temperature magnetization dependence with Co concentration for samples obtained from $RCo_2$ and $RCo_3$ targets.

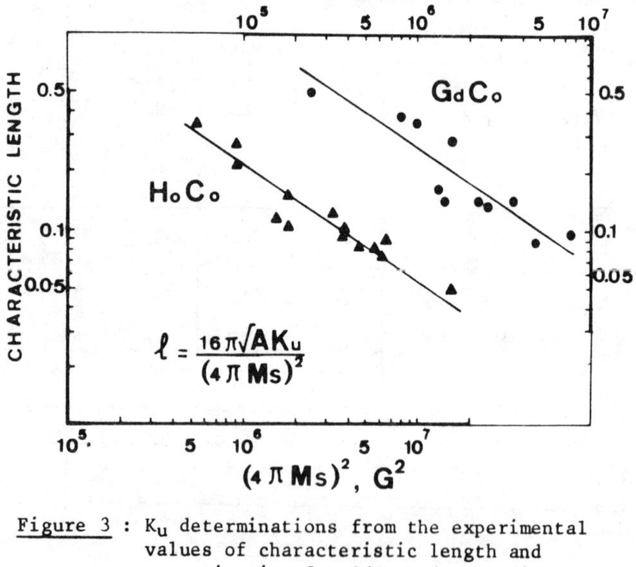


Figure 3 : $K_u$ determinations from the experimental values of characteristic length and magnetization for GdCo and HoCo films.

ROOM TEMPERATURE MAGNETIZATION

A great number of films with Co concentration ranging from x = 2.1 - 4.9 (67 - 83 At % Co) has been investigated at room temperature. An in-plane loop-tracer and the polar Kerr effect are used to determine the magnetization and the domain structure. The compositional dependence of the magnetization is illustrated on fig. 2 for samples obtained from $RCo_2$ and $RCo_3$ targets. The Co concentration of the room temperature compensated composition decreases regularly from Gd to Er due to the reduction of the molecular field $H_{Co-R}$. Another consequence of this physical property is the diminution in the compositional dependence of the magnetization from 800 G per At % Gd for GdCo to 400 G per At % Er for ErCo.

ANISOTROPY

From fig. 2, the following observations can be made concerning the existence of anisotropy perpendicular to the plane of the films :

1 - In the compositional range studied, the Gd, Dy, Ho films have been found to support bubble domains at room temperature while the Er films do not.

2 - This anisotropy seems to appear only for sufficiently high bias voltage except for $GdCo_3$.

3 - The HoCo film with about 75 At % Co can be deposited from the $HoCo_3$ target with a low bias voltage and from the $HoCo_2$ target with a high bias voltage. The first kind of samples do not support bubble domains while the second ones do.

In order to compare GdCo and HoCo as bubble materials, it is convenient to consider the characteristic length l as a function of $(4\Pi\ M_s)^2$ (fig.3). Similar $AK_u$ values are obtained for both kinds of films. Considering that the A constant is dominated by the Co - Co interactions[8], which are similar in both kinds of alloys, this leads to neighbouring values of $K_u$.

CONCLUSION

The main results on the $RCo_x$ amorphous films may be summarized as follows :

1 - The Co - Co interactions are high and independent of R giving confirmation that in the amorphous phase the transfer of rare earth conduction electrons to the 3d band does not take place unlike the case of the corresponding crystalline alloys.

2 - When the rare earth is not in an S state, the R contribution to the magnetization is reduced, due to a non-colinear spin arrangement, attributable to local anisotropy fluctuations.

3 - Bubble domains are observed in Gd, Dy, Ho compounds, without evidence of a specific action of the R on the origin of the anisotropy.

Similar anisotropy energies are obtained in Gd and Ho films subject to an important influence of the bias conditions during deposition.

The authors wish to thank T.S. Wang and A. Lienard for helpful contribution to magnetic measurements.

REFERENCES

1 - N. HEIMAN, A. ONTON, D.F. KYSER, K. LEE, C.R. GUARNERI, A.I.P. Conf. Proc., 24, 573-574 (1975).

2 - K. LEE, N. HEIMAN, A.I.P. Conf. Proc., 24, 108-109 (1975).

3 - N. HEIMAN, K. LEE, Phys. Rev. Lett., 33, 778-781 (1974).

4 - R. MEYER, H. JOUVE, J.P. REBOUILLAT, I.E.E.E. Trans. on Magn., Vol. Mag 11, 1335-1337 (1975).

5 - R. HARRIS, M. PLISCHKE, M.J. ZUCKERMANN, Phys. Rev., Lett., 31, 160 (1973).

6 - J.M.D. COEY, J. CHAPPERT, J.P. REBOUILLAT, T.S. WANG, Proc. Int. Conf. Mössbauer Spectroscopy, Cracow (Poland) 1975, Vol. 1, 347-348.

7 - R. LEMAIRE, R. PAUTHENET, J. SCHWEIZER, I.S. SILVERA, J. Phys. Chem. Solids, 28, 2471-2475 (1967).

8 - R. HASEGAWA, J. Appl. Phys., 45, 3109-3112 (1974).

# THE STATIC AND DYNAMIC PROPERTIES OF MAGNETIC BUBBLE DOMAINS IN Sm, Ca, Ge-SUBSTITUTED YTTRIUM IRON GARNET FILMS

P. F. Tumelty, R. Singh, and M. A. Gilleo
Materials Research Center, Allied Chemical Corporation, Morristown, New Jersey 07960

## ABSTRACT

The static and dynamic properties of magnetic bubble domains in garnet films of nominal composition $Y_{1.9}Sm_{0.1}Ca_1Fe_4Ge_1O_{12}$ grown by the LPE method on GGG substrates have been measured. The films supported 5-8 μm bubbles and collapse fields ranged from 60 to 110 Oe. The temperature coefficient $\alpha_{bc}$ of the collapse field became more negative with increasing $T_C$; at $T_C$ = 468 K, $\alpha_{bc}$ = -0.11%/°C at 50°C and $|\alpha_{bc}|$ increased by ∿ 8.4% for a 1 K increase in $T_C$. Pulsed field gradient bubble velocity measurements were made on films as grown and after ion implantation with 2.5 x $10^{14}$ Ne $cm^{-2}$ at 100 keV energy. In almost all cases, bubbles having rotation number $n_r$ = 1 were observed in ion-implanted films and $n_r$ = 0 bubbles were seen in as-grown samples. There was little evidence of dynamic conversion for $V < V_p$ for repetitive measurements with a given bubble. However, for randomly selected bubbles there was greater scatter in velocity. Mobility values for drive fields ≤ 4.5 Oe ranged from about 1100 to 1400 cm $sec^{-1}$ $Oe^{-1}$ with the dynamic coercive force always < 0.6 Oe.

## INTRODUCTION

The nominal magnetic film composition $Y_{1.9}Sm_{0.1}Ca_1Fe_4Ge_1O_{12}$ was first shown by Nielsen and coworkers[1] to have characteristics exceptionally well-suited to magnetic bubble domain device applications. The temperature dependence of bubble-collapse field and its relation to the Curie temperature, the effect of ion implantation on rotation number, and the question of velocity distribution in bubble translation measurements made on material of this nominal composition grown in this laboratory will be discussed.

## EXPERIMENTAL TECHNIQUES

The films were grown by liquid phase epitaxy (LPE) on (111) $Gd_3Ga_5O_{12}$ substrates in the manner described by Nielsen and coworkers.[1] The growth temperatures were in the range from 955°C to 985°C. The Curie temperature of the films was decreased by increasing the concentration of CaO + $GeO_2$ in the melt. The difference between the film and substrate lattice constants (corrected for strain) for all of the samples was less than 0.004 Å with the films under tension. Some of the samples were ion implanted with 2.5 x $10^{14}$ Ne ions/$cm^2$ at 100 keV energy to eliminate the formation of hard bubbles.

The static properties measured at 25°C for eight of the films include thickness h, demagnetized stripe width w, bubble-collapse field $H_{bc}$, mid-bias bubble diameter $d_{mb}$, and Curie temperature $T_C$; these data (Table I) were obtained by the usual optical techniques. The characteristic length ℓ and saturation magnetization $4\pi M_s$ were calculated from these data by the method of Fowlis and Copeland.[2]

The dynamic characteristics of magnetic bubble domains in this material were measured by means of the pulsed gradient field technique described by Vella-Coleiro and Tabor.[3] The bias field experienced by the bubble during translation was held constant by means of a compensating field[3] for measurements from which values of domain wall mobility were derived. The bubbles were generated by means of a large in-plane field[4,5] (2 to 3 kOe) and their displacements were kept to less than two bubble diameters. A value for the domain wall mobility μ was found by fitting the bubble velocity V vs bias field difference ΔH across the bubble diameter data to the relation[3]

TABLE I. Static properties at 25°C of 8 different $Y_{1.9}Sm_{0.1}Ca_1Fe_4Ge_1O_{12}$ films.

| Sample No. | h (μm) | w (μm) | $d_{mb}$ (μm) | $H_{bc}$ (Oe) | ℓ (μm) | $4\pi M_s$ (G) | $T_C$ (K) |
|---|---|---|---|---|---|---|---|
| 1 | 5.4 | 5.2 | 5.3 | 96 | 0.57 | 182 | 472 |
| 2 | 5.9 | 5.3 | 5.0 | 101 | 0.56 | 184 | 472 |
| 3 | 6.0 | 5.5 | 5.9 | 96 | 0.59 | 176 | 473 |
| 4 | 6.3 | 6.0 | 6.3 | 88 | 0.66 | 166 | 471 |
| 5 | 6.6 | 7.2 | 7.4 | 69 | 0.85 | 142 | 468 |
| 6 | 6.6 | 6.7 | 5.7 | 78 | 0.76 | 153 | 471 |
| 7 | 6.9 | 7.2 | 7.9 | 71 | 0.83 | 141 | 469 |
| 8 | 6.9 | 6.1 | 6.9 | 82 | 0.64 | 148 | 469 |

$$V = \frac{1}{2}\mu(\Delta H - \frac{8}{\pi} H_c), \qquad (1)$$

where $H_c$ is the dynamic coercive force. From the plot of V vs ΔH for Sample No. 8 (Fig. 1) a value of 1275 cm $sec^{-1}$ $Oe^{-1}$ was obtained for μ. For the other samples the mobility ranged from 1100 to 1400 cm $sec^{-1}$ $Oe^{-1}$ and $H_c$ was less than 0.6 Oe in all cases.

## RESULTS AND DISCUSSION

The temperature dependence of $H_{bc}$ has been measured from 20°C to 180°C by means of a Mettler microscope hot stage. For all samples $H_{bc}$ decreased nearly linearly with temperature as it increased to about 100°C, above which $H_{bc}$ fell off more rapidly. From these data the temperature coefficient of the collapse field at 50°C, which is defined by

$$\alpha_{bc}(50°C) = [(H_{bc})^{-1}(\Delta H_{bc}/\Delta T)]_{T=50°C} \qquad (2)$$

was determined. It was found that $\alpha_{bc}$ became more negative as $T_C$ increased (Fig. 2) so that some accommodation of the collapse-field temperature coefficient to that of the bias field is possible. Nielsen and coworkers[1] have pointed out that this is an important consideration and their data for one sample closely match the temperature dependence of a particular bias magnet. The value of $\alpha_{bc}$ derived from their results

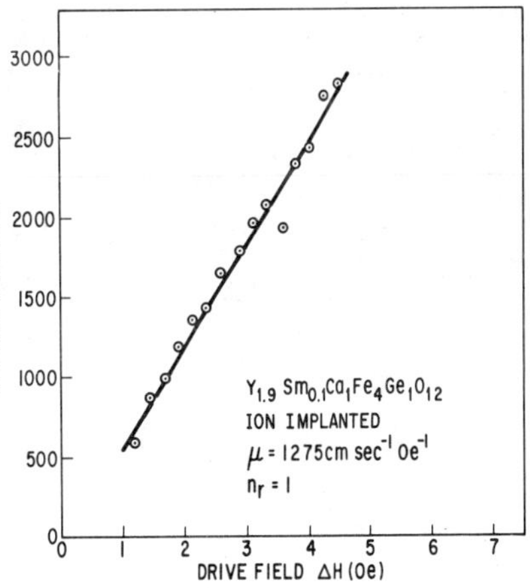


FIG. 1. Plot of velocity vs bias field difference for Sample No. 8 after ion implantation with 2.5 x $10^{14}$ Ne ions/$cm^2$ at 100 keV energy.

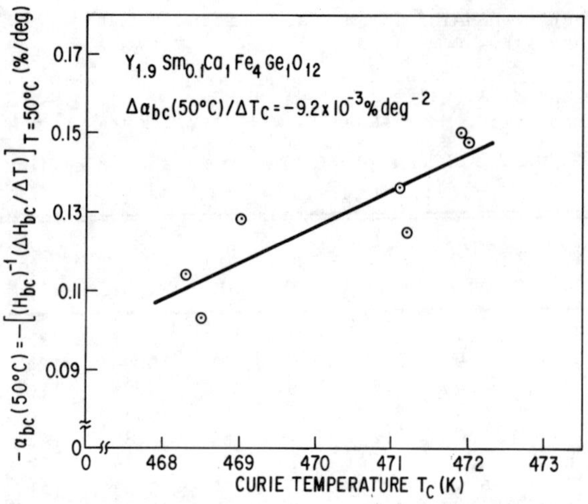


FIG. 2. Plot of temperature coefficient of bubble-collapse field at 50°C vs Curie temperature for seven films.

falls within the range shown in Fig. 2. The variation of $\alpha_{bc}$ is mainly a consequence of the increase in magnetic moment of the tetrahedral iron with an increase in Curie temperature achieved through a decrease in the Ge concentration on the tetrahedral sites. Very near the iron system compensation point the effects of the $Sm^{3+}$ ion, which has a moment opposite to that of the net iron moment in the present case, become important. The negative temperature coefficient of the oppositely directed $Sm^{3+}$ magnetization would steadily reduce the size of the negative temperature coefficient of the net magnetic moment of the material as $T_C$ is reduced from above that temperature corresponding to the compensation point for the net moment of the iron ions.

Ion implantation has a pronounced effect on the deflection angles observed in pulsed gradient field measurements on bubbles generated by a large in-plane field in this material (Fig. 3). Slonczewski and coworkers[6] relate the observed deflection angle to the rotation number $n_r$ which is defined by

$$n_r = (\gamma d|\Delta H|\sin\theta)/8V \qquad (3)$$

where $\gamma$ is the gyromagnetic ratio (taken to be $1.76 \times 10^7$ rad sec$^{-1}$ Oe$^{-1}$ for this composition[7]), d is the bubble diameter and $\theta$ is the deflection angle (taken as positive for counterclockwise deflection when the bias field points upwards toward the observer[8]). In this material as grown, bubbles generally had one pair of Bloch lines ($n_r = 0$) while bubbles generated in ion-implanted material generally had no Bloch lines ($n_r = 1$). Furthermore, in ion-implanted material the distribution of $n_r$ values about 1 was narrower than the distribution around 0 in as-grown material. However, $n_r = 1$ bubbles also occurred infrequently in as-grown material.

There is little evidence of dynamic conversion for velocities less than $V_p$ for repetitive pulsed gradient field measurements on a given bubble in these films (Fig. 4). The greater scatter in velocity observed in measurements made on randomly selected bubbles (Fig. 4) is most probably due to slight deviations from $n_r = 1$ (Fig. 3). Malozemoff[8] has shown that for constant $\Delta H$, velocity is dependent on revolution number with the maximum value occurring at $n_r = 1$.


ACKNOWLEDGMENTS

The authors are grateful to D. D. Badding for his expert technical assistance and to J. C. Walling for helpful discussions.


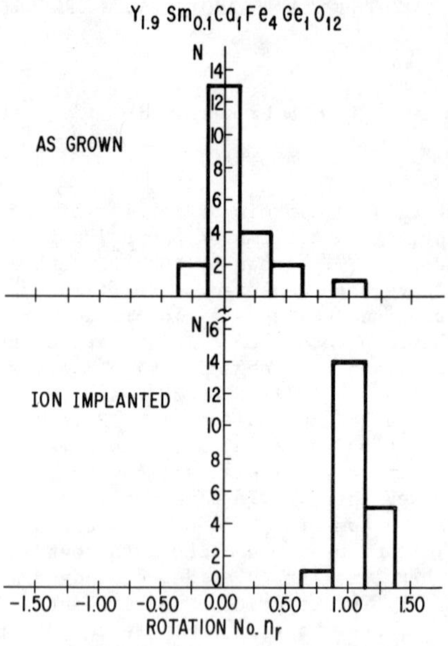


FIG. 3. Distribution of rotation number values at $\Delta H = 2.64$ Oe in as-grown and ion-implanted material. Bubble displacement was symmetrical about the centerline; a compensating field was not used.

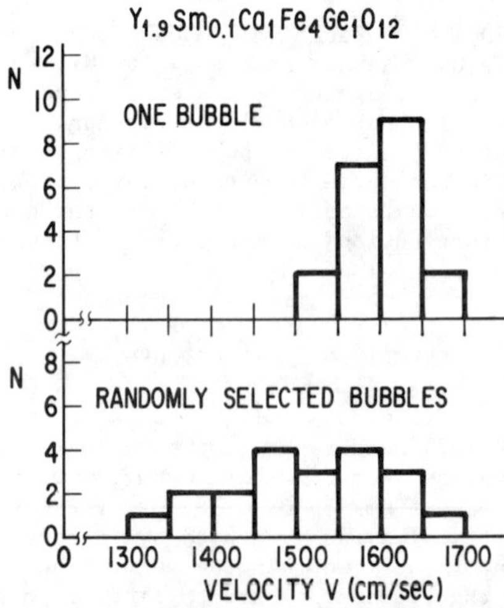


FIG. 4. Distribution of velocity values at $\Delta H = 2.64$ Oe for one bubble translated 20 times and for 20 randomly selected bubbles in an ion-implanted film. Bubble displacement was symmetrical about the centerline; a compensating field was not used.

REFERENCES


[1] J. W. Nielsen, S. L. Blank, D. H. Smith, G. P. Vella-Coleiro, F. B. Hagedorn, R. L. Barns, and W. A. Biolsi, J. Electron. Mat. 3, 693-707 (1974).

[2] D. C. Fowlis and J. A. Copeland, AIP Conf. Proc. 5, 240-243 (1972).

[3] G. P. Vella-Coleiro and W. J. Tabor, Appl. Phys. Lett. 21, 7-8 (1972).

[4] R. M. Josephs and B. F. Stein, AIP Conf. Proc. 18, 227-231 (1974).

[5] A. B. Smith, M. Kestigian, and W. R. Bekebrede, AIP Conf. Proc. 18, 167-171 (1974).

[6] J. C. Slonczewski, A. P. Malozemoff, and O. Voegeli, AIP Conf. Proc. 10, 458-477 (1973).

[7] F. B. Hagedorn, S. L. Blank, and R. J. Peirce, Appl. Phys. Lett. 26, 206-209 (1975).

[8] A. P. Malozemoff, J. Appl. Phys. 44, 5080-5089 (1973).

# GERMANIUM-SUBSTITUTED BUBBLE DOMAIN GARNET FILMS FOR USE OVER AN EXTENDED TEMPERATURE RANGE AND AT A HIGH DATA RATE*

D. M. Heinz, R. G. Warren and M. T. Elliott
Rockwell International, Electronics Research Division, Anaheim, CA 92803

ABSTRACT

Bubble domain material parameters required for device operation between -25 and 75°C at 1 MHz are reviewed and the potentials of four Ge-substituted garnet compositions for this use are presented. For a selected Ge composition, growth parameters, reproducibility of film properties and device operating margins are given.

## INTRODUCTION

The dual objectives of this investigation have been the development of a bubble garnet composition capable of device operation between -25 and 75°C at a data rate of 1 MHz and the means for its reproducible film growth. Bubble device performance is usually limited by different phenomena at high and low temperatures so that material parameters must be developed to meet performance goals across the temperature range.

Magnetic properties degrade rapidly as the Neel temperature $T_N$ is approached which requires the $T_N$ of a bubble composition to be about 100C° higher than the highest operating temperature. In this instance, for use at 75°C, $T_N$ should be $\geq$175°C. Satisfactory device performance also requires that bubble nucleation not occur due to the fields generated at the permalloy device elements by the in-plane drive field. Thus, the quality factor q and the wall energy $\sigma_w$ should be sufficiently large to prevent nucleation at the elevated temperature with the circuit spacing and the drive field normally employed. For operation of 4 to 6μm bubble devices at 75°C, room temperature values of $q \geq 4$ and $\sigma_w \geq 0.2$ ergs/cm$^2$ are necessary.

The wall mobility $\mu_w$ decreases and the coercivity $H_c$ increases on lowering the temperature, often by a factor of two on going from 25°C to -25°C. To realize satisfactory device performance at any temperature, the bubble velocity must be sufficiently high to permit the attainment of the desired data rate with the drive field normally employed. Thus the room temperature values of $\mu_w$ and $H_c$ should be of the proper magnitude to accommodate changes at lower temperatures. For 1 MHz operation of 4 to 6μm bubble devices at -25°C, room temperature values of $\mu_w \geq 1000$cm/sec-Oe and $H_c \leq 0.3$ Oe are necessary.

The values of q and $\sigma_w$ needed for high temperature device operation require a large value of the uniaxial anisotropy $K_u$, while the value of $\mu_w$ needed for low temperature device operation requires small values of $K_u$ and the damping parameter $\lambda$. Thus the optimum parameters result from a compromise between those which give best performance at high temperatures and best performance at low temperatures. The q, $\sigma_w$ and $\mu_w$ values given earlier are compromise values. Both high growth-induced $K_u$ and $\lambda$ are associated with the presence of Sm and Eu in magnetic garnets so that their concentrations must be critically controlled.

Since the permalloy device period is fixed in size, the bubble diameter must be within narrow limits for reliable device operation. Over the temperature range, this calls for a temperature coefficient of characteristic length $\ell_T (\equiv \Delta\ell/\ell\Delta T) \leq 0.3$ percent/C°. In addition, the temperature coefficient of magnetization $M_T (\equiv \Delta M/M\Delta T)$ must be negative over the temperature range in order that hexagonal ferrite permanent magnets, which have a negative $M_T$, may be used to provide the bias field.

The magnetic moment $4\pi M$ of a pure iron garnet is normally reduced to about 180G for 4 to 6μm bubble domains by substitution of non-magnetic ions for Fe ions. Ga has most often been employed for this purpose because about 90 percent substitutes on tetrahedral Fe sites[1], lowering the magnetization. On the other hand, about 99 percent of Ge substitutes on tetrahedral sites[2]. As a consequence, substitution of one Ga per garnet formula unit reduces the magnetization by about 1400G and substitution of one Ge per garnet formula unit, about 1700G. Thus less Ge is required and $T_N$ is at least 50C° higher than Ga compositions of the same $4\pi M$.

## EXPERIMENTAL INVESTIGATIONS

When Ge(IV) is substituted for Fe(III), charge neutrality may be maintained in the garnet formula unit by substituting an equal number of Ca(II) for Y(III). In this investigation, the molar ratios of CaO to $GeO_2$ were varied between 0.51 and 1.85 in melts of various compositions.

The nominal compositions of the Ge-substituted garnets grown in this study are listed in Table I along with typical film properties. All of the film growth was carried out by the horizontal dipping, liquid phase epitaxy method using equipment and techniques described previously[3]. For the normal range of growth conditions, the Ge content of a film increases with growth rate (which is opposite to the behavior of Ga). Thus the control of the growth rate and hence the growth temperature is critical in preparing films with specified bubble properties.

The first Ge-substituted garnet investigated was composition A, using the melt formulation given by Bonner et al[4]. In the initial attempt to deposit a film at the reported growth temperature of 980°C, the large degree of supercooling (about 50C°) produced surface nucleation and a very high film growth rate (>10μm/min). After a high temperature soak to redissolve the small crystallites, subsequent depositions were made using a more moderate degree of supercooling.

An estimate of the lattice parameter of composition A was made based on substitutions in $Y_3Fe_5O_{12}$(YIG). Thus the normal changes of lattice parameter[5] resulting from substituting Eu and Lu for Y, and Ca and Ge for Y and Fe (-0.004Å for the ion pair[2]) resulted in a calculated lattice parameter of 12.367Å for composition A. On a GGG substrate with a lattice parameter of 12.383Å, this film would be under such great tensile strain ($\Delta a$= 0.016Å) that it would crack. When the first films were characterized for lattice mismatch with the substrate by double crystal reflection rocking curve measurements,

TABLE I. CHARACTERISTICS OF GERMANIUM-SUBSTITUTED BUBBLE DOMAIN GARNET FILMS.

| | NOMINAL COMPOSITION | h μm | w μm | $H_{col}$ Oe | $4\pi M$ G | $\ell$ μm | $\ell_T$ -%/C° | $\sigma_w$ erg/cm$^2$ | q - | $T_N$ °C | $\mu_w$ cm/s-Oe | $H_c$ Oe |
|---|---|---|---|---|---|---|---|---|---|---|---|---|
| A | $Y_{1.64}Eu_{0.10}Lu_{0.30}Ca_{0.96}Fe_{4.04}Ge_{0.96}O_{12}$ | 7.1 | 6.2 | 92 | 168 | 0.64 | 0.31 | 0.14 | 4.5 | 184 | 2600 | 0.12 |
| B | $Y_{1.52}Eu_{0.30}Tm_{0.30}Ca_{0.88}Fe_{4.02}Ge_{0.88}O_{12}$ | 4.6 | 4.7 | 109 | 218 | 0.53 | 0.24 | 0.20 | 4.1 | 197 | 1400 | 0.24 |
| C | $Y_{1.88}Eu_{0.20}Ca_{0.92}Fe_{4.08}Ge_{0.92}O_{12}$ | 4.2 | 5.6 | 69 | 165 | 0.70 | 0.22 | 0.15 | 4.4 | 195 | 2300 | 0.17 |
| D | $Y_{1.95}Sm_{0.10}Ca_{0.95}Fe_{4.05}Ge_{0.95}O_{12}$ | 5.1 | 6.2 | 78 | 175 | 0.75 | 0.46 | 0.18 | 4.1 | 190 | 1520 | 0.20 |
| E | $Y_{1.75}Eu_{0.10}Tm_{0.30}Ca_{0.85}Fe_{4.15}Ge_{0.85}O_{12}$ | 3.8 | 4.0 | 133 | 272 | 0.46 | 0.20 | 0.27 | 5.2 | 210 | 1500 | 0.40 |

the films were found to be in slight compression ($\Delta a = -0.0005Å$). It was unlikely that Pb(II) expanded the film lattice because this melt composition utilized a Ca/Ge molar ratio of 1.86 and not much Pb is incorporated at film growth temperatures over 1000°C. A possible explanation for the disparity is that Lu had partially substituted on octahedral lattice sites where it expanded the lattice. Further evidence for octahedral site occupancy is that $T_N$ was about 10C° lower than estimated for this Ge content[2]. Using relations presented elsewhere[6], an analysis of the lattice expansion and low $T_N$ suggests that about 0.1 Lu per garnet formula unit is on octahedral sites. An addition of $Fe_2O_3$ was made to the melt with the expectation that some Lu would be forced out of the octahedral lattice sites. Films grown after this addition were in tension with higher $\sigma_w$ values. The data in Table I for composition A is for a sample with $\Delta a = 0.006Å$.

It was very difficult to grow films of composition A with reproducible properties from this melt formulation, in part due to the ease with which Lu was shifted by Fe from octahedral sites and in part due to the rapid loss of PbO at the high saturation temperature $T_{sat}$.

Despite the overall good properties exhibited by composition A, the problems of low $\sigma_w$ and growth of reproducible films were not resolved. The second melt composition was designed to increase the growth-induced $K_u$ (and hence increase $\sigma_w$) by raising the Eu content, and to reduce the rare earth occupancy of octahedral sites by utilizing larger Tm ions in place of Lu. Composition B films were grown without complications. Films were in slight tension and $T_N$ values were greater than those of composition A. The third Ge-containing melt was formulated to have a $\mu_w$ between the values of compositions A and B. Film growth of composition C was quite predictable but it suffered from low $\sigma_w$. The fourth melt was prepared to compare the properties of another attractive Ge-substituted composition described by Nielsen et al[7]. This composition exhibited a marginal $\sigma_w$ and a high $\ell_T$ for the intended application. Thus, based on the overall behavior of these four materials, composition B was selected for more extensive growth studies.

For the reproducibility study, the material performance goals were 4μm bubble operation over the -25 to 75°C temperature range at 0.5 MHz. Target tolerances for h and w were 4.0μm $\pm$5 percent and for $H_{col}$ was $\pm$5 Oe. In developing these properties, it was necessary to modify composition B, especially to obtain $q \geq 4$ for this higher $4\pi M$ material. The growth-induced $K_u$ was augmented by a stress-induced $K_u$ ($\Delta a = 0.008Å$). In the final formulation, the following anisotropy values were obtained from FMR and annealing studies (in units of $10^3$ ergs/cm$^3$): growth 6.4, stress 1.2 and cubic 1.5. This formulation is designated as composition E.

The reproducibility of growth of composition E was carried out using the melt formulation given in Table II. This formulation produces approximately 650 ml of melt with a solute concentration of 18.3 mole percent and a $T_{sat}$ of 1015°C. For a typical supercooling of 19C°, the film growth rate is 1.0μm/min. The film growth procedure was similar to that reported previously[3]; however automatic timing and sequencing features were added to the system for improved control of rotation and rotation reversal of the substrate, substrate insertion and withdrawal, and deposition duration control. Typical film properties are presented for composition E in Table I.

TABLE II. MELT FORMULATION FOR COMPOSITION E

| Oxide | Grams | Moles | Oxide | Grams | Moles |
|---|---|---|---|---|---|
| $Y_2O_3$ | 16.087 | 0.0712 | $GeO_2$ | 57.667 | 0.5513 |
| $Eu_2O_3$ | 2.042 | 0.0058 | $Fe_2O_3$ | 418.65 | 2.6215 |
| $Tm_2O_3$ | 6.325 | 0.0164 | $B_2O_3$ | 71.0 | 1.020 |
| $CaCO_3$ | 52.169 | 0.5212 | PbO | 3550.0 | 15.904 |

Films grown from this melt formulation were analyzed by electron microprobe and atomic absorption spectroscopy to yield the composition $Y_{1.75}Eu_{0.10}Tm_{0.30}Ca_{0.85}Fe_{4.15}Ge_{0.85}O_{12}$. ($T_N$, $\Delta a$ and $4\pi M$ values suggest that 0.01 Tm is present on octahedral lattice sites.) The analyses agree within the following tolerances per formula unit: Y $\pm$ 0.05, Eu $\pm$ 0.00, Tm $\pm$ 0.01, Ca $\pm$ 0.06, Fe $\pm$ 0.06 and Ge $\pm$ 0.21. This spread for Ge results from a low atomic absorption value which was caused by limited sensitivity and possible loss during sample digestion. The following distribution coefficients were obtained: Y 2.90, Eu 2.03, Tm 2.16, Ca 0.38, Fe 0.92 and Ge 1.79. Davies et al[8], reported the values Ca 0.15 and Ge 1.8 for films similar to composition D.

Fifty 37.5 mm films were grown from a single large volume melt. During this extended film growth series, PbO losses were replaced and solute depletion was compensated by decreasing the growth temperature and increasing the growth period as needed to maintain material parameters. Increments of 0.5C° and 10 sec were used. The total changes for this series were 3C° and 0.5 min. All of the films met the tolerances on h and w, and 88 percent met the tolerance on $H_{col}$. Values of $\sigma_w$ and q were 0.29 $\pm$ 0.02 ergs/cm$^2$ and 5.6 $\pm$ 0.5. The $\mu_w$ goal for 0.5 MHz operation of 500 cm/sec-Oe was bettered to the range of 700 $\pm$ 100 cm/sec-Oe. Although the $H_c$ goal of $\leq$0.3 Oe was not met, all films fell in the range of 0.4 $\pm$ 0.1 Oe, which is sufficiently low for good device operation.

The simultaneous growth of a number of 37.5 mm films was investigated. In the 10-at-a-time configuration described earlier[3], both the larger substrate diameter and greater melt viscosity limited our success in meeting the objectives of h uniformity of $\leq$ one interference fringe of Na light over 75 percent of the area and $H_{col}$ reproducibility of $\pm$5 Oe. A limited evaluation of 50 mm films grown one-at-a-time was also made for which these h and $H_{col}$ objectives were met.

The ultimate evaluation for this Ge-substituted garnet composition is in its performance at a data rate of 0.5 MHz between -25 and 75°C. Using a 20 K bit, T-bar circuit with a 16μm period and a 55 Oe drive field, the upper and lower bias fields (in Oe) were -25°C: 144.5, 133.5; 25°C: 137.5, 127.0; and 75°C: 124.5, 117.5. Device operation of this and other circuits at 0.5 MHz were extended over the range of -50 to 90°C.

In summary, the reasons for using Ge-substitution in bubble garnets has been discussed. A new composition has been described which has been grown with reproducible properties. This Ge-substituted composition has performed satisfactorily in devices which meet temperature range and data rate objectives.

## REFERENCES

*Work supported on U. S. Air Force Contracts F33615-73-C-5017 and F33615-75-C-5010.

1. S. G. Geller, J. A. Cape, G. P. Espinosa and D. H. Leslie, Phys. Rev. 148, 522-24(1966).
2. S. Geller, H. J. Williams, C. P. Espinosa and R. C. Sherwood, Bell System Tech. J. 43,565-623 (1964).
3. R. G. Warren, J. E. Mee, F. S. Stearns and E. C. Whitcomb, AIP Conf. Proc. 18, 63-67(1974).
4. W. A. Bonner, J. E. Geusic, D. H. Smith, L. G. Van Uitert and G. P. Vella-Coleiro, Mat. Res. Bull. 8, 1223-29 (1973).
5. L. G. Van Uitert, E. M. Gyorgy, W. A. Bonner, W. H. Grodkiewicz, E. J. Heilner and G. J. Zydzik, Mat. Res. Bull. 6, 1185-200 (1971).
6. H. L. Glass and M. T. Elliott, J. Cryst. Growth 27, 253-60 (1974).
7. J. W. Nielsen, S. L. Blank, D. H. Smith, G. P. Vella-Coleiro, F. B. Hagedorn, R. L. Barns and W. A. Biolsi, J. Electronic Mat'ls.3, 693-707 (1974).
8. J. E. Davies, E. A. Giess and J. D. Kuptsis, J. Mat. Sci. 10, 589-92 (1975).

# REPRODUCIBLE GROWTH AND BUBBLE PROPERTIES OF RARE-EARTH-SUBSTITUTED YCaGeIG FILMS

T. Obokata, H. Tominaga, T. Mori and H. Inoue
FUJITSU LABORATORIES LTD., Kawasaki, Japan


## ABSTRACT

Thermally stable bubble garnets $(YCaRE)_3(FeGe)_5O_{12}$ (RE: rare-earth ion) were studied as materials for high density and high speed bubble devices. Rare-earth ions used were EuYb, EuTm, EuEr, Sm, SmLu and SmTm. These garnets were particularly useful for bubble materials with bubble diameters of 6 μm to less than 1 μm. For example, the temperature stabilities of submicron bubble garnet YCaSmLuGeIG were very good up to 140°C, temperature coefficients of the stripe width and bubble collapse field being -0.01%/°C and -0.2%/°C, respectively. Typical values of 4πMs, uniaxial anisotropy field and wall mobility of this garnet were 518 Gauss, 1650 Oe and 600 cm/sec·Oe, respectively. For the highly reproducible growth the feedback control technique of the growth process was applied for YCaEuYbGeIG with 3 μm bubble diameters which was used for $10^5$ bits/chip bubble devices. By this technique the epitaxial growth of the garnet was well controlled resulting in an extremely high yield of films: for example, more than 60% for bubble collapse field Ho=140±3 Oe.


## INTRODUCTION

Bonner et al.[1] first reported that epitaxial films of $(Y,RE)_{3-x}Ca_xFe_{5-x}Ge_xO_{12}$ (RE: rare-earth ion) have excellent magnetic properties for bubble domain applications. We also studied independently the epitaxial growth and the magnetic properties of the rare-earth free iron garnets $Y_{3-2x}Ca_{2x}Fe_{5-x}V_xO_{12}$ (YCaVIG)[2] and $Y_{3-x}Ca_xFe_{5-x}Ge_xO_{12}$ (YCaGeIG)[3] for the purpose of obtaining a thermally stable garnet with a high domain wall mobility. However, it turned out to be very difficult to grow the high quality crystal of YCaVIG by the LPE technique. On the other hand the crystal growth of YCaGeIG was easy and this garnet was found to be very attractive for bubble domain devices. But the uniaxial magnetic anisotropy of YCaGeIG, which was induced by stress, was not large enough for stable bubble operation. We substituted, therefore, rare-earth ions for yttrium ions to provide high growth-induced anisotropy. In order to achieve the bit density of $10^5 \sim 10^6$ bits/cm$^2$, we have studied these garnets supporting 3 μm to submicron-diameter bubbles.

## EXPERIMENT AND RESULTS

Garnet films were grown on $Gd_3Ga_5O_{12}$ substrates by the usual LPE technique[4]. The adopted rare-earth ions were EuYb, EuTm, EuEr, Sm, SmLu and SmTm, which were chosen from the following criterions: (1) low damping, (2) large growth-induced anisotropy, (3) lattice constant matching. As to the criterion (2), Eu and Sm were selected according to the result of Gyorgy et al[5]. Table I shows typical bubble properties of these garnets. As a consequence of the characteristic length $\ell$ being adjusted in the range of 0.3 to 0.1 μm, the stripe width Sw was in the range of about 3 μm to 0.8 μm. The film thickness h was controlled to be nearly equal to Sw. As has been already reported[1,6], the temperature coefficients of Sw and the bubble collapse field Ho are smaller than those of $Y_{2.6}Sm_{0.4}Fe_{3.8}Ga_{1.2}O_{12}$ which is extensively used as a crystal for 6 μm-diameter bubbles. Figure 1 shows variations of the uniaxial anisotropy constant Ku and domain wall mobility $\mu_w$ with the increase of contents x of rare-earth ions. Ku increases linearly with x in the range $0.1 \leq x \leq 0.4$ and, as a result, a quality factor q ($=Ku/2\pi Ms^2$) also increases with x. In the case of YCaSmTmGeIG, for example, q increases from 2.4 to 3.6 as x increases from 0.1 to 0.3.

For mass memory applications a large number of films with similar properties (in particular Ho and h) are required. For this purpose, we investigated the method of the control for achieving reproducibility of the film properties using YCaEuYbGeIG which we chose as

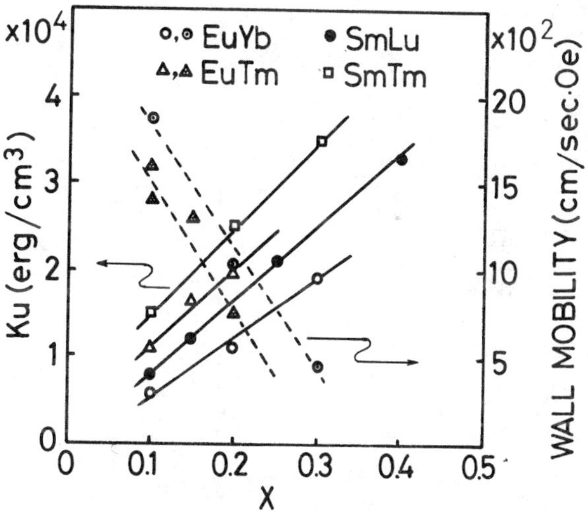


Fig. 1 Uniaxial anisotropy constant Ku and wall mobility as a function of x (x is shown in Table I).

Table I. Typical Film Properties

| MATERIAL | x * | STRIPE WIDTH Sw (μm) | FILM THICKNESS h (μm) | 4πMs (Gauss) | $\ell$ (μm) | $q(=\frac{Ku}{2\pi Ms^2})$ | μw (cm/sec·Oe) | TEMPERATURE COEFFICIENT (%/°C) $\frac{1}{Sw}\cdot\frac{\Delta Sw}{\Delta T}$ | $\frac{1}{Ho}\cdot\frac{\Delta Ho}{\Delta T}$ |
|---|---|---|---|---|---|---|---|---|---|
| YCaEuYbGeIG | 0.2 | 3.05 | 3.15 | 270 | 0.35 | 4.5 | 1,000 | - 0.1 | - 0.15 |
| YCaEuTmGeIG | 0.15 | 3.28 | 3.69 | 300 | 0.35 | 4.8 | 1,300 | - 0.07 | - 0.15 |
| YCaEuErGeIG | 0.16 | 3.0 | 3.3 | 295 | 0.34 | 3.3 | 700 | - 0.14 | - 0.13 |
| YCaSmGeIG | 0.09 | 3.3 | 2.7 | 183 | 0.42 | 4.5 | 1,800 | - 0.26 | - 0.1 |
| YCaSmLuGeIG | 0.4 | 1.66 | 2.32 | 518 | 0.12 | 3.2 | 600 | - 0.01 | - 0.2 |
| " | 0.4 | 0.81 | 0.63 | 434 | 0.10 | 3.0 | | | |
| YCaSmTmGeIG | 0.3 | 1.31 | 1.62 | 524 | 0.13 | 3.6 | 800 | - 0.01 | - 0.18 |
| " | 0.3 | 0.90 | 0.87 | 470 | 0.11 | 2.4 | | | |

* $Y_{2.1-8x/3}Ca_{0.9}Eu_xYb_{5x/3}Fe_{4.1}Ge_{0.9}O_{12}$, $Y_{2.1-x}Ca_{0.9}Sm_xFe_{4.1}Ge_{0.9}O_{12}$ x is the melt composition.

$Y_{2.1-10x/3}Ca_{0.9}Eu_xTm_{7x/3}Fe_{4.1}Ge_{0.9}O_{12}$, $Y_{2.25-12x/5}Ca_{0.75}Sm_xLu_{7x/5}Fe_{4.25}Ge_{0.75}O_{12}$

$Y_{2.1-5x}Ca_{0.9}Eu_xEr_{4x}Fe_{4.1}Ge_{0.9}O_{12}$, $Y_{2.25-7x/2}Ca_{0.75}Sm_xTm_{5x/2}Fe_{4.25}Ge_{0.75}O_{12}$

a crystal for $10^5$ bits/chip bubble devices. The melt composition used for the growth of $Y_{1.3}Ca_{0.9}Eu_{0.2}Yb_{0.6}Fe_{4.1}Ge_{0.9}O_{12}$ is as follows: 2.003 gms $Y_2O_3$, 2.058 $CaCO_3$, 0.4628 $Eu_2O_3$, 1.5144 $Yb_2O_3$, 22.97 $Fe_2O_3$, 13.99 $GeO_2$, 478.7 PbO and 9.574 $B_2O_3$. Device requirements are listed in Table II. A series of films grown from the same melt under constant conditions of the temperature and growth time exhibit a drift in their film parameters from run to run due to changes in the melt composition. As Hewitt et al.[7] showed, the simple empirically determined corrections of increasing the growth time by 10 second/run and decreasing the growth temperature by 0.1°C/run on the average can control the drift in YCaEuYbGeIG. However, this control scheme is not sufficient for our device requirements. We employed, therefore, a secondary control scheme or the feedback control technique. In this method, crystal properties are quickly measured after growth of each crystal and, from the deviations from the aims, the subsequent growth conditions are determined. Growth conditions are temperature, growth time and substrate rotation rate. Figure 2 shows the dependence of $4\pi Ms$, wall energy density $\sigma w$ and growth rate Vg on the supercooling at a constant rotation rate of 180 rpm. The temperature dependence of $4\pi Ms$ is quite different from that of Ga or Al doped iron garnets, i.e., $4\pi Ms$ decreases with decreasing temperature. From these data we empirically determined a control chart of Ho and h, which we adopted as the most significant film parameters, with regard to the temperature and growth time. By these control methods, more than forty runs have been made from the same melt. Figure 3 shows the cumulative percentage of Ho, h and Sw. It can be seen from this figure that the LPE growth of YCaEuYbGeIG is well controlled with an extremely high yield, 60%, of films satisfying all the device requirements as given in Table II.

Table II. Device Requirements for YCaEuYbGeIG

| PROPERTY | AIM | LIMITS |
|---|---|---|
| STRIPE WIDTH | 3.05 μm | 2.75 ~ 3.50 μm |
| FILM THICKNESS | 3.15 μm | 2.75 ~ 3.50 μm |
| COLLAPSE FIELD | 140 Oe | 137 ~ 143 Oe |
| WALL MOBILITY | 1000 cm/sec·Oe | |
| COERCIVITY | < 0.3 Oe | |
| $\ell$ | 0.35 μm | |
| $4\pi Ms$ | 270 Gauss | 240 ~ 300 Gauss |
| WALL ENERGY | 0.203 ergs/cm² | |

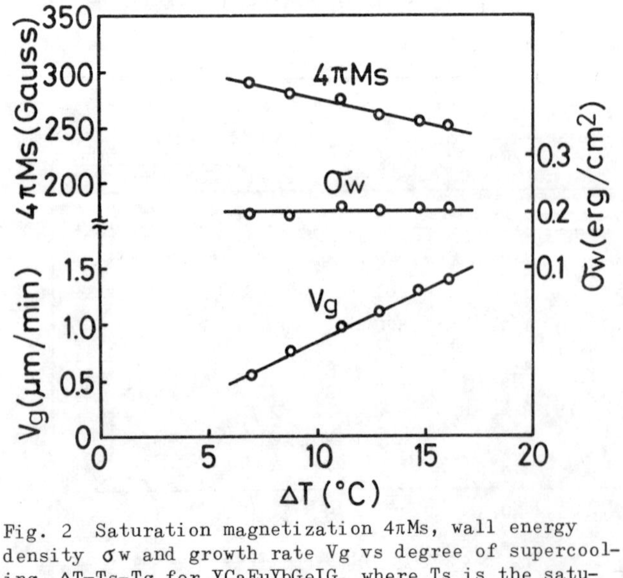


Fig. 2 Saturation magnetization $4\pi Ms$, wall energy density $\sigma w$ and growth rate Vg vs degree of supercooling $\Delta T = Ts - Tg$ for YCaEuYbGeIG, where Ts is the saturation temperature and Tg is the growth temperature.

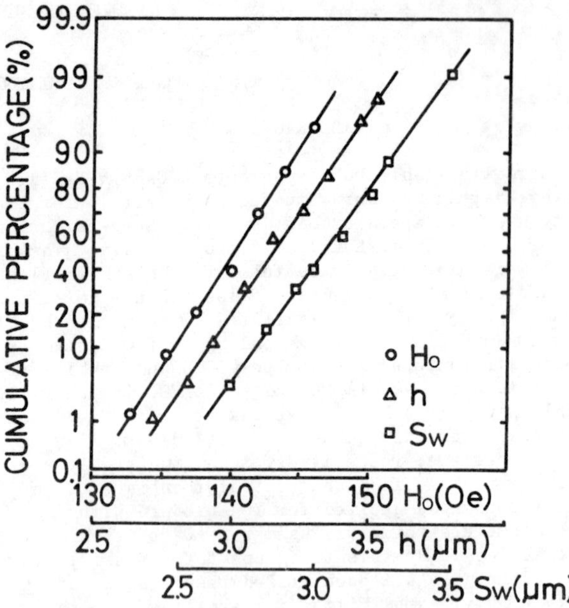


Fig. 3 Cumulative percentage of bubble collapse field Ho, film thickness h and stripe width Sw for YCaEuYbGeIG.

## DISCUSSION AND CONCLUSION

The reproducibility of Ho for YCaEuYbGeIG is very good, namely, 80% yield for Ho=140 Oe±3%, which is better than that of YSmGaIG: 40% yield for Ho=110 Oe ±3% obtained by using the same control method as mentioned above. This result is explained as follows: the increase of supercooling $\Delta T$ due to temperature fluctuation causes the increase of the film thickness and the decrease of $4\pi Ms$. As a result, the increase of Ho due to the increase of the film thickness is compensated by the decrease of Ho caused by the decrease of $4\pi Ms$, thus reducing the change of Ho caused by the temperature fluctuation.

In conclusion, the rare-earth substituted YCaGeIG has excellent bubble properties as materials for high density and high speed bubble devices. The controlled epitaxial growth of these garnets is easier than that of Ga or Al substituted iron garnets, and high quality epitaxial films of YCaEuYbGeIG with closely controlled magnetic properties are obtained with an extremely high yield.

## ACKNOWLEDGMENT

The authors are indebted to K. Asama for his guidance and helpful discussions, T. Namikata and K. Yamaguchi for their valuable criticism. They are pleased to thank M. Nakamura, F. Iwai and S. Kojima for the encouragement.

## REFERENCES

1. W. A. Bonner, J. E. Geusic, D. H. Smith, L. G. Van Uitert and G. P. Vella-Coleiro, Mat. Res. Bull 8, 1223 (1973). W. A. Bonner, AIP Conf. Proc. No. 18, 68 (1974). J. E. Geusic, D. H. Smith, L. G. Van Uitert and G. P. Vella-Coleiro, ibid, 69 (1974).
2. T. Obokata, K. Sokura and T. Namikata, IEEE Trans. Magn. MAG-9, 373 (1973).
3. T. Obokata, K. Sokura and T. Namikata, Annual Conf. of Japan Soc. of Appl. Phys. (1974) (in Japanese).
4. H. J. Levinstein, S. Licht, R. W. Landorf and S. L. Blank, Appl. Phys. Letters 19, 486 (1971).
5. E. M. Gyorgy, M. D. Sturge, L. G. Van Uitert, E. J. Heilner and W. H. Grodkiewicz, J. Appl. Phys. 44, 438 (1973).
6. A. H. Bobeck, P. I. Bonyhard and J. E. Geusic, Proc. of the IEEE 63, 1176 (1975).
7. B. S. Hewitt, R. D. Pierce, S. L. Blank and S. Knight, IEEE Trans. Magn. MAG-9, 366 (1973).

# $(LuSm)_3Fe_{5-x}Ga_xO_{12}$ GARNET FILMS FOR SMALL BUBBLE DIAMETERS*

J. T. Carlo, D. C. Bullock, R. E. Johnson, and S. G. Parker,
Texas Instruments Incorporated, P. O. Box 5936, M. S. 145,
Dallas, Texas 75222

ABSTRACT

The $(LuSm)_3Fe_{5-x}Ga_xO_{12}$ garnet system is investigated for bubble devices requiring a nominal bubble size from one to two microns. All film compositions are grown matched on GGG substrates by compensating for the change in lattice constant with different x by varying the Lu/Sm ratio. As the gallium concentration, x, is varied from 0 to 0.84; $\ell$ increases from 0.045 µm to 0.21 µm and $4\pi M$ decreases from 1750 Oe to 530 Oe. Temperature dependence of $\ell$ and Q shows the applicability of this material system for device operation up to 100°C. Quasistatic margins for T-bar propagation are given for two micron diameter bubbles in two films with different gallium concentrations.

## INTRODUCTION

The $(LuSm)_3Fe_5O_{12}$ garnet has been shown to possess high growth induced anisotropy[1,2] and submicron magnetic bubbles (0.8 µm) have been propagated in this material.[2] A drive field of 100 Oe was required for bubble propagation with Permalloy circuits[2] because of the high $4\pi M$ (1750 Oe) for this composition.[3] Giess[4] has shown the applicability of $4\pi M$ reduction in small bubble materials by the addition of gallium. In a similar manner, the $4\pi M$ for the $(LuSm)_3Fe_5O_{12}$ garnet composition has been reduced by gallium incorporation. For one micron bubbles $4\pi M$ is reduced to below 1000 Oe; for two micron bubbles $4\pi M$ is reduced to below 600 Oe.

## MATERIAL GROWTH

Standard LPE dipping techniques[5] were used to grow the material in an isothermal zone of a furnace with temperature control of ± 0.1°C. During growth a slice is rotated at 60 rpm with a rotation rate of 350 rpm after removal from the melt in order to remove flux droplets. A typical melt composition in mole percent consisted of: PbO(79.0), $B_2O_3$ (5.1), $V_2O_5$ (4.3), $Fe_2O_3$ (10.36), $Ga_2O_3$ (0.73), $Lu_2O_3$ (0.31) and $Sm_2O_3$ (0.20). The film composition grown from this melt is approximately $Lu_{1.8}Sm_{1.2}Fe_{4.32}Ga_{0.68}O_{12}$ where the gallium content was determined by Curie temperature data and the lutetium/samarium ratio from the film lattice constant[2] (12.383 Å).

The saturation temperature of the melt was 975°C with the epi films being grown 15°C - 25°C below the saturation temperature. A growth rate of 1.0 µm/min was obtained at 950°C, and 0.85 µm/min at 960°C. Film thickness was chosen to give bubble diameter of either one micron (x = 0.20, 0.31, 0.43, h = 2.1, 1.7, 1.2 µm) or two microns (x = 0.68, 0.84, h = 3.5, 1.9 µm). The segregation coefficient, $\alpha_{Ga}$, was 2.2 for growth at 960°C from the melt composition given above.

Film compositions with different x were grown on GGG substrates by compensating for the change in lattice constant with x by varying the lutecium/samarium ratio.

## MAGNETIC PROPERTIES

Variation of gallium concentration in the $(LuSm)_3Fe_{5-x}Ga_xO_{12}$ garnet system allows control of $\ell$ and $4\pi M$ to obtain the desired bubble material properties for a selected bubble diameter. Figure 1 shows the measured values of $4\pi M$ and $\ell$ versus x. $\ell$ was calculated from the measured thickness and static stripe domain period data at room temperature[6] (23°C). The value of $\ell$ was shown to be relatively independent of film thickness; e.g., for a given melt composition a film with thickness 5.4 µm has $\ell$ = 0.047 µm, while a film with thickness 4.3 µm has $\ell$ = 0.048 µm. This constancy in $\ell$ permits the selection of $\ell$ by choosing the melt composition and growth conditions, and the subsequent selection of film thickness by choosing growth time to give a desired bubble diameter.

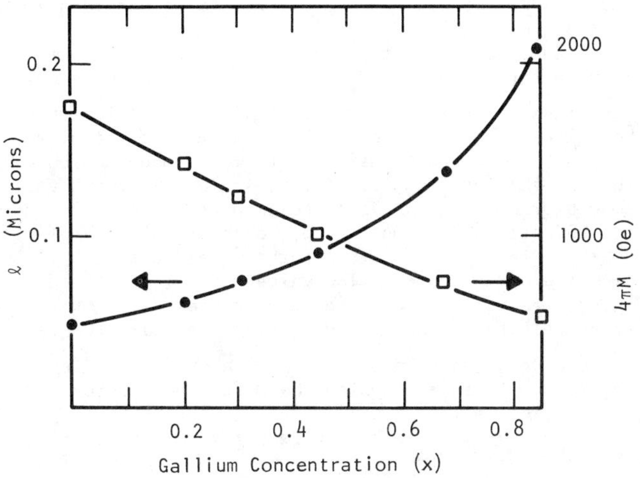


Fig. 1. $\ell$ and $4\pi M$ versus Gallium Concentration in $(LuSm)_3Fe_{5-x}Ga_xO_{12}$.

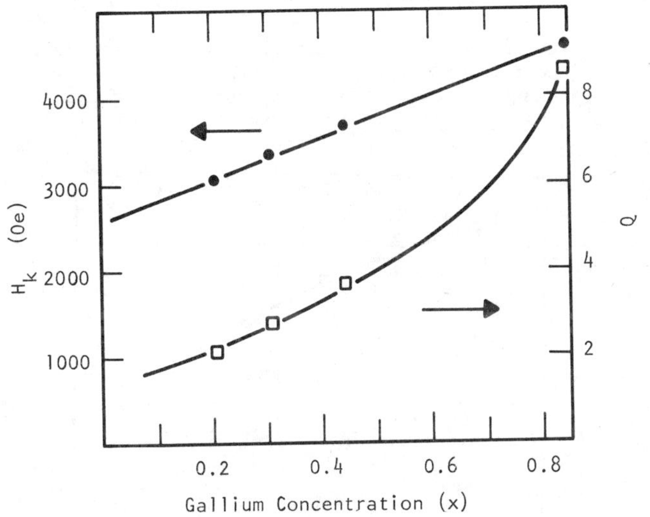


Fig. 2. $H_k$ and Q versus Gallium Concentration in $(LuSm)_3Fe_{5-x}Ga_xO_{12}$.

As illustrated in Fig. 1, by requiring an $h/\ell$ value of 10, a material for one micron diameter bubbles has x = 0.50, and a two micron material has x = 0.82. The values of $4\pi M$ for these values of x are also illustrated in Fig. 1. The values for $4\pi M$ are derived from measurements of h, $\ell$ and the bubble collapse field. For x = 0.50, $4\pi M$ = 900 Oe and for x = 0.82, $4\pi M$ = 530 Oe. These values of $4\pi M$ are important for determining bubble device drive field, which has been shown to be proportional to $4\pi M$.[3]

Figure 2 illustrates Q and $H_k$ versus x. The values for $H_k$ were determined by FMR and optical magnetometer measurements. For Q greater than three, the values derived from the two types of measurements are in agreement, while for low Q values, the optical magnetometer data gave higher values of $H_k$ and only the FMR data were utilized. We attribute this effect to the partial tilting of the magnetization into the plane in the case of low Q samples. As shown in Fig. 2, $H_k$ increases linearly with gallium additions. This increase in $H_k$, along with a decrease in $4\pi M$ permits Q to rise with increasing gallium concentration. By requiring $h/\ell = 10$ and $Q > 4$ at room temperature, this material system should provide useful bubble devices for a bubble size above one micron.

Figure 3 illustrates the temperature dependence of $H_k$ and A for x = 0.31. $H_k$ was measured in this temperature range by FMR. The value of $4\pi M$ used to interpret the FMR data was measured at 20°C and then extrapolated to higher temperatures using a Brillouin model calculation for the ferrimagnet. This calculation of $4\pi M$ versus temperature gave a -0.37%/°C temperature change in $4\pi M$ from 20°C to 100°C. Optical hysteresigraph[8] data of bubble strip out versus temperature yielded a -0.33%/°C temperature dependence. The values of A illustrated in Fig. 3 were derived from the experimental data for $\ell$ versus temperature and the $H_k$ and $4\pi M$ values. The temperature dependence for A agrees with the calculated temperature dependence using the model for A described by Carlo, et. al.[9]

Figure 4 illustrates the temperature dependence of $\ell$ for x = 0.31 and x = 0.84. The slow variation with temperature is desirable in order that the bubble diameter matches the device circuit over a wide operating range. Fig. 5 illustrates the temperature dependence of Q for x = 0.31 and x = 0.84. Only a slow decrease in Q ($\sim$ -.3%/°C) is observed with increasing temperature.

The coercivity of these materials ranges from less than 0.1 Oe for x = 0.2 up to 0.48 Oe for x = 0.84. Thus, although coercivity does increase with increasing gallium, the coercivity is low enough for good device operation.

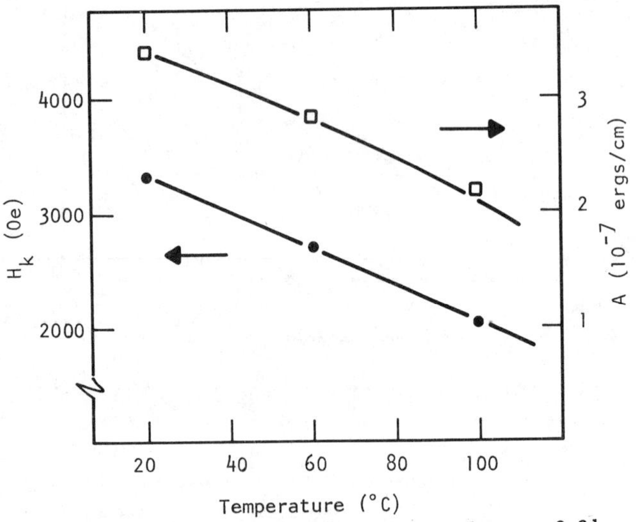


Fig. 3. $H_k$ and A versus Temperature for x = 0.31.

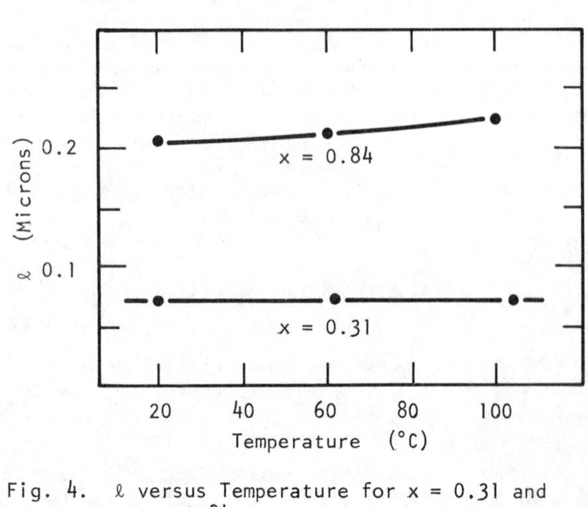


Fig. 4. $\ell$ versus Temperature for x = 0.31 and x = 0.84.

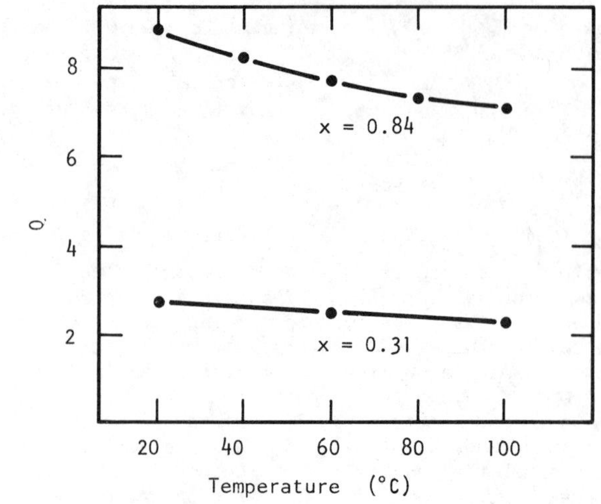


Fig. 5. Q versus Temperature for x = 0.31 and x = 0.84.

## DEVICE CHARACTERISTICS

Quasistatic circuit propagation was observed on a T-bar circuit (9 micron circuit period) at room temperature in two different compositions for two micron diameter bubbles. For the x = 0.68 material with 3.8 μm thickness a drive field of 48 Oe is required for propagation and the bias range of operation is from 427 Oe to 480 Oe (free bubble collapse field = 490 Oe). For the x = 0.84 material with 2.8 μm thickness, a drive field of 35 Oe is required and the bias range of operation is 256 Oe to 314 Oe (free bubble collapse field = 303 Oe). The latter material is superior to the first in required drive field which we attribute to a lower $4\pi M$ and a bubble diameter closer to film thickness. Note also that in the latter material, the bias range overlaps the free bubble collapse field, while for x = 0.68 the circuit does not operate at the collapse field.

## CONCLUSION

The $(LuSm)_3Fe_{5-x}Ga_xO_{12}$ garnet is a useful material system for bubble diameters from one to two microns. The ability to adjust the bubble diameter/film thickness ratio to match device parameters and circuit designs is made possible by adjusting the concentration of gallium in the film. If additional reduction in drive fields are required for practical device operation it is well known that the drive field can be further decreased by reducing the bubble thickness at the sacrifice of bubble sense signal. Thus for one micron bubble size, requiring $h/\ell = 10$ gives a $4\pi M$ of 900 Oe, while requiring $h/\ell = 4$ gives a $4\pi M$ of 770 Oe. The latter film, with one micron bubble size would have h = 0.5 μm and $h/\ell = 10$ film would have h = 1.0 μm.

The authors would like to thank N. E. Tidwell and W. L. Kriss for assistance in growing the epitaxial films. M. S. Shaikh, R. K. Watts, J. W. Taylor and R. L. Easley assisted in circuit processing.

## REFERENCES

*This work was partially supported by AFAL Contract No. F33615-75-C-1129.

1. F. M. Gyorgy, M. D. Sturge, L. G. Van Uitert, E. J. Heilner and W. H. Grodkiewicz, J. Appl. Phys. 44, 438 (1973).
2. D. C. Bullock, J. T. Carlo, D. W. Mueller and T. L. Brewer, AIP Conf. Proc. No. 24, Magnetism and Magnetic Materials, Ed. C. D. Graham, Jr. and J. J. Rhyne, p. 647 (1974).
3. M. H. Kryder, K. Y. Ahn, G. S. Almasi, G. E. Keefe and J. V. Powers, IEEE Trans. Magnetics, Vol. MAG-10, 825 (1974).
4. E. A. Giess, J. E. Davies, C. F. Guerci and H. L. Hu, Mat. Res. Bull. 10, 355 (1975).

5. H. J. Levinstein, S. J. Licht, R. W. Landorf and S. L. Blank, Appl. Phys. Letts. 19, 486 (1971).
6. D. C. Fowlis and J. A. Copeland, AIP Conf. Proc. No. 5, Magnetism and Magnetic Materials, Ed. C. D. Graham, Jr. and J. J. Rhyne, p. 240 (1971).
7. P. W. Shumate, D. H. Smith and F. B. Hagedorn, J. Appl. Phys. 44, 449 (1973).
8. C. D. Mee, IBM J. of Res. and Dev., p. 468 (July 1967).
9. J. T. Carlo, D. C. Bullock and F. G. West, IEEE Trans. Magnetics, Vol. MAG-10, 626 (1974).


# MAGNETIC PROPERTIES OF SPUTTER DEPOSITED GdCoCu AMORPHOUS BUBBLE FILMS


V. J. Minkiewicz, P. A. Albert, Robert I. Potter and C. R. Guarnieri
IBM Research Laboratory, San Jose, California 95193



## ABSTRACT

Amorphous films of the ternary alloys GdCoCu are attractive candidates for bubble device applications. We report the results of measurements on rf sputter deposited films of GdCoCu. The magnetic properties of the films were studied by polar Kerr rotation, stripe width and vibrating sample magnetometer measurements. The more promising compositions (nominally $Gd_{14}Co_{63}Cu_{23}$) support 1.3μm stripe domains, and generally had moderately high quality factors of Q = 7.9 with $4\pi M$ = 930 gauss and a characteristic length $\ell$ = 0.13μm. The stripe collapse field of these films was 750 Oe. At room temperature, the polar Kerr rotation hysteresis loop shows a minimum "residual" coercivity of $H_c \sim 2$ Oe, which may be process limited. The data indicate that the copper additive does not dilute the transition metal sublattice magnetization as rapidly as does molybdenum.


## INTRODUCTION

The initial experiments on sputter deposited films of amorphous alloys of GdCo[1] established that they were ferrimagnetic and could support bubble domains. The films were easily fabricated, and it was therefore clear that they were strong candidates for device applications. However, additional measurements showed that the temperature dependence of the magnetization was too severe to meet specifications. As a result, a great deal of attention has now been focused on the magnetic properties of ternary alloys of GdCoX. The constituent element X is generally chosen to dilute the transition metal sublattice with the hope that the temperature dependence of the magnetization can be reduced, and that the magnitude of the magnetization can be adjusted to accommodate a wide range of bubble diameters. There has been intense interest in the GdCoMo system.[2-4] Our effort is directed toward establishing and understanding the static magnetic properties of an alternative ternary alloy system GdCoCu. The dynamic properties are discussed in a companion paper.[5]

## EXPERIMENT

The amorphous alloy films were prepared by sputter deposition in an argon plasma with rf substrate bias. The argon gas pressure was 22.0 millitorr. The target potential was generally fixed at - 1400v, while the substrate bias varied between 0 and - 120v. The composition of the films from a given target could be varied, to a limited extent, in a systematic way by changing the bias on the substrates. In addition, as is known, biasing the substrates produced films with uniaxial anisotropy perpendicular to the film plane.[1] Deposition with zero bias generally gave films whose magnetization was in-plane. The deposition rate was approximately 8 Å/sec and the thickness of the films was 1.3 ± 0.1μm.

Two main techniques were used to measure the relevant magnetic properties. Polar Kerr rotation as a function of applied magnetic field for $\lambda$ = 6328 Å was used to monitor the collapse field $H_0$ and the static coercivity $H_c$. The polar Kerr rotation angle was typically 20 minutes. The magnetic compensation temperature $T_{comp}$ was identified with a reversal in the signature of the hysteresis loop. The magnetization and the quality factor Q were obtained from vibrating sample magnetometer measurements.

## RESULTS

Before presenting the results of the measurements, we first comment on the range of the composition of the films. In order to systematically study the effect of a diluent on the bubble properties of GdCoX, one must determine the compositions that are compensated at room temperature. This information is important because, for device applications, one generally wants to study those compositions for which the magnetization of the films, at room temperature, is transition metal dominated with both $T_{comp}$ and the Neel temperature $T_N$ well removed from room temperature. Fig. 1 is a ternary composition map that illustrates the compositions of the films that are compensated at room temperature. The compositions to the right of the compensation line are transition metal dominated. To study the effect of dilution, one increases the copper content and monitors the properties of films in this region. The cross-hatched area in the figure illustrates the portion of the map that was studied in our experiments. In this paper, we present representative results for films with 36 at. % and 23 at. % copper content.

The data given in Table I are representative results for films with these nominal copper compositions. The results clearly show that Cu and Mo dilute the transition metal sublattice in markedly different ways. In contrast to the strong dilution effects observed for Mo,[2-4] the addition of Cu has a much more gradual effect. Also included in Fig. 1 is the room temperature compensation line for GdCoMo. We note

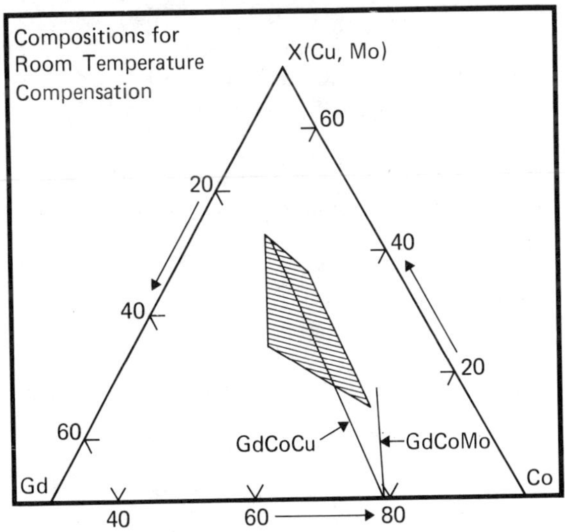


**Figure 1.** A ternary phase diagram that indicates the compositions for room temperature compensation for GdCoCu and GdCoMo.

that alloys in this system with Mo concentration greater than 20 atomic percent are paramagnetic at room temperature.(3) Amorphous films of GdCoCu, on the other hand, which have diluent contents of almost a factor of two times greater are still ferrimagnetic at room temperature. We also note that the quality factors of the films are moderately high.

The temperature dependence of $H_o$ and $H_c$ are illustrated in Fig. 2. The polar Kerr rotation hysteresis loop of a film with nominal composition $Gd_{16}Co_{56}Cu_{27}$ is shown as a function of the sample temperature. The results show that both parameters are very temperature sensitive. This sensitivity reflects the temperature dependence of the magnetization, which in turn implies that other basic film properties such as the characteristic length and domain stripe width are also temperature sensitive.(6) In view of the fact that $H_o$ continually increases, the results in Fig. 3 show that these films are not near the peak or the "flat" part of the magnetization curve at room temperature. The static coercivity is known to increase as the sample approaches magnetic compensation(7) (see Fig. 2 and also Table I). After a careful study of a great many films, deposited under a variety of sputtering conditions, it was found that the static coercivity

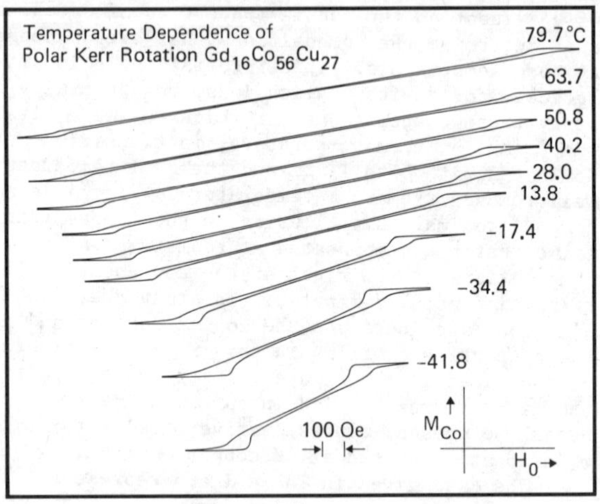


**Figure 2.** The temperature dependence of the Polar Kerr rotation angle for X = 6328Å. The vertical scale reflects the magnetization of the cobalt sublattice.

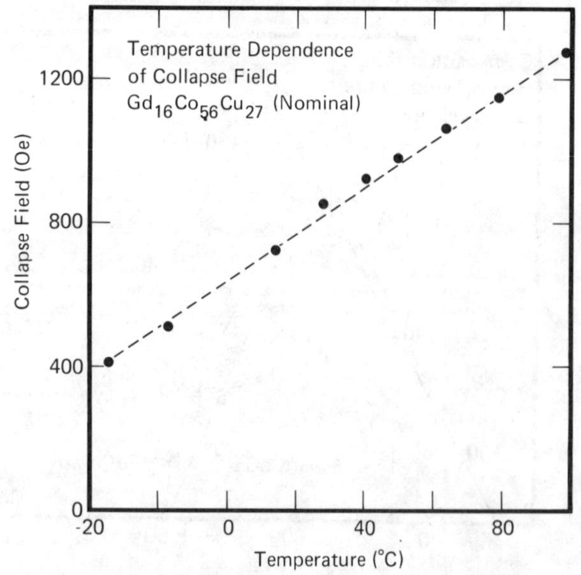


**Figure 3.** The temperature dependence of the bubble collapse field. The slope of the straight line is 6.3 Oe/ °C.

TABLE I. The table contains a list of typical properties of representative films. The films have a thickness and domain stripe width of 1.3 ± 0.1μm. The nominal compositions of films 1 to 3, and 4 and 5, are $Gd_{14}Co_{63}Cu_{23}$ and $Gd_{13}Co_{51}Cu_{36}$.

| Film # | 4πM (kGauss) | Q | $H_o$ (Oe) | $H_c$ (Oe) | $T_{comp}$ (°C) |
|---|---|---|---|---|---|
| 1 | 1.16 | 6.7 | 570 | 5.0 | - 103 |
| 2 | 0.96 | 8.3 | 640 | 5.0 | - 114 |
| 3 | 0.93 | 7.9 | 750 | 3.0 | - 130 |
| 4 | 0.76 | 7.0 | 420 | 7.0 | - 74 |
| 5 | 0.73 | 7.4 | 490 | 7.0 | - 78 |

generally had a lower limit of 2 Oe for as-deposited films. This result was somewhat surprising since the films should be "perfectly" homogeneous or completely disordered.

## CONCLUSIONS

The amorphous films of GdCoCu studied in these experiments do not provide answers to the problem of the temperature dependence observed for GdCo alloys. In view of our observation that Cu dilutes the transition metal sublattices much differently than Mo, we are attempting to increase the copper content of the films to further decrease the magnetization and to adjust the magnetization curve such that M(T) is relatively independent of T from 0°C to + 100°C. The observed minimum value of 2 Oe for the static coercivity in as-deposited films is larger than one would desire for device applications. The coercivity, however, has been found to be lowered by post-deposition treatments.(5)

## ACKNOWLEDGEMENTS

We would like to thank Drs. K. Lee, M. Lorenz, N. Heiman, and A. Onton for their advice and helpful discussions. We also thank C. Corpuz for his expert technical assistance.

## REFERENCES

1. Chaudhari, P., Gambino, R. J. and Cuomo, J. J., Appl. Phys. Lett. 22, 337 (1973).
2. Argyle, B. E., Gambino, R. J. and Ahn, K. Y., "Magnetism and Magnetic Materials - 1974," AIP Conference Proceedings No. 24, edited by C. D. Graham and J. J. Rhyne (AIP, New York 1975).
3. Chaudhari, P., Cuomo, J. J, Kirkpatrick, S. and Tao, L. J., "Magnetism and Magnetic Materials - 1974," AIP Conference Proceedings No. 24, edited by C. D. Graham and J. J. Rhyne (AIP, New York 1975).
4. Hasegawa, R., Argyle, B. E. and Tao, L. T., "Magnetism and Magnetic Materials - 1974," AIP Conference Proceedings No. 24, edited by C. D. Graham and J. J. Rhyne (AIP, New York, 1975).
5. Potter, Robert I., Minkiewicz, V. J., Lee, Kenneth, and Albert, P. A. (this Conference).
6. Thiele, A. A., J. Appl. Phys. <u>41</u>, no. 3, 1139 (1970).
7. Mee, C. D., IBM J. Res. & Dev. <u>11</u>, 468 (1967).

# LATTICE MISMATCH BETWEEN Pb-SUBSTITUTED EuIG LPE FILMS AND NdGG SUBSTRATES

T. S. Plaskett, E. Klokholm and D. C. Cronemeyer
IBM Thomas J. Watson Research Center, Yorktown Heights, New York 10598

## ABSTRACT

The lattice mismatch Δa as a function of Pb content of the film, for LPE grown EuIG films on (100) and (111) NdGG substrates, deviates from linearity at Pb contents greater than 1 wt % Pb. At these Pb contents the films are compressively strained. It is suggested that the deviation is caused by a change in valence state of Pb and Eu. Heat treating films of high Pb content above 900°C resulted in stress-relief.

## INTRODUCTION

The room temperature lattice mismatch (Δa) between an LPE garnet film and substrate has been reported[1,2] to be a linear function of Pb content of the film. In this paper we show that for EuIG on (100) and (111) NdGG, Δa is not a linear function of Pb for $\Delta a \lesssim 0$, i.e., when films at room temperature become compressively strained. As was previously reported[3,4], a very large induced uniaxial anisotropy (Ku) exists in films with high Pb contents. The relationship, if any, of Δa with Ku is discussed in an attempt to determine the source of Ku in EuIG. The effect of annealing on Δa and Ku is also described for high Pb content films.

## EXPERIMENTAL

The films were grown isothermally by the LPE "tipping" process from a $PbO:B_2O_3$ flux.[5] The films were grown on (100) and (111) chemically-mechanically polished NdGG substrates. The film thicknesses were all between 0.6 and 1.8 μm and were grown at rates between 0.1 and 0.4 μm/min. We had not sufficient thickness and growth rate data to separate their effects on Δa. The Pb content of the film was varied by changing the growth temperature. The empirical relationship of Pb content to the growth temperature is given by,

$$Pb = 2.18 \times 10^{-12} e^{52,400/RT} \quad (1)$$

where Pb is given as moles per garnet formula unit, R the gas constant in cal/mole -°K and T the growth temperature in °K. No significant difference was found in the constants of Eq. (1) for the (100) and (111) orientations. In this paper we report Pb content as wt % and for EuIG one wt % is approximately equal to 0.045 mole of Pb per garnet formula unit. The Pb content was measured by electron microprobe analysis.

The lattice mismatch (Δa) is defined as the lattice parameter of the substrate minus the undistorted lattice parameter of the film. A positive Δa indicates a film with tensile strain and a negative Δa compressive strain. From the lattice parameters published[6] for EuIG (12.498 Å) and NdGG (12.507 Å), a pure EuIG grown on NdGG is in tension. The Δa was calculated[7] from differences in the lattice spacing of the film and substrate as measured by the Bragg-Brentano x-ray diffraction technique. A Poisson's ratio of 0.3 was used to calculate Δa and it was assumed that the films are pseudomorphic.

The Ku was calculated from ferromagnetic resonance data taken parallel and perpendicular to the film plane. The reported[6] bulk magnetization value of 1172 gauss was used for calculating Ku and it was assumed to be independent of the Pb content. The Ku reported includes both growth and stress components, but does not include the crystalline anisotropy.

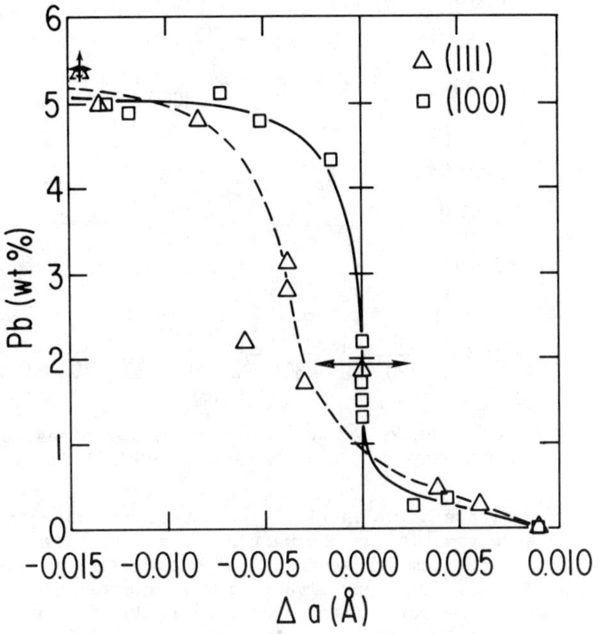


Fig. 1 Δa as a function of Pb for (100 and (111) films.

## RESULTS

The Δa as a function of Pb content is shown in Fig. 1 for (100) and (111) oriented films. For no Pb a Δa of 0.009 Å is expected.[6] The data shown in Fig. 1 extrapolates very closely to this value. As the Pb content increases the lattice parameter of the film increases linearly, thus decreasing Δa to zero. The curves of Δa with Pb for (100) and (111) films in tension are about the same; this is expected since the incorporation of Pb dilates the unit cell uniformly and garnets are elastically isotropic. At $\Delta a \lesssim 0$ there is an abrupt deviation from linearity especially for the (100) orientation. This deviation indicates that $a_f$ does not change with Pb content. Essentially the lattice is accepting more Pb without dilating the unit cell. Also, when $\Delta a \lesssim 0$, the curves for the (100) and (111) orientations separate. The change in Δa with Pb, for Pb contents between 1 to 5 wt % Pb is much smaller for (100) oriented films than for (111) films. At about 5 wt % Pb, the Δa increases without an increase in Pb content. There is, however, a slight decrease in the growth temperature as Δa increases. This means that equation (1) does not apply for low growth temperatures. We are unable to increase the Pb content above 6.0 wt %. This concentration was obtained by growing at about 760°C which is very near the eutectic temperature of the flux.

The effect of annealing high Pb content films on the Δa is shown in Fig. 2 as a function of annealing time for (100) and (111) orientation. At 900°C and above, the room temperature Δa decreases to zero with annealing time. Above 1000°C Δa became positive. Annealing at 800°C had no effect. There appears to be no significant difference between (100) and (111) orientations even though Δa prior to annealing was different; for the (100) it was -0.016 Å and for the (111) was -0.027 Å. Also annealing low Pb content films showed no change in Δa even at the high temperatures. In all instances, where there was a change in Δa, hillocks were visible on the film surface. Electron microprobe analyses showed that the hillocks were not a Pb-rich

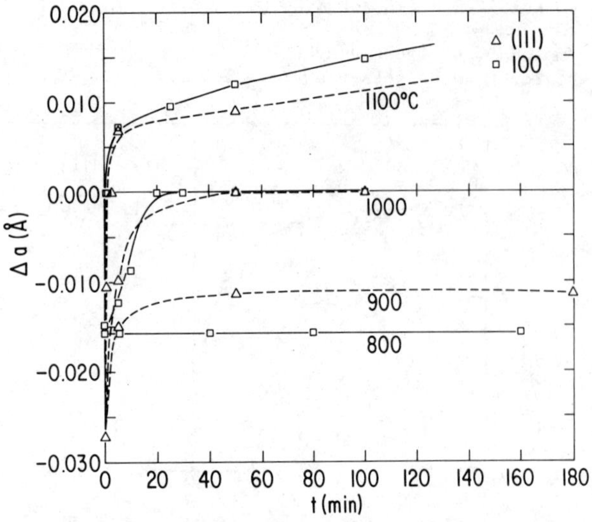


Fig. 2 Δa as a function of annealing times and temperatures for high Pb (100) and (111) films.

precipitate as we initially expected, but were the same composition as the matrix. The hillocks, therefore, are from stress-relief at the annealing temperatures.[8] The positive room temperature Δa which we measured after annealing is due to the difference in thermal expansion between the film and substrate. As Δa became positive the magnetic strip domains (which indicate a positive Ku) disappeared.

DISCUSSION

The behavior of Δa with Pb can be divided into three regions. In the region up to 1 wt % Pb, incorporation of Pb expands the lattice uniformly until the lattice parameter of the film matches the substrate. This is similar to most garnets and it is assumed that the Pb is present as $Pb^{2+}$ on the dodecahedral site. In the next region, from 1 to 5 wt % Pb, Δa does not greatly change by additions of Pb, i.e., Pb is incorporated into the lattice without dilating it. We propose that this can occur by a valence change according to the reaction,

| | $Pb^{2+}$ + | $2Eu^{3+} \rightleftarrows$ | $Pb^{4+}$ + | $2Eu^{2+}$ | (2) |
|---|---|---|---|---|---|
| ionic radii (site) | 1.29 Å (dodec) | 1.07 (dodec) | 0.77 (oct) | 1.17 (dodec) | |
| avg. | 1.14 | | 1.04 | | |

The net ionic radius of $Pb^{4+}:2Eu^{2+}$ is smaller than $Pb^{2+}:2Eu^{3+}$. If Pb is incorporated in the usual manner, then in this region the lattice would become compressively strained and it is this strain which may provide the driving force for the above reaction. Although oxidation potentials suggest that this reaction is not energetically favorable in the forward direction, these values are for room temperature and usually do not apply to the solid-state, where reactions occur at higher temperatures and oxidation potentials are modified by structure.[9] In the third region (greater than 5 wt %), Δa changes without further additions of Pb. This may be explained by reaction (2) being reversed at lower growth temperatures. As previously noted, the increase in Δa is accompanied by a decrease in the growth temperature even though the Pb content is constant.

The relationship of Δa with Ku is shown in Fig. 3. It is evident from this Figure that Ku is not magnetostrictive for the following reasons. At Δa ∿ 0 Ku increases without a change in Δa. Also Ku passes through a maximum at Δa ∿ -0.075 Å which is difficult to explain from magnetostrictive arguments.

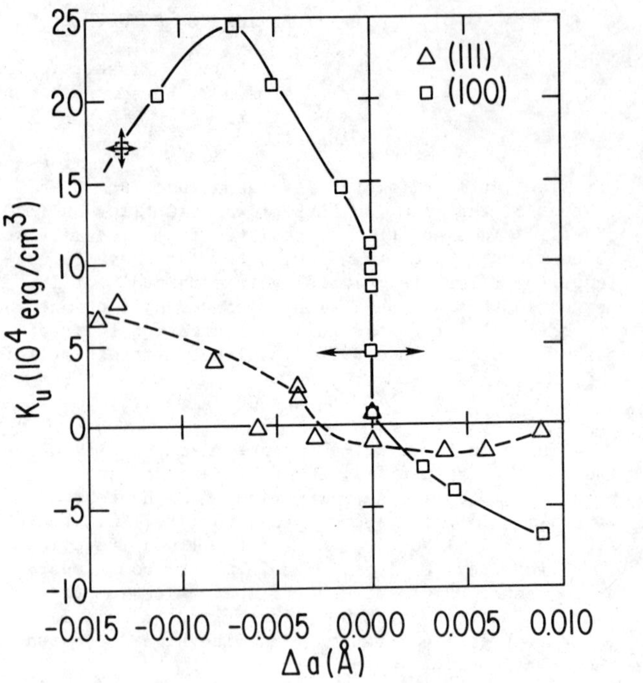


Fig. 3 Ku as a function of Δa for (100) and (111) films.

Only for (100) film in tension is Ku magnetostrictive; the magnetostrictive coefficient (λ) calculated agrees with published values.[6] The Ku values shown in Fig. 3 for Δa = 0.009 Å (no Pb) were calculated using published values of λ.

The source of Ku in this single rare-earth garnet is still not understood. Our experimental evidence suggests that Ku is related to the unusual behavior of Δa with Pb. Whether or not a change in valence state can account for the large Ku in (100) films is not clear at this time. Further studies of this problem are being conducted.


ACKNOWLEDGMENTS

The authors are grateful to M. W. Shafer for helpful discussions, A. H. Parsons for film preparation and characterization, R. Schad, S. O. Ellmann, F. Cardone and C. F. Aliotta for electron microprobe analysis and J. M. Karasinski for x-ray analysis.


REFERENCES


1. E. A. Giess, B. E. Argyle, D. C. Cronemeyer, E. Klokholm, T. R. McGuire, D. F. O'Kane, T. S. Plaskett and V. Sadagopan, AIP Conf. Proc. 5, 110 (1972).
2. J. C. Brice, J. M. Robertson, W. T. Stacy and J. C. Verplanke, J. Crystal Growth 30, 66 (1975).
3. T. S. Plaskett, E. Klokholm, D. C. Cronemeyer, P. C. Yin and S. E. Blum, Appl. Phys. Lett. 25, 357 (1974).
4. D. C. Cronemeyer, T. S. Plaskett and E. Klokholm, AIP Conf. Proc. 24, 586 (1975).
5. T. S. Plaskett and R. Ghez, AIP Conf. Proc. 24, 584 (1975).
6. L. G. Van Uitert, E. M. Gyorgy, W. A. Bonner, W. H. Godkiewicz, E. J. Heilner and G. J. Zydzik, Mat. Res. Bull. 6, 1185 (1971).
7. E. Klokholm, J. W. Matthews, A. F. Mayadas and J. Angilello, AIP Conf. Proc. 5, 105 (1972).
8. D. C. Miller and R. Caruso, J. Crystal Growth 27, 274 (1974).
9. M. W. Shafer, J. Appl. Phys. 36, 1145 (1965).

# VARIATIONS OF MAGNETIC ANISOTROPY WITHIN EPITAXIAL FILMS OF $Y_{2.85}La_{0.15}Fe_{3.75}Ga_{1.25}O_{12}$ OBTAINED FROM SPIN WAVE RESONANCE

B. Hoekstra and J.M. Robertson, Philips Research Laboratories Eindhoven, The Netherlands and G. Bartels, Philips Forschungslaboratorium Hamburg, Germany


## ABSTRACT

Spin wave resonance in epitaxial films of $Y_{2.85}La_{0.15}Fe_{3.75}Ga_{1.25}O_{12}$ is used to determine the variation of the magnetic anisotropy within these films. The difference between the magnetic anisotropy of the "transient layer", and the layers remote from the substrate has been investigated. It is shown that the anisotropy of the transient is not affected by "meltback" nor by interdiffusion between the film and the substrate but is related to the different growth conditions during the transient period compared with those during the "steady state" growth regime. The anisotropy of the transient layer of a vertically dipped film is shown to be same as the anisotropy of the bulk of a film obtained by horizontal dipping under the same conditions, using fast rotation.


## INTRODUCTION

The liquid-phase-epitaxy (LPE) method for growing single crystals of magnetic garnets on non-magnetic substrates gives rise to variations in the chemical composition within the magnetic film. In particular during the first transient period of rapid growth a thin transient layer[1] is formed which has been shown to be Pb-rich[2,3,4]. "Meltback", a partial dissolution of the substrate before the initiation of the growth, was reported in the growth of films at small supercooling[5]. We have investigated the variations of the uniaxial anisotropy in LPE La,Ga:YIG, a bubble material[6]. We have shown previously[7] that spin-wave-resonance experiments (SWR) provide evidence for a stratification of these films. We found that the anisotropy of the transient layer is different from the anisotropy of the bulk layers. In this paper we report further investigations into the origin of the anisotropy variations and the relation with the growth procedure.

## EXPERIMENTAL PROCEDURES

Films were grown by LPE on (111)-oriented GGG substrates. Two different melts (see Table I) yielded an approximate film composition $Y_{2.85}La_{0.15}Fe_{3.75}Ga_{1.25}O_{12}$. Melt A was used in a small crucible (500g. melt) only. From melt B films were grown using a small as well as a large crucible(2000g.melt). Typical growth temperatures are 850° C from melt A and 935° C from melt B. The growth from these melts, using the vertical mode of dipping, has been described previously[8,9]. From melt B we obtained films by dipping vertically as well as horizontally. In the latter case a constant axial rotation rate was used. Etching of the films was done in concentrated sulphuric acid at an etching rate of 1 µm/hr. Before etching, films were cleaned in a mixture of hydrogen peroxyde (30%) and sulphuric acid (1:3 by volume). Resonance experiments were made on small pieces of the films on a conventional EPR spectrometer operating at 10 GHz. The films were on one side of the substrate only and their lateral dimensions did not influence the spectrum. The dc field was applied in a (110)-plane perpendicular to the plane of the film, thus including the principal uniaxial and cubic axes.

## VARIATIONS OF THE UNIAXIAL ANISOTROPY

We have shown previously[7] that the perpendicular resonance spectrum of films from melt B in a large crucible exhibits a number of separate resonance lines, as

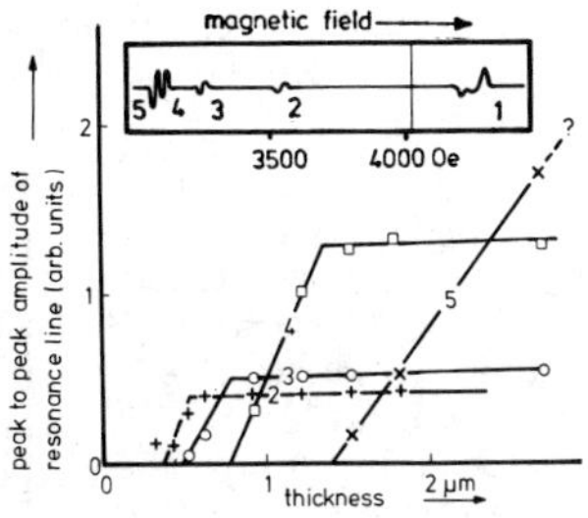


Fig. 1. Variation of the peak -to-peak amplitudes of the lines in the perpendicular resonance spectrum (insert in top) of a 2.7 µm La,Ga:YIG film with decreasing film thickness.

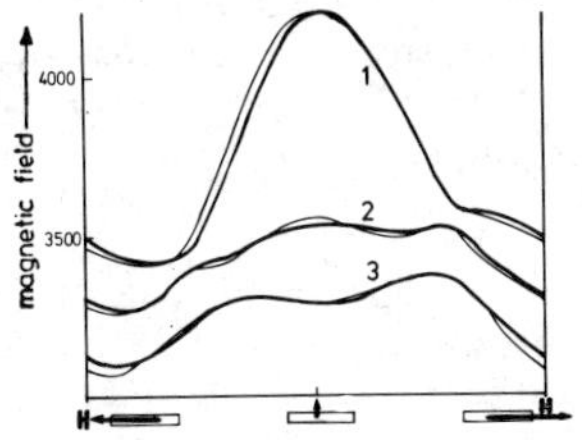


Fig. 2. Angular diagram of the experimental (heavy lines) and calculated (thin lines) field positions of the resonances of the film at 0.6 µm thickness. The numbers refer to the lines in fig. 1.

seen in the plot of the derivative of the absorption versus the applied field in the insert in fig. 1. In order to explain the successively linearly decreasing amplitudes of the lines with decreasing thickness of the film (fig. 1) and their constant position we have proposed[7] that the film consists of discrete strata which give rise to these apparently localised resonances.

We have calculated the spectra of a stratified single crystal film. The stratification is attributed to variations of the growth induced anisotropy which is, expectedly, sensitive for variations of the conditions during the growth process.
We have made a fit to the spectrum after only three lines remain, assuming three layers with different $K_u/M$ and stepwise changes in between. M is taken to be constant[2,4,10]. Since the garnet is a single crystal the exchange stiffness will be constant as well. The layers are coupled by the exchange interaction. We assume finally that there is no pinning[11] at the film-air and film-substrate interfaces, as appears from the vanishing of all modes except one at a critical angle[12]. We have solved the resonance equation[13] to find the resonance modes. In fig. 2 we compare the experimental positions of the lines (heavy curves) with the calculated positions (thin curves) as a function of the direction of the applied field. The set of parameters used is given in Table II, together with the parameters previously obtained[7] by assuming the resonances to be uniform modes within the layers, neglecting the coupling between the layers. We have taken $K_1/M$=-50 Oe (the cubic anisotropy). The exchange stiffness was $A=1.4 \times 10^{-7}$ erg/cm, as obtained from the low field spin waves in parallel resonance[14]. The good fit between the theoretical and experimental curves shows that this model can satisfactorily explain the observed mode positions.

An essential assumption in order to explain the peculiar dependence of the spectrum on film thickness is, in the above model, that with every resonance line

Table I Melt composition in % by weight

| melt | PbO | $B_2O_3$ | $Y_2O_3$ | $La_2O_3$ | $Fe_2O_3$ | $Ga_2O_3$ |
|---|---|---|---|---|---|---|
| A | 89.665 | 2.244 | 0.556 | 0.222 | 5.937 | 1.397 |
| B | 90.785 | 2.000 | 0.707 | 0.515 | 5.040 | 0.953 |

Table II Properties of the first three strata in a La, Ga-YIG film from melt B. $K_u/M$ includes a contribution $-2\pi M$ from demagnetization. a: data used in the calculation of the curves in fig. 2. b: data obtained previously.

| stratum | thickness (µm) | | $K_u/M$ (Oe) | |
|---|---|---|---|---|
| | a | b | a | b |
| 1 | 0.2 | 0.4 | -355 | -400 |
| 2 | 0.2 | 0.1 | - 15 | -145 |
| 3 | 0.4 | 0.3 | + 70 | +30 |

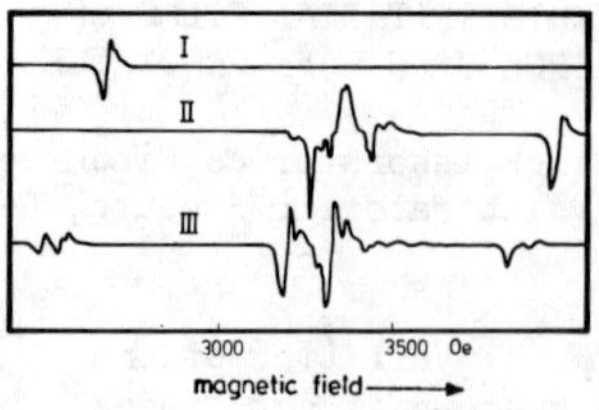


Fig. 3. Perpendicular resonance spectra of three films. I: a film from melt A, II: a film from melt B, III: a B-film grown on the A-film.

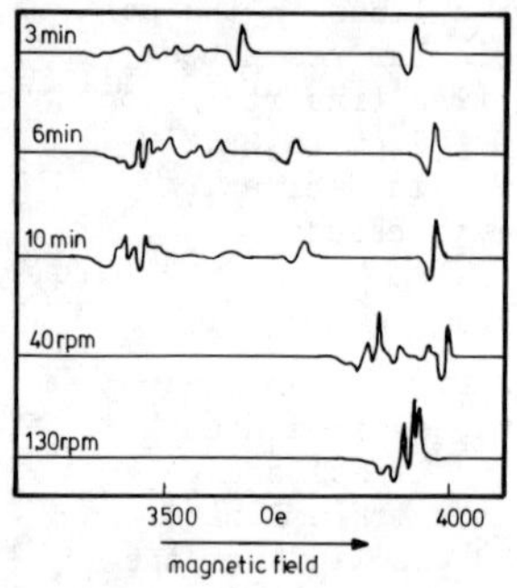


Fig. 4. Perpendicular resonance spectra of 5 B-films grown at the same temperature (≈45° C) supercooling) by vertical dipping for 3,6 and 10 min. and by horizontal dipping using a constant rotation of 40 rpm and 130 rpm.

in the spectrum there corresponds a stratrum where this resonance is nearly uniform and therefore excited by a uniform driving field. We have also made preliminary calculations assuming a more gradual variation of the anisotropy. These show that the SWR spectra obtained from such a model can be similar to those observed in our film. This would mean that the anisotropy does not necessarily change stepwise with the distance from the substrate but is possibly changing gradually, with different slopes in different regions.

ANISOTROPY VARIATIONS AND GROWTH CONDITIONS

Films obtained from melt B in a large crucible, as above, have layers remote from the substrate with positive anisotropy ($K_u \approx 2.10^3$ erg/cm$^3$). The transient layer (layer 1)[7] has a large negative $K_u$ ($\approx$-3.10$^3$ erg/cm$^3$). In small applied fields bubble domains are formed in the remote layers whereas next to the substrate the magnetization is still in-plane, thus preventing the formation of hard bubbles[9]. Films from melt A support hard bubbles. The SWR-experiments show that the transient layer has a positive $K_u$ which can be up to two times larger than $K_u$ of the remote layers which have $K_u \approx 3.10^3$ erg/cm$^3$. Due to the smaller growth temperature from melt A the latter films contain more Pb then the films from melt B[3]. Since $K_u$ increases with the Pb-concentration[15,16] the enhanced $K_u$ of the transient layer of an A-film may reflect its larger growth rate and therefore larger Pb-concentration[1]. From SWR one finds values for $K_u/M$. $K_u$ has been determined by taking M constant within the film. Alternatively one might suppose that M is smaller in the transient layer in accordance with observations of domain wall oscillations[7].

In order to investigate the origin of the negative $K_u$ of the transient layer in B-films we have grown films from melts A and B in small crucibles under different conditions. The upper plot in fig. 3 shows the perpendicular resonance spectrum of a 1.7 μm A-film, dipped vertically only half into the melt. The substrate was thereafter broken in two halves. One of the halves, only partly covered with the film, was subsequently dipped into melt B in a small crucible (supersaturation 25° C). The perpendicular spectrum (second trace in fig. 3) of the B-film, taken where the film grew directly on the substrate, shows the characteristics seen in fig. 1: at high field (3950 Oe) we find the resonance line of the transient layer due to its large negative $K_u/M$ (see Table II). Near 3500 Oe the resonances due to remote layers overlap. The spectrum taken where the B-film grew on the A-film, shown in the lower trace, is simply the sum of the spectra of the individual films, apart from some small shifts and different lineshapes. From the fact that the resonance of the transient layer occurs at nearly the same field position, whether film B is grown on the substrate or on another La, Ga:YIG film, we conclude that the anisotropy of the transient layer is not affected by the nature of the substrate on which it grows. Therefore we can conclude that "melt-back" or any deposit after polishing the substrate is not the origin of the anisotropy of the transient layer.

Subsequently we have grown B-films by vertical dipping at 45° C supersaturation. Dipping times were 3,6 and 10 minutes yielding 1.75, 3.1 and 4.4 μm thick films. Perpendicular resonance spectra of these films, (fig. 4) are again similar to the B-film spectrum of fig. 1. An important observation is that with increasing dipping time the high-field resonance of the transient layer scarcely changes position, indicating no appreciable effect from diffusion between film and substrate during growth. The resonances at 3450 Oe are due to bulk layers and change with longer dipping. The lower two traces in fig. 4 are of films grown under the same conditions but using horizontal dipping with rotation rates of 40 and 130 rpm. It is observed that the effect of the rotation is to move the bulk resonances towards the position of the transient layer in the vertically dipped films (The line at 4000 Oe in the spectrum of the 40 rpm film is not due to a transient but is due to a resonance localised in a 1.5 μm thick top layer). At 130 rpm the bulk has the same anisotropy as the transient and the film is fairly homogeneous. These observations are consistent with existing growth theory[1]. During the transient period the diffusion layer is established and is still very thin. After the transient period the presence of the diffusion layer affects the properties of the film which is deposited. The properties of the transient layer are therefore different from those of the film remote from the substrate. On using the horizontal dipping mode applying fast axial rotation the diffusion layer becomes very thin. The properties of a fast rotated film are thus expected to be close to those of the transient layer of a film obtained by vertical dipping. This is actually observed in fig. 4.

Summarizing, we conclude that the negative anisotropy of the transient layer in vertically dipped films is inherent in the growth process. Its anisotropy is not affected by either dissolution of the substrate or interdiffusion between the film and the substrate. The authors wish to acknowledge P.F. Bongers and W. Tolksdorf for many discussions and H.D. Jonker and R.P. van Stapele for comments on the manuscript.

REFERENCES

1) R. Ghez and E.A. Giess, Mat.Res. Bull, 8, 31 (1973)
2) J.E. Davies, E.A. Giess and J.D. Kuptsis, Mat. res. Bull. 10 (1975).
3) J.M. Robertson, M.J.G. van Hout and J.C. Verplanke, Mat. Res. Bull., 9, 555 (1974).
4) H.D. Jonker, A.E. Morgan and H.W. Werner, to be published in J.Cryst.Growth.
5) R.D. Henry and E.C. Whitcomb, Mat.Res.Bull. 10, 681 (1975).
6) W. Tolksdorf, G. Bartels, G.P. Espinosa, P. Holst, D. Mateika and F. Welz, J. Crystal Growth 17, 322 (1972).
7) B. Hoekstra, to be published in Solid State Comm.
8) J.M. Robertson, W. Tolksdorf, and H.D. Jonker, J. Cryst. Growth 27, 241, (1974).
9) J. Haisma, G. Bartels, and W. Tolksdorf, Philips Res. Repts. 29, 493 (1974).
10) C.S. Guenzer, C. Vittoria, and H. Lessoff, AIP Conf. Proc. 10, 1292 (1974).
11) P. Pincus , Phys. Rev. 118,658 (1960).
12) J.P. Omaggio and P.E. Wigen, AIP Conf. Proc. 24, 125 (1975).
13) A.M. Portis, Appl. Phys. Lett. 2, 69 (1963).
14) B. Hoekstra, to be published.
15) J.M. Robertson, unpublished results.
16) T.S. Plaskett, E. Klokholm, D.C. Cronemeyer, P.C. Yin and S.E. Blum, Appl. Phys. Lett. 25, 357 (1974).
17) F.H. de Leeuw and J.M. Robertson, J. Appl. Phys. 46, 3182 (1975).

# THE TEMPERATURE DEPENDENCE OF THE UNIAXIAL ANISOTROPY OF $Gd_{1-x-y}Co_xMo_y$ AMORPHOUS ALLOY FILMS ON GLASS SUBSTRATES

P. Chaudhari and D. C. Cronemeyer
IBM Research Center, Yorktown Heights, New York 10598

## ABSTRACT

The temperature dependence of the uniaxial anisotropy of $Gd_{1-x-y}Co_xMo_y$ films is described in this study as measured by ferromagnetic resonance. The magnetic moment of the films is obtained from a Faraday balance measurement. For this ferrimagnet, the sub-network moments and g-factors are related by the Wangsness equation which gives the net g-factor in terms of these sub-network moments and g-factors. Having these sub-network moments, the anisotropy energy can be described in terms of them, and it is found that the temperature dependence of the anisotropy constant can be fitted to a dipolar equation of the form:

$$K_u = \sum_{i,j=1}^{2} C_{ij} M_i M_j$$

## INTRODUCTION

$Gd_{1-x}Co_x$ metal alloy amorphous films have shown properties which are interesting for bubble domain devices[1, 2]. However, the net magnetization for $Gd_{1-x}$-$Co_x$[3] has such a steep temperature variation about the compensation temperature, that various diluents have been proposed. The sub-networks in this amorphous material, analogous to sub-lattices in crystalline ferrimagnets, which account for this magnetic moment compensation behavior are an interesting topic for investigation. Mo is one of the elements which form ternary alloys with Gd and Co, and is advantageous in decreasing the magnitude and temperature dependence of the magnetic moment. It was hoped that in the case of $Gd_{1-x-y}Co_xMo_y$ that the high uniaxial anisotropy typical of $Gd_{1-x}Co_x$ would not be diminshed too drastically to make the material unsuitable for bubble domain device operation. The exploration of the temperature variation of $K_u$ for $Gd_{1-x-y}Co_xMo_y$, and the investigation of the role of the sub-networks in this material were the goals of this particular study.

It might be expected according to simple molecular field theory that $K_u \sim M^n$, where n is between two and three, and M is the net magnetization. However, it seems more reasonable in the case of a ferrimagnet like $Gd_{1-x-y}Co_xMo_y$ to have $K_u \sim M_{Co}^n$, where $M_{Co}$ is the Co sub-network magnetization. The question is further complicated by the expected dipolar character of the dissimilar sub-networks, and the liklihood of Gd-Co pairs being active in causing the anisotropy. It was felt that the actual temperature behavior of $K_u$ would be of importance in addressing this question.

## EXPERIMENTAL PROCEDURE

Ferromagnetic resonance measurements were made at X-band (9.06 GHz) utilizing a Varian ESR spectrometer whose microwave cavity was fitted with a standard variable temperature apparatus utilizing a cooled nitrogen gas stream passing over a heater whose output was controlled by the amplified signal from a resistance thermometer. With this arrangement, measurements were possible covering 80 to 573°K. The samples utilized for this study were $Gd_{1-x-y}Co_xMo_y$ rf-sputtered onto glass substrates, and were about 1 µm in thickness. Most of the samples studied had compensation temperatures below room temperature. A strong surface mode was apparent with these samples, and it was found that such a mode could be removed by ion milling and overcoating the sample with SiO. In most cases, it was sufficient to obtain the parallel and perpendicular resonance fields, and cavity resonance frequency in order to calculate $K_u$ and g-values at each temperature. On occasion, a full rotation spectrum from parallel to perpendicular orientation was utilized in order to separate true bulk resonance from the surface resonances mentioned previously.

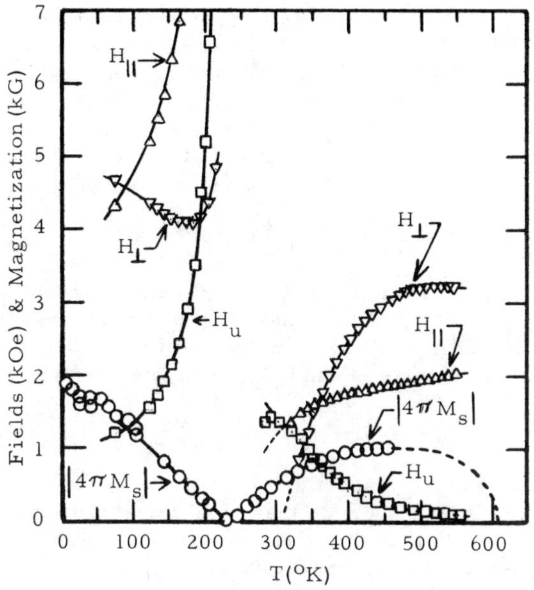


Fig. 1. The dependence of ferromagnetic resonance fields, $H_\parallel$, $H_\perp$, anisotropy field $H_u$, and magnetic moment $4\pi M_s$ upon temperature for a sample of $Gd_{0.15}Co_{0.74}Mo_{0.11}$.

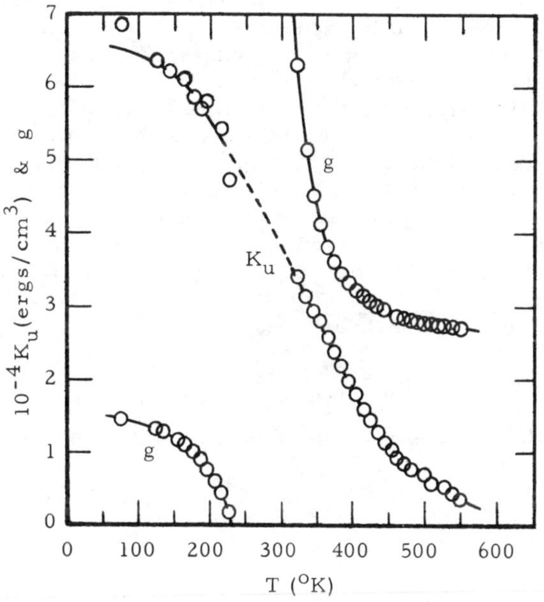


Fig. 2. The g-factor and uniaxial anisotropy $K_u$ (circles are experimental, solid line is drawn through the fit to eq. (5) ), for a $Gd_{0.15}Co_{0.74}Mo_{0.11}$ sample.

## RESULTS

The basic data for the resonance fields $H_\parallel$ and $H_\perp$ are shown (Fig. 1) for a particular sample of $Gd_{1-x-y}$-$Co_xMo_y$, along with the magnetic moment variation as obtained from Faraday balance measurements. From this data, and the cavity resonance frequency, one calculates $K_u$ and g by the usual equations combining parallel and perpendicular resonance[3]. This interpretation of the resonance fields is of course subject to the conditions that $H_\perp \geqslant 4\pi M_s$. The simultaneous solution of the parallel and perpendicular resonance fields is also subject to $H_\parallel \geq H_u = 2K_u/M_s$, so that by plotting $H_\parallel$ $H_\perp$, $H_u$, and $4\pi M_s$ (Fig. 1), the regions of valid solutions are clearly delineated. In Fig. 2, the temperature variations of $K_u$ and g are shown for this sample.

## ANALYSIS

If one assumes that 1 and 2 denote Gd and Co(Mo), respectively, and one assumes that the Wangsness equation[4] holds for g in terms of the sub-network magnetizations $M_1$ and $M_2$, then one may write:

$$g = (M_1 + M_2)/((M_1/g_1) + (M_2/g_2) \quad (1)$$

Since the net magnetization (algebraic) is: $M = M_1 + M_2$, then either $M_1$ or $M_2$ may be eliminated from eq. (1), and one obtains the following set of equations for the sub-network magnetizations:

$$M_1 = (Mg_1/g)(g_2-g)/(g_2 - g_1)$$
$$M_2 = (Mg_2/g)(g-g_1)/(g_2-g_1) \quad (2)$$

If one assigns appropriate values to $g_1$ and $g_2$, then the experimental data for g and M permit the derivation of the two sub-network magnetizations $M_1$ and $M_2$. Appropriate values are of course subject to some guidelines laid down by the known zero °K $4\pi M_s$ values of 26,520 and 18,160 G for crystalline Gd and Co, respectively. It was found that it was impossible to fit the observed $K_u(T)$ variations to be proportional to $M_2^n$ where n = 2 to 3, and although a fit for n = 4 is obtainable, there seems no justification for a quadrupolar type variation under the circumstances. Likewise, it proved to be impossible to fit $K_u(T)$ with a two-term expansion $C_{12}M_1M_2 + C_{22}M_2^2$; instead, it was necessary on the pseudo-dipolar hypothesis[5] to set:

$$K_u = \sum_{i,j=1}^{2} C_{ij}M_iM_j \quad (4)$$

With this three-term expansion, one can show that:

$$K_u = M^2 \sum_{n=0}^{2} a_n g^{-n} \quad (5)$$

where it can be readily ascertained that:

$$a_0 = \left[C_{11}g_1^2 - C_{12}g_1g_2 + C_{22}g_2^2\right]/(g_2-g_1)^2$$
$$a_1 = \left[C_{12}(g_1+g_2) - 2(C_{11}g_1 + C_{22}g_2)\right]g_1g_2/(g_2-g_1)^2$$
$$a_2 = \left[C_{11} - C_{12} + C_{22}\right]g_1^2g_2^2/(g_2-g_1)^2 \quad (6)$$

One may make a computer least squares fit of the $K_u$, M, and g data to eq. (5) in order to determine the best values for the $a_n$. Having these values for $a_n$ from the fit of eq. (5), an inversion of eq. (6) to obtain the $C_{ij}$ constants is readily accomplished by eq. (7):

$$C_{11} = a_0 + a_1g_1^{-1} + a_2g_1^{-2}$$
$$C_{12} = 2\left[a_0 + 0.5(g_1+g_2)g_1^{-1}g_2^{-1} + a_2g_1^{-1}g_2^{-1}\right]$$
$$C_{22} = a_0 + a_1g_2^{-1} + a_2g_2^{-2} \quad (7)$$

Two samples were chosen having fairly complete $K_u(T)$ and g(T) measured variations. The values found for

TABLE I

| S# | $a_0$ | $a_1$ | $a_2$ | $C_{11}$ | $C_{12}$ | $C_{22}$ | $g_1$ | $g_2$ |
|---|---|---|---|---|---|---|---|---|
| A | 30.0 | -130 | 137 | -0.558 | -1.43 | -0.617 | 2 | 2.19 |
| B | 41.8 | -182 | 196 | -0.041 | -0.486 | -0.288 | 2 | 2.12 |

Sample A: x = 0.74, y = 0.11, $T_N$=605°K, $T_c$= 232°K
Sample B: x = 0.71, y = 0.16, $T_N$=374°K, $T_c$= 238°K

these constants $a_n$ and $C_{ij}$, and the choices for $g_1$, $g_2$ to approximate the proper values of the sub-network magnetizations are given in Table I. In treating the data, one must realize that annealing effects enter in above 425°K[6]. If data too close to the compensation temperature is chosen, the spread in M, $K_u$, and g due to inhomogeneity in the sample may be responsible for an apparent dip in $K_u$[7], and spurious fluctuations in the sub-network magnetizations. The value of $g_1 = 2$ for Gd seems proper, although there is a measurement cited in the literature for Gd metal as 1.95[8].

## CONCLUSIONS

It has been found that the uniaxial anisotropy coefficient $K_u(T)$ as determined by FMR may be made to fit a pseudo-dipolar variation of the sub-network magnetizations according to eq. (4), or equivalently, the variation expressed by eq. (5) in terms of the net magnetization $M_s(T)$ and the g-factor g(T). The magnitudes of the pseudo-dipolar coupling constants $C_{ij}$ in eqs. (4, 7) should indicate the relative importance of such interactions between Gd-Gd, Gd-Co, and Co-Co, respectively. The determination of these constants shows that the Gd-Co coupling is dominant, and that even the Gd-Gd coupling is of considerable importance. The finding of the secondary importance of Co-Co coupling and the large variation of the constants $C_{ij}$ with Mo addition are surprising. These deductions from the magnitudes of the $C_{ij}$ suggest that although eqs. (4, 5) fit the $K_u(T)$ variation found experimentally, they do not really unveil the physical situation which is responsible for the anisotropy. However, the eqs. (4, 5) serve to fit the experimental data for $K_u(T)$ within a few percent.

## ACKNOWLEDGMENT

It is a pleasure to acknowledge the excellent magnetic moment measurements performed by H. Lilienthal.

## REFERENCES

1. P. Chaudhari, J. J. Cuomo, & R. J. Gambino, IBM J. Res. & Dev. 17, 66 (1973); Appl. Phys. Letters 22, 337 (1973)
2. P. Chaudhari, J. J. Cuomo, R. J. Gambino, S. Kirkpatrick, & L. J. Tao, AIP Conf. Proc. 24, 562 (1974); B. E. Argyle, R. J. Gambino, & K. Y. Ahn, AIP Conf. Proc. 24, 564 (1974)
3. D. C. Cronemeyer, AIP Conf. Proc. 18, 85 (1973)
4. R. K. Wangsness, Am. J. Phys. 24, 60 (1956)
5. J. H. Van Vleck, Phys. Rev. 52, 1178 (1937)
6. R. J. Kobliska, R. Ruf, J. J. Cuomo, AIP Conf. Proc. 24, 570 (1974)
7. T. Kobayashi, N. Imamura, & Y. Mimura, AIP Conf. Proc. 24, 566 (1974)
8. A. F. Kip, Rev. Mod. Phys. 25, 229 (1953)

# GROWTH-INDUCED ANISOTROPY IN YTTRIUM IRON GARNET FILMS GROWN BY LIQUID PHASE EPITAXY*

M. T. Elliott and H. L. Glass
Electronics Research Division, Rockwell International
Anaheim, California 92803

ABSTRACT

Growth-induced anisotropies were measured by ferromagnetic resonance techniques in LPE YIG films which contained appreciable concentrations of impurity Pb ions. Values of $K_u^G$ as large as $2.7 \times 10^4 erg/cm^3$ were found. The growth-induced anisotropy could be removed by annealing.

## INTRODUCTION

Growth-induced magnetic anisotropy is an important and extensively studied property of flux-grown magnetic garnet crystals. It is an especially important property in LPE (liquid phase epitaxy) garnet films which are used in magnetic bubble memory devices. For such applications the growth-induced contribution may be the major source of the uniaxial magnetic anisotropy that is required for the existence of bubble domains.

Bubble garnet compositions generally include Eu or Sm along with one or more smaller rare-earths (or Y). In these materials the large growth-induced anisotropy has been accounted for by models which invoke a partial ordering of the different rare-earth ions on dodecahedral c-sites along with magnetic interactions between the Eu or Sm and neighboring Fe ions[1-3]. Growth-induced anisotropy has also been observed in mixed garnets and substituted garnets with no magnetic rare-earths present. In these garnets the anisotropy has been attributed to ordered distortions of the crystal field which induce single ion anisotropy in Fe ions. The ordered distortions could arise from ordering of the rare-earth ions in mixed garnets[4] or of substituent ions in substituted garnets[5].

Growth-induced anisotropy has also been observed in single rare-earth (or Y) iron garnets. In these materials, the anisotropy has been associated with the incorporation of Pb, as an impurity, from the PbO-based fluxes employed in LPE growth. For Eu-iron garnet ($Eu_3Fe_5O_{12}$) the Pb-induced anisotropy is reported to be similar to the growth-induced anisotropy observed in mixed garnets containing Eu[6]. The same model, partial ordering of Eu on c-sites and magnetic interactions between Eu and Fe, has been invoked to explain the observations. For Y-iron garnet (YIG) there are contradictions in the literature. Some investigators report appreciable growth-induced anisotropy when Pb is incorporated[7], while other investigators say the effect is very small[6]. In this paper we present some results of our measurements of growth-induced anisotropy in LPE YIG films containing Pb. We will show that the Pb-induced uniaxial anisotropy can be as large as that in mixed garnets containing Eu.

## EXPERIMENT AND RESULTS

YIG films a few μm thick were grown on (111) oriented GGG (gadolinium gallium garnet) substrates by the isothermal dipping method of LPE. A $PbO-B_2O_3$ flux was used and film growth was carried out while the horizontally held substrates were rotated around a vertical axis. Melt compositions and growth conditions were similar to those described in our recent publication on Pb incorporation in LPE YIG[8]. The incorporation of Pb was controlled by selection of the film growth temperature. The Pb concentration in the films was determined indirectly by X-ray diffraction measurements of the film/substrate lattice misfit from which the film lattice parameter was calculated. The increase of the film lattice parameter ($\delta a_f$) relative to the value 12.376Å for pure YIG was used as the measure of Pb concentration; the increase being 0.013Å/wt % Pb[8].

Domain patterns in the films were observed using a polarizing microscope. With this technique, stripe domains were visible by the differential Faraday rotation when the films exhibited an appreciable perpendicular component of magnetization. (However, since the Faraday rotation decreases with increasing Pb concentration and changes sign at $\delta a_f \sim 0.02$Å[8,9], the absence of visible domains does not necessarily imply that the perpendicular component is negligible.) For small values of $\delta a_f$ the misfit stress in the films would be tensile and magnetostriction would produce an easy axis in the [111] direction of the film normal[10]; stripe domains would be visible. For films under small compressive stress, $\delta a_f > 0.007$Å, the stress-induced anisotropy was in-plane[10] and domains were not visible or were only faintly visible. However, for large compressive stresses, $\delta a_f \gtrsim 0.016$Å, intense stripe domain patterns were observed.

Quantitative measurements of the magnetic anisotropy were made by ferromagnetic resonance on small pieces cut from the samples. From the longitudinal and transverse resonance fields at 9.1 GHz the cubic and uniaxial anisotropy fields were determined[11]. Using the measured value of film/substrate misfit along with elasticity theory and the magnetostriction coefficient of YIG[10], the stress-induced contribution to the uniaxial anisotropy field was calculated and subtracted off to yield the growth-induced contribution. The growth-induced anisotropy $K_u^G$ was obtained from the corresponding anisotropy field ($2K_u^G/M$) using the magnetization M corresponding to the $\delta a_f$ value of the film[8]. These values of magnetization were confirmed, where possible, from the position of the bottom of the spin-wave band[12]. The results are tabulated below:

| Film No. | $\delta a_f$ (Å) | $K_u^G$ (erg/cm$^3$) |
|---|---|---|
| 1 | $8.3 \times 10^{-3}$ | $-0.58 \times 10^4$ |
| 2 | 15.6 | 1.33 |
| 3 | 17.4 | 1.91 |
| 4 | 23.9 | 2.70 |

For film 1 $K_u^G$ is negative, indicating an in-plane (non-uniaxial) anisotropy. For this film and for film 2, which has a positive (uniaxial) $K_u^G$, no stripe domains were observed in the polarizing microscope.

To supplement the above results, other pieces of the same films were annealed in oxygen for 8 hours at 1175°C to determine whether the growth-induced anisotropy would vanish as it does for mixed garnet films. The annealing resulted in a small amount of stress relief in film 1 and drastic stress relief in the other films. In fact, after annealing, films 2-4 were under tensile stress. From the shift in resonance fields brought about by annealing and taking into account the changes in film stress, the growth-induced anisotropies were redetermined[11]. For film 1, the values of $K_u^G$ before and after annealing agreed to within 3%; however, for the other films the discrepancies ranged from 15-30%.

## DISCUSSION

The results clearly demonstrated that LPE YIG films which contain Pb can exhibit a uniaxial growth-induced magnetic anisotropy which is comparable in magnitude to that observed in mixed iron garnets that contain magnetic rare-earths[1,13]. This growth-induced anisotropy increases with increasing Pb concentration. Using the value for film lattice expansion of 0.013Å/wt % Pb[8], $K_u^G$ has a value of $1 \times 10^4$ erg/cm$^3$ when the Pb concentration is about 1 wt % or 0.04 atoms per formula unit. If the growth-induced anisotropy is associated with ordering of Pb ions on dodecahedral c-sites, the relatively low concentration of Pb would imply a high degree of site selectivity compared to rare-earths such as Eu. This is not necessarily unreasonable, since Pb is a much larger ion than any rare-earth and since ionic radius mismatch (between Pb and Y in this case) appears to determine site preference[13].

Although the growth induced anisotropy correlates with Pb concentration, other factors may also be involved. For example, the incorporation of Pb is accompanied by the appearance of Y on octahedral a-sites which would normally be occupied only by Fe[8,14]. Since Y ions are much larger than Fe, ordering of Y on a-sites could produce large lattice distortions and single-ion anisotropies. Ordering of Y on a-sites could be linked to ordering of Pb on c-sites.

The incorporation of Pb also requires some charge compensation mechanism, since the Pb ions are divalent substituents for trivalent Y. Ferromagnetic resonance measurements have shown that for $\delta a_f \sim 0.007$Å the room temperature linewidth is a minimum[15]. This minimum, 0.15 Oe at 9.5 GHz, corresponds to the intrinsic linewidth of pure, perfect YIG. (For film 1, having $\delta a_f = 0.0083$Å, the linewidth at 9.1 GHz was 0.14 Oe.) For larger $\delta a_f$, the linewidth increases due to charge compensation effects. The dominant charge compensation mechanism appears to be the creation of tetravalent Fe[15,16]. Another possible mechanism is the presence of oxygen vacancies. The ordering of oxygen vacancies, or other defects, has been suggested as a source of growth induced anisotropy[17]. Since the charge compensating defects (tetravalent Fe or oxygen vacancies) are associated with Pb ions and since Y on a-sites may also be associated with Pb, the anisotropy could be due to ordering of some rather complex clusters. Whatever the ordered entity, the absence of magnetic species other than Fe favors those models which are based upon single-ion anisotropy induced by ordered distortions of the crystal field.

## REFERENCES

*Supported by the Air Force Office of Scientific Research (AFSC) under Contract F44620-75-C-0045.


1. A. Rosencwaig and W. J. Tabor, AIP Conf. Proc. 5, 57-70 (1972).
2. H. Callen, Appl. Phys. Lett. 18, 311-313 (1971).
3. M. D. Sturge, R. C. LeCraw, R. D. Pierce, S. J. Licht and L. K. Shick, Phys. Rev. B 7, 1070-1078 (1973).
4. H. Callen, J. Appl. Phys. 45, 2348-2350 (1974).
5. A. Akselrad and H. Callen, Appl. Phys. Lett. 19, 464-466 (1971).
6. T. S. Plaskett, E. Klokholm, D. C. Cronemeyer, P. C. Yin and S. E. Blum, Appl. Phys. Lett. 25, 357-359 (1974).
7. T. J. A. Popma, A. M. von Diepen and J. M. Robertson, Mat. Res. Bull. 9. 699-704 (1974).
8. H. L. Glass and M. T. Elliott, J. Crystal Growth 27, 253-260 (1974).
9. K. Shinagawa, H. Takeuchi and S. Taniguchi, Japan J. Appl. Phys. 12, 466 (1973).
10. P. J. Besser, J. E. Mee, P. E. Elkins and D. M. Heinz, Mat. Res. Bull. 6, 1111-1123 (1971).
11. M. T. Elliott, Mat. Res. Bull. 9, 1143-1150 (1974).
12. B. R. Tittman, AIP Conf. Proc. 5, 1128-1132 (1972).
13. E. M. Gyorgy, M. D. Sturge, L. G. Van Uitert, E. J. Heilner and W. H. Grodkiewicz, J. Appl. Phys. 44, 438-443 (1973).
14. J. M. Robertson, M. J. G. van Hout, J. C. Verplanke and J. C. Brice, Mat. Res. Bull. 9, 555-568 (1974).
15. H. L. Glass and M. T. Elliott, J. Crystal Growth, to be published.
16. M. T. Elliott, this conference.
17. W. T. Stacy and C. J. M. Rooymans, Solid State Communications 9, 2005-2008 (1971).


# EVIDENCE FOR NON-CUBIC MAGNETOSTRICTION IN BUBBLE GARNETS

M.H. Yang and M.W. Muller
Department of Electrical Engineering, Washington University, Saint Louis, MO 63130

## ABSTRACT

Several investigators have reported observations of homogeneous stripe domain nucleation in bubble garnet films (1,2,3). The domains formed under homogenous nucleation conditions tend to align at or near right angles with the in-plane component of the applied field. Both a magnetostatic effect associated with the garnets' cubic anisotropy (4) and magnetostriction (1,3) have been put forward to account for the observations, with partial success. We show that the experimental findings can be fully accounted for by postulating a non-cubic component of the magnetoelastic tensor. The sign and magnitude of this excess-uniaxial magnetoelastic constant can be estimated from the experiments. The non-cubic magnetoelastic interaction may have the same origin as the growth-induced uniaxial anisotropy of the film.

Supported by the National Science Foundation, Grant GH32000


1. A. Hubert, A.P. Malozemoff, J.C. DeLuca, J. Appl. Phys. 45, 3462 (1974).
2. S.K. Chung, M.W. Muller, J. Mag. Mag. Mat. (in press).
3. M.W. Muller, M.H. Yang, Trans. IEEE, MAG-11, (Sept. 1975).
4. M.W. Muller, J. Appl. Phys. 45, 5050 (1974).

# REDUCTION OF THE APPARENT ANISOTROPY OF BUBBLE GARNET FILMS UNDER ALUMINUM METALLIZATION

R. Wolfe and W. A. Johnson
Bell Laboratories, Murray Hill, New Jersey 07974

ABSTRACT

The magnetic field parallel to the surface of a garnet film which results in nucleation of a sea of bubbles, is lower for regions covered by an evaporated Al film than for the bare garnet. This field, which is related to the anisotropy field, $H_k$, decreases further under the Al when the specimen is annealed at temperatures up to 450°C (as much as 40% below the bare garnet value). The value of the in-plane bubble nucleation field returns to normal when the Al is removed. The effect is still present, but reduced in magnitude, when a spacer layer of $SiO_2$ up to 1.2 μm thick is deposited on the garnet beneath the Al film. The effect is therefore not related to a chemical interaction between the Al and the garnet. $H_k$ as determined from demagnetized domain stripe width and bubble collapse measurements is essentially unchanged under the Al film. This effect is interpreted in terms of stress gradients associated with granularity of the Al metallization.

INTRODUCTION

When a large magnetic field is applied to a bubble garnet film in an appropriate direction approximately parallel to the film surface and then removed, a sea of bubbles is nucleated. The minimum value of this field which gives rise to bubbles rather than stripe domains, is closely related to the anisotropy field, $H_k$, of the garnet.[1] In this paper, the changes in this in-plane bubble nucleation field observed under aluminum metallization are reported.

EXPERIMENTS

In these experiments, the metallization was typically an alloy of Al with 4.5 wt. % Cu, 0.5 μm thick, deposited by e-beam evaporation with the wafer at room temperature. The effects observed were similar when pure Al was used instead of this alloy. In most cases, the Al was patterned and covered with a 0.8 μm layer of RF sputtered $SiO_2$. The nominal substrate temperature during the $SiO_2$ deposition was 200°C. Only small quantitative differences were observed when the metallization was not covered with $SiO_2$. The LPE garnet films used were of various compositions with bubbles near 6 μm or 3 μm in diameter. These included compositions of the types:

$(YGdTm)_3(FeGa)_5O_{12}$

$(YSm)_3(FeGa)_5O_{12}$

$(YSmCa)_3(FeGe)_5O_{12}$

$(YLuSmCa)_3(FeGe)_5O_{12}$

This latter composition was recently described by Blank, et al.[2] The effects were qualitatively similar in all of these materials. The garnet films were implanted with $Ne^+$ ions ($2\times10^{14}$ per $cm^2$ at 100 keV) for hard bubble suppression.[3] Some experiments were performed without ion implantation and again the observed effects were essentially the same.

In a typical experiment, the "in-plane" field was applied close to a <110> direction in the (111) plane of the film. The film was then tilted a few degrees to find the angle for nucleation of all-black or all-white bubbles. The minimum field required to nucleate a number of bubbles was then measured for the bare garnet and an adjacent region under the metallization. This field was found to be lower under the Al than in the bare garnet. This effect is illustrated in Fig. 1.

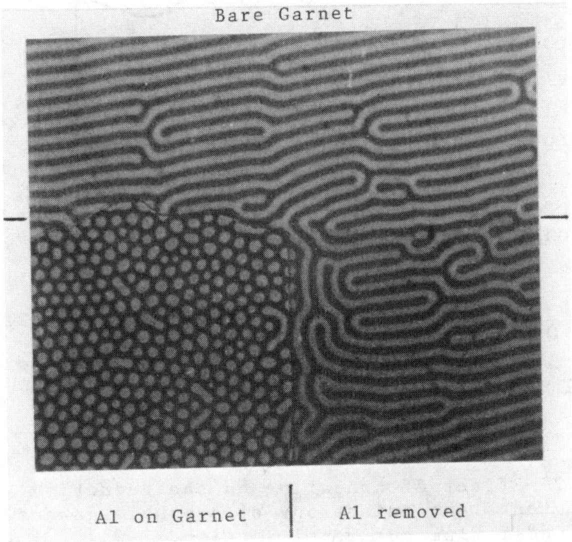


Fig. 1 Domains in a garnet film, viewed by reflection, show the reduction in apparent anisotropy under aluminum metallization after annealing for ½ h in $N_2$ at 450°C. Part of the Al film was removed by etching after annealing. A field of 1600 Oe was applied parallel to the film surface and then removed.

In this figure domains in a 6 μm thick film of nominal composition $Y_{1.44}Lu_{0.3}Sm_{0.3}Ca_{0.96}Ge_{0.96}Fe_{4.04}O_{12}$[2] are viewed by reflection. (The film is face down on an aluminized surface.) A field of 1600 Oe was applied close to the [110] direction in the film plane (horizontal in the photograph) and then removed. This field was sufficient to nucleate a sea of bubbles in the region covered with Al - 4.5% Cu, but not large enough to overcome the anisotropy field of the bare garnet, at the top of the photograph. In the lower right area of this figure, the Al film had been removed by an acid etch. In this region, except for slight darkening,[4] the properties are identical to those of the bare garnet which had never been covered with Al. This indicates that the effect is due to stress in the Al film and not to a chemical effect.

The reduction in the "apparent anisotropy field" under the aluminum is enhanced by annealing in an inert atmosphere, as shown in Fig. 2. The specimen was a 3 μm thick film of nominal composition $Y_{1.48}Lu_{0.3}Sm_{0.3}Ca_{0.92}Ge_{0.92}Fe_{4.08}O_{12}$.[2] The aluminum was covered with sputtered $SiO_2$ and had therefore been heated above 200°C. Separate chips from this wafer were annealed at the temperatures shown for ½ hour in nitrogen. The in-plane bubble nucleation field was measured at room temperature on adjacent bare and aluminized regions of each chip. This field dropped by 30% under the Al after the 400°C and 450°C anneals.

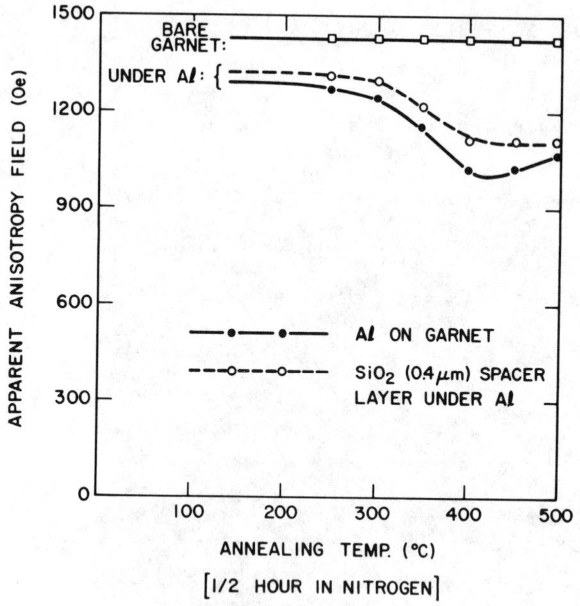


Fig. 2 Effect of annealing on the reduction of the apparent anisotropy of a bubble garnet film under aluminum metallization.

A similar drop to a minimum of 25 to 40% below the bare garnet value at 400°C or 450°C was found in all of the garnet compositions listed above.

At higher temperatures the apparent anisotropy field increased due to a decrease in the magnetization of the garnet. This effect is small in this Ca-Ge composition, but in the Ga-doped compositions it was so large that the apparent anisotropy field after annealing above 500°C was higher under the Al than in the bare garnet. This drop in magnetization is due to the rearrangement of Ga and Fe ions on the tetrahedral and octahedral sites accelerated by the chemical reduction of the garnet by the Al, as discussed elsewhere.[4] The strong tetrahedral site preference of Ge makes this effect smaller in the Ca-Ge materials.

The effect of depositing a layer of $SiO_2$ on the garnet before evaporating the Al film is shown in Fig. 2. With an RF sputtered spacer layer 0.4 μm thick, the drop in the apparent anisotropy is still present, but reduced in magnitude. A small effect was observed even with a spacer layer 1.2 μm thick, but the rise in anisotropy field at anneal temperatures above 450°C was not observed for spacer layers thicker than 0.2 μm. This again supports the conclusion that the reduction in anisotropy is associated with stress in the Al film and not with chemical interaction between the Al and the garnet.

Another method for determining the anisotropy field of a bubble garnet film is based on measurements of the bubble collapse field, the demagnetized domain stripe width and the film thickness.[5] Using this method, the anisotropy measured on several specimens was found to be the same within ±5% for areas under the Al metallization and for the bare garnet, even after annealing at temperatures up to 450°C. The stress in the Al film is therefore not affecting the "real" anisotropy field through the whole thickness of the garnet film. For anneals above 450°C the decrease in magnetization gives rise to an increase in the "real" as well as the "apparent" anisotropy field.

Visual observations of the domains while the in-plane field was applied showed that the stripe domains had disappeared in the bare garnet regions when the in-plane field was large enough to result in bubble nucleation. However, under the Al, stripe domains were still visible and were observed to break up into bubbles ("stripe domain pinching"[6]) at fields below the "real" anisotropy field.

## DISCUSSION

Stresses in Al or Al-Cu films as deposited on garnet substrates are typically in the range $-10^7$ to $-10^8$ dynes/cm$^2$. Measured values after annealing at 400°C are of the order of $-10^9$ (tension).[7] The uniform compressive strain transmitted to a garnet film on a relatively thick substrate would give rise to a reduction in the anisotropy field due to the negative magnetostrictive coefficients of the compositions listed above. However, the effect should be at least two orders of magnitude smaller than the observed effects. This is consistent with the absence of any change in the "real" anisotropy determined by bubble measurements. The stress buildup in this supersaturated Al-4.5 wt % Cu alloy is not due to a precipitation hardening since the effect is observed with pure Al as well. Large localized changes in wall energy (and anisotropy) occur at the edges of patterned metallization due to local stress fields.[8] These changes were shown to be reduced but not eliminated by the use of spacer layers up to 1 μm thick. The effects reported here for broad area aluminum metallization are probably similar localized anisotropy changes associated with increasing grain growth and resultant nonuniform stresses as the aluminum is annealed. This granularity is visible in regions of the aluminum film from which the covering layer of $SiO_2$ has been removed.

The fabrication of field access bubble devices involves all of the processes used in these experiments except that annealing in $N_2$ is replaced by heating in vacuum to at least 300°C during the Permalloy deposition step. The apparent anisotropy reduction under Al conductors can therefore influence the behavior of all such devices.

## ACKNOWLEDGMENTS

Valuable discussions with W. J. DeBonte, R. C. LeCraw, H. J. Levinstein, R. D. Pierce, B. J. Roman and G. P. Vella-Coleiro are gratefully acknowledged.

## REFERENCES

1. A. Hubert, A. P. Malozemoff and J. C. DeLuca, J. Appl. Phys. 45, 3562 (1974) and references therein.
2. S. L. Blank, J. W. Nielsen and W. A. Biolsi, presented at Meeting of Electrochem. Soc., Dallas, Texas, October 5, 1975. To be published.
3. R. Wolfe, J. C. North and Y. P. Lai, Appl. Phys. Lett. 22, 603 (1973).
4. R. C. LeCraw, E. M. Gyorgy and R. Wolfe, this conference.
5. D. C. Fowlis and J. A. Copeland, AIP Conference Proceedings No. 5, 240 (1971).
6. Y. Shimada, J. Appl. Phys. 45, 3154 (1974).
7. B. J. Roman and W. J. DeBonte, personal communications.
8. J. M. Dishman, R. D. Pierce and B. J. Roman, J. Appl. Phys. 45, 4076 (1974).

# EFFECTS OF ION IMPLANTATION ON AMORPHOUS Gd-Co*

E. L. Venturini, P. M. Richards, J. A. Borders, and E. P. EerNisse
Sandia Laboratories, Albuquerque, New Mexico 87115


## ABSTRACT

Studies of stress, composition, and magnetic resonance have been made on amorphous Gd-Co thin films implanted with nitrogen ions at fluences between $10^{10}$ and $10^{16}$ ions/cm$^2$. A new mode of magnetic resonance is produced in a small portion of the implanted region. In this portion $H_A - 4\pi M$ decreases with fluence and ultimately achieves a negative value considerably larger in magnitude than $4\pi M$ in the film prior to the implantation. This behavior is interpreted as arising from a decrease in the Gd-Co exchange interaction which reduces the alignment of the Gd spin system.


## INTRODUCTION

Ion implantation is a well known tool for altering the magnetic properties of rare earth iron garnets for bubble device applications. Its effect on amorphous metal alloy films is also of interest because of the possibility of "tuning" the net perpendicular anisotropy[1] and magnetization for optimum bubble characteristics. Studies of implanted amorphous films will hopefully lead to a better understanding of the origin of the perpendicular anisotropy field $H_A$.

High energy (1 to 3 MeV) argon ions have been shown to alter the anisotropy in Gd-Co films from Bitter pattern[2] and polar Kerr effect[3] studies.

We have studied the effects of somewhat lower energy nitrogen and argon ions implanted to higher fluence in Gd-Co. Using X-band magnetic resonance, we have identified a new mode introduced by the implantation and determined a room temperature value for the net anisotropy $H_A - 4\pi M$ associated with this mode for different fluences. An advantage of the resonance technique is that it, as opposed to the Kerr effect, can measure negative as well as positive values of $H_A - 4\pi M$ since it does not require the magnetization to be perpendicular to the film. A separate measurement of the lateral stress in a film versus fluence indicates that the net anisotropy change with implantation is not related to the stress.

## MEASUREMENTS

The lateral stress in a Gd-Co thin film prepared by rf sputtering onto a glass substrate has been monitored as a function of fluence for 500 keV $N^{++}$ ions by using a cantilever beam technique described previously.[4] It reached a peak of nearly $2\times10^{11}$ dynes/cm$^2$ at a fluence of $10^{14}$ ions/cm$^2$ and was relieved at $5\times10^{14}$ ions/cm$^2$. Such stress behavior at these low ion fluences generally is attributable to displacement damage caused by ion energy deposited into atomic collisions. The film was approximately $\frac{1}{2}$ µm thick, and most of the implanted ions stopped within the film.

Room temperature magnetic resonance was measured at 9.8 GHz on films whose composition was roughly $Gd_{.19}Co_{.81}$ and which showed compensation near 140°K. Most implantations for our resonance studies used 200 keV $N^+$ ions with a projected range of just over 2000 Å in amorphous Gd-Co. Since the films were nearly one µm thick, the $N^+$ ions stopped within the top 25% of the films. One implant was done with 500 keV $Ar^{++}$ ions at a fluence of $2\times10^{16}$ ions/cm$^2$. Their projected range is roughly equal to that of the 200 keV $N^+$ ions, and the ion energy deposited into damage is equivalent to that of $1.3\times10^{17}$ $N^+$ ions/cm$^2$.

Before implanting a given sample, a complete absorption versus applied field trace to 9 kOe was recorded to identify all X-band magnetic resonances. Following the implantation a similar trace to 9 kOe revealed a new resonance (at sufficiently high fluence) with a peak-to-peak width of only 150 Oe which was presumably located within the implanted region. By measuring the field for resonance of this new mode both parallel and perpendicular to the film and by using a ferromagnetic resonance approximation, a value for the net anisotropy $H_A - 4\pi M$ associated with this mode is obtained.

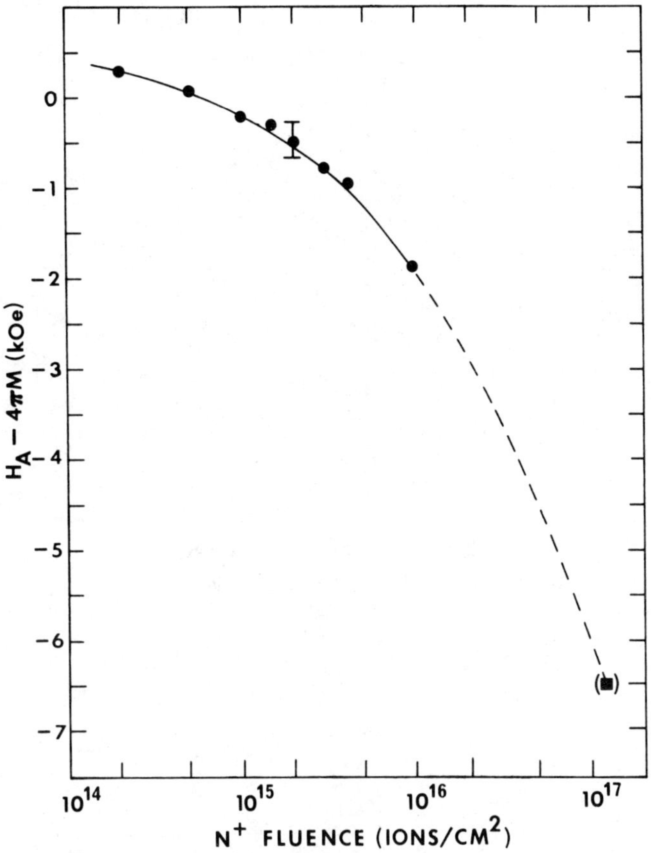


Fig. 1. Net anisotropy as determined from X-band resonance vs. fluence of 200 keV $N^+$ ions. Solid line is an aid to the eye only.

Figure 1 shows the resulting net anisotropy vs. $N^+$ fluence. The data point at the highest fluence shown in parentheses is from the $Ar^{++}$ implanted sample. The complete angular dependence at one fluence is shown in Fig. 2 together with the variation predicted for the uniform mode. There was essentially no angular variation of the intensity, and the resonance field varied by only 10% between room temperature and 100°K. The static magnetization was measured with a vibrating sample magnetometer and found to be $4\pi M$ = 1.8 kG before implantation. There was some evidence of an increase after implantation, but it was no more than 10% at the highest fluence.

Two MeV helium ion backscattering was also performed on the $Ar^{++}$ implanted film and an unimplanted film in order to determine composition vs. depth before and after implantation. The unimplanted film had a uniform composition of 19 at.% Gd, but the implanted one showed a reduced Gd concentration, presumably due to preferential sputtering, of 17 at.% in the first 600 Å of the 9700 Å thickness which then increased to the bulk 19 at.% by 1000 Å. X-ray diffraction gave no evidence of polycrystalline structure

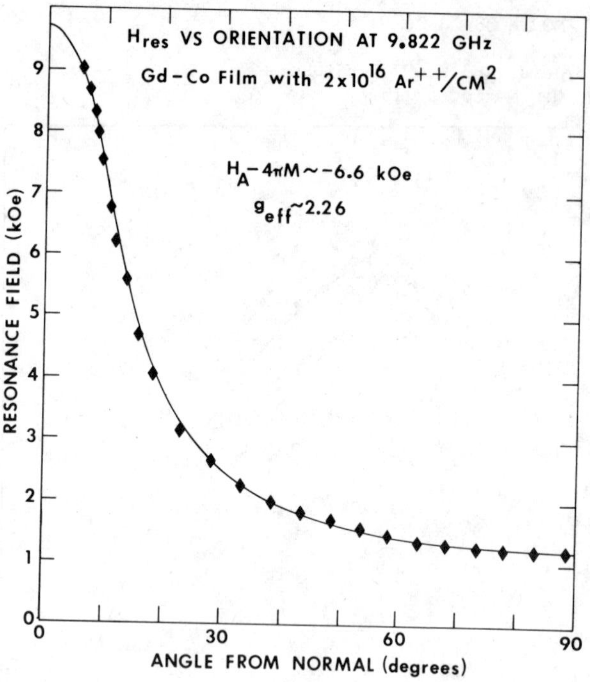


Fig. 2. Angular variation of the resonance field for the new mode in the most heavily implanted sample. The solid line is a uniform ferromagnetic resonance calculation using the parameters indicated above.

larger than 25 Å either before or after implantation.

DISCUSSION

The facts that the new resonance mode has an angular dependence consistent with a uniform mode and shows no significant angular dependence of intensity indicate that it is a volume rather than a surface mode.

The resonance data in Fig. 1 then show that $H_A - 4\pi M$ in a portion of the film decreases without apparent limit and reaches a large negative value at the highest fluence. Since it seems unlikely that ion bombardment would induce the ordering required to produce a negative $H_A$, we will take $H_A \geq 0$ in the most heavily damaged sample for the arguments which follow, which implies that $4\pi M$ has increased.

Consider the depletion of Gd as measured by the backscattering. A composition of 17 at.% Gd corresponds[1] to $4\pi M \cong 3$ kG, so that if the only effect of implantation were to change the Gd-to-Co ratio in amorphous GdCo, we should have a net in plane anisotropy of 3 kG or less in the $Ar^{++}$ implanted sample. Since the observed net anisotropy is 6.6 kG, other mechanisms must be operative. The new mode is essentially temperature independent, which means that it is primarily due to Co since the variation of M in Gd-Co below 300°K comes almost entirely from the temperature dependence of $M_{Gd}$. This argument is strengthened by the fact that the data of Fig. 2 fit g = 2.26, similar to the g-factor for Co metal. Hence we assume $M_{Co} \gg M_{Gd}$ and $4\pi M_{Co} \approx 6.6$ kG + $H_A$ in the region responsible for the new resonance mode. Magnetization curves of the unimplanted sample gave $H_A = 4.3$ kOe. If we assume that implantation could only decrease $H_A$, then $4\pi M_{Co} \lesssim 10.9$ kG. The Gd magnetization in $Gd_{.17}Co_{.83}$ is $4\pi M_{Gd} = 7.6$ kG. This large a value of $M_{Gd}$ would produce a sizeable temperature dependence which is not observed in the new resonance. Thus we conclude that $M_{Gd}$ has been greatly reduced in that part of the implanted region responsible for the resonance.

A decrease in $M_{Gd}$ is most readily explained by a decrease in the Gd-Co exchange interaction $J_{Gd-Co}$ which causes misalignment of the Gd spin system. (Since Gd has localized 4f magnetism we rule out the possibility that the atomic moment $\mu_{Gd}$ decreases; and since the Gd-Gd interaction is considerably weaker than the Gd-Co one,[5] ordering of the Gd spins is primarily determined by $J_{Gd-Co}$.)

Decrease of the atomic Co moment $\mu_{Co}$ upon implantation was reported in ref. 3. It is likely that $\mu_{Co}$ has decreased in our samples as well. Taking[5] $\mu_{Co} = 1.65\ \mu_B$ gives $4\pi M_{Co} = 10$ kG at 83 at.% Co in GdCo. Only in the unlikely event that $H_A$ is nearly unchanged by the implantation could we account for this large an $M_{Co}$. In the more probable event that $H_A$ has decreased considerably our result $4\pi M_{Co} \approx 6.6$ kG + $H_A$ shows that $\mu_{Co}$ has decreased due to structural rearrangement.

The bulk magnetization as measured by a vibrating sample magnetometer increased by no more than 10% with the heaviest implantation. This implies that the region in which the new resonance occurs with $4\pi M_{Co} \gtrsim 6.6$ kG is less than 400 Å thick, whereas the backscattering indicates that the composition varies to a depth of about 1000 Å, and the projected range of the implanted ions was about 2000 Å. This apparent discrepancy can be explained by assuming that the magnetization and/or $H_A$ is nonuniform throughout all but 400 Å of the implanted region. The nonuniformity will produce excessive broadening and make resonance from this major portion of the implanted depth unobservable.

In conclusion our data indicate that there is a relatively uniform section of about 400 Å or less within the implanted region for which $J_{Gd-Co}$ is greatly reduced so that the Gd spins are no longer aligned. It is also likely that $\mu_{Co}$ is reduced in this region. Both $\mu_{Co}$ and $J_{Gd-Co}$ depend upon the local environment and presumably can be reduced by intervening implanted ions and/or a destruction of the local atomic order. The fact that the largest net anisotropy changes occur at fluences above that corresponding to maximum lateral stress seems to rule out any strong effects due to local strains and magnetoelastic couplings.

ACKNOWLEDGMENTS

The authors thank R. E. Hampy for preparing the Gd-Co films, G. D. Peterson for the ion implantation, and D. H. Cooper for assistance with the measurements.

REFERENCES

*Work supported by the U. S. Energy Research and Development Administration.

1. P. Chaudhari, J. J. Cuomo, and R. J. Gambino, IBM, J. Res. Dev. 17, 66 (1973); P. Chaudhari, J. J. Cuomo, and R. J. Gambino, Appl. Phys. Lett. 22, 337 (1973).
2. R. J. Gambino, J. Ziegler, and J. J. Cuomo, Appl. Phys. Lett. 24, 99 (1974).
3. R. Hasegawa, R. J. Gambino, J. J. Cuomo, and J. F. Ziegler, J. Appl. Phys. 45, 4036 (1974).
4. E. P. EerNisse, Appl. Phys. Lett. 18, 581 (1971).
5. R. Hasegawa, B. E. Argyle, and L-J. Tao, A.I.P. Conf. Proceedings 24, 110 (1974).

# MAGNETIC ANISOTROPY OF $Ho^{3+}$ IN HOLMIUM-ALUMINIUM-GARNET

T. Egami [+*] and P. J. Flanders [+]
Laboratory for Research on the Structure of Matter,
University of Pennsylvania, Philadelphia, Pa. 19174
and
E. M. Gyorgy and L. G. Van Uitert
Bell Laboratories, Murray Hill, N. J. 07974

## ABSTRACT

The field induced magnetocrystalline anisotropy of $Ho^{3+}$ in Aluminium garnet was measured by the torque method at 4.2K in magnetic fields up to 70 kOe, and was compared to the magnetization. The exponent of the power law, $\mathcal{K}_4 \propto m^n$, was found to be 4 when m is small and to approach gradually to 10 as m was increased, as has been predicted by Callen and Callen; this measurement is the first confirmation of the power law of the field induced anisotropy.

## INTRODUCTION

The origin of the magnetocrystalline anisotropy of the rare earth-iron-garnets is known to be predominantly the crystalline electric field (CEF) on the rare earth ion and partly the anisotropic exchange between the rare earth spin and iron spin. In order to single out CEF effect, we studied the field induced anisotropy of Holmium-Aluminium garnet (HAG; $Ho_3Al_5O_{12}$) at low temperatures in high magnetic fields. The ground state multiplet of $Ho^{3+}$ ion with $(4f)^{10}$ electron configuration is $^5I_8$ which will split into several crystal field levels with overall splitting about 700K(1) when placed in garnets. The saturated moment of the state with $J_z = 8$ is $10\mu_B$, so that the application of magnetic fields up to 10 Tesla (1Tesla=10 kOe) would result in considerable change in the energy levels and the magnetic moment at sufficiently low temperatures. The results were compared to the magnetization for the purpose of obtaining the power law, and to the field dependent anisotropy of $Ho^{3+}$ in iron garnets.

## EXPERIMENTAL

Crystals with compositions $Ho_3Al_5O_{12}$, $Ho_3Fe_5O_{12}$ and $Ho_{0.5}Y_{2.5}Fe_5O_{12}$ were grown by the standard flux method (2), and were annealed at 1200°C for 16 hrs in oxygen to eliminate the growth induced anisotropy. Two cylinder shaped samples with the cylinder axis parallel to <100> and <110> crystallographic axis were cut out from the crystals of HAG. Also three cylinders with <110> axis were cut out, two from the HoIG crystals and one from the crystal of $Ho_{0.5}Y_{2.5}Fe_5O_{12}$. Each sample weighed about 40~50 mg.

The magnetocrystalline anisotropy was measured by the torque method described elsewhere (3), in applied magnetic fields up to 7 Tesla. The magnetization was measured by a low frequency vibrating sample magnetometer in fields up to 10 Tesla.

## MAGNETIC ANISOTROPY OF HAG

The magnetocrystalline anisotropy and magnetization of HAG at 4.2K were stongly field dependent as shown in Fig. 1. At the magnetic field of 7 Tesla, about 60% of the saturation magnetization of $Ho^{3+}$ ion was induced. At low fields it was found that the magnetization was linear with the field, but the anisotropy was proportional to $H^4$. By plotting $\mathcal{K}_4 = K_1 + \frac{1}{11}K_2$ and m, magnetization, on a log-log scale, we found that the differentially defined exponent of the power law

$$n = \frac{\partial \log \mathcal{K}_4}{\partial \log m} \quad (1)$$

changes from 4 for small m to 10 for large m, as shown in Figs. 2 and 3. The measurement made at 77K confirmed the $m^4$ law, with slightly different multiplying factor as will be discussed later. By extrapolation we estimate the anisotropy constant at saturation would be about $4 \times 10^8$ erg/g.

## MAGNETIC ANISOTROPY OF HoIG

We also studied the field dependent anisotropy of $Ho_3Fe_5O_{12}$ (HoIG) and $Ho_{0.5}Y_{2.5}Fe_5O_{12}$ (Ho-YIG) at 77K. The total magnetic moment of iron spins is about 4.7 $\mu B$ per formula at 77K, and the magnetic moment of holmium spin is about 2.6 $\mu B$ in HoIG and 3.2 $\mu B$ in Ho-YIG, and is always antiparallel to the total magnetic moment of iron spins. The moment of $Ho^{3+}$ is greater in Ho-YIG than in HoIG; so is the susceptibility. The smaller lattice constant of YIG than HoIG is thought to cause these differences. When a magnetic field is applied, the holmium moment becomes parallel to

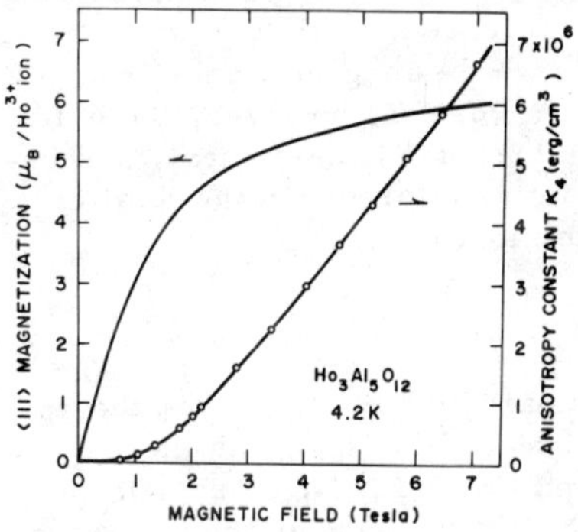


Fig. 1. Field dependence of anisotropy constant and magnetization along easy direction of $Ho_3Al_5O_{12}$ at 4.2 K

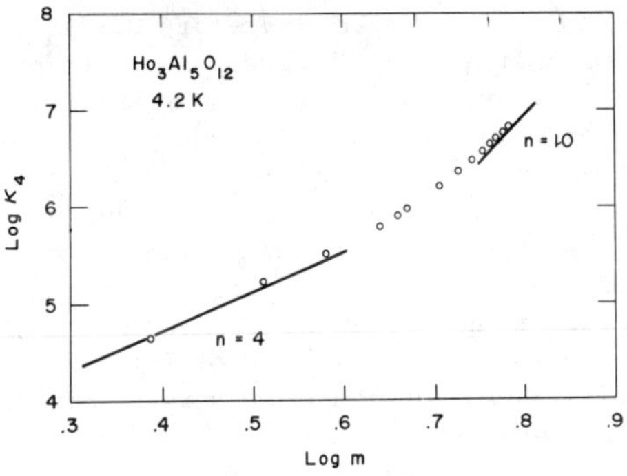


Fig. 2 Plot of $\mathcal{K}_4$ vs m in log-log scale.

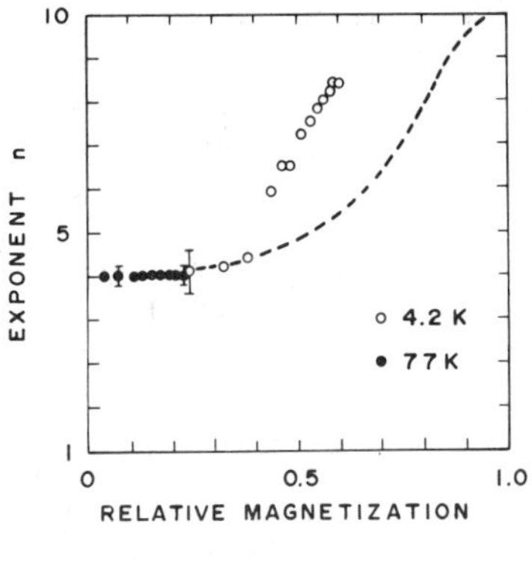


Fig. 3. Dependence of the exponent n defined by eq. (1) on magnetization. Broken line represents eq. (3).

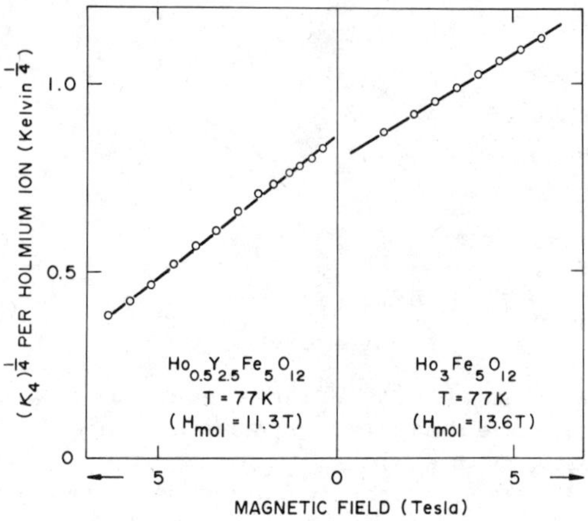


Fig. 4. Field dependence of $(\mathcal{K}_4)^{\frac{1}{4}}$ for HoIG and Ho-YIG. Contribution of YIG is subtracted from the data.

Table 1.

The anisotropy energy per $Ho^{3+}$ ion in the unit of Kelvin/atom. $\mu$ is the magnetization of $Ho^{3+}$ ion in $\mu_B$.

| Composition | T (K) | $\mathcal{K}_4$ $(K/\mu_B^4)$ |
|---|---|---|
| $Ho_3Al_5O_{12}$ | 4.2 | $0.816 \times 10^{-2} \times \mu^4$ |
| $Ho_3Al_5O_{12}$ | 77 | 0.440 " |
| $Ho_3Fe_5O_{12}$ | " | 0.908 " |
| $Ho_{0.5}Y_{2.5}Fe_5O_{12}$ | " | 0.521 " |

the field in HoIG, since the total holmium moment 7.8$\mu_B$ per formula is larger than the total iron moment. Consequently, as the field is increased, in HoIG the magnetic moment of holmium and hence the magnetic anisotropy of the crystal is increased. In Ho-YIG, however, the holmium moment is less than the total iron moment, and becomes antiparallel to the field. Moreover, the total effective field on the holmium ion,

$$H_{eff.} = H_{mol.} - H_{appl.} \qquad (2)$$

where $H_{mol.}$ is the molecular field and $H_{appl.}$ is the applied external field, is decreased as $H_{appl.}$ is increased. This is because the susceptibility of the iron system is small and the exchange interaction between the iron system and the holmium ion is small, and therefore the $H_{mol}$ is virtually unchanged as $H_{appl.}$ is changed. Thus, both the magnetization of holmium and the total anisotropy are decreased as the applied field is increased. $H_{mol.}$ may be estimated as the field at which the moment of HoIG or Ho-YIG becomes equal to the moment of YIG, in other words, as the field at which the moment of holmium vanishes. Such a field is about 13 Tesla for HoIG and about 10 Tesla for Ho-YIG. Fig. 4 shows that the $\frac{1}{4}$ power of the anisotropy attributed to a Ho ion (in the unit of Kelvin/atom) is proportional to the field, hence the magnetization. The intercept of the $K^{\frac{1}{4}}$ with the field axis is 13.6 T for Ho-IG and 11.4 T for Ho-YIG, in accordance with the estimates from magnetization. These results are in agreement with the measurement by Demidov et al[4]. The $m^4$ dependence of the anisotropy of three Ho-garnets are summarized in Table 1. There are some variations in the multiplication factor, which presumably is due to either different chemical composition or thermal expansion. However, considering the fact that the magnetic field dependence of the anisotropy constant was very much different in all of these four cases, the agreement in the m scaling is noteworthy.

## POWER LAW OF ANISOTROPY CONSTANT

The power law of the anisotropy constant first suggested by Akulov[5] and Zener[6], developed by Keffer[7] and generalized by Callen and Callen[8] states that the $\ell$-th order anisotropy constant $\mathcal{K}_\ell$ is related to m by

$$\mathcal{K}_\ell = \mathcal{K}_\ell^0 \cdot \hat{I}_{\ell+1/2}(\hat{I}_{3/2}^{-1}(m)) \qquad (3)$$

where

$$\hat{I}_{\ell+1/2}(x) = I_{\ell+1/2}(x)/I_{1/2}(x) \qquad (4)$$

and $I_k(x)$ is the hyperbolic Bessel function. Although the relation (3) was derived via the molecular field approximation and classical ($S \to \infty$) approximation, it appears to be widely applicable unless the anisotropy energy is very large compared to the exchange interaction. The reason behind this is that this relationship is based upon symmetry properties, rather than upon the details of the model, as has been pointed out by Callen and Shtrikman[9].

We calculated the exponent of the power law defined by (1), using (3) for $\ell = 4$. The result is shown in Fig. 3 with a broken line. It is seen that the exponent n is about 4 when m is small, and approaches 10 as m is increased. The celebrated $\ell(\ell+1)/2$ law applies only when m is larger than .95. This tendency of the transition of the power law from n=4 to n=10 is unambiguously reproduced for the case of HAG, although the transition to n=10 occurs at lower m than predicted from (3). Now the density matrices for HAG and for the classical model used to derive (3) are very different; in HAG the crystal field levels are unevenly distributed with overall splitting about 700K and the Zeeman splitting up to 100K at 70 kOe, whereas a continuous and constant density of states from -S to +S is assumed in the classical model. Therefore, here the argument by Callen and Shtrikman is under a severe test. The fact it predicts the tendency correctly indicates that the renormalizations of the anisotropy and magnetization, whether they are due to temperature or to the crystal field itself, are closely connected through symmetry.

As far as we are aware of, this measurement constitutes the first confirmation of the power law of the field induced anisotropy. It is, indeed, probably the better way to check the power law than by the temperature dependence since the measurements were done at a constant temperature, so that thermal expansion is not involved in the measurement.

## ACKNOWLEDGEMENTS

The authors are thankful to Dr. R. C. Sherwood for carrying out the low field magnetization measurement of HAG.

## REFERENCES

+ Supported in part by NSF Grant DMR72-03025.
* Also at Department of Metallurgy and Materials Science, University of Pennsylvania.

1. L. F. Johnson, J. F. Dillon, Jr., and J. P. Remeika, Phys. Rev. B1, 1935 (1970).
2. L. G. Van Uitert, W. H. Grodkiewicz and E. F. Dearborn, J. Amer. Ceram. Soc., 48, 105 (1965).
3. C. D. Graham, Jr., P. J. Flanders, and T. Egami, AIP Conf. Proc. 10, 759 (1973).
4. V. G. Demidov, A. K. Zvezdin, R. Z. Levitin, A. S. Markosyan, and A. I. Popov, Sov. Phys. Solid State, 16, 1379 (1975).
5. N. Akulov, Z. Phys., 100, 197 (1936).
6. C. Zener, Phys. Rev., 96, 1335 (1954).
7. F. Keffer, Phys. Rev., 100, 1692 (1955).
8. H. B. Callen and E. Callen, J. Phys. Chem. Solids, 27, 1271 (1966).
9. H. B. Callen and S. Shtrikman, Solid St. Commun., 3, 5 (1965).

# THE WEAK FERROMAGNETIC PROPERTIES OF DISORDERED-CONCENTRATED TRANSITION METAL ALLOYS

J. BEILLE, D. BLOCH and J. VOIRON
Laboratoire de Magnétisme, C.N.R.S.,
166-X, 38042-GRENOBLE-Cedex, France.

## ABSTRACT

The magnetic moments of transition atoms in disordered-concentrated metallic alloys depend on their immediate surrounding as well as on the mean electronic properties of the alloys. We discuss, using a few examples, the homogeneity of these alloys in the concentration range for the onset of ferromagnetism and of the effects of high magnetic fields, temperatures and pressures on their magnetic moment. Among the examples chosen are the $Ni_xPt_{1-x}$, $Ni_xPd_{1-x}$, $Fe_xGe_{1-x}$, $Fe_xCo_{1-x}Si$ and rare earth-yttrium-cobalt $Y_{1-x}T_xCo_2$ systems. Their physical properties are largely dependent on non-strongly magnetic transition atoms.

## INTRODUCTION

The transition atoms have or have not a magnetic moment depending on the mean electronic properties of the metallic alloy or compound in which they are located. However, in the compounds where the transition atoms are placed in different crystallographic sites, their moment can vary according to the particular site. In disordered-concentrated alloys, these moments can similarly depend on their immediate surroundings. One can then observe quite inhomogeneous spatial distributions of the magnetic moments. The models for the description of the magnetic properties of disordered-concentrated alloys are mainly phenomenologic ones. Among these, the rigid band model[1] or the virtual crystal model [2] are homogeneous models, in that they neglect the differences of local properties and surroundings. The models which are the most commonly used to take account of these differences are the Jaccarino-Walker model[3] which relates the existence of a magnetic moment on an atom to its immediate surrounding, and the superparamagnetic model which considers the magnetic moments as concentrated in magnetic clusters[4]. Somewhat less phenomenological, the C.P.A. method allows for different electronic structures on each atom. It can be extended to the disordered-concentrated alloys[5]. It leads to electronic distributions in qualitative agreement with results obtained from photoemission experiments[6]. However, it does not permit quantitative predictions of the ferromagnetic properties.

On the experimental side, one tries to specify not only the mean values of the magnetic moments, but eventually the local magnetic and electronic properties. To understand the existence or the stability of the magnetic moments, one can obtain help from experiments where one varies the chemical content, the temperature, the magnetic field and, or the pressure. In section II, we will consider the effects, at T = 0, of chemical substitutions on the magnetic moments of some disordered metallic alloys. In section III we will comment on the notion of homogeneity, especially in connection with the high-field magnetic behaviour. We will in section IV examine the thermal behaviour of these magnetic moments and in section V, their pressure dependence.

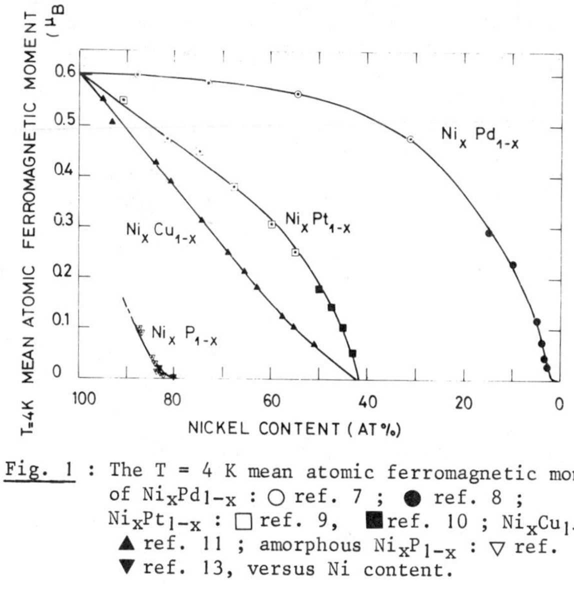


Fig. 1 : The T = 4 K mean atomic ferromagnetic moment of $Ni_xPd_{1-x}$ : ○ ref. 7 ; ● ref. 8 ; $Ni_xPt_{1-x}$ : □ ref. 9, ■ ref. 10 ; $Ni_xCu_{1-x}$ : ▲ ref. 11 ; amorphous $Ni_xP_{1-x}$ : ▽ ref. 12, ▼ ref. 13, versus Ni content.

## THE CHEMICAL PARAMETERS

One of the most common methods to study the stability of the magnetic moments consists of analyzing the magnetic properties of the solid solutions of a magnetic and a non-magnetic element. For this purpose, nickel is often chosen as the magnetic element ; Pd, Pt or Cu for instance can be mixed with nickel in any proportion to form disordered solid solutions having f.c.c. structure. At T = 0, an isolated nickel atom is non-magnetic. Ferromagnetism appears in a concentration range which depends on the non-magnetic element (Fig. 1). One can notice that the studies performed on amorphous alloys are often complementary to those undertaken on crystalline alloys. In some cases the two elements which have low solubility limits in the crystalline state can have a larger solubility, at room temperature, in the amorphous state. Such a situation occurs in $Ni_xP_{1-x}$[12, 13], $Fe_xGe_{1-x}$[14,15] or $Fe_xSi_{1-x}$[16]. Another situation, to which perhaps not enough attention has been paid, is that of alloys with two critical concentration ranges for the onset of ferromagnetism. This occurs in $(FeSi)_x(CoSi)_{1-x}$, for $0.2 \leq x \leq 0.8$[17-20] and $(FeTi)_x(CoTi)_{1-x}$ for $0.4 \leq x \leq 0.8$[20-22] (Fig. 2). Thus Fe, in FeSi, or Co in CoSi are not magnetic, but Fe or Co can have a magnetic moment in $Fe_xCo_{1-x}Si$ when some of their neighbours are Co or Fe.

Another typical system for studying the ferromagnetic moment on the transition atoms is the rare earth-transition metals systems. The cubic Laves phase $TCo_2$ have quite intriguing properties. The $YCo_2$[26] and $LuCo_2$[27] compounds are Pauli paramagnets. In the

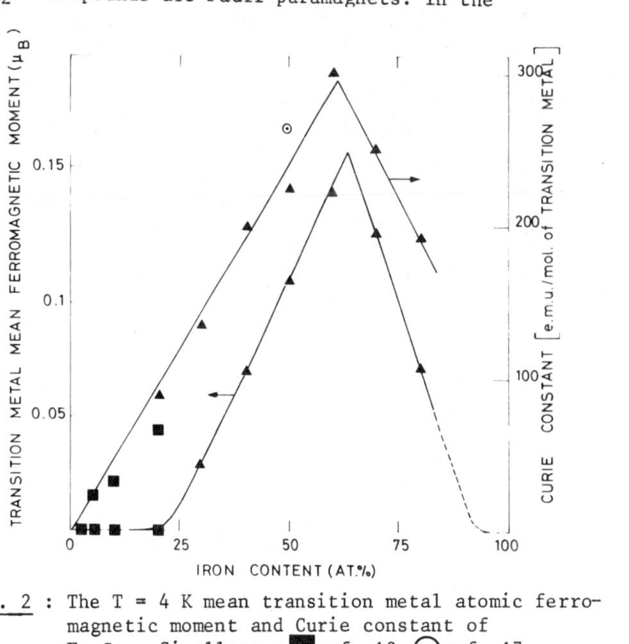


Fig. 2 : The T = 4 K mean transition metal atomic ferromagnetic moment and Curie constant of $Fe_xCo_{1-x}Si$ alloys : ■ ref. 13, ⊙ ref. 17, ▲ ref. 18, ● ref. 20, versus Fe content.

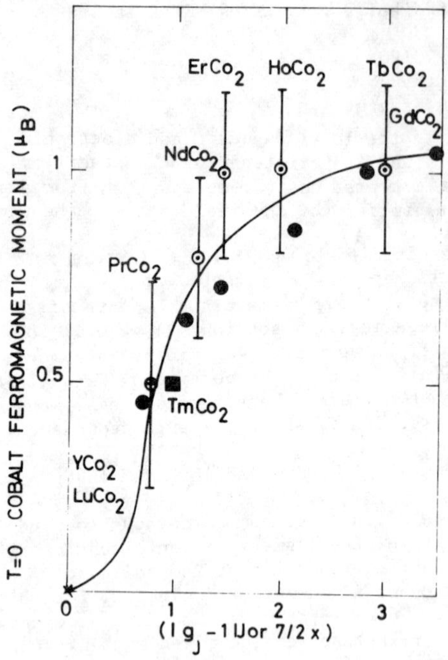


<u>Fig. 3</u> : T = 4 K mean ferromagnetic moment of cobalt in cubic Laves phases $TCo_2$ and $Gd_xY_{1-x}Co_2$ ; ● ref. 22, ⊙ ref. 23, ⊕ ref. 24, ■ ref. 25, X ref. 26 and 27, versus the mean spin. The full line is a guide for the eyes.

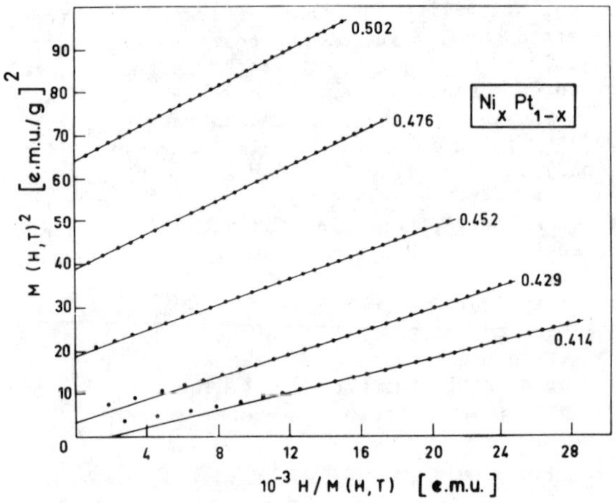


<u>Fig. 4</u> : $M^2(H,T)$ versus H/M at high magnetic field (150 kOe) and T = 4 K for $Ni_xPt_{1-x}$ alloys (ref. 10).

compounds containing a magnetic rare earth T, the cobalt atom possesses an induced[28] magnetic moment whose value, determined from magnetization measurements ($GdCo_2$) or neutron scattering experiments (Fig. 3), depends on the spin $|g_J-1|J$ of the neighbouring rare earth. Its maximum moment corresponds to that of one hole in the cobalt 3d-band. A similar dependence is observed for the solid solution $Gd_xY_{1-x}Co_2$, as a function of the mean spin (7/2)x. The comparison between the magnetic properties of these alloys in the crystalline and amorphous state leads to the conclusion that electronic transfers are of prime importance in determining the value of the maximum moment in the crystalline state. For instance cobalt in amorphous $YCo_2$ has a moment of $\simeq 1.6\ \mu_B$[29] ; this value is very similar to that of cobalt in amorphous $GdCo_2$ ($1.4\pm0.5\ \mu_B$)[30] or pure crystalline cobalt ($1.7\ \mu_B$). However the rigid band model does not give the key for the understanding of their magnetic properties.

## HOMOGENEITY AND HIGH FIELD BEHAVIOUR

In discussing the magnetic moment at T = 0, one can make an improvement on the rigid band model by introducing a Stoner type condition for the existence of ferromagnetism. This condition can be written $UN(E_F) > 1$, where U is an effective Coulomb intra-atomic interaction and $N(E_F)$ is the density of states at the Fermi level. When $UN(E_F)$ is smaller than one, the metal is non-magnetic ; when it is larger than one, it has a ferromagnetic moment. One of the most simple method to describe the magnetic properties of disordered-concentrated alloys is then the virtual crystal method, where the (mean) $UN(E_F)$ value can vary continuously with the chemical composition. The magnetic moment is then turned on at the critical concentration $c_F$ for the onset of ferromagnetism[2], when $UN(E_F) = 1$. Once this condition is fullfilled a magnetic moment occurs. It can be a strong ferromagnetic moment, or a weak one, associated with a slight polarization of the d electrons. Just above the critical concentration, the situation can in the latter case be described as a very weak homogeneous ferromagnet[31]. This picture is expected to have a reasonable degree of success in the case of alloys with a moderate critical concentration for the appearance of ferromagnetism (i.e. Ni-Pt but not Ni-Pd). Within this model, the difference $\Delta F$ between the free energy of the ferromagnetic and paramagnetic states is expressed as a Landau expansion of the magnetization M,

$$\Delta F = (1/2)AM^2(H) + (1/4)BM^4(H) - JM(H) \qquad (1)$$

where $A = A_1\ (c_F - c)$ changes its sign at the critical content. Minimizing $\Delta F$ versus M(H) leads to the equation :

$$M^2 = B^{-1}\ (H/M) + (c - c_F).A_1B^{-1} \qquad (2)$$

where $A_1$ and B are smooth functions of the concentration. We have shown that this kind of relationship is followed with a high accuracy for $Ni_xPt_{1-x}$ alloys (Fig. 4)[10,32], where the critical content is ≃42 at % Ni. More recent experiments, with fields up to 350 kOe[33] do not give evidence for any high field deviation. In agreement with this model, (see eq. 2), the H = 0 magnetic moment varies as $(c - c_F)^{1/2}$, in the vicinity of the critical content (Fig. 5). This relationship does not hold in the case of $Ni_xPd_{1-x}$ or $Ni_xP_{1-x}$ alloys. Furthermore in these cases (Fig. 6, 7) $M^2$ is not a linear function of H/M ; this can be associated[34] with inhomogeneities in the spatial distributions of the magnetic moments.

In the Jaccarino-Walker[3] model, an atom does or does not possess a magnetic moment, depending on its surroundings. The condition for the occurence of a magnetic moment has therefore some local aspects ; furthermore the transition to the magnetic state within this model is supposed to lead to a strong ferromagnetic moment. In the critical range for the onset of ferromagnetism (Fig. 6, 7) the high field experiments clearly demonstrate, after eventual saturation of the ferromagnetic regions or of the paramagnetic clusters a quite large susceptibility. It can be ascribed as due to weakly ferromagnetic or strongly paramagnetic regions in the sample. One should however notice that the high field susceptibility is not a constant, at a given temperature, independent of the applied field. Strongly paramagnetic domains may have a very large Stoner enhancement factor. For an enhancement factor of 100, the effective field is 10 MOe for an applied field of 100 kOe, and non-linear effects should be the rule. In fact, in the critical region for the onset of ferromagnetism, the high field susceptibility of $Ni_xPt_{1-x}$ and $Ni_xPd_{1-x}$ alloys decreases by a factor of about 2[35] from 40 to 150 kOe. One should notice that increasing values of the applied magnetic field do not always lead to a saturation of the magnetic susceptibility. We have shown in fact[36] that the susceptibility of $YCo_2$, a

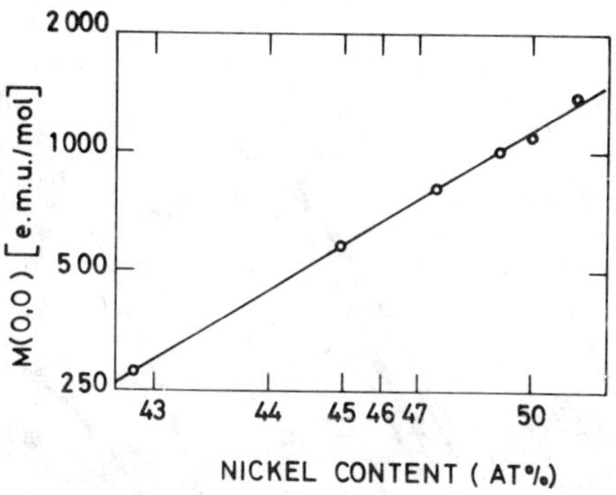


Fig. 5 : The spontaneous (T = 0, H = 0) moment of $Ni_xPt_{1-x}$ alloys as a function of Ni content.

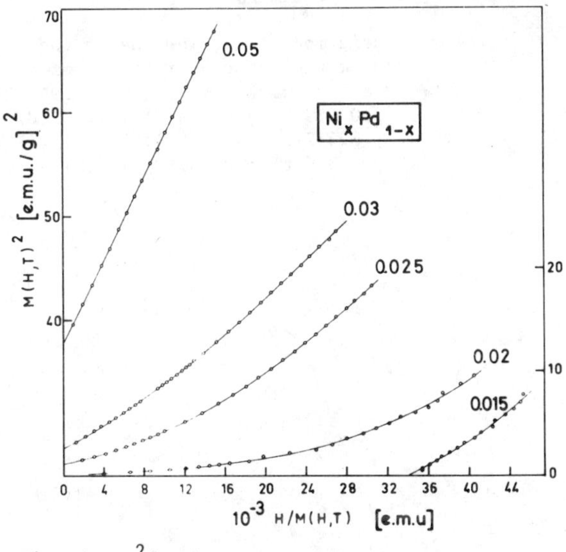


Fig. 6 : $M^2(H,T)$ versus H/M at high magnetic field (150 kOe) and T = 4 K of $Ni_xPd_{1-x}$ alloys (ref. 10).

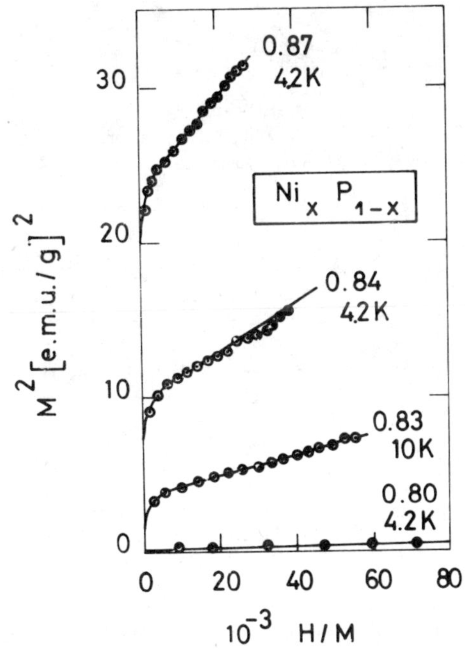


Fig. 7 : $M^2(H,T)$ versus H/M at high magnetic field (150 kOe) of amorphous $Ni_xP_{1-x}$ alloys (ref.13) (temperatures as indicated).

highly Stoner enhanced Pauli paramagnet, increases, at 4.2 K, of about 7 % between 0 and 150 kOe. It has been suggested[37] that a metamagnetic transition towards a large moment state could occur in some metals at sufficiently high fields. From the temperature and field dependence of the susceptibility, we estimated this threshold field to be about 1.4 MOe in $YCo_2$[36], which is out of the scope of actual high fields experiments. However an effective field of that order of magnitude can be achieved easily from a substitution of yttrium with a magnetic rare earth. For instance the molecular field acting on cobalt, in $GdCo_2$, is about 4.8 MOe. It exceeds the 1.4 MOe value in $ErCo_2$, $HoCo_2$, $DyCo_2$ and $TbCo_2$ as well as in $Gd_xY_{1-x}Co_2$ for cobalt atoms with at least two of its six yttrium neighbours substituted with gadolinium[36] (see Fig. 3). This allows, in these disordered solid solutions, for a discontinuous appearance of part of the magnetic moment on the cobalt atoms.

## HIGH TEMPERATURE BEHAVIOUR

The high temperature Curie-Weiss law for disordered concentrated transition metal alloys in the critical range for the onset of ferromagnetism is usually ascribed to the giant moments of the clusters[4]. From the Curie constant and from the low temperature moments, the mean magnetic moment per cluster and their concentration can be deduced. Typical values for such a magnetic moment are 10 - 30 $\mu_B$. The temperature independent component to the specific heat, as noticed in Ni-Cu[38], Ni-Rh[39] (but not Ni-Pd[40] or Ni-Pt[41]) is generally attributed to these clusters[42]. However, the Curie-Weiss constant, in simple metals or compounds, is often numerically quite different from what is expected from a local moment behaviour[43]. Furthermore, interactions between spin-fluctuation modes in homogeneous weak ferromagnets or strong paramagnets can lead to a Curie-Weiss behaviour[44,45], as well as to low temperature $T^2$ terms in the resistivity in the concentration range for the appearance of ferromagnetism[46]. It is therfore certainly useful to give some examples which can give more insight into the thermal behaviour of the disordered concentrated metal alloys. We first recall the thermal variation of the magnetization as deduced from the virtual weak ferromagnetic crystal approximation. The Landau coefficient A changes its sign at the Curie temperature. If one takes $A = A_2.[T^2 - T_c^2(c)]$ then eq. (2) can be written as :

$$M^2 = B^{-1} H/M + \left[T_c^2(c) - T^2\right] A_2 B^{-1} \qquad (2')$$

where $A_2$ and B are smooth functions of temperature and concentration ; $M^2$ is thus a linear function of H/M at any temperature. This is in quite good agreement with the experimental results for $Ni_xPt_{1-x}$[10,32,47] in the critical range for the onset of ferromagnetism (Fig. 8) or $Fe_{0.5}Co_{0.5}Si$[17]. One should note that in these alloys, nickel and platinum, as well as iron and cobalt possess a weak mean magnetic moment of similar order of magnitude. The $T^2$ dependence of A leads, in line with the Stoner theory, to a $T^2$ dependence of $M(0,T)^2$. This describes approximately the ferromagnetic behaviour below the Curie temperature (Fig. 9).

Spin fluctuations are to be considered as giving rise to a more realistic temperature dependence[35,45]. Similarly the low temperature specific heat of $Ni_xPt_{1-x}$ alloys[41] can be analyzed as $C = \gamma T + \beta T^3$, where the very large decrease of $\beta$ at the critical concentration cannot be explained as a softening of the lattice. It is then too more likely associated with spin-fluctuation effects. The high temperature susceptibility of $Ni_xPt_{1-x}$ alloys presents some interesting features (Fig. 10). These alloys follow a Curie-Weiss law at sufficiently high temperatures, with a Curie constant independent of the concentration and therefore independent of the T =0 spontaneous moment. The curves, Fig. 10b, for pure platinum and pure nickel have been drawn at scales 1/5 of those given Fig. 10a, in order to preserve identical

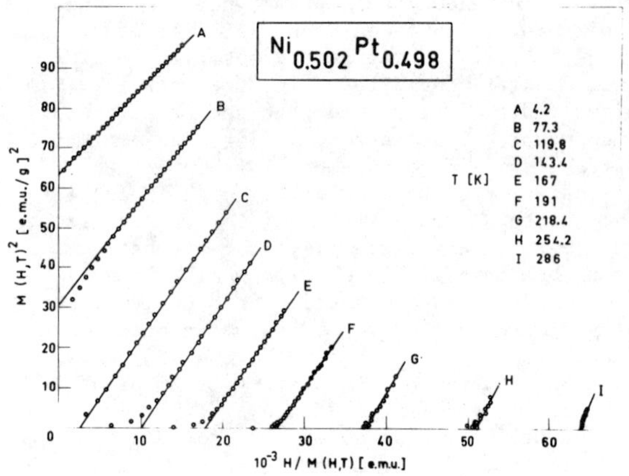


Fig. 8 : $M^2(H,T)$ versus H(T)/M for $Ni_{0.502}Pt_{0.498}$ at various temperatures from 4.2 to 286 K (ref.10).

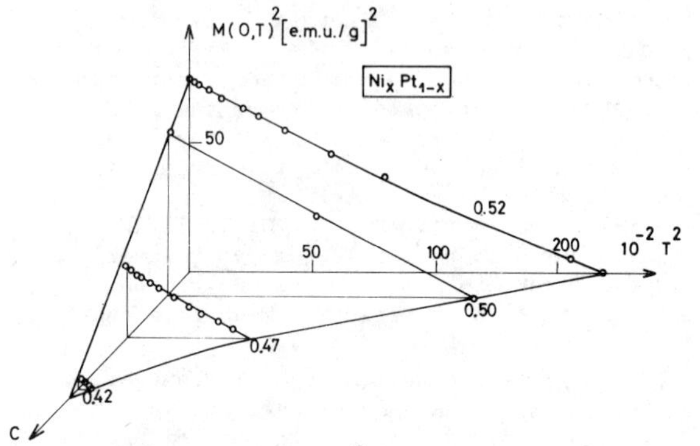


Fig. 9 : $M^2(0,T)$ versus $T^2$ for $Ni_xPt_{1-x}$ at various concentrations in the critical range for the appearance of ferromagnetism.

slopes. The Curie constant corresponds to an effective moment 2S $\mu_B$ = 0.9 $\mu_B$. This magnetic behaviour, with no variation of the Curie constant at the critical concentration, as well as the specific heat results suggest that $Ni_xPt_{1-x}$ alloys are rather homogeneous in the concentration range for the onset of ferromagnetism. Furthermore they suggest that Ni has a local moment at high temperature. The thermal dependence of the resistivity[48], specific heat[41,48] and the magnetic properties indicate that the spin fluctuating model for weak ferromagnets or strong paramagnets gives, at least qualitatively, a better description of their properties than the Stoner-Edwards-Wohlfarth approximation. Any quantitative comparison is however out of the range of actual theories.

Most of the alloys, whose T = 0 properties reflect an inhomogeneity, do preserve some of these properties at elevated temperatures. They appear in the temperature dependence of $M^2$ versus H/M as noticed for $Ni_{0.025}Pd_{0.975}$[10], amorphous $Fe_{0.3}Ge_{0.7}$ (Fig. 11) and amorphous $Ni_{0.87}P_{0.13}$ (Fig. 12). The inhomogeneous distribution of the magnetic moment in the critical range (≃ 2 at % Ni) for the onset of magnetism[50] disappears progressively in $Ni_xPd_{1-x}$ at higher nickel content (see Fig. 6) and higher temperature[35,61].

The Curie constant of $Fe_xCo_{1-x}Si$ (Fig. 2), $Fe_xGe_{1-x}$[15], $Ni_xP_{1-x}$[13] changes with the concentration. In fact CoSi is diamagnetic, with a full d-band, and FeSi is an anomalous Pauli paramagnet[49] ; Fe and Ni are metals whereas Ge and P are not metallic and such concentration dependence is to be expected. The $Fe_xCo_{1-x}$ system can be used in order to check the various models for the description of the onset of ferromagnetism in disordered-concentrated metal. Typical M(H) relationship for $Fe_xCo_{1-x}Si$ are given Fig. 13. The $M^2$ versus H/M relationship are linear (Fig. 14). The critical content for the appearance of ferromagnetism is, at 4 K, slightly larger than 20 at % Fe. Up to this concentra-

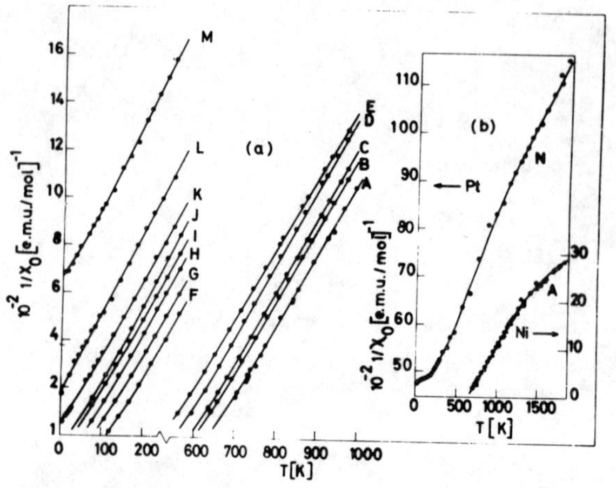


Fig. 10 : Thermal variation of the inverse of the initial paramagnetic susceptibility of $Ni_xPt_{1-x}$[9] alloys with : x = 1(A) ; 0.987(B) ; 0.977(C) ; 0.959(D) ; 0.909(E) ; 0.502(F) ; 0.476(G) ; 0.452(H) ; 0.429(I) ; 0.414(J) ; 0.368(K) ; 0.326(L) ; 0.241(M) ; 0(N).

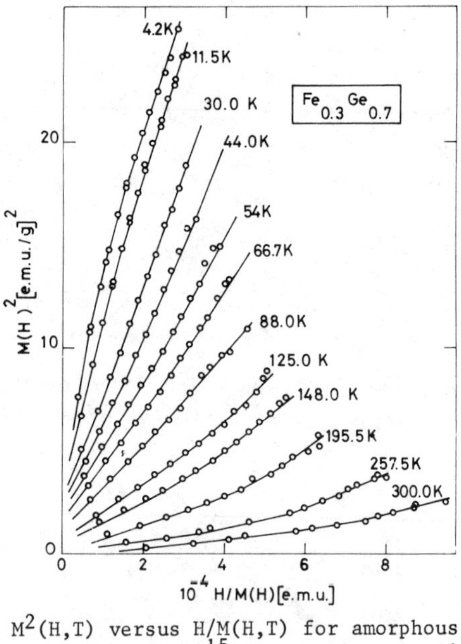


Fig. 11 : $M^2(H,T)$ versus H/M(H,T) for amorphous $Fe_{0.3}Ge_{0.7}$ alloy[15]. The maximum applied field is 150 kOe.

tion, Fe and Co atoms appear as non-magnetic. One can conversely, from M versus H/T, define the saturation moment as that of independent clusters. The saturation moment is, for x = 0.05 and x = 0.2, $0.7 \times 10^{-2}$ and $4 \times 10^{-2}$ $\mu_B$ per transition atom. From the Curie constant one can deduce the mean number $N^*$ of giant moments per transition atom, as well as the mean value $\mu^*$ of this giant moment for x = 0.05, $N^* = 3 \times 10^{-4}$ and $\mu^* = 24$ $\mu_B$, and for x = 0.2, $N^* = 20 \times 10^{-4}$ with $\mu^* = 16$ $\mu_B$. If we take these moments as due to strongly magnetized transition atoms, with a moment of ≃ 2 $\mu_B$ per atom, then respectively $3.5 \times 10^{-3}$ ans $2 \times 10^{-2}$ atoms have a magnetic moment. In fact, Mössbauer experiments indicate that Fe in CoSi has no field-induced moment larger than $5 \times 10^{-2}$ $\mu_B$ to be compared to a moment of about $10 \times 10^{-2}$ $\mu_B$[51] in $Fe_{0.5}Co_{0.5}Si$. Furthermore, from the decrease of the integrated spin echo intensities of $Co^{59}$ in $Fe_xCo_{1-x}Si$[52] the number of (field induced or spontaneous) magnetic cobalt atoms, equal to that of the cobalt atoms with Fe atoms in their nearest neighbour sites, is quite large, 80 and 17 $\times 10^{-2}$ for x = 0.2 and 0.05. Thus the magnetic moment is largely spread out in the crystal, and most of the transition atoms have a very weak magnetic moment. This allows one to understand the linear relationship $M^2$ versus H/M observed

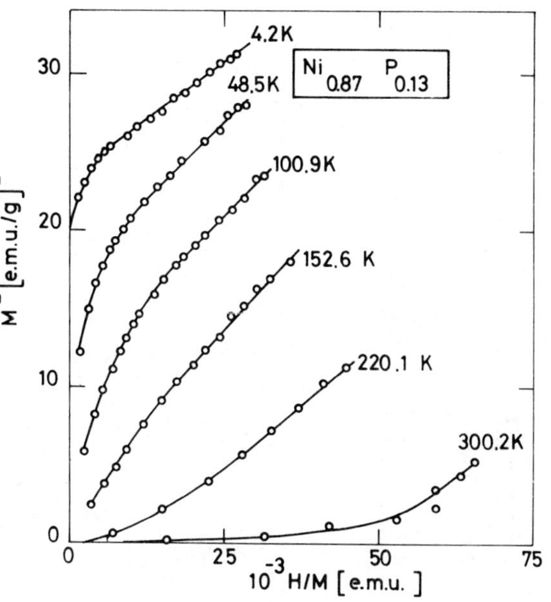


Fig. 12 : $M^2(H,T)$ versus $H/M(H,T)$ for amorphous $Ni_{0.87}P_{0.13}$[13]. The maximum applied field is 150 kOe.

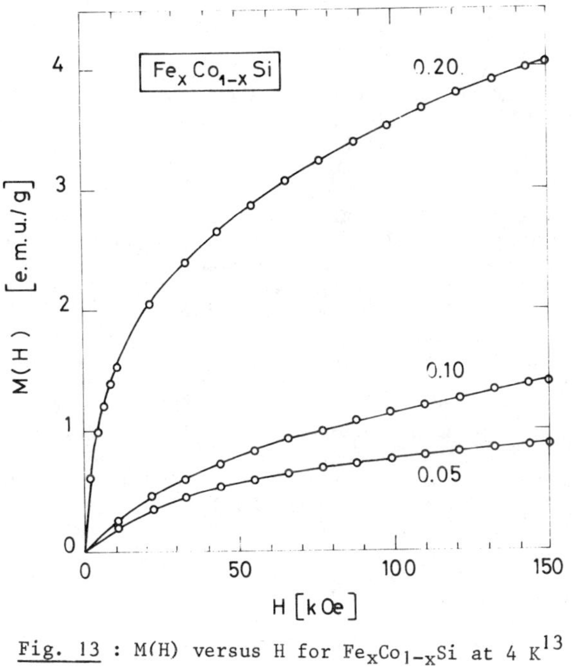


Fig. 13 : M(H) versus H for $Fe_xCo_{1-x}Si$ at 4 K[13]

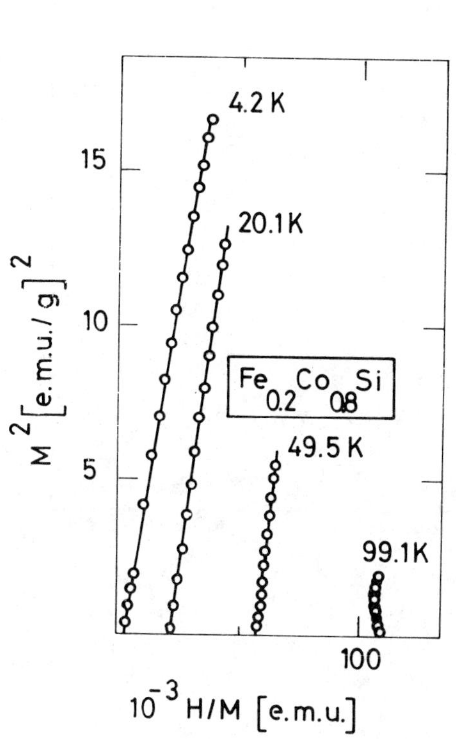


Fig. 14 : $M^2$ versus H/M for $Fe_{0.2}Co_{0.8}Si$ alloys at various temperatures.

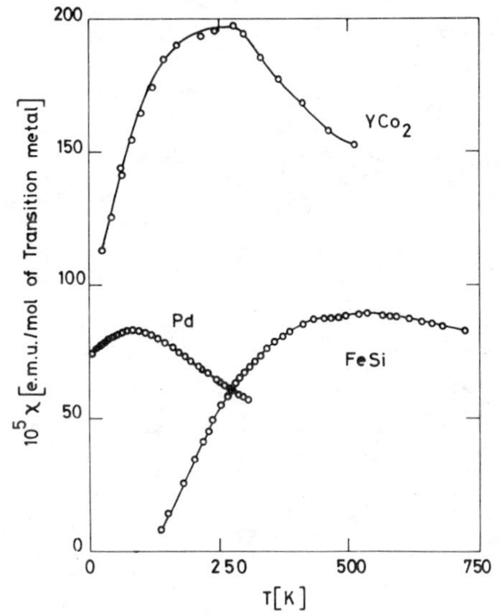


Fig. 15 : Thermal variation of the paramagnetic susceptibility of Pd[50], $YCo_2$[26] and FeSi[49].

for these alloys (Fig. 14). The large Curie constant is then partly associated with spin-fluctuation effects of the weakly magnetized transition metal atoms. In fact NMR spin echo spectrum has, from x = 0.4 to x = 0.7, a sharply defined peak[52] which confirms the rather good homogeneity of the weak ferromagnetic $Fe_xCo_{1-x}Si$ alloys in this concentration range.

It is important to notice however that the Curie-Weiss law for strong paramagnets is not, at least at moderate or low temperature, a completely general rule for strongly paramagnetic metals, as demonstrated, Fig. 15, for Pd, $YCo_2$ or FeSi, which are strongly enhanced paramagnets, with singularities in the density of states near the Fermi level.

In the magnetic $TCo_2$ compounds, the effective field on the cobalt atoms depends on temperature, to some extend through the temperature dependence of the matrix ($YCo_2$), but mostly through that of the rare earth T moment. Due to the metamagnetic process described in section II, this can give rise[36] to a first order transition instead of an usual Curie temperature, as observed in $HoCo_2$, $DyCo_2$ and $ErCo_2$. As the cobalt moment is induced, one should notice that its disappearance above the magnetic ordering temperature. This, in fact, is confirmed from the analysis of their paramagnetic susceptibility[53] and from the magnetic entropy[54] as well as polarized neutron scattering experiments[55].

## HIGH PRESSURE BEHAVIOUR

The magnetic moment can vary with temperature. It can similarly vary with pressure. Within the virtual crystal approximation, one can treat the effect of pressure P, through the consideration of its effect on the Landau coefficient in the free energy expansion (1). Then, with $A = A_3[P - P_c(c,T)]$, A changes its sign at the critical pressure $P_c$, and

$$M^2 = B^{-1}\, H/M + \left[P_c(c,T) - P\, A_3\right] B^{-1} \qquad (2'')$$

where $A_3$ and B are smooth functions of temperature, pressure and concentration. The intersection of the

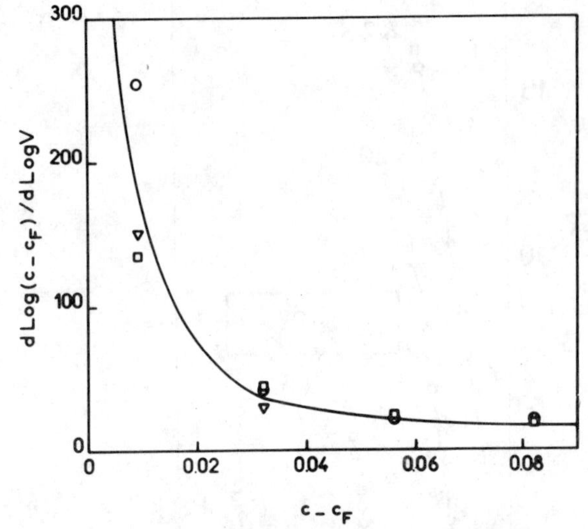


Fig. 16 : 2 dLog $T_c$/dLog V : ▽; 2 dLog M/dLog V : □; dLog $\chi_0$/dLog V : ○ as a function of c − $c_F$. The full line is obtained as dLog(c−$c_F$)/dLog V for $dc_F/dp = 10^{-3}$ $kbar^{-1}$ (from 32).

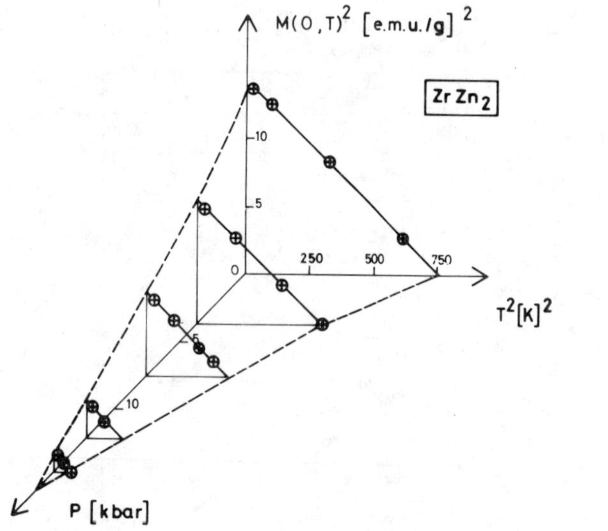


Fig. 17 : $M^2(H = 0,T)$ versus $T^2$ for $ZrZn_2$ at various pressures (from 56).

"surface" $A(T_1, P_1, c) = 0$ with a T plane thus gives the variation of the critical pressure as a function of the concentration ; similarly, the intersection with a P plane, that of the Curie temperature versus concentration and the intersection with a c plane, that of the Curie temperature with pressure. Within the virtual crystal approximation[32] :

$$2\,\frac{\text{dLog } M_0}{\text{dLog V}} = 2\,\frac{\text{dLog } T_c}{\text{d Log V}} = -\,\frac{\text{dLog } \chi_0}{\text{dLog V}} = \frac{\text{dLog }(c-c_F)}{\text{dLog V}} \qquad (3)$$

where $M_0$ is the T = 0, H = 0 spontaneous magnetization and $\chi_0$ the T = 0, H = 0 limit of the high field susceptibility. The experimental results for various $Ni_xPt_{1-x}$ alloys are given Fig. 16. They can be fitted with a single parameter $dc_F/dp = 10^{-3}$ $kbar^{-1}$, whose value can be understood semi-quantitatively[32].

Similar results have been obtained more recently for $ZrZn_2$ by Huber et al[56]. As expected from the weak ferromagnetic approximation $M_0$ and $T_c$ are approximately linear functions of the applied pressure (Fig. 17). The critical pressure is 18 kbar for the sample under consideration. At moderate pressure and at concentration close to the critical concentration, the relative pressure dependence $(1/T_c)(dT_c/dp)$ of homogeneous weakly ferromagnetic disordered alloys is thus expected

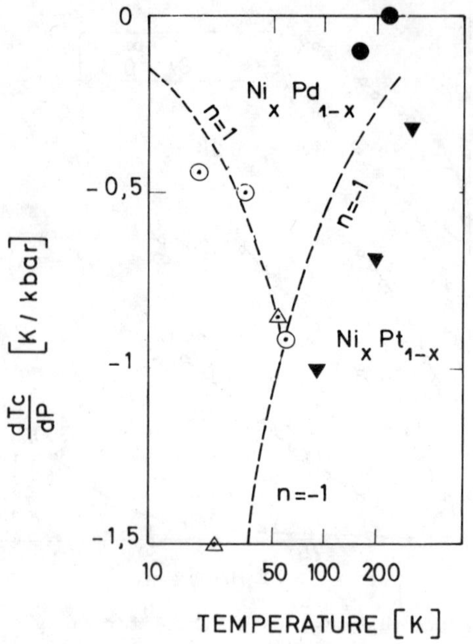


Fig. 18 : The pressure derivative $dT_c/dp$, versus $T_c$ for $Ni_xPt_{1-x}$ and $Ni_xPd_{1-x}$ alloys.
$Ni_xPt_{1-x}$ : △ ref. 32 ; ▼ ref. 64 ;
$Ni_xPd_{1-x}$ : ⊙ ref. 61 ; ● ref. 64. The dashed curves correspond to $dT_c/dp \simeq T_c^n$ with n = ± 1.

to be proportional to $(c - c_F)^{-1}$, that is to $T_c^{-2}$ itself. Thus $T_c dT_c/dp$ should be almost a constant[57]. Conversely, in the case of an indirect coupling between diluted local moments, it has been found that the Curie temperature followed a relationship $T_c \sim xJ\chi$ where x is the concentration of magnetic impurities, $\chi$ the matrix susceptibility and J the exchange interaction between localized spins and itinerant electrons[58]. For a typical enhanced paramagnetic metal, the volume dependence dLog $\chi$/dLog V of the susceptibility is quite large. It is for instance 6.5[8] for Pd at 40 K, 14[59] for $YCo_2$ at 40 K, and 10[60] for FeSi at 300 K. Thus a large pressure dependence of the Curie temperature is expected in the case of a strongly enhanced matrix. However, $dT_c/dp$ is then proportional to $T_c$, to be compared with $T_c^{-1}$ for the weak homogeneous model. In fact $Ni_xPd_{1-x}$ alloys in the critical range for the onset of ferromagnetism follow quite correctly such a local model behaviour[61-63].

We give, Fig. 18, in logarithmic scale, the pressure derivative $dT_c/dp$ as a function of $T_c$ for various disordered concentrated nickel alloys with low $T_c$. There is a striking difference between Ni-Pd alloys, where at low $T_c$, $dT_c/dp \sim T_c$ and Ni-Pt alloys where $dT_c/dp \sim T_c^{-1}$. However, this difference is reduced for large $T_c$, when Ni-Pd alloys recover some homogeneity.

## CONCLUSIONS

We have shown that partially magnetized transition atoms are of major importance to understand the chemical, temperature, field and pressure dependence of the physical properties of disordered-concentrated transition metal alloys. Although weak ferromagnets or largely Stoner enhanced paramagnets are often quite inhomogeneous due to the statistical disordering, large inhomogeneous effects are not always present, or of major importance, as demonstrated for $Ni_xPt_{1-x}$ or $Fe_xCo_{1-x}Si$ alloys.

## ACKNOWLEDGEMENTS

We are grateful to B.K. Chakraverty, G. Chouteau, D. Edwards, V. Jaccarino, R. Lemaire, M. Shimizu, R. Tournier and E.P. Wohlfarth for very helpful discussions and to A.S. Pavlovic for an accurate reading of the manuscript.

## REFERENCES


1 - See E.C. STONER, J. Phys. Rad., 12, 372 (1951).

2 - See J. MATHON, Proc. Roy. Soc. London, 306, 355 (1968).

3 - V. JACCARINO and L.R. WALKER, Phys. Rev. Letters, 15, 258 (1965).

4 - See J.S. KOUVEL, in Magnetism in Alloys, P.A. Beck and J.T. Wader Edt. AIME, N.Y., p. 165 (1972).

5 - J. VAN DER REST, F. GAUTIER and F. BROUERS, J. Phys. F, Metal Phys., 5, 995 (1975).

6 - See N.J. SHEVCHIK and D. BLOCH, This conference.

7 - C. SADRON, Ann. de Phys., 27, 417 (1932).

8 - J. BEILLE, Thesis, Grenoble (1975)(unpublished).

9 - V. MARIAN, Ann. de Phys., 7, 513 (1937).

10 - J. BEILLE, D. BLOCH and M.J. BESNUS, J. Phys. F, Metal Phys., 4, 1275 (1974).

11 - S.A. AHERN, M.J.C. MARTIN and W. SUCKSMITH, Proc. Roy. Soc. (London), 248, 145 (1958).

12 - D. PAN and J. TURNBULL, 19th Magn. and Magn. Mat. Conf., AIP Conf. Proc., 18, 643 (1974).

13 - To be published.

14 - O. MASSENET, H. DAVER and J. GENESTE, J. de Phys., 35, C4 279 (1974).

15 - Y. ENDOH, K. YAMADA, J. BEILLE, D. BLOCH, H. ENDO, K. TAMURA and J. FUKUSHIMA, Sol. Stat. Comm. (to be published).

16 - G. MARCHAL, P. MANGIN and C. JANOT, Sol. Stat. Comm. (to be published).

17 - D. BLOCH, J. VOIRON, V. JACCARINO and J.H. WERNICK Phys. Lett., 51A, 362 (1975).

18 - D. SHINODA, Phys. Stat. Sol., 11a, 129 (1972).

19 - J.H. WERNICK, G.K. WERTHEIM and R.C. SHERWOOD, Mat. Res. Bull., 7, 143 (1972).

20 - H. YASUOKA, R.C. SHERWOOD, J.H. WERNICK and G.K. WERTHEIM, Mat. Res. Bull., 9, 223 (1974).

21 - Y. ASADA and H. NOSE, J. Phys. Soc. Japan, 35, 409 (1973).

22 - R. LEMAIRE and J. SCHWEIZER, Phys. Lett., 21, 366, (1966).

23 - R.M. MOON, W.C. KOEHLER and J. FARREL, J. Appl. Phys., 36, 978 (1965).

24 - J. SCHWEIZER, Phys. Lett., 24A, 739 (1967).

25 - J. DEPORTES, D. GIGNOUX and F. GIVORD, Phys. Stat. Sol., 64b, 29 (1974).

26 - R. LEMAIRE, Cobalt, 33, 201 (1966).

27 - D. BLOCH, F. CHAISSE, F. GIVORD, J. VOIRON and E. BURZO, J. Phys., 32-C1, 659 (1971).

28 - B. BLEANEY, Proc. Roy. Soc., 276A, 19 (1963).

29 - K. LEE and N. HEIMAN, 20th Magn. and Magn. Mat. Conf., AIP Conf. Proc., 24, 108 (1975).

30 - L.J. TAO, S. KIRKPATRICK, R.J. GAMBINO and J.J. CUOMO, Sol. Stat. Comm. 13, 1491 (1973).

31 - D.M. EDWARDS and E.P. WOHLFARTH, Proc. Roy. Soc., A303, 127 (1968).

32 - H.L. ALBERTS, J. BEILLE, D. BLOCH and E.P. WOHLFARTH, Phys. Rev., B9, 2233 (1974).

33 - C.J. SCHINKEL, Col. intern. du C.N.R.S. sur la Physique sous champs magnétiques intenses, CNRS, 242, 25 (1975).

34 - See for instance D.M. EDWARDS, J. MATHON and E.P. Wohlfarth, J. Phys. F, Metal Phys., 5, 1619 (1975).

35 - J. BEILLE, D. BLOCH,and J. VOIRON, Col. Intern. du C.N.R.S. sur la Physique sous champs magnétiques intenses, CNRS, 242, 237 (1975).

36 - D. BLOCH, D.M. EDWARDS, M. SHIMIZU and J. VOIRON, J. Phys. F, Metal Phys., 5, 1217 (1975).

37 - E.P. WOHLFARTH and P. RHODES, Phil. Mag., 7, 1817 (1962).

38 - See K.P. GUPTA, C.H. CHENG and P.A. BECK, Phys. Rev., A133, 203 (1964).

39 - W.F. BRINKMAN, E. BUCHER, H.J. WILLIAMS and J.P. MAITA, J. Appl. Phys., 39, 547 (1968).

40 - See G. Chouteau, R. FOURNEAU, K. GOBRECHT and R. TOURNIER, Phys. Rev. Lett., 20, 143 (1968).

41 - See J. BEILLE, D. BLOCH and R. KUENTZLER, Sol. Stat. Comm., 14, 964 (1974).

42 - A. HAHN and E.P. WOHLFARTH, Helv. Phys. Acta, 41, 857 (1968).

43 - P.H. RHODES and E.P. WOHLFARTH, Proc. Roy. Soc., A273, 247 (1963).

44 - K.K. MURATA and S. DONIACH, Phys. Rev. Lett., 29, 285 (1972).

45 - T.O. MORIYA and A. KAWABATA, J. Phys. Soc. Japan, 34, 639 (1973).

46 - K. UEDA and T. MORIYA, J. Phys. Soc. Japan, 39, 605 (1975).

47 - M.J. BESNUS and A. HERR, Phys. Lett., A39, 83 (1972).

48 - D.J. GILLESPIE, C.A. MACKLIET and A.I. SCHINDLER, XIII Conf. Low Temp. Phys. (Boulder, 1972) Edt. by R.H. Kropschot and K.D. Timmerhaus, Univ. of Colorado, (1973).

49 - V. JACCARINO, G.K. WERTHEIM, J.H. WERNICK, L.R. WALKER and S. ARAJS, Phys. Rev., 160, 476 (1967).

50 - G. CHOUTEAU, R. TOURNIER and P. MOLLARD, J. de Phys., 5-C4, 185 (1974) and G. CHOUTEAU, Thesis, Grenoble (1973)(unpublished).

51 - G.K. WERTHEIM, J.H. WERNICK and D.W.E. BUCHANAN, J. Appl. Phys., 37, 3333 (1966).

52 - S. KAWARASAKI, H. YASUOKA and Y. NAKAMURA, AIP Conf. Proc., 10, 1632 (1973), Sol. Stat. Comm., 10, 919 (1972).

53 - D. BLOCH and R. LEMAIRE, Phys. Rev., B2, 2648 (1972).

54 - J. VOIRON, A. BERTON and J. CHAUSSY, Phys. Lett., 50A, 17 (1974).

55 - D. GIGNOUX, D. GIVORD, F. GIVORD, W.C. KOEHLER and R.M. MOON, (to be published).

56 - J.G. HUBER, M.B. MAPLE and D. WOHLLEBEN, Sol. Stat. Comm., 16, 211 (1975).

57 - E.P. WOHLFARTH, Phys. Lett., 31A, 525 (1970).

58 - D.J. KIM, Phys. Rev., 149, 434 (1966).

59 - J. VOIRON, J. BEILLE, D. BLOCH and C. VETTIER, Sol. Stat. Comm., 13, 201 (1973).

60 - A.S. PANFILOV, T.L. PIVOVAR, I.V.S. VECHKAREV, P.V. GELD and F.A. SIDORENKO, J.E.T.P., 68, 2134 (1975).

61 - J. BEILLE, Phys. Lett., 49A, 63 (1974).

62 - J. BEILLE and G. CHOUTEAU, J. Phys. F, Metal Phys., 5, 721 (1975).

63 - J. BEILLE and R. TOURNIER, J. Phys. F, Metal Phys., to be published.

64 - H. FUJIWARA, H. KADOMATSU and K. OHISHI, Proc. 4th Intern. Conf. on High Pressure, Kyoto 1974, The Physico-Chemical Society of Japan, 275 (1975).

# EXCHANGE COUPLING IN AMORPHOUS RARE EARTH-IRON ALLOYS

Neil Heiman, Kenneth Lee and Robert I. Potter
IBM Research Laboratory, San Jose, California 95193

## ABSTRACT

The technologically important amorphous rare earth (RE)-transition metal (TM) alloys provide an excellent system for investigating basic magnetic phenomena in amorphous magnetic materials. The sensitivity of the magnetic properties of the amorphous RE-TM alloys to slight compositional changes indicated the need for a rapid survey of the systematic variation of their properties. A brief review of this survey and the theoretical models will be presented. One result is to point out Fe as the most favorable TM for obtaining a more detailed understanding of the magnetic properties. We will present data obtained from $^{57}Fe$ Mössbauer effect, magnetometer and magneto-optic measurements on amorphous RE(Gd,Tb,Dy,Ho,Tm,Yb,Lu)-Fe amorphous alloy films. The Mössbauer effect results indicate that the Fe spin decreases with increasing RE content but the rate of decrease is dependent on the RE effective spin, (g-1)J. Using this information, we present a mean field theory model which reasonably reproduces the magnetic properties of the amorphous RE-Fe alloys, and which provides useful insights into their properties.

## INTRODUCTION

In addition to their interesting technological potential,[1,2,3,4] the amorphous rare earth (RE)-transition metal (TM) alloys present a convenient opportunity to study magnetic order in structurally amorphous materials. These alloys offer a number of advantages over other amorphous systems which exhibit magnetic ordering. They are easily produced in thin film form[1,5,6] without the need to maintain the substrates at cryogenic temperatures. They are amorphous at room temperatures and in general, crystallize only at temperatures in excess of 500K. They do not require the inclusion of "glass former" atoms to stabilize the amorphous phase, and they can be prepared over a very large range of compositions. The possible combinations of RE and TM constituents and the large range in x for $RE_{1-x}TM_x$ alloys include very diverse types of magnetic behavior. In addition, the crystalline RE-TM alloys exist in a number of different compositional phases. This allows the direct comparison of the magnetic behavior of amorphous alloys to the behavior of corresponding crystalline compounds.

In the early work on these materials, Rhyne et al.[7,8] demonstrated that the Curie temperature ($T_c$) for amorphous $Tb_{.33}Fe_{.67}$ (390K) is markedly lower than $T_c$ for crystalline $TbFe_2$ (711K). In contrast to this Chaudhari et al.[2] showed that $T_c$ for amorphous $Gd_{.33}Co_{.67}$ (>550K) is considerably higher than $T_c$ for crystalline $GdCo_2$(409K). Harris et al.[9] attributed these differences to the random orientation of the local anisotropy field. Gd, being an S state ion, does not couple to the local anisotropy field and is therefore unaffected while Tb, having the highest Stevens factor, couples very strongly with a resulting decrease in magnetic order. Tao et al.[10] while not ruling out anisotropy effects for $Tb_{.33}Fe_{.67}$ attributed the increased $T_c$ in $Gd_{.33}Co_{.67}$ to the fact that reduced density and increased disorder inhibit charge transfer from the RE to the TM.

In this paper, we present: (1) a survey of RE-TM systematics, (2) a discussion of the RE-Fe alloy system, (3) the problems associated with modeling the exchange coupling in RE-Fe alloys, (4) the results of $^{57}Fe$ Mössbauer Effect investigations, (5) our mean field model for $RE_{1-x}Fe_x$ alloys, and (6) our conclusions.

## SURVEY OF RE-TM SYSTEMATICS

Initially in order to test these models, we prepared samples of amorphous $Ho_{1-x}Co_x$ and $Ho_{1-x}Fe_x$ alloys, ($.40 \leq x \leq .85$). The results[6] show that, relative to the crystalline state, $T_c$ for the Ho Co alloys increases (similar to $Gd_{.33}Co_{.67}$) and $T_c$ for Ho Fe alloys decreases (similar to $Tb_{.33}Fe_{.67}$). These results suggest that the altered state of the TM and not local anisotropy is responsible for the differences in $T_c$ between the crystalline and amorphous materials. In order to more reliably determine the variation of the magnetic interactions, we undertook a systematic survey of $RE_{1-x}TM_x$ alloys.[11] We studied several different amorphous binary $RE_{1-x}TM_x$ alloy systems (RE = Gd, Tb, Dy, Ho, Er, Tm, Yb, Lu, Y), (TM = Ni, Co, Fe, Mn), ($0.40 \leq x \leq 0.90$). The samples were prepared in thin film form by thermal evaporation and examined with magnetometer and magneto-optic techniques. In the case of RE Co systems it was found that, for all RE, $T_c$ for the amorphous state was higher than $T_c$ for the crystalline material. For example, $T_c$ for crystalline RE $Co_2$ compounds vary between 0°K for $YCo_2$ and 409°K for Gd $Co_2$ whereas $T_c$ was >450K for all amorphous $RE_{.33}Co_{.67}$ alloys. In fact, one problem which inhibits the study of amorphous RE Co alloys is that $T_c$ is so high that crystallization or corrosion prevents the measurement of $T_c$. In contrast $T_c$ of <u>all</u> amorphous RE-Fe alloys is less than $T_c$ for <u>any</u> crystalline RE Fe alloy of the same Fe content. As shown in Fig. 1, $T_c$ of all amorphous RE Fe alloys are below 500K (and therefore can be conveniently measured). Furthermore, it is apparent from Fig. 1 that $T_c$ varies systematically with x and with the effective spin of the RE.

Amorphous RE-Mn and RE-Ni were also studied but over a more limited range of composition. No magnetic ordering was observed for the amorphous RE-Mn alloys. In fact, the addition of Mn to amorphous Gd Fe alloys radically decreases $T_c$. For example, while amorphous

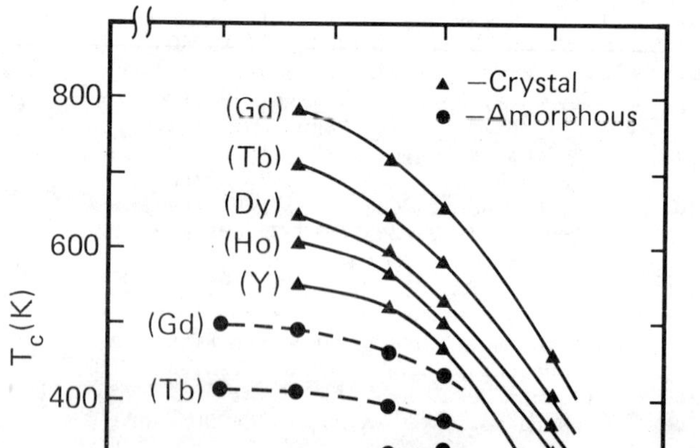

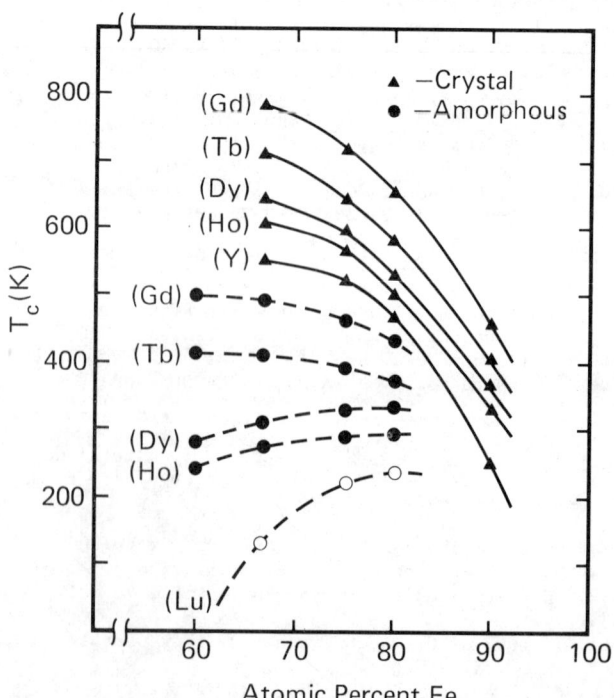

**Figure 1.** Curie Temperatures for both crystalline and amorphous RE-Fe alloys.

$Gd_{25}Fe_{75}$ has $T_c$ = 460K, $Gd_{.25}(Fe_{.68}Mn_{.07})$ has $T_c$ 290K. This is consistent with similar drastic reductions in $T_c$ which have been observed for the addition of Mn to amorphous $Fe_{.80}Si_{.20}$.[12] Amorphous $RE_{.33}Ni_{.67}$ alloys appear to have $T_c$ reduced by a factor of 2 to 3 compared to their crystalline counterparts. Since the Ni bands should be filled in $RE_{.33}Ni_{.67}$ alloys, Ni should not have a moment in these materials.[13] Thus the reduced $T_c$ indicates a decrease in the RE-RE exchange. More Ni rich alloys, where Ni bands should not be full, seem to behave more like Co, exhibiting an increased $T_c$ in the amorphous state.

## RE-Fe ALLOY SYSTEM

One result of the survey was to indicate that of all the RE-TM alloy systems, the RE-Fe alloys offer the best opportunity to obtain a more detailed understanding of the magnetic properties. The reasons for this are: 1) $T_c$ of all amorphous $RE_{1-x}Fe_x$ alloys are below the crystallization temperature; 2) $T_c$ appears to vary systematically with x and with the RE effective spin; 3) the Fe magnetic moment is sensitive to local coordination and atomic spacing; 4) the Fe sublattice moment can be investigated via the $^{57}Fe$ Mössbauer effect.

In spite of these advantages,a number of difficulties are encountered with the RE Fe alloys. For example, although $T_c$ varies systematically for these alloys, Rhyne et al.[14] have reported that the moment of amorphous $Gd_{.33}$ $Fe_{.67}$ is the same as the crystalline value whereas the moment of $Tb_{.33}$ $Fe_{.67}$ is reduced. This was interpreted as evidence that although local anisotropy effects do not play a significant role in the $T_c$ reduction, they do affect the moment. Taylor has measured the magnetization of a number of amorphous Gd Fe alloys,[15] and in contrast to Rhyne et al.,[14] report reduced moments for the amorphous alloys. A possible explanation of the conflicting data stems from the fact that the total net magnetization is the difference between two large sublattice magnetizations. Therefore, in addition to the usual experimental difficulties associated with complete saturation, sample size, etc., there is the need to accurately determine the composition of the amorphous alloys. The seriousness of the problem is evident when one realizes that an uncertainty in composition of only 1 atomic percent results in an uncertainty in $4\pi M$ of over 0.6 kilogauss. This is illustrated if one looks ahead to Fig. 7 where the effects of a 2% change in composition upon the mean field calculations are presented. Considering the difficulty of accurately determining the composition of amorphous alloys, we tend to attach little significance to small differences in the net magnetization. We therefore place more emphasis on the systematic variation of the magnetic behavior.

Even in the case of the systematic variation of $T_c$, a number of discrepancies appear. Rhyne et al.[16,17] have reported $T_c$ values for amorphous $RE_{.33}Fe_{.67}$ alloys which are generally in agreement with our values when the RE possesses high effective spin values, but their $T_c$ values for RE possessing low effective spins are generally lower than ours (see Fig. 2). In fact, they report[14] that $Y_{.33}Fe_{.67}$ does not possess long range magnetic order. This discrepancy is most likely due to the different methods used for determining $T_c$. Rhyne et al's.[16,17] values were obtained from Arrott plots of magnetization data, whereas our $T_c$ values were obtained from polar Faraday rotation measurements. The differences probably result from the fact that while the magnetic phase transition is very sharp for RE-TM alloys with high effective RE spin, the transition for low spin RE becomes more gradual. An extreme example of this can be seen in the temperature dependence of the Faraday rotation of $Lu_{.33}Fe_{.67}$ shown in Fig. 3. The data suggests the existence of magnetic order but it is difficult to assign any definite value for the ordering temperature, which explains to some extent why reported $T_c$ values for $Lu_{.33}Fe_{.67}$ and $Y_{.33}Fe_{.67}$ vary considerably. To make matters even more difficult, it appears that the magnetic behavior of RE-Fe alloys with low RE spin is quite complex. Fig. 3 shows the temperature dependence of the low field polar Faraday rotation signal for several compositions of $Lu_{1-x}Fe_x$. Since Lu does not have a moment, the Faraday rotation should be directly proportional to the magnetization. Thus the gradual decrease in rotation with increasing temperature reflects the expected decrease of M(T); however, in the case of $Lu_{.25}Fe_{.75}$, there is an unexpected decrease at low temperatures. In the case of $Lu_{.20}Fe_{.80}$, the rotation goes to zero at low temperatures. Could this be due to a wide distribution of exchange interaction-strengths, varying with composition, which causes spin glass type behavior and, at high Fe content, antiferromagnetism? It is clear that before this question can be answered the alloys with low RE spins

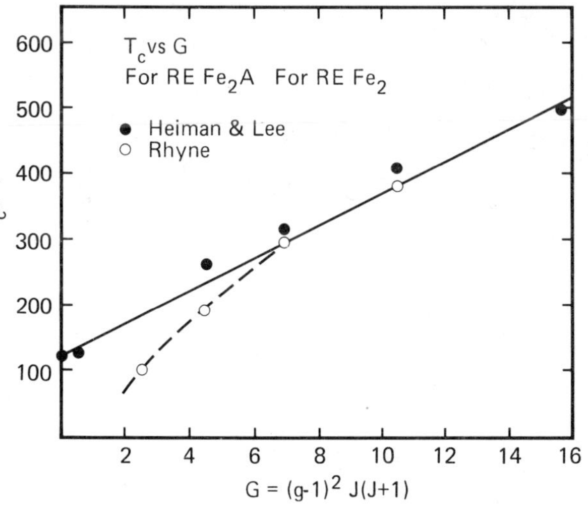


**Figure 2.** Reported $T_c$ values for $RE_{0.33}$-$Fe_{0.67}$ alloys as a function of the RE de Gennes factor, $(g-1)^2J(J+1)$.

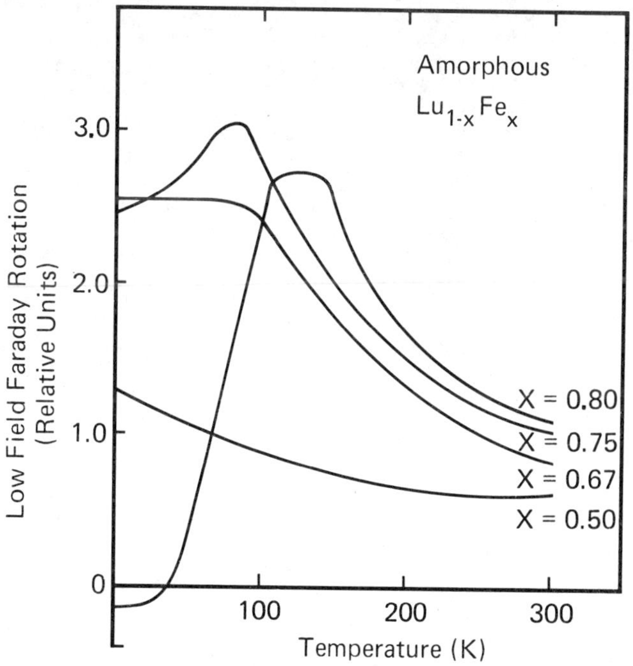


**Figure 3.** Polar Faraday Rotation as a function of temperature for a number of amorphous $Lu_{1-x}Fe_x$ alloys.

will require considerably more study.

## PROBLEMS WITH EXCHANGE MODELING

The combination of low $T_c$ for amorphous $RE_{.33}Fe_{.67}$ alloys, the absence of long range order in $Y_{.33}Fe_{.67}$, and the same magnetization for both crystalline and amorphous $Gd_{.33}Fe_{.67}$ has prompted Rhyne et al.[17] to suggest that the Fe-Fe exchange interaction is zero in the $RE_{.33}Fe_{.67}$ alloys and that the RE-Fe exchange is the major contribution to the total exchange. They have reported that with these assumptions mean field type amorphous calculations of $T_c$ for $RE_{.33}Fe_{.67}$ alloys produce approximate agreement with their $T_c$ data.

We have also attempted to fit both sets of $T_c$ data for amorphous $RE_{.33}Fe_{.67}$ with a simple molecular field model. We find that unless we use a different set of model parameters for each RE, the agreement with $T_c$ is only very approximate. More importantly, the calculated shape of the magnetization vs. temperature curve is completely incorrect.

Using a mean field model similar to that of Hasegawa,[18] we also attempted to determine the systematic variation of the exchange ineractions both as a function of x and of the RE species. We found that the mean field parameters were cross correlated and that the calculated result was very sensitive to the value of the Fe spin. Therefore an independent determination of the Fe spin is required in order to produce meaningful results from such a model. Fortunately, the results from Mössbauer effect studies[19] of a number of amorphous $RE_{1-x}Fe_x$ alloys were available.

## MÖSSBAUER EFFECT SPECTRA

Early Mössbauer effect studies of amorphous Re-Fe alloys were made using powdered samples obtained from grinding sputter deposited foils of $RE_{.33}Fe_{.67}$. The first study[20] failed to observe any effect below $T_c$ while the second study[21] observed a broad magnetically split pattern in $Dy_{.33}Fe_{.67}$ with an unusual asymmetry in the line intensities. We obtained room temperature and 4.2K Mössbauer spectra of as deposited evaporated amorphous films (thickness ≃ 5000 Å) of $RE_{(1-x)}Fe_x$ (RE = Gd, Tb, Dy, Ho) and ($0.60 < x < 0.90$). These films possessed uniaxial magnetic anisotropy allowing us to observe well resolved structure.

Because of the uniaxial magnetic anisotropy, Mössbauer spectra obtained with $T < T_c$ show a well resolved 4-line, rather than 6-line, pattern, as shown in Fig. 4 where the film was normal to the γ-ray beam (θ = 90°). The solid line in Fig. 4 is a fit to a single site hyperfine Hamiltonian. For simplicity, the distribution of hyperfine interaction parameters was assumed to be Lorentzian. The existence of more than a single site is readily apparent in the figure by the inability of fit to account for the shoulders on the right hand sides of the outer lines. These shoulders were also present but less apparent in other samples including as deposited sputtered films of Gd-Fe. When the films were tilted so that they were no longer normal to the γ-ray beam (Fig. 4, θ = 45°), a 6-line pattern appeared. The spectrum is similar to that reported by Forester[21] including the asymmetry in the intensity of the inner lines; however, in our case, the lines are better resolved because the spectrum of the oriented film is not broadened by the averaging over all orientations that occurs with powdered samples.

Figure 5 is a plot of the total internal magnetic field at 4.2K as determined by a single site Hamiltonian fit for several different $RE_{1-x}Fe_x$ samples. We observe the internal magnetic field to decrease with increasing RE concentration at a rate dependent on the effective RE spin, (g-1)J. The data can easily be fitted to a linear function of RE concentration (1-x); namely:

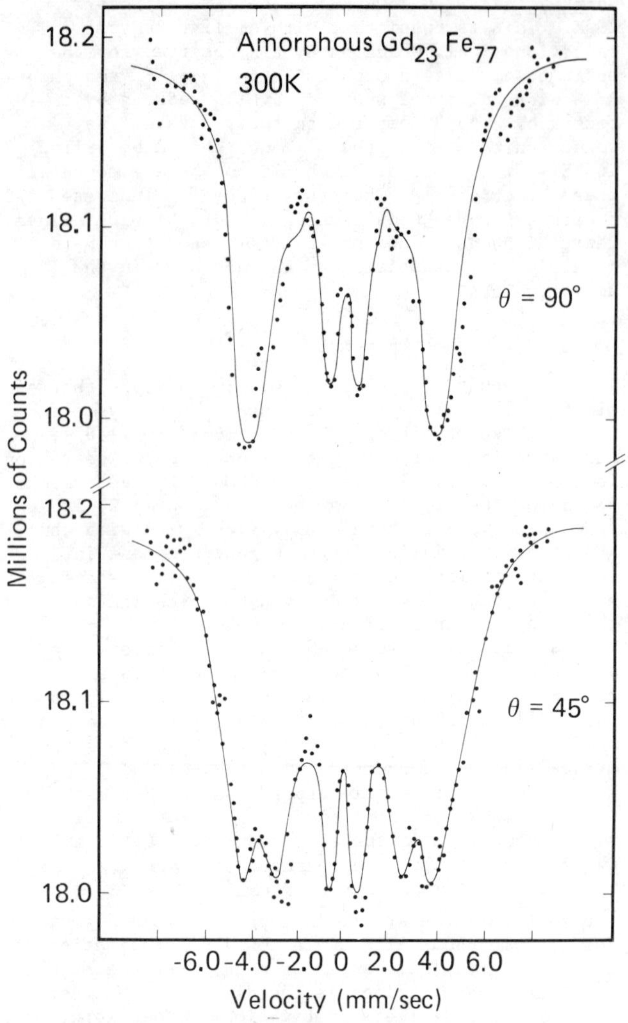


**Figure 4.** Room temperature Mössbauer spectra for amorphous $Gd_{0.23}Fe_{0.77}$. θ = 90° indicates the film was normal to the γ ray beam. θ = 45° indicates the film was tilted 45° to the beam.

$$H_{INT} = 348 - (1-x)\,[745 - 148(g-1)J] \qquad (1)$$

where the coefficient of (1-x) is dependent on (g-1)J. Although the linear fit was satisfactory, one expects the diluting effect of the RE to increase as the ratio of RE to Fe increases. The data in Fig. 5 is, therefore, fitted to the function:

$$H_{INT} = 348 - \left(\frac{1-x}{x}\right)\,[507 - 111(g-1)J] \qquad (2)$$

It is worth noting that Forester's data[21] for $Dy_{.33}Fe_{.67}$ can also be fitted by these expressions.

Due to the requirement of the mean field model for an independent determination of the variation of the Fe moment, we wished to extract Fe moment behavior from the $H_{INT}$ data. Unfortunately, establishing a correlation between the Fe moment and $H_{INT}$ in a magnetic alloy is quite complicated. The interpretation of Mössbauer spectra for the crystalline RE-Fe alloys is still being debated.[22] The problems in the crystalline materials involve multiple spectra due to the orientation of the easy axis relative to the electric field gradient as well as isotropic and anisotropic contributions from dipolar fields and core polarization arising from the RE spin. In spite of these considerations, there are a number of reasons to believe that for the amorphous materials, the variation of $H_{INT}$ with x and (g-1)J roughly approximates the variation in the Fe moment. First, while $H_{INT}$ for crystalline alloys also varies with x and (g-1)J, the variation is quite

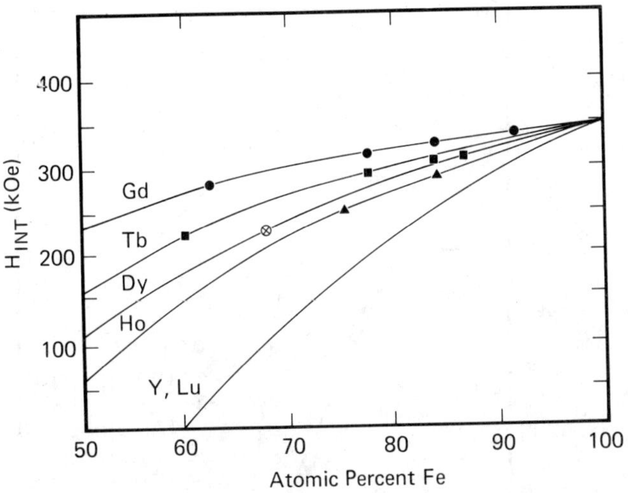


**Figure 5.** Internal magnetic field at 4.2K for several amorphous RE Fe alloys. The solid lines are a plot of Eq 1.

small compared with the very strong dependence on these parameters expressed in Eq. 1 or 2. This strong dependence in the case of the amorphous alloys indicates that most of the variation is due to different considerations. In fact, the variation is so large that it is difficult to attribute it to anything other than the variation of the Fe moment.

Second, one expects that for a truly random amorphous material, the contributions from dipolar fields and electric field gradient orientation would produce line broadening but not a shift in $H_{INT}$. Additional arguments in favor of viewing $H_{INT}$ in terms of the Fe moment are that Bleaney[23] and Lemaire,[24] have presented theoretical arguments and experimental data to support a model for crystalline RE Co alloys where the Co moments varies according to a formula very similar to Eq. 2. A similar idea has been proposed by Wallace[24] for crystalline RE-Fe alloys, but with a weaker dependence on (g-1)J. If we express Eq. 1 or 2 in terms of the Fe spin, we obtain

$$S_{Fe} = 1.1 - (1-x)\ [2.3 - 0.47\ (g-1)J] \qquad (3)$$

and

$$S_{Fe} = 1/2\ [2.2 - \left(\frac{(1-x)}{x}\right)(3.2 - 0.70\ (g-1)J)] \qquad (4)$$

Equation 4 is written to express the variation of the Fe spin in terms of a charge transfer model with the addition of a moment induced by the RE. An interesting feature of Eq. 4 is that it indicates that each RE transfers 3.2 electrons to the Fe bands. The number is very close to 3.0 which one would expect from simple valence electron arguments. 3.0 is also the number obtained by Lemaire[24] for the charge transfer model of crystalline RE Co alloys.

For these reasons and because an independent determination of the variation of the Fe spin is needed to proceed with mean field calculations of the exchange interactions, we have incorporated Eqs. 3 and 4 into the mean field model[26] and examined a number of procedures for obtaining exchange behavior.

## MEAN FIELD MODEL FOR $RE_{1-x}\ Fe_x$

Considering the strong dependence of $S_{FE}$ on x and (g-1)J, we wished to determine whether this feature by itself was sufficient to account for the systematic variation of $T_c$. For a mean field model the calculation of $T_c$ for two sublattices reduces to:

$$3kT_c = a_{11} + a_{22} + [(a_{11} - a_{22})^2 + 4a_{12}a_{21}]^{1/2}$$

where

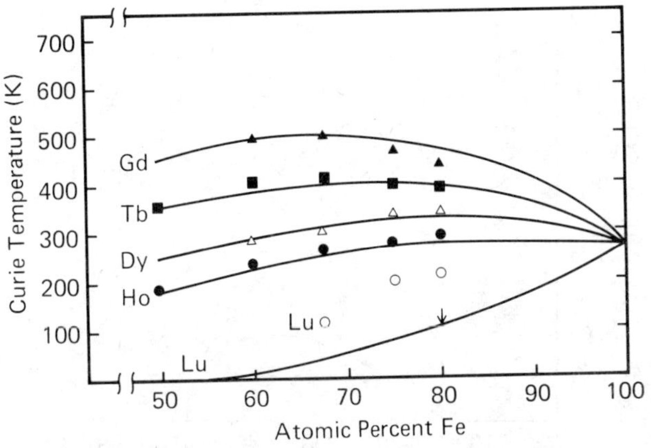


**Figure 6.** Curie temperatures for several amorphous RE-Fe alloys. The solid lines are from the fit obtained using Eq 3 for $S_{Fe}$ and fixed exchange constants.

$$a_{11} = (1 - x)\ Z_1\ J_{11}\ S_1\ (S_1 + 1) \qquad (5)$$

$$a_{22} = x\ Z_2\ J_{22}\ S_2\ (S_2 + 1)$$

$$a_{12}a_{21} = x\ (1 - x)\ Z_1 Z_2\ J_{12}^{\ 2}\ S_1\ (S_1 + 1)\ S_2\ (S_2 + 1)$$

and where the $Z_i$ are the coordination numbers (taken to be 12 in all cases), i = 1 refers to the RE atom, i = 2 refers to the Fe atom and $J_{ij}$ are the exchange constants. The deGennes factor for the RE is substituted for $S_1(S_1+1)$ and Eq. 3 or 4 is used to determine $S_2$. Extrapolating the $T_c$ values shown in Fig. 1 yields a $T_c$ for 100% (x=1.0) "amorphous Fe" of 270K; this extrapolation fixes $J_{22}$ at 20.5 x $10^{16}$ ergs. (Data of Alperin et al.[27] indicate a slightly lower extrapolated $T_c$ and work by Felsch[28,29,30] has indicated possible problems with extrapolating amorphous Fe alloy data; but for the purpose of fitting $T_c$ in Fig. 1, the extrapolation is valid.) $J_{11}$ is fixed at 1.6 x $10^{16}$ ergs by the $RE_{.33}Ni_{.67}$ data where it is reasonably assumed that $J_{11}$ is responsible for the entire exchange interaction. Thus $J_{12}$ is the only fully adjustable parameter, and was selected to produce the best fit to the data. We found that using Eq. 3 to determine $S_2$ yielded a better fit to published $T_c$ data than using Eq. 4. This is mainly due to the fact that Eq. 4 produces a more rapid decrease in $S_2$ at high RE content causing $T_c$ to decrease too rapidly for $x \leq 0.6$. When Eq. 3 is used, $J_{12}$ = 28.0 x $10^{16}$ ergs produced the best fit. The result, shown in Fig. 6, demonstrates that we were able to achieve a satisfactory fit to the data except for the case of $Lu_{1-x}Fe_x$; however, as mentioned above the magnetic properties of Lu Fe alloys are both complicated and different from high RE spin alloys.

It is certainly unrealistic to expect the exchange constants to be unaffected by compositional changes, and in fact, the above set of exchange constants, when used with a full mean field model, fail to reproduce the observed magnetization vs. temperature behavior. Nevertheless, we were encouraged by the fact that the approximate dependence of $T_c$ on composition could be entirely accounted for by the variation of the Fe spin. This suggests that it would be worthwhile to proceed with a more detailed model.

Because there is more data available for amorphous $Gd_{1-x}Fe_x$ and $Tb_{1-x}Fe_x$ than for other RE alloys, we decided to fit the magnetic data for these materials with the full mean field model in the hope of realistically determining the compositional dependence of $J_{ij}$. By inserting (g-1)J = 3.5 into Eqs. 3 and 4 we obtained the Fe spin as a function of concentration for $Gd_{1-x}Fe_x$ alloys. With the Fe spin thus determined, we adjusted the exchange constants

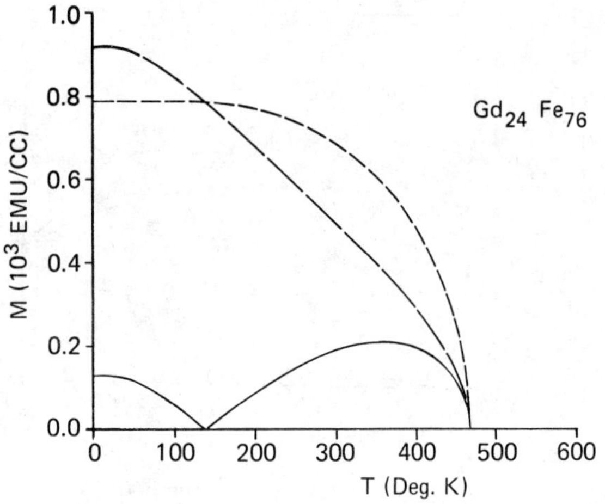


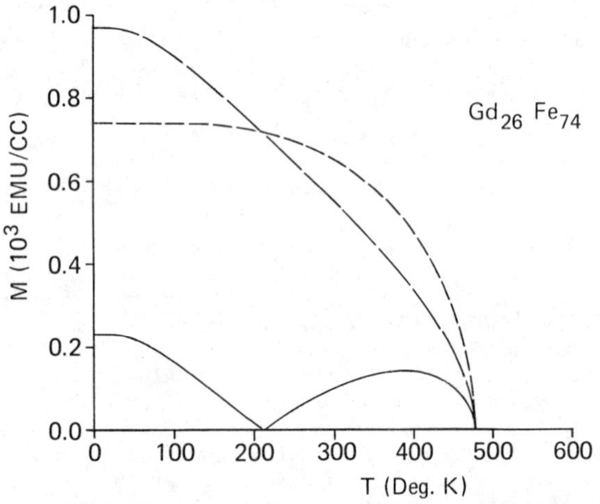


**Figure 7.** Examples of mean field calculations for amorphous $Gd_{1-x}Fe_x$ alloys. These figures of M vs T illustrate that a change of only 2 atomic percent in composition produces a large change in M(o) and $T_{comp}$, while $T_c$ and the sublattice magnetizations are altered only slightly.

to obtain the best fit to the magnetization vs. temperature curves for a number of concentrations. When these sets of exchange constants had been obtained we observed that the magnitude of both $J_{12}$ and $J_{22}$ decreased with increasing x. In fact, whether we used Eq. 3 or Eq. 4, the variations of $J_{12}$ and $J_{22}$ could be approximated by simple linear relations. In the case of Eq. 3:

$$J_{12} = 2.25 - 53.0\ (1-x)$$

$$J_{22} = 20.5 + 120.0\ (1-x) \qquad (6)$$

while for Eq. 4:

$$J_{12} = 1.4 - 46.8\ (1-x)$$

$$J_{22} = 20.5 + 115\ (1-x) \qquad (7)$$

All $J_{ij}$ are in units of $10^{16}$ ergs. In both cases $J_{11}$ was taken to be 1.6 x $10^{16}$ ergs as before. Note also that for x=1, $J_{22}$ is the same as for the previous $T_c$ calculations.

Examples of these mean field calculations are shown in Fig. 7 which show both sublattice magnetizations and the net magnetization as a function of temperature for two compositions which differ by only 2 atomic per cent. This figure illustrates the earlier point that we tend not to rely on small differences in the net magnetization or compensation temperature ($T_{comp}$) between crystalline and amorphous materials. Rather we emphasize the systematic variation of the magnetic properties, since these

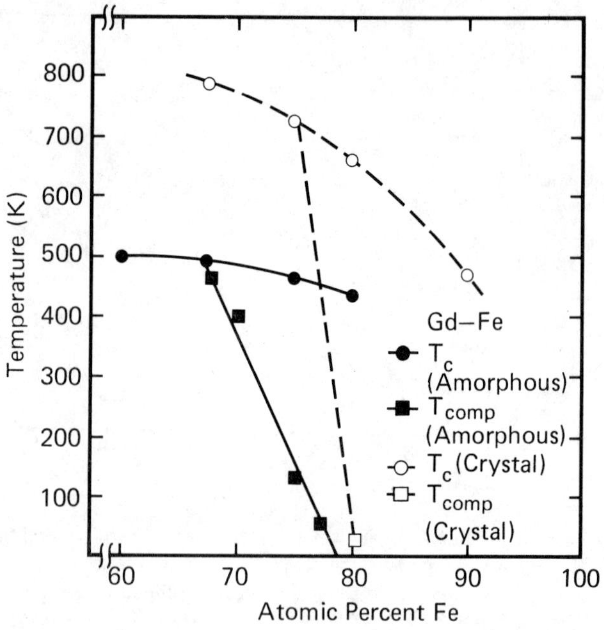


**Figure 8.** $T_c$ and $T_{comp}$ for both crystalline and amorphous $Gd_{1-x}Fe_x$ alloys. The solid lines are from a fit obtained using Eq 4 for $S_{Fe}$ and Eq 7 for the exchange constants.

systematic variations are not critically dependent upon the absolute composition.

That the simple linear approximations (Eqs. 6-7) succeed in producing the correct temperature dependence of the magnetization for amorphous $Gd_{1-x}Fe_x$ alloys can be seen in Fig. 8 which shows both $T_c$ and $T_{comp}$ for amorphous (and crystalline) $Gd_{1-x}Fe_x$ alloys. The solid lines are $T_c$ and $T_{comp}$ obtained from the mean field model using Eqs. 4 and 7. The linear approximation for $J_{12}$ and $J_{22}$ (Eq. 6) combined with Eq. 3 for the Fe spin does not produce as good a fit. This combination produces $T_c$ values which are slightly too large for x = 0.6. This should not be interpreted as demonstrating that Eq. 4 is more correct than Eq. 3 since there is no basis for requiring the $J_{ij}$ to be linear functions of x. Rather, the reasonably good results obtained with either set merely indicate that the $J_{ij}$ increase with increasing RE content in a manner that can be approximated by linear formulae such as Eqs. 5 and 7.

We next tested the extension of these same exchange constants (Eqs. 6 and 7) to other amorphous $RE_{1-x}Fe_x$ alloys. The only change in the model is to change the Fe spin according to Eq. 3 or 4. We found that, although the combination of Eq. 3 for the Fe spin and Eq. 6 for the $J_{ij}$ produces a slightly poorer fit to $Gd_{1-x}Fe_x$ data than the combination of Eq. 4 and 7, it reproduces $T_c$ for <u>all</u> amorphous $RE_{1-x}Fe_x$ alloys quite well as shown in Fig. 9. Furthermore, using the same equations, the model predicts the correct magnetization vs. temperature behavior for $Gd_{1-x}Fe_x$, $Tb_{1-x}Fe_x$ and $Dy_{1-x}Fe_x$ alloys; the calculated shape of the temperature dependence of the magnetization begins to deviate from the data only for alloys containing RE atoms with small (g-1)J values. In view of the complicated behavior of $Lu_{1-x}Fe_x$ alloys and the fact Eq. 6 was obtained from a fit to $Gd_{1-x}Fe_x$ data (the highest (g-1)J value), it is not surprising that some deviations occur in the regime of low (g-1)J.

It should be noted that the combination of Eqs. 4 and 7 did not fit the data well in the region x ≤ 0.6 (except for Gd Fe alloys) due to too rapid a decrease of $S_{Fe}$ with decreasing x.

## CONCLUSIONS

One should not use the success of Eqs. 4 and 7 as proof that the correct mean field model formulation for amorphous RE-TM alloys has been found. This is

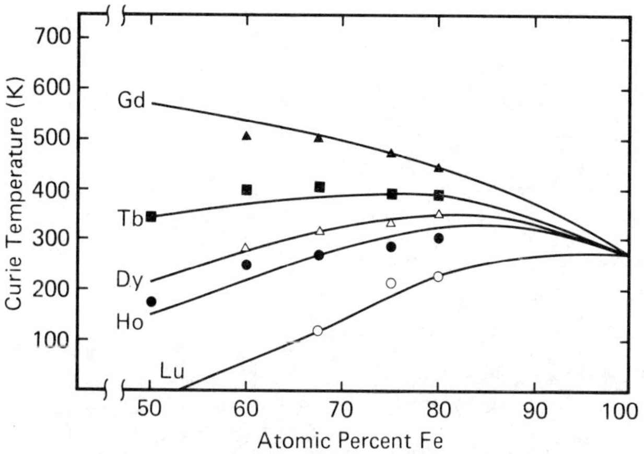


**Figure 9.** Plot of Curie temperatures for amorphous $RE_{1-x}Fe_x$ alloys demonstrating the fit (solid lines) obtained using Eq 3 for $S_{Fe}$ and Eq 6 for the exchange constants.

particularly true in view of the serious limitations of mean field theory applied to metallic systems. It is not our intention to interpret the mean field results quantitatively or to suggest that the model will accurately predict the magnetic characteristics of all $RE_{1-x}Fe_x$ alloys. On the other hand we believe the model correctly reproduces the systematic variation of the magnetic properties of the amorphous $RE_{1-x}Fe_x$ alloys. The success of the model combined with the Mössbauer effect results establish a number of features:

1.) The Fe moment is dependent on the concentration of RE and the value of (g-1)J for the RE. The dependence on both these parameters is much stronger than in the case of the crystalline alloys.
2.) The magnitude of both the Fe-Fe and the RE-Fe exchange constants increase with increasing RE concentration.
3.) Without the presence of an "induced Fe moment," the behavior of the Fe-Fe exchange is complex. It appears that a RE with a reasonably high effect spin (g-1)J, must be present in order to induce well defined ferromagnetic coupling between Fe atoms.

If one wishes to refine the mean field model, a number of items are worthy of consideration:

1.) Improved Mössbauer data are needed. The fit to present data is not sensitive to the choice of coefficient for Eq. 3 or 4. More data in range $x \leq 0.6$ should allow one to distinguish easily between the form of Eq. 3 and Eq. 4. More data for alloys containing RE species with small (g-1)J would make the fit much more sensitive to the data.
2.) Improved understanding of the various contributions to the internal field would allow better determination of the Fe spin. (Any other determination of the Fe spin would, of course, be just as useful.)
3.) Following Bleaney, one should replace (g-1)J with (g-1) $(\mu_{RE}/g)$ so that the correct temperature dependence of the induced moment is accounted for.
4.) There is very little data for the amorphous RE-TM alloys involving small (g-1)J. There is even less understanding of what little data exists. The behavior of these materials is very complex and should not properly be included in the current mean field model. A thorough investigation of the alloys of Y or Lu with the TM series seems to be called for.

While these and other improvements may lead to more refined models, it is unlikely that the improved models will prove the qualitative conclusions to be incorrect.

REFERENCES


1. P. Chaudhari, J. J. Cuomo and R. J. Gambino, IBM J. of Res. and Dev. 11, 66 (1973).
2. P. Chaudhari, J. J. Cuomo, and R. J. Gambino, Appl. Phys. Letters 22, 337 (1973).
3. R. J. Gambino, P. Chaudhari, and J. J. Cuomo in Magnetism and Magnetic Materials - 1973, AIP Conference Proceedings No. 18, edited by C. D. Graham Jr. and J. J. Rhyne (American Institute of Physics, New York - 1973), p. 578.
4. A. E. Clark, Appl. Phys. Letters 23, 642 (1973).
5. J. Orehotsky and K. Schröder, J. Appl. Phys. 43, 2413 (1972).
6. N. Heiman and K. Lee, Phys. Rev. Letters 33, 778 (1974). $T_c$ for crystalline alloys were obtained from K.N.R. Taylor Advan. Phys. 20, 551 (1971), and references cited therein.
7. J. J. Rhyne, S. J. Pickart and H. A. Alperin Phys. Rev. Letters 29, 1962 (1972).
8. J. J. Rhyne, S. J. Pickart and H. A. Alperin in Magnetism and Magnetic Materials - 1973, AIP Conference Proceeding No. 18, edited by C. D. Graham Jr. and and J. J. Rhyne (American Institute of Physics, New York, 1973).
9. R. Harris, M. Plischke, and M. J. Zuckermann, Phys. Rev. Letters 31, 160 (1973).
10. L. J. Tao, S. Kirkpatrick, R. J. Gambino, and J. J. Cuomo, Solid State Commun. 13, 1491 (1973).
11. K. Lee and N. Heiman, in Magnetism and Magnetic Materials - 1974, AIP Conference Proceeding No. 24, Edited by C. D. Graham, Jr. G. H. Lander, and J. J. Rhyne (American Institute of Physics, New York, 1974).
12. T. Mizoguchi, K. Yamauchi, and H. Miyajima in Amorphous Magnetizm, Edited by H. O. Hooper, and A. M. deGraaf (Plenum, New York - 1973), p. 325.
13. W. E. Wallace, Rare Earth Intermetallics (Academic Press, New York and London, 1973), p. 112ff.
14. J. J. Rhyne, J. H. Schelleng, and N. C. Koon, Phys. Rev. B10, 4672 (1974).
15. R. C. Taylor, J. Appl. Phys. (to be published) and private communication.
16. J. J. Rhyne and J. H. Schelleng, Bull. Am. Phys. Soc. 20, 458 (1975).
17. J. J. Rhyne, this conference.
18. R. Hasegawa, B. E. Argyle, and L. J. Tao, in Magnetism and Magnetic Materials - 1974, AIP Conference Proceedings No. 24, Edited by C. D. Graham, Jr., G. H. Lander and J. J. Rhyne (American Institute of Physics, New York, 1974), p. 110.
19. N. Heiman and K. Lee, Phys. Letters (to be published).
20. D. Sarkar, R. Segnan, E. K. Cornell, E. Callen, R. Harris, M. Plischke, and M. J. Zuckermann, Phys. Rev. Letters 32, 542 (1973).
21. D. W. Forester, R. Abbundi, R. Segnan, D. Sweger in Magnetism and Magnetic Materials - 1974, AIP Conference Proceeding No. 24, Edited by C. D. Graham, Jr., G. H. Lander and J. J. Rhyne (American Institute of Physics, New York, 1974).
22. See for example, M. P. Daniel, U. Atzmony, and D. Lebenaum, Phys. Stat. Solid. B59, 615 (1973).
23. B. Bleaney, Rare Earth Res., Proceedings of 3rd Rare Earth Conference, Edited by K. S. Vorres (Gordon and Breach, New York, 1964), Vol. 2, p. 499.
24. R. Lemaire, Cobalt 33, 201 (1966).
25. W. E. Wallace, op. cit. p. 186.
26. N. Heiman, K. Lee, R. I. Potter, and S. Kickpatrick (Submitted to Phys. Rev.).
27. H. J. Alperin, J. R. Cullen, and A. E. Clark, this conference.
28. W. Felsch, Z. Phys. 219, 280 (1969).
29. W. Felsch, Z. Agnew Phys. 29, 217 (1970).
30. W. Felsch, Z. Agnew Phys. 30, 275 (1970).

# SPIN WAVES IN A MODEL FOR AN AMORPHOUS FERROMAGNET*

R. Alben
Department of Engineering and Applied Science, Yale University, New Haven, Connecticut 06520

ABSTRACT

We present exact numerical calculations of linear spin wave excitations of ferromagnetically coupled spins in a random dense packing model for an amorphous ferromagnet. Results for both density of states and for inelastic neutron scattering are given. The neutron calculation shows a roton-like behavior which is explained in terms of static structural correlations rather than dynamic effects.

## I. INTRODUCTION

Much work on amorphous materials and disordered alloys has been based on direct numerical calculations for hand built or computer generated models which tell where the atoms, or at least a finite number of them, are.[1] This direct approach has the advantage that there is little room to adjust the theory, and the results clearly reflect the initial assumptions. Insights into the meaning of the calculation are generally much easier to come by after the results are in hand.

In this paper we present the first such direct calculations of spin wave density of states and neutron scattering intensities for relatively large (more than 1000 unit) model structures for amorphous metals.[2,3] We assume a simple constant nearest neighbors interaction Hamiltonian so as to be able to study the effects of structural disorder independently of quantitative disorder of the interactions. (The latter type of disorder is more conveniently studied in the context of crystalline alloys.) By contrast, past theories of spin waves in amorphous ferromagnets have relied on decorrelation approximations which often fail to distinguish between quantitative and structural disorder[4-8] and, as we shall see, miss out the essential behavior of the neutron scattering intensity for intermediate and large momentum transfer.

In doing a direct numerical calculation, it is desirable to treat models which are as large as possible. This reduces the effects of the surface and permits a significant sampling of different environments within the bulk. Unfortunately, however, the most straightforward techniques for describing magnetic as well as electronic and vibrational excitations involve the diagonalization of matrices. This rapidly becomes very costly for large models, particularly when, in addition to eigenvalues, eigenvectors are required as in calculating neutron scattering or projecting onto interior atoms to avoid surface modes. Therefore in this work we have employed a novel method for calculating excitation frequencies and intensities which can be applied to very large models. It is based on numerical solutions to equations of motion for one particle Green's functions.[9] The results are exact in that, although they are broadened with a resolution function, the function is known and its width can be chosen (See the Appendix).

This paper is divided into four sections. In Sec. II we describe the spin-wave Hamiltonian and obtain an expression for the neutron scattering law S(Q,E). In Sec. III we present results for density of states and S(Q,E) for two dense random packing (DRP) models. The results are compared with those calculated for a polycrystal of face-centered cubic (fcc) structure and for a disordered crystalline alloy. We also consider the low frequency hydrodynamic modes in terms of static resistance to changes in spin direction and obtain the spin wave stiffness constant as well as the contribution of these modes to the neutron scattering. Finally in Sec. IV we consider recent experimental results[10-12] in the light of the results of our model calculations.

## II. SPIN WAVE HAMILTONIAN

We consider the excitations of a system of spins $\vec{S}_i$ which are located at the sites i of a model structure. There is a nearest neighbor Heisenberg interaction given by the Hamiltonian:

$$H = - J \sum_{<i,j>} \vec{S}_i \cdot \vec{S}_j , \quad (1)$$

where the exchange constant J is the same for all pairs <i,j> close enough to be considered neighbors. This Hamiltonian is of course highly idealized, especially since the magnetic electrons in real amorphous magnets are at least partially itinerant in nature. However it is nonetheless useful for studying effects peculiar to structural disorder. We take J as positive (ferromagnetic) and assume a ground state with all spins aligned along z.

We next make the Holstein-Primakoff substitution:

$$S_i^+ \sim \sqrt{2S}\, a_i , \; S_i^- \sim \sqrt{2S}\, a_i^+ , \; S_i^z \sim S - a_i^+ a_i , \quad (2)$$

where S is the total spin quantum number and $a_i^+$ and $a_i$ are Bose operators which respectively create and destroy spin excitation at the site i. Higher order terms in the a's have been dropped. This approximation is valid for temperatures low compared with the ordering temperature. We will assume T = 0 K throughout our treatment. Equation (1) now takes the form:

$$H = - \frac{\bar{z} N J S^2}{2} + J S \sum_{<i,j>} (a_i^+ a_i + a_j^+ a_j - a_i^+ a_j - a_j^+ a_i), \quad (3)$$

where N is the number of sites in the model and $\bar{z}$ is the mean coordination number.

The neutron energy loss cross section at 0 K, averaged over all directions of the momentum transfer vector $\vec{Q}$ and of the neutron polarization, can be written as follows:[13]

$$\frac{d^2\sigma}{d\Omega\, dE} = \frac{2}{3} N \frac{k_f}{k_i} b^2 |F(Q)|^2 \, S'(Q,E) \quad (4)$$

where N is the number of spins, $k_i$ and $k_f$ are magnitudes of the incident and scattered wave vectors, b is the scattering length, F(Q) is the form factor and S(Q,E) is a "scattering law" which contains all of the information about the spin wave excitations. In terms of the a-operators. the scattering law is given by

$$S(Q,E) = - \frac{S}{N} \sum_{i,j} \mathrm{Im} \frac{1}{\pi\hbar} \int_0^\infty e^{i\frac{Et}{\hbar}} -i\langle a_i(t) a_j^+(0)\rangle e^{iQ\cdot(R_i - R_j)} dt, \quad (5)$$

where < > denotes the ground state expectation value. As shown in the Appendix, the products in the angular brackets define a set of Green's functions which may be straightforwardly calculated from their equations of motion yielding both S(Q,E) as well as a statistically accurate density of states.

## III. DENSE RANDOM PACKING MODELS

Dense random packing (DRP) models seem to accurately represent the geometrical arrangement of the atoms in various amorphous metals and alloys including such amorphous magnetic materials as $Co_{1-x}P_x$ and $GdFe_2$. The two models which we consider here represent the packing of hard spheres of a single size. The first model consists of the 996 sites closest to the center of the 7934 hand built model of J. Finney.[2] The second model consists of 890 sites surrounding the center of a 3999 computer generated model by C. Bennett.[3] We identify nearest neighbors as sites within 1.25 sphere diameters of each other, provided that neither atom of the pair has more than 12 other closer neighbors. The cut-off at 12 neighbors affected only 2% of the sites. With the above criteria the mean coordination number for interior sites is slightly less than 11 for both models.

In Fig. 1a, b we show results for $S(\vec{Q},E)$ for $\vec{Q}$ along two directions respectively in the 996 site model. In order to suppress contributions associated with surface sites, the spectral function was projected on the 400 sites nearest the center (See the Appendix). The limited size of the region of interior sites implies an intrinsic broadening in the Q dependence of about 1 $d_0^{-1}$, where $d_0$ is the sphere diameter. We take $d_0$ as 2.5 Å, a value roughly appropriate to CoP like materials. The width of the peaks at low Q are limited by this effect. The resolution in energy is limited by the grid size and is $\frac{1}{2}$JS.

It has recently become apparent that the elastic scattering $S(\vec{Q})$ associated with amorphous model structures is quite strongly dependent on the direction[14] as well as the magnitude of $\vec{Q}$. Our results show that the inelastic scattering also depends on the direction of $\vec{Q}$. In Fig. 1a we show results for the direction corresponding to the maximum $S(\vec{Q})$ for $|Q| = Q_1 \sim 3$ Å$^{-1}$. (The significance of $Q_1$ will be discussed after Eq. (11) below.) We note that there is a significant component of the inelastic scattering which may be regarded as showing roton-like behavior. That is, the energy peak for this component has a minimum near $Q_1$. One can see evidence of another minimum near $Q_2 \sim 5.7$ Å$^{-1}$. However the Q-dependent part of the scattering becomes smaller as $|Q|$ increases and for large Q, S(Q,E) becomes the same as the density of states. In Fig. 1b we show results for a direction for which $S(\vec{Q})$ for $|Q| = Q_1$ is essentially zero. The roton-like behavior is here less well defined, but the overall tendency of the scattering intensity to shift back and forth between high and low frequencies as $|Q|$ is increased is quite evident. The behavior of S(Q,E) averaged overall directions in the model is probably intermediate between the two cases shown.

In Fig. 2 we show results for density of states for the 996 site (DRP) model compared with that for a pure face centered cubic crystal and for a disordered crystalline alloy. Overall, the density of states for the DRP model appears to be a slightly broadened version of that for the fcc crystal. By contrast, the alloy result[15] shows a considerable broadening.

In Fig. 3 we show density of states and S(Q,E) for the 890 site model. Only the results for the direction of $\vec{Q}$ corresponding to the maximum $S(\vec{Q})$ for $|Q| = Q_1$ are shown. The roton-like peak for $Q = Q_1$ is sharper in this case than for the 996 atom model.

For comparison with the results for the amorphous models, we give in Fig. 4 the $S(\vec{Q},E)$, averaged over all directions of $\vec{Q}$, for an fcc crystal. These results

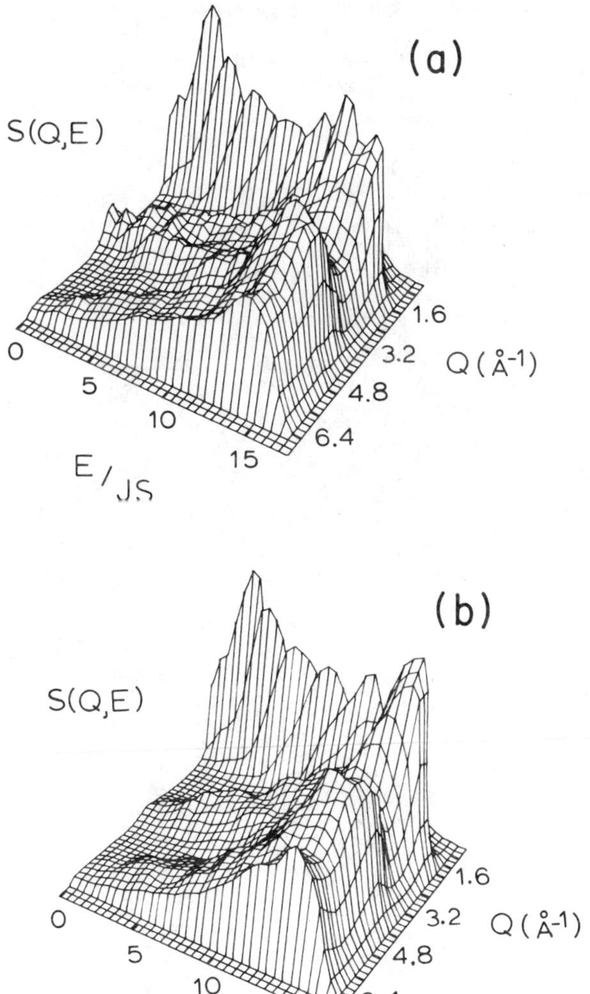


Fig. 1 Neutron scattering law S(Q,E) calculated by the equation of motion method for a 996 atom dense random packing model. (a) wave vector Q along a direction which has a strong peak in the elastic scattering at $\approx 3$ Å$^{-1}$. (b) wave vector along a direction for which the elastic scattering at 3 Å$^{-1}$ is zero for the model.

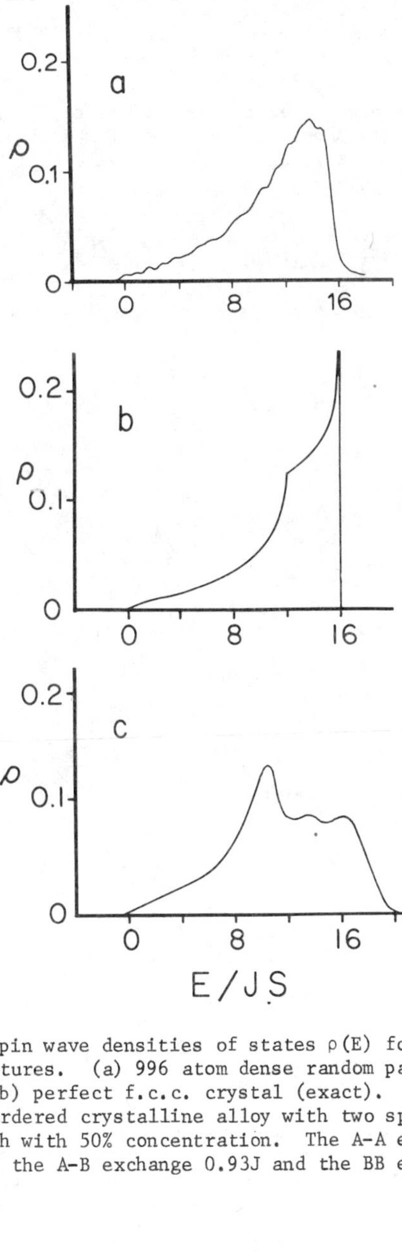


Fig. 2 Spin wave densities of states ρ(E) for various structures. (a) 996 atom dense random packing model. (b) perfect f.c.c. crystal (exact). (c) 5280 atom disordered crystalline alloy with two species A and B each with 50% concentration. The A-A exchange is 1.33J, the A-B exchange 0.93J and the BB exchange is 0.66J.

were obtained by calculating energies for $3 \times 10^6$ randomly chosen values for $\vec{Q}$ with $|Q| \leq 16\ d_0^{-1}$, and forming a histogram of the energy distribution for successive ranges of $|Q|$. The behavior bears considerable resemblance to that in Fig. 1 and Fig. 3. The principal differences are that the spectral features are sharper and the variation with $|Q|$ is more pronounced and more complex in the case of the crystal.

The principal features of S(Q,E) both for the amorphous and the crystal models are well explained by a simple theory analogous to that derived by Weaire and Alben[16] to describe neutron scattering from lattice vibrations in amorphous germanium. In this approach we consider correlations of the motion of nearest neighbors exactly, and neglect all other correlations. We first write the scattering law in terms of the normal mode operators $a_q^+$ for the Hamiltonian (3):

$$a_q^+ = \sum_i U_i^q\, a_i^+ \,, \tag{6}$$

where the $U_i^q$'s form a unitary transformation. Substituting in (5) we obtain:

$$S(\vec{Q},E) = \frac{S}{N}\sum_q \Big|\sum_i U_i^q\, e^{i\vec{Q}\cdot R_i}\Big|^2 \delta(E-E(q)), \tag{7}$$

where E(q) is the energy of the $q^{th}$ normal mode. Equation (7) can be rewritten in terms of a sum over all pairs <i,j>

$$S(\vec{Q},E) = \frac{S}{N}\sum_q \Big(1 + \sum_{<i,j>} 2\ \mathrm{Re}\ U_i^q\, \bar{U}_j^q\, e^{iQ(R_i\cdot R_j)}\Big)\delta(E-E_q). \tag{8}$$

We next average over all directions of $\vec{Q}$, take all neighbor distances as a constant R and neglect terms in the sum referring to pairs of sites which are not neighbors. This yields

$$S(\vec{Q},E) \sim \frac{S}{N}\sum_q \Big[1 + \frac{\sin QR}{QR} {\sum_{<i,j>}}' (U_i^q \bar{U}_j^q + U_j^q \bar{U}_i^q)\Big]\delta(E-E_q) \tag{9}$$

where the primed sum is over pairs of neighbors only.

On the other hand, by substituting (6) into the Hamiltonian (3) and noting that the U's are such that the coupling between modes drops out, we may obtain:

$$\frac{1}{N}\sum_{<i,j>} (U_i^q \bar{U}_j^q + U_j^q \bar{U}_i^q) = \bar{z} - \frac{E(q)}{J} \quad . \tag{10}$$

With the above relation, we can write the approximate scattering law, for a continuous distribution of modes as follows:

$$S(\vec{Q},E) \sim S\Big[1 + \Big(\bar{z} - \frac{E}{J}\Big)\frac{\sin QR}{QR}\Big]\rho(E), \tag{11}$$

where $\rho(E)$ is the density of states. In this approximation, the scattering law is just the density of states modulated by a factor which enhances the low energy scattering when sin QR is positive and enhances the high energy scattering when sin QR is negative. The wave vectors corresponding to enhancement of the low frequency scattering are given by: $Q_n R \approx (2n + \frac{1}{2}\pi)$, where n is a positive integer.

Of course, Eq. 11 is only an approximation, and indeed it has the unsatisfactory features of being negative when Q is small and E is large. (It does however preserve the sum rule: $\int dE S(Q,E) = S$). Nonetheless it does describe the overall behavior in Figs. 1, 2 and 4 as the reader may readily verify. In particular we find that $Q_1 R \sim \frac{5\pi}{2}$ and $Q_2 R \sim \frac{9\pi}{2}$ as predicted by (11). Also the high energy portion of the scattering for $\vec{Q} = Q_1$ is roughly a factor of two smaller than the density of states in this region. It might be noted that we have also calculated spin waves for

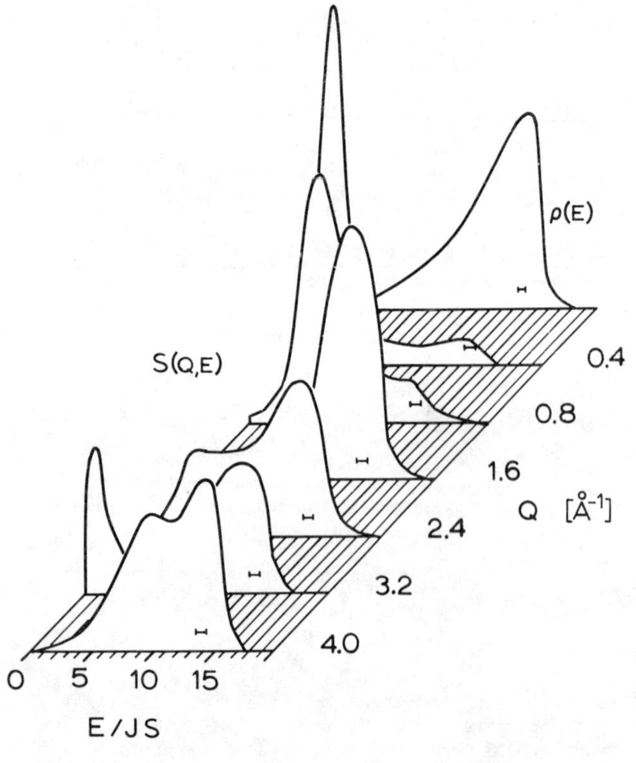


Fig. 3 Neutron scattering law S(Q,E) and density of states $\rho(E)$ calculated for an 890 atom dense random packing model. For S(Q,E), Q is along a direction for which the elastic scattering at $\approx 3$ Å$^{-1}$ is a maximum.

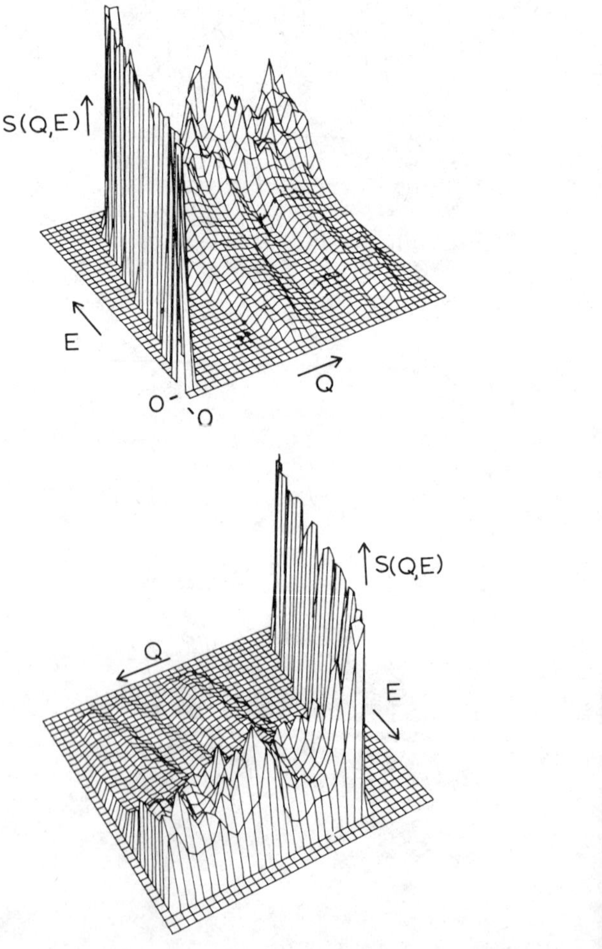


Fig. 4 Neutron scattering law S(Q,E) for a polycrystalline f.c.c. model. The results we obtained from averages over $3 \times 10^6$ randomly chosen wave vectors. Two views are shown. The ridges running down to zero energy correspond to the Q-values of the diffraction rings in the elastic scattering from a powder.

tetrahedrally coordinated models and found that (11) works quite well for those models also.

One case in which correlations beyond first neighbors are particularly important is the low energy regions where we expect hydrodynamic modes for which the phase of the spin precession changes slowly from site to site. This is also a region which is difficult to treat in a model calculation since a finite model cannot accommodate long wavelength excitations. Happily, both the energies and scattering intensities due to the long wavelength excitations can be well understood in terms of simple static properties of the models.

We first consider the energies of the long wavelength modes. The change in static free energy due to slow spatial variation in the average spin direction in a region centered at a position $\vec{x}$ can be written as

$$F = -\frac{1}{2S} D \int (\vec{\nabla}\vec{S}(\vec{x}))^2 \rho d^3x, \qquad (12)$$

where $\rho$ is the average number of spins per unit volume, and D is a constant related to the exchange interaction. For an f.c.c. crystal with nearest neighbor interactions: $D = 2JSd_0^2$ where $d_0$ is the neighbor distance. For model structures, D can be simply determined by a method[17] analogous to that used for determining elastic constants of amorphous models.[18] We enforce an average gradient of spin direction by making the boundary spins vary appropriately. The interior spins are then relaxed so as to minimize F. The result depends little on the direction in which the gradient is imposed and was essentially the same for the two DRP models: $D = (1.9 \pm 0.1)JSd_o^2$.

The spin wave energy E at long wavelength (small k) may be shown to depend simply on D:

$$E = D k^2, \qquad (13)$$

and thus D is referred to as the "spin wave stiffness" constant. We may conclude then that the spin wave stiffness of the DRP models is essentially the same as for fcc. The reason for this is that spin wave stiffness depends strongly on the range as well as on the number of exchange interaction. The lower coordination of models is thus compensated by the assumption that the exchange interacts exists with undiminished strength between sites with up to $1.25d_0$ separation. In a real insulator this would not be the case. In metals, of course, more serious modifications of the form of the interactions must be considered.

The scattering due to the long wavelength modes can be understood in terms of an argument derived by Axe[19] for a one dimensional model for vibrations in a disordered system. Axe assumed that near E = 0 the normal modes are plane waves, so that the $U_i^q$ 's in (6) become

$$U_i^q = \frac{1}{\sqrt{N}} e^{i\vec{k}(q)\cdot\vec{R}_i}, \qquad (14)$$

where $\vec{k}$ is the wave vector for the mode q. (This assumption is valid for all E in the case of the fcc crystal.) We relabel the low E modes according to their wave vector k, and rewrite the scattering law (5) as follows.

$$S(\vec{Q},E) = \frac{S}{N^2}\sum_k \Big|\sum_i e^{i(\vec{k}+\vec{Q})\cdot\vec{R}_i}\Big|^2 \delta(E-Dk^2)$$
$$= \frac{S}{N}\sum_k S'(\vec{k}+\vec{Q})\delta(E-Dk^2), \qquad (15)$$

where, as before, S' is the static structure function.

For a polycrystal S' is non zero only on spherical shells corresponding to the angular average of the Bragg spots. We consider the contribution near one such shell of radius $Q_B$. We then restrict the sum in (13) such that

$$|\vec{k} + \vec{Q}| = Q_B \qquad (16)$$

It is convenient to change the sum into an integral over the possible directions for $\vec{k}$ with respect to $\vec{Q}$ such that (16) is satisfied. We thus obtain for $|k| \ll Q_B$

$$S(Q,E) \sim S'(Q_B)\Delta Q_B(|Q|-Q_B)^2 \int \frac{\delta(E-D(|Q|-Q_B)^2\sec^2\theta)\,d\cos\theta}{\cos^3\theta} \qquad (17)$$

where $\Delta Q_B$ is the width of a Bragg spot and $\theta$ is the angle between $\vec{k}$ and $\vec{Q}$. The integral is easily performed yielding (for $Q_B \neq 0$):

$$S'(Q,E) \sim \frac{S'(Q_B)\Delta Q_B}{2D} \qquad E>D(|Q|-Q_B)^2$$
$$S'(Q,E) = 0 \qquad E<D(|Q|-Q_B)^2 \qquad (18)$$

Thus, near each $Q_B$ we expect scattering from the low energy modes. However the scattering takes the form of an abrupt rise to a constant value when $E>D(|Q|-Q_B)^2$. There is no peak as a function of E predicted by this theory for three dimensional systems. (A peak does result for the one dimensional case treated by Axe.) This is consistent with the numerical results in the very low E region of Fig. 4.

The above argument should also apply to the long wave length modes of amorphous materials, providing that there are reasonably sharp peaks in S'(Q). The effect of finite breadth of the peaks is just to broaden the abrupt rise in the scattering mentioned above.

## IV. RELATION TO EXPERIMENTS

Real amorphous magnets[10-12] are metallic alloys and differ from the model which we have studied in important ways. Firstly, the different environments of different sites lead us to expect some quantitative disorder in the exchange interactions, in addition to the topological disorder which we have treated. Similarly the size of the localized magnetic moments, and hence the interaction with neutrons, might vary randomly from site to site. Finally the metallic character should give a Rudderman-Kittel-like long ranged component of the effective exchange.

The effect of quantitative disorder in the exchange is potentially very important. This is because the exchange can be quite sensitive to the local environment and even relatively modest variations in the exchange can strongly affect the density of states (see Fig. 2c). However, since there is as yet no direct measure of exchange variations in real amorphous materials, the actual importance of quantitative disorder remains uncertain.

Random variation in magnetic moments will not in itself affect the excitations but can strongly affect S(Q,E). In particular, large moment variations will reduce coherent inelastic scattering and increase the Q-independent incoherent scattering. Indeed incoherent dominated behavior has been found for several rare-earth transition metal systems. Nonetheless, careful measurements of behavior near the special Q-vectors discussed in Sec. IV might yet reveal evidence of the structure dependent frequency enhancement oscillations predicted by the model calculations.

The presence of a long range oscillating component of the exchange should not alter our qualitative conclusions. The detailed form of the density of states and the spin wave stiffness would however no doubt be affected.

The changes from our model results which are expected for a real amorphous magnetic alloy are generally in the direction of reducing the Q-dependent structure in S(Q,E). It is therefore surprising the recent neutron results[12] for $Co_{1-x}P_x$ alloys have shown roton-like behavior which is sharper than expected from our model. Nonetheless, the overall behavior is qualitatively similar to what we have found. This is evidence then that the effect of static structural correlations in the neutron scattering, so evident in the model results, does occur in real materials.

We thank J. Axe, G. S. Cargill III and M. Blume for helpful discussions.

APPENDIX

The essence of the equation of motion method is to introduce Green's functions which may be Fourier analyzed to yield spectral densities from which both the neutron scattering intensity and a statistically accurate density of states can be obtained. The Green's functions can be efficiently calculated from numerical approximation to their equations of motion. Since the equations of motion are integrated for some finite number of time steps, the spectral functions and hence intensities and densities of states are broadened by a resolution function. We introduce the retarded Green's functions

$$G_{ij}(t) = -i < 0|a_i(t)\, a_j^+(0)|0>, \quad t \geq 0 \tag{A1}$$

where $<0|$ is the ground state with all spins aligned. These functions give the amplitude at site i and time t of a spin excitation created at time zero at site j. We next introduce the N transformed Green's functions

$$\tilde{G}_{i\vec{Q}}(t) = \frac{1}{N}\sum_j e^{i\vec{Q}\cdot\vec{R}_j} G_{ij}(t), \tag{A2}$$

which describe the amplitude at each site i at time t of an initial plane wave excitation with wave vector $\vec{Q}$. Here $\vec{R}_j$ is the position vector of site j. From (3) and (5) and the commutation relation for Bose operators we obtain the equations of motion for the $\tilde{G}_{i\vec{Q}}$'s:

$$i\hbar\frac{\partial}{\partial t}\tilde{G}_{i\vec{Q}}(t) = JS[Z_i\tilde{G}_{i\vec{Q}}(t) - \sum_j{}' G_{j\vec{Q}}(t)], \tag{A3}$$

where $Z_i$ is the number of neighbors of site i and the primed sum is over these neighbors. From (A1) and (A2) we obtain the initial conditions:

$$\tilde{G}_{i\vec{Q}}(0) = -i\, e^{i\vec{Q}\cdot R_i}. \tag{A4}$$

In our procedure we replace (6) by the difference equations

$$\tilde{G}_{i\vec{Q}}(t+\delta) = \tilde{G}_{i\vec{Q}}(t-\delta) + \frac{2\delta}{i\hbar}[JSz_i\tilde{G}_{iQ}(t) - JS\sum{}'\tilde{G}_{jQ}(t)] \tag{A5}$$

where $\delta$ is the time step for each iteration cycle. We have found that $\delta \leq \frac{1}{20}$ of the shortest time period of the system gives an accuracy of better than 1% in our final spectra. (The shortest time period is defined as Planck's constant divided by the maximum of the absolute value of the excitation energies.)

The scattering law (Eq. 5) can be rewritten:

$$S(Q,E) = \lim_{\lambda\to 0^+} -S\ \mathrm{Im}[\frac{1}{\pi\hbar}\int_0^\infty dt \sum_i e^{-i\vec{Q}\cdot R_i}\tilde{G}_{iQ}(t)e^{+i\frac{Et}{\hbar}}e^{-\lambda t}dt]. \tag{A6}$$

We replace (A6) by a sum of terms corresponding to an integration time T where typically $T \simeq 250\delta$. The principal effect of this approximation is to produce a broadening of the spectrum. (The value of $\lambda$ is chosen to effectively suppress termination oscillations without adding unnecessarily to the broadening.) The amount of broadening is inversely proportional to T and hence may be chosen at will. The choice is governed by the type of information obtainable from a given size model, and for very large models by the time penalty associated with high resolution.

The density of states is obtained indirectly by considering a spectral function for incoherent scattering. That is, the phases $\vec{Q}\cdot\vec{R}_i$ in (A2), (A4) and (A6) are replaced by random angles $\phi_i$. For a given set of angles $\{\phi\}$ we obtain a spectral function $\rho'$:

$$\rho'(\{\phi\}E) = \lim_{\lambda\to 0^+} -\frac{1}{\pi N}\mathrm{Im}[\frac{1}{\hbar}\int_0^\infty dt \sum_{i,j} G_{ij}(t)e^{i(\phi_i-\phi_j)}e^{+i\frac{Et}{\hbar}}e^{-\lambda t}] \tag{A7}$$

If we average $\rho'$ over many uncorrelated sets of angles, the terms in the double sum in (A7) referring to different sites drop out. The sum of the diagonal terms gives the true density of states. In practice, satisfactory results can be obtained by averaging results for a relatively small number of $\{\phi\}$'s. Alternatively we can choose random phases for a limited number of sites and replace the exponential factors in (A2), (A4) and (A6) by zero to obtain a density of states projected on just part of a model. Finally it should be noted that an exact projected density of states for a single site results when only one non-zero exponential factor is used.

The equation of motion method is particularly effective in treating large models since both the time and storage requirements for executing the calculation vary only linearly with model size. Memory requirements are 2N locations of random access memory and (Z+3)N locations of sequential memory. The number of arithmetic operations for a typical calculation of a spectral function with a resolution full width at half maximum of 3% of the spectral width is 250(4 + 2Z)N.

REFERENCES

* Work supported in part by the National Science Foundation.


1. See reviews by R. J. Bell, Rep. Prog. Phys. 35, 1315 (1972); and P. Dean, Rev. Mod. Phys. 44, 127 (1972).
2. J. L. Finney, Proc. Roy. Soc. (London) A 319, 479 (1970).
3. C. H. Bennett, J. Appl. Phys. 43, 2727 (1972).
4. C. G. Montgomery, J. I. Krugler and R. M. Stubbs, Phys. Rev. Letters 25, 669 (1970).
5. T. Kaneyoshi, J. Phys. C: Solid State 5, 3504 (1972).
6. E-Ni Foo and S. M. Bose, Phys. Rev. B9, 347 (1974).
7. J. E. Gubernatis and P. L. Taylor in *Amorphous Magnetism*, H. O. Hooper and A. M. de Graaf eds., (Plenum Press, New York, 1973) p. 405.
8. S. M. Bose, K. Moorjani, T. Tanaka and M. M. Sokoloski, ibid p. 421.
9. R. Alben and M. F. Thorpe, J. Phys. C: Solid State 8, L275 (1975); R. Alben, M. Blume, R. Krakauer and L. Schwartz, Phys. Rev. B (1975 - in press).
10. J. D. Axe, L. Passell and C. C. Tsuei in *Magnetism and Magnetic Materials-1974*, C. D. Graham, Jr., G. H. Lander, and J. J. Rhyne eds. (American Institute of Physics, New York, 1975) p. 119.
11. J. J. Rhyne, D. L. Price and H. A. Mook, ibid p. 121.
12. H. A. Mook, N. Wakabayashi and D. Pan, Phys. Rev. Letters 34, 1029 (1975).
13. See P. G. de Gennes in *Magnetism*, Vol. III, G. Rado and H. Suhl eds. (Academic Press, New York (1963) p. 123.
14. R. Alben, G. S. Cargill III and J. Wenzel, Phys. Rev. B, Feb. 1976.
15. See R. Tahir-Kheli, Phys. Rev. B6, 2838 (1972).
16. D. Weaire and R. Alben, J. Phys. C. 7, L189 (1974).
17. S. Kirkpatrick and A. B. Harris in *Magnetism and Magnetic Materials-1974*, C. D. Graham, Jr., G. H. Lander and J. J. Rhyne eds. (American Institute of Physics, New York 1975) p. 99.
18. R. Alben and D. Weaire, Phys. Rev. B9, 1975 (1974).
19. J. D. Axe in Proceedings of the Rhode Island Summer School on Amorphous Semiconductors (unpublished).

# SPIN EXCITATIONS IN RANDOM ANTIFERROMAGNETS

Scott Kirkpatrick
IBM T. J. Watson Research Center, Yorktown Heights, N.Y. 10598

***Abstract***

A first-principles description of the magnetic properties of the antiferromagnetic alloys $Rb_2Mn_pNi_{1-p}F_4$, $Rb_2Mn_pMg_{1-p}F_4$, and $RbMn_pMg_{1-p}F_3$, based on detailed calculations, is presented. The theory gives a good account of the existing neutron scattering data, as well as the temperature dependence of the sublattice magnetization and the nonlinear static susceptibility, and makes a number of predictions for future experiment. The calculations provide microscopic details of fluctuation phenomena due to disorder in antiferromagnets which do not occur in ferro- or ferri-magnets.

***Introduction***

This paper summarizes calculations of the low temperature properties of "ideal" random magnetic systems, those which can be described by a Heisenberg Hamiltonian, carried out in collaboration with A. B. Harris. Two long papers, which will appear elsewhere[1,2], describe the treatment of ferromagnets, ferrimagnets and antiferromagnets. In this report, only antiferromagnets will be considered.

Substitutional alloys of the form $RbA_pB_{1-p}F_3$ and $Rb_2A_pB_{1-p}F_4$, where A and B are (possibly) magnetic ions, are known to be ideal antiferromagnets in the sense described above. (Related compounds in which K replaces Rb are also studied, and have essentially identical properties.) In the former compounds, the magnetic ion sites constitute a three dimensional simple cubic (3D SC) lattice. In the latter, magnetic ions occur in well separated planes, the sites of which form a 2D square (SQ) lattice. De Jongh and Miedema[3] have recently given a thorough review of the experimental data on the pure "ideal" magnetic systems with these and similar structures.

We consider two alloy systems. The first, (A,B) = Mn-Ni, has both constituent ions magnetic. Such two component systems, in the 2D form, have recently been studied by inelastic neutron scattering[4]. In the second, (A,B) = Mn-Mg, and the Mn ions are diluted by non-magnetic Mg. In both cases, antiferromagnetic exchange interactions, primarily between spins which are nearest neighbors, result from superexchange, and the system's magnetic properties may be described by

$$H = \sum_{i<j} 2J_{ij}\, \boldsymbol{S}_i \cdot \boldsymbol{S}_j - \sum_i (\sigma_i \Delta_i S_i^z + g_i \mu_B\, \boldsymbol{h}_{ext} \cdot \boldsymbol{S}_i) \quad (1)$$

In (1), the exchange integrals, $J_{ij}$, the anisotropy, $\Delta_i$, the g-factor, $g_i$, and the magnitude of the spins, $S_i$, are random variables, since they depend upon the types of atoms occupying sites i and j. Interaction with an external field, $\boldsymbol{h}_{ext}$, which may be position and time dependent, is included, $\mu_B$ is the Bohr magneton, and $\sigma = \pm 1$ indicates in which of the two sublattices site i occurs. Values of $J_{ij}$ appropriate to the various pure species have been determined experimentally[3] and should be correct for the coupling between like spins in the alloys as well. The relation $J_{AB} = (J_{AA}J_{BB})^{1/2}$ which one expects from superexchange theory has been confirmed for Mn-Ni by studies of impurity resonances[5], and fixes the remaining coupling between unlike spins needed to describe the alloys.

We shall assume that the temperature is low enough and the magnitude of $S_i$ sufficiently great that a linearized, or classical, description of spin excitations from a Neel ground state is accurate. For $Mn^{2+}$, with S = 5/2, this should be a reasonable approximation. The anisotropy, $\delta$, defined in dimensionless form as $\Delta/2Jz$, where z is the number of nearest neighbors, is small in both 2D and 3D alloys, i.e. $\delta \approx 0.005$ for $Rb_2MnF_4$ and $\delta \lesssim 10^{-5}$ for the 3D systems[3]. However, we see below that because of special properties of two dimensional antiferromagnets, this anisotropy must be included for accurate treatment of the 2D alloys. The parameters we have used in calculations of both Mn-Ni and Mn-Mg alloys are given in Refs. 1 and 4. The considerations involved in their determination are discussed at greater length in those references.

It is convenient, in discussing the calculations which have been carried out, to distinguish calculations valid at arbitrary frequencies from those directed at studying properties of excitations at low energy and long wavelength. The imaginary part of the dynamic susceptibility, $\chi(q,\omega)$, provides a compact description of the excitations at arbitrary q and $\omega$, and can be calculated numerically by inverting the equations of motion for the spins in a large, computer-generated sample having the desired concentrations of the various magnetic ions[1,6,7]. Examination of microscopic details of the computer simulations then makes possible a fairly detailed characterization of the eigenmodes of the random system.

A continuum description, which reduces to results obtained through hydrodynamic arguments[8] for isotropic systems, is appropriate for modes at low q and $\omega$. The parameters in this description can all be expressed in terms of static response functions, e. g. the susceptibility or the exchange stiffness. These have been calculated for low concentrations of one species by exact expansion to lowest order in the concentration. Computer simulation was then used to obtain these response functions at arbitrary concentrations, by solving for the spin distributions which minimize the energy of the system in an appropriate external field. Again, the microscopic information which the computer simulations can provide proves useful in understanding the effects of fluctuations on the static response functions. Finally, an average medium theory can be applied to give simple interpolation formulae for the response functions as functions of concentration. Comparison with the low concentration and simulation results shows that the interpolation formulae are rather accurate for the two component system, but may be unreliable for dilute magnetic systems.

The principal results of these calculations, described in more detail below, are:

1) Extensive agreement has been obtained between theory and neutron scattering studies[4] on $Rb_2Mn_{.5}Ni_{.5}F_4$, a 2D two-component magnetic alloy. Two distinct branches of excitations are seen in this alloy, associated with modes in which the spin deviations occur predominantly on the Ni or Mn spins, respectively. The energies of the Ni modes, which occur at the higher frequencies, since $J_{NiNi} > J_{MnMn}$, can be estimated rather accurately from the first two odd moments of the susceptibility. Calculations of the dynamic susceptibility of the dilute system, $Rb_2Mn_pMg_{1-p}F_4$, for which the experiments have not yet been performed, are also presented below. So-called "Ising resonances" are prominent features in these at moderate to high dilutions.

2) The susceptibility, $\chi_\perp$, of an antiferromagnet in a uniform external field perpendicular to the sublattice magnetization, proves extremely sensitive to disorder. In isotropic dilute systems, $\chi_\perp$ diverges as the concentration of magnetic ions decreases towards a percolation threshold; it also diverges for arbitrary non-zero vacancy concentrations if the spatial dimensionality is two or less. Ferrimagnetic fluctuations, local regions in which the sublattice magnetizations do not cancel, are the cause of this divergence. This mechanism also explains the unusual nonlinear field dependence of the magnetization which Breed and co-workers[9] have observed in both the 2D and 3D dilute alloys.

***Excitations at Arbitrary Energies***

To calculate a dynamic susceptibility from (1) we introduce a periodic external field

$$\boldsymbol{h}_i^+(t) = h_i(\omega)\, (\boldsymbol{x} + i\, \boldsymbol{y})\, e^{-i\omega t} \quad (2)$$

and calculate the resulting spin deviation

$$\boldsymbol{S}_i^+(t) = S_i^+(\omega)\, (\boldsymbol{x} + i\, \boldsymbol{y})\, e^{-i\omega t}\ , \quad (3)$$

defining the two point susceptibility as the ratio

$$\chi^{+-}_{ij}(\omega) = g_i\, \mu_B\, S_i^+(\omega)/h_j(\omega) \quad (4)$$

The linearized equations of motion for spin deviations about the Neel state

$$\sigma_i \hbar\omega\, S_i^+(\omega) = 2 \sum_j J_{ij} (S_i S_j^+(\omega) + S_j S_i^+(\omega)) + \Delta_i S_i^+(\omega) - 2 g_i \mu_B h_i(\omega) S_i \,, \quad (5)$$

where $S_i$ is the magnitude of the spin on site i, can be put in a more convenient symmetrical form by introducing

$$s_i^+(\omega) = (S_i)^{-1/2} S_i^+(\omega) \quad (6)$$

to obtain

$$\sum_j (\sigma \hbar\omega - \boldsymbol{W})_{ij} s_j^+(\omega) = -2 g_i \mu_B h_i(\omega) (S_i)^{1/2} \quad (7)$$

where $\sigma$ without its site index denotes a diagonal matrix with elements $\sigma_i$ and

$$W_{ij} = \delta_{ij} (\Delta_i + 2 \sum_k J_{ik} S_k) + 2 J_{ij} (S_i S_j)^{1/2} \,. \quad (8)$$

The Fourier transformed susceptibility $\chi^{+-}(q,\omega)$, the imaginary part of which is observed experimentally, may now be expressed as

$$\chi^{+-}(q,\omega) = -2 < q \mid (\sigma \hbar\omega - \boldsymbol{W})^{-1} \mid q > \,, \quad (9)$$

where

$$< i \mid q > = g_i \mu_B (S_i/V)^{1/2} \exp(iq \cdot R_i) \,, \quad (10)$$

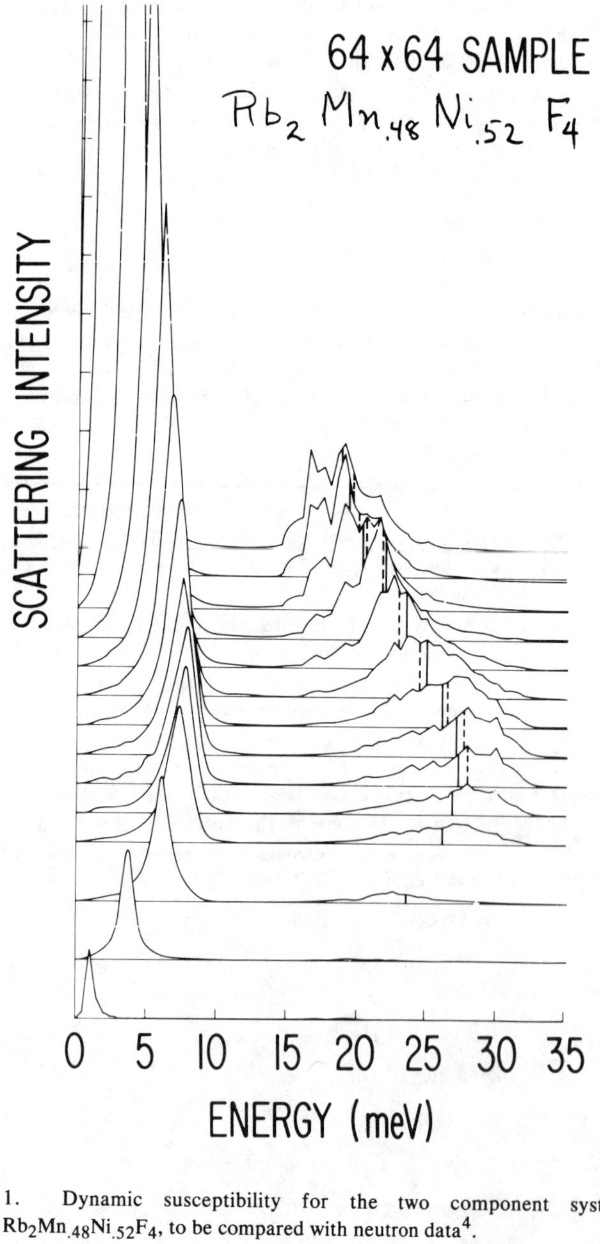


1. Dynamic susceptibility for the two component system, $Rb_2Mn_{.48}Ni_{.52}F_4$, to be compared with neutron data[4].

and V is the volume of the system.

Methods for calculating $\chi^{+-}(q,\omega)$ numerically for finite samples in which the disorder is modeled in detail have been introduced by Harris and Holcomb[6], Alben and Thorpe[7], and the author[1]. Appendix A of Ref. 1 describes the procedure followed in the calculations presented in this paper. Results for a two component system with equal concentrations of Ni and Mn are shown in Fig. 1. In Fig. 1 (and Figs. 2 as well) the scattering intensity (i.e. the imaginary part, $\chi''$) is plotted vertically as a function of frequency at constant $\boldsymbol{q}$ for a series of wave vectors $\boldsymbol{q}$ extending from (0,0), the center of the lattice Brillouin zone, to $\boldsymbol{Q} = (\pi/a, \pi/a)$, the magnetic zone center. Each curve is plotted to the same scale, and is shifted upwards in proportion to the magnitude of $\boldsymbol{q}$.

The most prominent feature in Fig. 1, also seen experimentally[4], is the presence of two branches of excitations, separated at all $\boldsymbol{q}$ by a gap of roughly 10 meV. The calculated susceptibilities at other concentrations[1] show that this separation between two branches persists for all values of p. By comparing the response $S_i^+(\omega)$ projected onto Ni or Mn sites we find[1] that, except near $\boldsymbol{q} = 0$ or $\boldsymbol{Q}$, the upper branch consists of modes in which only Ni spins are excited while the lower branch involves only Mn spins. The admixture of the other spin is typically less than 5 per cent of the response in either branch. Near q = 0 or $\boldsymbol{Q}$, continuum theory predicts excitations in which both types of spin precess in phase with comparable amplitudes. We find this to be true in the lower branch near these symmetry points. A greater admixture is also found in the upper branch close to 0 and $\boldsymbol{Q}$.

The presence of two branches with distinct character can be predicted rather simply by calculating appropriate moments of the spin wave density of states. A characteristic frequency for excitations involving A-atom spins is given by

$$[(\hbar\omega_A)^2_{avg}]^{1/2} = (N_A^{-1} \mathrm{Tr}\, P_A \,\sigma\, \boldsymbol{W} \,\sigma\, \boldsymbol{W})_{avg}^{1/2} \,, \quad (11)$$

where $P_A$ is a diagonal matrix which is 1 on A-atom sites and 0 elsewhere. For $Rb_2$ Mn-Ni $F_4$, the Ni excitation frequency defined in this way lies between 20 and 30 meV at all concentrations, while the Mn frequencies are always roughly 6 meV. These energies are characteristic of the two branches observed. We can extend this construction to define characteristic energies at a given wave vector by calculating projected moments of $\chi''(q,\omega)$. The ratio of third to first moments provides a useful definition[10]:

$$(\hbar\omega_A(q))^2_{avg} \equiv \frac{< q \mid P_A (\sigma \boldsymbol{W})^3 \sigma P_A \mid q >}{< q \mid P_A \,\sigma\, \boldsymbol{W} \,\sigma\, P_A \mid q >} \,. \quad (12)$$

The solid vertical lines on each of the upper branch peaks in Fig. 1 are the characteristic frequencies for Ni excitations as a function of $\boldsymbol{q}$ predicted by (12). These energies account for the Monte Carlo calculations and fall close to the experimental peak positions[4], which are indicated by dashed lines in Fig. 1.

The predicted line shapes and relative intensities in Fig. 1 are in good agreement with the experimental results of Ref. 4. This point is treated in more detail by Als-Nielsen (this conference).

In considering the dilute systems such as $Rb_2Mn_pMg_{1-p}F_4$, a common starting point is the so-called "Ising cluster approximation" [5,6], in which one calculates the energy required to raise or lower $S^z$ of a single spin by one unit. At low temperatures this will depend only upon the number of nearest neighbors present:

$$\Delta E_i = 2 z_i JS + \Delta_i \,. \quad (13)$$

For $Rb_2$Mn-$MgF_4$, these energies are 1.7, 3.3, 5.0, and 6.6 meV, for z = 1, 2, 3, and 4. Since excitation of a single spin is not an eigenstate of the random system, it is not obvious that distinctive modes with these energies will occur or, if they are seen, that their energies will not be shifted by the mixing by which excitations of greater spatial extent arise.

In Figs. 2a-c, we present scattering intensities calculated for the dilute $Rb_2$Mn-$MgF_4$ system at three concentrations. We find some evidence for the validity of the Ising cluster picture, as well as some idea of its limitations. Contributions to $\chi''$ from isolated paramagnetic

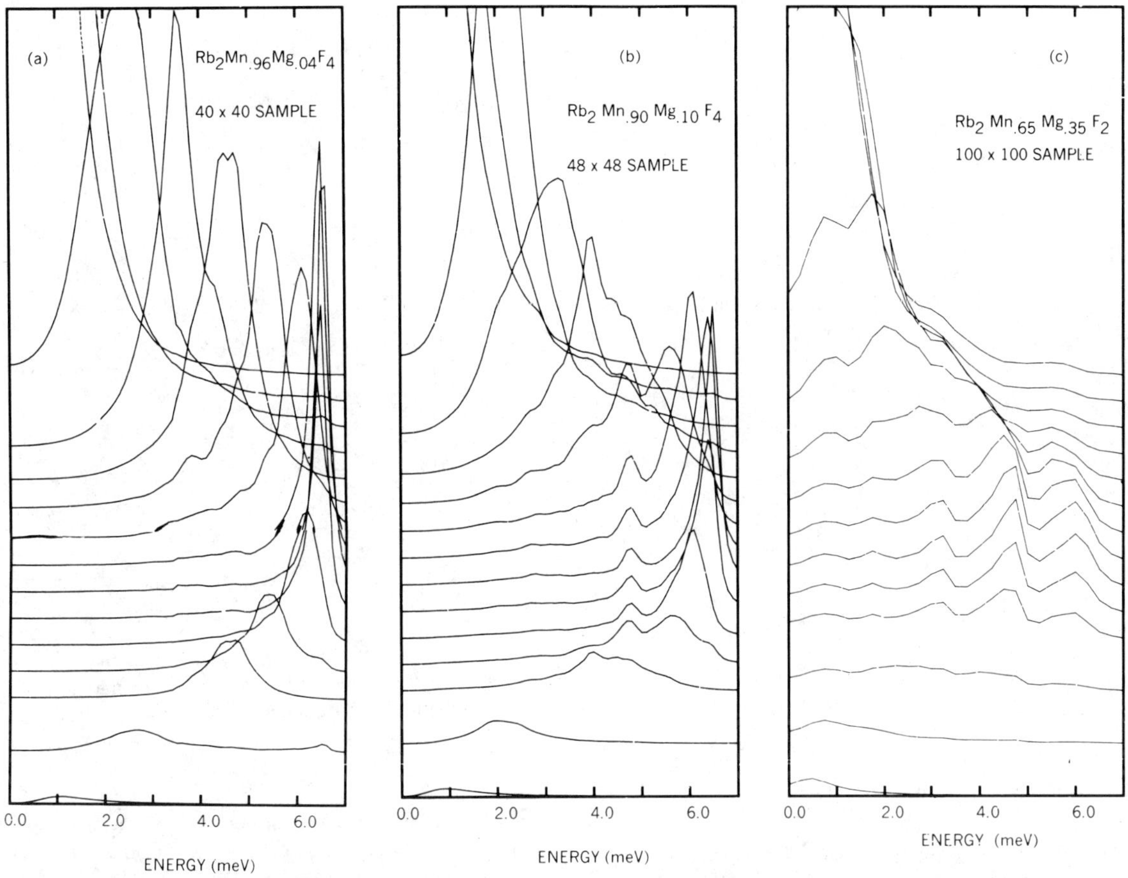


2. Dynamic susceptibility predicted for $Rb_2MnF_4$ diluted by (a) 4 per cent, (b) 10 per cent, and (c) 35 per cent of nonmagnetic Mg ions.

clusters are not included in Figs. 2a-c, since they can be distinguished experimentally by their temperature dependence, and can be shown to be small at the compositions considered.

In the most dilute case, with 4 per cent Mg replacing Mn (Fig. 2a), the only striking consequence of the disorder is a broadening of the lines. The intensities increase monotonically as $\boldsymbol{q}$ goes from 0 to $\boldsymbol{Q}$, as would occur in a pure system. Even though roughly one sixth of the Mn sites have three Mn neighbors at this concentration, there is no Ising peak at 5.0 meV in Fig. 2a. Removal of 10 per cent of the Mn atoms (Fig. 2b) broadens the lines further, and does produce a narrow resonance just below 5.0 meV. However this resonance extends only over a small range of $\boldsymbol{q}$, and its total weight, integrated over $\boldsymbol{q}$ and $\omega$, is a small part of the total scattering, even though roughly 30 per cent of the Mn atoms have 3 Mn neighbors. At a Mn concentration of .65 (Fig. 2c), however, Ising-like resonances provide all the scattering at the higher energies. These peaks were examined microscopically by projecting out the portion of $S_i^+(\omega)$ occurring on sites with 1,2,3, or 4 magnetic neighbors. We find that 80 per cent or more of the weight in the upper two peaks is due to excitation of spins with $z = 3$ or 4, respectively, while the character of the lowest peak is more evenly mixed. We suggest, therefore, that the Ising cluster approach will be useful in a system whose spins are not confined to isolated small clusters if and only if a) the system is strongly disordered, in this case requiring dilution nearly to the percolation threshold which occurs at $p = .59$, and b) the energies considered are high with respect to those at which a continuum picture is appropriate.

### *Continuum Theory*

Dispersion relations for the frequencies of long wavelength spin waves can be derived from the semiclassical magnetostatic energy associated with small deviations of the magnetization about its equilibrium value. A discussion of this procedure for ferromagnets, ferrimagnets, and antiferromagnets can be found in Refs. 2 and 11. For antiferromagnets, the continuum approximation gives

$$\omega_q^2 = \omega_0 + C^2 q^2 , \tag{14}$$

where the anisotropy gap frequency, $\omega_0$, is given by

$$\omega_0^2 = 4\gamma^2 K/\chi_\perp , \tag{15}$$

with $\gamma$ the gyromagnetic ratio, K the macroscopic anisotropy energy, and $\chi_\perp$ the perpendicular susceptibility, and

$$C^2 = 2\gamma^2 A/\chi_\perp , \tag{16}$$

where A is the exchange stiffness.

The response functions A, K, and $\chi_\perp$ must next be examined for a particular microscopic description and a specific lattice. In a pure system, with only nearest neighbor exchange, all three quantities can be expressed in terms of the two parameters, $\Delta$ and J. For example, for a SC lattice with lattice constant a, $K = \Delta S/a^3$, $\chi_\perp = (g\mu_B S/a^3)/(\Delta + 24JS)$, and (for small $\Delta$) $A \approx JS^2/a$. In the presence of disorder, the three quantities become independent and must be calculated separately.

To evaluate A and K for the antiferromagnet when $\Delta$ need not be small it is necessary to use the fact that[1]

$$\chi(\boldsymbol{Q} - \boldsymbol{q}^*, 0) \propto S_{avg}^2/(2K + A\,\boldsymbol{q}^{*2}) , \tag{17}$$

where $S_{avg}$ is the average spin density on one sublattice. Thus a microscopic calculation of the response to a staggered field determines K, and A is obtained by considering an external field of slightly shorter wavelength. Evaluation of $\chi_\perp$ for an antiferromagnet requires solving the microscopic equations

$$\textstyle\sum_j 2J_{ij}(S_i^x S_j + S_j^x S_i) + \Delta_i S_i^x = h_0 S_i \tag{18}$$

for the spin deviations $S_i^x$ induced by a uniform applied field $\boldsymbol{h} = h_0\boldsymbol{x}$. Then

$$\chi_\perp \propto \textstyle\sum_i S_i^x/(Nh_0) , \tag{19}$$

where N is the number of spins in the sample.

The simplest useful theory for A, K, and $\chi_\perp$ in an alloy is the average medium construction in which one calculates the response of the

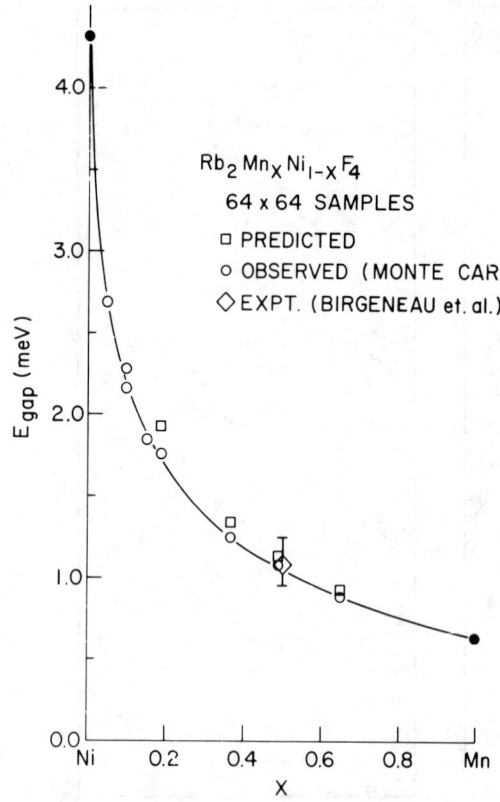


3. Comparison of the anisotropy gap energy for $Rb_2Mn_xNi_{1-x}F_4$ as predicted by the continuum theory using static response functions (square data points) with the actual energies observed in Monte Carlo simulations (circles) and by neutron scattering (diamond). The solid line is merely a guide to the eye

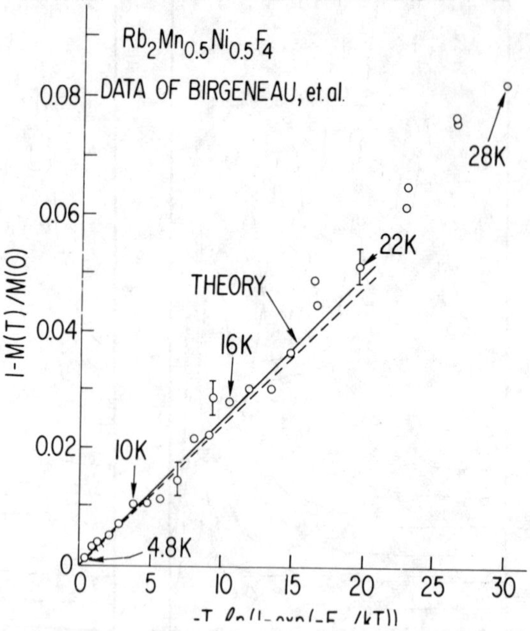


4. Sublattice magnetization as a function of temperature in $Rb_2Mn_{.5}Ni_{.5}F_4$, as obtained from neutron scattering intensities[4]. The solid and dashed lines are the predictions of continuum spinwave theory, using the exchange stiffness calculated as described in the text.

system to an external field such as $h_0\, \boldsymbol{x}$, or $h_0 \exp[i(\boldsymbol{Q} - \boldsymbol{q}^*)\cdot\boldsymbol{R}_i]$ by making the Ansatz that all spins of a given type of atom relax through the same angle, or with one Fourier component of uniform amplitude. In this way, the following interpolation formulae for response functions of an A-B alloy are obtained[1]:

$$K(p) = p K_A + (1 - p) K_B , \qquad (20)$$

$$A(p) = p^2 A_A + 2p(1-p)(A_A A_B)^{1/2} + (1-p)^2 A_B , \qquad (21)$$

neglecting anisotropy, and

$$\chi_\perp(p) = \frac{(pS_A-(1-p)S_B)^2 - 2p(1-p)S_A S_B(\alpha + \alpha^{-1})}{(p\, S_A\, \chi_{\perp A}{}^{-1/2} + (1-p)\, S_B\, \chi_{\perp B}{}^{-1/2}} , \qquad (22)$$

where $\alpha = (J_{AA}/J_{BB})^{1/2}$, and we have used $J_{AB} = (J_{AA}J_{BB})^{1/2}$ to simplify (21) and (22).

Comparison[1] of the predictions of (20-22) with exact expansions of A(p), K(p), and $\chi_\perp(p)$ for p close to 0 or 1 and with computer simulations for general values of p shows that (20) and (21) are excellent approximations in two component systems, while (22) is accurate as long as the spins of the two components are not too different. For $Rb_2Mn_pNi_{1-p}F_4$, with $S_{Ni} = 1$, $S_{Mn} = 5/2$, errors of roughly 20 per cent result from the use of (22) at intermediate concentrations[1]. For dilute magnetic systems, similar comparisons show[2] that (20) remains accurate and (21) is still qualitatively correct, but (22) is qualitatively incorrect. The unexpected consequences of dilution on $\chi_\perp$ are discussed in the following section.

Although our continuum theory is exact[8] in the long wavelength limit for an isotropic (classical) system, we must check its validity for applications in which the anisotropy makes all spin wave frequencies finite. The error terms can be evaluated exactly for p near 0 or 1, and are small[1,2], but a more thorough test of the use of static response functions to predict low energy dynamics is comparison with experiment. In Fig. 3 we compare the gap frequency, $E_g$, predicted by (15) with the one experimental measurement to date[4] and the actual lowest eigenvalues observed in computer simulations. In the worst case, the error appears to be less than 10 per cent.

Measurements of the temperature dependence of the sublattice magnetization in $Rb_2Mn_{.5}Ni_{.5}F_4$[4] provide a second test of the continuum theory. A semiclassical evaluation of the leading term in M(T) for an anisotropic antiferromagnet gives[1]

$$\Delta M/M \sim -(kT/4\pi cA) \ln[1 - \exp(-E_g/kT)] , \qquad (23)$$

where c is the lattice constant perpendicular to the planes of spins. In Fig. 4 we compare the prediction of (23) with data from Ref. 4, using two values for A(.5). The dashed line results from use of the interpolation formula (21), and is in good agreement with the data up to roughly 20 K. The solid line was calculated using A(.5) from computer simulation, and yields a slight improvement over the dashed line. In both cases, no free parameters beyond those known to describe the pure systems were employed in obtaining the fit to experiment.

### *Effects of Dilution on* $\chi_\perp$

Finally we shall consider the response functions as quantities of experimental interest in themselves. We single out the perpendicular susceptibility for discussion because it proves to be unusually sensitive to the presence of fluctuations on the microscopic scale.

If one makes the approximation of assuming a uniform response in $\chi_\perp$, as is done in (22), dilution has no effect on $\chi_\perp$:

$$\chi_\perp(p) = \chi_\perp(1) \qquad \text{(avg. medium)} . \qquad (24)$$

The exact result for an isotropic system with p close to 1, however, is[2,12]

$$\chi_\perp(p)^{-1}\, d\chi_\perp(p)/dp = 2 - 2P_0 , \qquad (25)$$

where

$$P_0 = N^{-1} \sum_q \omega_E{}^2/\omega_q{}^2 , \qquad (26)$$

with $\hbar\omega_E = 2zJ$. For the SC lattice this slope is −1.02, implying that the true nonuniform response causes an increase of $\chi_\perp$ upon dilution. In 2D, $P_0$ diverges as ln N, where N is the number of sites.

General arguments[2] show that the logarithmic divergence in $\chi_\perp$ occurs in 2D at all non-zero vacancy concentrations. We may obtain a lower bound, $\chi_{lb}$, to $\chi_\perp$, from the following approximate construction of

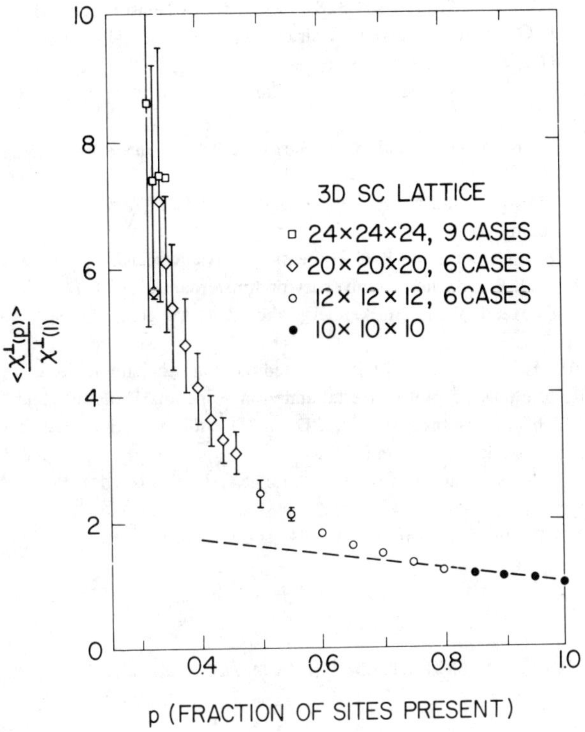


5. Perpendicular susceptibility, $\chi_\perp$, calculated by computer simulation for a dilute 3D simple cubic lattice. The dashed line indicates the prediction of the exact low concentration expansion, and furnishes a check on the numerical results.

the fluctuating response of the spins to a static transverse field, $h_0$. Divide the system into cells, denoted $v_i$, each L sites on a side. Approximate the response of the spins inside each cell by a single spin density wave,

$$m_\perp(x,y) = a_q S^z \sin qx \sin qy \ , \tag{27}$$

with the wave vector $q \sim L^{-1}$ chosen such that the spins on the walls of the cell do not rotate. The response in each cell can then be calculated independently. The exchange energy cost per cell of such a wave is proportional to $L^d q^2 a_q^2$, where d is the spatial dimensionality and $a_q$ is the amplitude of the response. The Zeeman energy in the ith cell is proportional to $S(v_i) a_q h_0$, where $S(v_i)$ is the net unbalanced spin in the cell. Minimizing the total energy, we find $a_q \sim S(v_i) L^{2-2d} h_0$, which leads to an induced spin density in each cell of order $h_0 S(v_i)^2 L^{2-2d}$. Since $S(v_i)^2$ is proportional to $L^d$, the average of the cell susceptibilities gives a lower bound of order $L^{2-d}$:

$$\chi_\perp \geq \chi_{lb} \propto L^{2-d} \ , \tag{28}$$

and $\chi_\perp$ must diverge in the presence of fluctuations in $S(v_i)$ if $d < 2$. The case $d = 2$ may be treated more accurately by further subdivision of those cells in which $S(v_i)$ is small. The result is that

$$\chi_\perp \geq \chi_{lb} \propto \ln L \qquad (d = 2) \ . \tag{29}$$

This divergence will have the effect, in an isotropic system, of modifying the dispersion relation for long wavelength spin waves so that

$$\omega_q \propto q/(\ln q)^2 \ .$$

In quasi-2D materials, however, the non-zero anisotropy field will keep $\chi_\perp$ finite.

Computer simulation of $\chi_\perp(p)$ in isotropic 3D lattices, as shown for the SC case in Fig. 5, reveals that $\chi_\perp$ also diverges when p approaches the percolation threshold from above. This divergence is also removed by anisotropy, since $\chi_\perp \leq p(g\mu_B S/\Delta a^3)$, but the qualitative feature that dilution increases $\chi_\perp$ sharply as $p \to p_c^+$ should remain true for physically reasonable values of $\Delta$. Calculations with the parameters appropriate to $Rb_2Mn_pMg_{1-p}F_4$ show that $\chi_\perp(p \gtrsim p_c)$ is enhanced by a factor of three over $\chi_\perp(1)$.

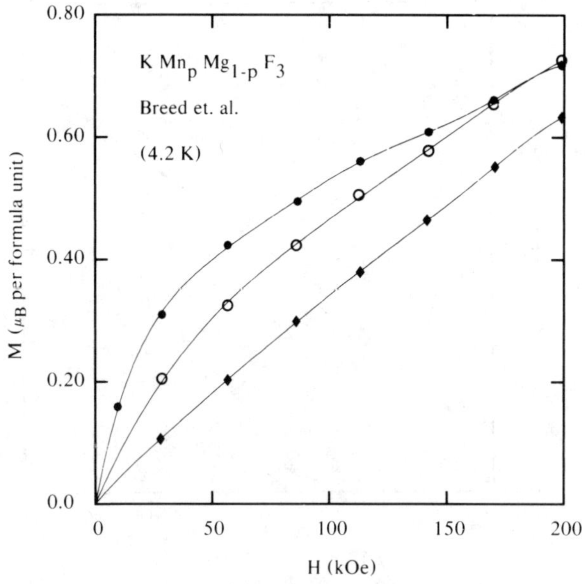


6. Magnetization as a function of field for dilute 3D $KMn_pMg_{1-p}F_3$ alloys at several concentrations (data of Ref. 8). The diamonds indicate data for p = .82, circles are p = .58, and the dots indicate p = .39.

To see if this behavior is observed in real materials we turn next to the experiments of Breed et. al.[9] on dilute 2D and 3D antiferromagnets. The obstacle to comparison with this work is that the magnetization proves to be a highly nonlinear function of the applied field. Since the effect is more dramatic in 3D we consider Breed's data for $KMn_pMg_{1-p}F_3$, plotted in Fig. 6. Although the susceptibility of pure $KMnF_3$ is linear in fields of 200 kOe or more, it is concave downward for each of the dilute samples. The susceptibility at low fields increases with dilution, but at fields of 100 kOe or more the apparent susceptibility remains constant or decreases with dilution. The same qualitative behavior occurs upon dilution in $K_2Mn_pMg_{1-p}F_4$ [9].

The presence of the nonlinearity, and the qualitative trends in both high and low-field susceptibilities can be explained from the same microscopic model (1) if we allow for the possibility that some of the spins may deviate through large angles. Then (18) becomes

$$\sum_j 2J_{ij} S_i S_j [\sin(\theta_i)\cos(\theta_j) + \cos(\theta_i)\sin(\theta_j)] + \Delta_i S_i \sin(\theta_i) = h_0 S_i \cos(\theta_i) \ . \tag{30}$$

These equations can be solved for large Monte Carlo samples by iteration. The low field susceptibility is the same as was given by the linear system (18), but at high fields $\chi_\perp$ is depressed. Results for three concentrations roughly comparable to Breed's cases are plotted in Fig. 7. We do not obtain quantitative agreement, possibly because the Monte Carlo samples were not big enough, but the qualitative features of the data are reproduced in the calculations.

Examination of the microscopic details of the nonlinear solution reveals that the mechanism which causes the susceptibility to saturate at unusually low fields is the same as the source of the increase in $\chi_\perp$ upon dilution. In constructing a lower bound to $\chi_\perp$ we found that the highly nonuniform response associated with locally ferrimagnetic regions could lead to divergences. One would expect large spin deviations to occur in these regions, and that these will approach saturation while the mean spin deviation is still small. To check this picture, we plot in Fig. 8 distributions of the transverse component of the spin for the three concentrations shown in Fig. 7, at a constant field of 30 kOe. In each case a broad distribution of $\sin(\theta)$ is observed, the width increasing with increasing dilution. Many of the spins are seen to relax in the direction opposite to the applied field. The distribution of local exchange fields which gives rise to the results of Fig. 8 should be measurable by the usual probes of internal fields, such as nmr and Mossbauer effect. Because of their unusual sensitivity to fluctuations on the microscopic

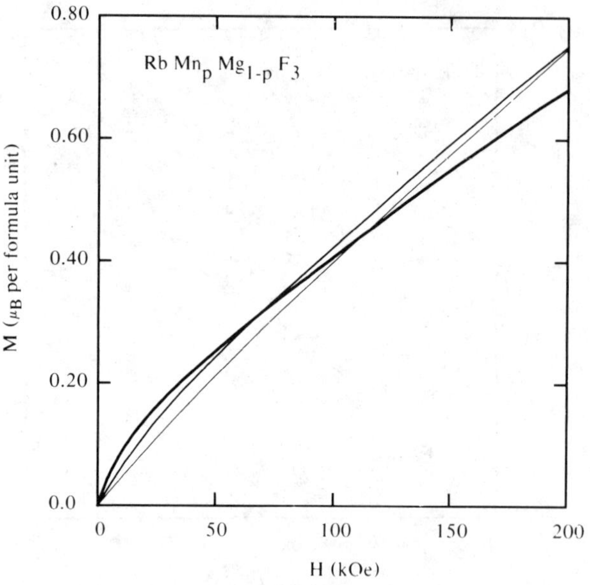


7. High-field magnetization, calculated by non-linear computer simulation as described in the text, for $RbMn_pMg_{1-p}F_3$ with p = .79 (light line), .65 (medium line) and .45 (thickest line).

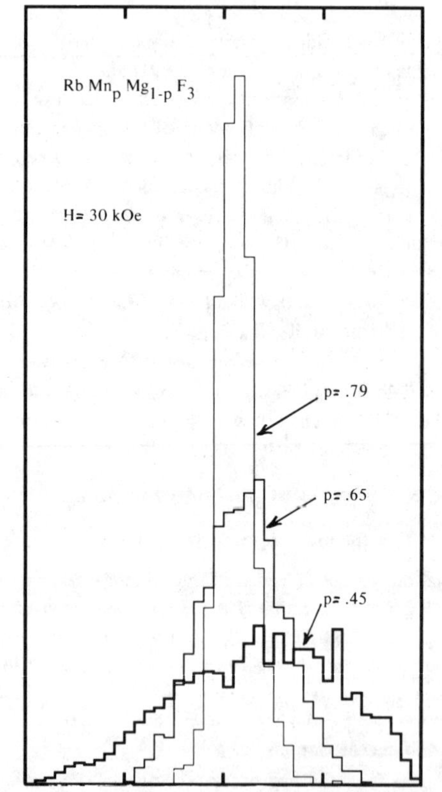


8. Probability distributions of the transverse spin deviations, $\sin(\theta_i)$, observed in the computer simulations of Fig. 7 at a field of 30 kOe.

scale, therefore, it appears that disordered antiferromagnets are worthy of continued experimental attention.

***References***

1. S. Kirkpatrick and A. B. Harris, Phys. Rev. ***B***, Nov. 15, 1975, to appear.

2. A. B. Harris and S. Kirkpatrick, in preparation.

3. L. J. de Jongh and A. R. Miedema, Advances in Physics ***23***, 1 (1974).

4. R. J. Birgeneau, L. R. Walker, R. J. Guggenheim, J. Als-Nielsen, and G. Shirane, J. Phys. ***C8***, L328 (1975); J. Als-Nielsen, R. J. Birgeneau, R. J. Guggenheim, and G. Shirane, Phys.Rev. ***B***, Nov. 15, 1975, to appear; J. Als-Nielsen, this conference.

5. R. A. Cowley and W. J. L. Buyers, Revs. Mod. Phys. ***44***, 406 (1972).

6. W. K. Holcomb and A. B. Harris, AIP Conference Proceedings ***24***, 102 (1974).

7. R. Alben and M. F. Thorpe, J. Phys. ***C8***, L275 (1975); R. Alben, this conference.

8. B. I. Halperin and P. C. Hohenberg, Phys. Rev. ***188***, 898 (1969).

9. D. J. Breed, thesis, University of Amsterdam (1969); D. J. Breed, K. Gilijamse, J. W. E. Sterkenberg, and A. R. Miedema, Physica ***68***, 303 (1973).

10. Evaluation of (12) is more tedious than the simple form suggests, but the terms required are tabulated in Appendix C of Ref. 1 in a form which permits their use for 2D or 3D cubic materials with arbitrary concentrations and parameters $J_{ij}$.

11. S. Kirkpatrick and A. B. Harris, AIP Conference Proc. ***24***, 99 (1974).

12. B. Dreyfus, J. Appl. Phys. ***34***, 1089 (1963).

## MAGNETIC EXCITATIONS IN AMORPHOUS METALLIC FERROMAGNETS*

J. D. Axe
Brookhaven National Laboratory, Upton, New York 11973

ABSTRACT

Inelastic neutron scattering experiments have beer carried out on several Fe-based alloy glasses in order to study the nature of the long wavelength magnetic excitations. Conventional triple-axis spectrometry was used but the scattering kinematics restricted the scattering to the near forward direction. Three different alloy systems have been studied: i) $Fe_{75}P_{15}C_{10}$ ($T_c$ = 597°K)[1]; ii) $(Fe_xMo_{1-x})_{80}B_{10}P_{10}$ with x = 0.93,0.8 and 0.81(170°K < $T_c$ < 450°K)[2]; and iii) $Fe_{33}Pd_{46}P_{21}$ ($T_c$ ∿ 190°K)[3]. For i) and the high $T_c$ sample of ii) spin waves consistent with a ferromagnetic dispersion relation, $\hbar\omega(q) = Dq^2 + Eq^4$, were observed. No intrinsic linewidth was detected over the entire range of momentum transfer for which the scattering triangle could be closed ($q \lesssim 0.2$ Å$^{-1}$). In both cases the values of D are too large to account for B in the relation $M(T)/M(0) = 1 - BT^{3/2} + \ldots$ and the large value of E suggests that higher order terms in M(T) should be- come important at relatively low temperatures, contrary to observations.[4] These discrepancies suggest that these propagating collective modes do not exhaust the low frequency spin fluctuations.

The temperature dependence of the spin waves is consistent with $\omega(T) = \omega(0)\ (1 - \lambda T^{5/2} + \ldots)$ for temperatures up to ∿2/3 $T_c$, and additional thermal broadening of the line shapes appears at elevated temperatures.

Less extensive observations of the sample ii) wit intermediate composition reveal broadened but still di- tinct spin waves at low temperatures. By contrast, in the most magnetically dilute samples, ii) with x = 0.8 and iii), no well-defined spin-wave scattering has beer observed. In the former case quasielastic temperature dependent "critical" scattering is observed near $T_c$.

* Work performed under the auspices of the U. S. Ener- Research and Development Administration.

1. J. D. Axe, L. Passell and C. C. Tsuei, AIP Conf. Proc. 24, 119 (1975).
2. J. D. Axe, T. Mizoguchi and G. Shirane, to be published.
3. J. D. Axe, L. Passell, S. M. Shapiro and C. C. Tsue unpublished.
4. C. C. Tsuei, to be published.

# PERPENDICULAR ANISOTROPY OF AMORPHOUS, ELECTRODEPOSITED COBALT-PHOSPHORUS ALLOY FILMS*

G. C. Chi and G. S. Cargill III+
Yale University, New Haven, Connecticut 06520

The in-plane magnetic saturation field and the intensity of small-angle x-ray scattering were observed for some amorphous, ferromagnetic Co-P alloy films (22.5 at. % P), 31 μm thick, produced by electrodeposition. Annealing the amorphous films reduces both the intensity of small-angle x-ray scattering and the in-plane magnetic saturation field. The x-ray scattering observations have been interpreted in terms of oriented, ellipsoidal scattering regions of high (or low) cobalt concentration, and therefore of high (or low) magnetization. Anisotropy energies calculated with a magnetostatic model, using observed x-ray scattering data, are consistent with measured in-plane saturation fields, both for as-deposited and for annealed films. The time-temperature dependences observed in annealing experiments suggest that the annealing effects are associated with reductions of composition inhomogeneities by atomic scale diffusion. Our results indicate that internal demagnetizing effects associated with anisotropic microstructure contribute to perpendicular easy-axis magnetic anisotropy in these amorphous ferromagnetic films.

## INTRODUCTION

A striking characteristic of amorphous Co-P alloys is magnetic softness, which has been attributed to an absence of magnetocrystalline anisotropy. However, in-plane saturation field and stripe domain observations indicate that the amorphous electrodeposited Co-P films have weak perpendicular easy-axis anisotropy.[1] Internal shape anisotropy resulting from composition fluctuations has been proposed as a possible origin for the experimentally observed magnetic anisotropy.[1] Small-angle x-ray scattering is a sensitive technique for detecting such composition fluctuations. We present in this paper parallel studies of structural and magnetic anisotropy in amorphous electrodeposited Co-P alloy films. The magnetic anisotropy data are compared with a magnetostatic model[2, 3] calculation, based on observed structural anisotropy parameters. The time and temperature dependences of observed annealing effects are compared with those expected for atomic scale diffusion.

## EXPERIMENTS

X-Ray Small-Angle Scattering.[4] The amorphous Co-P films were electrodeposited, using bath B of ref. 6. Specimens used for both x-ray and MH loop measurements contained 22.5 at. % P and were 31 μm thick. The x-ray scattering measurements ($0.01\,\text{Å}^{-1} < K < 5.0\,\text{Å}^{-1}$, where $K=4\pi\sin\theta/\lambda$) employed Mo $K_\alpha$ radiation and a Kratky camera, using a modified transmission geometry which allows the sample to be rotated with respect to the incident x-ray beam. This geometry and the dependence of observed scattered intensity on both rotation angle $\phi$ and K, for as-deposited and annealed samples, are shown in Fig. 1. The scattering is anisotropic, i.e. dependent on rotation angle $\phi$, for $K < 0.1\,\text{Å}^{-1}$. Annealing the Co-P films at 200°C for 6 h decreased the small-angle scattering (SAS) intensity by approximately 40% ($K < 0.05\,\text{Å}^{-1}$), but the intensity of the first diffuse peak at $3.0\,\text{Å}^{-1}$ remained unchanged.

The measured SAS could be well described as scattering by a collection of identical and well separated ellipsoids, with semiaxis lengths $A_1 \gg A_2 \geq A_3$. The $A_1$ axes were assumed to be nearly normal to the film plane. Comparison of model curves with experimental data for our as-deposited film indicates that $A_1 = 4400\,\text{Å}$, $A_2 = 165\,\text{Å}$, and $A_3 = 59\,\text{Å}$. Similar analysis of data for the annealed film gave only slightly different values for $A_2$ (164Å) and for $A_3$ (62Å). A schematic illustration of this anisotropic model structure is shown in Fig. 2.

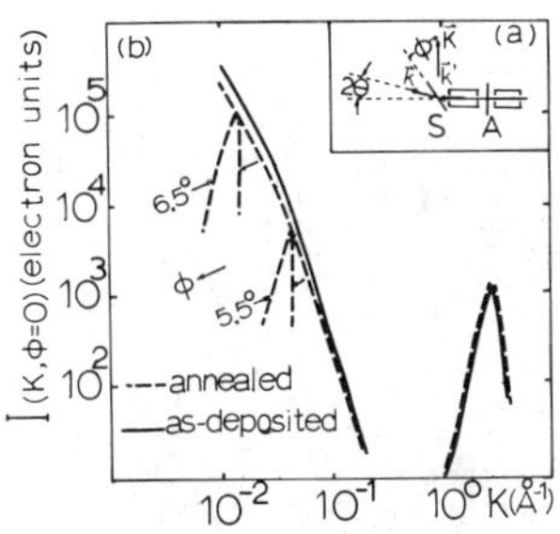


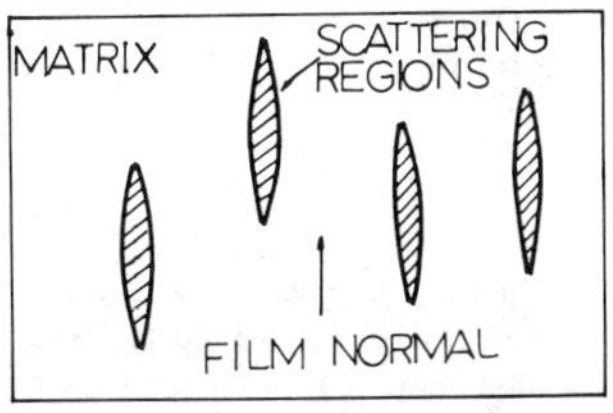


Fig. 2. (above) Proposed model for anisotropic microstructure (not drawn to scale).

Fig. 1. (left) (a) SAS geometry with scattering sample S, scattering angle 2θ, and rotation angle $\phi$; (b) observed dependence of SAS intensity on both $K = 4\pi\sin\theta/\lambda$ and $\phi$. Note log scales for I and K.

Magnetization Measurements and Annealing. The in-plane saturation field $H_s$ was measured with a 60 Hz MH loop tracer, using samples (2cm x 0.2 cm x 31 μm) for which external shape demagnetization effects could be neglected. Hysteresis loops for as-deposited, annealed (200°C for 6 h), and crystallized (340°C for 10 min) samples are shown in Fig. 3. $H_s$ was 70 Oe for the as-deposited sample and was 40 Oe for the annealed sample. The room temperature magnetization of these films was 69 emu/g ($4\pi M_s$ = 6800 G) and was unchanged by the annealing. The in-plane saturation field $H_s$ is simply related to the magnetic anisotropy energy $E_a$ as $E_a = -M_s H_s/2$.[1]

Kinetics of this annealing behavior was further investigated by annealing samples at 145°C and at 185°C for various periods of time. The in-plane saturation fields were measured at intervals during the annealing treatments. $H_s$ versus time curves are shown in Fig. 4 for both of the temperatures; $H_s$ decreases more rapidly at the higher temperature, as expected for a thermally activated process.

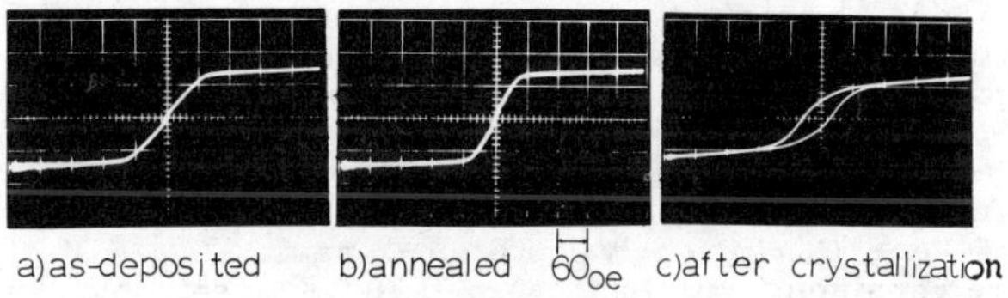


Fig. 3. In-plane MH loops for Co-P films.

## RESULTS AND DISCUSSION

The x-ray SAS intensity distribution for the proposed model of oriented ellipsoidal regions is proportional to the volume fraction $\alpha$ of the film occupied by these scattering regions, to the volume V of each region, and to the square of the difference, $\Delta\rho_{el}$, between the electron density in the scattering regions and the average electron density of the film. In the limiting case $K \to 0\,\text{Å}^{-1}$, the intensity I(K), in electron units per atom, can be expressed as[5]

$$\lim_{K\to 0} I(K) = \alpha\, V\, (\Delta\rho_{el})^2 / \rho_o \qquad (1)$$

where $\rho_o$ is the average bulk atomic density of the film.

Since this density (atoms per unit volume) is almost independent of composition for amorphous Co-P alloys of 19 - 24 at. % P,[6] it is most reasonable to attribute $\Delta\rho_{el}$ to differences in Co concentration $\Delta c_{Co}$ between the ellipsoidal scattering regions and their surroundings,[7] $\Delta\rho_{el} \approx 12\, \Delta c_{Co}\, \rho_o$. In this composition range the room temperature magnetization $M_s$ is nearly proportional to $c_{Co}$,[8] so Eq. (1) can be rewritten as

$$\lim_{K\to 0} I(K) = \text{const.}\ \alpha\ (\Delta M_s)^2, \qquad (2)$$

where $\Delta M_s$ is the difference in magnetization between the ellipsoidal scattering regions and their surroundings.

The magnetic anisotropy energy $E_a$ associated with such an inhomogeneous, anisotropic distribution of magnetization can be evaluated with a magnetostatic model[2, 3]

$$E_a = -\pi\ (\Delta M_s)^2\ \alpha(1-\alpha)\ \cos^2\theta \qquad (3)$$

where $\theta$ is the angle between the direction of magnetization and the film normal, and $E_a = -M_s H_s/2$. It follows that Eq. (3) can be rewritten as

$$H_s = 2\ \pi\ \alpha\ (\Delta M_s)^2/M_s$$
$$= \text{const.}\ \alpha\ (\Delta c_{Co})^2 \qquad (4)$$

if $\alpha \ll 1$. These considerations indicate that the in-plane saturation field $H_s$ should be proportional to the intensity of SAS, extrapolated $K\to 0$. Our measurements indicate that $H_s/I(0) = (41 \pm 3)\times 10^{-6}$ Oe/(eu/atom) for the as-deposited sample, and $(42 \pm 5)\times 10^{-6}$ Oe/(eu/atom) for the annealed sample, which is consistent with the proposed models for the structural and magnetic anisotropies.

The interpretation of our SAS observations in terms of identical, independently scattering, oriented ellipsoidal regions is certainly not a unique one. The experimental SAS results indicate that values of $A_1$ less than 2300 Å would be inconsistent with our observed SAS, but it is difficult to place an upper limit on $A_1$ from these measurements.[4] Furthermore, a collection of ellipsoids of revolution, with minor semiaxis lengths distributed between $A_2$ and $A_3$, could produce SAS like that calculated for identical general ellipsoids with minor semiaxes $A_2$ and $A_3$. However, the latter model is the simplest one which is consistent with the observed SAS. Another assumption of our model, a sharp interface between the scattering regions and the matrix, may be an over simplification for the as-deposited films, and it is almost certainly incorrect for the annealed films.

Only the product $V\alpha(\Delta\rho_{el})^2$ can be obtained from the SAS intensities, even if the "identical ellipsoids" scattering model is assumed to be correct. The volume $V = (4/3)\pi A_1 A_2 A_3$ cannot be accurately evaluated because of uncertainty in $A_1$. This lack of definiteness is illustrated in Fig. 5; the solid curves represent combinations of $\alpha$, $\Delta\rho_{el}$ and some of the $A_1$ values which are consistent with our SAS measurements on the as-deposited Co-P film. Also shown in Fig. 5 (dashed line) are the values of $\alpha$ and $\Delta M_s$ (or equivalently $\Delta c_{Co}$ or $\Delta\rho_{el}$) which are consistent with the observed 70 Oe anisotropy field for this sample, i.e. these values satisfy Eq. (4). This dashed line nearly coincides with the $\alpha$ vs $\Delta\rho_{el}$ curve for $A_1 = 13,000$ Å, but the $\alpha$ and $\Delta c_{Co}$ values remain undetermined. However, it seems to us unlikely that either $\alpha$ or $\Delta c_{Co}$ is beyond the range of values given in the figure.

The observed SAS intensities and in-plane saturation field $H_s$ decreased with annealing, as shown in Fig. 4. We propose that these annealing effects result from atomic scale diffusion, which reduces $\Delta c_{Co}$, $\Delta\rho_{el}$, and $\Delta M_s$. The T = 145°C annealing behavior shown in Fig. 4 agrees well with that predicted by the simple diffusion equation

$$H_s(t) = H_s(0)\ (\Delta c_{Co}(t)/\Delta c_{Co}(0))^2$$
$$= H_s(0)[\sum_{\nu=1}^{\infty} (4/\xi_\nu^2)\ \exp(-\xi_\nu^2 Dt/r_o^2)\ ]^2, \qquad (5)$$

with boundary conditions $\Delta c_{Co}(t)=0$ for $r \geq r_o = (A_2A_3)^{1/2}$, where $D = 7\times 10^{-19}$ cm²/sec and $\xi_\nu$ are the roots of the Bessel function of zero order.[9] This expression cannot be used to represent our experimental $H_s(t)$ observations for longer annealing times or higher annealing temperatures; this may be attributable to the breakdown of our assumption of a sharp interface between matrix and scattering regions and to the unrealistic boundary conditions required for this simple diffusion equation (5).

The diffusion coefficient $D = 7\times 10^{-19}$ cm²/sec is more than an order of magnitude larger than that reported for Ag tracer diffusion in amorphous $Pd_{80}Si_{20}$ at 205°C.[10] Further annealing experiments, together with investigations of more realistic diffusion equations and magnetostatic models, are needed to provide a more detailed interpretation of the annealing effects which we have described in this paper.

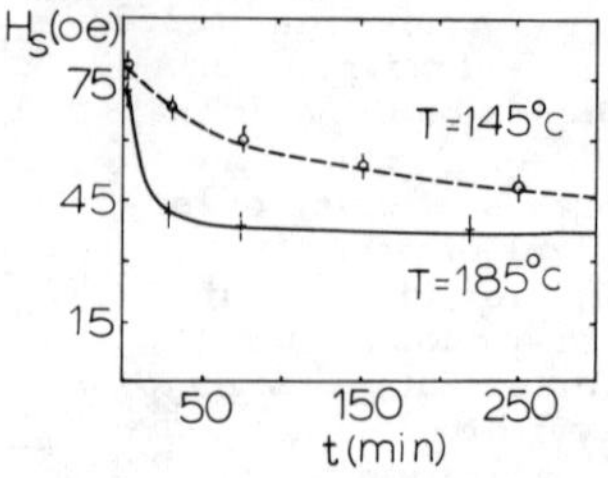


Fig. 4. (above) Effects of annealing on the in-plane saturation field $H_s$. Dashed line: calculation with Eq. (5).

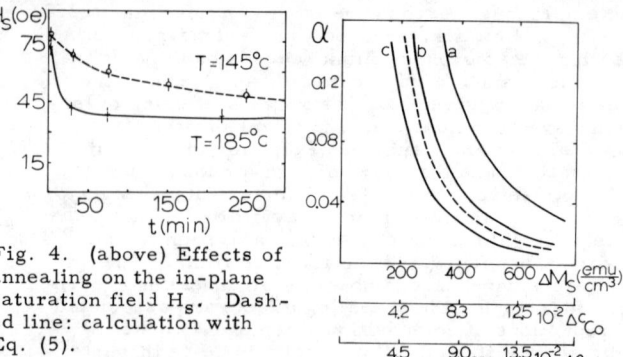

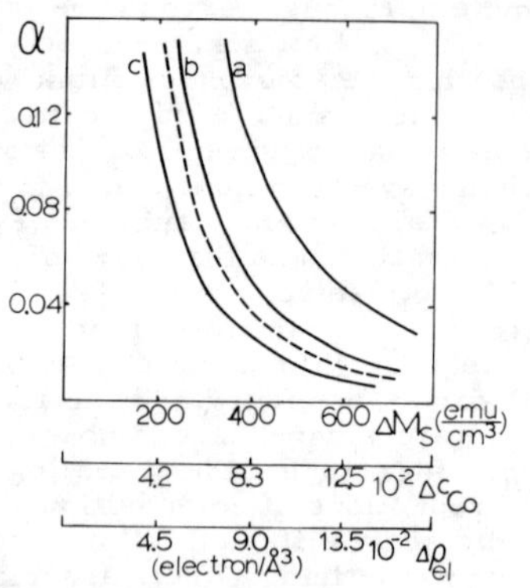


Fig. 5. (right) Solid lines: values of $\alpha$ and of $\Delta\rho_{el}$ for three choices of $A_1$ [(a) 4400 Å, (b) 10,000 Å, (c) 20,000 Å]. Broken line: values of $\alpha$ and $\Delta M_s$ consistent with observed $H_s$=70 Oe.

## CONCLUSIONS

Our x-ray SAS measurements have demonstrated that amorphous electrodeposited Co-P films with perpendicular easy axis magnetic anisotropy also have an anisotropic microstructure, which we describe in terms of oriented ellipsoidal regions of high (or low) Co concentration. Internal demagnetizing effects associated with this microstructure contribute to the observed magnetic anisotropy. Annealing these materials reduces both the SAS intensity and the magnetic anisotropy. We propose that these annealing effects result from atomic scale diffusion reducing anisotropic composition variations (microstructure) in the amorphous, electrodeposited films.

The authors are pleased to acknowledge helpful discussions with R. Alben and W. Kinney.

## REFERENCES

*Work supported by NSF.
+Present address: IBM T.J. Watson Research Center, Yorktown Heights, New York 10598.

1. G.S. Cargill III, R.J. Gambino, and J.J. Cuomo, IEEE Trans. Mag. MAG-10, 803 (1974).
2. T. Iwata, R.J. Prosen, and B.E. Gran, J. Appl. Phys. 37, 1285 (1966).
3. H. Maeda, J. Phys. Soc. (Japan) 29, 311 (1970).
4. G.C. Chi and G.S. Cargill III, "Annealing Effects in Amorphous Co-P Alloys," presented at Conference on Rapidly Quenched Metals, Cambridge, Mass., November, 1975; to be published in the Conference Proceedings.
5. G.S. Cargill III, Phys. Rev. Letters 28, 1372 (1972).
6. G.S. Cargill III and R.W. Cochrane, J. Phys. (Paris) 35, C4-269 (1974), and in Amorphous Magnetism, H.O. Hooper and A.M. deGraff, Eds. (Plenum Press, New York, 1973), p. 313.
7. We have not rigerously ruled out the possibility that the anisotropic scattering regions are voids within the films, which would be characterized by a very large $\Delta\rho_{el}$. However, the observed annealing effects, i.e. decreasing I(K) without changes of similar magnitude in $A_1$, $A_2$, or $A_3$, suggests that $\Delta\rho_{el}$ is more likely associated with composition inhomogeneities, i.e. with $\Delta c_{Co}$, than with actual voids or regions of low (or high) density (atoms per unit volume).
8. D. Pan and D. Turnbull, J. Appl. Phys. 45, 1406 (1974).
9. W. Jost, Diffusion in Solids, Liquids, Gases (Academic Press, New York, 1952), p. 45.
10. D. Gupta, K.N. Tu, and K.W. Asai, Phys. Rev. Letters 35, 796 (1975).

# COMMENT ON THE SPINWAVE SPECTRUM OF $Co_4P$ COMPARISON WITH A CONTINUUM THEORY OF AMORPHOUS FERROMAGNETISM

A. M. de Graaf,* Wayne State University, Detroit, MI 48202


## ABSTRACT

The continuum theory of amorphous ferromagnetism[1] which was developed in 1972, has been reevaluated. This theory generalizes the Landau-Lifshitz theory of ferromagnetic spinwaves by allowing the exchange constants to be spatially dependent. It is shown that this theory predicts long wavelength excitations of the form $E = D_m Q^2$ and $E = D_r(Q-Q_o)^2 + \Delta_Q$, where $D_m$ and $D_r$ are magnon and roton stiffness parameters, $Q_o$ the position of the roton minimum, and $\Delta_{Q_o}$ the roton gap. These features were observed in the amorphous ferromagnet $Co_4P$,[2] where $D_m$ = 185 meV Å$^2$, $D_r \simeq$ 185 meV Å$^2$, $Q_o$ = 3.13 Å$^{-1}$, and $\Delta_{Q_o}$ = 34 meV. It is shown that the roton branch is due to the randomness of the ferromagnet. It is also shown that the roton excitations are spatially damped magnon states. The experimental data indicate that the spatial extent of the damped states in $Co_4P$ is only a few atomic distances, rendering the continuum theory inapplicable.


## DISCUSSION

It was shown in Ref. 1 that integration of the energy density

$$\Delta E(\vec{r}) = \frac{1}{2} C_{ij}(\vec{r}) \frac{\partial M_\alpha}{\partial x_i} \frac{\partial M_\alpha}{\partial x_j} , \qquad (1)$$

associated with a long wavelength excitation above the ferromagnetic groundstate leads to the following equation of motion for the magnetization $\vec{M}(\vec{r})$

$$d\vec{M}(\vec{r})/dt = -\gamma \vec{M}(\vec{r}) \times \vec{H}(\vec{r}) ,$$

$$H_\alpha \equiv \left[\frac{\partial C_{ij}}{\partial x_i} \frac{\partial M_\alpha}{\partial x_j} + C_{ij} \frac{\partial^2 M_\alpha}{\partial x_i \partial x_j}\right] . \qquad (2)$$

The coefficients $C_{ij}$ are related to the exchange coupling constants, which in the amorphous ferromagnet are rapidly varying functions of the coordinate $\vec{r}$. Equation (2) is identical to the familiar Landau-Lifshitz equation, except for the first derivative term, which is due to the randomness of the amorphous ferromagnet.

We divide the volume of the sample in small volumes $\Omega = (2\pi/Q)^3$, where Q is the wavenumber of a spinwave excitation, and average Eq. 2 over $\Omega$. Since in the long wavelength limit the transverse magnetization varies little within $\Omega$, the averaged equation of motion for the transverse magnetization follows from Eq. 2 by replacing the rapidly varying coefficients by their averages over $\Omega$.

For values of Q close to zero, the average values of $(\partial C_{ij}/\partial x_i)$ and $C_{ij} (i \neq j)$ are zero, and the usual Landau-Lifshitz equation is obtained, giving rise to extended spinwaves with a quadratic dispersion relation. However, for values of Q close to $Q_o = 2\pi/a$ (a is an average lattice parameter) these averages are no longer zero, even though such Q values still represent long wavelength excitations. The modified Landau-Lifshitz equation can be solved approximately, and it is found that these spinwaves are spatially damped. The spatial extent d, and the dispersion relation of the damped spinwaves are, respectively[1]

$$d = 2C/\dot{C}$$
$$E = \gamma\hbar M_o [C(Q-Q_o)^2 + \dot{C}^2/2C] , \qquad (3)$$

where C is an average diagonal element of $C_{ij}$, $\dot{C}$ is the average spatial derivative of this element, and $M_o$ is the saturation magnetization. Comparison with the results of the $Co_4P$ experiment yields a = 2 Å and d = 3.3 Å. This implies that the transverse magnetization varies rapidly within $\Omega$, and the theory breaks down. In conclusion we have shown that the modified Landau-Lifshitz theory of spinwaves in amorphous ferromagnets predicts spatially damped roton-like excitations. The theory is only valid if the spatial extent of these excitations is large compared to the average interatomic distance.

A full account of this work together with a microscopic justification of the theory described here will be published elsewhere.


## ACKNOWLEDGMENT

The author wishes to thank R. Luzzi for helpful correspondence.


## REFERENCES

*Supported in part by the National Science Foundation.


[1]R.G. Henderson and A.M. de Graaf in Amorphous Magnetism, Plenum, 1973.

[2]H.A. Mook, et al. Phys. Rev. Letters 34, 1029 (1975).

# EFFECTIVE MEDIUM APPROXIMATION FOR FERROMAGNETIC SPIN WAVES IN AMORPHOUS SYSTEMS AND ALLOYS

L. M. Roth
SUNY/Albany, Albany, NY 12222

ABSTRACT

We have extended an effective medium approximation which was developed for the electronic structure of liquid metals and which is equivalent to the coherent potential approximation for alloys, to the problem of magnons in disordered ferromagnets. The magnon self-energy is obtained to first order in $1/\bar{z}$, where $\bar{z}$ is an effective number of nearest neighbors, and the propagator is renormalized in a self consistent manner. Comparisons are made with recent results on the diluted ferromagnet. The theory is readily applied to alloys and amorphous materials.

## INTRODUCTION

We have been studying the interesting problem of the magnon spectrum of disordered materials. This turns out to be a more difficult problem than the electronic structure, and is also related to the problem of phonons is disordered structures. Most work has been done on the diluted ferromagnet - the alloy in which one component is non-magnetic. Of interest also are amorphous structures and magnetic alloys.

There have been several attempts to apply the coherent potential approximation[1] (CPA) to magnons in the diluted ferromagnet. Despite the additivity of the simplest pseudopotential representing removal of a magnetic ion, the usual CPA arguments do not go through due apparently to the necessity of the magnon energy to vanish at zero wave number - a consequence of the Goldstone theorem. Harris et al[2] use a non-symmetric pesudopotential and do carry through a CPA for the simple cubic case. They find that the spectral function agrees reasonably well with moment calculations of Nickel.[3] Another important result is that of Tahir-Kheli[4] based on a T-matrix truncation shceme and which conserves frequency moments of the density of states through the third moment for the simple cubic lattice. There is also a recent effective medium calculation by Theumann & Tahir-Kheli.[4a]

There has been less work on amorphous systems - chiefly that of Gubernaites & Taylor[5] - using s somewhat different approach from the present work.

In this work we apply to this problem methods developed in dealing with the problem of electronic structure of liquid metals.[7,8] In that problem we obtained an effective medium approximation which is equivalent to the CPA.[7] In particular a perturbation approach showed that we have an expansion in $1/\bar{z}$, where $\bar{z}$ is an effective neighbor number. We emphasize the $1/\bar{z}$ aspect of the result in our treatment of the magnon problem.

## THEORY

The Hamiltonian we use for a single magnetic component is

$$H = \sum_{ij} H_{ij} a_i^+ a_j \quad (1)$$

$$H_{ij} = \sum_{\ell} J_{i\ell}\delta_{ij} - J_{ij} \quad (2)$$

where $a_i^+$ creates a spin deviation on site i. The sites are distributed in some amorphous manner, or correspond to a randomly chosen fraction of the sites of a lattice. We construct for the amorphous case the continuum Green's function[6,7]

$$G(R-R') = \langle \sum_{ij} \delta(R\ R_i)\, \mathcal{G}_{ij}\, \delta(R'-R_j) \rangle \quad (3)$$

where $\mathcal{G}_{ij}$ is the matrix magnon Green's function obeying the equation

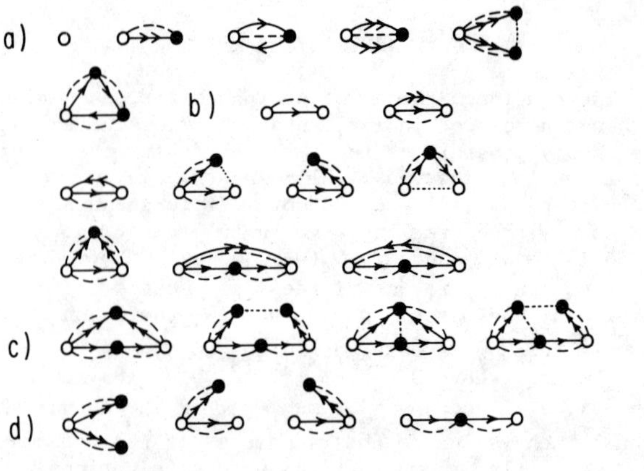

Figure 1. Diagrams in the expansion of G(R-R'). All diagrams through second order are included, and selected higher order diagrams.
(a) Closed irreducible diagrams.
(b) Open irreducible diagrams.
(c) Open irreducible diagrams in 4th order. The last two contribute to $M_{1k}$.
(d) Reducible diagrams in second order.

$$\omega \mathcal{G}_{ij} = \delta_{ij} + \sum_{\ell} H_{i\ell} \mathcal{G}_{\ell j} \quad (4)$$

and the brackets represent a configurational average. A perturbation expansion for G is depicted by the diagrams in Fig. 1, where the notation follows that of the liquid metal work.[7] Vertices represent distinct ions with each contributing a density factor n and some power of $1/\omega$ according to the perturbation sequence. Arrows represent -J(R-R'). The double arrow corresponds to J(R-R') from the first term of Eq.(2) and is new for the magnon case. Correlation between site positions is expressed in a Kirkwood approximation by including a pair distribution function g = 1 + h (g---, h...) for every pair of points (h-lines are used between points not connected by J-lines). Intermediate vertices (closed circles) are summed over.

The Green's function when Fourier transformed can be expressed in the form

$$G_k = \frac{n}{\omega - nM_0 + nM_k} \quad (5)$$

where $nM_0$ is a sum of irreducible diagonal self energy parts as shown in Fig.1a, and $-nM_k$ a sum of irreducible transfer parts as shown in Fig.1b, c.

We notice that if we integrate over R' in the diagrams of Fig.1b, (or equivalently set k=0) there is a term by term cancellation among diagrams, including these in Fig.1a, which have the same topological structure. This is a consequence of the Goldstone theorem and justifies our using $M_0 = M_{k=0}$ in Eq.(5). A similar cancellation occurs for all terms in the expansion of G - except for $1/\omega$.

We use a $1/\bar{z}$ argument and choose diagrams to sum with a maximum number of points in each order and furthermore renormalize vertices (i.e. $1/\omega \rightarrow 1/\omega - nM_0$) and transfer parts. This gives (if $\tilde{J}(R) = g(R)J(R)$)

$$M_k = \tilde{J}_k + \int \Big[ 2J_{k-k'} + 2nh_{k-k'}\tilde{J}_{k-k'} + \frac{n(1+nh_{k-k'})}{\omega - nM_0} J_{k-k'}^{\,2} \Big] \frac{M_{k'}}{\omega - nM_0 + nM_{k'}} \frac{dk'}{8\pi^3} - n \int \frac{h_{k-k'} M_{k'}^{\,2}}{\omega - nM_0 + nM_k} \frac{dk'}{8\pi^3} + M_{1k} \quad (6)$$

Here $M_{1k}$ corresponds to terms like the last two diagrams of Fig.1c, and has only been worked out in detail for the simple cubic case discussed next.

To compare with other work we consider the case of a lattice - the diluted ferromagnet - in which case we can use Eq.(6) with $h_k \to -1$, $n \to (1-x)$ where x is the concentration of non-magnetic ions. We use in particular the simple cubic nearest neighbor model. Then we find

$$(M_o - M_k) = 6J(1-\gamma_k)\left(1 - \frac{2x}{1-x}\frac{G_{100}}{G_{000}}\right) - \frac{x}{1-x}6J^2[(1-\gamma_{2k})(G_{200}+2G_{110}-3G_{100}{}^2/G_{000}) + 3(1+\gamma_{2k} - 2\gamma_k{}^2)(G_{110}-G_{100}{}^2/G_{000})] \tag{7}$$

where $\gamma_k$ = 1/3 (cos $k_xa$ + cos $k_ya$ + cos $k_za$) and the $G_R$ correspond to a site representation for G.

## DISCUSSION

In comparing this result with the above-mentioned theories we find: (1) To order x, Harris et.al. obtain the exact self energy, while the present result omits $1/z^2$ corrections - where here z=6. (2) The present result conserves moments of the spectral function through the second, while Tahir-Kheli's result conserves the third moment. This seems to be related also to a $1/z^2$ effect. Harris et.al. find that their result gives only one moment exactly. Tahir-Kheli's result contains only a term analagous to the first term of Eq.(7), representing first neighbors. We find that the second and third neighbor terms which contribute only to fourth or higher moments for the simple cubic lattice, are of the same order in 1/z as the first neighbor correction. The Theumann & Tahir-Kheli result, which is exact for $x \to o$, includes both types of terms and also conserves the second moment of the spectral function.

We have carried through the self-consistent calculation for the three independent parameters in Eq.(7)

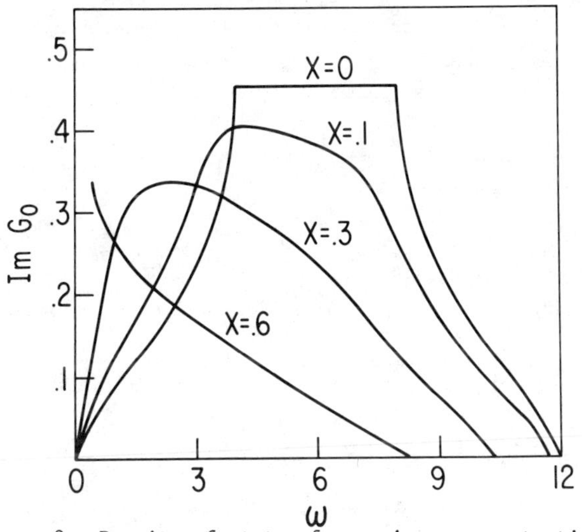


Figure 2. Density of states for various concentrations of non-magnetic ions for the diluted simple cubic ferromagnet.

and have used these to obtain the density of states ($n(\omega) = -1/\pi \, \mathrm{Im}G_{000}$) curves shown in Fig. 2. The results are similar to previous work. A detailed comparison of the k-dependent spectral function with the results of Nickel is in progress. We hope to find out whether our theory is accurate for this case, or whether other terms should be included (for example to give the correct result for $x \to o$). After this analysis we shall obtain results for amorphous systems and also study alloys by using methods worked out in the liquid metal work.[7]

## REFERENCES

1. An excellent review is R.J. Elliott, J.A. Krumhansl and P.L. Leath, Rev. Mod. Phys. 46, 465-543 (1974).
2. A.B. Harris, P.L. Leath, B.G. Nickel and R.J. Elliott, J. Phys. C 7, 1693-1719 (1974).
3. B.G. Nickel, J. Phys. C 7, 1719-1734 (1974).
4. R.A. Tahir-Kheli, Phys. Rev. B6, 2808-2825 (1972). See also A. Theumann, J. Phys. C 6,2822 (1973).
4a. A. Theumann and R.A. Tahir-Kheli, Phys. Rev. B12, 1796, (1975).
5. J.E. Gubernaitis & P.L. Taylor, Phys. Rev. B9, 3828 (1974).
6. L.M. Roth, Phys. Rev. B6, 2476-2484 (1974).
7. L.M. Roth, Phys. Rev. B11, 3769-3779 (1975), F. Yonezawa, L.M. Roth and M. Watabe, J. Phys. F 5,435-442 (1975).

# Bethe-Peierls-Weiss Approximation in Disordered Ferromagnets

K. Moorjani* and S. K. Ghatak
Groupe des Transitions de Phases
C.N.R.S., Grenoble 38042, France

ABSTRACT

The behavior of ferromagnetic atoms in a structurally disordered non-magnetic host is analyzed within the Bethe-Peierls-Weiss (BPW) approximation. The structural disorder, which is assumed to cause fluctuations in the exchange interactions, is found not to affect the value of the concentration of magnetic atoms ($C_m$) for the onset of ferromagnetism; however, above $C_m$, the Curie temperature relative to the Curie temperature of the averaged system is reduced by such fluctuations. The special case of structural disorder alone is studied within the same approximation and our conclusions are compared with earlier results in the literature.

## A. MODEL

In the spirit of the BPW approximation, we consider a cluster of (z+1) atoms; an atom at the origin with spin $S_o$ which interacts via the exchange integrals $J_{oi}$'s with its z nearest neighbor spins, $S_i$'s, located on the surface of the cluster. The rest of the medium is simulated by an effective internal field acting on the surface atoms. In the presence of an external applied field, $H_o$, the Heisenberg Hamiltonian can be written as,

$$\mathcal{H} = -2S_o \cdot \sum_{i=1}^{z} J_{oi}S_i - g\mu_B H_1 \cdot \sum_{i=1}^{z} S_i - g\mu_B H_o \cdot S_o \qquad (1)$$

$H_1$ being the sum of $H_o$ and the internal field due to the rest of the medium; g and $\mu_B$ are the g-factor and the Bohr magneton respectively. The translational invariance of a one component crystalline solid requires that all the interaction parameters $J_{oi}$'s be equal and the BPW analysis of Eq.(1) then leads to an expression for ordering temperature which is a function of the single exchange parameter, the value of the spin and the number of nearest neighbors. For a chemically and/or structurally disordered solid, however, $J_{oi}$'s will depend on the location and the species of atoms in the cluster and in general could take on z different values. The partition function needs to be averaged over all these possible configurations before the quantities of interest are derived from it. To avoid a great deal of cumbersome mathematics inherent in configuration averages, we make a simplifying assumption concerning the nature of structural disorder. We assume that the structural disorder causes the central magnetic atom to interact with some of its magnetic nearest neighbors via the exchange parameter $J_{oi}(=J_1)$ and via the interaction $J_{O2}(=J_2)$ with the rest of the magnetic n.ns. Such an assumption is not expected to change the general conclusions of the present investigation.

Lastly, since for the crystalline ferromagnetic lattice, the Curie temperatures obtained by treating the spins as classical vectors are within a few percent of those calculated quantum mechanically for $S \geqslant 1$ [1] we employ the classical approximation to calculate the partition function.

## B. METHOD

The procedure followed to obtain the Curie temperature for the system described above is quite straightforward. The classical partition function is given by $Z_{cluster} = \int e^{-\beta\mathcal{H}} d\Omega$, the integration being over the phase space available for the system; $d\Omega = \prod_{i=1}^{z+1} d\Omega_i$, where $d\Omega_i$ is the element of solid angle in the direction $S_i$. For a disordered solid, the partition function needs to be averaged over all possible configurations of the system to obtain $Z = \langle Z_{cluster}\rangle_{conf.avg.}$. The details of the configuration average are discussed in the next section.

The average magnetic moments, $M_o$ and $M_1$ of the central and surface atoms respectively are determined from

$$M_o = g\mu_B\langle S_o\rangle = \frac{1}{\beta}\frac{\partial \ln Z}{\partial H_o} \text{ and } M_1 = g\mu_B\langle S_1\rangle = \frac{1}{\beta z}\frac{\partial \ln Z}{\partial H_1} \qquad (2)$$

and by invoking the self-consistency condition $M_o = M_1$, one obtains an expression which determines $H_1$. The critical temperature $T_c$ is obtained by requiring that $H_1$ be finite in the absence of any external field.

## C. CALCULATION

In contrast to the crystalline case treated by Smart [2], not only must we consider the clusters with magnetic or non-magnetic atom at the center and k magnetic atoms on the surface, but, due to the presence of structural disorder, the probability $\rho(k;k_1k_2)$ that $k_1$ of the magnetic atoms have an interaction $J_1$ with the central magnetic atoms while $k_2 = k-k_1$ have an interaction $J_2$ must also be included in the evaluation of the partition function. If $Z^{(k)}_{mag:}(k_1,k_2)$ represents the partition function for such a configuration, then the partition function $Z^{(k)}_{mag:}$ corresponding to the configuration with a magnetic central atom and k magnetic surface atoms is simply given by

$$Z^{(k)}_{mag:} = \prod_{k_1=o}^{k} \left\{Z^{(k)}_{mag:}(k_1,k_2)\right\}^{\rho(k;k_1,k_2)} \qquad (3),$$

$$\text{where } \rho(k;k_1,k_2) = \frac{1}{2^k}\,\frac{k!}{k_1!(k-k_1)!} \qquad (4).$$

Introducing the dimensionless variables, $\lambda_i = \beta g\mu_B H_i$, (i=0,1); $p_i = \beta J_i$, (i=1,2); $\alpha_i = \mathrm{Sinh}\,2p_i S(S+1)/2p_i S(S+1)$, (i=1,2); one finds by straightforward calculation that up to $\lambda_1^2$

$$Z^{(k)}_{mag:}(k_1,k_2) = \alpha_1^{k_1}\alpha_2^{k_2}\Big[1 + \frac{1}{3}\lambda_o\lambda_1 S(S+1)(k_1L_1+k_2L_2) + \frac{1}{6}S(S+1)\lambda_o^2 + \frac{1}{3}\lambda_1^2 S(S+1)k_1k_2L_1L_2 + \frac{1}{6}S(S+1)\lambda_1^2 \times\left\{k_1[1+(k_1-1)L_1^2] + k_2[1+(k_2-1)L_2^2]\right\}\Big] \qquad (5),$$

where $L_i$'s(i = 1,2) are the Langevin functions,

$$L_i = \mathrm{Coth}[2\beta J_i S(S+1)] - \frac{1}{2\beta J_i S(S+1)} .$$

The corresponding expression for the configuration with non-magnetic central atom and k magnetic surface atoms is simply given by

$$Z^{(k)}_{n.mag:} = \text{const.}\left\{1 + \frac{S(S+1)}{6}k\lambda_1^2\right\} \qquad (6).$$

The rest of the arguments below for the calculation of

total partition function are analogous to those for the crystalline case (2). If c and (1-c) are the respective concentrations of magnetic and non-magnetic atoms, then the probability of $\xi_z^{(k)}$ of having k magnetic atoms on the surface is given by

$$\xi_z^{(k)} = \frac{z!}{k!(z-k)!} \quad c^k(1-c)^{z-k} \qquad (7)$$

and, since with k surface magnetic atoms there will be $c\,\xi_z^{(k)}$ clusters with a magnetic central atom and $(1-c)\xi_z^{(k)}$ clusters with a non-magnetic central atom, the total partition function is given by,

$$Z = \prod_{k=0}^{z} \left[Z_{mag:}^{(k)}\right]^{c\xi_z^{(k)}} \left[Z_{n.mag:}^{(k)}\right]^{(1-c)\xi_z^{(k)}}. \qquad (8)$$

The Eq$^{ns}$. (3) - (7) are now substituted in Eq.(8) and, following the procedure outlined in Sec.B, one obtains the self-consistency condition,

$$\frac{cz}{2}(L_1 + L_2) = 1 + (z-1)c^2 L_1 L_2 + (z-1)\frac{c^2}{4}\left(L_1 - L_2\right)^2. \quad (9)$$

## D. RESULTS AND DISCUSSION

We first note that above equation (9) has the correct limits; in the absence of structural disorder ($J_1 = J_2$, i.e., $L_1 = L_2$), the last term vanishes and one recovers the results discussed by Smart (2). If one further takes the limit of no chemical disorder, i.e., c = 1, one obtains the results of Brown and Luttinger (1). For the general case, Eq.(9) can be solved by letting $L_m = \frac{L_1 + L_2}{2}$ and $L_d = \frac{L_1 - L_2}{2}$. The resulting equation is a quadratic in $cL_m$ with only one physically acceptable solution $cL_m = (z-1)^{-1}$. Thus the value of the critical concentration, $C_m$, of magnetic atoms for the onset of ferromagnetism is given by $C_m = (z-1)^{-1}$ and is unaffected by fluctuations in J. For $c > c_m$, the behavior of the Curie temperature of the non-crystalline solid $T_c^{n.c.}$ as a function of the disorder parameter $\alpha = J_1 - J_2/J_1 + J_2$ is approximately given by the expression

$$T_c^{n.c.} \simeq T_c^c \left[1 - \frac{12}{5}\frac{1}{c^2(z-1)^2}\alpha^2\right], \quad c > c_m; \alpha << 1 \quad (10)$$

where $T_c^c = c(z-1)(J_1 + J_2)S(S+1)/3k$ is the Curie temperature of the crystalline solid with a mean exchange interaction $\frac{J_1 + J_2}{2}$. The fluctuations thus depress the value of $T_{c.}^{n.c.}$ relative to the mean-field value. Such a conclusion is in agreement with the exact results (3) based on the Ising model and other earlier work (4-7).

We have previously discussed the case of a structurally disordered Heisenberg ferromagnet (8). However, the earlier formulation contained a spurious periodicity in the sense that the disordered solid was composed of only one particular configuration of atoms in the cluster. In the present analysis, this restriction is removed by proper configurational average and the results are easily obtained from Eq.(10) by putting c=1. In this limit, our results are most easily compared with those of Brooks Harris (7) who finds an expression similar to Eq.(10) with $a = -\left.\frac{d\ell n T_c^{n.c.}}{d\alpha^2}\right|_{\alpha^2=0} = 0.073$ for the three dimensional simple-cubic Ising model in contrast to the mean-field value of $a = 0$. The present method gives the value $a = 0.096$ for the simple-cubic lattice.

Acknowledgments: It is our pleasure to thank Dr. B. Chakraverty and other members of the Phase Transition Group for their kind hospitality.

## REFERENCES

*Present address: The Johns Hopkins University, Applied Physics Laboratory, Laurel, Maryland 20810.
Work partially supported by Naval Sea Systems Command Contract.

1. H. A. Brown and J. M. Luttinger, Phys. Rev. 100, 685 (1955).
2. J. S. Smart, J. Phys. Chem. Solids 16, 169 (1960).
3. H. Falk and G. A. Gehring, J. Phys. C8, L298 (1975).
4. K. Handrich, Phys. Status Solidi 32, K55 (1969); ibid 44b, K17 (1971).
5. C. G. Montgomery, J. I. Krugel and R. M. Stubbs, Phys. Rev. Lett. 25, 669 (1970).
6. T. Kaneyoshi, J. Phys. C6, 3180 (1973).
7. A. Brooks Harris, J. Phys. C7, 1671 (1974).
8. S. K. Ghatak and K. Moorjani, Solid State Comm. 17, 923-925, (1975).

# SPONTANEOUS MAGNETIZATION OF AMORPHOUS FERROMAGNETS WITH ANISOTROPY

J. D. Patterson* and R. C. Weger
South Dakota School of Mines and Technology, Rapid City, South Dakota 57701

ABSTRACT

The mean field approximation was used to calculate by **iteration** (for simple cubic lattices of up to 1728 spins) the spontaneous magnetization of a spin one ferromagnet in which the amorphousness is simulated by fluctuating the magnitude of the uniaxial anisotropy term (of strength D and average value $\bar{D}$). **Fluctuations with $\bar{D}$ = 0 as well as about a positive** value of $\bar{D}$ were considered. Nearest neighbor (n.n.) **Ising exchange interactions** with periodic boundary conditions were used. In the approximation considered, each spin interacted with the local mean field created by its 6 n.n.'s and it also experienced a randomly fluctuating uniaxial anistropy. Generally, the fluctuations caused a decrease in the magnetization from the crystalline mean field case. For $\bar{D}$ =0 convergence problems were encountered near the Curie temperature. By examining both the free energy and the magnetization it is shown that increasing $\bar{D}$ suitably will cause a first order transition. For still higher values of $\bar{D}$, there is no thermodynamically stable magnetized state. The magnetization curve depends also on the amplitude of the D fluctuations.

## I. INTRODUCTION

Several workers have used a mean field approximation in discussing amorphous magnetism [1-4]. Handrich and Kobe [1] have done several mean field calculations in which amorphousness is simulated by allowing the exchange coupling between spins to fluctuate. In one calculation, they use an iterative mean field approximation on a finite lattice of spins (of spin 1/2). Our calculation is similar to that of Kobe and Handrich, but we emphasize the fluctuation of the strength of a uniaxial anisotropy term. We consider a simple cubic lattice of spins with spin one and with n.n. Ising exchange coupling and periodic boundary conditions. Each spin is also assumed to experience a single ion anisotropy of fixed direction but varying magnitude. In the mean field approximation the effect of the exchange coupling is to introduce, for each spin, a local molecular field due to the six n.n.'s. At very low temperatures (T) we initialized each of the quantum statistically averaged z-components of spin to have the physically reasonable (for our choice of parameters) value of 1. We then iterated throughout a finite lattice of spins until the values for the statistically averaged z-components, calculated at T, were self consistent. The temperature was increased to T + ΔT and the procedure was repeated with the initial values being the self consistent values at T. In order to test for thermodynamic stability at each step, the free energy was calculated and compared to the free energy of the unmagnetized solution. Our calculations were done on lattices of $10^3$(=1000) and $12^3$(=1728) spins. At least for temperatures fairly well below the Curie temperature, the lattice with 1000 spins gave essentially the same results for the overall average magnetization (defined below) as was obtained with the lattice of 1728 spins for the same physical situation. The magnetization in this temperature range was also essentially unchanged (for either lattice) by using different uniformly selected random sets of anisotropy constants (with fixed average value and fixed maximum fluctuation).

## II. MEAN FIELD APPROXIMATION

We write the Hamiltonian, for n spins, as

$$H_1 = \frac{-1}{2}\sum_{i,j=1}^{n} J_{ij} S_i^z \cdot S_j^z - H\sum_{i=1}^{n} S_i^z + \sum_{i=1}^{n} D_i (S_i^z)^2 \quad (1)$$

where $S_i^z$ denotes the spin operator at site i, and the exchange constants $J_{ij} \equiv J_{ji}$ are assumed to be zero unless i and j refer to nearest neighbor spins. Eq. (1) can be considered to result from a Heisenberg interaction with an appropriate large exchange anisotropy. The second term on the right hand side of this equation represents the Zeeman energy with the magnetic field H (in suitable units) along the z-axis. The third term is the uniaxial anisotropy term. In general, $D_i$ may vary from site to site and $J_{ij}$ may vary from nearest neighbor pair to nearest neighbor pair. Assuming the z-direction is the direction of magnetization, this Hamiltonian becomes in the mean field approximation,

$$H_2 = \sum_{i=1}^{n} h_i \quad (2a)$$

where

$$h_i = -(\lambda_i + H)S_i^z + D_i (S_i^z)^2 \quad (2b)$$

$$\lambda_i = \sum_{j=1}^{n} J_{ij} \langle S_j^z \rangle \quad (2c)$$

and $\langle S_j^z \rangle$ refers to the quantum statistical average of the operator $S_j^z$. We assume all spins are spin 1. Setting $\hbar$ = 1, the energy eigenvalues of Eq. (2b) are

$$E_i(m_i) = -(\lambda_i + H)m_i + D_i m_i^2 , \quad (3)$$

where $m_i$ = 1,0,-1.

The total partition function for the system ($Z_T$) is equal to the product of the $Z_i$'s, where

$$Z_i = \sum_{m=-1}^{1} e^{-\beta E_i(m)} , \quad (4)$$

$\beta = (kT)^{-1}$, k is Boltzmann's constant and T is the temperature. The local magnetization at the site j is given by

$$\langle S_j^z \rangle = \sum_{m=-1}^{1} e^{-\beta E_j(m)} m/Z_j , \quad (5a)$$

and the overall magnetization is defined by

$$M = \frac{1}{n}\sum_j \langle S_j^z \rangle . \quad (5b)$$

The free energy F can be calculated from the definition F = U - TS where U is the internal energy and S is the entropy. We obtain, in the mean field approximation,

$$F = -\frac{1}{\beta}\sum_j \ln Z_j + \frac{1}{2}\sum_{i,j} J_{i,j} \langle S_i^z \rangle \langle S_j^z \rangle . \quad (6)$$

The free energy is an extremum when Eq. (5a) is used to determine the local magnetization at each site.

We consider only solutions for the spontaneous magnetization when the magnetic field is zero. Notice that $\langle S_j^z \rangle = 0$ (for all j) is always a solution of (5a). By Eqs. (4) and (6), we can calculate the free energy $F_0$ for this case. It's worthwhile to examine if $\langle S_j^z \rangle = 0$, for all j, could be a stable solution at absolute zero. For simplicity, we assume that all the $J_{ij}$'s and $D_j$'s are positive. As $\beta \to \infty$, in the magnetized case, we assume $\langle S_j^z \rangle \to 1$ for all j. Stability of the magnetized solution, characterized by F, requires $F_0 > F$. For a simple cubic lattice we find that this implies $\bar{D} \leq 3\bar{J}$, where $\bar{D}$ and $\bar{J}$ are the averages over the lattice of $D_i$ and $J_{ij}$. For the crystalline case ($J_{ij}$(n.n.) = J, $D_j$ =D) it's easy to show for all D<3J (including negative D with J>0) that a magnetized solution will have lower free energy than the unmagnetized solution.

Even in the range $\bar{D} < 3\bar{J}$ we have the possibility of either a first or a second order transition. For the crystalline case, the critical ratio of D/J for which the system changes from first order to second order is readily obtainable [5]. Expanding Eq. (5a) in $\langle S_z \rangle$ and using that $\langle S_z \rangle \to 0$ as $\beta \to \beta_c$ (i.e. at Curie temperature) we find that the Curie temperature for a second order transition is determined by

$$e^{\beta_c D} + 2 = 12\beta_c J \ . \qquad (7)$$

The critical ratio of D/J will occur when the $\beta_c$ derivative on the right hand side of Eq. (7) equals the $\beta_c$ derivative on the left hand side. Using the resulting equation as well as Eq. (7) we find that the critical ratio is $(D/J) \approx 2.78$. Thus for $D/J < 2.78$ we have a second order transition with the Curie temperature determined by Eq. (7). For $2.78 < D/J < 3$ we have a first order transition with the Curie temperature obtained by setting $F = F_o$. For $D/J > 3$ no magnetized solution is stable.

III. RESULTS AND DISCUSSION

All our calculations were with the magnetic field set to zero. In Fig. 1, we have plotted the magnetization (determined by Eqs. (5)) versus the temperature for a case in which $J = 1/2$ and $D_i$ randomly and uniformly fluctuates between +1 and -1. The Curie temperature for the crystalline case also shown is (in our units) $T_c \equiv (1/\beta_c) = 4J = 2$. The free energy of the amorphous magnetized solution was calculated to be lower than the free energy of the unmagnetized amorphous solution for every temperature for which a point was plotted. This is also true for Fig. 2. The "error bars" on the magnetization for the amorphous case were obtained by plotting the results for three different sets of randomly selected $D_i$'s. Two of the sets were for a lattice of 1728 spins and one of the sets was for a lattice of 1000 spins. If only the two 1728 spin runs had been included then the bars would be somewhat shorter, but they still would show the property of increasing in length as the Curie temperature is approached. The calculated magnetization for the amorphous case ceased to converge after 100 iterations at a temperature just above the temperature of the last plotted point. For $T>0$, but fairly well below $T_c$, the magnetization of the amorphous case is clearly below that of the crystalline case. Due to the increasing uncertainty of our results as $T_c$ is approached from below we cannot extrapolate accurately to a Curie temperature for the amorphous case A smaller fluctuation of D led to a smaller reduction of the magnetization at all temperatures for which calculations were done.

Fig. 2 is for the $J = 1/2$ and $\bar{D} = 1.49$ ($.99 \leq D_i \leq 1.99$) case. The crystalline case, also shown, has $\bar{D}/J = 2.98$ and shows a first order transition. As

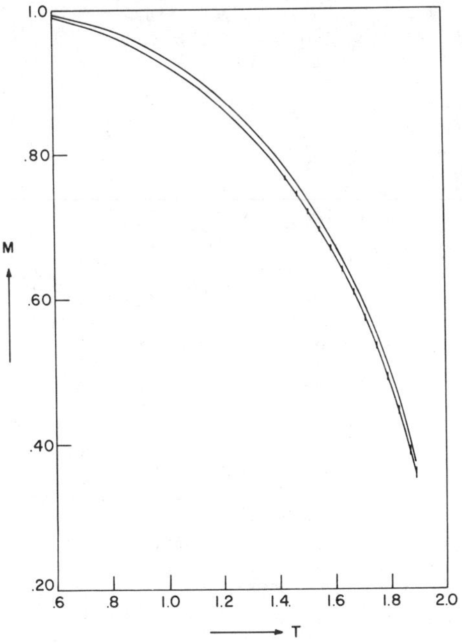


Fig. 1. M vs T. Upper curve is crystalline case ($J=1/2$, $D=0$) and lower curve is amorphous case ($J=1/2, -1 \leq D_i \leq 1$).

shown by an evaluation of the free energy, the amorphous case also undergoes a first order transition just above the last plotted point. This calculation indicated the necessity of evaluating the free energy. We stopped plotting at a temperature where the free energy of the magnetized solution became more than the free energy of the unmagnetized solution. However, the iterative procedure continued to produce magnetized solutions (presumably metastable) until considerably higher temperatures. We notice here that the fluctuations in D decrease the Curie temperature by about 20%. The uncertainty of our results is rather small at all temperatures, which we might expect for a well defined first order transition.

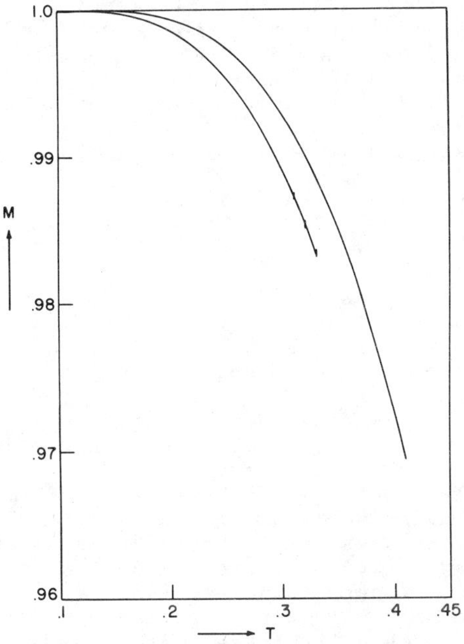


Fig. 2. M vs T. Upper curve is crystalline case ($J=1/2$, $D=1.49$) and lower curve is amorphous case ($J=1/2, .99 \leq D_i \leq 1.99$).

We should state our convergence criterion. After selecting initial values at some temperature, we iterated until the sum of the absolute values of the differences between $\langle S_j^z \rangle$ on one cycle and $\langle S_j^z \rangle$ on the next cycle was less than E. If this couldn't be accomplished in 100 or less iterations we said the method failed to converge. We tried different E values ($E = 10^{-7}$, $10^{-9}$) and the results for the magnetization did not differ at all except very slightly at temperatures very close to where the method ceased to converge entirely. We used $E = 10^{-7}$ for most of our runs. Also, in iterating to find the value of $\langle S_j^z \rangle$ for a particular j, we used the best available value for the other $\langle S_i^z \rangle$'s which were nearest neighbors of site j.

One author (JDP) would like to thank Prof. R. A. Tahir-Kheli for his support of this work.

REFERENCES

*Partially supported by NSF Grant #39023

1. S. Kobe and K. Handrich, Fiz. Tverd Tela 13, 887 (1971) [Soviet Phys.-Solid State 13, 734 (1971)] and references cited.
2. T. Kaneyoshi, J. Phys. C: Solid State Phys. 6, 3130 (1973).
3. G. B. Taggart, R. A. Tahir-Kheli, and E. Shiles, Physica 75, 234(1974) and references cited.
4. E. Shiles, AIP Conference Proceedings 24, 250 (1975) and references cited.
5. M. Blume, Phys. Rev. 141, 517(1966) does a similar calculation.

# A PHASE DIAGRAM FOR AMORPHOUS MAGNETIC ALLOYS*

R. Harris and D. Zobin, Eaton Electronics Laboratory,
McGill University, Montreal, Quebec, H3C 3G1, Canada

ABSTRACT

We have constructed a phase diagram for the magnetic behaviour of an amorphous metallic alloy using ideas suggested by recent work of Sherrington and Southern. We propose that a spin-glass state may be stabilized by the presence of a local anisotropy field of random orientation and constant magnitude, D, rather than by fluctuations of the exchange constant J. Calculations have been performed for S=1 and the classical case of S→∞ with several values of D, and over a wide range of temperature. The magnetization curves show the existence of a well-defined spin-glass transition temperature and a sizeable coercive field, which are both functions of D. The results are shown to be in qualitative agreement with experiments on amorphous $REFe_2$ ferromagnetic alloys.

## INTRODUCTION

Significant progress has recently been made in the study of ferromagnetism in amorphous materials [1]. One approach is based on Gubanov's ideas [2] and uses phenomenological fluctuation of the Heisenberg coupling constant J to explain values of Curie temperature and saturation magnetization obtained for various amorphous ferromagnets. A similar physical idea, expressed in much more elegant mathematical formulation, was used by Edwards and Anderson [3] and later by Sherrington and Southern [4] and Fischer [5] to explain the cusp found experimentally in the susceptibility of the so-called 'spin glasses'. In reference 4, an investigation is presented of the conditions for spin-glass or ferromagnetic ordering leading, for example, to predictions about the ranges of concentration within which each type of ordering should be found.

A different model, assuming the existence of a random 'crystal field' in amorphous materials [6], but taking no account of any fluctuations in exchange coupling, was quite successful in explaining some of the magnetic properties of amorphous rare-earth-iron ($REFe_2$) alloys. In particular, the model explained in a natural way the large decreases of Curie temperature and spontaneous magnetization compared with the values in the corresponding crystalline compounds. However, neither this nor any other of the existing models is capable of explaining the huge coercivity observed at low temperatures in amorphous $TbFe_2$ and $DyFe_2$ [7]. Because of its unusually high value and its temperature dependence [8], we believe this to be an intrinsic property of the amorphous compounds, and not a domain effect. The temperature dependence of the low-field susceptibility, $\chi(o)$, of amorphous $TbFe_2$ and $DyFe_2$ [8] is also rather unusual, and is reminiscent of that characteristic of the spin glasses. In particular, it has a maximum at some well-defined temperature, which by analogy with the spin glasses we might call the spin-glass temperature.

To explain this behaviour, we have generalized the random anisotropy model using ideas suggested by reference 4. Our results show magnetization curves and a magnetic phase diagram which are in qualitative agreement with experiment.

## MODEL

As in our earlier work [6,9] we assume that in amorphous rare-earth-iron alloys there exists a 'crystal field' which varies from point to point in the material, in a manner caused by the topologically disordered structure. The magnetic interaction is described by a simple Heisenberg model with the exchange coupling being of RKKY-type. The total Hamiltonian we use is then of the following form:

$$\mathcal{H} = \sum_i V_i - \frac{1}{2} J \cdot \sum_{i,j} \bar{S}(i) \cdot \bar{S}(j) \qquad (1)$$

Here, $V_i$ is a local single-ion anisotropy field at the site i, $\bar{S}(i)$ is the angular momentum operator for the magnetic ion on the site i, and J is a Heisenberg nearest-neighbour coupling constant. Various forms of $V_i$ have been considered in previous work [6,9]. In this paper we choose the simplest uniaxial form, $V_i = -D[S_{Z_i}(i)]^2$, where $Z_i$ refers to the local direction of anisotropy at site i. Previously [6] the Hamiltonian (1) was analyzed in the Hartree-Fock approximation with a molecular field in a unique Z direction, yielding values for the Curie temperature and the reduced magnetization in qualitative agreement with experiment, but failing to produce the correct shape of the magnetization curves. To make the treatment of (1) more general, we consider the molecular field at each site to be the vector sum of two fields, one which is again constant in magnitude and directed along the macroscopic magnetization vector, and the other, a local component of the exchange field which is allowed to vary from site to site. Introduction of such local components of the exchange field appears to be the simplest way to take into account the fluctuations undoubtedly present in a disordered material. The Hamiltonian can therefore be written as

$$\mathcal{H} = -\sum_i \Big\{ D[S_{Z_i}(i)]^2 + (H_Z^{mol} + H_Z^{appl}) \cdot S_Z(i) + \bar{H}_i^{loc} \cdot \bar{S}(i) \Big\} \qquad (2)$$

It should be emphasized here that the appearance of local components of the exchange field is entirely due to the misalignment of spins produced by the random anisotropy field, and consequently, they have the greatest magnitude at T=0°K.

## RESULTS AND DISCUSSION

A number of different procedures might be adopted to investigate magnetization as a function of applied field for the Hamiltonian (2). All involve some approximations to reduce the problem to a manageable form. We choose to assume that local components of the exchange field have the same magnitude throughout the amorphous sample, but are random in direction. Further, in an attempt to reduce the amount of computing time required, without significantly altering the qualitative features of the model, we assume that the random anisotropy field, constant molecular field, and the local components of the exchange field are in the same plane. We can then determine $|\bar{H}^{loc}|$ selfconsistently for fixed D, $H_Z = H_Z^{mol} + H_Z^{appl}$, and T, substitute its values back into the simplified version of (2), and calculate the magnetization in the macroscopic Z direction.

Calculations have been performed for S=1 and for the classical case of infinite S, and some typical magnetization curves for S=1 are shown in figure 1. Also shown is a

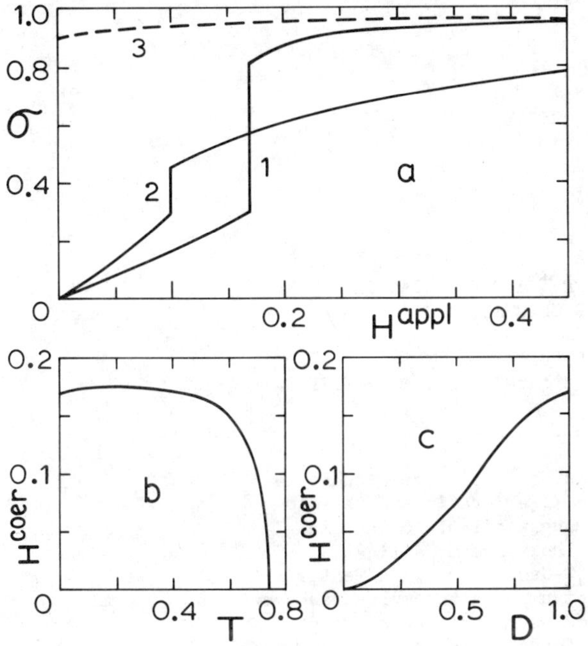


Figure 1. Reduced magnetization as a function of applied field for S=1 and D=1.0. Curve (a)(1) has T=0; (a)(2) T=0.7 and (a)(3) is taken from reference 6 (see text). Curve (b) shows $H^{coer}$ as a function of T for S=1 and D=1.0. Curve (c) shows $H^{coer}$ as a function of D for S=1 and T=0. All quantities are expressed in units of zJ.

magnetization curve corresponding to the simple model of reference 6. Both the S=1 and the classical curves exhibit a discontinuous change in the magnetization at a critical field which we identify as the coercive field, $H^{coer}$. Also shown in figure 1 are the temperature dependence of $H^{coer}$ for D=1.0 and its dependence at zero temperature on the value of D. The results for classical spins are qualitatively very similar.

There are several other important aspects of the effect of temperature on the magnetization curves. First of all, in the S=1 case, the low field susceptibility $\chi(o)$ shows an initial increase with temperature reaching a maximum at the spin-glass temperature. Also near this temperature the coercive field $H^{coer}$ vanishes, a fact which can be used as an alternative definition of the effective spin-glass temperature. In the classical case, which may be more appropriate to describe amorphous $REFe_2$ compounds because of the high values of angular momentum on the rare earth ions (S=6 for $Tb^{3+}$ or S=15/2 for $Dy^{3+}$), values for the spin glass temperature can still be given in terms of the vanishing coercive field, even though the peak in $\chi(o)$ is no longer observed.

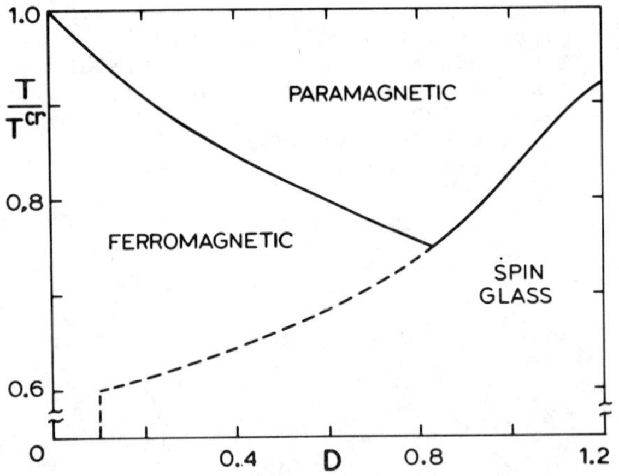


Figure 2. Magnetic phase diagram for S=1. The dashed line indicates the spin-glass transition temperature in the ferromagnetic phase.

We can thus construct a magnetic phase diagram, and distinguish three main regions, namely paramagnetic, ferromagnetic and spin-glass. Figure 2 shows such a diagram for S=1. The paramagnetic phase exists at high temperatures, and as T is lowered the system undergoes a phase transition to ferromagnetic state, if $D \lesssim 0.8$, or spin-glass state, if $D \gtrsim 0.8$. The ferromagnetic region can, in turn, be separated into two subregions. One, in which only ferromagnetic ordering is possible, corresponds to $D \lesssim 0.1$. In the other one, for $0.1 \lesssim D \lesssim 0.8$, both types of ordering are found, with the spin glass behaviour apparently stable at low temperatures. It is interesting to note that the phase diagram constructed here is qualitatively similar to that obtained by Harris [10] for the Heisenberg spin-glass model of reference 4 and by Sherrington and Kirkpatrick [11] for the exactly soluble infinite range Ising model.

Although our model is still not sufficiently realistic to provide a basis for quantitative comparison with experiment, we suggest that amorphous $TbFe_2$, for example, corresponds to D of about 0.7-0.8 in our phase diagram. In order to obtain quantitative agreement with the data, the model must be extended to include ferrimagnetic (as opposed to ferromagnetic) interactions between atoms of rare-earths and iron. We suggest, however, that magnetization measurement on amorphous rare-earth-noble metal or rare-earth-aluminum compounds would provide a simple test of the model.

REFERENCES

* Work supported by the National Research Council of Canada

1. G.S. Cargill III, AIP Conf. Proc. 24, 138-44 (1975).
2. A.I. Gubanov, Fiz. Tverd. Tela 2, 502-5 (1960) [Sov. Phys. Solid State 2, 468-70 (1961)].
3. S.F. Edwards and P.W. Anderson, J. Phys. F5, 965-74 (1975).
4. D. Sherrington and B.W. Southern, J. Phys. F5, L49-53 (1975).
5. K.H. Fischer, Phys. Rev. Lett. 34, 1438-41 (1975).
6. R. Harris, M. Plischke and M.J. Zuckermann, Phys. Rev. Lett. 31, 160-2 (1973).
7. A.E. Clark, Appl. Phys. Lett. 23, 642-4 (1973).

8. J.J. Rhyne, J.H. Schelleng and N.C. Koon, Phys. Rev. B10, 4672-9 (1974).
9. R.W. Cochrane, R.Harris, M. Plischke, D.Zobin and M.J. Zuckermann, J.Phys. F5, 763-73 (1975).
10. R. Harris, unpublished.
11. D. Sherrington and S. Kirkpatrick, to be published.

# MÖSSBAUER RESONANCE OF $FE^{2+}$ and $FE^{3+}$ CATIONS IN SILICATE GLASSES

R. A. Levy
Rensselaer Polytechnic Institute, Troy, N. Y. 12181

ABSTRACT

The Mössbauer effect has been used for a quantitative determination of the $Fe^{2+}/Fe^{3+}$ ratio and the distribution of $Fe^{3+}$ cations among available octahedral and tetrahedral sites in a wide range of silicate glasses of the type $SiO_2$-$Na_2O$-CaO-"$Fe_2O_3$". A feature common to the Mössbauer spectrum of all samples was the appearance of three broadened asymmetric peaks which were fitted to a sum of six lorenzian-shaped components: a doublet arising from the contribution of the $Fe^{2+}$ cations and two other pairs corresponding to the $Fe^{3+}$ cations with tetrahedral and octahedral coordination. The $Fe^{2+}$ doublet is observed to emerge with the largest isomer shift ($0.95 \pm 0.05$ mm/sec relative to $Co^{57}$ in Cr) and quadrupole splitting ($2.50 \pm 0.10$ mm/sec). The higher value of the isomer shift is due to the presence of an extra d-electron which tends to screen the s-electrons and lower in the process the electron density at the $Fe^{57}$ nuclear site. The larger quadrupole splitting is probably the result of axial or rhombic symmetry distortions generated by the electron outside the half-filled spherical shell. For the two types of ferric cations, the differences in the Mossbauer parameters are less pronounced. There appears to be a general tendency for the $Fe^{3+}$ cations with lower coordination number to have a lower isomer shift but the overlap in the range of values makes this a doubtful criterion by itself. The quadrupole splitting, shows a more pronounced difference for the types of sites. The splitting for the tetrahedral sites ranges from 1.00 to 1.55 mm/sec, while that for the octahedral sites ranges from 0.62 to 0.91 mm/sec indicating the deviation from spherical symmetry to be more pronounced in the case of octahedral sites. The $Fe^{2+}/Fe^{3+}$ ratio is observed to increase with calcium oxide additions, reduced basicity, lower oxygen partial pressures, and higher equilibration temperatures. Contrary to published reports, a large fraction (over 50%) of the $Fe^{3+}$ cations is found to be octahedrally coordinated in the alkaline silicate glasses $SiO_2$-$Na_2O$-"$Fe_2O_3$". No significant change in this fraction is detected by either calcium oxide additions or reduced basicity of the melt.

Work supported by the National Aeronautics and Space Administration.

# CORRELATION FUNCTION AND SUSCEPTIBILITY OF SITE AND BOND RANDOM HEISENBERG PARAMAGNET*

Kazuko Kawasaki
Department of Physics, Nara Women's University, Nara, 630 Japan

Raza A. Tahir-Kheli
Department of Physics, Temple University, Philadelphia, Pennsylvania 19122

ABSTRACT

The wave-vector dependent susceptibility and the correlation function for a Heisenberg paramagnet on a Bravais lattice with site or nearest neighbor exchange bond randomness is studied by high temperature series expansion technique. The first five coefficients for the spin correlation function $\langle S_k S_{-k} \rangle$ and susceptibility, $\chi(k)$, are calculated for arbitrary k, the sign of the exchange and the spin magnitude. [2, 1] Pade for $\chi(0)$ is used to calculate an approximate value for the magnetic transition temperature $T_c$ (random). A correlation length $\xi(T)$ is defined through the use of Ornstein-Zernike representation for the correlation function for ka << 1. The behavior of $\xi(T)$ and $\chi(k)$ is analyzed in terms of a reduced temperature $\varepsilon = [T - T_c(\text{random})]/T_c(\text{random})$. It is noted that when $\varepsilon \to 0$, the range $\xi(T)$ diverges and $\chi(k)$ reaches a maximum. The increase in these quantities is the more pronounced the larger the system randomness.

## FORMAL THEORY

Collins[1] has analyzed the behavior of the wave-vector dependent susceptibility, $\chi(k)$, and the correlation function, S(k), for a Heisenberg paramagnet with isotropic, nearest neighbor exchange in terms of a high temperature series expansion. In this paper we extend Collins' work to the case of site or bond random paramagnet. In other words, we work either with a system where the concentration of sites occupied by magnetic atoms is $c(0 < c \leq 1)$ or where the concentration of connected nearest neighbor bonds is p $(0 < p \leq 1)$.

The relevant Hamiltonian is

$$H = -\sum_{ij} p_{ij} J(i,j) c_i c_j \vec{S}_i \vec{S}_j \tag{1}$$

Here $c_i$ $(p_{ij})$ are the usual site (bond) occupation operators which are unity if site i is occupied by a magnetic atom (the exchange bond ij is connected); otherwise, they are zero. Also J(ij) = J only if

and j are neighboring sites. The magnitude of the pin S (h = 1) and the sign of J are arbitrary as are he concentrations c and p.

High temperature series expansions for the eneralized susceptibility

$$(k) = \beta \sum_{\{c,p\}} \sum_f e^{-ik.f} \int_0^1 d\lambda \left\langle c_0 S_0^z \; c_f S_f^z(i\beta\lambda) \right\rangle_{(\text{config})}$$

nd the correlation function

$$S(k) = \sum_{\{c,p\}} \sum_f e^{-ik.f} \left\langle c_0 S_0^z \; c_f S_f^z \right\rangle_{(\text{config})}$$

re computed by standard procedures with the generalizaion necessary for dealing with lattice randomness. ecause we are dealing with quenched disorder, the eighted configurational averaging, denoted by $\sum_{\{c,p\}}$ as to be performed last. Note that $\langle \ldots \rangle_{(\text{config})}$ enotesa canonical average over the spin variables of he system consistent with given configuration of ccupied sites or connected bonds.

For $k \to 0$, the correlation function S(k) is exanded in a Taylor series

$$S(k) = S(0)[1 - \xi^2(T)k^2 + O(k^4)] \,. \tag{2}$$

Recasting this in the Ornstein-Zernike form

$$S(k) \simeq S(0)\, \kappa^2(T)/[\kappa^2(T) + k^2] \tag{3}$$

we interpret $\xi(T) = 1/\kappa(T)$ as a measure of the correlation length (i.e. the short range order, SRO).

## RESULTS

We have evaluated the first five coefficients in the high temperature series expansion for arbitrary Bravais lattices. These results are checked by comparison with the previous results for several cases; the site dilution problem in the spin system with $k = 0$[2], the pure Heisenberg system (c = p = 1) with arbitrary k-values[1] and the bond dilution problem with $S = \frac{1}{2}$ and $k = 0$[3]. Our series are too short to reliably predict the system properties near the transition temperature. However, away from $T_c$, i.e. $T \gtrsim 1.5\, T_c$, our results should be reasonable.

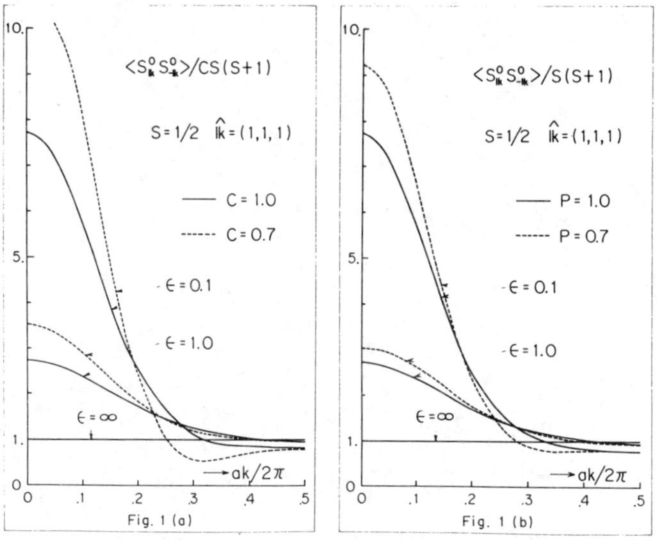


Fig. 1 The k-dependent spin correlation function along (1,1,1) direction for Heisenberg paramagnet on f.c.c. lattice with J > 0.
(a) with magnetic concentration c = 1.0 and 0.7
(b) with magnetic concentration c = 1.0 and bond concentration p = 1.0 and 0.7.

In Fig. 1 (a) and (b), the normalized k-dependent spin correlation function is plotted for both site and bond random paramagnet on f.c.c. lattice (J > 0). At the high temperature limit, the normalized spin correlation function is of course unity. However, with decreasing temperature, a peak about k = 0 appears in each curve. (It becomes peaked at $k = a\pi$ in an antiferromagnet.) The deviation from unity is more sensitive to the concentration c than to p. [as shown in Figs. 1 (a) and (b)]

The k-dependent susceptibilities are plotted against $k_BT/J$ (the inverse of the expansion parameter) for a site diluted paramagnet (J > 0) on b.c.c. lattice in Fig. 2. Although the susceptibility for finite k, $\chi(k)$, depends only weakly on the temperature, the uniform susceptibility $\chi(0)$ grows rapidly as the temperature approaches the critical point and diverges at $T_c$.

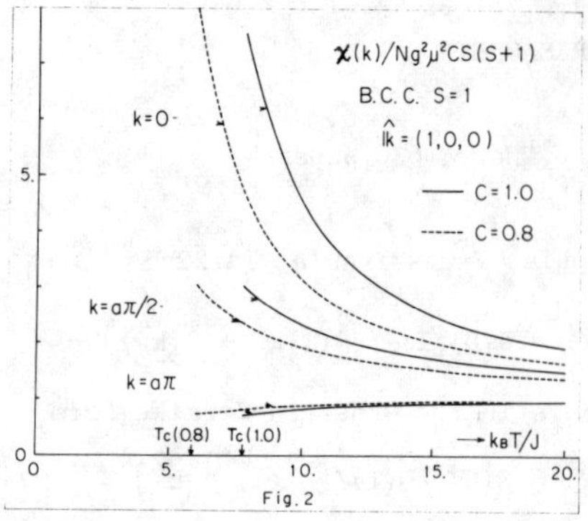


Fig. 2 The k-dependent susceptibility along x-direction, (1,0,0), versus $k_BT/J$ for a site diluted paramagnet with $J > 0$, on a b.c.c. lattice. The scale is normalized such that all curves are unity at infinite temperature. $T_c(1.0)$ and $T_c(0.8)$ denote the transition temperatures for $c = 1.0$ and 0.8, respectively.

In order to investigate the short range order effect for various spin systems, it is convenient to look at the correlation length $\xi(T)$ as a function of the reduced temperature $\varepsilon = (T-T_c)/T_c$. In Fig. 3 we show $\xi^2(\varepsilon)$ for s.c lattice ($S^c = 5/2$) with magnetic concentration c = 1.0, 0.8 and 0.6. The correlation length is found to increase monotonically with decreasing ε, and becomes large as the concentration decreases from 1 to 0.6.

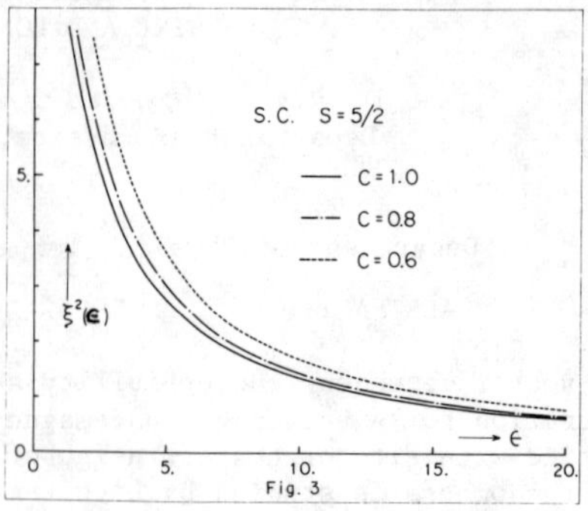


Fig. 3 The square of the effective correlation length $\xi^2(T)$ versus the reduced temperature $\varepsilon = (T-T_c)/T_c$ for a site diluted Heisenberg paramagnet with $J > 0$ on a s.c. lattice with various magnetic concentrations, i.e. c = 1.0, 0.8 and 0.6.

## REFERENCES

1. M. F. Collins, Phys. Rev. 2 4552 (1970); 4, 1588 (1971).
2. D. J. Morgan and G. S. Rushbrooke, Mol. Phys. 4, 291 (1961).
3. E. Brown, J. W. Essam and C. M. Place, J. Phys. C: Solid St. Phys. 8 321 (1975).

*Supported by the U. S. National Science Foundation DMR 73-07651-A02.

# MODELING SLATER-PAULING CURVES

R. C. O'Handley and D. S. Boudreaux, Materials Research Center, Allied Chemical Corporation, Morristown, New Jersey 07960.

The net magnetic moment per transition-metal (TM) atom in a series of alloys varies with TM content in a way that is readily interpreted in terms of the band model of magnetism. Such Slater-Pauling-like curves for glassy metallic alloys of the form $TM_{80}P_{10}B_{10}$[1] (at.%) and for crystalline TM borides,[2] $(TM)_2B$ and TM B, show characteristic departures from the standard Slater-Pauling curve for crystalline alloys. When metalloids are present, 1) the curves are shifted on a TM-content scale, 2) their peaks are depressed, and 3) a steeper slope is observed in the data for the metamagnetic alloys (V, Cr or Mn in Fe). To investigate these differences, the rigid band model is examined using three densities of states: $D(E)$=constant, $D(E)\propto E^{1/2}$, and $D(E)\propto E$. Assuming the Goodenough[3] form of d-band exchange splitting, $\Delta$, only the latter two densities of states allow for strong ferromagnetism ($\Delta > E_0 - E_F$, where $E_0$ is the energy at the top of the minority-spin sub-band). These models also show that the Goodenough splitting is inadequate when the data peaks much below (<90% of) 2.5 $\mu_B$/metal atom and the slope to the right of the peak remains approximately 1 $\mu_B$/valence electron. The shift in the data for the TM borides relative to the standard Slater-Pauling curve is interpreted on the basis of these models to show that the TM d-bands attract 1.6 electrons per boron atom present. The glassy $TM_{80}P_{10}B_{10}$ alloys show less charge transfer, an average of 1.5 electrons per metalloid atom. This hybridization of metalloid conduction bands with TM d-bands (or alternatively, the mediation by the metalloids of a greater degree of overlap between TM d-levels) increases the itinerant character of the d-bands, lowering their density of states and hence their exchange splitting. This accounts for the depressed peak in the data for metalloid-containing alloys. The broader d-bands also bias the alloys toward the antiferromagnetic state. Thus the addition of an anticoupling metal (*e.g.*, Mn, Cr or V in Fe) to an alloy more effectively reduces the net moment when a metalloid is present. This explains the steeper slopes observed in the data for the metamagnetic alloys.

1. T. Mizoguchi, K. Yamaguchi and H. Miyajima in Amorphous Magnetism, ed. H. O. Hooper and A. M. deGraaf (Plenum, NY, 1969) p. 325.
2. M. C. Cadeville and E. Daniel, J. de Phys. 27, 29 (1966).
3. J. B. Goodenough, J. Appl. Phys. 39, 403 (1968).

# MAGNETIC PROPERTIES OF AMORPHOUS Fe-Ge FILMS

G. SURAN, Laboratoire de Magnétisme, C.N.R.S., 92190 Meudon-Bellevue (France), H. DAVER and J.C. BRUYERE, Groupe de Transition de Phases, C.N.R.S., 38042 Grenoble (France).


## ABSTRACT

The temperature dependence of saturation magnetization and the critical concentration $x_c$ corresponding to the onset of ferromagnetism were determined by ferromagnetic resonance on amorphous $Fe_xGe_{1-x}$ thin films. In the concentration range $0,43 < x < 0,53$ Ms follows a linear law as a function of temperature. The critical concentration is close to $x_c = 0,4$ and were deduced from the temperature dependence of magnetic moment and from $T_c$ versus concentration. The results are discussed in term of a band model calculation and some pecularities are explained by the amorphous structure corresponding to the concentration range studied.


## INTRODUCTION

The influence of the structural fluctuation upon fundamental magnetic properties were studied in various amorphous ferromagnetic compounds |1|. Mostly transition metal (TM) - metalloïd (MA) alloys were investigated as pure TM is hard to obtain at room temperature in the amorphous state. Co and Ni alloys (CoP and NiP) have been mainly studied but only few results were reported on Fe based amorphous alloys. The influence of disorder on $M_s$ or on $T_c$ is not a resolved problem. The experimental data in CoP were explained by a rigid band model |2,3| and in NiP by itinerant magnetism |4| or by some ferrimagnetic structure |5|. Amorphous alloys are also well adapted for a determination of the critical concentration $x_c$ corresponding to the appearence of ferromagnetism as no singularities arise from the symmetry of the crystalline structure.

In the present paper, we report measurements on amorphous Fe alloys. We determined the saturation magnetization $M_s$ near 0°K, $M_s(T)$, $T_c$ and $x_c$ on $Fe_xGe_{1-x}$ thin films where x is in the range of 0.3 to 0.6.

## EXPERIMENTAL PROCEDURE

The $Fe_xGe_{1-x}$ films were prepared by codeposition of constituants under a pressure of $10^{-7}$ Torr onto glass substrates cooled down to 77°K |6|. We used a single evaporation source, and as the vapour pressure of Fe and Ge are quite different particular care was taken to avoid any compositional gradient along the thickness of the film : The evaporation rate is sufficiently slow that no concentration gradient is formed in the crucible, the charge of the melt is large (∿ 10g) and only a small fraction of the initial charge is evaporated. The results reported here were obtained on 1000 Å ± 200 Å thick films. The composition with an accuracy of 1 % were determined by a back scattering analysis |7|, the surface of the film being bombarded by $^4He^+$ ions. The analysis detected no other element (∿ 100 ppm) than Fe and Ge in the films. From measurements during abrasion we could verify that gradient along the thickness of the film is less than 2 %.

The electron diffraction studies of the films showed them to be amorphous for $0 < x < 0.7$ at 300°K. For higher Ge concentration ($x < 0.2$) the local structure is polytetrahedral, like in pure amorphous Ge films, and for higher Fe concentrations ($x \sim 0.6$) the dense random packing of hard spheres describe best the observed structure |6|.

The magnetic properties were studied by ferromagnetic resonance at 17,5 GHz. The microwave cavity was in a Dewar system which allowed measurements from 4.2 to 330°K the temperature stability being better than 0.2°K. The magnetic field corresponding to the resonance peak could be measured with an accuracy of ± 2 Oe.

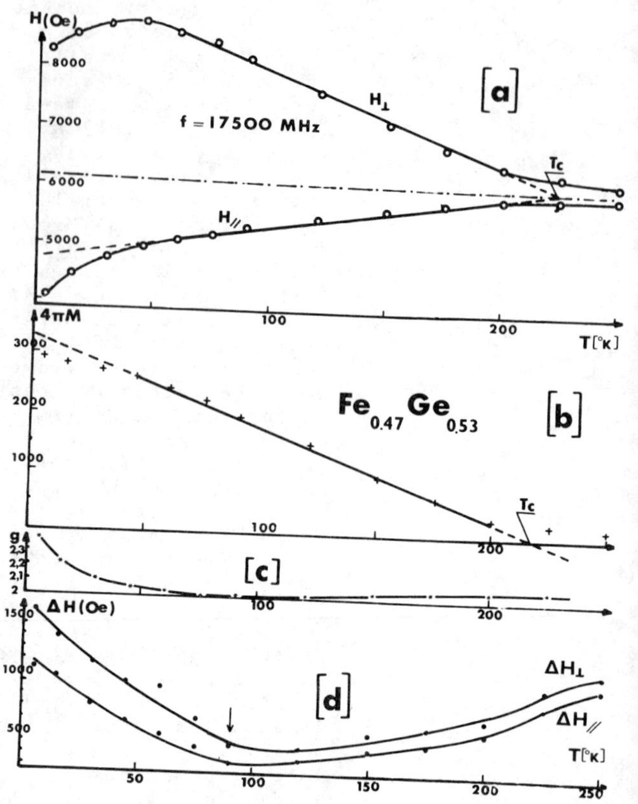


Fig. 1 - Temperature dependence of a) resonance fields b) saturation magnetization c) g factor d) derivative linewidth corresponding to perpendicular (ΔH) and parallel ($\Delta H_{//}$) orientatio

## RESULTS AND DISCUSSION

A typical set of measurements as a function of temperature is presented on fig. 1a-d. The demagnetizing field $4\pi M_{eff}$ and g factor (fig. 1c) were calculated from the two resonance fields $H_\perp$ and $H_{//}$ (fig. 1a) where $H_\perp // \vec{n}$ and $H_{//} \perp \vec{n}$ ($\vec{n}$ : normal to the film plan) using the Kittel resonance formula. In the present case $4\pi M_{eff}$ is equal to $4\pi M_s$ (fig. 1b) as we verified that the various anisotropy fields are negligible : crystalline anisotropy does not exist as the structure is amorphous, a study of domain structure showed that induced uniaxial anisotropy is negligible. The strain induced anisotropy field studied systematically by depositing simultaneously films on glass substrates with various thermal expansion coefficient α |7| was found to vary (< 2 - 3% of total resonance field) as a function of α. The present results are those obtained on substrates with $\alpha = 46.10^{-7}$ for which the strain is judged to be minimal. As it is shown on fig. 1b $M_s(T)$ follows a linear law in a large range of temperature except for $T < 50$°K and near the Curie temperature $T_c$. The large deviation from the linear law near $T_c$ could be partly due to the contribution of induced magnetization to $4\pi M_{eff}$ as the measurements are made with an applied d.c. field close to 6000 Oe. However the intensity of the resonance signal is much higher than that expected for a purely paramagnetic one, so ferromagnetic moment even here seems to be the dominant contribution. Paramagnetic resonance corresponding to $H_\perp = H_{//}$ could not be observed. In these conditions $T_c$ was determined, somewhat arbitrary, by extrapolating the experimental M(T) curve to M = 0. In the low temperature region ($T < 90$°K) both g factor (fig. 1c) and resonance linewidth (fig. 1d) present an anomalous behaviour : at

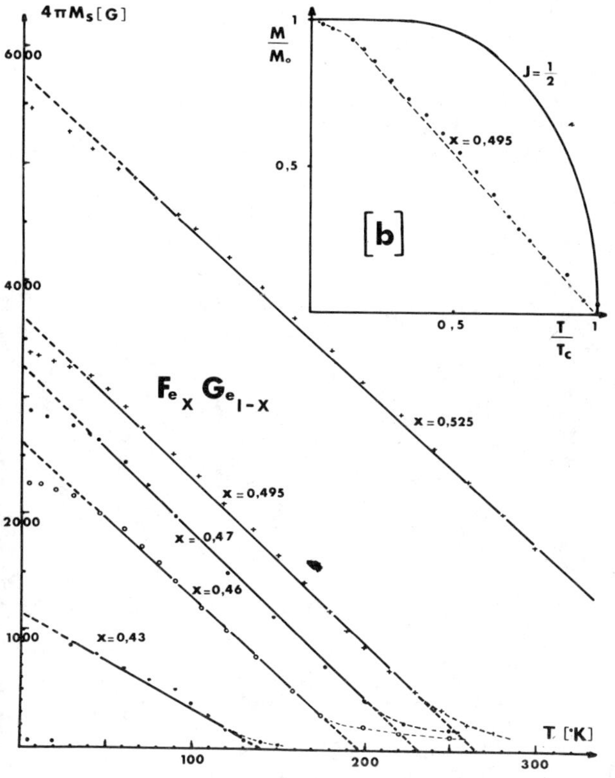


Fig. 2 - Temperature dependence of (a) the saturation magnetization for various Fe concentrations (b) reduced magnetization.

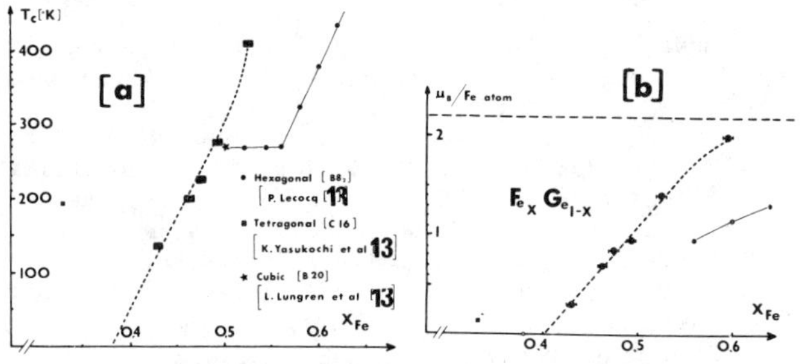


Fig. 3 - Compositional dependence of (a) magnetic moment at 4,2°K per Fe atom b) Curie temperature.

90°K, g = 2.017 and $\Delta H$ = 330 Oe while both values increase progressively when T is lowered so at T = 4.2°K, g = 2.33 and $\Delta H$ = 1600 Oe.

The temperature dependence of $M_s$ for various iron concentrations is shown in fig. 2.$M_s$ (0°K) decreases with decreasing Fe content, the linear variation of $M_s$ with T is observed for Fe concentrations as large as 0.525 and the deviations at low temperature become more marked particularly for compositions close to the critical one. The reduced magnetization curve (fig. 2b) falls far below the theoretical one (J = 1/2) corresponding to a crystalline compound. Our experimental results are quite different from that found in other TM-MA alloys : in CoP (20 % P) $M_s(T)$ follows the Bloch

law $M_s(T) = M_s(0)\ |1-BT^{3/2}|$ up to $T/T_c = 0.2$ (1)

Our results can be explained by a much larger fluctuation of exchange constant and lower value of $T_c$ than in preceeding case : as $T_c$ decreases the temperature region where Bloch law is assumed to be valid decreases. Computations based on CPA |9| or MF |10| theories show that with increasing exchange fluctuation the flattening of $M_s(T)$ is more marked and $M_s(T)$ approach a linear law. The large fluctuation of exchange integral is due to high MA concentration in the film : in the concentration range studied the structure is intermediate between that corresponding to Ge rich and Fe rich films so larger local disorder is expected than that observed in amorphous structure closed to hard packed spheres. Preliminary measurements showed that for higher Fe concentration ($x \sim 0.6$) $M_s$ approaches the Bloch law.

The critical concentration $x_c$ was deduced from the compositional dependence of the magnetic moment $\mu_B$(0°K) per Fe atom (fig. 3b) and from $T_c$ versus concentration (fig. 3a). The extrapolated value of both curves gives concordant values with $x_c = 0.4 \pm 0.02$ |8|. This value of $x_c$ could be explained by a rigid band model or by percolation theories. The percolation threshold is $x_c \sim 2/z$ |9| (z : number of nearest neighbour) and $x_c$ = 0.33 for z = 6. This result is not unrealistic but it is difficult to estimate its validity as z is not well known. A calculation based on rigid band model gives $x_c$ = 0.35 or $x_c$ = 0.43 depending upon either all four |2| valence electrons or only the two $4p^2$ electrons |11| of Ge is transferred to TM to fill up the 2,6 holes in 3d band of Fe atoms. However band model predicts a linear decrease of $\mu_B(0)$/Fe atom with increasing Ge content which is not satisfied in the present experiments. For example when x = 0.525 the experimental and calculated values are respectively 2.07$\mu_B$ and 0.7 to 0.8 $\mu_B$. These discrepancies could be also attributed to the structure of the film : as one increases Ge concentration the interatomic distance increases so electron transfer to TM becomes less effective.

Theory predicts both an increase or decrease of $T_c$ of amorphous state as compared to the crystalline one. Fig. 3a shows that in the amorphous state $T_c$ is higher than the corresponding hexagonal one, but close to that of the cubic phase. Mössbauer spectroscopy shows also a similarity of magnetic structure in cubic and amorphous state |12|.

Several interpretations can be proposed in order to explain the low temperature anomalies reported in fig. 1 and which were observed for the whole composition range studied.

One involves strain induced anisotropy due to some unavoidable difference between the thermal expansion of film and substrate. Such anisotropy could well explain the decrease of $H_\perp$ and increase of $\Delta H$ for low temperatures, the minimum at 90°K in $\Delta H(T)$ corresponding to the deposition temperature. One cannot exclude, the possibility of mixed magnetism : Fe can exist in an antiferromagnetic state depending upon the distance of nearest neighbours and this antiferromagnetism is predominant at low temperatures.

Some preliminary results on resonance linewidth is presented in fig. 1d and fig. 4. Fig. 1d shows $\Delta H(T)$ for a given composition and fig. 4 $\Delta H$ as a function of Fe content at T = 300°K, the results for other temperatures has the same trend. The relaxation mechanism between T = 90°K and $T_c$ can be explained by two contribution : An "intrinsic" relaxation $\lambda$ which increases as Fe content decreases and is related to exchange constant fluctuations and a temperature dependent term proportional to $\lambda/M_s$. The details will be described in a fuller paper.

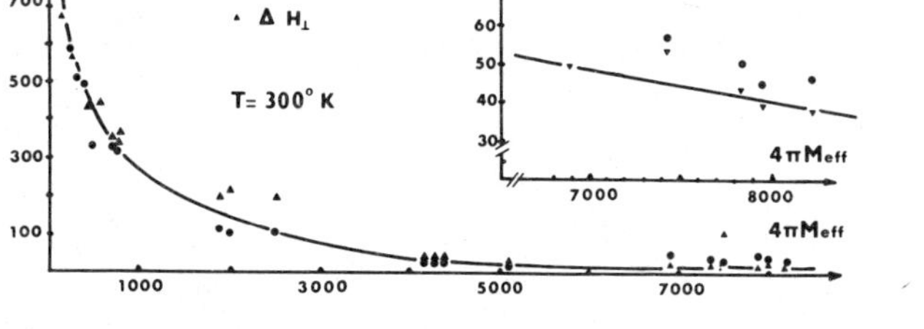


Fig. 4 - Resonance linewidth as a function of 4πM at room temperature.

In conclusion f.m.r. is shown to be well adapted for the determination of basic magnetic properties of amorphous thin films even near the critical concentration. We could determine $M_s(0)$, $M_s(T)$, $x_c$ in amorphous FeGe films. In order to confirm the reported data more information is needed in particular a better knowledge of the amorphous structure, and measurements on films with higher Fe concentration. The low temperature anomalies could be perhaps understood by Mössbauer spectroscopy. This work is actually in progress.

The authors are indebted to M. Tessier for technical assistance.

REFERENCES

|1| G.S. Cargill A.I.P. Conf. Proc. 24 138 (1975).
|2| A.W. Simpson, D.R. Brambley, Phys. Stat. Sol. (b) 43 291 (1971).
|3| D. Pan, D. Turnbull, J. Appl. Phys., 45 1406 (1974).
|4| A.W. Simpson, D.R. Brambley, Phys. Stat. Sol. (b) 49 685 (1972).
|5| D. Pan, D. Turnbull, A.I.P. Conf. Proc. 18 646 (1974)
|6| D. Massenet, H. Daver, J. Geneste, J. Phys. Paris, C4 35 279 (1974).
|7| G. Suran, H. Daver, Nguyen Van Dang, J.C. Bruyère, 7th Int. Coll. on Magnetic Films, Regensbourg, 1975, paper 8B-8.
|8| The present value of $x_c$ is correct. The one reported in ref. 6 was determined erronously, the measurements being rather approximative.
|9| R.A. Tahir-Kheli, Amorphous Magnetism (Plenum New York, 1973) p. 393.
|10| S. Kobe, K. Handrich, Sov. Phys. Solid State, 13 734 (1971).
|11| K. Yamauchi, T. Mizoguch, J. Phys. Soc. Japan, 39 541 (1975).
|12| C. Jeandey, P. Peretto, Phys. Stat. Sol. (a), 30 71 (1975).
|13| L. Lungren et al., Phys. Scripta, 1 69 (1970).
P. Lecocq, Ann. Chim., 8 106 (1963).
K. Yasuchochi, K. Kanematsu, J. Phys. Soc. Japan, 16 429 (1961).

# NEGATIVE MAGNETORESISTANCE IN UNIAXIAL AND AMORPHOUS FERROMAGNETS†

L. Berger
Physics Department, Carnegie-Mellon University, Pittsburgh, Pennsylvania 15213

ABSTRACT

When a d.c. current crosses a 180° domain wall, the current lines are sharply bent at the wall by twice the Hall angle of the material. This is a simple consequence of Maxwell's equations and of the existence of off-diagonal Hall components in the resistivity tensor. In a demagnetized sample, where the current crosses many walls, the current lines follow a zigzag pattern. This leads to an increase of the effective ohmic resistance of the sample. Then a negative magnetoresistance $\delta R/R_o$ should be observed if an external field parallel to the easy axis removes the walls. For a current running normal to stripe domains, we predict $\delta R/R_o = -\beta^2$, where $\beta$ is the tangent of the Hall angle. Other kinds of domains give similar results. The present theory may explain the negative transverse magnetoresistance $\delta R/R_o = -2 \times 10^{-4}$ associated with domains, observed recently in sputtered Gd - Co films where $|\beta| \approx 10^{-2}$. For very pure cobalt at 4 K ($|\beta| \approx 0.6$), $\delta R/R_o \approx -0.4$ is predicted.

## INTRODUCTION

Existing low-field magnetoresistance data[1,2] for ferromagnetic metals concern mostly cubic materials such as iron, nickel and their alloys. Because of the presence of several easy directions of magnetization, a modest external field is able to change the magnetization direction in the various domains. In turn, this affects the measured resistance since it depends on the cosine of the angle between magnetization and current. The magnetoresistance behavior is best characterized by the quantity $\Delta\rho/\rho_o = (\rho_\parallel - \rho_\perp)/\rho_o$, called ferromagnetic anisotropy of resistivity.[2] Here $\rho_\parallel$ and $\rho_\perp$ are the value of the resistivity when the saturation magnetization is respectively parallel and perpendicular to the current, and $\rho_o$ is the zero-field resistance.

In materials having only one easy magnetization axis (h.c.p. Co or Gd), small fields can only switch the magnetization between opposite directions. This does not affect the square of the cosine of the angle between magnetization and current. As a result, the ferromagnetic anisotropy of resistivity is ineffective as a cause of magnetoresistance. We describe below a different mechanism, which may be active in these uniaxial materials. It might also be important in amorphous ferromagnets since one easy magnetization axis, associated with the induced anisotropy energy, is often present in them.

## STRATIFIED MEDIUM MODEL

Recently,[3] we solved Maxwell's equations in the neighborhood of a 180° domain wall, in a material where the resistivity tensor has off-diagonal components associated with the existence of the Hall effect. It turns out that the current lines are sharply bent at the wall. In Fig. 1a, we show the resulting current distribution in the case[4] of an array of equidistant walls in a very large sample (i.e., "stripe" domains). The average current direction is assumed to be normal to the walls, and the walls to be at rest. Then the bending of the current lines at the walls is by twice the Hall angle of the material. This zigzag pattern of current lines leads to an increase of the average ohmic resistance of the sample, as measured with two potential probes on a line parallel to the average current. Then a negative magnetoresistance $\delta R/R_o$ should be observed if an external field parallel to the easy axis removes all the walls and saturates the sample. We find for arbitrary $\beta$:

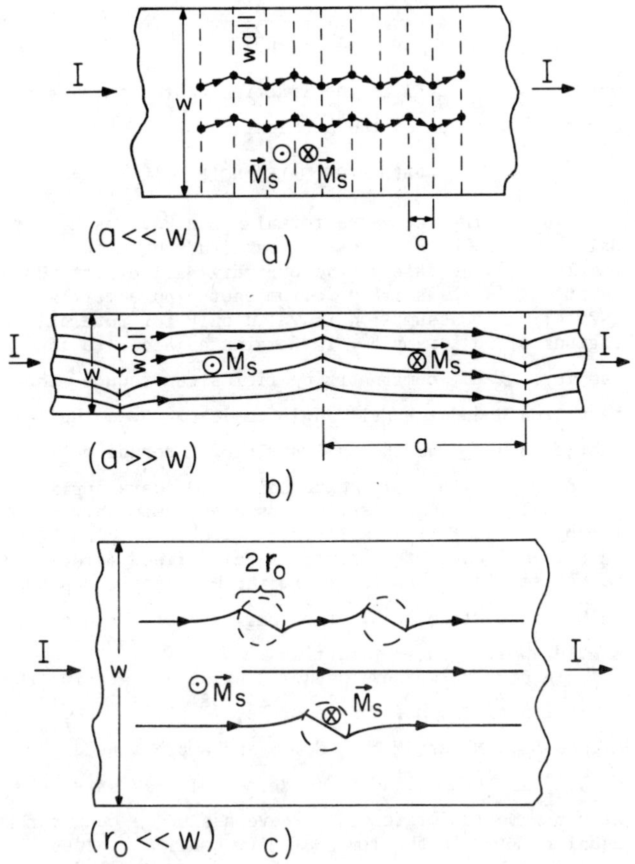


Fig. 1. Zigzag pattern of current lines for various simple models of domain structure.

$$\frac{\delta R}{R_o} = -\beta^2 \tag{1}$$

where $\beta$ is the tangent of the Hall angle, and where $R_o$ is the resistance in the absence of walls. Note that $\delta R/R_o$ is independent of the wall spacing a as long as a is much smaller than the sample width w. Note also that $\delta R/R_o = 0$ is predicted if the current runs parallel to the walls.

## SINGLE-WALL MODEL

We consider another model, where[3,4] the spacing a of the walls is much larger than the sample width w (Fig. 1b). We find in that case, for $|\beta| << 1$:

$$\frac{\delta R}{R_o} = -0.54\,\frac{w}{a}\,\beta^2 \tag{2}$$

and for $|\beta| >> 1$:

$$\frac{\delta R}{R_o} = -\,\frac{w}{a}\,|\beta| \tag{3}$$

## BUBBLE-DOMAIN MODEL

In this model,[5] the sample contains an array of cylindrical domains of reversed magnetization, all having the same radius $r_o$ (Fig. 1c). The average number of cylinders per unit area is $n_w$. We assume $w>>r_o$ and $n_w^{-1/2}>>r_o$. Also, the current is in a direction normal to the cylinder axes. We find for arbitrary $\beta$:

$$\frac{\delta R}{R_o} = -2f \frac{\beta^2}{(1+\beta^2)^{1/2}} \qquad (4)$$

where $f = \pi r_o^2 n_w << 1$ is the fraction of the total sample area occupied by the bubbles.

## VALUES OF HALL ANGLE

In writing the above formulas, we have assumed the Hall angle to have a constant magnitude over the whole sample. In the case of the ordinary Hall effect, dominant in low-resistivity ferromagnets (pure metals at T<20 K), this assumption is valid only for internal regions of bulk samples; in domains adjacent to the sample surface, demagnetizing fields can reduce[6] the induction $\vec{B}$ and the Hall angle to zero. Even the induction $\vec{B} = \vec{M}_s$ and the Hall angle of internal regions could vary if too large external fields were applied.

However, these restrictions are unnecessary for ferromagnets of high resistivity (i.e., at T>100 K or in alloys), where the anomalous Hall effect (side-jump mechanism) is dominant. There the Hall angle depends on $\vec{M}_s$ rather than on $\vec{B}$. Then our present formulas should apply even to thin films.

In Fig. 2, we have gathered existing data for the tangent $\beta$ of the Hall angle of Fe,[7,8] Ni,[9-11] Co,[12] Fe-Al, Fe-Si,[8] Ni-Cr, Ni-V,[9] Gd,[13] and amorphous $Gd_{26}Co_{74}$,[14] as a function of the zero-field resistivity $\rho_o$ at the same temperature. We have assumed a constant B equal to $M_s$. In the limit of very small $\rho_o$ values, where the ordinary Hall effect is dominant, the Hall angle of the compensated metal Fe decreases after a

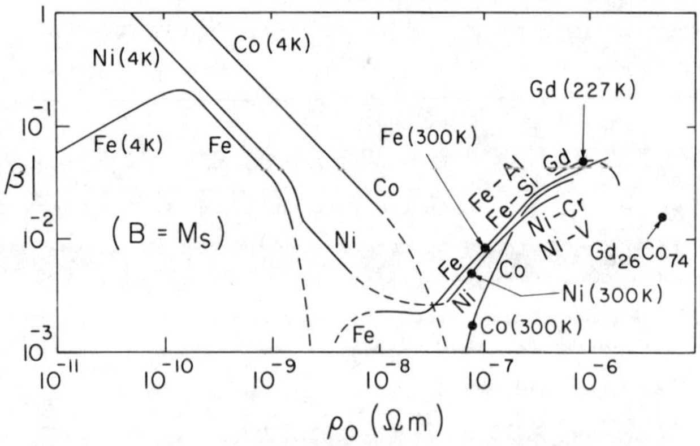


Fig. 2. Absolute value of the tangent of the Hall angle for various materials, at a field $B=M_s$. In abscissa is plotted the zero-field resistivity $\rho_o$ at the same temperature.

maximum, while the Hall angle of non-compensated Ni and Co is large ($|\beta| \gtrsim 1$) and steadily increasing. For large $\rho_o$ values, where the anomalous Hall effect is dominant, the Hall angle increases again but seems limited to $|\beta| \simeq 0.05$ by the incipient disappearance of ferromagnetism caused by alloying or high temperatures.

## NEGATIVE MAGNETORESISTANCE OF Gd-Co

A negative magnetoresistance $\delta R/R_o \simeq -2 \times 10^{-4}$ has been observed recently[15] in amorphous, sputtered, films of $Gd_{25}Co_{75}$ around room temperature. It seems associated with the removal by the field of domain walls crossed by the current.

Therefore, it might be caused by the mechanism proposed in the present paper. From the resistance and Hall data of their Fig. 5(A), we derive roughly $|\beta| \simeq 0.01$ for their samples. Then our Eq.(1) predicts $\delta R/R_o \simeq -1 \times 10^{-4}$, in rough agreement with their experimental value quoted above.

In the case of cobalt of presently attainable purity at 4.2 K, where $|\beta| \simeq 0.6$ is reached[4] for $B = M_s$, our Eq.(1) predicts $\delta R/R_o \simeq -0.36$. Note, however, that other magnetoresistance mechanisms, associated with cyclotron orbits which intersect the domain walls, have been recently proposed[16] for low-resistivity ferromagnets.

We are grateful to T. R. McGuire as well as to L. M. Falicov and to R. V. Coleman for helpful discussions on these topics.

## REFERENCES

† Work supported by the U.S. National Science Foundation.


1. H. C. Van Elst, Physica 25, 708 (1959).
2. T. R. McGuire and R. I. Potter, IEEE Transactions on Magnetics, MAG-11, 1018 (1975).
3. D. L. Partin, M. Karnezos, L. C. deMenezes, and L. Berger, J. Appl. Physics, 45, 1852 (1974).
4. L. Berger, J. Phys. Chem. Solids 35, 947 (1974).
5. L. Berger, Proc. 19th Annual Conf. on Magnetism and Magn. Materials, Boston, 1973 (Amer. Inst. of Physics, New York, 1974) p. 918.
6. The case of the ordinary Hall effect in thin films has been treated by S. H. Charap, J. Appl. Phys. 45, 397 (1974).
7. A. K. Majumdar and L. Berger, Phys. Rev. B7, 4203 (1973); R. W. Klaffky and R. V. Coleman, Phys. Rev. B10, 2915 (1974).
8. W. Jellinghaus and M. P. DeAndres, Ann. Physik, 7, 189 (1961).
9. W. Koster and W. Gmohling, Z. Metallkunde 52, 713 (1961).
10. R. Huguenin and D. Rivier, Helv. Phys. Acta 38, 900 (1965).
11. W. A. Reed and E. Fawcett, J. Appl. Phys. 35, 754 (1964).
12. N. V. Volkenshtein and G. V. Fedorov, Sov. Phys.-JETP 11, 48 (1960); I. A. Tsoukalas, Phys. Stat. Solidi 23, K41 (1974).
13. N. V. Volkenshtein, I. K. Grigorova, and G. V. Fedorov, Sov. Phys.-JETP 23, 1003 (1966); N. A. Babushkina, Sov. Phys.-Solid State, 7, 2450 (1966); N. V. Volkenshtein and G. V. Fedorov, Phys. Metals and Metallogr., 18, N°1, 26 (1964).
14. A. Ogawa, T. Katayama, M. Hirano, and T. Tsushima, Proc. 20th Annual Conf. on Magnetism and Magn. Materials, San Francisco, Dec. 1974 (Amer. Inst. Physics, New York, 1975) p. 575.
15. K. Okamoto, T. Shirakawa, S. Matsushita and Y. Sakurai, Proc. 20th Annual Conf. on Magnetism and Magn. Materials, San Francisco, Dec. 1974 (Amer. Inst. Physics, New York, 1975) p. 113.
16. Yu. I. Mankov, Sov. Phys.-Solid State 14, 62 (1972); R. I. Shekhter and A. S. Rozhavskii, Sov. Phys.-JETP 38, 383 (1974); G. G. Cabrera and L. M. Falicov, to be published.

# THE MAGNETIC PROPERTIES OF 3d IMPURITIES IN AMORPHOUS Cu-Zr ALLOYS

T. Mizoguchi* and T. Kudo
Faculty of Science, Gakushuin University, Mejiro, Tokyo, Japan

ABSTRACT

The magnetic susceptibility and electrical resistivity of amorphous $(Cu_{.57}Zr_{.43})_{1-x}M_x$ alloys, where M expresses 3d transition metal impurities of V, Cr, Mn, Fe, Co and Ni, were measured. The host $Cu_{.57}Zr_{.43}$ alloy shows temperature independent Pauli pramagnetic behavior. The addition of 3d impurities only slightly affects the value of the susceptibility which still remains temperature independent except for Mn, showing that these 3d impurities have no localized magnetic moment in this amorphous alloy. The susceptibility of amorphous Cu-Zr containing Mn can be expressed as the sum of a temperature independent part, and a temperature dependent part which obeys the Curie Weiss law. The effective magnetic moment per Mn atom is about 1.6 μB (gS $\sim$ 1.45 assuming g = 2).

## INTRODUCTION

Dilute additions of 3d group transition metals to a nonmagnetic amorphous host alloy has been studied by Hasagawa and Tsuei[1] and also Szofran, et al.[2]. The principal result of this initial work is that magnetic moments are found associated with the 3d metal. In this paper we report electric and magnetic measurements of 3d additions to amorphous Cu-Zr which show, with one exception, the opposite behavior with no magnetic moment for the 3d atoms. Hasagawa and Tsuei's alloys were of the form, $M_xPd_{.80-x}Si_{.20}$, where M was Cr, Mn, Fe or Co. These four impurities were found to have well defined effective moments from Curie Weiss law susceptibility with values 3.6, 5.7, 5.8 and 4.4 $\mu_B$, respectively, for a concentration at $x \approx 0.03$.

In the present work $M_x(Cu_{.57}Zr_{.43})_{1-x}$ was studied, where M represents the 3d metals previously listed as well as V and Ni with various concentrations $x < .06$. The host alloy $Cu_{.57}Zr_{.43}$ is a simple binary which Waseda and Masumoto[3] have studied with x-ray in detail. Szofran, et al.[2] were the first to use amorphous Cu-Zr as a host alloy with Fe as the added impurity over a concentration range $0 \leq x \leq .12$. They reported an effective magnetic moment for their alloys, a result we could not duplicate in our own work which indicated zero moment for the Fe impurity.

## EXPERIMENTAL PROCEDURE

The alloy specimens were prepared by rapid quenching from the melt in a plasma jet furnace, and confirmed to be amorphous by x-ray diffraction. The nominal alloy compositions were verified with electron microprobe analysis. Magnetic susceptibility was measured with a pendulum type magnetometer between 4.2 and 300K in a field of about 8 KOe. Some measurements in higher field up to 18 KOe between 2.1K and 300K were done with a force balance magnetometer. Electrical resistivity and magnetoresistance was measured between 4.2 and 300K using a standard four lead method.

## EXPERIMENTAL RESULTS AND DISCUSSION

The host $Cu_{.57}Zr_{.43}$ alloy showed temperature independent Pauli-paramagnetism. The addition of 3d impurities of V, Cr, Fe, Co or Ni for concentration of about 2% slightly affects the value of the susceptibility, which still remains temperature independent. As shown in Fig. 1, the change due to 2% Fe was only slightly above the host alloy and other impurities (except Mn) had values close to these two curves. This clearly indicates that these 3d impurities have no localized magnetic moment in this amorphous alloy. The susceptibility of the alloys containing up to 6% V or Ni shows no significant variation with changing its concentration.

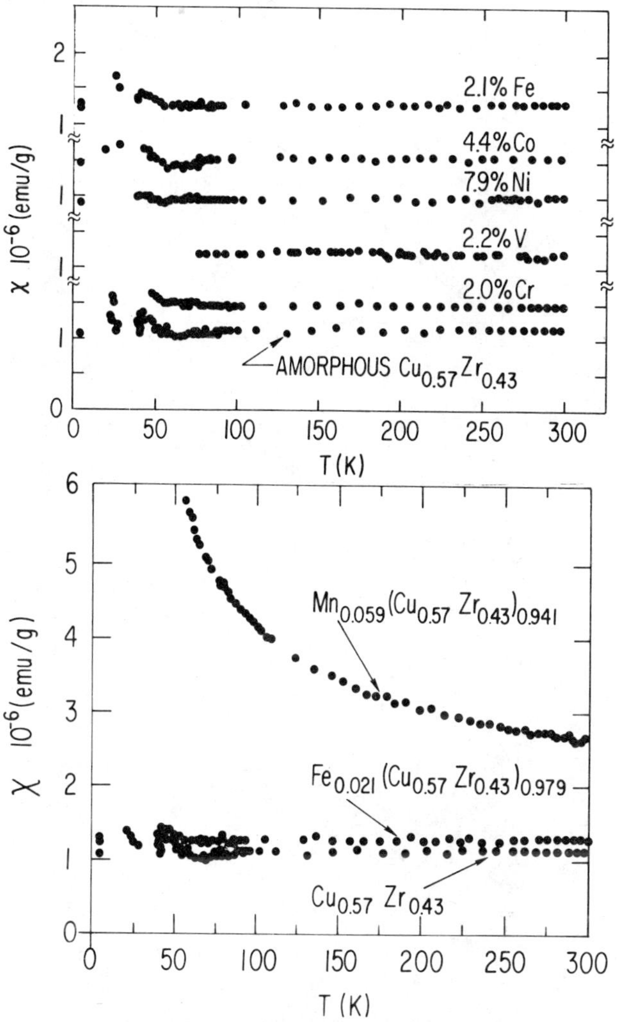


Fig. 1: The temperature dependence of the magnetic susceptibility of amorphous $Cu_{.57}Zr_{.43}$ and $M_x(Cu_{.57}Zr_{.43})_{.1-x}$, where M = V, Cr, Mn, Fe, Co or Ni.

In contrast to these 3d impurities, Mn does have a localized moment in the Cu-Zr amorphous alloy. Its susceptibility can be expressed as follows:

$$\chi = \chi_\circ + \frac{C}{T-\theta} \qquad (1)$$

that is, the sum of a temperature independent part and a temperature dependent part which obeys the Curie-Weiss law, as shown in Fig. 2. The effective magnetic moment per Mn atom is about 1.6 $\mu_B$ (g S $\sim$ 1.45 assuming g = 2). The asymptotic Curie temperature, θ, and temperature independent susceptibility, $\chi_\circ$, obtained from least square fitting of the data above 20K are -8.7K and $1.8 \times 10^{-6}$ emu/g for x = 0.059 Mn. As shown in Fig. 3, the induced magnetization at low temperatures (T < 20K) shows a nonlinear dependence with field and exceeds the value expected from Eq. (1) inspite of the negative θ. No characteristic maximum in the susceptibility due to spin freezing was observed down to 2.1K.

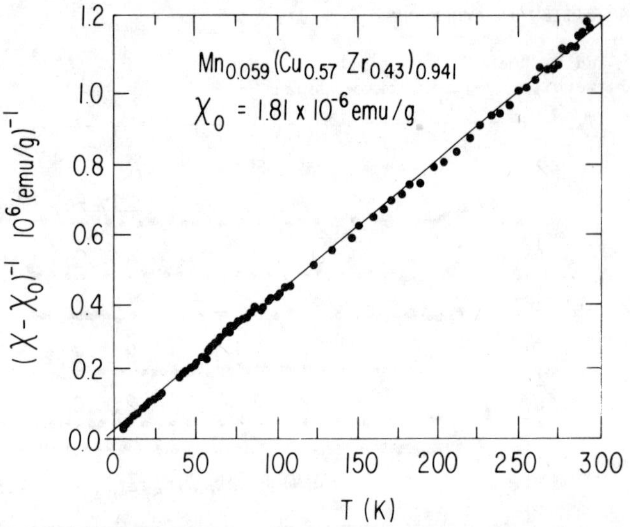


Fig. 2: The inverse of a temperature dependent part of susceptibility versus temperature for an amorphous $Mn_{.059}(Cu_{.57}Zr_{.43})_{.941}$ alloy.

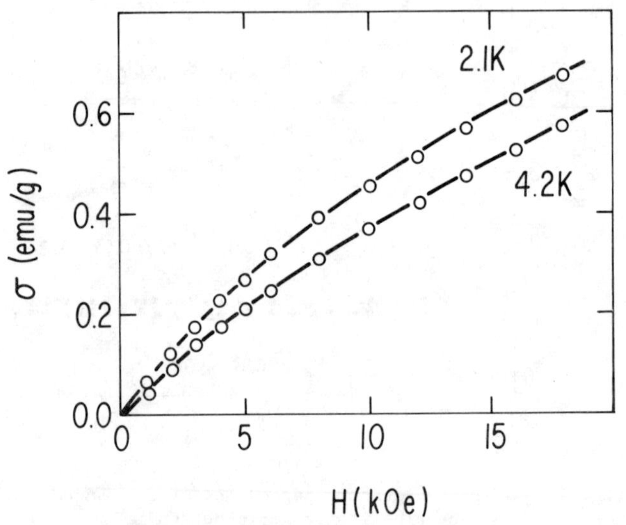


Fig. 3: The field dependence of magnetization at 4.2 and 2.1K for amorphous $Mn_{.059}(Cu_{.57}Zr_{.43})_{.941}$ alloy.

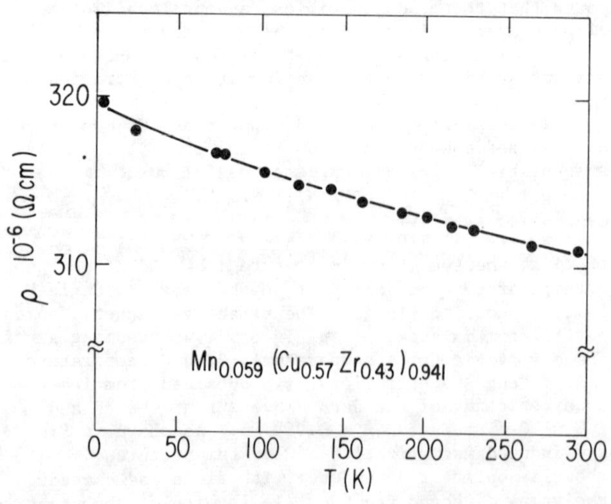


Fig. 4: The electrical resistivity of amorphous $Mn_{.059}(Cu_{.57}Zr_{.43})_{.941}$ alloy.

The electrical resistivity of amorphous $Cu_{.57}Zr_{.43}$ alloy is weakly temperature dependent and increases slightly with decreasing temperature ($\rho(4.2K)/\rho(300K) \simeq 1.036$). As shown in Fig. 4 the addition of 5.9 at.% of the magnetic impurity (Mn) causes no significant effect. No magneto-resistance effects in these alloys were observable.

Szofran, et al.[2] reported that the amorphous $Cu_{.60}Zr_{.40}$ alloy had a large paramagnetic susceptibility which was strongly temperature dependent. They also observed a local-moment-type contribution of the Fe impurity to the susceptibility. It would be difficult to attribute the drastic change in the susceptibility solely to the slight difference between the composition of the alloys.

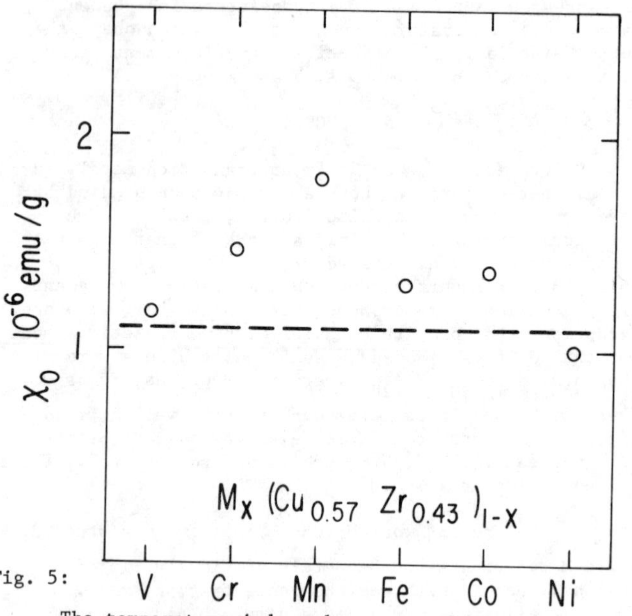


Fig. 5: The temperature independent part of susceptibility of amorphous $M_x(Cu_{.57}Zr_{.43})_{1-x}$ alloy, where M = V, Cr, Mn, Fe, Co or Ni, for $\chi \simeq 0.02$.

In Fig. 5 the temperature independent Pauli susceptibility of the alloys are plotted versus the atomic number of impurity elements. The increase of the Pauli susceptibility is considered to be due to the virtual level formed by 3d impurities near the Fermi level. The experimental result seems to support the theoretical prediction[4] that the impurity atoms which have a half filled d shell are most likely to be localized and split by the exchange interaction.

ACKNOWLEDGEMENTS

The authors wish to express their sincere thanks to Dr. T. McGuire and Dr. S. Sekizawa for helpful discussions, arrangements and encouragement. Thanks are due to H. Lilienthal for assistance of resistivity and low temperature susceptibility measurements.

REFERENCES

* Present address: IBM T. J. Watson Research Center Yorktown Heights, N. Y. 10598

1) R. Hasegawa and C. C. Tsuei, Phys. Rev. B2,1631, (1970);
R. Hasegawa, J. App. Phys. 41, 4096 (1970).

2) F. R. Szofran, et al. AIP Conf. Proc. 18, 282 (1973).

3) Y. Waseda and T. Masumoto, 8. Phys. B21, 235 (1975).

4) T. Moriya, Prog. Theor. Phys. 33, 157 (1965).

# A CALORIMETRIC STUDY OF $MnO\cdot Al_2O_3\cdot SiO_2$ GLASSES

R.W. Kline and A.M. de Graaf,* Wayne State University, Detroit, MI 48202;
L.E. Wenger and P.H. Keesom,† Purdue University, W. Lafayette, IN 47907.

## ABSTRACT

In order to further elucidate the magnetic properties of concentrated manganese aluminosilicate glasses, the specific heat of two samples containing 13.3 at.% and 14.3 at.% Mn was measured in the temperature range 2-38 K. Below ~50 K the susceptibility of these same samples exhibits superparamagnetic type behavior which terminates in a sharp peak at 2.95 K and 4.25 K, respectively, seeming to indicate the onset of long range antiferromagnetic order. However, the specific heat does not show a cooperative-type peak. Instead, at low temperatures, the specific heat exhibits a broad, smooth maximum approximately 10 K wide, superimposed on a nearly linear background. The specific heat result is consistent with the broad dip in the sound velocity which has been observed in these systems.[1] It has previously been shown that the susceptibility and the sound velocity can be accounted for by dividing the glass in small domains (anisotropy parameter KV) and applying Néel's theory of small magnetic particles.[2] It is shown that this theory accounts for the temperature dependence of the specific heat and leads to values of KV which are in reasonable agreement with those obtained from the susceptibility and the sound velocity. The size of the domains is ~12 Å. The net magnetic moment of the domains is about seven times smaller than the magnetic moment needed to fit the susceptibility data. No satisfactory explanation for this discrepancy is available.

## INTRODUCTION

In a recent series of papers attention has been drawn to the remarkable magnetic properties of cobalt and manganese aluminosilicate glasses.[1,2] These glasses exhibit a relatively sharp peak in the low field, low temperature susceptibility, very similar to the peak observed in the susceptibility of the so-called spin glass alloys.[3] There is also a broad dip in the sound velocity, which correlates with the susceptibility peak. These observations were successfully accounted for by dividing the glass in antiferromagnetically ordered domains whose net magnetic moments point in random directions, and subsequently applying Néel's theory of small antiferromagnetic particles. The peak in the susceptibility and the dip in the sound velocity are due to the freezing of the domain moments in random anisotropy directions below a temperature on the order of KV/k, where K is the anisotropy energy density of the domains and V their volume. These glasses show also an anomaly in the low temperature specific heat. This anomaly may roughly be described as a broad smooth maximum superimposed on a nearly linear background. This maximum approximately coincides with the dip in the sound velocity. The purpose of the present study is to further test the domain model.

## EXPERIMENTAL

The glass samples for this study were prepared by firing appropriate mixtures of reagent grade manganese carbonate and aluminum oxide with pure silica sand in an arc-image furnace. In the heating process the carbonate is converted to manganese oxide by dissociation. The molten material was quenched by pouring into an aluminum mold and allowing it to cool to room temperature. Portions of the samples were examined by powder X-ray diffraction and electron microscopy. No crystallinity was detected by these techniques. The two samples used for the specific heat measurements were in the form of cylinders about 2 cm long and 1.3 cm thick. Each sample weighed about 8 grams. The specific heat was measured using a standard heat pulse technique which has been described elsewhere.[4] The magnetic susceptibility of a small piece of glass cut from the end of each of the cylindrical samples was measured at 500 Hz using a low field (~5 G) mutual inductance bridge technique. The compositions of the samples studied and the corresponding magnetic susceptibility data [i.e., effective ionic moment $\mu_{eff}$, peak temperature $T_{max}$, the value of KV/k obtained from a least squares fit of the experimental data] are listed in Table I.

TABLE I

Compositions and Susceptibility Data

| Molar % | | | at.% | | | |
|---|---|---|---|---|---|---|
| MnO | $Al_2O_3$ | $SiO_2$ | Mn | $\mu_{eff}(\mu_B)$ | $T_{max}$(K) | KV/k(K) |
| 40 | 20 | 40 | 13.3 | 6.29[a] | 2.95 | 8.7[b] |
| 40 | 10 | 50 | 14.3 | 6.89[a] | 4.25 | 12.6 |

[a]R.W. Kline, Ph.D. Thesis, W.S.U., 1974, and Ref. 2.

[b]Extrapolated value.

The specific heat data for the sample containing 14.3 at.% Mn are displayed in Fig. 1. The magnetic susceptibility of the same sample is shown in the insert of this Figure. A least squares fit of the susceptibility using the expression given in Ref. 2 is also shown in the insert.

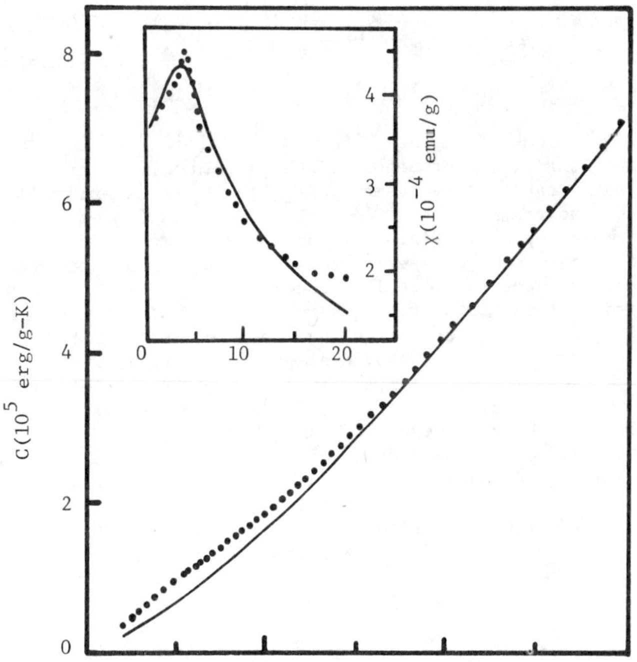


Fig. 1. Specific heat vs. temperature of the 14.3 at.% Mn glass. Dots are experimental points; solid curve is $\alpha T + \beta T^{3/2} + \gamma T^3$. Insert shows susceptibility of the same sample (dots experimental points, solid curve theoretical fit).

## DISCUSSION

We begin by dividing the glass in $\nu$ small domains of volume V and anisotropy energy density K. The domains are assumed to be antiferromagnetically ordered, but because of their smallness they possess a net magnetic moment equal to $\mu_{eff}\sqrt{N}$, where N is the number of magnetic ions per domain. These magnetic moments are oriented randomly in the glass. The specific heat of small noninteracting magnetic domains has been described by Livingston and Bean,[5] and we shall use their result to analyze the experimental data. Since the magnetic moments of the domains are several $\mu_B$, we may use the expression for the specific heat of a domain with infinite magnetic moment:

$$C_d = k[\tfrac{1}{2} - \frac{e^a}{4I}\cdot(1 - 2a + \frac{e^a}{I})] , \tag{1}$$

where

$$a \equiv \frac{KV}{kT} , \qquad I = \int_0^1 e^{ax^2}\, dx .$$

Since each domain carries a net magnetic moment, we speculate that, as in a ferrimagnet, long wavelength spinwave excitations give rise to a contribution to the specific heat proportional to $T^{3/2}$. Phonons contribute a $T^3$ term. The data also indicate the presence of a term linear in T, whose origin will be discussed below. The expression for the specific heat used to analyze the data is then

$$C = \alpha T + \beta T^{3/2} + \gamma T^3 + \nu C_d . \tag{2}$$

We used Eqs. (1) and (2) to obtain a least squares fit of the data. The best values of the constants $\alpha$, $\beta$, $\gamma$, $\nu$, and KV/k, which were adjusted in the fitting routine, are listed in Table II. The values of KV/k in Table I and Table II differ by a factor of about three, which considering the crudeness of the domain model represents reasonable agreement. In Fig. 1 we have plotted the background $\alpha T + \beta T^{3/2} + \gamma T^3$ for the 14.3 at.% Mn sample, while in Fig. 2 we compare $\nu C_d$ with the difference between the experimental data and the background for the same sample. The agreement is good. The agreement for the 13.3 at.% Mn sample (graph not shown here) is somewhat better. It should be noted that at 38 K $\gamma T^3$ is only about 10% of the measured specific heat (the negative sign for the 14.3 at.% glass is therefore not meaningful). Independently, measurements of the specific heat between 100 K and 400 K as well as a theoretical estimate using a sound velocity of 6 x $10^5$ cm/sec, also showed that at 38 K a Debye-like specific heat is about 10% of the measured specific heat.

TABLE II

Specific Heat Parameters

| at.% Mn | $\alpha\left(\frac{erg}{g\text{-}K^2}\right)$ | $\beta\left(\frac{erg}{g\text{-}K^{5/2}}\right)$ | $\gamma\left(\frac{erg}{g\text{-}K^4}\right)$ | $\nu$(#/g) | KV/k(K) |
|---|---|---|---|---|---|
| 13.3 | $8.4\times10^3$ | $2.3\times10^3$ | +1.3 | $2.2\times10^{20}$ | 31.5 |
| 14.3 | $4.0\times10^3$ | $4.0\times10^3$ | -2.3 | $1.7\times10^{20}$ | 36.6 |

The linear contribution to C requires comment. So far we have assumed that all domains have the same KV value. However, a narrow peaked distribution with relatively flat tails would be more realistic. Integration of $C_d(KV/kT)$ over such a distribution gives $C_d$ at its peak, while its tails lead to a contribution nearly linear in temperature. A maximum value of KV/k of about 100 K is required to explain the magnitude of $\alpha$.

From the values of $\beta$ and using spinwave theory, we can estimate the exchange constant for these glasses. We find a value of about 80 K, which is just about the temperature for which the susceptibility of these glasses begins to depart from linear Curie-Weiss behavior.[2] This is also consistent with the fact that the increase in magnetic entropy associated with Eq. (2) up to 38 K is about 90% of the maximum possible magnetic entropy.

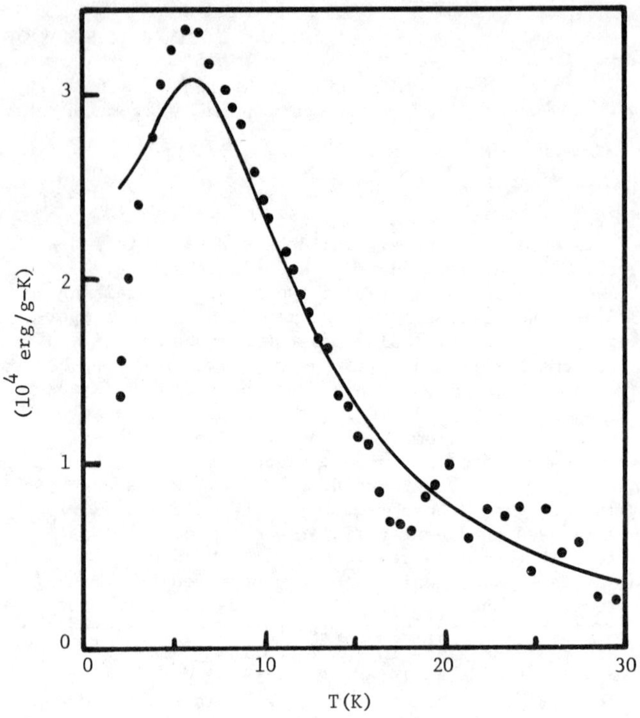


Fig. 2. Domain contribution $\nu C_d$ of the 14.3 at.% Mn glass (solid curve); $\alpha T + \beta T^{3/2} + \gamma T^3$ subtracted from experimental data (dots).

Using for the density a value of 3 g/cm$^3$, we have calculated K, V, N, and the magnetic moment of a domain. These are listed in Table III, together with the magnetic moment obtained from the least squares fit of the susceptibility. The values of K are typical of crystalline Mn compounds. At present no convincing explanation is available for the factor of about seven or eight difference between the values of the magnetic moment.[6] Additional experimental and theoretical work is needed to obtain a more detailed understanding of these glasses. Such work is currently in progress and will be reported on in a future publication.

TABLE III

Domain Parameters

| at.% Mn | K(erg/cm$^3$) | $V^{1/3}$(Å) | N | $\frac{\mu}{\mu_B}$(C) | $\frac{\mu}{\mu_B}(\chi)$ |
|---|---|---|---|---|---|
| 13.3 | $2.8\times10^6$ | 11.5 | 15.3 | 24.6 | 206[c] |
| 14.3 | $2.5\times10^6$ | 12.6 | 21.0 | 31.6 | 215 |

[c]Extrapolated value.


## ACKNOWLEDGMENTS

One of us (A.M. de Graaf) would like to thank G.F. Hawkins and F.S. Huang for assistance with the preparation of this paper.

REFERENCES

*Supported in part by the National Science Foundation.

†Work supported by NSF-MRL Grant DMR 72-03018 A03.

[1]T.J. Moran, et al., Phys. Rev. 11, 4436 (1975).

[2]R.A. Verhelst, R.W. Kline, A.M. de Graaf, and H.O. Hooper, Phys. Rev. 11, 4427 (1975).

[3]V. Cannella and J. Mydosh, Phys. Rev. B 6, 4220 (1972).

[4]G.M. Seidel and P.H. Keesom, Rev. Sci. Instrum. 29, 606 (1958).

[5]J.D. Livingston and C.P. Bean, J. Appl. Phys. 32, 1964 (1961).

[6]See however Ref. 2, p. 4434.

# SPINWAVE DISPERSION AND TEMPERATURE DEPENDENCE OF MAGNETIZATION IN AN AMORPHOUS Co-P ALLOY*

James R. McColl and D. Murphy
Yale University, New Haven, Connecticut 06520

G. S. Cargill III and T. Mitzoguchi+
IBM T. J. Watson Research Center, Yorktown Heights, N.Y. 10598

## ABSTRACT

We have used ferromagnetic resonance to measure the spinwave dispersion coefficient D and the temperature dependence of magnetization M(T) in thin films of an electrodeposited amorphous Co-P alloy of ∿25 at. % P. From spinwave dimensional resonance, D = 138 ± 5 mev Å$^2$ at 4.2°K, which is in substantial agreement with that expected from the coefficient of $T^{3/2}$ in the low temperature magnetization ($T < T_C/3$). These results indicate that long wavelength spinwave excitations are responsible for 90 ± 8 % of the observed decrease of magnetization at these low temperatures and that apparent discrepancies between results from inelastic neutron scattering and from low temperature magnetization measurements for amorphous ferromagnetic alloys should be reexamined.

## INTRODUCTION

Ferromagnetic resonance (FMR) has been widely used to investigate long wavelength magnetic excitations in thin films of crystalline and polycrystalline materials; it has proved to be useful in complementing the more direct technique of inelastic neutron scattering. It therefore seemed worthwhile to use FMR to study both spinwave dispersion and temperature dependence of magnetization in amorphous magnetic materials, and to this end, we have investigated FMR in thin films of an amorphous, ferromagnetic Co-P alloy.

This study was motivated in part by apparent discrepancies between results from inelastic neutron scattering and from low temperature magnetization measurements for two amorphous ferromagnetic alloys: $Co_{80}P_{20}$[1] and $Fe_{75}P_{15}C_{10}$[2]. The neutron scattering results are consistent with existence of long wavelength spinwave excitations with the usual dispersion relation

$$E(q) = Dq^2. \tag{1}$$

Thermal excitation of spinwaves satisfying Eq. (1) yields the well known result[3] that the magnetization at low temperatures follows the equation

$$M(T) = M(0)\,(1 - B\,T^{3/2} - C\,T^{5/2} + \dots), \tag{2}$$

where

$$B = \frac{2.612\ g\mu_B}{M(0)}\left(\frac{k}{4\pi D}\right)^{3/2}. \tag{3}$$

In these equations, M is the magnetization, $\mu_B$ is the Bohr magneton, and k is Boltzmann's constant. For the $Co_{80}P_{20}$, D = 185 mev Å$^2$ at room temperature from the neutron scattering measurements[1], and the value of B predicted[4] from Eq. (3) is $B_{sw} = 1.65 \times 10^{-5}$ $°K^{-3/2}$. This is significantly smaller than the value for B, $B_{mag} = 2.6 \times 10^{-5}$ $°K^{-3/2}$, obtained by Cochrane and Cargill from low temperature magnetization measurements on a film of similar composition;[5] $B_{sw}/B_{mag} = 0.63$. This may suggest that additional, non-wavelike excitations are also contributing to the magnetization reduction at low temperatures, i.e. to $B_{mag}$; such a direct comparison of these $B_{sw}$ and $B_{mag}$ values may not be proper, since different samples, prepared in different laboratories, were used in these two experiments. However, a similar discrepancy has also been reported for amorphous $Fe_{80}P_{15}C_{10}$: $B_{sw}/B_{mag} = 0.68$.[2]

## EXPERIMENTAL

Thin films of Co-P were prepared by electrodeposition onto pyrex substrates which had been coated first with 500 Å of Cr and then with 5000 Å of Au by rf sputtering. Film thicknesses were determined by optical interference techniques. Electron microprobe measurements indicated that the films contained 24.0 ± 1.5 at. % P, but magnetization and Curie temperature measurements are more consistent with previous data[6] if a value of 25 ± 1 is taken for the P content. The films were grown under conditions similar to those described in ref. 7, but the bath composition was modified to incorporate more phosphorus in the films.[8]

In the FMR experiment, thin films subjected to microwave magnetic fields at frequency ν, in an external magnetic field $H_q$ perpendicular to the film, display resonances satisfying

$$h\nu = g\mu_B\,(H_q + H_a - 4\pi M) + Dq^2 \tag{4}$$

where $H_a$ is the perpendicular uniaxial anisotropy field.[9] From M-H loop measurements on a 5 μm thick film, we found $H_a$ ∿ 100 Oe at room temperature, which is small compared to 4πM ∿ 4000 G; such measurements also suggest that $H_a(T)$ is proportional to M(T) for T < 300°K.[10]

For the "main mode," i.e. the mode of strongest intensity and smallest q, the resonance field $H_o$ should be given by $h\nu \simeq g\mu_B\,(H_o + H_a - 4\pi M)$. This mode is also easily observed with H parallel to the film, in which case the resonance condition is

$$h\nu \simeq g\mu_B\,(H_o\,(H_o - H_a + 4\pi M))^{1/2}, \tag{5}$$

so that $4\pi M - H_a$ and g are determined by observation of the main mode in the two orientations. We find a temperature independent (4.2°K, 77°K, 296°K) g-value of 2.125 ± 0.007 for our samples. At 77°K, $4\pi M - H_a$ values for the four films ranged from 5560 to 5740 G, with an average of 5650 G. We attribute this variation in $4\pi M - H_a$ to small (± 1 at. %) differences in P content.

FMR spectra for four films of thickness 1530 to 3110 Å, at 77°K, with H perpendicular to the films, are shown in Fig. 1. The data were obtained with ν = 9.22 GHz. In addition to the main mode at field $H_o$ (which is off scale at this sensitivity), additional weaker modes at lower fields $H_q$ are clearly demonstrated. These modes are shifted from $H_o$ by the $Dq^2$ term in Eq. (4) and are associated with allowed standing-wave values of q, i.e. wavelengths comparable to the film thicknesses t. The small variation in $H_o$ from film to film reflects the variation in $4\pi M - H_a$ mentioned above. Our main task is to extract a value for D from the positions of these resonances. This task reduces to assigning q values to each and every mode according to a physically reasonable, self-consistent procedure.

## DISCUSSION

In Fig. 2 we illustrate two procedures that have been used in the past for determining mode assignments. The first procedure utilizes the assumption that the magnetization is strongly pinned at the film surfaces,[11] so that standing-wave values of q satisfy q = pπ/t, where p is an odd integer. A self-consistent set of odd-integer assignments and the corresponding values of D for each of the four films are shown as the squares plotted in Fig. 2. It is apparent that a good consistency of D-values is obtained.

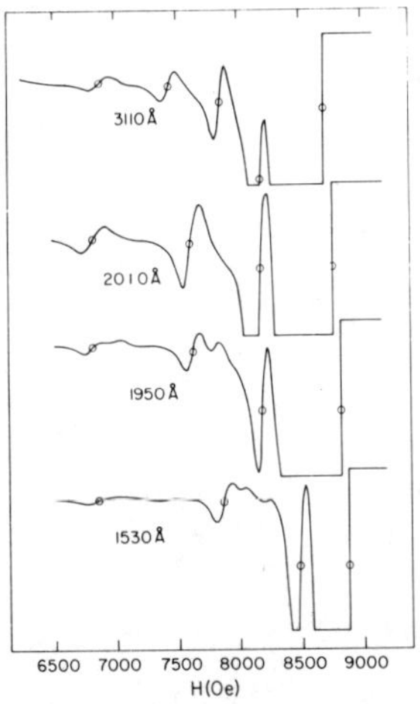


Fig. 1. FMR derivative spectra at 77°K. Mode positions $H_o$ and $H_q$ are indicated by circles. A few weaker, "forbidden" modes can also be seen.

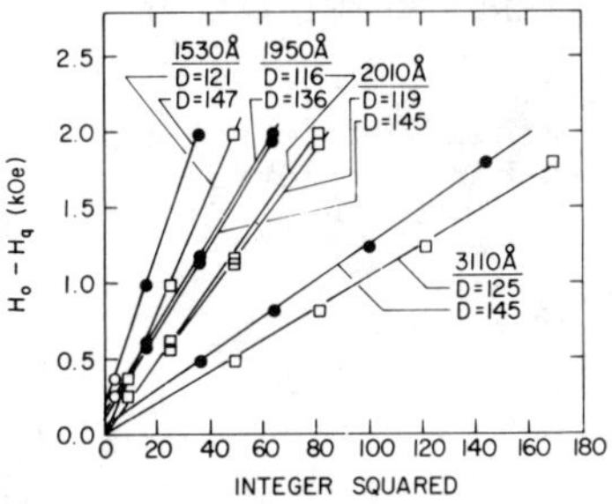


Fig. 2. Splitting of resonance fields $H_q$ from the main mode fields $H_o$ vs integers squared: squares: odd-integers circles: even integers.

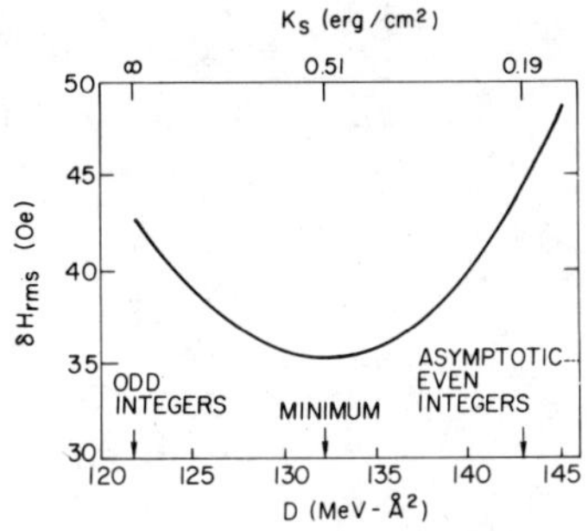


Fig. 3. RMS deviation $\delta H_{rms}$ vs D. Optimum values of $K_S$ are also indicated.

A second procedure follows from the observation that a wide variety of surface boundary conditions[12] all lead to the same transcendental equation for the standing-wave values of q:

$$\tan(qt/2) = \frac{2\ g\ K_S\ \mu_B}{M\ D\ q}, \qquad (6)$$

where $K_S$ is an effective surface anisotropy energy per unit area and is taken to have the same value for both film surfaces. If $K_S$ is not too large and D is not too small, it is found that $(qt/\pi)^2$ tends asymptotically to an even-integer-squared plus a constant. In this procedure, $H_o - H_q$ values are plotted vs even-integers-squared and the slopes of lines passing asymptotically through data points for the higher modes yield values for D.[12] This procedure, along with the resulting D values, is also illustrated (circles) in Fig. 2. The filled circles indicate the modes used in the asymptotic fitting. It is apparent that these values are about as self-consistent as those obtained by the first procedure, but that D values from the two procedures differ by about 17%.[13]

A more general procedure for obtaining D is to find values of D and $K_S$ which yield the best overall fit to the resonance field values for all films. Rather than use a standard, non-linear, two-parameter, least-squares algorithm to accomplish this, we have adopted the following somewhat more illuminating (and simpler!) procedure. We choose a value for $K_S/D$ and solve Eq. (6) for q for all films. We then solve the one-parameter, linear, least-squares equations for the value of D which minimizes the sum of squares of deviations

$$\delta H_{rms} = [\sum_{1}^{N} (H_{obs} - H_{calc})^2/N]^{1/2} \qquad (7)$$

and plot this vs D. Results of these calculations are shown in Fig. 3. The minimum deviation occurs for D = 132 mev Å$^2$; the corresponding value of $K_S$ is 0.51 erg/cm$^2$. Also indicated in Fig. 3 are the values of D obtained using the two procedures discussed above; the D value obtained by actually solving Eq. (6) yields a somewhat better overall fit to the experimental data. From our FMR data for 77°K, we conclude that D = 132 ± 5 mev Å$^2$.

We have also obtained FMR data on the 2010 and 3110 Å films at 4.2 and 296°K. Since this range of thicknesses is small, it seemed most reasonable not to attempt to deduce $K_S$ from the limited data, but to use the value of $K_S/MD$ found from the 77°K data, and simply to find the value of D which gives the best overall fit to the data at each of the two additional temperatures. The resulting values of D, and also the values of $4\pi M - H_a$, obtained from these two films are given in Table I. Corresponding values of the "exchange stiffness parameter" $A = MD/(2g\mu_B)$ are also given.

Table I

| | | | |
|---|---|---|---|
| T (°K) ......... | 4.2 | 77 | 296 |
| D (mev Å$^2$) ......... | 138 | 132 | 94 |
| A ($10^{-7}$ erg/cm) ... | 2.65 | 2.45 | 1.24 |
| $4\pi M$-$H_a$ (emu/cm$^3$)... | 5780 | 5590 | 3950 |

## CONCLUSIONS

The values of D found here may not be directly compared with the inelastic neutron scattering results,[1] since the sample compositions are different. The present values of D may be used, however, to evaluate B from Eq. (3). From the 4.2°K results, we calculate $B_{sw} = (3.9 \pm 0.2) \times 10^{-5}$ °K$^{-3/2}$. In a direct determination of B in the same samples by using FMR to measure M(T), to be described in more detail elsewhere,[14] we find $B_{mag} = (4.34 \pm 0.15) \times 10^{-5}$ °K$^{-3/2}$, i.e. $B_{sw}/B_{mag}$ = 0.90. From these results we conclude that 90 ± 8 % of the magnetization reduction at low temperatures in these samples is due to thermal excitation of long wavelength spinwaves. The previously mentioned, much larger discrepancy between neutron scattering and low temperature magnetization data for amorphous ferromagnetic alloys remains unresolved.

We acknowledge with pleasure helpful discussions with R. Barker, C. H. Wilts, C. H. Bajorek, and technical assistance of G. C. Chi.

## REFERENCES

*Work at Yale University supported in part by N.S.F.
+Permanent address: Faculty of Science, Gakushuin University, Mejiro, Tokyo, Japan.

1. H.A. Mook, N. Wakabayashi, and D. Pan, Phys. Rev. Letters 34, 1029 (1975).
2. J.D. Axe, L. Passell, and C.C. Tsuei, AIP Conf. Proc. 24, 119 (1975); C.C. Tsuei and H. Lilienthal, to be published; N. Kazama, T. Masumoto, and H. Watanabe, J. Phys. Soc. (Japan) 37, 1171 (1974).
3. C. Herring and C. Kittel, Phys. Rev. 81, 869 (1951).
4. Only room temperature neutron scattering data are are available. A larger value of D, and a smaller value for $B_{sw}$, would presumably have been obtained from measurements at lower temperatures. The $Co_{80}P_{20}$ sample was produced by D. Pan at Harvard University. The M(0) value is taken from ref. 6.
5. R.W. Cochrane and G.S. Cargill III, Phys. Rev. Letters 32, 476 (1974).
6. D. Pan and D. Turnbull, J. Appl. Phys. 45, 1406 (1974).
7. G.S. Cargill III and R.W. Cochrane, in Amorphous Magnetism, H.O. Hooper and A.M. deGraff, Eds. (Plenum Press, New York, 1973), p. 313; and J. Phys. (Paris) 35, C4-269 (1974).
8. The present plating bath consisted of $H_3PO_3$ (95 g/l), $CoCO_3$ (50.6 g/l), $CoCl_2 \cdot 6H_2O$ (139 g/l), and $H_3PO_4$ (50 g/l).
9. C. Kittel, Introduction to Solid State Physics, 4th ed. (Wiley, New York, 1971), Chap. 17.
10. G.S. Cargill III, R.J. Gambino, and J.J. Cuomo, IEEE Trans. Mag. MAG-10, 803 (1974).
11. M.H. Seavey Jr. and P.E. Tannenwald, Phys. Rev. Letters 1, 168 (1958).
12. C.H. Bajorek and C.H. Wilts, J. Appl. Phys. 42, 4324 (1971).
13. C.H. Wilts and S.K.C. Lai, IEEE Trans. Mag. MAG-8, 280 (1972).
14. J.R. McColl, D. Murphy, G.S. Cargill III, and T. Mitzoguchi, to be published.

# XPS STUDIES OF CO-SPUTTERED Gd-Fe ALLOYS

G. Güntherodt and N.J. Shevchik
Max-Planck-Institut für Festkörperforschung,
Stuttgart, Federal Republic of Germany

## ABSTRACT

The electronic structure of co-sputtered $Gd_xFe_{1-x}$ alloys ($0 \le x \le 1$) has been studied by means of monochromatized X-ray photoemission (XPS). For the amorphous Gd- and Fe-rich compositions, respectively, the conduction band spectra, including the 4f levels of Gd, are surprisingly very similar to those of crystalline Gd and Fe. These spectra of the alloys appear to be well described as a linear superposition of those of the pure elements. Small Gd→Fe charge transfer upon alloying is inferred from the unchanged Fe 2p core-level line shape asymmetry and the small shift in energy of the Gd- and Fe core levels. On the basis of a simple relationship between charge transfer and altered interatomic separation distances, the charge transfer in the amorphous alloys appears to be too small, compared to the crystalline phases, in order to account for the magnitude of the Fe moment.

## INTRODUCTION

During the past few years, amorphous rare earth (RE)-transition metal (TM) alloys, as for instance Gd-Co and Gd-Fe, have attracted strong attention because of their possible applications as bubble domain- and magneto-optical devices.[1,2] Because of the significantly different size of the TM and RE atoms, they do not form equilibrium solid solutions, but rather metastable, amorphous alloys. The latter are prepared by either sputtering[1,2] or thermal evaporation.[3,4] Up to now, only the magnetic, magneto-optical and structural properties of amorphous RE-TM alloys have been investigated. In this preliminary study, we have investigated for the first time the electronic structure of co-sputtered $Gd_xFe_{1-x}$ alloys over the entire composition range by means of X-ray photoemission (XPS).

## EXPERIMENT

The samples for our investigation were prepared by dc co-sputtering of the pure metal targets in a pure (99.999%) argon atmosphere at 30 mTorr onto room temperature substrates. It has been shown previously[1] by electron microscopy and electron diffraction that the as-prepared alloys are amorphous and structurally stable at room temperature.[3] However, the as-deposited films of pure Gd and Fe are crystalline.[3] The monochromatized X-ray photoemission spectra (XPS) of co-sputtered $Gd_xFe_{1-x}$ alloys have been measured by means of a Hewlett-Packard 5950 ESCA spectrometer, with a resolution of 0.6eV. The oxygen build-up at the sample surface during the course of the measurements has been controlled by frequently recording the intensity of the O 1s core level and not letting it exceed about one monolayer.

## RESULTS

Figure 1 shows representative room temperature spectra of the conduction band- and 4f level region of co-sputtered $Gd_xFe_{1-x}$ for $0 \le x \le 1.0$. The spectra of pure, polycrystalline Fe and Gd, respectively, are in good agreement with measurements of Ley et al.[5] and McFeely et al.[6] The s,d-derived conduction bands are seen in the vicinity of $E_F$, whereas the Gd 4f level is found around 9.4eV below $E_F$. Upon alloying in going from polycrystalline Fe and Gd, respectively, to amorphous Fe- and Gd-rich alloy compositions, we observe a surprisingly small change and striking similarity in the XPS spectra. As a function of increasing Gd concentration, we see in Fig.1 a continuous reduction in intensity of the Fe conduction band, but no significant change in its position in energy and width. At the same time, the density of states (DOS) at $E_F$ is not drastically changed. On the other hand, the Gd 4f level is continuously increasing in intensity, but not significantly shifting in energy with respect to pure Gd. The XPS spectra of the amorphous alloys can well be described by a linear superposition of those of the pure, crystalline metals. This is shown for $Gd_{0.5}Fe_{0.5}$ in Fig. 2, where the maxima of the conduction

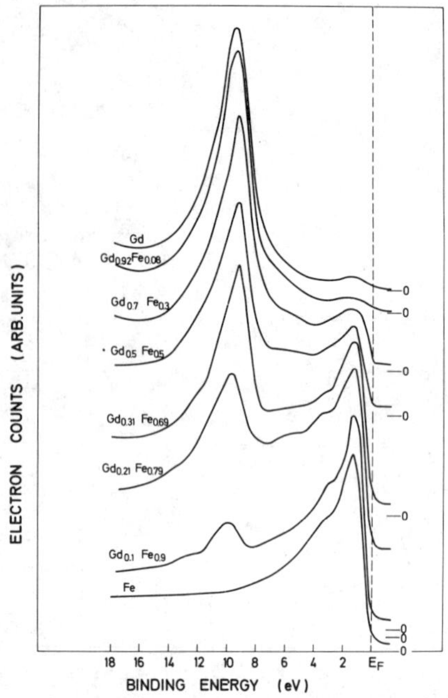


Fig. 1 XPS spectra of $Gd_xFe_{1-x}$ alloys for $0 \le x \le 1.0$

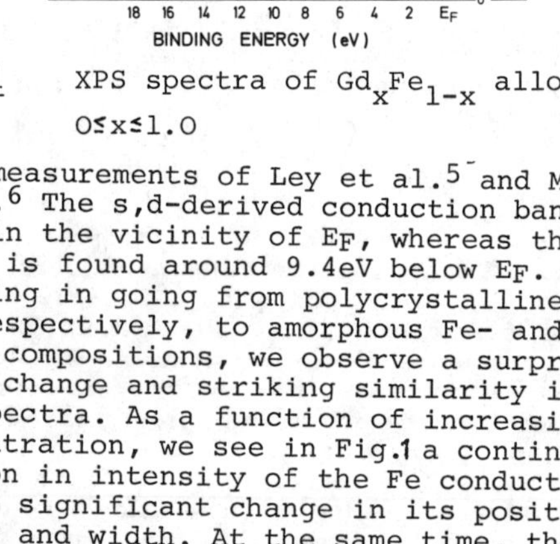

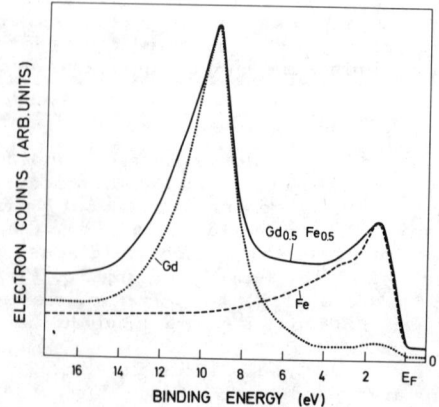

Fig. 2 Decomposition of the XPS spectrum of $Gd_{0.5}Fe_{0.5}$

band and of the 4f level, respectively, have been matched with those of pure Fe and Gd.

The Fe 2p and Gd 4d core levels have been measured over the entire composition range. With respect to other, corresponding core levels, they show fairly small shifts of less than about 1eV towards higher binding energy with decreasing Fe content. If a total charge transfer took place so that Gd and Fe atoms, respectively, had electronic configurations like Eu and Mn, the core levels would shift by several eV towards higher binding energy. Hence we conclude that, if the small core level shifts are solely due to charge transfer, the latter is fairly small. The $2p_{3/2}$ core level of pure Fe shows an asymmetric line shape as seen in Fig. 3. Upon alloying no change in the line-shape asymmetry can be seen in Fig. 3, where $Gd_{0.5}Fe_{0.5}$ is shown as a representative example. (The two core levels have been matched for comparison.)

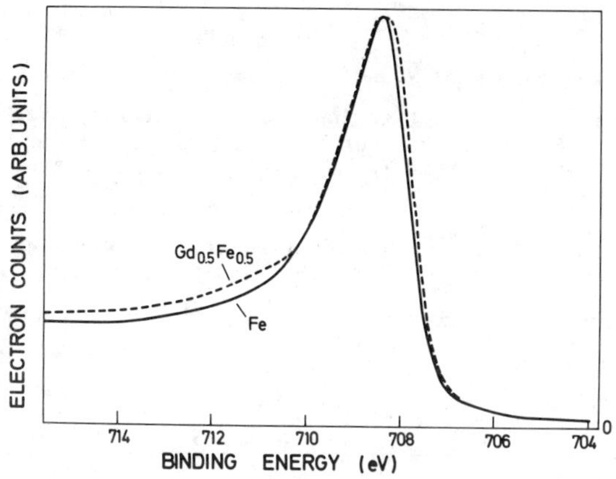


Fig. 3 Fe $2p_{3/2}$ core level of pure Fe and $Gd_{0.5}Fe_{0.5}$

DISCUSSION

Increasing charge transfer from Gd to the partially filled minority spin band of Co (majority band filled) with increasing Gd concentration in crystalline Gd-Co intermetallics accounts for the decrease of the Co magnetic moment and of the ordering temperature.[7] The higher Co moment and ordering temperature in amorphous Gd-Co alloys, compared to their crystalline phases, is explained by a reduced charge transfer due to the increased Gd-Gd separation.[7] On the other hand, an analogous interpretation for crystalline and amorphous Gd-Fe alloys in terms of charge transfer is much more complex since majority as well as minority spin bands of Fe are both partially filled. Hence, an attempt to account for the change in Fe moment upon charge transfer depends on a detailed knowledge of the DOS of both spin bands. Moreover, such an attempt meets with a big controversy about photoemission results on transition metals[5,8-10] and the predictions of the Stoner-Wohlfarth-Slater band model of ferromagnetism. Apart from this, however, certain trends in charge transfer can be deduced from our XPS data. In crystalline Gd-Fe phases, the decrease in Fe moment with decreasing Fe content[11] is ascribed to an increasing charge transfer from Gd to Fe atoms, analogous to crystalline Gd-Co phases.[7] The transfer in crystalline Gd-Fe phases is expected in the order of 1 electron per Gd atom, as estimated for crystalline $GdCo_2$.[7] A similarly large charge transfer is to be expected also for amorphous Gd-Fe alloys since in the crystalline as well as amorphous phase of e.g. $GdFe_2$ the Fe moment of $1.55\mu_B$ is the same,[11] compared to $2.22\mu_B$ in pure Fe. However, although in the amorphous Gd-Fe alloys, the shift of the Gd 4d core level indicates an increasing charge transfer with decreasing Fe concentration, analogous to the crystalline phases, its magnitude of at most 1eV can account only for a small transfer. Moreover, in the amorphous phase, these electrons must not necessarily be transferred completely onto the Fe d levels because of the extra space provided for them due to the smaller density. The Fe 2p core level shift indicates also a small charge transfer. If the Fe d level is filled up at the expense of the s level, no large charge transfer need take place, in agreement with the small core-level shift.

Although there is Gd→Fe and Fe s→d charge transfer taking place in amorphous Gd-Fe alloys, its magnitude is much smaller compared to what is expected for the crystalline phase. Hence, the reduced, but unchanged Fe moment in the crystalline as well as amorphous phase of e.g., $GdFe_2$ cannot be explained on the basis of a simple relationship between charge transfer and observed interatomic separation distances, as in the case of Gd-Co alloys. A further clue for this we believe is seen in the Fe 2p core-level line shape asymmetry. As has recently been shown for Pt and Ni,[13] the core-level asymmetry is a measure of the local DOS at $E_F$. Since we do not observe any change in the Fe 2p core-level asymmetry in going from pure, crystalline Fe over the entire composition range of amorphous Gd-Fe alloys, we conclude that the local DOS of Fe at $E_F$ is not signficantly changed. This might be considered as indicative of a small charge transfer to the Fe d level. However, it remains subject to future investigations how this unchanged local DOS at $E_F$ can be reconciled with the change in Fe moment from $2.22\mu_B$ for pure, crystalline Fe to $1.55\mu_B$ for crystalline as well as amorphous $GdFe_2$.

We thank L. Ley and P.C. Kemeny for valuable discussions.

+Present address: SUNY Stony Brook, Stony Brook,N.Y., U.S.A.

REFERENCES

1. P. Chaudhari, J.J. Cuomo and R.J. Gambino, IBM J.Res. Developm. 17, 66 (1973).
2. P. Chaudhari, J.J. Cuomo and R.J. Gambino, Appl. Phys. Lett. 22, 337 (1973).
3. J. Orehotsky and K. Schröder, J. Appl. Phys. 43, 2413 (1972).
4. K. Lee and N. Heiman, AIP Conf. Proc. No. 24, 108 (1974).
5. L. Ley, S.P. Kowalczyk, F.R. McFeely and D.A. Shirley, to be published.
6. F.R. McFeely, S.P. Kowalczyk, L. Ley and D.A. Shirley, Phys. Rev. 45A, 227 (1973).
7. L.J. Tao, S. Kirkpatrick, R.J. Gambino and J.J. Cuomo, Sol. State Comm. 13,1491 (1973).
8. D.T. Pierce and W.E. Spicer, Phys. Rev. B6, 1787 (1972).
9. G. Busch, M. Campagna and H.C. Siegmann, Phys. Rev. B4, 746 (1971).
10. J.E. Rowe and J.C. Tracy, Phys. Rev. Lett. 27, 799 (1971).
11. J.J. Rhyne, J.H. Schelleng and N.C. Koon, Phys. Rev. 10, 4672 (1974).
12. G.S. Cargill, AIP Conf. Proc. No. 18,p.631, (1974).
13. N.J. Shevchik,Phys.Rev.Lett. 33, 1336 (1974).

# MAGNETIC RESONANCES IN a-$TbFe_2$, a-$GdFe_2$ and a-$YFe_2$*

S. M. Bhagat** and D. K. Paul
University of Maryland, College Park, Maryland 20742

ABSTRACT

The resonance microwave absorption in amorphous rare-earth-iron compounds (e.g., $TbFe_2$, $GdFe_2$ and $YFe_2$) has been investigated at several frequencies and temperatures. The resonances showing a four-line ($S_1$ through $S_4$) spectra, though somewhat similar in all three systems, have many puzzling features. The strong resonance ($S_1$) represents the usual $q \cong 0$ mode maintained by thermally activated magnon hopping among clusters whose local magnetization is close to the saturation magnetization, $M_s$. The $S_4$-line, however, appears to have much weaker magnetization, and has been assigned to localized magnon mode. A preliminary discussion of the data is presented in terms of a spin-cluster model.

## INTRODUCTION

We report measurements on the spin resonances in a-$TbFe_2$, a-$GdFe_2$ and a-$YFe_2$ at several frequencies between 9 and 22 Ghz and at several temperatures between 77 K and 500 K. The samples were made available to us by Dr. Rhyne. Bulk samples of amorphous rare-earth-iron compounds have been prepared by d.c. triode-sputtering method. The details of the structural and compositional characterization of these materials have been described earlier.[1,2] The data were taken with a conventional microwave spectrometer, the sample temperature being monitored by a thermocouple attached to the cavity.[3] Typically, the samples were small chips (∿ 1 mm x 1 mm x 0.1 mm thick) or needles (∿ 0.1 mm dia x 2 mm long). However, in a-$YFe_2$ most of the data were taken with a parallelopiped (1 mm x 2 mm x 4 mm) specimen. The magnetic field was applied in the sample plane.[4] For convenience, we present the results in three parts.

## RESULTS

(i) Room Temperature: Fig. 1 shows a typical spectrum. This trace is for a sample of a-$GdFe_2$ at 10.8 Ghz. However, the spectra in the other two materials are formally identical, that is, in every case we observe a four line spectrum: a strong resonance ($S_1$) accompanied by three weak satellites ($S_2$ through $S_4$) on its high field side. The following points should be noted: (a) the relative intensities of the satellites are very sensitive to the position of the sample in the cavity; (b) in comparison with $S_1$, the satellites are weakest in a-$GdFe_2$ and strongest in a-$YFe_2$; (c) after due account for the demagnetizing field, the field for $S_1$ obeys the equation

$$(\omega/\gamma)^2 = H(H + 4\pi M_s) \tag{1}$$

where the values of $\gamma$ and $M_s$, along with the values of the saturation magnetization measured earlier,[1] are given in Table I. Here we point out the first surprising result. Even though a-$YFe_2$ does not exhibit

Table I. Parameters for Equation (1) in Amorphous $RFe_2$ (where R is Tb, Gd or Y) at 300 K.

| Material | $\gamma$ ($S^{-1}$ $Oe^{-1}$) | $4\pi M_s$ (Oe) (Resonance Data) | $4\pi M_s$ (Oe) Static Meas.[A] |
|---|---|---|---|
| a-$TbFe_2$ | $1.79 \times 10^7$ | 4570 | 4380 |
| a-$GdFe_2$ | $1.52 \times 10^7$ | 4750 | 4670 |
| a-$YFe_2$ | $1.85 \times 10^7$ | 5400 | 156[B] |

A. Experimental data from Rhyne, et al. (Ref. 1).
B. Shows short-range ordering only.

ordering it seems to show one spin resonance which looks very much like ordinary ferromagnetic resonance from a system with a sizeable $M_s$; (d) the second surprise comes when we examine $H_4$, the field for $S_4$, and

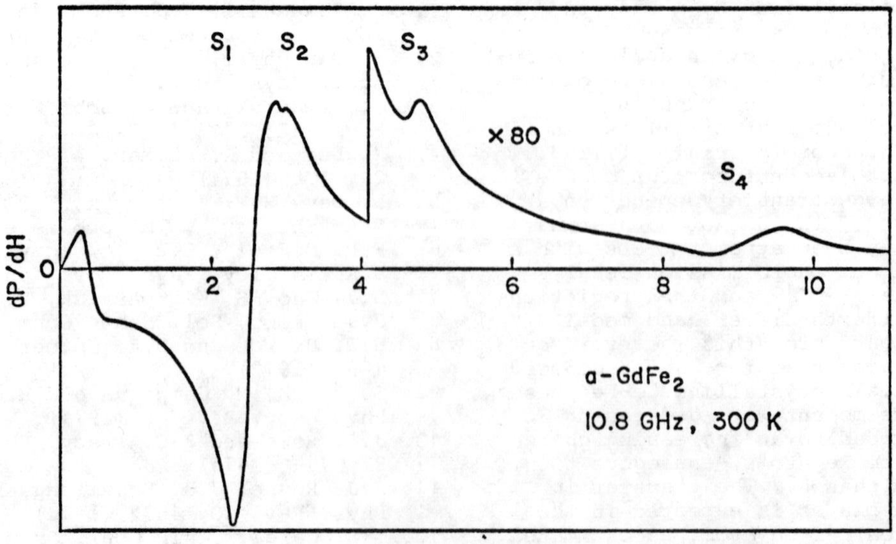


Figure 1. Field derivative of the power absorption.

find that it obeys the equation:

$$(\omega/\gamma') = H_4 - C \qquad (2)$$

with C = (4200 ± 100) Oe and $\gamma' \cong 1.2$ for all three systems. Further, $H_4$ is independent of sample, shape or size, i.e., does not reflect the changes in demagnetizing fields; (e) in a-$TbFe_2$ and a-$GdFe_2$, $S_2$ is sufficiently weak and sufficiently close to $S_1$ that so far we have not been able to measure it with any precision; (f) the next surprise is that in a-$YFe_2$ the position of $S_2$ is strongly dependent upon the angle between the applied field and the long axis of the parallelopiped (Fig. 2). Further, as shown in Fig. 2, this anisotropy reduces with increasing frequency.

(ii) Low Temperature: Several more surprising results emerge when the temperature is lowered; (a) the $S_1$ line in a-$TbFe_2$ first broadens and then essentially disappears. It is, in some sense, replaced by several weak resonances; (b) for all three systems the $S_4$ line is completely unaffected by the variation of temperature down to 77 K; (c) in a-$YFe_2$ the $S_1$ and $S_2$ lines shift lower in field and the anisotropy in $S_2$ noted earlier (Fig. 2) disappears at 77 K. On the other hand, the $S_1$ line in a-$GdFe_2$ lowers in field and behaves, by and large, as expected for ordinary ferromagnetic resonance.

(iii) High Temperature: So far we have looked mainly at a-$TbFe_2$. Once again, $S_4$ is found to be unaltered for $300 < T \lesssim 500$ K; this despite the fact that a-$TbFe_2$ has its Curie point at about 380 K. The variation of $S_1$ with temperature cannot be explained using the relation $(\omega/\gamma)^2 = [H(H + 4\pi M_s)]$ unless one is willing to make $\gamma$ (or g-value) a rather strong function of temperature. In order to account for the a-$TbFe_2$ data one will have to let g increase from 2.04 to 3 in the temperature range between 300 and 400 K.[3]

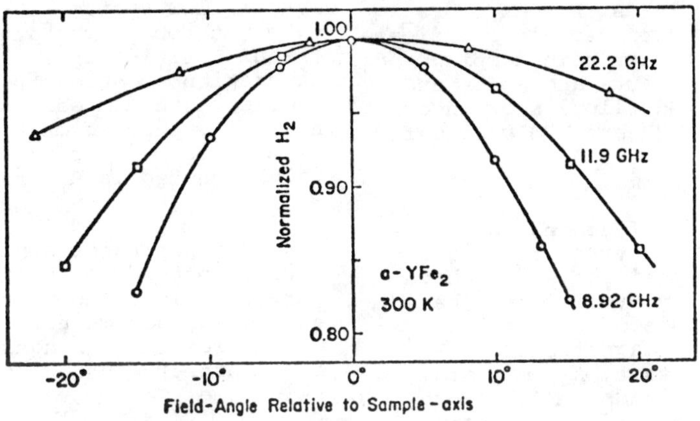


**Figure 2.** Dependence of resonance field for $S_2$ on field angle. At 0° the field is parallel to the longest edge of the parallelopiped.

## DISCUSSIONS

Some of the above results can be rendered plausible if, following Rhyne et al[1,2] these systems are regarded as composed of loosely coupled clusters of spins with randomly varying magnetizations (exchange) and/or anisotropies. The $S_1$ resonance in a-$TbFe_2$ and a-$GdFe_2$ can then be attributed to those clusters whose magnetization is close to the full saturation value. Since this is essentially a $q \simeq 0$ line (uniform mode) it must be maintained by magnon hopping between clusters. If the hopping energy is large (small) we expect lowering of temperature to have a strong (weak) effect on this line. It is conceivable that a-$TbFe_2$ and a-$GdFe_2$ are representative of these two situations. For instance, static measurements show that a-$TbFe_2$ develops a large coercive field and magnetic aftereffect at low T. It seems to us that the disappearance of $S_1$ is closely tied to this phenomenon. On the other hand, at this time we do not understand why non-magnetic a-$YFe_2$ has so many clusters with large $M_s$ to account for the $S_1$ observed in this material.

We suggest that the $S_4$ line comes from clusters whose magnetization is very small, hence the linear dependence on frequency. This low value of the magnetization is most likely a consequence of the combined effects of the varying exchange and dipolar fields and if this picture is right it is not surprising that $H_4$ is independent of demagnetizing fields and/or the static magnetization. The constant term in Eq. (2) may be related to an effective exchange frequency, $\omega_e$ (= $Dq^2$), and since it is negative it implies a negative value for $q^2$ or Im [q] >> Re [q]. This mode will therefore be highly localized.

We are deeply grateful to Dr. Rhyne for the loan of the specimens.

## REFERENCES

* Supported in part by the N.S.F. and the O.N.R.
** Address till July 1976, University of Uppsala, Institute of Physics, Uppsala, Sweden.
1. J. J. Rhyne, J. H. Schelleng and N. C. Koon, Phys. Rev. B10, 4672 (1974).
2. J. J. Rhyne, S. J. Pickart and H. A. Alperin, AIP Conf. Proc. 18, 563 (1974).
3. S. M. Bhagat and D. K. Paul, Phys. Rev. Lett 35, 1458 (1975).
4. Some measurements have now been made with magnetic field perpendicular to the sample surface and the same features were observed as with the parallel field.

# FMR IN SOME AMORPHOUS RE - 3-d TRANSITION METAL FILMS

P. Lubitz, J. Schelleng, C. Vittoria
Naval Research Laboratory, Washington, D.C. 20375

and

Kenneth Lee, IBM Research Laboratory, San Jose, Ca. 95193

## ABSTRACT

Ferrimagnetic Resonance (FMR) and vibrating sample magnetometer (VSM) measurements were made on amorphous $GdCo_3$ and Gd-, Ho-, Tb- and $DyFe_2$ for temperatures from 90 to 500K. For Gd alloys, the T dependence of the linewidth is in general agreement with that predicted for a T independent Landau-Lifshitz damping parameter of $\sim 10^8$/s. Ho- and $TbFe_2$ have progressively broader resonances indicating an increase in both intrinsic damping and inhomogeneous broadening. Dy- and $HoFe_2$ and to a lesser degree the Gd alloys show inconsistencies between the FMR and VSM values for the magnetization.

## INTRODUCTION

Magnetic properties of amorphous alloys of rare earth (RE) and 3-d transition metals (TM) of approximate composition $RE\text{-}TM_2$ have recently been extensively studied.[1-3] These measurements have emphasized such static properties as saturation magnetization, M, uniaxial anisotropy constant, $K_u$, magnetostriction and coercive fields. FMR has provided some information about dynamic parameters, such as g value, in addition.[4] To date, however, the dynamic quantities accessible to FMR, such as relaxation rates, exchange stiffness and g values are largely unstudied in RE-TM alloys. Knowledge of these properties may be useful, both in understanding relaxation processes and magnetic interactions in amorphous alloys and in developing practical devices incorporating high frequency and/or non-uniform excitations, e.g., transducers and bubble memories.

Using VSM, and FMR measurements at 9, 22 and 72 GHz and temperatures from 90 to about 500K, we have studied a number of thin film amorphous RE-TM alloys whose static properties have been previously characterized[1]. The variation of the linewidth with frequency allows the intrinsic contribution to be separated from most inhomogeneous sources of linewidth which are not strongly frequency dependent.

Based on a two sublattice model[5] with sublattice Landau-Lifshitz relaxation rates $\lambda_1$ and $\lambda_2$, moments $M_1$ and $M_2$ and gyromagnetic ratios $\gamma_1$ and $\gamma_2$ ($\gamma_i = g_i e/2mc$), the peak-to-peak linewidth obtained from the derivative of absorption, $\Delta H$, is related to the relaxation rates by:

$$\Delta H = 2\omega\lambda/\sqrt{3}\gamma_1\gamma_2(M_1+M_2) \qquad (1)$$

where $\lambda = \lambda_1(\gamma_2/\gamma_1) + \lambda_2(\gamma_1/\gamma_2) \simeq \lambda_1+\lambda_2$. g and $(4\pi M - 2K_u/M)$ were determined from the measured fields for in plane ($\parallel$) and perpendicular ($\perp$) uniform resonance[4]:

$$(\omega/\gamma) = H_\perp - (4\pi M - 2K_u/M) \qquad (2a)$$

$$(\omega/\gamma)^2 = H_\parallel(H_\parallel + 4\pi M - 2K_u/M) \qquad (2b)$$

where

$$\gamma = (M_1+M_2)/(M_1/\gamma_1 + M_2/\gamma_2) \qquad (3)$$

Separation of M and $2K_u/M$ requires VSM measurements.

## EXPERIMENTAL

The thin films were prepared by thermal evaporation onto glass substrates[1] except for the $GdCo_3$ films which were bias field sputtered to produce large $K_u$[4]. Film thicknesses were .3 to 1. μm except for $HoFe_2$ which was 750Å thick. FMR measurements were made with conventional reflection cavity spectrometers with field modulation to display the field derivative of absorption. Variable temperatures for the 9 GHz measurements were obtained using a $N_2$ gas flow system. Data at 22 and 72 GHz were taken at room temperature (RT). VSM data were taken at fields up to 18 KOe by which point saturation appeared to be complete in all samples. The compositions may vary from the nominal values as much as 2%, in which case the RT moments may differ considerably from the values quoted in Reference 1.

## RESULTS

The resonance spectra for the RE-TM films studied often showed magnetic excitation structure additional to that expected in a uniform saturated ferrimagnetic film. Nevertheless, in most cases the strongest absorption lines could be interpreted using Eq. 2. From this analysis using the $\parallel$ and $\perp$ resonance fields, the values of g and of $(4\pi M - 2K_u/M)$ were determined. We list below these quantities and the narrowest linewidths obtained at RT for the three frequencies: At the higher frequencies, some of the resonances could not be observed.

The $GdFe_2$, $GdCo_3$ and $HoFe_2$ resonances were observed in the temperature range from 90 to 500K at 9.23 GHz. In Fig. 1 we show the linewidths, g values and $(4\pi M - 2K_u/M)$ derived from FMR data using Eq. 2 as well as VSM data for $4\pi M$ and $2K_u/M$, as functions of temperature. The thinness of our $HoFe_2$ sample precluded accurate determination of its moment. The linewidth in $TbFe_2$ broadens rapidly below RT. For $DyFe_2$ very little temperature dependence was seen in any of the quantitites.

## DISCUSSION

For $GdFe_2$ only a small part of the width at RT appears to be non-intrinsic. The 9 and 22 GHz data indicate $\lambda \simeq 2.2 \times 10^8$/s near RT. Again, the non-intrinsic contribution is small. For $HoFe_2$ using the FMR values for M we find that $\lambda \simeq 3 \times 10^8$/s and that at least 200 Oe of width at RT is not intrinsic. Values of $\lambda$ for $DyFe_2$ appear to be of the same order of magnitude and are another order of magnitude larger in $TbFe_2$. However, since we lack data at high frequencies for these two alloys, these $\lambda$'s should be considered as upper bounds.

The temperature dependence of the $GdFe_2$ data at 9 GHz is relatively easy to interpret, since the FMR and VSM data for M are consistent. The magnetization curve is typical of an inhomogeneous system[6,7]. The change in g value with temperature can be interpreted as reflecting higher values of $M_{Fe}/M_{Gd}$ as the Neel temperature, $T_N$, is approached (see Fig. 1). The linewidth data follow approximately Eq. 2, i.e., $\Delta H \propto 1/M(T)$ above RT, although near $T_N$ the linewidth increases somewhat less rapidly. The origin of the slight rise in width below RT has not been determined. The temperature dependence of the width in $GdCo_3$ shows a similar qualitative dependence on $1/M(T)$. A quantitative comparison to M(T) is not possible

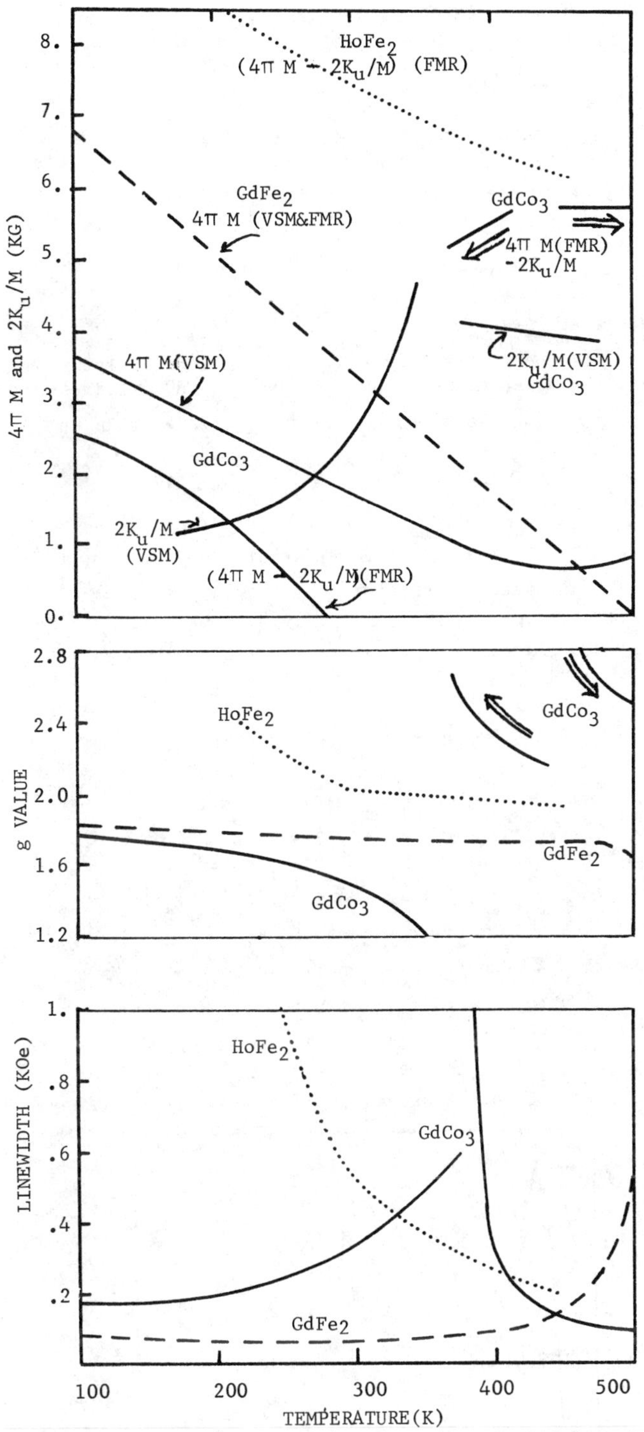


Fig. 1. $4\pi M$, $2K_u/M$, g and $\Delta H_\perp$ for $GdCo_3$ (solid lines) $GdFe_2$ (dashed lines) and Ho $Fe_2$ (dotted lines) as functions of temperature. The double arrows along the $GdCo_3$ data indicate the sequence of measurements. The FMR data were taken at 9.23 GHz.

using our present VSM data because the data are not sufficiently precise near the compensation point ($T_c$). Above $T_c$ the lines become very narrow which suggests that $\lambda$ is greatly reduced from its RT value. The FMR data for Ho (and Dy) alloys indicate that these systems are ordered well above RT. The increase in g in $HoFe_2$ at low T may reflect increasing Ho moment relative to the Fe moment, although in that case M should reduce. The large linewidths at low temperature in Ho- and $TbFe_2$ suggests inhomogeneous broadening with magnetostrictive origins.

| composition | g | $4\pi M$-$2K_u/M$(KG) | $\Delta H_{\|}$(Oe) | $\Delta H_\perp$(Oe) |
|---|---|---|---|---|
| | | 9.23 GHz | | |
| $GdFe_2$ | 1.74 | 3.8-4.3 | 45 | 70 |
| $GdCo_3$ | 1.43 | -.8 | 320 | 320 |
| $HoFe_2$ | 2.05 | 7.33 | 300 | 300 |
| $TbFe_2$ | >2 | ~6 | 1500 | 1500 |
| $DyFe_2$ | 2.1 | 18.-20. | 800 | 80 |
| | | 22.0 GHz | | |
| $GdFe_2$ | 1.75 | 4.83 | 160 | 170 |
| $GdCo_3$ | 1.38 | -1.0 | 750 | 800 |
| $HoFe_2$ | 2.03 | 7.8 | 450 | 450 |
| | | 72.5 GHz | | |
| $GdFe_2$ | 2.05 | 7.07 | 650 | 600 |
| $HoFe_2$ | 2.04 | 7.95 | 950 | 1350 |

## CONCLUSIONS

FMR and VSM observations of $GdFe_2$, $GdCo_3$ and Ho-$Fe_2$ indicate Landau-Lifshitz relaxation rates of about 1,2 and $3 \times 10^8$/s respectively near RT. These rates are comparable to those in pure TM's[8]. Additional contributions, especially in $HoFe_2$, which decrease with annealing, may be a result of non-uniform strains which couple to the magnetostriction.

In general, FMR indicates a considerably different state of ordering than the average over the sample as obtained from VSM. The $GdFe_2$ M(T) data, and other measurements[6,9] indicate that the amorphous RE-TM alloy systems may be microscopically inhomogeneous. In this case, and especially if the system is not well ordered magnetically, the FMR may be probing small (no VSM signal is seen), iron rich (g>2) or microcrystalline regions of the sample which have large M and/or $K_u$. In any case, a number of questions remain as to the interpretation of the FMR in these systems.

## REFERENCES


1. Kenneth Lee and Neil Heiman, AIP Conf. Proc. 24, 108 (1974).
2. J. Orehotsky and K. Schroder, J. Appl. Phys. 43, 2413 (1972).
3. J.J. Rhyne, J.H. Schelleng and N.C. Koon, Phys. Rev. B10, 4672 (1974).
4. D.C. Cronemeyer, AIP Conf. Proc. 18, 85 (1973).
5. S.V. Vonsovskii, Ed., Ferromagnetic Resonance, p. 66 (Pergamon Press, New York 1966).
6. C. Vittoria, P. Lubitz and J. Schelleng, AIP Conf. Proc. (to be published).
7. D.L. Huber and R.P. Siemann, Solid State Comm. 17, 769 (1975).
8. S.M. Bhagat and P. Lubitz, Phys. Rev. B10, 179 (1974).
9. S.M. Bhagat and D.K. Paul, AIP Conf. Proc. (to be published).

# NET ANISOTROPY AND FERROMAGNETIC RESONANCE FREQUENCY OF AN AMORPHOUS FERROMAGNET

Peter M. Richards
Sandia Laboratories, Albuquerque, NM 87115

ABSTRACT

The net anisotropy constants and the ferromagnetic resonance (FMR) frequency are calculated in a model of amorphous ferromagnetism for which the axes of strong local uniaxial anisotropy are nearly randomly oriented throughout the sample, and, consequently, the equilibrium directions of local magnetization are nonuniform. A slight preferential ordering produces net anisotropy perpendicular to the plane of a thin film sample as appropriate to sputtered Gd-Co films. The net anisotropy is a maximum for a given $H_A$ near $H_A/H_E = 1.5$, where $H_A$ and $H_E$ are the local anisotropy and exchange fields, and at $H_A/H_E = 10$ it is about 3 times less than its value for $H_A/H_E \rightarrow 0$, corresponding to uniform magnetization. The FMR resonance frequency $\omega$ would contain a term of the order of $H_A^2/H_E$, which can be comparable to the net anisotropy field $\langle H_A \rangle$, if the magnetization were uniform, but this term cancels out if proper account is taken of the nonuniformity. As a result, $\omega$ is the same as would be computed for a uniform anisotropy field $\langle H_A \rangle$.

## INTRODUCTION

Amorphous magnetic thin films have been studied[1-3] which show a net anisotropy favoring magnetization perpendicular to the plane of the film. This feature makes the materials interesting for magnetic bubble devices as well as for basic studies of amorphous magnetism. The net anisotropy, which has been measured by ferrimagnetic resonance[4] and other means, is believed to be the result of local ordering[1,5] which gives a slight overall preference for the perpendicular orientation of magnetization. In the most widely investigated Gd-Co films the local anisotropy has been proposed[5] to be that of $GdCo_5$ units which have an easy plane anisotropy constant[6] of $K = 4 \times 10^7$ ergs/cm$^3$. Since the observed[4] net anisotropy constant is of the order of $10^5$ ergs/cm$^3$ in the amorphous films, only about 1% alignment of the local easy planes is required

Thus, the atomic spin experiences a local anisotropy field far in excess of the net anisotropy field. and there can be spatial variations in the direction of magnetization due to the nearly random directions of the local anisotropy field. In this paper we calculate the net anisotropy and ferromagnetic resonance frequency taking into account the spatial variations.

## NET ANISOTROPY

We have recently[7] discussed a model in which the local magnetization $\vec{M}_i$ makes an angle $\theta_i$ with respect to the local uniaxial anisotropy axis and an angle $\Delta\theta_i$ with respect to the local exchange field $\vec{H}_E$. The exchange plus anisotropy energies are therefore

$$U = K \sum_i \sin^2 \theta_i - \lambda M_i \langle M \rangle \sum_i \cos \Delta\theta_i \tag{1}$$

where K is the uniaxial anisotropy constant, assumed to be constant in magnitude throughout the specimen; the local exchange field is taken to be $\vec{H}_E = \lambda \langle \vec{M} \rangle$ in a mean field approximation where the magnitude of $\langle \vec{M} \rangle$ is $\langle M \rangle = N^{-1} M_i \sum_i \cos \Delta\theta_i$ and $M_i$, the magnitude of local magnetization, is the same throughout. For a given angle $\alpha_i$ of $\vec{H}_E$ with respect to the local anisotropy axis, energy minimization requires that

$$\sin(2\alpha_i - 2\Delta\theta_i) = \frac{2H_E}{H_A} \sin \Delta\theta_i \tag{2}$$

where $H_A = 2K/M_i$.

If the probability $\rho(\beta)$ that the local anisotropy axis makes an angle between $\beta_i$ and $\beta_i + d\beta_i$ with respect to the c-axis of net anisotropy is written as

$$\rho(\beta) = \frac{1}{4\pi} [1 + \sum_{n=1}^{\infty} \delta_{2n} P_{2n}(\cos \beta)] \tag{3}$$

where $P_{2n}$ is the $2n^{th}$ order Legendre polynomial and $\delta_{2n} \ll 1$ (only even numbers are included since we expect $\rho$ to be an even function of $\cos \beta$), then the combination formula for spherical harmonics shows that the energy may be expressed as

$$U = \text{const} + \sum_{n=1}^{\infty} \delta_{2n} P_{2n}(\cos \theta_c) k_{2n} \tag{4}$$

to first order[8] in the $\delta_{2n}$ where $\theta_c$ is the angle of $\langle \vec{M} \rangle$ with respect to the c-axis. The reduced anisotropy constants $-k_2/|K|$ and $k_4/|K|$ are plotted in Fig. 1 as functions of $H_A/H_E$. From (4) it is evident that the net n = 1 anisotropy constant is $\langle K_1 \rangle = -\frac{3}{2} k_2 |\delta_2|$. Note that $k_2$ and $k_4$ have opposite signs and that $k_4$ is at most about 20% of $k_2$. With the sign convention $\delta_{2n} = |\delta_{2n}| K/|K|$ the value of $k_{2n}$ is independent of the sign of K. The microscopic origins of Eq. (3) which lead to a net anisotropy are not considered here.

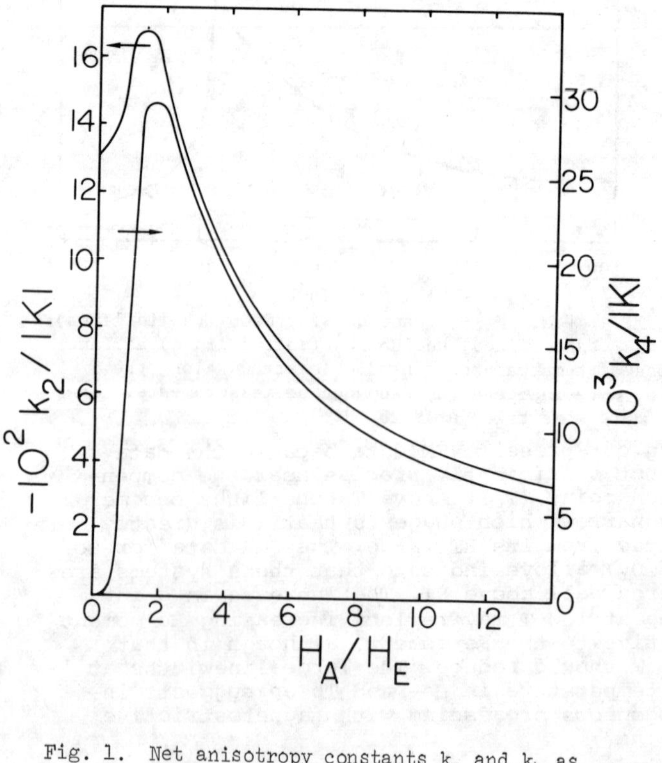


Fig. 1. Net anisotropy constants $k_2$ and $k_4$ as defined by Eq. (4).

## FERROMAGNETIC RESONANCE FREQUENCY

The ferromagnetic resonance (FMR) equations of motion are

$$\frac{d\vec{M}_i}{dt} = \gamma[M_i \times (\vec{H}_o + \vec{H}_{A_i}) + \lambda\vec{M}_i \times \langle\vec{M}\rangle] \qquad (5)$$

$$\frac{d\langle\vec{M}\rangle}{dt} = \gamma N^{-1} \sum_i [\vec{M}_i \times (\vec{H}_o + \vec{H}_{A_i})] \qquad (6)$$

where $\vec{H}_o$ is the external field; $\gamma$ is the gyromagnetic ratio; N is the number of spins; $\vec{H}_{A_i}$ is the local anisotropy field; the local exchange field has been approximated by $\lambda\langle\vec{M}\rangle$ as in the previous section; and $\langle\vec{M}\rangle = N^{-1}\sum_i \vec{M}_i$. Because of the fluctuation in local anisotropy directions, the equilibrium axes of $\vec{M}_i$ and $\langle\vec{M}\rangle$ do not coincide, but are at the angle $\Delta\theta_i$. The procedure therefore is to let $\vec{M}_i = (M_i^{x'}, M_i^{y'}, M_i)$ in the x′, y′, z′ coordinate system and $\langle\vec{M}\rangle = (M^x, M^y, \langle M\rangle)$ in the x, y, z system — where $\hat{z}'$ and $\hat{z}$ are equilibrium directions of $\vec{M}_i$ and $\langle\vec{M}\rangle$, respectively — and to keep only terms linear in $M_i^{x'}$, $M_i^{y'}$, $M^x$, $M^y$ in Eqs. (5) and (6). We then assume an $e^{i\omega t}$ time dependence and look for a solution with $\omega \ll \gamma\lambda\langle M\rangle = \gamma H_E$, in which case the left-hand side of Eq. (5) may be set equal to zero to obtain expressions for $M_i^{x'}$ and $M_i^{y'}$ in terms of $M^x$ and $M^y$ which may then be substituted into Eq. (6) to yield a secular equation for $\omega$.

For simplicity we make the following assumptions: (i) $H_o$ is sufficiently large compared with the net anisotropy field that the equilibrium z direction of $\langle\vec{M}\rangle$ is along $\vec{H}_o$; (ii) $H_o \ll H_A \ll H_E$ so that $H_o$ and $H_A^2/H_E$ are comparable in magnitude and both type terms must be included, but terms of the order of $H_A(H_A/H_E)^2$ may be neglected. Note that $\Delta\theta_i$ is of the order of $H_A/H_E$. For the large local anisotropies of the materials of interest here, only the case $H_A \ll H_E$ is of practical concern to FMR, since considerable exchange narrowing would be required in order to make the line narrow enough to be observed. (iii) terms of the order of $\delta_2 H_A^2/H_E$ are negligible, where $\delta_2$ is defined by Eq. (3). That is, we keep $H_A^2/H_E$ terms only when they are multiplied by angular factors which do not average to zero for a completely random distribution of anisotropy axes. Terms of the order of $\delta_2 H_A$ are included, however, since they give rise to the lowest order effects of net anisotropy. With assumptions (i)-(iii), the result is

$$\omega^2 = (H_o + \tfrac{1}{2} H_A\langle 3\cos^2\alpha - 1\rangle - \tfrac{2}{3} H_A^2/H_E + \tfrac{5}{2} H_A\langle\Delta\theta \sin 2\alpha\rangle)^2 - \tfrac{1}{4} H_A^2(\langle \sin 2\varphi \sin^2\alpha\rangle^2 + \langle \cos 2\varphi \sin^2\alpha\rangle^2) \qquad (7)$$

where $\varphi$ is the azimuthal angle of $\hat{z}'$ with respect to the x, y, z axes and $\alpha$, as before, is the angle of the local anisotropy axis with respect to $\langle\vec{M}\rangle$. In arriving at Eq. (7) we have used, consistent with assumption (i), $H_A \cos^2\theta_i = H_A \cos^2(\alpha_i - \Delta\theta_i) \approx H_A \cos^2\alpha_i + H_A\Delta\theta_i \sin 2\alpha_i$.

If the equilibrium directions of local magnetization did not fluctuate, so that $\Delta\theta_i = 0$, then there would be a contribution to $\omega$ of the order of $H_A^2/H_E$, which is likely to be comparable to $H_A\langle 3\cos^2\alpha - 1\rangle$. However, Eq. (2) gives $\Delta\theta_i \approx \frac{1}{2}(H_A/H_E)\sin 2\alpha_i$ for $H_A/H_E \ll 1$, so that the complete coefficient of $H_A^2/H_E$ in (7) is $-\frac{2}{3} + \frac{5}{2}\cdot\frac{1}{2}\langle\sin^2 2\alpha\rangle$, which is zero for a random distribution of anisotropy axes ($\langle\sin^2 2\alpha\rangle = 8/15$). In his original work on the subject,[9] Schlömann assumed all the local magnetizations to be aligned in the same direction and therefore found an $H_A^2/H_E$ correction to the resonance frequency. This study, however, shows that the lack of complete alignment produces another $H_A^2/H_E$ term which exactly cancels it to make the net correction identically zero. As a consequence, one may show that the distribution of anisotropy axes given by Eq. (3) (assuming only $\delta_2$ is important) produces a result for $\omega$ which is the same as would be obtained for a uniform magnetization and a uniform anisotropy field $\langle H_A\rangle = \delta_2 H_A/5 = 2\langle K_1\rangle/\langle M\rangle$, so that the resonance frequency may be calculated simply as if the spatial variations of the amorphous magnet were ignored.

## CONCLUSIONS

The bulk observed anisotropy constant $\langle K_1\rangle$ depends on the ratio of $H_A/H_E$ as well as on the local anisotropy field $H_A$ and the distribution of anisotropy axes. In particular, Fig. 1 shows that, for a given $H_A$, $\langle K_1\rangle$ goes through a maximum at $H_A/H_E \approx 1.5$ and for larger $H_A/H_E$ decreases to values considerably less than would be obtained if the magnetization were uniform ($H_E \to \infty$). The variation in the directions of local magnetization does not, however, influence the ferromagnetic resonance frequency which may be computed using standard formulae for uniform magnets and the same bulk $\langle K_1\rangle$. A further feature of note is that for finite $H_A/H_E$ there can be small components of fourth and higher order net anisotropy even though the local anisotropy is second order ($E_A \propto \cos^2\theta$).

## REFERENCES

*Work supported by the U. S. Energy Research and Development Administration.

1. R. J. Gambino, P. Chaudhari, and J. J. Cuomo, A. I. P. Conf. Proc. 18, 578 (1974).
2. N. Heiman and K. Lee, Phys. Rev. Lett. 33, 778 (1974).
3. N. Heiman, et al., A. I. P. Conf. Proc. 24, 573 (1975).
4. D. C. Cronemeyer, A. I. P. Conf. Proc. 18, 85 (1974).
5. A. Onton, N. Heiman, W. Parrish, and J. C. Suits, Bull. Am. Phys. Soc. 20, 458 (1975).
6. W. A. J. J. Velge and K. H. J. Buschow, J. Appl. Phys. 39, 1717 (1968).
7. P. M. Richards, Phys. Lett. 55A, 121 (1975).
8. Since the exchange part of the total energy involves $\langle\cos\Delta\theta\rangle^2$, there will be contributions of the order of $\delta_{2m}\delta_{2n}$ to the effective net anisotropy energy. Such terms are ignored here.
9. E. Schlömann, J. Phys. Chem. Solids 6, 242 (1958).

# CURIE TEMPERATURES OF AMORPHOUS RFe$_2$ ALLOYS

J. J. Rhyne

Institute for Materials Research, National Bureau of Standards*
Washington, DC 20234 and White Oak Laboratory, NSWC, White Oak, MD 20910

## ABSTRACT

Curie temperatures of the series of amorphous rare earth-iron alloys RFe$_2$ (where R = Gd, Tb, Dy, Ho, Er, and Y) have been determined from Belov-Goryaga plots of the magnetization isotherms. The Curie temperature of GdFe$_2$ is 500 K, and drops sharply (e.g. ErFe$_2$ T$_c$ = 105 K) as one proceeds to the right in the above series (decreasing R spin). The observed Curie temperatures exhibit a smooth variation with the DeGennes factor of the rare earth ion, with the zero spin limit YFe$_2$ exhibiting no long-range order. This is in marked contrast to the analogous crystalline Laves phase compounds for which YFe$_2$ has a 535 K Curie temperature and which show a much weaker dependence on the rare earth spin. These results imply a more significant effect of the structural disorder on the direct Fe-Fe exchange versus the RKKY Fe-R and R-R couplings. A molecular field model describing these interactions has been used to calculate the expected Curie temperature for both the amorphous and crystalline series. Overall agreement with the observed T$_c$'s was less satisfactory for the amorphous than for the crystalline materials.

## INTRODUCTION

This paper presents results for the Curie temperatures of the series of ferrimagnetically ordered heavy rare earth-iron (RFe$_2$) amorphous alloys prepared by Battelle Northwest Laboratories[1] using rapid dc sputtering. The Gd, Tb, and Ho alloys have been previously studied extensively by neutron[2] and X-ray scattering[3] and have been determined to be amorphous with a structure corresponding to a dense random packing of R and Fe atoms.[4] Previous magnetization studies[5] on the Tb, Gd, and Y alloys over the entire range of magnetic order have provided evidence for a strong local random direction anisotropy interaction in amorphous TbFe$_2$ of magnitude comparable to the R-Fe exchange. This produces a large high field susceptibility and a significant lowering of the measured 0 K spontaneous magnetization due to the non-colinear R spin alignment. At low temperatures coercive fields as large as 32 KOe are observed with a sharp temperature dependence qualitatively explained by an interacting spin-cluster model.[5] Evidence for this nonhomogeneous or "micro-domain" ordering of the spin system on a scale of order 100Å is provided also by neutron small angle scattering.[2]

## MAGNETIZATION

Magnetization data as a function of temperature for the RFe$_2$ series were taken using a PAR vibrating sample magnetometer and an 18 KOe electromagnet. Isothermal magnetization data were plotted as $\sigma^2$ ($\sigma$ = magnetic moment/g, units of $\sigma$: 1 emu/g = 1 erg/(gauss-g)) versus $H_i/\sigma$ ($H_i$ = applied field corrected for demagnetization) to determine the Curie temperature as shown in Figure 1a for HoFe$_2$. This method (Belov-Goryaga[6] plot) assumes the free energy near T$_c$ can be written as a sum of second and fourth order terms in the magnetization, and it provided the most accurate determination of T$_c$ for these amorphous alloys. Determinations of T$_c$ from the vanishing of the spontaneous moment in the isothermal magnetization curves presents considerable uncertainty due to the field-dependent high field susceptibility near T$_c$. Attempts to use such extrapolations will produce a higher T$_c$ than the 0 intercept curve on the Belov-Goryaga plot.

From the data of Figure 1a, T$_c$ for amorphous HoFe$_2$ is 194K. The susceptibility above T$_c$ taken from the H/$\sigma$ intercept exhibits a linear temperature dependence for T>215 K and extrapolates to a $\theta$ = 206 K as shown in Figure 1b. The slope corresponds to an effective moment of 6.7 $\mu_B$.

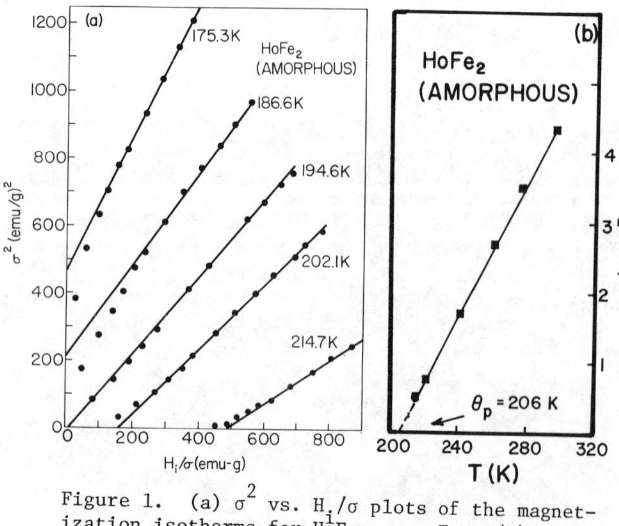

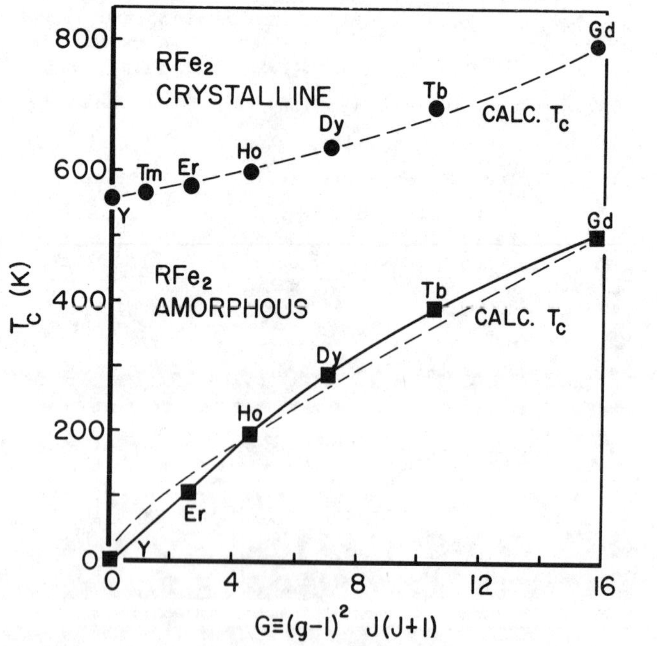

Figure 2. Curie temperatures of crystalline RFe$_2$ compounds and amorphous RFe$_2$ alloys versus DeGennes factor. The dashed line is a molecular field calculation (see text).

Figure 1. (a) $\sigma^2$ vs. $H_i/\sigma$ plots of the magnetization isotherms for HoFe$_2$ near T$_c$. (b) Reciprocal susceptibility obtained from H/$\sigma$ intercept of (a) versus temperature.

## CURIE TEMPERATURES

The transition temperatures for the RFe$_2$ series (R=Gd, Tb, Dy, Ho, Er and Y) in both crystalline and amorphous forms are shown in Figure 2 as a function of the effective rare earth spin squared given by the DeGennes factor $G=(g_J-1)^2 J(J+1)$. The crystalline compound data taken from Burzo[7] and Buchow and Van Stapele[8] show a large 535 K Curie temperature for

$YFe_2$ (G=0) which increases monotonically as the rare earth spin moment increases to 793 K for $GdFe_2$. In contrast, the amorphous alloys show a much stronger dependence of $T_c$ on rare earth spin, and anomalously show no long range magnetic order for $YFe_2$. The ordering temperature of amorphous $TmFe_2$ could only be determined to be less than 50 K due to the complicating large coercive field behavior seen in these alloys at low temperatures. The strong dependence of $T_c$ on rare earth spin and the anomalous vanishing of magnetic order as the rare earth spin goes to zero is quite puzzling, and suggests that the background Fe-Fe exchange (in $YFe_2$) is strongly quenched by the structural disorder, leaving only the R-Fe and R-R exchange to affect magnetic order. These latter two are of a long range electron polarization wave indirect exchange type which may be less effected by disorder than the Fe-Fe exchange interaction which is probably of much shorter range.

The dashed lines in Figure 2 show the results of applying a two sublattice molecular field calculation by Cullen[9] to the crystalline and amorphous systems. Due to its intrinsic nature, the molecular field model applied to determining $T_c$ is not expected to give accurate results for the exchange constants; however, it is useful in making comparisons between crystalline and amorphous systems. In this model the rare earth and iron sublattice magnetizations are written respectively as

$$M_R = N_R \mu_B J\, B_J \left[ \frac{J_{R-R}(g_J-1)^2 J M_R/N_R + J_{R-Fe}(g_J-1) J M_{Fe}/N_{Fe}}{kT} \right] \quad (1)$$

and

$$M_{Fe} = N_{Fe} \mu_B S B_S \left[ \frac{J_{Fe-Fe} S M_{Fe}/N_{Fe} + J_{R-Fe} S(g_J-1) M_R/N_R}{kT} \right] \quad (2)$$

where $B_J$ and $B_S$ are the Brillouin functions for rare earth total angular momentum J and for iron spin S; $J_{Fe-Fe}$, $J_{R-Fe}$ and $J_{R-R}$ are the exchange couplings, $\mu_B$ is the Bohr magneton and $N_R$ and $N_{Fe}$ are the number of rare earth and iron atoms per formula unit.

Solving expressions (1) and (2) for the transition temperature gives

$$\theta_c = \tfrac{1}{2}\theta_c^{Fe} + J_{R-R}G/6k + \sqrt{\left(\frac{\theta_c^{Fe}}{2} - \frac{J_{R-R}G}{6k}\right)^2 + \frac{J_{Fe-R}^2\, S(S+1)G}{9k^2}} \quad (3)$$

where G is the DeGennes factor and $\theta_c^{Fe} = \frac{S(S+1)}{3k} J_{Fe-Fe}$ is the ordering temperature for the compound with no rare earth spin ($YFe_2$). Expression (3) has been fit to the observed transition temperatures of Figure 2 and the results are given by the dashed lines. The exchange constants obtained for the crystalline case were $J_{Fe-Fe}$= 832K, $J_{R-Fe}$=-137K and $J_{R-R}$=98K which represents the data very adequately. The fit for the amorphous case is less satisfactory, and yielded exchange constants of $J_{Fe-Fe}$=0 (non-magnetic $YFe_2$) $J_{R-Fe}$ = -129K and $J_{R-R}$ = 73K, the latter two being only slightly reduced from their crystalline counterpart values. The departures from the molecular field calculation, and the rapid rise in $T_c$ for the amorphous alloys compared with the crystalline compounds, suggests that the Fe-Fe exchange energy may not vanish uniformly, particularly for the higher spin rare earth alloys (e.g. $GdFe_2$ for which $T_c$ = 500K), but may may be dependent on the presence of rare earth exchange to become effective.

The author appreciates many helpful discussions with J. R. Cullen and N. C. Koon.

REFERENCES


* Present address.

1. Prepared by R. Allen of Battelle Northwest.
2. J. J. Rhyne, S. J. Pickart and H. A. Alperin AIP Conference Proceedings 18, 563 (1974); S. J. Pickart, J. J. Rhyne and H. A. Alperin Phys. Rev. Letters 33, 424 (1974).
3. G. S. Cargill, AIP Conf. Proc. 18, 631 (1974).
4. For a review see G. S. Cargill, "Structure of Metallic Alloy Glasses," Solid State Phys., 30, edited by Seitz, Turnbull and Ehrenreich, Academic Press, New York 1975, pp 227-289.
5. J. J. Rhyne, J. H. Schelleng and N. C. Koon, Phys. Rev. B 10, 4672 (1974).
6. K. P. Belov and A. N. Goryaga, Fiz. Met. Metallov., 2, 3 (1956).
7. E. Burzo, Z. Angew Phys., 32 127 (1971).
8. K.H.J. Buchow and R. P. Van Stapele, J. Appl. Phys. 41, 4066 (1970).
9. J. R. Cullen, private communication. This result is similar to a calculation by R.Hasegawa, B. E. Argyle and L. J. Tao, AIP Conf. Proc. Series 24, 110 (1975).

# PRESSURE DEPENDENCE OF MAGNETIC PROPERTIES OF AMORPHOUS RE - TM THIN FILMS

J.W.M. Biesterbos, M. Brouha and A.G. Dirks,
Philips Research Laboratories, Eindhoven, The Netherlands.

ABSTRACT

The pressure dependence of the Curie temperature of various amorphous RE-TM alloys (prepared by electron beam coevaporation), having $100 < T_c < 400$ K, was measured. In all cases relatively large negative $dT_c/dP$ values were found, similar to those for crystalline RE-TM intermetallic compounds. This implies that the lower $T_c$ of the amorphous alloys (20-300 K lower than of the corresponding crystalline phases) cannot be simply ascribed to the lower densities of the amorphous alloys. Like we have found previously for the crystalline compounds, the $dT_c/dP$ values for the amorphous RE-TM alloys agree qualitatively with the itinerant model of band ferromagnetism.[6,7]

A number of amorphous rare earth (RE)-transition metal (TM) films have been prepared by simultaneous electron beam evaporation ($5 \times 10^{-10}$ torr) onto substrates at room temperature (RE ≡ Ho,Gd,Y; TM ≡ Fe,Co). The purity of the starting materials was 99.9% for the RE-metals and 99.99% for the TM. The films were removed from the substrates (KBr) prior to the measurements. The films having a thickness of 1 μm were sandwiched between protective layers. It was checked by X-ray diffraction that the films were amorphous.

The pressure dependence of the Curie temperature ($T_c$) of these films was measured up to 40 kbar in a piston-cylinder apparatus under hydrostatic conditions, using a mixture of alkanes (B.P. = 50-80 °C) as pressure transmitting medium in a teflon container placed inside the usual solid medium[1]. $T_c$ was determined from the initial susceptibility versus temperature curves obtained by using the sample as the core of one half of a differential transformer. The pressure dependence of the magnetization was measured in a BeCu vessel up to 15 kbar in a Faraday balance.

Roughly speaking three types of initial susceptibility versus temperature curves were found of which examples are presented in Fig. 1. For two types of curves $T_c$ is well defined as is clear from Figs. 1a and 1b.

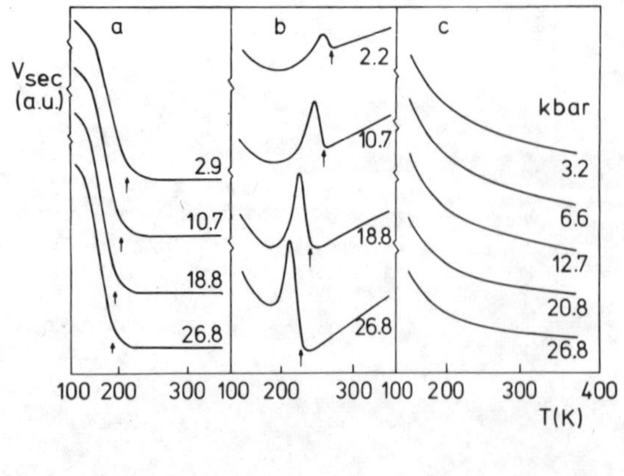


Fig. 1. Secondary voltage of the differential transformer, which holds the sample, versus temperature for amorphous $Gd_{55}Co_{45}$ (a), $Ho_{42}Fe_{58}$ (b) and $Y_{45}Co_{55}$ (c) at different pressures.

However, the systems Gd-Fe (50,60,70% Fe*), Y-Fe (45% Fe), Y-Co (50,55,67,75% Co) and Gd-Co (70,80% Co) showed smooth curves, like those for $Y_{45}Co_{55}$ presented in Fig. 1c, from which we were not able to determine $T_c$. It has to be remarked that for these alloys the magnetization curves in an applied field of 10 kOe show an identical behaviour.

The results collected in Figs. 2a and 2b were determined from curves like those in Figs. 1a and 1b. For all compositions $T_c$ is seen to decrease markedly with pressure at a rate of -1 to -5 K/kbar.

In the amorphous state these materials show a large difference in $T_c$ with respect to the corresponding crystalline state[2]. The Curie temperatures of all measured amorphous alloys ($100 < T_c < 400$ K) are lower than those of the crystalline compounds in the same range of composition for the Fe as well as for the Co alloys. For the investigated Gd-Co alloys the difference is about 20 K and for the Fe alloys up to about 300 K. If the change in $T_c$ was merely due to the difference in density, one would expect $T_c$ to increase with pressure, since the amorphous alloys have densities which are up to 15% lower[3].

In previous investigations concerning the pressure dependence of $T_c$ in crystalline RE-Co and RE-Fe compounds most $dT_c/dP$ values appeared to be strongly negative, except for compounds with relatively high $T_c$[4,5]. The variation of $dT_c/dP$ observed for the various compounds is in agreement with the itinerant electron model of ferromagnetism in metals[6,7]:

$$dT_c/dP = 5/3\,\kappa T_c - 5/6\,\kappa B T_c^{-1} \qquad (1)$$

Here $\kappa$ represents the compressibility and B is a parameter depending on the intra-atomic Coulomb repulsion and band parameters at the Fermi level. For compounds with a relatively high value of $T_c$ the second term is small and the first positive term will lead to a slight increase of $T_c$ with pressure, as observed for $Gd_2Co_{17}$ (Fig. 2c) and related $(RE)_2Co_{17}$ compounds. For compounds having relatively low $T_c$ values the second term will take over. We found that the parameter B is nearly constant for a series of compounds throughout the composition range up to $GdCo_3$[5b].

For the crystalline RE-Fe compounds[5a] a similar behaviour was observed, with a different value for B. Here the $(RE)Fe_2$ compounds have the highest value of $T_c$ (550-800 K) and correspondingly a small positive value of $dT_c/dP$.

As shown in the lower part of Fig. 2a the amorphous Y-Fe system behaves similarly. With increasing Fe content $T_c$ decreases and the slope of the $T_c$ versus P curves becomes increasingly negative. The average slope varies from -2.4 up to -4.8 K/kbar which is nearly equal to the value of -3.5 K/kbar for the crystalline $(RE)_2Fe_{17}$ compounds. For the Ho-Fe and Gd-Fe amorphous alloys $T_c$ and $dT_c/dP$ are nearly equal to the value for $Y_{45}Fe_{55}$.

Out of a whole series of Gd-Co and Y-Co alloys, only for the Gd-Co alloys containing less than 60% Co, plots of $T_c$ versus P could be made (Fig. 2b), because of difficulties in determining $T_c$ from curves like those presented in Fig. 1c.

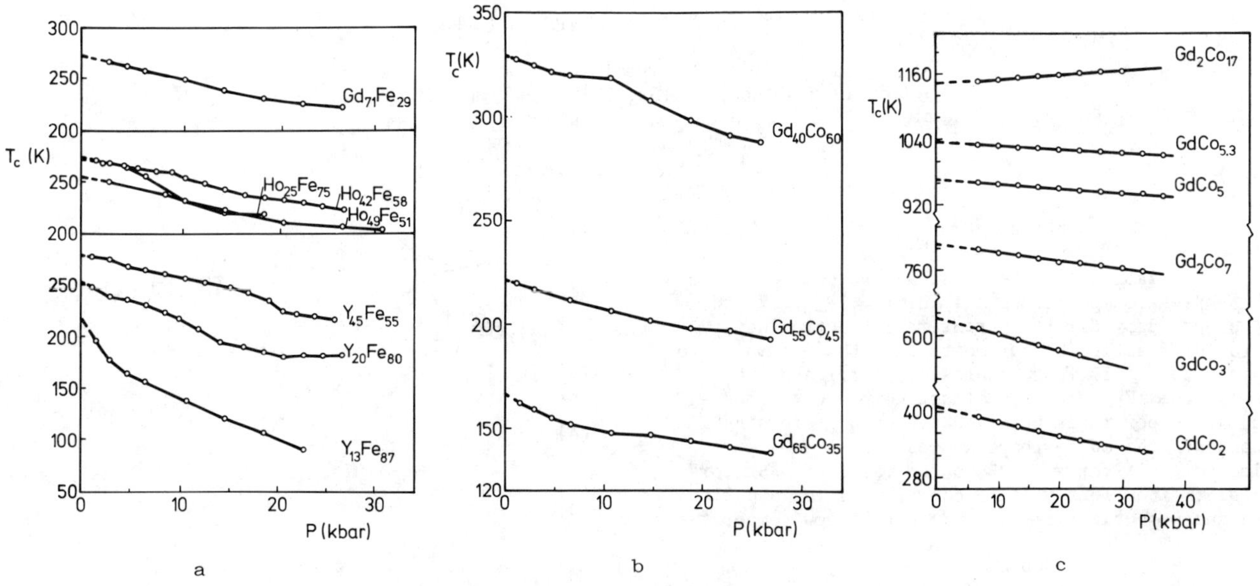


Fig. 2. Curie temperature versus pressure for various amorphous RE-Fe (Fig. 2a) and Gd-Co (Fig. 2b) alloys. For comparison Fig. 2c shows the pressure dependence of $T_c$ for several crystalline intermetallic Gd-Co compounds.

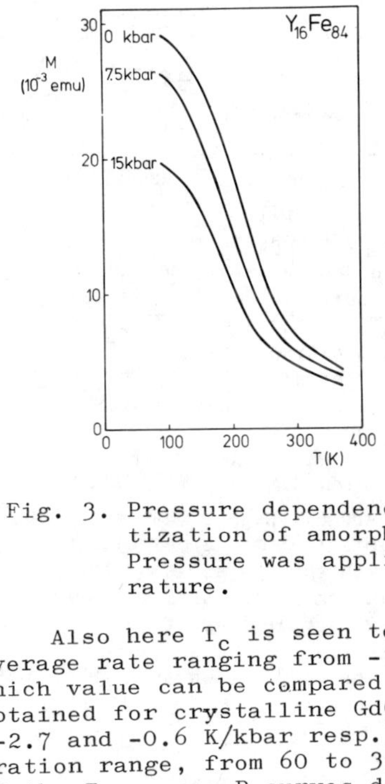


Fig. 3. Pressure dependence of the magnetization of amorphous $Y_{16}Fe_{84}$. Pressure was applied at room temperature.

Also here $T_c$ is seen to decrease at an average rate ranging from -1.0 to -1.7 K/kbar, which value can be compared with the values obtained for crystalline $GdCo_2$ and $Gd_4Co_3$ (-2.7 and -0.6 K/kbar resp.). In this concentration range, from 60 to 35% Co, the slope of the $T_c$ versus P curves decreases with decreasing Co concentration. The same effect was observed for crystalline Co compounds with relatively low $T_c$ (<400 K); here it was ascribed mainly to a decrease of the parameter B, arising from the fact that the Fermi level nearly reaches the top of the 3d band.

In special cases the effect of pressure on the magnetization of amorphous RE-TM alloys can be extremely large as is shown in Fig. 3 for $Y_{16}Fe_{84}$. From this figure one estimates $(1/M_o)dM_o/dP$ to be at least $4.10^{-2}$ 1/kbar at 0 K, which is one order of magnitude larger than for the crystalline RE-TM compounds[8] (the value of the comparable $Y_2Fe_{17}$ compound equals $(1/M_o)\ dM_o/dP = (0\pm0.5)10^{-3}\ kbar^{-1}$)

In the crystalline compounds the change of crystal structure through the series of RE-Fe and -Co compounds appears not to affect those parameters of the 3d band which determines $T_c$ and $dT_c/dP$[5]. In these compounds however, the different structures are closely related. Remarkably also in the extreme case of the absence of a regular structure, like in the investigated amorphous alloys, $dT_c/dP$ still behaves in close agreement with crystalline RE-TM systems of comparable $T_c$ values. This suggests that also in the amorphous RE-TM systems the bulk magnetic properties are governed mainly by the itinerant 3d electrons residing in relatively narrow energy bands. As in the crystalline state these 3d band will be gradually filled due to electron transfer from the RE atoms.

The authors would like to thank Mr. M. Farla for careful preparation of the samples and Mr. R. Kuipers for his skilful technical assistance.

REFERENCES


1. Brouha M., Rijnbeek A.G., Rev. Sci. Instrum. 44, 852 (1973).
2. Lee K., Heiman N., A.I.P. Conf. Proc. MMM-1974, Eds. Graham C.D. Jr., Landerand G.H., Rhyne J.J., p. 108.
3. Tao L.J., Kirkpatrick S., Gambino R.J., Cuomo J.J., Solid St. Comm. 13, 1491 (1973).
4. Brouha M., Buschow K.H.J., Miedema A.R., IEEE Trans. Magn. MAG-10, 182 (1974).
5. Brouha M., Buschow K.H.J., a) J. Appl. Phys. 44, 1813 (1973); b) J. Phys. F: Metal Phys. 3, 2218 (1973).
6. Wohlfarth E.P., J. Phys. C: Solid State Phys. 2, 68 (1969).
7. Edwards L.R., Bartel L.C., Phys. Rev. B5, 1064 (1972).
8. Brouha M., Buschow K.H.J., Europhysics Conference Abstracts 1A, 26 (1975).


*) All compositions are given in at.perc.

# MAGNETIC PROPERTIES OF BULK AMORPHOUS $Tb_xFe_{1-x}$*

H. A. Alperin
Naval Surface Weapons Center, White Oak, MD 20910
National Bureau of Standards, Washington, DC 20234
and
J. R. Cullen and A. E. Clark
Naval Surface Weapons Center, White Oak, MD 20910

ABSTRACT

Magnetization measurements have been made on four bulk samples (compositions x = .018, .118, .167 and .25) prepared by direct current rapid sputtering. Neutron diffraction measurements show these samples to be amorphous. All four samples are ferrimagnetic; however, the 17% sample is almost completely compensated at temperatures below 100K. With decreasing Tb-content, the Curie temperatures fall from a maximum of 405K for x = .25 to the value of 245K for x = .018. Anomalously large coercive fields and time dependent magnetizations are present at low temperatures.

## INTRODUCTION

Rare earth-transition-metal alloys with compositions corresponding to crystalline counterparts have been prepared in bulk form by d. c. rapid sputtering and by coevaporation as thin films.[1] Neutron scattering measurements[2] on the bulk samples of composition $R\text{-}Fe_2$ (with R = Tb, Gd, Ho and Y) have shown them to be amorphous both crystallographically and magnetically with a dense random packing of unequal-size spheres. Magnetization measurements have shown $TbFe_2$, $GdFe_2$ and $HoFe_2$ to be ferrimagnetic while $YFe_2$ exhibits only short range cluster order.[3]

At low temperatures and in zero applied field intense small-angle neutron scattering has been observed[4] for $TbFe_2$, $HoFe_2$ and to a lesser extent for $YFe_2$, indicative of magnetic inhomogeneities of size >90A, 40A and 8A respectively while no such inhomogeneities have been observed for $GdFe_2$. These in turn can be correlated with the very large coercive fields that have been observed.[3,5]

The work presented here extends the previous measurements to Tb-Fe compositions with much lower Tb content in order to observe the effect of the decreasing magnetic anisotropy as one approaches pure amorphous iron.

## EXPERIMENTAL

Samples of composition $Tb_xFe_{1-x}$ with x = .018, .118, .167 and .25 were prepared by rapid d. c. sputtering.[6] Typical sample size was 7 mm x 7 mm x 1 mm. Neutron diffraction measurements on these samples yielded amorphous patterns similar to those reported previously.[7] Magnetization measurements to a maximum field of 19 kOe were made using a vibrating sample magnetometer on small chips (∿30 mg to 60 mg) removed from the larger pieces. In no case was the field sufficient to achieve saturation.

The composition of the richest iron sample was determined by neutron activation analysis to be 0.049 ± .001 weight-% Tb and 0.95 ± .05 weight-% Fe which corresponds to 0.0177 ± .0004 atomic % Tb.

## RESULTS

Magnetization data (M vs H) were taken at discrete temperatures from 4.2K to 412K. The latter temperature is well below the recrystallization temperature for these materials. Below the Curie temperature ($T_c$) the curves show a strong high field susceptibility. Extrapolation of M from high fields back to the demagnetization field yields the spontaneous moments ($\sigma_0$) plotted as a function of temperature in Figure 1. The Curie temperatures

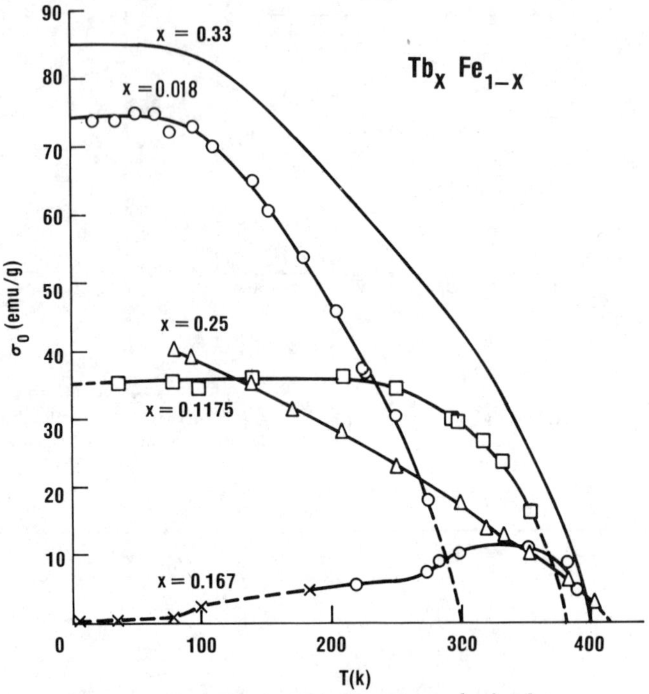


Spontaneous moment vs temperature derived from magnetization data as explained in the text. The smooth curve for x = .33 is taken from reference 3. For the x = .167 sample below 200K reliable extrapolations could not be made due to the anomalously large coercive fields. The points shown (x) are values of the observed remenance. The x = .1175 data is also extrapolated while the x = .25 sample was not measured below 77K.

obtained from Figure 1 are all higher than those obtained from "Arrott" plots (see table below). This difference correlates with the sharpness of the transition; the x = .018 sample transition is very diffuse while the x = .25 sample is most sharp.

Figure 2 shows the "Arrott" plots for the x = .018 composition. The isotherm whose high field

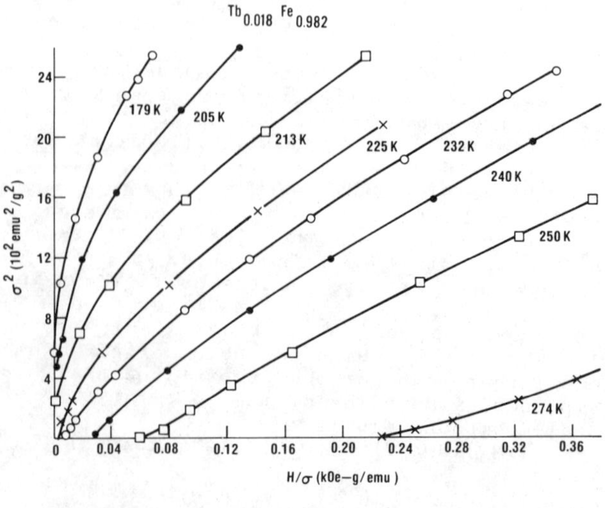


"Arrott" plot isotherms for $Tb_{.018}Fe_{.982}$. The Curie temperature is ∿245K as explained in the text.

portion extrapolates thru the origin defines $T_c$. In this case it is ∿245K. At low fields the deviation of these plots from straight lines is probably indicative of the nonuniformity (in this case due to the amorphous nature) of these materials.[8] Curie temperatures obtained in this way are given vs composition in the table. Extrapolation to x = 0 would give an expected $T_c$ = 200K for pure amorphous iron.

Upper limits to the iron moment can be determined by assuming anti-parallel alignment of Tb and Fe and the full Tb moment of $9\mu_B$. These are given below.

| x | 0.018 | .118 | .167 | .25 | .333(ref.3) |
|---|---|---|---|---|---|
| $\mu_{Fe\ max}$ $(\mu_B)$ | 0.95 | 1.68 | 1.81±.07 | 2.14 | 2.30 |
| $T_c$(°K) | 245 | 365 | 380 | 404 | 383 |

Figure 3 shows the anomalously large coercive fields that develop in these materials at low temperatures. These data were taken from hysteresis loops which exhibited time-dependent behavior (increasing with decreasing x) characteristic of small magnetic particles, although these amorphous materials are not a collection of physically distinct particles.

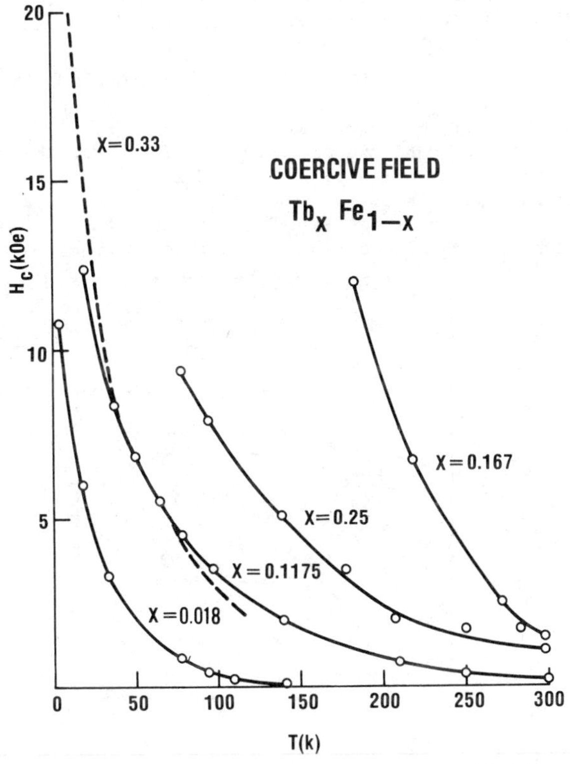


Temperature dependence of the coercive field from hysteresis loop measurements. The dashed curve is taken from reference 3. Data could not be taken below 183K for the x = .167 sample because of insufficient applied field to traverse the complete loop.

CONCLUSIONS

The importance of these data is that they reveal the role played by the rare-earth in the magnetic order and coercivity of these alloys. If one assumes the properties of the 1.8%-Tb sample (low $T_c$, high $H_c$ and low average Fe moment) are characteristic of pure amorphous iron then, since Fe atoms have little or no spin-orbit coupling, we are forced to consider sources other than anisotrophy to explain these proerties. The low average Fe moment for example, can be due to partial cancellation of larger individual moments, due in turn to non-collinear alignment of a large fraction of the Fe atoms. Since the coherence length associated with Fe atoms for any such configuration would be very small (between 10A and 30A) a change in moment direction due to an applied field in the "wrong" direction must take place by spin reversal of regions of volume $\sim 10^3 A^3$. These would be impeded from reversal by exchange forces coupling them to neighboring regions. A given region would flip when the gain in Zeeman energy exceeded the loss due to exchange decoupling on its surface. Then the coercive field $H_c = RJ/\mu_B$ where R is the ratio of the number of surface spins to the number in the volume of the cluster and $J/\mu_B$ is the exchange energy in oersteds. The large regions would reverse first. For $J/\mu_B \sim 10^5$Oe and $R \sim 10^{-1}$ we expect coercive fields of the observed magnitude.

We expect this effect to occur in the rare earth-richer samples as well, though masked by spin-orbit effects. The lack of long range ferromagnetism in $YFe_2$ is most likely due simply to the dilution of the Fe-Fe interaction by the non-magnetic Y atoms.

The thermal activation model described by Rhyne et al[3] to explain the dependence of $H_c$ on T at small T can operate here as well. The relatively small values of $H_c$ observed for the 1.8% sample is probably due to the small exchange while the large $H_c$ observed for the 16.79% sample is due to the small, almost compensated moment.

We are grateful for the assistance of W. Gilmor and for discussions with J. Rhyne.

REFERENCES

*Work supported by the Naval Surface Weapons Center Independent Research Fund and Office of Naval Research.

1. Lee, K., and Heiman, N., AIP Conf. Proc. 24, 108 (1975)
2. Rhyne, J. J., Pickart, S. J., Alperin, H. A., AIP Conf. Proc. 18, 563 (1974).
3. Rhyne, J. J., Schelleng, J. H., Koon, N. C., Phys. Rev. B, 10, 4672 (1974).
4. Pickart, S. J., Rhyne, J. J., Alperin, H. A., Phys. Rev. Lett. 33, 424 (1974); Pickart, S. J., Rhyne, J. J., Alperin, H. A., AIP Conf. Proc. 24, 117 (1975).
5. Clark, A. E., Appl. Phys. Lett. 23,642 (1973).
6. Samples prepared at Battelle Northwest Laboratories and furnished by R. Allen.
7. Pickart, S. J., et al, Phys. Lett. 47A, 73 (1974); Alperin, H. A. Proc. Int'l. Conf. Magnetism, Moscow, Aug. 1973, Vol. 4, pg. 358 (Aug. 1973).
8. Shtrikamn, S. and Wohlforth, E. P., Physics 60, 427 (1972).

# THE STRUCTURE OF AMORPHOUS Gd-Co ALLOY FILMS

C. N. J. Wagner,* N. Heiman, T. C. Huang, A. Onton, and W. Parrish
IBM Research Laboratory, San Jose, California 95193

## ABSTRACT

Co-Gd alloy films with 77-23 and 68-32 atomic ratios (no bias) and with 77-17-6, Co-Gd-Ar (-100 V bias on substrate) were prepared by bias sputter deposition on 50 μm thick Be substrates. Within experimental uncertainties, the interference function, I(K), obtained from reflection and transmission of Mo-Kα radiation by a given sample was identical. A comparison of G(r) for various compositions leads to an identification of the resolved components of the first maximum as Co-Co, Co-Gd, and Gd-Gd nearest neighbors at r=2.5, 2.95, and 3.45Å, respectively. At larger values of r ($\gtrsim$6Å), G(r) of the biased film has greater peak heights than G(r) of the unbiased films, suggesting longer range structural coherence in the biased film.

## I. INTRODUCTION

It was found by Chaudhari et al.[1] that amorphous Gd-Co films prepared by sputter deposition with bias applied to the substrates possess such uniaxial magnetic anisotropy. Subsequently, Heiman et al.[2,3] have reported magnetic uniaxial anisotropy in a number of other rare earth-transition metal amorphous alloys, prepared by thermal evaporation. However, no magnetic anisotropy was observed in Gd-Co films when they were prepared by sputter deposition without a bias voltage. It was thought that the magnetic anisotropy may have structural origin. In order to investigate this effect, Gd-Co films were prepared by sputtering with and without bias applied to the substrate, and the structures of biased and unbiased Gd-Co films were determined by both X-ray transmission and reflection techniques.[4]

## II. PREPARATION OF Gd-Co FILMS

Films of Gd-Co were sputtered in an argon atmosphere at a pressure of 25 μm with 0 and -100 V bias applied to the substrate [50 μm Be foils]. The film thickness, t, was made 20 μm to yield a μt-value of about 0.8 to 1.0 for MoKα radiation, where μ is the linear absorption coefficient. The compositions of the samples, as checked by electron microprobe analysis, were found to be 23.2% Gd, 76.6% Co, 0.2% Ar (0 bias); 16.5% Gd, 77.2% Co, 6.3% Ar (-100 V bias); and 32.2% Gd, 67.5% Co and 0.3% Ar (0 bias).

## III. X-RAY ANALYSIS

The X-ray scattering data of the Gd-Co films were obtained with MoKα radiation in reflection and in transmission using a bent graphite monochromator in the diffracted beam.

The elastically scattered intensity per atom $I_{eu}(K)$ expressed in electron units must be known in order to determine the interference function I(K), i.e.,[4]

$$I(K) = \frac{I_{eu}(K) - [\langle f^2\rangle - \langle f\rangle^2]}{\langle f\rangle^2}, \qquad (1)$$

where $K=4\pi\sin\theta/\lambda$, and $\langle f\rangle$ is the scattering factor of the alloy. The total measured X-ray scattering, $I_f(K)$, is the difference between the sample and substrate scattering, corrected for absorption in the sample.

Then $I_{eu}(K)$ is given by:

$$I_{eu}(K) = \beta\,\frac{I_f(K)}{PA_f} - QI_{in}(K) - I_{mu}(K), \qquad (2)$$

where β is a normalization constant, P and $A_f$ are polarization and absorption factors, $I_{in}(K)$ is the inelastic or Compton scattering, Q is the monochromator attenuation function,[5] and $I_{mu}(K)$ is the multiple scattering. Since the graphite monochromator in the diffracted beam partially removed the inelastic scattering from the sample, the K-dependence of Q must be known.[5]

The atomic distribution function G(r) was obtained by Fourier transformation of F(K)=K[I(K)-1]), i.e.,[4]

$$G(r) = 4\pi r[\rho(r)-\rho_o] = \frac{2}{\pi}\int_0^\infty F(K)\exp(-\alpha K^2)\sin Kr\, dK, \qquad (3)$$

where $\rho_o$ is the average atomic density of the alloy, ρ(r) is the total atomic distribution function, and α is an artificial damping factor, usually taken to be <0.01, but α was set equal to zero in our case.

Large modulations were usually visible in G(r) at small values of r (r<2Å) due to slowly varying errors in F(K), most likely arising from the errors in the monochromator attenuation function, Q(K), used in Eq. (2). Additionally, a short wavelength modulation of low amplitude was clearly visible at values of r>6Å. Both modulations were removed by the Kaplow procedure[6] yielding refined interference functions, I(K), and atomic distribution functions, G(r).

## IV. EXPERIMENTAL RESULTS AND DISCUSSION

The refined interference functions I(K) of the transmission data are presented in Fig. 1 for the unbiased alloys with 32.2 and 23.2 at.% Gd and the biased film with 16.5% Gd and 6.5% Ar. The reflection data yields identical interference functions. Also shown in this figure is the interference function of

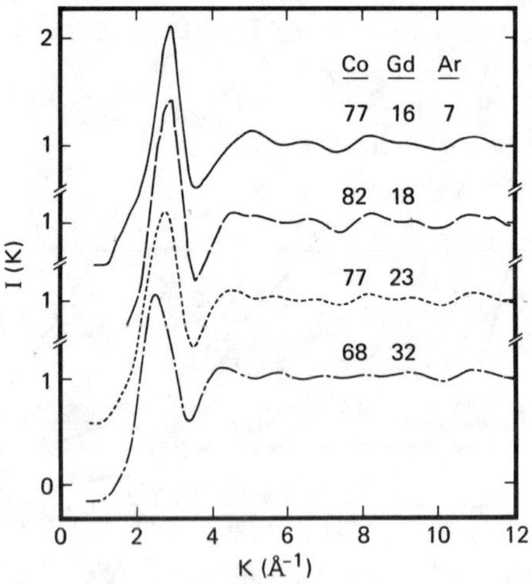


Fig. 1 Refined interference functions of Gd-Co alloys in MoKα transmission.

an unbiased alloy with 18 at.% Gd determined by Cargill.[7] The Fourier transforms of I(K) of these alloys, i.e., the reduced distribution functions G(r) are presented in Fig. 2. The first peak in G(r) is split into three maxima, which are most clearly visible in the alloy with 32.2 at.% Gd, at positions $r_1$=2.5, 2.95, and 3.45Å.

The diffuseness of the interference patterns of the Co-Gd films was quite surprising. The first peak is relatively broad with a width of ΔK=0.95Å$^{-1}$ at half-maximum height, which is much larger than that observed in liquid-quenched metal-metalloid alloys (ΔK~0.5Å$^{-1}$).[8] The heights of the first and second peaks in I(K) are also quite low. This could be a consequence of the lack of structural order, but might also be due to interference among the partial functions $I_{11}$, $I_{22}$, and $I_{12}$ of I(K):[4]

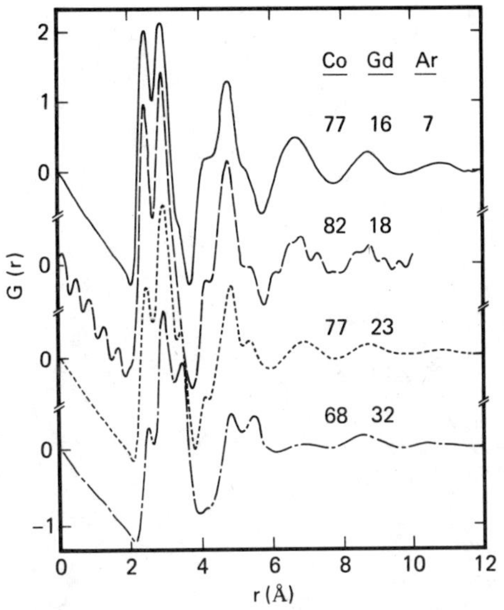


Fig. 2 Reduced radial distribution functions of Gd-Co amorphous alloys.

$$I(K) = W_{11}(K)I_{11}(K)+W_{22}(K)I_{22}(K)+2W_{12}(K)I_{12}(K), \quad (4)$$

where

$$W_{ij}(K) = c_i c_j f_i(K) f_j(K)/\langle f(K)\rangle^2 . \quad (5)$$

In these expressions $c_j$ is the atomic concentration of element j and $f_j(K)$ is the corresponding atomic scattering factor. The values of the weighting factors $W_{ij}(0)$ at K=0 are given in Table I.

The position of the first peak in I(K) decreases with increasing Gd concentration. This may be because the Goldschmidt diameter[7] of the Gd atom (3.60Å) is much larger than that of a Co atom (2.50Å).

The most surprising result, however, is the splitting of the first peak in the reduced distribution function G(r) of the Co-Gd alloys. Similar observation was made by Cargill[7] on Gd-Co and Gd-Fe alloys and by Rhyne, et al.[9] on Tb-Fe alloys. If we assume that $W_{ij}(K)$ is a slowly varying function of K, then we can approximate[4] G(r) as the weighted sum of the partial $G_{ij}(r)$ (i.e., Eq. (4) with G(r) substituted for I(K)). Thus, it is possible to associate the individual peaks at $r_1$=2.50Å, $r_1^2$=2.95Å and $r_1^3$=3.45Å with distances corresponding to the nearest neighbor separations between Co-Co, Co-Gd, and Gd-Gd atoms. The heights of these subsidiary peaks seem to be proportional to the weighting factors $W_{ij}(0)$. Since $2W_{12}(0) \approx 0.5$ for the three alloys considered, the height of the central peak is relatively unaffected by the alloy composition.

At the outset of this investigation, it was thought that the magnetic anisotropy might have structural origin. One possibility for such anisotropy might be a preferred atomic orientation in the film with respect to the substrate. Such orientation might be detectable if the X-ray intensities from the films are measured in transmission and in reflection. As has been observed in the unbiased films, the transmission and

Table I. Values of the weighting factors, $W_{ij}(0)$, for Co-Gd alloys of different Gd concentrations.

| Unbiased Film %Gd | $W_{GdGd}(0)$ | $2W_{GdCo}(0)$ | $W_{CoCo}(0)$ |
|---|---|---|---|
| 32.2 | 0.28 | 0.50 | 0.22 |
| 23.2 | 0.17 | 0.49 | 0.34 |
| 18 | 0.12 | 0.45 | 0.43 |
| **Biased Film** %Gd %Ar | | | |
| 16.5 6.5 | 0.13 | 0.46 | 0.41 |

the reflection data of the bias-sputtered alloy of composition 77-17-6 (Co-Gd-Ar) showed no structural differences. Therefore, no large scale preferred orientation appears to exist even in the biased film.

However, the total interference function I(K) of the Gd-Co-Ar film shows features which are not present in the unbiased films. We should compare the biased 77-17-6 film with the unbiased 82-18 film since they are closest in Gd/Co ratio as well as $W_{ij}(0)$ (see Table I). As shown in Fig. 1, the only differences observed are in the neighborhood of the first and second peaks. Both peaks are shifted slightly toward larger K values, and a shoulder is clearly visible below the first peak of the biased film.

It is tempting to correlate at every value of K the difference in height of I(K) between the biased and unbiased films possessing the same Gd/Co value with the modulation of the Laue monotonic scattering $(\langle f^2\rangle - \langle f\rangle^2)/\langle f\rangle^2$. Such modulations are observed in crystalline, ordered solid solutions, and are a consequence of short range compositional order.[10]

The reduced distribution functions, G(r), of the biased film and the unbiased 18% Gd film[7] are practically identical. The larger height of the first split peak in Cargill's G(r) curve is due to an increased resolution because of a larger $K_{max}$-value (16.4Å$^{-1}$) used in the Fourier integral than in ours ($K_{max}$=14.35Å$^{-1}$). However, the peaks in G(r) at larger values of r (i.e., r>6Å) seem to have larger heights in the biased film than in the unbiased films. This would seem to indicate that the biased film, the one with magnetic anisotropy, possesses structural order over larger distances than the unbiased films.

ACKNOWLEDGMENT

We would like to thank D. F. Kyser for electron microprobe analysis of film compositions, and D. L. Salazar for assistance in sample preparation.

REFERENCES

*Permanent address: Materials Department, UCLA, Los Angeles, California 90024.

1. P. Chaudhari, J. J. Cuomo, and R. J. Gambino, IBM J. Res. Develop. 17, 66 (1973).
2. N. Heiman, A. Onton, D. F. Kyser, K. Lee, and C. R. Guarnieri, in Magnetism and Magnetic Materials, 1974, AIP Conf. Proc. No. 18, C. D. Graham, Jr. and J. J. Rhyne, (eds.) (Am. Inst. Phys., New York, 1975) p. 573.
3. K. Lee and N. Heiman, in Magnetism and Magnetic Materials, 1974, AIP Conf. Proc. No. 18, C. D. Graham, Jr. and J. J. Rhyne (eds.) (Am. Inst. Phys., New York, 1975) p. 108.
4. C. N. J. Wagner, in Liquid Metals, Chemistry and Physics, S. Z. Beer (ed.) (Marcel Dekker, New York, 1972) Ch. 6, p. 257.
5. W. Ruland, Brit. J. Appl. Phys. 15, 1301 (1964).
6. R. Kaplow, S. L. Strong, and B. L. Averbach, Phys. Rev. 138, (1965).
7. G. S. Cargill, in Solid State Physics, F. Seitz, D. Turnbull, and H. Ehrenveich (eds.) (Academic Press, New York, 1975) Vol. 30, in press.
8. B. C. Giessen and C. N. J. Wagner, in Liquid Metals, Chemistry and Physics, S. Z. Beer (ed.) (Marcel Dekker, New York, 1972) Ch. 15, p. 633.
9. J. J. Rhyne, S. J. Pickart, and H. A. Alperin, in Magnetism and Magnetic Materials, 1973, C. D. Graham, Jr. and J. J. Rhyne (eds.) (Am. Inst. Phys., New York, 1974) p. 563.
10. B. E. Warren, X-Ray Diffraction, (Addison Wesley, Reading, Mass., 1969).

# THE EFFECT OF Co ON THE ANISOTROPY OF AMORPHOUS Gd-Fe FILMS

R. C. Taylor

IBM Research Center, Yorktown Heights, N.Y. 10598

Recent work on sputtered films of amorphous GdCo and GdFe alloys has shown that these films have large perpendicular anisotropy, which has been attributed principally to a pair-ordering mechanism involving transition metal atoms. Chaudhari and Cronemeyer,[1] however, have found Gd-Co pairs to be dominant in sputtered Gd-Co-Mo films. We have observed that evaporated amorphous films in the Gd-Fe binary system display strong perpendicular anisotropy over the composition range $Gd_8Fe_{92}$ to $Gd_{33}Fe_{67}$. In the evaporated Gd-Co system, however, anisotropy is in-plane when the Gd content exceeds 10%, with the direction of the easy axis of magnetization determined by the positions of the Gd and Co evaporation sources. Assuming that the deposition processes are the same in both alloys, it seems likely that different pairs are dominant in the two alloy systems. We have established that Fe-Fe is the dominant pair in GdFe alloys by fitting sublattice magnetizations from molecular field theory to the experimental anisotropy data.

In order to study the behavior of anisotropy in the pseudo-binary GdFe-GdCo system, a number of evaporated films of amorphous Gd-Fe-Co alloys were prepared. Four series of alloys were studied with 8, 21, 24 and 30 atom % Gd and varying Co/Fe ratios. It was found that while magnetization might change by a factor of two in a series, a much larger change took place in the perpendicular anisotropy constant ($K_u$) and the anisotropy field ($H_K$), which usually decreased monotonically with increasing Co concentration. For example, in alloys containing 24% Gd, room temperature saturation magnetization ($4\pi M_s$) remained between 800 and 1300 G as the Co/Fe ratio increased from 0 to 1, but $K_u$ decreased by almost an order of magnitude from $2.5 \times 10^5$ to $3.3 \times 10^4$ ergs/cm$^3$ and $H_K$ from 6000 to 1000 Oe. The higher the Gd content, the more rapidly $K_u$ and $H_K$ decreased with increasing Co. At 21, 24 and 30% Gd, anisotropy went in-plane when Co reached 45% of the transition metal content indicating the strong negative influence of Gd-Co pairs on perpendicular anisotropy. (Co-Co pairs would have a positive influence on perpendicular anisotropy as evidenced by the perpendicular anisotropy in GdCo films with less than 10% Gd). However, when little Gd was present in the ternary alloys, as in the 8% Gd series, anisotropy remained perpendicular, but was at a minimum at a Co/Fe ratio of 1. This finding, together with those of Hoffman and Miyasaki[2] for Fe-Co alloys, suggest the negative influence of Fe-Co pairs on perpendicular anisotropy.

The results indicate that like pairs (Fe-Fe, Co-Co and Gd-Gd) contribute to perpendicular anisotropy in these alloys, with Fe-Fe pairs dominant in GdFe alloys. Unlike pairs (Gd-Fe, Gd-Co and Fe-Co) tend to cause in-plane anisotropy with Gd-Co pairs dominant in GdCo alloys and Gd-Co and Fe-Co pairs dominant in ternary alloys with high Co content. This suggests the possibility of preferential atomic ordering of like species in the growth direction during deposition.

1. P. Chaudhari and D. C. Cronemeyer (this conference).
2. H. Hoffman and T. Miyasaki, IEEE Trans. Magn., MAG-10, 556 (1974).

# ROOM TEMPERATURE COMPOSITION DEPENDENCE OF THE ANISOTROPY ENERGY OF $Ho_xTb_{1-x}Fe_2$ SINGLE CRYSTALS

C. M. Williams, N. C. Koon, and J. B. Milstein
Naval Research Laboratory, Washington, D. C. 20375

The single ion model suggests that the anisotropy energy of the $Ho_xTb_{1-x}Fe_2$ compounds should scale linearly with holmium concentration. In order to verify this model the room temperature magnetocrystalline anisotropy energy was investigated for a series of $Ho_xTb_{1-x}Fe_2$ single crystals for $0.82 < x < 0.91$ using torque magnetometry techniques. These measurements revealed, for the first time, that anisotropy constants of higher order than $K_1$ and $K_2$, the first and second order magnetocrystalline anisotropy constants, are required to account for all of the observed anisotropy. The additional constants, $K_3$ and $K_4$, correspond to the coefficients of the eighth and tenth order direction cosines in the anisotropy energy expansion. A comparison of the composition dependence of the anisotropy constants with the single ion model showed only fair agreement for $K_1$ and very poor agreement for $K_2$, $K_3$, and $K_4$. The poor agreement is attributed to the fact that the anisotropy constants are not unique, i.e., they depend strongly on the number of Fourier harmonics included in the torque data analysis.

The problem of non-unique anisotropy constants is a direct consequence of non-orthogonality of the basis set used in the conventional expansion of the anisotropy energy. This problem is resolved by expanding the anisotropy energy in terms of a orthogonal set of basis functions, namely, the cubic harmonics [1]. Because of orthogonality of this basis set, the new anisotropy constants, $\kappa_\ell$, can be determined directly without the use of Fourier analysis as an intermediate step. The results obtained using the cubic harmonic expansion indicate that even harmonics up to $\ell = 10$ are required to fit the torque data. The composition dependence of the $\kappa_\ell$'s show that $\kappa_4$, $\kappa_6$, and $\kappa_{10}$ scale linearly with holmium concentration, and that $\kappa_4$ and $\kappa_{10}$ change signs somewhere in the neighborhood of 86.6%Ho, in agreement with the single ion model. $\kappa_8$, which is of the same order of magnitude as $\kappa_6$, did not scale linearly with concentration, but showed a pronounced minimum in the neighborhood of 86% holmium. This latter behavior suggests that the origin of $\kappa_8$ may be the results of interaction other than the crystal field.

1. F. M. Mueller and M. G. Priestly, Phys. Rev. <u>148</u>, 638 (1966).

# RHOMBOHEDRAL DISTORTION IN HIGHLY MAGNETOSTRICTIVE LAVES PHASE COMPOUNDS*

A. E. Clark and J. R. Cullen
Naval Surface Weapons Center, White Oak, MD 20910
and
O. D. McMasters
Iowa State University, Ames, Iowa 50010
and
E. R. Callen
American University, Washington, DC 20016

## ABSTRACT

A model is presented which produces a magnetostrictively driven rhombohedral distortion ($\lambda_{111}$) in cubic diamond-like structures. No tetragonal magnetostrictive distortion ($\lambda_{100}$) is allowed, which is consistent with the experimental observation, $\lambda_{111} >> \lambda_{100}$. From x-ray and single crystal measurements we find $|\lambda_{111}/\lambda_{100}| > 600$ for $DyFe_2$. Rhombohedral distortions ($\lambda_{111}$) are examined in the ternary $Tb_{1-x}Gd_xFe_2$, $Tb_{1-x}Y_xFe_2$ and $Tb_{1-x}Dy_xFe_2$ systems by x-ray methods. The single-ion nature of the giant Tb magnetostriction in $TbFe_2$ is verified.

## INTRODUCTION

This paper is divided into two parts. In the first part we introduce a "Double Tetrahedron" (DT) model of magnetostriction which has the correct form to account for the anomalously large $\lambda_{111}$ magnetostriction of the rare earth ions in the cubic Laves phase structure, while at the same time leaving $\lambda_{100}$ at conventional magnitudes.[1] It is now possible to understand why huge rhombohedral distortions are found only in the Laves phase compounds containing Tb, Sm, Er, and Tm; and no tetragonal distortion is observed in any compound. In the second part of this paper we report rhombohedral distortions in the $Tb_{1-x}Gd_xFe_2$, $Tb_{1-x}Y_xFe_2$ and $Tb_{1-x}Dy_xFe_2$ ternary systems. Careful x-ray lattice parameter measurements, similar to those employed by Dwight and Kimball[2] were used to obtain values of $\lambda_{111}$. This method is rapid, applicable to materials of large magnetostriction, and eliminates the tedium of single crystal growth, preparation, orientation, etc. Important features such as magnetization axis changes can be easily identified by this method.

## THE DOUBLE TETRAHEDRON MODEL

The cubic rare earth-iron Laves phase compounds are the only known compounds possessing huge magnetostrictions ($\simeq 2000 \times 10^{-6}$) at room temperature. It is surprising that compounds of such high symmetry exhibit strains of this magnitude. It is even more surprising that the magnetostriction arises almost exclusively from the $\lambda_{111}$ distortion.[3,4] We propose that this huge anisotropy, $\lambda_{111} >> \lambda_{100}$, arises from the existence of two tetrahedral sites in the C15 structure which allows the energy to reduce by internal distortion. This distortion lowers the symmetry and drives an external rhombohedral distortion ($\lambda_{111}$). No tetragonal distortion ($\lambda_{100}$) is driven by this mechanism which leaves only conventional (small) magnetostriction values for $\lambda_{100}$. In Figure 1a we illustrate the DT distortion for Tb compounds. The two equivalent sites are situated at 0,0,0, and 1/4, 1/4, 1/4 in the C15 lattice and labeled A (or A') and B (or B'). Here the [111] direction is singled out as the magnetization direction. The 4f electron cloud of Tb ( -e) is oblate in shape and lies perpendicular to the magnetization axis. Because of the closer proximity of the 4f electron cloud on A to B' than to B, a stretching of the A - B bond occurs, causing the resultant internal distortion denoted by the arrows. The increase in a exceeds the decrease in b, yielding a net expansion along [111] for Tb. For a prolate 4f charge density, i.e.

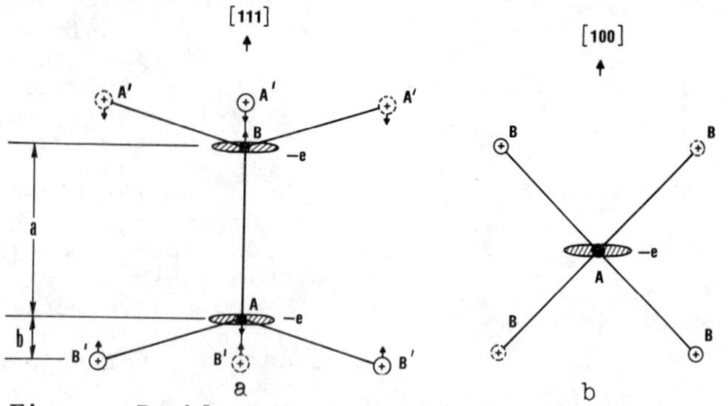


Fig. Double Tetrahedron Model of Magnetostriction; (a) M||[111], (b) M||[100]. Solid circles, ○, denote atoms above plane of figure; broken circles, ◌, denote atoms below.

for Sm, Er and Tm, a contraction of a occurs producing the observed negative magnetostriction ($\lambda_{111}$).[3] The inclusion of the iron ions tends to enhance this effect further.[1] Thus we conclude that a rhombohedral distortion exists whenever the magnetization points along [111] i.e. for $SmFe_2$, $ErFe_2$ and $TmFe_2$. The case of the magnetization parallel to [100] for $DyFe_2$ and $HoFe_2$ is depicted in Figure 1b. Here the 4f electron cloud is equidistant to all nearest neighbors and no internal distortion develops.

We present here the expression for the strain, $\varepsilon_{xy}$, as a function of the coefficients of the quadratic ($v_2$) and third order ($v_3$) terms in the crystalline potential. $C_{44}$ and k are the external and internal moduli. $v_3'$ is the strain derivative of $v_3$.

$$\varepsilon_{xy} \cong (v_3' v_3/k - v_2) < xy > /C_{44}.$$

For the extra-ordinary magnetostriction (first term) to dominate it is clear that the relevant internal mode must be rather soft, i.e. $\sim 10^{12}$ cps or less. Such frequencies are not uncommon in metals with more than one atom per cell. For details, see reference 1.

## X-RAY DETERMINATION OF $\lambda_{111}$ IN TERNARY COMPOUNDS

In this part of the paper we report values of $\lambda_{111}$ calculated from precise x-ray lattice parameter measurements on small polycrystalline samples. Spacing between the (620) and (440) planes were carefully measured using long wavelength Cr radiation. Under the magnetostrictive $\lambda_{111}$ distortion the (620) and (440) diffraction lines each split into two lines of equal intensity. The fractional difference in spacing between the planes corresponding to each doublet is calculated from: $\Delta d/d = 1 - \sin\theta_2/\sin\theta_1$. $\theta$ is the Bragg angle. This displacement is related

Table I. X-Ray Determination of $\lambda_{111}$

| Reflection | Magnetostriction ($\Delta d/d$) |
|---|---|
| 111, 222, 333, 444 | $\frac{4}{3}\lambda_{111}$ |
| 110, 220, 330, 440 | $\lambda_{111}$ |
| 310, 620 | $\frac{3}{5}\lambda_{111}$ |

to the magnetostriction $\lambda_{111}$ according to Table I.

$Tb_{1-x}Gd_xFe_2$ and $Tb_{1-x}Y_xFe_2$ Magnetostriction. The x-ray method cited above was used to determine the reduction of the magnetostriction of $TbFe_2$ as Tb is replaced by Gd and Y. Gd possesses no angular momentum and Y is nonmagnetic. If the magnetoelastic energy is single-ion in origin, the magnetostriction should be linear in concentration x (at 0°K). Modification of this linear dependence is expected at higher temperatures, becoming pronounced near the Curie temperature, $T_c$. In Figure 2 we plot the values of $\lambda_{111}$ calculated from the (620) and (440) line splittings for $Tb_{1-x}Gd_xFe_2$ and $Tb_{1-x}Y_xFe_2$ (x = 0, .4, .6). The uncertainty is approximately 100 x $10^{-6}$.

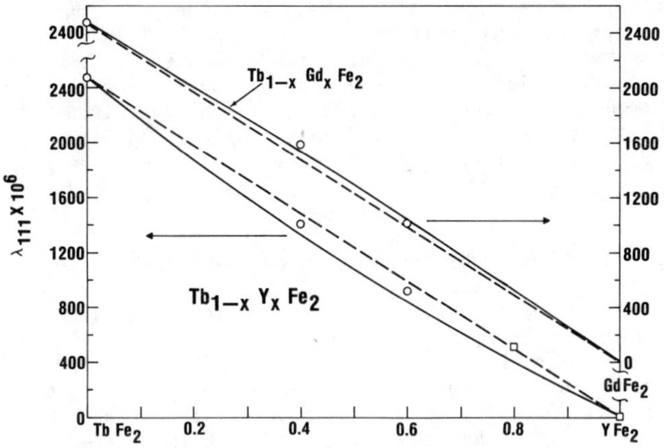


Fig. 2 $\lambda_{111}$ for $Tb_{1-x}Y_xFe_2$ and $Tb_{1-x}Gd_xFe_2$. See text.

Values for $YFe_2$ and for $Tb_{.2}Y_{.8}Fe_2$ denoted by squares were calculated from polycrystal samples assuming $\lambda_{111} = \lambda_p/.637$.[4] The dotted line is a linear relationship; the curved line is the corrected curve based upon the variation of the Tb sublattice magnetization[5] with x at T = 293°K. Because $YFe_2$ has a lower $T_c$ than $TbFe_2$, the theoretical curve for $Tb_{1-x}Y_xFe_2$ lies below the linear relationship. In $Tb_{1-x}Gd_xFe_2$, the theoretical curve is higher. Although the original data of Dwight and Kimball[2] for $Tb_{1-x}Y_xFe_2$ implies a dependence on x far from linear and a critical concentration (x ≅ .7) beyond which $\lambda_{111} = 0$, our measurements indicate that the magnetoelastic energy is truly single ion in origin. No critical concentration was found. Our measurement of magnetostriction using conventional strain gage methods clearly demonstrates the presence of a large $\lambda_{111}$ at x = .8.

$Tb_{1-x}Dy_xFe_2$ Magnetostriction. One of the most interesting rare earth-iron ternary systems is $Tb_{1-x}Dy_xFe_2$. $TbFe_2$ and $DyFe_2$ posses huge magnetic anisotropies (K's) of opposite sign. Near x = .7, the anisotropy compensates, yielding a technically important material with high $\lambda/K$ ratio. The x-ray technique is an excellent method to determine the precise compensating concentration easily. No single crystals are necessary. At the compensating concentration the magnetostriction falls abruptly from its large extra-ordinary DT value to near zero. This is illustrated in Figure 3. For x < .73, the magnetostriction is high, implying that the magnetization (M) is parallel to [111]. For x = .8 and 1.0, no splitting was perceptable, implying M parallel to [100]. This is consistent with Dwight and Kimball,[2] who pointed out that no tetragonal distortion was found in any $RFe_2$ compound.

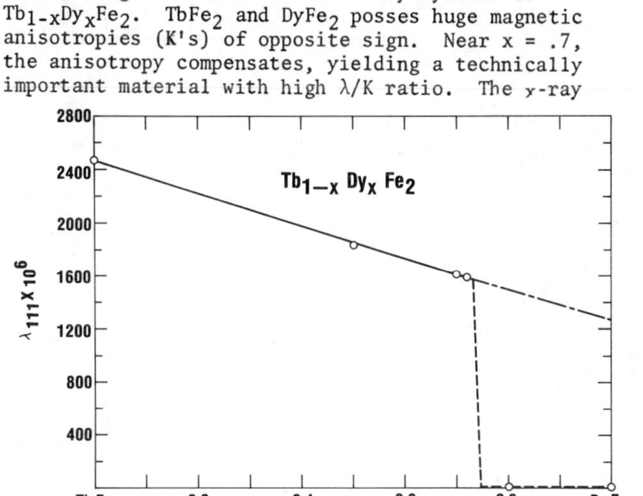


Fig. 3 $\lambda_{111}$ for $Tb_{1-x}Dy_xFe_2$.

From these data and single crystal $\lambda_{100}$ measurements in $DyFe_2$ we can infer the ratio of extra-ordinary to ordinary magnetostriction: $\lambda_{111}/\lambda_{100}$. The value for this ratio is difficult to obtain by direct measurement on a single compound. Extrapolating our $\lambda_{111}$ data to $DyFe_2$ utilizing the small temperature correction, we find $\lambda_{111}$ for $DyFe_2$ equal to 1260 x $10^{-6}$, about 1/2 that of $TbFe_2$. Conventional strain gage methods were used to obtain the value of $\lambda_{100}$ for single crystal $DyFe_2$. We find $\lambda_{100} = 0 \pm 4 \times 10^{-6}$, a a very small magnetostriction. The huge intrinsic magnetoelastic interaction characteristic of the rare earth elements is effectively shorted out in the Laves phase structure whenever M||[100]. See Fig. 1(b). Thus the ratio of the extra-ordinary magnetostriction to the ordinary magnetostriction in these compounds is huge: $|\lambda_{111}/\lambda_{100}| > 600$. In Table II we list values of $\lambda_{111}$ at room temperature in order of decreasing magnetostriction. The strains for polycrystal samples, defined by $\lambda_p \equiv 2/3\,(\lambda_{||} - \lambda_{\perp})$ at 25 kOe, are included for comparison.

Table II. Magnetostriction of $RFe_2$ Compounds

| | $\lambda_{111} \times 10^6$ | $\lambda_{100} \times 10^6$ | $\lambda_p \times 10^6$ |
|---|---|---|---|
| $TbFe_2$ | 2460 | - | 1750 |
| $SmFe_2$ | -2100 | - | -1560 |
| $DyFe_2$ | 1260 | 0±4 | 430 |
| $ErFe_2$ | - 300 | - | - 230 |

REFERENCES

* Supported by the Office of Naval Research, the Naval Surface Weapons Center Internal Research Fund and the Naval Sea Systems Command.


1. Cullen, J. R., and Clark, A. E., NSWC/WOL/TR 75-7111, White Oak Laboratory, NSWC, Silver Spring, MD 20910.
2. Dwight, A. E., and Kimball, C. W., Acta Cryst. B30, 2791 (1974).
3. Clark, A. E., Proc. 19th Conf. on Mag. & Magnetic Materials, AIP Conf. Proceedings No. 18, 1015 (1974).
4. Clark, A. E., Cullen, J. R., and Sato, K., Proc. 20th Conf. on Magnetism and Magnetic Materials, AIP Conf. Proc. No. 24, 670 (1975).
5. Rhyne, J. J., private communication.

# MOSSBAUER AND MAGNETIZATION STUDY OF $Tb(Fe_{1-x}Al_x)_2$ †

R.S. Preston, S.P. Taneja, S.M. Drensky,
A.E. Dwight, C.W. Kimball and L.R. Sill
Department of Physics, Northern Illinois University
DeKalb, Illinois 60115

## ABSTRACT

The magnetization and magnetic hyperfine field at $^{57}Fe$ nuclei have been measured in $Tb(Fe_{1-x}Al_x)_2$ Laves phase alloys to examine possible correlations of the magnetic properties with the concentration dependence of the rhombohedral distortion.[1] The magnetic moment per molecule at 300 K is almost constant for all concentrations for which the crystal is distorted ($x < 0.24$); the saturation moment at 300 K then begins to decrease smoothly. The maximum magnetic hyperfine field also varies smoothly over the concentration range which spans the onset of detectable distortion. We conclude that in this concentration range there are no real anomalies in the lattice magnetization nor in the magnetizations of either of the two oppositely magnetized sublattices.

## INTRODUCTION

Dwight and Kimball[1] determined from x-ray diffraction measurements at room temperature that the Laves phase $TbFe_2$, which is nominally of the cubic $MgCu_2$-type, exhibits a spontaneous rhombohedral distortion. Because the lattice is magnetically ordered (ferrimagnetic) at this temperature, they attributed this distortion to spontaneous magnetostriction. They found that this distortion can be decreased by substituting nonmagnetic Y atoms for some of the Tb atoms, and seems to disappear rather suddenly for values of Y concentration above a certain value. Preston, et al.[2], made a further study of the ternary system $Tb_xY_{1-x}Fe_2$ using the Mossbauer effect. From measurements of the isomer shift and of the angle between the magnetic dipole and electric quadrupole field axes they concluded that the iron tetrahedra within the unit "cube" suffer an additional displacement or distortion even in concentration regions where no rhombohedral distortion is seen. They noted that the rhombohedral distortion seems to disappear at about the value of x for which the spontaneous room-temperature lattice magnetization goes to zero when the rare-earth and the iron sublattice magnetizations happen to compensate each other[3]. Although this fact is noteworthy, it does not seem to have any theoretical significance since neither of the sublattice magnetizations disappears at this point. On the contrary, the iron sublattice magnetization, as measured by the Fe hyperfine field[2,4], varies smoothly throughout this region of Y concentration. Also, as far as can be seen from the published data of Buschow and van Stapele[3] (see curve a, Fig. 1 of Ref. 2), the total spontaneous magnetization has no anomaly in this region except for the change in sign of dM/dx at the compensation point. From this it appears likely that for both the iron and the rare-earth sublattices the spontaneous magnetization is a smoothly varying function of Y concentration. However, the magnetization data of reference 3 are not very detailed in the region of interest.

Dwight and Kimball reported, further, that the rhombohedral distortion can also be made to diminish and finally disappear if either Al or Co atoms are substituted for Fe atoms in $TbFe_2$ to form $Tb(Fe_{1-x}Al_x)_2$ or $Tb(Fe_{1-x}Co_x)_2$. Curve a of Figure 1 shows their data for $Tb(Fe_{1-x}Al_x)_2$.

We considered it important to make a more detailed study of the dependence of the rhombohedral distortion on the spontaneous magnetization for one of these three systems. We chose $Tb(Fe_{1-x}Al_x)_2$ so as to avoid any possible complications associated with magnetic compensation. The magnetization of the Tb sublattice already predominates in pure $TbFe_2$, so that decreasing the moment on the "iron" sublattice by substituting non-magnetic Al atoms can only further diminish the moment of the "iron" sublattice relative to that of the Tb sublattice. We made room-temperature magnetization and Mössbauer measurements over the same range of Al concentrations which had been used for the x-ray measurements of the distortion. The magnetization measurements give us the overall lattice magnetization, the Mössbauer measurements of the hyperfine field give us an indirect measure of the "iron" sublattice magnetization, and a comparison of the two gives us an indication of the behavior of the Tb sublattice magnetization.

## EXPERIMENT

The method of preparation of the alloy samples has been described in reference 1.

The magnetization measurements were made with a vibrating sample magnetometer using the saturation moment of nickel as a standard for calibration. The results of the magnetization measurements are shown in curve b of Figure 1. The magnetization first rises slightly and then falls off with increasing Al concentration. There are two reasons for this variation of the magnetization. One is that the magnetization of the Tb sublattice increases relative to that of the "iron" sublattice as iron atoms are removed. This would tend to increase the total lattice magnetization monotonically if it were not for the fact that the magnetic ordering temperature decreases with increasing Al concentration[5], and eventually the total magnetization must go to zero at 300 K. Although the spontaneous magnetization is falling smoothly in the concentration range where the distortion disappears, it is varying more rapidly than at lower Al concentrations.

The Mössbauer spectra are very complex except for that of $TbFe_2$. This is because the random substitution of Al for Fe on the "iron" sublattice

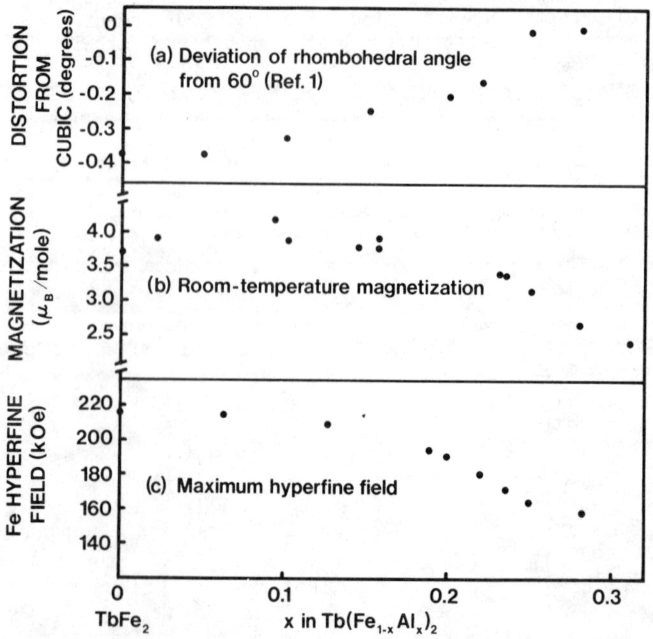


Figure 1. Rhombohedral distortion, magnetization, and Fe hyperfine field for $Tb(Fe_{1-x}Al_x)_2$.

produces a great variety of environments for the Fe atoms, and a corresponding variety of hyperfine-split spectra which are all superimposed to produce the spectrum observed at any particular concentration of Al. However, for pure $TbFe_2$ there are only two components[4], and the one with the greater intensity also has the larger magnetic hyperfine splitting. As the aluminum concentration is increased, the other spectral components begin to appear, and the relative intensity of this one (for which the near neighbors on the "iron" sublattice continue to be Fe atoms) begins a steady decline. It soon becomes a hopeless task to try to fit the observed spectrum uniquely by a superposition of spectra having varying intensities and hyperfine Hamiltonians. Nevertheless, the outer two lines of this component are still distinguishable from the rest of the spectrum, even at fairly high concentrations of Al. One such spectrum and the best-fit curve are shown in Figure 2. The parameters of the Hamiltonians for the different

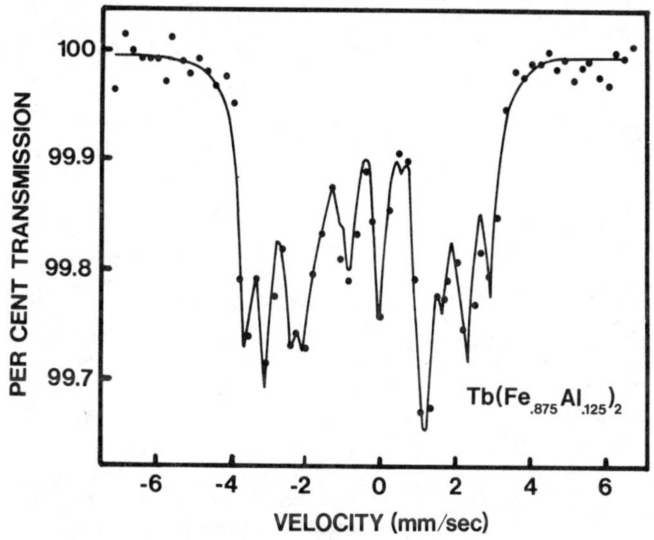


Figure 2. Spectrum for $TbFe_2$ with 12.5% of the Fe replaced by Al. The solid line represents a fit to six superimposed Fe spectra. Most of the details of this fit cannot be taken seriously, but the locations of the outermost lines of the spectrum for Fe with no Al nearest neighbors can be determined quite accurately, since they are the outermost lines of this complicated spectrum.

spectral components are not all determined equally well by the fitting process. However, we do obtain good approximate values of the magnetic hyperfine field for the component of interest. They are shown in curve c of Figure 1. It is apparent that there is nothing unusual about the concentration dependence of the hyperfine field for this component of the spectrum in the region where the distortion seems to vanish. Since the hyperfine field for Fe having any given near-neighbor configuration is probably proportional to the sublattice magnetization, these hyperfine field values should be nearly proportional to the spontaneous magnetization of the "iron" sublattice. Thus the "iron" sublattice magnetization appears to be a smooth function of the Al concentration.

The magnitude of the Tb sublattice magnetization is the difference between the total magnetization (curve b, Figure 1) and the magnitude of the "iron" sublattice magnetization (proportional to curve c, Figure 1). Since both these quantities decrease with increasing Al concentration, the Tb sublattice magnetization must also curve downward with increasing Al concentration, and somewhat more steeply than the total magnetization of curve b. Since both curve b and curve c are smooth in the region where the distortion disappears, the Tb sublattice magnetization must also be varying smoothly in this region.

## CONCLUSION

We conclude that the total and sublattice magnetizations of $Tb(Fe_{1-x}Al_x)_2$ do vary with x, and that the Tb sublattice magnetization, in particular, varies quite rapidly for values of x near the point where the rhombohedral distortion vanishes. But there are no real anomalies in the magnetic behavior at this point.

Note added in proof: Clark, et al.,[6] have concluded on the basis of both theory and experiment that the rhombohedral distortions in this and certain other $RFe_2$ alloys persist as long as the rare-earth sublattice is magnetized along a [111] direction. This would mean that small but non-zero rhombohedral distortions persist beyond the points where Dwight and Kimball[1] considered them to have vanished. The present results, as well as the previous finding that distortion of the Tb sublattice also does not vanish at this point,[2] are qualitatively consistent with the conclusions of Clark et al.

## ACKNOWLEDGEMENT

We wish to thank R. Biggers for making some of the magnetization measurements.

## REFERENCES:

†Based on work performed under the auspices of the National Science Foundation.

1. Dwight, A.E., and Kimball, C.W., Acta Cryst. B30, 2791-2793 (1974).
2. Preston, R.S., Dwight, A.E., Fedro, A.J., and Kimball, C.W., AIP Conference Proceedings 24, 660-661 (1975).
3. Buschow, K.H.J., and Van Stapele, R.P., J. Appl. Phys. 41, 4066-4069 (1970).
4. Dariel, M.P., Atzmony, U., and Lebenbaum, D., Proceedings of the Tenth Rare Earth Conference, U.S. Atomic Energy Commission, Oak Ridge, 1973, pp. 439-447.
5. Oesterreicher, H., J. Phys. Chem. Solids 34, 1267-1280 (1973).
6. A.E. Clark, J.R. Cullen, O.D. McMasters, and E.R. Callen, reported elsewhere in these Proceedings.

# MAGNETIC PROPERTIES OF $Gd_{1-x}Fe_x$ FILMS

C. Vittoria, P. Lubitz and J. Schelleng
Naval Research Laboratory, Washington, D. C. 20375

## ABSTRACT

The saturation magnetization, M, g-value, exchange and magnetic damping parameter were measured in films of $Gd_{1-x}Fe_x$, where $0<x<1$, as a function of temperature and frequency using ferromagnetic resonance (FMR) and vibrating sample magnetometer techniques. We find for $x \sim .9$ that the dominant exchange coupling is between the Fe-Fe ions. The exchange-conductivity linewidth broadening mechanism is minimal in this system.

## INTRODUCTION

The magnetic structure of rare earth-transition metal alloy system and the fundamental mechanisms by which these ions interact are of basic physical interest. These alloys systems also show promise for a variety of applications, e.g., permanent magnets ($SmCo_5$), "bubble" domain magnetic materials ($Gd_{1-x}Co_x$) and transducers ($TbFe_2$).

The $Gd_{1-x}Fe_x$ film system was recently studied[1] at 9 GHz. The magnetization and the g-factor were measured as a function of x; the exchange stiffness constant, A, was measured for the iron rich films ($x\sim.8-.96$) from the spin-wave resonance spectrum; the linewidth was fitted to the Landau-Lifshitz form for magnetic damping from ~100 to 500K. The compensation point form was found to be at $x\sim.78$. The exchange constant, A, decreased for increasing Gd concentration ($0<1-x<.2$). We surmised[1] that the dominant exchange contribution to A "comes" from the Fe-Fe exchange. We now extend this work by measuring the magnetization as a function of temperature. From these and previous[1] measurements of M we are able to estimate the relative quantitative importance of the Fe-Gd and Gd-Gd exchange with respect to the Fe-Fe exchange. We find that indeed the Fe-Fe exchange is predominant over any other combination of exchange, thus confirming our previous[1] ideas.

Since the exchange-conductivity mechanism[2] contribution to the measured linewidth was determined[1] to be minimal for Gd-Fe alloys, this should serve an ideal system in which to study intrinsic magnetic-relaxation-processes in metals. Thus, as a further test on the validity of the Landau-Lifshitz model for relaxation, we measured the linewidth as a function of frequency (9, 32, 72 GHz.) We find that the linewidth, $\Delta H$, is proportional to frequency as predicted from the Landau-Lifshitz mechanism.

## RESULTS

From FMR measurements the magnetization is deduced from the in-plane ($\parallel$) and perpendicular ($\perp$) resonance lines and M is given as

$$4\pi M = H_\perp + (H_\parallel/2) - (H_\parallel(H_\perp + 1.25H_\parallel))^{\frac{1}{2}} \qquad (1)$$

Also from the magnetometer measurements the uniaxial anisotropy contribution is negligible in comparison with $4\pi M$. In Fig. 1 M is plotted as a function of temperature for x = .83. The fact that M decreases for temperatures below room temperature implies that the Gd moment is directed opposite to the Fe moment, and that the Gd sublattice is only polarized extensively at low temperatures. In attempting to explain the temperature variation of the total moment, M, we have assumed M to be composed of two sublattice magnetizations, $M_{Gd}$ and $M_{Fe}$, coupled ferrimagnetically.

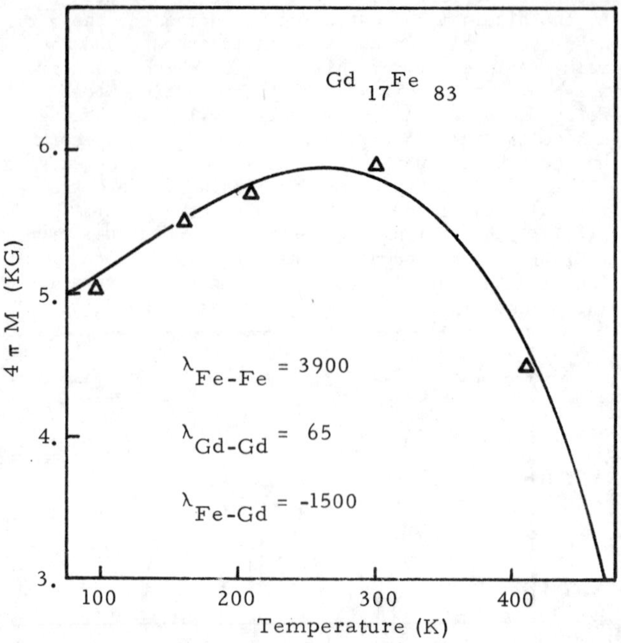


Fig. 1. Magnetization as a function of temperature for $Gd_{17}Fe_{83}$. The points were determined from Eq. 1. The curve is based on Eq. 2 with $M_{Fe}$ = 1.1 KG and $M_{Gd}$ = .71 KG.

The sublattice magnetizations can be expressed[3] in terms of a Brillouin function

$$M_J(T) = M_J(0)\, B_S[(gS\mu_\beta)_J(\lambda_{J-J}M_J - \lambda_{J-J'}M_{J'})/kT] \qquad (2)$$

where J≡Fe or Gd and J' denotes the other sublattice so that the total magnetization is $M=M_J-M_{J'}$. Also $S_{Fe}=1$; $(gS)_{Fe}=2.2$

$S_{Gd}=7/2$; $(gS)_{Gd}=7.3$

$\lambda_{J-J}$ is defined as the exchange constant within a sublattice. $\lambda_{J-J'}$ is the exchange coupling between sublattices. The signs of $\lambda_{J-J}$ and $\lambda_{J-J'}$ are opposite to each other, since the coupling is assumed ferrimagnetic. The values of $\lambda_{Gd-Gd}$, $\lambda_{Fe-Fe}$ and $\lambda_{Fe-Gd}$ were varied for best fit of M versus T. The values of $M_{Fe}(0)$ and $M_{Gd}(0)$ were also varied within reasonable range, since a priori, these values can be reasonably estimated. Similar fitting procedure was done for a number of films with nominal values of $.82< x < .96$. We find, as a rule, that $\lambda_{Fe-Fe} > \lambda_{Fe-Gd} > \lambda_{Gd-Gd}$ and in particular for x = .83 $\lambda_{Fe-Fe}=3900$, $\lambda_{Fe-Gd}=-1500$ and $\lambda_{Gd-Gd}=65$. This confirms our earlier results where[1] we surmised $\lambda_{Fe-Fe}$ is the dominant exchange.

For higher concentrations of Gd ($x<.65$) we find that M is nearly linearly dependent on T from 100K to the Neel temperature as shown in Fig. 2 for a typical sample. With reasonable parameters for this system, the preceding analysis cannot predict a linear relationship over a broad range of T. We propose that for these concentrations we

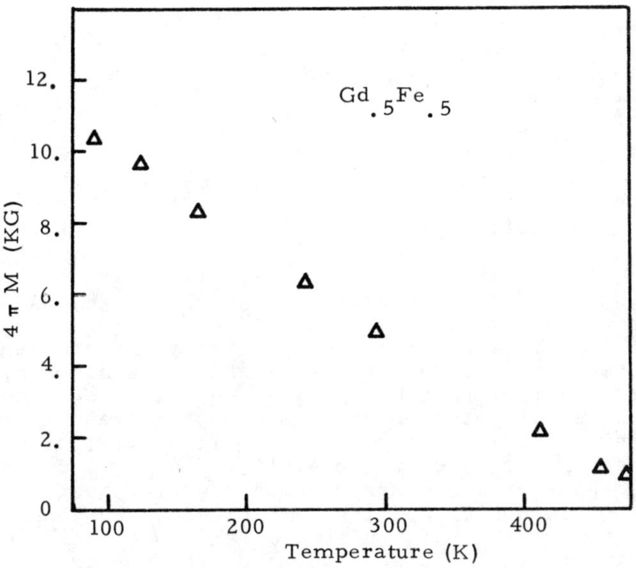


Fig. 2. Magnetization as a function of temperature for $Gd_{.5}Fe_{.5}$. The points were determined from resonance data using Eq. 1.

have a disordered ferromagnet whose M versus T is nearly linear as predicted by Huber and Siemann[4]. This notion is corroborated by the fact that there are no stable[5] crystal phases for $x < \frac{2}{3}$. However, it is not clear from our data whether the disorder is due to the films being amorphous or a combination of many mixed phases. Our X-ray data indicates that there are no well defined separate phases for $x < \frac{2}{3}$.

In fitting[1] the temperature variation of the linewidth we made some simplifying assumptions, namely: the increase in linewidth at low temperature was attributed to inhomogenous broadening and, therefore, is not intrinsic. Assuming a Landau-Lifshitz damping parameter of $\lambda = 1.15 \times 10^7$ Hz, deduced from fitting the temperature variation[1] of the linewidth for x= .86, we calculated the frequency dependence of the linewidth. In Table I the linewidths were measured at 9, 32 and 72 GHz and the predicted linewidth is given as

$$\Delta H \simeq 1.15 \frac{\lambda f}{\gamma^2 M}$$

where f is the frequency, $\gamma = g \frac{e}{2mc} \sim 2.84 \times 10^6 \frac{Hz}{Oe}$. We find that there is reasonable agreement between the measured and predicted linewidths and it appears that the Landau-Lifshitz model is adequate in explaining the relaxation model of these films.

## CONCLUSIONS

(1) For iron rich Gd-Fe films, the iron-iron exchange predominates over any other type of exchange. The analysis is based on our assumption of an ordered ferrimagnet with two sublattices. This assumption is not valid for low values of x. We must resort to an analysis[4] applicable to disordered ferrimagnets to explain our M versus T data.

(2) The Landau-Lifshitz model for magnetic relaxation is adequate in predicting the measured linewidth as a function of frequency and temperature.

| f(GHz) | (ΔH)(Oe) experimental | (ΔH)(Oe) calculated |
|---|---|---|
| 9.23 | 26 | 24 |
| 32. | 75 | 81 |
| 72.5 | 180 | 180 |

Table 1. Measured and predicted linewidth for $Gd_{14}Fe_{86}$ at a frequency of 9.23 GHz.

## REFERENCES


[1]P. Lubitz, J. Schelleng and C. Vittoria, Solid State Comm. (to be published.)

[2]G.T. Rado and J.R. Weertman, J. Phys. Chem. Solids 11, 315 (1959).

[3]L.R. Maxwell and S.J. Pickart, Phys. Rev. 92, 1120 (1953).

[4]D.L. Huber and R.P. Siemann, Solid State Comm. 17, 769 (1975).

[5]R. Elliott, ed. Constitution of Binary Alloys, First Dupplement (McGraw-Hill, New York, 1965).

# AMORPHOUS ALLOYS AS SOFT MAGNETIC MATERIALS II

E. M. Gyorgy, H. J. Leamy, R. C. Sherwood, & H. S. Chen
Bell Labs Murray Hill, New Jersey 07974

## ABSTRACT

A year ago C. D. Graham Jr. [1] reviewed the behavior and applications of ferromagnetic glasses. This paper will discuss the progress made since that time by a number of investigators in various laboratories. Substantial effort has been directed towards the study of the dynamic properties of these alloys. The dynamic properties examined include the complex permeability and the coercive force as a function of frequency, the rapid flux reversal time as a function of applied field, and the damping associated with domain wall motion. Essentially all of the dynamic properties can be understood by considering the ferromagnetic glasses to be just soft magnetic materials with a resistivity about three times that of permalloy. The details of the domain structure in the two materials differ, but not in a substantial or fundamental way. Because the glassy alloys are in general not isotropic, their domain structure is determined by a local anisotropy which arises due to strain or "pair ordering".

## INTRODUCTION

The discussion here will be limited to amorphous magnetic materials in ribbon form since it apears that this form will be most useful for practical applications. The ribbons are produced by directing a molten stream between two counter rotating rollers or onto the inside of a rotating drum [1]. The mechanical properties, details of the random atomic arrangement, etc. will not be discussed here since it is our aim only to review the recent investigations that relate to possible applications.

In order to consider possible uses of the ferromagnetic glasses it is necessary to discuss the static and dynamic properties in some detail. Many of the results obtained can be interpreted with a few models and assumptions based on the probable domain configurations present. These models are not to be taken as either totally accurate or definitive, but rather are used to set forth the pertinent material parameters. Different assumptions may change the details of the interpretations but is is unlikely that the overall trends demonstrated will be substantially different.

Much of the data reported is for Metglas 2826 (registered trademark of Allied Chemical) and for a zero magnetostrictive alloy [2]. The properties of these alloys are quite similar and the emphasis on their properties reflects, at least in part, their availability.

## MAGNETIZATION PROCESSES

As has been shown by a number of investigators [1,3,4,5,6,9] amorphous alloys are not magnetically isotropic but rather have a local anisotropy. We neglect for a moment the anisotropy produced by applying a unifrom stress to the material. This local anisotropy may be produced by non-uniform strains, field annealing or by some process that occurs during the initial cooling. The value of the induced anisotropy ($K_u$), of course depends upon the alloy composition and previous thermal-mechanical treatment. For the present it is sufficient to state that values of $K_u$ ranging from 800ergs/cc [7] for a magnetically annealed NiFe alloy to approximately $5x10^4$ ergs/cc for as quenched $Fe_{.8}P_{.13}C_{.07}$ [8] have been observed.

This local induced anisotropy determines the domain structure within the constraints of the sample shape. In Fig. 1 we show the domain structure of the as quenched, zero magnetostrictive alloy ($\lambda = 0$). This sample was prepared by the spin quenching process and the domain pattern clearly shows the flow that occurred during cooling. In the spin process a fine stream of molten metal, 0.04 cm in diameter, is directed onto the inside of a rapidly rotating drum. Thus the outer edges of the ribbon have experienced flow perpendicular to the roll direction. As seen in Fig. 1, the direction of the anisotropy induced by this flow is along the flow direction; that is, mostly parallel to the roll direction near the center of the tape and primarily perpendicular to the roll direction at the edges of the tape. As might be expected from an examination of the domain pattern, the B-H loop taken with relatively small applied fields is square but exhibits a relatively low remanence ratio. The square part of the loop arises from the motion of the 180 domain walls present in the sample, the low remanence results from the regions where the direction of local anisotropy is not parallel to the roll direction. For the zero magnetostrictive alloy discussed here, a field of approximately 6 Oe is required to saturate the sample, implying a value of $1.5x10^3$ ergs/cc for $K_u$. Similar results are obtained from roller quenched materials. In this case, the lateral flow of material between the rolls also produces a local anisotropy perpendicular

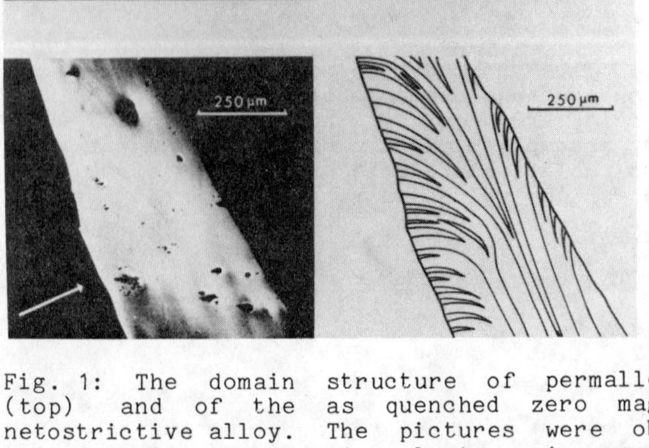

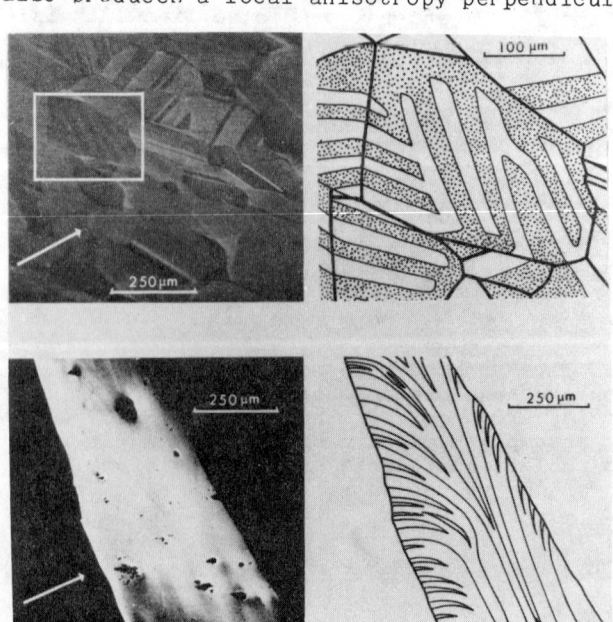

Fig. 1: The domain structure of permalloy (top) and of the as quenched zero magnetostrictive alloy. The pictures were obtained using the scanning electron microscope technique of Fathers [9].

to the roll direction [3,9]. The value of $B_r/B_s$ obtained for almost all samples investigated seems to be about 0.4 + 0.1. For the $\lambda = 0$ sample discussed here $B_R/B_s = 0.45$, for spin quenched $Fe_{.80}P_{.13}C_{.07}$ the remanence ratio is 0.42 and for a number of NiFe alloys $B_R/B_s$ lies between 0.3 and 0.5.

The domain pattern after a magnetic anneal, both parallel and perpendicular to the roll direction is shown in Fig. 2 [3] for the same alloy as was used for Fig. 1. These domain patterns clearly demonstrate that the easy direction of magnetization is determined by the magnetic field during annealing. As is obvious from Fig. 2, annealinging the sample with the applied field along the roll direction substantially increases the remanence ratio, and annealing perpendicular to the roll direction decreases the remanence ratio essentially to zero (Fig. 3). Values of $B_R/B_s$ up to 0.9 have been reported for $Ni_{.40}Fe_{.40}P_{.16}B_{.06}$ (Metglas 2826 - registered trademark) cooled with a field along the roll direction at about 0.1 deg/min from above the Curie temperature [10]. The B-H loops observed for field annealed samples are quite square and often show a few large Barkhausen jumps, indicating that as is suggested by Fig. 2, a small number of walls are nucleated and that these walls are then essentially free to sweep through the sample.

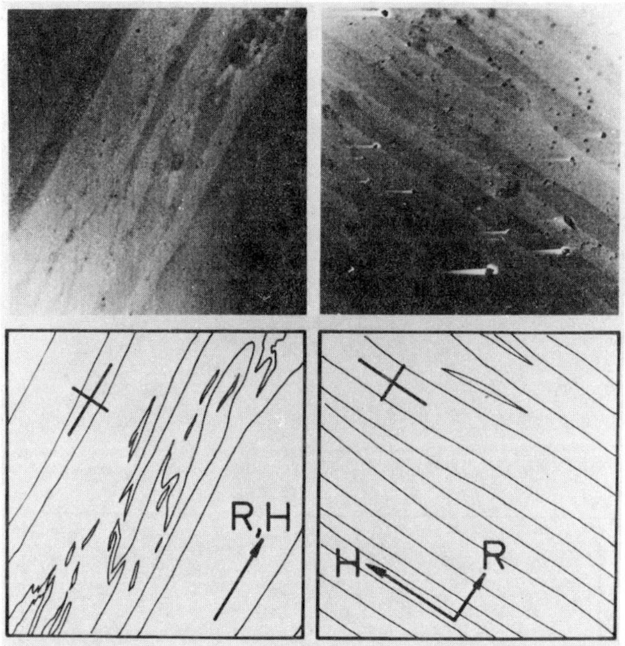


Fig. 2: The domain structure after magnetic annealing for the composition given in Fig. 1. On the left H is parallel to the roll direction (R), on the right H is perpendicular to R.

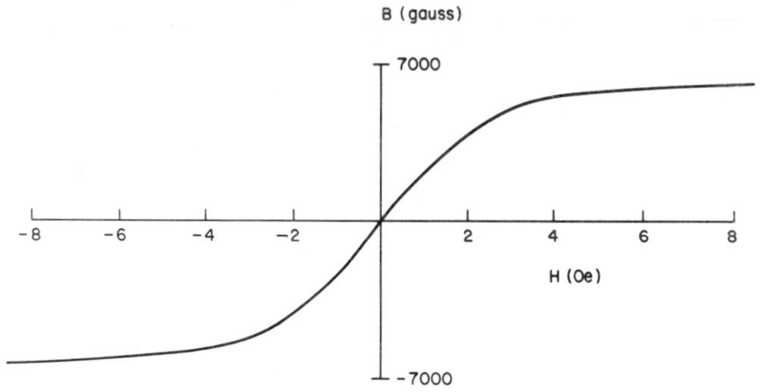


Fig. 3: The B-H loop for the alloy shown in Fig. 2 with H perpendicular to R.

The coercive force can be discussed if we make the not unreasonable assumption that, at least in well annealed samples, the domain walls are pinned only at the surface. The surface roughness shown in Fig. 4 helps to justify this assumption. For the case of 180 walls pinned at the surface, the coercive

Fig. 4: The surface of a spin quenched ribbon. The surface shown is the one that was in contact with the rim and is approximately 1mm wide.

force can be written, essentially from dimensional analysis, as .

$$H_c = \frac{4(AK)^{1/2}S}{M_s d}$$

where d is the sample thickness and S is a constant less than one which describes the details of the surface roughness. The case S = 1 approximately describes the situation when the totally pinned wall expands discontinuously [11]. For the $\lambda = 0$ composition with A estimated at $10^{-6}$ ergs/cm, $H_c = 0.1S$; indicating that S is about 0.1 for this sample. We re-emphasize that this model is to be regarded as illustrative and is used here primarily to set forth the pertinent material parameters. Since A varies as $M_s^2$ (at least for a crystalline material)[12] and, as will be shown later for annealed alloys, $K_u$ to a good approximation varies as $M_s^2$, the coercive force should vary as $M_s$. The only data available to date cover too small a temperature to make a convincing comparison of this simplified model possible.

The low field initial susceptibility can also be discussed in terms of a model similar to that applied to the coercive force. We again assume 180 walls pinned at the surface with a wall spacing of L and a tape thickness d. The usual equation of wall motion

$$mx + x + x = 2M_sH$$

is used [13]. The term containing the wall mass, m, is neglected in the frequency range to be considered here, the damping constant, is taken to arise from eddy currents only and the spring constant has its origin in the added energy required to bow out the walls. With these assumptions and at low frequencies

$$= M_s^2d^2/[18L(AK)^{1/2}]$$

As in the derivation of the coercive force, the cosine term that accounts for the angle between M and H is taken equal to one. For the $\lambda = 0$ composition with $d = 2.5\times10^{-3}$ at 100 Hz, was measured to be 330, which yields a value of L equal to approximately 0.1mm. This value is reasonable in view of the domain patterns shown in Fig. 2.

As the magnetic field is increased, increases showing that added walls are nucleated and/or that the walls are breaking free from their pinning points. An analysis of under these conditions is quite difficult and for our present purposes not very instructive. Rather, we will now consider the dynamic response of the domain walls when an ac field large enough to reverse the magnetization through the B-H loop is applied. Under these conditions the domain walls are assumed to be essentially free of surface pinning and their behavior is governed only by eddy current damping. The assumption that eddy current damping substantially accounts for the domain wall velocity is justified on the basis of measurements made in the single wall regime [14]. If the applied field is given by $H = H_0 \sin 2\pi ft$ the apparent coercive force is

$$H_{ac} = (\pi fLH_0 / M_s)^{1/2}$$

The experimental results for the as quenched $\lambda = 0$ alloy show that Hac is proportional to the cube rather than square root of f (Fig. 5). Thus, in terms of the model, the domain spacing (L) must be a function of frequency, a result previously found for single crystal Si-Fe [15] and a 50-50 Ni-Fe alloy [16]. Qualitatively this frequency dependence of L can be understood by considering that at low frequencies a wall nucleated at an energetically favorable site sweeps through other potential nucleation sites before the field has increased enough for these sites to nucleate additional walls. At higher frequencies the wall will not yet have reached these sites before the field has increased enough for them to nucleate a wall. The data in Fig. 5 imply that L varies as $f^{-.33}$. The slope of the three curves in Fig. 5 vary as $H_0^{.5}$ as predicted by the model. The domain spacing (L) at 100 Hz deduced in this way is 0.3 mm for the as quenched sample and 1.0 mm for a field annealed sample. While this latter value is high it is not unreasonable considering that the domain walls must be nucleated each cycle.

Applying the same free wall model to minor loops, the core loss is shown also to be proportional to L [16]. From the 50 kHz loss data on a NiFe alloy [10], L is found to be 0.1 mm. Again this is a reasonable value. The data in Ref. 10 are not for totally free

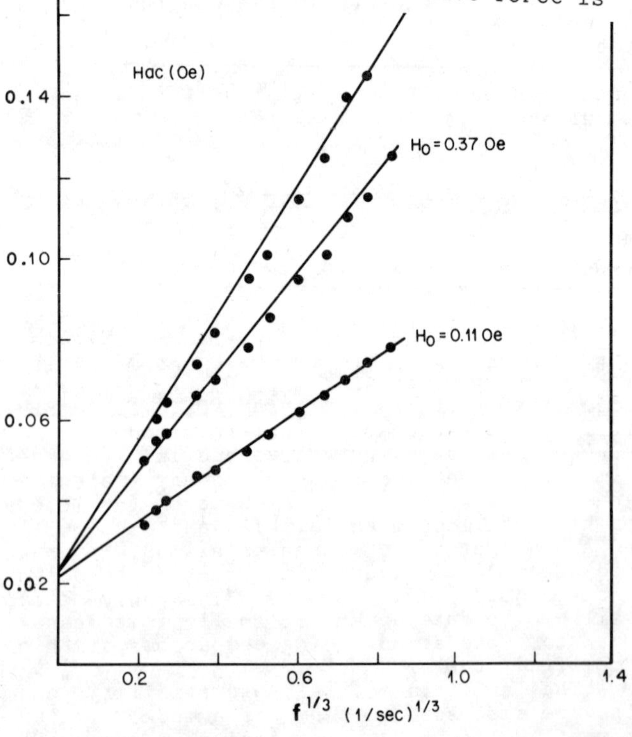


Fig. 5: The ac coercive force as a function of the cube root of the drive field frequency. Three values of the field amplitude are shown.

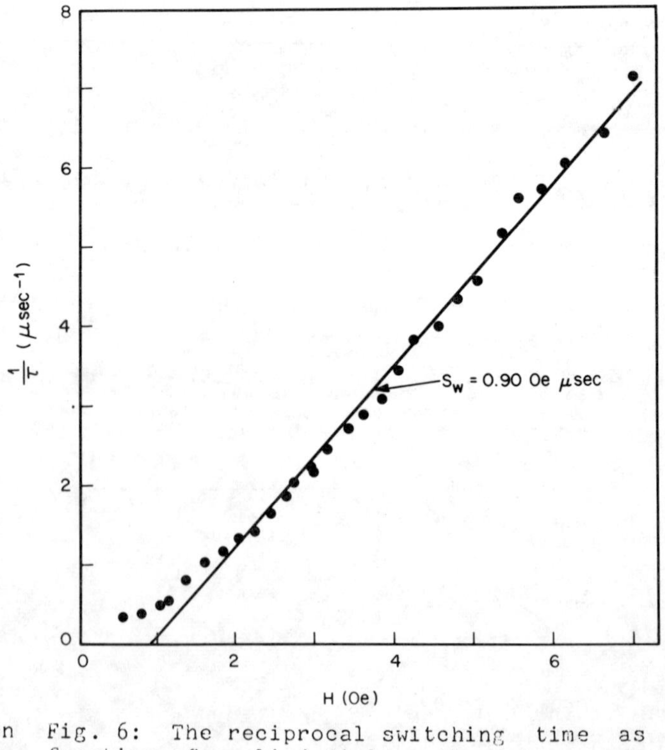


Fig. 6: The reciprocal switching time as a function of applied field. The data are for a field annealed zero magnetostrictive sample.

the time required for the integrated flux to change from 10% to 90% of the total flux switched. The results for the zero magnetostrictive alloy, field annealed at 300 C for 1 hr are shown in Fig. 6. Figure 6 is typical of the results obtained with a large class of magnetic materials [18]. They show the usual non-linear region at low drive fields that corresponds to domain wall motion, and a linear region at high drive fields due to non-uniform rotation of the magnetization. The non-linearity of the low field region arises because the number of domain walls increases with increasing drive field.

The inverse slope of the high field region is defined as the switching coefficient (Sw). The switching coefficient is the sum of a term due to intrinsic damping Sr(w) and a term that arises from eddy current losses Se(w) [19]. The non-uniform rotational model gives Sr(w) = 0.3 Oe microsec and for the tape core used here, Se(w) = 0.5 Oe microsec. Thus, Sw = Se(w) + Sr(w) = 0.8, which is in good agreement with the experimental value of 0.9 (Fig. 6). The threshold for rotation is 1 Oe, which is suprprisingly low compared to the expected value of 3 Oe for the anisotropy field. The value of the anisotropy field is taken from data obtained with nominally the same alloy annealed with H perpendicular to the roll direction (Fig. 3). In general, if a uniaxial anisotropy exists, the threshold for rotation is equal to the anisotropy field [20]. While we presently have no quantitative data to support an explaination of this discrepancy, we note that recent experiments [6,22] suggest that the field anneal induced anisotropy is very sensitive to the thermal history and sample preparation procedure.

The preceding discussion of the magnetization processes can be summarized by stating that the domain patterns are determined by the direction of the induced anisotropy. The response of these domain configurations to applied fields can be predicted with relatively simple expressions containing the intrinsic magnetic parameters A, $K_u$, $M_s$ and $\lambda$, structure sensitive quantities such as L and d, and the resistivity of the material. For all practical purposes, A is the same for all materials for which $T_c$ is well above room temperature. For most amorphous alloys, $M_s$ is only somewhat less than that of the corresponding crystalline alloy [21]. The magnetostrictive constant may be positive or negative and is typically of the order of $10^{-5}$ for both amorphous and crystalline materials. The resistivity of amorphous alloys is inherently higher (by about a factor of three) than that of crystalline materials. Thus it appears that, for the magnetic behavior considered here, amorphous materials will not differ in a fundamental way from the corresponding crystalline materials unless $K_u$ can be substantially reduced.

## MATERIAL PROPERTIES

It is not our purpose to discuss here the origin of ferromagnetism in amorphous alloys. Rather, we will review briefly the experimental results pertinent to possible technical application. Among the parameters that characterize a magnetic material, only $M_s$, $\lambda$, and the exchange constant A (and hence $T_c$) can be considered intrinsic in amorphous materials. The induced anisotropy ($K_u$) depends on the method of preparation and subsequent heat treatment.

The low temperature value of $M_s$ is fairly well represented by a Slater-Pauling curve [21]. The discontinuities observed between alloys of different crystal structure are, of course, absent in the data obtained with amorphous materials. The curve for amorphous materials is about 0.2 Bohr magnetons per atom lower than the equivalent crystalline curve. This decrease may be due to the randomness of the structure and to the dilution caused by the metalloid atoms. Iron has approximately 2 $\mu_B$ per atom, Co one and Ni zero. Mn, Cr and V have -3, -4, and -5 $\mu_B$ per atom respectively. Viewed from an applications standpoint, the principal conclusion to be drawn is that it is very improbable that an amorphous alloy with a larger magnetization than that of iron can be found in these alloy systems.

The dependence of $T_c$ on composition is quite smooth, having none of the discontinuities observed for crystalline materials. For Fe-Ni and Co-Ni, $T_c$ decreases monotonically from about 610 K to zero as the Ni content increases [21]. The Co-Fe series exhibits a maximum $T_c$ at the 50-50 alloy given as 950 K by Ref. [21] and 700 K by Ref. [2]. Whether this discrepancy arises because the glass forming additions are different for the two alloys or because different measurement techniques were employed cannot be determined at present. The values of $T_c$ that are above the glass transition temperature ($T_g$) are of course determined by extrapolation. In addition, the ferromagnetic transitions in at least some of the high concentration alloys are not sharp and therefore a clear determination of $T_c$ is not possible [22].

From a practical standpoint the Curie temperatures above the glass transition temperature are not of interest except inasmuch as $T_c$ effects the value of $M_s$. In fact, for some applications it may even be necessary to maintain temperatures below the embrittlement temperature ($T_B$). $T_B$ is obtained by heating the sample for 20 hrs at a fixed temperature, cooling to room temperature and then sharply bending the sample. The annealing temperature for which brittle failure is observed is defined as $T_B$ [23]. In Fig. 7 we show $M_s$, $T_c$, $T_g$ and $T_B$ as a function of composition for three alloy systems. The data in Fig. 7 should only be considered as semi-quantitative since they represent measurements on materials with only one of the many possible combinations of glass formers. Since the role of the glass formers in the embrittlement process is not well understood, alloys containing other glass formers may well exhibit different behavior.

The magnetostrictive constant ($\lambda$) has only been systematically studied in the CoFe system [24]. The value of $\lambda$ in this series varies from $20 \times 10^{-6}$ for Fe to $-5 \times 10^{-6}$ for Co. The zero magnetostrictive alloy discussed previously is $(Co_{.96}Fe_{.04})_{.75}$ PBAl. The value of $\lambda$ also depends on glass formers. For $Fe_{.77}P_{.16}B_{.04}Al_{.03}$, $\lambda$ changes from $26 \times 10^{-6}$ to $16 \times 10^{-6}$ as the ratio: P:B changes from 4 to 2.3. The addition of Ni to an Fe alloy decreases $\lambda$ [1]. Part of this decrease at least can be attriuted to the decrease in $T_c$ as the Ni content is increased (Fig. 7).

Since in general a $T_c$ well above room temperature and a large $M_s$ are required in most applications, the induced anisotropy ($K_u$) is the most significant magnetic property that is susceptible to control. Thus, an understanding of the origin of $K_u$ is most important in the technical development of amorphous alloys.

For an as quenched NiFe alloy with positive magnetostriction, Egami et al. [4] find that $K_u(T)$ is directly proportional to $\lambda(T)$. Thus it is reasonable to assume, as they do, that the origin of $K_u$ in this case is most likely the internal stress field.

For the $\lambda = 0$ alloy, internal stresses do not account for the induced anisotropy observed in as quenched samples. Rather it is necessary to postulate that pair ordering of some sort occurs during the initial cooling. As discussed in connection with Fig. 1, the direction of the pair ordering appears to be determined by the flow directions present at that time [3,9]. In this regard, our use of the term "pair ordering" is taken by analogy with the well known effect in crystalline materials. In glasses, this effect would be manifested by an axiosymmetric departure from complete isotropy of the pair correlation functions that describe the amorphous atomic arrangement.

The effect of field heat treatment has been shown in Fig. 2. Such effects are best observed in samples that have been totally stress relieved in order to eliminate magnetostrictive effects. Much of the experimental effort in the study of field induced anisotropy has been concentrated on the NiFe alloys. As a function of temperature, $K_u$ for these alloys has been shown to be proportional to $M_s^2$ [4]. This dependence suggests that the field induced anisotropy is due to some kind of pair ordering [26]. However, since other mechanisms may have an $M_s^2$ dependence, the pair ordering mechanism cannot be considered as having been definitely established. Furthermore unlike metal atoms are not required to produce a field induced anistropy since $Fe_{.75}P_{.15}C_{.10}$ responds to magnetic annealing treatment [5]. As expected from the model, the magnetic heat treatments are essentially reversible [6]. The kinetics involved however depend on the previous thermal history [6,22].

The field induced anisotropy cannot be eliminated by cooling from above $T_c$ in zero field since in this case the anisotropy direction is determined by the local direction of magnetization. However very rapid cooling from above $T_c$ should greatly decrease the value of $K_u$. Thus, from our model for it appears that the permeability should improve for a rapidly cooled sample. Encouraging results have been obtained with the $\lambda = 0$ alloy. For example, at 100Hz and a flux swing of 20 G, the permeability was increased by a factor of four.

Before leaving this topic we note that the optimum temperature for a totally effective stress relief anneal is about 600 K [7] and that magnetic annealing is most effective just below $T_c$. Thus for the alloys described in Fig. 7, it may be difficult if not impossible to obtain a non-brittle, stress relieved, field annealed sample.

## APPLICATIONS

We will not present a detailed discussion of device applications such as magnetic amplifiers, transformers, etc.; but rather will take a more general outlook. Possible applications of a magnetic material are determined by its static loop and dynamic properties such as permeability and losses. In addition, cost and stability must be considered.

The static properties of amorphous materials are most naturally compared to those of crystalline NiFe alloys (Permalloys). Common commercial Permalloys possess coercive forces of about 0.01 to 0.1 Oe and saturation magnetizations between 8000 and 16000 G. The amorphous materials at the present stage of development are very similar. The NiFe and $\lambda = 0$ amorphous alloys usually studied have $B_s$ values ranging from 7000 to 15000 G and values of $H_c$ approximately equal to 0.01 Oe. The remanence ratio ($B_R/B_S$) can be as large as 0.8 to 0.9 for both crystalline and amorphous materials. Also, both materials can, following appropriate field heat treatment, ex-

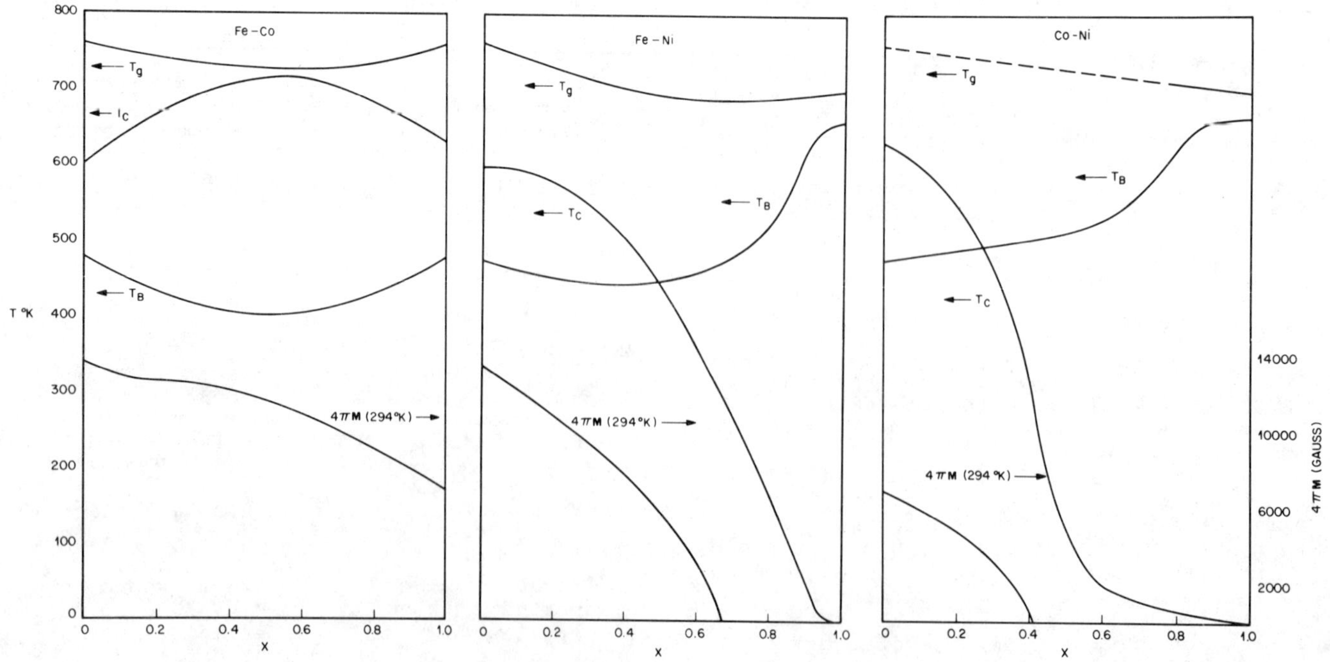


Fig. 7: The saturation magnetization ($M_s$), Curie temperaature ($T_c$), glass transition temperature ($T_g$), and embrittlement temperature ($T_B$) for CoNi, CoFe, and FeNi. The glass formers are $P_{.16}B_{.06}A_{.03}$.

hibit zero remanence (Fig. 3). Crystalline materials admit the added flexibility of a roll induced anisotropy, a technique that has not as yet been exploited in amorphous materials.

The situation can be summarized by the statement that the static behavior of amorphous alloys and that of the usual crystalline Permalloys is essentially identical. However, some crystalline alloys (Supermalloy) have been prepared with coercive forces as low as

TABLE I: A comparison of the permeabilities of Supermalloy and a = 0 Metallic Glass

| f(Hz) | B(G) | Metallic Glass (a) | Metallic Glass (b) | Supermalloy (c) | Supermalloy (c) |
|---|---|---|---|---|---|
| 100 | 20 | 3.2 | 4.2 | 67.5 | 7.0 |
| 100 | 100 | --- | 10.2 | 72.0 | 8.0 |
| 100 | 1000 | --- | 47.0 | 120.0 | 14.0 |
| 1000 | 20 | 3.1 | 4.0 | 63.0 | 7.0 |
| 1000 | 100 | --- | 10.8 | 65.0 | 7.0 |
| 1000 | 1000 | --- | 31.9 | 89.0 | 12.0 |
| 10000 | 20 | 2.8 | 4.0 | 42.0 | 6.0 |
| 10000 | 100 | --- | 5.7 | 43.0 | 7.0 |
| 10000 | 1000 | --- | --- | 45.0 | 8.0 |

(a) 0.001" thick, as quenched metallic glass
(b) metallic glass: field heat treated at 300 C
(c) Supermalloy results obtained on 0.001" ribbon by M. Sheppard at Arnold Engineering
(d) This material was worked by unwrapping and rewrapping the toroid.

All permeabilities are to be multiplied by 1000

0.002 Oe [27]. To reach this low coercive force required a substantial development effort. It would apear likely that in the future amorphous alloys with such extremely low coercive forces will be produced, given the appropriate developmental effort.

The situation concerning the permeability is about the same. The permeability of the amorphous alloys is comparable to that obtained from a number of magnetically soft crystalline alloys but it is not as high as the best results obtained from Permalloy type alloys. Table I lists the permeability at a number of frequencies and flux swings of 20, 100, 1000 G for the $\lambda$ = 0 alloy and Supermalloy. The values given for the $\lambda$ = 0 alloy are about the same as those mesasured for $Ni_{.40}Fe_{.40}P_{.14}B_{.06}$ [10]. They are substantially lower than those given for Supermalloy and even somewhat lower than those given for "damaged" Supermalloy. Again it seems reasonable to assume that future refinements will yield amorphous alloys as good as Supermalloy.

A major problem that will require extensive investigtion is the stability of amorphous materials with time. Luborsky et al. [10] have suggested that for some applications the stabilty is satisfactory and that for others it is not. Stability considerations depend, of course, upon the assumed temperature of operation and even of storage. While it is too early to reach any definite conclusions, it appears that the stability problem will always be more serious for amorphous materials than for crystalline materials.

Rather than conclude on a somewhat negative note it should be remarked that in a remarkably short time amorphous magnetic materials have been produced that in many aspects are competitive with crystalline materials. With expected future improvements glassy magnets may well play an important role in a number of applications.

ACKNOWLEDGMENTS

We are very greatful to F. Greenwald and M. Sheppard of Arnold Engineering for supplying the material and data used in Table I. Many discussions and measurements were facilitated by H. A. Brooks.

REFERENCES

[1] T. Egami, P. J. Flanders, and C. D. Graham, Jr., A.I.P. Conf. Proc. 24, 697 (1974).
[2] R. C. Sherwood, E. M. Gyorgy, H. S. Chen, S. D. Ferris, G. Norman and H. J. Leamy, A.I.P. Conf. Proc. 24, 745 (1974)
[3] H. S. Chen, S. D. Ferris, E. M. Gyorgy, H. J. Leamy and R. C. Sherwood, J. Appl. Phys. 26, 405 (1975).
[4] T. Egami and P. J. Flanders, this issue.
[5] B. S. Berry and W. C. Pritchet, Phys. Rev. Letters 34, 1022 (1975)
[6] F. E. Luborsky, this issue.
[7] F. E. Luborsky, J. J. Becker and R. O. McCary, IEEE Trans. Mag. MAG-II, 1644 (1975)
[8] H. Fujimori, T. Masumoto, Y. Ohi and M. Kikuchi, Japan J. Appl. Phys. 13, 1889 (1974
[9] H. J. Leamy, S. D. Ferris, G. Norman, D. C. Joy, R. C. Sherwood, and E. M. Gyorgy, J. Appl. Phys. 26, 259 (1975).
[10] F. E. Luborsky, R. O. McCary and J. J. Becker, to be published in the Proc. of Second Int. Conf. on Rapidly Quenched Metals.
[11] S. Chikazumi & S. Charap, Physics of Magnetism (Wiley, New York, 1964), p. 208.
[12] J. Smit & H. P. J. Wijn, Ferrites, (Wiley, New York, 1959), p. 68.
[13] E. M. Gyorgy in Treatise of Solid State Chemistry, Vol. 2, N. B. Hannay ed., Plenum Press, New York, 1975.
[14] R. C. O'Handly, J. Appl. Phys. 46, 4996 (1975)
[15] J. N. Sun, T. R. Haller and J. J. Kramer, J. Appl. Phys. 7, 2379 (1974).
[16] T. Higuchi, J. Phys. D,: Appl. Phys. 7, 2379 (1974).
[17] F. E. Luborsky, Personal Communication.
[18] E. M. Gyorgy in, Magnetism Vol. 3, G. T. Rado & H. Suhl, eds., (Academic Press, New York, 1964).,p 541
[19] N. Menyuk and J. B. Goodenough, J. Appl. Phys. 26, 8 (1955).
[20] E. M. Gyorgy and D. Treves, J. Appl. Phys. 33S, 1222 (1962).
[21] T. Mizoguchi, K. Yamauchi and H. Miyajima in Amorphous Magnetism H. O. Hooper and H. M. de Graaf, eds., (Plenum Press, New York, 1973), p. 325.
[22] H. J. Leamy, E. M. Gyorgy, R. C. Sherwood and H. S. Chen, this issue.
[23] H. S. Chen, to be pubished.
[24] H. A. Brooks, to be published in J. Appl. Phys.
[25] H. A. Brooks, this issue.
[26] S. Taniguichi, Sci. Rept. Inst. Tohoku Univ. A-7, 269 (1955).
[27] R. M. Bozorth, Ferromagnetism (P. van Nostrand Co., New York, 1951) p. 870.

# DOMAIN OBSERVATIONS IN AN AMORPHOUS IRON-NICKEL ALLOY

Joseph J. Becker
General Electric Research and Development Center
Schenectady, New York 12301

ABSTRACT

Domain structures were observed on Allied Chemical Metglas[TM] ribbons, $Fe_{40}Ni_{40}P_{14}B_6$, using colloid on one surface as received. Sharply defined patches with fine domain structure were seen. They would disappear in fields large enough to produce saturation. Increasing tension also caused their disappearance, as did annealing. Long mobile boundaries were occasionally seen between patches and near regions of high stress. These observations suggest regions with local strain-magnetostriction anisotropy which can be overcome by applied fields or stress, or eliminated by annealing.

Ribbons of Allied Chemical Metglas[TM], nominally $Fe_{40}Ni_{40}P_{14}B_6$, show among other features a remanence less than half of the saturation magnetization. An applied field on the order of 20 to 100 Oe is require to attain saturation[1,2]. Applied tensile stresses on the order of 100 MPa raise the magnetization in a field of 1 Oe to near its saturation value[1,2]. Annealing in the range of 200-300° C has the same effect[3].

The structural origin of these features is as yet obscure. There have been a few indications[1,4,5] that regions of quenched material may have uniaxial anisotropy normal to the ribbon.

In this investigation the classic ferromagnetic colloid technique was used. The Metglas[TM] ribbon as received has one surface that is much smoother than the other. This smooth surface needs only a little cleaning in acetone to be suitable for domain observation with colloid. Thus there is no uncertainty about the effects of surface preparation procedures. All pictures were made with dark-field illumination.

A number of striking features were observed. The colloid very characteristically forms patches, as shown in Fig. 1. These patches have a fine cellular domain structure, on a considerably smaller scale than the sample thickness. In a small applied field parallel to the ribbon axis, the cellular domains split into one of two complementary patterns, depending on the sense of the field. Fig. 2 illustrates this behavior. The light regions in 2b correspond to the dark regions in 2c. This reversal is shown even more strikingly in Fig. 3, in which two scratches are available as reference marks.

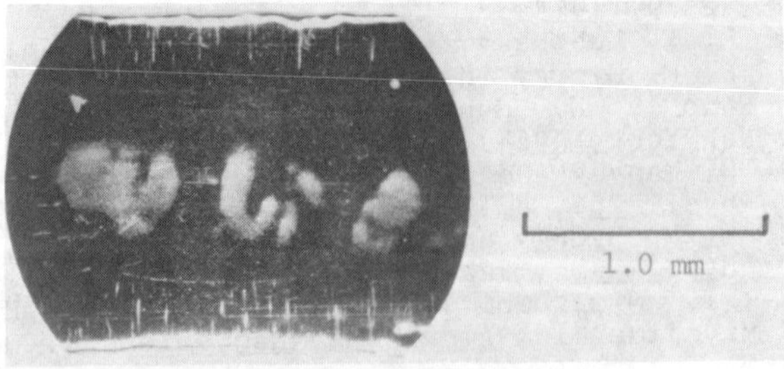


Fig. 1. Patches of fine domain structure on Metglas[TM] ribbon. No applied field. Ribbon axis horizontal.

This behavior indicates that the magnetization within the patches has a component coming out of the sample surface. The applied field aligns the aggregates of colloid particles. It probably also increases their magnetization, as the patterns tend to be easier to see in applied fields. The aligned particles then collect only on alternate domain boundaries. Fig. 4 is a sketch of this situation. The curved arrows indicate simply the local field distribution from the surface poles, which does not depend on the details of the wall structure. In the absence of an applied field, the particles will stick along both a and b. A field H as shown does not move the domain walls, but aligns the colloid particles, which will then stick along b but not a. Reversing H reverses the situation.

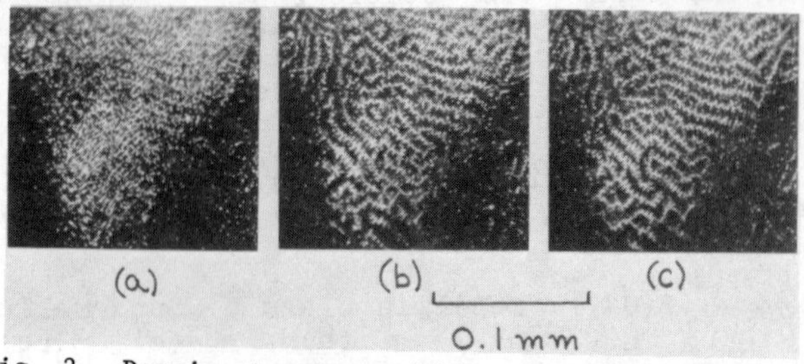


Fig. 2. Domain structure within patch with and without applied field. a) Zero applied field. b) and c) 8 Oe in opposite directions. Ribbon axis and applied field horizontal. Sample thickness 40 µm.

As H is further increased and the magnetization rotates, the wall spacing will remain roughly the same but the walls will disappear as the magnetization becomes parallel to the observation surface. This behavior can be seen in Fig. 5. Fields large enough to saturate the material cause the patches to

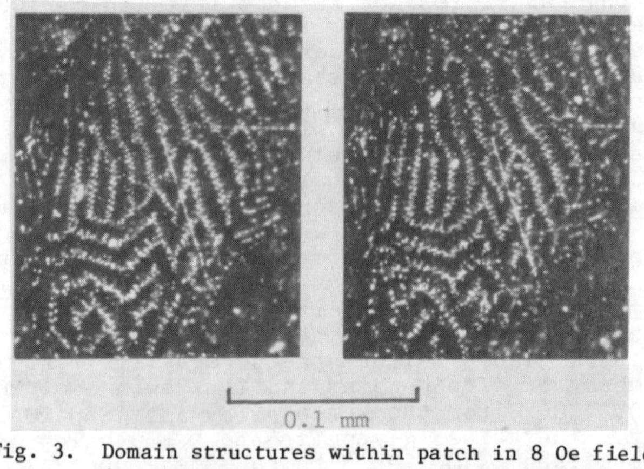


Fig. 3. Domain structures within patch in 8 Oe field in opposite senses. Straight lines are scratches. Ribbon axis and applied field horizontal.

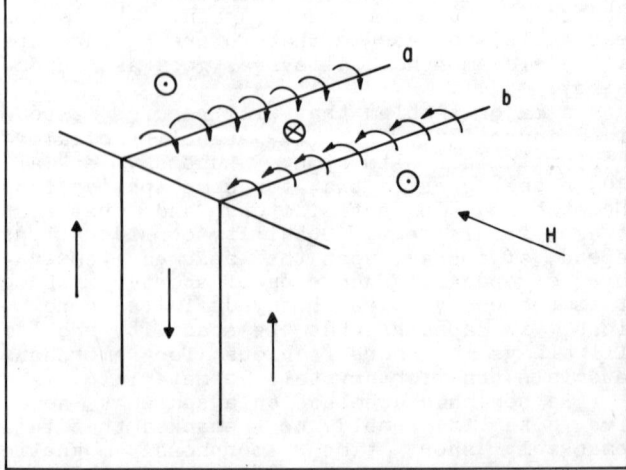

Fig. 4. Local field distribution on surface of sample with magnetization normal to surface.

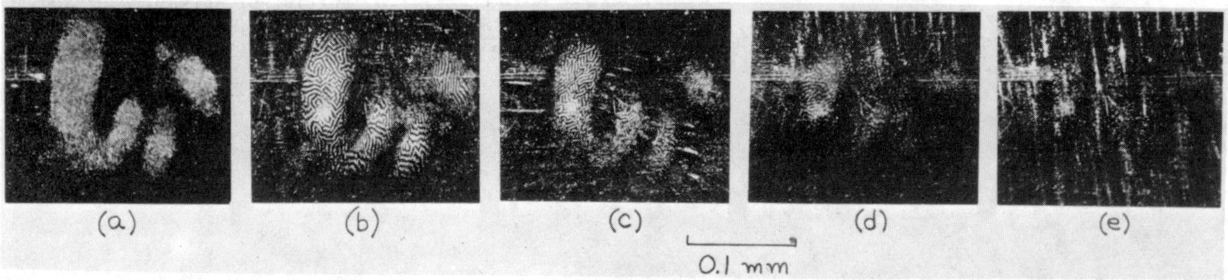


Fig. 5. Domain structures of patch in increasing field. a) Zero applied field. b) - e) Approximately 10, 20, 30, and 40 Oe. Ribbon axis and applied field horizontal.

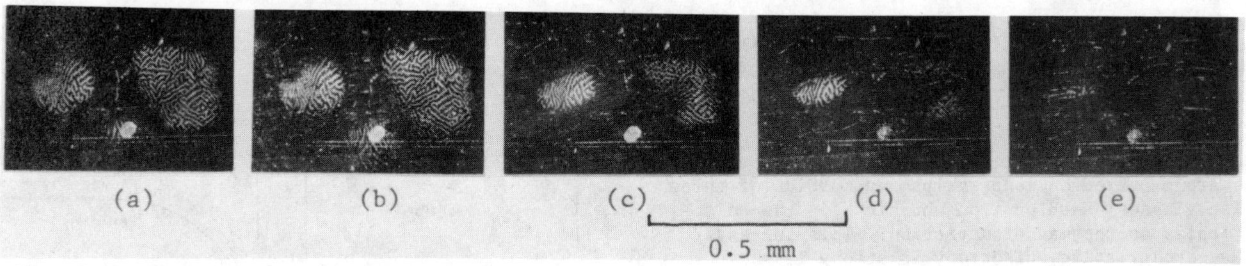


Fig. 6. Domain structures in patch with increasing tension. a) Zero tension. b) - e) Approximately 30, 60, 90, and 120 MPa. Applied field 8 Oe in all cases. Ribbon axis, tension axis, and applied field horizontal.

shrink and disappear.

Of course, the walls actually make various angles with H, the surface is irregular, and the detailed relationship to the cell structure has not been worked out. However, it seems clear that the patches are regions in which there is some type of anisotropy out of the sample plane. Their effect on the total domain structure is to cause the observed lowering of the remanence.

Next the domain structures were observed while tension was applied to the ribbon. Increasing tension also causes the patches to shrink and disappear, as shown in Fig. 6. The values of tension at which this happened are on the same order as those required to raise the magnetization to saturation in a small field[2]. Using $\lambda$=11 × $10^{-6}$ (ref. 1), a tension of 100 MPa produces an anisotropy field $3\lambda\sigma/M_s$ of about 60 Oe, the same order of magnitude as the applied field that causes the disappearance of the patches.

Annealing has also been shown[3] to make magnetization easier in low fields, and to bring about complete stress relief at about 300° C. The same patch before and after annealing at 200 and 250° C is shown in Fig. 7. The domain structure in an 8 Oe applied field becomes less convoluted as the anneal proceeds, suggesting that the local anisotropy is decreasing. After a 300° C anneal, the patch was no longer visible. This suggests that the patches may be associated in some way with local stresses.

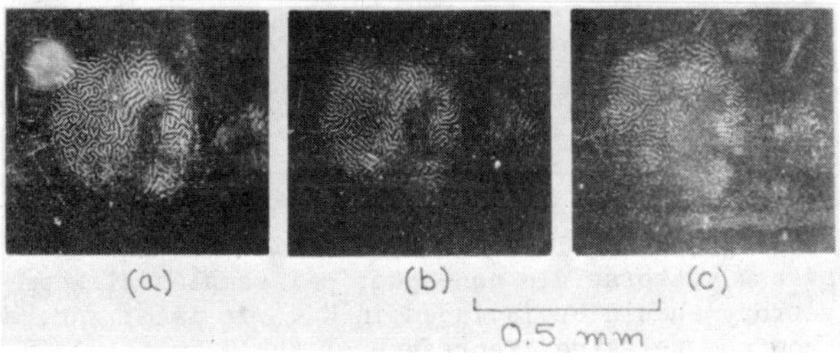


Fig. 7. Domain structure of patch before and after heat treatment. a) Untreated. b) After 15 minutes at 200° C. c) After 15 minutes at 250° C. Applied field 8 Oe in all cases. Ribbon axis and applied field horizontal.

Long curved mobile boundaries were occasionally seen in the regions between the patches. A few such walls can be seen running vertically in Fig. 5a, including one near the center in the reentrant portion of the large patch. They were quite numerous at the cut end of the sample, as in Fig. 8. Deliberate deformation of the sample resulted in many such walls, tending to run generally perpendicular to the visible slip lines. These walls tend to surround either regions of stress or patches. They have not been further analyzed, but their configuration also seems to support the suggested association of patches with stresses.

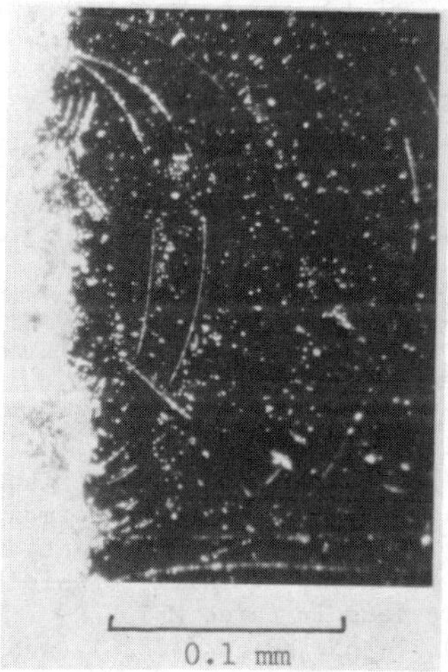


Fig. 8 Curved boundaries in vicinity of cut end of strip (cut end at left).

These domain observations indicate that the material as produced has magnetic inhomogeneities on a scale much larger than the thickness. These appear to be regions of local anisotropy associated with local stresses. The applied fields or stresses or annealing treatments that make the visible domain patches disappear are also those that increase the low-field magnetization.

## REFERENCES

1. T. Egami, P.J. Flanders, and C.D. Graham, Jr., AIP Conf. Proc. 24, 697-701 (1975).
2. J.J. Becker, IEEE Trans. MAG-11, 1326-7 (1975).

3. F.E. Luborsky, J.J. Becker, and R.O. McCary, IEEE Trans. MAG-11, (1975).
4. H. Fujimori, T. Matsumoto, Y. Obi, and M. Kikuchi, Japan. J. Appl. Phys. 13, 1889-1900 (1974).
5. H.J. Leamy, S.D. Ferris, G. Norman, D.C. Joy, R.C. Sherwood, E.M. Gyorgy, and H.S. Chen, Appl. Phys. Lett. 26, 259 (1975).

# MAGNETIC CORE LOSS AND INTERNAL STRESS IN METALLIC GLASSES

R. C. O'Handley
Allied Chemical Corporation, Materials Research Center
Morristown, NJ 07960

## ABSTRACT

Magnetic properties related to anisotropy and stress are measured on long strips and toroids of three metallic glasses based on iron and nickel. The anisotropy scales as the magnetostriction. This suggests that magnetostrictive anisotropy is strong in these materials and that their average internal stress is comparable, being on the order of $10^8$ dyne/cm$^2$. The initial stress sensitivity of the remanent magnetization is large and is well predicted by simple theory. Barkhausen structure observed upon magnetization reversal suggests a magnetostrictive interaction between domain walls and pinning centers. Stress-relieved toroids of these metallic glasses of finite magnetostriction show core loss decreasing with magnetostriction and approaching that of carefully annealed permalloys ($\lambda_s \approx 0$) at 1000 gauss and $10^4$ Hz.

## INTRODUCTION

The attractive soft-magnetic properties of continuous-cast metallic glasses based on iron, cobalt and/or nickel are the object of growing interest. Hitherto, most of the published magnetic data has focused on the d.c. properties of these materials.[1-3] The low anisotropy[4] and high resistivity characteristic of these materials suggests that their a.c. performance will also be superior.[5-7] In this paper we present and discuss some experimental results pertaining to anisotropy, stress and a.c. core loss in three METGLAS® alloys, $Fe_{80}P_{16}C_3B_1$(#2615), $Fe_{40}Ni_{40}P_{14}B_6$(#2826), and $Ni_{49}Fe_{29}P_{14}B_6Si_2$(#2826B). The continuous glassy ribbons studied here are formed by rapid quenching ($\sim 10^6$ K sec$^{-1}$) from a melt of the desired composition. Magnetostriction was determined by a static-magnetic-field, a.c.-gauge-voltage technique using semiconductor strain gauges. Standard hysteresis equipment is employed for the d.c. magnetic measurements. The power loss is determined from the complex product of undistorted voltage across, and current through an inductive element whose core is the test material.[8] Where annealing is indicated, it was done in vacuum for two hours at 225°C for alloy #2615 and 325°C for alloys #2826 and #2826B.[6,7]

## RESULTS AND DISCUSSION

### Magnetization

As quenched strips 30 cm in length of all three alloys show comparable anisotropy fields and remanence ratios $M_r/M_s$ (Figs. 1, 2 and Table 1).

A general expression for the anisotorpy should include a magnetostrictive term as well as a stress-independent, magnetostructural (glassy analogue of magnetocrystalline) anisotropy:

$$K = K_o + \lambda_s \sigma_i \qquad (1)$$

Here $\lambda_s$ is the saturation magnetostriction and $\sigma_i$ is

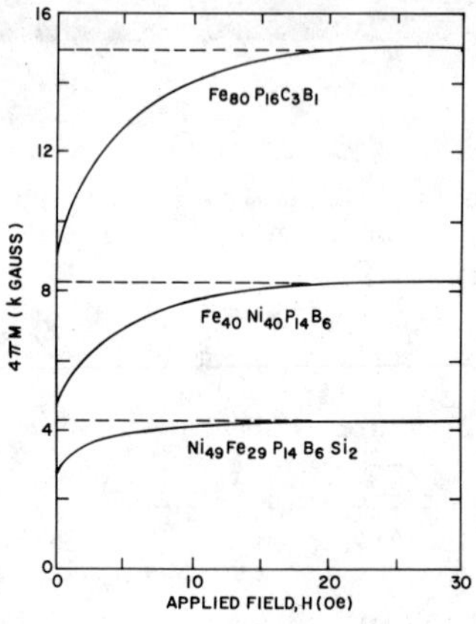


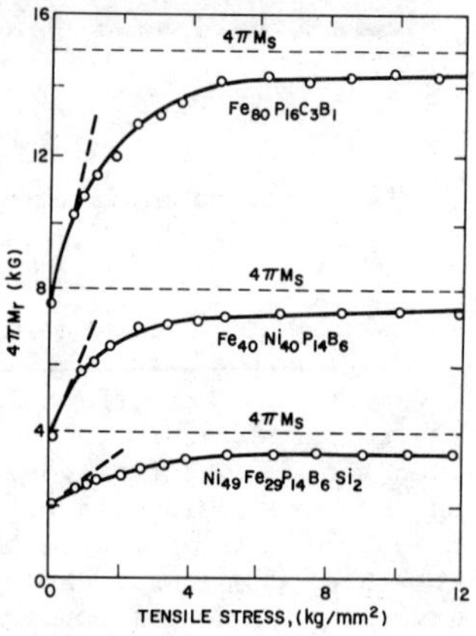


Fig. 1 Magnetization of straight strips.

Fig. 2 Stress sensitivity of remanence.

Table 1

SOME PROPERTIES OF METGLAS$^R$ ALLOYS

| GLASSY ALLOY | #2615 | #2826 | #2826B |
|---|---|---|---|
| $\rho$ ($\mu\Omega$cm) | 150 | 130 | 140 |
| $4\pi M_s$ (kG) | 14.9 | 8.2 | 4.2 |
| $M_r/M_s$ | .40 | .45 | .54 |
| Annealed | .42 | .58 | .70 |
| $K\times10^{-3}$(erg/cm$^3$), Eq. 2 | 3.6 | 1.4 | 0.5 |
| Annealed | 3.4 | 1.1 | 0.4 |
| $K\times10^{-3}$(erg/cm$^3$), Eq. 3 | 14.0 | 8.3 | 4.0 |
| $H_c$ (mOe) Strip | 80 | 17 | 13 |
| Annealed | 68 | 12 | 6 |
| Toroid | 62 | 50 | 57 |
| Annealed | 50 | 19 | 11 |
| $\lambda_s\times10^6$ | 30 | 11 | 3 |
| $\Lambda_r$ (kG mm$^2$/kg) | 4.0 | 2.2 | 0.7 |

the average internal stress for a uniform distribution of stress vectors. In non-spherical samples a shape anisotropy should be included in $K_o$. It is of interest to know the relative importance of these two contributions to the total anisotropy of these metallic glasses.

A convenient measure of the anisotropy is the area above the magnetization curve (Fig. 1)

$$K = f \int_{M_r}^{M_s} H dM \qquad (2)$$

where f is 1 for uniaxial anisotropy and 4 for cubic

anisotropy when the applied field direction bisects the angle between the easy directions of magnetization.[9] Another measure of anisotropy may be had by fitting the approach to saturation with the equation

$$M = M_s - \frac{8K^2}{105\ M_s H^2} - \ldots \qquad (3)$$

Values of the integral in Eq. 2 and K from Eq. (3) obtained for straight strips of the three glassy alloys are shown in Fig. 3 to increase with magnetostriction. Comparison of the two sets of results gives $f \approx 3.5 \pm 0.4$. One interpretation of Fig. 3 is that 1) magnetostrictive anisotropy is strong in these long strips, $\lambda_s \sigma_i > K_o$, and 2) the average internal stress, $\sigma_i = \partial K/\partial \lambda_s$, is of comparable magnitude in all three glassy alloys and equal to about $4 \times 10^8$ dyne/cm$^2$. The first statement is supported by the observed decrease in anisotropy upon stress-relief annealing (Table 1).[7] The second statement is not unreasonable because 1) these materials were prepared with the same technique and quench rate, 2) they presumably possess a common structure, and 3) this structure is free from grain boundaries.

The sensitivity of the magnetization to *applied* stress $\sigma_a$ is proportional to $\lambda_s B_s/K$ which for $K_o < \lambda_s \sigma_i$ is approximately $B_s/\sigma_i$. The initial stress sensitivities $\Lambda_r$ taken from Fig. 2 are listed in Table 1.

## Coercivity

Unintegrated, d.c. ($\sim 10^{-2}$ sec$^{-1}$), hysteresis loops (shown in Fig. 4 for alloy #2826) reveal barkhausen pulses resulting from the irregular motion of the domain walls. The same reduction in the number of

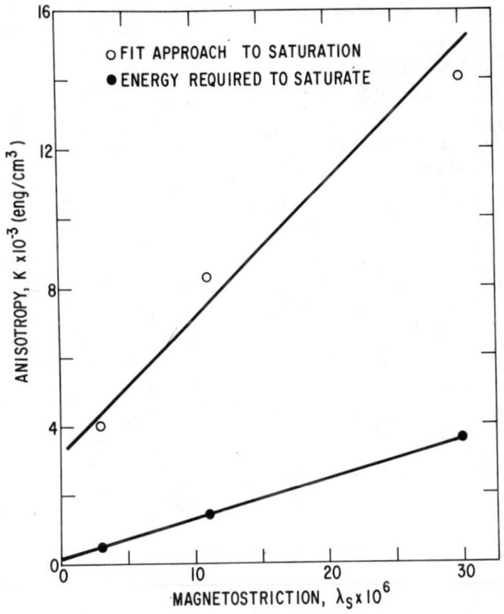


Fig. 3 Anisotropy in as-quenched, straight strips. Values of integral in Eq. 2 (dots) and K from Eq. 3 (circles).

barkhausen pulses shown in Fig. 4 for winding stress is also observed for applied tensile stress on a strip. Stress-relief annealing smooths out the envelope of pulses (eliminating those of largest magnitude) and shifts the distribution to lower fields. This suggests that the pinning centers impeding wall motion are largely magnetostrictive in nature.

A measure of the coercivity is the mean field over which the barkhausen pulses are distributed. Averages of these values of $H_c$ and those taken from B-H loops are listed in Table 1. The effects of applied stress on the coercivity are not as well defined as those on the remanence.

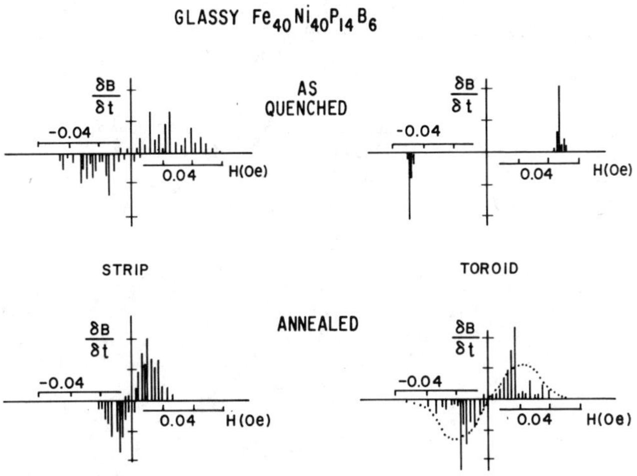


Fig. 4 Barkhausen pulses during magnetization reversal of glassy alloy #2826. Vertical scale arbitrary.

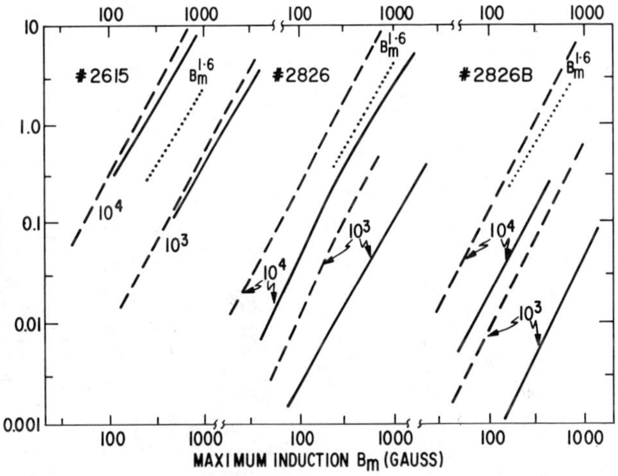


Fig. 5 Core loss before (dashed line) and after (solid line) stress relief as a function of maximum induction at $10^3$ and $10^4$ Hz. Dotted line shows the Steinmetz induction dependence.

## Core Loss

Power loss per cycle in a magnetic material may be expressed from Poynting's theorem as the volume integral of $H \cdot \partial B/\partial t$. It is expected, therefore, that the unintegrated hysteresis loops $\partial B/\partial t$ vs H shown in Fig. 4 should reveal something about core loss. Each pulse results from an irreversible process in the material as a domain wall moves from one pinning site to another. The dashed line superimposed on Fig. 4 shows the shape of the response observed for a fully annealed Mo-permalloy core under the same conditions. No pulses can be resolved for this core. Instead the magnetization reverses in a more uniform process. The behavior of the simply annealed metallic glass core begins to approach that of the permalloy core in this respect.

Fig. 5 compares the core-loss for the three metallic glasses. The trend in these glassy alloys, as with most conventional soft-magnetic materials, is toward lower core-loss in materials of lower magnetostriction.

It is worth noting that after a simple stress-relief anneal, these metallic glasses of *finite magnetostriction* exhibit low core losses which approaches those of the carefully treated commercial permalloys of zero magnetostriction. For example, at $10^4$ Hz and 1000 gauss, the core loss of alloy #2826B (annealed) is only a factor of 2 greater than the loss reported for 4-79 permalloy and Superperm 80[10] of comparable thickness. Even lower core losses than those shown in Fig. 5 have been achieved in METGLAS alloy 2826 by field annealing.[6,7]

The author gratefully acknowledges helpful discussions with R. Hasegawa, L. I. Mendelsohn, E. A. Nesbitt and P. J. Flanders, and the technical assistance of M. O. Sullivan.

REFERENCES

1. T. Egami, P. J. Flanders and C. D. Graham, Jr., Appl. Phys. Lett. 26, 128 (1975)
2. R. C. Sherwood, E. M. Gyorgy, H. S. Chen, S. D. Ferris, G. Norman and H. J. Leamy, AIP Conf. Proc. No. 24 (AIP, NY, 1975) p. 745
3. M. Kikuchi, H. Fujimori, Y. Obi and T. Masumoto, Japan. J. Appl. Phys. 14, 1077 (1975)
4. R. Hasegawa, this conference
5. R. C. O'Handley, J. Appl. Phys. 46, 4996 (1975)
6. E. A. Nesbitt, L. I. Mendelsohn and G. R. Bretts, to be published
7. F. E. Luborsky, J. J. Becker and R. O. McCary, IEEE Magnetics 11, 1644 (1975)
8. This method has been recommended to us with helpful suggestions by F. E. Luborsky and R. O. McCary, General Electric Corporate Research and Development, Schenectady, NY; IEEE Standard 106-1972
9. S. Chikazumi, Physics of Magnetism (John Wiley & Sons, NY, 1964) p. 137
10. Trade names of Arnold Engineering, Inc. and Magnetic Metals, Inc. Data taken from manufacturers handbooks.

# MAGNETOSTRICTION AS A FUNCTION OF GLASS FORMERS IN AMORPHOUS ALLOYS OF Fe-P-B-Al

H. A. Brooks and H. S. Chen
Bell Laboratories, Murray Hill, New Jersey 07974

ABSTRACT

Magnetostriction of amorphous-metal alloys has been found to display an unexpected dependence on the glass formers P and B in Fe-P-B-Al. This observation has been used to determine lower values for magnetostriction in the series, $Fe_x(P_{1-y}B_y)_{1-x-.03}Al_{.03}$ and for the cobalt-substituted series $(Fe_xCo_{1-x})_{.77}P_{.14}B_{.06}Al_{.03}$.

Samples of these series were produced by the centrifugal quenching technique, and have dimensions of 25 μ in thickness by 1 mm in width.

Magnetostriction data was generated using semiconductor gages. Magnetostrictive strain was produced using a stationary sample in the plane of rotation of a saturating H-field of 3 kG.

In the first series, $Fe_x(P_{1-y}B_y)_{1-x-.03}$, the x value was held to within five percent of .77; the y values were allowed to range from .1 to .33. Within the range of y values used a decrease by a factor of 1.5 in y is capable of producing an increase of 1.6 in the magnetostriction. This is exemplified by the +26 μs value for $Fe_{.77}P_{.16}B_{.04}Al_{.03}$ with y = .2 compared to +16 μs for $Fe_{.77}P_{.14}B_{.06}Al_{.03}$ with y = .3.

The composition $Fe_{.77}P_{.14}B_{.06}Al_{.03}$ was then used as the basis of a cobalt substituted series, $(Fe_xCo_{1-x})_{.77}P_{.14}B_{.06}Al_{.03}$. With the glass former proportions now held constant, the magnetostriction values as a function now of x were compared to those of a previously investigated series $(Fe_xCo_{1-x})_{.75}P_{.16}B_{.06}Al_{.03}$ with y = .27. The curve for the former series remains below that of the latter consistent with the initial observation that the x = 1 composition, $Fe_{.77}P_{.14}B_{.06}Al_{.03}$, with $\lambda$ = +16 μs has less magnetostriction than the x = 1 composition $Fe_{.75}P_{.16}B_{.06}Al_{.03}$ with $\lambda$ = +20 μs. While this implies an earlier zero crossing for the magnetostriction of the lower series, the present data seems to indicate that the Fe-Co amorphous alloys will cross zero at values of Fe of 4% and Co of 96% with only minor variations among various alloy systems.

# KINETICS OF REORIENTATION OF MAGNETICALLY INDUCED ANISOTROPY IN AMORPHOUS $Ni_{40}Fe_{40}P_{14}B_6$

F. E. Luborsky
General Electric Corporate Research and Development, Schenectady, New York 12301

## ABSTRACT

The kinetics of the reorientation of magnetically induced anisotropy in stress free toroids of $Ni_{40}Fe_{40}P_{14}B_6$ is reported. The anisotropy is deduced from measurements of the area between magnetization curves. Changes in direction of the induced anisotropy are assumed proportional to the changes in the remanence-to-saturation ratio. Isothermal anneals were performed in circumferential fields starting from the induced anisotropy perpendicular to the plane of the toroid. For a well annealed sample, simple first order kinetics were observed, suggesting a single rate process, with $\Delta E = 1.4$ eV and $\nu_o = 10^{11}$ $sec^{-1}$.

## INTRODUCTION

In a previous paper[1] we described the wide range of d-c and a-c properties achievable from amorphous metal tapes with the nominal composition of $Ni_{40}Fe_{40}P_{14}B_6$. These properties were obtained by first stress relieving and then magnetically annealing the toroid.[2] The properties then depended only on the influence of the magnetically induced anisotropy; contributions from other anisotropies were all non-existent or negligible.

Sherwood et al[3] first described the field induced anisotropy developed in some metallic glasses but gave no kinetic data. Berry and Pritchet[4] studied magnetic annealing in $Fe_{75}P_{15}C_{10}$. Stress induced ordering was found to be reversible with an activation energy, $\Delta E$, of 2.2 eV but no kinetics were given for the field induced ordering. We assume that the origin of the field induced anisotropy is the same as in crystalline bulk and thin film NiFe; namely atom pair ordering. Recent studies[5] on bulk alloys of various composition of NiFe showed that the removal of excess vacancies changed $\Delta E$ for the rotation of the induced anisotropy, from 0.8 eV to 2.4 eV and the frequency factor, $\nu_o$, from $10^4$ $sec^{-1}$ to $10^{13}$ $sec^{-1}$. Analysis[6] of annealing of zero magnetostrictive $Ni_{82}Fe_{18}$ thin films showed that $\Delta E$, in the presence of excess vacancies, was 0.85 eV increasing to 2.2 eV when excess vacancies were removed, while $\nu_o$ changed from $10^7$ $sec^{-1}$ to $10^{14}$ $sec^{-1}$. It may be difficult to quantitatively describe or define a vacancy in an amorphous metal but the amorphous metals will certainly have locations analogous to vacancies in crystals.

In this paper we examine the kinetics of reorientation of the field induced anisotropy in amorphous $Ni_{40}Fe_{40}P_{14}B_6$. These results are then compared to results reported for crystalline NiFe alloys.

## EXPERIMENTAL METHODS AND RESULTS

A toroidal sample was prepared from METGLAS 2826[R] ribbon, ∿0.005 cm thick and 0.16 cm wide, by winding about 12 turns of this ribbon into a tungsten cup. Fifty turns of copper wire were wound around the cup for drive and sense windings. The assembly was sealed in a Pyrex tube, evacuated, and back-filled with pure dry nitrogen. The sample was stress-relieved by heating for 2 hrs at successively increasing temperatures in ∿25° increments, ending at 360°C. This is about 5°C below the beginning of crystallization. The sample was cooled through its Curie temperature, $T_c$, of 250°C, at a rate slower than ∿1°C/min, in a large field perpendicular to the plane of the toroid, $H_\perp$, to develop a minimum remanence-to-saturation ratio, $M_r/M_s$, in the circumferential measurement direction. The toroid was then isothermally annealed in a field, $H_\parallel$ = 4.5 Oe, generated by one of the drive-sense windings. The wire size was sufficiently large that the temperature rise due to $H_\parallel$ was <1°C. At the end of the anneal the toroid was returned to its original magnetic state, for the next isothermal anneal by reheating to 280°C, i.e. above $T_c$, and cooling in $H_\perp$ at the same rate as before. D-C hysteresis loops were traced using an integrating fluxmeter.

The changes in reduced $M_r/M_s$ on annealing are shown in Fig. 1 by the solid lines, where the initial value of $M_r/M_s = M_o$ and its limiting final value is $M_\infty$. $M_s$ was determined from M at 10 Oe. The values of $M_o$ were measured at the beginning of each isothermal anneal after the cooling in $H_\perp$ ; the values of $M_\infty$ were obtained first from independent tests by cooling slowly through $T_c$ in $H_\parallel$ (Fig. 2). The actual values used, given in Table I, were derived from each isothermal annealing sequence to give the best fit to the first order kinetic equation. The values of $M_\infty$ in Table I are within the scatter associated with the direct determination of $M_\infty$ in Fig. 2. The values of both $M_o$ and $M_\infty$, measured at room temperature, after recooling through $T_c$, varied somewhat. These variations appeared to be due to changes in the cooling rate and to a slow irreversible metallurgical change in the alloy structure. However there was no observable change in the kinetics of reorientation, for this sample, during these anneals.

A second toroid was prepared, identical to the first, to determine if the anneal kinetics are differ-

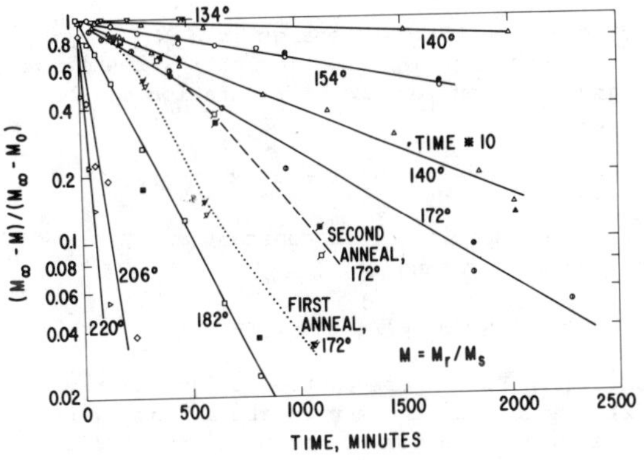


Fig. 1 Change in reduced remanence during isothermal annealing of toroids in a circumferential field of 4.5 Oe. Solid symbols for measurements at room temperature; open symbols for measurements at the anneal temperature. Dashed and dotted lines are for a new toroid annealed at 172°C.

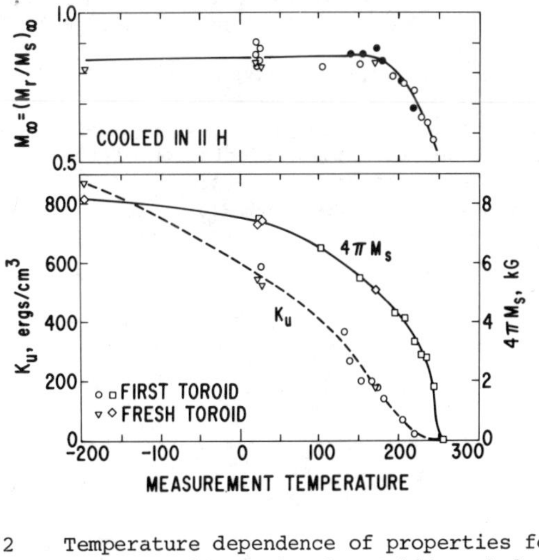


Fig. 2 Temperature dependence of properties for the stress relieved toroid of Fig. 1. Solid symbols obtained as best fit to isothermal anneal kinetics.

ent for a fresh sample, i.e. a sample not repeatedly exposed to high temperatures. This sample was given only a single stress relief anneal of 2 hrs at 325°C, cooled rapidly to 280°C and then slowly in H . The first isothermal reorientation results are shown by the dotted line in Fig. 1. The sample was then reheated to 280°C and cooled slowly in H to reset the toroid. The second isothermal anneal tests are shown by the dashed line. It is apparent that the annealing of the fresh sample does not follow first order kinetics, it anneals faster the first time and it appears that it may approach the kinetics of the first sample after further high temperature exposures.

The induced anisotropy, $K_u$, has been obtained by measuring the area between the initial and final magnetiziation curves. At room temperature, values between 400 and 850 ergs/cm$^3$ were obtained. The variations occur for the reasons indicated above. These values are in good agreement with the value of 800 ergs/cm$^3$ obtained[2] from our measurements on straight ribbons of the same material. The temperature dependence of $K_u$ was obtained by the same method. The results are shown in Table I and Fig. 2. From these we find, by least squares fit, $K_u$ proportional to $M_s^{3.7\pm.2}$. An exponent of 2 is expected for uniaxial anisotropy.

The change in $K_u$ during the isothermal annealing has also been evaluated but the measurement of the areas could not be made very precisely. The changes in $K_u$ did appear to follow the same kinetic relation as the changes in $M_r/M_s$.

## THE KINETICS

If we assume that the change in $M_r/M_s$ is proportional to the change in direction of $K_u$, the first order rate equation for the reorientation may be written as

$$-d(M_\infty - M)/dt = \nu(M_\infty - M) \qquad (1)$$

where $\nu$ is the specific rate constant and $M = M_r/M_s$ at time t. On integrating

$$\ln[(M_\infty - M)/(M_\infty - M_o)] = -\nu t. \qquad (2)$$

The results in Fig. 1 are plotted using the relation of eqn. (2). Note that the curves are linear over the entire range of time for all temperatures for the values of $M_\infty$ shown in Table I. The specific rate constants evaluated from these results are also listed in Table I.

Assuming a thermally activated process the temperature dependence of $\nu$ is given by the Arrhenius equation

$$\ln\nu - \ln\nu_o = -\Delta E/kT \qquad (3)$$

where k is the Boltzmann constant and T the absolute temperature. The Arrhenius plot is given in Fig. 3. From this we obtain $\Delta E = 1.4$ eV and $\nu_o = 0.9 \times 10^{11}$/sec.

## DISCUSSION AND CONCLUSIONS

We have made the assumption that the values of $M_r/M_s$ reflect the resultant orientation of the field induced magnetic anisotropy. This has been discussed[2]

TABLE I Rate Constants and Induced Anisotropy

| | measured at room temperature | | | measured at aging temperature | | | |
|---|---|---|---|---|---|---|---|
| °C | $M_o$ | $M_\infty$ | $\nu$ sec$^{-1}$ | $M_o$ | $M_\infty$ | $\nu$ sec$^{-1}$ | $K_u$ ergs/cm$^3$ |
| 27 | -- | -- | -- | -- | -- | -- | 600 |
| 140 | .090 | .90 | $1.5\times10^{-6}$ | .075 | .86 | $1.5\times10^{-6}$ | 280 |
| 154 | .14 | .86 | $7.0\times10^{-6}$ | .118 | .86 | $7.1\times10^{-6}$ | 200 |
| 172 | .105 | .88 | $2.5\times10^{-5}$ | .106 | .88 | $2.3\times10^{-5}$ | 180 |
| 182 | .12 | .82 | $7. \times10^{-5}$ | .12 | .84 | $7.4\times10^{-5}$ | 140 |
| 206 | .135 | -- | -- | .13 | .77 | $3.0\times10^{-4}$ | 70 |
| 220 | .091 | -- | -- | .12 | .68 | $6.4\times10^{-4}$ | 25 |

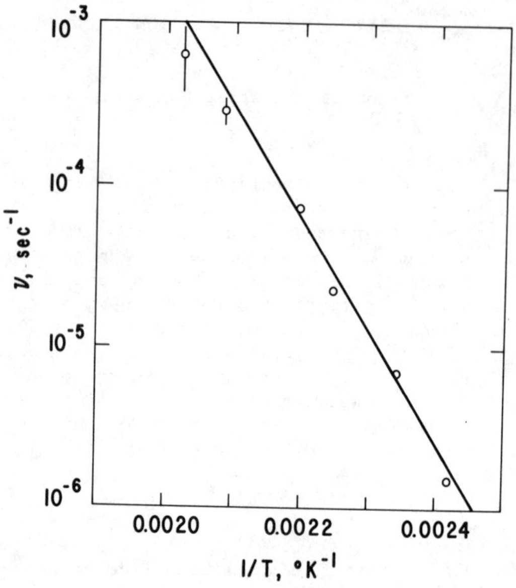


Fig. 3 Temperature dependence of the specific rate constant obtained from the isothermal anneal results of Fig. 1.

with reference to results on straight ribbons of this same material and appeared to be valid. Thus for slow cooling in $H_\perp$ we would expect to obtain $M_r/M_s = 0$ in the measurement direction. Similarily slow cooling in $H_\parallel$ should give $M_r/M_s = 1$. The deviations from 0 and 1 actually observed are due to contributions from other anisotropies. The values of $\Delta E = 1.4$ eV and $\nu_o = 10^{11}$ sec$^{-1}$ appear to fall between the extremes of excess vacancy, and vacancy deficient, controlled atom reorientation. Changing the quench rate during solidification or changing the preanneal temperature exposure could conceivably change the kinetics. The difference in kinetics of the fresh sample is reminiscent of the stabilization of crystalline alloys. This stabilization in crystalline alloys occurs by removal of vacancies, predominately by diffusion to grain boundaries. We suggest that in these amorphous alloys analogous voids on the atomic scale are present. These cooperate in the atom reorientation to produce the induced anisotropy and to control its reorientation kinetics. The stabilization, by removal of vacancies, can only occur by a densification process.

## ACKNOWLEDGMENTS

The experimental assistance of B. J. Drummond has been greatly appreciated.

## REFERENCES


R Registered trade name of Allied Chemical Corp.

1. F. E. Luborsky, R. O. McCary, J. J. Becker, The Second International Conf. on Rapidly Quenched Metals, Nov. 1975. To be published.
2. F. E. Luborsky, J. J. Becker and R. O. McCary, IEEE Trans. on Magnetics, MAG-11 XXX (1975).
3. R. C. Sherwood, E. M. Gyrogy, H. S. Chen, S. D. Ferris, G. Norman and H. J. Leamy, Magnetism and Magnetic Materials - 1974, AIP Conf. Proc. No. 24, p. 745.
4. B. S. Berry and W. C. Pritchet, Phys. Rev. Letters, 34, 1022 (1975).
5. A. Ferro, G. Griffa and G. Montalenti, IEEE Trans. on Magnetics, MAG-2, 764 (1966).
6. M. Takayasu, S. Uchuyama, K. Takahashi and T. Fujii, IEEE Trans. on Magnetics, MAG-10, 552 (1974).


Note added in proof: A reference which is earlier than Sherwood et al[3] demonstrated magnetic annealing in $Fe_{75}P_{15}C_{10}$. This reference is B. S. Berry and W. C. Pritchet, U.S. Patent 3,820,040, June 25, 1974.

# THE EFFECT OF HEAT TREATMENT ON THE CURIE TEMPERATURE OF A METALLIC GLASS

H. J. Leamy, E. M. Gyorgy, R. C. Sherwood, T. Wakiyama, & H. S. Chen
Bell Labs Murray Hill, New Jersey 07974

## ABSTRACT

Metallic glasses undergo structural relaxation at elevated temperatures and their mechanical properties are often dramatically altered as a consequence. We have discovered that the magnetization behavior of a $(Ni_{.5}Co_{.5})_{.75}P_{.16}B_{.06}Al_{.03}$ glass is similarly sensitive to heat treatment. The ferromagnetic transition in this material is not sharp, and heat treatment serves to broaden the transition further. Tc cannot be defined by conventional methods; e.g., $(M)^{1/\beta}$ vs $(H/M)^{1/\gamma}$ plots or a.c. permeability measurements. Both the initial permeability and the high field magnetizationincrease upon annealing, at rates that are identical to those of structural relaxation processes. While transmission electron microscopy examination of the glass failed to reveal either crystallization or phase separation, the magnetic properties suggest that a microscopically inhomogeneous distribution of the ferromagnetic species is developed during quenching and heat treatment.

## INTRODUCTION

Many amorphous materials are metastable with respect to crystallization. Glasses may also be unstable with respect to atomic rearrangements which collectively lower the free energy without producing crystallization. These rearrangements, or structural relaxations, may be activated thermally and typically exhibit a broad distribution of activation energies. The extent of the relaxation in glasses produced by undercooling the liquid phase depends upon the severity of the undercooling and the rate at which it is carried out [1]. In short, the structural state of a glass depends upon its thermal history. Metallic glasses are no exception to this general description of the glassy state. Broadly distributed relaxation activation energies have been experimentally determined [2,3], and mechanical and magnetic property changes have been observed to accompany annealing [4,5]. We report in this paper a similar sensitivity to heat treatment of the magnetization behavior of a $(Ni_{.5}Co_{.5})_{.75}P_{.16}B_{.06}Al_{.03}$ glass, which was selected from among compositions studied previously [6] to insure that measurements near Tc would not be complicated by simultaneous structural relaxation.

## EXPERIMENTAL

Ribbon shaped samples of 25μ thickness and 1 mm width were produced by cooling the molten alloy at $>10^5$ K/sec. The magnetization behavior of both quenched and annealed samples was determined by vibrating sample magnetometer measurements. The heat treatment, 300 C for 20 hours, was chosen to insure substantial relaxation and, at the same time, to avoid crystallization. The avoidance of crystallization was confirmed by x-ray and transmission electron microscopic examination of the annealed samples.

Selected magnetization curves for quenched and annealed samples are compared in Fig. 1, which shows that the heat treated material possesses a higher moment and initial permeability at all temperatures. The dramatic nature of this heat treatment effect is shown more clearly in Fig. 2, where M(T) for both samples in high and low external fields is plotted. The transition to the paramagnetic state is, by comparison with data reported for a $Co_{.7}B_{.2}P_{.1}$ glass [7], "smeared" over a range of temperature; the effect being most pronounced for the heat treated sample. The "smeared" nature of the transition is also evidenced by the failure of standard $(M)^{1/\beta}$ vs $(H/M)^{1/\gamma}$ plots to accurately define Tc. Such plots, whether constructed with mean field or Heisenberg exponents, yield a nonlinear relationship. For example, Fig. 3 shows $M^{2.5}$ vs $(H/M)^{.75}$ for the heat treated sample, from which we may conclude only that Tc lies somewhere between 230 and 400 K. Precise Tc determination by other methods also proved elusive. The permeability, μ, for example, does not exhibit a maximum at Tc in these samples. As Fig. 4 shows, the absence of a

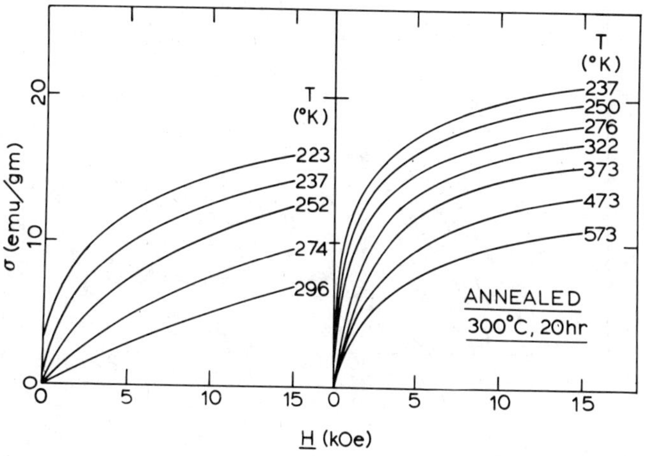


Fig. 1: M(H) for as quenched and annealed samples

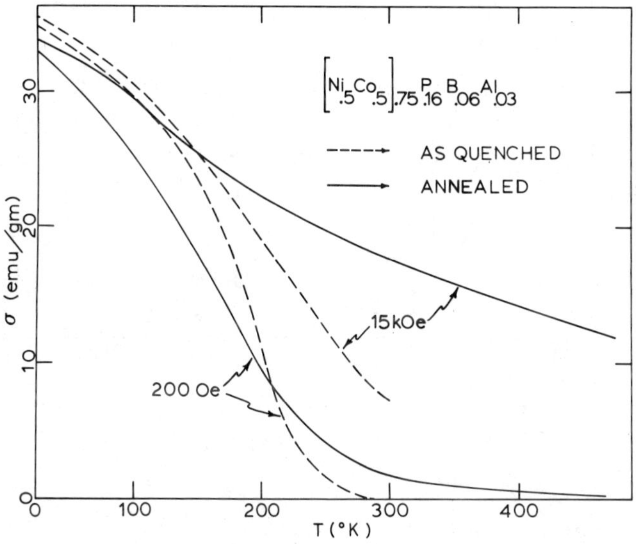


Fig. 2: The temperature dependence of M for H = 200 Oe and H = 15kOe

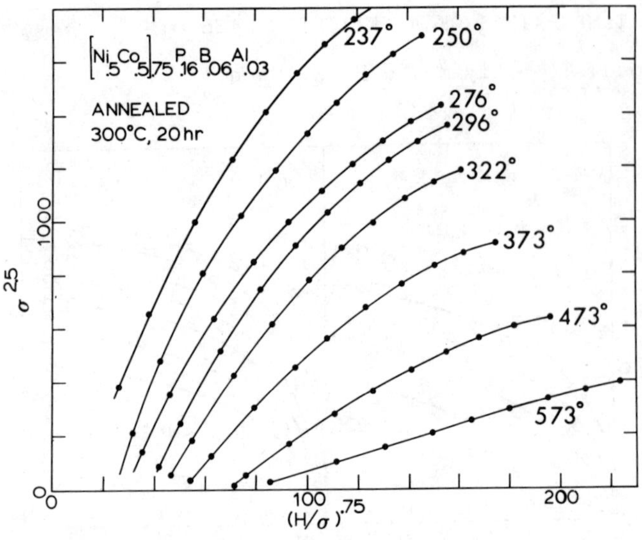


Fig. 3: $M^{2.5}$ vs $(H/M)^{.75}$ for an annealed sample.

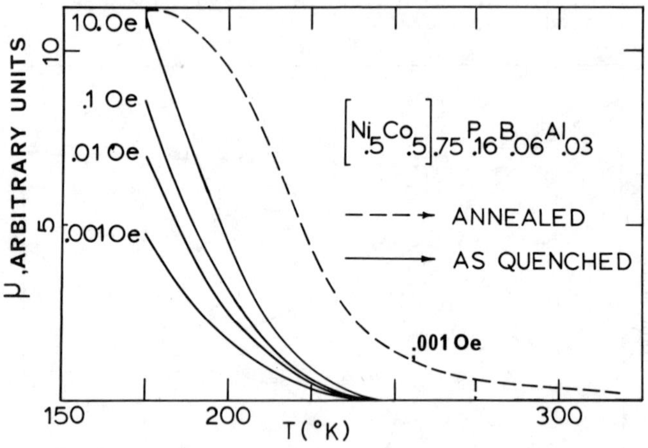


Fig. 4: The temperature dependence of the effective permeability for quenched and annealed samples. μ was measured at $4.5 \times 10^3$ Hz.

maximum in μ is independent of the magnitude of the external field. This result is, however, not suprising for this material since the anisotropy is uniaxial and is expected to vary as $M^2$ [8]. The primary contribution to μ is rotation against the anisotropy, which varies as $M^2/K_u$. Thus no maximum in μ is expected.

In order to examine the kinetics of the heat treatment effect, the room temperature magnetization behavior of alloys annealed for various times and at various temperatures was examined. As Fig. 5 shows, the effect increases with both time and temperature of annealing. If the magnetization at H =15kOe is taken as a measure of the magnitude of the effect, the time dependence may be compared with isothermal ageing curves obtained by differential scanning calorimetry. Such a comparison is shown in Fig. 6. It is obvious that the time rate of heat evolution parallels closely the change in high field magnetization. Both are initially large and fall to a small, nearly constant value after about 15 min. Ageing behavior of this sort is typical of reactions which proceed via steps whose activation energies are broadly distributed. The activation energy spectrum for structural

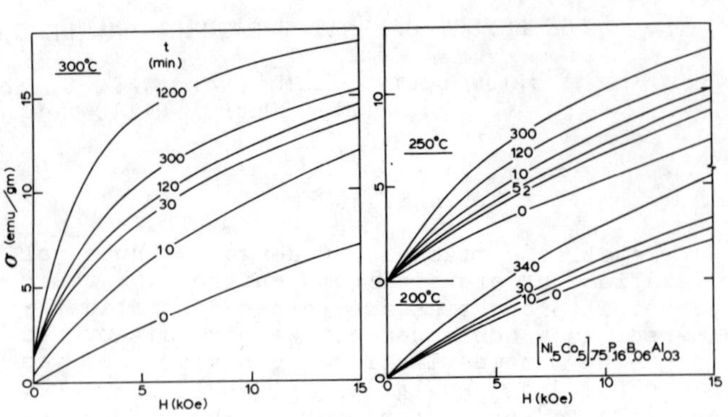


Fig. 5: M(H) of heat treated samples.

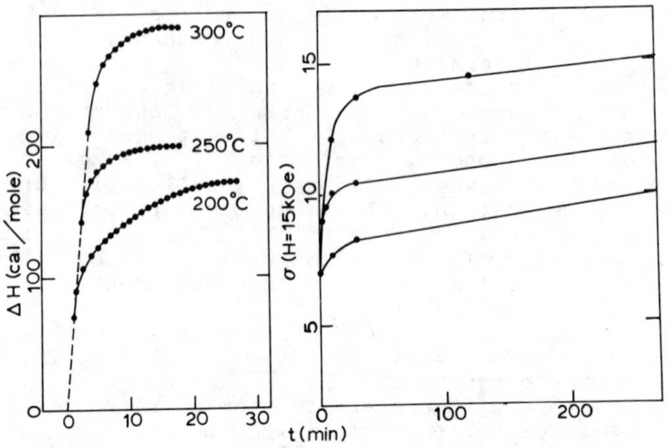


Fig. 6: A comparison of the heat evolved with the change in M(H = 15kOe) during isothermal annealing.

relaxation in similar metallic glasses, moreover, is distributed from 20 to 30 kcal/mole [3]. It seems certain, therefore, that the annealing effect described here is occasioned by the structural relaxation of the glass upon annealing.

## DISCUSSION

The results presented here possesses two distinguishing features. The ferromagnetic transition is smeared and cannot be described in terms of normal critical exponents. Second, the form of M(H,T) is a sensitive function of the structural state of the glass. Although the magnetization process in these glasses is obviously complex, these features may be qualitatively understood with reference to the structure of the alloy.

The metallic glass examined here contains only 37.5 at.% of ferromagnetic ions, (Ni is nonmagnetic in these P containing glasses [6]). Further, we expect in view of the rapid quenching treatment, that these will be randomly distributed in a Bernal type [9] random close packed arrangement. The environment of each ferromagnetic ion in a random array of this sort is variable with respect to both the number of neighboring ferromagnetic ions and their spatial disposition. The smaller clusters that must exist in such an array are expected to behave superparamagnetically and thus would account for the smeared transition. In this connection, the M(H) curves in Fig. 1 can be fit by the standard theory of superparamagnetism [10] if a suitable distribution of particle sizes is assumed. The effect of heat treatment is to weight more heavily the large particle size portion of the distribution.

Although the theory of randomly diluted ferromagnetic lattices near the percolation threshold [11] is not directly applicable to the amorphous state, many of the phenomena expected for diluted lattice magnets seem attractive for qualitative application here. Specifically, the onset of ferromagnetism as well as M(H,T) are predicted to depend upon structure; i.e. upon the connectivity of the lattice and the range of the exchange interaction. Such a sensitivity is required to explain the remarkable change in magnetic behavior with structural relaxation described here,and the strong composition dependence of Tc reported earlier [6].

REFERENCES


[1] B. G. Bagley, Amorphous and Liquid Semiconductors Plenum Press, New York, 1974 Chapt. 1

[2] H. S. Chen, H. J. Leamy, & M. Barmatz, J. Non-Cryst. Solids 5 444-448 (1971)

[3] H. S. Chen & E. Coleman, unpublished

[4] H. S. Chen, S. D. Ferris, E. M. Gyorgy, H. J. Leamy, & R. C. Sherwood, Appl. Phys. Lett. 26 405-406 (1975)

[5] F. E. Luborsky, this volume

[6] R. C. Sherwood, E. M. Gyorgy, H. S. Chen, S. D. Ferris, G. Norman, & H. J. Leamy, A. I. P. Conf. Proc. 24 , 745-746 (1974)

[7] T. Mizoguchi, N. Ueda,K. Yamauchi, &H. Miyajima, J. Phys. Soc. Japan 34 1691 (1973)

[8] E. M. Gyorgy, this volume

[9] J. D. Bernal, Nature 183 , 141-147 (1959)

[10] I. S. Jacobs & C. P. Bean, in Magnetism Vol. III Academic Press, New York 1963, Chapt. 6

[11] J. W. Essam in, Phase Transitions and Critical Phenomena; Vol. 2 Academic Press, New York, 1972 Chapt. 6

# MAGNETIC PROPERTIES OF AMORPHOUS $Fe_{40}Ni_{40}P_{14}B_6$*

C. L. Chien
The Johns Hopkins University, Baltimore, Maryland 21218

R. Hasegawa
Materials Research Center, Allied Chemical Corporation, Morristown, New Jersey 07960

## ABSTRACT

We have used Mössbauer spectroscopy to study the magnetic properties of amorphous $Fe_{40}Ni_{40}P_{14}B_6$ (Allied Chemical METGLAS 2826). Spectra were taken at temperatures from 4.2 K to 1000 K. The amorphous state is preserved up to 670 K at which temperature we observed a rapid transformation into the crystalline state.

For the as-quenched sample at 300 K, the spectra show that magnetization lies in the plane of the ribbon and predominantly along the ribbon direction. At temperatures between 240 K and 77 K the magnetization direction is a strong function of temperature - first tilting significantly out of the plane of the ribbon and then moving back closer to the ribbon plane.

The magnetic spectra consist of broadened 6-line patterns. The spontaneous magnetization shows a significantly more rapid decrease with temperature than does the magnetization of crystalline Fe or Ni. Above the sharply defined $T_c$ = 537 K, the spectra consist of doublets indicative of quadrupole splitting as expected. In the crystalline state the material consists of two distinct magnetic phases with ordering temperatures of 425 K and 750 K. This information plus x-ray measurements and the values of other parameters in the Mössbauer spectra allow us to identify the phases as $Fe_{.55}Ni_{.45}$ and $(Fe,Ni)_3P_{.7}B_{.3}$.

## RESULTS AND DISCUSSION

Amorphous $Fe_{40}Ni_{40}P_{14}B_6$(METGLAS$^R$ 2826) has been investigated from 4.2 K to 1000 K by $Fe^{57}$ Mössbauer spectroscopy. The sample was made by rapid quenching from melt. Its amorphous nature was verified by x-ray diffraction. The sample has the geometric form of a long ribbon with a nominal width of 3.4 mm and a thickness of 40 μm. Magnetically, 2826 is a very soft material.[1,2]

Absorbers were made by placing several ribbons parallel to each other, covering an area of about 1.5 cm x 1.5 cm. For measurements with T < 300 K, the sample was sandwiched between thin Al foils and held by vacuum grease. The Al foils were attached to the OFHC cold foot with no stress applied to the sample. All measurements were made in zero applied magnetic fields with the spectrometer oriented in the north-south direction so that the magnetic field of the earth is perpendicular to the ribbon plane. Fringe fields such as those due to the magnetic transducer have been carefully checked to have no effect. The sample was further surrounded by magnetic shielding materials.

In Fig. 1, three spectra are shown of the as-quenched sample taken of 300 K with different geometric arrangements, as indicated on the right of the figure. Fig. 1a shows a six-line pattern with a line area ratio of about 3:4:1:1:4:3. One immediately concludes that the easy axis of magnetization is in the ribbon plane. This is in apparent disagreement with the results of Egami et al.[1,2] who indicated that the easy axis is perpendicular to the ribbon face. If the easy axis were indeed perpendicular to the ribbon face, in Fig. 1a, one would have observed an area ratio of 3:0:1:1:0:3, i.e., two of the six peaks would disappear. Fig. 1b and 1c are spectra where the ribbon face was turned $30^0$

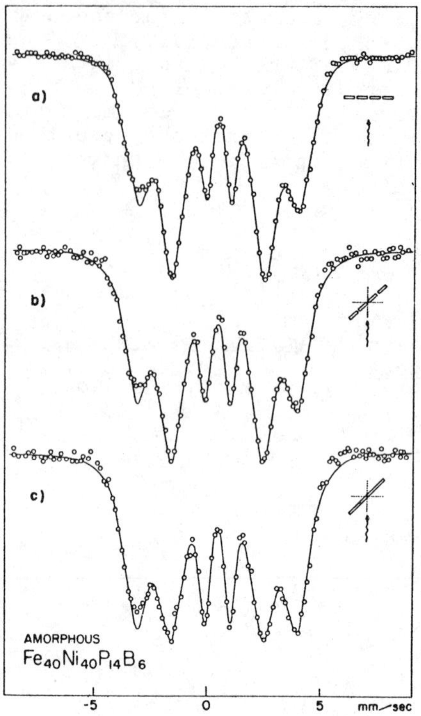


Fig. 1. Mössbauer spectra of amorphous $Fe_{40}Ni_{40}P_{14}B_6$ at 300 K under different geometries. On the right, top view of the experimental conditions are shown. Short and long rectangles represent the short and long edges of a single ribbon. The γ-ray is indicated by the wavy lines.

with respect to the γ-ray while keeping either the long or the short edge of the ribbon perpendicular to the γ-ray. Fig. 1b is practically the same as Fig. 1a whereas Fig. 1c shows very different intensities for the #2 and #5 peaks. This is possible only if the easy axis of the as-quenched sample is predominantly along the long edge of the ribbon. This conclusion was also reached by the magnetic resonance measurement.[4]

The reasons for the disagreement between the present results and those from earlier measurements[1,2] is not entirely clear. It may be, however, (as suggested by recent domain observations[3]) that the perpendicular direction of the easy axis seen by the previous workers occurs only in a small fraction of the ribbon while the majority portion has an in-plane magnetization as seen in the present work, in which the entire thickness of the ribbon is measured.

As shown in Fig. 2, below 240 K, the easy axis on the average begins to tilt out of the ribbon plane, achieving a maximum tilt angle of about $45^0$ at 110 K. Below 110 K, the easy axis begins to tilt back toward the ribbon plane. This tilting was found to be continuous and reproducible in both cooling and heating cycles, and was present in several samples from different batches. The mechanism of the tilting is not well understood and is being subjected to further investigation. Thermally produced stresses are a likely cause. The above results are pertinent to as-quenched samples, i.e., samples that have not been heated above 300 K. It was clearly shown by Egami et al.[1,2] that high temperature annealing with a magnetic field, or with stresses, can greatly affect the bulk magnetic properties of the material.

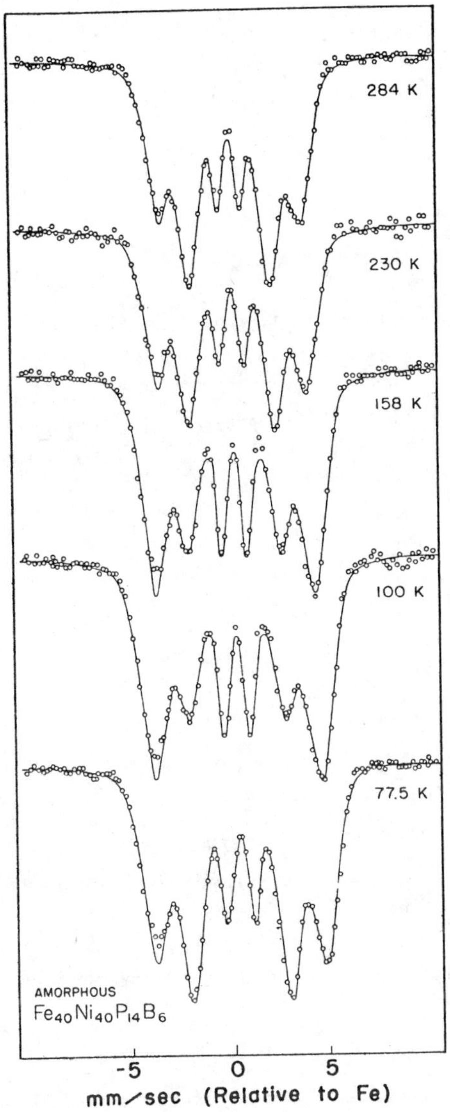


Fig. 2 Mössbauer spectra of amorphous $Fe_{40}Ni_{40}P_{14}B_6$ at low temperatures. γ-ray is perpendicular to the ribbon plane.

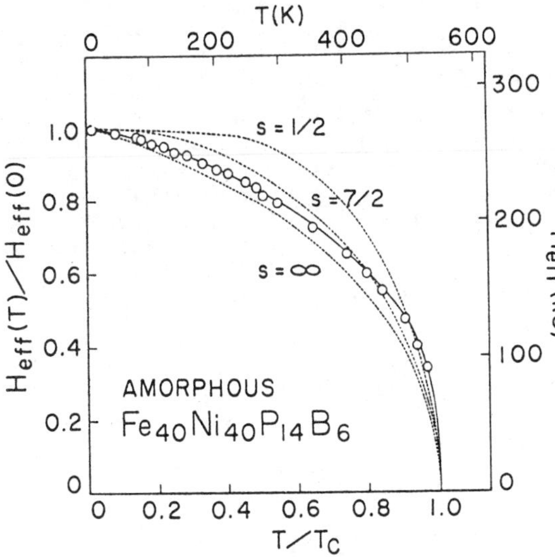


Fig. 3 Spontaneous magnetization of amorphous $Fe_{40}Ni_{40}P_{14}B_6$. The dashed curves are mean field results for S= 1/2, 7/2 and ∞.

It should be mentioned that Mössbauer spectroscopy separates out the magnitude change (line positions) and the directional change (line intensities) of the magnetization. The direction of the magnetization with respect to the gamma rays has a strong effect on line intensities, as mentioned above. The magnetization direction does not, however, affect the total hyperfine splitting, which measures the magnitude of the magnetization. In Fig. 3, the spontaneous magnetization vs. T is shown. It falls off with T much more rapidly than does the magnetization of crystalline Fe or Ni. At low temperatures, the magnetization satisfies the $T^{3/2}$ law expected from excitations of long wavelength spin waves, but with an anomalously large coefficient. The transition at $T_c$ = 537 ± 2 K is found to be sharp. These results are discussed elswhere.[5]

Doublet spectra due to quadrupole interaction are observed above 537 K. At about 670 K, which we identify as the glass-to-crystalline transition temperature, a sudden change of the spectrum occurs. A weak magnetic hyperfine pattern appears in addition to the doublet. Above 750 K and up to 1000 K, (the highest temperature reached) the sample is paramagnetic with the resultant spectra consisting of three peaks.

As the temperature is lowered from 1000 K, one 6-line hyperfine pattern with negligible quadrupole interaction begins to appear at 750 K. Another 6-line pattern with sizable quadrupole interaction appears at 425 K. The hyperfine field values of these two patterns are shown in Fig. 4. From the observed hyperfine parameters, the magnetic ordering temperature, and x-ray studies[4], we conclude that there are two dominant Fe-containing phases: $Fe_{.55}Ni_{.45}$ ($T_c$ ~ 750 K) and $(Fe,Ni)_3P_{0.7}B_{0.3}$ ($T_c$ ~ 425 K) in the crystalline sample.[4]

We are indebted to J. C. Walker, L. A. Davis, and J. J. Gilman for their interest in the present work.

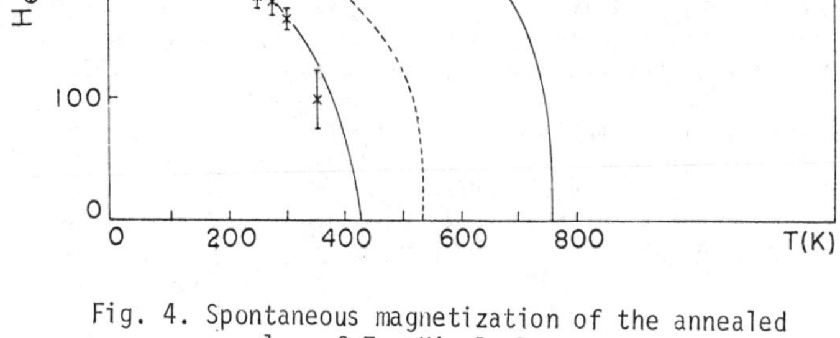

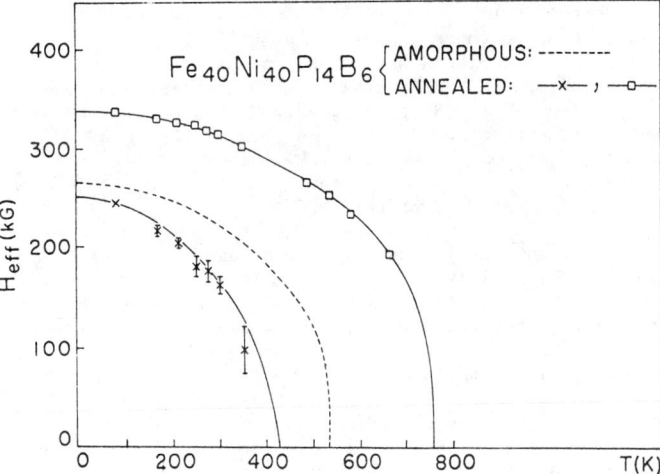

Fig. 4. Spontaneous magnetization of the annealed samples of $Fe_{40}Ni_{40}P_{14}B_6$.

REFERENCES

*Work partially suported by the National Science Foundation.

1. T. Egami, P.J. Flanders and C.D. Graham, Appl. Physics Letters 26, 128 (1975).

2. T. Egami, P.J. Flanders and C.D. Graham, AIP Conf. Proc. No. 24, 697 (1975).

3. J.J. Becker, this conference.

4. R. Hasegawa, this conference.

5. C. L. Chien and R. Hasegawa, to be presented at The International Topical Conference on Structure and Excitations of Amorphous Solids, (Williamsburg, 1976).

# MAGNETIZATION, ANISOTROPY AND COERCIVITY OF A GLASSY METALLIC ALLOY

R. Hasegawa
Materials Research Center, Allied Chemical Corporation, Morristown, NJ 07960

## ABSTRACT

A glassy alloy in ribbon form with the composition $Fe_{40}Ni_{40}P_{14}B_6$ (METGLAS[R] 2826), obtained by rapid quenching from the liquid, has been investigated using a ferromagnetic resonance technique (300-700 K) and a B-H hysteresigraph. The observed room temperature values for the saturation magnetization $M_s$, the g-factor and the coercivity $H_c$ are 630 G, 2.05 and 35 mOe respectively. A uniaxial magnetic anisotropy ($K_u = 1.5 \times 10^4$ erg/cm$^3$ at 300 K) is found to be directed along the ribbon axis and varies as $M_s^2$. The dc coercivity data are discussed in terms of a recent theory.

## INTRODUCTION

The remarkable magnetic and mechanical properties of amorphous metallic alloys were first observed in the Fe-P-C system.[1] With recent developments in materials synthesis[2] a number of metallic glasses containing iron and nickel can now be prepared in continuous ribbon form. Previous studies[3,4] made on these ribbons indicated their potential usefulness for various low field magnetic applications. However, the earlier work[3] contained ambiguities as to the direction of magnetic anisotropy and the saturation magnetization values. We have recently studied these quantities as a function of temperature for a ribbon having the composition $Fe_{40}Ni_{40}P_{14}B_6$ (METGLAS 2826) by Mössbauer[5] and ferromagnetic resonance (FMR) techniques. The purpose of this report is to present the results obtained by the FMR technique in conjunction with the coercivity data.

## EXPERIMENTAL

A ribbon sample about 2 mm wide and 35 μm thick was prepared by rapidly quenching the liquid and examined by X-ray diffraction to confirm the glassy state of the material. The ribbon was then polished mechanically and electrolytically to reduce the thickness to about 15 μm; circular disk samples about 1.5 mm in diameter were then cut from the ribbon. An EPR spectrometer (Varian E-12) was used to observe the ferromagnetic resonance at f = 9.335 GHz as a function of temperature (300-700 K) along the directions parallel and perpendicular to the plane of the disk. The sample was placed in a nitrogen atmosphere to minimize surface oxidation. Straight ribbons about 30 cm long were used to obtain the temperature dependence (300-500 K) of the coercivity $H_c$, the value of which was determined by observing the time derivative of the flux. The measurements were made by sweeping the field at f = 0.01 Hz.

## RESULTS AND DISCUSSION

To obtain information about the magnetic anisotropy of the present samples, the resonance field H was measured at 300 K as a function of the field direction with respect to the ribbon axis. An example of the results, shown in Fig. 1, indicates that the present ribbon has a

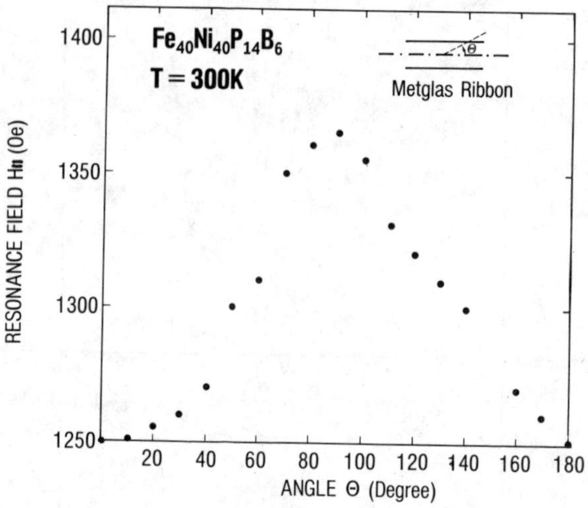


Fig. 1. Resonance field as a function of the external field direction.

uniaxial magnetic anisotropy along the ribbon axis. Thus the resonance condition may be written[6] as

$$\left(\frac{\omega}{\gamma}\right)^2 = H_{\|}^{o}(H_{\|}^{o} + 4\pi M_s) \qquad (1)$$

where conventional nomenclature is used. The resonance field $H_{\|}$ includes the effect of the anisotropy of the sample. For the two directions, $H_{\|}$ longitudinal (easy axis) and transverse (hard axis) to the ribbon direction, $H_{\|}$ may be given respectively by[7]

$$H_{\|} = H_{\|}^{o} - H_k \qquad (2a)$$
$$H_{\|} = H_{\|}^{o} + H_k \qquad (2b)$$

where $H_{\|}^{o}$ is the resonance field in the absence of the anisotropy field $H_k$. When $H_{\|}^{o} >> H_k$, the difference between the resonance fields for the easy and hard axis corresponds approximately to $2H_k$. In the present case, it is found that $H_k$ ranges between 45 and 60 Oe at 300 K for the several samples studied. A conclusion drawn from this experiment is that the direction of the anisotropy is on the average along the ribbon axis. Further evidence for this finding is provided by a Mössbauer study.[5]

The FMR signal for the case of perpendicular resonance has only one absorption line which is due to the uniform precession of spins. The present disk sample is too thick for the other spin wave modes to be excited. Thus the resonance condition for this case may be given as[6]

$$\frac{\omega}{\gamma} = H_{\perp} - 4\pi M_s \qquad (3)$$

where $H_{\perp}$ is the external field. Using Eqs. (1-3), it is therefore possible to determine quantities γ (or g-factor), $4\pi M_s$, $H_k$ and, hence, the uniaxial anisotropy energy $K_u$ (= $H_k M_s/2$). It is found that g = (2.05 ± .02) which is compared with g (= 2.1) for permalloy.[8] Other results are summarized in Fig. (2) and (3). It is noticed that the $4\pi M_s$ value at 300 K is about 10% lower than the previous value.[3] Since a more direct measurement[9] gives about the same value as the present one, we take the room temperature $4\pi M_s$ value as 7900 G. The anisotropy data [Fig. (3)] implies $T_c \sim 525$ K, which is compared with $T_c = 520$ K determined from a thermomagnetization study.[5] The FMR spectra above 580 K is very broad (linewidth $\Delta H \sim 1$kOe) and weak and is, therefore, not reliable. However, above $T \sim 640$ K well-defined spectra (with narrower $\Delta H$) begin to develop due to the crystallization of the glassy alloy. In the present case, crystallization took place rapidly around 643 K. It was found that the crystallization temperature ranges between 640 and 670 K for several samples studied. Upon transformation into the cyrstalline state, the present sample seems to have at least two magnetic phases, one of which has $T_c \sim 750$ K and the other $T_c \sim 430$ K. A recent Mössbauer study for the same alloy confirms this finding.[5] Furthermore, a detailed comparison between thermomagnetization and X-ray data show that the higher and lower $T_c$ phases correspond to fcc $Fe_{0.55}Ni_{0.45}$ and tetragonal $(Fe_{0.58}Ni_{0.42})_3P_{0.7}B_{0.3}$ respectively.[5]

The temperature dependence of the uniaxial anisotropy energy $K_u$ (Fig. 3) is best described by a relation $K_u = kM_s^2$ with $k = 4 \times 10^{-2}$ as shown by the inset of Fig. 3. This suggests that the coupling energy for the anisotropy is dipolar in origin. Since the present ribbon is magnetostrictive,[3] the internal stress introduced during rapid-quenching may give rise to the anisotropy. On the other hand, it has been reported[10] that a zero-magnetostriction glassy alloy ribbon exhibits evidence of directional magnetic ordering. It is likely therefore that these two mechanisms are combined to give rise to the observed anisotropy as suggested in Ref. 11.

It is found that the room temperature coercivity $H_c$ falls between 30 and 50 mOe for the various sample ribbons examined. If it is assumed that magnetization reversal takes place mainly by domain wall motion, then by analogy to a similar problem[12] examined previously one may write $H_c = (2K_u/M_s)f$ with $f = \sum_i \Delta X_i/X_i$ where $\Delta X_i$ stands for the fluctuation in the quantity $X_i$ within a region corresponding to the domain wall width parameter $\delta[=(A/K_u)^{1/2}]$. Here $X_i$ includes exchange stiffness constand A, $K_u$ and the dimensions of the ribbon. Using $4\pi M_s$ and $K_u$ data [Figs. (2) and (3)], an excellent fit of the $H_c$ vs T data to the above formula was obtained with $f = 6.8 \times 10^{-4}$ as shown in Fig. 4. This analysis assumes a gradual change of the quantity $X_i$ across the width $\delta$. In light of the recent theory of coercivity[13]

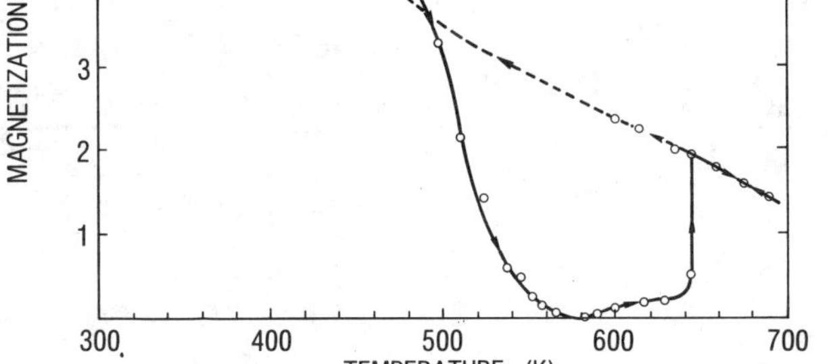

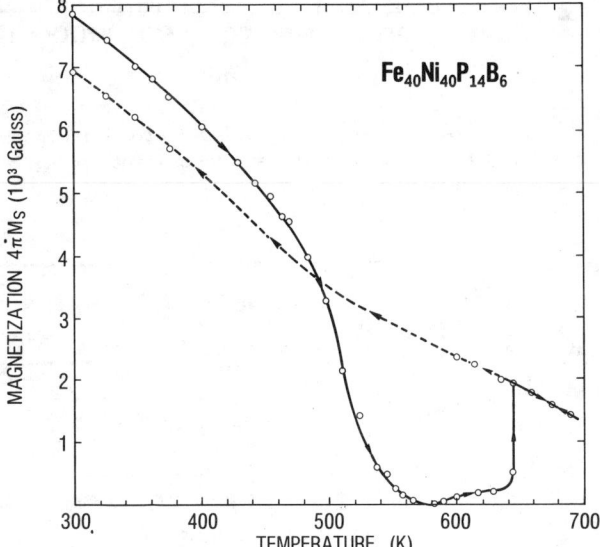

Fig. 2. Saturation magnetization versus temperature. Arrows indicate the heating cycle starting from T = 300 K.

which assumes a change in $X_i$ confined within the barrier width W, f is given by $0.19(W/\delta)(A/A'-K'/K)$ where A' and K' are the exchange and anisotropy constants in the barrier. Thus the value of f obtained above corresponds to an effective pinning width $W(A/A'-K'/K)$ of $5 \times 10^{-9}$ cm. If we assume furthermore the change in A and K is 2∼3%, the barrier width W is on an order of 10 Å.

The author acknowledges D. I. Paul, J. J. Gilman, L. A. Davis, R. C. O'Handley and H. Yue for their interest in the present work.

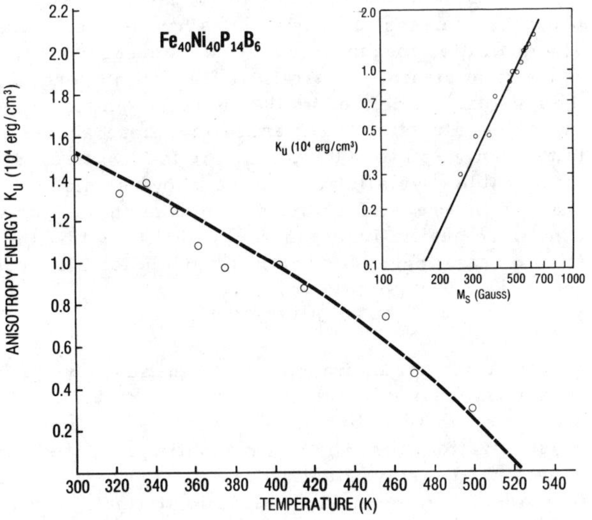


Fig. 3. Anisotorpy energy as a function of temperature. The inset shows the quadratic dependence of $K_u$ on $M_s$.

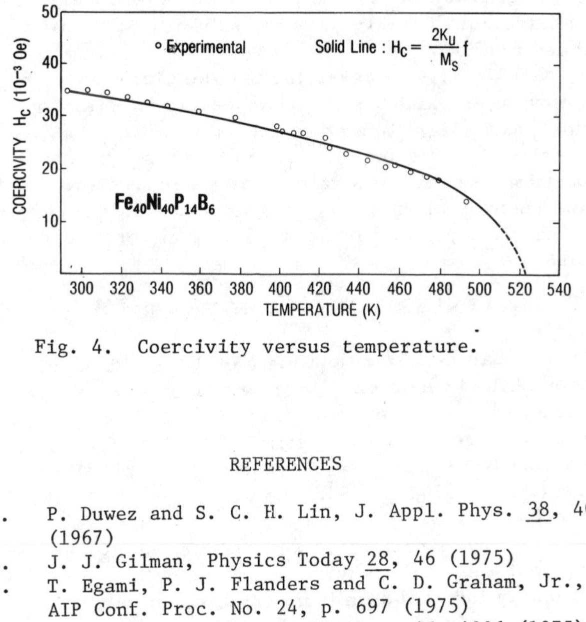


Fig. 4. Coercivity versus temperature.

REFERENCES

1. P. Duwez and S. C. H. Lin, J. Appl. Phys. 38, 4096 (1967)
2. J. J. Gilman, Physics Today 28, 46 (1975)
3. T. Egami, P. J. Flanders and C. D. Graham, Jr., AIP Conf. Proc. No. 24, p. 697 (1975)
4. R. C. O'Handley, J. Appl. Phys. 46, 4996 (1975)
5. C.-L. Chien and R. Hasegawa (this conference)
6. C. Kittel, Phys. Rev. 71, 270 (1947); 73, 155 (1948)
7. R. L. Conger and F. C. Essig, Phys. Rev. 104, 915 (1956)
8. G. G. Scott, Rev. Mod. Phys. 34, 102 (1962)
9. R. Hasegawa and R. C. O'Handley, Proc. 2nd Int. Conf. on Rapidly Quenched Metals (Boston, 1975)
10. H. S. Chen, S. D. Ferris, E. M. Gyorgy, H. J. Leamy and R. C. Sherwood, Appl. Phys. Lett. 26, 405 (1975)
11. R. C. O'Handley (this conference)
12. R. J. Gambino, P. Chaudhari and J. J. Cuomo, A.I.P. Conf. Proc. 18, 578 (1974)
13. R. Friedberg and D. I. Paul, Phys. Rev. Lett. 34, 1234 (1975) and D. I. Paul (this conference)

# ANNEALING EFFECTS IN AMORPHOUS MAGNETIC ALLOYS

C.D. Graham, Jr., T. Egami, Robert S. Williams, and Y. Takei*
Dept. of Metallurgy and Materials Science and
Laboratory for Research on the Structure of Matter
University of Pennsylvania, Philadelphia, Pa. 19174

## ABSTRACT

The isothermal annealing behavior of some amorphous ferromagnetic alloy ribbons has been determined. The quantities measured were the macroscopic stress at constant strain, the strain at fracture in bending, and the anisotropy, defined as the energy required to reach magnetic saturation. The annealing temperatures were in the range 150 to 300°C, which is too low to cause measurable crystallization in the alloys studied. The changes in stress, fracture strain, and anisotropy with time were qualitatively similar, suggesting that the same atomic rearrangements are responsible for all three.

## INTRODUCTION

A number of amorphous magnetic alloys undergo substantial changes in their mechanical and magnetic properties on annealing at temperatures too low to cause crystallization.[1,2] The soft magnetic properties improve: the coercive field decreases, the permeability increases, and the field required to attain saturation decreases. The mechanical properties deteriorate: in particular, the alloys become increasingly brittle, although the hardness increases slightly or not at all. Surprisingly, these property changes are not reflected in normal x-ray diffraction or electron microscope observations; the structural changes are too subtle for detection by these methods.

We have attacked the problem indirectly, by following the kinetics of the annealing effects, hoping in this way to learn something of the mechanisms involved. Specifically, we have measured macroscopic stress relaxation at constant strain, changes in fracture strain, and changes in the energy required to attain magnetic saturation, all as a function of time at constant temperature.

## EXPERIMENTAL PROCEDURE

Samples were mainly Metglas ribbons purchased from Allied Chemical Corp., although some measurements have been made on similar ribbons of our own manufacture. The compositions were Metglas 2826, nominally $Fe_{40}Ni_{40}P_{14}B_6$, and Metglas 2826B, $Fe_{29}Ni_{49}P_{14}B_6Si_2$. Ribbons were about 50μm thick and 2mm wide.

Stress relaxation was followed using a method suggested by Luborsky.[3] Lengths of ribbon were wound into a cylindrical aluminum container (a magnetic tape core box) of 6.5 mm inside radius. The ribbon was held against the inside of the cylinder by its own elasticity. The surface strain was about 0.003, positive on the outside or tension side, and negative on the inside. The corresponding stress is about $4x10^9$ dyne/$cm^2$=$50x10^3$ lb/$in^2$=350MPa, which is less than half of the fracture stress in tension. The sample in the core box was annealed for various time intervals at constant temperature in a silicone oil bath. After each annealing period, the ribbon was removed from the core box and its equilibrium radius of curvature was measured. As stress-relaxation proceeds, the ribbon increasingly conforms to the diameter of the core box. The stress relaxation is plotted on a scale from 1.0, corresponding to the initial stress, to 0, corresponding to complete stress relaxation or zero spring-back.

Fracture strain was determined by slowly compressing a U-shaped length of ribbon between the parallel faces of a small vise, and noting the separation between the jaws at the time of fracture. The surface strain is then calculated on the assumption that the ribbon bends in a circular arc, since there is little or no plastic deformation before fracture. The fracture strain in the as-prepared ribbons is difficult to determine, since the ribbons can often be folded sharply back on themselves without fracture; only the values in Fig. 1 have been normalized at zero time. The test becomes unreliable as the ribbons become very brittle. Annealing was carried out on straight lengths of ribbon sealed into pyrex tubes under vacuum, or immersed directly into silicone oil.

As a measure of the magnetic effects of annealing, we use the energy required to attain saturation. This is given by the area to the left of the magnetization curve,

i.e., $\int_0^{M_s} H dM$, which we call the anisotropy.

We find measurable changes in both the Curie temperature and in the room-temperature saturation magnetization on annealing. In the alloys studied, the Curie temperature increases sometimes 20°C or more, and the saturation magnetization increases by several percent. We have not yet investigated this phenomenon in detail.

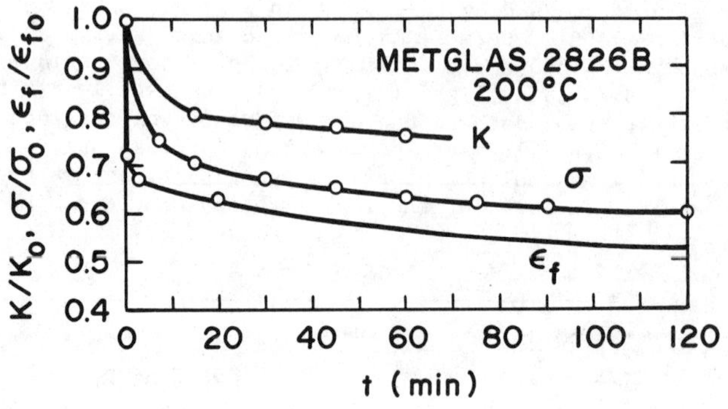


Fig. 1. Effect of annealing on magnetic anisotropy, strain at fracture, and stress at constant strain, all normalized.

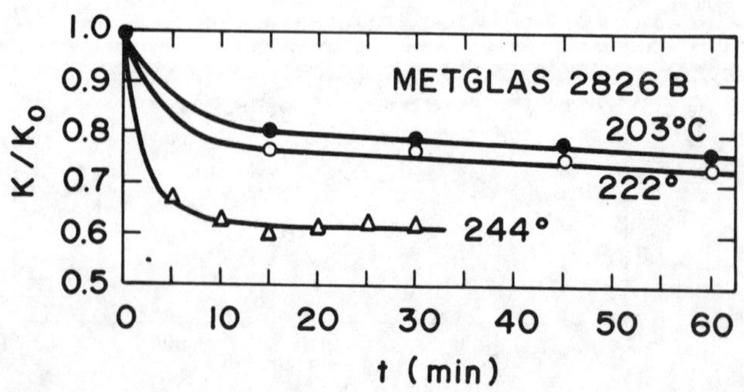


Fig. 2. Effect of annealing at various temperatures on magnetic anisotropy.

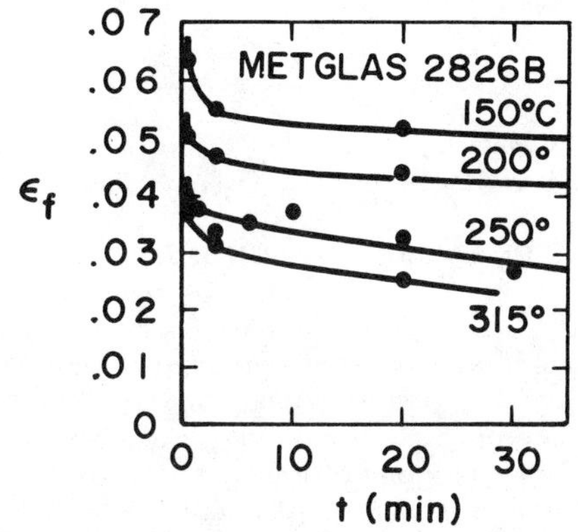


Fig. 3. Effect of annealing at various temperatures on the fracture strain. Additional data points at longer times not shown.

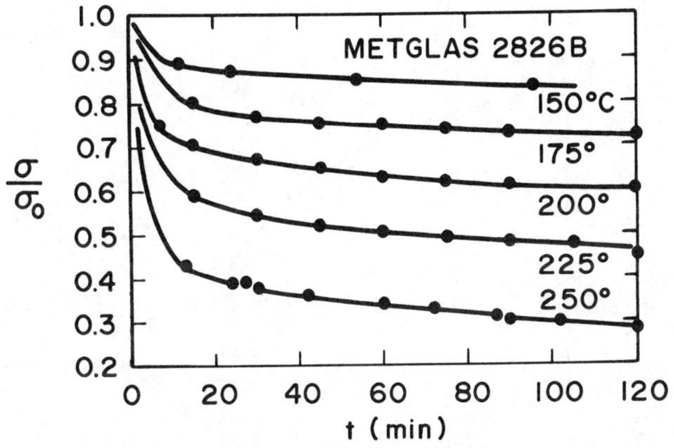


Fig. 4. Effect of annealing at various temperatures on stress at constant strain. Additional data points at longer times not shown.

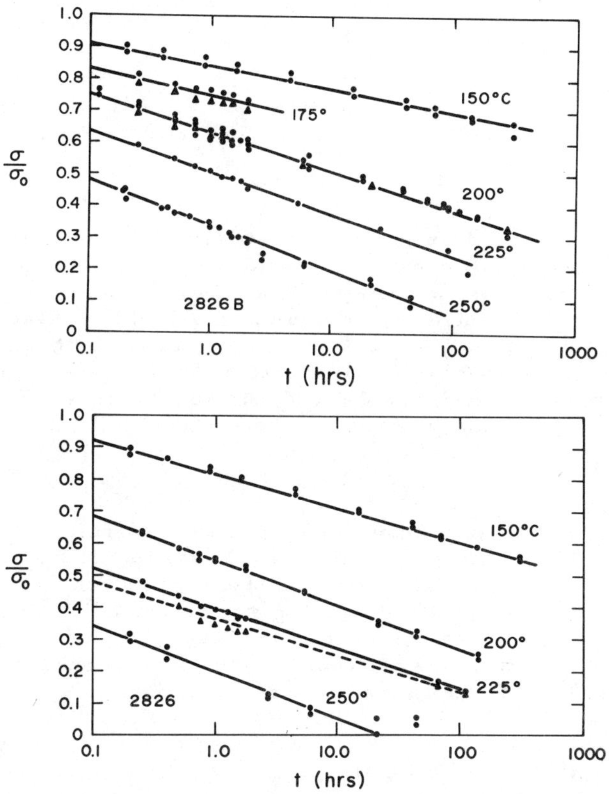


Fig. 5. Stress relaxation at longer times for two alloy compositions. Dotted line is for an increased radius of curvature.

## RESULTS AND DISCUSSION

Because the three measurements were made separately, the times and temperatures were not generally the same. Fig. 1 shows the best comparison of the three measurements under similar conditions. Fig. 2 shows additional data for the anisotropy, Fig. 3 for the fracture strain, and Fig. 4 for the stress relaxation. The behavior is qualitatively similar for all three: there is a rapid drop in the measured quantity, largely complete in about 10 minutes, followed by a much slower, long-term drop. The initial rapid drop is strongly temperature dependent, but the slow drop is very weakly temperature dependent. Attempts to deduce activation energies from the data, or to fit the curves to conventional kinetic equations, have been unsuccessful. However, the stress-relaxation data for long time periods are approximately linear when plotted as $\sigma/\sigma_0$ vs log time. (See Figs. 5a and b.) Note that at 250°C, almost complete stress relaxation takes place at about 100 hours.

If after substantial annealing time, a ribbon is removed from its core box and re-coiled with reverse curvature, the initial rapid drop in stress is no longer observed, but the slow long-term process is not significantly changed.

It appears from these results that the same (or closely similar) mechanisms are responsible for all three of the observed effects. Other evidence points to the origin of the anisotropy in the interaction between the magnetostriction and the non-uniform stresses produced in the ribbon by the manufacturing operation.[4, 5] This would account for the similar time dependence of the stress-relaxation and the anisotropy decay.

The fact that significant changes in properties occur in very short times at low annealing temperatures implies that short-range atomic motions are involved.

## ACKNOWLEDGEMENTS

This work was supported by the National Science Foundation through the Laboratory for Research on the Structure of Matter under Contract DMR 72-03025, and by the Sony Corp. P.J. Flanders gave valuable assistance with the measurements.

## REFERENCES

*on leave from Sony Corp., Japan

1. T. Egami, P.J. Flanders, and C.D. Graham, Jr., AIP Conf. Proc. 24 697 (1975).
2. H. Fujimori, T. Masumoto, Y. Obi, and M. Kikuchi, Japan. J. Appl. Phys. 13 1889 (1974).
3. F.E. Luborsky, J.J. Becker, and R.O. McCary, IEEE Trans. on Magnetics MAG-11 1644 (1975).
4. T. Egami and P.J. Flanders, this Conference, 5C9.
5. J.J. Becker, this Conference, 5C3.

# TEMPERATURE DEPENDENCE OF "MAGNETIC ANISOTROPY" IN AMORPHOUS ALLOYS

T. Egami +* and P.J. Flanders +
Laboratory for Research on the Structure of Matter,
University of Pennsylvania, Philadelphia, Pa. 19174

## ABSTRACT

For the purpose of investigating the origin of the magnetic anisotropy in the Fe-Ni based amorphous alloys, we studied the temperature dependence of the magnetic anisotropy, defined as the energy required to saturate the magnetization. It is concluded that the anisotropy of the as-quenched alloy is most likely due to the internal stress field, and the field induced anisotropy is presumably due to short range atomic pair ordering.

## INTRODUCTION

Recent studies of the magnetic properties of the rapidly quenched transition metal based amorphous alloys prepared in the form of ribbon[1-3] showed that they can be excellent soft magnetic materials. In their as-quenched state, however, they have some anisotropy which reduces the magnetization at low fields substantially below the saturation value, typically by about 60%. Domain observations (Fig. 1) indicates that the alloy contains regions with different directions of local easy axis[4-5]. In particular, the presence of the fine domain pattern in some region of the sample suggests that the local easy axis of that region has a component out of plane of the ribbon. Also it is observed that the dimensions of the region in which the local anisotropy is reasonably homogeneous is of the order of few hundred microns. To investigate the origin of this local anisotropy, we studied its temperature dependence.

The magnetization curve of a typical amorphous alloy is shown in Fig. 2 (solid line). It resembles that of a polycrystalline solid in that the remanence is smaller than the saturation magnetization, and a finite field is necessary to achieve magnetic saturation. In the absence of anisotropy, magnetization should saturate at an infinitesimal field (dotted line in Fig. 2). We define the anisotropy energy in amorphous alloys by the area enclosed by these two curves (shadowed area in Fig. 2).

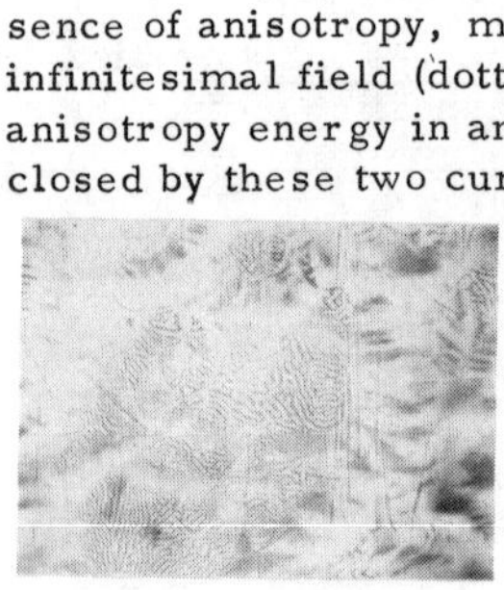

Fig. 1. Domain pattern obtained by colloid technique. Small field (< 100 Oe) perpendicular to the surface is applied. (x 100)

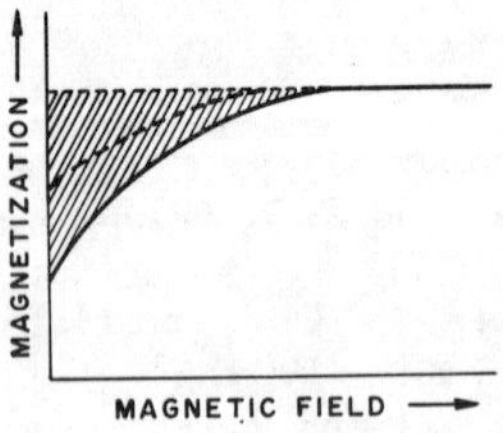


Fig. 2. Definition of the anisotropy energy (shadowed area). 2826 B.

It represents an average of the local anisotropy energies rather than the macroscopic anisotropy of the specimen as a whole. A typical value of the anisotropy ranges from $10^3$ to $10^4$ erg/cm$^3$ at room temperature for the Ni-Fe based amorphous alloys.

## TEMPERATURE DEPENDENCE

The magnetic anisotropy thus defined, K, and the magnetization, m, were measured at temperatures above -196°C for Metglas[6] alloys 2826 and 2826B, whose nominal compositions are $Ni_{40}\,Fe_{40}\,P_{14}\,B_6$ and $Ni_{49}\,Fe_{29}\,P_{14}\,B_6\,Si_2$ respectively. The results for 2826B alloy are shown in Figs. 3 and 4. For 2826 alloy the measurement was carried out only up to 100°C, while the Curie temperature is 240°C, because the irreversible annealing effect takes place during the measurement. The logarithmic plot of K vs m (Fig. 5) shows that except for the temperature regions close to

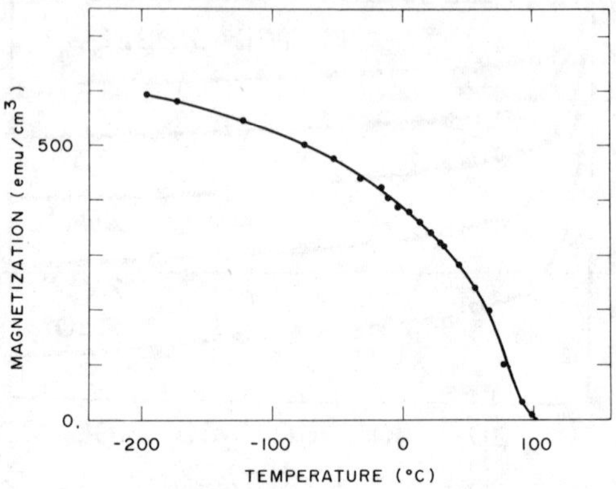


Fig. 3. Temperature dependence of magnetization. Metglas 2826 B.

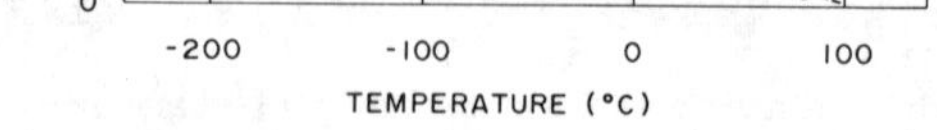

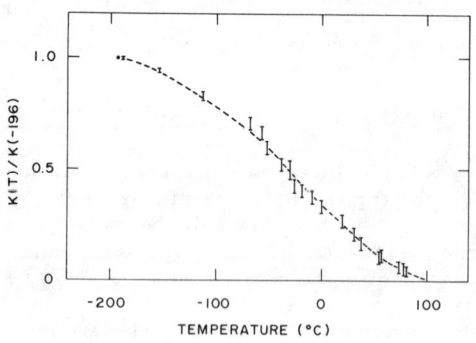

Fig. 4. Temperature dependence of K, Metglas 2826B. In this particular case, K(-196) = $8 \times 10^3$ erg/cm$^3$.

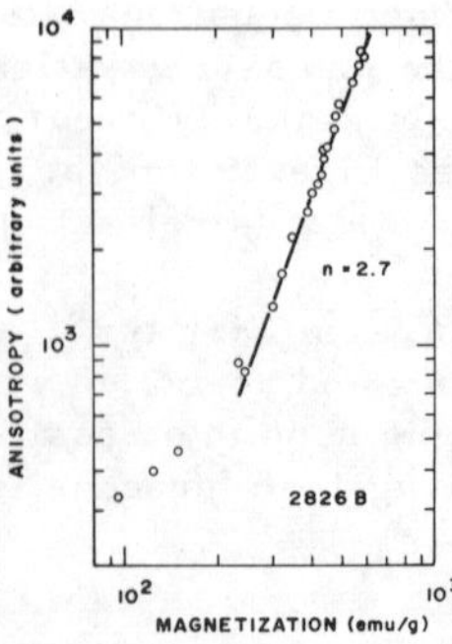


Fig. 5 Log-log plot of K vs M.

$T_c$, the power law

$$K \propto m^{2.7\ (\pm .2)} \quad (2826B)$$

holds. The similar plot for 2826 alloy yielded the power of 2.6 ± 0.2. This result clearly indicates that the origin of the anisotropy is neither compositional fluctuations nor microvoids, since anisotropies with these origins are proportional to the demagnetizing energy, therefore, to $m^2$.

## STRESS EFFECT

Since the alloys studied here have positive magnetostriction [1], the application of tensile stress to the specimen increases the low field magnetization, as shown by the dashed line in Fig. 2. Therefore, the value of K, defined as the area in Fig. 2, decreases with stress. This occurs because the applied stress makes the direction of the stress magnetically easy, not because it decreases the anisotropy. Under a constant stress, therefore, the reduced area is proportional to the magnetostriction. An important finding here is that K decreases with the stress $\sigma$, at a rate which is independent of temperature. In Fig. 6, the ratio of the value of K under a stress $\sigma$ to K without stress, $K(\sigma)/K(0)$, is shown for various values of stress and

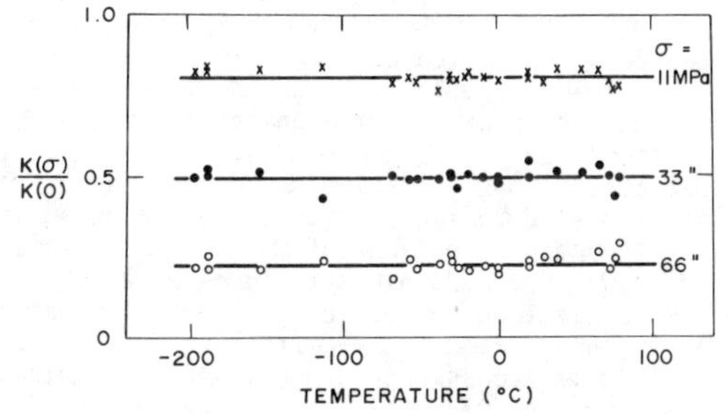


Fig. 6. Temperature dependence of $K(\sigma)/K(0)$ for three different stresses. The sample is 2826 alloy; similar results were obtained for 2826 alloy.

temperature. In this temperature range, the value of K (0) changes by more than a factor of 30, and yet the ratio $K(\sigma)/K(0)$ stays independent of temperature. This indicates that the anisotropy is exactly proportional to magnetostriction regardless of temperature.

We consider this experiment is a more direct and accurate method to correlate the anisotropy and the magnetostriction, than to compare independent measurements of two quantities, for instance by measuring the magnetostriction with strain gauge.

## FIELD INDUCED ANISOTROPY

It is known that an anisotropy can be induced by annealing the amorphous alloy in a magnetic field. [2, 7] Applying the field perpendicular to the length of the ribbon of 2826 alloy, we created an anisotropy which had an associated anisotropy field of 3.2 Oe. The magnetization was linear up to the saturation as was described in ref. 7. It was found that the slope of the magnetization with respect to the field was unchanged above -196°C, up to about 100°C where an irreversible effect of annealing begins. It is, therefore, concluded that the field induced anisotropy is proportional to $m^2$, distinctly different from the as-quenched anisotropy.

## ORIGIN OF ANISOTROPY

The principal results of the present measurements are as follows:

1) As-quenched anisotropy is proportional to magnetostriction in its temperature dependence, and is not due to microscopic shape anisotropy.

2) Field induced anisotropy is proportional to $m^2$, and its origin is different from the as-quenched anisotropy.

The result 1) strongly suggests that the as-quenched anisotropy is likely to be due to inhomogenious internal strain built in during the rapid quench. This view is not decisive, since there is a finite possibility that the anisotropy is of another origin but has the same temperature dependence as magnetostriction. The case as such could indeed happen in crystalline material, since the temperature dependence is determined by the symmetry regardless of the origin [8]. However, in amorphous alloys where the symmetry is purely statistical, such a possibility must be judged small. Also, the idea of magnetostrictive anisotropy is supported by the annealing behavior[7], and by the fact that the anisotropy can be drastically increased by rolling deformation, and can be reduced by subsequent annealing. The result 2) is less informative, but suggests that the field induced anisotropy is, like the crystalline case [9], due to some atomic pair ordering, since the pair correlation function $\langle \vec{S}_i \vec{S}_j \rangle$ scales with $m^2$ except whem m is very close to unity [10]. The likelihood that this anisotropy is due to some shape anisotropy is again minute.

We conclude, therefore, the most likely origins of the anisotropies are:

As-quenched anisotropy → residual internal strain

Field-induced anisotropy → atomic pair ordering

+supported in part by NSF Grant DMR72-03025
*also at Department of Metallurgy and Materials Science, University of Pennsylvania

## REFERENCES


1. T. Egami, P.J. Flanders, and C.D. Graham, Jr., AIP Con. Proc. 24, 697 (1975).
2. R.C. Sherwood, E.M. Gyorgy, H.S. Chen, S.D. Ferris, G. Norman, and H.J. Leamy, AIP Conf. Proc. 24, 745 (1975).
3. H. Fujimori, T. Masumoto, Y. Obi, and M. Kikuchi, Japan, J. Appl. Phys. 13, 1889 (1974).
4. J.J. Becker, in this conference.
5. H. Kronmüller, Max-Planck-Institut für Metallforschung, Stuttgart, private communication.
6. Trade Mark of Allied Chemical Corp., Morristown, N.J.
7. F.E. Luborsky, J.J. Becker, and R.O. McCary, IEEE Trans. Mag., 11, 1644 (1975).
8. H.B. Callen and E. Callen, J. Phys. Chem. Solids, 27, 1271 (1966).
9. S. Chikazumi and C.D. Graham, Jr., in "Magnetism and Metallurgy," ed. A.E. Berkowitz and E. Kneller (Academic Press, New York, 1969).
10. E. Callen and H.B. Callen, Phys. Rev. 139, A455 (1965).

# DOMAIN STRUCTURE DURING MAGNETIZATION OF AN ANNEALED AND ELASTICALLY STRAINED AMORPHOUS Ni-Fe-P-B ALLOY

J. W. Shilling
Research Center, Allegheny Ludlum Steel Corp., Brackenridge, Pa. 15014

## ABSTRACT

The magnetic domain structure was studied in a 0.17 cm wide x 32 μm thick ribbon of a $Ni_{40}Fe_{40}P_{14}B_6$ amorphous alloy (Metglas Alloy 2826 manufactured by Allied Chemical) during dc and ac magnetization after stress relief annealing and under various levels of applied tensile stress. The longitudinal Kerr effect was used for domain observation. Under an applied tensile stress, a simple antiparallel 180° wall structure was observed and the flux density of the sample correlated well with the observed 180° wall displacement. The domain wall spacing/sheet thickness ratio (2L/d) was extremely large (∿48) and decreased with increasing magnetizing frequency and applied tensile stress; the dependence of 2L/d on magnetizing frequency is characteristic of systems in which domain wall motion is eddy current limited. Significant improvements in losses and permeability during magnetization at power frequencies will occur if methods can be found to reduce these rather large domain wall spacings.

## INTRODUCTION

The magnetic properties of amorphous alloys have been extensively studied[1-7], and these investigations have shown that amorphous alloys are extremely soft magnetically having dc coercivities less than 0.01 Oe. This fact is consistent with the absence of a large magnetocrystalline anisotropy. The alloys do have a finite magnetostriction depending on composition[2] and thus the magnetic properties are quite stress-sensitive. Amorphous alloys also have electrical resistivities 2 to 3 times larger than crystalline alloys with similar Ni/Fe ratios. Despite extremely low coercivities and high electrical resistivities, the 60 to 400 $H_z$ losses and permeabilities of amorphous alloys have been somewhat disappointing since values only approaching those of 4-79 permalloy have been obtained[5-6]. The presence of residual stresses and rather large domain wall spacings have been suggested as being responsible for the relatively high losses of amorphous materials. The present study was initiated in order to determine the domain structure present in a well annealed amorphous material having a finite and positive magnetostriction as a function of magnetizing frequency and applied tensile stress. It must be remembered that the domain structures observed in the samples studied here are probably not representative of those existing in as-manufactured ribbons due to the presence, in the latter case, of large residual stresses resulting from rapid cooling during manufacture. The material chosen for study was an amorphous ribbon sample, Metglas 2826, manufactured by Allied Chemical. The nominal composition and physical properties of this material are as follows: [3-5]composition

$Fe_{40}Ni_{40}P_{14}B_6$; $4\pi Ms$ 8.2kG; $\lambda s$ 11 x $10^{-6}$; $\rho$ 130 μΩ-cm.

## PROCEDURE

Domains were observed using the longitudinal Kerr effect. The apparatus has been described previously.[8] Samples for domain observation ∿2 cm long were cut from longer lengths of 0.17 cm wide amorphous ribbon and the ribbon surface polished using conventional metallographic polishing procedures. The polished ribbon samples were then placed between glass plates and annealed at 288°C for 4 hrs in $H_2$ in order to flatten the ribbon samples and remove residual stresses produced by mechanical polishing. The polished surfaces of the annealed samples were then coated with ZnS for minimum reflectivity. Attempts to observe domain structure using the Kerr effect on as-manufactured ribbon surfaces were unsuccessful even when ZnS coatings were applied. The samples after mechanical polishing were ∿0.0032 cm (1.25 mil) thick.

During domain observation, the ∿2 cm long samples were magnetized along their lengths using Helmholtz coils with no flux closure. Due to the absence of flux closure and the rather small aspect ratio of these samples, significant effects of demagnetizing fields originating at the sample ends on domain structure and the shape of the B-H loop were observed. The Helmholtz coils were driven by a bipolar power supply/amplifier which in turn was driven by a low frequency function generator. A 1000 turn secondary coil was placed around the sample and integrated using a Walker MF-3D integrating fluxmeter. For observations under applied tensile stress, the sample was glued to a copper plate, and the plate was then bent with the sample mounted on the tension side of the plate. Strain gages mounted on the plate were used to estimate the strain in the ribbon sample which was converted to stress using a modulus for the ribbon sample[4] of 15 x $10^4$N/mm².

Even after the above sample preparation procedure, domain contrast was relatively weak (compared to 3% Si-Fe). The domain structure was easy to observe but difficult to photograph. Movies of the domain structure during 0.1 $H_z$ magnetization were taken at 8 frames/sec and shown at the Magnetism Conference. These movies more clearly illustrate the observed domain structures than the micrographs included in this manuscript.

## RESULTS AND DISCUSSION

Examples of the domain structures observed in a sample of amorphous ribbon are shown in Fig. 1. The domain structures have been drawn to the right of the micrographs in order to aid the reader in recognizing the structure in the micrograph. In the as-annealed condition (no external stress) rather non-uniform structures were observed which varied with the frequency of the demagnetizing field (Fig. 1a). Since the material is amorphous and has no crystalline anisotropy yet does have a rather large positive magnetostriction[1-5], local easy directions of magnetization are determined by rather small variations in residual stress. In the as-annealed condition, the easy directions of magnetization were not simply defined and the domain structures were somewhat random. Magnetization curves in this condition were not particularly square even after correcting for the demagnetizing field of the sample, and changes in sample magnetization were observed which did not correlate with changes in surface domain structure. This indicated that the domain structure observed on the surface of as-annealed samples (not stressed in tension) was not representative of the bulk domain structure.

The surface domain structure in samples under tensile stress was much simpler, reproducible, and based on measurements of the bulk flux density, was representative of the bulk domain structure. Examples of this structure are shown in Figs. 1b,c. The structure is shown under an applied tensile stress of 4kg/mm² in the demagnetized state and after magnetization in an applied field of 3.7 Oe. A B-H curve for this stressed sample taken at 0.1 $H_z$ is shown in Fig. 2. The low remanence value and slope of the B-H curve were primarily due to demagnetizing effects occurring at the ends of the rather short sample which was not flux closed as discussed above. The ascending and decending branches are superimposed in Fig. 2 since the coercivity was <0.02 Oe.

Magnetization at 0.1 $H_z$ under applied tensile stress occurred by uniform motion of two 180° walls separating three domains of antiparallel magnetization. If the sense of the applied field was such that growth of the "white" domains in Fig. 1b was favored, magnetization occurred by movement of the 180° walls toward the center (as viewed edge-to-edge) of the ribbon. The walls moved furthest along the center line (X-X' in Fig. 1b) due to demagnetizing effects at the sample ends. Impingement occurred first at the center of the sample leaving two reverse spikes at each sample end. Such a structure is shown in Fig. 1c in an applied field of 3.7 Oe. A sharp break in the B-H loop occurred at initial impingement (H=2.70e) as shown in Fig. 2. The slope of the B-H loop for H >2.70e was finite and associated with gradual removal of the reverse spikes at each sample end. In applied fields of the opposite sense, the center domain in Fig. 1b grew in size

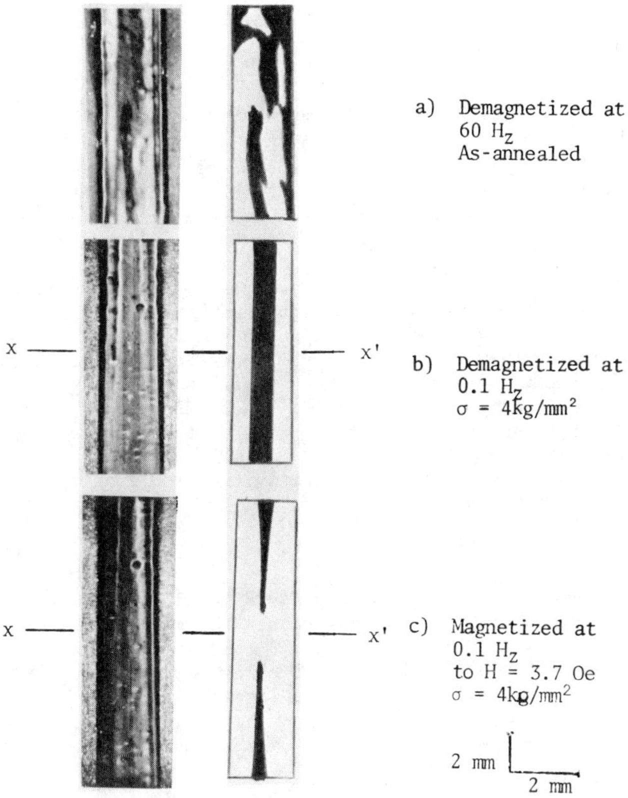


Figure 1. Domain structure in an amorphous ribbon sample.

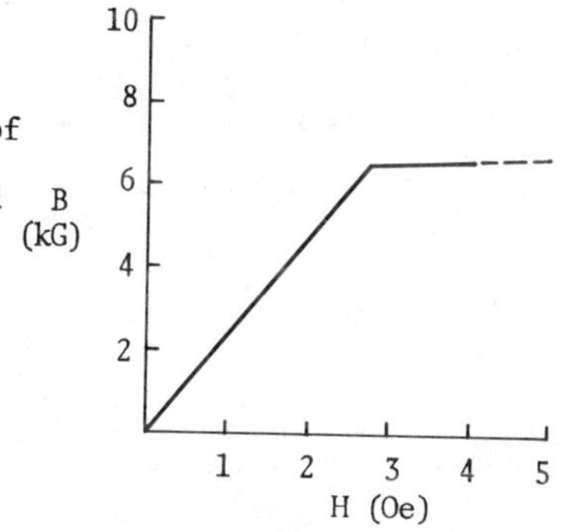


Figure 2: B-H loop of an amorphous sample under $4 kg/mm^2$ applied tensile stress.

and eventually reached the sample edge. The motion was largest along the center-line resulting in four triangular domains, one at each corner of the sample, after the two domain walls had run to the edges of the sample along the center-line. This domain process thus explains the fact that the induction at the knee of the B-H loop shown in Fig. 2 was somewhat less than saturation.

Domain structures during magnetization at 60 and 600 $H_z$ were observed by strobing the wall motion. The fact that the wall motion could be strobed indicates that it was relatively reproducible. Simple 180° wall motion was observed, as at 0.1 $H_z$, except that a refinement in domain structure was noted with increasing frequency. These data are shown in Fig. 3 where the domain wall spacing/sheet thickness ratio is plotted as a function of magnetizing frequency. In the "unstressed" condition, the sample was still glued to the copper plate but the stress relaxed to near-zero. Some small tensile stress most likely existed in the "unstressed" condition since simple 180° wall structures were observed. As shown in Fig. 3, the domain wall spacing decreased with increasing frequency and also decreased with applied tensile stress. The frequency dependence is characteristic of materials undergoing eddy current limited domain wall motion and this general phenomenon has been discussed previously [8-10]. Measurements of wall spacing using indirect techniques have indicated similar reductions in wall spacing with increasing frequency in amorphous alloys.[7] The effects of stress on domain wall spacing are most likely due to removal of transverse flux closure structure at the sample ends which, in turn, forces a domain spacing refinement. The domain spacing refinement produced by tension may thus be unique to the short samples studied here and may not occur in a flux closed toroidal sample. Tension in the latter case would be expected to produce a simple 180° domain structure with a large 2L/d spacing. A domain spacing refinement with increasing frequency should also occur in a toroidal sample providing there are sufficient reverse nuclei of critical size present which can grow in the rather small fields present during ac magnetization.[9] In the present sample, these nuclei most likely originated at the sample ends.

The domain wall spacing has been plotted in Fig. 3 as a ratio to the sheet thickness since eddy current losses are generally proportional to this quantity in the limit of simple planar 180° wall motion.[11] The values for 2L/d observed in this amorphous alloy are extremely large and similar to those observed in single crystals of 3% Si-Fe.[9] Such large values will unavoidably lead to large eddy current losses despite the large electrical resistivity of the amorphous alloys.

Thus significant improvements in losses and permeability during magnetization at frequencies from 60 to 400 $H_z$ can be expected in amorphous alloys if ways to reduce the domain wall spacing can be found. A natural reduction in domain wall spacing with increasing frequency will occur if a large number of critically sized metastable domain nuclei can be produced in these alloys. Such nuclei occur naturally in crystalline materials with finite crystalline anisotropy whenever easy directions of magnetization intersect sample surfaces. For example, in slightly misoriented single crystals and polycrystalline samples of 3% Si-Fe, 2L/d is typically less than 2 and can be less than 1 under some circumstances[8-9].

Figure 3: Domain wall spacing to sheet thickness ratio (2L/d) "unstressed" and under $4 kg/mm^2$ tensile stress as a function of magnetizing frequency (log scale).

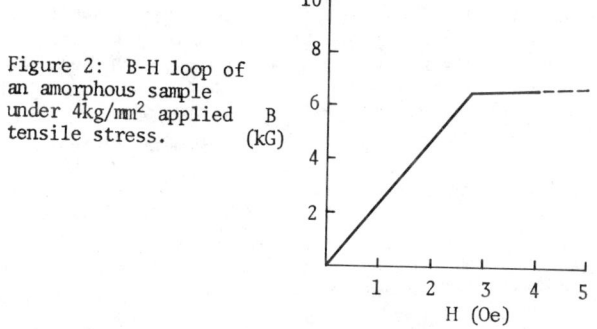

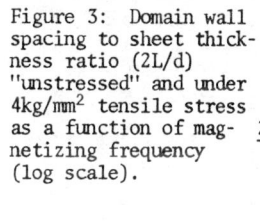

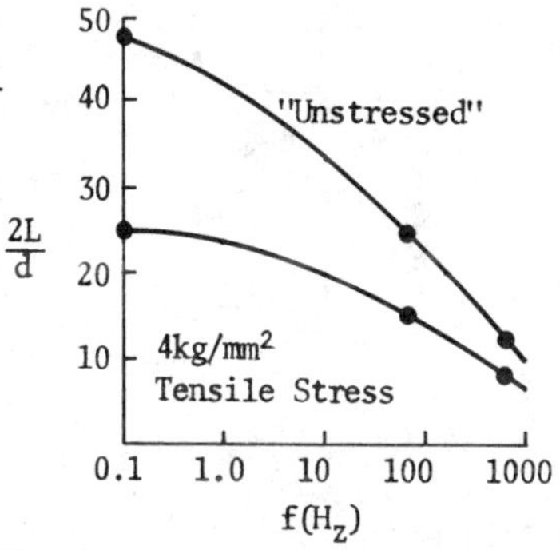


ACKNOWLEDGEMENTS

This work was initiated at the suggestion of C. D. Graham, Jr. and J. J. Becker, whose guidance and continued interest are gratefully acknowledged.

REFERENCES

1. T. Egami, P. J. Flanders and C. D. Graham, Jr., AIP Conf. Proc. #24, 697-701 (1975).
2. R. C. Sherwood, E. M. Gyorgy, H. S. Chen, S. D. Ferris, G. Norman, and H. J. Leamy, AIP Conf. Proc. #24, 745-746 (1975).
3. P. J. Flanders, C. D. Graham, Jr., and T. Egami, IEEE Trans. Mag., Mag-11, 1323-1325 (1975).
4. J. J. Becker, IEEE Trans. Mag., Mag-11, 1326-1328(1975).
5. R. C. O'Handley, This Conference.
6. F. E. Luborsky, J. J. Becker and R. O. McCary, IEEE Trans. Mag., Mag-11, 1644-1649 (1975).
7. R. C. O'Handley, J. Appl. Phys., 46, 4996-5001 (1975).
8. J. W. Shilling and G. L. Houze, Jr., IEEE Trans. Mag., Mag-10, 195-223 (1974).
9. J. W. Shilling, IEEE Trans. Mag., Mag-9, 351-356 (1973).
10. J. W. Shilling, AIP Conf. Proc. #5., 1504-1508 (1972).
11. R. H. Pry and C. D. Bean, J. Appl. Phys., 29, 532-533 (1958).

# RECENT DEVELOPMENTS IN THE THEORY OF SPIN GLASSES

D. Sherrington
IBM, Thomas J. Watson Research Center, Yorktown Heights, N.Y. 10598
and Physics Dept., Imperial College, London SW7 2BZ, England*

## ABSTRACT

The spin glass is a magnetic arrangement found in systems with randomly competing exchange interactions. It is characterized by local moments frozen in direction but with no long-range order. A review is presented of recent theoretical formulations of the spin glass as a new magnetic phase. An integral part of the development is the recognition that a new type of order parameter is necessary for the spin glass and plays a non-negligible role even for more conventional phases in disordered systems.

Results are also presented of computer simulations which help to illustrate the physics and complement the experiments.

## INTRODUCTION

The expression "spin glass" has its origin as a label to describe the states of certain substitutional magnetic alloys at low to intermediate concentrations, such as <u>CuMn</u> or <u>AuFe</u> in the concentration range between about 0.1 and 15at.%. In these states the local moments appear to be frozen, in the sense that the thermally averaged spin at any site is non-zero, but exhibit no long range order. Experimental evidence leading to this picture has been reviewed by Mydosh[1] and by Fiory[2].

Recent theoretical studies[3-9] have been concerned with the characterization of the spin glass as a new magnetic phase and with attempting to examine its thermodynamics in the simplest non-trivial approximations. The picture which has emerged is of a state likely to be much more widespread than that which led to the coining of the name. As alluded to earlier <u>two</u> minimal properties are necessary for a phase to be described as a spin glass; (i) thermal freezing of spins, (ii) no long-range order. A convenient way to formalize these requirements is to require (i) a non-zero value of an order parameter $q = \overline{| < S_i > |^2}$, where < > refers to a thermal average and the bar to a spatial average, (ii) $\overline{<S_i><S_j>} \to 0$ as $(R_i - R_j) \to \infty$. It is evident that only a disordered system will exhibit this phase.

In disordered systems even if condition (ii) does not hold, say the state is overall ferromagnetic or periodically antiferromagnetic, the order parameter q is still useful to supplement the more usual order parameter $m = \overline{<S_i> \cos(Q.R_i)}$, where $Q$ is the wavevector of the average periodicity. It is only in pure systems that $q = m^2$.

This paper is concerned primarily with this aspect of spin glasses; i.e., characterization as a new thermodynamic phase of disordered magnets. Also reported are results of computer simulations of spin glasses. Elementary excitations are not discussed, nor are transport properties which are intimately related to such excitations.

## THEORETICAL ANALYSIS

In the canonical examples of spin glasses, the substitutional alloys <u>AuFe</u> and <u>CuMn</u>, the spins are randomly distributed on a lattice. They interact via exchange of RKKY type, long-ranged ($R^{-3}$) and oscillatory ($\cos 2k_F R$). A pair of magnetic ions chosen randomly in an alloy may thus exhibit either ferromagnetic or antiferromagnetic exchange interaction. It is this random competition between the two types of exchange which is believed to be responsible for the existence of the spin glass phase and thus one might reasonably expect such a phase to be possible in any system where randomly competing exchange information occurs. Many other experimental situations can be envisaged which lead to such competition (e.g. amorphousness, random superexchange, topological disorder etc.). Much of the recent analysis has therefore been performed with models simpler than that appropriate to the substitutional alloys, particularly Heisenberg or Ising models with exchange interactions $J_{ij}$ distributed randomly and independently of one another over both positive and negative values. It has been suggested that anisotropy effects may be important in characterizing relaxation effects in spin glasses[10] and that in certain amorphous systems they play the role of a random spin director[11]. Such effects are ignored in this paper which concentrates on the effects of exchange disorder.

Let us consider explicitly the simple example of a nearest neighbor Ising model

$$H = \frac{1}{2}\sum_{ij} J_{ij} S_i S_j ; \; S = \pm 1 \tag{1}$$

with the $J_{ij}$ independently distributed with the same distribution. Before describing the formal mathematical tricks which have been applied to this system let us consider the problem heuristically in molecular field theory starting from the usual equation

$$<S_i> = \tanh (\sum_j J_{ij} <S_j>/kT) \tag{2}$$

To exclude absolutely the possibility of normal magnetic order let us take $\overline{J_{ij}} = 0$. The standard procedure to find an ordering temperature is to linearize (2) and look for non-trivial solutions. One clearly finds such a solution at kT equal to the largest eigenvalue of the $J_{ij}$ matrix. This however corresponds to a localized solution, i.e. a superparamagnetic cluster, while the true thermodynamic phase transition is a cooperative process involving an extensive number of spins. It is better approximated by the temperature corresponding to the highest extended-state eigenvalue of the $J_{ij}$ matrix[10]. For the special case of symmetric distribution this temperature is thus given approximately by $T = z^{1/2}\Delta J/k$ where $\Delta J$ is the standard deviation of the J distribution and z is the number of nearest neighbors. This result, as also the relevance of the q order parameter, may be seen by averaging the square of the linearized mean field equation

$$<S_i>^2 = (kT)^{-2}[\sum_j J_{ij}{}^2 <S_j>^2 + \sum_{jl}{}' J_{ij} J_{il} <S_j><S_l>] \tag{3}$$

$$\text{to give } \overline{<S_i>^2} = z(\Delta J/kT)^2 \; \overline{<S_j>^2} \tag{4}$$

On the other hand averaging the linearized m.f. equation itself gives $\overline{<S_i>} = 0$ in this case; that is, there is no ferromagnetism, this is a spin glass.

A formal procedure to treat this problem has been introduced by Edwards and Anderson (EA)[4]. It starts by averaging the free energy over the disorder,

$$F = -kT \int \prod_{(ij)} dJ_{ij} \; P\{J_{ij}\} \; \ln Z\{J_{ij}\}, \tag{5}$$

where $Z\{J_{ij}\}$ is the partition function for a particular set of $J_{ij}$, and $P\{J_{ij}\}$ is their probability distribution function, which for simplicity is taken as Gaussian,[12]

$$P\{J_{ij}\} = \prod_{(ij)} ( (2\pi)^{1/2}\Delta J)^{-1} \exp[-(J_{ij} - J_\circ)^2/2(\Delta J)^2]. \tag{6}$$

The averaging is carried out over the free energy rather than the partition function since the bonds are thermodynamically immobile (quenched)[13]. Mathematically it is however easier to average a partition function (since it is a sum of exponentials rather than a logarithm of such a sum). The desired averaging may be achieved formally by use of the identity

$$\ln Z = \lim_{n\to 0} \frac{1}{n}(Z^n - 1) \qquad (7)$$

Thus if $Z^n$ is averaged and the resulting quantity analytically continued to $n \to 0$ the physical free energy may be obtained. For integral n we may write

$$Z^n = \mathrm{Tr}_n \exp\{1/2\Sigma_{ij} J_{ij} \Sigma_{\alpha=1}^{n} S_i^\alpha S_j^\alpha / kT\} \qquad (8)$$

where α is a dummy label. Spin operators with different labels commute. Since $J_{ij}$ enters only as the argument of an exponential, (8) is readily averaged against $P\{J_{ij}\}$. Assuming that analytic continuation from integral to continuous n can be made, F is thus given by

$$F = -kT \lim_{n\to 0} 1/n\{\mathrm{Tr}_n \exp[1/2\Sigma_{ij}(\frac{J_o}{kT}\sum_\alpha S_i^\alpha S_j^\alpha +$$

$$\frac{(\Delta J)^2}{2(kT)^2}\sum_{\alpha\beta} S_i^\alpha S_j^\alpha S_i^\beta S_j^\beta)]-1\}. \qquad (9)$$

The randomness has been removed in favour of taking the limiting behaviour of a system in which replicas α interact through a quartic interaction; we shall refer to this system as the n-system. Analogous limit representations for quenched random systems have been used in various studies of critical behaviour[14-16] which have not allowed for a spin glass phase.

The mean magnetization m and the order parameter q introduced above may also be expressed in terms of the n-system as[17]

$$m = \overline{\langle S_i \rangle} = \lim_{n\to 0} \langle S_i^\alpha \rangle_n, \qquad (10)$$

$$q = \overline{\langle S_i \rangle^2} = \lim_{n\to 0} \langle S_i^\alpha S_i^\beta \rangle_n; \ \alpha \neq \beta, \qquad (11)$$

where the subscript n refers to a thermal average over the n-system. Thus q is related to the correlation between spins at the same site in different replicas which results from their effective interactive interaction. To generalize self-consistent field theory to the present problem one therefore introduces into (9) the variational order parameters

$$m^\alpha = \langle S_i^\alpha \rangle_n, \qquad (12a)$$

$$q^{\alpha\beta} = \langle S_i^\alpha S_i^\beta \rangle_n \ ; \ \alpha \neq \beta, \qquad (12b)$$

using the substitutions

$$1/2\Sigma_{ij} S_i^\alpha S_j^\alpha \to zm^\alpha \sum_i (S_i^\alpha - m^\alpha/2), \qquad (13a)$$

$$1/2\Sigma_{ij} S_i^\alpha S_j^\alpha S_i^\beta S_j^\beta \to zq^{\alpha\beta}\Sigma (S_i^\alpha S_i^\beta - q^{\alpha\beta}/2). \qquad (13b)$$

The molecular field free energy is obtained by minimizing the substituted free energy of the n-system with respect to $m^\alpha$, $q^{\alpha\beta}$ and taking the $n\to 0$ limit. Making the ansatz that the physical extremum has all $q^{\alpha\beta}$ equal, the quadratic term in (13b) may be further simplified using the identity

$$\exp\lambda(\sum_\alpha S_i^\alpha)^2 = (2\pi)^{-1/2}\int dx \exp(-x^2/2 + \lambda^{1/2} x \sum_\alpha S_i^\alpha). \qquad (14)$$

This leads finally to the coupled equations for q and m:[9,18]

$$q = 1-(2\pi)^{-1/2}\int_{-\infty}^{\infty} dx\, e^{-x^2/2}(kT/\tilde{J}q^{1/2})x$$

$$\tanh(\tilde{J}q^{1/2}x/kT + \tilde{J}_o m/kT), \qquad (15a)$$

$$m = (2\pi)^{-1/2}\int_{-\infty}^{\infty} dx\, e^{-x^2/2}\tanh(\tilde{J}q^{1/2}x/kT + \tilde{J}_o m/kT), \qquad (15b)$$

where

$$\tilde{J}_o = zJ_o, \ \tilde{J} = z^{1/2}\Delta J. \qquad (16)$$

These equations lead to the phase diagram of Fig. 1, which is expressed in terms of the dimensionless ratios $\tilde{J}_o/\tilde{J}$ and $kT/\tilde{J}$; it is trivially rescaled to describe models

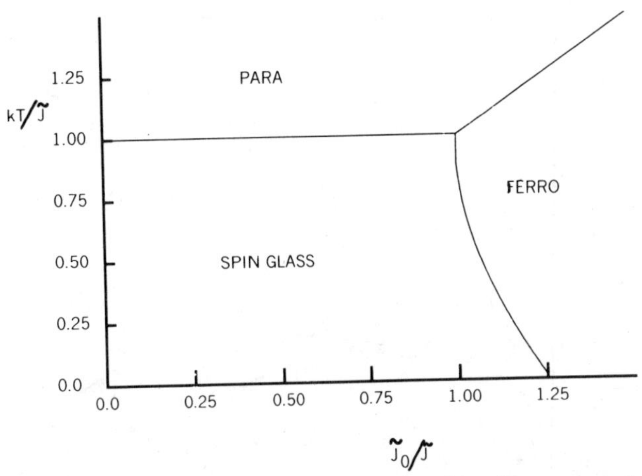


Fig. 1 Phase diagram of Ising spin glass/ferromagnet

in which $\tilde{J}_o$ and $\tilde{J}$ are known functions of external parameters. Both spin-glass ($q \neq 0$, $m = 0$) and ferromagnetic ($q \neq 0$, $m \neq 0$) regions occur, as well as a paramagnetic ($q = 0$, $m = 0$) region. All the transitions are second order. Above we have been taking $\tilde{J}$ to be positive. If it is negative, m is defined as the sub-lattice magnetization, and the phase diagram is simply a reflection of that shown about $\tilde{J}_o = 0$.

If $\tilde{J}/\tilde{J}_o \ll 1$ one finds $q \simeq m^2$ as expected physically. Fig. 2 shows values of $m(T)$ and $q^{1/2}(T)$ as obtained from (15) for various values of $\tilde{J}/\tilde{J}_o$. Note that $q^{1/2}(T)$ is consistently higher than $m(T)$ for finite $\tilde{J}$ and tends to

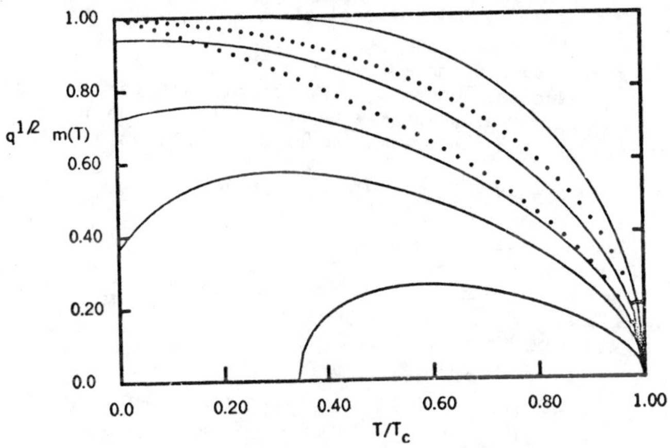


**Fig. 2** n(T), $q^{1/2}(T)$ for exchange disordered Ising model. Solid lines denote m(T) for ratios $\tilde{J}_o/\tilde{J}$ (top to bottom) ∞, 2.0, 1.5, 1.3, 1.1. Dotted lines show $q^{1/2}(T)$ for $\tilde{J}_o/\tilde{J} = 2.0$ (upper points) and 0.0 (lower points).

unity as $T\to 0$. Deviations in $q^{1/2}$ (or q) at low temperature are linear in T while both m and $q^{1/2}$ approach their critical points as $(T_c - T)^{1/2}$. The effect of fluctuations appears to be strongest at low temperatures. At T=0 m is suppressed for $\tilde{J}_o/\tilde{J} < \sqrt{\pi/2}$ but for larger $\tilde{J}_o$ rises rapidly, initially as

$$m \sim (18\pi)^{1/4}(\tilde{J}_o/\tilde{J})^2[(2/\pi)^{1/2} - \tilde{J}/\tilde{J}_o]^{1/2}, \qquad (17)$$

becoming essentially unity by $\tilde{J}_o/\tilde{J} \sim 3$.

The susceptibility is readily obtained from the modification of (15) to include the effects of an external field; a contribution H/kT is added to the argument of the tanh. The zero-field susceptibility is given by

$$\chi(T) = (1-q)/(kT - \tilde{J}_o(1-q)) \qquad (18a)$$

$$= \chi^{(o)}/(1-\tilde{J}_o\chi^{(o)}) \qquad (18b)$$

where $\chi^{(o)}$ is the result for $\tilde{J}_o = 0$. Results for two

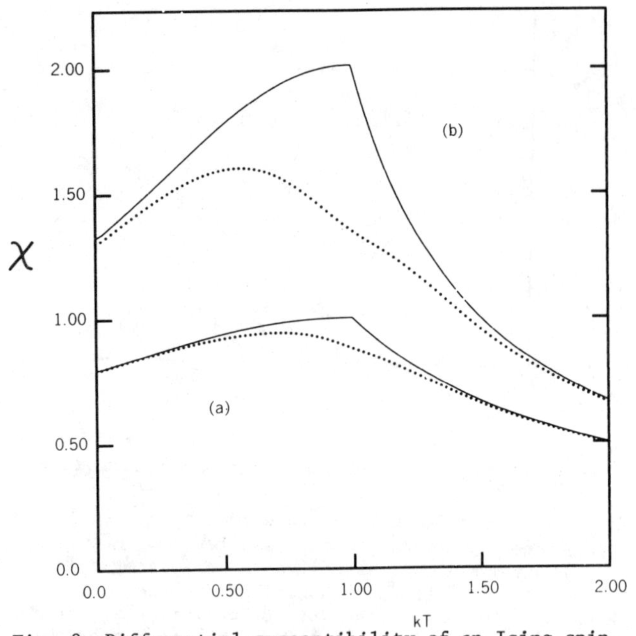


Fig. 3 Differential susceptibility of an Ising spin glass without external field (solid lines) and with a field $H = 0.1\tilde{J}$ (dotted lines) for (a) $\tilde{J}_o/\tilde{J} = 0$ (b) $\tilde{J}_o/\tilde{J} = 0.5$

spin glass systems are shown in Fig. 3. The susceptibility has a cusp at the ordering temperature which is smoothed out bv an external field. $\chi$ remains finite as $T\to 0$. Real substitutional alloy spin glasses exhibit qualitatively identical behavior although the experimental cusp is more symmetric[19].

Another quantity of thermodynamic interest is the specific heat which may be obtained from the internal energy. In the present model this is given by

$$U = -N[m^2\tilde{J}_o/2 + \tilde{J}^2(1-q^2)/2kT] \qquad (19)$$

and leads in the spin glass phase to the specific heat curve given in Fig. 4. At low temperatures the specific heat is linear in T, as found experimentally both in spin glasses[20,1] and in ordinary glasses[21]. In contrast to

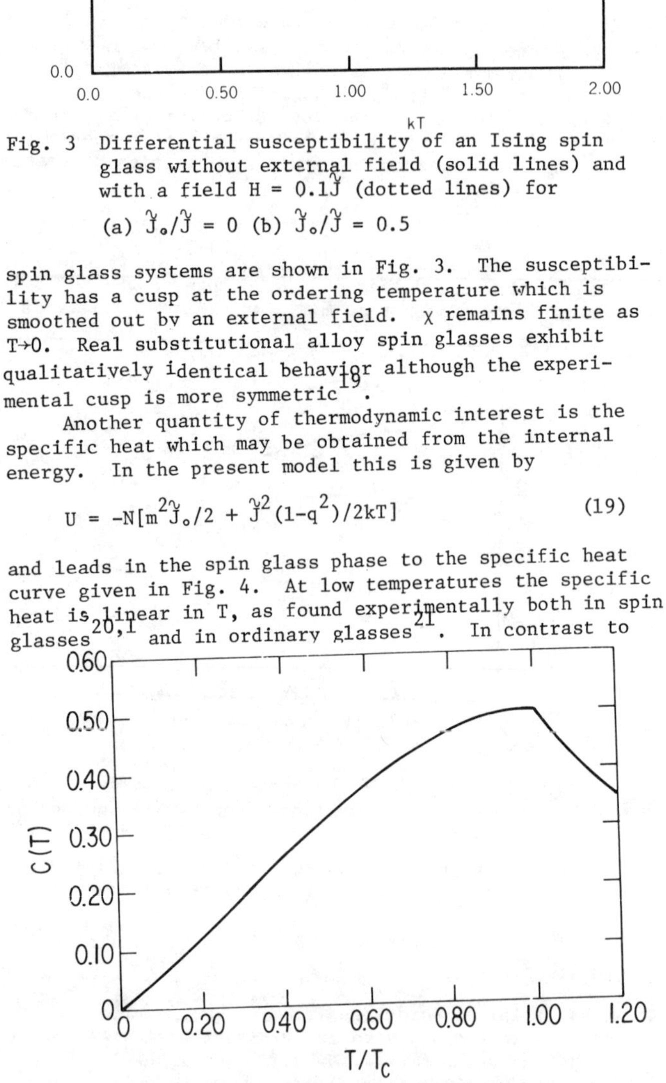


Fig. 4 Specific heat of an Ising spin glass

more conventional mean field theories the specific heat obtained here does not vanish above the ordering temperature; rather it decays as $T^{-2}$. Experimentally, cusps such as obtained here have not been observed: a rounded peak is usually seen with a maximum at a temperature a little higher than that of the susceptibility cusp.

The above analysis has been given for only the simplest mean field theory which allows for a difference in $\langle S_i\rangle$ and $(\langle S_i\rangle^2)^{1/2}$, that is one with no short-range order. Several improvements analagous to those for pure systems can be envisaged. Nevertheless it clearly demonstrates the relevance of a q-like order parameter for disordered magnetic systems. The method can readily be extended to include Heisenberg[4-6] or intermediate models, longer range interactions[5,6] and other exchange distributions. Furthermore one may generalize beyond bond disorder. For example, Aharony[16] has shown how higher order interactions of the type found in (9) result from a site disorder model.

The restriction of the effective n- Hamiltonian to terms of quartic order in S is a special consequence of the use of the Gaussian exchange model. In general, terms of all orders are expected. The qualitative feature of a phase diagram exhibiting three phases (paramagnet, ferromagnet, spin glass) for $\tilde{J}_o \geq 0$ is unlikely to be altered by such higher order terms; although the location of phase transition lines would be improved by the inclusion of further order parameters such as

$$\overline{\langle S_i\rangle^3} = \lim_{n\to o} \langle S_i^\alpha S_i^\beta S_i^\gamma\rangle_{\alpha\neq\beta\neq\gamma}, \qquad (20)$$

no new phases are to be expected. It has however recently been predicted by Harris, Lubensky and Chen[22] that a sufficiently skew $J_{ij}$ distribution could lead to a first order spin glass transition. Using renormalization group analysis on a system with $\tilde{J}_o = 0$ these authors find a new stable fixed point associated with $q \neq 0$ having mean field exponents in 6 or more dimensions. With skewness as non-ordering parameter a tricritical point is found with mean field properties for 4 or more dimensions.

## COMPUTER SIMULATION

A further insight into the microscopic nature of spin glasses may be obtained by computer simulation. Such studies complement the more conventional experiments both in providing microscopic information and in permitting comparison with analytical theory for Hamiltonians simpler than those found in nature. At the present time only zero-temperature simulations are available[30]; these probe only ground state properties.

One group of simulations has been performed by Kirkpatrick and Sherrington[23] on nearest neighbor Ising models with exchange disorder both of Gaussian and of random ±J character and with a variety of applied magnetic fields. For simplicity a simple cubic lattice was used. After choosing a set of random bonds (with the appropriate distribution) ground state searches were made starting from a random spin configuration. The search procedure consisted of picking bonds randomly, checking the energetic consequences of turning over either or both the spins joined by that bond and flipping the spin(s) if so doing either lowers the energy or increases it by an amount less than a cut-off which is reduced with each iteration. The resulting spin distribution was used as starting point for a second sweep sequence with a reset initial cut-off. The procedure was repeated until no further improvement in the ground state energy resulted.

The results reported here are for Gaussian exchange

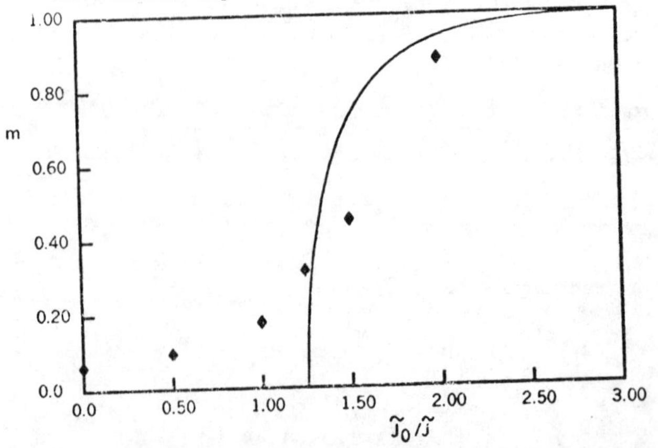


Fig. 5 $m(\tilde{J}_o/\tilde{J})$ at T = 0 for Ising model with Gaussian random exchange interaction. Curve is from molecular field theory. Points are computer simulations in the presence of a small external field.

distribution on a 20 x 20 x 20 lattice with only a small or zero external field. For small enough $\tilde{J}_0/\tilde{J}$ different random initial spin configurations converged to different "ground states" although their energies were very close. This is in accord with recent pictures of normal glasses[25, 26] which attribute their linear specific heat to tunneling between local minima. Fig. 5 compares $m(\tilde{J}_0/\tilde{J})$ given by simulations in the presence of a small field ($0.05\ \tilde{J}$) with that obtained from mean field theory at $T = 0$. The simulations are consistent with the m.f.t. prediction of a second order phase transition from a spin glass to a ferromagnet at $\tilde{J}_0/\tilde{J}$ of order unity. Fig. 6

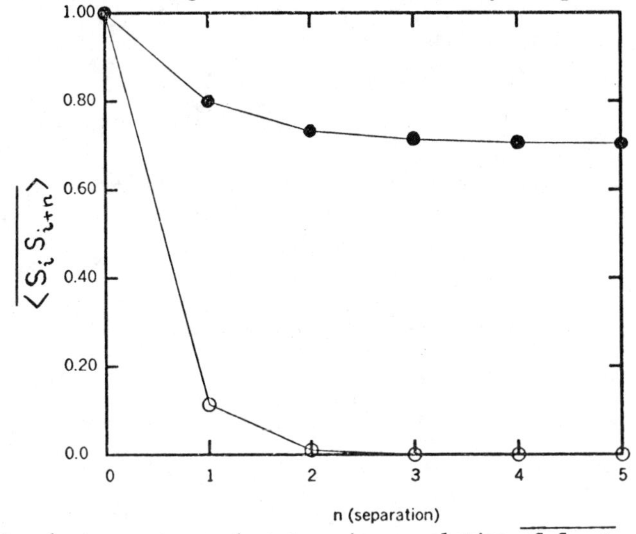


Fig. 6 Average ground state spin correlation $\overline{\langle S_i S_{i+n} \rangle}$ along principal axes for random Ising simulation. Solid circles are for $\tilde{J}_0/\tilde{J} = 2.0$ (ferromagnet), open circles for $\tilde{J}_0/\tilde{J} = 0.5$ (spin glass).

shows average spin correlations found along principal axes for a typical spin glass ($\tilde{J}_0/\tilde{J} = 0.5$) and ferromagnet ($\tilde{J}_0/\tilde{J} = 2.0$) away from the extremes of $\tilde{J}_0/\tilde{J}$: for $\tilde{J}_0 = 0$ there is no average correlation while for $\tilde{J} = 0$ it is complete.

Simulations on a model closer to a real substitutional alloy have been performed by Walker and Walstedt[26]. They have considered a model with classical Heisenberg spins randomly distributed on an f.c.c. lattice and interacting through a truncated RKKY exchange.[27] The RKKY parameters used were those corresponding to pure Cu. The simulations reported here employed an array of 30 unit cells per side, with periodic boundary conditions, at a spin concentration of 0.3% (324 spins). Again random initial spin configurations were employed but thereafter iteration proceeded by systematic spin rotation towards iteratively instantaneous local molecular fields. Self-consistent states were found in which the spins were

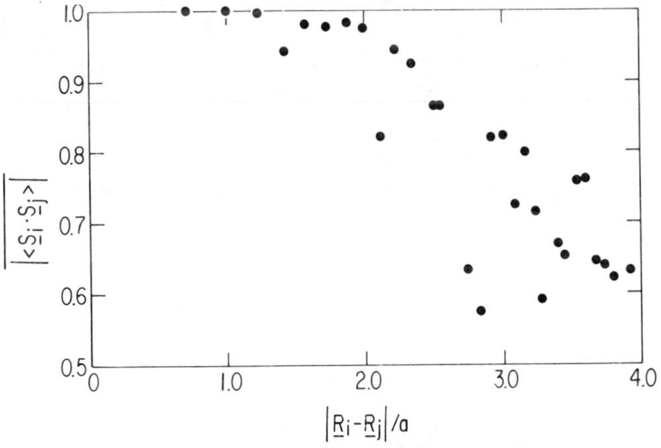


Fig. 7 Average spin correlation $|\overline{\langle S_i . S_j \rangle}|$ obtained for five runs on simulation of 0.3% spins in Cu as a function of spatial separation. A purely random distribution would lead to a value 0.5.

locally aligned but which on average were distributed isotropically with no long-range order. Fig. 7 shows spin correlation results for this model.

Most early spin-glass theory was formulated in terms of the distribution of local molecular field strengths $p(H)$. Fig. 8 shows this distribution as obtained in

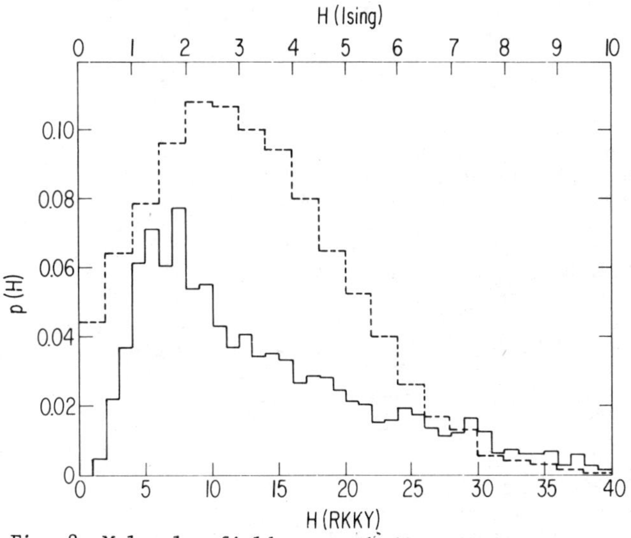


Fig. 8 Molecular field strength distribution as obtained for $J_0 = 0$ Ising (dashed histogram) and CuX simulations (solid histogram). Both plots are normalized to range exhibited. In the Ising case H is expressed in units of $\Delta J$. In the CuX case it is defined as $H = |\sum_j J_{ij} \underline{S}_j|$ where $|\underline{S}| = 1$ and $J_{ij} = 176.8 \cos(9.822 R_{ij})/R_{ij}^3$ where $R_{ij}$ is measured in units of the lattice constant $a = 3.61$ Å.

the two simulation studies reported above. Note that in the Heisenberg problem $p(H=0)=0$, indicating that the linear specific heat found in spin glasses cannot have its origin in isolated spin flips but must involve larger reorientations.

## CONCLUSION

We have seen that in the description of magnetic ordering in spatially disordered systems order parameters describing average moment periodicity should be supplemented, or even supplanted, by one (or more) characterizing local spin freezing effects which lack periodicity. The discussion presented here has been concerned only with static properties of spin glasses. Further study is needed of dynamic effects[28].

Computer simulations bear out the interpretation of a spin glass as having no long range order and a glassy ground state character.

From the analysis presented above there appears to be no reason to believe spin glasses are restricted to substitutional metallic alloys and the occurence of states exhibiting local spin freezing but no long range order can be widely anticipated.[29]

## ACKNOWLEDGEMENTS

Many of the results reported here were obtained in collaboration with Dr. S. Kirkpatrick to whom the author is deeply indebted. He would also like to thank Drs. L. R. Walker and R. E. Walstedt for making available their unpublished work.

## REFERENCES

* Permanent address

1. J. A. Mydosh, A.I.P. Conf. Proc. 24, 131 (1974).
2. A. T. Fiory, this conference.
3. K. Adkins and N. Rivier, J. de Physique 35, C4-237 (1974).

4. S. F. Edwards and P. W. Anderson, J. Phys. F5, 965, (1975).
5. K. H. Fischer, Phys. Rev. Lett. 34, 1438 (1975).
6. D. Sherrington and B. W. Southern, J. Phys. F5, L49 (1975).
7. D. Sherrington, J. Phys. C8, L208 (1975).
8. B. W. Southern, J. Phys. C8, L213 (1975).
9. D. Sherrington and S. Kirkpatrick, to be published
10. A philosophically related model for an RKKY - coupled spin glass has been introduced by D. A. Smith (J. Phys. F4, L266 (1974)). The system is considered in terms of clusters within which all the spins are coupled with exchange bonds of strength greater than order kT but with intercluster coupling only by bonds weaker than this. The spin glass temperature is that at which infinite clusters occur.
11. R. Harris, M. Plischke and M. J. Zuckermann, Phys. Rev. Lett. 31, 160 (1975); R. Harris and D. Zobin, This conference.
12. Gaussian distributions are known from renormalization group studies to have stability disadvantages but are used here only for mathematical convenience and only within a mean field theory.
13. R. Brout, Phys. Rev. 115, 824 (1959).
14. V. J. Emery, Phys. Rev. B11, 239 (1975).
15. G. Grinstein, Ph.D. thesis, Harvard University (1974 unpublished).
G. Grinstein and A. H. Luther, to be published.
16. A. Aharony, Phys. Rev. Lett. 34, 590 (1975).
17. These identifications follow from the observations that

$$\langle S_i\rangle = \partial/\partial h \ \ln \mathrm{Tr} \exp \left[\Sigma_{i\neq j} J_{ij} S_i^\alpha S_j^\alpha/kT + hS_i^\alpha\right]_{h=0}$$

$$\langle S_i\rangle^2 = \partial/\partial h' \ln \mathrm{Tr}_{\alpha\beta} \exp \left[\Sigma_{i\neq j} J_{ij} (S_i^\alpha S_j^\alpha + S_i^\beta S_j^\beta)/kT + h' S_i^\alpha S_i^\beta\right]_{h'=0}$$

where $\alpha \neq \beta$ are dummy labels.
18. The $m^\alpha$ are also taken all equal at the extremum.
19. V. Cannella and J. A. Mydosh, Phys. Rev. B6, 4220 (1972).
20. J. E. Zimmerman and F. E. Hoe, J. Phys. Chem. Solids 17, 52 (1960).
21. R. C. Zeller and R. O. Pohl, Phys. Rev. B4, 2029 (1971).
22. A. B. Harris, T. C. Lubensky, and J-H Chen, to be published.
23. S. Kirkpatrick and D. Sherrington (unpublished).
24. P. W. Anderson, B. I. Halperin and C. M. Varma, Phil. Mag. 25, 1 (1972).
25. W. A. Phillips, J. Low Temp. Phys. 7, 351 (1972).
26. L. R. Walker and R. E. Walstedt (unpublished).
27. Similar simulations have also been performed by A. de Rozario and D. A. Smith (umpublished). and by R. Harris and D. Zobin (unpublished).
28. Some progress in this direction has been made by D. A. Smith (dynamic relaxation of finite clusters, ref. 10 and to be published) and by N. Y. Rivier (elementary excitations, Wiss. Z. Tech. Univ. Dresden (to be published) and J. Phys. F5, 1745 (1975)).
29. See also for example J.M.D. Coey and P. W. Readman Nature 246, 476 (1973).
30. Since this paper was written a preprint has appeared (K. Binder and K. Schröder, to be published) presenting results of a finite temperature Monte Carlo study of the Ising model with nearest-neighbor interactions distributed with a zero-mean Gaussian probability distribution. The susceptibility found exhibits a cusp, while the specific heat has a broad peak.


# ONSET OF MAGNETIC ORDER IN V-Fe AND Cu-Ni


Helmut Claus†
University of Illinois, Chicago, Illinois 60680



ABSTRACT

In the spin cluster alloys V-Fe and Cu-Ni, near the critical concentration to the onset of ferromagnetism, the a.c. susceptibility displays a fairly sharp peak at the magnetic ordering temperature, $T_c$.[1,2] Measurements of this susceptibility peak for alloys in states of different heat treatments reveal a large but continuous change of $T_c$, as the degree of short range order is increased. Changes in $T_c$ of more than a factor of 10 are observed. In V-Fe, $T_c$ decreases with increasing short range order; whereas in Cu-Ni, $T_c$ is increasing with increasing short range order, as expected from the different sign of the short range order parameter in the two alloy systems. Low field a.c. susceptibility measurements are thus very convenient in studying the kinetics of atomic short range order. For both V-Fe and Cu-Ni, there is a sudden pronounced decrease in the height of the susceptibility peaks as $T_c$ drops below a temperature of 3K. This behavior suggests that alloys, with a susceptibility peak below 3K, are no longer ferromagnetic but, probably, show a transition into a state where the spin cluster moments are frozen in a local anisotropy field, without long range order. The occurance of ferromagnetism in spin cluster alloys is thus preceded by a spin glass like state. Transitions into a long range ferromagnetic state only occur above a threshold temperature of about 3K.


†Work performed while a guest scientist at the Institut für Festkörperforschung, KFA Jülich, Germany.


1. H. Claus, Phys. Rev. Lett. 34, 26 (1975).
2. H. Claus, Phys. Lett. 51A 283 (1975).

# $\mu^+$ SPIN PRECESSION EXPERIMENTS IN SPIN GLASSES

A. T. Fiory
Bell Laboratories, Murray Hill, New Jersey 07974

ABSTRACT

The precession of polarized muons was used to measure the effect of ordering of the local moments in spin glass alloys, CuMn and AuFe. A distribution in the local fields seen by the muons, of dipolar origin, cause an inhomogeneous broadening that relaxes the muons' polarization. An onset of such broadening occurs near the ordering temperature, and the temperature width of the transition is an increasing function of applied field. Results extrapolated to zero temperature are consistent with a static or frozen spin system. Marked hysteresis effects observed below the ordering temperature are associated with similar behavior in the magnetization.

Measurements of the spin precession of polarized positive muons implanted in the spin glass alloys CuMn and AuFe were used to observe the static local magnetic field distribution associated with magnetic ordering. The technique, used previously to measure local fields in a number of materials, was reviewed by Kossler last year.[1] In studies of this type the muon can be considered a radioactive proton occupying random interstitial sites in the alloy. Muon spin precession, as observed via oscillations in the emitted positron intensity ($\mu^+ \rightarrow e^+ + \nu + \bar{\nu}$, $\tau$=2.2 μsec), at a fixed angle, is a sensitive measure of the local magnetic field. Previous techniques used to measure internal fields in spin glasses include Mössbauer effect[2] and nuclear[3] and electron spin[4] resonance studies.

Confirmation a few years ago by Cannella, Mydosh and Budnick[5] of the existence of a sharp phase transition in the magnetic susceptibility of alloys of a few percent concentration of magnetic impurities, first observed in the Mössbauer effect hyperfine field, led to renewed theoretical interest[6-12] in these unique materials. It is believed that below a certain freezing temperature, the magnetic impurities, which interact via the oscillatory long range Ruderman-Kittel-Kasuya-Yosida (RKKY) indirect exchange, order into configurations which are neither ferromagnetic nor anti-ferromagnetic.

Among the first theoretical treatments of local moment interactions[13-16] were random molecular field theories in which the fields experienced by individual spins are expected to be broadly distributed, as short range magnetic ordering sets in at low temperatures. In its original formulation, theory did not predict a phase transition.[16] Rivier and Adkins developed a modification allowing for a short-range order phase transition, however.[6] A more recent theoretical proposal of Edwards and Anderson,[7] developed further by other authors,[8-11] shows that short-range order in itself is not necessary for a spin glass transition. There is a finite expectation that a spin $S_i$ will freeze below the ordering temperature $T_0$ in the absence of short-range order. A mean field order parameter q is defined as an average of $S_i^2$ over the system for a long time span, and behaves like $(1-T/T_0)$ near $T_0$. One could expect $\sqrt{q}$ to correspond to the mean polarization of the individual spins which in turn dictates the breadth of the internal field distribution.

In the muon precession experiment, as in a $^{63}$Cu NMR experiment, the Larmor precession frequency is influenced by both the direct magnetic dipolar fields from the magnetic impurities as well as the contact field from conduction electrons polarized by RKKY interactions. These fields are expressed as:

$$H^{(\mu)}_{\text{dipolar}} = \frac{g\mu_B\langle S_i\rangle}{r^3}\,[3\cos^2\theta - 1]\,, \tag{1}$$

$$H^{(\mu)}_{\text{RKKY}} = \frac{g\mu_B J_{sd}\langle S_i\rangle}{2E_F r^3}\,[\cos(2k_F r)]\,, \tag{2}$$

where Yosida's approximation has been used to represent the RKKY field, and $J_{sd}$ is the exchange energy at zero wave-vector.[17] The dipolar field tends to be stronger by a factor $2E_F/J_{sd}$, which for CuMn is about 10.[18] The situation is analogous to the static inhomogeneous line broadening seen in host NMR,[3,18,19] although for NMR in $^{63}$Cu the RKKY field dominates. Using NMR terminology, the muon depolarization time, $T_2^*$, is equivalent to an effective linewidth, defined by

$$\Delta = (\gamma_\mu T_2^*)^{-1}, \tag{3}$$

where $\gamma_\mu$ is the muon's gyromagnetic ratio, $8.5\times10^4$/Gs.

Samples for these experiments were prepared by quenching the melt and cold rolling, followed by an anneal and a second quench. Two specimens, $CuMn_{0.7\%}$ of 40 cm$^2$ area ×1 cm thick and $AuFe_{1.5\%}$ 30 cm$^2$ × 1 cm, were studied in detail. Ordering temperatures of 7.7K and 11.6K, respectively, were determined by low field ac susceptibility measurements on several small pieces cut from each sample: Variations of several tenths of a degree in the position of the susceptibility cusp were noted. Samples were clamped to the cold finger of a helium cryostat. Aluminum foil windows in the thermal radiation shield and the aluminum vacuum jacket had a combined thickness of 0.48 g/cm . The magnetic field was applied transverse to the beam direction, which is also the direction of the muon's initial polarization.

In a local field of arbitrary direction, the muon's spin precesses in a cone coaxial with the local field. After a brief residence time t the decay positron is emitted preferentially along the muon's spin, and the positron detection rate is proportional to the cosine of the angle between the spin and the detector direction. Positron detectors were positioned at 0° and 90° relative to the beam, and data were simultaneously collected with 0.5 and 4 ns per channel resolutions using time to amplitude converters and 2048 channel ADCs. The muon stopping rate was $7\times10^3$sec$^{-1}$. Raw data of positron counts as a function of time show asymmetry oscillations characteristic of the muon's precession frequencies. Data are analyzed in accordance with the following formula:

$$N(t) = N_0 e^{-t/\tau}[1+A(t)] + \text{background}\,, \tag{4}$$

i.e., the total number of events N obeys the radioactive decay law modulated by an angular asymmetry represented by the term A(t). The accidental background amounted to about 15%. A(t), extracted from the data by a linear fit to Eq. (4), represents an amplitude averaged over the specimen. Results for the 0° detector are shown in Fig. 1 for $CuMn_{0.7\%}$ at two temperatures straddling $T_0$. The signal appears as a damped oscillation due to dephasing in an inhomogeneous field, an effect referred to as "slow" depolarization.

Non-linear least squares fits to the A(t) results were made, in which the precession frequency, asymmetry decay time ($T_2^*$), asymmetry amplitude and phase were fitted parameters. Typical fits for an exponential decay are also shown in Fig. 1. Reduced chi-squares averaged 1.2: Fits to a gaussian depolarization produced values about 5% higher. The

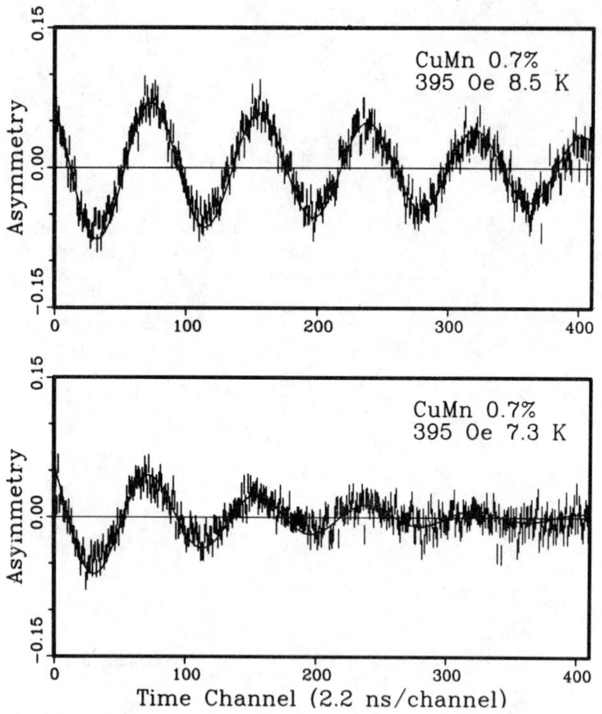


Fig. 1 The asymmetry in the positron detection signal (denoted by $A(t)$ in Eq. 4) is plotted against the muon's residence time in the $CuMn_{0.7\%}$ sample. The transition temperature $T_o$ is about 7.7K. The solid line is the result of a non-linear least squares fit to the data for exponentially damped cosine function. Error bars denote the standard statistical error of the data. The samples were field-cooled in 395 Oe from 20K for the measuring temperatures indicated.

chi-squares ratio of 1.05 has a 20% chance of being random. Therfore, the exponential depolarization time was used in Eq. (3) to calculate experimental values for Δ.

Most of the data were taken by cooling in the measuring field, such as the example of Fig. 1. Values of Δ obtained from fits to data collected by the two positron detectors were combined, and the results are shown in Figs. 2 and 3 where $\Delta/T_o$ is plotted against $T/T_o$ for several applied fields. To within an experimental uncertainty of about 10G there are no shifts in the precession frequencies in the low temperature phase. It is expected from magnetization measurements on similar specimens that Lorentz and depolarization fields be small.[20]

Additional experiments were performed, testing the magnetic remanence effect[20,21] of cooling in a strong (2500 Oe) field and taking precession data in a weaker (275 Oe) field. Specimens of

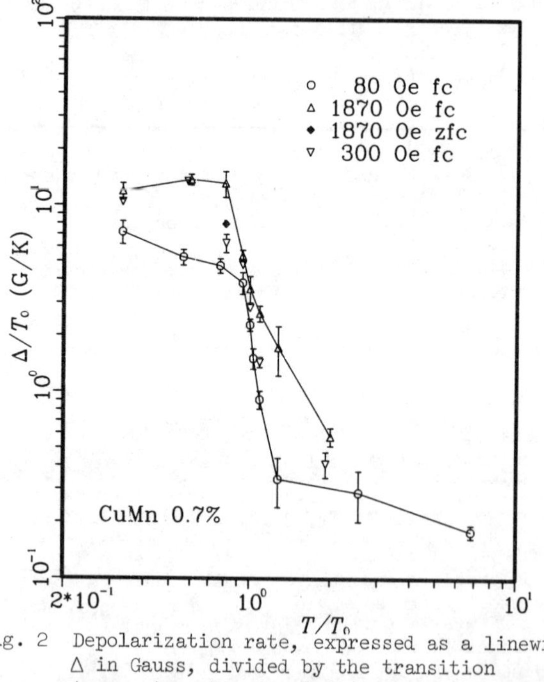


Fig. 2 Depolarization rate, expressed as a linewidth Δ in Gauss, divided by the transition temperature $T_o$, is plotted against reduced sample temperature $T/T_o$ for several values of the applied magnetic field. $T_o$=7.7K. fc corresponds to field cooling, zfc to zero field cooling (one data point is shown). Error bars are combined statistical standard error and standard error of the mean of several measurements. The solid lines connect points taken under similar conditions. Note log-log scales.

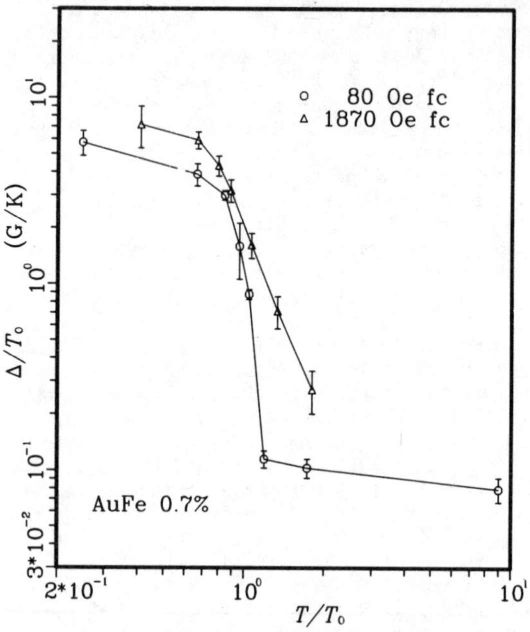


Fig. 3 Reduced temperature dependence of $\Delta/T_o$ for $AuFe_{1.5\%}$ where $T_o$=11.6K, at two fixed fields.

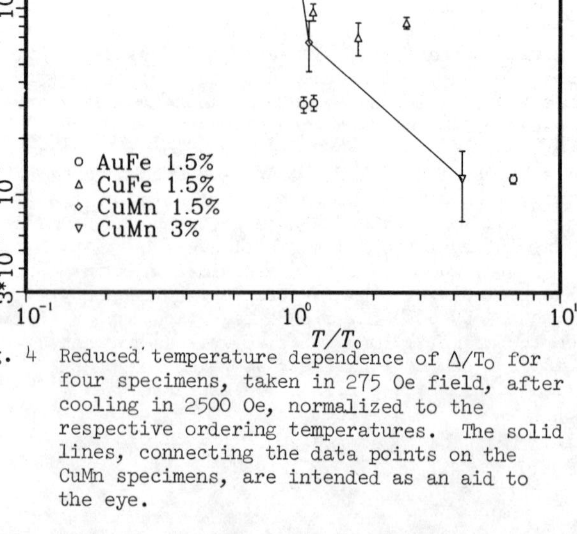

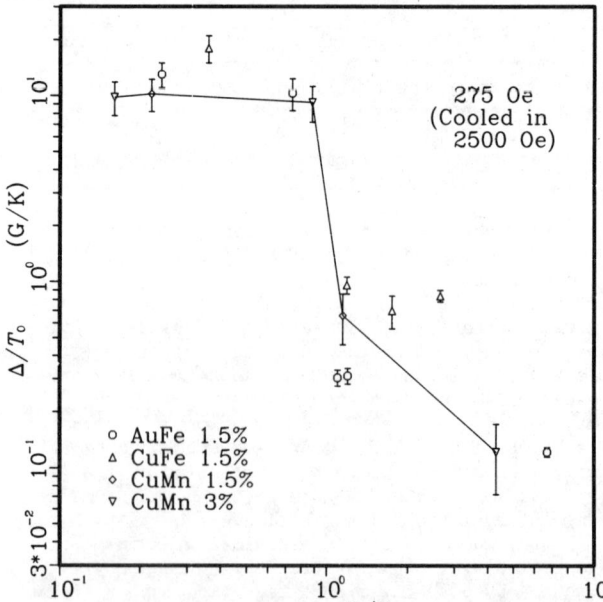

Fig. 4 Reduced temperature dependence of $\Delta/T_o$ for four specimens, taken in 275 Oe field, after cooling in 2500 Oe, normalized to the respective ordering temperatures. The solid lines, connecting the data points on the CuMn specimens, are intended as an aid to the eye.

$CuMn_{1.5\%}$ and $CuMn_{3.0\%}$, $CuFe_{1.5\%}$ and $AuFe_{1.5\%}$ were surveyed. A large $\Delta$ was observed for all data taken below the $T_0$'s, and a small $\Delta$ above. The results for this second set of experiments are combined in a reduced plot shown in Fig. 4. Further data, taken in zero measuring field, showed no precession oscillations. A Fourier transform spectrum was studied in each case. In the absence of an external field, an isotropic internal field distribution of unique magnitude would result in oscillations in the 0° detector with an amplitude reduced by a factor of 2/3. Maximum apparent polarization, or A(0), is expected for uniaxial transverse internal fields, with muons precessing in a plane.

Linewidth calculations for the NMR case have been made by Walstedt and Walker for the fields of dipolar and RKKY sources.[19] Applying their results for a random frozen spin configuration, the dipolar and RKKY contributions to the muon linewidth in $CuMn_{0.7\%}$, considered separately, are found to be 84G and 6.4G, respectively. The calculated dipolar lineshape is very close to a Lorentzian. For $AuFe_{1.5\%}$ the calculated dipolar linewidth is 71G. These estimates should be insensitive to the details of the ordering in the spin glass phase,[22] and should scale with the average value of the local moment. These estimates for complete spin freezing agree fairly well with a zero temperature extrapolation of the measured values of $\Delta$.

The following conclusions have been drawn from the experimental data:

(a) At the ordering temperature, there is a rather abrupt appearance of local fields that cause muon depolarization. The steep slope for $\Delta$ near $T_c$ at low fields suggests agreement with the mean field theory of freezing.[7] (There is a background contribution to $\Delta$ from nuclear dipolar fields, which in Cu is on the order of 1G[23]). It should be noted that if the local moments were simply non-interacting paramagnets, their dipolar fields would be negligible and could not explain the substantial increase at 1890 Oe observed in $\Delta$ for $T > T_0$.

The data taken in weak fields do not show a gradual onset of local ordering, predicted by the random molecular field theories.[16] A small amount of local ordering for $T > T_0$ would lead to appreciable dipolar broadening in $\Delta$. However, fluctuations on the time scale of local moment precession frequencies would be averaged out in our experiments: i.e., fluctuation rates greater than $(T_2^*)^{-1}$.

(b) The smearing of the transition in an applied field is due to the creation of larger local fields, both above and below $T_0$. Possibly, this is why resonance experiments done in a strong field fail to see a sharp transition point.[3,4] Field cooling produces stronger fields than zero-field cooling, Fig. 2. Previous observations of a field effect on the magnetic susceptibility[5] and on the remanent magnetization[20,21] are associated with induced spin freezing. The induced effects of strong field cooling persist only for temperatures below $T_0$.

(c) The results are consistent with a random selection of interstitial sites by the muons.

(d) The local field distribution contains no strong peaks or singularities. Although such an effect might be expected for a static conduction electron spin density wave, a picture proposed by Overhauser in an early theory on this subject,[15] much larger dipolar fields rule out their possible observation in this experiment.

The apparent observation of a well defined Mossbauer hyperfine field has been attributed to a saturation effect,[24] and therefore does not contradict the muon and NMR observations of a broad internal field. The data do not show evidence of the existence of a finite probability density of zero molecular field as had been proposed.[14,16,25] If a substantial fraction of the magnetic impurities experience weak fields, in turn producing weak dipolar fields, the observed linewidths would be much smaller. On the other hand the muon technique does not appear to be particularly sensitive in distinguishing between the possible presence of clusters of short-range correlations and purely random orientations of the frozen spins.

Recent NMR data of MacLaughlin and Alloul,[3] reported to this conference, seem to indicate that about half of the $^{63}Cu$ nuclei experience relatively weak RKKY fields, concomittant with dynamical effects, in the ordered phase. The disagreement with the present data owes no easy resolution, although their experiments are probing the RKKY field (though on a similar time scale) in specimens of lower impurity concentration, and in relatively higher applied fields. The absence of structure at $T_0$ in NMR and specific heat[26] remain unexplained by present theory.

The experiments were done in collaboration with W. J. Kossler and D. E. Murnick. The valuable assistance of L. V. Medford, R. P. Minnich, M. F. Robbins and W. Vulcan is also acknowledged. Conversations with P. W. Anderson, P. Hohenberg, M. W. Klein, L. R. Walker, R. E. Walstedt, H. Alloul and D. E. MacLaughlin have been fruitful. The experimental data were taken at N.A.S.A.'s Space Radiation Effects Laboratory, Newport News, Virginia.

REFERENCES

1. W. J. Kossler, Magnetism and Magnetic Materials - 1974, AIP Conference Proceedings No. 24 (1975).
2. C. E. Violet and R.J. Borg, Phys. Rev. 149, 540 (1966). W. Marshall, T. E. Cranshaw, C. E. Johnson and M. S. Ridout, Rev. Mod. Phys. 36, 399 (1964); U. Gonser, R. W. Grant, C. J. Meechan, A. H. Muir, and H. Wiederisch, J. Appl. Phys. 36, 2124 (1965).
3. D. E. MacLaughlin and H. Alloul, to be published.
4. J. Owen, M. E. Browne, V. Arp, and A. F. Kip, J. Phys. Chem. Solids 2, 85 (1957); D. Griffiths, Proc. Roy Soc. 90, 707 (1967); K. Okuda and M. Date, J. Phys. Soc. Japan 27, 839 (1969).
5. V. Cannella, J. A. Mydosh, and J. I. Budnick, J. Appl. Phys. 42, 1689 (1971); V. Cannella and J. A. Mydosh, Phys. Rev. B 6, 4220 (1972); V. Cannella, Amorphous Magnetism, edited by H. O. Hooper and A. M. de Graaf (Plenum, New York, 1973), p. 195.
6. K. Adkins and N. Rivier, J. Phys. 35, C4-237 (1974).
7. S. F. Edwards and P. W. Anderson, J. Phys. F 5, 965 (1975).
8. D. Sherrington and B. W. Southern, J. Phys. F 5, L49 (1975).
9. D. Sherrington, J. Phys. C 8, L208 (1975).
10. K. H. Fischer, Phys. Rev. Letters 34, 1438 (1975).
11. D. Sherrington and S. Kirkpatrick, Phys. Rev. Letters 35, 1792 (1975).
12. D. A. Smith, J. Phys. F. 5, 2148 (1975).
13. A. Blandin and J. Friedel, J. Phys. Radium 20, 160 (1959).
14. W. Marshall, Phys. Rev. 118, 1519 (1960).
15. A. W. Overhauser, J. Phys. Chem. Solids 13, 71 (1960).
16. M. W. Klein and R. Brout, Phys. Rev. 132, 2412 (1963); M. W. Klein, Low Temperature Physics-LT13, edited by K. D. Timmerhaus, W. J. O'Sullivan and E. F. Hammel (Plenum, New York, 1974), p. 397.
17. K. Yosida, Phys. Rev. 106, 893 (1957).
18. $J_{sd}$=1.3eV, quoted from satellite NMR work of N. Karnezos and J. A. Gardner, Phys. Rev. B 9, 3106 (1974).
19. R. E. Walstedt and L. R. Walker, Phys. Rev. B 9, 4857 (1974).
20. J. Soultie and R. Tournier, J. Low Temp. Phys. 1, 95 (1969).
21. J. S. Kouvel, J. Phys. Chem. Solids 21, 57 (1961); O. S. Lutes and J. L. Schmit, Phys. Rev. 125, 433 (1962); 134, A676 (1964).

22. A computer simulation of spin glass ordering was performed by R. E. Walstedt and L. R. Walker - private communication. No difference in the field distribution at the unoccupied sites between the ordered and random spin configuration was found.
23. I. I. Gurevich, E. A. Mel'eshko, I. A. Muratova, B. A. Nikol'sky, V. S. Roganov, V. I. Selivanov, and B. V. Sokolov, Phys. Lett. 40A, 143 (1972).
24. M. W. Klein, Phys. Rev. 136, A1156 (1964).
25. C. Held and M. W. Klein, Phys. Rev. Letters 35, 1783 (1975).
26. L. E. Wenger and P. H. Keesom, Phys. Rev. B11, 3497 (1975), and to be published.

## THEORY OF SYSTEMS WITH RANDOM BOND ISING INTERACTIONS IN THE PARAMAGNETIC PHASE

R. A. Tahir-Kheli*† and R. J. Elliott,
Department of Theoretical Physics,
University of Oxford, Oxford OX1 3PQ, England

A theory giving the spin correlation functions, susceptibilities, and transition temperatures of systems containing random distributions of bonds with varying Ising exchange interactions is developed by generalising an idea due to Lines[(1)]. The coherent potential approximation is used in terms of two parameters which depend on temperature but not on frequency. Detailed results are presented for the b.c.c. lattice with arbitrary bond dilution and with a gaussian distribution of bond strengths. For the quenched dilute bond case the critical concentrations below which magnetic long range order disappears are found to be 22.34%, 16.03% and 10.47% for s.c., b.c.c., and f.c.c. respectively. For the Gaussian distribution it is found that the physics of the problem depends largely on the parameter $\varepsilon^2$ which specifies the mean square deviation of the exchange (from its average value which we take to be unity). We find that the paramagnetic state goes monotonically into the ferromagnetic state (this transition is specified by divergent susceptibility, i.e. $\chi^{-1} = 0$) as long as $(\varepsilon^2)_{max} = \varepsilon_o^2 \leq 1$. However, for $\varepsilon_o^2 > 1$, the paramagnetic state does not change into ferromagnetic state and $\chi^{-1} = 0$ does not provide a description of the new phase. We conjecture that this phase may be of the spin glass variety. This conjecture helps us extend the phase diagram between paramagnetic and ferromagnetic phases to the interface (i.e. $\varepsilon_o^2 > 1$) between paramagnetic and spin glass phases.

*Supported in part by the S.R.C. and the N.S.F.
†Permanent address: Department of Physics, Temple University, Philadelphia, Pennsylvania 19122

## MÖSSBAUER AND MAGNETIZATION MEASUREMENTS IN FeCr SOLID SOLUTIONS*

B. D. Dunlap and A. T. Aldred
Argonne National Laboratory, Argonne, Illinois 60439

R. J. Nemanich and C. W. Kimball
Northern Illinois University, DeKalb, Illinois 60115

ABSTRACT

Mössbauer effect and magnetization measurements have been obtained for Fe-Cr alloys having Cr concentrations c = 0.20, 0.25 and 0.30. The magnetically split Mössbauer spectra have been analyzed by the procedure of Window[1] to give the probability distribution of hyperfine fields P(H) as a function of temperature and concentration. This distribution is closely approximated by two Gaussians: one centered near H = 0 and the other at a value of $H_1$ which is concentration and temperature dependent. The halfwidths ΔH of these Gaussians are independent of temperature and concentration. Both $H_1$ and the zero field magnetization are proportional to $(T - T_c)^{1/2}$ with $T_c = 160 \pm 3$ and $255 \pm 3$ K for c = 0.25 and 0.30, respectively. For c = 0.20, $H_1$ goes to zero at $T_c = 60 \pm 3$ K. The magnetization for this sample is strongly field-dependent and the spontaneous magnetization difficult to obtain. These results are in general agreement with a suggestion of Shull and Beck[2] that the Fe-Cr solid solutions are mictomagnets for $0.09 \leq c \leq 0.23$ and are ferromagnetic for higher concentrations.

*Based on work performed under the auspices of the U.S. Energy Research and Development Administration.
[1]B. Window, J. Phys. E: Sci. Inst. 4, 401 (1971).
[2]R. D. Shull and P. A. Beck, AIP Conf. Proc. 24, 95 (1975).

# A CALORIMETRIC INVESTIGATION OF THE SPIN GLASS: CuMn*

L. E. Wenger and P. H. Keesom
Purdue University, West Lafayette, Indiana 47907

## ABSTRACT

In a continuing effort to further understand the magnetic ordering found in spin glasses, the specific heats of polycrystalline samples of $Cu_{.988}Mn_{.012}$ and $Cu_{.976}Mn_{.024}$ were measured in the temperature range 2-40 K. The low-field susceptibility measurements on the same samples showed the characteristic sharp cusp-like peaks at 11.1 and 19 K, respectively. However, there is no indication of a discontinuity nor of a co-operative-type peak in the magnetic specific heats at these ordering temperatures. Instead, the magnetic specific heats increase smoothly with the temperature and display rounded maxima centered at 14 and 26 K, respectively. The magnetic entropy changes from absolute zero to the ordering temperature are only 0.33 of $cR \ln (2J+1)$ and to 30 K are approximately 0.66 of $cR \ln (2J+1)$. These results on CuMn are in agreement with a previous calorimetric investigation of AuFe and appear to be in disagreement with the most recent theories on spin glasses used to explain the susceptibility peak.

## INTRODUCTION

These measurements of the low-temperature specific heat of the spin glass alloy: CuMn, were initiated as a result of a prior low-temperature calorimetric investigation on AuFe alloys.[1] The cusplike peak seen in the susceptibility suggests a magnetic ordering at a well-defined temperature $T_0$. By extending the temperature range of the calorimetric measurements above the ordering temperature, it was hoped that these heat capacity data would help in characterizing the magnetic ordering in CuMn alloys.

The completely reversible, low-field susceptibility results on CuMn[2,3] show sharp, cusplike peaks at $T_0$. A magnetic field of a hundred oersteds destroys the sharp peaks and produces broad maxima. The susceptibility work of Mukhopadhyay and Beck[3] between 200 and 300 K shows a Curie-Weiss dependence with an effective magnetic moment $p_{eff}$ that approaches the single impurity limit. At temperatures below 200 K, a gradual diminishing of the $1/\chi$ vs T slope is found which apparently indicates an increase in $p_{eff}$. The susceptibility again follows a Curie-Weiss law as $T \to T_0$ from above but with a different Curie temperature. Below $T_0$, the susceptibility remains finite even at temperatures approaching absolute zero.

From magnetization data,[3] the cluster concentration and average magnetic moment are estimated by fitting the data to a Brillouin function for temperatures above $T_0$. The magnetization measurements for temperatures below $T_0$ depend very strongly on the external field present while cooling the alloy. Hysteresis losses and isothermal remanent magnetization are then found.

An early analysis of neutron scattering[4] pointed to the absence of long-range magnetic ordering in CuMn. More recent neutron diffraction measurements[5] in the temperature range 77 to 400 K, detected the existence of short range order over approximately three fcc unit cells. For Mn concentrations of 15 to 25 at.%, the short range coupling induces clustering and giant ferromagnetic moments. This ferromagnetic clustering is consistent with the mictomagnetic interpretation[6] of the susceptibility and magnetization results. When the Mn concentration is reduced, the tendency towards complete randomness is increased, and then the statistical fluctuations in the Mn density become responsible for cluster formation rather than the short range order.

The first low-temperature specific heat measurements on magnetic alloys were done on CuMn ($.004 \le c \le .10$)[7] between 1.5 and 5 K and indicate a magnetic contribution which is linear temperature dependent and essentially concentration independent. Later specific heat measurements[8] on dilute CuMn alloys ($.0015 \le c \le .0115$) were expanded to a temperature range of 0.3 to 10 K. The magnetic specific heat of the very dilute alloys increases linearly with the temperature rising to a rounded maximum and then gradually falls off at higher temperatures. Until now, no calorimetric data were available for the higher Mn concentrations over a temperature range that included the temperature $T_0$ and on samples which demonstrated the susceptibility cusp.

## EXPERIMENTAL

The present calorimetric and susceptibility investigation studied two polycrystalline samples, $Cu_{.988}Mn_{.012}$ and $Cu_{.976}Mn_{.024}$, in the temperature range 2-40 K. In order to ensure a fairly uniform distribution of Mn ions, both samples were annealed in a vacuum for 120 hours at 1000°C and then allowed to cool slowly to room temperature. The susceptibility was measured by an ac technique (~ 17 Hz) in a field of less than one oersted. The heat capacity measurements were made by using a standard heat-pulse technique and a $^3He$ cryostat.[1] The experimental error associated with these measurements is estimated to be approximately one percent below 20 K and increasing to a maximum of 2% at higher temperatures.

The difference between the measured specific heat and the specific heat of the same alloy as if the impurity atoms were nonmagnetic is defined as the magnetic specific heat. The nonmagnetic specific heat can only be estimated from the specific heat of pure copper. An appreciable change in the lattice specific heat is anticipated when the atomic mass of the host matrix differs substantially from the transition-metal solute. For the CuMn alloys, this change is expected to be small. Their elastic constants[9] exhibit a small linear decrease with manganese concentration while the temperature dependences are similar to those of pure copper. The variation in the Debye temperature $\Theta_0$ with concentration was calculated from these elastic moduli and is approximately -0.6°K/at. percent Mn. The nonmagnetic specific heat for the CuMn alloys was then estimated by utilizing the equation given by Martin[10] for copper with the appropriate change in the cubic term due to the Mn concentration. This estimation may still leave the nonmagnetic specific heat with substantial error above 15 K as it completely neglects the influence of the Mn impurities on the shape of the phonon dispersion curves.

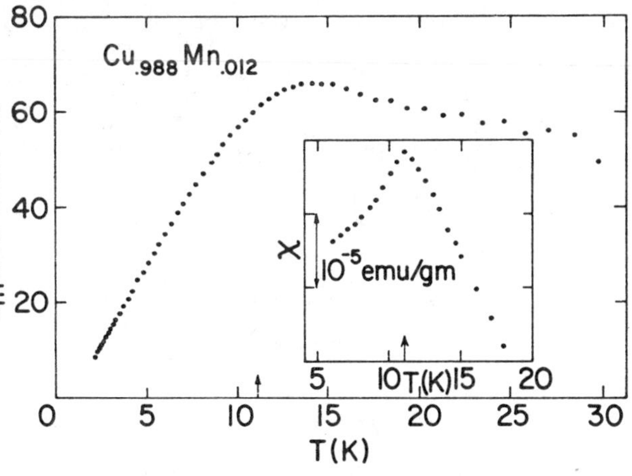


Figure 1. The magnetic specific heat and susceptibility of $Cu_{.988}Mn_{.012}$.

## RESULTS

The low-field susceptibility results for the $Cu_{.988}Mn_{.012}$ and $Cu_{.976}Mn_{.024}$ samples are displayed in the inserts of Figs. 1 and 2. The sharp, cusplike peaks occur at 11.1 and 19 K, respectively, and are similar to prior susceptibility results.

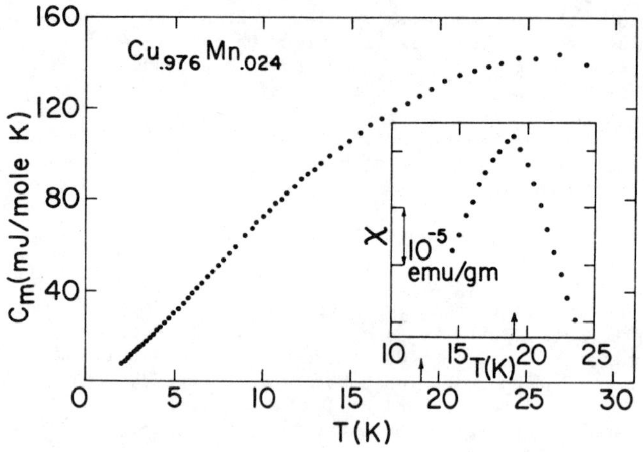


Figure 2. The magnetic specific heat and susceptibility of $Cu_{.976}Mn_{.024}$.

In Figs. 1 and 2, the magnetic specific heats $C_m$ of $Cu_{.988}Mn_{.012}$ and $Cu_{.976}Mn_{.024}$ are displayed as a function of the temperature. The low temperature magnetic specific heats show an approximate linear temperature dependence and begin to flatten out at higher temperatures. Subsequently, the magnetic specific heats display broad, rounded maxima centered at 14 and 26 K, respectively. At the ordering temperatures as indicated by the arrows, $C_m$ shows no evidence of any peak, discontinuity, or anomaly that could be associated with their susceptibility peaks. Taking into account the scatter of the data, an anomaly larger than 1½% of the magnetic specific heat for the CuMn alloys should have been readily observable at their respective ordering temperatures. Because of the large scatter in $C_m$, the magnetic specific heats are not shown to the highest measured temperatures as the nonmagnetic specific heat becomes a large fraction of the total measured specific heat at these high temperatures. In addition, the increased uncertainty in the lattice specific heat contribution at the highest temperatures will lead to further errors in the magnetic specific heat.

The maximum entropy change associated with the magnetic system should be equal to $cR \ln(2J+1)$, where c is the concentration of the magnetic ions, R is the gas constant, and J is the electron spin of the ions. The values of the spin J for the various alloys are determined from the low-field susceptibility results[4] just above $T_o$. The measured increases of magnetic entropies between 0 K and $T_o$ of $Cu_{.988}Mn_{.012}$ and $Cu_{.976}Mn_{.024}$ are 57 and 123 mJ/mole·K, respectively. This increase is 0.33 of $cR \ln(2J+1)$ for each alloy. Extending to 30 K, the entropy increases are measured to be 118 and 185 mJ/mole·K, respectively (~ 0.66 $cR \ln(2J+1)$). Therefore these entropy considerations argue against the freezing of a majority of the spin system as well as against a strong Weiss field-type long-range cooperative ordering occurring at $T_o$.

## DISCUSSION

In the most recent attempts to treat the magnetic properties of the spin glasses, in particular the sharp cusplike susceptibility peak, three theoretical approaches have been developed. In principle, the magnetic specific heat can be predicted from each of these theoretical treatments and can be compared to the experimental results. The first is that of Adkins and Rivier[11] which generalizes a local molecular model by calculating P(H) using an analogue to the random walk problem. This theory assumes a short range order parameter, m(T), which orients the impurity spins in a preferred but random direction within a correlation length of a few unit cells. The temperature dependence for m(T) goes to zero at the ordering temperature in a similar fashion as the molecular field theories, i.e. like $(T_o-T)^{\frac{1}{2}}$. With this sudden disappearance of the short range order, there corresponds a cusp in the susceptibility. However, the magnetic specific heat will include a contribution that is a function of the temperature derivative of m(T). Consequently at $T_o$, the magnetic specific heat appears to approach an infinite value.

In a second approach, Edwards and Anderson[12] have considered a classical Heisenberg model with a random distribution of exchange interactions. The basis of this approach is a time dependent correlation function q of the solute spins. The quantity q is somewhat similar to an ordering parameter and leads to discontinuities in the slope of the susceptibility and the magnetic specific heat, as would be expected for a third-order phase transition. Recently, this theory has been modified for the quantum mechanical case[13] which leads to a cusp with a finite slope just below $T_o$ for both the susceptibility and for the specific heat.

The final approach of D. A. Smith[14] utilizes the competition between the RKKY interaction and the thermal disorder energy. The susceptibility shows a cusp at $T_o$ where an infinite cluster first forms and freezes in. A small number of loose spins in the infinite cluster are thought to be responsible for most of the low temperature magnetic specific heat and susceptibility properties, i.e. $C_m \propto T$ and a finite $\chi$ as $T \rightarrow 0$. However, to agree quantitatively with the experimental results, a majority of the impurity spins are required to be loose spins. In addition, this approach lacks a quantitative prediction of $C_m$ at $T_o$.

All of these approaches appear not to agree with the experimental specific heat results and consequently no satisfactory theory exists which explains the alloy problem.

## REFERENCES

* Work supported by NSF Grant No. DMR71-01821 A01 and by NSF-MRL Grant No. DMR72-03018 A04.
1. L. E. Wenger and P. H. Keesom, Phys. Rev. B11, 3497 (1975).
2. V. Cannella, in Amorphous Magnetism, edited by H. O. Hooper and A. M. de Graaf (Plenum, New York, 1972), p. 195.
3. A. Mukhopadhyay and P. A. Beck, Solid State Comm. 16, 1067 (1975).
4. A. Arrott, J. Appl. Phys. 36, 1093 (1965).
5. H. Sato, S. A. Werner, and R. Kikuchi, J. Phys. (Paris) 35-C4, 23 (1974).
6. P. A. Beck, Met. Trans. 2, 2015 (1971).
7. J. E. Zimmerman and F. E. Hoare, J. Phys. Chem. Solids 17, 52 (1960); L. T. Crane and J. E. Zimmerman, J. Phys. Chem. Solids 21, 310 (1961).
8. F. J. du Chatenier and J. de Nobel, Physica 28, 181 (1962).
9. D. L. Waldorf, J. Phys. Chem. Solids 16, 90 (1960).
10. D. L. Martin, Phys. Rev. B8, 5357 (1974).
11. K. Adkins and N. Rivier, J. Phys. (Paris) 35-C4, 237 (1974); J. Phys. F: Metal Phys. 5, 1745 (1975).
12. S. F. Edwards and P. W. Anderson, J. Phys. F: Metal Phys. 5, 965 (1975).
13. K. H. Fischer, Phys. Rev. Letters 34, 1438 (1975); D. Sherrington and B. W. Southern, J. Phys. F: Metal Phys. 5, L49 (1975).
14. D. A. Smith, J. Phys. F: Metal Phys. 4, L266 (1974).

# MAGNETIC ORDERING OF $Au_{0.92}Fe_{0.08}$: AN ULTRASONIC INVESTIGATION

G. F. Hawkins, T. J. Moran,* and R.L. Thomas
Wayne State University, Detroit, Michigan 48202

## ABSTRACT

The cusplike peak seen in the low field susceptibility of $Au_{0.92}Fe_{0.08}$ is suggestive of magnetic ordering at 29 K. Recent measurements[1] of the specific heat and entropy on a single crystal between 3 and 50 K do not, however, show a cooperative-type peak. In an attempt to further understand this discrepancy we have made high resolution measurements of the longitudinal wave sound velocity, which is expected to show a critical anomaly at a magnetic phase transition. Over a frequency range from 5 to 55 MHz and between 4 and 150 K no such anomaly has been observed. These results appear to rule out long range ordering in this alloy. A satisfactory explanation of these phenomena has not yet been made.

## INTRODUCTION

The nature of the magnetic structure of AuFe alloys in the concentration range $1 < c < 12$ at.% Fe has been an unresolved problem of considerable interest for several years. Measurements of the low-field ac magnetic susceptibility[2] show sharp peaks at temperatures which scale systematically with c, suggestive of the onset of some kind of magnetic ordering. However, recent low temperature heat capacity data[1] for these alloys do not show cooperative-type behavior. The low temperature heat capacity data can be represented over a wide temperature range by an expression of the form $C = aT + bT^3$, where the coefficient of the linear term is anomalously large, in agreement with the results of previous experiments[3] at lower temperatures.

During the past two years several molecular field theories[4-8] have been proposed to describe the magnetic behavior of such alloys. All these theories predict either a peak or a discontinuity in the heat capacity accompanying the predicted sharp peak in the susceptibility, in disagreement with experimental evidence[1] for AuFe. In addition, it has been known for some time that a spin density wave state[9] could account for the susceptibility peak. A spin density wave state, however, also implies a cusp in the heat capacity.

In view of the discrepancy between current theories for the susceptibility peak and the negative result of the heat capacity experiment we have undertaken the present ultrasonic investigation to test furhter the nature of the magnetic ordering of this alloy. It can be shown from general thermodynamic arguments[10] that if an anomaly is present in the heat capacity at the susceptibility peak temperature, there must also be an anomaly in the temperature dependent sound propagation. Such ultrasonic measurements near magnetic phase transitions[10] have been fruitful for a variety of solids. In particular, measurements of sound velocity can be carried out with very high resolution using phase comparison techniques. For example, large anomalies are observed in the sound velocity for temperatures close to cooperative phase transitions, and from such data one can often extract values for critical experiments.[10] Thus on the basis of existing theories one would expect to observe an anomaly in the temperature dependence of the sound velocity near the susceptibility peak temperature.

## EXPERIMENTAL

The $Au_{0.92}Fe_{0.08}$ single crystal[11] was cut from the same boule as the heat capacity[1] and susceptibility[11] samples. Faces were spark planed parallel to (110) crystallographic planes, yielding an ultrasonic sample 0.80 cm long. Velocity measurements were made in the range 5-55 MHz using a phase comparison technique[10] capable of detecting transit time changes of a part in $10^7$. Commercially calibrated Ge and Pt thermometers were used for the ranges $1.5 \text{ K} < T < 20$ K, and $20 \text{ K} < T < 150$ K, respectively. For purposes of comparison, similar measurements were made for a pure gold single crystal.

Figure 1 shows the temperature dependence of the longitudinal wave sound velocity at 30 MHz in Au and $Au_{0.92}Fe_{0.08}$, for $20 \text{ K} < T < 80$ K, together with that of the susceptibility[11] for the same $Au_{0.92}Fe_{0.08}$ crystal. The measured sound velocity in the alloy at 77 K is $3.46 \times 10^5$ cm/sec. No evidence of cooperative-type behavior is seen. These data are representative of the behavior observed for the entire range of frequencies. Measurements of the shear wave sound velocity were similarly smooth over this temperature range. These data, together with the heat capacity results,[1] appear to rule out long range ordering in this alloy.

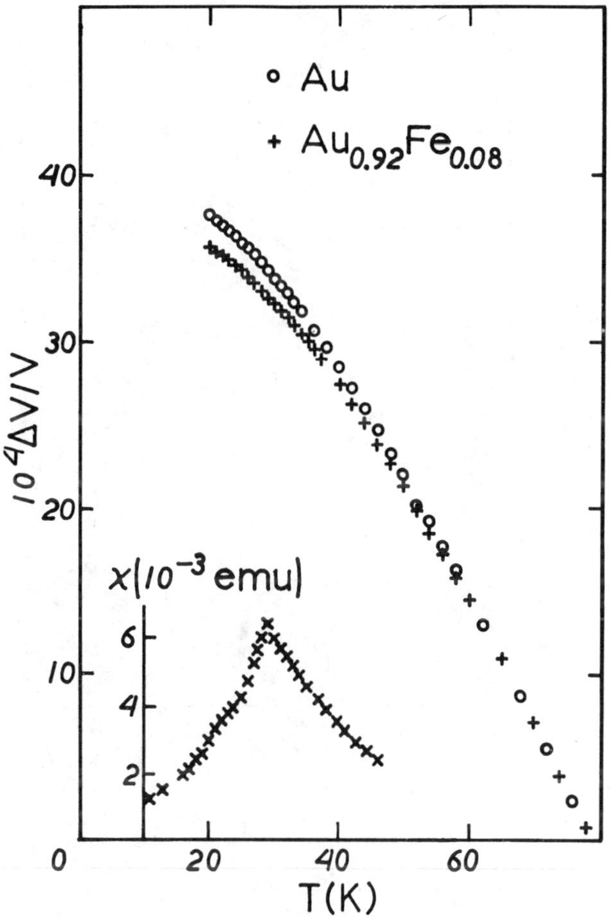


Fig. 1. Temperature dependence of the 30 MHz longitudinal wave sound velocity for propagation along [110] for $Au_{0.92}Fe_{0.08}$ and Au, together with that of the magnetic susceptibility for $Au_{0.92}Fe_{0.08}$.

## DISCUSSION

A thermodynamic description of the low temperature behavior of an elastic modulus of a metal has been given by Alers.[12] In the case of pure Cu, for example, he finds a predicted bulk modulus B, of the form

$B = B_0(1 - 9.7 \times 10^{-10}\, T^2 - 5.6 \times 10^{-11}\, T^4)$. Due to the possible effects of dislocation damping, comparison of this expression to experiment was made by Alers[12] for $Cu_{0.9}Al_{0.1}$, confirming the qualitative correctness of the description for the non-magnetic alloy. We have performed a least squares fit of the present data to expressions of the form $\Delta v/v = A_1 + A_2T^2 + A_3T^4$, and $\Delta v/v = A_1 + A_3T^4$ (since the coefficient $A_2$ was found[12] to be negligible at these temperatures for the non-magnetic alloy). Since there was a slight mismatch of the Ge and Pt thermometers near the ends of their ranges (∿20 K), we have smoothed the data in this region prior to performing the polynomial fits. The results are shown in Fig. 2, and it will be noted that the coefficient of the $T^2$ term is about two orders of magnitude larger than might be expected from a non-magnetic alloy. This result is also consistent with the observation[1,3] of an anomalously large linear term in the heat capacity. In view of the fact that neither the present ultrasonic experiment nor the previous heat capacity experiment[1] has yielded an anomaly near the susceptibility peak temperature, the current theory of the magnetic behavior of this alloy is incomplete.

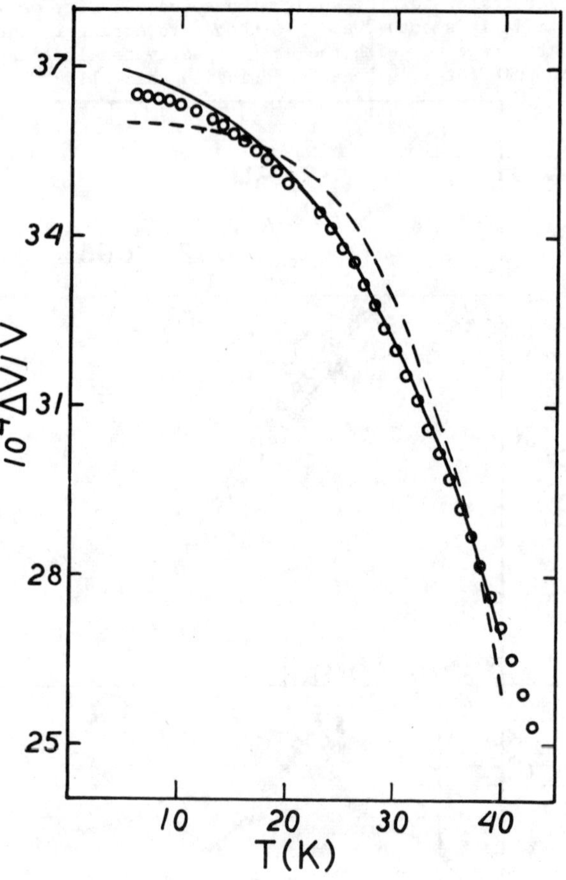


Fig. 2. Comparison of temperature dependence of the 50 MHz longitudinal wave sound velocity with expressions $\Delta v/v = 37 \times 10^{-4} - 4.3 \times 10^{-7}\ T^2 - 1.2 \times 10^{-10}\ T^4$ (solid curve), and $\Delta v/v = 36 \times 10^{-4} - 3.96 \times 10^{-10}\ T^4$ (dashed curve). The apparent small jump in the experimental points at 20 K is due to a slight mismatch in thermometers.

ACKNOWLEDGMENTS

The authors would like to thank Professor S.A. Werner for supplying the AuFe sample and the magnetic susceptibility data, and Professor A.M. de Graaf for helpful discussions.

REFERENCES

*Present address: AFML/LLP, Wright-Patterson AFB, Dayton, Ohio 45433

[1] L.E. Wenger and P.H. Keesom, Phys. Rev. B 11, 3497 (1975).

[2] V. Cannella and J.A. Mydosh, Phys. Rev. B 6, 4220 (1972).

[3] See for example J.E. Zimmerman and F.E. Hoare, J. Phys. Chem. Solids 17, 52 (1960).

[4] K.A. Adkins and N.Y. Rivier, J. Phys. Paris 5 C4, 237 (1974).

[5] B. Southern, J. Phys. C 8, L213 (1975).

[6] D. Sherrington, J. Phys. C 8, L208 (1975).

[7] D. Sherrington and B.W. Southern, J. Phys. F 5, L49 (1975).

[8] S.F. Edwards and P.W. Anderson, J. Phys. F 5, 965 (1975).

[9] A.W. Overhauser, J. Phys. Chem. Solids 13, 71 (1960).

[10] B. Lüthi, T.J. Moran, and R.J. Pollina, J. Phys. Chem. Solids 31, 1741 (1970).

[11] The sample and susceptibility data were kindly provided by S.A. Werner, U. of Missouri (Columbia).

[12] G.A. Alers in "Physical Acoustics", W.P. Mason, ed. (Academic Press, New York, 1966) Vol. IV A, p. 277).

## MICTOMAGNETISM IN Fe-Al ALLOYS*

R. D. Shull, H. Okamoto, and Paul A. Beck
University of Illinois, Urbana, Illinois, 61801

Scattered indications of mictomagnetism in Fe-Al alloys have been available in the literature[1,2,3]. A systematic study with metallurgically well defined specimens (heat treatment based on an X-ray diffraction study[4]) shows that alloys with [$Fe_3Al$]-type atomic order and with 27 to 30 at.% Al, which are ferromagnetic at room temperature, become mictomagnetic on cooling to 90K, or lower. Alloys with 30.4 to 32 at.% Al, ordered in the same way, change from paramagnetic to mictomagnetic at temperatures between 82 and 62K. For $Fe_{68}Al_{32}$ the transition temperature remains the same if the structure changes to [FeAl]-type. Alloys of the latter structure become mictomagnetic on cooling at least up to 45 at.% Al, where the transition temperature is 10K. $Fe_{52}Al_{48}$ does not become mictomagnetic above 2.2K. A more detailed account of the study will be published elsewhere.

*This work was supported by a grant from the U. S. Energy Research and Development Administration to the Materials Research Laboratory of the University of Illinois.

1. Kouvel, J. S., Appl. Phys. 30, 4, 313S-314S (1959).
2. Arrott, A., and H. Sato, Phys. Rev. 114, 6, 1420-1426 (1959).
3. Danan, H., and H. Gengnagel, J. Appl. Phys. 39, 2, 678-679 (1968).
4. Okamoto, H., and P. A. Beck, Met. Trans. 2, 569-574 (1971).

# LOW TEMPERATURE MAGNETIC PROPERTIES OF THE SPIN GLASS SYSTEM PtMn

E.F. Wassermann and J.L. Tholence
2.Physikalisches Institut der RWTH Aachen, D-5100 Aachen, Germany
and C.R.T.B.T., B.P. 166, F-38042 Grenoble, France

## ABSTRACT

The reversible susceptibility and the thermoremanent (TRM) and isothermal remanent magnetization (IRM) of three PtMn alloys ( 0.5; 1.0; 2.5at%) have been measured in the temperature range from 0.05 to 4.2K. The reversible susceptibility shows pronounced peaks at the spin glass freezing temperature $T_M \propto c$ for the two most dilute alloys. The peaks are connected to the appearance of a remanent magnetization below $T_M$, as e.g. in the AuFe system. With increasing impurity concentration a tendency towards antiferromagnetism is observed. The spin value for Mn in Pt, as taken from a Curie-Weiss law is slightly decreasing from S = 2.5 to S = 2.2 as the concentration increases. A comparison in terms of scaling law (specific heat, susceptibility, freezing temperature) shows that in PtMn the interactions between the impurity atoms are much weaker than in other spin glass systems, e.g. CuMn.

## INTRODUCTION

Since the discovery of the sharp cusp in the low field a.c. susceptibility of non dilute AuFe alloys by Cannella and Mydosh [1]) extensive work has been done on the low temperature properties of magnetic alloy systems showing this kind of anomaly. These magnetic materials were called the "Spin Glasses" or "Mictomagnets", depending on the concentration range of the impurities [2)3)]. Spin glass refers mainly to the lower concentrated alloys, where the long range indirect RKKY-interaction is responsible for the freezing of the spins, and where the scaling law for magnetization and specific heat holds [4)].

Typical systems, now already widely studied, are the noble metal matrix alloys containing the 3d transition metals Cr, Fe and Mn in concentrations below about 3 at%. The spin glass behaviour of transition metal - transition metal alloys are not so widely studied, especially those having Pt as a matrix. Within the Pt matrix systems, however, spin glass behaviour is only expected for the PtMn system [2)].

The properties of the PtMn system are relatively unknown. In the work of Nieuwenhuys et al.[5)] on the specific heat of PtMn alloys (0.36%; 0.88%; 1.64%) maxima due to magnetic ordering have been observed, the position of the maximum changing with concentration and applied field. Sacli et al.[6)] recently remeasured the specific heat of four PtMn alloys (between 0.26 and 1.35 at%), and found scaling of $T_{Max}$ with concentration for the three more dilute alloys. At low temperatures $\Delta C$ is proportional to T and concentration independent. A spin value of S = 2.5 was obtained by extrapolating the impurity specific heat at temperatures above 5K with a 1/T-dependence. Such an extrapolation is justified since a dependence $\Delta C = A/T - B/T^2$, with A proportional to $c^2$ and B proportional to $c^3$, is expected and observed in systems, where the impurity spins are coupled by an interaction $1/r^3$ [7)]. Resistivity measurements by Sarkissian and Taylor [8)] (0.09 - 18 at%Mn) also indicate that PtMn is a spin glass system. A characteristic $T^{3/2}$ dependence of $\Delta\rho$ is observed in the temperature range below the resistance minimum for alloys with impurity concentrations above 1%.

In the present work we show that PtMn is indeed a spin glass system. The results on the susceptibility behaviour of this system will clearly exhibit the occurrence of a sharp maximum at the freezing temperature $T_M$.

## RESULTS AND DISCUSSION

The magnetization of three samples (0.5%; 1.0; 2.5 at% Mn) has been measured by an extraction method in a low d.c. field (max. 3 kOe) in a temperature range between 0.04 and 4.2K. The impurity concentration has been determined with an activation method.

Fig.1 shows the reversible susceptibility $\chi_{rev}$ due to the Mn impurities versus the reduced

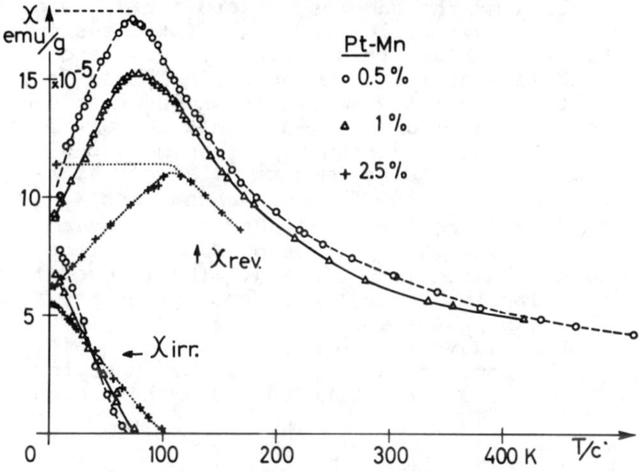


Fig.1 Impurity contribution of the susceptibility as measured in low field $\chi_{rev}$ versus the normalized temperature T/c for three PtMn alloys. Cooling of the samples in a weak field from above $T_M$ (the spin glass freezing temperature) results in the occurrence of an irreversible part of the susceptibility $\chi_{irr}$. $\chi_{total}$ (dashed lines) is the sum of both parts, $\chi_{rev}+\chi_{irr}$.

temperature T/c. A maximum is observed for all the three alloys (0.5%: $T_M$ = 0.4K; 1.0%: $T_M$ = 0.8K; 2.5%: $T_M$ = 2.5K), and scaling holds for the two more dilute samples. For the more concentrated alloy the freezing temperature rises more rapidly than the concentration. A similar nonproportional increase of $T_M$ with c has been observed by Sacli et al.[6)] for PtMn alloys above 1% Mn. Sacli and coworkers believe that this behaviour is due to the onset of ferromagnetic ordering between the Mn atoms. Our measurements, however, indicate that there is always a tendency to antiferromagnetic ordering. Moreover, as one can see in Fig.1, the absolute $\chi$ value for the 2.5% sample is even lower than in the two less concentrated ones. The effective magnetic moment as taken from the Curie-Weiss law above $T_M$ in Fig.2 (0.5%: $p_{eff}$ = 6.0; 1%: $p_{eff}$ = 5.65; 2.5%: $p_{eff}$ = 5.3) decreases with increasing concentration. This corresponds to a reduction of the spin value from S = 2.5 to S = 2.2. The paramagnetic Curie temperature

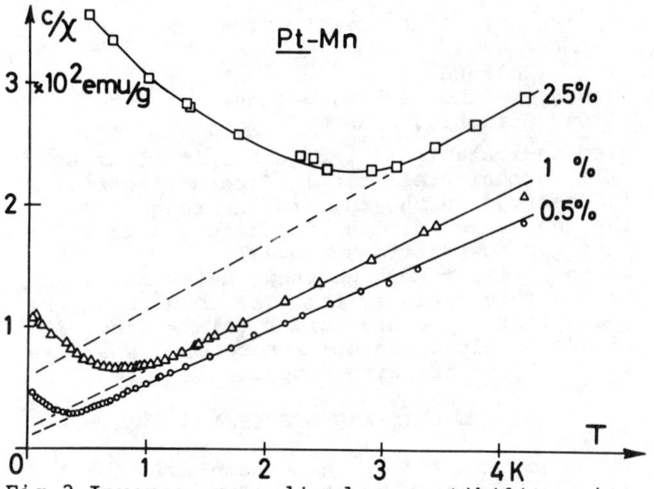


Fig.2 Inverse, normalized susceptibility c/χ versus temperature for three PtMn alloys.

is always negative. It rises from 0.2K for the 0.5% alloy to about 1K for the sample containing 2.5% Mn.
Cooling of the samples in zero field, as done for the reversible part $\chi_{rev}$ of the susceptibility in Fig.1, results in the occurrence of the peaks at $T_M$ if the measuring field is kept sufficiently small. If, however, the alloys are cooled in a weak field h from above $T_M$, the magnetization $M_{total}$ as measured in that field contains the reversible part $\chi_{rev} \cdot h$ and a second, the thermoremanent part $\chi_{irr} \cdot h$. The sum of both parts leads to a constant susceptibility $\chi_{total}$ below $T_M$ as indicated by the dashed lines in Fig.1. The irreversible contribution to the total susceptibility $\chi_{irr}$ is shown in the lower part of Fig.1. It vanishes always at the corresponding $T_M$.
The determination of $\chi_{irr}$ can be understood from Fig. 3, where the thermoremanent magne-

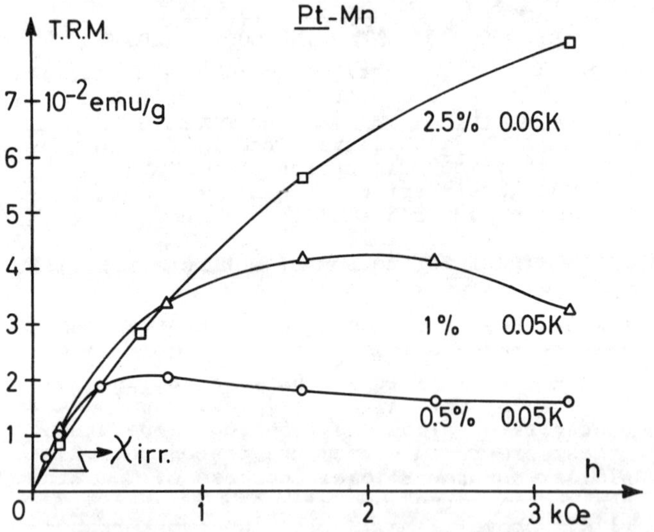


Fig.3 Thermoremanent magnetization (TRM) versus applied field h for three PtMn alloys. The field h is applied at temperatures $T > T_M$ and the samples are then cooled to 0.05K. The TRM is the remanent magnetization after removal of h.

tization TMR is plotted versus the field h. h is the field applied during cool down from $T > T_M$ to 0.05K. The TRM values shown are the magnetization which rests, after h is reduced to zero. $\chi_{irr}$ as plotted in Fig.1 is the temperature dependence of the initial slope $\Delta$(TRM)/$\Delta$h of the curves in Fig. 3.
The remanent magnetization in spin glasses can be understood in the following way. For T approaching zero, the randomly frozen in moments form magnetic domains. The resulting moment of such domain lies in one direction of the anisotropy axis of the domain. Its mean value is $M = \sqrt{n}\,\mu$, where n is the number of the frozen single moments $\mu$ within the domain [3].
From the measured value of the initial susceptibility as taken from Fig.1 and the linear term of the specific heat as taken from reference [6], one can calculate [4] a value for the spin of the Mn, S = 3.2, for T approaching zero. This value is in resonable agreement with the value S = 2.5, taken from the Curie-Weiss plot above $T_M$.
Finally, a comparison of the PtMn system with other spin glass systems containing Mn shows that the interaction between the Mn atoms in Pt are much weaker than in other systems. The absolute values of the freezing temperatures per unit concentration are an order of magnitude smaller in PtMn than in CuMn (or AuFe) [2] and about 5 times smaller than in AgMn and AuMn. At the same time the absolute values of the susceptibility for $T \to 0$, $\chi(0)$ are an order of magnitude higher in PtMn ($\chi(0) = 1.9 \cdot 10^{-2}$emu/mole) than in CuMn ($\chi(0) = 1 \cdot 10^{-3}$ emu/mole). This indicates that the density P(H=0,T=0) for the molecular field in PtMn has a high absolute value but a low half width, corresponding to a high absolute susceptibility at T = 0 and a small $T_M$. It also shows that the amplitude of the RKKY interaction at large distances is small in PtMn compared to e.g. CuMn or AuFe. Our results also demonstrate that short range interactions in PtMn have more antiferromagnetic character.

ACKNOWLEDGMENTS

The authors are very much indepted to Dr. O. Bethoux for preparation and analyzation of the alloys.

REFERENCES

1) Cannella, V. and Mydosh, J.A., Phys.Rev. B6, 4220 (1972)
2) Mydosh, J.A., Proc. 20th Annual Conf. on Magnetism and Mag. Mat., San Francisco 1974 AIP Conf. Proc. 24, Ed. H.C. Wolfe, New York 1975, p.3
3) Tholence, J.L. and Tournier, R., J. de Phys. 35, C4-229 (1974)
4) Souletie, J. and Tournier, R., J.Low Temp. Phys. 1, 95 (1969)
5) Nieuwenhuys, G.J., Pikart, M.F., Zwart, J.J. Boestoel, B.M., and Van Den Berg, G.J., Physics 69, 119 (1973)
6) Sacli, Ö.A., Emerson, D.J., and Brewer, D. F., J. Low Temp. Phys. 17, 425 (1974)
7) Chouteau, G. Thesis University of Grenoble, France 1973, unpublished.
8) Sarkissian, B.V.B. and Taylor, R.H., J. Phys. F (GB) 4, L243 (1974)

# THE HYDROGENATION OF PdCr ALLOYS

J.A. Mydosh[x]

Institut für Festkörperforschung der Kernforschungsanlage, 517 Jülich, Germany

## ABSTRACT

In contrast to the "giant moment" ferromagnetic PdFe system, PdCr alloys exhibit "weak moment" or spin fluctuation behavior ($T_{sf} \simeq 200K$). It is only for concentrations above $\simeq 7$ at.% Cr that such alloys possess sufficient impurity-impurity interactions to stabilize the moments and thereby permit a spin glass type of impurity freezing to occur at low temperatures. We report here electrical resistivity measurements between $\simeq 1.3$ and 50K for a series of PdCr alloys (0.24 to 3.5 at.% Cr) which have been charged with hydrogen (H/Metal $\simeq 0.6$) such that the exchange enhancement of the Pd matrix is destroyed. For the (PdCr)H alloys, a maximum appears in the low temperature resistivity when the Cr concentration is greater than 0.6 at.%. This marked decrease in the resistivity as the temperature is lowered indicates the onset of the moment freezing processes and takes place in the hydrogenated samples for at least a factor of 10 less Cr than in the PdCr system. The maximum progressively shifts to higher temperatures as the concentration of Cr is further increased. Our results suggest a strong similarity between (PdCr)H and (PdFe)H alloys in that both systems show typical spin glass properties at the same concentrations, and thus behave comparable to the noble metal based spin glasses.

## INTRODUCTION

We have recently found that the hydrogenation of PdFe alloys[1] causes a destruction of the giant moment ferromagnetism, and resistivity behavior similar to that observed in spin glasses[2] manifests itself. Very new low field ac susceptibility measurements[3] have confirmed the existence of the spin glass freezing for (PdFe)H and (PdCo)H. Previously, spin glass behavior was reported[4,5] in the ternary system (PdAg)Fe, and a parallel[1,5], based upon the reduction of the exchange enhanced susceptibility, was drawn between the addition of Ag and H to the Pd matrix.

We now turn our attention to the antithetical (weak moment) PdCr system. Earlier resistivity[6-9] and susceptibility[6,10] experiments can be interpreted in terms of the following model. Cr impurities, when introduced into the Pd host, locally blot out the uniform exchange enhancement, but in this process the Cr losses its low temperature magnetic moment ($T_{sf} \simeq 200K$), and thus the PdCr systems represent a favorable candidate with which to test the ideas of the localized spin fluctuation theory[11]. Only when the Cr concentration is sufficiently increased to provide a suitable Cr local environment, does a stable Cr moment appear, now however, in a non-enhanced Pd matrix, and oscillating impurity-impurity interactions can occur, leading to a spin glass type[9] of state at low temperatures. This requires a Cr concentration of at least 7 at.%. We have hydrogenated a series of PdCr alloys and find that, once the H depresses the Pd susceptibility to zero, strong and randomly interacting Cr moments determine the magnetic behavior of the (PdCr)H system, even for Cr concentration $\simeq 0.5$ at.%. Here spin glass behavior should be present, as in the Pd(Fe, Co or Mn) H alloys, and all of these systems are directly comparable to the noble metal - 3d solute spin glasses.

## EXPERIMENTAL

The PdCr alloys (0, 0.24, 0.65, 1.7, 3.5, 5.4 and 10.5 at.% Cr) were fabricated by arc melting in ultra pure argon the appropriate weights of 6-9 Pd and Marz-grade Cr. The nominal by weight concentration agreed very well with an (absorption spectroscopy) analysis which also showed the presence of other impurities to be less than $\simeq 10$ ppm. After rolling the alloys into foils 0.1-0.2 mm thick, they were strain annealed at 1000°C for 24 hours and quenched into water. Hydrogenation was carried out in a special furnace at a temperature of 400°C under a hydrogen pressure of 3100 Torr for a period of about 4 h. The samples were then slow cooled (0.25°C/min) under this hydrogen pressure, and H to metal ratios (H/M) (determined by weighing) of about 0.6 were obtained. This ratio decreased very strongly with Cr concentrations above 3.5 at.%, so that only H/M = 0.08 for Pd + 5.4 at.% Cr and H/M $\simeq$ 0 for Pd + 10.5 at.% Cr could be acquired. No difference in the resistivity behavior of these two alloys were detected when comparing our data with those of similarly concentrated hydrogen free specimens[7-9]. Rather than employing a more complicated H charging procedure for the higher concentrations, e.g. a low temperature electrolysis or a few kbars of hydrogen gas pressure, we focus in the present paper on $c \leq 3.5$ at.% Cr where H/M was always greater than 0.5. The resistivity was measured from 1.35 to 100K by using the standard V/I four-point probe method. A relative accuracy of a few parts in 100000 was attained with a combination photoamplifier, digital voltmeter, voltage compensation technique. The resistivity data above 50K is severely affected by a phase transformation (or migration) among the H interstitial sites, and this precludes an accurate separation of the magnetic resistivity for higher temperatures.

## RESULTS

In Fig. 1 we present the overall resistivity behavior of the (PdCr)H alloys. For comparison "pure" PdH is shown in curve (a). Its resistivity is relatively constant up to about 20K which means that subtracting this $\rho(c=o,T)$ from the $\rho(T)$ for the Cr alloys has no effect in the temperature dependences of $\Delta\rho(c,T)$ for $T \leq 20K$. Also, as reported earlier[1], a very shallow resistivity minimum ($\simeq 50$ ppm) at about 7K was again observed in PdH. Curve (b) is our most dilute (c=0.24 at.%) of the Cr alloys. Here $\rho(T)$ rises sharply with decreasing temperature down to 1.35K. When plotted as $\rho(T)$ versus log T, a logarithmic dependence is found between approximately 4.5 and 20K. However, when we attempted to take into account the phonon resistivity and subtract the "pure" PdH resistivity, a stronger than logarithmic fall off in $\Delta\rho$ occurred as $T \rightarrow 50K$. This difficulty at the higher temperatures left us unsure as to describing the dilute (PdCr)H system within a simple Kondo-resistivity framework. Without more dilute alloys and a low T —analysis, we can now only roughly estimate a Kondo temperature of order 10K - considerably larger than that for AgCr of $10^{-2}$K.[12] In curve (c), we note

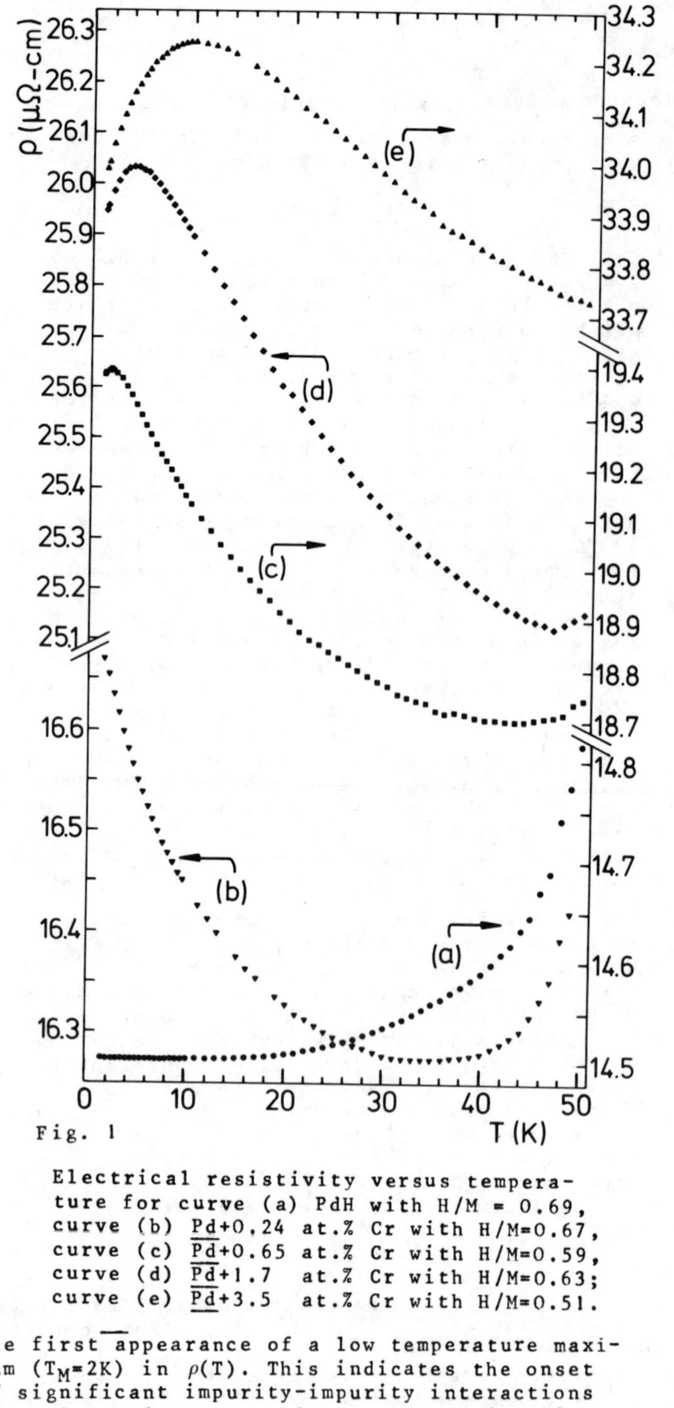


Fig. 1

Electrical resistivity versus temperature for curve (a) PdH with H/M = 0.69, curve (b) Pd+0.24 at.% Cr with H/M=0.67, curve (c) Pd+0.65 at.% Cr with H/M=0.59, curve (d) Pd+1.7 at.% Cr with H/M=0.63; curve (e) Pd+3.5 at.% Cr with H/M=0.51.

the first appearance of a low temperature maximum ($T_M$=2K) in $\rho$(T). This indicates the onset of significant impurity-impurity interactions in analogy with the noble metal - 3d impurity alloy systems. The fact that the maximum in the hydrogenated alloys is already present for 0.65 at.% Cr - a factor of at least 10 less Cr than for PdCr - suggests that Cr represents a strong or good moment impurity when introduced into the PdH matrix. Above the maximum there is a similar $\rho$(T) behavior as found for the lower concentration, but with the Cr-Cr interactions causing a wider deviation at the low temperature end from the Kondo-log T dependence. The maximum in $\rho$(T) is progressively shifted to higher temperatures as the Cr - c is further increased - see curves (d) and (e). This signifies the growing strength of the impurity-impurity interactions as the Cr atoms come, on average, closer together. For the 3.5 at.% Cr sample, there seems to be a linear increase in $\rho$ at the lowest temperatures. Presently, there is little experimental or theoretical guidance on how to relate the resistivity behavior of a spin glass to its freezing temperature, $T_0$.[13] This temperature certainly lies below the maximum in $\rho$, somewhere within a linear in T region and above the very low temperature characteristic $T^{3/2}$ dependence. Therefore, without a direct magnetic measurement, $T_0$ can only crudely be determined as around 2K for the 3.5 at.% Cr-PdH alloy. This value may be compared with $T_0$=10K for (Pd + 3.5 at.% Fe)H and $T_0$ = 4K for (Pd + 3.5 at.% Co)H.[3]

## CONCLUSIONS

The hydrogen charging of PdCr alloys has greatly altered the $\rho$(T) behavior from that of PdCr. A comparison with (PdFe)H shows much the same $\rho$(T)-properties. Thus, these two quite diverse magnetic systems converge upon hydrogenation into a similar type of good moment, randomly interacting alloy. However, based upon contrasting the values of $T_M$ and $T_0$ per at.% Fe or Cr, the interaction strength between Fe impurities is somewhat larger than between Cr impurities. Susceptibility measurements of the (PdCr)H would be of definite interest.

A further parallel may be drawn with the noble metal - 3d solute Kondo/spin glasses. There is also here an analogous $\rho$(T) behavior, but, for the hydrogenated alloys, a greater concentration of magnetic impurity ($\simeq$5 times) is required to observe similar effects. The reduced electronic mean free path of the PdH systems could be an important factor in dimishing the inter-impurity coupling. Consequently, the hydrogen charging process offers a new variable for future ingestigations of Kondo and spin glass alloys.

## REFERENCES

x Present address: Kamerlingh Onnes Laboratorium der Rijksuniversiteit, Leiden, The Netherlands.

1. J.A.Mydosh,Phys.Rev.Letters 33,1562(1974).
2. For a review of this area see J.A.Mydosh, AIP Conf.Proc. 24,131(1975).
3. J.P.Burger et al. in Proceedings of the 14th International Conference on Low Temperature Physics LT-14,edited by M.Krusius and M.Vuorio (North-Holland,Amsterdam,1975) Vol.3,pg.278.
4. J.I.Budnick et al.,AIP Conf.Proc. 18,307 (1974).
5. R.A.Levy and J.A.Rayne, Phys.Letters 53A, 329(1975) and in LT-14,Vol.3,pg.262;R.P. Bittner et al., AIP Conf.Proc. 24,481(1975)
6. H. Nagasawa,Journ.Phys.Soc.Japan 28,1171 (1970).
7. W.M.Star, thesis, University of Leiden (1971) and Physica 59,128(1972).
8. F.C.C.Kao and G.Williams,Phys.Rev. B7,267 (1973).
9. R.M.Roshko and G.Williams in LT-14,Vol.3, pg.274.
10. J.E.van Dam, thesis, University of Leiden (1973).
11. N.Rivier and M.J.Zuckermann,Phys.Rev.Lett. 21,904(1968);N.Rivier and V.Zlatič,J.Phys. F 2,L87 and L99(1972).
12. I.R.Williams et al.,Solid State Commun. 8, 125(1970).
13. P.J.Ford and J.A.Mydosh, to be published.

# MÖSSBAUER MEASUREMENTS IN SPIN-GLASS ALLOY $(Pd_{0.5}Ag_{0.5})_{0.99}Fe_{0.01}$ *

S. De Benedetti, J. A. Rayne and A. Zangwill
Carnegie-Mellon University, Pittsburgh, Pennsylvania 15213

and

R. A. Levy
Rensselaer Polytechnic Institute, Troy, New York 12181

## ABSTRACT

Mössbauer measurements from 1.6 K to 300 K have been made in the spin-glass alloy $(Pd_{0.5}Ag_{0.5})_{0.99}Fe_{0.01}$. The room temperature spectrum yields a quadrupole splitting of 0.237±0.006 mm $sec^{-1}$. Data below 3.85 K show the existence of a magnetic hyperfine field, which has an average value of 270±1 kG at 1.6 K. The low temperature results cannot be satisfactorily represented with a distribution of hyperfine fields. Preliminary fits to a relaxation model are discussed.

## INTRODUCTION

The properties of the $(Pd_{1-x}Ag_x)_{0.99}Fe_{0.01}$ ternary alloy system have been studied by a number of techniques.[1-4] For $0 \leq x \leq 0.25$ ferromagnetic ordering is observed, the Curie temperature being determined by the spin paramagnetic susceptibility of the host matrix as predicted by theory.[5] For higher silver concentrations both susceptibility and resistivity measurements indicate spin-glass behaviour,[6] similar to that recently reported[7] in the Pd-H(1% Fe) system with hydrogen-to-metal ratio near 0.6. In this paper we report Mössbauer measurements on the ternary alloy with x=0.50 over a temperature range from 1.6 K to 300 K.

## EXPERIMENT

Measurements were made on a stress-relieved foil, approximately 25 μm thick, used in previous work.[1] The room temperature spectrum was obtained with a constant acceleration spectrometer operating in conjunction with a backscatter detector.[8] Low temperature data were obtained in the standard transmission mode. Above 4.2 K, temperatures were maintained to an accuracy of better than 0.1 K, using a calibrated carbon thermometer in an electrothermal feedback circuit. Temperatures below 4.2 K were achieved by controlled pumping of the liquid helium bath.

* Work supported by National Science Foundation.

## RESULTS AND DISCUSSION

Figure 1 shows the room temperature spectrum examined in a velocity range of ±1.50 mm $sec^{-1}$ with velocity increments of 0.01 mm $sec^{-1}$. The data reveal a partially resolved doublet with approximately equal linewidths, but amplitudes differing in the ratio 1.20 to 1.00. A no-constraint Lorentzian fit to the two lines yields a quadrupole splitting of 0.237±0.006 mm $sec^{-1}$ and an isomer shift relative to metallic iron of 0.223±0.006 mm $sec^{-1}$.

This behaviour differs from that at lower silver concentrations ($0 \leq x \leq 0.25$), where the spectra have the appearance of broadened single lines with an approximately constant isomer shift of 0.185±0.010 mm $sec^{-1}$. For these concentrations, the line broadening is possibly caused by the presence of a combination of quadrupole doublets with different values of splitting and isomer shifts due to the different local environments for each iron atom. Evidently, the quadrupole splitting must be much less than the natural linewidth, since a partial resolution of the doublet would otherwise occur.

For x=0.50, the appearance of the partially resolved doublet and the higher value of isomer shift are believed to be associated with the filling of the alloy

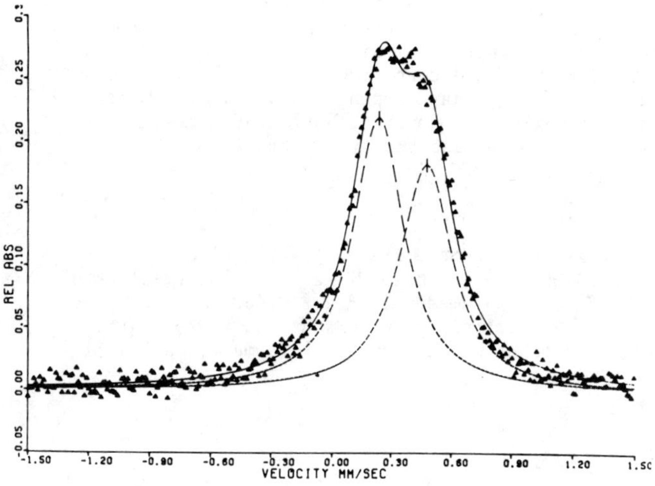


Figure 1. Room temperature spectrum for $(Pd_{0.5}Ag_{0.5})_{0.99}Fe_{0.01}$ showing partially resolved doublet. Dashed curves are computed components of doublet. Source is $Co^{57}$ in chromium.

4d-band. The observed asymmetry of the doublet may be due to an anisotropic recoil-free fraction in the alloy. However, it is more likely to be caused by a distribution of quadrupole splittings and isomer shifts, as discussed above.

Figure 2 shows the low temperature Mössbauer data with an expanded velocity range of ±6.00 mm $sec^{-1}$. Above 4.0 K the spectra appear as broadened lines centered at 0.334±0.006 mm $sec^{-1}$. The shift relative to room temperature reflects the second-order Doppler effect. There is a loss in resolution both from the expanded range and the use of non-ideal specimen thick-

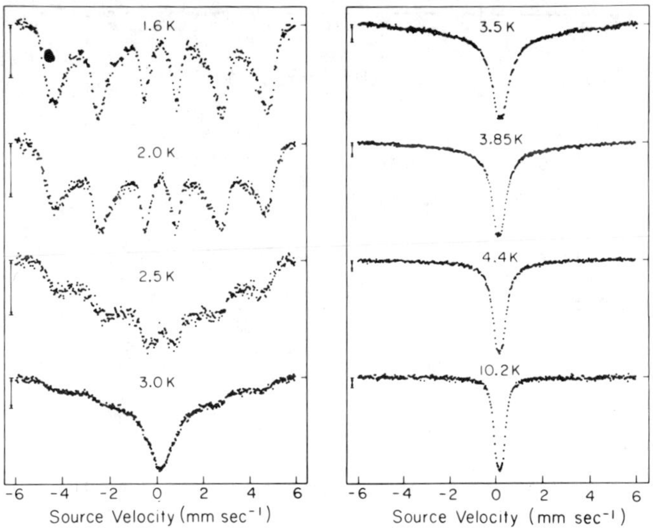


Figure 2. Low temperature Mössbauer spectra for $(Pd_{0.5}Ag_{0.5})_{0.99}Fe_{0.01}$. Vertical lines correspond to two percent change in transmission. Source is $Co^{57}$ in palladium.

ness in transmission, as evidenced by the absence of the previously observed quadrupole doublet. Close to the glassy transition temperature near 3.8 K, as determined by susceptibility[2] and resistivity data[3,4], the central peak shows significant additional broadening, while at 3 K there is evidence of magnetic hyperfine splitting. The six-line pattern is clearly resolved at 1.6 K, at least part of the observed asymmetry being due to the distribution of hyperfine fields in the alloy.

A computer fit to the latter spectrum has been attempted to investigate the latter effect. A distribution of pairwise symmetric six-line patterns is assumed, each of the constituent Lorentzians being constrained only to have a constant width of 0.6 mm $sec^{-1}$ at half maximum. The distribution of fields is given by a normalized probability function. A suitable choice of the latter is given by

$$P(x) \sim x^{\alpha} \exp(-x/\beta) \quad , \tag{1}$$

where x is a reduced field variable given by (312.3-H)/28.97, H being in kG. The best values of the fitting parameters $\alpha$,$\beta$ are 0.657 and 0.908, respectively. It is assumed that the isomer shift, which depends on the local environment, varies linearly with H.

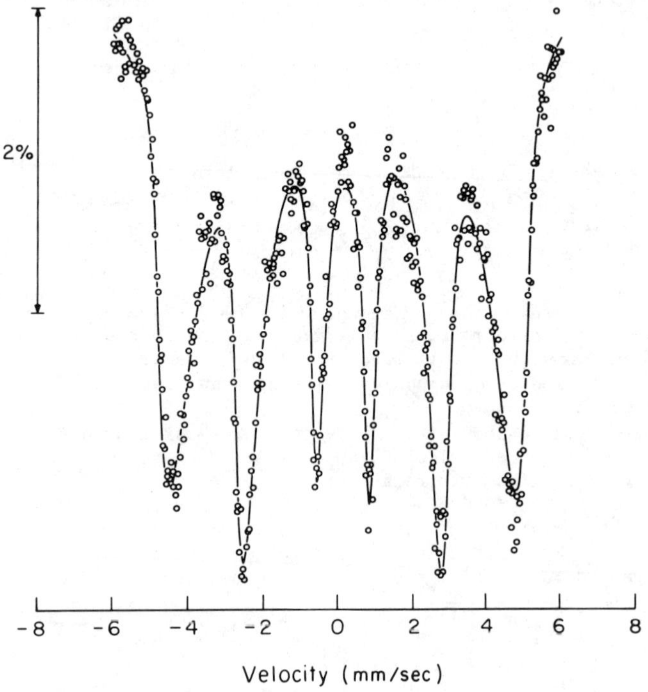


Figure 3. Computer fit to Mössbauer data for $(Pd_{0.5}Ag_{0.5})_{0.99}Fe_{0.01}$ at 1.6 K.

Figure 3 shows the best fit to the data using this procedure. The average hyperfine field is 270±1 kG, and the corresponding isomer shift is 0.340±.005 mm $sec^{-1}$, which is close to the value in the paramagnetic regime. There is a systematic divergence from the experimental data in the region of the central trough; for higher temperatures the fit becomes progressively less satisfactory. The failure of such a general fitting procedure implies that some additional factors must be considered.

Preliminary attempts have been made to fit the data with a relaxation model.[9] In this model, the hyperfine field is allowed to relax isotropically at a rate given by the parameter $\lambda$. If $\omega_0$ is the nuclear Larmor frequency, the collapse of the six-line pattern to a single line corresponds to the transition from the static case $\lambda \ll \omega_0$ to the limit $\lambda \gg \omega_0$, where the average hyperfine field is zero. The parameter $\lambda$ is assumed to be thermally activated.

$$\lambda = \lambda_0 \exp(-U/k_B T) \quad , \tag{2}$$

where in a random alloy different iron sites presumably have different values of $\lambda_0$ and U.

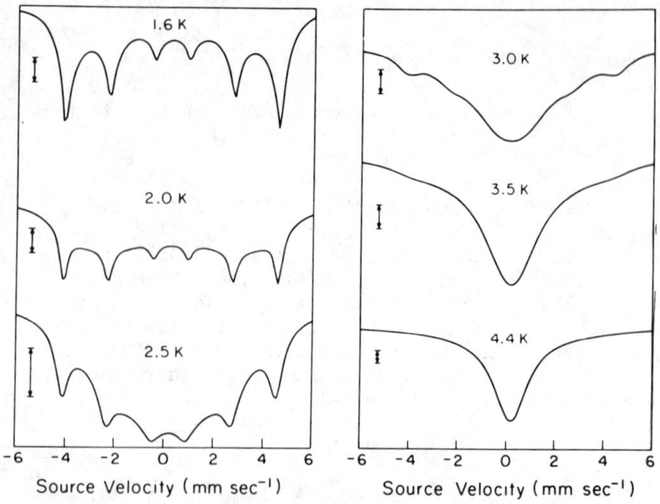


Figure 4. Representative Mössbauer spectrum for relaxation model discussed in text. Activation temperatures ($U_i=k_B T_i$) are $T_1$=15.3 K, $T_2$=8.0 K with weights $P_1$=0.2, $P_2$=0.8. Vertical bars represent two percent change in transmission.

Figure 4 shows a computed spectrum, assuming only two different sites with activation energies $U_1$, $U_2$, weighting factors $P_1$, $P_2$ and the same value of $\lambda_0$. A value of 270 kG is taken for the static value of the hyperfine field. Both components are assumed to collapse to the same Lorentzian at high temperatures. Clearly the general features of Fig. 4 conform rather well to the overall behaviour of the experimental spectra. In a more realistic calculation both the activation temperatures and weights would become fitting parameters, while a distribution of static hyperfine fields would also be assumed. The effects of the latter could be determined from the lowest temperature spectrum where the static limit has been presumably reached. Fits of this type are currently being attempted and will be reported in a future communication.

Thanks are due to Professor P. A. Flinn for advice on the computer minimization routine and also to Dr. S. Dattagupta for helpful discussion concerning the relaxation model. The authors wish to acknowledge the continued interest of Professor J. Budnick in this work.

REFERENCES

1. R. A. Levy, J. J. Burton, D. I. Paul and J. I. Budnick, Phys. Rev. B9, 1085 (1974).
2. J. I. Budnick, V. Cannella and T. J. Burch, Proceedings of the 19th Conference on Magnetism and Magnetic Materials (1973), p. 307.
3. K. V. Rao, O. Rapp, C. L. Johansson, J. Budnick, T. J. Burch and V. Cannella, Proceedings of the 20th Conference on Magnetism and Magnetic Materials (1974), p. 474.
4. R. P. Bittner, R. A. Levy and J. A. Rayne, Proceedings of the 20th Conference on Magnetism and Magnetic Materials (1974), p. 481; R. A. Levy and J. A. Rayne, Phys. Letters A53, 329 (1975).
5. T. Takahashi and M. Shimizu, J. Phys. Soc. Jap. 20, 26 (1965).
6. J. A. Mydosh, Proceedings of the 20th Conference on Magnetism and Magnetic Materials (1974), p. 131.
7. J. A. Mydosh, Phys. Rev. Letters 33, 1562 (1974).
8. R. A. Levy, P. A. Flinn and R. A. Hartzell, (to be published).
9. S. Dattagupta and M. Blume, Phys. Rev. B10, 4540 (1974).

# EXCITATIONS OF SUBSTITUTIONALLY DISORDERED ANTIFERROMAGNETS*

R. A. Cowley†
Brookhaven National Laboratory, Upton, New York 11973

## ABSTRACT

Measurements by neutron inelastic scattering techniques have provided detailed information about the magnetic excitations of substitutionally disordered antiferromagnets. In the dilute systems, such as $MnF_2/ZnF_2$, the excitations broaden and decrease in energy with increasing Zn concentration, while in mixed magnetic systems, such as $KMnF_3/KCoF_3$ there are usually two branches of the excitations which correspond to excitations propagating largely on one or other of the magnetic ions. Both of these results are in reasonable accord with the predictions of theories based on the coherent potential approximation and with the results of computer simulations. Recent high resolution experiments on the $MnF_2/ZnF_2$ system have shown that there is structure in the line shape of the excitations arising from the different excitation energies of Mn ions surrounded by different numbers of Zn neighbors.

## INTRODUCTION

The transition metal fluorides of the rutile and perovskite structures provide ideal systems with which to study the excitations in disordered systems. This is because although these metal ions are chemically very similar they are magnetically very different and single crystals may be grown in which magnetically very different ions are distributed randomly throughout the crystal. The magnetic excitations in these crystals can then be studied in detail with neutron inelastic scattering techniques.

Until 1971 most of the work[1] on these systems was on examples in which one of the constituents was present in low concentration. Examples were found of systems in which the impurities introduced localized excitations and others in which they created a resonant perturbation of the host spin waves, and the results were found to be in excellent accord with theory when it was developed as a series in the concentration of the impurities. More recently, both experimental and theoretical work has been performed on more concentrated systems, and it is the objective of this review to briefly describe the experimental results and to outline the theoretical developments. Throughout this paper details of both the experimental techniques and of the theoretical calculations which are described in the original papers will be omitted and only a broad overall picture described.

Much of the theoretical work described later has been concerned with extending the coherent potential approximation (CPA), originally developed to describe electronic and phonon systems, to apply to these magnetic systems. It is possible, however, to understand many of the results with the aid of the nearest neighbor Ising model of magnetic excitations. If the system consists of two magnetic species, A and B, with nearest neighbor exchange interactions $J_{AA}$, $J_{AB}$, and $J_{BB}$, the Ising frequency of an ion of type $\lambda$ surrounded by $\mathbf{r}$ ions of the same type and (Z-r) ions of the other type $\lambda'$ is

$$E_\lambda(r) = r\, J_{\lambda\lambda} S_\lambda + (Z-r)\, J_{\lambda\lambda'} S_{\lambda'} \quad .$$

If the probability of r neighbors of type $\lambda$ is $p_\lambda(r)$ the mean frequency is given by $E(\lambda) = \sum_r p_\lambda(r) E_\lambda(r)$ .

Using this simple model and assuming that the exchange constants $J_{AB}$ are given by the geometric mean of the pure crystal exchange constants, the mean frequencies of the different ions may be calculated, and are list-

| System | Conc. | Calc. | Expt. | Ref. |
|---|---|---|---|---|
| $MnF_2/CoF_2$ | 0.95 | 3.46 | 3.57±0.05 | 2 |
| | | 1.51 | 1.49±0.02 | |
| | 0.30 | 2.42 | 2.32±0.10 | 3 |
| | | 1.20 | 1.20±0.06 | |
| | 0.10 | 2.05 | 2.02±0.08 | 4 |
| | | 1.05 | 1.09±0.08 | |
| | 0.05 | 1.92 | 1.86±0.06 | 4 |
| | | 1.00 | 0.90±0.10 | |
| $KMnF_3/KCoF_3$ | 0.80 | 6.96 | 6.55±0.15 | 5 |
| | | 2.27 | 2.26±0.02 | |
| | 0.29 | 7.04 | 6.80±0.10 | 3 |
| | | 2.30 | 2.30±0.10 | |
| $KMnF_3/KNiF_3$ | 0.97 | 8.51 | 7.68±0.10 | 6 |
| | | 2.27 | 2.17±0.05 | |
| | 0.25 | 11.7 | 11.8 ±0.50 | 7 |
| | | 3.12 | 3.1±0.15 | |
| $Rb_2MnF_4/Rb_2NiF_4$ | 0.5 | 6.81 | 6.48 | 8 |
| | | 1.92 | 1.79 | |
| $MnF_2/ZnF_2$ | 0.78 | 1.18 | 1.19±0.02 | 9,10 |
| | 0.69 | 1.04 | 1.05±0.03 | 11 |
| $CoF_2/ZnF_2$ | 0.86 | 1.60 | 1.59±0.02 | 12 |
| | 0.69 | 1.28 | 1.12±0.05 | 12 |

TABLE I

The Ising frequency compared with measured zone boundary frequencies (THz). The concentration is that of the first component.

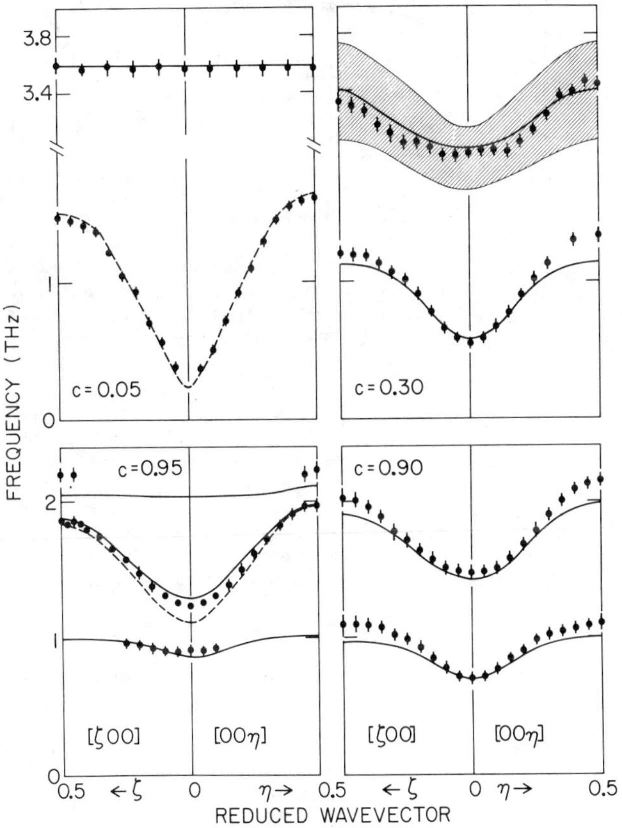


Fig. 1. Dispersion relations of the $Mn_{1-c}Co_cF_2$ system.[2-4] The solid lines are CPA calculations[13,4] and the dotted lines, the dispersion relations of the pure material The shaded region for c = 0.30 gives the width predicted by the calculations.

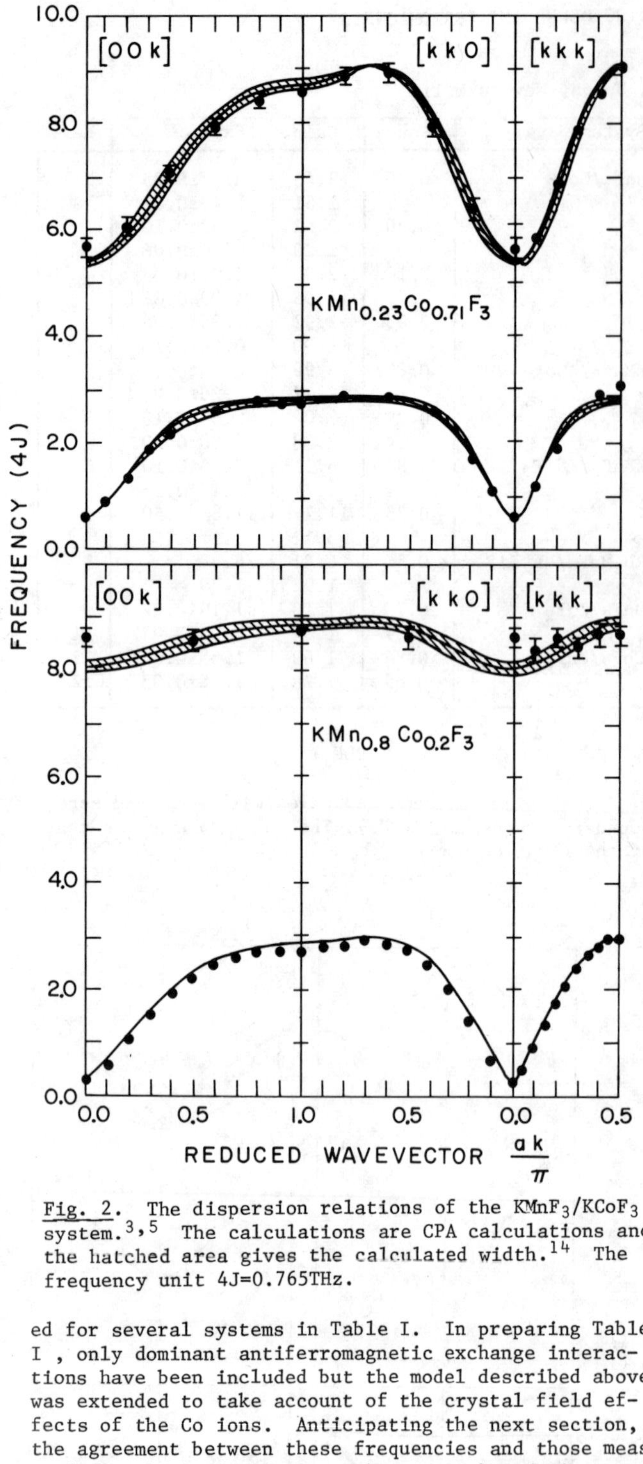


Fig. 2. The dispersion relations of the $KMnF_3/KCoF_3$ system.[3,5] The calculations are CPA calculations and the hatched area gives the calculated width.[14] The frequency unit 4J=0.765THz.

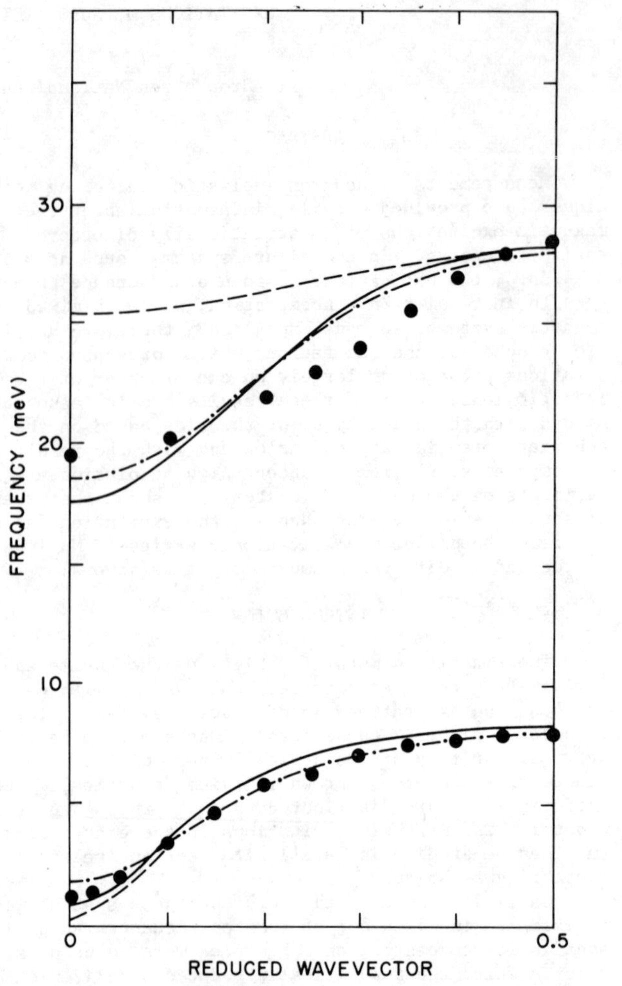


Fig. 3. The dispersion relation[8] for $Rb_2Ni_{0.5}Mn_{0.5}F_4$ compared with CPA calculations using the BPE(dot-dash) VCA(dot) and SSC(solid) approximations.[15]

ed for several systems in Table I. In preparing Table I , only dominant antiferromagnetic exchange interactions have been included but the model described above was extended to take account of the crystal field effects of the Co ions. Anticipating the next section, the agreement between these frequencies and those measured for the excitations at those zone boundaries, which in the pure crystals give these Ising frequencies, is surprisingly good for such a simple model.

## EXPERIMENTAL RESULTS

In systems containing two magnetic ions, the energy distributions of the scattered neutrons mostly have two peaks for each wave vector in agreement with the Ising model described above. Measurements of the frequencies of these two peaks have been made in the $MnF_2/CoF_2$ system[2-4] and are shown in Fig. 1. The introduction of a few Co ions into $MnF_2$ produces a mode localized on the Co ions with a frequency well above that of the $MnF_2$ band. With increasing concentration of Co this excitation broadens, decreases in frequency and shows dispersion, while the branch of lower frequency becomes less dispersive. Rather surprisingly, this band exhibits dispersion even when the Mn concentration is only 0.05 unlike the results at the Mn-rich end. At this concentration scattering is also observed above the Co band, which is associated with the Co ions that are neighbors of a Mn ion. This scattering gives a distinct peak only for low concentrations of Mn ions.

Fig. 2 shows similar results in the $KMnF_3/KCoF_3$ system.[3,5] There are two bands of excitations; the one of lower frequency is very similar to that of pure $KMnF_3$ even when the Mn concentration is only 0.29 while the branch of higher frequency shows an increasing amount of dispersion as the Mn concentration decreases. Qualitatively similar results have been obtained in the $KMnF_3/KNiF_3$ system[6,7] and in the analogous two-dimensional magnetic system $Rb_2MnF_4/Rb_2NiF_4$ as shown in Fig. 3. In several of these measurements the intensity of the excitations was also determined and some broadening of the excitations observed. In most cases, however, this broadening was only slightly larger than the experimental resolution and detailed measurements could not be made.

Measurements have also been reported in the $MnF_2/ZnF_2$ system[9,10] and in the $CoF_2/ZnF_2$ system.[12] In both cases when the Zn concentration is increased, the excitations decrease in frequency and broaden as shown in Figs. 4 and 5. In these cases, the broadening is considerably larger than in the mixed magnetic systems discussed above. When the concentration of magnetic ions becomes close to the percolation limit in both systems the excitations are very broad, as shown for the $MnF_2/ZnF_2$ system in Fig. 6. In the $MnF_2/ZnF_2$ system no significant difference in the inelastic scattering was found between a specimen with a concentration just above the percolation limit from one below that limit.

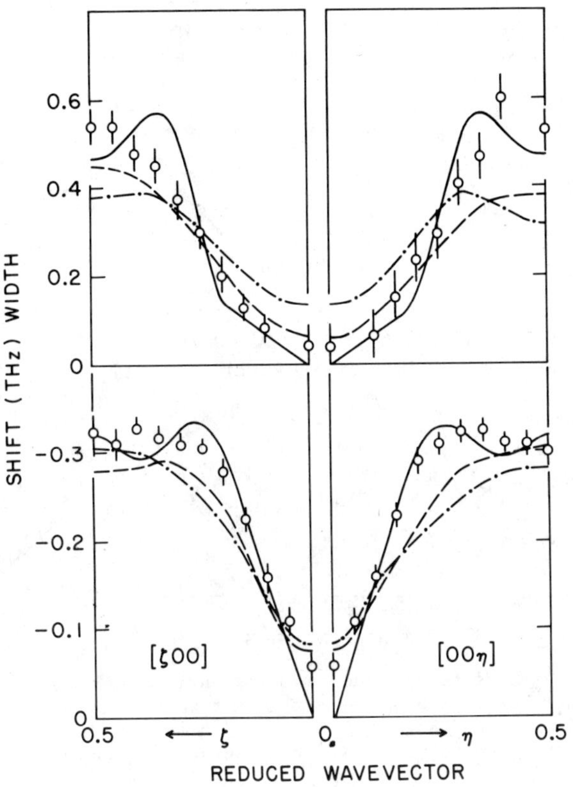


Fig. 4. The frequency shifts from pure $MnF_2$ and widths for $Mn_{0.78}Zn_{0.22}F_2$ compared with CPA calculations of theories by Holcomb(solid), BPE(dotted) and VCA(dot-dash).[11]

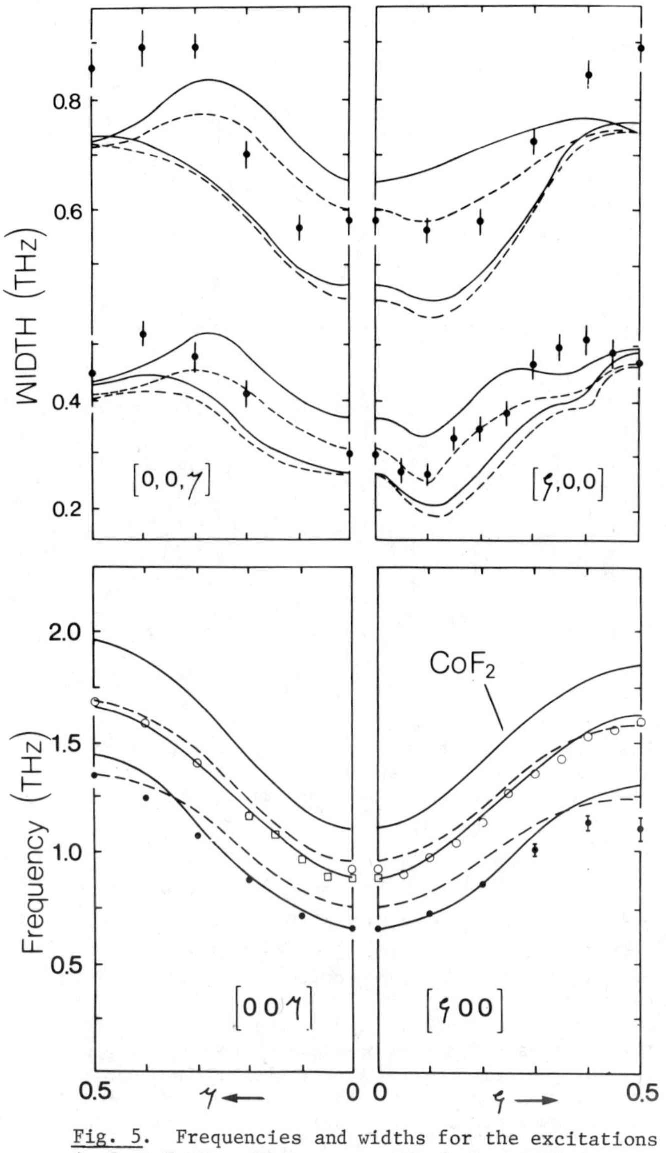


Fig. 5. Frequencies and widths for the excitations in $Co_{1-c}Zn_cF_2$. The upper part of the width diagram and the lower frequency results in the lower part of the diagram are for c=0.31. The concentration in the other sample is 0.14. The solid curves for the dispersion relation give the results of CPA calculations while for the width the CPA theories are in order of decreasing width VCAH, VCAE, BPEH, BPEE.

In the experiments described above the results could be described by listing the position, width, and intensity of the neutron groups. Recently[11], experiments have been performed with improved resolution which show that there is structure in the line shape of the distributions which was obscured in the earlier experiments. These results are shown in Fig. 7 and the structure arises from the different frequencies, $E_\lambda(r)$, of Mn ions surrounded by different numbers, r, of neighboring Mn ions. This result and the agreement shown in Table I demonstrates the extent to which the simple Ising model provides at least a qualitative understanding of these systems.

## COHERENT POTENTIAL APPROXIMATION

The coherent potential approximation is the most used technique for describing the excitations in disordered systems. Several different ways have been developed to apply this approach to these magnetic systems but the most generally useful[13] starts from the Ising model described in the Introduction and which has already been seen to provide a great deal of understanding of these systems. The Ising frequencies vary from site to site so that the magnetic excitations will be scattered by these fluctuating frequencies and this scattering may be calculated using the single-site CPA approximation. In this approach a complex frequency dependent potential is placed at each site in the crystal and then determined by requiring that on average this potential gives the same effect as that of all the randomly fluctuating Ising frequencies .

There are two difficulties with the application of this theory to real magnetic systems; firstly unlike the analogous alloy problems the site frequencies depend upon the local environment and so are at least partially correlated even when the distribution of the ions is completely random. This effect should be most pronounced when there are only a small number of neighbors, but no attempt has as yet been made to investigate the effect of these correlations. Secondly, the interactions between the spins are not of the Ising form but also include transverse parts. The effect of these transverse interactions in antiferromagnets is to make the self-energy a complex 2x2 matrix which is both frequency and wave vector dependent. Inclusion of the transverse parts in the CPA is very difficult in practice and various different approximations have been adopted to overcome this difficulty.

This last difficulty is least important in the $MnF_2/CoF_2$ system because by chance the transverse parts of the exchange interactions in the pure materials are very similar. In this case, therefore, the single-site CPA can be applied directly and the results[13,4] which are shown in Fig. 1 are in very satisfactory agreement with experiment.

It is also possible to treat the transverse parts of the exchange interactions by exploiting the fact that the Ising frequencies are unchanged if the spin of a particular species is changed by a factor,k, provided that the exchange constants are scaled appropriately. Tonegawa[14] has exploited this spin scaling approach (SSC) and shown that the transverse interac-

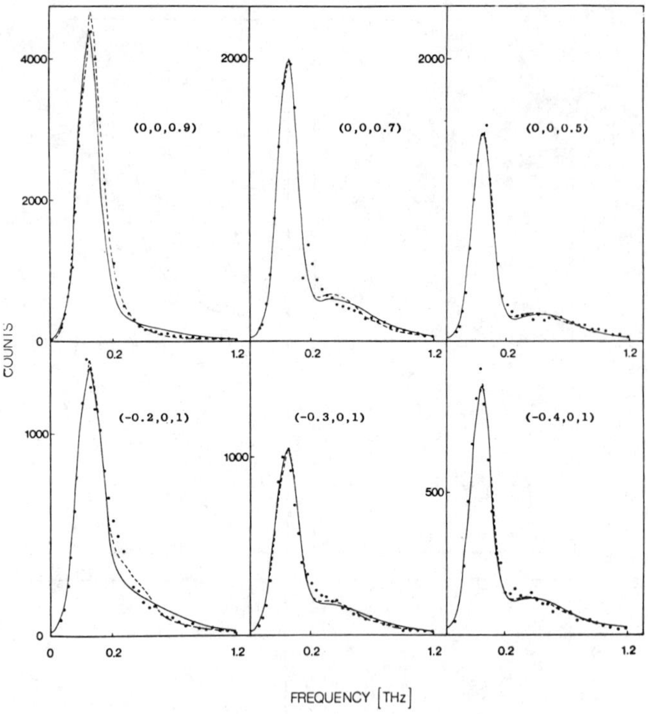


Fig. 6. Scattered neutron distributions from $Mn_{0.32}Zn_{0.68}F_2$ compared with CPA calculations based on the BPE (broken line) and VCA (solid line) approximations broadened with the resolution and added to a quasi-elastic Gaussian.[10]

tions may be made identical if $S_AJ_{AB} = S_BJ_{BB}$ and $S_BJ_{AB} = S_AJ_{AA}$ which is closely satisfied for $KMnF_3/KCoF_3$. The results of the calculations which are shown in Fig. 2 give excellent agreement with experiment.

In the case of the Ni/Mn systems, neither of these approaches may be used and so the off-diagonal part of the self-energy must be treated less satisfactorily. In all of the approximations, this term is treated as a product of a wave vector dependent term, of the same form as it has in the pure crystals, and of a frequency dependent term. If the interactions are Heisenberg-like, as is closely the case in the Ni/Mn system, symmetry requires that at zero frequency and wave vector the diagonal and off-diagonal parts of the self-energy matrix are identical. Buyers et al[13](BPE) then assumed that the frequency dependence of the off-diagonal part of the self-energy was the same as that of the diagonal part. The results of calculations using this approach[15] are shown in Fig. 3. Alternatively, we may use symmetry to fix the magnitude of the off-diagonal self-energy and assume it to be frequency independent (VCA). The results in Fig. 3 show that the form assumed for the off-diagonal part of the self-energy does influence the results, and that the BPE approach appears to be the more satisfactory. Finally, the spin scaling approach was used to minimize the fluctuation in the transverse parts of the interaction, and the remainder of their effect treated using the VCA approximation. Fig. 3 shows that this procedure gives very similar results both to the BPE approximation and to experiment.

In systems containing non-magnetic ions the single-site CPA incorrectly permits the magnetic excitations to spread onto the non-magnetic ions. This unphysical behavior is most appropriately prevented by placing a large potential on the non-magnetic sites (hard core, H). Alternatively, the non-magnetic sites may be omitted from the calculations except for the effect they have on the Ising frequencies of the magnetic ions (empty core, E). In Figs. 4-6 the results of some of the calculations[16,10] are compared with experimental

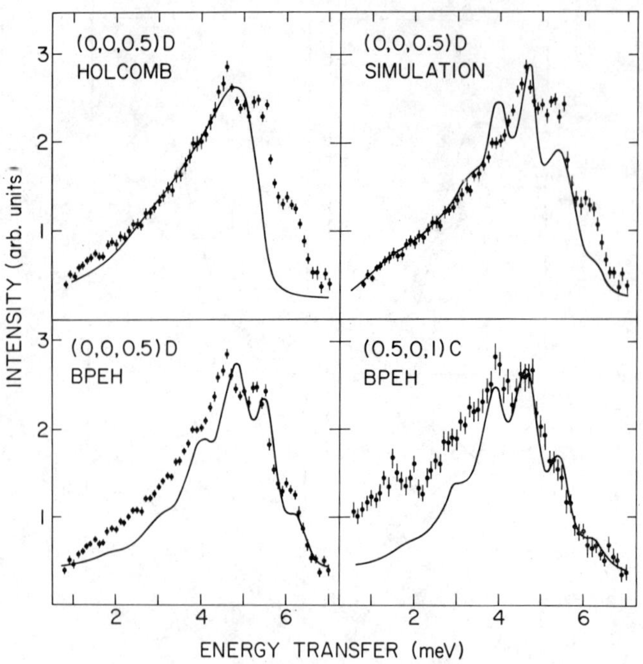


Fig. 7. Scattered neutron distributions[11] from $Mn_{0.68}Zn_{0.32}F_2$ compared with CPA theories due to Holcomb[17] and Buyers et al[16], and computer simulations[21], convoluted with the resolution function.[11]

results. Overall the theories are in reasonable accord with experiment except that the calculations mostly predict a width somewhat smaller than that observed. These theories also predict structure in the line shape of the excitations, which as shown in Fig. 7, is in very reasonable agreement with that observed in $MnF_2/ZnF_2$.

An alternative development of the CPA for these systems has been given by Holcomb.[17] When one of the species is non-magnetic the transverse parts of the exchange interactions may be treated exactly at the expense of treating the magnetic ions as having an average environment for all but one of their neighbors. The results of this theory are compared with experiment in Figs. 4 and 7, and the overall agreement is possibly better than with the single-site CPA theory. The theory does, however, fail to predict structure in the line shape, Fig. 7, and also predicts a fairly well-defined resonance in the widths, Fig. 4, which is not observed. It seems reasonable that both of these deficiencies are a result of treating the magnetic ions as having a largely average environment.

Various other variants of the CPA have been developed for these systems. Krey and Schtilling[18]have proposed a theory very similar to that of Holcomb's except that the cluster around each ion is treated less satisfactorily. Several workers[19] have attempted to include the transverse interactions by treating each bond self-consistently, but the results are not even in qualitative accord with experiment.

A relatively simple cluster theory of these materials has been developed by Dzyub.[20] This theory has the advantage that it does not use a self-consistent procedure like the CPA, but the results are quantitatively less satisfactory than for the single-site CPA theories. Finally, computer simulations have been performed for the $MnF_2/ZnF_2$ system[21] and for the $Rb_2NiF_4/Rb_2MnF_4$ system[22] and as shown in Fig. 7 give a very satisfactory amount of the measurements.

## CONCLUSIONS

Measurements of the magnetic excitations in disordered antiferromagnets have provided detailed information about the excitations in disordered systems. The results are in agreement with the predictions of the single-site CPA provided that the appropriate Ising energies are included for each ion. There is then difficulty in treating the transverse parts of the interactions satisfactorily but the use of the spin-scaling approach or the BPE approximation that the frequency dependence of the diagonal and off-diagonal parts of the self-energy is the same appears to yield satisfactory results. A less empirical approach to the inclusion of the transverse exchange is desirable.

ACKNOWLEDGMENTS

Many people have contributed to my understanding of the work described in this paper. In particular, I am grateful to my collaborators: W. J. L. Buyers, G. C. Coombs, O. W. Dietrich, T. M. Holden, D. A. Jones, G. Meyer, G. Shirane, R. W. H. Stevenson, and E. C. Svensson.

REFERENCES

* Supported by the U. S. Energy Research and Development Administration.

† Permanent address: Department of Physics, University of Edinburgh, Scotland.

1. R. A. Cowley and W. J. L. Buyers, Rev. Mod. Phys. 44, 406-450 (1972).
2. T. M. Holden, R. A. Cowley, W. J. L. Buyers, and R. W. H. Stevenson, Solid State Commun. 6,145-7(1968)
3. W. J. L. Buyers, T. M. Holden, E. C. Svensson, R. A. Cowley, and R. W. H. Stevenson, Phys. Rev. Lett. 27, 1442-5 (1972).
4. E. C. Svensson, S. M. Kim, W. J. L. Buyers, S. Rolandson, R. A. Cowley, and D. A. Jones, AIP Conf. Proc. 24, 161-2 (1975).
5. E. C. Svensson, W.J. L. Buyers, T. M. Holden, R. A. Cowley, and R. W. H. Stevenson, Can. J. Phys. 47, 1983-1985 (1969).
6. T. M. Holden, R. A. Cowley, W. J. L. Buyers, E. C. Svensson and R. W. H. Stevenson, J. de Phys. 32 C1, 1184-5 (1971).
7. G. J. Coombs, R. A. Cowley, D. A. Jones, G. Parisot, and D. Tocchetti (this conference).
8. J. Als-Nielsen, R. J. Birgeneau, H. J. Guggenheim, and G. Shirane (to be published).
9. E. C. Svensson, T. M. Holden, W. J. Buyers, R. A. Cowley, G. J. Coombs, and D. A. Jones, Proc. Int. Conf. on Magnetism, Moscow, 1973.
10. G. J. Coombs, R. A. Cowley, W. J. L. Buyers, E. C. Svensson, T. M. Holden, and D. A. Jones (to be published).
11. O. W. Dietrich, G. Meyer, R. A. Cowley, and G. Shirane(to be published)and W. J. L. Buyers(private communication).
12. R. A. Cowley, O. W. Dietrich, and D. A. Jones, J. Phys. C-Solid State 8, 3023-3035(1975).
13. W. J. L.Buyers, D. E. Pepper, and R. J. Elliott, J. Phys. C-Solid State 5, 2611-28 (1972).
14. T. Tonegawa, Prog. Theor. Phys. 51, 1293-1311(1974).
15. G.J.Coombs and R. A.Cowley, J. Phys.C 8,1889-1900 (1975).
16. W. J. L. Buyers, D. E. Pepper, and R. J. Elliott, J. Phys. C.-Solid State 6, 1933-52(1973).
17. W.K.Holcomb, J.Phys.C-Solid State 7,4299-4313(1974).
18. U.Krey and H.J.Schlichting,Zeit.fur Physik B21,157-164 (1975).
19. J.B.Parkinson,J.Phys.C-Solid State 6,2337-49(1973).
20. I.P.Dyzub,Physica Status Solidi 61,383-392(1973); ibid 66,339-347(1974).
21. W.K.Holcomb& A.B.Harris,AIP Conf.Proc.24,102-3(1975).
22. R.Alben & M.F.Thorpe,J.Phys.C 8,L275-9(1975).

## THE COHERENT POTENTIAL APPROXIMATION FOR THE DILUTED ANISOTROPIC ANTIFERROMAGNET*

William K. Holcomb, Department of Physics, The University of Alabama, University, Alabama 35486

ABSTRACT

The excitations of the diluted anisotropic Heisenberg antiferromagnet are studied by extending the previous work[1] for the isotropic case. The scattering of a vacancy and its shell of neighbors is treated coherently. The system is described in terms of an effective medium determined self-consistently using the CPA condition. Results of calculations for the density of states and the spectral weight function are presented for several impurity concentrations with values of the exchange constants appropriate to the system $(Co_{1-x}Zn_x)F_2$. Comparison with neutron scattering data[2] for this system is made and reasonable agreement is found. It must be noted, however, that the agreement is not as good as the excellent results that were found[1] between the isotropic theory and neutron scattering data for $(Mn_{1-x}Zn_x)F_2$. This is not understood. It has also been found that, contrary to speculations based on qualitative arguments concerning the larger gap between host and impurity bands, non-analytical and unphysical results are obtained if the hard core psuedo-potential is not included in the calculations. A brief discussion of these analytical difficulties is made and their physical origin is explained.

[1]William K. Holcomb, J. Phys. C7, 4299-4313, 1974.
[2]R. A. Cowley, O. W. Dietrich, D. A. Jones, J. Phys. C8, 3023-3033, 1975.

*A detailed account of this work will appear in Journal of Physics C9, 1976.

# THE OBSERVATION OF STRUCTURE IN THE NEUTRON SCATTERING FROM ANTIFERROMAGNETIC $Mn_{0.32}Zn_{0.68}F_2$

E.C. Svensson, W.J.L. Buyers and T.M. Holden
Atomic Energy of Canada Limited, Chalk River, Ontario, Canada, K0J 1J0
D.A. Jones
University of Aberdeen, Aberdeen, Scotland

## ABSTRACT

On the basis of Ising theory one would expect the distributions of neutrons scattered from magnetic excitations in substitutionally disordered antiferromagnets to exhibit a multiple-peaked structure because of the different Ising frequencies associated with the different compositions of the clusters of ions surrounding the magnetic ions. The calculations of Buyers et al. also predict such structure, but other theories predict no structure in the neutron scattering. To resolve the question, we have carried out high-resolution neutron-scattering measurements on a single crystal of antiferromagnetic $Mn_{0.32}Zn_{0.68}F_2$ using a triple-axis crystal spectrometer. The measurements were carried out at 5.05 K at a reduced wave vector $\zeta = 0.45$ in each of the [00ζ] and [ζ00] directions. Definite structure is observed at both positions. For the [00ζ] direction, peaks are observed at 0.23 and 0.47 THz and there is a shoulder at ≈0.65 THz. For the [ζ00] direction, there are peaks at 0.30 and 0.50 THz and a shoulder at ≈0.70 THz. No existing theory gives an adequate description of the results.

## INTRODUCTION

Although much theoretical[1-7] and experimental[8-14] effort has been devoted to the problem, the behavior of spin waves in randomly disordered crystals is not yet clearly understood. For example, the available theories disagree on whether or not the spin-wave lineshapes for a pseudo-binary mixed crystal should exhibit an Ising-resonance structure. Two-peaked scattered-neutron distributions are readily observed[8,10] for specimens containing two species of magnetic ions which give rise to separate bands of excitations, but we are referring to the question of substructure within the scattering from a given band. In this paper we will be concerned, in particular, with the existence of structure within the single band of excitations for a specimen, $Mn_{1-c}Zn_cF_2$, containing a high concentration of non-magnetic impurities.

The theory of Buyers, Pepper and Elliott[4], which is a cluster version of the coherent-potential approximation (CPA), predicts a multiple-peaked lineshape arising from the Ising resonances of a Mn ion surrounded by varying numbers of Mn neighbors. In this theory the transverse part of the exchange interaction is treated by an approximation and a repulsive hard-core pseudopotential is introduced to keep the spin waves off the Zn ions. (We shall refer to this theory as BPEH.)

The theory of Holcomb[6], in which the CPA is applied in a different way, predicts no structure in the spin-wave lineshape. The transverse exchange interactions within a cluster are treated exactly but no account is taken of the dependence of the longitudinal part of the exchange on the number of neighboring Zn ions.

Monte Carlo computer calculations have been carried out by Holcomb and Harris[7] (HH) by populating randomly a three dimensional lattice of 8192 sites with Zn and Mn ions. These calculations predict that there should be structure in the spin-wave lineshape.

It has been assumed in the theories that the spin structure is unchanged by dilution with zinc and that the spin waves can be described by linear quasi-boson theory. Both assumptions may be unwarranted at large zinc concentrations especially near the critical concentration, c ~ 0.72, where long-range magnetic order disappears.

Predictions of the theories for a zone-boundary

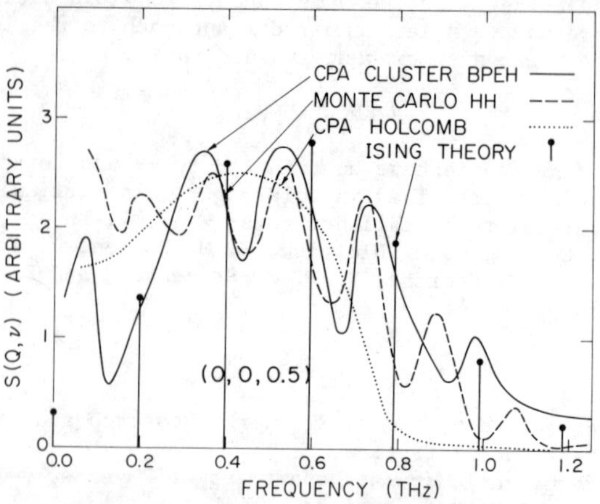


Fig. 1. Comparison of theoretical lineshapes for the c-axis zone-boundary spin-wave response in $Mn_{1-c}Zn_cF_2$. The results of the CPA cluster theory of Buyers et al. (Ref. 4), the Monte-Carlo calculations of Holcomb and Harris (Ref. 7), and Ising theory are for c = 0.65 while the results of the CPA theory of Holcomb (Ref. 6) are for c = 0.68. The heights of the flags at the Ising frequencies are proportional to the probabilities of occurrence of the corresponding clusters.

wave vector in $Mn_{1-c}Zn_cF_2$ are shown in Fig. 1. The CPA-cluster[4] and Monte Carlo[7] calculations give qualitatively similar results with good agreement for the three main resonances but not elsewhere. In contrast, the CPA theory of Holcomb[6] gives a smooth lineshape which falls off rapidly on the high-frequency side. Note that the resonances in the BPEH and HH theories occur at lower frequencies than the corresponding Ising resonances which are also shown in Fig. 1.

In previous neutron-scattering measurements on crystals containing non-magnetic impurities[8,9,11] the resolution has been too low to observe any Ising-resonance structure. In this paper we report high-resolution measurements on a single crystal of $Mn_{0.32}Zn_{0.68}F_2$. Definite structure in the lineshape is observed.

## EXPERIMENT AND RESULTS

The measurements were made with a triple-axis spectrometer operating at constant momentum transfer. Ge(111) and pyrolytic graphite (002) planes were used for the monochromator and analyser, respectively, and the scattered-neutron energy was fixed at 1.81 THz.

The neutron scattering for wave vectors close to the a-axis and c-axis zone-boundary positions is shown in Fig. 2. Structure is apparent in both distributions, with two or possibly three features being identifiable. The observed positions are, however, significantly different from the Ising-resonance frequencies $\omega_r$ (r = 0,...z) associated with either a-axis or c-axis modes, $\omega_r = (r/z)\omega_{max}^{a,c}$. Lower frequencies are observed in the c-axis measurements in contrast to the higher frequencies for c-axis spin waves in pure $MnF_2$.

The observed scattering with background subtracted is compared with the predictions of the theories in Fig. 3. The theoretical lineshapes have been folded with the experimental resolution and scaled to experiment.

We see that Holcomb's theory[6] gives a fair description of the envelope of the scattering at intermediate frequencies, but fails to predict the resonance structure. It also does not account for the rapid in-

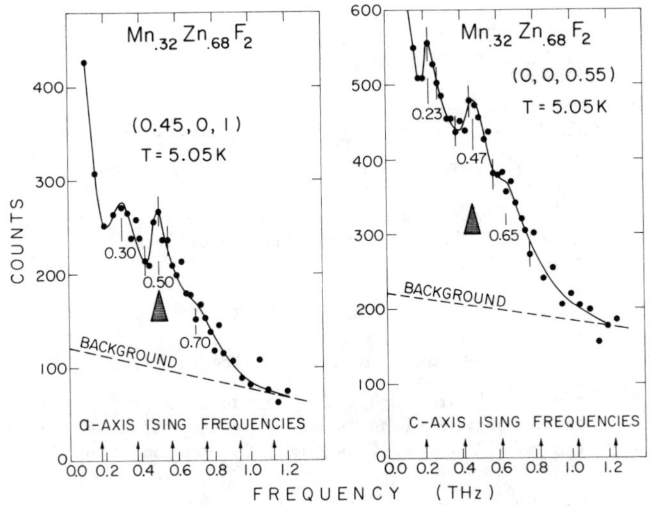


Fig. 2. Scattered-neutron distributions for wave vectors near zone boundaries in $Mn_{0.32}Zn_{0.68}F_2$. The instrumental resolution (FWHM) is given by the base of the solid triangle.

crease in scattering at low frequencies and cuts off somewhat too rapidly at high frequencies.

The BPEH and HH theories succeed in predicting structure qualitatively similar to that observed, but give slightly too much scattering at large frequencies and too little at low frequencies. The predicted positions of the two main resonances are approximately correct for the a-axis but not for the c-axis.

The measurements have not been corrected for elastic incoherent nuclear scattering, and hence the discrepancies between the theories and experiment at low frequencies may be less than indicated.

## CONCLUSION

We have shown that the spin-wave lineshapes in the dilute antiferromagnet $Mn_{0.32}Zn_{0.68}F_2$ exhibit structure ascribable to two or possibly three Ising resonances. Similar structure has also recently been observed by Dietrich et al.[14] for $Mn_{0.68}Zn_{0.32}F_2$ and by March et al.[13] for $KMn_{0.96}Zn_{0.04}F_3$. The BPEH theory and Monte Carlo method give results qualitatively similar to experiment, but significant discrepancies remain. The observation of structure in the lineshape indicates that it is necessary to include the random environment of an ion in any defect theory.

The present study shows that no existing theory for spin waves in substitutionally disordered systems is completely satisfactory. At low frequencies the difficulties may, at least in part, be associated with the neglect in the theories of the randomization of the spin direction produced by the varying direction of the local magnetic-dipole force. Near the critical concentration for the existence of long-range magnetic order, finite temperature effects may also be important, although they are expected to be small for all but the lowest resonances.

## ACKNOWLEDGMENTS

We are grateful to H. Nieman and R. Campbell for expert technical assistance.

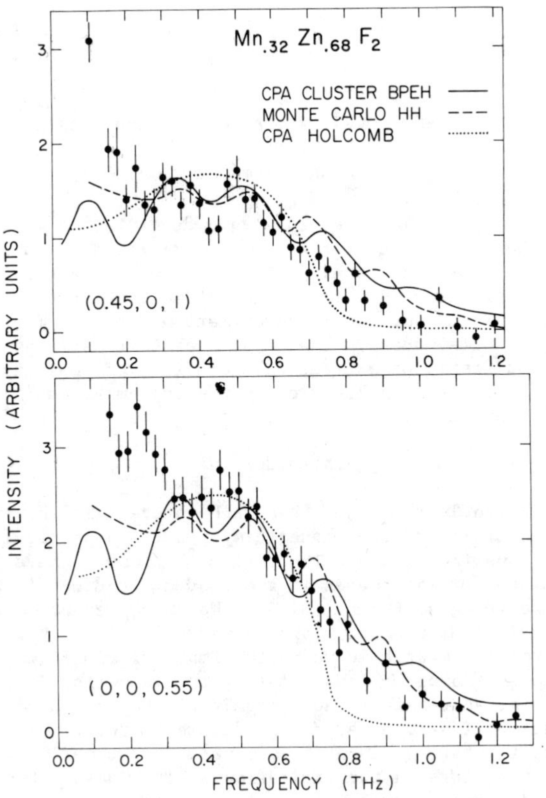


Fig. 3. Comparison of experiment with the CPA theory of Holcomb (Ref. 6), the Monte-Carlo calculations of Holcomb and Harris (Ref. 7), and the CPA cluster theory of Buyers et al. (Ref. 4). The Monte-Carlo calculations are actually for $Mn_{0.35}Zn_{0.65}F_2$.

## REFERENCES

1. R.A. Cowley and W.J.L. Buyers, Rev. Mod. Phys. 44, 406-50 (1972).
2. A.B. Harris, P.L. Leath, B.G. Nickel and R.J. Elliott, J. Phys. C 7, 1693-1718 (1974).
3. R.J. Elliott, J.A. Krumhansl and P.L. Leath, Rev. Mod. Phys. 46, 465-543 (1974).
4. W.J.L. Buyers, D.E. Pepper and R.J. Elliott, J. Phys. C 5, 2611-28 (1972) and 6, 1933-52 (1973).
5. G.J. Coombs and R.A. Cowley, J. Phys. C 8, 1889-1900 (1975).
6. W.K. Holcomb, J. Phys. C 7, 4299-4313 (1974).
7. W.K. Holcomb and A.B. Harris, AIP Conf. Proc. 24, 102-3 (1975).
8. For a summary of much of the earlier experimental work see E.C. Svensson, W.J.L. Buyers, T.M. Holden, R.A. Cowley and R.W.H. Stevenson, AIP Conf. Proc. No. 5, pp. 1315-33 (1972).
9. E.C. Svensson, T.M. Holden, W.J.L. Buyers, R.A. Cowley, G.J. Coombs and D.A. Jones, Proc. Int. Conf. on Magnetism, ICM-73 (Nauka, Moscow, 1974) Vol. V, pp. 567-70, and G.J. Coombs, R.A. Cowley, W.J.L. Buyers, E.C. Svensson, T.M. Holden and D.A. Jones (to be published).
10. E.C. Svensson, S.M. Kim, W.J.L. Buyers, S. Rolandson and R.A. Cowley, AIP Conf. Proc. No. 24, pp. 161-2 (1975).
11. R.A. Cowley, O.W. Dietrich and D.A. Jones, J. Phys. C (in press).
12. J. Als-Nielsen, R.J. Birgeneau, H.J. Guggenheim and G. Shirane, Phys. Rev. (in press).
13. R.H. March, E.C. Svensson, T.M. Holden, R. Stedman and D.A. Jones (these proceedings).
14. O.W. Dietrich, G. Mayer, R.A. Cowley and G. Shirane (private communication).

# CALCULATIONS OF SPIN WAVES IN A RANDOMLY DILUTED TWO-DIMENSIONAL ANTIFERROMAGNET*†

R. Alben and M. F. Thorpe
Department of Engineering and Applied Science, Yale University, New Haven, Connecticut 06520

## ABSTRACT

We have calculated spin wave densities of states and the neutron scattering law S(Q,E) for a randomly diluted two-dimensional antiferromagnet. The calculations were done by the equation of motion method within the linear spin wave approximation for relatively large finite models. The results exhibit sharp peaks characteristic of excitations localized on small numbers of sites. Application to the system $Rb_2Mn_cZn_{1-c}F_4$ is discussed.

## INTRODUCTION

Recently there has been much interest in the properties of disordered magnetic systems. In particular, both experimental[1] and theoretical[2-4] results have been obtained for spin waves in the random two-dimensional antiferromagnet $Rb_2Mn_{1-c}Ni_cF_4$. We here present theoretical results which might be expected to apply to the related system in which the impurities are (nonmagnetic) Zn or Mg ions rather than (magnetic) Ni ions. That is, we consider the randomly diluted two-dimensional antiferromagnet.[5] The dilute system is of particular interest since it is expected to clearly display structure in the excitation spectra associated with different surroundings of the magnetic ions.[6]

Our approach is to compute the spin wave density of states and the neutron scattering law S(Q,E) within the linear spin wave approximation for finite models with periodic boundary conditions. We restrict ourselves to T = OK and assume the ground state to be the Néel state. The latter assumption might be expected to break down particularly badly for small isolated clusters of magnetic ions such as are prevalent at low impurity concentrations. Therefore we examine the exact quantum mechanical results for two and three spin clusters and find that the qualitative predictions of the spin wave analysis continue to hold even in these cases.

In order to deal with relatively large models we have used the equation of motion method for the spin wave problem.[2,7]

## MODEL AND RESULTS

We consider a square lattice of sites, a fraction c of which are occupied by magnetic ions. The interaction between ions is given by the following Hamiltonian which describes the interactions in pure $Rb_2MnF_4$:[8]

$$H = J \sum_{<i,j>} S_i S_j + \sum_i H_i^A S_i^z , \qquad (1)$$

The first sum is over pairs of occupied sites and J is the nearest neighbor exchange which we take +0.655 meV (antiferromagnetic). The second sum is over individual occupied sites and $H_i^A$ is +0.03 meV for sites on the down sublattice and -0.03 meV for sites on the up sublattice. The spin s is 5/2.

In Fig. 1 we show the density of states, calculated in the linear spin wave approximation for three concentrations. As c is reduced from 1.0 to below

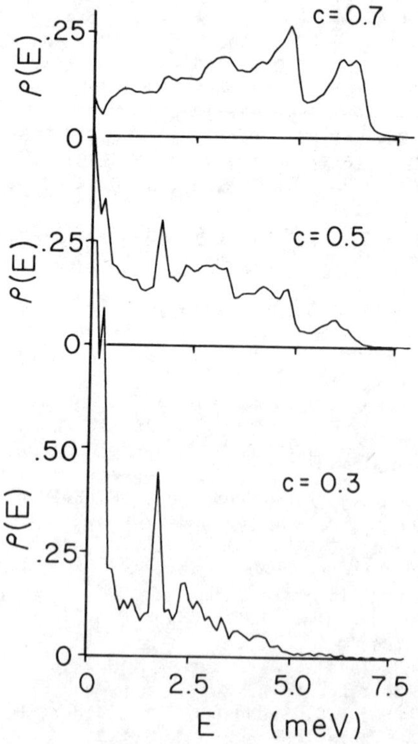


Fig. 1 Density of spin wave states for three concentrations c of magnetic sites. (3600 site models. The resolution width Δ is 0.6 meV.)

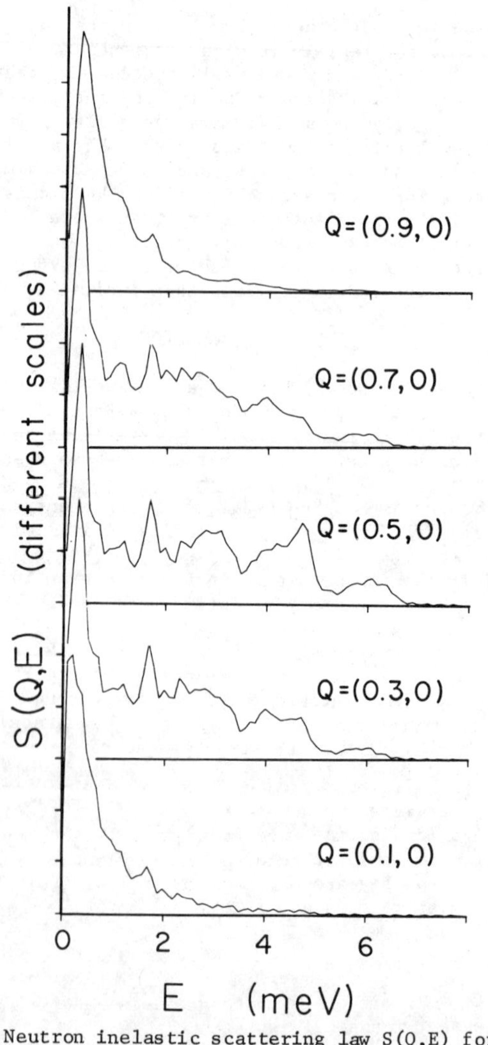


Fig. 2 Neutron inelastic scattering law S(Q,E) for c = 0.5. (3600 site models; Δ = 0.6 meV). The curves for each value of Q are normalized differently for ease of presentation (see Fig. 3).

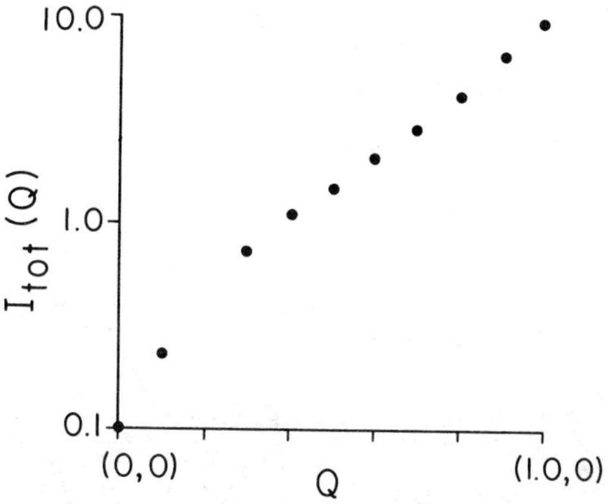


Fig. 3 Total integrated scattering intensity as a function of wave vector Q along the magnetic [1,0] direction for c = 0.5. (3600 site models; $\Delta$ = 0.6 meV).

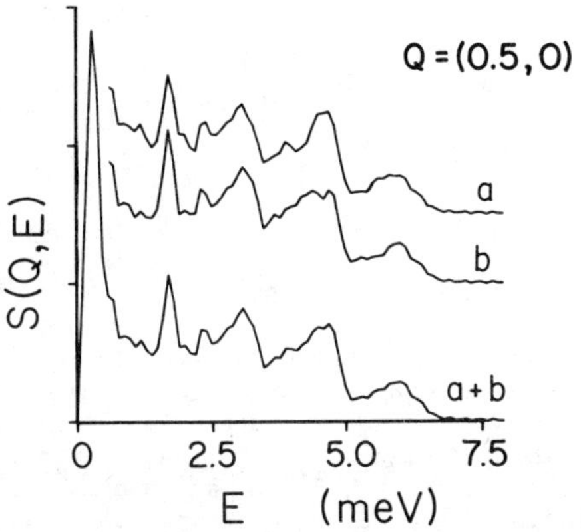


Fig. 4 Structure in S(Q,E) for c = 0.5 with Q at a magnetic zone boundary point. (10000 site models; $\Delta$ = 0.6 meV).

the percolation concentration, which is $\sim$0.6 for the square lattice, the density of states changes character from a smooth function to a distribution of sharp features from different types of small isolated clusters. Of particular note is the feature at E = JS = 1.64 meV. This may be shown to arise from modes associated with "triads" in which two spins share a third spin as their sole magnetic neighbor. The third spin may have no additional neighbors (isolated triads) or may connect to a larger cluster.

In Fig. 2 we show S(Q,E) for one concentration c = 0.5 and Q along the (1,0) direction. (The wave vector Q is indexed to the magnetic Brillouin zone as in Refs. 1,3.) It should be noted that in the pure system, S(Q,E) shows a single sharp mode whose energy first increases as Q increases from (0,0) to (0.5,0) and then decreases as Q increases from (0.5,0) to (1,0). The overall tendency of the spectrum to be weighted toward higher energies for Q near the (0.5,0) remains even for the dilute case.

The scales in Fig. 2 were chosen differently for each Q in order to display the change in shape of S(Q,E). In Fig. 3 we show the behavior of the total integrated intensity with Q. The intensity changes by about a factor of 100 in going from the zone center to the zone edge. A similar behavior is found for the pure antiferromagnet.

## APPLICATION TO $Rb_2Mn_cZn_{1-c}F_4$

In Fig. 4 we show S(Q,E) for c = 0.5 obtained by averaging results for several different large models. The peaks and subpeaks in the spectrum are evidently reproducible features of the linear spin wave solution for the dilute system. It would be interesti~g to find such structure in a real system. However before these results can be applied to a real system we must consider two questions: A. Does the Hamiltonian (1) describe the interactions? and B. Does the linear spin wave solution make sense, particularly for concentrations below the percolation threshold where the sharp features are most evident?

For $Rb_2Mn_{1-c}Zn_cF_4$ the Hamiltonian (1) almost certainly describes the interactions within the layers. There are however also significant interactions between layers which need not cancel as effectively in the dilute system as they do for pure $Rb_2MnF_4$.[8] Some deviations from the predicted behavior are to be expected from this effect. However, judging by the essentially two dimensional character which has been found for the mixed system $Rb_2Mn_cNi_{1-c}F_4$,[1] we may hope that the deviations will be small.

Deviations from the linear spin wave approximation are also expected to increase for the dilute system. We consider then the exact results for two spin and three spin clusters. We neglect the small anisotropy term. For the two spin case the exact energies are given by

$$E^{(2)} = \frac{1}{2} J\ [j(j+1)-2s(s+1)], \tag{2}$$

where the total spin j varies from 0 to 2s. The ground state is j = 0 and allowed transitions associated with S(Q,E) connect only to the state j = 1. The allowed excitation energy is $\Delta E^{(2)}$ = J = 0.655 meV. The linear spin wave analysis predicts an excitation energy of 0, while the maximum energy of the pure system is 4JS = 6.55 meV. For system with a high spin, the spin wave result for a two spin cluster is a relatively good approximation.

For three spins, the exact energies are given by

$$E^{(3)} = \frac{1}{2} J\ [j(j+1)-j'(j'+1)-s(s+1)], \tag{3}$$

where $2s \geq j' \geq 0$ and $j' + s \geq j \geq |j'-s|$. The allowed transitions are summarized by selection rules $\Delta j'=0,\pm1$ and $\Delta j=0,\pm1$. For the present case the transition energies are: $\Delta E^{(3)}$ = 0 meV, 1.64 meV, 2.29 meV, 3.28 meV, 5.57 meV. The linear spin wave result gives transition at 0 and 1.64 meV in agreement with the two lowest allowed excitations.

Although the excitation spectra of larger isolated clusters become considerably more complex, these results suggest that the linear spin wave predictions of a multipeak structure in S(Q,E) might well be an observable in $Rb_2Mg_cZn_{1-c}F_4$ and related materials.

We thank M. Blume and R. J. Birgeneau for helpful discussions.

## REFERENCES

* Work supported in part by NSF.
† Calculations at Brookhaven National Laboratory supported by E.R.D.A.

1. R. J. Birgeneau, L. R. Walker, H. J. Guggenheim, J. Als Nielsen and G. Shirane, J. Phys. C <u>8</u>, L328 (1975); J. Als Nielsen, R. J. Birgeneau, H. J. Guggenheim and G. Shirane, Phys. Rev. B (in press).
2. D. L. Huber, Solid St. Commun. 14, 1153 (1974).
3. R. Alben and M. F. Thorpe, J. Phys. C8, L275 (1975).
4. S. Kirkpatrick and A. B. Harris, Phys. Rev. B (in press).
5. D. L. Huber, Phys. Rev. B10, 4621 (1974).
6. W. K. Holcomb and A. B. Harris in Magnetism and Magnetic Materials, 1974, C. Graham, Jr., J. Rhyne and G. Lander, eds. (A.I.P., New York 1975), p.102.
7. see R. Alben, M. Blume, H. Krakauer and L. Schwartz, Phys. Rev. B (in press).
8. R. J. Birgeneau, H. J. Guggenheim and G. Shirane, Phys. Rev. B1, 2211 (1970).

# RESONANT PERTURBATION OF SPIN WAVES IN $KMn_{0.99}Mg_{0.01}F_3$ AND $KMn_{0.96}Zn_{0.04}F_3$

R.H. March*, E.C. Svensson and T.M. Holden
Atomic Energy of Canada Limited, Chalk River, Ontario, Canada, K0J 1J0
R. Stedman
AB Atomenergi, Studsvik, FACK, S-61101 Nyköping 1, Sweden
D.A. Jones
University of Aberdeen, Aberdeen, Scotland

## ABSTRACT

The perturbation of spin waves in the perovskite antiferromagnet $KMnF_3$ by non-magnetic Mg and Zn impurities has been studied by neutron inelastic scattering. On the basis of Ising theory, a resonant perturbation is expected at about 1.9 THz, and a marked resonant perturbation centered in the range 1.8 - 1.9 THz is observed for both materials. For $KMn_{0.99}Mg_{0.01}F_3$, single-peaked lineshapes are observed with a maximum impurity-induced width Γ of $0.107 \pm 0.012$ THz. For $KMn_{0.96}Zn_{0.04}F_3$, double-peaked lineshapes are observed near resonance and the maximum Γ is $0.354 \pm 0.015$ THz. The effect of the impurities on the frequencies is small relative to the effect on the widths.

## INTRODUCTION

The case in which non-magnetic ions are substituted for magnetic ions in antiferromagnetic insulators provides one of the simplest examples for the study of the effects of impurities on elementary excitations. At low impurity concentration, one expects, on the basis of Ising theory, a resonant perturbation of the spin waves while at higher impurity concentrations one expects broad multiple-peaked lineshapes because of the different Ising energies associated with the different clusters of neighbors that then occur with comparable probability. (A cluster consisting of $r = 0,1,...n$ impurity ions and $n - r$ host ions, where n is the number of nearest antiferromagnetically-coupled neighbors, has an associated Ising energy $\nu_m(n - r)/n$ where $\nu_m$ is the maximum frequency of the host spin waves.)

The resonant perturbation of spin waves in $Mn_{0.95}Zn_{0.05}F_2$ has been studied[1] in detail by neutron scattering, and the results are adequately described[1,2] by first-order Green-function theory[2]. Neutron-scattering results for specimens of $Mn_{1-c}Zn_cF_2$ having $c = 0.22$, 0.68 and 0.75 have also been reported[3] and are reasonably well described[3,4] by the theory of Buyers et al.[4] which is based on the coherent-potential approximation (CPA). Higher-resolution measurements on $Mn_{0.32}Zn_{0.68}F_2$ which show a multiple-peaked structure are reported[5] in these proceedings, and a similar study has also recently been carried out[6] for $Mn_{0.68}Zn_{0.32}F_2$. In this paper we report the results of neutron-scattering measurements at 4.3 K on single-crystal specimens of $KMn_{0.99}Mg_{0.01}F_3$ and $KMn_{0.96}Zn_{0.04}F_3$.

## EXPERIMENT AND DISCUSSION

The measurements on $KMn_{0.99}Mg_{0.01}F_3$ and $KMn_{0.96}Zn_{0.04}F_3$ were carried out, respectively, at Studsvik and Chalk River using triple-axis crystal spectrometers operated in the constant-momentum-transfer (constant-$\vec{Q}$) mode with fixed scattered-neutron energy.

A selection of scattered-neutron distributions for $KMn_{0.96}Zn_{0.04}F_3$ is shown in Fig. 1. A marked double-peaked structure is clearly resolved in the measurements near the principal ($r = 1$) resonance which is predicted by Ising theory to occur at $\frac{5}{6}\nu_m = 1.91$ THz. Distributions D-H show how rapidly the lineshape changes as one passes through the resonance wave vector. In distributions B and C the resonance mode is seen as a separate weak peak or shoulder. The shifts from the corresponding frequencies[7] for $KMnF_3$ (arrows in Fig. 1) are $\lesssim 0.1$ THz.

The observed widths for $KMn_{0.96}Zn_{0.04}F_3$ are shown in Fig. 2 together with the calculated resolution

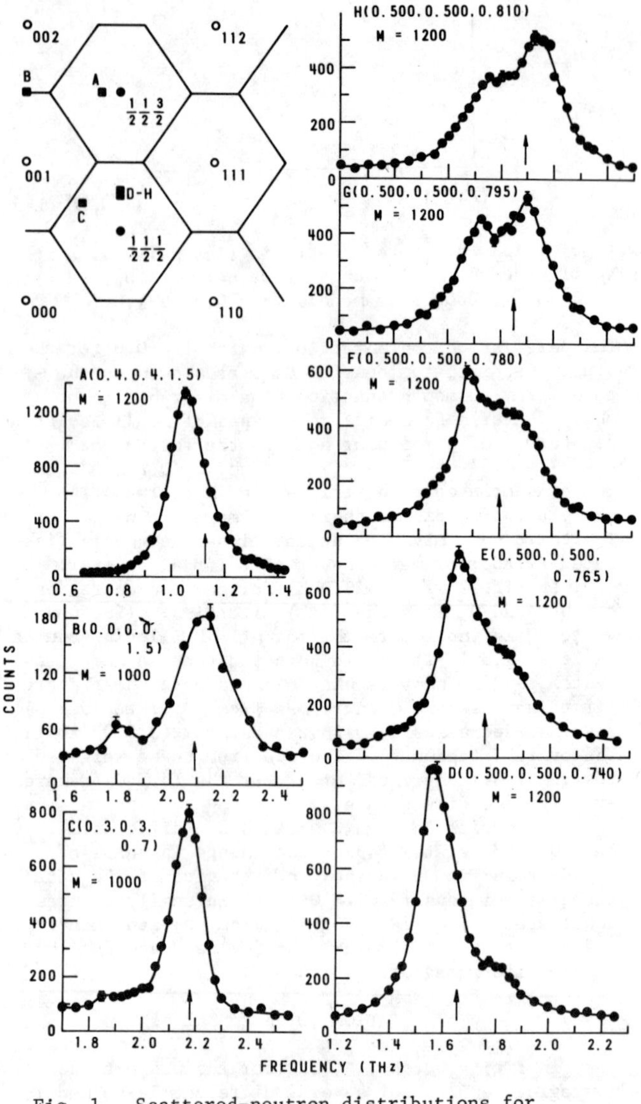


Fig. 1. Scattered-neutron distributions for $KMn_{0.96}Zn_{0.04}F_3$ obtained in constant-$\vec{Q}$ scans at the positions (A-H) indicated on the reciprocal-lattice diagram. Each distribution is labelled with the values of $a\vec{Q}/2\pi$ (a is the lattice constant) and the monitor count M. Arrows indicate the corresponding frequencies for $KMnF_3$ (Ref. 7).

widths. One sees that the Zn impurities cause a large "resonant" perturbation at intermediate wave vectors. The positions of the maxima in the [00ζ], [ζζ0] and [ζζζ] directions correspond respectively to resonance frequencies of 1.87, 1.88 and 1.86 THz which are very close to the Ising-theory value of 1.91 THz. In each direction there is also a weak subsidiary maximum at a lower wave vector than the main maximum. These correspond to frequencies of 1.40, 1.36 and 1.42 THz for the [00ζ], [ζζ0] and [ζζζ] directions and are probably attributable to the second ($r = 2$) Ising resonance expected at $\frac{4}{6}\nu_m = 1.53$ THz. Weak shoulders corresponding to these modes were also observed in several distributions.

The results for $KMn_{0.96}Zn_{0.04}F_3$ exhibit much more structure than observed[1] for $Mn_{0.95}Zn_{0.05}F_2$. This is probably attributable to the larger spacing between the Ising levels for the $K(Mn,Zn)F_3$ system (0.38 vs. 0.21

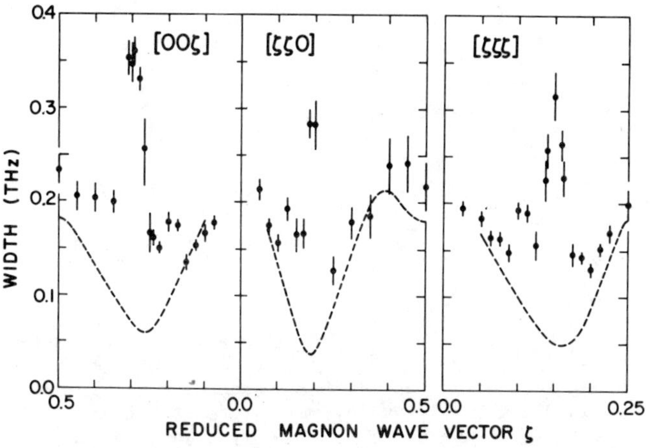


Fig. 2. Observed widths (FWHM) of scattered-neutron distributions for $KMn_{0.96}Zn_{0.04}F_3$ and the calculated widths of the experimental resolution (dashed lines).

THz) and suggests that measurements on $KMn_{1-c}Zn_cF_3$ specimens having large c would give more conclusive evidence for multiple-peaked lineshapes than obtained in the recent measurements[5,6] on $(Mn,Zn)F_2$ specimens.

A selection of scattered-neutron distributions for $KMn_{0.99}Mg_{0.01}F_3$ is shown in Fig. 3. No marked structure in the lineshapes was observed, but at certain wave vectors there was a suggestion of a weak "resonance-mode" peak or shoulder on the side of the main peak as illustrated by distribution B. The frequency shifts for $KMn_{0.99}Mg_{0.01}F_3$ are barely significant, but the impurity-induced widths Γ exhibit a characteristic resonance behavior as shown in Fig. 4. The values of Γ were obtained by deconvolution assuming Lorentzian intrinsic lineshapes and Gaussian experimental resolution. Deviations of the actual lineshapes from the assumed forms will not significantly change the results. The maxima correspond to "resonance" frequencies of 1.79, 1.92 and 1.86 THz for the [00ζ], [ζζ0] and [ζζζ] directions in satisfactory agreement with the results for $KMn_{0.96}Zn_{0.04}F_3$ and the prediction of Ising theory. Within the accuracy to which the values of c are known (about 20%), the maximum values of Γ for the Mg-doped and Zn-doped specimens scale with impurity concentration. The present measurements therefore indicate that Zn and Mg impurities have essentially identical effects on the spin waves as expected.

Frequency shifts and widths for $KMn_{0.95}Zn_{0.05}F_3$, calculated on the basis of first-order Green-function theory, were given by Cowley and Buyers[2] (see their Fig. 24). These are not at all in agreement with the present measurements, but this is believed (Cowley and Buyers, private communication) to be the result of an error in the numerical calculations. Preliminary calculations using CPA theory[4] have been carried out in collaboration with W.J.L. Buyers and a detailed comparison of theory and experiment will be published later.

ACKNOWLEDGMENTS

We would like to acknowledge helpful discussions with W.J.L. Buyers and R.A. Cowley and to thank K.-O. Isaxon, S. Sandell, R. Campbell and H. Nieman for valuable technical assistance. One of us (ECS) is grateful to the staff of AB Atomenergi for their hospitality and assistance during his stay at Studsvik.

REFERENCES

*Permanent Address: Dalhousie University, Halifax, Nova Scotia.

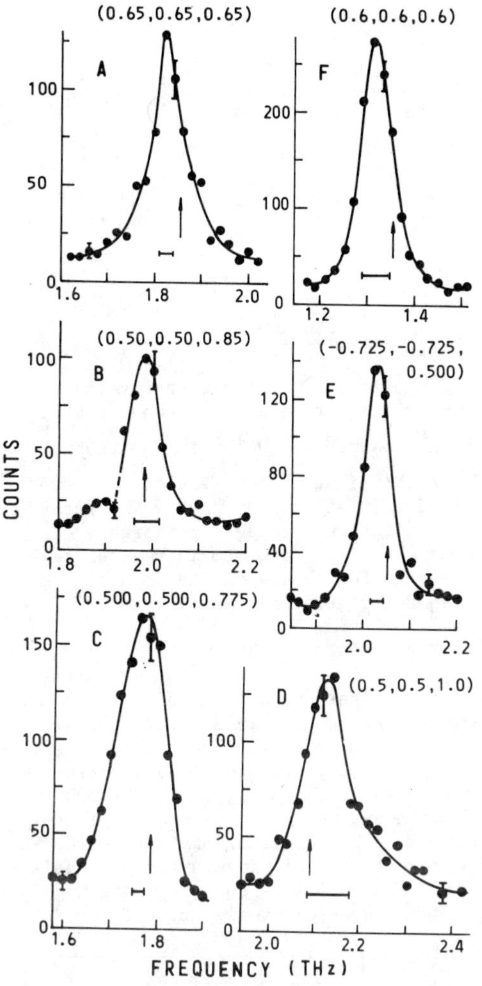


Fig. 3. Scattered-neutron distributions for $KMn_{0.99}Mg_{0.01}F_3$. Each distribution is labelled with the value of $a\vec{Q}/2\pi$. Horizontal bars and arrows indicate respectively the calculated resolution widths and the corresponding frequencies for $KMnF_3$ (Ref. 7).

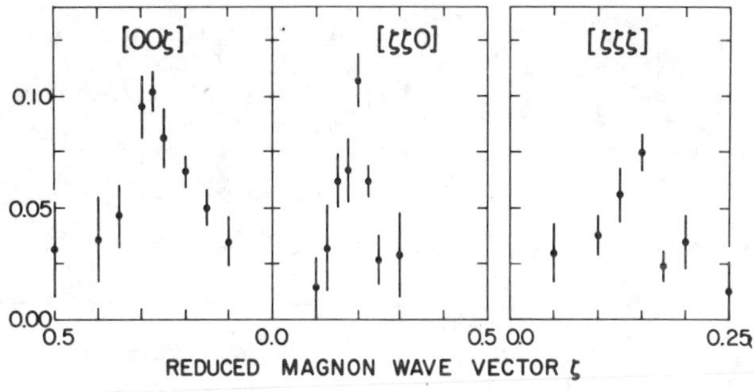


Fig. 4. Impurity-induced widths for $KMn_{0.99}Mg_{0.01}F_3$.

1. E.C. Svensson, T.M. Holden, W.J.L. Buyers, R.A. Cowley and R.W.H. Stevenson, Solid State Commun. 7, 1693-6 (1969); E.C. Svensson, W.J.L. Buyers, T.M. Holden, R.A. Cowley and R.W.H. Stevenson, AIP Conf. Proc. No. 5, pp. 1315-33 (1972).
2. R.A. Cowley and W.J.L. Buyers, Rev. Mod. Phys. 44, 406-50 (1972).
3. E.C. Svensson, T.M. Holden, W.J.L. Buyers, R.A. Cowley, G.J. Coombs and D.A. Jones, Proc. Int. Conf. on Magnetism ICM-73 (Nauka, Moscow, 1974) Vol. V, pp. 567-70.
4. W.J.L. Buyers, D.E. Pepper and R.J. Elliott, J. Phys. C 6, 1933-52 (1973).
5. E.C. Svensson, W.J.L. Buyers, T.M. Holden and D.A. Jones (these proceedings).
6. O.W. Dietrich, G. Meyer, R.A. Cowley and G. Shirane (private communication).
7. S.J. Pickart, M.F. Collins and C.G. Windsor, J. Appl. Phys. 37, 1054-55 (1966).

# SPIN WAVES IN $KNi_{0.75}Mn_{0.25}F_3$

G. J. Coombs and R. A. Cowley
Department of Physics, University of Edinburgh
Scotland

D. A. Jones
Department of Natural Philosophy
University of Aberdeen, Scotland

G. Parisot and D. Tocchetti
Institut Laue-Langevin, Grenoble, France

## ABSTRACT

The magnetic excitations in a single crystal of the substitutionally disordered antiferromagnet, $KMn_{0.25}Ni_{0.75}F_3$ have been measured at 10 K using neutron inelastic scattering techniques. The results show two fairly well-defined excitations for each wave vector corresponding to excitations propagating largely on either the Mn or the Ni ions, and are in reasonable agreement with calculations based on the coherent potential approximation for disordered systems.

## EXPERIMENTAL RESULTS

This experiment is part of a continuing program to measure the magnetic excitations in substitutionally disordered crystals and so by comparing the results with the predictions of theory to test the theories of the excitations in disordered systems.[1] $KMnF_3$ and $KNiF_3$ are both simple antiferromagnets and simple cubic at high temperatures. The substitutionally disordered single crystal was grown by the Stockbarger process by cooling a melt of the appropriate quantities of highly purified $KMnF_3$ and $KNiF_3$. The single crystal obtained had a simple cubic structure with a lattice parameter of 4.063 Å. The temperature dependence of a magnetic Bragg reflection was determined and found to be zero above 229 K. Below this temperature it increased with decreasing temperature more slowly than would be expected for a pure antiferromagnet, possibly due to a concentration gradient in the crystal. From these latter results, and assuming a linear dependence of $T_N$ on concentration, we estimate the concentration of Mn ions to be 0.25 ± 0.05.

The neutron inelastic scattering measurements were performed at the Institut Laue-Langevin using triple-axis crystal spectrometers situated both on a thermal neutron beam and on a beam from the hot source. In all of the measurements the spectrometers were controlled so as to obtain energy distributions while keeping the momentum transfer fixed. In some of the experiments the incident wave vector was held constant throughout each scan at a wave vector between 3.06 $Å^{-1}$ and 4.1 $Å^{-1}$ while in other experiments the scattered wave vector was fixed at a value between 4.1 $Å^{-1}$ and 5.1 $Å^{-1}$.

Measurements were made of the excitations with wave vectors propagating along the three principal symmetry axes at 10 K. The results of 4 scans are shown in Fig. 1. They show that there are two fairly well-defined branches of the excitations whose dispersion relations are shown in Fig. 2. The width of the neutron groups shown in Fig. 1, is somewhat larger than the experimental resolution especially close to the zone boundary of the branch with lowest frequency. The intensity of the scattering of both branches decreased as the wave vector increased towards the zone boundary. Unfortunately, the experimental difficulties encountered in making the measurements were such as to preclude a detailed and quantitative study of the widths and intensities of the excitations.

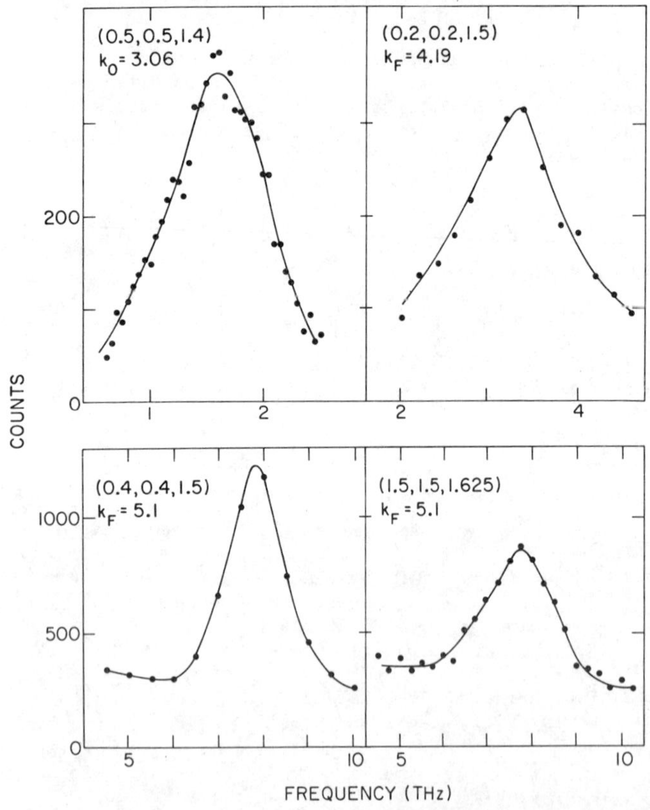


Fig. 1. Typical scattered neutron distributions for $KNi_{0.75}Mn_{0.25}F_3$ at 10 K observed for various wave-vector transfers (given in reciprocal lattice units). The distributions were obtained with either $k_0$ or $k_F$ fixed at the values ($Å^{-1}$) indicated.

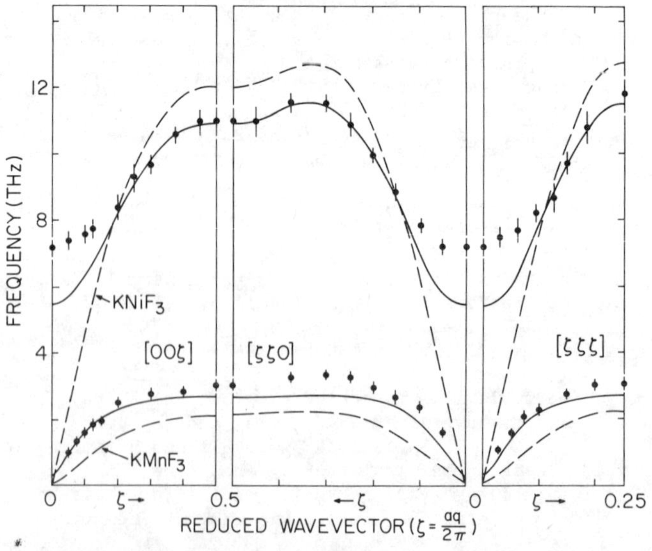


Fig. 2. The frequencies of the excitations in $KNi_{0.75}Mn_{0.25}F_3$ at 10 K. The dotted lines give the corresponding frequencies of the pure materials. The solid lines are calculations based on the coherent potential approximation with scaling of the spins to take account of the transverse part of the exchange interactions.

## CALCULATIONS AND DISCUSSION

The experimental results shown in Fig. 2 show that the branch of lower frequency is very similar in shape to that of pure $KMnF_3$ although the frequencies are almost 50% higher than those of the pure material[2]. The frequency of the upper branch at the zone boundary is close to the maximum frequency found in pure $KNiF_3$.[3] Qualitatively these results and the overall shape of the dispersion relations is very similar to those found in the $KMnF_3/KCoF_3$ system[4] and in the two-dimensional system $Rb_2MnF_4/Rb_2NiF_4$ system.[5]

The frequencies of the excitations at the zone boundary may be calculated using the Ising model for the excitations. When the Ni-Ni and Mn-Mn exchange constants are taken from measurements on the pure materials and the Mn-Ni constant obtained from measurements on dilute Ni in $KMnF_3$[6], the model predicts frequencies of 11.5 THz and 2.87 THz in reasonable accord with the results given for the [111] zone boundary in Fig. 2.

Calculations have also been performed using various approximations based on the coherent potential approximation. The calculations were performed in a similar manner to those reported[7] for the $Rb_2Mn_{0.5}Ni_{0.5}F_4$ system. In Fig. 2 we show the results obtained when the spin scaling approach was used to include the transverse parts of the exchange interactions. Very similar results were also obtained when the transverse interactions were included by use of the approximation developed by Buyers, Pepper, and Elliott.[8] The theory clearly gives a very reasonable description of the experimental results. Some of the discrepancies may arise from the uncertainty in the concentration of the specimen or from changes in the exchange constants with lattice parameter.

Financial support for this work was provided by the Science Research Council.

## REFERENCES

1. For a review see R. A. Cowley(this conference).
2. S. J. Pickart, M. F. Collins, and C. G. Windsor, J. Appl. Phys. 37, 1054-1055 (1966).
3. S. R. Chinn, H. J. Zeiger, and J. R. O'Connor, J. Appl. Phys. 41, 894-895 (1970).
4. W. J. L. Buyers, T. M. Holden, E. C. Svensson, R. A. Cowley, and R. W. H. Stevenson, Phys. Rev. Letters 27, 1442-1445 (1971).
5. R. J. Birgeneau, L. R. Walker, H. J. Guggenheim, J. Als-Nielsen, and G. Shirane, J. Phys. C: Solid State Phys. 8, L328-333 (1975).
6. T. M. Holden, R. A. Cowley, W. J. L. Buyers, E. C. Svensson, and R. W. H. Stevenson, J. de Physique 32, C1, 1184-1185 (1971).
7. G. J. Coombs and R. A. Cowley, J. Phys. C: Solid State Phys. 8, 1889-1900 (1975).
8. W. J. L. Buyers, D. E. Pepper,and R. J. Elliott, J. Phys. C: Solid State Phys. 5, 2611-2628 (1972).

# MAGNETIC EXCITATIONS IN ANTIFERROMAGNETIC $Ni_{1-x}Co_xO$

V. Wagner[+)], D. Tocchetti[*)] and B. Hennion[°)]
Institut Laue-Langevin, B.P. 156, Grenoble 38042, France

## ABSTRACT

The dispersion curves of the magnetic excitations were determined by inelastic neutron scattering techniques for a Ni-rich sample (x=0.30) at 4.2 K. This is the first alloy of fcc antiferromagnets of type II for which the spinwave dispersion has been studied. Two branches of well-defined excitations were observed along the [ζζζ] and [ζζ0] directions. The lower branch is split at $q \approx 0$ due to anisotropy and domain effects, the energy gap being $\varepsilon(0)=3.2$ THz. The upper branch with $\varepsilon(0)=12.0$ THz becomes very close to the magnon dispersion in pure NiO for increasing wavevector.

The results have been analysed in terms of a molecular field model assuming isotropic Heisenberg exchange and accounting for the exchange mixing of the spin orbit levels of the ground state multiplet of $Co^{2+}$. The dominating exchange interaction between Ni and Co was found to be $2J_2=2.4 \pm 0.2$ THz.

## INTRODUCTION

NiO and CoO are collinear antiferromagnets which have NaCl structure in the paramagnetic phase. In the antiferromagnetic phase the spins order in a slightly distorted fcc arrangement of type II. The spin waves in NiO ($T_N$=523 K) are well known from inelastic neutron scattering and can be described in terms of a Heisenberg model including a small anisotropy[1]. In the case of CoO ($T_N$=290 K) the situation is more complex as the orbital angular momentum of $Co^{2+}$ is not completely quenched by the crystal field. The predominant exchange is between next nearest neighbors as in NiO, but spin-orbit coupling and single-ion anisotropy of $Co^{2+}$ are important for the magnetic properties[2,3].

Despite experiments with optical[4,5] and neutron scattering techniques[6,7] the magnetic excitations in CoO arising from the ground state multiplet have not yet been clearly established. This is a drawback for a study of the magnetic excitations in the mixed crystal, which is very interesting because both the exchange interaction and anisotropy of the pure materials are quite different.

We have concentrated on a Ni-rich sample (x=0.30) where the excitations of the host are well known. The concentration of Co was chosen above the percolation limit, so that the resonant modes arising from the Co levels could be expected to spread out. No localised magnetic mode can be expected for a Co impurity in NiO except for single-ion like longitudinal excitons of $Co^{2+}$. The dispersion of the magnetic excitation was measured by neutron inelastic scattering and analysed within a molecular field model[8]. From this the predominant exchange interaction between Ni and Co was deduced, as will be discussed in more detail.

## EXPERIMENTAL RESULTS

The experiment was carried out at the high flux beam reactor at Grenoble. Most of the data were collected with the triple-axis spectrometer IN1 at the hot source using incoming energies up to 250 meV. Some additional measurements at small energy transfer were made at the triple-axis spectrometer IN3 at a thermal neutron guide. The sample was a multi-domain crystal[9] at 4.2 K and 4 $cm^3$ in size. It was cut from a flame-grown single crystal containing 30% Co and oriented with a [1$\bar{1}$0] axis perpendicular to the scattering plane. Magnetic Bragg peaks of domains A and B, which arise from the spin alignment in planes (111) and (11$\bar{1}$) respectively, were of same intensity.

The results of scans in different magnetic zones are summarized in Fig. 1. The bars indicate whether the scattered neutron peaks were observed in a scan

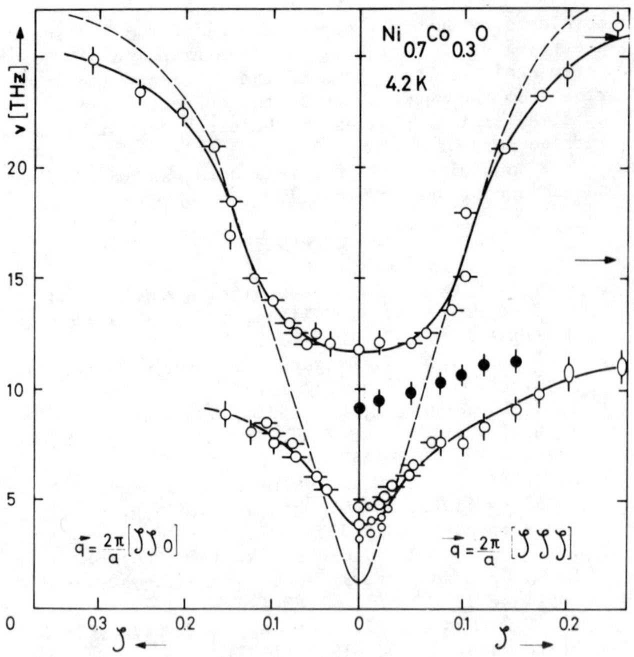


Fig. 1

Dispersion curves in antiferromagnetic $Ni_{0.7}Co_{0.3}O$. Open circles: magnetic excitations, full circles: TO phonon. The full line is a guide for the eye. The magnon dispersion in NiO[1] is indicated by the dashed line.

with constant energy transfer $\varepsilon$ or constant momentum transfer $\hbar\vec{Q}$. The wavevector $\vec{q}$ is reduced with respect to magnetic lattice points and given in units of the chemical cell, however. The lattice parameter of the pseudocubic chemical unit cell was found to be 4.197 Å as compared to 4.177Å for pure NiO. Two branches of well defined magnetic excitations were observed within the energy range of the magnons in pure NiO. The magnetic character was established by the Q-dependence of the scattered intensity. Most of the scattered neutron groups of the upper branch were broader than expected from instrumental resolution. Comparison with a pure NiO sample showed that the broadening is related to disorder in the mixed crystal. At low energies line broadening was less evident and it was possible to resolve domain effects[1] near $q \approx 0$. The higher branch starts at 12.0 THz in the zone center and becomes very similar to the magnon branch in pure NiO. The lower branch is split near q=0 with an energy gap at 3.2 THz. It could not be followed completely along the [110] direction. There was a third branch observed in the main symmetry direction [111] which is presumably the TO phonon and not a magnetic excitation. Its frequency at a nuclear lattice point is 11.8 THz and agrees with ir-data of a mixed crystal of similar composition[10].

## DISCUSSION

It is well known that in a simple antiferromagnet a molecular field model is a good approximation for the spinwave energies at the zone boundary of a symmetry direction. This also holds for a mixed crystal if the different clusters are considered[8]. In the present case, the simplest Hamiltonian accounting for the major interactions is

$$H(i) = \lambda(i)\vec{\ell}_i\vec{S}_i + 2\sum_d J_{i,i+\delta}\vec{S}_i\langle\vec{S}_{i+\delta}\rangle \qquad (1)$$

where the first order spin-orbit interaction is nonzero only for Co. Only contributions of the 6 next nearest neighbors to the molecular field will be considered. Nearest neighbor exchange and anisotropy are neglected as of minor importance. For a cluster with a Ni ion at the center (S=1,ℓ=0) and (6-r) Ni and r Co ions at the n.n.n. sites, the excitation energy for a spin deviation at the central Ni ion is

$$\varepsilon(Ni) = 2(6-r)J_2(NiNi)\langle S^z_{Ni}\rangle + 2rJ_2(NiCo)\langle S^z_{Co}\rangle \qquad (2)$$

with $2J_2(NiNi) = 4.6$ THz[1] and where $\langle S^z_{Co}\rangle$ has to be calculated with allowance for the mixing of the Co levels by the echange field. After averaging over the clusters we find $2J_2(NiCo) = 2.4 \pm 0.2$ THz and $\langle S^z_{Co}\rangle = 1.4$ from a fit to the frequency of the upper branch at the zone boundary $\vec{q} = \frac{2\pi}{a}\left[\frac{1}{4}\frac{1}{4}\frac{1}{4}\right]$ (cf. arrow in Fig. 1). The exact value of $J_2(NiCo)$ depends on the value of $\lambda$, which has been taken to be 6.09 THz[11].

For a cluster with a Co ion at the central site we have to diagonalize (1) within the space of the ground state multiplet of Co($S = \frac{3}{2}, \ell=1$) in order to get its single ion energies in the mixed crystal. With our value for $J_2(NiCo)$ and $J_2(CoCo) = 0.9$ THz[2,7], this gives the zone boundary frequency of the lower magnetic branch near 16 THz, if it arises from Co spin fluctuations transforming as $S^-$. This is not confirmed by the experimental result, although the neglect of n.n. interaction and some uncertainty of the spin-orbit coupling and $J_2(CoCo)$ may partially account for the discrepancy. Therefore we suppose the lower branch to mix with another excitation near the zone boundary. So far it is not clear whether there is mixing by transverse exchange with the lowest spin-orbit transition transforming as $S^+$ at 9.2 THz or whether the spin fluctuations of Co interact with the TO phonon.

The lower branch is split near the zone center due to anisotropy and domain effects as in NiO[1]. The lowest frequency mode is considered as an AFMR mode which is shifted by the large anisotropy of $Co^{2+}$ with respect to $Ni^{2+}$. Within an average model we find the single ion anisotropy of Co close to 1 THz in agreement with ir-measurements on mixed crystals with $x \leqslant 0.016$[12]. A detailed study of the dependence of the AFMR modes on the Co concentration is in progress.

In conclusion we should note that the value of n.n.n. exchange $2J_2(NiCo) = 2.4$ THz is larger than would be expected from the geometric mean of the respective exchange in the pure materials. A larger n.n.n. exchange in CoO, as claimed on theoretical grounds[3], would reconcile the geometric mean rule and our finding.

One of the authors (V.W.) gratefully acknowledges financial support from the Bundesministerium für Forschung und Technologie, Bonn.

## REFERENCES


+) Guest scientist from Universität Würzburg, Germany
*) Present address: EMBL Grenoble, France
°) Present address: CEN Saclay, France
1) M.T. Hutchings and E.J. Samuelson, Phys. Rev. B6, 3447 (1972)
2) J. Kanamori, Progr. Theoret. Phys. 17, 177 (1957), 17, 197 (1957)
3) T. Yamada and O. Nakanishi, J. Phys. Soc. Jap. 36, 1304 (1974)
4) I.G. Austin and E.S. Garbett, J. Phys. C3, 1605 (1970)
5) R.R. Hayes and C.H. Perry, Sol. State Comm. 14, 173 (1974)
6) J. Sakurai, W.J.L. Buyers, R.A. Cowley and G. Dolling, Phys. Rev. 167, 510 (1968)
7) V. Wagner and W. Drexel, J. Magnetism and Magnetic Materials, in press
8) R.A. Cowley and W.J.L. Buyers, Rev. Mod. Phys. 44, 406 (1972)
9) G.A. Slack, J. Appl. Phys. 32, 1571 (1960)
10) P.J. Gielisse, J.N. Plendl, L.C. Mansur, R. Marshal S.S. Mitra, R. Mykolajewycz and A. Smakula, J. Appl. Phys. 36, 2446 (1965)
11) P. Cossee, Mol. Phys. 3, 125 (1960)
12) C.R. Becker, Ph. Lau, R. Geick and V. Wagner, phys. stat. sol. (b) 67, 653 (1975)

# CRYSTAL FIELD IN LIQUID CERIUM AND PRASEODYMIUM*

A.H. Millhouse and A. Furrer
Institut für Reaktortechnik ETHZ
EIR, CH-5303 Würenlingen, Switzerland

ABSTRACT

By means of inelastic neutron scattering experiments on liquid La, Ce, and Pr it is demonstrated that a well defined local crystal field of uniaxial symmetry exists in liquid rare earths. The results indicate that the effective charge distribution in liquid rare earth metals is independent of the particular rare earth element.

## INTRODUCTION

The static model of a liquid, in the form of a list of atomic coordinates, has been used for some years to provide a basis for the description of many of the equilibrium properties of a liquid. Within the framework of the static model of a liquid there is no significant difference between a liquid and an amorphous material. Various methods have been used to calculate such a list of atomic coordinates by considering a disordered cluster of hard spheres[1]. The most recent of these calculations was carried out for an amorphous rare earth alloy with the specific aim of examining the local crystal field within such a cluster[2]. The calculation revealed a dominant and well defined uniaxial component with the local axes varying randomly from site to site. Thus, to a first approximation, the crystal field Hamiltonian in a liquid would be expected to be of the form

$$\mathcal{H}_{cf} = -DJ_z^2 \qquad (1)$$

where D is the uniaxial crystal field parameter and z varies randomly from ion to ion. An obvious test of the above model is to perform an inelastic neutron scattering experiment using a liquid rare earth. Elastic neutron scattering measurements on liquid La, Ce, and Pr[3,4,5] have shown that the major difference between the liquid structure factors, S(Q), is an extra magnetic contribution for Ce and Pr which has a Q-dependence in agreement with the dipole approximation to the magnetic form factors.

In a previous publication[6] we presented results which indicate that the crystal field in liquid rare earths is well defined and in agreement with the uniaxial model. These results are extended here by considering the Q-dependence of the scattering and a quantitative comparison is made between the results for Ce and Pr.

## EXPERIMENTAL RESULTS

The data were obtained using the triple-axis spectrometer at reactor Saphir operating in the "constant-Q" mode. The incident neutron energy was fixed at 14.96 meV and the beam passed through a pyrolitic graphite filter. The approximately 100 gr samples of La, Ce, and Pr were sealed in evacuated Ta cylindrical containers of 2 cm diameter and maintained at a sample temperature of 1012°C. Johnson-Matthey 99.9 % vacuum melted La, Ce, and Pr were used.

The data for liquid Ce are shown in Fig. 1. The dashed curves were obtained by passing a smooth curve through the liquid La data and scaling it by assuming that the intensity is proportional to

$$I \sim Nb^2 A\, S(Q) \qquad (2)$$

where N is the total number of ions in the sample, b is the coherent scattering amplitude, A is the transmission factor, and S(Q) is the liquid structure factor. This is a valid assumption since La and Ce are both almost purely coherent scatterers. We observe an extra scattering for Ce which is not present for non-magnetic La. Although the La scattering decreases smoothly with neutron energy gain, there is an indication of some structure in the scattering from Ce. Thus, the difference between the two scaled data sets was formed and an attempt was made to fit this difference with Gaussian peaks. The best fit was obtained for three Gaussian

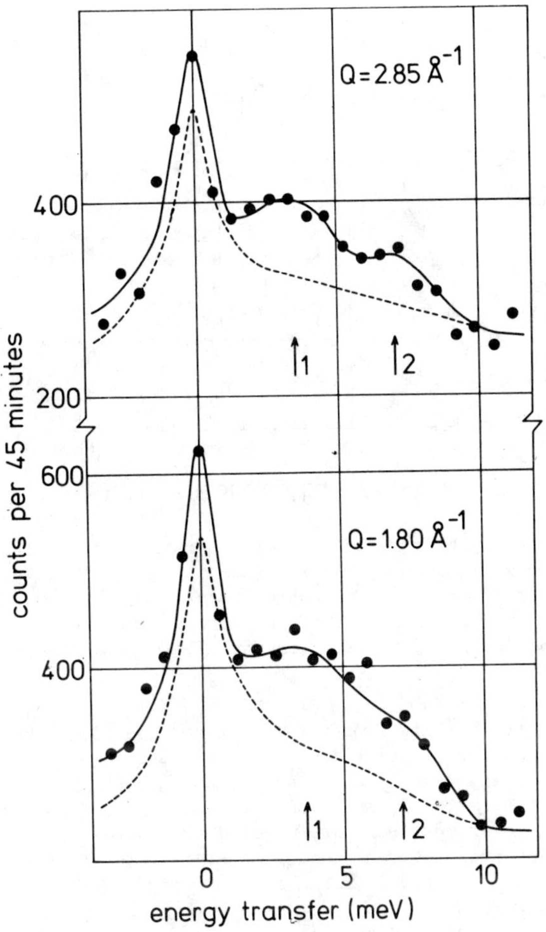


Fig. 1. Energy spectra of neutrons scattered from liquid cerium (points) at 1012° C. The solid and dashed curves are explained in the text.

Table I. Fitting parameters for the inelastic peaks observed in liquid Ce: positions (Δ), half widths (Γ), and intensities (I).

| Q (Å$^{-1}$) | 1.80 | 2.85 |
|---|---|---|
| $\Delta_1$ (meV) | 3.7±0.6 | 3.5±0.3 |
| $\Gamma_1$ (meV) | 3.1±0.6 | 3.0±0.3 |
| $I_1$ | 540±110 | 400±40 |
| $\Delta_2$ (meV) | 7.2±0.6 | 7.1±0.3 |
| $\Gamma_2$ (meV) | 2.6±0.6 | 2.2±0.3 |
| $I_2$ | 340±80 | 210±30 |

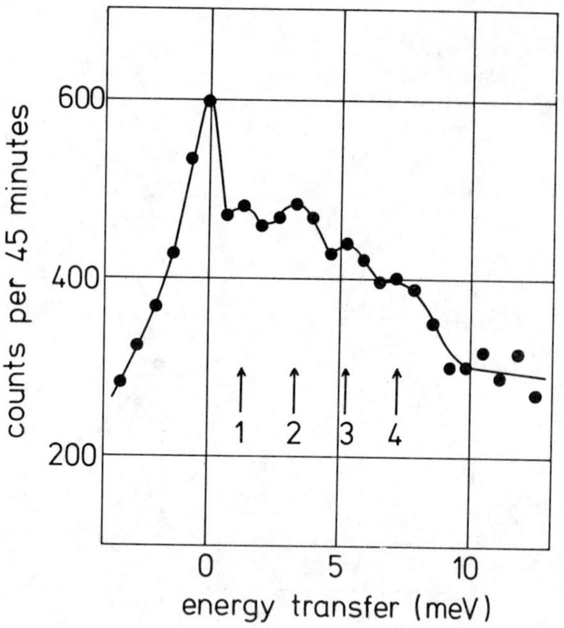


Fig. 2. Energy spectrum of neutrons scattered from liquid praseodymium (points) at 1012° C and for Q=2.50 $Å^{-1}$. The solid curve is merely drawn as a guide to the eye. The arrows indicate the positions of the four crystal field peaks as described in the text.

peaks. In the fitting procedure the positions ($\Delta$), half-widths ($\Gamma$), and peak heights were varied. The sum of this fitted intensity and the scaled La intensity is shown as the solid lines in Fig. 1. In addition to the peak at zero energy transfer, we find two inelastic peaks which are numbered 1 and 2 in Fig. 1. The fitting parameters are listed in Table I.

In the case of liquid Pr shown in Fig. 2 one cannot compare directly with liquid La since Pr has a non-negligible incoherent scattering cross section. However, there is certainly extra scattering which is of magnetic origin and this scattering has some structure. It is just possible to see traces of four inelastic peaks at 1.2, 3.2, 5.2, and 7.2 meV. Due to the fact that the half widths are somewhat larger than the separation of the peaks, they are not well resolved in this case.

## ANALYSIS OF RESULTS

The magnetic origin of the extra intensity for Ce is confirmed by our measurements in the sense that this intensity decreases with Q roughly in accordance with the magnetic form factor, f(Q), for $Ce^{3+}$. This Q-dependence is given by

$$I_i(Q) \sim f^2(Q)\,\exp(-BQ^2/8\pi^2) \qquad (3)$$

where B is the Debye-Waller temperature parameter. In a separate experiment we found B=6.3 $Å^2$ at 740°C in the bcc δ-phase for Ce. Using this along with the dipole value for the form factor gives

$$I_i(Q=1.80\ Å^{-1})/I_i(Q=2.85\ Å^{-1}) = 1.89\ ,$$

or roughly what is observed.

Under the action of the Hamiltonian given in Eq. (1) the $^2F_{5/2}$ ground state multiplet will split into three doublets and the transitions between these levels would give rise to two inelastic peaks at energies $\Delta_2/\Delta_1=2$ which agrees with what is observed. The uniaxial model gives for the intensity ratio $I_1/I_2=1.6$ which is also in agreement with the observations. We obtain for $|D|$ the value 1.74±0.08 meV.

For Pr the uniaxial model predicts four inelastic peaks at energies D, 3D, 5D, and 7D. This is also in good agreement with the observations. For $|D|$ the value 1.1±0.1 meV is obtained.

## DISCUSSION AND CONCLUSIONS

Our measurements indicate that for liquid Ce and Pr an extra magnetic intensity occurs as inelastic peaks in the observed energy spectra. This demonstrates that the crystal field is well defined in liquid rare earth metals. The widths of the crystal field transitions are of the same magnitude as those in the solid phase[7,8]. This indicates that no further line broadening mechanisms exist in the liquid, such as valence transitions due to promotion of a 4f electron into the conduction band.

It is interesting to note that the overall crystal field splitting for liquid Ce and Pr are both about 10% lower than those in the solid phase below room temperature[8,9].

A comparison of the uniaxial crystal field parameters, D, for Ce and Pr can be made by using the point charge model result

$$D \sim \langle r^2\rangle\, \alpha_J/R^3\ , \qquad (4)$$

where $\langle r^2\rangle$ is a radial integral[10], $\alpha_J$ is a reduced matrix element, and R should be chosen as the position of the first maximum in the radial distribution function[4]. By using the experimental value of $|D_{Ce}|$ and Eq. (4) we obtain $|D_{Pr}|$=1.2 meV, which is in excellent agreement with what is observed. This means that the effective charge distribution in liquid rare earth metals does not depend on the particular rare earth element.

## REFERENCES

* Supported by the Schweizerische Nationalfonds zur Förderung der wissenschaftlichen Forschung

1. G.S. Cargill III, Solid State Physics, ed. H. Ehrenreich, F. Seitz, and D. Turnbull (Academic Press New York 1975), vol. 30, p.227.
2. D. Sarkar, R. Segnan, E.K. Cornell, E. Callen, R. Harris, M. Plischke, and M.J. Zuckermann, Phys. Rev. Lett. 32, 542 (1974). R.W. Cochrane, R. Harris, and M. Plischke, J. Non-cryst. Solids 15, 239 (1974).
3. M. Breuil and G. Tourand, Phys. Letters 29A , 506 (1969).
4. R. Bellissent and G. Tourand, J. Physique 36, 97 (1975).
5. J.E. Enderby and V.T. Nguyen, J. Phys.C (GB) 8,L112 (1975).
6. A.H. Millhouse and A. Furrer, Phys. Rev. Lett. 35, 1231 (1975).
7. A.H. Millhouse and A. Furrer, Solid State Commun. 15, 1303 (1974).
8. A.H. Millhouse and A. Furrer, Proceedings of the First Conference on Crystalline Electric Field Effects in Metals and Alloys, ed. R.A.B. Devine (University of Montreal Press, Montreal 1974),p.494.
9. B.D. Rainford, in Magnetism and Magnetic Materials - 1971 A.I.P. Conference Proceedings 5, 591 (1972).
10. A.J. Freeman and R.E. Watson, Phys. Rev. 127, 2058 (1962).

# GROUND- AND EXCITED-STATE SPIN WAVES IN $PrAl_2$

H.-G. Purwins
Département de Physique, Université de Genève, Geneva, Switzerland
W.J.L. Buyers, T.M. Holden and E.C. Svensson
Atomic Energy of Canada Limited, Chalk River, Ontario, Canada, K0J 1J0

ABSTRACT

The magnetic excitations in ferromagnetic $PrAl_2$ have been studied by means of neutron inelastic scattering. Branches of excitations corresponding to four of the six allowed magnetic-dipole transitions have been observed. At the magnetic zone boundary, where the excitation spectrum most closely approximates a single-ion spectrum, the observed frequencies are $1.54 \pm 0.04$, $2.05 \pm 0.05$ and $3.76 \pm 0.06$ THz. The results are analysed in terms of exchange and crystal-field parameters using the pseudoboson theory for all states of the $J = 4$ manifold. The crystal-field-only ground state is found to be a non-magnetic doublet but the exchange interaction is strong enough to induce almost the full moment at low temperatures. At elevated temperatures in the ordered phase, excited-state spin waves have been observed at $0.55 \pm 0.10$ THz. These correspond to a transition between the upper member of the exchange-split ground doublet, and the second member of the exchange-split triplet state.

## INTRODUCTION

$PrAl_2$ is a metallic ferromagnet[1] ($T_c = 33$ K) having the cubic Laves-phase crystal structure. The crystal field at the Pr sites causes the [001] direction to be an easy axis of magnetization and partly reduces the measured magnetic moment per Pr ion[2], $2.88 \pm 0.05\ \mu_B$, below the free-ion moment of 3.2 $\mu_B$. These magnetization experiments have been interpreted[2] in terms of a model involving the crystal-field and a molecular field. The predicted ground state of the Pr ion in the crystal field is a $\Gamma_3$ doublet, both components of which are non-magnetic in the absence of a molecular field. In this paper we report neutron-scattering measurements on $PrAl_2$ which confirm that the non-magnetic doublet lies lowest and consequently that the ordering is of the induced-moment type.

## EXPERIMENT AND RESULTS

The measurements were performed with a triple-axis crystal spectrometer at the NRU reactor, Chalk River, operated in the constant-momentum-transfer mode with fixed scattered-neutron energy. Scattered-neutron distributions observed at 4.4 K at the centre of the "acoustic" zone, (220), the zone boundary, (003), and the centre of the "optic" zone, (002), are shown in Fig. 1. Because of the magnon structure factors, only "acoustic" modes are observed in the Brillouin zone containing (220) and only "optic" modes are observed in the Brillouin zone containing (002). At (003) where the spectrum is approximately the same as the single-ion spectrum, three distinct magnetic peaks and a phonon peak are observed. The frequencies of the magnetic peaks are $1.54 \pm 0.04$, $2.05 \pm 0.05$ and $3.76 \pm 0.06$ THz and their intensities are in the ratio 0.1 : 0.5 : 1.0 . An additional weak mode at $2.2 \pm 0.1$ THz is observed near the acoustic zone centre. The mode assignments in Fig. 1 were made partly from measurements at different wave vectors and partly on the basis of a model fitted to the experimental results (see below).

The dispersion relations for excitations propagating along the [00ζ] direction are shown in Fig. 2. Near $\zeta = 0.5$ in the "acoustic" zone the [00ζ] TA phonon branch interacts with the lower branches of magnetic excitations. The dashed curve shows the dispersion relation for this phonon branch at 300 K. The groups at 4.4 K associated with this phonon branch are observed largely via the magnetic cross-section induced by the magnon-phonon interactions between the phonon mode and the two lowest magnetic-dipole modes (2 and 3). The

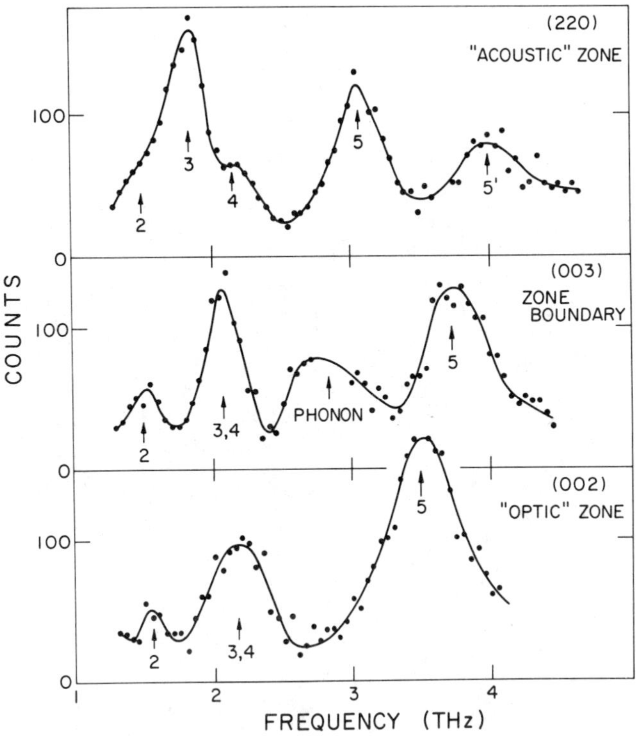


Fig. 1. Scattered-neutron distributions for $PrAl_2$ at 4.4 K observed in measurements at the centres of the "optic" and "acoustic" zones and at the zone boundary. Each distribution is normalized to the same number of incident neutrons. The branch labels are assigned on the basis of model calculations described in the text.

results also indicate that an interaction splits the strongest branch (5) of the spectrum. The splitting is large in the "acoustic" zone, but in the "optic" zone only one mode is observed, suggesting that the interaction is much weaker there. The nature of this interaction is not yet understood. It may, however, be a magnon-phonon interaction. Branches (1) and (7) correspond to quadrupole excitations which are not directly observable in the neutron scattering.

As the temperature is raised within the ordered phase (Fig. 3) a new peak appears at $0.55 \pm 0.1$ THz which corresponds to a transition between the first and third excited states of $PrAl_2$. Excited-state spin waves have previously been observed[3] in $TbAl_2$ where the intensity is believed to arise almost entirely through an interaction with the intense ground-state spin wave. In $PrAl_2$, on the other hand, the excited-state spin-wave intensity results directly from the large matrix element for the transition and the population factor.

## DISCUSSION

The solid lines in Fig. 2 represent the best fit of the pseudoboson model[4] to the experimental data, allowing the crystal-field parameters[5] X and W and the nearest-neighbor (nn) and next-nearest-neighbor (nnn) exchange parameters to vary. The model was fitted to those experimental frequencies unaffected by interactions, to the intensity ratios of the three branches observed at (003), and to the low temperature moment[2]. The pseudoboson-model parameters and the resulting molecular-field $T_c$ and ordered moment are given in Table I. The difference between the calculated local f-moment and the bulk moment can be caused by conduc-

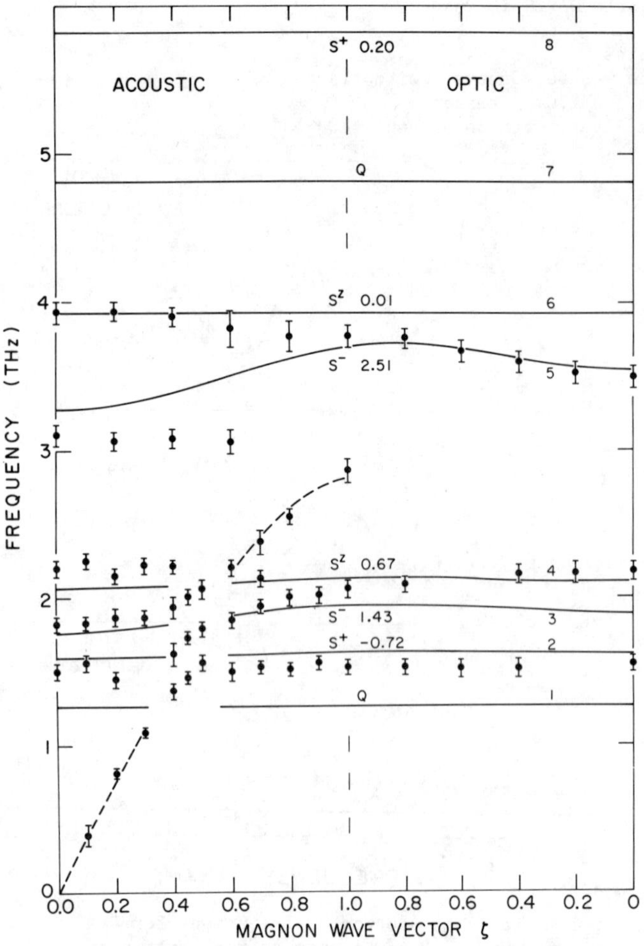


Fig. 2. Dispersion relation for excitations propagating along the [00ζ] direction in $PrAl_2$ at 4.4 K. The "acoustic" modes were obtained between (220) and (221) while the "optic" modes were obtained between (002) and (003). ζ has units 2π/a where a is the lattice parameter. The solid lines represent the best fit to the pseudoboson model and the mode labels (1-8) and transition matrix elements are assigned on the basis of the model. The dashed line represents the [00ζ] TA phonon branch at 300 K.

tion electron polarization. With the single-ion parameters of Purwins et al.[2], also shown in Table I, the model predicts frequencies which are systematically about 15% higher than experiment. Also, the intensity of the third branch is predicted to be less than that of the second and fourth branches in contradiction to experiment. While the pseudoboson model gives a qualitatively correct description of the frequencies and intensities of the ground-state spin waves and also the frequency (predicted value 0.63 THz) of the excited-state spin wave, the fit is not entirely satisfactory as can be seen from Fig. 2. The generalization to an anisotropic exchange interaction and the inclusion of

TABLE I

Model Parameters and Calculated Properties for $PrAl_2$

| | Present Work | Purwins et al.[2] |
|---|---|---|
| X | 0.77 ± 0.02 | 0.70 |
| W ($10^{-3}$ THz) | −71 ± 4 | −91 |
| $J_{nn}$ ($10^{-3}$ THz) | −6.2 ± 0.3 | |
| $J_{nnn}$ ($10^{-3}$ THz) | −3.0 ± 0.2 | |
| $\mathcal{J}(0)$ ($10^{-3}$ THz) | −61 ± 3 | −72 |
| $T_c$ (K) | 33 | 33 |
| $\langle\mu\rangle$ ($\mu_B$) | 3.11 | 2.88 |

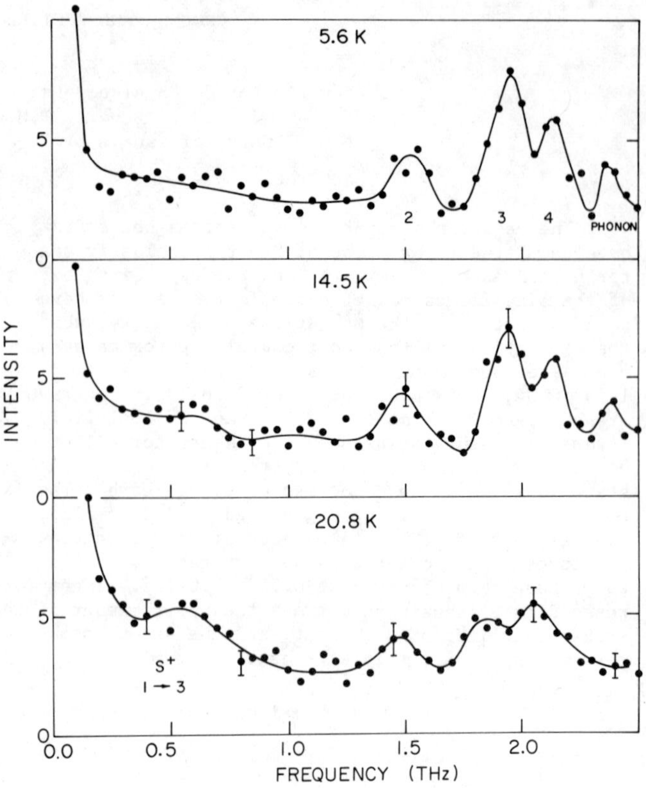


Fig. 3. Temperature dependence of the neutron scattering from $PrAl_2$ at a wave-vector transfer, 2π/a (2,2,0.7). Each distribution is normalized to the same number of incident neutrons.

exchange to more distant neighbors may lead to improvements in the model.

The ratio $J_{nnn}/J_{nn}$ is much larger for $PrAl_2$ than for $TbAl_2$[6] or $NdAl_2$[7] and this feature is related to the observed decrease in frequency of the fifth branch between the zone boundary and the centre of the optic zone. The present work substantiates the prediction[2] that the $\Gamma_3$ doublet is the crystal-field-only ground state but gives model parameters that are somewhat different from those obtained previously. The exchange is found to be three times greater than that required for magnetic order at $T = 0$ and is sufficient to induce almost the full moment in the ground state.

ACKNOWLEDGMENTS

We would like to acknowledge the expert technical assistance of H. Nieman, D. Tennant, M. Potter and A. Hewitt.

REFERENCES

1. H.J. Williams, J.H. Wernick, E.A. Nesbitt and R.C. Sherwood, J. Phys. Soc. Japan 17, suppl. B2, 91-95 (1962).
2. H.-G. Purwins, E. Walker, B. Barbara, M.F. Rossignol and P. Bak, J. Phys. C 7, 3573-82 (1974).
3. H.-G. Purwins, J.G. Houmann, P. Bak and E. Walker, Phys. Rev. Lett. 31, 1585-7 (1973).
4. W.J.L. Buyers, T.M. Holden, E.C. Svensson, R.A. Cowley and M.T. Hutchings, J. Phys. C 4, 2139-59 (1971). For the analysis of a similar problem, see also T.M. Holden, E.C. Svensson, W.J.L. Buyers and O. Vogt, Phys. Rev. B 10, 3864-76 (1974).
5. K.R. Lea, M.J.M. Leask and W.P. Wolf, J. Phys. Chem. Solids 23, 1381-1405 (1962).
6. W. Buhrer, M. Godet, H.-G. Purwins and E. Walker, S.S. Comm. 13, 881-4 (1973).
7. J.G. Houmann, P. Bak, H.-G. Purwins and E. Walker, J. Phys. C 7, 2691-6 (1974).

# THEORY OF SINGLE-DOUBLET EXCITATIONS IN PRASEODYMIUM*

P. Bak
Brookhaven National Laboratory, Upton, New York 11973

ABSTRACT

The magnetic excitation spectrum in a paramagnetic singlet-doublet system is calculated using a diagrammatic high density expansion technique. The lowest order diagrams, which correspond to the random phase approximation (RPA), give a detailed description of the wavevector and temperature dependence of the four exciton modes in Praseodymium in terms of a Hamiltonian including isotropic Heisenberg exchange interactions and anisotropic, dipolar-like interactions. The leading contributions to the linewidths of the excitations are obtained by extending the 1/Z expansion of the generalised susceptibility propagators one order beyond the random phase approximation. This damping corresponds to spin wave scattering on single-site fluctuations. The theoretical spectral functions are in detailed agreement with experiment.

## I. INTRODUCTION

For the last several years there has been a great interest in the properties of localized magnetic systems which possess a nonmagnetic singlet ground state.[1] The ordering in such systems occurs as an exchange polarization of the ground state, provided the exchange interaction between the magnetic ions exceeds a certain threshold value.

Most of the theoretical work on singlet-ground state systems has been based on the random phase approximation (RPA) using a variety of different representations of the single-ion states, or on pseudoboson theories which essentially give the zero temperature limit of the RPA. More elaborate theories based on various higher-order decouplings of equations of motions of spin operators give corrections to the exciton energies. However, inconsistencies and ambiguities are introduced through the decoupling procedures and it is difficult to estimate the errors introduced via the truncations. Another shortcoming of most of these theories is that they give infinite lifetimes of the excitations and thus completely neglect damping effects. An alternative type of theory is the diagrammatic Green's function expansion method developed by Vaks, Larkin and Pikin (VLP) for spin-operators to calculate correlation functions to any order of perturbation theory.[2] The formalism has later been generalized by Kashchenko et al.[3] and by Izyumov and Kassan-Ogly.[4] Yang and Wang[5] extended the method to any multilevel magnetic system using a standard basis operator representation. The VLP formalism corresponds to the semi-invariant theory introduced by Stinchcombe et al. which has been applied to the Ising model in a transverse field.[6] In contrast to the other techniques, this kind of theory gives a well defined high-density expansion parameter (1/Z) or $(1/r_0^3)$, allowing us to perform systematic selfconsistent calculations, and to recognize the physical processes involved. The approximations are based on physical, not technical reasons. Praseodymium is an example of a singlet ground state magnet in which the exchange is only slightly undercritical with respect to magnetic ordering. The ground state on the hexagonal sites is the pure $|J^z = 0\rangle$ singlet, and the first excited state is the doublet $|x\rangle$, $|y\rangle$, where $|x\rangle = 1/\sqrt{2}\,(|1\rangle + |-1\rangle)$ and $|y\rangle = -i/\sqrt{2}\,(|1\rangle - |-1\rangle)$. The paramagnetic excitation spectrum has recently been measured by Houmann et al.[7] using inelastic neutron scattering technique. The temperature dependence of the excitation energies were found to be in substantial agreement with a random phase theory (RPA), and the lowest lying mode shows a clear tendency towards softening as the temperature is lowered towards 0 K. In addition to strong temperature dependence of the exciton energies, a dramatic increase of the intrinsic linewidth was observed as the temperature was raised from 6K to 30K, where well-defined modes cease to exist. dhcp Praseodymium seems to be the simplest real singlet-ground-state magnet, and significantly more information is now available on the excitations in this material than in any other paramagnetic system. This makes the element Pr almost ideal for a confrontation between experiment and theoretical model calculations.

## GREEN'S FUNCTION FORMALISM AND RPA THEORY

The effective Hamiltonian describing the magnetic ions on the hexagonal sites in Pr may be written

$$\mathcal{H} = \sum_i \Delta (S_i^z)^2 - \sum_{ij,\alpha\beta} \mathcal{J}_{ij}^{\alpha\beta} S_i^\alpha S_j^\beta, \quad \alpha,\beta = x,y \qquad (1)$$

$\Delta$ is the crystal field splitting (= 3.2 MeV[7]). We introduce the Green's functions

$$G^{\alpha\beta}(r_1,\tau_1;r_2,\tau_2) = \langle T_\tau S^\alpha(r_1,\tau_1) S^\beta(r_2,\tau_2) \rangle , \qquad (2)$$

where $\langle T_\tau \ldots \rangle$ denotes the thermal average of the $\tau$ ordered product of operators in the interaction representation. The Green's functions are defined for $\alpha,\beta = x,y$ only. It is a standard procedure[8] to expand the Green's functions in the form

$$G^{\alpha\beta}(\vec{r}_1,\tau_1;\vec{r}_2,\tau_2) = \langle T_\tau S^\alpha(\vec{r}_1,\tau_1) S^\beta(\vec{r}_2,\tau_2) \mathcal{S}(\beta) \rangle_0 \Big/ \langle \mathcal{S}(\beta) \rangle_0$$

$$\mathcal{S}(\beta) = T_\tau \exp\left(-\int_0^\beta \mathcal{H}_{int}(\tau)\,d\tau\right) \qquad (3)$$

where the averages are taken with respect to the single-ion part of the Hamiltonian.

In general, this expansion can be represented by a sum of connected diagrams representing an aggregate of single-cell blocks joined by interaction lines. The non-interacting Green's functions, $g_0(i\omega_n) = 2\Delta(\Delta^2 - i\omega_n)^2)^{-1}$, are represented by full lines, and the interactions by wavy lines. $\omega_n$ are the Matsubara frequencies. The propagators $G^{\alpha\beta}(\vec{q}, i\omega_n)$ are represented by double lines. The single-cell blocks are surrounded by broken lines to denote that the propagators involved are restricted to the same site. A diagram containing N independent wavevector labels, which are eventually summed over, is of order $(1/Z)^N$. The zeroth order diagrams simply consist of non-interacting Green's function connected with interaction lines (see the RPA-diagrams in Fig. 2a). The poles of the RPA-propagators, which give the energy spectrum are[9]

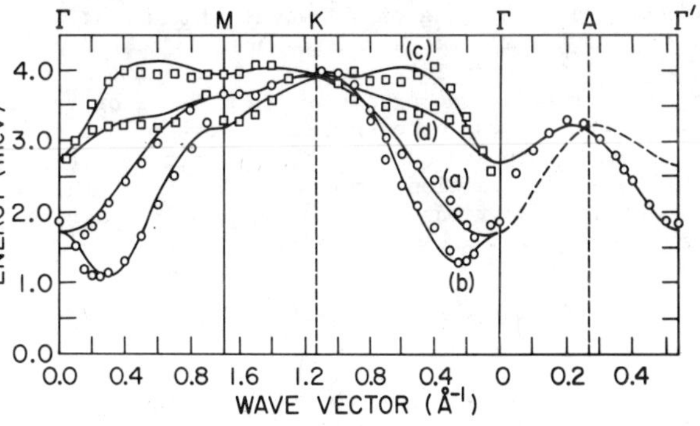


Fig. 1 Dispersion relations for magnetic excitations propagating on the hexagonal sites in dhcp Praseodymium. The experimental data is taken from Ref. 7. The full lines represent a least squares fit as described in the text.

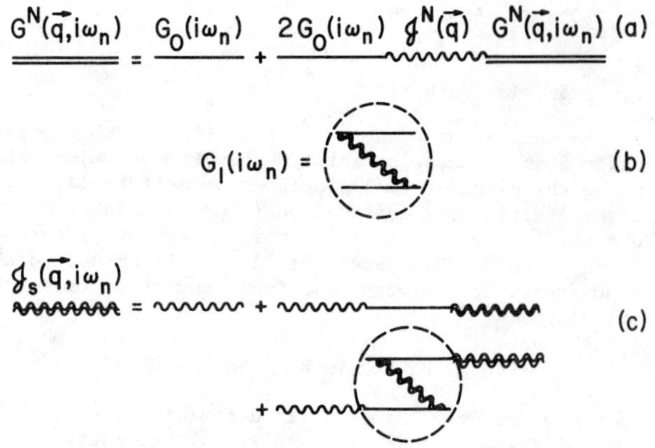


Fig. 2 a) RPA diagram representing the zeroth order term in the (1/Z) expansion. b) First order single-cell block representing fluctuation scattering of excitons. c) "Screened" interaction.

$$(\omega_{\vec{q}}^{N})^2 = \Delta^2 - 4\Delta R\,\mathcal{J}^N(\vec{q}) \tag{4}$$

where $R = \{1-\exp(-\beta\Delta)\}\{1+2\exp(-\beta\Delta)\}^{-1}$ and $J^N(\vec{q})$ are eigenvalues of a $4\times4$ matrix describing the exchange interactions between the ions. The measured dispersion relations were fitted to this expression (see Fig. 1) and interatomic exchange parameters were derived. The polarisations of the exciton modes are in general elliptic or linear, and the polarisation vector changes rapidly as a function of $\vec{q}$, in particular at large wavevectors.

The RPA theory has thus allowed us to set up a theory of the dispersion relations, and to construct a complete model to describe the magnetic properties of the hexagonal ions, and we can now calculate higher order diagrams without introducing any adjustable parameters.

## FLUCTUATION DAMPING OF EXCITATIONS

To find the first order corrections to the Green's functions we collect all the single-cell blocks with one interaction-loop. The real part of these diagrams give rise to a small shift of the excitation energies.[10] It turns out, however, that only the diagram in Fig. 2b gives a contribution to the damping of the excitations. The imaginary part, which is of interest here, has the following analytic expression in $\omega_n,\vec{q}$ space:

$$G_1(i\omega_n) = 2g_0(i\omega_n)^2\, b\, i\, \mathrm{Im}\,\frac{1}{N}\sum_{\vec{q}'} \mathcal{J}_s(\vec{q}', i\omega_n) \tag{5}$$

where $b = \{5\exp(-\beta\Delta)+\exp(-2\beta\Delta)\}\{1+2\exp(-\beta\Delta)\}^{-2}$. The diagram corresponds to scattering of excitation waves on single-site fluctuations of the population difference (or quadrupole moment) R. The intermediate states are magnetic excitations with wavevector $\vec{q}'$. This effect corresponds to scattering of spinwaves on fluctuations of $< S^z >$ for a simple ideal ferromagnet as discussed by Vaks et al.[2] It is interesting that the damping occurs to first order in the expansion. For boson or fermion systems, the lowest order imaginary part of the self energy occurs in second order in the high density expansion due to interaction between excitations. The resulting Green's functions including all chain diagrams involving $G_0(i\omega_n)$ and $G_1(i\omega_n)$ are

$$G^N(\vec{q},i\omega_n) = \{G_0(i\omega_n)+G_1(i\omega_n)\}\{1-2\mathcal{J}^N(\vec{q})(G_0(i\omega_n)+G_1(i\omega_n))\}^{-1} \tag{6}$$

The spectral functions, which are proportional to the neutron scattering cross-section, can easily be expressed in terms of the Green's functions using the fluctuation-dissipation theorem:

$$S^N(\vec{q},\omega) = \frac{1}{\pi}\{1-\exp(-\beta\omega)\}^{-1}\, \mathrm{Im}\, G^N(\vec{q},\omega)$$

$$= \frac{1}{\pi}\{1-\exp(-\beta\omega)\}^{-1}\,\frac{\gamma(\omega)(\Delta^2-\omega^2)}{((\omega_{\vec{q}}^{N})^2-\omega^2)^2 + (2\gamma(\omega)\mathcal{J}^N(\vec{q}))^2} \tag{7}$$

where the damping parameter $\gamma(\omega)$ must be determined selfconsistently by solving the integral equation

$$\gamma(\omega) = (\Delta^2-\omega^2)\,\mathrm{Im}G_1(\omega)$$

$$= 16\Delta^2 b \int_{\omega'} d\omega' \,\frac{\left(\frac{1}{4}\sum_N N_T^{\,N}(\omega')^2\,\mathcal{J}^2(\omega')\gamma(w)\right)}{(\omega'^2-\omega^2)^2 + (2\gamma(\omega)\mathcal{J}(\omega'))^2} \tag{8}$$

The summation over $\vec{q}'$ space has been changed to an integration over $\omega'$ space. $N_T^N(\omega)$ is the density of states for the N'th exciton mode at temperature T, and $\mathcal{J}(\omega')$ is the value of $\mathcal{J}^N(\vec{q}')$ for a mode with energy $\omega'$ determined by (4). The theoretical lineshapes convoluted with the experimental resolution function are compared with experiment in Fig. 3. Almost complete agreement between positions of peaks, intensities and lineshapes is observed at any temperature. The agreement is least perfect at the highest temperatures, where the linewidth, as calculated to first order in the (1/Z) expansion, is comparable to the energy as calculated to zeroth order. Hence, damping effects due to other effects, such as interactions between excitations occurring to order $(1/Z)^2$ in the expansion, scattering on phonons, or direct Coulomb scattering with conduction electrons, can be entirely neglected at least at moderate temperatures. This fundamental understanding of the excitations in Pr may be valuable for the understanding not only of other singlet ground state systems of magnetic or non-magnetic nature, such as the hydrogen bonded ferroelectrics, but also of more complicated systems, since the formalism can easily be extended to arbitrary level schemes.

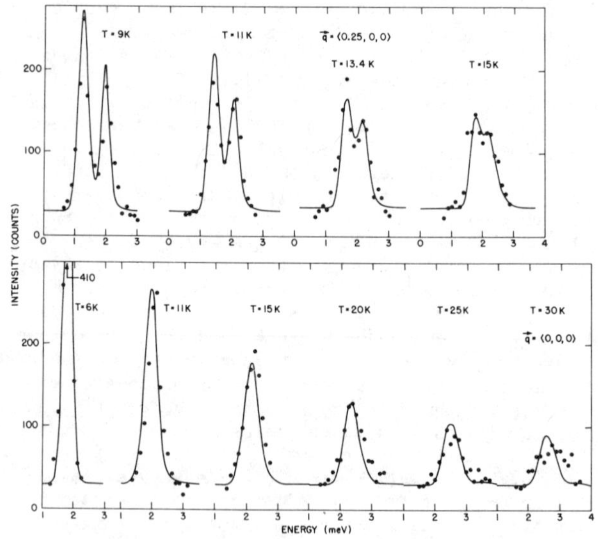


Fig. 3 Spectral functions for magnetic excitations in Pr. Points: neutron measurements (Ref. 11), lines: selfconsistently calculated lineshapes convoluted with the experimental resolution (Gaussian, full width at half maximum = 0.36 meV).

## REFERENCES

* Work supported by Energy Research and Development Administration.


1. R.J. Birgeneau, AIP Conf. Proc. 10, 1664 (1973).
2. V.G. Vaks, A.I. Larkin and S.A. Pikin, Sov. Phys. JETP 26, 188 (1968); ibid. 26, 647 (1968).
3. M.P. Kashchenko, N.F. Balakhonov and L.V. Kurbatov, Sov. Phys. JETP 37, 201 (1973).
4. Yu.A. Izyumov and F.A. Kassan-Ogly, Fiz. metal. metalloved 30, 225 (1970).
5. F.H-Y. Yang and Y-L. Wang, Phys. Rev. B10, 4714 (1974).

6. R.B. Stinchcombe, G. Horwitz, F. Englert and R. Brout, Phys. Rev. 130, 155 (1963); R.B. Stinchcombe, J. Phys. C6, 2484 (1973).
7. J.G. Houmann, M. Chappelier, A.R. Mackintosh, P. Bak, O.D. McMasters and K.A. Gschneider, Jr., Phys. Rev. Lett. 34, 587 (1975).
8. A.H. Abrikosov, L.P. Gorkov and I.E. Dzyaloshinski, Methods of Quantum Field Theory in Statistical Physics, (Prentice-Hall, Englewood Cliffs, N.J., 1963).
9. P. Bak, Phys. Rev. Lett. 34, 1230 (1975) and Phys. Rev. (to be published).
10. D. Hsing-Yen Yang and Y-L. Wang, Phys. Rev. B12, 1057 (1975).
11. J.G. Houmann and A.R. Mackintosh, private communications.

## A STUDY OF THE FREQUENCY DEPENDENT SUSCEPTIBILITY OF THE COMPOUND, $CeAl_3$, BY INELASTIC NEUTRON SCATTERING*

T. Brun and S. Sinha
Argonne National Laboratory, Argonne, Illinois 60439
A. S. Edelstein and R. Majewski
University of Illinois, Chicago, Illinois 60680 and
Argonne National Laboratory, Argonne, Illinois 60439
H. R. Child
Oak Ridge National Laboratory, Oak Ridge, Tennessee 37830

ABSTRACT

Using neutron time-of-flight techniques the $Im\chi(q,\omega)$ for polycrystalline sample of the compound $CeAl_3$ has been measured for $0.5\ \text{Å}^{-1} < q < 2.5\ \text{Å}^{-1}$ at 2.3, 7, 40, and 78 K. The symmetrized scattering function has been least squares fitted to a sum of two Lorentzians at the upper two temperatures and to a single Lorentzian at the lower two temperatures. The crystal field splitting that we determine by this procedure is 4.7 meV. The widths of the central quasi-elastic peak are 1.3, 1.1, 3.9, and 4.2 meV at 2.3, 7, 40 and 78 K, respectively, and are independent of q. The intensities are approximately correct since when they are inserted in the appropriate integral relation the result is approximately equal to the susceptibility. The results are consistent with a crystal field model with the 3/2-level as groundstate and the 1/2-level and 5/2-level as first and second excited states. Over the range of q investigated the amplitude variation is that expected for the form factor of atomic $Ce^{3+}$. The quasi-elastic peak at 78 and 40 K has contributions from scattering from predominantly the $\pm$ 3/2 and $\pm$ 1/2 levels. At 7 K this peak is due only to scattering from the $\pm$ 3/2 levels. The fractional occupancy of the localized $4f_1$ configuration is $0.9 \pm 0.1$ at 11 K. Hence most of the reduction in $\mu^2_{eff} \cong 3\ kT\chi/A_0$ occurs because of a short spin correlation time and crystal field effects. We are taking further measurements to improve the statistics and thus enable us to better compare our widths with the theory of Gotze and Schlottmann.[1]

[1] W. Gotze and P. Schlottmann, J. Low Temp. Phys. 16, 87 (1974).

* Research sponsored by the U. S. Energy Research and Development Administration under contract with Union Carbide Corporation.

## STRUCTURE, MOMENT, AND MAGNETIC EXCITATIONS OF THE TRANSPARENT FERROMAGNET $Rb_2CrCl_4$

M. T. Hutchings, Materials Physics Division, Harwell, Didcot, England and P. Day, M. Fair and A. K. Gregson, Inorganic Chemistry Laboratory, Oxford, England.

The crystal and magnetic structure, the moment variation with temperature, and the magnon dispersion at 4.5K, of ferromagnetic $Rb_2CrCl_4$ have been investigated using neutron scattering techniques. At 295K the tetragonal crystal structure is confirmed to be that of the layer compound $K_2NiF_4$(1) (I4/mmm), and we have observed no indication of an orthorhombic distortion such as that found in $K_2CuF_4$(2). The $Cr^{2+}$ $(3d)^4$ $^5A_1$ ions' spins order ferromagnetically in three dimensions [3d] at 63 $\pm$ 2K, and lie in the a-b plane. The observed moment at 4.3K is $3.8 \pm 0.4\ \mu_B$ and varies with temperature in a manner characteristic of a [2d] ferromagnet with weak anisotropy or interlayer exchange. The magnon dispersion shows the excitations to be two dimensional in character with no measurable dispersion perpendicular to the layers. The dispersion in the a-b plane can be fitted using a single exchange parameter $J_{nn} = 7.66 \pm 0.15$K (where $H_{12} = -2J_{12}\underline{S}_1 \cdot \underline{S}_2$). The spin dynamics should thus follow closely those of a [2d] easy-plane, ferromagnet. Their temperature dependence will be the subject of future investigations.

(1) H. J. Seifert and K. Klatyk, Z. Anorg. allg. Chem. 334, 113 (1964).

(2) R. Haegele and D. Babel, Z. Anorg. allg. Chem. 409, 11 (1974).

# MAGNETIC EXCITATIONS IN HOLMIUM PHOSPHIDE

A. Furrer
Institut für Reaktortechnik ETHZ, EIR, 5303 Würenlingen, Switzerland

E. Kaldis
Laboratorium für Festkörperphysik, ETH, 8049 Zürich, Switzerland

ABSTRACT

Using a large single crystal the magnetic excitation spectrum of HoP has been studied by neutron inelastic scattering at various temperatures below and above $T_c$=5.5 K. Below $T_c$ nearly all the spin wave strength resides in the transition to the 6th excited state which originates from the exchange split $\Gamma_3^{(2)}$ doublet ground state and $\Gamma_4^{(1)}$ triplet excited state of the $Ho^{3+}$ ion in an octahedral crystal field. Weaker ground state transitions up to the 9th excited state have also been observed. The spin wave branches turn out to be almost independent of wave vector. The data have been analysed on the basis of a Hamiltonian containing crystal field, bilinear exchange and quadrupole interactions. The stability of the anomalous flopside spin structure of HoP is discussed in terms of these model parameters.

## INTRODUCTION

Holmium phosphide belongs to the large class of NaCl-structured rare earth monopnictides. Below $T_c$=5.5K a magnetically ordered flopside spin structure has been observed[1], i.e. the moments in (111) planes are in the x- or y-directions, and the moment directions in adjacent (111) planes are mutually perpendicular as shown in Figure 1. Various models have been suggested in order to explain this anomalous spin arrangement. However, it is still an open question whether the flopside spin structure is stabilized by dipole forces[2], by quadrupole interactions[3], or by potential minima effects based on the tunneling model[4]. This ambiguity is partly due to the absence of experiments on reliable single crystals. Recently the growth of large single crystals of rare earth phosphides was possible[5], so that a study of the magnetic excitation spectrum of HoP became feasible.

In the present work we report neutron inelastic scattering measurements of magnetic excitations in single crystal HoP and give a theoretical interpretation of the observed spectrum. The analysis of the data is based upon a single-ion Hamiltonian, since the dispersion of the observed spin wave branches is very small.

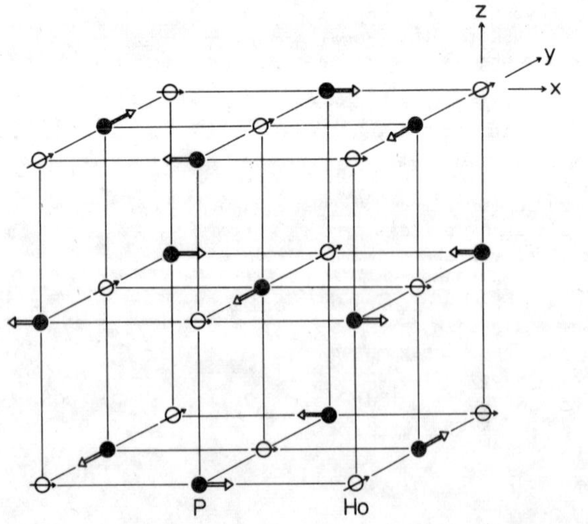


Fig. 1. Crystal structure of HoP. The arrows on the Ho ions point to the moment directions. The arrows attached at the P ions show the proposed phosphorus displacements.

From the resulting crystal field, bilinear exchange and quadrupolar parameters we can explain the appearance of the flopside spin structure.

Table I. Matrix elements of $\vec{S}$ for $Ho^{3+}$ in HoP. $\alpha$ denotes the symmetry of the state $|p\rangle$.

| p | $E_p$(meV) | $\alpha$ | $\langle p\|S_\alpha\|0\rangle$ | p | $E_p$(meV) | $\alpha$ | $\langle p\|S_\alpha\|0\rangle$ |
|---|---|---|---|---|---|---|---|
| 0 | 0 | z | 7.84 | 9 | 12.00 | + | -0.65 |
| 1 | 1.55 | (++--) | - | 10 | 12.39 | (++--) | - |
| 2 | 1.76 | - | -1.40 | 11 | 13.77 | (++--) | - |
| 3 | 1.86 | + | 1.05 | 12 | 14.36 | + | -0.23 |
| 4 | 2.14 | z | -1.05 | 13 | 14.82 | + | -0.11 |
| 5 | 4.72 | z | -0.01 | 14 | 15.40 | z | -0.09 |
| 6 | 10.43 | - | -3.85 | 15 | 16.52 | - | 0.06 |
| 7 | 11.65 | z | 0.01 | 16 | 17.11 | (++--) | - |
| 8 | 12.00 | - | 0.71 | | | | |

## THEORY

In a single-ion picture the magnetic behaviour of HoP can be described by the Hamiltonian

$$\mathcal{H}_1 = B_4(O_4^0+5O_4^4)+B_6(O_6^0-21O_6^4)-\sum_i g^2\mu_B^2\lambda\langle\vec{S}\rangle\cdot\vec{S}_i-\lambda_Q\langle O_2^0\rangle O_2^0 \quad , \qquad (1)$$

where the $B_n$ are the crystal field parameters, the $O_n^m$ are the Stevens operator equivalents, $\vec{S}$ is the total angular momentum operator, $\lambda$ is the molecular field parameter and $\lambda_Q$ is the quadrupolar parameter. The molecular field energies $E_p$ and wavefunctions $|p\rangle$ of $Ho^{3+}$ in HoP, p=0,1,...,16, have been obtained by diagonalizing $\mathcal{H}_1$. The energies $E_p$ and the matrix elements for transitions between the ground state $|0\rangle$ and the excited states $|p\rangle$ are listed in Table I for the best fitted parameters resulting from the present experiments. The cross section of the magnetic excitations is approximately proportional to the squares of the matrix elements $\langle p|S_\alpha|0\rangle$ ($\alpha$=+,-,z). Therefore we expect to observe a strong transverse $S_-$-transition to the 6th excited state at an energy of 10.4 meV, as well as weaker transitions to the 2nd, 3rd, and 4th excited states at energies of about 2 meV and transitions to the 8th and 9th excited states at an energy of 12 meV.

## CRYSTAL GROWTH, EXPERIMENTAL METHOD AND ANALYSIS

Single crystals of HoP (0.5 cm$^3$ volume) were grown for the first time using a novel high temperature phosphorus pressure transport[5] in evacuated and sealed tungsten crucibles[6]. The growth conditions have been similar to those already reported for GdP[7]. The temperature conditions were $\Delta T$=2180 (bottom)-2050 (lid) = 130°C. The necessary phosphorus pressure to avoid decomposition of HoP has been created in the crucible by thermal decomposition of WP. The lattice constant, a, of HoP was found to be 5.623 Å.

The experiments were carried out on a triple-axis spectrometer at the reactor Diorit, Würenlingen. The measurements have been performed in the constant-Q mode of operation with neutron energy loss. The analyser energy was kept fixed at 14.96 meV and the scattered beam was passed through a pyrolytic graphite filter to avoid higher order contamination. The measurements have been made at T=1.5 K, 4.2 K, 20 K, 50 K, and 80 K for wave-vectors along the three principal symmetry directions around various reciprocal lattice points. From the energy spectra measured in the paramagnetic state the crystal field parameters have been determined to be

$$B_4 = (-3.1 \pm 0.1) \times 10^{-4} \text{ meV} ,$$
$$B_6 = (-4.5 \pm 0.9) \times 10^{-7} \text{ meV} ,$$

which are in good agreement with the data obtained from neutron spectroscopy on a polycrystalline sample[8].

Figure 2 shows the distribution of scattered neutrons for $\vec{Q}=2\pi/a(1.5,1.5,1.5)$ at T=1.5 K. The energy spectrum exhibits besides the elastic line an intense peak at 10.4 meV which is a transverse magnon-like excitation corresponding to a $S_-$-transition from the ground state to the 6th excited state. Unresolved shoulders at the high energy sides of these peaks can be identified as the transitions to the 2nd, 3rd, 4th, 8th, and 9th excited states as mentioned in the previous section. The observed energy spectra turned out to be surprisingly independent of the scattering vector $\vec{Q}$. In particular, the energy of the intense $S_-$-transition did not change by more than 0.3 meV for different wave-vectors. The results obtained at T=4.2 K are very similar to the energy spectra observed at T=1.5 K; the average decrease of the energy of the intense $S_-$-mode is only 0.2 meV.

The energy spectra measured in the ordered state have been analysed on the basis of the single-ion Hamiltonian (1). The crystal field parameters were held constant at the values found from the paramagnetic excitation spectra, and only the molecular field parameter $\lambda$ and the quadrupolar parameter $\lambda_Q$ were varied. From the least squares fitting procedure of the data at 1.5 K and 4.2 K we obtained

$$\lambda = ( 7.1 \pm 0.5) \times 10^8 \text{ Oe}^2/\text{meV} ,$$
$$\lambda_Q = (-2.5 \pm 0.6) \times 10^{-5} \text{ meV} .$$

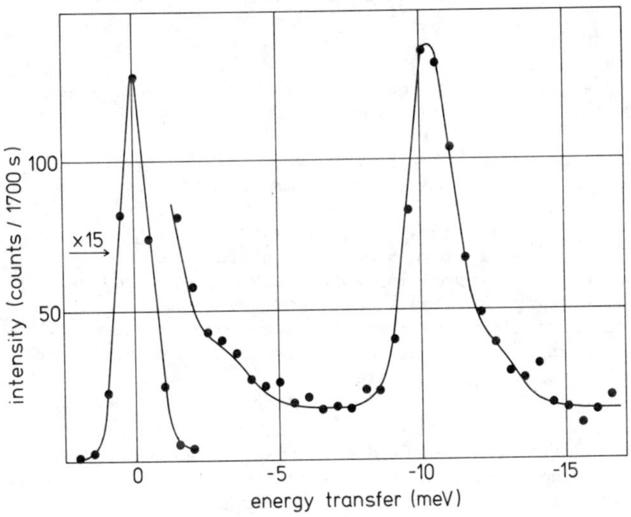


Fig. 2. Energy distribution of neutrons scattered from HoP at 1.5 K for $\vec{Q}=2\pi/a(1.5,1.5,1.5)$. The solid line corresponds to the values calculated from the model parameters.

## DISCUSSION

In the rare earth monopnictides appreciable exchange interactions exist only between nearest and next-nearest neighbours. The exchange $J_1$ is between a Ho ion and its 12 nearest neighbours, six on the same sublattice and six on the adjacent sublattices. All six second-nearest neighbours lie on the adjacent sublattices. In the Heisenberg type of exchange, however, the exchange energy between the spins on adjacent sublattices vanishes, since the moment directions are mutually perpendicular. Therefore, the dominant bilinear exchange for HoP is the ferromagnetic exchange $J_1$ within the same sublattice:

$$J_1 = (1/12)\, g^2\mu_B^2\lambda = (3.1\pm0.2)\times10^{-3} \text{ meV} .$$

The quadrupolar coupling, $\lambda_Q$, may arise from magneto-elastic couplings to strains, acoustic phonons or optic phonons, and is analogous to the coupling responsible for Jahn-Teller transitions. However, Mullen et al.[9] found for several (rare earth)-Sb compounds, that only strain coupling was important, and we assume that this is also true for HoP.

We now proceed to explain the appearance of the flopside spin structure of HoP on the basis of the model parameters determined from the present experiments. Trammell[2] has shown that the flopside spin structure can be stabilized by dipole forces. His calculations, however, are based on a wrong exchange energy between a Ho ion and its neighbours on the same sublattice which has been estimated to be 0.7 meV, while the present experiments predict 2.3 meV. Stevens and Pytte[4] suggest a model in which the rare earth moments tunnel between equivalent potential minima. They suppose that the phosphorus ions are alternately displaced along the +x and -x direction, so that half the Ho ions are in compressed octahedra of neighbours and half in extended octahedra. Consequently the resulting quadrupole interaction should be averaged to zero, which is in contradiction with the present experiments.

Levy and Chen[3] have pointed out that the unusual flopside spin structure of HoP can be explained if $\lambda_Q<0$. This conclusion is based on a fundamental description of the crystal distortion in compounds containing $Cu^{2+}$ or $Mn^{3+}$ at octahedral sites due to the Jahn-Teller effect[10]. If the distortions occur alternately along the x and y axis, the lattice of the magnetic ions is divided into two sublattices with mutually perpendicular moment directions. For HoP the displacement of the phosphorus ions has to produce a net extension of the octahedra along the moment directions, since $\lambda_Q<0$. This is realized if the phosphorus ions are displaced as shown in Figure 1. It should be possible to detect these internal distortions by high resolution x-ray or neutron diffraction experiments.

A detailed report of this work will be published elsewhere[11].

The authors are grateful to Professors G. Busch and W. Hälg for their continuous encouragement and to V. Hildebrandt for valuable assistance in the crystal growth experiments.

## REFERENCES

1. H.R. Child, M.K. Wilkinson, J.W. Cable W.C. Koehler, and E.O. Wollan, Phys. Rev. 131, 922 (1963).
2. G.T. Trammell, Phys. Rev. 131, 932 (1963).
3. P.M. Levy and H.H. Chen, Phys. Rev. Lett. 27, 1358 (1971).
4. K.W.H. Stevens and E. Pytte, Solid State Commun. 13, 101 (1973).
5. E. Kaldis, J. Cryst. Growth 24/25, 53 (1974).
6. E. Kaldis, Principles of the vapour growth of single crystals, in: Crystal Growth, theory and techniques, ed. by C.H.L. Goodman (Plenum Press, 1974).
7. W. Beckenbaugh, J. Evers, E. Kaldis, and V. Hildebrandt, Helv. Phys. Acta 47, 423 (1974).
8. R.J. Birgeneau, E. Bucher, J.P. Maita, L. Passell, and K.C. Tuberfield, Phys. Rev. B8, 5345 (1973).
9. M.E. Mullen, B. Lüthi, P.S. Wang, E. Bucher, L.D. Longinotti, J.P. Maita, and H.R. Ott, Phys. Rev. B10, 186 (1974).
10. J. Kanamori, J. Appl. Phys. 31, 14S (1960).
11. A. Furrer and E. Kaldis, to be published in J.Phys.C.

# SPIN-WAVE AND CRITICAL SCATTERING OF NEUTRONS FROM COBALT DISULFIDE

M. Iizumi* and J. W. Lynn
Brookhaven National Laboratory†, Upton, New York 11973
A. Ohsawa
Tohoku University, Sendai, Japan
H. Ito
Hokkaido University, Sapporo, Japan

ABSTRACT

The triple-axis neutron scattering technique has been used to measure the temperature dependence of the spin-wave and critical scattering below $T_c$(=120 K) in the metallic ferromagnet $CoS_2$. The spin-wave dispersion relations at small propagation vectors($|q|\leq 0.3A^{-1}$) can be described by $\hbar\omega = Dq^2 - Eq^4$ with D of about 110 meV-$A^2$ and E/D of $1\sim 2A^2$ at 78 K, with a small anisotropy observed in the different symmetry directions of the crystal. At small q, D is found to renormalize as $(1-T/T_c)^\beta$ with $\beta$=0.24. Some indication of a triple-peaked structure in the scattering spectrum was observed immediately below $T_c$.

The properties of the metallic conductor $CoS_2$ and its solid solutions with $FeS_2$ and $NiS_2$ are considerably less complicated to interpret than the 3-d ferromagnetic elements and alloys because the former's electrical and magnetic characteristics both arise from the same single d-band which is free from hybridization with s or p bands. For this reason they have been extensively studied in recent years[1] and many of their properties have been successfully interpreted in terms of the tight-binding Hubbard model. These systems therefore appear to be favorable to study the influence of the itineracy of the d-electrons on the magnetic excitation spectrum.

In the present study we report measurements of the spin-wave spectrum and critical scattering for $CoS_2$, which has the pyrite crystal structure and is ferromagnetic below 120 K.[1] The spin-wave dispersion relations were measured on a single crystal which was grown by the chemical transport technique. Unfortunately, the neutron scattering measurements are hindered by the relatively large absorption and incoherent scattering cross sections for cobalt. Moreover, the magnetic moment is only 0.84 $\mu_B$ per cobalt atom, and this coupled with the small sample size (0.2 $cm^3$) available has limited the present spin-wave measurements to wave vectors less than $0.3A^{-1}$. Since the spin-wave dispersion was found to be approximately isotropic in $\vec{q}$, however, the measurements of the critical scattering and the renormalization of the spin waves at small wave vectors could be taken in the forward direction on a powder sample of considerably larger volume(5 $cm^3$). This sample was in the form of a plate 8 mm thick.

The neutron scattering measurements were made with the triple-axis spectrometers at the Brookhaven High Flux Beam Reactor. Most of the spin-wave measurements on the single crystal were taken with the constant-E technique with an incident neutron energy of 41 meV. Because of the inherently weak inelastic intensity, we were obliged to use the relaxed horizontal collimation of 40' full width at half maximum (FWHM) both before and after the pyrolytic graphite (PG) monochromator and analyzer. The nominal FWHM energy resolution under these conditions was 4 meV. The measurements on the powder were made with the constant-Q technique. The incident energy in this case was 13.7 meV and 10' collimation was used before and after the monochromator and analyzer. The nominal FWHM energy resolution was 0.2 meV, and useful data could be obtained for scattering angles larger than 0.5°. A PG filter was used for both 41 and 13.7 meV incident neutrons to reduce unwanted higher order wavelengths. All the spin-wave energies presented have been corrected for the finite resolution of the instrument.

The dispersion relations for the spin waves in the high-symmetry directions at 78 K are shown in Fig. 1.

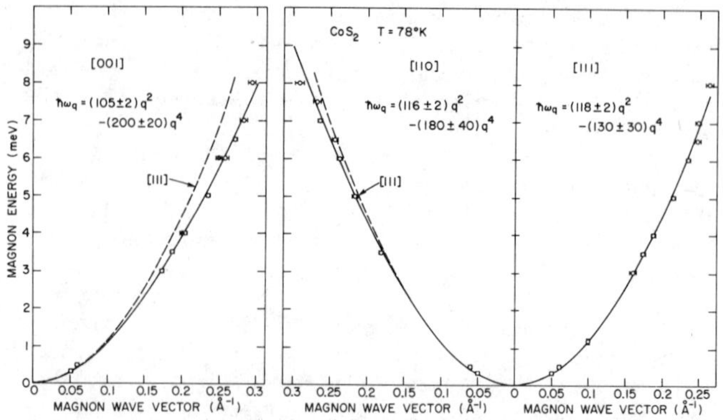


Fig. 1. Dispersion relations along the principal symmetry directions of the crystal. Open and closed circles indicate the data obtained with the single crystal sample, while the open squares show the data obtained with the powder sample. The solid curves are the best fits to the data for each direction.

Our data extend only to 0.3 $A^{-1}$, which is about 1/4 of the way to the zone boundary (1.14 $A^{-1}$) in the [001] direction. Over this wave-vector range the spin-wave dispersion relation can be expanded in a power series in $q^2$. The data have therefore been least-squares fitted to the expression $\hbar\omega_q = Dq^2 - Eq^4$. The resultant fits are shown in the figure as solid curves. For the three high-symmetry directions [001], [110], and [111], we obtained values for the spin-wave stiffness parameter D of 105, 116 and 118 meV-$A^2$, respectively. The smaller value of D in the [001] direction is outside our experimental error, so there appears to be a small amount of anisotropy present. On the other hand, measurements of the bulk magnetization show only a very small anisotropy.[2] The ratios E/D of the fourth-order coefficients to the second-order coefficients are 1.9 ± 0.2, 1.6 ± 0.3, and 1.1 ± 0.3 $A^2$ for the [001], [110], and [111] directions, respectively.

The temperature dependence of the spin-wave scattering was observed in two different energy regions. For energies between 4 and 8 meV the magnetic scattering as measured in constant-E scans shows well-defined peaks at low temperatures, which then increase in width with increasing temperature without any significant shift of the peak positions. Although a very broad peak is observed even above $T_c$, one cannot attribute a propagating character to this peak, because the width of the peak suggests that there are no side peaks in the spectral weight function. In order to determine if there is any resemblance to the propagating behavior observed in iron and nickel[3], it is necessary to measure the magnetic scattering at higher energies and wave vectors, which are unfortunately experimentally inaccessible at present because of the small sample size.

At small values of q the spin-wave energies change rapidly near $T_c$. It is convenient to present the renormalization of the spin-wave energies by the temperature dependence of the stiffness parameter D(T). The results are shown in Fig. 2. We have fit these data to the expression $D(T) = D(0)\ (1 - T/T_c)^\beta$. With $T_c$ fixed to the measured value, the least-squares fit gave a value of $\beta$=0.24±0.02, which agrees well with the exponent of 0.25±0.01 obtained by Jibu et al[4] from measurements of the magnetization.

The observed temperature dependence of the magnetic scattering below $T_c$ is shown in Fig. 3 for a wave vector of 0.06 $A^{-1}$. To obtain this data a large back-

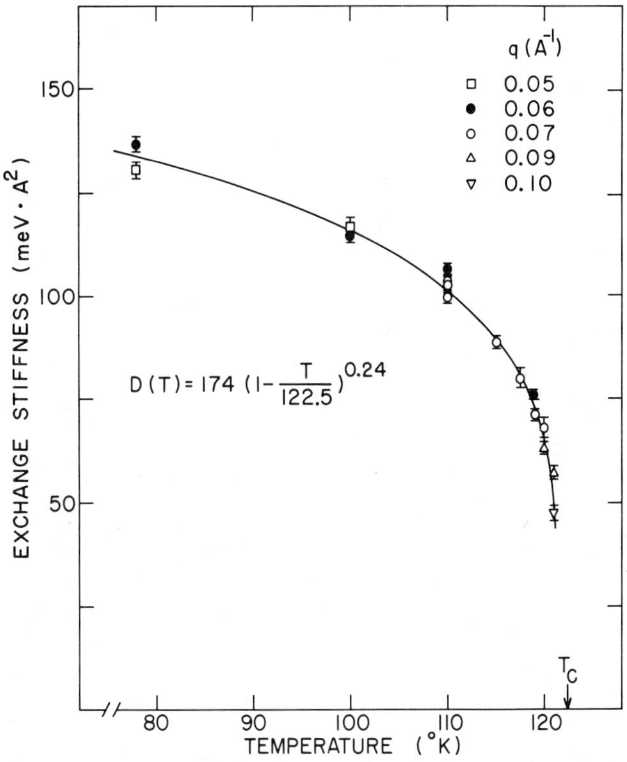


Fig. 2. Renormalization of the spin-wave stiffness parameter, D, with temperature. The solid curve is the best fit to the data.

ground correction has been made to the data at energies around E=0. This background originates primarily from air scattering and small angle scattering from the cryostat and sample. The observed scattering around E=0 at 78 K and room temperature coincided within experimental error and were taken as background. This amounted to 1245±15 counts/23 minutes at E=0. The absence of a central component at 110 K assures that the background subtraction has been performed satisfactorily. The spectra at temperatures closer to $T_c$ suggest that there may be a triple-peaked structure, the two side bands being the spin waves at $\pm\hbar\omega_q$, and the central peak being due to longitudinal correlations. Such a central diffusive mode is not present in other nearly isotropic ferromagnets such as Fe, Ni[5], Co[6], EuS or EuO[7], whereas this component has been clearly seen in isotropic antiferromagnets such as $RbMnF_3$[0] and in very anisotropic ferromagnets such as MnP.[9] Although the present data suggest that there is a central mode in $CoS_2$, clearly more effort will be required in order to unambiguously decide this question.

The authors are indebted to G. Shirane for suggesting this study and for many stimulating discussions. They have also benefited greatly by enlightening discussions with S. Ogawa, Y. Yamaguchi, and L. Passell. We would also like to thank C. J. Glinka for kindly providing us with one of the data processing programs.

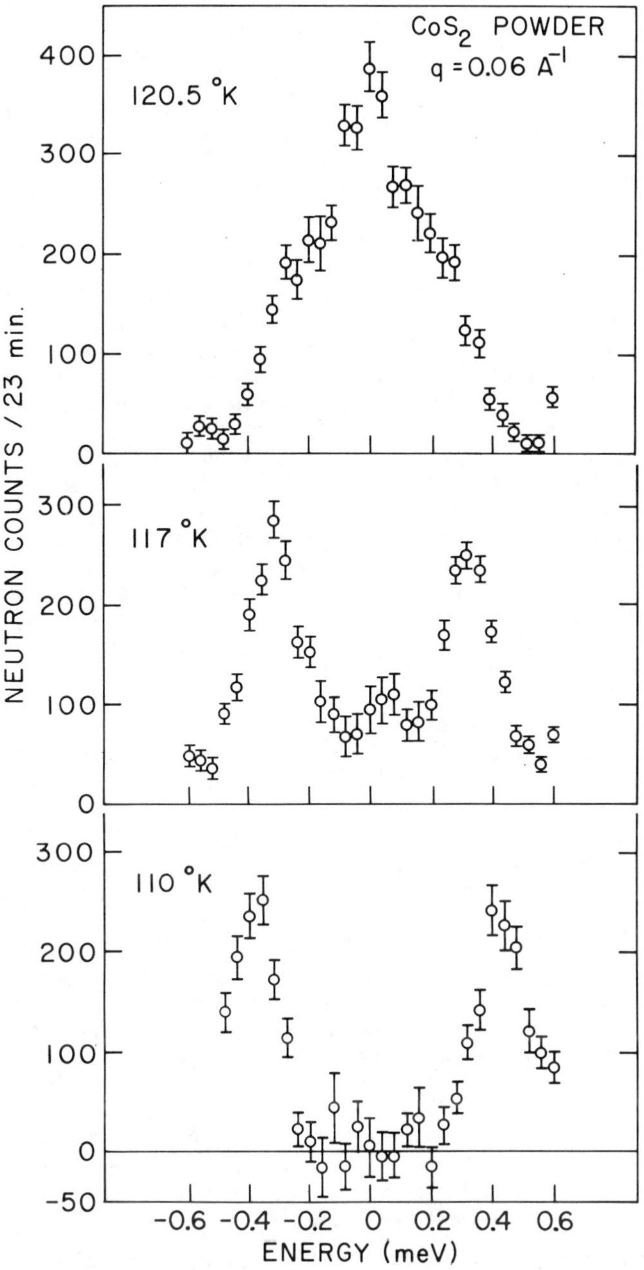


Fig. 3. Constant-Q profiles of the magnetic scattering below $T_c$(= 122.5 K). Background has been subtracted. The data suggest that there may be a central component to the scattering below $T_c$.

REFERENCES

* On leave from Japan Atomic Energy Research Institute, Tokai, Japan.

† Work at Brookhaven performed under the auspices of the U. S. Energy Research and Development Administration.

1. K. Adachi, K. Sato and M. Takeda, J. Phys. Soc. Japan 26, 631-638 (1969); H. S. Jarrett, W. H. Cloud, R. J. Bouchard, S. R. Butler, C. G. Frederick, and J. L. Gillson, Phys. Rev. Lett. 21, 617-620 (1968); S. Ogawa, S. Waki, and T. Teranishi, Int. J. Magnetism 5, 349-360 (1974).
2. K. Adachi, K. Sato, M. Okimori, G. Yamauchi, H. Yasuoka, and Y. Nakamura, J. Phys. Soc. Japan 38, 81-86 (1975).
3. J. W. Lynn, Phys. Rev. B 11, 2624-2637 (1975); H. A. Mook, J. W. Lynn, and R. M. Nicklow, Phys. Rev. Lett. 30, 556-559 (1973).
4. M. Jibu, Y. Ishikawa, and K. Tajima, Phys. Lett. 45A, 235-236 (1973).
5. M. F. Collins, V. J. Minkiewicz, R. Nathans, L. Passell, and G. Shirane, Phys. Rev. 179, 417-430 (1969); V. J. Minkiewicz, M. F. Collins, R. Nathans and G. Shirane, Phys. Rev. 182, 624-631 (1969).
6. C. J. Glinka, V. J. Minkiewicz, and L. Passell, Proc. 20th Annual Conf. on Magnetism & Magnetic Materials, AIP Conf. Proc. 20, 283-284 (1974).
7. L. Passell, J. Als-Nielsen, and O. W. Dietrich, Proc. Fifth IAEA Symposium on Neutron Inelastic Scattering (International Atomic Energy Agency, Vienna, 1972) pp. 619-629.
8. A. Tucciarone, H. Y. Lau, L. M. Corliss, A. Delapalme, and J. M. Hastings, Phys. Rev. B 4, 3206-3245 (1971).
9. V. J. Minkiewicz, K. Gesi, and E. Hirakawa, J. Appl. Phys. 42, 1374-1375 (1971).

# INVESTIGATION OF $FeBO_3$ BY BRILLOUIN SCATTERING FROM THERMAL MAGNONS

W. Jantz, J. R. Sandercock[+] and W. Wettling
Institut für Angewandte Festkörperphysik der Fraunhofergesellschaft
D 78 Freiburg, Eckerstr. 4, West Germany
[+] Laboratories RCA, Badenerstr. 569, CH 8048 Zürich, Switzerland

ABSTRACT

Brillouin scattering from thermal magnons of the low frequency branch is used to determine with high precision all relevant magnetic parameters of iron borate. Vapor transport grown crystals of high optical quality are employed. Using the backscattering configuration the symmetric and antisymmetric exchange constants are determined for temperatures $77 < T < T_N$. The magnon energy exhibits a remarkable anisotropy which is mainly due to the trigonal lattice distortion. The forward scattering geometry measures magnons with small exchange energy, therefore giving more detailed information about the spin wave energy gap. In addition, ferromagnetic resonance data at various microwave frequencies are reported, comparing favorably with the results of forward scattering.

Iron borate, $FeBO_3$, is one of the rare materials that exhibit a spontaneous ferromagnetic moment at room temperature and are transparent at optical frequencies [1]. This combination of properties is essential for magnetooptic applications.

$FeBO_3$ has the rhomohedral calcite structure. The refined structure data [2] are lattice constant $d = 5.520$ Å and rhombohedral angle $\delta = 49.54°$. The magnetic sublattices are (111) layers of alternating direction of magnetisation. The coupling is essentially antiferromagnetic, however a small canting of the sublattice magnetisations results in a weak ferromagnetic moment ($4\pi M = 115$ Gauss at room temperature) constrained to lie in the (111) plane. Thus $FeBO_3$ is an easy-plane weak ferromagnet [3].

The well established methods used to study magnetic properties, of which the most important are ferromagnetic resonance [4], Raman [5] and inelastic neutron scattering [6], have fairly recently been augmented by the method of light scattering from long wavelength acoustic magnons [7]. Such measurements have been performed using a multipass Fabry Perot interferometer with the green argon laser line, for which wavelength the material is fairly transparent.

The energy of spin waves of the low frequency acoustic branch in $FeBO_3$ is given by [4]

$$\omega = ( \omega_H^2 + \omega_k^2 + \omega_\Delta^2 )^{1/2} \qquad (1)$$

where

$$\omega_H = \gamma ( H ( H + H_{DM} ) )^{1/2} ; \; H \perp \langle 111 \rangle \qquad (2a)$$

$$\omega_k = \gamma \cdot H_E \cdot d \cdot k \cdot \cos\varepsilon \; ; \; \text{wavevector } k \ll \pi/d \qquad (2b)$$

H is the applied static field, $\gamma$ is the gyromagnetic ratio, $H_{DM}$ is the Dzialoshinski-Moriya field of the antisymmetric exchange interaction that produces the canting of the sublattice magnetic moments, $H_E$ is the symmetric exchange field and $\varepsilon$ is the angle between a direct and the associated reciprocal lattice vector. The small contribution $\omega_\Delta$ measures

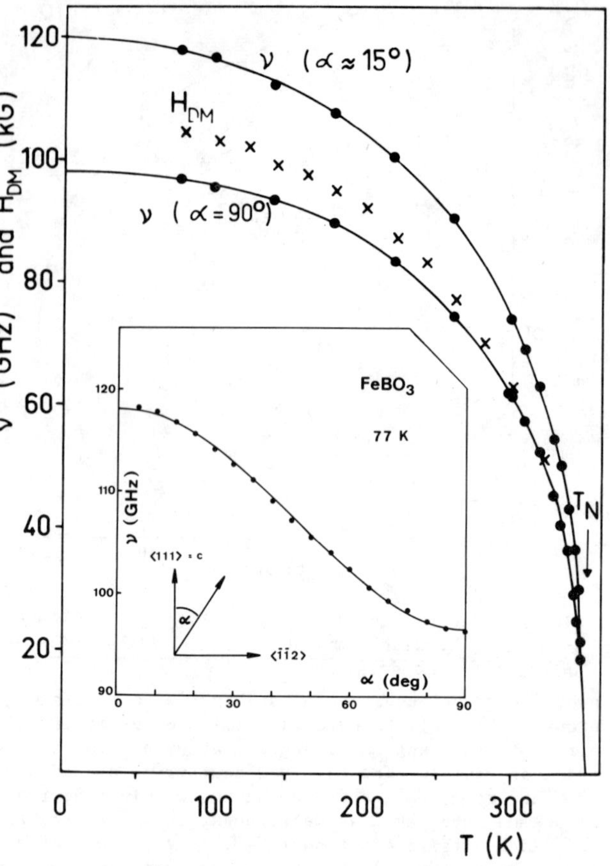


Fig. 1 Temperature dependence of the magnon frequency $\nu = \omega/2\pi$ measured in $H = 0$ for $\alpha = 0$ and 15° and of the field $H_{DM}$. The insert shows the directional dependence of $\nu$, at 77 K.

the zero field, zero wavevector energy gap.

From eq. (1) it is seen that an accurate determination of one of the terms is possible only if the others are kept small. Thus, to measure $\omega_\Delta$, both H and k must be small, whereas to determine $\omega_k$ a larger k is desirable In order to select the magnitude of $\vec{k}$, both the backscattering ($k \approx 5 \times 10^5$ cm$^{-1}$) and the low angle scattering ($k \approx 5 \times 10^4$ cm$^{-1}$) configurations were used. The measurements could be made at variable temperature $77 < T < T_N = 348$ K and applying external fields up to 6 kG for a simultaneous determination of $H_{DM}$. For the backscattering experiments a spherical sample of 2 mm diameter was used to allow a measurement of the directional dependence of $\omega$ on $\alpha$, the angle between $\vec{k}$ and the ⟨111⟩ c-axis.

Fig. 1 shows the temperature dependence of $\omega$ for $H = 0$, which equals $\omega_k$, because for $k = 5.4 \times 10^5$ cm$^{-1}$ $\omega_\Delta$ in eq. (1) may be neglected. Two orientations of $\vec{k}$ with respect to c are shown, revealing an anisotropy of the exchange energy, of which the full angular dependence, for 77 K, is given in the insert. The anisotropy is, primarily, due to the trigonal distortion of the $FeBO_3$ lattice, but also contains information about the exchange constants. $\omega_k$ is mainly determined by $J_1$, the nearest neighbour intersublattice exchange, to a smaller extent by $J_1'$, the nearest neighbour intra- and $J_2$, the next nearest neighbour intersublattice exchange. The complete formulae describing the anisotropy are lengthy and shall be given elsewhere[8]. Using $J_1$, $J_2$ and $J_1'$ as variable parameters, an excellent fit to the directional dependence may be obtained (solid line), giving $J_1 + J_2 = -297$ GHz and $J_2 - J_1' = -12$ GHz. Using the result of two magnon Raman scattering[5], viz. $\omega_k/2\pi = 8.055 \times 10^3$ GHz at 80 K for $k = (\pi/d, \pi/d, 0)$, and the expression

$$\omega_{(\pi/d,\,\pi/d,\,0)} = 2\pi \cdot 20\left(2(2J_1' - J_1 - 3J_2) \times (J_1' - J_1)\right)^{1/2} \qquad (3)$$

the exchange constants may be separately evaluated:

$$J_1 = -268 \text{ GHz}$$
$$J_2 = -29 \text{ GHz} \quad (T = 77 \text{ K}) \qquad (4)$$
$$J_1' = -17 \text{ GHz}$$

$J_1$ agrees well with known values[5, 9], whereas $J_1'$ and $J_2$, accurately determined for the first time, are markedly different from data estimated previously[5]. The effective exchange field $H_E$ to be inserted in eq. (2b) for 77 K is calculated from (4) to be $1.99 \times 10^3$ kG for $k \parallel c$ and $1.65 \times 10^3$ kG for $k \perp c$.

To good approximation

$$\frac{(J_1 + J_2)_{77\,K}}{(J_1 + J_2)_T} \simeq \frac{(J_1)_{77\,K}}{(J_1)_T} \qquad (5)$$

thus from the values (4) together with the data in Fig. 1 all three exchange constants can be evaluated for any $T$ between 77 K and $T_N$.

Repeating the backscattering measurements with an applied external field, the magnitude and the temperature dependence of $H_{DM}$ were obtained. The data are also given in Fig. 1.

To obtain scattering from spin waves with small $k$, necessary for measuring $\omega_\Delta$, a low angle scattering geometry was used. The sample was a 0.2 mm thick plate containing the c-axis. The scattering process rotates the light polarization by 90 degrees, therefore $k$ must be calculated according to the theory of light scattering in birefringent media[10].

In Fig. 2 the measured room temperature magnon frequencies as a function of the magnetic field between 0 and 200 Gauss are given for different wavevectors $k$. All solid lines have been calculated using a single set of parameters, viz. $J_1 + J_2 = -189$ GHz, $H_{DM} = 61$ kG and $\omega_\Delta = 1.5$ GHz. The first two results agree well with the backscattering data, which may be obtained for room temperature using Fig. 1 and eq. (5).

Unfortunately, it turned out that $\omega_\Delta$ cannot be determined very accurately ($\pm 1$ GHz), because only the smallest values $k \leqslant 3 \times 10^4$ cm$^{-1}$ make $\omega_k$ small enough not to dominate in eq. (1). These small wavevectors, however, most sensitively depend on the scattering angle, therefore their accuracy is degraded by the finite aperture (2 degrees) of the optical system collecting the scattered light.

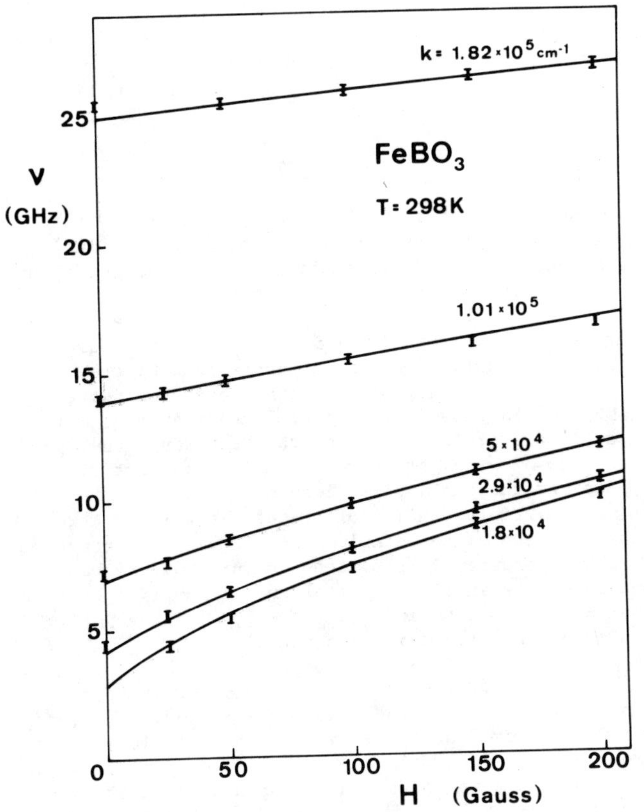


Fig. 2 Results of forward scattering measurements showing magnon frequency as a function of applied field H and wavevector $k \parallel c$.

In order to avoid this difficulty measurements of the ferromagnetic resonance ($k = 0$) were performed between 5 and 250 Gauss using a wideband microwave system (2 - 12 GHz). A very good fit to the data measured at 294 K has been obtained with $H_{DM} = 64$ kG and $\omega_\Delta = 3.0 \pm 0.2$ GHz. The demagnetisation, though small (2.4 Gauss), was taken into account.

REFERENCES


1. R. C. LeCraw, R. Wolfe and J. W. Nielsen, Appl. Phys. Lett. 14, 352-54 (1969).
2. R. Diehl, Sol. State Comm. 17, 743-45 (1975).
3. I. E. Dzialoshinski, Sov. Phys. JETP 6, 1120-33 (1958).
4. L. V. Velikov, A. S. Prokhorov, E. G. Rudashevski and V. N. Seleznev, Sov. Phys. JETP 39, 909-15 (74).
5. N. Koshizuka, T. Okuda, M. Udagawa and T. Tsushima, J. Phys. Soc. Jap. 37, 354-62 (1974).
6. T. M. Holden, E. C. Svensson and P. Martel, Can. J. Phys. 50, 687-91 (1972).
7. J. R. Sandercock and W. Wettling, Sol. Stat. Comm. 13, 1729-32 (1973).
8. W. Jantz, J. R. Sandercock and W. Wettling, to be published
9. A. E. Meixner, R. E. Dietz and D. L. Rousseau, Phys. Rev. B7, 3134-41 (1973).
10. R. W. Dixon, IEEE J. Quant. Electr. QE-3, 85-93 (1967).

# VIRTUAL SPIN WAVES IN ITINERANT FERROMAGNETS ABOVE $T_c$*

M.A. Klenin,[+] SUNY, Stony Brook, New York 11794 and
J.A. Hertz[++], University of Chicago, Chicago, Illinois 60637

ABSTRACT

We carry out a microscopic calculation of the dynamic susceptibility in itinerant (Hubbard-like) paramagnets using a reformulation of an earlier theory[1] which retains rotational invariance explicitly. Because the fluctuating order parameter has the character of a three-dimensional vector field, the susceptibility calculated within a generalized Hartree theory is found to include both transverse (spinwave-like) and longitudinal (Stoner-like) response. Our results differ from those of the RPAthrough the appearance of propagating spinwave-like collective modes and through a virtual spin splitting of the single particle density of states, as well as through the smearing of the single-particle band structure which we previously reported. Damping occurs because of the analytic form of the Stoner-like part and because of fluctuations of the effective gap parameter entering into the spin wave part.

The recent neutron scattering experiments of Mook, Lynn and Nicklow[2] point up a serious deficiency in the RPA or Stoner theory of itinerant electron ferromagnetism. The data appear as if the splitting of up and down spin bands and the spin wave excitations out of this broken-symmetry state exist above as well as below $T_c$, except at very long wavelengths. It is not possible within simple Stoner theory to account for why such excitations live as long as they apparently do, and why they exist for such long wavelengths. The natural inference from the pattern of the data is that local magnetic order exists over relatively long distances, so that magnons of wavelengths short compared to such distances can propagate almost as if the system were ordered. This picture is taken as a hypothesis in several recent discussions of the problem.[3] In this work we use the microscopic functional integral formulation of the interacting paramagnon problem to show that such a picture emerges in a natural way when the non-linear character of the fluctuation effects is taken into account. We used such a formal description earlier to discuss virtual local moment formation and the static susceptibility of these systems[1]; the phenomena we examine here are the dynamic manifestations of the physics described there.

In order to retain the symmetry of the problem most intimately connected with the spin wave excitations, we write the Hubbard interaction Hamiltonian in a manifestly rotationally invariant way.[4]

$$H' = -\frac{2}{3}U\sum_i \vec{S}_i\cdot\vec{S}_i = -\frac{2}{3}U\sum_i(\psi_i^+\frac{\vec{\sigma}}{2}\psi_i)\cdot(\psi_i^+\frac{\vec{\sigma}}{2}\psi_i) \quad (1)$$

where $\vec{\sigma}$ is the vector of Pauli matrices. The Stratonovich-Hubbard[4,5] transformation leads to the following expression for the partition function

$$Z = Z_0\int\delta\vec{\xi}\exp\left\{-\frac{1}{\beta}\int_0^\beta d\tau\,\xi^2(\tau)\right\}\cdot\left\langle T\exp\left\{-\int_0^\beta d\tau\sum_i c\,\vec{\xi}_i(\tau)\cdot\vec{S}_i(\tau)\right\}\right\rangle_0 \quad (2)$$

where $c=(4U/3\beta)^{\frac{1}{2}}$. The brackets indicate the expectation value taken with respect to the non-interacting system. Writing the result formally as $Z[\Psi] = \int\delta\vec{\xi}\exp(-\Psi[\xi])$, one sees that $\Psi[\xi]$ is a generalized Landau-Ginzburg functional. The terms in which are quadratic in $\xi$ describe free paramagnons and those of higher order, couplings between them. Viewed in terms of this picture, the approximation we make below amounts to approximating all these higher-order coupling by local interactions, and treating them in Hartree-Fock approximation.

One way to get a Hartree-Fock approximation is to ask the question: What single particle Hamiltonian, or in this case, what quadratic free energy functional best approximates the true functional $\Psi[\xi]$? In Ref. 1 we posed and solved this problem for the scalar field case; the generalization to the vector field is trivial. The required functional is

$$\Psi_0[\xi] = \frac{1}{2}\sum_{\vec{q}\omega,\alpha\beta}\left\langle\frac{\partial^2\Psi[\xi]}{\partial\xi^\alpha_{\vec{q}\omega,\alpha\beta}\partial\xi^\beta_{-\vec{q}-\omega}}\right\rangle_{\Psi_0}\xi^\alpha_{\vec{q}\omega}\xi^\beta_{-\vec{q}-\omega} \quad (3)$$

where the average is over the approximate distribution; i.e., $\langle f[\xi]\rangle_{\Psi_0} = 1/Z[\Psi_0]\int\delta\vec{\xi}f[\xi]\exp(-\Psi_0[\xi])$. Carrying out the differentiation,

$$\Psi_0[\xi] = \sum_{\vec{q}\omega,\alpha\beta}\left[\delta_{\alpha\beta} - \frac{2}{3}U\bar{\chi}^0(\vec{q},\omega)\right] \quad (4)$$

where $\bar{\chi}^0$ is the susceptibility of a non interacting system in the presence of the fields $c\ \xi(\tau)$ averaged over the distribution $\exp(-\Psi_0)$. The structure of the partition function, susceptibilities, etc., is therefore like that in RPA, except that the free electron susceptibility is replaced by $\bar{\chi}^0$.

The evaluation of $\bar{\chi}^0$, of course, involves approximations. Here we will ignore the spatial inhomogeneity and time dependence of the fields $\xi_i(\tau)$ and evaluate the average as if the $\xi_i(\tau)$ were uniform and static. We consequently obtain a susceptibility which depends only on the average magnitude of the local vector field. We have discussed the formal aspects of the problem, including its diagrammatic formulation, in some detail in Ref. 1; the only additional feature of the approximation in the present case is that it imposes a local spherical symmetry in the calculation of $\bar{\chi}^0$. In particular one derives the result

$$\bar{\chi}^0(q,\omega) = 4\pi\left(\frac{3}{2\pi\sigma^2}\right)^{3/2}\int_0^\infty\xi^2 d\xi\,\exp(-3\xi^2/2\sigma^2)\left(\frac{1}{3}\chi^0_\parallel(c\,\xi)+\frac{2}{3}\chi^0_\perp(c\,\xi)\right) \quad (5)$$

where now $\chi^0$ is functionally dependent on only the single variable $\xi$ with variance

$$c^2\sigma^2 = \frac{4U}{3N\beta}\sum_{\vec{q}\omega}\langle|\xi_{\vec{q}\omega}|^2\rangle \quad (6)$$

$\chi^0_\parallel$ and $\chi^0_\perp$ are the parallel and transverse susceptibilities of the non interacting system in the presence of a static uniform field of magnitude $c\,\xi$

$$\chi^0_\parallel(q,\omega;\xi) = -\frac{1}{2}\sum_{\substack{K\\ s=\pm1}}\frac{f(E_k+s\,c\,\xi) - f(E_{k+q}+s\,c\,\xi)}{E_k - E_{k+q} - \omega} \quad (7)$$

$$\chi^0_\perp(q,\omega;\xi) = -\frac{1}{2}\sum_{\substack{K\\ s=\pm1}}\frac{f(E_k+s\,c\,\xi) - f(E_{k+q}-s\,c\,\xi)}{E_k - E_{k+q} - \omega + 2s\,c\,\xi} \quad (8)$$

The transverse ($\perp\xi$) contribution has a Stoner gap 2c, while the longitudinal part ($\parallel\xi$) is the source of a Fermi-level shift of the sort we reported in our earlier work. The weights of 1/3 and 2/3 which appear in Eqn. 5 are the result of averaging over the orientation of $\vec{\xi}$, so that we are left with the average over the distribution of magnitudes of $\vec{\xi}$. Since this distribution takes on its maximum value at $\xi\simeq\sigma>0$, and since the contribution from $\chi^0_\perp$ would contain well-defined spinwave-like excitations if the distribution were very sharply peaked about this maximum, we expect the excitation spectrum to contain at least in principle some remnants of the spin wave structure characteristic of a gap approximately given by $c\xi$.

There remains, however, the important quantitative question: Under what conditions will the excitations remain well defined and when will they be overdamped? Obviously, any serious attempt to deal with the experiments in real systems ought to take into account the full complexities of their band structure. To see how the band parameters enter into such considerations, however, we turn now to an approximate calculation of the spectrum.

The susceptibility is given by

$$\chi(q,\omega) = \bar{\chi}^\circ(q,\omega)/(1 - 2/3\, U\bar{\chi}^\circ(q,\omega)) \quad (9)$$

We therefore search for the zeroes of the real part of the denominator of (9) to find approximate excitation energies and then compute the imaginary part of these energies to determine how strongly damped the excitations are. Expanding Eqns. 7 and 8 to lowest order in $q$ and $\omega$, and averaging with respect to $\xi$, we find that the first order term in $\omega$ vanishes, as it must above the transition. Defining the parameters $r, \lambda, \tau$,

$$1 - \frac{2}{3} U \bar{\chi}^\circ(q,\omega) = r - \tau^2\omega^2 + \lambda^2 q^2 \quad (10)$$

we find after some lengthy but straightforward calculation the expressions

$$r = 1 - \frac{2}{3} UN(E_F) + \frac{5}{27} U N''(E_F) \cdot c^2\sigma^2$$
$$\tau^2 = \frac{UN(E_F)}{3c^2\sigma^2} - \frac{1}{54} UN''(E_F) \quad (11)$$
$$\lambda^2 = \frac{U}{54} \frac{d^2}{dE^2}\left[N(E) v_F^2(E)\right]\Big|_{E=E_F}$$

In order to simplify our expressions we have assumed that the density of states $N(E)$ is symmetric about the Fermi energy $E_F$, that $N''(E) < 0$, and that we may ignore derivatives of higher than second order of $N(E)$ and $v_F(E)$. The expressions (11) are low temperature limits; that is, we ignore the temperature dependence they acquire from the finite-T smearing of the Fermi surface and assume (as in Ref. 1) that the dominant temperature dependence comes from $c^2\sigma^2$. $v_F$ has its usual definition, the average magnitude of the gradient of $E(k)$ at the Fermi surface.

The dispersion relation

$$\omega^2(q) = \frac{1}{\tau^2}(r + \lambda^2 q^2) \quad (12)$$

can be simplified. As we have previously noted[1], when T exceeds typical spinwave energies, the finite-terms in (6) can be ignored, and the principal T-dependence of $c^2\sigma^2$ comes from the $c^2$ itself:

$$c^2\sigma^2 = \frac{UT}{N}\sum_q \left(1 - \frac{2}{3} U \bar{\chi}^\circ(q,\omega)\right)^{-1} \quad (13)$$
$$\approx UT$$

Thus r becomes

$$r(T) = 1 - \frac{2}{3} UN(E_F) + \frac{5}{27} UN''(E_F) \cdot UT = \frac{3(T-T_c)}{2UC} \quad (14)$$

where the effective Curie constant is

$$C = \frac{3T}{2U}\left(\frac{2}{3} UN(E_F) - 1\right)^{-1}, \quad (15)$$

and the dispersion relation

$$\omega^2(q) = \frac{3T(T-T_c)}{C}\left[1 - \frac{1}{18}\frac{N''(E_F)}{N(E_F)} UT\right]^{-1} + \lambda^2 q^2/\tau^2 \quad (16)$$

The Hartree theory is at best limited to values of q greater than the inverse correlation length. (For smaller q, collisional damping dominates the dynamics.) Hence

$$\lambda^2 q^2 \gg r \quad (17)$$

or

$$qv_F \gg \sqrt{U(T-T_c)} \quad (18)$$

Damping within the theory comes in principle from two sources, the smearing out of the gap appearing in $\chi_\perp^\circ$, and the Landau damping always present in $\chi_\parallel^\circ$. One finds, however, that the lowest order contributions to the imaginary part of $\omega$ from $\chi_\perp^\circ$ are proportional to $\omega^3$ and $q^2\omega$, so damping from this source does not become important until $\omega(q)$ is of the order of the rms gap $c\sigma$; that is when $\lambda q/\tau \gtrsim c$, which reduces to

$$(qv_F)^2 \gtrsim 27/UN''(E_F) \quad (19)$$

$\mathrm{Im}\,\chi_\parallel^\circ$ makes a contribution approximately given by

$$\mathrm{Im}\,\omega(q) \simeq \frac{\pi}{12}\frac{1}{q\bar{v}_F^2} \quad (20)$$

Here $\bar{v}_F$ differs from $v_F$. It is defined (again in terms of the band structure) by the expression

$$\frac{N(E_F)}{\bar{v}_F} = \int \frac{d^{(2)}S}{|\nabla_k E(k)|}\,\delta(\hat{q}\cdot\nabla_k E(k)) \quad (21)$$

The requirement that this contribution to the damping not wash out the spectrum completely is just that $\mathrm{Re}\,\omega(q) \gg \mathrm{Im}\,\omega(q)$ or that

$$(qv_F)^2 \gg \sqrt{T/N''(E_F)} \quad (22)$$

Using Eqns. 14 and 15, this can be written

$$\frac{q\bar{v}_F}{c\sigma} \gg \left(\frac{2}{3} UN(E_F) - 1\right)^{-\frac{1}{4}} \quad (23)$$

That is, we expect the sloppy spin waves to be completely overdamped in very weak ferromagnets.

In summary, we have derived an approximate expression for the dispersion relation of spinwave-like excitations above the Curie temperature in strong ferromagnets. We are able to describe both characteristic frequencies and damping in terms of various band parameters, and to determine a regime in which these excitations should be observable. Direct comparison with experiment, however, awaits more precise numerical computation.

## REFERENCES

*Supported by N.S.F. and N.S.F.-M.R.L.
+also Guest Scientist, Brookhaven National Laboratory, Upton, New York 11973, under auspices of U.S. E.R.D.A.
++Alfred P. Sloan Foundation Fellow

1) J.A. Hertz and M.A. Klenin, Phys. Rev. B10, 1084 (1974)
2) H.A. Mook, J.W. Lynn and R.M. Nicklow, Phys. Rev. Letters 30, 556 (1973)
3) R.E. Prange and V. Korenman, AIP Conf. Proc. 24, 325 (1974); J.B. Sokoloff, preprint.
4) J.R. Schrieffer, CAP Summer School Lectures, 1969 (unpublished)
5) R.L. Stratonovich, Soviet Physics - Doklady 2, 416 (1958); J. Hubbard, Phys. Rev. Letters 3, 77 (1959)

# MAGNON MODES OF A GYROTROPIC MAGNETIC BAR

T. M. Sharon and A. A. Maradudin
Department of Physics, University of California, Irvine, California 92664

## ABSTRACT

We present a theory of magnetostatic modes which are wavelike in the direction parallel to the axis of an infinitely long, ferromagnetic bar of square cross-section, and localized at its surface. We solve second order partial differential equations for the magnetic scalar potential by assuming a solution for $\varphi$ of the form $\varphi(r,\theta,z)=e^{iqz}f(r,\theta)$. We expand $f(r,\theta)$ in terms of basis functions which transform according to the irreducible representations of $C_4$, the proper point group of the bar. The boundary conditions of continuity of $\varphi$ and $\vec{B}_n$ across the surface of the bar therefore need to be applied at only one of the four faces of the bar and are automatically satisfied at the remaining faces. This is done by making $\varphi$ and $B_n$ continuous at a discrete set of points along this face. The solvability condition for the resulting set of homogeneous equations for the expansion coefficients in the expressions for $\varphi$ yields the dispersion curves for the magnetostatic surface modes. We also present the dispersion curves for the magnetostatic surface modes for a gyrotropic right circular cylinder.

## INTRODUCTION

We have recently studied surface[1] and edge[2] localized magnons or spin waves for a simple cubic Heisenberg ferromagnet with exchange interactions between nearest and next nearest neighbor spins. Such modes may play a role in phase transitions[3] and chemical reactions[4] at magnetic surfaces, and may have technological applications in surface or topographic wave guides for magnetic excitations. Inasmuch as the spins in a real ferromagnet interact through long range dipole interactions, as well as through short range exchange interactions, the former of which dominate in the long wavelength limit, it is of interest to see how the results of Refs. 1 and 2 are modified by the inclusion of dipolar interactions between spins.

In a recent study[5] the spin wave spectrum was obtained for a discrete, infinite bar of a simple cubic ferromagnet of square cross section, assuming nearest and next nearest neighbor short-range exchange interactions as well as long-range dipolar interactions. In the presence of the dipolar interactions structure was found in this spectrum which was absent when only the short range exchange interactions were taken into account. The nature of the structure is depicted schematically in Fig. 1, and appears to be of the type, observed in other, similar, contexts,[6] of the repulsion of levels which occurs when the dispersion curve for a localized excitation (a resonance mode) falls in the region of a continuous spectrum. Since this structure appears in the long wavelength limit, in the presence of the dipolar interactions, in the absence of retardation, it is natural to assume that the localized excitation is a magnetostatic mode of the ferromagnetic bar, localized at its surface or edges.

To ascertain the reasonableness of this assumption we obtained the dispersion relations for magnetostatic modes localized at the surface of a ferromagnet in the form of an infinite, right circular cylinder and for magnetostatic modes in a ferromagnet in the form of an infinite bar possessing a square cross section.

## MAGNETOSTATIC SURFACE MODES ON A GYROTROPIC CYLINDER

We consider a right circular cylinder of radius a whose axis is parallel to the z-axis. In the magnetostatic approximation we introduce a scalar potential $\varphi(\vec{x},\omega)\exp(-i\omega t)$, in terms of which $\vec{H}=-\nabla\varphi$ and $\vec{B}$ is related to $\vec{H}$ by the magnetic permeability tensor appropriate to a ferromagnet in the presence of a dc magnetic field along the z-axis.

We find that $\varphi(\vec{x},\omega)$ obeys the equation

$$\frac{\partial^2\varphi}{\partial x^2}+\frac{\partial^2\varphi}{\partial y^2}+\nu(\omega)\frac{\partial^2\varphi}{\partial z^2}=0\ , \qquad (2.1)$$

where $\nu(\omega)=\dfrac{\omega_0^2-\omega^2}{\omega_v^2-\omega^2}$ for $0\le r<a$, and $\nu(\omega)\equiv 1$ for $r>a$; $\omega_o=\gamma H_0$, and $\omega_v=\gamma[H_0(H_0+4\pi M_0)]^{\frac{1}{2}}$, where $H_0$ is the dc magnetic field strength and $M_0$ is the saturation magnetization, while $\gamma$ is the gyromagnetic ratio.

We assume solutions of Eq. (2.1) which are wavelike along the axis of the cylinder. Inside the cylinder the solution is finite at the origin, and increases exponentially with increasing r; outside the solution decreases exponentially with increasing r.

The continuity of $\varphi$ and of $B_r$ at the boundary r=a yields the pair of dispersion equations whose solutions give the relation between $\omega$ and q:

$$(\nu(\omega))^{\frac{1}{2}}\mu_1(\omega)\frac{I_n'((\nu(\omega))^{\frac{1}{2}}qa)}{I_n((\nu(\omega))^{\frac{1}{2}}qa)}-\frac{K_n'(qa)}{K_n(qa)}\pm\frac{n\mu_3(\omega)}{qa}=0, \qquad (2.2)$$

where the upper sign obtains for $\omega>\omega_v$, and the lower for $0<\omega<\omega_0$. The $I_n(x)$ and $K_n(x)$ are the modified Bessel functions of the first and second kinds, respectively. It is found that there are no solutions in the frequency range $0<\omega<\omega_0$. Solutions of Eq. (2.2) for $\omega>\omega_v$ have been obtained numerically for several values of n, and are displayed in Fig.

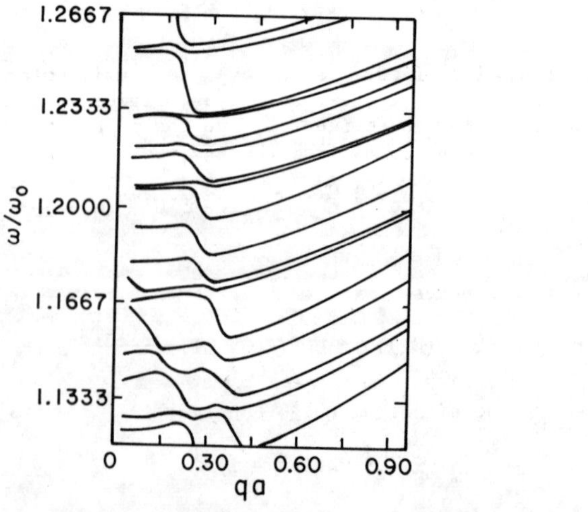


Fig. 1. Schematic depiction of the dispersion relations for spin waves in a discrete bar of a ferromagnet with square cross section, in the long wavelength limit. ($\omega_v/\omega_0$=1.00167)

2 for $n = 1,2,3,4$. (There are no solutions for $n = 0$.) It should be noted that they have the qualitative form suggested by the results shown in Fig. 1.

## MAGNETOSTATIC SURFACE AND EDGE MODES ON A FERROMAGNETIC BAR OF SQUARE CROSS SECTION

We now turn to a solution of Eq. (2.1) for a bar of square cross section, whose axis is parallel to the z-axis. The edge of the square cross section of the bar is 2a, and its center is at x=y=0. The interior and exterior solutions for $\varphi(r,\theta,z;\omega)$ can be written as

$$\varphi_{in}^{(j)}(r,\theta,x;\omega) = e^{iqz} \sum_{n=0}^{\infty} \sum_{\lambda=1}^{2} a_{n\lambda}^{(j)} I_n((\nu(\omega))^{\frac{1}{2}} qr) \times$$
$$\times f_{n\lambda}^{(j)}(\theta) \qquad 0 \le r \le a \qquad (3.1a)$$

$$\varphi_{out}^{(j)}(r,\theta,z;\omega) = e^{iqz} \sum_{n=0}^{\infty} \sum_{\lambda=1}^{2} b_{n\lambda}^{(j)} K_n(qr) \times$$
$$\times f_{n\lambda}^{(j)}(\theta) \qquad r \ge a \ , \qquad (3.1b)$$

where $j = \Gamma_1, \Gamma_2, \Gamma_3, \Gamma_4$ labels the irreducible representations of the point group $C_4$ of the bar, and $f_{n1}^{(j)}(\theta)$ and $f_{n2}^{(j)}(\theta)$ are the parts of $\cos n\theta$ and $\sin n\theta$, respectively, which belong to each of these irreducible representations, for example:

$$f_{n1}^{(\Gamma_1)}(\theta) = \cos\frac{n\pi}{2} \cos^2\frac{n\pi}{4} \cos n\theta$$

$$f_{n1}^{(\Gamma_2)}(\theta) = \cos\frac{n\pi}{2} \sin^2\frac{n\pi}{4} \cos n\theta \qquad (3.2)$$

$$f_{n2}^{(\Gamma_1)}(\theta) = \cos\frac{n\pi}{2} \cos^2\frac{n\pi}{4} \sin n\theta$$

$$f_{n2}^{(\Gamma_2)}(\theta) = \cos\frac{n\pi}{2} \sin^2\frac{n\pi}{4} \sin n\theta$$

With the interior and exterior potentials expanded in terms of these symmetry adapted functions it suffices to satisfy the boundary conditions at only one face of the square bar, e.g., the face defined by $x = +a, -a \le y \le +a$: the boundary conditions are then automatically satisfied at the three remaining faces. In practice we do this by making $\varphi$ and $B_x$ continuous at a discrete set of points along this face. These points are chosen so as to divide the face x=+a, -a≦y≦+a into segments of equal length but not to include the corners, which are shared by two adjacent faces. An even number of points was chosen in the case of modes of $\Gamma_2, \Gamma_3, \Gamma_4$ symmetry because here n>0, and for each value of n the index $\lambda$ assumes the two values $\lambda$=1,2. Thus the first N nonzero values of n in the expansions (3.1) give rise to 2N terms in these expansions, which requires 2N points at which the boundary conditions are satisfied. For modes of $\Gamma_1$ symmetry, n can equal zero. In this case the $f_{n2}^{(\Gamma_1)}(\theta)$ vanishes identically. Thus the first N nonzero values of n in this case give rise to 2N-1 terms in the expansions (3.1), which requires 2N-1 points at which the boundary conditions are satisfied.

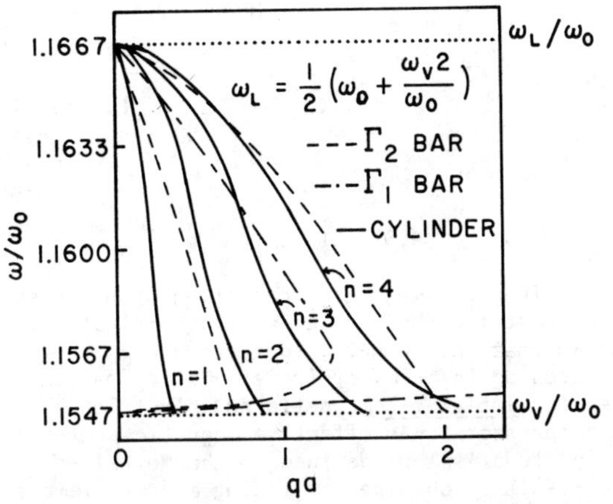


Fig. 2. The dispersion relations for magnetostatic surface modes on a gyrotropic right circular cylinder, and on a gyrotropic bar of square cross section. ($\mu_0$=1, $\omega_v/\omega_0$=1.15467)

In Fig. 2 we present dispersion curves for modes of $\Gamma_1$ and $\Gamma_2$ symmetry, obtained with N=4 in each case. The convergence of this method seems to be quite good. When these curves were recalculated with N=5 the lower frequency mode of $\Gamma_2$ symmetry shifted a maximum of 0.7%, while the higher frequency mode of this symmetry shifted a maximum of 1.1%. The higher frequency mode of $\Gamma_1$ symmetry shifted a maximum of 2.5%, while the lower frequency mode shifted by 10%. In each case the shift was in the direction of lower frequency. The higher frequency mode of $\Gamma_1$ symmetry and the two modes of $\Gamma_2$ symmetry have the form to explain the results shown in Fig. 1, which lends support to the latter advanced in the Introduction. Results for modes of $\Gamma_3$ and $\Gamma_4$ symmetry are now being obtained.

## REFERENCES

* Work supported in part by AFOSR Grant No. 76-2887.

1. Ipatova, I. P., Klochikhin, A. A., Maradudin, A. A., and Wallis, R. F., in Elementary Excitations in Solids, (Plenum Press, New York, 1969), p. 476.

2. Sharon, T. M., and Maradudin, A. A., Solid State Communications 13, 187 (1973).

3. Trullinger, S. E., and Mills, D. L., Solid State Communications 12, 819 (1973).

4. Petzinger, K. G., and Scalapino, D. J., Phys. Rev. B8, 266 (1973).

5. Sharon, T. M., and Maradudin, A. A., Bull. Am. Phys. Soc. 20, 89 (1975).

6. See, for example, Chen, T. S., Alldredge, G. P., de Wette, F. W., and Allen, R. E., Phys. Rev. Letters 26, 1543 (1971).

# Magnetic Scattering of Neutrons by a Relativistic Atom in the $\ell^N$ Configuration

C. Stassis and H. W. Deckman
Physics Department
Iowa State University
Ames, Iowa 50010


## ABSTRACT

We examined the magnetic scattering of neutrons by an atom described by a relativistic Hamiltonian. It is shown that the magnetic scattering amplitude can be expressed in terms of (relativistic) matrix elements of relativistic magnetic and electric multipole operators. An effective magnetic scattering amplitude is then defined so that a relativistic result is obtained by taking matrix elements of this operator between non-relativistic states of the atom. Quite generally the effective magnetic scattering amplitude can be expressed in terms of relativistic radial integrals and the Racah tensors $W^{(0,k)k}$ and $W^{(1,k')k}$ ($k' = k, k\pm1$). For simplicity, explicit expressions for the relativistic radial integrals are given for the experimentally important case of neutron scattering by an atom in the $\ell^N$ configuration.


## INTRODUCTION

We examine the magnetic scattering of neutrons by an atom described by a relativistic Hamiltonian and we express the magnetic scattering amplitude in terms of relativistic radial integrals which can be evaluated using Hartree-Fock-Dirac (D-F) radial wavefunctions. We will assume that the atomic Hamiltonian can be written as

$$H = \sum_i H_D(i) + V \tag{1}$$

where the sum is over the electrons of the atom, and

$$H_D = c\vec{\alpha}\cdot\vec{p} + \beta mc^2 + \frac{Ze^2}{r} \tag{2}$$

is an one-electron Dirac Hamiltonian. The interaction, V, between the electrons of the atom is usually approximated by the Breit interaction. Most of the approximation methods used in obtaining solutions of Eq. (1) are based on the central field approximation. In this case the atomic states can be expressed in terms of one-electron states, whose large and small radial wavefunctions will be denoted by $F_{n\ell j}(r)$ and $G_{n\ell j}(r)$ respectively. Throughout this paper we will use the common convention of denoting by a rounded ket $|)$ a relativistic wavefunction, and by an angular ket $|>$ a nonrelativistic wavefunction.

## THE MAGNETIC SCATTERING AMPLITUDE

We consider the scattering of neutrons arising from the coupling of the neutron magnetic moment to the atomic current density

$$\vec{j}(\vec{r}) = \sum_i ce\vec{\alpha}(i)\delta(\vec{r}-\vec{r}_i) . \tag{3}$$

By the same manipulations as those used in the non-relativistic treatment of the problem,[1] the magnetic scattering amplitude can be expressed in terms of relativistic multipole operators:

$$f(\vec{q}) = |\gamma| r_0 \vec{\sigma}\cdot(f| \sum_{k,m} \left(\frac{8\pi}{2k+1}\right)^{\frac{1}{2}} \{\vec{X}^{*}_{km}(\hat{q})T^{(e)}_{km} - i(\hat{q}\times\vec{X}^{*}_{km}(\hat{q}))T^{(m)}_{km}\}|i), \tag{4}$$

where $\vec{X}_{km}(\vec{q})$ is a vector spherical harmonic and $T^{(m)}_{km}$ and $T^{(e)}_{km}$ are relativistic magnetic and electric multipole operators defined by

$$T^{(e)}_{km} = \frac{1}{\sqrt{2}}\left(\frac{m}{e\hbar q}\right)\int \vec{A}^{(e)}_{km}\cdot\vec{j}\,d\vec{r} \tag{5a}$$

and

$$T^{(m)}_{km} = \frac{1}{\sqrt{2}}\left(\frac{m}{e\hbar q}\right)\int \vec{A}^{(m)}_{km}\cdot\vec{j}\,d\vec{r} \tag{5b}$$

In these equations,

$$\vec{A}^{(m)}_{km} = i^k\,(4\pi(2k+1))^{\frac{1}{2}}\,j_k(qr)\vec{X}_{km}(\hat{r}) , \tag{6a}$$

$$\vec{A}^{(e)}_{km} = i^{k+1}\left(\frac{4\pi(2k+1)}{k(k+1)}\right)^{\frac{1}{2}} [q\vec{r}j_k(qr)Y_{km}(\hat{r}) + \frac{1}{q}\nabla\{Y_{km}(r)\frac{d}{dr}(r\,j_k(qr))\}], \tag{6b}$$

$j_k(qr)$ is a spherical Bessel function and the current density operator is defined by Eq. (3). The magnetic and electric multipole operators defined by Eqs. (5) have parities $(-)^{k+1}$ and $(-)^k$ respectively. These operators are irreducible tensor operators of rank k in the sense that they are represented by 4x4 matrices whose non-zero elements are irreducible tensor operators of rank k.

The evaluation of the relativistic matrix elements of these operators between relativistic states of the atom is complicated by two factors. First, traditionally one uses a non-relativistic description for the atomic states. Second, one must evaluate matrix elements between relativistic j-j coupled atomic states, whereas most of the techniques of evaluating matrix elements are adapted for non-relativistic LS coupled states. These problems can be avoided by using the effective operator approach first introduced by Sanders and Beck[2] in the study of hyperfine interactions. In this method, for any relativistic operator A one defines an effective operator $A_e$ by requiring that

$$(\Phi|A|\Phi') = <\Phi|A_e|\Phi'> , \tag{7}$$

where $|\Phi)$ and $|\Phi')$ are relativistic atomic states which in the non-relativistic limit go into the states $|\Phi>$ and $|\Phi'>$ respectively.

Using this approach the multipole operators and the magnetic scattering amplitude can be written in an effective operator form. Taking matrix elements of these operators between non-relativistic states of the atom produces a relativistic result. Since the multipole operators are one-electron operators it is easily shown that the magnetic scattering amplitude can be expressed in terms of relativistic radial integrals and the generalized Racah tensors $W^{(0,k)k}$ and $W^{(1,k')k}$ ($k' = k, k\pm1$), whose matrix elements are evaluated by standard techniques. Thus the problem of calculating the magnetic scattering amplitude reduces to evaluating a certain number of relativistic radial integrals using some type of D-F calculation.

Since there is no point in reporting detailed expressions for the general case, we illustrate the formalism by considering the important case of elastic scattering by an atom in a single Russell-Saunders state of the $\ell^N$ electronic configuration. This is a good approximation for the magnetic scattering of neutrons by rare earth ions, in their ground state. In this case by parity and angular momentum conservation only odd magnetic multipoles whose order of multipolarity is less than or equal to $2\ell+1$ contribute to the magnetic scattering amplitude. In addition the orbital rank k' of these multipoles can take only the values $k\pm1$ and must be less than or equal to $2\ell$. In this case the magnetic scattering amplitude can be written in the conventional form $p(\vec{q})\vec{q}_m\cdot\vec{\sigma}$, where $\vec{q}_m$ is the so called magnetic scattering vector and

$$p(\vec{q}) = |\gamma| r_o \sum_k \left(\frac{2}{k(k+1)}\right)^{\frac{1}{2}} P_k'(\cos\Theta)\langle\theta JM|T^{(m)}_{k0}|\theta JM\rangle \quad (8)$$
$$k=1,3\ldots\ 2\ell+1$$

where $\theta\equiv\alpha SL$, $P_k'$ is the derivative of a Legendre polynomial, and $\Theta$ is the angle between the axis of quantitation of the atom and the scattering vector. In this equation, $T^{(m)}_{ko}$ is an <u>effective</u> magnetic multipole operator,

$$T^{(m)}_{k0} = i^{k+1}[R^{(m)}(0,k,k)W_o^{(0,k)k} + \sum_{k'=k\pm1} R^{(m)}(1,k',k)W_o^{(1,k')k}] \quad (9)$$
$$k = 1,3\ldots2\ell+1;\ k' \le 2\ell ,$$

where we defined the relativistic radial integrals by

$$R^{(m)}(0,k,k) = (-)^{k+\ell}(2\ell+1)\left[\frac{\ell(\ell+1)(2\ell+1)(2k+3)}{k}\right]^{\frac{1}{2}} \times \begin{pmatrix}\ell & k+1 & \ell\\ 0 & 0 & 0\end{pmatrix}\begin{Bmatrix}k+1 & 1 & k\\ \ell & \ell & \ell\end{Bmatrix} j^{(m)}(0,k,k) \quad (10a)$$

and

$$R^{(m)}(1,k',k) = (-)^{\ell} i^{k-k'-1}(2\ell+1)\left[\frac{(2k+1)(2k'+1)}{2}\right]^{\frac{1}{2}} \times \begin{pmatrix}k' & 1 & k\\ 0 & 1 & -1\end{pmatrix}\begin{pmatrix}\ell & k' & \ell\\ 0 & 0 & 0\end{pmatrix} j^{(m)}(1,k',k). \quad (10b)$$

In these equations $j^{(m)}(0,k,k)$ and $j^{(m)}(1,k',k)$ are given by

$$j^{(m)}(0,k,k) = \frac{4mc}{\hbar(2\ell+1)^2}\int r(j_{k+1}+j_{k-1})\,[(\ell+1)F_+G_+ - \ell F_-G_- - \tfrac{1}{2}(F_+G_- + F_-G_+)]\,dr , \quad (11a)$$

$$j^{(m)}(1,k+1,k) = \frac{4mc}{\hbar(2\ell+1)^2}\int \frac{r(j_{k+1}+j_{k-1})}{k(2k+1)}[-(\ell+1)\times(k+1)(2\ell-k)F_+G_+ - \ell(k+1)(2\ell+k+2)F_-G_- + \tfrac{1}{2}(2\ell-k)(2\ell+k+2)(F_+G_- + F_-G_+)]dr , \quad (11b)$$

$$j^{(m)}(1,k-1,k) = \frac{4mc}{\hbar(2\ell+1)^2}\int \frac{r(j_{k+1}+j_{k-1})}{(k+1)(2k+1)}[k(\ell+1)\times(2\ell+k+1)F_+G_+ + k\ell(2\ell-k+1)F_-G_- + \tfrac{1}{2}(2\ell+k+1)(2\ell-k+1)(F_+G_- + F_-G_+)]dr , \quad (11c)$$

where $F_\pm = F_{n\ell j=\ell\pm\frac{1}{2}}$ and $G_\pm = G_{n\ell j=\ell\pm\frac{1}{2}}$ are one electron radial wavefunctions obtained by some type of D-F calculation.

It can be seen (Eqs. (8) and (9)) that in this simple case the magnetic scattering amplitude can be obtained from the non-relativistic expressions by simply replacing the non-relativistic radial integrals R by their relativistic counterparts. It can easily be shown that this statement is correct for any elastic scattering process. For an inelastic process, on the other hand, the magnetic scattering amplitude contains angular matrix elements of the tensors $W^{(1,k')k}$ where k' is an odd integer whereas these tensors are absent in the non-relativistic scattering amplitude. It is easily seen that in the non-relativistic limit

$$j^{(m)}(0,k,k) = \bar{j}_{k+1} + \bar{j}_{k-1} , \quad (12a)$$

$$j^{(m)}(1,k\pm1,k) = \bar{j}_{k\pm1} , \quad (12b)$$

where

$$\bar{j}_k = \int_0^\infty R^2_{n\ell}(r)\, j_k(qr)dr , \quad (13)$$

and $R_{n\ell}(r)$ is the non-relativistic radial wavefunction. Using these relations the multipole operators and the magnetic scattering amplitude become identical to those of the non-relativistic formulation.[1]

For the 4f-electrons of the rare earth metals the non-relativistic limits given by Eqs. (12) are good approximations for the integrals $j^{(m)}$, provided that the definition of the $j_k$ (Eq. (13)) is replaced by its relativistic generalization

$$\bar{j}_k = \frac{\ell+1}{(2\ell+1)}\int (F_+^2 + G_+^2)j_k(qr)dr + \frac{\ell}{2\ell+1}\times\int (F_-^2 + G_-^2)j_k(qr)dr , \quad (14)$$

an equation first proposed by Freeman and Desclaux[3]. In this case the $j^{(m)}$ integrals calculated using this approximation (Eqs. (12) and (14)) do not differ significantly from those obtained using Eqs. (11).

In conclusion, we showed that the scattering of neutrons by an atom described by a relativistic Hamiltonian can be formulated along lines similar to those of the non-relativistic problem[1], by using the effective operator approach. Matrix elements between non-relativistic atomic states of the magnetic scattering amplitude, written as an effective operator, produce a relativistic result. In the important case of elastic scattering, the magnetic scattering amplitude can be obtained from the non-relativistic expressions by simply replacing the non-relativistic radial integrals R by their relativistic counterparts. The relativistic radial integrals can be calculated using radial wavefunctions obtained by some type of D-F calculation.

The authors are grateful to Dr. B. N. Harmon for many useful discussions during the course of this work.

REFERENCES


1. C. Stassis and H. W. Deckman, Phys. Rev. B 12, 1885 (1975); J. Phys. C: Solid State Physics (to be published).
2. P. G. H. Saunders and J. Beck, Proc. Roy. Soc., Ser. A 289, 97 (1965).
3. A. J. Freeman and J. P. Desclaux, Int. J. Magnetism 3, 311 (1972).

# ELECTRON SPIN POLARIZATION IN FERROMAGNETIC METALS

R. Meservey, P.M. Tedrow, and D. Paraskevopoulos,*
Francis Bitter National Laboratory,†
Massachusetts Institute of Technology, Cambridge, Ma. 02139

ABSTRACT

Various types of experiments which have been designed to measure the electron spin polarization of ferromagnets are described. Results obtained for the 3d ferromagnetic metals are given. The experimental results are briefly discussed in relation to proposed theoretical interpretations and related experiments, with particular emphasis on tunneling. A phenomenological correlation between magnetic moment and electron spin polarization, which is observed in Fe, Co, and Ni, appears to be valid for Ni-Mn and Ni-Ti ferromagnetic alloys.

## INTRODUCTION

The 3d ferromagnetic metals have been studied intensively for years, and the amount of information collected about Fe, Co, and Ni is astonishing. In spite of this great activity, the exact electronic behavior which causes the ferromagnetism is uncertain. A recent review of experimental work on the emission of spin-polarized electrons from magnetic materials by Siegmann[1] has presented a strong case for the inadequacy of the Stoner-Wohlfarth-Slater (SWS) theory of ferromagnetism. On the other hand, the great difficulty of improving this theory has been described recently by Gutzwiller.[2] In addition the interpretation of the various experiments is still uncertain and in some cases apparently conflicting. In this paper we list and briefly describe various types of experiments which have given, or may give shortly, data on electron spin polarization in ferromagnetic metals with special emphasis on tunneling. We try to indicate the capabilities and particular characteristics of the methods and give results where available. We also briefly discuss the relationship of current results with theoretical interpretations and with other experiments. Finally we describe some recent results that we have obtained on ferromagnetic alloys.

## EXPERIMENTAL TECHNIQUES

### Photoemission

The spin polarization P of electrons photoemitted from ferromagnets was first demonstrated in Gd[3] and Dy[4] and in Eu compounds.[5,6] Shortly thereafter, the spin polarization of photoelectrons from Ni was measured.[7] In previous attempts[8-10] on the 3d ferromagnetic metals the failure to observe polarization is perhaps to be explained by the use of crossed E and H fields rather than the parallel E and H field arrangement used by Bänninger et al.[7] In these latter experiments a thin evaporated film of the ferromagnetic metal was irradiated with a high pressure Hg-χe arc with a high frequency limit of 6 eV. A magnetic field was applied perpendicular to the film surface to align the ferromagnetic domains, and the electrons were extracted with an electric field also perpendicular to the film. The electrons were deflected by a cylindrical condenser to transform the longitudinal spin direction into a transverse one, and then accelerated by a voltage of 100 keV into a Mott detector. The Mott detector consists of a thin gold foil and symmetrically placed counters, and, because of the spin-orbit interaction of the electrons with the Au atoms, the asymmetry in the electron count measures the spin polarization of the electron beam. Further details of the technique are found in Ref. 6; see also Campagna et al.[11]

The films were deposited at 400 K, 78 K, or 4.2 K in a vacuum of $10^{-8}$ Torr and subsequently meassured at 4.2 K in a vacuum of $2 \times 10^{-10}$ Torr. The results are summarized in Fig. 1. The designations I and II refer to films deposited at 400 K and 4.2 K respectively. Films annealed at 400 K after deposition were of Type I; films deposited at 78 K were no different from those deposited at 4.2 K. Films made by low temperature deposition (Type II) are known to have reduced crystalline order approaching an amorphous state and a corresponding reduction in saturation magnetic moment. In fact the reductions in P for Fe and Ni shown in Fig. 1 are approximately proportional to the known reductions in the magnetic moment for such films.[12] In the case of Co the results have considerable scatter, particularly for the Type II films.

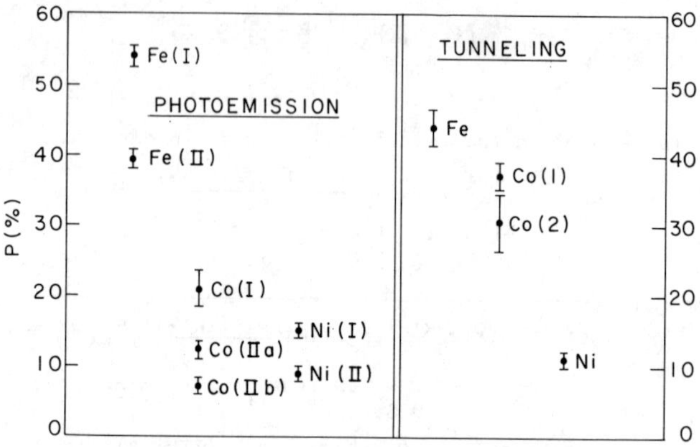


Fig. 1. Chart comparing electron spin polarization P as measured in photoemission and tunneling.

The energy level below $E_F$ from which the photoelectrons are excited could be controlled to some extent. When the full spectrum of the χe-Hg arc was used, the work function $\phi = 5.5$ eV of Ni allowed a band about 0.8 eV below $E_F$ to be excited. A filter could reduce this to 0.4 eV and made no noticeable effect on P. Further investigation[13] of the energy dependence of P was done by coating Fe, Co, and Ni with up to about one monolayer of Cs. In this case the energy of the incident photons was controlled by a monochromator. The result in each case is a monotonic decrease in P as energy levels farther below $E_F$ are probed.

Another important result obtained by photoemission was a measurement of the escape length of the photoelectrons. From the measured polarization of a Cu film coated with increasing amounts of Ni and a Ni film coated with increasing amounts of Cu, the depth of origin of the polarized photoelectrons was shown to be about 11 Å.[14]

### Electron Tunneling

The method of determining the spin polarization of a tunnel current from a ferromagnetic metal relies on the splitting of the quasi-particle states of a very thin film of Al in a magnetic field. In Fig. 2(a) we show the superconducting density of states plotted as a function of voltage applied to the tunnel junction with the separate spin desnities of states shown displaced by $2\mu_B H$.[15] In Fig. 2(b) the measured conductance corresponding to this density of states is shown for a junction made with Al, $Al_2O_3$, and a non-ferromagnetic material, Ag, measured at finite temperature (0.4 K for $T_c = 2.5$ K). The conductance σ shows peaks corresponding to the peaks in the density of states in the

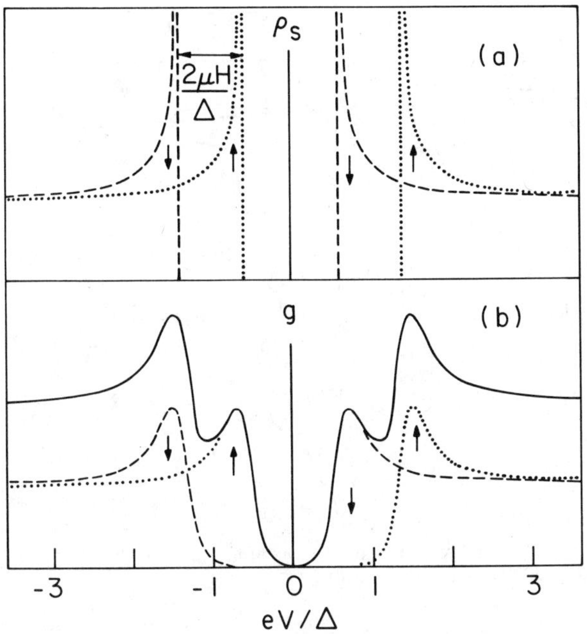


Fig. 2. (a) Theoretical quasiparticle density of states for a very thin spin-paired superconductor in a parallel magnetic field showing the Zeeman splitting. (b) Differential conductance of an Al-$Al_2O_3$-Ag tunnel junction as a function of voltage at $T/T_c = 0.2$.

two spin directions. When a junction is made with Al, $Al_2O_3$, and a ferromagnetic metal, the asymmetry of the tunneling curves (shown in Fig. 3 for Co) can be used to determine the spin polarization of the current from the ferromagnet.[16] The measured values of P for Fe, Co and Ni are shown in Fig. 1.[17] The results are very similar to those obtained by photoemission. P is in all cases positive. The values for Fe and Ni are between the values for Type I and Type II films by photoemission. For Co the average value of P is between the value for Fe and Ni and in each experiment shows great variability. The energy region probed by the tunneling method is within $10^{-3}$ eV of $E_F$. Studies of ultra-thin layers of Co on a normal metal imply a characteristic escape length for spin-polarized electrons in the ferromagnet of about 3 Å.[18]

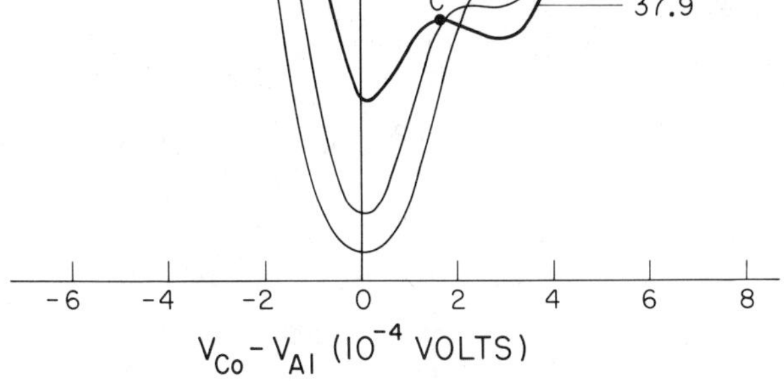

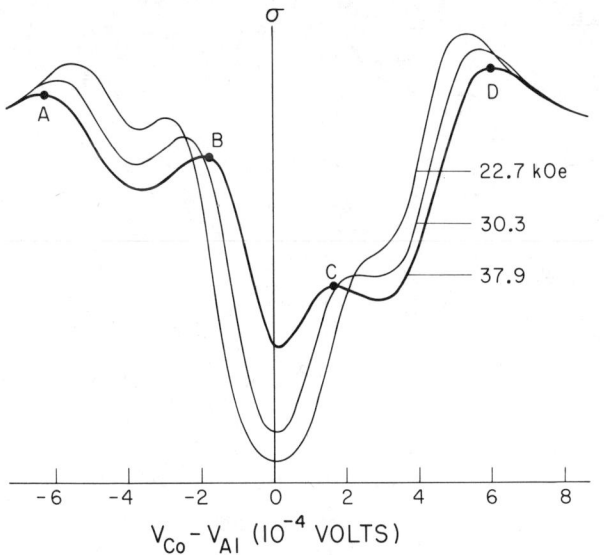

Fig. 3. Measured differential conductance of an Al-$Al_2O_3$-Co junction for 3 applied fields. The polarization can be determined from the conductance differences between points A, B, C, and D.

## Field Emission

Spin polarization has been reported for field emitted electrons from ferromagnetic metal point emission tips.[19,20] This technique is very similar to the technique of photoemission except for the emission process itself. A EuS-coated tungsten tip has been shown to give a very high value of polarization.[21] For Ni, positive and negative values of P have been obtained in different crystal directions.[20] However, the field emission from 3d ferromagnetic metals is complicated and, it appears, as yet not entirely understood. Ni gives different signs of polarization from different lattice sites, and the results are extremely sensitive to the surface conditions.[22] Measurements[23] on W, which would not be expected to give polarization, did give a polarization and suggest that some of the earlier work needs confirmation. At present the results of field emission are obscure, but this technique holds much promise. One interesting capability is that by using a viewing screen with an aperture in it one can determine the lattice site from which the measured electrons originate. The energy region probed by field emission is about 0.1 eV below $E_F$. The process of necessity takes place in a region of high electric field in which the surface layer of the metal plays a decisive role.

## Magneto-Optical Kerr Effect

The magneto-optical Kerr effect (MOKE) has been measured in Ni, Co, and Fe and gives information on electron spin polarization.[24,25] MOKE measures the off-diagonal elements of the optical conductivity tensor $\sigma_{xy}^{(2)}(\omega)$, the absorptive component of which is proportional to the difference in absorption of right- and left-circularly polarized light. The sign of $\sigma_{xy}^{(2)}(\omega)$ is related to the spin polarization of the electron states which contribute to the MOKE absorption. Figure 4 shows $\omega\sigma_{xy}^{(2)}(\omega)$ for Ni as a function of energy below $E_F$. According to Erskine and Stern the negative sign of the conductivity implies a negative spin polarization closer than 0.5 eV to $E_F$ (where $\omega\,\sigma_{xy}^{(2)}(\omega)$ crosses the estimated intraband contribution shown as a vertical line on Fig. 4). At larger energy differences the results imply a positive polarization. The sign of the result is claimed to be model independent for Ni and depends essentially on the sign

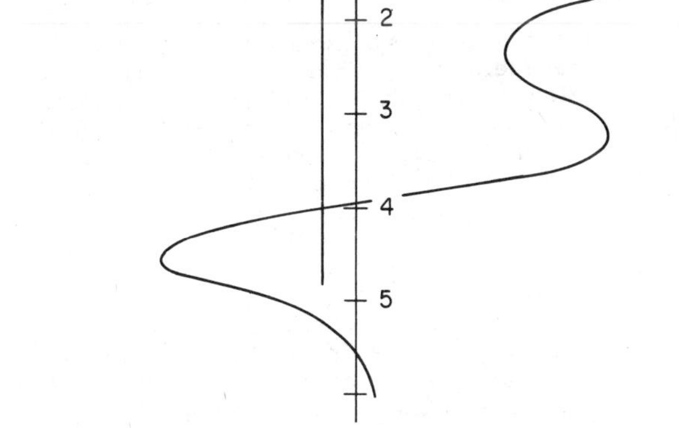

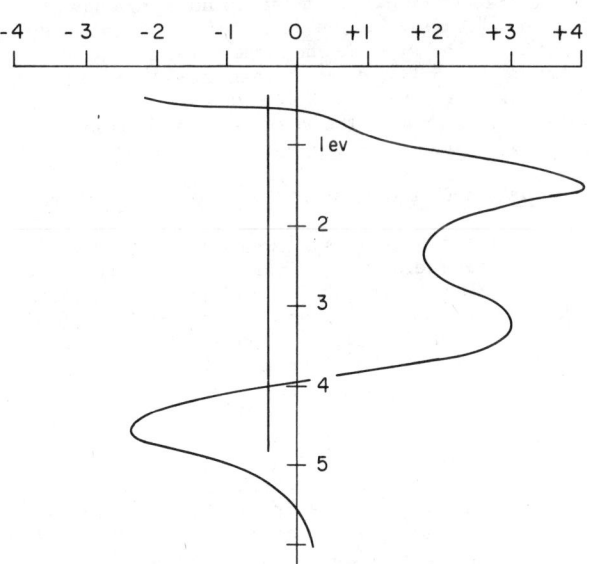

Fig. 4. Magneto-optical Kerr effect in Ni showing change in sign of the absorptive component of the optical conductivity tensor (adapted from reference 24).

of spin-orbit coupling which is known from atomic calculations. For Fe and Co the interpretation is apparently more model dependent and less clear. MOKE spectra are relatively insensitive to surface effects because the optical penetration is typically hundreds of angstroms. The optical method eliminates the uncertainty in the energy of the states studied caused by the electronic work function. With MOKE the coupling in Fe, Co, and Ni is presumably to the d electrons and the s-p electrons contribute very little.

### Electron Capture in Deuterium Ion Reflection

A charged deuteron beam reflected from or passing through a ferromagnetic crystal can capture electrons.[26] The spin polarization of the electrons is transferred to the deuterium nucleus by the hyperfine interaction and can thus be determined in a subsequent nuclear reaction. Using this method Rau and Sizman[26] obtained for Ni a value of P = - 96% for a [110] surface and negative values for all other directions measured except the [120] surface where P = + 16%. This method samples only the electron density at the surface. At 150 keV and an angle of 0.3° from the surface the deuterons probe the region of closest approach, which for an ideally flat surface is about 1 Å from the surface. Naturally this method demands high perfection in the crystal surfaces and a lack of surface contamination.

### Thermal Energy Atomic Beam Scattering

A somewhat different technique using the diffraction of a thermal atomic beam of hydrogen from a magnetic surface has been described by Thompson and Felcher.[27]

### Mössbauer Measurement of Conduction Electron Polarization

A Mössbauer absorption method designed to determine the spin polarization of conduction electrons in Fe has been described by Gamlitsky et al.[28] In this experiment a 60 Å film of Sn lay between two layers of Fe each 150 Å thick. The Mössbauer absorption spectrum of 23.8 keV γ rays from $Sn^{119}$ nuclei in this sandwich were measured at low temperature with and without a magnetic field H = 25 kOe. The observed splitting of the $Sn^{119}$ line for H = 0 was interpreted as caused by the conduction electrons in the iron diffusing into the Sn layer and, through the contact interaction with the Sn nuclei causing a hyperfine splitting. The fact that this hyperfine splitting did not increase at H = 25 kOe was interpreted by the authors as showing that the conduction electrons of Fe have a negative spin polarization. The special characteristic of this experiment is that only electrons with an s-like wave function can enter into the hyperfine splitting, but as yet the interpretation of the results is somewhat uncertain.

### Recombination Radiation in P-Type Semiconductors

An interesting proposal has been made by Scifres et al[29] to detect the spin polarization of electrons injected from a ferromagnet into a p-type semiconductor. Provided the semiconductor has a valence band spin-orbit splitting >> kT the spin polarization of the injected electrons could be determined by the polarization of the recombination radiation.

## DISCUSSION

The band structure of ferromagnetic Ni has been extensively studied,[30,31] and the density of states as calculated on the basis of the Stoner-Wohlfarth-Slater (SWS) theory is somewhat as shown in Fig. 5. According to this view the majority spin d-electron band is completely filled, its top below $E_F$. At $E_F$ the density of minority-spin states predominates. Thus, it was expected that photoemission from close to $E_F$ would give negative (that is, minority) electron spin polarization and that positive polarization would be observed at energies further below $E_F$. Actually, no difference was seen for energy regions 0.4 eV and 0.8 eV below $E_F$; this result is not easy to explain. However, Wohlfarth[32] has shown that the photoemission results do not necessarily conflict with SWS theory with plausible values of the band splitting and the energy gap between the top of the d band and $E_F$. However, when such a model is applied to the density of states of Co or Fe[33] (see Fig. 6) the results can not be made to agree with experiment. Smith and Traum[34] have assumed momentum conservation in the photoemission process and calculated a complicated dependence of the sign and magnitude of P on the initial electron energy and the photon energy. This view appears to conflict with the very simple dependence of P on photon and electron energies obtained in cesium-coated nickel.[13] Magneto-optical Kerr effect measurements in Ni appear to be consistent with the density of states of the SWS theory although for Co and Fe the situation is more obscure.

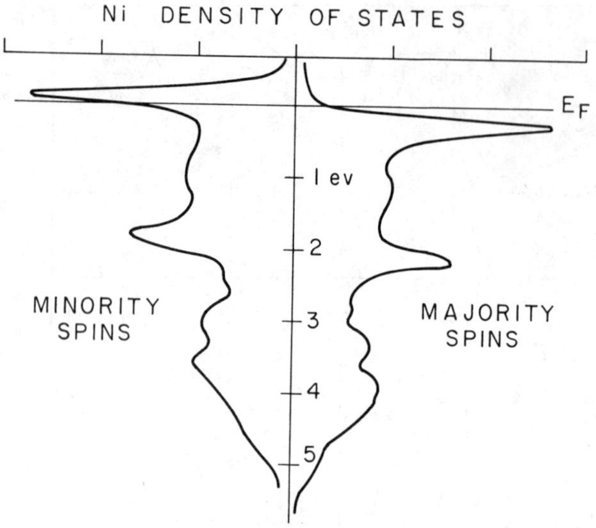


Fig. 5. Density of states of ferromagnetic nickel (adapted from reference 30).

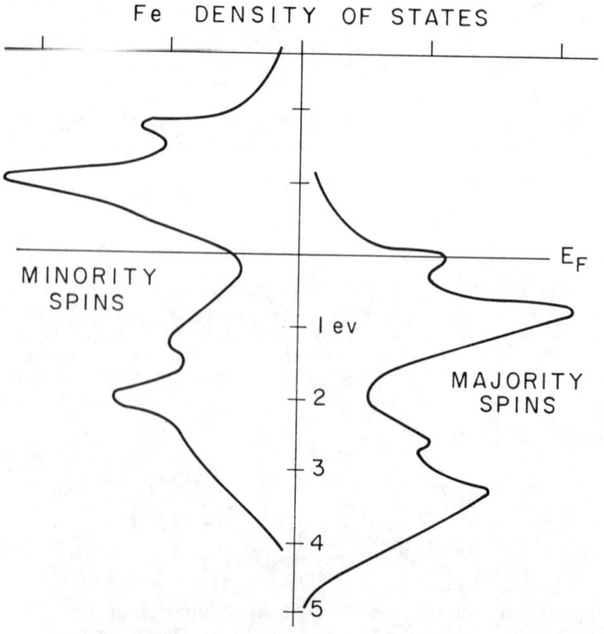


Fig. 6. Density of states of ferromagnetic iron (adapted from Wakoh and Yamashita, reference 33).

A calculation of the field emission from the [100] plane of Ni by Politzer and Cutler[35] giving P = -4% agrees approximately with P = - 10% obtained in field emission experiments by Gleich et al.[20] Since the photoelectric and tunneling results were in polycrystalline samples they do not necessarily conflict with this calculation. More recent field emission work has shown the interpretation of these experiments may be complicated and further confirmation is needed. One conclusion of the Politzer-Cutler calculation is that field emission in the [100] direction consists largely of electrons in the 4 s-p band, and that the high polarization of the d band (- 80%) only contributes sufficiently to give a net P = - 4%. This result seems to agree with X-ray measurements which show a negative polarization in the outer parts of the unit cell and with the deuteron reflection measurements which also give negative polarization outside the metal surface.

Concerning the tunneling experiments, it was originally assumed that the electron polarization would be negative in Ni because of the high density of states in the minority direction at $E_F$. However, there seems to be no firm basis for such an expectation. For a simple one-dimensional band model Harrison[36] has shown that the density of states does not enter into the magnitude of the tunneling current. This proof certainly has a limited range of validity, but it shows that the observed positive polarization is not necessarily in conflict with band calculations. On the other hand the agreement of tunneling and photoemission does appear to conflict with a simple interpretation of the experiments on the basis of the SWS band theory. One way to explain the tunneling results has been advanced by Luther and Fulde.[37] They calculate that the surface layer of ferromagnetic Ni is positively polarized. Kautz and Schwartz also calculated the variations caused by a surface,in the relative spin density in a spin-polarized electron gas.[38] It is reasonable that tunneling should be determined mainly by the surface, but again the agreement with photoemission, which has a characteristic probing depth of 10 Å, makes this view difficult to maintain.

Although there is no clear fundamental understanding of all of the experimental results, there are certain striking patterns. One of these is the remarkable similarity of the photoemission and tunneling results. For Fe and Ni the values of P from tunneling are bracketed by the photoemission results (it is probable that the disorder in the tunneling films is intermediate between the Type I and Type II photoemission samples). For Co the results for tunneling and the Type I photoemission samples are intermediate between Fe and Ni. In addition, for Co the value of P in both methods varies from sample to sample in a manner implying that some important variable is not being kept constant during the preparation. The polarization of Gd is the least in each case. Both types of experiments show saturation in P at a magnetic field which should align the domains in the films.

TABLE I. Correlation between saturation magnetic moment $n_B$ and measured spin polarization of Fe, Co, and Ni normalized to P(Fe).

| $n_B(\mu_B/at)$ | $P[n_B(Fe)/P(Fe)]$ | |
|---|---|---|
| | Tunneling | Photoemission |
| Fe 2.22 | 2.22 | 2.22 |
| Co 1.72 | 1.72 | 0.86 |
| Ni 0.62 | 0.55 | 0.62 |

A principal result is that with the 3d metals, the values of P scale approximately with the saturation magnetization. Table I shows that this correlation is indeed very striking. Here we present effective values of the magnetic moment per atom $n_B$ assuming that they are proportional to P and using the value of P for Fe to normalize the result. The fact that Gd with $n_B = 7.12$ and P = 4.3% (from tunneling) does not fit into this scheme is hardly surprising since the 4f electrons in the rare earth ions are effectively localized and their ferromagnetism is well described by an indirect Heisenberg interaction.

A number of attempts have been made to include many-body effects in ferromagnetic band calculations.[39-43] In particular Hertz and Aoi[42] have presented a theory which applies to tunneling. They assume, on the basis of a plausible estimate and the calculation of Cutler and Politzer for Ni, that the tunneling current consists predominantly of s electrons. According to these authors s-d hybridization leads to a positive polarization which is of the proper size if spin-wave emission self-energy effects are included. The values estimated for Fe, Co, and Ni are 37%, 26% and 8% respectively. This theoretical picture implies that the conduction properties of the two-spin directions are of primary importance in determining P. Although the estimated values of polarization are of the right sign and magnitude to agree with experiment the question remains how to interpret the photoemission results in this picture.

## ALLOY MEASUREMENTS

On the basis of the above discussion there are two conjectures which we can investigate by tunneling measurements in ferromagnetic alloys. One hypothesis is that electron spin polarization is proportional to the saturation magnetic moment of 3d metal alloys as is the case for the three pure elements. Figure 7 shows the well-known Slater-Pauling curve[44] in which saturation magnetic moment/atom is plotted against electron concentration for the 3-d metals. If the correlation between saturation magnetic moment and electron spin polarization is valid, the measured values of P properly normalized should fall on the Slater-Pauling curve.

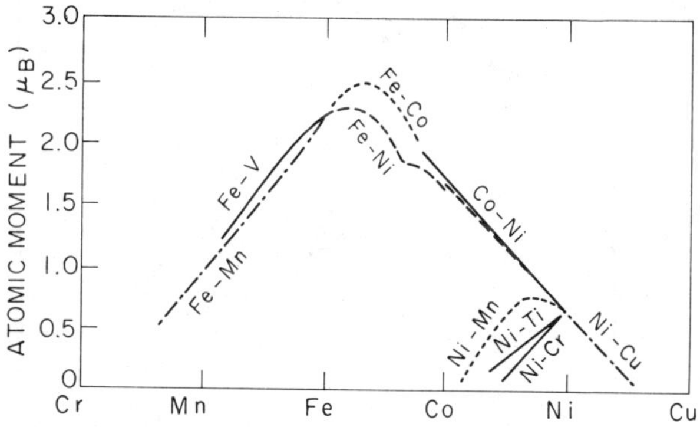


Fig. 7. Slater-Pauling curve showing the saturation magnetic moment/atom of various ferromagnetic metals as a function of electron concentration.

A second hypothesis assumes that the tunneling current is carried by the mobile s-p electrons. The introduction into Ni of various transition metal impurities which are believed to scatter preferentially either majority or minority spin electrons may possibly increase or decrease P. According to Fert and Campbell[45] the residual resistance of Ni with various transition metal impurities as is given in Fig. 8. From this point of view Cr impurities scatter majority electrons more strongly, whereas Co, Mn, and Ti impurities scatter minority electrons more strongly. If this scattering decreases the tunneling probability, then we would expect Cr to decrease P, whereas Co, Mn, and Ti should increase P. For Mn in Ni the magnetic moment at first increases and then decreases beyond about 8 at. %. Since the scattering hypothesis assumes small amounts of the scattering element, 8% is probably beyond the range of its validity. With Co in Ni the magnetic moment increases linearly with the Co con-

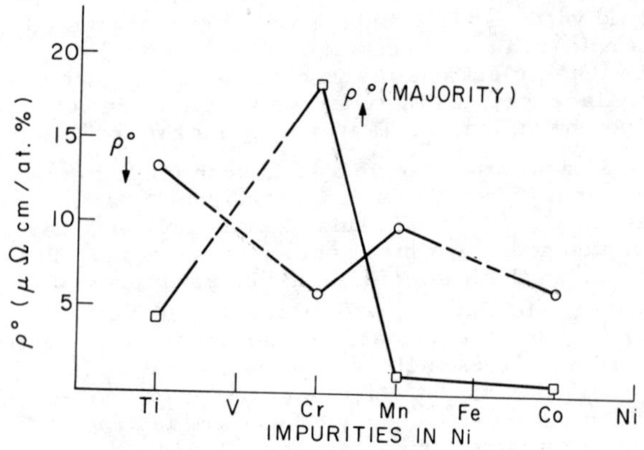


Fig. 8. Residual resistivity of transition element impurities in Ni (adapted from reference 45).

tent. Only in the case of Ti should the scattering model be qualitatively different from the magnetic moment hypothesis, since the magnetic moment decreases with Ti content.

The results of recent tunneling measurements of Ni films with various amounts Mn and Ti are plotted in

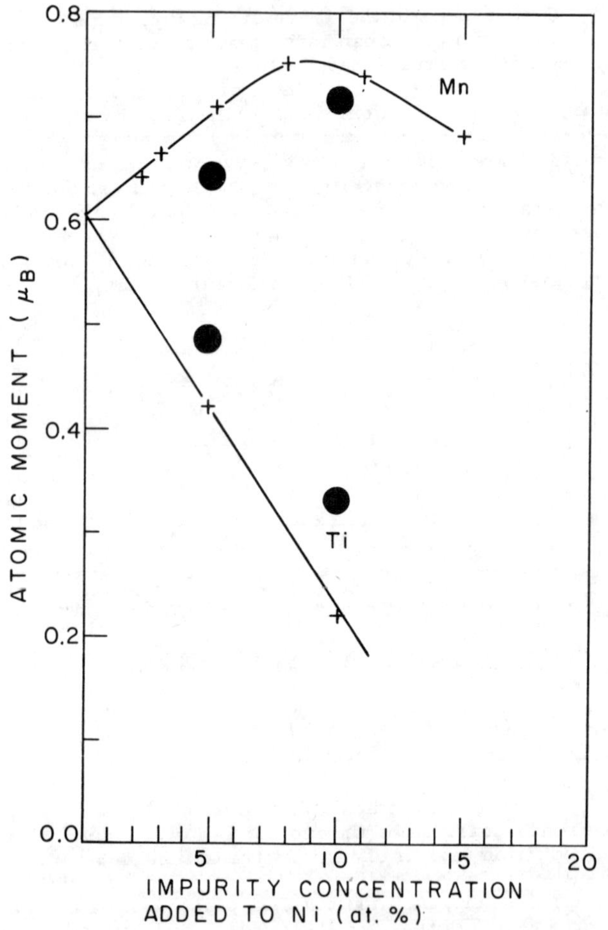


Fig. 9. Saturation magnetization of Ni with Mn and Ti impurities with polarization measurements normalized to pure Ni shown as superposed (circles).

Fig. 9. In this figure the crosses and solid lines are measurements of the saturation magnetic moment of Ni-Mn and Ni-Ti alloys as a function of impurity concentration. The circles are measurements of the polarization of the alloys films in terms of $n_B$, the saturated magnetic moment/atom:

$$n_B(\text{alloy}) = n_B(\text{Ni}) \cdot \frac{P(\text{alloy})}{P(\text{Ni})}.$$

The alloys films were prepared by flash evaporation of intimately mixed powders of the proper concentration. The value of P(Ni) used in the normalization was that obtained from a Ni film deposited on the same substrate as the alloy film and having the same Al counterelectrode. This procedure tends to correct for variations in junction preparation as well as for the fact that flash evaporation gives lower values of P for the Ni films. This lowering of $P_{Ni}$ by as much as 30% is associated with increased disorder in the films, which is caused by the relatively poor vacuum conditions during flash evaporation of powders. Because of the lowered values of $P_{Ni}$ we consider these measurements preliminary, but there is little doubt that the results strongly support the correlation of P with $n_B$, and in the case of Ti do not support the hypothesis depending on preferential minority spin scattering in Ni-Ti. Experiments are underway to investigate the full region of the Slater-Pauling curve.

## REFERENCES


∗ Also Physics Department, Boston University, Boston, Ma.

† Supported by the National Science Foundation.

1. H.C. Siegmann, Phys. Reports 17C, 37 (1975).
2. M.C. Gutzwiller, A.I.P. Conf. Proc. Magnetism and Magnetic Materials, 1972, p. 1197.
3. G. Busch, M. Campagna, P. Cotti, and H.C. Siegmann, Phys. Rev. Lett. 22, 597 (1969).
4. U. Bänninger, G. Busch, M. Campagna, and H.C. Siegmann, Conf. Internat. de Magnetisme, Grenoble, France, 1970 (unpublished).
5. G. Busch, M. Campagna, and H.C. Siegmann, Solid State Commun. 7, 775 (1969).
6. G. Busch, M. Campagna, and H.C. Siegmann, J. Appl. Phys. 41, 1044 (1970).
7. U. Bänninger, G. Busch, M. Campagna, and H.C. Siegmann, Phys. Rev. Lett. 25, 585 (1970).
8. H.A. Fowler and L. Marton, Bull. Amer. Phys. Soc. 4, 235 (1959).
9. R.L. Long, Jr., V.W. Hughes, J.S. Greenberg, I. Ames, and R.L. Christensen, Phys. Rev. A138. 1630 (1965).
10. A.B. Baganov and D.B. Diatroptov, Sov. Phys. JETP 27, 713 (1968).
11. M. Campagna, D.T. Pierce, K. Sattler, and H.C. Siegmann, J. Appl. Phys. 34, 6 (1973).
12. W. Felsch, Z. Angew. Phys. 30, 275 (1970); K. Tamura and H. Endo, Phys. Lett. A29, 52 (1969).
13. H. Alder, M. Campagna, and H.C. Siegmann, Phys. Rev. B8, 2075 (1973).
14. D.T. Pierce and H.C. Siegmann, Phys. Rev. B9, 4035 (1974).
15. R. Meservey, P.M. Tedrow, P. Fulde, Phys. Rev. Lett. 25, 1270 (1970).
16. P.M. Tedrow and R. Meservey, Phys. Rev. Lett. 26, 192 (1971).
17. P.M. Tedrow and R. Meservey, Phys. Rev. B7, 318 (1973).
18. P.M. Tedrow and R. Meservey, Solid State Commun. 16, 71 (1975).
19. M. Hofmann, G. Regenfus, O. Schärpf, and P.J. Kennedy, Phys. Lett. A25, 270 (1967).
20. W. Gleich, G. Regenfus, and R. Sizmann, Phys. Rev. Lett. 27, 1066 (1971).
21. N. Müller, W. Eckstein, W. Heiland and W. Zinn, Phys. Rev. Lett. 29, 1651 (1972).
22. N. Müller, Physics Letters 54A, 415 (1975); M. Campagna, T. Utsumi, and D.N.E. Buchanan. J. Vac. Sci Technol. Jan. 1976.
23. G. Regenfus and P. Sütsch, Z. Physik 266, 319 (1974).
24. J.L. Erskine and E.A. Stern, Phys. Rev. Lett. 30, 1329 (1973).
25. G.S. Krinchik and A.V. Artemiev, Sov. Phys. JETP 26, 1080 (1968).

26. C. Rau and R. Sizmann, Proc. 5th Intl. Conf. on Atomic Coll. in Solids, Gattlinburg, Tenn. (1973), ed. S. Datz; W. Brandt and R. Sizmann, Phys. Lett. A37, 115 (1971).
27. E. D. Thompson and G. Felcher, A.I.P. Conf. Proc. Magnetism and Magnetic Materials, 1974, p. 394.
28. V.Y. Gamlitsky, O.A. Gurskovsky, V.I. Nikolayev, I.N. Nikolayev, V.M. Cherepanov, and S.S. Yakimov, Zh.Exper. Teor. Fiz. 67, 756 (1974).
29. D.R. Scifres, B.A. Huberman, R.M. White, and R.S. Bauer, Solid State Commun. 13, 1615 (1973).
30. L. Hodges, H. Ehrenreich, and N.D. Lang, Phys. Rev. 152, 505 (1966).
31. E.I. Zornberg, Phys. Rev. B1, 244 (1970), and references given in this paper.
32. E.P. Wohlfarth, Phys. Lett. A36, 131 (1971).
33. E.P. Wohlfarth, J. Appl. Phys. 41, (1970); S. Wakoh and J. Yamashita, J. Phys. Soc. (Japan) 21, 1712 (1966).
34. N.V. Smith and M.M. Traum, Phys. Rev. Lett. 27, 1388 (1971).
35. B.A. Politzer and P.H. Cutler, Phys. Rev. Lett. 28, 1330 (1972).
36. W.A. Harrison, Phys. Rev. 123, 85 (1961).
37. P. Fulde, A. Luther, and R.E. Watson, Phys. Rev. B8, 440 (1973).
38. R.L. Kautz, MIT Thesis (1975) unpublished.
39. P.W. Anderson. Philos. Mag. 24, 203 (1971).
40. S. Doniach, A.I.P. Conf. Proc. Magnetism and Magnetic Materials (1971) p. 549.
41. J.A. Hertz and D.M. Edwards, Phys. Rev. Lett. 28, 1334 (1972).
42. J.A. Hertz and Koya Aoi, Phys. Rev. B8, 3252 (1973).
43. D.J. Kim, unpublished.
44. This curve was adapted from Ferromagnetism by R.M. Bozorth, D. Van Nostrand, New York (1951), p. 441.
45. A. Fert and I.A. Campbell, Phys. Rev. Lett. 21, 1190 (1968); O. Jaoul and I.A. Campbell, J. Phys. F: Metal Physics 5, L69 (1975). Other references are contained in these papers.

# PHOTOELECTRON-SPIN POLARIZATION,-SPIN FLUCTUATION AND - PARAMAGON INTERACTIONS AND ENERGY SPECTRA IN PHOTOEMISSION FROM ITINERANT, NARROW d-BAND MAGNETIC METALS: INCLUSION OF TEN-FOLD, SPIN-ORBITAL, d-ELECTRON DEGENERACY

Edward Siegel
24 Fifth Avenue
New York, New York 10011

## ABSTRACT

The non-degenerate Doniach theory of photoemission electron spin polarization and energy spectra via photo-electron-spin polarization, photoelectron - spin fluctuation and photoelectron-paramagnon interactions for narrow-band, itinerant ferromagnetic metals is extended to include true d-electron (hole) ten-fold spin-orbital degeneracy. The one-electron state renormalization of the energy shift, Doniach's "relaxed orbital correction," is rederived in this Siegel-Kemeny ten-fold degenerate Hubbard model (T.D.H.M.) and naturally embodies Hund's rule, enhanced coulomb repulsive interaction and enhanced attractive exchange interactions due to the degeneracy. The Doniach explanation of the positive electron spin polarization, in agreement with experiment, and as opposed to the non-degenerate Stoner band theory of ferromagnetism, is shown to still hold when spin-orbital degeneracy is included for both maximum and minimum allowable exchange energy matrix elements and within even a Hartree-Fock, one (degenerate)-electron theory.

# A LOCALIZED MOMENT MODEL FOR PARAMAGNETISM IN $Ni_3Al$

A. Parthasarathi and Paul A. Beck
University of Illinois, Urbana, Illinois, 61801

## ABSTRACT

Analysis of magnetization vs. field isotherms at 45° to 65°K to 12.7 kOe in terms of two Brillouin functions (taking into account interaction between dipoles in the molecular field approximation) and the use of susceptibility data up to 300°K led to a localized moment model for $Ni_3Al$, involving a Ni moment of 0.37 $\mu_B$ and a low concentration of magnetic clusters with average moments of 12 to 16 $\mu_B$. Magnetization values extrapolated by means of the model to 200 kOe agree with published data[1a] for a different specimen to within 7.4%.

## INTRODUCTION

$Ni_3Al$ and $Ni_3Ga$ (the latter with a slightly hyperstoichiometric Ni content) have been described[1a,2] as weak Stoner-Wohlfarth type band ferromagnets. The magnetoresistance of $Ni_3Al$ specimens of various compositions was studied[3] at 4.2°K in fields up to 340 kOe with results consistent with a weak band-ferromagnetic model. However, for the ferromagnetic specimens the agreement was good only at very high fields, where the ordinary positive magnetoresistance, due to the conduction electrons, is large as compared with the negative magnetoresistance. In a more recent study[4] magnetoresistance measurements were made at temperatures up to 90°K and in fields not higher than 23 kOe. Under such conditions of relatively high temperatures and low fields the results emphasize the negative magnetoresistance, which was found in various specimens to be 2 to 20 times larger than that predicted by the theory.[3] It was pointed out[4] that the temperature dependence of these negative magnetoresistance values is similar to that observed with localized moment ferromagnets. In the latter case the negative magnetoresistance is considered to be a result of the field-induced suppression of spin disorder scattering. A clear implication of this work[4] is that a localized moment model may be appropriate for $Ni_3Al$.

Although the presence of magnetic clusters in paramagnetic $Ni_3Ga$ and $Ni_3Al$ with slightly hypocritical Ni contents has been quite satisfactorily established,[5,6] it was decided to reanalyze the paramagnetic data for stoichiometric $Ni_{75}Al_{25}$ above the Curie temperature by taking into account the interaction between magnetic dipoles in the molecular field approximation. This analysis was planned so as to detect the possible presence of smaller magnetic moments, in addition to the giant moments of the magnetic clusters and, thus, the possibility of accounting for the entire magnetization in terms of a localized moment model.

## EXPERIMENTAL PROCEDURES

An alloy was prepared from 99.99% pure nickel and 99.997% pure aluminum by arc melting in argon atmosphere on a water-cooled copper plate, using a tungsten electrode. The specimen was homogenized by annealing at 1200°C for seven days. After homogenizing, the specimen was ground to the required size and annealed at 1200°C for 15 minutes to relieve the stress and the disorder in the surface layer, caused by grinding. After a fast quench, the specimen was chemically etched to remove the slight surface contamination due to the last, short anneal. Mass spectrographic analysis revealed the presence of 10 ppm W, 6 ppm Cr, 5 ppm Fe, and 2 ppm Cu. Quantitative analysis gave 74.49 at. % Ni. The magnetization was measured by the Faraday method.

## RESULTS

It is seen in Fig. 1 that at 40°K there is a distinct remanence. No remanence was found at 45°K. At 78°K and at higher temperatures the isotherm is a straight line up to 12.6 kOe. The reciprocal initial susceptibility vs. temperature, Fig. 2, shows Curie-Weiss behavior at temperatures above about 180°K. The Curie constant calculated from the high temperature data gives an effective moment per Ni atom of 0.937 $\mu_B$, in reasonable agreement with the 1 $\mu_B$ value obtained by de Boer.[1a] The average Ni moment $\mu = gS = 0.37$ $\mu_B$ was calculated from the effective moment, assuming g=2. Since this value is only 23% higher than that found recently[7] for antistructure Ni atoms in the CsCl-type alloy NiAl, which have eight Ni nearest neighbors with no moment, it seems reasonable to attribute the value obtained in the present work to single Ni atoms in the $Ni_3Al$ structure, each with eight Ni nearest neighbors also having a moment of $\mu = 0.37$ $\mu_B$ and with four Al nearest neighbors. This interpretation means that the contribution to the magnetization of both the temperature-independent susceptibility $\chi_0$ and the magnetic clusters is considered to be negligible in the temperature range 180° to 300°K.

The deviation from the straight line in Fig. 2 at lower temperatures suggests the appearance of magnetic clusters below 180°K. Information about the giant moment of the magnetic clusters and their concentration was obtained by least-squares fitting the parameters in Eq. (1) to the data in the temperature range of 45° to 65°K, Fig. 1.

$$\sigma = Nc_1\mu_1 B[\mu_1, \frac{H+\lambda_1\sigma}{T}] + Nc_2\mu_2 B[\mu_2, \frac{H+\lambda_2\sigma}{T}] \quad (1)$$

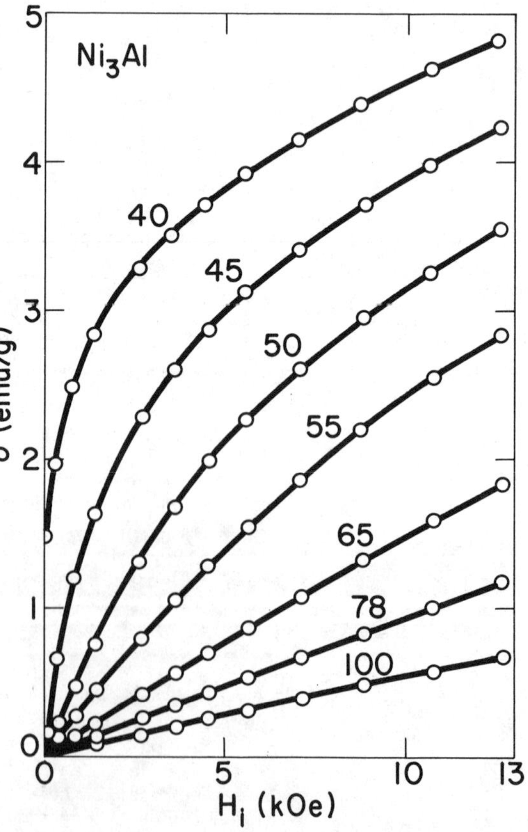


Fig. 1 Magnetization vs. internal field at temperatures indicated for each graph in °K.

with the Brillouin function:

$$B[\mu, \frac{H+\lambda\sigma}{T}] = \frac{(\mu+1)}{\mu}\text{ctnh}[(\mu+1)\frac{\mu_B(H+\lambda\sigma)}{kT}]$$

$$-\frac{1}{\mu}\text{ctnh}\frac{\mu_B(H+\lambda\sigma)}{kT}.$$

Here $\mu_1$ is the average giant moment and $c_1$ is its concentration. The Ni atomic moment of 0.37 $\mu_B$, as determined from the Curie constant, is used for $\mu_2$. The corresponding concentration $c_2$ is calculated by subtracting from the overall Ni concentration (0.75 at. fraction) the total concentration of Ni atoms participating in magnetic clusters, with the average Ni moment in the magnetic clusters estimated at 0.5 $\mu_B$:

$$c_2 = 0.75 - c_1 \frac{\mu_1}{0.5}. \qquad (2)$$

Eq. (1) assumes that g=2 in $\mu$=gS and that S is not required to be an integer multiple of 1/2. The parameters determined by least-squares fitting are $\mu_1$, $c_1$, $\lambda_1$ and $\lambda_2$, where $\lambda_1$ and $\lambda_2$ are the Weiss molecular field coefficients, expressing the interaction of $\mu_1$ and $\mu_2$, respectively, with other dipoles in their neighborhood. In view of the observation that, even at the highest temperatures of measurement where the Curie-Weiss susceptibility is lowest the value of the temperature-independent susceptibility may be considered negligible in comparison, no term was introduced into Eq. (1) based on such a susceptibility. Expecting the parameters to be obtained by least-squares fitting to vary with the temperature, the least-squares fitting was done separately for pairs of adjacent isotherms, as seen in Table I. The root mean square relative deviation (RMSRD) values are 0.1%, or less, so that all three fits may be considered excellent. The parameter values listed in Table I turned out to be only slightly dependent on the temperature between 45° and 65°K. However, using the low temperature parameter values in Eq. (1) to calculate the magnetization at higher temperatures results in calculated values much higher than the measured ones. This discrepancy is due to the gradual disappearance of the magnetic clusters in the temperature range between 55° and 180°K, where the $1/\chi_i$ vs. T graph is distinctly curved. However,

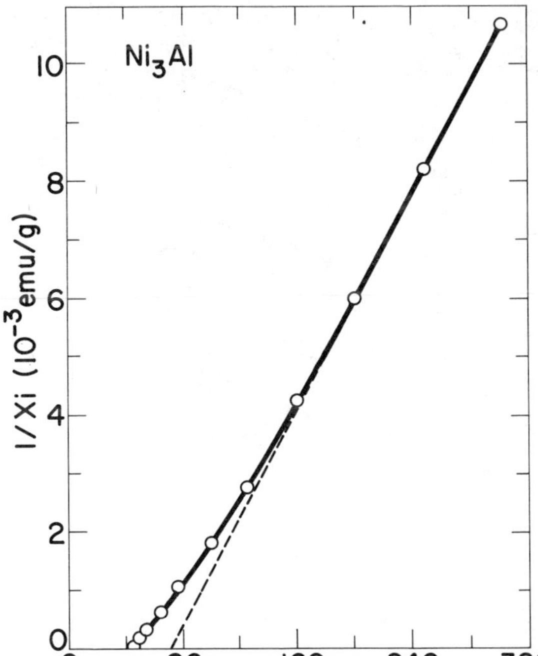


Fig. 2 Reciprocal initial susceptibility vs. temperature.

Table I. PARAMETERS FOR $Ni_{75}Al_{25}$ DERIVED FROM THE DATA

| Temp. range (°K) | $\mu_1$ ($\mu_B$) | $c_1$ ($10^{-2}$ at. fraction) | $\lambda_1$ ($\frac{Oe.g}{emu}$) | $\mu_2$ ($\mu_B$) | $c_2$ (atomic fraction) | $\lambda_2$ ($\frac{Oe.g}{emu}$) | RMSRD (%) |
|---|---|---|---|---|---|---|---|
| 45.1 50 | 15.1 | 0.369 | 16983 | 0.37 | 0.639 | 2569 | 0.10 |
| 50 55.1 | 16.2 | 0.267 | 16871 | 0.37 | 0.663 | 7797 | 0.065 |
| 55.1 64.9 | 12.6 | 0.380 | 20727 | 0.37 | 0.654 | 6333 | 0.065 |

because the $\sigma$ vs. $H_i$ isotherms are here straight lines, the data in this range cannot be analyzed by the method used at lower temperatures.

## DISCUSSION

A localized moment model for paramagnetism in $Ni_3Al$ was developed above on the basis of data extending only to H=12.7 kOe. As a test for this model, its predictions for high fields may be compared with the measurements of de Boer.[1a] Unfortunately, high field paramagnetic data for $Ni_3Al$ have been published[1a] only for 77.3°K, and they are given in graphical form. Nevertheless, a rough comparison is certainly possible. The magnetization values for 77.3°K, taken from Fig. 3.7 of de Boer's thesis[1b] are approximately 4.1, 5.8, and 8.5 emu/g for 50, 100, and 200 kOe, respectively. The corresponding values calculated from Eq. (1), with the parameters listed in Table I for 45°-50°K, are 4.2, 6.4, and 9.1 emu/g. At 50 kOe (3.8 times the maximum field used in determining the parameters) the agreement is quite good, but at 200 kOe (15.7 times higher than the maximum field used for deriving the parameters) the calculated value is 7.4% higher than the measured one. It was found that by a slight change of a single parameter, $\mu_2$, from 0.37 to 0.35, the difference at 50 kOe is eliminated and at 200 kOe it is reduced to 4.3% Clearly, the model may be expected to allow fitting the parameters to high field data to obtain excellent agreement. Furthermore, it seems likely that this fitting would result in parameter values not much different from those determined in the present work. The ability of the present localized moment model to reproduce the lack of paramagnetic saturation even at very high fields is undoubtedly due to the fact that a large fraction of the magnetization is attributed by the model to very small atomic moments acted upon by a very low Weiss molecular field. The lack of saturation up to 200 kOe in the ferromagnetic state could be accounted for in essentially the same way.

## ACKNOWLEDGEMENTS

This work was performed with the help of a grant from the U. S. National Science Foundation. We wish to extend our thanks to the International Nickel Company for the grant of the high purity nickel metal and to the Aluminum Company of America for making available the high purity aluminum used.

## REFERENCES

1a. F. R. de Boer, C. J. Schinkel, J. Biesterbos and S. Proost, J. Appl. Phys. 40, 1049 (1969). See also
1b. F. R. de Boer, Thesis Amsterdam, 1969.
2. P. F. de Chatel and F. R. de Boer, Physica 48, 331 (1970).
3. K. H. Chang, R. H. Van der Linde and E. G. Sieverts, Physica 69, 467 (1973).
4. P. D. Hambourger, R. J. Olwert and C. W. Chu, Phys. Rev. B, 11, 3501 (1975).
5. W. de Dood and P. F. de Chatel, J. Physics F: Metal Physics 3, 1039 (1973).
6. T. F. M. Kortekaas, Thesis, Amsterdam, 1975.
7. A. Parthasarathi and Paul A. Beck, Sol. St. Comm. (1975).

# MAGNETIC SUSCEPTIBILITY OF Ni-Aℓ ALLOYS AT HIGH TEMPERATURES

Sigurds Arajs and J.R. Kelly
Department of Physics
Clarkson College of Technology
Potsdam, New York 13676

## ABSTRACT

The mass magnetic susceptibility ($\chi$) has been determined as a function of temperature (T) of binary Ni-Aℓ alloys, containing 2.1, 4.2, 6.5, 8.7, 11.5 and 16.2 at. % Aℓ, between their ferromagnetic Curie temperatures ($T_C$) and 1500 K. The $\chi$ data satisfy the modified Curie-Weiss law, $\chi = C/(T-\theta)+\chi_0$, where $C,\theta$, and $\chi_0$ are constants, between T' and 1500 K. The temperature T' is some minimum temperature which decreases with increasing Aℓ content. The above constants give additional support to the recent interpretation of paramagnetism by Beck and Flynn, postulating the existence of localized magnetic moments up to 1500 K.

## INTRODUCTION

At the present time there are no satisfactory theories for the paramagnetic state of ferromagnetic materials involving the transition elements. Neither Fe or Ni or Co satisfy the Curie-Weiss law,

$$\chi = C/(T-\theta) \ , \qquad (1)$$

or the modified Curie-Weiss law,

$$\chi = C/(T-\theta) + \chi_0, \qquad (2)$$

where $\chi$ is the magnetic susceptibility, T the absolute temperature, and $C,\theta$, and $\chi_0$ are constants over the whole paramagnetic solid state region. Recently Beck and Flynn[1] have suggested that the paramagnetic behavior of Ni above the ferromagnetic Curie temperature ($T_C$) is strongly influenced by the short-range magnetic order effects which can exist up to at least 1000K.[2] Furthermore, they have found that $\chi$ data of Ni[3] can be fitted very satisfactorily to Eq. 2 between 1000 and 1500 K. From the value of C and the assumption that the g-factor is equal to 2, they conclude that the magnetic moment on Ni atoms is about 0.7 $\mu_B$ which they compared with the ferromagnetic moment of 0.6 $\mu_B$. The major implication of this analysis is that there are localized moments on Ni atoms in the paramagnetic state at elevated temperatures. The purpose of the present investigations is to present new high temperature $\chi$ measurements on Ni-Aℓ alloys and to analyze them from the viewpoint mentioned above.

## EXPERIMENTAL CONSIDERATIONS

The Ni-Aℓ alloys used in this study were prepared by levitation melting using Johnson and Matthey Ni and Aluminum Corporation of America Aℓ. The metallurgical procedures used for making these alloys are very similar to those described elsewhere.[4] The $\chi$ data were measured by means of the apparatus, based on the Faraday method, similar to that used before.[3]

## RESULTS AND DISCUSSION

The behavior of $1/\chi$ of Ni and Ni-Aℓ alloys, containing 2.1, 4.2, 6.5, 8.7, 11.5, and 16.2 at. % Aℓ, as a function of T up to 1500 K is shown in Fig. 1. All curves exhibit a downward curvature with respect to the T axis characteristic to Ni alloys. We have found that it is impossible to fit the $\chi$ data to Eq. 2 above $T_C$ (neglecting the magnetic effects just above $T_C$ associated with the critical aspects of the transition) and 1500 K with $\chi_0$ as a constant. However, if we assume that magnetic short-range effects extend to high temperatures to some limiting temperature T', then above T' the experimental $\chi$ data for all Ni-Aℓ alloys satisfy Eq. 2. Because $T_C$ decreases with increasing Aℓ concentration, T' also decreasing in a similar fashion. We have found from our computer analysis on pure Ni that when T' = 1000 K, C = 3.98 $10^{-3}cm^3Kg^{-1}$. This value of C implies that the effective paramagnetic moment $\mu^{Ni}_{eff} = g[S(S+1)]^{1/2} = 1.36\ \mu_B$, where $\mu_B$ is the Bohr magneton and S the spin quantum number. If g = 2 (as was assumed by Beck and Flynn[1]), then $\mu^{Ni} = gS = 0.698\ \mu_B$. This value is in good agreement with that

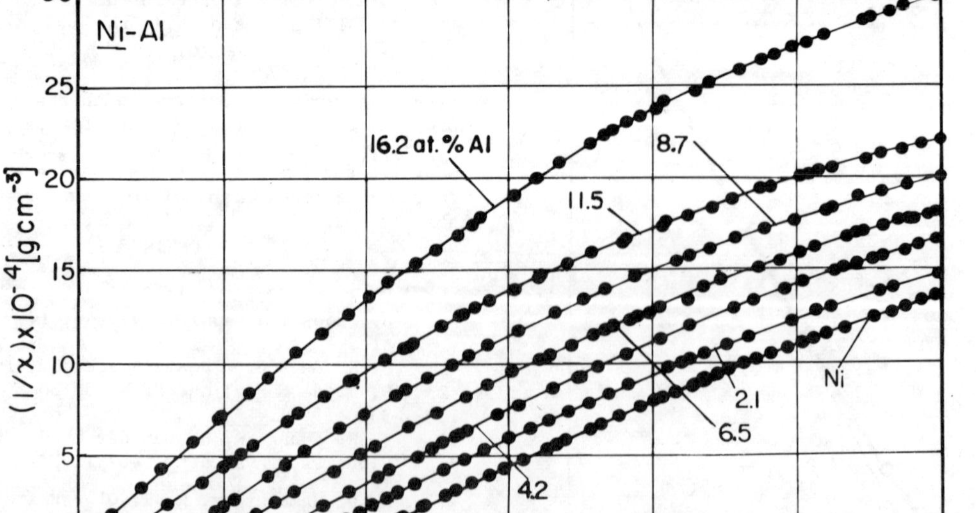


Fig. 1. Inverse Magnetic Susceptibility of Ni-Aℓ Alloys

calculated by Beck and Flynn.[1] A similar computer analysis has been carried out on all our Ni-Aℓ alloys, with T' gradually decreasing to almost 800 K for the 16.2 at. % Aℓ sample. Figure 2 shows the computer plots of $1/(\chi-\chi_0)$ as a function of T which clearly illustrate the linearity at elevated temperatures. If the quantities C, determined from the linear portions of Fig. 2, are analyzed in a manner similar to the Ni case, we find that the paramagnetic moments per Ni atom $\mu^{Ni}_{eff}$, and hence (assuming g = 2) also the paramagnetic $\mu^{Ni}$ gradually decrease with increasing Aℓ content. For pure Ni the quantity $\mu^{Ni}$, obtained from the paramagnetic high T data, is about 0.1 $\mu_B$ larger than the corresponding ferromagnetic moment. The reasons for this discrepancy are not clear at the present time. It is interesting to note that our values of $\mu^{Ni}$ (which for space purposes are not presented in detail here), calculated from the $\chi$ data at high T, form a straight line with respect to the Aℓ concentration, which is approximately parallel to the $\mu^{Ni}$ line obtained from the ferromagnetic data[5-8]. The difference of about 0.1 $\mu_B$, observed for pure Ni, appears to remain the same also for the Ni-Aℓ alloys.

In summary, our $\chi$ results on Ni-Aℓ alloys can be fitted to Eq. 2 at elevated temperatures. The values of the constants support the ideas of Beck and Flynn[1] that localized magnetic moments exist in Ni-Aℓ alloys up to 1500 K.

ACKNOWLEDGMENTS

This work has been supported by the Office of Naval Research under Grant Number N0014-70-A-0311-001. The authors are grateful to Professor P.A. Beck for his preprint of Reference 1 and for reopening our interest in the high temperature behavior of ferromagnetic transition metals.

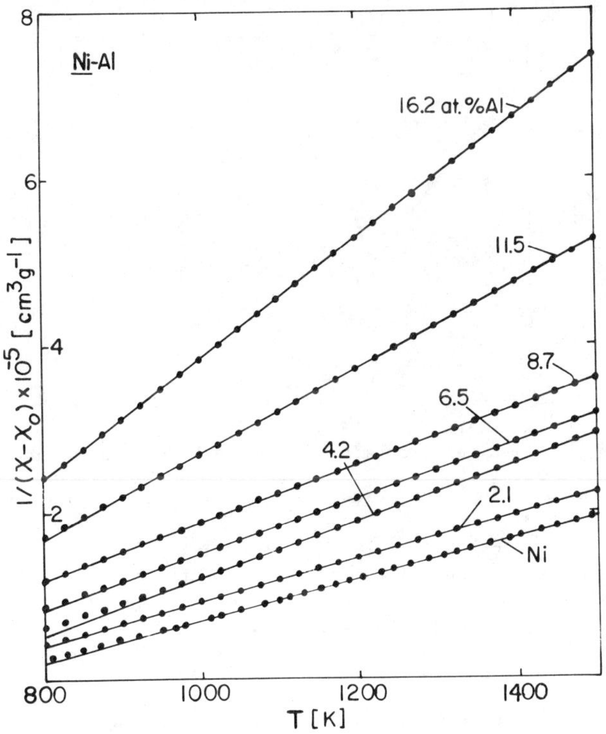


Fig. 2. $1/(\chi-\chi_0)$ as a function of temperature for Ni-Aℓ alloys

REFERENCES

1. P.A. Beck and C.P. Flynn (to be published in Sol. State Commun).

2. M. Braun and R. Kohlhaas, phys. stat. sol. 12, 429 (1965).

3. S. Arajs and R.V. Colvin, J. Phys. Chem. Solids 24, 1233 (1963).

4. H. Chessin, S. Arajs, and R.V. Colvin, J. Appl. Phys. 35, 2419 (1964).

5. C. Sadron, Ann. phys. 17, 371 (1932).

6. V. Marian, Ann. phys. 7, 459 (1937).

7. W. Sucksmith, Proc. Roy. Soc. (London) A171, 525 (1939).

8. J. Crangle and M.J.C. Martin, Phil. Mag. 4, 1006 (1959).

## MAGNETIC MOMENT DISTRIBUTION IN $Ni_3Al$*

G. P. Felcher
Argonne National Laboratory, Argonne, Illinois 60439
J. S. Kouvel†
University of Illinois, Chicago, Illinois 60680, and
A. E. Miller
University of Notre Dame, Notre, Dame, Indiana 46556

ABSTRACT

A polarized-neutron diffraction study was made of the unpaired-spin density distribution in the $Cu_3Au$-type ordered alloy $Ni_3Al$ at 4.2°K in a 10 kOe field. The system $Ni_{3+x}Al_{1-x}$ is weakly ferromagnetic for $x > -0.02$; for our single-crystal specimen ($x \approx 0.035$) the average Ni moment is 0.125 $\mu_B$ and $T_c \approx 70$°K. The measured fundamental (fcc) reflections describe a 3d magnetic form factor whose radial dependence and asphericity (82% $T_{2g}$, 18% $E_g$) are similar to those for pure nickel. Consistent with the tetragonal symmetry at each Ni site, the superlattice reflection measurements show that the cubic $T_{2g}$ and $E_g$ distributions split into 3% (xy), 79% (yz,zx), 6% $(3z^2-r^2)$, and 12% $(x^2-y^2)$, z being the tetragonal axis. Hence, much less unpaired-spin moment is associated with the xy lobes directed toward nearest-neighbor Al atoms than with the yz and zx lobes directed toward nearest-neighbor Ni atoms. These results appear to agree with recent band calculations on $Ni_3Al$, which indicate that the electrons in xy-type orbitals have far fewer states near the Fermi energy than those in yz- and xz-type orbitals.[1]

*Performed under the auspices of the U. S. Energy Research and Development Administration.
†Supported in part by the National Science Foundation and the Office of Naval Research.

[1] G. C. Fletcher, Physica 62, 41 (1972) and private communication.

# DETECTION OF LOCAL MOMENTS IN THE 3d SERIES BY ELECTRON ENERGY-LOSS SPECTRA OF 3p→3d TRANSITIONS

Mary Beth Stearns and Lee A. Feldkamp
Research Staff, Ford Motor Company, Dearborn, Michigan 48121

ABSTRACT

We have measured the 3p→3d core transition spectra of Cr, Mn, Fe, Co and Ni by the energy loss of ∿ 20 KeV electrons with energy resolution of 0.1 eV. As is well known from atomic physics, a discrete or localized state which is hybridized with a continuum gives rise to a distinctive dispersive lineshape. We generalize this lineshape theory to apply to the solid state. The spectral shape of Cr shows the usual Fermi edge and extended structure of normal metals, indicating that it has no local states while those of Mn, Fe, Co and Ni have a dispersive shape showing that they have local states. This confirms that a purely itinerant (Stoner) model of ferromagnetism is incorrect, even for Ni, and agrees with the model that ferromagnetism arises through the coupling of localized $d_\ell$ electrons by exchange polarized itinerant $d_i$ electrons. The width of the localized states gives a measure of the hybridization of the $d_\ell$ states with the itinerant $d_i$ and 4s states and continuum εf states through Coster-Kronig transitions.

Spectra with improved resolution over previous measurements[1,2] have been obtained for the 3p→3d core transitions of Cr, Mn, Fe, Co and Ni. The energy loss of ∿ 20 KeV electrons transmitted through single crystal thin films of ∿ 500Å was measured with an energy resolution of 0.1 eV and an angular resolution of 1-2 mrad. The spectra were measured varying the scattering angle and thus momentum transfer from 0.1 to 2Å$^{-1}$

We show that the widths and variation of the lineshapes are mainly due to solid state effects and a proper interpretation reflects the band structure of the 3d series. In particular, one of us (MBS), proposes that a major contribution to the widths comes from the hybridization of the localized $d_\ell$ states above the Fermi level with the itinerant $d_i$ states of similar energy. There is much previous evidence[3] which shows that: All the d electrons of the elements Sc through Cr are itinerant. At Mn some electrons have become localized and the degree of localization then increases across the series with Ni having the largest fraction of $d_\ell$ electrons. This behavior is the underlying feature which determines the ferromagnetic behavior of the 3d series and we show that it also accounts for the salient behavior of the lineshapes, namely Sc through Cr having very different lineshapes than Mn through Ni. As is well known in atomic physics a discrete or localized state hybridized with a continuum gives rise to a dispersive lineshape. We see in Fig. 1 that the lineshape of Cr (which should also be typical of elements to the left of Cr) shows the usual Fermi edge and extended structure of normal metals, indicating that Cr has no localized states. In contrast the lineshapes of Mn, Fe, Co and Ni are seen to have dispersive shapes, indicating that they have localized d states. Thus, the lineshapes show that a purely itinerant (Stoner) model[4] of ferromagnetism is incorrect, even for Ni, and gives added confirmation to the model[3] where the $d_\ell$ moments are aligned by a small fraction of $d_i$ electrons which have been polarized through Coulomb exchange interactions with $d_\ell$ electrons.

We analyze the lineshapes of Fe, Co and Ni in terms of their solid state structure as follows: The atomic Fano model[5] for one discrete state and several continua can be easily modified to apply to a solid. Using the notation typical of the closely related hybridization problem of dilute local moments in a nonmagnetic host , the linewidth of the hybridized localized state φ is

$$\Delta = \Gamma/2 = \pi(|V_E|^2 \rho_{d_i} + |W_E|^2 \rho_s) + \Gamma_{3p}/2 \quad , \qquad (1)$$

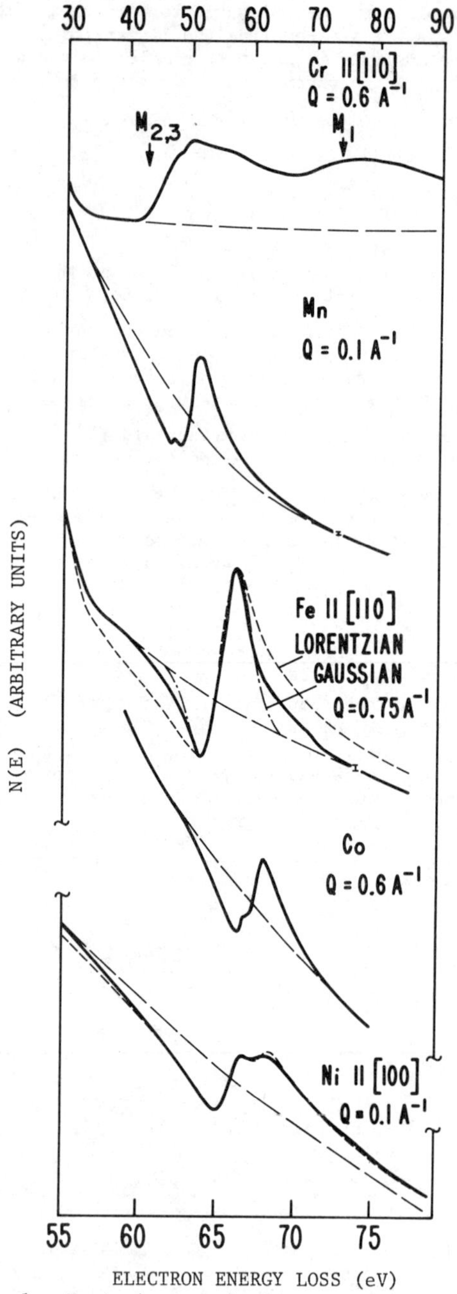


Fig. 1. Typical transmission energy-loss spectra for the 3p→3d excitations of Cr, Mn, Fe, Co and Ni. Note that the energy scale of Ni is expanded over that for the other elements.

where Δ is the halfwidth of the localized state after hybridizing with all the continuum states and $\rho_{d_i,s}$ is the density of states for itinerate $d_i$ and 4s electrons. $V_E$ and $W_E$ are the interaction matrix elements of the empty $d_\ell$ state with the $d_i$ and 4s states respectively. The electrons interact through the Coulomb potential. $\Gamma_{3p}$ is the 3p hole width which comes about mainly from super Coster-Kronig (C-K) transitions, i.e. a 3d electron falling into the 3p hole and promoting another 3d electron into the continuum; usually εf states[6]. The lineshape is given by[5]

$$N(E) - N(E_o) \sim (q^2 + 2\varepsilon q - 1)/(1+\varepsilon)^2 = f(\varepsilon,q) \quad , \qquad (2)$$

where $\varepsilon = (E - E_o)/\Delta$ is the electron energy measured relative to the center of the hybridized-localized state in units of Δ. $N_o(E)$ is the smooth background obtained by extrapolating from the regions far from $E_o$ as shown by the long-dashed curves in Fig. 1. The lineshape parameter q is given by

$$q = M_\phi/\pi(V_E \rho_{d_i} M_{d_i} + W_E \rho_s M_s + U_E \rho_f M_f) \quad , \qquad (3)$$

where $M_\phi$ is the matrix element for the transition operator T from the initial state ($3p^6 3d^n$) to the hybridized localized state, i.e. $M_\phi = \langle 3d_\ell|T|3p\rangle$ plus other terms like $\mathcal{P}\int V_{E'} M_{di} dE'/(E-E')$, which are expected to be small since $V_{E'}$ and $M_{di}$ are not strongly energy dependent. The other M's are transition matrix elements between the initial state and the unperturbed continuum states, i.e. $M_{di} = \langle 3d_i|T|3p\rangle$, $M_s = \langle 4s|T|3p\rangle$ and $M_f = \langle \varepsilon f|T|3d\rangle$. $U_E$ is the sum of the interaction matrix elements of the localized state with the $\varepsilon f$ states, the C-K matrix elements. Only those C-K matrix elements contribute to the dispersive character which involve the $d_\ell$ state.

A slight modification of Eq. (2) is needed since the 3p core state has two components, $3p_{3/2}$ and $3p_{1/2}$ in an intensity ratio of 2:1. They are expected to be split by about 1.5-2 eV for the latter part of the 3d series. Taking this into account we have

$$N(E) - N_0(E) \sim f(\varepsilon,q) + \tfrac{1}{2}f(\varepsilon - \delta,q) \quad , \qquad (4)$$

where $\delta$ is the spin splitting in units of $\Delta$.

Since Fe has the best known band structure of the 3d series let us estimate its various parameters. Band structure calculations[7] give the effective mass of the 4s and $d_i$ electrons as about $2m_e$, so $\rho_{d_i,s} \simeq 1/6\ eV^{-1}$. Band structure calculations or a comparison of this data with that of soft x-ray emission spectra (XSX),[8,9] as given later, gives the intra-atomic exchange energy for the 3d series to be $\sim 1.7$ eV per electron spin. $V_E$ is expected to be about the same strength, perhaps a little smaller since it is a very similar matrix element, thus we take $V_E \simeq 1.4$ eV as a reasonable value. $W_E$ is expected to be smaller than $V_E$; from a variety of data[4,10] we estimate it to be $\sim 0.7$ eV. Thus we get partial widths $\Delta_{di} \simeq 1.0$ eV and $\Delta_s \simeq 0.3$ eV.

We evaluate q by assuming that $M_\phi$ and $M_{di}$ are comparable since they involve transitions of the same energy and symmetry. In the dipole approximation, p to s transitions are generally much less likely than p to d transitions, so $M_s$ is expected to be about four times smaller than $M_{di}$.[6] Thus we obtain $q \simeq 1.2$.

Fe has two discrete states above $E_F$ which are very close in energy. From Fano's derivation of the case of two discrete states with one continuum[5] we see that two discrete nearly degenerate states give a linewidth which is twice that of a single state while q remains unchanged. Thus we estimate $\Delta_{Fe}$ from $d_i$ and s hybridization to be about 2.6 eV and $q \simeq 1.2$.

The fit to Eq. 4 for Fe (at Q=0.6 $A^{-1}$) is shown as the dotted curve in Fig. 1. We see this fit is not very good in the wings so we have also fit Eq. 4 with the Lorentzian distribution replaced by a Gaussian. This is the dashed curve in Fig. 1. We see that the real lineshape is somewhere in between. The deviation from a Lorentzian may be due to having multiple localized states in Fe. The Lorentzian fit gave q=1.5, $\Delta$= 3.0 eV whereas the Gaussian fit gave q=1.3, $\Delta$=4.0 eV. For both fits $E_0$=54.6±0.2 eV and $\delta$ was taken as 1.5 eV. Thus we shall take q≈1.4 and $\Delta$≈3.6 eV. The values obtained for $\Delta$ were independent of Q but q varied systematically with Q from ~1.05 at Q=0.1 $A^{-1}$ to ~1.5 at Q=1.2 $A^{-1}$. Thus the estimated value of $\Delta_{Fe}$ ≈2.6 eV is smaller than the measured value by about 1 eV or 0.5 eV per local state. The estimated value of q is in good agreement with the measured value.

We expect that $\Delta_{di}$ and $\Delta_s$ are about the same or slightly larger for Ni as for Fe. However since Ni has about one half of a localized level above the Fermi level, $\Delta_{Ni}$ should be smaller than 1.3, i.e. $\sim 0.7$ eV, and q should remain about the same as for Fe. A fit to the Ni data is shown in Fig. 1 by the short dashed curve. In this case a Lorentzian fits very well. The parameters used for this fit were q = 1.2 ± 0.1, $\Delta_{Ni}$ = 1.1 ± 0.1 eV, $\delta$ = 1.9 ± 0.1 eV and $E_0$ = 66.0 ± 0.2 eV. The spin-orbit splitting of the $3p_{3/2}$ and $3p_{1/2}$ levels is clearly seen. It is clear from Fig. 1 that a better fit to the data could be made by letting the $3p_{1/2}$ state be broader than the $3p_{3/2}$ state. An increased broadening of this type is expected due to super C-K transitions of the $M_2$-$M_3$ $M_{4,5}$ type. Introducing such an increased broadening gave an almost perfect fit to the data for $\Delta_{1/2}$ = 1.5 eV at $E_0$ = 67.7 eV and $\Delta_{3/2}$ = 1.1 eV at $E_0$ = 65.8 eV with $\delta$ = 1.9 eV and q = 1.2 as before. Thus the full width due to $M_2$-$M_3$ $M_{4,5}$ transition is about 0.8 eV.

We conclude that the $d_i$ and s hybridization appear to account for all but about 0.5 eV/per localized level of the measured half width; we attribute the remainder to C-K transitions. This agrees well with the value for the super C-K transition found above. The atomic calculations of McGuire[6] which were used to explain the linewidth of Ni by Dietz et al.[11] appear to be about a factor of two to four too large.

As seen in Fig. 1, Mn and Co both show structure on the lower energy lobe. Not much is known about the moments and band structure of antiferromagnetic Mn, so as yet we have not tried to interpret its spectra. The structure of Co is consistent with there being discrete states at two different energies above $E_F$, the farthest from $E_F$ having a large density of states and the other being very near $E_F$ and having only a small portion above $E_F$. This would not seem unreasonable.

A measure of the intra-atomic exchange interaction for Fe, Co and Ni can be obtained from comparing the position of $E_0$, which measures the position of the spin-down level $E^\downarrow$, with the peak in the SXS data of Refs. 8 and 9. This peak should be very close to the equivalent level $E^\uparrow$ in the spin-up band. These values are listed in Table I. This type comparison should be very good for Fe and Ni but may be a oversimplification for Co since it is mainly hcp but contained some fcc. In column 4 we give the moment due to unpaired spin (corrected by the g factor) and column 5 gives the level splitting per spin which should be a measure of the intra-atomic exchange, $U^{d-d}$. We expect this quantity to be quite constant over a given series and we see that indeed it is. The only band calculation that gives exchange splitting of this size for Fe is that of Duff and Das[7]; all other calculations give about half the value measured here. Three band calculations[12] for Ni give about the splitting observed here. The fact that the comparison of electron energy-loss spectra and SXS seem to be valid would indicate that either relaxation effects are negligible or they have shifted the bands to be nearly the same over the effective time of these two types of measurements.

Table I. Intra-atomic exchange interaction from SXS and electron energy-loss spectra.

| | $E^\uparrow$(eV) | $E^\downarrow$(eV) | $\mu_s(\mu_B)$ | $U^{d-d}$ |
|---|---|---|---|---|
| Co | 57.8 | 60.6 | 1.6 | 1.8 |
| Fe | 51.3 | 54.6 | 2.1 | 1.6 |
| Ni($3p_{3/2}$) | 64.8 | 65.8 | 0.55 | 1.8 |

## REFERENCES


1. J. L. Robins and J. B. Swan; Proc. Phys. Soc., (London) 76, 857 (1960).
2. B. Sonntag, R. Haensel and C. Kunz, Sol. St. Commun. 7, 597 (1969).
3. M. B. Stearns, Phys. Rev. B4, 4081 (1971); B8, 4383 (1973); in press (1975).
4. E. C. Stoner, Rept.s Progr. in Phys. 11, 43 (1947).
5. U. Fano, Nuovo Cemento 12, 156 (1935); Phys. Rev. 124, 1866 (1961).
6. E. J. McGuire, J. Phys. Chem. Sol. 33, 577 (1972); Sandia Lab. Rept. No. SC-RR-71 0835.
7. K. Duff and T. P. Das, Phys. Rev. B3, 192 (1971).
8. D. H. Tomboulian and D. E. Bedo, Phys. Rev. 121, 146 (1961).
9. J. R. Cuthill, A. J. McAlister, M. L. Williams and R. E. Watson, Phys. Rev. 164, 1006 (1967).

10. S. Hufner, G. K. Wertheim and J. H. Wernick, Phys. Rev. 8B, 4511 (1973) and G. Gruner, Adv. in Phys. 6, 941 (1974).
11. R.E. Dietz, E.G. McRae, Y.Yafet and C.W. Caldwell, Phys. Rev. Lett. 33, 1372 (1974).
12. S. Wakoh, J. of Phys. Soc. of Japan, 20, 1894 (1965) J.W.D. Connolly, Phys. Rev. 159, 415 (1967); J. Langlinais and J. Callaway, Phys. Rev. B5, 124 (1972).


# THERMAL EXPANSION, VOLUME MAGNETOSTRICTION AND SHIFT OF THE CURIE TEMPERATURE WITH PRESSURE OF $Ni_3Al$ AND Ni-Pt ALLOYS


J.J.M.Franse, T.F.M.Kortekaas and N.Buis
Natuurkundig Laboratorium der Universiteit van Amsterdam, The Netherlands



## ABSTRACT

In alloys of the $Ni_3Al$ phase and in disordered Ni-Pt alloys a transition occurs from enhanced paramagnetism to weak ferromagnetism at 74.6 and 42.5 at % Ni, respectively. The alloys in the direct neighbourhood of the critical concentration for ferromagnetism show large and positive magneto-volume effects. At passing the critical concentration a sharp and drastic change in the low temperature thermal expansion can be observed. The ferromagnetic alloys start at low temperatures with a negative thermal expansion coefficient. From thermal expansion and specific heat data the electronic Grüneisen parameter has been determined for both systems. This parameter shows a maximum at the critical concentration.
The experimental results of our magneto-volume studies can be described with one single magneto-elastic coupling parameter. In the model of weak itinerant ferromagnetism this parameter can be expressed in the volume derivatives of several band parameters. The experimental values for these volume derivatives are discussed.


## INTRODUCTION

Magneto-volume studies on strongly paramagnetic and weakly ferromagnetic alloys provide additional information about the model that is applicable for a description of the magnetic properties of these systems. Large and positive magneto-volume effects are to be expected in the model of weak itinerant ferromagnetism.
In describing our magneto-volume experiments we use the following expression for the thermodynamic potential per gram G:

$$G = G_o + \frac{1}{2} A\sigma^2 + \frac{1}{4} B\sigma^4 + \frac{1}{2\rho\kappa}\omega^2 + \frac{p\omega}{\rho} - H\sigma \quad (1)$$

where $G_o$ is independent of $\sigma$ and $\omega$, $\sigma$ is the magnetization per gram, $\omega$ the relative change in volume, $\rho$ the density, $\kappa$ the compressibility, p the external pressure, H the external magnetic field and A and B material constants. The parameter B is independent of the volume in both systems. For the derivatives of A we introduce the notations:

$$A' = (\partial A/\partial T)_{T=T_c} \quad \text{and} \quad C = -\frac{1}{2}(\partial A/\partial\omega)_{p=0,H=0}$$

The magnetic contribution to $\omega$ can be written as:

$$\omega_M(H,T) = \rho\kappa C\sigma^2(H,T) \quad (2)$$

The shift of $T_c$ with pressure is given by:

$$\partial T_c/\partial p = -2\kappa C/A' \quad (3)$$

All our experimental data fit quite well to eqs.(2) and (3). This gives a justification for the use of eq.(1). This equation has the form of Belov's expression for the thermodynamic potential near $T_c$ [1].

## THERMAL EXPANSION

The ferromagnetic alloys start with a negative thermal expansion coefficient and show a typical invar behaviour. The paramagnetic alloys can be described between 4.2 K and 12 K by the expression:

$$\frac{\Delta l}{l} = a\,T^2 + b\,T^4 \quad (4)$$

where a and b represent the electron and phonon contributions to the thermal expansion, respectively. Following eq.(2) we expect an additional contribution to $\Delta l/l$ in eq.(4) equal to $\rho\kappa C\sigma^2(T)/3$ for the ferromagnetic alloys. With a quadratic temperature dependence of $\sigma^2(T)$ we can use eq.(4) also for the ferromagnetic alloys with $a = a_e + a_M$ ; $a_e$ is the electronic, $a_M$ the magnetic contribution to a.
The parameter a is plotted in Fig. 1 as a function of the alloy concentration. The full curves connect the experimental a values. The points connected by the broken curves give $a_e$; $a_M$ has been calculated from volume magnetostriction and magnetization data.
From values for $a_e$ and from specific heat data the electronic Grüneisen parameter has been found to be: about 4 for $Ni_3Al$ and about 6 for Ni-Pt [2].

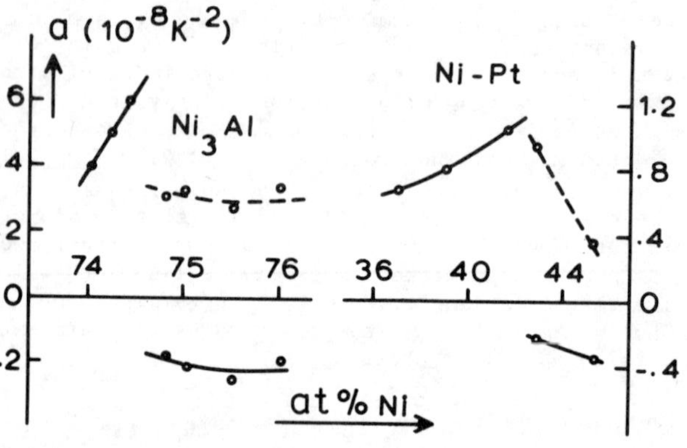


Figure 1.
The parameter a in eq.4 as a function of the alloy concentration.

TABLE I

| system | at% Ni | $\rho\kappa C$ ($10^{-6}$ gauss$^{-2}$g$^2$cm$^{-6}$) Volume magnetostriction T = 4.2 K | Volume magnetostriction $T = T_c$ | Pressure |
|---|---|---|---|---|
| $Ni_3Al$ | 74 | 0.63 | | |
| | 74.2 | 0.56 | | |
| | 74.4 | 0.54 | | |
| | 74.8 | 0.63 | 0.42 | 0.45 |
| | 75 | 0.64 | 0.36 | 0.50 |
| | 75.5 | 0.67 | 0.39 | 0.43 |
| | 76 | 0.76 | 0.40 | 0.38 |
| Ni-Pt | 36.9 | 4.50 | | |
| | 39 | 4.05 | | |
| | 41.4 | 3.92 | | |
| | 42.9 | 3.47 | 3.37 | 3.6 |
| | 45.2 | 3.32 | 2.18 | 2.2 |

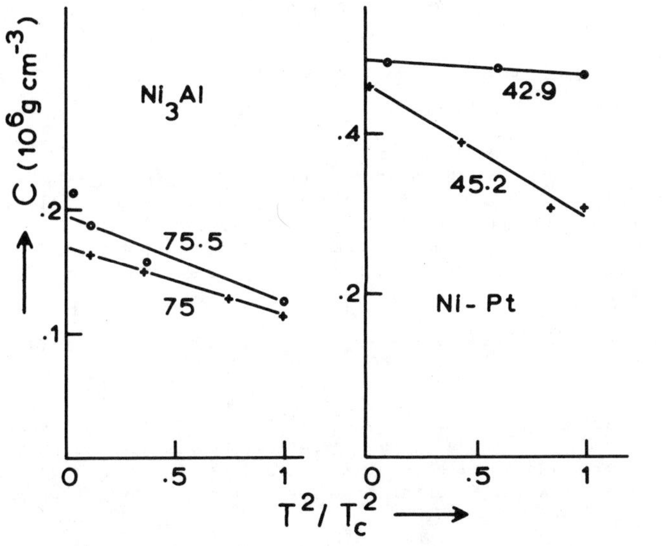


Figure 2.
Temperature dependence of the magneto-elastic coupling parameter C. The curves are labelled with the nickel content in at %.

TABLE II

| at% Ni | $\partial T_c/\partial p$ (K/kbar) | $T_c$ (K) | $\alpha$ ($K^2$/kbar) |
|---|---|---|---|
| 74.8 | -0.58 | 30 | 17.4 |
| 75 | -0.50 | 43 | 21.5 |
| 75.5 | -0.42 | 59 | 24.8 |
| 76 | -0.36 | 72 | 25.9 |

VOLUME MAGNETOSTRICTION

Values for the parameter $\rho\kappa C$ in eq.(2), as determined from volume magnetostriction measurements, are given in Table I.

The magneto-elastic coupling parameter C is plotted in Fig. 2 as a function of $(T/T_c)^2$ for some ferromagnetic $Ni_3Al$ and Ni-Pt alloys ( $\rho \simeq 7.4$ and 17 g cm$^{-3}$, $\kappa = 4.2$ and $4.3 \cdot 10^{-13}$ dyne$^{-1}$cm$^2$ for $Ni_3Al$ and Ni-Pt, respectively).

SHIFT OF THE CURIE TEMPERATURE WITH PRESSURE

Magnetization measurements under pressure offer another, independent way of determining the parameter $\rho\kappa C$. Such measurements have been performed by Buis et al.[3] on ferromagnetic $Ni_3Al$ compounds. Values for $\partial T_c/\partial p$ and for Wohlfarth's[4] phenomenological parameter $\alpha(= - T_c\ \partial T_c/\partial p)$ are given in Table II.

With eq.(3) the corresponding $\rho\kappa C$ values have been calculated, see Table I. For the Ni-Pt alloys we used data reported by Alberts et al.[5] in order to obtain $\rho\kappa C$ from the pressure experiments.

DISCUSSION

The positive sign and the order of magnitude of the magneto-elastic coupling parameter C in these alloy systems are in agreement with Wohlfarth's expression for this parameter in the band model[6]. After some rearrangements this expression can be written as

$$C = \frac{1}{4NN(E_F)\mu_B^2}\left[\frac{\ln N(E_F)}{\partial\omega} + \bar{I}\,\frac{\ln I}{\partial\omega} + f(\bar{I}-1)\,\frac{T^2}{T_c^2}\right] \quad (5)$$

where I is the effective interaction between the itinerant electrons, $\bar{I} = I\ N(E_F)$, $f = \partial \ln\{N(E_F)T_F^2\}/\partial\omega$ and $T_F$ the effective degeneracy temperature, expressed in the first and second derivatives of N(E) to the energy at the Fermi level.
Eq.(5) predicts a quadratic temperature dependence for C and a slope of the C versus $(T/T_c)^2$ curves proportional to $(\bar{I}-1)$, the inverse of the Stoner enhancement factor. This agrees very well with the experimental observations, given in Fig. 2.
From values for C at 0 K we deduce for $\partial \ln \bar{I}/\partial\omega$: 0.8 for $Ni_3Al$ and 1.45 for Ni-Pt.
Taking $\partial \ln N(E_F)/\partial\omega = 5/3$ (uniform deformation of the d band with pressure in combination with Heine's result for the volume dependence of the bandwidth[7]) we calculate for $\partial \ln I/\partial\omega$: -0.8 for $Ni_3Al$ and -0.2 for Ni-Pt. A negative value for the volume derivative of I is in agreement with Kanamori's expression[8] for the effective interaction between the itinerant electrons. With the data, given above, we obtain for the ratio of the effective and bare interactions $I/I_b$: 0.5 for $Ni_3Al$ and 0.85 for Ni-Pt.
The range of possible $I/I_b$ values in Kanamori's expression is between 0 and 1. It is certainly not a trivial case that our $I/I_b$ values fall into this interval. Wohlfarth and Bartell report $I/I_b$ values between 0.5 and 1 for several other systems ($ZrZn_2$, Fe-Ni Invar, etc.).
From the temperature dependence of C we derive for $\partial \ln T_F/\partial\omega$: -6 for $Ni_3Al$ and -11 for Ni-Pt. The errors in these numbers are in the order of 30%.
At a uniform deformation of the d band with pressure this volume derivative should be: -5/3. For higher derivatives of N(E) to the energy the uniform deformation approximation is apparently less accurate.
Some doubt about the validity of this approximation arises also for $N(E_F)$ from the large values of the electronic Grüneisen parameter. It is, however, difficult to estimate the effects of electron-phonon and other magnetic interactions on this parameter. An indication that such interactions are present comes from the peak in the Grüneisen parameter at the critical concentration for ferromagnetism[2].

REFERENCES

1. K.P.Belov, Magnetic Transitions, New York (1961).
2. T.F.M.Kortekaas, Thesis Amsterdam (1975).
3. N.Buis et al., to be published.
4. E.P.Wohlfarth, ICM Moscow 2, 28 (1973).
5. H.L.Alberts, J.Beille, D.Bloch and E.P.Wohlfarth, Phys.Rev. B9, 2233 (1974).
6. E.P.Wohlfarth, J.Phys.C 2, 68 (1969).
7. E.P.Wohlfarth and L.C.Bartell, Phys.Letters 34A, 303 (1971).
8. J.Kanamori, Prog.Theor.Phys. 30, 275 (1963).

# HYPERFINE FIELDS AT INTERSTITIAL POSTIVE MUON SITES IN FERROMAGNETIC MATERIALS*

P. Jena
Physics Department, Northwestern University, Evanston, Ill. 60201

## ABSTRACT

A simple theory of hyperfine field seen by a positive muon at an interstitial site in ferromagnetic materials is presented. The ferromagnetic host is characterized by a free electron conduction band of sp-electrons and localized d-electrons. The theory is extended to study hyperfine fields at interstitial atomic impurities in ferromagnetic hosts. The field systematics are seen to arise from the screening of the impurity charge. Hyperfine fields at interstitial muon sites in Fe, Co and Ni and interstitial Li, Be, B, C, N, O, F and Ne sites in Ni are calculated and compared with most recent experimental data.

## INTRODUCTION

Recently hyperfine fields at interstitial positive muon sites in Fe,[1,2,3] Co,[2] and Ni[1,3,4] have been measured by studying the angular distribution of decay positrons. The fields are negative in all cases and increase in magnitude from Ni to Fe. In addition to being an excellent probe of the interstitial magnetization in ferromagnetic materials, the muons have the additional unique advantage of shedding light on the explicit role of core electrons on the hyperfine field.

The hyperfine field, $B_{hf}$ at the muon site can be written as

$$B_{hf} = - \frac{8\pi}{3} \mu_B P(0) \quad , \qquad (1)$$

where the spin density, P(0) is given by the difference between the electron density for spin ↑ and spin ↓ ,

$$P(0) = P^{\uparrow}(0) - P^{\downarrow}(0) \qquad (2)$$

Since the muons do not have a core in the conventional sense, the hyperfine field that arises due to the interaction of the muon spin with the spin polarized conduction band would be primarily due to Fermi contact interaction. However, the hyperfine field at an atomic impurity of charge 1, say Li, in the same host would be significantly different due to the role of core-electrons. This can be brought about in three ways. (1) The potential seen by an electron due to the muon would be Coulomb type $e^2/r$ that goes to infinity at r = o. For an atomic impurity, however, the effective potential is a pseudopotential that is finite at small r. (2) The spin density at the atomic impurity site would be enhanced due to the orthogonality of the core electrons to the plane waves. (3) The contribution to the $B_{hf}$ due to the exchange interaction between the core electrons and polarized conduction electrons would not exist in the case of the muon.

There have so far been two attempts to explain the hyperfine fields at muon sites in Fe and Ni. Stearns[5] has extended the conduction electron polarization (CEP) curve in Fe to the interstitial region by using the measured value of $B_{hf}$ at $\mu^+$ site in Fe. The value of $B_{hf}$ used was -3kG whereas the recent experiments reveal the field to be -10.7 kG. This procedure does not allow an abinitio calculation of the hyperfine field. Furthermore, the CEP curve is not well known for Ni and Co. Patterson and Falicov[6] have used the local field approximation to explain the $B_{hf}$ seen by $\mu^+$ in Ni. Such an approximation, however, is not valid for interstitial impurities and not much confidence can be put on the quantitative aspect of their calculation.

We present here a simple theory of the hyperfine field seen by an interstitial impurity in a ferromagnetic host. The theory is based on the screening of the impurity charge and is an extension of earlier theories by Jena and Geldart[7] for Heusler alloys and by Daniel and Friedel[8] for ferromagnetic Fe. We apply the theory to compute $B_{hf}$ at both interstitial muon and atomic impurities (Li, Be, B , C , N, O, F , Ne) in ferromagnetic materials.

## MODEL

To compute the electron density per spin $\sigma$ at the origin, $P^{\sigma}(0)$, we define the local impurity potential by a spin dependent potential, $V^{\sigma}(r)$ relative to the uniformly polarized background. The impurity potential is then approximated by a square well of range "a" and depth "$V_o$", namely,

$$V^{\sigma}(r) = \begin{cases} -(V_o + \sigma\Delta) & r < a \\ 0 & r > a \end{cases} \quad , \qquad (3)$$

where σ = +1 for spin ↑ and -1 for spin ↓ electron. For a muon, the range of the potential, "a", is assumed to be the Thomas-Fermi screening length (a = .6A° for $\mu^+$ in Ni). The depth, $V_o$ is obtained self-consistently by satisfying the Friedel sum rule,

$$Z = \frac{2}{\pi} \sum_{\ell} (2\ell + 1) \, \delta_{\ell}(E_F, V_o, a) \quad , \qquad (4)$$

where the partial wave phase shifts, $\delta_{\ell}$ depend upon the Fermi energy, $E_F$, $V_o$, and a. The splitting of the well, $\Delta/E_F$ is determined from the measured interstitial moment density in Ni ($\Delta/E_F \simeq .11$) and is then allowed to scale linearly with the magnetic moment per molecule of the host in question. The screening charge, Z of an interstitial muon is unity. Having determined all the parameters, the wave functions for each spin, hence the spin density, were computed by solving the Schrodinger equation exactly and by requiring that the wave functions inside and outside the well match continuously and differentially at the cell boundary, r = a.

## RESULTS

Our computed hyperfine field, (normalized to the host magnetic moment) $B_{hf}/\mu_n$ at an interstial muon site is plotted in Fig. 1 as a function of $k_F$. We have used a = .6A° throughout. Using the number of sp-electrons per atom in Fe, Co, and Ni to be respectively 1.2, .7 and .6 and the known magnetic moment $\mu_n$ , we obtain $B_{hf}$ values to be -5.6 KG in Fe, -2.5 kG in Co and -.6 kG in Ni. These compare satisfactorily with experimental values of -10.7 kG in Fe, -5 to -6.7 kG in Co and -0.66 kG in Ni. The sign of the field is the result of

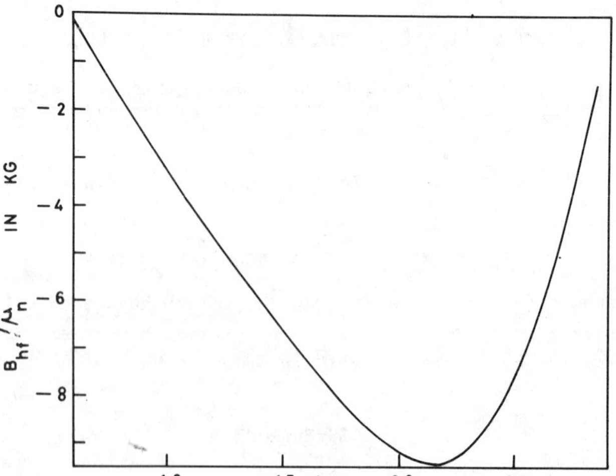


Fig. 1. Hyperfine field (normalized to the magnetic moment per molecule of the host) in kG seen by an interstitial positive muon as a function of $k_F$, free-electron Fermi wave number of the host material.

interference between background spin polarization and local potentials and, as has been discussed earlier,[7] is a direct consequence of the screening of the impurity charge. In commenting on the remaining discrepancy between theory and experiment in Fe and Co, it should be borne in mind that the dipolar field at the interstitial muon site is assumed to be zero while extracting $B_{hf}$ values from the measured internal fields. Although the dipolar field at a static interstitial muon site in Ni vanishes from symmetry considerations, it is non-zero in Fe and Co. It is usually argued that the fast diffusing muon averages out this field by hopping from one site to another several times during its life span. Secondly our theory does not take into account the band structure effects of the ferromagnetic host.

To compute $B_{hf}$ at Li, Be, B, C, N, O, F , and Ne sites in Ni, we have approximated the range of the potential, "a" by the Wigner-Seitz radius of the impurity ion. The depth of the potential well, $V_o$ for a given atomic impurity was determined selfconsistently from Eq. (4) by specifying the appropriate screening charge, Z (eg. Z=3 for B and Z = 8 for Ne). The hyperfine field, $B_{hf}$ in this case can be written by,

$$B_{hf} = - \frac{8\pi}{3} \mu_B \, \alpha \, (k_F) \, P(0) \qquad (5)$$

where $\alpha(k_F)$ is the core enhancement factor[9] and originates from the orthogonality of the plane waves to the impurity core. It can be given by,

$$\alpha \, (k_F) = \frac{1}{N(k_F)} \left[ 1 - \sum_{ns} R_{ns}(0) \, T_{ns}(k_F) \right]^2 \qquad (6)$$

where $N(k_F)$ is the OPW normalization constant and $R_{ns}(0)$ is the value of the radial part of the "ns" core orbital at the origin (r=o). $T_{ns}(k_F)$ is the Bessel transform of the core wave function,

$$T_{ns}(k_F) = \int_0^\infty j_o \, (k_F r) \, R_{ns}(r) \, r^2 \, dr \quad . \qquad (7)$$

For a muon, $\alpha(k_F) = 1$, while for atomic impurities, Li, Be, --, Ne we have computed $\alpha$'s using Hartree-Fock wave functions. The value of $\alpha$ for Li in Ni is 58 and varies little for elements in the same row of the periodic table. The results are compared with the experimental data[10] in Fig. 2. Again, the agreement with experiment

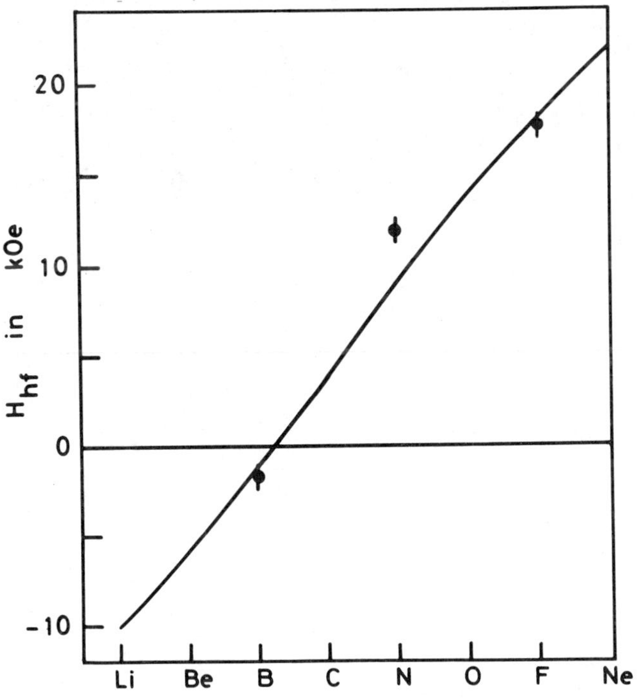


Fig. 2. Hyperfine fields in kG at interstitial atomic impurities in ferromagnetic Ni.

is satisfactory. Further experimental work is necessary to substantiate the model discussed here.

The author takes great pleasure in thanking Professors D.J.W. Geldart and K.S. Singwi for stimulating duscussions.

REFERENCES

*Work supported by ARPA through Northwestern University Materials Research Center.

1 M.L.G. Foy, N. Heiman and W.J. Kossler, Phys. Rev. Lett. 30, 1064(1973).

2 I.I. Gurevich, A.I. Klimov, V.N. Maiorov, E.A. Meleshko, I.A. Muratora, B.A. Nikolsky, V.S. Roganov, V.I. Selivanov and V.A. Suyetin, Zh. Eksp. Teor. Fiz. 66, 374 (1974).

3 J.H. Brewer, D.G. Fleming, K.M. Crowe, R.F. Johnson, B.D. Patterson, A.M. Portis, F.N. Gygax and A. Schenk, Physica Scripta, 11, 144 (1975).

4 B.D. Patterson, K.M. Crowe, F.M. Gygax, R.F. Johnson, A.M. Portis and J.H. Brewer, Phys. Letters 46A, 453 (1974).

5 M.B. Sterns, Phys. Letters 47A, 397 (1974).

6 B.D. Patterson and L.M. Falicov, Solid State Comm. 15, 1509 (1974).

7 P. Jena and D.J.W. Geldart, Solid State Comm. 15, 139 (1974).

8 E. Daniel and J. Friedel, J. Phys. Chem. Solids 24, 1601 (1963).

9 P. Jena and D.J.W. Geldart, Phys. Rev. B7, 439 (1973).

10 H. Hamagaki, K. Nakai, Y. Nojiri, I. Tanihata and K. Sugimoto (to be published).

# MOMENT DISTURBANCES IN Ni-Cu ALLOYS*

R. A. Medina† and J. W. Cable
Solid State Division, Oak Ridge National Laboratory
Oak Ridge, Tennessee 37830

## ABSTRACT

The magnetic diffuse scattering of polarized neutrons from ferromagnetic Ni-Cu alloys with 19.8, 29.6 and 53.5 at.% Cu was measured at 4.2 K. The data show that the Cu atoms are not polarized and that the negative polarization exists only around the Ni sites.

## INTRODUCTION

The moment disturbances in $Ni_{1-c}Cu_c$ alloys were measured with the technique of diffuse magnetic scattering of unpolarized neutrons by Cable et al.[1] (c=.2) and by Aldred et al.[2] for a range of concentrations. These experiments seem to imply the existence of a uniform moment and/or a local moment on Cu atoms totaling -0.1 $\mu_B$ over the range 0-40 at.% Cu. This is a surprising result because in the same range the average moment changes from 0.616 $\mu_B$ to 0.166 $\mu_B$. On the other hand, the result is inconsistent with the diffraction data of Ito and Akimitsu[3] which show that the uniform moment decreases monotonically with increasing Cu content. This inconsistency suggested a new measurement of the moment disturbances with the diffuse scattering of polarized neutrons which gives more easily interpretable results.

## CROSS SECTION AND SOME EXACT RELATIONS

It can be easily proved that the cross section per atom for diffuse scattering of polarized neutrons from a binary alloy with the magnetization perpendicular to the scattering plane is:

$$\left(\frac{d\sigma}{d\Omega}\right)_\sigma = c(1-c)\left[(\Delta b)^2 S(\vec{K}) + 0.54\sigma\Delta b M(\vec{K}) + (.27)^2 T(\vec{K})\right] \quad (1)$$

$$c(1-c)S(\vec{K}) = \sum_{\vec{R}} e^{i\vec{K}\cdot\vec{R}} \langle (p_{\vec{R}+\vec{T}} - c)(p_{\vec{T}} - c)\rangle \quad (2)$$

$$c(1-c)M(\vec{K}) = \sum_{\vec{R}} e^{i\vec{K}\cdot\vec{R}} \langle (p_{\vec{R}+\vec{T}} - c)\, f_{\vec{T}}\mu_{\vec{T}}\rangle \quad (3)$$

$$c(1-c)T(\vec{K}) = \sum_{\vec{R}} e^{i\vec{K}\cdot\vec{R}} \langle f_{\vec{T}+\vec{R}}\mu_{\vec{T}+\vec{R}}(f_{\vec{T}}\mu_{\vec{T}} - \langle f\mu\rangle)\rangle \quad (4)$$

where $\Delta b$ is the difference between the nuclear scattering amplitudes of the components, $\sigma = \pm 1$ is the neutron spin state, $p_{\vec{R}}$ is a site occupation operator, $\mu_{\vec{R}}$ is the magnetic moment at site R, $f_R(\vec{K})$ is the form factor of that moment, and the brackets $\langle \cdots \rangle$ stand for the configurational average. The quantity $S(\vec{K})$ is the short range order (SRO) scattering function, $M(K)$ is the scope of this experiment, and $T(\vec{K})$ is the function measured with unpolarized neutrons. Simple relations hold for those quantities, i.e. it can be proved[4] that:

$$\frac{d\langle\mu\rangle}{dc} = M(0)/S(0). \quad (5)$$

When polycrystalline samples are used, only spherical averages (i.e. $M(\vec{K}) = M(K)$) can be measured. From Eqs. (2), (3) and (4) we obtain the following large K limits. For simplicity we have assumed that all atoms have the same spherically averaged form factor $\langle f(K)\rangle$.

$$\lim_{K\to\infty} M(K)/\langle f(K)\rangle = \Delta\langle\mu\rangle = \langle\mu_i\rangle - \langle\mu_h\rangle \quad (6)$$

$$\lim_{K\to\infty} T(K)/\langle f(K)\rangle^2 = \langle(\Delta\mu)^2\rangle/c(1-c)$$

$$= (\Delta\langle\mu\rangle)^2 + \frac{1}{(1-c)c}\left[(1-c)\langle(\Delta\mu_h)^2\rangle + c\langle(\Delta\mu_i)^2\rangle\right] \quad (7)$$

where i and h stand for impurity and host atoms. As Eq. (6) shows, the large K limit of $M(K)/\langle f(K)\rangle$ is $\Delta\langle\mu\rangle$, the difference of the average moments of the two components. On the other hand, the large K limit of $c(1-c)T(K)/\langle f(K)\rangle^2$ is the total moment fluctuation $\langle(\Delta\mu)^2\rangle$, which consists of $c(1-c)(\Delta\langle\mu\rangle)^2$ and the average of the moment fluctuations of each kind of atom (see Eq. (7)).

Finally, Marshall[5] gives the following approximate expression for $T(\vec{K})$:

$$M(\vec{K}) = \langle f(K)\rangle S(\vec{K}) M(\vec{K}) \quad (8)$$

$$T(\vec{K}) = \langle f(K)\rangle^2 S(\vec{K}) M(\vec{K})^2 + \text{small terms}. \quad (9)$$

## EXPERIMENTAL PROCEDURES

The samples used were polycrystalline plates with 19.8, 29.6 and 53.5 at.% of Cu in $^{62}Ni$. This particular Ni isotope was chosen in order to have the maximum signal. From the isotopic composition, we calculated $\Delta b = 1.621 \pm 0.02\ 10^{-12}$ cm. The samples were prepared by melting in a berylia crucible, rolling, and annealing at 1050 C for 16 hours. The measurements were carried out at the polarized neutron facility of the HFIR at ORNL. The wavelength of the neutrons used was $\lambda = 1.067$ Å. All measurements were done at 4.2 K. For the 20% and 30% alloys a field of 25 kOe was used, for the 53.5% alloy a field of 10 kOe was used. In order to avoid the main beam at smaller angles (<2.5°), a beryllium analyzer was used as in a triple-axis spectrometer with $\Delta E = 0$. The cross sections of both spin states were measured. $M(K)$ was obtained from their difference while the nuclear cross section was obtained from their sum. Equation (9) was used to estimate the small magnetic contribution to the sum cross section. The multiple scattering contributions to $M(K)$ for the 20, 30, and 53.5 at.% Cu alloys were estimated to be -0.055, -0.026, 0.0 $\mu_B$ with a representative error of ± 0.015 $\mu_B$. In all the calculations the following lattice parameters were used: 20% a = 3.532 Å, 30% a = 3.54 Å, 53.5% a = 3.56 Å. These were obtained by interpolation of published data.[6]

## RESULT AND ANALYSIS

The nuclear cross section was least squares fitted to the following form:

$$S(K) = \overline{S(\vec{K})} = 1 + \Sigma_\lambda\, \alpha_\lambda\, Z_\lambda\, j_o(KR_\lambda) \quad (10)$$

where $\alpha_\lambda$ is the SRO parameter, $Z_\lambda$ is the coordination number and $R_\lambda$ is the radius of the shell $\lambda$. The results of the fitting are given in Table I. For the 30% alloy it was necessary to use up to nine shells in order to get a good fit with reasonably small SRO parameters; in this case the parameters were constrained to be small.

The diffraction experiments in pure Ni[7] and in Ni-Cu alloys[3] have shown that the average moment density is composed of a local atomic-like moment density and a uniform negative polarization between sites. It has been previously assumed that this negative magnetization is truly uniform for which a δ-function form factor has been used. If this were really the

ORNL DWG. 75-15871

TABLE I. SRO PARAMETERS

| c | $\alpha_1$ | $\alpha_2$ | $\alpha_3$ | $\alpha_4$ | $\alpha_5$ | $\alpha_6$ | $\alpha_7$ | $\alpha_8$ | $\alpha_9$ | $x^2/N$ |
|---|---|---|---|---|---|---|---|---|---|---|
| 0.198 | 0.1175(19) | −0.0614 | 0.0445 | −0.0584 | 0.0263 | | | | | 1.3 |
| 0.296 | 0.143(82) | −0.0328 | 0.0432 | −0.0098 | 0.0108 | 0.0225 | 0.0037 | −0.0211 | 0.0065 | 0.97 |
| 0.535 | 0.1339(9) | −0.0748 | 0.0438 | −0.0480 | 0.0148 | | | | | 0.80 |

case, the K = 0 limit of the diffuse scattering data should agree with the derivative of the local moment (-1.35 $\mu_B$/Cu atom) and not with the derivative of the bulk moment as it does (see Table II). We therefore think that some other kind of form factor should be associated with the negative polarization; this form factor should be negligible at the Bragg peaks but should be important for the K values characteristic of the diffuse scattering experiment.

The negative magnetization has been attributed to sp-band polarization. On the other hand, Moon[8] has suggested that it is part of the d-electron moment. We have followed this last assumption as explained in detail elsewhere.[9] The Ni form factor used was:

$$\langle f(K)\rangle = [1 + \alpha(1-c)]f_\ell(K) - \alpha(1-c)f_{ov}(K) \qquad (11)$$

where $f_{ov}(K)$ is the "overlap" form factor to be associated with the negative polarization, $f_\ell(K)$ is the form factor of the local moment, and $\alpha$ is a constant equal to 0.154. Both $f_\ell(K)$ and $\alpha$ were chosen to be consistent with the diffraction data. For the analysis of the moment disturbances we have neglected the difference between the form factors of Ni and Cu. This introduces no error if the copper moment is negligible.

We have fitted $M(K)/\langle f(K)\rangle$, with an expression of the form:

$$M(K)/\langle f(K)\rangle = \Delta\langle\mu\rangle + \Sigma_\lambda Z_\lambda \gamma_\lambda j_0(KR_\lambda) \qquad (12)$$

in order to obtain $\Delta\langle\mu\rangle = \langle\mu_{Cu}\rangle - \langle\mu_{Ni}\rangle$. We have also fitted $M(K)/\{\langle f(K)\rangle S(K)\}$ with a similar expression for obtaining the extrapolated value to K = 0.

The results of those fittings are shown in Table II and Fig. 1. The arrows in the figure stand for the values of $d\langle\mu\rangle/dc$ obtained from the magnetization data.

The error quoted for $\langle\mu_{Ni}\rangle - \langle\mu_{Cu}\rangle$ is the statistical error from the fitting. An uncertainty in the multiple scattering correction of about 0.015 $\mu_B$ should be added to this. Our data are clearly consistent with the assumption that the Cu atoms have no moment. The $\Delta\langle\mu\rangle$ values we have obtained are smaller than those obtained with the unpolarized neutron experiment. For example, at 20% Cu we obtain $\Delta\langle\mu\rangle = 0.478\ \mu_B$ compared with the previous result[1,2] of 0.60 $\mu_B$. A large part of this discrepancy comes from the improper spherical average approximation of Eq. (9) used in these earlier analyses, which was

$$\overline{S(\vec{K})M(\vec{K})^2} \simeq S(K)M(K)^2 \ . \qquad (13)$$

This approximation is correct at small K values but wrong for large K. We calculated that about half of this discrepancy comes from this approximation. The remaining difference may arise from the different form factor assumptions used and from the small terms of Eq. (9) that were neglected in the previous analyses. The beauty of the polarized neutron method is that $\Delta\langle\mu\rangle$ is obtained directly from the data without any model assumption. The good agreement between $M(0)/S(0)$ and the derivative of the average moment shows that the moments are determined by their local environment.

In summary the polarized neutron data show that the magnetic moments in Ni-Cu alloys are determined by their local environment and are associated only with the Ni atoms.

ORNL DWG. 75-15872

**TABLE II**

| c | $\langle\mu_{Ni}\rangle - \langle\mu_{Cu}\rangle$ | $\langle\mu\rangle/(1-c)$ | $\mathfrak{M}(0)/S(0)$ | $d\langle\mu\rangle/dc$ | $\langle\mu_{Cu}\rangle$ |
|---|---|---|---|---|---|
| 0.198 | 0.478(5) | 0.486(3) | 1.125(10) | 1.140(10) | 0.006(13) |
| 0.296 | 0.413(6) | 0.397(6) | 1.128(10) | 1.120(10) | − 0.011(12) |
| 0.535 | 0.091(3) | 0.068(9) | 0.65(2) | 0.56(5) | −0.011(8) |

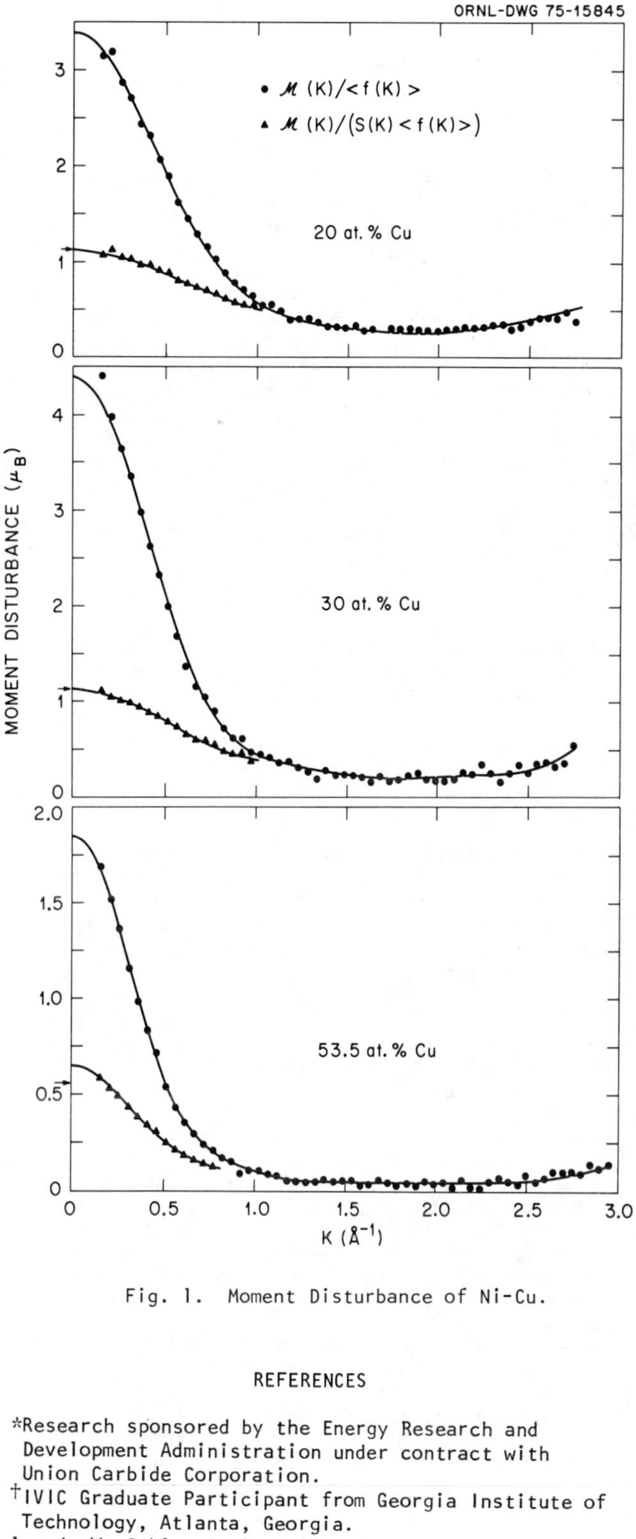


Fig. 1. Moment Disturbance of Ni-Cu.

REFERENCES

*Research sponsored by the Energy Research and Development Administration under contract with Union Carbide Corporation.

†IVIC Graduate Participant from Georgia Institute of Technology, Atlanta, Georgia.

1. J. W. Cable, E. O. Wollan and H. R. Child, Phys. Rev. Lett. 22, 1256 (1969).
2. A. T. Aldred, B. D. Rainford, T. J. Hicks and J. S. Kouvel, Phys. Rev. B 7, 218 (1973).
3. Y. Ito and J. Akimitsu, J. Phys. Soc. Japan 35, 1000 (1973).
4. R. A. Medina and J. W. Garland, to be published.
5. W. Marshall, J. Phys. C 1, 88 (1968).
6. M. Hayase, M. Shiga and Y. Nakamura, J. Phys. Soc. Japan 34, 925 (1973).
7. H. A. Mook, Phys. Rev. 148, 495 (1966).
8. R. M. Moon, AIP Conf. Proc. 24, 425 (1975).
9. G. P. Felcher, J. W. Garland, J. W. Cable and R. Medina, this Conference.

# TEMPERATURE (FIELD) DEPENDENT PHASE SHIFTS : A USEFUL TOOL IN THE APPROACH TO THE KONDO PROBLEM

J. SOULETIE
Centre de Recherches sur les Très Basses Températures, C.N.R.S., B.P. 166 Centre de Tri, 38042 Grenoble-Cedex, France.

ABSTRACT

A number of illustrations are given which experimentally support the affirmation contained in the title and underline the ability of the model to provide easy predictions for a variety of effects (resistance, magnetoresistance, susceptibility, magnetization, specific heat). The approximation involved is equivalent to one which has been, to some extent, justified in sophisticated extensions of the sd model and amounts to the definition of an effective temperature dependent exchange integral ; thus Abrikosov's approximate solution, $2 J_{eff}(T/T_K)\,\rho \sim 2 J \rho/[1 - 2 J \rho \ln(kT/D)]$ implies, in our model, a temperature dependence of the phase shift which leads to the Hamann Fischer solution for the resistivity. Other properties are also discussed (susceptibility, re-entrant $T_c(c)$ curves, etc ...) with emphasis put on the interesting physical picture which the model in addition provides.

## INTRODUCTION

Any solution of the Schrödinger equation in a central field can be written as the sum of products of spherical functions by radial functions $P(r)/r$ such that :

$$\frac{d^2P(r)}{dr^2} + \left[k^2 - V(r) - \frac{\ell(\ell+1)}{r^2}\right]P(r) = 0 \qquad (1)$$

the asymptotic form of which is the stationary wave :

$$P(r) \sim a_\ell \sin\left[kr - \frac{\ell\pi}{2} + \delta_\ell(k)\right]. \qquad (2)$$

The effect at long distances of the potential $V(r)$ can be expressed in terms of $\delta_\ell(k)$ only ; the determination of many properties implies an integration over the energies after which, in a Fermi gas, only the value of the phase shift at $k_F$ will ultimately remain : this is true for the scattering power $g[\delta_\ell(k_F)] \sim (2\ell+1)\sin^2\delta_\ell(k_F)$ and for the Friedel sum rule [1] which relates to $\delta_\ell(k_F)$ the total number $Z_\ell$ of conduction electrons which have been subtracted from the band and confined in a virtual bound state of $\ell$ symmetry : $\delta_\ell(k_F)/\pi = Z_\ell/\gamma_\ell$ where $\gamma_\ell$ is the degeneracy of the considered state. The functions of $\delta_\ell(k_F)$ which we thus obtain are determined in a very general way and do not depend of the structure of the potential $V(r)$. If this remains true in the case of the Kondo [2] effect (or of the local spin fluctuations [3]) then the general solution of the problem would reduce to the determination of $\delta_\ell(H,T)$ for given field and temperature. The generality of the relations between various properties that one obtains by eliminating $\delta_\ell(H,T)$ between theoretical expressions has motived our first attempts [4] and those, simultaneous and independent, of Nagasawa [5] and Loram et al. [6] A common approximation (not necessary in principle but which we will adopt) consists in neglecting all but one phase shift : the $\ell = 2$ phase shift if $V(r)$ is the potential due to impurities of the iron series in a noble matrix such as Cu, Ag or Au.

## THE PHENOMENOLOGICAL MODEL : RESISTANCE vs. SUSCEPTIBILITY AND MAGNETORESISTANCE vs. MAGNETIZATION.

If $Z^\uparrow$ and $Z^\downarrow$ electrons of each spin are confined on $V(r)$ we have from the Friedel sum rule :

$$\frac{Z^\uparrow}{2\ell+1} + \frac{Z^\downarrow}{2\ell+1} = \frac{Z}{2\ell+1} = \frac{\delta^\uparrow}{\pi} + \frac{\delta^\downarrow}{\pi} = \frac{2\delta_v}{\pi}\ , \qquad (3)$$

where $Z$ is the charge difference between the matrix and the impurity.

$$\frac{Z^\uparrow}{2\ell+1} - \frac{Z^\downarrow}{2\ell+1} = \frac{2S}{2\ell+1} + \frac{\delta^\uparrow}{\pi} - \frac{\delta^\downarrow}{\pi} = \frac{2\delta_m}{\pi}\ . \qquad (4)$$

The advance of phase $2\delta_m$ of one spin direction over the other is related to an excess of charge, i.e. a moment $2S$ along the direction $\uparrow$ if we take the convention that $Z^\uparrow \geqslant Z^\downarrow$ ($\delta^\uparrow \geqslant \delta^\downarrow$). This moment points in the plus direction with respect to an external field if $\delta^\uparrow$ refers to the electrons of plus spin. Let there be $N^+$ impurity moments pointing along the field direction, $N^-$ along the reverse ; the current $\sigma^+$ carried by the electrons of plus spin will be proportional to $[N^+g(\delta^\uparrow) + N^-g(\delta^\downarrow)]^{-1}$ where $g(\delta^\uparrow)$ and $g(\delta^\downarrow)$ are the scattering powers associated to majority and minority electrons on the impurity site. There exists a similar term (with $\delta^\uparrow$ and $\delta^\downarrow$ interchanged) for the conductance of the electrons of minus spin. The resistivity $\rho = (\sigma^+ + \sigma^-)^{-1}$ in atomic units ($e = \hbar = m = 1$) is, if $p$ is the number of conduction electrons per atom : [7,8]

$$R(H,T) = (2\ell+1)\frac{2\pi c}{k_F p}\left[(1 - \cos 2\delta_v \cos 2\delta_m) - \frac{(\sin 2\delta_v \sin 2\delta_m)^2}{1 - \cos 2\delta_v \cos 2\delta_m}\left(\frac{N^+-N^-}{N^++N^-}\right)^2\right]$$

with

$$\frac{N^+-N^-}{N^++N^-} = \frac{M(H,T)}{2S} = \frac{M(H,T)\pi}{(2\ell+1)2\delta_m}\ . \qquad (5)$$

For well defined moments we find a first term giving in zero field the constant resistivity already determined by Blandin and Friedel [9] and, in the presence of an external field, a negative magnetoresistance proportional to the square of the magnetization as first shown by Korringa and Gerritsen.[10] The susceptibility $\chi$ follows a Curie law and the magnetization $M(H,T)$ is given by the Brillouin function.

Experimentally, in a dilute alloy, the resistivity depends on the temperature and the susceptibility deviates from the Curie law and reaches a constant value for vanishing temperatures.[11] The logic of our argument is to test whether the same thermal dependence of the phase shift on the temperature (i.e. $\delta_m(T)$ decreasing from $\delta_m(T_F) = \delta_m$ to $\delta_m(0) = 0$) would simultaneously account for the anomalies in the resistivity and the apparent cancellation of the effective moment. From (5) the resistivity would appear as a linear function of $\cos 2\delta_m(T/T_K)$ where $\delta_m(T/T_K)$ can be deduced from the experimental susceptibility. We have :

$$R(T) = (2\ell+1)\frac{2\pi c}{k_F p}\left[1 - \cos 2\delta_v \cos 2\delta_m\left(\frac{T}{T_K}\right)\right] \qquad (6)$$

$$\frac{S}{S+1}\chi(T) = \frac{g^2\mu_B^2}{3kT}S^2(T) = \frac{(2\ell+1)^2}{\pi^2}\frac{g^2\mu_B^2}{3k_T}\delta_m^2(T)\ . \qquad (7)$$

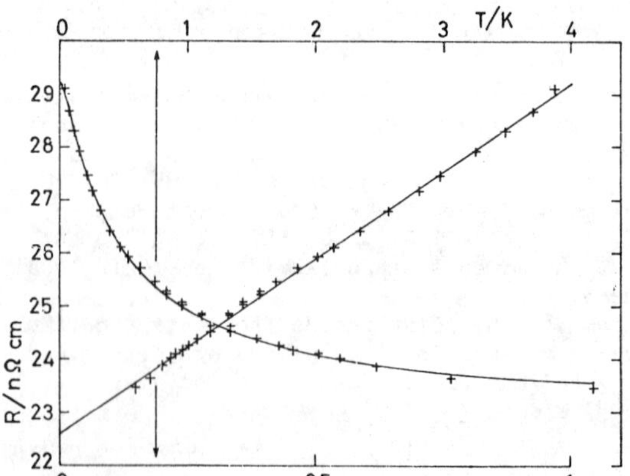


Fig. 1 : Resistance of a 17 p.p.m. Au Fe Alloys vs. temperature between 50 mK and 4.2 K. The same data are also shown vs. cos $2\delta_m(T)$ with $\delta_m(T)$ being determined from the experimental susceptibility (from ref. 7).

Fig. 2 : The resistance in the magnetic (curve 1) and in the non magnetic limits (curve 2) according to Friedel, respectively the high and low temperature limits in our model. Also shown are experimental estimates for the low temperature limit in alloys of the iron series with Au (•) and Cu (x). (From ref. 38, 39 and 40).

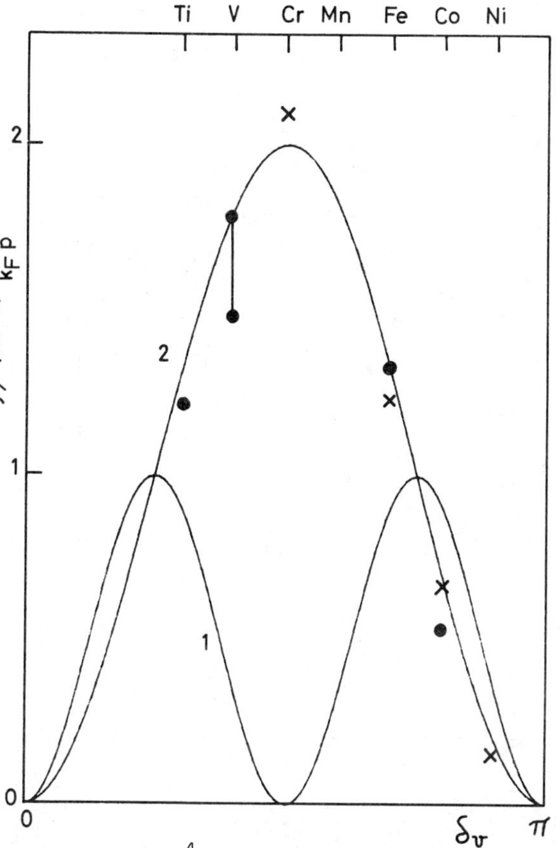


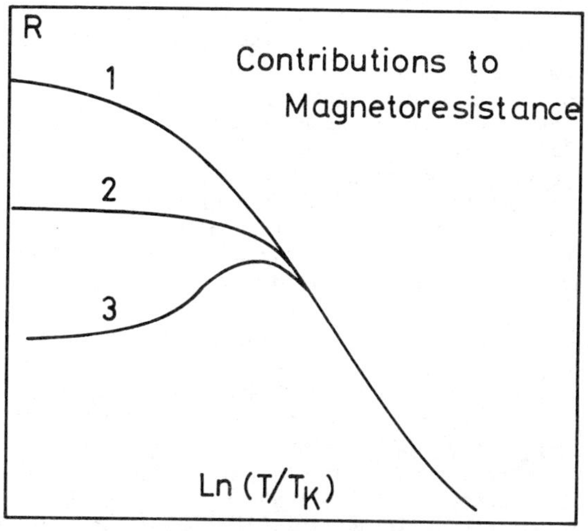


Fig. 3 : Curve 1 schematically shows the temperature dependence of the resistivity in zero field corresponding to a complete cancellation of the moment. For $H \neq 0$ this term saturates (curve 2) to the value corresponding to the frozen moment $\sim\delta_m(T=0,H)$. The negative magnetoresistance $\sim M^2$ is responsible for a maximum at $T \sim H$ (curve 3).

We have shown [4] that this linearity is well obeyed in alloys where an impurity of the iron series is diluted in a noble matrix : for ex. Au V, Au Fe, Cu Cr and Cu Fe (see Fig. 1). Simultaneously Nagasawa [5] has obtained a very good verification of the same relation in Rh Fe and Rh Mn. In those last two systems the resistivity is an increasing function of the temperature and the matrix is a transition metal. Notice, in this occasion, that we consider both systems which are generally refered to as Kondo systems (with $J \gtrless 0$) as well as systems which are considered as illustrations of the theory of local spin fluctuations. A particularly interesting feature of the model is precisely its ability to predict, without any calculation, whether the resistivity will be an increasing or a decreasing function of the temperature : by making $\delta_m = 0$ and $\delta_m = \delta_v$ we obtain the values predicted by Blandin and Friedel [9] in the non-magnetic and in the magnetic case respectively, which, in our model, are the two limits of low and high temperature of the resistivity. From Fig. 2 the resistivity would thus be an increasing function of the temperature for systems like Co or Ni in Au or Cu where $\delta_v$ is either larger than $3\pi/4$ or smaller than $\pi/4$. Loram et al. [6] have developed a similar argument for alloys of Pd with the iron series. All these predictions receive impressive experimental confirmation [4,5,6,12] for values of $\delta_v$ always very close to those which can be derived from the consideration of the Mendeleev table. The $\rho$ vs. $\cos 2\delta_m(T)$ diagram of Fig. 1 permits a direct determination of $\cos 2\delta_v$ (the slope of the linear variation) and of the unitarity limit $\rho_u$ which in the case of noble metals compares with the result $\rho_u = (2\ell+1)\frac{2\pi c}{k_F \rho}$ of the free electron theory (in the case of a transition metal matrix one should also take account of the presence of a d band). [6]

Thus the temperature dependence of the resistivity (curve 1 of Fig. 3) appears experimentally related to a temperature dependence of the moment. Introducing $\delta_m(T,H)$ to account for the effect of an external field we expect only small modifications of the resistivity term (Eq. 6) as long as $T \gg H$. [7,8] In contrast when the temperature becomes negligible the resistivity should saturate to a constant value corresponding to the phase difference $2\delta_m(T=0,H)$ (curve 2, Fig. 3). A contribution to the magnetoresistance is thus introduced which is associated with the freezing of the spin amplitude (f.s.a.): this term, negative in the example of Fig. 3, would be positive in the case of a resistance which increases with T. In addition to this contribution the term associated with the freezing of the spin order (f.s.o.), which was justified in Eq. (5), should be considered :

$$-\frac{\left[\sin 2\delta_v \sin 2\delta_m(T,H)\right]^2}{1-\cos 2\delta_v \cos 2\delta_m}\left[\frac{M(H,T)\pi}{2(2\ell+1)\,\delta_m(T,H)}\right]^2 . \qquad (8)$$

This term, relatively unsensitive to the variation of $\delta_m(H,T)$, introduces a negative contribution, nearly proportional to the square of the magnetization, which qualitatively explains the presence of a maximum [13] for $\mu H \sim kT$ in the magnetoresistance of those systems where the resistivity increases for decreasing temperatures (curve 3, Fig. 3). The f.s.o. term is zero in the particular case of a virtual bound state symmetric at the Fermi level ($Z_\ell = 2\ell+1$, $\delta_v = \pi/2$), which in principle corresponds to the case of Cr in a noble matrix : no maximum was observed in the magnetoresistance of the Cu Cr system. [14] The f.s.o. term is also zero in isoelectronic systems such as Pd Ni where $\delta_v = 0$. The f.s.a. magnetoresistance is then positive and at absolute zero its expression is particularly simple :

$$R(H) = 2\,\rho_u \sin^2 \delta_m(H,T=0) . \qquad (9)$$

From (4), at T = 0, we have $\frac{2\ell+1}{\pi}\,2\delta_m(H,T=0) \equiv M(H,T=0)$ and if the magnetization remains small enough (this is the case for Pd Ni in fields up to 100 kOe) we have :

$$R(H) \sim \frac{\pi^2}{25}\,\frac{\rho_u}{2}\,M^2(H,T=0) \qquad (10)$$

which is well verified qualitatively and quantitatively in Pd Ni (Fig. 4). [6]

## COMMENTS

This phenomenological model suffers from some ambiguities at the level of the definitions : Nagasawa [3] associated the variations of $\delta_m(T)$ with the variations of the effective moment as is done in this paper (eq. 7).

This definition agrees with what is done in Hamann's formula. We, on the other hand [2], had determined $\delta_m(T)$ from a susceptibility written $\sim [S(T)][S(T)+1]/T$, an expression which seems difficult to sustain when S(T) takes non integer values (but which associates at low temperature a resistance proportional to $T^2$ with a constant susceptibility). Experimentally, good linearity is obtained over a large temperature range with either definition with the cost of minor modifications of the parameters $\delta_v$ and $\rho_u$. This, incidentally, shows the limits of the model ; a problem of accuracy leaves a certain choice at the level of the definitions of the properties in terms of the phase shifts. This may to some extent explain why good agreement is obtained with definitions of the susceptibility and resistivity which are open to criticism ; in particular, for Blandin and Friedel [9] the expression (5) would not account for spin flip effects in the resistivity which are essential, in principle, to the Kondo effect. [2] Our phenomenological approach, which apparently supposes just the reverse, ignores these difficulties and finds its justification in the good agreement observed and in the number of the predictions which it makes possible. We will see that Hamann's formula [15] provides, a posteriori, an indirect theoretical support for our assumption.

Let us stress some major features of our argument : the Eq. (5) accounts for the signs and magnitudes of the observed variations of the resistance in dilute systems if the phase difference $2\delta_m$ depends on the temperature T and on the external field H. This dependence appears consistent with the apparent variations of the moment which can be deduced from susceptibility and magnetization measurements. The strength of the model and its interest, from the point of view of the predictions which it makes possible, come from the fact that no fitting parameter is involved since $\delta_m(T)$ is eliminated and $\delta_v$ is determined from the Friedel sum rule ; the values obtained for $\delta_v$ in all cases reasonably agree with this criterion.

PHENOMENOLOGICAL APPROACH vs. sd MODEL

The absence of any fitting parameter is not a new feature : it constitutes the advantage that Blandin and Friedel [9] got, from the beginning, from the application of the sum rule to determine $\delta_v$ and $\delta_m$ in Eq. (5). This Eq. with $\delta_m$ constant should be compared to the sd approach in first Born approximation where two parameters V and J are introduced to fit a single constant (the resistivity in this approximation). For $\ell = 0$ and $S = 1/2$ we have :

$$R = \frac{2\pi c}{k_F \rho}\left[2\pi^2 V^2 \rho^2 + 2\pi^2 J^2 \rho^2 S(S+1)\right]. \qquad (11)$$

The first significant contributions to the Kondo problem however were performed within the framework of the sd model and of particular interest is the approach of Abrikosov [16] who, using parquet graph techniques, showed that the summation of the most divergent terms amounts in the utilisation of an effective exchange integral $J_{eff}(T/T_K)$ expressed in terms of J as :

$$2\,J_{eff}\left(\frac{T}{T_K}\right)\rho \simeq \frac{2\,J\,\rho}{1 - 2\,J\rho\,\mathrm{Ln}\,\frac{T}{T_F}} \simeq \left[\mathrm{Ln}T - \mathrm{Ln}T_K\right]^{-1} \quad \text{with } T_K = T_F \exp\left(\frac{1}{2J\rho}\right). \qquad (12)$$

The same expression was more directly derived by Anderson [17] using a scaling procedure. The variation of $J_{eff}$ in terms of LnT (Fig. 5) exhibits a divergency at $T = T_K$ which would imply a divergency of R when $J_{eff}(T)$ is substituted for J in Eq. (11). If $T_F > T_K$, i.e. when $J < 0$, this divergency appears at a temperature which can be reached in an experiment, which is unacceptable. Of course Eq. (12) ceases to be valid when $J_{eff}$ becomes too large and hopefully a correct calculation would eliminate the possibility of a divergency occuring at a finite temperature. We want to stress that, from our point of view, it is not essential that $J_{eff}$ should remain finite at any finite temperature, and that Hamann's solution [15], which is considered correct for $T \gtrless T_K$, supposes precisely the reverse. At $T = T_F$ this solution (one of the most advanced developments of the s.d. model) should reduce to Eq. (11). Hamann's solution (modified by Fischer [18] to account for potential effects) at $T = T_F$ reduces in fact to :

$$R = \frac{2\pi c}{k_F \rho}(2\ell+1)\left[1 - \cos 2\delta_v \left(1 + 4\pi^2 J^2 \rho^2 S(S+1)\right)^{-1/2}\right] \qquad (13)$$

which is equivalent to Eq. (11) for the small values of $J\rho S$ but would not diverge whatever might be J. Indeed, the introduction of the $J_{eff}(T/T_K)$ of Abrikosov (Eq. 12) in the expression (13) recovers Hamann's complete solu-

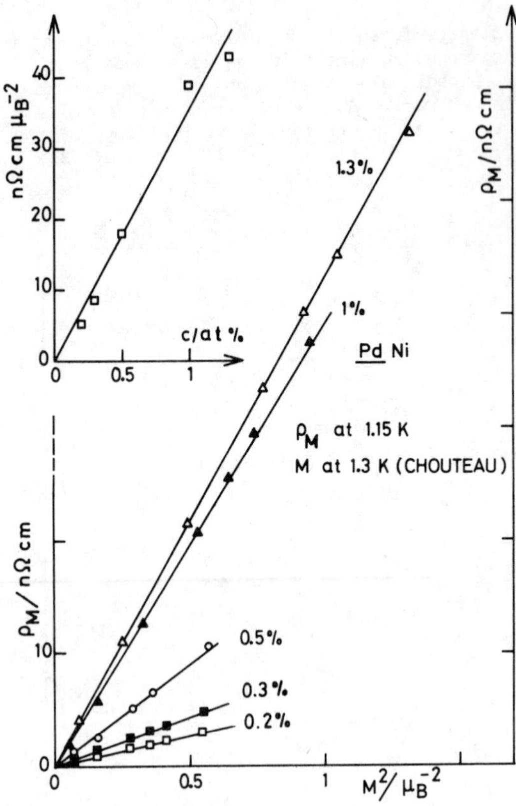


Fig. 4 : The positive impurity magnetoresistance $\rho_M$ of Pd Ni alloys at 1.15 K versus the square of their magnetization M at 1.3 K. The insert shows the slope $\rho_M M^{-2}$ versus the concentration in at. %. (From ref. 8).

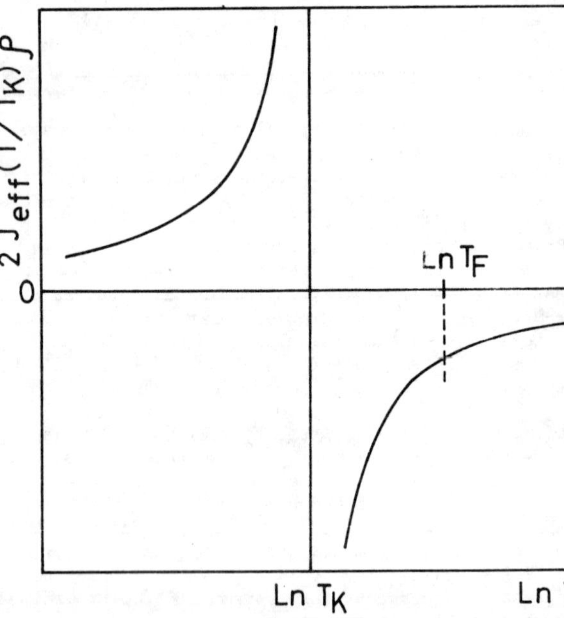


Fig. 5 : $J_{eff}(T/T_K)$ according to Abrikosov. The sign of J is fixed by the respective positions of $T_K$ and $T_F$. For $J < 0$ a divergency occurs at $T_K < T_F$.

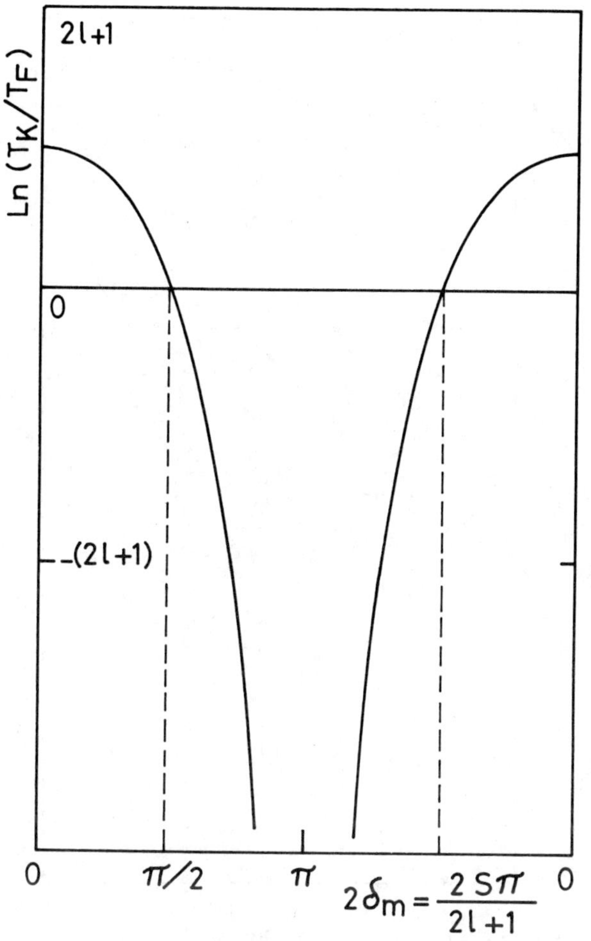


Fig. 6 : $\mathrm{Ln}\,\frac{T_K}{T_F} = \frac{(2\ell+1)\delta_m(T_F)}{\mathrm{tg}\,2\delta_m(T_F)}$ vs. $2\,\delta_m(T_F)$.

The main features of the experimental variations of $T_K$ in terms of the high temperature moment are correctly described.

tion at any temperature. Thus the superiority, if any, of Hamann's over Abrikosov's result is not in a better approximation to the $J_{eff}(T/T_K)$ variation, which is the same, but in the utilisation of an expression (13) which cannot exceed the unitarity limit for any J. Moreover Eq. (13) can be expressed as the Blandin-Friedel result (5) with the same definition of $\delta_v$ and with

$$\mathrm{tg}\,2\delta_m = 2\,\pi\,J\rho\,\sqrt{S(S+1)}\,. \qquad (14)$$

This relation expresses the correspondence between the model of Friedel and the s.d. model, i.e. what the Shrieffer-Wolff [19] transformation describes. Indeed for small values of J S and in the approximation of a Lorentzian virtual bound state [20], the Shrieffer-Wolff transformation yields, in terms of phase shifts [21] :

$$2\pi j_{SW}\rho S \simeq (\mathrm{tg}\,\delta^{\uparrow} - \mathrm{tg}\delta^{\downarrow}) \sim \frac{\mathrm{tg}2\delta_m}{\cos^2\delta_v} = \frac{2\pi J\rho S}{\cos^2\delta_v}\,; \qquad (15)$$

we find between $j_{SW}$ and J the factor $\cos^2\delta_v$ which should be introduced when potential effects are present.[22] Notice that since Eq. (13) is expected to account for spin flip effects so should Eq. (5).

The derivation by Abrikosov and Anderson of a $J_{eff}(T/T_K)$ is equivalent to the utilisation of a temperature dependent phase difference $2\delta_m(T/T_K)$, i.e. the fundamental assumption of our phenomenological model. Using the approximate expression (12) in Eq. (14) we have :[21]

$$\mathrm{tg}2\delta_m(T/T_K) = \frac{\pi\sqrt{S(S+1)}}{\mathrm{Ln}T_K - \mathrm{Ln}T}\,, \qquad (16)$$

or alternatively

$$\delta_m(T/T_K) = \left[\frac{\pi}{2} + \mathrm{arctg}\,\frac{\mathrm{Ln}(T/T_K)}{\pi\sqrt{S(S+1)}}\right], \qquad (17)$$

which when introduced in Eq. (6), gives Hamann's equation :

$$R = (2\ell+1)\,\frac{2\pi c}{k_F\rho}\left[1-\cos 2\delta_v\,\frac{\mathrm{Ln}(T/T_K)}{\left(\pi^2 S(S+1)+(\mathrm{Ln}T/T_K)^2\right)^{1/2}}\right]. \qquad (18)$$

Finally Eq. (14) applied at $T = T_F$ determines $T_K$ in terms of $T_F$ for all values of the phase difference $2\delta_m(T_F)$, i.e. of the moment. We have

$$\mathrm{Ln}(T_K/T_F) = \frac{\pi\sqrt{S(S+1)}}{\mathrm{tg}2\delta_m(T_F)} = (2\ell+1)\,\frac{\delta_m(T_F)}{\mathrm{tg}2\delta_m(T_F)}\,. \qquad (19)$$

The predicted $T_K$ dependence (Fig. 6) reproduces some features which have been experimentally observed.[23] Notice also that the expression (14) associates negative values of J with the cases where our phenomenological model predicts (in agreement with the experiment) a resistivity which increases when T decreases (i.e. for $\pi/4 < \delta_v < 3\pi/4$ see Fig. 2) and positive values of J in the other cases.

In short the assumption of a temperature dependent moment, which constitutes the basis of our phenomenological approach, amounts to the utilisation, justified by Abrikosov and Anderson, of an effective exchange integral $J_{eff}(T)$ depending on the temperature. Their expression for $J_{eff}(T/T_K)$ yields the $\delta_m(T/T_K)$ which makes our model identical to Hamann's. The application of the Friedel sum rule determines $\delta_v$ and $\delta_m$ (thus J) and consequently $T_K$.

## EXTENSION TO OTHER PROPERTIES

In the line of our phenomenological model we may now use the theoretical expression (17) to calculate the susceptibility :

$$\frac{S}{S+1}\,\chi = \frac{g^2\mu_B^2}{3k_T}\,\frac{(2\ell+1)^2}{4}\left[\frac{1}{2} + \frac{1}{\pi}\,\mathrm{arctg}\,\frac{\mathrm{Ln}(T/T_K)}{\pi\sqrt{S(S+1)}}\right]^2\,. \qquad (20)$$

This expression does not diverge, but its asymptotic development for large T leads to Scalapino's result [24] which diverges at $T_K$. One may also write the entropy $E(T) = Nk\,\mathrm{Log}\left(2S_{eff}(T/T_K)+1\right)$ from which an expression for the specific heat is easily derived.

Let us now, in a particular case, show how our model allows recovery of a spectacular result, providing furthermore an interesting phenomenological picture. Abrikosov and Gorkov [25] (A.G) have expressed the depression of the superconducting transition as :

$$\mathrm{Ln}\,\frac{T_c}{T_{c_0}} = \Psi(\tfrac{1}{2}) - \Psi(\tfrac{1}{2} + \frac{1}{4\pi\tau_K k_B T_c})\,, \qquad (21)$$

where $\tau_K$ is the electronic lifetime due to exchange scattering. In the resistivity, additional potential effects enter which interfere with exchange terms but would not affect the pairing since they shift by the same amount the wave functions for the two spin directions. In the usual relation between electronic lifetime and resistivity $\rho R = \tau^{-1}$, we have $\tau \neq \tau_K$ in the general case. But $\tau = \tau_K$ in those cases where potential effects are absent, i.e., with our conventions, for $\delta^{\downarrow} = 0$ and $\delta^{\uparrow} = 2\delta_v = 2\delta_m$. Thus :

$$\tau_K^{-1} = \frac{c(2\ell+1)\,\sin^2 2\delta_m}{\pi\rho}\,, \qquad (22)$$

and the dependence of $\tau_K^{-1}$ on $\delta_m$ is similar to curve 1 of Fig. 3 which represents the resistivity in the magnetic case in Friedel's scheme. We have a different A.G curve for each well-defined moment ; the maximum depression of $T_c$ is obtained for $2\delta_m = \pi/2$, i.e. for $S = \frac{2\ell+1}{4}$ which makes $\tau^{-1}$ maximum corresponding to the unitary limit, see Fig. 7. The A.G. curves corresponding to the other constant values of the moment deduce from this one in an affinity of ratio $(\sin^2 2\delta_m)^{-1}$ along the c axis : thus the same curve is obtained for complementary values of $\delta_m$ (spin S and $\frac{2\ell+1}{2} - S$) and no depression is expected when S is either zero or a maximum of $(2\ell+1)/2$.

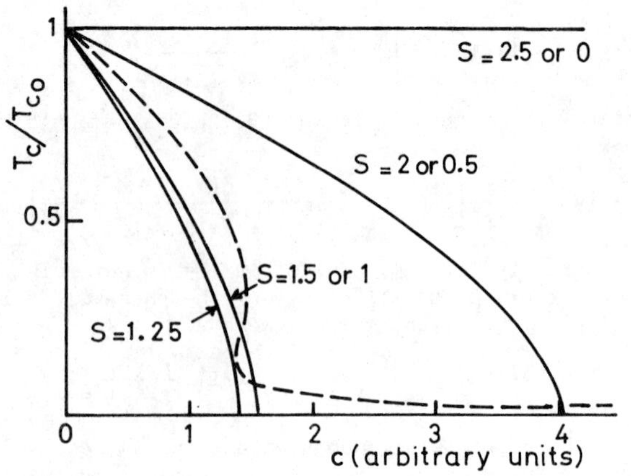


Fig. 7 : The A.G. curves for $T_c(c)$ are shown for different well-defined values of the impurity moment (with $\ell = 2$). The dotted curve schematizes qualitatively what happens for a temperature dependent moment with $T_K \ll T_{c_o}$.

If we now let $\delta_m$ depend on T to account for a Kondo effect, the $T_c(\tau_K(T))$ point will describe a continuous path across the different A.G. curves (the ordinate of curve (1) is multiplied by $(\sin^2 2\delta_m(T/T_K)^{-1})$: for decreasing moments this path will approach the limiting unitarity curve, reach and contact it for $\delta_m(T)=\pi/4$, then depart from it when the temperature and the moment are further decreased. If $\delta_m(T_{c_o}) \gg \pi/4$ we observe the re-entrant behaviour first predicted by Muller Hartmann and Zittartz [26] (M.H.Z.) and observed by Riblet et al.[27] in La $Al_2$ Ce. Indeed, using $\delta_m(T/T_K)$ from Eq. (17) and $\tau_K$ from Eq. (22), the expression (21) provides immediately the M.H.Z. result :

$$\mathrm{Ln}\,\frac{T_c}{T_{c_o}} = \Psi(\tfrac{1}{2}) - \Psi\left[\frac{1}{2} + \frac{(2\ell+1)c}{4\pi^2\rho\, k_B T_c}\;\frac{\pi^2 S(S+1)}{\pi^2 S(S+1)+(\mathrm{Ln}\,\frac{T}{T_K})^2}\right] \quad (23)$$

(notice the factor $(2\ell+1)$ in the right hand side).

## CONCLUSION

Our phenomenological model is capable of new predictions which agree with experiment qualitatively and quantitatively. Besides being consistent with sophisticated theoretical developments of the s.d. model, it provides in addition a simple physical picture and the determination of the parameters J and V. It also applies to a wide variety of alloys without making an experimentally arbitrary distinction, between spin fluctuation and Kondo systems.

To have a completely coherent picture one would desire a more satisfactory expression than Eq. (17) for the thermal dependence of $\delta_m(T/T_K)$, which is wrong, strictly speaking, since Abrikosov's result (12) is not valid for large values of $J_{eff}$. Now, taking as granted the fact that $2\delta_m(T/T_K)$ will vary from above to below the value $\pi/2$, i.e. that $J_{eff}$ as defined by Eq. (14) diverges at some temperature, the approximation provided by Eq. (12) (Fig. 5) may remain a good description for the variations of $J_{eff}$ in a considerable range of temperatures around the divergence, if the parameters are fitted to experiment. Such a description would be reminiscent of the Lorentzian approximation used by Anderson [20] to describe a virtual bound state in terms of the energy : notice the similarity of Eq. (16) with the expression $\mathrm{tg}\delta_v = \Gamma/(E-E_F)$ for the phase shift in the Anderson model. Indeed a good fit to experimental data around $T_K$ is obtained with Hamann's formula, when $\pi\sqrt{S(S+1)}$ is considered as a fitting parameter much as what is generally done with $\Gamma$ ; experimentalists have thus obtained ad hoc $\pi\sqrt{S(S+1)}$ values which are too small by a factor of order 5 but which make formula (18) fit the experiment with considerable accuracy [28,29] (comparable to that of our phenomenological plots like in Fig. 2) in a large range around and above $T_K$. (Notice that the factor $\pi\sqrt{S(S+1)}$ in Eq. (17) affects only the width of the temperature dependence ; the magnitude of the moment, as given by the magnitude of $\delta_m(T/T_K)$ is of the correct size).

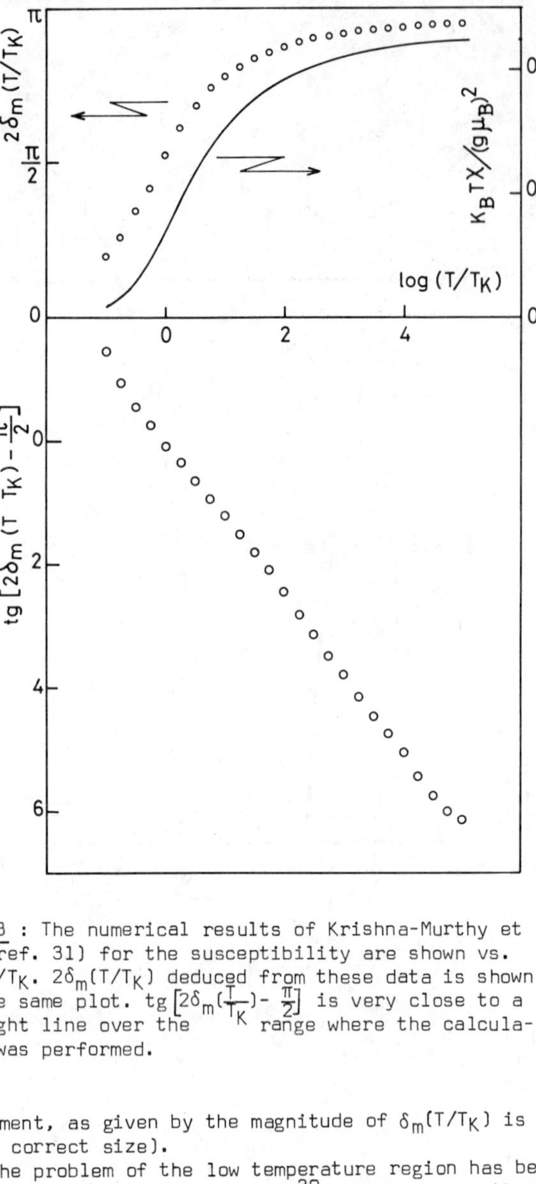


Fig. 8 : The numerical results of Krishna-Murthy et al. (ref. 31) for the susceptibility are shown vs. Log $T/T_K$. $2\delta_m(T/T_K)$ deduced from these data is shown in the same plot. $\mathrm{tg}\left[2\delta_m(\frac{T}{T_K}) - \frac{\pi}{2}\right]$ is very close to a straight line over the range where the calculation was performed.

The problem of the low temperature region has been recently clearified by Nozières [30] who has shown the usefulness in this range of the notion of T and H dependent phase shifts to obtain exact expressions for the first terms of the susceptibility and specific heat. For increasing T the phase reaches large values but developments are again available at high temperatures where $2\delta_m \sim (\pi-\varepsilon)$ with $\varepsilon$ small. The present utilisation of Eq. (17) (with $\pi\sqrt{S(S+1)}$ fitted to the experiment) deals fairly well with the cross-over region ($2\delta_m(T/T_K) \sim \pi/2$). A rather convincing illustration of this point is given in Fig. 8. Krishna-Murthy et al.[31] have used the numerical method (exact in principle). recently developed by Wilson to calculate $T\chi/g^2\,\mu_B^2$, i.e. in our notations, $\pi^{-2}\delta_m^2(T/T_K)$. We have determined $2\delta_m(T/T_K)$ and plotted $\mathrm{tg}(2\delta_m(T/T_K)-\pi/2)$ from their data vs. $\log(T/T_K)$. The deviation of this last plot from a straight line does not appear very meaningful in the range of T where their calculation was performed.

Many further developments of the above ideas can be imagined. For example the Fig. 2 shows that the resistivity of Au Fe [32] in the dilute limit can be expressed in terms of a $\delta_m(T)$ deduced from the susceptibility.[33] A pair approximation developed by Matho and Beal Monod [34] would account for high temperature high concentrations deviations and the presence of a maximum. The evolution and the progressive disappearance of this maximum when the concentration is decreased are well described if we use, for the interaction energy, the expression derived by Caroli [35] in terms of the phase shifts and introduce furthermore in this expression the same $\delta_m(T)$ dependence.[32]

Finally we consider a few experimental objections ; most can be easily ruled out (for example we have neglected s and p phase shifts which can account for minor differences of Z from its theoretical value). It is then of interest to note the objections raised by Alloul [36] on the basis of the recent striking N.M.R. data on Cu nuclei in Cu Fe. Both the distance of the satellites [37] from the central peak (which corresponds to the value of the spin polarization on a well defined shell of close neighbours) as well as the width of the central peak (once the effects attributed to interactions have been subtracted) appear proportional to the susceptibility, according to Alloul. The width of the central peak is proportional to the spin polarization at long distances. In a simple minded application of our ideas, at long distances, (when we can define a $\delta_m$) the spin polarization should reflect the thermal dependence of the phase shift and the thermal dependence of the width of the central peak should sizeably differ from plain proportionality to the susceptibility.

REFERENCES


1. J. Friedel, J. Phys. Radium 23, 692 (1962).
2. J. Kondo, Progr. Theoret. Phys. (Kyoto) 32, 37 (1964).
3. P. Lederer and D.L. Mills, Phys. Rev. 165, 837 (1968).
4. J. Souletie, J. of Low Temp. Phys. 7, 141 (1972).
5. H. Nagasawa, Solid State Commun. 10, 33 (1972).
6. J.W. Loram, R.J. White and A.C.D. Grassie, Phys. Rev. B 5, 3652 (1972).
7. J. Souletie, in Proceedings of the L.T. 13 Conf., Boulder (Colorado) p. 479, vol. 2 (1973).
8. J.L. Genicon, F. Lapierre and J. Souletie, Phys. Rev. B 10, 3976 (1974).
9. A. Blandin and J. Friedel, J. Phys. Radium 20, 160 (1959).
10. J. Korringa and A.N. Gerritsen, Physica 19, 457 (1953).
11. J.L. Tholence and R. Tournier, Phys. Rev. Letters 25, 867 (1970).
12. T. Sugawara, M. Takano and S. Takayanagi, J. Phys. Soc. Japan.
13. P. Monod, Phys. Rev. Letters 19, 1113 (1967).
14. M.D. Daybell and W.A. Steyert, Phys. Rev. Lett. 20, 195 (1968).
15. D.R. Hamann, Phys. Rev. 158, 570 (1967).
16. A.A. Abrikosov, Physics 2, 21 (1965).
17. P.W. Anderson, J. Phys. C 3, 2346 (1970).
18. K. Fischer, Z. Physik 225, 444 (1969).
19. J.R. Schrieffer and P.A. Wolff, Phys. Rev. 149, 491 (1966).
20. P.W. Anderson, Phys. Rev. 124, 41 (1961).
21. J. Souletie, J. Phys. F (Metal Phys.) 5, 342 (1975).
22. K.D. Schotte, Z. Physik 212, 467 (1968).
23. M.D. Daybell and W.A. Steyert, Rev. Mod. Phys. 40, 380 (1968).
24. D.J. Scalapino, Phys. Rev. Lett. 16, 937 (1966).
25. A.A. Abrikosov and L.P. Gor'kov, Zh. Eksp. Teor. Fiz. 39, 1781 (1960) ; Sov. Phys. J.E.T.P. 12, 1241 (1961).
26. E. Müller-Hartmann and J. Zittartz, Z. Phys. 234, 58 (1970).
27. G. Riblet and K. Winzer, Solid State Commun. 9, 1663 (1971).
28. J.W. Loram, T.E. Whall and P.J. Ford, Phys. Rev. B 2, 857 (1970).
29. M.D. Daybell in Magnetism edited by T. Rado and H. Suhl (Academic Press New York and London) vol. 5 p. 121-147 (1973).
30. P. Nozières, J. Low Temp. Phys. 17, 31 (1974).
31. H.R. Krishna-Murthy, K.G. Wilson and J.W. Wilkins in Proceedings of the L.T. 14 Conference, vol. 3, p. 310 (1975).
32. O. Laborde and J. Souletie in Proceedings of the L.T. 14, Conference, vol. 3, p. 366 (1975).
33. J.L. Tholence and R. Tournier, 7ème Conf. Intern. de Magnétisme, 1970, J. de Physique, Suppl. 32 (2-3), C1-211 (1971).
34. K. Matho and M.T. Béal Monod, Phys. Rev. B 5, 1899 (1972).
35. B. Caroli, J. Phys. Chem. Solids 28, 1427 (1967).
36. H. Alloul (this issue).
37. C.P. Slichter (this issue).
38. C. Rizzuto, Rep. Prog. Phys. 37, 147 (1974).
39. M.D. Daybell, D.L. Kohlstedt, and W.A. Steyert, Solid State Commun. 5, 871 (1967).
40. D.K.C. Mac Donald, W.B. Pearson and I.M. Templeton Proc. Roy. Soc. London A 266, 161 (1962).

# $^{63}$Cu NMR SATELLITE STUDIES OF THE ELECTRON SPIN POLARIZATION AND IMPURITY SUSCEPTIBILITY IN Cu-Fe

H. Alloul
Laboratoire de Physique des Solides*, Université Paris-Sud, 91405 - Orsay (France)

ABSTRACT

NMR measurements on both bulk and near neighbour nuclei of the impurity have been performed in dilute Cu-Fe alloys (c ranging from 20 to 4000 ppm) in external fields of about 60 and 30 kGauss. Interaction between impurities broaden the satellite resonances, but their differential NMR shift $\Delta K$ with respect to the pure metal was found independent of field and concentration. For several different satellite resonances, the differential NMR shifts $\Delta K$, all have the same temperature dependence, showing the absence of any change of the spatial dependence of the spin polarization from $T << T_K$ to $T >> T_K$. The data for $\Delta K(T)$ scale with the Mossbauër hyperfine field (≡ Knight shift) data on $^{57}$Fe, if a value $K_{orb} = (5 + 0.5)$ % is taken for the orbital contribution to $K(^{57}Fe)$. These results confirm that the impurity susceptibility has a $(1 - (T/17)^2)$ temperature dependence below $T_K$. The spin polarization and the T dependence of the local susceptibility $\chi(T)$ are discussed in terms of recent theoretical calculations for both the sd and Anderson model.

## INTRODUCTION

A great number of experiments have been aimed at understanding the local properties of Kondo systems[1]. Great efforts have been made in particular in order to compare the temperature dependences of the macroscopic susceptibility $\chi_{mac}$, the localized impurity susceptibility $\chi_{loc}$ and the spin polarization n(r) at large distances from the impurity, through the broadening $\Delta H$ and the shape of the host NMR line[2]. A direct study of n(r) in the vicinity of the impurity can be performed when discrete satellite resonances, associated with near neighbour nuclei of the impurity, can be resolved from the bulk NMR line[3,4].

It has been debated whether the electron electron correlations which build up in the Kondo state would be associated with the appearance of an extended electron cloud ferromagnetically polarized with respect to the localized d spins . The idea of such a polarized electron cloud found apparent experimental support in the observation in Cu-Fe of an anomalous increase below $T_K$ (30 K) of the measured susceptibility $\chi_{mac}$ and the $^{63}$Cu NMR linewidth $\Delta H$[6] with respect to $\chi_{loc}$ as given by Mossbauër effect measurements. Both $^{27}$Al satellite NMR data [3,4] in Al-Mn, and $^{107}$Ag NMR width data in (Au-Ag)V alloys[7], below the respective values of $T_K$(∿ 1000 K and 300 K respectively) did not support this idea of a negative spin polarization of the electron cloud. This was confirmed for the particular case of Cu-Fe by Tholence and Tournier[8], through careful magnetization measurements. These authors demonstrated that Fe impurities which are not too far apart might be more magnetic than isolated impurities. They could then extrapolate a value for $\chi_{mac}$ of isolated impurities, which scales with $\chi_{loc}$. Although such interactions between impurities could also explain the NMR width data, Potts and Welsh found that $\Delta H$ increases linearly with c at any temperature[9]. This allowed them to conclude that an anomalously large spin polarization builds up below $T_K$, and is not associated with an overall significant macroscopic susceptibility. Recently progress in the understanding of Cu-Fe was made by Boyce and Slichter through their observation of satellite $^{63}$Cu resonances from room temperature to below $T_K$, for some of them. The major feature of their results is that $\chi_{mac}$, $\chi_{loc}$ and the spin polarization n(r) as given by the differential Knight shift $\Delta K$ of some shells of neighbours of the impurity, all quantities follow within experimental accuracy a $(T+T_K)^{-1}$ temperature dependence. Finally careful measurements of the width $\Delta H$ of the bulk NMR line, showed that its anomalous increase below $T_K$ might be associated with the magnetization of interacting impurities or chemical clusters of Fe. Together these results imply that no major anomaly of n(r) occurs below $T_K$.

Although previous major ambiguities have been cleared up, it is still quite important to understand whether changes in the detailed shape of n(r) do occur over broadest range of temperatures. Also, both experimental and theoretical efforts are still needed in order to determine the range of validity of the linear relationship between n(r) and $\chi_{loc}$. In the present work, we report observation by transient NMR techniques of the satellite resonances associated with near neighbour nuclei of the impurity in Cu-Fe. The great sensitivity achieved in these experiments allowed us to observe some satellites which were lost by Boyce and Slichter below $T_K$, and to measure with great accuracy the T dependence of $\Delta K$ for other satellite resonances. This experimental technique also allowed us to measure the spin lattice relaxation time $T_1$ of some satellite resonances, which yields the Fe electron spin relaxation[12]. Here we shall focus essentially on the results concerning the static susceptibilities, while more details on the dynamic susceptibility measurements will be given in a further publication. The experimental results are described in section II, while the most important experimental data are given in section III. Comparison between NMR and recent Mossbauër data[13] for $\chi_{loc}$ are presented in section IV. A discussion of the results in comparison with theoretical calculations is then performed in section V.

## EXPERIMENTAL TECHNIQUES

The purpose of these experiments was to observe very weak satellite resonances on the wing of the huge $^{63}$Cu NMR line. As the NMR spectra are usually very broad (a few kGauss), it is obviously impossible to realize ideal pulse conditions for all $^{63}$Cu nuclei in the sample, because of the limited magnitude of the available r-f field $H_1$ (In our experimental case $H_1 \lesssim 80$ Gauss). At low temperatures, the width of some satellite resonances was much greater than $H_1$. In such conditions observation of the free precession decay of the satellite nuclei is impossible. The satellite NMR lineshape was obtained by monitoring the variation with external field of the amplitude of a spin echo formed with a couple of r-f pulses. A finite value of the r-f field $H_1$ is required for such measurements, which leads to an intrinsic limitation of the resolution. Indeed even for $H_1$ as small as 15 G, it was impossible to resolve satellites at less than 100 kHertz from the bulk $^{63}$Cu NMR line. Such a limitation does not occur in the c. w. experiments performed by Boyce and Slichter[10]. In order to overcome this difficulty the present NMR data has been taken in high external field ($H \simeq 60$ kGauss and 30 kGauss for some data points), to split the satellites from the bulk $^{63}$Cu resonance as much as possible. The data were taken in a superconducting magnet, with two insert cryostats allowing to span respectively the temperature ranges (1.2 ÷ 20K) and (15 - 300 K). A classical pulse spectrometer[14] was used, the higher frequency $\omega \sim 65$ MHz being obtained by frequency doubling at the output of the transmitter. Only for a few experimental conditions was the signal to noise ratio high enough to allow direct observation on an oscilloscope screen. It was usually necessary to integrate either through a boxcar integrator or by using a transient digital memory and a digital averager. In such conditions the sensitivity was found much greater than for c. w. experiments as, at

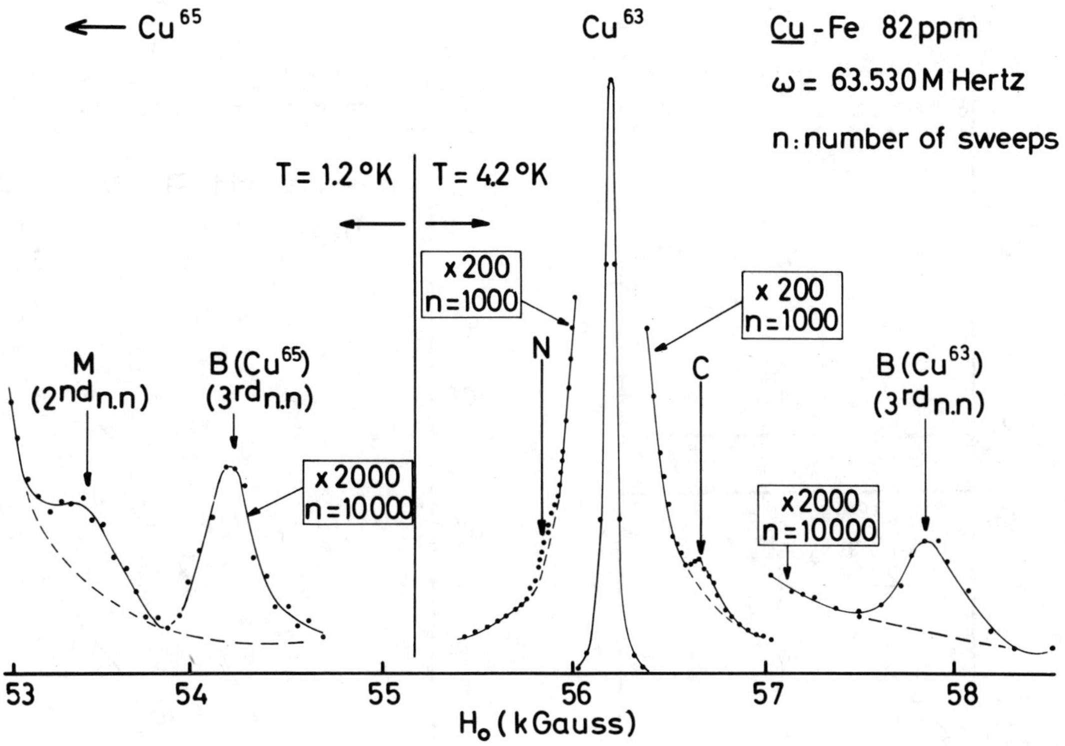


Fig 1 : Variation of the spin echo intensity(arbitrary vertical scale) versus external field for Cu Fe 82ppm. The vertical scale is magnified in order to reveal the satellite resonances which are very weak with respect to the bulk $^{63}$Cu line. The signal to noise ratio is correspondingly increased by digital averaging over n observations of the signal. Note that repetition rates are much greater for observation of the satellite resonances which allows one to reduce the amplitude of the background signal with respect to the satellite resonances as their $T_1$ considerably differ at these temperatures.

1.2°K, we were able to observe all five satellites previously observed by Boyce and Slichter at room temperature (they could only detect one of them at 1.2°K), and that on an 80 ppm sample. The general features of the spectra are apparent on Fig. 1, in which satellite resonances from four shells of near neighbour nuclei can be seen. The total time required to observe such a spectrum did not exceed four hours. A greater accuracy could be obtained for limited field sweeps on a given satellite line. At higher temperatures $4.2 < T < 77$K the sensitivity was somewhat reduced as high repetition rates could not be used without appreciable heating of the sample. Finally at 77K, and higher, the limited resolution did not allow the observation of the two inner satellite resonances C and N. From these points it can be seen that c. w. and pulse experiments are complementary as the pulse NMR allows to detect satellites when they are broad and far from the bulk $Cu^{63}$ NMR, while c. w. NMR allows to resolve satellites which are near the bulk $^{63}$Cu line. Finally a great number of powder samples have been used, covering a wide range of Fe concentrations (20, 82, 275, 348, 880, 1500, 4000 ppm)

EXPERIMENTAL RESULTS

1 - Interactions between impurities

Interactions between impurities in Cu-Fe are known to be important especially in the low T range. Being aware of this point, we have carefully investigated the effect of increasing concentration on the observed satellite resonances, which is illustrated on Fig.2 by results obtained on the 3rd nearest neighbours(satellite B). The position of the satellite does not change with concentration, within 1% accuracy while its width increases. For the 880 ppm sample the satellite NMR becomes poorly resolved from the bulk NMR line. Various sources for the broadening of a satellite resonance can be invoked.

(i) a concentration independent part associated with the non scalar part of the interaction between the impurity moment and the nuclear spins. This might be either dipolar or pseudodipolar interactions through the conduction electrons

(ii) A distribution of susceptibilities of the impurities associated with interactions between impurities.

Both these two components appear on Fig.2.It can be seen that a concentration as low as 20 ppm is required in order to obtain a situation representative of the single impurity spectrum. The concentration dependent part of the linewidth increases slightly slower than linearly in c, and becomes large for moderate concentrations at low temperature. The satellite NMR lines observed at low T were symmetrical, and their shift $\Delta K$ was independent of concentration. Then the interactions between impurities which yield the broadening of the satellite resonances correspond to a symmetric distribution of width $\Delta\chi$ of the impurity susceptibilities, centered at the isolated impurity value $\chi_{loc}$. This implies that the near neighbour $^{63}$Cu nuclei of those Fe atoms which bear a quite different magnetic moment than isolated impurities are wiped out in such experiments. Such a result appears quite natural as the spin polarization around such impurities should greatly differ from that around isolated impurities. The long range interactions between impurities which yield the distribution of susceptibilities around $\chi_{loc}$ have already been observed through impurity NMR in dilute Al-Mn and Au-V alloys , and correspond to a width

$$\Delta\chi \propto c\,\chi^2_{loc} \qquad (1)$$

Extensive work was not performed here in order to obtain accurate measurements of the satellite widths (e. g. we did not systematically extrapolate the satellite width

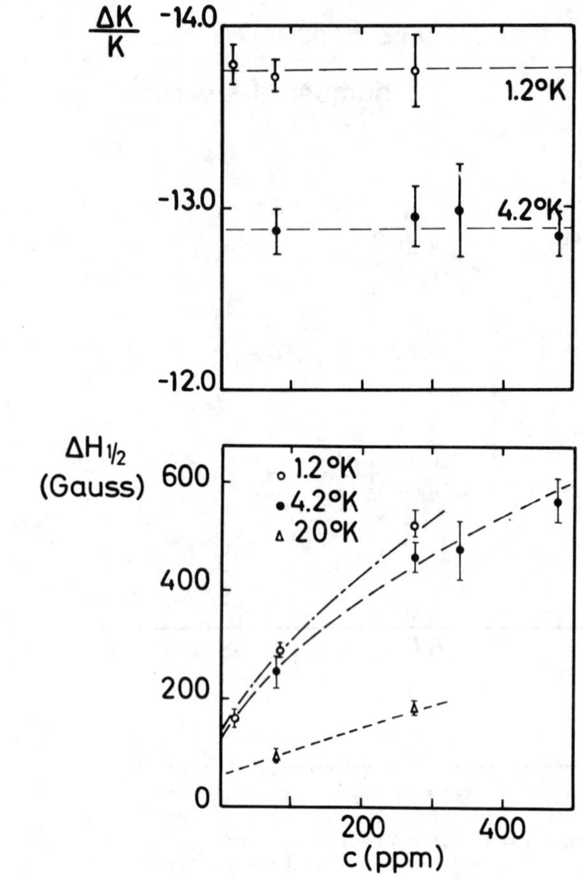


Fig.2 :
Concentration dependence of $\Delta K/K$ and the satellite linewidth $\Delta H_{1/2}$, for the 3rd nearest neighbour shell at 63 MHertz.

down to zero $H_1$ value). It can nevertheless be seen that the concentration dependent part of the broadening is increased by a factor of about 4 between 1.2 and 20°K while $\chi_{loc}$ only increases by a factor 1.5. This higher increase of $\Delta\chi$ at low temperatures with respect to Eq. (1) is of course identical in origin with the excess broadening of the bulk $Cu^{63}$ NMR linewidth below $T_K$. Although the nuclear spins in the vicinity of magnetic Fe groups are not directly observed, these impurities do contribute to the width of the distribution $\Delta\chi$ through their interactions with the isolated Fe atoms. We should then at this point retain as an essential point that interactions between impurities mainly yield an increase of width of the satellite resonances. Consequently $\Delta K$ directly gives a measure of the spin polarization for isolated impurities. It should on the other hand be noticed that the shapes of the impurity, the bulk $^{63}Cu$ or the satellite resonances, are somewhat related with the distribution of impurity magnetization, and are affected by interactions between impurities.

2 - Spin polarization.

As mentionned above, we were able to observe all five satellite resonances at T = 1.2°K. Two of them (B and M) could be followed up to 250 K. C and N could be seen up to 20 K. We only attempted to observe A at 1.2 and 250 K, as the spin lattice relaxation time $T_1$ was expected to be prohibitively small in the intermediate temperature range. Within an experimental accuracy of about 5%, $\Delta K/K$ was found independent of external field, on satellite B. Results on satellite N will not be presented here as the accuracy on $\Delta K(T)$ is much poorer than in the other cases. The results on satellites B, C and M are summarized in fig. 3 where $\Delta K/K$ is plotted versus $(T + 27.6)^{-1}$, which appears to be a convenient representation for $\chi(T)$[13]. The agreement between our results and those of Boyce and Slichter is quite good showing the high absolute accuracy which can be reached for

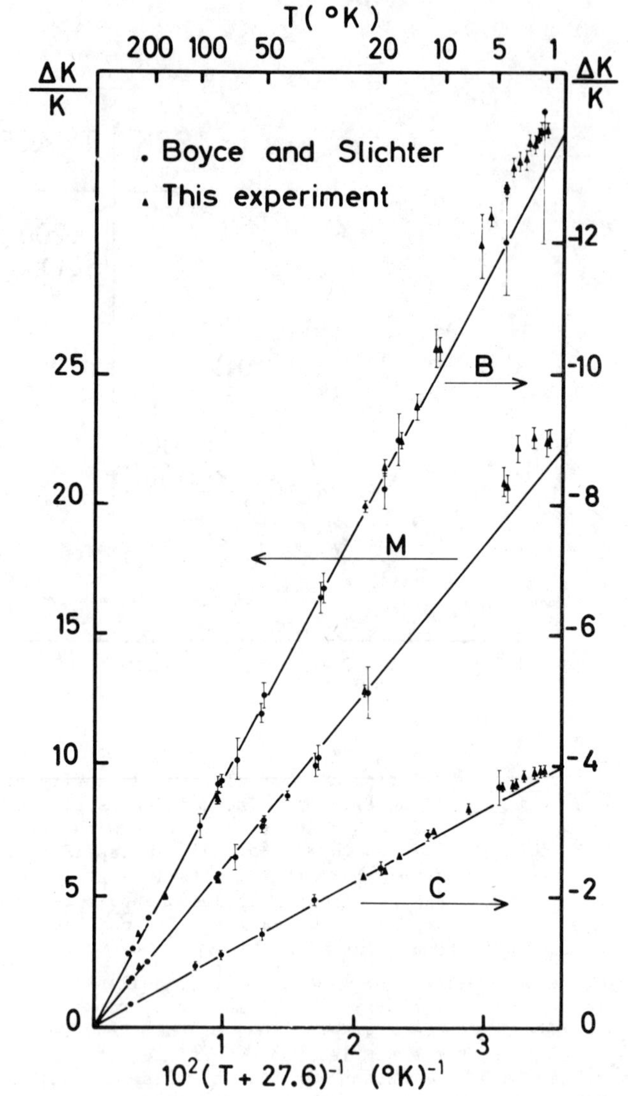


Fig.3 : NMR data for the excess Knight shift $\Delta K$, normalized to the pure copper shift K = 0.232%, plotted versus $(T+27.6)^{-1}$ for the three satellites B, M and C.

such spectroscopic measurements.

It can be seen that for all three cases it is impossible to draw a single straight line through all the data points. This fact is totally independent of the numerical value taken for the Weiss temperature in the abscissae. A slight deviation of about 7 % of $\Delta K(T)$ with respect to the linear dependence is observed below about 20 K. From these temperature dependence studies, it can be seen that $\Delta K$ has a similar temperature dependence for all three satellite resonances. Thus, we can conclude that the spin polarization shape, at least near the impurity does not depend on temperature. This result is valid to within a very good accuracy over the whole temperature range, as can be seen by numerical comparisons of accurate data obtained, at given temperatures (Table I). This data allows us to conclude that the spatial and T dependence of the spin polarization can be factorized, that is

$$n(r, T) = \phi(r)\ A(T) \qquad (2)$$

## COMPARISON WITH LOCAL SUSCEPTIBILITY DATA

An important point to be understood now is whether the slight deviation with respect to a $(T + 27.6)^{-1}$ law below $T_K$ corresponds to a non linear dependence of $\Delta K$ on $\chi_{loc}$, or to a T dependence of $\chi_{loc}$. We therefore have to examine in detail the data for $\chi_{loc}$, which can be obtained either by Mossbauër effect measurements[13] of the impurity Knight shift $K(^{57}Fe)$ or by neutron scattering

| T (°K) Shell | 1.3 | 20.4 | 77 | 245 |
|---|---|---|---|---|
| A ($1^{rst}$ n.n) | -62.25±2.5 | | | -6.91±0.13 (9.01±0.53) |
| B ($3^{rd}$ n.n) | -13.73±0.1 | -7.98±0.08 (1.72±0.03) | -3.46±0.04 (3.97±0.08) | -1.43±0.02 (9.60±0.20) |
| C | - 3.94±0.05 | -2.32±0.04 (1.70±0.05) | | -0.433±0.026* (9.1±0.66) |
| M ($2^{nd}$ n.n) | 22.4±0.2 | 12.85±0.25 (1.74±0.05) | 5.61±0.11 (3.99±0.12) | +2.30±0.05 (9.74±0.31) |

Table I ΔK/K values for four shells of neighbours at four temperatures. The numbers between brackets at a given T are ΔK(1.3)/ΔK(T). They are the same within experimental accuracy for all shells of neighbours.
* This result is deduced by extrapolation from ref (10).

experiments[16,17]. Accurate Mossbauër data have been obtained at low T on a very dilute Cu-Fe sample (c=10 ppm), the accuracy being reduced to about ± 5 % on $K(^{57}Fe)$ at 100 K. It is known from impurity NMR measurements that the Knight shift on a 3d impurity is the sum of a d spin and an orbital term

$$K(^{57}Fe) = K_{(d)} + K_{(orb)} \qquad (3)$$

where

$$K_{(i)} = H_{hf}^{(i)} \mu_B^{-1} \chi^{(i)} \qquad (4)$$

where $H_{hf}^{(i)}$ and $\chi^{(i)}$ are the (i) hyperfine field per Bohr magneton $\mu_B$, and $\chi^{(i)}$ the corresponding susceptibility [(i) = (d) or (orb)]. It is usually assumed that $K_{orb}$ and $\chi_{orb}$ are temperature independent, while the T dependences of K and χ originate from $K_{(d)}$ and $\chi_{(d)}$. In early interpretations of their data $K_{(orb)}$ was assumed to be negligible by Steiner et al[18], who concluded that $\chi_{loc}$ was different from $\chi_{mac}$. Nevertheless, after careful reexaminations of their data[19,13], it could be shown that $K_{(orb)}$ can vary from (5.7 ± 0.2) to (5.6 ± 0.9) % depending upon the extrapolation procedure employed to deduce the T → ∞ limit for $K(^{57}Fe)$. We have fitted the NMR data for ΔK/K for satellite B, with that for $K(^{57}Fe)$, and a surprisingly good fit is found between these two quantities for $K_{orb}$ = (5 ± 0.5) % (fig. 4). Such a fitting procedure does not imply any assumption on the temperature dependence of $\chi_{loc}(T)$ but is somewhat dependent upon the exact value of $K_{(orb)}$. The fact that the obtained value for $K_{(orb)}$ is in quite good agreement with that expected from direct comparisons of $K(^{57}Fe)$ with $\chi_{mac}$ allows one to deduce that within experimental accuracy

$$n(r, T) = \phi(r)\, \chi_{loc}(T) \qquad (5)$$

For T < 6 K, the exact value of $K_{(orb)}$ has no real importance, and it can be seen, in the inset of Fig. (3) that both Mossbauër and NMR data show that

$$\chi_{loc}(T) = \chi_{loc}(o)\, \left[1 - (T/\theta)^2\right] \qquad (6)$$

with θ = (17 ± 1)°K.
It is then clear that the validity of Eq. (5) implies that all these quantities $\chi_{loc}$, $\chi_{mac}$ and n(r) are proportional to each other.

Recent neutron diffraction experiments[17] seem to contradict this point as an increase of $\chi_{loc}$ of about

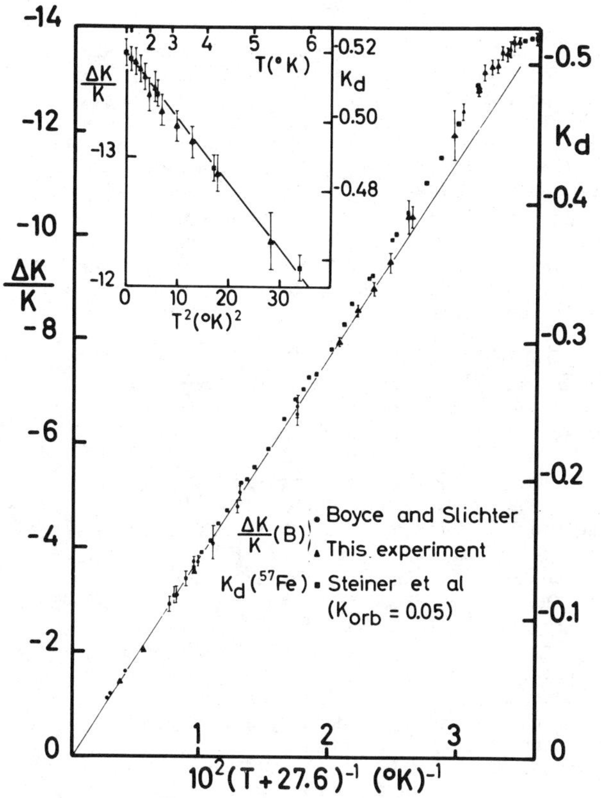


Fig.4 :

Comparison between ΔK/K for satellite B, and the d contribution $K_d$ to $K(^{57}Fe)$. A constant orbital shift $K_{orb}$ = 0.05 has been assumed. Typical error bars for $K_d$, when not given are the same as for the lowest T points. For ΔK/K, the error bars at high T are smaller than the size of the data points. In the insert the low T points are plotted versus $T^2$.

a factor of 3 from 10 to 1.5 K could be detected. As the sensitivity of neutron scattering experiments is very small for such dilute alloy problems, rather large concentrations of impurities have been taken (c > 400 ppm) and extrapolation to c = 0 has been performed in order to obtain $\chi_{loc}$ for isolated impurities. Such an approach might not be right for these type of experiments as the authors had to consider that the magnetic scattering amplitude is the same for all types of paired impurities. Such a simplified approach might not be exact as we really do not know the exact microscopic structure of these impurity pairs. Such simplified approaches have indeed been at the origin of misinterpretations of the bulk $^{63}$Cu NMR width data, as it was considered that ΔH was associated with single impurities because it increased linearly with c, (at low field)[9,11]. Another point mentioned by the authors is that in neutron experiments the viewing time scale might be less than the intrinsic fluctuation time τ of the magnetic center. Although such a possibility should not be discarded, nuclear spin lattice relaxation measurements tell us that τ is nearly independent of temperature up to 20 K. We cannot at present imagine how such an increase of the measured "susceptibility" might then occur between 10 and 1.2 K. The neutron experiments are not fully understood yet, but in view of the Mossbauër and NMR results there is no doubt that the susceptibility is fully localized on the impurity site. Let us finally mention that although the data for $\chi_{mac}$ could be fitted by Hüfner and Steiner with a $(T + 27.6)^{-1}$ law down to 1.2°K, the low T points were obtained by extrapolating the data to vanishing concentration . Such a procedure was not accurate enough to allow the observation of the low T behaviour shown in Fig. (4).

## DISCUSSION

During the last few years, real progress has been done towards a theoretical solution of the Kondo problem[20]. In particular an exact calculation of $\chi(T)$ for the sd hamiltonian[21] as well as for the symmetric Anderson hamiltonian[22] has been performed, for S = 1/2. Consequently, comparison of experiments with exact theories will hopefully become possible in the future for more realistic experimental cases. But, for the time being, it can be considered that at least the qualitative features of the present experimental results should be verified by the S = 1/2 solutions.

1 - Spin polarization within the sd model : T dependence

In the sd limit, we know from perturbation theory that, at $T \gg T_K$, the asymptotic form for the spin polarization is given by

$$n(r) = -\frac{1}{4\pi} J\rho(\varepsilon_F) \frac{\cos 2k_F r}{r^3} < S_z > \quad (7)$$

Both the NMR width data, and the present experimental results indicate that this type of linear relationship between n(r) and $< S_z >$ should also hold at T = 0°K. The confirmation of this experimental result only requires then a theoretical demonstration of the validity of Eq. (7) at T = 0. In such a case this relationship would hold in scaling procedures between strong and small coupling regimes. Indeed a demonstration along the lines developed by Nozières[24] or by Yamada and Yosida[25] might allow this question to be answered. The only calculation done up to now which might bear some resemblance with the exact solution is probably that given by Ishii[26], in which the ground state at T = 0°K is built as a singlet, as done by Yosida and Yoshimori[26]. In this calculation Ishii shows that Eq. (7) holds for r smaller than a critical radius $r_c$ such as

$$k_F r_c \sim - J\rho(\varepsilon_F) \exp \left\{ - \frac{1}{J\rho(\varepsilon_F)} \right\} \sim \frac{D}{k_B T_K} \Big/ \text{Log} \left(\frac{D}{k_B T_K}\right) \quad (8)$$

where $D \sim E_F$ is the bandwidth. For $r \gg r_c/(-J\rho(\varepsilon_F))$ n(r) has a different form

$$n(r) = -\frac{1}{2\pi} < S_z > \frac{\cos 2k_F r}{r^3} \quad (9)$$

It can be immediately seen that for $T_K \sim 30$ K and $D/k_B \sim 10^5$ K, the distance at which n(r) should be modified with respect to Eq. (7) would correspond to a sphere of about $10^7$ nuclei around the impurity, and the variation of n(r) would be impossible to observe experimentally. Although the singlet ground state theory of Yosida Yoshimori and Ishii [26] is not proved to be rigourous, it is clear that their results are in rather good agreement with experiment for many physical quantities at T = 0°K.

2 - Shape of the spin polarization

Of course, even for $T \gg T_K$, Eq. (7) does not give the exact shape of the spin polarization. For instance, solutions of the Anderson model within Hartree-Fock approximation[27] show that the asymptotic limit is only valid for

$$k_F r \gg \frac{\varepsilon_F}{\Delta} \sin \eta^\sigma \quad (10)$$

where $\eta^\uparrow$ and $\eta^\downarrow$ are the phase shifts at the Fermi level and $\Delta$ the width of the virtual bound states, the asymptotic limits being

$$n(r) = +\frac{1}{2\pi} \frac{\sin(\eta^\uparrow - \eta^\downarrow)}{\eta^\uparrow - \eta^\downarrow} \frac{\cos(2k_F r + \eta^\uparrow + \eta^\downarrow)}{r^3} <S_z> \quad (11)$$

in the magnetic limit, and

$$n(r) = +\frac{1}{2\pi} \frac{\cos(2k_F r + 2\eta)}{r^3} < S_z > \quad (12)$$

in the non magnetic limit. When a virtual bound state lies in the vicinity of the Fermi level a preasymptotic reduction of the amplitude of the spin polarization is found for $r \ll r_c$, which may be greater than a few atomic distances. Such preasymptotic deviations, which also appear for charge density oscillations, have been evidenced by comparing the satellite NMR data with the bulk NMR linewidth, in Al-Mn[27]. In Cu-Fe the bulk NMR linewidth measurements, when compared with computations of the NMR lineshape associated with the asymptotic form of the spin polarization, allow one to derive

$$|J_{eff}\, \rho(\varepsilon_F)| \equiv 2 \frac{\sin(\eta^\uparrow - \eta^\downarrow)}{(\eta^\uparrow - \eta^\downarrow)} = 0.74 \pm 0.1 \quad (13)$$

Calculating then the expected values for $\frac{\Delta K}{K}(r)$ from Eq (11) for the various near neighbours of the impurity allows to conclude that 7 shells of impurities should be observed in our experimental range if the asymptotic limit was valid up to the nearest neighbours of the impurity. Experimentally 5 shells (even 6 as for instance satellite N might be the superposition of the contribution of 2 shells) of neighbours of the impurity are observed. Then the spin polarization does not deviate much in amplitude with respect to the asymptotic limit. This points out that the particular structure of d density of states which occurs in CuFe does not drastically influence the amplitude of the spin polarization near the impurity.

Some authors[29] have made a correspondence between the susceptibility $\chi_{mac}$, the resistivity $\rho$ and magnetoresistance $\rho(H)$ in Kondo systems, by using a Hartree-Fock formulation, with T dependent phase shifts deduced from an effective spin

$$\eta^\uparrow(T) - \eta^\downarrow(T) = \frac{2\pi}{5} S(T) \quad (15)$$

Here S(T) was deduced from the temperature dependence of the susceptibility, assuming that

$$\chi(T) = g^2 \mu_B^2 \frac{S(T)\,[S(T) + 1]}{3k_B T} \quad (16)$$

Such an approach gives good agreement with experiment for $\rho$ and $\rho(H)$. Nevertheless, there is no physical reason for giving to these phase shifts the same meaning as they have for Hartree-Fock theory. For instance, considering that Kondo effect is merely a change from a degenerate virtual bound state (vbs) at T = 0 to a split vbs at $T = \infty$, would correspond for n(r) to a progressive change from Eq. (12) to Eq. (11). As S(T) goes from S(o) = 0 to S(300 K) $\sim$ 1.25 from Eq. (16), the phase shift dependent coefficient in equation (11) should change progressively from 1 to 0.6 from T = 0 to T = 300 K which is not observed experimentally. Also such a change between spin split to degenerate vbs, would, within HF theory correspond to a change of the preasymptotic shape of the spin polarization with temperature.

It is clear that the Kondo problem is somewhat more complicated, and that the change of phase shifts which is required in order to explain the increase of resistivity is rather linked with the appearance of a triple peaked structure for the density of states $\rho_d(E)$ at T = 0, as was suggested by Grüner and Zawadowski[30]. Yamada[25] does indeed obtain such a shape for $\rho_d(E)$ at T = 0 using a perturbation expansion in U of the Anderson Hamiltonian. The present experimental results indicate that such a central peak in the structure of $\rho_d(E)$ does not contribute appreciably to the spin polarization, as this central peak should progressively be wiped out with increasing temperature. This is indeed what Ishii's result of Eq. (8) and (9) tells us in the case of large U/$\Delta$, for which the sd model applies, as the effect of the central peak in $\rho_d(E)$ is only to induce important changes of n(r) for $r \gg r_c$. It can then be thought that deviation from a linear relationship between n(r) and $\chi(T)$ will only be observable either at large distance from the impurity for large $U/\Delta$, or for high $T_K$ systems, such as Al-Mn.

3 - T dependence of $\chi_{loc}$

The conclusion drawn above, although not presently based on a rigourous theoretical proof is quite important on the experimental side. In view of the quite good accuracy which can be obtained in measurements of n(r), and of the lack of sensitivity to interactions between impurities, this experimental technique appears to be a rather accurate way for obtaining measurements of $\chi(T)$. Fig. (4) shows us indeed a good experimental plot for $\chi(T)$ in a Kondo system. At low T, as was already deduced indirectly from specific heat measurements[31], $\chi(o) - \chi(T)$ has a quadratic T dependence. Unfortunately for $T \ll T_K$ the results of Wilson's numerical calculation do not have the required accuracy to allow a distinction between a T and a $T^2$ law. Although various authors demonstrate that the leading temperature correction to $\chi_{loc}(o)$ has a $T^2$ form[25,32], no accurate calculation of the $T^2$ coefficient has been performed. It is not consequently possible at the time being to compare the $T^2$ coefficients of $\Delta\chi(T)$ and $\Delta\rho(T)$ at low temperature. In the crossover regime, small deviations from a Curie Weiss law are observed as indicated by Wilson's numerical results[21,22]. Krishna Murthy et al.[22] point out that for $T_K < T < 20\ T_K$ the best Curie Weiss law which fits their numerical data is

$$\chi(T) \simeq \frac{0.17(g\mu_B)^2}{k_B(T + 2T_K)} \qquad (17)$$

the T = 0 limit being

$$\chi(o) \sim 0.1(g\mu_B)^2/k_B\,T_K \qquad (18)$$

Such a result implies that, for $T < T_K$, $\chi(T)$ increases with respect to the high T Curie Weiss law (17). This is exactly the tendency shown by the low T deviation of $\Delta K(T)/K$ in fig. (4). Although the qualitative features of the data are well explained by present calculations for S = 1/2, more complete quantitative comparisons still require further refinements of the theory, namely calculations for S > 1/2, and for a non symmetric Anderson model. The experimental knowledge of Cu-Fe is presently such that it will be a very good candidate for future comparisons with theory.

ACKNOWLEDGEMENTS

The author should like to thank P. Bernier, J. Friedel, D. E. Mac Laughlin and J. Szeftel for constant interest and stimulating discussions.

REFERENCES

* Laboratoire associé au CNRS.

1 For a review article of experimental data-taken with hyperfine techniques up to 1973 see A.Narath Magnetism vol.V chapter 5;GT.Rado and H.Suhl ed.(Academic Press, New York 1973).

2 T.Sugawara, J.Phys.Soc.Japan 14,643 (1959).

3 H.Launois and H.Alloul,Sol.State Comm.7,525 (1969)

4 H.Alloul,P.Bernier,H.Launois and J.P.Pouget J.Phys. Soc.Japan 30,101 (1971).

5 A.J.Heeger,Solid State Physics Vol 23(ed.F.Seitz,D. 1969)Turnbull and H.Ehrenreich, New York Academic Press p'283.

6 D.C.Golibersuch and A.J.Heeger, Phys.Rev.182,584(1969)

7 A.Narath and A.C.Gossard,Phys.Rev.183,391 (1969)

8 J.L.Tholence and R.Tournier,Phys.Rev.Lett 25,867(1970)

9 J.E.Potts and L.B.Welsh,Phys.Rev.B 5,3421 (1972)

10 J.B.Boyce and C.P.Slichter,Phys.Rev.Lett.22,401(1969)

11 H.Alloul,J.Darville and P.Bernier,J.Phys.F Metal Phys. 4, 2050(1974).

12 A preliminary report of this work is given in Phys.Rev. letters 35,460 (1975)

13 P.Steiner,S.Hüfner and W.V.Zdrojewski,Phys.Rev.B 10, 4704 (1974)

14 W.G.Clark,Rev.Sci.Instrum.35,316(1964)

15 Such a broadening has been evidenced on single crystal work by T.S.Stakelon:Ph D thesis,University of Illinois (1974) unpublished.

16 C.Stassis and C.G.Shull,Phys.Rev.B 5,1040 (1972)

17 M.H.Dickens,C.G.Shull,W.C.Koehler and R.M.Moon,Phys.Rev. Lett 35, 595 (1975)

18 P.Steiner,W.V.Zdrojewski,D.Gumprecht and S.Hüfner,Phys. Rev.Lett.31,355 (1973).

19 I.A.Campbell,Phys.Rev.B 10,4036 (1974).

20 A review article of these recent developments has been given by P.Nozières Proceedings of LT14,vol 5 M.Krusius and M.Vuorio ed.(North Holland,American Elsevier 1975) p 339.

21 K.G.Wilson,Rev.Mod.Phys.47,773(1975)

22 H.R.Krishna-murthy,K.G.Wilson and J.W.Wilkins,J.Phys.Rev. Letters 35,1101(1975)

23 K.Yosida,Phys.Rev.106,893 (1957)

24 P.Nozières,J.Low T.Physics 17,31(1974)

25 K.Yamada Prog.Theor.Phys.53,970(1975),54(1975) K.Yosida and K.Yamada Ibid.53(1975)

26 Ishii private communication,a report of these results is given in K.Yosida and Yoshimori Magnetism Vol V chapter 9 G.T.Rado and H.Suhl ed (Academic Press New York 1973)

27 H.Alloul,J.Phys.F.Metal Phys.4,1501(1974)

28 F.Mezei and G.Grüner,Phys.Rev.Lett.29 1465(1972)

29 J.Souletie,J.Low Temp.Phys.7,141(1972) Similar viewpoints have been developed by others:A.M.Stewart Lett.el nuovo cimento 2,1236(1971),H.Nagasawa Solid Stat.Commun.10,33 (1972) J.W.Loram,R.J.White and ADC.Grassie,Phys.Rev.B 5 3659(1972).

30 G.Grüner and A.Zawadowski.Reports on Progress in Physics 37,1497 (1974).

31 B.B.Triplett and N.E.Phillips,Phys.Rev.Lett.27,1001(1971)

32 M.T.Beal Monod and D.L.Mills,Solid State Commun. 14,1157 (1974)

# NMR STUDIES OF THE KONDO EFFECT REVISITED*

C. P. Slichter
Materials Research Laboratory and Department of Physics
University of Illinois
Urbana, Illinois 61801

## ABSTRACT

Both theoretical and experimental aspects of the Kondo effect have at times been highly controversial, but work in the last two years appears to have led to unanimity.

Early work on the system CuFe showed that the width of the $Cu^{63}$ NMR line had the same temperature dependence as the magnetic susceptibility for temperatures, T, well above the Kondo temperature, $T_K$, but not well below. As an explanation it was proposed that the shape of the magnetization density $\sigma(\vec{r},T)$ at distance $\vec{r}$ from the impurity obeyed a law $\sigma(\vec{r},T) = \chi(T)g(\vec{r},T)H$ where H is the applied field, $\chi(T)$ the magnetic susceptibility, and $g(\vec{r},T)$ a temperature dependent shape function. Boyce and Slichter measured $\sigma(\vec{r},T)$ at $Cu^{63}$ nuclei which were the near neighbors of the Fe atom. They found that these nuclei give weak resonances resolved from the main $Cu^{63}$ resonance. The shape function for $\sigma$ does not vary with temperature. They explained why. They showed that the linewidth results were spurious since they were strongly affected by the presence of clusters of magnetic atoms. Alloul has extended this work obtaining important new results. Single crystal studies by Stakelon identify several satellites, and show that crystal fields act on the magnetic ions.

## INTRODUCTION

Considerable experimental and theoretical[1-4] effort has been devoted to determining the nature of the electronic spin state of dilute magnetic alloys that exhibit the Kondo effect and much controversy has resulted. One would like to know the state of the impurity d electrons and the form of the correlations of their spins with the spins of the conduction electron gas, both above and below the Kondo temperature, $T_K$. A quantity that can yield information on the former is the conduction electron spin density, $\sigma(\vec{r},T)$. It is the response of the spins of the conduction electron gas at position $\vec{r}$ relative to the impurity, and at temperature T, to the polarization of the magnetic atom by an external magnetic field H. The operator representing it is $\sum_i s_{zi}\delta(\vec{r}_i - \vec{r})$ where $\vec{r}_i$ and $s_{zi}$ are the position and z-component of spin of the $i^{th}$ electron.

The controversy concerns two forms of the spin polarization which have been proposed to describe the linear response to H

$$\sigma(\vec{r},T)/H = \chi(T)f(\vec{r}) \qquad (1)$$

versus

$$\sigma(\vec{r},T)/H = \chi(T)g(\vec{r},T) \qquad (2)$$

where $\chi(T)$ is the spin susceptibility of a single impurity, often closely fitted by a Curie-Weiss law, $\chi(T) \sim 1/(T + T_K)$.[4,5] For Eq. (1) the spatial shape of the spin density, given by $f(\vec{r})$, does not change with temperature both above and below $T_K$. Only the magnitude of $\sigma(\vec{r},T)$ varies with T. It follows $\chi(T)$. The Kondo effect then manifests itself only through $\chi(T)$. For Eq. (2) the spatial shape function $g(\vec{r},T)$ assumes a function of T which changes when T is lowered below $T_K$ with $g(\vec{r},T) = f(\vec{r})$ for $T \gg T_K$, but not for $T < T_K$. This shape change below $T_K$ is said to be a manifestation of the formation of the "Kondo condensed state."[4]

Both forms of $\sigma(\vec{r},T)$ have been predicted theoretically. Above the Kondo temperature all calculations yield expressions of the form of Eq. (1). Far from the impurity $\sigma(\vec{r},T)$ is given by the Ruderman-Kittel-Kasuya-Yosida (RKKY)[3] oscillations or, equivalently, the Friedel oscillations.[1] Close to the impurity, the magnetic ion structure and the wave vector dependence of the scattering are important as shown by Jena and Geldart[6] and by Alloul.[7]

Experimentally there is also disagreement as to whether Eq. (1) or Eq. (2) describes the spin density. The classic Kondo system is CuFe, which has a convenient Kondo temperature of 29 K.[5] Golibersuch and Heeger,[8] comparing Mössbauer data with their NMR linewidth results, conclude that the spin polarization contains a spatially extended term which is aligned ferromagnetically with the local d-spin on the Fe and accounts for one half the total bulk susceptibility. This spin polarization term goes to zero above about 16 K and 60 kG. The NMR linewidth studies were extended by Potts and Welsh[9] who conclude that below $T_K$ there is either an enhancement of the spin polarization or a formation of an additional long range spin polarization. The original Mössbauer results of Steiner et al.[10] indicated that a small antiferromagnetic polarization of the electron gas existed and was destroyed above 30 K and 100 kG. Thus these authors supported Eq. (2).

Boyce and Slichter[11] realized that it is possible to measure $\sigma(\vec{r},T)$ by means of nuclear magnetic resonance and thus decide which equation is correct. They obtained the spin density at Cu neighboring sites to single Fe impurities by observing the shift of their weak satellite resonances from the main Cu resonance both above and below $T_K$. Alloul has refined the measurements obtaining important new results at the lowest temperatures as reported in the next talk. The satellite NMR technique has been useful in obtaining $\sigma(\vec{r},T)$ in other dilute magnetic alloys. Studies done below the characteristic temperature only have been performed by Alloul et al. on AlMn[13] and by the group at Illinois on CuNi,[14] CuCo,[15] CuV,[16] CuTi,[17] and CuSc.[18] Satellite studies above $T_K$ only have been done by Karnezos and Gardner[19] and Tompa[20] on CuMn and by Boyce, Aton, and Slichter[21] on CuMn and CuCr. Single crystal studies have been done on CuNi, CuCo, CuFe by Stakelon, and on CuMn by Aton in not yet published theses. However, only for the system CuFe have measurements been made both above and below the Kondo temperature so that is the system I discuss in this paper.

## MEASUREMENT OF THE SPIN DENSITY IN CuFe

The host NMR spectrum consists of a large main line resonance signal due to atoms far from the impurity plus weak satellite signals due to atoms near the impurity. The main line resonance of CuFe has been thoroughly studied.[8,9,22,23,24,25] Studies of its width led to proposing Eq. (2). Boyce and Slichter showed these conclusions were in error owing to the presence of pairs or higher clusters which affect the main line width.

The satellites have their NMR field shifted from that of the main line due to changes in the conduction electron spin density, $\sigma(\vec{r},T)$, and due to changes in the conduction electron charge density, $\rho(\vec{r},T)$, resulting from the presence of the impurity. Since these changes fall off rapidly with distance from the impurity, only the near shells of neighbors are shifted sufficiently in field to be observed as satellites to the main line. Each shell produces a distinct resonance. The near neighbor satellite shift can be

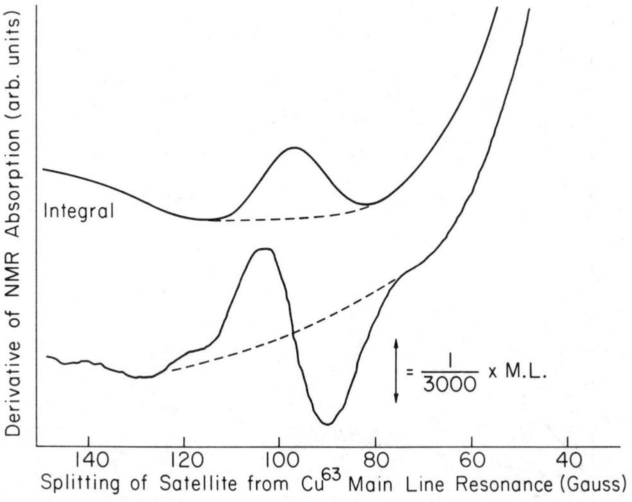


FIGURE 1

Satellite B of the $Cu^{63}$ main line resonance in Cu-0.05 at. % Fe at 298 K and 34.3 kG. The lower trace is the average of 109 sweeps of the derivative of the absorption signal. The arrows show the scale relative to the main line (M.L.) peak-to-peak intensity. The upper trace is the integral of this signal performed digitally with the signal averager.

separated into two parts: the magnetic shift due to $\sigma(\vec{r},T)$ which is linear in the external magnetic field H and the electric quadrupole shift due to $\rho(\vec{r},T)$. The quadrupole shift will be field independent if it is first order and will vary as 1/H if it is second order. This assumes H is small enough so as not to cause a change of the Kondo state. Boyce and Slichter find that the magnetic shift dominates in CuFe as is expected for this magnetic alloy.

Figure 1 shows a typical satellite to the $Cu^{63}$ main line in CuFe.[11] The lower trace is the derivative of the absorption signal and the upper trace is the integral of this experimental curve. The dashed line is the estimated baseline.

Boyce and Slichter observed five such satellite resonances in CuFe and measured positions as a function of external magnetic field. (See Fig. 2) All the splittings are linear in external magnetic field for all the temperatures and fields studied, showing that the electric quadrupole effects are small and that there are changes of state induced by the application of H.

Since the magnetic shift dominates, the experimental shift is a measure of $\sigma(\vec{r},T)$. The additional field at the nucleus at positions $\vec{r}$ relative to the impurity and at temperature T, $\delta H(\vec{r},T)$, is related to $\sigma(\vec{r},T)$ by the equation

$$\delta H(\vec{r},T)/H \equiv \Delta K(\vec{r},T) = -\frac{8\pi}{3}\gamma_e \hbar\,\sigma(\vec{r},T)/H . \quad (3)$$

The minus sign in the above equation occurs because the spin and the moment of the electron have the opposite sign. It is convenient to quote the ratio ΔK/K where K is the Knight shift of the pure copper since ΔK/K is proportional to $\sigma(\vec{r},T)/H$ with constants of proportionality which are properties of pure copper and thus are independent of $\vec{r}$ and T. ΔK/K at 300 K for each of the satellites are listed in Fig. 2. For satellite A a direct dipole-dipole contribution is included so that the isotropic part to be compared with Eq. (3) is $\Delta K/K|_A = -5.24 \pm 0.3$ at 300 K.

Boyce and Slichter found that the splittings are independent of concentration for the concentrations studied: 500, 830, 1000, and 5000 ppm. The intensity of the satellites varied nearly linearly with concentration showing that they arose from single Fe impurities, not pairs or larger clusters. Boyce and Slichter explain why satellites due to pairs or larger clusters were not observed.

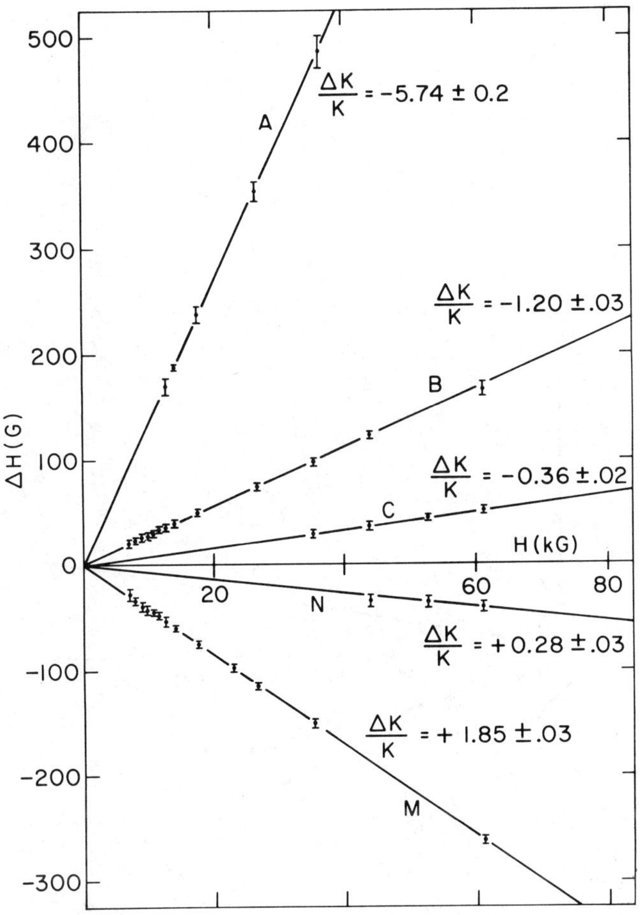


FIGURE 2

Magnetic field dependence of satellite separations from main $Cu^{63}$ resonance at 300 K. Shift of satel lite, H, in gauss from the $Cu^{63}$ resonance versus applied field H, in kilogauss.

## SATELLITE NMR RESULTS

The temperature dependence of ΔK/K tells how $\sigma(\vec{r},T)/H$ varies with temperature at each of the neighboring lattice sites $\vec{r}$ for which a satellite was observed. If ΔK/K fits a Curie-Weiss law, $1/(T+\theta)$, a plot of the reciprocal of ΔK/K versus T would yield a straight line with a temperature intercept of $-\theta$, where θ is the Curie-Weiss temperature. Figure 3 is such a plot. The straight lines, which are least-squares fits to the data, obey a Curie-Weiss law with θ's of 29.2 ± 2.4 for B; 27.6 ± 4.0 for C; 29.2 ± 2.3 for M; 29.3 ± 7.1 for N. Each of the θ's is the same within experimental error, and they are the same as the Curie-Weiss θ of 29 ± 1 K determined from the bulk susceptibiltiy measurements of Tholence and Tournier.[5] ΔK/K is plotted versus $1/(T+29)$ in Fig. 4 to emphasize the low temperature ($\lesssim T_K$) points. The fact that the straight lines fit the data shows that ΔK/K has the same temperature dependence as $\chi(T)$ both above and below $T_K$. Alloul will discuss more precise low temperature points in his talk.

Boyce and Slichter therefore concluded that Eq. (1) describes the results, and that the entire manifestation of the Kondo effect on $\sigma(\vec{r},T)$ is through its effect on the susceptibility, causing the magnetic to non-magnetic transition of $\chi(T)$ as T is lowered below $T_K$. There is no shape change of $\sigma(\vec{r},T)$, no additional spin polarization below $T_K$, and there is no additional spin compensating electron cloud below $T_K$.

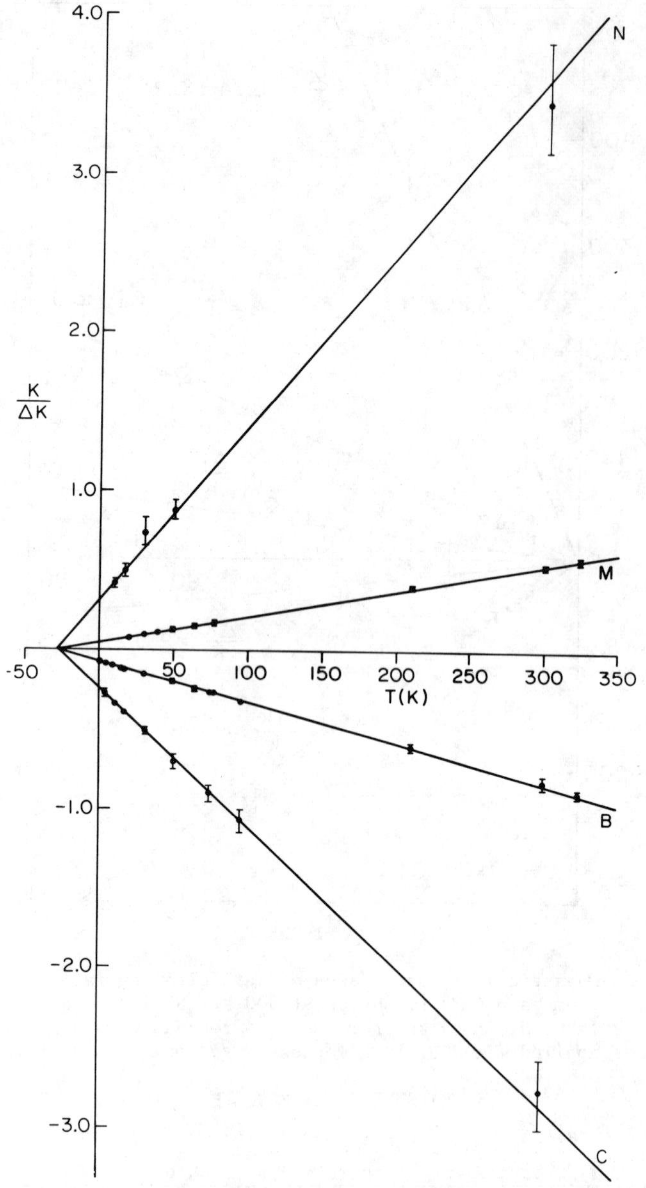


FIGURE 3

$[\Delta K/K]^{-1}$ versus temperature for four of the satellite resonances.

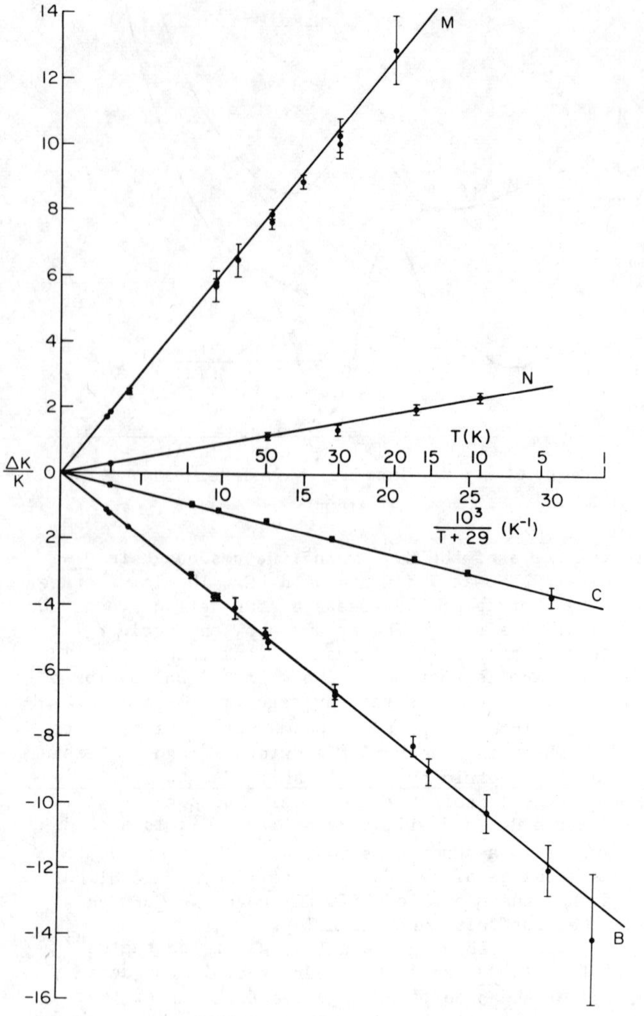


FIGURE 4

The Knight shifts versus $1/(T + 29)$ for four of the satellites. No additional polarization is seen to form below $T_K$ = 29 K.

## A SIMPLE EXPLANATION FOR THE CORRECTNESS OF EQUATION (1)

The question arises can one understand why Eq. (1) is correct rather than Eq. (2)? Boyce and Slichter present the following argument.

Consider an impurity which has a permanent magnetic moment (e.g., Fe in Cu) in the Anderson model. The density of states is shown in Fig. 5. The diagram labelled "state a" shows the parabolic density of states of the conduction electron plus the Lorentzian density of states associated with the impurity atom. Since the down spin resonance occurs below the Fermi energy $E_F$ those states are nearly full, whereas the up spin resonance, located above $E_F$, contains only a few electrons. Thus "state a" corresponds to a net down spin ("up" magnetic moment) on the impurity atom. "State b" corresponds to a net up spin ("down" magnetic moment).

States a and b are degenerate in the absence of an external magnetic field, $H_o$. Application of a magnetic field produces two effects: (i) it changes state "a" and state "b" and (ii) it changes the relative likelihood of finding an impurity in the two states. Boyce and Slichter show that (i) is

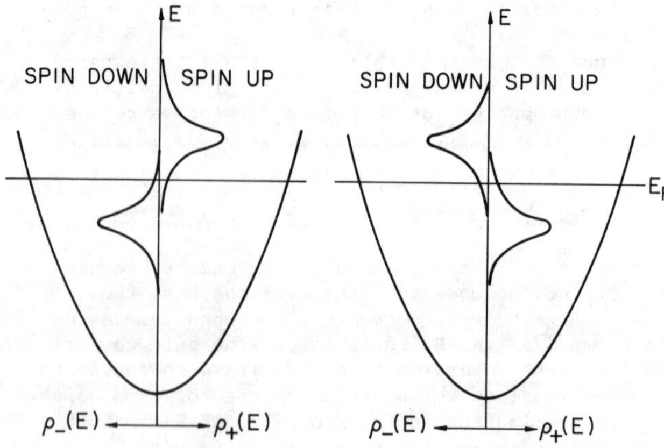


FIGURE 5

Density of states in the Anderson model. Showing the densities $\rho_+(E)$ and $\rho_-(E)$ of up and down spin electrons versus energy, E. The parabolic curves of a free electron are shown as are the extra Lorentzian densities contributed by the magnetic impurity.

unimportant for a strongly magnetic atom such as Fe in Cu.

The second effect represents the fact that state "a" has a lower magnetic energy than state "b" for a magnetic field pointing in the +z direction, hence a larger Boltzmann factor. The resultant magnetization obeys Curie's law in the range of magnetic fields giving linear response.

Since the Anderson model uses the Hartree-Fock approximation, it does not include spin-flip scattering except in terms of the self-consistent populations of the up and down spins. The scattering of a particular spin orientation may be strong since the scattering resonances are close to the Fermi energy.

In the usual s-d Hamiltonian used in the theory of the Kondo effect, the spin-flip scattering leads to the divergences. One must go beyond the Hartree-Fock treatment of Anderson to get the effect. Taking the Hartree-Fock solutions as a starting point, one must include terms which couple together solutions such as the states "a" and "b". State "a" is really a family of states, as can be seen by starting with "a" at absolute zero. "a" is a many electron state. At T = 0 K it is composed of one-electron states which are full below $E_F$ and empty above. At T > 0 K, some of the one-electron states above $E_F$ are occupied, some below empty. We can designate a many electron state then not only by "a" or "b" but also by which one-electron states are occupied. For convenience we label these states: $a_0$, $a_1$, $a_2$,...etc. with $a_0$ being the Anderson ground state. These states and the states $b_0$, $b_1$, $b_2$,... contain many which are either degenerate or nearly degenerate. They are coupled together by the exact Hamiltonian. Resolution of that degeneracy is the heart of the Kondo problem. Presumably the ground state will no longer be $a_0$ or $b_0$, but some linear combination of the low lying $a_i$'s and $b_i$'s -- mostly from within about $k_BT_K$ of the ground state, much as in the BCS theory. In contrast, with BCS, however, the ground state is not split off by a gap from the excited states. The Kondo effect will not have a significant effect on the position or width of the d-wave scattering resonances, since these properties are determined by chemical considerations (the number of d-electrons the atom should have), energies much larger than the Kondo energies.

At temperatures well above $T_K$, we can neglect the Kondo effect. Then the Anderson model should describe things. Application of a magnetic field will cause a preferential alignment corresponding to inducing a preponderance of a-like states.

If $T \ll T_K$, there will also be a net magnetization. Whether that arises from a repopulation effect (i.e., occupation of previously unoccupied low lying states as in the normal spin susceptibility of conduction electrons) or an induced moment effect (admixture of excited states into the ground state as in Van Vleck temperature independent paramagnetism) we cannot say.

As one goes up in temperature, thermal excitation will produce repopulations which progressively break up the Kondo state, leading to the break-down of the "resolution degeneracy" effects, perhaps in analogy to the way thermal hopping takes over from tunelling in the problem of diffusion of hydrogen in solids.

A calculation of spin density at neighbor sites valid for $T \gg T_K$ can be made in the Anderson model by recognizing that the impurity scatters electrons. The scattering can be described in terms of a phase shift $\delta_{\ell\sigma}(k)$ where $\ell$ denotes the angular momentum, and $\sigma$ the spin orientation (up or down) of the scattered wave of vector k. It gives $\Delta K/K$ as

$$\frac{\Delta K(\vec{r})}{K} = \frac{\chi}{\chi_s} \frac{2\pi\Sigma_\sigma m_\sigma}{[\delta_{2+}(k_F)-\delta_{2-}(k_F)]} \int \rho_1(W_k)\,dW_k\, f(W_k,T)$$
$$[n_2{}^2(kr)-j_2{}^2(kr)]\,\sin^2\delta_{2\sigma}{}^a(k)-2n_2(kr)j_2(kr)$$
$$\cdot\, \sin\delta_{2\sigma}{}^a(k)\cos\delta_{2\sigma}{}^a(k)] \qquad (4)$$

where the superscript "a" refers to the "a" configuration of Fig. 5, $n_2$ and $j_2$ are the usual $\ell = 2$ spherical Bessel functions, $\rho_1(W_k)$ the density of conduction electron states of one spin, $f(W_k,T)$ the Fermi function, of energy $W_k$ and temperature T and zero magnetic field, $\chi$ is the spin susceptibility of the impurity, and $\chi_s$ that of the conduction electrons (both on a per atom basis), and $m_\sigma$ is 1/2 or -1/2 spin up or down respectively. In deriving Eq. (4) we have assumed that only the d-wave scattering is important in polarizing the spins owing to the fact that the d-wave phase shifts must be substantial if one is to satisfy the Friedel sum rule for an impurity atom with a partially filled d-shell. We have neglected the repopulation effects within state "a" which result from the change in relative position of the up-spin and down-spin resonances. The factor $\chi$ represents the degree to which the applied field has polarized the impurity and is given by the Boltzmann factor of states a and b.

The formula was computed assuming the Anderson model to be valid, but we know that model does not apply below $T_K$. Boyce and Slichter suggest a reasonable approximation would be (1) replace Curie's law $\chi$ by the Kondo $\chi$ and (2) replace $f(W_k,T)$ by a function $F(W_k,T)$ which describes the admixture of excited states, similarly to the BCS theory. $F(W_k,0)$ should then go from 1 to 0 over a width in energy of order $k_BT_k$.

The effect of the f or F is to cut off the upper limit of integration near $E_F$. Boyce and Slichter then show that Eq. (4) implies that near the impurity $\Delta K/K$ is insensitive to the form of the cut-off (i.e., whether one replaces f by F) and that far from the impurity $\Delta K/K$ is so small that it contributes little to the magnetization.

They conclude, then, that the manifestation of the Kondo effect on the shape of the spin polarization should be through the multiplicative factor of the spin susceptibility.

## IDENTIFICATION OF THE SATELLITES

Identification of which shell of neighbors gives rise to a given satellite is possible in principle if one uses single crystals by observing the variation of satellite position and intensity with orientation of the magnetic field relative to the crystal axes. In this way Stakelon showed that satellite B is the third neighbor shell. He has shown that a satellite similar to A is the first neighbor in CuCo. By "similar" we mean (1) $\Delta K/K$ scales nearly in the ratio of the susceptibilities of Co vs Fe, (2) the satellite possesses an observable structure in the powder, the only satellite for either magnetic atom to do so. Thus satellite A is almost surely the first neighbor.

Identification of other satellites can be done using the lineshape, width, and relative intensity of the satellite resonances. It shows that satellite M is the second neighbor, and C is most likely the fourth.

## CRYSTAL FIELD EFFECTS ON MAGNETIC ATOMS IN Cu

The NMR studies give the first proof that the crystal field produces important effects on the iron group atoms in Cu. Unequivocal evidence comes from the single crystal NMR experiments of T. Stakelon on CuCo. He finds there is a pseudodipolar interaction between the Co and the first neighbor Cu nuclei which is not axially symmetric about the Cu-Co internuclear vector. This result fits the symmetry of the site about this axis (it is a two-fold axis), however axial symmetry would result if one could neglect all atoms other than the Co and the Cu whose nucleus is under study. Thus the crystal potential must play a role. A qualitative but not a quantitative explanation of the asymmetry has been completed.

Boyce and Slichter[11,26] have shown how the crystal field can affect the satellite splitting. The cubic crystal potential at the Fe site will split the five-fold angular degeneracy of the iron d-electrons. This effect can readily be included in the theory of Jena-Geldart on the spatial dependence of $\Delta K$ as described by Boyce and Slichter in another paper at this conference. Addition of a crystal potential will give both diagonal and off-diagonal matrix elements to the Jena-Geldart Hamiltonian. The diagonal elements split and shift the d-electron energy levels. The off-diagonal elements would be present even in pure Cu, for which they would be responsible for converting plane waves into Bloch waves. Boyce and Slichter assume the same result occurs with an impurity atom. Using the Bloch waves enhances the theoretical Knight shift by several orders of magnitude over the free electron value. One expects a similar enhancement will occur with an impurity atom, and that in computing the ratio $\Delta K/K$ the enhancement will cancel out. Boyce and Slichter then show that a cubic crystal splitting of the Fe d-states can produce agreement within 15% of all the measured $\Delta K/K$'s while giving the proper spin and number of electrons.

REFERENCES

*This research was supported in part by the Energy Research and Development Administration under Contract AT(11-1)-1198.


1. A thorough review of the magnetic properties of metallic alloys has appeared in Magnetism, edited by H. Suhl (Academic Press, New York, 1973), Vol. 5.
2. J. Kondo, in Solid State Physics, edited by H. Ehrenreich, F. Seitz, and D. Turnbull (Academic, New York, 1969), Vol. 23.
3. A. Narath, Crit. Rev. Solid State Sci. 3, 1-37 (1972).
4. A. J. Heeger, in Solid State Physics, edited by H. Ehrenreich, F. Seitz, and D. Turnbull (Academic, New York, 1969), Vol. 23.
5. J. L. Tholence and R. Tournier, Phys. Rev. Lett. 25, 867-871 (1970).
6. P. Jena and D. J. W. Geldart, Phys. Rev. B 7, 439-450 (1973).
7. H. Alloul, J. of Phys. F 4, 1501-1516 (1974).
8. D. C. Golibersuch and A. J. Heeger, Phys. Rev. 182, 584-586 (1969).
9. J. E. Potts and L. B. Welsh, Phys. Rev. B 5, 3421-3441 (1972).
10. P. Steiner, W. V. Zerojewski, D. Gumprecht, and S. Hüfner, Phys. Rev. Lett. 31, 355-359 (1973).
11. J. B. Boyce and C. P. Slichter, Phys. Rev. Lett. 32, 61-64 (1974).
    Also a longer version to be published in Phys. Rev.
12. H. Alloul, Phys. Rev. Lett. 35, 460-463 (1975).
13. H. Alloul, P. Bernier, H. Launois, and J. P. Pouget, J. Phys. Soc. Japan 30, 101-116 (1971).
14. D. C. Lo, D. V. Lang, J. B. Boyce, and C. P. Slichter, Phys. Rev. B 8, 973-979 (1973).
15. D. V. Lang, J. B. Boyce, D. C. Lo, and C. P. Slichter, Phys. Rev. Lett 29, 776-779 (1972); D. V. Lang, D. C. Lo, J. B. Boyce, and C. P. Slichter, Phys. Rev B 9, 3077-3085 (1974).
16. D. Follstaedt. Thesis, University of Illinois, unpublished (1975).
17. D. Follstaedt, D. Abbas, T. Stakelon, and C. P. Slichter, to be published.
18. D. Abbas, to be published.
19. N. Karnezos and J. A. Gardner, Phys. Rev B 9, 3106-3112 (1974).
20. K. Tompa, Phys. Stat. Sol 62, 265-269 (1974).
21. J. B. Boyce, T. J. Aton, and C. P. Slichter, in AIP Conference Proceedings on Magnetism and Magnetic Materials, edited by D. G. Graham and J. J. Rhyme , (AIP, New York, 1974); No. 18, Part 1, p. 252.
22. A. J. Heeger, L. B. Welsh, M. A. Jensen, and G. Gladstone, Phys. Rev. 172, 302-319 (1968).
23. H. Alloul, J. Darville, and P. Bernier, J. of Phys. F 4 2050-2066 (1974).
24. H. Takenaka, Y. Oda, and K. Asayama, J. Phys. Soc. Japan 37, 961-964 (1974).
25. T. Sugawara, J. Phys. Soc. Japan 14, 643-652 (1959).
26. J. B. Boyce and C. P. Slichter, AIP Conference Proceedings in Magnetism and Magnetic Materials (1975).


## SUSCEPTIBILITY OF THE ANDERSON AND KONDO MODELS OF DILUTE MAGNETIC ALLOYS.

H. R. Krishna-murthy*, Laboratory of Atomic and Solid State Physics, Cornell University, Ithaca, N. Y. 14853.

ABSTRACT

Recent calculations[1] of the susceptibility of the symmetric Anderson model, done with K. G. Wilson and J. W. Wilkins, are presented. The results map nicely on to the susceptibility of the Kondo model. The approach, based on the numerical renormalisation group techniques developed by Wilson for the Spin-$\frac{1}{2}$ Kondo Hamiltonian[2], is also suitable for calculating the susceptibility of the asymmetric Anderson model, and preliminary results are reported. Work on the possible extension of the techniques to calculate transport properties, and on the connection with phenomenological treatments is being pursued.

References:

* I.B.M. Graduate Fellow.


1. H. R. Krishna-murthy, K. G. Wilson and J. W. Wilkins, Phys. Rev. Lett. 35,1101 (1975).
2. K. G. Wilson, Rev. Mod. Phys. 47, 773 (1975).

# MAGNETIC STUDIES OF ACTINIDES--EVIDENCE FOR LOCALIZED *5f* ELECTRONS*

G. H. Lander
Argonne National Laboratory, Argonne, Ill. 60439

## ABSTRACT

The position of the *5f* elements in the periodic table suggests that the electronic properties of these elements (and their compounds) will resemble those of the lanthanide series. However, the extended nature of the *5f* wave functions leads to fundamental differences between *4f* and *5f* systems. In this review the evidence for "localized" magnetism will be presented. Results of magnetization, Mössbauer, neutron and low-temperature x-ray experiments on Np, Pu, and Am compounds will be used to illustrate both the similarities to and differences from lanthanide magnetism.

A determination of the ground-state *5f* electron wave functions is, in principle, possible by measuring the neutron magnetic cross section. The interpretation of such experiments on $UO_2$, USb, and PuP requires a knowledge of the radial extent of the *5f* electrons, which we obtain from relativistic Dirac-Fock calculations, and the use of the tensor-operator formalism to treat the spin-orbit interaction. This interaction in $Pu^{3+}$ (a $5f^5$ configuration) results in a magnetic form factor that initially increases with increasing scattering angle. For USb the experimental magnetic scattering is used to determine the ground-state wave function of the $U^{3+}$ ion. In addition the temperature dependence of the quadrupole moment in USb has been measured and yields information both on the size of the crystal-field and exchange interactions, and on the interplay between them.

## INTRODUCTION

The unique problems associated with describing the electronic structure of the actinide elements and compounds were recognized as soon as the first investigations were performed some 35 years ago. A consequence of the spatial extent of the outermost electrons is that the actinide ions are sensitive to their environment, i.e., large crystal-field interactions are present.[1] On the other hand, the *5f* electrons have a high angular momentum state, and the term $\ell(\ell + 1)/r^2$ in the Schrödinger equation then acts in combination with the atomic potential to produce a centrifugal barrier confining the electron to an annular region around the nucleus, i.e., the spin-orbit coupling is large.[2] The presence of both large spin-orbit and crystal-field interactions leads to complications that are rarely found in either the *3d* transition series, in which the crystal field dominates, or the lanthanide (*4f*) series, in which the spin-orbit term dominates.

What are those properties of the system that suggest the presence of localized *5f* electrons? (1) A large spin-orbit coupling, i.e., the tendency for the spin and angular momenta to combine together in such a way as to produce a good quantum number J. Of course, the presence of a large crystal-field interaction may break down Russell-Saunders coupling, necessitating the use of intermediate-coupling g factors and even J mixing.[2] (2) The high-temperature susceptibility should reflect the localized moment behavior giving an effective moment $\mu_{eff} = g\sqrt{J(J + 1)}$. (3) If conduction electrons are present to mediate the signal between nonmagnetic and actinide ions the nuclear spin-lattice relaxation rate $1/T_1$ at the anion will sense the induced moment (i.e., the susceptibility) at the actinide site. (4) As a consequence of the large orbital moment (excluding S-state ions) (a) an appreciable spin-lattice interaction should exist. In turn, this leads to magnetic anisotropy and the probable crystallographic distortion of the chemical unit cell in order to minimize the magnetoelastic energy. (b) A large orbital contribution to the hyperfine field will be present at the actinide nucleus. (5) The crystal-field interaction will lift the degeneracy of the J multiplets and result in wave functions that reflect the symmetry of the actinide ion. Such a representation implies the presence of well defined spin-wave excitations, which reflect both the crystal-field and exchange interactions in the material.

By attempting to characterize the *5f* electrons as localized we are, of course, assuming that the *5f* bands have a narrow energy distribution in momentum space and that their energies lie well below the Fermi energy $E_F$. Extensive band-structure calculations have been performed on the actinide elements,[1] all showing that the *5f* states are neither narrow nor well below $E_F$. Davis has presented a preliminary calculation for the NaCl compounds.[3] The presence of f electrons near $E_F$ contributes to the peak in the density of states. Such a peak is compatible with the high values for the electronic specific heat, which, in mJ/mole $K^2$, are 23.3, 9.6, and 49.0 for US, UP, and UN, respectively. Although hybridization makes it difficult to define the *5f* bandwidth, the average value suggested by the band calculations is 2 - 3 eV, as opposed to $\sim$ 0.3 eV for the localized *4f* electrons. uv photoemission spectra[4] on US have been interpreted as supporting the itinerant picture of the *5f* electrons in this compound, although Veal has argued for a more cautious approach in connecting theory and experiment.[5] We will return to US later; the important point of introducing the band-structure ideas is that we do not expect any one model to correctly describe <u>all</u> the properties of these actinide compounds.

In the elemental metals, at least for the first half of the series, the spatial extent and corresponding wave function overlap causes a situation which requires a description in terms of itinerant *5f* electrons. The concept of spin fluctuations (i.e., incipient magnetic behavior) appears to explain the resistivity, specific heat, and susceptibility.[6] Attempts to provide more direct evidence of such phenomena with microscopic techniques, such as nuclear-gamma-ray resonance or neutron scattering are difficult, but worthwhile, endeavors for future research.

In this article we will concentrate almost exclusively on the actinide compounds that form with elements of Group VA and VIA of the periodic table and crystallize in the NaCl structure. In the first part we will discuss magnetization, nuclear-gamma-ray resonance (Mössbauer), neutron and low-temperature x-ray experiments. In the second part we will present results from recent "second generation" neutron experiments on single crystals of uranium compounds.

## ACTINIDE COMPOUNDS WITH THE NaCl CRYSTAL STRUCTURE

Some properties of actinide compounds with the NaCl structure[7-9] are given in Table I. All these compounds are metallic with resistivities ranging between 100 and 2000 μΩ cm. The presence of conduction electrons and their possible interaction with the *5f* electrons must therefore be kept in mind. The object in studying the trends in such a table is to see if unique *5f* electron configurations can be assigned, as is the case for example, in similar lanthanide (*4f*) compounds. If the actinide ions are tripositive, the configurations will be $U^{3+} - f^3$, $Np^{3+} - f^4$, $Pu^{3+} - f^5$, and $Am^{3+} - f^6$. The actinide contraction will favor these configurations as one proceeds to the heavier ions and this is confirmed by the large effective moments of Cm and Bk ($\sim 8\mu_B$), which suggest $f^7$ and $f^8$ configurations, respectively.[6] The spatial extent of the *5f* wave functions means that those compounds with the smallest lattice spacings may have the most complicated behavior. Indeed Hill[10] proposed a direct correlation between the An - An separation and the occurrence of magnetic order, with the critical separation being about 3.4 Å, i.e., $a_o \sim 5.1$ Å in the NaCl structure.

Table I. Magnetic properties of NaCl actinide compounds. For a rhombohedral distortion, easy axis <111>, c and a are defined here as distances measured parallel and perpendicular to the trigonal axis such that c/a = 1 in the cubic phase.

| | $a_o$ (Å) | Magnetism | Temp. (°K) | $\mu_{eff}$ ($\mu_B$) | $\mu_{sat}$ ($\mu_B$) | Easy Axis | 1-c/a ($\times10^4$) | $f^n$ |
|---|---|---|---|---|---|---|---|---|
| UC | 4.960 | TIP | --- | --- | --- | --- | --- | 2 |
| UN | 4.890 | AF I | 53 | 3.1 | 0.75 | <100> | -6 | 2? |
| UP | 5.589 | AF I | 125 | 3.2 | 1.9 | <100> | <5 | 3 |
| UAs | 5.779 | AF I<br>AF IA | 127<br>63 | 3.4 | 1.9<br>2.2 | <100> | <5 | 3 |
| USb | 6.197 | AF I | 241 | ∿3.8 | 2.8 | <100> | ? | 3 |
| US | 5.489 | F | 178 | 2.25 | 1.70 | <111> | +105 | 2? |
| USe | 5.744 | F | 160 | ∿2.4 | 2.0 | <111> | +81 | 3? |
| NpC | 5.000 | AF I<br>F | 310<br>220 | 3.4 | 2.1 | <100><br><111> | <5<br>+23 | 4? |
| NpN | 4.897 | F | 87 | 2.4 | 1.4 | <111> | -52 | 4? |
| NpP | 5.615 | AF 3+3- | 130 | 2.8 | 1.8<br>2.3 | <100> | -42 | 4 |
| NpAs | 5.838 | AF 4+4-<br>AF I | 175<br>142 | ∿2.6 | 2.5 | <100><br><100> | -8<br>≤3 | 4 |
| NpSb | 6.254 | AF I | 207 | ∿2.3 | 2.5 | <100> | <15 | 4 |
| NpS | 5.527 | AF II | 20 | 2.2 | 0.9 | <100> | <5 | 4? |
| PuC | 4.977 | AF I | ∿100 | ∿1 | 0.8 | <100> | ? | 5 |
| PuN | 4.905 | AF ? | 13 | 1.1 | ? | ? | ? | 5 |
| PuP | 5.659 | F | 126 | 1.1 | ∿0.5 | <100> | ? | 5 |
| PuSb | 6.240 | F | 85 | 1.0 | 0.6 | ? | ? | 5 |
| PuS | 5.537 | TIP | --- | --- | --- | --- | --- | 6 |
| AmSb | 6.239 | TIP | --- | --- | --- | --- | --- | 6 |

TIP--temperature independent paramagnetism.
AF --antiferromagnetism, I - type I + -; IA - type IA 2+ 2-; II - type II; F - ferromagnetism.

Starting with the heavier ions in Table I, AmSb exhibits temperature independent susceptibility. This behavior is consistent with, but does not necessarily prove, the existence of a $5f^6:{}^7F_0$ ground state. The situation for PuS is similar. All the remaining Pu compounds exhibit low effective magnetic moments (∿ 1 $\mu_B$) and ordered moments of between 0.4 and 0.8 $\mu_B$. These values are consistent with a $5f^5:{}^6H_{5/2}$ configuration. As we discuss in more detail below, the magnetic form factor of this configuration is sufficiently unique that measurements of the neutron magnetic cross section, even from polycrystalline samples, unambiguously defines the ground state. However, for $f^5$ states (both Sm and Pu) the excited J multiplet is fairly close in energy to the ground state. This means that the effects of intermediate coupling and J mixing may be appreciable.

Assignments for the neptunium and uranium compounds are not nearly as straightforward. For example, the effective magnetic moment $\mu_{eff}$ of $f^2$, $f^3$, and $f^4$ free-ion states are 3.58, 3.62 and 2.68 $\mu_B$, respectively. The ordered moments $\mu_{sat}$ are 3.20, 3.27, and 2.40 $\mu_B$, respectively. With the added complications of crystal-field and exchange effects the identification of the configuration from either $\mu_{eff}$ or $\mu_{sat}$ may be ambiguous. The Np compounds have ordered moments ranging between 1.4 and 2.5 $\mu_B$, and the values of $\mu_{eff}$ are also consistent with a $5f^4$ configuration. For Np compounds additional information is available from Mössbauer-effects measurements. The hyperfine field $H_{hf}$ consists of a core contribution, taken in the first approximation to be proportional to $\langle S_z\rangle = \chi(g_J - 1)\langle J_z\rangle$, and an orbital contribution proportional to $\langle J||N||J\rangle\langle r^{-3}\rangle\langle J_z\rangle$, where $\langle J||N||J\rangle$ is a reduced matrix element dependent on the ionic configuration, and $\langle r^{-3}\rangle$ is the average value of $1/r^3$ for the open-shell electrons. Thus, $H_{hf} \propto \langle J_z\rangle = \mu_{sat}/g_J$. In Fig. 1 we show the correlation[11] between the hyperfine field and the ordered moment as determined by neutron diffraction. Of particular interest is that the linear relationship, $H_{hf} = (1915 \pm 50)\ \mu_{sat}$ kOe/$\mu_B$, extrapolates through the $Np^{3+}$ value (corresponding to a $5f^4$ configuration) calculated with intermediate-coupling wave functions and relativistic values for $\langle r^{-3}\rangle$. Another feature of the correlation is that it extends to values of $\mu_{sat}$ less than 1 $\mu_B$, as found in the Laves phase compounds.[12] Such a result indicates that the value of $\langle r^{-3}\rangle$ for the $5f$ electrons remains constant in these neptunium compounds.

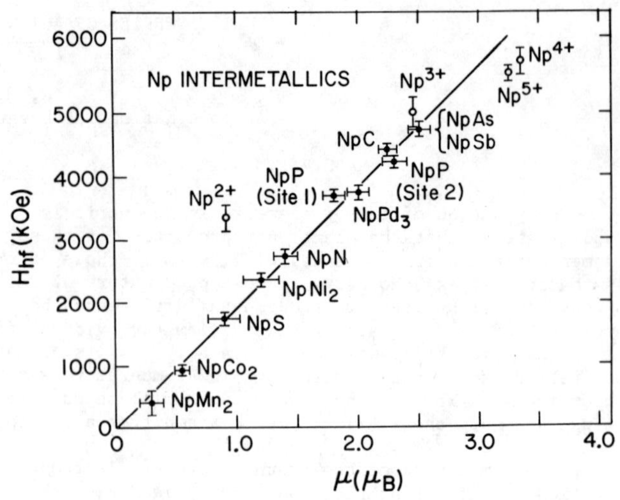


Fig. 1. Linear relationship between the hyperfine field $H_{hf}$ and the magnetic moment $\mu_{sat}$ in neptunium intermetallics. Experimental points are shown by closed circles and calculated free-ion values by open circles.

The uranium compounds in Table I have been examined with a variety of experimental techniques and have been discussed with both a localized f electron formalism,[2] and an itinerant band-structure approach.[3] As stated above, we expect to need both models to understand all the magnetic properties. For the compounds UP, UAs, and USb, which all have reasonably large U - U spacings, the localized $5f^3$ configuration appears appropriate. The situation is more complicated for UC, UN, and US. With a $f^3$ configuration and the octahedral coordination of the NaCl structure the crystal-field interaction leads to a quartet $\Gamma_8$, or a magnetic doublet $\Gamma_6$ as the ground state. Assuming the predominance of the fourth-order term $V_4 = A_4\langle r^4\rangle$, in comparison to the sixth-order term $V_6$, the ordered moments $\mu_{sat}$ should lie between 1.33 and 2.55 $\mu_B$. The intermediate coupling g factor is 0.7595 as compared to the ${}^4I_{9/2}$ value of 8/11 = 0.7273, and we do not expect J mixing to perturb $\mu_{sat}$ by more than ± 10%. However, UC is nonmagnetic and UN, although it has a large $\mu_{eff}$ of 3.1 $\mu_B$, orders with a moment of only 0.75 $\mu_B$. Alternatively, both materials may be considered $5f^2$ states, in which the ground state may be a nonmagnetic singlet. The magnetic ordering in such a system is then induced by an exchange field large enough to mix the ground and excited (magnetic) states. This model has achieved quantitative success in Pr, Tb, and Tm compounds.[13] The crystal-field splitting $E(\Gamma_4) - E(\Gamma_1)$ necessary to produce the temperature-independent susceptibility in UC is ∿ 1500 K (130 meV). This is too large an energy separation to be observed directly with present neutron spectrometers, and the determination of the ground state by fitting the magnetic susceptibility should be treated with caution (see below). In the case of UN attempts to measure the spin-wave spectra are currently in progress.

## SPIN LATTICE INTERACTIONS

In the discussions above we have not shown explicitly how the determination of the easy axis of magnetization and the occurrence of lattice deformations relates to the localization of 5*f* electrons, except in the sense that these effects are a consequence of a strong spin-lattice interaction mediated by the orbital moment. With the exception of Ce compounds, both the easy axis and lattice distortions are understood in the lanthanide series in terms of the crystal-field ground states. Such descriptions present problems for actinide compounds. For example, in Table I we note that in both U and Np compounds the easy axis are <100> for antiferromagnets and <111> for ferromagnets. The correlations thus appear

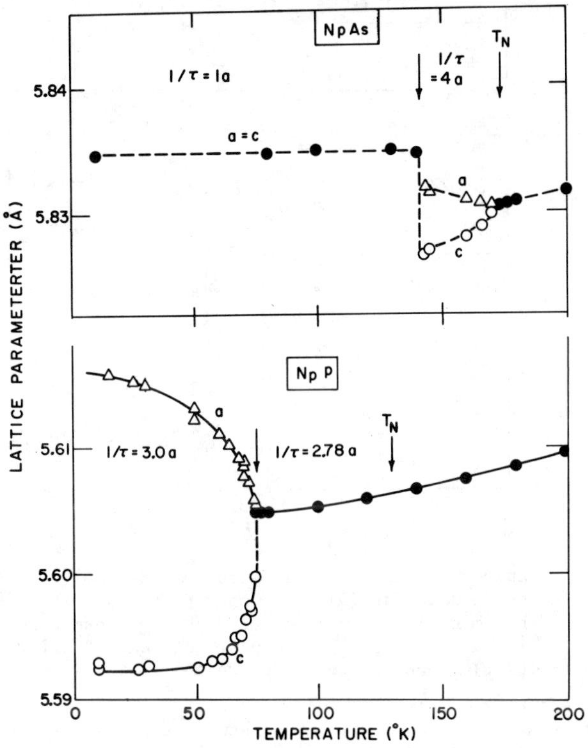


Fig. 2. Lattice parameters of NpP and NpAs as a function of temperature. The quantity $1/\tau$ gives the repeat distance of the magnetic structure.

with the magnetic structures rather than with the ground-state configuration. The temperature dependence of the lattice parameters of NpP and NpAs (Fig. 2) present evidence for very dramatic and unusual spin-lattice interactions.[9] For the commensurate long-range magnetic structures ($T \leq 74$ K in NpP with a 3 +, 3 - layered structure, and $142 < T < 175$ K in NpAs with a 4 +, 4 - structure) tetragonal distortions are observed. However, for magnetic ordering incommensurate[8] with the lattice ($74 < T < 130$ K in NpP), or the simple type I +, - structure in NpAs, the lattice is apparently cubic. The effect in NpAs, in which the symmetry is raised (from tetragonal to cubic) in the ordered state, is so unusual that one is tempted to suggest a valence change occurs at 142 K. The electrical resistivity changes by an order of magnitude at this transition.[8] Notice, however, that distortions in all type I antiferromagnets are absent, or at least very small,[9,14] $(c-a)/a \leq 10^{-3}$. At present no explanation for these spin-lattice interactions has been advanced. Recently, neutron experiments at Argonne on antiferromagnetic $UO_2$ have shown that an internal shear deformation of the oxygen sublattice occurs below $T_N$, whereas the uranium atoms do not move.[15] The magnetoelastic energy is minimized by this internal strain, which dominates the magnetic behavior of $UO_2$ through strong spin-lattice coupling, and the <u>dimensions of the overall unit cell remain unchanged</u>.[16] Although $UO_2$ is an insulator and has the fluorite structure, the general formulation of this new effect may also apply to the NaCl compounds. Indeed, similar effects have been proposed for certain lanthanide compounds.[17]

## ELECTRON WAVE FUNCTIONS DETERMINED BY NEUTRON SCATTERING

The scattering of thermal neutrons yields, in principle, information about the radial and angular distributions of the unpaired electrons in a solid. In practice, achieving this goal is difficult. The experiments require accurate measurements from single crystals, and comparison with theory requires the use of the tensor-operator method[18] together with relativistic values for the one-electron radial integrals.[19]

The magnetic scattering length is defined as a vector $\vec{E}$, with spherical components $E_Q$ given by

$$(2\pi\hbar/m)\, E_Q = \langle \psi_e | T_Q^K (e,\vec{\kappa}) | \psi_e \rangle \tag{1}$$

The electron wave functions are represented by $\psi_e$ and $T_Q^K(e,\vec{\kappa})$ defines a tensor operator. The neutron-electron interaction is expressed as a tensor of rank one, so that we need to evaluate three terms, $Q = 0, \pm 1$. The presence of an unquenched orbital moment means that the magnetization density is a vector quantity[20,21] with three components, $M_x$, $M_y$, and $M_z$. We may associate the components $E_Q$ as follows, $E_0 \to M_z$, $E_{+1} \to -(M_x + i M_y)/\sqrt{2}$, and $E_{-1} \to (M_x - i M_y)/\sqrt{2}$. The magnetic moment density is therefore obtained by Fourier transforming $M_x$, $M_y$, and $M_z$. In certain cases $\vec{M}$, the total magnetization, may be perpendicular to the scattering vector $\vec{\kappa}$ (this is common practice in experiments with polarized neutrons), and then $E_{+1} = E_{-1} = 0$, and the moment density is a scalar quantity as is found in most transition metal compounds. We may write $E_Q = -2\, p\, q^2$, where $p = (0.27 \times 10^{-12})\, \mu f(\vec{\kappa})$ cm is the conventional magnetic scattering length, $\mu$ is the magnetic moment in Bohr magnetons, $q^2$ is the square of the magnetic interaction vector, and $f(\vec{\kappa})$ is the form factor, which is related to the magnetization density through the Fourier transform

$$\mu f(\vec{\kappa}) = \int M(\vec{r}) e^{i\vec{\kappa}\cdot\vec{r}}\, dr. \tag{2}$$

In the tensor-operator formalism

$$f(\vec{\kappa}) = \langle j_0 \rangle + c_2 \langle j_2 \rangle + c_4 \langle j_4 \rangle + c_6 \langle j_6 \rangle \tag{3}$$

where

$$\langle j_i \rangle = \int_0^\infty U^2(r)\, j_i(\kappa r) dr \tag{4}$$

are the radial integrals of the one-electron wave functions and $j_i(\kappa r)$ are spherical Bessel functions. The coefficients $c_i$ are defined by the electronic configuration of the magnetic ion and the experimental conditions. From these equations, assuming we know the magnetic moment $\mu$ (obtained by extrapolating the cross section to $\kappa = 0$), and the geometric term $q^2$, an <u>effective</u> magnetic form factor $f'(\vec{\kappa})$ can always be deduced from both the observed and calculated cross sections. If $q^2 = 1$ then $E_{+1} = E_{-1} = 0$ and $f' = f$ as in Eq. (2); but, in general, the situation is more complex.

Accurate form-factor measurements have been reported for $UO_2$ (both in the ordered and paramagnetic states) and US. For $UO_2$ in the ordered state the ambiguity in the magnetic moment direction requires that the coefficients $c_i$ in Eq. (3) be averaged over the (001) plane before comparing with experiment; thus losing a great deal of information about the anisotropy of the ground state. In the paramagnetic state[22] the induced moment of $0.0374\, \mu_B$ is too small to be able to compare $f(\vec{\kappa})$ values calculated with different models. However, this experiment does show that the radial integrals $\langle j_i \rangle$ of Eq. (4) derived from relativistic Dirac-Fock calculations[19] are a good representation of the spatial extent of the 5*f* electrons in $UO_2$.

Polarized neutrons have been used to measure the magnetic form factor[23] in ferromagnetic US. The neutron

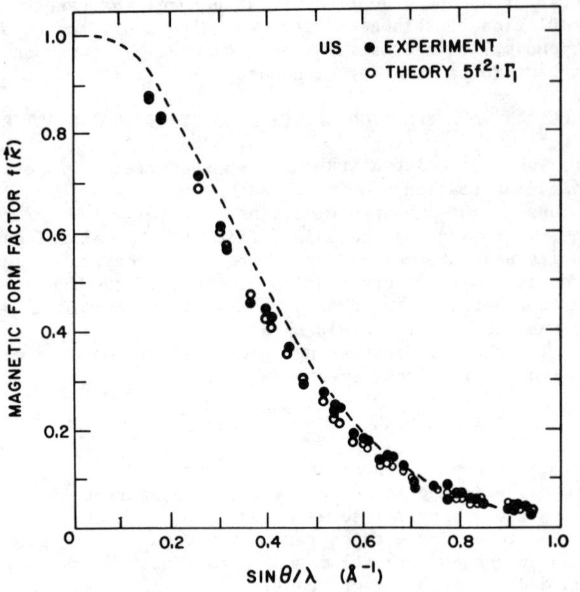


Fig. 3. Magnetic form factor for US. The solid points are from Wedgwood (Ref. 23) using a magnetic moment of 1.70 $\mu_B$. The open points are obtained with a $5f^2$:$^3H_4$:$\Gamma_1$ plus exchange model and the relativistic $\langle j_i \rangle$ functions of Ref. 19. The broken line is a smooth curve drawn through the theoretical form factor of the $5f^4$:$^5I_4$ configuration.

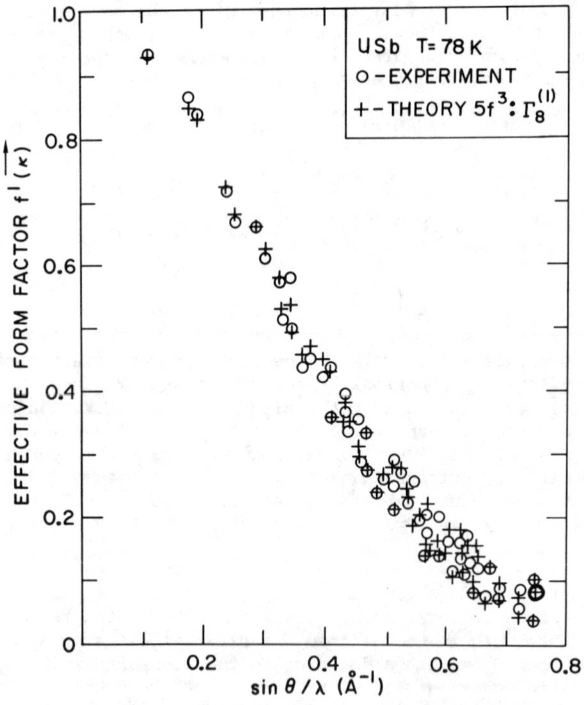


Fig. 4. Effective magnetic form factor of USb. The relativistic integrals of Ref. 19 have been used together with the $5f^3$:$\Gamma_8^{(1)}$ configuration.

data suggests a magnetic moment of 1.70 $\mu_B$, whereas magnetization experiments[23] give only 1.55 $\mu_B$. Similar discrepancies occur in $3d$ transition metals and compounds and are ascribed to conduction-electron polarization. In US, it probably arises from hybridization of the $5f$, 6d, and 7s bands and, because of spatial delocalization of these wave functions, contributes to $f(\vec{\kappa})$ only for $\sin\theta/\lambda < 0.1$ Å$^{-1}$. The form factor of the localized part is shown in Fig. 3. The data do not depart much from a smooth curve, and we would expect considerable more anisotropy for $5f^3$ and $5f^4$ states. We have reanalyzed this data with the relativistic $\langle j_i \rangle$ integrals and, in agreement with Wedgwood's original analysis, obtain the best fit (open points in Fig. 3) for the $5f^2$:$\Gamma_1$ singlet configuration.

Considerably more anisotropy is present in the magnetic form factor of USb, see Fig. 4. The overall fit with a $5f^3$:$\Gamma_8^{(1)}$ ground state is excellent, again showing

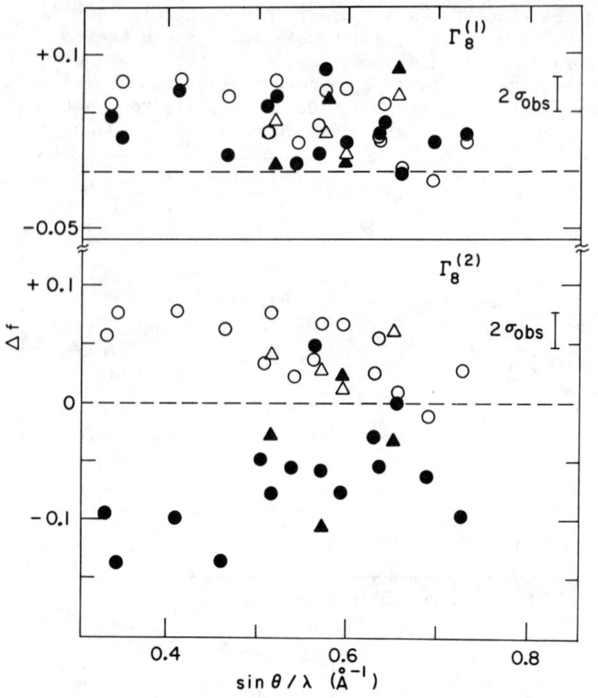


Fig. 5. Anisotropy in the form factor of USb. $\Delta f$ is defined in Eq. (5). The open points are the experimental values, the solid points those calculated with the $\Gamma_8^{(1)}$ (upper plot) and $\Gamma^{(2)}$ (lower plot) ground state eigenfunctions.

the accuracy of the $\langle j_i \rangle$ integrals. The anisotropy in Fig. 4 allows us to distinguish between a number of possible ground states. We define a term

$$\Delta f = f(\vec{\kappa}_1) - f(\vec{\kappa}_2) \tag{5}$$

where $|\vec{\kappa}_1| = |\vec{\kappa}_2|$. From Eq. (3) we see that this eliminates the (spherically symmetric) term in $\langle j_0 \rangle$ and focusses on higher-order integrals that reflect the magnetic quadrupole, octapole, and higher moments. In Fig. 5 $\Delta f$ for a number of pairs of reflections is plotted versus $\sin\theta/\lambda$. Assuming the LLW parameter x = 0.8, (x reflects the ratio between $V_4$ and $V_6$, and the analyses are independent of x for $0.7 < x \leq 1.0$. The sign of x depends on the coordination only) the ground-state wave functions are

$$W<0 \quad \Gamma_8^{(2)} \quad \psi_e = 0.79|9/2\rangle - 0.59|1/2\rangle - 0.14|-7/2\rangle,$$

$$\mu_{sat} = 2.12\ \mu_B$$

$$W>0 \quad \Gamma_8^{(1)} \quad \psi_e = 0.97|7/2\rangle - 0.25|-1/2\rangle - 0.01|-9/2\rangle,$$

$$\mu_{sat} = 2.36\ \mu_B\ .$$

Now from Fig. 5 the $\Gamma_8^{(2)}$ state is clearly an incorrect assignment. Note that this has the $|9/2\rangle$ state as the major term. For $0.3 < \sin\theta/\lambda < 0.5$Å$^{-1}$ the anisotropy comes primarily from the magnetic quadrupole moment, which is sensed through the $\langle j_2 \rangle$ function. The shape of the magnetization density observed in USb is incompatible with both the $\Gamma_8^{(2)}$ or free-ion $|9/2\rangle$ states. Point-charge calculations give x > 0 and W < 0 with a $\Gamma_8^{(2)}$ ground state. This is found in all Nd compounds[24] ($4f^3$).

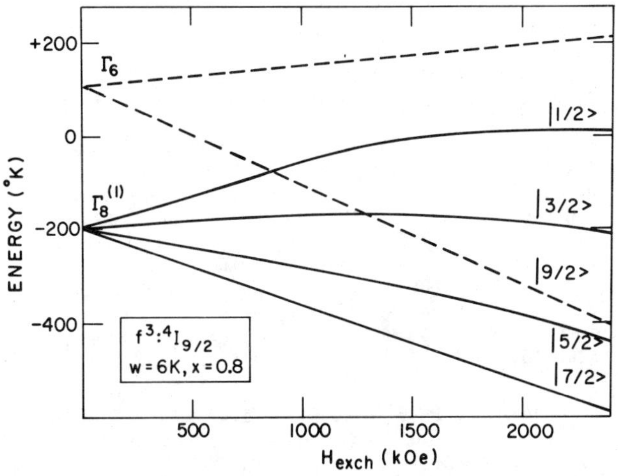


Fig. 6. Variation of the crystal-field levels with an internal exchange field. The $\Gamma_6 - \Gamma_8^{(1)}$ splitting at H = 0 represents that proposed for USb. As $H_{exch} \to \infty$ the eigenfunctions have the character indicated on the right-hand-side of the diagram. (The higher $\Gamma_8^{(2)}$ crystal-field level is not shown.)

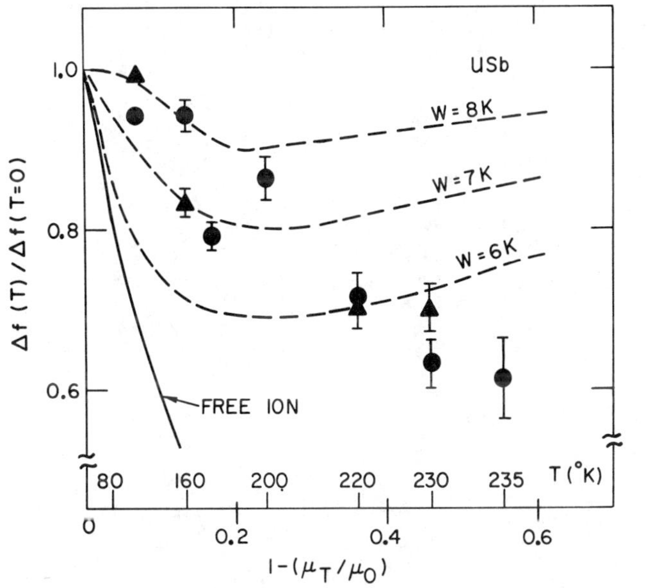


Fig. 7. Variation of the form factor anisotropy as a function of the dipole moment (temperature is an implicit parameter). The circles and triangles represent different pairs of Bragg reflections. The solid line is the free-ion result, the broken curves calculations using the crystal-field levels of Fig. 6 and a simple molecular-field model.

Our assignment of the ground state is also opposite to that suggested by Troc and Lam[25] from an analysis of the high-temperature susceptibility of UP and UAs. However, the present neutron experiments represent the first direct identification of a crystal-field eigenstate in an actinide intermetallic compound.

We have also measured the temperature dependence of Δf. This quantity is very sensitive to the occupation of states with different symmetry from the ground state. In Fig. 6 we show the suggested crystal-field level scheme for USb in the presence of an internal exchange field (the molecular field model). In the absence of $H_{exch}$ the $\Gamma_8^{(1)} - \Gamma_6$ energy separation is ∿ 300 K. As $H_{exch}$ is increased the degeneracy of the $\Gamma_8$ state is raised, and the gap between the lowest level of the excited $\Gamma_6$ state is narrowed. If the $\Gamma_8$ state is separated by more than about 500 K from $\Gamma_6$ the anisotropy will be temperature independent. On the other hand, in the free-ion picture (which is approached on the extreme right of Fig. 6) the temperature dependence of the anisotropy is very abrupt, being approximately proportional to the fifth power of the dipole moment at low temperatures. The measured quantity Δf(T) is related to the effective quadrupole moment $<O_2^0>$ by normalizing, thus $<O_2^0> \propto \Delta f(T)/\Delta f(T = 0)$ and this is plotted as a function of the reduced dipole moment in Fig. 7. Note that the experimental $<O_2^0>$ shows some temperature dependence, placing an upper limit of ∿ 400 K on the $\Gamma_8^{(1)} - \Gamma_6$ energy separation. Using a molecular field model as illustrated in Fig. 6 we have calculated Δf(T), and the results for W = 6, 7, and 8 K are shown in Fig. 7. Near $T_N$ $H_{exch}$ rapidly diminishes, leading to an increase in the $\Gamma_8^{(1)} - \Gamma_6$ separation, with a corresponding increase in Δf. The data do not reflect this upturn near $T_N$, but do suggest that W ∿ 7 K is a reasonable value. The present elastic scattering measurements show the need for single-crystal inelastic neutron experiments because the exchange interactions will undoubtedly produce considerable dispersion in the crystal-field levels. This may well account for the failure to detect crystal-field levels using polycrystalline samples and time-of-flight techniques.[26] However, the concept of a well-defined crystal-field state does appear to be valid for antiferromagnetic USb.

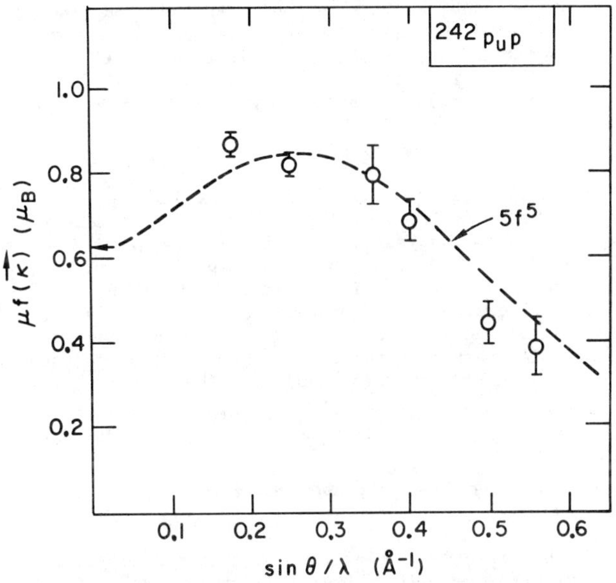


Fig. 8. Results of polarized neutron experiment on $^{242}$PuP.

As suggested by Table I the Pu ions usually belong to the $f^5$ configuration. The magnetic cross section for this state is most unusual, having a maximum at $\sin\theta/\lambda$ ∿ 0.3 $Å^{-1}$ rather than at zero. The reason is that L = 5, S = 5/2, and these oppose each other, J = L - S = 5/2. The localized magnetization density from the orbital moment is opposed by a large, but diffuse, negative spin density. The total magnetization therefore changes sign as one proceeds away from the nucleus. The Fourier transform of the magnetization density then has a maximum at $\kappa \neq 0$. Since this form factor is unique to the $f^5$ configuration, neutron scattering provides a simple method of establishing the ground state. Unfortunately, single crystals of Pu compounds have not yet been produced, but in Fig. 8 we show results obtained with polarized neutrons from a polycrystalline sample of ferromagnetic $^{242}$PuP. Preferred orientation effects, which establish the easy axis as <100>, limit the number of reflections that can be measured. The smooth curve is a best fit to the data using relativistic $<j_i>$ integrals and the $5f^5{:}{}^6H_{5/2}{:}\Gamma_8$ ground state. This form factor extrapolates to give $\mu_{sat}$ = 0.62 $\mu_B$, the magnetization value[27] is 0.42 $\mu_B$. In these configurations we will certainly have to take into account the effects of J mixing since estimates[2] show that the $^6H_{5/2}$ state may not make up more than ∿ 66% of the ground state. Calculations of the magnetic cross section for these complicated situations are in progress.

## SUMMARY

We have concentrated almost exclusively here on the properties of the NaCl actinide compounds. As Table I illustrates, we are able to interpret most of the magnetic properties in terms of localized *5f* electrons behavior. However, direct proof of the crystal-field level-picture remains hard to obtain. Even in the so-called simple systems the microscopic exchange and spin-lattice interactions remain totally unresolved. For example, the sudden increase in the ordered magnetic moment in UP at low temperature has been analyzed in terms of both a crystal-field effect[28] and a change in the valence state.[29] We have discussed the spin-lattice interaction with reference to recent observations[15] on the insulating actinide antiferromagnet $UO_2$, but the application of these ideas to systems that exhibit effects such as in Fig. 2 will require a substantial theoretical effort. In some cases, e.g., UN, US, and possibly UC and UP, the *5f* band is probably broad enough that a simple localized model will not predict the correct magnetic behavior. X-ray photoemission and de Haas van Alphen experiments on these systems should prove most rewarding.

As discussed in the final section, the measurement of the elastic magnetic cross section is providing quantitative information on the radial and angular wave functions of the unpaired *5f* electrons. Further experiments of this nature, together with inelastic neutron experiments to measure the elementary excitations, will be most valuable. To accomplish these goals, a much greater effort should be made to grow single crystals.

## ACKNOWLEDGMENTS

I am indebted to many colleagues at Argonne National Laboratory for stimulating discussions and collaboration over the years, in particular, A. T. Aldred, M. B. Brodsky, T. O. Brun, B. D. Dunlap, J. Faber, Jr., A. J. Freeman, F. Y. Fradin, D. J. Lam, and M. H. Mueller. The technical assistance of R. L. Hitterman, H. W. Knott, and J. F. Reddy in solving a variety of problems is greatly appreciated. I am grateful to Oscar Vogt of the ETH, Zurich for supplying the USb crystal.

## REFERENCES

*Work performed under the auspices of USERDA.

1. "The Actinides: Electronic Structure and Related Properties," edited by A. J. Freeman and J. B. Darby (Academic Press, New York, 1974), A. J. Freeman and D. D. Koelling, Vol. I, ch. 2.
2. S. K. Chan and D. J. Lam, Ref. 1, Vol. I, ch. 1.
3. H. L. Davis, Ref. 1, Vol. II, ch. 1.
4. D. E. Eastman and M. Kuznietz, Phys. Rev. Letters, 26, 846 (1971), J. Appl. Physics, 42, 1396 (1971).
5. B. Veal, Ref. 1, Vol. II, ch. 3, pp. 101-107.
6. W. J. Nellis and M. B. Brodsky, Ref. 1, Vol. II, ch. 6. M. B. Brodsky, AIP Conf. Proc. 5, 611 (1972).
7. A detailed survey up to 1972 is given by D. J. Lam and A. T. Aldred, Ref. 1, Vol. I, ch. 3. An excellent literature survey of uranium compounds is given by J. Grunzweig-Genossar, M. Kuznietz, and F. Friedman, Phys. Rev. 173, 562 (1968).
8. For the Np compounds see A. T. Aldred, B. D. Dunlap, A. R. Harvey, D. J. Lam, G. H. Lander, and M. H. Mueller, Phys. Rev. B 9, 3766 (1974).
9. For lattice distortions see G. H. Lander and M. H. Mueller, Phys. Rev. B 10, 1994 (1974).
10. H. Hill in "Plutonium 1970 and Other Actinides," edited by W. N. Miner (AIME, New York, 1971) p. 2.
11. B. D. Dunlap and G. H. Lander, Phys. Rev. Letters, 33, 1046 (1974).
12. A. T. Aldred, B. D. Dunlap, D. J. Lam, G. H. Lander, M. H. Mueller and I. Nowik, Phys. Rev. B 11, 530 (1975).
13. See, for example, B. R. Cooper and O. Vogt, Phys. Rev. B 1, 1218 (1970).
14. J. A. Marples, C. F. Sampson, F. A. Wedgwood, and M. Kuznietz, J. Phys. C 8, 708 (1975).
15. J. Faber, Jr., G. H. Lander, and B. R. Cooper, see paper in this Conference.
16. S. J. Allen, Phys. Rev. 166, 530 (1968); 167, 492 (1968); R. A. Cowley and G. Dolling, Phys. Rev. 167, 464 (1968).
17. K. W. H. Stevens and E. Pytte, Solid State Comm. 13, 101 (1973).
18. W. Marshall and S. W. Lovesey, "Theory of Thermal Neutron Scattering," (Oxford U.P., London, 1971); G. H. Lander, T. O. Brun and O. Vogt, Phys. Rev. B 7, 1988 (1973) and references therein.
19. A. J. Freeman, J. P. Desclaux, G. H. Lander, and J. Faber, Jr., Phys. Rev. B (in press).
20. O. Steinsvoll, G. Shirane, R. Nathans, M. Blume, H. A. Alperin, and S. J. Pickart, Phys. Rev. 161, 499 (1967)
21. E. Balcar, J. Phys. C 8, 1581 (1975).
22. G. H. Lander, J. Faber, Jr., A. J. Freeman and J. P. Desclaux, Phys. Rev. B (in press).
23. F. A. Wedgwood, J. Phys. C 5, 2427 (1972).
24. A. Furrer, J. Kjems, and O. Vogt, J. Phys. C 5, 2246 (1972); J. Phys. C 7, 3365 (1974); J. Phys. C 8, 1054 (1975).
25. R. Troc and D. J. Lam, Phys. Stat. Solidi B 65, 317 (1974).
26. F. A. Wedgwood, J. Phys. C 7, 3203 (1974).
27. D. J. Lam, F. Y. Fradin, and O. L. Kruger, Phys. Rev. 187, 606 (1969).
28. C. Long and Y. L. Wang, Phys. Rev. B 3, 1656 (1971).
29. J. M. Robinson and P. Erdos, Phys. Rev. B 8, 4333 (1973); Phys. Rev. B 9, 2187 (1974).

# BULK PROPERTIES OF $UIr_2$ AND $UIr_3$*

M. B. Brodsky, R. J. Trainor,
A. J. Arko, and H. V. Culbert
Argonne National Laboratory, Argonne, Illinois 60439

ABSTRACT

The electrical resistivity and magnetic susceptibility of the cubic intermetallic compounds $UIr_2$ ($MgCu_2$-type) and $UIr_3$ (ordered, $AuCu_3$-type) have been measured between 2-300K. Low temperature specific heats have been measured between 1.5-4K for $UIr_2$ and 2.5-10K for $UIr_3$. The susceptibility of $UIr_3$ is temperature independent and substantially lower than that of isoelectronic $URh_3$ being 0.57 vs 0.97 x $10^{-3}$ emu/mole at room temperature. The susceptibility of $UIr_2$ is very weakly temperature dependent above 100K, and has a value of 1.18 x $10^{-3}$ emu/mole at room temperature. The electrical resistivities of both compounds follow power law dependences at low temperatures, of the form $\rho-\rho_0 = AT^n$, with n = 1.9 for $UIr_2$ and n = 3.7 for $UIr_3$. These results indicate that $UIr_2$ may be a spin fluctuation compound, while $UIr_3$ behaves as a simple transition metal compound, with even less d-f character than $URh_3$ (n = 3.0). The specific heats of both $UIr_2$ and $UIr_3$ may be fit to $C = \gamma T + \beta T^3$. A large value of $\gamma$, 62.5 mJ/(mole-$K^2$), is found for $UIr_2$, consistent with a narrow 5f band at the Fermi level. The lack of any magnetic phenomena in $UIr_3$ is explained by the hybridization of the 5f electrons into f-d bands, which mostly lie below the Fermi level.

## INTRODUCTION

In view of the large variety of magnetic phenomena found in metallic actinides, it is desirable to systemmatically study actinide systems to help unravel the physics underlying the various phenomena. This paper presents the results of a study of the electrical resistivity, magnetic susceptibility, and low temperature specific heat of $UIr_2$ and $UIr_3$. $UIr_3$ forms peritectically in the $MgCu_2$-type cubic Laves phase, and $UIr_3$ crystallizes congruently in the ordered $AuCu_3$ structure. The measurements on $UIr_3$ are of special interest since they may be compared with preliminary deHaas-vanAlphen (dHvA) results for this compound,[1] and with the fairly complete dHvA and band structure results for isoelectronic $URh_3$.[2]

## EXPERIMENTAL

Samples were prepared by arc-melting the constituents, followed by electrolytic machining where necessary. The correct single phase structures were verified by x-ray diffraction methods. The $UIr_2$ was annealed at 1040°C for ten days. Resistivity and susceptibility measurements were made between 2 and 300K, while the specific heat measurements cover the 1.5-4 and 2.5-10K intervals for $UIr_2$ and $UIr_3$, respectively. Experimental techniques used in this work have been described previously.[3,4,5]

## RESULTS

The temperature dependence of the susceptibility is given in Fig. 1 for both compounds. Although both sets of data are essentially temperature-independent, there is a slight maximum in the $UIr_2$ data at 60 ± 2K. The room temperature value for $UIr_3$ (0.570 x $10^{-3}$ emu/mole) is about 0.6 as large as the value for $URh_3$ (0.973 x $10^{-3}$ emu/mole),[6] and the $UIr_2$ value is slightly higher (1.176 x $10^{-3}$ emu/mole) than the $URh_3$ value.

The electrical resistivity data up to 300K are shown in Fig. 2. Above about 100K, $\rho$ is linear for $UIr_3$, but exhibits a tendency toward saturation for $UIr_2$. The values of $\rho_{300}-\rho_0$ are 122 μΩcm for $UIr_2$ and 34 μΩcm for $UIr_3$ (vs 54 for $URh_3$). There is a small bump in the $\rho$-T curve for $UIr_2$ with a height of only 0.5 out of 60 μΩcm, and centered at 57K, with a width of 2-3K. The

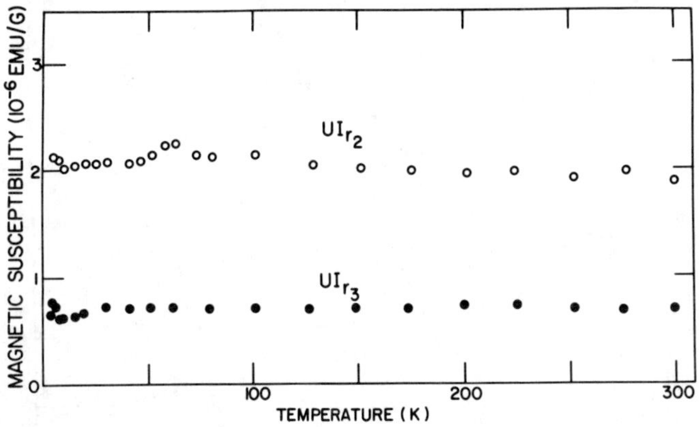


Fig. 1. Magnetic susceptibility vs temperature for $UIr_2$ and $UIr_3$.

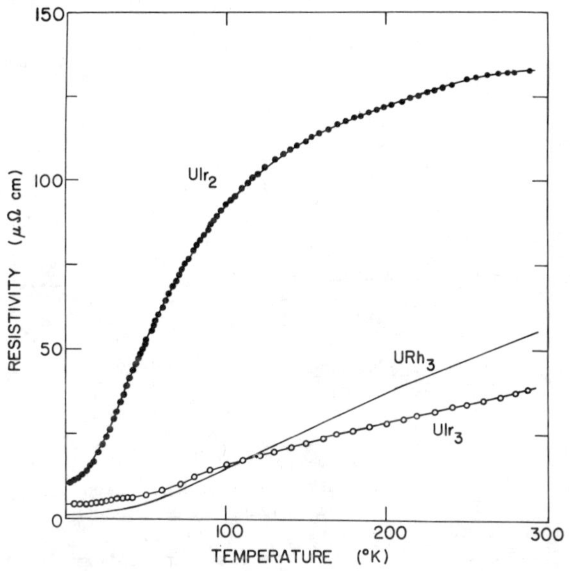


Fig. 2. Electrical resistivity of $UIr_2$ and $UIr_3$ vs temperature. Also shown is the resistivity-temperature curve for $URh_3$.

low temperature resistivities are plotted as $\log(\rho-\rho_0)$ vs log T in Fig. 3. Both sets of data may be represented by $\rho-\rho_0 = AT^n$, with n = 1.9 for $UIr_2$ and n = 3.7 for $UIr_3$. The latter value is to be compared to n = 3.0 for isoelectronic $URh_3$.[6]

The specific heats of both $UIr_2$ and $UIr_3$ are well described by $C = \gamma T + \beta T^3$ over the entire measured temperature ranges (1.5-4 and 2.5-10K, respectively). For $UIr_2$ $\gamma$ = 62.5 ± 1.0 mJ/(mole-$K^2$) and $\beta$ = 0.50 ± 0.03 mJ/(mole-$K^4$), corresponding to a Debye temperature, $\Theta_D$ = 227 ± 5K, whereas for $UIr_3$ $\gamma$ = 19.5 ± 1 mJ/(mole-$K^2$) and $\beta$ = 0.31 ± 0.03 mJ/(mole-$K^4$), corresponding to $\Theta_D$ = 293 ± 10K.

## DISCUSSION

The higher power law exponent for the $UIr_3$ resistivity data versus $URh_3$ indicates significant s-s or p-p scattering in addition to s-d scattering. The much smaller $\rho_{300}-\rho_0$ for $UIr_3$ vs $URh_3$ is in agreement with this proposal. This conclusion is supported further by the dHvA data, which show more s-like orbits than are found in $URh_3$. The tentative band structure for $UIr_3$ (based strongly on the URh$_3$ results) explains the lack of magnetism in UIr$_3$, also, since the f-electrons are all strongly hybridized into f-d bands, which mostly lie well below the Fermi level.

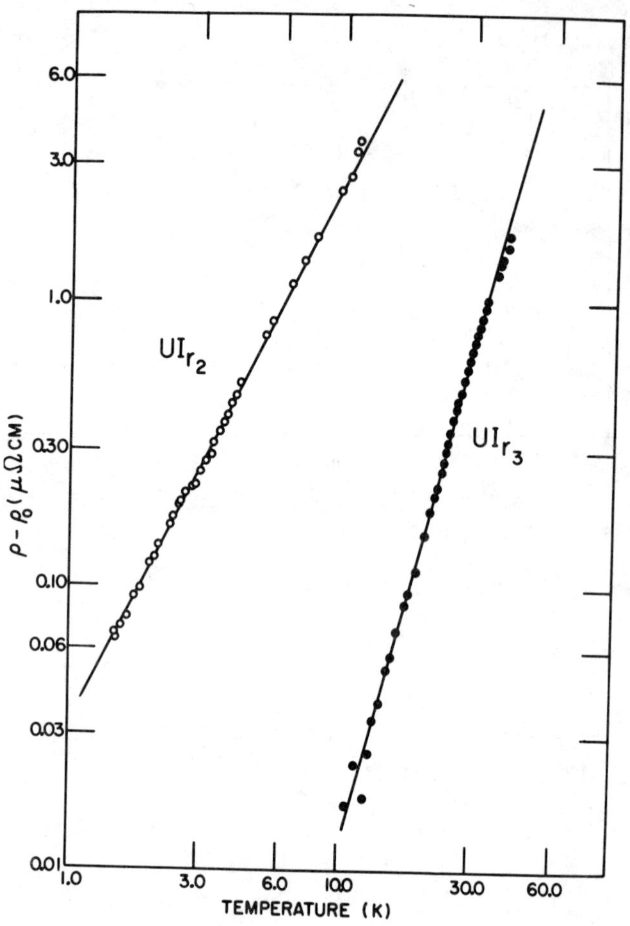


Fig. 3. Log ($\rho$-$\rho_o$) vs Log T for UIr$_2$ and UIr$_3$ at low temperatures.

The small maximum in the UIr$_2$ susceptibility accompanied by the small resistivity anomaly, is probably not associated with a magnetic transition. Most likely it is due to a slight cubic to tetragonal distortion as is found in other actinide cubic Laves phase compounds with transition metals.[7] The lack of magnetic ordering is supported further by an almost trivial temperature dependent susceptibility between 150-300K. Additional measurements on UIr$_2$ in the temperature region near 60K are necessary to determine the cause of the susceptibility maximum. Among these are x-ray diffraction and specific heat.

However, the low-temperature $T^2$ resistivity for UIr$_2$ indicates a magnetic phenomenon. By analogy with many other actinide compounds, it is likely that UIr$_2$ is a spin fluctuation compound.[8,9] The slope of the $T^2$ regime yields a spin fluctuation temperature,[10] $T_{sf}$, of 200K, while the limit of the $T^2$ regime only yields $T_{sf} \sim 60K$. However, for a $T_{sf}$ this large an upper limit to the $T^2$ dependence becomes hard to separate from the total resistivity. We point out that since the spin fluctuation contribution to the specific heat, $T^3 \ln T/T_{sf}$, goes to $T = 0$ as $T^3$, a spin fluctuation contribution would be inseparable from the lattice contribution for $T << T_{sf}$. For UAl$_2$, a $T_{sf}$ = 23K permitted the observation of a low temperature upturn in C/T due to the spin fluctuation term.[9] The large value for $\gamma$ is in agreement with a spin fluctuation model which requires a narrow band at or near the Fermi level. A rough estimate of the exchange enhancement factor, $S \approx \chi/\gamma$, of 2.3 is obtained from the experimental data. This does not allow for the unknown electron-phonon and spin-fluctuation enhancements of the electronic specific heat (expected to be small relative to S) or the orbital contribution to the susceptibility.

One mechanism which has been widely used to explain the lack of magnetism in many actinide compounds is broadening due to 5f-5f direct overlap as a consequence of relatively short interactinide distances.[11] This mechanism is certainly operable in the cubic Laves phase structure where the U-U distance in UIr$_2$, for example, is only 3.25A. In the case of AuCu$_3$-type compounds, the larger U-U distance, e.g., 4.023A for UIr$_3$, should lead to local moment behavior if 5f-5f overlap is the only mechanism for broadening. However, in some actinide-transition metal compounds additional broadening occurs via 5f-6d hybridization which is favorable in this structure, and is stronger than 5f-5f overlap. Hence, there may be no magnetic behavior. By adding more 5f electrons as in going from URh$_3$ to PuRh$_3$, one may obtain a situation with some unhybridized, localized 5f electrons. Thus, PuRh$_3$ is a good example of 5f local moment behavior.[6,12]

REFERENCES

*Work supported by the U.S. Energy Research and Development Administration.


1. A. J. Arko, M. B. Brodsky, G. W. Crabtree, D. Karim, J. B. Ketterson, D. D. Koelling, and L. R. Windmiller, 5th International Conference on Plutonium and Other Actinides, Baden-Baden, September, 1975, (to be published by North-Holland Publ. Co., Amsterdam).
2. A. J. Arko, M. B. Brodsky, G. W. Crabtree, D. Karim, D. D. Koelling, L. R. Windmiller, and J. B. Ketterson, Phys. Rev. B2, 4102 (1975).
3. M. B. Brodsky, N. J. Griffin, and M. D. Odie, J. Appl. Phys. 40, 895 (1969).
4. J. W. Ross and D. J. Lam, Phys. Rev. 165, 617 (1968).
5. H. V. Culbert, D. E. Farrell, and B. S. Chandrasekhar, Phys. Rev. B3, 794 (1971).
6. W. J. Nellis, A. R. Harvey, and M. B. Brodsky, A.I.P. Conf. Proc. 10, 1076 (1973), and A. R. Harvey, M. B. Brodsky, and W. J. Nellis, Phys. Rev. B7, 4137 (1973).
7. A. T. Aldred, B. D. Dunlap, D. J. Lam, and I. Nowik, Phys. Rev. B10, 1011 (1974).
8. M. B. Brodsky, Phys. Rev. B9, 1381 (1974).
9. R. J. Trainor, M. B. Brodsky, and H. V. Culbert, Phys. Rev. Lett. 34, 1019 (1975).
10. Determined from $\rho \propto (T/T_{sf})^2$. See for example, R. Jullien, M. T. Béal-Monod, and B. Coqblin, Phys. Rev. B9, 1441 (1975).
11. H. H. Hill, in Pu 1970 and Other Actinides, ed., W. N. Miner, (Am. Inst. of Min., Met., Pet. Eng., New York, 1970), pp. 2-19.
12. R. J. Trainor and M. B. Brodsky, to be published.

# APPLICATION OF STANDARD-BASIS OPERATORS TO THE THEORY OF ELECTRONIC AND MAGNETIC PROPERTIES OF ACTINIDE COMPOUNDS*

J. M. Robinson
Indiana University-Purdue University at Fort Wayne,
Fort Wayne, Indiana, 46805

ABSTRACT

We derive the electronic energy bands of a cubic metallic crystal, starting from localized s (or d) and f orbitals in Hubbard's atomic representation. The standard-basis operator technique of Haley and Erdös is applied to this model and leads to electronic and magnetic properties intermediate between those of the localized and band descriptions. The paramagnetic susceptibility and s-f hybridization follow the Curie-Weiss law at high temperatures and level off with decreasing temperature. These results agree qualitatively with the behavior observed in many actinide intermetallic compounds, as does also a simple calculation of the electrical resistivity of the model.

## I. INTRODUCTION

The theory of the electronic and magnetic properties of metallic actinide compounds is complicated by the 5f electrons, which exhibit behavior[1,2] intermediate between that predicted by the localized (or crystal field) model and the band model. Most calculations of energy bands have employed operators which "create" and "destroy" single electrons in certain energy levels. These operators are not appropriate for describing the localized or atomic-like levels, because the latter levels do not in general represent the energies of single electrons. In contrast, the standard-basis[3] (or atomic)[4] operators cause a transition of an atom from one atomic state to another. The two states may or may not differ in the number of electrons on the atom. Hubbard[4] showed how a theory involving these operators can interpolate between the Boltzmann statistics of the localized model and the Fermi -Dirac statistics of the band model. Haley and Erdös[3] then developed a systematic Green's function method for the standard-basis operators in the case of Boson excitations, and their results are easily extended to the Fermion case. In what follows, we apply this method to a simple model designed to contain the basic physics of the metallic actinides.

## II. THE THEORETICAL MODEL

We consider a simple cubic crystal of N atoms, each having exterior spin-degenerate s (or d) and f orbitals with energies $t_o$ and e, respectively. The Hamiltonian H is given by the expression

$$H=\sum_p \sum_i (\varepsilon_p-\zeta n_p)L^i_{pp}+\sum_{pqrs}\sum_{ij} B^{pqrs}_{ij} L^i_{pq} L^j_{rs} \quad , \qquad (1)$$

where $\zeta$ denotes the Fermi level. The sums over i and j run over the N lattice sites. The indices p, q, r, s denote the atomic states, depicted with their energies $\varepsilon_p$ in Fig. 1, and the number of electrons in the state p is $n_p$. The standard-basis[3] operators $L^i_{pq}$ cause a transition of the atom at site i from the state q to the state p. The hopping matrix $B^{pqrs}_{ij}$ is equal to $t_{ij}$ for s-s hopping (e.g. [pqrs]=[6512] and equals $V_{ij}$ for s-f hopping (e.g. [pqrs]=[6515]). Thus, $t_{ij}$ leads to a tight-binding s band, and $V_{ij}$ yields the s-f hybridization. We neglect f-f overlap and consider only Fermion excitations, i.e. $n_s-n_r=n_p-n_q=1$. In Fig. 1, the short-range Coulomb repulsion between an f and an s electron on the same site is denoted by G, the external magnetic field by h, and the magnetic moments of the s and f orbitals by $\mu$ and $\bar{\mu}$, respectively. The f-f Coulomb repulsion is assumed to be so strong that no more than one f orbital may be occupied at a given site, whereas the s-s replusion is neglected. Furthermore, we shall consider only the lowest-energy configurations with zero, one, or two electrons per atom.

| ORBITALS f | ORBITALS S | P | $\varepsilon_P$ |
|---|---|---|---|
| — | — | 1 | 0 |
| — | ↑ | 2 | $t_o^+$ |
| — | ↓ | 3 | $t_o^-$ |
| — | ↑↓ | 4 | $2t_o$ |
| ↑ | — | 5 | $e^+$ |
| ↑ | ↑ | 6 | $e^+ + t_o^+ + G$ |
| ↑ | ↓ | 7 | $e^+ + t_o^- + G$ |
| ↓ | — | 8 | $e^-$ |
| ↓ | ↑ | 9 | $e^- + t_o^+ + G$ |
| ↓ | ↓ | 10 | $e^- + t_o^- + G$ |

Fig. 1. The ten allowed configurations of a single atom. Each row gives the electronic occupation of the f and s orbitals, the label p, and the energy $\varepsilon_p$ of a particular configuration. For example, the configuration p=7 has a spin-up f electron, a spin-down s electron, and an energy $\varepsilon_p=e^++t_o^-+G$. Here, $e^\pm=e\pm\bar{\mu}h$ and $t_o^\pm=t_o\pm\mu h$. The other symbols are explained in the text.

The equations of motion for the Green's functions $G^{ij}_{pqrs} = \langle\langle L^i_{pq};L^j_{rs}\rangle\rangle_E$ (in the notation of Ref. 3) yield in the RPA approximation the following:

$$[E-\varepsilon_q+\varepsilon_p+\zeta]G^{ij}_{pqrs}=i\,\delta_{ij}\delta_{ps}\delta_{qr}D_{pq}/2\pi+D_{pq}\sum_{\ell mn} B^{qpmn}_{i\ell}G^{\ell j}_{mnrs} , \qquad (2)$$

where $D_{pq}=D_p+D_q$ and $D_p = \langle L^i_{pp}\rangle$. We restrict ourselves to the paramagnetic zero-field (h=o) case, for which the occupation probabilities $D_p$ are subject to the constraints $D_2=D_3$, $D_6=D_7=D_9=D_{10}$, and $D_5=D_8$. Conservation of particle number and probability yield the additional conditions $\Sigma D_p=\Sigma n_p D_p=1$ for the case of one electron per atom. These constraints reduce the number of independent $D_p$ to three. The poles of the Green's functions give the quasiparticle excitation spectrum and are solutions of the equations

$$\mathcal{D}(E)=E(E-t_o)(E-t_o-G)(E-G)-\mathcal{f}\xi=0 \quad , \qquad (3)$$

$$\mathcal{f}=(D_{12}+D_{24})(E-t_o-G)+2D_{56}(E-t_o) \quad , \qquad (4)$$

$$\xi=E(E-G)t_k+V_k^2(D_{15}(E-G)+2D_{26}E) \quad . \qquad (5)$$

Here $t_k$ and $V_k$ are Fourier transforms of $t_{ij}$ and $V_{ij}$, respectively. The quantities $D_{pq}$ in Egs. (2-5) must be determined self-consistently from the Green's functions in the usual manner[3], and the details are omitted here. In the paramagnetic case this procedure yields four non-linear equations in the variables $D_2$, $D_4$, $D_6$, and $\zeta$. A computer program was written which solves these equations by iteration with a damping factor.[6] The calculations were carried out on the CDC 6600 computer of Indiana University.

The paramagnetic susceptibility $\chi(T)$ is calculated by considering the first-order effects of a small external magnetic field h (See Fig. (1)). For example, $D_2=D_2^\circ-\delta$ and $D_3=D_3^\circ+\delta$ , where $D_2^\circ$ and $D_3^\circ$ are zero-field solutions and $\delta$ is a small positive number proportional to h. Unfortunately, the resulting expression for $\chi(T)$ is too complicated to be reproduced here.

## III. THERMODYNAMICS OF THE MODEL

Under the conditions $t_{ij}=V_{ij}=0$, the solutions of Eq. (2) yield

$$D_p = Z^{-1}\exp(-(\varepsilon_p - n_p\zeta)/k_BT) \quad . \tag{6}$$

where Z is the partition function. Eq. (6) describes the atomic limit, i.e. non-interacting atoms with discrete energy levels.

The model has also been studied for the case e=o, $V_k=V$, and

$$t_k = -t(\cos k_x a + \cos k_y a + \cos k_z a). \tag{7}$$

Equation (7) describes a tight-binding s band in a cubic lattice with lattice constant a. The calculations shown in Figs. 2-4 are for the case $\mu=\bar{\mu}=\mu_B$, G=0.5W, and t=0.25W, where W is arbitrary energy.

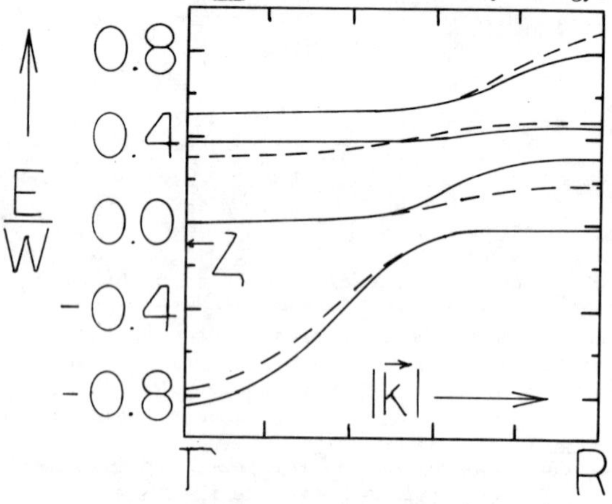


Fig. 2. The quasiparticle energy bands in the [111] direction in the simple cubic lattice for V=0.2W, $t_0$=-0.1W. Solid curves: T=0.03W. Dashed curves: T=0.25W.

Figure 2 depicts typical quasiparticle energy bands for V=0.2W and $-t_0$=0.1W (Eqs. (3-5)). At a low temperature T (Fig. 2, solid curves) the electrons occupy the mostly itinerant states of the lower s-like band. At a higher T (Fig. 2, dashed curves) the lower band narrows and increases in energy, leading to increasing localization. This results in a change from a nearly T-independent Pauli-type susceptibility (Fig. 3, solid curve ) to a Curie-Weiss behavior, a change which is observed experimentally in many actinide intermetallics.[1,2] The correlation function $d=|\langle L^{1}_{25}\rangle|$ is a measure of the admixture of the s and f orbitals due to the hybridization. The rapid decline in d with increasing T (Fig. 3, dashed curve ) is similar to the "thermal dehybridization" effect proposed by Doniach.[7] We estimate the electrical resistivity ρ by assuming that only electrons in s orbitals contribute to the conductivity and that the localized electrons act as scattering centers. One finds

$$\rho \cong \rho_o (D_5+2D_6)/(D_2+D_4+2D_6) \quad . \tag{8}$$

Equation (8) is plotted in Fig. 4 for $\rho_o$=100μΩ-cm and compared to data on $UA\ell_4$, a typical intermetallic actinide compound. Because of the more complex crystal and band structure of this compound, a quantitative fit to the data is not expected here.

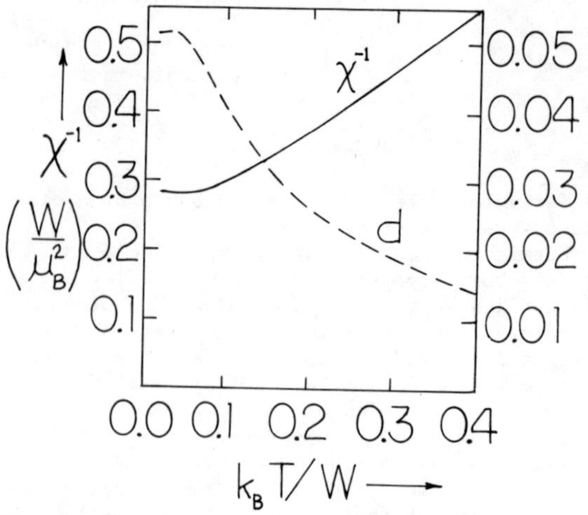


Fig. 3. The inverse paramagnetic susceptibility $\chi^{-1}$(T)(solid curve) and the hybridization parameter d(T)(dashed curve) for $t_o$=-0.1W, V = 0.2W.

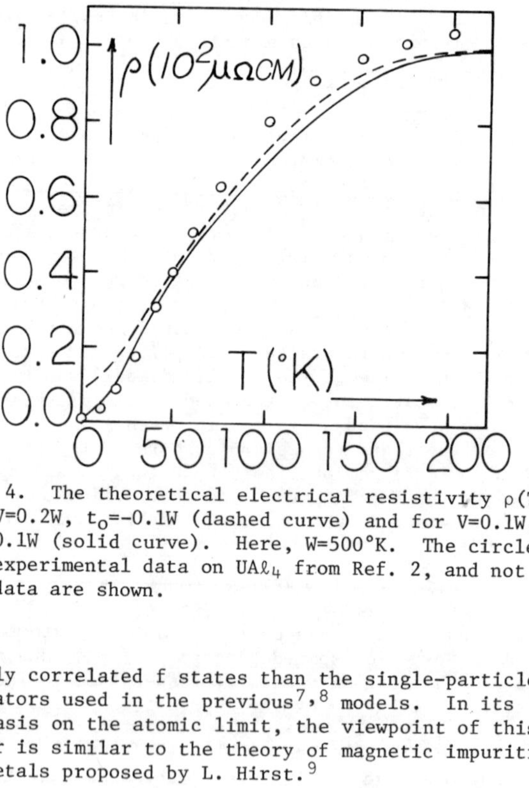


Fig. 4. The theoretical electrical resistivity ρ(T) for V=0.2W, $t_0$=-0.1W (dashed curve) and for V=0.1W, $t_0$=-0.1W (solid curve). Here, W=500°K. The circles are experimental data on $UA\ell_4$ from Ref. 2, and not all data are shown.

## IV. DISCUSSION

The calculated thermodynamic properties of the model indicated that it contains much of the basic physics of the metallic actinide compounds. In order to make more quantitative applications to experiments, the intraatomic Coulomb, spin-orbit, and crystal field interactions will be included in the Hamiltonian in future work.

In the spin-fluctuation model of Doniach[7], the thermal dehybridization and resistivity behavior results from incorporating in the quasiparticle energy an imaginary part having a $T^2$ temperature dependence. It is interesting that we obtain similar behavior with real quasiparticles. The standard-basis operators seem more appropriate for treating the highly correlated f states than the single-particle operators used in the previous[7,8] models. In its emphasis on the atomic limit, the viewpoint of this paper is similar to the theory of magnetic impurities in metals proposed by L. Hirst.[9]

## REFERENCES

*This work was supported by a 1975 Summer Faculty Grant from Indiana University-Purdue University at Fort Wayne.

1. B. Dunlap et al, A.I.P. Conf. Proc. 24 351(1975).
2. K. Buschow and H. van Daal, A.I.P. Conf. Proc. 5, 1464(1972).
3. S. Haley and P. Erdös, Phys. Rev. B5, 1106(1972).
4. J. Hubbard, Proc. Phys. Soc. A, 277 , 237(1964); 285, 542(1965).
5. J. Robinson, A.I.P. Conf. Proc. 24, 246(1975).
6. The damping factor was suggested by G. Bendixen, I.U.-Purdue Department of Mathematics.
7. S. Doniach, in The Actinides: Electronic Structure and Related Properties, A. J. Freeman and J. B. Darby, Jr., eds. (Academic Press, Inc., New York 1974) vol. 2, ch. 2, p. 51.
8. R. Jullien et al, Phys. Rev. B9, 1441(1974).
9. L. Hirst, Phys. Kondens. Materie 11, 255(1970).

# MAGNETIZATION FLUCTUATION RENORMALIZATION OF SINGLE PARTICLE AND MAGNON ENERGIES*

Victor Korenman, Joanne L. Murray and R. E. Prange
Department of Physics and Astronomy, and Center for Theoretical Physics,
University of Maryland, College Park, Maryland 20742

ABSTRACT

We have suggested[1] that the dependence on magnetic disorder of certain properties of itinerant ferromagnets is best studied by first computing them in the presence of a spatially slowly varying magnetization density, and then functionally averaging the magnetization. We report self-consistent rpa calculations of single particle and spin wave energies and damping in the presence of such a magnetization modulation, of arbitrary amplitude. The average energy gap is proportional to the population difference of suitably defined up- and down-spin bands. The gap, and also the local magnetization magnitude, do not decrease in proportion to the average magnetization, but as the mean square gradient of the magnetization direction. At wave vectors large compared to those dominant in the background, the spin wave stiffness also contains a term proportional to this gradient squared. The coefficient is positive for parabolic bands in the strong limit. These results are consistent with the observed insensitivity of band structure to the average magnetization, and with the persistence of short wavelength magnons above the Curie temperature.

One usually treats magnetic electrons in terms of states with spin parallel to some field direction. In the presence of a spatially varying magnetization it is convenient to use states with spin everywhere parallel or antiparallel to the <u>local</u> magnetization direction. In terms of these states the many body hamiltonian density is $H = H_o + H_1$, with $H_o$ the original expression. In the absence of spin-orbit forces and external fields $H_1$ comes entirely from kinetic energy terms and is given by

$$H_1 = [|\vec{g}(r)|^2 + |\vec{a}(r)|^2]\rho(r)/2m + j_{\mu\nu}(r)\cdot [\tfrac{1}{2}\,\vec{a}(r)\sigma^- + \tfrac{1}{2}\,\vec{a}^*(r)\sigma^+ - \vec{g}(r)\sigma^z]_{\mu\nu} \quad (1)$$

We have defined

$$\vec{a}(r) = (\sin\theta\ \nabla\phi - i\nabla\theta)e^{-ib}/2$$
$$\vec{g}(r) = (\nabla b + \cos\theta\ \nabla\phi)/2. \quad (2)$$

Here $\theta(r)$ and $\phi(r)$ are polar and azimuthal angles describing the magnetization direction relative to a fixed axis, while $b(r)$ is the third Euler angle which fixes the orientation of the local spin coordinate frame. This is arbitrary, and we fix $b(r)$ by requiring $\vec{g}(r)$ to be transverse. The spin current $\vec{j}_{\mu\nu} = [\psi^+_\mu \nabla\psi_\nu - (\nabla\psi^+_\mu)\ \psi_\nu]/2mi$ where the indices refer to the states (+, -), with spin (parallel, antiparallel) to the local magnetization direction.

By construction we must solve this Hamiltonian with the constraint of uniform magnetization, i.e., in the absence of spin waves. More precisely, this is in the absence of spin waves with wavevector $q \lesssim Q$ where $(2\pi/Q)^3$ is the averaging volume used to define the local magnetization. All the effects of longer wavelength magnetization fluctuations are included in $H_1$. They appear in a rotationally invariant manner since only gradients are involved. We will compute these to second order in $\vec{a}$ and $\vec{g}$, second order in magnetization gradients.

We here consider the self-consistent Hartree-Fock solution, in a one band model with a local interaction of strength U. We list and comment on the more salient results. A fuller discussion will be given elsewhere.

## Single Particle Properties

1. The local magnetization magnitude (spin per unit volume) is expressed in terms of occupation densities of the +, - states as $|\vec{M}(r)| = n_+ - n_- - 2\Sigma'|\vec{v}(k)\cdot\vec{a}(r)|^2/\Delta^2$. Here $\vec{v}$ is the velocity, $\Delta$ is the zero order spin splitting, and the primed sum is only over singly occupied states. $|\vec{M}|$ differs from the population difference because the perturbed eigenstates are slightly tilted from the local magnetization direction.

2. The band energies are $\varepsilon_{\pm}(k)=\varepsilon_o(k) \mp \frac{1}{2}U(n_+-n_-)+ \langle|a|^2\rangle/2m \mp \{\langle|\vec{a}\cdot\vec{v}(k)|^2\rangle - \langle|a\cdot v(k)|^2\rangle_{avg}\}/\Delta$.

The angular brackets denote a spatial average, while the subscript avg means also an average over the singly occupied states. Note that the average direct gap for these states is given by $n_+ - n_-$ rather than by $|\vec{M}|$.

3. The total energy change for this system is

$$\delta E = D\ (n_+ - n_-)\ V \langle|a|^2\rangle = D \int d\vec{r}\ |\vec{M}|\ |\nabla\hat{M}|^2. \quad (3)$$

where D is the usual rpa expression for the spin wave stiffness and $\hat{M} = \vec{M}/|\vec{M}|$. This expression implies the Landau-Lifshitz equation for magnetization motion, and the usual expression for the energy of a single spin wave.

4. Although the average direct gap is given by $n_+ - n_-$, the change in the indirect gap between the + and - fermi surfaces, $\varepsilon_-(k_F^-) - \varepsilon_+(k_F^+)$ drives a repopulation of the two sub bands. This is

$$\delta n_+ = \{(\Gamma_+/N_+)+(\Gamma_-/N_-)-2(\bar{\Gamma}/\bar{N})\}/\{(1/N_+)+(1/N_-)-2U\}$$

Here $N_+$ is the density of states at the up fermi surface, $\Gamma_+$ the integral of $\langle|\vec{a}\cdot\vec{v}(k)|^2\rangle/\Delta$ over the + fermi surface, $\bar{N} = n_+ - n_-$ and $\bar{\Gamma} = \Sigma'\langle|\vec{a}\cdot\vec{v}(k)|^2\rangle/\Delta$. In the strong limit $N_-$ vanishes, but $\Gamma_-$ vanishes as well to give $\delta n_+ = 0$. At finite temperature there is some repopulation, proportional to $\exp(-\Delta_1/kT)$ where $\Delta_1$ is the gap between the top of the minority spin band and the fermi energy.

5. To evaluate the above expressions we need the thermodynamic average of $|\nabla\hat{M}|^2$. This is generally given[1] as a functional average, which we cannot perform. At low temperatures, however, where a representation in terms of independent spin waves is appropriate, $\langle|\nabla\hat{M}|^2\rangle$ can be written as $\Sigma_q q^2 N_q$ where $N_q$ is the thermal occupation of spin waves with wave vector q. This gives $\delta\Delta \propto \delta n \propto T^{5/2}$ in the weak case, $e^{-\Delta_1/kT}T^{5/2}$ in the strong. The reduction in the local magnetization magnitude is proportional to $T^{5/2}$ in either case. These results are consistent with de Haas Van Alphen measurements[2] in Fe, where repopulation proportional to the bulk magnetization change, as predicted by Stoner theory is not found. We remark that Edwards[3] predicts $T^{5/2}$ behavior, using a fermi liquid argument. Even in the weak case his coefficient differs from ours.

6. The fact that the sub band populations, energy splitting and local magnetization magnitude are not directly given by the bulk magnetization, allows for the persistence of split bands well above the Curie temperature. We interpret the high temperature neutron experiments[4] in iron and nickel as

demonstrating that the band splitting does in fact persist in these cases. Note that the Curie temperature does not correspond to the Stoner transition, where $|\vec{M}|$ goes to zero. The transition is caused by the disordering of the small regions of finite magnetization, as in Heisenberg systems. The size of the ordered regions implied by the neutron experiments is such that our perturbation parameter $vQ/\Delta$ may remain quite small above $T_c$.

7. The $\vec{g}$ terms do not affect the single particle energies to the order considered. They enter the Hamiltonian as the vector potential of a magnetic field acting only on the orbital motion. Energy shifts go as $H^2 \propto (\nabla \times \vec{g})^2$ which is fourth order in gradients. These terms do, however, lead to real electron scattering. An expression for the scattering rate is easily written down but, due to the "softness" of the scattering, may not be experimentally significant. It may be most useful to note that the scattering is precisely that produced by a random magnetic field with local value $\vec{h}(r) = (\hbar c/e)\ \nabla \times \vec{g}(r) = (\hbar c/2e)(\nabla \cos\theta \times \nabla\phi)$. Note that $\vec{h}(r)$ vanishes unless there are at least two distinct wave vectors present in the magnetization fluctuation spectrum. This corresponds to the fact that electron scattering is a two magnon process in the usual treatment. At low temperatures where a spin wave representation is valid we may compute the rms value of $\vec{h}$ as proportional to $(kT/D)^{5/2}(\hbar c/e|M|)$. This gives approximately 10 gauss $T^{5/2}$ for nickel in zero external field.

## Magnon Properties

The collective excitations of our system, in the presence of the fluctuating magnetization, are given by the usual rpa expression, for example as a sum of particle-hole bubble graphs. The evaluation of these expressions is complicated, however, by the need to use the single particle energies and wave functions as perturbed by the fluctuating background field. We have evaluated the kernel of the integral equation for a magnon excitation, to second order in $\vec{a}$ and $\vec{g}$. We then used an extension of the perturbation scheme of Korenman and Prange[5] to find the energy shift and width and perturbed wave function for the magnon, to this order. First consider the $\vec{a}$ terms.

8. The energy of an excitation with wave vector q is given by $\hbar\omega = Dq^2 - 2D\langle|a|^2\rangle + Aq^2\langle|a|^2\rangle + B\langle|\vec{q}\cdot\vec{a}|^2\rangle$. The terms A and B, which are corrections to the stiffness D, are similar to, but not the same as, those found by Izuyama[6]. In particular this correction is always positive for parabolic bands in the strong limit. The quantity D reduces as $T^2$ with temperature because of the temperature dependence of the Fermi factors, so that the net effect on the stiffness is a reduction with temperature. Now, it should be noted that the validity of this expression is more limited than that of our previous results. If the background contains fluctuations with wave vector larger than q, it will be a serious error to neglect the time dependence of these fluctuations. We must set $Q \ll q$ and separately treat the influence of shorter wave length fluctuations. We have not yet carried out this calculation.

9. The presence of the term $-\ 2D\langle|a|^2\rangle$ is at first surprising and seems to violate rotational invariance. However it is a real physical effect. It reflects the fact that the physical magnon is not an excitation which rotates the magnetization away from the z-direction, but an excitation rotating the magnetization away from its local orientation. Note that this implies that a physical magnon does not generally represent a change of one unit in the bulk magnetization. There is no contradiction with the q = 0 limit since, because $Q \ll q$ is required, $\langle|\vec{a}|^2\rangle$ goes to zero in this limit. In general this term is partially cancelled by the explicit time dependence of the background fluctuation. In the case of a single spin-wave background the cancellation is complete, giving a magnon-magnon interaction just involving A and B.

10. The $\vec{a}$ terms give a magnon decay rate which corresponds to the interaction with a single spin wave and a particle-hole pair of one spin. In the limit of long electron mean free path the width is $\Gamma(q)/\omega(q) \approx \langle|\vec{a}xq|^2\rangle/kq^3$ where k is the fermi momentum. Again this is only valid for $Q \ll q$.

11. To the order considered, the effect of the $\vec{g}$ terms is to replace the gradient in the equation of motion for $\vec{M}$ by $\nabla + 2i\ \vec{g}(r)$. The term $D\ \nabla^2$ then leads to a magnon decay rate which corresponds to two-magnon scattering processes. At low temperatures the scattering vertex is proportional to $(D\vec{q}/|M|)\cdot[(\vec{q}_1-\vec{q}_2) \times (\vec{q}_1 x \vec{q}_2)]/|\vec{q}_1-\vec{q}_2|^2$ where $\vec{q}_1$ and $\vec{q}_2$ are the wave vectors of the adsorbed and emitted magnons. The term proportional to $\nabla^4$ gives a correction to D proportional to $\langle|\vec{g}|^2\rangle$. In rpa the coefficient seems to be negative in the weak case, and may be of either sign in the strong. We cannot compare it directly with the A and B terms above since the relative size of $\langle|\vec{a}|^2$ and $\langle|\vec{g}|^2\rangle$ is not known.

## REFERENCES

*Supported in part by the National Science Foundation.


1. R. E. Prange and V. Korenman, AIP Conf. Proc. 24, 325-326 (1975).
2. G. Lonzarich and A. V. Gold, Can. J. Phys. 52, 694 (1974).
3. D. M. Edwards, Can. J. Phys. 52, 704 (1974).
4. H. A. Mook, J. W. Lynn, and R. M. Nicklow, Phys. Rev. Letters 30, 556 (1973). J. W. Lynn, Phys. Rev. B. 11, 2624 (1975).
5. V. Korenman and R. E. Prange, Phys. Rev. B6, 2769 (1972).
6. T. Izuyama, Phys. Letters 9, 293 (1964).

# THE IMPORTANCE OF VERTEX CORRECTIONS IN PARAMAGNON THEORIES*

K. Levin and J. Hertz+
James Franck Institute, University of Chicago, Chicago, Illinois

M. T. Beal-Monod
Physique des Solides, Orsay, France

## ABSTRACT

We show by explicit calculation of the first order diagrams that vertex corrections to the fermion-paramagnon interaction are of relative order unity. The consequence of this result is that current paramagnon theories can at best be only qualitatively correct. A full theory of paramagnon effects must take systematic account of vertex corrections. Our result also serves to underline the significant differences between the electron-phonon and fermion-paramagnon systems. There is no Migdal theorem for the latter, in contrast to the former case. This happens essentially because the bare fermion-paramagnon interaction is the short range fermion-fermion repulsion $I = O(E_F)$, whereas in the phonon case the electron-phonon coupling $g << E_F$. The absence of a Migdal theorem for paramagnons has important implications for superfluid $He^3$.

Studies of paramagnon effects in nearly ferromagnetic Fermi systems have generally been based on the random phase approximation and therefore ignore all vertex corrections to the fermion-paramagnon interaction. These investigations include the early work of Doniach and Engelsberg on normal liquid $He^3$ and Pd[1] and of Berk and Schrieffer on the suppression of superconductivity in transition metals,[2] as well as more recent work on superfluid $He^3$ by a number of authors.[3] In all these studies, the fermion-paramagnon interaction is treated in direct analogy to the electron-phonon interaction in metals. While such approaches no doubt contain much of the correct physics, the validity of the analogy rests on the establishment of a Migdal theorem for paramagnons. In the electron-phonon problem, the Migdal theorem allows one to ignore vertex corrections.[4] In this note, however, we show, by calculating the first few vertex corrections explicitly, that there is no Migdal theorem for paramagnons. We also discuss the importance of these corrections in several problems.

In the general interacting Fermi system, nothing like a Migdal theorem exists to simplify the description of the coupling of collective and single-particle excitations. However, if such a system is near a ferromagnetic instability, one may try to apply the arguments used in the electron-phonon problem[5] to conclude that in strongly enhanced systems, vertex corrections to the RPA result are $O(\omega_{sf}/E_F) << 1$ where $\omega_{sf}$ is a typical paramagnon energy and $E_F$ is the Fermi energy. This argument is incorrect, however, as we shall show below.[6]

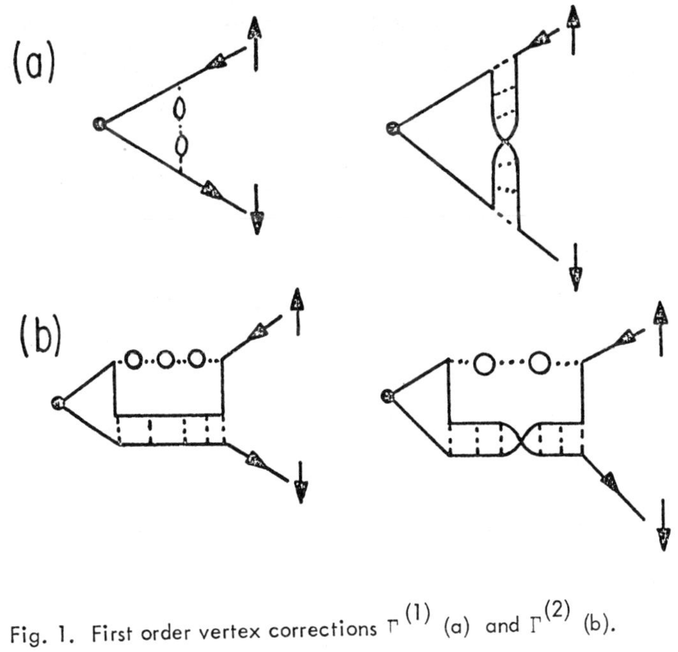

Fig. 1. First order vertex corrections $\Gamma^{(1)}$ (a) and $\Gamma^{(2)}$ (b).

The first order vertex corrections shown in Fig. 1a give

$$\Gamma^{(1)}(k,k+q) = iI^3 \int \frac{d^4p}{(2\pi)^4} \chi(p-k)G(p)G(p+k) \tag{1}$$

where $I$ is the Hubbard contact interaction and $\chi$ is the susceptibility. The reason for the breakdown of the Migdal theorem can be seen in the following way: We approximate $\chi$ by $\omega_{sf}^{-1}$ for $|p_0| < \omega_{sf}$, and take it to be zero outside this range. Then

$$\Gamma^{(1)} \approx \frac{iI^3}{\omega_{sf}} \int_{-\omega_{sf}}^{\omega_{sf}} \frac{dp_0}{2\pi} \int \frac{d^3\vec{p}}{(2\pi)^3} G(p)G(p+q) \approx \frac{I^3 N(0)}{\omega_{sf}} \left(\frac{\omega_{sf}}{E_F}\right) \tag{2}$$

The factors of $\omega_{sf}$ cancel, so that $\Gamma^{(1)} = O(I)$. In the phonon case, the vertex correction $\Gamma^{(1)} \approx g_{ep}(g_{ep}N(0))^2$ is small because the electron-phonon coupling constant $g_{ep}$ is small relative to electron energies. This important difference between the paramagnon and phonon problems results from the fact that the paramagnons (unlike the phonons) are excitations of the fermions themselves and are therefore strongly coupled to them.

We can make a more explicit evaluation of (1) in the limit that the four-vector $q \to 0$. We examine the case where the frequency is set equal to zero before the wavevector. This is the limit in which the Migdal theorem applies in the phonon case. Using a specral representation of $\chi$, the $p_0$ integration can be done exactly within an effective mass approximation with the result

$$\Gamma^{(1)} \equiv \Gamma^{(1)}(k_F, 0; k_F, 0) =$$

$$\frac{I^3 m}{(2\pi)^2 k_F} \int_0^{2k_F} pdp \int_0^{\infty} \frac{d\omega}{\pi} \chi''(p,\omega) \left[\frac{2(\omega + v_F p)}{(\omega + v_F p)^2 - (p^2/2m)^2}\right] \tag{3}$$

where $\chi''$ is the paramagnon spectral weight function. Details of the algebra leading to (3) will be published elsewhere.[7] For $\chi''$ we use the Doniach-Engelsberg model:

$$\chi''(p,\omega) = \frac{\chi(p)\,\omega\,\omega_p}{\omega^2 + \omega_p^2}\,\theta(v_F p - |\omega|) \tag{4}$$

with the static susceptibility given by ($\bar{I} = IN(0)$)

$$\chi(p) = N(0)\,[1 - \bar{I} + \tfrac{1}{3}(p/2k_F)^2]^{-1} \tag{5}$$

The cutoff in (4) is consistent with the long wavelength form of the full RPA $\chi(q,\omega)$.[8] Because the paramagnon spectral weight is concentrated at low $\omega$ (note that $\omega$ can never exceed $v_F p$), the factor in brackets in (3) is just $2/v_F p$, for $p << k_F$. Then, doing the $\omega$ integration

$$\Gamma^{(1)} \approx \frac{2N(0)I^3 m}{(2\pi)^3 k_F} \int_0^{2k_F} 2p\,dp \ln\left(\frac{N(0)}{\chi(p)}\right) \tag{6a}$$

$$\approx -\frac{4I}{\pi}[1 + \ln 3 + 3(1-\bar{I})\ln(1-\bar{I})] \tag{6b}$$

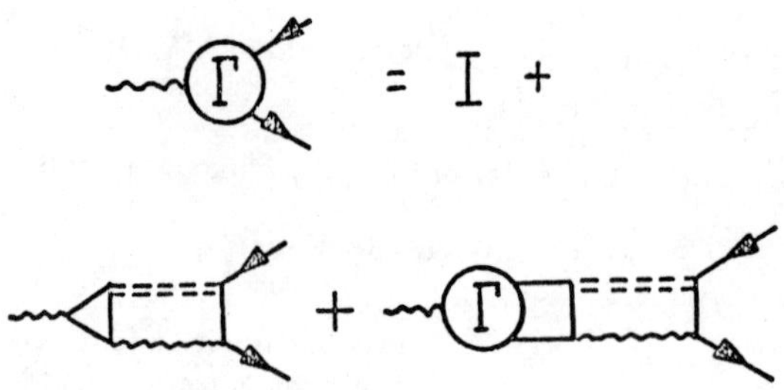


Fig. 2. Integral equation for the full vertex obtained by iterating the corrections of Fig. 1b. A wavy line stands for a paramagnon and a double dotted line for the density fluctuation propagator which occurs in Eqn. (7).

$\Gamma^{(1)}$ thus is finite and of relative order unity. Thus there is no Migdal theorem for paramagnons.

The two diagrams of Figs. 1b and 1c, while they appear to be of second order in $\chi$, can be combined to yield terms which are actually of first order[9]

$$\frac{I^2\chi_o}{1 - I^2\chi_o^{\,2}} - \frac{I}{1 - I^2\chi_o^{\,2}} = \frac{-I}{1 + I\chi_o} \approx -\frac{1}{2} I \tag{7}$$

(Here $\chi_o$ is the susceptibility of the noninteracting system.) Hence the sum of these diagrams is

$$\Gamma^{(2)}(k, k+q) = \tag{8}$$

$$-\frac{1}{2} I^4 \int \frac{d^4p}{(2\pi)^4 i} \left[ \int \frac{d^4k'}{(2\pi)^4 i} G(k'+p)G(k')G(k'+q) \right] G(k+p)\chi(p)$$

It is a reasonable approximation here to replace $\Lambda$, the quantity in brackets in (8), by its value when the 4-vectors q, p, and p-q are zero, with the frequencies set equal to zero before the wavevectors, since $\chi(p)$ is large only for $|\vec{p}| << 2k_F$ and $p_o << v_F |\vec{p}|$. We find $\Lambda = \frac{1}{2} N'(0)$ and

$$\Gamma^{(2)} \equiv \Gamma^{(2)}(k_F, 0; k_F, 0) = \frac{1}{4} I^2 N'(0) \Sigma_{DE}(k_F) \tag{9}$$

Here $\Sigma_{DE}$ is the self energy as evaluated by Doniach and Englesberg.[1] It is of order $I^2 N'(0) E_F$ and is finite at $I = 1$, so $\Gamma^{(2)}/I$, like $\Gamma^{(1)}/I$, is of order unity.

A major effect of both $\Gamma^{(1)}$ and $\Gamma^{(2)}$ corrections is to shift the instability point to an value of $I \neq 1$. It is difficult to estimate this effect very quantitatively because evaluating the integral in eqn. (3) requires knowing the spectral weight function well for $p \approx k_F$, which is beyond the region of validity of the simple model of eqns. (4)-(5).

Vertex corrections also affect the mass enhancement. To see this, we use the full vertex in place of one of the I's in the Doniach-Engelsberg calculation of $\Sigma$. The mass enhancement depends on $\partial\Sigma(k)/\partial k_o$ and thus on $\partial\Gamma(k, k+q)/\partial k_o$. One can check, by differentiating (1) with respect to $k_o$ and proceeding with calculations analogous to those that led to (3), that $\Gamma^{(1)}$ does not have any singular behavior in its energy derivative. However, $\Gamma^{(2)}$ clearly does, since it is proportional to $\ln(1 - I)$.[1] The iteration of the $\Gamma^{(2)}$ vertex correction (fig. 2) leads to an approximate expression for $\Gamma$:

$$\Gamma = I + \frac{\Gamma^{(2)}}{1 - \Gamma^{(2)}/I} = \frac{I}{1 - \frac{1}{2}\Lambda I \Sigma_{DE}} \tag{10}$$

Consequently, the differentiation of $\Sigma = (\Gamma/I)\Sigma_{DE}$ gives a mass enhancement

$$\frac{m^*}{m} = 1 - \left.\frac{\partial\Sigma}{\partial k_o}\right|_{k_o=0} = 1 - \frac{\partial\Sigma_{DE}/\partial k_o}{(1 - \frac{1}{2}\Lambda I \Sigma_{DE}(k_F, 0))^2} \tag{11}$$

The correction to the DE result vanishes in the case of particle-hole symmetry, since then both $\Lambda$ and $\Sigma_{DE}(k_F, 0)$ vanish. On the other hand, in systems like Pd, where N'(0) is quite large, this effect may be very significant.

As a final illustration of the importance of vertex corrections, we note that the absence of a Migdal theorem for paramagnons makes it impossible to construct an Eliashberg-like theory for paramagnon-induced pairing in superfluid $He^3$. In contrast to the situation in ordinary superconductors, the same paramagnon effects which are essential for the superfluidity become the major obstacle to building a first principles theory of the pairing.

## REFERENCES

*Work supported by NSF, AFOSR and CNRS.
+Alfred P. Sloan Foundation Fellow.

1. S. Doniach and S. Engelsberg, Phys. Rev. Lett. 17, 750 (1966).
2. N. F. Berk and J. R. Schrieffer, Phys. Rev. Lett. 17, 433 (1966).
3. P. W. Anderson and W. F. Brinkman, Phys. Rev. Lett. 30, 1108 (1973), W. F. Brinkman, J. Serene and P. W. Anderson, Phys. Rev. A 10, 2386 (1975), Y. Kuroda, Progr. Theor. Phys. 53, 349 (1975), L. Tewordt, D. Fay, P. Dorre and D. Einzel, Phys. Rev. B 11, 1914 (1975).
4. A. B. Migdal, Soviet Physics JETP 7, 496 (1958).
5. D. J. Scalapino, in Superconductivity, edited by R. D. Parks (Dekker, 1960), pp. 473-77.
6. To the best of our knowledge, the only explicit discussion of this question in the literature was given by C. Pethick in Lectures in Theoretical Physics, Vol. 9B (K. Mahanthappa and W. Brittin, eds., Gordon and Breach, 1969), pp. 245-49. He did not reach any firm conclusions.
7. J. Hertz, K. Levin and M. Beal-Monod, Solid State Communications, to be published.
8. See, e.g., J. R. Schrieffer, Superconductivity (Benjamin, 1964), pp. 137-141.
9. S. Ma, M. T. Beal-Monod and D. R. Fredkin, Phys. Rev. 174, 227 (1968).

# SUPERCONDUCTING AND MAGNETIC PROPERTIES OF $V_{3-x}Fe_xSi$ AND $V_{3-x}Mn_xSi$

R. L. Bergner, V. U. S. Rao and S. G. Sankar
Department of Chemistry
University of Pittsburgh
Pittsburgh, Pennsylvania 15260

## ABSTRACT

The superconducting, magnetic and structural properties of intermetallic compounds $V_{3-x}Fe_xSi$ and $V_{3-x}Mn_xSi$ have been investigated. The crystal structure is of the β-W type up to about x=0.5 in both systems, with the lattice parameter decreasing with increasing x. The superconducting transition temperature, $T_S$, and the upper critical field, $H_{c2}$, were determined from resistivity measurements. For x=0.015 in $V_{3-x}Fe_xSi$, $H_{c2}$ was enhanced over that of $V_3Si$. However, no such enhancement was found upon addition of Mn impurities to $V_3Si$. Susceptibility measurements indicate a local moment on Fe in $V_{3-x}Fe_xSi$ with $\mu_{eff} \sim 1.2\ \mu_B$. No detectable moment was associated with Mn in $V_{3-x}Mn_xSi$. These results indicate an antiferromagnetic coupling between Fe and conduction electron spins. On the Fe-rich side (x=2.0 to 3.0), the crystal structure was of the cubic $BiF_3$ type. In this region the Curie temperature decreased with increasing vanadium content.

## INTRODUCTION

The β-W compounds have received a great deal of attention owing to their high upper critical fields $H_{c2}$ and high superconducting transition temperatures $T_S$. However, to our knowledge, no detailed examination of the influence of paramagnetic impurities on the superconducting properties of β-W compounds has been performed.[1] The systems $V_{3-x}Fe_xSi$ and $V_{3-x}Mn_xSi$ seem ideally suited for such investigations owing to the existence of appreciable solid solubility (up to about x=1.0) found by us in the present study. Furthermore, since $Fe_3Si$, which has the cubic $BiF_3$ structure, is ferromagnetic with a Curie temperature $T_M \sim 800$ K, there exists the interesting possibility of examining both the superconducting and magnetic properties over a wide range of composition. Another aspect of this work is to observe the presence or otherwise of local moments on impurities such as Fe and Mn, and their influence on superconducting behavior.

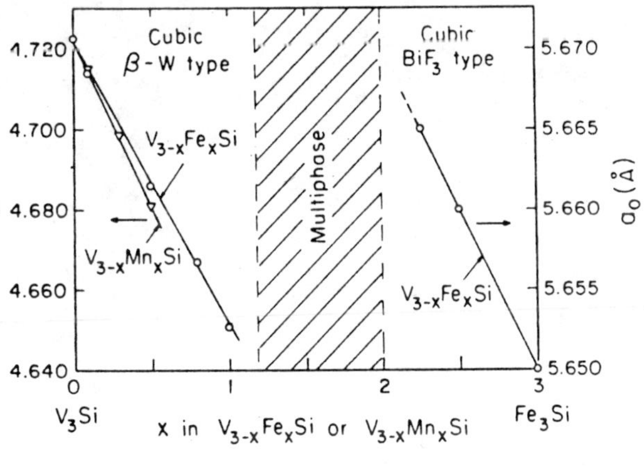


Figure 1. Variation of lattice constant with composition in $V_{3-x}Fe_xSi$ and $V_{3-x}Mn_xSi$.

## EXPERIMENTAL

The purity of the starting materials was 99.99% for V, Fe and Mn and 9N for Si. The compounds were prepared by induction melting the elements together in a water-cooled copper boat. The compounds were remelted several times to achieve homogeneity, annealed for a week at 700°C and quenched in ice water. The samples were then spark cut into elongated rods for resistivity measurements.

The lattice constants were determined by standard powder diffraction methods. The magnetic measurements were made by a Faraday technique. The superconducting transition temperatures $T_S$ and the resistive critical fields $H^*_{c2}$ were determined in applied fields up to 120 kOe employing an Intermagnetic General superconducting magnet.

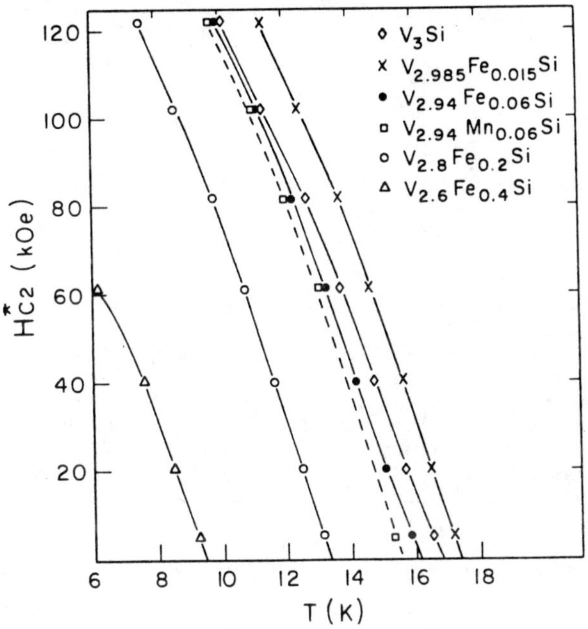


Figure 2. Resistive critical field $H^*_{C2}$ vs T curves for $V_{3-x}Fe_xSi$ and $V_{3-x}Mn_xSi$.

## RESULTS AND DISCUSSION

It was found that substantial amounts of V could be replaced by Fe and Mn in $V_3Si$ without altering the crystal structure. Single phase materials were obtained up to about x=1.0 in both $V_{3-x}Fe_xSi$ and $V_{3-x}Mn_xSi$. In this region the lattice constant decreased with increasing x as shown in Figure 1. On the Fe rich side of $V_{3-x}Fe_xSi$, the crystal structure was of the cubic $BiF_3$ type between x ∿ 2 and x = 3. The lattice constant increased slightly with V content (Fig. 1). Between x ∿ 1.0 and x ∿ 2.0 the alloys were found to be multiphase. Similar investigation of the Mn rich side in $V_{3-x}Mn_xSi$ has not been performed since the crystal structure of $Mn_3Si$ itself has not yet been established.

The critical field $H^*_{c2}$ vs T curves for some representative compositions are shown in Fig. 2. It is seen that $H^*_{c2}$ for $V_{2.985}Fe_{.015}Si$ is higher than that of the parent compound $V_3Si$. However, introduction of larger amounts of Fe depresses both $T_S$ and $H^*_{c2}$ as seen from the figure. Intoduction of dilute Mn impurities in $V_3Si$ did not result in enhancement of $H^*_{c2}$.

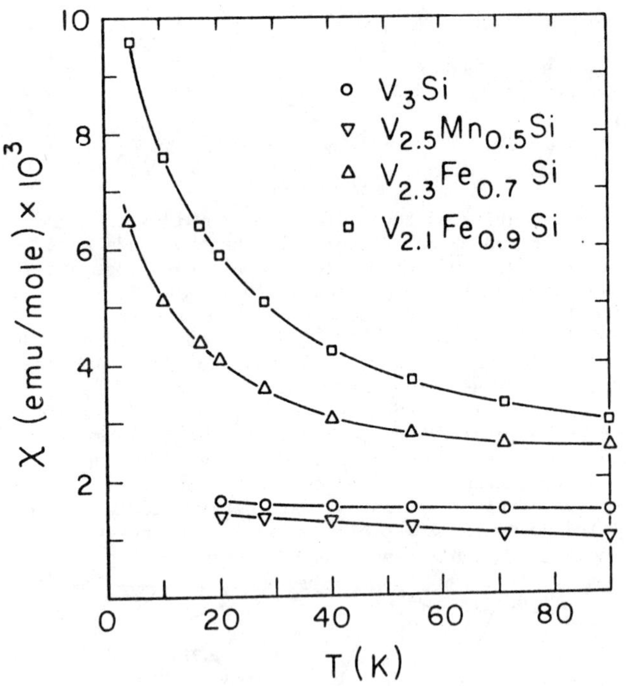


Figure 3. Magnetic susceptibility vs. T for $V_{3-x}Fe_xSi$ and $V_{3-x}Fe_xSi$.

To understand the magnetic states of the Fe and Mn impurities in $V_3Si$, susceptibility ($\chi$) measurements were performed above $T_S$. The results are shown in Fig. 3. $V_3Si$ has a weakly temperature dependent susceptibility, arising mainly from the Pauli paramagnetism of the 3d band. The composition $V_{2.5}Mn_{0.5}Si$ also shows similar behavior (Fig. 3), indicating the absence of local moments on Mn atoms. The slight decrease in $\chi$ of $V_{2.5}Mn_{.5}Si$ with respect to $V_3Si$ suggests a decrease in the 3d density of states at $E_F$. On the other hand, the compositions $V_{2.3}Fe_{.7}Si$ and $V_{2.1}Fe_{0.9}Si$ show strongly temperature dependent susceptibilities. By subtracting the $\chi$ of $V_3Si$ from those of the above compounds, the contribution to $\chi$ from the Fe atoms was estimated. This contribution was found to obey the Curie-Weiss law with $\mu_{eff}^{Fe} \sim 1.2\ \mu_B$. Susceptibility results therefore enable one to conclude that Fe impurities in $V_3Si$ possess local moments whereas Mn impurities are nearly nonmagnetic.

It is thus possible to understand the origin of the enhancement of $H_{c2}^*$ on the basis of the Schwartz and Gruenberg model[2] if one postulates an antiferromagnetic coupling between the Fe and conduction electron spins. Similar behavior has been found for Mn impurities in $MoGa_4$ by Fischer *et al.*,[3] and for Fe impurities in $ZrV_2$ and $HfV_2$ by Duffer *et al.*[4] The depression of $T_S$ and $H_{c2}^*$ upon introduction of Mn into $V_3Si$ may be explained as a consequence of the decrease in the density of states at $E_F$.

The Curie temperature, $T_M$, of the Fe rich regions of $V_{3-x}Fe_xSi$ having the cubic $BiF_3$ structure are shown in Fig. 4. It is seen that increasing V content results in a marked reduction in $T_M$. For comparison, the variation of $T_C$ with x is also shown for $V_{3-x}Fe_xSi$ in Fig. 4. Further work, including Mossbauer measurements are being planned to examine the coexistence of superconducting and magnetic order in this system.

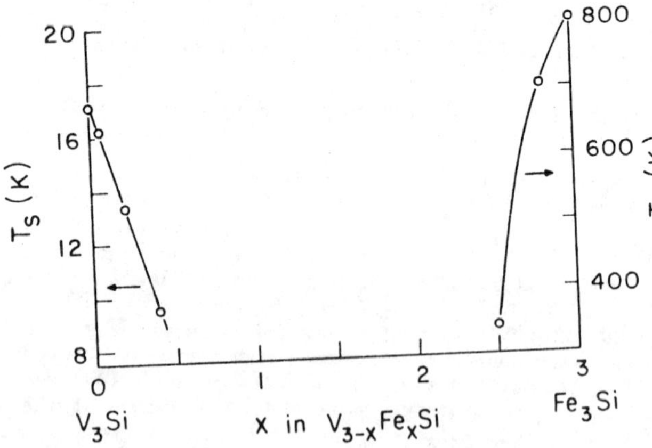


Figure 4. Variation of superconducting transition temperature $T_S$ and Curie temperature $T_M$ with composition for $V_{3-x}Fe_xSi$.

ACKNOWLEDGEMENT

The authors wish to acknowledge the assistance of Dr. R. Obermyer, Dr. D. M. Gualtieri and Mr. P. Duffer in the experimental work.

REFERENCES

1. K. H. Bennemann and J. W. Garland, Intern. J. Magnetism 1, 97-122 (1971).
2. B. B. Schwartz and L. W. Gruenberg, Phys. Rev. 177, 747-748 (1969).
3. O. Fischer, H. Jones, G. Bongi, C. Frei and A. Treyvand, Phys. Rev. Letters 26, 305 (1971).
4. P. Duffer, S. G. Sankar, V. U. S. Rao, R. L. Bergner and R. Obermyer, Phys. Stat. Solidi (in press).

# CANONICAL BAND THEORY OF THE VOLUME AND STRUCTURE DEPENDENCE OF THE IRON MAGNETIC MOMENT

J. Madsen, O.K. Andersen,
Electrophysics Department, Technical University, DK-2800 Lyngby, Denmark
U.K. Poulsen and O. Jepsen
AEK Research Establishment Risø, DK-4000 Roskilde, Denmark

ABSTRACT

The ferromagnetic moment at 0°K of bcc, hcp and fcc Fe has been obtained as a function of atomic volume from self-consistent, spin-polarized band calculations employing the atomic sphere approximation and the local spin density approximation for exchange and correlation. Our results are in agreement with the measured bcc moment and its volume dependence, they indicate a very strong volume dependence of the hcp and fcc moments near the normal atomic volume, and they explain the pressure induced (bcc, ferromagnetic) → (hcp, paramagnetic) transition around 100 kb. These results have a simple interpretation in terms of the Stoner criterion and the bcc, hcp and fcc canonical d state densities.

The prediction of the ground state properties, such as magnetic moments, magnetic structures and crystal structures, is the first step towards a detailed understanding of the magnetism of the transition metals. Asano and Yamashita[1] obtained the correct magnetic moments and relative stability of the ferro- and antiferromagnetic structures for the 3d transition metals using one adjustable parameter, I, for the effective exchange interaction, together with paramagnetic energy bands corresponding to the observed atomic volumes and crystal structures. For ferromagnetism, their model is the Stoner model and, within the uncertainty of the energy bands, I turned out to be the same (55 mRy) for all 3d metals considered. The local spin density (LSD) approximation[2] to the theory of ground state properties of Hohenberg, Kohn and Sham removes the arbitrariness of the band structure and obviates the use of an adjustable exchange parameter. Moreover, if the exchange splitting is treated to first order in the magnetization, if the contribution from the sp electrons is neglected, and if the charge and spin densities are spherically symmetric in the atomic sphere the LSD approximation leads to the model of Asano and Yamashita[3,4]. Values of I, computed by Gunnarsson[3] under these approximations, range from 59-74 mRy for V through Ni. We have performed self-consistent, spin-polarized band calculations[4] with the Barth-Hedin version of the LSD approximation, only invoking the Stoner model when interpreting the results. This model turns out to be quite successful, and we deduce values of I in close agreement with those of Gunnarsson. The conjecture, that it is mainly the band structure which determines the variation in magnetic moments, and crystal structures of transition metals, is further supported by considering the properties of a single metal as functions of pressure, and in this paper we present such calculations for Fe.

Experimentally it is well known that, at low temperatures and atmospheric pressure, Fe is bcc with a ferromagnetically ordered moment of 2.2 bohr magnetons per atom. The atomic radius, $S_o$, is 2.662 bohr radii. With the application of pressure, the moment, m, and the atomic radius, S, are both reduced with initial slopes[5,6] of respectively $\partial \ln m/\partial P = -2.5\cdot 10^{-4}$/kb and $\partial \ln S/\partial P = -2.0\cdot 10^{-4}$/kb. At about 100 kb, when these quantities are reduced by only a few per cent, there is a transition to the hcp structure accompanied by a complete loss of magnetic moment and with no change of atomic volume[6]. At high pressures, Fe is therefore similar to the isoelectronic 4d and 5d transition metals, Ru and Os, with 8 valence electrons.

We have performed self-consistent, spin-polarized calculations for bcc, hcp and fcc Fe as a function of atomic volume. We did not consider the possibilities of antiferromagnetic structures or other crystal structures, and we did not compute the pressure. Even this program requires several hundred individual band calculations and we therefore used the atomic sphere approximation (ASA)[7] to the KKR method whereby the structure constants become canonical, that is, independent of energy and atomic volume. Moreover, we neglected the sp-d hybridization, except in a few cases mentioned below, but we retained the much stronger s-p hybridization. With these approximations, the crystal structure merely enters a self-consistent calculation through the canonical sp and d number-of-states functions[7,8], $n(p_s,p_p,\infty)$ and $n(\infty,\infty,p_d)$, and these were obtained once and for all for the three different crystal structures through diagonalizations and summations in the respective Brillouin zones. In Fig. 1 we show the canonical d state densities, $N_d(p_d) \equiv \partial n(\infty,\infty,p_d)/\partial p_d$. The relation between the canonical "energy" $p_{\ell\alpha}$, referring to spin $\alpha$, and the true energy E is $p_{\ell\alpha}(E) = (4\ell+2)[D_{\ell\alpha}(E)+\ell+1]/[D_{\ell\alpha}(E)-\ell]$, where $D_{\ell\alpha}(E) = \partial \ln \phi_{\ell\alpha}(E,r)/\partial \ln r|_S$, and $\phi_{\ell\alpha}$ is the solution of the $\ell$'th radial Schrödinger equation for the spin dependent potential in the atomic sphere.

The results for the volume dependence of the magnetization are shown in Fig. 2. They demonstrate how, under compression, the "kinetic" energy overcomes the "exchange" energy with a resulting loss of magnetic moment, and they predict that hcp Fe is not ferromagnetic when the atomic radius is reduced by a few per cent. It is, however, beyond the accuracy of these calculations to reliably predict whether hcp Fe is ferromagnetic at normal atomic volume since, for the bcc structure at normal volume, the predictions, m = 2.7 bohr magnetons per atom and $\partial \ln S$ = 2.9, are fairly crude. In order to establish the role of the sp-d hybridization, and hence the accuracy of the LSD approximation, we performed two fully hybridized calculations for the bcc structure using the potentials obtained

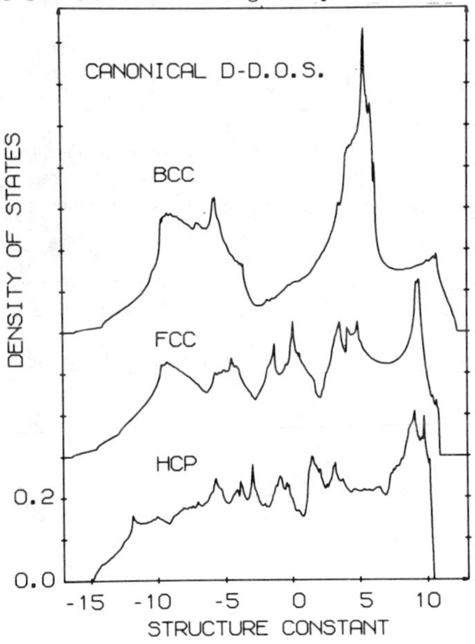


FIG. 1. Canonical d state densities. The abscissa is $p_d$ and the paramagnetic Fermi level for Fe is at 4.8 and 4.3 for the bcc and the close-packed structures respectively. For the hcp structure, the ideal c/a ratio was used.

self-consistently from the non-hybridized calculations. The results, m = 2.17 magnetons per atom and $\partial \ln m/\partial \ln S$ = 1.9, are in good agreement with the experimentally observed moment and in fair agreement with its

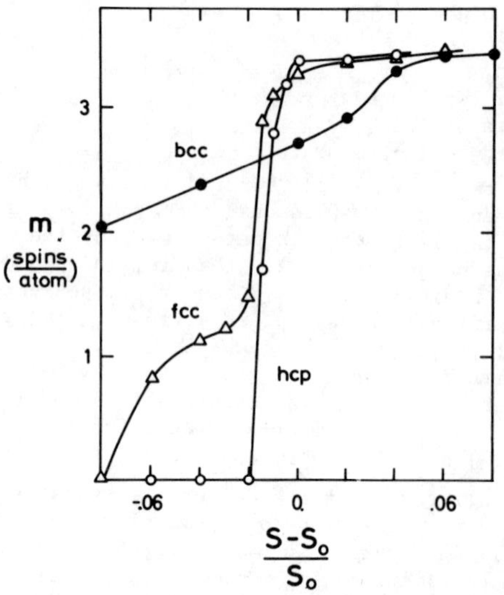


FIG. 2. The ferromagnetic moment vs. atomic radius for Fe, obtained neglecting the sp-d hybridization and using a frozen charge density for the core.

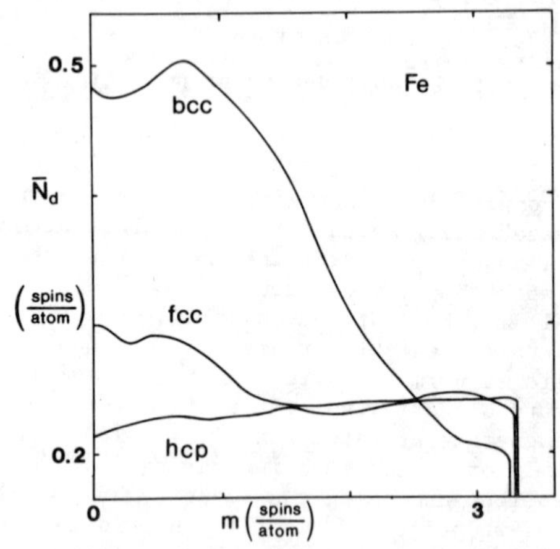


FIG. 3. Average, canonical d state densities for the number of d states, n, given in TABLE I.

volume derivative. Even for transition metals, the LSD approximation therefore seems to warrant fully hybridized calculations.

We now interpret Fig. 2 and estimate the volume reduction needed for the occurrence of the (bcc, ferromagnetic) → (hcp, paramagnetic) phase transition using the Stoner model. With this approximation, the only role of the sp band is to determine the number of d electrons and, since in the following all quantities refer to the unhybridized d band, we shall omit the subscript d. The spin-polarized band structure, E↑ and E↓, is given in terms of the paramagnetic band structure, E, as

$$E_{\downarrow}^{\uparrow} = E \mp \tfrac{1}{2}mI(E) \quad , \qquad (1)$$

where the energy dependence of I is nearly linear and derives from the fact[3,4] that I is a matrix element involving $\phi^4(E,r)$. For 3d bands, the function inverse to $p_\alpha(E)$ may be parametrized simply as

$$E_\alpha(p) = C_\alpha + p/\mu_\alpha S^2 \quad , \qquad (2)$$

where C is the band centre and μ the band mass[7]. If we insert (2) in (1), we see how the parameters C↑ and μ↑, C↓ and μ↓, characterizing the spin-polarized bands, may be transformed into the paramagnetic band parameters C and μ plus the Stoner parameters I(C) and dI/dE. For a fixed number, 2n, of d electrons, the moment is now determined by the condition that $E{\uparrow}(p{\uparrow}) = E{\downarrow}(p{\downarrow}) = E_F$, where $p_{\downarrow}^{\uparrow} \equiv p(n \pm \tfrac{1}{2}m)$, and p(n) is the function inverse to n(∞,∞,p). This leads to the Stoner criterion $(p{\uparrow}-p{\downarrow})/\mu S^2 = mI$, with $I \equiv \tfrac{1}{2}[I(E(p{\uparrow})) + I(E(p{\downarrow}))]$, or more conveniently

$$\mu S^2 \; I \; \overline{N}(n,m) = 1 \qquad (3)$$

Here, $\overline{N}(n,m)$ is the canonical d state density, averaged about the paramagnetic Fermi level for n states, over a range corresponding to m states.

Parameters obtained from the spin-polarized calculations are shown in Table I and, within the limits indicated, the number of d electrons and the Stoner parameters are constant, independent of the crystal structure and independent of the atomic radius over the range shown in Fig. 2. Moreover, the paramagnetic d band mass and its volume derivative are independent of the crystal structure. In (3), therefore, the structure dependence is exclusively in the function $\overline{N}$ of m, the volume dependence is exclusively in $\mu S^2$, and I is a constant. The average, canonical d state densities obtained from Fig. 1, for the values of n given in Table I, are shown in Fig. 3, and they reflect that, for Fe, the paramagnetic Fermi level is situated on a peak of the state density in the bcc structure , but not in the close-packed structures. Together with (3) and Table I, this figure explains the magnetization curves in Fig. 2.

Since, for a given atomic volume, the paramagnetic potential parameters were found to be independent of the crystal structure, it is reasonable to assume that this also holds for the spherically averaged charge density and hence for the non-magnetic interaction term of the total energy. Moreover, the energy associated with the sp electrons is assumed to be independent of the crystal structure since, in all cases, the occupied part of the sp band is nearly isotropic. Within the Stoner model, the magnetization at which the total energy of bcc Fe equals that of paramagnetic hcp Fe is therefore determined by

$$\int_0^m \frac{m'dm'}{2\overline{N}_{bcc}(n,m')} + 2\int_0^n [p_{bcc}(n') - p_{hcp}(n')]dn' = \frac{\mu S^2 m^2 I}{4} \qquad (4)$$

which predicts that the phase transition takes place at a 4 per cent reduction of atomic radius.

We are grateful to Dr. D.G. Pettifor for drawing our attention to the properties of hcp Fe.

TABLE I. d Band Parameters

| | |
|---|---|
| 2n (electrons/atom) | 6.73 ± .08 |
| I (mRy) | 70 ± 2 |
| dI/dE | .073 ± .010 |
| $\mu(S_o)$ | 8.55 ± .20 |
| $\partial \ln \mu / \partial \ln S$ | 2.17 |

REFERENCES

[1]S. Asano and J. Yamashita, Prog. Teor. Phys. 49, 373 (1973)

[2]U. von Barth and L. Hedin, J. Phys. C5, 1629 (1972) O. Gunnarsson, B.I. Lundqvist and S. Lundqvist, Solid St. Commun. 11, 149 (1972), see also J.C. Slater, Quantum Theory of Molecules and Solids, Vol. IV. (McGraw-Hill, 1974)

[3]O. Gunnarsson, J. Phys. F (to be published)

[4]U.K. Poulsen, J. Madsen and O.K. Andersen (to be published)

[5]D. Bloch and A.S. Pavlovic, Adv. in High Pressure Research, R.S. Bradley, ed., (Academic, London-New York, 1969), 125.

[6]D.L. Williamson, S. Bukshpan and R. Ingalls, Phys. Rev. B 6, 4194 (1972)

[7]O.K. Andersen, Solid St. Commun. 13, 133 (1973) and Phys. Rev. B (Oct 15, 1975)

[8]O. Jepsen, U.K. Poulsen and O.K. Andersen (to be published)

# MAGNETIC MOMENTS OF FERROMAGNETIC GADOLINIUM ALLOYS*

H. W. White[†], B. J. Beaudry, P. Burgardt, S. Legvold and B. N. Harmon
Ames Laboratory-ERDA and Department of Physics, Iowa State University, Ames, Iowa 50010

## ABSTRACT

Saturation magnetization and lattice parameter measurements have been made on samples of polycrystalline and single crystal c-axis gadolinium and on samples of polycrystalline gadolinium alloyed with scandium, yttrium, lutetium, lanthanum, magnesium, ytterbium and thorium. The magnetic measurements were made near 4.2 K in fields up to 30 kOe. Alloy concentrations ranged up to 30 atomic per cent. For pure single crystal gadolinium the saturation moment was measured to be $7.63 \pm .01 \mu_B$ per gadolinium atom. This result indicates the conduction electron contribution is about 13% higher than the value obtained from earlier measurements. For polycrystalline gadolinium the saturation moment per atom was found to be $7.555 \pm .015 \mu_B$. The excess moment per Gd atom for the polycrystalline alloys fell near a straight line when plotted as a function of the unit cell axes ratio c/a. The c/a ratios ranged from 1.583 to 1.595 and the excess moments were between 0.78 and $0.44 \mu_B$. Surprisingly, there was no correlation of excess moment with atomic volume or with the valence of the solute atoms.

## INTRODUCTION

Nigh, Legvold and Spedding[1] found the saturation magnetic moment for pure gadolinium to be $7.55 \mu_B$ per Gd atom. The source of the excess moment of $0.55 \mu_B$ per Gd atom over the value of $gJ = 7.0 \mu_B$ arising from the seven unpaired 4f electrons has since attracted significant theoretical attention. It is now attributed to the polarization of the conduction band electrons by the localized 4f electrons via the exchange interaction. The interaction between the localized 4f moments is mediated indirectly by this 4f-conduction electron interaction as described by RKKY theory[2].

Recent investigations of the conduction electron polarization of ferromagnetic Gd metal showed the conduction electron spin density is mostly of d character. The d-f exchange was found to be dominant[2,3].

The present work was undertaken to measure the magnetic moment per Gd atom and the lattice parameters in pure Gd and several Gd alloys to gain insight into the nature of the exchange mechanism. Measurements are reported on pure Gd and Gd alloyed with scandium, yttrium, lutetium, thorium, lanthanum, magnesium and ytterbium in concentrations ranging up to about 30 atomic per cent.

## EXPERIMENTAL

The La, Y, Sc and Lu alloys were prepared by arc melting high purity constituents in an argon atmosphere. The Mg, Yb and Th alloys were prepared by heating in sealed Ta crucibles. Spectrographic and fusion analyses were made on all samples. The major impurities, listed in parts per million, in the Gd used were as follows: Fe≈4, Ni≈4, W≈6, Ta≈40, La≈6, Ce≈8, O≈610, N≈80 and H≈470. Magnetization measurements were made on a polycrystalline sample from this mother material and will be referenced as Gd(Polycrystal Mother). In view of the interest in knowing the saturation magnetic moment for pure Gd, measurements were also made on a second and a third sample. The second was a polycrystalline sample which had been electrotransported to reduce oxygen content. It will be referred to as Gd(Polycrystal). It had a 300 K to 4.2 K resistivity ratio of 353. The third sample was a single crystal oriented such that the magnetization was measured along the c-axis. It had a corresponding resistivity ratio of 284 and was labeled Gd(Single Crystal c-axis).

The saturation magnetization of each sample was measured near 4.2 K in magnetic fields up to 30 kOe using a vibrating sample magnetometer. High purity iron and nickel were used for the calibration of the apparatus. The results were analyzed in a manner similar to that used in Ref. 1. The lattice parameters were measured at room temperature. Details of the sample preparation, experimental apparatus and technique plus additional experimental results will be reported elsewhere.

## RESULTS AND DISCUSSION

Figure 1 shows representative data points for the magnetization versus 1/H, the inverse internal field, for Gd(Single Crystal c-axis), Gd(Polycrystal Mother), the Fe standard and one representative alloy $Gd_{.852}Y_{.148}$. These data show the approach to saturation values for these four samples.

For the single crystal Gd the saturation moment per atom along the c-axis was measured to be $7.63 \pm .01 \mu_B$. This result implies that the excess moment above $7.0 \mu_B$ per Gd atom is about 13% higher than the value obtained from earlier measurements[1].

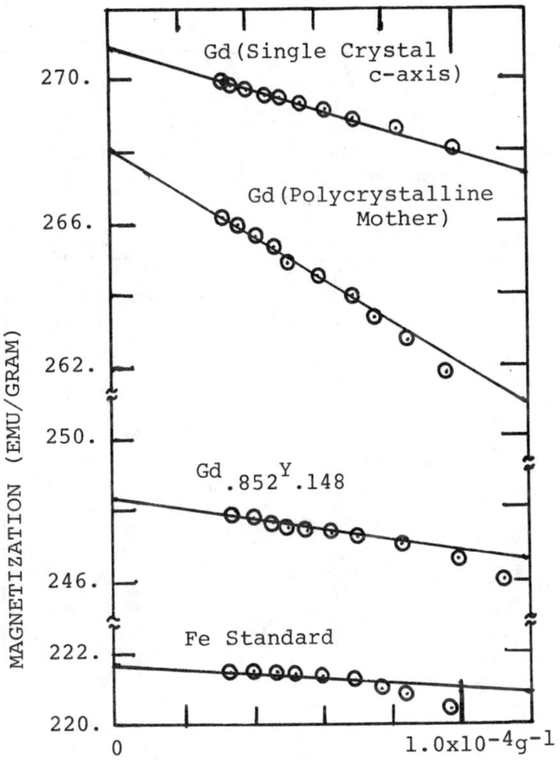


Figure 1. A plot showing representative data points for the magnetization near 4.2 K versus 1/H, the inverse of the internal field, for the samples Gd(Single Crystal c-axis), Gd(Polycrystal Mother), the alloy $Gd_{.852}Y_{.148}$ and the Fe standard.

This c-axis moment for the single crystal sample is within the experimental uncertainty of the c-axis saturation moment for single crystal Gd as reported to us by Roeland and co-workers[4].

For both polycrystalline Gd samples a corresponding value of $7.555 \pm .015 \mu_B$ was measured. Low field data, not shown in FIG. 1, indicated that the approach to saturation occurred at much lower fields for the single crystal than for any of the polycrystalline samples. The relatively larger magnetic hardness

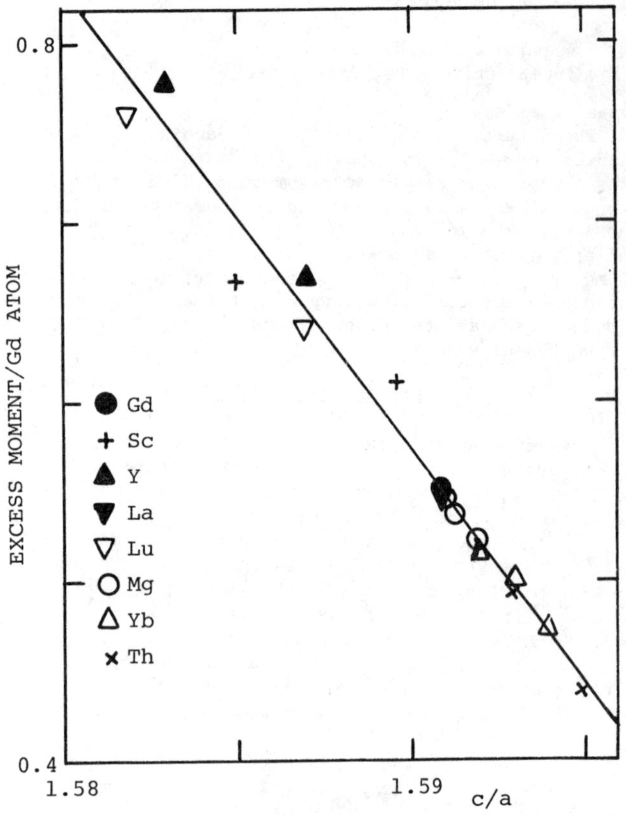


Figure 2. The excess magnetic moment per Gd atom in units of $\mu_B$ versus the unit cell axes ratio c/a for Gd(Polycrystal Mother) and the Gd alloys. The symbols indicate the solute atom.

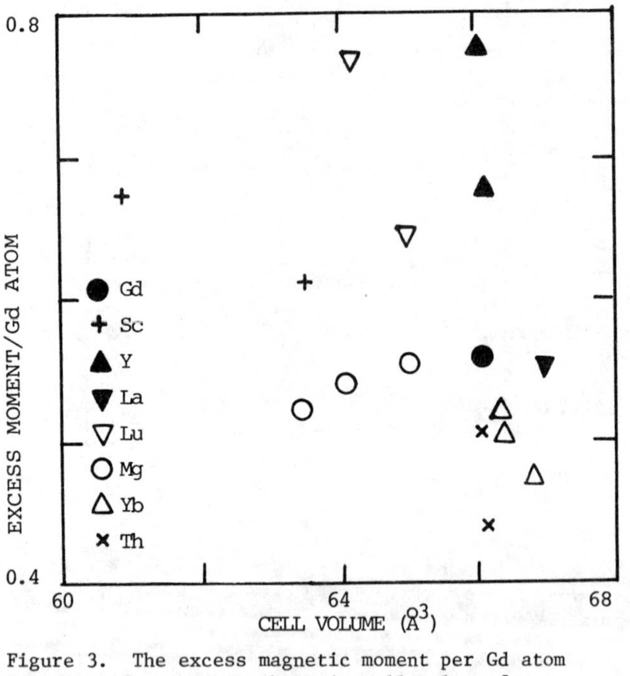


Figure 3. The excess magnetic moment per Gd atom in units of $\mu_B$ versus the unit cell volume for Gd(Polycrystal Mother) and the Gd alloys. The symbols indicate the solute atom.

and the lower saturation values for the two polycrystalline Gd samples can probably be attributed to either higher oxygen content or lattice strains, or possibly both.

Since the approaches to saturation for all polycrystalline samples were similar to those shown in FIG. 1 a uniform magnetic behavior was assumed for all of the polycrystalline samples. To be consistent the measured saturation magnetic moments of the polycrystalline alloys were compared with the saturation magnetic moment of polycrystalline Gd.

Figure 2 shows the excess moment per Gd atom for the polycrystalline mother Gd and for the alloys plotted as a function of the unit cell axes ratio c/a as determined by room temperature measurements. The excess moment is defined to be the measured moment per Gd atom minus $7.0\mu_B$. The points lie near a straight line which has been drawn through the data. FIG. 3 shows that there is essentially no correlation of the excess moment with atomic volume. Neither does there seem to be any apparent correlation with the valence of the solute atoms.

The dependence of the saturation moment on the c/a ratio as reported here has not been previously reported. One can only speculate about the mechanism responsible for this behavior. A conduction electron-conduction electron exchange enhancement could lead to a c/a dependence; however, calculations for pure Gd show very little sensitivity in the conduction electron wavefunctions as c/a varies from 1.63 to 1.57. On the other hand, it is also possible that in the polycrystalline samples all of the 4f moments are not being completely aligned, and that the degree of alignment is modulated by the c/a ratio. It may be necessary to obtain a more complete characterization of the 4f electron states in the rare earth metals and alloys in order to explain our results. It also appears plausible that any oxygen content or lattice strain might cause a change in the c/a ratio and simultaneously cause a decrease in the measured saturation moment[4].

References:

*Work supported in part by the U.S. Energy Research and Development Administration.

†Permanent address: Physics Department, University of Missouri, Columbia, MO 65201.

The authors wish to acknowledge helpful discussions with P.-A. Lindgård who communicated to us the results of Roeland and co-workers.

1. H. E. Nigh, S. Legvold and F. H. Spedding, Phys. Rev. 132, 1092-97 (1963).
2. B. N. Harmon and A. J. Freeman, Phys. Rev B 10, 1979-93 (1974).
3. R. M. Moon, W. C. Koehler, J. W. Cable and H. R. Child, Phys. Rev. B 5, 997 (1972).
4. L. W. Roeland, G. J. Cock, F. A. Muller, A. C. Moleman, R. G. Jordan and K. A. McEwen, (Private Communication).

# THE ANALYSIS OF MAGNETIC NEUTRON SCATTERING DATA*

G. Felcher
Argonne National Laboratory, Argonne, Illinois 60439
J. W. Garland
University of Illinois, Chicago, Illinois 60680
J. W. Cable and R. Medina†
Oak Ridge National Laboratory,‡ Oak Ridge, TN 37830

## ABSTRACT

The determination of the proper magnetic form factors for use in the analysis of diffuse-scattering data is discussed, and that information which is immediately available from such data is considered. Apparent discrepancies between the results of diffuse scattering and diffraction experiments on ferromagnetic Ni-Cu alloys are resolved. It is shown that the data indicate that the negative spin density usually attributed to conduction electrons instead arises largely from the overlap of localized wavefunctions, as was first suggested by Moon.

## INTRODUCTION

In principle, elastic neutron scattering offers by far the most complete experimental information on the spatial distribution of the magnetization in magnetic materials. Yet, despite the pioneering work of Marshall, *et al.*,[1] the extraction of theoretically meaningful information from diffuse elastic neutron-scattering data has remained a difficult, not completely solved problem. This is illustrated by the recent appearance of apparent major inconsistencies between the results of diffuse neutron-scattering experiments[2,3] and other experiments[3,4] for ferromagnetic Ni-Cu alloys. We discuss here the determination of the proper magnetic form factors, $f(\vec{K})$, for use in the analysis of diffuse-scattering data, and consider that information which is immediately available from such data in the limits of small and large momentum transfers. Upon explicit consideration of the diffuse elastic scattering[2,3,5] and diffraction[4] results for Ni-Cu alloys, the apparent inconsistencies are found to disappear under proper analysis. Space limitations prevent our giving here more than a few simple results; derivations and a complete set of formulas will appear elsewhere.

## NEUTRON DIFFRACTION

Neutron diffraction measures the Fourier transform,

$$\tilde{M}(\vec{G}) = N^{-1} \int M(\vec{r}) \exp(i\vec{G}\cdot\vec{r})\, d^3\vec{r}, \qquad (1)$$

of the magnetization density at the reciprocal lattice vectors, $\vec{G}$, where N is the number of unit cells. By the definition of $\vec{G}$, this is just the Fourier transform of the average magnetization density of a unit cell, which is usually assumed to be given by

$$\langle M_n(\vec{x})\rangle \simeq \langle \mu_n^{loc}(\vec{x})\rangle + \mu^{cond}. \qquad (2)$$

Here, $\mu_n^{loc}(\vec{x})$ is the localized atomic magnetization density of the atom(s) in the $n^{th}$ cell, and $\mu^{cond}$ is the average conduction-electron magnetization density. One can define an average magnetic form factor,

$$f(\vec{G}) = \langle \tilde{M}_n(\vec{G})\rangle / \langle M_n\rangle, \qquad (3)$$

where $M_n = \tilde{M}_n(0)$ is the total magnetization of the $n^{th}$ cell. Then, from Eq. (2) one finds

$$f(\vec{G}) \simeq (1+\alpha) f^{loc}(\vec{G}) - \alpha\delta_{\vec{G}\vec{0}}, \qquad (4)$$

where

$$f^{loc}(\vec{G}) = \langle \tilde{\mu}_n^{loc}(\vec{G})\rangle / \langle \mu_n^{loc}\rangle. \qquad (5)$$

Equation (4) allows one to determine $\alpha$ and $f^{loc}(\vec{G})$ from the experimentally determined $f(\vec{G})$. For elements and for alloys in which only one chemical species is magnetic, theoretical calculations generally are in excellent agreement with the values of $f^{loc}(\vec{G})$ obtained from Eq. (4), thus supporting the use of Eqs. (2) and (4) and allowing one to determine $f^{loc}(\vec{K})$ for all $\vec{K}$.

## DIFFUSE ELASTIC SCATTERING

The diffuse elastic scattering of neutrons offers information on the Fourier components of the local magnetization of disordered magnetic materials for all wavevectors away from the Bragg peaks. The use of spin-polarized neutrons enables one to measure directly the correlation function $M(\vec{K})$, which for a binary alloy is given by

$$M(\vec{K}) = [Nc(1-c)]^{-1} \sum_{m,n} \langle (p_m - c)\mu_n(\vec{K})\rangle \exp(i\vec{K}\cdot\vec{R}_{nm}), \qquad (6)$$

where c is the impurity concentration, $p_m$ is the impurity occupation number of site m, and $\vec{R}_{nm} = \vec{R}_n - \vec{R}_m$.

The analysis of $M(\vec{K})$ is especially simple in the limits of small and large momentum transfers. An exact limiting point for $M(\vec{K})$ is provided by the relationship

$$\lim_{K\to 0} \{M(\vec{K})/S(\vec{K})\} = d\langle\mu\rangle/dc, \qquad (7)$$

where $S(\vec{K})$ is the Fourier transform of the Cowley short-range-order parameters. In the opposite limit

$$M(\vec{K}) = \langle\mu(\vec{K})\rangle_{imp} - \langle\mu(\vec{K})\rangle_{host} + \delta M(K), \qquad (8)$$

where the subscripts "imp" and "host" denote averages over impurity atoms and host atoms, respectively, and where $\delta M(\vec{K})$ is an oscillatory, rapidly decreasing function of K.

Unfortunately, until recently diffuse elastic scattering experiments have been performed only with unpolarized neutrons and multicrystalline scatterers. Thus, they have measured directly only the spherical averages of $S(\vec{K})$ and of the function

$$T(\vec{K}) = [Nc(1-c)]^{-1} \sum_{m,n} [\langle \mu_m(\vec{K})\mu_n(\vec{K})\rangle - \langle\mu(\vec{K})\rangle^2] \exp(i\vec{K}\cdot\vec{R}_{nm}), \qquad (9)$$

not $M(\vec{K})$. In general, one may write

$$T(\vec{K}) = M^2(\vec{K})/S(\vec{K}) + \delta T(\vec{K}), \qquad (10)$$

where $\delta T(\vec{K})$ contains small linear terms plus complex nonlinear terms which are very difficult to evaluate numerically. These terms render the accurate determination of $M(\vec{K})$ next to impossible for alloys in which $\mu_n$ depends very nonlinearly on its chemical environment. The usual procedure for finding $M(\vec{K})$ is poor for large K, as is discussed in reference 5, and for some systems is poor even for small K.

## APPARENT INCONSISTENCIES

In previous analyses of diffuse elastic-scattering data the form factor $f^{loc}(\vec{K})$ has been used, and it has been assumed that some part of the total magnetization arises from a conduction-electron spin polarization not seen by the neutrons. Thus, the extrapolated K=0 limit of $M(K)/S(K)$ should have given $d\langle\mu^{loc}\rangle/dc$ rather than the bulk magnetization values $d\langle\mu\rangle/dc$ typically found. For the case of ferromagnetic Ni-Cu alloys such an analysis[3] gave the surprising result that $\mu^{cond} = \langle\mu\rangle - \langle\mu^{loc}\rangle$ is very nearly equal to -0.10, independent of Cu concentration over the range $0 \leq c \leq 0.4$, assuming $\langle\mu\rangle_{Cu} = 0$. Thus, the data appeared to be internally self-consistent. However, more recent neutron-diffraction studies indicate that $\mu^{cond}$ decreases with increasing Cu concentration even more rapidly than does $\langle\mu\rangle$. This is in substantial disagreement with the result inferred from both the low- and high-K limits of the diffuse-scattering data. The disagreement found is too large to be explained either by possible experimental errors or uncertainties or by nonlinear terms not included in the Marshall[1] formalism.

RESOLUTION OF INCONSISTENCIES--THE PROPER FORM FACTOR

Although only the local atomic magnetization, $\langle\mu^{loc}\rangle$, is seen in neutron diffraction, it is clear that the total moment associated with the site n in an alloy must be fairly well localized and should be seen in small-angle diffuse scattering. It is improper to ascribe any part of the bulk magnetization of a magnetically inhomogeneous alloy to a uniform conduction-electron spin polarization not seen by small-angle diffuse scattering or to use a purely local atomic form factor in analyzing diffuse-scattering data. Either a conduction-electron spin polarization induced by the local atomic part of $\mu_n$ or, as was first suggested by Moon,[6] overlap bonding or antibonding spin densities can give rise to a "nonlocal" contribution to $\mu_n$ having a form factor, $f^{nonloc}(\vec{K})$, which is large for small values of K but is very small at the Bragg peaks. This is confirmed by rough calculations for Ni alloys, which also indicate that induced conduction-electron spin densities and d-band overlap spin densities yield very similar form factors, as is shown[7] in Fig. 1.

The proper form factor for use in the analysis of diffuse-scattering data can be written as

$$f(\vec{K}) = (1+\alpha)f^{loc}(\vec{K}) - \alpha f^{nonloc}(\vec{K}), \qquad (11)$$

where $\alpha$ is determined by Eq. (4), as is discussed above in the section on neutron diffraction. Since $\alpha$ typically is small and since $f^{nonloc}(\vec{K})$ is rather insensitive to its method of calculation, the proper $f(\vec{K})$ is easily determined to order one or two per cent. The proper f(K) for Ni-Cu alloys is compared in Fig. 2 with $f^{loc}(K)$ as determined by Mook[8] and with that f(K) which would be produced by a consistent or very long-range conduction-electron spin polarization.

The use of the correct form factor for Ni-Cu alloys should lead to the satisfaction of Eq. (7) with $\langle\mu\rangle$ given by the bulk magnetization value and to the result $(1-c)M(\vec{K})/f(\vec{K}) \to \langle\mu\rangle$ in the limit of large momentum transfers. Since $f(\vec{K})$ approaches $f^{loc}(\vec{K})$ as K approaches zero, Eq. (7) is properly satisfied. An

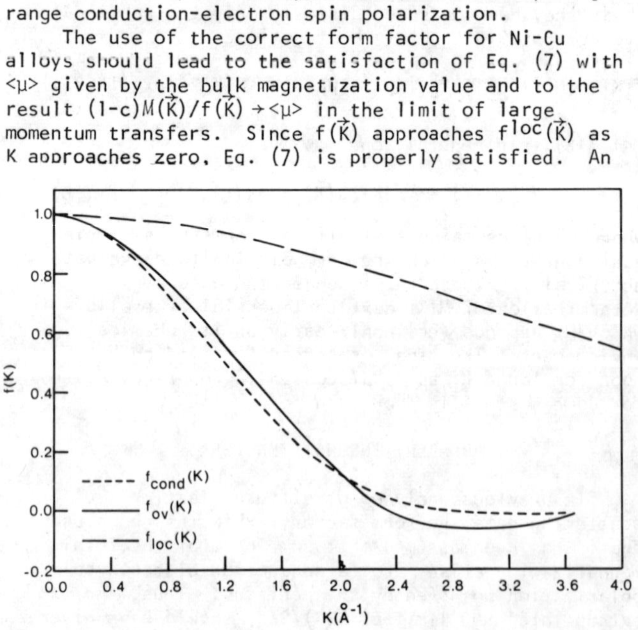


Figure 1. Partial form factors for Ni-Cu alloys: f(K) for an induced conduction-electron polarization, for the Ni-Ni overlap spin density, and for the local spin.

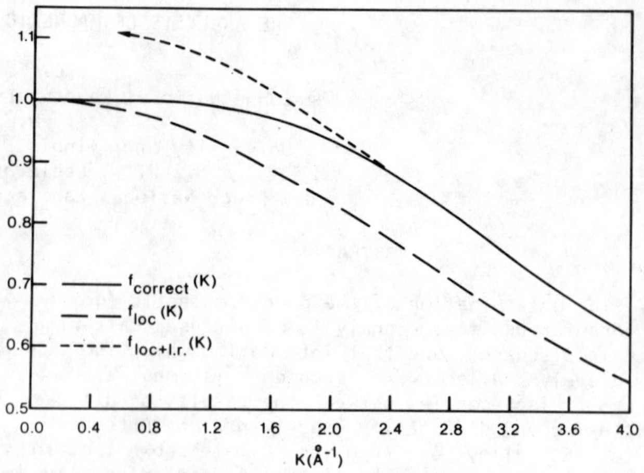


Figure 2. Form factors for Ni-Cu alloys: the correct f(K), the local or Mook f(K), and the f(K) appropriate to a local plus a long-range conduction-electron spin.

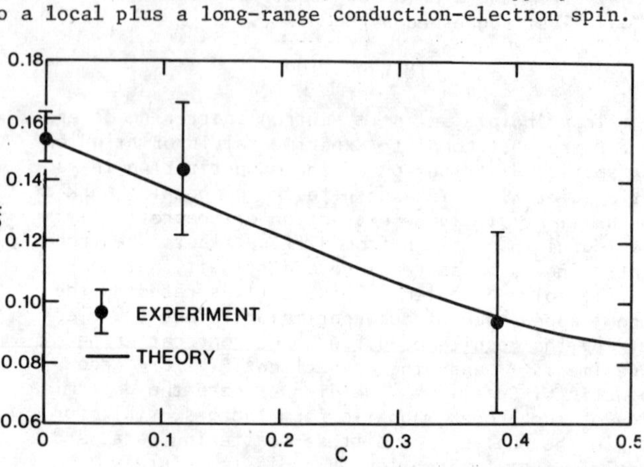


Figure 3. The ratio, $\alpha_{eff}$, of the delocalized part of $\langle\mu\rangle$ to the total $\langle\mu\rangle$ for Ni-Cu alloys from diffraction experiments and from Eq. (12), based on ref. 7.

improved analysis of the old diffuse-scattering data[2,3] gives a satisfactory value of $\langle\mu\rangle$. The results of recent polarized-scattering experiments[5] give excellent agreement with the bulk magnetization values of both $\langle\mu\rangle$ and $d\langle\mu\rangle/dc$; the data is sufficiently good to lend quantitative support to our choice of $f(\vec{K})$. The correct choice of form factor and correct analysis of large-angle scattering data removes all inconsistencies

In principle, the concentration dependence of $\alpha$ can be used to judge the relative importance of conduction-electron and d-d overlap contributions to the nonlocal part of $\langle\mu\rangle$. If $\langle\mu^{nonloc}\rangle$ were due to conduction-electron spin polarization, to first order $\alpha$ should be independent of c. On the other hand, if it were due to d-d overlap it should depend on the number of host-host bonds and hence on c according to the formula

$$\alpha_{eff} = \alpha_1 - c(\alpha_1 - \alpha_2)[1 + (1-c)g_1/\langle\mu\rangle_{host}], \qquad (12)$$

where $\langle\mu\rangle_{imp}$ is assumed to be zero, $\alpha_1$ and $\alpha_2$ are independent of c and measure host-host and impurity-host bonding, respectively, and $g_1 \simeq -.035$ is defined by Marshall[1] and is known experimentally. For Ni-Cu alloys, one may assume $\alpha_2 \simeq 0$; Fig. 3 compares the prediction of Eq. (12) for Ni-Cu alloys with the values of $\alpha$ given by diffraction experiments.[4] The experimental values support the point of view of Moon, but better experimental results clearly are desirable.

REFERENCES

*Supported in part by the U. S. Energy Research and Development Administration.

†IVIC Graduate participant from Georgia Institute of Technology, Atlanta, Georgia 30332.

‡Operated by Union Carbide Corporation for ERDA.

1. W. Marshall, J. Phys. C 1, 88 (1968); E. Balcar and W. Marshall, ibid., 966 (1968).
2. J. W. Cable, E. O. Wollan, and H. R. Child, Phys. Rev. Lett. 22, 1256 (1969).
3. A. T. Aldred, B. D. Rainford, T. J. Hicks, and J. S. Kouvel, Phys. Rev. B 7, 218 (1973).
4. Y. Ito and J. Akimitsu, J. Phys. Soc. Japan 35, 1000 (1973).
5. R. Medina and J. Cable, in this volume.
6. R. M. Moon, AIP Conf. Proc. 24, 425 (1975).
7. The results shown are typical of several calculations, including two from R. E. Watson and A. J. Freeman, Phys. Rev. 178, 725 (1969).
8. H. A. Mook, Phys. Rev. 148, 495 (1966).

## CRYSTAL FIELD EFFECTS ON THERMAL EXPANSION AND MAGNETOSTRICTION OF TmSb

H.R. Ott
Laboratorium für Festkörperphysik, ETH Zürich,
8049 Zürich, Switzerland
B. Lüthi
Physics Department, Rutgers University
Now Brunswick, N.J. 00903

ABSTRACT

The Schottky-anomaly of the low temperature specific heat $c_p$ is a well known crystal field effect. Recently it has been shown that the splitting of crystal field levels also leads to anomalies in the elastic constants [1]. According to the Grüneisen-relation, the linear thermal expansion coefficient α should show similar anomalies as well. We have measured the thermal expansion along the [100]-axis of a TmSb single crystal between 1.5 K and 20 K and we observe a negative Schottky-type anomaly of the order of $2 \times 10^{-6}$ which corresponds nicely with the anomaly in $c_p$ [1]. The Grüneisen-constant $\gamma = (3\alpha\Omega/\kappa c_p)$ is a measure of the volume dependence of the crystal field energy level splitting. From our measurements below 10 K we deduce a reduction of the splitting of the lowest energy levels with decreasing volume, being in contrast to point charge model predictions. We obtain (with $\Omega = 33.89\ cm^3$, $\kappa = 15 \times 10^{-13}$ cgs) $\gamma = -1.4$. Above 10 K the measured thermal expansion strongly deviates from the Schottky-type anomaly, indicating other than pure crystal field contributions.

In addition we have measured the magnetostriction ε along the [100]-axis at constant temperatures in magnetic fields up to 15 kOe. These measurements allow in principle a determination of the magnetic ion-lattice coupling-constant $g_2$ [1] in magnitude and sign and a comparison with values obtained from elastic constant experiments can be made. Below 4 K we observe $\varepsilon = 10.8 \times 10^{-6}$ in 15 KOe. Above 4 K ε rises to $14 \times 10^{-6}$ at 10 K and slowly decreases thereafter. The crystal field contribution to the magnetostriction is mainly due to quadrupolar effects (coupling constant $g_2$ in ref.1 or $B^{\gamma}_{02}$ in ref.2). The higher order term ($B^{\alpha}_{04}$) which determines α is negligibly small. In order to obtain agreement between theory and experiment, exchange-striction effects have to be taken into account. Recent measurements of the pressure dependence of the magnetic susceptibility have led to similar conclusions [3]. A more detailed account will be published elsewhere.

REFERENCES

1.) M.E. Mullen, B. Lüthi, P.S. Wang, E. Bucher, L.D. Longinotti, J.P. Maita and H.R. Ott, Phys. Rev. B10, 186 (1974)
2.) E.R. Callen and H.B. Callen, Phys.Rev.129, 578 (1963)
3.) R.P. Guertin, J.E. Crow, L.D. Longinotti, E. Bucher, L. Kupferberg, S. Foner, Phys. Rev. B12, 1005 (1975)

# MAGNETIC PROPERTIES OF $ScFe_2$*

S. G. Sankar and W. E. Wallace
Department of Chemistry
University of Pittsburgh
Pittsburgh, Pennsylvania 15260

ABSTRACT

$ScFe_2$ crystallizes in the hexagonal C-14 Laves phase. Magnetic measurements performed in the temperature range 4-600 K reveal that the sample orders ferromagnetically with a Curie temperature of 545 K. Saturation magnetization measurements performed at 4.2 K yield a value of 2.9 $\mu_B$ per formula unit of $ScFe_2$. The easy axis of magnetization was determined by examining the changes in the intensities of x-ray pattern on aligning a powder at room temperature in a field of 20 kOe. This was found to be along the c-axis. Mössbauer pattern at room temperature exhibits a magnetic hyperfine interaction with a complex spectrum.

## INTRODUCTION

$RFe_2$ compounds (where R = heavy rare earth) have been extensively studied during the past decade with a view to understand their structural and magnetic characteristics.[1] In general, they crystallize in cubic C-15 type Laves structure. However, their analogue $ScFe_2$ exhibits a hexagonal structure. Early work on the structural aspect of $ScFe_2$ indicated that it possesses a $MgNi_2$-type structure.[2] Recent re-examination of the x-ray work revealed that the material crystallizes in $MgZn_2$-type structure.[3] In view of the desire to understand the magnetic properties of the entire family of $RFe_2$ compounds, magnetic measurements and Mössbauer studies of $ScFe_2$ have been undertaken.

## EXPERIMENTAL

The sample of $ScFe_2$ was prepared by melting together stoichiometric quantities of Sc (purity 99.9%) and Fe (purity 99.99%). The resultant alloy button was annealed under vacuum at 1000 C for one week. X-ray diffraction pattern taken using Cu-$K_\alpha$ radiation confirmed that the sample was single phase and the lines could be indexed on the basis of a hexagonal C-14 type structure. The lattice constants calculated are (a = 4.953 Å, c = 8.093 Å) in reasonable accord with literature. Magnetization was measured by the Faraday technique. The temperature dependence of the magnetization ($\sigma$) was recorded in the range 4.2 to 600 K at an applied field of 5 kOe. The Curie temperature was estimated from a $\sigma^2$ vs T curve. Saturation magnetization measurements were performed at 4.2 K with the applied field varying from 0 to 20 kOe. These results are shown in figure 1.

## RESULTS AND DISCUSSION

The Curie temperature obtained from the magnetic measurements is 545 K. This value is in close agreement with that reported by Ikeda et al.[4] However, these authors have not reported the saturation magnetic moment in $ScFe_2$. Saturation magnetization value calculated from our experiment is 2.90 $\mu_B$ per formula unit. This value is considerably smaller than that measured in metallic iron and is closer to those of $YFe_2$ and other $RFe_2$ compounds. Further, the slow development of saturation of magnetic moment up to about 10 kOe is suggestive of the fact that the compound may exhibit a high magnetic anisotropy. Anisotropy field was measured by casting an oriented powder into a sphere and measuring the magnetization parallel and perpendicular to the easy direction. The anisotropy field was estimated to be 45 kOe.

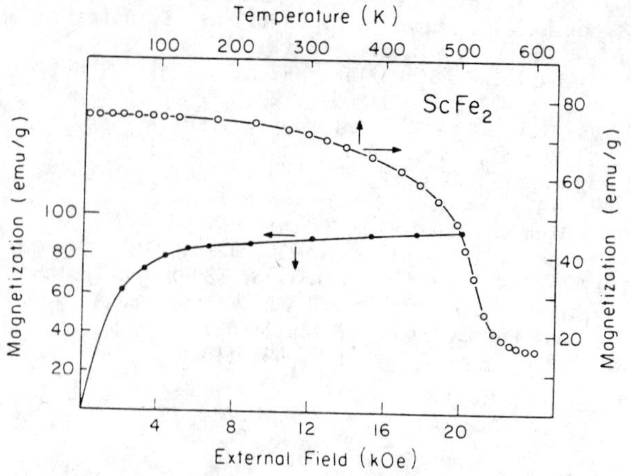


Figure 1. Variation of magnetization as a function of field and of temperature.

$ScFe_2$ is a unique member of the $RFe_2$ series by virtue of its tendency to crystallize in the hexagonal Laves phase. Therefore, it was considered desirable to establish its easy direction of magnetization. This was accomplished by examining the x-ray diffraction pattern of an aligned sample. The results are indicative of the fact that $ScFe_2$ possesses an easy c-axis at room temperature.

$Fe^{57}$ Mössbauer spectra were obtained using an Elron unit at room temperature and at 78 K. Two six-line spectra are observed, which may be associated with the two crystallographically distinguishable types of iron atoms. The unit cell of $ScFe_2$ contains eight iron atoms - six equivalent atoms in the 6h sites and two equivalent atoms in 2a sites. However, because of the complex nature of the spectrum, detailed analysis was not made. The hyperfine fields calculated from a fit to a six-line spectrum is 188 kOe at room temperature and 197 kOe at 78 K. These values are remarkably lower than that of metallic iron but are close to the other members of the $RFe_2$ series.[5]

In conclusion, it may be pointed out that although $ScFe_2$ is unique crystallographically, its magnetic properties resemble closely those of $YFe_2$.[6]

## REFERENCES

* This work is supported by a grant from ERDA.

1. W. E. Wallace, Rare Earth Intermetallics, Chapter 11, Academic Press Inc., New York (1973)

2. A. E. Dwight, Trans. of the A.S.M. 53, 479-500 (1961)

3. A. E. Dwight, C. W. Kimball, R. S. Preston, S. P. Taneja and L. Weber, J. Less Common Metals 40, 285-291 (1975)

4. K. Ikeda, T. Nakamichi, T. Yamade and M. Yamamoto, J. Phys. Soc. Japan 36, 611 (1974).

5. W. E. Wallace, J. Chem. Phys. 41, 3857-3863 (1964)

6. K. H. J. Buschow, J. Less Common Metals 40, 361-363 (1975)

# DETERMINATION OF THE ENERGY LEVEL PARAMETERS AND CRYSTAL FIELD SPLITTING FOR Fe IN Cu

J. B. Boyce
Xerox Palo Alto Research Center
Palo Alto, California 94304

C. P. Slichter*
Materials Research Laboratory and Department of Physics
University of Illinois
Urbana, Illinois 61801

## ABSTRACT

Values for the Friedel-Anderson energy level parameters for Fe in Cu have been obtained using the measured conduction electron spin density around the Fe impurity. The expression for the spin density due to Jena and Geldart was applied to the NMR data on the spectral shifts of the Cu near neighbor resonances. A good explanation for this data could not be obtained unless the crystal field splitting of the Fe d-levels was taken into account. Direct experimental evidence for such a non-negligible crystal field splitting has been obtained in the single crystal NMR studies of T. Stakelon. The energy level parameters we obtain are: t - e crystal field splitting, $\Delta$ = 0.5eV; virtual level width, $\Gamma_d$ = 0.7eV; Coulomb splitting of spin-up and spin-down virtual levels, U = 5.6eV. These results are consistent with CuFe being magnetic in the Anderson sense ($U/\Gamma_d = 8 > 1$) and yield, via the Friedel sum rule, a total of seven d-electrons with a net spin of 1.25.

## INTRODUCTION

The electronic structure of transition element impurities in simple metal hosts is usually discussed in terms of virtual bound states described by the Friedel-Anderson model.[1,2] In this model the density of states of the host is modified by the addition of the d-levels of the transition metal impurity. These d-states are broadened by mixing with the host conduction electrons, giving rise to Lorentzian peaks in the density of states for each direction of spin, $\sigma$, centered at $E_d^\sigma$ and having width $\Gamma_d$. This picture of the electronic structure of dilute alloys is consistent with the direct measurements of optical and photoemission studies[1] and also with the indirect probes provided by susceptibility, resistivity, specific heat, etc.[2]

The effect of the cubic crystal field on the above band structure has been considered by Yafet.[3] He concludes, however, that the crystal field splitting is small compared with the virtual level width, at least in the case of CuNi. Evidence that this may not be the case for CuFe and CuCo is given by the single crystal NMR studies of T. Stakelon.[4] He observes that in CuCo there is a pseudodipolar interaction between the Co and the first neighbor Cu nuclei which is not axially symmetric about the Cu-Co internuclear vector. This result fits the symmetry of the site about this axis (it is a two-fold axis); however an axially symmetric result would be found if one could neglect all atoms other than the Co and the Cu whose nucleus is under study. Thus the crystal potential must play a role. Below we consider the effect of this crystal field splitting on the satellite NMR spectrum of CuFe and, using the theory of Jena and Geldart,[5] obtain an approximate set of electronic structure parameters for the Fe impurity.

## RESULTS AND DISCUSSION

The NMR spectrum of CuFe consists of the large main line signal due to $Cu^{63}$ nuclei far from the Fe impurity plus five small satellite resonances, labeled A, B, C, M, and N, due to $Cu^{63}$ nuclei near the Fe atom.[6] The change in the Knight shift of a satellite at position r relative to the Fe impurity, $\Delta K(r)$, divided by the pure Cu Knight shift, K, is given by

TABLE I

Comparison of the experimental and theoretical results for $\Delta K/K$ at 300K.

| | | $\Delta K/K$ at 300K | |
|---|---|---|---|
| Shell | r (A°) | Experiment (Satellite) | Jena and Geldart with crystal field splitting |
| 1 | 2.55 | -5.24±0.3 (A) | -5.28 |
| 2 | 3.61 | 1.85±0.03(M) | 1.78 |
| 3 | 4.42 | -1.20±0.03(B) | -1.41 |
| 4 | 5.10 | -0.36±0.02(C) | -0.34 |
| 5 | 5.71 | [0.28±0.03(N)] | 0.21 |
| 6 | 6.25 | | 0.10 |
| 7 | 6.75 | | -0.12 |
| 8 | 7.22 | | 0.12 |
| 9 | 7.66 | | 0.09 |
| 10 | 8.07 | | 0.02 |

$$\Delta K(r,T)/K = C\,\sigma(r,T)/H, \qquad (1)$$

where $\sigma(r,T)$ is the conduction electron spin density at position r and temperature T, H is the external magnetic field and C is a constant for a given host. The experimental values for $\Delta K/K$ for the five satellites observed are listed in Table I. Satellites A, M, B, and C are due to the first, second, third, and fourth shells of neighbors, respectively, as determined by considerations of satellite lineshape, width, and intensity.[7] This identification is in agreement with T. Stakelon's single crystal studies[4] where M is determined to be the second neighbor and B the third. The identification of N is uncertain.

Jena and Geldart (JG)[5] have obtained an expression for $\sigma(r,T)$ that is based on the Friedel-Anderson model, and, since it takes into consideration the wave number dependence of the scattering and the magnetic ion structure, it is valid close to the impurity. They obtain for the change in the number density of electrons of spin $\sigma$

$$\delta n^\sigma(r) = -\frac{5}{4\pi^2}\frac{\Lambda^\sigma(r)}{r^3}\sin\delta_d^\sigma(E_F)\cos[2k_F r + \delta_d{}^\sigma(E_F) + \theta^\sigma(r)]. \qquad (2)$$

$\delta_d^\sigma(E_F)$ is the scattering phase shift evaluated at the Fermi energy, $E_F$, and $k_F$ is the Fermi wavevector. $\Lambda^\sigma(r)$ and $\theta^\sigma(r)$, which are functions of $\Gamma_d$ and $\delta_d^\sigma(E_F)$ given by JG, are corrections in close to the impurity. $\Lambda^\sigma(\infty) = 1$ and $\theta^\sigma(\infty) = 0$ so that for r large Eq. (2) reduces to the usual asymptotic expression. Using the fact that

$$2\sigma(r,T) = [\delta n^\uparrow(r) - \delta n^\downarrow(r)]\;|\langle S_z(T)\rangle|/S, \qquad (3)$$

Eqs. (1)-(3) specify $\Delta K/K$. In Eq. (3), $\langle S_z(T)\rangle$ is the thermal average of the z-component of the Fe spin, S. Both are given by the measured susceptibility if one assumes g=2, i.e., S=1.25 and $\langle S_z(T)\rangle \sim H/(T+29)$.[8] A good fit of the Knight shift data of Table I to these expressions could not be obtained. This might suggest that crystal field effects should be important, as suggested by the single crystal experiments of T. Stakelon.[4]

We modify the above expressions to include the effects of the crystal potential in the following two ways.

First, instead of having two five-fold degenerate virtual levels centered at $E_d^\sigma$, one has four virtual levels centered at $E_{d(t)}^\sigma$ and $E_{d(e)}^\sigma$. $E_{d(t)}^\sigma$ refers to the threefold degenerate $t_{2g}$ orbitals (xy, xz, yz) and $E_{d(e)}^\sigma$ refers to the twofold degenerate $e_g$ orbitals $(x^2 - y^2, 3z^2 - r^2)$. We assume that the $e_g$ orbitals lie above the $t_{2g}$ orbitals in energy and their splitting is $\Delta$, the crystal field splitting. (This ordering is to be expected since the nearest neighbor atoms are in the (110) direction. We tried unsuccessfully to fit the data with the crystal field levels inverted.) We take the widths of all four virtual levels to be the same and the Coulomb splittings of the $t_{2g}$ and $e_g$ orbitals to be the same.

The second modification of the JG theory due to the crystal field splitting is to weight the spin density due to scattering from the $t_{2g}$ and $e_g$ orbitals differently in different crystal directions.[8] These weighting factors, $W_e(\vec{r})$ and $W_t(\vec{r})$, take into account the different directional dependence of the $t_{2g}$ and $e_g$ orbitals and give rise to the experimentally observed anisotropy in the single crystal experiments. $W_e(\vec{r})$ and $W_t(\vec{r})$ are readily calculated for the neighboring shells of atoms with the known directional dependence of the $t_{2g}$ and $e_g$ orbitals and with the normalization condition that $W_e(\vec{r}) + W_t(\vec{r}) = 1$.

With the above modifications due to the crystal field splitting Eq. (1) becomes

$$\frac{\Delta K}{K}(\vec{r},T) = C\,\frac{1}{2}\,\frac{|\langle S_z(T)\rangle|}{HS}\Big\{W_t(\vec{r})[\delta n_t^\uparrow(r) - \delta n_t^\downarrow(r)] + W_e(\vec{r})[\delta n_e^\uparrow(r) - \delta n_e^\downarrow(r)]\Big\} \qquad (4)$$

where $\delta n_t^\sigma(r)$ and $\delta n_e^\sigma(r)$ are given by Eq. (2) with t and e subscripts on the phase shifts. The Friedel sum rule puts the following constraints on the energy level parameters:

$$\begin{Bmatrix}3\\2\end{Bmatrix}\left[\delta_{d\begin{Bmatrix}t\\e\end{Bmatrix}}^{\uparrow}(E_F) + \delta_{d\begin{Bmatrix}t\\e\end{Bmatrix}}^{\downarrow}(E_F)\right] = \pi\, N_{d\begin{Bmatrix}t\\e\end{Bmatrix}}$$

$$\begin{Bmatrix}3\\2\end{Bmatrix}\left[\delta_{d\begin{Bmatrix}t\\e\end{Bmatrix}}^{\uparrow}(E_F) - \delta_{d\begin{Bmatrix}t\\e\end{Bmatrix}}^{\downarrow}(E_F)\right] = \pi\, 2S_{\begin{Bmatrix}t\\e\end{Bmatrix}} \qquad (5)$$

$$N_d = N_{d(t)} + N_{d(c)}$$

$$S = S_t + S_e$$

$N_d$ is the total number of electrons on the Fe d-level and S = 1.25.[8] With these constraints, three parameters, $\Gamma_d$, $S_e$, and $N_{d(e)}$, determine $\Delta K/K$. These parameters were varied over a physically reasonable range and a good fit to the satellite Knight shifts at 300K was obtained. Satellite N was excluded from the fit since its identification is uncertain. The parameters obtained are: $\Gamma_d$=0.71eV, $\Delta$=0.52eV, U=exchange splitting=5.6eV, $E_{d(t)}^\uparrow=0.19E_F$, $E_{d(t)}^\downarrow=0.99E_F$, $E_{d(e)}^\uparrow=0.26E_F$, $E_{d(e)}^\downarrow=1.06E_F$, and $N_d$=7 electrons. The virtual bound states that correspond to these parameters are shown in Fig. 1. The fit yielded a total of 7 d-electrons on the Fe atom corresponding to one of the two s-electrons on the isolated Fe atom going into the Cu conduction band and the other into the Fe d-band. This agrees with $N_d$ determined from resistivity measurements.[2] The value of $U/\Gamma_d = 8 > 1$ is consistent with CuFe being magnetic in the Anderson sense and a width of 0.7eV and an exchange splitting of 5.6eV are in rough agreement with estimates made from other experiments.[1,2] For example, optical experiments on AgMn yield $\Gamma_d \simeq 0.5$eV and U $\simeq$ 5eV.[1] The value of the crystal field splitting for CuFe is comparable to the virtual level width, as it should be for crystal field effects to be important.

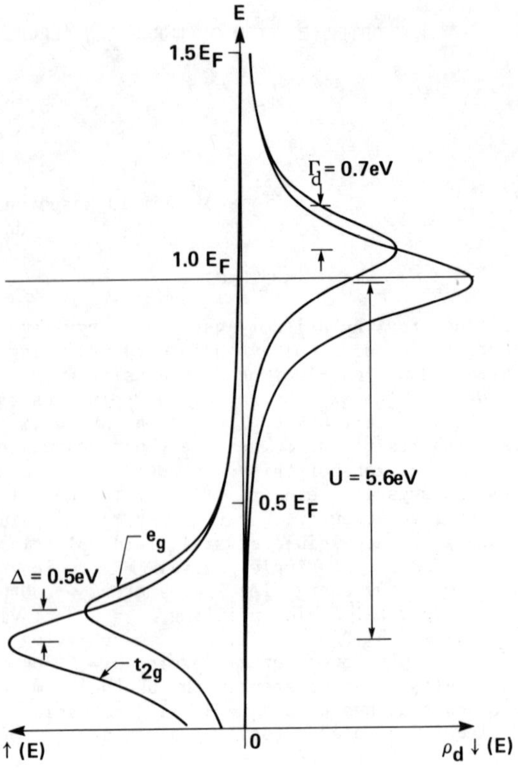


Fig. 1: Extra density of d-states due to an Fe impurity in Cu.

The values of $\Delta K/K$ determined in the above manner are listed in Table I. These results with crystal field splitting are seen to fit the data quite well, deviating at most by 15% at the third neighbor. In addition, it is seen that the experimental shift of satellite N is best fit by the calculated value for the fifth neighbor. Since the calculated values of $\Delta K/K$ were obtained by fitting only to the first four neighbor shifts, one may possibly conclude that satellite N is the fifth neighbor. Also, it should be noted that the calculated $\Delta K/K$'s for the sixth through tenth neighbors are $\leq 0.12$. This is not within the experimental resolution and so is consistent with the fact that only the first five shells of neighbors were observed as resolved satellites to the main line.

We believe that our results illustrate physically the way in which a crystalline potential introduces angular dependence into the spin density. A better theoretical grounding is needed before one can have an estimate of how seriously to take the numerical results.

*This research was supported in part by the Energy Research and Development Administration under Contract AT(11-1)-1198.

REFERENCES

1. G. Gruner, Adv. in Phys. 23, 941 (1974).
2. C. Rizzuto, Rep. Prog. Phys. 37, 147 (1974).
3. Y. Yafet, Phys. Lett. 26A, 481 (1968); Y. Yafet, J. Appl. Phys. 39, 853 (1968).
4. T. Stakelon, Thesis, University of Illinois, 1974 (unpublished).
5. P. Jena and D. J. W. Geldart, Phys. Rev. B7, 439 (1973).
6. J. B. Boyce and C. P. Slichter, Phys. Rev. Lett. 32, 61 (1974).
7. J. B. Boyce and C. P. Slichter, Phys. Rev., to be published.
8. J. L. Tholence and R. Tournier, Phys. Rev. Lett. 25, 867 (1970).

# EFFECTS OF A NON MAGNETIC IMPURITY ON THE APPEARANCE OF MAGNETISM IN A KONDO SYSTEM

R. Guérinot and R. Tournier
Centre de Recherches sur les Très Basses Températures, C.N.R.S.,
B.P. 166 Centre de Tri, 38042 Grenoble-Cédex, France.

ABSTRACT

In order to verify the effect of one impurity on the variation of local density of states and the appearance of magnetism, we have introduced Co atoms at low concentration c into $\underline{Cu}Fe_x$ dilute alloys.

Some previous results, namely magnetization and initial susceptibility measurements are briefly recalled. We have observed a large enhancement of the number of magnetic carriers and we calculate their ratio to the number of Fe atoms.

We present new results on susceptibility measurements extended to the temperature range $0.005 \leqslant T \leqslant 1.3$ K.

The system exhibits spin glass properties : an ordering temperature $T_M(c,x)$ evolving from $x^2$ to cx law and an ordering susceptibility $\chi_M(c)$ weakly c-dependent. We deduce the magnetic ratio and we compare to $\underline{Cu}Fe_x$ properties.

## INTRODUCTION - PREVIOUS RESULTS

Starting from the local Friedel-Anderson condition of magnetism, it was suggested that the oscillations of charges induced by other impurities generate a local variation in the density of states of the virtual bound state at the Fermi level and are able to induce magnetism[1].

We have inserted Co atoms in very dilute $\underline{Cu}Fe_x$ alloys ($x \leqslant 2.10^{-4}$) for it has been proved that in the low concentration limit ($c \leqslant 5.10^{-3}$) Co atoms are in a non magnetic state[2].

The 13 $\underline{Cu}Co_cFe_x$ samples, the $\underline{Cu}Co_c$ and $\underline{Cu}Fe_x$ matrix have been elaborated in the same conditions. The atomic concentrations c and x given here are nominal ; a spectroscopic analysis of excess melts does not show large differences between nominal and real concentrations.

Magnetization measurements at T=1.3 K in the field range (0,90 kOe) for $0 \leqslant c \leqslant 3.10^{-3}$ show a saturation magnetization $\sigma_s(c,x)$ proprotional to cx in the field $H_s$=60 kOe[3]. We observe a large enhancement of $\sigma_s$ (fig. 1)

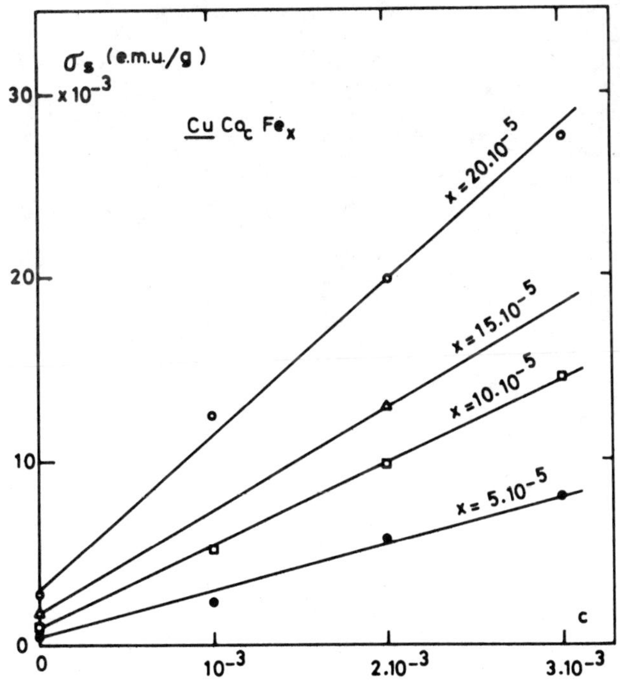


Fig. 1 : Saturation magnetization $\sigma_s$ vs c. The values for c = 0 are those of ref. 4. The c axis corresponds to x = 0.

and we conclude that the effect of Fe-Co interactions is stronger than Fe-Fe or Co-Co interactions. The calculation of $\alpha_1 = \frac{\sigma_s}{\sigma_\circ}$, where $\sigma_\circ$ is the total moment of all Fe atoms with a saturation moment $\mu_{Fe}=2.4\ \mu_B$ at the concentration x gives a first evaluation of the ratio of the number of magnetic carriers to the number of Fe atoms. $\alpha_1$ is reported in the Table.

In a first approach, the initial susceptibility $\chi_i = \frac{\partial M}{\partial H}\Big|_{H<1\ kOe}$ measured in the range $1.3\ K \leqslant T \leqslant 4.2$ K may be fitted according to the law $\chi_i = \chi_\circ + \frac{C}{T+\theta}$, where C is the Curie constant, with the following results :

i) $\chi_\circ(c,x) \simeq \frac{\partial M(c,x)}{\partial H}\Big|_{H \geqslant H_s}$ with $H_s$ = 60 kOe.

ii) $\theta(c,x) = \langle T_K \rangle \simeq 0.4$ K with $\frac{\partial\theta}{\partial c}\Big|_x < 0$
$\theta(c,x) \ll \theta(0,x) < 29$ K, the Kondo temperature of Fe in Cu.

iii) $C(c,x) \propto cx$ for $c \geqslant 2.10^{-3}$ and $C(c,x) \gg C(0,x)$

The evolution of $\theta(c,x)$ and $C(c,x)$ with c indicates the increase in the number of magnetic Fe atoms.

Assuming that the apparent spin S' entering into the Curie constant C has the same value as in the expression of $\sigma_s$, we can get both S' and the concentration n of carriers. S'(c,x) is found to be 2.35±0.20. $\alpha_2$=n/x is reported in the table. The order of magnitude of $\alpha_2 \simeq \alpha_1/2$ indicates that the critical radius $R_c$ of the $\underline{Cu}Co_cFe_x$ system is c-independent for $0 \leqslant c \leqslant 3.10^{-3}$ and the critical sphere includes about z=120 sites[3,4]. Thus a Co atom in the vicinity of a Fe atom plays the same role as an Fe near another Fe.

## INITIAL SUSCEPTIBILITY AT LOW TEMPERATURE SPIN GLASS PROPERTIES

The results mentioned above have been completed by initial susceptibility measurements in a field less than 160 Oe in the temperature range $0.05\ K \leqslant T \leqslant 1.3$ K by adiabatic demagnetization technique and we have extended the measurements to the concentration c = $4.10^{-3}$.

We have determinated the order temperature $T_M$ and the corresponding initial susceptibility $\chi_M$ for 6 samples ($T_M \gtrsim 0.05$ K for $cx \gtrsim 3.10^{-7}$).

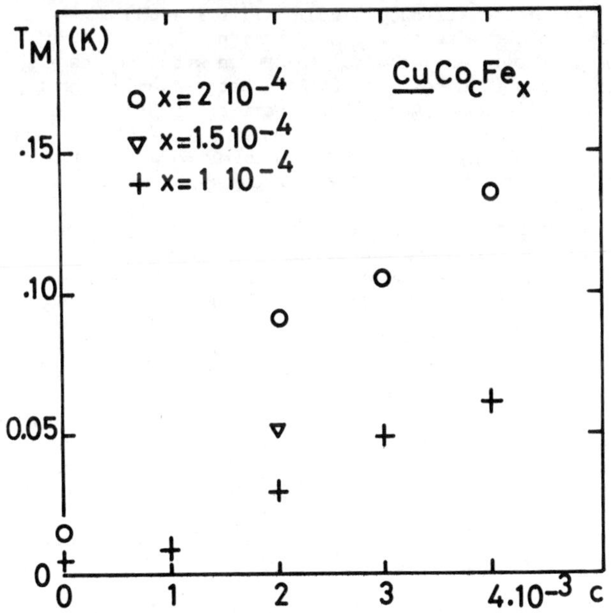


Fig. 2 : Ordering temperature $T_M$ vs c. The values for c = 0 are those of ref. 6.

Other measurements have been performed in the range $0.005\ K \leq T \leq 0.05\ K$ using SQUID techniques and a dilution $He^3$-$He^4$ refrigerator[5]. At the present time we have the results concerning two samples ($c=10^{-3}$ ; $x=10^{-4}$) ($c=2.10^{-3}$ ; $x=10^{-4}$) and the results given for c=0 by Hirschkoff et al.[6] (fig. 2).

$T_M$ is shown to evolve gradually from a $x^2$ law (c=0) to a cx law ($c \gtrsim 3.10^{-3}$). The effect of Co atoms on Fe atoms is important in the transition region at low Co concentration.

These properties are reminiscent of $\underline{Cu}Fe_x$ and $\underline{Au}Fe_x$ and we can expect scaling laws for $\underline{Cu}Co_cFe_x$.

The ordering temperatures $T_M$ and susceptibilities $\chi_M$ are respectively :

$T_M \simeq 3.5\ 10^5\ x^2$ $\quad \chi_M \simeq 10.10^{-6}$ uem/g for $\underline{Cu}Fe_x$[5],

$T_M \simeq 6\ 10^6\ x^2$ $\quad \chi_M \simeq 2.10^{-6}$ uem/g for $\underline{Au}Fe_x$[6].

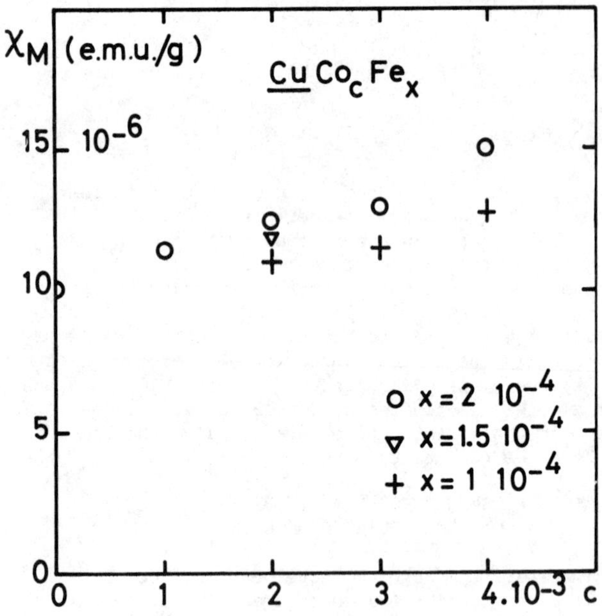


Fig. 3 : Ordering susceptibility $\chi_M$ vs c. The value for c = 0, $x = 2.10^{-5}$ is that of ref. 6.

According to (5), we could write $T_M \simeq 5800\ c^*$ where $c^* = zx^2$ is the effective concentration of pairs of Fe atoms. In our case we write $T_M \simeq 2900\ n$ with n=concentration of magnetic Fe atoms ($n=2c^*$) because a Co atom replaces an Fe atom in a pair. Referring to previous results on $\underline{Au}Co_cFe_x$[7], it has been already observed that the variation of c did not affect $T_M$ while $\chi_M$ was modified linearly with c.

From our values of $T_M$ we deduce n and we estimate the ratio $\alpha_3=n/c$ which is close to $\alpha_2/2$. We conclude that only half the number of magnetic Fe atoms -of which the moment is saturated in 60 kOe- contribute to magnetic ordering. This is a property of $\underline{Cu}Fe$ systems.

The presence of magnetic clusters in $\underline{Cu}Fe_x$ alloys has been suggested[8] and the question is now if short range clustering is not a dominant effect.

The results reported figure 1 are obtained by difference between the saturation magnetization of $\underline{Cu}Co_cFe_x$ alloys and the corresponding $\underline{Cu}Co_c$ alloys. The effect of the clustering of Co atoms between them is eliminated (and it has been shown[2] that this clustering is weak for $c<5.10^{-3}$). On the other hand, the magnitude of the observed effect cannot be explained by Co-Fe clustering alone.

We have observed on another sample ($c=5.10^{-3}$, $x=2.10^{-4}$) some deviations to the present simple results which could be attributed to such a clustering.

TABLE OF MAGNETIC RATIOS

| c | $1.10^{-3}$ | 1 | 1 | 2 | 2 | 2 | 2 | 3 | 3 | 3 | 4 | 4 |
|---|---|---|---|---|---|---|---|---|---|---|---|---|
| x | $5.10^{-5}$ | 10 | 20 | 5 | 10 | 15 | 20 | 5 | 10 | 20 | 10 | 20 |
| $\alpha_1$ | .21 | .25 | .30 | .54 | .46 | .41 | .47 | .77 | .69 | .66 | | |
| $\alpha_2$ | .15 | | .11 | .27 | .27 | .20 | | .40 | .37 | .31 | | |
| $\alpha_3$ | | .03 | | | .10 | .10 | .15 | | .16 | .18 | .22 | .23 |

CONCLUSION

The properties of $\underline{Cu}Co_cFe_x$ and $\underline{Cu}Fe_x$ systems are analogous but the number of magnetic Fe atoms becomes greater in the first system.

If we compare the weak magnitude of RKKY interactions between Fe atoms and between Co atoms to the Co-Fe interaction, we think the appearance of magnetism in Fe atoms depends on the variation of local density of states.

REFERENCES

1 - R. Tournier, LT 13, Vol. 2 (1973).
2 - B. Tissier and R. Tournier, Proc. 14th Int. Conf. Low Temp. Physics, edited by M. Krusius and M. Vuorio (Helsinki, Finland, 1975) (North-Holland Publ. Co., Amsterdam, Oxford - American Elsevier Publ. Co., New York) Vol. 3, p. 378.
3 - G. Chouteau, R. Guérinot, R. Tournier, Proc. 14th Int. Conf. Low Temp. Physics, edited by M. Krusius and M. Vuorio (Helsinki, Finland, 1975) (North-Holland Publ. Co., New York) Vol. 3, p. 346.
4 - J.L. Tholence and R. Tournier, Phys. Rev. Letters 25, 13, 867 (1970).
5 - G. Frossati et al. (to be published).
6 - E.C. Hirschkoff et al., J. of Low Temp. Phys. 5, 155 (1971).
7 - P. Costa-Ribeiro, Thesis Grenoble 1973 (unpublished).
8 - S. Mishra and P. Beck, Phys. Stat. Sol. (a) 19, 267 (1973).

# THE KONDO EFFECT IN THE ANDERSON MODEL*

Tadashi Arai
Argonne National Laboratory, Argonne, Illinois 60439

## ABSTRACT

We show that the Green's function solution of the Anderson model calculated by the functional derivative method developed earlier yields a density of states having the three-peak structure proposed by Grüner and Zawadowski and that the central peak indeed introduces the ℓnT dependence.

## INTRODUCTION

It is generally accepted that the problem of localized moments in metals may be discussed more naturally by the Anderson model[1] than by the s-d model, although the physics involved in the Kondo effect[2] can be explained more clearly by the s-d model. Of course, there are several treatments[3,4] of the Kondo effect based on the Anderson model, but it is difficult to understand from their results why spin-flip processes involved in the Kondo effect are not suppressed drastically when the Hartree-Fock density of states for d electrons at the Fermi level is negligibly small for a magnetic case. The Schrieffer-Wolff transformation[5] can convert the Anderson model to the s-d model, but this is possible only in the limit where the intra-atomic interaction U is larger than the width Δ of the localized levels. To circumvent the difficulties, Grüner and Zawadowski[6] have proposed a density of states exhibiting a three-peak structure. At high temperatures where many-body corrections are not strong enough to modify the Hartree-Fock results, two broad but well-separated peaks corresponding to the majority- and minority-spin electrons appear. The conduction-electron scattering is then described with two phase shifts. At zero temperature, on the other hand, the system is in the singlet ground state with a well-defined single phase shift, introducing a narrow third peak. They have assumed that this narrow resonance is responsible for the Kondo effect. Zlatic et al[7] have constructed a scattering amplitude by adding this narrow resonance but did not find the ℓnT dependence.

## CALCULATION OF THE GREEN'S FUNCTION

In this paper, we shall show that a Green's function solution of the nondegenerate Anderson model calculated by the functional derivative method developed earlier[6] yields a density of states having a three-peak structure and that the central peak indeed introduces a ℓnT dependence. Let us consider the equations of motion for the two types of Green's functions $\Gamma^{(\pm)}_{k\lambda\sigma} \equiv \langle\langle C_{d\sigma}(t)N^{(\pm)}_{n\bar\sigma}(t)C^{\dagger}_{\lambda\sigma}(t')\rangle\rangle$, where $N^{(+)}_{d\bar\sigma}(t) \equiv N_{d\bar\sigma}(t) = C^{\dagger}_{d\bar\sigma}(t)C_{d\bar\sigma}(t)$, $N^{(-)}_{d\bar\sigma}(t) = 1 - N_{d\bar\sigma}(t)$, and $\lambda = d$ or $k$. Here, in line with the Hubbard treatment[9] of narrow bands, the projection operators $N^{(\pm)}$ are inserted so that the intra-atomic interaction U is correctly calculated whenever an electron with opposite spin $\bar\sigma$ appears at the impurity site. The one-electron Green's function is then given by $G = \Gamma^{(+)} + \Gamma^{(-)}$.

The equations of motion involve the following two types of terms

$$\mathcal{G}_1=\sum_k V_{dk} \langle\langle C_{k\sigma}(t)[N_{d\bar\sigma}(t)-\langle N_{d\bar\sigma}\rangle]C^{\dagger}_{\lambda\sigma}(t')\rangle\rangle \quad , \tag{1}$$

$$\mathcal{G}_2=\sum_k\{V_{dk}\langle\langle C_{d\sigma}(t)[C^{\dagger}_{d\bar\sigma}(t)C_{k\bar\sigma}(t)-\langle C^{\dagger}_{d\bar\sigma}C_{k\bar\sigma}\rangle]C^{\dagger}_{\lambda\sigma}(t')\rangle\rangle$$
$$-V_{kd}\langle\langle C_{d\sigma}(t)[C^{\dagger}_{k\bar\sigma}(t)C_{d\bar\sigma}(t)-\langle C^{\dagger}_{k\bar\sigma}C_{d\bar\sigma}\rangle]C^{\dagger}_{\lambda\sigma}(t')\rangle\rangle\}, \tag{2}$$

where $V_{kd}$ is the hopping matrix element between conduction electron k and localized electron d. If one neglects $\mathcal{G}_1$ and $\mathcal{G}_2$, the equations can be solved immediately, yielding $G_0$. Since $\mathcal{G}_1$ and $\mathcal{G}_2$ can be reduced to functional derivatives of G, they can be calculated iteratively by inserting $G_0$. The reason why, instead of the Hartree-Fock solution, $G_0$ is used as the zeroth order approximation is that the intra-atomic interaction U then is included rigorously in each step of the perturbation calculation, making our result distinctly different from any other existing calculation. The result is

$$2\pi G^{-1}_{dd\sigma}(\omega) = \frac{(\omega-\varepsilon_d)(\omega-\varepsilon_d-U)-iG_{dd\sigma}(\omega)Y(\omega)}{\omega-\varepsilon_d-(1-n_{d\bar\sigma})U}$$
$$-\sum_k \frac{|V_{dk}|^2}{\omega-\varepsilon_k}\frac{[\omega-\varepsilon_d-(1-n_{d\bar\sigma})U]^2}{(\omega-\omega_a)(\omega-\omega_b)-iG_{dd\sigma}(\omega)Y(\omega)}, \tag{3}$$

where $\varepsilon_d$ and $\varepsilon_k$ are the energies of electrons k and d; $n_{d\sigma} = \langle N_{d\sigma}\rangle$, and

$$Y(\omega) = (U/2)^2 M_{\bar\sigma}[\omega-\varepsilon_d-(1-n_{d\bar\sigma})U](\omega-\omega_{-x})^{-1}(\omega-\omega_x)^{-1},$$
$$\omega_{a,b}=\varepsilon_d + (1-n_{d\bar\sigma})U \mp \sqrt{n_{d\bar\sigma}(1-n_{d\bar\sigma})}U,$$
$$\omega_{\pm x} = \varepsilon_d+(\tfrac{3}{4}-\tfrac{n_{d\bar\sigma}}{2})U \pm [iG_{dd\sigma}(\omega)](U/4), \tag{4}$$
$$M_{\bar\sigma} = \frac{1}{2}\sum_k[V_{dk}\langle C^{\dagger}_{d\bar\sigma}C_{k\bar\sigma}\rangle + V_{kd}\langle C^{\dagger}_{k\bar\sigma}C_{d\bar\sigma}\rangle] .$$

If $\sum_k|V_{dk}|^2/(\omega-\varepsilon_k)$ is replaced by $\varepsilon_k$, the structure of the above expression becomes exactly the same as that of the Hubbard model obtained in Ref. 10. The value of $\sum_k|V_{dk}|^2/(\omega-\varepsilon_k)$ may be evaluated as $\Gamma - i\Delta$, where $\Gamma(<0)$ and $\Delta$ are the effective energy shift and the width parameter of the d states, respectively. However, the term involving $iG_{dd\sigma}(\omega)Y(\omega)$ in the summation $\Sigma_k\ldots$ should be handled with care, since $iG_{dd\sigma}(\omega) = [1-f(\omega)]A(\omega)$, where $f(\omega)$ is the Fermi-Dirac distribution function and $A(\omega)$ the spectral function. In this particular term, $\sum_k|V_{dk}|^2/(\omega-\varepsilon_k)$ cannot be reduced to $\Gamma - i\Delta$, and instead the relation

$$\sum_k|V_{dk}|^2 f(\omega)/(\omega-\varepsilon_k) = \kappa(\omega)-i\mathcal{K}(\omega) \quad , \tag{5}$$

should be used where $\kappa(\omega)$ is the Kondo integral responsible for the ℓnT dependence of the resistivity and $\mathcal{K}(\omega) = f(\omega)\Delta$. Then

$$2\pi G^{-1}_{dd\sigma}(\omega) = \frac{(\omega-\varepsilon_d)(\varepsilon-\varepsilon_d-U) - [1-f(\omega)]A(\omega)Y(\omega)}{\omega-\varepsilon_d-(1-n_{d\bar\sigma})U}$$
$$\frac{(\Gamma-i\Delta)[\omega-\varepsilon_d-(1-n_{d\bar\sigma})U]^2}{(\omega-\omega_a)(\omega-\omega_b)-(1-k+i\ell)A(\omega)Y(\omega)+\ldots}\quad , \tag{6}$$

where $k = (\kappa\Gamma+\mathcal{K}\Delta)/(\Gamma^2+\Delta^2)$ and $\ell = (\mathcal{K}\Gamma-\kappa\Delta)/(\Gamma^2+\Delta^2)$.

## RESULTS AND DISCUSSION

Since $M_{\bar\sigma}$, Γ and Δ are all proportional to $|V^2|$, the Kondo term $(k-i\ell)$ is of order $|V^4|$ in $G_{dd\sigma}(\omega)$ and of order $|V^6|$ in the conduction electron Green's function $G_{kk\sigma}(\omega)$ in agreement with the result of the Schrieffer-Wolff transformation. However, the ℓnT dependence will appear only if the density of d electrons, $A(\omega)$, is finite at the Fermi level. We shall show this by solving the polynomial equation $G^{-1}_{dd\sigma}(\omega) = 0$ graphically. Since the method is completely parallel to that used in the Hubbard model,[10] we shall only outline the procedure and the results.

In the limit of $\Gamma=\Delta=0$, the poles of G are calculated by solving the quartic equation in $\omega$:

$$(\omega-\varepsilon_d)(\omega-\varepsilon_d-U) - [1-f(\omega)]A(\omega)Y(\omega) = 0 \quad , \tag{7}$$

where $f(\omega) \approx 0$ since we are calculating the scattering amplitude of conduction electron k traveling right above the Fermi level; $G_{dd\sigma}(\omega)$ is a part of $G_{kk\sigma}(\omega)$ for this electron. The term $Y(\omega)$ is generated by the correction term $\mathcal{G}_2$ and tends to introduce magnetism.

In fact, the solution calculated by Eq. (7) satisfies the condition for magnetism

$$\delta n_{d\sigma}(n_{d\bar{\sigma}})/\delta n_{d\bar{\sigma}} < -1 \quad . \tag{8}$$

As long as $M_{\bar{\sigma}}$ is small, however, the poles $\omega_1'$ and $\omega_2'$ will not deviate drastically from the isolated atomic levels $\varepsilon_d$ and $\varepsilon_d + U$ while the additional poles $\omega_{\pm x}'$ are physically unimportant since they are nearly equal to $\omega_{\pm x}$ and the spectral weights are vanishingly small.

The solutions of

$$(\omega-\omega_a)(\omega-\omega_b)-(1-k+i\ell)A(\omega)Y(\omega) = 0 \tag{9}$$

can be calculated similarly. At high temperatures where $\kappa(\omega)$ is small, $k \approx f(\omega)$ and $\ell \approx 0$ and the two solutions $\omega_a'$ and $\omega_b'$ will not be very different from $\omega_a$ and $\omega_b$, while the additional solutions $\omega_{\pm x}''$ are again virtually equal to $\omega_{\pm x}$ and their spectral weights are vanishingly small. We note that, for $1/2 < n_{d\sigma} \leq 1$ and $0 \leq n_{d\bar{\sigma}} < 1/2$, the value of $\omega_a$ changes drastically depending upon the value of $n_{d\bar{\sigma}}$ but $\omega_b$ remains more or less constant. In the fully magnetic limit $n_{d\sigma} = 1$ and $n_{d\bar{\sigma}} = 0$, $\omega_a$ is equal to $\varepsilon_d + U$ but the value decreases rapidly as the magnetization $n_{d\sigma} - n_{d\bar{\sigma}}$ decreases and, in the nonmagnetic limit $n_{d\sigma} = n_{d\bar{\sigma}} \approx 1/2$, $\omega_a$ becomes equal to $\varepsilon_d$. Meanwhile, $\omega_b$ increases from $\varepsilon_d + U$ to $\varepsilon_d + 1.2\ U$ and then decreases to $\varepsilon_d + U$. As the temperature is lowered, $\kappa(\omega)$ and $(1 - k)$ increase, making $\omega_a$ larger.

The real parts of the complete solutions $\bar{\omega}_1$, $\bar{\omega}_a$, etc., of Eq. (6) can then be calculated from intersections of the following two curves.

$$h = (\omega-\omega_1')(\omega-\omega_a')(\omega-\omega_{-x}')(\omega-\omega_{-x}'')(\omega-\omega_x')(\omega-\omega_x'')(\omega-\omega_2')(\omega-\omega_b') \tag{10}$$

$$g = \Gamma[\omega-\varepsilon_d-(1-n_{d\bar{\sigma}})U](\omega-\omega_{-x})^2(\omega-\omega_x)^2 \quad , \tag{11}$$

as is illustrated in Fig. 1. The contribution from the imaginary parts of the equations can be estimated perturbationally. For the nonmagnetic case, $\omega_1' = \omega_a'$ and $\omega_2' = \omega_b'$ and hence the difference between the two solutions $\bar{\omega}_1$ and $\bar{\omega}_a$ (or $\bar{\omega}_2$ and $\bar{\omega}_b$) is small as compared with their widths. Therefore, the two peaks $\bar{\omega}_1$ and $\bar{\omega}_a$ (or $\bar{\omega}_2$ and $\bar{\omega}_b$) will look like a broad single peak. As the magnetization $n_{d\sigma} - n_{d\bar{\sigma}}$ increases from zero to one, the difference $\omega_a' - \omega_1'$ increases, yielding two distinct peaks $\bar{\omega}_1$ and $\bar{\omega}_a$. Although $\bar{\omega}_1$ remains more or less constant, $\bar{\omega}_a$ increases from $\sim \varepsilon_d$ to $\sim \varepsilon_d + U$, while $\bar{\omega}_a$ and $\bar{\omega}_b$ will continue to exhibit a single peak

As the temperature is lowered, $\kappa(\omega)$ and hence $(1-k)$ and $\ell$ increase, making $\omega_a'$ and $\bar{\omega}_a$ larger. The calculated value of $n_{d\sigma}$ will then decrease and, to maintain the self-consistent requirement, Eq. (8), $n_{d\bar{\sigma}}$ has to increase, resulting in the reduction of the magnetization $n_{d\sigma} - n_{d\bar{\sigma}}$ as well as $\omega_a$.

Let us now summarize the foregoing results. At high temperature where the Kondo integral is negligible, a magnetic solution with the three-peak structure appears as long as the term $Y(\omega)$ which enhances magnetism dominates over the term proportional to $\Gamma - i\Delta$, which is generated from the correction term $\mathcal{G}_1$ and which tends to suppress magnetism. As long as there exists a small density of d states at the Fermi level $\varepsilon_F$, the value of $(1-k)$ increases as T decreases, tending to make $\omega_a$ larger. If the third peak $\omega_a$ lies below $\varepsilon_F$, the peak $\bar{\omega}_a$ will be shifted towards $\varepsilon_F$, increasing the density of d states and enhancing the $\ell nT$ dependence. As T is lowered further, the magnetization $n_{d\sigma} - n_{d\bar{\sigma}}$ will decrease, making $\omega_a$ smaller. Consequently $\omega_a$ will no longer increase and instead remain constant at $\sim \varepsilon_F$. If, on the other hand, the third peak $\bar{\omega}_a$ appears above the Fermi level, the tendency to increase $\bar{\omega}_a$ again introduces two competing procedures. One is the tendency to reduce the density of d states at $\varepsilon_F$ and hence the tendency to increase $\bar{\omega}_a$ will be suppressed greatly. The second effect is to reduce the magnetization $n_{d\sigma} - n_{d\bar{\sigma}}$, making $\omega_a'$ and hence $\bar{\omega}_a$ smaller. If $\bar{\omega}_a$ becomes smaller and approaches to $\varepsilon_F$, the density of d states at $\varepsilon_F$ increases, enhancing the $\ell nT$ dependence in $(1-k)$.

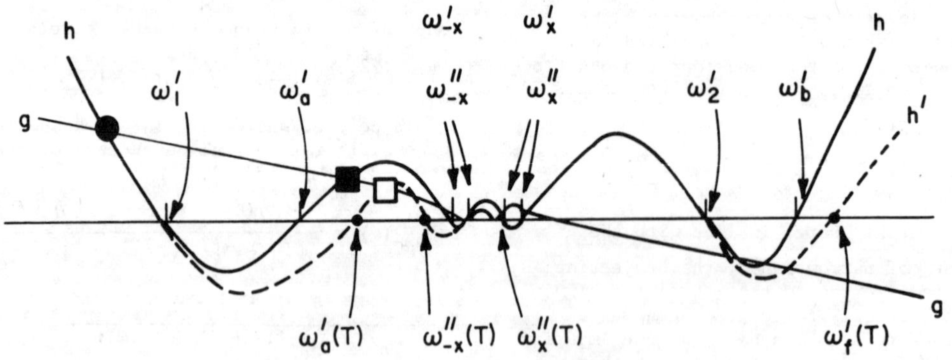


Fig. 1 Plot of h and g. The broken lines as well as $\omega_a'(T)$, $\omega_{\pm x}''(T)$ and $\omega_b'(T)$ illustrate deviations at lower temperature. •, ■, and □ denote $\bar{\omega}_1$, $\bar{\omega}_a$, and $\bar{\omega}_a(T)$, respectively.

Hence the third peak $\bar{\omega}_a$ will be shifted towards $\varepsilon_F$ and locked in at $\sim \varepsilon_F$.

As the magnetization $n_{d\sigma} - n_{d\bar{\sigma}}$ becomes small, however, $\omega_a'$ becomes nearly equal to $\omega_1'$ and will not decrease further. Then the mechanism to keep $\omega_a$ around $\varepsilon_F$ will disppear and $\omega_a$ starts to increase, thus reducing the density of d states at $\varepsilon_F$ and hence reducing the $\ell nT$ dependence. The system will then behave like spin fluctuations. This is possible because the imaginary parts of Eq. (6) are proportional to $[\omega-\varepsilon_d-(1-n_{d\bar{\sigma}})U]^n$, with n = 1 or 3, and hence the width of the third peak remains narrow.

REFERENCES

*Work performed under the auspices of the U. S. Energy Research and Development Administration.

1. Anderson, P. W., Phys. Rev. 124, 41 (1961).
2. Kondo, J., Progr. Theoret. Phys. (Kyoto) 32, 37 (1964).
3. Suhl, H., Phys. Rev. Letters 19, 442 (1967); Levine, M., and Suhl, H., Phys. Rev. 171, 567 (1968); Schrieffer, J. R., J. Appl. Phys. 38, 1143 (1967); Wang, S. Q., Evenson, W. E., and Schrieffer, J. R., Phys. Rev. Letters 23, 92 (1969); Hamann, D. R., Phys. Rev. Letters 23, 95 (1969); Phys. Rev. B2, 1373 (1970).
4. Also see the renormalization group theoretical treatment by Krishna-Murthy, H. R., Wilson, K.G., and Wilkins, J. W., PHys. Rev. Letters 16, 1101 (1975).
5. Schrieffer, J. R. and Wolff, P. A., Phys. Rev. 149, 491 (1966).
6. Grüner, G and Zawadowski, A., Solid State Comm. 11, 663 (1972).
7. Zlatic, V., Grüner, G. and Rivier, N., Solid State Comm. 14, 639 (1974).
8. Arai, T. and Tosi, M., Solid State Comm. 14, 947 (1974); Arai, T., Phys. Rev. Letters 33, 486 (1974); Arai, T and Tosi, M. P., submitted to Phys. Rev.
9. Hubbard, J., Proc. Roy. Soc. A276, 238 (1963); A281, 401 (1964).
10. Arai, T., AIP Conference Proc. 24, 341 (1974).

## THE POSSIBILITY OF INTERCONFIGURATION FLUCTUATIONS IN DILUTE Eu ALLOYS

Ron G. Pirich and C. R. Burr, Physics Department
State University of New York at Binghamton,
Binghamton, NY 13901

ABSTRACT

We originally conducted some preliminary magnetic susceptibility, electrical resistivity and X-ray diffraction measurements on a series of dilute alloys of Eu in Mg with concentrations of 2.60, 0.073 and 0.019 atomic percent Eu in Mg. At room temperature, the magnetic susceptibility suggests that the Eu ions are in the divalent $^8S_{7/2}$ state. However, below 10K, the magnitude and temperature independence of the magnetic susceptibility together with the electrical resistivity and X-ray diffraction results led us to consider the possibility that we may be observing an interconfiguration crossover or magnetic ordering of an intermetallic phase. Since there appears to be little metallurgical information about this system and in order to identify the mechanism producing this behavior, we have grown more concentrated MgEu compounds with concentrations of 5.70 and 8.80 atomic percent Eu in Mg.

Our measurements of all our alloys, both dilute and concentrated, indicate that they are homogeneous. The magnetic susceptibility of our more concentrated alloys, 5.70 at % and 8.80 at %, seems to be neither divalent or trivalent and is less than pure Mg and roughly temperature independent. X-ray analysis indicates the presence of a new structure which does not appear in our more dilute alloys.

# HYPERFINE STUDIES OF A VERY DILUTE ALLOY BY NUCLEAR ORIENTATION: $Pd^{54}Mn$

J. O. Thomson and J. R. Thompson
Department of Physics, University of Tennessee, Knoxville, Tennessee 37916

## ABSTRACT

Nuclear Orientation measurements have been made on a very dilute alloy of $^{54}Mn$ in Pd. The measurements were made in applied magnetic fields from 0 to 9000 Gauss and at temperatures from 11.5 to 100 mK.

Our results in low magnetic fields ($\lesssim$ 430 Gauss) are not describable by models based on a free paramagnetic Mn spin ($A\vec{I}\cdot\vec{S}$ coupling, slow electronic relaxation) or based on a simple hyperfine field approximation ($AI_zS_z$ coupling, rapid electronic relaxation). The data, however, can be fit with a model in which the local field axis is not always parallel to the applied field. This may arise from long range impurity interactions or relaxation/fluctuation effects. In high applied fields the data can be adequately described in terms of a hyperfine field model. We find that the magnitude of the saturation hyperfine field is (415 ± 20)kG for $Pd^{54}Mn$.

## INTRODUCTION

The variety of interesting properties of dilute magnetic impurities in metallic hosts[1] arises in large measure from the interaction of conduction electron spins $\vec{s}$ with the impurity spin $\vec{S}$. This interaction with the Hamiltonian of the form $H = J_{sd}\vec{s}\cdot\vec{S}$ leads in some cases to a behavior of the susceptibility $\chi$ of a completely isolated magnetic impurity, $\chi = C/(T + T_K)$, where $T_K$ is a temperature characteristic of the interaction. This interaction also leads to the long range RKKY indirect exchange interaction between impurity spins. Scaling considerations[2] indicate that no matter how dilute the alloy, there will be a temperature $T_i$ below which the RKKY interactions become important. Interpretation of experimental results is often complicated by the uncertainty as to which interaction, that characterized by $T_K$, or that characterized by $T_i$ is dominant under the experimental conditions.

We present here initial measurements on the nuclear orientation (NO) of $^{54}Mn$ dissolved in PD. The measurement consists of determining the angular distribution $W(\theta,T)$ of $\gamma$ radiation from the $^{54}Mn$ nucleus oriented (dominantly) by a hyperfine interaction with the magnetic electrons.

Novel information may be obtained on the magnitude and direction of the impurity magnetization and, perhaps, on the structure of the impurity in extremely dilute alloys. For $^{54}Mn$, we have

$$W(\theta,T) = 1 - A_2P_2(\cos\theta) - A_4P_4(\cos\theta) \qquad (1)$$

The anisotropy depends on the angle $\theta$ between the laboratory quantization axis and the direction of $\gamma$ ray emission through the Legendre polynomials $P_2$ and $P_4$. The temperature and hyperfine interaction dependences are contained in the factors $A_2$ and $A_4$ which are determined in part by the thermal population of nuclear substates.

## EXPERIMENTAL

We determine the anisotropy by measuring the $^{54}Mn$ count rate at temperature T at $\theta=0$ relative to the isotropic count rate at $T \to \infty$. Since appreciable anisotropy occurs only for very low temperatures, the measurements were performed in a $^3He$ - $^4He$ dilution refrigerator. The Pd-Mn sample was mounted on one end of a post of 99.999% Cu by a pressure clamp. Primary thermometry was done by nuclear orientation of $^{60}Co$ in a single crystal of hcp Co which was mounted with a pressure fit in a Cu holder and screwed into the Cu post, just above the Pd-Mn sample.

The sample was prepared from Pd sponge and foil with ~20 μCi of carrier free $^{54}MnCl_2$. These were placed in a high purity $ZrO_2$ crucible and sealed in a quartz tube. The chloride was reduced with $H_2$, which was then gettered with a hot Zr getter which also eliminated any $O_2$. The Pd was then melted in an rf induction furnace. Analysis of the sample indicated <1 ppm Mn, 9 ppm of Fe, and ~0.1% Zr.

The anisotropy $W(0,T)$ was measured in various applied magnetic fields, $H_a$. The $\gamma$ counts from the $^{54}Mn$ and $^{60}Co$ were registered in a NaI(Tl) detector, amplified by a gain stabilized amplifier, and then fed to separate counters. Thus the primary thermometry was done simultaneously with the Mn anisotropy measurements. After measuring $W(0,T)$ by sweeping T, the applied field was changed to a new value and the process repeated.

## RESULTS AND DISCUSSION

Our data for $W(0,T)$, corrected for $^{60}Co$ counts in the $^{54}Mn$ window, are shown in Fig. 1. The points shown have been read at equal intervals of $T^{-1}$ from smooth curves drawn through the raw data (which were taken at much smaller intervals of $T^{-1}$) and corrected for the $^{60}Co$ radiation which was Compton scattered into the $^{54}Mn$ window. The size of the plotted symbols is substantially larger than our statistical errors which are the order of 0.2% for the $^{54}Mn$. Also shown for reference in Fig. 1 are curves calculated assuming there is a hyperfine interaction describable by a temperature independent field at the nucleus $\vec{H}_n = \vec{H}_{hf} + \vec{H}_a$, where $\vec{H}_{hf}$ is a hyperfine magnetic field. Such a constant $\vec{H}_{hf}$ representation is expected to be valid only when the electronic and nuclear spins are decoupled, and when the magnetization is constant over the whole temperature range.

This constant hyperfine field representation fits our data at 1720 G and above very well when $T^{-1} > 25K^{-1}$, i.e., $T < 0.04K$. From these high field data we find the saturation hyperfine field is -415 ± 20 kG where the error includes counting errors, uncertainties in the correction for $^{60}Co$ counts and thermometry errors.

We have compared the behavior of our data with that predicted by several more realistic models including[3]

a. free spin, $\omega_n\tau_e >> 1$ ($A\vec{I}\cdot\vec{S}$ coupling)
b. free spin, $\omega_n\tau_e << 1$ ($AI_z < S_z >$ coupling)
c. the semiempirical model[4] of Kitchens and Taylor $H/H_{sat} = B_S[\mu H_a/k_B(T + T_K)$.

Here $\omega_n$ is the nuclear Larmor frequency, $\tau_e$ is the electronic spin relaxation time, $A\vec{I}\cdot\vec{S}$ is the interaction Hamiltonian between nuclear spin $\vec{I}$ and electronic spin $\vec{S}$ and $B_S$ is the Brillouin function for spin S.

As expected, all models satisfactorily fit the data for high applied fields, since the electronic spin is saturated in the temperature range of this study. At fields of 430 Gauss and below, however, the data and model predictions disagree in two important respects: (1) as the temperature is raised $W(0,T)$ approaches unity more slowly than the models predict. (2) More importantly, for $T^{-1} \gtrsim 50K^{-1}$, the anisotropy given by the models is greater than that observed. This may be seen in Fig. 1, where it is evident that (for $H_a \lesssim$ 430 Gauss) the apparent hyperfine field decreases with decreasing temperature.

If we adopt the position that the hyperfine interaction must approach a constant value as the temperature is lowered, then it becomes evident from our data that there must be an angular averaging which prevents us from measuring the full anisotropy $W(0,T)$ associated with a given hyperfine interaction as given in Eq. 1. This averaging attenuates the $P_4$ term more

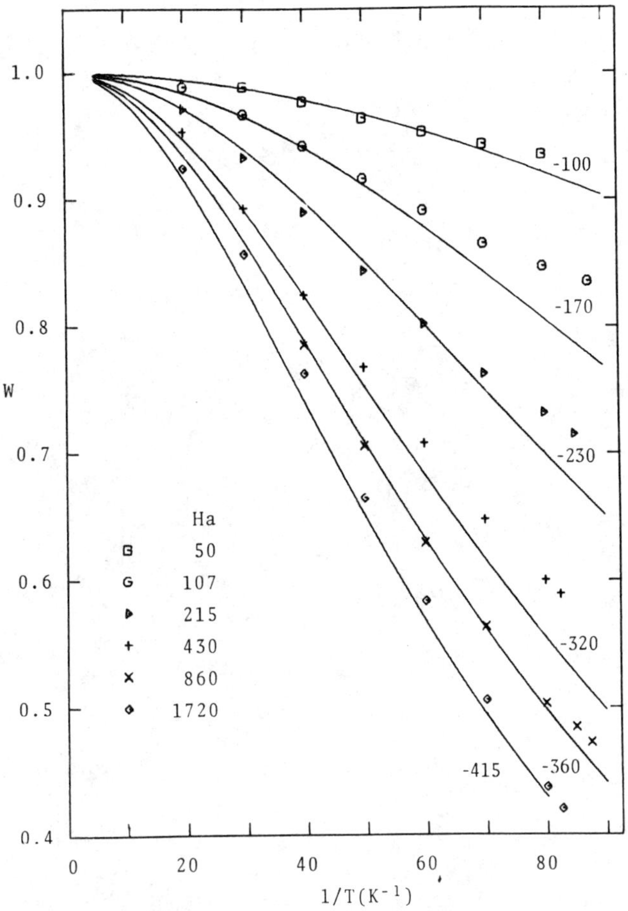


Fig. 1 The $^{54}$Mn Anisotropy as a function of 1/T for Applied Fields $H_a$ in Gauss as indicated. The curves are calculated assuming a constant hyperfine field acts on the $^{54}$Mn nuclei, and the value of this field in kiloGauss is indicated. Note that the data disagree with this model for $T^{-1} > 50K^{-1}$.

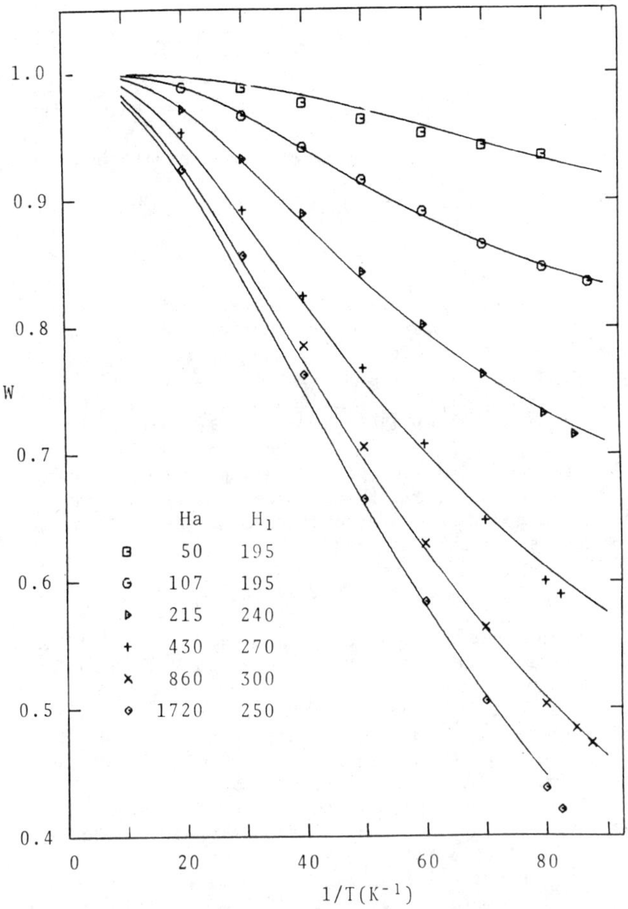


Fig. 2 The plotted points are as in Fig. 1. The curves are calculated using a distribution of canting fields acting on the Mn spin. The width of this field distribution, $H_1$, is given in Gauss.

severely then the $P_2$ term and accounts for the falloff of the apparent hyperfine field below 20mK where the $A_4P_4$ term first becomes appreciable.

Thus the local axis of the hyperfine interaction must have a distribution of angles with respect to the laboratory axis, defined by the applied field and the position of the $\gamma$ ray detector.

To introduce such an effect we have calculated W(0,T) using a model which introduces an angular averaging via field, $H_2$, which is random in direction with a Gaussian distribution in magnitude $P(H_2) \sim \exp(-H_2^2/2H_1^2)$, and which acts on the Mn electronic spin. The results of this model are compared with our data in Fig. 2. In this calculation we have used $-H_{sat}$ = 415kG as determined above and S = 5/2 and g = 3 from specific heat and susceptibility results.[5]

We have chosen the optimum value for $H_1$ for each applied field. We see from Fig. 2 that the data agree with the curves drawn assuming $H_1$ is independent of temperature. The magnitude of $H_1$ is mildly dependent upon applied field, although all of the data can be represented by a canting field of $H_1$ = 245 ± 55 Gauss.

This model although oversimplified, may represent the situation better than one might suspect at first. Omitted, of course, are the detailed spin-spin interactions which in an average sense are accounted for by the self consistent molecular field in the Weiss theory. However, one expects, for the RKKY interaction between a given impurity and its distant neighbors, an equal number of ferromagnetic and antiferromagnetic couplings. Thus the Weiss field will be substantially weakened and to first order may be neglected at very low concentrations.

The origin of this canting interaction may be long range RKKY interactions, perhaps combined with a magnetic anisotropy. Thus $H_1$ parameterizes the interaction strength and should in this case scale with impurity concentration. An alternate disaligning mechanism may be localized spin fluctuations.[1]

In summary, we have presented evidence that there is a distribution of the spin direction for Mn impurities in Pd which lasts for times greater than the nuclear Larmor precession time. Further study is underway to understand the interaction responsible for this disalignment.

REFERENCES

1. Magnetism, Vol. V Ed. by H. Suhl, (Academic, New York, 1973), p. 1-416.
2. Soulitie, J. and Tournier, R., J. Low Temp. Phys. 1, 95-108 (1969).
3. Alloul, H. and Bernier, P., Ann. Phys. (Paris) 8, 169-232 (1974).
4. Kitchens, T. A. and Taylor, R. D., Phys. Rev. B9, 344-346 (1974).
5. Nieuwenhuys, G. J., Boerstoel, B. M., and Star, W. M., Low Temperature Physics: LT13 Ed. Timmerhaus, O'Sullivan, Hammel (Plenum, New York, 1974), p. 506-519.

# NMR SPIN LATTICE RELAXATION RATE IN NiA$\ell$ DUE TO 3-d IMPURITIES*

J. R. Willhite and J. O. Brittain
Northwestern University, Evanston, Illinois 60201
L. B. Welsh
Northwestern University and UOP Inc., Des Plaines, Illinois 60016

ABSTRACT

The $A\ell^{27}$ NMR spin lattice relaxation in stoichiometric NiA$\ell$ due to the magnetic 3-d impurities Cr, Mn and Fe have been studied. The recovery of the NMR signal following a saturation train is exponential in time. The impurity induced relaxation rate ($T_{1\,imp}^{-1}$) is linear in impurity concentration for $c \leq 500$ppm. The magnetic field dependence of the rate is $T_{1\,imp}^{-1} = (A+BH)^{-1}$ for $1.5 \leq H \leq 15$kOe and $T_{1\,imp}^{-1}$ is temperature independent for $1.5 \leq T \leq 4.2$K. Measured values of the impurity magnetization and of the impurity broadening of the host NMR linewidth are used to interpret $T_{1\,imp}^{-1}$ using the model of energy diffusion in the nuclear spin system to the impurity.

## INTRODUCTION

The properties of magnetic impurity states in metallic systems have been extensively studied, both in regards to their bulk properties, such as resistivity and magnetization, and their NMR and other microscopic properties. The impurity induced host NMR relaxation offers the chance of obtaining information regarding the dynamic susceptibility of the impurity.[1-6] However the effects of energy diffusion in the nuclear spin system has been discussed for few systems: $La_{1-c}Gd_cA\ell_2$ [2], CuMn[3], and $La_{1-c}R_cX_3$ [4] (R = Gd or Ce and X = In, Sn or Pb). We report NMR results on 3-d impurities in the ordered intermetallic compound NiA$\ell$ where the impurity-impurity interactions are not as strong as for Gd in $LaA\ell_2$. The weaker host NMR relaxation rate $T_{1p}^{-1}$ in NiA$\ell$, as compared to Cu, will show different properties of the impurity induced rate.

The intrinsic magnetic and NMR properties of pure stoichiometric NiA$\ell$ have been reported[7,8] and are those of a simple Pauli paramagnet. The susceptibility ($\chi$) of $(0.202 \pm 0.005) \times 10^{-6}$ emu/G is temperature independent ($1.3 \leq T \leq 300$K) and the $A\ell^{27}$ NMR relaxation rate is $T_{1p}^{-1} = (0.035 \pm 0.006)T\,sec^{-1}$ and is magnetic field independent ($1.5kOe \leq H \leq 15kOe$). Limited atomic disorder caused by the introduction of 1 at.% excess Ni increases $\chi$ by 15%. However $\chi$ remains temperature independent, indicating that excess Ni atoms have no local magnetic moment.

The addition of Cr, Mn or Fe to stoichiometric NiA$\ell$ introduces a temperature dependence to $\chi$.[8] For impurity concentrations (c) less than 500ppm, the excess magnetization (M) is proportional to c and the Curie-Weiss $\theta$ are independent of c. The impurity magnetizations are well fit by Brillouin functions. Values for the impurity spin (S) and $\theta$ are listed in Table I.

## NMR MEASUREMENTS AND RESULTS

$A\ell^{27}$ NMR spin lattice relaxation measurements were performed in the temperature range 1.5 to 300K in magnetic fields from 1.5 to 15kOe. A non-commercial phase coherent pulsed spectrometer was used. The signal observed was the amplitude of the spin echo following a saturation train - t-90° - $\tau$-43° pulse sequence. Samples were prepared from 5-9's starting material and were formed by melting in an arc furnace in argon. NMR data was taken on powders which had been comminuted in a porcelain mortar and then strain annealed.

Table I. MAGNETIZATION PARAMETERS

| | Cr | Mn | Fe |
|---|---|---|---|
| S | 1 | 1 | 3/2 |
| $\theta$(k) | 1.2 | 1.0 | 1.0 |

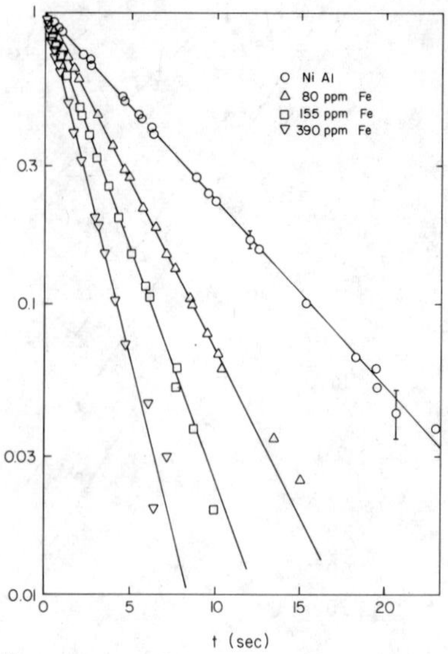


Figure 1. $A\ell^{27}$ spin echo recovery function in pure NiA$\ell$ doped with Fe.

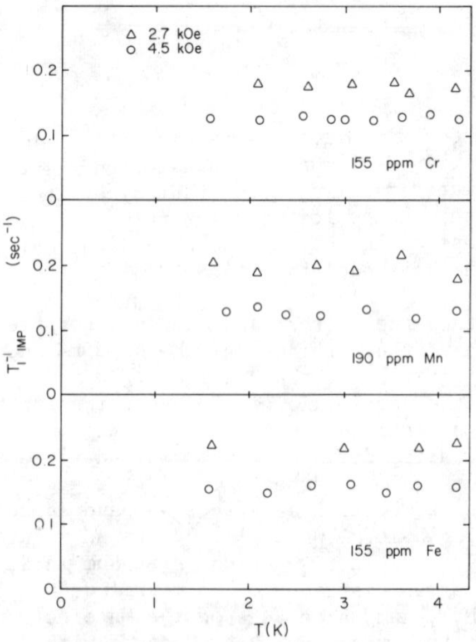


Figure 2. Impurity contribution to the $A\ell^{27}T_1^{-1}$ vs temperature for samples of NiA$\ell$ doped with Cu, Mn or Fe.

The $A\ell^{27}$ spin echo relaxation signal at 4.2K and 7kOe is shown in Fig. 1 for pure NiA$\ell$ doped with dilute amounts of Fe. For impurity concentrations $50 \leq c \leq 1000$ ppm the relaxation function is purely exponential. The impurity contribution to the relaxation rate, $T_{1\,imp}^{-1} = T_1^{-1} - T_{1p}^{-1}$, is proportional to the impurity concentration for $c \leq 500$ppm.

The magnetic field dependence of $T_{1\,imp}^{-1}$ for all dilute impurity samples is $T_{1\,imp}^{-1} = A+BH^{-1}$.

Figure 2 shows the temperature dependence of $T_{1\,imp}^{-1}$ at 2.7 and 4.5kOe for three samples containing dilute Cr, Mn or Fe impurities. The impurity contribution to the relaxation rate is temperature independent from 1.5 to 4.2K. At 77K $T_{1\,imp}^{-1}$ is masked by the larger value of the pure host rate and thus we will only consider values for $T_{1\,imp}^{-1}$ at temperatures below 4.2K.

## ANALYSIS

The relaxation of the host nuclei in a metal by impurity moments is generally assumed to proceed by one of four mechanisms;[1] direct coupling to the transverse (TD) or longitudinal (LD) fluctuations of the impurity dipole moment or indirect coupling via real excitations of the conduction electrons to the transverse fluctuations (BGS) or via virtual excitations (GH). The first three mechanisms (TD, LD, BGS) will produce a local relaxation rate for nuclei a distance r from an impurity that can be written $T_1{}^{-1}_{imp}(r) = \xi/r^6$. The GH mechanism will produce a local rate proportional to $r^{-5}$. This rate is expected to be very weak and would have a field dependence unlike that observed and thus will not be considered.

The long time relaxation rate due to impurities for the case of NiAℓ becomes:[3]

$$T_1{}^{-1}_{imp} = \frac{32}{9}\pi^2 \frac{No^2 c}{n_d}\xi \qquad (1)$$

where No is the density of Aℓ (or Ni) in NiAℓ and $n_d$ is the number of Aℓ nuclei about an impurity which cannot exchange energy among themselves due to the large local fields. The number $n_d$ should increase with the impurity magnetization.

The BGS rate is given (for $\gamma_e H\tau \ll 1$) by:[1]

$$T_1^{-1}(r) = r^{-6}\pi \frac{[JN(E_F)E_F]^2}{4\hbar K_F^6 (T_1T)_p K_B} \frac{T}{(T+\Theta)} \frac{S^2 B_s(x)}{x}\tau \qquad (2)$$

where $\gamma_e$ and $\tau$ are the impurity electron spin gyromagnetic moment and relaxation time, $(T_1T)_p$ is from the pure host relaxation and is constant and the impurity is assumed to obey a Brillouin function of spin S and argument $x = g\mu_B SH/k(T+\Theta)$. The quantity J is the impurity s-d exchange constant and $N(E_F)$ is the host electronic density of states at the Fermi energy, $E_F$. For small values of x, $\xi$ has the temperature dependence of $\tau T/(T+\Theta)$.

The impurity electron spin relaxation time can be estimated from the Korringa rate[9] due to interaction with the conduction electrons. This rate is:

$$\tau_K^{-1} = \frac{\pi}{\hbar}[JN(E_F)]^2 kT. \qquad (3)$$

Another possible contribution to $\tau^{-1}$ is the impurity-impurity interaction rate given by[2]

$$\tau_{RKKY}^{-1} = \frac{4}{3}\pi^2\left[\frac{1}{6}\pi S(S+1)\right]^{\frac{1}{2}}[JN(E_F)]^2 Noc/\hbar k_F^3 \qquad (4)$$

where S is the impurity spin. However an impurity electron relaxation rate of this form would lead to a concentration independent $T_1{}^{-1}_{imp}$ when substituted into eqs. 1 and 2. Thus although eq. 4 predicts a rate comparable to eq. 3 for the concentrations used in our experiment, it cannot give the dominant impurity relaxation rate.

The quantity $JN(E_F)$ can be[8,10] estimated from the impurity broadening of the host NMR linewidth using a free electron approximation for the RKKY oscillations. The values obtained in this manner are 0.75, 0.80 and 0.72 for Cr, Mn and Fe. Using these values the BGS rate is predicted to be a factor of twenty larger than either the LD or TD rate in the temperature and field range in which we have data. The conclusions of this paper would not change if the BGS rate were not dominant, since the TD and LD rates have very similar temperature and field dependences in the region in which we took data.

For low magnetic fields, assuming $\tau$ is field independent, the field dependence of $T_1{}^{-1}_{imp}$ should be determined by $n_d$. The observed field dependence of $T_1{}^{-1}_{imp}$ implies that $n_d = \alpha + \beta B_s(x)$. The magnetic field independent contribution to $n_d$ is attributed to the large electric field gradients near an impurity. The nuclei inside the nuclear spin diffusion barrier which are observed in an NMR experiment produce a non-exponential nuclear recovery. We did not observe this form of recovery and from careful study of the recovery of a sample containing 1000ppm Fe we estimate $n_d \leq 150$ at 4.2K and 10kOe.

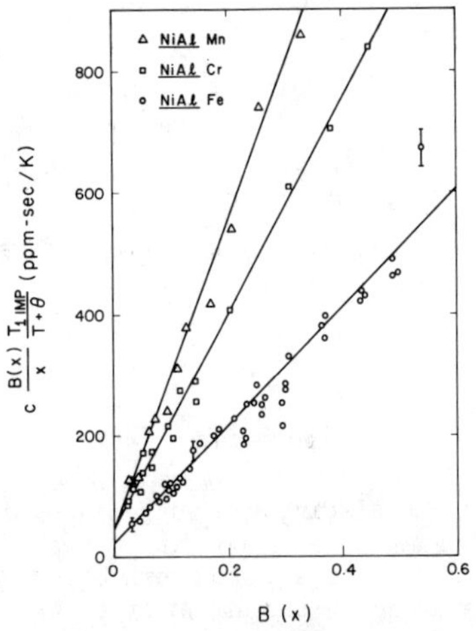


Figure 3. The quantity $c\,\frac{B(x)}{x}\,\frac{T_{1\ imp}}{T+\Theta}$ vs B(x) for Cu, Mn and Fe impurities in NiAℓ. B(x) and Θ are from magnetization measurements.

Table II

| | Cr | Mn | Fe |
|---|---|---|---|
| a (ppm-sec/K) | 50±15 | 50±15 | 20±5 |
| b (ppm-sec/K) | 1750±150 | 2500±250 | 980±100 |

The values of $T_1^{-1}$ are observed to be temperature independent. Using this fact, eqs. 1 and 2, and assuming the functional form of $n_d$ given above, the temperature dependence of $\tau^{-1}$ is determined to be $\tau^{-1} \propto T$. With a temperature dependence of this form for $\tau^{-1}$, $cB_s(x)T_{1\ imp}/(x(T+\Theta))$ should be equal to $a+bB_s(x)$. The data for Cr, Mn or Fe impurities in NiAℓ are plotted in this manner in Fig. 3. The values for a and b are given in Table II.

Since values for $n_d$ could not be determined in our work, the magnitude of $\tau$ could not be established. We would estimate the actual value of $\tau^{-1}$ to be larger than or equal to that given by eq. 3.

REFERENCES

* This work supported by the NU Materials Research Center.

1. B. Giovanni, P. Pincus, G. Gladstone and A. J. Heeger, J. Phys. (Paris) 32, C1-163 (1971).
2. M. R. McHenry, B. G. Silbernagel and J. H. Wernick, Phys. Rev. B5, 2958 (1972).
3. H. Alloul and P. Bernier, J. Phys. F:Metal Physics 4, 870 (1974).
4. L. B. Welsh, C. L. Wiley and F. Y. Fradin, Phys. Rev. B11, 4156 (1975).
5. D. S. Schreiber and R. F. Wade, AIP Conf. Proc. 24, 472 (1974).
6. F. Y. Fradin, Phys. Rev. B5, 1119 (1972).
7. J. R. Willhite, L. B. Welsh, T. Yoshitomi and J. O. Brittain, Solid State Comm 13, 1907 (1973).
8. J. R. Willhite, thesis Northwestern University (unpublished).
9. J. Korringa, Physica 16, 601 (1950).
10. R. E. Walstedt and L. R. Walker, Phys. Rev. 139, 4857 (1974).

# ELECTRICAL RESISTIVITY AND MAGNETIC SUSCEPTIBILITY OF $Pt_{1-x}Co_x$

K. V. Rao,*+ O. Rapp and Ch. Johannesson
Royal Institute of Technology
Stockholm, Sweden

J. I. Budnick and T. J. Burch
University of Connecticut**
Storrs, Connecticut 06268

V. Cannella
Wayne State University
Detroit, Michigan 48202

ABSTRACT

Electrical resistivity measurements have been carried out on 4 samples of PtCo with between 0.6 and 5 at % Co. This resistivity data exhibits a strong temperature dependence in the impurity contribution. The overall temperature and concentration dependent characteristics of the data bear a strong resemblance to features attributed to scattering from localized spin fluctuations. Below the ordering temperatures low field susceptibility experiments show a pronounced deviation from the behavior expected of a ferromagnet for the samples with Co concentration equal to 0.8 and 2 at %.

## INTRODUCTION

Pt based alloys with the transition metal impurities Co,[1] Fe,[2] Ni[3] and Cr[4] show a wide range of interesting and challenging temperature dependent resistivity behaviors. The Pt matrix, with an exchange enhancement significantly lower than that of Pd, can allow studies of the interesting regions of magnetic behavior between those systems which show a dominant long range ferromagnetic spin polarization and those which exhibit Kondo like spin compensation.[5] Early evidence of the interesting behavior of the Co moment in the PtCo system was contained in the magnetization work of Crangle and Scott,[6] the NMR studies of Graham and Schreiber,[7] and the resistivity measurements of Shen, Schreiber and Arko.[8] From their experiments, clear evidence is presented for a strong decrease in the Co moment with temperature. Two studies of the electrical resistivity at single dilute concentrations of Co have been reported.[1,8] The most extensively characterized work has been that of Williams et al. and Swallow et al. who have studied in detail the susceptibility, resistivity and magnetoresistance of a single sample of PtCo with a concentration equal to 0.06 at % Co. They confirm a strong reduction in the local Co moment at about 1.6 K. This conclusion was also reached through the nuclear orientation measurements of Gallop and Campbell[9] and in the detailed magnetization studies of Tissier and Tournier.[10] We report some measurements of both the electrical resistivity and low field susceptibility of the PtCo alloy system.

$Pt_{1-x}Co_x$ samples were prepared with $x = 0.05, 0.02, 0.008$ and $0.006$. The 0.05 ingot was prepared by arc melting 99.995% Pt and 99.999% Co in an argon atmosphere. Lower concentration ingots were made by dilution of this ingot. Samples were prepared for susceptibility measurements by cutting cubes from the ingots which were then ground into spheres and annealed. Bars were cut from the ingots, annealed, drawn into wires, and reannealed for resistivity measurements. All heat treatments were done in evacuated quartz tubes at about 1000 C for 2 to 4 days and then the tubes and their contents were air quenched. The $x=0.05$ and $x=0.008$ samples were analysed to determine composition.

## RESISTIVITY

Fig. 1 shows the temperature dependence of the impurity contribution to the resistivity, $\Delta\rho(T)$, for $Pt_{1-x}Co_x$ alloys with $x = 0.05, 0.02, 0.008$ and $0.006$. $\Delta\rho(T) = \rho_{alloy}(T) - \rho_{Pt}(T)$. In the $x = 0.008$, $0.02$ and $0.05$ alloys expected changes in the slopes of $\Delta\rho(T)$,

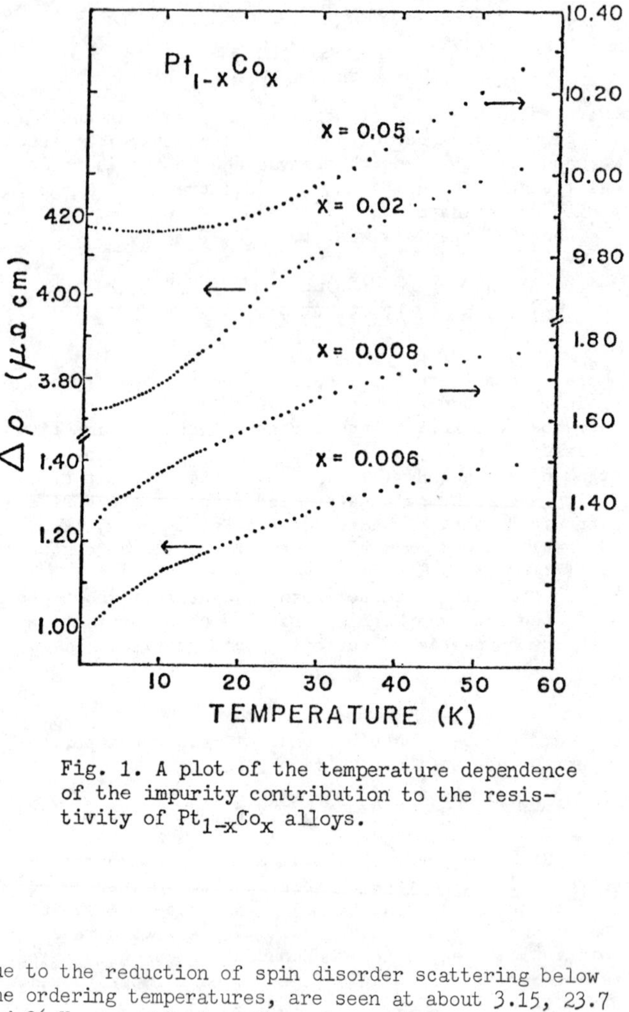


**Fig. 1. A plot of the temperature dependence of the impurity contribution to the resistivity of $Pt_{1-x}Co_x$ alloys.**

due to the reduction of spin disorder scattering below the ordering temperatures, are seen at about 3.15, 23.7 and 96 K respectively. These temperatures are close to the ordering temperatures observed in our susceptibility measurements. The ordering temperature of the $x=0.006$ alloy falls outside the temperature range investigated. In addition, in the $x=0.05$ sample a pronounced minimum in the resistivity is observed at about 7 K, well below the ordering temperature ($T_c \approx 96$ K) of that alloy. No simple explanation is proposed for this behavior.

$\Delta\rho(T)$ for the PtCo system for $T<50$ K shows a marked similarity to the earlier reported results on RhFe[12,13] and to the model proposed by Kaiser and Doniach[11] for behavior in spin fluctuation systems. Kaiser and Doniach concluded that the temperature dependence at low temperatures ($T < \theta/4$) should go as $T^2$ while for $T > \theta/4$ they predicted a temperature dependence which should go as T, where $\theta$ is the spin fluctuation temperature. In their study of the resistivity of the PtFe system Loram et al.[2] based their analysis on the Kaiser and Doniach model and fitted their data with an expression of the form $\Delta\rho(T) = A + B\ln(T^2+\theta^2)^{\frac{1}{2}}$. For dilute samples they found that A and B were constants which scaled with concentration and that $\theta$ was constant. Williams et al.[1] have analyzed $\Delta\rho(T)$ for one sample of $Pt_{.9994}Co_{.0006}$ using the same expression and found a value of 0.7 K for the spin fluctuation temperature, $\theta$. Our data for the 0.006 and 0.008 alloys can be fitted to such an expression between the magnetic ordering temperatures and about 10 K. The analysis yields values of 0.860 and 1.12 $\mu\Omega$cm for A, 0.112, 0.142 $\mu\Omega$cm for B and about 3 and 5 K for $\theta$ for the 0.006 and 0.008 alloys respectively. Our values for A normalized to the Co concentration are in good agreement with that found by Williams et al. for their dilute sample. Our values for B/x, on the other hand, are about a factor of three larger than that found for the dilute PtCo alloy.

This analysis of our data leads us to conclude that the data can be described with a spin fluctuation temperature not associated with a single impurity but which varies with Co concentration. Such a conclusion has been deduced from studies of the PtFe system by Loram et al.[2] In fact, our data are fitted by spin fluctuation temperatures which are very close to those obtained for PtFe at similar concentrations.

## SUSCEPTIBILITY

In addition to the resistivity experiments we have measured the magnetic susceptibility, $\chi(T)$, of the 0.05, 0.02, 0.012, 0.008 and 0.006 samples, using a low frequency (50 Hz) low field (5 gauss) mutual inductance technique. The data for the 0.008, 0.02 and 0.05 samples are shown in Fig. 2, where $\chi/D$ is plotted vs $T/T_c$ for each sample. D is the demagnetizing factor for the nearly spherical samples and $T_c$ is the magnetic ordering temperature. The divergence of the measured susceptibility to a value 1/D as the temperature is lowered to $T_c$ indicates the onset of long range magnetic order with a bulk magnetization. This is consistent with what is expected for a ferromagnet at the Curie temperature. Below $T_c$ we find a pronounced reduction in the measured susceptibility. This reduction becomes larger with decreasing Co concentration. A reduction in susceptibility below $T_c$ has been previously[4] found for magnetic alloy systems in which a transition from long range ferromagnetic order to a more complex (spin glass or mictomagnetic) type of order is observed as the matrix susceptibility or the concentration of the magnetic component is reduced. Examples include the $(Pd_xAg_{1-x})_{.99}Fe_{.01}$ system[14] and the AuFe system.[15] If we define the normalized reduction in $\chi$ at T = 0 by $\Delta\chi = [\chi(T_c) - \chi(0)]/\chi(T_c)$ we find that $\Delta\chi = 1 - Kx$ for $0.008 \leq x \leq 0.05$. Here K is a constant 0.156/at %.

The values of $T_c$ obtained from the susceptibility measurements are 2.75, 8.4, 22.8 and 96 K for the 0.008, 0.012, 0.02 and 0.05 alloys respectively. $T_c$ for the 0.006 sample is below 1.5 K. This data is in excellent agreement with the work of Crangle and Scott[6] for concentrations above x = 0.01. $T_c$ for the 0.008 sample falls well below the Crangle and Scott curve and approaches the curve of ordering temperatures measured by Tissier and Tournier[10] in dilute samples (x = .00662 to .00163). The total variation of the transition temperature with concentration suggests a transition from a high concentration to a low concentration behavior at about 0.6 at % Co. A sample of this concentration would exhibit an ordering temperature somewhere between 1 and 2 K, which spans the moment reduction temperature of 1.6 K reported by Gallop and Campbell[9] and Tissier and Tournier.[10] This transition in the concentration dependence of the ordering temperature could be a direct result of the drastic reduction in Co moment observed for the low concentration samples at temperatures less than 1.6 K.

## ACKNOWLEDGEMENTS

We acknowledge helpful discussions with Dr. S. Skalski of Fordham University and Gwyn Williams of The University of Manitoba and the support of Svenska Natur Veten Skapliga Sorskningsrad.

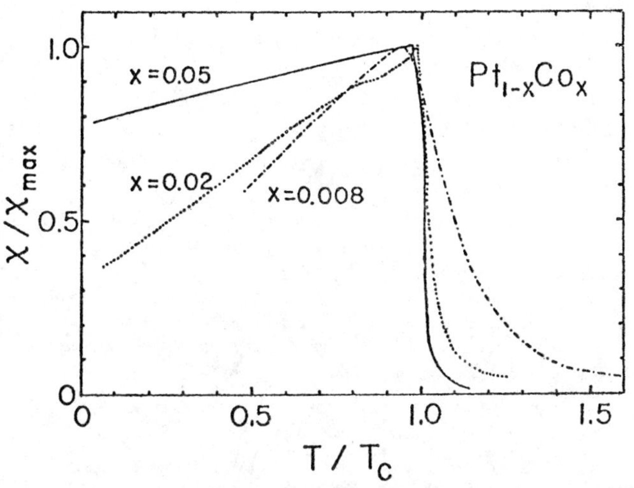


**Fig. 2. A plot of the reduced susceptibility versus the reduced temperature for $Pt_{1-x}Co_x$ alloys.**

## REFERENCES

* Permanent address: Clarkson College of Technology, Potsdam, New York 13676.

\+ Supported in part by The National Science Foundation.

** Supported in part by The University of Connecticut Research Foundation.

1. G. Williams, G. A. Swallow and J. W. Loram, Phys. Rev. B11, 344 (1975); G. A. Swallow, G. Williams and A. D. C. Grassie, Phys. Rev. B11, 337 (1975).
2. J. W. Loram, R. J. White and A. D. C. Grassie, Phys. Rev. B5, 3659 (1972).
3. C. A. Mackliet, A. I. Schindler and D. J. Gillespie, Phys. Rev. B1, 3283 (1970).
4. R. M. Roshko and G. Williams, Phys. Rev. B9, 4945 (1974).
5. F. C. C. Kao and G. Williams, Phys. Rev. B7, 263 (1973).
6. J. Crangle and W. R. Scott, J. Appl. Phys. 36, 921 (1965).
7. L. D. Graham and D. S. Schreiber, J. Appl. Phys. 39, 963 (1968).
8. L. Shen, D. S. Schreiber, and A. J. Arko, Phys. Rev. 179, 512 (1969).
9. J. C. Gallop and I. A. Campbell, Solid State Comm. 6, 831 (1968).
10. B. Tissier and R. Tournier, Solid State Commun. 11, 895 (1972).
11. A. B. Kaiser and S. Doniach, Int. J. Magn. 1, 11 (1970).
12. A. P. Murani and R. B. Coles, J. Phys. C. Suppl. 2, S159 (1970).
13. R. J. White and G. A. Swallow, Physics Letters 35A, 427 (1971).
14. J. I. Budnick, T. J. Burch and V. Cannella, AIP Conf. Proc. 18, 307 (1974).
15. V. Cannella and J. A. Mydosh, Phys. Rev. B6, 4220 (1972).

# INVESTIGATION OF CONDUCTION ELECTRON POLARIZATION EFFECTS IN $Fe_3Si$ BASED TERNARY SYSTEMS+

K. Raj, V. Niculescu* and J. I. Budnick
University of Connecticut, Storrs, Connecticut 06268

S. Skalski
Fordham University, Bronx, New York 10458

## ABSTRACT

Spin-echo spectra of transition elements in the ternary alloys $Fe_{3-x}T_xSi$ where T = V and Mn show clear satellite structure for small x. For these alloys, it is concluded that changes in the "4s" CEP alone produce the satellite structure. Studies made on some $Fe_3Si$ based quarternary alloys are consistent with the above conclusions.

## INTRODUCTION

It has been shown[1] that in the binary transition metal alloys, Fe-Mn and Fe-V, similar hyperfine field distributions are obtained for the impurity and host nuclei. In fact, it has been claimed that if CEP is dominant, the field shifts for impurity and host nuclei located a given distance from an impurity, should be equal in these alloys. However, due to the general tendency toward random distribution of the impurities into various sites, the CEP contribution is difficult to sort out experimentally.[2]

The system $Fe_{3-x}T_xSi$, where T is a 3d transition metal, provides an excellent opportunity to study CEP effects because of the extreme site selectivity of T into the $Fe_3Si$ matrix.[3] The substitution of T into distant specific neighbor shell (e.g. 3rd for Mn and V) eliminates chemical and direct environment effects normally present in dilute alloy studies where a 1nn impurity can drastically affect the host atom local moment.[4]

We report well resolved spin-echo spectra indicating a clear contribution from CEP for the transition metals in $Fe_{3-x}T_xSi$ in the low composition range. The CEP effects for both magnetic and non-magnetic solutes entering the same site are compared and interpreted. The effect of intervening atoms on the spin-density oscillations is examined by extending measurements to some $Fe_3Si$ based quarternary systems.

The $Fe_3Si$ system and the effect of 3d-impurity substitution: $Fe_3Si$ is ferromagnetic and shows a very high degree of Crystallographic ordering. Its crystal structure (Fig. 1) exhibits four sites A, B, C and D with sites A,C being equivalent. Near neighbor configurations are listed in Table I. The sites A, C and B

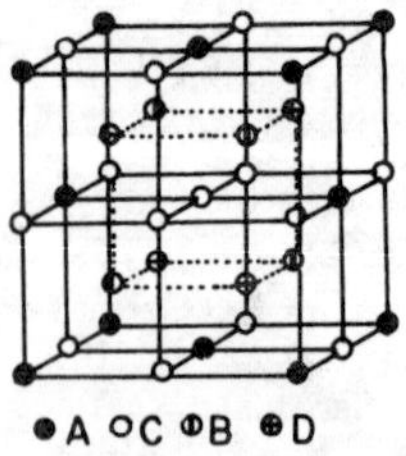

●A ○C ⊕B ⊛D

Fig. 1. Crystal structure of $Fe_3Si$. Site occupation A,C:Fe, B:Fe, D:Si

TABLE I

$Fe_3Si$ Neighbor Configurations

| Site | 1nn | 2nn | 3nn |
|---|---|---|---|
| A,C | 4B,4D | 6A,C | 12A,C |
| B | 8A,C | 6D | 12B |
| D | 8A,C | 6D | 12D |

are normally occupied by Fe and the site D by Si. Fe in B sites is denoted by Fe(B); Si in D sites by Si(D) etc.

In the $Fe_3Si$ system 3d impurities to the left of Fe enter the B sites and those below and to the right of Fe enter the A, C sites.[3] In the present article we present results for V and Mn both of which enter the B sites. Mn will have third and higher near neighbor (nn) Mn - Mn interactions and so will V. None of these impurities have first nn interactions. Thus direct overlap effects are minimized and we can clearly consider the effects of distant shells.

We have also prepared and studied well ordered quarternaries containing the same 3d elements in $Fe_3Si$ namely (FeMnV)Si, (FeVCo)Si and (FeMnCo)Si. The NMR studies made on the quarternaries clearly indicate that the selectivity of the elements is the same as is found for the ternary $Fe_3Si$ based alloys e.g. in (FeMnCo)Si, Mn enters the B sites and Co the A, C sites.

## EXPERIMENTAL RESULTS

NMR Spectra of $Fe_3Si$ based Ternary and Quarternary Systems: In ordered $Fe_3Si$, one observes three separate resonances; Fe(B) at 46.6 MHz, Fe(A,C) at 30.0 MHz and Si(D) at 31.5 MHz. The addition of small amounts of 3d elements produce well resolved satellites with intensities consistent with the random distribution of the impurity entering a certain shell selectively. The spectral features of the impurity are very similar to the host atom it replaces. For example the Mn spectrum shows the same hyperfine field distribution as the Fe(B). A detailed discussion of Mn substituted $Fe_3Si$ over a wide composition range is available elsewhere.[5] Only those spectra will be presented here which are relevant to the discussion.

(FeMn)Si: The results are shown in Fig. 2. Since Mn enters the B site, the Fe(B) and Mn spectra are superposed. The Mn NMR at 274.6 MHz is due to no Mn in the third neighbor shell and the low frequency satellites appear as a result of Mn with 1 Mn, 2 Mn and so on in the 3 nn shell. The intensities of both the Fe(B) and Mn resonances are consistent with Mn entering the 3 nn shell randomly. The scaling of the field shifts for the first satellite of the two spectra gives $\Delta H^{Mn}/\Delta H^{Fe(B)} = 0.78$ and indicates a predominant CEP effect as will be discussed later. This ratio can be compared with that for the Fe - Mn binary alloys i.e., $\Delta H^{Mn}/\Delta H^{Fe} = 1.26$ where the effect was considered due to only CEP.[2]

(FeV)Si: The main $V^{51}$ resonance in this alloy appears at 53.0 MHz. In Fig. 3, we show the overlap of the V and Fe(B) spectrum in $Fe_{2.98}V_{0.02}Si$.

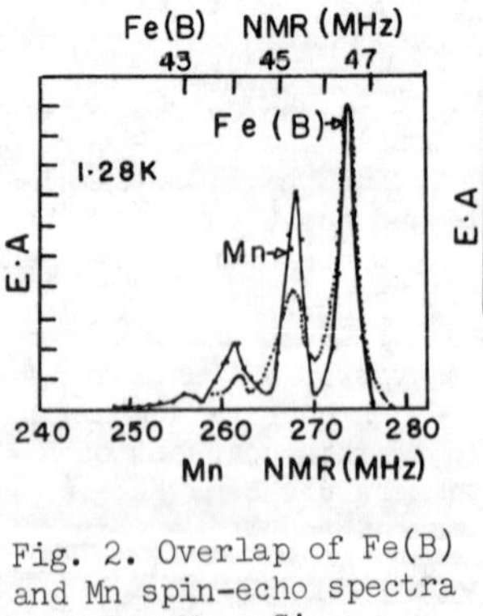


Fig. 2. Overlap of Fe(B) and Mn spin-echo spectra in $Fe_{2.92}Mn_{.08}Si$

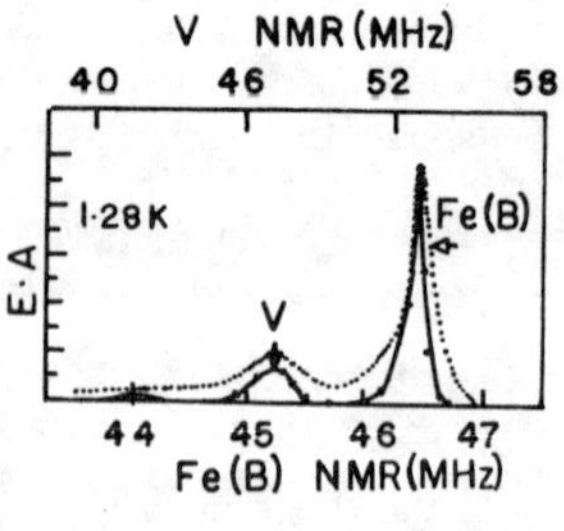


Fig. 3. Overlap of Fe(B) and V spin-echo spectra in $Fe_{2.98}V_{0.02}Si$

Like the Mn, V also enters the B site, the difference however, being that whereas Mn carries a moment of about 2.2 $\mu_B$,[6] V has a very much smaller and possibly a zero moment.[7] The spectral features for the V substitution are the same as for the Mn. Almost identical field shifts are observed. From Fig. 3, we obtain $\Delta H^{V}/\Delta H^{Fe(B)} = 0.57$ and, as we will discuss later, indicates a dominant CEP effect. It is interesting to note that exactly the same ratio $\Delta H^{V}/\Delta H^{Fe} = 0.57$ was found for the Fe-V binary alloy. The agreement between these two values strongly suggests similar mechanisms for the hyperfine field in the Fe-V binary alloy and the (FeV)Si system.

$Fe_3Si$ Based Quaternary Systems: The studies made on (FeV)Si and (FeMn)Si indicate that conduction electron scattering is fairly independent of the solute magnetic moment. To explore this idea further and to study the effect of intervening atoms on the spin density, some $Fe_3Si$ based quarternary systems were measured. In $Fe_{2.84}Mn_{0.08}V_{0.08}Si$ (Fig. 4), both Mn and V enter the B site and the hyperfine field distributions for these elements are consistent with a random substitution of x = 0.16 impurity in a shell of 12. This result further supports the premise that the spin density is sensitive only to the total number of impurity atoms. The value of $\Delta H^{Mn}/\Delta H^{V} = 1.42$ is obtained as expected from our ternary results. Because of the dominance of V in the low frequency range, the Fe NMR is not observed.

In the alloy $Fe_{2.84}Mn_{0.08}Co_{0.08}Si$, the spin density at the Mn is expected to be affected due to the presence of some Co in the 1 nn to Mn. The Mn and Fe(B) spectra, on the other hand, show a clear scaling (Fig. 5) with $\Delta H^{Mn}/\Delta H^{Fe(B)} = 0.73$ unaffected by the Co substitution. Approximately the same field shift is served as for the ternary $Fe_3Si$ containing Mn.

The sample $Fe_{2.84}Co_{0.08}V_{0.08}Si$ was found to contain a small amount of second phase. The preliminary NMR results indicated that the V hyperfine field shifts were the same as in (FeV)Si.

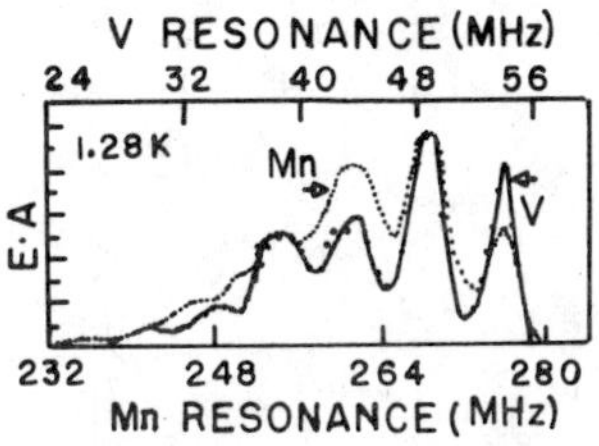


Fig. 4. Overlap of Mn and V spin-echo spectra in $Fe_{2.84}Mn_{.08}V_{.08}Si$

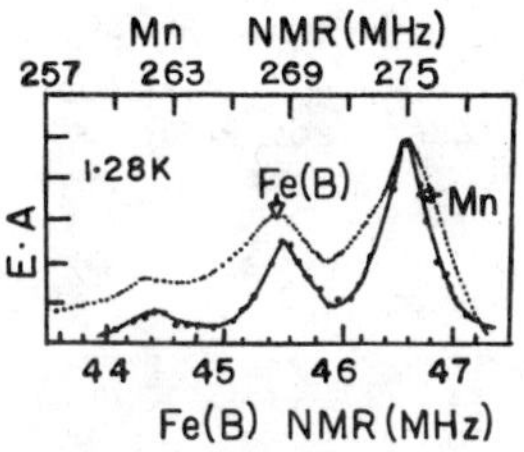


Fig. 5. Overlap of Fe(B) and Mn spin-echo spectra in $Fe_{2.84}Mn_{.08}Co_{.08}Si$

## CONCLUSION

In an effort to understand the origin of the field shifts, we assume that the hyperfine field at the nucleus is made up of two parts: Core polarization and the CEP of 4s electrons. One can then write

$$H(r) = -a\sigma_{4s}(r) + H_{cp}(r)$$

$$H(\infty) = -a\sigma_{4s}(\infty) + H_{cp}(\infty)$$

where H(r) and $H(\infty)$ are hyperfine fields at a nucleus distant r and infinity from the impurity respectively, a is the atomic hyperfine coupling constant, $\sigma_{4s}$ is the "4s" spin density and $H_{cp}$ is the core polarization. The ratio of the field shifts $[\Delta H(r) = H(r) - H(\infty)]$ for the impurity (i) and the host (Fe) becomes

$$\frac{\Delta H^{i}(r)}{\Delta H^{Fe}(r)} = \frac{-a_i\delta\sigma_{4s}(r) + \delta H^{i}_{cp}(r)}{-a_{Fe}\delta\sigma_{4s}(r) + \delta H^{Fe}_{cp}(r)}$$

where $\delta_{4s}$ and $\delta H_{cp}$ represent changes in the spin density and core polarization respectively.

The above expression indicates that if $\Delta H^{i}/\Delta H^{Fe}$ is the same as the ratio of the coupling parameters, the core polarization term should be negligible. On the other hand if $\Delta H^{i}/\Delta H^{Fe}$ is different from the ratio of the coupling parameters, then the core polarization term should be important. We use the ratio of the atomic hyperfine coupling constants in these alloys. This procedure is supported by our results on the quarternaries.

The results of the present studies are summarized in Table II along with that for the binary alloys. It is to be noted that the field shift ratios, assuming signs of hyperfine fields to be negative as is true for the binary alloys, agrees well with the ratios of the coupling parameters. From the agreement one concludes that for the Mn and V substituted ternaries (and also some quarternaries) "4s" polarization is primarily responsible for $\Delta H$.

TABLE II

Experimental ratios of hyperfine field shifts in various alloys and comparison with the ratio of coupling parameters (a).

| SAMPLE | EXPTL. RATIO | RATIO OF a's (8) | IDENTIFICATION |
|---|---|---|---|
| $Fe_{2.92}Mn_{.08}Si$ | 0.78 | 0.86 | $\Delta H^{Mn}/\Delta H^{Fe(B)}$ |
| $Fe_{2.98}V_{.02}Si$ | 0.57 | 0.62 | $\Delta H^{V}/\Delta H^{Fe(B)}$ |
| Fe - Mn(1) | 1.26 | 0.86 | $\Delta H^{Mn}/\Delta H^{Fe}$ |
| Fe - V(1) | 0.57 | 0.62 | $\Delta H^{V}/\Delta H^{Fe}$ |
| $Fe_{2.84}Mn_{.08}Co_{.08}Si$ | 0.73 | 0.86 | $\Delta H^{Mn}/\Delta H^{Fe(B)}$ |
| $Fe_{2.84}Mn_{.08}V_{.08}Si$ | 1.42 | 1.39 | $\Delta H^{Mn}/\Delta H^{V}$ |

It must be remarked that the model used here is different from the one used by Rubinstein et al.[1] and thus the conclusions drawn for the binary alloys are different. The present model suggests, unlike the conclusions drawn by Rubinstein et al., that in Fe-V alloys, the satellites are produced largely due to CEP and in Fe - Mn alloys core polarization must be considered.

Our conclusion that "4s" polarization is responsible for the observed field shifts for both the magnetic (Mn) and nonmagnetic (V) impurities is further supported by the good agreement between the field shift ratio and the atomic 4s coupling parameter ratio for the Mn and V fields in the quarternary system ($Fe_{2.84}Mn_{.08}V_{.08}Si$, see Table II). Furthermore the introduction of an impurity into the 1 nn shell (Co in $Fe_{2.84}Mn_{.08}Co_{.08}Si$, see Table II), has a negligible effect on the observed field shifts.

The effect on the field shifts due to a change in the 2 nn to Mn and V was examined by making off stoichiometric samples. Replacing a Si by a Fe in these shells produces high frequency satellites without any change in the field shift ratios.

ACKNOWLEDGEMENTS

We wish to thank Dr. T. J. Burch for helpful discussions and the Institute of Materials Science of the University of Connecticut for the use of X-ray facilities.

REFERENCES

+ Supported in part by the University of Connecticut Research Foundation.
* Visiting Fulbright - Hays scholar from Univ. of Cluj, Romania.

1. M. Rubinstein, G. H. Stauss and J. Dweck, Phys. Rev. Letters 17, 1001 (1966).
2. M. B. Stearns, Phys. Rev. B9, 2311 (1974).
3. T. J. Burch, T. Litrenta and J. I. Budnick, Phys. Rev. Letters 33, 421 (1974).
4. G. K. Wertheim, V. Jaccarino, J. H. Wernick and D. N. E. Buchanan, Phys. Rev. Letters 12, 24 (1964).
5. V. Niculescu, K. Raj, T. J. Burch and J. I. Budnick. To be published in the Phys. Rev. B.
6. S. Yoon and J. G. Booth, Phys. Letters 48A, 381 (1974).
7. W. A. Hines, A. H. Menotti, J. I. Budnick, T. J. Burch, T. Litrenta, V. Niculescu and K. Raj, to be published in the Phys. Rev. B.
8. I. A. Campbell, J. Phys. C, 2, 1338 (1969).

# N.M.R. SHIFT AND RELAXATION STUDIES OF Mn IMPURITIES IN LIQUID BISMUTH

R. Dupree*† and R. E. Walstedt†
*University of Warwick, Coventry, United Kingdom, and †Bell Laboratories, Murray Hill, N. J. 07974

ABSTRACT

The static and dynamic behavior of dilute Mn localized moments in liquid Bi have been investigated by NMR measurements in both $^{55}$Mn and $^{209}$Bi nuclei. A K,χ plot for $^{55}$Mn NMR reveals a temperature independent (orbital) contribution to the susceptibility; such a contribution is absent in CuMn. The $^{55}$Mn spin-lattice relaxation apparently consists of two terms, one due to the fluctuating hyperfine field of the local moment with $T_2/T = 6.4\times10^{-9}$ sec $K^{-1}$ and the other Korringa like with $T_2T$ of 18 msec K. The Mn impurities induce a large, temperature dependent shift of the $^{209}$Bi NMR line and cause the associated relaxation rate to increase rapidly with impurity concentration. The correlation time deduced from the Mn relaxation is an order of magnitude too short, however, to explain the Bi relaxation.

Recent experiments on CuMn alloys[1] have shown that it is possible to observe the nuclear magnetic resonance (NMR) of $^{55}$Mn in liquid alloys where it has a localized magnetic moment and thus obtain, in principle, much useful microscopic information in such systems. Of the liquid "simple" metal-Mn alloys whose magnetic susceptibility has been investigated BiMn has the lowest liquidus temperature for a reasonable concentration ($\geq$ 1%) of Mn. This, together with a measured $p_{eff} \sim 5(\mu_B)^2$, apparently makes it an ideal system to study in detail. In the following we report and discuss measurements of shift and spin-lattice relaxation time for both $^{55}$Mn and $^{209}$Bi.

The $^{55}$Mn NMR shift is plotted vs a Curie-law susceptibility in Fig. 1. Note that both K and χ vary by more than a factor two over the temperature range investigated. The straight-line behavior found here is interpreted in the conventional way[3] by taking $K_{Mn} = K_s+K_{orb}+K_d$ and $\chi_{Mn} = \chi_s+\chi_d+\chi_{orb}$, where $K_d = \alpha\chi_d$ and $K_{orb} = \beta\chi_{orb}$. We estimate $K_s \sim 0.2\%$ and neglect $\chi_s << \chi_{orb},\chi_d$. $K_s$ and the orbital quantities are assumed to be independent of temperature, so that the temperature-dependences of K and χ arise solely from the d-spin contributions. β is estimated to be[1,4] 65 $(\text{emu/mole})^{-1}$. Drawing a line of slope β from the point K = 0.2%, χ = 0, the intersection of this line with a line drawn through the data partitions $K_{Mn}$ and $\chi_{Mn}$ into d-spin and orbital parts. We find that $K_{orb}$ = 0.45%, indicating some orbital susceptibility despite the apparently normal $p_{eff}$ for the manganese. This is not the case for CuMn where $K_{orb}$ and hence $\chi_{orb} \to 0$.[5] The hyperfine field of 63 kG is very nearly the same as that of CuMn.[1]

The inverse of the $^{209}$Bi shift (relative to Bi metal) is plotted vs temperature for several Mn concentrations in Fig. 2. The shift caused by the addition of Mn is seen to be large, positive, and strongly temperature-dependent. However, it does not follow $\chi_{Mn}$, appearing to become zero at about 200K for these concentrations. This behavior is consistent with a hypothetical negative shift contribution which is independent of temperature and linear in concentration. The positive shift is clearly caused by the RKKY spin-density oscillations generated by the Mn moments. Interestingly, it is opposite in sign to that of $^{63}$Cu in CuMn and opposite to that predicted by current theories[6] of the spin density oscillations around Mn impurities in simple metals. The origin of the suggested negative contribution to the shift is not clear because it is extremely large having $\Delta K/cK_{Bi} = -9\pm1$ whereas a typical value for a valence difference of 4 would be about 1.

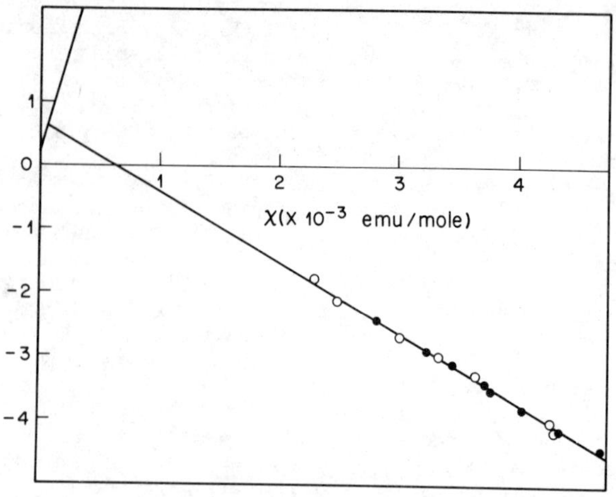


Fig. 1. Plot of $K_{Mn}$ versus $\chi_{Mn}$ for different concentrations of Mn showing analysis into orbital and d spin contributions (● 14.7% Mn, ○ 6.0% Mn). $\chi_{Mn}$ taken from Ref. 2.

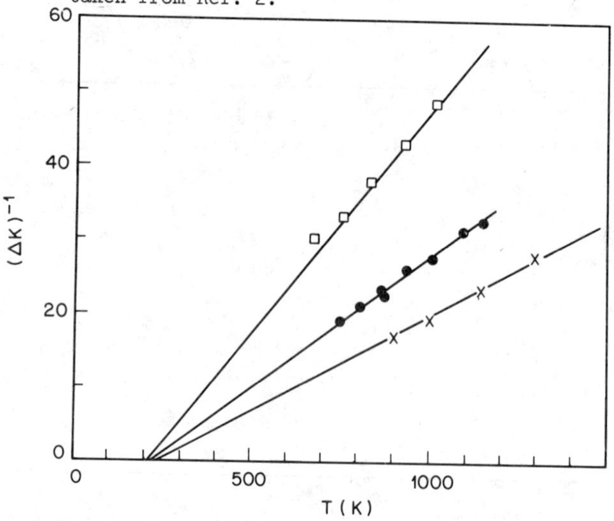


Fig. 2. Inverse NMR shift of $^{209}$Bi relative to Bi metal versus temperature for several concentrations of manganese (× 14.7% Mn, + 10.3% Mn, □ 6.0% Mn.)

The longitudinal relaxation time $T_1$ for $^{55}$Mn is too short in these alloys to be measured directly. Owing to the motional narrowing effect of the liquid state, $T_2$ measurements are assumed to yield the same result. $T_2$ is plotted vs T for various concentrations of Mn in Fig. 3. It is clear that the most concentrated alloy has the longest $T_2$. Further, $T_2$ for each alloy increases nonlinearly with temperature, appearing to flatten at the highest temperatures. We interpret the observed behavior as follows. The dominant $T_1$ mechanism is presumed to be the fluctuating hyperfine field of the local moment, which in the limit of zero concentration is given by[7]

$$1/T_1 = 1/T_2 = 2A_d^2S(S+1)T_{1e}/3 \qquad (1)$$

where $A_d$ is the d-spin hyperfine coupling constant (in frequency units), and the role of the correlation time

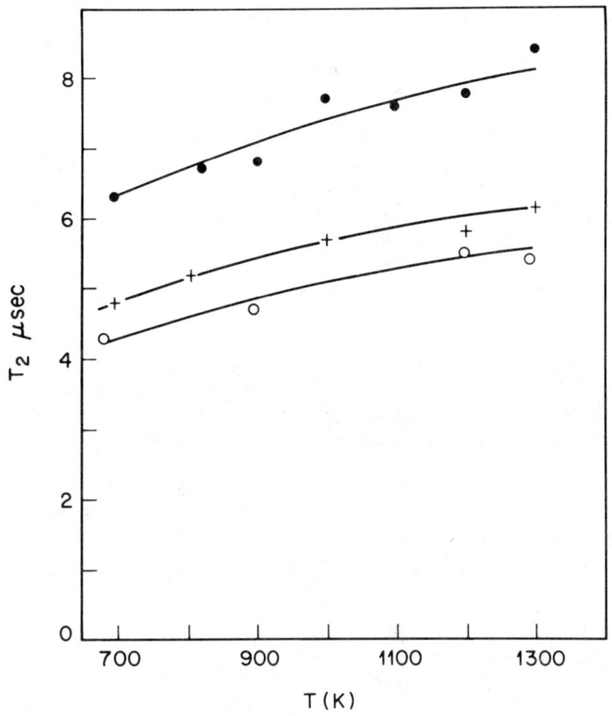


Fig. 3. Relaxation time $T_2$ of $^{55}$Mn in Bi (● 14.7% Mn, + 6.7% Mn, ○ 1.8% Mn).

is played by the local moment spin-lattice relaxation time $T_{1e}$. Since $T_{1e}^{-1} \propto T$, one finds $T_2 \propto T$ from Eq. (1), in qualitative accord with the data of Fig. 3. At finite concentrations the fluctuation rate of the local moment will be augmented by the RKKY indirect exchange coupling, thereby decreasing the rate in Eq. (1) and making $T_2$ longer at higher concentrations, as is found. Combining these effects, however, predicts straight-line behavior for the plot of Fig. 3. Although the scatter in the data might permit such a fit, a much better fit is made by adding a relatively strong Korringa term, i.e. $T_2T$ = 18 msec K to the rate of Eq. (1). The curves drawn correspond to these three combined effects, then, with $T_{1e}T = 3.6\times10^{-11}$ sec K and the RKKY fluctuation rate $1/\tau_{RKKY} = 1.2\times10^{14}$ c $sec^{-1}$, where the combined inverse correlation time is $1/\tau_c = 1/\tau_{RKKY} + 1/T_{1e}$. From $T_{1e}T$ we deduce an $\ell = 2$ s-d exchange parameter[8] of $|J_2| = 0.59$ eV where the effect of electron-electron interactions has been estimated from the Korringa product of pure Bi.[9] It is assumed here that the mixing contribution to exchange predominates. The quantitative values must be considered tentative here in view of the large and unaccountable Korringa term which we have adduced.

Lastly, we have found that the relaxation rate of the $^{209}$Bi is strongly increased by the addition of Mn. One mechanism for this would be the fluctuating transferred hyperfine field at the nearest-neighbor position to a Mn moment. Neglecting more long-range effects one can write the transferred hyperfine coupling parameter $A_t$ in terms of the $^{209}$Bi shift ΔK caused by the addition of Mn impurities as

$$A_t = \frac{3\Delta K \gamma k_B T}{\nu c g \mu_B (S+1) S} \qquad (2)$$

where ν is the nearest-neighbor coordination number in the liquid. Then the corresponding relaxation rate is simply that of Eq. (1) modified by the occupation probability factor cν:

$$(1/T_1)_{Bi} = \frac{2}{3} A_t^2 S(S+1) T_{1e} c\nu / \qquad (3)$$

Substitution of appropriate numbers into (2) and (3) yields a rate which is too small by more than an order of magnitude. Presumably, therefore, there is some other, even stronger, relaxation mechanism present. Since bismuth has a large quadrupole moment, quadrupolar relaxation is a possibility. However, this too is usually an order of magnitude smaller than that observed. Further experiments are under way to resolve this question.

REFERENCES

1. R. E. Walstedt and W. W. Warren, Jr., Phys. Rev. Lett. 31, 365-368 (1973).
2. S. Tamaki and S. Takeuchi, J. Phys. Soc. Japan 22, 1042-1045 (1967).
3. A. M. Clogston, V. Jaccarino and Y. Yafet, Phys. Rev. 134 A 650-661 (1964).
4. A. Freeman and R. E. Watson in Magnetism, ed. H. Suhl and G. Rado (Academic, New York, 1965) Vol. IIA p. 167.
5. A similar situation is found, for example, for $Cu_{.90}Al_{.10}$:Mn, where $\chi_{orb} > 0$ but $\chi_d$ is unchanged from that of CuMn.
6. P. Jena and D. J. W. Geldart, Phys. Rev. 7B, 439-450 (1973).
7. R. E. Walstedt and A. Narath, Phys. Rev. B6, 4118-4125 (1972).
8. R. E. Walstedt and L. R. Walker, Phys. Rev. B11, 3280-3291 (1974).
9. R. Dupree and E. F. W. Seymour in Liquid Metals, ed. S. Z. Beer (Marcel Dekker, New York), p. 461, (1972).

# HYPERFINE FIELDS IN THE ABSENCE OF MAGNETIC ORDER IN Dy-Sc ALLOYS

R. Abbundi and R. Segnan, American University, Washington, D.C. 20016;
and
J.J. Rhyne and D.M. Sweger, Natl. Bureau of Standards, Wash. D.C. 20234

ABSTRACT

$Tb_xSc_{1-x}$ and $Gd_xSc_{1-x}$ alloys have been shown to exhibit no long-range magnetic order for $x < .25$ (in Tb-Sc) from neutron scattering and magnetization experiments in contrast to Y based alloys and conventional theory which predicts a vanishing of $T_N$ only as $x \to 0$. We have investigated the Dy hyperfine fields in a series of $Dy_xSc_{1-x}$ alloys in a similar range in which no $T_N$ is found (above 4K). Although no long-range order is present, the alloys do show a well-defined hyperfine splitting corresponding to a field approximately equal to that in pure Dy metal and independent of Dy concentration. This is present even in a 2% Dy alloy which is the lowest concentration measured. This surprising result suggests a spin-relaxation mechanism is operative which produces the observed hyperfine fields, but which is not accompanied by static long-range magnetic order.

## INTRODUCTION

Scandium possesses chemical and electronic properties similar to the heavy rare earth elements but without the 4 f magnetic electrons. It is thus of interest to investigate what changes, if any, occur in the magnetic properties of the rare earths in diluting them with this element.

Previous work on many heavy rare earth--rare earth alloys have indicated that a linear relationship exists between the ordering temperature and the two-thirds power of the average DeGennes factor[1] which is given by

$$\overline{G} = \sum_i C_i G_i \qquad (1)$$

$$G_i = (g_i - 1)^2 J_i(J_i + 1) \qquad (2)$$

where $C_i$ is the concentration of that magnetic constituent with a DeGennes factor of $G_i$. It is also found that the ordering temperature drops to zero only for zero concentration.

Measurements on rare earth - Sc alloys however have given anomalous results. Nigh et. al.[2] have observed in Gd-Sc a vanishing of the ordering temperature at the rather large Gd concentration of 15%. Neutron experiments[3] on Tb-Sc show that for this system the critical concentration for a nonzero $T_N$ is 25% Tb.

The present work investigates a series of Dy-Sc alloys in an effort to further pursue the rather anomalous behavior of the RE-Sc system. In this study we examined the hyperfine interaction at $^{161}$Dy nuclei in a series of Dy-Sc alloys.

## EXPERIMENTAL DETAILS

A series of alloys containing 2, 5, 10, 25, 35, 50 and 75 atomic percent dysprosium were prepared for this experiment. The constituents were carefully weighed and then arc-melted in an argon atmosphere several times to ensure uniformity. The samples were also weighed after melting to determine that any mass loss was minimal. Following this, the alloys were filed in a "dry box" and mounted on thin plastic discs to be used as Mössbauer absorbers. The density of $^{161}$Dy contained in the absorbers varied from 1.7 $^{mg}/cm^2$ to 13.1 $^{mg}/cm^2$.

The resonance of interest was the 25.6 keV transition in $^{161}$Dy. To produce a gamma ray of this energy a 131 mg $GdF_3$ source enriched to 99.99% $^{160}$Gd was employed. By means of thermal neutron irradiation $^{160}$Gd produces $^{161}$Tb which possesses a 6.9 day half-life and decays by electron emission to $^{161}$Dy. Periodic 24 hour irradiations at the NBS Research Reactor results in a source strength of approximately 13 millicuries. This source yielded an experimentally measured linewidth of .53 $^{cm}/sec$. The $\gamma$ - rays were detected by a high resolution Si(Li) crystal operated at 115K and the absorption spectra were stored in a 400 channel multichannel analyzer.

Operating in a constant acceleration mode required a spectrometer capable of handling velocities of $\pm$ 26 $^{cm}/sec$. This was accomplished by the use of a commercially available velocity transducer. The drive is equipped with a Moiré grating of known spacing which was used to check not only the linearity of the drive but also served as a means for obtaining an absolute velocity calibration.

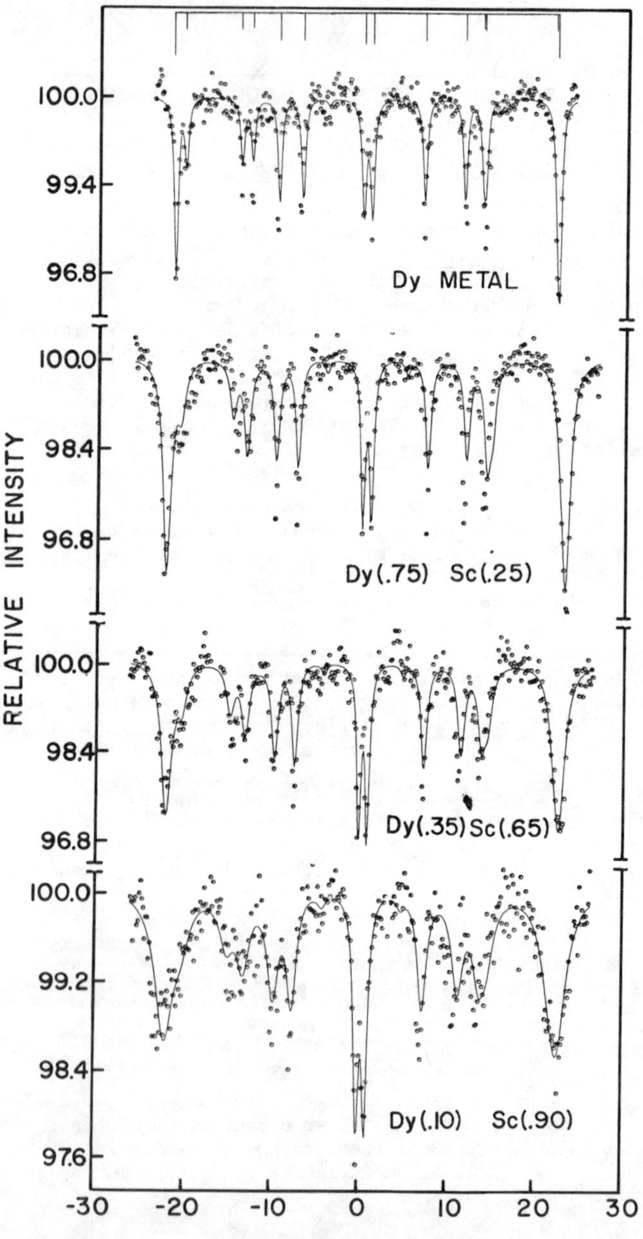


Fig. 1. $^{161}$Dy Mössbauer spectra at T = 4.2 K for Dy metal, $Dy_{.75}Sc_{.25}$ alloy, $Dy_{.35}Sc_{.65}$ alloy and $Dy_{.10}Sc_{.90}$ alloy. The bars at the top indicate the position and relative amplitude of the hyperfine lines in Dy.

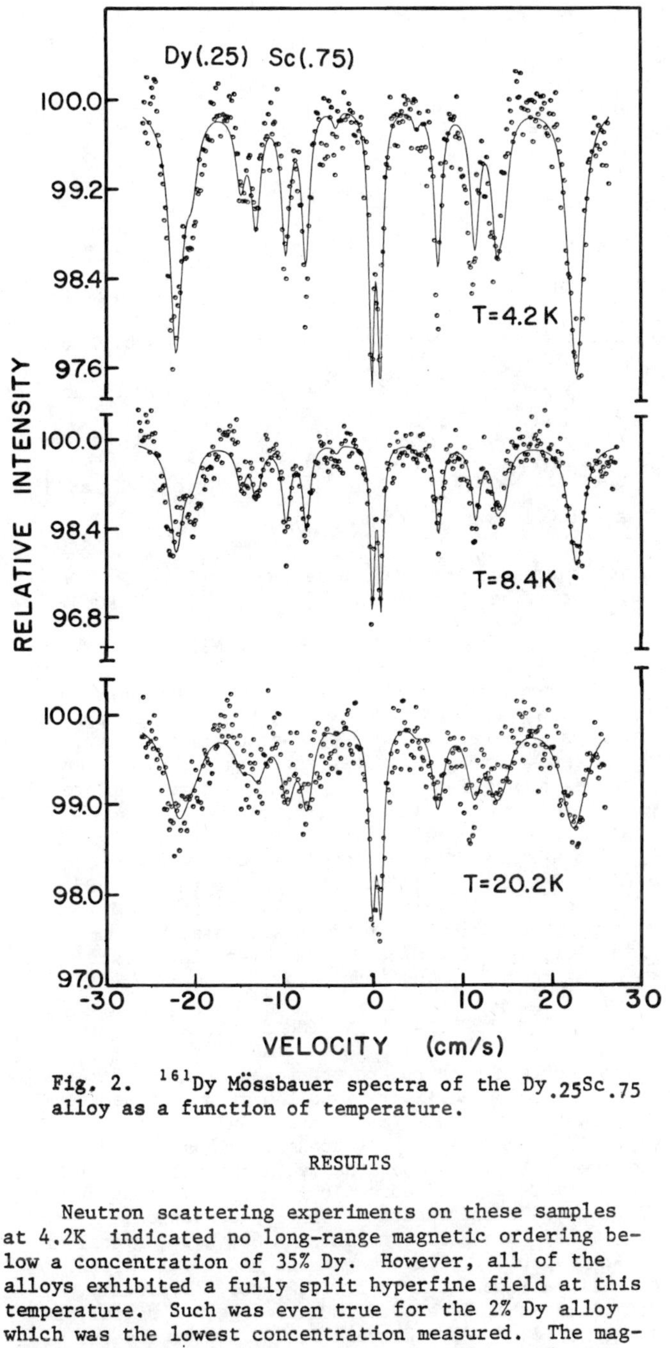


**Fig. 2.** $^{161}$Dy Mössbauer spectra of the $Dy_{.25}Sc_{.75}$ alloy as a function of temperature.

## RESULTS

Neutron scattering experiments on these samples at 4.2K indicated no long-range magnetic ordering below a concentration of 35% Dy. However, all of the alloys exhibited a fully split hyperfine field at this temperature. Such was even true for the 2% Dy alloy which was the lowest concentration measured. The magnitude of the splitting was approximately equal to that found in pure Dy metal and is independent of Dy concentration. However, as can be seen in figure 1, the width of the lines, in particular the outer peaks, is very much concentration dependent as a result of the distribution of neighboring Dy sites. A comparison of the 75%, 35% and 10% alloys to Dy metal clearly indicates broadening of the lines as the concentration of Dy decreases.

The observed hyperfine splitting as a function of temperature is shown in figure 2 for the $Dy_{.25}Sc_{.75}$ alloy. The data for this alloy, which exhibits no long-range magnetic order, indicate that the hyperfine splitting is essentially independent of temperature until it collapses into a single line near 27 K as shown in figure 3. Shown for comparison in this figure is the hyperfine field for Dy metal[4] which closely follows a Brillouin function temperature dependence. (The temperature axis for the Dy has been normalized to reflect the same critical temperature as observed for the alloy.) The qualitative features of the pattern for the alloy as a function of temperature closely resemble paramagnetic $Dy_2O_3$. In the oxide the full splitting is maintained until a very sharp transition to a single line occurs within a narrow temperature range.[5] This splitting has been attributed to a spin lattice relaxation mechanism in which $\tau > 10^{-9}$ sec. The relative amplitudes and sharpness of the individual lines are a function of the relaxation time with short relaxation times yielding only a single line due to the fact that the spins change direction so rapidly that the field averages to zero.

The temperature dependence of the hyperfine interaction in the Dy-Sc system and the lack of static long-range order suggests a slow relaxation mechanism involving the Dy 4 f spin system in which the full hyperfine splitting is observed for temperatures low enough such that the relaxation frequency is slower than the Larmor precession frequency. The large single ion anisotropy interaction of the Dy ion could provide a possible source for this relaxation effect.

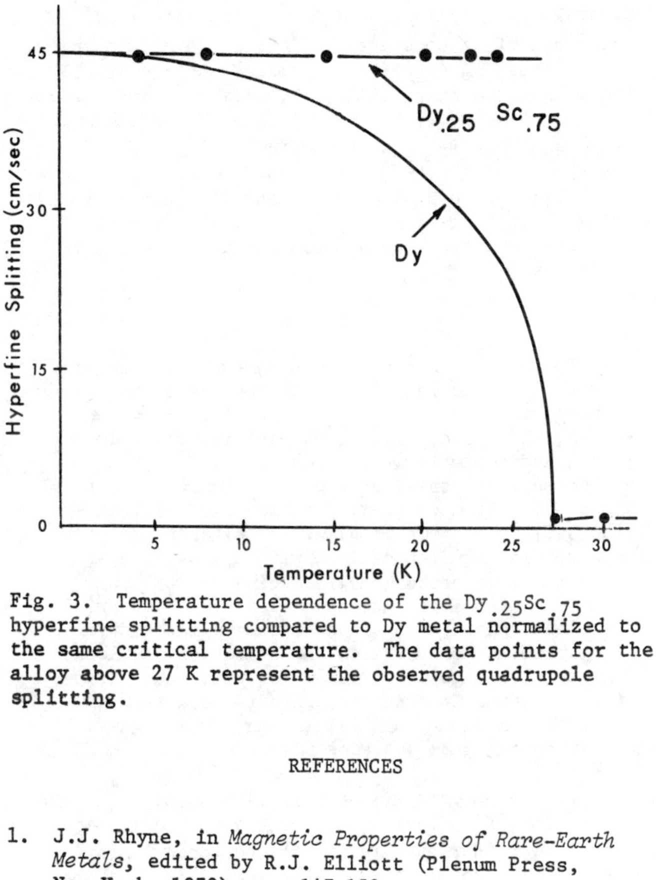


**Fig. 3.** Temperature dependence of the $Dy_{.25}Sc_{.75}$ hyperfine splitting compared to Dy metal normalized to the same critical temperature. The data points for the alloy above 27 K represent the observed quadrupole splitting.

## REFERENCES


1. J.J. Rhyne, in *Magnetic Properties of Rare-Earth Metals*, edited by R.J. Elliott (Plenum Press, New York, 1972), pp. 147-152.

2. H.E. Nigh, S. Legvold, F.H. Spedding and B.J. Beaudry, J. Chem. Phys. 41, 3799 (1964).

3. H.R. Child and W.C. Koehler, J. Appl. Phys. 37, 1353 (1966).

4. D.M. Sweger, R. Segnan and J.J. Rhyne, Phys. Rev. B 9, 3864 (1974).

5. S. Ofer, B. Khurgin, M. Rakavy and I. Nowik, Phys. Lett. 11, 205 (1964).

# NUCLEAR RESONANCE OF YTTERBIUM LOCAL MOMENTS IN GOLD†

D. M. Follstaedt, W. J. Meyer,
D. C. Barham, and A. Narath
Sandia Laboratories
Albuquerque, New Mexico 87115

ABSTRACT

We have observed the $^{173}Yb$ nuclear resonance of dilute (0.1 at. %) ytterbium impurities in gold using a phase-coherent, spin-echo spectrometer. The experiments were performed at ~500 MHz in external fields of ~100 kOe at pumped helium temperatures. At these fields and temperature the electronic moment is completely polarized and the local moment relaxation regime $\omega_e T_{2e} \gg 1$ applies. The resonance has been studied in AuYb powders and single crystals. The data obtained with the external field in the [100] direction between 67 and 110 kOe can be fit with a crystal field splitting of $\Delta(\Gamma_8 - \Gamma_7) = 84.4(3.5)$ K and a hyperfine constant $A = 160.4(0.9) \times 10^{-20}$ ergs. These values agree well with those obtained in ESR and magnetic susceptibility experiments. Nuclear relaxation time measurements show that $T_2 \propto H_o^2$. This field dependence and the magnitude [147(8) μsec at 109.4 kOe] agree well with predictions for the $\frac{1}{2} \leftrightarrow -\frac{1}{2}$ transition based on a local moment fluctuation model using an exchange constant $\langle \mathcal{J}^2 \rangle = 0.72$ eV$^2$ derived from the reported thermal broadening of the ytterbium ESR.

## INTRODUCTION

A localized moment in a non-magnetic metal interacts with the metal's conduction electrons via an exchange interaction, the simplest form of which is $-\mathcal{J}\vec{S}\cdot\vec{s}$, where $\vec{S}$ and $\vec{s}$ are, respectively, the local moment and conduction-electron spins. This interaction perturbs the local-moment eigenstates causing transitions between them, both real and virtual. Fluctuations between local-moment eigenstates have been studied theoretically,[1] with a low-temperature and a high-temperature regime being identified.

Since the local moment nucleus interacts with the electronic moment via a hyperfine interaction $A\vec{I}\cdot\vec{J}$, fluctuations between the states of the local-moment total angular momentum ($\vec{J}$) will cause transitions between the nuclear spin ($\vec{I}$) states and affect the nuclear spin relaxation. The high-temperature regime has been investigated with NMR in the local-moment system CuMn[2]. A study of nuclear relaxation rates in the low-temperature regime $\omega_e T_{2e} \gg 1$ (the subscript e refers to the local moment) is needed for a complete understanding of the exchange perturbation on the local-moment states. NMR relaxation studies of the WCo system[3] show evidence for low-temperature local-moment fluctuations, but these are partially obscured by other nuclear relaxation processes.

To study further the low-temperature regime we have chosen the local-moment system AuYb. This choice was based on several criteria. Ytterbium dissolves in gold as a trivalent (J = 7/2) ion. Its crystal-field ground state is an isotropic $\Gamma_7$ doublet,[4,5] which is needed for a narrow nuclear resonance in a powder sample. Crystal-field studies show that the doublet is well isolated by ≳ 80 K from the higher $\Gamma_8$ and $\Gamma_6$ states, so that anisotropy due to field-induced admixture of higher states is minimized. The $\Gamma_7$ ESR has been observed in weak fields,[6] showing that the $\omega_e T_{2e} \gg 1$ requirement is satisfied in our experiments at 1 K. The ESR hyperfine structure gave a value for A from which the zero-field resonance frequency could be calculated; the ESR thermal broadening yielded a value for $\langle \mathcal{J}^2 \rangle$ which suggested that favorable nuclear $T_2$'s (≳ 100 μsec) could be obtained at high fields (≈ 100 kOe).

Table I

| ν | H (exp.) | H (theo.)* |
|---|---|---|
| 362.4 MHz | --- | 0.00 kOe |
| 450.0 MHz | 66.98(0.10) kOe | 66.97 kOe |
| 475.0 MHz | 91.23(0.10) kOe | 91.16 kOe |
| 492.0 MHz | 109.41(0.10) kOe | 109.46 kOe |

* Obtained with K = +1.0%, $\Delta(\Gamma_8 - \Gamma_7) = 83.6$ K, and $A = 160.10 \times 10^{-20}$ ergs.

## $^{173}Yb$ RESONANCE IN POWDERED SAMPLES

We were able to detect the $^{173}Yb$ resonances in (0.1 at. %) powdered samples at 1 K. The NMR apparatus is a phase-coherent, spin-echo spectrometer, which repetitively digitizes the signal in the time interval appropriate for the echo. A more complete account of experimental details will be published elsewhere.[7]

At 490 MHz, the resonance was centered at 102(4) kOe. Its full inhomogeneous width was ~25 kOe, compared with a width of ~14 kOe expected from admixture anisotropy. At the position of maximum intensity, $T_2 \sim 100$ μsec; on the high (low) field side, $T_2$ became longer (shorter).

## $^{173}Yb$ RESONANCE IN SINGLE CRYSTALS

To obtain a narrower resonance for a quantitative study of the $\Gamma_8$ admixture into the ground state, we studied the resonance in 0.1 at. % AuYb single crystals. Our best data (strongest signal and narrowest line) were obtained with the external field in the [100] crystal direction. We shall limit our discussion to these data in this presentation.[7]

Three data points taken with our best crystal are shown in Table I. The resonances had a full width at half maximum equal to 1.0(0.2) kOe. Typically, digitizing the signal 200,000 times gave a signal-to-noise ratio of ~8. We were able to determine the center of the line to within ± 0.1 kOe.

A detailed examination of the resonance positions shows a nonlinear field dependence, indicating that the lowest order perturbation treatment is inadequate. To extract the crystal-field splitting of the $\Gamma_8$ quartet from the $\Gamma_7$ ground-state doublet, we have used a computer routine which diagonalizes the crystal field and Zeeman Hamiltonians. From the ground-state eigenvector, $\langle J_z \rangle_{g.s.}$ is obtained, which combined with the hyperfine constant A gives the nuclear resonance frequency as a function of applied field.

We must also include the interaction of the nucleus with the external field ($\gamma_{173} = 2\pi \times 0.205$ MHz/kOe). This must be corrected for the Knight shift of the conduction electrons screening the $Yb^{3+}$ ion. The Knight shift can contain contributions which are negative (due to the spin moment of 5d screening electrons) as well as positive (due to the orbital moment of 5d screening electrons and the direct hyperfine interaction of 6s electrons).[8] Examination of Knight shifts in 5d metals suggests that a reasonable range of values is $-1\% \leqslant K \leqslant +3\%$. With

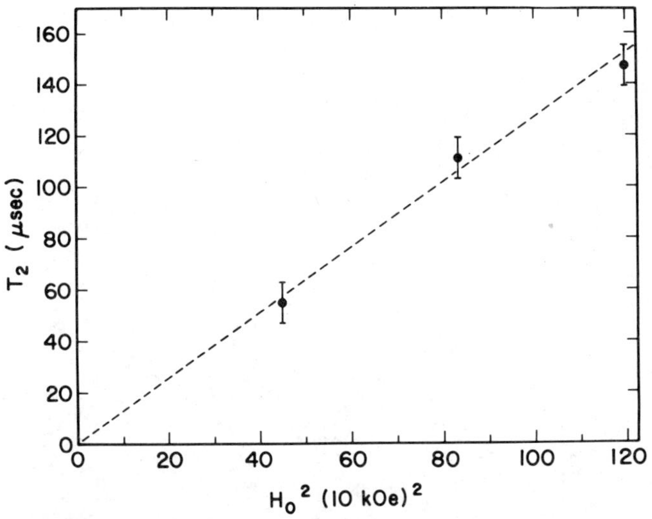


Figure 1. $T_2$ values obtained for the resonance positions in Table I plotted versus $H_0^2$. The dashed line is the theoretical prediction of the model discussed in the text using $\langle \mathcal{J}^2 \rangle = 0.72\ eV^2$.

this range of uncertainty and the uncertainty in resonance position we obtain $\Delta(\Gamma_8 - \Gamma_7) = 84.4(3.5)$ K and $A = 160.4(0.9) \times 10^{-20}$ ergs. These values agree well with crystal-field splittings given by Williams and Hirst[4] (≥ 79 K) and Murani (94 K) and the hyperfine constant of Tao et al.,[6] scaled to $^{173}Yb$ $[159.0(4.4) \times 10^{-20}$ ergs].

TRANSVERSE NUCLEAR RELAXATION TIMES ($T_2$)

Shown in Figure 1 are the $T_2$ data for the [100] resonances versus $H_0^2$. We were unable to obtain data below 67 kOe, evidently because $T_2$ becomes too short.

We wish to examine the $T_2$ data in terms of theoretical expressions given for a local-moment nucleus whose relaxation rate is determined by fluctuations of a highly polarized local moment $(g\mu_B H)/(k_B T) \gg 1$. Transverse fluctuations of the moment due to virtual excitations to higher levels give a nuclear relaxation rate for S-state ions[1]

$$T_2^{-1} = \frac{\eta \pi k_B\ T\ A^2 S^2 (\mathcal{J}\rho)^2}{2\hbar (g\mu_B H_o)^2} ,$$

where $\rho$ is the density of conduction electron states for one spin at the Fermi energy and $\eta (\equiv \frac{1}{2}[4I(I+1) - 1])$ is an enhancement factor which is needed when only the $\frac{1}{2} \longleftrightarrow -\frac{1}{2}$ nuclear transition is being observed.[3,9] We assume this is the case for $^{173}Yb$ ($I = 5/2$, $\eta = 17$) because of strain-induced first-order quadrupole splitting of the satellite transitions. The other symbols have their usual meanings. Since the ground state is well isolated, we attempt to explain the $T_2$'s with the above expression for an effective $S_{eff} \equiv \frac{1}{2}$ moment. We use g for the $\Gamma_7$ doublet, scale the exchange interaction as $[g_{eff}(g_J - 1)/g_J]\mathcal{J}\vec{S}_{eff} \cdot \vec{s}$, and the hyperfine interaction as $A_{eff}\vec{S}_{eff} \cdot \vec{I}$. Using the low field values for $A_{eff}$ and $g_{eff}$ (appropriate for $\langle J_Z \rangle_{g.s.} = 3/2$), the value $\langle \mathcal{J}^2 \rangle = 0.72$ $eV^2$ from Tao et al.,[6] and $\rho = 0.15\ eV^{-1}\ spin^{-1}$,[10] we obtain the dotted line in Fig. 1. The model agrees well with the data, even at fields as high as 110 kOe. The continued agreement at higher fields even in the presence of appreciable VanVleck admixture, is attributable to $A_{eff}$ and $g_{eff}$ both being proportional to $\langle J_Z \rangle_{g.s.}$, so that the field dependences due to admixture cancel.

SUMMARY

We believe this to be the first spin-echo study of a low-temperature local-moment nucleus whose relaxation is determined by fluctuations of the moment without serious competition from other mechanisms. The crystal-field model explains the resonance positions quite well. The magnitude of $T_2$ and its $H_0^2$ field dependence agree well with theoretical relaxation rates due to transverse fluctuations of the local moment, when the $Yb^{3+}$ ground state is appropriately treated as an $S_{eff} = \frac{1}{2}$ ion. Longitudinal fluctuations are evidently "frozen out" as expected for the temperatures and fields used in our experiments.[1] Virtual transitions of the moment evidently dominate the fluctuation spectrum in this regime.

REFERENCES

† This work supported by the U. S. Energy Research and Development Administration, ERDA.


1. R. E. Walstedt and A. Narath, Phys. Rev. B6, 4118 (1972).
2. R. E. Walstedt and W. W. Warren, Jr., Phys. Rev. Lett. 31, 365 (1973).
3. A. Narath, Physica Scripta 11, 237 (1975); Phys. Rev. (to be published).
4. G. Williams and L. L. Hirst, Phys. Rev. 185, 407 (1969).
5. A. P. Murani, J. Phys. C: Metal Phys. Suppl., No. 2, S153 (1970).
6. L. J. Tao, D. Davidov, R. Orbach, and E. P. Chock, Phys. Rev. B4, 5 (1971).
7. We expect to publish a complete and detailed report of our work in Phys. Rev.
8. A. C. Gossard, V. Jaccarino, and J. H. Wernick, Phys. Rev. 133, A881 (1964).
9. R. E. Walstedt, Phys. Rev. Lett. 19, 146 (1967); 19, 816 (1967).
10. A. Narath, Phys. Rev. 163, 232 (1967).

# OBSERVATION OF AN ELECTRO-NUCLEAR SINGLET GROUND-STATE IN Au $^{171}$Yb

G. Frossati, J.M. Mignot, D. Thoulouze and R. Tournier
Centre de Recherches sur les Très Basses Températures, C.N.R.S.,
B.P. 166 Centre de Tri, 38042 Grenoble-Cédex, France.

## ABSTRACT

We describe SQUID susceptibility measurements on several dilute natural and monoisotopic Au Yb alloys down to 7 mK. Whereas Au $^{174}$Yb obeys a Curie-law in the whole range of temperature, a very unusual case of a Van-Vleck-like behavior due to an electro-nuclear singlet ground-state is reported for Au $^{171}$Yb. This peculiar situation also results in a Schottky anomaly centered at 45 mK, which is observed in specific heat measurements down to 25 mK.

## INTRODUCTION

In the last few years, a number of experimental studies down to the millikelvin range have been devoted to the hyperfine-enhanced nuclear susceptibilities of Van-Vleck compounds[1]. However, there has been less interest in the very low temperature hyperfine properties of rare earth ions of which the crystal-field ground-state is a Kramers'doublet.

The purpose of this paper is to show how, in such a case, the strong hyperfine coupling between the effective spin one-half[2] $\vec{S}_{eff}$ defined inside the electronic doublet ground-state with a finite nuclear spin $\vec{I}$ strongly alters the magnetic properties of the ion below 100 mK. The maximum effect is obtained with a nuclear spin one-half for an antiparallel coupling, which leads to complete compensation of the ionic moment. This striking phenomenon actually occurs in the isotope $^{171}$Yb as demonstrated by a comparison of the low temperature susceptibilities of Au Yb, Au $^{171}$Yb and Au $^{174}$Yb. The same effect is observed in the specific heat of Au $^{171}$Yb which exhibits a Schottky anomaly centered at 45 mK.

## EXPERIMENTS

Samples of Au Yb (95 and 610 at.ppm) and Au $^{171}$Yb (42 and 448 at.ppm) were prepared by melting appropriate amounts of components in a high-vacuum induction furnace (Au : 5N Johnson Matthey ; Yb : 3N LEICO ; 171 Yb : 2N4 Oak-Ridge, 90.60 % enriched) and casting the alloy in a water cooled copper mold. The Au $^{174}$Yb sample (c ∿ 30 at.ppm) was prepared by collecting the Ytterbium on gold foils in an isotope separator, the enrichement being more than 99 %.

The susceptibility measurements were performed by a d.c. SQUID method[3] in the mixing chamber of a dilution refrigerator[4] in continuous operation down to 7 mK. A constant field of 0.5 Oe trapped in a Niobium cylinder was applied to the sample. The temperature was deduced from the a.c. susceptibility of a powdered CMN with a diameter equal to the height. This scale was in agreement with that obtained from the static susceptibility of another CMN measured with the SQUID. Silicon and Carbon resistors were used as secondary thermometers.

The specific heat measurements were carried out in an adiabatic demagnetization cryostat[5] by a quasi-adiabatic method.

Susceptibilities are plotted as functions of 1/T in figures 1 and 2. Previous measurements[6] have shown that the Au-Yb system obeys a Curie-law down to 100 mK. In the framework of an ionic model, this linear variation is attributed to the magnetization of the $\Gamma_7$ crystal-field doublet ground-state, governed by the effective-spin Zeeman hamiltonian :

$$\mathcal{H}_{Zee} = g_s \mu_B \vec{S}.\vec{H} \quad (\vec{S}\text{=spin one-half, } g_s = 3.4). \qquad (1)$$

It can be seen on figures 1 and 2 that only $^{174}$Yb follows this simple Curie behavior down to 7 mK. Both natural Au Yb samples exhibit deviations below 40 mK.

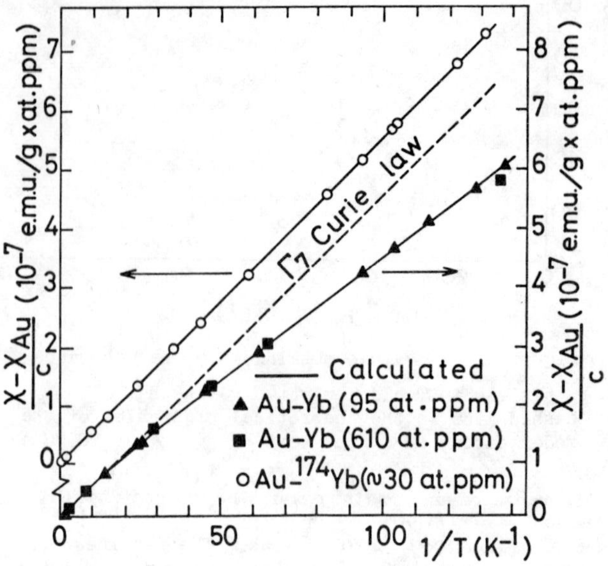


Fig. 1 : Reduced susceptibilities of Au Yb and Au $^{174}$Yb alloys versus 1/T. The "arbitrary units" of the SQUID magnetometer have been calibrated from the high temperature Curie slope which was previously measured in a classical extraction apparatus (see Ref. 6).

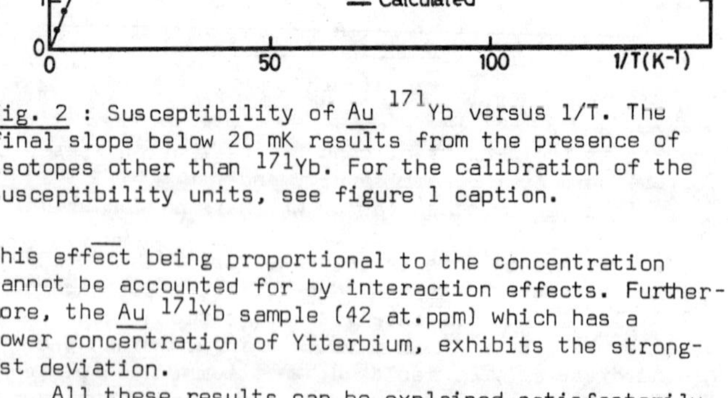

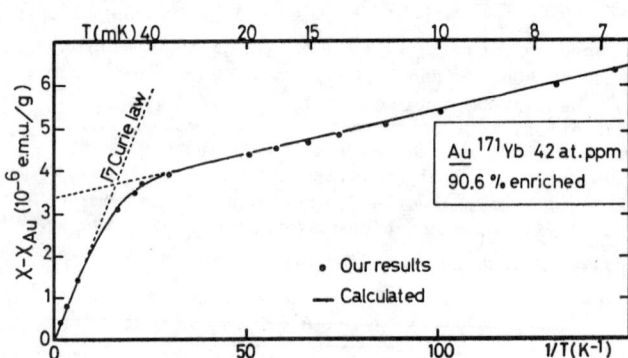

Fig. 2 : Susceptibility of Au $^{171}$Yb versus 1/T. The final slope below 20 mK results from the presence of isotopes other than $^{171}$Yb. For the calibration of the susceptibility units, see figure 1 caption.

This effect being proportional to the concentration cannot be accounted for by interaction effects. Furthermore, the Au $^{171}$Yb sample (42 at.ppm) which has a lower concentration of Ytterbium, exhibits the strongest deviation.

All these results can be explained satisfactorily in terms of hyperfine interactions occuring in the Ytterbium isotopes which have finite nuclear spins (table I). To take this effect into account, one must diagonalize the total hamiltonian

$$\mathcal{H}_T = \mathcal{H}_{Zee} + \mathcal{H}_{hf} = g_s \mu_B \vec{S}.\vec{H} + A_s \vec{S}.\vec{I} \qquad (2)$$

where $A_s$ is the hyperfine constant reported in table I. It is then easy to calculate the total magnetization as a function of H and T. In $^{171}$Yb, the antiparallel coupling of $\vec{I}$ and $\vec{S}$ (both spins one-half) creates a non-magnetic singlet ground-state[7] of the electro-nuclear system.

Table I

Hyperfine parameters of the isotopes of Ytterbium

| isotope | I | $A_S$ (mK) | Abundance in natural Yb (at. percent) | Abundance in commercial $^{171}Yb$ (at. percent) |
|---|---|---|---|---|
| $^{171}Yb$ | 1/2 | +126.6 | 14.3 | 90.60 |
| $^{173}Yb$ | 5/2 | - 34.9 | 16.1 | 1.15 |
| $^{174}Yb$ | 0 | — | 31.8 | 1.48 |
| others | 0 | — | 37.8 | 6.77 |

This situation results in a temperature-independent Van-Vleck-like susceptibility below 40 mK. In $^{173}Yb$ the high temperature electronic Curie slope is only reduced by a factor of 4/9 below 20 mK, whereas no reduction occurs in $^{174}Yb$ for which I = 0. The variation of $\chi$ in natural Yb is a mixture of the contributions of the various isotopes with appropriate coefficients of abundance. The calculated curves have been fitted to our results on figures 1 and 2, leading to good agreement in the entire temperature range.

The specific heat of the more concentrated Au $^{171}Yb$ sample has been measured down to 25 mK. The results presented on figure 3, after substracting the contribution of the gold matrix ($C_{Au}$ = 37 T erg/g.K), show a Schottky anomaly centered at 45 mK. This is in agreement with the level-scheme calculated from equ. 2 for which the maximum specific heat is at $T_M = 0.35\ A_S$. Moreover, a very good fit can be obtained in the whole temperature range as shown on fig. 3.

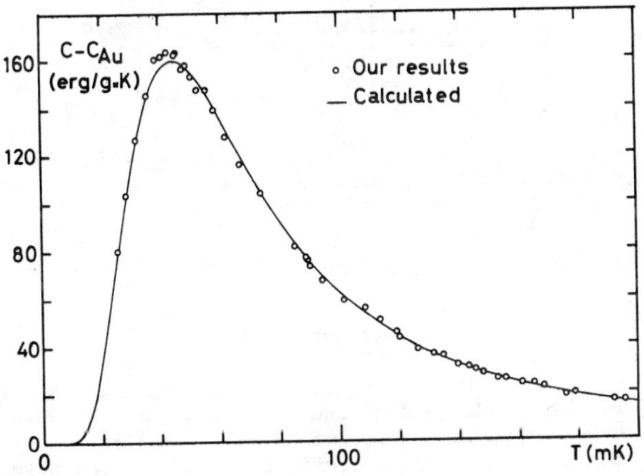


Fig. 3 : Electronuclear specific heat of $^{171}Yb$ below 200 mK. The best fit of our results is obtained for a concentration c = 407 at.ppm which is in good agreement with the analysed value of the excess melt (448 ± 10 at.ppm).

A more detailed account of this experiment will be given elsewhere but it must be emphasized that this effect is observed in the absence of electronic ordering, unlike usual hyperfine specific heats[8] which presuppose the saturation of the electronic moment. Whereas hyperfine specific heats normally result from the population effect of the nuclear levels split by an hyperfine field, our measurements reflect the population of the electro-nuclear levels of $^{171}Yb$.

CONCLUSION

In conclusion, this very low temperature study of monoisotopic Au Yb alloys allowed us to observe two unusual behaviours resulting from the existence of an electro-nuclear singlet ground-state in $^{171}Yb$ : a Van-Vleck-like temperature-independent susceptibility below 40 mK and an electro-nuclear specific heat Schottky anomaly centered at 45 mK. As for the Kondo effect reported previously by several authors[6,9] in the Au Yb system, the Curie-law followed by $^{174}Yb$ down to 7 mK rules out a value of $T_K$ higher than some tenths of a millikelvin. Further work is being performed on this system, in connection with Kondo and interaction effects and will be reported in due course.

We are very grateful to Dr. BOURIANT from the Institut des Sciences Nucléaires de Grenoble for collecting the isotope 174. We thank Drs. J.L. GENICON, J.C. LASJAUNIAS, A. RAVEX and M. VANDORPE for help in the experiments.

REFERENCES

1 - K. Andres and E. Bucher, J. Low Temp. Phys. 9, 267 (1972).
J.L. Genicon and R. Tournier in Proc. 14th Int. Conf. Low Temp. Phys., 1975, North Holland - American Elsevier, Amsterdam - New York (1975), Vol III, p. 200.
2 - A. Abragam and B. Bleaney, Electron Paramagnetic Resonance, Clarendon Press (Oxford, 1969).
3 - R.P. Giffard, R.A. Webb and J.C. Wheatley, J. Low Temp. Phys. 6, 533 (1972).
4 - G. Frossati and D. Thoulouze, Proceedings of 5th Int. Conf. Cryo. Eng., p. 229 (Kyoto) 1974 - IPC Science and Techn. Press Ltd.
5 - J.C. Lasjaunias, A. Ravex, D. Thoulouze, M. Vandorpe (to be published).
6 - B. Cornut and B. Coqblin in Proc. 14th Int; Conf. Low Temp. Phys., 1975, North Holland - American Elsevier, Amsterdam - New York (1975), Vol. III, p. 410.
7 - J. Flouquet, Ann. Phys. 8, 5 (1974).
8 - B. Holmström, A.C. Anderson, and M. Krusius, Phys. Rev. 188, 888 (1969) and references therein.
9 - A.P. Murani, Solid State Commun. 12, 295 (1973).

Y. Talmor and J. Sierro, Phys. Rev. B 11, 300 (1975).
F. Gonzalez-Jimenez, F. Hartmann-Boutron, P. Imbert, B. Cornut and B. Coqblin, J. Physique Colloq. 35, C6-421 (1974).

# KONDO RESISTIVITY IN THE SINGLET GROUND STATE SYSTEM $La_{1-x}Pr_xSn_3$*

A. I. Abou-Aly, S. Bakanowski, N. F. Berk[+], J. E. Crow, and
T. Mihalisin, Physics Department, Temple University
Philadelphia, Pennsylvania 19122

ABSTRACT

We have observed a minimum followed by a maximum in the resistivity of the singlet ground state system $La_{1-x}Pr_xSn_3$. In sufficiently dilute samples ($x < .2$), resistivity, magneto-resistance and magnetization measurements indicate that this striking resistance behavior is not attributable to magnetic ordering. Also, this local maximum in the magnetic contribution to the resistivity cannot be explained by a first Born approximation calculation of the inelastic scattering from the crystalline electric field (CEF) levels. The observed behavior is strongly suggestive of "Kondo" scattering from the excited magnetic CEF levels followed by a precipitative drop in the resistivity ascribable to the thermal deactivation of the excited CEF levels.

## INTRODUCTION

Within recent years there has emerged an increasing awareness of the significance of crystalline electric fields (CEF) in metallic systems and of their manifestations on both equilibrium and transport properties. Systems which have been of particular interest are those containing non-Kramers rare earth ions ($J$ = even) where the CEF-ground state of the ion in the absence of exchange interaction may be non-magnetic. However, little attention has been devoted to examining "Kondo" behavior in such systems. The observation and analysis of Kondo behavior in rare earth alloys and intermetallic compounds has been restricted to systems with the Kramers ions- Ce and Yb. Recently, measurements of several equilibrium and transport properties in $(\underline{La},Pr)Sn_3$ have indicated that the CEF-ground state of $Pr^{3+}$ is a non-magnetic singlet and have shown anomalous behavior indicative of Kondo or resonant scattering from the excited CEF levels.[1,2] In particular, $PrSn_3$ has been shown to order anti-ferromagnetically ($T_n \simeq 9$ K) with a local resistivity maximum in the vicinity of $T_n$ followed by a rapid decrease of the resistivity at lower temperatures.[3] This behavior persists as $PrSn_3$ is alloyed with $LaSn_3$ even though $T_n$ is depressed to zero. We have measured the resistivity, magneto-resistance and superconducting transition temperature, $T_s$, in $(La,RE)Sn_3$ with RE = Ce,Pr,Nd,Sm,Tb and Tm to ascertain to what extent this behavior can be attributed to long or short range magnetic ordering, CEF-only effects and/or "Kondo" behavior.

## RESULTS AND DISCUSSION

Shown in Fig. 1 (a) and (b) is $\rho_x(T)-\rho_0(T)$ (i.e. magnetic plus residual resistivity) versus T for $(\underline{La},Pr)Sn_3$ for x = 0.02 to 1.0. The arrows indicate $T_n$ - defined by the maximum in the low field magnetization. The low field magnetization for samples with $x < 0.3$ did not indicate any ordering above 2 K. Sample preparation and experimental techniques have been described previously.[2] The salient features of the resistivity are (1) the resistivity due to magnetic scattering has a local maximum followed by an abrupt decrease, (2) this local maximum persists even for samples that do not show any long range ordering in the magnetization, (3) the temperature at which the maximum occurs, $T_{max}$, is independent of concentration for $x < 0.6$, and (4) the resistivity for x = 1.0 has a substantial negative temperature coefficient of resistivity (NTCR) for temperatures much larger than $T_n$ (i.e. $T >> T_n$). In contrast; note the behavior of $La_{0.96}Nd_{0.04}Sn_3$ in Fig. 1 (a).

We have discussed these features in greater detail in a previous publication[2] and have shown that these

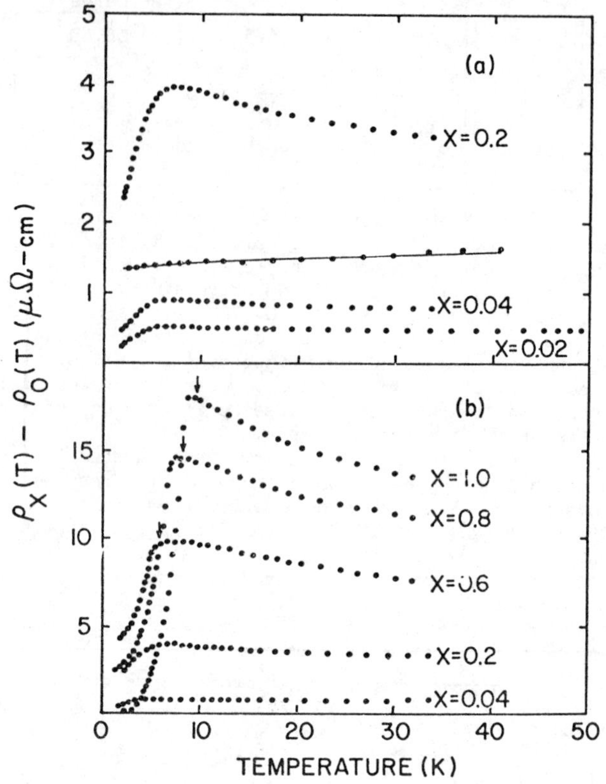


Fig. 1 Resistivity of $La_{1-x}Pr_xSn_3$ minus the phonon contribution vs. temperature for various concentrations, x, of Pr. $La_{.96}Nd_{.04}Sn_3$ is shown by the solid curve which has been shifted up by 1μΩ-cm. The arrows indicate the Neel temperatures determined by low field magnetization.

features cannot be exclusively attributed to long range magnetic ordering. The resistive behavior for the concentrated samples ($x \geq 0.6$) is qualitatively similar to the predictions of Suezaki and Mori[7]; however, systems such as $NdSn_3$, which has a similar antiferromagnetic structure and ordering temperature, have[3] characteristically different resistive behavior-typifying, the prediction of the Fisher-Langer theory.[5] Also, the features in the resistivity of $(\underline{La},Pr)Sn_3$ remain even when $T_n \rightarrow 0$ K. Thus the principal features are not ascribable to precursor behavior associated with the magnetic transition and one must examine other possible contributing interactions such as CEF and/or short range magnetic ordering.

In the absence of spin-spin exchange interactions and CEF, the spin disorder resistivity calculated in a first Born approximation (FBA) due to the s-f interaction is temperature independent. The presence of a cubic CEF partially removes the (2J+1) degeneracy of the angular momentum states and gives rise to a temperature dependent spin-disorder resistivity due to the inelastic scattering from the CEF split ions. Within a FBA and with the singlet (i.e., non-magnetic) ground state, one obtains the precipitious drop in the resistivity as observed but no allowable CEF parameters lead to the NTCR which is observed for $T > T_{max}$. The rapid drop in the resistivity in $(\underline{La},Pr)Sn_3$ is semi-quantitatively consistent with the calculation of the spin-disorder resistivity using the CEF parameters deduced from specific heat and susceptibility measurements.[1]

Van Peski-Tinbergen and Dekker (VD)[6] have calculated within a FBA the temperature and field dependence of the spin-disorder resistivity due to single ions and

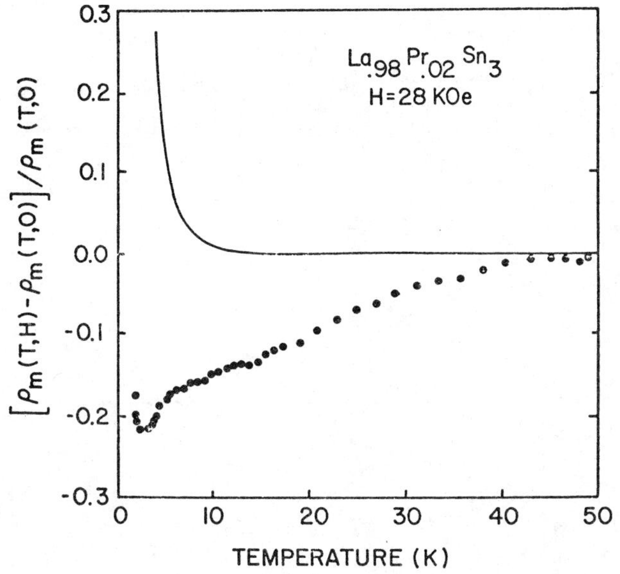


Fig. 2 The magnetic contribution to the magneto-resistance vs. temperature. Solid curve is CEF-only magneto-resistance with H parallel to (100) axis, and CEF parameters from Ref. (1).

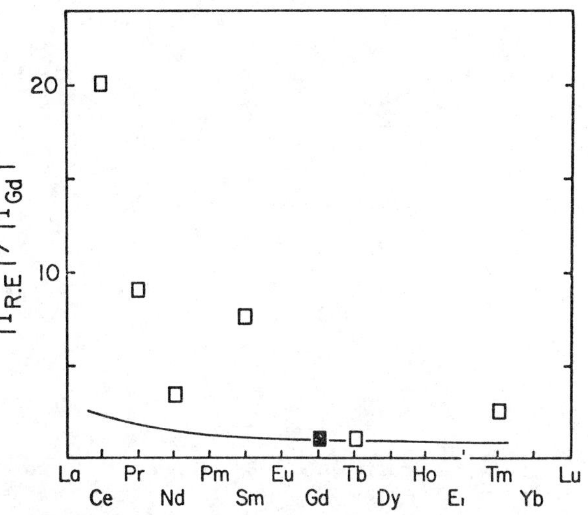


Fig. 3 Normalized s-f exchange constant for the rare earths in $La_{1-x}R.E._xSn_3$ obtained from depression of $T_s$. (Solid Curve-normalized exchange constant for $La_{1-x}R.E._xAl_2$ obtained from ref. 13, and ■ obtained from Ref. (12)).

exchange coupled pairs. Under certain circumstances, they obtain a NTCR due to inelastic scattering from the exchange coupled pairs. They require that the quantity $WF(k_fR)$ must be positive where W is the exchange constant for spin-spin interactions and $F(k_fR)$ is a decreasing oscillatory function of the Fermi momentum, $k_f$, and the interparticle distance, R. Quantitative comparison between their results and the measured resistivity is not appropriate since they have indicated that their calculations are only qualitatively correct. However, a comparative analysis with the resistivity versus temperature of a system such as (La,Nd)$Sn_3$ which is "magnetically similar" to (La,Pr)$Sn_3$ should indicate whether or not the NTCR observed from (La,Pr)$Sn_3$ is due to pair-wise interaction.[7] Based on a simple molecular field consideration and the similarity of their magnetic structures and ordering temperatures, we would expect that their pair-wise interactions would be comparable. As long as the valance of the constituents in (La,RE)$Sn_3$ remains unchanged, $k_fR$ should be insensitive to small variations in R. Thus, assuming the VD model is relevant to the Pr-results, one would expect the resistive behavior of these two systems to be similar. This is contrary to what is observed as shown in Fig. 1 (a).

The temperature dependence of the magnetoresistance in a longitudinal magnetic field of 28 kOe for $La_{.98}Pr_{.02}Sn_3$ has been measure and is shown in Fig. 2. The quantity plotted is the magnetic contribution to the magnetoresistance if the difference in the field dependence of the impurity potential scattering for the pure and doped sample is negligible. The solid curve is a calculation of the magnetic contribution to the magnetoresistance for a field along the (100) direction for the CEF parameters previously mentioned. This calculation was done using the procedure described previously for the zero field case[2] except now the ion states and energies are recalculated exactly in the applied field. The agreement is poor. Estimates of the magnetoresistance due to Kondo behavior in the absence of CEF effects using the results of Kinzer et et al[8] are susbstantially different from the experimental results. It therefore seems that higher order calculations including the effects of CEF are needed to explain the experimental results.

Shown in Fig. 3 is the exchange constant, $\Gamma_{RE}$ normalized to $\Gamma_{Gd}$[9] versus the RE. There values were obtained from $dT_s/dx|_{x\to 0}$ using the s-f interaction, $(g-1)\Gamma$ J.s, and neglecting the effects due to CEF. The solid curve represents and corresponding values of $\Gamma_{RE}/\Gamma_{Gd}$ for the (La,RE)$Al_2$ system.[10] The anomalously large values of $\Gamma_{RE}$ for the lighter RE in (La,RE)$Sn_3$ is partially a reflection of the fact that $\Gamma_{Gd}$ is lower in this system as compared to (La,RE)$Al_2$. This strong resonant behavior has been recently associated with the occurrence of the Kondo effect in Ce-alloys with $g < 1$ and $\Gamma$ negative.[11]

CONCLUSION

We have shown that the prominent features of the resistive behavior in (La,Pr)$Sn_3$ cannot be attributed to long or short range magnetic ordering. Our data for the resistivity is consistent with a log T dependence at higher temperatures ($T > 25K$) where CEF effects are unimportant. However, the temperature range is limited and the rough estimates of the residual resistivity we must use make quantitative analysis difficult. We are presently extending the range of our measurements to correct these difficulties and to check for the expected exponential dependence of the resistivity at low temperatures. However, the data is strongly suggestive of "Kondo" behavior in the presence of CEF.

REFERENCES

*Work supported by the National Science Foundation and Research Corporation

+Present address: Physics Dept., New York University, New York, New York 10003

1) R. W. McCallum, W. A. Fertig, C. A. Luengo, M. B. Maple, E. Bucher, J. P. Maita, A. B. Sweedler, L. Mattix, P. Fulde, and J. Keller, Phys. Rev. Letters 34, 1620 (1975).
2) A. I. Abou-Aly, S. Bakanowski, N. F. Berk, J. E. Crow and T. Mihalisin, Phys. Rev. Letters 35, 1387 (1975). (See also P. Lethuillier and P. Haen, Phys. Rev. Letters 35, 1391 (1975).)
3) B. Stalinski, Z. Kletowski and Z. Henkie, Phys. Stat. Sol. (a) 19, K165 (1973).
4) Y. Suezaki and M. Mori, Prog. Theor. Phys. 41, 1177 (1969).
5) M. E. Fisher and J. S. Langer, Phys. Rev. Letters 20, 6635 (1968).
6) T. Van Peski-Tinbergen and A. J. Dekker, Physics 29, 917 (1963).

7) P. Lethuillier, J. Pierre, G. Fillron and B. Barbara, Phys. Stat. Sol. (a) 15, 613 (1973).
8) H. Keiter, E. Muller-Hartmann and J. Zittartz, Solid St. Comm. 16, 1247 (1975).
9) A. M. Toxen, P. C. Kwok and R. J. Gambino, Phys. Rev. Letters 21, 792 (1968).
10) M. B. Maple, Magnetism V, (ed. H. Suhl), Academic Press, New York (1973) Pg. 289.
11) B. Cornut and B. Coqblin, Phys. Rev. B5, 4541 (1972).


## MAGNETIZATION OF DILUTE Au-Fe ALLOYS AT VERY LOW TEMPERATURES*


T. Steelhammer and O. G. Symko
Department of Physics, University of Utah, Salt Lake City, Utah 84112



### ABSTRACT

The magnetization of Au-Fe alloys with Fe concentrations ranging from 52 ppm to 2 ppm has been measured from 2 K down to 10 mK in a dilution refrigerator. The measurements performed with a SQUID magnetometer in a field of 10 Oe show that the alloys exhibit a Curie-Weiss behavior at high temperatures with strong departures from this law below 0.1 K. The onset of these deviations occurs at temperatures which vary only slightly with concentration. As the concentration of impurities is reduced the magnetization also is reduced and appears to be approaching a non-magnetic state, which could be indicative of single-impurity effects.


### INTRODUCTION

The study of the behavior of a transition metal impurity in a non-magnetic metal host at low temperatures can provide understanding about the interactions of the impurity with its surroundings. We have measured the magnetization of dilute Au-Fe alloys at temperatures from 2 K down to 10 mK. This system has been studied extensively at high temperatures[1,2] and from such studies it is claimed to have a Kondo temperature $T_K$ of ~0.1 K characterizing the onset of spin-compensation of the impurity. Recent very low temperature measurements of the magnetization of Au-Fe[3] were interpreted in terms of spin-compensated Fe impurities and Fe pairs; the latter dominate the behavior of the alloys at the lowest temperatures. The measurements presented here are aimed at showing the magnetic behavior of Au-Fe alloys even more dilute than in the above studies. Since impurity-impurity effects tend to be quite strong, it is important to have very dilute alloys if single-impurity effects are to be observed as demonstrated in the Ag-Mn system.[4] Although Kondo spin compensation is usually thought to be associated with single-impurity effects (from high temperature extrapolations), it should be kept in mind that there can be crystal field effects and hyperfine interactions as well.

In order to see single-impurity effects and to assess the importance of impurity-impurity interactions, the magnetization of a series of Au-Fe alloys was measured with a SQUID magnetometer as a function of temperature in a $^3He$-$^4He$ dilution refrigerator. The Fe concentrations of the alloys varied from 52 ppm to 2 ppm. The temperature was measured with a CMN thermometer as described in Ref. 4.

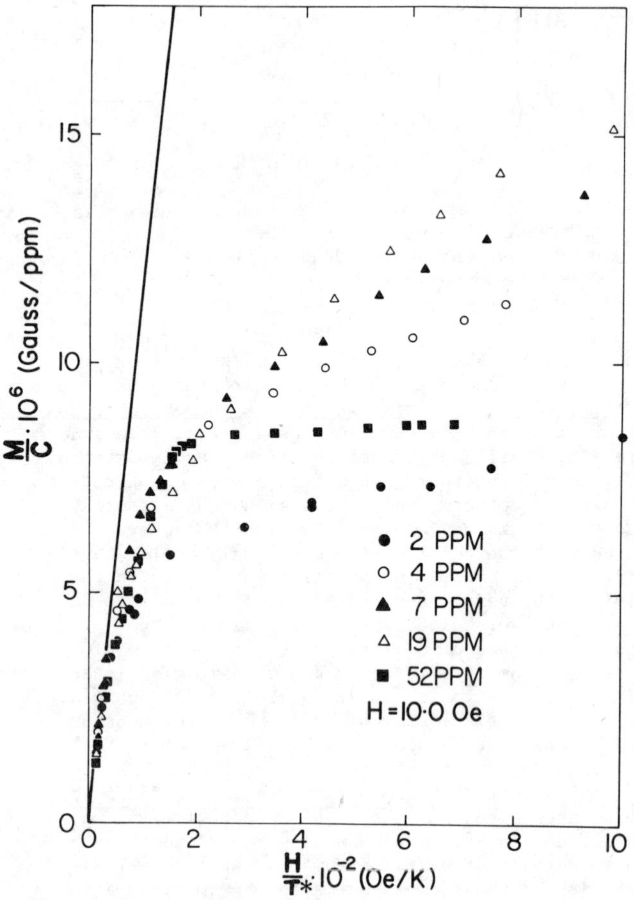


Fig. 1. Magnetization of Fe impurities in Au as a function of H/T for various concentrations. The solid line is a Brillouin function for a "free spin" of 1.

### RESULTS AND DISCUSSION

The magnetization of the Au-Fe alloys follows a Curie-Weiss law at high temperatures with an effective moment of approximately $3\mu_B$ and with a Curie-Weiss temperature of approximately 80 mK. In the vicinity of 0.1 K all the samples show strong departures from this law. This behavior is shown in Fig. 1 where the magnetization per Fe impurity is plotted as a function of H/T for the concentrations studied. The solid line corresponds to a Brillouin function for an effective spin of 1 and it is used only as a guide for showing the strong departure of Au-Fe from "free spin" behavior. The results in Fig. 1 show that there are two effects present. As the impurity concentration is reduced impurity-impurity interactions become smaller and the magnetization would be expected to approach "free spin" behavior; however this does not occur. At very low concentration the magnetization departs even more from the "free spin" curve; this implies that another interaction must be responsible for such a trend. The magnetization seems to be approaching one that would correspond to a singlet ground state.

The results show that impurity-impurity effects are extremely important; they persist even for the most dilute sample studied here. The kind of impurity-impurity interactions present here can be complicated in view of the fact that there can be Fe pairs interacting with each other or with individual Fe impurities. The results of Ref. 3 were analyzed in terms of Fe pairs. In Cu-Fe alloys, where similar possibilities exist, the results were analyzed in terms of pairs of impurities[5,6] and also in terms of clusters of as many as four impurities acting as units.[7] We do not have any reason for applying such an analysis at this stage, although it is possible that small clusters do exist in spite of the fact that we have such high dilution. Previous measurements at high temperatures of more concentrated alloys[8] were interpreted in terms of concentration dependent impurity-impurity interactions increasing as the concentration decreased. Such a model would cause the ordering temperature to vary with concentration, however we observe that the departures from Curie-Weiss law occur over a narrow temperature range for our concentrations. This behavior favors the idea of another interaction causing the strong departure from the free spin curve.

The strong concentration dependence in Fig. 1 indicates that we have not reached the limit where only single-impurity effects exist. Nevertheless the data show an approach to a non-magnetic state as the concentration is reduced. Our high concentration data are in agreement with the measurements of Ref. 3 in that as the concentration is reduced the magnetization per impurity goes from saturation toward the free spin curve. On further reduction of the concentration, the magnetization per impurity shows a drastic departure away from the free spin curve. This implies competing interactions. It is possible that this is evidence of a Kondo spin-compensated state. It may be also that the behavior just observed is a limiting case of spin glass.[9,10] However it could also be crystal field splitting of the impurity states and the approach to a singlet ground state. This hypothesis can be tested by measurements on a single crystal of Au-Fe. More data are needed on samples with Fe concentrations of less than 1 ppm.

ACKNOWLEDGMENTS

The samples were prepared by Dr. F. Rosenberger. We are grateful to Dr. L. Moberly for help with some of the measurements.

REFERENCES

*Supported by NSF.

1W. Loram, A. D. C. Grassie, and G. A. Swallow, Phys. Rev. B2, 2760 (1970).

2J. L. Tholence and R. Tournier, J. Phys. 32, C1-211 (1971).

3G. Frossati, J. L. Tholence, D. Thoulouze and R. Tournier, Proc. of Low Temp. Conf. LT14, 3, 370 (1975), Helsinki.

4J. C. Doran and O. G. Symko, Proc. of 19th Annual Magnetism Conference, Boston, 980 (1973).

5E. C. Hirschkoff, M. R. Shanabarger, O. G. Symko, J. C. Wheatley, J. Low Temp. Phys. 5, 545 (1971).

6J. Tholence and R. Tournier, Phys. Rev. Letters 25, 867 (1970).

7S. Mishra and P. A. Beck, Phys. Stat. Sol. (2) 19, 267 (1973).

8J. C. Liu and F. W. Smith, Solid St. Comm. 17, 595 (1975).

9R. J. Borg and T. A. Kitchens, J. Phys. Chem. Solids 34, 1323 (1974).

10P. Steiner, G. N. Beloserskij, D. Gumprecht, W. v. Zdojewski and S. Hüfner, Solid State Comm. 14, 157 (1974).

# EVIDENCE OF CRYSTAL FIELD AND ORBITAL ENERGY LEVEL SPLITTINGS FOR TRANSITIONAL IMPURITIES IN METALS

John Gardner
Department of Physics, Oregon State University, Corvallis, OR 97331

ABSTRACT

A comparison of the Knight shift and susceptibility of dilute Co impurities in a wide variety of nonmagnetic metal hosts demonstrates that crystal field and orbital splitting of the Co ionic levels cannot be neglected, as is done in most theoretical treatments of the subject. The crystal field level splitting of the $^4F$ Hund's rule ground state is at least a few tenths of an eV in low temperature noble metal hosts for which the characteristic coupling energy $kT_K$ to conduction electrons is 0.1 eV or smaller. In liquid metal hosts, crystal field effects appear to be small. Data on polyvalent liquid metals indicate that one or more ionic orbitals lie about 0.3 eV above the $^4F$ level and are populated when $kT_K$ is large. The level could be either a $^4P$ or $^5D$ ionic level, corresponding respectively to intra- or interconfiguration mixing in the polyvalent host metals.

For transitional impurities in metals, direct experimental information about ionic energy levels is difficult to obtain because of the relatively large d-conduction electron interaction. Unlike rare earths there is little unambiguous evidence that crystal field and orbital splittings of d states are important. In fact it has been customary to use very simple approximations for the d states, like the orbitally degenerate Anderson model or special cases like the Kondo and LSF models.[1,2] We will discuss below indirect but compelling evidence that such simple approximations are not valid in general for transitional impurities.

We will limit the present discussion to magnetic properties of Co impurities and show schematically in Fig. 1 the CEF ionic energy level diagram suggested by Hirst[3] for Co. We assume an intermediate cubic crystalline electric field, neglect spin-orbit coupling, and show only a few of the lowest-lying energy levels. In the Anderson model most of the finer details are ignored. For example, all the quartet spin levels of the $(3d)^7$ configuration would be degenerate and separated from the doublet levels by $4\bar{J}$.

Fortunately, it is not difficult to write down a sensible approximation for the magnetic susceptibility of an impurity without oversimplifying the ionic energy levels so drastically. The major difficulties are identifying the ionic ground state and including conduction electron effects. Independent of the details of the ion, it is obvious that the conduction electron interaction $V_{kd}$ profoundly alters the magnetic properties of the ion. The otherwise degenerate Zeeman levels of the ionic ground state are coupled and split by $V_{kd}$ so that the magnetic susceptibility ceases to follow a Curie law below some characteristic temperature $T_K$ and approaches a constant. This phenomenon has been discussed extensively in the literature[2] for nondegenerate impurities. Generalizing the susceptibility to the degenerate case, we assume that the susceptibility of an impurity ion of spin S and effective orbital quantum number $\tilde{L}$ can be written as

$$\chi = \chi_{spin} + \chi_{orb} \qquad (1)$$

$$\chi_{spin} = g_S^2 \mu_B^2 S(S+1)/F(T,T_K) \qquad (2)$$

$$\chi_{orb} = g_L^2 \mu_B^2 \tilde{L}(\tilde{L}+1)/F(T,T_K) + \chi_{VanVleck}.$$

$F(T,T_K)$ is usually approximated by $3(T+T_K)$, but the exact form is not important for this discussion. The only other major approximation is in neglecting the susceptibility due to conduction electron polarization by the impurity. This polarization is known to be small beyond the impurity site and is estimated to contribute no more than a few percent to $\chi$.[4] If the energy separation between levels is less than the characteristic energy $kT_K$, several orbitals may be coupled into the ionic "ground state". If so an appropriate average is taken over each ionic level. Since all low-lying levels are orbitally degenerate, the Van Vleck susceptibility should be small for Co, and we shall neglect it. A more complete discussion is to be published elsewhere.[5]

The Knight shift $\kappa$ of Co impurities in a number of liquid and solid metal hosts is shown in Fig. 2 vs. the Co susceptibility $\chi$. The qualitative characteristics include a linear dependence of $\kappa$ on $\chi$ in the nonmagnetic region ($\chi<10^{-3}cm^3$/mole), a sudden increase for liquid CuCo, and a relatively much smaller Knight shift in the low temperature noble metal hosts. These features are difficult to understand unless the ionic character is properly taken into account.

If 8Dq is larger than $kT_K$, the ionic ground state for solid AuCo and CuCo is the ground $T_{1g}$ triplet shown in Fig. 1. Since liquid "crystal fields" are zero on the average, we assume that the liquid CuCo ground state is an unsplit $^4F$ level, and that in polyvalent hosts, higher-lying orbital(s) are coupled in as $T_K$ becomes large and Co "nonmagnetic". With these assumptions we find that $kT_K$ is approximately 0.02 eV for AuCo, 0.1 eV for CuCo, and rises to about 2 eV for AlCo. We note that our neglect of spin-orbit coupling is probably not quantitatively valid in noble metal hosts.

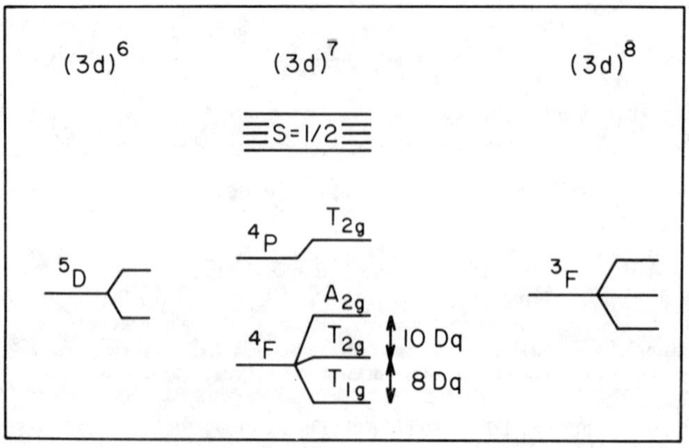


Figure 1; Schematic energy level diagram of Co in a cubic crystalline electric field. Spin-orbit splitting is neglected.

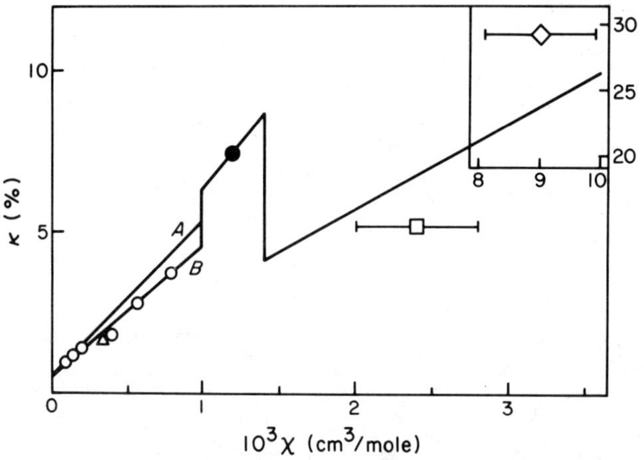


Figure 2: Co Knight shift vs. Co. susceptibility for dilute Co impurities in liquid Al and $Al_xCu_{1-x}$ (open circles, ref. 6), liquid Cu (closed circle, ref. 6), liquid Sn (triangle, ref. 7,8), Cu at 4K (square, ref. 9,10), and Au at 4K (diamond, ref. 11,12).

The Co Knight shift is given by

$$\kappa = \kappa_o + \alpha\chi_{orb} + \beta\chi_{spin} \quad (4)$$

$$= \kappa_o + \chi[\alpha(1+\chi_{spin}/\chi_{orb})^{-1} + \beta(1+\chi_{orb}/\chi_{spin})^{-1}]. \quad (5)$$

The coefficient of $\chi$ in Eq. 5 is independent of $F(T,T_K)$ if $\chi_{VanVleck}$ is neglected. Here $\kappa_o$ (=0.5%) is the constant contact term, $\alpha$ and $\beta$ are respectively the orbital and spin hyperfine coupling constants.

The spin and orbital susceptibilites of Co are comparable but $|\beta|$ is much smaller than $\alpha$, so the exact value of $\beta$ is not important. In Eq. (5), we take $\beta$ = -10 mole/cm$^3$, (-55 kG/$\mu_B$), a typical experimental value[13] and find $\alpha$ by fitting the above formula to the precisely determined liquid CuCo data. $\alpha$ is found to be 145 mole/cm$^3$, in good agreement with the theorectical value of 135 mole/cm$^3$ for $Co^{++}$ ions.[13] The solid lines in Fig. 2 show the computed Knight shift for the above ionic ground state assignments. In the non-magnetic region the "ground state" of the ion is assumed to be a linear superposition of the $^4F$ state with a (approx 0.3 eV) higher lying $^4P$ (line A) or $^5D$ (line B) level. Either are in acceptable agreement with experiment. A superposition of the $^4F$ and $^3F$ orbitals is not in agreement with experiment. The fit to the solid noble metal host data is also quite satisfactory.

In summary, we have shown that the features of the NMR data can be explained semi-quantitatively by reasonable assumptions about the proper ionic state. Both crystal field and orbital ionic energy level splittings must be included for a proper description. We also note that only a small number of distinct ionic orbitals appear to be populated even for "non-magnetic" impurity states. This appears to invalidate the use of band-type models such as the LSF or virtual bound state model which require a large distribution of populated ionic levels and configurations.

REFERENCES

* Supported by the Oregon State University Foundation General Research Fund

1. P. W. Anderson, Phys Rev 124, 41 (1961).
2. Magnetism (G. Rado and H. Suhl, eds.) Vol V. Academic Press, N.Y. (1973). Several review articles on the impurity problem are included in this volume.
3. L.L. Hirst, Z. Physik 241, 9(1971).
4. D.V. Lang, J.R. Boyce, D.C.Lo, and C.P. Slichter, Phys Rev Letters 29, 776 (1972).
5. J.A. Gardner, Phys Rev B, to be published.
6. A. L. Ritter, J.C. Bremer, and J.A. Gardner, Phys Rev B10, 3246 (1974).
7. R. Dupree, Phys Lett 44A, 435 (1973).
8. J.A. Gardner and C. Ardary, to be published.
9. S. Wada and K. Asayama, J. Phys Soc Japan 30, 1337 (1971)
10. R. Tournier and A. Blandin, Phys Rev Letters 24, 397 (1970).
11. A. Narath and D.C. Barham, Phys Rev B7, 2195 (1973).
12. E. Boucai, B. Lecoanet, J. Philon, J.L. Toulence, and R. Tournier, Phys Rev B3, 3834 (1971).
13. A.J. Freeman and R.E. Watson, Magnetism, (G. Rada and H. Suhl, eds.) Vol IIA, Academic Press NY 1965.

# THERMOPOWER OF NEARLY-MAGNETIC ALLOYS AND METALS*

A. B. Kaiser
Physics Dept., Victoria University of Wellington,
Private Bag, Wellington, New Zealand.

## ABSTRACT

In nearly-magnetic alloys in which the host and impurity are transition metals of similar electronic structure, e.g. Pd-Ni, Rh-Fe and Ir-Fe, the thermopower shows a peak at about the spin fluctuation temperature deduced from resistivity measurements. We suggest that a 'spin-fluctuation drag' thermopower component, produced by the disequilibrium of spin fluctuations in the presence of a thermal gradient, could cause a peak in the thermopower of nearly-magnetic metals and alloys. We outline the expected characteristics of this thermopower component and make a comparison with experimental measurements.

## NEARLY-MAGNETIC ALLOYS

The alloys which we consider are nearly-magnetic alloys such as Pd-Ni which can be described by the Lederer-Mills[1] model. In this model, the Coulomb interaction between opposite-spin d electrons is increased from a value U in the host to a value $(U + \delta U)$ in the impurity cells, so that enhanced spin fluctuations occur in the d-band at impurity sites. Lederer and Mills showed that scattering of conduction electrons off these spin fluctuations led to a $T^2$ term in the resistivity at low temperatures.

We have shown[2] that the Lederer-Mills model can be extended to other dilute transition-metal alloys in which the host and impurity have similar electronic structure, e.g. Ir-Fe and Rh-Fe. Since the spin fluctuations behave like bosons, the $T^2$ resistivity dependence changes to a linear T law as temperature T increases, in analogy to the Bloch-Gruneisen law for the electron-phonon resistivity. Fitting our universal curve for the spin-fluctuation resistivity[2] to the experimental data yields the value of $T_{sf}$, the spin fluctuation temperature representing the peak in the spin fluctuation excitation spectrum. The values of $T_{sf}$ listed in Table I below were derived in this way.

## EXPERIMENTAL THERMOPOWER BEHAVIOUR

Foiles and Schindler[3] found a large peak in the thermopower of nearly-magnetic Pd-Ni alloys. This peak is absent in pure Pd and in a more concentrated ferromagnetic Pd-Ni alloy, suggesting that it is due to localized spin fluctuations. The additional thermopower due to the Ni impurities is shown in Fig. 1.

Measurements on other nearly-magnetic alloys have been made by Touger and Sarachik[5] (Ir-Fe) and by Graebner et al.[6] (Rh-Fe). From these data we can summarize the characteristic behaviour of the thermopower of nearly-magnetic alloys as follows:

1) A large peak occurs at a temperature $T_p$ which is of the order of the spin fluctuation temperature $T_{sf}$ (see Table I).

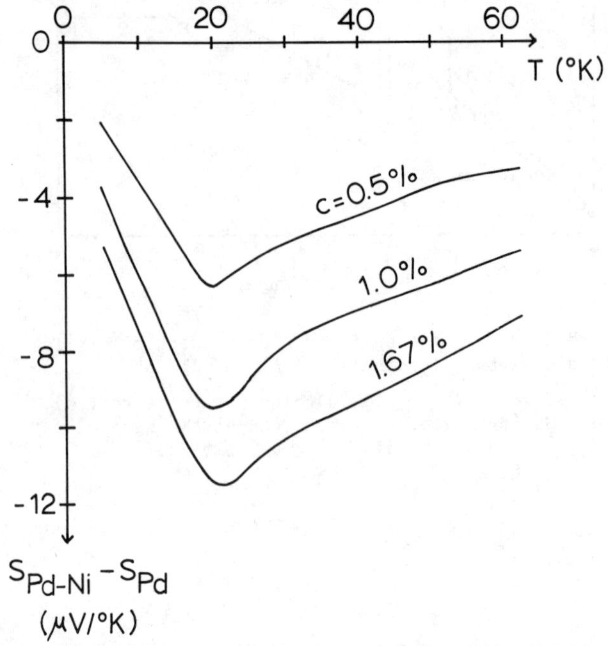


Fig. 1 Experimental values of the thermopower of $S_{Pd-Ni}$ of Pd-Ni alloys (Foiles and Schindler[3]) minus the thermopower $S_{Pd}$ of pure Pd (Fletcher and Grieg[4]); c is the impurity concentration in atomic %.

| | $T_{sf}$(°K) | $T_p$(°K) |
|---|---|---|
| Pd-Ni | 23[7] | 20[3] |
| Ir-Fe | 28[2,8] | 30[5] |
| Rh-Fe | 2[9,6] | 2.5[6] |

Table I Experimental spin fluctuation temperatures $T_{sf}$ and thermopower peak temperatures $T_p$; the superscripts give references to the source of each value.

2) The sign of this peak is negative for Pd-Ni and Rh-Fe, but positive for Ir-Fe.

3) At the lowest temperatures, the thermopower appears to vary linearly with temperature.

4) The magnitude of the peak increases with concentration in Pd-Ni and Ir-Fe; the behaviour of Rh-Fe may be affected by interactions.[6]

## EFFECT OF SPIN FLUCTUATIONS ON DIFFUSION THERMOPOWER

As a first attempt to account for the large thermopower peaks, detailed calculations were made of the effect of spin fluctuations on the diffusion thermopower assuming an equilibrium spin fluctuation distribution as in our calculation of electrical resistivity. However, the effect on the diffusion thermopower turns out to be trivial. The essential reason for this is that in the Lederer-Mills model, the spin fluctuations scatter the conduction electrons at a rate depending on the energy difference of the initial and final conduction electron states, but not on their energy relative to the Fermi energy. There is therefore no large asymmetric variation of the effective conduction-electron relaxation time across the Fermi surface, as required for a greatly enhanced diffusion thermopower. Hence to account for the large thermopower peaks, we need either to go beyond the Lederer-Mills model, e.g. to consider higher-order processes, or to consider drag effects.

For Kondo alloys, theory[10] and experiment[11] indicate that the diffusion thermopower shows a peak at about the Kondo temperature $T_K$, the magnitude of the peak being insensitive to impurity concentration. This insensitivity to impurity concentration is characteristic of the diffusion thermopower when impurity scattering is dominant, so the observed strong increase in magnitude of the peaks as impurity concentration increases in Pd-Ni and Ir-Fe suggests that the principal origin of these peaks may not be a diffusion effect. We therefore turn our attention to drag effects.

## SPIN-FLUCTUATION DRAG THERMOPOWER

In nearly-magnetic materials, the presence of a thermal gradient will produce disequilibrium in the spin-fluctuation distribution, since more spin fluctuations are excited at higher temperatures. Consequently, there will be a bias of spin-fluctuation wavevectors down the thermal gradient. This bias will tend to be transmitted to the conduction electrons when they scatter off the spin fluctuations, leading to a 'disequilibrium' or 'drag' component in the thermopower. Analogous drag effects due to phonons and to magnons[12] are already well known. A drag effect should occur whenever conduction electrons are scattered by excitations in disequilibrium, including localized spin fluctuations which are confined to impurity cells rather than propagating throughout the crystal.

Suppose that the net rate (per unit volume) at which spin fluctuations of wave vector $\vec{q}$ are absorbed in collisions with conduction electrons is $\partial n(\vec{q})/\partial t$; this rate represents the degree of disequilibrium in the spin fluctuation spectrum due to the thermal gradient, and would be zero for thermal equilibrium. Then as for the case of phonons[13] or magnons[14], we can consider the resulting momentum transfer to the electrons as balanced by a force due to an electric field $\varepsilon_x$ which prevents current flow:

$$-N|e|\ \varepsilon_x = \sum_{\vec{q}} \frac{\partial n(\vec{q})}{\partial t}\ \hbar(\vec{q} + \vec{G}) \qquad (1)$$

where $-N|e|$ is the total charge of conduction electrons per unit volume and $\vec{G}$ is a reciprocal lattice vector. The resulting spin-fluctuation drag thermopower is given by

$$S_{sf} = \varepsilon_x \left(\frac{\partial T}{\partial x}\right)^{-1} \qquad (2)$$

where $\partial T/\partial x$ is the temperature gradient.

This simple picture reveals some of the characteristics of $S_{sf}$:

1) For Normal conduction electron-spin fluctuation scattering processes ($\vec{G} = 0$), $S_{sf}$ is negative, since electrons are dragged down the thermal gradient. However, if Umklapp processes dominate, the sign of the momentum transfer $\hbar(\vec{q} + \vec{G})$ is likely to be reversed, i.e. electrons are dragged up the thermal gradient, and $S_{sf}$ could be positive.

2) In the case of localized spin fluctuations, $\partial n(\vec{q})/\partial t$ and hence $S_{sf}$ will proportional to the impurity concentration c if interactions between impurities are absent.

3) At higher temperatures, above the characteristic spin-fluctuation temperature $T_{sf}$, the thermopower $S_{sf}$ will be greatly reduced. At these temperatures, the spin-fluctuation spectrum becomes blurred out[2,15] and the effect of the spin fluctuations is reduced. In addition, if spin-fluctuation interactions other than collisions with conduction electrons become important, not all the spin-fluctuation wavevector bias will be transmitted to the conduction electrons. This reduction in $S_{sf}$ above $T_{sf}$ means that $S_{sf}$ will show a peak at a temperature roughly of the order of $T_{sf}$. (This peak is somewhat analogous to the phonon drag peak often observed at about 0.2 $\theta_D$, where $\theta_D$ is the Debye temperature).

4) At very low temperatures, one might expect the gradient in the spin-fluctuation energy density E(T) to yield a drag thermopower proportional to $\partial E/\partial T$ as for phonon and magnon drag[13], i.e. proportional to the specific heat $C_{sf}$ of the spin fluctuations. Since $C_{sf}$ is linear in temperature at low temperatures[16], this would give a linear T dependence for $S_{sf}$ in analogy to the $T^3$ dependence for phonon drag and the $T^{3/2}$ dependence for magnon drag.

5) Since different sources of thermoelectricity are additive, a spin-fluctuation drag component $S_{sf}$ will add to the diffusion component $S_d$ and phonon drag component $S_g$ to give a total thermopower

$$S = S_d + S_g + S_{sf}. \qquad (3)$$

For a dilute alloy in which $S_d$ and $S_g$ are approximately the same as in the host, $S_{sf}$ due to localized spin fluctuations is the difference between the alloy and host thermopowers, as plotted in Fig. 1. (There may also be a drag thermopower due to uniform spin fluctuations in the host, but this will be much smaller with a peak at much higher temperature).

Thus spin-fluctuation drag does appear to give a qualitative explanation of the thermopower of nearly-magnetic alloys, in particular the strong increase of the peak magnitude with impurity concentration c, and further work should prove interesting. Theoretically an explicit calculation of $\partial n(\vec{q})/\partial t$ is needed and experimentally it would be of interest to check whether $S_{sf}$ is linear in c for smaller values of c.

### REFERENCES

*Supported by the University Grants Committee, New Zealand.


1. P. Lederer and D.L. Mills, Phys. Rev. 165, 837-44 (1968).
2. A.B. Kaiser and S. Doniach, Int. J. Magnetism 1, 11-22 (1970).
3. C.L. Foiles and A.I. Schindler, Phys. Lett. 26A, 154-5 (1968).
4. R. Fletcher and D. Grieg, Phys. Lett. 17,6-7 (1965).
5. J. Touger and M.P. Sarachik, Bull. Am. Phys. Soc, 19, 304 (1974).
6. J.E. Graebner, J.J. Rubin, R.J. Schutz, F.S.L. Hsu, W.A. Reed and R.J. Higgins, AIP Conf. Proceedings 24, 445-6 (1975).
7. H.G. Purwins, Y. Talmor, J. Sierro and F.T. Hedgecock, Solid State Comm. 11, 361-5 (1972).
8. M.P. Sarachik, Phys. Rev. 170, 679-82 (1968).
9. R.L. Rusby, J. Phys. F: Metal Phys. 4, 1265-74 (1974). Rusby fits his Rh-Fe data to the alternative spin fluctuation theory of N. Rivier and V. Zlatic. J. Phys. F: Metal Phys. 2, L99-104 (1972).
10. H. Suhl and D. Wong, Physics 3, 17-36 (1967); see also V. Zlatic and N. Rivier, J. Phys. F: Metal Phys. 4, 732-8 (1974).
11. E.g. M.D. Daybell, in Magnetism V, ed. H. Suhl, Academic, New York, (1973), 121-47.
12. M. Bailyn, Phys. Rev. 126, 2040-54 (1962); F.J. Blatt, D.J. Flood, V. Rowe, P.A. Schroeder and J.E. Cox, Phys. Rev. Lett. 18, 395-6 (1967).
13. E.g. R.D. Barnard, Thermoelectricity in metals and alloys, Taylor and Francis, London (1972), Chs. 1 and 5.
14. C.M. Bhandari and G.S. Verma, Nuovo Cim. 60B, 249-252 (1969).
15. R. Jullien, M.T. Beal-Monod and B. Coqblin, Phys. Rev. B 9, 1441-57 (1974).
16. G. Chouteau, R. Fourneaux, R. Tournier and P. Lederer, Phys. Rev. Lett. 21, 1082-5 (1968).

# KONDO EFFECT IN Y-RICH ALLOYS CONTAINING A Ce IMPURITY

T. S. Petersen, S. Legvold and P. Burgardt
Ames Laboratory-USERDA and Department of Physics
Iowa State University, Ames, Iowa 50011

ABSTRACT

Sugawara et al. have found that Ce impurities in some nonmagnetic rare earth hosts exhibit the Kondo effect. Maple and Wittig have shown that when pressure is applied to Y-Ce, the Kondo effect disappears, assumedly because the 4f level is driven from below to above the Fermi level. In the work reported here, alloy hosts of $Y_xLa_{1-x}$ are used to get the slope of the Kondo resistivity as a function of atomic volume for samples containing 2 atomic percent Ce impurity. The slope of the $\rho$ vs. ln T, ($d\rho/d\ \ln T$) varies markedly with atomic volume and shows a sharp peak near x = 0.95. Furthermore, by using Y-Lu host alloys the Kondo effect will disappear at the atomic volume calculated for the Maple and Wittig result.

## INTRODUCTION

Sugawara et al.[1,2] found that Ce impurities in the nonmagnetic rare earths Y and La produce a minimum in the low temperature resistivity that is not found in the nonmagnetic rare earths such as Sc and Lu[3]. The additional resistivity at low temperatures giving rise to this minimum in these dilute Ce alloys has been attributed to the s-f exchange interaction proposed by Kondo. Coqblin and Schrieffer[4] have used an Anderson type model for Ce and have shown that the Kondo effect (resistivity minimum) in dilute Ce alloys can be explained by a resonant exchange interaction between the conduction electrons and the Ce impurity moment, taking into account the orbital angular momentum of the 4f electron. The resistivity due to this mechanism is proportional to (- ln T) except at very low temperatures where a plateau is reached.

The magnitude of this exchange scattering resistivity (or the magnitude of the spin-spin interaction constant) depends on the relative position of the 4f level to the Fermi level in Ce. Maple and Wittig have shown[5] that when pressure is applied to Y-Ce it will diminish the size of the resistivity minimum, making the minimum disappear near 80 kbar. Also, pressure on La-Ce causes the slope of the low temperature logarithmic resistivity to go through a maximum at a pressure of about 16 kbar as reported by Kim and Maple.[6] The pressure results on La-Ce have been explained by assuming that the 4f level approaches the Fermi level with increasing pressure.[7] This causes a larger coupling to the local moment and increased Kondo resistivity slope. The Y-Ce pressure results suggest that the Ce atoms are becoming nonmagnetic under pressure and that the 4f level moves above the Fermi level at 80 kbar.

These pressure results prompted the present investigation wherein a pseudo-pressure approach is used. Y was alloyed into La-Ce to simulate the effect of pressure on La-Ce and Lu was alloyed into Y-Ce. The atomic volumes of pure Lu, Y and La are 17.77, 19.86 and 22.5 cc/mole respectively, and this made the alloying technique feasible.

## EXPERIMENTAL PROCEDURE AND RESULTS

The samples were prepared by arc melting together weighed amounts of the constituents. After the final melt, the buttons were arc cast into rods from which the actual samples were cut by hand. An x-ray analysis was done to determine the atomic volume of each alloy. All of the samples were polycrystalline hcp, except for $La_{0.98}Ce_{0.02}$ which was dhcp. The electrical resistivity was measured using a four probe dc technique and temperatures from 1.8 to 40 K were reached by pumping on a $^4$He reservoir and by using a manganin heater on the reservoir.

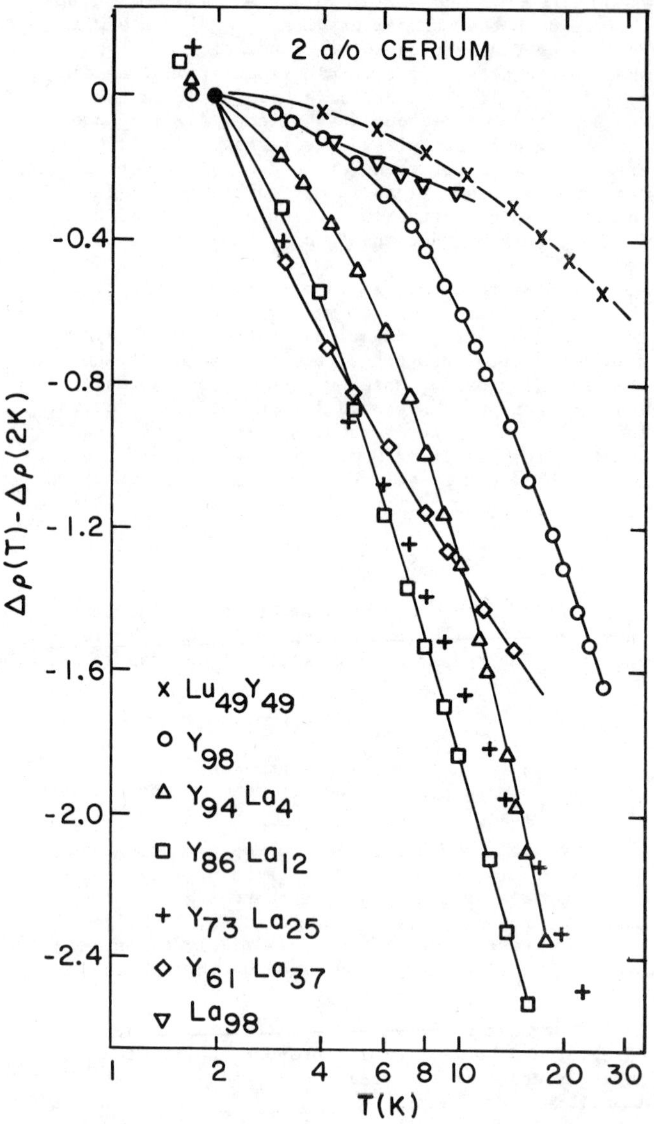


Fig. 1 Resistivity of 2% cerium in alloys of Lu-Y and Y-La. The base alloy resistivities have been subtracted. A constant has been subtracted from each curve to obtain a common origin at 2 K.

Samples with and without 2 atomic percent cerium were prepared in order that the phonon contribution to the resistivity could be subtracted. The resistivities, $\Delta\rho = \rho$(with cerium) - $\rho$(base alloy), are plotted in Fig. 1. Without subtracting the base alloy contribution, the resistivities of the samples with 2% cerium generally show a minimum in the temperature range 15 to 20 K. In a theoretical study, Matho and Beal-Monod[8] derived the following expression for the resistivity:

$$\rho = A + B \ln 1/(T^2 + C^2)^{1/2}. \qquad (1)$$

It was gratifying to find that in most instances this expression gave an excellent fit to the Kondo resistivity data over a considerable temperature range.

The slopes of the resistivities, $d\rho/d \log T$, taken from Fig. 1 in the vicinity of the linear behavior (5 to 10 K) are plotted in Fig. 2 as a function of the atomic volume of the alloy. The addition of Y to La-Ce decreases the atomic volume and an enhancement of the Kondo effect can be seen, reaching a maxi-

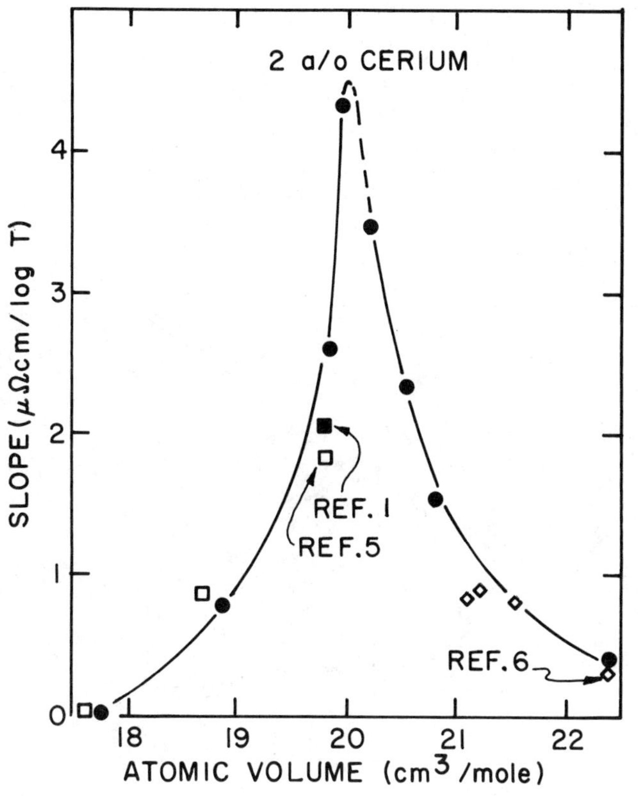


Fig. 2 Slopes of the resistivities shown in Fig. 1, $d\rho/d\ \log_{10} T$ vs. the atomic volume of the alloy.

mum for the sample $Y_{0.94}\ La_{0.04}\ Ce_{0.02}$ (atomic volume = 20.0 $cm^3$/mole). By progressing to lower atomic volume by adding Lu to Y-Ce, the slope of the resistivity goes to zero. From the pressure work on Y-Ce[5], the resistance anomaly had nearly disappeared at 57 kbar which corresponds to an atomic volume of 17.4 cc/mole,[9] which is slightly less than the atomic volume where no resistivity minimum was seen; the results, then, show excellent agreement. For the sample $Y_{98}Ce_2$ the present result is higher than the slope of the resistivity from Sugawara[1] and Maple and Wittig;[5] the latter used a sample with 1% of Ce. The pressure results on La-Ce[6] are also plotted in Fig. 2. This sample contained the dhcp phase and contained 3% Ce; we find the numerical agreement is good at zero pressure. The maximum slope in that work occurred at a pressure of 15.9 kbar (21.2 cc/mole as calculated from the compressibility of La).[9]

These results may be understood if one assumes that a decrease in the atomic volume causes a shift upwards of the 4f level with respect to the Fermi level. Since the 4f level has been estimated[4] to lie about 0.1 eV below the Fermi level in La-Ce, this shifting of the 4f level causes a strong increase in the interaction between the conduction electrons and the local moments. However, when the 4f level goes above the Fermi level the cerium atom loses its local moment and the resistivity minimum has disappeared. The data indicate that the 4f level has completed this transition for the $Lu_{0.98}\ Ce_{0.02}$ alloy.

Helpful discussions with B. N. Harmon and the aid of J. Schwartz and B. J. Beaudry in obtaining the data are gratefully acknowledged.

REFERENCES

1. T. Sugawara, J. Phys. Soc. Japan 20, 2252 (1965).
2. T. Sugawara and H. Eguchi, J. Phys. Soc. Japan 21, 725 (1966).
3. T. Sugawara, I. Yamase and R. Soga, J. Phys. Soc. Japan 20, 618 (1965).
4. B. Coqblin and J. R. Schrieffer, Phys. Rev. 185, 847 (1969).
5. M. B. Maple and J. Wittig, Solid State Comm. 9, 1611 (1971).
6. K. S. Kim and M. B. Maple, Phys. Rev. B 2, 4696 (1970).
7. B. Coqblin, M. B. Maple and G. Toulouse, Intern. J. Mag. 1, 333 (1971).
8. K. Matho and M. T. Beal-Monod, Phys. Rev. B 5, 1899 (1972).
9. K. Gschneidner, Rare Earth Alloys (D. Van Nostrand & Co., New York, 1961).

# THE EFFECT OF INTERCONFIGURATIONAL EXCITATION ENERGIES LYING CLOSE TO THE FERMI ENERGY IN THE DILUTE MAGNETIC ALLOY

Samuel P. Bowen
Dept. of Physics, Virginia Polytechnic Institute
Blacksburg, Va. 24061

ABSTRACT

The case of an impurity in a metal for which low-lying term energies of two configurations are close together is considered. It is shown that the type of interaction with the conduction electrons as in the Anderson model can induce a temperature dependent shift of the effective configuration energies. It is also shown by second order perturbation theory that the interconfiguration excitation energies are shifted as log T for high temperatures and with a $1-T^2$ dependence at low temperatures.

This temperature dependent shift of the interconfigurational energies can describe quite well many of the properties of the dilute magnetic alloy and of some f-shell intermetallic compounds.

Calculations showing the resistivity and heat capacity at zero magnetic field and for differing field values are compared with experiment.

The detailed description of the calculations of this thermal shifting of small interconfigurational energies and the application of this effect to various experimental systems will be discussed in detail elsewhere. This effect would appear to offer an alternative description for many systems to the usual Kondo effect.

# MAGNETIC FIELD DEPENDENCE OF THE SPECIFIC HEAT OF PdDy ALLOYS

H.A. Zweers, G.J. Nieuwenhuys, H.W.M. van der Linden,
Kamerlingh Onnes Laboratorium der Rijksuniversiteit, Leiden, The Netherlands.
and
J.A. Mydosh*,
Institut für Festkörperforschung der Kernforschungsanlage, Jülich, Germany.

## ABSTRACT

We have measured the specific heat of PdDy alloys between 1.2 and 20 K in external fields ranging from 0 to 20 kOe. The data are analysed with respect to the crystal field splitting CFS of the Dy-ion in a cubic symmetry. Previously, different crystal field level schemes had been proposed for this system based upon the interpretation of EPR, magnetization and specific heat measurements. The application of a straightforward CFS analysis still shows systematic deviations from our experimental data. We present here some different CFS possibilities, which, when combined with a weak internal field, form a reasonable describtion for the specific heat data.

## INTRODUCTION

The ground state of Dy ions ($4f^9$ electrons, $^6H15/2$), when diluted into Pd, experiences a weak crystal field potential of cubic symmetry induced by the fcc Pd matrix. Lea et al.[1] have calculated the possible level schemes as a function of two parameters, W and X. W is an overall energy scaling parameter, while X concerns the relative importance of the fourth and sixth order potential terms.

EPR results[2] clearly indicate a Γ8 quartet to be the lowest level, but the X-parameter is not uniquely resolved[2-6], with possible values around -0.55 and -0.42. On the other hand, magnetization data[7,8] do not give a suitable fit assuming X=-0.54 and X=+0.75. In all cases W is found to be negative and of such magnitude that there is a few degrees Kelvin separation between the lowest Γ8 and the first excited level. Specific heat measurements[9,10] between 0.5 and 6 K were interpreted in terms of two contributions: one from crystal fields, and an additional source coming from superparamagnetic clustering of the Dy atoms. The salient features of these excess specific heat data were a broad minimum at about 2 K, with the lowest temperature points still increasing in magnitude. We have extended the specific heat experiments on PdDy alloys to include the effects of an external field and higher temperatures. As a result of our analysis we found inconsistencies among the previously suggested level schemes and we propose new possibilities for resolving these difficulties.

## EXPERIMENTAL

Three PdDy (0.3, 0.6 and 1.0 at.% Dy) were melted and remelted several times after turning over. An X-ray scanning microanalysis showed the formation of some dendrites, which were removed by an additional heat treatment (200 hours, 1000° C) for the 0.3 and 0.6 at.% alloys. The X-ray scanning microanalysis then showed these alloys to contain a homogeneous Dy content in the bulk. Only some Dy clusters, presumably $Dy_2O_3$, were found at the surface, which were removed, as far as possible, by etching with nitric acid.

The heat capacity was measured using the standard adiabatic method with an overall accuracy better than 1%[11] between 1.3 and 20 K in external magnetic fields from 0 to 20 kOe. The data for PdDy 0.3 at.% are omitted due to a significant scatter in the excess specific heat, which is caused by the large specific heat of Pd itself. The excess specific heats, ΔC, for the homogeneous PdDy 0.6 at.% and "inhomogeneous" (some dendrite formation) PdDy 1 at.% alloy show only slight deviations at low temperature (figure 1), which clearly indicate that any large concentration or precipitation dependence is rather small. Due to a

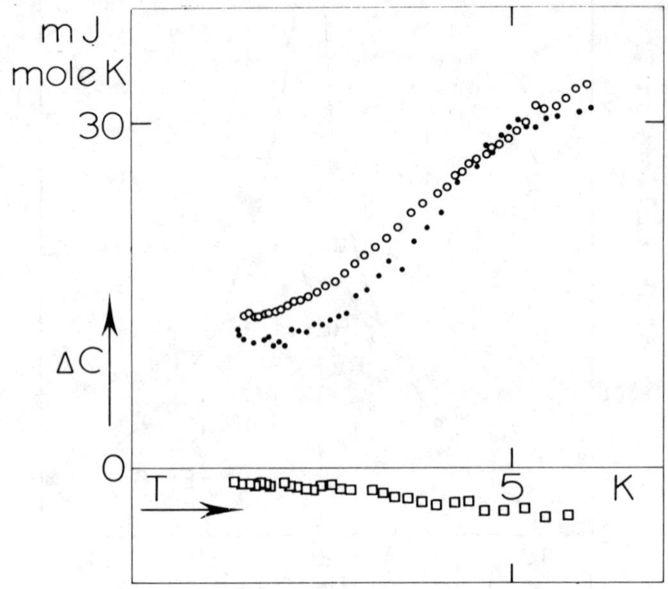


The excess specific heat ΔC (in zero magnetic field) due to alloying for PdDy 1 at.% (○), PdDy 0.6 at.% (●), and PdLu 0.3 at.% ( ). All data are normalized to 1 at.%.

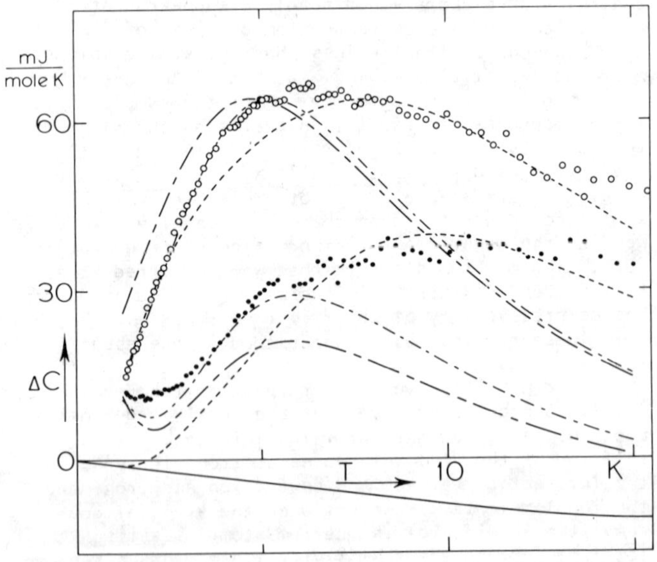


The excess specific heat ΔC due to alloying for PdDy 0.6 at.% in zero field (●); in 20 kOe (○). The full line ——— represents PdLu. All data are normalized to 1 at.%. — - — calculated for X = + 0.64; — - — calculated for X = - 0.54 (also approximately X = -0.40); - - - - - - calculated for X=+0.40; the additional parameters W and c are given in the text. The calculated values are given with respect to the PdLu base line.

possible Dy-Dy interaction, which strongly influences magnetic data for concentrations of ≈ 3 at.%[12], the PdDy 0.6 at.% alloy gives the most favourable results with which to test a CFS scheme. The lowest points in figure 1 and line in figure 2 deal with ΔC (independent of magnetic field) for PdLu, which is thought to be a good approximation of the change of electronic and lattice specific heat due to alloying Pd with rare-earths.

## CALCULATIONS

We made several attemps to fit the difference in specific heat between PdLu and PdDy alloys to crystal field theory using Lea's parameters X and W. The procedure is as follows: (a) A Γ8 is found to be the lowest level (EPR), (b) Specific heat data suggest the presence of two maxima (one at $T < 1$ K, one at $T > 5$K) with a minimum in between ($T \approx 2$K), (c) For PdDy 0.6 at.% the ratio between the high temperature maximum and the minimum is used to obtain X. The specific heats for $-1 < X < +1$ (the total range for X, and for W positive and negative) in steps of $\Delta X = 0.2$ are computer calculated starting from the partition function. Only three possibilities result: $X = -0.54\pm0.01$, $X = -0.40\pm0.01$ and $X = +0.64\pm0.03$ (W is always negative), (d) W is obtained by scaling the temperature scale (for the above three X values, $W = -0.05$K). In all cases the experimental data give a broader structure than predicted with these parameters (see figure 2), which suggests that interactions cannot be completely ignored, (e) So we rely on the data in external magnetic fields of 10 and 20 kOe. These are calculated for the three sets of parameters by numerical intergration over all field directions (we have used polycrystalline samples). The maxima of the curves in a field of 20 kOe are fitted by adoption of an effective concentration, c. In this way we obtained three possible set of parameters: $X = -0.54$, $W = -0.05$K, $c = 0.412$ at.%; $X = -0.40$, $W = -0.05$K, $c = 0.418$ at.%; $X = +0.64$, $W = -0.05$K, $c = 0.360$ at.%. Calculated curves for magnetic fields of 0 and 20 kOe are shown in figure 2. Recognizing the relatively larger importance of the low temperature data (entropy considerations), it is clear that the best fit is obtained with $X = +0.64$.

Some remarks should now be made. (I) All data at high temperatures (10 to 20 K) are systematically too large. (II) The values obtained for the effective concentrations, c, suggest (apart from possibly not seen cluster contributions) a loss of calculated entropy at high temperatures according to (I). (III) We did not consider the possibility of a splitting of the ground state Γ8, causing a maximum at $T < 1$K, due to some type of interaction effects. We are continuing our studies of the specific heat below 1K. Therefore, at present, no straightforward analysis can be made. W and X should be chosen comparatively in order to get a nearly temperature independent contribution between 7 and 20 K. In a first attempt, we found a good combination: $X = -0.22$, $W = -0.15$K with a $c = 0.55$ at.% being the analysed concentration. This nicely fits the data in zero field between 7 and 15K. On the other hand the calculated data in a magnetic field of 20 kOe are much too high at 4K, not at 1.5K. So the first excited level, a Γ8, gives too large a contribution when it is split up by a field. The next try had to do with a doublet, a Γ6, as the first excited level. We then took $X = +0.40$, $W = -0.10$K and $c = 0.420$ at.%. Now the fit, especially in the field and at low temperatures, is much better. However, we feel a better agreement can be obtained, simply by an extensive trial and error method.

## CONCLUSION

A straightforward CFS analysis does not give satisfactory results. Our specific heat data suggest the low temperature maximum ($T < 1$K) to be the result of a splitting of the ground state Γ8 by an internal magnetic field due to interaction effects. The first excited level (at approximately 12K seperation), most probably, will be a Γ6, which, together with the other levels, build up the high temperature (5 to 20K) specific heat. Data for $T < 1$K should give additional information, which is very important with respect to entropy considerations.

## ACKNOWLEDGMENT

This work is part of the research programme of FOM and has been made possible by financial support from ZWO and from the Metaalinstituut TNO. The samples were skillfully prepared through cooperation of the Leiden FOM-MT IV group and the Amsterdam FOM-MT V group. We are grateful to Dr. G.J. van den Berg for his continuous interest in this work.

## REFERENCES

*Present address: Kamerlingh Onnes Laboratorium der Rijksuniversiteit, Leiden, The Netherlands.


1. K.R. Lea, M.J.M. Leask and W.P. Wolf, J.Phys.Chem. Solids 23, 1381 (1962).
2. R.A. Devine, J.M. Moret , J. Ortelli, D. Shaltiel, W. Zingg and M. Peter, Solid State Commun. 10, 575 (1975).
3. H.C. Praddaude, Phys. Lett. 42A, 97 (1972).
4. R.A.B. Devine, Solid State Commun. 13, 935 (1973).
5. H.C. Praddaude, Solid State Commun. 16, 1019 (1975).
6. Y. Yang, N.L. Huang Liu and R. Orbach, Conference on Crystalline Electric Field Effects in Metals and Alloys. University of Montreal, Canada, June 26-29 (1974).
7. H.C. Praddaude, S. Foner, E.J. McNiff Jr. and R.P. Guertin, AIP Conf. Proc. 10, 1115 (1973).
8. R.P. Guertin, H.C. Praddaude, S. Foner, E.J. McNiff Jr. and B. Barsoumian, Phys.Rev. B 7, 274 (1973).
9. L.L. Isaacs, Phys. Rev. B 9, 2228 (1974).
10. R.A.B. Devine, P. Jacques and M. Poirier, Phys.Rev. B 11, 563 (1975).
11. B.M. Boerstoel, W.J.J. van Dissel and M.B.M. Jacobs, Physica 38, 287 (1968), Comm. Kam. Onnes Lab., Leiden No. 363a.
12. H.A. Zweers, Ph.D. thesis (University of Leiden, 1976) (unplublished), available upon request.

# MODEL FOR MAGNETIC PROPERTIES OF Pd(Gd) ALLOYS

H.C. Praddaude, S. Foner, E.J. McNiff, Jr.
Francis Bitter National Magnet Laboratory,† Massachusetts Institute of Technology, Cambridge, Massachusetts 02139
R.P. Guertin*
Tufts University, Medford, Massachusetts 02155

## ABSTRACT

A model is presented which describes the paramagnetic properties of polycrystalline $Pd_{1-x}Gd_x$ alloys for $0.0003 \le x \le 0.02$, $3\ kG \le H \le 215\ kG$ and $1.3\ K \le T \le 4.2\ K$. The effective field acting on the Gd is the resultant of the applied field plus an anisotropy field which is postulated to be transverse to the external field. This transverse field is responsible for the slow approach to magnetic saturation. The fit of the magnetic data to this model yields a saturation magnetization of $7\mu_B$/Gd-atom (equal to the $Gd^{3+}$ free ion value) for all concentrations.

## INTRODUCTION

Magnetic moment measurements in polycrystalline paramagnetic $Pd_{1-x}Gd_x$ alloys for $0.0003 \le x \le 0.02$, applied magnetic field $3\ kG \le H \le 215\ kG$, and $1.3\ K \le T \le 4.2\ K$, show that the field dependence of the magnetic moment $\sigma$ is not described by a Brillouin function with J = 7/2. As previously reported, a semiempirical formula(1) for $\sigma(H)$ gave a good fit for data at T = 4.2 K and fields above 10 kG. However, fits of measurements at the lowest concentrations with this formula showed an increased Gd saturation moment. Because $\sigma(H)$ is in general a slowly varying function of H (with no sharp anomalies) it is possible to describe $\sigma(H)$ by several functional dependences on H; clearly, a suitable physical model will help to resolve this situation.

In this paper we present a model which leads to a very good description of $\sigma(H, T)$ for all the paramagnetic $Pd_{1-x}Gd_x$ alloys studied. In order to test this model we have extended the earlier measurements over a wider range of H and T, particularly for the more dilute alloys where the Gd moment is a small fraction of the total measured moment. A detailed analysis of the data is being published elsewhere.(2)

## PHYSICAL MODEL

One of the striking features of the $\sigma(H)$ data(3) was the observed slow approach to magnetic saturation. This is surprising because some of our Pd(Gd) alloys become ferromagnetic at low temperatures, and the ferromagnetic Gd-Gd interaction would suggest an approach to saturation which is more rapid than the Brillouin function $B_{7/2}$. However if we assume an effective field component perpendicular to the applied field, then the slow approach to saturation would be understandable. The fields acting on the Gd moment are illustrated in Fig. 1. The applied field $\vec{H}$ is along the z-axis, and the average anisotropy field $\langle H^2_{an}\rangle^{1/2}$ is in the x-y plane. The thermal average of the Gd moment is aligned along the effective field, $\vec{H}_{eff}$, which is the resultant of $\vec{H}$ and $\langle H^2_{an}\rangle^{1/2}$. Because the anisotropy field can be along an arbitrary and variable direction in the x-y plane, the macroscopically measured Gd magnetization, $\sigma(H,T)$, will be the projection of the thermal average of the Gd magnetization along the z-axis. The total magnetization is

$$\sigma = (\alpha H/H_{eff}) B_{7/2}(J\beta H_{eff}) + \chi_{mat} H, \qquad (1)$$

where $\alpha$ and $\beta$ are fitted parameters, $B_{7/2}(\ )$ is the Brillouin function for the angular momentum J = 7/2, and $\chi_{mat}$ is the matrix susceptibility.

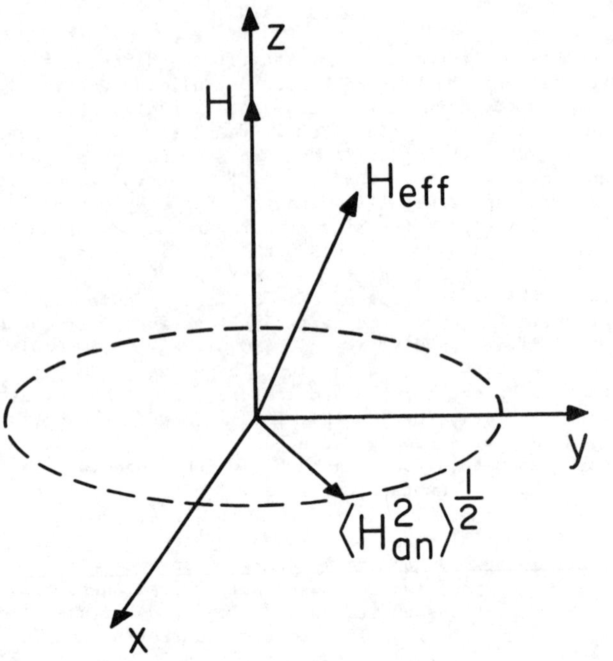


Fig. 1. Illustration of the applied magnetic field H, transverse anisotropy field $\langle H^2_{an}\rangle^{1/2}$, and resultant effective field $H_{eff}$. The average Gd moment is assumed to be aligned along the instantaneous $H_{eff}$. Because $\langle H^2_{an}\rangle^{1/2}$ can have an arbitrary direction in the xy-plane, the measured time-averaged Gd moment is the z-component of the effective moment along $H_{eff}$.

## DISCUSSION

Equation (1) describes the experimentally observed weak temperature dependence of the slow approach to saturation of the Gd moment. To attain the saturation value of the Gd moment two conditions must be satisfied: 1) $B_{7/2}(\ ) \approx 1$; and 2) the applied field H must be much larger than the anisotropy field $\langle H^2_{an}\rangle^{1/2}$, i.e., $H/H_{eff} \approx 1$. Once T is reduced to satisfy condition 1), then $H/H_{eff}$ dominates the field dependence of $\sigma$; further reduction of T has a negligible influence on the approach to saturation. At present the origin of $\langle H^2_{an}\rangle^{1/2}$ is not understood from first principles. It is possible that the effective anisotropy field arises from Gd-Gd interactions. It may also result from the s-f interaction between the Pd conduction electrons and the 4f electrons of the Gd. This exchange interaction may be written as $J\vec{S}\cdot\vec{s}$ where $\vec{S}$ is the Gd spin and $\vec{s}$ is the total conduction electron spin in the neighborhood of the Gd. Only the z-component of $\vec{s}$ has an average value different from zero for $H \neq 0$. This term gives rise, to first order, to the exchange correction to the applied field. This correction is written as $(1-\delta)$ and it is

Table I – Parameters $\beta$ and $H_0$ derived from fits of Eq. (1) to the magnetization data at 4.2 K for $Pd_{1-x}Gd_x$ alloys. The parameter $\theta$ was calculated as indicated in the text.

| x | 0.0003 | 0.0005 | 0.0010 | 0.0025 | 0.0035 | 0.0050 | 0.0065 | 0.01 | 0.02 |
|---|---|---|---|---|---|---|---|---|---|
| $\beta$(1/kG) | 0.028 | 0.028 | 0.031 | 0.037 | 0.039 | 0.043 | 0.049 | 0.061 | 0.127 |
| $\theta$ (K) | - | - | 0.3 | 0.9 | 1.1 | 1.4 | 1.7 | 2.2 | 3.2 |
| $H_0$ (kG) | 7 | 8 | 3 | 11 | 13 | 14 | 14 | 17 | 16 |

included in the coefficient $\beta$. The transverse ($s_x$, $s_y$) components of $\vec{s}$ average to zero and, to first order, will not contribute to the Gd spin energy. However, higher order perturbations will not necessarily average to zero. In particular, terms of the form $\langle s_x^2 + s_y^2 \rangle \neq 0$ will make a contribution. Here, we assume that $\langle H_{an}^2 \rangle = |H| H_0$, where $H_0$ is a parameter, i.e., the transverse anisotropy squared is proportional to the applied field or, equivalently, to $\langle s_z \rangle$. The qualitative reason for this assumption is that a certain degree of coherence in the spin precession of the conduction electrons is needed in order to produce an anisotropy field. If $H_0/|H| \ll 1$ then $H/H_{eff} \approx H/(|H| + H_0/2)$, which reduces to the approximation used in Ref. 1.

The parameters in Eq. (1) are $\alpha = Ng_J \mu_B J$ and $\beta = (1-\delta) g_J \mu_B / k(T-\theta)$, where N is the number of Gd atoms, $\mu_B$ is the Bohr magneton, $g_J$ is the Landé g-factor, k is the Boltzmann constant and $\theta$ is the paramagnetic Curie temperature. Here we have assumed that the Gd-Gd interaction that gives rise to ferromagnetism in Pd(Gd) can be approximated by the replacement of T by $(T-\theta)$. In effect we assume that a Curie-Weiss law is valid.

We have fitted Eq. (1) to our magnetic data for a wide range of Gd concentrations for which the alloys were paramagnetic, and over a wide range of H and T. The results are summarized in Table I. Note that the parameter $H_0$ varies from 3 to 17 kG as a function of x. These values of $H_0$ strongly affect the approach to saturation. For example, for $H_0$ = 10 kG, $B_{7/2}(\ ) = 1$, and H = 100 kG, the Gd moment is only 95% saturated. Because $H_0$ increases with increasing x, it is more difficult to magnetize the higher Gd concentration alloys despite the fact that they are closer to being ferromagnetic. From the values of $\beta$ in Table I and the values of $(1-\delta)$ obtained from EPR[4] we obtain the tabulated values for $\theta$ which increase with increasing x as expected. The values of $\theta$ are not to be confused with $T_c$; values of $\theta$ derived based on the molecular field approximation are expected to be larger than measured values of $T_c$.

Within experimental uncertainty the fitted values for $\alpha$ give a moment of $7\mu_B$/Gd-atom for all Gd concentrations, which is equal to the $Gd^{3+}$ free ion value. This is consistent with $\alpha = Ng_J \mu_B J$.

The present results and the model are satisfying because they give an explanation for: a) the high field characteristics which appear to be nearly temperature independent; b) the low field characteristics which are weakly temperature dependent; c) the difficulty of magnetizing the alloys; and d) the reason that the semiempirical formula[1,2] fits the high field data so well. The saturation Gd moment of $7\mu_B$/Gd-atom is obtained for all concentrations (within experimental error). Furthermore the model permits us to determine the paramagnetic Curie temperature $\theta$ for the alloy series.

## REFERENCES

† Supported by the National Science Foundation.
* Work supported by National Science Foundation Grant No. DMR 75-09494.

1. H.C. Praddaude, Phys. Letters 34A, 281 (1971).
2. H.C. Praddaude, S. Foner, E.J. McNiff, Jr. and R.P. Guertin, To be published in Phys. Rev. B1, Jan. 1, 1976.
3. R.P. Guertin, H.C. Praddaude, S. Foner, E.J. McNiff, Jr. and B. Barsoumian, Phys. Rev. 7B, 274 (1973).
4. H. Cottet and M. Peter, Solid State Commun. 8, 1601 (1970).
H.C. Praddaude and H. Gärtner, Phys. Letters 34A, 217 (1971).

# NEUTRON SCATTERING INVESTIGATION OF THE CENTRAL MODE AND ACOUSTIC PHONON ANOMALY ARISING FROM THE JAHN-TELLER PHASE TRANSITION IN $TbVO_4$

M. T. Hutchings
Materials Physics Division, Harwell, Didcot, Oxon. OX11 ORA, England,
R. Scherm
Institute Laue-Langevin, Ave. des Martyrs, B.P.156, 38042 Grenoble, France, and
S. R. P. Smith
Department of Physics, University of Essex, Colchester, Essex CO4 3SQ, England.

## ABSTRACT

Neutron scattering techniques have been used to study the low-lying excitations in terbium vanadate at and above the structural phase transition, a co-operative Jahn-Teller phase transition caused by the electron-phonon coupling between the $Tb^{3+}$ ions and the lattice. Above the transition temperature at 33K, $T_D$, the lowest $Tb^{3+}$ ion levels are two singlets separated by ~ 0.49 THz, and a doublet midway between them. The principal interaction is with the <100> transverse acoustic phonon mode and its effect increases as $T \rightarrow T_D$. The phonon couples almost equally with the electronic transitions between the degenerate levels of the doublet and between the singlet levels. The neutron scattering cross-section from the system is dominated by the nuclear scattering, and we find the phase transition is characterised by a sharp anisotropic central (zero frequency) mode due to the coupling of the phonon to the electronic doublet, whose intensity diverges as $T \rightarrow T_D$, $q \rightarrow 0$. The coupling with the singlet-singlet transition gives rise to an anticrossing with the <100> phonon which increases as $T \rightarrow T_D$. Using linear response theory techniques an expression for the complete scattering function $\mathcal{S}(\vec{Q},\omega)$ has been derived, and in a simplified form this has been used to interpret the data on the central mode, confirming its origin.

## 1. INTRODUCTION

The rare earth vanadates have proved to be ideal systems for studying co-operative Jahn-Teller phase transitions[1]. $TbVO_4$ exhibits such a structural phase transition, from the tetragonal $D_{4h}^{19}$ structure to an orthorhombic $D_{2h}$ structure at $T_D \sim 33K$, due to the electron-phonon coupling between the $Tb^{3+}$ ion and the lattice. This system has been extensively studied by a variety of experimental methods[1], of which the most relevant here are the Raman scattering studies of Elliott et al[2] and the Brillouin scattering and ultrasonic studies of Sandercock et al[3]. We present here the results of neutron scattering experiments on $TbVO_4$, which show that the phase transition in $TbVO_4$ exhibits characteristics both of a structural and a magnetic phase transition.

The analyses of Elliott et al and Sandercock et al may be summarized as follows. The lattice phonons couple predominantly to the lowest four levels of the $Tb^{3+}$ ion, which comprise (for $T > T_D$) two singlets separated by 16.5 $cm^{-1}$ (0.49 THz) and a doublet midway in energy between the singlets. (Below $T_D$, the doublet is split and the singlet separation is increased, to about 50 $cm^{-1}$ in each case, by the low-symmetry distortion). The phonon coupling is of roughly equal strength to an excitation between the singlets and to an excitation (ie a zero-frequency excitation) between the levels of the doublet. The full effect of the doublet will not be observed for $T > T_D$ unless one carries out experiments at frequencies less than the linewidth of the zero-frequency excitation within the doublet, as illustrated by the difference between the ultrasonic (low-frequency) and Brillouin scattering (intermediate-frequency) elastic constant measurements. In neutron scattering experiments, one anticipates that the effect of the doublet will be to provide scattering centred at zero energy transfer (central mode) and the effect of the singlets to form a coupled mode with the acoustic phonon.

The basic properties of the system are thus well understood, and mean field theories can be applied very successfully. The interaction is linear in the electronic and phonon co-ordinates, and therefore the central mode arises in first order in the theory[4] as in magnetic systems[5] in contrast to the second order interactions which are believed to cause central modes in structural phase transitions such as $SrTiO_3$.[6] We use linear response theory here to derive expressions for the scattering functions at and above the transition, and compare our experimental data[7] with calculations based on a simplified form of the theory. We show the variation of the central peak intensity with wavevector is a much more sensitive probe of the wavevector dependence of the total interaction between the ions than the anticrossing of the coupled mode energy dispersion[8]. However experiments aimed at investigating such a central peak are plagued by all the usual difficulties encountered when measuring scattering intensity adjacent to that from a Bragg peak.

We describe the experimental results in the next section, and the theory for the coupled response functions in section 3. The application of a simplified form of the theory to the data is described in section 4, and a summary and conclusion given in section 5.

## 2. EXPERIMENTS

### 2.1 Apparatus

The measurements reported here were almost all obtained using the IN3 and IN2 triple-axis spectrometers at the Institute Laue-Langevin, Grenoble. The triple-axis spectrometer in PLUTO, Harwell, was used to obtain the highest phonon energies. The IN3 spectrometer is sited on a thermal neutron guide which reduces the higher energy content of the beam incident on the monochromator, and therefore the $\lambda/2$ component on the specimen. The IN2 spectrometer was used with a double pyrolytic graphite (P.G.) crystal monochromator to give $\lambda = 4.046Å$, higher orders being eliminated by a cooled beryllium filter. In Table I we summarize the monochromator and analyser planes used and some measured dimensions (full width at half height) of the Gaussian resolution function[9] of each instrument. The seven distinct elements of the resolution matrix for IN2 were determined experimentally and used directly in the convolutions described below. The corresponding components for IN3 were determined partly by measurement and partly by calculation. The resolution ellipsoid slope $(\Delta\nu/\Delta q)_{Res.}$ pointed predominantly along [010] relative to $\vec{Q} \sim \vec{\tau}(400)$ for the IN3 measurements, and along $[0\bar{1}0]$ relative to $\vec{Q} \sim \vec{\tau}(200)$ for those using IN2. Here $\vec{Q} = \vec{k}-\vec{k}'$, where $\vec{k}$ (E) and $\vec{k}'$ (E') are the initial and final neutron wavevector (energy). We define a reduced excitation wavevector by $\vec{q} = \vec{Q}-\vec{\tau}$, and energy transfer by $h\nu = \hbar\omega = E-E'$.

Two samples were used, both grown by pulling from a flux[10]. Sample A ~ 4 x 4 x 6 $mm^3$ was used for the first measurements on IN3, and sample B ~ 5 x 5 x 11 $mm^3$ was investigated on IN2. Each sample was mounted with the c-axis perpendicular to the scattering plane in a helium-filled aluminium can attached to the cold finger of a Cryogenics Associates CT14 cryostat, and its temperature was controlled using a Thor Cryogenics 3010 controller. The temperature was measured with a platinum resistance themometer, and the stability was ± 6 mdeg.K per day at ~ 33K, with reproducibility within 25 mdeg.K.

TABLE I

Monochromator (Mon.) and Analyser (An.) crystals, and F.W.H.M. of scans through the Bragg peak $\underline{\tau}$ at wavelength λ for the two main instruments used. Units are THz and $Å^{-1}$. $a_0 = 7.179Å$, $c_0 = 6.329Å$.

| | Mon./An. | λ(Å) | $\underline{\tau}$ | $\Delta\nu^{FWHM}_{incoh.}$ | $\Delta\nu^{FWHM}_{elastic}$ | Approx. slope $(\Delta\nu/\Delta q)_{Res.}$ THz Å | $\Delta Q_x^{FWHM}$ | $\Delta Q_y^{FWHM}$ | $\Delta Q_z^{FWHM}$ |
|---|---|---|---|---|---|---|---|---|---|
| IN3 | Cu(111)/Ge(220) | 2.417 | (400) | 0.145 | 0.022 | 2.1 | 0.0106 | 0.0109 | 0.067 |
| IN2 | Bent P.G.(002)/P.G.(002) | 4.046 | (200) | 0.044 | 0.012 | 1.2 | 0.0137 | 0.0088 | 0.053 |

2.2 Experimental Results

The transition $T_D$ was determined for each crystal by monitoring the elastic intensity of the (400) Bragg peak of sample A(IN3), and (200) or (220) Bragg peak for sample B (IN2). The intensities varied markedly with temperature particularly near $T_D$, for example the (200) peak intensity increased by over a factor of four between 109K and $T_D$. The full reason for the increase is not clear, but it must in part be due to critical scattering as the transition is approached. There is no doubt that the intensities of the peaks suffer from extinction, and for this reason a detailed examination of their variation with T and $\vec{Q}$ was not carried out. Indeed some of the increase at $T_D$ is possibly due to changes in extinction. $T_D$ was found to be 33.3 ± 0.2K for sample A, and 32.8 ± 0.2K for sample B. Below $T_D$ we observed splitting of the Bragg peaks along [010] and [100] consistent with the expected domain formation. All of our measurements were confined to temperatures at and above the transition, with the sample in the tetragonal phase.

The initial data were taken on the IN3 spectrometer using sample A[7]. As the electronic coupling is strongest to the transverse acoustic phonon propagating along the [010] direction and polarised along [100] we have confined our attention to this phonon mode. Its energy dispersion was measured at low wavevectors $\vec{q}$ at temperatures 153K, 73K, 53K, 43K, 35K and at $T_D$ = 33.3K. Typical energy scans at constant-Q are shown for four temperatures in Fig.1. The data shown are the measured counts uncorrected for background, which rises to about 40 counts at $\nu$ = 0. The growth of a central peak and reduction,

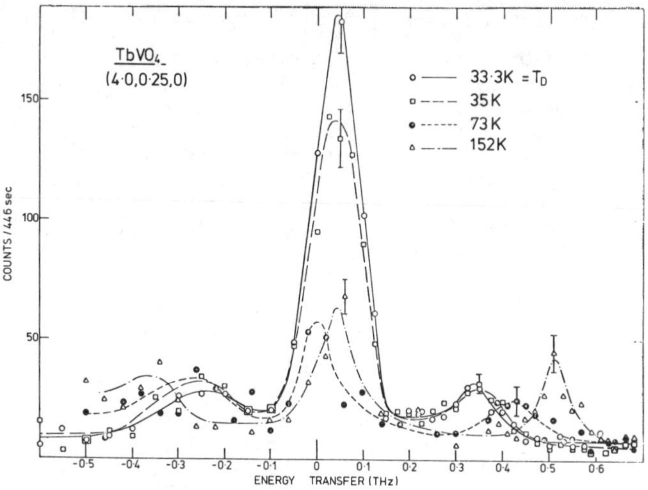


Fig. 1. Typical neutron groups observed on IN3 scanning energy transfer at the point (4.0,0.25,0). The softening of the transverse acoustic mode, and growth of the central peak as T ⟶ $T_D$ is clearly seen.

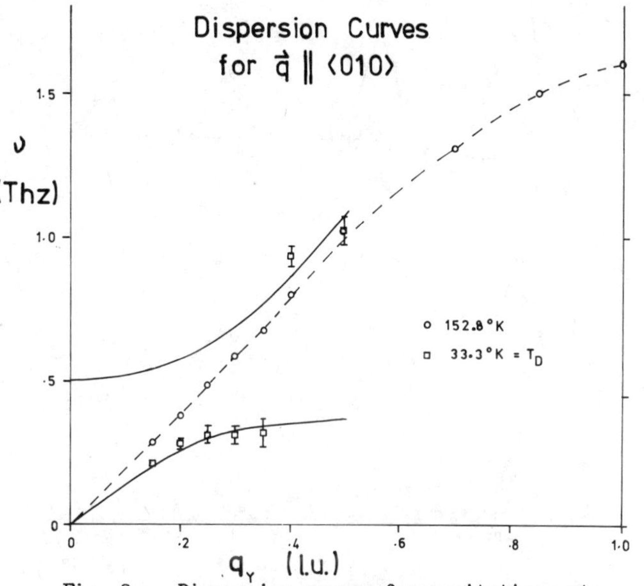


Fig. 2. Dispersion curves for excitations at 152.8K and 33.3K = $T_D$ for $\vec{q}$ along [010] measured relative to (400). The broken curve is drawn through the measured points. The solid curves are the theoretical best fit to the data at $T_D$, see §4.1.

or softening, of the phonon energy-gain and loss peaks from the lower branch of the coupled exciton-phonon modes can be clearly seen. The small shift of the central peak from zero energy transfer is due in part to the slope of the resolution function of the instrument in ν-q space, and in part to a small mis-set in the zero of the spectrometer of ~ 0.025 THz.

The excitation dispersion is plotted in Fig. 2 at 153K when the phonon is almost decoupled from the exciton, and at $T_D$ = 33.3K where the anticrossing of the phonon and singlet-singlet exciton mode is apparent. The data points have been corrected for the instrumental zero error but not for small resolution shifts, expected to be less than 0.01 THz. Both the intensity of the central peak, and the effect of the coupling between the exciton and phonon modes, increase as T approaches $T_D$. As expected from the form of the electronic wavefunctions the neutron scattering cross section appears to come only from the nuclear interaction part of the total scattering cross section, so that it is not possible to observe the branch which is predominantly excitonic in character away from the crossover region. Note the gradient of the lower branch at small q has decreased considerably at T = $T_D$, but does not fall to zero as only half of the total Jahn-Teller coupling involves the singlet-singlet excitation, the other half involving the doublet. The theoretical fit to these excitation data will be discussed in §4.1.

The most striking feature of the experimental results is undoubtedly the growth of the central

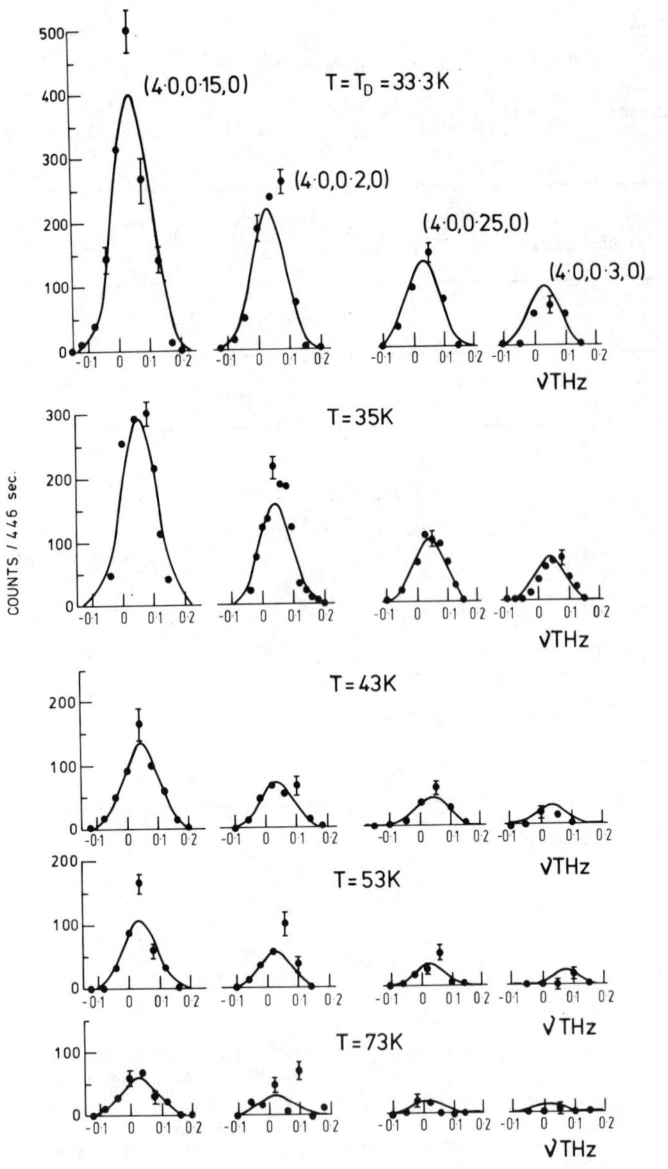


Fig. 3. Central peak observed on IN3 at four values of $\vec{Q}$, and at five temperatures. The points are the experimental data. The solid line is the best fit of the cross section convoluted with the resolution function of the instrument, see §4.2.

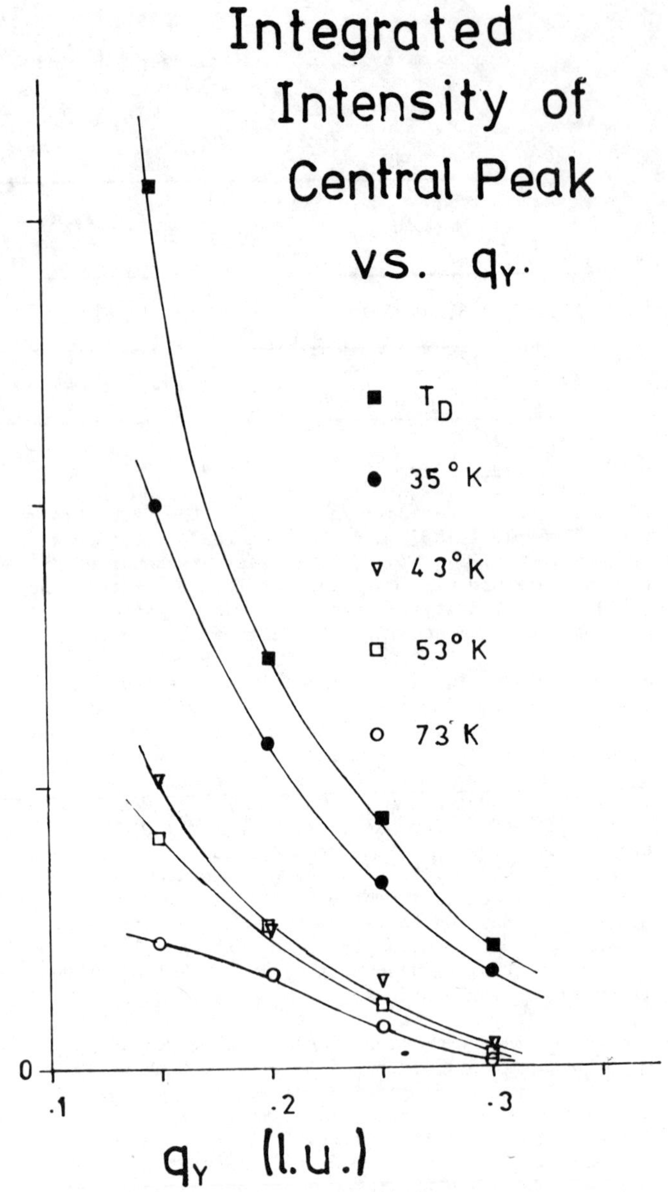


Fig. 4. Variation of the experimental central peak intensity, integrated over $\nu$, with wavevector along [010] at various temperatures above $T_D$.

component at $\nu = 0$ as T approaches $T_D$, and in Fig.3 the experimental profiles, with the estimated background and excitation scattering subtracted, are shown at four values of $\vec{Q}$, and at five temperatures. The energy width of the peaks are all resolution limited. The experimental peak intensities, integrated over $\nu$, are summarized in Fig. 4, and clearly show a divergence as $T \rightarrow T_D$ and $q \rightarrow 0$.

In order to examine the behaviour of the central peak at lower values of $\vec{q}$ without contamination from the Bragg peak, it was necessary to increase the resolution by decreasing the energy of the incident beam on the sample. This was accomplished using the IN2 spectrometer with $\lambda = 4.046$Å, but the loss in beam intensity was such that the scattering from the central mode was reduced by over 1/100 and the coupled excitation could no longer be easily observed. The central peak scattering from sample B was examined at $T_D$, and at 2, 4 and 9K above $T_D$. At $T = T_D$, it was confirmed that the scattering was symmetric about the lattice point (2,0,0) and its anisotropy in $\vec{Q}$ was examined. The intensity was found to fall off far more rapidly with $q_x$ than with $q_y$ and below $|q_y| < 0.05$ the Bragg contamination became prohibitively large.

(We take Ox along [100], and Oy along [010]). The experimentally observed central peak at $T = T_D$, with background subtracted, is shown for different values of $\vec{q}_y$ along [0$\bar{1}$0] in the left hand side of Fig. 5. Again in all the scans the peak intensity appears at slightly positive energy transfers due to the slope of the resolution function in $\nu$-q space. The two most striking changes as $|q|$ approaches the lower values of the range, < 0.060, are the rapid rise in intensity of the peak, and the occurrence of a sharp second peak in the scattering at higher $\nu$.

## 3. THEORY

In this section, we derive expressions for the scattering response functions for $T > T_D$ using linear response function theory[11]. The basic ideas of the electron phonon coupling were given by Elliott et al[2], and developed by Sandercock et al[3].

### 3.1 Single ion response

Above $T_D$, the $Tb^{3+}$ site symmetry is $D_{2d}$, and the lowest four electronic energy levels are two singlets ($A_1$ and $B_1$) separated by $2\varepsilon = 16.5$ cm$^{-1}$[4], and an E doublet midway between them. The Jahn-Teller electron-

Fig. 5. Central peak observed at $T = T_D$ on IN2 at various values of $\vec{q}$ along $[0\bar{1}0]$ relative to (200). The points on the left are experimental with background subtracted, and the lines through them are guides to the eye. The curves on the right are the best theory fit to the data as described in §4.2.

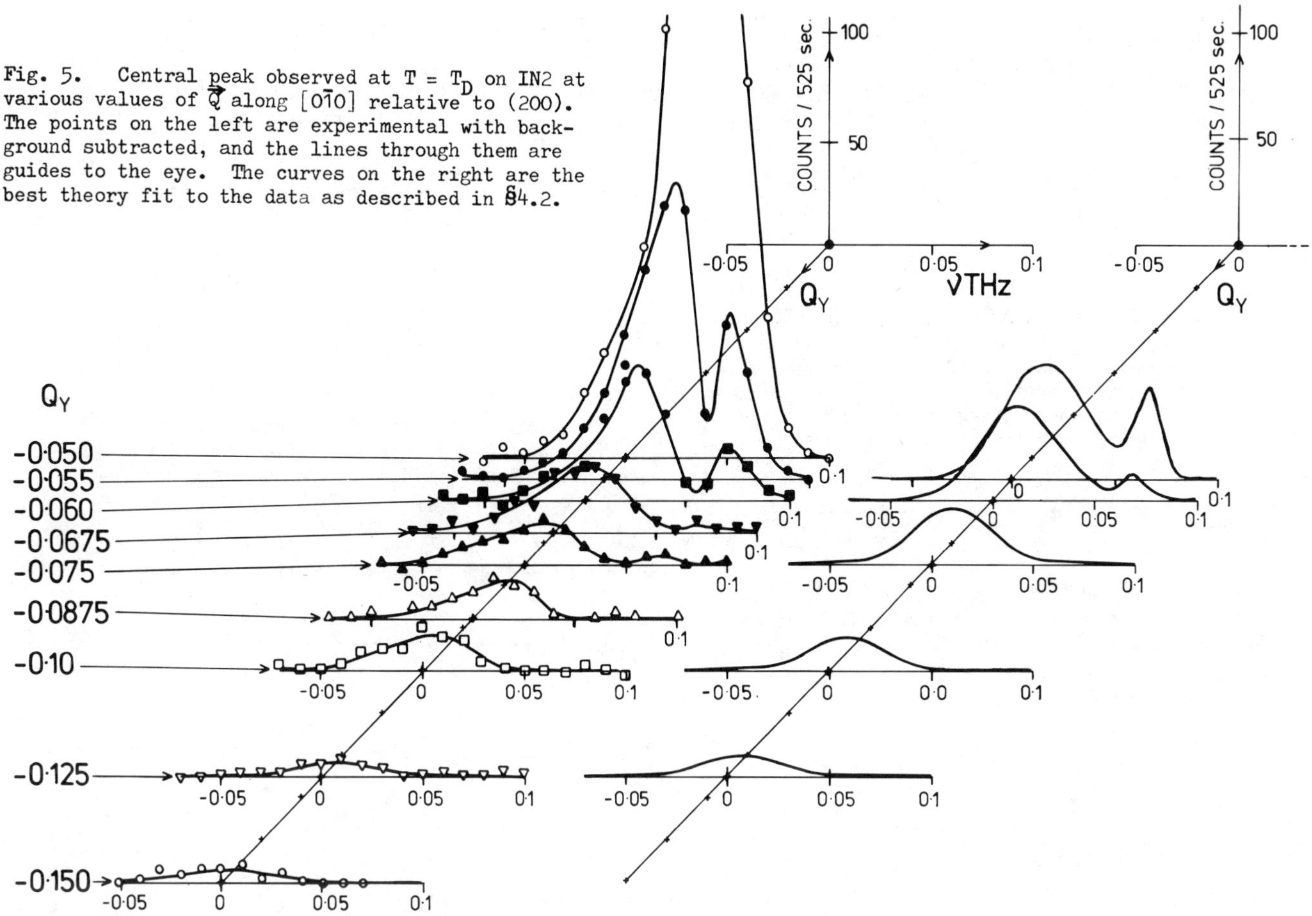


phonon coupling involves a $\bar{B}_1$ symmetry electronic operator $\sigma^z = \sigma_1{}^z + \sigma_2{}^z$, where $\sigma_1{}^z$ is an operator connecting the singlets, and $\sigma_2{}^z$ an operator between the doublet levels [in the notation of Elliott et al,

$\sigma_1{}^z = \frac{1}{2}(1 + \tau^z)\sigma^z$ and $\sigma_2{}^z = \frac{1}{2}(1 - \tau^z)\sigma^z$]. The static susceptibilities of these operators are

$$g_1 = \varepsilon^{-1} \tanh(\beta\varepsilon)P_1; \quad g_2 = \beta P_2, \tag{1}$$

where $P_1$ and $P_2$ are the relative populations of the singlet and doublet levels, i.e.

$$P_1 = \cosh(\beta\varepsilon) \quad [1 + \cosh(\beta\varepsilon)]^{-1};$$

$$P_2 = [1 + \cosh(\beta\varepsilon)]^{-1} \tag{2}$$

$\sigma_1{}^z$ creates an excitation of frequency $2\varepsilon$, and $\sigma_2{}^z$ excites a zero frequency (relaxational) response, so the appropriate single particle (uncoupled) response functions are

$$G^s_{11}(\omega) = g_1 \frac{4\varepsilon^2}{4\varepsilon^2 - \omega^2 - i\omega\Gamma_1} \quad ; \quad G^s_{22}(\omega) = g_2 \frac{\Gamma_2}{\Gamma_2 - i\omega}, \tag{3}$$

where $\Gamma_1$ and $\Gamma_2$ are phenomenological relaxation times, and $G^s_{ij}(\omega)$ is the susceptibility Green function $\langle\langle\sigma_i{}^z;\ \sigma_j{}^z\rangle\rangle_\omega$.

3.2 Jahn-Teller coupling

The electron-phonon (Jahn-Teller) interaction is assumed[2] to be equal for both types of electronic operator $\sigma_1{}^z$ and $\sigma_2{}^z$

$$\mathcal{H}_{JT} = \sum_{n\vec{q}} \xi_n(\vec{q})\sigma^z(\vec{q})a_n(-\vec{q}),$$

where $a_n(\vec{q}) = c_n^+(\vec{q}) + c_n(-\vec{q})$ for the $n^{th}$ phonon branch of wavevector $\vec{q}$ and frequency $\omega_n(\vec{q})$. By transforming to displaced phonon co-ordinates, the total effective electronic coupling is (setting $\hbar=1$)

$$-\tfrac{1}{2}\sum_{\vec{q}} J(\vec{q})\sigma^z(\vec{q})\sigma^z(-\vec{q}), \tag{4}$$

where $$J(\vec{q}) = \sum_n [K_n(\vec{q}) - N^{-1}\sum_{\vec{q}'} K_n(\vec{q}')];$$

$$K_n(\vec{q}) = 2|\xi_n(\vec{q})|^2/\omega_n(\vec{q}).$$

The second term in the expression for $J(\vec{q})$ is the self-interaction which must be subtracted to avoid interactions $\sigma^z(\ell)\sigma^z(\ell')$ for which $\ell' = \ell$. We shall consider separately the coupling to only one phonon branch n = a, and let the other phonons be included in the total interaction (4). Thus, we shall consider the Jahn-Teller Hamiltonian

$$\mathcal{H}_{JT} = \mathcal{H}_{ee} + \mathcal{H}_{ea}, \tag{5}$$

where $$\mathcal{H}_{ee} = -\tfrac{1}{2}\sum_{\vec{q}} \mathcal{J}_a(\vec{q})\sigma^z(\vec{q})\sigma^z(-\vec{q}) \tag{6}$$

and $$\mathcal{H}_{ea} = \sum_{\vec{q}} \xi_a(\vec{q})\sigma^z(\vec{q})a_a(-\vec{q}). \tag{7}$$

$\mathcal{J}_a(\vec{q})$ is the total interaction (4) less the coupling to the branch n = a;

$$\mathcal{J}_a(\vec{q}) = J(\vec{q}) - K_a(\vec{q}) \tag{8}$$

3. Coupled response functions

The single ion uncoupled response function for $\sigma^z$ is

$$G^s_{zz}(\omega) = G^s_{11}(\omega) + G^s_{22}(\omega).$$

The single ion operators $\sigma^z(\ell)$ are coupled to each other by $\mathcal{H}_{ee}(6)$, giving a wavevector-dependent response $G^o_{zz}(\vec{q},\omega)$ for $\sigma^z(\vec{q})$. Since the effective internal field acting on $\sigma^z(\vec{q})$ is $\mathcal{J}_a(\vec{q})\sigma^z(\vec{q})$, we can write down the response of $\sigma^z(\vec{q})$ to an external field $h^z(\vec{q})$ as

$$\sigma^z(\vec{q}) = G^s_{zz}(\omega)[J(\vec{q})\sigma^z(\vec{q}) + h^z(\vec{q})] = G^o_{zz}(\vec{q},\omega)h^z(\vec{q}). \tag{9}$$

Thus, $G^s_{zz}(\omega)$ and $G^o_{zz}(\vec{q},\omega)$ are related by

$$G^o_{zz}(\vec{q},\omega) = G^s_{zz}(\omega)\,[1 - \mathcal{J}_a(\vec{q})G^s_{zz}(\omega)]^{-1}. \tag{10}$$

We now consider the coupled mode response functions when the operators $\sigma^z(\vec{q})$ and $a_a(\vec{q})$ are coupled by $\mathcal{H}_{ea}$ of (7). The appropriate uncoupled responses are $G^o_{zz}(\vec{q},\omega)$ of (10) and the phonon response

$$G^o_{aa}(\vec{q},\omega) = \frac{2}{\omega_a(\vec{q})}\,\frac{\omega_a(\vec{q})^2}{\omega_a(\vec{q})^2 - \omega^2 - i\omega\gamma_a(\vec{q})}, \tag{11}$$

where $\gamma_a(\vec{q})$ is a damping parameter. By a technique similar to that used in (9), one finds that the coupled response functions are

$$G_{zz}(\vec{q},\omega) = G^o_{zz}(\vec{q},\omega)D^{-1}$$

$$G_{az}(\vec{q},\omega) = G_{za}(\vec{q},\omega) = \xi_a(\vec{q})G^o_{aa}(\vec{q},\omega)\times G^o_{zz}(\vec{q},\omega)D^{-1}$$

$$G_{aa}(\vec{q},\omega) = G^o_{aa}(\vec{q},\omega)D^{-1},$$

where $$D = 1 - |\xi_a(\vec{q})|^2\, G^o_{aa}(\vec{q},\omega)G^o_{zz}(\vec{q},\omega). \tag{12}$$

The response functions (12) have a pole at $\omega = 0$, corresponding to the response $G^s_{22}(\omega)$ of the E doublet. The zero frequency responses are

$$G_{zz}(\vec{q},0) = \frac{g_1+g_2}{1 - J(\vec{q})(g_1+g_2)};$$

$$G_{aa}(\vec{q},0) = \frac{2}{\omega_a(\vec{q})}\left\{\frac{1 - [J(\vec{q})-K_a(\vec{q})](g_1+g_2)}{1 - J(\vec{q})(g_1+g_2)}\right\}. \tag{13}$$

The phase transition involves a zone centre $\vec{q} = 0$ displacement, and one sees from (13) that the susceptibilities diverge at $\omega = 0$ where $J(0)(g_1+g_2) = 1$, which defines the transition temperature $T_D$. At frequencies $\omega > \Gamma_2$, the linewidth of the single-particle zero frequency excitation, the E doublet no longer contributes to the response, and in the limit of no damping the response functions have finite frequency poles at $\omega^2 = \omega_\pm(\vec{q})^2$, where $\omega_\pm(\vec{q})^2$ are solutions of the coupled mode equations

$$[\omega_a(\vec{q})^2 - \omega^2]\,[4\varepsilon^2\{1 - g_1J(\vec{q})\} - \omega^2] = 4\varepsilon^2 g_1\omega^2 K_a(\vec{q}) \tag{14}$$

as in equation (5.9) of Elliott et al[2], with $\langle\sigma^x\rangle = g_1\varepsilon$.

## 3.4 Scattering response functions

The quantity observed in a scattering experiment is the scattering response function $\mathcal{S}_{ij}(\vec{q},\omega)$, given by

$$\mathcal{S}_{ij}(\vec{q},\omega) = \pi^{-1}\,\mathfrak{Im}\{G_{ij}(\vec{q},\omega)\}\,[1 - e^{-\beta\omega}]^{-1} \tag{15}$$

and we shall concern ourselves here with $\mathcal{S}_{aa}$ only, since the observed neutron scattering intensity appears to be due entirely to the phonon contribution. In the limit of zero damping, $\mathcal{S}_{aa}$ can be represented by a sum of delta functions

$$\mathcal{S}_{aa}(\vec{q},\omega) = \sum_\alpha A_\alpha(\vec{q})\delta(\omega-\omega_\alpha(\vec{q}))\exp[\beta\omega_\alpha(\vec{q})/2] \tag{16}$$

with the five delta functions located at $\omega_\alpha(\vec{q}) = 0$ and $\omega_\alpha(\vec{q})^2 = \omega_\pm(\vec{q})^2$. The intensities $A_\alpha(\vec{q})$ are given by

$$A_0(\vec{q}) = \frac{2K_a(\vec{q})g_2}{\beta\omega_a(\vec{q})[1 - J(\vec{q})g_1][1 - J(\vec{q})(g_1+g_2)]} \tag{17}$$

$$A_\pm(\vec{q}) = \frac{\omega_a(\vec{q})[\omega_\pm(\vec{q})^2 - 4\varepsilon^2\{1 - \mathcal{J}_a(\vec{q})g_1\}]}{\omega_\pm(\vec{q})\sinh(\beta\omega_\pm(\vec{q})/2)[\omega_\pm(\vec{q})^2 - \omega_\mp(\vec{q})^2]}. \tag{18}$$

Note that the intensity of the central peak, $A_0(\vec{q})$ diverges at $\vec{q} = 0$ at the transition, whereas $A_\pm(\vec{q})$ does not. This critical divergence is associated with a critical narrowing of the central mode, as can be seen if one evaluates $\mathcal{S}_{aa}$ as $\omega \rightarrow 0$ by replacing $G^o_{aa}$ and $G^s_{11}(\omega)$ by their values at $\omega = 0$, leaving the response of the central peak as

$$\mathcal{S}^o_{aa}(\vec{q},\omega) = \frac{2\Gamma_2K_a(\vec{q})g_2\omega}{\pi[1-\exp(-\beta\omega)]\omega_a(\vec{q})[1-J(\vec{q})g_1]^2[\omega^2+\Gamma_2(\vec{q})^2]}. \tag{19}$$

This has the form of a Debye relaxational response with a linewidth

$$\Gamma_2(\vec{q}) = \Gamma_2[1 - J(\vec{q})(g_1+g_2)][1 - J(\vec{q})g_1]^{-1} \tag{20}$$

which vanishes at $\vec{q} = 0$ at $T = T_D$.

## 3.4 Neutron scattering cross-section

The theory developed above involves the coupled response of the electronic system and a single phonon branch, and is generally applicable so long as there is no coupling to other phonon branches with frequencies close to the excitations involved. In applying the theory to the experimental results we take the branch n = a to be an acoustic phonon propagating along the y-axis of the crystal, transversely polarised along the x-axis.

We discuss the neutron scattering in terms of the function $S(\vec{Q},\omega)$, which is related to the partial differential cross-section by

$$\frac{d^2\sigma}{d\Omega dE'} = \frac{k'}{k}S(\vec{Q},\omega) \tag{21}$$

Following Marshall and Lovesey[12], the scattering due to the acoustic phonon branch a close to the reciprocal lattice point $\vec{\tau}$ can be written

$$S(\vec{Q},\omega) = L_{\vec{\tau}}\,\frac{|\vec{Q}\cdot\sigma_a(\vec{q})|^2}{\omega_a(\vec{q})}\,\mathcal{S}_{aa}(\vec{q},\omega)\,(\vec{Q}-\vec{q}-\vec{\tau}) \tag{22}$$

where $L_{\vec{\tau}}$ is a factor which includes the Debye-Waller factor, $\vec{\sigma}_a(\vec{q})$ is the polarisation vector, and $\mathcal{S}_{aa}(\vec{q},\omega)$ can be taken to be the scattering response function obtained in §3.3. The detector current obtained when the spectrometer is set for a wavevector transfer $\vec{Q}_o$ and energy transfer $\omega_o$ is given by the convolution of the instrumental response and the scattering function (22)[9]:

$$I(\vec{Q}_o,\omega_o) = B_{\vec{\tau}}\int d\vec{Q}d\omega R(\Delta\vec{Q},\Delta\omega)S(\vec{Q},\omega), \tag{23}$$

where $R(\Delta\vec{Q},\Delta\omega)$ is the resolution function, and $\Delta\vec{Q} = \vec{Q} - \vec{Q}_o$, $\Delta\omega = \omega - \omega_o$.

## 4. INTERPRETATION OF RESULTS

### 4.1 The coupled mode dispersion

Expressions for the coupled mode frequencies are given by the solutions of (14), but since it is never possible to observe both branches simultaneously, we cannot determine $J(\vec{q})$ and $K_a(\vec{q})$ independently. The solid curves in Fig. 2 represent a fit to the data points assuming that $J(\vec{q})$ and $K_a(\vec{q})$ are independent of $q_y$. Using the 152.8K values as the unperturbed phonon frequencies $\omega_a(\vec{q})$, we find $J(0) = 0.71$ THz and $K_a(0) = 0.46$ THz from a fit to the data using a least squares routine in which we require that $g_1$ and $g_2$ are given by (1) and that $J(0)(g_1+g_2) = 1$ at $T = T_D$. This determines the unperturbed crystal field splitting $\varepsilon$ as 0.28 THz. The agreement between theory and experiment is quite good, but the most important feature of the analysis is that the fit is not markedly improved by letting $J(\vec{q})$ vary with $\vec{q}$ as in (24) below, showing that the parameter $\alpha(\vec{e}\cdot)$ is small. This is similar to the result obtained by Kjems et al[8], who found no variation of $J(\vec{q})$ and $K_a(\vec{q})$ with the magnitude of $\vec{q}$ in their experiments on $TmVO_4$. There is a puzzling discrepancy between our value for $K_a(0)$ and the value 35.7 $cm^{-1}$ (= 1.07 THz) obtained by Sandercock et al (principally from ultrasonic data).

### 4.2 The Central Mode

In order to confirm that the observed scattering is of the dynamic origin discussed in §3 we have made a preliminary comparison of the experimental data with the theory of §3 using several simplifying approximations. We hope to carry out a more complete comparison later. Sandercock et al estimate $\Gamma_2 \approx 30$ GHz, and one therefore expects that the linewidth of the central peak ($\Gamma_2(\vec{q})$ of (20)) will be less than the triple axis resolution. Accordingly, we assume throughout a delta function response (17) for the central peak. In order to perform a convolution of the form of (23), we need to know the $\vec{q}$-dependence of $J(\vec{q})$ and $K_a(\vec{q})$.

These functions will depend primarily on the direction of $\vec{q}$ which we represent by the unit vector $\vec{e}$. We assume that for small $q$, $J(\vec{q})$ can be written

$$J(\vec{q}) \;=\; J(\vec{e})[1-\alpha(\vec{e})q^2] \qquad (24)$$

(with a similar expression for $K_a(\vec{q})$, where $J(\vec{e}) = \mathcal{L}t_{q\to 0}\, J(\vec{e}q)$ . The amplitude $A_o(\vec{q})$ of the central peak in (17) depends principally on the last term in the denominator. Ignoring the $\vec{q}$-dependence of the remaining terms, we can write

$$A_o(\vec{q}) \;=\; \frac{1}{\omega_a(\vec{q})}\,\frac{X(\vec{e})}{\kappa(\vec{e})^2+q^2}\,, \qquad (25)$$

where $X(\vec{e}) = 2K_a(\vec{e})g_2\{\beta[1-J(\vec{e})g_1]\alpha(\vec{e})J(\vec{e})(g_1+g_2)\}^{-1}$

$$\kappa(\vec{e})^2 \;=\; [1-J(\vec{e})(g_1+g_2)]\,[\alpha(\vec{e})J(\vec{e})(g_1+g_2)]^{-1}$$

The maximum value $J(0)$ for $J(\vec{e})$ occurs when $\vec{e}$ is along $\vec{e}_o$ = [100] or [010], in which case $\kappa(\vec{e})$ vanishes at the transition temperature; thus $A_o(\vec{q})$ has the form usually assumed for critical scattering. $\kappa(\vec{e})$, the inverse correlation length, will increase rapidly as $\vec{e}$ moves away from $\vec{e}_o$, and one therefore expects that the central peak response $A_o(\vec{q})$ has a maximum for $\vec{e} = \vec{e}_o$, and falls off rapidly away from $\vec{e}_o$, as is indeed observed experimentally. This effect will be most pronounced at $T_D$.

We have made use of this property of $A_o(\vec{q})$ in order to obtain a first approximation to the convolution (23). We assume that $A_o(\vec{q})$ is zero except when $\vec{e}$ lies within a small angle of the [010] axis, when we put $A_o(\vec{q}) = A_o(q\vec{e}_y)$. Then (23) reduces simply to an integration over $q_y$, of the form

$$I(\vec{Q}_o,\omega_o) = B'\int dq_y\,\frac{X(\vec{e}_y)}{\kappa(\vec{e}_y)^2+q_y^2}\,R(\Delta\vec{Q},\Delta\omega)\,\delta(\vec{Q}-q_y\vec{e}_y-\vec{\tau}), \qquad (26)$$

where $\vec{e}_y$ = [010], and $\vec{\tau}$ is taken to be the (2,0,0) or (4,0,0) point . B' contains the $|\vec{Q}.\vec{\sigma}_a(\vec{q})|^2$ term in (22) which allows scattering to be observed for $\vec{q}$ along $\vec{e}_y$ but not along $\vec{e}_x$, and is assumed to be constant for small $\vec{q}$. Note that the $q^2_y$ factor in the volume element of integration in (26) has cancelled with a similar factor from the $\omega_a(\vec{q})^{-1}$ terms in (22) and (25). We see that the $\vec{q}$ dependence of the central peak is sensitive to that of $J(\vec{q})$ and that the form of (26) results from the assumption made in (24).

We have calculated the form of (26) assuming that the resolution function is Gaussian[9] using a simple stepwise numerical integration method and omitting the q = 0 point as this corresponds to a static strain. As the resolution function for both instruments slopes in $\nu$-q space away from $\vec{\tau}$ along $\pm$Oy the maximum intensity in a constant-Q energy scan will occur at $\nu>0$ if the central mode intensity increases as $|q|$ decreases. The extreme case of this effect is when the extremity of the resolution aperture passes through a reciprocal lattice point, with intensity ~const $\delta(\vec{Q}-\vec{\tau})$, in which case a sharp peak at $\nu = (\Delta\nu/\Delta q_y)_{Res.}\times q_y$ is observed.

The expression (26) was fitted to the data shown in Figs. 3 and 5 using a least squares routine. The variable parameters were taken to be a small residual background, a small shift in the energy zero setting of the spectrometer, the inverse correlation range parameter $\kappa$, and the scaling factor at each temperature B'X which was taken to be independent of $\vec{q}$. It proved not possible to obtain a value of $\kappa$ from the data, indicating $\kappa \ll Q_{oy}$ in which case the calculated main peak profile is independent of $\kappa$. However at the lowest values of $Q_{oy}$ studied on IN2, the calculations do in fact predict the small sharp additional peak at higher $\nu$ observed in the experiments as is seen from Fig.5. This additional sharp peak occurs at a $\nu$ setting very close to that given by the intersection of the resolution function with Bragg intensity at the lattice point, from which it must therefore be distinguished. Its origin lies in the fact that at low $\kappa$ the cross section builds up very rapidly as $q_y \to 0$ in a manner which 'overtakes' the closing of the resolution aperture. Calculations show that the intensity of this sharp peak is very dependent on the value of $\kappa$, as expected, and it disappears as $\kappa$ approaches the value of $Q_{oy}$. Furthermore it is clear that the steps in the numerical integration of (26) must be finely spaced in order to give the true intensity of such a sharp peak. This computational difficulty is however matched by the experimental problems caused by extinction which are equally difficult to correct for! Providing any Bragg contribution can be subtracted, the variation of such a sharp second peak at different values of $Q_{oy}$ can in principle lead to a direct estimate of the value of $\kappa$. The fact that it is present, in addition to the broader peak, clearly indicates the latter is not due to some change in the static behaviour of the Bragg peak.

The fits to the IN3 data were made with $\kappa$ fixed at 0.001 at all T. The energy zero shift was found to be +0.02$\pm$0.01 THz and this adds to the true resolution shift of the peak centre which, for example, is calculated to be 0.025 THz at $Q_{oy}$ = 0.15. The agreement of this data, and that taken on IN2 at $|Q_{oy}|$ > 0.075 is sufficiently good for us to be confident that the general features of the analysis are correct, and verifies the origin of the central mode scattering and the form of $J(\vec{q})$ given in (24). However there are discrepancies at the smallest $Q_{oy}$ in Fig. 3 and particularly in Fig. 5. Fits to the

IN2 data shown in Fig. 5 for $T=T_D$ gave $\kappa = 0$ within the error, and negligible energy zero shift and background, but they fail badly at $Q_{oy} \leqslant 0.06$ where there appears to be additional intensity present. This discrepancy was also present, but by no means so severe, at the higher temperatures investigated on IN2.

At present we can only speculate as to the origin of this discrepancy at low $Q_{oy}$. It may be due to the 'linear' approximation used being too severe, or a more complex variation of $J(\vec{q})$ with $\vec{q}$. Also, we have not included in our analysis the contribution of the coupled excitation modes, although they are expected to have energies ~ 0.07 THz at $|Q_{oy}| = 0.05$, somewhat higher in energy than that at which the extra intensity is observed. We feel that until these differences are resolved the variation of the scale parameters B'X with temperature will be difficult to interpret.

## 5. SUMMARY AND CONCLUSION

We have observed neutron scattering intensity from $TbVO_4$ near zero energy transfer and at wave vectors along ⟨100⟩ directions which diverges as $q \rightarrow 0$ and $T \rightarrow T_D$. We believe that this central mode is caused by the characteristic "soft mode" of the co-operative Jahn-Teller phase transition, namely the zero-frequency response of the electronic doublet whose phonon-coupled intensity diverges at $T_D$. Linear response theory has been used to calculate the frequencies and intensities of the modes of the coupled system. The data has been shown to be qualitatively consistent at the larger wavevectors investigated with a simplified model in which the interactions are only present along [010] and [100] and in which the total effective ion-ion coupling $J(q) \sim (1 - \alpha q^2)$. However the scattering observed at the lowest values of q has not been fully explained, nor has the apparently small value of effective inverse correlation length at all temperatures investigated. These anomalies may well be due to the simplifying assumptions we have made in the analysis, and it is possible that they may provide information on the nature of the finite low frequency excitations in the system. It is also possible that there may be a contribution to the scattering from purely static effects[6].

The acoustic phonon also couples equally to the singlet-singlet exciton, and the effect is to produce an anticrossing in their coupled modes which is maximum at $T_D$ and has been analysed to yield values for $J(0)$ and $K_a(0)$.

The observation of the central mode is a consequence of the condition that the symmetry of the soft mode of the system is the same as the symmetry of an excitation within a degenerate level of the electronic system. One may therefore anticipate that this type of central mode will occur in all systems in which this condition is satisfied. In fact this is usually the case in systems showing magnetic phase transitions including those of the induced moment type, where the lowest electronic state is a singlet. Our measurements have therefore not only verified to a large extent our understanding and interpretation of this particular system[2,3,4], but the general principles may also be expected to apply to the behaviour of singlet ground state magnetic compounds which have recently excited considerable interest[5].

We would like to thank Dr. S. H. Smith of the Clarendon Laboratory, Oxford for growing the crystals used in these experiments.

## REFERENCES

+Work supported by the Science Research Council.

1. G. A. Gehring and K. A. Gehring, Rep. Prog. Phys. 38, 1-90 (1975).
2. R. J. Elliott, R. T. Harley, W. Hayes and S. R. P. Smith, Proc. R. Soc. Lond. A328, 217-266 (1972).
3. J. R. Sandercock, S. B. Palmer, R. J. Elliott, W. Hayes, S. R. P. Smith and A. P. Young, J. Phys. C5, 3126-46 (1972).
4. E. Pytte, Phys. Rev. B8, 3954-9 (1973).
5. See for example, W. J. L. Buyers, "Magnetism and Magnetic Materials 1974, Proc. 20th Ann. Conf.", Eds. C. D. Graham, G. H. Lander, and J. J. Rhyne, (A.I.P. 1975) p27-32.
6. J. D. Axe, S. M. Shapiro and G. Shirane, "Anharmonic Lattices, Structural Transitions and Melting" Ed. T. Riste, (Noordhoff-Leiden, 1974), 23-37.
   F. Schwabl, ibid., p.87-112.
7. A brief report of some of the data is given by M. T. Hutchings, R. Scherm, S. H. Smith and S. R. P. Smith, J. Phys. C8, L393-6 (1975).
8. J. K. Kjems, W. Hayes and S. H. Smith, Phys. Rev. Letters 35, 1089-92 (1975).
9. See B. Dorner, Acta Crys. A28, 319-27 (1972), and references therein.
10. S. H. Smith, G. Garton and B. K. Tanner, J. Cryst. Growth 23, 335-40 (1974).
11. S. R. P. Smith, "Proc. of 3rd Int. Conf. on Light Scattering from Solids, Campinas, 1975", Ed. M. Balkanski et al, (to be published).
12. See W. Marshall and S. W. Lovesey, Theory of Thermal Neutron Scattering, Oxford 1971, p.64

# MAGNETOELASTIC INTERACTIONS IN $UO_2$*

J. Faber, Jr. and G. H. Lander
Argonne National Laboratory, Argonne, Ill. 60439
and
B. R. Cooper
Department of Physics, West Virginia University
Morgantown, W. Virginia 26506

ABSTRACT

Neutron diffraction measurements of the elastic magnetic scattering cross section from antiferromagnetic $UO_2$ show additional nuclear intensity below $T_N$ = 30.8 K. We have examined the possibility of analyzing the additional scattering in terms of homogeneous distortions, which involve shifts of the oxygen atoms from their fluorite lattice sites. The behavior arising from the presence of these homogeneous distortion modes formed the basis for Allen's theory of a cooperative Jahn-Teller effect in $UO_2$. However, an analysis in terms of these homogeneous distortions cannot explain the neutron data. But, by extending Allen's concepts to include *inhomogeneous* deformations, corresponding to a zone boundary $q = (\pi/a)(1,0,0)$ phonon, excellent agreement is obtained between theory and experiment. The oxygen displacement is 0.014(1) Å from the fluorite lattice positions, and the inhomogeneous deformation ($T_{2g}(Q^1)-T_{1g}$) does not require a change in the dimensions of the unit cell. The essential features of Allen's theory for $UO_2$ can still be maintained.

- - - - - - - - -

The first-order antiferromagnetic transition in $UO_2$ aroused considerable interest when it was first reported by Frazer et al.[1] Perhaps the most unusual aspect of this transition is that it appears to occur without a *macroscopic* lattice distortion. In this paper we start from the fact that the orbital momentum of the two 5*f* electrons for $U^{4+}$ is strongly coupled to the lattice modes.[2] Here our interest is in the manifestation of the magnetoelastic coupling on the static distortion. These interactions in $UO_2$ were recognized by Allen,[3] and, by considering modes coupling to the electronic $\Gamma_5$ ground state,[4] he showed how an *internal* distortion could minimize the magnetoelastic energy without an appreciable lattice deformation.

Recent neutron measurements[5] of the elastic scattering cross section from $UO_2$ have shown that unexpectedly large anisotropy occurs at high scattering angles. A careful analysis[6] of the data led to the conclusion that additional nuclear scattering is present below $T_N$ in $UO_2$. (By "additional" we mean scattering that is not allowed by the symmetry of the ideal fluorite structure.) Regular nuclear Bragg reflections for the fcc structure occur for $h,k,\ell$ all even or odd; whereas the type I antiferromagnetic reflections have mixed indices with $h+k$ = even, and $\ell$ even or odd.[1] In Fig. 1 we show the difference cross section (observed-calculated magnetic). Full details of the magnetic cross section calculation will be published elsewhere. From the figure, however, it is clear that additional elastic scattering occurs at the $\ell$ odd positions (note that these intensities are $<10^{-3}$ of the nuclear Bragg reflections). The question is whether these additional intensities can be understood in terms of internal (static) displacements of the atoms that arise as a consequence of the spin-lattice interaction.

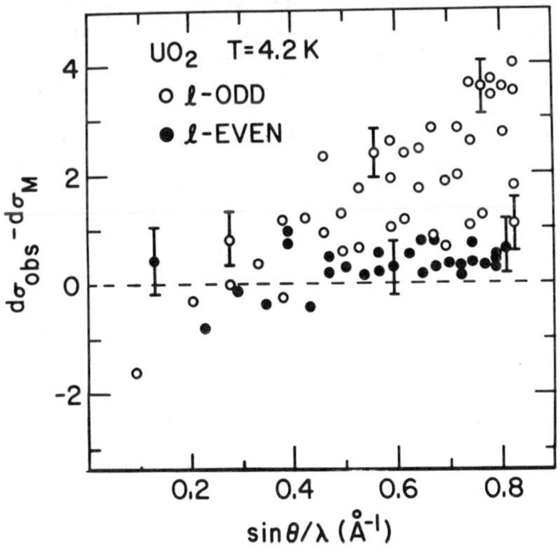


Fig. 1. The difference cross section $d\sigma_{obs}$ (observed) - $d\sigma_M$ (calculated magnetic) plotted for each Bragg reflection. The open circles are magnetic reflections with *h*, *k*, even and $\ell$ odd; the closed circles are reflections with *h*, *k*, odd and $\ell$ even.

Following Allen[3] we consider the local modes of the oxygen cube surrounding a $U^{4+}$ ion that transform as the representations contained in the symmetric square $[\Gamma_5^2]$,

$$\Gamma_5 \times \Gamma_5 = A_{1g} + T_{1g} + E_g + T_{2g} \quad . \tag{1}$$

The modes of these types present for the oxygen cube are one $A_{1g}$ mode, one $T_{1g}$, one $E_g$, and two $T_{2g}$ modes. The $A_{1g}$ (breathing) mode represents a uniform dilation of the oxygen cube and cannot by itself give rise to additional scattering. There are no $T_{1g}$ distortion modes present. The only $T_{1g}$ mode is the rigid rotation of the unit cell and, as with the $A_{1g}$ mode, cannot give rise to additional scattering. The two components of the $E_g$ mode, $Q_\varepsilon$ and $Q_\theta$, correspond to orthorhombic and tetragonal deformations, respectively. The $T_{2g}$ modes are (in Allen's notation[3]) $Q^1$, which gives a shear deformation, and $Q^2$, which is the *internal* distortion. All these modes (Fig. 2) are homogeneous deformations (corresponding to q = 0 phonons) and, except for the $Q^2$ mode, which corresponds to an optic phonon, lead to a change in dimensions of the unit cell.

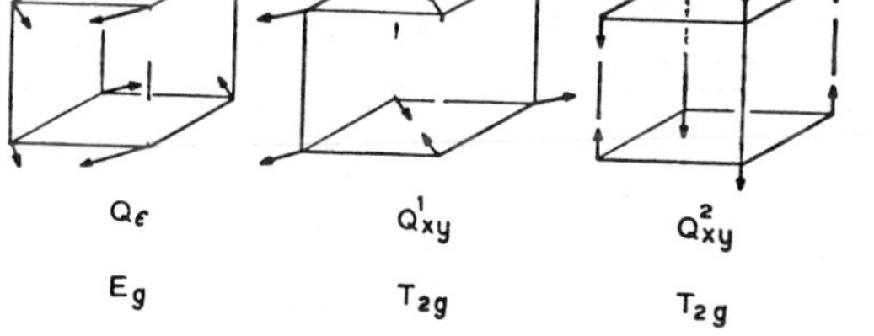

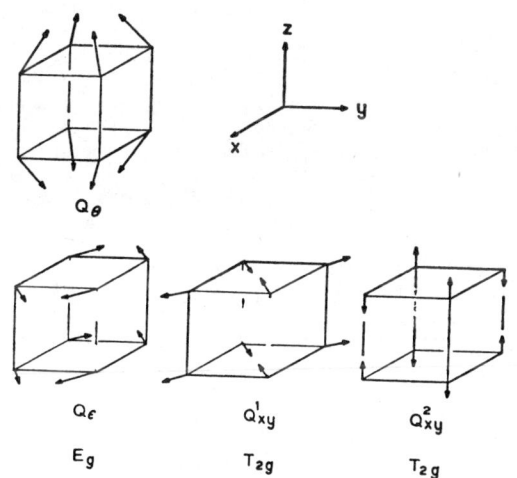

Fig. 2. Homogeneous deformations with $E_g$ and $T_{2g}$ symmetry that mix $\Gamma_5$ states. One component of each $T_{2g}$ mode is shown.

The lowering of the overall symmetry of the unit cell has two consequences in the diffraction pattern. First, the diffraction from planes of equivalent spacing in the cubic phase occurs at slightly different positions in the distorted phase. This splitting of the diffraction profile is best observed with high resolution at high scattering angles since $\Delta\theta = -(\Delta d/d)\tan\theta$, where $\theta$ is the Bragg angle and d is the cubic lattice spacing. In $UO_2$, low-temperature x-ray experiments[7] have failed to detect any such splitting of the diffraction profiles and experiments recently performed at ANL confirm the earlier results.[8] Second, the distortion will modify the Bragg intensities. For those reciprocal lattice points that have zero intensity in the fcc diffraction pattern, we have calculated the cross section from a monoclinic unit cell that is distorted from cubic by the $T_{2g}(Q^1)$ mode. Intensities of between 2 and 4 mb/mole at $\sin\theta/\lambda \sim 0.7$ Å$^{-1}$ can be achieved if the angle between the $\vec{a}$ and $\vec{b}$ axes is increased to 90.5°. However, this deformation also gives contributions to the $\ell$ even reflections, and would result in a splitting of the high-angle diffraction lines at least 50 times greater than the upper limit given by the x-ray experiments. Thus, this deformation is inconsistent with results from both neutron and x-ray diffraction experiments. Similar arguments can be made for $E_g(Q_\varepsilon)$ and $E_g(Q_\theta)$ modes, in which the $\vec{k}$ dependence calculated for the $\ell$ odd reflections is also incorrect. In the context of Allen's model our results should therefore be compatible with the internal $T_{2g}(Q^2)$ mode. Unfortunately, this mode, in which the two oxygen tetrahedra move against one another, does not give intensity at the magnetic reciprocal lattice points. Instead, the intensities of the fundamental fluorite reflections are modified.

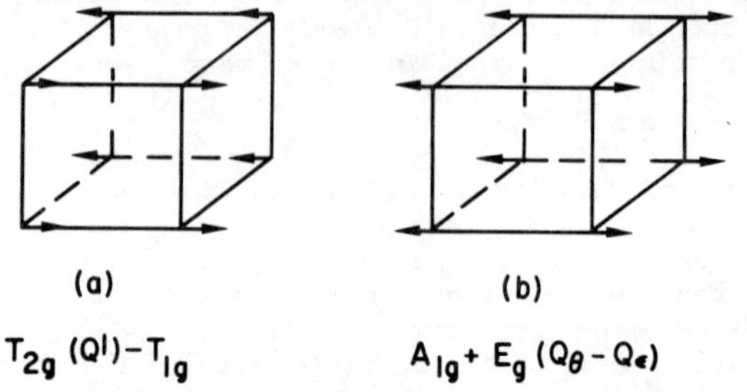


Fig. 3. Inhomogeneous deformations (linear combinations of the normal modes) of the cube of oxygen atoms surrounding each metal atom that mix $\Gamma_5$ states. One component of each mode is shown.

Thus, none of the four modes contained in the symmetric square $[\Gamma_5^2]$ can account for the neutron intensities of Fig. 1. These modes all correspond to q = 0 phonons, i.e., to homogeneous deformations of the unit cell. The next step is to consider *inhomogeneous* deformations. By this, we mean deformations that are not identical as one moves from uranium to uranium in the lattice. Two simple inhomogeneous deformations, which correspond to zone boundary $q = (\pi/a)(1,0,0)$ phonons, are shown in Fig. 3. In Fig. 3(a) a $T_{2g}(Q^1)$ shear deformation is combined with the $T_{1g}$ pure rotation, but with the sense of the atomic displacements of nearest neighbor oxygen atoms rotated by $\pi$ between adjacent uranium atoms. In Fig. 3(b) an $E_g$ mode is combined with a $A_{1g}$ uniform dilation. Since these modes are degenerate, the combinations can be made in a variety of ways. Thus, the $(Q^1 + T_{1g})$ mode is six-fold degenerate, and the $(E_g + A_{1g})$ mode is three-fold degenerate.

To test the compatibility of the various mode configurations with the neutron diffraction results of Fig. 1, we have performed a least-squares refinement with the quantities $(d\sigma_{obs} - d\sigma_M)$ of Fig. 1 as experimental input. The calculated cross section $d\sigma_D$ arises from oxygen displacements. The only parameter in this calculation is $\Delta$, the fractional coordinate shift of the oxygen atom from its equilibrium position. The results are illustrated by plotting the point-by-point residual in Fig. 4. In Fig. 4(a), which considers the $(Q^1 - T_{1g})$ mode, the fit is clearly excellent ($\chi^2 = 1$), whereas the $(E_g + A_{1g})$ mode (Fig. 4(b)) cannot account for the additional neutron scattering. Both types of modes do not contribute to the $\ell$ even reflections (solid points in Fig. 1) so these are purely magnetic in origin. The least-squares fit of $d\sigma_D$ and $(d\sigma_{obs} - d\sigma_M)$ in Fig. 4(a) gives $\Delta = (2.6 \pm 0.2) \times 10^{-3}$, which implies a shift in the oxygen position of 0.014 Å. It is worth pointing out that at low temperatures the oxygen temperature factor[5] $B_O = 0.22$ Å$^2$ so that the mean thermal vibrational amplitude $\langle u_0 \rangle = [B/(8\pi^2)]^{1/2} = 0.053$ Å. The static displacements are therefore quite small in terms of the lattice vibrations.

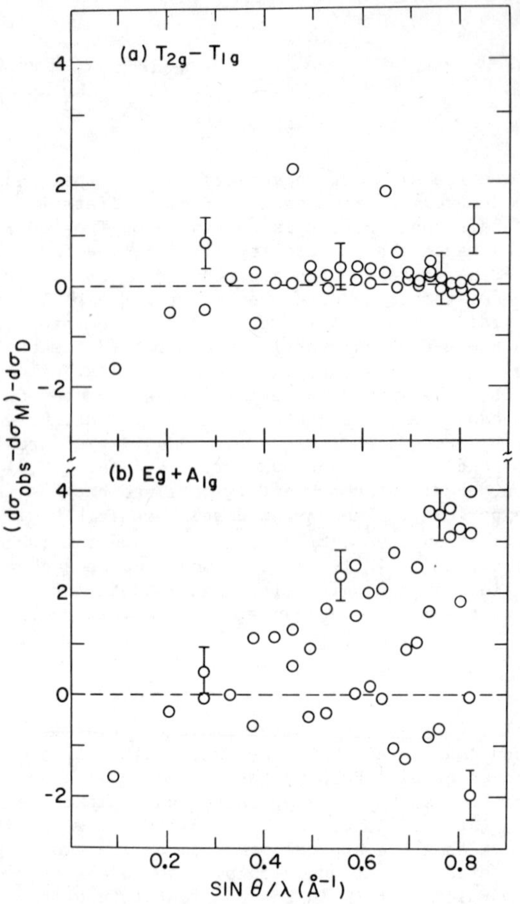


Fig. 4. Residual cross sections plotted for reflections with $\ell$ odd after subtracting $d\sigma_D$ (arising from the inhomogeneous deformation of the oxygen sublattice) from the $d\sigma_{obs} - d\sigma_M$ values of Fig. 1. The calculations of $d\sigma_D$ in (a) and (b) use modes from Fig. 3(a) and Fig. 3(b), respectively.

In summary, we have presented a brief survey of possible distortion modes involved in the spin-lattice interactions for the actinide antiferromagnet $UO_2$. Our object has been to develop a theory in which the magnetoelastic energy is minimized by an internal rearrangement in such a way as to maintain the cubic (i.e., $a=b=c, \alpha=\beta=\gamma=\pi/2$) parameters of the overall unit cell. The oxygen motion is not allowed within the space group Fm3m, and a lower symmetry has to be considered. The introduction of $q \neq 0$ modes does not affect the coupling between $\Gamma_5$ states and the essential features of Allen's theory still apply. In addition, the static displacement of the oxygen sublattice gives rise to anisotropy at the uranium site that suggests a noncollinear magnetic configuration.[6] We hope soon to study the dynamics of this process by measuring the temperature dependence of the transverse acoustic phonons in $UO_2$ near $T_N$.

We are pleased to acknowledge a number of informative discussions with S. J. Allen.

REFERENCES

*Work supported by the U. S. Energy Research and Development Administration.

1. B. C. Frazer, G. Shirane, D. E. Cox, and C. E. Olsen, Phys. Rev. 140, A 1449 (1965).
2. R. A. Cowley and G. Dolling, Phys. Rev. 167, 464 (1968).
3. S. J. Allen, Phys. Rev. 166, 530 (1968; *ibid.* 167, 492 (1968).
4. H. U. Rahman and W. A. Runciman, J. Phys. Chem. Solids 27, 1833 (1966).
5. J. Faber, Jr., Amer. Inst. Physics. Conf. Proc. 24, 51 (1975).
6. J. Faber, Jr., G. H. Lander, and B. R. Cooper, Phys. Rev. Letters 35, 1770 (1975).
7. See Ref. 12 of Brandt and Walker [Phys. Rev., 170, 528 (1968)]. See also Ref. 31 of White and Sheard [J. Low Temp. Physics, 14, 445 (1974)]. J. D. Pirie and T. Smith, Phys. Status Solidi, 41, 221 (1970).
8. J. Faber, Jr. and M. H. Mueller (to be published).

# CHARGE ORDERING IN $Fe_3O_4$*

J. B. Sokoloff
Department of Physics
Northeastern University
Boston, Massachusetts 02115

ABSTRACT

A year ago, a theory of charge ordering in $Fe_3O_4$ based on the cooperative Jahn-Teller effect, with $\Delta 5$ symmetry phonons of wave vector $(0,0,\frac{1}{2})$ as condensing modes, was proposed by Yamada.[1] This theory explains the critical neutron scattering above the ordering temperature $(T_v)$[2] at half integer, but not at mixed integer scattering vectors. It also explains the existence of half integer wave vector diffraction peaks below $T_v$,[3,4] but not the existence of mixed integer peaks.

In Yamada's charge ordering, only electrons on every other plane of B-site ions perpendicular to the c-axis order, with the remaining planes having their charge disordered. It is proposed here that if the remaining planes had a charge ordering of wave vector (1,0,0), (0,1,0) or (0,0,1), the existence of mixed integer diffraction peaks below $T_v$ could be explained (since such wave vectors when added to reciprocal lattice vectors for the primitive fcc lattice give "mixed integer" wave vector). Mean field theory on a model Hamiltonian having both Coulomb and lattice distortion induced electron-electron interactions gives a different transition temperature for each type of ordering, in agreement with Mossbauer effect[5] but not neutron scattering[2,3,4] experiments.

Ewald method[6] calculations of the Coulomb interaction energy have been performed for the Verwey ordering, the AB ordering with wave vectors $(0,0,\frac{1}{2})$ and $(0,0,1)$[7], the Yamada ordering,[1] and a combination of $\Delta 5^{(1)}$ Yamada ordering of wave vector $(0,0,\frac{1}{2})$ and $\Delta 5^{(2)}$ Yamada ordering of wave vector (0,0,1). The Verwey ordering gives the lowest electrostatic energy. Since the ground state is not the Verwey ordering, the Jahn-Teller effect must be important in stabilizing the actual ground state. A more complete report of this work is given in reference 8.

REFERENCES

* Work supported by NSF grant under DMR-72-03282A02

1. Y. Yamada, A.I.P. Conference Proceedings, N.24, 79(1975).
2. Y. Fujii, G. Shirane, and Y. Yamada, Phys. Rev. B10, 2036 (1975).
3. G. Shirane, S. Chikazumi, J. Akimitsu, K. Chiba, M. Matsui, and Y. Fujii, J. Phys. Soc. Japan (in press); see also E. J. Samuelson, E. J. Bleekov, L. Dobrzynski, and T. Riste, J. Appl. Phys. 39, 1114 (1968); Y. Yamada, K. Suzuki, S. Chikazumi, Appl. Phys. Lett. 13, 172 (1968).
4. M. Iizumi and G. Shirane, Solid State Communications (in press).
5. R. A. Buckwald and A. A. Hirsch, Bulletin of the A.I.P. 20, 383 (1975); Solid State Communications 17, 621 (1975).
6. J. M. Ziman, Principles of the Theory of Solids, second edition (Cambridge University Press, 1972), p.39.
7. The AB ordering is a linear combination of $\Delta 5^{(1)}$ and $\Delta 5^{(2)}$ ordering, such that each plane of B-site ions is ordered. The term AB ordering was introduced in reference 3.
8. J. B. Sokoloff, Physical Review (in press).

# CURRENT UNDERSTANDING OF LOW TEMPERATURE PHASE TRANSITION OF MAGNETITE, PARTICULARLY IN RELATION TO THE BEHAVIOR OF MAGNETOCRYSTALLINE ANISOTROPY

Sōshin Chikazumi
Institute for Solid State Physics
University of Tokyo
Roppongi, Tokyo, Japan

## ABSTRACT

Recent experiments have revised old concepts of the low temperature phase of magnetite, such as the Verwey order scheme, orthorhombic crystal symmetry etc. Neutron diffraction has revealed that doubly degenerate $\Delta_5$ phonon modes tend to condense towards a point 0.3 K below the transition point, $T_V$. At $T < T_V$, however, one of the degenerate modes, $\Delta_5^{(1)}$, which has ionic shifts mainly parallel to the monoclinic $b$-axis, appears. It seems that there is no ionic ordering (charge ordering) of the wave-length of $2c$, associated with the $\Delta_5^{(1)}$ mode.

Below $T_V$, the crystal symmetry is monoclinic, and its single phase was realized by the application of a magnetic field parallel to [001] and at the same time by squeezing the crystal about [$\bar{1}$11] while cooling through $T_V$. The magnetocrystalline anisotropy was measured on a single-phase specimen by using a computerized fully-automatic torque magnetometer. The anisotropy energy is expressed by

$$E_a = K_a\alpha_a^2 + K_{aa}\alpha_a^4 + K_b\alpha_b^2 + K_{bb}\alpha_b^4 + K_{ab}\alpha_a^2\alpha_b^2 - K_u\alpha_{\bar{1}01}^2$$

where $\alpha_a$, $\alpha_b$ and $\alpha_{\bar{1}01}$ are direction cosines of the angle of spin from monoclinic $a$- and $b$-axes and [$\bar{1}$01], respectively. It was found that $K_a$, $K_b$ and $K_u$ exhibit temperature dependences of the activation-type with activation energies of about 0.02 eV.

## INTRODUCTION

Magnetite, or $Fe_3O_4$, is a typical ferrimagnetic substance. The crystal structure at room temperature is the inverse spinel (cubic), in which one $Fe^{3+}$ ion in a chemical formula occupies a tetrahedral or an A site, while other $Fe^{3+}$ and $Fe^{2+}$ ions occupy two octahedral or B sites. Unusually high conductivity of this oxide has been considered to be due to the hopping of electrons between these $Fe^{2+}$ and $Fe^{3+}$ ions on B sites. This material undergoes a phase transition at a low temperature from cubic to a crystal form of lower symmetry. At the same time, the electrical conductivity drops tremendously as this hopping ceases. Verwey[1] proposed an ordering scheme of $Fe^{2+}$ and $Fe^{3+}$ ions on B sites, in which $Fe^{2+}$ layers and $Fe^{3+}$ layers laminate alternately along the $c$-axis. It was believed that this ordering scheme was verified by the observation of the 0 0 2 magnetic scattering of neutrons by Hamilton[2].

Recently, however, many experiments disproved this old concept, so that we are obliged to construct a new charge ordering scheme. Moreover, it seems that this charge ordering is coupled with some kinds of phonon modes.

The purpose of this paper is to describe recent experimental data concerning this phase transition, and to try to understand the nature of this phenomenon. Magnetocrystalline anisotropy was measured for a specimen squeezed about one of the <111> axes and also cooled in a magnetic field along one of the <100> axes to remove all kinds of twin structures. The characteristic temperature change of the magnetocrystalline anisotropy suggests the presence of some charge ordering on B sites.

## QUALITY AND TWINNING OF SPECIMEN

Occurrence of the low temperature phase transition in $Fe_3O_4$ is most sensitively detected by measuring a discontinuous change in electrical conductivity (Fig. 1)[3]. The conductivity just above and just below the transition point, or Verwey temperature ($T_V$), is plotted as a function of transition temperature in Fig. 2[3]. There is a tendency that the higher the transition temperature, the larger the discontinuous change in conductivity. In the best specimen which undergoes transition at 124.7 K the conductivity changes[5] by a factor of 230.

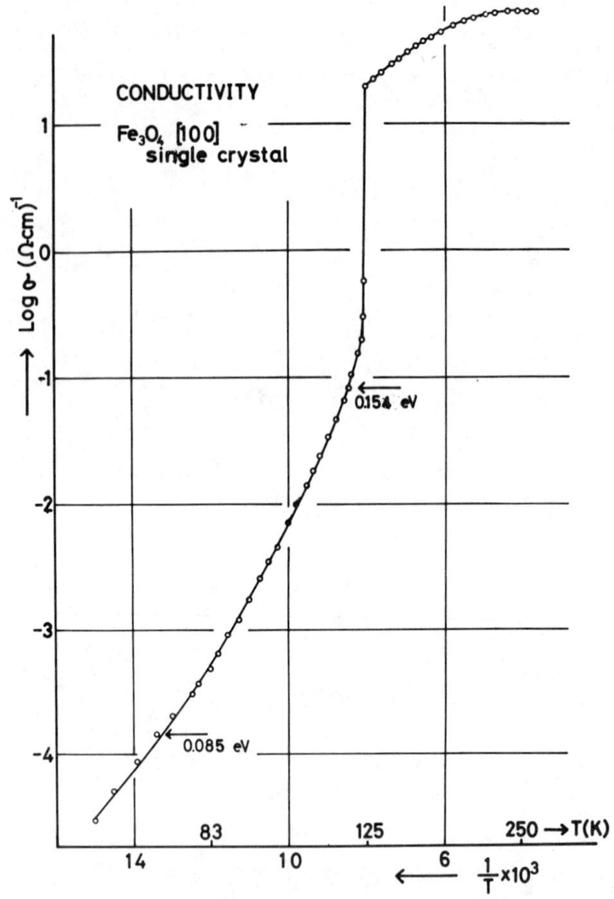


Fig. 1 Temperature dependence of conductivity for single crystal of $Fe_3O_4$ measured along the cubic [100]. No field applied during the process of cooling and measurement. (Ref. 3)

These specimens are all single crystals which were grown from the melt in Pt crucibles in the atmosphere of $CO_2$, using the raw materials, 99.9 to 99.99 % pure[3]. According to the chemical analysis[4], the content of vacancies in B sites is less than 0.7 %, and the content of Pt is less than 0.08 %. It seems that the quality of specimens is mainly determined by residual stresses in the crystal, rather than by the purity of the specimen.

Previously, the crystal structure below $T_V$ was believed to be orthorhombic. It is now obvious from various experiments that the crystal deformation is nearly rhombohedral.[5] Usually such a rhombohedrally deformed crystal forms a twin structure[6], as shown in Fig. 3, with the unique $c$-axis (magnetically easiest axis), almost perpendicular to the twin boundaries. Thus the crystal symmetry may be monoclinic. The tilt of the $c$-axis from the original cubic [001] is 0.23°.

In order to avoid such a twin structure, a single crystal of cylindrical form with its axis parallel to cubic [$\bar{1}$11], is squeezed about its axis by the thermal contraction of several separated aluminium rings (Fig. 3) during cooling through $T_V$. At the same time, a magnetic field is applied parallel to one of the cubic <100>, in order to establish the $c$-axis. It was confirmed by neutron diffraction[7] that this procedure can really produce a single monoclinic phase. Such single-phase specimens have been used in our neutron diffraction, magnetic anisotropy, magnetization and conductivity measurements.

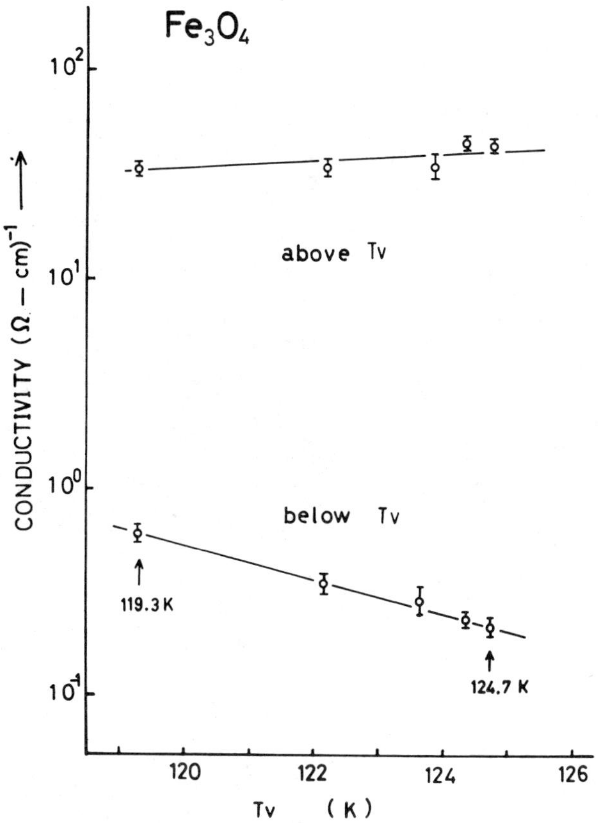


Fig. 2 Conductivity of $Fe_3O_4$ just above and just below the transition temperature, $T_v$, for several specimens as a function of $T_v$. (Ref. 3)

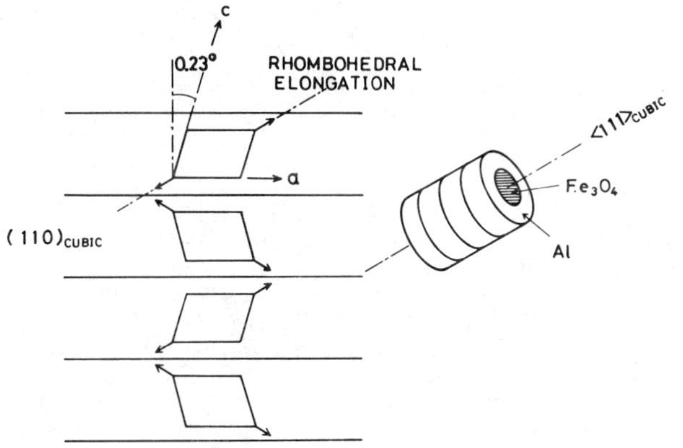


Fig 3 Twin structure of $Fe_3O_4$ below $T_v$ caused by rhombohedral elongation along the cubic <111>, and the illustration of a squeezed crystal in which a complete phase was obtained.

## NEUTRON AND ELECTRON DIFFRACTION EXPERIMENTS

Discovery of the $h\ k\ \ell+\frac{1}{2}$ spots by electron[8] and neutron[9] diffraction experiments have revealed the presence of the complexity in the crystal structure of low temperature phase. Moreover, a detailed investigation[5] of the elastic scattering of neutrons with 0 0 2 and 2 0 0 cubic indices has revealed that the finite intensities reported so far for these reflections[2,9] are all caused by simultaneous reflection (or double scattering), so that these lines are all non-existent. This fact disproved the Verwey order[1] as well as the C-C model[10] which was constructed by modifying the Verwey order. This means that even if there exists some ordering arrangement of $Fe^{2+}$ and $Fe^{3+}$ ions on B sites, the average numbers of $Fe^{2+}$ ions and $Fe^{3+}$ ions on the same cubic {100} plane must be equal.

The appearance of extra spots with $h\ k\ \ell+\frac{1}{2}$ type indices signifies that the unit cell is doubled along the $c$-axis. Based on the fact that the intensities of these extra spots increase with an increase of scattering angle, Samuelsen et al.[9] suggested that such extra spots are caused by ionic shifts rather than the ordered arrangement of $Fe^{2+}$ and $Fe^{3+}$ ions.

In recent neutron diffraction experiments, Fujii et al.[11] observed sharp critical scatterings at $4\ 0\ \frac{1}{2}$ and $8\ 0\ \frac{1}{2}$ etc. just above $T_v$. The intensity of these lines starts to emerge several degrees above $T_v$ and the reciprocal of the intensity tends to vanish at a point 0.3 K below $T_v$. Y. Yamada[12] interpreted this critical phenomenon in terms of the $\Delta_5$ phonon mode, in which ions oscillate mainly in the $a$-$b$ plane with the wave-length of $2c$ ($c = a$ in the cubic phase).

Detailed investigations[5] have been made on $2\ 0\ \ell+\frac{1}{2}$ type ($\ell = 0,1,2,3,4$) neutron reflections from the low temperature phase. The magnetic intensities were separated by applying a weak magnetic field of 1.1 kOe perpendicular to the $c$-axis, and by observing the change in scattering intensity. The result shows that both the nuclear intensity, $F_N^2$, and the magnetic intensity normalized by magnetic form factor squared, $F_M^2/f_{mag}^2$, increase with an increase of scattering vector. None of the postulated models of charge ordering with a wave-length $2c$ explained this fact. The impression is that both $F_N^2$ and $F_M^2/f_{mag}^2$ are caused simply by ionic shift in the $\Delta_5$ mode. If there is any ordered charge arrangement, it would have a cell length of $c$ rather than $2c$, as in the case of the $\Delta_5$ mode.

Recent study of neutron scattering[13] of $h\ h\ \ell+\frac{1}{2}$ type has revealed that only one $\Delta_5$ mode ($\Delta_5^{(1)}$ mode) out of two degenerate $\Delta_5$ modes ($\Delta_5^{(1)}$ and $\Delta_5^{(2)}$ modes) is condensed at $T < T_v$. If we define the monoclinic axes as shown in Fig. 4, the condensed $\Delta_5^{(1)}$ mode has the ionic shifts parallel to the $b$-axis on every other ($c$-plane) layer. The $\Delta_5^{(2)}$ mode with the ionic shifts parallel to the $a$-axis is nonexistent.

Besides the critical scattering due to $\Delta_5$ phonon modes, diffuse streaks were observed at 4 0 0 type reciprocal lattice points by electron[14] and neutron scattering[11]. These streaks appear in a temperature range much wider than the critical scattering. Although their origin has not yet been clarified, it seems to be caused by some kind of ionic oscillation in the lattice. This behavior may be closely related to the softening of the $c_{44}$ elastic modulus observed for $Fe_3O_4$ at $T > T_v$[15,16], as will be discussed later.

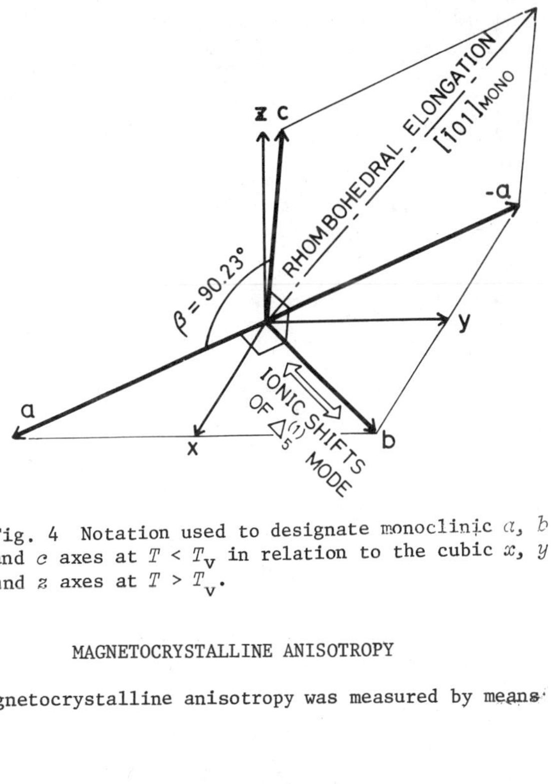


Fig. 4 Notation used to designate monoclinic $a$, $b$ and $c$ axes at $T < T_v$ in relation to the cubic $x$, $y$ and $z$ axes at $T > T_v$.

## MAGNETOCRYSTALLINE ANISOTROPY

Magnetocrystalline anisotropy was measured by means

of a computer-controlled automatic torque magnetometer[17] in which the specimen can be rotated stereographically after establishing the $c$-axis by cooling in a magnetic field through $T_V$. At the same time, the specimen is squeezed with aluminum rings (Cf. Fig. 3), so as to make the crystal of a single-phase. The shape effect was corrected by computer, referring to the anisotropy measurement at $T > T_V$.

The magnetocrystalline anisotropy per unit volume is expressed by the formula

$$E_a = K_a\alpha_a^2 + K_{aa}\alpha_a^4 + K_b\alpha_b^2 + K_{bb}\alpha_b^4 + K_{ab}\alpha_a^2\alpha_b^2 - K_u\alpha_{\bar{1}01}^2 , \qquad (1)$$

where $\alpha_a$, $\alpha_b$, and $\alpha_{\bar{1}01}$ are the direction cosines of the spontaneous magnetization, referring to the monoclinic $a$-, $b$-axes and [$\bar{1}$01] which is the longest cubic <111>, respectively (Cf. Fig. 4). This formula is the same as that first given by Calhoun[18], except for the last term which arises from the rhombohedral deformation of the crystal and the associated ionic arrangement. All $K$'s are the anisotropy constants which take the following values in $10^5$ erg/cc at 4.2 K.[19]

$$K_a = 25.53,\ K_b = 3.66,\ K_u = 2.13,\ K_{aa} = 1.76,\ K_{bb} = 2.42,\ K_{ab} = 7.00 . \qquad (2)$$

A large value of $K_a$ makes the $c$-axis easiest, and the $a$-axis hardest. Although the $K_u$ value is rather small, its temperature change is as large as those of $K_a$ and $K_b$ as shown in Fig. 5. The temperature dependences of these constants were carefully measured by rotating the spontaneous magnetization in a range of angle about the $c$-axis within which occurrence of any twinning was avoided. It is interesting to note that the temperature

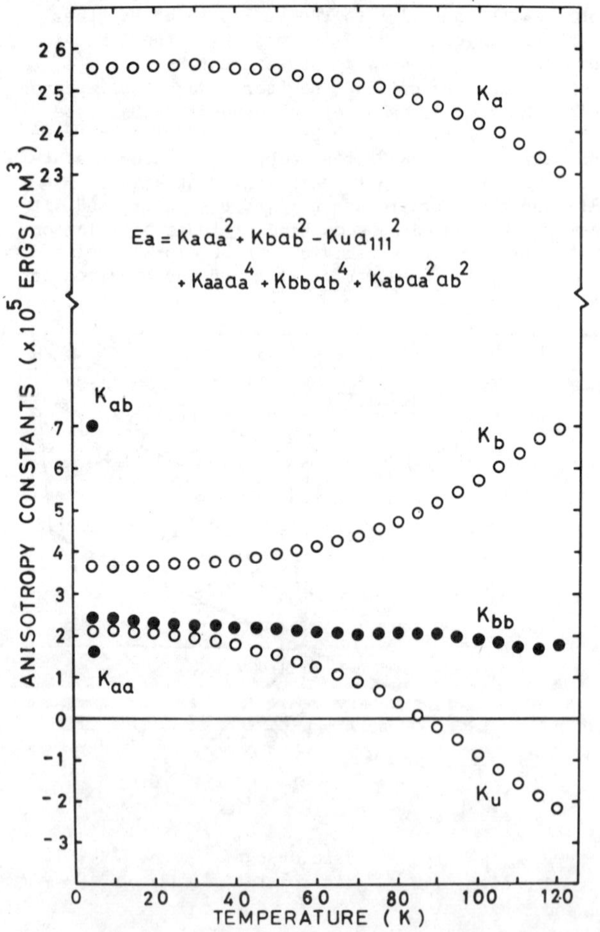


Fig. 5 Temperature dependence of anisotropy constants $K_a$, $K_b$ and $K_u$ (Cf. Eq. (1)) for $Fe_3O_4$ at $T < T_V$. (Ref. 19)

dependence of $K_a$, $K_b$ and $K_u$ can be expressed approximately by the formula

$$K = K_0 + \Delta K\, e^{-\frac{Q}{kT}} , \qquad (3)$$

where $K_0$ is the value at 0 K, $\Delta K$ is approximately $30 \times 10^5$ erg/cc for all these $K$'s except for signs (− for $K_a$ and $K_u$; + for $K_b$), and $Q \simeq 0.020 \sim 0.025$ eV.

The ratios of $K_{aa}$, $K_{bb}$ and $K_{ab}$ given in Eq. (1) are very close to 3 : 3 : 10 which is calculated by a coordinate transformation from cubic to monoclinic by assuming the $K_1$ term in the cubic anisotropy is preserved.

Fig. 6 shows the temperature dependence of $K_1$ at $T > T_V$, together with the $K_1$ values which are approximated by averaging the values of $K_{aa}$, $K_{bb}$ and $K_{ab}$ at $T < T_V$ by using the relationship

$$K_1 = \frac{1}{3}\left(-\frac{4}{3}K_{aa} - \frac{4}{3}K_{bb} - \frac{2}{5}K_{ab}\right) . \qquad (4)$$

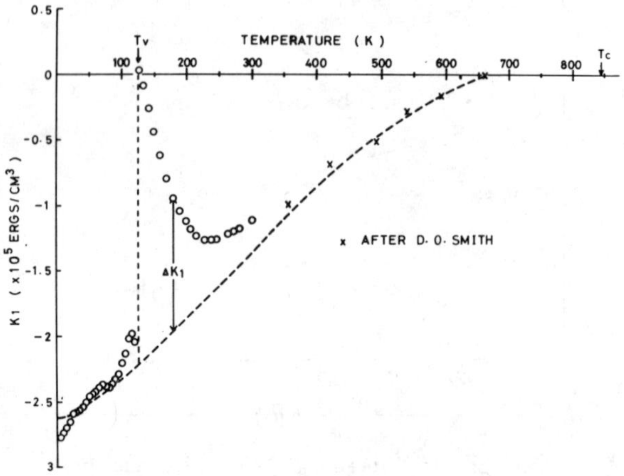


Fig. 6 Temperature dependence of cubic anisotropy constants $K_1$ for $Fe_3O_4$ (Ref. 19). The data for $T > 300$ K are taken from Ref. 20. Data for $T < T_V$ were calculated from $K_{aa}$, $K_{bb}$ and $K_{ab}$ by using Eq. (4).

It is interesting to note that the $K_1$-$T$ curve for $T < T_V$ is in good agreement with the smooth extrapolation of the curve for $T > 300$ K. The deviation of $K_1$ from this basic curve (broken curve in Fig. 6), denoted by $\Delta K_1$, increases with a decrease of temperature towards $T_V$. Similar, but much more exaggerated, behavior was observed for Co-substituted magnetite by Bickford et al.[21], and interpreted by Slonczewski[22] in terms of one-ion anisotropy of $Co^{2+}$ ions on B sites. We found that the temperature dependence of $\Delta K_1$ in Fig. 6 can be well reproduced by the Slonczewski mechanism by assuming the presence of $Co^{2+}$-like anisotropic ions, such as $Fe^{1+}$ or $Fe^{2+}$ ions with a doublet as the ground states caused by a lattice distortion or some other reasons. One difficulty is that the population of such anisotropic ions must be only about 0.012 of the total number of $Fe^{2+}$ ions.

Characteristic temperature dependence of $K_a$, $K_b$ and $K_u$, as shown in Fig. 5, can also be well-explained in terms of the Slonczewski mechanism by assuming a certain distribution of anisotropic ions among four kinds of B sites. That is $N_1 = 0.011N_{2+}$, $N_2 = 0$, $N_3 = N_4 = 0.019N_{2+}$, where $N_{2+}$ is the total number of $Fe^{2+}$ ions and $N_1$, $N_2$, $N_3$ and $N_4$ refer to the B site 1, 2, 3 and 4 indicated in Fig. 7. The site 1 has the trigonal axis parallel to the monoclinic [$\bar{1}$01] which is the longest cube diagonal. Thus the symmetry of the assumed ionic distribution conforms with that of the lattice deformation. Mystery, however, still remains concerning the small populations of anisotropic ions as also indicated by $\Delta K_1$ at $T > T_V$.

Even after subtracting the effect of $K_1$ and $\Delta K$ in Eq. (3) from the experimental values of $K_a$, $K_b$ and $K_u$ given in Eq. (2), there still remain the following finite values ( $10^5$ erg/cc ) at 0 K:

$$K_{a0} = 6.58,\quad K_{b0} = 27.58,\quad K_{u0} = -19.49. \qquad (5)$$

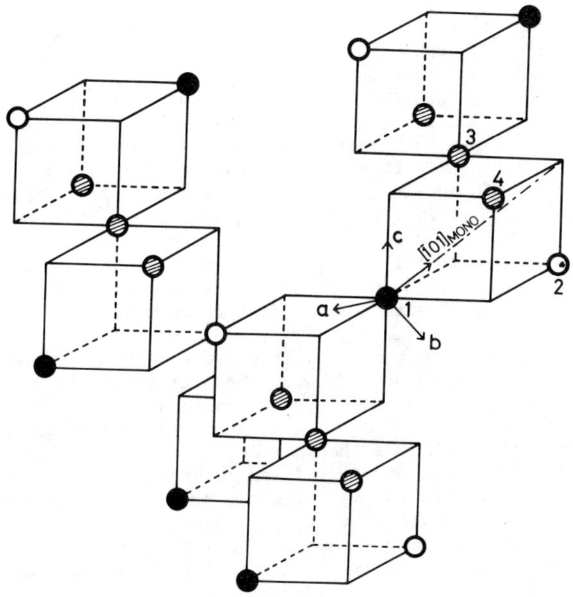


Fig. 7 Distribution of anisotropic ions among four kinds of B sites 1, 2, 3 and 4. The site 1 has the trigonal axis parallel to the monoclinic [$\bar{1}$01], which is the longest cube diagonal. The axes $a$, $b$ and $c$ refer to the monoclinic lattice.

Since these values should include the magnetoelastic energy caused by a large spontaneous deformation of the monoclinic lattice, it is rather hard to interprete the values given in Eq. (5), before knowing the values of magnetostriction constants.

OTHER DATA FOR SINGLE-PHASE CRYSTALS

Saturation magnetization was measured[23] as a function of temperature for a single-phase specimen, the

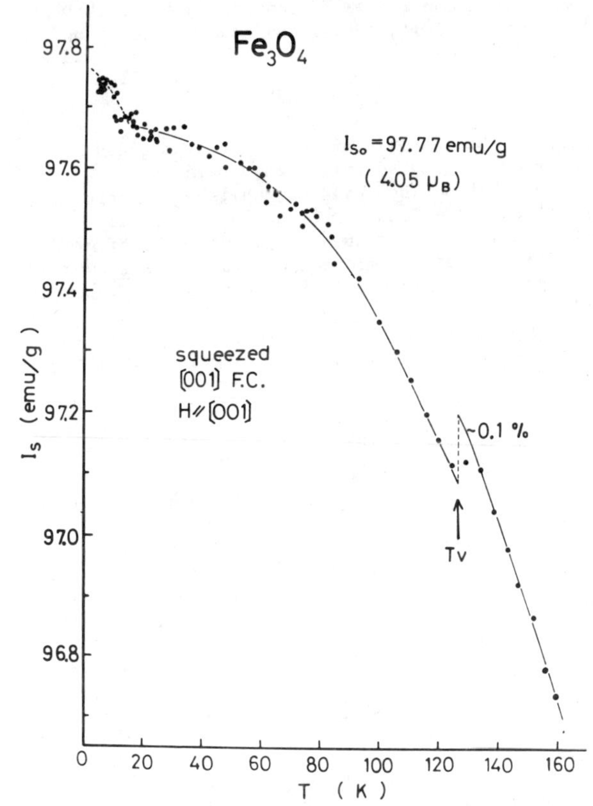


Fig. 8 Saturation magnetization as a function of temperature measured for single-phase magnetite. (Ref. 23)

result of which is shown in Fig. 8. In this case the field was applied parallel to the cubic [001] which slightly deviates from the easiest axis because of the presence of $K_u$. The saturation value $I_s$ was calculated by extrapolating the field, $H$, to infinity by using the relationship.

$$I = I_s\ (1 - \frac{b}{H^2}) \quad . \tag{6}$$

It is interesting to note that the saturation magnetization drops about 0.1 % at $T_V$ during the course of cooling, even after correcting the effect of the magnetocrystalline anisotropy. One more striking feature is that saturation magnetization increases about 0.1 % below 15 K. A corresponding anomaly is also found in $K_1$ - $T$ curve below 15 K (Cf. Fig. 6). The reasons for these anomalous changes have not yet been clarified.

The anisotropy of the electrical conductivity was measured[3] for the specimen in which orthorhombic twins were removed by the well-known field cooling technique[18], but still contains rhombohedral twins. As seen in Fig. 9, just below $T_V$ the $c$-axis is most conductive, while the $a$-axis is least conductive. Below about 60 K, however, this order is reversed, so that the $b$-axis becomes most conductive, while the $c$-axis becomes least

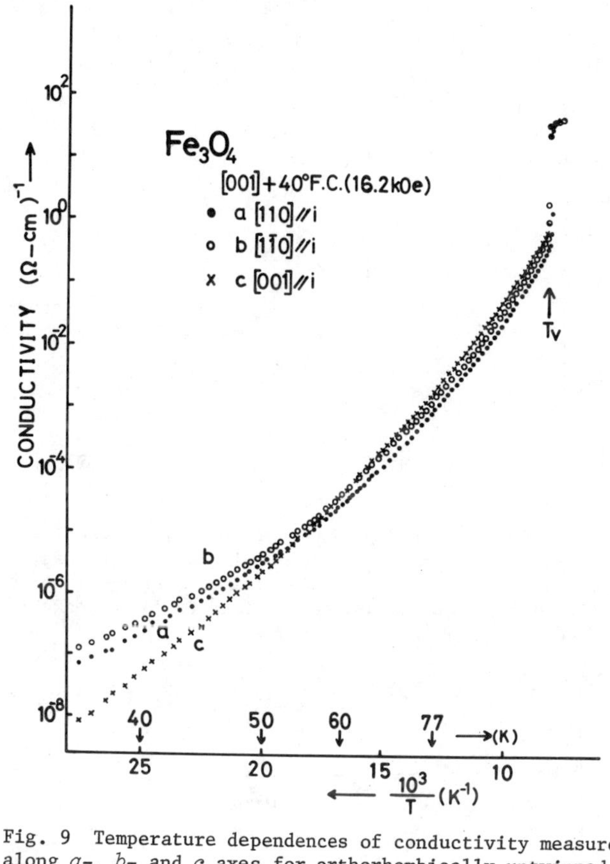


Fig. 9 Temperature dependences of conductivity measured along $a$-, $b$- and $c$ axes for orthorhombically untwinned, but rhombohedrally twinned crystal of $Fe_3O_4$. (Ref. 3)

conductive. For the completely single-phase specimen which was squeezed by insulated aluminum rings, the conductivity is measurable only along the squeezed axis, i.e. monoclinic [$\bar{1}$01] or cubic [$\bar{1}$11] (Cf. Fig. 4). It was found that in the whole temperature range below $T_V$, the conductivity measured parallel to this specified [$\bar{1}$11] was less conductive by a factor of 1.3 than that for the average <111> measured for the completely twinned crystal.

DISCUSSION

It was first observed by Moran et al.[15] that the elastic modulus $c_{44}$ partially softens during cooling towards $T_V$. Kino[16] measured the temperature dependence of $c_{44}$ for our specimen which exhibits a discontinuous increase at $T_V$ as shown in Fig. 10. If we define the softening part of $c_{44}$, i.e. $\Delta c_{44}$ as the difference from the value of $c_{44}$ just below $T_V$, we found that the reciprocal of $\Delta c_{44}$ tends to vanish at 81 K.

It is interesting to note that this temperature dependence of $\Delta c_{44}$ is quite similar to that of $\Delta K_1$ as shown in Fig. 11. As mentioned above, $\Delta K_1$ is well-explained by the Slonczewski mechanism which gives a more complicated formula for $\Delta K_1$ - $T$ curve, so that no significant meaning can be given for the extrapolated point, 81 K. However, the similarity of the temperature dependence of $\Delta c_{44}$ and $\Delta K_1$ suggests some close connection between the two quantities. As mentioned before, diffuse streaks show a similar temperature dependence. If $\Delta K_1$ is to be explained in terms of an anisotropic ion, the softening of $c_{44}$ and accordingly the final condensed rhombohedral deformation might be caused by the orbital deformation of anisotropic ions. The condensation of this rhombohedral distortion seems to be suddenly triggered by the condensation of the $\Delta_5{}^{(1)}$ phonon mode. The mechanism of this process is still unknown.

Neutron diffraction from single-phase $Fe_3O_4$ is now being performed. We hope that this result gives the final solution on the ordering scheme below $T_V$.

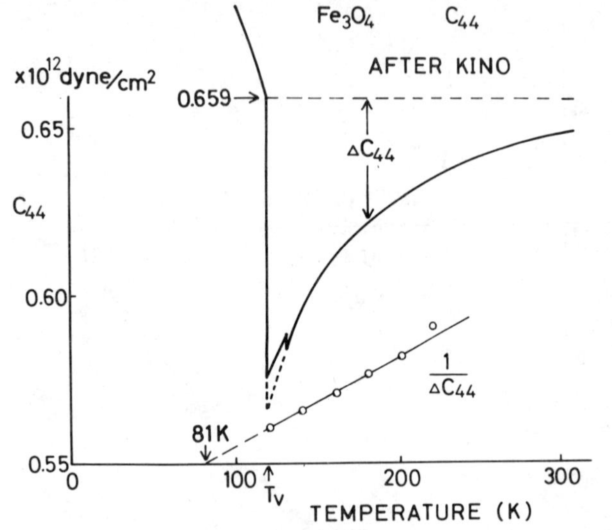


Fig. 10 Temperature dependence of $c_{44}$ elastic modulus of $Fe_3O_4$ measured by Kino[16] using ultrasonic $c_{44}$ shear wave. $\Delta c_{44}$ is defined by a change in $c_{44}$ from the value just below $T_V$. The reciprocal of $\Delta c_{44}$ tends to vanish at 81 K.

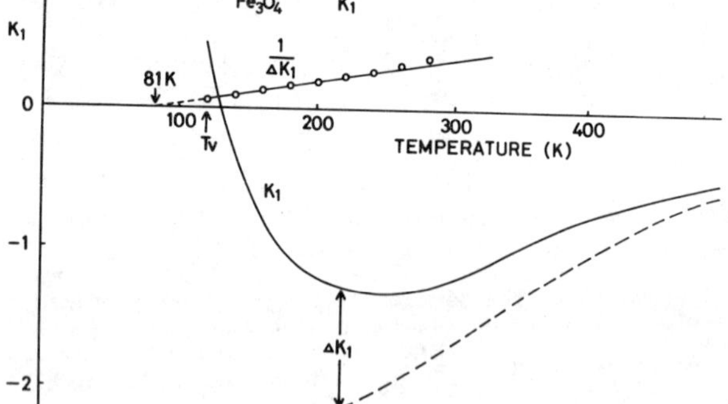


Fig. 11 Temperature dependence of $\Delta K_1$ above $T_V$ for $Fe_3O_4$. Data are taken from Refs. 20 and 21. The $1/\Delta K_1$- $T$ curve extrapolates to 81 K.

## SUMMARY

Concerning the low temperature phase transition of magnetite, $Fe_3O_4$, the following facts have been clarified:

1. The transition temperature or Verwey temperature $T_V$, ranges from 119 to 125 K, depending on the quality of the specimen. The residual stresses seem to be the main factor to influence $T_V$.
2. At $T < T_V$, no finite intensities were assigned for the 0 0 2 and 2 0 0 neutron scatterings. This fact disproved the Verwey ordering scheme as well as the C-C model[10]. If there exists some ionic ordering on B sites, it should have equal numbers of $Fe^{2+}$ and $Fe^{3+}$ on any cubic {100} planes.
3. At $T < T_V$, the deformation of the crystal is nearly rhombohedral. The crystal symmetry is monoclinic with the $c$-axis (magnetically easiest axis) making an angle of 0.23° with the original cubic $z$-axis ($a > c > b$). (Cf. Fig. 4)
4. At $T < T_V$, neutron diffraction revealed the presence of the $\Delta_5{}^{(1)}$ phonon mode with the periodic ionic shift mainly parallel to the monoclinic $b$-axis with the wave-length of $2c$. No charge ordering seems to be associated with this phonon mode.
5. At $T < T_V$, magnetocrystalline anisotropy is expressed by the formula with the monoclinic symmetry. The anisotropy constants $K_a$, $K_b$ and $K_u$ exhibit a characteristic temperature dependence. The constants $K_{aa}$, $K_{bb}$ and $K_{ab}$ are well expressed in terms of cubic $K_1$.
6. At $T_V$, the spontaneous magnetization drops about 0.1 %, and this drop is recovered below 15 K.
7. At $T < T_V$, the electrical conductivity becomes anisotropic. Namely the $c$-axis is most conductive just below $T_V$, whereas it becomes least conductive below 60 K. The monoclinic [$\bar{1}$01] or the longest cubic <111> seems to be least conductive among four cubic <111>.
8. In a narrow temperature range above $T_V$, neutron critical scatterings corresponding to the $\Delta_5$ modes were observed.
9. In a wider temperature range at $T > T_V$, characteristic and similar temperature dependences were observed for various physical quantities such as magnetocrystalline anisotropy, the $c_{44}$ elastic modulus and diffuse electron and neutron scattering.

In conclusion, the condensation of an ordered arrangement of Fe ions which include some anisotropic ions seems to be coupled with some phonon modes through orbital strain, and its actual condensation seems to be triggered by the condensation of the $\Delta_5$ phonon mode.


## ACKNOWLEDGMENTS

Most of the magnetic and electric data cited in this paper were measured by colleagues in my laboratory.[3,19,23] They will be published in more details elsewhere. Thanks are due to Dr. Y. Kino of Matsushita Research Institute Tokyo Inc., who offered his unpublished data of elastic modulus. The author is much indebted to Professor A. Yoshimori of the Institute for Solid State Physics, University of Tokyo, and Professor M. Tachiki of Institute for Iron, Steel and Other Metals, Tohoku University, for their valuable discussion.


## REFERENCES


1. E. J. W. Verwey and P. W. Haayman: Physica 8 979 (1941); E. J. W. Verwey, P. W. Haayman and F. C. Romeijn: J. chem. Phys. 15 181 (1947).
2. W. C. Hamilton: Phys. Rev. 110 1050 (1958).
3. M. Matsui, S. Todo and S. Chikazumi: unpublished.
4. We are indebted to Professor T. Katsura of Tokyo Institute of Technology for this chemical analysis.
5. G. Shirane, S. Chikazumi, J. Akimitsu, K. Chiba, M. Matsui and Y. Fujii: J. Phys. Soc. Japan 39 947 (1975).
6. S. Chikazumi, K, Chiba, K, Suzuki and T. Yamada: Ferrites (Proc. Int. Conf.). (University of Tokyo Press, Tokyo,1971) p. 595.
7. Private communication from G. Shirane.
8. T. Yamada, K. Suzuki and S. Chikazumi: Appl. Phys. Letters 13 172 (1968).
9. E. J. Samuelsen, E. J. Bleeker, L. Dobrzynski and

T. Riste: J. appl. Phys. 39 1114 (1968), Kjeller Rep. (Inst. for Atomenergi, Kjeller, Norway, 1967) KR-122.
10. S. Chikazumi, K. Chiba, M. Matsui, J. Akimitsu and S. Todo: Proc. Int. Conf. Mag., Moscow (Publishing House ≪NAUKA≫, Moscow, 1974) Vol. 1-(1) 137.
11. Y. Fujii, G. Shirane and Y. Yamada: Phys. Rev. B11 2036 (1975).
12. Y. Yamada: AIP Conf. Proc. No.24 79 (1975).
13. M. Iizumi and G. Shirane: Solid State Comm. 17 433 (1975).
14. K. Chiba, K. Suzuki and S. Chikazumi: J. Phys. Soc. Japan 39 839 (1975).
15. T. J. Moran and B. Lüthi: Phys. Rev. 187 710 (1969).
16. Y. Kino: unpublished.
17. K. Abe and S. Chikazumi: to be published in Japan. J. Appl. Phys.
18. B. A. Calhoun: Phys. Rev. 94 1577 (1954).
19. K. Abe, Y. Miyamoto and S. Chikazumi: unpublished.
20. D. O. Smith: Phys. Rev. 102 959 (1956).
21. L. R. Bickford, Jr., M. S. Brownlow and R. F. Penoyer: Proc. Instrum. Elec. Eng. 104B 238 (1957); R. F. Penoyer and L. R. Bickford, Jr.: Phys. Rev. 108 217 (1957).
22. J. C. Slonczewski: J. Appl. Phys. 29 448 (1958); Phys. Rev. 110 1341 (1958); J. Appl. Phys. 32 253S (1961).
23. M. Matsui, S. Todo and S. Chikazumi: unpublished.

## INDUCED MOMENT MAGNETISM IN $TbAsO_4$: CONSTANT COUPLING APPROXIMATION

J. W. McPherson† and Yung-Li Wang
Department of Physics, Florida State University, Tallahassee, Florida 32306

ABSTRACT

The constant-coupling approximation (CCA) is applied to the Ising model with a transverse field which is then used to study the induced-moment magnetism in $TbAsO_4$. We first show that in contrast to previous work,[1] the CCA is valid throughout the entire temperature range; there exist no anomalies at low temperatures. A comparison of the phase boundaries in the Δ(transverse field)-T plane as predicted by the molecular-field approximation (MFA), CCA, and the high temperature series expansion (HTE), shows that the CCA predicts a transition temperature which is much closer to the HTE value for all Δ. The temperature dependence of the sublattice magnetization is also shown for the CCA and the MFA. A fitting to the experimentally observed energy level splitting is shown for $TbAsO_4$. By using a single parameter, the exchange integral, a good agreement is found between the theory and the experiment for the whole range of temperature.

*Supported in part by National Science Foundation, under Grant No. GH-40174.

†Present address: Department of Physics, University of North Carolina at Greensboro, Greensboro, North Carolina 27412.

[1]R. Blinc and S. Svetina, Phys. Rev. 147, 423 (1966).

## LINEAR AND BILINEAR MAGNETOELECTRIC EFFECTS IN MAGNETICALLY BIASED MAGNETITE

G. T. Rado and J. M. Ferrari
Naval Research Laboratory, Washington, D.C. 20375

ABSTRACT

In a synthetic magnetite ($Fe_3O_4$) crystal at 4.2°K the macroscopic anisotropy coefficients depend[1] on an external electric field $\vec{E}$. We now show that this dependence gives rise to linear as well as to two kinds of bilinear magnetoelectric (ME) effects provided the crystal is biased by means of an external static magnetic field $\vec{H}_0$. Upon expanding the previously proposed[1] free energy density $\tilde{F}$ to third order in the perturbing fields $\vec{E}$ and $\vec{h} = \vec{H}-\vec{H}_0$, we obtain terms (among others) of the form $-P_{oi}E_i$, $-\alpha_{ij}E_ih_j$, $-\frac{1}{2}\beta_{ijk}E_ih_jh_k$ and $-\frac{1}{2}\gamma_{ijk}E_iE_jh_k$, where i,j,k denote a,b or c. By requiring the magnetization $\vec{M}$ to be in equilibrium in each order of the perturbation, we calculate, without adjustable parameters, the dependences on $\vec{H}_0$ of $P_{oi}$ (obtained previously[1]) and of the ME susceptibilities $\alpha_{ij}$, $\beta_{ijk}$ and $\gamma_{ijk}$. We measure the dependence on $\vec{H}_0$ of these three susceptibilities by using 1 kHz fields for $\vec{E}$ and $\vec{h}$ and observing the 2 kHz as well as the 1 kHz parts of both $\vec{P}(=-\partial\tilde{F}/\partial\vec{E})$ and $\vec{M}(=-\partial\tilde{F}/\partial\vec{H})$. For $\vec{h}\parallel\vec{c}$ (= easy axis) and $\vec{H}_0$ oriented arbitrarily in the bc plane, the measured $\vec{H}_0$-dependence of $\alpha_{ac}$ agrees very well with the single-domain-rotation version of the theory whenever $H_0$ is sufficiently large and $\vec{H}_0$ is not parallel to $\vec{b}$. For $\vec{h}\parallel\vec{H}_0\parallel\vec{b}$ (= intermediate axis), the measured $H_{ob}$-dependences of $\alpha_{ab}$ and $\beta_{abb}$ all agree reasonably well with the two-domain-rotation version of the theory at all values of $H_{ob}$. Our measured $|\beta_{abb}|$ at $H_0$=0 is also calculable and is the largest $|\beta_{ijk}|$ ever observed in any material.

[1]G.T. Rado and J.M. Ferrari, AIP Conf. Proc. 24, 661 (1975), and a detailed paper to appear in Phys. Rev. B.

# MÖSSBAUER, NMR, AND X-RAY STUDIES AND A NEW ORDERING MODEL OF $Fe_3O_4$

S. Iida, K. Mizushima, M. Mizoguchi, J. Mada, S. Umemura, K. Nakao and J. Yoshida
Department of Physics, University of Tokyo, Bunkyo-ku, Tokyo, Japan

ABSTRACT

We propose a new ordering model of $Fe_3O_4$. Saturation magnetization measurements detected a decrease of only 0.1% at the Verwey transition in cooling. Mössbauer study gave the directions of the electric field gradient of $Fe^{2+}$ ions and NMR spin echo observation separated 8 $Fe^{3+}$ ions on A sites and 8 $Fe^{2+}$ and 8 $Fe^{3+}$ ions on B sites. X-ray diffraction confirmed the c-doubling of the unit cell. We believe that electrostatic interaction and stabilization of electron orbital of $Fe^{2+}$ ions are most important, which create a Verwey-like electron ordering with a rhombohedral distortion, being followed by a correlated buckling with the **k** vector of magnitude $\pi/c$. In these displacements the A site $Fe^{3+}$ ions play a decisive role to stabilize $d\varepsilon_{xy}$ orbital of $Fe^{2+}$-II with an important choice of the displacements, changing the crystal symmetry into magnetoelectric.

Very careful saturation magnetization measurements were made at the Verwey transition temperature on cooling and on heating for single crystals of $Fe_3O_4$ with and without squeezing. Umemura[1] detected the decrease of just 0.1% in cooling. Cylindrical compression gave results that are consistent with the crystal symmetry being rhombohedral[2,3] with an orthorhombic magnetic symmetry.[4]

Observation of Mössbauer spectra of single crystal $Fe_3O_4$ had been reported already.[5,6,7] Essential advancement of Mada's study[8] is in its asignment and in taking into account the effect of the angle between the hyperfine field $H_{hf}$ and the principal direction of the electric field gradient, EFG. One of our results is shown in Fig. 1 and Table 1. Our grouping of the signal is similar to that of Hargrove and Kündig.[5] The fact that $Fe^{2+}$ II has the largest negative FFG along [001] is regarded as a key to solving the structure.

An extra high quality large single crystal of $Fe_3O_4$ was furnished by Mr. I. Sasaki of Fuji Electrochemical Co. through Dr. M. Sugimoto of the Institute for the Physico-Chemical Research. A cylindrical rod of diameter 18 mm and length 40 mm was studied at 4.2°K with NMR spin echo technique. The results for the specimen cooled without magnetic field are shown in Figs. 2 and 3. An essential advance from the previous studies by Rubinstein et al[6] and Kovtun et al[9] is in intensity. We have observed 24 lines which are well separated.

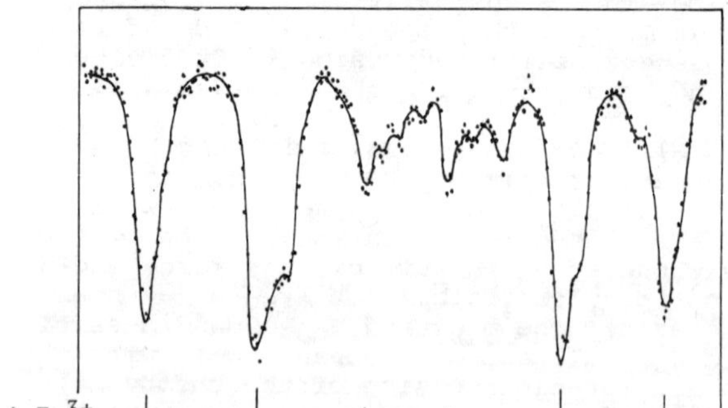

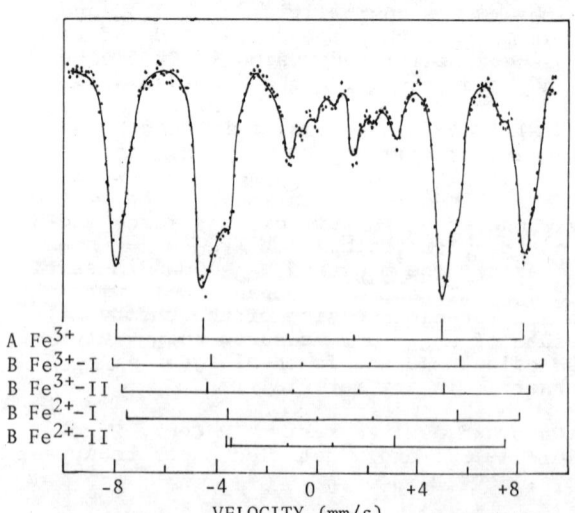

Fig. 1. An example of the Mössbauer Spectra of $Fe_3O_4$ at 78°K. Specimen is a $(1\bar{1}0)$ plate cooled in a magnetic field along [001].

| Site | | $H_{int}$(KOe) | I.S.(mm/s) | Q.S.(mm/s) | EFG Axis |
|---|---|---|---|---|---|
| Tet | $Fe^{3+}$ | 504 | 0.38 | — | — |
| Oct | $Fe^{3+}$(I) | 535 | 0.78 | 0.5 | [110] |
| | $Fe^{3+}$(II) | 511 | 0.58 | — | — |
| | $Fe^{2+}$(I) | 485 | 0.87 | 1.5 | $[1\bar{1}0]$ |
| | $Fe^{2+}$(II) | 355 | 1.10 | 2.2 | [001] |

Table 1. An example of the numerical data of Mössbauer study of $Fe_3O_4$ at 78°K, which corresponds to Fig. 1.

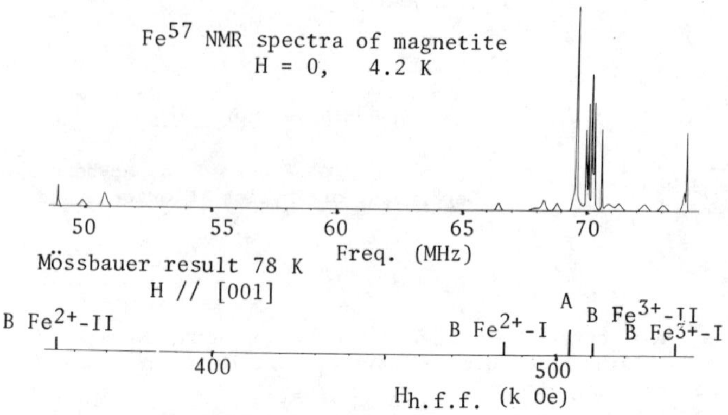


Fig. 2. NMR spectra of zero field cooled $Fe_3O_4$ at 4.2°K and an example of the Mössbauer spectrum at 78°K, shown with the same scale.

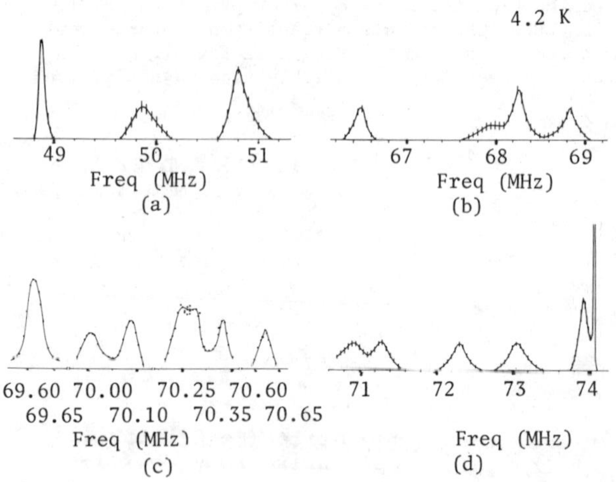


Fig. 3, Intensity comparison of the NMR spectra of Fig. 2.

The line at 50.75 MHz is a doublet and the line at 69.61 MHz are triplets and we have three overlapped singlet lines at about 60.26 MHz. As is shown in Fig. 2, the lines of Fig. 3(a), (b), (c) and (d) should correspond to four $Fe^{2+}$-II lines, four $Fe^{2+}$-I lines, eight A site $Fe^{3+}$ lines plus two $Fe^{3+}$-II lines, and two $Fe^{3+}$-II lines plus four $Fe^{3+}$-I lines.

Using a single crystal of $Fe_3O_4$, we have obtained clear signals of $\{4, -4, \pm\frac{1}{2}\}$, $\{8, 0, \pm\frac{1}{2}\}$, $\{4, 0, 4\pm\frac{1}{2}\}$ and $\{4, 4, 4\pm\frac{1}{2}\}$ lines by X- ray at 78°K.

Substantial evidence[10-12] suggests that Verwey's original model for charge ordering in $Fe_3O_4$ is incorrect. Taking into consideration our new results of Mössbauer, NMR and X-ray, we propose a new ordering model for $Fe_3O_4$ at low temperatures. Let us assume the a, b, c axes and the rhombohedral elongation along $[1\bar{1}0]$, [110],

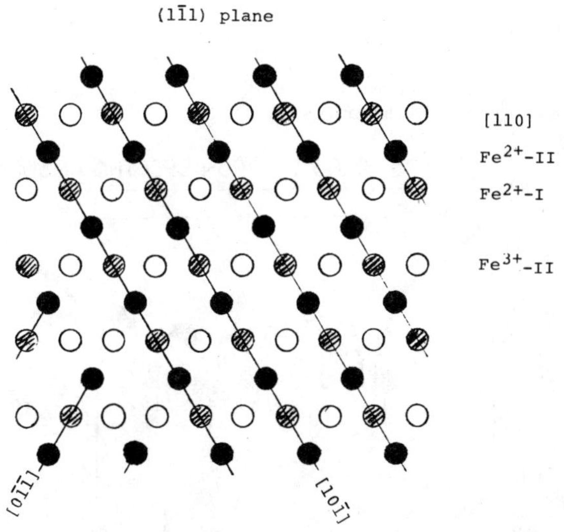


Fig. 4. Proposed electronic ordering on (1$\bar{1}$1) plane. a, b, c, axes and the elongation are assumed along [1$\bar{1}$0], [110], [001] and [$\bar{1}$11] respectively. In the lower left corner, a possible twinning is illustrated.

[001] and [$\bar{1}$11] respectively. We regard the electrostatic Coulomb interaction as most important and that part of Verwey's idea is still effective. The straight line order of $Fe^{2+}$, however, is along [10$\bar{1}$] or [011]. The choice of [10$\bar{1}$] or [011] will be important, creating a microscopic twinning and adding an additional extinction factor for the diffraction experiments.[12-14] For simplicity, however, let us assume that the choice is along [10$\bar{1}$], then the structure on a (1$\bar{1}$1) plane is as represented in Fig. 4. Here, $Fe^{2+}$ ions are classified into two groups and the $Fe^{2+}$ ions, illustrated by black circles in the figure are plainly distinguished, being well separated on ($\bar{1}$11) planes. We assume that these ions are $Fe^{2+}$-II ions, whose extra electron forms a $d\varepsilon_{xy}$ type orbital. Then the remaining $Fe^{2+}$ is $Fe^{2+}$-I and the $Fe^{3+}$ in the figure is $Fe^{3+}$-II. $Fe^{3+}$-I are located at both sides from $Fe^{2+}$-II along [1$\bar{1}$0]. We assume that, in order to further decrease the elctronic orbital energy of the extra electron and the electrostatic interaction energy of $Fe^{2+}$ ions, the crystal lattice exhibits shifts of $Fe^{3+}$ ions on A sites and then a correlated buckling of the lattice whose **k** vector is along c axis with the magnitude of $\pi/c$, as briefly shown in Fig. 5. In the figure the cations on a same (001) plane are on the same line. Now in order to stabilize the $d\varepsilon_{xy}$ orbital of $Fe^{2+}$-II further, half of the $Fe^{3+}$ ions on A sites exhibit a shift along [1$\bar{1}$0]. There is an important choice, whether these ions are A II with the shift along [$\bar{1}$10] downwards or they are A I with the shift along [1$\bar{1}$0] upwards. In the figure we have assumed the former and it triggers all the shifts shown. In the figure arrows indicate the direction of the shift in xy-plane. In this way, the crystal loses center of symmetry and it becomes magnetoelectric, as was found by Rado.et al.[15] In fig. 5, the chained line represents the line [10$\bar{1}$] in Fig. 4, and we have assumed that nodal and maximum buckling planes are the planes with $Fe^{3+}$-I and $Fe^{2+}$-II, and planes with $Fe^{3+}$-II and $Fe^{2+}$-I are at equivalent phase locations of the buckling. Then, geometrically, this buckling immediately differentiates 4 $Fe^{3+}$-I and 4 $Fe^{3+}$-II on A sites and 4 $Fe^{3+}$-I, 4 $Fe^{3+}$-II, 4 $Fe^{2+}$-I and 4 $Fe^{2+}$-II on B sites. Our model agrees with the Mössbauer results precisely up to the sign and the direction of the EFG.

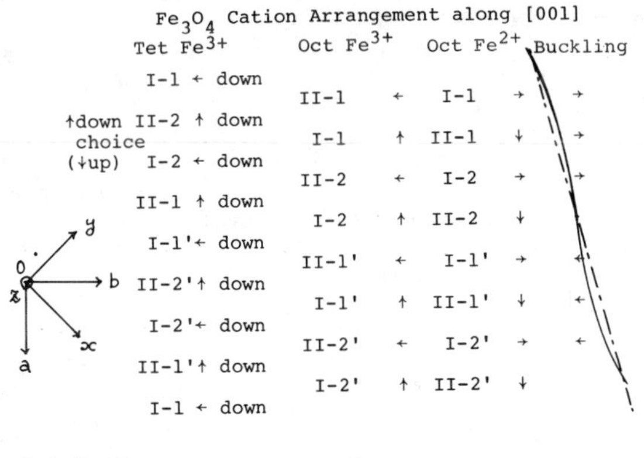


Fig. 5. Cation arrangement of ordered $Fe_3O_4$ along [001]. Arrows indicate the expected direction of the displacement of each ion. The sinsoidal curve indicates the expected buckling, whose precise directions of the displacements are not yet known.

Accurate agreement with the neutron diffraction experiments will be a problem in future, because there are so many adjustable parameters. Magnetic intensities, however, must be weak because in addition to the aforementioned twinning complexity, the shifts assumed in Fig. 5 should be decorated in such a way as to reduce the charge difference between $Fe^{2+}$ and $Fe^{3+}$.

REFERENCE

1), S. Umemura and S. Iida, J. Phys. Soc. Japan 40, No. 3 (1976)
2), H. P. Rooksby and B. T. M. Wills, Acta Cryst. 6, 565 (1953)
3), S. Iida, M. Yamamoto and S. Umemura, AIP Conf. Proc. 18, 913 (1974)
4), L. R. Bickford, Jr., Rev. Mod. Phys. 25, 75 (1953), S. C. Abrahams and B, A. Calhoun, Acta. Cryst. 8, 257 (1955)
5), R. S. Hargrove and W. Kündig, Solid State commun. 8, 303 (1970)
6), M. Rubinstein and D. W. Forester, Solid State Commun. 9, 1675 (1971)
7), A. Ito, J. de Phys. C6, 35, 325 (1974)
8), J. Mada and S Iida, J. Phys. Soc. Japan 39, 1627 (1975)
9), N. M. Kovtun and A. A. Shamyakov, Solid State Commun. 13, 1345 (1973)
10), S. Chikazumi, K. Chiba, K. Suzuki and T. Yamada, Proc. Int. Conf. Ferrites, Univ. Tokyo Press, 595 (1971)
11), Y. Yamada, AIP Con. Proc. 24, 79 (1975)
12), Y. Fujii, G. Shirane and Y. Yamada, Phys. Rev. B 11, 2036 (1975)
13), G. Shirane, S. Chikazumi, J. Akimitsu, K. Chiba, M. Matsui and Y. Fujii, J. Phys. Soc. Japan 39, 949 (1975)
14), M. Iizumi and G. Shirane, Solid State Commun. 17, 433 (1975)
15), G. T. Rado and J. M. Ferrari, AIP Conf. Proc. 24, 661 (1975)

# CRYSTAL CHEMISTRY AND ELECTRON LOCALIZATION IN Sn-DOPED $Fe_3O_4$ +

B. J. Evans, Lu San Pan and R. H. Vogel, Department of Chemistry, University of Michigan, Ann Arbor, Michigan 48104


## ABSTRACT

For $Fe_{3-x}Sn_xO_4$ at 300 K and x > 0.1, a paramagnetic quadrupole doublet pattern, in addition to the usual two magnetic patterns, is observed in the $^{57}Fe$ NGR spectra. This doublet is due to A site $Fe^{2+}$ whose concentration for $0.1 \leq x \leq 0.5$ is such that approximately equal amounts of $Fe^{2+}$ and $Fe^{3+}$ are on the B site. The 200 kG field at the $Sn^{4+}$ site at 300 K is smaller than that in $NiFe_2O_4$. At 80 K a paramagnetic quadrupole doublet is observed in the $^{119}Sn$ NGR spectrum and is due to $Sn^{2+}$.


## INTRODUCTION

Most nuclear gamma ray resonance (NGR) studies of doped $Fe_3O_4$ have been concerned with materials for which there is a decrease in the $Fe^{2+}$ content.[1] We have, therefore, synthesized Sn-doped $Fe_3O_4$, $Fe_{3-x}Sn_xO_4$ for which $Fe^{2+}$ increases with increasing x and obtained their $^{57}Fe$ and $^{119}Sn$ NGR spectra at 300 K and 80 K. Questions of particular interests are: (1) effect of a $Fe^{2+}Fe^{3+}$ ratio > 1 on the conduction mechanism and the role of $Sn^{4+}$, (2) confirmation of A site $Fe^{2+}$, hereafter noted as $Fe^{2+}(A)$, for x < 0.5, and (3) perturbation by Sn of the electron localization at the Verwey transition.

## EXPERIMENTAL

$Fe_{3-x}Sn_xO_4$ with x = 0, .05, .1, .2, and .5 were synthesized by firing mixtures of spectroscopic grade $Fe_2O_3$, Fe, and $^{119}Sn$ enriched $SnO_2$ in evacuated and sealed silica tubes at 1273 K. X-Ray diffraction analysis with a Guinier focusing camera showed the samples to be single phase and with lattice parameters for x = 0, 0.05, 0.1, and 0.2 of 0.8395 nm, 0.8411 nm, 0.8241 nm, and 0.8453 nm, respectively, in good agreement with earlier measurements.[2]

Standard transmission NGR techniques and least-squares curve fitting techniques were employed. 25 mCi $^{57}Co/Rh$ and 0.5 mCi $^{119}Sn/CaSnO_3$ sources were used and were always at the same temperature as the absorber. The lines were assumed to be lorentzian for both nuclei and the $-1/2 \rightarrow -1/2$ transition was used to obtain the relative occupancies of the A and B sites by Fe since the corresponding lines were the most well resolved.

Table I

$^{57}Fe$ NGR Parameters of $Fe_{3-x}Sn_xO_4$ at 300 K

| x | Pattern | $H_{eff}$ (kG) | $IS_{II}-IS_I$ ($mm\ s^{-1}$) | Area Ratio II/I |
|---|---|---|---|---|
| 0 | I | 493 | 0.40 | 1.80 |
| | II | 463 | | |
| 0.05 | I | 399 | 0.36 | 2.20 |
| | II | 379 | | |
| 0.1 | I | 400 | 0.31 | 2.04 |
| | II | 375 | | |
| 0.2 | I | 395 | 0.32 | 2.8 |
| | II | 366 | | |
| 0.5 | I | 389 | 0.33 | 3.2 |
| | II | 358 | | |

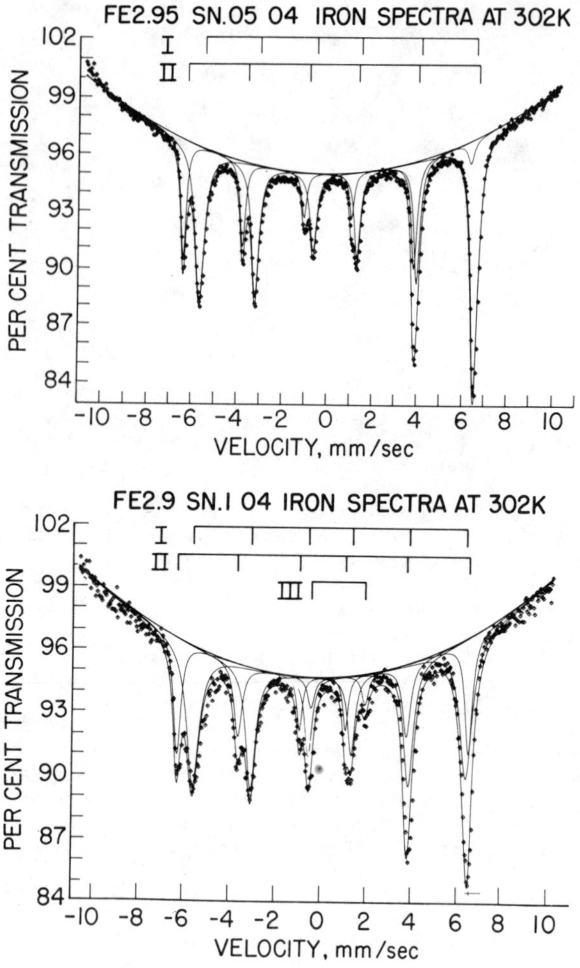


Fig. 1. $^{57}Fe$ NGR spectra of $Fe_{3-x}Sn_xO_4$, x = 0.05, and 0.1, at 302 K. Pattern III for x = 0.1 is to be noted.

## RESULTS

$^{57}Fe$ spectra at 300 K for x = 0.05 and 0.01 are given in Fig. 1 and those for $^{119}Sn$ at 300 K and 80 K for x = 0.1 are given in Fig. 2. In order to account for the $^{57}Fe$ spectra for x > 0.1, it is necessary to introduce a quadrupole doublet (Pattern III) with an isomer shift, δ, of 0.83 mm $s^{-1}$ w.r.t. Fe and a quadrupole splitting, $\Delta E_Q$, of 2.35 mm $s^{-1}$. The NGR parameters of patterns I and II (cf. Fig. 1) for all samples are given in Table I. The III/I area ratio is 0.12 for x = 0.1 and 0.37 for x = 0.2. An accurate fit of pattern III for x = 0.5 has not been possible. The Verwey transition has been determined using $^{57}Fe$ NGR and is above 100 K and rather sharp for x = 0.05. For x > 0.05, the transition commences above 120 K and is smeared over a temperature interval of at least 40 K. The $^{119}Sn$ NGR parameters of the magnetic hyperfine pattern at 300 K and 80 K for x < 0.5 are given in Table II. The absorption line at 0 mm $s^{-1}$ is believed to be due to a $SnO_2$ impurity undetected in the X-ray diffraction analysis. For x > 0.05, a quadrupole doublet (Pattern II of Fig. 2) appears in the 80 K spectra and is to be associated with the low temperature phase of $Fe_3O_4$. This pattern has a δ of 3.0 mm $s^{-1}$ w.r.t. $CaSnO_3$ and a $\Delta E_Q$ of 2.06 mm $s^{-1}$ and is, therefore, characteristic of $Sn^{2+}$. The II/I area ratio is 0.06 and 0.25 for x = 0.1 and 0.2, respectively.

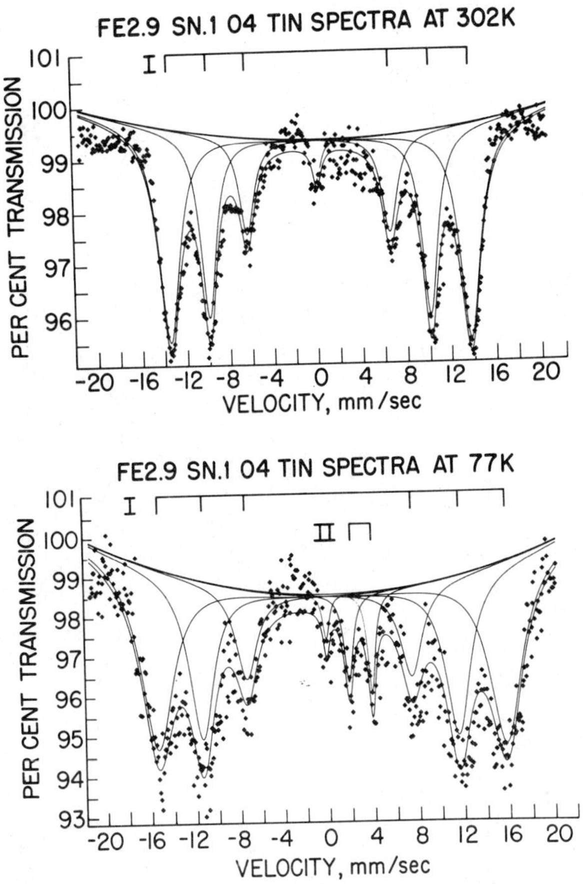


Fig. 2. $^{119}Sn$ NGR spectrum of $Fe_{2.9}Sn_{0.1}O_4$ at 302 K and 77 K (below the Verwey transition). The paramagnetic pattern II and its large isomer shift in the 77 K spectrum is to be noted.

Table II

$^{119}Sn$ NGR Parameters of $Fe_{3-x}Sn_xO_4$ at 300 K and 80 K

| x | $H_{eff}$ at 300 K (kG) | $H_{eff}$ at 80 K (kG) | Isomer shift at 80 K (mm $s^{-1}$) | Area Ratio II/I |
|---|---|---|---|---|
| 0.05 | 195 | 243 | +0.14 | ---- |
| 0.1 | 197 | 229 | +0.17 | 0.06 |
| 0.2 | 184 | 232 | +0.22 | 0.25 |

DISCUSSION

The significant findings of this study are: (1) the large II/I, apparently (A site)/(B site), area ratios of the $^{57}Fe$ spectrum, (2) the presence of paramagnetic $Fe^{2+}$ at 300 K, and (3) the existence of $Sn^{2+}$ in the low-temperature phase. If all the A and B site cations contributed to patterns I and II, respectively, the II/I area ratio should decrease with increasing x since $Sn^{4+}$ occupy the B site;[2] this is not the case. The increase in the II/I area ratio with x is due to the partial occupation of the A site by $Fe^{2+}$, with the fractional occupation increasing with increasing x[2] *and* to the fact that these $Fe^{2+}$(A) ions contribute to neither pattern I nor II. The $Fe^{2+}$(A) gives rise to pattern III. If it is assumed that all $Fe^{2+}$(A) ions contribute to pattern III, then good agreement is obtained between the II/I area ratio and the fraction of A sites occupied by $Fe^{2+}$ for x = 0.1 and 0.2. The cation distributions for these two cases are $(Fe^{2+}_{.88}Fe^{2+}_{.12})$ $[Fe^{3+}_{.92}Fe^{2+}_{.98}Sn_{.1}]$ and $(Fe^{3+}_{.63}Fe^{2+}_{.37})$ $[Fe^{3+}_{.97}Fe^{2+}_{.83}Sn_{.2}]$. For x = 0.5, the cation distribution determined from the II/I area ratio is $(Fe^{3+}_{.3}Fe^{2+}_{.7})$ $[Fe^{3+}_{.8}Fe^{2+}_{.7}Sn_{.5}]$. The remarkable feature of these results is the apparent adjustment in the A site $Fe^{2+}$ concentration to give approximately equal amounts of $Fe^{2+}$ and $Fe^{3+}$ on the octahedral site. A similar phenomenon was observed in $Fe_{3-x}Cr_xO_4$.[3] The paramagnetic character of $Fe^{2+}$(A) is due perhaps to clustering of B site $Sn^{4+}$ ions about such A site ions.

The magnetic $^{119}Sn$ pattern exhibits somewhat larger linewidths and hyperfine fields at 80 K than at 300 K. The larger fields are due simply to the increase in the magnetization but the larger linewidths may be due to the lower symmetry of the low-temperature phase, as expected. The quadrupole doublet due to $Sn^{2+}$ is unexpected and suggests that the conduction electrons are localized on the Sn sites below the Verwey transition. Further support for this suggestion is found in the increase in the intensity of the $Sn^{2+}$ pattern with increasing x.

While the $^{57}Fe$ area ratio results are suggestive of pair-wise hopping at 300 K, the localization of electrons on the $Sn^{4+}$ sites at low temperatures is more readily explicable in terms of band conduction.

REFERENCES

+ Support of this study by the National Science Foundation, Grant GH-41419 is gratefully acknowledged.

1. B. J. Evans, AIP Conf. Proc. 24, 73 (1975) and references therein.
2. F. Basile, C. Djega-Mariadassou and P. Poix, J. Phys. Chem. Solids 35, 1067 (1974).
3. M. Robbins, G. K. Wertheim, R. C. Sherwood, and D. N. E. Buchanan, J. Phys. Chem. Solids 32, 717 (1971).

# SOFT BULK MODULUS AT THE CONFIGURATIONAL PHASE TRANSITION IN $Sm_{1-x}Y_xS$

T. Penney, R. L. Melcher, F. Holtzberg and G. Güntherodt*
IBM Watson Research Center, Yorktown Heights, New York 10598

## ABSTRACT

Solid solutions of $Sm_{1-x}Y_xS$ undergo a configurational phase transition with increasing x, similar to the transition in SmS under pressure. We have measured the sound velocities and bulk modulus of $Sm_{1-x}Y_xS$ and find that the bulk modulus is soft in the gold, mixed valent phase. We consider a simple model for SmS under pressure to illustrate the relevant physical processes involved.

## DISCUSSION

SmS undergoes a first order phase transition under pressure at 6.5 kbar.[1,2,3] Its lattice collapses by ~12%, changing from a black semiconducting phase with Sm $f^6$ configuration to a gold metallic phase with mixed $f^6$ and $f^5d$ configuration. $Sm_{1-x}Y_xS$ shows a similar collapse with increasing x at $x_c = 0.15$ and zero pressure.[4,5] We have measured the sound velocities at room temperature and 30 MHz for single crystals of the $Sm_{1-x}Y_xS$ system.[6] Although we find discontinuities in the propagating modes at $x_c$, none is soft. The bulk modulus, B, can be calculated from the measured sound velocities and is given in Fig. 1. Note that in the black phase, $x \leq 0.15$, B is relatively constant, but in the gold phase $B \rightarrow 0$ as $x \rightarrow x_c$ from above. We believe that this softness throughout the gold phase ($x \rightarrow x_c$) results from the fact that the Sm is in a mixed configuration where the $f^6$ and $f^5d$ configurations are degenerate at the Fermi level.

In order to illustrate the physics of the problem, we will consider the case of SmS under pressure and assume extremely simple forms for the lattice and electronic energies. We treat each Sm equivalently, as if each had a valence 2+d and an average ionic size. The lattice energy per volume, $V_o$, is taken to be

$$E_l = 1/2B_o(e-e_d)^2 - pe \quad (1)$$

Here p is the pressure, $e=(V_o-V)/V_o$ is the volume strain with respect to a reference volume, $V_o$. The average strain, $e_d$, is defined by $e_d=(1-d)e_2 + de_3$. The strains $e_2$ and $e_3$ give the equilibrium volumes of $Sm^{2+}S$ and $Sm^{3+}S$, relative to $V_o$ and neglecting the electronic terms. It is convenient to set $V_o=V_2$ so that $e_2=0$ and $e_d=de_3$.

The model band structure for SmS in the semiconducting phase is given in Fig. 2a. The $f^6$ states form a band with density of states, $N_f$, and width, $1/N_f$. This band contains only one state per Sm, since only one electron is removed in forming the $f^5d$ state. The d electrons form a band with density of states, $N_d$. The bottom of this d band is separated from the f band by a gap, $\Delta$, when $e=e_2$. A major aspect of the model is that the d band position is dependent through the deformation potential, $\eta$, on the volume. The zero temperature electronic energy in the black phase (Fig. 2a) where the d band lies above the f band, is just $E_{el}=-1/(2N_f)$. The equilibrium strain is given by $\partial E/\partial e = 0 = B_oe-p$ where $E=E_l+E_{el}$. The bulk modulus, B, is given by $B= \partial^2E/\partial e^2 = B_o$. The gap closes to zero when $E_d=\Delta - \eta e$ is 0, that is at $p=B_o\Delta/\eta$.

Once the bands overlap the situation is as given in Fig. 2b and the electronic energy is

$$E_{el}=N_f(E_F{}^2-N_f{}^{-2})/2 + N_d(E_F{}^2-E_d{}^2)/2 \quad (2)$$

with $E_d=\Delta - \eta e$

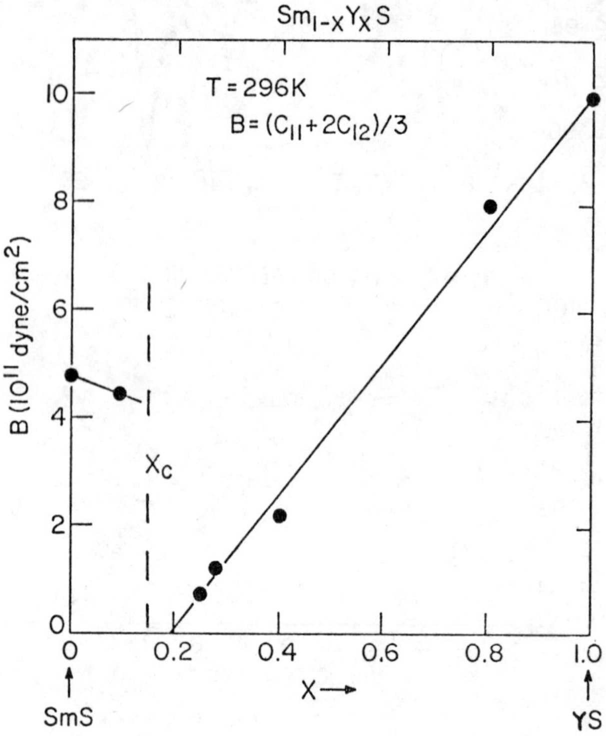


Fig. 1 Bulk modulus of $Sm_{1-x}Y_xS$ vs. composition, x, at 296K as computed from measured sound velocities. The point at x=0.4 is calculated using an interpolated value for one of the shear modes.[6]

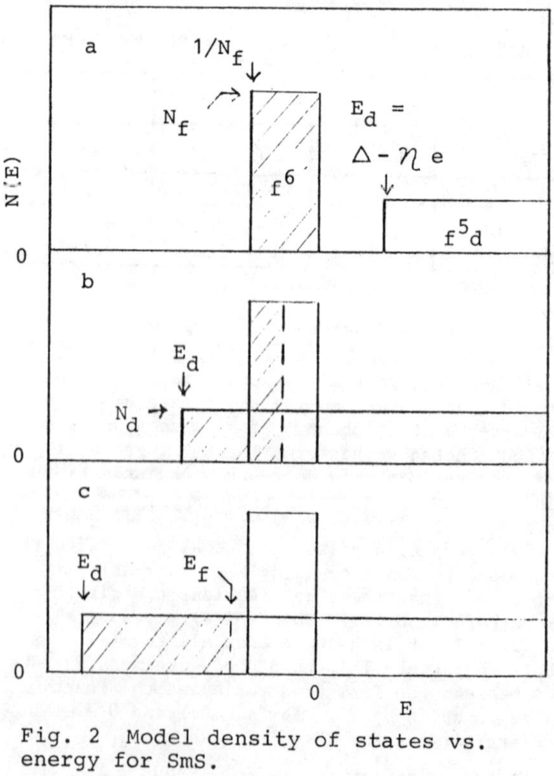


Fig. 2 Model density of states vs. energy for SmS.

The fraction, d, of $f^5d$ configuration is given by the electrons in the d band $d=N_d(E_F-E_d)$ and by the holes in the f band $d=-N_fE_F$. These expressions give $E_F=E_dN/N_f$ and $d=-E_dN$ where $1/N=1/N_d+1/N_f$. The electronic energy is then $E_{el}=-E_d^2N/2-1/(2N_f)$. The electronic contribution to the B is $B_{el}=\partial^2E_{el}/\partial e^2=-\eta^2N$. The lattice contributes an additional term, since d is now strain dependent and must be determined self-consistantly. We have from Eq. 1 $B_1=\partial^2E_1/\partial e^2=B_o(1-e_3\eta N)^2$. So the bulk modulus in the mixed valent phase is

$$B/B_o=(1-e_3\eta N)^2-\eta^2N/B_o \qquad (3)$$

This simple analytic form is possible due to the simplicity of the model and should be considered an illustrative rather than a quantitive result. There are two terms which soften B. The purely electronic term $-\eta^2N/B_o$ arises from a redistribution of electrons between levels as their relative energy is changed due to strain. The lattice term $-\eta Ne_3$ results from a renormalization of the lattice which is coupled to the electronic energy by $\eta$. That is when a transition $f^6$ to $f^5d$ is made, the lattice shrinks because $f^5$ is smaller than $f^6$ ($e_3>0$). Since this strain results in a d band lowering, the energy for a d excitation is reduced.

The magnitude of these terms may be estimated. From this work, $B_o$ for SmS is $4.75\times10^{11}$ erg/cm$^3$ which is 16 eV per SmS. From $B_o$ and the pressure dependence of the band edge[2], $\eta$ is 5 eV. N is about $N_d$ for $N_f\gg N_d$ and $N_d\approx0.5$ eV$^{-1}$ from XPS results on YS.[7] Since $Sm^{2+}S$ and $Sm^{3+}S$ differ in volume by 0.16, $e_3$ will be less than that, since there is also an electronic term, $\eta/B_o$, contributing to this volume difference. We find that the renormalization term ($\eta Ne_3\leq0.4$) and the electronic term ($\eta^2N/B_o\approx0.75$) are of the order unity. These values are suggestive and are given only to show that they are large enough to cause a significant reduction in B. If the calculated B is negative, then the phase is unstable and a first order transition will take place.

For sufficiently high pressure, $E_f$ lies below the f band as in Fig. 2c. Since $d=1=(E_F-E_d)N_d$ the electronic energy is $E_{el}=E_d+1/(2N_d)$ and $B_{el}=\partial^2E/\partial e^2=0$. Because $e_d=e_3$ is constant in Eq. 1, we have $\partial^2E/\partial e^2=B=B_o$. That is there is no softening of B in this phase. Softening occurs only when the two bands overlap at $E_F$ as in Fig. 2b.

We next discuss the sharpness of the transition and the stability of the intermediate valence. In this model, if $B>0$, a smooth transition from $f^6$ to $f^5d$ will occur with pressure as in the case of SmSe and SmTe. If, however, $B<0$ then a first order transition will take place from pure $f^6$ to pure $f^5d$. Although SmS shows a first order transition, it goes to a mixed valent combination of $f^6$ and $f^5d$, not to pure $f^5d$. We have assumed for simplicity that $\eta$ and $B_o$ are the same for $Sm^{2+}S$ and $Sm^{3+}S$. However, $B_o$ is about twice as large for $Sm^{3+}S$ as it is for $Sm^{2+}S$. Therefore, from Eq. 3, B will be larger for $Sm^{3+}S$. If $B<0$ for $Sm^{2+}S$ at $p_c=6.5$ kbar the first order transition will begin. But if $B>0$ for $Sm^{3+}S$ at $p_c$, the transition will not go all the way to pure $Sm^{3+}$ but will stop at some intermediate value of d.

Similarly, the intermediate valence could be stabilized if $\eta$ is smaller for $Sm^{3+}$.

Previously Wio, Alascio and Lopez[8,9] (WAL) as well as Varma and Heine[10] (VH) have considered this phase transition including lattice renormalization and deformation potential effects. At T=0, WAL find that the intermediate valence state is not stable if there is a first order transition. VH add a nonlinear term $e_4d(1-d)$ to the average strain, $e_d$, in Eq. 1. They find that $e_4$ must be very large, of the order of 1, to stabilize the intermediate phase. Batlogg Schoenes and Wachter[11] have discussed the narrowing of the band gap.

The situation for $Sm_{1-x}Y_xS$ with increasing x is similar to SmS with increasing p.[2,3] For $x<x_c$, B (Fig. 1) is nearly constant as in SmS for $p<p_c$, and at $x=x_c$, the lattice collapses.[5] The reason is that partial substitution of metallic YS for SmS lowers the d band while adding one electron per YS to the d band. When $E_F$ drops to the top of the f band the situation will be similar to Fig. 2b and B will be given by an expression like Eq. 3. If $B>0$ a continuous transition would take place. The occurrence at room temperature of a sharp transition at $x_c$ implies that $B<0$ at the onset of the transition. Just as discussed for SmS, if at $x=x_c$, $B_o$ increases or $\eta$ decreases with increasing d, the transition may stop at some intermediate value of d.

The fact that, for $x>x_c$, B (Fig. 1) lies well below the straight line between SmS and YS indicates that there is some softening over this entire range. Since softening occurs only when the $f^6$ and $f^5d$ overlap at $E_F$, (Fig. 2b) the data shows that this gold phase has intermediate valence.

Recently, Von Molnar and Holtzberg[12] have measured the low temperature specific heat of $Sm_{1-x}Y_xS$. In the gold phase they find that $\gamma$ is an order of magnitude larger than in the black phase, reflecting the large density of states at $E_F$ in Fig. 2b for the gold phase. In the black phase, $E_F$ cuts the d band above the f band where the density of states is lower.

We thank E. Pytte and D. Sherrington for useful discussions, J.M. Rigotty for making the measurements and P.M. Lockwood for the crystals.

REFERENCES

* Current address, Max-Planck Institut fur Festkorperforshung, Stuttgart, Germany

1 A. Jayaraman, V. Nayaranamurti, E. Bucher and R.G. Maines, Phys, Rev. Letters 25, 1430 (1970).
2 A. Chatterjee, A.K. Singh and A. Jayaraman, Phys. Rev. B6, 2285 (1972).
3 A. Jayaraman, A.K. Singh, A. Chatterjee and U. Devi, Phys. Rev. B9, 2513 (1974).
4 L.J. Tao and F. Holtzberg, Phys. Rev. B11, 3842 (1975).
5 T. Penney and F. Holtzberg, Phys. Rev. Letters 34, 322 (1975).
6 R.L. Melcher, G. Guntherodt, T. Penney and F. Holtzberg, Proc. 1975 IEEE Ultrasonics Symposium, Los Angeles.
7 R.A. Pollak, J.L. Freeouf, F. Holtzberg and D.E. Eastman, Phys. Rev. Letters 33, 820 (1974).
8 B. Alascio and A. Lopez, Solid State Comm. 14, 321 (1974).
9 H.S. Wio, B. Alascio and A. Lopez, Solid State Comm. 15, 1933 (1974).
10 C.M. Varma and V. Heine, Phys. Rev. B11, 4763 (1975).
11 B. Batlogg, J. Schoenes and P. Wachter, Physics Letters 49A, 13 (1974).
12 S. von Molnar and F. Holtzberg, this conf.

# LOW TEMPERATURE SPECIFIC HEAT OF $Sm_{1-x}Y_xS$

S. von Molnar and F. Holtzberg
IBM Research Laboratory, Yorktown Heights, N.Y. 10598

ABSTRACT

The $Sm_{1-x}Y_xS$ system, in analogy to pure SmS under pressure, exhibits a large volume decrease and color change, from black to gold, with increasing x. Both effects are related to the proximity in energy of the Sm $f^5d$ configuration to the $f^6$. Transport and lattice constant measurements by Penney and Holtzberg[1] indicate that there is a much larger density of states at the Fermi energy in the gold phase than in the black phase. We have measured the low temperature (1.3°K to ∿ 12°K) specific heat of small ($\lesssim$ 20 mg) single crystal samples cleaved from the same crystals used in preparing the transport samples. The results indicate an order of magnitude greater density of states in the gold phase than in the black, in agreement with the predictions of the model.

## INTRODUCTION

In 1970 Jayaraman, et al.,[2] discovered that SmS undergoes an electronic phase transition, from semiconductor to metal, with the application of 6.5 kbar pressure. Concomittant with the electronic transition, the crystal volume decreases by 13.5%[3] without change in the face centered cubic structure. Measurements by Maple and Wohlleben[4] indicate that even in the high pressure phase, where the Sm ions are expected to have substantial trivalent character with total angular momentum J ≠ 0, the susceptibility is temperature independent and small below approximately 100°K. Bader, Phillips and McWhan[5] demonstrated that the low temperature specific heat, C, increases dramatically when SmS is transformed to the metallic state.

Penney and Holtzberg[1] have performed detailed studies of the transport and structural properties of the alloy system $Sm_{1-x}Y_xS$. This system behaves in most respects like SmS under pressure, but with the advantage that studies do not have to be performed while the sample is contained in a cumbersome high pressure cell. The major disadvantage is that each Y atom (configuration $4d^15s^2$) contributes a conduction electron to the alloy and acts as a diluent to the Sm sublattice. Their results of resistivity and Hall measurements led them to propose the density of states shown in Fig. 1. It was the purpose of the present study of the low temperature specific heat to test this model. As will be shown, limiting coefficient of the linear term in the specific heat, γ, which for normal metals is proportional to the electron density of states, is consistent with the model. In addition, this note reports an anomalous increase in the specific heat below 5K for SmS and samples containing large (23 and 33%) concentrations of Y and a superconducting transition temperature for YS of ∿ 2.8K.

## RESULTS AND DISCUSSION

The microcalorimetric technique developed by Bachmann, et al.[6] was used in the measurements reported here. This is a thermal relaxation method in which the decay time, τ, of the specimen and bolometer through a known thermal leak is measured. The heat capacity at constant pressure, C, is then extracted through the relation $\tau = \frac{C}{k}$, where k is the known thermal conductance of the leak. Tens of milligrams of material are sufficent for most experiments. Thus the study of samples cleaved from the same crystal used in preparing the transport samples was possible. The absolute error in C from all sources is estimated to be $\pm$ 10%. Our measurements on ∿90 mgms of 99.999% pure Cu give results which obey the equation

$$C = \gamma T + \beta T^3 \qquad (1)$$

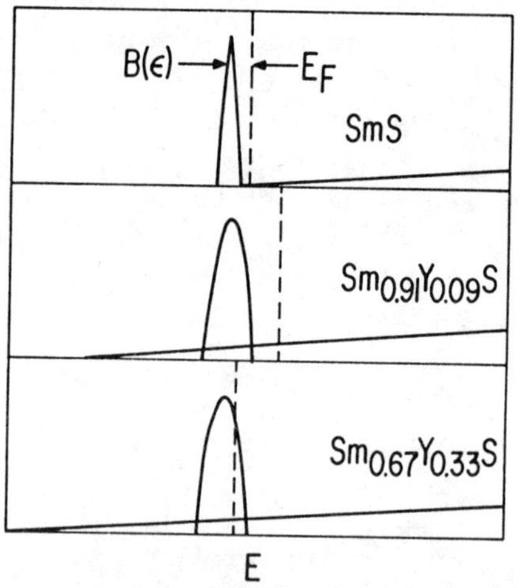


Fig. 1: Schematic electronic density of states vs. energy. B(ε) represents states made up predominantly of Sm f electrons. After Penney and Holtzberg, Ref. 1.

with $\gamma = 6.9 \times 10^{-4}$ J/mole $K^2$ and $\beta = 5.38 \times 10^{-5}$ J/mole $K^4$. These values are in reasonable agreement with those reported by Ahlers[7].

In writing Eq. 1 it has implicitly been assumed that the heat capacity can be interpreted as the sum of separable electronic, i.e. γT, and lattice contributions. For normal metals, γ is proportional to the electron density of states at the Fermi Energy and β is related to the Debye characteristic temperature, $\theta_D$, by

$$\beta = 1944\, n\, \theta_D^{-3} \text{ J/mole } K^4, \qquad (2)$$

where n is the number of atoms per molecule.

These same assumptions are made in analyzing the $Sm_{1-x}Y_xS$ alloys. Fig. 2 depicts the temperature variation of the specific heat between 1.3 and ∿ 12K. The contribution to C from the bolometer has been subtracted from all measurements and the curves have been plotted as $\frac{C}{T}$ vs $T^2$ in order to display (see Eq. 1) the variation in γ between samples. The limiting values are listed in Table I, as are the $\theta_D$ values for those alloys where a reasonable fit could be made. For comparison we list Debye temperatures calculated from velocity of sound measurements[8] using de Launay's tables.[9] The comparison, where possible, between the $\theta_D$ value is satisfying and lends credence to our estimates of γ. The small systematic difference between experimental and calculated values is under investigation. Sample $Sm_{.77}Y_{.23}S$ was included in order to show the trend in γ. The observed scatter in the data is due to the fact that this material experiences a phase transition at ∿ 200K with a sharp lattice contraction. This degrades the thermal link between bolometer and sample.

Obviously the most striking effect in C is the large change in γ between the various samples. YS, a metal with one d electron, has a suitably large C compared with Cu (with one s electron) corresponding to the large density of states in a d band. The γ value for the $Sm_{1-x}Y_xS$ sample with x = .33 is one order of magnitude larger than YS and two orders of magnitude larger than Cu, indicating that in the gold phase there is a very large density of states contributing to γ as indicated in Fig. 1. On the other hand, for the black phase, x = 0

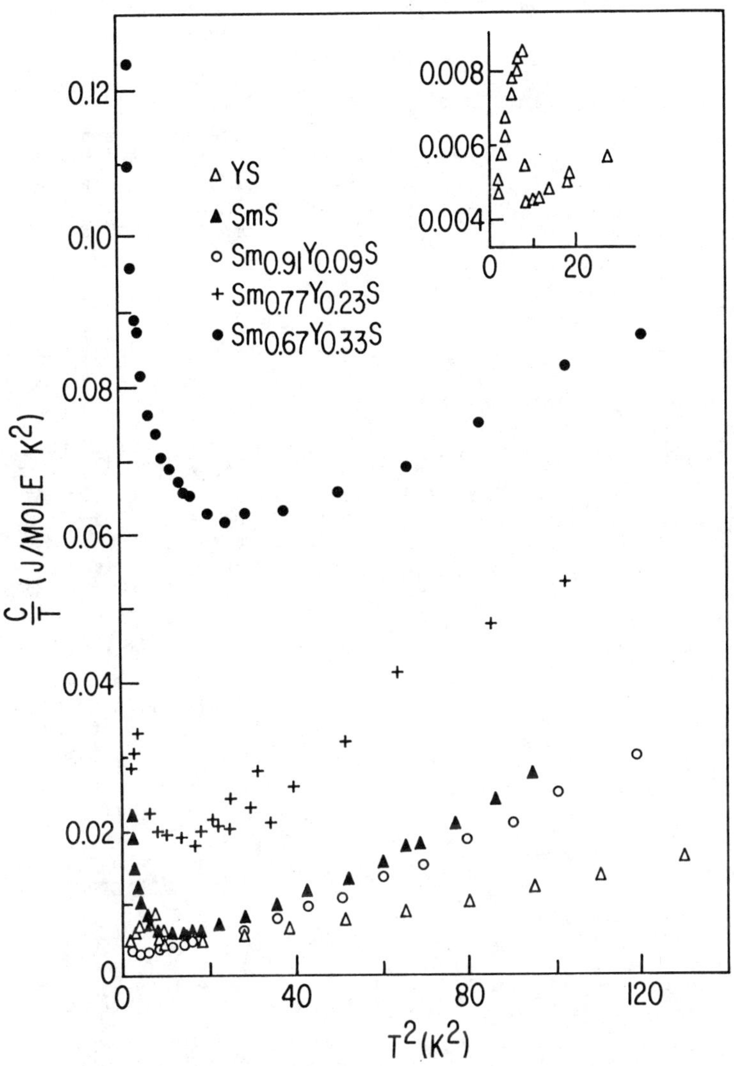


Fig. 2: Temperature variation of the specific heat for various samples of the alloy system $Sm_{1-x}Y_xS$. The insert plots $\frac{C}{T}$ vs $T^2$ for YS.

TABLE I

| Sample | $\gamma$(J/mole $K^2$) | $\theta_D$(exp)(K) | $\theta_D$(calc)(K) |
|---|---|---|---|
| SmS | .001 ± .001 | 247 | 267 |
| $Sm_{.91}Y_{.09}S$ | .003 ± .001 | 268 | 282 |
| $Sm_{.77}Y_{.23}S$ | .014 ± .005 | | |
| $Sm_{.67}Y_{.33}S$ | .05 ± .005 | | |
| YS | .003 ± .001 | 358 | 370 |
| Cu | .00069 ± .00003 | 331 | 345 |

and x = .09, the $\gamma$ values are not well determined but lie in the range 0 to .004 J/mole $K^2$, at least an order of magnitude smaller than the gold samples.

It is also noteworthy that the plots for SmS and for the 23 and 33% samples contain an anomalous low temperature contribution. Low temperature deviations were also observed by Bader, et al.,[5] who attributed them to impurity effects. Although this hypothesis appears reasonable in this case, especially since neither of two samples of $Sm_{.91}Y_{.09}S$ showed the effect, an intrinsic cause due to $Sm^{3+}$ cannot be ruled out. Studies of other Y concentrations as well as the $SmAs_{1-x}S_x$ system are planned to shed light on this problem.

Finally, we have observed superconductivity in YS (see insert, Fig. 2), with a well defined specific heat jump at 2.8°K.

SUMMARY AND ACKNOWLEDGEMENTS

In summary, the specific heat of small (10-25 mgm) single crystals of $Sm_{1-x}Y_xS$ has been measured. The estimates of $\gamma$ confirm the general features of the model for the density of states proposed by Penney and Holtzberg.[1]

We gratefully acknowledge the excellent technical assistance of S. Hanrahan, R. Linn, P. Lockwood and J. Rigotty. We also wish to thank T. Penney for many clarifying discussions concerning this subject.

REFERENCES


1) T. Penney and F. Holtzberg, P.R.L. 34, 322 (1975).
2) A. Jayaraman, V. Narayanamurti, E. Bucher, and R. G. Maines, P.R.L. 25, 1430 (1970).
3) E. Kaldis and P. Wachter, Solid State Comm. 11, 907 (1972).
4) M. B. Maple and D. Wohlleben, P.R.L. 27, 511 (1971).
5) S. D. Bader, N. E. Phillips, and D. B. McWhan, P.R. B7, 4686 (1973).
6) R. Bachmann, et al., Rev. Sci. Instruments 43, 205 (1972).
7) G. Ahlers, Rev. Sci, Instr. 37, 477 (1966).
8) R. Melcher, G. Güntherodt, T. Penney and F. Holtzberg, Proc. IEEE Symposium on Ultrasonics 1975, Los Angeles.
9) G. A. Alers, in Physical Acoustics, edited by W. P. Mason (Academic Press, Inc., New York, 1965, Vol. III, Part B).

# MIXED CONFIGURATION GROUND STATE IN Sm COMPOUNDS

J. W. Schweitzer
University of Iowa, Iowa City, Iowa 52242* and
Groupe des Transitions de Phases, C.N.R.S., Grenoble, France

## ABSTRACT

As a model for the electronic properties of Sm compounds exhibiting mixing of the $4f^6$ and $4f^5$ configurations we have studied a Hubbard model for the f and d bands generalized to include the inter-atomic Coulomb mechanism for configuration mixing that has been suggested by Kaplan and Mahanti.

## INTRODUCTION

In Sm compounds such as $SmB_6$, SmS, SmSe, and SmTe the $4f^6$, $4f^5 5d^1$, and perhaps $4f^5 6s^1$ configurations of the Sm ion are nearly degenerate giving rise to so-called intermediate valence behavior. Many experiments[1] on SmS under pressure have established a semiconductor to metal transition at 6.5 kbar where the ionic configuration of Sm goes from $4f^6$ to a configuration which is a mixture of $4f^6$ and $4f^5$ with electrons delocalized into the conduction band. Similar behavior is observed[1] in SmSe and SmTe at higher pressures except for the transition being continuous. In contrast experiments[2] on $SmB_6$ at normal pressure indicate that the Sm ions exist in a configuration which is a mixture of $4f^6$ and $4f^5$ plus a localized electron of predominantly 5d character. Another example[3] where 5d electrons seem to be localized in a state of mixed configuration is $Sm_{1-x}Y_xS$ for $x \leq 0.15$. In this range of composition the Sm ion configuration appears to be a mixture of $4f^6$ and $4f^5 5d^1$ where the weight of $4f^5 5d^1$ increases with incrasing Y concentration. At x = 0.15 there is a transition similar to that of SmS at 6.5 kbar where electrons are presumably delocalized. These examples illustrate that the intermediate valence state can exhibit localization as well as itinerancy of the electrons. Here we discuss a model that can yield a localized mixed configuration ground state and we study the stability of this state against delocalization of the electrons.

## MODEL

We consider a two-band Hubbard model generalized to include the mechanism suggested by Kaplan and Mahanti[4] for configuration mixing. This mechanism is the part of the inter-atomic Coulomb interaction which produces a simultaneous mixing of the configurations on two sites. The Hamiltonian is

$$H = \sum_{i\sigma} \Big( \epsilon_f f^\dagger_{i\sigma} f_{i\sigma} + \tfrac{1}{2} I f^\dagger_{i\sigma} f_{i\sigma} f^\dagger_{i-\sigma} f_{i-\sigma} + \epsilon_d d^\dagger_{i\sigma} d_{i\sigma}$$

$$+ G \sum_{\sigma'} f^\dagger_{i\sigma'} f_{i\sigma'} d^\dagger_{i\sigma} d_{i\sigma} \Big) + \sum_{ij\sigma} t_{ij} d^\dagger_{i\sigma} d_{j\sigma}$$

$$+ \frac{1}{2} \sum_{\substack{i\sigma \\ (i \neq j)}} \sum_{j\sigma'} \xi_{ij} \left( f^\dagger_{i\sigma} d_{i\sigma} + d^\dagger_{i\sigma} f_{i\sigma} \right) \left( f^\dagger_{j\sigma'} d_{j\sigma'} + d^\dagger_{j\sigma'} f_{j\sigma'} \right) .$$

Here $f^\dagger_{i\sigma}$ and $d^\dagger_{i\sigma}$ create an electron of spin $\sigma$ at the i-th site in the Wannier "f-orbital" and "d-orbital", respectively. The parameters $\epsilon_f$ and $\epsilon_d$ give the atomic level energies while I and G denote the f-f and f-d intra-atomic Coulomb repulsion matrix elements. The elements $\xi_{ij}$ are the inter-atomic Coulomb matrix elements $\xi_{ij} = \langle w_{di} w_{dj} | v | w_{fi} w_{fj} \rangle$ where $w_{di}$ and $w_{fi}$ are the Wannier orbitals on the i-th site. The elements $t_{ij}$ are the transfer integrals for the d band.

The major simplication of this model is the neglect of the orbital degeneracy associated with f and d electrons. Also it should be noted that we have included only those parts of the Coulomb interaction described by I, G, and $\xi$, and have assumed the f band has zero width. Thus we view the f states as highly correlated and the d states uncorrelated as in the Falicov-Kimball model[5] for metal-insulator transitions which is just the $\xi = 0$ limit of the model considered here.

We consider the case where the electron density is two electrons per site and where the energy difference between the atomic configurations $f^1 d^1$ and $f^2$, $D = \epsilon_f + \epsilon_d + G - 2\epsilon_f - I$, is of similar magnitude as G and the d bandwidth $W_D$. Furthermore we assume that the f-f intra-atomic Coulomb repulsion I is very large compared with these parameters and hence only $f^2$ and $f^1$ are possible. Clearly $f^2$ and $f^1 d^1$ are intended to simulate the Sm $4f^6$ and $4f^5 5d^1$ configurations. In this regime of parameters one expects mixing between the $f^2$ and $f^1 d^1$ configurations.

If $\xi$ is sufficiently small compared with the bandwidth $W_D$, the effect of $\xi$ is only to break the conservation of f-electron number on each site. Goncalves da Silva and Falicov[6] have given a treatment of this case based on the coherent potential approximation which adequately describes the physics. In the opposite extreme where $W_D = 0$, Kaplan and Mahanti[4] find that a sufficiently large negative $\xi$ will give a mixed configuration state as the ground state in a mean field approximation. It is of interest to investigate the effect of $W_D \neq 0$ on the occurrence of this localized mixed configuration ground state.

## RESULTS AND DISCUSSION

Following Kaplan and Mahanti[4] we take as a variational trial ground state $\Psi = \prod_i \varphi_i$, where $\varphi_i$ are two-electron functions for the i-th site. Note that this choice restricts the state to the subspace of exactly two electrons per site and consequently the d-transfer term in the Hamiltonian is rigorously zero in this subspace. Furthermore, it neglects correlations between sites. One obtains the following result for the state of this form which minimizes $\langle \Psi | H | \Psi \rangle$.

$$\Psi_{MGS} = \prod_i \left[ A f^\dagger_{i\uparrow} f^\dagger_{i\downarrow} - B \frac{1}{\sqrt{2}} \left( f^\dagger_{i\uparrow} d^\dagger_{i\downarrow} - f^\dagger_{i\downarrow} d^\dagger_{i\uparrow} \right) \right] |0\rangle$$

where $B^2 = 1 - A^2 = (\psi - \delta)/(2\psi)$ with $\psi = -4 \sum_j \xi_{ij}/W_D$ and $\delta = D/W_D$ provided $\psi \geq \delta$. For $\psi < \delta$ the state is pure $f^2$ or pure $f^1 d^1$ depending on whether $\delta$ is greater than or less than zero. This result for $\Psi$, which defines what we shall call the localized mixed configuration ground state (MGS), is of course just the result obtained by Kaplan and Mahanti[4] for the ground state of the model with $W_D = 0$.

In order to investigate the stability of this state against electron-hole excitations which delocalize electrons we have considered one-hole and one-electron states relative to $\Psi_{MGS}$. The "one-hole" states belong to the subspace spanned by the states $f_{i\sigma}\Psi_{MGS}$ and $d_{i\sigma}\Psi_{MGS}$, and the one-hole energies are the eigenvalues of the Hamiltonian in this subspace. Likewise the "one-electron" states are found by diagonalizing the Hamiltonian in the subspace spanned by the states $f^\dagger_{i\sigma}\Psi_{MGS}$ and $d^\dagger_{i\sigma}\Psi_{MGS}$. Details of the calculation and the results will be presented elsewhere.

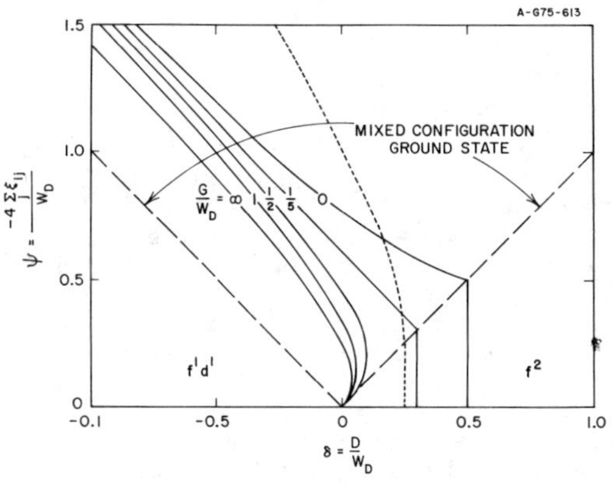


Fig. 1. "Phase diagram" for the ground state. The localized mixed configuration ground state (MGS) requires $\psi > |\delta|$ and is stable against single delocalized electron-hole excitations to the right of solid curve for particular $G/W_D$ and stable against delocalization to metallic state to right of dotted curve.

If the minimum single delocalized electron-hole excitation energy $E_{exc}$ is negative the localized mixed ground state is unstable against this type of excitation. In Fig. 1 the solid curves in the parameter space of $\psi$ and $\delta$ are defined by $E_{exc} = 0$ for various values of $G/W_D$. Here we have taken the minimum d-band energy to be $-W_D/2$. The region to the left of the curve corresponds to $E_{exc} < 0$ and therefore to a region where there is this electron-hole instability toward delocalizing the electrons. The dashed curves $\psi = |\delta|$ divide the parameter space into regions of pure $f^2$ $(B = 0)$, MGS $(0 < B < 1)$, and pure $f^1d^1$ $(B = 1)$.

Now clearly stability of the localized mixed configuration ground state against a single electron-hole excitation is only a necessary condition for this state to be the ground state within our variational treatment. In the Falicov-Kimball model $(\xi = 0)$ one knows that the $f^2$ state can be metastable while the "metallic state" of an $f^1$ configuration at each site and one itinerant d electron per site is the true ground state. Hence we have calculated for comparison the energy of this metallic state. Assuming a rectangular density of states for the d band one finds that $E(f^1, \text{d band}) < E_{MGS}$ if $\delta < -\psi + \sqrt{\psi}$. The dotted curve in Fig. 1 gives the boundary on which these two states have the same energy.

Figure 1 can be viewed as a phase diagram for the ground state of SmS and related systems. One can associate $\epsilon_d$ with the position of the $t_{2g}$ branch of the crystal field split d band as found in these compounds. Then $\epsilon_d$ will clearly decrease with lattice compression. If changes in $W_D$, $\xi$, and G are neglected, the effect of lattice compression is to decrease $\delta$ at fixed $\psi$ and $G/W_D$.

If the system is initially in a pure $f^2$ state and the lattice is compressed causing $\delta$ to decrease, there are several possibilities for the evolution of the electronic ground state of the system. For $\psi$ sufficiently large the system will cross the $\psi = \delta$ boundary into the localized mixed configuration ground state phase where the number of d electrons per site increases as $B^2/2 = (\psi - \delta)/4\psi$, yet there are no itinerant d electrons until an instability curve is reached. For $G/W_D > 1/4$ this is the boundary with the metallic state where there is one d electron per site delocalized. At finite temperatures there should be a first-order phase transition as in the Falicov-Kimball model for some region to the right of this boundary curve. If $G/W_D < 1/4$ it is possible for a restricted range of $\psi$ to cross the single electron-hole instability curve. Since creating more electron-hole excitations takes increased energy, the d electrons will be delocalized in a gradual manner. For a small number of delocalized electrons these states are described by the single electron-hole excitation spectrum we have calculated. For sufficiently small $\psi$ there can be no localized mixed configuration state.

Clearly the behavior shown in Fig. 1 is very well suited for a qualitative discussion of Sm compounds and solid solutions. Reasonable estimates of the parameters G, $W_D$, D, and $(-4\sum_j \xi_{ij})$ is SmS yield values of the order of 0.1 eV, and therefore one can expect to see all the various possibilities indicated by Fig. 1 in the SmS alloys. The only significant approximation that has entered into obtaining Fig. 1 is the neglect of correlations between sites in $\Psi_{MGS}$. However one expects this mean field treatment to yield a reliable phase diagram for considerations of the localization of the d electrons in the ground state.

For a quantitative discussion of Sm systems one of course can not neglect, as we have in the treatment presented here, the orbital character of the f and d electrons. Hund's rule, spin-orbit interactions, and crystal field effects must be taken into account to adequately describe the ground state of these systems. Since our treatment starts with a localized state it is straightforward to include these effects, however it would involve a considerable increase in complexity.

REFERENCES

* Permanent address

1. For recent reviews of experiments see M. B. Maple and D. K. Wohlleben, AIP Conf. Proc. 18, 447 (1974), and M. Campagna and G. K. Wertheim, AIP Conf. Proc. 24, 22 (1975).
2. R. L. Cohen, M. Eibshutz, and K. W. West, Phys. Rev. Lett. 24, 383 (1970); J. C. Nickerson, R. M. White, K. N. Lee, R. Bachmann, T. H. Geballe, and G. W. Hull, Phys. Rev. B 3, 2030 (1971); M. Aono, S. Kawai, S. Kono, M. Okusawa, T. Sagawa, and Y. Takehana, Solid State Commun. 16, 13 (1975).
3. T. Penney and F. Holtzberg, Phys. Rev. Lett. 34, 322 (1975).
4. T. A. Kaplan and S. D. Mahanti, Phys. Lett. 51A, 265 (1975).
5. L. M. Falicov and J. C. Kimball, Phys. Rev. Lett. 22, 997 (1969).
6. C. E. T. Goncalves da Silva and L. M. Falicov, J. Phys. C 5, 906 (1972).

# THERMOELECTRIC PROPERTIES OF EuSe

J. Heleskivi[†], Technical Research Centre of Finland, SF-02150 Espoo 15, Finland and T. Shiosaki, Kyoto University, Kyoto, Japan


## ABSTRACT

The thermoelectric power and the electrical conductivity of the semiconducting magnetic EuSe has been measured in the temperature range 7-300 K and in external magnetic fields between 0-13 kG in order to study the reasons for the conductivity minimum and the strong negative magnetoresistance found near the magnetic ordering temperature. At room temperature the thermoelectric power was of the order of -100 μV/K decreasing monotonously to -10 μV/K at 20 K. At external magnetic fields exceeding 10 kG the monotonous decrease continued below 20 K reading a value of 5 μV/K at 7 K. This shows that the changes in conductivity at fields above 10 kG are caused by changes in the mobility of the carriers. At external fields below 10 kG a positive peak was found in the thermoelectric power the peak value being of the order of 1 mV/K at magnetic fields below 4 kG. The thermoelectric power peak temperature at a certain field coincided approximately with the conductivity minimum at the same field. These findings indicate that the conductivity minimum and the negative magnetoresistance are accompanied by large changes in the carrier concentration at magnetic fields below 10 kG. The p-type state points to conduction in the 4f level or in a band formed by overlap between the 4f level and the anion $p^6$ band as proposed earlier by band structure calculations.


## INTRODUCTION

EuSe is a magnetic semiconductor with a Neel temperature of 4.6 K. The magnetic properties of EuSe arise from the well-localized spins of the $Eu^{2+}$ ions whereas the origin of the charge carriers is not fully understood. Pure and stoichiometric EuSe seems to be a rather good insulator even at room temperature. Doping with other rare earth metals and non-stoichiometry always leads to an increase of the n-type conduction in EuSe.

Very strong influence of magnetic ordering on the electrical properties has been observed experimentally in EuSe[1,2]. The most pronounced feature is the deep minimum in the electrical conductivity found near the magnetic ordering temperature. The conductivity decreases in some cases by approximately 12 decades. In the presence of a sufficiently large external magnetic field the minimum almost disappears indicating a very strong negative magnetoresistance.

The reasons for the conductivity minimum and the negative magnetoresistance are still unknown to some extent especially at zero and small external magnetic field conditions. Hall measurements performed in external magnetic fields exceeding 10 kG show that changes in the mobility are mainly responsible for the changes in the conductivity above 10 kG[1]. However, a field of 10 kG changes the conductivity in some EuSe crystals by 10 decades. To get information about the carrier concentration at zero field conditions magnetic fields below 0.1 kG must be used[2]. This on the other hand is impossible because of experimental reasons.

Thermoelectric power measurements offer another method to detect changes in the carrier concentration without the need of any external magnetic field. In EuO the deep minimum in the conductivity is accompanied by a peak in the thermoelectric power indicating a minimum in the carrier concentration[3]. In this paper we present the results of thermoelectric power measurements in EuSe in order to provide information about the reasons for the conductivity minimum and the strong magnetoresistance in magnetic fields below 10 kG.

## EXPERIMENTAL RESULTS

The thermoelectric power and the conductivity of EuSe were measured in the temperature range 7-300 K and in external magnetic fields between 0-13 kG. The magnetic field was perpendicular to the thermal gradient. The measurements were performed on unintentionally doped EuSe crystals with a n-type room temperature carrier concentration of $3 \cdot 10^{18}$ 1/cm$^3$ according to Hall effect measurements. The corresponding room temperature thermoelectric power was -160 μV/K. The crystals were provided with soldered Indium contacts.

The results of conductivity measurements are shown in Fig. 1 and corresponding thermoelectric power in Fig. 2.

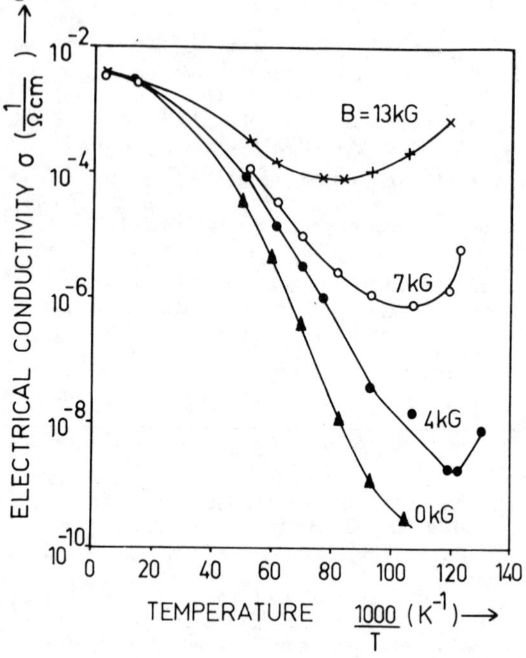


Fig.1. Conductivity vs. inverse temperature in EuSe at different external magnetic

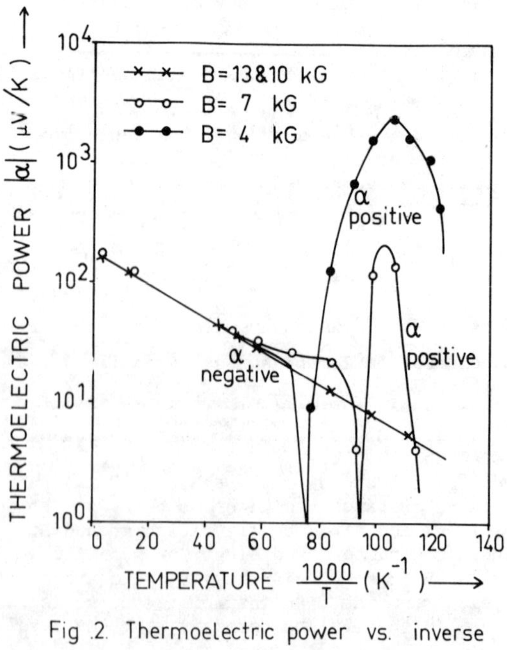


Fig 2. Thermoelectric power vs. inverse temperature in EuSe at different external magnetic fields. Observe the sign reversal

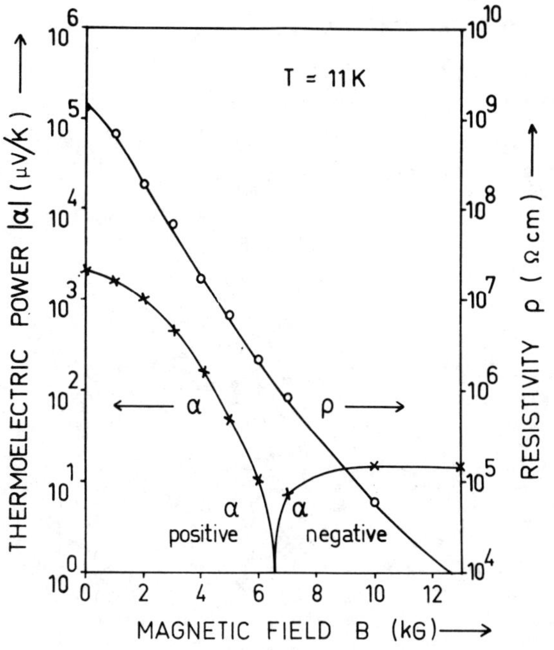


Fig.3. Thermoelectric power and resistivity vs. external magnetic field in EuSe at a fixed temperature of 11K.

The conductivity minimum changes to higher temperatures and becomes broader with increasing magnetic field.

The results obtained at a fixed temperature of 11 K are shown in Fig. 3.

DISCUSSION

At high temperatures the thermoelectric power in Fig. 2 is negative in consistency with Hall measurements. At magnetic fields above 10 kG the thermoelectric power decreases monotonously without any further magnetic field dependence at higher fields. Comparing this result with the conductivity behaviour above 10 kG shown in Fig. 1 one can conclude that the changes in the conductivity as a function of temperature and magnetic field are mainly caused by changes in the mobility and not in the carrier concentration. The same can be seen in Fig. 3 where the thermoelectric power stabilizes to a constant negative value of -15 μV/K at fields above 10 kG even if the resistivity still continues its decrease with increasing magnetic field. These findings are in consistency with high field Hall effect measurements[1].

At fields below 10 kG and temperatures below 20 K the thermoelectric power changes its sign and has a very strong positive peak approximately at the same temperature where the conductivity has its minimum at a specified field. The sign reversal and the strong positive peak points to the possibility that the conductivity changes below 10 kG are accompanied by large changes in the carrier concentrations. At room temperature the conductivity type in EuSe seems to change from n-type to p-type below a conductivity level of about $10^{-6}$ 1/Ωcm[4]. The sign reversal in the thermoelectric power happens approximately at the same conductivity level as can be seen when comparing Figs. 1 Cho[5] has proposed that p-type conduction could be possible in a band combined by the 4f level and the anion $p^6$ band in EuSe. However, the energy difference between the 4f level and the conduction band seems to be to large to allow hole conduction in such a combined band in the measured crystal. Consequently it must be assumed that impurity effects are responsible for the p-type state.

ACKNOWLEDGEMENTS

The authors are grateful to Mr H. Stubb for kindly supplying us with the samples and to Prof. T. Stubb, Mr M. Mäenpää and Mr J. Lavin for helpful discussions.

REFERENCES

† The author has been a senior scientist at the Academy of Finland during the work.

1 Y. Shapira, S. Foner & N.F. Oliveira, Phys.Rev. B 10, 4765-4780,1974.

2 J. Heleskivi, M. Mäenpää & T. Stubb, Negative Magnetoresistance in EuSe, Technical Research Centre of Finland, Semiconductor Laboratory, Report 11, Helsinki 1975.

3 A.A. Samokhvalov, A.Ya. Afanas´ev, B.A. Gizhevskii, N.N. Loshkareva & M.I. Simonova, Sov.Phys. Solid State 16, 365-366, 1974.

4 T. Penney & T. Kasuya, J.Appl.Phys. 42, 1403-1409, 1971.

5 S.J. Cho, Phys.Rev. B 1, 4589-4603, 1970.

## LOW SPIN HIGH SPIN TRANSITION OF IRON IN 1T-$Fe_xTa_{1-x}S_2$; $0 < X \leq 1/3$

M. Eibschütz and F. J. DiSalvo
Bell Laboratories, Murray Hill, New Jersey 07974

ABSTRACT

Magnetic susceptibility (χ) and Mössbauer effect measurements of iron substituted in 1T-$TaS_2$ are reported. The susceptibility χ in $Fe_xTa_{1-x}S_2$ for $0 < X \leq 1/3$ has an anomalous behavior. For X = 0.05, χ below 150 K shows that Fe is divalent low spin state (S=0). Above 150 K, χ increases with temperature indicating a smooth transition in the spin configuration of iron from low spin to a magnetic high spin configuration. For X > 0.05 the transition is moved to higher temperature. Mössbauer effect measurements of $Fe^{57}$ in $Fe_xTa_{1-x}S_2$ indicates that below the transition the Fe spin configuration is low spin $Fe^{2+}$ state ($^1A_{1g}$) which transforms gradually increasing temperature to high spin $Fe^{2+}$ state ($^5T_{2g}$). At high temperatures T > 800 K almost all the iron ions are in $Fe^{2+}$ high spin state. The transition occurs when the trigonal crystal field energy and the Hund's rule exchange energy are comparable in magnitude. The isomer shift data show that the transition is dynamical and that the rate constants for the processes indicated by $^1A_{1g} \rightleftarrows {}^5T_{2g}$ is faster than $10^7$ sec$^{-1}$.

# LOCALIZED MOMENTS AND MAGNETIC INTERACTIONS IN Fe-DOPED LAYER COMPOUNDS $NbSe_2$ AND $TaSe_2$ *†

R. V. Coleman, R. M. Fleming, D. A. Whitney
Physics Department, University of Virginia, Charlottesville, Virginia 22901

E. R. Domb and D. J. Sellmyer
Behlen Laboratory of Physics, University of Nebraska, Lincoln, Nebraska 68588

## ABSTRACT

Localized moments and magnetic interactions have been studied in the Fe-doped layer compounds $Fe_xTaSe_2$ and $Fe_xNbSe_2$ for $.01 < x < .10$. For small x the Fe impurities carry a localized magnetic moment as indicated by magnetic susceptibility measurements and, in a number of cases, Kondo-like behavior is observed in the transport properties. At larger values of x there is evidence for an ordered magnetic state at low temperatures.

## EXPERIMENTAL RESULTS

Dilute iron alloys of the form $Fe_xTaSe_2$ and $Fe_xNbSe_2$ have been studied by measuring magnetoresistance, Hall effect, and magnetic susceptibility for temperatures between 1.4 and 300 K. Single crystals have been grown from sintered powders by iodine vapor transport. The powders were sintered at 910°C while the single crystals were grown near 710°C. The $2H\text{-}Fe_xNbSe_2$ crystals show a resistance minimum[1] in the range 8 - 20 K. $T_{min}$ scales with concentration and $\rho_{min}$ shows anisotropy in a magnetic field for x as low as .0025. In the case of $Fe_xTaSe_2$ a similar minimum is observed only for $x \geqq .03$. This transition in behavior for x between .02 and .03 is connected with the stabilization of a different phase of $TaSe_2$ by iron at the growth temperature of 710°C. For $x \geqq .03$ the 2H phase is produced and is completely stable. For $x \leqq .02$ the 4Hb phase is produced at the growth temperature and can be maintained by quenching. The phases of the single crystals have been checked by x-ray precession photographs and by x-ray powder patterns from ground up single crystals.

The resistance minima observed for crystals of $Fe_xNbSe_2$ and $Fe_xTaSe_2$ are shown in Fig. 1. As shown for $Fe_xTaSe_2$ a parallel field of 150 kG is sufficient to quench the minimum while the perpendicular field reduces the temperature of $\rho_{min}$ by $\sim 5$ K. In addition the positive to negative Hall transition[2,3] observed in pure crystals at the onset of the CDW is rapidly quenched by addition of iron. An anomalous positive Hall resistivity is observed at low temperatures as evidenced by strong non-linearities in the field and temperature dependence. For $4Hb\text{-}Fe_xTaSe_2$ where x = .02 and .01, the Hall coefficient remains negative at 4.2 K, but is reduced in magnitude. The general behavior is demonstrated in Fig. 2.

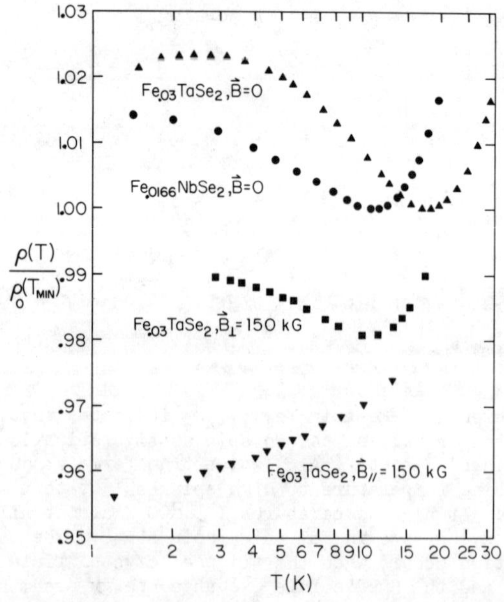


Fig. 1. Temperature dependence of resistivity for $Fe_{.03}TaS_2$ at ($\bar{B}=0$, $\bar{B}_\perp = 150$ kG, $\bar{B}_{//} = 150$ kG) and $Fe_{.0166}NbSe_2$ at $\bar{B} = 0$.

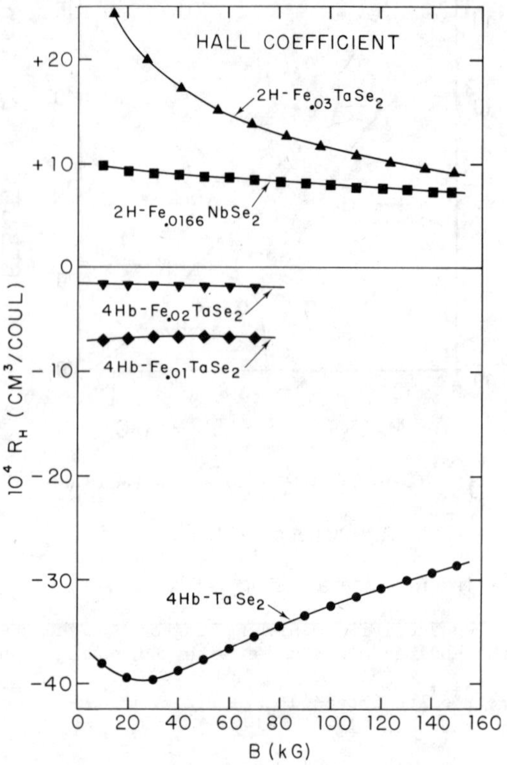


Fig. 2. Field dependence of the Hall coefficient for $4Hb\text{-}TaSe_2$, $4Hb\text{-}Fe_xTaSe_2$, $2H\text{-}Fe_xTaSe_2$ and $2H\text{-}Fe_xNbSe_2$.

The magnetic susceptibility has been measured in both powders and single-crystals and results are given in Figs. 3 and 4 and Table I. The susceptibility can be fit to the Curie-Weiss expression $\chi(T) = \chi_0 + C/(T-\theta)$ for temperatures greater than $T_{max}$, the temperature at which $\chi(T)$ is a maximum. An increase in B generally decreases $T_{max}$. In addition $\chi$ is extremely anisotropic for the $4Hb\text{-}Fe_{.02}TaSe_2$ crystal, as seen in Fig. 4.

## DISCUSSION

The Curie-Weiss behavior of $\chi(T)$ clearly indicates that the Fe impurities carry a local moment. Except for the $Fe_{.10}TaSe_2$ sample all the $\mu_{eff}$ values are consistent with spin values between 1 and 2. It is interesting that $\mu_{eff}$ for the $2H\text{-}Fe_{.05}TaSe_2$ crystal is equal to that for the multiphase powder of the same iron concentration. This suggests that there is relatively little difference in the local environment of an Fe atom in the 2H, 4Hb, and 3R phases. However, the strong anisotropy seen in the $4Hb\text{-}Fe_{.02}TaSe_2$ crystal indicates highly anisotropic crystalline fields at the Fe sites in this structure.

The resistance minima are consistent with spin-flip, conduction-electron scattering in the usual Kondo mechanism while the strong non-linear behavior of the Hall effect is characteristic of dilute magnetic alloys where skew scattering is present. A negative magnetoresistance which increases rapidly at low temperature is also present and is predicted in such

Table I. Summary of Susceptibility Data

| Material | Phase | $\mu_{eff}(\mu_B)$ | $\theta$(K) | $T_{max}$(K) |
|---|---|---|---|---|
| $Fe_{.02}TaSe_2$ | 3R-powder | 4.9 | -9 | |
| $Fe_{.02}TaSe_2$ | 4Hb-xtal | 4.4⊥ | -5 | |
| | | 0.4// | ∿0 | |
| $Fe_{.03}TaSe_2$ | 2H, 4Hb, 3R-powder | 3.3 | 11 | 8(1.5 kG) |
| $Fe_{.05}TaSe_2$ | 2H, 4Hb, 3R-powder | 3.2 | 15 | 9(1.5 kG) |
| | 2H-xtal | 3.2// | -3 | 2.7(1.5 kG) |
| | | | | 1.5(10.4 kG) |
| $Fe_{.10}TaSe_2$ | 2H-xtal | 7.8⊥ | 2.3 | 15(1.6 kG) |
| | | | | 10(10.4 kG) |
| $Fe_{.05}NbSe_2$ | 2H-powder | 3.2 | -7 | |

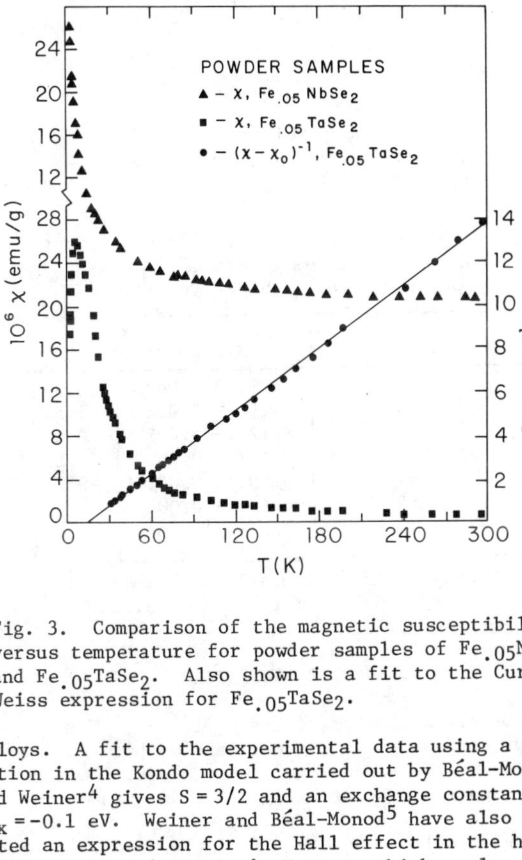


Fig. 3. Comparison of the magnetic susceptibility versus temperature for powder samples of $Fe_{.05}NbSe_2$ and $Fe_{.05}TaSe_2$. Also shown is a fit to the Curie-Weiss expression for $Fe_{.05}TaSe_2$.

alloys. A fit to the experimental data using a calculation in the Kondo model carried out by Béal-Monod and Weiner[4] gives $S = 3/2$ and an exchange constant $J_{ex} = -0.1$ eV. Weiner and Béal-Monod[5] have also calculated an expression for the Hall effect in the high field regime and obtain $\ell n$ H terms which replace the $\ell n$ T terms. Hall data on $Fe_xTaSe_2$ up to 150 kG show this logarithmic dependence at high fields and can be fit with $J_{ex} = -0.2$ eV.

An interesting feature of 4Hb-$Fe_{.02}TaSe_2$ is the presence of a local moment in the susceptibility data while no resistance minimum is observed. The 4Hb phase is highly anisotropic due to the alternating coordination of the layers. The two dimensional character is clearly observed in both the electronic conduction and the magnetic susceptibility and may play some role in quenching the spin-flip scattering.

The final point to be discussed is the origin of the susceptibility maxima in $Fe_xTaSe_2$ with $x \geq .03$. These maxima indicate the onset of some type of magnetically ordered state which might be due either to a two[6]-or-three[7] dimensional spin-glass (SG) transition, a spin-density-wave (SDW) in the electron gas,

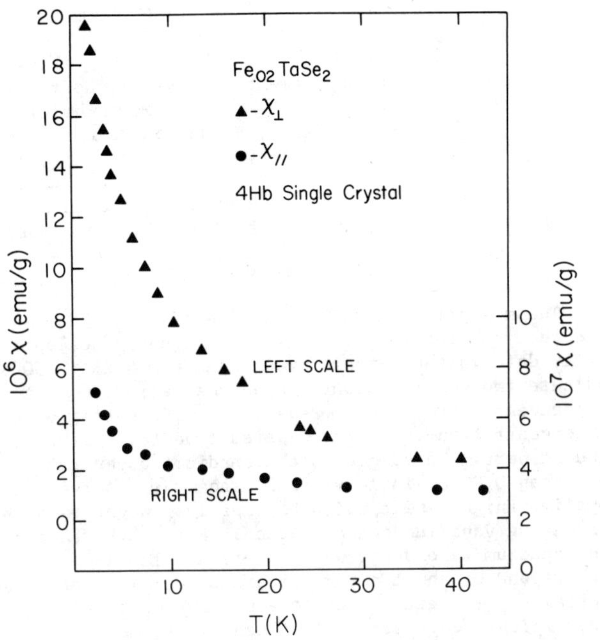


Fig. 4. Magnetic susceptibility versus temperature for 4Hb-$Fe_{.02}TaSe_2$ for fields oriented parallel and perpendicular to the basal plane.

or a local-antiferromagnetic ordering (LAO) in atomic clusters of Fe atoms. It is difficult to decide definitely which of these mechanisms is present. However, the $\theta$ values for the samples showing maxima in $\chi(T)$ are predominantly positive (suggesting ferromagnetic interactions), which is inconsistent with the LAO hypothesis. It is also well known[7] in spin-glass alloys that increasing the field does not shift $T_{max}$ as we observe, but only broadens the peak. The concentration dependence of $T_{max}$ in both the SG and SDW theories is approximately $T_{max} \propto x^n$ where $n \leq 1$. In order to prove that a SDW state exists below $T_{max}$, neutron diffraction measurements will probably be required. We are pursuing high-field magnetization measurements in order to try to elucidate the nature of the ordered magnetic state in these materials.

ACKNOWLEDGEMENTS

The authors wish to thank Professors V. Celli, M. Fowler and L. M. Falicov for useful discussions. Professors K. Lawless and R. H. Hanscom have provided valuable help with the x-ray work. Estelle Phillips has made major contributions to the sample preparation.

REFERENCES

*Work supported by the U. S. Atomic Energy Commission Contract No. AT-(40-1)-3105 and NSF No. DMR72-03208A01.

†Work above 80 kOe was performed while the authors were Guest Scientists at the Francis Bitter National Magnet Laboratory, which is supported at the Massachusetts Institute of Technology by the National Science Foundation.

1. R. C. Morris, B. W. Young, and R. V. Coleman, AIP Conf. Proc. No. 18, 292-296 (1974).
2. D. A. Whitney, R. M. Fleming, and R. V. Coleman, Solid State Communications, to be published.
3. R. C. Morris, Phys. Rev. Letters 34, 1164-1166 (1975).
4. M. T. Béal-Monod and R. A. Weiner, Phys. Rev. 170, 552-559 (1968).
5. R. A. Weiner and M. T. Béal-Monod, Phys. Rev. B3, 145-147 (1971).
6. B. Fisher and M. W. Klein, Phys. Letters 48A, 329-330 (1974).
7. J. Mydosh, AIP Conf. Proc. No. 24, 131-137 (1975).

# MAGNETIC SUSCEPTIBILITY AND NMR IN $Ta_{1-x}V_xS_2$*

N. Karnezos
Department of Physics, Northwestern University, Evanston, Ill. 60201
M.W. Shafer and G.V. Subba Rao †
IBM T. J. Watson Research Center, Yorktown Heights, N.Y. 10598
and
L.B. Welsh
Corporate Research Center, UPO Inc., Ten UOP Plaza
Des Plaines, Ill. 60016 and Northwestern University, Evanston, Ill. 60201

## ABSTRACT

The magnetic and NMR properties of $Ta_{1-x}V_xS_2$ alloys have been studied for $0.25 \leq x \leq 0.90$. Magnetic susceptibility data in the temperature range from 4.2K to 300K indicate two contributions. A paramagnetic temperature independent part which decreases with increasing vanadium concentration and a Curie Weiss type temperature dependent part. In samples with vanadium concentration less than 0.80 a single $V^{51}$ resonance was observed with negative Knight shift which becomes less negative with increasing vanadium concentration. For x=0.90 sample two vanadium resonances were observed. The second vanadium resonance and the temperature dependent part of susceptibility are attributed to about 20% of vanadium atoms which occupy sites with localized moments.

## INTRODUCTION

The 1T polytype of $TaS_2$ has a layered structure with one formula unit per unit cell. The layers are formed from sandwiches three atoms thick. The middle sheet accomodates the metal atoms which occupy octahedral holes formed by the top and bottom sheets of close packed chalcogen atoms. Pure 1T-$TaS_2$ undergoes phase transition at the temperatures of 352K and 200K. It has a small magnetic susceptibility[1], of the order of $10^{-7}$ emu/mole above room temperature, and becomes diamagnetic below the phase transition with a susceptibility of about $-50 \times 10^{-6}$ emu/mole.

Various attempts in forming $VS_2$, which is the x=1 limit of $Ta_{1-x}V_xS_2$, were unsuccessful. DeVries and Haas[2] and recently Silbernagel[3] prepared and studied the magnetic susceptibility and NMR of $V^{51}$ resonance in $V_5S_8$. The magnetic susceptibility of $V_5S_8$ consists of a temperature independent band contribution and a Curie Weiss type temperature dependent part which is attributed to vanadium atoms between the layers with localized d-states. The pressence of vanadium atoms with localized moments causes the $V^{51}$ resonance to split into three lines[3]; Resonance I with a large negative Knight shift due to vanadium sites with localized moments, resonance II with a positive Knight shift due to hyperfine fields induced by the localized moment at a vandium site which is in immediate proximity to the localized moment and resonance III with small and negative Knight shift due to a third vanadium site further apart from the localized moment.

Recent studies[4] of the effect of cation doping on charge density waves indicate that superlattice formation is destroyed in $Ta_{1-x}V_xS_2$ compounds even for small vanadium concentrations. The purpose of this work is to study the crystal structure, magnetic susceptibility and NMR properties of $V^{51}$ resonance in $Ta_{1-x}V_xS_2$ compounds as a function of vanadium concentration and compare them in the limit of $x \to 1$ to those of $V_5S_8$.

Solid solutions of $Ta_{1-x}V_xS_2$ were prepared by mixing appropriate amounts of $TaS_2$, V metal and S. The mixture was pressed into a pellet and heated in evacuated sealed quartz tube slowly to 500 C and held at 650 C for 4 days and finally at 900 C for one week and quenched. The product was ground to powder and reheated in quartz ampoules at 900-1000°C for periods ranging from 4 days to one week and then water quenched. Powders obtained in this way were examined by x-rays. Only lines due to 1T phase were noted throughout the solid solution. The lattice parameters a and c in $Ta_{1-x}V_xS_2$ decrease rather linearly with increasing x from a=3.369Å and c=5.894Å for x=0 to a=3.282Å and c=5.694Å for x=0.95.

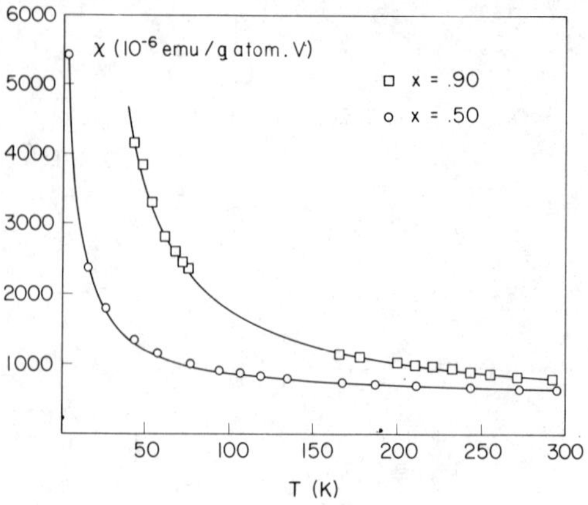


Fig. 1. Magnetic susceptibility versus temperature, solid line is a least square fit of the data to Eq. (1).

TABLE I. $\chi_o$, Curie Constant C and Curie temperature θ obtained from least square fit of susceptibility data in $Ta_{1-x}V_xS_2$ to Eq. (1).

| x | χ ($10^{-6}$ emu/g°atom.V) | C (emu.K/g atom.V) | θ (K) |
|---|---|---|---|
| 0.50 | 513 ± 10 | 0.037 ± 0.002 | 3.4 |
| 0.70 | 347 ± 23 | 0.163 ± 0.002 | 3.5 |
| 0.75 | 349 ± 19 | 0.129 ± 0.002 | 0.9 |
| 0.80 | 324 ± 33 | 0.107 ± 0.002 | -5.8 |
| 0.85 | 295 ± 35 | 0.150 ± 0.002 | -0.6 |
| 0.90 | 296 ± 16 | 0.136 ± 0.002 | -9.2 |
| $V_5S_8$ | 370 | 0.133 | -21.0 |

## MAGNETIC PROPERTIES

The magnetic susceptibility of the $Ta_{1-x}V_xS_2$ alloys for $0.50 \leq x \leq 0.90$ has been measured between 4.2 and 300K using the Faraday method. In Fig. 1 we present the magnetic susceptibility data as a function of temperature for the samples x=0.50 and x=0.90. The solid line is a least square fit of the data to a form

$$\chi = \chi_o + C/(T+\theta). \qquad (1)$$

The experimental data suggest that the magnetic susceptibility of $Ta_{1-x}V_xS_2$ compounds consists of a temperature independent part $\chi_o$ and a Curie Weiss type temperature dependent part. In Table I we summarize the $\chi_o$, the Curie constant C and Curie-Weiss temperature θ, results from our samples and $V_5S_8$[2]. $\chi_o$ shows a monotonic decrease with x, while C increases by about an order of magnitude from x=0.50 to 0.90 with a substantial scatter from sample to sample around x=0.80. It is important to notice that the C for x=0.90 sample is the same as in $V_5S_8$[2]. Since the magnetic susceptibility of $TaS_2$ is very small and temperature independent, we attribute the measured susceptibility of the $Ta_{1-x}V_xS_2$ alloys to the vanadium atoms. The Curie constant is proportional to the amplitude and number of the localized moments, $C = N\mu^2_{eff}/3k$. Assuming a vanadium site with localized moment similar to the one found in $V_5S_8$[2] (g =1.63,S=1), we find that in the x=0.90 sample 20%

of the vanadium atoms have a localized moment. For the other samples with $0.70 \leq x < 0.90$, this number fluctuates between 16% and 25%.

## NUCLEAR MAGNETIC RESONANCE

The $V^{51}$ NMR has been studied using a C.W. wide line spectrometer at 8 and 16MHz on 325 mesh powder samples of $Ta_{1-x}V_xS_2$ solid solutions where $0.25 \leq x \leq 0.90$. Due to the strong electric field gradients, all the quadrupole satellites were wiped out and only the central transition was observed. In samples with $x<0.90$, a sinlge $V^{51}$ resonance was observed from 100 to 300K. The $V^{51}$ Knight shift, measured with respect to the $V^{51}$ resonance in an aqueus solution of $NaVO_3$, is temperature independent, negative, and it becomes less negative with increasing x. In Fig. 2 we plot the Knight shift of $V^{51}$ as a function of x.

In the x=0.90 samples two $V^{51}$ resonances were observed. Resonance A, with a negative and temperature independent Knight shift of $K_A$=-0.04%, and resonance B, with a positive and temperature dependent Knight shift. The two resonances are shown in Fig. 3 after subtracting the background due to the aluminum resonance in the probe. The line widths of the two resonances A and B are inversely proportional to the temperature and increase from about 30 Gauss to 75 Gauss in the temperature range from 300K to 90K. The Knight shift of resonance B is of the form $K=K_0+A_0/(T+\theta)$ where $A_0$=0.77K and $K_0$=-0.02%. The value of $K_0$ differs from the value of $K_0$(II)=-0.254% found[3] for resonance II in $V_5S_8$. In Fig. 4, we plot the Knight shift of resonance B for the x=0.90 sample versus $\chi$. The effective field at the vanadium site which is responsible for resonance B can be estimated from the relation

$$H_{eff} = (N\mu_B)(dK/d\chi), \qquad (2)$$

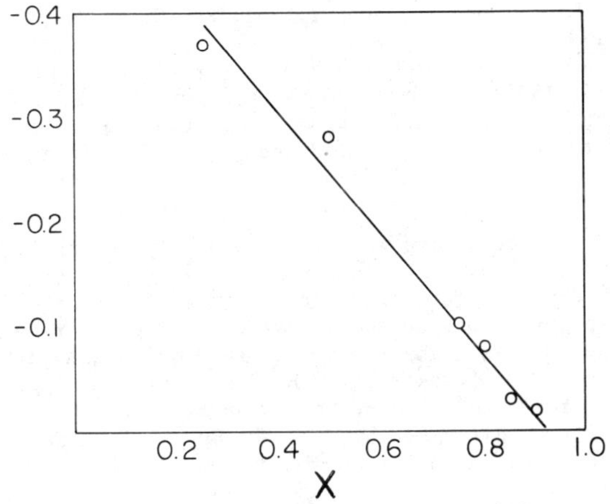


Fig. 2. Knight shift of resonance A at 300 K versus x in samples $Ta_{1-x}V_xS_2$.

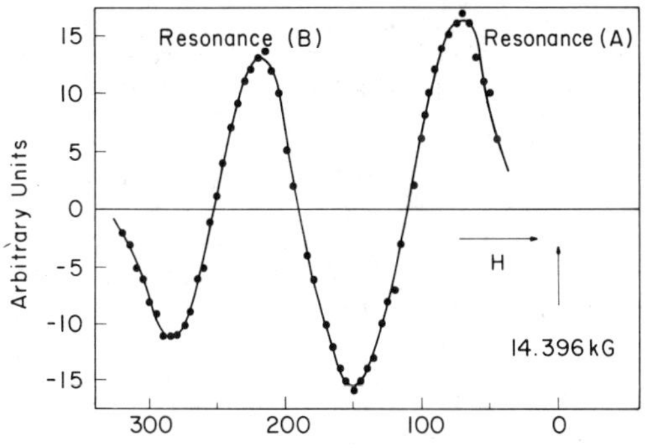


Fig. 3. Two vanadium resonances obtained from x=0.9 sample at 90 K and at the frequency of 16 MHz.

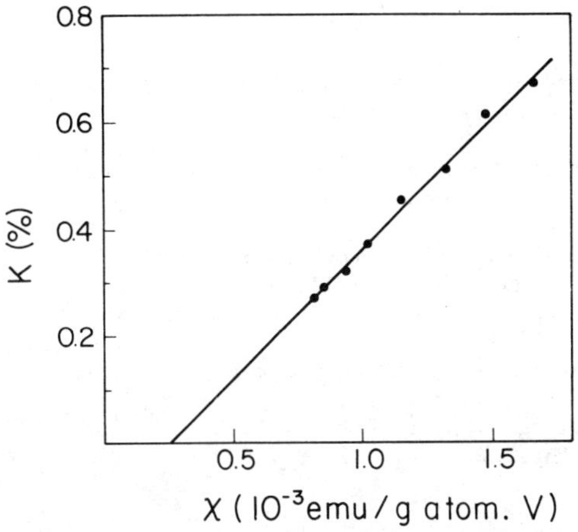


Fig. 4. Knight shift of resonance B in $Ta_{0.1}V_{0.9}S_2$ versus susceptibility.

where N is the number of localized moments and $\mu_B$ the Bohr magneton. Assuming 20% of the vanadium atoms occupy sites with localized moment and from the slope of the linear relation between K versus $\chi$, $dK/d\chi$ = 4.8(g atom V/emu), we estimate $H_{eff}$=5.4kG. This value is close to $H_{eff}$=6.7kG obtained from resonance II by Silbornagel et al.,[3] suggesting that the vanadium atoms which are responsible for resonance B, occupy sites similar to the type II atoms in $V_5S_8$. Resonance A arises from vanadium atoms further apart from the localized moments and is probably similar to resonance III observed in $V_5S_8$. In $V_5S_8$ the resonance of the paramagnetic V atoms has been observed. This resonance has not been observed here presumably due to the poor signal to noise ratio. The absence of resonance B for $0.7 \leq x < 0.90$ can be justified if the vanadium atoms between the layers are randomly distributed, instead of occupying well defined sites as in $V_5S_8$. Such a model is consistent with the results of the x-ray studies.

## CONCLUSIONS

From magnetic susceptibility results we conclude that for small vanadium concentrations the vanadium is mainly substitutional. For concentrations $x \approx 0.70$ and up, about 20% of the vanadium atoms go into a different site which we assume is an intergap site with a structural arrangement similar to $V_5S_8$. However, our powder x-ray measurements have not been able to unequivocally verify a monoclinic phase of the $V_5S_8$ type although we did observe a few extra lines indicating a distortion from the hexagonal symmetry of the 1T-phase. The fact that we see a continuous decrease in the lattice parameters of the 1T-phase as the vanadium is increased from $x \approx 0.75$ to $x \approx 0.95$ is consistent with the absence of a two phase region i.e., a 1T-phase and a $V_5S_8$ phase, and indicates a continuous single phase solid solution which becomes increasingly distorted as the vanadium concentration becomes greater than $x \approx 0.75$-0.80. It is likely that it distorts to a phase structurally similar to $V_5S_8$, because when $x \approx 0.95$ a $V_5S_8$ is actually seen.

The substitutional vanadium atoms affect the conduction band which is reflected in the change of the temperature independent part of the Knight shift and the magnetic susceptibility. The 20% nonsubstitutional vanadium atoms have localized moments which are responsible for the temperature dependent susceptibility and for the type B $V^{51}$ resonance.

REFERENCES

*Work supported by AFOSR and NSF through Northwestern University Materials Research Center.

†Present Address, Department of Chemistry, Indian Institute of Technology, Madras, India.


1) J. A. Wilson, F. J. DiSalvo, and S. Mahajan, Advances in Physics 24, 117 (1975).
2) A. B. DeVries and C. Haas, J. Phys. Chem. Solids 34, 651 (1973).
3) B. G. Silbernagel, R. B. Levy and F. R. Gambel, Phys. Rev. B11, 4563 (1975).
4) F. J. DiSalvo, B. G. Bagley, J. A. Wilson and J. V. Waszczak, Phys. Rev. B12, 2220 (1975).


## THEORY OF ELECTRONIC PROPERTIES OF SmS, SmSe, SmTe, $SmB_6$*

S.D. Mahanti, T.A. Kaplan and Mustansir Barma
Department of Physics, Michigan State University, East Lansing, MI 48824

ABSTRACT

Optical absorption data in semiconducting Sm-chalcogenides suggest[1] the existence of low energy excitations from $f^6$ to $f^5(d\text{-}t_{2g})$ localized configurations (Frenkel excitons) with a weak $f^6$ to $f^5$ (ds)-band excitation[2] starting from a lower energy. Based on this picture, Kaplan and Mahanti[3,4] suggested that such excitons are created in pairs via an intersite Coulomb interaction W. Within a mean field theory they found a second order phase transition from an unmixed $f^6$ to a coherently mixed ($Af^6+Bf^5d$) localized state (CMLS) as a function of decreasing volume. We have now investigated the effect of lattice energy on the transition. With reasonable values of W we can semiquantitatively reproduce the PV curves for SmS, (discontinuity in V vs. P), SmSe and SmTe (smooth behavior). We can also reproduce the amount of mixing of $f^6$ and $f^5d$ configurations in collapsed SmS. In addition, the case of $SmB_6$ fits naturally into our general picture.

The metallic behavior in collapsed SmS is interpreted in terms of partial lowering of a broad conduction band at the first order transition.[5] In our model, the number of electrons n occupying this band is small (≃ 0.1/Sm atom). This picture is based on the following. (a) The effective mass at the band bottom (which occurs at the X-point[6]) was given[7] as $m^* \simeq .8m$; this agrees approximately with band calculations.[6] (b) The band lowering as a result of the lattice collapse in SmS is about 0.6 ev.[8] (c) The chemical potential which would then appear to be at .6 ev above the band bottom will be lowered within our model because it costs an additional energy to transfer an electron from the CMLS to the band. Combining (a), (b) and (c) we find that about .15 electron per Sm atom occupies the band.

Choosing n=0.1/Sm atom and a mobility[9] of $10cm^2$/V-sec. (at room temperature) one obtains a value of the electrical conductivity in agreement with experiment.[10]

Our picture of the CMLS clearly gives agreement with the Mössbauer experiments,[11] and can be seen to be in accord with the XPS results.[12] We may also be able to account for the large linear specific heat at low temperatures, the nonmagnetic nature of the CMLS, the temperature dependence of the resistivity, and the plasma edge[2] although our consideration of these points has been only preliminary.

REFERENCES


1. F. Holtzberg and J.B. Torrance, AIP Conf. Proc. 5, 860 (1971).
2. E. Kaldis and R. Wachter, Sol. St. Comm. 11, 907 (1972).
3. T.A. Kaplan and S.D. Mahanti, Phys. Lett. 51A, 265 (1975), referred to as KM below.
4. The basic physical mechanism discussed by KM was first proposed by E.T. Jaynes and E.P. Wigner, Phys. Rev. 79, 213 (1950), as a mechanism for the ferroelectricity in $BaTiO_3$. Also see "Ferroelectricity" by E.T. Jaynes, Princeton U. Press, 1953. As noted by KM the actual ordering need not be ferroelectric; in fact, due to the dipolar nature of the interaction, the ordering due to this mechanism is probably quite complicated. We thank Prof. M.H. Cohen for pointing out the work of Jaynes and Wigner.
5. This was discussed by T.A. Kaplan, S.D. Mahanti and M. Barma, Bull. Am. Phys. Soc. 20, 383 (1975).
6. K. Lendi, Phy. Cond. Matt. 17, 215 (1974); S.T. Cho, Phys. Rev. B1, 4589 (1970).
7. A.V. Golubkov, E.V. Goncharova, V.P. Zhuze and I.G. Manoilova, Sov. Phys. Sol. St. 7, 1963 (1966). If instead of considering the conduction band bottom to be at Γ point, one takes it at the X point,[6] then the value of the effective mass obtained from this experiment will be reduced to $\simeq .8m/3^{2/3}$.
8. T. Penney and F. Holtzberg, Phys. Rev. Lett. 34, 322 (1975). (We used $\Delta V/V = 12\%$ at the transition).
9. S. Methfessel and D.C. Mattis, Handbuch Der Physik XVIII-1, Sect. 45, (Springer-Verlag, Berlin, 1968).
10. S.D. Bader, N.E. Phillips and D.B. McWhan, Phys. Rev. B7, 4686 (1973).
11. J.M.D. Coey, S.K. Ghatak and F. Holtzberg, AIP Conf. Proc. 24, 38 (1974).
12. M. Campagna, E. Bucher, G.K. Wertheim and L.D. Longinotti, Phys. Rev. Lett. 33, 165 (1974).


* Supported by NSF Grant GH-34565.

# MAGNETOCRYSTALLINE ANISOTROPY IN $FeCr_2S_{4-x}Se_x$

L. GOLDSTEIN, P. GIBART, L. BROSSARD
Laboratoire de Magnétisme, CNRS - 92190-BELLEVUE (FRANCE)

ABSTRACT :

$K_1(T)$ in spinels $FeCr_2S_{4-x}Se_x$ was deduced from high field magnetization measurements. $K_1(5.10^6 erg/cm^{-3})$ increases with x. For the highest selenium content ($x \sim 1.1$), $K_1(T)$ exhibits a maximum around 15°K. These features are explained in term of single ion magnetocrystalline anisotropy arising from $Fe^{II}$ in $T_d$, $C_{3v}$ and $C_{2v}$ symmetry. Special emphasis is given in $Fe^{II}$ with 2S-2Se near neighbours.

## 1 - INTRODUCTION

The thiospinel $FeCr_2S_4$ is a ferrimagnet with a large magnetocrystalline anisotropy ($K_1$ at 4.2°K = $5.10^6 erg/cm^{-3}$). Van Stapele (1) has shown that the anisotropy arises from $Fe^{II}$ in A sites. Hoekstra et al (2) deduced the anisotropy of $Fe^{2+}$ ions in A sites in Fe doped $CdCr_2S_4$ crystals by FMR and torque measurements. The $^5E$ ground doublet of $Fe^{II}$ is split by the exchange field and the second-order spin-orbit coupling. (2)

The coumpounds $FeCr_2S_{4-x}Se_x$ with $0 < x < 1.25$ were reported to be spinel ferrimagnets. For $x > 1.25$, $FeCr_2S_{4-x}Se_x$ crystallizes in the ordered defect NiAs structure type $Cr_3S_4$. By studying the magnetocrystalline anisotropy on $FeCr_2S_{4-x}Se_x$ it is possible to get information about the influence of the deformation of $Fe^{II}$ site. Furthermore assuming the local distortions of tetrahedral sites are distributed at random, the influence of random deformation on the magnetocrystalline anisotropy was investigated.

## 2 - MEASUREMENTS

Single crystals spheres were polished (diameter ∿ 1-3 mm)) and X Ray oriented either in <110> or <100> direction. The measurements were carried out between 4.2°K and 70°K. To saturate the sample along <110> it was necessary to apply high magnetic field. The measurements were done in Grenoble using SNCI facilities. In this equipment, a Bitter coil produced a stable field in the range 0-150 KOe ; the extraction method was used to get the data. $K_1$ is deduced from the surface between the <100> and <110> curves.

## 3 - RESULTS

Typical example on magnetization of curves at 4.2°K is given on figure 1 for $FeCr_2S_4$,

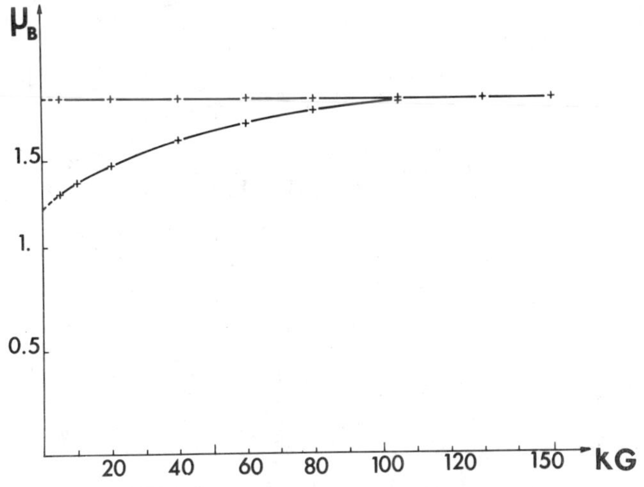


Fig. 1 - Magnetization at 4.2°K of $FeCr_2S_4$ along <100> (upper) and <110> (lower) directions.

$K_1$ vs T for 4 different x values (x = 0,0.4,0.8,1.1) are plotted in fig. 2.

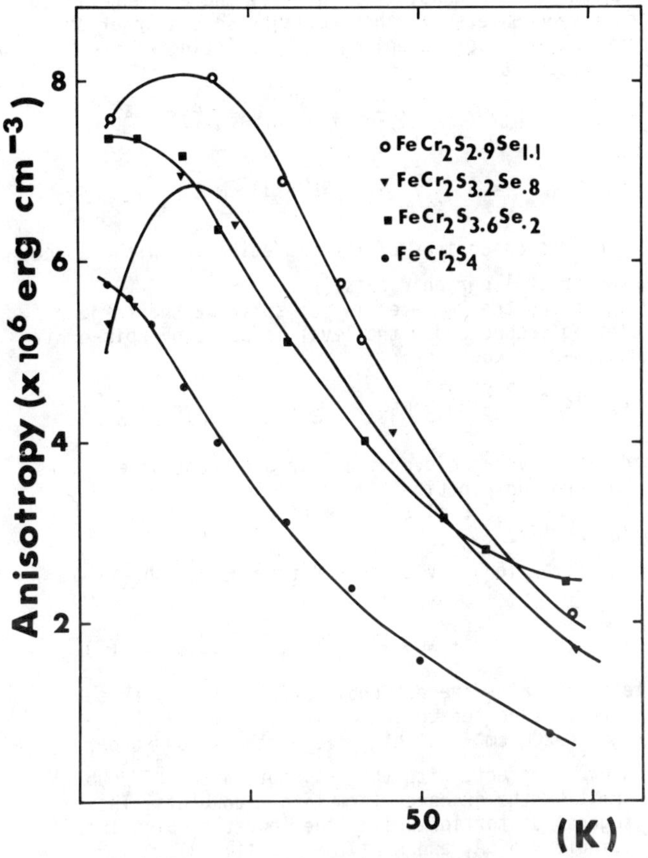


Fig. 2 - Magnetocrystalline anisotropy vs T for $FeCr_2S_{4-x}Se_x$ for x = 0, .2, .8, 1.1.

Two main features appear in these curves :
- $K_1$ at 4.1°K increases for x increasing up to x = 1
- $K_1$ (T) shows a maximum vs T for the compositions x = .8 and x = 1.1.

## 4 - INTERPRETATION

4-a - Crystal chemistry :

The substitution of selenium atoms for sulfur atoms in the closed packed anion arrangements gives four types of tetrahedral anion neighbors for iron atoms : 4 sulfur atoms ($T_d$ symmetry, 3 sulfur, one selenium ($C_{3v}$), 2 sulfur, 2 selenium ($C_{2v}$) and even 1 sulfur, 3 selenium ($C_{3v}$). Mössbauer spectra on paramagnetic samples (3) at room temperature clearly show that the substitution of selenium atoms for sulfur follows a statistical law. In other words the partition of the different tetrahedral sites follows the law :

$$p^m = C_4^m \left(\frac{x}{4}\right)^m \left(1-\frac{x}{4}\right)^{4-m} \qquad (1)$$

where p is the probability of having a tetrahedral site with m selenium atoms ; x is the proportion of selenium in the crystal. For instance in the limit composition x = 1.25, P°(4S-$T_d$ site) =.22, $P^1$ = .4, $P^2$ = .28, $P^3$ = .09. In Mössbauer paramagnetic spectra, each site gives rise to a doublet, which shows that the tetrahedral sites are locally distorted.

4-b - Magnetocrystalline anisotropy :

The results are explained in terms of one ion model theory of magnetocrystalline anisotropy. The total anisotropy energy of the system is given by adding the individual contribution of each kind of Fe atom : Fe in $T_d$, $C_{3v}$ or $C_{2v}$ symmetry. The number of each type of site follows the statistical law. The anisotropy of $Fe^{II}$ in $T_d$ site has been studied by Hoekstra. The $^5E$ doublet is split by the exchange field into five doublets, further split by second order spin-orbit and spin-spin coupling. The splitting of the lowest doublet is :

$$\Delta E = 2\delta\,|1-3(\alpha_1^2\alpha_2^2 + \alpha_2^2\alpha_3^2 + \alpha_3^2\alpha_1^2)|^{-1/2}$$

where $\delta = 6\,(\lambda^2/\Delta_c + \rho)$ $\delta = 11\ cm^{-1}$

$\alpha_i$ = direction cosines of the magnetization with respect to the crystallographic axes.

In $C_{3v}$ sites the $^5E$ level is not split by the trigonal field. To second order the level is split by spin-orbit and spin-spin coupling :

$$\Delta E=2\{\delta^2|1-3(\alpha_1^2\alpha_2^2+\alpha_2^2\alpha_3^2+\alpha_3^2\alpha_1^2| + \frac{32}{3}(\frac{\lambda\Delta'}{\Delta_c})^2\ (\alpha_1+\alpha_2+\alpha_3)^2\}^{1/2}$$

which when averaged over the four different sites gives a non zero contribution (3).

In $C_{2v}$ sites (2S-2Se) the effective hamiltonian is the following (5).

$$H = H0\ (\text{free ion}) + V\ \text{cubic} + V\ \text{tetragonal} + \Delta_{ex}\ (\frac{\vec{M}}{M_0})\vec{S}$$

$$+ \lambda\vec{L}\vec{S} + V_{ss} + \mu_B H\ (\vec{L}+2\vec{S}) \qquad (2)$$

V tetra contains the non cubic crystal field terms. V tetra is expected to be smaller than the exchange term, $\Delta_{ex}\approx 200\ cm^{-1}$ but higher than the second order spin-orbit effect. With this assumption the $^5E$ doublet is split by the exchange term into 5 doublets. The tetragonal distortion splits the lowest doublet into two singlets of $A_1$ and $A_2$ symmetry (fig. 3).

The second order spin-orbit interaction shifts the lowest states. This shifting is a function of the direction cosines of the magnetization and is responsable for the magnetocrystalline anisotropy. A detailed calculation of the energy in the ground doublet gives :

$$E_{A1} = E_{5E}-\Delta_{ex}\ (\frac{M}{M_0})m-\Delta' - \frac{\lambda^2}{\Delta_c}\ |18-3 < S_z^2 >_m| \qquad (3)$$

$$E_{A2} = E_{5E}-\Delta_{ex}\ (\frac{M}{M_0})m-\frac{\lambda^2}{\Delta_c}\ |6+3 < S_z^2 >_m|$$

where $A_1$ or $A_2$ is the lowest level depending in the sign of $\Delta'$ (first order energy of the tetragonal field): low temperature Mössbauer measurements now in process will give further information on the local distortions.

We have $K_1 = 2.5\ 10^{-15}$ erg/ion instead of $9.2\ 10^{-16}$ erg/ion in the perfect cubic symmetry.

Taking into account the statistical distribution of selenium atoms among anion sites, $K_1(T)$ must be averaged over all tetragonal distortion axes. In other words the ox, oy, oz axis have the same probability of being the tetragonal deformation axis. The contribution energy of $C_{2v}$ site is given by :

$$E = \frac{N}{3}\ P^2\ |Ex+Ey+Ez| \qquad (4)$$

where N = Fe atom/$cm^3$

$P^2$ is the probability of having an Fe atom with 2S-2Se. Ex, Ey, Ez are the anisotropy energies for tetrahedra distorted along ox, oy, oz respectively. Fig. 4 gives the value $K_1(T)$ deduced from relation (4) for different values of $\Delta'$ and $\lambda$.

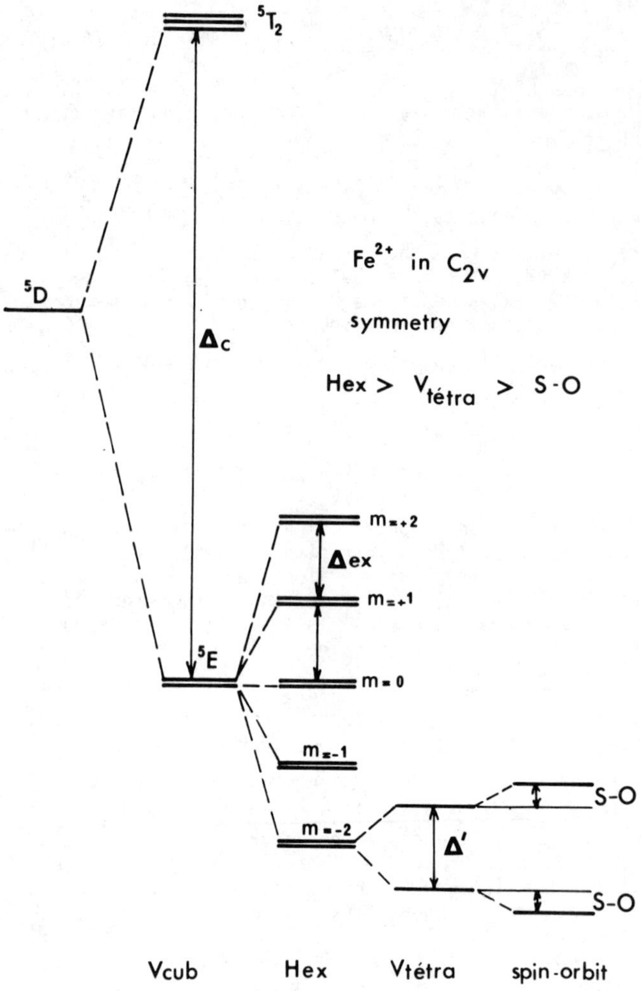


Fig. 3 - Energy levels of $Fe^{II}$ ($3d^5$) in $C_{2v}$ symmetry.

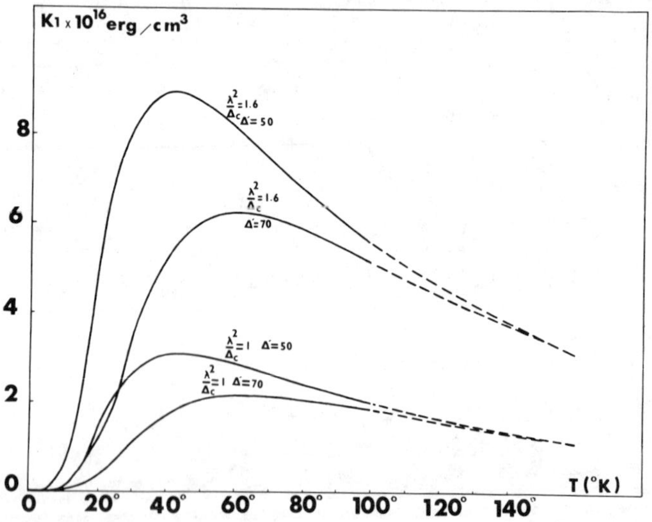


Fig. 4 - Calculated $K_1$ vs T for $Fe^{II}$ in $C_{2v}$ symmetry averaged over the three directions of tetragonal distortions for different values of $\delta$ and $\Delta'$.

At 0°K, $K_1$ is 0, $K_1(T)$ exhibits a non zero contribution for $T \neq 0$ with a maximum around 20°K. The $K_1$ value in this region is of the same order of magnitude as in the $T_d$ site.

These calculations show that for high concentration of Se, where the probability for 2S-2Se is high, a maximum of the curve $K_1(T)$ should appear. This qualitatively explains the shape of the curve observed. A rigorous quantitative fit is almost impossible ; values of $\lambda$, $\Delta_c$ and $\Delta'$ are missing, and the one ion model is well adapted to dilute system but is not so effective in concentrated one. Furthermore substitution of selenium causes local distortions of the tetrahedral sites even in the case of 4 sulfur neighbors as shown by Mössbauer spectra.

The increase of $K_1$ at 0°K for low concentration of selenium (Fig. 4) can be also explained in terms of the variation of the crystalline cubic field and spin orbit interaction. In a tetrahedral site, $\Delta_c$ and $\lambda$ both decrease from oxides to sulfides and so they are expected to decrease further in selenides. But the ratio $\delta = \frac{6\lambda^2}{\Delta_c}$, which is directly related to $K_1$, is expected to increase due to the more rapid decreasing of $\Delta$ :

$FeCr_2O_4$ $\lambda = 80\ cm^{-1}$ $\Delta = 4000\ cm^{-1}$ $\delta = 9.6\ cm^{-1}$

(5)

$FeCr_2S_4$ $\lambda = 76\ cm^{-1}$ $\Delta = 2500\ cm^{-1}$ $\delta = 14.4\ cm^{-1}$

On the other hand the statistical tetragonal deformation of tetrahedral sites due to selenium induces a decrease of $K_1$ at very low temperature.

Acknowledgements : the authors wish to express there gratitute to Pr R. Pauthenet and J.C. Picoche for the magnetization measurements at SNCI Grenoble.

Ref. : 1) R.P. Van STAPELE, J.S. Van WIERINGEN and P.F. BONGERS, J. Phys. (Paris) 32, CI-53 (1971).

2) B. HOEKSTRA and R.P. Van STAPELE, Phys. Stat. Sol.(b) 55, 607 (1973).

3) Unpublished results.

4) B. HOEKSTRA, R.P. Van STAPELE and A.B. VOERMANS, Phys. Rev. 6, 1762 (1972).

5) L. GOLDSTEIN, Thesis, Paris 1974.

# ELECTRON-PHONON COUPLING AND THE LOSS OF MAGNETISM IN NiS

G. Parisot
Institut Laue-Langevin, B.P. 156, Grenoble 38042 France

J.M.D. Coey and R. Brussetti
Groupe des Transitions de Phases-C.N.R.S., B.P. 166, Grenoble 38042, France

F. Gompf, G. Czjzek and J. Fink
Institut für Angewandte Kernphysik, K.F.A. Karlsruhe, Germany

## ABSTRACT

The changes in the phonon spectrum of NiS at its first-order transition from itinerant-electron antiferromagnet to Pauli paramagnet have been studied by triple-axis and time-of-flight neutron techniques.

Three of the elastic constants were derived from the initial slopes of the acoustic phonon dispersion relations at five temperatures between 180 and 300 K. $C_{44}$ and $C_{11}$ soften discontinuously in the metallic phase, despite the volume decrease. $C_{33}$ is unchanged within experimental error. The values of these constants are 0.41(3), 2.4(2), and 1.9(2) x $10^{-12}$ cgs in the antiferromagnetic semimetal and 0.24(2), 2.0(2), and 1.9(2) x $10^{-12}$ cgs in the paramagnetic metal.

The weighted density of states was measured in the semimetallic phase at 245°K and in the metallic phase at 269°K. In the low frequency spectrum, the acoustic modes are appreciably lower in the metallic phase than in the semimetallic one, in agreement with the elastic constants data. Furthermore a sharp drop near 25 meV in the 245K spectrum is considerably broadened at 269K, indicating a stronger dispersion of the optical modes above the transition or possible modifications of longitudinal acoustic branches near the zone boundaries. These results agree with $^{61}Ni$ Mössbauer data. The recoilless fraction (Debye-Waller factor) and second-order Doppler shift analysed using a Debye+Einstein model give a partial density of states for nickel which is in reasonable agreement with the neutron data.

It is clear from the ensemble of our measurements that the transition in NiS occurs because of the greater entropy of the metallic phase. This entropy is neither magnetic nor purely electronic in origin. Most of it is associated with the lattice, and it may be attributed essentially to the electron-phonon coupling.

# ELECTRICAL PROPERTIES OF $CdCr_2Se_4$ THIN FILMS

D. I. Tchernev and A. J. Syllaios
Department of Electrical Engineering, The University of Texas at Austin, Austin, Texas 78712

ABSTRACT

Measurements of the electrical conductivity σ and the Seebeck coefficient Q were performed on a series of In doped $CdCr_2Se_4$ thin films grown by RF sputtering of stoichiometric mixtures of CdSe, $Cr_2Se_3$ and InSe and subsequent heat treatment in Se atmosphere. The Seebeck coefficient changes from positive to negative upon annealing in vacuum at 430°C and thereby introducing Se defects. The conductivity and the absolute value of Q increase monotonically between 77°K and room temperature. No resistivity peak or any anomaly of Q around the Curie temperature as observed in bulk samples were present in either n or p-type films.

## INTRODUCTION

The chalcogenide spinel $CdCr_2Se_4$ is a ferromagnetic semiconductor. The study of the electrical properties of $CdCr_2Se_4$ films was undertaken for two reasons. First, because of its high mobility[1], $CdCr_2Se_4$ is particularly suitable for the experimental study of the exchange interaction of the band electrons with the localized magnetic moments. We have concluded earlier[2] that for electrons traveling at high velocities in a ferromagnetic semiconductor without too much scattering it should be possible to observe changes in the trajectory dependent properties when standing spin waves or 180° domain walls are present. The resonance effect of the standing spin waves and the fact that the whole volume of the sample will be occupied by the magnetic structure promises a greatly increased effect. The measured films, however, have very small mobilities, $\mu < 10^{-2}$ $cm^2$/V-sec, and no such effects were observed. The second reason is that there have been both theoretical and experimental discussions on the temperature dependence of the electrical conductivity and the Seebeck coefficient of this compound[1,3].

In n-type doped $CdCr_2Se_4$ bulk samples a conductivity minimum is found near the Curie temperature $T_C$ = 130°K. The Seebeck coefficient is anomalous near $T_C$ showing two local extrema. To account for this behavior a mixed conduction model has been proposed[1] with donor levels arising from Se deficiencies. A magnetic impurity state model has also been proposed[3].

We have found $CdCr_2Se_4$ thin films to be ferromagnetic[4] with Curie temperature $T_C$ = 120°K. In this paper we report the electrical transport properties of these films. The films were studied before and after an annealing treatment in vacuum.

## EXPERIMENTAL

Electrical measurements were made on thin films grown[4] on single crystal sapphire substrates by RF sputtering of stoichiometric mixtures of CdSe, $Cr_2Se_3$ and InSe. An amorphous mixture of the constituent elements, highly deficient in Se, is thus deposited on the substrate. A subsequent heat treatment in Se atmosphere is then carried out and $Cd_{1-x}In_xCr_2Se_4$ films are grown according to the reaction

$$(1-x)Cd + xIn + 2Cr + ySe + (4-y)Se \xrightarrow[1h]{500^\circ C} Cd_{1-x}Cr_2Se_4 \quad (1)$$

with x = 0.1. It was found that the number of moles of Se, y, deposited on the substrate during sputtering must be

$$y \geqslant 3.3 \quad (2)$$

so that the reaction (1) proceeds to the formation of $Cd_{1-x}In_xCr_2Se_4$. The balance 4-y is then provided by Se vapors condensing on the substrate. The films have the spinel structure, and the lattice constant is found to be a = 10.75 Å in agreement with the bulk value[5]. Subsequently the films were annealed in a dynamic vacuum at different temperatures. Reduction is thus achieved by introducing Se vacancies. All films were 4500 Å thick.

In-Hg amalgam (~60% In) proved to make good, noise-free ohmic contacts to $CdCr_2Se_4$ down to 77°K. The I-V characteristics are linear up to (20 mA, 100 V). The dc resistivity was measured by passing a constant current through the film and monitoring the voltage across it (2-probe method). Because of the finite size of the contacts the distance between them could not be measured accurately. Therefore there was a corresponding uncertainty in the exact value of the resistivity. The relative variation with T, however, was measured more accurately.

## RESULTS

When the films are annealed in vacuum they transform from p to n-type. This is associated with loss of Se as seen in Table 1. The ratio $I_{Se}/I_{Cr}$ decreases with annealing at higher temperatures. $I_{Se}/I_{Cr}$ represents the ratio of the intensities of the $K\alpha_1$ emission lines of Se and Cr as measured in the electron microprobe, which is proportional to the actual concentration ratio within a correction factor for X-ray absorption and fluorescence effects. An estimate of the actual concentration ratio [Se]/[Cr] is made by measuring $I_{Se}/I_{Cr}$ for stoichiometric $CdCr_2Se_4$ single crystals used as a standard. It is seen from Table 1 that films as grown and with no vacuum annealing treatment contain about 60% excess Se above the stoichiometric concentration. Because the films were prepared in Se atmosphere, a layer of Se is condensed on the film surface as seen in the SEM photograph in Figure 1a. When the films are annealed above 230°C the surface Se layer is removed as seen in Figure 1b.

Curve a in Figure 2 shows the change in the resistivity of a film during annealing in vacuum. The resistivity decreases in a non-exponential manner during heating at a rate of 200°C/hr, and there is a minimum at 430°C. When the temperature is stabilized at 450°C the resistivity decreases again (vertical part of curve a) and then remains constant for the rest of the annealing time of about 20 hours. At subsequent coolings and heatings the resistivity follows curve b. At slower heating rates the derivative of the corresponding heat-

TABLE 1
Electrical Properties of Thin Films

| Sample | Annealing Temperature | $\frac{I_{Se}}{I_{Cr}}$ | $\frac{[Se]}{[Cr]}$ | Activation Energy (eV) | Type | Resistivity Ω-cm |
|---|---|---|---|---|---|---|
| Film A | No vacuum annealing | 0.59 | 4.843 | 0.02 | p | $10^{-1}$ |
| Stoichiometric $CdCr_2Se_4$ crystals | -- | 0.37 | 3.037 | 0.20 | p | $10^3$ |
| Film B | 400°C | 0.31 | 2.545 | 0.04 | p | 1 |
| Film C | 450°C | 0.23 | 1.888 | 0.09 | n | 10 |

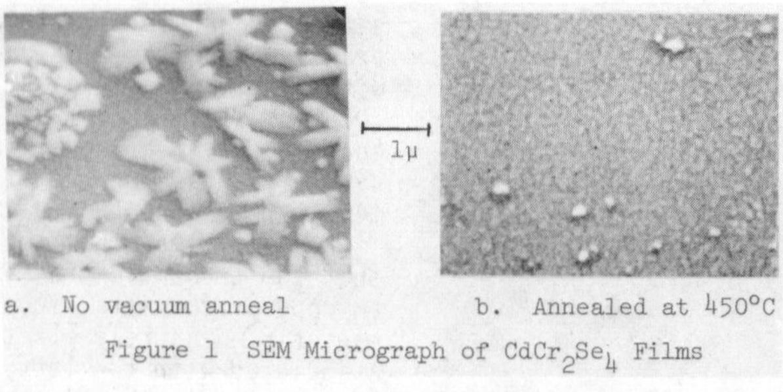


a. No vacuum anneal b. Annealed at 450°C

Figure 1 SEM Micrograph of $CdCr_2Se_4$ Films

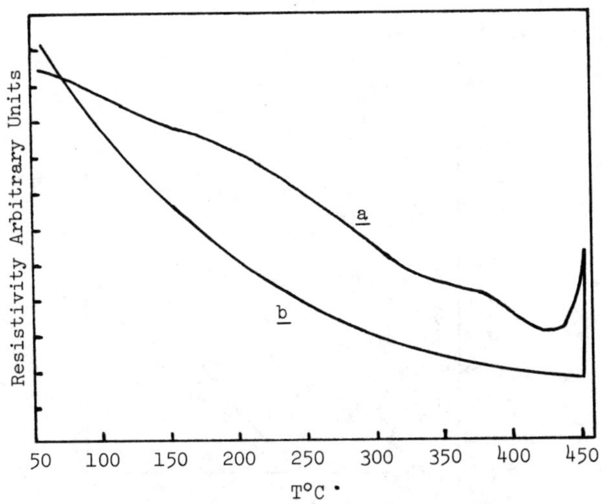


Figure 2 Change of the Resistivity During Annealing in Vacuum

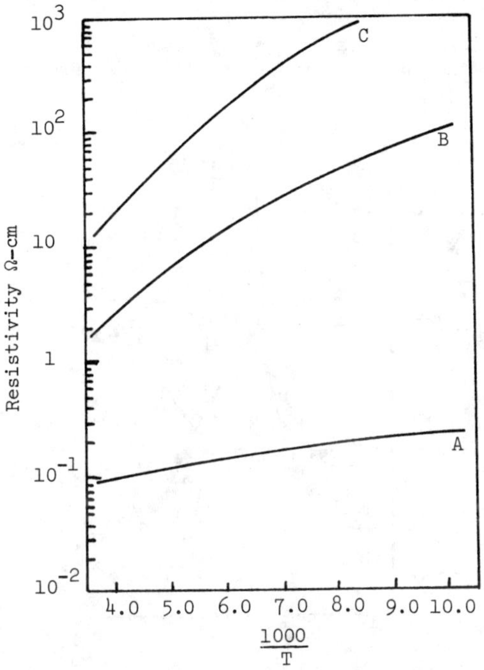


Figure 3 Temperature Dependence of the Resistivity

ing curve is smaller and the minimum is not so pronounced. The cooling curve b, however, does not change.

Figure 3 depicts the variation of the resistivity between 77°K and room temperature as a function of 1/T for films identified in Table 1. The activation energy as calculated from these curves increases when the films are annealed in vacuum at higher temperatures. The highest value of the activation energy at 0.09 eV for the n-type film is about half of that reported for n-type single crystals[1]. Reduction of the activation energy with annealing in Se atmosphere has been reported for bulk samples[3]. Whereas the order of magnitude of the room temperature resistivity of the n-type films agrees with that of n-type single crystals[1] at about 10Ω cm, the temperature variation of the film and bulk resistivities differ substantially. Namely, the resistivity peak observed near $T_C$ in the bulk is absent in the films at T = 120°C, the measured film Curie temperature. For all films the slope of the $\ell n\rho$ vs 1/T curve decreases with decreasing temperature, indicating a decrease in the carrier mobility with temperature.

In Figure 4 the temperature dependence of the thermoelectric power Q is shown for the p and n-type films. Q varies monotonically with T for all films. The anomaly of Q near $T_C$ observed in n-type bulk material was not present.

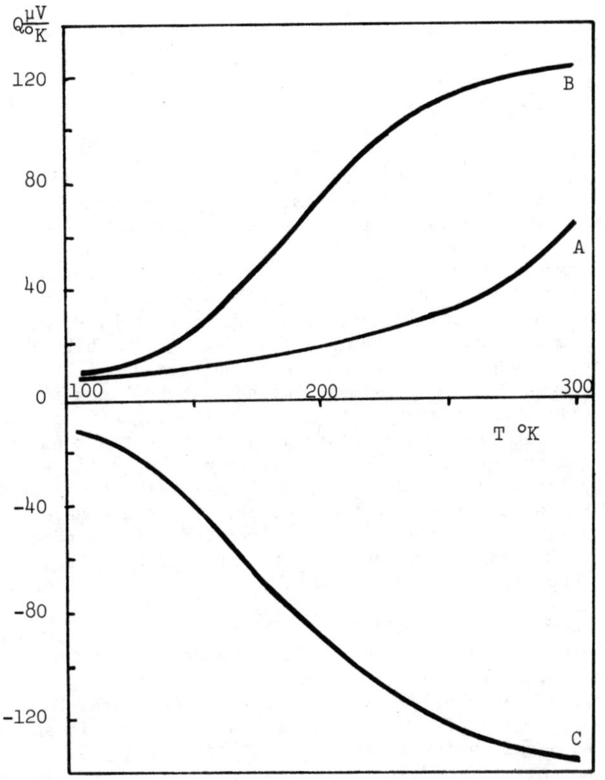


Figure 4 Temperature Dependence of the Thermoelectric Power

Neither magnetoresistance nor Hall effect was observable even though the experimental set-up permitted measurement of Hall mobilities as low as $10^{-2}$ cm²/V sec.

## DISCUSSION

As seen in Table 1, removal of Se reduces the electrical conductivity of the films. This is in contrast with the bulk behavior[1] where the conductivity of stoichiometric samples is small and increases considerably with Se reduction. The large conductivity of the unsufficiently reduced films is the result of the large conductivity of the Se surface layer as seen in Figure 1a and Se between the $CdCr_2Se_4$ grain boundaries. This also accounts for the small Hall effect because of the metallic behavior of Se.

In the case of the reduced films (n-type films) the unusually small Hall effect would indicate a rather large carrier concentration, $n > 10^{19}$ cm$^{-3}$, which, combined with low mobility, $\mu < 10^{-2}$ cm²/V-sec, would result in resistivities of the order of magnitude shown in Table 1, i.e., $\rho \sim 10$ Ω-cm.

A current reexamination of the sputtering conditions indicates that bulk-like $CdCr_2Se_4$ films may be grown by RF sputtering from CdSe + $Cr_2Se_3$ targets enriched in Se and/or CdSe onto heated sapphire substrates without a subsequent heat treatment in Se atmosphere.

*Supported by NSF Grant DMR 73-02416

## REFERENCES

1. Amith, A. and L. Friedman, Phys. Rev. B, 2, 434 (1970).
2. Tchernev, D. I. and A. J. Syllaios, AIP Conf. Proc. No. 10, Part I, 515 (1973).
3. Larsen, P. K. and A. B. Voermans, J. Phys. Chem. Solids 34, 645 (1973).
4. Tchernev, D. I. and A. J. Syllaios, AIP Conf. Proc. 24, 376 (1975).
5. Baltzer, P. K., P. J. Wojtowicz, M. Robbins and E. Lopatin, Phys. Rev. 151, 151 (1966).

# OBSERVATION OF DOMAINS FIRST ORDER TRANSITIONS IN NORMAL SPINEL $ZnCr_2Se_4$

R. Plumier, M. Sougi, A. Miedan-Gros, M. Lecomte
Service de Physique du Solide et de Résonance Magnétique
CEN-Saclay, BP n°2, 91190 Gif-sur-Yvette
France

ABSTRACT

Specific heat, neutron diffraction, magnetization and susceptibility measurements have been performed in an extended range of temperatures on powder $ZnCr_2Se_4$. They indicate that in this compound the usually admitted Néel temperature $T_N = 21\,K$ is actually a first order transition temperature at which helimagnetic macrodomains spontaneously break into metamagnetic microdomains. Most of these domains split again at $T = 45.5\,K$. The high temperature magnetic structure is centred tetragonal and a true $T_N = 105\,K$ is determined. Entropy due to Bloch walls flexibility explains the stability of the magnetic phase. A(T,H) phase diagram is obtained which indicates that in the presence of a magnetic field, the micromagnetic phase may be stable at $T < 21\,K$; this also explains the observed kink of the magnetization curves. From this phase diagram, an entropy value is obtained which is in good agreement with the one obtained from direct calorimetric measurements.

Normal spinel $ZnCr_2Se_4$* has been extensively studied in the past[1,7] and, as the magnetic susceptibility has a sharp maximum around $T_c = 21\,K$, this temperature has generally been considered as the Néel temperature. However as specific heat measurements indicate that, in addition to a fairly large peak (from $T = 17\,K$ to $T = 23\,K$) culminating at $T = 20.5\,K$, two other narrower peaks exist at $T = 45.5\,K$ and $105\,K$, careful neutron work has been recently performed at EL3 Saclay on a powder specimen in an extended range of temperatures and magnetic fields, the preliminary results of which have just been published[8]. The main results and conclusions are as follows :

1) Up to a so-called Néel temperature (referred as $T_c$ from now on), diffraction patterns are very similar to the one obtained at 4.2 K [3,4] (fig.2). Well defined magnetic satellites are observed with little change in intensities and angular positions. The helix angle $\varphi$ varies from 41° at 4.2 K to 38° at 21 K (fig.1) ; this $\varphi$ variation, also observed in other chalcogenide spinels[9,10], may be explained by the existence of a small biquadratic term[10]. However, starting at about 17 K and with an intensity increasing as a function of temperature, we notice that a broad peak appears on the left side of the $(000)^{\pm}$ magnetic satellite.

2) At $T_c$ and slightly above $T_c$, the opposite situation occurs. The well defined $(000)^{+}$ satellite practically disappears (it actually does so at $T = 21.2\,K$) whereas the broad peak reaches its full intensity. A careful search on both sides of the (111) nuclear reflection (which remains narrow) shows the existence of both the $(111)^{\pm}$ broad satellites. Thus, all mechanisms other than small particles (here small magnetic domains) scattering effects may be discarded to explain the anomalous width of the peaks. From the angular positions of $(000)^{\pm}$ and $(111)^{\pm}$, a pitch angle $\pi/5$ is deduced. This is an interesting result as it indicates that at $T > T_c$, $ZnCr_2Se_4$ no longer is a helimagnet but a metamagnet with a centred tetragonal cell $(a/\sqrt{2}, a/\sqrt{2}, 5a)$. This situation maintains itself up to $T'_c \simeq 45.5\,K$ where a decrease of the peak height together with an increase of the line width are observed (fig.2). However in the temperature range $21\,K < T < 45\,K$ and from the observed line width corrected for experimental line width[11], we obtain $d \simeq 53$ Å for the size of the reflecting domains (microdomains). It is quite remarkable that this is just 5a ($\simeq 52.42$ Å), the height of the metamagnetic unit cell and also twice the minimum size for a $2\pi$ spin rotation. No magnetic interaction will thus exist between microdomains in the metamagnetic state (no poles). It is interesting to notice that this would not have been the case for helimagnetic domains of the same size. In the temperature range

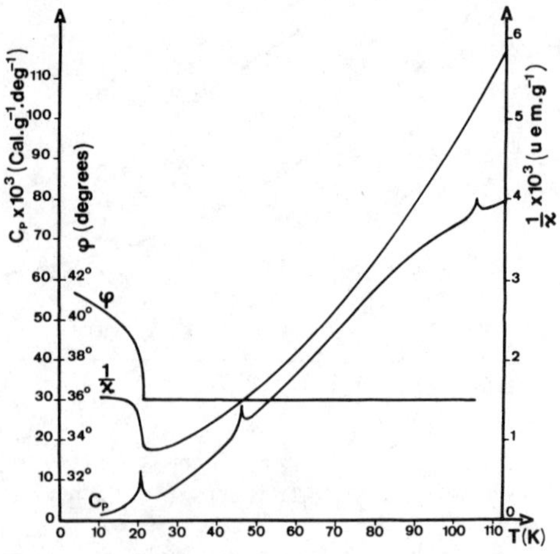


Fig.1 - Specific heat, magnetic susceptibility and pitch angle variations as a function of temperature.

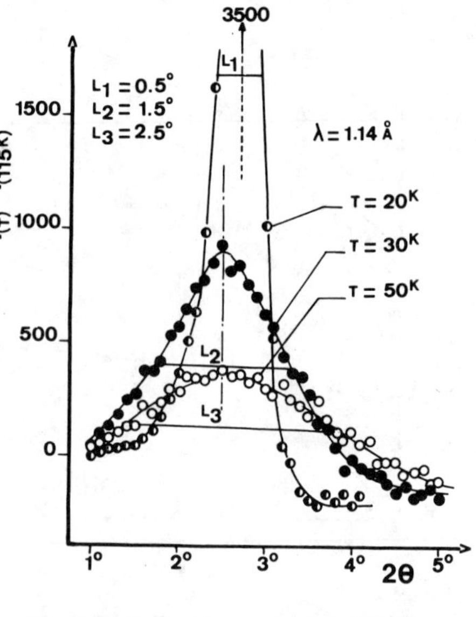


Fig.2 - Part of neutron spectrum

$45.5\,K < T < 105\,K$, from the observed line width, we obtain in the same way $d \simeq 32$ Å for the average size of the reflective domains. This indicates that at $T'_c = 45\,K$, a fairly large number of microdomains break into two smaller microdomains. These will be 5a/2 high. This is the minimum size for a single $2\pi$ spin rotation and no magnetic interaction.

3) Up to 105 K, whereas no further change in line width nor angular position are observed, the intensities of the broad magnetic peaks regularly decrease until they completely disappear. When due account is taken of line widths and Lorentz factors, we observe that, within experimental limits, there is no change in magnetization neither at $T_c$ nor at $T'_c$. This set of results lead us to consider 105 K as the true Néel temperature. It is worth noticing that the relative magnetic exchange energy $\Delta w$ between the helimagnetic and metamagnetic phases is $< 1\%$ and that this reduction is the smallest which may be achieved by a magnetic cell commensurate to the nuclear one.

4) However the most striking results are those obtained when at $T < T_c$, a magnetic field is applied along the scattering vector of the $(000)^{\pm}$ magnetic reflection.

a) At moderate field intensities, we first look for the variation as a function of temperatures of the switching field $H_s$ for Bloch walls (called B.W. from now on) motion[4]. The result depicted on fig.3 indicates that the pinning energy of B.W. gets smaller and smaller as we approach $T_c$, from the low temperature side. This is an important result as it shows that at $T \geq T_c$, B.W. may be nucleated at a very low cost ; in addition, these B.W. will have great flexibility and, as a result, large entropy sets in[12,13]. At finite temperatures, the entropy term TS will compete with both $\Delta w$ and surface energy at B.W. in such a way that as soon as its free energy becomes lower, the microdomain phase (phase II) will take over the macrodomain phase (phase I). Our experimental results indicate that this is indeed what happens in $ZnCr_2Se_4$ at $T_c = 21\,K$. In the same spirit, we consider the transition occurring at $T'_c = 45\,K$ from phase II to phase III (fig.3) as being due to a balance between an increase B.W. surface energy and an entropy increase arising from additional split of the domains.

b) If after the wall switching has occurred, we go on increasing the magnetic field intensity, we observe that at all temperatures a critical field $H_c(T)$ exists (fig.3) such that $H < H_c$, the well defined $(000)^{\pm}$ satellite is still there although reduced as a consequence of the canting of the moments towards the field direction[4]), where as at $H > H_c$, a broad peak appears with the same line width and angular position as the one previously observed at $T > T_c$ and zero field. In addition, we notice some remanence effects around $H_c(T)$ while coming back to zero magnetic field ; this is a typical result for first order transitions. From fig.3, where the $H_p(T)$ line gives the magnetic field for transition to the paramagnetic state, it follows that the experimental points $H_c(T)$ nicely fall on a transition line described by the parabolic curve $10^{-8}\,H^2 = 2.16(21 - T)$. This relation not only confirms the first order character of the transition we have just described but also leads to a direct determination of the entropy change at the transition as may be seen from the following. From fig.1, we notice that at $T < T_c$, the magnetic susceptibility $\chi_I$ (independent of T and H as it should be[14]) is about one half the extrapolated susceptibility $\chi_{II}$. Remembering that in the same range of temperatures magnetic saturation is practically reached in both phases, we obtain by equating the free energy of both phases: $H^2(\chi_{II} - \chi_I) \simeq 2S(T_c - T)$ which has the expected form. This relation gives $S \simeq 0.74$ cal.mol.deg$^{-1}$ whereas direct calorimetric measurements led to $S =$ 0.42 cal.mol.deg$^{-1}$. This last type of measurements gives $S = 0.07$ cal.mol.deg$^{-1}$ at 45.5 K where we have seen that the number of microdomains is increased by a factor $\alpha$ with $\alpha \lesssim 2$. The latent heats at $T_c$ and $T'_c$ are respectively 8.4 cal/mole and 3 cal/mole. The agreement between these experimental values is quite satisfactory and could

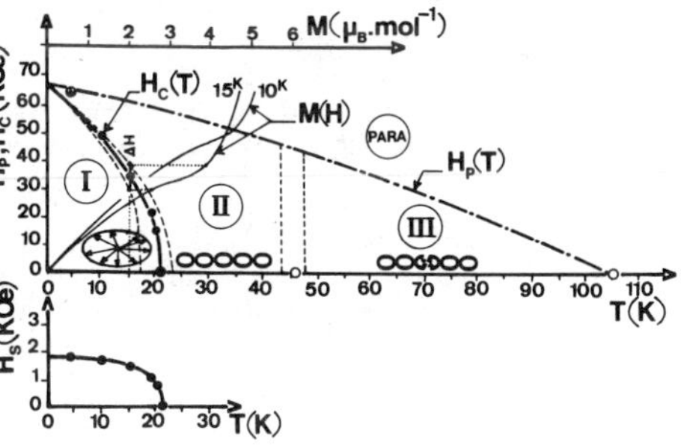


Fig.3 - Variation of $H_s$, $H_c$, $H_p$ and $\Delta H$ as a function of temperature ; M(H) are typical magnetization curves $T < T_c$ (• neutron work, o specific heat, ◐ pulsed magnetic field)

be improved if the increased pinning energy and reduced flexibility of B.W. at $T < T_c$ were taken into account.

The first order transition also offers an explanation of the kink observed in the magnetization curves at $T < T_c$ (fig.3). The field value at the centre of the kink is just $H_c(T)$, the spread being due to the unavoidable distribution of size, shape, internal stresses and imperfections in the various macrodomains which forbids them to suffer the transition at the same time. From fig.3, we see why this spread $\Delta H$ which also manifests itself in specific heat and neutron diffraction measurements is an increasing function of temperature.

The microdomains in phase II are, in some way, reminiscent of the superantiferromagnets extensively studied by Néel[15]. However Néel was dealing with the magnetic properties of small isolated grains whereas in $ZnCr_2Se_4$ the microdomains appear spontaneously in an otherwise massive sample. We may thus understand why certain properties like the susceptibility enhancement are the same in both cases. Others, however, like the phase transition observed at $T_c$ are typical of cooperative phenomenon occurring in an infinite medium.

We do not see any reason why the phase transition observed in $ZnCr_2Se_4$ could not also occur in other magnetic compounds. The unusual magnetic properties of $HgCr_2S_4$ [6,9] and $MnCl_2$ [16] seem to indicate that these compounds also suffer such a transition. From the foregoing, it is obvious that careful neutron scattering around the admitted transition temperatures together with detailed specific heat measurements will be of great help to determine such transitions. Such a work is now under way in our laboratory on normal spinels $Ag_{1/2}In_{1/2}Cr_2S_4$ [10] and $Cu_{1/2}In_{1/2}Cr_2Se_4$ [17]. Preliminary results indicate in these compounds a behaviour similar to the one observed in $ZnCr_2Se_4$ and reported here.

References

1. F.K. Lotgering, Proc. Int. Conf. on Magnetism, Nottingham, Sept.1964, p.533.
2. F.K. Lotgering, Solid State Commun. 3, 347 (1965).
3. R. Plumier, C.R.Acad.Sci.Paris, 260, 3348 (1965).
4. R. Plumier, J. Physique 27, 213 (1966).
5. Y. Allain et al., C.R.Acad.Sci.Paris, 260, 4677 (1965).
6. P.K. Baltzer et al., Phys. Rev. 151, 367 (1966).
7. K. Dwight, N. Menyuk, Phys. Rev. 163, 435 (1967).
8. R. Plumier et al., Phys. Lett. 54A, (1975).
9. J.M. Hastings, L.M. Corliss, J. Chem. Phys. Solids, 29, 9 (1968).
10. R. Plumier, M. Sougi, Solid State Commun. 9, 417 (1971).
11. R. Plumier, J. Physique, 24, 741 (1963).
12. Y.Y. Li, Phys. Rev. 101, 1450 (1956).
13. M.M. Farztdinov, Sov. Phys. Uspekhi, 7, 855 (1965).
14. A. Herpin, "Théorie du Magnétisme", P.U.F. (1968).
15. L. Néel, Phys. des Basses Températures, Les Houches (1961), p.413, C. DeWitt Edit. ; C.R.Acad.Sci.Paris 253, 9 (1961) ; 253, 203 (1961).
16. M.K. Wilkinson et al., Phys. Semian Progr. Rep., March 10, 1958, ORNL 2430, p.37.
17. R. Plumier et al., J. Physique C-1, 324 (1971).

* Compound prepared by Dr. F.K. Lotgering, Philips Research Lab.

# MAGNETIC INTERACTIONS IN EUROPIUM HEXABORIDE

W. S. GLAUNSINGER*
Department of Chemistry
Arizona State University
Tempe, Arizona 85281

## ABSTRACT

$EuB_6$ samples of widely separated electron density have been studied by EPR. In bulk samples, narrow signals have been observed and attributed to conduction electrons coupled to the europium moments. The exchange field increases linearly with temperature above the temperature range where critical spin fluctuations occur. The europium resonance splits into 2 lines below 21 K, and the splitting increases with decreasing temperature, with the low-field line being considerably more intense than the high-field line, which is unobservable near 10 K. Several models for the exchange interaction between the europium moments have been examined, and it is found that the RKKY interaction can account for the magnitude of the high-temperature exchange field in bulk $EuB_6$, but that another interaction, quite possibly the BR interaction, makes the dominant contribution to its temperature dependence. The splitting of the europium resonance, which cannot be attributed to crystalline-electric-field effects, may result from a magnetic transition, possibly involving magnetic-polaron formation.

## INTRODUCTION

The rare earth hexaborides crystallize in a cesium chloride structure, with both rare earth ions and boron octahedra located on simple cubic lattice sites.[1] Due to the rigidity of the boron framework, the metal-metal spacing is approximately the same in each compound. Their simple structure, coupled with the fact that most of the rare earth ions are paramagnetic, make these materials important model systems for fundamental studies of magnetic interactions in the solid state.

$EuB_6$ is unique because it is the only semiconductor,[2] as well as being the only ferromagnet,[3] among the rare earth hexaborides. The electron density in $EuB_6$ varies from about 0.09 el/Eu-site in bulk material[2] to about $10^{-3}$ el/Eu-site for thin films grown on a boron substrate.[4] Although several exchange mechanisms have been suggested for the exchange interaction between the europium moments,[5-7] the observation of a temperature-dependent paramagnetic Curie temperature in both bulk[4,8] and thin-film preparations,[4] which can be correlated with electron-density changes, and relatively large g-shift[9] suggests that the europium moments are coupled by mobile electrons. On the basis of a very approximate calculation of the Weiss constant, Fisk[10] has suggested that the Bloembergen-Rowland (BR) interaction is dominant. Recently, Wood[11] has pointed out that the Ruderman-Kittel-Kasuya -Yosida (RKKY) interaction should dominate in bulk $EuB_6$. However, both Fisk and Wood assumed that nearest-neighbor exchange predominates, which is in general a questionable approximation in view of the long-range nature of indirect-exchange interactions.

In this paper the results of a variable-temperature EPR study of $EuB_6$ at two widely separated electron densities are reported and interpreted, with the purpose of elucidating the important exchange mechanisms[5] coupling the europium moments.

## EXPERIMENTAL

Bulk samples were prepared by reacting $Eu_2O_3$ with crystalline boron in a zirconium diboride crucible at 1800°C. Samples having a lower electron density than bulk samples were obtained by keeping the sample at 2400°C for an extended period and then removing the crystallites of $EuB_6$ from the crucible lid.

Samples were prepared for EPR study by reducing the particle size to less than 45 μ and pressing the powder into a 1×4 mm tube. EPR spectra were recorded at X-band using second derivative signal detection.[12] The field sweep was calibrated with an adjustable single-crystal ruby standard.[13] A liquid helium flow system was used to produce temperatures down to 10 K which were stable to within 0.5 K.

Accurate g-factors were measured using charred dextrose as an internal standard,[14] and resonant fields were corrected for demagnetization.[9]

## RESULTS AND DISCUSSION

Typical second-derivative EPR spectra above 25 K in bulk samples and samples having a lower electron density than bulk samples, hereafter referred to as I and II, respectively, are displayed in Fig. 1. A weak, narrow signal near g = 2 is superimposed on the europium resonance in I. This signal is distorted under the experimental conditions used to record the europium resonance, but by decreasing the modulation amplitude and field-sweep range and rate and increasing the signal amplification, undistorted signals were recorded with a signal-to-noise ratio of 25:1 at 296 K. The lineshape is approximately Lorentzian. The g-factor is independent of temperature and equal to

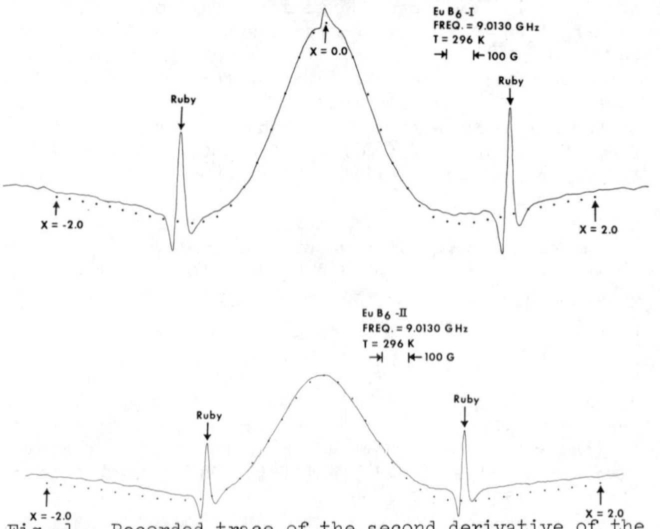


Fig. 1. Recorded trace of the second derivative of the EPR absorption in bulk (I) and low-electron-density (II) $EuB_6$. The squares are values computed from a Lorentzian lineshape function normalized to the maximum in the experimental spectrum. The resonance lines of the ruby standard are used to calibrate the field sweep.

2.000 ± 0.002. The linewidth is about 8 G at 360 K and increases linearly with decreasing temperature between 360 and 150 K at a rate of 0.8 G/100 K and then increases more rapidly with decreasing temperature below 150 K. The resonance line was too broad to be detected below about 40 K. The narrowness of this signal, as well as its appearance at g = 2.000, indicates that it arises from mobile spins. The fact that this signal is only observed in I suggests that it is due to conduction electrons. From the measured signal-to-noise ratio, it is estimated that I and II differ in electron density by at least an order of magnitude. As shown below, the temperature dependence of the conduction-electron linewidth is similar to that of the europium moments, which suggests a strong coupling between these two spin systems. The increase in conduction-electron linewidth with decreasing temperature between 360 and 150 K implies that the relaxation rate increases with decreasing temperature. This is reasonable because the magnetic susceptibility of $EuB_6$ increases with decreasing temperature, thereby enhancing the magnetic strength of the europium resevoir to which the conduction electrons are coupled. The rapid increase in linewidth below 150 K signals the onset of critical spin fluctuations. In such case the conduction electrons experience local fields from the europium moments which fluctuate more slowly as the temperature decreases, thereby causing the relaxation rate, and hence linewidth, to increase rapidly with decreasing temperature and eventually broadening the signal beyond observation.

The europium resonance can be fitted to a Lorentzian lineshape function in the central region and an exponential lineshape function in the wings. The g-factor above 25 K in both I and II is independent of temperature and equal to 2.000 ± 0.003. Above 150 and 90 K the linewidth increases linearly with decreasing temperature in I and II, respectively, whereas below these temperatures the linewidth increases more rapidly. The rapid increase in linewidth below 150 K in I and below 90 K in II is characteristic of the occurrence of critical spin fluctuations. Above 150 and 90 K, respectively, critical spin fluctuations are unimportant, as evidenced by the relatively weak temperature dependence of the linewidth. Above the temperature range where critical spin fluctuations occur, to a good approximation the exchange field is given by

$$H_e = 10\, M_2^d / 3\Delta H, \qquad (1)$$

where $M_2^d$ is the dipolar contribution to the second moment and $\Delta H$ is the half-width at half-maximum absorption. The temperature dependence of the exchange

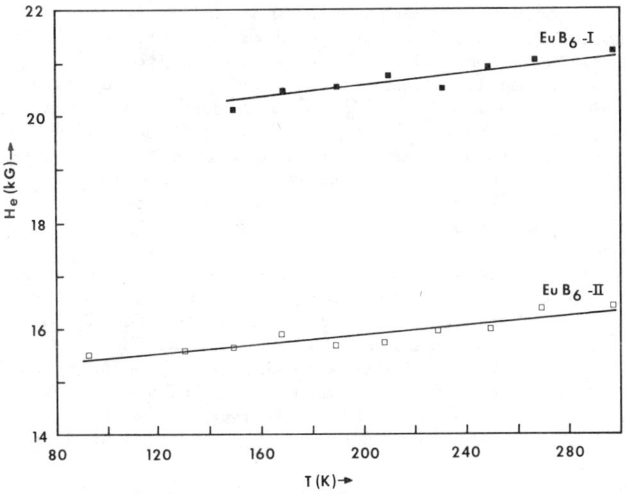


Fig. 2. Temperature dependence of the exchange field in bulk (I) and low-electron-density (II) $EuB_6$.

field is shown in Fig. 2. The exchange field in I is larger than in II, but in both samples the exchange field increases linearly with temperature at a rate of 5.4 G/K in I and 4.2 G/K in II. The larger exchange field in I implies a stronger magnetic coupling between the europium moments, which in turn implies a larger Weiss constant. Considering only nearest-neighbor exchange interactions, the predicted Weiss constants are 16.0 and 12.4 K in I and II, respectively,[15] which are not in very good agreement with that determined from magnetic susceptibility measurements (about 9 K[3,4]). This discrepancy suggests the presence of exchange interactions between nonnearest-neighbor europium moments, which are larger in I than II. Furthermore, the linear increase of exchange field with temperature in both samples indicates that the magnetic coupling increases with temperature. Hence dipolar, direct exchange, or superexchange interactions cannot be dominant. On the RKKY model, the magnetic behavior in I only can be estimated, because the electron density in II is too low to apply the RKKY model numerically. Taking the sf exchange integral to be 0.036 eV and the conduction-electron density to be 0.09 el/Eu-site, the exchange field predicted on the RKKY model is 17.4 kG,[16] which is about 80% of the experimental value. Assuming a 10% increase in electron density between 165 and 300 K in I, it is estimated that the exchange field should increase with temperature by about 2 G/K,[15] which is only about 40% of the experimental value. If a possible difference between the squares of the valence and conduction-band wave functions and the correction to the BR interaction resulting from the unavailability of the lowest states in the conduction band are neglected, then at a given temperature the exchange field due to the BR interaction is the same in both I and II. Taking the valence-electron density to be 2 el/Eu-site and the band gap to be 0.38 eV,[10] the exchange field predicted on the BR model is 29.4 kG,[15] which is about 40% larger than the experimental value. A 1% decrease in temperature dependence of the valence-electron density in I between 30 and 300 K would cause the exchange field to increase with temperature by about 3 G/K,[15] which is about 60% of the experimental value, and, in conjunction with the RKKY prediction, would account for the temperature dependence of the exchange field in I. Hence it appears that the RKKY interaction can account for the magnitude of the high-temperature exchange field in I, but that another interaction, quite possibly the BR interaction, makes the dominant contribution to its temperature dependence. The g-shift provides further evidence for the dominance of the RKKY interaction at high temperatures. In the absence of a magnetic-resonance bottleneck, the g-shift predicted on the RKKY model is 0.006 in I,[16] which is in good agreement with the experimental value (0.007 ± 0.003). Although the RKKY interaction can account for the high-temperature exchange field and g-shift in I, it fails to predict the correct low-temperature behavior. In I, the Weiss constant predicted on the RKKY model is 24 K,[16] which is in poor agreement with experiment, suggesting that the RKKY model is inadequate at low temperatures.

In order to investigate the origin of the inadequacy of the RKKY model at low temperatures, low-temperature EPR measurements were undertaken. Just below 25 K the europium resonance becomes distorted, and below 21 K it was evident that the resonance was splitting into two lines. The splitting increases with decreasing temperature, and below 15 K the splitting was clearly resolved, with the low-field line being considerably more intense than the high-field line, which is unobservable near 10 K. The splitting of the europium resonance is greater than 1 kG at 15 K. Both I and II exhibited this behavior.[15] A possible explanation of the splitting of the europium resonance is that it might arise from the splitting of the free-ion europium levels by the crystalline-electric field. However, an estimate of the fine-structure anisotropy ($b_4 < 250$ G[17-19]) indicates that it is too small to explain the large splitting. Furthermore, the temperature-independent fine structure cannot be attributed to crystalline-electric-field effects. Another possibility is that the splitting results from a magnetic transition near 25 K. Such a transition may involve magnetic-polaron formation;[5] however, the experimental data are insufficient to establish the existence of a magnetic polaron in $EuB_6$. In order to resolve this question, EPR studies of samples of controlled electron density, prepared by incorporating known amounts of gadolinium into $EuB_6$, are being undertaken.

REFERENCES

*Research supported by the Research Corporation and Arizona State University

1. Kissling R., Acta Chem. Scand., 4, 209-227 (1950).
2. Samsanov G.V., High Temperature Compounds of Rare Earth Metals with Nonmetals, Consultants Bureau, New York, 1965, 1-100.
3. Krause J.L and Sienko M.J., J.Solid State Chem., 6, 590-594 (1973).

4. Bachmann R., Lee K.N., Geballe T.H., and Menth A., J.Appl. Phys., 41, 1431-1432 (1970).
5. Matthias B.T., Phys. Lett., 27A, 511 (1968).
6. Ruderman M.A. and Kittel C., Phys. Rev., 96, 99-102 (1954).
7. Bloembergen N. and Rowland T.J., Phys. Rev., 97, 1679-1698 (1955).
8. Hacker H., J.Mag. Res., 18, 175-178 (1972).
9. Glaunsinger W.S., J.Mag. Res., 18, 265-275 (1975).
10. Fisk Z., Phys. Lett., 34A, 261-262 (1971).
11. Wood V.E., Phys. Lett., 37A, 357-358 (1971).
12. Glaunsinger W.S. and Sienko M.J., J.Mag. Res., 10, 253-262 (1973).
13. Glaunsinger W.S. and Sienko M.J., J.Chem. Phys., 62, 1883-1889 (1975).
14. Vana N.J. and Unfried E., J.Mag. Res., 6, 655-659 (1972).
15. Glaunsinger W.S., phys. stat. sol (b), to be published.
16. Glaunsinger W.S., phys. stat. sol (b), 70, K151-K157 (1975).
17. Cheema S.U. and Smith M.J.A., J.Phys. C: Solid St. Phys., 4, 1231-1241 (1971).
18. Low W. and Rosenberger V., Phys. Rev., 116, 621-623 (1959).
19. Shuskus, Phys. Rev., 127, 2022-2024 (1962).

## MAGNON-MAGNON AND MAGNON-PHONON RELAXATION IN $FeF_2$†

E.F. Sarmento
Universidade Federal de Alagoas, Maceio, Brazil

B. Zeks
Universidade Federal de Pernambuco, Recife, Brazil and
Institute J. Stefan, Ljubljana, Yugoslavia

S.M. Rezende††
Universidade Federal de Pernambuco, Recife, Brazil and
Department of Physics, University of California, Santa Barbara, California 93106

ABSTRACT

The relaxation of the $\vec{k} = 0$ magnon in the antiferromagnet $FeF_2$ is studied theoretically. Several magnon-magnon and magnon-phonon interaction processes are investigated. It is shown that despite the large magnetoelastic interaction in $FeF_2$, magnon-phonon processes give negligible contribution to the linewidth. The existing temperature dependent linewidth measurements can be accounted for by four-magnon processes.

The antiferromagnetic resonance in $FeF_2$ was studied several years ago by far-infrared spectroscopy techniques[1], yielding information on material parameters and on characteristics of the $\vec{k} = 0$ magnon. More recently, data on magnon properties have been added by Raman[2] and neutron scattering[3,4] experiments. Despite the fact that ferrous fluoride is a simple two-sublattice antiferromagnet, representative of a class of very high anisotropy materials, with all of its important parameters known, its $\vec{k} = 0$ magnon linewidth has not been explained microscopically. Such an explanation should provide an important test of our understanding of the magnon interactions in this material.

Fig. 1 shows the experimental temperature dependent part of the linewidth of the $\vec{k} = 0$ magnon, obtained by FIR[1] and Raman[2] techniques. At temperatures above $T_N/4$ they give linewidth measurements in reasonably good agreement with each other which cannot be readily explained by the existing analytic linewidth calculations[5,6]. Due to this fact we were led to investigate several different interaction processes that could relax the zone-center magnon.

Since the $Fe^{2+}$ ion in $FeF_2$ has a non-zero orbital angular momentum, the spin-orbit-crystal field couplings give rise to a strong magnetoelastic interaction. As a consequence magnon-phonon processes should be considered as possible relaxation mechanisms of much more importance than in crystals with S-state magnetic ions, such as the well studied $MnF_2$. Let us consider these first. Take for the magnetoelastic interaction Hamiltonian

$$\mathcal{H}_{me} = \sum_i b_{\alpha\beta\gamma\delta} S_i^\alpha S_i^\beta \frac{\partial R_i^\gamma}{\partial x_\delta} \quad (1)$$

where $S_i^\alpha$ is the $\alpha$- component of the spin operator at lattice site i, $R_i^\gamma$ is the $\gamma$-component of the displacement operator and $b_{\alpha\beta\gamma\delta}$ are the lowest order magnetoelastic constants. By expressing $S_i$ and $R_i$ in (1) in terms of magnon and phonon operators one obtains the several N magnon-1 phonon interaction Hamiltonians. The linear magnon-phonon interaction leads to a splitting in the magnon-phonon dispersion curves which has been observed in $FeF_2$[4]. Measurement of this frequency splitting allows us to determine some of the non-vanishing $b_{\alpha\beta\gamma\delta}$. At T = 4.2 K they turn out to be of the order of $10^{-15}$ erg. Higher order magnon-phonon interactions provide possible magnon relaxation channels. Due to the forms of the magnon and phonon dispersion curves, one can see that 2 magnon-1 phonon processes cannot conserve both energy and momentum to relax the k = 0 mode. This is not the case for 3 magnon-1 phonon and 4 magnon-1 phonon processes, which can contribute to the relaxation. We have calculated the corresponding relaxation rates using the procedures of Ref. 6, with the integrals evaluated numerically. We have found that 3 magnon-1 phonon contribute negligibly ($\sim 10^{-3}$ $cm^{-1}$ at 30 K). 4m-1ph processes, in which the magnon mode to be relaxed combines with one thermal magnon and one phonon to give two output magnons, give a much larger contribution, which is plotted in Fig. 1. Despite the fact that in the latter case there is one more magnon involved in the process, contributing with a relatively small thermal occupation number, many more magnon and phonon states are allowed by the conservation rules to participate in the process, giving rise to a larger relaxation rate. Here we note that our first evaluation of the magnon-phonon relaxation rate, done analytically under the high temperature approximation of Ref. 6, led to an approximately correct temperature dependence for the linewidth but grossly overestimated. The numerical results show actually that magnon-phonon processes cannot account for the measured AFMR linewidth of $FeF_2$.

Among the possible magnon interaction mechanisms, four magnon processes have been found to give the largest contribution to the linewidth over most of the ordered state temperature range. The relaxation rate for a magnon mode $k_1 = 0$ due to a process in which it combines with a thermal magnon $k_2$ giving two output magnons, is

$$\eta_1 = \frac{2\pi}{\hbar^2}(e^{\beta\hbar\omega_1}-1)\sum_{k_2k_3}|C|^2 e^{\beta\hbar\omega_2}\bar{n}_2\bar{n}_3\bar{n}_{2-3}\delta(\omega_1+\omega_2-\omega_3-\omega_4) \qquad (2)$$

where C is the appropriate coefficient which takes into account all possible processes. The 4-magnon processes[5,7] from both the exchange and the anisotropy interactions can either involve magnons of the same type (2α-2α) or can be of the intermode type, in which one α mode combines with a β mode to give another αβ mode pair. Since $k_1 = 0$, energy-momentum conservation makes the joint density of states largest when $k_2$ and one of the output magnons have large momenta. This simplifies the coupling coefficients for the two types of processes which can be written approximately as

$$C^{\alpha\alpha} = 2z_2J_2(u_0v_0-v_0^2) - 4u_0^2K$$

$$C^{\alpha\beta} = z_2J_2(u_0v_0-2u_0^2) - 4v_0^2K \qquad (3)$$

where $J_2$ is the exchange constant, K the anisotropy constant ($H_E \equiv 2Sz_2J_2/g\mu_B$, $H_A = (2S-1)/g\mu_B$) and $u_0$ and $v_0$ are the k = 0 spin wave operator transformation coefficients[7]. In $FeF_2$, $H_A$(200 kOe) is comparable to $H_E$(580 kOe), $u_0 = 1.1$ and $v_0 = 0.5$, leading to an intramode scattering vertex $C^{\alpha\alpha}$ which is dominated by the anisotropy interaction. Furthermore, we note that the exchange and anisotropy scattering amplitudes interfere destructively in the intramode case, whereas in the intermode case $C^{\alpha\beta}$ they interfere constructively. As a result the two types of processes have comparable contributions to the relaxation rate, as opposed to the situation in $MnF_2$[7].

The relaxation rate (2) was calculated with only one major approximation, namely that the magnon dispersion relation is linear $\omega = \omega_0 + vk$. This leads the energy delta function to $\delta(\omega) = (k_2-k_3)\,\delta(\cos\theta - 1/vk_2k_3$, where θ is the angle between $k_2$ and $k_3$, and $k_3$ is bound by $k_2 \geq k_3$. The 4 magnon relaxation rate for mode 1 then becomes

$$\eta_1^{(4)} = \frac{9\pi(e^{\beta\hbar\omega_0}-1)}{2\hbar^2(\omega_{zB}-\omega_0)k_m^6}\int_0^{k_m}dk_2\int_0^{k_2}dk_3\frac{|C|^2k_2k_3(k_2-k_3)e^{\beta\hbar\omega_2}}{(e^{\beta\hbar\omega_2}-1)(e^{\beta\hbar\omega_3}-1)(e^{\beta\hbar\omega_{2-3}}-1)} \qquad (4)$$

where

$$|C|^2 = \frac{1}{2}|C^{\alpha\alpha}|^2 + |C^{\alpha\beta}|^2 \qquad (5)$$

and we have made $v = (\omega_{zB}-\omega_0)/k_m$, where $k_m = (6\pi^2/\Omega)^{1/3}$ is the maximum value of k in the spherical approximation, $\omega_{zB}$ and $\omega_0$ are the magnon frequencies at the edge and at the center of the Brillouin zone. The factor 1/2 in $C^{\alpha\alpha}$ in (5) takes account of the fact that the two output magnons are indistinguishable when all the magnons are of the same type. The relaxation rate (4) was evaluated as a function of temperature for $FeF_2$, assuming that the dependence of v on the temperature is the same as for $\omega_0$[1,2]. Since in $FeF_2$, at T = 0 K, $\omega_0$ = 75 K and $\omega_{zB}$ = 112.5 K, and at T = 60 K they are renormalized by about 30%, we see that in the range 20 - 60 K one is not allowed to use the approximations which are valid for the $MnF_2$ case[6], namely that $K_BT \ll \hbar\omega$ (and the upper limit of the integral extended to ∞) or $K_BT \gg \hbar\omega$. Therefore the integrals in (4) were calculated numerically, leading to the results shown in Fig. 1. The results are in satisfactory agreement with

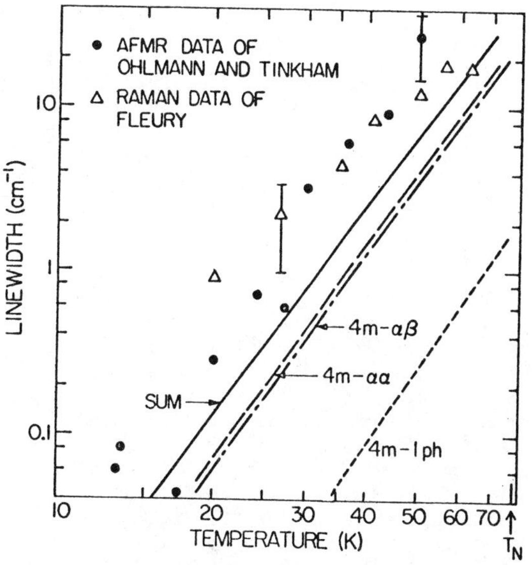


Fig. 1. Comparison Between Experimental Data and Linewidth Calculations For $FeF_2$

the experimental data if one considers the approximations used in the calculation and the fact that the experimental errors (the vertical bars in the figure indicate the instrumental width) are quite significant at lower temperatures, due to the fact that zero-temperature linewidth is subtracted from the measured values.

We have also considered 6 magnon interaction, which have been shown to be important in the AFMR relaxation of $MnF_2$. Here, due to the large magnon frequencies of $FeF_2$, the magnon population is small even at high temperatures. As a consequence, six or more magnon processes have negligible contribution compared to the 4 magnon processes.

The authors express their gratitude to Prof. R.M. White of the Xerox Research Center for helpful discussions.

REFERENCES

†Work partially supported by CNPq, BNDE and CAPES (Brazilian Government), and The National Science Foundation (USA).

††J.S. Guggenheim Foundation Fellow.

1. R.C. Ohlmann and M. Tinkham, Phys. Rev. 123, 425 (1961).
2. P.A. Fleury, Proc. 2nd Int. Conf. Light Scattering (ed. M. Balkanski, Flamarion, Paris, 1971). p. 151.
3. M.T. Hutchings, B.D. Rainford, and H.J. Guggenheim, J. Phys. C: Solid St. Phys. 3, 307 (1970).
4. B.D. Rainford, J.G. Houmann, and H.J. Guggenheim, Proc. Inelastic Scatt. Neutrons in Solids and Liquids (IAE Agency, Vienna, 1972), p. 655.
5. A.B. Harris, D. Kumar, B.I. Halperin, and D.C. Hohenberg, Phys. Rev. B3, 961 (1971).
6. R.M. White, S.M. Rezende, and L.C.M. Miranda, 20th Conf. Magn. Magnetic Mat. AIP Conf. Proceedings No. 241 (San Francisco, 1974), p. 172.
7. R.M. White, R. Freedman, and R.B. Woolsey, Phys. Rev. B10, 1039 (1974).

# HYPERFINE INTERACTIONS IN ANTIFERROMAGNETIC EuTe USING THE Te-125 MOSSBAUER RESONANCE

N. A. Blum
The Johns Hopkins University, Applied Physics Laboratory,* Laurel, Maryland 20810
and
R. B. Frankel
Francis Bitter National Magnet Laboratory, † MIT, Cambridge, Massachusetts, 02139

ABSTRACT

Europium telluride crystallizes in the NaCl structure and is antiferromagnetic with a Néel temperature $T_N$ of 9.6 K. We have used the 35.5 keV Mössbauer transition in Te-125 to examine the nature of the spontaneous magnetic moment which has been observed in conducting samples below $T_N$. At 80 K the Mössbauer resonance spectrum consisted of a single absorption line comparable in width with the spectrum of nonmagnetic cubic ZnTe, indicating the absence of a significant quadrupole splitting at this temperature. At 4.2 K the line-width increased by a factor of 1.5, which is equivalent to a single magnetic hyperfine field at the Te sites of about 73 kOe. The line showed gradual broadening with increasing external longitudinal magnetic field up to 70 kOe at 4.2 K. These observations are consistent with the magnetization measurements as well as with recent spin-echo NMR experiments showing resonances which may be attributed to Te-125. We conclude from the shape of the Mössbauer line that the broadening observed at 4.2 K is most likely due to a single magnetic hyperfine field of magnitude 73 kOe acting at the Te sites. An applied magnetic field $H < 65$ kOe induces an internal field component of opposite sign and nearly equal magnitude at the tellurium site, but near the canted to paramagnetic transition ($H \approx 66$ kOe at 4.2 K) the observed hyperfine field increases more rapidly with H, reaching a value of 112 kOe when $H = 70$ kOe.

## INTRODUCTION

The divalent europium monochalcogenides (EuO, EuS, EuSe, and EuTe) crystallize in the f.c.c. NaCl structure and are magnetically ordered at low temperature as a result of the localized spins on the $Eu^{++}$ ions. At sufficiently low temperature EuO, EuS, and EuSe are ferromagnetic, while EuTe is antiferromagnetic. The relatively simple lattice structure and variety of magnetic phenomena observed in this class of materials have resulted in their properties being extensively investigated in recent years. [1,2]

Samples of EuTe prepared by two different methods were previously examined by magnetization [3] and spin-echo NMR [4] techniques. Insulating samples ($\rho \approx 10^6$ ohm-cm) behaved like pure stoichiometric EuTe, exhibiting ideal antiferromagnetic behavior at an external magnetic field $H = 0$ for $T < T_N$. [3] Conducting samples ($\rho \approx 10^{-2}$ ohm-cm) were found to have a spontaneous magnetic moment at $H = 0$ for $T < T_N$. [3] Both types exhibited NMR spectrum echoes which were tentatively identified as Te-125 and Te-123 resonances. [4] The magnetization and NMR results are consistent in that they both suggest, for the conducting sample, and possibly for the non-conducting sample also, a departure from the magnetic symmetry imposed by the NaCl structure.

Oliveira, et al. [3] and Shapira, et al. [5] have speculated that the anomalous magnetic behavior of the conducting samples is due to the magnetic effects of conduction electrons which are present because of lack of perfect stoichiometry in the sample, voids, impurities, etc. These effects should be nonlocalized and uniformly characteristic of the entire sample. It is the purpose of the present experiment to investigate this hypothesis and to make contact with the NMR results of Raj, et al. [4].

## EXPERIMENTAL RESULTS

The EuTe Mössbauer absorber was prepared from pieces of the same crystal (#102) used in the magnetization [3] and NMR [4] experiments by grinding the EuTe together with fine powdered alumina under dry nitrogen, and hermetically encapsulating the mixture in a plastic disc. The absorber thickness was approximately 3 mg/cm$^2$ Te-125. The reference single-line absorber, containing 1.5 mg/cm$^2$ Te-125, was similarly prepared, using chips from a single crystal of cubic ZnTe. [6]

Mössbauer spectra [7] of EuTe (#102) and cubic ZnTe were obtained at 80 K for an external magnetic field $H = 0$, and at 4.2 K for several values of H from 0 to 70 kOe. Examples of the data are shown in Fig. 1, where least squares fit single Lorentzian solid lines are shown superimposed on the data points. All the spectra have the appearance of single lines of various widths, showing no resolvable structure, and having a nearly Lorentzian shape. Single line least square fits were used for the purpose of comparing ZnTe spectra in a known magnetic field with the EuTe spectra. Typical least square fits are shown in Fig. 2 and the results are summarized in Table I. The ZnTe and EuTe absorbers are not quite the same effective thickness and the calculation of the internal magnetic field, $H_{int}$, at the Te site in EuTe takes this into account. The line shape of the broadened EuTe sample (4.2 K, H = 0) matches that observed in ZnTe (4.2 K, H = 70) with an accuracy which strongly favors an interpretation of the broadening in EuTe as due to a single internal field, $H_{int} = 73$ kOe.

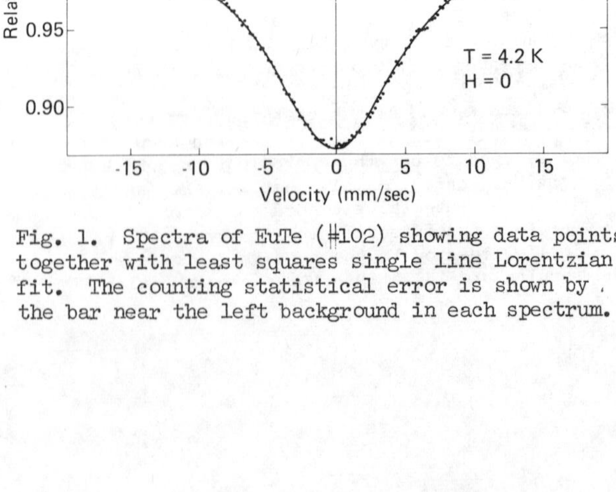

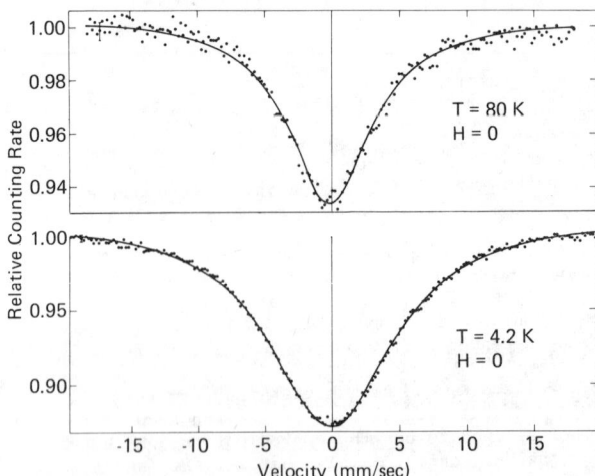

Fig. 1. Spectra of EuTe (#102) showing data points together with least squares single line Lorentzian fit. The counting statistical error is shown by the bar near the left background in each spectrum.

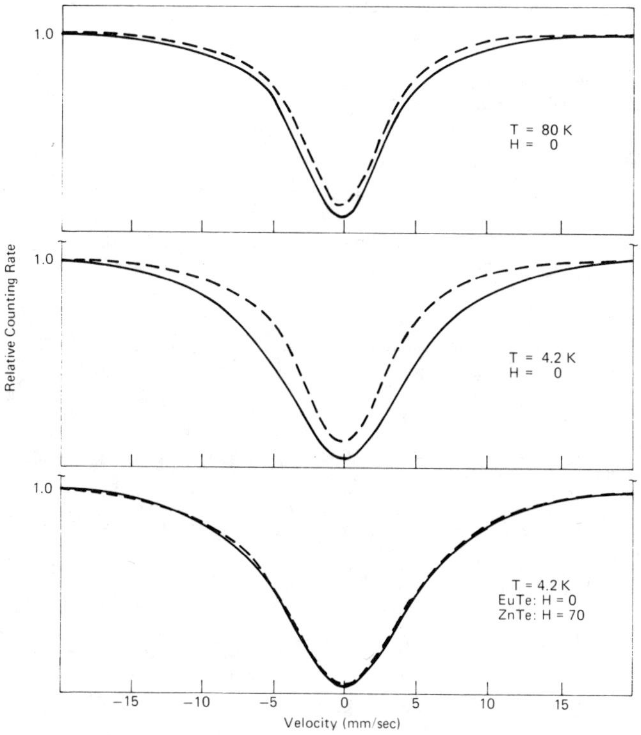


Fig. 2. Spectra showing superimposed least squares fits of EuTe (#102)(solid line) and ZnTe (dashed line) for various temperatures and magnetic fields. The data points were omitted to avoid cluttering the figure.

Table I. Summary of experimental results. I and Γ are, respectively, intensity and full width at half maximum intensity of least square fit Lorentzian lines.

| | | ZnTe | | | EuTe # 102 | | |
|---|---|---|---|---|---|---|---|
| T Kelvin | H kOe | I % | Γ mm/sec | $H_n$ kOe | I % | Γ mm/sec | $H_n$ kOe |
| 80 | 0 | 3.9 | 6.80 | 0 | 6.8 | 7.50 | 0 |
| 4.2 | 0 | 13.0 | 7.70 | 0 | 17.0 | 11.20 | 73 |
| 4.2 | 50 | 12.5 | 9.60 | 50 | 18.9 | 11.40 | 78 |
| 4.2 | 65 | 12.2 | 10.10 | 65 | 18.9 | 11.50 | 81 |
| 4.2 | 70 | 12.0 | 10.30 | 70 | 18.9 | 12.70 | 113 |

## DISCUSSION

The NMR results of Raj, et al.[4] identify two echoes near 115.1 and 98.0 MHz, with the sample at 1.4K, as having the proper frequencies and intensity ratio to be consistent with Te-125 and Te-123 resonances in a field $H_{int}$ = 86 kOe. The width and shape of the EuTe Mössbauer line in zero external field at 4.2 K (see Table I) is consistent with a single hyperfine field $H_{int}$ = 73 kOe at the Te sites. For $T_N$ = 9.6 K a sublattice magnetization curve, approximated by a Brillouin function with S = 7/2, indicates that M(4.2K) ≈ 0.85 M(1.4 K).[8] The NMR result therefore corresponds to $H_{int}$ ≈ 73 kOe by the Mössbauer measurements, if we make the reasonable assumption that the net internal magnetic field (everywhere, and at the Te site in particular) follows the sublattice magnetization. Because the Mössbauer line components are unresolved, it is not possible to eliminate the possibility of a small quadrupole interaction or a non-unique value of $H_{int}$. However, the observed line shape, compared with ZnTe in an external field, discriminates against a field distribution that is weighted towards low values. The excellent agreement between the Mössbauer and the NMR results supports the hypothesis that the field is unique, or at least that it has a major component that is unique. The NMR results taken together with the Mössbauer results indicate that all, or nearly all, the Te nuclei in the sample experience the hyperfine field, and not just nuclei in or near domain walls.

It is possible that there may be a small distortion of the crystal structure coincident with the with the magnetic phase transition at T = $T_N$. Such behavior has been observed in other rare-earth monochalcogenides [9]; although in this case $Eu^{++}$ is a S-state ion which is believed to couple only weakly to the lattice. Our inability to obtain a satisfactory fit to the T = 4.2 K, H = 0 spectrum with a quadrupole doublet, together with the NMR results, supports the view that a possible crystal distortion does not produce a significant quadrupole broadening compared with the magnetic hyperfine broadening.

In an applied magnetic field, the observed field, $H_n$, increases in magnitude, but by much less than H, until H ≈ 65 kOe, where the AF → P magnetic phase transition occurs.[3] $\vec{H}_n = \vec{H} + \vec{H}_{DM} + \vec{H}_{int}$ where $H_{DM}$ is the demagnetization field. For the sample geometry used (a thick disc filled with separated, irregularly-shaped EuTe polycrystals), $H_{DM}$ ≈ 3.0 kOe. The results (see Table I) support a model in which there is an initial contribution to $H_{int}$, negative with respect to H, which changes more rapidly in the vicinity of the AF → P transition.

There is still no direct evidence for a lattice distortion which would reduce the Te site symmetry. The origin of $H_{int}$ at the Te sites remains unknown; the NMR results indicate that it is present in both conducting and non-conducting samples.

One of us (NAB) wishes to acknowledge helpful discussions with J.I. Budnick, N. Bykovets, and K. Raj.

REFERENCES

* Work supported by the Naval Sea Systems Command,
† Work supported by the National Science Foundation.

1. S. Methfessel and D. C. Mattis, in Handbuch der Physik, edited by S. Flugge (Springer-Verlag, Berlin, 1968), Vol. 18, Part I.
2. C. Haas, CRC Crit. Rev. Solid State Sci. 1, 47(1970).
3. N. F. Oliveira, Jr., S. Foner, Y. Shapira, and T.B. Reed, Phys. Rev. B5, 2634 (1972).
4. K. Raj, J. I. Budnick, and T. J. Burch, AIP Conf. Proc. 24 (1974 Conference on Magnetism and Magnetic Materials), 44 (1975).
5. Y. Shapira, S. Foner, and N. F. Oliveira, Jr., Phys. Rev. B5, 2647 (1972).
6. J. Oberschmidt and P. Boolchand, Phys. Rev. B8, 4953 (1973).
7. V. I. Goldanskii and E. F. Makarov, in Chemical Applications of Mössbauer Spectroscopy, edited by V. I. Goldanskii and R. H. Herber (Academic Press, New York, 1968), Chap. 1.
8. G. Will, S. J. Pickart, H. A. Alperin, and R. Nathan, J. Phys. Chem. Solids 24, 1979 (1963).
9. F. Hulliger, M. Landolt, R. Schmelczer, and I. Zarbach, Solid State Commun. 17, 751 (1975).

# DOUBLE RESONANCE EXCHANGE INTERACTIONS AND MAGNETIC ORDER

M.I.Darby and P.J.Webster,
Department of Pure and Applied Physics,
University of Salford, Salford M5 4WT, England.

## ABSTRACT

The results of calculations of exchange interactions are presented for f.c.c. and simple cubic lattices assuming an RKKY or Caroli and Blandin type double resonance exchange mechanism. Magnetic ordering temperatures have been determined for ferromagnetic and anti-ferromagnetic types 1, 2 and 3 arrangements of magnetic moments.

The theoretical predictions are correlated with experimental results obtained from measurements on palladium based Heusler alloys, in which the electron concentration may be continuously varied and the magnetic sites may be ordered on an f.c.c. or s.c. lattice depending on heat-treatment.

## INTRODUCTION

Indirect exchange interactions exhibit a long range oscillatory behaviour and in consequence can give rise to a variety of antiferromagnetically ordered structures in addition to ferromagnetism. The detailed form of the exchange coupling depends sensitively upon the conduction electron concentration. It should therefore be possible in principle to change the type of magnetic order in a material by altering the electron concentration, for example by substitution of a proportion of one element with another, whilst retaining the same crystallographic order. In the present work calculations have been made using the RKKY indirect exchange interaction and a Caroli-Blandin[1] (C-B) type of double resonance exchange mechanism, in order to predict the various types of magnetic order that should occur in f.c.c. and s.c. lattices as the electron concentration is varied. Magnetic ordering temperatures have been determined for ferromagnetic and antiferromagnetic types 1, 2 and 3[2] (designated F, A1, A2 and A3 below) for each lattice and the theoretical predictions have been correlated with experimental results for some palladium based Heusler alloys.

## EXCHANGE INTERACTIONS

According to the molecular field theory, the magnetic ordering temperature $T_C$ is given by

$$k_B T_C = \frac{2}{3} S(S+1) \sum_{i=1}^{N} \varepsilon_i J_{ij} \quad . . . (1)$$

where $J_{ij}$ is the exchange interaction between an arbitrarily selected magnetic atom j and its $i^{th}$ neighbour. $\varepsilon_i$ takes the values +1 or -1 depending on whether the moments on atoms i and j are parallel or antiparallel.

Values of $\gamma_i$, the subsums of $\varepsilon_i$ over all atoms i equidistant from j, are given in table I for both f.c.c. and s.c. lattices for all four types of order considered.

The RKKY theory and the C-B model give the following expressions for $J_{ij}$ for atoms i and j separated by a distance $R_{ij}$:

$$\text{RKKY:} \quad J_{ij} = -\frac{9\pi}{8} \frac{n^2\Gamma^2}{E_f} \frac{\cos(2k_f R_{ij})}{(k_f R_{ij})^3} \quad . . . (2)$$

$$\text{C-B :} \quad J_{ij} = -\frac{25}{4\pi} \frac{E_f}{S^2} \sin^2\phi \frac{\cos(2k_f R_{ij} + 2\phi)}{(k_f R)^3} \quad . . (3)$$

Here $k_f$ is the Fermi wavenumber, $E_f$ the Fermi energy, $\phi$ a phase factor, $\Gamma$ the s-d exchange constant, and n is the number of conduction electrons per atom.

Using the values of $\gamma_i$ in table I and expressions (2) and (3), $T_C$ was calculated as a function of $k_f$ and $\phi$ for each of the 8 cases considered. In so doing it was assumed that the lattice parameters (a) and (2a) respectively for the s.c. and f.c.c. lattices remained constant and that a free electron relationship $E_f \propto k_f^2$ obtains. The first assumption is appropriate for the

TABLE I. Distances ($R_{ij}$, in units of a), subsums ($\gamma_i$) of parallel (+) and antiparallel (-) neighbours (i) for simple cubic and f.c.c. magnetic structures, F, A1, A2 and A3. (It is assumed that the lattice parameter of the f.c.c. unit cell, 2a, is just that of the simple cubic unit cell a.)

| $R_{ij}$ | Simple cubic | | | | | Face-centred cubic | | | | |
|---|---|---|---|---|---|---|---|---|---|---|
| | i | F | A1 | A2 | A3 | i | F | A1 | A2 | A3 |
| $\sqrt{1}$ | 1 | 6 | -6 | -2 | 2 | | | | | |
| $\sqrt{2}$ | 2 | 12 | 12 | -4 | -4 | 1 | 12 | -4 | 0 | -4 |
| $\sqrt{3}$ | 3 | 8 | -8 | 8 | -8 | | | | | |
| $\sqrt{4}$ | 4 | 6 | 6 | 6 | 6 | 2 | 6 | 6 | -6 | 2 |
| $\sqrt{5}$ | 5 | 24 | -24 | -8 | 8 | | | | | |
| $\sqrt{6}$ | 6 | 24 | 24 | -8 | -8 | 3 | 24 | -8 | 0 | 8 |
| $\sqrt{8}$ | 7 | 12 | 12 | 12 | 12 | 4 | 12 | 12 | 12 | -4 |
| $\sqrt{9}$ | 8 | 30 | -30 | -10 | 10 | | | | | |
| $\sqrt{10}$ | 9 | 24 | 24 | -8 | -8 | 5 | 24 | -8 | 0 | -8 |
| $\sqrt{11}$ | 10 | 24 | -24 | 24 | -24 | | | | | |
| $\sqrt{12}$ | 11 | 8 | 8 | 8 | 8 | 6 | 8 | 8 | -8 | -8 |
| $\sqrt{13}$ | 12 | 24 | -24 | -8 | 8 | | | | | |
| $\sqrt{14}$ | 13 | 48 | 48 | -16 | 16 | 7 | 48 | -16 | 0 | 16 |
| $\sqrt{16}$ | 14 | 6 | 6 | 6 | 6 | 8 | 6 | 6 | 6 | 6 |
| $\sqrt{17}$ | 15 | 48 | -48 | -16 | 16 | | | | | |
| $\sqrt{18}$ | 16 | 36 | 36 | -12 | -12 | 9 | 36 | -12 | 0 | -12 |
| $\sqrt{19}$ | 17 | 24 | -24 | 24 | -24 | | | | | |
| $\sqrt{20}$ | 18 | 24 | 24 | 24 | 24 | 10 | 24 | 24 | -24 | 8 |
| $\sqrt{21}$ | 19 | 48 | -48 | -16 | 16 | | | | | |
| $\sqrt{22}$ | 20 | 24 | 24 | -8 | -8 | 11 | 24 | -8 | 0 | 8 |
| $\sqrt{24}$ | 21 | 24 | 24 | 24 | 24 | 12 | 24 | 24 | 24 | 8 |

Heusler alloys discussed below, and the second should not affect any qualitative conclusions.

## MAGNETIC ORDER

For specified values of $k_f$ and $\phi$ the type of order which is energetically favoured is that with the highest $T_C$. Hence it can be seen that it is the oscillatory function $\Sigma\, \varepsilon_i \cos(2k_f R_{ij}+2\phi)/(k_f R_{ij})^3$ which is important in determining which magnetic structure is the more stable. The type of magnetic order favoured for varying $k_f$ and $\phi$ is shown in figures 1a and 1b for the f.c.c. and s.c. lattices respectively. Lattice parameters, 2a = 6.38 Å for the f.c.c. and a = 3.19 Å for the s.c. were used. The overall sign of $J_{ij}$ has been changed from that given in (2) and (3). The necessity for this change has been noted before[5].

It can be seen from figures 1a and 1b that all four types of magnetic order are theoretically possible for each lattice. Additionally, in some regions a sequence of several different magnetic orderings can be produced by only a small change in $k_f$. A comparison of figures 1a and 1b indicates that whilst several diagonal trends are similar, the f.c.c. case is dominated by large ferromagnetic regions whereas the s.c. case is dominated by type 1 antiferromagnetism.

## APPLICATION TO HEUSLER ALLOYS

Webster and Tebble[3] investigated the magnetic ordering of the Heusler alloys, $Pd_2MnIn$, $Pd_2MnSn$ and

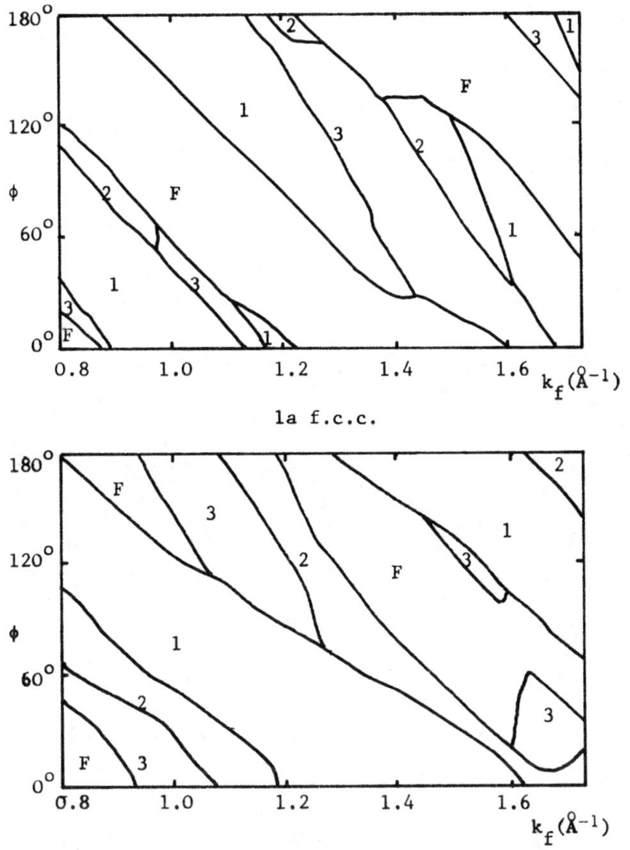


Figures 1a, 1b. Regions of magnetic order for f.c.c. and s.c. lattices as a function of $\phi$ and $k_f$.

Pd$_2$MnSb, and Ramadan[4] the isostructural series of alloys at the intermediate compositions Pd$_2$Mn(In$_{1-x}$Sn$_x$) and Pd$_2$Mn(Sn$_{1-x}$Sb$_x$). In these alloys the Pd atoms form a simple cube with the Mn and the (In/Sn/Sb) atoms ordered on 2 interpenetrating f.c.c. sub-lattices at alternate body centres. The distances between the magnetic Mn atoms is sufficiently large (> 4.5 Å ) for it to be assumed that indirect exchange mechanisms are dominant. By varying x (0<x<1) it is possible to effect a continuous change in the conduction electron concentration, and to alter the type of magnetic order. The experimentally observed types of magnetic order are illustrated as a function of composition in figure 2.

It is possible, but only at high In concentrations, to produce disorder between the Mn and (In/Sn) sites by rapid quenching from high temperatures. This leads to a random arrangement of Mn and (In/Sn) atoms each half occupying a simple cubic lattice of lattice parameter a = 3.19 Å, instead of the two interpenetrating f.c.c. sub-lattices of side 2a = 6.38 Å. The resulting magnetic order is found to be simple cubic type A1.

The theoretical predictions illustrated in figure 1b indicate two broad bands of s.c. antiferromagnetism type A1. For the f.c.c. lattice figure 1a predicts only one region where there is agreement with the experimentally observed sequence of f.c.c. antiferromagnetism type A2, followed by type A3 then ferromagnetism as the electron concentration is increased. This is in the region $\phi \simeq 60°$ and $k_f \sim 0.96$. This region also corresponds to antiferromagnetism type A1 for the s.c. case.

## DISCUSSION

It has been shown that the C-B double resonance exchange mechanism, with suitable choice of parameters, is capable of explaining the observed complicated variations in magnetic order, A2 → A3 → F, and A1 when disordered, of some palladium based Heusler alloys provided

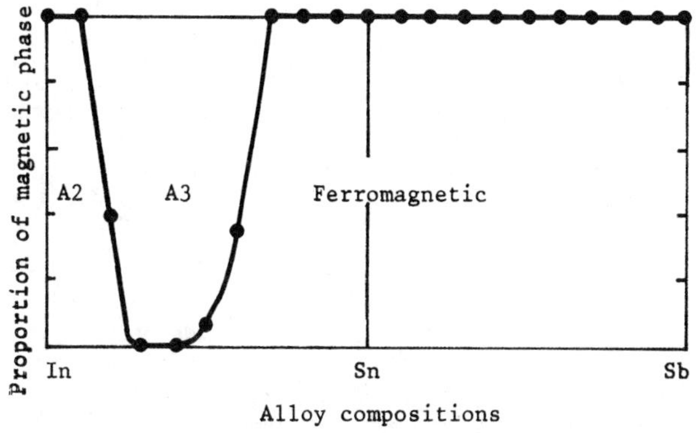


Figure 2. Proportion and type of magnetic phase as a function of composition for alloys Pd$_2$MnIn$_x$Sn$_{1-x}$ and Pd$_2$MnSn$_x$Sb$_{1-x}$.

that the sign of the exchange constants $J_{ij}$ (2) and (3) are changed to positive. The necessity for this sign change has previously been noted [5].

Assuming[6] that the number of s-p electrons per formula unit varies between 4 and 6 for the alloys of interest, and that the lattice constants are essentially independent of composition, estimates of $k_f$ from the free electron expression for the electron density indicate that it lies within the range $1.2 \leq k_f \leq 1.4$ (Å$^{-1}$). Considering the limitations of the free electron estimate, the value of $k_f \sim 0.96$ suggested by the present work is in reasonable agreement with it. However, the value for the phase angle $\phi = 60°$ is significantly higher than the value of 35° predicted on theoretical grounds[1] for other Heusler alloys.

For both the s.c. and f.c.c. lattices the exchange interactions between atoms at distances up to 31.26(Å), corresponding to 24th and 12th nearest neighbours respectively, have been included in the summations. Calculations made with fewer interactions indicated that the qualitative results are not very sensitive to the number chosen, and that the first three interactions dominate the situation.

## REFERENCES


1. B Caroli and A Blandin, J.Phys.Chem.Solids, 27, 503-508, 1966.
2. D ter Haar and M E Lines, Phil.Trans.Roy.Soc., 254 A, 521-555, 1962.
3. P J Webster and R S Tebble, Phil.Mag., 16, 347-361, 1967.
4. M R I Ramadan, Magnetic order in some palladium based Heusler alloys, Thesis, University of Salford, 1975.
5. Y Ishikawa and Y Noda, AIP Conf.Proc.No 24, "Magnetism and Magnetic Materials - 1974", 145-151, 1975.
6. T Kasuya, Solid State Commun. 15, 1119-1122, 1974.

# THE SPINFLOP BICRITICAL POINT IN $MnF_2$*

A.R. King
Physics Department, University of California
Santa Barbara, California 93106

H. Rohrer †
IBM Research Laboratory, Saumerstr. 4
8803 Ruschlikon, Switzerland

ABSTRACT

The competition of different kinds of orderings near the spinflop bicritical point of antiferromagnets gives rise to a new type of critical behavior governed by the anisotropy crossover exponents. We have measured the magnetic phase diagram of $MnF_2$ in the vicinity of the spinflop bicritical point using susceptibility and NMR. For a carefully aligned sample (misalignement of the applied field with respect to the easy axis of magnetization $< 2 \times 10^{-6}$ rad) we obtained for the crossover exponent $\phi = 1.26 \pm 0.03$, in good agreement with theory. The experiment further yields the orientation of the field scaling axis and the universal critical-line amplitude ratio Q both in only fair agreement with theory.

## EXPERIMENTAL

The measurements were made on a single crystal of $MnF_2$ grown in our laboratory, oriented, and cut into a cylinder with the cylinder axis parallel to the crystalline c axis. The locations of the phase boundaries were determined by the parallel susceptibility $\tilde{\chi}_{\parallel}$. Lock-in amplification and chart recorder display were used.

Temperatures were measured with a $N_2$ vapor pressure bulb driving a pressure transducer, whose output was measured with a differential voltmeter. The overall system was calibrated each day in zero field against a platinum resistance thermometer. Thermal contact between vapor pressure bulb and sample was maintained by a sapphire rod, giving a thermal lag between the vapor pressure and sample temperature of a few seconds. Surrounding the sample and vapor pressure bulb was a copper shield, wound uniformly with a resistance wire heater.

Temperature stabilization was accomplished with a standard temperature controller driving the resistance heater. The great temperature sensitivity of the vapor pressure allowed the temperature to be stabalized to within a part in $10^5$.

Alignment of the sample was accomplished first by mechanically tilting the sample and its holder to maximize the susceptibility peak at the spinflop transition. Fine adjustments were made by tilting the field at the sample with a small field from a pair of Helmholtz coils mounted outside the magnet dewar, transverse to the large field. The alignment accomplished this way was within $\sim 2 \times 10^{-6}$ rad.

Field measurements were made with nmr of a proton sample in a small coil mounted near the sample.

NMR of the $^{19}F$ nuclei was also studied in the SF configuration. Additional information on the SF-PM phase boundary and the dynamics of the transition were obtained with this method. This work is described in another contribution at this conference by the same authors, hereafter referred to as II.

The first order antiferromagnetic (AF) to spinflop (SF) transition gave by far the sharpest peak in $\tilde{\chi}_{\parallel}$, of the three phase boundaries. The width of the nearly symmetric peak, of about 100 Oe, was constant over the temperature region investigated, about 1 K below the bicritical point $T_b$. These measurements were made with field scans at constant temperature: the phase boundary was assumed to lie at the maximum of $\tilde{\chi}_{\parallel}$.

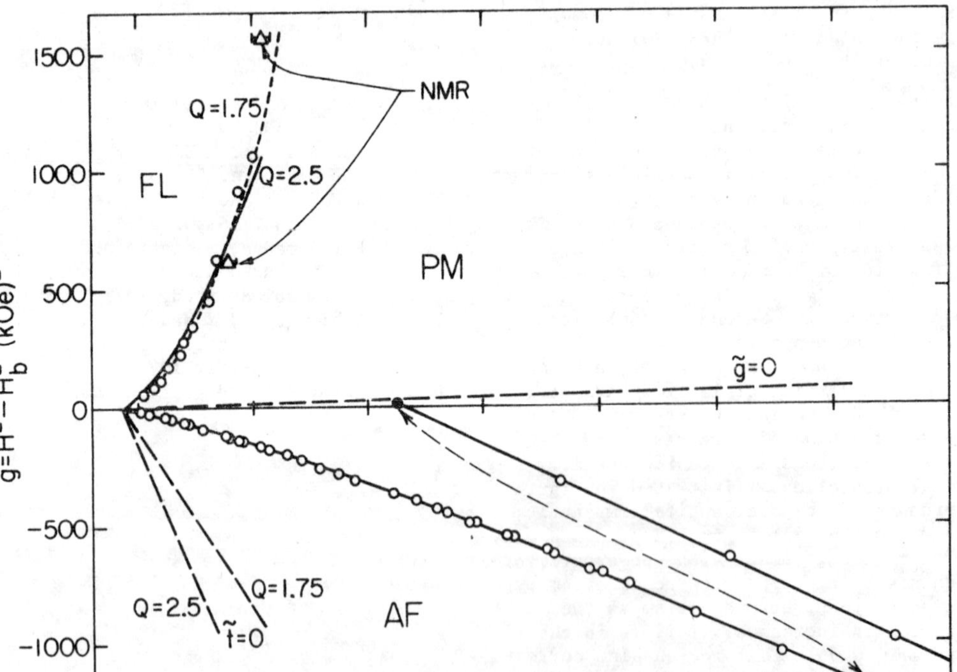


Fig. 1. Phase diagram of $MnF_2$ in the vicinity of the bicritical point ($T_b = 64.793$ K, $H_b = 118.355$ kOe). The AF-PM boundary continues to higher temperature and lower field as indicated by the insert. The two nmr points are explained in II.

| | $Q$ | $\phi$ | $q(10^{-6}kOe^{-2})$ | $\bar{\sigma}_{AF}$(mK) | $\bar{\sigma}_{FL}$(mK) |
|---|---|---|---|---|---|
| Experiment first attempt, Q fixed | 2.5 | 1.21 | 0.6070 | 1.6 | 4.5 |
| Experiment first attempt, Q varied | 1.25 | 1.29 | 1.3838 | 1.9 | 1.7 |
| Experiment second attempt, Q fixed | 2.5 | 1.23 | 0.6874 | 1.4 | 2.6 |
| Experiment second attempt, Q varied | 1.75 | 1.26 | 1.0637 | 1.6 | 1.6 |
| Theory[1,2] | 2.5 | 1.25 | 1.35 | | |

Table I. Parameters obtained from least-squares fit of data on first attempt and second attempt (AF-PM transition adjusted upward). Standard deviations are given separately for the AF-PM boundary ($\bar{\sigma}_{AF}$) and SF-PM boundary ($\bar{\sigma}_{FL}$). Theoretical values are shown for comparison.

The peak amplitude went to zero with increasing temperature in a nearly linear fashion, but surprisingly, at a temperature well above $T_b$ (by about 50 mK). In fact, this susceptibility peak is observed along the extension of the spinflop line out into the paramagnetic phase, at temperatures where the AF-PM boundary had clearly dropped to lower fields. Both $\tilde{\chi}_{\parallel}$ peaks were seen on the same field scans.

The second-order AF-paramagnetic (PM) boundary gave a much broader, asymmetric peak in $\tilde{\chi}_{\parallel}$. Again, most of these data points were measured with field scans at constant temperature. On the low field side, $\tilde{\chi}_{\parallel}$ increased nearly linearly toward the transition: on the high field side, $\tilde{\chi}_{\parallel}$ had the appearance of a broadened divergence. Two attempts were made at fitting the data on this phase boundary. First the peak in $\tilde{\chi}_{\parallel}$ was identified with the phase boundary. Second, since the effect of broadening would be to shift the peak in $\tilde{\chi}_{\parallel}$ toward the side with smaller slope, a point was chosen roughly halfway between the peak and the point of maximum slope. This point was about 50 Oe higher in field than the maximum of $\tilde{\chi}_{\parallel}$. These points are the ones plotted in Fig. 1.

The second-order SF-PM transition presented the greatest experimental difficulties. No change in $\tilde{\chi}_{\parallel}$ could be detected by field scans across this boundary: all data were taken with temperature scans at constant field. This was done at rates slow enough that no significant difference was observed between up and down scans. The shape of the peak in $\tilde{\chi}_{\parallel}$ was similar to that observed at the AF-PM boundary: a much steeper slope on the disordered side, with a broadened peak. The peak amplitude was considerably smaller than the AF-PM peak, and decreased with increasing field, which limited the maximum field at which the boundary cou ld be located.

The AF-PM(||) and SF-PM(⊥) critical lines were fit to the expression[1,2]

$$\tilde{g}/\tilde{t}^{\phi} = -W_{\parallel} \;,\; + W_{\perp} \qquad (1)$$

where $\tilde{g} = g - pt$, $\tilde{t} = t + qg$, $t = (T-T_b)/T_b$, $g = H^2-H_b^2$. The $\tilde{g} = 0$ scaling axis, which is the extension of the spinflop transition, was determined experimentally, while the crossover exponent $\phi$, $Q \equiv W_{\perp}/W_{\parallel}$, $T_b$ and $q$ were varied.

A least-squares curve-fitting procedure was used, with the standard deviation $\bar{\sigma}$ given in mK, computed for each branch separately. On the first attempt, the theoretical value of $Q = 2.5$ gave a poor fit to the SF-PM boundary, and also gave too large a slope for the $\tilde{t} = 0$ scaling axis. (See Table 1). When Q was varied, the best fit was obtained with $Q = 1.25$, close to the orthorhombic case[3], but with much better agreement in the $\tilde{t} = 0$ scaling axis.

On the second attempt, with the AF-PM transition shifted higher by 50 Oe, the fit for $Q = 2.5$ was not unreasonable, but again gave too steep a $\tilde{t} = 0$ scaling axis. (See Fig. 1 and Table I). When Q was again varied, the best overall fit was obtained, with $Q = 1.75$, but neither Q nor q agreed so well with the theory. It appears that demanding Q match the theory gives poor agreement in q; relaxing the restriction on $\phi$ results in a better overall fit, with better agreement in q at the expense of poorer agreement in Q.

During the whole fitting procedure, the best value of $\phi$ varied only a few percent from the theoretical value of 1.25, indicating very good agreement with at least this aspect of the theory.

REFERENCES

*Work supported in part by the National Science Foundation.

†Visiting scientist, Department of Physics, University of California, Santa Barbara, CA 93106, while experimental work was done.

1. M.E. Fisher and D.R. Nelson, Phys. Rev. Lett. 32, 1550 (1974).
2. M.E. Fisher, Phys. Rev. Lett. 34, 1638 (1975).
3. H. Rohrer, Phys. Rev. Lett. 34, 1638 (1975).

# ELECTRONIC MOMENTS INDUCED BY NUCLEAR MOMENTS IN VAN VLECK COMPOUNDS

J.L. Genicon and R. Tournier
Centre de Recherches sur les Très Basses Températures, C.N.R.S.,
B.P. 166 Centre de Tri, 38042 Grenoble-Cédex, France.

ABSTRACT

We have measured, at low temperatures and in high magnetic fields, the nuclear magnetization of several Van-Vleck compounds of Praseodymium. In the $(Pr_xLa_{1-x})Sn_3$ system, we reached saturation of the nuclear magnetization. This saturation value decreases with increasing field. To account for this behaviour, we have proposed a simple "nuclear field" model. This model can also be used for simple systems such as $PrIn_3$ or for intermediate cases such as $PrCu_6$.

## INTRODUCTION

Singlet ground-state compounds of rare earths have been extensively studied in recent years. In Van-Vleck compounds of Praseodymium, due to the existence of a large hyperfine interaction, the local field at the Pr nuclei is highly enhanced over the external magnetic field. A complete study[1,2] of many such compounds has been made, in low fields by Andres, Bucher and co--workers. By adiabatic enhanced nuclear demagnetization, they have obtained temperatures in the mK range. They have also shown that, in low fields, the nuclear susceptibility is enhanced by a factor $(1+K)^2$ where K is the Knight shift of the ion being considered.

## EXPERIMENTS

Our measurements were performed in a conventional magnetization measurement apparatus. Two types of technique can be used on the same apparatus. First, with the extraction technique, we measure the total magnetization of the sample from 50 mK to 4 K and in fields up to 100 kOe. Second, without moving the sample, we detect the change in magnetization produced by heating the sample from an initial temperature $T_i$ to a final temperature $T_f$ always close to 1 K. In that case, we measure only the nuclear part of the magnetization as the electronic contribution, essentially the Van-Vleck one, is temperature independent.

To test our method, we have measured the nuclear magnetization of ferromagnetic cobalt[3] in three different fields. In each field, the nuclear magnetization follows a Curie-law. From these measurements, we have deduced a value of -220±5 kOe for the hyperfine field in good agreement with other results[4].

## NUCLEAR FIELD MODEL

The local field at the Pr nuclei is enhanced by a factor (1+K), through hyperfine interactions, over the external field. This gives, in a classical way, for the nuclear magnetization

$$M_n = N g_n \mu_n I \mathcal{B}_I \left[ g_n \mu_n I H(1+K)/k_B T \right] \quad (1)$$

K is related to the electronic magnetization M induced by the external field by

$$K = h_f \cdot (M/H) = (A/N g_J \mu_B g_n \mu_n)\,(M/H) \quad (2)$$

A is the hyperfine constant and other symbols have their usual meaning. The "nuclear field" created by $M_n$ is

$$H_n = h_f \cdot M_n \quad (3)$$

which acts on the electronic ground-state. The maximum permissible value for $H_n$ can be calculated when $M_n$ is saturated ; in the case of a $Pr^{3+}$ ion, we find 2.4 kOe (we have assumed, following Ref. 1, that $g_n = 1.71$, $g_J = 0.8$, $I = 5/2$, and $A/k_B = 52.5$ mK). The electronic moment generated by this nuclear field is

$$M'_n = M(H+H_n) - M(H) \text{ or } M'_n = H_n \frac{dM}{dH} \text{ when } H_n \ll H \quad (4)$$

This gives, in first approximation, for the enhanced nuclear magnetization

$$M_{en} = M_n + M'_n = M_n + H_n\,(dM/dH) \quad (5)$$

or

$$M_{en} = (1+\eta)\, N g_n \mu_n I \mathcal{B}_I \left[ g_n \mu_n I\, H(1+K)/k_B T \right] \quad (6)$$

with $\eta = h_f\, dM/dH$ and $K = h_f\, M/H$. If the electronic susceptibility is field independent $\eta = K$ and one obtains a nuclear susceptibility enhanced by a factor $(1+K)^2$.

## RESULTS ON $PrIn_3$, $(Pr_xLa_{1-x})Sn_3$ AND $PrCu_6$

A simple example of a Van-Vleck compound is $PrIn_3$[5,6] which crystallizes in the cubic $Cu_3Au$ structure. The susceptibility of a polycrystalline sample, measured at 4 K up to 100 kOe, is nearly field independent and has a value of $2.6\ 10^{-2}$ emu/mole giving a value of K equal to 4.9 using Eq. (2). The nuclear magnetization[7], measured in three different fields (33.3, 66.2, 93.3 kOe) follows each time a Curie-law with the same enhanced nuclear Curie constant. From these measurements, we deduce a value of K equal to 5.2±0.1. Thus, in this case, the nuclear moment is enhanced by an electronic moment which is also field independent.

A more complicated case is the $(Pr_xLa_{1-x})Sn_3$[7,8,9] system. The magnetization M per ion for x=0.1 and x=0.16 (Fig. 1), measured at 1.15 K, is very large and not proportional to the external field. The nuclear magnetization $M_{en}$ (Fig. 2) does not follow a Curie-law over all the temperature range. Instead, in each applied field, the saturation is reached. The saturation value decreases with increasing field and the saturation is reached at higher temperature in higher field.

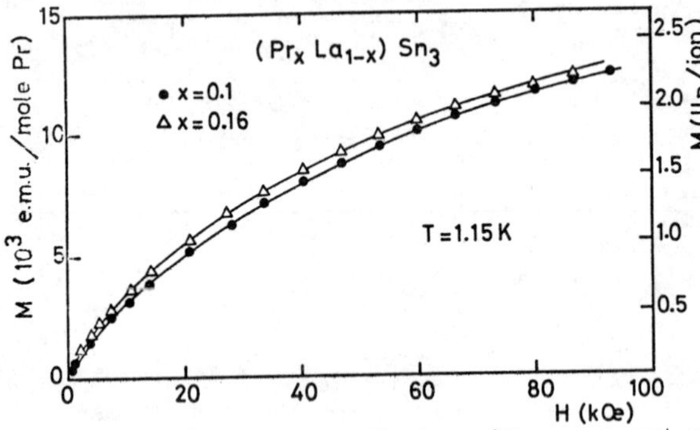


Fig. 1 : Magnetization M of $(Pr_xLa_{1-x})Sn_3$ measured at 1.15 K versus applied field.

TABLE I

| H (kOe) | $h_f(M/H)$ | $h_f(dM/dH)$ | $\eta$ |
|---|---|---|---|
| 36.4 | 39 | 25.5 | 24.5 |
| 65.8 | 30.5 | 15.4 | 14.5 |
| 93 | 25.25 | 10.9 | 10.5 |

Applied field H, values of Knight shift and enhancement factor deduced from the magnetization measured at 1.15 K, and values of the enhancement factor η deduced from the nuclear magnetization in $(Pr_{0.1}La_{0.9})Sn_3$.

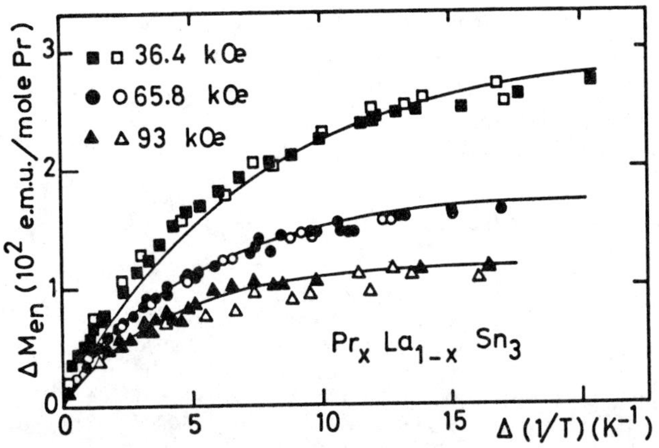


Fig. 2 : Change of nuclear magnetization $\Delta M_{en}$, measured in different fields, between the temperature $T_i$ and the temperature $T_f$ versus $\Delta(1/T)=1/T_i-1/T_f$ for $(Pr_xLa_{1-x})Sn_3$. Closed symbols : x=0.1 ; open symbols : x=0.16. For clarity, only a few points for the sample with x=0.16 have been shown. The solid curves are calculated using Eq. (6) and values of $h_f(M/H)$ and $h_f(dM/dH)$ of Table I.

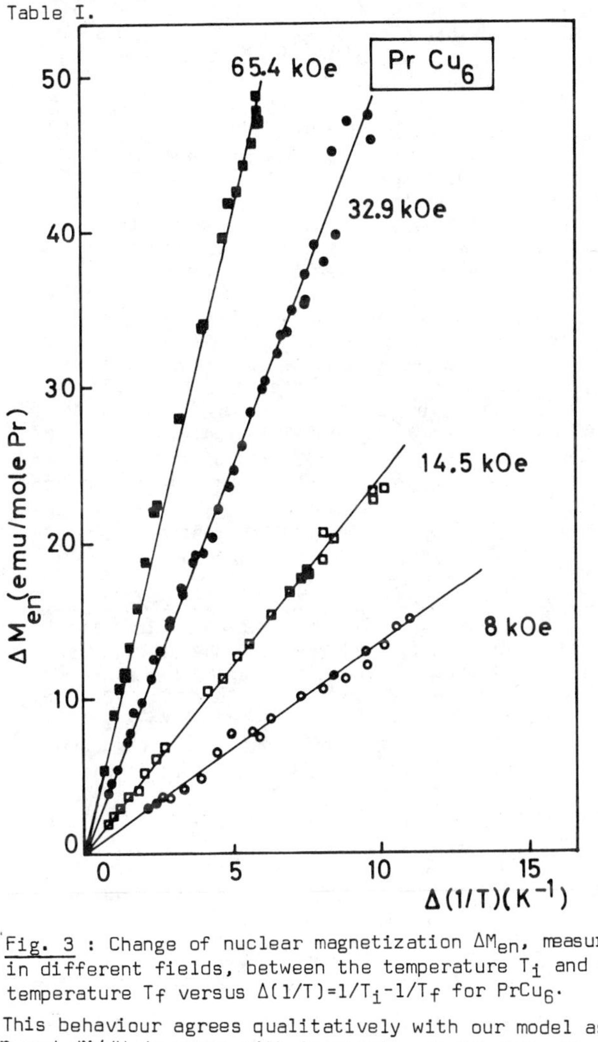


Fig. 3 : Change of nuclear magnetization $\Delta M_{en}$, measured in different fields, between the temperature $T_i$ and the temperature $T_f$ versus $\Delta(1/T)=1/T_i-1/T_f$ for $PrCu_6$.

This behaviour agrees qualitatively with our model as $\eta$ and dM/dH decrease with increasing field while the local field H(1+K) at the Pr nuclei increases. Quantitative measured and calculated values of $\eta$ and K are summarized in Table I and Fig. 2. Within experimental accuracy, the values of $\eta$ and of $h_f$(dM/dH) are the same for the two samples. It can be seen that the measured values agree well with values calculated using our simple model.

TABLE II

| H (kOe) | M/H (emu/mole) | dM/dH (emu/mole) | enh. factor | 1+K |
|---|---|---|---|---|
| 8 | 0.062 | 0.062 | 12.65 | 13.5 |
| 14.5 | 0.061 | 0.061 | 12.4 | 13.3 |
| 32.9 | 0.061 | 0.059 | 12.25 | 12.5 |
| 65.4 | 0.059 | 0.057 | 12 | 11.7 |

Applied field H, values of M/H, dM/dH and enhancement factor $[(1+h_f\ M/H)\ (1+h_f\ dM/dH)]^{1/2}$ deduced from the electronic magnetization, and values of the enhancement factor (1+K) deduced from the nuclear magnetization in Pr $Cu_6$.

An intermediate case is $PrCu_6$ which crystallizes in the orthorhombic $CeCu_6$ structure[2]. The magnetization of a polycrystalline sample, measured at 1.2 K, shows a very weak curvature as a function of applied field (due to the scale of figures, this curve would have been a straight line ; so we do not present it here). The nuclear magnetization (fig. 3) follows a Curie law in each applied field with different enhanced nuclear Curie constants. Using our model, we can compare the values of (1+K) deduced from the nuclear magnetization to the values of the enhancement factor deduced from the magnetization measured at 1.2 K. Results are presented in Table II. Within experimental accuracy, the agreement observed is good.

## CONCLUSION

We have shown, by direct measurements in high fields of the nuclear magnetization, the effect of a nuclear moment upon a non magnetic electronic ground state. Our model, which takes into account the possible curvature of the electronic magnetization, agrees well with the observed properties of enhanced nuclear magnetization. The nuclear moment is enhanced with an electronic moment equal to the field derivative of the electronic magnetization times the nuclear field.

## REFERENCES

1 - K. Andres and S. Darack, Phys. Rev. B 10, 1967 (1974).
2 - K. Andres and E. Bucher, J. Low Temp. Phys. 9, 267 (1972).
3 - J.L. Genicon and R. Tournier, Proc. 14th Int. Conf. Low Temp. Physics, edited by M. Krusius and M. Vuorio (Helsinki, Finland, 1975) (North-Holland Publ. Co., Amsterdam. Oxford - American Elsevier Publ. Co., New York) Vol. 3, p. 200.
4 - A.J. Freeman and R.E. Watson in Magnetism, edited by G.T. Rado and H. Suhl (Academic Press, New York and London, 1965) Vol. II A, p. 167.
5 - K. Andres and E. Bucher, J. Appl. Phys. 42, 1522 (1971).
6 - A.M. Van Diepen, R.S. Craig and W.E. Wallace, J. Phys. Chem. Solids 32, 1867 (1971).
7 - J.L. Genicon and R. Tournier, to be published.
8 - P. Lethuillier and J. Chaussy, to be published in Phys. Rev. .
9 - R.W. Mc Callum, W.A. Fertig, C.A. Luengo, M.B. Maple and E. Bucher and J.P. Maita and A.R. Sweedler and L. Mattix and P. Fulde and J. Keller, Phys. Rev. Lett. 34, 1620 (1975).

# ROTATIONAL INVARIANCE AND THE COUPLING OF ACOUSTIC WAVES TO ZERO FREQUENCY PSEUDOSPIN MODES IN PARAMAGNETS*

P. A. Fedders
Department of Physics, Washington University,
St. Louis, Missouri 63130
and
R. L. Melcher
IBM Thomas J. Watson Research Center,
Yorktown Heights, New York 10598

ABSTRACT

Non-resonant changes in acoustic velocities due to pseudospin-phonon interactions in anisotropic paramagnets are investigated with particular emphasis on the additional contributions to the interaction which arise from the requirement of rotational invariance and the coupling of the acoustic waves to the pseudospin normal modes. In general one does not know the pseudospin rotational properties and thus the pseudospin-phonon interaction can only be written in the crystal coordinate system rather than the laboratory system. On the other hand, since the crystal coordinate system is a rotating reference frame, one cannot write equations of motion for the pseudospins. These restrictions limit acoustic velocity change calculations to the thermodynamic limit of $\omega\tau \ll 1$ where $\tau$ is a typical pseudospin relaxation time and $\omega$ is the frequency of an acoustic wave. However, we show that equations for the "zero frequency" pseudospin modes (like the pseudomagnetization) can be derived in either the crystal or laboratory coordinate systems. Thus the thermodynamic calculation can be extended from $\omega\tau \ll 1$ to all frequencies less than pseudospin resonant frequencies. Simple calculations for pseudospin $\frac{1}{2}$ systems elucidate these points.

## I. INTRODUCTION

Recently there has been considerable interest in the consequences of rotational invariance and finite deformation theory on magnetoelastic or spin-phonon phenomena in ordered systems.[1-4] Explicit discussions of the rotational contributions to the spin-phonon coupling in paramagnetic systems have been given by Kumar et al.[5] and by Melcher.[6] However, these treatment are only applicable when the rotational properties of the spin operators are known. In many cases the spin system is described by pseudospin operators whose rotational properties in real space are not those of simple vectors and, in fact, may be unknown. In more recent work Bonsall and Melcher[7] have shown a means of circumventing this lack of knowledge of the rotational properties of the pseudospins by performing the calculation for the acoustic velocity change entirely in the coordinate system of the crystal. In this paper we shall show that although this type of calculation is in general valid only for frequencies much less than spin relaxation rates, the results can be rather easily extended to a less restricted range of frequencies that are much less than spin resonance frequencies.

The rest of this section will be devoted to a discussion of the pseudospin-phonon problem in different frequency regimes. In Section II the thermodynamic calculation of Ref. 7 is extended to all magnetic field directions in the x-z plane and in Section III we shall show how to extend the calculation to all frequencies much less than spin resonant frequencies when the pseudospin rotation properties are not known.

For our purpose we shall assume a simple two-level pseudospin system with a splitting or resonant frequency $\omega_0$ and a decay rate $\Gamma$ or lifetime $\tau = 1/\Gamma$ such that $\omega_0 \gg \Gamma$. In particular, as we shall see, $\Gamma$ refers to the decay of the pseudospin $m = 0$ mode or what might be called the pseudomagnetization in a two level system. In an earlier paper Fedders[8] described the dynamical effects of spins (or pseudospins) on acoustic waves in terms of mode-mode coupling of spin modes to acoustic modes. From this work it is easily seen that there are three distinct and increasingly less restrictive frequency regimes for spin-phonon coupling. We shall classify them as (i) the adiabatic regime $\omega \ll \Gamma$, (ii) the below resonance regime of $\omega \ll \omega_0$, and (iii) the unrestricted regime where frequencies are only restricted by the continuum limit where wavelengths are much greater than interatomic spacings.

The thermodynamic calculation of Ref. 7 is valid only in the adiabatic regime of $\omega \ll \Gamma$. In this regime the pseudospins follow the lattice waves and are in instantaneous local equilibrium with the lattice which exactly fits the assumptions of the thermodynamic calculation. Even excluding resonant effects, Ref. 8 shows that spins or pseudospins introduce frequency dependent terms into the velocity that are different in the regimes $\omega \ll \Gamma$ and $\Gamma \ll \omega \ll \omega_0$. These terms are of the same order of magnitude as the $v_{xz} - v_{zx}$ terms of Ref. 7 and thus cannot be neglected except at the lowest frequencies.

## II. THERMODYNAMIC CALCULATION

In this section we consider two simple but nontrivial systems and calculate the changes in acoustic velocities due to the spin-phonon interaction including contributions due to rotational motion of the lattice. The calculation is a generalization of the results of Ref. 7 to the case of an external magnetic field with an arbitrary orientation in the x-z plane of the laboratory coordinate system. Thus we consider the case of a pseudospin $\frac{1}{2}$ system doubly degenerate in the absence of an external magnetic field at a site of $D_{2d}$ symmetry. The externally applied magnetic field $\vec{H}$ lies in the laboratory x-z plane at an angle $\theta$ with the laboratory z-axis. We restrict the calculation to elastic strains and rotations in the x-z plane, i.e., $e_{xz} = (u_{xz} + u_{zx})/2$ and $\omega_{xz} = (u_{xz} - u_{zx})/2$ where $u_{ij} = \partial u_i/\partial x_j$ and $\vec{u}$ is the lattice displacement. For simplicity we neglect contributions to the pseudospin-phonon interaction arising from the quadratic displacement gradients. These terms do not effect the frequency dependent contributions to $(\Delta v/v)$ which are of interest here.

A straightforward extension of the argument of Ref. 7 for a non-Kramers doublet yields the following Hamiltonian.

$$\mathcal{H} = g_{\parallel}\mu_0\sigma_z^c H_z^c + g_{\parallel}\mu_0 F_{44}\sigma_z^c H_x^c e_{xz} + 2C_{44}e_{xz}^2 , \qquad (1a)$$

where the subscript c denotes the crystal coordinate system and thus

$$\vec{H}^c = H[\sin(\theta + \omega_{xz}),\ 0,\ \cos(\theta + \omega_{xz})] . \qquad (2)$$

For a Kramers doublet the most general Hamiltonian is

$$\mathcal{H} = g_{\parallel}\mu_0\sigma_z^c H_z^c + g_{\perp}\mu_0\sigma_x^c H_x^c + 2C_{44}e_{xz}^2 + g_{\parallel}\mu_0 F_{44}\sigma_z^c H_x^c e_{xz} + g_{\perp}\mu_0 F'_{44}\sigma_x^c H_z^c e_{xz} . \qquad (1b)$$

We emphasize that Eqs. (1) cannot be transformed to the laboratory coordinate system unless the pseudospin rotational properties are known.

The calculation of the acoustic velocity in the thermodynamic limit is a straightforward generalization of Ref. 7. We use the notation $(\Delta v/v)_{th}$ for the change in velocity in the thermodynamic limit. The results for the non-Kramers doublet are

$$(\Delta v/v)_{th} = (\beta n_s(g_{\parallel}\mu_0 H)^2/8\rho v^2)[\cos^2\theta(1\mp 2F_{44}) - \sin^2\theta(1\mp F_{44})^2], \tag{3a}$$

in the high temperature limit $\beta\mu_0 g_{\parallel} H << 1$ where $\beta = 1/kT$, $n_s$ is the spin density and $\rho$ is the crystal mass density. The upper and lower signs refer to $v_{xz}$ and $v_{zx}$ respectively where $v_{ij}$ is the transverse acoustic velocity for phonons polarized along the i direction and traveling along the j direction. Semilarily, for the Kramers doublet,

$$(\Delta v/v)_{th} = -(\beta n_s(\mu_0 H)^2/8\rho v^2)[g_{\parallel}^2\sin^2\theta(1\mp F_{44})^2 + g_{\perp}^2\cos^2\theta(1\pm F'_{44})^2 - g_{\parallel}^2\cos^2\theta(1\mp 2F_{44}) - g_{\perp}^2\sin^2\theta(1\pm 2F'_{44})]. \tag{3b}$$

## III. RESULTS

If the rotational properties of the pseudospin operators are known, the thermodynamic calculation can easily be extended to all frequencies consistent with the elastic continuum limit. One way to do this is to transform the Hamiltonian to the laboratory coordinate system and then calculate the acoustic velocity change due to the pseudospin normal modes m = +1,0,-1. This calculation is done in Ref. 8 and one can express the results in terms of $(\Delta v/v)_m(\omega)$, the frequency dependent velocity change due to the mode m. The total velocity change $(\Delta v/v)$ due to the pseudospin-phonon interaction is then

$$(\Delta v/v) = (\Delta v/v)_{th} + \Sigma_m[(\Delta v/v)_m(\omega) - (\Delta v/v)_m(0)] \tag{4}$$

where $(\Delta v/v)_{th}$ is the differential velocity change in the thermodynamic limit. The reason for the last term in Eq. (4) is the $(\Delta v/v)_{th}$ includes the effects from the spin normal modes at zero frequency, and thus this term must be subtracted out.

If the rotational properties of the pseudospin operators are not known, one cannot transform the Hamiltonian to the laboratory coordinates. In addition, the pseudospin normal mode problem cannot be done in the crystal coordinate system without knowledge of the pseudospin rotation properties. The reason is simply that the crystal coordinate system is a rotating coordinate system and not an inertial frame. However, at $\omega = 0$ the rotating field terms vanish and thus the thermodynamic calculation is correct in this limit.

In spite of this, it turns out that the thermodynamic calculation can be rather easily modified to lift the restriction $\omega << \Gamma$ although the restriction $\omega << \omega_0$ cannot be circumvented without knowledge of pseudospin rotation operators. The reason is that the m = 0 pseudospin mode or the pseudomagnetization is not affected to first order in the rotating frame. In the rest of this section we shall first show how to obtain the velocity change due to the m = 0 mode in the laboratory coordinate system. We shall then show that this is identical to the calculation in the crystal coordinate system and explicitly show that knowledge of the pseudospin rotational properties are not necessary.

The calculation of $(\Delta v/v)_0(\omega)$, the change in the velocity due to the m = 0 pseudospin mode, is a minor extension of the treatment in Ref. 8. Since only first order terms in $e_{xz}$ and $\omega_{xz}$ are necessary, we consider the general first order Hamiltonian in the laboratory coordinate system

$$H = \vec{a}\cdot\vec{\sigma} + e_{xz}\vec{b}\cdot\vec{\sigma} + \omega_{xz}\vec{c}\cdot\vec{\sigma} \tag{5}$$

where $\vec{a}$, $\vec{b}$, and $\vec{c}$ do not depend on $e_{xz}$ and $\omega_{xz}$. The z-component of $\vec{\sigma}$ in the spin coordinate system that diagonalizes the zero order part of the Hamiltonian in Eq. (5) is $\tilde{\sigma}_z = \vec{a}\cdot\vec{\sigma}/a$, where $a = |\vec{a}|$. Further, the only part of the interaction term of Eq. (5) that couples to the m = 0 pseudospin normal mode is the projection of the interaction term on $\tilde{\sigma}_z$. That is, we can use the Hamiltonian

$$H = a\,\tilde{\sigma}_z + G\,e_{xz}\tilde{\sigma}_z$$
$$G = \vec{a}\cdot(\vec{b} \pm \vec{c})/a\ , \tag{6}$$

where we have set $\omega_{xz} = \pm e_{xz}$. The rest of the calculation follows Ref. 8 and yields

$$(\Delta v/v)_0(\omega) = -n_s\beta G^2\chi'_0(\omega)/8\rho v^2\ . \tag{7}$$

The phenomenological form for $\chi'_0(\omega)$ in the high temperature limit is $\Gamma^2/(\omega^2+\Gamma^2)$.

In the crystal coordinate system, the Hamiltonian can be written as

$$H = \vec{a}^c\cdot\vec{\sigma}^c + e_{xz}\vec{b}^c\cdot\vec{\sigma}^c + \omega_{xz}\vec{c}^c\cdot\vec{\sigma}^c \tag{8}$$

where $\vec{\sigma}^c$ is the Pauli pseudospin vector in the crystal. The transformation between the laboratory and crystal coordinate system for $\omega << 1$ can be written as

$$\sigma_i^c = \sigma_i + \omega_{xz}\,\Sigma_j\,\gamma_{ij} \tag{9}$$

where $\gamma_{ij} = -\gamma_{ji}$ to lowest order in $\omega_{xz}$. By substitution and a direct comparison of Eqs. (5) and (8), one obtains $\vec{a} = \vec{a}^c$, $\vec{b} = \vec{b}^c$ and

$$c_i = \sigma_i^c + \Sigma_j\,a_j\,\gamma_{ji}\ . \tag{10}$$

However, it is easily seen that this additional term in c does not alter Eqs. (6) because

$$tr[(\vec{c}-\vec{c}^c)\cdot\vec{\sigma}\tilde{\sigma}_z] \sim (\vec{c}-\vec{c}^c)\cdot\vec{a} = 0\ . \tag{11}$$

This proves that $(\Delta v/v)_0(\omega)$ can be calculated in either the laboratory or crystal coordinates and that a knowledge of the rotational properties of the pseudospin operators is not necessary. There is no problem with the $m \neq 0$ modes since the thermodynamic calculation treats them correctly for $\omega << \omega_0$.

For the non Kramers case we obtain a correction to Eq. (3a)

$$(\Delta v/v)_c = (\beta(g_{\parallel}\mu_0 H)^2/8\rho v^2)\sin^2\theta(1\mp F_{44})(\omega^2/(\omega^2+\Gamma^2)), \tag{12a}$$

and for the Kramers case the correction to Eq. (3b) is

$$(\Delta v/v)_c = [\beta(\mu_0 H\sin\theta\cos\theta)^2/8\rho v^2(g_{\parallel}^2\cos\theta + g_{\perp}^2\sin^2\theta)] \times [g_{\parallel}^2(F_{44}\mp 1) + g_{\perp}^2(F'_{44}\pm 1)]^2 \times (\omega^2/\omega^2+\Gamma^2)\ . \tag{12b}$$

## REFERENCES

*Work supported in part by NSF.

[1]H. F. Tierston, J. Math. Phys. 5, 1298 (1964).
[2]W. F. Brown Jr., J. Appl. Phys. 36, 944 (1965).
[3]D. E. Eastman, Phys. Rev. 148, 530 (1966).
[4]R. L. Melcher, Phys. Rev. Letters 25, 1201 (1970).
[5]S. Kumar, T. Ray, and D. K. Ray, Phys. Stat. Sol. 37, 165 (1970).
[6]R. L. Melcher, Phys. Rev. Letters 28, 165 (1972).
[7]L. Bonsall and R. L. Melcher, Bull. Am. Phys. Soc. 20, 349 (1975). L. Bonsall and R. L. Melcher, submitted to Phys. Rev.
[8]Peter A. Fedders, Phys. Rev. B 12, 2045 (1975).

# PREDICTED NEW COMPONENTS OF MAGNETIC FORCE ON A FERROMAGNET UNDERGOING RESONANCE

F.R. Morgenthaler
Department of Electrical Engineering and Center for Materials Science and Engineering, Massachusetts Institute of Technology, Cambridge, Mass. 02139

## ABSTRACT

The present author has recently postulated for a ferromagnet a new energy-momentum tensor that includes non-electromagnetic linear momentum in the rest frame of the lattice. Such momentum persists - indeed, is most important - in the magnetostatic limit.

In that limit, and neglecting exchange and anisotropy for simplicity, we show the leading terms of the force density are given by $\bar{f} = \mu_o \bar{M}^{\circ}\cdot\nabla\bar{H} \pm 2\,\omega_{\Delta H}\bar{G}$ where $\bar{M}^{\circ}$ is the equilibrium magnetization vector, and $\bar{G}$ the momentum density; the sign corresponds to the sense of precession. The second term, previously omitted by the present author when assuming a lossless material, is shown to be essential in preventing force paradoxes under steady-state resonance conditions. Remarkably, this is true even when one considers the Kittel "k=0" uniform precession mode of a small ellipsoid for which $\bar{G}$ might be expected to be negligible even when a magnetic field gradient is present. In comparison with the commonly accepted theory, our tensor implies differences not only in the magnetostatic force density, but also in the net force whenever transient conditions produce a time rate of change of the total material momentum.

## STATIONARY MATTER WITHOUT ANISOTROPY

We begin by setting forth both the new and conventional theories[1] In order to focus on the essential points with a minimum of complication, we here restrict ourselves to reviewing the case of stationary rigid matter without anisotropy or polarization.

In addition to Maxwell's Equations, the torque equation governing the magnetization vector $\bar{M}$ is, in MKS units,

$$\frac{\partial\bar{M}}{\partial t} = \gamma\mu_o\bar{M} \times [\bar{H} + \bar{H}^{ex} + \bar{H}^{loss}] \qquad (1)$$

where $\gamma$ is the gyromagnetic ratio (negative); $\bar{H}$ the magnetic field; $\bar{H}^{ex}$ the exchange field

$$\bar{H}^{ex} = \frac{\partial}{\partial x_j}\left(\lambda \frac{\partial\bar{M}}{\partial x_j}\right), \qquad (2)$$

with $\lambda$ the isotropic exchange constant and summation over repeated indices assumed; $\bar{H}^{loss}$ is the phenomenological damping field

$$\bar{H}^{loss} = \pm \Delta H \frac{\bar{M} \times \bar{u}}{\bar{M}\cdot\bar{u}} \qquad (3)$$

with $\Delta H$ the resonance half-linewidth and $\bar{u}$ the unit vector parallel to the equilibrium magnetization (the choice of sign must correspond to the sense of precession of $\bar{M}$).

As is well known, the four dimensional energy momentum tensor is made up of the stress tensor $\bar{\bar{T}}$, momentum density $\bar{G}$, power flux vector $\bar{S}$ and the energy density w. In the new theory:

$$T_{ij}^{new} = -\mu_o\lambda\frac{\partial\bar{M}}{\partial x_i}\cdot\frac{\partial\bar{M}}{\partial x_j} + \varepsilon_o E_i E_j + \mu_o H_i(H_j + M_j)$$
$$- \frac{1}{2}[\mu_o \frac{M^2}{\bar{M}\cdot\bar{u}}\ \bar{u}\cdot\bar{H}^{ex} + \varepsilon_o E^2 + \mu_o\bar{H}\cdot(\bar{H}+\bar{M}-\bar{u}\bar{M}\cdot\bar{u})]\delta_{ij} \qquad (4)$$

$$G^{new} = \frac{1}{2\gamma\bar{M}\cdot\bar{u}}\left(\bar{M} \times \frac{\partial\bar{M}}{\partial x_i}\right)\cdot\bar{u} + \mu_o\varepsilon_o[\bar{E}\times(\bar{H}+\bar{M}-\bar{M}^{\circ})]_i \qquad (5)$$

$$S_j = -\mu_o\lambda\frac{\partial\bar{M}}{\partial x_j}\cdot\frac{\partial\bar{M}}{\partial t} + (\bar{E} \times \bar{H})_j \qquad (6)$$

and

$$w = \frac{1}{2}\mu_o\lambda\frac{\partial\bar{M}}{\partial x_j}\cdot\frac{\partial\bar{M}}{\partial x_j} + \frac{1}{2}\varepsilon_o E^2 + \frac{1}{2}\mu_o H^2 \qquad (7)$$

where $\bar{M}^{\circ} = \frac{1}{2}(M^2/\bar{M}\cdot\bar{u} + \bar{M}\cdot\bar{u})\bar{u} \simeq M\bar{u}$, $\bar{E}$ the electric field, and $\delta_{ij}$ the Kronecker delta.

The conventional theory has the same expressions for $\bar{S}$ and W but $\bar{\bar{T}}$ and $\bar{G}$ are given by

$$T_{ij} = -\mu_o\lambda\frac{\partial\bar{M}}{\partial x_i}\cdot\frac{\partial\bar{M}}{\partial x_j} + \varepsilon_o E_i E_j + \mu_o H_i(H_j + M_j)$$
$$- \frac{1}{2}(\varepsilon_o E^2 + \mu_o H^2)\delta_{ij} \qquad (8)$$

and

$$\bar{G} = \mu_o\varepsilon_o\ \bar{E} \times \bar{H} \qquad (9)$$

Notice that if $\bar{M} = M\bar{u}$ (no magnetic resonance excitation) and $\bar{u}$ is uniform (single domain), the corresponding new and conventional values of $\bar{\bar{T}}$ and $\bar{G}$ are identical. In the magnetostatic limit all terms proportional to $\varepsilon_o$, the permittivity of free space, become negligible.

## PREDICTED NEW COMPONENTS OF MAGNETIC FORCE ON A SPHERE UNDERGOING RESONANCE

In the magnetostatic limit, and neglecting magnetic anisotropy and quantum mechanical exchange for simplicity, the conventional theory has $\bar{G}=0$; therefore from the divergence of Eq.(8),

$$\bar{f} = \mu_o(\bar{M}\cdot\nabla)\bar{H} \qquad (10)$$

In the same limit, also without anisotropy and exchange, the new theory gives

$$\bar{f}^{new} = \mu_o(\bar{M}^{\circ}\cdot\nabla)\bar{H} \pm 2\omega_{\Delta H}\bar{G}^{new} + \ldots \qquad (11)$$

Here $\omega_{\Delta H} = -\gamma\mu_o\Delta H$ is the relaxation frequency and again the choice of sign corresponds to the sense of precession of $\bar{M}$. The total force acting upon a rigid body ferromagnet can be calculated in either model from

$$\bar{F} = \oint_{S>S_o} \bar{\bar{T}}\cdot\bar{n}\,da - \frac{d}{dt}\int_{V_o} \bar{G}\,dV \qquad (12)$$

where $S_o$ represents the surface of the material enclosing a volume $V_o$.

If there is no time-varying momentum within the volume, the net force will be independent of material details as long as they do not alter the external field[2] However, when there is time-varying momentum the situation is quite different.

Consider a small spherical sample of radius R and saturation magnetization M, restrained from moving in an external d.c. magnetic field having a constant gradient H' in the z-direction; assume further that a spatially uniform positive circularly polarized rf magnetic field of frequency $\omega$ and small amplitude $h_a$ is also present. If the field gradient is not too large ($H'R << \Delta H$), the sphere will everywhere be in resonance and the magnetic field inside of the sphere approximately given by

$$\bar{H} \simeq \bar{i}_z(H_o + H'z) - \bar{i}_\rho\frac{1}{2}H'\rho - \frac{1}{3}\bar{M} + \bar{h}_{rf} \qquad (13)$$

where $\rho = r\sin\theta$ and $z = r\cos\theta$ are cylindrical coordinates,

$$\bar{h}_{rf} = h_a(t)(\bar{i}_x\cos\omega t + \bar{i}_y\sin\omega t) \quad (\omega>0) \qquad (14)$$

and

$$\bar{M} \simeq M[\sin\theta_o(\bar{i}_x\cos\phi_o + \bar{i}_y\sin\phi_o) + \bar{i}_z\cos\theta_o] \quad (15)$$

If prior to $t = 0$, the midplane of the sphere ($z=0$) is exactly in steady state resonance with $h_a(t) = h$, Eq. (1) has the approximate solution

$$\tan\theta_o \simeq h/\Delta H \ll 1 \quad (16)$$

and

$$\phi_o \simeq \omega t - \pi/2 + H'z/\Delta H \quad (17)$$

irrespective of whether the new or conventional theory applies. From Eq.(10), the conventional force density is

$$\bar{f} \simeq \bar{i}_z\mu_o MH' \cos\theta_o \quad (18)$$

From Eqs.(5) and (11),

$$\bar{G}^{new} \simeq \bar{i}_z \frac{M}{2\gamma} \frac{\sin^2\theta_o}{\cos\theta_o} \frac{H'}{\Delta H} \quad (19)$$

and

$$\bar{f}^{new} \simeq \bar{i}_z\mu_o MH'[\tfrac{1}{2}(\cos\theta_o + 1/\cos\theta_o) - \sin^2\theta_o/\cos\theta_o] \quad (20)$$

The steady-state rigid-body force acting upon the sphere can be calculated from

$$\bar{F} = \int_V \bar{f}\, dv + \oint_{S_o} [\bar{\bar{T}}_{(1)} - \bar{\bar{T}}_{(2)}]\cdot\bar{i}_r\, da \quad (21)$$

where the subscripts (1) and (2) refer respectively to boundary values just outside and just inside the surface of the sphere. According to the conventional theory, the surface integral of (21) vanishes and

$$\bar{F} = \bar{i}_z\mu_o MH' V\cos\theta_o \quad (22)$$

where $V = 4/3\pi R^3$.
According to the new theory, the corresponding integral reduces to

$$\oint_{S_o} \frac{1}{2}\mu_o \bar{h}_{rf}\cdot(\bar{M} - \bar{i}_z M\cos\theta_o)\,\bar{i}_r da \simeq \bar{i}_z \frac{1}{2}\mu_o MH'V\frac{\sin^2\theta_o}{\cos\theta_o} \quad (23)$$

Therefore the new theory while giving a very different force distribution, agrees as it must with the total force. The three components of the distribution can be interpreted as: (a) the reaction force upon the lattice due to acceleration of the magnon wavepacket; (b) the interaction force at the surface between the magnons and the driving field and (c) the dissipation force produced as magnons are annihilated. Note that the (c) is essential in maintaining agreement and that the wave number $k_z = -H'/\Delta H$, although small, cannot be set equal to zero.

Now if for $t>0$, $h_a(t)=0$, the resonance cone angle $\theta_o$ will decay and the forced steady-state response of $\bar{M}$ at $\omega$ will be replaced with a frequency of precession that is a function of z owing to the field gradient. In particular, Eq.(1) now has the approximate solution

$$\tan\theta_o \simeq (h/\Delta H)e^{-\omega_{\Delta H}t} \quad (24)$$

and

$$\phi_o \simeq \omega t - \pi/2 + H'z/\Delta H - \gamma\mu_o H'zt \quad (25)$$

again irrespective of which model applies. The conventional force density remains equal to Eq.(18) with $\cos\theta_o$ simply changing according to Eq.(24); although Eq.(11) still applies, the new force density is altered from Eq.(20) because

$$\bar{G}^{new} \simeq \bar{i}_z \frac{M}{2\gamma} \frac{\sin^2\theta_o}{\cos\theta_o}\left(\frac{H'}{\Delta H} - \gamma\mu_o H't\right) \quad (26)$$

contains a term due to the now time-dependent wave vector,

$$k_z = -\left(\frac{H'}{\Delta H} - \gamma\mu_o H't\right) \quad (27)$$

Owing to the disappearance of $\bar{h}_{rf}$, the surface integral in Eq.(21) vanishes (magnons can no longer interact with the driving field at the surface). Therefore according to the new theory, the total force on the sphere is for $t>0$ and $\theta_o \ll 1$,

$$\bar{F}^{new} \simeq \bar{i}_z\mu_o MH'V[1 - \theta_o^2(0)(1 + \omega_{\Delta H}t)e^{-2\omega_{\Delta H}t}] \quad (28)$$

and

$$\bar{F} \simeq \bar{i}_z\mu_o MH'V[1 - \frac{1}{2}\theta_o^2(0)e^{-2\omega_{\Delta H}t}] \quad (29)$$

Notice that because of Eq.(12), $\bar{F}^{new}$ must satisfy

$$\bar{F}^{new} = \bar{F} - \frac{d\bar{G}^{new}}{dt} V \quad (30)$$

as indeed it does.

When $H'R \gg \Delta H$, the sphere is no longer all in resonance at the same time. Detailed calculations similar to those here given, then indicate that for $t>0$, the conventional and new theories predict respectively,

$$F_z \simeq \mu_o MH'V - \frac{H'}{\Delta H}\frac{P}{\omega}e^{-2\omega_{\Delta H}t} \quad (31)$$

and

$$F_z^{new} \simeq \mu_o MH'V - \frac{H'}{\Delta H}\frac{P}{\omega}(\omega_{\Delta H}t + C)e^{-2\omega_{\Delta H}t} \quad (32)$$

where P is the rf power absorbed by the sphere and

$$C = \frac{1}{2}\,\frac{(u_o^2 - 1)\tan^{-1}u_o + u_o}{(u_o^2 + 1)\tan^{-1}u_o - u_o} \quad (33)$$

for $u_o = H'R/\Delta H$. The approximations are most valid provided $H'R \ll \omega/(-\gamma\mu_o)$ and $u_o$ is either small or large. Plots of Eqs. (31) and (32), as a function of $\omega_{\Delta H}t$, are shown in Fig. 1; the latter equation for the limiting values of C=1 and C=1/2.

NUMERICAL EXAMPLE

For $H'R/\Delta H = 2$, $R = 5 \times 10^{-3}$m., $\omega = 6\pi \times 10^9$ rad/sec. and P = 100 watts peak, $\Delta F = H'P(2\Delta H\omega) = 10^{-6}$ newtons. Naturally, increased power will make the force larger but parametric instabilities will eventually occur unless the radius is also increased; however, if R is too large, the magnetostatic approximation can fail.

In order to measure the transient force, it may be desirable to transmit the force to a nonresonant piezoelectric transducer in the form of a parallel plate capacitor. The open-circuit voltage produced by the force would be on the order of

$$V = \frac{k}{1-k^2}\sqrt{\frac{\varepsilon}{C_{elas}}}\,\frac{\Delta F}{C_o}$$

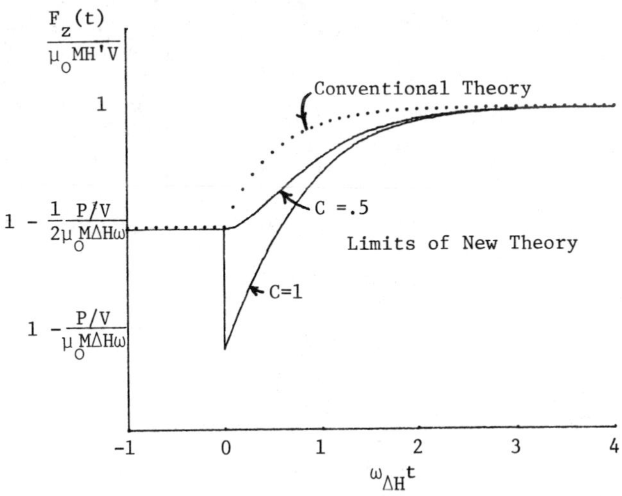


Fig. 1 Plots of the predicted rigid-body force acting upon a ferromagnetic sphere both when it is in steady-state resonance ($t<0$) and after removal of the rf driving field at $t=0$.

where k is the coupling coefficient, $\varepsilon$ the permittivity, $C_{elas}$ the appropriate elastic constant and $C_o$ the "clamped" capacitance of the transducer. For ZnO, and $C_o$ = 1/2 pfd., a force of $10^{-6}$ newtons would produce approximately 15 μvolts.

REFERENCES

1. F.R. Morgenthaler, "Linear Momentum in the Rest Frame of a Ferromagnet and the Electrodynamic Consequences", AIP Conference Proceedings, No.18, Magnetism and Magnetic Materials - 1973, p.720. References to the conventional theory are also included.
2. Outside of the material, $\bar{\bar{T}}$ is of course independent of the choice of model; it is simply the Maxwell stress-tensor.

## EXCITED STATE EPR AND EXCHANGE INTERACTIONS IN PARAMAGNETIC SINGLET-GROUND-STATE SYSTEMS

B. R. Cooper
Dept. of Physics, West Virginia University, Morgantown, W. Va. 26506
C. Y. Huang
Los Alamos Scientific Laboratory*, Los Alamos, N. M. 87545
and K. Sugawara
Dept. of Physics, Case Western Reserve University†, Cleveland, Ohio 44106

ABSTRACT

We have performed EPR in excited crystal-field states of singlet-ground-state systems, and relate these new results, and earlier results in TmN and TmP, to the expected behavior including anisotropic exchange effects. The $Tm^{3+}$ excited $\Gamma_5$ EPR in TmAs is qualitatively similar to that in TmP. The g-factor approaches from below the predominantly fourth-order crystal-field-only value of approximately 2.2 at high temperatures; while below 100°K the g-factor shows anomalous behavior like that in TmP. For $Tm_{0.5}La_{0.5}As$ the g-factor between 200°K and 500°K shows no temperature dependence within experimental uncertainties. However, the experimental uncertainty in the g-factor for $Tm_{0.5}La_{0.5}As$ is very substantial, ∿0.4; and the fact that the lattice constant differs from that of pure TmAs makes it likely that effects besides simple dilution of exchange are going on. For TmSb, the g-factor measured between 100 and 300°K smoothly approaches a value of about 2.1 from below. In $Tb_{0.1}La_{0.9}P$, above about 30°K the resonance g-factor for the excited $\Gamma_5$ state is near the fourth-order crystal-field value, while below 30°K the g value increases smoothly to a value 2 or 3% greater at 4.2°K. (TbP is a singlet-ground-state antiferromagnet; while below the critical value of x the behavior of $Tb_xLa_{1-x}P$ is purely paramagnetic even at 0°K. So $Tb_{0.1}La_{0.9}P$ is a singlet-ground-state paramagnet.) We have extended our theory for the temperature dependence of the excited state g-factor to include anisotropic exchange effects, and relate this to the anomalous behavior in TmN, TmP, and TmAs. From the high temperature g-factor behavior giving the shift from the crystal-field-only asymptotic limit, we deduce the Tm-Tm exchange, $J(0)$, for the thulium monopnictides. These values of $J(0)$ and their overall variation going from the nitride to the antimonide compare well with the behavior for the same quantity found by comparison of the g-shift of Gd in the thulium monopnictides with the temperature dependence of the susceptibility for the pure host.

*Work performed under the auspices of the U.S.E.R.D.A.
†Supported by the National Science Foundation

## $Gd^{3+}$ EPR AS A PROBE OF THE ANTIFERROMAGNETIC CRITICAL BEHAVIOR IN RARE EARTH SYSTEMS WITH SUBSTANTIAL CRYSTAL-FIELD SPLITTINGS

K. Sugawara
Dept. of Physics, Case Western Reserve University†, Cleveland, Ohio 44106
C. Y. Huang
Los Alamos Scientific Laboratory*, Los Alamos, N. M. 87545
and B. R. Cooper
Dept. of Physics, West Virginia University, Morgantown, W. Va. 26506

ABSTRACT

We have followed the EPR of dilute Gd impurities through $T_N$ for TbP, a crystal-field singlet-ground-state antiferromagnet. This compliments our EPR study of the excited $\Gamma_5{}^{(2)}$ level of $Tb^{3+}$ in TbVA(VA=P,As,Sb) on approaching $T_N$ from above. In addition we have studied the Gd EPR in $Tb_xY_{1-x}P$ and $Tb_xLa_{1-x}P$ with x close to and well below the threshold value for antiferromagnetism. Also we have studied the Gd EPR on going through $T_N$ for CeP, CeAs, and CeSb where the crystal-field splitting from $\Gamma_7$ to $\Gamma_8$ is about 160°K for CeP, 140°K for CeAs, and 25°K for CeSb. The Gd EPR has also been studied in $Ce_{0.05}La_{0.95}P$,$Ce_{0.2}La_{0.8}P$ and $Ce_{0.2}La_{0.8}Sb$. Our analysis includes positive and negative frequency effects, to deal with the situation where line width is substantial compared to resonance field, and also considers effects of the coupling of the Gd impurity probe with the antiferromagnetic host. For Gd in TbP, the resonance g goes through a well defined maximum at $T_N$, while the line width increases as temperature increases above $T_N$. For Gd in the cerium monopnictides the g-factor also has a maximum at $T_N$, and the line width increases above $T_N$. We relate this behavior to the critical susceptibility behavior and the crystal-field interaction.

*Work performed under the auspices of the U.S.E.R.D.A.
†Supported by the National Science Foundation

# EQUATION OF STATE FOR THE $Ce_{1-x}Th_x$ VALENCE TRANSITION*

J. M. Lawrence, M. C. Croft, J. M. Markovics, and R. D. Parks
University of Rochester, Rochester, New York 14627


## ABSTRACT

The valence transition in $Ce_{1-x}Th_x$ is isomorphic, and, hence, analogous to a liquid-vapor transition (the order parameter being the volume V); and it exhibits mean-field behavior. We have found that in the critical region the x-T phase boundary is nearly parallel to the x-axis, and that the data for resistivity R (which couples linearly to the order parameter) fits a Landau equation of state of the approximate form

$$0.35\left(\frac{\Delta R}{R_o}\right)^3 + 0.67\left(\frac{\Delta x}{x_o}\right)\left(\frac{\Delta R}{R_o}\right) = \frac{\Delta T}{T_o} - 0.07\frac{\Delta x}{x_o}$$

This is in analogy to a mean-field ferromagnet with $\{M,H,T\} \sim \{V,T,x\}$; consequently $dR/dT|_{max} \sim (\Delta x)^\gamma$ with $\gamma = 1$, as demonstrated by the data. The mean field nature is due to the electroelastic coupling associated with the large volume change. This coupling enters the free energy as the $\Delta x(\Delta V)^2$ term, which may reflect the dominance of anharmonic strain with increased alloying. Magnetic susceptibility and specific heat measurements reveal that both the average magnetic moment and entropy couple linearly to the order parameter. The latter finding is consistent with the existence of the term $-\Delta V\Delta T$ in the free energy as implied by the above equation of state.


## INTRODUCTION

In recent work[1,2,3] the phase diagram for SmS has been examined. One can look at, say, the phase boundary in the x-T plane for $Sm_{1-x}Gd_xS$ and give an intelligent commentary concerning the electron system-- e.g., at what concentration the f level enters the band, where the entropy difference between the states disappears, and the role of the lattice coupling. We have recently shown[4] that cerium, when alloyed with thorium, has a second-order valence transition of mean-field character. We have now extended this result to obtain not only the phase diagram, but the equation of state for the transition; here we present these results and discuss the significance of the various terms which enter the free energy. In particular we will show that the lattice coupling and the large spin entropy change play a dominant role in the $Ce_{1-x}Th_x$ transition.

## EXPERIMENTAL RESULTS

To obtain the equation of state we have utilized the fact that the electrical resistivity R couples linearly to the order parameter[4] (which for the isomorphic transitions is the volume V). We then obtained resistance data on a number of samples of differing concentrations in the critical regime. By measuring the temperature of the midpoint of the hysteresis loop on first-order samples, and the inflection point of the S-shape curves for second-order samples we obtain the phase diagram of Fig. 1a. The salient feature is the very weak variation of $T_o$ with thorium dilution.

In our earlier work[4] we proposed an equation of state of the form

$$A\left(\frac{\Delta R}{R_o}\right)^3 + B\left(\frac{\Delta x}{x_o}\right)\left(\frac{\Delta R}{R_o}\right) = \frac{\Delta T}{T_o} - C\frac{\Delta x}{x_o} \qquad (1)$$

$T_o$ and C are obtained from the phase diagram, given $x_o = 0.265$ from the earlier work. To further test this we least-squares fitted the data for each separate concentration to Eq. (1), obtaining best values for A(x) and B(x). The values so obtained exhibited

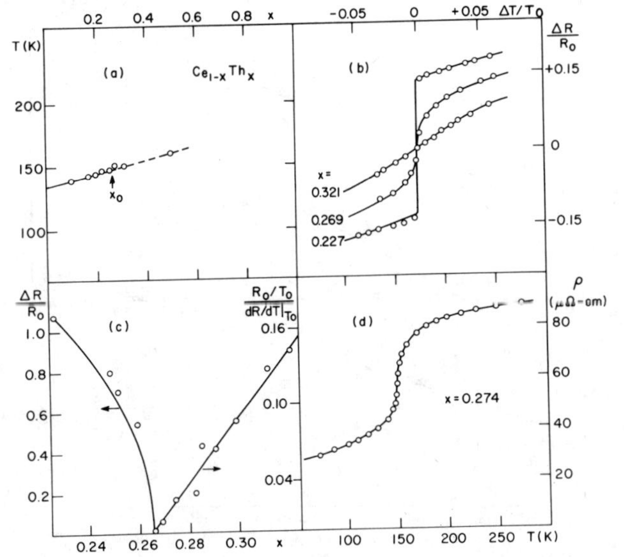


Fig. 1. (a) Phase boundary for the $Ce_{1-x}Th_x$ α-γ transition, $x_o$ being the critical concentration. (b) Resistance data for three concentrations, plotted with the predictions of the equation of state [Eq. (1)]. (c) The first order jumps and generalized inverse susceptibility, with the predictions of Eqs. (2) and (3). (d) The resistivity of a near-critical sample, exhibiting magnitudes.

10% scatter but no trend in x for concentrations $0.2 \lesssim x \lesssim 0.3$, i.e., they are essentially independent of x; and, we find A = 0.35, B = 0.67, $x_o$ = 0.265, and $T_o$ = 148 K, C = +0.07.

In order to make this procedure more convincing we present in Fig. 1 several plots of the data together with the theoretical predictions of Eq. (1). Figure 1b shows directly the ΔR vs ΔT curves for first order, near critical, and continuous transitions. This is perhaps more transparent if one notes that for the above equation of state, assuming $T_o$ vs x to be perfectly flat (i.e., neglecting the $C\Delta x/x_o$ term) one can draw an exact analogy to a mean-field ferromagnet if we let $\{\Delta R, \Delta T, \Delta x\}_{valence} \sim \{\Delta M, \Delta H, \Delta T\}_{mag}$. Given this analogy one recognizes in Fig. 1b the usual M(H) curves for $T<T_o$, $=T_o$, $>T_o$. Furthermore

$$\frac{R_o/T_o}{(dR/dT)_{T_o}} = B\left(\frac{x}{x_o}\right)^{+\gamma} \qquad (2)$$

represents the inverse susceptibility with $\gamma = 1$ for the mean field case. This is plotted in Fig. 1c, which serves both to exhibit the goodness of the B parameter, and that $\gamma = 1$. Also given is a plot of the first order jumps; i.e.

$$2\left.\frac{\Delta R}{R_o}\right|_{T=T_o} = \left(4\frac{B}{A}\frac{\Delta x}{x_o}\right)^{1/2} \qquad (3)$$

Given B from above this serves as an independent check of A and β = 1/2 exponent.

The equation of state as it stands represents only that part of the resistivity that reflects the phase transition. Since the coupling to the volume is linear, and the equation is normalized to the critical values, the parameters A and B should, in principle, remain unchanged were $\Delta V/V_o$ substituted for $\Delta R/R_o$;

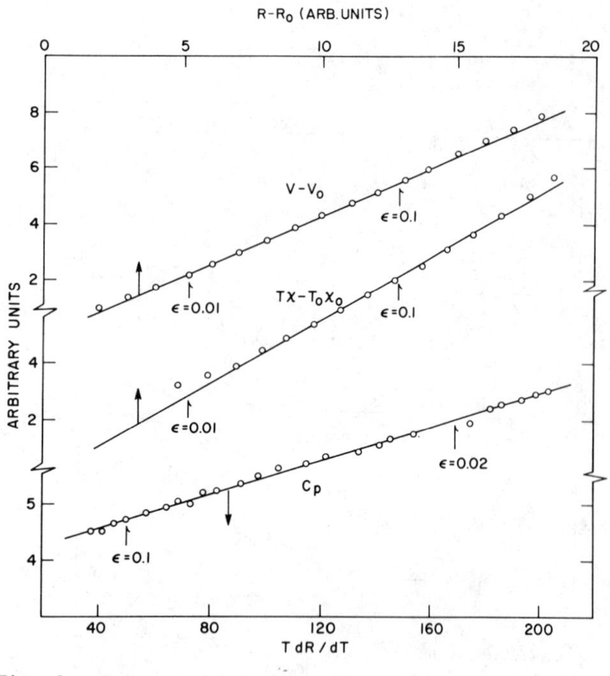


Fig. 2. Demonstrating linear coupling of V, R, $\chi T$ and S. $\Delta V$ and $\Delta\chi T$ are plotted versus $\Delta R$, while $C_p$ is plotted versus T dR/dT. All data shown for $T > T_o$

however this would only work if the backgrounds in each case were correctly subtracted out. In Fig. 1d we note $\rho$(300 K) ~ 90 $\mu\Omega$cm, $\rho(T_o)$ ~ 60 $\mu\Omega$cm and $\rho$(4.2 K) ~ 30 $\mu\Omega$cm. These values are, to within the 10% scatter, independent of concentration in the region 20-30 at % Th.[5] We have, in the above analysis, subtracted off $\rho$(4.2 K). The remaining large values for $\rho$ and $d\rho/dT$ indicate that the dominant contribution to $\rho$ comes from spin-disorder scattering (see the discussion below) and that the large variation in $d\rho/dT$ is primarily due to the phase transition. In fact, the phonon background should also be subtracted out, but this is rather more difficult, and we haven't done it. We estimate that the quoted values for A, B and C are correct to 10 or 15%.

## DISCUSSION

In the above we have strongly utilized the linear coupling of the resistivity to the order parameter. This is a consequence of the large spin-disorder scattering that occurs in cerium, which in turn is undoubtedly connected with the proximity of the f-level to the Fermi surface. One expects that as the cerium spins demagnetize, the volume will couple to the effective number of moments [$\chi T \sim N_{eff}(g\mu_B)^2$ where $\chi$ is the magnetic susceptibility]. The scattering cross-section for the conduction electrons, presumably also linear in $N_{eff}$, e.g., $1/\tau \sim N_{eff}\, J^2 g^2 \mu_B{}^2$, will reduce correspondingly. One then would have linear coupling of $\chi T$, R and V. In Fig. 2 we plot our preliminary data for $\chi T$, and the data for V against that for R at the same values of $\varepsilon = (T-T_o)/T_o$; this indeed establishes the linear coupling.

If we integrate Eq. (1) we obtain for the critical contribution to the free-energy

$$F \propto \frac{A}{4}\left(\frac{\Delta V}{V_o}\right)^4 + \frac{B}{2}\left(\frac{\Delta x}{x_o}\right)\left(\frac{\Delta V}{V_o}\right)^2 - \left(\frac{\Delta T}{T_o}\right)\left(\frac{\Delta V}{V_o}\right) + C\left(\frac{\Delta x}{x_o}\right)\left(\frac{\Delta V}{V_o}\right) \qquad (4)$$

Here we have reverted to using $\Delta V$ notation, á la the discussion at the end of the last section; an over-all proportionality factor is still missing. We now want to give an educated guess as to what each term represents.

The last term, giving the weak temperature variation of $T_o$, has the form of a $-\mu N$ term; one expects such a term to arise from a microscopic hamiltonian containing a $+(\Delta\varepsilon)n_f$ term, where $\Delta\varepsilon(x)$ is $E_f-\varepsilon_F$, the distance of the f-level from the fermi level $\varepsilon_F$, and $n_f$ gives the occupancy of the f level. Since presumably $n_f$ is coupled to V, the existence of the $\Delta x\Delta V$ term is consistent with a picture where $\Delta\varepsilon \sim \Delta x$ which could be explained by a downward shifting of the bottom of the conduction band accompanied by a downward shifting in $\varepsilon_F$ with increasing conduction electron density due to thorium (thorium being tetravalent, cerium trivalent).

The $-\Delta T\Delta V$ term arises from entropy. There is a large spin entropy ~$R\ell n6$ associated with the J = 5/2 moment in $\gamma$-cerium, most of which should be lost on transformation to the $\alpha$ state. The large magnitude of this term relative to the $\Delta x\Delta V$ valency term is what determines the relative flatness of the phase diagram. (It also could be used to determine the overall multiplicative factor for F, which should be slightly less than unity.) It hinges, however, on the entropy being linear in the volume. To substantiate this, we exhibit in Fig. 2 preliminary results for the specific heat; assuming $C_p = C_o + C_1\, T\, dS/dT$, we have plotted $C_p$ vs T dR/dT, demonstrating the coupling.

The $x(\Delta V)^2$ interaction term represents the coupling to the lattice. Anderson and Chui suggest, on the basis of elastic continuum theory, that due to the different sizes of the ions of the two valence states, there will be an infinite ranged attractive interaction between ions of the same size, $-E_h(\Delta V)^2$; however, any field such as P or x which strains the crystal far enough will cause it to enter an anharmonic regime where the interaction becomes repulsive; i.e. we should write $[E_a(x) - E_h(x)](\Delta V)^2$. This would explain not only the existence of such a term in the equation of state, but why the transition is mean-field.

At this point we are seeking only a zero-order description. The numerical values are not meant to be taken too seriously; similarly we do not mean to exclude the contribution of other mechanisms, such as a size-effect contribution to the $\Delta x\Delta V$ term, or an alternative explanation of the $\Delta x(\Delta V)^2$ bootstrapping.

## CONCLUSION

By obtaining sufficient data in the critical region, we have been able to derive an equation of state for the $Ce_{1-x}Th_x$ valence transition, good to 10 or 15%. The essential features reflect the dominance of the lattice coupling and the spin entropy. It should be possible to extend this result in the sense of pinpointing which mechanisms could or could not give rise to the various terms.

## REFERENCES

*Supported by ONR and ARO(D).

1. A. Jayaraman, P. Dernier and L. D. Longinotti, Phys. Rev. B 11, 2783 (1975).
2. T. Penney and F. Holtzberg, Phys. Rev. Letters 34, 322 (1975).
3. C. M. Varma, Preprint to be published in Rev. Mod. Physics, see also works cited therein.
4. J. M. Lawrence, M. C. Croft and R. D. Parks, Phys. Rev. Letters 35, 289 (1975).
5. M. Nicolas-Francillon, thesis, Faculte des Sciences d'Orsay, 1973 (unpublished).
6. P. W. Anderson and S. T. Chui, Phys. Rev. B 9, 3229 (1974).

# SPIN GLASS PROPERTIES AND METAL-INSULATOR TRANSITIONS IN $(Ti_{1-x}V_x)_2O_3$

J. DUMAS, C. SCHLENKER, J.L. THOLENCE, R. TOURNIER
Groupe des Transitions de Phases et Centre de Recherches sur les Très Basses Températures, Centre National de la Recherche Scientifique
B.P. 166X, 38042 Grenoble-Cedex, France.

ABSTRACT

The initial magnetic susceptibility $\chi$, the thermoremanent (TRM) and isothermal remanent (IRM) magnetizations have been measured below 10 K on single crystals of $(Ti_{1-x}V_x)_2O_3$ for $0.007<x<0.076$. The data show a spin glass behavior in the whole concentration range, with peaks in the curve $\chi(T)$. A strong anisotropy is found for all magnetic properties. The anisotropy found in the height of the susceptibility peak, and the anisotropy of the TRM and IRM curves, indicate that the peak is closely related to irreversible magnetic phenomena. It is also shown from the concentration dependence of the magnetic properties that the system is metallic or semimetallic throughout the explored concentration range. These data establish that the anomaly in the low temperature specific heat of these compounds, reported elsewhere, is mainly due to the spin glass properties.

## INTRODUCTION

Titanium sesquioxide $Ti_2O_3$ shows a semiconductor-metal transition around 450 K, due to a gradual reduction of the bandgap caused by an anomalous increase of the c/a ratio of the lattice parameters of the corundum structure.[1] The magnetic susceptibility below 400 K is small and temperature independent.[2] The incorporation of Vanadium in $Ti_2O_3$ both increases the c/a ratio and reduces the size of the electrical transition without changing its temperature range.[3,4] $(Ti_{1-x}V_x)_2O_3$ is reported to be a metal at all temperatures for $x>0.02$.[3] For $0.02<x<0.10$, the specific heat shows at low temperature an anomalously large linear term.[5] This linear term had previously been attributed to a narrow V impurity band falling at the Fermi level and responsible for a high density of states.[6]

We have reported that the magnetic susceptibility of $(Ti_{1-x}V_x)_2O_3$ shows between 10 K and 300 K a Curie-Weiss behavior in the concentration range $0.001<x<0.10$, with ferromagnetic Curie temperatures.[7] The effective magnetic moment per V atom was found to be 1.8 $\mu_B$ for $0.04<x<0.10$ and to depend on x for smaller concentrations. These data established the presence of strong magnetic moments localized on the V ions, even for large x when the matrix is metallic. Furthermore, a preliminary magnetic study at low temperatures showed for the initial reversible susceptibility plotted versus T a peak at a critical temperature $T_1$ and thermoremanent (TRM) and isothermal remanent (IRM) magnetizations for $T<T_1$.[8] [The TRM is obtained when a magnetic field h is applied during the cooling process from a temperature > $T_1$ down to the measurement temperature T and then suppressed. The IRM is obtained when h is applied at the temperature T and then suppressed .

The purpose of the present work was to study the low temperature magnetic properties versus V concentration, and to discuss them versus the specific heat anomaly. Evidence for an anisotropy in spin glass properties is also reported for the first time and discussed.

## EXPERIMENTAL RESULTS

The $(Ti_{1-x}V_x)_2O_3$ single crystals with $0.007<x<0.076$ were grown by the Czochralsky method in a commercial triarc furnace ; the V concentration was determined by atomic absorption spectroscopy. The initial reversible susceptibility $\chi$ measured in low dc fields (H<20 Oe) down to 50 mK, as described elsewhere [9], is plotted as a function of temperature on figure 1, for several concentrations $x\geqslant0.007$. The curves showing peaks are characteristic of a spin glass, even for the smallest concentration x=0.007. The temperature $T_1$ of the peak

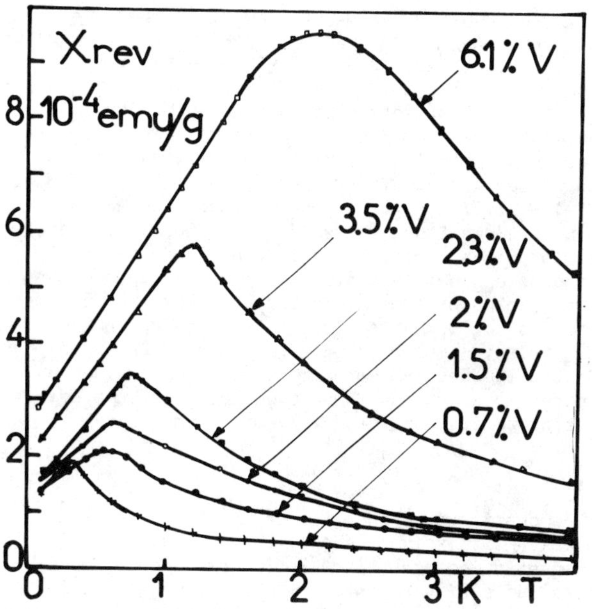


Figure 1. Initial reversible susceptibility versus temperature for $(Ti_{1-x}V_x)_2O_3$ single crystals. The V concentrations are given on the curves in at.%.

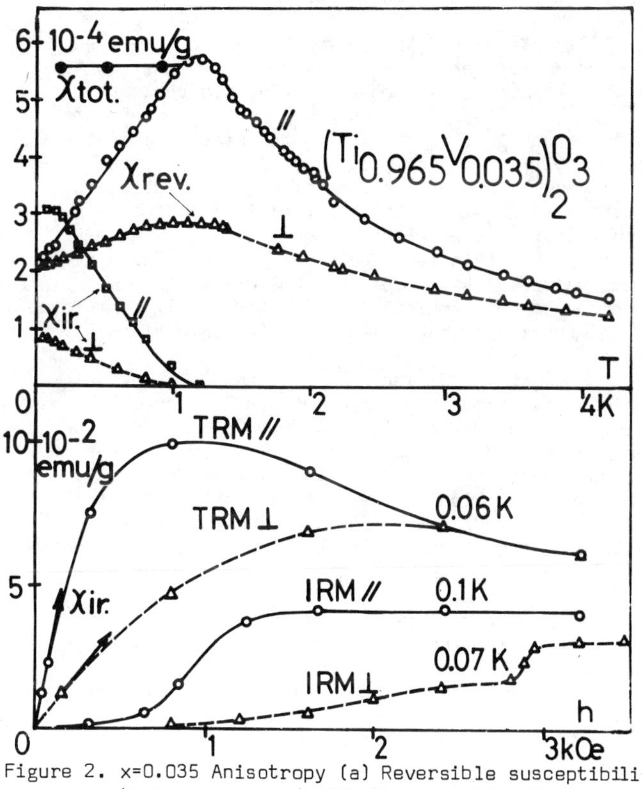


Figure 2. x=0.035 Anisotropy (a) Reversible susceptibility $\chi_{rev}$ measured with the magnetic field parallel and perpendicular to the $[1\bar{3}1]$ rhomboedral axis. The irreversible susceptibilities $\chi_{ir}$ deduced from the initial slope of the TRM versus h (figure 2b) are also shown. (b) TRM and IRM versus h in the parallel and perpendicular case. The $IRM_{//}$, being measured at a higher temperature than the TRM, saturates to a smaller value.

is found to depend approximately linearly on x in the whole concentration range. The values $\chi(0)$ extrapolated to 0 K are almost concentration independent for $x<0.035$ and increase with x for higher concentrations. Figure 2a shows for the sample $x=0.035$ the initial susceptibilities $\chi_{/\!/}$ and $\chi_{\perp}$ measured with a magnetic field respectively parallel and perpendicular to the $[1\bar{3}1]$ rhomboedral axis.

$\chi_{/\!/}$ is found to be larger than $\chi_{\perp}$ and the peak is much higher for $\chi_{/\!/}$ than for $\chi_{\perp}$ ; However the temperatures of the peaks and the values $\chi(0)$ are the same in both cases. The TRM and IRM are plotted versus magnetic field h for both orientations on figure 2b. The initial slopes of the curves of the TRM versus h give the irreversible susceptibilities, $\chi_{ir/\!/}$ and $\chi_{ir\perp}$ as plotted versus temperature on figure 2a. $\chi_{ir}(T)$ should be equal to $\chi(T_1) - \chi(T)$. [9] Figure 2a shows that good agreement is found between these values for both parallel and perpendicular cases. The TRM and IRM in both directions show the same general behavior as in other spin glass systems [9], plus a strong anisotropy in low fields. The IRM remains negligible at low temperature as long as the dc field is smaller than a few hundred Oe. In high fields the IRM and TRM saturate to the same value in both directions but the $TRM_{/\!/}$ shows a large bump before decreasing to its saturated value in high field.

## DISCUSSION

The quasi proportionality of the temperature $T_1$ of the peak with x implies that the interactions between magnetic moments follow a $r^{-3}$ law. Such a law is characteristic of RKKY interactions through conduction electrons. [10,11] It thus indicates that the $(Ti_{1-x}V_x)_2O_3$ crystals are in a metallic state at low temperatures for all $x>0.007$. As the present data show no steep anomaly with x for $x>0.007$, one may estimate that the density of states at the Fermi level $g(E_F)$ does not change abruptly with x in the explored concentration range.

It is clear from the present data that the specific heat anomaly at low temperature is at least partly due to the spin glass properties. One may evaluate the electronic contribution to the excess low temperature specific heat $\Delta Cp$, from the high temperature magnetic susceptibility $\chi_o$. Above 300 K, $\chi_o$ is temperature independent and is due at least partly to the Pauli paramagnetism. As $\chi_o$ is of the order of $10^{-4}$ emu/mole [7], $\gamma$ calculated in a free electron model is approximately 10 mJ/mole.K$^2$. This is an upper estimation. The experimental values of the slope $\alpha=d(\Delta Cp)/dT_{T\to 0}$ decrease versus x from 78 to 56 mJ/mole.K$^2$ for $0.039<x<0.081$. Therefore, $\Delta Cp$ has to be attributed mainly to the spin glass properties.

The experimental decrease of $\alpha$ versus x may be attributed to some enhancement of the RKKY interactions, through an increase of $g(E_F)$. Such an effect would be expected to lead to a decrease of $\chi(o)$ versus x. The observed independence of $\chi(o)$ for $x<0.035$ may result from the competition between this effect and the following mechanism : when x is increased, ferromagnetic couplings between the V moments enhance $\chi(o)$. The experimental data show that for $x>0.035$, the second mechanism is predominant.

The anisotropy shown on figure 2 establishes that there is a hard direction for the magnetic moments. For $T>T_1$ the curves $\chi_{/\!/}(T)$ and $\chi_{\perp}(T)$ correspond respectively to an enhancement and a reduction of the susceptibility compared to the high temperature behavior. This result may be attributed to the presence of superparamagnetic regions with the same average hard axis. As the anisotropy energy of these regions is not much smaller than kT, the macroscopic parallel susceptibility is found larger than the perpendicular one. For $T<T_1$, the relaxation times of these regions become larger than the experimental measurement times and they give rise to remanent magnetizations. The saturated $TRM_{/\!/}$ and $TRM_{\perp}$ are thus due to the same regions and are expected to be the same, in agreement with the experimental data (Figure 2b).

The origin of the anisotropy is not clear. The value of $\chi(o)$ is determined by the density of magnetic impurities in zero molecular field [11] : the absence of any detectable anisotropy in $\chi(o)$ indicates that the local anisotropy is not the dominant mechanism . One may also notice that the anisotropy of the matrix 3d electronic bands should lead to some anisotropy in the long distance magnetic interactions. The dipolar coupling could then induce some easy direction for the magnetic moments.

## ACKNOWLEDGEMENTS

The authors wish to thank J. Devenyi and J. Mercier for the crystal growth and J. Palleau and P. Amiot for the chemical analysis. They are also grateful to J. Friedel, Sir Nevill Mott and J. Souletie for helpful discussions.

## REFERENCES

1. Honig J.M. and Reed T.B., Phys. Rev. 174, 1020 (1968).
2. Schlenker C., Dumas J., Buder R, Waksmann B., Adler D., Shin S.H., Reed T.B., Proc. ICM 73 (Publ. House Nauka Moscow, 1974) Vol. V, p. 134.
3. Chandrashekhar G.V., Won Choi Q, Moyo J., Honig J.M. Mat. Res. Bull. 5, 999 (1970).
4. Mott N.F., Friedman L., Phil. Mag. 30, 389 (1974).
5. Sjöstrand M.E. and Keesom P.H., Phys. Rev. B7, 3558 (1973).
6. Van Zandt L.L., Phys. Rev. Lett. 31, 598 (1973).
7. Dumas J., Schlenker C., Natoli R.C., Sol. State Commun. 16, 493 (1975).
8. Dumas J., Schlenker C., Tholence J.L., Tournier R., Sol. State Commun. 17, 1215 (1975).
9. Tholence J.L., Tournier R., J. Phys. 35,C4-229 (1974).
10. Blandin A., Friedel J., J. Phys. Rad. 20, 160 (1959).
11. Souletie J., Tournier R., J. Low Temp. Phys. 1, 95 (1969).
12. Neel L. Low Temperature Physics (Les Houches, 1961), edited by C. de Witt, Dreyfus B. and De Gennes P.G. (Gordon and Breach, New York, 1962).

# PARAMAGNETIC-ANTIFERROMAGNETIC TRANSITION IN HUBBARD MODEL†

J. Florencio Jr. and K. A. Chao§
Departamento de Fisica, Universidade Federal de Pernambuco,
Recife-Pe., Brazil

## ABSTRACT

The Gutzwiller's variational scheme is used to obtained a first order para-antiferromagnetic transition in a s-band Hubbard model including the virtual electron hoppings. The AFM ground state in the phase diagram is restricted between the electron densities $n_1<1$ and $n_2=2-n_1$. It is also bounded from below by a critical value of U/W. The complete AFM ordering appears only for n=1. As n approaches to $n_1$ or $n_2$ along the phase boundary, the AFM ordering gradually disappears. The AFM ordering is essentially due to the virtual electron hoppings. We also found an area in the phase diagram where the para- and the AF-magnetic states coexist. As the temperature is raised, the present theory predicts a transition from the AFM insulating to the PM metallic state with the possibility of having an intermediate PM insulating state.

## INTRODUCTION

Various models have been proposed to investigate the magnetic ordering and the metal to insulator transition in a narrow band due to the electron correlation. The simplest and most tractable one is the s-band Hubbard model[1] which incorporates the intraatomic correlation energy U to the bare bandwidth W. In this model the ferromagnetic ordering appears only at the atomic limit, i.e. U/W approaches infinity, as one of the degenerate states. As U/W increases one expects a paramagnetic (PM) metallic to an antiferromagnetic (AFM) insulating transition, with a possible PM insulating intermediate state. The conjecture is in accordance with experiments. However, Mott[2] and Herring[3] have pointed out that the magnetic transition and the metal-insulator transition (MIT) are due to different mechanisms. The former is caused by the electron virtual hoppings which are of the order of $W^2/U$. On the other hand, the metallic conductivity requires an activation energy of the order U. Therefore, for sufficiently large U/W the MIT temperature can not be lower than the Néel temperature. This is generally observed, for example, in $V_2O_3$[4].

In the insulating state far from the MIT the effect of electron virtual hoppings has been discussed by Caron and Pratt[5] and Esterling and Lange[6]. Near the MIT the importance of the virtual hopping in regard to the possible transitions between the AFM insulating, the PM insulating and the PM metallic states will be investigated in this paper. The Gutzwiller's variational scheme[7] is particularly useful for the present problem because the trial function is constructed with emphasis on the localized properties of electrons.

## MODEL AND ANALYSIS

Consider N electrons in a lattice of L sites with N≤L (If N>L we consider the hole). Half of them are with up spins and half with down spins. We separate the lattice into two sublattices L(↑) and L(↓). For k within the inner half of the first Brillouin zone, we define the single-particle creation operators

$$d^{\dagger}_{k\sigma}=S\{\sum_{g\in L(\sigma)} e^{ikg}a^{\dagger}_{g\sigma}+\zeta\sum_{g\in L(-\sigma)} e^{ikg}a^{\dagger}_{g\sigma}\} \qquad (1)$$

where $a^{\dagger}_{g\sigma}$ is the creation operator associated to the Wannier state at site g and spin σ. S is a normalization factor and ζ is the variational parameter.

Using Gutzwiller's projection operator[7], the ground state trial function for a correlated system can be expressed as

$$\Psi=\prod_{g}\{1-(1-\xi)n_{g\uparrow}n_{g\downarrow}\}\prod_{k\in K} d^{\dagger}_{k\uparrow}d^{\dagger}_{k\downarrow}|0> \qquad (2)$$

where $n_{g\sigma}=a^{\dagger}_{g\sigma}a_{g\sigma}$, ξ is the second variational parameter, |0> the vacuum and K is the Fermi sea of the uncorrelated system. The physical significance of the variational parameters is obvious: ζ measures the AFM ordering (ζ=0 for complete AFM state and ζ=1 for PM state), and ξ indicates the strength of correlation.

To calculate the energy, we directly incorporate the electron virtual hopping in the Hubbard Hamiltonian instead of using the second order perturbation formalism. If an up spin electron is adjacent to a down spin electron, the virtual hopping will lower the energy of each electron by V(U). From the following (5) we see that the ground state has an optimum number ν of doubly occupied atoms. Therefore, if a doubly occupied atom is next to an empty one the energy should increase by V(U). For the region of large U which we are interested in, ν is very small and V(U) approaches asymptotically to $W^2/U$. Consequently the contribution from the second case is negligible.

We then have the effective Hamiltonian

$$H=\sum_{gf\sigma} t_{gf}a^{\dagger}_{g\sigma}a_{f\sigma}+U\sum_{g} n_{g\uparrow}n_{g\downarrow} - V(U)\sum_{gf\sigma}{}'\{n_{g\sigma}(1-n_{g-\sigma})n_{f-\sigma}(1-n_{f\sigma})- n_{g\sigma}n_{g-\sigma}(1-n_{f\sigma})(1-n_{f-\sigma})\} \qquad (3)$$

where the primed sum is restricted to the nearest neighbors. The energy $E=<\Psi|H|\Psi>/<\Psi|\Psi>$ can be calculated by quasi-chemical approximation[8]. The reader is referred to the original works for the details which will not be given here. Let us define $\bar{\varepsilon}(n)$ as the average band energy per electron for electron density n, and η=ζξ be a new variational parameter. The value of η is restricted between the band limit η=1 and the atomic limit η=0. In terms of the dimensionless quantities $\alpha=U/4|\bar{\varepsilon}(n=1)|$ and $\beta=ZV(U)/4|\bar{\varepsilon}(n=1)|$ where Z is the coordination number, the normalized energy per electron ε is obtained from

$$n\varepsilon=-(1+\zeta)^2(n-2\nu)\{\nu^{\frac{1}{2}}+(1-n+\nu)^{\frac{1}{2}}\}^2/4(1+\zeta^2)+\alpha\nu -\beta\{(1+\zeta^4)(n-2\nu)^2(1+\zeta^2)^{-2}-2\nu(1-n+\nu)\}. \qquad (4)$$

ν is the number of doubly occupied atoms normalized with respect to the total number of lattice site. Eq. (4) is valid for |1-n|<<1.

ε is then minimized with respect to ν and η to determine the ground state. We will first consider the case n=1 to illustrate the essential features of the present theory. In this case the condition ∂ε/∂η=0 yields

$$\nu=\{(1+\zeta)^2-\alpha(1+\zeta^2)-2\beta\}/4(1+\zeta)^2, \qquad (5)$$

$$\varepsilon=-\{2(1+\zeta)^2\nu^2+\beta\}/(1+\zeta^2). \qquad (6)$$

For given value of $\zeta$, $\nu$ decreases monotonically to zero as U increases to the critical value

$$U_o(\zeta)=\{4(1+\zeta)^2|\bar{\varepsilon}(n=1)|-2ZV(U_o)\}/(1+\zeta)^2. \qquad (7)$$

This indicates a MIT. Under the condition $\nu=0$ it is easy to see the minimum energy occurs at $\zeta=0$. Therefore, the ground state is AFM-insulating for large U.

For small U V(U) can be neglected. Then we see from Eqs. (5) and (6) that $\varepsilon$ is less for larger value of $\zeta$. Consequently in this region the ground state is PM-metallic.

## RESULTS AND DISCUSSIONS

The exact transition from the PM metallic to the AFM insulating as U increases depends on the form of V(U). $V(U)=t^2/U$ derived from the second order perturbation is valid only for large U. Since for small U the ground state is PM, the exact form of V(U) is no longer important as long as it is small. In this region we assume the following form for numerical calculation

$$V(U)=W^2/4Z^2U \qquad ;U\geq U_o=8|\bar{\varepsilon}(n=1)| \qquad (8)$$

$$=W^2U\{12|\bar{\varepsilon}(n=1)|-U\}/1024Z^2|\bar{\varepsilon}(n=1)| \qquad ;U\leq U_o$$

Note that $V(U)\to 0$ as $U\to 0$, and around the area of interest $U\simeq 8|\bar{\varepsilon}(n=1)|$ the V(U) is continuous with continuous first derivative.

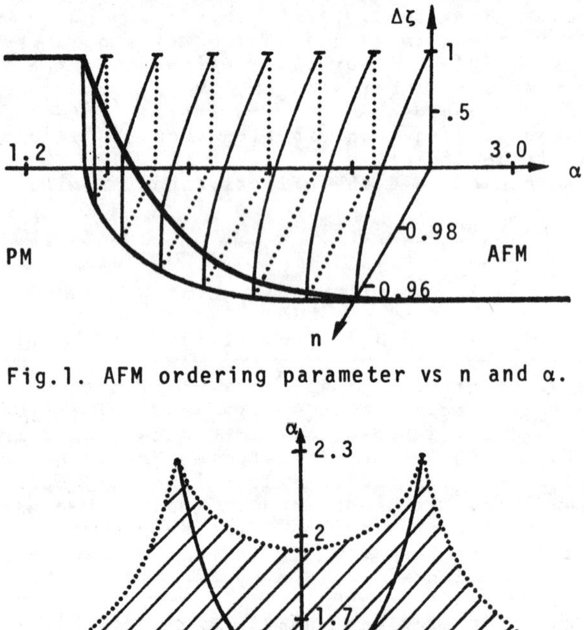


Fig.1. AFM ordering parameter vs n and α.

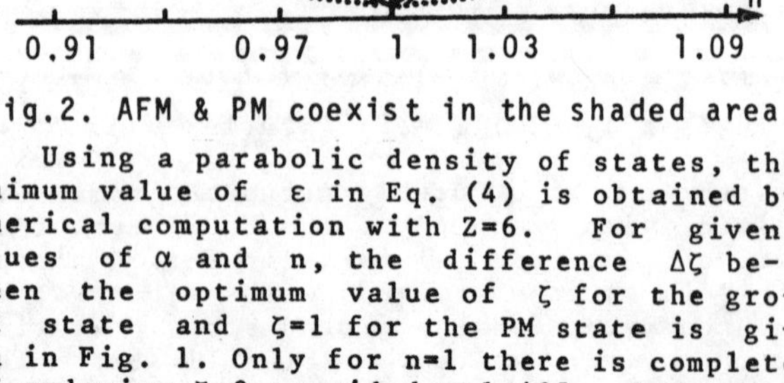


Fig.2. AFM & PM coexist in the shaded area.

Using a parabolic density of states, the minimum value of $\varepsilon$ in Eq. (4) is obtained by numerical computation with Z=6. For given values of $\alpha$ and n, the difference $\Delta\zeta$ between the optimum value of $\zeta$ for the ground state and $\zeta=1$ for the PM state is given in Fig. 1. Only for n=1 there is complete AFM ordering $\zeta=0$ provided $\alpha>1.425$. As n gets smaller along the phase boundary in the n-$\alpha$ plane, the AFM ordering becomes weaker and disappears entirely at n=0.9548. Across the phase boundary the transition is first order. The complete phase diagram is shown in Fig. 2 by the solid curve. In the shaded area we see that the AFM and the PM states coexist.

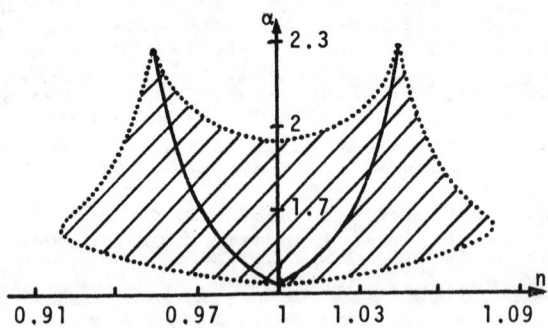

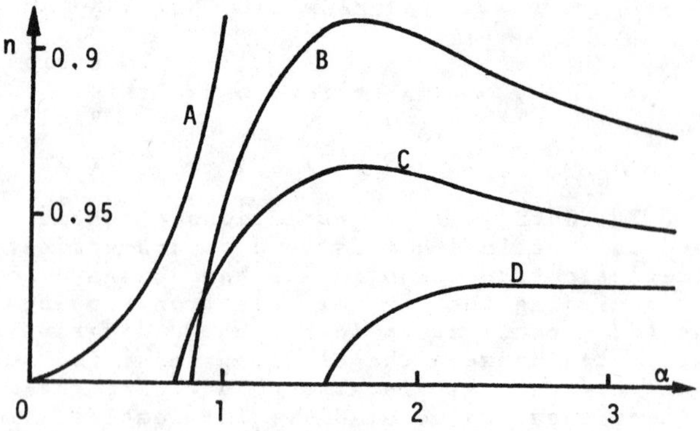


Fig.3. A comparison of the AFM-PM phase boundaries from different authors.

Recently Ogawa, Kanda and Matsubara[9] and Takano and Uchinami[10] have extended the Gutzwiller method with different approximations to study the AFM state. In Fig. 3 we compare the AFM-PM phase diagrams from various works: Curve A from Penn[11], curve B from Takano and Uchinami, curve C from Ogawa et.al. and curve D from the present theory. For certain values of n curves B and C show an AFM to PM transition as $\alpha$ increases. This is unphysical.

For special case n=1, we found from (5) that at T=0 the PM state is metallic if $\alpha<1.98$ but insulating if $\alpha>1.98$, while the AFM state is always insulating. Since the AFM state has less entropy than the PM state an AFM insulating to PM metallic transition occurs at the Néel temperature if $1.425<\alpha<1.98$. On the other hand if $\alpha>1.98$, the transition should be from the AFM insulating to the PM insulating, and then to the PM metallic as the temperature is raised. This behavior is observed in $V_2O_3$[4].

## REFERENCES

† Work partially supported by BNDE, CNPq and CAPES (Brasilian government).

§ Leave of absence from The Department of Physics, University of Linköping, Linköping, Sweden. Present address.

1. J. Hubbard, Proc. R. Soc. A276 238-254, (1963); A281 401-419, (1964).
2. N. F. Mott, Metal-Insulator Transition (Taylor & Frances Ltd., London 1974).
3. C. Herring, Magnetism IV, Eds. G. T. Rado and H. Suhl (Academic Press, N. Y. 1966).
4. D. B. McWhan and J. P. Remeika, Phys.Rev. B2 3734-3750 (1970).
5. L. G. Caron and G. W. Pratt, Rev. Mod. Phys. 40 802-806, (1968).
6. D.M. Esterling and R. V. Lange, Rev. Mod. Phys. 40 796-799, (1968).
7. M. C. Gutzwiller, Phys. Rev. Lett. 10 159-164 (1963); Phys. Rev. 137A 1726-1735, (1965).
8. For the details of QCA, see K. A. Chao, Phys. Rev. B4 4034-4046 (1971); B8 1088-1098; J. Phys. C7 127-145, (1974).
9. T. Ogawa, K. Kanda and T. Matsubara, Prog Theoret. Phys. 53 614-633, (1975).
10. F. Takano and M. Uchinami, Prog. Theoret. Phys. 53 1267-1285, (1975).
11. D. R. Penn, Phys. Rev. 142 350-365 (1966)

# MAGNETOSTRICTION NEAR THE NÉEL TEMPERATURE OF $MnF_2$

Y. Shapira and R.D. Yacovitch, Francis Bitter National Magnet Laboratory,*
Massachusetts Institute of Technology, Cambridge, Massachusetts 02139

ABSTRACT

The longitudinal and transverse magnetostriction in $MnF_2$ were measured near the Néel temperature. The low-field data are compared with a model in which the magnetostriction is largely caused by the strain dependence of the dominant exchange interaction. The magnetic phase diagram for $\vec{H}\parallel[001]$, obtained from the magnetostriction data, is analyzed in terms of recent theories.

## INTRODUCTION

$MnF_2$ is a uniaxial easy-axis antiferromagnet with a Néel temperature $T_N = 67.3$K. The $Mn^{++}$ions (S state, spin 5/2) form a body-centered tetragonal structure. The dominant exchange interaction of an ion at (0,0,0) is with its neighbors at ($\pm\frac{1}{2}$, $\pm\frac{1}{2}$, $\pm\frac{1}{2}$). The anisotropy favors the [001] axis and is weak compared to the exchange interaction. Here we present a study of the longitudinal and transverse magnetostriction along both the [001] and [110] directions, for temperatures $65 < T < 78$K and in magnetic fields H up to 140 kOe. The experimental technique is the same as in ref.1.

Figure 1 shows the H-dependence of $\Delta\ell/\ell = [\ell(H) - \ell(0)]/\ell(0)$, where $\ell$ is the sample's length along the direction of measurement. These data are for one temperature above $T_N$ and another somewhat below it. In both cases $\Delta\ell/\ell$ is proportional to $H^2$, at low H. At 66.79K, $\Delta\ell(H)$ exhibits an anomaly near the antiferromagnetic (AF) to paramagnetic (P) transition.[2] Below we focus on: 1) the quantitative behavior of $\Delta\ell$ at low H, and 2) the magnetic phase diagram for $\vec{H}\parallel[001]$, determined from the magnetostriction measurements. Some data for $\vec{\ell}\parallel\vec{H}\parallel[001]$ were presented earlier.[1] The discussion below is based also on new data for the configurations: $\vec{\ell}\parallel[001]$ with $\vec{H}\parallel[110]$; $\vec{\ell}\parallel\vec{H}\parallel[110]$; $\vec{\ell}\parallel[110]$ with $\vec{H}\parallel[1\bar{1}0]$; and $\vec{\ell}\parallel[110]$ with $\vec{H}\parallel[001]$.

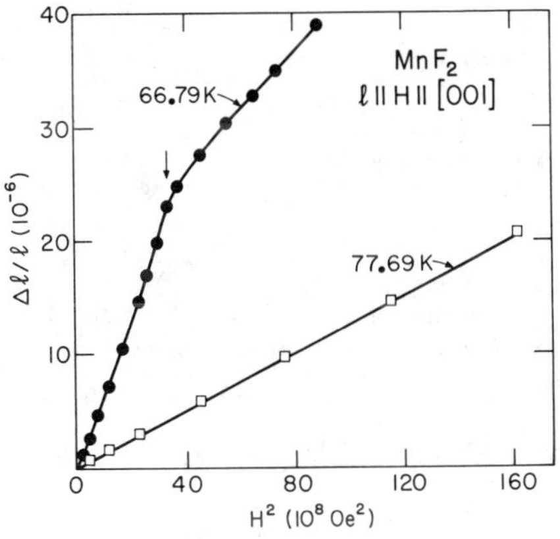


Fig. 1. Longitudinal magnetostriction along the [001] axis. The arrow for the curve at 66.79K marks the transition from the AF phase to the P phase (see ref. 2 and Fig. 4).

## LOW-FIELD BEHAVIOR

For all configurations of $\vec{\ell}$ and $\vec{H}$, $\Delta\ell$ is proportional to $H^2$ at low H. The ratio $[(\Delta\ell/\ell)/H^2]_0$ at low H was measured as a function of T from 65 to 78K. With $\vec{H}\parallel[001]$ this ratio exhibited a large lambda anomaly near $T_N$. The anomaly was considerably smaller for configurations with $\vec{H}\parallel[110]$ (see Figs. 2 and 3).

To interpret the results, consider a sample which at H = 0 occupies a volume of 1 $cm^3$. The dependence of $\ell$ on H and the dependence of the magnetic moment I on uniaxial pressure p (applied along $\vec{\ell}$) are related by the equation

$$(1/\ell)(\partial\ell/\partial H)_{T,p} = -(\partial I/\partial p)_{T,H} \quad (1)$$

For low H the susceptibility $\chi = I/H$ is independent of H, so that

$$\Delta\ell/\ell \equiv [\ell(H) - \ell(0)]/\ell(0) = -(H^2/2)(\partial\chi/\partial p). \quad (2)$$

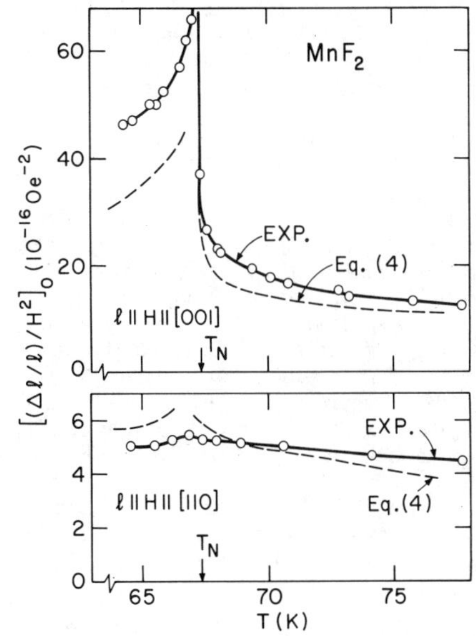


Fig. 2. Temperature dependence of the ratio $[(\Delta\ell/\ell)/H^2]_o$ for the longitudinal magnetostriction along the [001] and [110] directions.

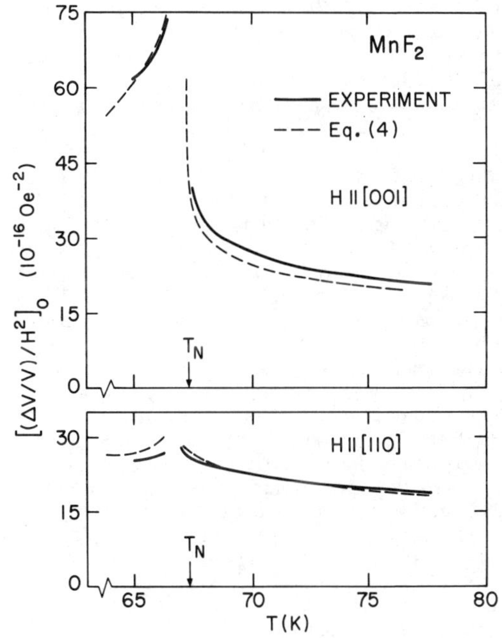


Fig. 3. Temperature dependence of the ratio $[(\Delta V/V)/H^2]_o$ for $\vec{H}\parallel[001]$ and $\vec{H}\parallel[110]$.

We now make the assumption that the only magnetic interactions of an ion at (0,0,0) are a Heisenberg exchange interaction with its neighbors at ($\pm\frac{1}{2}$, $\pm\frac{1}{2}$, $\pm\frac{1}{2}$) and a direct interaction with $\vec{H}$, i.e., we neglect the anisotropy and all exchange interactions other than with next-nearest-neighbors.[3] One can show that under these assumptions $(T\chi)$ is a function of J/T only (J = exchange constant). Therefore,

$$(\partial\chi/\partial p) = (\partial J/\partial p)(\partial\chi/\partial J) = -(1/J)(\partial J/\partial p)[\partial(T\chi)/\partial T], \quad (3)$$

so that

$$\Delta\ell/\ell = H^2(1/2J)(\partial J/\partial p)[\partial(T\chi)/\partial T]. \quad (4)$$

Concerning the above, note the following: a) Eqs. (1) –(4) also apply to the volume magnetostriction $\Delta V/V$ provided that the derivatives with respect to uniaxial pressure p are replaced by derivatives with respect to the hydrostatic pressure $P_h$. b) Eq. (4) also holds if the Heisenberg interaction between next-nearest-neighbors is replaced by an Ising interaction. However, in this case $\chi$ depends on the direction of $\vec{H}$. c) Eq. (4) is usually not valid if nearest- as well as next-nearest neighbor exchange interactions are both included, or if the anisotropy is included, because then $(T\chi)$ is not a function of $J/T$ only. The general case is discussed by Callen and Callen[4] and is reviewed by Callen.[5] Under the simplifying assumptions introduced above, the results of Callen and Callen are equivalent to Eq. (4).[1]

Well above $T_N$ the effect of the anisotropy on the susceptibility of $MnF_2$ is small so that Eq. (4) is expected to give a good description of the data. Even at 75 K this equation should hold reasonably well. However, near and below $T_N$ the anisotropy cannot be ignored, as is evident from the difference between the susceptibilities $\chi_{\parallel}$ and $\chi_{\perp}$ parallel and perpendicular to [001], respectively.[6] Therefore, deviations from Eq. (4) are expected at $T \leqslant T_N$. Nevertheless, as we shall see, Eq.(4) gives a fair description of the experimental data even near $T_N$ provided that one uses $\chi_{\parallel}$ whenever $\vec{H}\parallel[001]$, and $\chi_{\perp}$ whenever $\vec{H} \perp [001]$. As expected theoretically,[7] $\partial(T\chi_{\parallel})/\partial T$ exhibits a large lambda anomaly near $T_N$, whereas the anomaly for $\partial(T\chi_{\perp})/\partial T$ is much smaller.[6]

Since the T-dependence of both $\chi_{\parallel}$ and $\chi_{\perp}$ have been measured,[6] the only remaining parameter necessary to evaluate the right-hand side of Eq. (4) is $(1/J)(\partial J/\partial p)$. In the analogous expression for $\Delta V/V$ one needs $(1/J)(\partial J/\partial P_h)$. Assuming that $T_N \propto J$, the results of Benedek and Kushida[8] give $(1/J)(\partial J/\partial P_h) = (4.49 \pm 0.3) \times 10^{-12}\ cm^2/dyne$. For p along [001] the thermal expansion data of Gibbons[9] give $(1/J)(\partial J/\partial p) \cong 2.6 \times 10^{-12}\ cm^2/dyne$, as discussed in Ref. 1 (see also Ref. 10). Since equal uniaxial pressures along [001], [110] and [$1\bar{1}0$] are equivalent to a hydrostatic pressure, we have $(1/J)(\partial J/\partial p) \cong 0.95 \times 10^{-12}\ cm^2/dyne$ for p along [110]. This leaves no adjustable parameters in Eq. (4) or its analog for $\Delta V/V$.

Figure 2 compares the T-dependence of $[(\Delta \ell/\ell)/H^2]_0$ for the longitudinal magnetostriction along [001] and [110] with values calculated from Eq. (4). The overall agreement is reasonably good, although the observed lambda anomaly for $\vec{\ell}\parallel\vec{H}\parallel[001]$ is larger than calculated. The lambda anomaly for the configuration $\vec{\ell}\parallel[110]$ with $\vec{H}\parallel[001]$ (not shown here) is smaller than that calculated from Eq. (4). These discrepancies are attributed to the neglect of the anisotropy. Physically, we expect that the change in anisotropy caused by a compression in the basal plane will be of opposite sign to that caused by a compression along [001], because the two change the c/a ratio in opposite ways. In the volume magnetostriction, which is the sum of $\Delta\ell/\ell$ for $\vec{\ell}\parallel[001]$, [110], and [$1\bar{1}0$], the effects of the anisotropy partially cancel. As may be seen in Fig. 3, $\Delta V/V$ for both $\vec{H}\parallel[001]$ and $\vec{H}\parallel[110]$ are in good agreement with the volume-analog of Eq. (4). The lambda anomaly is much larger for $\vec{H}\parallel[001]$ than for $\vec{H}\parallel[110]$, as expected from the relative magnitude of the anomalies in $\partial(T\chi_{\parallel})/\partial T$ and $\partial(T\chi_{\perp})/\partial T$.

Finally we note that: a) the agreement of the transverse magnetostriction (not shown here) with Eq.(4) is approximately the same as for the longitudinal magnetostriction in Fig. 2; b) at 78 K, where $\chi$ is nearly isotropic, the magnetostriction for a given direction of $\vec{\ell}$ is approximately (within ~20%) independent of the direction of $\vec{H}$; and c) for all T, the data for $\vec{\ell}\parallel\vec{H}\parallel[110]$ are within ~10% of those for $\vec{\ell}\parallel[110]$ with $\vec{H}\parallel[1\bar{1}0]$.

## MAGNETIC PHASE DIAGRAM

The magnetic phase diagram of $MnF_2$ was measured earlier using ultrasonic and susceptibility techniques.[2] In the present work the phase boundaries for $\vec{H}\parallel[001]$ were determined from traces of $\ell$ vs H at constant T, and $\ell$ vs T at constant H. The second-order transitions (AF-P and SF-P in the notation of Ref. 2) appeared as lambda anomalies in $\partial\ell/\partial H$ and $\partial\ell/\partial T$. The phase boundaries are shown in Fig. 4, and are in general agreement with the data in Ref. 2. Comparison was made with recent theories of Fisher et al.[11] Using the notation of Ref. 11b our data give: $T_N = (67.32 \pm 0.03)$K; $H_b = (119.3 \pm 0.8)$ kOe; $dT_c^{\parallel}/dH^2 = (1.53 \pm 0.05) \times 10^{-10}\ K/Oe^2$ at $H=0$; and $T_b^* = (65.13 \pm 0.1)$K. A fit of the data above 90 kOe to Eq. (5) of Ref. 11b, using the theoretical values of Q and $\phi$ but letting $T_b$, q and $w_{\parallel}$ vary, gave $\tilde{T} - T_b = (0.78 \pm 0.03)$ K (compared to the theoretical estimate 1.22 K), and $T_b = (64.80 \pm 0.05)$ K. A more definitive comparison with theory will require better alignment of $\vec{H}$ along [001](presently within 2°), and a more precise determination of $T_c$ (H) than the present precision of 0.04 K at high H.

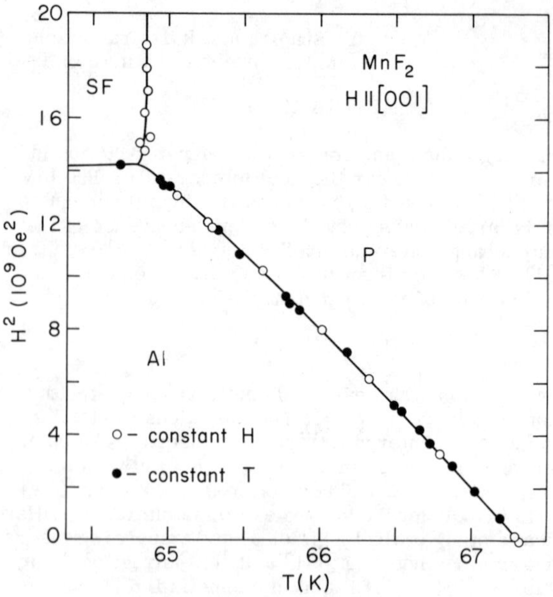


Fig. 4. Magnetic phase diagram of $MnF_2$, in the $H^2$-T plane, for $\vec{H}\parallel[001]$.

## REFERENCES


* Supported by the National Science Foundation.

1. Y. Shapira, R.D. Yacovitch and D.R. Nelson, Solid State Commun. 17, 175 (1975) and Phys. Letters 53A, 19 (1975).
2. Y. Shapira and S. Foner, Phys.Rev. B1, 3083 (1970).
3. Due to the c/a ratio of $MnF_2$ the ions at (0,0,0) and ($\frac{1}{2}$, $\frac{1}{2}$, $\frac{1}{2}$) are next-nearest neighbors. In Ref. 1 they are erroneously called nearest neighbors.
4. E. Callen and H.B. Callen, Phys. Rev. A139, 455 (1965).
5. E. Callen, J. Appl. Phys. 39, 519 (1968).
6. C. Trapp, Ph.D. thesis, U. of Chicago, 1963 (unpublished); S. Foner in Magnetism, Ed. by G.T. Rado and H. Suhl, (Academic Press, New York, 1963) Vol. 1, p. 383.
7. M.E. Fisher, Phil. Mag. 7, 1731 (1962), Physica 26, 618 (1960); M.F. Sykes and M.E. Fisher, Physica 28, 919 and 939 (1962).
8. G.B. Benedek and T. Kushida, Phys. Rev. 118, 46 (1960). See also K.C. Johnson and A.J. Sievers, Solid State Commun. 10, 829 (1972).
9. D.F. Gibbons, Phys. Rev. 115, 1194 (1959).
10. Dudko et al. [Soviet Physics-Solid State 12, 65 (1970)] give $(1/J)(\partial J/\partial p) = 1.9 \times 10^{-12}\ cm^2/dyne$ for p along [001]. However, their data analysis involves a "classical" correction which in our view is largely unwarranted. Neglecting this correction, $(1/J)(\partial J/\partial p) = 2.4 \times 10^{-12}\ cm^2/dyne$, in reasonable agreement with the value obtained from Gibsons' data.
11. M.E. Fisher, a) AIP Conf. Proc. 24, 273 (1975); b) Phys. Rev. Letters 34, 1634 (1975).

# MAGNETIZATION AND NEUTRON DIFFRACTION STUDIES ON $Mn_3Si$*

J. I. Budnick, V. Niculescu,** W. A. Hines, A. H. Menotti, K. Raj and T. J. Burch
Department of Physics and Institute of Materials Science,
University of Connecticut, Storrs, Connecticut 06268

S. J. Pickart
Department of Physics, University of Rhode Island
Kingston, Rhode Island 02881

ABSTRACT

Bulk magnetization and neutron diffraction measurements on $Mn_3Si$ are reported. Preliminary conclusions indicate a complex magnetic structure at low temperatures. The influence of dilute iron substitution on the magnetic behavior of $Mn_3Si$ is discussed.

## INTRODUCTION

It has been demonstrated that the $Fe_3Si$-$Mn_3Si$ system forms a continuous range of solid solutions with the f.c.c. $DO_3$-like structure.[1] Previous studies on the role of the selective substitution of Mn into the $Fe_3Si$ lattice[2-4] pointed out the direct correlation between a specific near neighbor environment and the magnetic behavior of the $Fe_{3-x}Mn_xSi$ system.

As suggested by neutron diffraction and bulk magnetization measurements[2] and confirmed by NMR work,[4] the presence of a small amount of Mn in ferromagnetic $Fe_3Si$ decreases the average magnetic moment of the Fe atoms which are first near neighbors to the Mn. High Mn concentrations ($x > 0.75$) lead to a rhombohedral antiferromagnetic arrangement of the magnetic moments[2] at low temperatures. For $x > 2.0$, Booth et al.[2] observe evidence for antiferromagnetism in their material which was two phase.

It thus appears that a detailed exploration of the magnetic properties of $Mn_3Si$ could extend these studies and provide valuable information on the Mn spin behavior.

In this paper we report preliminary results on the bulk magnetization and neutron diffraction studies in a well-ordered $Mn_3Si$ sample. In addition we have studied the magnetic behavior of a single sample, $Mn_{2.99}Fe_{0.01}Si$, in order to explore possible effects of magnetic impurities on the magnetic ordering.

## EXPERIMENTAL RESULTS AND DISCUSSION

### 1. Sample preparation and crystalline structure

The alloys $Mn_3Si$ and $Mn_{2.99}Fe_{0.01}Si$ were formed by arc melting the appropriate proportions of high purity Mn, Si and Fe in an argon atmosphere. The weight loss in the melting process was approximately 1% for samples of about 10 g. Analysis showed less than 100 ppm Fe impurity for the $Mn_3Si$ ingot.

The ingots were annealed in evacuated sealed tubes at 800°C for 24 hours and slowly cooled to 600°C at a rate of 40°C/hour. The samples were then water quenched. The ingots were powdered to #400 mesh and the powder was annealed at 600°C for one hour under vacuum ($2 \times 10^{-6}$ torr) and cooled to room temperature in about two hours.

X-ray and neutron diffraction analyses showed that the powdered samples were single phase and well-ordered with a cubic $DO_3$ type structure.

The lattice parameter for $Mn_3Si$ from the X-ray data was a = 5.726 Å, in good agreement with the value reported by Gladijsevskij[1] (a = 5.728 Å). The neutron diffraction data at room temperature indicate that $Mn_3Si$ is essentially well-ordered crystallographically with only (4 ± 1)% of the Si atoms distributed randomly on Mn sites.

### 2. Magnetization results

The magnetic measurements were carried out on $Mn_3Si$ and $Mn_{2.99}Fe_{0.01}Si$ over the temperature range 3 - 77K and room temperature, for magnetic fields up to 20 kOe, with a P.A.R. vibrating sample magnetometer.

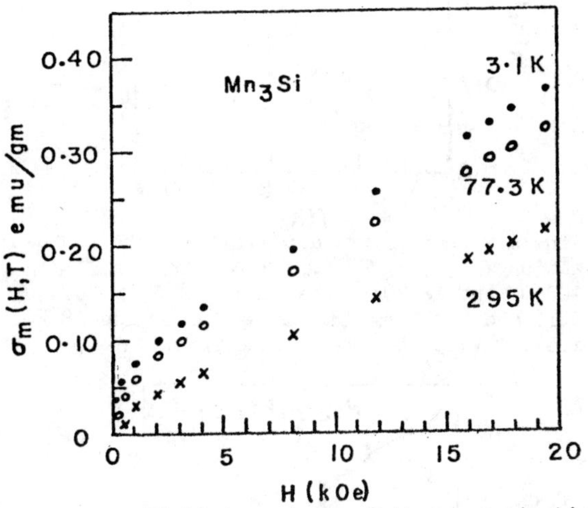


Fig. 1. The field dependence of the magnetization $\sigma_m(H,T)$ in emu/gm for $Mn_3Si$ at the temperatures 3.1, 77.3 and 295K.

TABLE I

The values of the parameters $\chi_m(T)$ and $\sigma_m^o(T)$ for $Mn_3Si$ and $Mn_{2.99}Fe_{0.01}Si$.

| Sample | T(K) | $\chi_m(T) \times 10^5$ (emu/g) | $\sigma_m^o(T)$ (emu/g) |
|---|---|---|---|
| $Mn_3Si$ | 3.1 | 1.40 ± 0.05 | 0.090 ± 0.004 |
| | 9.3 | 1.38 ± 0.05 | 0.086 ± 0.004 |
| | 44.2 | 1.32 ± 0.05 | 0.072 ± 0.004 |
| | 47.1 | 1.30 ± 0.05 | 0.071 ± 0.004 |
| | 77.3 | 1.32 ± 0.05 | 0.066 ± 0.004 |
| | 295 | 1.00 ± 0.05 | 0.026 ± 0.004 |
| $Mn_{2.99}Fe_{0.01}Si$ | 10.2 | 1.49 ± 0.05 | 0.196 ± 0.004 |
| | 35.6 | 1.43 ± 0.05 | 0.172 ± 0.004 |
| | 77.0 | 1.29 ± 0.05 | 0.156 ± 0.004 |
| | 295 | 0.93 ± 0.05 | 0.067 ± 0.004 |

Figure 1 shows the measured bulk magnetization per unit mass for $Mn_3Si$ as a function of magnetic field for temperatures of 3.1, 77.3 and 295K. At high fields, the dependence of the $Mn_3Si$ magnetization can be approximately described by the form

$$\sigma_m(H,T) = \chi_m(T) \cdot H + \sigma_m^o(T)$$

where $\chi_m(T)$ is the paramagnetic susceptibility per unit mass and $\sigma_m^o(T)$ represents a spontaneous magnetization. The values of $\chi_m(T)$ are obtained from the slope of the straight lines at high fields while the $\sigma_m^o(T)$ are taken from the intercepts with the $\sigma_m$-axis. A summary of $\chi_m(T)$ and $\sigma_m^o(T)$ deduced from the data are given in Table I. Figure 2 shows the low temperature dependence of the reduced magnetization $\sigma_m(H,T)/\sigma_m(H,10K)$ for magnetic fields of 100 Oe and 18 kOe. A broad bump at ~30K is observed which becomes more pronounced in low fields and which we tentatively associate with a change in spin structure.

In the $Mn_{2.99}Fe_{0.01}Si$, the field dependence of the magnetization has the same behavior as for $Mn_3Si$. The Fe substitution in the lattice produces an increase of the magnetization $\sigma_m^o(T)$, while the values of $\chi_m(T)$ remain practically unchanged (Table I). At the same time, the temperature dependence of the magnetization at low temperatures becomes slightly more pronounced

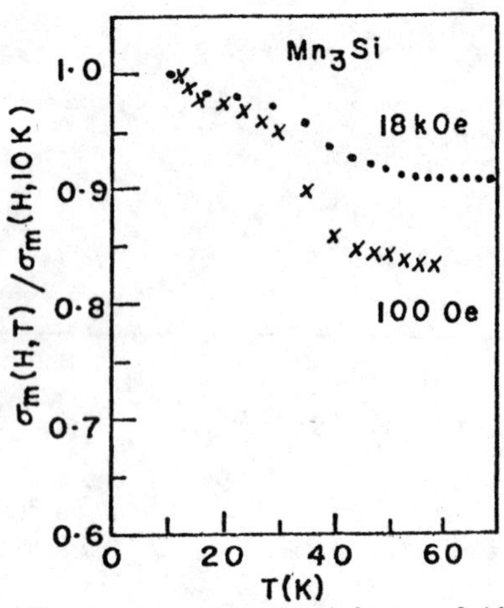


Fig. 2. The low temperature dependence of the reduced magnetization $\sigma_m(H,T)/\sigma_m(H,10K)$ for $Mn_3Si$ at the fields 100 Oe and 18 kOe. The values of $\sigma_m(H,10K)$ are 0.00416 and 0.333 emu/gm for H = 100 Oe and 18 kOe respectively. (Note the vertical origin.)

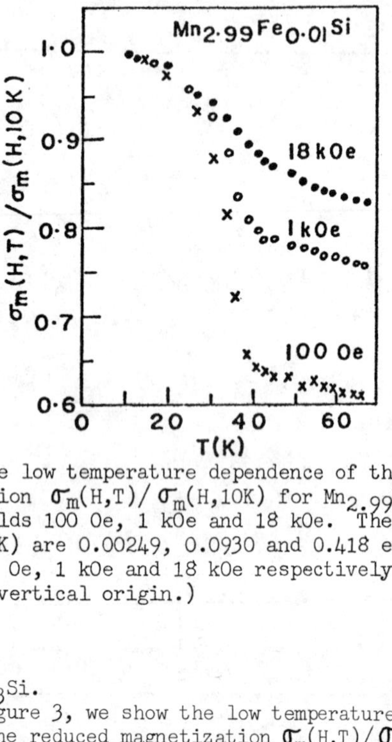


Fig. 3. The low temperature dependence of the reduced magnetization $\sigma_m(H,T)/\sigma_m(H,10K)$ for $Mn_{2.99}Fe_{0.01}Si$ at the fields 100 Oe, 1 kOe and 18 kOe. The values of $\sigma_m(H,10K)$ are 0.00249, 0.0930 and 0.418 emu/gm for H = 100 Oe, 1 kOe and 18 kOe respectively. (Note the vertical origin.)

than in $Mn_3Si$.

In Figure 3, we show the low temperature dependence of the reduced magnetization $\sigma_m(H,T)/\sigma_m(H,10K)$ for $Mn_{2.99}Fe_{0.01}Si$ in which is indicated a relatively larger change at ~30K, which increases for lower fields.

3. Neutron measurements

The neutron diffraction data taken on $Mn_3Si$ at 4K show, besides the lines corresponding to the cubic $DO_3$ type structure, several weak low-angle superlattice reflections that could not be indexed on the cubic unit cell. In addition, increases of about 60% were noted in the lowest angle $DO_3$ type superlattice lines. These effects, if of magnetic origin as the magnetization experiments suggest, indicate a magnetic structure with components both commensurate and incommensurate with the lattice. The low angle superlattice lines fit an incommensurate component propagating along [111] with a wavelength very nearly $\sqrt{2}\, a_0$ and a magnitude of 0.75-1.0 $\mu_B$ depending on whether it is spiral or sinusoidal. No fit could be found assuming any other principal cubic axis for the propagation direction.

A collinear model for the in-phase component is definitely ruled out by the data. The simplest non-collinear antiferromagnetic model, with equal Mn moments, is a triangular arrangement of the three face-centered Mn sublattices (i.e., each sublattice pointing along a face diagonal). This model results in approximately the experimentally observed intensity ratio of the first two superlattice lines for a Mn moment equal to 1.82 $\mu_B$.

Preliminary calculations varying the magnitude of the moments and the angles of the sublattices simultaneously indicate that the intensities, as might be expected for powders, are not very sensitive to such changes. Single crystal measurements will have to be performed to distinguish between such models and to index the satellite peaks unambiguously.+

CONCLUSIONS

The magnetization data on $Mn_3Si$ suggest some change in spin structure, possibly a magnetic phase transition at about 30K. The existence of an ordered state at low temperatures is supported by our observation of some hysteresis in the magnetic behavior.

The neutron diffraction data taken at 4K suggest a complex, possibly antiferromagnetic arrangement of the Mn sublattices with components both commensurate and incommensurate with the lattice. The weak moment deduced from magnetization $\sigma_m^o$ data $\sim 3\times10^{-3}\mu_B$/formula unit at low temperature is consistent with an antiferromagnetic ordering.

The substitution of 0.25 at.% Fe increases the moment to a value $\sim 6\times10^{-3}\mu_B$/formula unit.

Measurements of the temperature dependence of the neutron scattering as well as further studies of the magnetization will be made to produce a more detailed picture of the spin structure of $Mn_3Si$ at low temperatures and of the nature of the transition observed at ~30K.

ACKNOWLEDGEMENTS

We acknowledge the helpful assistance of J. J. Rhynne in taking the neutron data. The data were taken at the National Bureau of Standards reactor at Gaithersburg, Maryland. Analysis of the samples was kindly provided by Cameca Inst. Inc., 101 Executive Blvd., Elmsford, New York 10523.

REFERENCES


* Supported in part by the University of Connecticut Research Foundation and by NSWC White Oak Laboratory Independent Research Funds.

** Visiting Fulbright Hays Scholar (University of Cluj) Romania.

\+ After this paper was written, we learned of recent work of Tomiyoshi and Watanabe (J. Phys. Soc. Jap. 39, 295 (1975)), who report a helical spin structure for $Mn_3Si$ based on single crystal TOF measurements. While agreeing with the magnitude and direction of the propagation vector found above, these authors report no change in the nuclear superlattice reflections, over an unspecified temperature range. We plan to repeat our powder measurements as a function of temperature in an attempt to resolve this discrepancy.

1. E. I. Gladijsevskij, Fix. Metal i Metalloved, 2, 454 (1956).
2. S. Yoon and J. G. Booth, Phys. Letters 48A, 381 (1974); J. G. Booth, J. E. Clark, J. D. Ellis, P. J. Webster and S. Yoon, Proc. Intern. Conference on Magnetism, Moscow (1973), Vol IV, p. 577.
3. T. J. Burch, T. Litrenta and J. I. Budnick, Phys. Rev. Letters 33, 421 (1974); T. J. Burch, S. Pickart, T. Litrenta and J. I. Budnick, Phys. Letters, 53A, 321 (1975).
4. V. Niculescu, K. Raj, T. J. Burch and J. I. Budnick, to be published in the Phys. Rev. B; W. A. Hines, A. H. Menotti, J. I. Budnick, T. J. Burch, T. Litrenta, V. Niculescu and K. Raj, to be published in Phys. Rev. B.

# FERROMAGNETIC ORDERING IN SOLID $H_e^3$ DUE TO GROUND STATE VACANCIES*

A. Widom and J. B. Sokoloff
Department of Physics
Northeastern University
Boston, Massachusetts 02115

ABSTRACT

Recent measurements by Kummer et al., give strong experimental evidence that solid $H_e^3$ orders ferromagnetically. The data are discussed on the basis of a mean field theory of ferromagnetic ordering due to ground state vacancies proposed by the authors. A few other possible models are also discussed.

Experimentally observed magnetic properties of solid $H_e^3$ cannot be described by a simple Heisenberg model[1] which previously appeared to be adequate.[2] Recently, the authors have pointed out that the presence of a small concentration of ground state vacancies ($x \sim 10^{-4}$ vacancies / site) give results in qualitative agreement with known thermal experimental data near melting for T above the spin ordering temperature $T_c$.[3] Similarly, Greywall[4] has discussed specific heat anomalies in solid $H_e^3$ and finds $\sim 10^{-4}$ Landau-Fermi "quasiparticles" per site which we interpret as vacancies. The motion of vacancies in solid $H_e^3$ will lead to an energy lowering of the totally ferromangetic state with respect to other spin configurations, as pointed out by Thoules[5] and Nagaoka[6] and as discussed recently in more detail by Sokoloff.[7] In a recent publication, the authors put forward a mean field theory of ferromagnetic ordering in a model for solid $H_e^3$ having both mobile vacancies and a Heisenberg model antiferromagnetic exchange interaction.[8] It is possible in our model to have a first order ferromagnetic phase transition and still have an inverse susceptibility above the Curie temperature with a negative temperature intercept. Here we will discuss the experimental evidence for ferromagnetic ordering, and how our mean field theory compares with the data.

The most direct experimental evidence for ferromagnetic ordering to date was provided by Kummer et al,[9] who find that the magnetic ordering temperature in solid $H_e^3$ shows a general increase with increasing external magnetic field. Such behavior appears not to be consistent with antiferromagnetic or spin flipped antiferromagnetic orderings[10] because a uniform external field tends to increase the tendency towards ferromagnetism and decrease the tendency towards antiferromagnetism (and hence decrease the Néel temperature). Furthermore, the anisotropy due to magnetic dipole interaction is too small to give a spin-flop transition at the correct temperature to explain the data.

If the phase transition is first order, we may use the magnetic Clausius-Clapeyron equation

$$dT_c/dH = -\Delta m/\Delta s \quad , \qquad (1)$$

where $\Delta s$ and $\Delta m$ are respectively the discontinuities in the entropy and magnetization. Since $\Delta s = s(\text{disordered}) - s(\text{ordered}) > 0$ , and $\Delta m = m(\text{disordered}) - m(\text{ordered})$ must be opposite in sign to $dT_c/dH$ , $dT_c/dH > 0$ vigorously implies ferromagnetism! If the usual second order ferromagnetic transition occurs, there are no singularities, but there is still a peak in the specific heat for small H, whose temperature does not increase with increasing H.[9]
In the experiments of Kummer, et al.[9] ,

$$\frac{\partial T_M}{\partial H} \approx 4.4 \qquad (H \text{ in mK}) \qquad (2)$$

for fields greater than 0.42 T.

Our molecular field theory for vacancy induced ferromagnetism[8] gives a first order transition to a totally ferromagnetically ordered state. Thus, substituting $\Delta s = \ln 2$ and $\Delta m = -1$ (i.e. m = 1 is total saturation) in equation (1) gives

$$dT_c/dH \approx (\ln 2)^{-1} \approx 1.4 \quad , \qquad (3)$$

which is close to equation (2) but still a little low. In reality, since the system is never completely disordered just above $T_c$ (as it is assumed to be in our mean field theory), $\Delta s$ is less than ln2 . In fact from figure 2 in Kummer, et al.,[9] $\Delta s$ could easily be half this value. Furthermore, if m < 1 just below $T_c$, $\Delta s/\Delta m$ could be reduced further, as seen from the s vs. m curve in reference 8.

The initial decrease in the transition temperature observed by Kummer, et al.,[9] for magnetic fields less than 0.4T can be explained in the following way, which is not inconsistant with the vacancy model: Let us postulate that the F vs. m curve for a given field has three minima corresponding to an antiferromagnetic, ferromagnetic, and paramagnetic state. If for H = 0, the antiferromagnetic minimum lies lowest for sufficiently low temperature, the system could be antiferromagnetic for very low fields but ferromagnetic for higher fields (because H lowers the ferromagnetic relative to the antiferromagnetic state energy). Of course, since the apparent existence of the initial decrease of the transition temperature with increasing H is only based on measurements for a single value of H, this speculation must be considered tentative.

It must be admitted that there exist other mechanisms which could qualitatively explain the data. For example, an effective four spin exchange induced by phonons could conceivably be long range, which is necessary (see reference 11) to reconcile the various zero field data. It should be pointed out that a Hamiltonian containing Ising and biquadratic Ising model terms with appropriately chosen parameters can give antiferromagnetic ordering for low magnetic fields and ferromagnetic ordering for higher fields.[12] As mentioned previously, this situation would give the correct behavior of the transition temperature as a function of magnetic field.

REFERENCES

* Work supported by National Science Foundation Grant under DMR72-03282-A02.

1. W. P. Kirk and E. D. Adams, Phys. Rev. Lett. 17, 392 (1972); J. M. Dunson and J. M. Goodkind, Phys. Rev. Lett. 32, 1343 (1974); W. P. Halperin, C. N. Archie, F. B. Rasmussen, R. A. Buhrman, and R. C. Richardson, Phys. Rev. Lett. 32, 927 (1974).
2. R. A. Guyer, R. C. Richardson, and L. I. Zane, Rev. Mod. Phys. 43, 532 (1971); A. Landesman, Ann. Phys. (Paris) 8, 53 (1974); S. B. Trickey, W. P. Kirk, and E. D. Adams, Rev. Mod. Phys. 44, 668 (1972).
3. A. Widom and J. B. Sokoloff, J. Low Temp. Phys. (in press).
4. D. S. Greywall, Phys. Rev. B11, 4717 (1975).
5. D. J. Thoules, Proc. Phys. Soc. (London) 86, 893 (1965).

6. Y. Nagaoka, Phys. Rev. 147, 393 (1966); Solid State Communications 3, 409 (1965).
7. J. B. Sokoloff, Phys. Rev. B2, 3707 (1970); B3, 3826 (1971); and B4, 232 (1971).
8. J. B. Sokoloff and A. Widom, Phys. Rev. Lett. 35, 673 (1975).
9. R. B. Kummer, E. D. Adams, W. P. Kirk, A. S. Greenberg, R. M. Mueller, C. V. Britton, and D. M. Lee, Phys. Rev. Lett. 34, 517 (1975).
10. A. Landesman, J. Phys. (Paris), Coll. 31, C3 (1970); L. J. de Jongh and A. R. Miedema, Advan. Phys. 33, 1 (1974).
11. R. A. Guyer, Phys. Rev. A9, 1452 (1974); see also L. H. Nosanow and C. M. Varma, Phys. Rev. 187, 660 (1969).
12. F. Y. Wu (private communication). See also D. Merlini and C. Gruber, J. Math. Phys. 13, 1814 (1972).

## MAGNETOSTRICTION IN $MnF_2$ ABOVE AND BELOW $T_N$*†

J. Matolyak‡, A. S. Pavlovic and M. S. Seehra**
Physics Department, West Virginia University, Morgantown, WV 26506

A detailed study of the magnetostriction in $MnF_2$ has been made. The measurements were made using a capacitive bridge in conjunction with a capacitive dilatometer. The data were taken as a function of temperature (4.2 - 115 °K), magnetic field strength H (upto 15 kOe), magnetic field orientation and the measuring direction. Above $T_N$ (67.3 °K), the data is anisotropic with respect to the measuring direction viz. the magnetostriction along the [001] direction is larger than that along [010]. In both directions however, the data varies as $H^2$. Below $T_N$, the observed magnetostriction $\lambda$ in the [011] direction can be described by $\lambda = \lambda_o(H) + a(T)H^2\sin 2\theta_H$, where $\theta_H$ is the field angle from [001], the easy axis. Calculations for the linear magnetostriction are done using the known magnetic, elastic and magnetoelastic interactions for $MnF_2$. In agreement with the observations, the calculations predict that (1) the linear magnetostriction along the principal directions is zero, and (2) $\lambda \propto H^2\sin 2\theta_H$. An order of magnitude increase in $\lambda$ is observed between $T_N$ and 4.2 °K. Details of the experimental results and calculations will be published elsewhere.

*Supported in part by the NSF Grant GH-41126.
†Based on the Ph.D. dissertation of J. Matolyak, West Virginia University 1975 (unpublished). An initial report appeared in Phys. Letters A <u>49</u>, 333 (1974).
‡Presently at Indiana University of PA., Indiana, PA 15701
**A. P. Sloan Research Fellow.

# THEORY OF RANDOM ANISOTROPIC MAGNETIC ALLOYS+

Per-Anker Lindgård
Physics Department
A.E.C. Risø
DK-4000 Roskilde, Denmark

## ABSTRACT

A mean field - crystal field theory is developed for random, multi-component, anisotropic magnetic alloys. It is specially applicable to rare earth alloys. A discussion is given of multicritical points and phase transitions between various states characterized by order parameters with different spatial directions or different ordering wave vectors. Theoretical predictions for the phase diagrams and magnetic moments, based on known parameters for the rare earth alloys Nd-Pr, pure dhcp Nd, TbEr and TbTm alloys, agree with experimental observations. A simple procedure to include fluctuation corrections in the mean field results is also discussed.

## INTRODUCTION

The physics of anisotropic mixtures have several interesting aspects. Multicritical points (bi-, tri-, tetra-critical points, etc.) may be realized for simple model systems. We shall discuss these within the context of mean-field theory taking the crystal field, i.e., the anisotropy, exactly into account at any temperature. The theory is a generalization of the mean-field theory for the spin flop transition for an antiferromagnet in a magnetic field,[1] and also of the theory by Wegner[2] for an antiferromagnetic alloy.

Another aspect of anisotropic magnetic alloys that is of interest is their significance for the understanding of the rare earth metals. A number of experiments on rare earth alloys have been made giving phase diagrams and magnetization curves, e.g., Er-Tb and Dy by Millhouse and Koehler,[3] Nd-Pr by Lebech et al.[4] To a large extent these data have not been analyzed and fully utilized to extract information about the crystal fields and exchange interactions in these materials. The present theory may provide a basis for doing so. In the ordered phase the free energy, the susceptibility and the moment components are calculated self-consistently.

## MOLECULAR FIELD THEORY FOR PHASE TRANSITIONS

Let us consider the phase diagram for magnetic phases of an alloy of two elements with different susceptibilities; for example an alloy of two rare earth metals. We shall assume a perfect random alloy of ions of types 1 and 2 for which the Hamiltonian can be written

$$\mathcal{H} = \sum_i \{c_1(V^i_{c_1} - c_1\sum_j \mathcal{J}_{11}(ij)\,\vec{J}_{1i}\cdot\vec{J}_{1j}) + c_2(V^i_{c_2} - c_2\sum_j \mathcal{J}_{22}(ij)\,\vec{J}_{2i}\cdot\vec{J}_{2j}) - c_1c_2\sum_j 2\mathcal{J}_{12}(ij)\,\vec{J}_{1i}\cdot\vec{J}_{2j}\}, \qquad (1)$$

where $c_n$ are the concentrations, $V^i_{cn}$ the crystal fields, and $\mathcal{J}_{nm}(ij)$ the exchange interaction between the angular momenta $\vec{J}_{ni}$ and $\vec{J}_{mj}$. In the molecular field approximation (1) reduces to a single site Hamiltonian for which we can obtain the free energy exactly. For simplicity we shall first discuss the case where $<\vec{J}_1>$ and $<\vec{J}_2>$ are parallel in the ordered phase. We introduce the Fourier transformed interaction constants and angular momenta defined as $\mathcal{J}_{nm} = \sum_R 2\mathcal{J}_{nm}(R)\exp(i\vec{Q}\cdot\vec{R})$ and $J_m = \sum_R J_{mR}\exp(i\vec{Q}\cdot\vec{R})$, where $\vec{Q}$ is the ordering wave vector.

Near a second order transition the molecular fields, $H_n$, are small, and the magnetic moment induced at a site is equal to the molecular field times the paramagnetic susceptibility.

$$<J_1> = H_1\chi^0_1 = (\mathcal{J}_{11}c_1<J_1> + \mathcal{J}_{12}c_2<J_2>)\chi^0_1$$
$$<J_2> = H_2\chi^0_2 = (\mathcal{J}_{21}c_1<J_1> + \mathcal{J}_{22}c_2<J_2>)\chi^0_2 \qquad (2)$$

This gives the condition for a finite moment:

$$\frac{1}{\chi_1}\frac{1}{\chi_2} = (\mathcal{J}_{12})^2 c_1c_2\ , \quad \frac{1}{\chi_n} = \frac{1}{\chi^0_n} - c_n\mathcal{J}_{nn}\ , \qquad (3)$$

where $\chi_n$ is the enhanced susceptibility. If we express the concentrations in terms of $c = c_1$ and $(1-c) = c_2$, then the condition for the ordering temperature $T_N$ is an equation of second order in c: $Ac^2 + Bc + C = 0$, B and C depend on T through the susceptibilities $\chi^0_n(T)$. By considering the free energy, one can show that (3) is generally true also for second order phase transitions between different ordered phases. $\chi_n$ has then to be calculated in the ordered phase. It is convenient and physically much more meaningful for an alloy to consider the molecular fields to be the order parameters rather than the individual moments. The competing ordered phases may have order parameters which differ with respect to moment directions, ordering wave vectors or both. Since the discussion is analogous we consider the former case. In a coordinate system, where the enhanced susceptibilities are diagonal, the condition for having a multi-critical (bi- or tetra-) point is simply that (3) is fulfilled for the two components α and β. That is when

$$\chi_1^{\alpha\alpha}(c,T)\chi_2^{\alpha\alpha}(c,T) = \chi_1^{\beta\beta}(c,T)\chi_2^{\beta\beta}(c,T). \qquad (4)$$

The condition for having a bi-, tri or tetra-critical point depends on whether the fourth derivitive of the free energy vs. order parameter, in the ordered phase, is less, equal or greater than zero, respectively. The ordered phases are separated by a first order transition at a bi-critical point, and a mixed phase at a tetra-critical point.

A simple improvement of the molecular field results can be obtained by considering the remaining exchange interaction as a perturbation. The free energy which includes the fluctuation to second order can be written approximately.

$$F = F_{MF} - kT\sum_{R,\alpha}(\mathcal{J}_R^{\alpha\alpha})^2(\chi_0^{\alpha\alpha})^2\ , \qquad (5)$$

where $F_{MF}$ is the molecular field free energy.

## APPLICATIONS

The Tb-Er alloys were measured by Millhouse and Koehler[3] and the Tb-Tm alloys are presently being investigated.[5] Tb orders with spiral ordering with the moments in the basal hexagonal plane and Er and Tm order with a c-axis modulated (CAM) structure with the moments along the hexagonal c-axis. No detailed analysis has yet been made of these data, we shall therefore, as a first calculation, only include the dominant physical features in the basic model and no adjustable parameters.

a)

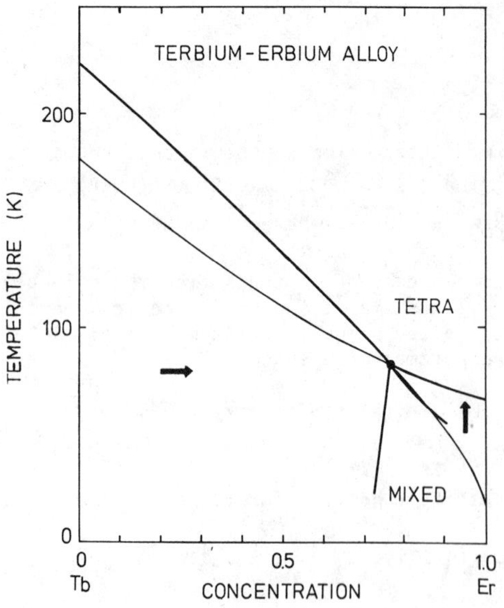


b)

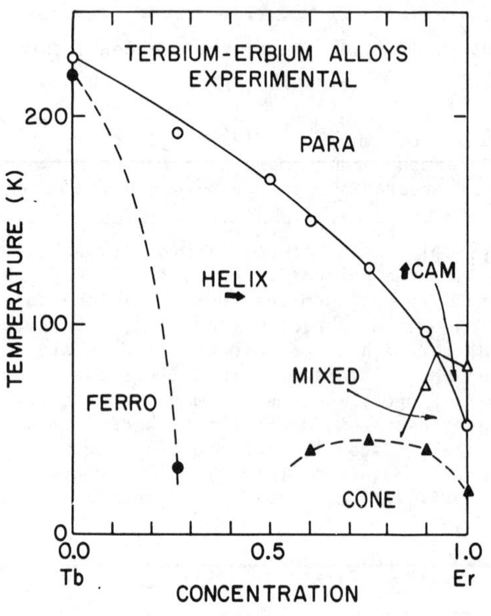


Fig. 1 a) The calculated phase diagram for the TbEr alloys. It is very similar for the TbTm alloys. The heavy full lines show the second order phase transition lines. The thin full lines show where the ordering would occur if the system was not perturbed by the different order, already present. b) The experimental data for the TbEr alloys from Ref. 3.

The Hamiltonian is a Heisenberg exchange interaction plus a crystal field, which is assumed for simplicity to contain only the axial $B_2^0 O_2^0$ term and the hexagonal $B_6^0(O_6^0 + \frac{77}{8} O_6^6)$. One exchange constant is obtained from the experimental $T_N$ for Tb. The remaining exchange interactions are obtained using the scaling by the de Gennes factors. $B_2$ is obtained from the measured difference between the paramagnetic Curie temperatures. $B_6$ is for Tb obtained from the spinwave energy gap and $B_2$(Tb). For Er and Tm, it is obtained by scaling the $B_6$ (Tb) by the appropriate Steven's factors. The calculation is done self-consistently in the magnetization components and using the complete level schemes. The resulting, calculated phase diagrams are shown in Fig. 1a. A comparison with the experimental results on Fig. 1b for Tb-Er shows that the simplified model describes the experiments quite well. A large region of helical ordering is found and a small pocket near the Er or Tm end with CAM structure, separated from the helical ordering by a mixed phase. The critical point is in agreement with experiment found to be a tetracritical point. The calculation does not consider the other observed structure changes such as for example to the ferromagnetic order. In a more detailed analysis of the phase diagrams, it is clear that one has to include other crystal field terms as well as magnetostriction, which is known to play an important role at the ferromagnetic transition. The results for Tb-Tm are similar. It is apparent from Fig. 1 that the interaction between the different elements is larger than expected using the de Gennes scaling, a feature also found for Tb-Tm. The pure dhcp Nd crystal may be considered as a 50-50% alloy of cubic- and hexagonal-site Nd. The hexagonal sites order at ~20 K. Using known parameters, the calculation shows that the moments at the hexagonal sites induce a weak magnetic order on the cubic sites in the same direction and with the same ordering vector $Q_h$. At ~8 K a second order phase transition makes the cubic sites order with the moments in essentially the same direction. At lower temperatures a perpendicular component develops, which turns the moments on the cubic sites to an angle of approximately 30° from the hexagonal sites with a strongly quenched moment. This is in agreement with a preliminary analysis of neutron scattering measurements on pure Nd single crystals.[6]

## CONCLUSION

It was shown that the mean-field random alloy theory agrees with observations for the rare earth alloys for all measured concentrations. The reason for the success of the simple theory is presumably 1) that the two-ion interaction in the rare earth metals is of long range, and 2) that the real order parameters in the theory are the mean-fields which to a much greater degree of accuracy are site independent than the individual moments. The rare earth alloys may be good candidates for investigating multi-critical points.

## REFERENCES

+Supported in part by the U.S. Energy Research and Development Administration.

1. H. Thomas, Proc. Conf. on Magnetism, Chania Crete (1969).
2. F. Wegner, Sol. State Com. 12, 785 (1973).
3. A. H. Millhouse and W. C. Koehler, Les Elements des Terres Rares, Colloques Intern. C.N.R.S. No. 180, Paris, Vol II, 214 (1970).
4. B. Lebech, K. A. McEwen and P. A. Lindgård, J. Phys. C 8, 1684 (1975).
5. P. A. Hansen and B. Lebech, Acta Cryst. A 31 (1975) to be published.
6. B. Lebech and B. D. Rainford, Proc. Int. Conf. on Magnetism 2, 248 (1973).

Acknowledgement: It is a pleasure to thank members of the solid state physics group and the Iowa State University for their hospitality during my visit when this work was completed.

# LATENT HEAT AT THE SPIN-REORIENTATION OF THE WEAK FERROMAGNET $Fe_3BO_6$

C. Voigt and N. Manderla
Institut für Werkstoffe der Elektrotechnik, Rhein.Westf.
Technische Hochschule Aachen, 51 Aachen, Germany

## ABSTRACT

The latent heat L at the spontaneous spin-reorientation of the weak ferromagnet $Fe_3BO_6$ has been measured with a sensitive differential scanning calorimeter (dsc). For single crystals we find L = 5.1 J/mol. The transition is endothermic on heating. We have also calculated L from the temperature dependence of the energy necessary to rotate the spins with an applied field from the easy to the hard axis. We obtain L = 5.2 J/mol from magnetization curves near $T_{sr}$. The spontaneous transition is found to be a continuous one with a width of about 1 K. The field dependence of the transition temperature as determined with the dsc agrees well with the one determined by magnetization measurements.

## INTRODUCTION

$Fe_3BO_6$ is an orthorhombic, weak ferromagnet with an ordering temperature $T_N$ = 508 K[1]. At $T_{sr}$ = 415 K iron borate undergoes a spin - reorientation with the resulting weak moment rotating in the [ 010 ] plane from [ 100 ] to [ 001 ] as the temperature is increased. Susceptibility and Mössbauer measurements[2,3,4] demonstrate that the sublattice moments also are always in the [ 010 ] plane. The spin-reorientation is similar to the one observed in some of the rare earth orthoferrites except that the transition in $Fe_3BO_6$ is much sharper[1]. No experiments have been reported which would allow to decide whether at $T_{sr}$ the weak moment rotates continuously from [ 100 ] to [ 001 ] or changes its direction abruptly in a first order phase transition. We have performed detailed magnetization and differential thermal analysis experiments in the vicinity of $T_{sr}$ to obtain information on the nature of the transition. All experiments were performed on single crystals with masses between 5 and 25 mg.

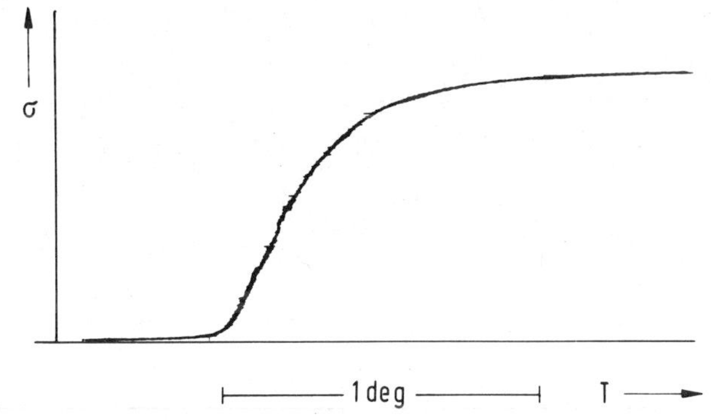


Fig. 1 Temperature dependence of the magnetic moment $\sigma \parallel [001]$ near $T_{sr}$. H = 4 A/cm. H || [ 001 ].

## MAGNETIZATION MEASUREMENTS

Fig. 1 shows, near $T_{sr}$, the temperature dependence of the component $\sigma$ of the weak moment per mass parallel to [ 001 ] . A small field parallel to [ 001] had to be applied to avoid domain splitting. The curve has been obtained by monitoring $\sigma$ as the temperature is slowly increased through the transition and is then decreased again. It can be seen that there is no temperature hysteresis. The transition is a continuous one with a width of about 1 K. We conclude that the spins rotate continuously as it is also found in the rare earth orthoferrites.

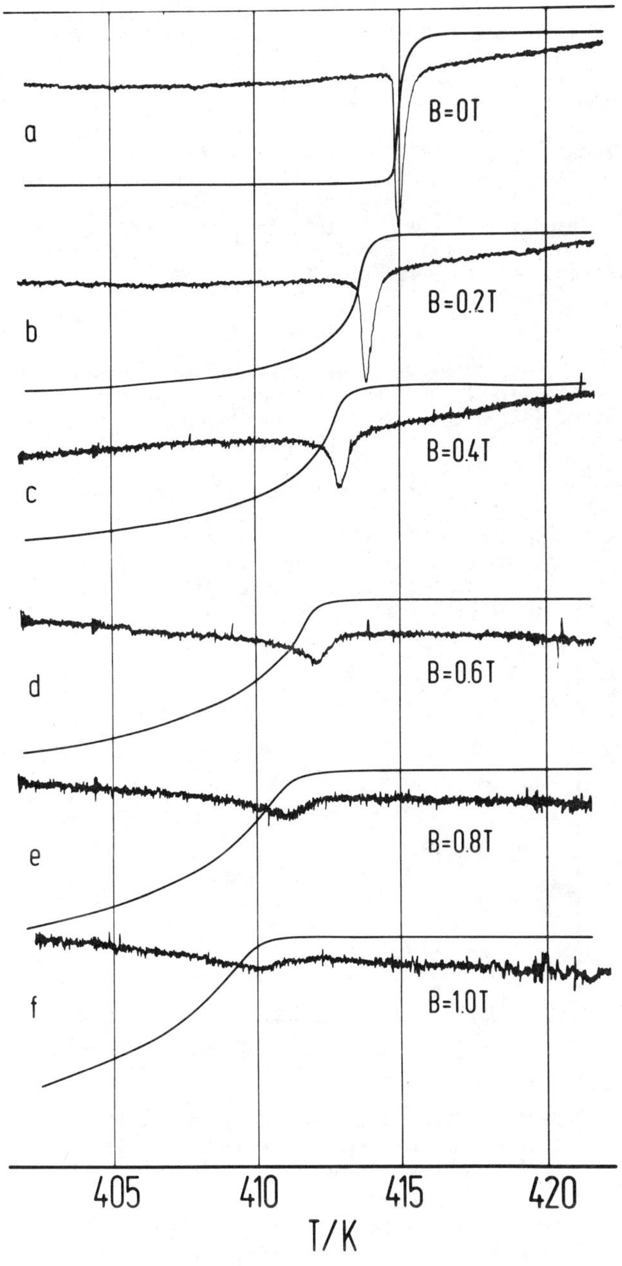


Fig. 2 Magnetic moment $\sigma \parallel [001]$ and dta-output $\Delta T$ as a function of temperature for $B \parallel [001]$ . T increasing. (noisy line) : dta, ——— : $\sigma$

## DIFFERENTIAL THERMAL ANALYSIS (DTA)

In a dta experiment the sample is placed in a temperature chamber which is heated so that the temperature increases linearily with time. The temperature difference $\Delta T$ between the sample and a reference sample also in the chamber is monitored as a function of temperature (or time). In the stationary state $\Delta T$ will be constant if no reaction takes place in the sample. Any reaction connected with absorption or development of heat will mark itself in a sudden change of $\Delta T$ (peak in the dta diagram).

We have built a sensitive dta apparatus for the investigation of iron borate single crystals. Fig. 2a shows the dta curve in zero field together with the temperature dependence of the weak moment parallel to [ 001 ] . The endothermic peak observed at $T_{sr}$ de - monstrates that the spontaneous spin-reorientation is accompanied by an absorption of heat as the temperature is increased. The peak width is about the same as the width found magnetically.

In an external field the reorientation occurs at $T < T_{sr}$ for B || [ 001 ] and at $T > T_{sr}$ for B || [ 100 ] . It can be seen from Fig.2b-2f that the dta peak coincides with the temperature where $\sigma$ has its strongest increase,i.e. where the velocity of the spin-rotation is near its maximum. Thus we obtain for both methods the same field dependence of the transition temperature. As B is increased the transition becomes less sharp for both curves on the low temperature side. This results from the fact that the field supports the rotation of the weak moment to the [ 001 ] direction. Analogous results were found for B|| [ 100 ] .

## LATENT HEAT

In a dta experiment one measures the temperature at some point of the crystal, in our case at the surface. In order to determine the amount of heat de - veloped one has to use a different arrangement, a dif-ferential scanning calorimeter (dsc). In a dsc the temperature is usually measured at the suspension (sample holder) of the sample in such a way that the area of the dsc peak is proportional to the heat de-veloped. We have built a dsc and have used $KNO_3$ for calibration which has a transition at 402 K. Our result for the latent heat L at $T_{sr}$ is L = 5.1 J/mol.

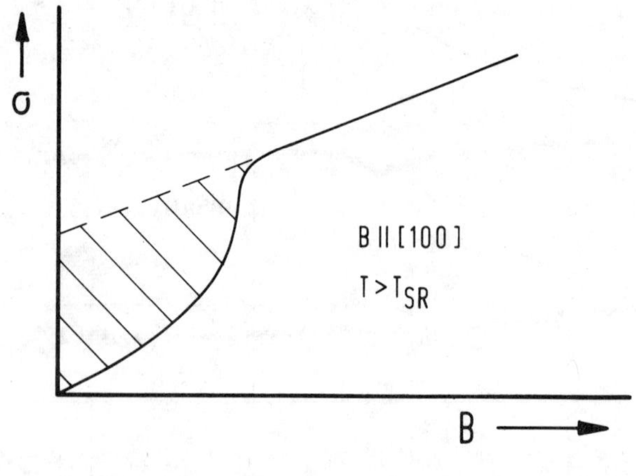


Fig. 3 Determination of the energy necessary to rotate the weak magnetic moment from the easy to the hard direction.

Following the ideas published previously [5,6,7] one can determine L also from magnetic data using the magnetic analog of the Clausius - Clapeyron equation :

$$L = T_{sr} \left( \frac{\mu_0 \sigma \, dH}{dT} \right)_{T_{sr}} = T_{sr} \left( \frac{dF}{dT} \right)_{T_{sr}} \qquad (1)$$

F is the energy difference between the states in which the weak moment is parallel to [ 100 ] and [ 001 ] . The evaluation of F from the magnetization curves with B || [100] for $T > T_{sr}$ (hatched area in Fig. 3) yields L = 5.2 J/mol. This value is in rather good agreement with the dsc value. It should be pointed out, however, that the error for both measurements is larger than the good agreement might suggest. dF/dT can be de-termined with an error of about 3 %, the inaccuracy of the dsc value could be as high as 15 %. Nevertheless our results demonstrate that the calculation of the latent heat using the Clausius-Clapeyron-like equation is well justified.

## ACKNOWLEDGMENT

The authors would like to thank Prof. Hempel for his kind supervision of this work.

## REFERENCES

(1) R. Wolfe, R.D. Pierce, M. Eibschütz and J.W.Nielsen Solid State Commun. 7 , 949-952 (1969) .
(2) M.Hirano, T. Okuda, T. Tsushima, S. Umemura, K. Kohn and S. Nakamura, Solid State Commun. 15 , 1129 - 1133 (1974) .
(3) C. Voigt and D. Bonnenberg, Physica 80 B , 439 - 443 (1975) .
(4) O.A. Bayukov, V.M. Buznik, V.P. Ikonnikov and M.I. Petrov, Proc. Intern. Conf. Mössbauer spectr. Poland - Cracow 1975, Vol. 1, 263 - 264 .
(5) Y. Shapira, Phys. Rev. 184 , 589 - 600 (1969) .
(6) L.M. Levinson and S. Shtrikman, Solid State Commun. 8 , 209 - 211 (1970) .
(7) J.W. Allen, Phys. Rev. B 8 , 3224 - 3228 (1973) .

# $MnWO_4$, CALORIMETRIC STUDY OF THE BIFURCATED ANTIFERROMAGNETIC ANOMALY*

Christopher P. Landee[§] and Edgar F. Westrum, Jr., Department of Chemistry, University of Michigan, Ann Arbor, Michigan 48104.

ABSTRACT

The heat capacity of a powdered $MnWO_4$ sample has been measured from 5 - 350 K. The data show three anomalies below 20 K: a small peak at 6.8 ± 0.1 K, then two large sharp peaks at 12.57 ± 0.05 and 13.36 ± 0.05 K. The magnetic entropy was measured as R ln 6. The data between 5 and 11.5 K obeys a power law dependence $C_{mag} = AT^B$ where B = 1.73. The sharp double peak is similar to the bifurcated anomaly in $MnCl_2$ which originates from two distinct antiferromagnetic phases, reported by R. B. Murray et al. The double anomaly is discussed in terms of the superexchange properties of $MnWO_4$.

## INTRODUCTION

In the course of determination on the heat capacities of the antiferromagnetic first-row transition metal tungstates,[1,2] the heat capacity of $MnWO_4$ has been measured adiabatically from 4 - 350 K.

The structure of manganese tungstate (isostructural with ferrous, cobalt, nickel, and zinc tungstates[3]) is characterized by zigzag chains of metal filled oxygen octahedra aligned along the c-axis. The crystal structure of the isomorph $FeWO_4$ is shown in figure 1.

While ferrous, cobalt, and nickel tungstates possess a common magnetic structure with ferromagnetic ordering within the chains and antiferromagnetic alignment of layers,[4,5] the magnetic structure of manganese tungstate is considerably more complicated with 32 manganese ions in the magnetic unit cell.[6] Half of the $MnWO_4$ magnetic unit cell is depicted in figure 2.

## EXPERIMENTAL

The sample of $MnWO_4$ purchased from Rocky Mountain Research was nearly amorphous but after compression into a pellet and firing at 1100 °C for one day X-ray diffraction analysis revealed a sharp, strong pattern with only $MnWO_4$ lines. Hence, the calorimetric sample was prepared by pressing the pellet, removing the top and bottom surfaces and firing in a platinum crucible for five days at 1100 °C. The surface was again removed after firing and the pellet itself broken into pieces small enough to fit into the calorimeter. No impurity lines were detected in the diffraction analysis and the lattice parameters were in excellent accord with literature values. Chemical analysis for major constituents was in good accord with the theoretical values; less than 40 ppm iron was detected as an impurity. The 85 g sample was prepared in a single batch.

Measurements were made in the Mark II adiabatic cryostat which utilized a gold-plated OFHS copper calorimeter vessel. All calibrations of this instrument are ultimately referred to measurements made at the National Bureau of Standards.

In terms of the Lindemann melting formula, zinc tungstate would be expected to be a good estimate of the lattice contribution for manganese tungstate. A comparison of the respective Debye thetas and their ratios confirm this prediction, for above 120 K the ratio is very constant and near unity out to the highest temperatures measured for the manganese compound. The ratio begins to rise as the temperature drops below 125 K and then drops even more sharply at 65 K as magnetic heat capacity affects the theta for $MnWO_4$. Since Debye theory predicts a ratio of the thetas equal to the square root of the inverse atomic weights of the cations (= 1.091), this value was chosen as the low-temperature limit of the ratio curve and a smooth interpolation between this lower value and the experimental ratios above 65 K adopted.

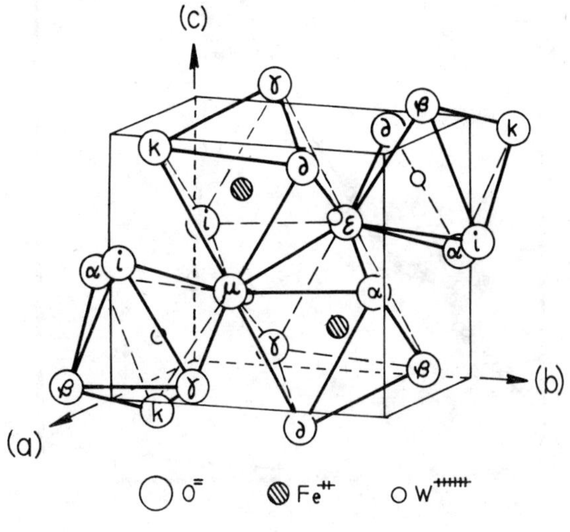


Fig. 1. Crystallographic unit cell of $NiWO_4$-type tungstates (after Ulku[5]).

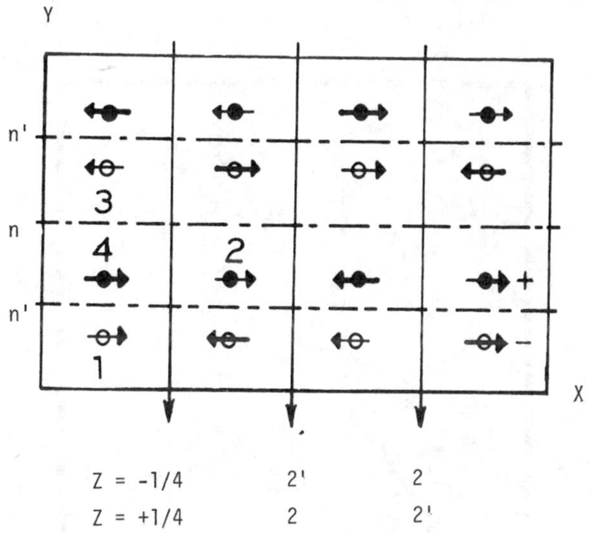


Fig. 2. Magnetic structure of manganese tungstates (cf. Duchs[7] -- only one-half of the magnetic cell is shown).

## RESULTS AND DISCUSSION

Low-temperature heat capacity data on a granular sample of manganese tungstate is shown in figure 3. There is a small peak at (6.8 ± 0.1) K and a large, bifurcated anomaly with peaks at (12.57 ± 0.05) K and (13.36 ± 0.05) K.

Using the heat capacity data of $ZnWO_4$ as the basis for a lattice contribution, the magnetic entropy was evaluated for $MnWO_4$ and found to be R ln 6, characteristic of spin disordering for an S = 5/2 system.

A logarithmic plot of the data (figure 4) reveals a power-law dependence of the magnetic heat capacity upon temperature over the range 4 - 11.5 K, excluding the region immediately about the 6.8 K peak. The

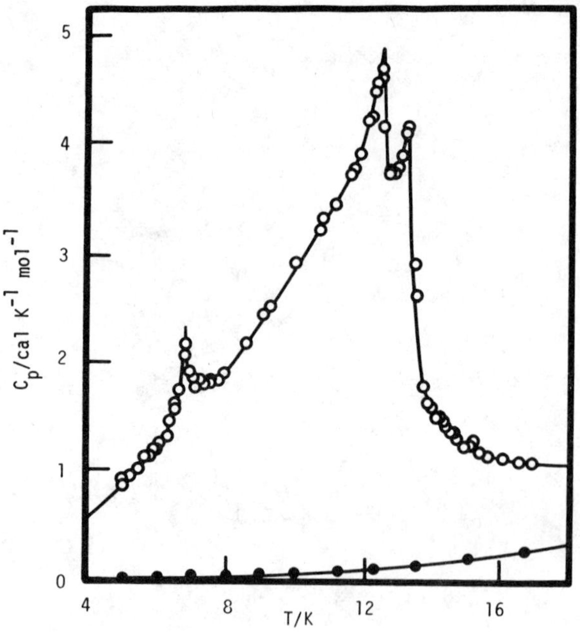


Fig. 3. Low-temperature heat capacity of manganese and zinc tungstates. $MnWO_4$ ——o—— and $ZnWO_4$ ——●——

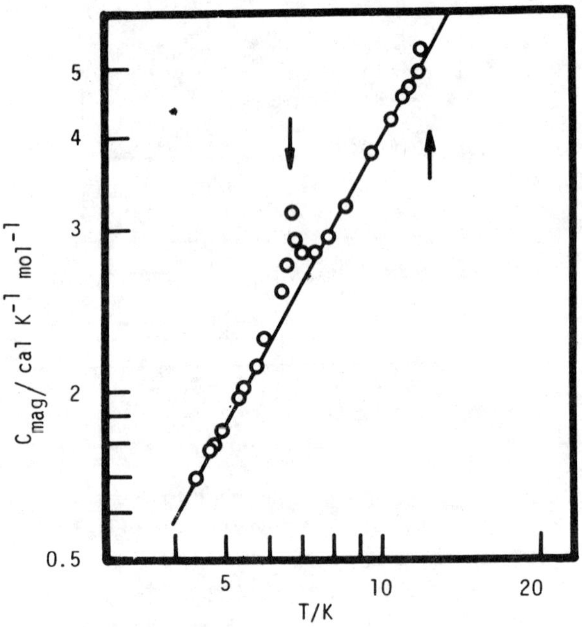


Fig. 4. Ln $C_{mag}$ vs. Ln T for $MnWO_4$.

relationship found was

$$C_{mag} = 0.052\ T^{1.73} .$$

The bifurcation of the main peak is considered to be characteristic of manganese tungstate and not particular to the sample measured. General material defects have the property of lowering and broadening the heat capacity of a magnetic transition, but the 12.57 K peak is higher and as sharp as the 13.36 peak. Weitzel's discoveries of two simultaneous types of magnetic order in mixed crystals of (Mn, Fe)$WO_4$[7,9] made it important to check for iron contamination of the sample but a spectroscopic examination revealed only 40 ppm iron impurity, three orders of magnitude less than would be required to separate the peaks by the amount observed.

The tungstate series, $MWO_4$, shares structural and magnetic similarities with the series of transition metal dichlorides, $MCl_2$. In both series the metal ions occupy octahedral holes in closest-packed lattices of anions.[10] The nn exchange is over intervening anions with 90° angles. For such arrangements, the Goodenough-Kanamori rules predict ferromagnetic exchange for $Fe^{2+}$-$Fe^{2+}$, $Co^{2+}$-$Co^{2+}$, and $Ni^{2+}$-$Ni^{2+}$ but antiferromagnetic exchange for $Mn^{2+}$-$Mn^{2+}$.[11] The magnetic structures of the dichlorides are similar to those of the tungstates mentioned above; the iron, cobalt, and nickel compounds have ferromagnetic layers coupled antiferromagnetically to adjacent layers. In $MnCl_2$, the spins are ordered in a more complicated manner within each layer.[12]

In $MnCl_2$, the low-temperature maxima in the heat capacity is also bifurcated with the twin peaks occuring at 1.81 and 1.96 K.[13] Neutron diffraction experiments revealed the presence of two distinct AFM-phases at low temperatures;[12] the peaks in the heat capacity correspond to the phase transitions AFMII → AFMI → PM.

The similarity of the 90° superexchange and the bifurcation of the heat capacity anomalies in $MnCl_2$ and $MnWO_4$ suggest several low-temperature magnetic phases in manganese tungstate.

A more complete account of this work will appear in the Journal of Chemical Thermodynamics.

REFERENCES

* Supported by the National Science Foundation.
§ Present address: Department of Chemistry, Washington State University, Pullman, Washington 99163.


1. W. G. Lyon and E. F. Westrum, Jr., J. Chem. Thermodynamics 6, 763 (1974).
2. C. P. Landee and E. F. Westrum, Jr., Ibid. [in press].
3. A. W. Sleight, Acta. Crystallogr, Section B, 28, 2899 (1972).
4. D. Ülkü, Zeit. Fur Krist., 124, 192 (1967).
5. H. Weitzel, Solid State Commun., 8, 2071 (1970).
6. H. Dachs, Solid State Commun., 7, 1015 (1969).
7. H. Weitzel, Solid State Commun., 7, 1249 (1969).
8. H. Weitzel, Neus Jahrb. Mineral., Abhand. 113, 13 (1970).
9. H. Weitzel, Zh. Fur Krist., 131, 289 (1970).
10. H. Grime and J. A. Santos, Zeit. Krist., 88, 136 (1934).
11. J. B. Goodenough, Magnetism and the Chemical Bond, Interscience, New York, (1963).
12. M. K. Wilkinson, J. W. Cable, E. O. Wollan, and W. C. Koehler, Oak Ridge National Laboratory Report ORNL-2501, unpublished (1958); Oak Ridge National Laboratory Report ORNL-2430, unpublished (1958).
13. R. B. Murray, Phys. Rev., 128, 1570 (1962).

# MAGNETO-CALORIC EFFECT STUDIES ON SINGLE CRYSTAL $MnCl_2 \cdot 4H_2O$ WITH MAGNETIC FIELDS APPLIED ALONG THE c CRYSTALLOGRAPHIC AXIS†

R. A. Butera, T. A. Reichert* and E. J. Schiller
Department of Chemistry, University of Pittsburgh
Pittsburgh, Pennsylvania 15260

## ABSTRACT

The magneto-caloric effect has been determined over the temperature range from 0.4 K to 1.7 K with applied magnetic field up to 20 kG. Correlation of the variation of $(\partial T/\partial H)_S$ with the transition boundaries have been determined and evidence that the adiabats do not cross the spin flopped to paramagnetic boundary tagentially is discussed. Results suggesting that an additional transition is occuring in the heretofore considered paramagnetic region at fields greater than 9 kG are presented.

## INTRODUCTION

Reichert, Butera and Schiller (1) have developed a method utilizing a derivative of the adiabatic variation of temperature vs. applied magnetic field as a sensitive probe to determine changes in the nature of the spin interactions of an antiferromagnetic system as a function of temperature and applied magnetic field. They applied this method to an investigation of the thermal and magnetic behavior of a spherical single crystal of $MnCl_2 \cdot 4H_2O$ with magnetic fields up to 9 kG applied along the c crystal axis in the temperature region from 1.3 K to 1.7 K. The use of the variation of the magneto-caloric effect, $(\partial T/\partial H)_S$, as a probe of the magnetic interactions appeared very promising. Further investigations were carried out in this laboratory to correlate the behavior of $(\partial T/\partial H)_S$ in the antiferromagnetic-spin flopped and the spin flopped-paramagnetic regions.

## RESULTS AND DISCUSSION

The variation of the magneto-caloric effect, as derived from the experimental T vs. H adiabatic data, is shown in figures 1 and 2. The data in the region of the spin flopped-paramagnetic lambda line exhibit negative "cusps" in $(\partial T/\partial H)_S$ vs. T at constant H. This result is similar to that reported by Reichert, et al. (1), however, the magnitudes of these "cusps" do not correspond to the slope of the lambda line and are consistantly lower in magnitude. This is an indication that, in contrast with the behavior at the antiferromagnetic-paramagnetic lambda line, the adiabats do not cross the spin flopped-paramagnetic lambda line tangentially. This result, and the fact that $(\partial T/\partial H)_S$ is finite at the "cusps", is a strong indication that neither $C_H$ nor $(\partial M/\partial T)_S$ diverge to infinity as the spin flopped-paramagnetic lambda line is crossed. This result is in agreement with the fact that Reichert (2) found that negative exponents were required to fit the heat capacity data for fields greater than 8 kG. Referring to Griffith's equation (3) for the slope of an isentrope,

$$(\partial T/\partial H)_S = (\partial T/\partial H)_s - S_b'(H)(\partial T/\partial S)_H \qquad (1)$$

Reichert et al. (1) have shown that this may be rewritten as

$$(\partial T/\partial H)_S = (dT_N/dH) - S_b'(H)(T/C_H). \qquad (2)$$

The apparent discrepancy between $(\partial T/\partial H)_S$ and the slope of the lambda lines shown in figure 1 can now be explained. The second term in eq. 1 reduces to zero when the heat capacity diverges to infinity as the boundary is crossed. However, this term becomes quite important if the heat capacity remains finite as the lambda line is crossed and must be considered. The derivative of the entropy at the boundary, $S_b'(H)$, has been obtained from a fit of the data of Reichert (2) and also Giauque et al. (4). This has been used to determine the magnitude of $C_H$ at the boundary which

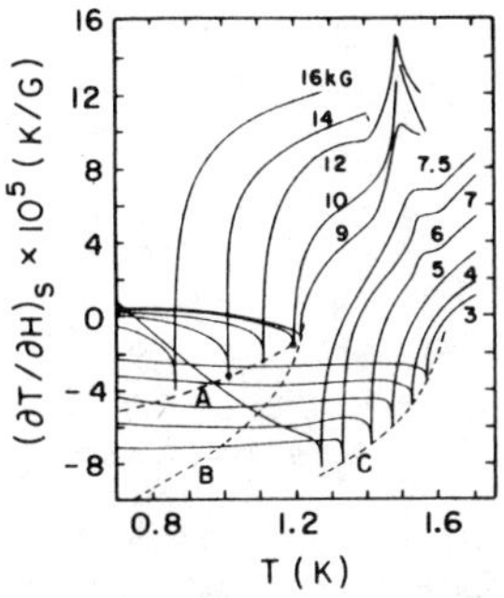


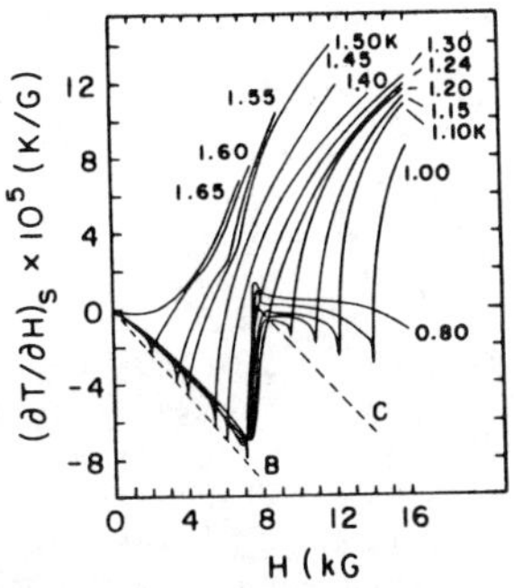


**Curve A.** $(\partial T/\partial H)_S = (dT_N/dH) - S_b'(H)[T/C_H]$.
**Curve B.** $(dT_N/dH) = -10.64 \times 10^{-9}$ (H-7800)
**Curve C.** $(dT_N/dH) = -11.6 \times 10^{-9}$ (H)

**Figure 1.** $(\partial T/\partial H)_S$ vs T at constant H.
**Figure 2.** $(\partial T/\partial H)_S$ vs H at constant T.

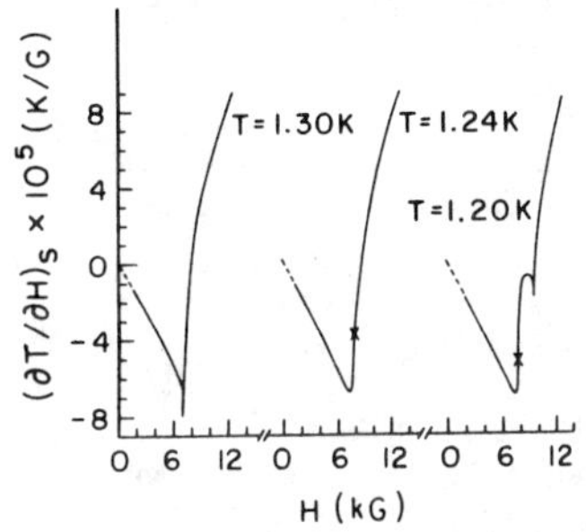


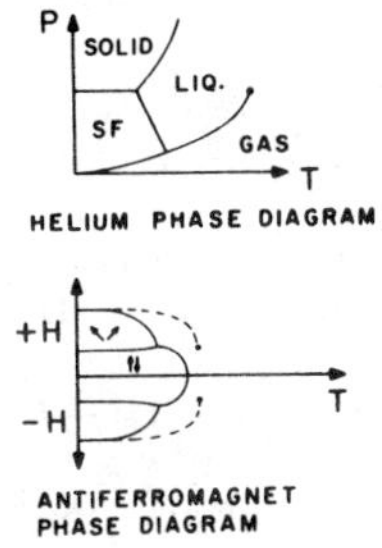


**Figure 3.** $(\partial T/\partial H)_S$ vs H at constant T in the region of the bicritical point
**Figure 4.** He and AF phase diagrams

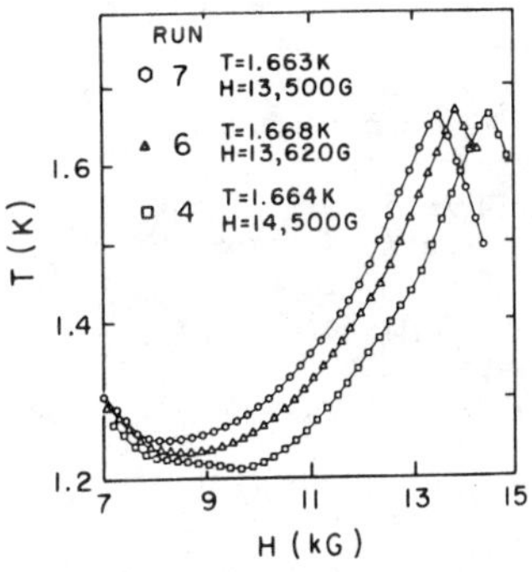


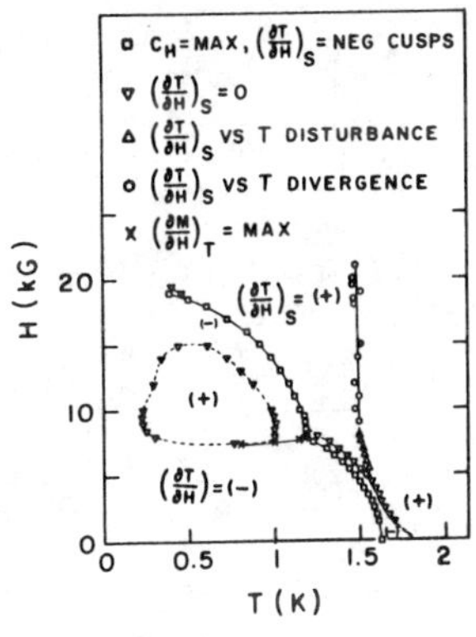


**Figure 5.** T vs. H adiabat data
**Figure 6.** Magneto-caloric effect diagram in the H-T plane

compares well with the experimental values reported by Reichert (2) and also Giauque et al. (4). The results of this treatment are summarized in Table I and also indicated by the dashed line (A) in Fig. 1. This result is thus conclusive evidence that the heat capacity and $(\partial M/\partial T)_H$ both remain finite at the spin flopped-paramagnetic transition boundary.

In order to correlate the magneto-caloric effect data with the antiferromagnetic-spin flopped transition, we found it necessary to derive values of $(\partial T/\partial H)_S$ vs. H at constant T (Figure 2). This data exhibits a dispersive type of behavior in the region centered about 7.8 kG, which sharpens as the temperature is lowered. As the temperature is increased toward that of the bicritical point, one observes a temperature region in which no cusps occur and

Table I

| H kG | $T_N$ K | $\frac{dT_N}{dH}$ x$10^5$ K/G | $\frac{\partial T}{\partial H}$ x$10^5$ K/G | $S_b'(H)$x$10^5$ gibbs/mole | $C_H$ calc. cal/mole-K |
|---|---|---|---|---|---|
| 3 | 1.568 | -3.48 | -3.48 | -3.78 | ∞ |
| 4 | 1.527 | -4.64 | -4.64 | -5.04 | ∞ |
| 5 | 1.475 | -5.80 | -5.80 | -6.30 | ∞ |
| 6 | 1.412 | -6.96 | -6.96 | -7.56 | ∞ |
| 7.5 | 1.293 | -8.70 | -8.20 | -9.45 | 24.4 |
| 10 | 1.194 | -2.34 | -1.50 | -12.67 | 18.0 |
| 14 | 1.015 | -6.60 | -2.91 | -17.64 | 4.9 |
| 18 | 0.677 | -10.85 | -5.93 | -22.68 | 3.1 |

$(\partial T/\partial H)_S$ varies smoothly with applied magnetic field (Figure 3). Thus, the antiferromagnetic-spin flopped transition appears to be characterized by a point of inflection in $(\partial T/\partial H)_S$ vs. H at constant T and not by the appearance of a "cusp". It is evident that the behavior of the magneto-caloric effect, and thus the nature of this transition, is quite different from that of either the antiferromagnetic-paramagnetic or the spin flopped-paramagnetic regions. Fisher (5) has proposed that the spin flopped-antiferromagnetic transition is first order and thus the properties of the system should vary in a discontinuous manner as the boundary is crossed. Giauque et al. (4) have reported the equilibrium magnetization vs. temperature data for $MnCl_2 \cdot 4H_2O$ with H parallel to the c crystal axis. Their data indicate that the magnetization varies continuously as the antiferromagnetic-spin flopped region is crossed. However, they also report the adiabatic variation of temperature vs. applied magnetic field at field sweep rates of 20 and 100 G/sec. Reference to Figure 6 of their paper indicates that a small but apparent discontinuity occurs (and is enhanced) at the lower rate of field change. They also report entropy production occurring in this region at temperatures near 0.3°K.

We have converted the published T vs. H adiabatic data of Giauque et al. (4) to $(\partial T/\partial H)_S$ vs. T at constant H for comparison with our data and have obtained a very good correspondence. Both the values of $(\partial T/\partial H)_S$ derived from the data of Giauque et al. and those obtained in this laboratory exhibit hysterisis in the antiferromagnetic-spin flopped regions.

Reichert et al. (1) observed an anomalous region in $(\partial T/\partial H)_S$ vs. T at constant H which was of unexplained origin. Careful consideration of our $(\partial T/\partial H)_S$ data as well as that derived from the T vs. H data of Giauque et al. (4) revealed that this anomalous region was evident in all cases. Since different samples were used by the two laboratories, one can, therefore, attribute the anomalous region to an intrinsic property of the material. It is unfortunate that, at the time of these investigations, the primary thrust of the work in both laboratories was to investigation the lambda lines and spin flop boundaries, and therefore the only data available covering the anomalous region is that which was incidentally obtained near the starting points of the adiabats in the paramagnetic region, heretofore considered "uninteresting".

M. E. Fisher (6) has suggested that this may be experimental evidence of the existence of the magnetic first-order boundary ending in a monocritical point which would be analogous to the liquid-gas boundary line observed in helium. Figure 4 shows the comparison of the helium phase diagram with that of an uniaxial antiferromagnet as given by Fisher (5). However, this diagram has been modified by the addition of the first-order boundary indicated by the dashed line as suggested by Fisher (6).

R. B. Griffiths (7) has pointed out that the adiabats should exhibit a shift in entropy if the boundary does in fact exist, is first order and corresponds to the anomalous region in question.

In light of the suggestions of Fisher (6) and Griffiths (7), we have conducted an intense review of our adiabat data. The results are given in Figure 5 which shows a series of adiabats which pass nearly through the anomalous region. Two salient features are immediately evident from Figure 5, these are: 1) $(\partial T/\partial H)_S$ along a given adiabat changes sign in a temperature-field region where the sample is presumed to be paramagnetic and 2) more importantly, the adiabats cross each other. It is thus evident that entropy is produced and the system is not in thermodynamic equilibrium in this region.

The $(\partial T/\partial H)_S$ data of Reichert et al. (1) and this work, as well as that derived from the T vs. H data of Giauque et al. (4), exhibit anomalies which appear to become sharper in the region between 5 and 9 kG (Figure 1). Consideration of this behavior with respect to the thermal characteristics of the sample (i.e., the locus of changes in sign and of extrema in $(\partial T/\partial H)_S$), leads to the observation that the curvature of the $(\partial T/\partial H)_S = 0$ line in the region above the zero field Neel temperature (1.62°K) is concave upward while that below 1.62°K is concave downward (Figure 6). The point at which this curvature change occurs also corresponds to the juncture with the line connecting the anomalies in $(\partial T/\partial H)_S$ occurring between 5 and 9 kG. Thus, it is reasonable to conclude that there may be a relationship between the zero field transition temperature and the occurrence of this anomaly. The existence and exact nature of this relationship is a subject for further investigation.

Let us now summarize the salient features of the magnetocaloric effect as they appear in the H-T plane (Figure 6). These features consist of: 1) the locus of $(\partial T/\partial H)_S = 0$; 2) the locus of negative cusps in $(\partial T/\partial H)_S$; 3) the locus of the divergence in $(\partial T/\partial H)_S$; and 4) the region of the development of the disturbance in $(\partial T/\partial H)_S$.

Reichert et al. (1) have shown that the locus of $(\partial T/\partial H)_S = 0$ lies at higher temperatures than the lambda line (i.e., the locus of negative cusps in $(\partial T/\partial H)_S$) for all fields up to the region of the bicritical point. Giauque et al. (4) have reported that the locus of $(\partial T/\partial H)_S = 0$ is coincident with the lambda line for fields greater than that of the bicritical point. Our data are consistent with the previous work; however, we find that in the region above 18 kG the locus of $(\partial T/\partial H)_S = 0$ is no longer coincident with the lambda line as shown in Figure 6. In addition, we determined that a second locus of $(\partial T/\partial H)_S = 0$ exists within the region normally associated with the spin-flopped state. The nature of this second region is of interest in that it forms a closed region wherein the magneto-caloric effect is positive. The high temperature edge of this locus is parallel to the lambda line while the low field edge is parallel to the spin flopped-antiferromagnetic boundary. Apparently, the magnetic interactions within this region are different than those in the surrounding regions. The exact nature of this difference is, at present, unknown. However, it may be that the occurrence of this region is simply due to the existence of the transition boundaries. The region between 1.00 and 1.20°K is apparently unique in that the system can pass from the antiferromagnetic region to the spin-flopped region without a change in the sign of the magneto-caloric effect. This effect has been reported for $MnCl_2 \cdot 4H_2O$ and also for $CoCl_2 \cdot 6H_2O$ by McElearney et al. (8).

Magnetization data obtained during adiabatic demagnetization have been reported by Giauque et al. (4). Comparison of this data with the magneto-caloric effect obtained in the same experiment, shows clearly that the sample exhibits what appears to be an infinite positive slope in the magnetization as well as a corresponding negative cusp in the magneto-caloric effect in the region near 0.4°K for fields of 7, 7.1 and 7.2 kilogauss. For fields greater than 7.2 kG the behavior changes abruptly. The slope of the magnetization, corresponding to a point of inflection in the magnetization vs. temperature, becomes large and negative producing a smooth and finite positive maximum in the magneto-caloric effect. These effects correlate very well with the anomalous behavior of $C_H/T$ that is evident in the only heat capacity data available for this material over this field-temperature region (4). Coupling this information with the field-temperature locus of the change in the sign of the magneto-caloric effect, Figure 6, leads us to the conclusion that there are modifications in the magnetic order in $MnCl_2 \cdot 4H_2O$ occurring in the region below 0.5°K. The origin and exact nature of these changes are unknown and, thus, warrant further study.

REFERENCES


* Present address - Biotechnology Program, Carnegie-Mellon University, Pittsburgh, Pennyslvania 15213.

† Work partially supported by the USAEC.

1. T. A. Reichert, R. A. Butera and E. J. Schiller, Phys. Rev. 1, (11) 4466 (1970).
2. T. A. Reichert, Ph.D. Thesis, University of California (Berkeley), (1968).
3. J. Skalyo, Jr., A. F. Cohen, S. A. Friedberg and R. B. Griffiths, Phys. Rev. 164, 705 (1967).
4. W. F. Giauque, R. A. Fisher, E. W. Hornung and G. E. Brodale, J. Chem. Phys. 53, 1474 (1970).
5. M. E. Fisher, Reviews of Modern Physics, 39, (2) (1967).
6. M. E. Fisher, private communication.
7. R. B. Griffiths, private communication.
8. J. N. McElearney, H. Forstat and P. T. Bailey, Phys. Rev. 181, 887 (1969).


## NEUTRON SCATTERING MEASUREMENTS OF THE α-γ TRANSFORMATION IN CeTh ALLOYS*


A. S. Edelstein
University of Illinois at Chicago Circle and Oak Ridge National Laboratory
H. R. Child
Oak Ridge National Laboratory, Oak Ridge, Tennessee 37830


ABSTRACT


Quasi-elastic neutron scattering measurements using 1.0 and 4.4 Å neutrons have been made on polycrystalline fcc alloys of $Ce_{1-x}Th_x$ (x = .20, .266, .29). The first alloy undergoes a first order transformation at approximately 140 K in which the lattice parameter decreases from 5.13 to 4.98 A. The second alloy undergoes a second order transition while the Bragg peaks of the third shift in position between 140 and 170 K. For x = .2 at 296 K the q dependence for $0.4 \leq q \leq 6$ Å$^{-1}$ of the diffuse scattering is in approximate agreement with that of the square of the atomic 4f form factor of Ce. The extrapolation of the diffuse cross-section to q = 0 gives a value of 260 millibarns which is consistent with a value of μ = 2.56 $\mu_B$. With decreasing temperature the magnitude of the diffuse scattering $d\sigma/d\Omega$ for x = 0.2 decreases about 30 millibarns as one approaches the transition from above, decreases about 100 millibarns at the transition and 30 millibarns below the transition. The fractional occupancy of the localized $4f_1$ configuration η is 0.4 ± 0.1 at 11 K. There is not any evidence of a compensating conduction electron polarization. The spatial extent of the moment is not affected by the fluctuations. The square of the effective moment $\mu_{eff}^2 = 3\,kT\chi$ is reduced at 11 K by the small value of η, a small value for the spin correlation life-time and possibly crystal field effects.



*Research sponsored by the Energy Research and Development Administration under contract with Union Carbide Corporation.


## STUDY OF THE SPIN-ORIENTATION TYPE PHASE TRANSITION IN $SmCrO_3$*


G. Gorodetsky, S. Shaft and A. Shaulof
Ben Gurion University of the Negev, Beer Sheva, Israel
B.M. Wanklyn
Clarendon Laboratory, Oxford, England
and
B. Sharon and I. Yaeger**
Weizmann Institute, Rehovot, Israel


ABSTRACT


The spin orientation type phase transition in $SmCrO_3$ at T ≃ 34°K has been studied by means of bulk magnetization, sound wave propagation and specific heat measurements. The high and low temperature phases belong to $\Gamma_4(Cr{:}\ G_xF_z)$ and $\Gamma_2(Cr{:}\ G_zF_x)$ respectively. We find a reorientation region of ∿1°K with the beginning and the end points corresponding to second order phase transitions. Though common in the orthoferrites, this is the first clear evidence of a continuous spin reorientation in a $RCrO_3$ compound. The continuous nature of the reorientation process is evident from peaks in the a and c axis susceptibilities and the softening of the elastic constant $C_{55}$ at both ends of the transition region due to a resonant magnetoelastic coupling. From the anomaly in the specific heat, the entropy associated with the spin reorientation is found to be ΔE/R = 0.21±0.01°K.
A complete account of this work will be published elsewhere.



* Supported in part by the U.S.-Israel Binational Science Foundation through Research Grant No. 452.



** Present address: Dept. of Physics, The University of Manitoba, Winnipeg, Canada.

# SCALING FUNCTIONS FOR MULTICRITICAL PHENOMENA

David R. Nelson*
Physics Department, Harvard University, Cambridge, Massachusetts 02138

ABSTRACT

Recent theoretical work has led to closed form expressions for the thermodynamic functions which characterize complicated multicritical phenomena. Renormalization group recursion relations are used to map multicritical systems out of the critical regime, at which point Landau theory with fluctuation corrections can be employed. We review here applications of this approach to tricritical points, spin flop transitions, dipolar systems, and Heisenberg ferromagnets below $T_c$. Although the calculations have presently been carried out only to first order in $\varepsilon = 4 - d$, they nevertheless give a good qualitative understanding of multicritical behavior. In particular, they provide a detailed picture of the crossover between competing kinds of critical behavior, and allow treatment of ordered phases with spontaneously broken symmetries.

## I. INTRODUCTION

Tremendous progress has been made in recent years with the renormalization group theory of critical phenomena.[1] It has become a rather mechanical procedure to construct recursion relations and produce the associated fixed points and eigenvalues when a new variety of critical behavior is encountered. This program, when carried out with the methods described in Reference 1, or using, say, a field-theoretic approach,[2] leads rather straightforwardly to critical exponents.

More ambitiously, one can also attempt to determine the universal scaling functions associated with complicated varieties of critical behavior. This has been done for ordinary critical points[3] using a Feynman graph technique involving exponentiation of logarithms to produce power law singularities.[4] Unfortunately, this procedure has been unsuccessful in determining the scaling functions associated with complex multicritical behavior. The singularity structure of these functions is sufficiently complicated to thwart a naive exponentiation of logarithms.

Consider the phase diagrams for a metamagnet (with a tricritical point) and a uniaxial antiferromagnet (with a bicritical point) shown in Fig. 1. Both systems have critical lines as well as singled-out multicritical points in the (T,H)-plane. In general, it is believed[1] each critical point or line is described by a distinct fixed point, so one is forced to deal simultaneously with a variety of different sorts of critical behavior. A scaling description is conveniently formulated in terms of variables t and g which are linear combinations of $(T - T_t)$ and $(H - H_t)$[5] or $(T - T_b)$ and $(H - H_b)$.[6] In particular, the lines $g = 0$ should correspond to the tangent of the first order lines at the multicritical point.[5,6a] An extended crossover scaling assumption in terms of these variables requires that the susceptibility, for example, be expressible as

$$\chi \approx t^{-\gamma}\Phi(g/t^{\phi}) \qquad (1.1)$$

for t and g sufficiently small. Here, $\gamma$ is the susceptibility exponent along the locus $g = 0$, and $\phi$ is the crossover exponent introduced by Riedel and Wegner.[7] Equation (1.1) is designed to display explicitly singularities in the neighborhood of the multicritical point. To describe a particular critical line, it is assumed in addition[7] that the scaling function $\Phi(z)$ is singular at $\dot{z}$, with a divergence given by

$$\Phi(z) \approx A|z - \dot{z}|^{-\dot{\gamma}} , \qquad (1.2)$$

where $\dot{\gamma}$ is the critical line susceptibility index.

Standard renormalization group methods[1,2] work well in determining $\gamma$, $\phi$, and $\dot{\gamma}$. However, because of singularities like (1.2) in $\Phi(z)$, an actual determination of scaling functions is more complicated. Recently, explicit calculations of functions like $\Phi(z)$ have become possible by using recursion relations to map multicritical Hamiltonians out of the critical regime, and then employing Landau theory with fluctuation corrections.[8] Closed form expressions for functions like $\Phi(z)$ have been obtained for the tricritical[8] and bicritical[9] systems shown in Fig. 1, as well as for dipolar systems[10] and isotropic ferromagnets below $T_c$.[11] In this paper we describe briefly the "matching" procedure which leads to explicit results for crossover scaling functions to first order in $\varepsilon$, and summarize the various results which have been obtained. Wherever possible, we shall try to indicate possible experiments suggested by the new results.

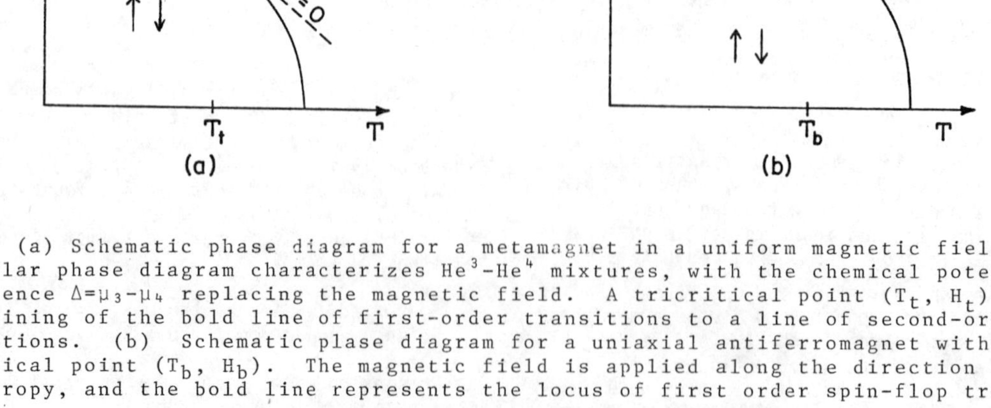

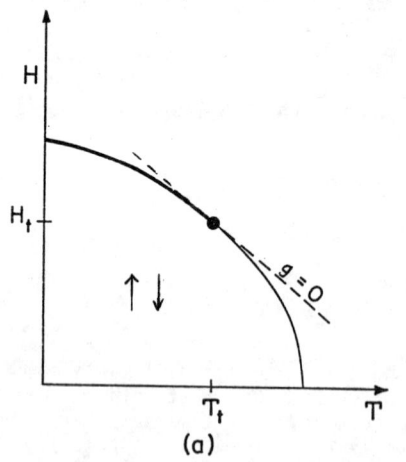

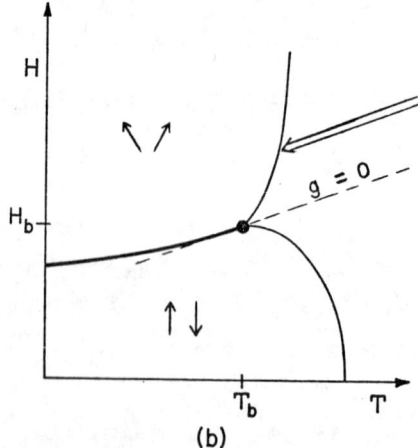

Fig. 1 (a) Schematic phase diagram for a metamagnet in a uniform magnetic field H. A similar phase diagram characterizes $He^3$-$He^4$ mixtures, with the chemical potential difference $\Delta = \mu_3 - \mu_4$ replacing the magnetic field. A tricritical point $(T_t, H_t)$ marks the joining of the bold line of first-order transitions to a line of second-order transitions. (b) Schematic plase diagram for a uniaxial antiferromagnet with a bicritical point $(T_b, H_b)$. The magnetic field is applied along the direction of anisotropy, and the bold line represents the locus of first order spin-flop transitions.

## II. THERMODYNAMIC FUNCTIONS FROM RECURSION RELATIONS

Consider an isotropic Landau-Ginzburg-Wilson Hamiltonian for Ising-like continuous spins $S(\vec{R})$ in d-dimensions, namely

$$\mathcal{H} \equiv \mathcal{H}/k_B T = -\int d\vec{R}[\tfrac{1}{2}(\vec{\nabla}S)^2 + \tfrac{1}{2}rS^2 + uS^4]. \quad (2.1)$$

When this Hamiltonian is transformed into momentum space, it will be assumed, as usual, that the momentum integrals are cut off by a Brillouin zone of unit radius. A direct attempt to calculate, say, the susceptibility using exact one-particle propagators will not work. Since the "mass" of the renormalized propagators goes to zero at the critical point,[1] there are well-known problems with such a direct graphical approach at small momenta, associated with the existence of a large correlation length. These difficulties can be circumvented by first using recursion relations to map the Hamiltonian (2.1) out of the critical regime so that the correlation length of the resulting "block-spin" system is of order unity. At this point, a direct graphical calculation presents no problems. Renormalization group theory can be used to relate non-critical thermodynamic functions obtained in this way to the corresponding quantities deep within the critical region. These relationships, together with solutions of the recursion relations, lead directly to crossover scaling functions.[8]

The approach sketched above is in the spirit of Wilson's early ideas about applications of the renormalization group to critical phenomena.[12] Riedel and Wegner have studied these ideas using model recursion relations,[13] while Nauenberg and Nienhuis have explored their utility in obtaining equations of state for the two-dimensional Ising model.[14] We will sketch here how to implement the program described above for the continuous spin Hamiltonians already extensively investigated via the ε-expansion.

As a simple example, consider first the recursion relations generated from the application of a differential renormalization group[15] to (2.1), namely

$$dr(\ell)/d\ell = 2r(\ell) + Au(\ell)/[1+r(\ell)] \quad (2.2a)$$

$$du(\ell)/d\ell = \varepsilon u(\ell) - Bu^2(\ell)/[1+r(\ell)]^2 \quad (2.2b)$$

where $A = 3/2\pi^2$ and $B = 9/2\pi^2$. These equations describe correctly to $O(\varepsilon)$ the evolution of the "partially dressed" parameters entering the renormalized Hamiltonian $\mathcal{H}(\ell)$ after a fraction $1-e^{-d\ell}$ of the degrees of freedom have been integrated out. We assume that the initial values r and u are themselves of order ε. Nicoll et al.[16] have discussed solutions of the system (2.2) which are accurate even in the limit $r(\ell) \to \infty$. Since we are only interested in integrating until $r(\ell) = O(1)$, substantial simplifications can be made. (This condition suffices to make the correlation length $\xi(\ell)$ of order unity.) In particular, it is easily verified that the functions

$$r(\ell) = t(\ell) - \tfrac{1}{2}Au(\ell) + \tfrac{1}{2}Au(\ell)t(\ell)\ln[1+t(\ell)] \quad (2.3a)$$

$$u(\ell) = ue^{\varepsilon\ell}/Q(\ell), \quad (2.3b)$$

with

$$Q(\ell) = 1 + Bu(e^{\varepsilon\ell}-1)/\varepsilon, \quad t(\ell) = te^{2\ell}/Q^{\frac{1}{3}}(\ell) \quad (2.4)$$

solve (2.2) to order ε <u>even when $r(\ell)$ is of order unity</u>.

With these solutions of the recursion relations at hand, it is straightforward to proceed to the calculation of thermodynamic functions. The susceptibility χ is particularly simple. The susceptibility $\chi(\ell)$ associated with the Hamiltonian $\mathcal{H}(\ell)$ is related to the desired susceptibility of the initial system (2.1) by a multiplicative factor,[1]

$$\chi = e^{2\ell - \int_0^\ell \eta(\ell')d\ell'}\chi(\ell), \quad (2.5)$$

where $\eta(\ell)$ is a function which reduces to the critical exponent η near a fixed point.[1,15] To order ε, we can take $\eta(\ell) \equiv 0$. The multiplying factor in (2.5) arises from a spin rescaling[1] and is closely related to the field theoretic wave function renormalization constant. Evaluating (2.5) at $\ell = \ell^*$ such that $r(\ell^*) \approx 1$ insures that the correlation length involved in the calculation of $\chi(\ell^*)$ is of order unity. The function $\chi(\ell^*)$ can then be determined using Landau theory with fluctuation corrections. The end result of this procedure is the susceptibility associated with (2.1) to first order in ε for an arbitrary four-spin coupling u of $O(\varepsilon)$,[8]

$$\chi = t^{-\gamma}[(1-Bu/\varepsilon)t^{\frac{1}{2}\varepsilon} + Bu/\varepsilon]^{\frac{1}{3}}, \quad (2.6)$$

where $t = t(\ell=0)$ and $\gamma = 1 + \tfrac{1}{6}\varepsilon$.

The free energy obeys a modified homogeneity relation, namely[17-19]

$$F = \int_0^\ell e^{-d\ell'}G_0(\ell')d\ell' + e^{-d\ell}F(\ell), \quad (2.7)$$

which contains an inhomogeneous term in constrast to (2.5). This contribution can be thought of as a line integral along a renormalization group trajectory, where the kernel $G_0(\ell)$ is just[17,19]

$$G_0(\ell) = (16\pi^2)^{-1}\{\ln[1+r(\ell)] - \tfrac{1}{2}\} \quad (2.8)$$

to leading order. Evaluating (2.7) at the same $\ell^*$ as was used in the susceptibility calculation gives the leading order result for the free energy, namely

$$F = (-t^2/48u)\{[1+Bu(t^{-\frac{1}{2}\varepsilon}-1)/\varepsilon]^{\frac{1}{3}} - 1\}. \quad (2.9)$$

Determinations of other thermodynamic quantities proceed in an analogous fashion. The equations (2.2) are inappropriate in the ordered phase where $r(\ell) \to -\infty$. However, recursion relations suitable for this region are easily constructed by first shifting the spin field by the exact magnetization M, $S \to S + M$.[8]

To test the consistency of the procedure sketched above, one can differentiate the basic equation (2.5) and (2.7) with respect to $\ell = \ell^*$. The results are independent of the precise choice of $\ell^*$,[8] as should be expected. Indeed, explicitly differentiating (2.5) with respect to $\ell$ gives the requirement

$$\left[\frac{\partial}{\partial\ell} + \frac{du(\ell)}{d\ell}\frac{\partial}{\partial u(\ell)} + \frac{dr(\ell)}{d\ell}\frac{\partial}{\partial r(\ell)} + 2 - \eta(\ell)\right]\chi(\ell) = 0, \quad (2.10)$$

which is a close relative of the Callen-Symanzik equation.[2] It should not be difficult to fit the ideas outlined here into the framework of a field theoretic approach.

Of course, expressions like (2.5) and (2.7) can also be evaluated in the limit $\ell \to \infty$. It is not hard to show[17] that the susceptibility (inverse renormalized mass squared) is

$$\chi = \lim_{\ell\to\infty} e^{2\ell}/r(\ell), \quad (2.11)$$

while the renormalized four-spin coupling constant is simply

$$u_R = \lim_{\ell\to\infty} e^{-\varepsilon\ell}u(\ell). \quad (2.12)$$

The free energy can be determined as a power series in the exact magnetization,[20]

$$F(M) = F(0) + \tfrac{1}{2}\chi^{-1}M^2 + u_R M^4 + \ldots \quad (2.13)$$

Such an approach has been taken by Stephen et al.[21] who have analyzed tricritical points using a parquet summation procedure, and Nicoll et al.[22] who have discussed a compressible Ising system by studying recursion relations in the limit $\ell \to \infty$. Natterman and Trimper[23] have studied the crossover from asymptotic critical to mean field critical behavior using a related technique.

We hope the above discussion has conveyed at

least the essential ideas involved in determining thermodynamic functions from recursion relations. The reader is referred elsewhere[8] for details.

## III. APPLICATIONS TO SPECIFIC SYSTEMS

We summarize here, graphically whenever possible, particular results for a number of multicritical systems. Although the list is not exhaustive, most of the predictions discussed here have direct experimental consequences.

### A. Tricritical Points

This is the most straightforward application of the "matching" approach, the calculation having been essentially outlined in the previous section. The continuous spin Hamiltonian describing tricritical behavior proposed by Riedel and Wegner is simply the generalization of (1.1) to n-component spins,

$$\overline{\mathcal{H}} \to \overline{\mathcal{H}}_0 = -\int d\vec{R}[\tfrac{1}{2}(\vec{\nabla}\vec{S})^2 + \tfrac{1}{2}r|\vec{S}|^2 + u|\vec{S}|^4], \qquad (3.1)$$

together with the addition of a six-spin interaction, namely

$$\overline{\mathcal{H}}_v = -v\int d\vec{R}|\vec{S}|^6 \qquad (3.2)$$

Here, $n = 2$ should correspond to $He^3$-$He^4$ mixtures, while $n = 1$ should describe metamagnetic tricritical points. With the stabilizing additional term, one can now imagine varying both r and u through negative values. (This might be achieved by adjusting T and H in a metamagnet.) The usual Landau analysis,[25] neglecting the effect of the fluctuation term in (3.1), gives the phase diagram shown in Fig. 1a with the first order line given by negative values of u. Between 3 and 4 dimensions, a tricritical point should occur for all $v > 0$ at $r = 0$, $u = 0$.[24]

If one is interested only in the behavior of thermodynamic functions near the critical line, it is possible to set v to zero at the start in an $O(\varepsilon)$ calculation. The reason is that v is an irrelevant variable, and should contribute only corrections to asymptotic scaling of the kind first discussed by Wegner.[26] Thus, it is necessary only to repeat the calculations sketched in Sec. II for both r and u small (in particular, $u \ll \varepsilon$). This leads immediately to the tricritical crossover scaling appropriate to the critical line to first order in $\varepsilon$.[8]

The results for the $He^3$-$He^4$ mixtures concentration X (one derivative of the free energy with respect to chemical potential difference $\Delta = \mu_3 - \mu_4$) to $O(\varepsilon)$ are compared with experiment in Fig. 2. A scaled deviation of the concentration from its tricritical value is plotted against a scaled combination of the special variables t and g discussed in Sec. II. The solid lines represent the scaling function determined by Riedel et al.[27] from vapor pressure and calorimetric data on $He^3$-$He^4$ mixtures.[28] Data taken at a variety of different temperatures seemed to fall approximately on the two universal curves shown.[27] The upper curve ($g > 0$) corresponds entirely to normal phase data, while the lower curve is derived from normal phase data on the right and superfluid phase data on the left. The lambda line is given by $t/|g|^{\frac{1}{2}} \approx 1.1$.

The dashed line shows the prediction of a Landau theory which includes a $v|\vec{S}|^6$ term, while the dotted line summarizes the renormalization group calculations[8] to $\theta(\varepsilon)$. Renormalization group theory clearly does a better job than the Landau result near the lambda line and in the normal phase. The theory breaks down as indicated for $g < 0$ in the superfluid phase due to the neglect of the $v|\vec{S}|^6$ term. However, a direct graphical analysis in three dimensions by Stephen et al.[21] of the Hamiltonian (3.1)-(3.2) in this region indicates that Landau theory should suffice here up to logarithmic corrections for the scaling functions, as found earlier for the tricritical exponents.[24] In this sense, these calculations are complementary to the $\varepsilon$-expansion results. Fisher[29] has pointed out

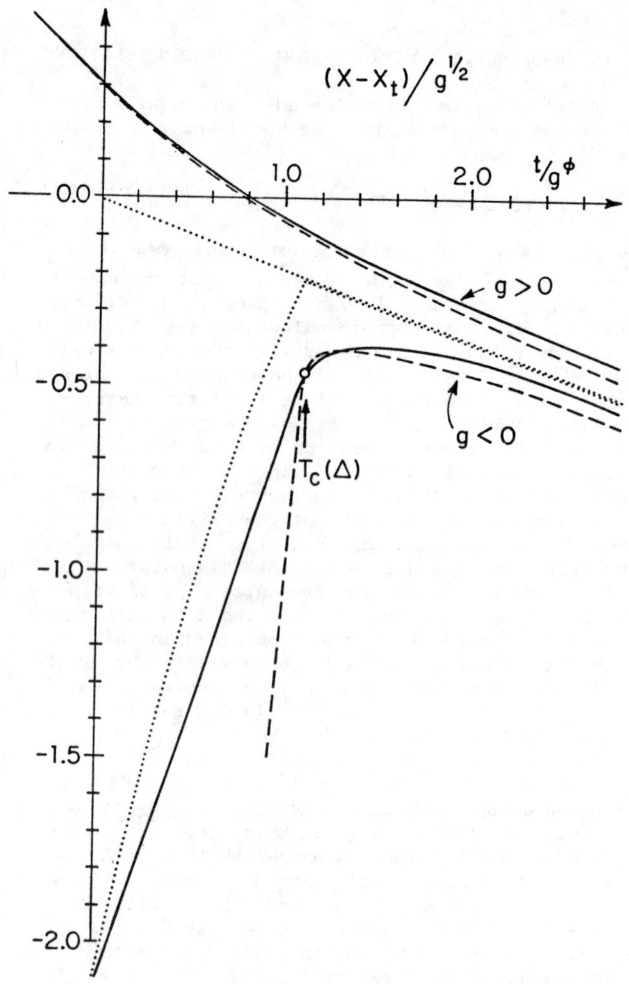


Fig. 2 Scaled deviation of the $He^3$-$He^4$ mixtures concentration from its tricritical value plotted against the scaling parameter $t/g^{\phi} = t/g^{1/2}$. The solid lines represent an experimentally determined scaling function,[27] while the dashed lines show the results of renormalization group theory. The dotted line is the prediction of a Landau analysis.

that the maximum occurring above $T_c$ in the lower branch of the scaling function signals the presence of a $t^{1-\alpha}$ singularity in X. This maximum also occurs in the theoretical expression for X to $O(\varepsilon)$, and the analysis explicitly confirms the presence of the $t^{1-\alpha}$ singularity.[8]

Further experimental work on tricritical scaling functions would certainly be of interest. (Other functions besides X are discussed in Reference 27.) Unfortunately however, it is unlikely that current experiments could actually study the detailed changeover of one effective critical exponent[30] into another. Both experiment[27] and theory[27,8] indicate that experiments studying regions several orders of magnitude closer to the critical line would be necessary to observe the crossover of an effective tricritical exponent into an asymptotic lambda line exponent.

### B. Bicritical Points

Phase diagrams with bicritical points[6] should occur in uniaxial antiferromagnets with spin-flop transitions[31a] and in flop-like displacive transitions in perovskite crystals.[31b] Renormalization group calculations have been carried out[9] on a model Hamiltonian believed to describe bicritical points,[31] which consists of (3.1) with an additional term,

$$\overline{\mathcal{H}} = \mathcal{H}_0 + \mathcal{H}_b\ ,$$

$$\mathcal{H}_b = \tfrac{1}{2}g\int d\vec{R}\left[\frac{n-m}{n}\sum_{i=1}^{m} s_i^2 - \frac{m}{n}\sum_{i=m+1}^{n} s_i^2\right] . \qquad (3.3)$$

For $GdAlO_3$, which was recently investigated in detail by Rohrer,[32] (3.3) applies with $n = 2$, $m = 1$, while for $MnF_2$, $n = 3$ and $m = 1,2$ should be appropriate.[6a] The analysis is complicated by the existence of two distinct length scales, one associated with the longitudinal spin field $\vec{S}_{\parallel} \equiv (S_i,\ i = 1, \ldots, m)$, and one associated with the transverse spin field $\vec{S}_{\perp} \equiv (S_i,\ i = m+1, \ldots, n)$. One approach[9a] (which works when $g > 0$) is to integrate the recursion relations until the transverse correlation length is of order unity. At this point, the transverse spins can be integrated entirely out of the problem. One is left with an m-component isotropic effective Hamiltonian whose properties are easily calculable,[9a] and the various bicritical thermodynamic functions follow immediately in scaling form. Setting the four-spin coupling constant to its critical value, $u = u_c = 2\pi^2\varepsilon/(n+8)$, the susceptibility, for example, can be expressed in the scaling form (1.1) with[9a]

$$\Phi(z) = (1-x)^{-\dot{\gamma}}R_A^{(m+2)/(m+8)}\left[1+\varepsilon\Delta_1(x)+\tfrac{1}{2}\varepsilon\frac{m+2}{n+2}\ln\left(1+\frac{m}{n-m}x\right)/R_A\right] \qquad (3.4)$$

where $x = z/\dot{z}$, $\dot{\gamma} = 1 + \tfrac{1}{2}(m+2)\varepsilon/(m+8)$, and

$$\Delta_1(x) = \frac{1}{2}\left[\frac{2x-n-2}{n+8}\ln\left(1+\frac{m}{n-m}x\right) + \frac{n}{n+8}x\ln\left(\frac{n}{n-m}\right)\right]/(1-x) , \qquad (3.5)$$

$$R_A(x) = \frac{n-m}{n+8}(1-x)^{\frac{1}{2}\varepsilon} + \frac{m+8}{n+8} . \qquad (3.6)$$

It is interesting that (3.4) - (3.6) reveal another singularity in $\Phi(x)$ proportional to $(1-x)^{\frac{1}{2}\varepsilon}$, in addition to the $(1-x)^{-\dot{\gamma}}$ factor predicted by the phenomenological theories.[7]

Results for the bicritical susceptibility (with $n = 2$, $m = 1$) for an arbitrary u of order $\varepsilon$ [9a] are compared with the results of high temperature expansions[33] in Fig. 3. We have plotted the effective exponent $\gamma_{eff}$, defined as[30]

$$\gamma_{eff} = d\ln\chi/d\ln\dot{t} , \qquad (3.7)$$

on the path of approach to the critical line at fixed g shown in Fig. 1b. The horizontal axis is measured in units of $\log_{10}\dot{t}$, where $\dot{t} = [T - T_c(g)]/T_c(g)$. Both curves remain initially at an Ising value of $\gamma_{eff}$, and then crossover in a range of about one decade, leveling off at bicritical XY values of $\gamma_{eff}$. The curves then exhibit a final nonuniversal crossover to the mean field prediction $\gamma_{eff} = 1$. The series work, of course, should be considered more reliable, but the qualitative similarities between the two curves are quite encouraging. The prediction of Landau theory is given by the horizontal line at $\gamma_{eff} = 1$.

The bicritical nonordering susceptibility $\tilde{\chi}(t,g)$, defined as two derivatives of the free energy with respect to magnetic field, should display rather interesting universal crossover scaling behavior. The scaling prediction for $\tilde{\chi}$ is[6]

$$\tilde{\chi}(t,g) \approx t^{2-\alpha-\phi}\Psi(g/t^{\phi}) . \qquad (3.8)$$

A plot of the $\Psi(x)$ to first order in $\varepsilon$[9a] is shown in Fig. 4 for $n = 2$ and $n = 3$. The behavior at $z = \dot{z}^{\pm}$ is governed by specific heat-like singularities in $\Psi(z)$,[6]

$$\Psi(z) \sim (z-\dot{z}_{\pm})^{-\dot{\alpha}^{\pm}} \qquad (3.9)$$

across the upper and lower critical lines. The plot of $\Psi(\chi)$ for Heisenberg-to-Ising and Heisenberg-to-XY crossovers shows an asymetric minimum, whose location and depth should be universal constants. Although series results for $\tilde{\chi}(t,g)$ are not presently available, an account of such work is in preparation.[32]

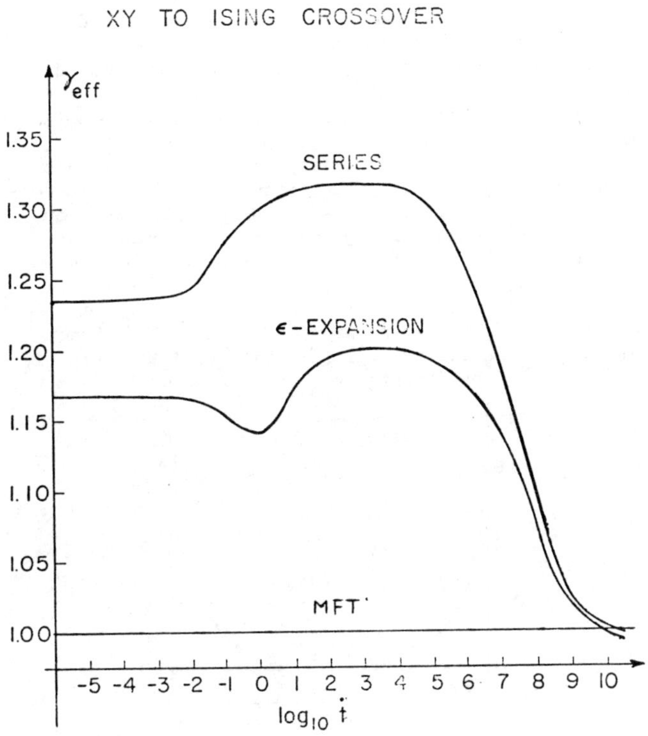


Fig. 3 Plot of $\gamma_{eff} = -d\ln\chi/d\ln\dot{t}$ to $O(\varepsilon)$ compared with the results of series espansions. The prediction of Landau theory is the horizontal line $\gamma_{eff} = 1$. An irrelevant variable has been adjusted in the renormalization group calculation to make the crossover to $\gamma_{eff} = 1$ occur at the same place on the $\log_{10}\dot{t}$ axis as the series results. The shape of both curves before the crossover to mean field theory sets in should be universal.

There is considerable hope that experiments will be able to determine a number of bicritical crossover scaling functions. Because of the relatively large crossover exponent ($\phi \approx 1.18 - 1.25$), bicritical crossovers should occur rather more rapidly than in tricritical systems. Indeed, Rohrer[32] has apparently observed both a critical line and a bicritical point exponent for the nonordering susceptibility of $GdAlO_3$ by making small variations of magnetic field near the spin flop point. The effective anisotropy entering (3.3) is expected to go from a few percent at $H = 0$ to zero at the bicritical point,[6a] which usually occurs in flop fields of about 10 - 100 KOe. Thus, by varying magnetic field it should be possible to "fine tune" the anisotropy g as well as the temperature t in a detailed experimental study of the bicritical region.

### C. Dipolar Systems

Fisher and Aharony[35] have analyzed the effect of adding a new perturbation to the Hamiltonian (3.1), namely

$$\mathcal{H}_d = \tfrac{1}{2}g\sum_{i,j=1}^{n=d}\int_{\vec{q}}\frac{1}{(2\pi)^d}\frac{q_iq_j}{q^2}\hat{S}_i(\vec{q})\hat{S}_j(\vec{q}) \qquad (3.10)$$

where the integral runs over a Brillouin zone of unit radius, and the $\hat{S}_i(\vec{q})$ are Fourier coefficients of the original spin field $\vec{S}(\vec{R})$. Such a term is generated by the presence of dipolar interactions, and leads to dipolar critical behavior in contrast to the usual, n-component Heisenberg behavior.[35] For small g, one can study the crossover between these two competing varieties of critical phenomena. As $T \to T_c(g)$ with g fixed, the effective critical exponents should be initially Heisenberg in character, and then change over to their dipolar values in the asymptotic critical regime.

Although it is doubtful that g can be varied appreciably in a given experiment, one can examine the behavior of substances with dipolar interactions of varying strength.[36] Thus, it may be possible to obtain at least a rough picture of crossover in dipolar systems.

The dipolar susceptibility crossover scaling function has been treated by Natterman.[37] Both the susceptibility and specific heat scaling functions were subsequently calculated using the method described in Sec. II.[10] The resulting plot of $\gamma_{eff}$ for the susceptibility [with $T \to T_c(g)$ at fixed g] is shown in Fig. 5. The horizontal axis is now $\log_{10}[\dot{t}(\dot{z}/g)^{1/\phi_D}]$, where $\phi_D$ is the dipolar crossover exponent ($\phi_D = \gamma_D$). Although there are some differences in detail with the work of Natterman,[37] the qualitative pictures agree. As is

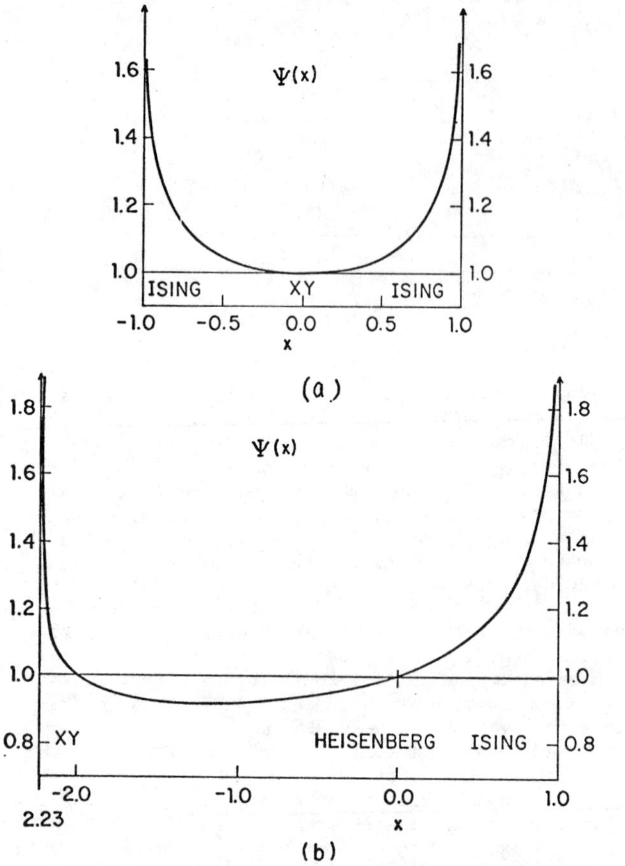


Fig. 4 Plots of the scaling functions $\Psi(x)$ for the nonordering susceptibility $\tilde{\chi}$. Fig. (a) shows the symetric plot appropriate to cross-over from n=2 to m=1 (XY to Ising), while Fig. (b) shows $\Psi(x)$ for n=3, m=1 (Heisenberg to Ising, x>0) and n=3, m=2 (Heisenberg to XY, x<0)

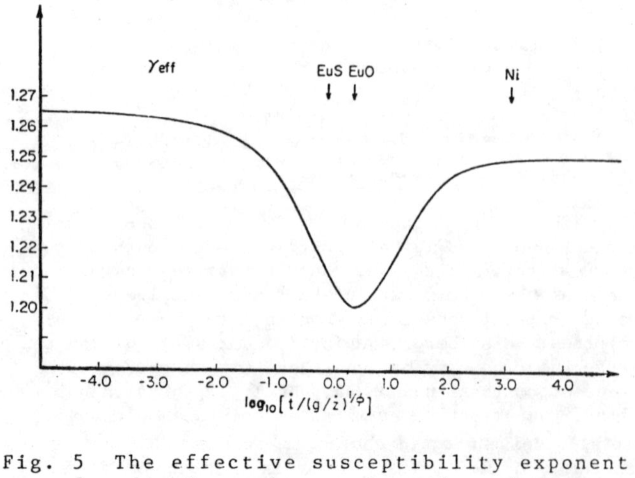


Fig. 5 The effective susceptibility exponent $\gamma_{eff}$ for dipolar systems plotted against $\log_{10}[\dot{t}/(g/\dot{z})^{1/\phi}]$.

well-known, the numerical difference between the Heisenberg and dipolar critical exponents is rather small.[35] Surprisingly, there is a large dip in $\gamma_{eff}$ which almost completely obscures this difference! The existence of the dip seems to be related to additional singularities in $\Phi(z)$ of the sort discussed for bicritical systems. Similar conclusions can be drawn for the specific heat.[10] The arrows in Fig. 5 indicate roughly the minimum values of z which have been obtained experimentally for various substances.[10] It seems probable that the best experiments have only begun to explore the asymptotic dipolar region, and that one should not expect a monotonic variation of the critical properties of dipolar systems with increasing dipolar coupling strength.

### D. Coexistence Curve Singularities in Isotropic Ferromagnets

Isotropic magnetic systems with continuous rotational symmetry behave rather differently than simple Ising-like magnets below the Curie temperature. As the magnetic field H tends to zero for $T < T_c$, it is expected that the longitudinal susceptibility $\chi_L$ diverges,

$$\chi_L \sim H^{-\frac{1}{2}\varepsilon} \tag{3.11}$$

as $H \to 0^+$, where, as usual, $\varepsilon = 4 - d$. At $T_c$, on the other hand, the longitudinal susceptibility diverges according to

$$\chi_L \sim H^{-(\delta-1)/\delta} \tag{3.12}$$

as $H \to 0^+$, a result which follows from the usual scaling description of a critical point.[39,40] This state of affairs, summarized in Fig. 6, can be viewed as an example of multicritical crossover behavior.[11] Given a ferromagnet with magnetization M in a field H, the Griffiths[40] scaling form for $\chi_L$ is

$$\chi_L = M^{1-\delta}\,\Phi(t/M^{1/\beta}), \tag{3.13}$$

where the coexistence curve should be described by a particular value of the argument of $\Phi(x)$, $x = \dot{x} < 0$. To describe the coexistence curve singularity, we assume in addition[11] that $\Phi(x)$ is singular at $\dot{x}$,

$$\Phi(x) \sim (x - \dot{x})^{-\frac{1}{2}\varepsilon}, \tag{3.14}$$

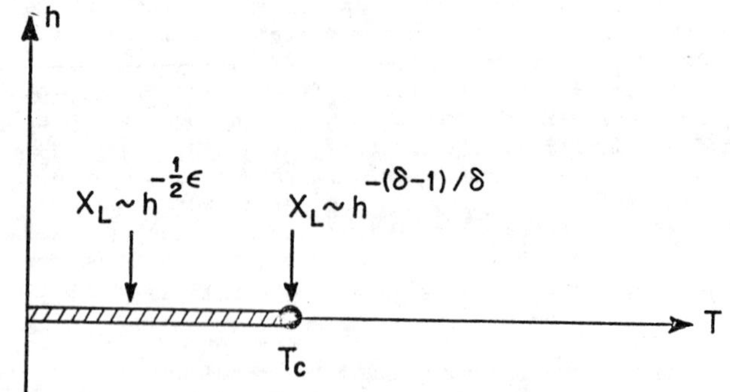


Fig. 6 Different kinds of divergences in a Heisenberg ferromagnet.

where the choice of exponent is dictated by (3.11).

Calculations of $\Phi(x)$ were not possible using the Feynman graph analysis that originally produced the equation of state for an ordinary ferromagnet.[3] Using the recursion relation "matching" approach described here, together with a parquet graph summation procedure, it is in fact possible to determine $\Phi(x)$ to first order in $\varepsilon$. The result is[11]

$$\Phi(x) = \frac{1}{(x+12u_c)} - \frac{3\varepsilon}{2(n+8)} \frac{x+36u_c}{(x+12u_c)^2} \ln(x+12u_c) - \frac{4\varepsilon u_c}{(x+12u_c)^2} + \frac{n-1}{5-n} \frac{1}{(x+12u_c)} \left\{ \left[ 1 + \frac{n+7}{n+8} \frac{x+4u_c}{x+12u_c} \left( (x+4u_c)^{-\frac{1}{2}\varepsilon} - 1 \right) \right]^{\frac{5-n}{n+7}} - 1 \right\} \tag{3.15}$$

where $\dot{x} = -4u_c$.

Real magnetic systems, of course, represent some space group and do not display complete rotational symmetry. There are always easy axes along which the spins prefer to align in the ordered state. It turns out that symmetry-breaking terms in the Hamiltonian clamp the divergence described above, so it is reasonable to inquire if there are any experimental consequences of the coexistence curve singularities discussed in this section. A rather large effect should occur in the longitudinal susceptibility of hexagonal layered metamagnetic crystals such as $CrCl_3$, $CoBr_2$, and $NiCl_2$. In particular, it is predicted[11] that the divergence which would be present in the absence of symmetry-breaking perturbations leads to an enhanced longitudinal susceptibility as $T \to T_c$ from below. Specifically,

$$\chi_L \sim |t|^{-\gamma'} = |t|^{-\gamma - \frac{1}{2}\varepsilon|\phi_W|}, \tag{3.16}$$

where $\gamma$ is the susceptibility exponent measured above $T_c$, and $\phi_W$ is the (irrelevant) crossover exponent associated with hexagonal perturbations. Results from the epsilon expansion indicate that[11] $\gamma' = 2.12$, which would be a strikingly large susceptibility exponent. Needless to say, experiments on hexagonal layered crystals would be invaluable.

## IV. SUMMARY

We have sketched developments which allow crossover scaling functions to be constructed for Landau-Ginzburg-Wilson continuous spin models of critical behavior. The results, which are now available to first order in $\varepsilon$ for a variety of multicritical problems, appear to give a good qualitative picture of crossover phenomena.

The calculational techniques discussed here are not restricted to multicritical phenomena. They should be useful in any problem for which recursion relations can be constructed. Rudnick[41] has used them to give a careful analysis of the "runaway" renormalization group flows which occur in the Baxter-like model considered by Wilson and Fisher,[42] while Aharony[43] has employed them in analyzing a new kind of singularity which occurs in random systems. One might hope that they would be of some utility in studying problems with a spatially varying order parameter.

versations and collaboration with Dr. Joseph Rudnick. Numerous conversations with Professor Michael E. Fisher on all aspects of the work were invaluable. I have also profited from interactions with Mr. E. Domany, Dr. A. D. Bruce, Dr. J. M. Kosterlitz, and Dr. T. S. Chang. E. Domany kindly pointed out an error in a preliminary version of Fig. 4.

The research reported here was supported in part by the National Science Foundation under Grant No. DMR72-02977 A03 and through the Materials Science Center at Cornell.

## REFERENCES

* Junior Fellow, Harvard Society of Fellows

1. For Reviews, see K. G. Wilson and J. Kogut, Physics Reports 12c, 77 (1974), S.-K. Ma, Rev. Mod. Phys. 45, 589 (1973) and M. E. Fisher, Rev. Mod. Phys. 42, 597 (1974).
2. E. Brézin, J. C. LeGuillou, and J. Zinn-Justin, Phys. Rev. D 8, 434, 2418 (1974).
3. E. Brézin, D. J. Wallace, and K. G. Wilson, Phys. Rev. Lett. 29, 591 (1972), Phys. Rev. B 7, 232 (1973).
4. K. G. Wilson, Phys. Rev. Lett. 28, 548 (1972).
5. E. K. Riedel, Phys. Rev. Lett. 28, 675 (1972).

6a. M. E. Fisher and D. R. Nelson, Phys. Rev. Lett. 32, 1350 (1974).

b. M. E. Fisher, Phys. Rev. Lett. 34, 1634 (1975).

7. E. K. Riedel and F. Wegner, Z. Phys. 225, 195 (1969). See also, e.g., P. Pfeuty, D. Jasnow, and M. E. Fisher, Phys. Rev. B 10, 2088 (1974).
8. D. R. Nelson and J. Rudnick, Phys. Rev. Lett. 35, 178 (1975), J. Rudnick and D. R. Nelson, Phys. Rev. B (in press).

9a. D. R. Nelson and E. Domany, Phys. Rev. B (in press).

b. J. M. Kosterlitz, J. Phys. C (in press).

10. A. D. Bruce, J. M. Kosterlitz, and D. R. Nelson, submitted to J. Phys. C.
11. D. R. Nelson, Phys. Rev. B (in press).
12. See Sec. 1.1 of K. G. Wilson and J. Kogut in Ref.1.
13. E. K. Riedel and F. J. Wegner, Phys. Rev. B 9, 294 (1974).
14. M. Nauenberg and B. Nienhuis, Phys. Rev. B 11, 4152 (1975).
15. F. J. Wegner and A. Houghton, Phys. Rev. A 8, 401 (1973).
16. J. F. Nicoll, T. S. Chang, and H. E. Stanley, Phys. Rev. Lett. 32, 1446 (1974).
17. J. Rudnick, Phys. Rev. B 11, 363 (1975).
18. M. Nauenberg and B. Nienhuis, Phys. Rev. Lett. 33, 1598 (1974).
19. D. R. Nelson, Phys. Rev. B 11, 3504 (1975).
20. G. Jona-Lasinio, Nuovo Cim. 34, 1790 (1964).
21. M. J. Stephen, E. Abrahams, and J. P. Straley, Phys. Rev. B 12, 256 (1975).
22. J. F. Nicoll, T. S. Chang, and H. E. Stanley, M.I.T. preprint.
23. Th. Natterman and S. Trimper, Phys. Lett. 50A, 307 (1974).
24. E. K. Riedel and F. J. Wegner, Phys. Rev. Lett. 29, 349 (1972).
25. See, e.g., R. Bausch, Z. Phys. 254, 81 (1972).
26. F. J. Wegner, Phys. Rev. B 5, 4529 (1972).
27. E. K. Riedel, H. Meyer, and R. P. Behringer, J. Low Temp. Phys., to be published.
28. G. Goellner, R. Behringer, and H. Meyer, J. Low Temp. Phys. 13, 113 (1973).
29. M. E. Fisher, private communication.
30. For a discussion of effective critical exponents, see Ref. 13.

31a. D. R. Nelson, J. M. Kosterlitz, and M. E. Fisher, Phys. Rev. Lett. 33, 813 (1974).

b. A. Aharony and A. D. Bruce, Phys. Rev. Lett. 33, 427 (1974).

32. H. Rohrer, Phys. Rev. Lett. 34, 1638 (1974).
33. S. Singh and D. Jasnow, Phys. Rev. B 11, 3445 (1975). See also P. Pfeuty, D. Jasnow, and M. E. Fisher,

Phys. Rev. B 10, 2088 (1972), and S. Singh and D. Jasnow, Phys. Rev. B 12, 493 (1975).
34. P. Gerber and M. E. Fisher (unpublished).
35. M. E. Fisher and A. Aharony, Phys. Rev. Lett. 30, 559 (1973); A. Aharony and M. E. Fisher, Phys. Rev. B 8, 3323 (1973).
36. A summary and analysis of a variety of different experiments is given in G. Ahler and A. Kornblit, Phys. Rev. B 12, 1938 (1975).
37. Th. Natterman, Leipzig preprint.
38. See, eg., M. E. Fisher, M. N. Barber, and D. Jasnow, Phys. Rev. A 8, 1111 (1973) and references therein.
39. B. Widom, J. Chem. Phys. 43, 3898 (1965).
40. R. B. Griffiths, Phys. Rev. 158, 176 (1967).
41. J. Rudnick, Case Western Reserve preprint.
42. K. G. Wilson and M. E. Fisher, Phys. Rev. Lett. 28, 240 (1972).
43. A. Aharony, Bell Laboratories preprint.

# RESONANT $^{19}F$ nmr ENHANCEMENT NEAR THE BICRITICAL POINT IN $MnF_2$*

A.R. King
Physics Department, University of California
Santa Barbara, California 93106

H. Rohrer †
IBM Research Laboratory, Saumerstr. 4
8803 Ruschlikon, Switzerland

## ABSTRACT

We have investigated the $^{19}F$ nmr in the spinflop (SF) phase of $MnF_2$ near the SF-paramagnetic (PM) phase boundary. We found a very broad (500 Oe) resonance, strongly enhanced with $h_1 \parallel H_0$, whose phase undergoes a continuous change through nearly 360°, including both normal and inverted absorption and dispersion, as the temperature is lowered below the phase boundary at $T_c$. The amplitude shows a broad maximum at about 1 K below $T_c$ and drops rapidly to zero at $T_c$, while the linewidth, surprisingly, is nearly constant.

The data are fit to a resonant enhancement model in which the frequency $\omega_0$ of a low-lying electronic mode coupling to $h_1 \parallel H_0$ is driven from above the nuclear resonance frequency $\omega_n$, down through $\omega_n$, toward zero as T approaches $T_c$ from below. Using the Bloch equations and enhancement coupling constant from molecular field theory, the nmr amplitude and phase give sufficient information to determine both the electronic frequency $\omega_0$ and damping $T_2$. The electronic mode is not seen directly since $\omega_0 T_2 \sim 1$.

## INTRODUCTION

During the course of an experiment to determine the magnetic phase boundaries near the bicritical point $(T_b, H_b)$ in $MnF_2$ (to be presented at this conference in another paper by the authors, hereafter referred to as I), we attempted to observe the $^{19}F$ nmr to get information on the spin dynamics in the crossover region. We expected a large enhancement of the nmr signal, because of the following argument. In molecular field theory[1] the sublattice magnetizations in the flopped state make an angle θ with the x-axis when a field H is applied parallel to the z-axis. The component of the sublattice magnetization M along the field is

$$M^z = M_0 H/2H_e \qquad (1)$$

independent of temperature, and the magnitude of **M** is given by

$$M = M_0 B_s[M\, g\mu_B S H_e/kT] \qquad (2)$$

independent of field. In the above equations, $H_e$ is the exchange field, $M_0$ the saturated sublattice magnetization, $B_s$ the Brillouin function, $\mu_B$ the Bohr magneton, and k the Boltzmann constant. At the transition temperature, $\theta = \pi/2$, and $M^z = M$. Thus the nmr enhancement for $h_1$ parallel to H should diverge near the critical field $H_c$ since $d(M_x)/dH$ diverges like $[1 - (H/H_c)^2]^{-\frac{1}{2}}$. The experiments did indeed show a large enhancement of the nmr signal, but showed a much more complicated behavior than predicted by this simple model.

## EXPERIMENTAL RESULTS

The $^{19}F$ nmr was observed at two frequencies (524 and 5[illegible] MHz), shifted downward by about 8% from the free $^{19}F$ frequency. This shift agrees well with the shift calculated in the paramagnetic state. A bridge-type nmr spectrometer was used, which was in all cases

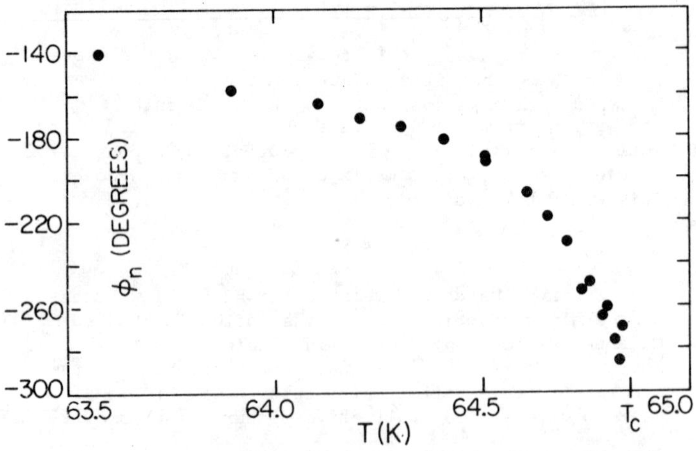


Fig. 1. Phase of the nmr signal. Absorption corresponds to $\phi = 0$.

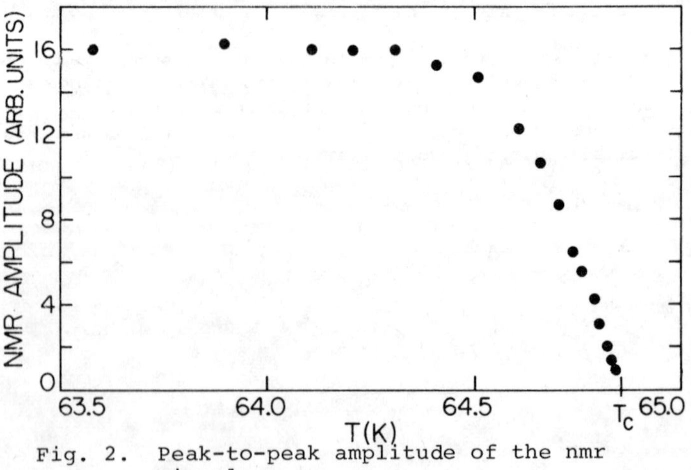


Fig. 2. Peak-to-peak amplitude of the nmr signal.

tuned to observe purely absorptive signals. Field modulation and lock-in detection were used, so derivative signals were obtained. The rest of the experimental details are discussed in I.

Although the spectrometer was tuned for absorption, the $^{19}F$ nmr phase (Fig. 1) was observed to change with temperature through nearly 360°, from absorption at the lowest temperature to dispersion, inverted absorption, and the opposite sign of dispersion near the phase boundary. Accompanying this phase change was a change in amplitude (Fig. 2) which was relatively constant over a range of temperature but which dropped to zero at the phase boundary. The extrapolation to zero of the amplitude was taken as the phase boundary, which made possible measurements in a region where the susceptibility signals became too small to be useful.

Throughout the entire temperature range, the nmr linewidth remained constant to within 5% at about 500 Oe. No indication was seen of a critical divergence in the linewidth[2]. This effect remains not understood.

## ANALYSIS

The strength of the nmr signal indicates a large enhancement[3] due to the electronic spin system. Therefore we attribute the nmr behavior to the electronic response. The nmr signal is proportional to the square of the amplitude of the effective field at the nucleus,

$$h_1^2(\text{eff}) = h_1^2(1 + \eta)^2 \qquad (3)$$

where $h_1$ is the rf field and $\eta$ is the enhancement factor due to the hyperfine interaction. For large enhancements, we neglect 1 with respect to $\eta$. If the electronic system response is complex (exhibits phase lag $\phi_e$) the nmr response will also be complex, with phase lag $\phi_n = 2\phi_e$. Since $\phi_n$ changes by nearly $2\pi$, $\phi_e$ must change by nearly $\pi$. This would occur if the enhancement were due to a resonant mode whose frequency $\omega_0$ were driven from above the nmr frequency $\omega_n$, down through $\omega_n$, toward zero at the phase transition. Taking the effective rf field transverse to M as $h_1 \sin\theta$, we find from the Bloch equations,

$$|\eta|^2 = (A/\gamma_n\hbar)^2(\gamma_e\langle S\rangle T_2 \sin\theta)^2/[1+(\omega_n-\omega_0)^2T_2^2] \qquad (4)$$

and

$$\tan\phi_e = 1/(\omega_0-\omega_n)T_2 \qquad (5)$$

Solving for $\omega_0-\omega_n$ and $T_2$, we find

$$\omega_0-\omega_n = \eta^*\langle S\rangle \cos\phi_e \sin\theta/\eta\langle S\rangle^* T_2^* \sin\theta^* \qquad (6)$$

and

$$T_2 = T_2^*\eta\langle S\rangle^* \sin\theta^*/\eta^*\langle S\rangle \sin\phi \sin\theta \quad , \qquad (7)$$

where

$$T_2^* = \gamma_n\hbar\eta^*/A\gamma_e\langle S\rangle^* \sin\theta^* \qquad (8)$$

Here all starred quantities refer to the point where inverted absorption is observed ($\phi_e = -\pi/2$). The magnitude of $T_2$ and $\omega_0-\omega_n$ depends on an estimate of the enhancement $(\eta^*)^2 \approx 710$, which was made by comparing the $^{19}F$ nmr amplitude and width with the proton resonance in the plastic sample holder. When $\langle S\rangle$ and $\theta$ are calculated from Eqs. (1) and (2), we find $\omega_0-\omega_n$ and $T_2$ as shown in Figs. 3 and 4. We see that $\omega_0-\omega_n$ does become negative as assumed, and that it appears to approach $-\omega_n$; i.e., $\omega_0 \to 0$, at the phase transition. However, the scale of $\omega_0-\omega_n$ is determined by only a very rough guess of $\eta^*$, so if the mode in fact goes soft at the transition, this result must be considered fortuitous. $T_2$, on the other hand, shows only a gradual increase near the phase transition. Since $\omega_0 T_2 \sim 1$, the electronic mode is not seen directly.

It should be emphasized that the above calculation is intended only as a somewhat qualitative analysis of the data. Molecular field theory is not expected to give the correct critical exponents, and other factors, such as ellipticity of the precession modes have been omitted.

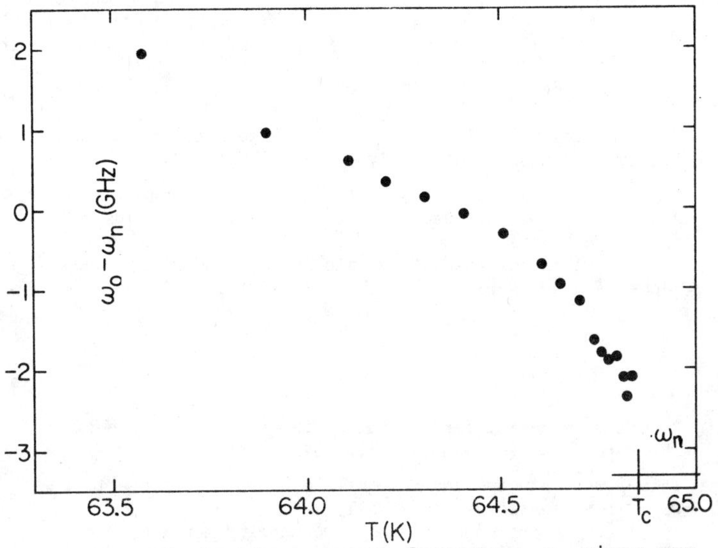


Fig. 3. Electronic resonant frequency $\omega_0$ minus nmr frequency $\omega_n$, from Eqs. (6) and (8).

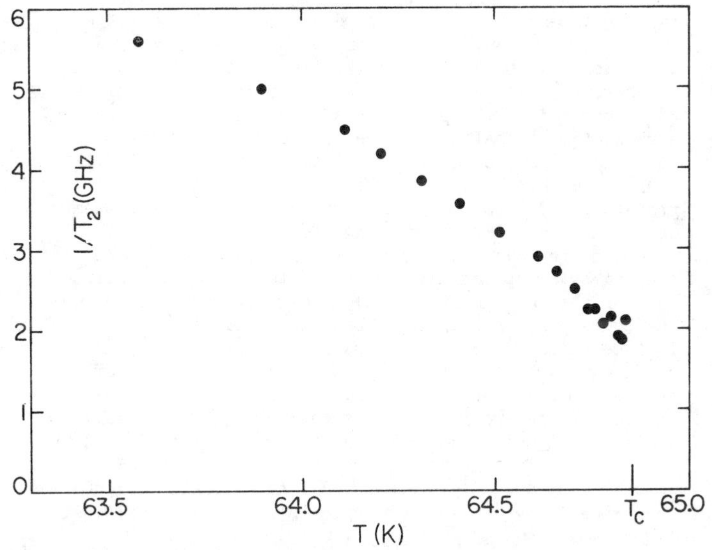


Fig. 4. Electronic $1/T_2$ from Eqs. (7) and (8).

## REFERENCES

*Work supported in part by the National Science Foundation.

†Visiting scientist, Department of Physics, University of California, Santa Barbara, CA 93106, while experimental work was done.


1. T. Nagamiya, K. Yosida, R. Kubo, Adv. in Phys. 4, 1 (1955).
2. P. Heller and G. Benedek, Phys. Rev. Lett. 8, 428 (1962).
3. See, for example, A.C. Gossard and A.M. Portis, Phys. Rev. Lett. 3, 164 (1959) and A.R. King, D. Paquette, and V. Jaccarino, Phys. Lett. 50A, 229 (1974).

# SPIN FLOP TRANSITION AND BICRITICAL BEHAVIOR OF ANTIFERROMAGNETIC LAYER COMPOUNDS

Franz S. Rys
Institut für Theoretische Physik, Freie Universität Berlin
1 Berlin 33 (-West), Arnimallee 3

## ABSTRACT

The observed spin flop transition in some antiferromagnetic layer compounds of the type $(CH_2)_n(NH_3)_2MnCl_4$ (n = 1,...5) is studied in view of the quasi-twodimensional nature of the spin system. The behavior of the system in the neighborhood of the bicritical point is studied and the bicritical exponents are calculated. A possible occurence of a tetracritical phase is discussed.

## INTRODUCTION

Recently, some antiferromagnetic biammonium compounds of the type $(CH_2)_n(NH_3)_2MnCl_4$ with values of n ranging from 1 to 5 were studied and some of their magnetic properties were investigated experimentally[1]. In several respects, these compounds resemble the corresponding monoammonium salts $(C_nH_{2n+1})_2(NH_3)_2MnCl_4$, studied some time ago[2]. Below the Néel temperature $T_N$ (which is typically of the order of 40 to 50 K) an antiferromagnetic spin order sets inalong an easy axis which is perpendicular to the layer plane. In addition, in some compounds a weak ferromagnetic moment has been observed parallel to the plane[3]. Furthermore, a remarkably sharp spin flop transition has been measured in the propyl (n=3) compound[3]. The sharpness of this transition is due to the layered nature of the compound on the one hand and to the availability of a single domain crystal on the other hand. The magnetic properties of these compounds are determined by a system of spins localised at the sites of the $Mn^{++}$-ions (with a ground state $^6S$); the spins having a value S = 5/2 are coupled by a superexchange coupling to its nearest neighbors on the same layer as well as on neighboring layers. The corresponding coupling constants J and J' show a stron lattice anisotropy; indeed, the ratio $\eta$ = J'/J is very small ($\eta = 10^{-8}$ or less, cf. Ref. 2). From the existence of the spin flop transition, the presence of an anisotropic spin interaction must be concluded (thus modifying a simple Heisenberg coupling ansatz between two coupled spins). Therefore, we expect that the spin system can be described by a quasi-twodimensional **anisotropic** Heisenberg model. Several interesting questions arise now: What is the reason for a non-vanishing long range spin order at finite temperatures ? What are the values of the critical exponents describing the magnetic critical behavior ? What kind of cross-over behavior will occur near the critical point ? Is there a possibility for a tetracritical point to show up in a magnetic phase diagram ? In the following, some of these questions will be discussed using the preliminary experimental data available up to date.

## MAGNETIC ORDER IN QUASI-TWODIMENSIONAL SPIN SYSTEMS

According to the Mermin-Wagner (M-W) theorem[5], no long-ranged spin order can occur in a twodimensional isotropic Heisenberg system. However, the presence of a small interlayer spin coupling or the occurence of an anisotropic term in the spin coupling energy invalidates the proof of the M-W theorem. We expect one of these two effects to cause a spin order in the present compounds. Using the methods for the derivation of the M-W theorem, one can easily generalise the results; upper limits for the sublattice magnetisation as a function of the lattice and/or the spin anisotropy parameter can be derived. As an example, for the case of a vanishing spinanisotropy, and for small values of $\eta$, the sublattice magnetisation is bounded from above by:

$$|m| < m_u(\eta) = c\,/\log \eta/^{-1} \quad (\eta \to 0).$$

A detailed analysis of the experimental data is expected to show which of the two reasons mentionned above does cause a spin order to occur in the layered compounds.Correspondingly, various possibilities for cross-over phenomena in the critical region have to be studied.

## SPIN FLOP TRANSITION

Preliminary experimental results on the spin flop transition for the propyl compound are in qualitative agreement with the theoretical results. Recently, the critical behavior of a spin flop transition was investigated[4], a possible phase diagram was proposed and the critical exponents at the bicritical point were calculated. Assuming a twodimensional critical behavior (in some region near the bicritical point) for the present systems, the corresponding twodimensional values of the critical exponents have been evaluated (up to third order in the $\epsilon$-expansion) and are listed with the threedimensional values in Table I. Typically, the twodimensional values are larger than the corresponding threedimensional ones.According to the preliminary evaluation of the experimental data, a power law behavior was found for the increase of the magnetisation at the spin flop transition,as a function of the reduced temperature, with the corresponding exponent $\tilde{\beta}$ agreeing well with the calculated value. It is hoped that, soon, more experimental data will be available for the present compounds.

| Exponent | $\phi$ | $\tilde{\beta}$ | $\tilde{\gamma}$ | $\tilde{\delta}$ |
|---|---|---|---|---|
| d = 2 | 1,61 | 0,88 | 0,72 | 1,80 |
| d = 3 | 1,22 | 0,85 | 0,40 | 1,47 |

TABLE I

Some bicritical exponents in two and three dimensions (for the definitions,see Ref.4)

## TETRACRITICAL POINT

An alternative phase diagram for the spin flop system was proposed in addition[4]. A new intermediate phase with two different coexisting order parameters is expected to occur between the antiferromagnetic and the spin flopped phase, provided that the number of spin components exceeds a lower limit.In two and three dimensions, this lower limit amounts to: $n^x(d=2) = 2,88$ , $n^x(d=3) = 3,12$;(up to $\epsilon^3$)
Thus, the inequality

$$n > n^x$$

which is a necessary condition for the existence of a tetracritical point, is fullfilled in two (but not in three) dimensions in the case of an ordinary three-component Heisenberg model. Assuming, as it was done before, a quasi-twodimensional behavior of the layer compounds, this would give a possibility for a magnetic tetracritical point to exist. Experimentally, the observed weak ferromagnetic moment perpendicular to tha antiferromagnetic spin order, observed in some compounds[3], could possibly be regarded as an evidence for the existence of a mixed phase with two perpendicular spin order parameters. Of course, more experimental data are required to confirm this speculation.

## REFERENCES

1. H.Arend, K.Tichy, K.Baberschke and F.Rys, to appear in Sol. State Comm. (1976)

2. L.J.de Jongh and A.R. Miedema, Adv. in Phys.23,1 (1974)

3. Franz Rys and K.Baberschke, to appear in Helv.Phys.Acta

4. M.E.Fisher and D.R.Nelson, Phys. Rev. Letters 32, 1350 (1974)

5. N.D.Mermin and H.Wagner, Phys. Rev. Letters 17,1133(1966)

# TRICRITICAL BEHAVIOR OF DYSPROSIUM ALUMINUM GARNET*

N. Giordano† and W.P. Wolf
Department of Engineering and Applied Science, Yale University
New Haven, Connecticut 06520

## ABSTRACT

Recent measurements of some of the magnetic properties of dysprosium aluminum garnet near one of its tricritical points are analyzed in the light of current theoretical predictions. In particular we consider the form of the internal field - temperature phase diagram, scaling of the magnetization and the behavior of the non-ordering susceptibility near the λ-line. All results are found to be in good agreement with the principal features of the theory, although the present accuracy is not high enough to detect any logarithmic corrections or to check the predicted value of the crossover exponent.

## INTRODUCTION

In a recent publication[1] we reported a study of the magnetic properties of dysprosium aluminum garnet (DAG)[2] near its tricritical point for fields applied along {110}. Using two novel methods based on hysteresis associated with domain effects, we were able to determine the first order phase boundaries and hence the location of the tricritical point (TCP). In addition, we measured the magnetization (M) as a function of applied field ($H_o$) and temperature (T) in the vicinity of the TCP. From these data, a wide variety of information can be obtained. In our previous paper[1] we determined the TCP exponents $\beta_u$, $\delta_\pm$ and $\gamma_u$,[3] and the shape of the M-T phase diagram, all of which were found to be in good agreement with present theory. In this paper we shall give results for the internal field ($H_i$) - temperature phase diagram, describe the results of a scaling analysis of the magnetization data, and discuss the behavior of the non-ordering susceptibility near the λ-line.

## RESULTS

Figure 1 shows the results for the $H_i$-T phase diagram over the reduced temperature range $|t| \equiv |T-T_t|/T_t \lessgtr 0.1$, where $T_t$ is the tricritical temperature.

There are three principal sources of uncertainty in determining $H_i$ at the phase boundary. One arises from the demagnetizing correction, which requires a knowledge of the absolute value of the magnetization. With our present experimental setup this contributes an uncertainty of ±1 Oe in $H_i$. A second error comes from trapped flux in our superconducting magnet. This was minimized by careful cycling, but it may still amount to some ±5 Oe. The third source of error arises from the uncertainty in the location of the onset of the phase transition. This error is minimized with our new method and amounts to ±1 Oe.

The results in Fig. 1 can be seen to fall well within these uncertainties. The data were fitted to an equation of the theoretically predicted form[4]: $(H_{i\pm} - H_t) = A_\pm(T - T_t) + B_\pm(T - T_t)^2$, where $H_t$ is the internal field at the TCP and the subscripts ± refer to $T \gtrless T_t$. Values for $H_t$ = 3173 Oe and $T_t$ = 1.808 K were taken from Ref. 1, and the corresponding fit, shown by the solid line in Fig. 1, gave $A_+ = -1.53 \pm 0.04$ kOe/K, $A_- = -1.61 \pm 0.03$ kOe/K; $B_+ = -1.99 \pm 0.30$ kOe/K$^2$, $B_- = -0.51 \pm 0.11$ kOe/K$^2$. This is consistent with the theory[4] which predicts a continuous first derivative and a discontinuous second derivative at the TCP. Similar results were obtained using other values for $H_t$ and $T_t$ within the error limits given in Ref. 1.

Fig. 2 shows scaling plots for the magnetization. Following previous work,[5] we plot $m/|t|^{\beta_u}$ versus $g/|t|^{\beta_u\delta_u}$ where $g = h + pt$, and $h = (H-H_t)/H_t$; $M_t$ is the tricritical magnetization and p is the slope of the phase boundary at the TCP in h - t space.[6,7] Values for $T_t$ = 1.808 K, $H_t$ = 3173 Oe, and $M_t$ = 250.6 emu/cm$^3$ were taken from Ref. 1, p = 0.866 was derived from Fig. 1, while for the exponents we used the theoretically predicted values $\beta_u = 1$, $\delta_u = 2$, which are also in good agreement with the experimental estimates.[1]

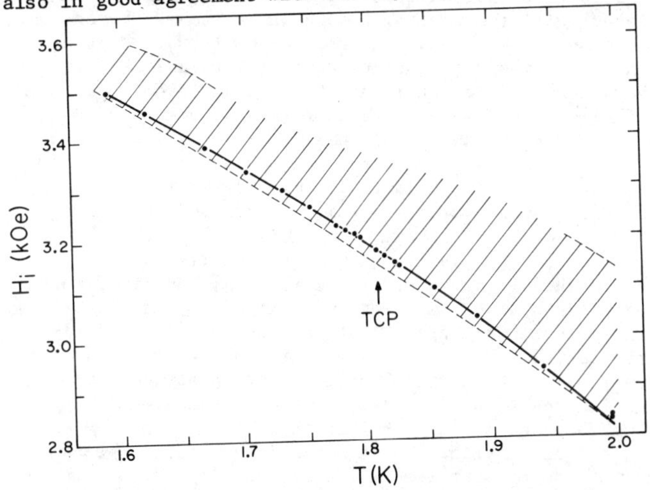


Fig. 1. $H_i$ - T phase diagram. The points are the experimental data. The solid curve is the fit to the data described in the text. The shaded area is the approximate scaling region.

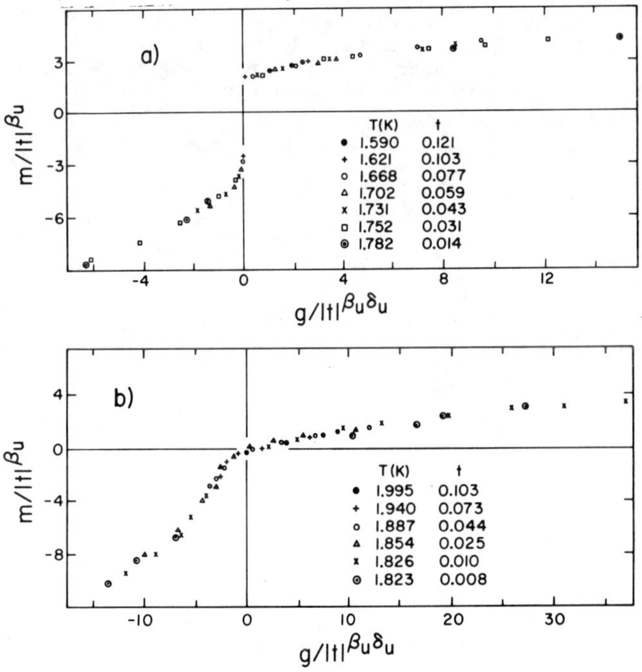


Fig. 2. Scaling plot of $m/|t|^{\beta_u}$ versus $g/|t|^{\beta_u\delta_u}$. a: $T < T_t$; b: $T > T_t$.

When we first plotted our data in this scaling form , we found that all the data did not, in fact, fall on the same curve. Rather, we found that there were systematic deviations which would correspond to uniform shifts in the value for $H_i$ for an entire isotherm. Such systematic errors could be due to changes in trapped flux which may vary from run to run, and we therefore shifted $H_i$ for each isotherm by a small amount to collapse the data. The required shifts were all less than 5 Oe, corresponding to the uncertainty in the trapped flux. We note that these shifts only affect the horizontal positions and not the shapes of the curves in Fig. 2.

We see from Fig. 2 that our data are quite consistent with scaling. The fact that the data scale near the first order phase boundary, i.e., near g = 0 for $T < T_t$ indicates that they are also consistent with the theoretical prediction $\gamma_\pm = 1$.

In Fig. 3 we show the same data plotted in a second scaling form[5]: $m/|g|^{1/\delta_u}$ versus $t/|g|^{1/\beta_u\delta_u}$. The same small shifts in $H_i$ described above were used in these plots. We see that the data are once again consistent with the scaling hypothesis. The two hatch marks on the vertical axis are the amplitudes corresponding to the exponents $\delta_\pm$ along the tricritical isotherm, which were determined from data presented in Ref. 1. We see that the data in Fig. 3 are in good agreement with these amplitudes.

We have also determined the approximate region over which scaling holds, by simply determining where the data deviate from the common curve in plots like Fig. 2. The results are shown in Fig. 1 as the shaded region. The boundaries of the scaling region are shown by a dashed line. Where there is no dashed line, this indicates that our data did not extend beyond the scaling region. We caution that this is only an approximate estimate of the scaling region and that our data are really not accurate enough to reach any definitive conclusions about a preferred second scaling axis.[8] Nevertheless, we can conclude that the scaling region is very asymmetrical, being much larger on the paramagnetic side than on the antiferromagnetic side of the phase boundary. This same asymmetry has also been found in $^3He$ - $^4He$ mixtures,[7] and it would be interesting to see if it is present also in other systems.

Fig. 4 shows our results for the non-ordering susceptibility, $\chi \equiv (\partial M/\partial H_i)_T$ near the λ-line for four different temperatures. These results were obtained by smoothing and differentiating the M-$H_i$-T data. The smoothing was necessary only very near the λ-line ($H_i - H_\lambda$ < 10 Oe) where χ changes most rapidly. We estimate the uncertainty in χ to be no more than 3% and that the smoothing affects the resolution only for $H_i - H_\lambda$ < 2 Oe. From Fig. 4 we see that the data strongly suggest that χ diverges at the λ-line, as expected theoretically.[9]

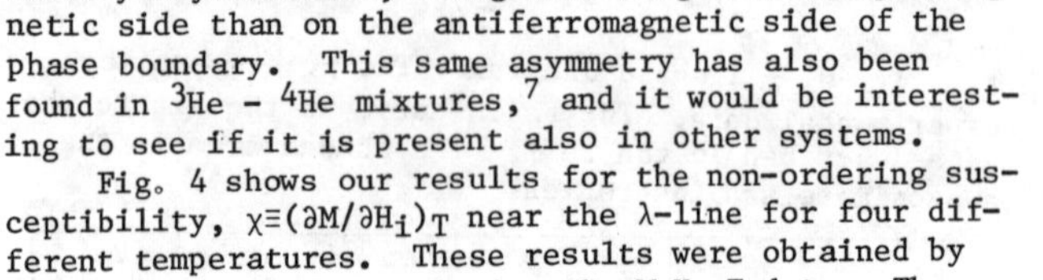

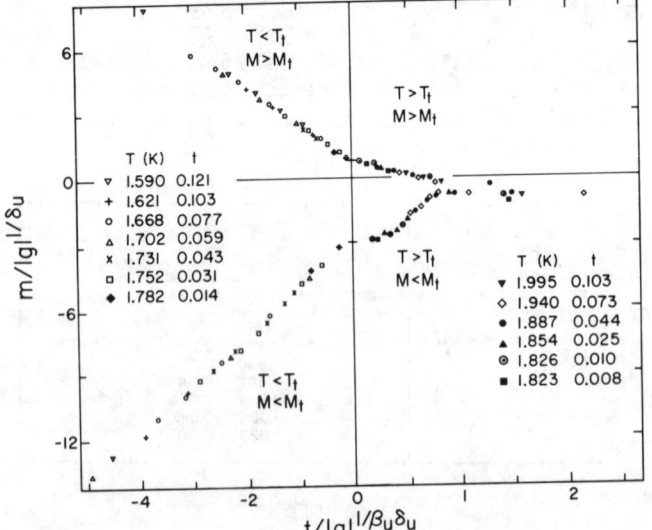

Fig. 3. Scaling plot of $m/|g|^{1/\delta_u}$ versus $t/|g|^{1/\beta_u\delta_u}$. Note the change of scale for t > 0 ($T > T_t$).

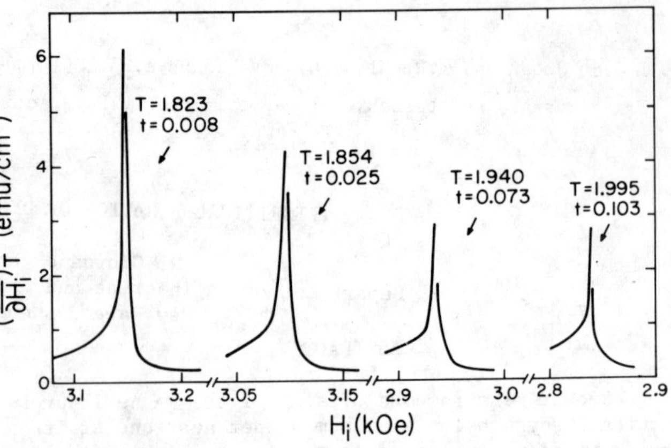


Fig. 4. Non-ordering susceptibility, $(\partial M/\partial H_i)_T$ versus $H_i$ for different values of $T > T_t$, near the λ-line. The reduced temperatures, t are also indicated.

## SUMMARY AND CONCLUSIONS

From the results shown in Fig. 1-4 and also those reported in Ref. 1, we conclude that the M-T and $H_i$-T phase diagrams, all of the exponents associated with the magnetization ($\beta_u$, $\delta_\pm$, $\gamma_u$ and $\gamma_\pm$), the equation of state near the TCP, and the behavior of χ near the λ-line are all in good agreement with the present theory. We have discovered no evidence for any logarithmic corrections in that all of our data are quite consistent with simple power law singularities. Qualitatively this implies that the theoretically predicted logarithmic corrections[10] must be quite small. Upper limits for these corrections will be discussed elsewhere. Our results are very similar to those found for $^3He$ - $^4He$ mixtures. These similarities include not only the values of the exponents, but also the asymmetry in the scaling region and the small size of the logarithmic corrections.

We would like to thank R. Alben, R.B. Griffiths, D.R. Nelson and M.J. Stephen for several enlightening discussions.

## REFERENCES


* Supported in part by NSF and the the U.S. Army Research Office.

† NSF Graduate Fellow.

1. N. Giordano and W.P. Wolf, Phys. Rev. Lett 35, 799 (1975).
2. For a summary of previous work on DAG see D.P. Landau, B.E. Keen, B. Schneider and W.P. Wolf, Phys. Rev. B3, 2310 (1971); W.P. Wolf, AIP Conf. Proc. 24, 255 (1975).
3. We use here the notation proposed by R.B. Griffiths, Phys. Rev. B7, 545 (1973).
4. E.K. Riedel, Phys. Rev. Lett. 28, 675 (1972).
5. G.F. Tuthill, F. Harbus and H.E. Stanley, Phys. Rev. Lett. 31, 527 (1973).
6. The scaling hypothesis strictly applies only to the singular part of the magnetization. The regular part of M will in general not be of the scaling form. However, it can be shown (N. Giordano, to be published) that in this particular case, the regular contribution to M will cause only negligible deviations from the scaling form. This same simplification has been found to occur in $^3He$ - $^4He$ mixtures (Ref. 7).
7. E.K. Riedel, H. Meyer and R. Behringer (to be published).
8. M.E. Fisher, Phys. Rev. Lett. 34, 1634 (1975).
9. R.B. Griffiths and J.C. Wheeler, Phys. Rev. A2, 1047 (1970).
10. E.K. Riedel and F.J. Wegner, Phys. Rev. B7, 248 (1973).

# BICRITICAL BEHAVIOR IN AN ANISOTROPIC HEISENBERG ANTIFERROMAGNET

D. P. Landau*
University of Georgia, Athens, Ga. 30602

K. Binder
Universität des Saarlandes, Saarbrücken, W. Germany

## ABSTRACT

A Monte Carlo method has been used to study the field induced transitions in a classical, anisotropic Heisenberg antiferromagnet with nearest-neighbor interactions. The spins were arrayed on simple cubic lattices as large as 20x20x20 with periodic boundary conditions and anisotropy constant $\Delta=0.2$. The spin-flop (SF) phase line as well as the AF→P and SF→P phase boundaries were studied over a wide range of field and temperature. A bicritical point was found at $kT_b/J=1.07\pm0.01$ and $\mu H_b/J=3.32\pm0.01$. The shape of the phase boundaries and critical magnetization near $(T_b,H_b)$ agree well with theoretical predictions.

## INTRODUCTION

Recent interest in multicritical phenomena has produced a resurgence in the study of magnetic field induced transitions in anisotropic antiferromagnets. Systems of interest fall broadly into two distinct classes: (a) those which as the applied magnetic field increases undergo a transition from the AF (antiferromagnetic) state directly to the P (paramagnetic) state; (b) those which for some range of low temperatures first undergo a transition to an intermediate SF (spin flop) state and enter the P state only at much higher fields. The systems in class (a) sometimes possess tricritical points with a new set of tricritical exponents which differ from the usual critical exponents. The critical field diagrams for systems in class (b) are considerably more complicated. The high temperature terminus of the spin flop phase boundary has been named the bicritical point since two phases, the AF and SF states, become critical simultaneously at that point. It has also been predicted[1-5] that the exponents at each point on the critical "surfaces" are determined by the number of spin components going critical at that point. The exponents for a point with n-critical spin components should be identical to the usual critical exponents for a system of spin dimensionality n. The AF→P boundary should therefore have Ising-like exponents, the SF→P line XY-like exponents, and the bicritical point itself should be Heisenberg-like. The shape of the 2nd order lines coming out of the bicritical point should be described by the "crossover" exponent $\phi$. The behavior of such systems has been fully predicted by renormalization group theory; however, relatively little is known experimentally[6] and most of these predictions are untested. We have therefore carried out computer simulations of a simple model which exhibits a bicritical point in order to compare the results with theory.

## MONTE CARLO PROCEDURE

An importance sampling Monte Carlo method was used to study a model with the Hamiltonian:

$$\mathcal{H} = J \sum_{(ij)} (1-\Delta)(S_{ix}S_{jx}+S_{iy}S_{jy})+S_{iz}S_{jz} \quad +\mu H \sum_i S_{iz} \qquad (1)$$

where $\vec{S}_i$ is a unit vector in the direction of the classical magnetic moment at site i, and the summation (ij) is over nearest neighbor pairs only. All of our computer experiments were carried out with the uniaxial exchange anisotropy fixed at $\Delta=0.2$. The spins were arrayed on a 20x20x20 simple cubic lattice with periodic boundary conditions. For each value of field and temperature Markov chains of spin configurations were generated using the usual technique.[9] Since only the static properties were being determined, it was possible to save computing time by considering the spins in turn rather than via random choice. Computer time was further reduced by carrying out the same procedure using the same random numbers for four different lattices simultaneously. (These modifications have already been proven effective in our studies[10] of two-dimensional anisotropic Heisenberg lattices.) A minimum of 80 MC steps/spin (and sometimes as many as 300 MC steps/spin) were discarded initially to allow the system to reach equilibrium. Then from 200 to 1200 MC steps/spin were generated and these configurations were used for calculating thermodynamic averages. (In order to reduce the effect of correlations[11] on error estimates only every 5th or 10th configuration was actually used.)

Data were taken along several different paths. The AF→P transition was studied keeping h=H/T constant. (Such paths remain roughly perpendicular to the phase boundary.) The AF→SF phase boundary was examined by isothermal field sweeps and the SF→P phase line by sweeping T at constant field. Data were generally taken in both directions along each path.

## RESULTS AND DISCUSSION

The behavior of the internal energy U, magnetization $M_z$, and order parameter $\psi$ was studied near the phase boundaries. (In the AF state the order parameter $\psi_{AF}=\sqrt{<m^2>}$ where $\vec{m}$ is the sublattice magnetization whereas $\psi_{SF}=\sqrt{<m_x^2+m_y^2>}$ in the spin flop phase.) Because of the finite size of the lattice, $\psi$ does not go to zero at $T_c(H)$ but shows a small finite size "tail" extending to high temperatures. Thus, although $\psi$ can be used to determine the approximate location of $T_c(H)$, accurate estimates can be made only from the inflection points in U and $M_z$. (For the SF→P transition $M_z$ remains virtually constant through $T_c(H)$ and only the temperature variation of U is useful. This inflection point in U corresponds to a singularity in the specific heat in the infinite system.) The critical field results are plotted in Fig. 1. The start of the spin flop boundary was indicated by the onset of hysteresis. We estimate that the bicritical point occurs for $h_b=\mu H_b/kT_b=3.1\pm0.1$ and $kT_b/J=1.07\pm0.01$. From Fig. 1 it is obvious that $T_b$ lies well below $T_N$. For comparison, $T_b/T_N\sim0.96$ in $MnF_2$ whereas here the ratio is ~0.70. Far from the bicritical point the mean field critical lines, labeled $T_c^{\perp}$ and $T_c^{\parallel}$ in Fig. 1, describe the data well. Near $(T_b,H_b)$, however, the data clearly deviate from mean field behavior. The mean field bicritical temperature $T_b$* is almost 3% greater than the actual $T_b$. (In $MnF_2$ the difference is less than 0.5%.)

Log-log plots of $\psi$ vs $\varepsilon=|1-T/T_c(H)|$ were used to estimate the order parameter exponent $\beta$ along the 2nd order phase boundaries. Because of the finite size rounding the last several points for smallest $\varepsilon$ fall systematically high. Additional data on other lattice sizes will allow us to correct accurately for finite size rounding, but for the moment we estimate that along the AF→P phase line $\beta=0.30\pm0.05$ and for the SF→P transition $\beta=0.35\pm0.05$. Our estimate for $\beta$ at the bicritical point is $\simeq0.3$, however, this value is affected by the uncertainty in the location $(T_b,H_b)$ as well by finite size. More precise estimates of critical exponents cannot be given until the effect of finite size on both the ordering temperatures as well as bulk properties can be quantitatively determined.

The proper scaling axes for studying bicritical behavior are not the H,T axes[3,4] but rather a skew set of axes which are related to the physical axes by

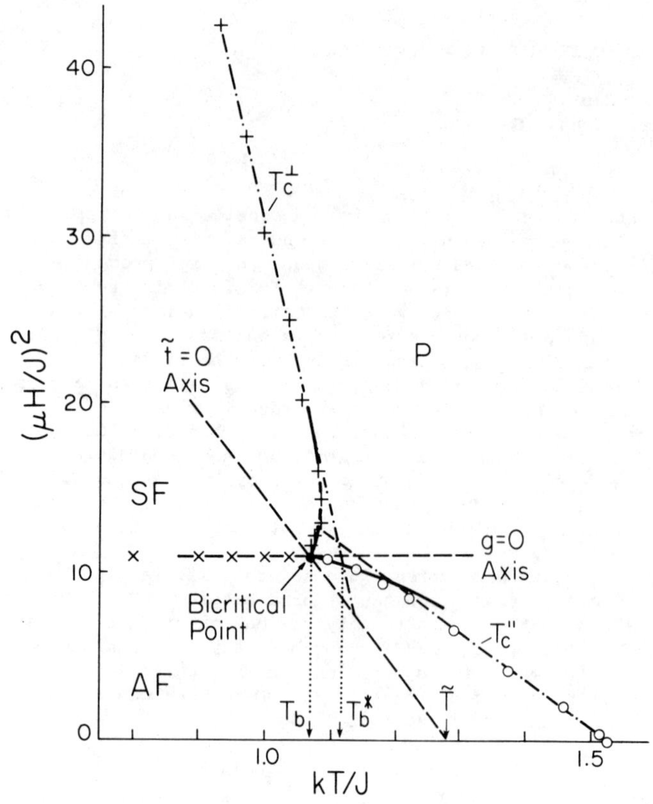


Fig. 1 Phase diagram showing asymptotic phase lines—— fitted to data using Eqn. 3 and the theoretical scaling axes---. The mean field bicritical temperature $T_b$* and phase lines —— • —— are shown for comparison.

$$g = \Delta(H^2) - pt \tag{2a}$$

$$\tilde{t} = t + q\Delta(H^2) \tag{2b}$$

where $\Delta(H^2)=H^2-H_b^2$ and $t=(T/T_b-1)$. The constant p is determined by the slope of the spin flop line at $(T_b,H_b^2)$. The slope q in Eqn. (2b) is such that

$$g/\tilde{t}^{\phi} = +w_{\perp}, -w_{\parallel} \tag{3}$$

where $w_{\perp}$ and $w_{\parallel}$ are constants referring to the SF→P and AF→P phase lines respectively. The crossover exponent $\phi$ has been estimated[1] as $\phi \simeq 1.25$. The ratio $w_{\perp}/w_{\parallel}$ has also been estimated[3,4] at ~2.5. With the accuracy and number of data points available it was not possible to independently fit the critical field curves for all these parameters. Taking the slope $p=(0.4\pm0.4)kOe^2/K$ (from the data), assuming $\phi=1.25$ and using the renormalization group estimate[4] for q, we fitted the asymptotic behavior of the critical field curves near $(T_b,H_b)$ with $w_{\perp}=84\pm10kOe^2$ and $w_{\parallel}=35\pm8kOe^2$. (Note that the "experimental" ratio = 2.4±0.8 easily includes the theoretical estimate.) The scaling axes used as well as the fitted phase lines are shown in Fig. 1.

The behavior of the critical magnetization $M_z^c$ near the bicritical point has also been predicted. The experimental results are shown in Fig. 2. The discontinuity in the critical magnetization as the bicritical point is approached should[1] disappear as $t^{\tilde{\beta}}$ where $\tilde{\beta} \approx 0.85$. Although our data are consistent with this behavior, no stronger statement can be made. Along the 2nd order phase lines $(M_z^c-M_z^b)=a_{\perp,\parallel}t^{\tilde{\beta}}$ where $M_z^b$ is the value at the bicritical point. These data can be fitted by this form with $a_{\perp}>>a_{\parallel}$ as shown in Fig. 2.

In summary, computer simulations of the field induced phase boundaries show good agreement with several theoretical predictions about bicritical behavior. The deviation from mean field phase boundaries is much greater here than in the case of $MnF_2$, (the only real uniaxial bicritical system studied to date).

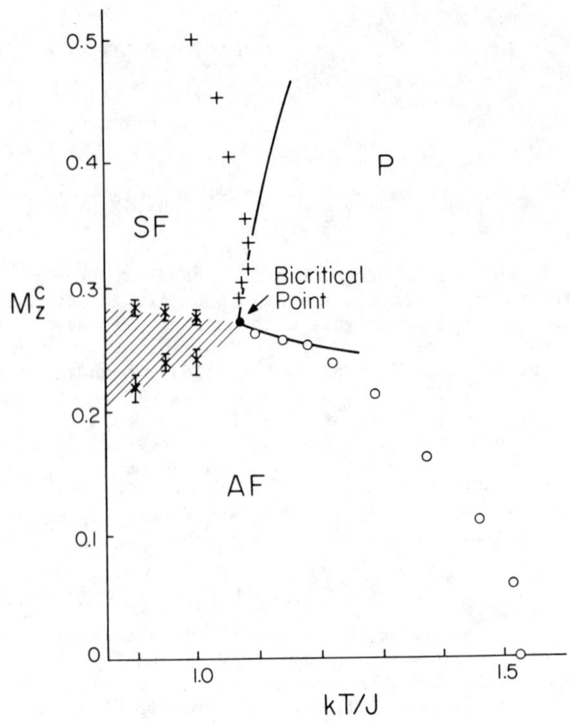


Fig. 2 Temperature dependence of the critical magnetization. The AF-SF phase separation is shown by the ruled lines. The solid lines are fitted to the data near $T_b$ assuming $\Delta M_z^c \alpha t^{\tilde{\beta}}$.

Further study should produce a very useful test of theoretical predictions.

One of us (D.P.L) wishes to thank the Institut für Festkörperforschung der KFA Jülich and the Universität des Saarlandes for their hospitality during a portion of the time this work was carried out.

*Supported in part by the National Science Foundation and the Alexander von Humboldt Stiftung.

REFERENCES

1. M.E. Fisher and D.R. Nelson, Phys. Rev. Lett. 32, 1350(1974).
2. D.R. Nelson, J.M. Kosterlitz, and M.E. Fisher, Phys. Rev. Lett. 33, 813(1974).
3. M.E. Fisher, AIP Conf. Proc. 24, 273(1975).
4. M.E. Fisher, Phys. Rev. Lett. 34, 1634(1975).
5. J.M. Kosterlitz, D.R. Nelson, and M.E. Fisher, (to be published).
6. The only high resolution data near a bicritical point are for uniaxial $MnF_2$ (Ref. 7) and biaxial $GdAlO_3$ (Ref. 8).
7. Y. Shapira, S. Foner, and A. Misetich, Phys. Rev. Lett. 23, 98(1969); Y. Shapira and S. Foner, Phys. Rev. B1, 3083(1970); and A. R. King and H. Rohrer, paper presented at this conference.
8. H. Rohrer, AIP Conf. Proc. 24, 268(1975); H. Rohrer, Phys. Rev. Lett. 34, 1638(1975).
9. For a general review see K. Binder, Advanc. Phys. 23, 917(1974); for applications to tricritical phenomena see D.P. Landau, Phys. Rev. Lett. 28, 449(1972), and B.L. Arora and D.P. Landau, AIP Conf. Proc. 10, 870(1973).
10. K. Binder and D.P. Landau, Physical Review B (in press).
11. H. Müller-Krumbhaar and K. Binder, J. Statist. Phys. 8, 1(1973).

# CANTED-PARAMAGNETIC PHASE BOUNDARY AND BICRITICAL POINT OF $NiCl_26H_2O$*

N.F. Oliveira, Jr., A. Paduan Filho and S.R. Salinas
Instituto de Física da Universidade de São Paulo, C.P. 20516, S. Paulo, S.P., Brasil

ABSTRACT

The canted - paramagnetic phase boundary and the bicritical point of $NiCl_26H_2O$ were investigated down to $T = 0.45$ K by differential magnetization measurements. The canted - paramagnetic boundary obeys a $T^{3/2}$ law below $T \sim 1.3$ K and yields $H_c(0) = 144 \pm 1$ kOe. The boundaries near the bicritical point are well described by the scaling theory with n=2.

## INTRODUCTION

$NiCl_26H_2O$ is a monoclinic salt which becomes antiferromagnetic below 5.34 K. A considerable number of investigations have been carried out of its magnetic and thermal properties.[1-6] In the present paper we show new experimental data concerning its magnetic phase diagram, focusing mainly on two points: the canted-paramagnetic phase boundary and the region near the bicritical point. The canted--paramagnetic boundary had never been determined before and we succeeded in determining it down to 0.45 K by differential magnetization (dM/dH) measurements, in magnetic fields of up to 145 kOe. These measurements, together with the spin-flop fields, provide an independent estimate of the exchange and anisotropy fields which can be compared with other reported values obtained from different techniques. The bicritical point and the spin--flop boundary had been determined previously by a technique similar to ours[5] but these measurements were not precise enough to allow comparison with the theories recently developed.[7] We have accurately determined all three boundaries near the bicritical point, taking special care with the problem of alignment between the easy-axis (from hereon called $a_{\|}$) and the magnetic field.

## EXPERIMENTAL

dM/dH was measured with a bridge similar to that described by Maxwell[8] operating at 155 Hz and with a modulation field of ∿20 Oe peak to peak. The variable temperature cryostat was similar to that described by Oliveira and Quadros[9] except for the fact that in the present measurements the pick-up coils were placed inside the sample bath.

The external magnetic field (H) was provided by two different superconducting magnets: - one, able to produce 145 kOe with a central homogeneity of about 0.1% in 1 cm; - and another able to produce 75 kOe , but with an homogeneity better than 0.01% in 2.5 cm. This last magnet was used for the precise measurements near the bicritical point where very good alignment between $a_{\|}$ and H was needed. To obtain this alignment, the magnet was suspended from the head of the cryostat by three posts. The lenght of these posts could be varied from the outside by means of a fine thread screw, thus allowing adjustments of less than 0.05° in the direction of H .

The samples were single crystals grown from solution and whose parallelepiped shape was elongated in the $a$ direction. Dimensions of a typical sample were 1cm x 0.2cm x 0.2cm. For measurements with $H \| a_{\|}$ the crystals were supported inside the pick-up coils so that the coil axis would lie in the $a$-$c$ crystallographic plane and make an angle of 9°±2° with the $a$-axis. This is roughly the position of $a_{\|}$ as reported by different authors.[3,4] For the measurements near the bicritical point, subsequent adjustments of the external field direction were made to obtain optimal alignment. These adjustments were monitored by the height of the dM/dH peak at the spin-flop transition just below the bicritical point and checked by the frequency dependence of the peak height, a procedure that has been described by Blazey et al.[10] Temperatures were determined by vapor pressure techniques.

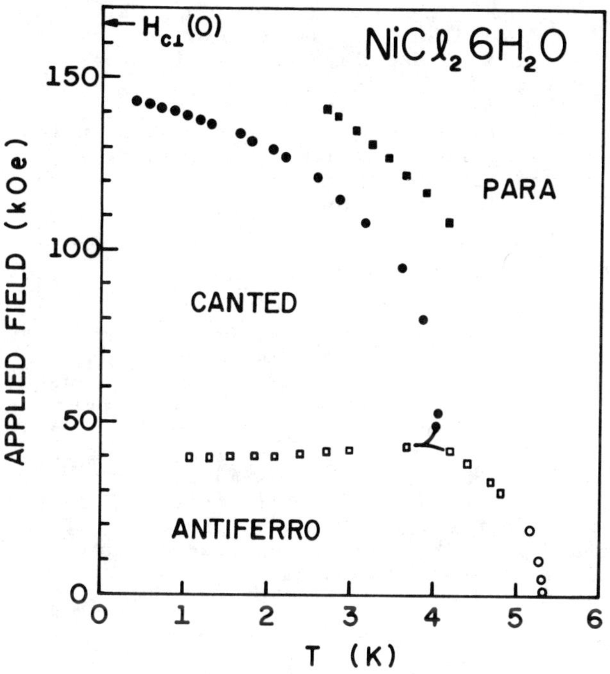


Fig. 1 - Magnetic phase boundaries. Details are explained in the text.

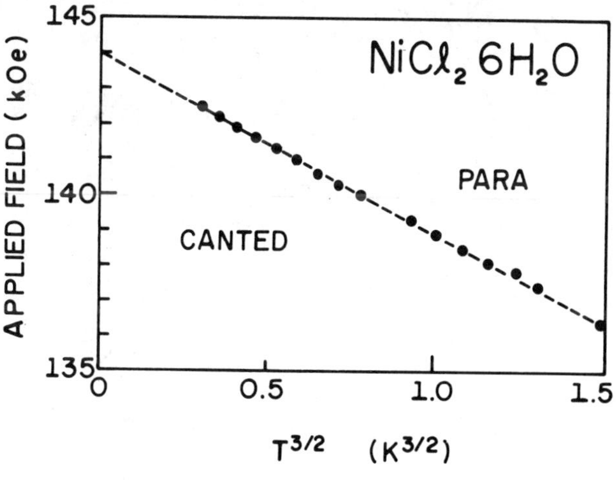


Fig. 2 - Canted-paramagnetic boundary for H parallel to the easy-axis.

## RESULTS AND DISCUSSION

Fig. 1 shows the magnetic phase diagram of $NiCl_2 6H_2O$ including data obtained in the present work (solid dots, solid squares and solid line), data from ref. 2 (open circles) and data from ref. 5 (open squares). The transitions to the paramagnetic phase at high fields are seen for $H \| a_{\|}$ (solid dots) and $H \| b$ (solid squares). For $H \| a_{\|}$, a $T^{3/2}$ law is closely obeyed below $T \sim 1.3$ K as seen in fig. 2. The extrapolation to $T = 0$ yields $H_{c\|}(0) = 144 \pm 1.0$ kOe, where the large error quoted is due to the uncertainty in field calibration and the fact that no attempt was made to correct for the demagnetization factor of the sample. The coefficient of $T^{3/2}$, determined from a least squares fit, is $-5.07 \pm 0.03$ kOe/$K^{3/2}$. Spin wave calculations of this boundary for a monoclinic structure and including one-ion anisotropy are in progress and will be compared to this experimental value.

The value of $H_{c\|}(0)$ together with a suitable extrapolation to T=0 of the spin-flop line can be used to calculate the exchange and anisotropy fields by means of the relations: $H_{c\|}(0) = 2H_E - H_A$ and $H^2_{SF}(0) = H_A \cdot H_{c\|}(0)$. Taking $H_{SF}(0)$ as $39.3 \pm 0.2$ kOe [5] one obtains: $H_E = 77.4 \pm 1.0$ kOe and $H_A = 10.7 \pm 0.4$ kOe. This value of $H_E$ is 10% lower than the values deduced from antiferromagnetic resonance[1] and zero field susceptibility[4] ($H_E = 86 \pm 1$ kOe). More recently however, Kimura[6] estimated the exchange constants by fitting calculations made in a pair model approximation to the susceptibility data of Hamburger and Friedberg.[4] His parameters result in $H_E = 75 \pm 5$ kOe, in reasonable agreement with the present value. The value of $H_A$ compares well with other reported data.[1,4] It predicts a $T = 0$ critical field for $H \perp a_{\|}$ of $H_{c\perp}(0) = 2H_E + H_A = 165$ kOe. This value is marked by an arrow in fig. 1 and is seen to be quite consistent with the position of the boundary measured with $H \| b$.

The data near the bicritical point is seen in fig. 3. The solid dots mark the position of the peaks observed in dM/dH by sweeping H at constant T. The open dots were obtained by varying T at constant H and in this case the transition was identified with a kink in dM/dH. The misalignment between H and $a_{\|}$, estimated as described in ref. 10, was less than $0.1^{o}$. The diameter of the dots in fig. 3 is compatible with the resolution of our apparatus in temperature ($\sim 0.003$ K) and field ($\sim 30$ Oe). A computer fit of the theory of spin-flop bicritical phase boundaries[7] was made to the experimental points. The parameters for the optimal scaling axes were obtained from the present spin-flop data, and the antiferro - paramagnetic boundary from Johnson and Reese.[2] The solid line in fig. 3 corresponds to the best fit obtained for $n = 2$ (the dimension of the order parameter) and the dashed line to that for n=3. A reasonably good fit was obtained for n=2 but a good fit was not possible for n=3. The dashed line is seen to deviate sistematically from the experimental points (although it represents a best fit) because of the imposed relation $(\omega_{\perp}/\omega_{\|}) = 2.51$ ($\omega_{\perp}$ and $\omega_{\|}$ as defined in ref. 7) characteristic of n=3. For n=2, the final values of the adjusted parameters were: $T_b = 3.940$ K and $\omega_{\perp} = \omega_{\|} = 9962 (kOe)^2$. This result will be the subject of further investigation and discussion, and presented in forthcoming publications.

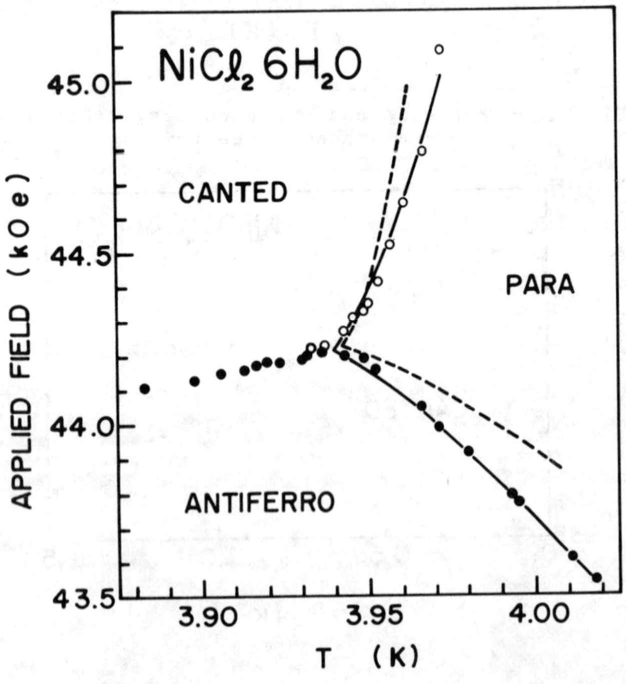


Fig. 3 - Phase boundaries near the bicritical point. Solid and dashed lines represent theoretical fits. Details are explained in the text.

## REFERENCES


* Work partially supported by FINEP and CNPq.



1. M. Date and M. Motokawa, J. Phys. Soc. Japan 22, 165 (1967).
2. W.L. Johnson and W. Reese, Phys. Rev. B 2, 1355 (1970).
3. A. Nakanishi, K. Okuda and M. Date, J. Phys. Soc. Japan 32, 282 (1972).
4. A.I. Hamburger and S.A. Friedberg, Physica 69, 67 (1973).
5. C.C. Becerra and A. Paduan Filho, Phys. Letters A 44, 13 (1973).
6. I. Kimura, J.Phys.Soc.Japan 37, 946 (1974).
7. M.E. Fisher, Phys. Rev. Letters 34, 1634 (1975).
8. E. Maxwell, Rev.Sci.Instrum. 36, 553 (1965).
9. N.F. Oliveira Jr. and C. Quadros, J.Phys. E 2, 967 (1969).
10. K.W. Blazey, H. Rohrer and R. Webster, Phys. Rev. B 4, 2287 (1971).

# CRITICAL BEHAVIOUR OF MAGNETIC SYSTEMS WITH HELICAL STATE

M. Droz
Department of Physics, Temple University
Philadelphia, Pennsylvania 19122

and

Laboratory of Atomic and Solid State Physics
Cornell University, Ithaca, New York 14853

and

M. D. Coutinho-Filho
Departmento de Fisica, Universidade Federal de Pernambuco, 50000 Recife, Brasil and Laboratory of Atomic and Solid State Physiçs, Cornell University
Ithaca, New York 14853

ABSTRACT

We consider a metamagnet in which the nearest neighbour $J_1$ and next nearest neighbour interlayers interaction $J_2$, compete in such a way that the system reaches a helical-like ordering. A renormalization group analysis of this model (to all order in $\varepsilon = d - 4$) shows that the critical exponents are Heisenberg-like, with n = 2, 4 for the para ↔ sinusoidal and para ↔ planar helical transitions respectively. These results differ from the ones obtained when the helical ordering is due to a Dzyaloshinskii-Moriya interaction. Boundaries of the phase diagram around the bicritical points are studied by scaling and renormalization group analysis.

## INTRODUCTION

Many materials undergo a second-order transition from a paramagnetic phase to a phase having a magnetic superstructure[1]. One of the most widely studied classes is that of the rare earth metals. The ordered phase may, for example, exhibit a sinusoidal (S), planar helical (H), or conical (C) ordering, for which the pitch parameter can be large compared to the dimension of the chemical unit cell.

Various physical mechanisms can lead to a superstructure. For example, an indirect isotropic exchange interaction between the localized moments, supplemented by a Dzyaloshinskii-Moriya interaction[2], leads to a superstructure. Due to the fact that the interactions are not parity-invariant the superstructure will be "oriented" (left-handed or right handed). This problem has been studied recently by Liu[3] within the framework of the renormalization group technique and the ε expansion. He found that such a system belongs to the same universality class as the simple ferromagnet, namely that the critical exponents are Ising-like (n = 1), xy-like (n = 2) and Heisenberg like (n = 3) for the sinusoidal, planar-helical and conical ordering, respectively. Symmetry arguments[2] show that the Dzyaloshinskii-Moriya interaction cannot be present in crystals for which the paramagnetic phase has a high symmetry.

Superstructures can also result from the competition between ferromagnetic and antiferromagnetic interactions. Such a competition can be present in crystal with high symmetry where the Dzyaloshinskii-Moriya interaction is forbidden. Such is the case, for example, in a simple tetragonal metamagnet composed of many (d-1) - dimensional layers with nearest neighbor ferromagnetic-like interaction $J_0$. The layers are coupled together via a nearest neighbor interaction $J_1$ and an antiferromagnetic next-nearest neighbor interaction ($J_2 < 0$). As shown by the phase diagram (Fig. 1), the helical state appears for $J_1 < 4|J_2|$, regardless of the sign of $J_1$. Note that such a superstructure is different from the preceding one in the sense that, since the interaction is parity invariant, the superstructures will grow right- or left-handed at random.

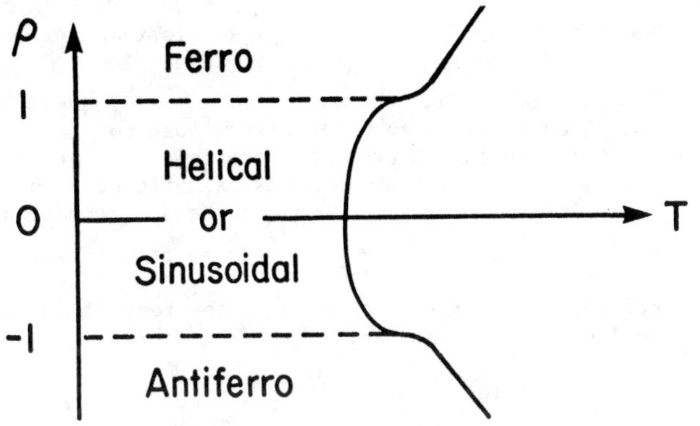


**Fig.1: Shematic phase diagram.**

The phase diagram exhibits two bicritical points where two λ-lines (related to different kinds of ordering) meet together with a first order line.

While similar model have been studied thoroughly in mean field approximation[4], little has been done beyond this approximation. The purpose of this note is to report results of a renormalization group analysis[5] of the critical behavior of the metamagnet (with single ion field anisotropy D), when $J_1$ and $J_2$ compete in such a way that a sinusoidal or planar helical state is reached. For $J_1$ and $J_2$ such that the ordered state is ferromagnetic or antiferromagnetic, the analysis of Nelson and Fisher[6] has shown that the exponents are Ising-like or xy-like for uniaxial or two-component planar spin, respectively. For the helical state we find that the critical exponents belong to a different universality class. Namely the critical behavior is described by a 2n Heisenberg model, and it leads to Heisenberg-like n =2,4 exponents for the transitions para ↔ sinusoidal and para ↔ planar helical respectively.

## MODEL AND RESULTS

The model is the following. We consider a d-dimensional metamagnet. In each (d-1) dimensional layer the spins are located on the sites of a square lattice. Nearest neighbor sites are separated by a distance $a_{/\!/}$. The spins in the layers interact through a nearest neighbor ferromagnetic interaction $J_0$. The layers, separated by a distance $a_\perp$, are coupled through nearest neighbor interaction $J_1$ and an antiferromagnetic next nearest neighbor interaction $J_2 < 0$. A single ion field anistropy D favors the alignment of the spin $\vec{S}(\vec{R}) = \{S^\alpha(\vec{R})\ \alpha = 1..\ n\}$ in the ⊥-direction (D > 0) or in the layers (D < 0). The hamiltonian reads:

$$H = -\frac{1}{2}\sum_{\vec{R}\vec{R}'} J(\vec{R} - \vec{R}')\vec{S}(\vec{R}).\vec{S}(R') - \frac{D}{2}\sum_{\vec{R}} S_\perp^2(\vec{R}) \qquad (1)$$

Going to a continuous spin system by adding for each spin a weight factor with O(n) symmetry[7], we have in the Fourier space.

$$H = -\frac{1}{2}\sum_{\vec{q}}\sum_{\alpha} u_2^{\alpha}(\vec{q})S^{\alpha}(\vec{q})S^{\alpha}(-\vec{q})$$

$$- u \sum_{\vec{q}_1 .. \vec{q}_4} S^{\alpha}(\vec{q}_1)S^{\alpha}(\vec{q}_2)S^{\beta}(\vec{q}_3)S^{\beta}(\vec{q}_4)\delta^{(d)}(\vec{q}_1+\vec{q}_2+\vec{q}_3+\vec{q}_4) \qquad (2)$$

where $u_2^{\alpha}(\vec{q}_1) = k_B T - D\delta_{\alpha\perp} - J(\vec{q})$ (3)

and $\frac{1}{2} J(\vec{q}) = \sum_{j=1}^{d-1} J_0 \cos(q_j a_{/\!/}) + J_1 \cos(q_\perp a_\perp)$

$$+ J_2 \cos(2q_\perp a_\perp) \qquad (4)$$

The particular feature of the model is that for $-1 \leq \rho = J_1/4|J_2| \leq 1$, $J(\vec{q})$ has a maximum for $\pm\vec{Q}$, where $\vec{Q} = (0\ 0, .. a_\perp^{-1} \text{Arc} \cos \rho)$. Thus for the helical state the fluctuations with wave vector close to $\pm\vec{Q}$ will play a crucial role. In order to take this fact properly into account we introduce a new magnetic Brillouin zone by shrinking the original one[11] in the $\perp$ direction from $\left\{-\frac{\pi}{a_\perp}; \frac{\pi}{a_\perp}\right\}$ to $\left\{-\frac{Q}{2}; \frac{Q}{2}\right\}$. To go from the extended Brillouin zone representation to the reduced one, we define the new spin variables $S^{\alpha\ell}(\vec{q})$ by:

$$S^{\alpha}(\vec{q}) = \sum_{\ell} S^{\alpha\ell}(\vec{q})\Delta_{\ell}(\vec{q}) \qquad (5)$$

$$\text{where } \Delta_{\ell}(\vec{q}) = \begin{cases} 1 & \text{if } (\ell-1)Q/_2 \leq |q_\perp| \leq \ell\, Q/_2 \\ 0 & \text{otherwise} \end{cases} \qquad (6)$$

The renormalization group analyes proceeds as follows:[5]

i) One expresses the Hamiltonian in terms of the new variables $S^{\alpha\ell}$ by the folding process sketched on Fig. 2.

ii) One rescales the spin variables in such a way that the non-critical modes can be integrated out.

Thus after a sufficient number of iterations the initial Hamiltonians flows to a new one corresponding to a 2n Heisenberg model, namely

$$H = -\frac{1}{2}\sum_{\vec{q}}{}'\sum_{\alpha}\sum_{\ell=1,2} u_2^{\alpha\ell}(-\vec{q})S^{\alpha\ell}(\vec{q})S^{\alpha\ell}(-\vec{q})$$

$$- \sum_{\vec{q}_1 \; \vec{q}_4}{}'\sum_{\alpha\beta}\sum_{\ell m=1,2} \delta^{(d)}(\vec{q}_1+\vec{q}_2+\vec{q}_3+\vec{q}_4)u.$$

$$S^{\alpha\ell}(\vec{q}_1)S^{\alpha\ell}(\vec{q}_2)S^{\beta m}(\vec{q}_3)S^{\beta m}(\vec{q}_4). \qquad (7)$$

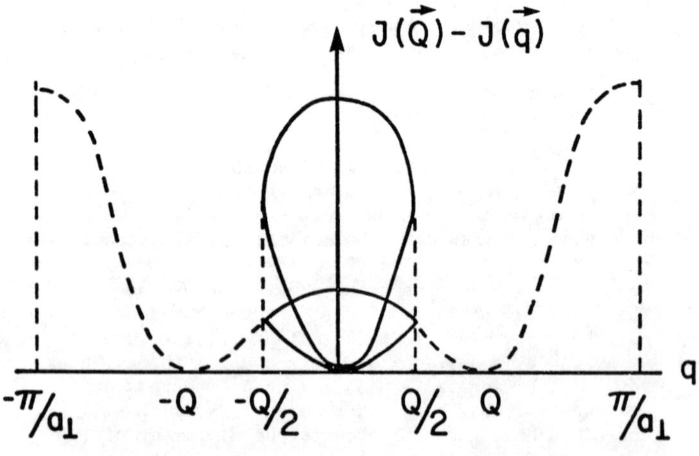


**Fig. 2: Illustration of the folding process.**

$u_2^{\alpha\ell}(\vec{q})$ is obtained from $u_2^{\alpha}(\vec{q})$ by appropriate momentum shifts, and $\sum'$ means summation over the new Brillouin zone.

The above result can be simply understood from a physical point of view. In the helical state the ordering is characterized by an amplitude and a phase and not just by an amplitude as it is the case for the simple (anti-or) ferromagnetic case. Note that the above calculation has been done using an O(n) symmetry weight factor. If one takes into account the particular symmetry of a crystal by adding perturbations with lower symmetry (cubic, hexagonal, tetragonal,..) the stable fixed point could be a new one.[9] The behavior of the system around the bicritical points is particularly interesting, as the usual $\varepsilon = 4 - d$ expansion breaks down[10] at those bicritical points. Nevertheless, a scaling analysis can be done[5] around the bicritical points, the reduced temperature $t = \frac{T-T_B}{T_B}$ is related to the scaling field $g = 1 - |\rho|$, through the relation: $t \sim g^{1/\phi_B}$ where the crossover exponent $\phi_B$ is equal to the susceptibility exponent $\gamma_B$. For d = 3, $\gamma_B$ is greated than one,[10] Consequently the critical second order lines and the first order line will meet tangentially at the bicritical points (Fig. I).

The behavior of such systems in an external magnetic field[10] where tricritical behavior is expected is also quite interesting.

Experimental studies of these predictions would be a valuable test of the theory. Different values of $\rho$ can be achieved in suitable materials by applying pressure or in alloys by varying the composition.

<u>Note</u>: During the preparation of this note we learned that same problem has been studied independently by T. Garel and P. Pfeuty.[8]

## REFERENCES


(1) D. E. Cox IEEE Trans. Mag., MAG-8, 161 (1972).

(2) I. Dzyaloshinskii, J. Phys. Chem. Solids 4, 241 (1958); T. Moriya, Phys. Rev. 120, 91 (1960).

(3) L. L. Liu, Phys. Rev. Lett. 31, 459 (1973).

(4) See T. Nagamiya in Solid State Physics, ed. by H. Ehrenierch, F. Seitz and D. Turnbull (Academic Press, New York, 1967) Vol. 20, p. 305. and references therein.

(5) See M. D. Coutinho-Filho and M. Droz (to be published) for details of the calculations.

(6) D. Nelson and M. E. Fisher, Phys. Ref. B11, 1030 (1975).

(7) See e.g. K. G. Wilson and T. Kogut, Phys. Rep. 126, 75 (1975).

(8) T. Garel and P. Pfeuty, to be published.

(9) D. Mukamel, Phys. Rev. Lett. 34, 481 (1975).

(10) M. Droz, to be published.

(11) For any value of Q, it exists a $\tilde{Q}$ (such that $\tilde{Q}a_\perp/2\pi = p/q$ with q and q integers) arbitrarily close to Q. If $p \neq 1$, one folds the modes of p consecutive Brillouin zones ($[-p\pi/a_\perp\ ;\ p\pi/a_\perp]$) in the reduced one $[-\tilde{Q}/2; \tilde{Q}/2]$.

# DIFFERENTIAL GENERATORS FOR MAGNETIC HAMILTONIANS WITH PAIRED SPIN-MOMENTA AND OTHER SPIN GROUPINGS AND FOR APPROXIMATE EQUATIONS OF STATE OF SYSTEMS WITH COMPETING FIXED POINTS

J.F. Nicoll, T.S. Chang and H.E. Stanley
Massachusetts Institute of Technology, Cambridge, Massachusetts 02139

## ABSTRACT

In treating critical phenomena of system Hamiltonians simulating complex realistic materials, the full structure of the renormalization group is usually unnecessarily complicated and cumbersome to work with. We point out that if the special features of the individual Hamiltonians are incorporated into the renormalization group equations from the beginning, considerable simplification can be achieved in the working forms of the generators. We give two simple examples. The first is an exact generator for Hamiltonians with paired spin-momenta. This generator admits an exact nontrivial fixed point which in the isotropic case is related to the spherical model; it also admits other perturbative higher order nontrivial fixed points. The second is an approximate generator for Hamiltonians with arbitrary spin-momentum groupings. Both generators are expressible as nonlinear second order partial differential equations. We further point out that generators for special purposes can be derived with or without specification of the Hamiltonian. As an example we develop a special differential generator which automatically yeilds the equations of state for complex systems with competing fixed points.

## INTRODUCTION: PHILOSOPHY OF THIS APPROACH

In applying the renormalization group to critical phenomena, one approach is to utilize the full and exact machinery at the beginning and only later to discard those elements which are not needed for the system under consideration or which may be neglected in the approximation employed. The possible complexity of the full renormalized group apparatus is enormous whether expressed in functional differential equations such as those of Refs. 1-2 or the field-theoretic, Callan-Symanzik method described, e.g., in Ref. 3.

In the former case, the Hamiltonians which can be described and which must be considered in perturbational analysis include terms which have no immediate connection to the system of interest. For example, a Hamiltonian intended to describe a material with a uniform magnetization develops "momentum dependence" when "renormalized" by the differential equation. Thus, although we are interested only in spatially uniform quantities, we are forced to consider spatially varying ones.

In the latter case, we must employ the full panoply of field theoretic perturbation theory to calculate simple functions of the parameters of the "renormalized" Hamiltonian. While this method avoids the introduction of intermediate spatially varying quantities (and thus can be carried to higher order in perturbation theory more easily) it appears that the field theory must be constructed anew for each system. That is, while the differential equation approach applies definite operators which are the same for every system, the field theoretic approach requires the calculation of whatever diagrams are found to be appropriate. There is no operational definition of the results of these diagramatic expansions.

A third approach is the construction of specialized renormalization group generators. This allows us to define the renormalization procedure as operations on the Hamiltonian and still treat only the desired spatially uniform (or otherwise specified) quantities. Since the resultant renormalization group generators are far simpler than their full counterparts, we must restrict ourselves in some fashion. We have considered three such restrictions: (i) specification of the type of Hamiltonian to be considered; (ii) approximate generators; and (iii) generators for special purposes.

## RESTRICTION ON THE HAMILTONIAN

From the full renormalization group equation we have derived an exact generator suitable for systems with Hamiltonians which are functionals of momentum paired spins. That is, we consider $H(z^{\alpha\beta})$ where

$$z^{\alpha\beta}=\int d\vec{x}\, S^{\alpha}(\vec{x})S^{\beta}(\vec{x}) - CONSTANT. \tag{1}$$

Such a restriction is "closed under renormalization". The generator for such systems

$$\frac{\partial H}{\partial \ell}=dH+(\tilde{\sigma}-d)z^{\alpha\beta}\frac{\partial H}{\partial z^{\alpha\beta}}+\frac{\partial^2 H}{\partial z^{\alpha\beta}\partial z^{\alpha\beta}}-\frac{\partial H}{\partial z^{\alpha\beta}}\frac{\partial H}{\partial z^{\alpha\beta}}. \tag{2}$$

is identical whether we begin with the full Wilson[1] or Wegner-Houghton[2] equations. Here $\ell$ is the renormalization parameter and $\tilde{\sigma}$ is the "propagator exponent" (cf. Ref. 4). It is extremely interesting that this exact simple generator is identical for the two very different generators of Refs. 1-2 since this identity shows the equivalence of the full generators in this limit without recourse to perturbation theory.

This class of systems forms the classical analog of BCS pairing in superconductivity and in anisotropic superfluidity where complex tensor order parameters are obtained through momentum pairing. It offers even more interest through the observation that for even order critical points, the isotropic version of (2) is the exact generator for the generalized spherical limit (number of spin components $n\to\infty$). For the ordinary critical case, we can obtain an exact fixed point of

$$H^{*}=\frac{2\tilde{\sigma}-d}{4}\left(z^2-\frac{d}{4}\right), \tag{3a}$$

where we use the isotropic variable $z=z^{\alpha\alpha}$. Linearizing about this fixed point we obtain the exact spectrum of eigenvalues for eigenfunctions (the Hermite polynomial $H_p$) associated with $z^p$,

$$\lambda_p=d-\tilde{\sigma}p. \tag{3b}$$

For $\tilde{\sigma}=2$, this reduces to the usual spectrum of the spherical model.

## APPROXIMATE GENERATORS

We can admit a larger class of Hamiltonians and still retain simplicity in the generator if we need only an approximate calculation. One interesting class of Hamiltonians is that defined by

$$H=H(\{\langle S^{\alpha_1}\cdots\cdots S^{\alpha_m}\rangle\}), \tag{4a}$$

where

$$\langle S^{\alpha_1}\cdots\cdots S^{\alpha_m}\rangle=\int d\vec{x}\, S^{\alpha_1}(\vec{x})\cdots\cdots S^{\alpha_m}(\vec{x}), \tag{4b}$$

and $(\alpha_1\ldots\alpha_m)$ runs over all possible sets of spin component indices. This corresponds to the admission of any number of delta functions in the momentum space representation of the Hamiltonian. Each delta function defines a grouping of some subset of the spins. Approximate generators dealing with such Hamiltonians can be derived from both the Wilson and Wegner-Houghton generators. In the latter case, we can express it as

$$\frac{\partial H}{\partial \ell}=dH+\left\langle\frac{\tilde{\sigma}-d}{2}\vec{S}\cdot\vec{\nabla}H\right\rangle+\left\langle \mathrm{tr}\ln\left[\delta_{\alpha\beta}+\frac{\partial^2 H}{\partial S^{\alpha}\partial S^{\beta}}\right]\right\rangle. \tag{5}$$

Here we define $\partial/\partial S^{\beta}\langle S^{\alpha_1}\cdots S^{\alpha_m}\rangle\equiv\partial/\partial S^{\beta}(S^{\alpha_1}\cdots S^{\alpha_m})$, $\partial/\partial S^{\beta}\langle\cdots\rangle_1\langle\cdots\rangle_2\equiv\langle\cdots\rangle_1\partial/\partial S^{\beta}\langle\cdots\rangle_2+\langle\cdots\rangle_2\partial/\partial S^{\beta}\langle\cdots\rangle_1$, *etc.*, and $\langle 1\rangle=\Omega$, where $\Omega$ is the volume of the system.

The generator (5) is applicable to the study of the simple compressible ferromagnet model of Sak[5]:

$$H=\frac{r}{2}\langle\vec{s}^2\rangle+\frac{u}{4}\langle\vec{s}^2\vec{s}^2\rangle+\frac{v}{4\Omega}\langle\vec{s}^2\rangle\langle\vec{s}^2\rangle. \tag{6}$$

We have used (5) to solve the complete nonlinear renormalization trajectories for the parameters (r,u,v) of the Hamiltonian (6) (see the equation of state given below and Ref. 4).

The generator (5) and the similar generator which can be derived from the Wilson generator both have (2) as their common limit in the spin pairing limit or for the generalized spherical limit ($n\to\infty$) for even order critical points. This demonstrates that the approximate generators become exact in these limits.

GENERATOR FOR THE EQUATION OF STATE

The usual construction of renormalization group generators has as its purpose the location of fixed points and computation of the critical point exponents which characterize each fixed point. For a fixed point to exist, specific scaling transformations are made on the Hamiltonian and on the spin vectors themselves. This means that the spins in the renormalized Hamiltonian are not, strictly speaking, equal to the spins of a real physical system. If we wish to deal with physical free energies and magnetizations, we must revise the generator by undoing the scaling of the spins and Hamiltonian. Having done so, it is straight-forward to show that the Helmholtz free energy is the limit of the unscaled but renormalized Hamiltonian as the renormalization parameter $\ell\to\infty$. This is so because as $\ell$ becomes infinite, all the spin fluctuations have been incorporated into the Hamiltonian, and we have, in fact, evaluated the partition function. Therefore, we can consider the renormalization group equations as a generator for the free energy and hence the equation of state. Using the spin grouping approximate generator (5), for example, we have

$$\frac{\partial A(M,\ell)}{\partial \ell}=\exp(-d\ell)\langle \mathrm{tr}\ln[\delta_{\alpha\beta}+\exp(\delta\ell)\frac{\partial^2 A}{\partial M^\alpha \partial M^\beta}]\rangle, \tag{7a}$$

where the physical free energy is given by

$$A(M)=\lim_{\ell\to\infty} A(M,\ell). \tag{7b}$$

The equation of state is then determined by

$$h_\alpha=\partial A/\partial M^\alpha. \tag{7c}$$

Although (7) is equivalent to (5) (differing only by scale changes), it cannot be used to find fixed points. Each fixed point is simply a critical point of A(M) at which it has some value; the equation of state simply reduces to 0=0 at such a point. The proper procedure is to use a generator such as (5) to study the renormalization properties of the Hamiltonian parameters. This information can then be translated into the equivalent information about the Helmholtz free energy and used to solve (7). As an example, we calculate using (5) the renormalization trajectories $r(\ell)$, $u(\ell)$, and $v(\ell)$ for the Sak Hamiltonian and obtain a first or "zero-loop" approximation to $A(M,\ell)$

$$A_0(M,\ell)=\frac{r(\ell)}{2}\exp(-2\ell)M^2+\frac{u(\ell)+v(\ell)}{4}\exp(-\epsilon\ell)M^4. \tag{8}$$

where we have set $\delta=2$ and $\varepsilon=4-d$. This gives for the free energy

$$A_0(M)=\frac{r_\infty}{2}M^2+\frac{u_\infty+v_\infty}{4}M^4, \tag{9a}$$

where we define

$$(r_\infty,u_\infty,v_\infty)=\lim_{\ell\to\infty}(r(\ell)e^{-2\ell},u(\ell)e^{-\epsilon\ell},v(\ell)e^{-\epsilon\ell}). \tag{9b}$$

The limits $r_\infty$, $u_\infty$, and $v_\infty$ are nonlinear scaling fields which are functions of the initial Hamiltonian parameters and which can be calculated from the nonlinear solutions. Equation (8), however, does not give a complete solution to the generator equation (7), even to the order of accuracy of the generator itself.

Using (8) as a first guess we can obtain a refinement or "one-loop" approximation. The simplest result to display is the equation of state which even in the one-loop approximation depends only on $r_\infty$, $u_\infty$, and $v_\infty$:

$$\begin{aligned}h={}&r_\infty M+(u_\infty+v_\infty)M^3+Mr_\infty|r_\infty|^{-\epsilon/2}\\&\cdot\Big\{(n-1)(u_\infty+v_\infty)\Big[1+\frac{M^2(u_\infty+v_\infty)}{r_\infty}\ln\Big|1+\frac{M^2(u_\infty+v_\infty)}{r_\infty}\Big|-\frac{M^2(u_\infty+v_\infty)}{r_\infty}\Big]\\&+(3u_\infty+v_\infty)\Big[1+\frac{M^2(3u_\infty+v_\infty)}{r_\infty}\ln\Big|1+\frac{M^2(3u_\infty+v_\infty)}{r_\infty}\Big|-\frac{M^2(3u_\infty+v_\infty)}{r_\infty}\Big]\Big\}\end{aligned} \tag{10}$$

The detailed derivation of (6) and the expression for $r_\infty$, $u_\infty$, and $v_\infty$ are given in Ref. 6.

Equation (10) incorporates all the nonlinear crossover effects of the Sak model. It is interesting to note that the form of (10) can be derived from the generator (7) even if the functional character of the nonlinear scaling fields has not been determined.

SUMMARY

We have introduced three generators for the renormalization group which are restricted in some way. We suggest that they may represent the lowest levels of three hierarchies of limited generators each of which is capable of describing more general systems than the previous one. Second. we may also imagine a sequence of approximate generators which have increasing degrees of accuracy. The improvements in the generators might be achieved by increasing the nonlinearity of the equations. Finally, as exemplified by the generator for the free energy and the equation of state, we may also recast generators in different forms in order to render them useful for a particular aspect of the renormalization group study of a particular physical system.

Thus, the generators described here are not only useful for the solution and study of an extremely broad and interesting class of systems, but they also should be regarded as paradigms for future generators which have even more extensive applicability.

REFERENCES

*Work supported by the National Science Foundation and the Air Force Office of Scientific Research.

1. Wilson, K.G. and Kogut, J. Phys. Repts. 12, 76-199 (1974).

2. Wegner, F.J. and Houghton, A. Phys. Rev. A8, 401-412 (1972).

3. Brézin, E., Le Guillou, J.C., and Zinn-Justin, J., in Phase Transitions and Critical Phenomena Vol 6. Ed. C. Domb and M.S. Green (Academic Press, London, in press).

4. Nicoll, J.F., Chang, T.S. and Stanley, H.E., Phys. Rev. A (in press).

5. Sak, J., Phys. Rev. B10, 3957-3960 (1974).

6. Nicoll, J.F., Chang, T.S., and Stanley, H.E. (M.I.T. preprint.)

# LOGARITHMIC FREQUENCY DEPENDENCE OF THE PHASE IN THE A.C. SUSCEPTIBILITY OF IRON WHISKERS WITHIN MILLIDEGREES OF THE CRITICAL TEMPERATURE

B. Heinrich and A.S. Arrott,
Simon Fraser University, Burnaby, British Columbia,
Canada V5A 1S6

ABSTRACT

The frequency dependence of the a.c. susceptibility of an iron whisker is analyzed in the region of nonlinear behavior within 20 millidegrees of $T_c$. A phenomenological equation of the form $\phi = \varphi_f \cdot f_o \ell n (1 + f/f_o)$ fits the data for the frequency dependence of the phase. The parameter $\varphi_f$ reaches a peak and $f_o$ becomes a minimum near or at $T_c$. The characteristic frequency $f_o$ may be related to critical slowing down reaching the millisecond region.

## INTRODUCTION

Iron whiskers exhibit rapid changes in the out phase component of the a.c. susceptibility in the immediate vicinity of the magnetic critical point $T_c$. Certain measures of the loss decrease by more than an order of magnitude on going away from $T_c$ whether that departure is made by increasing or decreasing the temperature by 10's of millidegrees, by applying d.c. fields of tenths of a gauss or by increasing the frequency of measurement into the range of a few kHz. These effects have been shown in a series of previous reports on this continuing investigation[1]. This report focuses upon an attempt to become more quantative concerning the behavior very close to $T_c$ as a function of frequency in the range where the in phase component is not affected by the out phase component, that is the square of the phase is much, much less than unity. For the whisker reported on here this restricted the measurements to 10kHz and less to maintain $\varphi^2 < 0.006$. Lack of sensitivity set the lower limit at 0.5kHz. The functional form of the results could be fit with a two parameter expression from which one can abstract the behavior in the limit of low frequencies and deduce a characteristic frequency. By some coincidence the characteristic frequencies fall right in the range of frequency measurement for the range of temperature where the effects are most interesting. It may be conjectured that these frequencies are associated in some way with critical slowing down. But no theoretical contact is established as yet. The phenomenological expression used to fit the data for the phase as a function of frequency is

$$\varphi = \varphi_f f_o \, \ell n \, (1 + f/f_o) \qquad (1)$$

In the limit of $f_o >> f$ this is just a proportionality between $\varphi$ and $f$ with $\varphi = \varphi_f f$, but near $T_c$ the values of $f_o$ fall below 1kHz and the argument of the $\ell n$ goes from 2 to 20 over the range of frequencies used. This range is sufficient and the fit such that it appears that one might take the phenomenology seriously.

## MEASUREMENTS

The whisker is maintained at a temperature which remains constant to 1 mdeg. Gradients in temperature along the whisker have not been detectable. The driving coil produces uniform magnetic fields (in the absence of the whisker) from 0.01 to 0.1 oersted, typically .05 Oe. The pickup coil surrounds the sample (and driving coil) in such a way that the coupling is closer to magnetometric than ballistic. Signals are normalized to the low frequency in phase signal below $T_c$ where the signal is completely determined by geometry, that is where $\chi' = 1/4\pi\bar{D}$, where the bar on the demagnetizing factor indicates the average appropriate for the detector coil geometry.

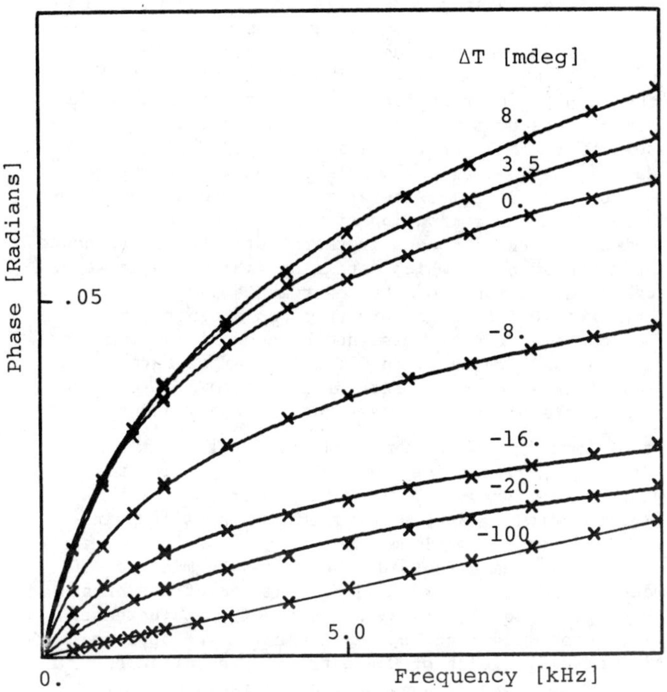


Fig. 1 Frequency dependence of the phase for several temperatures near $T_c$. The lowest curve is in the temperature range of the curling pattern. The curves through the data points are fits of Eq. 1.

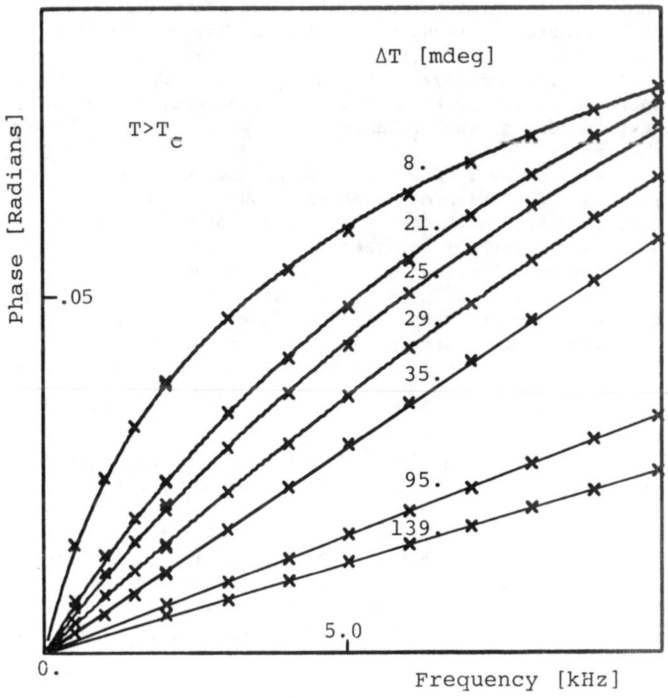


Fig. 2 Frequency dependence of the phase for several temperatures above $T_c$. The curves through the data points are fits of Eq. 1.

## RESULTS

The low frequency in phase component of the a.c. susceptibility remains constant with increasing temperature to within millidegrees of $T_c$. With increasing temperature from below $T_c$ the out phase component shows a region of low loss which has been previously ascribed to a magnetization curling pattern[2]. The low loss region ends abruptly $\sim 20$ mdeg below $T_c$. From there and from $\sim 20$ mdeg above $T_c$ the phase deviates more and more from linear dependence upon frequency as one approaches $T_c$. The phase at 10kHz reaches a maximum somewhat above $T_c$. Figure 1 shows the results for temperatures up to and including the temperature of that maximum and Figure 2 shows the results for that temperature and above. The fit to the data is as good as could be expected considering the constraints of temperature stability and the problems of absolute determination of phase. The pair of points at 2kHz for each isotherm reflects the sequence of measurement in which data from 2kHz to 10kHz were taken on one range of the instruments and then the data from 0.5 to 2kHz were taken on another range. The temperatures are assigned by comparison with runs at constant frequency in which the temperature range was swept at constant rate of 10 mdeg/min. The zero of temperature has been assigned to the isotherm of minimum $f_o$. While uncertainties as to what to take for $T_c$ remain they are no larger than $\pm$ 5 mdeg whether one looks at peak losses, extrapolates $M_s$ or $1/\chi_i$ to zero, or adopts a criteria such as the minimum in $f_o$ or the maximum in $\varphi_f$.

The isotherms outside the range -20 mdeg to +30 mdeg are quite linear. In the language of the phenomenology of Eq. 1, it may be stated $f_o$ has increased well beyond the range of measurement. The application of d.c. fields of 0.4 oersteds are sufficient also to increase $f_o$ well beyond the range of detection. An applied field of 0.4 oersted produces an internal field of 0.1 oersted in this range of temperature.

## TEMPERATURE DEPENDENCE OF THE PARAMETERS

From the fits of the phenomenological expression one can obtain values for the initial derivative of phase with respect to frequency. Rather than $\varphi_f$, the parameter plotted in Fig. 3 is $\beta \equiv \varphi_f/\chi' 4\pi\bar{D}$ inasmuch as the quantity has a direct interpretation outside the anomalous region about $T_c$. For eddy current losses $\beta$ should be temperature and frequency independent as long as the mechanism of magnetization remains the same.

Well below $T_c$ where the Landau domain structure is found, $\beta$ depends on temperature only through the conductivity. Within less than 1 deg of $T_c$ the Landau structure gives way to what has been termed the curling configuration. The value of $\beta$ for the Landau structure close to $T_c$ is designated by $\beta_L$, the value in the curling pattern region by $\beta_C$ and the value at higher temperatures by $\beta_H$. Part of the present program of research is to account for these three values of $\beta$ in terms of the magnetization process in each of the regions[3]. A general rule is that the more inhomogeneous the magnetization process the higher is the value of $\beta$. For this reason one might associate the region of very high $\beta$ around $T_c$ with magnetic fluctuations.

The uniqueness of the function $f_o \ln (1 + f/f_o)$ should be a significant clue in determining the nature of the statistical fluctuations dominating the dynamic behavior at $T_c$. The temperature dependence of the parameter $\tau_c = 1/f_o$ is shown in Fig. 3. The temperature dependence does not follow theoretically expected behavior [4]

$$\tau = \tau_o \left( \frac{T-T_c}{T_c} \right)^{-5/3} \qquad (2)$$

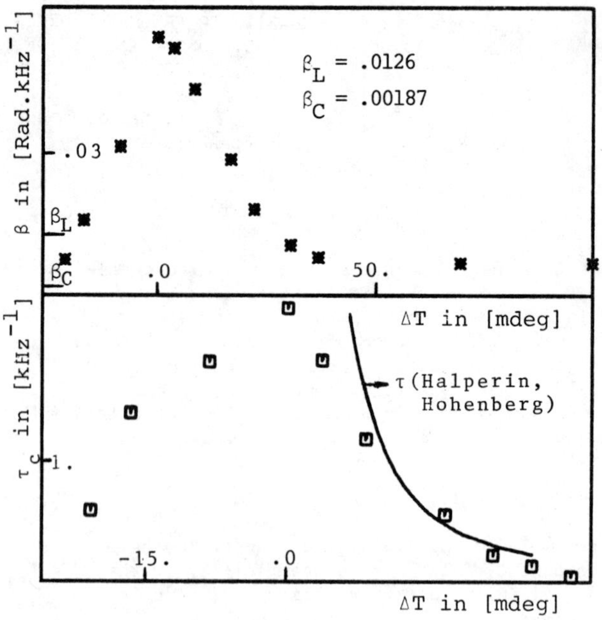


Fig. 3 The temperature dependence of a) $\beta \equiv \varphi_f/\chi' 4\pi \bar{D}$; b) $\tau_c \equiv 1/f_o$; and c) Eq. 2 with $\tau_o = 5{\times}10^{-12}$ sec.

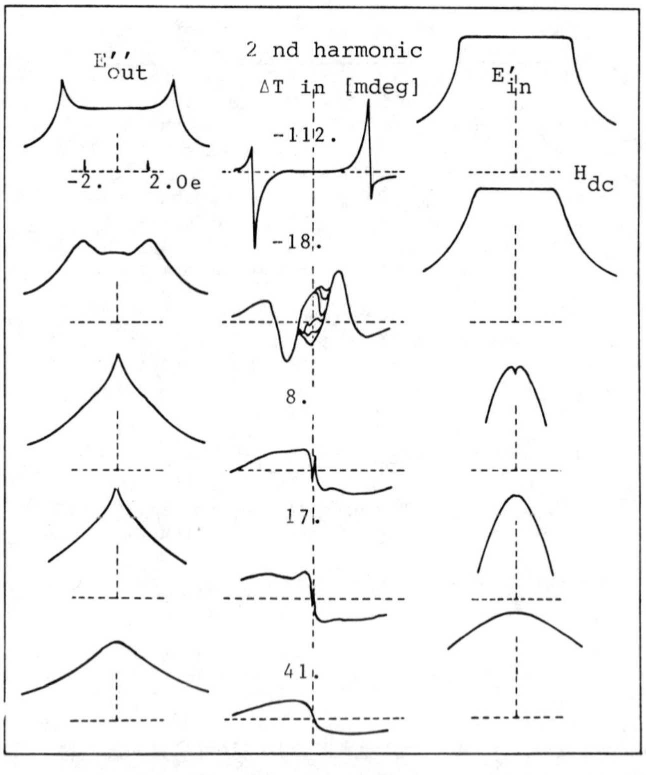


Fig. 4 The dependence of the out phase signal, the second harmonic signal and the in phase signal upon d.c. applied field. The data are uncorrected for phase errors. The temperatures of the isotherms are indicated. The in phase signals for $\Delta T = 8$ and $\Delta T = 14$ mdeg are shown with the base line at 80% of the signal.

however its magnitude requires reasonable value for $\tau_o$ ($\tau_o = 5{\times}10^{-12}$ sec). $(T-T_c) = 10$ mdeg from $T_c$ one obtains 1 msec for $\tau$. Inasmuch as this is the right order of magnitude to associate critical slowing down with $\tau_c \equiv 1/f_o$, it seems not inappropriate to show in Fig. 3 the expression of Eq. 2 along side the values of $\tau_c$ extracted from the data by using Eq. 1. This is done only for the high temperature side of the peak. The behavior on the low temperature side is clearly more complicated than this.

Though theoretical interpretations of these results has not yet been carried out, a framework for treating the role of magnetic dipole interaction in damping near $T_c$ has been given by Maleev[5]. The fluctuations are constrained by magnetostatics. Frequency dependent loss coefficient has also been observed in YI6[6].

FIELD DEPENDENCE OF LOSSES

The measurement of the a.c. response as a function of the d.c. bias field has been the main tool in the continuing study of the behavior of iron whiskers. Only in the region very near $T_c$ has the frequency dependence been of more value. For example, the field at which the susceptibility deviates from a constant value is a good measure of the spontaneous magnetization. This departure field $H_D$ is clearly reflected in the out phase component as peaks. This behavior is shown in the first line of Fig. 4 for temperature corresponding to the region of the curling patterns. The second harmonic shows the departure field even more strikingly. Within 20 mdeg of $T_c$ it becomes difficult to follow $H_D$, partly because of the broadening of the peaks but also because of the appearance of hysteresis which is most evident in the 2nd harmonic response. The out phase component near to $T_c$ shows clearly that there is an extra loss appearing in the last few tenths of an oersted on approaching zero field. This loss is slightly reflected in the in phase component which is shown magnified 4 times in both $\Delta T = 8$ and $\Delta T = 17$ mdeg. By $\Delta T = + 35$ mdeg the behavior is regular.

REFERENCES

1. A.S. Arrott, B. Heinrich, J.E. Noakes, Conf. Proc., AIP, Magnetism and Magnetic Materials, 6, 822, 1973, 6, 941 (1973), 24, 287 (1974).

2. B. Heinrich and A.S. Arrott, Proc. Int. Conf. of Magnetism, ICM-73, Vol. IV, 556 NAUKA Moscow 1974.

3. A.S. Arrott, B. Heinrich and D.S. Bloomberg, IEEE Transactions on Magnetism, Vol-Mag 10, No. 3, 950, 1974. B. Heinrich and A.S. Arrott, AIP Conf. Proc., 24, 702, 1974.

4. B.I. Halperin and P.C. Hohenberg, Phys. Rev. 177, 952, 1969, and private communication.

5. G.A. Vugal'ter and S.V. Maleev, Fiz. Tvord. Tela 14, 3428, 1972 (Sov. Phys. - Solid State 14, 2764, 1973). S.V. Maleev, Zh. Eksp. Teor. Fiz. 66, 1809, 1974 (Sov. Phys. - JETP 39, 889, 1974).

6. S.D. Luzyanin, P.D. Dobyekin and V.P. Khabronin, Zh. Eksp. Teor. Fiz. 66, 1079, 1974, (Sov. Phys. - JETP 39, 528, 1974).

# CRITICAL SPIN DYNAMICS IN UNIAXIAL FERROMAGNETS*

R. Raghavan and D. L. Huber
Department of Physics, Univ. of Wisconsin, Madison, Wi. 53706

ABSTRACT

We report the results of a study of the critical hydrodynamics of an easy axis ferromagnet with weak anisotropic exchange. Self-consistent equations based on the mode coupling hypothesis are derived for the decay rates of the longitudinal and transverse spin fluctuations in the region below the Curie point. Numerical solutions are obtained as a function of anisotropy, and the crossover from quasi-isotropic to fully anisotropic behavior is discussed. We also consider dipolar interactions. Over a limited range of wave vector and temperature they are shown to lead to behavior analogous to that arising from anisotropic exchange. Calculations of the spin wave linewidth in EuO at the reduced temperature $1-T/T_c=0.05$ are in good agreement with experiment. Details of the work will be presented elsewhere.

*Research supported by the National Science Foundation.

# DYNAMICAL CRITICAL BEHAVIOR OF 3D S=½ XY MODEL

M. Howard Lee[+]
Department of Physics, University of Georgia, Athens, Georgia 30602

ABSTRACT

In an earlier report, a mean-field version of the 3D S=½ XY model was presented. Corrections to the mean-field-like description are provided here via the relaxation function of Mori. This relaxation function in energy space is expressible in a continued fraction whose coefficients are certain equilibrium correlation functions. We show that the analytic structure of the relaxation function is related to solutions of an eigenvalue problem, in which the eigenvalues are energies associated with various excitations of the system. Dynamical critical behavior of our system is described on the basis of the behavior of these eigenvalues and the analyticity of the relaxation function.

In an earlier report (referred to as I)[1], we presented a mean-field version of the dynamical behavior of the 3D S=½ XY model above the critical temperature. The resulting picture is one in which a certain collective excitation becomes dominant over all other possible modes of excitations. This mean-field-like description is exact in the high-frequency limit.

Our approach to the dynamical behavior of the XY model is, as described in I, formally perturbative. The mean-field-like description is our zeroth order version of the dynamical behavior. Here we shall show corrections to the mean-field-like description by including certain factors neglected in I. The corrections are brought about (i) by considering the commutativity of the order of the system with the Hamiltonian and (ii) by determining the analytic structure of the relaxation function.

Our principal results are given by certain eigenvectors which we have established for our problem. One can associate the dynamical modes of the system with these eigenvectors as functions of temperature and energy. In terms of these eigenvectors one can describe the onset of the dynamical critical behavior.

Our model is defined by the XY Hamiltonian

$$H = -2J\Sigma(S_i^x S_j^x + S_i^y S_j^y), \tag{1}$$

where the sum is over n.n. pairs and J is the exchange integral. The time evolution of the transverse total spin at a given wavevector k may be given as[1]

$$S_k^x(t) = \sum_n (it)^n/n!.\ c_n(k). \tag{2}$$

Thus if $c_n$ can be term-by-term evaluated via the time derivatives or via the commutativity with H, then the time behavior of the spin operator may be obtained from the resulting operator expansion. The first nontrivial coefficient is $c_2$ which, with $\omega_o = q_o ½J$, where $q_o$ is the coordination number, we write as

$$c_2(k) = \omega_o^2\ S_k^x + \psi_k + \theta_k, \tag{3}$$

where the last two terms of (3) are defined in I. The next coefficient $c_3$ can be generated from $c_2$, and then others in this manner.

Our approach to this problem of extracting the time evolution from (2) is to consider the RHS of (3) in a formally perturbative manner. We observe first that the commutativity of $\theta_k$ with H generates operators which are all proportional to $\theta_k$. Since $\theta_k$ is proportional to a difference of two essentially similar operators, those quantities containing $\theta_k$ when ensemble-averaged must be small and unimportant. Thus $\theta_k$ is neglected in our present consideration, but it is to be added as higher order corrections to the present work.

Proceeding in this vein, we observe that $\psi_k$ contains operators which are tantamount to the longitudinal correlations. Since for our system the longitudinal correlations are much weaker than the transverse correlations, it is reasonable to regard $\psi_k$ as a kind of formal perturbation to $S_k^x$. The zeroth order of our perturbation is one in which we set $\psi_k=0$ and it gives rise to a mean-field description for the system. In this present work we shall consider corrections to zeroth order by taking into account the full effect of $\psi_k$. The commutativity of $\psi_k$ with H, unlike that of $\theta_k$, generates nonlinear terms. As a result, there is added richness in the excitation spectrum of our system. We shall refer to this procedure of including the nonlinear terms as first order.

It is possible to obtain $c_n$ explicitly to all n, valid to first order in our approximation. In particular, we show below two examples from which all others can be generated. Using (3) we obtain, valid to first order,

$$c_3 = (\omega_o^2+\omega_1^2)\ y_k + U^{(1)} y_k \tag{4}$$

$$c_4 = \omega_o^2(\omega_o^2+\omega_1^2) S_k^x + (\omega_o^2+\omega_1^2+\omega_2^2) U^{(1)} S_k^x + U^{(2)} S_k^x \tag{5}$$

where $y_k = c_1$, $\omega_n = q_n ½J$, where $q_n$ is the end-point sum of a chain of n+1 nearest-neighbor links embedded in a given lattice (e.g., $q_o$ is the coordination number, $q_1 = q_o - 1$ if the lattice is a simple cubic lattice), and

$$U^{(n)} S_k^x = \Sigma\ e^{ik.r_1} \prod_{m=1}^{2n} J_{m\ m+1}\ S_m^z \cdot S_{2n+1}^x \tag{6}$$

$$U^{(n)} y_k = i\ \Sigma\ e^{ik.r_1} \prod_{m=1}^{2n} J_{m\ m+1}\ S_m^z\ S_{2n+1}^x\ S_{2n+2}^x \tag{7}$$

where the lattice sums are over all allowed n.n. lattice points.

Our results (4 and 5) show that, valid to first order, $c_n$ can be evaluated once the commutativities of $U^{(n)}S^x$ and $U^{(n)}y$ with H are determined. We show below these commutativities:

$$[H, U^{(n)} S_k^x] = \omega_{2n+1}^2\ U^{(n-1)} y_k + U^{(n)} y_k \tag{8}$$

$$[H, U^{(n)} y_k] = \omega_{2n}^2\ U^{(n)} S_k^x + U^{(n+1)} S_k^x. \tag{9}$$

With the above commutativities we have evaluated $c_n$ up to n = 10. Unlike the case of zeroth order (where one can readily recognize the nature of the resulting series), in this case of first order it is not a simple matter to find an analytic expression for the series expansion representing $S_k^x(t)$. There is, however, an indirect way the summability can be studied by utilizing the results of $c_n$ we now have at our disposal.

The apparent difficulty associated with the summability of the series expansion (2) may be overcome by a method due to Mori[2]. Mori considers a dynamical variable f as a vector in a Hilbert space in which the time evolution of f(t) is represented by its changing projection onto an axis defined in this space. Thus, one can write with f=f(o)

$$f(t) = R(t)\ f + \tilde{f}(t) \tag{10}$$

where R(t) is a function defined in terms of Kubo inner product

$$R(t) = (f(t), f)\ (f,f)^{-1} \tag{11}$$

and $\tilde{f}(t)$, an orthogonal projection in the Hilbert space, is a convolution of **R**(t) and a secondary vector evolved from (exp iLt) f, where L is the Liouville operator. The new function defined above (11), called the relaxation function of Mori, may be used to obtain other dynamical quantities. Mori has shown that

$$R(t) = \int_c dz\, e^{zt} R(z) \quad (12)$$

where R(z) is a continued fraction in terms of certain equilibrium correlation functions $\Delta_n$,

$$R(z) = \frac{1}{z+} \frac{\Delta_1}{z+} \frac{\Delta_2}{z+} + \ldots \quad (13)$$

Since our spin operator $S_k^x$ is a dynamical variable which satisfies all the required properties of f, we may deduce the relaxation function for our system from (11) provided that (i) $\Delta_n$ can be evaluated and (ii) the continued fraction (13) can be made tractable. If we let $f = S_k^x$, we find that $\Delta_n$ is given in terms of $c_n$ as follows:

$$\Delta_1 = (c_1, c_1)\,(c_0, c_0)^{-1} \quad (14a)$$

$$\Delta_2 = \frac{(c_2, c_2) - (c_2, c_0)\,(c_0, c_2)\,(c_0, c_0)^{-1}}{(c_1, c_1)} \quad (14b)$$

and so on. Using our results for $c_n$, we obtain

$$\Delta_1 = \omega_0^2 + \phi_{01} \quad (15a)$$

$$\Delta_2 = \omega_1^2 + \lambda_{01} - \phi_{01} \quad (15b)$$

and so on, where

$$\phi_{nm}(k) = (U^{(n)} S_k^x, U^{(m)} S_k^x)\,(S_k^x, S_k^x)^{-1} \quad (16)$$

$$\lambda_{nm}(k) = (U^{(n)} y_k, U^{(m)} y_k)\,(y_k, y_k)^{-1}. \quad (17)$$

Thus it is in principle possible to know $\Delta_n$ once $\lambda$ and $\phi$ are given. At present only $\Delta_1$ has been studied in detail whose critical behavior in 3 dimensions is believed to be $|T-T_c|^\gamma$.

The second difficulty of the method of Mori is posed by the intractability of the continued fraction (13) which, as a result, has limited the application of this general method. The customary procedure is to terminate the continued fraction which is tantamount to making a multi-pole approximation to the continued fraction. This procedure, while dubious mathematically, has been argued on physical grounds.[2] We shall approach this problem from a point of view of matrix theory.

The continued fraction representation for R(z) may be expressed as

$$R_k(z) = \lim_{n\to\infty} P_n(z)/Q_n(z) \quad (18)$$

where $P_n(z)$ and $Q_n(z)$ are polynomials of nth degree. They are directly obtainable from the above continued fraction. It is interesting to note that these polynomials may be Hurwitz polynomials[3] depending on $\Delta_n$, which are themselves functions of temperature and wave-vector. The polynomials can be further related to determinants

$$\lim_{n\to\infty} P_n(z) = \det N_k(z) \quad (19)$$

$$\lim_{n\to\infty} Q_n(z) = \det D_k(z) \quad (20)$$

where $N_k(z)$ and $D_k(z)$ are infinite tridiagonal matrices shown below:

$$D_k(z) = \begin{bmatrix} z & -a_1 & & & \\ -a_1 & z & -a_2 & & \\ & -a_2 & z & -a_3 & \\ & & -a_3 & z & \cdot \\ & & & \cdot & \cdot \end{bmatrix} \quad (21)$$

where $a_j \equiv i\Delta_j^{1/2}$ and $N_k(z)$ is obtained from $D_k(z)$ by deleting the first row and first column. That is, det $N_k(z)$ is the cofactor of the leading element of $D_k(z)$.

From (12) we observe that $R_k(t)$ may be determined if the singularities of $R_k(z)$ are known. The singularities are given entirely by zeros of det $D_k(z)$ if det $N_k(z)$ is well behaved (which we shall assume here). We can rewrite (21) as

$$D_k(z) = z\,I - \mathcal{D}_k \quad (22)$$

where I is the unit matrix of $\infty$ rank and $\mathcal{D}_k$ is obtained from $D_k(z)$ by deleting the principal diagonal elements (z,z,z,...) and by changing the signs.

The requirement that det $D_k(z) = 0$ is also the requirement for nontrivial solutions of a system of linear homogeneous equations. That is, there exist eigenvectors $X_k = \text{col}\,(x_1\ x_2\ x_3\ \ldots)$ such that we have an eigenvalue equation

$$\mathcal{D}_k X_k = z_k X_k\,. \quad (23)$$

Hence, zeros of det $D_k(z)$ are eigenvalues of $\mathcal{D}_k$ and to obtain the eigenvalues we need merely to diagonalize $\mathcal{D}_k$. The eigenvectors $X_k$ are in energy space. Hence, their physical significance is somewhat obscured. Nevertheless, they are expected to describe in some appropriate way the nature of the dynamic modes of the system.

If $\mathcal{D}_k$ satisfies the Hurwitz stability criteria,[3] then $z_k$ must be complex numbers with nonpositive real parts. Also since $\mathcal{D}_k$ is a traceless matrix, the time reversal symmetry requires that the eigenvalues must occur in pairs (e.g., $z_1$ and $-z_1$). In particular, if $\mathcal{D}_k$ represents a Hurwitz polynomial, then all eigenvalues must lie on the imaginary axis.

Our matrix $\mathcal{D}_k$ has been diagonalized in two limits: (i) mean-field-like limit and (ii) uniform correlation limit. In a mean-field-like (trivial) limit, $a_1 = i\omega_0$ and $a_j = 0$ for all $j \geq 2$. Hence, our matrix is readily diagonalizable. The resultant eigenvalues are $z_k = \pm i\omega_0$ (independent of k) whose eigenvectors have but two non-zero elements. In a uniform correlation limit, $a_j = ia$ for all j. Although our matrix is infinite-dimensional here, it is still diagonalizable. The resultant eigenvalues are an infinite-dimensional version of a Chebyshev polynomial of the second kind. These eigenvalues form a continuum (a cut) on the imaginary axis from + i 2a to -i 2a. The behavior of our matrix is expected to be somewhere in between the two limits. If in particular, $|\Delta_j| > |\Delta_{j+1}|$ for all j, then the leading eigenvalues may be extracted by limiting the dimensions of our matrix. The convergence of the eigenvalues are expected to depend on the relative strengths of the correlation functions.

When certain correlation functions are dominant in some physical regime, $\mathcal{D}_k$ may be reduced to a finite-dimensional case. It is then possible to provide, for example, a description for the dynamical critical behavior. Assume that, away from the critical region, the correlation functions satisfy $|\Delta_1| > |\Delta_2| > |\Delta_3| \ldots$. Then, we obtain the following leading eigenvalues:

$$z_k(1,2) = \pm i(\Delta_1 + \Delta_2 + \Delta_3 + \ldots)^{1/2} \quad (25)$$

$$z_k(3,4) = \pm i(\Delta_1\Delta_3 \ldots)(\Delta_1 + \Delta_2 + \ldots)^{-1/2} \quad (26)$$

Now as $T \to T_c$, only $\Delta_1$ is expected to show strong critical behavior[4]. Hence, the contributions from z(1,2) can show only small changes as $T \to T_c$. However, as $z(3,4) \to 0$ as $T \to T_c$, these modes enhance the central peak in the diffuse scattering cross section. That is, they exhibit critical slow-down.

+Supported by Research Corporation and NATO.

1. M. H. Lee, AIP Conf. Proc. 18, 714(1973) and Phys. Rev. B8, 3290(1973).
2. H. Mori, Prog. Theor. Phys. 34, 399(1965).
3. R. Yarlagadda, SIAM J. Appl. Math. 16, 1146(1968).
4. M. H. Lee, Phys. Rev. B 12, 276(1975).

# FIRST-ORDER TRANSITIONS, SYMMETRY, AND THE ε-EXPANSION

D. Mukamel*, S. Krinsky, and P. Bak
Brookhaven National Laboratory†, Upton, New York 11973

ABSTRACT

We have used the group theoretical method of Landau and Lifshitz to derive effective Hamiltonians for certain paramagnetic to antiferromagnetic transitions having order-parameters with $n \geq 4$ components, and we have performed a renormalization group analysis in $4-\varepsilon$ dimensions. The first-order nature of the order-disorder transitions in Cr(n=12), Eu(n=12), $UO_2$(n=6), and MnO (n=8) can be explained by noting that the corresponding Hamiltonians possess no stable fixed points in $4-\varepsilon$ dimensions. We predict that all fcc type I($\vec{m}\perp\vec{k}$), type II and type III($\vec{m}\perp[100]$,$\vec{k}=[\frac{1}{2}01]$) antiferromagnetic transitions are first-order. We hope our work will serve as a guide in an experimental search for new examples of first-order transitions. We have also considered a 2m-component Hamiltonian which possesses a unique, non-isotropic, stable fixed point for each value of $2m \geq 4$. When 2m=4, the Hamiltonian describes the paramagnetic to antiferromagnetic transitions in $TbAu_2$, $DyC_2$, Tb, Ho, Dy, and the structural transition in $NbO_2$. If these transitions are second-order, we predict they all belong to the same universality class. For 2m=6, the Hamiltonian describes the antiferromagnetic transitions in $TbD_2$, Nd, $K_2IrCl_6$, and $MnS_2$. We predict these transitions belong to a single universality class, and suggest that experiments be performed to test our results.

## I. SYMMETRY AND FIRST-ORDER TRANSITIONS

In recent years the importance of symmetry in classifying second-order transitions has been recognized[1], and much progress has been made in grouping second-order transitions into universality classes. According to universality critical behavior should depend only upon a small number of a system's properties, such as the spatial dimensionality d, the number of components of the order-parameter n, and the symmetry of the Hamiltonian. On the other hand, symmetry considerations have been ignored in many recent theoretical treatments of first-order transitions. In this work, we wish to emphasize the importance of symmetry considerations in classifying first-order transitions. The relation between a system's symmetry properties and its phase transitions can be studied using the phenomenological theory developed by Landau.[2] Wilson[3] has extended the Landau theory to include the effects of fluctuations by using an expansion in $\varepsilon=4-d$. We have used the group theoretical method of Landau and Lifshitz[2] to construct effective Hamiltonians describing certain paramagnetic to antiferromagnetic transitions having order-parameters with $n \geq 4$ components[4], and we have performed a renormalization group analysis in $4-\varepsilon$ dimensions. We have found systems for which the renormalization group equations possess no stable fixed points[5] to lowest nontrivial order in ε, and suggest this lack of a stable fixed point within the ε-expansion is indicative of a first-order transition. We present support for this rule and propose specific experiments to test it.

According to the theory of Landau and Lifshitz[2], the symmetry breaking order-parameter associated with a second-order transition transforms according to an irreducible representation R of the symmetry group $G_0$ of the disordered phase. The number of independent components of the order-parameter is equal to the dimensionality n of the representation according to which it transforms. For transitions not involving a change in the unit cell, the order-parameter transforms according to the point group of the high symmetry phase, and the dimensionality of the representations satisfy $n \leq 3$. It has been emphasized[4] that when the unit cell is increased in one or more directions the order-parameter transforms according to the space group of the high symmetry phase, which may have representations with dimensionality $n \geq 4$. The renormalization group equations in $4-\varepsilon$ dimensions are dramatically different for systems with $n \leq 3$ than for those with $n \geq 4$. Brezin, Le Guillou, and Zinn-Justin[6] have shown that for $n \leq 3$, the isotropic fixed point is always stable, while for $n \geq 4$ it is always unstable. Hence, $n \geq 4$ systems may have either no stable fixed point as discussed in this section, or a stable anisotropic fixed point as discussed in the next section.

Using Landau symmetry arguments, one can immediately predict three classes of order-disorder transitions expected to be first-order:

(1) If the symmetric part of $R^3$, denoted $[R^3]$, contains the identity representation, i.e., $[R^3] \supset E$, then the Landau expansion includes third-order invariants and the transition is predicted to be first-order.

(2) Transitions for which the order-parameter transforms according to a reducible representation of $G_0$ are expected to be first-order.

(3) If the direct product of the antisymmetric part of $R^2$, denoted $\{R^2\}$, and the vector representation contains the identity representation, i.e., $\{R^2\} \times V \supset E$, and if the propagation vector $\vec{k}$ of the low temperature phase is temperature independent, then the transition is predicted to be first-order. Such transitions might be expected to result in a commensurate low temperature phase.

When none of these three criteria is satisfied, then the transition may be either first-order or second order, depending on the values of the coefficients in the Landau expansion. We wish to present experimental support for a fourth phenomenological rule which will be predictive in certain cases when the above Landau symmetry criteria are not.

(4) If there is no stable fixed point within the ε-expansion then the transition is first-order.

Symmetry does not allow the existence of third-order invariants for the magnetic systems we consider, so rule (1) is not predictive. Rule (2) predicts first order transitions in the type II antiferromagnets DySb[7] and HoSb[8] where the magnetic moment lies along the [001] direction, with components both parallel and perpendicular to the propagation vector $\vec{k}$. The order-parameter for these systems has 12 independent components and belongs to two irreducible representations. The transition in DySb is known to be first-order[9], while the nature of the transition in HoSb is not well-established. Experiments[10] on HoSb are consistent with the interpretation of a first-order transition.[38]

The antiferromagnetic transition in $MnBr_2$[11] is associated with a temperature independent propagation vector $\vec{k} = [\frac{1}{4}0\frac{1}{4}]$. Since the representation R satisfies $\{R^2\} \times V \supset E$, rule (3) predicts the transition to be first-order in agreement with experiment.

Let us note that if one was to observe a first-order transition in a ferromagnet, then rule (4) could never be invoked to explain it. Since the unit cell is not changed, it follows that $n \leq 3$, so the isotropic fixed point is always stable.[6] Despite a comprehensive search through the literature, we have found no well-established paramagnetic to ferromagnetic first-order transition, although a possible candidate is that in MnAs.[12] Rule (4) can be applicable to certain antiferromagnetic systems, since the unit cell is increased and one can have $n \geq 4$. We have found five different $n \geq 4$ component Hamiltonians for which there exist no stable fixed points in $4-\varepsilon$ dimensions, and rule (4) suggests the corresponding transitions are first-order. We now describe the physical systems corresponding to these Hamiltonians:

Type I antiferromagnets $\vec{m}\perp\vec{k}$(e.g.$UO_2$[13]): These are fcc crystals with space group Fm3m exhibiting an antiferromagnetic structure with propagation vector $\vec{k}_1=[100]$ and magnetization perpendicular to $\vec{k}_1$. The star of $\vec{k}_1$

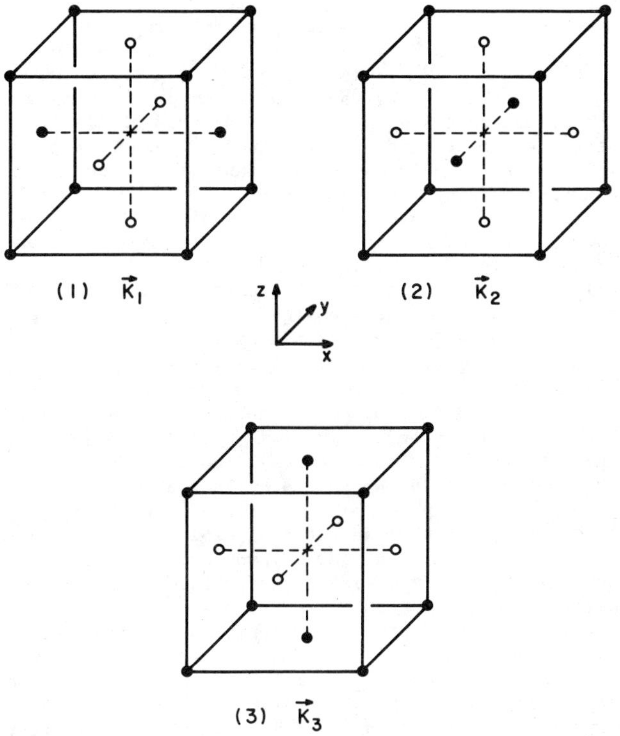


Fig. 1. The three magnetic lattices of type I antiferromagnets. The lattices 1,2,3 belong to the wave vector $\bar{k}_1=[100]$, $\bar{k}_2=[010]$, and $\bar{k}_3=[001]$, respectively.

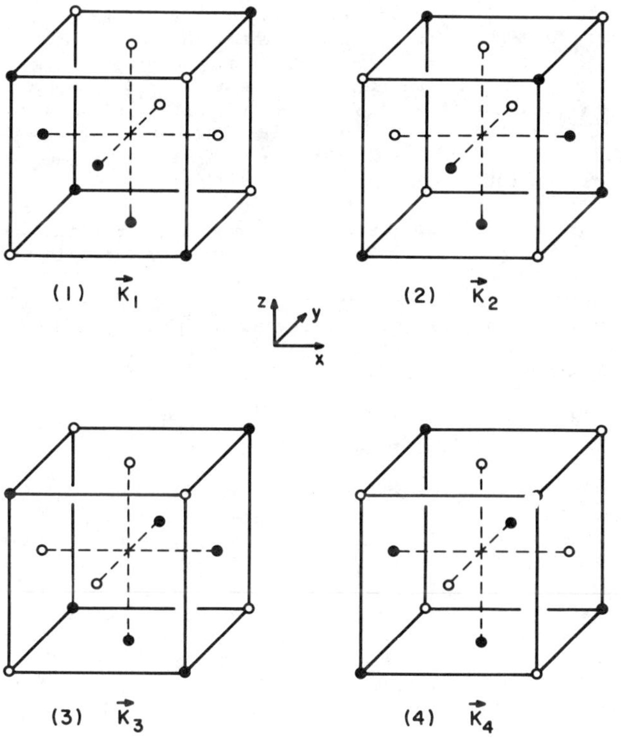


Fig. 2. The four magnetic lattices of type II antiferromagnets. The lattices 1,2,3,4 belong to the wave vectors $\bar{k}_1=[\frac{1}{2}\frac{1}{2}\frac{1}{2}]$, $\bar{k}_2=[-\frac{1}{2}\frac{1}{2}\frac{1}{2}]$, $\bar{k}_3=[\frac{1}{2}\frac{1}{2}-\frac{1}{2}]$, and $\bar{k}_4=[\frac{1}{2}-\frac{1}{2}\frac{1}{2}]$, respectively.

consists of the three vectors: $\bar{k}_1=[100]$, $\bar{k}_2=[010]$, and $\bar{k}_3=[001]$.(See Fig. 1.) Two components of the order-parameter are associated with each $\bar{k}_i(i=1,2,3)$, corresponding to the two independent directions of the magnetization in the plane perpendicular to $\bar{k}_i$. The transition is described by a 6-component Hamiltonian constructed from the five independent fourth-order invariants which can be formed from the order-parameter. The transition in $UO_2$ is known to be first-order.[13]

Type II antiferromagnets $\bar{m}\perp\bar{k}$ (e.g.,MnO[14],NiO[14], ErSb[8], EuTe[15]): These are fcc crystals with space group Fm3m whose ordered state is composed of ferromagnetic (111) planes coupled antiferromagnetically. (See Fig. 2.) The magnetization is perpendicular to the propagation vector $\bar{k}_1=[\frac{1}{2}\frac{1}{2}\frac{1}{2}]$ and the star of $\bar{k}_1$ consists of the four vectors $\bar{k}_1=[\frac{1}{2}\frac{1}{2}\frac{1}{2}]$, $\bar{k}_2=[-\frac{1}{2}\frac{1}{2}\frac{1}{2}]$, $\bar{k}_3=[\frac{1}{2}\frac{1}{2}-\frac{1}{2}]$, and $\bar{k}_4=[\frac{1}{2}-\frac{1}{2},\frac{1}{2}]$. Two components of the order-parameter are associated with each $\bar{k}_i$ (i=1,...,4), corresponding to the two independent directions of the magnetization in the plane perpendicular to $\bar{k}_i$. The 8-component Hamiltonian is constructed from the six fourth-order invariants which can be formed from the order-parameter.[39]It has recently been shown[16] that MnO indeed has a first-order transition. The nature of the transitions in NiO and MnSe is not yet well established. Measurements of the order-parameter[17] of ErSb indicate the transition is second-order, but specific heat measurements[18] indicate there may exist a latent heat. We urge that further experimental work be done on ErSb.

Type II antiferromagnets $\bar{m}||\bar{k}$(e.g., TbAs, TbSb,TbP[8]) Since the magnetization is parallel to the propagation vector, the order-parameter has four independent components corresponding to $\bar{k}_i(i=1,...,4)$, and the effective Hamiltonian is constructed from the three fourth-order invariants which can be formed from the order-parameter. Previous measurements[8] on these systems have been performed far below the transition temperature with the objective of determining the ordered state. We suggest it is worthwhile to study these substances very close to $T_c$ to see if the transition is first-order.

Type III antiferromagnets $\bar{m}\perp[100]$,$\bar{k}=[\frac{1}{2}01]$. These are fcc crystals with space group Fm3m,exhibiting an antiferromagnetic order in which the non-primitive unit cell is doubled in the x-direction(corresponds to $\bar{k}_1=[\frac{1}{2}01]$)and in this case the magnetization is perpendicular to the x-axis. The star of $\bar{k}_1$ consists of the six vectors $\pm\bar{k}_1=\pm[\frac{1}{2}01]$,$\pm\bar{k}_2=\pm[1\frac{1}{2}0]$, and $\pm\bar{k}_3=\pm[01\frac{1}{2}]$. (See Fig.3) There is only one compound,βMnS[19],which exhibits this type III structure, but its paramagnetic space group is F$\bar{4}$3m,a subgroup of Fm3m. The corresponding n=12 representation is reducible(12=6+6),so the transition is predicted to be first-order by Landau rule (2).

Cr[20,21]: This is a bcc crystal with space group Im3m exhibiting[20] a transverse sinusoidal magnetic structure with propagation vector $\bar{k}_1=[k00]$. The star of $\bar{k}_1$ consists of the six vectors $\pm\bar{k}_1=\pm[k00]$,$\pm$ $\bar{k}_2=\pm[0k0]$, and $\pm\bar{k}_3=\pm[00k]$. Two components of the order-parameter are associated with each $\bar{k}$-vector, corresponding to the two independent directions of the magnetization. The 12-component Hamiltonian is constructed from the seven fourth-order invariants which can be formed from the order-parameter. The phase transition in Cr is known to be first-order.[21]

Eu[22,23,24]: This is also a bcc crystal and it exhibits[22] a spiral magnetic order corresponding to a propagation vector $\bar{k}_1=[k00]$. (See Fig. 4.) The order-parameter and Hamiltonian are the same as for Cr. Mössbauer effect[23] and specific heat measurements[24] show the transition in Eu is first-order.

Our main point is that even if the Landau theory (or mean field theory) predicts a second-order transition, the ε-expansion may imply the transition is first-order.[40]This is of significance because the first-order transition in Cr has puzzled theorists for years. The discontinuous nature of the transition was discovered by Arrott[21] over a decade ago, and despite numerous attempts to clarify the origin of the transition using various microscopic models together with mean-field-like theories, no satisfactory explanation has yet been reported. Similarly, the first-order transition in $UO_2$[13] has defied explanation. Blume[25] tentatively proposed that the transition could be explained by a model assuming a singlet ground state, but it was later found the ground state was not a singlet. In mean field theory one is trying to find a physically reasonable mechanism for making the fourth-order coefficient in the corresponding Landau-type expansion nega-

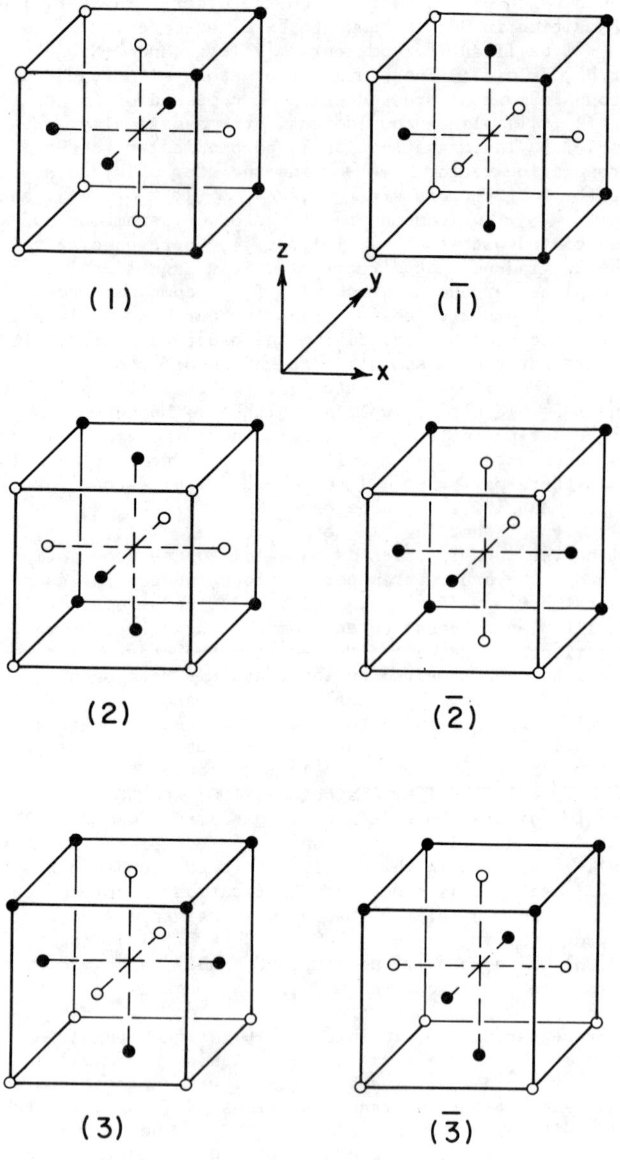


Fig. 3. The six magnetic lattices of type III antiferromagnets.

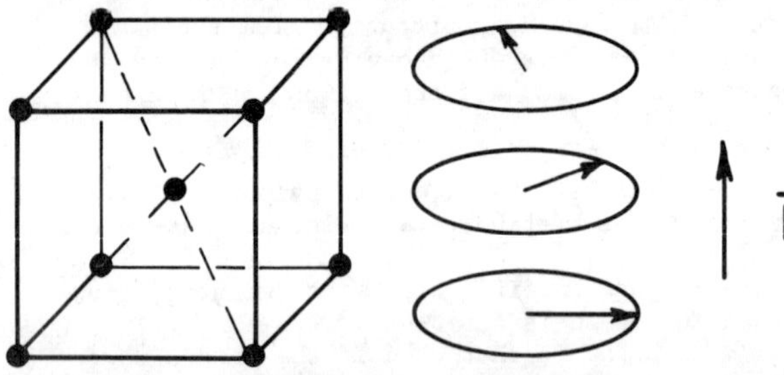


Fig.4. The helical magnetic structure of Eu. The arrows indicate the spin alignment within the hexagonal planes.

tive. This has proved difficult and may not be possible for these systems, so we suggest that our rule (4) is the appropriate explanation for their first-order nature.

There has been a very interesting recent experiment[26] using neutron scattering measurements to study the antiferromagnetic order-parameter in MnO as a function of temperature and applied uniaxial stress. It was found that a large [111] stress can change the transition from first-order to second-order. This is consistent with our ideas, since the [111] stress breaks the cubic symmetry and the transition in the system under stress is described by an n=2 order-parameter, so the isotropic n=2 fixed point is stable. The observed value of $\beta \sim 0.3$ is consistent with the value $\beta = 0.33$ expected for a continuous transition in a three-dimensional XY-type model. We would similarly expect that an applied direct magnetic field in the [111] direction would cause the transition to become second-order, since once again the cubic symmetry would be broken, and the order-parameter would again have n=2. We urge that similar experiments be performed on other systems predicted to be first-order due to a lack of a stable fixed point. Applying an external field which reduces the symmetry of the high temperature phase can change an $n \geq 4$ system into an $n \leq 3$ system.

We are certainly not trying to say that mean field theory never describes a first-order transition. For example, the tricritical point and the first-order transition in $FeCl_2$[27] can be explained quite satisfactorily within the framework of mean field theory. The fourth-order coefficient in the Landau expansion can be made negative by varying the external direct magnetic field. When one does not have an external experimentally variable field, it may be harder to provide a physically reasonable mechanism making a fourth-order coefficient negative. For this reason, we believe our rule (4) may be of value in understanding and finding first-order transitions.

First-order magnetic transitions from a disordered state to an ordered state which are not explained by Landau's arguments (2) and (3) are quite rare. In fact, the physical systems we have found corresponding to Hamiltonians with no stable fixed points exhaust many, and possibly, all of such known first-order transitions. In Table I, we list the materials corresponding to Hamiltonians with no stable fixed points and we distinguish those systems for which the transitions are known to be first-order from those in which the nature of the transitions has yet to be determined. We predict these systems will all have first-order transitions and urge that experiments be performed to test our results. It would be interesting to investigate whether the systems which are predicted to be first-order because of the lack of a stable fixed point behave differently from systems predicted to be first-order by Landau arguments. The agreement of our results with existing experiments lends support to the idea that the symmetry criteria for the absence of a stable fixed point in $4-\varepsilon$ dimensions may be the same as in three dimensions. It would be of great interest to find a group theoretical criterion for the absence of a stable fixed point in $4-\varepsilon$ dimensions.

## II. PHYSICAL REALIZATIONS OF A 2m-COMPONENT HAMILTONIAN

The $\varepsilon$-expansion indicates that universality is weaker for systems with order-parameters having $n \geq 4$ components than for systems with order-parameters having $n \leq 3$ components. To leading order in $\varepsilon$ the isotropic fixed point is always stable[6] for $n \leq 3$ systems, which suggests that the critical exponents are independent of the system's anisotropy. When $n \geq 4$, the isotropic fixed point is no longer stable[6] and critical exponents may depend on the anisotropy of the system. The total absence[5] of a stable fixed point is an extreme case of the difference between $n \leq 3$ and $n \geq 4$ systems. In the preceding section we discussed five different $n \geq 4$ Hamiltonians having no stable fixed point, and we suggested this to be indicative of a first-order transition. Now we wish to consider the following 2m-component Hamiltonian[5,29],

$$H = -\frac{1}{2}\sum_{i=1}^{m} r(\phi_i^2+\bar{\phi}_i^2) + (\nabla\phi_i)^2 + (\nabla\bar{\phi}_i)^2$$

$$- u\Big(\sum_{i=1}^{m}\phi_i^2+\bar{\phi}_i^2\Big)^2 - v\sum_{i=1}^{m}(\phi_i^2+\bar{\phi}_i^2)^2 - w\sum_{i=1}^{m}\phi_i^2\bar{\phi}_i^2 ,$$

Table I

Physical Systems Corresponding to Hamiltonians with no Stable Fixed Points

| Antiferromagnetic Order | Number of Components of Order-Parameter | Systems Known to be First-Order | Systems Predicted to be First-Order |
|---|---|---|---|
| Type I $(\bar{m}\perp\bar{k})$ | n = 6 | $UO_2$[13] | |
| Type II $(\bar{m}\parallel\bar{k})$ | n = 4 | | TbAs, TbSb, TbP,[8] CeS, TbSe, NdSe, NdTe[28] |
| Type II $(\bar{m}\perp\bar{k})$ | n = 8 | MnO[14,16] | NiO[14], MnSe[14], $\alpha$-MnS[19], ErP, ErSb[8], EuTe[15] |
| Type III $(\bar{m}\perp[100], \bar{k}=[\frac{1}{2}01])$ | n = 12 | | |
| Sinusoidal $(\bar{m}\perp\bar{k}=[k00])$ | n = 12 | Cr[20,21] | |
| Screw Spiral $(\bar{m}\perp\bar{k}=[k00])$ | n = 12 | Eu[22,23,24] | |

which for each value of $2m \geq 4$ possesses a unique, non-isotropic, stable fixed point, corresponding to $u^*$, $v^* \neq 0, w^*=0$. The critical behavior of systems described by this Hamiltonian in general belongs to a different universality class than the isotropic 2m-component system ($v=w=0$). The critical exponents corresponding to the stable fixed point are:

$$\nu = \frac{1}{2} + \frac{3(m-1)}{4(5m-4)}\varepsilon + \frac{(m-1)(20m^2 + 253m-334)}{16(5m-4)^3}\varepsilon^2 + O(\varepsilon^3),$$

$$\eta = \frac{(m-1)(2m-1)}{4(5m-4)^2}\varepsilon^2 + O(\varepsilon^3) \quad .$$

Expansions for the exponents $\beta$ and $\gamma$ can easily be obtained by using the scaling relations.

When $2m=4$, this Hamiltonian describes the antiferromagnetic transitions in $TbAu_2$, $DyC_2$, Tb, Ho, Dy, and the structural transition in $NbO_2$. Since we have found there exists a unique stable fixed point, we predict these different phase transitions are described by the same critical exponents and we suggest this be tested experimentally. If we set $\varepsilon=1$ in the expansion for the exponents, neglecting $O(\varepsilon^3)$, we find

$$\beta = 0.39, \quad \nu = 0.70 \quad, \gamma = 1.39 \quad .$$

These are the same as for the isotropic n=4 model.

When $2m=6$, this Hamiltonian describes the antiferromagnetic transitions in $TbD_2$, Nd, $K_2IrCl_6$, and $MnS_2$. Since there exists a unique stable fixed point, we predict these different systems have the same critical exponents, and we suggest this be checked experimentally. Inserting $\varepsilon=1$ into the expansion for the exponents, neglecting $O(\varepsilon^3)$, we obtain

$$\beta = 0.38, \ \nu = 0.69, \ \gamma = 1.38 \quad .$$

These values are significantly different from those obtained by setting $\varepsilon=1$ into the expansion for the isotropic n=6 model,

$$\beta = 0.41, \ \nu = 0.73, \ \gamma = 1.45 \quad .$$

This indicates that anisotropy is affecting the critical exponents in a non-negligible manner. The exponents for H with 2m=4 and 2m=6 are very close to those of the isotropic n=3 Heisenberg model

$$\beta = 0.38, \ \nu = 0.68, \gamma = 1.35 \quad .$$

We should emphasize that there was no more reason, a priori, to have expected n=3 Heisenberg values than n=1 Ising values.

Some of the physical systems which we have found to correspond to H may lie outside the domain of attraction of the stable fixed point, in which case the transition is not expected to be second-order. If the system exhibits a second-order transition, we predict the critical exponents to be as given above.

Let us now describe the physical systems corresponding to H.

$TbAu_2$, $DyC_2$: These are tetragonal crystals whose paramagnetic space group is I4/mmm. They exhibit[30] a transverse sinusoidal magnetic structure, with the propagation vector $\bar{k}_1$ in the basal plane and the magnetization along the z-axis. In the case of $TbAu_2$ the star of the vector $\bar{k}_1$ consists of the four vectors: $\pm\bar{k}_1 = \pm[kk0]$ and $\pm[k\text{-}k0]$, while in the case of $DyC_2$ the star consists of the vectors $\pm\bar{k}_1 = \pm[k00]$ and $\pm\bar{k}_2=[0k0]$. The order-parameter has in both cases four independent components, and the Hamiltonian which describes the system is H with 2m=4 and w=0.

Tb, Dy, Ho: These rare earth elements crystallize in the hcp structure with space group $P6_3$/mmc. They exhibit[31] a spiral magnetic strucrure with magnetic moments in the basal plane and propagation vector along the c-direction. The star of the propagation vector consists of the two vectors $\pm\bar{k}=\pm(0,0,k)$. For each of these vectors there are two equivalent directions of the spins in the basal plane, so the order-parameter has four independent components. These systems are also described by H with 2m=4 and w=0.

$NbO_2$: This is a tetragonal crystal whose space group is $P4_2$/mnm. At ∿ 800°C it undergoes a structural transition[32] in which its symmetry is reduced to $I4_1/a$. The structure below the transition is associated with a reciprocal vector $\bar{k}_1 = [\frac{1}{4}\frac{1}{4}\frac{1}{2}]$. The star of $\bar{k}_1$ consists of the four vectors $\pm\bar{k}_1=\pm[\frac{1}{4}\frac{1}{4}\frac{1}{2}]$ and $\pm\bar{k}_2=\pm[-\frac{1}{4}\frac{1}{4}\frac{1}{2}]$. The order-parameter has four components and the transition is described by H with 2m=4 and $w \neq 0$.

$TbD_2$: This is an fcc crystal whose paramagnetic space group is Fm3m. It exhibits[33] a longitudinal sinusoidal magnetic structure, with the propagation vector $\bar{k}_1$ being parallel to the x-axis. The star of the vector $\bar{k}_1$ consists of the six vectors: $\pm\bar{k}_1=\pm[k00]$, $\pm\bar{k}_2=\pm[0k0]$, $\pm\bar{k}_3=\pm[00k]$. The order-parameter has six independent components and the system is described by H with 2m=6 and w=0.

Nd: This is a double hexagonal close packed crystal whose space group is $P6_3$/mmc. It exhibits[34] a longitudinal sinusoidal magnetic structure, with propagation vector $\bar{k}_1$ in the basal plane. The star of $\bar{k}_1$ consists of six vectors: $\pm\bar{k}_1=\pm$ [k000], $\pm\ \bar{k}_2=[0k00]$, and $\pm\bar{k}_3=$

$\pm[00k0]$. The order-parameter has six independent components and the system is described by H with 2m=6 and w=0.

$K_2IrCl_6$: This is a fcc crystal whose space group is Fm3m. It exhibits[35] an antiferromagnetic structure of the third kind, (see Fig. 3) where the non-primitive unit cell is doubled in the x-direction (corresponds to a reciprocal vector $\bar{k}_1=[\frac{1}{2}01]$).The magnetization, in this case, is along the x-axis. The star of $\bar{k}_1$ consists of six vectors: $\pm\bar{k}_1=\pm[\frac{1}{2}01]$, $\pm\bar{k}_2=\pm[1\frac{1}{2}0]$, and $\pm\bar{k}_3=\pm[01\frac{1}{2}]$. The order-parameter has six independent components and the system is described by H with 2m=6 and w≠0.

$MnS_2$: This crystal has the pyrite structure which is a NaCl-like arrangement of Mn and $S_2$ groups, with the axes of the $S_2$ groups along the body diagonals. The paramagnetic space group is Pa3. The magnetic structure[36] below the transition temperature is type III antiferromagnetic, as in $K_2IrCl_6$. The paramagnetic space group of $MnS_2$(Pa3) is a subgroup of the paramagnetic space group of $K_2IrCl_6$(Fm3m). This might have given rise to additional fourth-order invariants, but it turns out that $MnS_2$ is also described by H with 2m=6 and w≠0.

## III. SUMMARY

Certain antiferromagnetic transitions are described by order-parameters having $n\geq 4$ components.[4] We have found physical examples of transitions having n=4,6,8, and 12 components, and we have used the group theoretical method of Landau and Lifshitz[2] to derive effective Hamiltonians for these systems. Renormalization group equations in 4-ε dimensions[3] are dramatically different[6] for systems with $n\leq 3$ component order parameters, and those for systems with $n\geq 4$. When $n\leq 3$ the isotropic fixed point is always stable, while for $n\geq 4$ it is always unstable. Hence it is possible for $n\geq 4$ systems to have <u>no</u> stable fixed point[5], and we believe this lack of a stable fixed point may provide the explanation for the first-order nature of the transitions in Cr(n=12), Eu(n=12), $UO_2$(n=6), and MnO(n=8). In Table I, we list some physical systems which we predict have first-order transitions.

We have also considered a 2m-component Hamiltonian which possesses a unique, non-isotropic, stable fixed point for each value of $2m\geq 4$. When 2m=4, this Hamiltonian describes the paramagnetic to antiferromagnetic transitions in $TbAu_2$, $DyC_2$, Tb, Ho, Dy, and the structural transition in $NbO_2$. We predict these transitions belong to the same universality class. For 2m=6, this Hamiltonian describes the transitions in $TbD_2$, Nd, $K_2IrCl_6$, and $MnS_2$. We predict these transitions belong to a single universality class.

The $n\geq 4$ component systems we have been discussing provide an exciting subject for experimental investigation. In particular, it would be of great interest to find new examples of first-order paramagnetic to antiferromagnetic transitions. We hope a comprehensive study of these $n\geq 4$ systems will be undertaken.

An excellent review of neutron scattering determinations of magnetic structures is given by D. E. Cox[37] and we have found his work to be an invaluable reference.

## IV. ACKNOWLEDGMENTS

We wish to thank J. D. Axe, G. A. Baker, Jr., M. Blume, K. Carneiro, L. M. Corliss, D. E. Cox, V. J. Emery, J. M. Hastings, Y. Imry, R. Pynn, S. M. Shapiro, G. Shirane, and S. Shtrikman for illuminating discussions.

## REFERENCES


* Present address: Baker Laboratory, Cornell University, Ithaca, New York 14850.

† Work performed under the auspices of the U. S. Energy Research and Development Administration.

1. M. E. Fisher, Rep. Progr. Phys. <u>30</u>, 615 (1967); L. P. Kadanoff, W. Götze, D. Hamblen, R. Hecht, E. Lewis, V. Palciauskas, M. Rayl, J. Swift, D. Aspines, and J. Kane, Rev. Mod. Phys. <u>39</u>, 395 (1967).
2. L. D. Landau and E. M. Lifshitz, <u>Statistical Physics</u> (Pergamon, New York 1968) 2nd Ed.Ch. XIV.
3. K. G. Wilson and J. Kogut, Phys. Rept. <u>12C</u>, 75 (1974); M. E. Fisher, Rev. Mod. Phys. <u>46</u>, 597 (1974).
4. D. Mukamel, Phys. Rev. Lett. <u>34</u>, 481-5 (1975). See also, R. Alben, C. R. Acad. Sci. B <u>279</u>, 111 (1974).
5. D. Mukamel and S. Krinsky, J. Phys. C. Letters (to be published); P. Bak, S. Krinsky, and D. Mukamel, Phys. Rev. Lett. (submitted); D. Mukamel and S. Krinsky, Phys. Rev. B (submitted); D. Mukamel and S. Krinsky, Phys. Rev. B (submitted); P. Bak and D. Mukamel, Phys. Rev. B (submitted).
6. E. Brezin, J. C. Le Guillou, and J. Zinn-Justin, Phys. Rev. B <u>10</u>, 892-900 (1974).
7. G. P. Felcher, T. O. Brun, R. J. Gambino, and M. Kuznietz, Phys. Rev. B <u>8</u>, 260 (1973).
8. H. R. Child, M. K. Wilkinson, J. W. Cable, W. C. Koehler, and E. O. Wollan, Phys. Rev. <u>131</u>, 922 (1963).
9. E. Bucher, R. J. Birgeneau, J. P. Maita, G.O. Felcher, and T.O. Brun, Phys. Rev. Lett. <u>28</u>, 746 (1972).
10. H. Taub, S. J. Williamson, W. A. Reed, and F. S. L. Hsu, Solid State Commun. <u>15</u>, 185 (1974).
11. E. O. Wollan, W. C. Koehler, and M. K. Wilkinson, Phys. Rev. <u>110</u>, 638 (1958).
12. N. N. Sirota, E. A. Vasilev, and G. A. Govor, J. Phys. <u>32</u>, C987(1971); L. H. Schwartz, E. L. Hall, and G. P. Felcher, J. Appl. Phys. <u>42</u>, 1621(1971).
13. B. C. Frazer, G. Shirane, D. E. Cox, and C. E. Olsen, Phys. Rev. <u>140</u>, A1448 (1965).
14. C. G. Shull, W. A. Strauser, and E. O. Wollan, Phys. Rev. <u>83</u>, 333 (1951).
15. G. Will, S. J. Pickart, H. A. Alperin, and R. Nathans, J. Phys. Chem. Solids <u>24</u>, 1679 (1963).
16. D. Bloch, R. Maury, C.Vettier, and W. B. Yelon, Phys. Lett. <u>49A</u>, 354 (1974).
17. S. M. Shapiro and P. Bak, J. Phys. Chem. Solids <u>36</u>, 579 (1975).
18. F. Hulliger and B. Natterer, Solid State Commun. <u>13</u>, 221 (1973).
19. L. M. Corliss, N. Elliott, and J. M. Hastings, Phys. Rev. <u>104</u>, 924 (1956).
20. G. Shirane and W. J. Takei, J. Phys. Soc. Japan <u>17</u> - B III, 35 (1962).
21. A. Arrott, S. A. Werner, and H. Kendrick, Phys. Rev. Lett. <u>14</u>, 1022 (1965).
22. N. G. Nereson, C. E. Olsen, and G. P. Arnold, Phys. Rev. <u>135</u>, A176 (1964).
23. R. L. Cohen, S. Hüfner, and K. W. West, Phys. Rev. <u>184</u>, 263 (1969).
24. B. C. Gerstein, F. J. Jelinek, J. R. Mullaly, W. D. Shickell, and F. H. Spedding, J. Chem. Phys. <u>47</u>, 5194 (1967).
25. M. Blume, Phys. Rev. <u>141</u>, 517 (1966).
26. D. Bloch, D. Hermann-Ronzaud, C. Vettier, W. B. Yelon, and R. Alben, Phys. Rev. Lett. <u>35</u>, 963 (1975).
27. R. J. Birgeneau, G. Shirane, M. Blume, and W. C. Koehler, Phys. Rev. Lett. <u>33</u>, 1098 (1974).
28. P. Schobinger-Papamantellos, P. Fisher, A. Niggli, E. Kaldis, and V. Hildebrandt, J. Phys. C <u>7</u>, 2023 (1974).
29. This Hamiltonian with w=0 has previously been studied by Brezin et al in Ref. 6, and also by G. Grinstein, AIP Conf. Proc. <u>24</u>, 313 (1974).
30. M. Atoji, J. Chem. Phys. <u>48</u>, 560, 3384 (1968).
31. O. W. Dietrich and J. Als-Nielsen, Phys. Rev. <u>162</u>, 315 (1967); W. C. Koehler, J. Appl. Phys. <u>36</u>, 1078 (1965); M. K. Wilkinson, W. C. Koehler, E. O. Wollan, and J. W. Cable, J. Appl. Phys. <u>32</u>, 485 (1961); W. C. Koehler, J. W. Cable, E. O. Wollan, and M. K. Wilkinson, J. Phys. Soc. Japan <u>17</u>, Suppl. B-III, 32 (1962); W. C. Koehler, H. R. Child, E. O. Wollan, and J. W. Cable, J. Appl. Phys. Suppl. <u>34</u>,

1335 (1963).

32. B. O. Marinder, Ark. Kemi. 19, 435 (1963); S. M. Shapiro, J. D. Axe, G. Shirane, and P. M. Raccah, Solid State Commun. 15, 377 (1974); R. Pynn, J. D. Axe, and R. Thomas (to be published).

33. D. E. Cox, G. Shirane, W. J. Takei, and W. E. Wallace, J. Appl. Phys. 34, 1352 (1963).

34. R. M. Moon, J. W. Cable, and W. C. Koehler, J. Appl. Phys. 35, 1041 (1964).

35. V. J. Minkiewicz, G. Shirane, B. C. Frazer, R. C. Wheeler, and P. B. Dorain, J. Phys. Chem. Solids 29, 881 (1968).

36. J. M. Hastings, N. Elliott, and L. M. Corliss, Phys. Rev. 115, 13 (1959).

37. D. E. Cox, BNL Report 13822 (1969); D. E. Cox, IEEE Trans. on Magnetics MAG-8, 161 (1972).

38. It has been suggested that CoO is type II with magnetization in a non-symmetry direction (W. Roth, Phys. Rev. 111, 772 (1958)). In this case rule 2 predicts that the transition is first order. However, the magnetic structure is still controversial. (See, for instance, M. D. Rechtin and B. L. Averbach, Phys. Rev. B 6, 4294 (1972)).

39. S. A. Brazovskii and I. E. Dzyaloshinskii, JETP Lett. 21, 164 (1975) described phase transitions in type II antiferromagnets using an n=4 model. Their Hamiltonian is the Hamiltonian we have found describes the case $\vec{m}||\vec{k}$. They studied this model to first order in ε and found no stable fixed point by studying the flow diagrams. In our analysis of this Hamiltonian we found marginal operators to first order in ε, and we determined the stability by studying the recursion relations to second order in ε.

40. ε-expansion technique has been used to predict first order transitions in type II superconductors by B. I. Halperin, T. C. Lubensky, and S. K. Ma, Phys. Rev. Lett. 32, 292 (1974). The transition in superfluid $He^3$ liquid is described by an 18 component order-parameter and has been predicted to be first order due to lack of stable fixed point by D. R. T. Jones, A. Love, and M. A. Moore, preprint, August 1975.

## THERMODYNAMIC BEHAVIOR NEAR VALENCE INSTABILITIES*

R. D. Parks and J. M. Lawrence
University of Rochester, Rochester, New York 14627

### ABSTRACT

Recent work on the characterization of the second-order, mean-field, valence transition in $Ce_{1-x}Th_x$ is briefly reviewed. Extended work has culminated in the quantification of the Landau free energy functional which describes the transition. The understanding of the microscopic origin of the different terms in the free energy is illuminated by measurements of the specific heat and the effects of solutes of varying valence and atomic size. Results for thermopower and magnetic susceptibility are also presented and are utilized to characterize the integral and non-integral valence states. Throughout, comparisons are made with the SmS system.

### INTRODUCTION

Cerium undergoes a valence transition which is isomorphic as in the Sm-chalcogenides, and which terminates in a critical point in the P-T plane. It is characterized by large decreases in the fcc lattice parameter, the magnetic susceptibility and the electrical resistivity upon going from the high temperature (γ) phase to the low temperature (α) phase. These phenomena are thought to be associated with a delocalization of the 4f electrons with decreasing temperature or increasing pressure.

Recently[1] we have located the second-order critical point in the x-T plane of the alloy system $Ce_{1-x}Th_x$. Critical exponent studies reveal that the transition is of mean field type. This has made it possible to extend the result by obtaining a Landau equation of state for the critical region. We discuss herein the microscopic origin of the corresponding terms in the free energy - in particular, the lattice coupling (which causes the transition to be mean-field), the importance of spin entropy and the effect of alloying. Comparisons are made with the SmS system which has been widely studied (see, e.g. Refs. 2 - 4). In addition, resistivity, susceptibility, thermopower and specific heat results are given which provide information about the pure integral and non-integral valence states. We show that, in contrast to the SmS system, the integral valence (γ) state has a moment; and exhibits behavior characteristic of a spin fluctuation system (as in the Kondo effect) due to the proximity of the f-level to the Fermi surface.

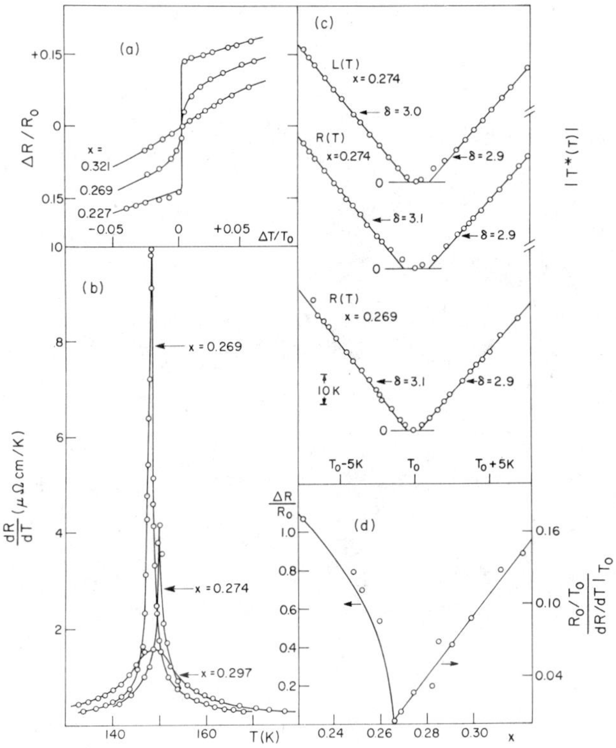


Fig. 1. a) Resistivity R(T) for three concentrations of $Ce_{1-x}Th_x$, plotted together with the prediction of the equation of state. b) dR/dT for three samples exhibiting continuous transitions, showing the onset of the divergence. c) Plots of $T^*(T) = (Y(T) - Y(T_o))/(dY/dT)$ for Y = L(T) (sample length) and Y = R(T) demonstrating the linear coupling of R to L and the mean-field exponent. d) The first-order jumps in R(T) and the inverse susceptibility $(dR/dT|_{max})^{-1}$ plotted with the prediction of the equation of state.

### THERMODYNAMICS IN THE CRITICAL REGION

In Fig.1. we give a succinct summary of our

earlier work on $Ce_{1-x}Th_x$ alloys.[1,5] The order parameter for the isomorphic transition is the volume. It couples linearly to the resistivity (Fig. 1.c.). The large first order jumps which occur in these quantities diminish on alloying and vanish at the critical concentration $x_o$, above which one sees continuous transitions (Fig. 1.a). As $x \to x_o^+$ the maximum in the derivative of the s-shaped curves diverges (Fig. 1.b). The phase boundary of first-order transitions and its extension along the line of $\frac{dR}{dT}\Big|_{max}$ is shown in Fig. 2. In order to obtain the critical indices for the transition one has to take into account which first or second derivative of the free energy is being measured, and also the pathway of approach to the critical point. This has been discussed in the general case by Griffiths and Wheeler[6] who point out that the only uniquely defined direction is the one asymptotically parallel to the coexistence curve. The basic ideas can be understood by analogy to the magnetic case. Since the $Ce_{1-x}Th_x$ phase boundary is nearly parallel to the x-axis we can draw the analogy that $\{\Delta V, \Delta x, \Delta T\}_{Th\text{-}Ce} \sim \{\Delta M, \Delta T, \Delta H\}_{magnet}$. In the magnetic case

$$M(H, T = T_o) \sim H^{1/\delta}$$

$$\frac{\partial M}{\partial H}\Big|_{H=0} \sim (T-T_o)^{-\gamma}$$

so that for the valence transition we expect

$$\Delta V(T, x = x_o) \sim (\Delta T)^{1/\delta}$$

$$\frac{\partial V}{\partial T}\Big|_{\Delta T=0} \sim (x-x_o)^{-\gamma}$$

By plotting $T^* = \Delta V/(dV/dT)$ we obtain $\delta = 3.0$ (Fig. 1c.). This is the value for mean-field behavior. It then becomes possible to locate the critical concentration [Fig. 1.d.] and to obtain the equation

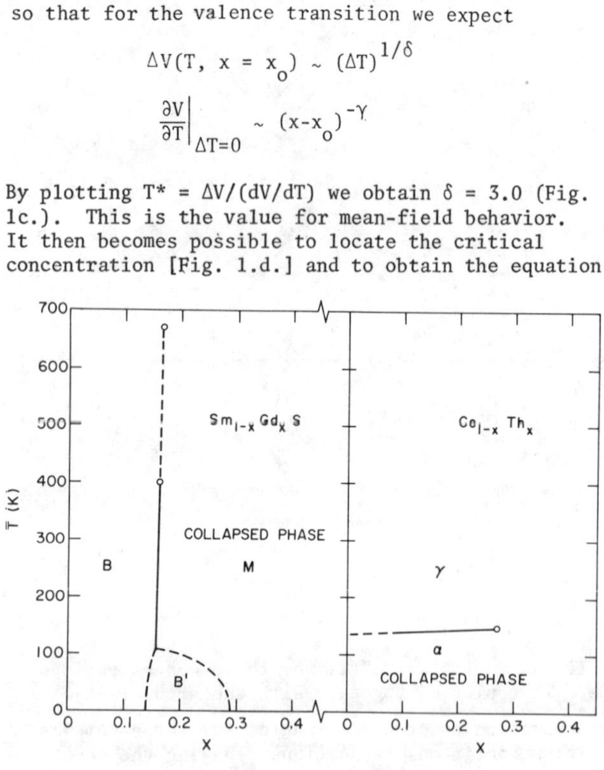


Fig. 2. Comparison of the phase diagrams for the valence tranistions in $Sm_{1-x}Gd_xS$ and $Ce_{1-x}Th_x$.

of state for the whole critical region. The numerology is carried out in Ref. 5, where we find

$$A\left(\frac{\Delta V}{V_o}\right)^3 + B\left(\frac{\Delta x}{x_o}\right)\left(\frac{\Delta V}{V_o}\right) + C\frac{\Delta x}{x_o}\frac{\Delta V}{V_o} = \frac{\Delta T}{T_o}$$

where $A = 0.67$, $B = 0.35$, $C = 0.07$, $x_o = 0.265$, $T_o = 148K$.

As we discuss below, the mean-field character should also occur in the SmS transition. However, since the phase boundary there is nearly parallel to the T-axis, one will measure a different set of critical indices. In particular, if one had the analogy $\{\Delta V, \Delta x, \Delta T\}\ Sm_{1-x}Gd_xS \sim \{\Delta M, \Delta H, \Delta T\}_{magnet}$, then for $x = x_o$ one would measure

$$\Delta V \sim (T - T_o)^{\beta} \quad T<T_o$$

$$\Delta V \equiv 0 \text{ for } T>T_o.$$

In the work of Jayaraman et. al.,[1] the curve for $Sm_{0.85}Gd_{0.15}S$ has this appearance, with $T_o \sim 600$ K. For x slightly greater than $x_o$ one would have analogs of the M(H,T) curves with $H \gtrsim 0$. These have inflection points for $T \sim T_o$. In this light the inflection points at ~600 K in $a_o(T)$ for x = 0.17, 0.18 suggest one is passing the point of closest approach to the critical point for those concentrations of Gd. (To make this more convincing and be consistent with the positive dT/dx for that system, one has to argue that the x = 0.15 room temperature lattice constant is that of the metastable insulating phase.)

## DISCUSSION OF THE FREE ENERGY

One obtains mean-field behavior when the range of the interaction responsible for the transition becomes large compared to the distance between particles. In most (though not all) real systems it is due to a coupling with the lattice (e.g., superconductivity, Jahn-Teller transitions or the hydrogen-niobium "liquid-gas" transition). Since the difference in ionic sizes of the two 4f valence states is quite large, the system will be strongly tied to the lattice[7] and one can obtain effectively infinite force ranges[8] and hence mean-field behavior. The interaction enters the free energy as a $\Delta x\, \Delta V^2$ term. Anderson and Chui[8] propose that when the crystal becomes strained enough (by application of P, or by alloying) the lattice-mediated interactions between ions of the same size (γ-γ or α-α) will change from harmonic and attractive to anharmonic and repulsive, causing the coefficient of the $\Delta V^2$ term to change sign and hence drive the transition second order. Whether or not this is the correct explanation of the lattice coupling, it is <u>highly</u> probable that the SmS transition will turn out to be mean-field, since in that case also the 15% volume change should dominate the transition.

In Fig. 2 we compare the phase diagram describing the valence transition in $Ce_{1-x}Th_x$ with that determined[1] for the chalcogenide system $Sm_{1-x}Gd_xS$. For the present discussion we focus only on the region $T \gtrsim 100$ K of the latter phase diagram. The solid line is a line of first order valence transitions which terminates presumably in a critical point which lies somewhere between the two circle data points. The phase labelled B is semiconducting in the region just below the critical point, whereas the phase labelled M; which is the higher valence, higher density phase, is metallic. The striking difference between the two phase diagrams, if one focuses on the region near the critical point, is that the phase boundary is nearly vertical for the Sm system, whereas it is nearly horizontal for the Ce system. This difference reflects the relative importance of entropy in the two cases, the entropy term being an extremely important term in the free energy functional describing the transition in the Ce system, whereas it plays a minor role in the Sm system The role of entropy in determining the relative flatness of the phase diagram is immedi-

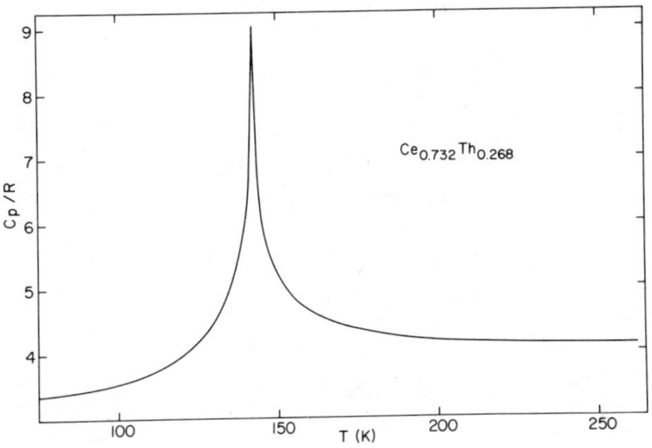


Fig. 3. The specific heat $C_p$ of $Ce_{0.732}Th_{0.268}$.

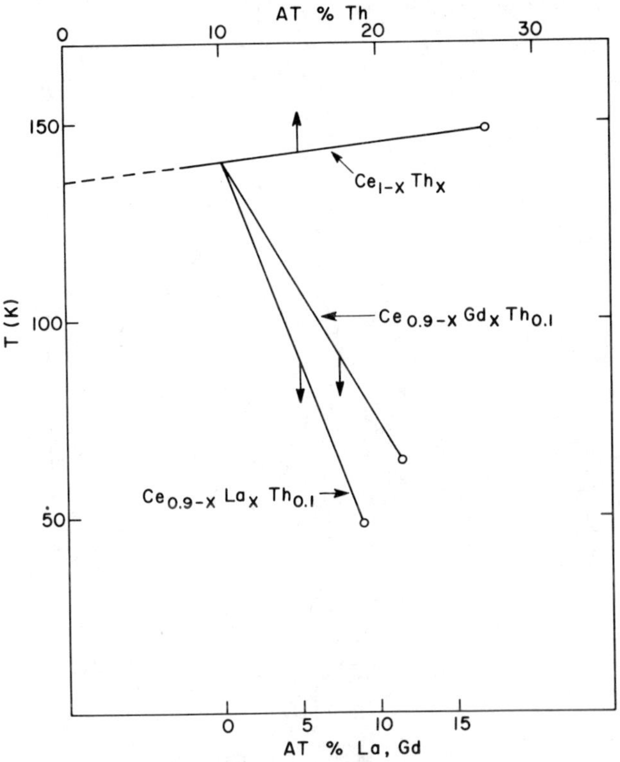


Fig. 4. The phase boundaries, with critical end-points, for different RE solutes in the alloys $Ce_{0.9-x}RE_xTh_{0.1}$.

ately apparent from the equation of state. Note that the entropy term may be written as $\Delta T\Delta V$ only because it was experimentally demonstrated that the volume couples linearly to the entropy. The source of the large entropy in the Ce system is the spin entropy, which for temperatures just above the phase boundary (Fig. 2) is a large fraction of that calculated for isolated Ce atoms (see discussion below). The spin entropy decreases upon going through the transition because of the delocalization of the 4f electrons. The absence of a large spin entropy term in the Sm system reflects the fact that the magnetic ground state of the non-collapsed phase ($Sm^{2+}$) has J = 0 and, hence, $Sm^{2+}$ displays only a weak Van Vleck susceptibility.

To further quantify this we have measured specific heat for a near-critical sample [Fig. 3]. To obtain the entropy change we have integrated under the specific heat curve, choosing a background which smoothly joins the low temperature portion of the curve to $C_p \simeq 4R$ above $T_o$. One obtains $\Delta S = \int C_p dT/T \sim 0.4R$. This is indeed a substantial fraction of the total spin entropy and is confirming evidence for our argument. It is apparent, however, that the full entropy $R\ell n6$ of the J = 5/2 $\gamma$-state is not removed at the transition. This is for two reasons. First of all, insofar as the low temperature state is mixed-valent, with a moderately narrow (~1000 K) level degenerate with the Fermi level one expects an enhanced specific heat coefficient.[9] This is indeed observed[10] in pure $\alpha$-Ce and is found to be $1.54 \times 10^{-3}$ R/K. Integrated up to 150 K this accounts for $\Delta S \sim 0.2R$. The point is that part of the spin entropy of the high temperature state appears as electronic entropy in the low temperature state. Secondly the data above 200 K shows $C_p$ rather larger than the 3R expected for the lattice entropy. This reflects the fact (to be discussed below) that in the $\gamma$-state the moment $\langle\mu^2\rangle \sim \chi T$ continues to set in well outside the region of the phase transition, which will be reflected as well in entropy production.

The slope of the phase diagram reflects not only the $\Delta V\Delta T$ entropy term but also the chemical potential term $C\Delta x\Delta T$. It has been argued that the main effect of alloying is to create an effective pressure field due to size mismatch between solute and host. On the other hand it seems likely that the valency of the solute could play an important rôle independent of size effect. To test this we have undertaken a study of $dT_o/dx$ for different dopants (Fig. 4). The reason for using $Ce_{0.9}Th_{0.1}$ as the starting point is to eliminate the metastable dhcp $\beta$-phase known to exist in pure cerium.

From the data one can make the following table for various dopants:

| | valence | $a_o$(fcc) | $dT_o/dx$ |
|---|---|---|---|
| Sc | 3 | 4.54 Å | ~0 K/at % |
| Gd | 3 | 5.12 | -6.6 |
| La | 3 | 5.30 | -9.4 |
| Th | 4 | 5.03 | +0.4 |

The Sc data is from Gschneidner et. al.[11] and the $a_o$ for Gd is extrapolated from the metallic radius. If one plots $dT_o/dx$ vs the fcc lattice constant one finds that the points for trivalent solutes fall on a line, the slope of which is ~11 K/at % - Å. This gives the magnitude of the size effect. One might expect the size effect to vanish for a dopant with $a_o \simeq 5.00$ Å, (which is half way between $a_o(\alpha) = 4.85$ Å and $a_o(\gamma) = 5.15$ Å). Extrapolating from the data, a trivalent atom of this size will have $dT_o/dx \sim -5.5$°K/at %. This much of $dT_o/dx$ for La and Gd dilution must be due to valency. Similarly, if one assumes the size effect is independent of valency, and draws a parallel line through the point for thorium one finds that valency should give $dT_o/dx \sim +0.8$ K/at %, which is reduced to 0.4 K/at % because of size effects. Such a procedure depends on making the correct choice of $a_o$ for which the size-effect should vanish. In any case it clearly shows that size mismatch (or valence mismatch), taken alone, is not enough to explain the various slopes, but that both effects must be considered simultaneously. For thorium dilution the two effects are nearly equally weighted. (Note also that, although the phase boundaries for La or Gd solutes are not as flat as for thorium there is still at least an order of magnitude difference in the slope of the $Sm_{1-x}Gd_xS$

diagram.)

## THE INTEGRAL VALENCE STATE

Considerable attention has been focused on the intermediate valence state in both Ce and SmS; it is pictured as a situation where the f electrons fluctuate into the band with characteristic frequencies $\sim 10^{-13} - 10^{-12}$ sec; these fluctuations (corresponding to the $V_{sf}$ mixing in the Anderson Hamiltonian) are tied to the lattice, since they involve removal of an electron from the localized orbital. It seems not to be so widely appreciated that if the integral valence phase is magnetic one has the possibility of spin fluctuations of the kind that arise in the Kondo effect; these do not have to be tied to the lattice in that they invoke only a reversal of the sign of the local spin. The condition for the valence transition, viz., that the f-level be near the Fermi energy, makes it extremely likely that such fluctuations should occur. (The effect can be ignored in SmS, where the integral valence phase is non-magnetic.)

To demonstrate this, we consider first the electrical resistivity. [See Fig. 1.d. of Ref. 6] In the γ-state it is large (~90μΩcm) and essentially temperature independent. Of this, perhaps 40μΩcm can be explained away by phonons and impurity scattering. The remainder is undoubtedly connected with the cerium spin; however, it is too large (for spin 1/2) to be explained by ordinary exchange. The implication is that the exchange constant is large and negative, as in the Kondo effect. Since the lattice constant of γ-Ce essentially implies it is trivalent, the level must be fully occupied and below the Fermi level, but close enough to give rise to the large scattering.

As further substantiation of the above picture we present susceptibility measurements for a near-critical sample [Fig. 5.]. Above the phase transition region one sees an asymptotic $1/(T + \theta)$ dependence with $\theta \sim 150$K. Since crystal field splittings of cerium in cubic environments are typically small (e.g. ~ 50 K), it seems probable that this temperature dependence is at least in part due to an inherent spin fluctuation. It has been shown[12] for the (dilute) Anderson model that $\chi \sim N_{eff}/(T + 2T_K)$ for $T_K < T < 20\, T_K$. This fit is a good approximation in the regime where $\chi T = N\langle\mu^2\rangle$ is not saturated. One expects entropy to be produced over the same region, and in fact the specific heat well above the valence transition, shown in Fig. 3, is slightly enhanced over what one would expect from lattice and normal electronic specific heats. [As mentioned above, this also helps explain why the full Nkℓn6 spin entropy is not liberated at the phase transition.]

Lastly, we exhibit the thermopower S for a near-critical sample (Fig. 6.). In the γ state, S is increasing rapidly with decreasing temperature; it gets cut off by the phase transition. [This is clearer in first-order samples, which don't exhibit the rounded maximum, but increase right up to the transition.] The peak value is large and positive (~30μV/K). This is the type of behavior one observes in Kondo systems, and also spin-fluctuation systems (see Kaiser[13], these proceedings).

Our interpretation is that the spin fluctuation temperature is smaller than the phase transition temperature. The susceptibility reflects both this fluctuation and crystal field effects. The thermopower would peak at $T_{SF}$, but is cut off by the phase transition. The resistivity shows the effect of the large negative exchange constant, but only in the

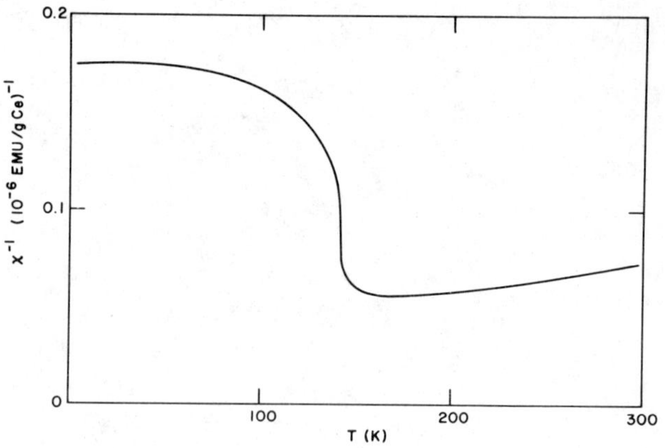


Fig. 5. The inverse magnetic susceptibility of $Ce_{0.732}Th_{0.268}$.

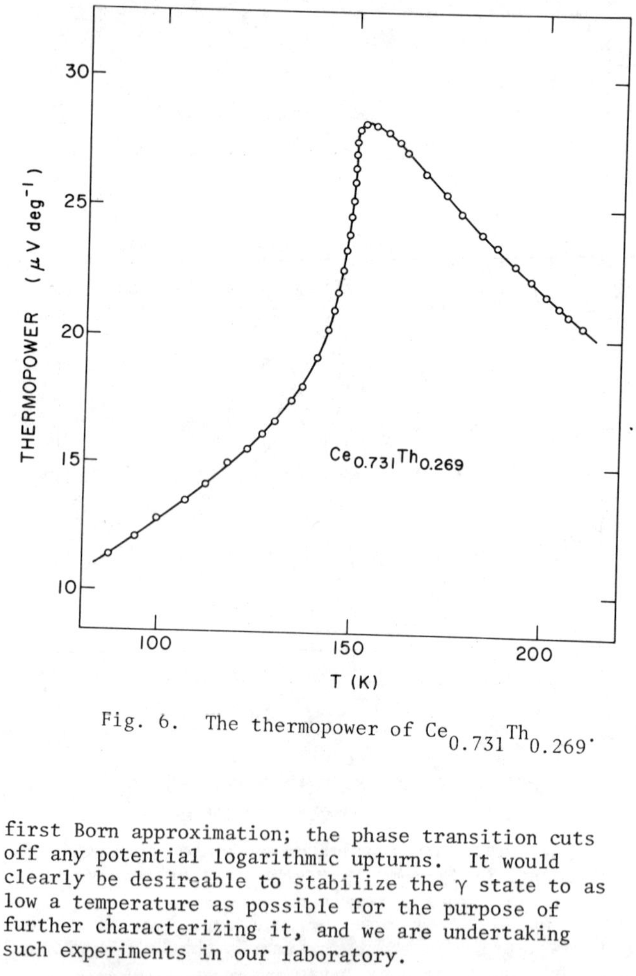


Fig. 6. The thermopower of $Ce_{0.731}Th_{0.269}$.

first Born approximation; the phase transition cuts off any potential logarithmic upturns. It would clearly be desireable to stabilize the γ state to as low a temperature as possible for the purpose of further characterizing it, and we are undertaking such experiments in our laboratory.

## THE INTERMEDIATE VALENCE STATE

We can also obtain quantitative information concerning the low temperature state from our susceptibility measurements. The low temperature results given in Fig. 5 represent the susceptibility after a magnetic impurity contribution was subtracted out. This subtraction was straightforward: by taking a high density of data points in the 2-20 K region, excellent fits to the function $\chi \sim A+B/T$ were obtained. The magnitude of B is 0.5% of the high temperature Curie constant. Since the extrinsic magnetic impurity level in our starting materials was too small to account for such a large value of B, it is reasonable to assume that it is due primarily to the presence of ~1/2% untransformed cerium atoms (perhaps those residing on grain boundaries). (We also suspect that the anamolous low temperature tail seen by Koskimaki and Gschneidner[10] in the susceptibility of α-cerium is due to the same phenomenon). Given this, the susceptibility of α-Ce or α-$Ce_{1-x}Th_x$ ($x \lesssim 0.3$) is essentially constant, with the value $5.7 \times 10^{-6}$ emμ/g-Ce.

This provides a clue to the width of the 4f level in α-Ce. Recently Varma and Yafet[9] have obtained the result $\chi \sim \mu^2/\Delta$ for $T \ll \Delta$ by examining the Anderson model for an isolated impurity spin for the case when the 4f level is pinned to the Fermi level. Analyzing our susceptibility results on $Ce_{0.73}Th_{0.27}$ (Fig. 5) according to this result gives $\Delta \sim 0.06$ eV. This is 3-6 times larger than values typically quoted for 4f level widths. Varma and Yafet obtain a similar formula for the case where the rare earth atoms form a lattice, which is more applicable to the case at hand. In the latter case the parameter Δ takes on a somewhat different significance: it corresponds to the level width of the narrow f-s (or f-d) band which results from the f-s hybridization.

Since the enhancement factor is quite large for α-Ce, we have searched for a paramagnon $T^2$ contribution to the low temperature resistivity; however none was found. Within experimental error the low temperature resistivity is that of a decent metal, with a large temperature independent contribution from the thorium impurities. This is, of course, consistent with the picture that the f-level forms a band, whose width is large enough that any paramagnon effects are too small to detect (which does not rule out their presence.)

## CONCLUDING REMARKS

Having a compact thermodynamic description of the valence transition in $Ce_{1-x}Th_x$ sets the stage for further studies aimed at unraveling the microscopic physics of the mixed valence problem. In terms of the transition itself, it is probable that there are at least two important boot-strapping mechanisms, viz., (1) the acoustic strain effect discussed in some detail above and (2) an excitonic effect first proposed by Ramirez and Falicov,[14] the latter arising from the Coulomb interaction between the f-holes and the conduction electrons. Examining this interaction as an additional term in the Anderson Hamiltonian, Khomskii and Kocharjan[15] have shown that it leads to a broadening of the f-level as it approaches the Fermi level - a new element in the mixed valence problem. These two mechanisms are expected to be important in the Sm-chalcogenides as well as the cerium-based systems; however, in the latter case spin entropy also plays an important role in the valence transition. Clearly, a major question which remains open concerns the relative importance of the various driving terms. Such information will likely come from experiments which are sensitive to the width of the 4f level (e.g., neutron scattering, photoemission, and thermopower).

## ACKNOWLEDGMENTS

We are grateful to T. Penney, C. Varma and Y. Yafet for illuminating discussions and we give special thanks to M. Croft, M. Manheimer, J. Markovics and I. Zoric for permitting us to cite their unpublished experimental results.

## REFERENCES

*Supported in part by the U. S. Army Research Office and by the U. S. Office of Naval Research.

1. J. M. Lawrence, M. C. Croft and R. D. Parks, Phys. Rev. Lett. 35, 289 (1975).
2. A. Jayaraman, P. Dernier and L. D. Longinotti, Phys. Rev. B 11, 2783 (1975).
3. T. Penney and F. Holtzberg, Phys. Rev. Lett. 34, 322 (1975).
4. M. B. Maple, in Magnetism and Magnetic Materials - 1973, AIP Conference Proceedings No. 18, edited by C. D. Graham and J. J. Rhyne (American Institute of Physics, New York, 1974), p. 447.
5. J. M. Lawrence, M. C. Croft, J. M. Markovics and R. D. Parks, these proceedings.
6. R. B. Griffiths and J. C. Wheeler, Phys. Rev. A 2, 1047 (1970).
7. C. M. Varma and V. Heine, Phys. Rev. 11, 4763 (1975).
8. P. W. Anderson and S. T. Chui, Phys. Rev. B 9, 3229 (1974).
9. C. M. Varma and Y. Yafet, to be published in Phys. Rev. (1975).
10. D. C. Koskimaki and K. A. Gschneidner, Jr., Phys. Rev. B 11, 4463 (1975).
11. K. A. Gschneidner, Jr., R. O. Elliott and R. R. McDonald, J. Phys. Chem. -Solids 23, 1191 (1962).
12. H. R. Krishna-murthy, K. G. Wilson and J. W. Wilkins, Phys. Rev. Lett. 35, 1101 (1975).
13. A. B. Kaiser, These proceedings
14. R. Ramirez and L. M. Falicov, Phys. Rev. B 3, 2425 (1971).
15. D. I. Khomskii and A. N. Kocharjan, Preprint, Lebedev Physical Institute, Moscow, 1975.

# CRITICAL SCATTERING SCALING FUNCTIONS AND THE MEASUREMENT OF η

Craig A. Tracy†

Institute for Theoretical Physics
SUNY at Stony Brook
Stony Brook, N. Y. 11794

## ABSTRACT

In the past decade a number of authors have reported the measurement of a non-zero value for the critical exponent η. We analyze the method of data analysis in these critical scattering experiments and conclude that no experiment to date unambiguously and directly establishes that η is greater than zero. We then discuss what we feel would be an unambiguous determination of η. These conclusions are a result of our computation of the k-dependent susceptibility $\chi(\vec{k},T)$ for the two-dimensional Ising model in zero magnetic field. In the scaling limit $k \to 0$, $\xi \to \infty$ such that $y = k\xi$ is fixed, $\chi(\vec{k},T) = \xi^{\gamma/\nu} X_{\pm}(y) + o(\xi^{\gamma/\nu})$ (ξ is the correlation length). We compare these exact results with the various phenomenological scattering approximates (Ornstein-Zernike, Fisher-Burford, etc.). This comparison provides insight into those regions of $y = k\xi$ where these approximates are applicable. Such insight is important since the region of experimentally accessible y is rather limited.

## INTRODUCTION

In the past decade a great amount of work has gone into the study of critical magnetic scattering. If the inelasticity effects are negligible and multiple scattering may be neglected, then the neutron scattering cross section is proportional to the $\vec{k}$-dependent susceptibility $\chi(\vec{k},T)$, where $\vec{k}$ is the momentum transfer. This is the so-called quasielastic approximation.[1] For a simple spin system the $\vec{k}$-dependent susceptibility $\chi(\vec{k},T)$ is related to the static spin-spin correlation function $<\sigma_0\sigma_{\vec{R}}>$ by

$$\chi(\vec{k},T) = \sum_{\vec{R}} e^{i\vec{k}\cdot\vec{R}}[<\sigma_0\sigma_{\vec{R}}>-M_s^2] \quad (1)$$

where $M_s$ is the spontaneous magnetization. For $\vec{k}=0$ $\chi(\vec{k},T)$ reduces to the thermodynamic susceptibility $\chi(T) \equiv\chi(0,T)$.

As T approaches the critical temperature $T_C$, the correlation length $\xi(T)$ and the thermodynamic susceptibility $\chi(T)$ diverge and are usually parametrized by[2]

$$\xi \sim \xi_0^{\pm}|1-T/T_C|^{-\nu} \qquad (T\to T_C^{\pm}) \quad (2)$$

and

$$\chi(T) \sim C_{0\pm}|1-T/T_c|^{-\gamma} \qquad (T\to T_C^{\pm}). \quad (3)$$

The critical exponent $\eta$ is defined[3] by

$$\chi(\vec{k},T_C) \sim \bar{C}_1 k^{-2+\eta} \qquad (T=T_C, k\to 0). \quad (4)$$

One notes that the exponent $\eta$ is defined by a property of the system for T exactly equal to $T_C$, whereas $\gamma$ and $\nu$ are defined by an approach to $T_C$. In Table I se summarize the experimental results for $\eta$ (for magnetic systems).

The scaling region is defined by the limit

$$T\to T_C, \quad k\to 0 \quad (5a)$$

such that the scaled variable

$$y=k\xi \qquad \text{is fixed.} \quad (5b)$$

In this scaling limit Kadanoff[4] and Fisher[2,3] assume that $\chi(\vec{k},T)$ which is a function of two variables $\vec{k}$ and T reduces to essentially a function of one variable

$$\chi(\vec{k},T) = \xi^{\gamma/\nu}X_{\pm}(y)+o(\xi^{\gamma/\nu}). \quad (6)$$

The functions $X_{\pm}(y)$ are referred to as scaling functions.

As $y\to\infty$ the large-y behavior of $X_{\pm}(y)$ defines an exponent $\hat{\eta}$

$$X_{\pm}(y) \sim C_1 y^{-2+\hat{\eta}} \qquad (y\to\infty). \quad (7)$$

The one-length scaling hypothesis states that $\eta$ defined by (4) and $\hat{\eta}$ defined by (7) are the same. Furthermore it is implicity assumed that $C_1$ is non-zero and it reproduces the constant $\bar{C}_1$ in (4). Thus the one-length scaling hypothesis states that

$$\hat{\eta} = \eta \qquad \text{and } C_1\neq 0. \quad (8)$$

Table I. Experimental results for the critical exponent $\eta$ determined by direct measurements of the neutron scattering cross section. The last column gives the reference to this work.

| System | $\eta$ | Ref. |
|---|---|---|
| DAG[a] | 0.12±0.1 | 5 |
| $MnF_2$ | 0.05±0.02 | 6 |
| $RbMnF_3$ | 0.055±0.01 | 7 |
| $K_2NiF_4$ | ***[b] | 8 |
| $K_2CoF_4$ | 0.2±0.1 | 9 |
| $MnTiO_3$ | 0.2±0.15 | 10 |

[a] Dysprosium Aluminum garnet

[b] Recent re-analysis of the data now gives the scattering cross section consistent with the Ornstein-Zernike pole approximation in the range $y\lesssim 10$. R.J. Birgeneau, private communication.

If there were more than one length scale present in the problem, then we would find (8) being violated. From here on we assume $\hat{\eta}=\eta$ and no longer distinguish between $\eta$ and $\hat{\eta}$.

Once one has (8) and demands that the scaling function connect onto (3) in the $y\to 0$ limit and onto (4) in the $y\to\infty$ limit, then one obtains the relationship[2,3,4]

$$(2-\eta)\nu=\gamma. \quad (9)$$

## PHENOMENOLOGICAL SCALING FUNCTIONS

We now discuss the problem of the measurement of $\eta$. First of all, the mere definition (4) of $\eta$ makes it impossible to directly measure for the simple reason that no experiment is performed exactly at $T=T_c$. From an experimental point of view, the definition (7) is more appropriate. Thus a direct measurement of $\eta$ involves measuring cross sections for large y. If one assumes the one-length scaling hypothesis, then small-y measurements allow one to determine $\gamma$ and $\nu$ and by (9) to deduce $\eta$. Unfortunately the mathematical limit $y\to\infty$ in (7) cannot be performed in the laboratory. Experimental values of y range up to 65 but perhaps 20 to 30 is more typical. Thus one has the problem of extrapolating the data into the regime where (7) is a valid approximation to the scaling functions $X_{\pm}(y)$. This extrapolation involves assuming a phenomenological formula for $X_{\pm}(y)$ and using this in a least squares fitting program. Thus to discuss further the measurement of $\eta$ we must discuss various phenomenological approximates for $X_{\pm}(y)$.

In general $X_{\pm}(y)$ are functions which depend on the system under consideration. However, for physically realistic systems no exact calculation of $X_{\pm}(y)$ has ever been carried out. Therefore the phenomenological formulas for $X_{\pm}(y)$ have been of an approximate nature, and over the years a large number of approximates[3,11-18] have been proposed. The most famous of these phenomenological approximates is that of Ornstein and Zernike[11]

$$X_{OZ}(y) = x_0(1+y^2)^{-1} \quad (10)$$

while some of the more recent approximates are those of Fisher,[3]

$$X_F(y) = A(1+y^2)^{-1+\eta/2}, \quad (11)$$

and those of Fisher and Burford,[12]

$$X_{FB}(y) = A\frac{(1+\phi_c^2 y^2)^{\eta/2}}{1+y^2}. \quad (12)$$

One notes that for $y\to\infty$ both (11) and (12) incorporate the expected large-y behavior (7) of $X_{\pm}(y)$. Because of this the "$\eta$" emerging from the least-squares fitting of these formulas to the data is interpreted as a determination of $\eta$. We discuss the validity of this procedure below.

Recently Wu, McCoy, Tracy, and Barouch[19-23] have computed exactly in the scaling limit the spin-spin correlation function $<\sigma_{0,0}\sigma_{M,N}>$ for the two-dimensional Ising model in zero magnetic field. It is possible therefore to compare the above phenomenological approximates with the exact scaling functions $X_{\pm}(y)$ for the two-dimensional Ising model. This has been done by Tracy and McCoy[21] and in Table II we present some of their results.

## DETERMINING $\eta$

Furthermore one can ask how well do the commonly used phenomenological approximates extract $\eta$ ($\eta=\frac{1}{4}$ for the two-dimesional Ising model) from the exact $X_{\pm}(y)$. That is to s we can use the exact values of $X_{\pm}(y)$ as "data" over variou ranges of y and try by using various phenomenological form to extract the exponent $\eta$. From Table II we can see that the Ornstein-Zernike pole dominates $X_{\pm}(y)$ in the region $y<10$. Thus since the Ornstein-Zernike pole term has zero $\eta$ one should not use data from this region in attempting to extract $\eta$. For instance, if one uses the Fisher ap-

Table II. This table compares the exact scaling function $X_+(y)$ for the two-dimensional Ising model with various approximate scaling functions. The approximates $X_{OZ}(y)$, $X_F(y)$, and $X_{FB}(y)$ are all normalized to unity at y=0, and hence, are compared with the scaling function $X_+(y)/X_+(0)$. The quantity $X_{FL}(y)=C_1y^{-7/4}[1+C_2 \times \ell ny+C_3)y^{-1}]$ is the Fisher-Langer approximate. The error in percent is defined as the exact value minus the approximate value, divided by the exact value. The upper row gives the error. Thus, for instance, the Ornstein-Zernike pole approximate $X_{OZ}(y)$ agrees with the exact $X_+(y)/X_+(0)$ to within 5% over the range $0\le y\le 11.2$.

| Approx. | 1% | 5% | 10% |
|---|---|---|---|
| $X_{OZ}(y)$ | $0\le y\le 4.0$ | $0\le y\le 11.2$ | $0\le y\le 21$ |
| $X_F(y)$ | $1170<y<\infty$ | $163<y<\infty$ | $64<y<\infty$ |
| $X_{FB}(y)$ | $0\le y<4.2$ | $0\le y<15$ | $0\le y<\infty$ |
| | $1145<y<\infty$ | $138<y<\infty$ | |
| $C_1y^{-7/4}$ | $1170<y<\infty$ | $163<y<\infty$ | $64<y<\infty$ |
| $X_{FL}(y)$ | $5.4<y<\infty$ | $2.5<y<\infty$ | $1.7<y<\infty$ |
| $X_{FB}(y)$ with LSV[a] | $8.8<y<103$ | $0\le y<310$ | $0\le y<800$ |

[a]Least-squares value.

Table III. This table gives the predicted critical exponent η from a least-squares fit to a phenomenological formula. The formulas are given in the left-hand column. The upper row gives the interval over which the formula was fitted. The number of data points, which are equally spaced and equally weighted, is also given. All data is for $T>T_c$. Recall $\eta=\frac{1}{4}$ is the exact result.

| Approx. | (10,30) 50 pts. | (20,40) 50 pts. | (20,60) 100 pts. |
|---|---|---|---|
| $X_F(y)$ | 0.086 | 0.123 | 0.131 |
| $X_{FB}(y)$ | 0.143 | 0.168 | 0.176 |
| $C_1y^{-2+\eta}$ | 0.095 | 0.125 | 0.133 |
| Fisher-Langer $C_1y^{-2+\eta}+C_2(\ell ny)y^{-3+\eta}$ | 0.299 | 0.264 | 0.260 |
| Fisher-Langer $C_1y^{-2+\eta}+[C_2\ell ny+C_3]y^{-3+\eta}$ | 0.256 | 0.2510 | 0.2507 |
| $C_1y^{-2+\eta}+By^{-3+\eta}\ \frac{y^{\lambda-1}}{\lambda}$ | 0.248 | 0.247 | 0.247 |

oximate $X_F(y)$ [see (11)] to fit the data in the range y≤10,then a least-squares fitting program gives $\eta\approx 0.02$. e results are presented in Table III. One notes there no real difference between the Fisher approximate and e $y^{-2+\eta}$ approximate. The Fisher-Burford approximate ts the data quite well in the sense that the computed lues from the fitted $X_{FB}(y)$ reproduce the exact input lues of $X_+(y)$ to three significant figures. In Table, , the last row, the Fisher-Burford approximate $X_{FB}(y)$ th values $\eta=0.1756$, $\phi_c=0.0571$ and A=2.6687 [those parameters were determined from a least-squares fit of $X_{FB}(y)$ 100 values of $X_+(y)$ equally spaced and equally weighted er the interval $20<y<60$] is compared with the exact $X_+(y)$. e fit is excellent over experimentally accessible y, but e predicted η is in approximately 30% error.

The reason for these large errors is that $X_F(y)$ and $_{FB}(y)$ are not really valid in the range of experimentally accessible y. The "η" appearing in these formulas is the true η from a least-squares-fitting point of view only when $y^{-2+\eta}$ is a good approximation. But as can be seen from Table II this is not the case in the region y<1000.

When $y^{-2+\eta}$ is not a good approximation to $X_\pm(y)$ one must consider the correction terms to (7). Fisher[24] and Fisher and Langer[25] have argued that for $y\to\infty$

$$X_\pm(y) \sim C_1y^{-2+\eta}[1\pm C_2y^{-(1-\alpha)/\nu}+C_3^\pm\, y^{-1/\nu}], \quad (13)$$

with α the critical exponent describing the divergence of the specific heat. For the two-dimensional Ising model ν=1 and $y^{\alpha/\nu}$ is replaced by ℓny (see Ref. 21 for the constants $C_1$, $C_2$, and $C_3$). The form (13), which we call the Fisher-Langer approximate, has more recently been discussed by Stell,[26] Tracy and McCoy,[20,21] Stell and Hochen,[27] Fisher and Aharony,[14] and Brézin,et.al.[28]

We now use the same "data" as above but this time we use the Fisher-Langer approximate as the fitting function. As one sees from Table III the results are remarkably improved. The experimental results listed in Table I used either the Fisher or the Fisher-Burford of the $y^{-2+\eta}$ approximates in analyzing the data. From our least-squares experiment we have shown that these three approximates are not trustworthy in the present experimental range of y values. Furthermore, our "data" suffer no problems of resolution[29] or inelasticity corrections,[30] and still these approximates are unable to extract the exponent η. Also the value $\eta=\frac{1}{4}$ is a large number for η, and thus the two-dimensional Ising model should provide the easiest test for these phenomenological formulas. Thus we must conclude that any analysis of critical scattering data that makes use of the Fisher, Fisher-Burford, or $y^{-2+\eta}$ approximates must be seriously questioned when it comes to extracting the exponent η from the scattering data. Recently Birgeneau[31] has reanalyzed his neutron scattering data for $K_2NiF_4$ and has concluded that his data is consistent with the Ornstein-Zernike pole approximate (1) in the range y≤10. Basically the pole dominates the cross-section to such an extent in the range of experimental y that it is impossible to distinguish between the simple Ornstein-Zernike pole approximate and the more mathematically complicated forms that one expects on theoretical grounds.[14,20-23,28]

We would like to give a series of steps that we feel will lead to an unambiguous measurement of η (this assumes, of course, that the resolution and inelasticity corrections are also make). We consider the case $T>T_c$, and only at the end remark about the case $T<T_c$.

(i) Data must exist in both the large- and small- kξ regions (a priori one doesn't know large kξ from small kξ, but in practice, some estimate for ξ is usually available). The data in the small-kξ region in conjunction with the Ornstein-Zernike pole approximation allows one to determine ξ, and hence the scaled variable y=kξ.

(ii) Test the data to determine if it scales.

(iii) Determine the value of y at which deviations from the Ornstein-Zernike pole first become significant (we denote this value by $y_{OZ}$).

(iv) For data which satisfy $y>y_{OZ}$ (and y>>1) use the Fisher-Langer approximate (13) as a fitting function. To not have too many fitting parameters one might first try setting $C_3^+=0$. As a check on the Fisher-Langer approximate, the value of the exponent α obtained from the least-squares fit should be compared with independent measurements of α.

(v) If the data are good enough to have seen the exponent α, then fixing α to the best known value is perhaps wise. A final fit with the Fisher-Langer approximate with this fixed α then gives an improved estimate for η. Furthermore if the data warrant it, one can include the $C_3^+y^{-1/\nu}$

term to get a better fit. This term may prove important for small $\alpha$.

(vi) The value of $\eta$ obtained should be independent of the cutoff $y_{OZ}$.

(vii) If data exist below $T_C$, then this can provide additional checks on the Fisher-Langer approximate. For instance, the only difference in the second term in (13) above and below $T_C$ is the sign.[25] Furthermore, Hocken and Stell[27] and Brézin, et.al[28] have shown that the ratio $C_3^+/C_3^-$ is the negative of the ratio of the specific-heat amplitudes above and below $T_C$.

## SCALING FUNCTIONS for the TWO-DIMENSIONAL ISING MODEL

In conclusion we give the analytical results of Wu, McCoy, Tracy, and Barouch[20-23] for the scaling functions for the two-dimensional Ising model in zero magnetic field. For a derivation of these results along with a discussion of their analytic properties, the interested reader is referred to Ref. 20-23.

$$X_+(y)=2^{9/4}\pi(\sinh\beta_c E_1+\sinh 2\beta_c E_2)^{1/8}\int_0^\infty d\theta\ \theta[1-\eta(\theta)]$$
$$\exp\left(\int_\theta^\infty dx\ x\ \ln x[1-\eta^2(x)]-h(\theta)\right) J_0(2\theta y), \quad (14)$$

$$X_-(y)=2^{9/4}\pi(\sinh 2\beta_c E_1+\sinh 2\beta_c E_2)^{1/8}\int_0^\infty d\theta\theta\Bigg[[1+\eta(\theta)]$$
$$\exp\left(\int_\theta^\infty dx\quad x\ln x[1-\eta^2(x)]-h(\theta)\right)-2\Bigg]J_0(2\theta y), \quad (15)$$

$$h(\theta)=\left(\frac{\theta\eta'(\theta)}{2\eta(\theta)}+\frac{\theta^2}{4\eta^2(\theta)}\quad[(1-\eta^2(\theta))^2-(\eta'(\theta))^2]\right)\ln\theta, \quad (16)$$

$\eta(\theta)$ is a Painlevé funciton of the third kind[32] and satisfies the differential equation

$$\frac{d^2\eta}{d\theta^2}=\frac{1}{\eta}\left(\frac{d\eta}{d\theta}\right)^2-\frac{1}{\eta}+\eta^3-\frac{1}{\theta}\frac{d\eta}{d\theta} \quad (17)$$

with the boundary conditions

$$\eta(\theta)=-\theta[\ln(\theta/4)+\gamma_E]+O(\theta^5\ln^3\theta) \quad (18)$$

as $\theta\to 0$, $\gamma_E=0.577215\cdots$ is Euler's constant, and

$$\eta(\theta)=1-2\pi^{-1}K_0(2\theta)+O(e^{-4\theta}) \quad (19)$$

as $\theta\to\infty$ where $K_0(x)$ and $J_0(x)$ are Bessel functions. Furthermore it is known[22,23] that $\eta(\theta)$ can be written for sufficiently large $\theta$ as ($\theta=\frac{1}{2}t$)

$$\frac{1-\eta(t/2)}{1+\eta(t/2)}=\sum_{k=0}^{\infty} g^{(2k+1)}(t), \quad (20)$$

$$g^{(2k+1)}(t)=(-1)^k\pi^{-(2k+1)}\int_1^\infty dy_1\cdots\int_1^\infty dy_{2k+1}\prod_{j=1}^{2k+1}\frac{e^{-ty_j}}{(y_j^2-1)^{\frac{1}{2}}}$$
$$\prod_{j=1}^{2k}(y_j+y_{j+1})^{-1}\prod_{j=1}^{k}(y_{2j}^2-1) \quad (21)$$

(for k=0 the last two products are replaced by unity). The analytic structure of $X_\pm(y)$ is shown in Fig. 1.

## ACKNOWLEDGMENTS

The author wishes to thank Professor B.M.McCoy for the many fruitful discussions and Professor T.T.Wu, Professor C.N.Yang, Professor M.Blume and Professor E. W.Montroll for much needed encouragement. We also acknowledge useful discussions with Professor M.E.Fisher and Dr. B. Mozer.

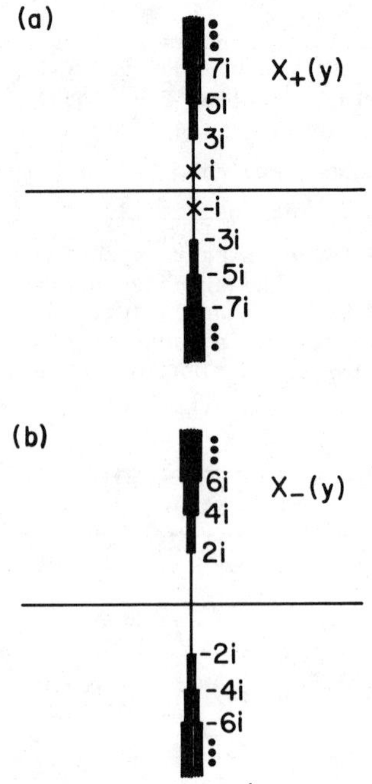


Fig. 1.a) Analytic structure of $X_+(y)$ in the complex y plane. The branch points at $\pm(2n+1)i$, n=1,2,3,..., are square-root-type singularities, and the symbol × denotes a simple pole. The poles are the Ornstein-Zernike poles b) Analytic structure $X_-(y)$ in the complex y plane. $X_-($ has only branch points which are located at $\pm 2ni$, n=1,2, 3,... .

## REFERENCES

†Supported in part by NSF Grant #32998X

1. See, for example, W. Marshall and S. W. Lovesey, The Theory of Thermal Neutron Scattering (Oxford U.P. (1971).
2. M.E.Fisher, Rep. Prog. Phys. 30, 615-730 (1967).
3. M.E.Fisher, J.Math. Phys. 5, 944-962 (1964).
4. L.P.Kadanoff, Physics 2, 263-272 (1966); L.P.Kadanoff, W.Gotze, D.Hamblen, R.Hecht, E.A.S.Lewis, V.V.Palciau M.Rayl, J.Swift, D.Aspnes, and J.Kane, Rev. Mod. Phys 39, 395-431 (1967).
5. J.C.Norvell, W.P.Wolf, L.M.Corliss, J.M.Hastings, and R.Nathans, Phys. Rev. 186, 567-576 (1969).
6. M.P.Schulhof, P.Heller, R.Nathans, and A.Linz, Phys. Rev. B1, 2304-2311 (1970).
7. A. Tucciarone, H.Y.Lau, L.M.Corliss, A.Delapalme, and J.M.Hastings,Phys.Rev. B4,3206-3245 (1971).
8. R.J.Birgeneau, J.Skalyo,Jr., and G.Shirane, Phys. Rev B3, 1736-1749 (1971).
9. H.Ikeda and K.Hirakawa, Solid State Commun. 14,529-53
10. J.Akimitsu and Y. Ishikawa, Solid State Commun. 15, 1123-1127 (1974).
11. L.S.Ornstein and F.Zernike, Proc. Acad. Sci. Amsterda 17, 793(1914); Physik Z. 19, 134 (1918); 27,761(1926)
12. M.E.Fisher and R.J.Burford, Phys. Rev. 156, 583-622 (1967).
13. D.S.Ritchie and M.E.Fisher, Phys. Rev. B5, 2668-269 2693 (1972).
14. M.E.Fisher and A.Aharony, Phys. Rev. Lett. 31, 1238-1241 (1973); Phys. Rev. B10, 2818-2833 (1974).
15. A. Aharony, Phys. Rev. B10, 2834-2844 (1974).
16. H.B.Tarko and M.E.Fisher, Phys. Rev. Lett. 31, 926-930 (1973); Phys. Rev. B11, 1217-1253 (1975).
17. M. Ferer, M.A.Moore, and M.Wortis, Phys. Rev. Lett. 22, 1382-1385 (1969).
18. M.Ferer ane M.Wortis. Phys. Rev. B6, 3426-3444(1972)
19. E.Barouch, B.M.McCoy, and T.T.Wu, Phys. Rev. Lett. 31, 1409-1411 (1973).
20. C.A.Tracy and B.M.McCoy, Phys.Rev. Lett. 31, 1500-1504 (1973).

21. C.A.Tracy and B.M.McCoy, Phys. Rev. B12, 368-387 (1975).
22. T.T.Wu, B.M.McCoy, C.A.Tracy and E.Barouch, Phys. Rev. B13, 316 (1976).
23. B.M.McCoy and C.A.Tracy, unpublished results.
24. M.E.Fisher in Critical Phenomena, edited by M.S.Green and J.V.Sengers, Natl. Bur. Stand. (U.S.)Misc.Publ. No. 273 (U.S.GP),Washington, D.C.,1966),pg.108.
25. M.E.Fisher and J.S.Langer, Phys. Rev. Lett. 20, 665-668 (1968).
26. G.Stell, Phys. Lett. A27, 550-551 (1968).
27. R. Hocken and G.Stell, Phys. Rev. A8, 887-898(1973).
28. E. Brézin, D.J.Amit, and J.Zinn-Justin, Phys. Rev. Lett. 32, 151-154 (1974); E. Brézin, J.C.le Guillou, and J.Zinn-Justin in Phase Transitions and Critical Phenomena, edited by C.Domb and M.S.Green (Academic, N.Y., to be published), Vol. 6.
29. For a discussion of the importance of resolution corrections see J.Als-Nielsen and O.W.Dietrich, Phys. Rev. 153, 706-717(1967), and the discussion in Ref.7.
30. See the discussion in Ref. 7.
31. R.J.Birgeneau, private communication.
32. See E.Ince, Ordinary Differential Equations (Dover, New York, 1945), Chapt. 14.

## SPIN DYNAMICS AND CRITICAL FLUCTUATIONS OF A SITE-RANDOM, TWO-DIMENSIONAL ANTIFERROMAGNET: $Rb_2Mn_{0.5}Ni_{0.5}F_4$

J. Als-Nielsen
AEC Research Establishment Risø, 4000 Roskilde, Denmark

R.J. Birgeneau
Dept. of Physics, M.I.T.*, Cambridge, Mass 02139 and
Bell Laboratories, Murray Hill, N.J. 07974

G. Shirane
Brookhaven National Laboratory+, Upton, N.Y. 11973

ABSTRACT

In $Rb_2Mn_{.5}Ni_{.5}F_4$ the Mn and Ni ions are randomly distributed on a plane square lattice. The exchange energy between neighboring spins is of the Heisenberg form $J\vec{S}_i\cdot\vec{S}_j$ with J=7.4 K, 25.7 K and 89 K for pairs of Mn-Mn, Mn-Ni, and Ni-Ni spins. The mean anisotropy field for the Mn and Ni spins is about 0.7% of the exchange field. The spin dynamics studied by inelastic neutron scattering has been described elsewhere (Phys. Rev. B12, 4963, (1975)) and only the results are summarized here. The dynamic susceptibility $\chi''(q,\omega)$ exhibits two well-defined peaks in frequency $\omega$ for a fixed wavevector q. The dispersion relations of the two modes shown in fig. 1 are well accounted for in a simple 4-sublattice model (J. Phys. C,8, L320 (1975)), and the width of the zone boundary peaks corresponds to the quintet of Ising frequencies of a spin with different configurations of neighboring spins. The results are in good agreement with $\chi''(q,\omega)$ as calculated by the CPA method (J. Phys. C,8, 1889 (1975)) and also with Monte Carlo calculations (Phys. Rev. B12, 4980 (1975)), the only discrepancy being on the intensity of the lower branch for $\xi<0.2$, cf fig. 10 p 4971 and fig. 12 p 4998 in Phys. Rev. B12 (1975).

Our study of the phase transition is described in a Letter of J. Phys. C9, 1976. The critical exponent $\beta$ of the order parameter is 0.16 to be compared with $\beta$ = 0.14 and 0.15 found in $K_2NiF_4$ and $K_2MnF_4$. The small anisotropy in these materials implied that the transverse fluctuations contribute significantly to the critical scattering. Below $T_N$ we found that the transverse correlation length $\kappa_t^{-1}$ is independent of temperature. Above $T_N$ we make the mean field assumption of $\kappa_t \propto (T-T_N^*)^\nu$ with $\nu$ being the critical exponent of the diverging longitudinal correlation range $\kappa_\ell \propto (T-T_N)^\nu$. With this assumption we can unambiguously separate the total scattering into its transverse and longitudinal components. Using this procedure for our data on the alloy and on $K_2MnF_4$, as well as in a reanalysis of the $K_2NiF_4$ data (Phys. Rev. B3, 1736 (1971)), we find the results shown in fig. 2. The pure systems and the random system appear to have common critical exponents close to those of the (2d) Ising model. It is interesting to note that as the anisotropy field relative to the exchange field increases from 0.2% to 0.7% the correlation length moves continuously down towards the (2d) Ising value.

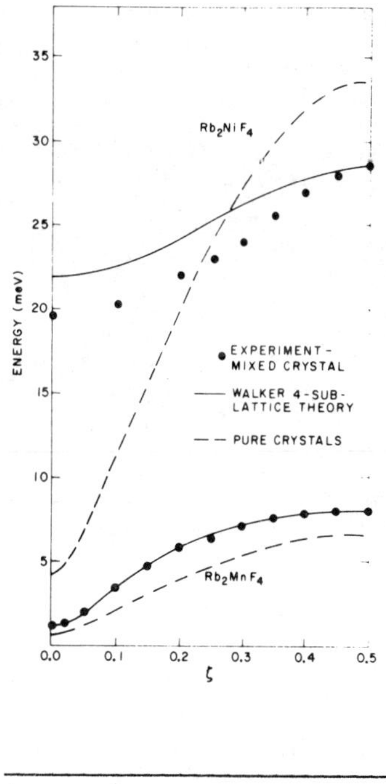


Fig.1 Dispersion relations along the (ζ,0,0) direction.

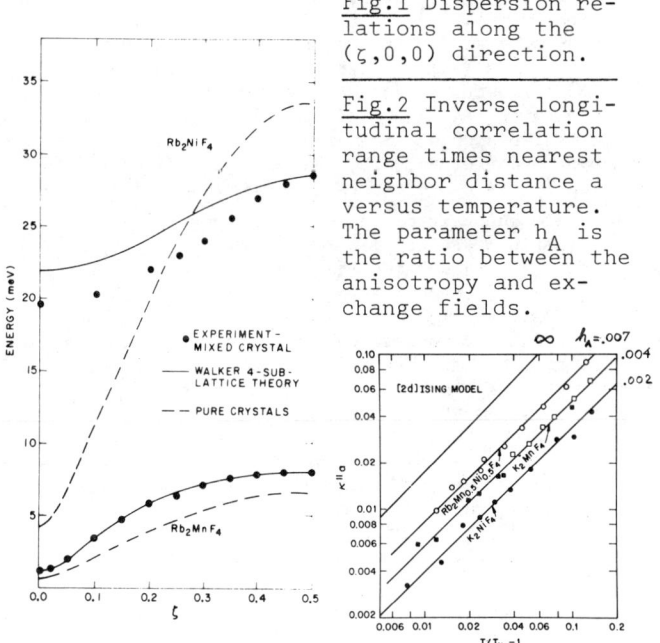

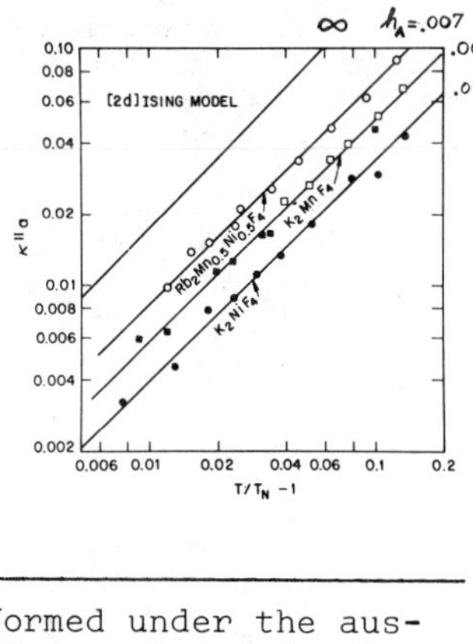


Fig.2 Inverse longitudinal correlation range times nearest neighbor distance a versus temperature. The parameter $h_A$ is the ratio between the anisotropy and exchange fields.

+Work at Brookhaven performed under the auspices of the US Energy Research and development Agency.

*Work at M.I.T. supported by NSF-MRL Grant No. DMR12-03027-ADS

# RANDOMLY DILUTE TWO DIMENSIONAL ISING MODELS*

R. Fisch and A. B. Harris
Department of Physics, University of Pennsylvania
Philadelphia, Pennsylvania 19174

ABSTRACT

Calculations of the specific heat and magnetization of quenched, site-diluted, N x N square and triangular Ising lattices have been carried out by a Monte Carlo method. For spin concentrations x of 0.8 and 0.9, lattices of size N = 64 did not give sharp transitions. For a triangular lattice with N = 128 and x = 0.904, we found a well-defined peak in the specific heat and an abrupt change in the magnetization at $T = 0.865\ T_c(1)$. Linear interpolation gives $s \equiv \frac{d}{dx}\left[\frac{T_c(x)}{T_c(1)}\right]_{x=1} = 1.40 \pm 0.05$, in excellent agreement with the high temperature series calculations of Rushbrooke et al. For the square lattice we calculate $s = 1.5 \pm 0.1$. We also determined site magnetization as a function of the number of "live" nearest neighbors.

## I. INTRODUCTION

For many years there has been great interest in the properties of randomly diluted magnetic systems. A simple theoretical model for such systems, which has been widely studied, is the randomly diluted Ising model, which we write as

$$\mathcal{H} = -J \sum_{\langle ij \rangle} p_i p_j \sigma_i \sigma_j, \qquad (1)$$

where $\langle ij \rangle$ indicates that the sum is over pairs of nearest neighbors, $\sigma_i = \pm 1$ is an Ising variable, and $p_i$ is a random variable equal to unity if the site i is occupied and is zero otherwise. In the "quenched" model which we consider the free energy is taken to be the configurational average, denoted by $\langle\langle\ \rangle\rangle$, of the free energy over the occupation variables. We assume the $p_i$'s to be uncorrelated with $\langle\langle p_i \rangle\rangle = x$.

In early work,[1] the thermodynamic functions of this model were studied via series expansions in powers of either (J/kT), where T is the temperature, or x. More recently, renormalization group treatments have been given[2,3] which indicate that for $2 < d < 4$, where d is the spatial dimensionality, one expects cross-over behavior such that for $\epsilon \equiv [T - T_c(x)]/T_c(1)$ sufficiently small one enters a randomness dominated regime where the critical exponents assume values characteristic of local disorder. Here $T_c(x)$ is the transition temperature (defined by the onset of long range order) for concentration x. In this paper we present a Monte-Carlo study of this model in the hope of elucidating the noncritical behavior. Heretofore, such information has only been obtained by series methods or via approximations[4] wherein fluctuations in local environment are treated in an average way. By using a simulation method we are able to study the importance of these fluctuations and also to give a qualitative picture of the thermodynamic properties over the entire range of temperature.

This paper is organized as follows. In Sec. II we describe briefly the numerical approach used, and we discuss our results in Sec. III.

## II. MONTE-CARLO TECHNIQUE

The calculations were made by a method similar to that used by Ogita et al[5] and by Stoll et al.[6] Toroidal periodic boundary conditions were used. Pseudo-random numbers were obtained using two well-tested generators.[7] One sequence was used to choose the jth spin. The other determined the new value, $\sigma_j$, of the jth spin according to the Boltzmann distribution:

$$P(\sigma_j) = \{1 - \tanh[E_j/kT]\}/2, \qquad (2)$$

where $E_j$ is the energy of the jth spin in the field of its neighbors. The program was checked by performing calculations on an undiluted system and comparing the results with the exact results for a finite lattice given by Ferdinand and Fisher.[8] There were no statistically significant errors in the Monte-Carlo results for x = 1 at any temperature.

For each lattice a distribution of impurities was chosen by using the random number generator. The same impurity distribution was used at all temperatures for a given concentration. On each run, the lattice was allowed to reach thermal equilibrium and then its behavior was averaged over 1000 times steps in order to simulate the thermal average. (On the average, each spin is examined once per time step.) Temperatures of interest were run two or three times.

## III. DISCUSSION

In Fig. 1 we show results for the specific heat obtained by graphically differentiating our Monte Carlo results for the internal energy for two sizes of triangular lattices. The apparent divergences seen in Fig. 1 occur at the same temperature, $T_c(x)$, determined by the onset of long range order. The smaller peak in the specific heat for x = 0.904 at a temperature below the main transition is attributed to the effects of metastable states. It is clear that our data does not permit us to discuss the possible modification of the critical exponents due to random dilution. For the transition temperature we obtained the value $T_c(0.904) = (0.865 \pm 0.005)\ T_c(1)$. Judging from the value of the transition temperature for x = 0.8 as indicated by our results shown in Fig. 1, we conclude that $T_c(x)$ is a nearly linear function of x for $0.9 \lesssim x \leq 1$. Thus, we deduce the limiting slope $s \equiv T_c(1)^{-1}\, dT_c/dx|_{x=1} = 1.40 \pm 0.05$, in excellent agreement with the high temperature series calculations of Rushbrooke et al.[1] We have performed similar calculations for the square lattice which give $s = 1.5 \pm 0.1$.

In Fig. 2 we show $\sigma_z(T)$, the magnetization of sites classified according to the number, z, of "live"

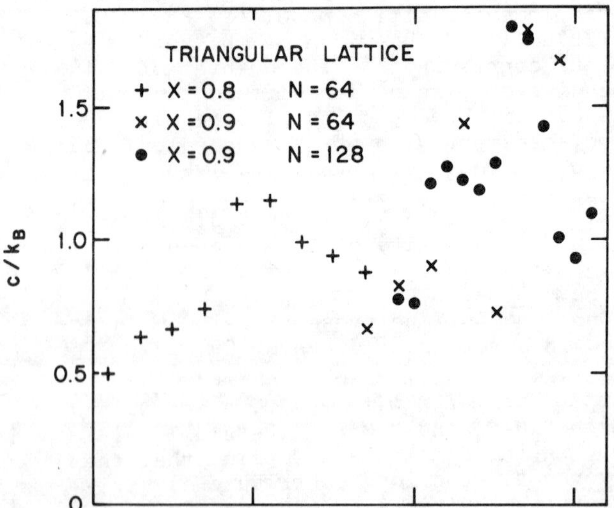


Figure 1

Specific heat per spin, c, of N x N randomly diluted triangular lattices with spin concentration x subject to periodic boundary conditions. The points were obtained by a finite difference method from a direct calculation of the energy.

nearest neighbors. Thus, if $z_i$ is the number of live nearest neighbors of the site i, we define

$$\sigma_z(T) = \sum_{z_i = z} \langle\sigma_i\rangle_T / \sum_{z_i = z} p_i, \qquad (3)$$

where $\langle\ \rangle_T$ indicates a thermal average. In the simplest theories(4) one assumes that $\langle\sigma_i\rangle_T$ is independent of i. However, as our data shows, this is not even approximately true. In fact, we see that the effective field acting on a site with z neighbors is approximately proportional to z. A difficulty in making this idea quantitative is caused by the inaccuracy inherent in the use of mean field theory for a two dimensional model. One can define an effective field via the relation

$$\sigma_z(T) = \tanh[H_{eff}(z)/k_B T]. \qquad (4)$$

For x = 0.904 and T = 0.80 $T_c(1)$ the data of Fig. 2 inserted in Eq. (4) gives $H_{eff}/k_BT$ = 0.38, 0.56, 0.84, 1.13, and 1.45 for z = 2, 3, 4, 5, and 6, respectively. Qualitatively, one has a linear relation between $H_{eff}(z)$ and z. As one would expect, deviations from linearity are such as to cause $H_{eff}(z)/z$ to increase slightly as z increases. Physically, we expect that $\langle\sigma_i\rangle_T$ is almost completely characterized by the value of $z_i$.

In Fig. 2 we see a regime for $0 \le \epsilon \le 0.05$ where magnetization occurs in our finite sample using a limited time average, but probably would not occur for an infinite sample. The size of this interval is the same as that below $T_c(x)$ where the specific heat results (for x = 0.904) are seen to display anomalous behavior.

Domain wall excitations are known(9) to play a role in the dynamics of pure systems. These effects become even stronger in the presence of dilution, since the energy of formation of a domain wall is proportional to the number of live bonds in the wall. In fact, it is clear that in the dilute system, in contrast to the pure case, domain walls in such low energy configurations are metastable and give rise to thermally activated relaxation mechanisms. This domain wall pinning below the transition causes the shape of the fluctuation regions of reversed spin to be more irregular in the dilute system than in the pure case. As a consequence of their irregular shape, large scale fluctuations tend to be unstable with respect to breaking up. Above the transition temperature the energy of the fluctuations is less important than their entropy, and irregularly shaped domains are rarer than regularly shaped ones. For these more circular domains piecewise decay is less likely and a longer relaxation time results. These domains with "giant moments" may explain the occurrence of apparent magnetization in our finite samples for $0 \le \epsilon \le 0.05$ for x = 0.904.

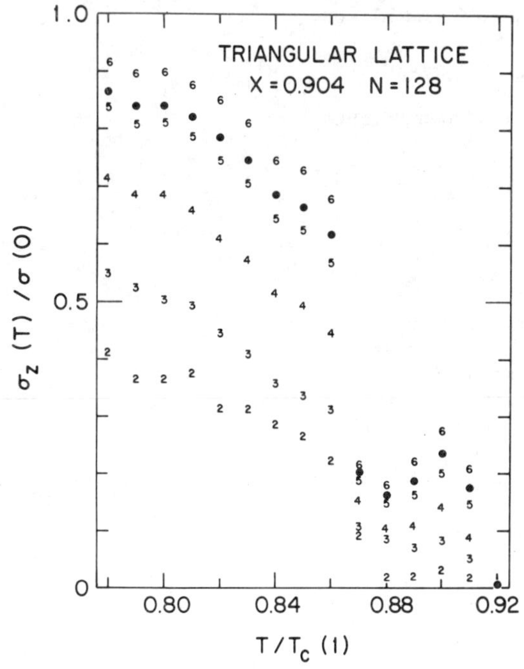


Figure 2

Thermal average magnetization per spin, $\sigma_z(T)$, as a function of temperature, T, for various values of z, the number of "live" nearest neighbors. The dots • show the average magnetization per spin of the entire lattice.

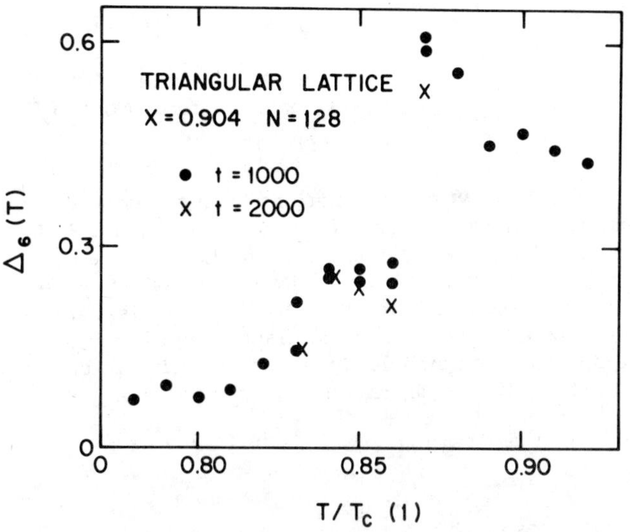


Figure 3

Standard deviation, $\Delta_6(T)$, of the average magnetization of sites with 6 "live" nearest neighbors as defined by Eq. (5) versus temperature.

Numerically, we may study these fluctuations by studying the way they influence the finite time averages we calculate. As in Eq. (3) we may define the dispersion, $\Delta_z(T)$, as

$$\Delta_z^2(T) = \left\{ \sum_{z_i = z} \langle\sigma_i\rangle_T^2 / \sum_{z_i = z} p_i \right\} - \sigma_z(T)^2. \qquad (5)$$

By the central limit theorem, we expect

$$\Delta_z^2(T,t) = \Delta_z^2(T,\infty) + \tau_k/t, \qquad (6)$$

where $\tau_k$ is a kinetic relaxation time. Large values of $\Delta_z(T,\infty)$ indicate that our system is trapped in a metastable configuration. Data for t = 1000 and 2000 time steps in shown in Fig. 3.

## REFERENCES

*Work supported in part by the Office of Naval Research and the National Science Foundation.

1. Rushbrooke, G. S., Muse, R. A., Stephenson, R. L., and Pirnie, K., J. Phys. C 5, 3371-86 (1972), and references therein.
2. Lubensky, T. C., Phys. Rev. B 11, 3573-80 (1974) and references therin.
3. Grinstein, G. and Luther, A., to be published.
4. Handrich, K., Phys. Stat. Sol. 32, K55-58 (1969).
5. Ogita, N., Ueda, A., Matsubara, T., Matsuda, H., and Yonezawa, F., J. Phys. Soc. Jpn. 26, Suppl. 145-9 (1969).
6. Stoll, E., Binder, K., and Schneider, T. Phys. Rev. B 8, 3266-89 (1973).
7. Lewis, P. A. W., Goodman, A. S., and Miller, J. M., IBM Sys. J. 8, 136-146 (1969); Hutchinson, D. W., Comm. ACM 9, 432-3 (1966).
8. Ferdinand, A. E. and Fisher, M. E., Phys. Rev. 185, 832-46 (1969).
9. Schneider, T. and Stoll, E. Phys. Rev. Lett. 31, 1254-9 (1973).

# CRITICAL EFFECTS OF REGULARLY SPACED POINT DEFECTS

M. E. Fisher and Helen Au-Yang
Baker Laboratory, Cornell University, Ithaca, New York 14853

ABSTRACT

Exact analytical calculations[1,2] of the critical behavior of square lattice Ising models with point defects regularly spaced on an $m \times n$ grid, may be related to a suitable scaling theory.[3] Defects considered include: missing magnetic sites, modified or missing exchange bonds, and bent and straight missing bond pairs. The specific heat singularity remains logarithmic but at a shifted critical temperature

$$T_c(x) = T_c(0)[1 - Q_1 x + Q_2 x^2 \ln x - Q_3(\tau)x^2 - Q_4 x^3 (\ln x)^2 + O(x^3 \ln x)] ,$$

where $x = 1/mn$ is the defect concentration and $\tau = n/m$. The amplitude of the logarithmic singularity varies[2] as

$$A(x) = A_0 + A_1 x[\ln x - A_2(\tau)] - A_3 x^2 \ln^2 x + O(x^2 \ln x).$$

The coefficients $Q_1$, $Q_2$, $A_0$ and $A_1$ may be predicted correctly via a scaling theory[3] in terms of the critical amplitudes of the incremental specific heat due to a single, isolated defect. Generalizations can be surmized and the results have some relevance for the theory of random systems.

REFERENCES

1. H. Au-Yang, M.E. Fisher and A.E. Ferdinand, Phys. Rev. B (1976).
2. H. Au-Yang, Phys. Rev. B (1976).
3. M.E. Fisher and H. Au-Yang, J. Phys. C 8, L 418 (1975).

# ISING FERROMAGNET WITH QUENCHED IMPURITIES

E. Stoll
IBM Zurich Research Laboratory, 8803 Rüschlikon, Switzerland

T. Schneider*
IBM T.J. Watson Research Center, Yorktown Heights, New York 10598

Using the Monte Carlo technique, we have studied a two-dimensional Ising model with quenched non-magnetic impurities. In particular, we calculated the temperature dependence of the order parameter of the zero-field susceptibility and of the specific heat. These results indicate that a second-order phase transition is preserved for $p \geq 0.9$ where $1 - p$ denotes the concentration of randomly distributed non-magnetic sites. Moreover, they yield $\beta \approx 1/8$ and $\gamma \approx 7/4$ indicating that these exponents equal the corresponding ones of the pure Ising model. The specific heat, however, exhibits a different temperature dependence. It is found to be smooth at $T_c(p)$ and reaches a maximum at $T > T_c(p)$.

*On leave from IBM Zurich Research Laboratory, 8803 Rüschlikon, Switzerland.

# CRITICAL EXPONENTS ASSOCIATED WITH ISOLATED IMPURITIES IN FERROMAGNETS

T. K. Bergstresser
Physics Department, Clark University, Worcester, MA 01610

ABSTRACT

We have investigated the spin-½ simple cubic Ising model with isolated impurities. This work does not make use of a mean-field approximation, but does use exponents and correlation functions deduced by series expansion methods. Denote by $\sigma$ the bulk magnetization, and denote by $\sigma_i$ either the statistically averaged spin on the impurity or on a neighbor to the impurity. We have obtained detailed numerical results of $\sigma_i/\sigma$ and $\sigma_i$ for $0 \leq T \leq T_c$, and of $(\partial\sigma_i/\partial H)/(\partial\sigma/\partial H)$ for $T \geq T_c$. The quantity $\sigma_i$ has the same critical exponents $\beta$, $\gamma$, $\gamma'$ and $\delta$ as does the bulk magnetization $\sigma$. The leading singular correction terms introduced by the impurity have also been found. This work supports the hypothesis that hyperfine fields on probe atoms which are dilute impurities in the magnetic lattice do indeed measure the bulk exponents for T sufficiently close to $T_c$. We provide quantitative estimates of the distortion in $\beta$ and $\gamma$ produced by an impurity when T is not close to $T_c$.

We present some results of a detailed calculation on the behavior of the spins on lattice sites in the spatial region of an isolated impurity in an Ising-model ferromagnet. These calculations do not use a mean-field approximation. We are motivated to consider this problem by the fact that measurements of hyperfine fields are used to deduce critical exponents in magnetic materials, and the fields are often measured at a probe nucleus which is itself an impurity in the magnetic host. The fact that the impurity probe disturbs the magnetic environment in the region around it, the probe disturbance effect, is the problem that we consider.[1] This model calculation is not directly comparable to experimental results, so that we compare our results to previous theories, all of which make use of a mean-field approximation.

For the results presented here, the host is modeled by a spin-½ Ising model on the simple cubic lattice. To define conventions, the interaction between a pair of lattice sites is

$$H_{ij} = J_{ij}\, \sigma_i \sigma_j \quad \text{and} \quad \sigma_i = S_i^z/S_i \tag{1}$$

Here $S_i^z$ is the z-component of the spin operator on site i, and $S_i$ is the spin (quantum number). The exchange interaction $J_{ij}$ is limited to nearest neighbors but depends on the presence of the impurity. Let J denote the ratio J(host-impurity)/J(host-host), and let S denote the spin on the impurity. The quantities $\sigma_c$, $\sigma_1$ and $\sigma$ denote the canonical ensemble averaged spin on the impurity site, a nearest neighbor site, and the bulk host spin, respectively, in the limit $H \to 0^+$, where H is a uniform external magnetic field. The method of calculation is described elsewhere.[2] The essence of the method is to relate quantities in the lattice-with-defect to perfect-lattice correlation functions. The relations are exact, but to evaluate them one uses approximations for the needed correlation functions.[3]

Fig. 1 shows results for a diamagnetic impurity (J=0). It plots $\sigma_1/\sigma$ as a function of reduced temperature $t=|1-T/T_c|$. Curve 1 shows our results, and curve 2 shows the results of the mean-field (MF) theory of Lovesey and Marshall (LM)[4], using the same hamiltonian. The quantity $\sigma_1/\sigma$ asymptotically approaches a constant in the critical region as a function of t, and the next term in the expansion, the leading singular term, has exponent $(1-\alpha')$ in our result and exponent $\beta$ in the theory of LM. This discrepancy partly explains the greater degree of constancy of curve 1, since $(1-\alpha') \simeq 0.875$ and $\beta \approx 0.312$ or $\beta = 0.5$,

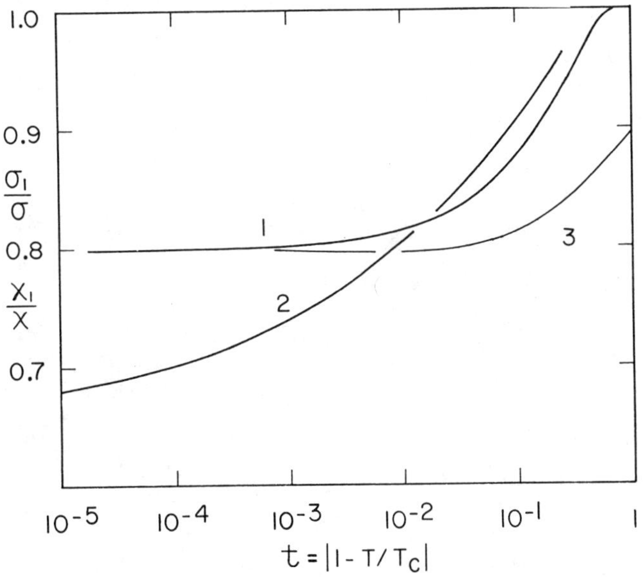


Figure 1. Curves 1 and 2, the spin of the nearest neighbor to a zero-spin defect in the simple cubic lattice, relative to the bulk spin, for $T<T_c$. Curve 1, result for the spin-½ Ising model. Curve 2, result of mean-field theory. Curve 3, the susceptibility of the nearest neighbor to the defect, relative to the bulk susceptibility, for $T>T_c$.

depending on whether the exact or MF bulk-magnetization, $\sigma$, is used in the evaluation of LM's theory. An explanation of the appearance of the exponent $(1-\alpha')$ can be given using an imperfect analogy with a simpler problem. Consider a lattice with undisturbed host exchange constants, but with a localized magnetic field acting on the spin $\sigma_c$ with magnitude adjusted to impose the condition $\sigma_c=0$. It is easily demonstrated for $T \leq T_c$ that

$$\sigma_1/\sigma = (1-\Gamma_1)/(1-\sigma^2) \qquad \text{with } \Gamma_1 = \langle\sigma_c\sigma_1\rangle \tag{2}$$

Here $\Gamma_1$ is a correlation function defined in the perfect lattice[3] and has exponent $(1-\alpha')$.

The quantity $\sigma_1$ is approximately the reduced hyperfine field measured at the impurity nucleus. (Farther-than-nearest neighbors and other effects also play a part.) An estimation of the probe-induced disturbance of the measurement of the exponent $\beta$ is given by

$$\Delta\beta = -\Delta\log(\sigma_1/\sigma)/\Delta\log t \tag{3}$$

For display we choose the temperature range $t=10^{-3}$ to $10^{-2}$ and tabulate $\Delta\beta/\beta$ for our results and the MF results in Table I. The results for other ranges of t may be seen in Fig. 1. Note that the MF result greatly underestimates the degree of constancy of $\sigma_1/\sigma$ in the critical region.

Consider now the exponent $\gamma$. Here we have $T>T_c$. Define

$$\chi = \lim_{H\to 0^+} \partial\sigma/\partial H, \qquad \chi_1 = \lim_{H\to 0^+} \partial\sigma_1/\partial H \tag{4}$$

Curve 3 of Fig. 1 plots $\chi_1/\chi$ v.s. log t. Here our result is even more nearly a constant in the critical region, which implies even less probe disturbance in the measurement of the exponent $\gamma$; see Table I. This is partly explained by the fact that $\chi_1/\chi$ does not go to 1 at t=1, but rather attains the value 1 at $t=\infty$. The leading singular term has exponent $(1-\alpha)$.

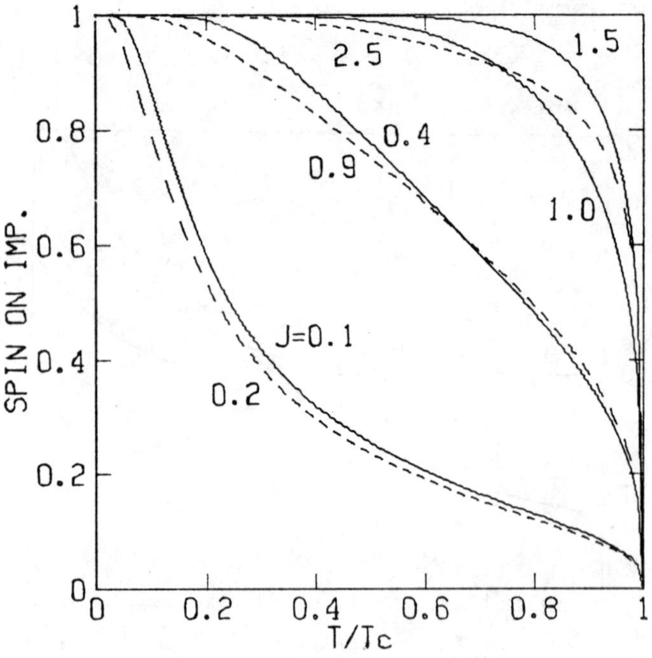


**Figure 2.** Spin on a magnetic impurity in the spin-½, simple cubic Ising model. Solid lines, the spin of the impurity is S=½. Dashed lines, the spin of the impurity is S=5/2. The numbers give the impurity-host exchange constant, relative to the host exchange. S=½, J=1 gives the bulk magnetization, σ.

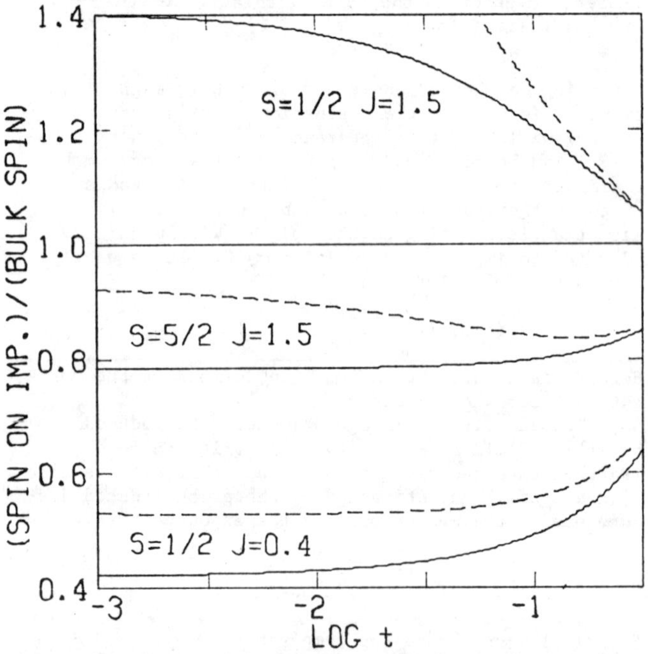


**Figure 3.** Spin on a magnetic impurity, as in Fig. 2. The solid lines give the present results; the dashed lines give the mean-field results.

Fig. 2 shows some of our results for the magnetic impurity. We plot $\sigma_c$ v.s. $T/T_c$ for a selection of values of J and S. Here the reduced hyperfine field is given mainly by $\sigma_c$. (Neighbors do contribute a portion to the field.) The probe-induced distortion of the measurement of β is given in Table I for the case S=½, J=0.4, and S=5/2, J=0.9. The behavior of $\sigma_c$ for these values approximates the case of the $Ru^{99}$ probe in Ni (See Ref. 5). Partly because of our convention in Eqn. 1, a larger J is required for larger S. After J is chosen this way the results are not greatly sensitive to S.

In Fig. 3 we compare our results for the magnetic impurity with MF results.[5,6] The MF calculations were intended to apply to the Heisenberg Model, but they apply with equal or greater validity to the Ising model. We plot $\sigma_c/\sigma$ v.s. log t. For small J the MF calculation does predict the high degree of constancy of $\sigma_c/\sigma$ in the critical region, although the exponent of the leading singular term is wrong; we find $(1-\alpha')$ v.s. $2\beta$ for MF in this case. The MF result fails for large J, as expected.

We summarize our results on exponents in Table II, which gives information of asymptotic probe-disturbance effects, although detailed calculations as in Figs. 1-3 are necessary for a full discussion. The results imply that the numerator and denominator of each line have the same critical exponent; β,γ,γ' and δ respectively. The last two lines have not been discussed in this text, but their interpretation should be clear.

It is usually assumed that a measurement of the hyperfine field on a probe nucleus, even if it is a magnetic impurity, does measure the desired exponents, provided that the temperature is sufficiently close to $T_c$. Our results give no reason to doubt this assumption, and in fact strengthen confidence in it since our results are more favorable than the less-accurate mean-field results previously known.

Table I Distortion of exponents induced by the impurity probe in the reduced temperature range $t=10^{-3}$ to $10^{-2}$. See Eqn. 3. The first three lines refer to the diamagnetic impurity, see also Fig. 1. The last two lines refer to the magnetic impurity, see Figs. 2 and 3.

| | |
|---|---|
| $\Delta\beta/\beta$ | -0.022 |
| $\Delta\beta/\beta$ (MF) | -0.117 |
| $\Delta\gamma/\gamma$ | 0.000 |
| $\Delta\beta/\beta$ (S=½, J=0.4) | -0.009 |
| $\Delta\beta/\beta$ (S=5/2, J=0.9) | -0.007 |

Table II Asymptotic expansions at the critical point. $\sigma_i$ denotes either $\sigma_c$ or $\sigma_1$. For $\sigma_1$, the constant K = 0.7986±0.0013, and $A_1$ = 1.88±0.15. ψ is the exponent of the entropy on the critical isotherm.

$$\lim_{H\to 0^+} [\sigma_i/\sigma] \approx K + A_1 t^{(1-\alpha')} \qquad \text{for } T\to T_c^-$$

$$\lim_{H\to 0^+} \left[\frac{\partial\sigma_i}{\partial H} \bigg/ \frac{\partial\sigma}{\partial H}\right] \approx K + A_2 t^{(1-\alpha)} \qquad \text{for } T\to T_c^+$$

$$\lim_{H\to 0^+} \left[\frac{\partial\sigma_i}{\partial H} \bigg/ \frac{\partial\sigma}{\partial H}\right] \approx K + A_3 t^{(1-\alpha')} \qquad \text{for } T\to T_c^-$$

$$\lim_{T\to T_c} [\sigma_i/\sigma] \approx K + A_4 H^{\psi} \qquad \text{for } H\to 0^+$$

REFERENCES

*Partially supported by the NSF

1. For a discussion of Hyperfine techniques and probe disturbance effects, see C. Hohenemser, in Hyperfine Interactions Studied by Nuclear Reactions and Decay, Contributed Papers, E. Karlsson and R. Wäppling, eds. (University of Uppsala, Sweden, 1974), p. 166, and references therein. Also, for example, C. Hohenemser, T. Kachnowski, and T.K. Bergstresser, to be published; G. K. Shenoy, in Perspectives in Mössbauer Spectroscopy, S.G. Cohen and M. Pasternak, eds. Plunum Publishing Co. (1973); G.K. Wertheim, H.J. Guggenheim and D.N.E. Buchanan, Phys. Rev. Letters 20, 1158 (1968).
2. T.K. Bergstresser, submitted to Phys. Rev. B.
3. H.B. Tarko and M.E. Fisher, Phys. Rev. B11, 1217 (1975); C. Domb, in Phase Transitions and Critical Phenomena, C. Domb and M.S. Green, eds., Academic Press (1974), p. 357.
4. S.W. Lovesey and W. Marshall, Proc. Phys. Soc. (London) 89, 613 (1966).
5. D.A. Shirley, S.S. Rosenblum, and E. Matthias, Phys. Rev. 170, 363 (1968).
6. V. Jaccarino, L.R. Walker and G.K. Wertheim, Phys. Rev. Letters 13, 752 (1964).

# EFFECT OF IMPURITIES ON HYPERFINE CRITICAL EXPONENTS

R. M. Suter and C. Hohenemser
Physics Department, Clark University, Worcester, Massachusetts 01610

ABSTRACT

The critical exponent β has been measured in $Ni_{1-x}Cu_x$ for $0.001 \leq x \leq 0.017$. To determine the average magnetization, we measured the average hyperfine field at dilute impurities of $^{111}Cd$ via perturbed angular correlations. The exponent β was found to be sensitive to x and appears to shift from 0.38 for x = 0 to 0.48 at x = 0.017. The results are in qualitative agreement with renormalization group theory; they also suggest an explanation of exponent shifts reported by Oddou et al. in $\underline{Ni}^{181}Ta$.

The effect of impurities on critical behavior in ferro and antiferromagnets is a problem of general interest to which a number of contributions have been made. Hyperfine experiments by Wertheim[1], and more recently in our laboratory[2], have shown that very dilute or isolated impurities do not disturb the critical exponent β. Using probes which are impurities, these experiments show that $\beta_{hf}$ is probe independent and is equal to the bulk exponent, provided that the dilution requirement is met and an appropriately asymptotic temperature range is used. Similar conclusions have been reached theoretically by Bergstresser[3] in calculations of the spin disturbance due to isolated nonmagnetic impurities in an Ising model. In contrast, Aharony[4] has treated finite concentrations of nonmagnetic impurities via the renormalization group approach, and obtains from this a multitude of fixed points[5], and some cases in which the sharpness of the phase transition is destroyed.

In the present study, we are concerned with the predictions of Aharony; in particular we attempt to see if any of the anticipated fixed points can be isolated, and whether the effective values of β are a continuous function of a parameter such as impurity concentration. Our experiments were carried out on $Ni_{1-x}Cu_x$, with $0.001 \leq x \leq 0.017$. The Cu concentrations were determined by neutron activation analysis through use of the positron emitting $^{64}Cu$ isotope. The critical exponent β was deduced from the temperature variation of the hyperfine field measured at $^{111}Cd$ impurities via perturbed angular correlations (PAC). The 173-247 keV gamma-gamma cascade in $^{111}Cd$ was populated by 2.8 day $^{111}In$ that had been diffused into samples at concentrations of less than $10^{-4}$.

Earlier experiments[2,6] have shown that for a pure Ni host, the hyperfine field at isolated $^{111}Cd$ nuclei yields β = 0.383(4) over a reduced temperature range of $3\times10^{-4} \leq x \leq 2\times10^{-2}$. This result agrees with bulk experiments and with predictions for the universality class (n,d) = (3,3)[5].

The PAC measurements on $Ni_{1-x}Cu_x$ were done in zero applied field on polycrystalline samples in a three counter geometry described previously[6]. For isolated impurity probes in a homogeneous environment, we expect a static hyperfine interaction characterized by a hyperfine field $H_{hf} = \omega_L/\mu$. In this case, the background corrected PAC signal is

$$R(t) = 0.2A_2(1 + 2\cos\omega_L t + 2\cos 2\omega_L t) \qquad (1)$$

where $A_2$ is the cascade anisotropy, and t the delay time between emission of the cascade gamma rays. In a disordered alloy we expect that the probe atoms will experience a distribution of hyperfine fields. Assuming that this a Gaussian, the background corrected PAC signal is

$$R'(t) = 0.2A_2(1 + 2e^{-t^2/4\Delta^2}\cos\bar{\omega}_L t + 2e^{-t^2/\Delta^2}\cos 2\bar{\omega}_L t) \qquad (2)$$

where Δ and $\bar{\omega}_L$ are the distribution width and mean, respectively. The justification for the symmetric nature of the distribution function lies in the fact that near $T_c$, where the damping is evident, the correlation length is large enough to include many impurities. Hence there exists a mean environment, and environments which produce hyperfine fields both smaller and larger than that of the mean. This picture is born out by the symmetric nature of almost all the Fourier transforms of the background corrected data.

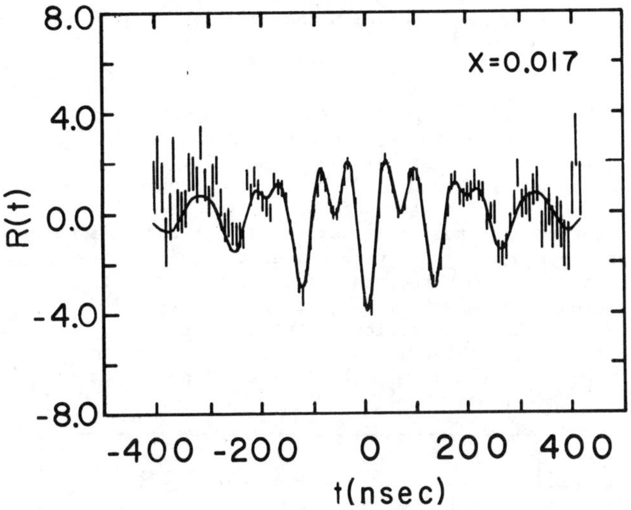


Fig. 1. Background corrected PAC data for the sample with x = 0.017.

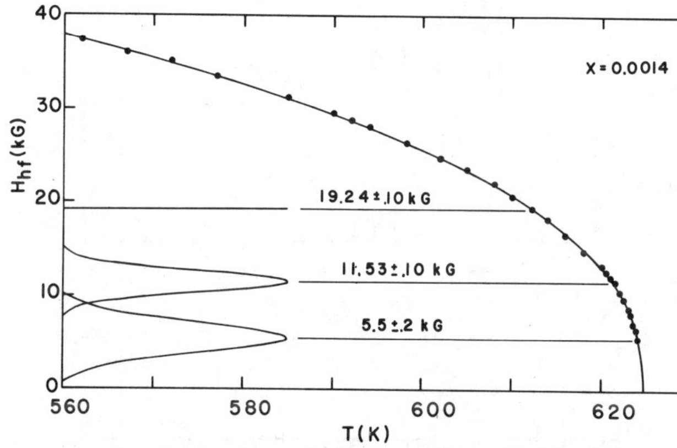


Fig. 2. Variation of the hyperfine field with temperature for the sample with x = 0.0014. At the left are shown three of the distribution functions obtained in the fit of the background corrected data. Errors for the hyperfine field points are less than .2kG and are not visible on the scale of the plot.

In analyzing our data, multi-parameter, nonlinear least squares fits of the background corrected PAC signal were made to Eq. (2), with Δ, $A_2$ and $\bar{\omega}_L$ as independent parameters. In general, the quality of the fits was excellent; an example is shown in Fig. 1. As $T_c$ was approached, the distribution width increased; at the same time the mean frequency remained well defined. The variation of $\bar{\omega}_L$ and Δ are illustrated in Fig. 2 for a sample with x = 0.0014.

The average implicitly taken in deducing $\bar{\omega}_L$ is in principle equivalent to the average performed in making a bulk measurement on an impure magnet; in this sense, determination of $\bar{\omega}_L$ has the character of a bulk measurement. The fit of the spectra with a Gaussian is, of course, somewhat arbitrary, but is not expected to strongly influence the result since the field distribution appears to be rather narrow. An alternative method for deducing $\bar{\omega}_L$ (which is currently being tried) consists of averaging the Fourier transform of the PAC signal. Preliminary analysis shows that deduced mean frequencies will not be substantially altered.

To extract the critical exponent β, the mean frequencies were fitted with the equation

$$\bar{\omega}_L = Dt^{\beta}, \quad t = 1-T/T_c \qquad (3)$$

With D, $T_c$ and β as parameters. Points were weighted with the inverse square of the probable error in $\bar{\omega}_L$. To determine the range of the asymptotic region, a series of fits to Eq. (3) were made in which the maximum value of t was successively reduced. For all concentrations studied it was found that the critical parameters D, $T_c$ and were constant to within statistical error when the reduced temperature was restricted to $t \leq 2\times10^{-2}$, suggesting that $t \leq 2\times10^{-2}$ is a good characterization of the asymptotic region. For the case x = 0.0025, a plot of β as a function of $t_{max}$ appears in Fig. 3, and a logarithmic plot of $\bar{\omega}_L$ vs. t in Fig 4.

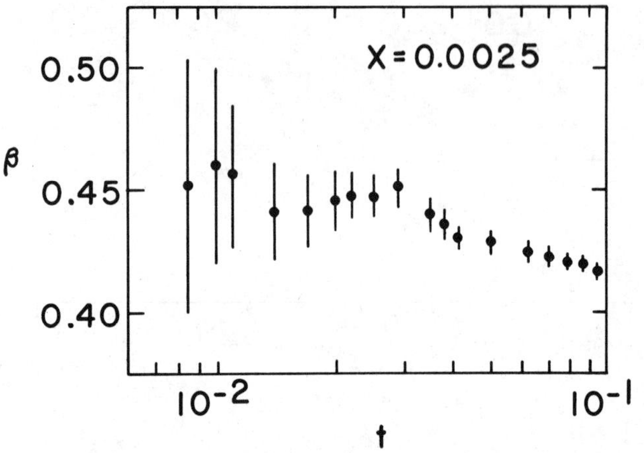


Fig. 3. Range of fit variation of β for the sample with x = 0.0025. The minimum reduced temperature for each fit is 8.6 x $10^{-4}$.

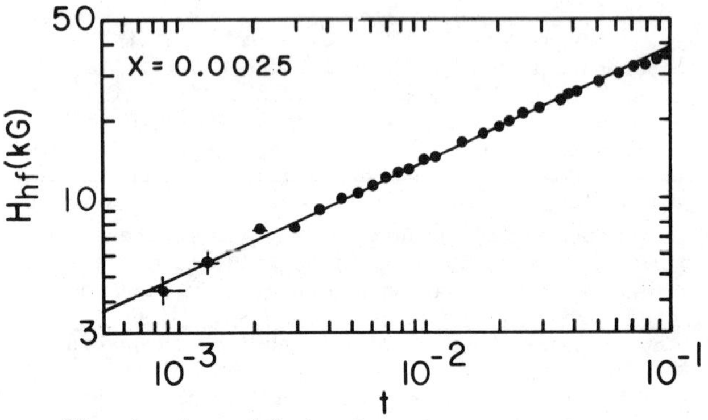


Fig. 4. Logarithmic plot of data for the sample with x = 0.0025

The results of experiments on pure Ni and three alloy systems are shown in Fig. 5. It is apparent that even for quite small Cu concentrations the exponent β is substantially shifted with respect to pure Ni. The case of x = 0.017 suggests a saturation of the β shift at β ≃ 0.48. A tentative interpretation of the observed behavior is that β varies continuously with impurity concentration until a new fixed point with β ≃ 0.48 is reached.

Aside from implications for general theory, the results suggest a new interpretation of results on $Ni^{181}Ta$ reported in 1973 by Oddou et al[7]. In these experiments, sources containing 3ppm Hf were reported to yield a hyperfine exponent of β = 0.417(10). This result is not only in disagreement with hyperfine measurements on Ni with very dilute impurity probes, but also with the previously mentioned principle that hyperfine exponents for very dilute impurity probes are undisturbed. Recently, Oddou et al.[8] have reported that the 1973 sources had other impurities (mostly Al and Si) previously not recognized, in total concentration of $1\text{-}3\times10^{-3}$. This suggests that the β value reported earlier does not violate probe independence for isolated probes as thought, but merely reflects a β shift in accord with our current results.

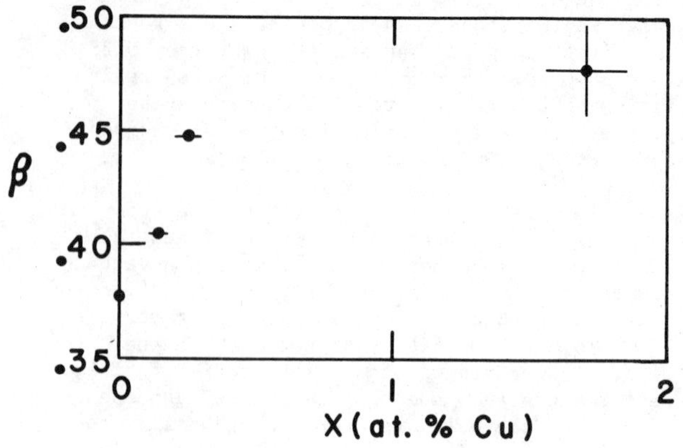


Fig. 5. β vs. x in $Ni_{1-x}Cu_x$.

References

† Research supported in part by the National Science Foundation

1. G.K. Wertheim, H.J. Guggenheim and D.N.E. Buchanan, Phys. Rev. Letters 20, 1158 (1968).
2. C. Hohenemser, T.A. Kachnowski and T.K. Bergstresser, Phys. Rev., to be published.
3. T.K. Bergstresser, Phys. Rev., to be published.
4. A. Aharony, Phys. Rev. B12, 1038 (1975).
5. M.E. Fisher, Rev. Mod. Phys. 12, 76 (1974).
6. T.A. Kachnowski, A.M. Gottlieb and C. Hohenemser, in Hyperfine Interactions Studied in Nuclear Reactions and Decay, Contributed Papers, E. Karlsson and R. Wappling, eds. (Uppsala, 1974), p. 168.
7. J.L. Oddou, J. Berthier and P. Peretto, Phys. Lett. 45A, 445 (1973).
8. J.L. Oddou, P. Peretto and J. Berthier, contribution to International Meeting on Hyperfine Interactions, Leuven, Belgium, September 10-12, 1975; also J.L. Oddou, private communication.

# CRITICAL BEHAVIOR OF THE BAND GAP IN FERROMAGNETIC SEMICONDUCTORS*

J. S. Helman and I. Balberg+
Centro de Investigación del I.P.N., Mexico 14, D.F., Mexico

and

S. Alexander
Racah Institute of Physics, Hebrew University, Jerusalem 91000 Israel

ABSTRACT

The shift of the band gap in ferromagnetic semiconductors due to magnetization and critical fluctuations is determined in detail using recent results of the theory of critical phenomena. The band shift $\Delta E$ is expressed in terms of the reduced magnetic field h and the reduced temperature t. The dominant contributions to $\Delta E$ are found to be the following: In the critical region, both above and below the critical temperature $T_c$, $\Delta E \propto -h^{1/\delta} - B_0 h^{(1-\alpha)/\beta\delta}$ for large h. For small h, $\Delta E \propto (-|t|^{\beta} - B_0|t|^{1-\alpha})$ for $T < T_c$ and $\Delta E \propto -1.7\, t^{-\gamma} h + B_0 t^{1-\alpha}$ for $T > T_c$. Outside the critical region, below $T_c$, $\Delta E \propto (-|t|^{1/2})$ for small h and $\Delta E \propto -h^{1/3}$ for large h; above $T_c$, $\Delta E \propto -h/t + B_0 t^{1/2}$ for small h and $\Delta E \propto -h^{1/3} + B_0 h^{1/3}$ for large h. Here $\alpha$, $\beta$ and $\delta$ are the usual critical exponents and $B_0 \approx 1$.

## INTRODUCTION

The temperature and magnetic field dependence of the energy gap in ferromagnetic semiconductors has been studied for ten years already,[1-5] and the general features of these dependences are well understood today.[6,7] On the other hand the behavior of these dependences near the critical point $T_c$ were not established in detail neither experimentally[8] nor theoretically. Using recent results of the theory of critical phenomena we have determined, in a previous study[9], the temperature dependence of the energy gap when no magnetic field is applied. In the present work an extension of this study is made to include the temperature dependence of the band shift in the presence of a magnetic field.

The calculation is done by using the Ising model result for the band shift[6,9]

$$\Delta E = -\frac{J}{2} S m + \frac{J^2}{4} \frac{\Omega}{(2\pi)^3} \int_0^{\pi/a} d\vec{q}^{\,3} \frac{\Gamma_{\vec{q}}}{E_{\vec{k}} - E_{\vec{k}-\vec{q}}} . \qquad (1)$$

Here, J is the exchange coupling constant between a conduction electron in the state $\vec{k}$ and the localized spin $\vec{S}$ of a magnetic ion, $\vec{q}$ is the electron momentum transfer, $E_k = \hbar^2 k^2/2m^*$ is the electron energy, m is the reduced magnetization, $\Gamma_q$ is the spin correlation function, a is the lattice constant and $\Omega$ is the volume per spin.

The calculations are performed for two regions. For the critical region we incorporate group renormalization results[10,11] for the magnetization and for the correlation function into Eq. (1). For the classical region the Ornstein-Zernike correlation function with the magnetization dependent correlation length, and the classical equation of state, are used.[12]

The results obtained are analyzed for the cases of "small" and "large" magnetic fields and are presented in a form which enables easy comparison with experiment.

## THE CORRELATION FUNCTIONS AND THE EQUATIONS OF STATE

The renormalization group result for the asymptotic high momentum (x>>1) correlation function for an Ising ferromagnet in the critical region is given by[11]

$$\Gamma(x,t,m) = G_0 \frac{t^{-\gamma}}{x^{2-\eta}} \left\{ A - \frac{B}{x^{1/\nu}} - \frac{\Phi(z)\, z^{\alpha-1}}{x^{(1-\alpha)/\nu}} \right\} . \qquad (2)$$

Here, $t=(T-T_c)/T_c$, $z=t/m^{1/\beta}$, $x=q/\kappa$ and $\kappa=\kappa_0 t^{\nu}$ is the reciprocal of the correlation length. The exponents $\alpha$, $\gamma$, $\delta$, $\nu$, $\eta$ are the usual critical exponents. Up to first order in the dimensionality parameter $\varepsilon$,

$$\Phi(z) = D\left[C - (\beta/18\alpha) z + O(z^2)\right] \quad \text{for } |z| \ll 3 \qquad (3)$$

and

$$\Phi(z) = D\left[-\frac{\beta}{18\alpha} z^{1-\alpha} + 0.29\,\gamma\, z^{\gamma-1}\right] \quad \text{for } z \gg 1 \qquad (4)$$

where $C=1.03\beta/(1-\alpha)$ and $F=0.46$ . The values $A=0.962$, $B=3$ and $D=11.13$ are determined by using Eqs. (2) and (4) in the limit $z \to \infty$ and by comparing with the corresponding result[10] for m = o.

In the classical region we use the Ornstein-Zernike approximation[12,13]

$$\Gamma_q = \kappa_0^2\, S(S+1)/(\kappa_1^2 + q^2) \qquad (5)$$

where

$$\kappa_1^2 = \kappa_0^2 (t + m^2)/(1 - m^2) . \qquad (6)$$

The t and h dependences are easily determined by the proper equation of state. In the critical region and up to order zero in $\varepsilon$ the equation of state is given by[11]

$$h/m^{\delta} = 1 + t/m^{1/\beta} \quad \text{for } t/m^{1/\beta} \ll 1$$

and

$$h/m^{\delta} = 0.58 (t/m^{1/\beta})^{\gamma} \quad \text{for } t/m^{1/\beta} \gg 1 \qquad (7)$$

while in the classical region it is given by[12]

$$h/m = t + m^2/3 \qquad (8)$$

where h is the reduced magnetic field. For determination of the dominant behavior it is enough to consider the magnetization to the lowest order in t. The results of the developments to this order are shown in Table I. In the table "large" h in the critical region corresponds to z<<3 for t>0 and to $|z| \ll 1$ for t<0. "Small" h corresponds to z>>3 for t>0 and to $|z| \lesssim 1$ for t<0. In the classical region for both t>0 and t<0 "large" h is simply $h \gg |t|$ and vice-versa.

TABLE I. $m(t,h)$

| region | $T > T_c$ | $T < T_c$ |
|---|---|---|
| critical large $h$ | $h^{1/\delta}\left(1 - t/\delta h^{1/\beta\delta}\right)$ | $h^{1/\delta}\left(1 + \lvert t\rvert/\delta h^{1/\beta\delta}\right)$ |
| critical small $h$ | $1.7\, t^{-\gamma} h$ | $\lvert t\rvert^{\beta}\left(1 + \beta h/\lvert t\rvert^{\beta\delta}\right)$ |
| classical large $h$ | $(3h)^{1/3}\left[1 - t/(3h)^{2/3}\right]$ | $(3h)^{1/3}\left[1 + \lvert t\rvert/(3h)^{2/3}\right]$ |
| classical small $h$ | $h/t$ | $\sqrt{3}\lvert t\rvert^{1/2}\left(1 + h/2\sqrt{3}\lvert t\rvert^{3/2}\right)$ |

TABLE II. $\Psi(t,h)$

| region | small $h$ | large $h$ |
|---|---|---|
| critical $T < T_c$ | $-\lvert t\rvert^{1-\alpha}$ $-\frac{18\alpha(1-\alpha)}{19\alpha-1} h \lvert t\rvert^{1-\alpha-\beta\delta}$ | $-h^{(1-\alpha)/\beta\delta}$ $+(1-\alpha)\left(\frac{1}{\beta\delta}+\frac{1}{18\alpha}\right) t\, h^{-\alpha/\beta\delta}$ |
| critical $T > T_c$ | $t^{1-\alpha}$ $-\frac{15\alpha\gamma}{\beta} t^{-\gamma-1} h^2$ | $-h^{(1-\alpha)/\beta\delta}$ $+(1-\alpha)\left(\frac{1}{\beta\delta}+\frac{1}{18\alpha}\right) t\, h^{-\alpha/\beta\delta}$ |
| classical $T > T_c$ $2k \gg \kappa_1$ | $t^{1/2} + \frac{1}{2} h^2 / t^{5/2}$ | $(3h)^{1/3} - \frac{1}{2} t/(3h)^{1/3}$ |
| classical $T > T_c$ $2k \ll \kappa_1$ | $-t^{-1/2} + \frac{1}{2} h^2 / t^{5/2}$ | $-(3h)^{-1/3} - \frac{1}{2} t/(3h)^{1/3}$ |

CRITICAL BAND SHIFT IN FERROMAGNETIC SEMICONDUCTORS

*In the neighborhood of $T_c$ one has to consider both terms of Eq.(1). This is because the first term (the contribution of the first Born approximation) which is proportional to the magnetization may vanish. Thus, the critical behavior of $\Delta E$ is determined both, by the $m(t,h)$ and by the $\Gamma_q(t,h)$ given above. One can follow then the procedure used previously[9] to show that in the critical region the second term in Eq.(1) is proportional to $-\Phi(z)m^{(1-\alpha)/\beta}$. Using this, the behavior of $\Delta E$ below and above $T_c$ is determined from Eqs.(3) and (4) and Table I. This behavior is summarized in the first two rows of Table II.*

*In the classical region for $T>T_c$ the use of Eqs.(5) and (6) in the integral of Eq.(1) leads to:*

$$\Delta E = -\frac{J}{2} S m - \frac{3J^2 S(S+1) m^* a}{8\pi\hbar^2 k}\left[\operatorname{arctg}\frac{2k}{\kappa_1} + O(\kappa_1 a)\right]. \quad (9)$$

*When $2k \gg \kappa_1$ the second term of Eq.(9) will be proportional to $\kappa_1$ while for $2k \ll \kappa_1$ this term will be proportional to $-\kappa_1^{-1}$. These results are summarized in the last two rows of Table II.*

*Outside the critical region, for $T<T_c$, the magnetization plays the dominant role. This classical temperature region will be considered in detail elsewhere.*

DISCUSSION

*The above calculation was made for $k>0$ since the second Born approximation diverges when $k=0$. In this case an infinite number of diagrams have to be considered.[6] However in the experiments[3] the optical absorption edge at a finite absorption is measured and thus $k>0$.*

*To assess the relative contribution of the two terms of Eq.(1) let us estimate their coefficients. If we write*

$$\Delta E \propto -m(t,h) + B_0 \Psi(t,h) + \text{constant} \quad (10)$$

*where m is one of the elements in Table I and $\Psi$ is the corresponding element in Table II, then, for $2k \gg \kappa_1$, $B_0 \simeq 0.146\ JS/E_k$[6,9]. Using typical values such as $E_k=0.1eV$ and $JS=0.5eV$ we get $B \simeq 0.7$, i.e., the coefficients of the two terms in Eq.(10) are of the same order of magnitude. Thus, their relative contributions can be directly determined by substituting the experimental h and t into the corresponding elements in Tables I and II.*

REFERENCES

*This work was done within the framework of the Joint Program of Scientific Collaboration between the Israel National Council for Research and Development and the Mexican National Council for Science and Technology. One of us (J.S.H.) would like to thank Research Corporation for partially supporting this work.

+On leave of absence from The Hebrew University, Jerusalem, Israel.


1. P.Wachter, Crit.Rev.Solid State Sciences 3, 189-241 (1972).
2. G.Busch and P.Wachter, Phys. kondens. Materie 5, 232-242 (1966).
3. G.Harbeke and H.Pinch, Phys.Rev.Letters 17, 1090-1092 (1966).
4. H.W.Lehmann and F.P.Emmenegger, Solid State Commun. 7, 965-968 (1969).
5. T.Mitani and T.Koda, Proc. XII Int. Conf. on the Physics of Semiconductors (B.G.Teubner, Stuttgart 1974) pp. 889-893.
6. F.Rys, J.S.Helman and W.Baltensperger, Phys. kondens. Materie 6, 105-125 (1967).
7. C.Haas, Phys. Rev. 168, 531-538 (1968).
8. The authors know only about one attempt to determine the temperature dependence of the band gap close to $T_c$. The attempt however was indirect (the measurement was of the conductivity-activation-energy) and limited to $T/T_c<0.99$. This is the work on EuO of F.Holtzberg, T.R.McGuire, T.Penney, M.W.Shafer and S.Von Molnar, IBM Corporation Final Technical Report, Contract report DAAH01-70-C-1309, June 1971. Unpublished.
9. S.Alexander, J.S.Helman and I.Balberg, Phys. Rev B, January 1st (1976).
10. M.E.Fisher and A.Aharony, Phys. Rev. Letters 31, 1238-1241 (1973).
11. E.Brezin, J.C.Le Guillou and J.Zinn-Justin, Phys. Rev.Letters 32, 473-475 (1974), and in Phase Transitions and Critical Phenomena, Vol.IV (C.Domb and M.S.Green Eds.) to be published by Academic Press.
12. H.B.Tarko and M.E.Fisher, Phys. Rev. B11, 1217-1253 (1975).
13. I.Balberg, S.Alexander and J.S.Helman, Phys. Rev. Letters 33, 836-839 (1974).

# NEW MÖSSBAUER STUDY OF THE CURIE POINT IN IRON†

M. A. Kobeissi and C. Hohenemser
Department of Physics, Clark University, Worcester, Mass. 01610

ABSTRACT

Using a highly controlled two stage furnace, we have done careful new measurements in the Fe critical region using the Mössbauer Effect. The value of $\beta$ is found to be 0.379 (4) for the reduced temperature range $10^{-3} < t < 2\times10^{-2}$. This agrees with theory, bulk measurements on Fe, and most data on other ferromagnets in the universality class (N,d) = (3,3), but disagrees with an earlier Mössbauer measurement. In addition we have searched for the line shift anomaly reported by Preston and do not find any to within the statistical uncertainty of our data.

According to the universality principle, the values of static critical exponents in magnets are primarily determined by effective spin dimensionality, N, and lattice dimensionality, d.[1] Recently, renormalization group methods have provided expansions in $\varepsilon = 4-d$ for static critical exponents in all physically interesting universality classes. The results by and large agree with previously derived high temperature series values, and for the first time give a complete picture of the variation of critical exponents with universality class.[2] From an experimental point of view one would like to test this picture.

In this paper we are concerned with Fe, an isotropic ferromagnet with (n,d) = (3,3). Its critical exponents should (a) be equal to predictions for the class (N,d) = (3,3); and (b) be equal to measured exponents for other members of the class. For the exponent $\beta$ in Fe neither of these conditions appear to apply to earlier Mössbauer measurements, though there appears to be no problem with bulk results (see below). We have therefore done a new measurement of $\beta$ in Fe, and report on it in the following.

As in previous work by Preston[3,4] the Mössbauer Effect (ME) was used, and the assumption is made that in the critical region the reduced hyperfine field, h, is equal to the reduced spontaneous magnetization, $\sigma_s$. Thus, for the present purposes, the exponent $\beta$ is defined by the asymptotic relation

$$\lim_{\substack{T\to T_c \\ H_a\to 0}} h = Dt^{\beta}; \quad t = 1 - T/T_c \tag{1}$$

where $T_c$, $H_a$ and D are the Curie temperature, the applied field and a dimensionless scale parameter, respectively. ME determinations of h in ferromagnets have the advantage of permitting measurements with $H_a = 0$ and thereby avoid any assumptions about the magnetic equation of state near $T_c$. That $\sigma_s = h$ in the critical region has been demonstrated through extensive comparisons between hyperfine and bulk measurement in various materials.[5]

Unlike Preston, who did an absorber experiment, our ME results were obtained with an Fe source mounted inside a furnace. Our choice of a source experiment was dictated by our judgment that temperature control and uniformity is the most important requirement for an improved experiment, and that success in this could be more easily achieved for a relatively small source "spot" than for a much larger absorber surface.

The source was made by electroplating 2.8 mC of $^{57}Co$ on a 3 mm diameter area of 99.99% pure Fe foil, and diffusing the activity in an ultra-pure $H_2$ atmosphere for 15 hours. This produced a nearly uniform distribution in the 0.0015 cm thick foil with an estimated concentration of radioactive $^{57}Co$ of less than $4\times10^{-4}$. At this concentration, the effects of impurity interactions can be expected to be negligible even in the critical region. The truth of this assertion is demonstrated by the absence of line broadening for $t > 10^{-3}$.

As absorber we used a 25 mm diameter matrix of $K_4Fe(CN)_6\cdot3H_2O$ enriched to 90% in $^{57}Fe$, and containing 0.25 mg/cm$^2$ of $^{57}Fe$. The absorber, which was at room temperature, was driven by a constant acceleration drive. For an emission spectrum of natural width an experimental linewidth of 0.35 mm/sec is expected.[6] The difference between this value and our observed linewidth of 0.398 (5) mm/sec at room temperature can be accounted for by a small amount of resonant self-absorption in the source.

Our specially designed vacuum furnace, details of which will be published,[7] consisted of a tubular outer heater and a disk shaped inner heater, both with bifilar windings. Near the Curie point of Fe the temperature gradient across the source was found to be less than 0.05 K. The furnace stability was 0.02 K per day. To create an environment nearly free of magnetic fields, the furnace was shielded with mu-metal from stray fields of the Mössbauer drive, with the result that at the source the magnetic field was less than 0.15 G. Regular calibration procedures indicated that the velocity calibration and velocity zero were stable to within 0.1%. To fit the spectra a standard multi-parameter non-linear least squares program was used in which line intensity, position and width were all treated as free parameters. In this way the hyperfine field was determined to an accuracy of better than 0.5 kG over $10^{-2} < t < 0.7$, and to an accuracy of better than 1 kG for $10^{-3} < t < 10^{-2}$. Though data were obtained for $t < 10^{-3}$ spectra from this region were excluded because they showed no discernable structure other than a single broad line.

Hyperfine fields were fitted with Eq. (1) with D, $T_c$ and $\beta$ as parameters. In order to examine the extent of the critical region, a series of fits were made in which the maximum falue of t was successively reduced. The result was that for all values of $t_{max} < 2\times10^{-2}$ the values of D, $T_c$ and $\beta$ obtained as a function of $t_{max}$ appear in Fig. 1.

Our conclusion from the above analysis is that the asymptotic region for Fe is defined by $t < 2\times10^{-2}$, in good agreement with expectations.[5] For this region our best estimate of all three critical parameters is

$$\beta = 0.379\ (4); \quad D = 1.66\ (3); \quad T_c = 1042.91\ (4)$$

As a check on the above procedure, $T_c$ was also determined by thermal scanning: i.e. the transmission at the centroid of the spectra was determined in the

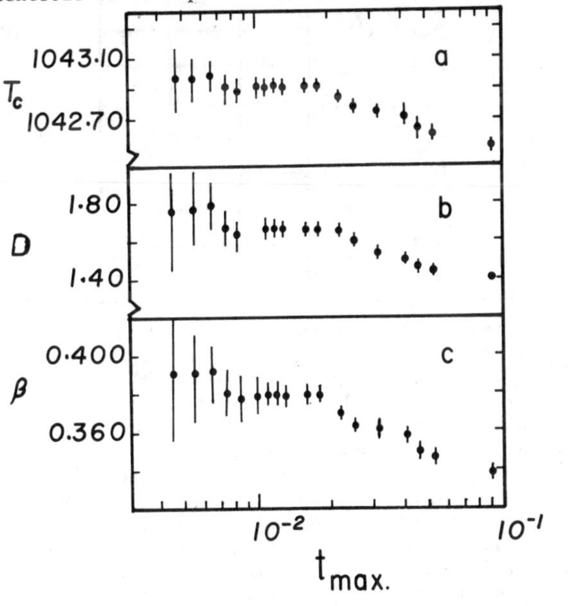


FIG. 1. Variation of $\beta$, D and $T_c$ with range of fit.

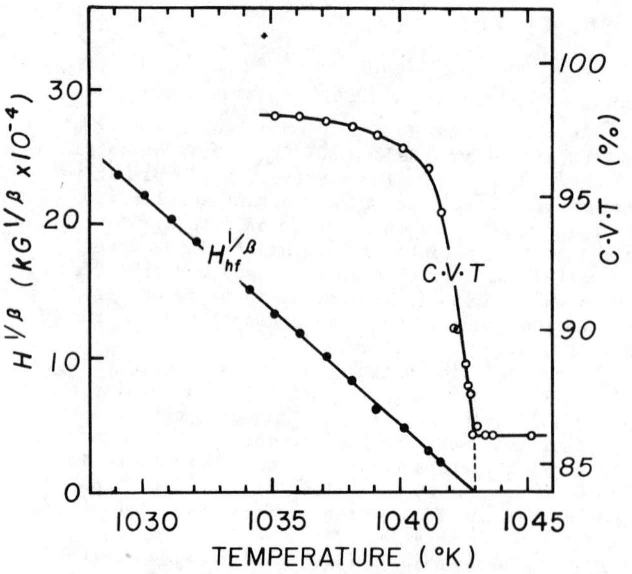


FIG. 2. Thermal scan and power law determination of $T_c$.

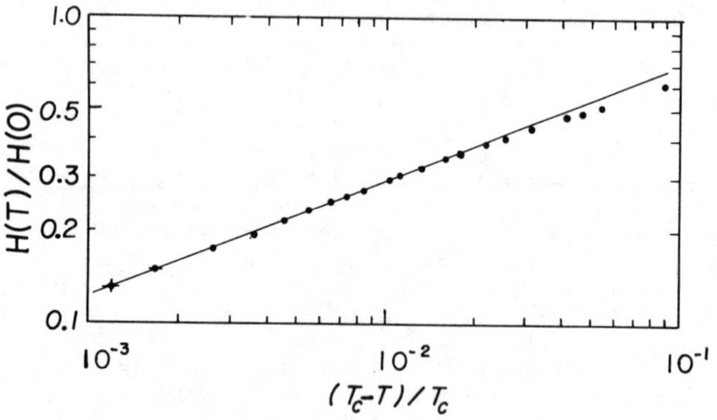


FIG. 3. Logarithmic plot of the reduced hyperfine field.

TABLE I. β-VALUES FOR (N,d) = (3,3) FERROMAGNETS

| Material | | Range in t ($10^{-4}$ $10^{-3}$ $10^{-2}$ $10^{-1}$) | Ref. |
|---|---|---|---|
| $CrBr_3$ | 0.368 ( 5) | | 8 |
| EuO | 0.368 ( 2) | | 9 |
| EuS | 0.335 (10) | | 10 |
| YIG | 0.370 ( 5) | | 10 |
| Ni | 0.383 ( 4) | | 5 |
| Fe | 0.342 ( 4) | | 4 |
| | 0.37 (2 ) | | 4 |
| | 0.368 | | 11 |
| | 0.379 ( 4) | | This work |
| Theory | 0.38 | | 2 |

neighborhood of the transition, with the result illustrated in Fig. 2. From this it is concluded that $T_c = 1043.0$ (1), in good agreement with the above. Also shown in Fig. 2 is a plot of $H_{hf}$ raised to the $1/\beta$ power, for $\beta = 0.379$. This illustrates graphically the degree of agreement between the two $T_c$ estimates. A conventional logarithmic plot of h appears in Fig. 3.

A comparison to theory and other (N,d) = (3,3) ferromagnets appears in Table I. In constructing this table we have selected only those results with significant data in the region $t \leq 10^{-2}$. The exclusion of results based only on data in $t \geq 10^{-2}$ is justified by the strong likelihood that asymptotic power law behavior fails there.[5] The effect of the exclusion is to eliminate some, though not all of the scatter from the table. Under the constraint applied in constructing Table I, we conclude that our value of β in Fe is in good agreement with other (N,d) = (3,3) ferromagnets, with the exception of EuS. Our result also agrees well with theory. The significance between our result and bulk measurements of Arrott et al[11] is difficult to evaluate, since no error is given for the bulk value. The discrepancy with Preston's Mössbauer results[4] remains unexplained.

Since our drive stability was more than adequate, we also checked for the discontinuity in the line shift reported by Preston at $T_c$.[3] As can be seen from Fig. 4 no anomaly appears in our data to within statistics, indicating that at least for our source there is no basis for believing the transition to be even marginally of first order. At the same time, explanation of Preston's anomaly on instrumental grounds is difficult since Preston reported greater sharpness of his discontinuity as the temperature regulation and uniformity improved.

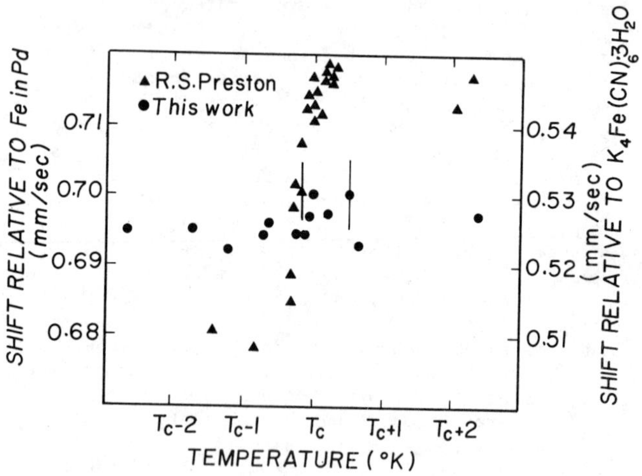


FIG. 4. Line shift near $T_c$ as observed by Preston and in this work.

REFERENCES

† Supported in part by the National Science Foundation.

1. R. B. Griffiths, Phys. Rev. Lett. 24, 1479 (1970).
2. M. E. Fisher, Rev. Mod. Phys. 46, 597 (1974).
3. R. S. Preston, Phys. Rev. Lett. 19, 75 (1967).
4. R. S. Preston, J. Appl. Phys. 39, 1231 (1968).
5. C. Hohenemser, T. A. Kachnowski and T. K. Bergstresser, Phys. Rev. B, to be published.
6. S. Margulies and J. R. Ehrmann, Nucl. Instr. 12, 131 (1961).
7. A. M. Kobeissi, to be published.
8. J. T. Ho and J. D. Litster, Phys. Rev. Lett. 22, 603 (1969).
9. N. Menyuk, K. Dwight and T. B. Reed, Phys. Rev. B3, 1689 (1971).
10. D. D. Berkner and J. D. Litster, to be published. See also D. D. Berkner, Ph.D. thesis, MIT (1974).
11. A. S. Arrott, B. Heinrich and D. S. Bloomberg, Magnetism and Magnetic Materials in 1972, p. 941.

# ABSOLUTE MEASUREMENT OF THE CRITICAL SCATTERING CROSS SECTION IN COBALT

C. J. Glinka*†≠ and V. J. Minkiewicz*†≠
University of Maryland, College Park, Maryland 20742
and
L. Passell**
Brookhaven National Laboratory, Upton, New York 11973

## ABSTRACT

Small-angle neutron scattering techniques have been used to study the angular distribution of the critical scattering from cobalt above $T_c$. These measurements have been put on an absolute scale by calibrating the critical scattering directly against the nuclear incoherent scattering from cobalt. In this way the interaction range $r_1$, which appears in the classical and modified Ornstein-Zernike expressions for the asymptotic form of the spin pair correlation function and is related to the strength of the spin correlations, has been determined. We obtain $r_1/a = 0.46 \pm 0.03$ for the ratio of the interaction range to the nearest-neighbor distance in cobalt. This result is in good agreement with theoretical predictions. Lack of agreement among previous determinations of the ratio $r_1/a$ made in iron failed to provide a definitive comparison with theory.

## INTRODUCTION

From the angular distribution of neutrons critically scattered from magnetic systems, considerable information can be obtained regarding the long range spin correlations which develop near the critical point. In the quasi-static limit[1] the scattered neutron intensity is directly proportional to the wave-vector-dependent susceptibility $\chi(q)$ which in the Ornstein-Zernike approximation has the simple form

$$\chi(q)=\frac{\chi_o}{r_1^2(\kappa_1^2+q^2)}, \quad \chi_o=(g\mu_B)^2 S(S+1)/3k_B T \qquad (1)$$

for small scattering vectors q. $\chi(q)$ is related to the spin pair correlation function $\gamma(r)$ by Fourier inversion, which, from Eq.(1), gives the well-known asymptotic dependence

$$\gamma(r) \propto \frac{1}{r_1^2} \quad \frac{e^{-\kappa_1 r}}{r} \qquad (2)$$

for large distances r. The correlation range of the spin fluctuations, $\xi=\kappa_1^{-1}$, is obtained directly from the width of the Lorentzian shaped critical scattering and has been extensively investigated in numerous materials. However, to determine the parameter $r_1$, the so-called interaction range which is related to the strength of the spin correlations, measurements of $\chi(q)$ must be put on an absolute scale. For this reason there have been relatively few experimental determinations of $r_1$ to afford comparison with theoretical calculations.

In the course of detailed measurements of the small-angle scattering from polycrystalline, face-centered cubic cobalt, it was realized that cobalt is a particularly favorable system for an accurate measurement of the critical scattering cross section, and hence of the interaction range $r_1$. Absolute cross section measurements are usually carried out by normalizing the scattered intensity against that measured under identical experimental conditions from a sample whose cross section is well known. One difficulty with such a procedure, however, is that the scattering volumes of the two samples must be precisely known and taken into account when comparing intensities. Additional complications arise if the data must be corrected for multiple scattering which is often the case.[2] For cobalt, however, many of these difficulties may be circumvented by calibrating the critical scattering directly against the rather large and accurately measured[3] nuclear, spin-incoherent cross section ($\sigma_{incoh}$ = 5.9 b) of cobalt itself. In this way factors of sample volume exactly cancel when forming intensity ratios. Furthermore, for the thin (2mm) slab sample geometry used in our experiments (necessitated by the moderately large absorption cross section of cobalt), analytic[2] and Monte Carlo calculations indicated that multiple scattering corrections were completely negligible for both the incoherent and critical scattering.

## DESCRIPTION OF MEASUREMENT AND RESULTS

Our small-angle scattering measurements were made with a triple-axis neutron spectrometer using a mono-energetic incident beam of 13.5 meV. The angular dependence of the scattering was studied by operating the spectrometer in the double-axis mode in which all neutrons scattered through a fixed angle are detected, regardless of their energies. Energy analysis of the scattering was carried out in a separate series of triple-axis, constant-q scans. The wave vector and frequency dependence of the small-angle scattering in the vicinity of the critical point, obtained by correlating these two types of measurements, have been briefly described in a previous paper.[4]

Our absolute cross section measurement was predicated on being able to separate the intensity recorded in a double-axis scan into contributions due to magnetic scattering, nuclear incoherent scattering and that due to all other sources. This is partially accomplished quite simply by measuring the forward scattering at temperatures well below the critical point ($T_c$=1115°C), for example room temperature, where there is no magnetic scattering over the angular range of our double axis scans. At such temperatures the elastic magnetic scattering is concentrated entirely in Bragg peaks away from the forward direction. Inelastic magnetic scattering, i.e. spin wave scattering, also gives no contribution to the observed intensity as a fortuitous consequence of the extremely steep spin wave dispersion in cobalt. For a neutron to scatter from a spin wave through an angle $\theta$, it must transfer energy $\hbar\omega$ given by the dispersion relation $\hbar\omega=\pm Dq^2(\omega,\theta)$ and also satisfy the usual momentum and energy conservation conditions. Near the forward direction the dispersion relation and conservation conditions can be satisfied simultaneously only for scattering angles $\theta<(\hbar^2/m_n)/D$. For cobalt at room temperature, the stiffness constant $D\sim500$ meV-Å$^2$.[5] The spin wave scattering is therefore confined to angles $\theta<0.25°$ which lay outside the range of our angular scans.

The forward scattering which is observed at low temperatures is predominantly the elastic, nuclear, incoherent scattering. Inelastic nuclear (i.e. phonon) scattering is suppressed near the forward direction by the $q^2$ factor in the phonon cross section, and other possible sources of scattering, such as multiple Bragg and domain-wall scattering, are also expected to be small compared to the strong incoherent scattering. There is, of course, a contribution to the recorded intensity from sources other than the sample, such as scattering from the furnace which surrounded the sample, room background, detector noise, etc. To estimate this extrinsic background scattering, the following procedure was adopted. First, the transmission T of the cobalt sample was carefully measured to be 0.404 ($T = \exp(\rho\sigma_t\tau)$, where $\rho$ is the number density, $\tau$ is the sample thickness, and $\sigma_t$ is the total cross section which includes the absorption and incoherent cross sections, for example). Then knowing the incoherent cross section we could compute $T' = \exp[-\rho(\sigma_t - \sigma_{incoh})\tau] = 0.440$, which would be the transmission of the sample were there no inco-

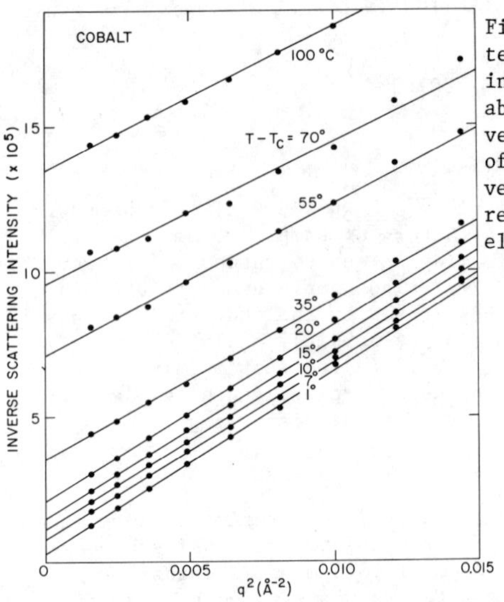


Fig. 1. Inverse intensities measured in angular scans above $T_c$ plotted versus the square of the scattering vector after correcting for inelasticity.

herent scattering. A surrogate sample was then prepared by stacking foils of indium, having the same lateral dimensions as the cobalt sample, to a thickness which gave a measured transmission equal to T'. Indium was chosen for this purpose because it produces essentially no incoherent scattering but does have an appreciable absorption cross section. The room temperature, small-angle scattering observed from such an indium sample should then differ from that recorded for the cobalt sample only as a result of the incoherent scattering from the cobalt. By subtracting two such scans a a resultant uniform level of scattering was obtained which could then be associated entirely with the incoherent scattering.

Since the Debye-Waller factor in the neutron cross section is nearly unity in the forward direction, the small-angle incoherent scattering is effectively temperature independent. Hence the difference between identical angular scans taken at room temperature and at temperatures near $T_c$ could be attributed entirely to magnetic critical scattering. Even very close to $T_c$, however, the critical scattering was observed to be inelastic to an appreciable degree. Because of this inelasticity, the total intensity measured at a fixed scattering angle represents a spread of scattering vectors q. Extensive triple-axis measurements of the energy distribution of the critical scattering at and above $T_c$ did provide the necessary information to enable the effects of the inelasticity to be removed from our double-axis data in a manner which has been described in Ref. 4. When corrected for inelasticity, the angular dependence of our double-axis data is quite adequately described by the Ornstein-Zernike expression given in Eq.(1). This can be seen in Fig. 1 in which the reciprocal of the corrected double-axis intensity is plotted versus $q^2$ for several temperatures up to 100°C above $T_c$. The intercepts of the straight lines in Fig. 1 with the $q^2$-axis determine the value of $\kappa_1^2$ for each temperature.

After correcting for inelasticity, the critical scattering intensity measured in double-axis scans above $T_c$ is directly proportional to the differential cross section,[6]

$$\left(\frac{d\sigma}{d\Omega}\right)_{crit} = \frac{2}{3} S(S+1)\left(\frac{\gamma e^2}{m_e c^2}\right)^2 |f(q)|^2 \frac{1}{r_1^2} \frac{1}{\kappa_1^2+q^2} \quad (3)$$

where S is the effective spin of the scatterer, $\gamma$ is the neutron magnetic moment in nuclear magnetons, and f(q) is the magnetic form factor. The same constant of proportionality relates the level of incoherent scattering deduced from the room temperature scans to the cross section

$$\left(\frac{d\sigma}{d\Omega}\right)_{incoh} = \frac{\sigma_{incoh}}{4\pi} \quad (4)$$

Hence, ratios of the critical to incoherent intensities can be equated to those obtained by dividing Eqs.(3) and (4), evaluated at the corresponding wave vector q. Since the value of $\kappa_1^2$ in Eq.(3) is known from plots like those shown in Fig. 1 and $|f(q)|^2 = 1$ over the small-angle range of our data, these ratios yield absolute values for the interaction range $r_1$. Using for the value of the effective spin appropriate for face-centered cubic cobalt,[7] S=0.88, we obtain from our measured intensity ratios the result $r_1$=1.16±0.04Å for temperatures within 15°C of $T_c$. While $r_1$ varies from material to material, the ratio of $r_1$ to the nearest neighbor distance $a_{nn}$ is expected to have universal significance. For cobalt near $T_c$, $a_{nn}$=2.55Å so that

$$r_1/a_{nn} = 0.46\pm0.03. \quad (5)$$

Table I. Results of measurements of the ratio of the interaction range $r_1$ to the nearest neighbor distance near $T_c$. A=2.52Å for iron and 2.55Å for cobalt

| $r_1/a$ | Material | Reference |
|---|---|---|
| 0.42±0.02 | Iron | [11] Gersch, Shull, Wilkinson 1956 |
| 0.58±0.07 | Iron | [12] Ericson and Jacrot, 1960 |
| 0.38 | Iron | [13] Spooner and Averbach, 1966 |
| 0.43±0.04 | Cobalt | [14] Bally et al., 1968 |
| 0.46±0.03 | Cobalt | This work. |

Our value for the ratio $r_1/a_{nn}$ is in good agreement with theoretical predictions, which have proven to be rather insensitive both to methods of calculation and to the crystal lattice. For example, the mean field approximation leads to the result[8] that $r_1/a_{nn}$ = 0.408. High temperature series expansion calculations[9] based on the Heisenberg model give values for $r_1/a_{nn}$ ranging from 0.45 to 0.52 depending on the spin and type of lattice; for a spin-1 system on a fcc lattice, which is the case most applicable to cobalt, $r_1/a_{nn}$ = 0.462. Similar calculations[10] for the spin-1/2 Ising model yield values for $r_1/a_{nn}$ from 0.44 to 0.46, the former number applying to an fcc lattice.

The few previous measurements of the ratio $r_1/a$ are listed in Table 1. The lack of agreement among the results obtained for iron was a motivating factor for our measurement. The only previous determination of the interaction range in cobalt was made by Bally et al.[14] who normalized their neutron scattering data to bulk susceptibility measurements. Their result is in good agreement with our own, obtained by a completely different method. In view of our result, there now appears to be a discernable trend in the available data which indicates that experiment and theory are in basic accord.

* Work supported by the National Science Foundation
† Guest scientist at Brookhaven National Laboratory.
‡ Present address; NBS, Washington, DC 20234
≠ Present address: IBM, San Jose, CA 95193.
** Work at Brookhaven supported by AEC.

1. In the quasi-static limit the energy transferred by the scattered neutron is small compared to its incident energy.


2. B. N. Brockhouse, L. M. Corliss and J. M. Hastings, Phys. Rev. 98, 1721 (1955).
3. R. I. Schermer, Phys. Rev. 130, 1907 (1963).
4. C. J. Glinka, V. J. Minkiewicz and L. Passell, AIP Conf. Proc. No. 24, 283 (1975).
5. G. Shirane, V. J. Minkiewicz and R. Nathans, J. Appl. Phys. 39, 383 (1968).
6. W. Marshall and S. W. Lovesey, Theory of Thermal Neutron Scattering, Oxford Press (1971) p. 469.
7. H. P. Myers and W. Sucksmith, Proc. R. Soc. 207A, 427 (1951).
8. P. G. DeGennes and A. Herpin, Comptes Rendus 243, 1611 (1956).
9. D. S. Ritchie and M. E. Fisher, Phys. Rev. B5, 2668 (1972).
10. M. E. Fisher and R. J. Burford, Phys. Rev. 156, 583 (1967).

11. H. A. Gersch, C. G. Shull and M. K. Wilkinson, Phys. Rev. 103, 525 (1956).
12. M. Ericson and B. Jacrot, Phys. Chem. Solids 13, 235 (1960).
13. S. Spooner and B. L. Averbach, Phys. Rev. 142, 291 (1966).
14. D. Bally, M. Popovici, M. Totia, B. Grabcev and A. M. Lungu, Neutron Inelastic Scattering, (IAEA, Copenhagen, 1968), Vol II p. 75.

# CRITICAL BEHAVIOR OF DILUTE MAGNETS WITH SITE-RANDOMNESS

Tatuo Kawasaki
Physics Department, College of Liberal Arts
Kyoto University, Kyoto 606, JAPAN

The renormalization transformation is examined in a random magnet with site-randomness. ($H=-\sum J_{ij}\sigma_i\sigma_j\xi_i\xi_j$ ; $\sigma$-spin and $\xi$-occupation operator) A transformation function of random-site variables was introduced as well as that of spin variables $T^S$ and the free energy is expressed as

$$F=\langle F_{\xi'}\rangle_{\xi'}=\left\langle \ln \mathrm{Tr}_{\sigma'}\, e^{H'(\sigma';\xi')}\right\rangle_{\xi'}$$

$$=\langle F_{\xi}\rangle_{\xi}=\left\langle\left\langle T^R(\xi';\xi)\ln \mathrm{Tr}_{\sigma\sigma'}T^S(\sigma';\sigma)\times e^{H(\sigma;\xi)}\right\rangle_{\xi}\right\rangle_{\xi'},$$

where relations $\mathrm{Tr}_{\sigma}T^S(\sigma';\sigma)=\mathrm{Tr}_{\sigma}T^S(\sigma';\sigma)=1$ and $\langle T^R(\xi';\xi)\rangle_{\xi}=\langle T^R(\xi';\xi)\rangle_{\xi}=1$ should be satisfied.

By setting the function $T^R$ be

$$T^R(\xi;\xi')=\prod_{\text{block}}\{1+[p(1-p)]^{-1}(\xi_i'-p)\Sigma(\xi_{ij}-p)\},$$

the transformation equation is found to be

$$(1-2q)K'=(1-2q)[2f_1^2K+4f_1^2(1+f_2-2f_1^2)K^2+3f_1^2L+2f_1^2M]$$
$$-4qf_1^2(1+2f_2-3f_1^2)K^2$$

$$(1-2q)L'=(1-2q)[f_1^2(1+7f_2-8f_1^2)K^2+f_1^2M]$$
$$-qf_1^2(1+14f_2-15f_1^2)K^2$$

$$(1-2q)M'=4(1-2q)[f_1^2(f_2-f_1^2)K^2]-8qf_1^2(f_2-f_1^2)K^2$$

up to the second order of the cumulant expansion method, where the expression inside the square blackets are completely the same as Niemeyer-VanLeeuwen gave for the pu pure system. This set of equation suggests that the critical exponent may change by dilution of magnetic ions. Further investigation will be published elsewhere.

#This work is partially supported by US NSF Grant #GH39 023 and DMR73-07651-A02 and Ministry of Education of Japan.

# SERIES STUDIES OF CRITICAL EXPONENTS IN CONTINUOUS DIMENSIONS†

J. P. Van Dyke and William J. Camp
Sandia Laboratories, Albuquerque, New Mexico 87115

ABSTRACT

Using results we have previously derived for the high-temperature susceptibility expansion of classical models as a closed-form function of lattice dimension, we study the dimensional dependence of the critical exponent $\gamma$ and critical temperature, as well as the correction-to-scaling exponent $\Delta_1$, for Ising-like ($n = 1$) models. The numerical results are obtained by extrapolation of 10-th order series on loose-packed hypercubical lattices, and 8-th order series on close-packed hypertriangular lattices. The critical exponent $\gamma$ increases monotonically with decreasing dimension, d, for $d < 4$, and apparently tends to infinity at $d = 1$; while the critical temperature decreases monotonically and smoothly to zero at $d = 1$. Detailed contact is made with the $\epsilon$-expansion estimates for critical exponents obtained in the context of renormalization group theory.

A number of years ago, Fisher and Gaunt[1] carried out high-temperature susceptibility expansions through $T^{-11}$ for the spin-half Ising model on the family of (loose-packed) simple hypercubic d-dimensional lattices of coordination number $q(d) = 2d$. This enabled them to obtain an expansion for the Ising critical point $T_c(d)$ through fifth order in $d^{-1}$. Their results for the critical temperature have since been extended to isotropic n-vector models by Gerber and Fisher.[2]

We have used formal results for the susceptibility expansion of a much more general class of models[3] to reduce the problem of obtaining the susceptibility expansion through 10-th order (for general models) to a lattice counting problem. We thus extend the hypercubic-lattice results of Fisher and Gaunt[1] (through 10-th order) to all models of our general class.[4] In addition, we have done the necessary lattice counting to obtain 8-th order series for the susceptibility of all such models on a family of (closepacked!) hypertriangular lattices with coordination number $q_t(d) = d(d + 1)$. So, on the family of loosepacked lattices we can obtain expressions -- for any given model of our class - through $d^{-5}$ for $T_c(d)$; while, on the family of closepacked lattices, our expressions for $T_c(d)$ would extend through $d^{-8}$. [It is worth noting that on the hypertriangular lattices the n-th order expansion coefficient is a polynomial of degree 2n in d -- the coefficients do not have simple polynomial expressions in $q_t$.]

In the light of renormalization-group-theory (RGT) predictions that the susceptibility behave, for $d < 4$, as

$$\chi \approx \chi_0(1 - T_c/T)^{-\gamma(\epsilon)} + \chi_1(1 - T_c/T)^{-\gamma(\epsilon) + \Delta_1(\epsilon)} + \cdots, \qquad (1)$$

it is interesting to analytically continue our series coefficients to continuous dimension in the range $1 < d \le 4$, and make detailed comparisons with RGT results. Note that although the amplitudes $\chi_0$ and $\chi_1$ depend not only on $\epsilon(= 4 - d)$, but also on "irrelevant" variables (such as spin), the susceptibility exponent, $\gamma$, and correction-to-scaling exponent, $\Delta_1$, are universal and have been calculated within RGT through 3rd order in $\epsilon$.[5] In this paper we report on such comparisons for the $S = \frac{1}{2}$ and $S = \infty$ Ising models.[6]

In Fig. 1 we compare the "best" values for $\gamma(d)$ obtained -- on the hypertriangular lattices -- using standard ratio and Padé methods (which do not properly account for the confluent singularity in Eq. (1)) with the results of second-order $\epsilon$-expansions. The smooth curve is the $\epsilon$-expansion result

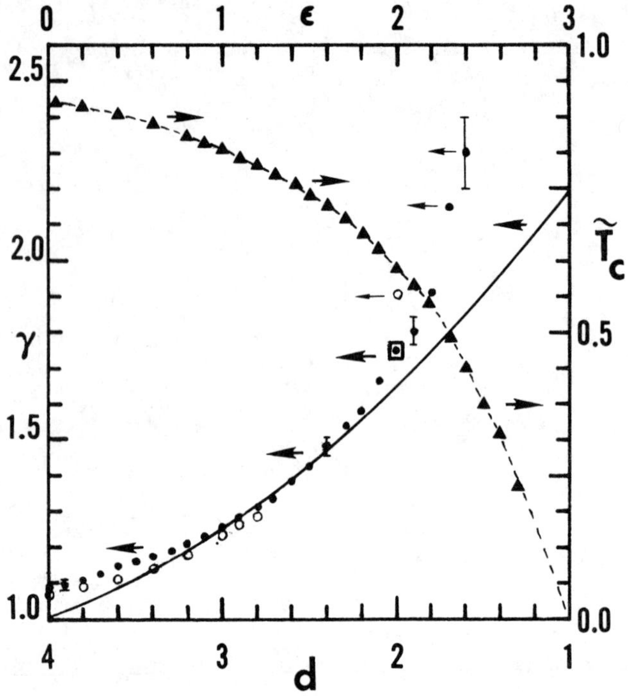


FIGURE CAPTION

Figure 1. The dimensional dependence of critical parameters on hypertriangular lattices. The solid triangles represent the scaled critical temperature $\widetilde{T}_c = kT/[d(d + 1)J]$. The solid curve is the 2nd order $\epsilon$-expansion result for $\gamma(n = 1)$. The solid and open circles are the apparent values of $\gamma$ deduced from straightforward series analysis for the $S = \frac{1}{2}$ and $S = \infty$ Ising models, respectively. The open square is the exact $S = \frac{1}{2}$, $d = 2$ result, $\gamma = 7/4$. The error brackets indicate typical apparent uncertainties in the series estimates for $\gamma$ in various parts of the region $0 < \epsilon < 3$.

(exact near $d = 4$). The solid dots are the results for the $S = \frac{1}{2}$ Ising model, the open square at $d = 2$ is the exact result $\gamma = 1.75$, and the open circles are the results for the $S = \infty$ Ising model. Several aspects of this figure warrant comment: (i) the deviation of series results from $\epsilon$-expansion results in the range $3 < d \le 4$ is due to the failure of the series methods employed to allow for the strong confluent correction term in Eq. (1). (ii) In the range $2.6 < d < 3.1$ series results and $\epsilon$-expansion results are in quite good agreement. (iii) Below $d = 2.6$ the results from series extrapolation increase with decreasing dimension much more rapidly than do the $\epsilon$-expansion results. (iv) There is a marked difference between the spin-half and spin-infinity series estimates at $d = 2$. As discussed by Camp and Van Dyke[7] this difference may be removed by allowing for strong confluent corrections to scaling at $S = \infty$ (there are only analytic corrections at $d = 2$ for $S = \frac{1}{2}$).

The difference between series and $\epsilon$-expansion results near $d = 4$ is removed when the series are fit to the reasonable crossover formula

$$\chi(K) = [A + (B/K\Delta_1(\epsilon))(\tau^{\Delta_1} - 1)]^{\theta}/\tau^{\gamma(\epsilon)} \qquad (2)$$

where $\tau = 1 - K/K_c$, $K = J/kT$, $\Delta_1(\epsilon) \approx \frac{1}{2}\epsilon$, and $\theta[\equiv (n + 2)/(n + 8)] = \frac{1}{3}$ for Ising-like ($n = 1$) models.[4,5] This formula reduces to Eq. (1) away from d = 4, and yields expected[5] logarithmic corrections at d = 4. Using it, we find $\gamma(d = 4) = 1.00 \pm 0.02$ for both $S = \frac{1}{2}$ and $S = \infty$.

In the range $2.7 < d < 3.3$, corrections to scaling are rather small for $S = \infty$, and very small for $S = \frac{1}{2}$. We are able to handle them quite successfully with, by now, standard methods.[7] We estimate that $\Delta_1$ is relatively constant near $\Delta_1 = 0.5 \pm 0.1$ in this range, while $\gamma$ varies smoothly from $\gamma \approx 1.15$ at d = 3.3 to $\gamma \approx 1.33$ at d = 2.8. Outside this range the corrections to the dominant behavior $\chi \sim \tau^{-\gamma}$ are so great that even methods which allow for confluent corrections[7] encounter severe difficulties. However, we have previously noted that $\Delta_1 = 0.5$ yields sensible results for the spin-S Ising model at d = 2, and as noted above, $\Delta_1$ is quite close to 0.5 near d = 3. We thus have fit the series coefficients to the form

$$\chi \approx \chi_0\tau^{-\gamma} + \chi_1\tau^{-\gamma + \Delta_1} + \chi_2\tau^{-\gamma + 1} \qquad (3)$$

with $\Delta_1$ given by [1/1] Padé approximant to the 3-rd order $\epsilon$-expansion result for $\Delta_1$ when $d > 3$, and by $\Delta_1 = \frac{1}{2}$ when $d \le 3$. The exponent $\gamma$ has been set equal to $\bar{\gamma} = (11\,d - d^2 - 4)/8(d - 1)$, which produces the known results $\gamma(d = 4) = 1$, $\gamma(d = 3) = 1.25$, $\gamma(d = 2) = 1.75$, and $\gamma(d = 1) = \infty$, and is close to the $\epsilon$-expansion result near d = 4.

The results for $\chi_1/\chi_0$ and $\chi_2/\chi_0$, for both $S = \frac{1}{2}$ and $S = \infty$ are displayed in Table I. First, examine the $S = \infty$ results. At d = 2, there are very large corrections to the dominant behavior $\tau^{-7/4}$, and the leading correction amplitude is negative. The large negative amplitude accounts reasonably for the large apparent exponent $\gamma(\infty) \simeq 1.9$ at d = 2 noted in the discussion of Fig. 1. Between d = 2 and d = 2.6, no method of analysis was successful for $S = \infty$. However, evidence <u>was</u> found that $\chi_1/\chi_0$ seemed to pass through zero in this range. This agrees with our results at d = 2.7, where the ratio is again positive. The value of $\chi_1/\chi_0$ increases to about 0.25 at d = 3, which accounts for the slightly low exponent $\gamma(\infty) \simeq 1.23$ previously found using standard ratio methods (see Ref. (7)). The ratio then decreases smoothly, goes through zero near d = 3.4 and becomes negative and increasingly large as d increases toward 4 dimensions. This latter fact accounts well for the increase of $S = \infty$ ratio and Padé estimates for $\gamma(\infty)$ over those predicted by $\epsilon$-expansion methods. Further, from Eq. (2), we see that both the negative sign and the increase in magnitude are requisite to produce the expected logarithmic corrections[5] at d = 4.

The $S = \frac{1}{2}$ results are somewhat different. For dimensions in the range $2.0 \le d \leqslant 3.1$ the estimated amplitude of the leading correction is very small or zero, consistent with the fact that it is zero with high precision at d = 2 and d = 3.[7] Above $d \approx 3.2$ $\chi_1/\chi_0$ becomes negative and grows rapidly in value. Indeed near d = 4 it is apparently considerably larger than the respective $S = \infty$ ratio, which accounts for the fact that the apparent series value of $\gamma$ is larger for $S = \frac{1}{2}$ than for $S = \infty$ in the range $3 \le d \le 4$.

In summary, series analysis is consistent with $\epsilon$-expansion results (for n = 1 models) when expected confluent singularities are taken into account. However, below $d \simeq 2.5$ dimensions series estimates for $\gamma$ become increasingly larger than $\epsilon$-expansion estimates (which tend to $\gamma \simeq 2.2$ at d = 1), and evidently tend smoothly to the known limit $\gamma(d = 1) = \infty$, although series analysis is very difficult below two dimensions. In this regard note in Fig. 1 that the correctly scaled critical temperature, $kT_c/q_tJ$, of the $S = \frac{1}{2}$ model is predicted by series analysis to approach zero continuously as d tends to one, which may be a significant result as regards the possibility of expanding scaling functions away from d = 1.

A more detailed account, including expansions for $T_c$ in powers of $d^{-1}$, complete series, and additional analysis is the subject of a subsequent publication.[6]

Table I. Variation of the parameters of Eq. (3) with dimension, d, for $S = \frac{1}{2}$ and $S = \infty$.

| d | $S = \infty$ $\chi_1/\chi_0$ | $S = \infty$ $\chi_2/\chi_0$ | $S = \frac{1}{2}$ $\chi_1/\chi_0$ | $S = \frac{1}{2}$ $\chi_2/\chi_0$ |
|---|---|---|---|---|
| 2.0 | -0.54 ± 4 | 1.17 ± 4 | +0.11 ± 20[a] | 0.20 ± 6 |
| 2.7 | +0.13 ± 8 | 0.58 ± 30 | —— | —— |
| 2.8 | +0.21 ± 8 | 0.33 ± 16 | —— | —— |
| 3.0 | +0.25 ± 1 | 0.12 ± 1 | +0.07 ± 4 | -0.13 ± 9 |
| 3.1 | +0.19 ± 2 | 0.17 ± 3 | +0.03 ± 1 | -0.13 ± 2 |
| 3.2 | +0.16 ± 2 | 0.07 ± 3 | -0.02 ± 1 | -0.10 ± 1 |
| 3.4 | +0.01 ± 1 | 0.19 ± 3 | -0.14 ± 2 | -0.06 ± 3 |
| 3.5 | -0.07 ± 0 | 0.21 ± 3 | -0.20 ± 1 | -0.03 ± 5 |
| 3.6 | -0.15 ± 0 | 0.26 ± 0 | -0.28 ± 1 | -0.00 ± 10 |
| 3.8 | -0.39 ± 0 | 0.30 ± 1 | -0.51 ± 5 | -0.00 ± 20 |
| 3.9 | -0.60 ± 0 | 0.29 ± 3 | -0.67 ± 1 | -0.00 ± 20 |
| 3.95 | -0.76 ± 0 | 0.26 ± 4 | -0.81 ± 0 | -0.23 ± 15 |

[a] The exact result is identically zero.

REFERENCES

† This work supported by the U. S. Energy Research and Development Administration, ERDA.

1. M. E. Fisher and D. S. Gaunt, Phys. Rev. <u>133</u>, A224 (1964).
2. P. R. Gerber and M. E. Fisher, Phys. Rev. <u>B10</u>, 4697 (1974).
3. J. P. Van Dyke and W. J. Camp, Phys. Rev. Lett. <u>35</u>, 323 (1975), and refs. cit.
4. W. J. Camp and J. P. Van Dyke, Bull. Am. Phys. Soc. <u>20</u>, 500 (1975).
5. K. G. Wilson and J. Kogut, Phys. Rept. <u>12C</u>, 248 (1972). Note that for $\gamma$ the 2nd order $\epsilon$ expansion result is better than 3rd order (e.g., for d = 3, 1.244 versus 1.195).
6. Full details, including series coefficients, will be published elsewhere.
7. W. J. Camp and J. P. Van Dyke, Phys. Rev. <u>B11</u>, 2579 (1975).

# SPIN-PEIERLS TRANSITIONS IN HEISENBERG ANTIFERROMAGNETIC LINEAR CHAIN SYSTEMS*

J. W. Bray, H. R. Hart, Jr., L. V. Interrante, I. S. Jacobs, J. S. Kasper, G. D. Watkins,** and S. H. Wee***
Corporate Research and Development, General Electric Company
Schenectady, New York 12301

and

J. C. Bonner
Applied Mathematics Dept.
Brookhaven National Laboratory
Upton, New York 11973

SUMMARY

We have recently reported[1,2] evidence for a spin-Peierls transition, i.e. a progressive spin-lattice dimerization phenomenon occurring below a transition temperature in a system of one dimensional (1-D) antiferromagnetic (AF) Heisenberg chains. The transition is second order, and at the lowest temperature the system is in a singlet ground state with a magnetic gap. This phenomenon has its historical origins in the background areas of metal-nonmetal transitions, particularly the Peierls instability[3] of a 1-D conductor; and of spin insulator magnetic model systems, especially uniform[4,5] and alternating[6,7] AF, 1-D, ($S = 1/2$) chains. A decade of theoretical development[8-12] in describing the behavior expected has developed our physical understanding without important qualitative changes appearing. Our experiments were carried out on newly synthesized donor-acceptor compounds of tetrathiafulvalene (TTF) with bisdithiolene metal complexes[13], e.g. $TTF^{+}\cdot MS_4C_4(CF_3)^{-}{}_4$, M = Cu, Au, which crystallize in a NaCl-like structure. The planar organic molecular units of these insulating compounds put them into the broader class of molecular crystals[14] which has been attracting increasing attention from solid state scientists.

The measurements are of magnetic susceptibility between 2.5K and 300K and of EPR at 20 GHz down to 1.5K. Only the $TTF^{+}$ cations have spins, and the anisotropic g-tensor values are within 1% of the free electron value. The static susceptibility for the Cu compound, displayed in Fig. 1, is identical for fields both parallel and perpendicular to the c-axis at all temperatures. Between 12K and 250K it is modeled very accurately by the susceptibility

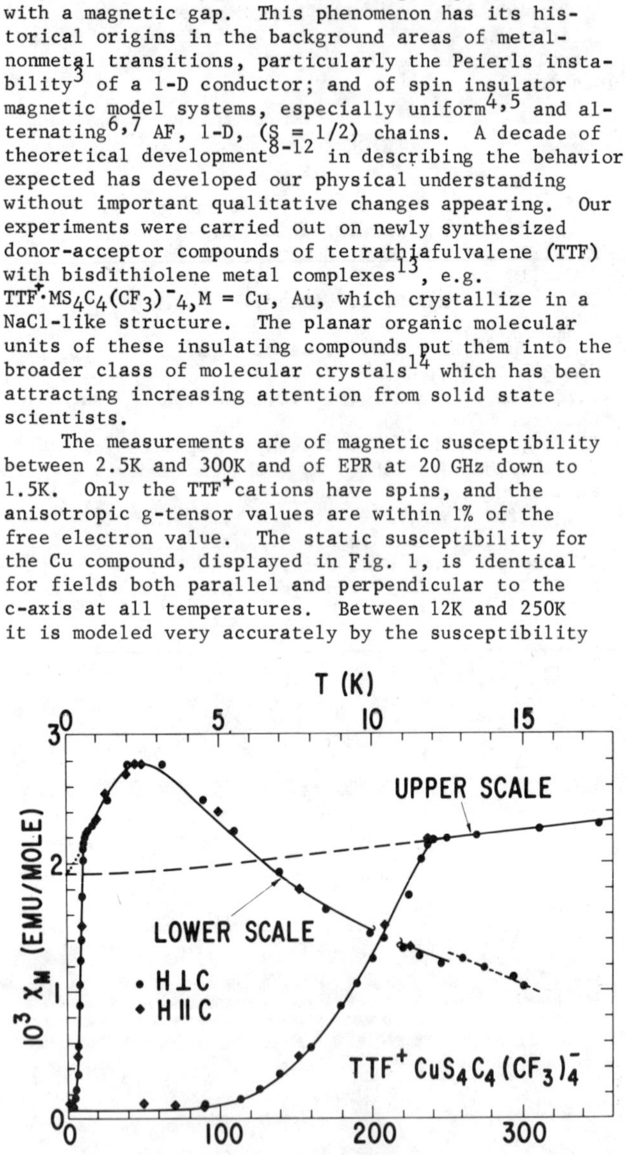


Fig. 1. Magnetic susceptibility of $TTF\cdot CuS_4C_4(CF_3)_4$ along two directions. Solid lines are calculated from a spin-Peierls theory which contains AF chains with uniform exchange above 12K and temperature-dependent alternating exchange below.

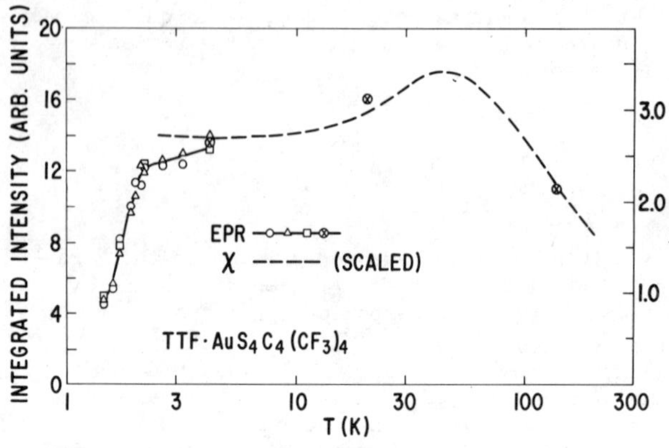


Fig. 2. Integrated EPR intensity vs. temperature for $TTF\cdot AuS_4C_4(CF_3)_4$ with H along two major crystal axes. Spectrometer sensitivity calibrated for all data shown. Dashed curve shows measured powder static susceptibility scaled to match at 4K.

for AF Heisenberg chains of spins ($S=1/2$) and uniform n.n. exchange coupling $J/k_B = 77K$ where $H_{ij} = JS_i \cdot S_j$. Below $T_c = 12K$, the susceptibility decreases sharply toward zero. The isotropy of this decrease, which is mirrored by the integrated EPR intensity, is incompatible with 3-D ordering of the spin system. In Fig. 2, we show similar data for the Au compound, susceptibility and integrated EPR intensity, for which $J/k_B = 68K$ and $T_c = 2.1K$.

The behavior is in very good agreement with a mean-field theory of the spin-Peierls transition (after Pytte[12]) embracing the 1-D uniform Heisenberg AF chain, 3-D lattice dynamics (in the harmonic approximation) and spin-phonon coupling. This formulation results in a Fröhlich-type Hamiltonian. Its solution leads to a BCS-like temperature-dependent magnetic gap and a progressive dimerization of the lattice below $T_c$. The susceptibility of this progressively alternating chain is calculated using the Bulaevskii model.[6] Only two experimental quantities are needed in the full description, J and $T_c$. The measurements demonstrate directly the characteristic properties with a textbook-model simplicity.

Additional effects in the transition region should be observable. Examples are an anomaly in the magnetic specific heat, a distortion of some lattice coordinate below $T_c$, a soft mode in the phonon spectrum above $T_c$, neutron scattering evidence of the 1-D character, etc. On the theoretical side, rigorous magneto-elastic theories which might explain the apparent success of the mean-field spin-Peierls theory are lacking, as is an understanding of the conditions under which the transition could become first order.

Several other materials have properties which have been suggested as examples of a spin-Peierls transition. They include Wurster's blue perchlorate, alkali-TCNQ salts, and doped or stressed $VO_2$. Objections, ranging from mild to severe, to identifying these materials with spin-Peierls transitions have been raised and are discussed in Ref. 2.

REFERENCES

*Work supported in part by a grant from the U.S. Air Force Office of Scientific Research under contract No. F-44620-71-C-0129, and (for J. C. B.) by the U.S. Energy Research and Development Administration.

**Present address: Physics Department, Lehigh University, Bethlehem, Pa. 18015

***Present address: Department of Laboratory Medicine, Microbiology Section, Mayo Clinic, Rochester, Minn. 55901

1. J. W. Bray, H. R. Hart, Jr., L. V. Interrante, I. S. Jacobs, J. S. Kasper, G. D. Watkins, S. H. Wee and J. C. Bonner, Phys. Rev. Letters, 35, 744-747 (1975).
2. A full paper by the authors is to be submitted for publication. The report in this Proceedings is a digest thereof.
3. R. E. Peierls, Quantum Theory of Solids (Oxford Univ. Press, London, England, 1955), p. 108.
4. J. C. Bonner and M. E. Fisher, Phys. Rev. 135, A640-658 (1964).
5. L. N. Bulaevskii, Zh. Eksp. Teor. Fiz. 43, 968-73 (1962), [Sov. Phys. JETP 16, 685-88 (1963)]
6. L. N. Bulaevskii, Zh. Eksp. Teor. Fiz. 44, 1008-14 (1963), [Sov. Phys. JETP 17, 684-87 (1963)] Fiz. Tverd. Tela 11, 1132-35 (1969), [Sov. Phys.-Solid State, 11, 921-24 (1969)].
7. W. Duffy, Jr., and K. P. Barr, Phys. Rev. 165 647-54 (1968).
8. H. M. McConnell and R. Lynden-Bell, J. Chem. Phys. 36, 2393-7 (1962); D. D. Thomas, H. Keller and H. M. McConnell, J. Chem. Phys. 39, 2321-29 (1963).
9. D. B. Chesnut, J. Chem. Phys. 45, 4677-81 (1966).
10. P. Pincus, Solid State Commun. 9, 1971-73 (1971); G. Beni and P. Pincus, J. Chem. Phys. 57, 3531-34 (1972); G. Beni, J. Chem. Phys. 58, 3200-2 (1973).
11. J. Y. Dubois and J. P. Carton, J. Phys. (Paris) 35, 371-6 (1974).
12. E. Pytte, Phys. Rev. B10, 4637-42 (1974).
13. L. V. Interrante, K. W. Browall, H. R. Hart, Jr., I. S. Jacobs, G. D. Watkins and S. H. Wee, J. Am. Chem. Soc. 97, 889-90 (1975).
14. Z. Soos, Ann. Rev. Phys. Chem. 27, 121-153 (1974).

## MAGNETIC ORDERING IN $NiZrF_6 \cdot 6H_2O$*

M. Karnezos, D. Meier and S. A. Friedberg
Carnegie-Mellon University, Pittsburgh, Pa., 15213

ABSTRACT

$NiZrF_6 \cdot 6H_2O$ crystallizes in the trigonal $NiSnCl_6 \cdot 6H_2O$ structure with one axially distorted $[Ni(H_2O)_6]^{++}$ complex per unit cell. Low-field susceptibilities, $\chi_{\parallel}$ and $\chi_{\perp}$, and $C_p$ for several values of $H_{\parallel}$ have been measured above ~0.05 K in a dilution cryostat. $\chi_{\parallel}$ and $C_p(H=0)$ exhibit large, sharp anomalies at $T_c=0.164$ K. The behavior of $\chi_{\parallel}$ and $\chi_{\perp}$ suggests that the salt is a uniaxial ferromagnet with the easy direction along the trigonal axis. $C_p(H=0)$ also shows a resolved Schottky anomaly with $C_{max} \simeq 0.24$ R at $T_{max} \sim 1.5$ K. This indicates that the $^3A_{2g}$ ground state of $[Ni(H_2O)_6]^{++}$ is split into a lower doublet and an upper singlet separated by ~3k. The spins are thus Ising-like at $T_c$. For $H_{\parallel} \neq 0$, the lower $C_p$ anomaly is rounded and shifted to higher T. These observations are quite well described by an Oguchi model in which pairs of spins are treated exactly and their interaction with the rest of the crystal is represented by a mean field. The pair Hamiltonian is

$$\mathcal{H}=D(S_{1z}^2+S_{2z}^2)-2J\underline{S}_1\underline{S}_2+g\mu_B(\underline{S}_1+\underline{S}_2)\cdot(\underline{H}+\gamma\underline{M})$$ where

$S_1=S_2=1$ and $\gamma=2(z-1)J/Ng^2\mu_B^2$ with $z=6$. We find $D/k=-3.1$ K, $g=2.33$, and $J/k=+0.014$ K. The effective interaction J/k contains a significant dipolar contribution which, in combination with the Ising character of the spins, may explain the mean field-like behavior of this salt.

A full account of this work will be published elsewhere.

*Work supported by the National Science Foundation and Office of Naval Research.

## ULTRASONIC ATTENUATION IN LINEAR ANTIFERROMAGNET $CsNiCl_3$*

D. P. Almond and J. A. Rayne
Carnegie-Mellon University, Pittsburgh, Pa., 15213

ABSTRACT

Ultrasonic attenuation measurements have been made in the linear antiferromagnet $CsNiCl_3$. Data were obtained from 2 to 180 K for longitudinal waves propagating along [0001] at frequencies up to 330 MHz. There is a broad maximum in the attenuation near 30 K, which appears to be related to the one-dimensional chain-like properties of $CsNiCl_3$. The behaviour is consistent with the temperature dependence of the heat capacity for a linear antiferromagnet with S=1. An attenuation anomaly is observed at $T_N = 4.36 \pm 0.04$ K, which agrees with the temperature corresponding to the lower of two transitions observed in NMR[1] and neutron diffraction experiments.[2] Near this anomaly the attenuation is of the form $\alpha \propto \omega^2 \exp(-\eta)$, where $\omega$ is the sound frequency and $\varepsilon = (T-T_c)/T_c$. The critical exponent $\eta = 1.35 \pm 0.3$ is close to the value 4/3 for an anisotropic three-dimensional system.[3]

Attenuation measurements have also been made in magnetic fields up to approximately 31 kG. There is a quadratic increase in the Neel temperature with increasing field. This behaviour resembles that recently reported in two other linear antiferromagnets, $CsMnCl_3 \cdot 2H_2O$[4] and α-CuNSal[5]. For fields in excess of 22 kG the attenuation anomaly increases dramatically possibly as the result of a spin-flop transition.

1. R. H. Clark and W. G. Moulton, Phys. Rev. B5, 788 (1972).
2. W. B. Yelon and D. E. Cox, Phys. Rev. B7, 2024 (1973).
3. G. E. Laramore and L. P. Kadanof, Phys. Rev. 187, 619 (1969).
4. H. Kobayashi, I. Tsujikawa and S. A. Friedberg, J. Low Temp. Phys. 10, 621 (1973).
5. W. G. Clark, L. J. Azevedo and E. O. McLean, Proc. 14th Int. Conf. Low Temp. Phys., (Helsinki, 1975).

*Work supported by National Science Foundation.

# POLY(METAL PHOSPHINATES): ANTIFERROMAGNETISM IN DISORDERED LINEAR POLYMERS

J. C. Scott,* T. S. Wei, A. F. Garito, and A. J. Heeger
Laboratory for Research on the Structure of Matter and Department of Physics,
University of Pennsylvania, Philadelphia, PA 19174+
and
H. D. Gillman and Piero Nannelli
Pennwatt Corporation Technological Center, King of Prussia, PA 19406‡

ABSTRACT

Poly(metal phosphinates) comprise a class of inorganic polymers in which the metal ions, which can be magnetic, form an integral part of the polymer backbone. Phosphinate groups bridge between metal centers and provide a path of antiferromagnetic superexchange J/k ~ 0-5K. A survey of the magnetic properties of a number of these materials is presented. It is found that disorder plays an important role in determining the magnetic properties, limiting the range of spin correlation. The susceptibility data of three poly(chromium phosphinates) are analysed in terms of a model of antiferromagnetic chains of classical spins with exchange varying randomly as a function of position. The width of the distribution of exchange correlates with the structural disorder as determined by x-ray and birefringence studies. The magnetic specific heat data are also shown to be qualitatively consistent with the proposed model.

## I. INTRODUCTION

The mathematical simplicity of one-dimensional systems has encouraged much theoretical interest in this area, particularly since the synthesis of suitable materials for experimental investigation.[1] The problem of disorder is another active field of experimental and theoretical interest.[2,3] In this paper we report a class of polymers, poly(metal phosphinates) (PMP),[4] which can be described in terms of disordered, one-dimensional magnetic models.

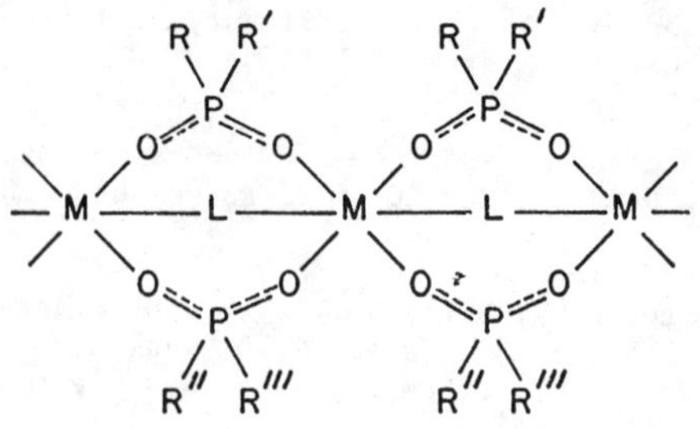


Fig. 1. General structure of poly(metal phosphinates). R, R', R'', R''' are alkyl groups or phenyl and may be the same or different. L is a bidentate singly charged ligand anion. M is a metal, the oxidation state of which determines the number of phosphinate bridges and ligands.

The polymeric nature of these compounds comes from the ability of the phosphinate anions (-OPRR'O-) (see Fig. 1) to bridge between metal centers. The structures and properties of these materials depend on the substituents R, R', etc., the type and oxidation state of the metal species, and the nature of any other ligands which may be present in the system.[5-15] Our present interest arises since the metal ion can be of a transition series and therefore a magnetic site in the linear polymer chain. Disorder may be introduced either in the fabric of the chain itself or in its relationship with other chains. Thus, we have the possibility of a magnetic system which is both one-dimensional and (in an appropriate sense) amorphous.

| Metal(s) | Phosphinate Bridges | Crosslinked or not (Estimate from Solubility) | Crystallinity from X-rays (and Birefringence-b) | Comments |
|---|---|---|---|---|
| Cobalt (II) | Bis-Dioctyl (I) | Not crosslinked | Weakly crystalline | Tetrahedral Co(7) |
| | Bis-Dioctyl (ii) | Crosslinked | Very crystalline | Octahedral Co(7) |
| | Bis-Diphenyl | Crosslinked | Weakly crystalline | |
| | Bis-Methyl Phenyl | Crosslinked | Moderately crystalline | |
| | Bis-Di-n-Butyl | Not crosslinked | Very crystalline (b) | |
| | Bis-Di-t-Butyl | Crosslinked | Moderately crystalline (b) | |
| | Bis-Di-iso-Pentyl | Crosslinked | Amorphous | |
| $Co_{.59}Zn_{.41}$ | Bis-Dioctyl | Not crosslinked | Weakly crystalline | |
| $Co_{.91}Zn_{.09}$ | Bis-Dioctyl | Not crosslinked | Weakly crystalline | |
| Chromium (III) | (Hydroxy) Bis-Methyl Phenyl | Not crosslinked | Amorphous (b) | Described more fully in Sec.IV |
| | (Hydroxy) Bis-Diphenyl | Not crosslinked | Amorphous (b) | |
| | Dioctyl Bis-Methyl Phenyl (B) | Not crosslinked | Highly disordered (b) | Described more fully in Sec.IV |
| | Dioctyl Bis-Methyl Phenyl (C) | Not crosslinked | Very crystalline (b) | Described more fully in Sec.IV |

Table I. Structural information on some typical poly(metal phosphinates).

We have measured the susceptibilities of many systems, based on metals of the first transition series: Cr, Mn, Fe, Co, Ni, Cu. To this list should be added Zn, which plays the role of a nonmagnetic control. Mixed systems based on Cu:Fe and on Co:Zn have also been examined. The number of R groups incorporated has been equally large: methyl, n-butyl, isopentyl, octyl, phenyl, and combinations thereof. In the trivalent Cr compounds other ligands may be present: for example, hydroxyl (OH), the halides (Cl, Br, etc.) or TCNQ. Syntheses are reported elsewhere.[5-11]

The paper is arranged as follows: in Sections II and III we present data on a wide variety of poly (metal phosphinates), and in Section IV we concentrate on the role of disorder in three particularly suitable examples. Section V examines what further steps will be necessary for a more complete understanding of these versatile materials.

## II. STRUCTURAL DETERMINATION

Analysis of the molecular structure in a polymeric material depends, in general, on inferences made from various indirect methods. X-ray analysis can, of course, be used to determine the structures of ordered polymers and, even in disordered materials, the presence of broad Bragg reflections gives some idea of the nature and extent of the disorder. In most PMP where x-ray structural analysis has been performed, only the backbone arrangement can be determined, the side groups being still randomly arranged.[12-15]

Birefringence studies can be used to examine the extent of optically anisotropic regions. Solubility measurements determine the degree of cross-linking between polymer chains.[16] Even mechanical properties may provide a clue to the degree of disordering.[9]

The application of these methods to the PMPs shows wide variations in cross-linking, crystallinity and optical activity (see Table I). Clearly, this is a very rich class of solids and much more information is needed for unambiguous assignment of structure. It will be shown that the magnetic properties can be used to help elucidate the details of the molecular arrangement.

## III. MAGNETIC PROPERTIES

The static susceptibility of the samples was measured as a function of temperature between 4K and 300K using a Faraday balance,[17] and corrected for core diamagnetism using Pascal's constants.[18] The resulting inverse susceptibility ($\chi_p^{-1}$) is plotted as a function of temperature, (Fig. 2) for some typical Co- and Ni-based materials. The high temperature data for these and several other materials were fitted by a nonlinear least-squares procedure to the form

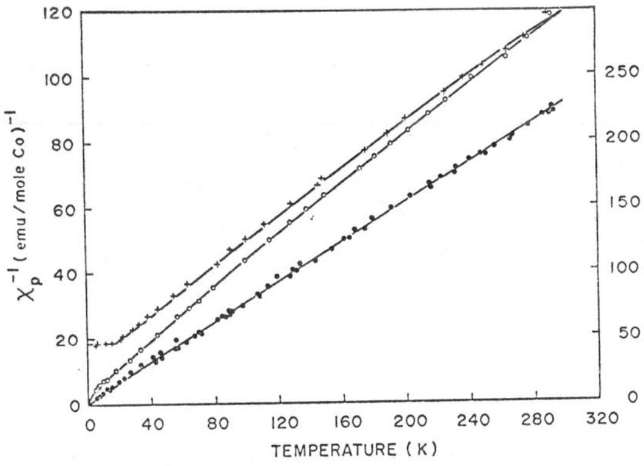


Fig. 2. Inverse susceptibility versus temperature poly(nickel dioctylphosphinate)(·), poly(cobalt di-n-butylphosphinate)(+), and poly(cobalt di-t-butylphosphinate)(O). Ordinate for Co materials is on the left, for Ni on the right. Lines are merely a guide for the eye.

$$\chi_p = \frac{C}{T + \Theta} + \chi_{VV}^{exp} \qquad (1)$$

where C (= $g^2\beta^2 S(S+1)N_A/3k$), $\Theta$ (the Curie-Weiss temperature) and $\chi_{VV}^{exp}$ (the temperature-independent Van Vleck contribution) were treated as fitting parameters. The results are summarized in Table II.

| Metal, Oxidation State and Spin | Θ(K) | exp $\chi_{VV}$ $10^{-3}$emu/mole | g (static) | g ESR | Linewidth Gauss |
|---|---|---|---|---|---|
| Ni(II),1 | 0-16 | 0 | 2.30-2.50 | not observed | |
| Co(II), 3/2 | 4-25 | 0-1 | 2.15-2.45 | not observed | |
| Mn(II), 5/2 | 0-10 | 1-2 | 1.80-1.90 | 2.00 | 135-810 |
| Cu(II) 1/2 | 0 | 0 | 2.26 | multiline spectrum $g_{AV}$=2.17 | |
| $Zn_xCo_{1-x}$ | 4-30 | 0 | 1.1-1.9 | not observed | |
| Cr(III), 3/2 | 8-24 | ~.3 | 1.85-2.00 | 1.97 | 200-600 |

Table II. Magnetic parameters in the poly(metal phosphinates).

There is no obvious systematic variation of Θ either as a function of the metal species or the alkyl substituents. Rather it seems that the effect of the R group is upon the molecular structure of the polymer and only indirectly on the magnetism.

For all metal species the values of $g\sqrt{S(S+1)}$ are consistent with absence of spin-pairing. Then S depends simply on the number of d electrons and we can obtain a value for g. This number is in good agreement with that from ESR in the cases where a spectrum has been observed. Ligand field theory predicts g shifts and values of $\chi_{VV}$, which are, in most cases, in good qualitative agreement with experiment.

It was hoped that some light might be cast on the percolation problem by examining systems: $Zn_xCo_{1-x}$. Unfortunately, nonmonatonic behavior of Θ, and unusual values of g lead to the conclusion that "clustering" of the metal species occurs and that there is not, in fact, dilution on a microscopic scale.

The mixed spin system, $[Cu_{0.5}Fe_{0.5}(OPOct_2O)_2]_n$, was examined in the hope of observing the noncancellation of moments. It is also possible that the different nature of the magnetic orbitals in Cu and Fe might lead to a different magnitude and/or sign of superexchange. The results for this material are qualitatively similar to many of the others.

ESR linewidths were measured at 10 and 30 GHZ. They show a wide range of values, but little frequency dependence (see Table II). We conclude that the broadening mechanism is not a distribution of g values, but must originate from dipolar interactions and fine structure. In general, the lines are Lorentzian in shape, indicating that exchange narrowing has occurred, and we expect such narrowing to remove any structure due to initial splittings, which indeed are never observed.

The copper spectra show structure typical of powder specimens with an axially anisotropic g value ($g_x = g_y = g_\perp$, $g_z = g_{//}$) or even having a g tensor of orthorhombic symmetry ($g_x \neq g_y \neq g_z$). This is consistent with the square-planar nature of the ligation in the copper polymers.[7,19]

The linewidth of poly(manganese di-t-butylphosphinate)[7,19] shows a frequency dependence. Of all the spin species observed, $Mn^{++}(d^5)$ is the most stable against fine structure splitting. Therefore, since the exchange frequency ($\omega_e \sim 1.4 \times 10^{11}$ Hz) is of the same order as the Larmour frequency ($\omega_0 \sim 1.8 \times 10^{11}$ Hz), one should expect to see evidence of nonsecular frequency-dependent broadening (the so-called "10/3 effect").[20,21]

The temperature dependence of the peak-to-peak linewidths of two chromium materials [Cr(OH)(OPMe$\varphi$O)$_2$][5] and Cr(OPMe$\varphi$O)$_2$(OPOct$_2$O)[9]] was measured between 4 and 300K. They show large increases below about 50K, and at even lower temperatures, the lines become highly asymmetric with a great deal of the intensity shifting to higher fields.

## IV. DISORDER IN ANTIFERROMAGNETIC CHAINS

The low temperature susceptibility behaves in a manner which falls between two limits: a broad maximum at some nonzero temperature and a divergence at T = 0 (see Fig. 2). This is particularly evident in the case of three chromium-based polymers[22] whose relative degrees of structural disorder can be unambiguously assigned. They are labeled mnemonically A, B, and C for the respectively amorphous, intermediate, and crystalline cases.

Sample A has a structure as shown in Fig. 3a.[5] The material is a powder, shows totally diffuse x-ray scattering, and virtually no birefringence. Materials B and C nominally have the same formula, [Cr(OPMe$\varphi$O)$_2$(OPOct$_2$O)]$_n$[9] (see Fig. 3b), but B is brittle, gives diffuse x-ray scattering, and shows only slight optical activity, whereas C is flexible, displays a number of broad Bragg reflections in addition to diffuse background scattering, and is highly birefringent. The solubility of the samples indicates minimal crosslinking in samples B and C, and only a little in sample A. Chain lengths are typically $10^3$-$10^4$ units.

a.

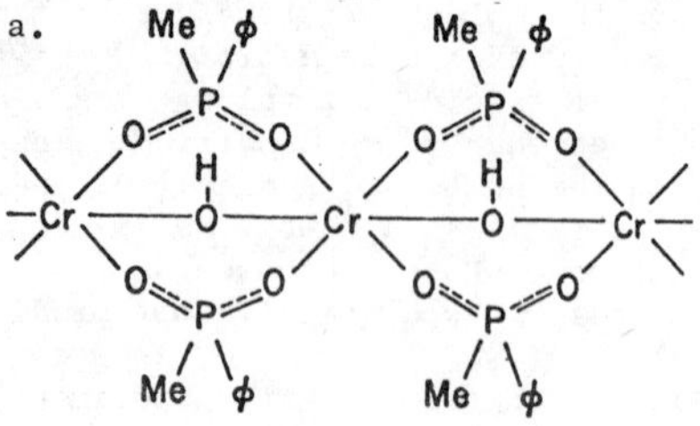


b.

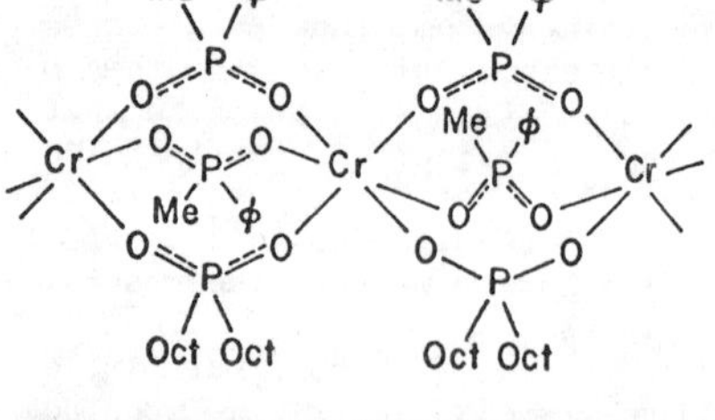


Fig. 3a. Most probable bridging structure for sample A.

Fig. 3b. Stoichiometric structure of samples B and C. Me ≡ $CH_3$-; $\phi$ ≡ $C_6H_5$-; Oct ≡ $C_8H_{17}$-

The magnetic susceptibilities below 70K are shown in Fig. 4 ($\chi_p^{-1}$ vs. T) and reveal the range of behavior described above. The differences can be related to the relative disorder by means of a model[22] in which disorder in the chain is introduced via an exchange, $J_i$, varying randomly along the chain according to a distribution $p(J_i)$. In particular, the square distribution function

$$p(J_i) = 1/\lambda \quad : \quad J - \lambda/2 \leq J_i \leq J + \lambda/2$$
$$= 0 \quad : \quad \text{otherwise} \qquad (2)$$

was applied to a line of classical spins and solved after the manner of Fisher.[23] The resulting expression for the susceptibility was found to be[22]

$$\chi = \frac{N_A g^2 \beta^2 S(S+1)}{3kT} \frac{1-\langle u \rangle}{1+\langle u \rangle} \qquad (3)$$

where the average of the transfer function, u, is given by

$$\langle u \rangle = \frac{kT}{2\lambda S(S+1)} \ln \left\{ \frac{(2J+\lambda)\ \sinh\ [(2J-\lambda)S(S+1)/kT]}{(2J-\lambda)\ \sinh\ [(2J+\lambda)S(S+1)/kT]} \right\}. \qquad (4)$$

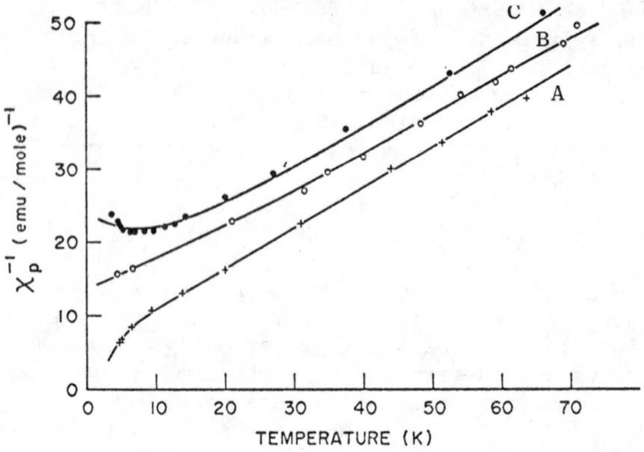


Fig. 4. Inverse susceptibility of samples A (+), B (o), and C (·).

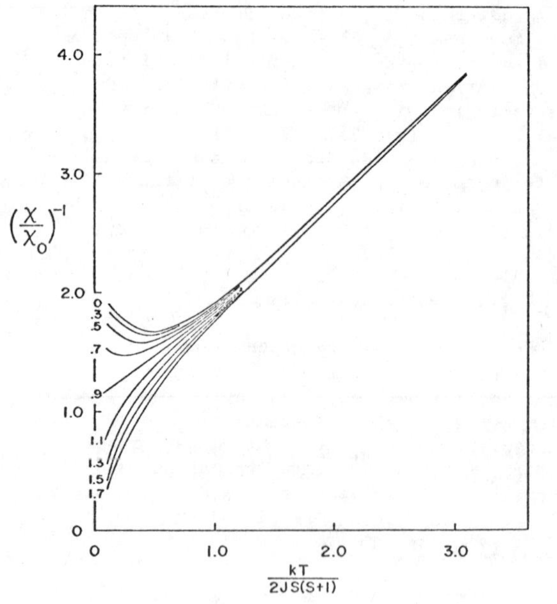


Fig. 5. Susceptibility for various values of the relative width of the distribution of exchange ($0 \leqslant \lambda/2J \leqslant 1.7$).

The behavior of $\chi(T)$ for various values of $\lambda/2J$ is shown in Fig. 5. It is obvious that the divergence is associated with a nonzero probability of ferromagnetic interactions; an interpretation which is physically very reasonable.

The data, corrected for $\chi_{VV}^{exp}$ as obtained from the high temperature analysis, were fitted in this form of $\chi$ for the parameters, g, J, and $\lambda$ to give the curves shown in Fig. 4. The values of the expressions g, 2JS(S+1)/k, and $\lambda/2J$ (the relative width of the square distribution) are given in Table III.

The width of the distribution parameterizes, in a unique manner, the magnetic disorder which can be seen to relate directly to the structural disorder. Even in the case of the most "crystalline" sample, the value of $\lambda$ is not negligible. This is as one would expect, since even the relatively well-ordered polymer is far from being a good crystal.

The specific heats of samples A, B, and C have also been measured, data from several zinc isomorphs being used in the separation of magnetic and vibrational contributions.[24] The magnetic specific heats thus obtained accounted for the expected magnetic entropy and showed a maximum which shifted to lower temperatures as the degree of disorder increased (see Fig. 6).

The low temperature specific heat cannot be properly described by the classical model described above, and to our knowledge no numerical computations have been done on the nonuniform chain. Our conclusions therefore stand as a purely experimental result. It should be noted that the behavior of the specific heat is at least qualitatively what one would expect of a disordered chain. In sample C there is a linear term suggestive of spin waves.[25,26] Scattering of these spin waves will lead to a broadening of the spectral density, to lower energy excitations and, when some neighbors are ferromagnetically coupled, to excitations of vanishingly small energy. Thus, the maximum in the specific heat is expected to shift to lower temperatures as the disorder increases.

| SAMPLE | g | 2JS(S+1)/k | λ/2J |
|---|---|---|---|
| A | 1.95 ± .05 | 13 ± 1 | 1.9 ± .2 |
| B | 1.95 ± .05 | 22 ± 1 | .90 ± .08 |
| C | 1.89 ± .06 | 25 ± 1 | .66 ± .06 |

Table III. Values of parameters in least-squares fit to model of randomized classical Heisenberg chain of data on the poly(chromium phosphinates): A, B, and C.

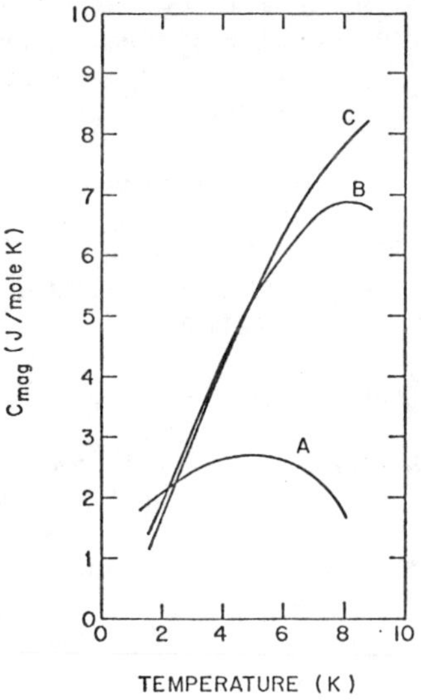


Fig. 6. Magnetic contribution to specific heat for samples A, B, and C.

## V. CONCLUSION

We have presented a broad initial study of a class of solids which promises to exemplify many interesting magnetic problems. In order to understand all the mechanisms involved in these polymers, it is necessary to go back to simpler systems: for example, dimers and more highly crystalline materials. The richness of PMP is such that these materials can be synthesised, and their examination is presently in progress. It is hoped that by doing so we shall determine the nature of the exchange mechanism and whether indeed it is possible that it changes sign. By examining the role of cross-linking we will, by extension, determine the properties of random three-dimensional magnets.

We end by emphasising the infinite variety of possible molecular structures of poly(metal phosphinates). In particular, it is hoped to use the polymerizing properties of the phosphinate groups and to vary the exchange by appropriate choice of the ligand L (Fig. 1). For example, the halides (X = Cl, Br, I), thiocyanate (SCN), thiophenyl ($SC_6H_5$), or nitrate ($NO_3$) groups might be used.

This paper is a summary of a further report which will appear elsewhere.

References

* Present Address: Laboratory of Atomic and Solid State Physics, Cornell University, Ithaca, New York 14853.

+ Supported in part by the NSF through the Laboratory for Research on the Structure of Matter, and grant #DMR-74-22923.

‡ Supported in part by ARPA through Order #2043, and by ONR through contract #N00014-69-C-0122.

1. L. J. deJongh and A. R. Miedema, Adv. in Phys. 23, 1 (1974).
2. H. O. Hooper and A. M. deGraff, eds., Amorphous Magnetism (Plenum, New York, 1973).
3. R. J. Elliott, J. A. Krumhansl, and P. L. Leath, Rev. Mod. Phys. 46, 465 (1974).
4. B. P. Block, Inorg. Macromol. Rev. 1, 115 (1970), and references therein.
5. P. Nannelli, M. D. Gillman, and B. P. Block, J. Polym. Sci., Part A-1, 9, 3027 (1971).
6. P. Nannelli, B. P. Block, J. P. King, A. J. Saraceno, O. S. Sprout, Jr., N. D. Peschko, and G. H. Dahl, J. Polym. Sci., Polym. Chem. Ed. 11, 2691 (1973).
7. H. D. Gillman, Inorg. Chem. 11, 3124 (1972); ibid. 13, 1922 (1974).
8. H. D. Gillman, P. Nannelli, and B. P. Block, J. Inorg. Nucl. Chem. 35, 4053 (1973).
9. P. Nannelli, H. D. Gillman, and B. P. Block, J. Polym. Sci., Polym. Chem. Ed. (in press).
10. S. H. Rose and B. P. Block, J. Polym. Sci., Part A-1, 4, 573 (1966).
11. S. H. Rose and B. P. Block, J. Am. Chem. Soc. 87, 2076 (1965).
12. V. Giancotti, F. Giordano, and A. Ripamonti, Die Makromol. Chem. 154, 271 (1972).
13. V. Giancotti and A. Ripamonti, J. Chem. Soc. (A), 706 (1969).
14. V. Giancotti, F. Giordano, L. Randaccio, and A. Ripamonti, J. Chem. Soc. (A), 757 (1968).
15. F. Giordano, L. Randaccio, and A. Ripamonti, Acta Cryst. B25, 1057 (1969).
16. P. J. Flory, Principles of Polymer Chemistry (Cornell University Press, Ithaca, New York, 1953).
17. J. C. Scott, Ph.D. Thesis, University of Pennsylvania.
18. J. Lewis and R. G. Wilkins, eds., Modern Coordination Chemistry (Interscience, New York, 1960) p. 403.
19. H. D. Gillman and J. L. Eichelberger, Inorg. Chem. (in press).
20. J. H. Van Vleck, Phys. Rev. 74, 1168 (1948).
21. P. W. Anderson and P. R. Weiss, Rev. Mod. Phys. 28, 269 (1953).
22. J. C. Scott, A. F. Garito, A. J. Heeger, P. Nannelli, and H. D. Gillman, Phys. Rev. B12, 356, (1975).
23. M. E. Fisher, Am. J. Phys. 32, 343 (1964).
24. T. S. Wei, J. C. Scott, A. F. Garito, A. J. Heeger, H. D. Gillman, and P. Nannelli, Phys. Rev. B (in press).
25. P. W. Anderson, Phys. Rev. 86, 694 (1952).
26. R. Kubo, Phys. Rev. 87, 568 (1952).

# METAMAGNETISM IN SINGLE CRYSTAL $\gamma$-Co(PYRIDINE)$_2$Cl$_2$

S. Foner and R.B. Frankel
Francis Bitter National Magnet Laboratory,† MIT, Cambridge, Mass. 02139
W.M. Reiff* and H. Wong
Northeastern University, Boston, Mass. 02115
G.J. Long**
University of Missouri-Rolla, Rolla, Missouri 65401

ABSTRACT

Small ($\lesssim$ 1 mg) single crystals of $\gamma$-Co(pyr)$_2$Cl$_2$ have been grown and measurements of magnetic moment vs applied field H and temperature have been made for $0 \le H \le 54$ kG and $1.25\,K \le T \le 4.2\,K$. At 1.25 K, metamagnetic transitions are observed along the a-axis at ~700 G; along the b-axis at both ~800 and ~1500 G; and along the c-axis at ~4 kG. The low field susceptibilities show a maximum at T ≃ 3.4 K for each orientation. Magnetic saturation is not achieved at 54 kG; about 2.5 $\mu_B$ is achieved along the a*- or b-axes, whereas a lower moment is achieved along the c-axis. Along the b-axis, the low field moment change is 1/2 the higher field moment change, suggesting a six-sublattice magnetic structure.

## INTRODUCTION

Previously[1,2] we reported on magnetic measurements on polycrystalline samples of Co(pyr)$_2$Cl$_2$, Fe(pyr)$_2$Cl$_2$, Fe(pyr)$_2$(NCS)$_2$ and Ni(pyr)$_2$Cl$_2$ where (pyr) is pyridine. These materials all have linear chain structures with ferromagnetic interactions along the chain and relatively weak interactions between chains. The temperature dependence of the low field data indicated metamagnetic behavior for all these linear chain systems. Another feature of the data was that the magnetic moment at high field was not saturated in fields to 200 kG. At that time we noted that detailed characteristics of the relatively low field transitions required measurements in single crystals. In this paper we present magnetic data obtained on single crystal $\gamma$-Co(pyr)$_2$Cl$_2$ which is isomorphous[3] with Cu(pyr)$_2$Cl$_2$. The results confirm the metamagnetic behavior at low field observed in the powder material.

## EXPERIMENTAL DETAILS

The experimental results were obtained with a vibrating sample magnetometer adapted to a superconducting solenoid. The Co(pyr)$_2$Cl$_2$ crystals were grown from absolute ethanol solution.

The single crystals were each nominally < 1 mg in weight. In order to obtain relatively large signals, samples composed of 2 to 3 oriented crystals were used. Individual crystal measurements were also made to guarantee that the composite samples reflected the single crystal behavior. There was a tendency for the crystallites to deteriorate or fracture with repeated measurements because the high temperature $\alpha$ phase transforms to the low temperature $\gamma$ phase and so the major part of the data was obtained for each orientation by cooling each batch of crystals only once.

## EXPERIMENTAL RESULTS

### General Features

The main features of the magnetic moment vs applied field, $\sigma$(H), for single crystal Co(pyr)$_2$Cl$_2$ are illustrated in Fig. 1. Figure 1a shows $\sigma$ (H) at 1.25 K for H parallel to the a*-axis. There is a single low-field transition and saturation is not achieved in fields of 50 kG. A more complex

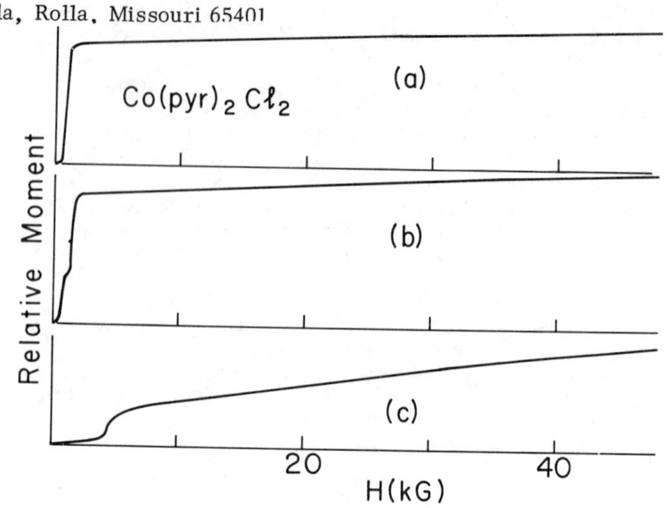


Fig. 1. Relative magnetic moment $\sigma$ versus applied field H at 1.25 K: a) H||a*-axis; b) H||b-axis; c) H||c-axis. The moments at H = 52 kG are ~ 2.5 $\mu_B$/Co atom for (a) and (b), and ~1.1 $\mu_B$/Co atom for (c).

behavior is observed for H parallel to the b-axis (Fig. 1b); $\sigma$(H) shows two sharp metamagnetic transitions with a rapid rise at low field which is approximately one-half the rapid rise at higher field. Again the high-field region shows a gradual increase of $\sigma$ up to 50 kG. The $\sigma$(H) for H along the c-axis is shown in Fig. 1c; for this orientation there is a single transition, but the field at which this transition takes place is considerably higher than those for the a*- or b-axes. Furthermore, the magnetic moment at ~50 kG is much less than for H along the a*- or b-axis.

### Low Field Transitions

For H along the a*-, b- or c-axis, the initial susceptibility versus T shows a maximum at T ≃ 3.4 K for each orientation, and a rapid decrease below 3.4 K. For each orientation there is a maximum in susceptibility in the range of 3.2 to 3.4 K, and the maximum change in susceptibility occurs in the region of about 3.2 K. The values of $T_c$ are therefore in good agreement with earlier measurements in polycrystalline samples.[2] The average susceptibilities for intermediate field regions are strongly dependent on H and T. As an example, $\sigma$(H) for different fixed temperatures with H along the b-axis is shown in Fig. 2. Measurements for other orientations will be presented elsewhere.

The critical field $H_c$ was determined as in Refs. 1 and 2. We find that $H_c$ increases as T is decreased for all the low field transitions. At 1.25 K, $H_c$ = ~700, ~800 and ~1500, and ~4 kG for the a*-, b-, and c-axis respectively, and in some cases $H_c$ exhibits hysteresis.

### High Field Results

Saturation is not achieved by 50 kG, but the magnetic moment for H along the c-axis is considerably smaller than

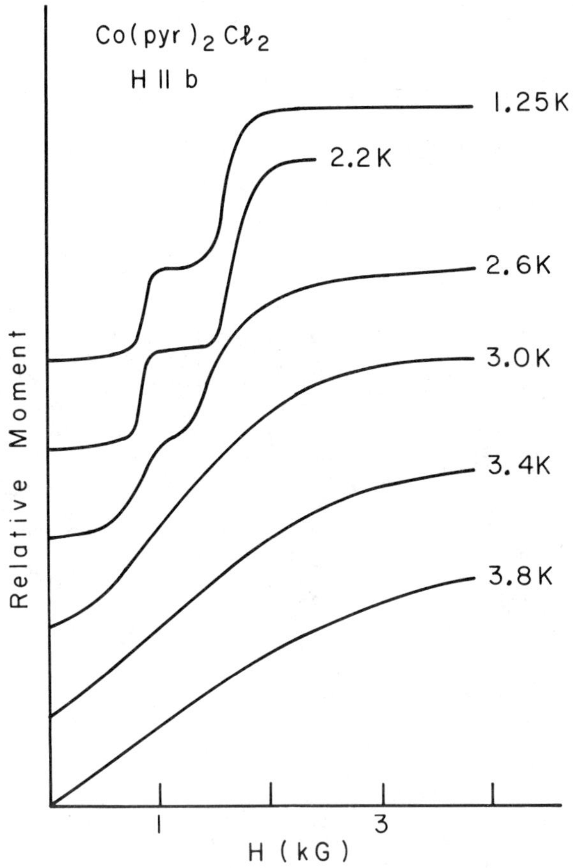


Fig. 2. Relative moment σ versus applied field H for $1.25 \leq T \leq 3.8$ K. The curves are displaced along the ordinate for clarity.

that along the other two axes. These results are consistent with the data presented earlier[2] for polycrystalline samples for fields of over 200 kG. The earlier polycrystalline data showed only one transition, which can be understood in terms of broadening introduced by the randomly oriented crystallites; the a*- and b-axis transitions occur at almost the same field and the second transition for the b-axis occurs over a relatively narrow angular orientation.

## DISCUSSION OF RESULTS

The σ(H, T) data for single crystal $Co(pyr)_2Cl_2$ presented here may be compared to the results reported by Narath[4] for $Co(H_2O)_2Cl_2$. Because of the bulky pyridine ligands, the interchain interactions are considerably reduced compared to the dihydrate and the transitions occur at fields more than one order of magnitude lower. There is no possibility for exchange via hydrogen bonding as in the case of the dihydrate.[5]

In the dihydrate case, the spins are ferromagnetically coupled along the c-axis, the b-axis is the easy magnetic axis, and external fields along b induce metamagnetic transitions at ~31 kG and ~46 kG. A six sublattice model describes the intermediate spin structure.[4] In the present case, the spins, which also are ferromagnetically coupled along the c-axis, appear to order antiferromagnetically along the b-axis as well. For H||b, one again observes two transitions. For H||a* a single metamagnetic transition occurs to a state with the spins aligned parallel to a*. The slightly lower $H_c$ for H||a* compared to H||b is consistent with our observed susceptibilities $\chi_{a}^{*} > \chi_b$.

## REFERENCES


Supported by the National Science Foundation.
Work supported by NSF Grant No. GH39010 and in part by Grant No. PP07143, Department of Health, Education and Welfare.
Supported by National Science Foundation Grant GP8653.



1. S. Foner, R.B. Frankel, W.M. Reiff, B.F. Little and G.J. Long, Solid State Commun. 16, 159 (1975).
2. S. Foner, R.B. Frankel, E.J. McNiff, Jr., W.M. Reiff, B.F. Little and G.J. Long, AIP Conf. Proc. 24, 363 (1975).
3. P.J. Clarke and H.J. Milledge Acta. Cryst. B31, 1543 1554 (1975).
4. A. Narath, J. Phys. Soc. Japan 19, 2244 (1964); Phys. Rev. 136, A766 (1964).
5. J.N. McElearney, S. Merchant and R.L. Carlin, Inorganic Chem. 12, 906 (1973).


# THEORY OF TWO SPIN INFRARED ABSORPTION IN ONE DIMENSIONAL HEISENBERG ANTIFERROMAGNETS


M. Drawid and J. W. Halley
School of Physics and Astronomy
University of Minnesota
Minneapolis, Minnesota, 55455


## ABSTRACT


Recent experimental work on one-dimensional paramagnets with antiferromagnetic Heisenberg interaction showing the existence of spin-wave-like collective excitations has generated interest in the study of two spin optical processes in such systems. No experiments have been performed primarily due to rather stringent selection rules imposed by the high symmetry of the magnetic crystal. A uniformly spaced chain will be optically inactive for Raman scattering but will give rise to infrared absorption to the extent of the reflection asymmetry in the crystallographic structure. Following we study the problem of infrared absorption in a uniform system which we hope to extend to non-uniform (dimerized) systems[1] where the symmetry will allow substantial infrared absorption. The simplest interaction leading to light absorption in a one-dimensional spin-system is given by $H_A = \sum_j (-1)^j S_j \cdot S_{j+1}$ For the lack of an obvious expansion technique the problem was treated in a fourth-order Green's function decoupling theory by one of us (JWH) and S. K. Lo[2] with the conjecture that the absorption will peak at twice the frequency of the zone-boundary spin waves with little interaction at T = 0 and substantial interaction at T = ∞. In the present paper we study that G. F. theory more thoroughly to find that the above conjecture was very well justified. In particular we observe that at T = 0 a fourth-order G. F. decoupling yields identical results to those obtained from a first principle non-interacting spin-wave theory. We display the results of the G. F. theory at various finite temperatures above T = 0 for a chain of 40 spins. We also find the exact classical frequency moments of the absorption and compare them with similar results obtained in a straightforward manner from the G. F. theory. We then obtain the absorption using the classical moments in the 3-pole approximation. The moments in the two theories agree quite well at all temperatures but the absorption starts differing markedly for $T/T_{MF} \geq 1$ where $T_{MF}$ is the mean field transition temperature. We argue that the 3-pole approximation is more reliable at higher temperatures.



1. J. W. Bray et.al., Phys. Rev. Lett. 35, 11 (1975).
2. S. K. Lo and J. W. Halley, Phys. Rev. B 8, 11 (1973).

# QUANTITATIVE STUDIES OF MAGNETIC COOLING ON A MAGNETIC SYSTEM WHICH OBEYS THE THIRD LAW.*

J. D. Johnson
Los Alamos Scientific Laboratory, Los Alamos, N. M. 87545
J. C. Bonner
Applied Mathematics Department, Brookhaven National Laboratory, Upton, L. I., N. Y. 11973

ABSTRACT

Quantitative estimates of magnetic cooling particularly in antiferromagnets have not been feasible in the past because the standard theoretical models and approaches disobey the third law of thermodynamics and are consequently unreliable at low temperatures.

Recently, an essentially exact treatment has been developed for the linear anisotropic Heisenberg-Ising chain, the only solvable model which does obey the third law. Novel and complex features in the cooling behavior of the antiferromagnet are observed, and arguments are presented to indicate that quasi-(1-D) magnetic systems may turn out to be serious rivals to the classic cooling paramagnets like CMN.

## INTRODUCTION

Magnetic cooling has not seen much theoretical development in recent years because standard theoretical models, for example, the classical paramagnet, the Ising and mean field antiferromagnets all disobey the third law of thermodynamics. This clearly limits their usefulness in a quantitative (or even qualitative) description of low temperature phenomena. In principle, series expansion treatments of more realistic models could be further pursued, but are hard to handle in, say, the low temperature, high field regime for antiferromagnets (AFs), which is the region of interest. Recently, however, there has been renewed experimental attention to AFs resulting from the now widespread availability of high field superconducting magnets. Very recent theoretical developments based on the renormalization-group (RG) approach have stimulated considerable interest in the uniaxial AF phase boundary and its bicritical point[1]. The phase behavior (and consequent magnetic cooling behavior) of the nuclear spin-½ AF, solid $He^3$, remains under intensive study[2], and the development of improved cryogenic techniques in relation to superconductivity has considerable current technological importance.

We present here the first exact treatment of a reasonably realistic magnetic model which behaves in accordance with the third law. This system is the one-dimensional (1-D), spin-½, Heisenberg-Ising magnet[3]. Since a comprehensive investigation with respect both to temperature and magnetic field is required, exact magnetic cooling studies of higher-D models seem out of the question at present. Quasi-(1-D) magnetic systems have been studied extensively experimentally in recent years[4] and interest in 1-D organic charge-transfer compounds has been enormously stimulated by the discovery of super conductivity in the quasi-(1-D), non-metallic, polysulfur nitride[5]. An amusing, and perhaps unexpected, feature of dimensionality and magnetic cooling is that 1-D systems, far from being rather artificial apologies for 'real' 3-D systems, actually surpass their higher-D counterparts in cooling potential. Therefore one should look first in the ranks of quasi-(1-D) systems for good magnetic coolers. Standard Wilson-type RG approaches are not suitable for studies of this nature because they assume a classical vector form for the spin which, of course, violates the third law. This disadvantage applies also to the comprehensive numerical solution by Blume et al. of the classical (∞ spin) Heisenberg chain in a field[6].

## SOLUTION OF THE MODEL

We consider the very familiar Heisenberg-Ising, 1-D, spin-½ H Hamiltonian with uniaxial anisotropy, $\gamma$, and periodic boundary conditions

$$H = -2J \sum_{i=1}^{n} \{S_i^z S_{i+1}^z + \gamma(S_i^x S_{i+1}^x + S_i^y S_{i+1}^y)\}$$
$$-g\,\beta\, H \sum_{i=1}^{n} S_i^z,$$

where the first term is the exchange term and the second the Zeeman term due to the applied field H.

This system has recently been studied analytically for temperature $T > 0$, primarily by McCoy and co-workers following work by Gaudin on the distribution of the zeroes of a certain set of functions, and their behavior as the anisotropy varies. The solution of the problem is not obtainable in closed form but in terms of an infinite set of coupled nonlinear integral equations. One of us[7] has developed a numerical iterative technique for solving these equations. The integrals themselves can be evaluated fairly expeditiously by use of an n-point trapezoidal rule, since it has been shown that the errors involved are exponentially small[7]. The results for the thermodynamic properties are in excellent agreement with earlier numerical extrapolations, confirming the validity of the basic Gaudin prescription. Extensive computations have been performed, primarily of the entropy and specific heat as a function of T and H, for various values of the anisotropy parameter, $\gamma$, ($\gamma > 0$), and for both the antiferromagnetic case ($J < 0$) and the corresponding ferromagnetic (paramagnetic) case ($J > 0$). This short report presents selected highlights of these calculations and a full account will be published in due course.

## COOLING CURVES (ISENTROPES)

In Fig. 1, we present the cooling curves (plots of reduced temperature, kT/J, versus reduced magnetic field, $g\beta H/J$, at constant entropy, S, for the antiferromagnet at a $\gamma$ value of 0.2. This plot is much more complex than its counterpart for the ferromagnet (single critical field at H=0) and for the isotropic ($\gamma$=1) AF (single critical field given exactly by the expression $H_c=4J/g\beta$). The reason is, of course, that the anisotropic AF system, $0 < \gamma < 1$, has <u>two</u> critical fields, a lower one located at $H_{c1} = 1.248J/g\beta$, where the magnetization first becomes non-zero, and an upper, located at $H_{c2}=2(1+\gamma)J/g\beta = 2.4J/g\beta$, where the magnetization saturates. (For 1-D, short-range systems, of course, critical singularities occur only at T=0.)

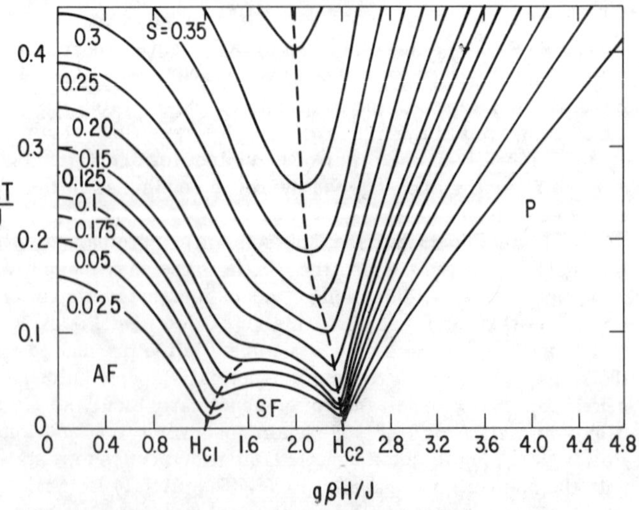


Fig. 1. Cooling isentropes for the 1-D AF for $\gamma$ = 0.2.

At low temperatures there is a maximum in the entropy isotherms in the vicinity of each critical field, reflecting a high density of states as, for example, the Zeeman energy becomes comparable with the AF exchange energy. By the magnetocaloric equation, the entropy maxima give rise to minima in the isentropes at corresponding H and T values, and hence regions of maximum cooling. Clearly, the cooling behavior is novel and complicated. Four different types of magnetic cooling are possible, involving both adiabatic magnetization and adiabatic demagnetization to either $H_{c1}$ or $H_{c2}$, depending on which of the three regions, $0 < H < H_{c1}$, $H_{c1} < H < H_{c2}$, and $H_{c2} < H$, the initial magnetic field value lies. The process of cooling an AF by adiabatic magnetization is reasonably familiar[8,9]. It is obvious from Fig. 1 that a potentially more powerful cooling process involves cooling by adiabatic demagnetization to $H_{c2}$ from an initial field as large as possible in comparison with $H_{c2}$[10]. There is an obvious analogy between this process and the cooling of a paramagnet by demagnetization to its critical field H=0.

Our exact treatment gives a microscopic statistical mechanical explanation of these phenomena in terms of the low-lying densities of states, and also analytic forms for the asymptotic functional dependence on T as $T \to 0$, but space limitations prevent discussion here[11].

There is a feeling that the loci of maximum entropy (isentrope minima) of a 1-D AF should bear some relation to the actual phase boundaries in, say, a 3-D AF. Hence these loci are shown dashed in Fig. 1. The 3-D analogue of our 1-D system is a uniaxial spin-flop AF, like $MnF_2$, and the shape of the three phase boundaries near the bicritical point is currently under scrutiny[1]. In the 1-D system we can loosely speak of a predominantly AF (short-range) ordered region, a (short-range) transverse ordered (SF) region and a high T, high field paramagnetic region (P). The dashed line emanating from $H_{c1}$ is then the presumed analogue of the first order 3-D spin-flop line. However, instead of remaining roughly parallel to the T axis and eventually joining up with the other loci, it bends sharply over and appears to terminate, though slight uncertainties inherent in our numerical procedures make this result difficult to determine conclusively.

## COMPARISON OF ANTIFERROMAGNET AND PARAMAGNET

The feeling is that demagnetizing a paramagnet down to H=0 is a more efficient cooling process than one involving AFs. However, the proper comparison to make with a paramagnet is to demagnetize an AF to the isentrope minima locus emanating from $H_{c2}$, and our exact solution allows such comparisons to be made. The determining factor is the entropy of the AF along the $H_{c2}$ locus compared with the entropy in zero field for the paramagnet.

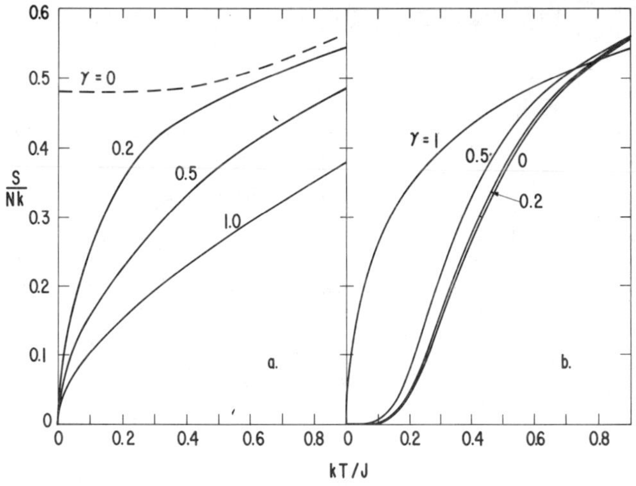


Fig. 2.(a) Entropy along the $H_{c2}$ locus for the 1-D AF for various $\gamma$ values. (b) Zero-field entropy for the 1-D ferromagnet for various $\gamma$ values.

These entropy curves are shown as a function of T for various $\gamma$ values for the antiferromagnet (Fig. 2a) and the paramagnet (Fig. 2b). The optimal system has as high an entropy as possible down to the lowest temperature possible, and by this criterion the AF beats the P, since the AF entropy rises from T=0, $H=H_{c2}$, as a power law $T^{1/2}$, while the P entropy rises exponentially slowly, owing to the existence of an energy gap between ground state and lowest excited states for $\gamma < 1$. $\gamma = 1$ is a special case, difficult to realize in practice.

## PRACTICAL APPLICATIONS

Our theoretical studies indicate that a 1-D antiferromagnet of fairly large anisotropy ($\gamma \sim 0.2$-$0.5$) would be a novel and efficient magnetic cooler. Accordingly, experiments are planned when a quasi- (1-D) system of this nature, with a J value such that the upper critical field is no larger than about 50 kOe, becomes available. Ultimately, 3-D effects would become important in a real system, but with minimal disturbance of the arguments presented above. Cooling would still result and the existence of a phase transition would have negligible effect[9]. From general arguments we expect the entropy at the (upper) critical field for AFs to go as $T^{d/2}$, where d is the dimensionality. Since the function $T^{1/2}$ rises more rapidly from zero than the function $T^{3/2}$, we have justification for the claim that a 1-D magnet is a better potential cooler than a uniform 3-D magnet. By using an undiluted linear system, the problem of lowering the magnetic specific heat by dilution in, say, the classic cooling paramagnet, cerium magnesium nitrate, would be avoided.

## ACKNOWLEDGEMENTS

We wish to acknowledge informative discussions with Professor S. A. Friedberg, whose extensive experimental investigations on the antiferromagnet $Cu(NO_3)_2 \cdot 2.5\ H_2O$ provided inspiration for this work.

*Supported by the U. S. Energy Research and Development Administration.

## REFERENCES

1. M. E. Fisher, Invited Paper, Proc. 20th Ann. Conf. on Magnetism and Magnetic Materials, p. 273 (1974).
2. J. D. Johnson and E. G. D. Cohen, Phys. Rev. B 12, 297 (1975).
3. J. C. Bonner and M. E. Fisher, Proc. Phys. Soc. (now J. Phys. C) 80, 508 (1962) showed that the Heisenberg-Ising AF obeys the third law for $\gamma > 0$.
4. L. J. de Jongh and A. R. Miedema, Adv. Phys. 23, 1 (1974).
5. See the review in Physics Today 28, 17 (June,1975).
6. M. Blume, P. Heller and N. A. Lurie, Phys. Rev. B 11, 4483 (1975).
7. J. D. Johnson, Phys. Rev. A 9, 1743 (1974).
8. W. P. Wolf, Phys. Rev. 115,1196 (1959).
9. J. C. Bonner and J. F. Nagle, Phys. Rev. A 5, 2293 (1972).
10. J. C. Bonner and S. A. Friedberg, Proceedings of the Conference on Phase Transitions and Their Applications, Eds. H. K. Henisch, R. Roy, and L. E. Cross, Pergamon Press, 1973, p 429.
11. J. C. Bonner and J. D. Johnson, work in preparation for publication.

# THE EXTRAORDINARY HALL EFFECT: INTUITIVE THEORY AND EXPERIMENTAL VERIFICATION

J.-N. Chazalviel and I. Solomon
Laboratoire de Physique de la Matière Condensée*
Ecole Polytechnique, 91120 Palaiseau, France

ABSTRACT

We present a review of the theoretical literature on the extraordinary Hall effect. By using the formulation of Nozières and Lewiner, we give an elementary presentation, stressing the physical aspects of the problem. Insight into the role of the spin-orbit interaction in Bloch functions and in the extraordinary Hall effect is provided by the concept of the transverse displacement of a spin-polarized electron. From this the different terms of Luttinger are found and discussed on intuitive grounds. Two types of effects are distinguished: the "side-displacement" contributions and the skew-scattering effect. The conflicting results of various theories in the literature are compared and criticized. In the very simple case of a III-V semiconductor the wavefunctions are well known and the calculations can be carried out exactly. We have measured the extraordinary Hall effect in such a system; by using a spin resonance method, the much larger ordinary Hall effect is circumvented as it gives no contribution to the signal. Our results provide an unambiguous test of the theory.

## I. INTRODUCTION

The early measurements in iron[1] gave an unusually large value for the Hall coefficient. Very soon it appeared that the Hall voltage in such magnetized media can be separated into two components, $V_H \propto R_H B + R_{an}(\mu_0 M)$, called respectively the normal and anomalous (or extraordinary) Hall effect. Many experimentalists[2-4] investigated the extraordinary Hall effect in various magnetic metals and alloys, plotting $R_{an}$ versus resistivity when the temperature or impurity content is varied, but no quantitative theory was attempted before the years 1950. Since then, there has been a great renewal of interest among theorists. Although some spin-orbit interaction was generally invoked, many conflicting papers were published;[3-15] nevertheless, the agreement with experiment remained rather unsatisfactory.

The failing of most of these theories was that they started from a free-electron model, resulting in a calculated effect several orders of magnitude too small. On the other hand, those theories taking into account the band structure could hardly be compared with experiment, due to the lack of knowledge of the wavefunctions in the investigated materials. Recently, the theory of the anomalous Hall effect has been worked out for semiconductors, where both the band structure and wavefunctions are well-known.[16-20] This has resulted in a much better understanding of the problem, and an experimental check of the theory has been provided.[21]

We first give an elementary presentation, closely following the theory of Nozières and Lewiner.[20] It gives a physical insight to the problem and makes clear the origin of the various terms found by Luttinger.[7] Then we discuss some recent controversial papers. Finally we report an experimental measurement in a semiconductor, which gives an unambiguous verification of the theory.

## II. SPIN-ORBIT INTERACTION AND BLOCH-FUNCTIONS

A simple illustration of the effect of spin-orbit interaction in a crystal is provided by III-V compound semiconductors. In these materials the gap occurs at the center of the first Brillouin zone (Γ point). At this point the conduction band is s like; the valence band is p like (wavefunctions X,Y,Z) and is split by spin-orbit interaction, just like the corresponding

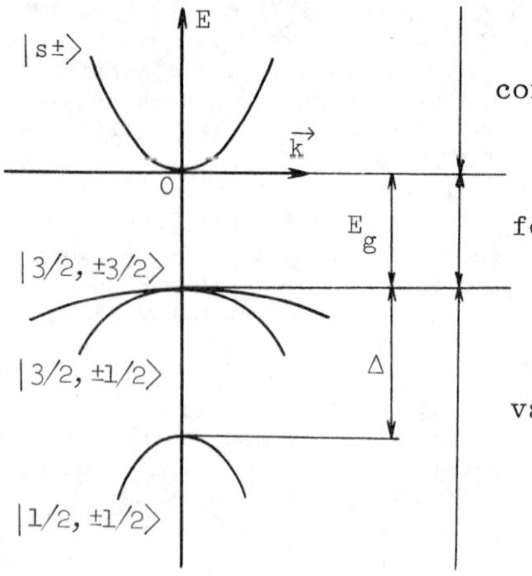


Fig. 1

Band-structure of a III-V compound semiconductor near the center of the first Brillouin zone (Γ point).

atomic levels. We call $E_g$ the band gap and Δ the spin-orbit splitting of the valence band (see Fig. 1).

In the vicinity of the Γ point, the wavefunctions and energy levels can be computed by $\vec{k}\cdot\vec{p}$ perturbation theory.[22] The eigenfunctions are of the Bloch type

$$\Psi_{n\vec{k}\sigma} = u_{n\vec{k}\sigma}\, e^{i\vec{k}\cdot\vec{r}}$$

where n and σ are respectively the band and spin index. For small gap materials, the periodic part of the Bloch functions for the conduction and valence bands can be expressed to a good approximation as linear combinations of $|s\sigma\rangle$ and $|\vec{R}\sigma\rangle$, where $|\vec{R}\rangle$ is the vector of components $|X\rangle$, $|Y\rangle$, $|Z\rangle$. The result for the conduction band has been given by Zawadzki[23] and may be written:

$$|u_{\vec{k}\sigma}\rangle = a|s\sigma\rangle - \frac{i\hbar P}{3m}\left[\left(\frac{2}{E_g} + \frac{1}{E_g+\Delta}\right)\vec{k}\cdot|\vec{R}\sigma\rangle + i\left(\frac{1}{E_g} - \frac{1}{E_g+\Delta}\right)2\vec{S}\cdot(\vec{k}\times|\vec{R}\sigma\rangle)\right] \quad (1)$$

with

$$a = 1 - \frac{1}{2}\frac{E}{E_g}\frac{3E_g^2+4E_g\Delta+2\Delta^2}{(3E_g+2\Delta)(E_g+\Delta)}$$

Here $\vec{S}$ is the spin operator and $iP = \langle s|p_x|X\rangle$. We have taken the terms only up to second order in $\vec{k}$, since $E_F$ is usually much smaller than $E_g$, and we have omitted the band index in $u_{\vec{k}\sigma}$, since we are concerned only with the conduction band wavefunctions.

The main effect of the spin-orbit interaction is confined to the last term in the square bracket. As pointed out by Adams and Blount,[8] this term is quite similar to the small components of the free electron Dirac wavefunction, the only difference being the order of magnitude: $(E/2mc^2)^{1/2}$ in that case, $(E/E_g)^{1/2}$ here. (here we use the fact that $\Delta \sim E_g$). The reason comes obviously from the different values of the interband spacing.

## III. MATRIX ELEMENTS OF $\vec{r}$

The matrix elements of $\vec{r}$ between Bloch states can be separated by performing an integration by parts.[24,25,5]

$$\langle\vec{k}\sigma|\vec{r}\,|\vec{k}'\sigma\rangle = \frac{1}{i}\vec{\nabla}_{\vec{k}'}\,\delta(\vec{k}-\vec{k}')$$

$$-\frac{1}{i}\,\delta(\vec{k}-\vec{k}')\iiint_{\Omega} u^{*}_{\vec{k}\sigma}\,\vec{\nabla}_{\vec{k}'}\,u_{\vec{k}'\sigma}\,\frac{d^3r}{\Omega}$$

where $\Omega$ is the volume of a unit cell. The first term, which would remain alone in the case of plane-wave states, is the "plane-wave part" of $\vec{r}$. The second term, hereafter noted as $\vec{R}$,[26] arises from the periodic part of the Bloch function and is usually called "periodic part" of $\vec{R}$.

Interpreting this term as a "polarization" of $u_{\vec{k}\sigma}$ inside a unit cell is in our view an unsatisfactory picture; its physical content can be made more evident by considering first a wave-packet, then taking the limit of infinite spatial extent.

$$\Psi = \iiint \frac{d^3k}{(2\pi)^3}\, u_{\vec{k}\sigma}(\vec{r})\,e^{i\vec{k}\cdot\vec{r}}\,c(\vec{k})$$

In the infinite spatial extent limit, $c(\vec{k})$ is localized in the Brillouin zone around some mean wavevector $\vec{k}_0$; developing $u_{\vec{k}\sigma}$ up to first order in $(\vec{k}-\vec{k}_0)$ gives:

$$\Psi = G(\vec{r})\,u_{\vec{k}_0\sigma}(\vec{r})\,e^{i\vec{k}_0\cdot\vec{r}} + \frac{1}{i}\left[\vec{\nabla}_{\vec{k}}u_{\vec{k}\sigma}(\vec{r})\right]_0\cdot\left[\vec{\nabla}_{\vec{r}}G(\vec{r})\right]e^{i\vec{k}_0\cdot\vec{r}} = F(\vec{r})+\varepsilon(\vec{r})$$

where $|\varepsilon| \ll |F|$ and $G(\vec{r})$ is the wave-packet envelope function (Fourier transform of $c(\vec{k}_0+\vec{k})$). In the case of free-electron states, $\Psi$ would reduce to $G(\vec{r})\exp(i\vec{k}_0\cdot\vec{r})$. The influence of $u_{\vec{k}\sigma}$ is twofold: first $\Psi$ is multiplied by $u_{\vec{k}_0\sigma}(\vec{r})$, a trivial result which yields the first part $F(\vec{r})$; since generally $u_{\vec{k}\sigma}$ has no inversion symmetry, this does give a "polarization" of the wavefunction inside each unit cell; however the envelope function is unchanged, and as can be seen from a calculation of $\iiint |F(\vec{r})|^2\vec{r}\,d^3r$, the average position of the wave-packet is not affected in the limit of infinite extent. A second effect of $u_{\vec{k}\sigma}$ is the appearance of an additional term $\varepsilon(\vec{r})$, which is proportional to $\vec{\nabla}_{\vec{r}}G(\vec{r})$, and thus results in a displacement of the envelope function. This contribution obviously remains non-null in the infinite extent limit; the corresponding displacement of the wave-packet is:

$$\vec{R} = \langle F|\vec{r}|\varepsilon\rangle + \langle\varepsilon|\vec{r}|F\rangle \approx$$

$$\iiint \vec{r}\left\{\left(G\vec{\nabla}G^{*}-G^{*}\vec{\nabla}G\right)\cdot\left(i\iiint_{\Omega} u^{*}_{\vec{k}_0\sigma}\left[\vec{\nabla}_{\vec{k}}u_{\vec{k}\sigma}\right]_0\frac{d^3r}{\Omega}\right)\right\}d^3r$$

where we have used the fact that $G(\vec{r})$ is a slowly varying function on the scale of a unit cell. Integrating by parts and using $\iiint |G(\vec{r})|^2\,d^3r=1$, one obtains:

$$\vec{R} = i\iiint_{\Omega} u^{*}_{\vec{k}_0\sigma}\,(\vec{\nabla}_{\vec{k}}u_{\vec{k}\sigma})_0\,\frac{d^3r}{\Omega} \qquad (2)$$

which corresponds exactly to the previously mentioned periodic part of $\vec{r}$. Remark: $\vec{R}$ is not itself an intrinsic quantity but depends on the choice of the basis Bloch functions and their relative phases: Namely if a basis $b = \{u_{\vec{k}\sigma}\exp(i\vec{k}\cdot\vec{r})\}$ is replaced by a basis $b' = \{u_{\vec{k}\sigma}\exp(-i\alpha(\vec{k})+i\vec{k}\cdot\vec{r})\}$, then $\vec{R}$ becomes $\vec{R}' = \vec{R}+\vec{\nabla}_{\vec{k}}\alpha(\vec{k})$. The physical (invariant) parameter for the displacement is thus $\vec{\nabla}_{\vec{k}}\times\vec{R}$ and one should verify that the physical results depend on $\vec{R}$ only through this quantity.

In the case of a III-V semiconductor, the value of $\vec{R}$ can be obtained by substituting into Eq. (2) the expression of $u_{\vec{k}\sigma}$ [Eq. (1)]. The exact result is

$$\vec{R} = \lambda\vec{S}\times\vec{k} \quad \text{or} \quad \vec{\nabla}_{\vec{k}}\times\vec{R} = 2\lambda\vec{S}, \qquad (3)$$

with

$$\lambda = \frac{2\hbar^2P^2}{3m^2}\left[\frac{1}{E_g^2}-\frac{1}{(E_g+\Delta)^2}\right] = \left(1-\frac{g^{*}}{2}\right)\frac{\hbar^2}{mE_g}\,\frac{2E_g+\Delta}{E_g+\Delta}$$

The algebra to get this general result is more tedious than instructive, but one can take as a simple example the case ($\vec{k}\,\|\,\vec{x}$, $\vec{S}\,\|\,\vec{z}$) which is very demonstrative; in this case

$$|u_{\vec{k}+}\rangle = |s+\rangle - \frac{i\hbar P}{3m}\left[\left(\frac{2}{E_g}+\frac{1}{E_g+\Delta}\right)k_x|X+\rangle + i\left(\frac{1}{E_g}-\frac{1}{E_g+\Delta}\right)(k_x|Y+\rangle - ik_x|Z-\rangle)\right]$$

From Eq. (1) and up to first order in $k_x$, we obtain:

$$\frac{\partial}{\partial k_x}|u_{\vec{k}+}\rangle = \frac{\partial a}{\partial k_x}|s+\rangle - \frac{i\hbar P}{3m}\left[\left(\frac{2}{E_g}+\frac{1}{E_g+\Delta}\right)|X+\rangle + i\left(\frac{1}{E_g}-\frac{1}{E_g+\Delta}\right)(|Y+\rangle - i|Z-\rangle)\right]$$

$$\frac{\partial}{\partial k_y}|u_{\vec{k}+}\rangle = -\frac{i\hbar P}{3m}\left[\left(\frac{2}{E_g}+\frac{1}{E_g+\Delta}\right)|Y+\rangle + i\left(\frac{1}{E_g}-\frac{1}{E_g+\Delta}\right)(-|X+\rangle + |Z-\rangle)\right]$$

$$\frac{\partial}{\partial k_z}|u_{\vec{k}+}\rangle = -\frac{i\hbar P}{3m}\left[\left(\frac{2}{E_g}+\frac{1}{E_g+\Delta}\right)|Z+\rangle + i\left(\frac{1}{E_g}-\frac{1}{E_g+\Delta}\right)(-|Y-\rangle + i|X-\rangle)\right]$$

Using $\dfrac{\partial a}{\partial k_x} = \dfrac{\partial a}{\partial E}\times\dfrac{\hbar^2k_x}{m^{*}} = -\dfrac{\hbar^2P^2}{3m^2}\times\dfrac{3E_g^2+4E_g\Delta+2\Delta^2}{E_g^2(E_g+\Delta)^2}\times k_x$,

the calculation of $\vec{R}(X,Y,Z)$ follows from Eq. (2); Z is obviously zero and the various terms in X cancel each other; finally $\vec{R}$ reduces to $Y=\lambda k_x/2=\lambda S_zk_x$, which corresponds exactly to Eq. (3). The displacement is thus proportional to the spin and vanishes with the spin-orbit interaction, which is stressed by the factor $(g^{*}-2)$ in Eq. (3).

This equation remains valid as long as $E \ll E_g$ (i.e. $k^2 \ll 2m^{*}E_g/\hbar^2 \sim 1/\lambda$); for larger values of $\vec{k}$, $|\vec{R}|$ no longer increases. The maximum value of $|\vec{R}|$ is thus of order $\sqrt{\lambda}$; this may be very large in the accidental case of very small band gap. Namely, $|\vec{R}|$ can become much larger than the dimension of a unit cell (which would not be the case if it arose from a "polarization" inside $\Omega$). For example $|\vec{R}_{max}| \sim \sqrt{\lambda} \approx 30$ Å in indium antimonide.

The displacement concept is useful for the physical understanding of the extraordinary Hall effect, and furthermore it provides a rigorous way to do the calculations: As shown by Nozières and Lewiner in their effective Hamiltonian formulation,[20] the exact form of $u_{\vec{k}\sigma}$ [Eq. (1)] can be disregarded, provided one replaces $\vec{r}$ by the effective operator $\vec{r}_{eff} = \vec{r}+\vec{R}$. Then a potential $U(\vec{r})$ becomes: $U(\vec{r}_{eff}) \approx U(\vec{r}) + \vec{\nabla}U(\vec{r})\cdot\vec{R} = U(\vec{r}) + \vec{\nabla}U\cdot\lambda\vec{S}\times\vec{k}$; which holds whenever $U(\vec{r})$ is slowly varying on the scale of a unit cell. The additional term has exactly the same form as a usual spin-orbit interaction; this is not surprising since the derivation of the spin-orbit interaction from Dirac's equation[27] bears a close resemblance to the effective Hamiltonian formulation of Nozières and Lewiner. The only difference is the value of $\lambda$; ($\lambda=-\hbar^2/2m^2c^2$ in usual spin-orbit theory; and $\lambda \sim \hbar^2/m^{*}E_g$ here).

## IV. SPIN-DEPENDENT HALL EFFECT: DISPLACEMENT CONTRIBUTION ($\omega\tau \gg 1$)

In the hypothetical case of an applied electric field of frequency $\omega/2\pi$ much larger than the momentum relaxation rate $1/\tau$ ($\omega\tau \gg 1$), the collisions can be neglected and the transverse conductivity can be obtained from the motion of an electron, taken as a wave-packet with components $c(\vec{k})$. This motion is described by the time-dependence of $c(\vec{k})$.

$$i\hbar \frac{\partial c(\vec{k})}{\partial t} = \frac{\hbar^2 \vec{k}^2}{2m^*} c(\vec{k}) + \sum_{\vec{k}'} e\vec{E} \langle \vec{k}\sigma | \vec{r} | \vec{k}'\sigma \rangle c(\vec{k}')$$

Here (-e) is the electron charge (e > o). Taking $\vec{E} \| \vec{x}$ and using for $\vec{r}$ the effective operator $i\vec{\nabla}_{\vec{k}} + \vec{R}$, it gives

$$i\hbar \frac{\partial c(\vec{k})}{\partial t} = \frac{\hbar^2 \vec{k}^2}{2m^*} c(\vec{k}) + eE_x \left( X + i \frac{\partial}{\partial k_x} \right) c(\vec{k}) \quad (4)$$

The transverse velocity of the wave-packet is a function of $c(\vec{k})$ and $\partial c(\vec{k})/\partial t$;

$$\frac{dy}{dt} = \frac{d}{dt} \iiint c^*(\vec{k}) \left( Y + i \frac{\partial}{\partial k_y} \right) c(\vec{k}) \frac{d^3k}{(2\pi)^3} = \iiint \left[ \frac{\partial c^*(\vec{k})}{\partial t} \left( Y + i \frac{\partial}{\partial k_y} \right) c(\vec{k}) + c^*(\vec{k}) \left( Y + i \frac{\partial}{\partial k_y} \right) \frac{\partial c(\vec{k})}{\partial t} \right] \frac{d^3k}{(2\pi)^3} \quad (5)$$

Substitution of the expression of $\partial c(\vec{k})/\partial t$ [Eq. (4)] into Eq. (5) gives:

$$\frac{dy}{dt} = - \frac{eE_x}{\hbar} \iiint c^*(\vec{k}) \left( \frac{\partial Y}{\partial k_x} - \frac{\partial X}{\partial k_y} \right) c(\vec{k}) \frac{d^3k}{(2\pi)^3}$$

If $c(\vec{k})$ is localized in $\vec{k}$-space, this reduces to:

$$\frac{dy}{dt} \approx - \frac{eE_x}{\hbar} \left( \frac{\partial Y}{\partial k_x} - \frac{\partial X}{\partial k_y} \right)$$

Therefore the application of an external electric field results in a non-null transverse velocity of the electron. This yields a transverse conductivity:

$$\sigma_{yx} = - \frac{ne}{E_x} \frac{dy}{dt} = \frac{ne^2}{\hbar} \left( \frac{\partial Y}{\partial k_x} - \frac{\partial X}{\partial k_y} \right) \quad (6)$$

The first term in Eq. (6) can be identified with the time variation of Y when the electron is accelerated by the electric field. The classical equation $\hbar dk_x/dt = -eE_x$ would indeed yield $dy/dt = dY/dt = (-eE_x/\hbar)\partial Y/\partial k_x$, hence $\sigma^a_{yx} = (ne^2/\hbar)\partial Y/\partial k_x$. In fact, the occurrence of X in the matrix elements of the perturbation $e\vec{E}\cdot\vec{r}$ provides an extra contribution $\sigma^b_{yx} = (-ne^2/\hbar)\partial X/\partial k_y$, leading to the final Eq. (6). This result is physically satisfactory, since it depends only on $\vec{\nabla}_{\vec{k}} \times \vec{R}$, and is thus phase invariant (see remark part III above).

In a III-V semiconductor, using the expression for $\vec{R}$ [Eq. (3)], the transverse conductivity [Eq. (6)] becomes:

$$\sigma_{yx} = \frac{ne^2}{\hbar} \lambda \langle 2S_z \rangle$$

If one defines a transverse mobility by $\sigma_{yx} = ne\mu_{yx}$, then

$$\mu_{yx} = \frac{e\lambda}{\hbar} = \left( 1 - \frac{g^*}{2} \right) \frac{\hbar e}{mE_g} \frac{2E_g+\Delta}{E_g+\Delta} \langle 2S_z \rangle$$

This is equivalent to Eq. (1.8) of ref. 16. With the band parameters of InSb ($E_g$=0.235eV, $\Delta$=0.9eV, $g^*$=-51.4) and for a spin-up electron, this gives $\mu_{yx}$=160 $cm^2$/Vsec.

## V. DISPLACEMENT EFFECT: ROLE OF THE COLLISIONS ($\omega\tau \ll 1$)

For a d.c. electric field, the problem is far more difficult because scattering can no longer be neglected. Since the pioneering paper of Karplus and Luttinger,[5] this has been the area of many controversies.[4-14] The fundamental article seems to be that of Luttinger.[7] However, his calculations are rather involved, and several attempts were made to put them in a more understandable form.[8-13] Doniach[11] and Fivaz[13] pointed out the effect of the transverse shift of the distribution function and Berger[14] introduced the side-jump concept, which was later confirmed by Lyo and Holstein.[28] However, discrepancies were still present until the more recent general synthesis of Nozières and Lewiner.[20] We first give an elementary presentation, following closely the very simple theory of these authors, then we discuss some of the last conflicting papers on the subject.

The transverse conductivity, as given by Nozières and Lewiner, is the sum of six terms:

$$\sigma^D_{yx} = \sigma^a_{yx} + \sigma^b_{yx} - \sigma'^a_{yx} - \sigma'^b_{yx} - \sigma''^a_{yx} - \sigma''^b_{yx} \quad (7)$$

Here again the a and b superscripts refer to the terms involving $\partial Y/\partial k_x$ and $-\partial X/\partial k_y$, respectively. In the limiting case $\omega\tau \ll 1$, one has $\sigma^a_{yx} = \sigma'^a_{yx} = \sigma''^a_{yx}$ and $\sigma^b_{yx} = \sigma'^b_{yx} = \sigma''^b_{yx}$; thus four among the six terms cancel each other, leading to a transverse conductivity opposite to the collisionless result [Eq. (6)].

The first two terms in Eq. (7) are those previously found in the case $\omega\tau \gg 1$; they are due to the transverse drift of the polarized electron when accelerated by the external electric field.

The two following terms in Eq. (7) are physically similar: namely, they describe the transverse motion of the electron which arises from its slowing down during a scattering event; this can be interpreted as a sudden side-jump of the wave-packet, as was emphasized by Berger[14]. The corresponding contribution to the transverse conductivity appears as the sum of two terms because the side-jump upon scattering can be written $\Delta\vec{r} + \Delta\vec{R}$;[20] this is in close relation with the occurrence of two terms for the side-drift contribution.[29] Since the system is stationary, one might expect the transverse motion due to the deceleration upon scattering (side-jump) to cancel exactly the transverse motion due to the acceleration by the electric field (side-drift). This is actually verified by the calculation[20] and to this level the final transverse conductivity should vanish.

This misleading result is obtained because the above reasoning is somewhat naïve: It implies that the electron was initially at rest and that its velocity arises only from the acceleration by the applied electric field. In fact, the electrons are distributed in $\vec{k}$-space and each of them undergoes a side-jump $2\lambda\vec{S} \times (\vec{k}'-\vec{k})$ upon a scattering event from state $|\vec{k}\rangle$ to state $|\vec{k}'\rangle$.[29] In the absence of external electric field, the distribution function (either Fermi or Boltzmann) has spherical symmetry around $\vec{k}$=0, and the sum of these side-jumps gives zero electrical current. When an external electric field is applied, the spherical symmetry is broken and a twofold contribution results:

First, because of the longitudinal shift $-e\vec{E}\tau/\hbar$ of the distribution function, the sum of the side-jumps of all the electrons no longer vanishes; this yields the above-mentioned contribution $-\sigma'^a_{yx}-\sigma'^b_{yx}$ to the transverse conductivity.

Second, because of the energy term $e\vec{E}\cdot\vec{r}$ in the Hamiltonian, a transverse shifting of the distribution function occurs, as a consequence of energy conservation upon scattering (see Fig. 2). As shown below, this transverse shift of the distribution function is responsible for the last two terms in Eq. (7).

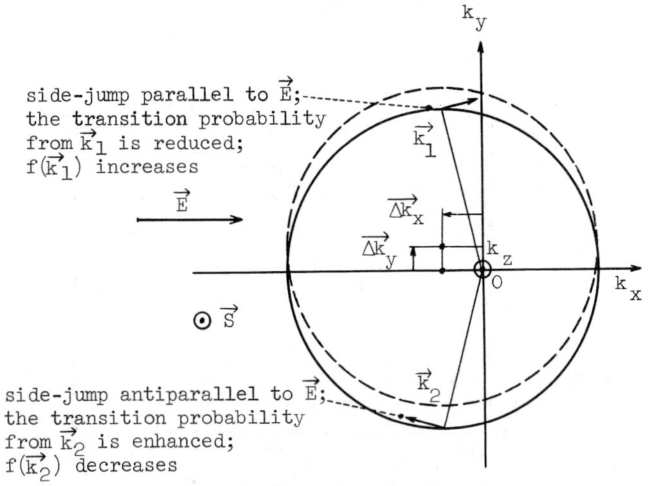


Fig. 2

Transverse shift of the distribution function and the displacement effect. The circle drawn in full line schematizes the distribution function $f(\vec{k}\,)$, shifted from equilibrium by the trivial quantity $\overrightarrow{\Delta k_x}=-e\vec{E}\tau/\hbar$. As shown in the figure, the electron side-jump upon scattering causes a change in potential energy, which results in a further shift of the distribution function $\overrightarrow{\Delta k_y}=2\lambda e(\vec{S}\times\vec{E})m^*/\hbar^2$. The dashed circle schematizes the final (stationary) position of $f(\vec{k}\,)$.

Proof: The amount of the shift can be found by writing that the energy change upon a scattering event is null in the stationary state. The potential energy change is a consequence of the side-jump in the electric field:

$$E_{pot}(\vec{k}\,') - E_{pot}(\vec{k}) = e\vec{E}\cdot 2\lambda[\vec{S}\times(\vec{k}\,'-\vec{k})] = 2\lambda e(\vec{S}\times\vec{E})\cdot(\vec{k}-\vec{k}\,')$$

For a small shift $\delta\vec{k}$ of the distribution function, the kinetic energy change upon a scattering event is:

$$E_{kin}(\vec{k}\,') - E_{kin}(\vec{k}) = \frac{\hbar^2\delta\vec{k}}{m^*}\cdot(\vec{k}\,'-\vec{k})$$

The stationary value of $\delta\vec{k}$ occurs when these two expressions cancel each other, which gives:

$$\delta\vec{k} = 2\lambda e(\vec{S}\times\vec{E})m^*/\hbar^2$$

The corresponding contribution to the transverse conductivity is

$$-\frac{ne}{E_x}\frac{\hbar\delta k_y}{m^*} = -\frac{ne^2}{\hbar}\lambda\langle 2S_z\rangle$$

As pointed out above, this corresponds to the last contribution $-\sigma''^{a}_{yx}-\sigma''^{b}_{yx}$ in Eq. (7). Here again the appearance of this contribution as the sum of two terms is due to the double nature of the side-jump.[29]

A classical picture can be obtained by looking at the time variation of $\langle y\rangle$ (see Fig. 3). Figure 3a corresponds to the incomplete result and zero transverse conductivity. Figure 3b corresponds to the final result. The side-drift during the acceleration by the electric field is compensated by the non-zero average value of $k_y$ due to the transverse shift of the distribution function. In this picture it might appear that only the side-jump contribution remains. Actually at $\omega=0$ the six terms in Eq. (7) are identical and one may argue that the cancellations occur in different manners. However the first two terms are frequency independent, in contrast to the last four which vanish as $1/(1+i\omega\tau)$ at high frequency; thus one cannot ignore the side-drift contribution, since it becomes dominant in the limiting case $\omega\tau \gg 1$[20,30] and is responsible for the intraband Faraday effect in ferromagnets.[3]

The final result at zero frequency is therefore the same as in the opposite limit [Eq. (6)] except for the change in sign:

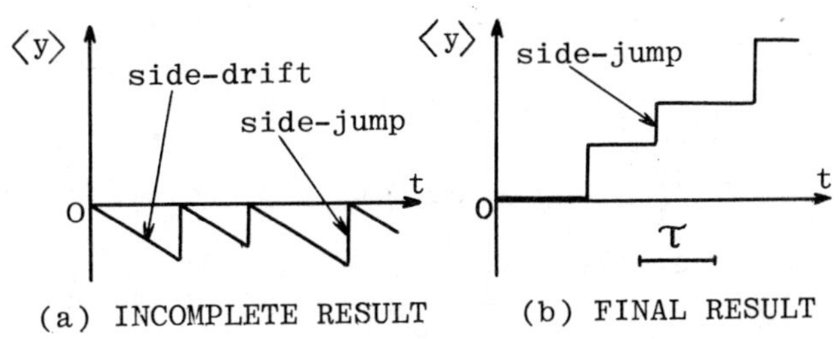


Fig. 3

Motion of an electron (schematic). The figure shows the time variation of the transverse position of the wave-packet. The curve in (a) corresponds to the incomplete result (Smit, see Ref. 4); the side drift of the wave-packet is compensated by the fast side jump during the scattering event. When the transverse shift of $f(\vec{k}\,)$ is considered (see text), a further contribution results and an "average" electron behaves as shown in (b).

$$\sigma^D_{yx} = -(ne^2\lambda/\hbar)\,\langle 2S_z\rangle \quad\text{and}\quad \mu^D_{yx} = -(e\lambda/\hbar)\,\langle 2S_z\rangle$$

This final result is equivalent to our Eq. (7) and Eq. (4.21) of Luttinger.[7] The transverse conductivity $\sigma^D_{yx}$ appears to be insensitive to the shape of the scattering potential. This explains that theories dealing with phonon scattering essentially yield the same result.[31,32] For a given material, if the resistivity $\rho$ is changed by varying the temperature or impurity content, the transverse mobility should remain constant, and if n is unchanged, the anomalous Hall constant $R_{an}$ should be proportional to $\rho^2$, a behavior often observed in ferromagnetic metals.[2-4]

Discussion: However the whole theory rests on the assumption that the potentials are slowly varying on the scale of the unit cell. If this approximation fails, two among the six contributions in Eq. (7) must be reconsidered, namely $-\sigma'^{b}_{yx}$ and $-\sigma''^{a}_{yx}$, because they arise both from the $\Delta\vec{r}$ part of the side-jump, which involves the correction to the scattering potential $\vec{\nabla}V\cdot\vec{R}$.[29] On the contrary the four other terms in Eq. (7) do not depend on this correction, and they would cancel out even for potentials violating the slowly varying condition.

In this view, some authors[14,28] have focused their attention on the $\Delta\vec{r}$ part of the side-jump, taking as granted the cancellation of the four other terms. (As they say, the terms originating directly from $\vec{R}$ cancel out). Although this point of view is legitimate, it has aroused some confusion recently[33-35] because in practice all the six terms originate from $\vec{R}$, and it is not always explicitly stated which of these terms are left out in those papers.[36] In a recent comment,[33] Smit claimed that the different contributions to the Hall effect cancel each other and lead to a null result; the terms considered by Smit in his last equation correspond to $[-\sigma''^{b}_{yx}+(\sigma^{b}_{yx}+\sigma^{a}_{yx})-\sigma'^{a}_{yx}]$, the second halves of the side-jump and distribution shift are missed. On the other hand, in the reply by Berger[34], only the b-type terms $[-\sigma''^{b}_{yx}+\sigma^{b}_{yx}-\sigma'^{b}_{yx}]$ appear. Ref. 35 is also a discussion of $\sigma^{b}_{yx}$.

Most of these papers are written from a free electron point of view, that is putting $\lambda = -\hbar^2/2m^2c^2 \simeq 10^{-6}\text{Å}^2$ in our formulas. As first pointed out by Smit[6], and as it is evident from our Eq. (3), the actual value of $\lambda$ for an electron in a solid is much larger ($10^3\text{Å}^2$ in InSb, and typically $1\text{Å}^2$ for 3d-metals). But formally, the free-electron approach and the band approach are identical; namely the free-electron approach does yield the same six terms as in Eq. (7).

## VI. SKEW SCATTERING CONTRIBUTION TO THE EXTRAORDINARY HALL EFFECT

Up to now in the transition probability $W_{\vec{k}\sigma\to\vec{k}'\sigma}$

we have only considered the spin-orbit corrections arising from the energy conservation term $\delta(E_{\vec{k}}-E_{\vec{k}'})$. A second type of correction is due to the matrix element of the scattering potential.

$$\langle \vec{k}\sigma|V|\vec{k}'\sigma\rangle = V_{\vec{k}\vec{k}'}\,[1-i\lambda(\vec{k}\times\vec{k}')\cdot\vec{S}]$$

Here again V has been assumed to be slowly varying on the scale of a unit cell and the spin-dependent terms have been kept only to lowest order in $\vec{k}$ and $\vec{k}'$. Using this expression for the matrix elements of V, the transition probability $W_{\vec{k}\sigma\rightarrow\vec{k}'\sigma}$, when calculated to the second Born approximation, involves spin-dependent terms of the form $(\vec{k}\times\vec{k}')\cdot\vec{S}$. The net result is a left-right asymmetry of the scattering cross-section (skew-scattering). This effect, which is analogous to the Mott-scattering of relativistic electrons in atomic physics, was first discussed by Smit.[6]

The relevant parameter in this case is the Hall angle, since it does not depend on the impurity content (which should lead, in principle, to a behavior $R_{an}\propto\rho$). Detailed calculations for this effect can be found in the literature.[17,37]

## VII. EXPERIMENTAL CHECK OF THE THEORY

A III-V compound semiconductor like indium antimonide provides an ideal case to test the theory of the extraordinary Hall effect. All the band parameters are accurately known and the deduced theoretical transverse mobility is $\mu^D_{yx}=-160$ cm$^2$/Vsec for unit spin polarization. The difficulty is that in such a material there is no spontaneous magnetization and the measurement of the anomalous Hall effect is hindered by the existence of a much larger ordinary Hall effect. However, the separation of this smaller effect has been performed unambiguously by using a spin-resonance method.[21,38] The principle of the experiment is to use as the active magnetization the rotating magnetization which appears under spin-resonance conditions, and which can be turned on and off by only a small change in the external magnetic field.

For small impurity concentrations ($n=N_D-N_A \lesssim 10^{14}$ cm$^{-3}$) the measured value of $\mu_{yx}$, referred to unit spin polarization, agrees within the experimental uncertainties with the theoretical value of the displacement effect, which provides an unambiguous test of this theory. For higher doping levels, the measured values of $\mu_{yx}$ increase and change sign for $n\approx1.5\times10^{14}$cm$^{-3}$. We have discussed this effect (see ref. 38) and shown it to be attributable to the increase in skew-scattering as both the Fermi energy and the longitudinal mobility increase. Furthermore in this regime multiple-scattering cannot be neglected.

The transverse mobility corresponding to the displacement effect is very large in InSb; this is due to the unusually small value of the band gap and to the location of electrons at $\Gamma$ point, whose high symmetry favors the spin-orbit effects. We have calculated the displacement contribution for germanium which has a smaller spin-orbit splitting and a larger gap; the effect is inhibited by the multivalley character of the conduction band and on the whole one should expect a transverse mobility of the same order of magnitude as for ferromagnetic metals, which involve narrow bands and where a typical interband spacing of 1 eV is not unreasonable. We have plotted in Fig. 4, as a function of atomic spin-orbit $\zeta$, the values of $|\mu_{yx}|$ for several ferromagnetic metals, where a $R_{an}\propto\rho^2$ component has been clearly identified. The value of $|\mu_{yx}|$ for germanium is plotted on the same scale, together with a typical curve showing the expected order of magnitude as a function of $\zeta$. As might have been expected, the values for the rare earth metals fall well off this line, which confirms the ineffectiveness of the present process in the case of Heisenberg systems. The values for the 3d metals, however are quite close to the theoretical curve and to the germanium value. As a further check the latter has been measured and found in good agreement with theory.[38]

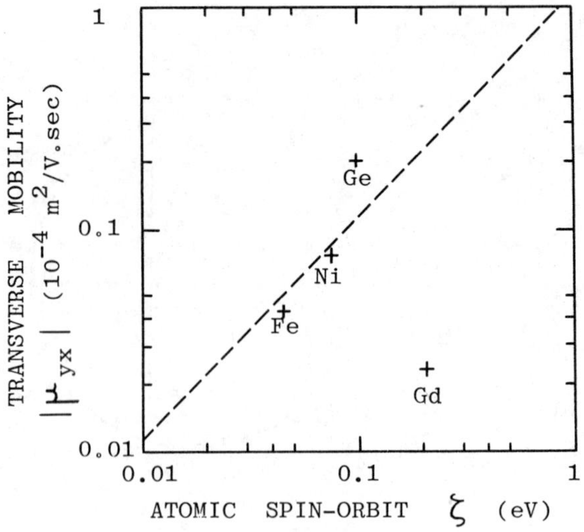


Fig. 4

Transverse mobility corresponding to the displacement effect as a function of atomic spin-orbit $\zeta$. The dashed line is an indication of the expected order of magnitude of the displacement effect. Its equation is $|\mu_{yx}|=\hbar e\zeta/mE_0^2$, where $E_0$ has been arbitrarily given the typical value 1 eV. The transverse mobility for ferromagnetic metals has been deduced from the extraordinary Hall effect data according to the formula $\mu_{yx}=-R_{an}\mu_0M_0/\rho^2ne=(R_{an}\mu_0/\rho^2)(\hbar^2/2m)(g^*/2)$. The data for Fe are the values quoted in Ref. 39; Ni is from Ref. 40, Ge from 38, and Gd (polycrystalline) from 41. The atomic spin-orbits were estimated from the calculated values for corresponding ions in Refs. 42 and 43.

These facts give strong evidence that the displacement mechanism is relevant to the extraordinary Hall effect in 3d ferromagnetic metals. Direct quantitative comparisons of theory with experiment have been attempted, starting from theoretical band structures of these materials[9,31] but until now the results were not much better than our rough order of magnitude evaluation. Now the effect seems to be understood, and the extraordinary Hall effect data might serve as a useful test for present band structure calculations.

## VIII. CONCLUSION AND PROSPECTS

We have presented a review of the theoretical literature on the extraordinary Hall effect. Two contributions have been distinguished: the displacement and the skew-scattering. The displacement effect has been presented from an intuitive viewpoint and the literature discussed in connection with our results. Our feeling is that several directions remain open for theorists; among these are the use of band-structure calculations of ferromagnetic metals, also the problem of the skew-scattering effect beyond the second Born approximation and in the multiple scattering regime. However, the most important step now is probably the extension of the theory to the case of localized scattering potentials, where the concept of the effective Hamiltonian fails. This includes the problem of the additional spin-orbit interaction brought by the scatterer, and possibly the interaction with a magnetic scatterer.[15] Recent experimental data[39,40] have shown the need for such a theory.


## ACKNOWLEDGMENTS

The authors are indebted to Dr. Y. Yafet for very useful comments on the manuscript. They also acknowledge a stimulating correspondence with Prof. J. Smit.

References

*Equipe de Recherche du Centre National de la Recherche scientifique.


1. E. H. Hall, Phil. Mag. 10, 301 (1880); H. A. Rowland, Am. J. Math. 2, 354 (1879).
2. For Fe, see J. P. Jan, Helv. Phys. Acta. 25, 677 (1952); C. Kooi, Phys. Rev. 95, 84 (1954); W. Jellinghaus and M. P. DeAndres, Ann. Physik 7, 189 (1961). For Ni, see W. Köster and O. Römer, Z. Metallk. 55, 805 (1964); R. Huguenin and D. Rivier, Helv. Phys. Acta. 38, 900 (1965). For Fe, Co, Ni, see N. V. Volkenshtein and G. V. Fedorov, Zh. Eksp. Teor. Fiz. 38, 64 (1960) [Soviet Phys.-JETP 11, 48 (1960)].
3. For a review of the literature on the extraordinary Hall effect, and also on the Faraday effect in ferromagnetics, see S. V. Vonsovskii, in Magnetism (John Wiley & Sons, New York, 1974), Vol. 2, pp. 1146-1164.
4. J. Smit, Physica 21, 877 (1955).
5. R. Karplus and J. M. Luttinger, Phys. Rev. 95, 1154 (1954).
6. J. Smit, Physica 24, 39 (1958).
7. J. M. Luttinger, Phys. Rev. 112, 739 (1958).
8. E. N. Adams and E. I. Blount, J. Phys. Chem. Solids 10, 286 (1959).
9. C. Strachan and A. M. Murray, Proc. Phys. Soc. 73, 435 (1959).
10. Y. P. Irkhin and V. G. Shavrov, Zh. Eksp. Teor. Fiz. 42, 1233 (1962) [Soviet Phys.-JETP 15, 854 (1962)]; L. E. Gurevitch and I. N. Yassievich, Fiz. Tverd. Tela. 5, 2620 (1963) [Soviet Phys. Solid State 5, 1914 (1964)].
11. S. Doniach, in Optical and Electronic Structure of Metals and Alloys, edited by F. Abeles (North Holland, Amsterdam, 1966), p. 471.
12. V. Christoph, Phys. Status Solidi 26, K17 (1968).
13. R. C. Fivaz, Phys. Rev. 183, 586 (1969).
14. L. Berger, Phys. Rev. B2, 4559 (1970); ibid. B5, 1862 (1972).
15. Here we are concerned with the theories starting from the band-magnetism model, i.e. one electronic system responsible for both the electric and magnetic properties. An alternative approach has been initiated by J. Kondo [Progr. Theor. Phys. 27, 772 (1962)] and F. E. Maranzana [Phys. Rev. 160, 421 (1967)] for the case of Heisenberg systems with non-magnetic conduction electrons.
16. A. J. F. Bastin, C. Lewiner, and N. Fayet, J. Phys. Chem. Solids 31, 817 (1970).
17. P. Leroux-Hugon and A. Ghazali, J. Phys. C5, 1072 (1972).
18. V. N. Abakumov and I. N. Yassievich, Zh. Eksp. Teor. Fiz. 61, 2571 (1972) [Soviet Phys.-JETP 34, 1375 (1972)].
19. C. Lewiner, O. Betbeder-Matibet and P. Nozières, J. Phys. Chem. Solids 34, 765 (1973).
20. P. Nozières and C. Lewiner, J. Phys. (Paris) 34, 901 (1973).
21. J.-N. Chazalviel and I. Solomon, Phys. Rev. Lett. 29, 1676 (1972).
22. J. C. Slater, in Quantum Theory of Molecules and Solids (Mc Graw-Hill, New York) Vol. 2, App. 5, p.455.
23. W. Zawadzki and W. Szymańska, Phys. Status Solidi B45, 415 (1971).
24. E. N. Adams, J. Chem. Phys. 21,2013 (1953).
25. Y. Yafet, in Solid State Physics, edited by F. Seitz and D. Turnbull (Academic, New York,1963), Vol 14, pp. 1-98.
26. Our notation $\vec{R}$ is from refs. 24 and 25. This $\vec{R}$ corresponds to $\vec{\rho}$ in ref. 20, $i\vec{J}$ in ref. 7, and $\vec{q}$ in refs. 4, 13 and 14.
27. L. I. Schiff, Quantum Mechanics (Mc Graw-Hill, New York, 1949), pp.311-328.
28. S. K. Lyo and T. Holstein, Phys. Rev. Lett. 29, 423 (1972).
29. The side-jump upon scattering is not simply $\Delta\vec{R} = \lambda\vec{S} \times \Delta\vec{k}$ as could be inferred from the expression of $\vec{R}$ [Eq. (3)]. A further side-jump $\Delta\vec{r}$ arises from the change in the phases of the scattered waves, due to the correction to the scattering potential $V_{eff} = V + \vec{\nabla}V \cdot \vec{R}$. This further contribution corresponds to the calculations in Refs. 14 and 28. Both contributions are discussed in Ref. 20. The origin of these two terms, $\Delta\vec{r}$ and $\Delta\vec{R}$, is exactly the same as for the $\sigma_{yx}^{b}$ and $\sigma_{yx}^{a}$ contributions in the electric field problem (collisionless case). Namely the first term is due to the correction to the perturbing Hamiltonian; the second term is due to the correction to $\vec{r}$ when the final position is calculated. With our particular choice of Bloch functions, one finds $\Delta\vec{r} = \Delta\vec{R}$, hence the total value of the side-jump $2\lambda\vec{S} \times \Delta\vec{k}$. The terms $-\sigma'^{a}_{yx}$ and $-\sigma'^{b}_{yx}$ in Eq. (7) are the direct consequence of the side jumps $\Delta Y$ and $\Delta y$ respectively; in the same manner, the last two terms $-\sigma''^{a}_{yx}$ and $-\sigma''^{b}_{yx}$, which we show to arise from a transverse shift of the distribution function, are a further consequence of the side-jumps $\Delta x$ and $\Delta X$.
30. S. K. Lyo and T. Holstein, Phys. Rev. B9, 2412 (1974).
31. H. R. Leribaux, Phys. Rev. 150, 384 (1966);
32. S. K. Lyo, Phys. Rev. B8, 1185 (1973).
33. J. Smit, Phys. Rev. B8, 2349 (1973).
34. L. Berger, Phys. Rev. B8, 2351 (1973)
35. S. K. Lyo, Phys. Rev. B11, 1260 (1975).
36. Splitting the side-jump into two parts and handling $\Delta\vec{r}$ separately results in a further drawback for a physical presentation, because $\Delta\vec{r}$ is an unphysical quantity, which depends on the choice of the relative phases of the basis Bloch functions. The true side-jump undergone by an electron upon scattering is $\Delta\vec{r} + \Delta\vec{R}$, which is of course "phase-invariant".
37. J.-N. Chazalviel, Phys. Rev. B10, 3018 (1974).
38. J.-N. Chazalviel, Phys. Rev. B11, 3918 (1975).
39. A. K. Majumdar and L. Berger, Phys. Rev. B7, 4203 (1973).
40. P. Jaoul, These de Doctorat d ' Etat, Université de Paris-Sud (1974) (unpublished).
41. R. S. Lee and S. Legvold, Phys. Rev. 162,431(1967).
42. M. Blume and R. E. Watson, Proc. Roy. Soc. (London) A271, 565 (1963).
43. M. Blume, A. J. Freeman and R. E. Watson, Phys. Rev. 134, A320 (1964).

# HALL EFFECT AND TRANSPORT IN 3d FERROMAGNETIC METALS*†

R. V. Coleman
Physics Department, University of Virginia, Charlottesville, Virginia 22901

ABSTRACT

Experimental data on the ordinary and extraordinary Hall effect observed in the 3d ferromagnetic metals iron, nickel and cobalt is reviewed. Emphasis is placed on data and analysis applicable to the high-field limit obtained at low temperatures and high magnetic fields. Complete data on iron single crystals in the range 1 - 300 K and in magnetic fields to 230 kG is presented. Results are discussed in terms of mobility transitions, magnetic breakdown, and intersheet scattering of electrons. Kohler plots of Hall resistivity and Hall angle are included with reference to deviations from Kohler's rule.

## I. INTRODUCTION

The Hall effect in ferromagnetic metals has been the subject of extensive study for many years and has generally been understood in terms of an ordinary Hall coefficient $R_o$ which is characteristic of a non-magnetic metal with the same band structure and an extraordinary Hall coefficient $R_s$ connected with the saturation magnetization of the ferromagnetic metal. The extraordinary Hall effect has been treated in many theoretical papers[1-5] and all agree that the spin-orbit coupling interaction is responsible for a transverse voltage which depends on the magnetization of the metal. In the presence of the spin-orbit interaction electrons with spin polarizations parallel and antiparallel to the magnetization are deflected in opposite directions at right angles to the electric current. If the two spin populations are unequal then a new transverse current appears which is cancelled by the resulting Hall voltage.

Karplus and Luttinger[1] were the first to treat the effect in detail using the force due to the spin-orbit part of the periodic potential which gave rise to a nonvanishing effect only if the carriers in a given band were coupled to other bands through the spin-orbit interaction. Many authors have considered the problem in subsequent papers and have developed many refinements of the theory a brief summary of which can be found in ref. 15. The original Karplus and Luttinger theory as well as all subsequent theories generally arrive at a dependence of $R_s$ on $\rho^2$ although the magnitude of the coefficient of the $\rho^2$ term can vary considerably with the details of the calculation.

The general expression used to represent the Hall resistivity is given by,

$$\rho_H(\bar{B}) = R_o\bar{B} + 4\pi\bar{M}_s R_s \qquad (1)$$

Under conditions where the electron mean free path is short the expression in formula (1) is generally verified experimentally as well as the relation $R_s \sim \rho^2$. This includes measurements on pure ferromagnetic metals in the range above 100 K and on ferromagnetic alloys. When expression (1) holds then separation of the constants $R_o$ and $R_s$ can be accomplished by simple extrapolation of the results above magnetic saturation in order to obtain an intercept at the point $\bar{B} = 0$. This intercept is associated with $R_s$ while the slope of the Hall resistivity above saturation is associated with the ordinary coefficient $R_o$.

For a complete understanding of the electronic behavior of metals it is always desirable to carry out experiments on very pure single crystals with very long electron mean free paths. The combination of low temperature and high magnetic fields can then attain conditions where $\omega_c\tau >> 1$ and theories can be tested in the high-field limit. For high purity single crystals of the ferromagnetic transition metals such experiments introduce complexities since expression (1) may no longer hold due to a nonlinear dependence on $\bar{B}$ above saturation and in addition $R_s$ is observed to depart from a dependence on $\rho^2$. This makes the analysis more difficult, but promises to reveal information about the complex electronic properties of ferromagnetic metals. In view of limited space this paper will concentrate on a review of data on high purity samples of the 3d ferromagnetic metals iron, nickel and cobalt.

## II. LOW FIELD LIMIT

Before discussing the more recent data at high magnetic fields it is useful to review the behavior of $R_o$ and $R_s$ as measured in previous experiments under conditions where the high-field limit was generally not attained. For polycrystalline samples of purity in the range 99.99% + and in magnetic fields not exceeding 30 kOe most investigators have used formula (1) in order to separate $R_o$ and $R_s$. Representative data obtained on the temperature dependence of the coefficients $R_o$ and $R_s$ is shown in Fig.1 & 2. This data was taken from a paper by Volkenshtein and Federov[6] where the residual resistance ratios were Fe(RRR = 11.5), Ni(RRR = 57.2), and Co(RRR = 66.3). In the case of $R_o$ iron shows a positive value over the temperature range from room temperature to 4.2 K while cobalt and nickel both show negative values of $R_o$. The behavior of $R_s$ is somewhat more complicated with iron and nickel showing positive and negative values respectively while for cobalt $R_s$ changes sign from positive to negative as the temperature is decreased. These general trends are representative of many of the experiments in the field, however the low temperature behavior can vary considerably from the above results as the samples become purer and are measured under conditions approaching the high-field limit. For example iron specimens with resistance ratios RRR $\geq$ 500 show a change in sign of $R_o$ at $\sim$ 80 K which is absent in the data of Fig. 1.

For pure crystals large contributions to the resistivity near H = 0 are observed at low temperatures due to the internal field within the individual domains. These effects have been studied in detail[7] and are understood in terms of the specific domain structure near H = 0. The resistance ratios quoted for high purity crystals of the ferromagnetic metals are therefore calculated on the basis of a measurement

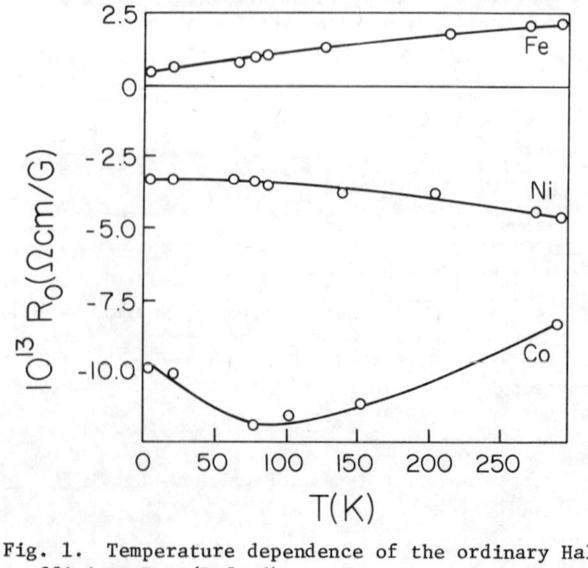


Fig. 1. Temperature dependence of the ordinary Hall coefficient $R_o$. (Ref. 6).

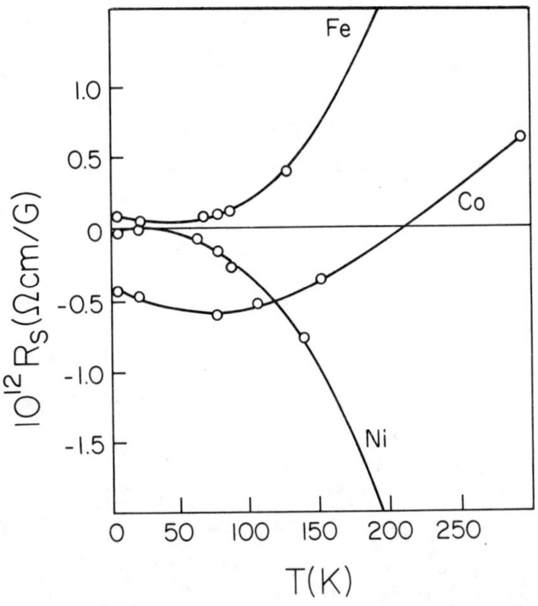


Fig. 2. Temperature dependence of the extraordinary Hall coefficient $R_s$. (Ref. 6).

of the low temperature resistivity in a longitudinal field sufficient to saturate the crystal as a single domain. Ratios quoted in this paper are calculated on this basis.

## III. HIGH-FIELD LIMIT

The behavior of $R_o$ at low temperatures and high field would be expected to exhibit transitions as $\omega_c\tau$ varies from the low-field to the high-field limit and these can be associated with a number of different mechanisms. These include mobility transitions[8] due to widely different effective masses and relaxation times for the various bands, open orbits[9] which play a dominant role for specific field directions at high fields, magnetic breakdown[10] due to tunneling between different sheets of the Fermi surface, and small angle intersheet scattering[11]. When possible effects of the ferromagnetic band structure are taken into account along with the need to separate $R_o$ and $R_s$ the complete analysis of any one ferromagnetic metal requires a formidable amount of data extending well into the high-field region. The data for iron[12] have been extended to the point where high-field limit effects are clearly dominant, nickel data in the high-field limit have been obtained in several experiments,[13,14] while very little data is available for cobalt in the high-field limit.

We will first review the data on iron which is fairly extensive and complex and then review the data on nickel with particular emphasis on several experiments which give information on the high-field limit behavior.

### A. Iron

As the purity of iron is increased, a major feature which develops is a sign change in the Hall resistivity below 80 K as shown in Fig. 3 for a single crystal with a RRR = 4000. This has been observed by a number of authors[8,15,16] working with both polycrystals and single crystals of high purity. In addition below 80 K the Hall resistivity becomes field dependent for magnetic fields above saturation and makes the use of formula (1) to separate $R_o$ and $R_s$ questionable. The effect is demonstrated in Fig. 4 where the derivative $d\rho_H/dB$ is plotted as a function of field and temperature. In the case of a linear field dependence the derivative is equal to a constant $R_o$ while in Fig. 4 $R_o(B)$ is clearly not a constant below 80 K.

A plot of $R_s$ obtained by use of formula (1) using the reasonably linear region just above saturation or

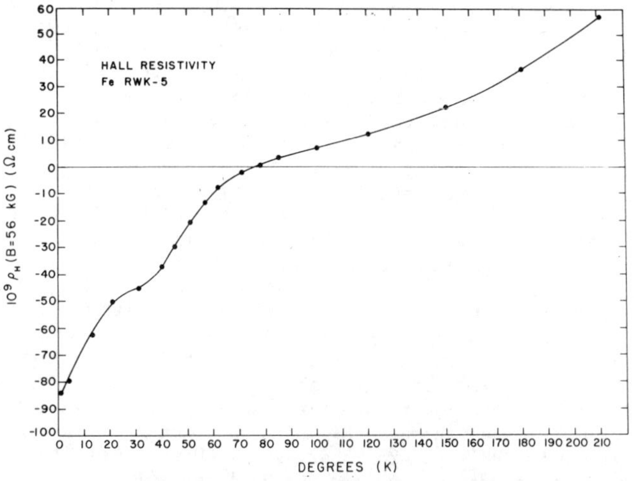


Fig. 3. Hall resistivity $\rho_H$ measured at $\bar{B} = 56$ kG as a function of temperature. RRR = 4000 (Ref. 15).

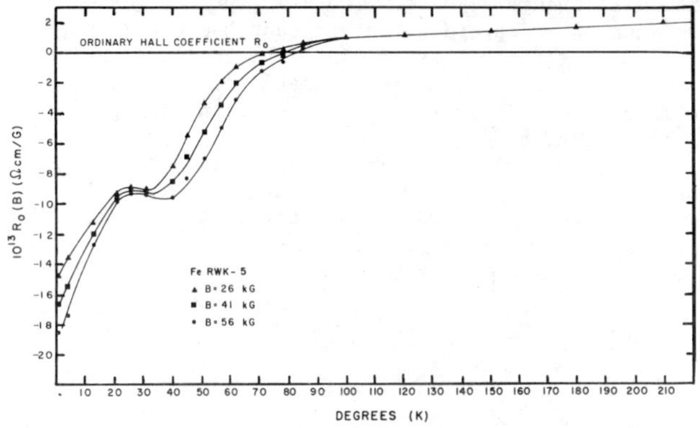


Fig. 4. Derivative of the Hall resistivity, $d\rho_H/dB = R_o(B)$ as function of temperature. (Ref. 15).

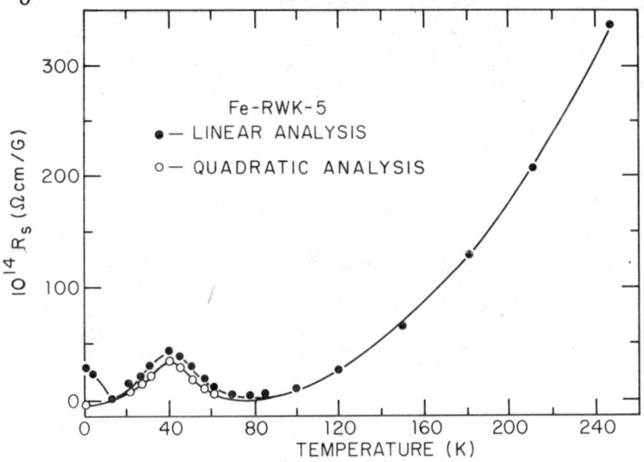


Fig. 5. Extraordinary Hall coefficient, $R_s$, as a function of temperature. (Ref. 15).

using the entire high field data with a term quadratic in B added to formula (1) is shown in Fig. 5. In both cases the high temperature behavior is consistent with $R_s \sim \rho^2$ while below 80 K a strong deviation is observed with an anomalous maximum at 40 K. This observation should not be taken as conclusively showing a low temperature anomaly in $R_s$ since $R_o$ and $R_s$ may not be clearly separable[14] as the high-field limit is approached at higher temperatures where $R_s$ has appreciable magnitude.

The data reviewed above were taken on a <111> axial single crystal with the magnetic field in a <112> direction and the Hall voltage measured in a <110> direction. The data were taken in fields up to $\bar{B} = 56$ kG and for RRR = 4000 this corresponds to $\omega_c\tau = 1$

at approximately 40 K. In order to obtain data over the range corresponding to much higher $\omega_c\tau$ values we have extended the measurements to $\bar{B} = 230$ kG and to temperatures of 1.1 K. For the <111> current direction with field and Hall voltage in <112> and <110> directions respectively the data up to $\bar{B} = 160$ kG at 4.2 K are shown in Fig. 6. The nonlinear field dependence is observed up to $\bar{B} \approx 100$ kG followed by a reasonably linear behavior above this field. In addition a linear extrapolation of the high field data to $B = 0$ gives a zero intercept indicating $R_s = 0$ at this temperature as expected and suggesting that the high-field limit for $R_o$ has been reached. However, other combinations of current, field, and Hall voltage direction in single crystals show a more complex behavior. For example in the case of the <100> current orientation and with field and Hall voltage in <100> directions the low temperature data are non-linear at high fields as shown in Fig. 7. Data for three specimens with RRR's of 425, 5842, and 7320 are shown. For the highest ratio specimen the Hall resistivity changes sign at $\sim$ 150 kG while the specimen of ratio 5842 will change sign at $\sim$ 200 kG. The low ratio specimen shows upward curvature, but has not reached a minimum at the highest field. The ordinary Hall coefficient $R_o$ for the highest ratio <100> crystal defined as $d\rho_H/dB$ is shown as a function of magnetic field in Fig. 8.

It is useful to plot the high-field data in a Kohler plot where the quantities are normalized by dividing by the zero-field resistivity $\rho_o$ and plotted against $B/\rho_o$ which is proportional to $\omega_c\tau$. Data can then be compared at the same $\omega_c\tau$ values and departures from Kohler's rule can be examined. Kohler's rule states that the quantity $(\rho_{ij}(\bar{B}) - \delta_{ij}\rho_o)/\rho_o$ should be a function of $B/\rho_o$ only. Samples of different purity or of the same purity but measured a different temperature should then follow a single curve on such a plot if Kohler's rule is obeyed. Such plots for the <111> and <100> current directions are shown in Figs. 9 and 10 for data obtained at 4.2 K. Previous data obtained on less pure polycrystals and on a less pure <111> single crystal is also included in Fig. 9.

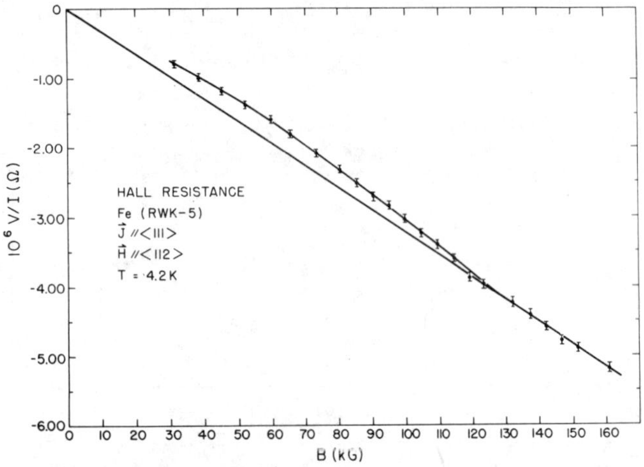


Fig. 6. Hall resistance as a function of $\bar{B}$ up to 160 kG at 4.2 K. $\bar{B}$//<112>, $V_H$//<112>, and $\bar{J}$//<111>. (Ref. 15)

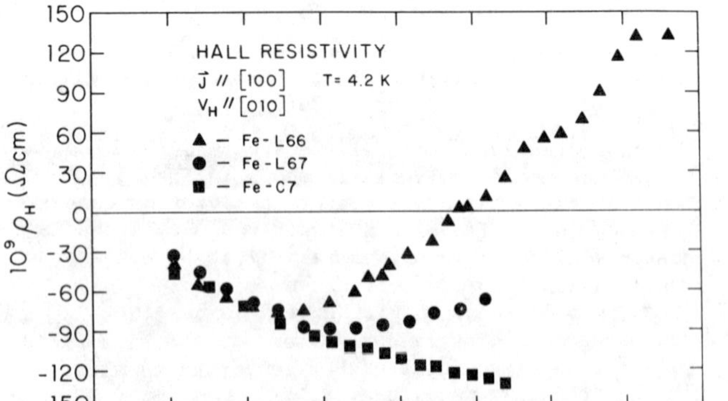


Fig. 7. Hall resistivity as a function of $\bar{B}$ up to 230 kG at 4.2 K. $\bar{B}$//<100>, $V_H$//<100>, and $\bar{J}$//<100>. RRR = 425, 5842, and 7520.

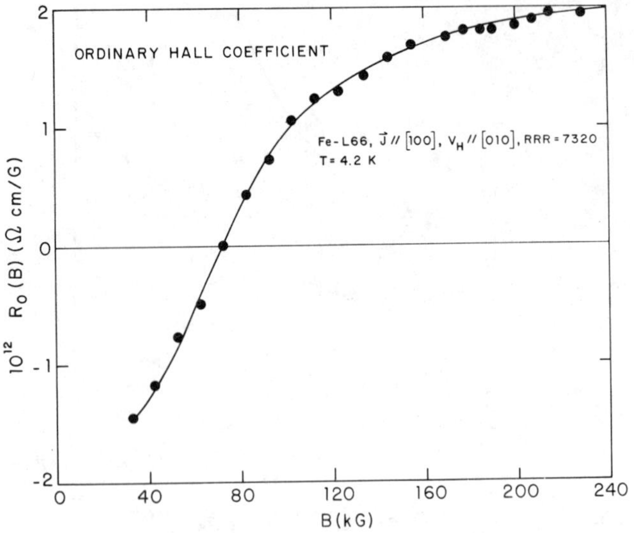


Fig. 8. Orindary Hall coefficient, $R_o = d\rho_H/dB$ as a function of $\bar{B}$ up to 230 kG.

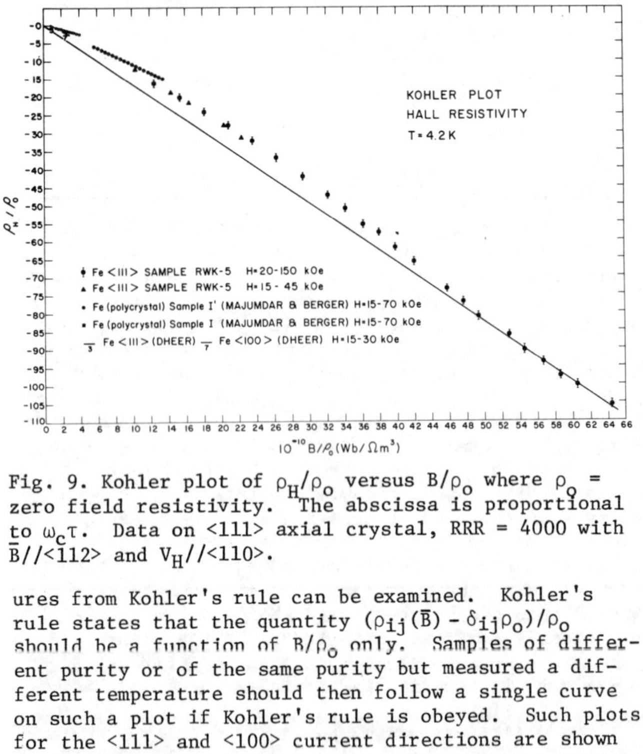


Fig. 9. Kohler plot of $\rho_H/\rho_o$ versus $B/\rho_o$ where $\rho_o$ = zero field resistivity. The abscissa is proportional to $\omega_c\tau$. Data on <111> axial crystal, RRR = 4000 with $\bar{B}$//<112> and $V_H$//<110>.

For further analysis of the high-field data the Hall angle defined as the ratio of the Hall resistivity to the magnetoresistance can be used to estimate the behavior in the ultimate high-field limit. Fig. 11 shows $\tan\phi = \rho_H/\rho(B)$ plotted as a function of $\rho_o/\bar{B}$. This is an inverse Kohler plot as first introduced by Majumdar and Berger[8] in an analysis of their data on high purity polycrystalline iron specimens. Extrapolation to infinite magnetic field can be indicated on such a plot and a value of the intercept can be estimated. The behavior is clearly highly dependent on crystal orientation and this will be discussed in Section IV.

B. Nickel

In the case of nickel measurements of the Hall effect have been carried out by Reed and Fawcett[17] on a single crystal with a RRR = 2700. This is the best

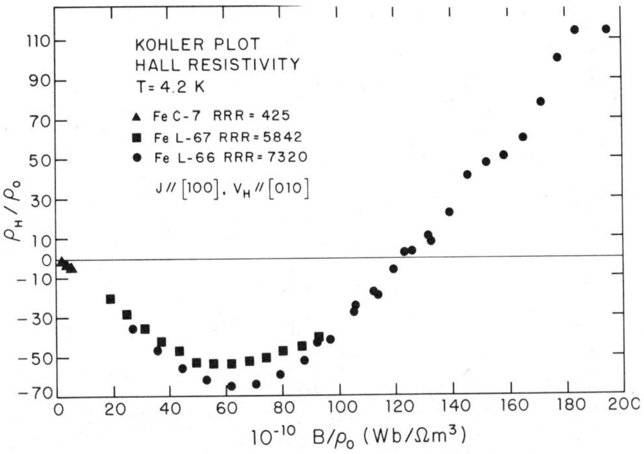


Fig. 10. Kohler plot of $\rho_H/\rho_o$ versus $B/\rho_o$. Data on <100> axial crystals. RRR = 425, 5842, and 7520. $\bar{B}$//<100> and $V_H$//<100>.

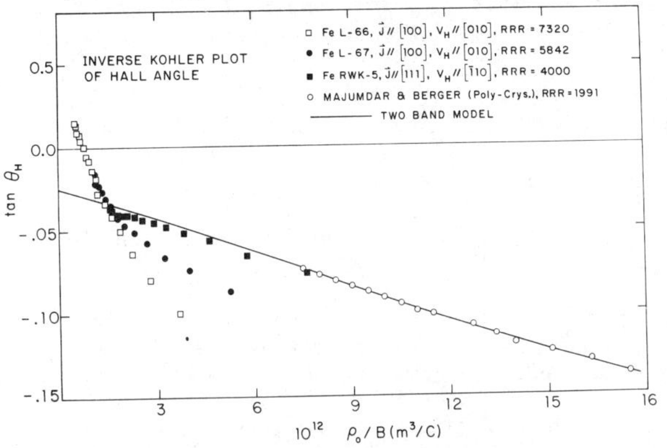


Fig. 11. Inverse Kohler plot for Hall angle, tan $\phi$ = $\rho_H/\rho(B)$ versus $\rho_o/B$.

case reported for an approach to the high-field limit, but even at 4.2 K and in fields to 80 kOe there is still some question about the existence of low-mobility d-band holes. Reed and Fawcett[17] have measured the Hall effect in a non-symmetry direction for where there are no open orbits and have observed a strictly linear field dependence to 80 kOe. The Hall voltage is negative indicating a dominant electron sheet and the Hall constant corresponds to a carrier concentration of 1.06 ± 0.03 electrons per atom. This result indicating approximately one electron per atom and the fact that the magnetoresistance shows saturation for closed orbit field directions suggests that nickel behaves as an uncompensated metal in the high-field limit. This lead Reed and Fawcett[17] to conclude that although nickel has an even number of electrons, the lifting of the spin degeneracy by the exchange interaction and the resulting spin-zone picture gives rise to the uncompensated behavior. This will be discussed further in Section IV.

Ehrlich, et al[14] have also made measurements of the Hall effect in high purity nickel using polycrystals of RRR equal to 2000 (NiIII), 490 (NiII), and 170 (NiI). A Kohler plot of the relative Hall resistivity versus $B/\rho_o$ is shown in Fig. 12 for these three specimens. The Hall resistivity in these experiments obeys Kohler's rule fairly well for $\omega_c\tau$ values up to approximately 10.

C. Cobalt

Data on the Hall effect in cobalt above $\omega_c\tau$ = 1 has not been obtained due to the difficulty of preparing high ratio crystals. In addition to the data reported in Ref. 6 and shown in Fig. 1, Volkenshtein, et al[18] have measured the Hall effect in single crystals of cobalt with current oriented both parallel and perpendicular to the c-axis. They measure a sub-

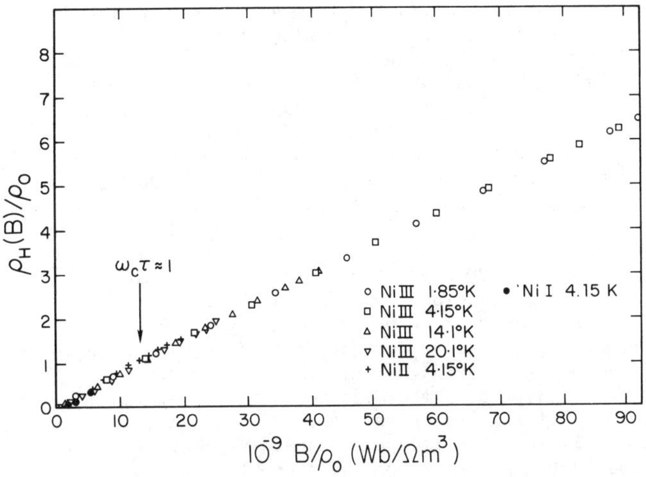


Fig. 12. Kohler plot for nickel polycrystals. RRR = 2000, 490, and 170.

stantial anisotropy in both $R_o$ and $R_s$ as might be expected since other transport properties in cobalt show a large anisotropy. The RRR's of these crystals were not reported in Ref. 18, but appear to be comparable to those of Ref. 6 and the data does not therefore extend into the high-field region. Coleman, et al[12] have measured magnetoresistance in cobalt crystals with RRR ∿ 400 and in fields to 150 kOe at 1.1 K. This data does extend to $\omega_c\tau > 1$, but no Hall resistivity data have yet been obtained.

## IV. DISCUSSION

The Hall effect in 3d ferromagnetic metals as measured in the high-field limit presents a number of interesting problems concerning the theory of galvanomagnetic properties. The general features of the high-field behavior have been formulated in the theory of Lifshitz, Azbel, and Kaganov[19] and this should certainly be the starting point for ferromagnetic metals. This theory generally relates the high-field behavior to features of the Fermi surface topology such as open and closed orbits and to the state of compensation as determined by an even or odd number of electrons per atom. Within the scope of this theory Kohler's rule would be expected to hold and analysis should proceed as in a non-magnetic metal except for the more complicated topology introduced by the spin-split Fermi surface and the need to consider the compensation condition more carefully. The compensation criterion for ferromagnetic metals was formulated by Reed and Fawcett[17] in terms of the parameter $n_A$ defined below.

$$-n_A = sZ - [F(\uparrow) + F(\downarrow) + J(\uparrow) + J(\downarrow)] \qquad (2)$$

The product sZ represents the number of electrons per unit cell, $F(\uparrow)$, $F(\downarrow)$ are the number of full spin-zones and $J(\uparrow)$, $J(\downarrow)$ are the number of spin-zones containing hole sheets with opposite signs of spin. $n_A$ must be an integer, but it can assume an even, odd or zero value in a ferromagnetic metal independent of the evenness or oddness of sZ. Therefore one cannot predict the compensation in advance from just a knowledge of sZ as in nonmagnetic metals. In fact the Hall effect data for nickel show $n_A = -1$ eventhough sZ = 28 and the above argument appears to be a logical explanation of the uncompensated behavior observed in nickel. The Fermi surface of nickel contains a one sheet electron surface making contact with the <111> faces of the Brillouin zone. By comparing the neck size measured from open orbit features in the magnetoresistance and the neck size calculated from the increase in the Hall coefficient for field along the symmetry axes Reed and Fawcett[17] obtain confirmation that the sheet contains a single sign of spin. It would be useful to carry nickel further into the high-field limit as a check on the correctness of the above conclusion, but at present this seems to be the accepted view.

Ehrlich, et al[14] have studied the detailed behavior of the Hall effect with respect to deviations from Kohler's rule. As indicated in Fig. 12 they found that the Hall resistivity obeyed Kohler's rule quite well for $\omega_c\tau$ values up to $\sim 10$, the highest value measured. On the other hand the Hall conductivity does not obey Kohler's rule. Samples of various purities or samples measured at different temperatures in the range 1 - 20 K all show conductivity maxima of different magnitude near $\omega_c\tau = 1$. At higher $\omega_c\tau$ values these conductivities show a trend toward a single function, but the data does not extend to high enough fields to check the high-field limit behavior in detail. Analysis of the magnetoresistance and magnetoconductivity of the same specimens is also included in Ref. 14 and this reference should be consulted for further details. Based on the present high-field data nickel would appear to be described adequately by the theory of Lifshitz, et al[19].

In the case of iron, the data obtained at high fields and low temperature clearly show that additional mechanisms outside the framework of the theory of Lifshitz, et al[19] must be taken into account. The most striking feature to be explained is the sign reversal of the Hall effect occurring in the <100> axial specimens for fields between 100 - 200 kOe corresponding to $\omega_c\tau$ in the range 100 - 200 and shown in Figs. 7 and 8. Similar sign changes have been observed in non-magnetic metals and can be explained either by magnetic-breakdown[10] or by intersheet scattering[11] occurring at "hot spots" on the Fermi surface. In the case of magnetic breakdown the electron makes an interband transition by tunneling through a small energy gap with a transition probability

$$P = \exp[-H_o/H] \qquad (3)$$

where the critical breakdown field $H_o$ is related to the energy gap $\Delta$ by

$$H_o = K\,\Delta^2 mc/\varepsilon_F eh \qquad (4)$$

This can result in a change of orbit topology for fields above $H_o$ and a consequent change in the hole or electron character of the orbit as well as in the compensation. For example compensated closed orbits can breakdown to form uncompensated closed orbits either of the electron or hole type. The magnetoresistance and Hall resistance calculated by Falicov and Sievert[10] for a transition from closed compensated orbits to closed uncompensated electron orbits are shown in Fig. 13. A reversal of the Hall resistivity occurs as a function of $\omega\tau$ and cases are given for various values of $\omega_0\tau = (|e|H_o\tau/mc)$. If $\tau$ is changed due to sample purity and $H_o$ remains the same then the expected sign changes will occur at different values of $\omega\tau$ since the onset of the transition occurs in the neighborhood

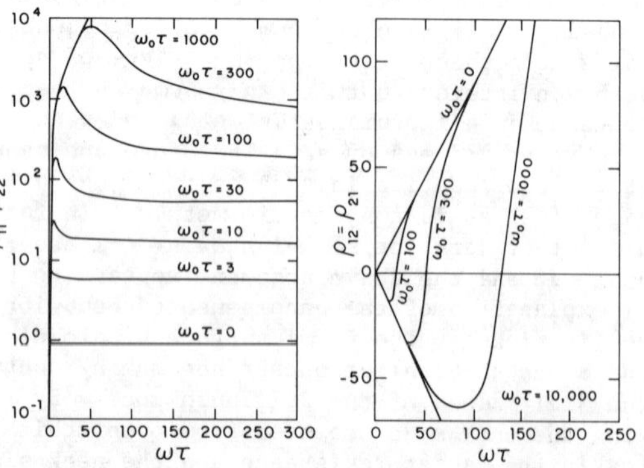


Fig. 13. Magnetoresistance and Hall resistance for a transition from closed compensated orbits to closed uncompensated electron orbits. (Ref. 10).

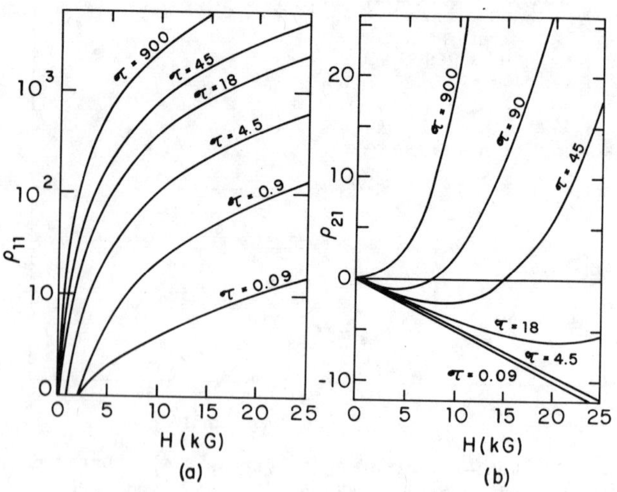


Fig. 14. Calculated transverse magnetoresistance (a) and Hall resistivity (b) for compensated orbits. $\mathcal{T}$ is measured in units of $10^{-11}$ sec. Case shown is for $\tau = 9.2 \times 10^{-10}$ sec.

of the critical field $H_o$. This is not the case for the data on the Hall resistivity of iron as shown in the Kohler plot of Fig. 10 where the minimum and crossing point will clearly occur at the same $\omega\tau$ value for the two samples of different purity.

A second mechanism which can produce a sign reversal in the Hall resistivity is localized intersheet scattering at points where two sheets of the Fermi surface are separated by a small gap. This scattering is characterized by a local scattering time $\mathcal{T}$ and represents a tunneling between the two sheets. However, in contrast to magnetic breakdown the probability of scattering decreases with increase of magnetic field since the time an electron spends in the region of the "hot spot" is decreased. In addition intersheet scatterimg can occur within the same band whereas magnetic breakdown is the result of interband transitions. Various possibilities exist for the behavior of the Hall resistivity, but it can be extremely sensitive to the presence of intersheet scattering and can exhibit a transition from electronlike to holelike behavior, or vice versa. Young, Ruvalds, and Falicov[11] have calculated typical curves based on a simple two channel model and obtain curves as shown in Fig. 14. These were calculated with particular reference to cadmium, but a similar regime could apply to iron with the introduction of appropriate values of $\mathcal{T}$ and $\tau$ and a more complicated Fermi surface. The sign change occurs at magnetic fields which scale with the value of $\mathcal{T}$ and can therefore account for the iron data on different purity crystals as shown in Fig. 7. More details of this mechanism as well as the magnetic breakdown mechanism within a detailed Fermi surface model for iron will have to be worked out, however.

For the <111> data as shown in Fig. 6 and 9 the behavior of the Hall resistivity is less anomalous. The Kohler plot as shown in Fig. 9 shows a relatively smooth behavior and the data fit smoothly with data on less pure specimens as shown. The curvature observed up to $\sim 100$ kG has been explained by Madjundar and Berger[8] as due to a mobility transition effecting $R_o$. This curvature would have an S-shape as a function of field and the behavior would return to a straight line at extremely high fields. Madjundar and Berger[8] have fitted their data on high purity polycrystals to a four band model and obtain a reasonably good fit. These authors have also used an inverse Kohler plot as shown in Fig. 15 in order to extrapolate the Hall angle to infinite field based on a two band model. They obtain a value of $\tan\phi_H = -0.027$ at point A and $\tan\phi_H = -0.025$ at point B of Fig. 15. The higher purity <111> data follow this same two band extrapolation as shown in Fig. 11 up to $\omega_c\tau \approx 70$. However, the <100> data clearly give a

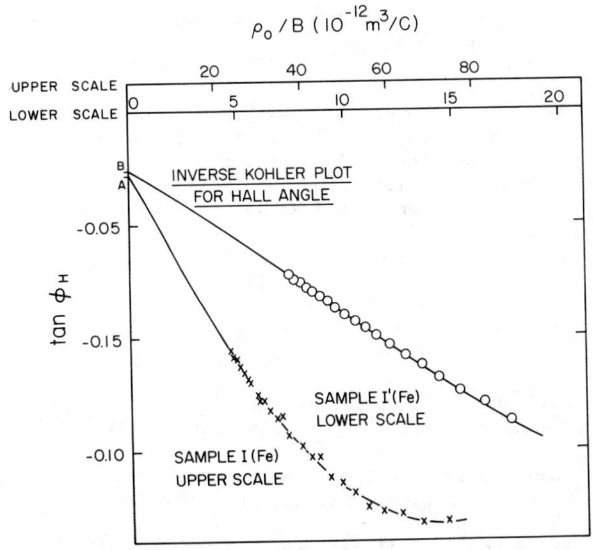


Fig. 15. Inverse Kohler plot for the Hall angle of two polycrystalline iron samples of different purity. Solid lines represent a two band model extrapolation to $\bar{B} = \infty$. (Ref. 8).

positive intercept and do not follow the same Kohler plot as also shown in Fig. 11.

At this point it is necessary to consider the more detailed topology of the Fermi surface particularly with respect to possible open orbit behavior. The primary evidence for open orbits comes from high field magnetoresistance data obtained by Reed and Fawcett[20,21] and by Coleman, et al.[12] For general field directions the magnetoresistance approaches a quadratic behavior in fields up to $\sim$ 90 kOe and indicates that iron is compensated with $n_A = 0$ consistent with formula (2) and with sZ = 26. The rotation diagrams show deep minima for fields in <100> and <110> directions and as first suggested by Reed and Fawcett[20] probably correspond to narrow bands of open orbits running parallel to <100> and <110> directions. Angadi, et al[22] have recently elucidated more details of the possible open orbit behavior by measuring the effects of stress on single crystals in fields up to 100 kOe. A more detailed review of the magnetoresistance data can be found in Refs. 12, 20, and 21.

The Fermi surface of iron has been studied in detail using de Haas van Alphen measurements first by Gold, et al[23] and more recently by Baraff[24]. These measurements along with comparisons to a number of band structure calculations[25-28] give a pretty complete picture of the many sheeted Fermi surface of iron and suggest a Fermi surface topology consistent with open orbits. A detailed review of the work in this field has been given by Gold[29].

Above 100 kOe the magnetoresistance begins to show structure in the field dependence which has features periodic in $1/\bar{B}$ as well as rather abrupt changes in slope particularly for field directions near <100>. An example is shown in Fig. 16 for a <100> axial crystal with a RRR = 9820 and measured in fields up to 220 kOe. The nonoscillatory structure in Fig. 16 suggests the possibility of magnetic breakdown and one such mechanism has been explored by Cabrera and Falicov[30]. They have examined the case of breakdown between two orbit regimes giving quadratic behavior, but with two different high-field coefficients. This mechanism produces changes in slope similar to those in Fig. 16 and must be considered as a possibility.

In considering the full complexity of the Fermi surface of iron it is necessary to consider not only the exchange splitting of the Fermi surface, but also the hybridization of the majority and minority bands under the action of spin-orbit coupling. Tawil and Callaway[31] have carried out such a calculation and the resulting changes in the Fermi surface

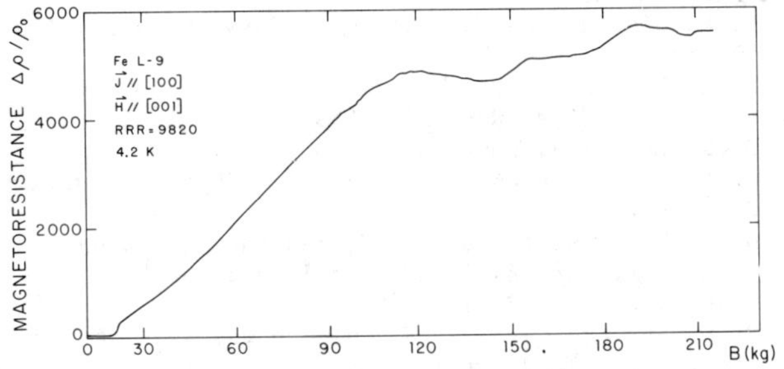


Fig. 16. Transverse magnetoresistance for <100> axial iron crystal with RRR = 9820. $\bar{B}$//<100> and $\bar{J}$//<100>. $\bar{B}$ = 20 - 230 kG.

topology allow the existence of many hybridized orbits. Many new energy gaps are also introduced which are capable of undergoing magnetic breakdown in fields accessible to experiment.

The elastic collision processes considered in most transport problems are spin-conserving and are usually characterized by a relatively short relaxation time $\tau$. In the spin-orbit coupling regime collisions with spin-flip are allowed and are characterized by a relatively long relaxation time $\tau_s$. Cabrera and Falicov[32] have formulated this problem in general and have applied it to the case in which two different scattering mechanisms combine in such a way that the decay to equilibrium goes through an intermediate quasiequilibrium state characterized by $\tau_s/\tau >> 1$. Magnetic breakdown effects have also been introduced and model calculations have been carried out for the expected behavior of the magnetoresistance for different values of $\tau/\tau_s$. The results clearly can simulate the characteristic features observed in Fig. 16, however, the detailed behavior of the Hall resistivity has not yet been given and this offers a possibility for bringing both the high-field magnetoresistance and Hall effect observations into a range of understanding.

## CONCLUSION

The high-field limit data as reviewed above has opened up a number of questions concerning transport mechanisms in ferromagnetic metals particularly with respect to iron. The theoretical framework required for analysis is probably available, but the complexity of the Fermi surface and the variety of behavior require both more data and analysis. Lowrey and Coleman[33] have extended the experimental work on iron above $\omega_c\tau = 100$ to additional orientations and higher fields and this data will be available soon. Nickel seems to be relatively well behaved as compared to iron, but checks at higher $\omega_c\tau$ values would be useful. The Fermi surface of cobalt[34] is rather complex and may suggest the possibility of high-field behavior of the type observed in iron. In addition the very strong anisotropy in cobalt might play an interesting role in the transport at high $\omega_c\tau$.

## ACKNOWLEDGEMENTS

The author acknowledges the contributions of Professors D. J. Sellmyer and R. C. Morris, Dr. R. W. Klaffky and Mr. W. H. Lowrey to various phases of the experimental work on iron. Valuable discussions have been held with Professors L. M. Falicov, J. Ruvalds and V. Celli.

## REFERENCES

*Work supported by the U. S. Atomic Energy Commission Contract No. AT-(40-1)-3105.
†Work above 80 kOe was performed while the author was a Guest Scientist at the Francis Bitter National Magnet Laboratory, which is supported at the Massachusetts Institute of Technology by the National

Science Foundation.


1. R. Karplus and J. M. Luttinger, Phys. Rev. 95, 1154 (1954).
2. J. M. Luttinger, Phys. Rev. 112, 739 (1958).
3. L. Berger, Phys. Rev. B2, 4559 (1970).
4. H. R. Leribaux, Phys. Rev. 150, 384 (1966).
5. P. Leroux-Hugon and A. Ghazali, J. Phys. C: Solid State Physics 5, 1072 (1972).
6. N. V. Volkenshtein and G. V. Fedorov, Zh. Eksperim. i Teor. Fiz 38, 64 (1960) [English transl: Soviet Phys.-JETP 11, 48 (1960)].
7. P. W. Shumate, Jr., R. V. Coleman and R. C. Fivaz, Phys. Rev. B1, 394 (1970).
8. A. K. Majumdar and L. Berger, Phys. Rev. B7, 4203 (1973).
9. E. Fawcett, Advan. Phys. 13, 139 (1964).
10. L. M. Falicov and Paul R. Sievert, Phys. Rev. 138, A88 (1965).
11. Richard A. Young, J. Ruvalds, and L. M. Falicov, Phys. Rev. 178, 1043 (1969).
12. R. V. Coleman, R. C. Morris and D. J. Sellmyer, Phys. Rev. 8, 317 (1973).
13. E. Fawcett and W. A. Reed, Phys. Rev. Lett. 9, 336 (1962).
14. A. C. Ehrlich, R. Hugenin, and D. Rivier, Phys. Chem. Solids 28, 253 (1967); A. C. Ehrlich and D. Rivier, Phys. Chem. Solids 29, 1293 (1968).
15. R. W. Klaffky and R. V. Coleman, Phys. Rev. 10, 2915 (1974).
16. P. N. Dheer, Phys. Rev. 156, 637 (1967).
17. W. A. Reed and E. Fawcett, Journ. Appl. Phys. 35, 754 (1964).
18. N. V. Volkenshtein, G. V. Fedrov and V. P. Shirokovskii, Phys. Metals and Metallography 11, 152 (1961).
19. I. M. Lifshitz, M. YaAzbel and M. I. Kaganov, Zh. Eksperim. i Teor. Fiz. 30, 220 (1955); 31, 63 (1956) [English transls: Soviet Phys.-JETP 3, 143 (1956); 4, 41 (1957)].
20. W. A. Reed and E. Fawcett, Phys. Rev. 136, A422 (1964).
21. W. A. Reed and E. Fawcett, International Conference on Magnetism, Nottingham (The Institute of Physics and the Physical Society, London, 1965), p. 120.
22. M. A. Angadi, E. Fawcett and M. Rasolt, Phys. Rev. Lett., 32, 613 (1974).
23. A. V. Gold, L. Hodges, P. T. Panousis and D. R. Stone, Int. J. Magn. 2, 357 (1971).
24. David R. Baraff, Phys. Rev. B8, 3439 (1973).
25. J. H. Wood, Phys. Rev. 126, 517 (1962).
26. S. Wakoh and J. Yamashita, J. Phys. Soc. Japan 21, 1712 (1966).
27. K. J. Duff and T. P. Das, Phys. Rev. B1, 192 (1971).
28. E. Abate and M. Asdente, Phys. Rev. 140, A1303 (1965).
29. A. V. Gold, Journ. Low Temp. Phys. 16, 3 (1974).
30. G. G. Cabrera and L. M. Falicov, Phys. Rev. 10, 4803 (1974).
31. R. A. Tawil and J. Callaway, Phys. Rev. B7, 4242 (1973).
32. G. G. Cabrera and L. M. Falicov, Phys. Rev. 11, 2651 (1975).
33. W. H. Lowrey and R. V. Coleman (to be published).
34. S. Wakoh and J. Yamashita, J. Phys. Soc. of Japan 28, 1151 (1970).


## ANISOTROPIC MAGNETORESISTANCE IN FERROMAGNETIC 3d TERNARY ALLOYS

T. R. McGuire, R. D. Hempstead and S. Krongelb
IBM Research Center, Yorktown Heights, N.Y. 10598

ABSTRACT

The electrical resistivity of ferromagnetic 3d transition metal alloys depends on the angle between the current and the magnetization. We have studied this anisotropic effect for selected compositions of ternary alloys containing Ni-Fe or Ni-Co with additions of Mn, Cr, Co, Cu and Zn. Both films and bulk samples were prepared. The alloy compositions were chosen to adjust the 3d + 4s electron concentrations (e.c.) in a continuous manner from 9.3 to 10.0. We find that the anisotropic ratio $\Delta\rho/\rho_{avg}$ (where $\Delta\rho = \rho_{\parallel} - \rho_{\perp}$, and $\rho_{\parallel}$ and $\rho_{\perp}$ are the resistivities parallel and perpendicular to the magnetization) has scattered values as a function of e.c. However, $\Delta\rho$ shows a continuous change with a maximum vlaue of $\sim .9 \times 10^{-6}\ \Omega$ cm for e.c. between 9.7 and 9.8. This agrees well with the behavoir of $\Delta\rho$ for the binary alloys Ni-Fe and Ni-Co[1,2]. The similarities of the binary and ternary systems with respect to $\Delta\rho$ gives support that d band filling is an important consideration in explaining the anisotropic magnetoresistance. Rigorous interpretation of the ternary data is complicated by the fact that a common d band is probably not formed which causes the wide variation found for the average resistivity ($\rho_{avg}$).


1) J. Smit, Physica XVI 612 (1951).
2) H. C. Van Elst, Physica 25, 708 (1959).

# MAGNETO-RESISTIVITY OF BETA-CERIUM[*]

P. Burgardt, S. Legvold, J. H. Queen and K. A. Gschneidner, Jr.
Ames Laboratory-USERDA and Departments of Physics and Materials Science and Engineering
Iowa State University, Ames, Iowa 50011

ABSTRACT

Longitudinal magneto-resistivity measurements of allotropically pure samples of dhcp cerium were made at temperatures from 4.2 to 90 K in fields up to 80 kOe. The results show negative magneto-resistivity which is typical of field suppression of spin flip scattering. In the vicinity of the Néel temperature complex results were obtained which are interpreted as the interaction of internal fields with the applied field.

## INTRODUCTION

Cerium metal is known to exist in three low temperature allotropes: β(dhcp, $\underline{a}$ = 3.6810 and $\underline{c}$ = 11.857 Å), γ(fcc, $\underline{a}$ = 5.1610 Å) and α(fcc, $\underline{a}$ = 4.85 Å). The β(dhcp) form of cerium, which has been known for some 30 years,[1] has a single occupied $\underline{4f}$ level that exhibits anti-ferromagnetic ordering.[2,3] The physical properties of β-Ce are expected to be interesting due to the proximity of the localized $\underline{4f}$ level to the Fermi level. Only in recent years have Gschneidner and co-workers[4] been able to produce allotropically pure samples of β-Ce and thus make physical measurements practical. In previous resistivity experiments[5] on samples of β-Ce we found an anomalous resistivity of the form $\rho \propto \log T$ and an unusual rise in resistivity with rising temperature in the vicinity of the ordering temperature (0<T<20 K). In this paper we present magneto-resistivity measurements of allotropically pure samples of β-Ce.

## EXPERIMENTAL PROCEDURE AND RESULTS

Samples of β-Ce were prepared from 99.9% cerium metal by a thermal cycling (298 to 4.2 to 298 K) and annealing (348 K) process.[4] The samples were spark cut to the final dimensions and were electro-polished prior to the experiment. X-ray investigations of the samples indicated that they were polycrystalline and were better than 98% pure dhcp.

The magneto-resistivity measurements were made by a four probe dc technique. By using sample currents of 100 mA and voltage sensitivity of $10^{-8}$V we obtained a relative precision in the resistance measurements of better than 0.05%. The sample temperature could be varied from 4.2 to 100 K and was determined by a calibrated gold-0.03% iron vs. copper thermocouple. The sample temperature was controlled to within 0.02 K. Magnetic fields from 0 to 85 kOe could be applied to the sample and in all cases the field was parallel to the sample current.

In our previous work on the electrical resistivity and magnetic susceptibility of β-Ce[5] we found that one must carefully avoid the temperature range from 15 to 50 K because β-Ce undergoes a partial conversion to α-Ce in that region. However, if the sample is cooled through the critical region in less than five minutes, experiments can safely be performed below 15 K. Therefore, we performed longitudinal magneto-resistivity measurements on β-Ce at T>60 K and T<15 K. Because of the importance of ordinary lattice magneto-resistivity effects, we also performed experiments on dhcp lanthanum at similar values of temperature and field such that the normal magneto-resistivity could be eliminated through proper application of Kohler's rule.[6] The results of the magneto-resistivity experiment on dhcp lanthanum are shown in Fig. 1. In accordance with Kohler's rule, the results can be written as $\Delta\rho/\rho_0 = 0.00092(H/\rho_0)^{1.35}$ for fields less than $H/\rho_0$ = 50 kOe/μΩcm and as $\Delta\rho/\rho_0 = 0.00362(H/\rho_0)$ for fields of $H/\rho_0$ = 50 to 350 kOe/μΩcm. The application of Kohler's rule to β-Ce is complicated because all processes that affect the conduction electron relaxation time contribute to the normal magneto-resistivity. Therefore, the spin resistivity must be incorporated in Kohler's rule. We assume that the temperature dependence of the phonon part of the resistivities of dhcp La and Ce are identical. This assumption is appropriate since the two materials are isostructural and have similar Debye temperatures. The residual resistivity of β-Ce is taken to be its resistivity at 2 K where the spins are essentially completely ordered and will not contribute to the resistivity. We may then write the spin resistivity of β-Ce as:

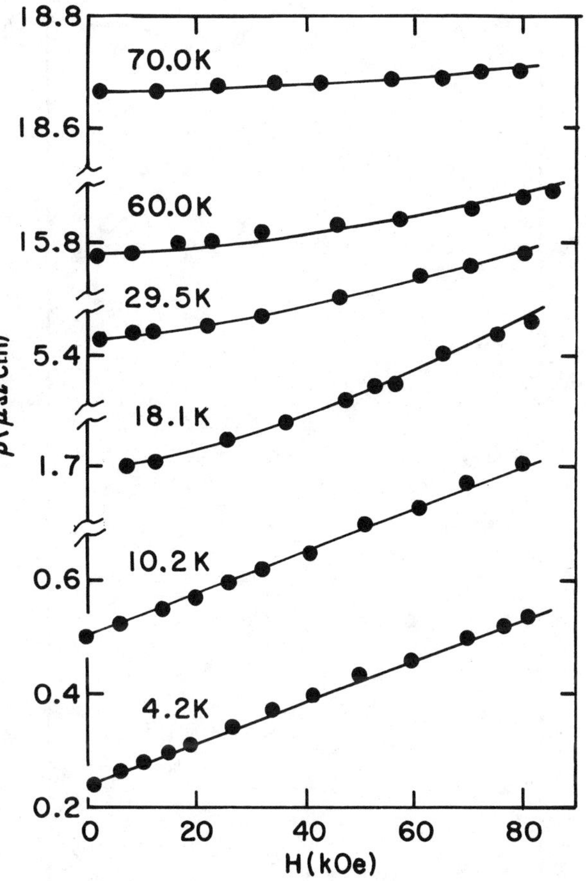


Fig. 1 Resistivity of dhcp lanthanum as a function of field at a constant temperature. The solid lines represent the Kohler's rule fit given in the text.

$$\rho_{SPIN}(T,H) = \rho_{experimental}(T,H) - \rho_{res} - \rho_{La}(T,0) - \rho_{NM}(T,H)$$

where $\rho_{NM}(T,H)$ is the normal magneto-resistivity. We assume that $\rho_{NM}(T,H)$ for β-Ce can be written by way of Kohler's rule derived from dhcp La to yield:

$$\rho_{NM}(T,H) = 0.00092\cdot(\rho_{res}+\rho_{La}(T,0) + \rho_{SPIN}(T,H)) \cdot (H/\rho_{res} + \rho_{La}(T,0) + \rho_{SPIN}(T,H))^{1.35}.$$

We can then solve for $\rho_{SPIN}(T,H)$ and obtain $\Delta\rho_{SPIN}(T,H) = \rho_{SPIN}(T,H) - \rho_{SPIN}(T,0)$. The results of the subtraction process are shown in Fig. 2. The magneto-resistivity of the cerium spin system is in general negative and can be described by the empirical equation: $\Delta\rho_{SPIN}(T,H) = -A(T)H^n$ where the values of the parameters A and n are given in Table 1. The salient features of the data are the large negative magneto-resistivity just above $T_N$ = 12.7 K and the positive upswing in the 4.2 K data at fields above 70 kOe.

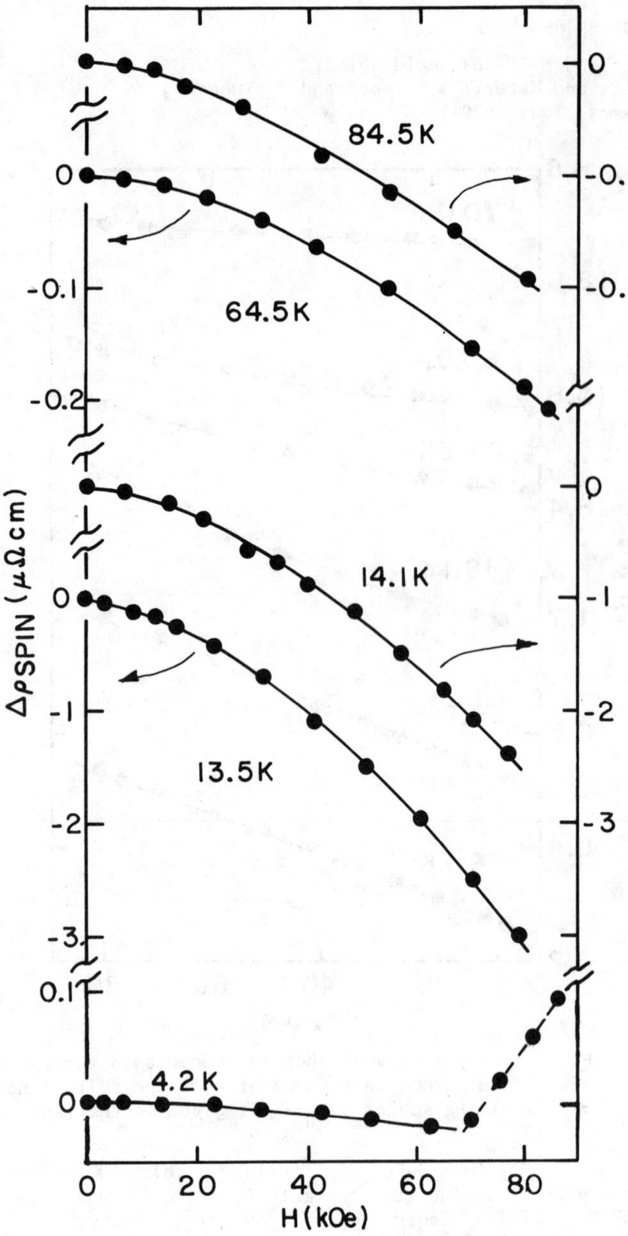


Fig. 2 The spin resistivity of β(dhcp) cerium evaluated at constant temperature as a function of field. The solid lines represent the empirical fit given in the text.

Table 1. Temperature (K), ρ(H=0) in μΩcm, the parameters A and n from the empirical equation $\rho_{SPIN} = -A\ H^n$

| T | $\rho_0$ | A(X $10^{-4}$) | n |
|---|---|---|---|
| 84.5 | 69.65 | 1.74 | 1.60 |
| 64.5 | 67.27 | 1.40 | 1.64 |
| 14.1 | 44.92 | 18.4 | 1.65 |
| 13.5 | 43.60 | 32.2 | 1.56 |
| 4.2 | 6.62 | 0.13 | 1.8 |

## DISCUSSION

Data that we obtained in the past showed that the zero field spin resistivity of β-Ce, obtained by subtraction of the zero field resistivity of dhcp La, exhibits a log T dependence at temperatures above the Néel point. Below the Néel temperature, $T_N$ = 12.7 K, the zero field resistivity decreases rapidly apparently as a result of the large internal field associated with magnetic ordering. This behavior suggests that the resistivity may be a result of Kondo scattering[7] between conduction electrons and the localized 4f levels which are near to the Fermi level. A resistivity behavior of this nature has been observed for various cerium alloys and has been interpreted[8,9] as scattering from crystal field split levels. On the basis of the observed crystal field splittings[10] one would expect a resistivity maximum at about 100 K which is not observed. On the basis of this we believe that the crystalline field models are not wholly satisfactory and that a simpler Kondo model is at this time more useful. A full discussion and justification for this interpretation of the data is given in Ref. 5.

The observed negative magneto-resistivities are consistent with the Kondo scattering hypothesis since Kondo scattering diminishes as certain spin flip processes are "frozen out" by the applied field. The success of the simple empirical fit to the negative magneto-resistivity and the relatively constant value of n in Table 1 suggests that a simple Kondo model may be adequate.[11] Immediately above the ordering temperature we observe strong negative magneto-resistivity. We interpret this result as field induced local moment alignment immediately above the ordering temperature such that the effective field would be the sum of the induced internal field and the applied field. The interesting positive upturn observed at 4.2 K is understood as an additional scattering due to disruption of anti-ferromagnetic order by the applied field. It is expected that the field necessary to flip the ordered spins at T = 0 is approximated by $H \sim k_B T_N / g J \mu_B$. For trivalent cerium with $T_N$ = 12.7 K we obtain H = 88 kOe. The correspondence between the predicted onset of positive magneto-resistivity and the observed value at 4.2 K as read from Fig. 2 is acceptable in light of the temperature at which the measurement was performed.

In conclusion, the negative magneto-resistivity of β-Ce is suggestive of Kondo scattering in a pure metal. The complicated nature of the results near to and below the ordering temperature can be understood as an interaction of applied and internal fields.

Helpful discussions with S. H. Liu and B. N. Harmon and the aid of J. O. Moorman with sample preparation are gratefully acknowledged.

## REFERENCES

*Prepared for the U.S. Energy Research and Development Administration under Contract No. W-7405-eng-82.

1. F. Trombe and M. Foex, Compt. Rend. 217, 501 (1943).
2. M. K. Wilkinson, H. R. Child, C. J. McHargue, W. C. Koehler and E. O. Wollan, Phys. Rev. 122, 1409 (1961).
3. J. M. Lock, Proc. Phys. Soc. London B 70, 566 (1957).
4. D. C. Koskimaki, K. A. Gschneidner, Jr., and N. T. Panousis, J. Crystal Growth 22, 225 (1974).
5. K. A. Gschneidner, Jr., P. Burgardt, S. Legvold, J. O. Moorman, T. A. Vyrostek and C. Stassis, to be published, J. Phys. F: Metal Phys., Vol. 6, No. 2 (1976).
6. M. Kohler, Ann. Phys. 32, 211 (1938).
7. B. Coqblin and J. R. Schrieffer, Phys. Rev. 185, 847 (1969).
8. F. E. Maranzana, Phys. Rev. Lett. 25, 239 (1970).
9. B. Cornut and B. Coqblin, Phys. Rev. B 5, 4541 (1972).
10. D. C. Koskimaki and K. A. Gschneidner, Jr., Phys. Rev. B 5, 2055 (1974).
11. T. Van Peski-Tinbergen and A. J. Dekker, Physica 29, 917 (1963).

# PRESSURE DEPENDENCE OF THE ELECTRONIC STRUCTURE OF NICKEL*

J. R. Anderson and Peter Heimann
Dept. of Physics, University of Maryland, College Park, Maryland 20742
J. E. Schirber
Sandia Laboratories, Albuquerque, New Mexico 87115
D. R. Stone
Laboratory for Physical Sciences, College Park, Md. 20740

## ABSTRACT

The influence of pressure on de Haas-van Alphen frequencies in nickel has been studied by means of both fluid helium and solid helium phase shift techniques. Two pieces of Fermi surface have been investigated, the copper-like majority-spin necks and the minority spin hole pockets. The changes are small; for example the effect of pressure on the nickel necks is much less than for the necks in copper.

## INTRODUCTION

In recent years there have been many studies, both experimental and theoretical, of the electronic structure of nickel. Experimental investigations of the Fermi surface by means of the de Haas-van Alphen (dHvA) effect have been especially useful in checking theoretical models of the nickel band structure.

We have been studying the dHvA effect in nickel at both atmospheric and higher pressures as part of our program to investigate the relationship between the band structures and magnetic properties of ferromagnetic metals. We believe that an understanding of itinerant ferromagnetism will depend in part upon learning the details of electronic structure.

Two small portions of the Fermi surface of nickel have been studied in detail at atmospheric pressure by means of the de Haas-van Alphen effect.[1,2] These are the copper-like, majority-spin necks centered at the L points and d-band hole pockets centered at the X points. Here we report some preliminary measurements of the effects of pressure on two dHvA frequencies, the frequency due to the necks for applied magnetic field H parallel to the easy direction of magnetization [111] ($F_n = 2.72 \times 10^6$ Gauss) and the frequency resulting from the hole pockets at X for H parallel to [001] ($F_p = 10.12 \times 10^6$ Gauss).

Two types of phase shift experiments were carried out:

1. Measurement of the shift in dHvA phase due to small hydrostatic pressures on the sample produced in fluid helium,[3] and

2. Measurement of the shift due to application of pressures of the order of kilobars with solid helium as the pressure transmitting medium.[4,5]

## EXPERIMENTAL TECHNIQUE

De Haas-van Alphen oscillations were measured by the conventional low-frequency field modulation technique in NbZr 55 kOe superconducting magnets. Field modulations up to 2 kOe peak at frequencies less than 100 Hz were obtained by means of small superconducting modulation coils.

The main sample used for the fluid He phase-shift experiments was a 4.5 mm diameter sphere. The sample used for the solid helium measurements was a smaller, nearly spherical sample of about 3 mm diameter.

Details of the fluid helium phase-shift technique are well known. The only modification here was the addition of on-line computing with a PDP-9 computer in order to control the experiment and to analyze the results.

The shift in phase $\Delta\theta$ of the dHvA oscillations with pressure $\Delta P$ can be related to the changes in both dHvA frequency F and saturation magnetization M as follows:

$$\frac{B}{F}\left(\frac{1}{2\pi}\frac{\Delta\theta}{\Delta P}\right) = \frac{d\ln F}{dP} - \frac{4\pi\,(1-D)M}{B}\frac{d\ln M}{dP} . \qquad (1)$$

Here B is the magnetic induction ($B = H + \frac{8\pi M}{3}$ for a sphere) and D is the demagnetizing factor. Note that the larger F/B the greater the observed phase shift, at least if one neglects the variation of M with pressure. (At magnetic fields typical for this experiment $\frac{F_n}{B} \sim 500$ and $\frac{F_p}{B} \sim 2500$.)

In order to determine this shift in dHvA phase with fluid helium pressures it was necessary to reproduce field sweeps over a few dHvA cycles very precisely. To this end the 55 kOe superconducting magnet was placed in a stable mode (stability of about 3 parts in $10^6$). The magnetic field variation was obtained by adding a computer controlled incremental dc sweep signal to the ac modulation signal at the summing junction of the high power current pump driving the modulation coil. Such a sweep could be repeated many times in order to enhance signal to noise by signal averaging. Since the sweep field was incremented in steps, each measurement point on a dHvA cycle was obtained at constant magnetic field.

The expected phase shifts in nickel due to pressures of 25 bars are quite small because the compressibility of nickel, $K_T$ (= $5.5 \times 10^{-4}$ $kbar^{-1}$),[6] is not large and because sample quality is such that large values of B must be used for satisfactory signal to noise, making F/B a rather small quantity. For example the change of phase of the neck frequency $F_n$ due to 25 bar pressure at an applied field of 20 kOe would be only about .001 cycle if one assumes that the only effect of pressure is to change the lattice spacing.

In order to study such small shifts we have used the computer to least squares fit a single sine wave to a few dHvA cycles, typically 3. We have made some tests to show that the exponential dependence upon magnetic induction which occurs in the expression for the dHvA amplitude can be omitted from the fitting function without significant loss of precision. The phase of the sine wave obtained from the least squares fit for atmospheric pressure was compared with the phase obtained at higher pressure; shifts of the order of $10^{-3}$ of a cycle could be detected providing the signal to noise ratio was greater than 10.

The solid helium phase-shift technique[5] is similar to the fluid helium technique, except for the fact that the sample plus pressure bomb must be raised above the liquid helium in the cryostat and allowed to warm up in order to change the pressure. Because much higher pressures are possible, the phase shifts are much greater and therefore the field sweeps over a few dHvA cycles were made with the 55 kOe magnet itself.

## RESULTS

By means of the fluid helium experiment we have obtained a value of $d\ln F_n/dP = (-4\pm2)\times 10^{-4}$ $kbar^{-1}$ for the neck oscillations. These measurements, made at applied magnetic fields between 16 kOe and 40 kOe, were independent of H within experimental error. At lower fields this value decreases in magnitude and even appears to change sign below about 10 kOe.

On the other hand, the solid helium technique gives a positive value of $d\ln F_n/dP$ $(6\pm1)\times 10^{-4}\,kbar^{-1}$. Measurements were made at 1, 2, and 3 kbars and at three field values, 10 kOe, 20 kOe, and 32 kOe. The phase shifts at different pressures were consistent and within experimental error the same shifts were observed at all values of H; consequently no value for dℓn M/dP could be determined.

The solid helium experiment gives a very small value for the [100] hole pockets,

$$\frac{d\ell n\ F_p}{dP} = (1.2\pm0.3) \times 10^{-4}\ \text{kbar}^{-1}.$$

Up to now, we have been unable to obtain a phase shift for the hole pockets using the fluid helium technique.

We are not able to account for the difference between the phase shifts of $F_n$ obtained by the two techniques. The shift in magnetic field corresponding to the phase shift for a pressure of 3 kbars is about 40 Oe. The field can be reproduced to at least one Oe by means of measurement of the magnetoresistance of the copper pickup coil. It is possible that the sample changes orientation when the pressure is changed. This is unlikely to be the cause of a major error in the solid helium experiments because a rather large shift in angle would be required. (If one assumes that the fluid helium shift is correct, a tilt of 6° would be required.) On the other hand, this tilt effect could be significant for fluid helium pressures of 25 bars. A shift of only about 0.6° could produce the negative shift that we have found in the fliud helium experiments. However it is also difficult to believe that such a tilt occurred since we consistently obtained the same phase shifts although the errors were large. In addition at higher fields, the sample would be more exactly lined up along [111], so that better agreement with the solid helium result would be expected at higher fields, which was not the case.

One might also note that the sphere used for the fluid helium experiments was significantly larger than that used at high pressures. However, we do not know any reason why this should affect our results. Domains should not be a problem since we are well above technical saturation.

## CONCLUSIONS

We have studied the effects of pressure on two dHvA frequencies, $F_n$ and $F_p$. In both cases the effect of pressure is quite small. We are inclined to believe the solid helium results rather than the fluid helium measurements since the experimental phase shifts are larger in the former and therefore sample tilting would be much less critical. Because of the discrepancy we are planning to repeat the fluid helium experiments.

Such small shifts are quite surprising when one remembers that for copper, which has a Fermi surface with similar [111] necks and similar compressibility, the change in dHvA frequency with pressure is 3 to 4 times larger. We have begun augmented-plane-wave calculations to investigate these experimental results.

## ACKNOWLEDGEMENTS

We wish to thank R. L. White and K. Neimiller for help with these experiments. The support of the University of Maryland Computer Science Center is gratefully acknowledged. Helpful discussions with J. K. Pinkston are gratefully acknowledged.

## REFERENCES

* Work supported by the N.S.F. under Grant No. DMR72-03007 A04 and by the O.N.R. under Contract No. N00014-67-A-0239-0024.

1. D. C. Tsui, Phys. Rev. 164, 669 (1967).
2. L. Hodges, D. R. Stone, and A. V. Gold, Phys. Rev. Lett. 19, 655 (1967).
3. I. M. Templeton, Proc. Roy. Soc. (London) A292, 413 (1966).
   J. E. Schirber and W. J. O'Sullivan, Phys. Rev. 184, 628 (1969).
4. J. E. Schirber, Cryogenics 10, 418 (1970).
5. J. E. Schirber and R. L. White, Jour. of Low Temp. Phys. - to be published.
6. G. A. Alers, J. R. Neighbours, and H. Sato, J. Phys. Chem. Sol. 13, 40 (1960).

# ELASTIC ANISOTROPY IN THE TSDW PHASE OF CHROMIUM

M.O. Steinitz
St. Francis Xavier University
Antigonish, Nova Scotia, Canada.
BOH 1CO

D.J. Stanley and E. Fawcett
University of Toronto
Toronto, Ontario, Canada.
M5S 1A7

## ABSTRACT

Sound velocity measurements in the transverse spin-density-wave (TSDW) phase of antiferromagnetic chromium reveal that, in contrast to some predictions, the polarization direction, $\vec{s}$, is elastically just as stiff as the direction normal to both $\vec{s}$ and the wave-vector, $\vec{Q}$, of the SDW.

## INTRODUCTION

Motivated by the prediction of Fletcher and Osborne[1] that the elastic response along the polarization direction, $\vec{s}$, in the transverse spin-density-wave (TSDW) phase of antiferromagnetic chromium should be elastically soft relative to other directions in the plane perpendicular to the wave vector, $\vec{Q}$, of the SDW, and wishing to complement the previous work of Munday and Street[2] on this problem, we have undertaken sound velocity measurements in the plane normal to $\vec{Q}$ as a function of the orientation of the sample with respect to an applied field.

## EXPERIMENT

Measurements were made on a single crystal of chromium, cooled through the Néel temperature ($T_N$ = 312°K) in a field of 107 kOe along the [001] axis, so that $\vec{Q}$ lay along [001]. The cooling field was then removed. The crystal had previously been shown to be essentially single-$\vec{Q}$ for cooling fields in excess of 50 kOe[3]. A pulse superposition technique[4] was employed, in which pulses of 30 MHz longitudinal waves were transmitted along the [010] axis in resonance with the echoes. A phase-tracking technique allowed the continuous observation of very small changes in velocity. An external field, $\vec{H}$, was applied in the plane normal to $\vec{Q}$ and the sample was rotated about $\vec{Q}$. Plots of relative change in sound velocity as a function of the angle

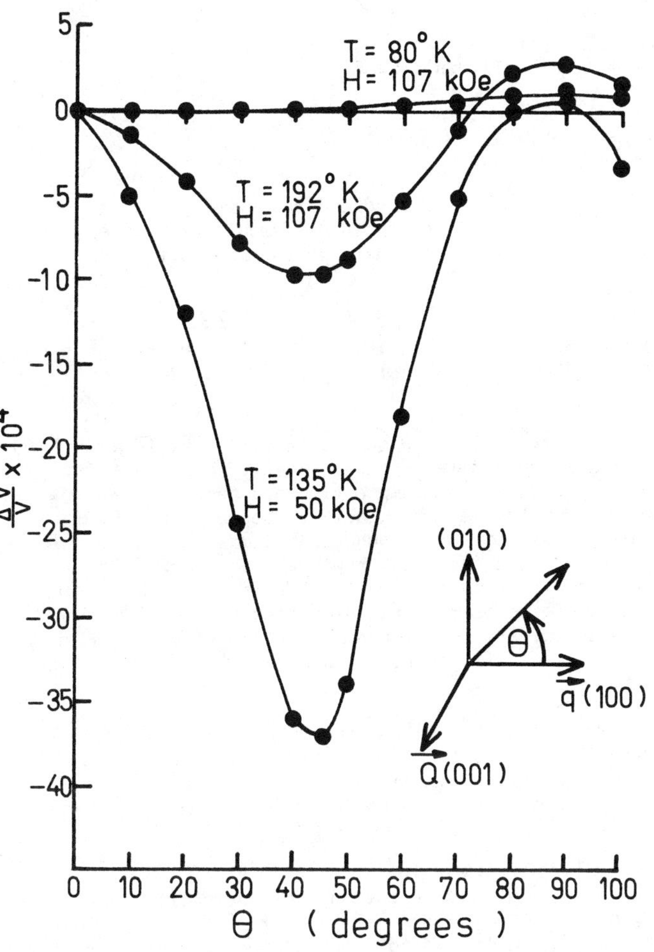


Figure 1. Dependence of 30 MHz sound velocity on orientation of Cr crystal with respect to a field applied in the plane normal to $\vec{Q}$.

of the field with respect to the sound propagation direction, $\vec{q}$, are shown in Figure 1 for several values of the temperature and applied field.

RESULTS

In the TSDW phase, for fields over about 30 kOe along [100] essentially all the polarization, $\vec{s}$, lies along [010][3]. Similarly, when $\vec{H}$ is along [010] all the polarization lies along [100] and none lies along [010]. This is due to the fact that for all antiferromagnets the perpendicular susceptibility is much larger than the parallel susceptibility. Thus, from Figure 1, one can see that the sound velocity is essentially the same, regardless of whether $\vec{s}$ is parallel or perpendicular to $\vec{q}$. The slight asymmetry of the curves is accounted for, in part, by the Alpher-Rubin effect[5], as shown by the curve in Figure 1 made at 80°K in the low temperature, longitudinal SDW phase, where $\vec{s}$ is along $\vec{Q}$. The sound velocity is, however, significantly reduced when the field lies at 45° between the two <100> axes normal to $\vec{Q}$. The anisotropy energy favours these directions[6], making them local energy minima, although $\vec{H}$ may make a <110> type direction the absolute minimum energy direction[7].

These results may be qualitatively understood in terms of the thermal activation model used earlier in the interpretation of the ultrasonic attenuation[8]. It was introduced by Werner et al.[9] in 1967. This model postulates small polarization domains that are thermally activated. The ultrasonic attenuation and velocity changes are then due to the relaxation process associated with the reorientation of the polarization of the domains. The order of magnitude of the velocity change is in agreement with the real part of the response calculated for the relaxation process in the attenuation work. A detailed analysis and calculation of the dependence of the velocity on the angle of the applied field, allowing a better estimate of the size and shape of the anisotropy energy is in progress and will be published elsewhere upon completion. This will involve an extension of the simplified two-state thermal activation model previously used by some of us[8], to include all possible polarization orientations in the plane perpendicular to $\vec{Q}$, rather than only those along [100] and [010], under the influence of an anisotropy energy with an assumed angular dependence.

REFERENCES


1. G.C. Fletcher and C.F. Osborne, J. Phys. F: Metal Phys., 3, L22-L25 (1973)
2. B.C. Munday and R. Street, J. Phys. F: Metal Phys., 1, 498-510 (1971)
3. M.O. Steinitz, L.H. Schwartz, J.A. Marcus, E. Fawcett and W.A. Reed, Phys. Rev. Lett., 23 979-982 (1969).
4. L.R. Testardi and J.H. Condon, Phys. Rev. B 1, 3928-3942 (1970)
5. R.A. Alpher and R.J. Rubin, J. Acoust. Soc. Am., 26, 452-453 (1954)
6. R.A. Montalvo and J.A. Marcus, Phys. Lett., 8, 151-153 (1964)
7. S.A. Werner, A. Arrott and M. Atoji, J. of Appl. Phys., 40, 1447-1449 (1969)
8. M.O. Steinitz, J.P. Kalejs, J.M. Perz and E. Fawcett, J. Phys. F: Metal Phys., 3, 617-622 (1973)
9. S.A. Werner, A. Arrott and H. Kendrick, Phys. Rev., 155, 528-539 (1967)


## ELECTRONIC STRUCTURE AND MAGNETIC PROPERTIES OF SCANDIUM*


Shashikala G. Das
Argonne National Laboratory, Argonne, Illinois 60439


ABSTRACT


The Van Vleck orbital and Pauli spin paramagnetic contributions to the magnetic susceptibility of single crystal and polycrystalline scandium are calculated using the APW method in conjunction with an LCAO interpolation scheme, which uses s, p and d type functions in the tight-binding representation. The warped muffin-tin APW potential was obtained from overlapping charge densities which were derived from the atomic configuration $3d^2 4s$; the exchange interaction was included in full Slater $\rho^{1/3}$ approximation. The Fermi energy is found to lie in a sharp peak in the density of states curve. The anisotropy in the magnetic susceptibility and the low temperature variation of the susceptibility of scandium, particularly the hump around 25°K observed by Spedding and Croat,[1] are successfully explained. The large enhancement of the low-temperature specific heat due to spin fluctuations calculated from the molecular field parameter in RPA approximation is shown to account for the absence of superconductivity in scandium.



*Work performed under the auspices of the U. S. Energy Research and Development Administration.


[1]F. H. Spedding and J. J. Croat, J. Chem. Phys. 58, 5514 (1973).

# MAGNETIC AND ELECTRIC PROPERTIES OF MnSb

Tu Chen, W. Stutius, and J. W. Allen
Xerox Palo Alto Research Center, Palo Alto, California 94304

G. R. Stewart
W. W. Hansen Laboratory of Physics
Stanford University, Stanford, California 94305

ABSTRACT

Transport, bulk and optical properties have been measured on $Mn_{1+x}Sb$ single crystals with x ranging between 0.013 and 0.152. Based on interpretation of the observed composition dependence of the physical properties, a simple metallic band model is proposed.

## I. INTRODUCTION

MnSb is a nickel arsenide structure compound. Similar to MnBi, this compound is a ferromagnet and has metallic conductivity.[1,2] However, unlike MnBi, this compound does not exhibit any phase transformation below its peritectic decomposition temperature and it has a large excess solubility of Mn beyond the stoichiometric composition. The excess Mn atoms in the compound occupy principally the bipyramidal interstitial sites, and they have significant effect on the magnetic properties of the compound.[3] In this work we have studied further the effects of the excess Mn on physical properties of the compound by nonstoichiometric MnSb (i.e. $Mn_{1+x}Sb$ with x = 0.013 to 0.152) single crystals. A model representing the electronic structure of MnSb as deduced from composition dependent physical properties will be presented.

## II. SAMPLE PREPARATION

Single crystals of different Mn concentration were grown by a pulling method from a molten mixture consisting of different Mn and Sb concentrations in a hydrogen and argon mixture atmosphere.[4] Crystal boules up to 1 cm in diameter and several cm long were grown. The crystals were cut and analyzed by a wet chemical method to determine the composition. It was determined that the concentration gradient along the length of the crystal boule is about 0.1 at. % Mn per cm or less. We have measured the Curie temperature of the chemically analyzed samples and established a curve representing the relationship between the Curie temperature and composition of the compound. This curve was used to determine the exact composition of the samples used in the various measurements. By the x-ray Laue method, the crystals were oriented and cut to a specific dimension for physical property measurements.

## III. TRANSPORT, BULK AND OPTICAL PROPERTIES

### A. Electrical Resistivity

The electrical resistivity of single crystals $Mn_{1+x}Sb$ has been measured between 4.2K and 680K.

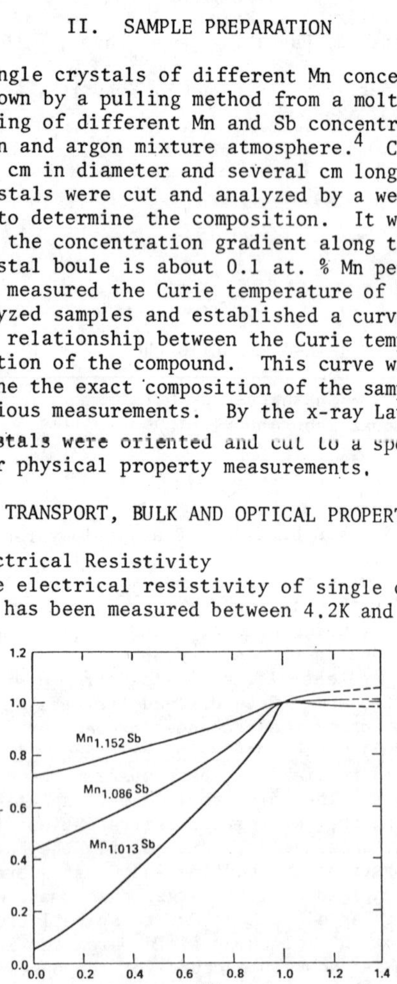


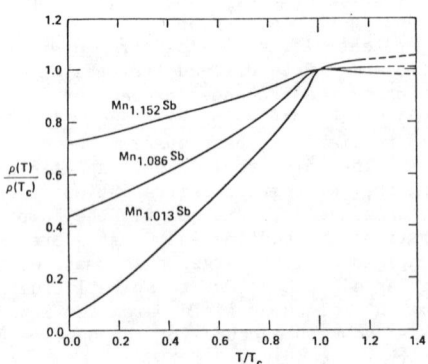

Fig. 1 Reduced resistivity of $Mn_{1+x}Sb$ vs. reduced temperatures

Fig. 1 shows the reduced resistivity versus reduced temperature for three different compositions of $Mn_{1+x}Sb$ crystals. The data pertaining to the graph are:

| Sample | $T_c$ in °K | ρ in Ω - cm at $T_c$ |
|---|---|---|
| $Mn_{1.013}Sb$ | 578 | $1.87 \times 10^{-4}$ |
| $Mn_{1.075}Sb$ | 516 | $2.52 \times 10^{-4}$ |
| $Mn_{1.152}Sb$ | 411 | $2.33 \times 10^{-4}$ |

The resistivity obtained in this work is an order of magnitude smaller than previously reported,[2] and depends strongly on composition and temperature. The resistivity curves are generally reproducible from sample to sample. We found that a maximum in the resistivity near the Curie temperature occurs only for samples with a large excess of Mn as shown in Fig. 1. For samples closer to the stoichiometric composition, the resistivity as a function of temperature is similar to that generally observed in a pure transition metal such as Ni or Fe.

For samples with a composition close to stoichiometric MnSb, the low temperature resistivity can be fitted by an expression $\rho=\rho_0+AT^2$ with $A=3x10^{-9}\Omega$-cm/deg$^2$. A $T^2$ - dependence of ρ is generally attributed to strong (s-p) to d electron scattering[5].

### B. Hall Effect and Thermoelectric Power

The Hall coefficient and thermoelectric power were measured on two crystals with different compositions, namely $Mn_{1.013}Sb$ and $Mn_{1.116}Sb$. Both the ordinary and the spontaneous Hall coefficient ($R_0$ and $R_s$) were found to be positive as is the case for MnBi. $R_0$ determined from the slope of Hall resistivity ($\rho_H$) versus magnetic field above 16 KOe was close to $10^{-10}m^3$/A-sec.

### C. Magnetic

The magnetic properties measured on four crystals of $Mn_{1+x}Sb$ with different compositions were essentially identical to those published by Okita and Makino.[3] This includes the temperature dependence of the magnetization, paramagnetic susceptibility and the magnetic anisotropy. In general, the magnetic moment decreases as the amount of excess Mn increases as shown in Fig. 2. We found, however, that the Curie temperatures do not depend linearly on composition as previously reported by Guillaud[1] and Okita and Makino.[3]

### D. Specific Heat

The heat capacity was measured on four samples with different compositions in the temperature range between 1.2K and 10K using a calorimeter described elsewhere.[6] The experimental curves were fitted by an expression of the form $C_p = A/T^2 + \gamma T + \beta T^3 + C_{mag}$, where the four terms represent the nuclear, electronic, lattice and magnon contributions, respectively. It was found that the coefficient γ increases almost linearly with x and values for γ determined for the four samples are shown in Fig. 2. The high values for γ point to a high density of states at the Fermi surface, and they are comparable to the γ of 18 mJ/mole-deg$^2$ obtained for Mn metal.

### E. Optical Properties

The room temperature optical conductivity and other optical constants were found from a Kramers - Kronig analysis of the reflectivity measured between .05 and 5 eV for four samples with x between 0.013 and 0.146. In contrast to the thermal, magnetic, and electrical properties, the optical results are essen-

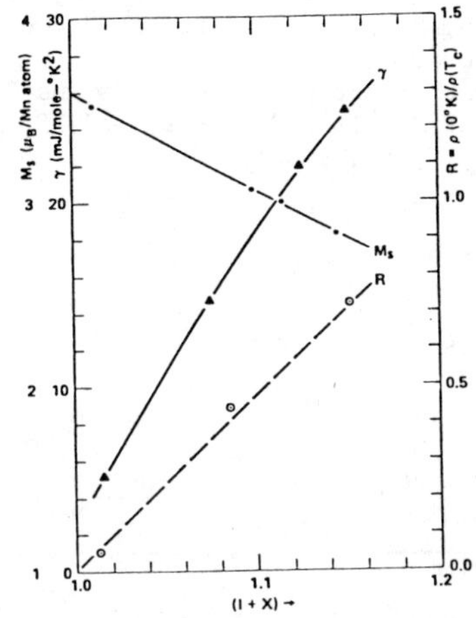


Fig. 2 Composition dependence of electronic specific heat coefficient γ, spontaneous magnetic moment $M_s$(T=0 K) and reduced residual resistivity $R=\rho(0)/\rho(T_c)$ of Mn.

tially independent of x. At low energies, the reflectivity falls smoothly with increasing energy from nearly 100% and a plasma edge is not observed. Thus the infrared optical conductivity does not display Drude behavior. There is a peak in the optical conductivity from 2 to 3.5 eV. As explained further below these optical properties are reminiscent of a transition metal and would be expected if narrow but itinerant 3d-bands overlap broad (s-p) bonding bands, with the Fermi level in the overlap region.

## IV. DISCUSSION

Although it is quite possible to account for the decrease in the magnetic moment in $Mn_{1+x}Sb$ with increasing excess Mn concentration on the basis of a ferrimagnetic arrangement of localized spins, the high values of the electronic specific heat coefficient γ and the strong composition dependence of γ suggest an explanation within a d-band model. Furthermore, the formation of a d-band over localized d states is favored by the small c/a ratio in MnSb (c/a ≃ 1.4).

In the following we will propose a straightforward and simple band model for the MnSb compounds which yields a coherent and consistent explanation of all the experimental data (Fig. 3).

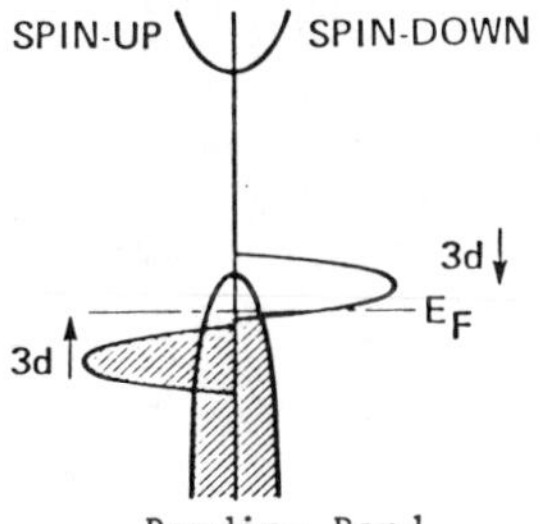


Fig. 3 Electron band model proposed for $Mn_{1+x}Sb$ schematic.

MnSb is a metal with hole conductivity and the Fermi level falls into the antimony 5(s-p)-derived bonding band. The manganese 3 d-bands are exchange split. The majority spin states are completely filled with 5 electrons and the Fermi level goes through the minority spin states which are filled with between 1.5 and 2.2 electrons. With the antibonding band empty and lying above the Fermi level, this leaves some holes in the (s-p) bonding band. Bouwma and Haas[7] have also suggested a model of this form to correlate the magnetization data and NMR results.

The consequences of our proposed band model become immediately transparent. If we increase the composition x in $Mn_{1+x}Sb$, we simply add more electrons and raise the Fermi level (we assume a rigid band model and also that the exchange splitting $\Delta_{ex}$ remains more or less unchanged). Since these electrons occupy the minority d-states, the magnetization decreases and the electronic specific heat coefficient γ, which is proportional to the density-of-states at the Fermi surface, increases. This is exactly what is observed experimentally.

The positive ordinary Hall constant is characteristic for hole conduction coming from the (s-p) bonding band. Also the positive sign of the spontaneous Hall constant is, according to a calculation by Kondorskiĭ[8], consistent with a scattering of conduction holes by minority-spin d electrons.

It is also easily seen that within our model the thermoelectric power Q will be negative despite the fact that holes carry the current. Q is proportional to $1/e \cdot [\partial \log \sigma(E)/\partial E]_{E=E_F}$. The major scattering process for the conduction holes is (s-p) to d scattering via the manganese 3 d-electrons and therefore the conductivity is proportional to $n_h/N_d(E_F)$[9], where $n_h$ is the number of holes and $N_d(E)$ the number of d states at the Fermi surface. An increase in the position of the Fermi level because of increase in x will leave $n_h$ near unchanged, but $N_d(E_F)$ increases significantly. Therefore $[\partial \log \sigma(E)/\partial E]_{E=E_F}$ is negative and a negative Q implies that e is positive (holes). We can also see that we are able to account for the rapid increase in the resistivity in $Mn_{1+x}Sb$ with increasing x solely by considering the increase in (s-p) to d scattering based on the same tendency of change in $n_h$ and $N_d(E)$ as a function of composition x described in the above. Of course, the increase in scattering by the excess Mn atoms distributed randomly in the bypyramidal interstitial sites will also lead to an increase in the resistivity.

The failure to observe a plasma edge is easily attributed to numerous low energy interband transitions, as would occur if the Fermi level passes through states where narrow 3 d-bands are strongly hybridized with (s-p) bonding bands. The resultant band structure should resemble that of a transition metal, where 3 d-bands overlap and hybridize with 4 s-bands, resulting in low energy interband transitions that dominate the infrared optical conductivity.[10] The lack of x-dependence of the optical properties is consistent with very little movement of the Fermi level because of the high density of states.

This simple band model allows us to give a coherent explanation of most phenomena observed experimentally in the MnSb compounds. However, a definite verification of this model requires a more direct determination of the electronic density-of-states, e.g., by photoemission experiments, which are planned in the near future.

ACKNOWLEDGEMENT

We would like to thank G. B. Charlan, P. L. Wolford and R. Allen for their assistance with the work and Y. Liang for his help in performing the infrared optical reflectance measurements, and R. Bauer for help with the Kramers - Kroning analysis. We also benefited from discussions with R. M. White.

## REFERENCES

1. C. Guillaud, Ann. Phys. 4, 671 (1949).
2. G. Fisher and W. B. Pearson, Can. J. Phys. 36, 1010 (1958).
3. T. Okita and Y. Makino, J. Phys. Soc. Japan 25, 120 (1968).
4. T. Chen, G. B. Charlan and R. C. Keezer, "Proc. 3rd Am. Conf. Cryst. Growth," 214, Stanford Univ. Ca., July 1975.

5. R. Bachmann et.al., Rev. Sci. Inst. 43, 205 (1972).
6. N. V. Volkenshtein, V. P. Dyakina and V. E. Startsev, Phys. Stat. Sol. 57, 9 (1973).
7. J. Bouwma and C. Haas, Phys. Stat. Sol. (b) 56, 299 (1973).
8. E. I. Kondorskiǐ, Sov. Phys. JETP 28, 291 (1969).
9. J. M. Ziman, "Electron and Phonon," 400 (Clarendon Press, Oxford, 1967).
10. H. Ehrenreich, "Optical Properties and Electronic Structure of Metals and Alloys," ed. F. Abeles, 109 (Wiley, N. Y., 1966).


# CRYSTAL ELECTRIC FIELD EFFECTS ON THE TRANSPORT PROPERTIES OF INTERMETALLIC RARE EARTH SYSTEMS†


Y. H. Wong and B. Lüthi
Department of Physics, Rutgers University, New Brunswick, N.J. 08903



ABSTRACT

We report on measurements at low temperatures (1.5 to 11K) of the electrical resistivity and thermal conductivity of pnictide compounds LnSb (Ln=La, Er, Tm) as well as the induced moment system $PrCu_5$. The anomalous temperature dependence of these transport properties provide strong evidence of an increased scattering of the carriers through exchange and aspherical Coulomb interaction with the localized 4f electrons.


Recently, there has been much interest in the effects of crystalline electric field (CEF) on various physical properties of metallic systems containing magnetic ions. For instance neutron scattering experiments, specific heat, susceptibility, EPR and elastic constant measurements show clearly the effects of CEF[1]. There have been also some reports on CEF effects on the transport properties (e.g. electrical resistivity[2,3], thermoelectric power[4]).

We present here thermal conductivity measurements of the pnictides LnSb (Ln=La, Er, Tm) and $PrCu_5$, which is a singlet ground state system. These measurements show the effect of CEF. All the samples used here are polycrystalline[5]. Of each compound we made at least two temperature runs with various temperature gradients which gave identical results. We first discuss the pnictides and then $PrCu_5$.

LnSb: In Fig 1 we show thermal conductivity measurements $\kappa/T$ versus T for LaSb, ErSb, TmSb. All these compounds are metallic as exhibited by electrical resistivity measurements, although the specific resistances $\rho$ at 2K are rather high. They are listed together with the resistance ratio in Table 1. LaSb shows a thermal conductivity typical of a metal[6], i.e.

$$\frac{\kappa}{T} = (A + BT^3)^{-1} \qquad (1)$$

where the phonon scattering term $BT^3$ should be small for low T. The result for LaSb in Fig. 1 shows for $\kappa/T$ a broad maximum around 3K. The temperature behaviour for T<3K is not explained by equation (1) but it may presumably be due to scattering at crystallite boundaries. The Lorenz number $L=\rho\kappa/T=1.91 \times 10^{-8}$ $\Omega$-W/K$^2$ at 2K (see Table I) is close to the theoretical value of $2.45 \times 10^{-8}$. It indicates that the thermal conductivity in this temperature range is governed by electrons. The decrease for T>4K could be due to the electron-phonon scattering $BT^3$ term. A numerical estimate of B gives a value of 0.01 cm-K/W which is an order of magnitude smaller than the experimental value[7]. We conclude therefore that the phonon channel in the thermal conductivity in this case is finite though small.

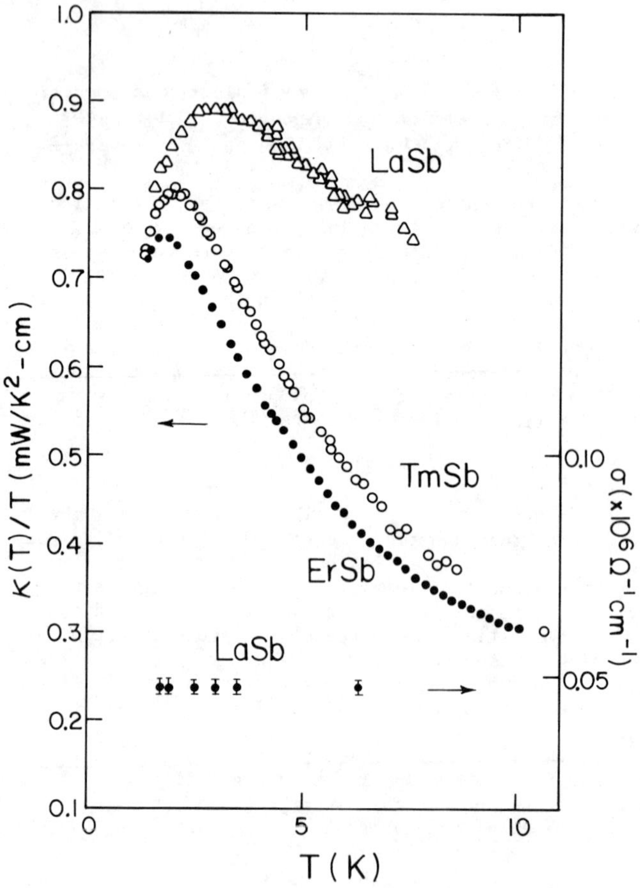


Figure 1

$\kappa/T$ versus T for the pnictides ($\Delta$ LaSb, o TmSb, • ErSb) and $\sigma(T)$ for LaSb. Notice the temperature independence of $\sigma(T)$ for LaSb.

Table I

| | $\rho(2K)^\dagger$ ($\mu\Omega$ cm) | $\rho(300K)/\rho(2K)$ | $L(2K)\ 10^{+8}$ [$(\Omega W)/K^2$] | $\theta_D$ (K) | $\Delta^*$ (K) |
|---|---|---|---|---|---|
| LaSb | 20 | 20 | 2.03 | 211 | - |
| TmSb | 32.6 | 11 | 2.52 | 237 | 27 |
| ErSb | 220 | 7.5 | | 237 | 34 |
| $PrCu_5$ | 1.3 | 50 | 3.00 | 262 | 28 |

† $\rho(T)$ for LaSb is temperature independent as shown in Fig. 1. For TmSb and ErSb, CEF effects and spin-disorder effect at $T_N$=3.5K (ErSb) are observed.

* Energy splitting of the CEF ground state and first excited state

TmSb and ErSb show a much stronger decrease for T>3K with subsequent flattening for T>8K. While the decrease in LaSb was about 10% of its maximum value, it is 53% and 50% for TmSb and ErSb respectively. We interpret this as due to an additional thermal resistance mechanism: scattering of the carriers by the magnetic ions. As discussed theoretically this scattering can take different forms, i.e. exchange and aspherical Coulomb[3,8,9],

$$H_{in} = \sum_{\vec{k},\vec{k}'} (J_{ex}\, \vec{s} \cdot \vec{J} + V_Q)\, a^{+}_{\vec{k}'} a_{\vec{k}} \qquad (2)$$

where the first term denotes the exchange interaction of a conduction electron of wavevector $\vec{k}$ with a localized moment of angular momentum $\vec{J}$ and the second term denotes the corresponding quadrupolar coupling. Electrical and thermal resistances due to these scattering mechanisms have been calculated previously for magnetic impurities[8,9]. In concentrated systems, as we deal here, the increase in electrical and thermal resistance arises from spin and quadrupole excitation at finite temperatures. Electrical resistivity calculations for such models have been performed before[9, 10].

$PrCu_5$: While the pnictides did not show ideal CEF effects because of their low electrical conductivity, $PrCu_5$ gives clearcut evidence of such effects as shown in Fig. 2. In this figure, we give $\sigma(T)$ and $\kappa(T)/T$ for the temperature range 1.5-11K. The values for $\rho(2K)$ and $\rho(300K)/\rho(2K)$ listed in Table I show that $PrCu_5$ is a good metal. Therefore any contribution of phonons to the thermal conductivity can be neglected in our temperature region. One notices in Fig. 2 a strong anomalous T dependence in both $\sigma(T)$ and $\kappa(T)/T$, indicating strong CEF effects. $PrCu_5$ does not order[11] magnetically down to 50 mK. The energy splitting of the two lowest states is listed in Table I. One notices a 60% effects for $\kappa/T$ and 40% for $\sigma(T)$ in the temperature range of 2-11K. Note that B from equation (1) in this case is $1.4 \times 10^{-5}$ cm-K/W which is much smaller than that of the pnictides. Therefore the anomalous behaviour of $\kappa(T)$ for T>4K is predominantly due to CEF effects. In the case of $\sigma(T)$, the Grüneisen contribution is likewise negligible in this temperature region, thus the temperature dependence can again be attributed to CEF effects.

It should be mentioned that the behaviour of $\kappa(T)/T$ for T<2K in all the samples measured, LnSb and $PrCu_5$, is not understood so far. It could be due to scattering at the crystallite boundaries as mentioned above. A quantitative account of the CEF effects on the thermal and electrical conductivity, taking into consideration the full CEF level scheme, has to await theoretical treatment. Measurements on other substances and a quantitative discussion will be reported elsewhere.

We would like to express our thanks to Dr. R. Filler and Dr. P. Lindenfeld for lending us their thermal conductivity apparatus and many advices. The substances used in this investigation were kindly provided by Dr. K. Andres, Dr. E. Bucher and Mr. L. D. Longinotti.

REFERENCES


† Research supported by NSF

1. See e.g. Conference Proceedings on "Crystalline Electric Field Effects in Metals and Alloys" June 1974, Universite de Montreal, R. A. B. Devine, editor.
2. B. Lüthi, M. E. Mullen, K. Andres, E. Bucher and J. P. Maita, Phys. Rev. B8, 2639 (1973)


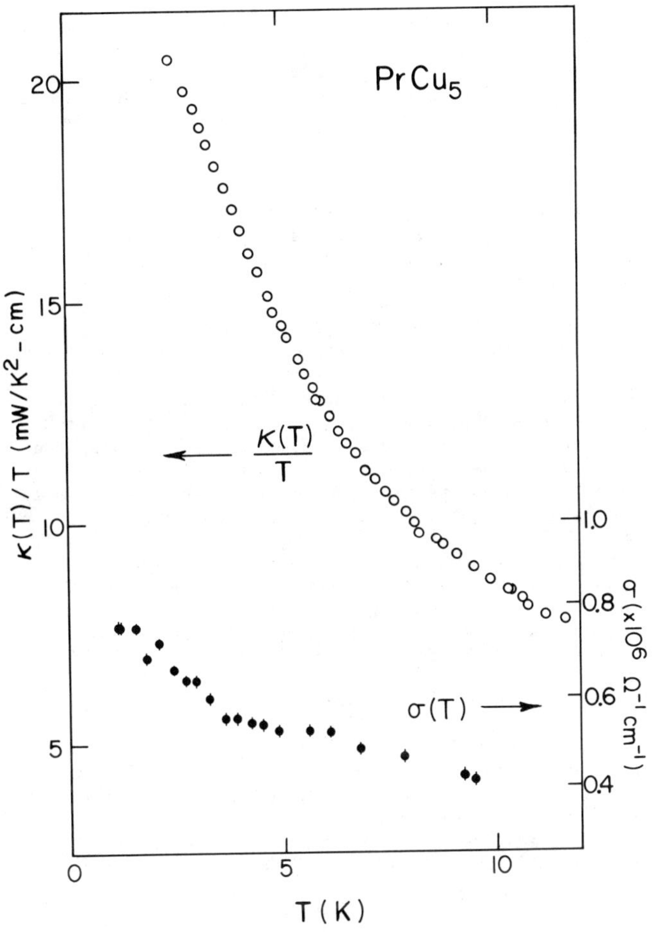


Figure 2

$\kappa/T$ and $\sigma$ as a function of temperature for $PrCu_5$.


3. A. Friederich and A. Fert, Phys. Rev. Letters 33, 1214 (1974)
4. J. Sierro, E. Bucher, L. D. Longinotti, H. Takayama and P. Fulde, Sol. St. Comm. 17, 79 (1975)
5. The pnictides samples are from the same batch as the ones used previously: M. E. Mullen et.al., Phys. Rev. B10, 186 (1974)
6. J. L. Olsen, "Electron Transport in Metals", Wiley-Interscience (1962)
7. For this estimate we took[6] $B = \frac{64\,(n/n_a)^{2/3}}{\kappa_\infty\, \theta_D^2}$ with $\theta_D$ from Table I, $\kappa_\infty$ using the Lorenz number and $\rho(300K)$ (Table I) and $n/n_a \sim 0.05$ (G. Güntherodt and P. Wachter, to be published)
8. P. Fulde and I. Peschel, Adv. in Phys. 21, 1 (1972) and references in it.
9. M. J. Sablik, H. H. Teitelbaum and P. M. Levy, AIP Conf. Proc. #10 p. 548 (1972)
10. R. J. Elliott, Phys. Rev. 94, 564 (1954)
11. K. Andres, E. Bucher, P. H. Schmidt, J. P. Maita and S. Darack, Phys. Rev. B11, 4364 (1975)

# MAGNETIC CHARACTERIZATION OF SEMI-AMORPHOUS NICKEL CATALYSTS AND THEIR METHANATION ACTIVITY

L.N. Mulay*, R.C. Everson*†, O.P. Mahajan, and P.L. Walker, Jr.
Material Sciences Department
The Pennsylvania State University, University Park, PA 16802

## ABSTRACT

Commercial catalysts (A) and (B) containing 43 and 67% wt. Ni supported on alumina have been successfully characterized by using magnetization ($\sigma$) versus H (to 20 k Oe) and T (77 - 600°K) type curves and by electron microscopy. Hysteresis curves yielded coercive force, remanence and $\sigma$(saturation) following heat treatment of the catalysts between 400-700°C. These parameters are interpreted in terms of the following properties of constituent particles: (a) superparamagnetic (b) single-domain-anisotropic and (c) multi-domain. Needlelike 'b' and 'c' type particles are formed when A is heated to about 600° and above 650°C respectively. The formation of 'c' type particles coincides with a decrease in the catalytic activity of A for the reaction $CO + 3H_2 \rightarrow CH_4 + H_2O$. B consisted of a larger fraction of 'a' type particles which on heat treatment increased $H_c$ and $\sigma$(sat) as expected, but was more thermally resistant to the formation of 'c' type particles. Unlike A, catalyst B did not show a significant decrease in catalytic activity upon the formation of 'c' type particles.

## INTRODUCTION

Magnetic properties of amorphous materials, including metal dispersions on oxides are fascinating indeed and have attracted considerable attention during recent years. One of us (L.N.M.) previously reported on the magnetic properties of dispersion of $\alpha$-$Fe_2O_3$ in zeolites[1] and of $Fe^{3+}$ ions in silicates[2,3] and glassy carbons.[4] Amongst dispersed magnetic systems the very fine dispersions of 3d metals such as Fe, Co, and Ni on various (diamagnetic) substrates (eg. $SiO_2$, $Al_2O_3$) are technologically most significant because of their widespread use as industrial catalysts in tonnage quantities. Significant fractions of such dispersions display superparamagnetic properties and are considered to be in part amorphous.

Correlations between the (superpara)magnetic properties of the 3d transition metal dispersions and chemisorption have been well recognized and widely investigated over the past three decades.[5-11] Most of the work has been on materials containing very low amounts (usually below 5%) of the metal which was dispersed on substrates to yield, so to speak ideal, "single phase" superparamagnetic systems.[12,13] While such studies have undoubtedly enhanced our understanding of the mechanism of chemisorption of electron donor and acceptor type molecules, very few attempts have been made to thoroughly characterize commercial catalysts containing large amounts of metal (up to 70%) and to correlate their magnetic properties with catalytic activity.

In this paper we present typical results on two commercial grade nickel on alumina catalysts containing 43 and 67% Ni with special reference to (i) the delineation of the properties of their magnetic components (superparamagnetic,[12,13] single-domain anisotropic[14] and multidomain ferromagnetic), (ii) the change in magnetic properties upon heat treatment to elevated temperatures and (iii) correlation of such changes with corresponding changes in catalytic activity for the reaction:
$CO + 3H_2 \rightarrow CH_4 + 2H_2O$.

## EXPERIMENTAL

Two nickel on $Al_2O_3$ catalysts A and B with the following physical characteristics were investigated. Nickel, wt.% (A) 43, (B) 67; BET surface area, $m^2/g$ (A) 51, (B) 117; average crystallite size, Å, as determined by x-ray line broadening, (A) 185, (B) 74.

Magnetization ($\sigma$) measurements were performed on a vibrating sample magnetometer (made by Princeton Applied Research Labs., Princeton, NJ) as a function of the field (up to 20 k Oe) and over a range of temperature (77° - 600°K) to yield especially values for saturation magnetization per gram ($\sigma$ sat) of nickel in the catalyst, the coercive force ($H_c$) and the remanence ($I_r$). Special quartz sample holders were designed to accommodate in vacuum the as received samples, and those after reaction and/or heat treatment as described below.

Heat treatment of the catalyst was conducted by (a) exposure to $H_2$ for 1 hour at selected heat treatment temperatures (HTT) in the range, 400 - 700°C followed by (b) exposure to the synthesis $H_2$/CO mixture for an additional 1 hour at HTT.

Catalytic activities were measured in terms of the methanation reaction which was carried out by passing a mixture of 17 mole percent of carbon monoxide (CO) in hydrogen ($H_2$) at atmospheric pressure through a vertical reactor packed with a known weight of catalyst pellets. These pellets were crushed to give fine powders (100 - 170 mesh) and low residence times, (0.003 - 0.019 secs) were used so that activities could be measured unaffected by diffusion effects.[15] Methane content in the product stream was monitored using a Hewlett Packard chromatograph with a column of 100 - 120 mesh fraction of Carbosieve B, obtained from Supelco Inc. The column was operated at 50°C. Methane production at 300°C, after reduction of the catalyst for 1 hour in $H_2$ and reaction for 1 hour, was used as the basis for comparison of activities of catalysts taken to different HTT.

## RESULTS AND DISCUSSION

Typical plots of the coercive force ($H_c$) measured at three temperatures ($T_m$ = 77°, 298°, 423°K) as a function of the HTT for catalysts A and B are shown in Fig. 1. Maxima in $H_c$ are observed around HTT's of 600°C and 650°C respectively for catalysts A and B. A comparison of these curves with Kneller and Luborsky's results[13] suggests that the maxima are probably due to two competing processes: first, conversion of 'a' superparamagnetic to 'b' single-domain anisotropic particles producing an increase in $H_c$ and then a conversion of such 'b' particles to 'c' type multidomain particles with zero $H_c$. These two conversions are believed to be characteristic of what occurs during the sintering of nickel supported on alumina. It should be noted that a partial conversion of 'a' to 'c', due to formation of closed multidomain configurations is not ruled out during heat treatment.

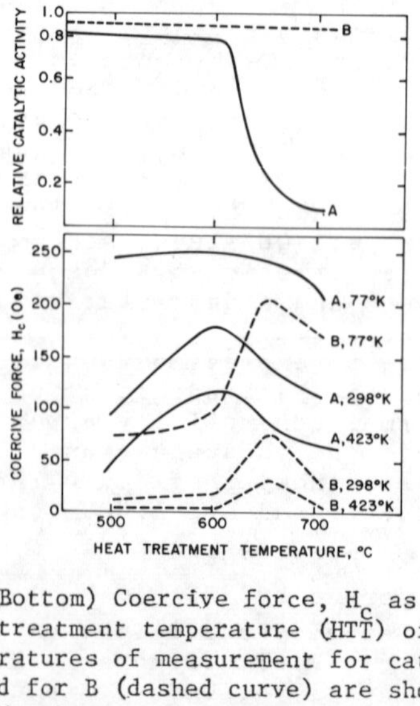


Fig. 1. (Bottom) Coercive force, $H_c$ as a function of the heat treatment temperature (HTT) of the sample. The temperatures of measurement for catalyst A (solid curve) and for B (dashed curve) are shown next to the curves. (TOP) Relative catalytic activity as a function of HTT.

Table I. Effects of Heat Treatment on Catalysts A and B.

| Heat Treatment Temperature, (HTT)°C | Wt. Fraction of Superparamagnetic Particles | | Mean Particle Size, Å (diam) of Superparamagnetic Fraction | |
|---|---|---|---|---|
| | A | B | A | B |
| As received | 0.36 | 0.58 | 25 | 24 |
| 500 | 0.27 | 0.48 | 29 | 25 |
| 600 | 0.21 | 0.40 | 54 | 28 |
| 700 | 0.12 | 0.26 | 37 | 26 |

Curves for catalyst A are indicative of the formation of large particles in the multidomain region corresponding to HTT's in the range 600 - 700°C. The relative catalytic activity for 'A' shown at the top of Figure 1 is seen to decrease sharply in this region. The temperature at which the activity of A commences to decrease sharply is seen to coincide with the HTT, at which the maxima in $H_c$ occurs. By contrast there is no such drop in the activity of B at its corresponding maxima in $H_c$. This aspect calls for further investigation. The important point to be noted is that catalyst B shows no loss in catalytic activity even up to a HTT of 700°C. We have estimated the fraction of superparamagnetic particles present in various heat treated samples and their mean diameters, using essentially the approaches and approximations described in the literature[10]. These are outlined in the appendix and the results are given in Table I. It is seen from Table I that for catalyst B, (despite its high concentration of nickel) the size of the superparamagnetic particles remains nearly constant and the fraction of such particles decreases less over the range of HTT's as compared to catalyst A.

Thus the magnetic technique provides a better characterization of the particles in terms of the critical sizes of superparamagnetic particles as well as their abundance than the gross estimates of crystallite sizes often reported in the catalysis literature. The small fraction (0.12) of superparamagnetic particles of A at HTT of 700°C indeed represents the least active catalyst. Since the chemisorptive properties of CO and $H_2$ have been shown to be dependent on the d-band characteristics of such particles[5,6] the magnetic technique may be said to provide additional characterization of the particles at the microscopic level.

The shape anisotropy of the particles is believed to be responsible for the increase in $H_c$ up to the transition temperature ($T_t$), and their disappearance is expected to cause the subsequent decrease in $H_c$; this conclusion is supported by evidence from electron microscopy, which clearly show the formation of needles or chain like clusters up to $T_t$ and their conversion to more regularly shaped particles, resembling those with multidomain properties, above $T_t$. The average crystallite size increased from 185 to 256Å for catalyst A, and from 74 to 160Å for B over the range of HTT.

Curves for the remanence ($I_r$) measured at the same temperatures as $H_c$ and as a function of HTT resemble those of $H_c$ and are not shown here. Their unique feature is that the peaks appeared sharper and at somewhat lower values of HTT corresponding to smaller particles than in the corresponding $H_c$ curves as one would expect from the theoretical treatment.[13] Thus, the parameters $H_c$ and $I_r$ are indeed useful in characterizing the presence of single-domain anisotropic particles, which are formed at least partially during the transition from the essentially superparamagnetic to the essentially multidomain regions. Furthermore, the HTT at which the single-domain anisotropic particles are formed appears to be characteristic of the "dispersive" state of the metal on the catalyst support.

Further details of studies on the chemisorption of $H_2$ and CO on the supported nickel and the activity of the catalysts will be published separately along with a discussion of other magnetic parameters.

## APPENDIX

The weight fractions $x_i$ of the superparamagnetic ("a" type) and hence of the multidomain ("c" type) particles in the various heat treated samples were estimated from the observed curves of the relative magnetization, $\sigma/\sigma_\infty$ versus H/T (at 573°C) by assuming Romanowski's relation[10] for a multi-component system given below.

$$\frac{\sigma}{\sigma_\infty} = \left(1 - \sum_i x_i\right) + \sum_i x_i L\left(\frac{I_{sp} v_i H}{kT}\right)$$

The expressions under the summation signs refer to superparamagnetic particles; L denotes the Langevin function and $v_i$ the mean volumes of the particles within the $x_i$ fraction. This expression is applicable at high fields such that saturation magnetization of multidomain particles show no change and superparamagnetic particles obey the well-known Langevin function. The average moment $\mu_i$ of the "a" type particle is defined by, $\mu_i = I_{sp} \cdot V_i$.

By assuming an overall mean volume $\bar{v}$ for all superparamagnetic particles, as in Selwood[5], a non-linear regression procedure was developed for estimating the weight fraction (x) of the "a" type particles, and the mean particle diameter of this fraction based on spherically shaped particles. The regression program used accomplished convergence very easily with a conventional hill climbing subroutine.

---

*Inquiries should be addressed to these authors.

†Present address: Chem. Engn. Dept., University of Natal, Durban, South Africa, 4001.

## REFERENCES

Only key references to selected reviews and papers are given.

1. L.N. Mulay and D.W. Collins, in "Amorphous Magnetism," M. Hooper and A.M. deGraff, Eds. Plenum Press, N.Y. (1973).
2. D.W. Collins and L.N. Mulay, J. Am. Ceram. Soc. 53, 74 (1970); 54, 52, 69 (1971).
3. D.W. Collins and L.N. Mulay (Proc. Intermag Conf 1968), IEEE Trans. Magnetics 4, 470 (1968).
4. L.N. Mulay, A. Thompson, P.L. Walker, Jr. in "Amorphous Magnetism" (See ref. 1).
5. P.W. Selwood, "Adsorption and Collective Paramagetism," Academic Press, N.Y. (1962).
6. J.H. Sinfelt, in "Annual Revs in Materials Science," Part 2, R.H. Bube and R.W. Roberts, Eds. Annual Reviews, Inc. Palo Alto, Calif. (1972).
7. T.E. Whyte, Jr., Catal. Revs. 8, 117 (1973).
8. J.I. McNab and R.B. Anderson, J. Catal. 29, 328 (1973).
9. J.L. Carter and J.H. Sinfelt, J. Catal. 10, 134 (1968); J. Phys. Chem. 70, 3003 (1966).
10. W. Romanowski, Zeits. anorg. allgem. Chem. 351, 180 (1967); V.W. Romanowski, H. Dryer and D. Nehring, ibid. 310, 286 (1961).
11. S.D. Robertson, S.C. Kloet, and W.M.H. Sachtler, J. Catal. 39, 234 (1975).
12. I.S. Jacobs and C.P. Bean in "Magnetism" Vol. III, Academic Press, N.Y. (1963).
13. E.F. Kneller and F.E. Luborsky, J. Appl. Phys., 34(3), 656 (1963).
14. C. Kittel and J.K. Galt in "Solid State Physics", F. Seitz and D. Turnbull, Vol. 3, 508 (1956).
15. R.A. Dalla-Betta, A.G. Piken and M. Shelef, J. Catal. 35, 54 (1974).

# TEMPERATURE STUDIES OF THE HYPERFINE MAGNETIC FIELD IN THIN IRON FILMS*

Robert J. Semper
St. Olaf College, Northfield, Minnesota 55057

C. L. Chien and J. C. Walker
The Johns Hopkins University, Baltimore, Maryland 21218

ABSTRACT

The hyperfine magnetic field of ultrathin single crystal iron films has been measured as a function of temperature using Mössbauer spectroscopy. The films were grown epitaxially by vacuum deposition of iron onto single crystal silver with a mica substrate. The samples showed a homogeneous magnetic structure for average thicknesses as small as 30 Å. The single crystal nature of the silver substrate grown epitaxially on synthetic mica has been verified by electron diffraction. The purity of the films has been tested by Auger electron analysis accompanying layer-by-layer argon sputtering of the film surface. The Mössbauer spectra of the resulting films show narrow lines, the relative intensities of which indicate that the magnetization tends to have a unique direction in the plane of the film.

The hyperfine structure measurements show further that the thin films have a smaller magnetic hyperfine field than that of bulk iron from about 100 K to 650 K. Below about 50 K the field appears to be slightly larger than that of bulk iron. The reduction of the thin film field at higher temperatures is greater for the thinner films. Work is in progress on high temperature measurements up to the Curie point in an effort to determine the magnetization curve completely.

## I. INTRODUCTION

To investigate the detailed nature of ferromagnetic ordering, experiments with thin single crystal and polycrystalline magnetic films have been undertaken.[1,2] One method of measuring the magnetism of these films involves the use of Mössbauer spectroscopy to sample the magnetic field at the nucleus.[3,4,5] Our work has involved Mössbauer measurements on ultrathin single crystal iron films.[6] The hyperfine magnetic field of the iron in these films has been studied as a function of both thickness and temperature.

## II. THIN FILM PREPARATION AND ANALYSIS

The thin films were grown by vacuum deposition of iron enriched to 90% $Fe^{57}$ onto substrates in an ultra-high ($2\times10^{-8}$ to $5\times10^{-9}$ torr) vacuum.[7] The base of the film was cleaved, iron-free (synthetic) mica. A layer of silver was then epitaxially grown on the mica to a thickness of about 600 Å. From previous epitaxial studies it is expected that the silver film is preferentially ordered with the (111) silver surface parallel to the (001) mica surface.[8] The iron was then epitaxially grown onto the silver where at least partial ordering of the (110) iron surface parallel to the (111) silver surface is expected.[9,10] Finally an overcoat of about 2000 Å of silver was added on top of the iron to prevent oxidation.

Electron diffraction studies of the silver substrate show a high percentage of epitaxial growth with the (111) plane parallel to the surface and indicate that the lower limit of silver platelet size is more than five microns.

The film thicknesses were measured using x-ray fluorescence techniques with the germanium K x-ray line (9.8 keV). Since this technique assumes uniform film coverage, the thickness measurements should be considered lower limits due to the possible presence of some island structure in the films. Therefore, errors as great as 25% may exist in the thickness measurements. Room temperature hyperfine magnetic field measurements were taken using scattering Mössbauer geometry (detecting the 6.3 keV x-ray) to increase the signal-to-noise ratio. Low and high temperature data were obtained with standard transmission geometry using special high count rate detectors.[11] For the transmission experiments the sample films were cut into pieces 1 cm$^2$ which were stacked together to increase the effective thickness of the absorber.

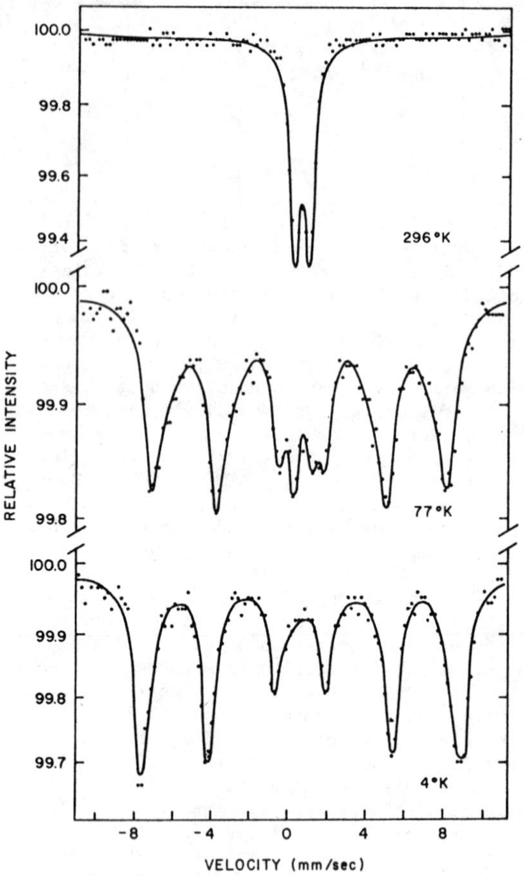


Fig. 1. Spectra of 30 Å Thick Film at 4.2°K. 77°K and 296°K Indicating Presence of Iron Oxide

## III. MÖSSBAUER SPECTRA OF THE THIN FILMS

The resulting room temperature spectra of a large number of film samples fell into two distinct categories. One group showed a hyperfine structure resembling that of bulk iron (the standard six-line spectra) while the other presented a doublet structure. There was only a very slight tendency toward a composite spectra, i.e., one showing both features.

Figure 1 shows the spectrum of a 30 Å thick film at 296°K, 77°K and 4.2°K. Note that while the spectrum is a doublet at room temperature, it is split out at low temperatures. This was true to a greater or lesser degree of all the doublet films we studied. For this film the hyperfine magnetic field strength was ~470 kG at 77°K and ~510 kG at 4.2°K. These low temperature spectra indicate the presence of iron oxide (either $Fe_3O_4$ or $Fe_2O_3$).

To further verify the origin of these doublets, the films were subjected to Auger electron spectroscopy with argon sputtering to allow a compositional depth profile. Through this techqnique the presence of oxygen in the iron layer was confirmed. Spectra taken periodically from 4.2°K to room temperature show

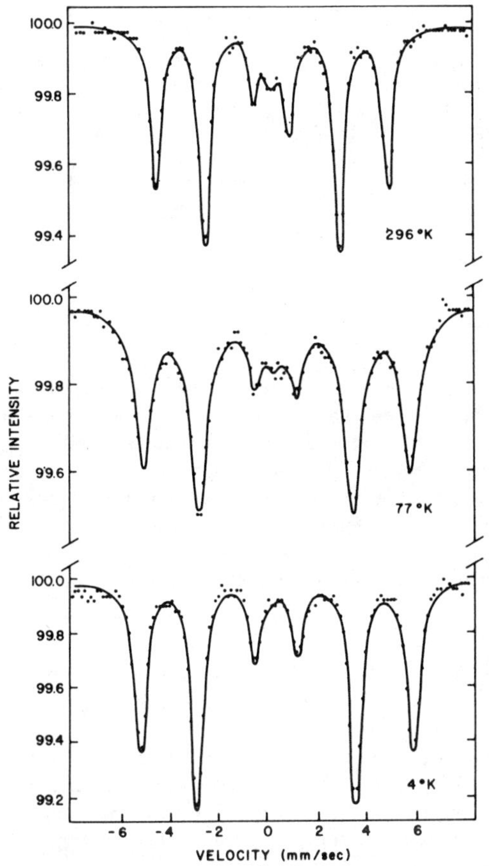


Fig. 2. Spectra of 33 Å Thick Film at 4.2°K, 77°K and 296°K Indicating Pure Iron (with a trace of superparamagnetic iron)

a progressive collapse of the hyperfine spectra similar to that obtained from superparamagnetic microparticles of iron oxide.[12]

It therefore seems that the doublet spectrum observed at room temperature is due to superparamagnetic relaxation of small clusters of iron oxide.[13] The Auger electron spectroscopy indicates that the oxidation is due to faulty silver cover layers. The thin films with a six-line pure iron spectrum showed both an adequate silver protection layer as well as no evidence of oxygen in the iron layer.

Returning to the thin films which exhibit a six-line spectra at room temperature, Figure 2 shows spectra for a 33 Å thick film at 296°K, 77°K and 4.2°K. At room temperature there is evidence of a very small singlet or doublet at the position of the inner two lines of the hyperfine spectrum. This anomaly disappears at low temperatures. Since there is no evidence of iron oxide being present at 4.2°K, this seems to be evidence of superparamagnetic pure iron present in very small quantities. Recent measurements indicate typical linewidths at 4.2°K of about .22 mm/sec for the films showing six lines.

Analysis of the relative intensity of the outer and middle lines of the spectra in both transmission and scattering indicates that the magnetization vector lies almost completely in the plane of the film. For the transmission case, with the outer line intensity normalized at 3, the middle line intensity measures about 3.8 (±5%) at room temperature, and 3.6 at 4.2°K.

In addition, analysis of the relative intensities for cases when a film was rotated about its normal in scattering geometry indicate that the magnetization vector is not randomly directed in the plane but has a preferred direction. This is indicative of epitaxial growth for both the iron and the silver, and is a further argument for the existence of single crystal iron films.

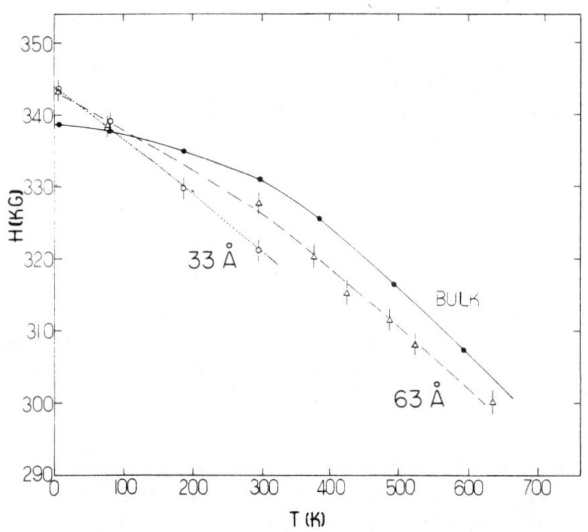


Fig. 3 Magnetic Hyperfine Field as a Function of Temperature for Bulk Fe, a 33 Å Film, and a 63 Å Film

## IV. TEMPERATURE AND THICKNESS DEPENDENCE OF THE HYPERFINE MAGNETIC FIELD

A general decrease in field strength as the thickness of the film is reduced is evident, for films less than 100 Å thick, with the reduction being about 4% for the 33 Å thick film. In Figure 3, the magnetic field of two films (33 Å and 63 Å thick) is plotted as a function of temperature. Both exhibit a magnetic field smaller than that of bulk iron at room temperature. However their fields tend to increase faster than that of bulk iron as the temperature is reduced and they actually have a larger field than bulk iron at 4.2°K by about 1.5%. This increase in the magnetic field at 4.2°K has been seen in other thin films less than 30 Å in thickness.

Above room temperature the magnetic field for the 63 Å thick film remains less than the bulk field. Due to difficulties in maintaining the silver cover layer at high temperatures, data were taken only up to 625°K. Further measurements in the high temperature region will be made after suitable experimental techniques are developed to maintain film integrity.

## V. CONCLUSIONS

The reduction of the hyperfine field at room temperature is less than that predicted by various spin wave theories. It appears to be more in line with molecular field theories, although the reason for the increase in the field magnitude above bulk field at low temperatures remains unclear.[1,2]

In general it has proved difficult to fabricate films with simple hyperfine spectra below 30 Å. Films thinner than this show doublet spectra mixed with six-line hyperfine spectra. We are presently developing techniques for measuring the magnetic field in thin films up to the Curie point as well as for the production of films of less than 30 Å thickness with good hyperfine spectra.

## REFERENCES

*Work supported by the National Science Foundation
1. U. Gradmann, Appl. Phys. 3, 161 (1974).
2. A. Corciovei, G. Costache, D. Vamanu, S. S. Phys. 27, 237 (1972) and references therein.
3. W. Zinn, Czech. J. Phys. B21, 391 (1971).
4. E. L. Lee, P. L. Balduc and C. E. Violet, Phys. Rev. Lett. 13, 800 (1964).
5. M. N. Varma and R. W. Hoffman, J. Appl. Phys. 42, 1727 (1971).

6. J. C. Walker, C. R. Guarnieri, R. J. Semper, AIP Conf. Proc. 10, 1539 (1972).
7. C. R. Guarnieri, Dissertation (The Johns Hopkins University, 1972) unpublished.
8. H. Jaeger, P. D. Mercer, and R. G. Sherwood, Surf. Sci. 6, 307 (1967).
9. H. C. Snyman and G. H. Olsen, J. Appl. Phys. 44, 887 (1973).
10. G. H. Olsen and H. C. Snyman, Acta Metallurgica 21, 769 (1973).
11. Robert J. Semper, C. Richard Guarnieri and J. C. Walker, Nuc. Instr. and Methods 128, 349 (1975).
12. T. K. McNab, R. A. Fox and A. J. F. Boyle, J. Appl. Phys. 39, 523 (1968).
13. A. M. Afanas'ev, I. P. Suzdalev, M. Ya. Gen, V. I. Gol'danskii, V. P. Kurneev, and E. A. Manykin, Zh. Eksp. Teor. Fiz. 58, 115 (1970).


# LOCAL SPIN FLUCTUATIONS IN CHEMISORPTION+


J. Handler and J. A. Hertz*
James Franck Institute, University of Chicago, Chicago, Illinois



## ABSTRACT

We introduce local spin fluctuations into the Newns-Anderson model of chemisorption. As in the bulk (wide band) magnetic impurity problem, this picture is correct in its essential features in both weak and strong correlation limits, provided that the correlation energy U is taken to be a phenomenologically determined parameter. Our focus here is on the differences between the chemisorption problem and the bulk impurity problem generated by the finite substrate bandwidth and the shape of its density of states. We show how a doubling of structure in the spectrum and new (Kondo) resonances near the Fermi level can arise, and discuss the microscopic origin of the parameters in this theory.


The purpose of this short paper is to show how a local spin fluctuation (LSF) picture like that used in describing magnetic impurities in bulk[1] can be used to describe correlations beyond the Hartree-Fock approximation in the theory of chemisorption. Such a description should be useful in understanding the spectroscopy of chemisorbed atoms, since in many situations of interest (e.g., H on transition metals) correlations are neither negligible nor dominant. Consequently neither the Anderson-Newns Hartree-Fock theory[2] nor the Gomer-Schrieffer induced bond picture[3] is applicable. In general, of course, an intermediate-coupling situation is calculationally intractable, but sometimes one can construct a phenomenological theory (in the spirit of Fermi liquid theory) which gives the correct structure of various response functions. In doing so, one gives up any hope of being able actually to calculate the parameters from perturbation theory. This is the sort of theory described below.

We start with the Anderson Hamiltonian:[2]

$$H = \sum_{k\sigma} \epsilon_k n_{k\sigma} + \sum_{\sigma} \epsilon_a n_{a\sigma} + U n_{a\uparrow} n_{a\downarrow} + \sum_k (V c_{k\sigma}^+ c_{a\sigma} + V^* c_{a\sigma}^+ c_{k\sigma}) \qquad (1)$$

Here k (not wavevector) labels the eigenstates in the substrate, and the $\epsilon_k$ are the eigenvalues. The subscript a denotes the adatom, U is the intra-atomic Coulomb energy on the adatom, and V is the adatom-substrate transfer matrix element. The strong and weak coupling limits are characterized by large and small values of U/V, respectively. We focus in this paper on calculating $G_a(E)$, the adatom Green's function, whose imaginary part gives the adatom density of states. (Although photoemission spectroscopy does not measure this function directly, structure in the adatom density of states is generally correlated with structure in the measured spectrum.)

In Hartree-Fock theory, the adatom self-energy consists of two pieces:

$$\Sigma = \Sigma_V + \Sigma_H = |V|^2 G_s(E) + U \langle n_a \rangle \qquad (2)$$

Here $G_s$ is the substrate surface electron Green's function and $\langle n_a \rangle$ is the mean adatom occupancy. $\Sigma_V$ takes into account (exactly) the hopping of electrons on or off the adatom, and $\Sigma_H$ is the Hartree energy shift due to the occupancy of the adatom. In this paper, we deal for simplicity with the so-called symmetric case, in which we assume that $\mathrm{Im}G_s(E)$ is an even function (relative to the Fermi level) and that the Hartree shift is such as to make the adatom density of states symmetric around $E_F$ as well. We take $E_F = 0$. The adatom energy level structure is ascertained by solving $G_a^{-1}(E) = 0$, that is

$$E = \Sigma_V(E) \qquad (3)$$

Bonding and antibonding levels below and above the substrate band limits can occur; Fig. 1a shows schematically the resulting adatom density of states.

If the adatom correlation is moderately strong, the local susceptibility there will be enhanced and the local spin fluctuations will relax very slowly. The electrons couple to these spin fluctuations, as represented by the self-energy [1,4]

$$\Sigma_{sf}(E) = U^2 \int \frac{d\omega}{2\pi i} G_a^{\,o}(E+\omega)\, \chi(\omega) \qquad (4)$$

Thus the imaginary part of $\Sigma_{sf}$ is a convolution of the spectral weight functions for $G_a^{\,o}$ and $\chi$:

$$\mathrm{Im}\,\Sigma_{sf} = -U^2 \int dE'\, N_a(E')\, \mathrm{Im}\,\chi(E-E') \,. \\ [\theta(E')\theta(E-E') - \theta(-E')\theta(E'-E)] \qquad (5)$$

$N_a$ is the adatom density of states, and $\theta$ is the step function. To the extent that the local enhancement is strong, $\mathrm{Im}\,\chi$ will be large only for quite small $\omega$. For purposes of this paper, we need only that its characteristic width $\omega_s$ be small compared to the energy scale on which $N_a(E)$ varies. If this is true, $|\mathrm{Im}\,\Sigma_{sf}|$ will be a slightly smeared and shifted image of the adatom density of states obtained in Hartree-Fock approximation. The only tricky point is that the step function (i.e., Pauli principle) constraints in (5) ensure that $|\mathrm{Im}\Sigma|$ vanish at E = 0 and that it be proportional to $E^2$ for small E. Figure 1b shows the resulting structure in the imaginary part of $\Sigma_{sf}(E)$ when the adatom density of states has the shape of Fig 1a.

To find the new resonance levels, we solve

$$G^{-1} = (G_a^{\,o})^{-1} - \Sigma_{sf} = E - \Sigma_V(E) - \Sigma_{sf}(E) \qquad (6)$$

by looking for places where the 45° line crosses the curve $\Sigma_V(E) + \Sigma_{sf}(E)$ (Fig. 1c). If $\mathrm{Im}\Sigma$ is small at such an E, we have a well-defined resonance. (No completely sharp levels survive, because $\mathrm{Im}\Sigma$ is nonzero everywhere except right at E = 0.) Each of the former bonding or antibonding levels can be split into two resonances, and the shape of $\Sigma_{sf}$ near $E_F$ can tend to pin a resonance near this energy. The level splitting can be thought of as a consequence of the fluctuating local field at the adatom. The Fermi level resonance is a Kondo effect. Its width is roughly $\omega_s$, so it is washed out for temperatures $\gtrsim \omega_s$.

The foregoing description shows the sort of new structure

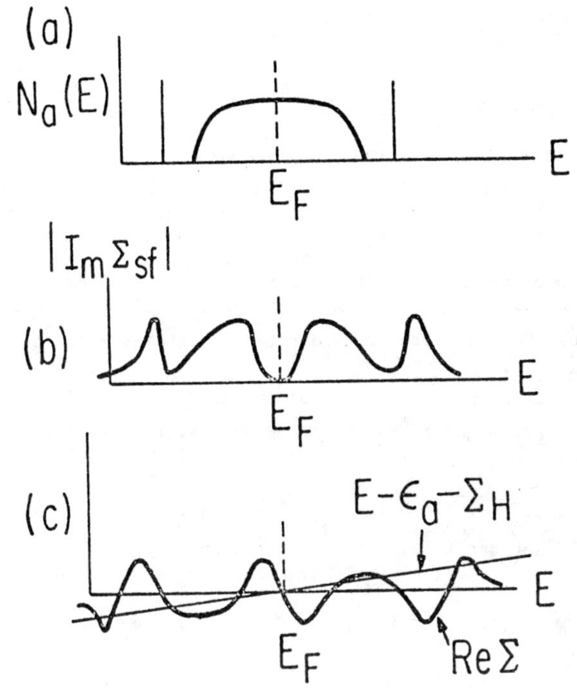


Fig. 1 (a) Hypothesized adatom density of states in Hartree-Fock approximation, for moderately strong bonding.
(b) Resulting structure of the imaginary part of the spin fluctuation self energy $\Sigma_{sf}$.
(c) The corresponding real part of $\Sigma$ and the solution of the Dyson equation (6).

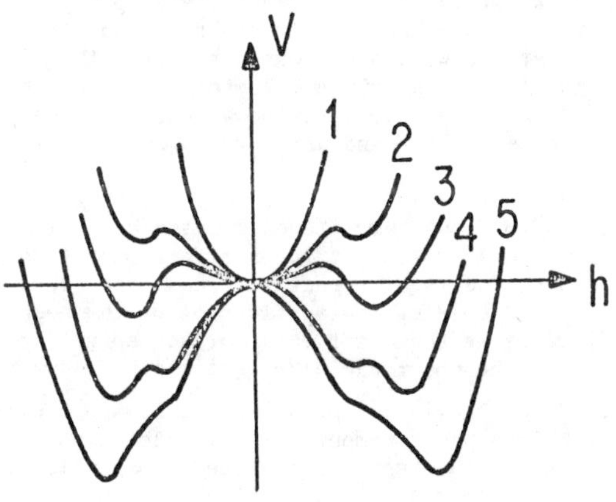


Fig. 2. Possible shapes of the free energy functional V(h) for different values of U. Curve I has the smallest U; curve 5, the largest.

which spin fluctuations can induce in the adatom spectral weight. The ingredients in the phenenomelogical model are just the surface Green's function $G_s(E)$ used to find $G_a(E)$ and $N_a(E)$, the coupling U, and whatever one needs to parametrize $\chi$, in particular $\omega_s$. In RPA the U of eqn. (4) is just the U of the Anderson Hamiltonian (1) and $\chi = \chi_0/(1 - U\chi_0)$ where $\chi_0$ is the susceptibility of the uncorrelated system. However, the cases where one wants to use this theory generally lie beyond the regime of validity of RPA, where vertex corrections which no one knows how to calculate reduce the coupling in (4) and keep $\chi$ from diverging. Accordingly, we cannot calculate the parameters of this theory from first principles.

Some insight into the parametrization of $\chi$ and the electron-spin-fluctuating coupling can be gained, however, through functional integral methods. We discuss these more fully elsewhere;[5] for this paper it is sufficient to examine the "effective potential" for spin fluctuations. It is obtained as follows. The standard tricks of the Stratonovich-Hubbard transformation express the partition function as a functional average:[6]

$$Z = Z_0 \int \delta x \exp\left[ -\frac{\pi}{\beta}\int_0^\beta dt\, x^2(t) - \int_0^\beta dt \int_0^1 d\lambda \sum_\sigma \sigma h(t) \langle n_a(t)\rangle_\lambda \right] \qquad (7)$$

where $\sigma h(t) = \sigma(\pi U/\beta)^{1/2} x(t)$ is a field acting on the adatom electron of spin $\sigma$, and the notation $\langle\ \rangle_\lambda$ means the expectation is to be taken in the presence of a field $\lambda h\sigma$. The effective potential is the argument of the exponential in static approximation, that is, if only time-independent x's are integrated over. The average $\langle n_a\rangle$ is simple to evaluate if the field is static; we find $Z_{st} \propto \int dh \exp[-\beta V(h)]$, where the potential V is a simple functional of $N_a(E)$:

$$V(h) = \frac{h^2}{U} - \sum_\sigma \int_{-\infty}^{h\sigma} d\epsilon \int_{-\infty}^{\epsilon} d\epsilon'\, N_a(\epsilon') \qquad (8)$$

That is, the second derivative of V is, except for a constant, just the adatom density of states, symmetrized around $E_F$.

Looking at the shape of V then tells us about the size of the spin fluctuations. For example, if V has two minima at $\pm h_0$, the fluctuating local Zeeman splitting of the bonding level should be about $2h_0$. This gives us a way of estimating the electron-paramagnon coupling U in (4)--it must be such as to produce a level splitting of $2h_0$. One can also estimate $\omega_s$ in a similar way. This requires going beyond the static approximation, so we leave this for a later publication.

The potential V(h) can have much more complex structure in the chemisorption problem than in the bulk magnetic impurity problem, since $N_a(E)$ is not a simple Lorentzian. An example of this new structure is shown in Fig. 2, for a case in which (as in Fig. 1) there are bonding and antibonding levels outside the band. Any local stationary point of V corresponds to a Hartree-Fock solution; there may clearly be many more such solutions here than in the bulk (wide band) problem. Of course, local spontaneous symmetry breaking is not permitted, so these solutions are unstable against fluctuations, but they do give us an idea of what values the local fluctuating field is most likely to take on.

In summary, then, the LSF picture provides a useful framework for interpreting structure in chemisorption spectra when the adatom correlation is intermediate to large. Furthermore, functional integral techniques give us some handle on the parameters of such a theory.

REFERENCES

+Supported by NSF and NSF-MRL.
*Alfred P. Sloan Foundation Fellow.

1. D. L. Mills, M. T. Beal-Monod and P. Lederer, Ch. 3 in Magnetism, Vol. 5 (G. Rado and H. Suhl, eds.) (Academic Press, 1973), pp. 89-117.
2. P. W. Anderson, Phys. Rev. 124, 41 (1961); D. M. Edwards and D. M. Newns, Phys. Letters 24A, 236 (1967); D. M. Newns, Phys. Rev. 178, 1123 (1969).
3. J. R. Schrieffer and R. Gomer, Surf. Sci. 25, 315 (1971); R. H. Paulson and J. R. Schrieffer, to be published.
4. S. Doniach and S. Engelsberg, Phys. Rev. Letters 17, 750 (1966).
5. J. A. Hertz and J. Handler, to be published.
6. S. Q. Wang, W. E. Evenson and J. R. Schrieffer, Phys. Rev. Letters 23, 92 (1969); J. R. Schrieffer, W. E. Evenson and S. Q. Wang, J. Phys. (Paris) 32, C1-19 (1971); D. R. Hamann and J. R. Schrieffer, Ch. 8 in Magnetism, Vol. 5 (G. Rado and H. Suhl, eds.) (Academic Press, 1973), pp. 237-252.

# SURFACE ACOUSTIC ATTENUATION DUE TO SURFACE SPIN WAVE IN FERRO- AND ANTIFERROMAGNETS

S. Maekawa
IBM Research Center, Yorktown Heights, N.Y. 10598
and M. Tachiki
The Research Institute for Iron, Steel and Other Metals, Tohoku University, Sendai 980 Japan

ABSTRACT

In spite of a number of theoretical predictions of interesting properties of surface magnons, observations have been severely restricted because of the smallness of the surface-to-volume ratio and the fact that only the magnetic resonance method has been available. In this paper, we propose theoretically a surface-acoustic-wave attenuation technique to observe a surface magnon with wave number equal to that of the acoustic wave. The surface acoustic wave gives a rotational motion to the lattice near the surface and induces a local deviation of the axis of magnetic anisotropy. Although shear strains responsible for the usual bulk magnon excitation vanish at the free surface, this rotational motion in anisotropic magnets leads to the same effect as that of the rotating magnetic field in the magnetic resonance method. Thus, the surface acoustic wave can excite a surface magnon. As an example, it is shown how a surface spin-flop instability proposed by Mills and Saslow may be easily observed in $MnF_2$ by using this method. The attenuation coefficient in anisotropic ferromagnets with surface anisotropy is also calculated.

We consider a magnetically anisotropic crystal with a flat surface. The z-axis is perpendicular to the surface and points into the crystal. The x-axis is parallel to the propagation vector $\vec{q}$ of a surface acoustic (SA) wave having frequency $\omega$. The displacement vector $\vec{u}$ accompanied by an SA wave is expressed by[1]

$$\left.\begin{aligned} u_x &= A\kappa_t[\exp(-\kappa_t z)-2q^2(q^2+\kappa_t^2)^{-1}\exp(-\kappa_\ell z)]\cos(qx-\omega t),\\ u_z &= -Aq[\exp(-\kappa_t z)-2\kappa_t\kappa_\ell(q^2+\kappa_t^2)^{-1}\exp(-\kappa_\ell z)]\sin(qx-\omega t),\\ u_y &= 0, \end{aligned}\right\} \quad (1)$$

with penetration components for longitudinal and transverse waves

$$\left.\begin{aligned} \kappa_\ell &= \sqrt{q^2-(\omega/v_\ell)^2}\ ,\\ \kappa_t &= \sqrt{q^2-(\omega/v_t)^2}\ , \end{aligned}\right\} \quad (2)$$

where $A$ is a constant proportional to the amplitude of SA wave, and $v_\ell$ and $v_t$ are velocities of bulk longitudinal and transverse waves, respectively. At the free surface all shear (symmetric) strains vanish, but the rotational (antisymmetric) deformation

$$\omega_{xz} = \frac{1}{2}(\partial u_x/\partial z - \partial u_z/\partial x)\ , \quad (3)$$

does not. Inserting (1) into (3), we obtain

$$\omega_{xz} = \frac{1}{2}Aq^2\xi^2\exp(-\kappa_t z)\cos(qx-\omega t), \quad (4)$$

where $\xi$ is given by a ratio of the velocity of SA wave $v$ to $v_t$ and has a value between 0.874 and 0.956.[1]

The spin-phonon interaction responsible for a one phonon-one magnon resonance process may be given in the usual form[2]

$$\varepsilon_{xz}(S_zS_x + S_xS_z), \quad (5)$$

where $\varepsilon_{xz}$ is the shear strain accompanied by SA wave. However, $\varepsilon_{xz}$ is almost zero in the penetration region of surface spin waves, because the penetration length of a surface spin (SS) wave is usually much shorter than that of an SA wave as will be shown later. Therefore, we include the interaction between rotational deformation (3) and spins, rather than the usual bulk magnetoelastic coupling (5).[2] Let the uniaxial crystal anisotropy energy associated with a spin $\vec{S}_i$ at the i-th site be expressed as $-DS_{iz}^2$, where $z$ indicates the z-axis fixed to the crystal. When the crystal lattice around the site-i is rotated by $\omega_{xz}$, the anisotropy energy is rewritten using spin components in the coordinate system fixed in space as

$$-D[S_{iz}^2 + \omega_{xz}(S_{iz}S_{ix} + S_{ix}S_{iz})]. \quad (6)$$

If we insert (3) into the second term of (6) we obtain the interaction Hamiltonian between SA wave and spins as

$$\mathcal{H}' = (1/2)q^2\xi^2D\sum_i\exp(-\kappa_t\ell_{iz})(S_{iz}S_{ix}+S_{ix}S_{iz})\cos(q\ell_{ix}-\omega t), \quad (7)$$

where $\ell_{iz}$ and $\ell_{ix}$ are the z and x components of the coordinate of site-i. This Hamiltonian may be considered equivalent to that of magnetic resonance with spacially varying magnetic field if one of the spin operators is replaced with the spin value. Melcher[3] has observed the rotational component of the shearing deformation accompanied by bulk transverse phonons. Because the rotational deformation of the SA wave is the main cause of SS wave excitations and because the coupling constant between the SA wave and spins must be the magnetic anisotropy constant, we can estimate the strength of the interaction and the absolute value of the attenuation. In the following we calculate the attenuation coefficient for typical ferro- and antiferromagnets.

Mills and Saslow[4] have proposed a surface spin-flop instability in a two-sublattice antiferromagnet having a body-centered crystal structure and a free (100) surface. As an example of the application of the mechanism derived above, we show that this surface spin-flop instability may be observed in $MnF_2$ utilizing SA waves. Manganese fluoride is a typical antiferromagnet with two sublattices whose moments are directed along the c-axis of its rutile crystal structure. A magnetic field $H$ is applied along the c-axis which is taken for the z-direction. When we take the c-plane for the free surface with the spin moment parallel to the z-direction, the SS-wave frequency $\omega_s$ and the inverse penetration length $p_a$ at zero wave number are expressed as[4]

$$\hbar\omega_s = \sqrt{\omega_A(\omega_E+\omega_A)} - \omega_H, \quad (8)$$

$$p_a = 2a^{-1}\sqrt{\omega_A/\omega_E} \quad \text{for} \quad \omega_A \ll \omega_E\ , \quad (9)$$

where $a$ is the distance between nearest neighbor $Mn^{2+}$ ions in the same sublattice and $\omega_H \equiv g\mu_B H$. $\omega_E$ and $\omega_A$ are defined as $\omega_E \equiv 8JS$ and $\omega_A \equiv D(2S-1)$ where $J$ represents inter-sublattice exchange. The surface spin-flop transition occurs when $\omega_s = 0$. Adopting $\omega_E = 1.04 \times 10^{-14}$ ergs/$Mn^{2+}$ ion, $\omega_A = 1.45 \times 10^{-16}$ ergs/$Mn^{2+}$ ion and $S = 5/2$,[5] we expect the surface spin-flop will occur at the critical field $H = 65.8$ kOe whereas the bulk spin-flop is known to occur at 93 kOe.[5] We also note that the value $p_a^{-1} = 14$Å obtained using (9) is very small compared with the penetration length of SA waves (∿ 10,000Å) obtained from (2).[6]

The attenuation of an SA wave, described by the coefficient $\alpha_q$, caused by SS wave excitation is obtained from (7) using the following procedure:
(i) Calculate the amplitude of the SS wave with wave number $q$ at each lattice site, where we may assume $q = 0$ since $q$ is very small.
(ii) Calculate the energy flow $P$ from the SA wave to the spin system per unit time and unit surface area using (7).
(iii) Obtain the coefficient $\alpha_q$ from the definition[1]

$$\alpha_q \equiv P/\nu E \quad , \qquad (10)$$

where $E$ is the mean energy of SA wave per unit surface area and $\nu$ is the velocity. Following this procedure[6], $\alpha_q$ is obtained as

$$\alpha_q = \frac{S\omega_A\omega_E\xi^4 q^2}{16\nu^2\rho a^2 G\hbar}\,\tau \quad \text{at} \quad \omega = \omega_s \, , \qquad (11)$$

where $\tau$ is the life time of the SA wave, $\rho$ is the density of the crystal and [6]

$$G = \frac{\kappa_t^2 + q^2}{2\kappa_t + q} + \frac{2q\kappa_t^2(q^2 + \kappa_t^2)^2}{\kappa\ell(q^2 + \kappa_t^2)^2} - \frac{4q\kappa_t}{q^2 + \kappa_t^2} \, . \qquad (12)$$

Utilizing elastic constants of $MnF_2$ at 4°K and the antiferromagnetic resonance line width observed by Shapira and Zak,[5] we estimate $\alpha_q$ in the SA wave at 1GHz as

$\alpha_q = 0.91$ dB/cm for $\tau = 3.6 \times 10^{-9}$ sec,

$\alpha_q = 9.1$ dB/cm for $\tau = 3.6 \times 10^{-8}$ sec.

These $\alpha_q$-values show that SA wave attenuation will be observable near the surface spin-flop transition point in $MnF_2$.

As an example of the application to ferromagnets, consider a simple cubic ferromagnet with nearest neighbor exchange interaction and a (001) plane as the free surface. In addition to the uniaxial anisotroy energy $-DSi_z^2$, the surface anisotropy energy $+ES_{iz}^2$, where $E > 0$, is acting on spins at the free surface. This model gives SS wave frequency $\omega_s$ and the increase penetration length $p_a$ at zero wave number as

$$\hbar\omega_s = \omega_H + \omega_A - e^2/(\omega_K + e), \qquad (13)$$

$$p_a = -a^{-1}\log(\omega_K/(\omega_K + e)), \qquad (14)$$

where $\omega_K \equiv 2JS$ and $e \equiv E(2S-1)$ with the nearest neighbor exchange constant $J$ and the lattice constant $a$. Neglecting magnetostatic effects for simplicity, and following the procedure mentioned above, we obtain the expression for $\alpha_q$ as[6]

$$\alpha_q = \frac{S\omega_A^2\xi^4 q^2}{8\rho\nu^2 G\hbar a^2}\,[1-\exp(-(\kappa_t+p_a)a)][1-(\frac{\omega_K}{\omega_K+e})^2]\,\tau \qquad (15)$$

at $\omega = \omega s$.

The derivations and the implications of (11) and (15) will be published elsewhere.[6]

One of the authors (S.M.) would like to acknowledge J. C. Slonczewski and B. E. Argyle for their helpful discussions.

REFERENCES

1. L. D. Landau and E. M. Lifshitz: "Theory of Elasticity" (translated by J. E. Sykes and W. H. Reid, Pergamon N.Y. 1970).
2. See, for example, E. E. Callen: J. Appl. Phys. 39, 519 (1968).
3. R. L. Melcher: Phys. Rev. Letters 25, 1201 (1970).
4. D. L. Mills and W. M. Saslow; Phys. Rev. 171, 488 (1968).
5. Y. Shapira and J. Zak: Phys. Rev. 170, 503 (1968).
6. S. Maekawa and M. Tachiki: in preparation.

## MAGNETIC BEHAVIOR OF ORGANIC COATED FERRITES

A. E. Berkowitz, J. A. Lahut, Lionel M. Levinson, I. S. Jacobs
General Electric Corporate Research & Development
Schenectady, New York 12301
D. W. Forester
Naval Research Lab.
Washington, D.C. 20375

ABSTRACT

$NiFe_2O_4$ micropowders (∿100Å dia.) coated with organic molecules such as oleic acid exhibit some remarkable properties[1]. (1) A large fraction of the spins are pinned in extremely high anisotropy fields; at 4.2K, <75% of the magnetization of bulk $NiFe_2O_4$ was achieved in fields of 200 kOe. Mössbauer data taken in high fields confirmed the spin pinning; otherwise the spectra were closely similar to those of ordered $NiFe_2O_4$. (2) When the coated particles were cooled below ∿80K in a field, large hysteresis loop shifts were present and the coercive forces were anomalously high for $NiFe_2O_4$. (3) As the organic coating was progressively removed, the degree of spin pinning decreased and the moment increased towards bulk values.

Coated $CoFe_2O_4$ micropowders also exhibited substantial spin pinning as evidenced by magnetization and Mössbauer measurements. The pinning disappeared when the coating was removed. The existence of shifted loops could not be determined due to the very high intrinsic magnetocrystalline anisotropy of $CoFe_2O_4$. Coated $Fe_3O_4$ micropowders did not show any significant amount of spin pinning nor any hysteresis loop shifts.

It was possible to eliminate two models that might explain this behavior. The Mössbauer spectra contained no evidence for any paramagnetism; all Fe spins were ordered. Therefore, it is unlikely that a magnetically "dead" layer exists in these coated micropowders. Another possibility is that features of the surface morphology, or defects introduced in the particles during preparation, might be responsible for the observed behavior. This hypothesis was tested by preparing ferrite micropowders by the same technique used to produce the coated particles, but without the organic coating. The average size of these uncoated particles was the same as that of the coated ones. Measurements on these uncoated ferrite micropowders did not indicate any significant amount of spin pinning or anomalous hysteresis behavior. Therefore, it seems reasonable to infer that the unusual properties of the coated $NiFe_2O_4$ and $CoFe_2O_4$ micropowders are due to an interaction between the organic molecules and the magnetic cations to which they are bonded. Presumably, a crystal field perturbation is involved. Evidently, the divalent Fe ion is not as sensitive to this interaction as are the Ni and Co ions in the other ferrites.

REFERENCES

1. A. E. Berkowitz, J. A. Lahut, I. S. Jacobs, Lionel M. Levinson and D. W. Forester, Phys. Rev. Lett. 34, 594 (1975).

*SPIN RELAXATION OF CONDUCTION ELECTRONS DUE TO SURFACE PARAMAGNETIC CENTERS.*[*]

*Adán R. Rodríguez and J.S. Helman*
*Centro de Investigación del I.P.N., México 14, D.F., México*

*ABSTRACT*

*We calculated the probability of spin flip per surface collision of a conduction electron due to the exchange interaction with paramagnetic centers outside the metal. This probability* $\varepsilon$*, is approximately given by*

$$\varepsilon = \frac{N k_F^4 \Omega^2 J^2}{36\pi E_F^{\,2}} \exp\{-4k_F\delta \left[\frac{m^*}{m}\,\frac{\Phi}{E_F}\right]^{1/2}\}$$

*where N is the surface density of paramagnetic centers and* $\delta$ *is their distance from the metal surface. m is the electron mass within the metal and m* is the electron effective mass in the medium containing the paramagnetic centers.* $\Omega$ *is the volume of the unit cell of this medium, J is the exchange coupling constant,* $\Phi$ *is the work function at the interface and* $E_F=\hbar^2 k_F^{\,2}/2m$ *is the Fermi energy.*

*The values of* $\varepsilon$ *calculated by using the above equation are consistent with those obtained on light metals from corresponding CESR*[1-3] *and Knight shift*[4] *measurements.*

*REFERENCES*

[*] *Work partially supported by ANUIES and CONACYT (Mexico) and Research Corporation (USA).*

1. *S.K.Wang and R.T.Schumacher, Phys. Rev. B8, 4119 (1973).*
2. *C.Taupin, J. Phys. Chem. Solids 28, 41 (1967).*
3. *M.A.Smithard, Solid State Comm. 14, 411 (1974).*
4. *R.H.Hammond and G.M.Kelly, Rev. Mod. Phys. 36, 185 (1964).*

ANOMALOUS SUPERPARAMAGNETISM

Itamar Eisenstein and Amikam Aharoni
Department of Electronics, The Weizmann Institute of Science, Rehovoth, Israel

ABSTRACT

The relaxation time for the magnetization reversal by thermal agitation, usually increases with increasing particle size, or with decreasing temperature. However, in certain cases this is not so, and the relaxation time is found experimentally[1,2] to decrease with increasing particle size. The explanation[1,3] of this anomaly is in terms of a switch-over to the magnetization curling mode. Under the rough approximations used so far for this mode, the anomaly turned out to be possible for cubic materials only. Using a more realistic Ritz model[4] for the curling, which emphasizes its non-linear nature, we find a reasonably good agreement with the experiment[2] on Fe, in that the general shape of the experimental curve is reproduced, even though the theoretical curve is shifted towards larger particle sizes. We also find a similar anomaly in uniaxial materials, which should be observed if the anistropy is low enough. In particular, we predict such an anomaly around room temperature, for particles of $BaCo_xTi_xFe_{12-2x}O_{19}$ with a radius of about 200 Å, and with x of about 1.3.

1. A.M. Afanas'ev et al., Sov. Phys. JETP, 31, 65 (1970).
2. A.P. Amulyavichus et al., Sov. Phys. JETP, 37, 859 (1973).
3. Afanas'ev et al., Sov. Phys. - Solid State, 14, 2175 (1973).
4. I. Eisenstein and A. Aharoni, J. Appl. Phys. (in press).

# THE COERCIVE FORCE AND THE THEORY OF FERROMAGNETIC DOMAIN WALL PINNING*

D. I. Paul
Columbia University, New York, New York, 10027

## ABSTRACT

We discuss the theory of domain wall pinning in bulk ferromagnetic materials as applied to the coercive force. A new theory which includes the wave nature of the ferromagnetic domain walls is explained. Specifically, an explicit algebraic expression for the magnetic field required to move a 180° ferromagnetic domain wall across a planar defect such as a grain boundary is obtained. The results show that the coercive force is proportional, amongst other factors, to the ratio of the barrier width to the domain wall width and to the fractional changes in the exchange and anisotropy constants upon entering the defect region. Using this expression, one can obtain the correct order of magnitude for many of the low and high coercive force materials whose magnetic properties are subject to the mechanism of domain wall motion. Applications to practical physical problems are suggested.

## INTRODUCTION

The coercive force, $H_c$, is the magnetic field required to produce zero magnetization in an initially saturated ferromagnetic specimen. Its value varies widely from material to material and is of great practical importance as well as theoretical interest. The energy dissipated in going around the hysteresis loop is of the order of $B_sH_c$ where $B_s$ is the saturation value of the magnetic induction. Hence, the coercive force determines to a large extent the hysteresis losses. In a good permanent magnet it may be of the order of $10^4$ Oersteds (CoPt or $Co_5Sm$) whereas very soft transformers (supermalloy) can go as low as 0.002 Oersteds for this parameter. In this paper, we shall limit our discussion to the effect of domain walls on this force. Further, except as otherwise stated, we shall consider bulk materials.

## DOMAIN WALL DISPLACEMENT

Given two magnetic regions or domains magnetized in opposite directions in an ideal crystal, Landau and Lifshitz[1] showed that the energy of the transition layer or ferromagnetic domain wall is not a function of position. Thus, wall displacement becomes an efficient mechanism for magnetic moment reversal, and the limiting obstacles to such motion are the crystalline defects. The subsequent discussion is predicated on the assumption that domain walls are present. The energetics of the formation of such walls represents yet another problem and the reader is referred to the work of Aharoni[2] and of Brown[3] for discussion of this topic.

Thus, one wants to determine the threshold magnetic field $H_c$ necessary for translational motion of the ferromagnetic transition layer. The coercive force, if not dominated by nucleation effects, is a measure of the field strength needed to carry a wall from one potential well to another - past the intervening energy barriers. Traditionally, it is determined by simply postulating the existence of a magnetic moment distribution for a given domain wall configuration and equating the change in the magnetostatic field energy to the maximum change in the configuration energy with respect to position. If the wall energy per unit area must increase by $\Delta\gamma$ for the wall to move a distance $\Delta z$, the procedure is to equate this energy to the gain in magnetostatic energy by reversal of the magnetic moment, $2HM\Delta z$. The coercive force is then given as

$$H_c = (d\gamma/dz)_{max.}/2M. \tag{1}$$

We recognize that such a simple approach is not equivalent to the solution of the differential equations governing the dynamics of the problem. Thus, it neglects the wave nature of the magnetic moment distribution which defines a domain wall. Such concepts as tunneling of the domain wall through a barrier, which would naturally appear in a solution to the differential equations, are just not taken into account. In a subsequent section, we shall present a solution to these differential equations and thus obtain results which do contain these factors for domain wall pinning in materials. In the next section, however, we briefly present some of the early formalisms, the results for which are governed by the above procedure. These formalisms were initially applied to the soft magnetic materials where they gave reasonable qualitative results. Essentially, we show that they cannot be extended to the high coercive force materials. Thus, we make no attempt to review the extensive theoretical developments of the past fifteen years which are based on the above and which are concerned with the quantitative explanations of experiments due to dislocation structure, internal stresses, plastic deformation, and temperature dependence of the soft magnetic materials. These topics are summarized in review articles by Trauble[4] and by Kronmuller[5].

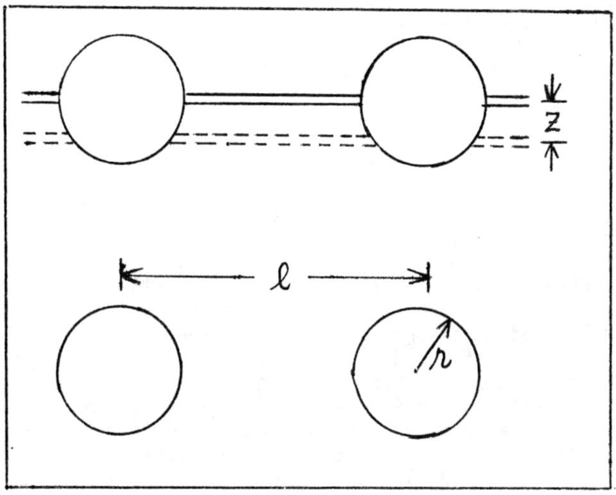


Fig. 1. Diagram of periodic inclusions as proposed by Kersten showing change in domain wall area for change z in location of wall position.

## SOFT MAGNETIC MATERIALS

Initial mechanisms considered for the coercive force of soft magnetic materials included

a) The reduction in wall area and thus wall energy thru the intersection of domain walls by inclusions proposed by Kersten[6]. The physical idea is illustrated in Fig. 1 where the assumption is made that the impurities are arranged in a periodic three dimensional array and have radii large compared to the wall width. At $z = 0$, the area of the wall prior to inclusions, $\ell^2$, is reduced to $\ell^2 - \pi r^2$. As the domain wall moves to the position given by the dashed line in Fig. 1, the change in area is given by $\ell - \pi(r^2 - z^2)$. Thus, the fractional reduction in wall energy as a function of z is

$$\sigma/\sigma_0 = [\ell^2 - \pi(r^2 - z^2)]/\ell^2 \tag{2}$$

This formula is inserted into Eq. (7) to yield

$$H_c = K\delta\alpha^{2/3}/2Mr \tag{3}$$

where $\alpha$, the packing fraction, is equal to $4\pi r^3/3\ell^3$ and $\delta$, the parameter governing the half-width of the domain wall is defined as $(A/K)^{1/2}$.

b) The reduction in wall energy due to periodic variations in the stress and thus in the magnetoelastic energy proposed by Kersten[6] and by Kondorsky[7]. The approach is similar to that given in a) above and the resultant formulae are

$$H_c = \pi\lambda S\delta/Md \ , \quad d >> \delta \qquad (4a)$$

and

$$H_c = \pi\lambda Sd/M\delta \ , \quad \delta >> d \qquad (4b)$$

where d is the periodic stress length, S is the magnitude of the stress variation, and $\lambda$ is the magnetoelastic energy constant.

Equations (3) and (4) have maximum values for $\delta = 2r$ and $\delta = d$ respectively and would at first sight appear to give reasonable values for the coercive force as a function of the density of defects. Neel[8] has pointed out, however, that both the periodicity and wall rigidity assumptions are not realistic. Random impurity positions and strain locations imply that one wall position is of approximately the same energy as another - thus reducing the maximum value obtainable for the coercive force from these mechanisms. The relaxation of the rigidity condition allows the domain walls to seek minimum positions but even with this allowance, maximum values of the coercive force are small. The formulas obtained by Neel are

$$H_c = \frac{\pi^2}{240}\frac{N}{L}\frac{\overline{(\gamma-\gamma_m)^2}}{\gamma_m}\left(\frac{\ln 4N}{3}\right)^{1/2} \qquad (5)$$

where L is of the order of the width of a domain, $\gamma_m$ is the wall energy and is equal to

$$2[A(K + \frac{3}{2}\lambda\sigma)]^{1/2}$$

for strains and

$$2[AK]^{1/2}$$

for inclusions, and $N = \pi L/2\eta$ where $\eta$ is the scale of the perturbation. Further,

$$\overline{(\gamma-\gamma_m)^2} = \begin{cases} \frac{9}{4}(\lambda\sigma\delta)^2 v & \text{for strains} \\ v'(1-v')\gamma_m^2 & \text{for inclusions} \end{cases}$$

where $\sigma$ represents the stress while v and v′ are the fractional volume of strains and inclusions, respectively.

Neel then calculated the result of fluctuations in the magnitude and direction of the magnetization caused by such randomly distributed irregularities and using Eq. (1) derived the formulas

$$H_c = \begin{cases} \frac{0.19 v\lambda^2\sigma^2}{KM}\left[1.4 + \frac{1}{2}\ln\left(\frac{2\pi M^2}{K}\right)\right], \ \frac{3}{2}\lambda\sigma << K & (6a) \\ \frac{0.46 v\lambda\sigma}{M}\left[1.4 + \frac{1}{2}\ln\left(\frac{4.5M^2}{\lambda\sigma}\right)\right], \ \frac{3}{2}\lambda\sigma >> K & (6b) \end{cases}$$

for the effect of the variation in the strain energy and

$$H_c = \frac{2Kv'}{\pi M_{av.}}\left[0.4 + \frac{1}{2}\ln\left(\frac{2\pi M_{av.}^2}{K}\right)\right] \qquad (6c)$$

for the effect of non-magnetic inclusions on the coercive force.

Using Eqs.(5) and (6) with v=1.0 and v′=0.01 for the coercive force, one obtains the values given in Table I below. We see that both the strain variation theories fall far short of any explanation for these materials - i.e., not only the magnitudes are too small but the trend is in the wrong direction. The modified Kersten inclusion theory also gives values that are too small. The demagnetizing effects of inclusions appears to contribute values of the correct order of magnitude for the coercive force of iron and cobalt and, in some forms of these materials, may be the major factor in determining the coercive force. However, this effect falls an order of magnitude short in explaining the coercive force for $Co_5Sm$.

Table I. - Values for the Coercive Force in Oersteds

| Material | Modified Kersten and Konsorsky theories for | | Neel Theory of Demagnetization Effects for | |
|---|---|---|---|---|
| | Strain | Inclusions | Strain | Inclusions |
| Iron | 10. | 0.0035 | 2.7 | 3.4 |
| Cobalt | 0.3 | 0.04 | 1.2 | 18. |
| $Co_5Sm$ | 0.25 | 0.15 | 1.0 | 1200. |

HARD MAGNETIC MATERIALS

Although the models described in the previous section have had some success in accounting at least qualitatively for the properties of low coercive force materials, they do not give any satisfactory values for the higher coercivity materials where $H_c$ is of the order of hundreds and sometimes thousands of Oersteds. As Carey and Isaac[9] point out, "at first sight the stress variation mechanism might seem a very plausible model inasmuch as many high coercivity materials are hard and brittle - properties associated with states of high localized internal strain. However, it turns out that the minimum value of the internal stress required to give rise to the observed coercivity is for some materials of the same order of magnitude as the breaking stress; moreover, some of the alloys are malleable and ductile. The magnetic properties of these materials clearly cannot be accounted for solely by the effects of stress centers on the movements of domain walls."

Major attempts to explain the hard permanent magnets have centered about the single domain particle concept. I.e., a major portion of the material (either due to precipitate interaction or use of pressed powders) consists of regions which are too small to contain domain walls and therefore require magnetic moment rotation to change the direction of magnetization. The energy barriers to be overcome may be due to shape anisotropy such as in alnico, lodex, cunife, etc, or to crystal anisotropy such as in CoPt and MnBi. The high theoretical values of coercivity obtained for coherent rotation of the single domain particles are not attained in practice. One reason given is that, for those materials whose rotation barriers are caused by crystal anisotropy, there exist modes (magnetization buckling and curling[10]) which produce magnetization reversal thru means other than magnetic moment rotation in unison. A second reason, put forth for materials whose energy barriers are due to shape anisotropy, is that magnetostatic interactions take place between the closely packed particles of a real material (fanning reversal mechanism[11]). Again, this collective mode reduces the energy barrier experienced by the rotation of a single particle.

In any case, the causes and explanations concerned with single domain type hard permanent magnets do not involve domain walls and thus fall outside the topic of this paper. We mention it above in order to give more perspective as to where the domain wall work is applicable. Rather, we shall show that in those hard magnetic materials which do contain ferromagnetic domain walls, such as $Co_5Sm$ and MnAl, their presence does not necessarily preclude the existence of very high coercive forces.

## NEW THEORY OF THE COERCIVE FORCE

A closer examination of the theories and methods described in the previous sections reveals that they do not represent a general solution to the non-linear differential equations governing the mechanism of domain wall motion in the presence of some physical barrier. Thus, the wave nature of the magnetic moment distribution both within and outside of the barrier and the consequences thereof are neglected in approaches such as given above. The resultant solutions do not contain effects of the width of the barrier (only of the barrier height) relative to the effective wave length of the magnetic moment distribution as represented by the domain wall width. Effects equivalent to tunneling do not appear.

In this section we review the work of Friedberg and Paul[12] who, by considering the non-linear differential equations for the problem of a domain wall in the presence of a planar barrier have been able to obtain explicit algebraic solutions which 1) reflect the concepts stated above, and 2) provide a framework for the coercive force of both soft magnetic materials and those hard magnetic materials which contain domain walls. The theory explains how different materials (possessing the same type of defects such as grain boundaries) can have widely different coercive forces. It also shows that such planar defects give the correct order of magnitude for the very high coercive force materials such as $Co_5Sm$ as well as for those lower coercive force materials whose planar defects are the major contribution. Similar solutions have been obtained for the effects of anti-phase boundaries by Hilzinger and Kronmuller[13] using discrete lattice theory and by Mitsek and Semyannikov[14] who postulated a linear variation in the magnetic moment distribution within the defect region.

We consider specifically the problem of a $180^0$ domain wall in an infinite medium divided into three regions as illustrated in Fig. 2. Regions 1 and 3 are identical and represent the homogeneous material while region 2 represents the energy barrier characterized by an abrupt change in the properties of the magnetic materials and caused by some defect such as a grain boundary. We examine the static equilibrium solutions to determine whether, for specific properties of the barrier and of the material, the domain wall is pinned. For a given barrier, there exists a continuous set of solutions of pinned domain walls corresponding to a range of external magnetic fields. We postulate that the maximum external field for which there exists a static domain wall solution corresponds to the coercive force for that particular obstacle[15]. Conversely, for a given external magnetic field, one can find the minimum width of region 2, (holding the other characteristics of the medium constant), for which there exists static domain wall solutions and obtain that barrier for which the given applied field is the coercive force.

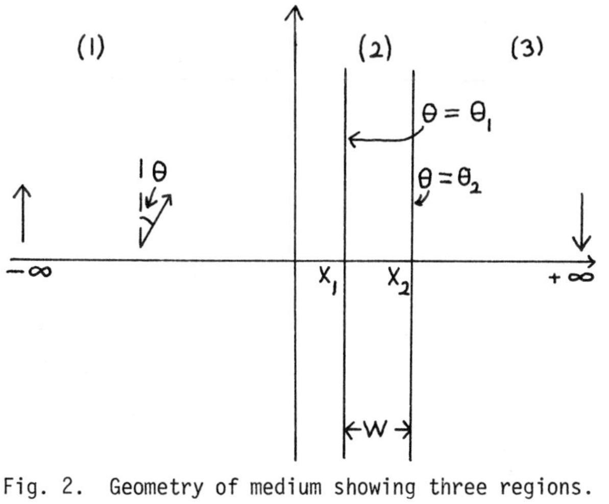


Fig. 2. Geometry of medium showing three regions.

We first obtain an implicit equiation for the magnetic field required to move a ferromagnetic domain wall across the barrier. We solve this under the restraint that the magnetostatic energy due to the external field is much less than the anisotropy energy of region 2. This restraint is not a limiting factor for most practical materials.

Assume that the magnetic moments lie in the plane of the coordinates perpendicular to the x axis and are functions of the x axis only, (i.e., neglect any variation in anisotropy direction between regions). Denote the angle between the magnetic moment and the fixed axis, z, by $\theta$. Then, using the notation as given previously, we write a) the energy density due to inhomogeneity in the distribution of the directions of the magnetic moments (i.e., the exchange energy) as $A\theta'^2$ where $\theta' = d\theta/dx$, b) the magnetic anisotropy energy density due to the presence of an axis of easy magnetization as $K\sin^2\theta$, and c) the magnetostatic energy density as $-HM\cos\theta$, where H is the applied field strength. Note that the contribution to the field strength due to the interaction between the magnetic moments is zero when all the magnetic moments are distributed perpendicular to the x axis and when $\theta$ depends only on x.

As discussed above, we look for static equilibrium solutions, i.e., the energy of the crystal,

$$\int (A\theta'^2 + K\sin^2\theta - HM\cos\theta)dv,$$

must have a minimum. Integrating the Euler equation for this integral yields the three non-linear equations,

$$- A_i\theta'^2 + K_i\sin^2\theta - HM_i\cos\theta = C_i \ , \qquad (7)$$

where the subscript i denotes the region of the material and where the quantities, A, K, and M have the same values in regions 1 and 3. Imposing the boundary conditions, $\theta(-\infty) = \theta'(-\infty) = \theta'(+\infty) = 0$, and $\theta(+\infty) = \pi$, yields the values $-HM_1$ for the constant $C_1$ and $+HM_1$ for $C_3$. Further, imposing continuity conditions for $\theta$ and $A_i\theta'$ at the two interfaces, $x_1(\theta_1)$ and $x_2(\theta_2)$, yields

$$[(ha/2b) + \cos\theta_1]^2 - [(ha/2b) + \cos\theta_2]^2 = 2h(a + 1)/b, \qquad (8)$$

and

$$W = (A_2/K_2)^{\frac{1}{2}} \int_{\theta_2}^{\theta_2} d\theta[\sin^2\theta - h\cos\theta + b\sin^2\theta_1 - ah\cos\theta_1 + h(a + 1)]^{-\frac{1}{2}}, \qquad (9)$$

where $W = x_2 - x_1$, $a = (A_1M_1/A_2M_2)-1$, $b = (A_1K_1/A_2K_2)-1$, and $h = M_2H/K_2$. Equation (2) is sketched in Fig. 3. We note that the lower branch, labelled B, of this curve corresponds to values of $\theta_1 > \theta_2$ and is therefore not considered here. Equations (8) and (9) determine h implicitly in terms of W and $\theta_1$. Maximizing h with respect to $\theta_1$, we can in principle obtain the coercive h for a given W.

As an illustration, we consider first the trivial case where the applied field, H, is zero. Then, from Eq. (8), $\cos\theta_1 = \cos\theta_2$, i.e., when $\theta_2$ ranges from 0 to $\pi/2$ radians, $\theta_1 = \theta_2$ and $W = 0$, while from $\pi/2$ to $\pi$ radians, $\theta_1 = \pi - \theta_2$, and $W > 0$. Thus, there exist solutions $W(\theta_1,\theta_2)$ for $H = 0$ with the minimum value being $W = 0$. This shows that any barrier, no matter how small, is sufficient to pin the wall in the absence of a magnetic field.

We next consider the more general case corresponding to h small but finite. As is shown in Table I, most physical cases of interest fall within this range.

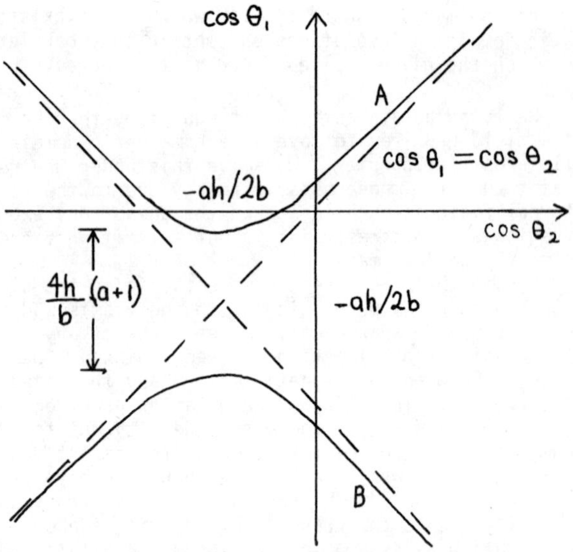


Fig. 3. Sketch of relationship between the angles $\theta_1$ and $\theta_2$ for arbitrary positive values of a, b, and h. Note: Curve A need not cross the abscissa.

Using this approximation, we rewrite Eq. (9) in the form

$$W = \left(\frac{A_2}{K_2}\right)^{\frac{1}{2}} \int_{\cos^2\theta_1}^{\cos^2\theta_2} \frac{d(-\cos^2\theta)}{\sin 2\theta(\sin^2\theta + b\sin^2\theta_1)^{\frac{1}{2}}}, \quad (10)$$

where we have dropped those terms in the integrand proportional to h. To evaluate the integral, we note from Eq. (8) that

$$\cos^2\theta_1 \approx \cos^2\theta_2 + 2h(a+1)/b, \quad (11)$$

and we approximate the integrand by substituting $\theta_1$ for $\theta$. We obtain the result,

$$H \approx \frac{K_1}{M_1}\frac{W}{\delta_1}\left(\frac{A_1}{A_2} - \frac{K_2}{K_1}\right)\sin^2\theta\cos\theta, \quad (12)$$

where $\delta_1 = (A_1/K_1)^{\frac{1}{2}}$ is the usual parameter governing the ferromagnetic domain wall width for the material. In accordinace with our prescription for the coercive force, we maximize the quantity $\sin^2\theta\cos\theta$ and get $\tan\theta_c = 2^{\frac{1}{2}}$, and the coercive force,

$$H_c = 0.38\frac{K_1}{M_1}\frac{W}{\delta_1}\left(\frac{A_1}{A_2} - \frac{K_2}{K_1}\right) \quad (13)$$

The quantity $K_1/M_1\delta_1$ in Eq. (13) represents the resistance of the material to the coercive force. We note the ratio, $W/\delta_1$, indicating the importance of the defect width to the domain wall width. For high anisotropy materials such as $Co_5Sm$, this ratio increases to order one even for such narrow barriers as grain boundaries. The factor $W[(A_1/A_2) - (K_2/K_1)]$ characterizes the planar defect or barrier and is a measure of its pinning strength. It appears naturally as a width, W, times an amplitude, $(A_1/A_2) - (K_2/K_1)$, and can be represented by a single number - making this effectively a one parameter theory. Further, by choosing a single fixed value for this parameter, one can obtain a range of $\sim 10^8$ in the values of the coercive force, depending on the particular material considered.

Inasmuch as a detailed calculation of the exchange and anisotropy constants within a grain boundary is not feasible, the value we chose for the defect parameter was got by simply assuming a 10% change in these constants (i.e., $A_1/A_2$=1.1 and $K_2/K_1$=0.9) and a reasonable value (i.e., 7.5A.) for the width W of the grain boundary. These numbers give agreement with the experimental value for the coercive force of Fe-Si4%. Keeping this characterization of the defect constant, we get the results tabulated in Table II for a number of ferromagnetic materials - both soft and hard. We see from the last two columns of this table that our theoretical results agree remarkably well with the experimental values.

It is perhaps of interest to mention at this point that the use of 7.5A. as the obstacle width does not negate the continuum approximation of micromagnetics. Several aspects may be mentioned including a) that the equations themselves do not break down as W becomes small. Thus, we may regard our value of W as an effective value to account for the sharp discontinuity in A and K at the boundaries and realize that actually W extends slightly further; b) that our calculation is one dimensional, but our material is three dimensional. For a close packed arrangement of atoms, we span many more than the nominal 3 atoms given by W = 7.5A.; and c) that the magnetic moment is a function of the electron spin rather than the atoms. Thus, 7.5A. may span a considerable number of magnetic moments.

We note of course that not all lamellar defects or even grain boundaries have the value chosen above and that there will be variations in the coercive force for different preparations of the same material. What is

Table II. Values of the coercive force due to grain boundaries. We assume the values W = $7.5\times10^{-8}$ cm, $A_1/A_2$ = 1.1 $K_2/K_1$ = 0.90, and $M_1/M_2$ = 1.

| Material | $M_1$ Magnetization | $K_1$ Anisotropy (ergs/cm$^3$x10$^{-5}$) | $A_1$ Exchange (ergs/cmx10$^6$) | $2\delta_1$ - Domain Wall Width (cmx10$^6$) | h Expansion Parameter | $H_c$-Coercive Force (Oersteds) | Observed Coercive Force |
|---|---|---|---|---|---|---|---|
| Supermalloy | 630. | 0.015 | 1.5 | 64. | 0.0002 | [a]0.0004 | 0.002 |
| Permalloy | 860. | 0.02 | 2.0 | 64. | 0.0002 | [a]0.0006 | 0.05 |
| [b]Iron-Si3% | 1590. | 3.7 | 2.2 | 4.8 | 0.0009 | 0.2 | 0.1 |
| Iron-Si4% | 1570. | 3.2 | 2.1 | 5.2 | 0.003 | 0.5 | 0.5 |
| Nickel | 485. | 0.7 | 0.5 | 5.2 | 0.003 | 0.3 | 0.7 |
| Iron | 1707. | 4.8 | 2.4 | 4.4 | 0.003 | 0.7 | 1. |
| Cobalt | 1400. | 45. | 4.7 | 2.0 | 0.007 | 20. | 10. |
| $Co_5Sm$ | 800. | 1500. | 2.0 | 0.24 | 0.056 | 9000. | 10000. |

[a]For these materials, the coercive force may be dominated by magnetostatic effects. In particular, for Permalloy, the rapid quenching of the disordered state should produce high stress fields and a Kondorsky-Néel-type contribution to the coercive force.
[b]This material is grain oriented. Therefore, we use W = $2.5\times10^{-8}$ cm.

interesting is that one can get the correct results to first order by choosing one average value for this parameter regardless of whether we look at iron, cobalt, $Co_5Sm$, etc. We discuss the implications of this in the next section.

We might point out that recent experiments on $Co_5Sm$ by A. Riley[16] using the 1 Mev electron microscope in England and by J. Livingston[17] in this country using the Kerr optical effect give experimental support to the pinning of $Co_5Sm$ domain walls by grain boundaries in fields of from 5 kilo-Oersteds to 10 kilo-Oersteds.

## DISCUSSION OF NEW THEORY

Certain concepts emerge from the theory. As we have noted, the width of the domain wall in the homogeneous material is of importance in establishing the nature of the obstacle needed to impede its motion. This is due to the fact that this width implies an "effective wave length" for the magnetic moment disturbance that constitutes a domain wall. One should be clear here. Although the disturbance in the magnetic moment or spin distribution does not start from $-\infty$ and go to $+\infty$, (i.e., the excitation is bounded and cannot be represented by a plane wave), this does not negate the wave nature of the phenomenon. Perhaps one can think of the magnetic moment distribution of the domain wall as a pulse which propagates within the material under the influence of external forces. The fact that it is a pulse implies the usual non-linearity of the dynamics of the phenomenon and thus the inherent mathematical difficulty of rigorously solving the three media boundary value problem. Nevertheless, the physical concept remains. If the effective wave length of the disturbance, i.e., the domain wall width, is small compared to the width of that part of the material containing the change in the magnetic properties of the medium, then the magnetic disturbance will have difficulty in traversing that area without an "effective dispersion" setting in. This is similar to optical pulses in a material media. As in optics, or in standard quantum mechanics for electrons, we note that this dispersion and spreading of the pulse increases as a function of its duration within the second medium. As long as the external field is not sufficiently strong to pull it completely out of the medium, complete magnetic reversal cannot occur.

It is for this reason that I wrote the symbol $\delta$ rather than $(A/K)^{\frac{1}{2}}$. The specialization to bulk materials where the intrinsic parameters A and K are the only ones present may perhaps obscure a more general statement that (and I have no rigorous proof) it is the "physical wall width" that appears in the denominator of the expression for domain wall pinning regardless of whether we are in bulk materials or not. Further, from the above, it appears that any changes in the medium which affect the original pulse of domain wall shape will act as obstacles to its propagation. Thus, for thin materials wherein magnetostatic effects are important in the domain wall magnetic moment distribution, any change in film thickness, d, could act as an obstacle. Although this parameter does not appear in the factor $(A_1/A_2) - (K_2/K_1)$ representing the amplitude of the disturbance in Eq. (12), it probably enters as an additional term $d_1/d_2$. Also, the width W of the barrier in this equation could be thought of as a measure of the extent of this thickness deviation. Noting that domain walls in amorphous material have a different structure than those in the crystalline phase of the same material[18], any local crystalline region in amorphous $Fe_{75}P_{15}C_{10}$ or GdCo may serve as a defect. From the above, bounds could be placed on the spatial extent of such defects before they seriously affect the coercive force. But one should really show all this - i.e., there remains much work to be done.

## APPLICATIONS

We recognize that, in real materials, grain boundaries and lamellar precipitates are usually not parallel to domain walls. In addition, these planar defects usually possess curvature. The increased complexity for this geometry is such that a closed form algebraic solution for the coercive force has not as yet been attained. This is due to the fact that the magnetic moment distribution is now a function of more than just the one coordinate z. At present, there are also no computer type calculations for these cases. Nevertheless, one can at this point suggest applications of the results of the previous sections to practical physical problems. In particular, for the high coercive force materials, special attention should be given to that metallurgical preparation which will produce "two dimensional" defects with their planes approximately parallel to the direction of magnetization. They should have the following desirable characteristics:

a) The thickness of the defect or precipiate should be large relative to the ratio $(k_BT_c/Ka)^{\frac{1}{2}}$ which governs the ferromagnetic domain wall width. Here, $k_B$ is the Boltzmann constant and $T_c$ is the Curie temperature.

b) The defect should be of such a nature as to markedly reduce the exchange bonding and/or the anisotropy within it. Thus, if lamellar precipitates are induced, their composition should be such that they make a poor ferromagnet re: Curie temperature considerations or have a structural anisotropy resembling that of an amorphous material. In this connection, it is of interest to consider the ferromagnetic material MnAl and other similar materials which show good magnetic properties when they are in the partially disordered state. In this state one may suppose that there exist aluminum dense lamellar regions several atomic layers thick. Such layers would have a drastically reduced exchange energy. (One can easily envisage a drop in the exchange constant to one-half of the ordered matrix value and a similar drop for the anisotropy constant.) Using this figure and the values M = 495, $T_c = 653^oK$, $A = 10^{-6}$ ergs/cm, and $K = 10^7$ ergs/cm$^3$, appropriate for MnAl, and the value $W = 12 \times 10^{-8}$cm for the width of the precipitate region, the coercive force is given by Eq. (12) as 4400 Oersteds in agreement with the observed value for this material.

c) The density of these planar defects should be sufficiently high such that domain walls attempting to sweep over large areas are pinned at these barriers. Thus, unreversed grains will require the nucleation of new domain walls.

d) The material should be cut such that there is a maxium change in the magnetic anisotropy in the direction of the magnetic moment rotation within the ferromagnetic domain wall - thus producing as narrow a wall as possible.

Many of these characteristics are of course already known from practical experience.

For the low coercive force materials, the anisotropy is small and therefore the domain wall width or variation of the magnetic moment distribution is large compared to the defect width. The consequence is that this distribution or disturbance passes relatively unscattered through the defect region and magnetization reversal occurs even for small external fields. Thus, it is exceedingly difficult to pin domain walls in low anisotropy material. For a higher coercive force in this type media, one must either prevent the formation of domain walls (the nucleation problem or the single domain concept) or make the barriers unusually wide, (i.e., large precipitates, large areas of strain variation, high angle grain boundaries, etc.).

## REFERENCES

*Supported in part by NSF Grant DMR72-03118 A02.


1. L. Landau and E. Lifshitz, Phys. Zeits Sowjetunion 8, 153 (1935)
2. A Aharoni, Phys. Rev. 119, 127-131 (1960); J. Appl. Phys. 32, 245S-246S (1961); Rev. Mod. Phys. 34, 227-238 (1962b)
3. W. F. Brown in Micromagnetics, Interscience Publishers (1963).
4. H. Trauble, in Magnetism and Metallurgy, Vol. II, 622-87 (Ed. by E. Kneller and A. Berkowitz, Academic Press) 1969.
5. H. Kronmuller, Int. J. Nondestr. Testing 3, 315 (1972).
6. M. Kersten, in Grundlagen einer Theorie der Ferromagnetischen Hysterese und der Koerzitskraft (S. Hirzel, Leipzig, 1943).
7. E. Kondorsky, Phys. Z. Sowjetunion 11, 597 (1937).
8. L. Neel, Physica 15, 225-234 (1949).
9. R. Carey and E. Isaac, in Magnetic Domains and Techniques for their Observation (Academic Press, New York, 1966).
10. E. H. Frei, S. Shtrikman, and D. Treves, Phys. Rev. 106, 446 (1957).
11. I. S. Jacobs and C. P. Bean, Phys. Rev. 100, 1060 (1955).
12. R. Friedberg and D. I. Paul, Phys. Rev. Ltrs. 34, 1234-1237 (1975); 1415 (1975).
13. H. R. Hilzinger and H. Kronmuller, Phys. Ltrs. 51A, 59 (1975).
14. A. I. Mitsek and S. S. Semyannikov, Sov. Phys. Sol. State 11, 899 (1969).
15. F. B. Hagedorn, J. Appl. Phys. 41,2491 (1970) used a similar type definition when considering two exchange coupled magnetic films.
16. A. Riley, Jrnl of the Less Common Metals 35, 305-313 (1974).
17. J. D. Livingston, Phys. Stat. Sol. (a) 18, 579-588 (1973).
18. D. I. Paul, J. Marti, and L. Valadez, A.I.P. Conf. Proc. 18, 1377-1381 (1973).


# ON THE QUANTUM THEORY OF BLOCH WALLS

R. Schilling
Institut für Theoretische Physik, ETH-Hönggerberg
CH-8049 Zürich, Switzerland

## ABSTRACT


A $180^\circ$ Bloch wall in a Heisenberg ferromagnet is described by the Hamiltonian :

$$H= -J \sum_{n,\delta} \vec{S}_n \cdot \vec{S}_{n+\delta} - A \sum_n (S_n^z)^2 + \alpha(\sum_{n\in F_N} S_n^z - \sum_{n\in F_1} S_n^z)$$

The two first terms are the Heisenberg exchange- and anisotropy energy and the third term takes into account the interaction of the spins in the boundary layers $F_1$ resp. $F_N$ with fictitious adjoining spins pointing up and down respectively.

A quantum theoretical treatment of H leads to results which differ qualitatively from those of classical studies in which the spin operators are considered as vectors with fixed length S. Since H commutes with the z-component of the total magnetisation $S^z = \sum_n S_n^z$, there exists a complete set of eigenstates for which the transverse components $\langle S_n^x \rangle$ and $\langle S_n^y \rangle$ vanish for all sites n. Selecting the eigenstates with further symmetry of H leads in some cases(for the lowest eigenstate in the subspace with $S^z$ =0 and an odd number of layers) to $\langle \vec{S}_n \rangle = 0$ in the middle layer of the wall although $\langle \vec{S}_n^2 \rangle = S(S+1) \neq 0$. Thus the quantum fluctuations concentrate at the middle of the wall. These fluctuations lead to more freedom for possible spin structures in a quantum spin system than in a classical one.

For a one dimensional chain with S=1/2 the properties of the groundstate are investigated. For the infinite chain the lowest eigenstates and eigenvalues $E(\alpha,M)$ in each subspace with $S^z$=M are calculated exactly. The same was carried out numerically for 2,3,4 and 5 sites. For all these cases the $E(\alpha,M)$ have the properties: $E(\alpha,M)= E(\alpha,-M)$ and $E(\alpha,M+1) > E(\alpha,M)$ for all M> 0 and all $\alpha$. There is much evidence that these properties are true for any number of sites. From this follws that the groundstate of the wall is not degenerate for an even number of sites and therefore has vanishing transverse components of the magnetisation, while the z-component has opposite sign at sites reflected at the middle of the wall. For an odd number of sites the groundstate is twofold degenerate and for their superposition the transverse components may not vanish. These superpositions are however energetically indifferent

## THEORETICAL MOTION OF A REALISTIC DOMAIN WALL

Amikam Aharoni
Weizmann Institute of Science, Rehovoth, Israel

All the theories of domain wall motion published so far, assume a one-dimensional wall structure, at least for a first order approximation. This is not justified, since the approximation of one-dimensionality is known to be a very poor one for both the theoretical and experimental studies of a stationary wall. In this work, a recent two-dimensional Ritz model for a wall is modified, by adding a third Ritz parameter, to contain the extra asymmetry which is always necessary for a moving wall. The modification is such that the calculations of the various energy terms are not more complicated than in the case of the stationary wall. Detailed numerical study is reported for the case of a wall moving at $10^2$ to $10^3$ m/sec, in a 1,000 Å thick permalloy film, with a uniaxial anisotropy in the plane of the film, at zero applied field. No "critical" velocity is found, in the sense encountered by Schlömann for his one-dimensional approximation, namely a velocity at which the wall structure is drastically changed. However, there is a rather sharp change in the wall mass at a velocity of about 400 m/sec, and this may be taken as an abrupt change which experimentalists might call a critical velocity. The wall mass is defined for this purpose in the conventional way, from the difference in energy between a moving and a stationary wall, and should affect the measured wall mobility. This mass decreases monotonically from its low-velocity value of 7 x $10^{-11}$ grams/cm$^2$ to 4 x $10^{-11}$ grams/cm$^2$ at a velocity of 1,100 m/sec.

## MAGNETIC DOMAIN WALL BOWING IN A PERFECT METALLIC CRYSTAL

W.J. Carr, Jr.
Westinghouse Research Laboratories, Pittsburgh, Pennsylvania 15235

ABSTRACT

From the equation of motion for a domain wall in a perfect crystalline sheet, expressions for the steady state mobility, power loss and wall profile are calculated in terms of the magnetic eddy current field acting on the wall. An exact expression for the eddy current field for a given wall shape is derived, and an iteration procedure is used to obtain approximate results. At low wall velocity ($0.16\, M_s H_A d/\varepsilon$ somewhat smaller than unity, where $M_s$ is the saturation magnetization, $H_A$ the applied field, d the sheet thickness and $\varepsilon$ the wall energy per unit area) the equation for the wall profile, $x = x_w(y)$, is approximately

$$x_w = \frac{M_s H_A d^2}{\pi\varepsilon}\left[\left(\sum_{m=1,3,5\cdots} 1/m^3\right)^{-1} \sum_{m=1,3,5\cdots} \frac{\sin\frac{m\pi}{2}}{m^4}\left(1-\cos\frac{m\pi y}{d}\right) - \frac{\pi y^2}{d^2}\right]$$

where the x axis is normal to the wall and y is along the sheet thickness, with the origin at the middle of the sheet. At some critical velocity, or applied field, which in rough approximation is given by $0.16\, M_s H_A d/\varepsilon$ equal to unity, no steady state solution exists. Severe wall bowing cannot occur except near this critical field, just before instability occurs.

## VARIATION OF THE FERROMAGNETIC DOMAIN WALL WIDTH AS A FUNCTION OF THE MAGNETIC ENERGY CONSTANTS*

H. Mohtadi and D.I. Paul
Columbia University, New York, New York 10027

The dependence of the width, $\delta$, of a 180° ferromagnetic domain wall on the magnetic exchange energy, A, anisotropy energy, K, and specimen thickness, D, have been determined by an approximate theoretical method in the region where the specimen thickness is of the same order of magnitude as the domain wall width. We show that there exists an asymptotic region defined by $(A/KD^2)^{1/2} > 0.7$ where the fractional ratio of domain wall width to specimen thickness is constant for all magnetic materials, and has the value $\delta \sim 0.5$ D, independent of the exchange and anisotropy energies. This is in accord with the value obtained by Hubert[1] for the special case of permalloy. Further, our results show that there exists a large jump in the value of the ratio $\delta/D$ in the region where $(A/KD^2)^{1/2} \sim 0.07$.

*Work supported by NSF Grant No. GH-34451.

[1]A. Hubert, Phys. Stat. Sol. 38, 699 (1970).

# BLOCH, NÉEL AND HEAD-TO-HEAD DOMAIN WALL MOBILITIES IN $YFeO_3$*

Ching H. Tsang
and
Robert L. White
Stanford Electronics Laboratory
Stanford University
Stanford, California 94305

and

Robert M. White
Xerox Palo Alto Research Center
3333 Coyote Hill Road
Palo Alto, California 94304

ABSTRACT

Velocities and mobilities of Bloch and Néel domain walls in single crystal $YFeO_3$ bars were measured using a Sixtus and Tonks transit time technique. At room temperature Bloch and Néel wall mobilities were measured to be $\mu_B = 6.16 \times 10^3$ cm/sec-Oe and $\mu_N = 5.8 \times 10^3$ cm/sec-Oe respectively. Mobilities were found to decrease monotonically and roughly linearly with increasing temperature in the temperature range 400-600°K. These results are to be contrasted with the much higher mobilities and exponential temperature dependence of mobility for Head-to-Head walls where the magnetic moments on both sides of the wall are pointing Head-to-Head against each other. The mobilities of these three types of walls were calculated on a Walker-like model of the moving domain wall. The results predict nearly the same mobility in all cases, assuming the same Gilbert damping parameter $\alpha$. In an effort to reconcile these predictions with the experimental results we have explored the possibility of the variation of $\alpha$ with domain wall type by calculating the spin wave spectra of a canted antiferromagnet containing the three kinds of domain walls, but again find the results very similar for the three kinds of walls.

Using a Sixtus and Tonks transit time technique[1], we have measured velocities and mobilities for Bloch and Néel domain walls in single crystal bars of $YFeO_3$, of dimension 1 $mm^2$ x 50 mm, with the long axis of the bar along the crystallographic b and a axes respectively.

The experimental set-up is similar to that used in our measurement of Head-to-Head domain wall velocities and reported on at this conference last year[2] except that the external field, applied parallel or anti-parallel to the magnetization, must now be directed normal to the bar axis. The pick-up loops are now narrow rectangular loops on the face of the bar along the field direction. A uniform reversed field is then applied to the whole bar, and a small electromagnet at one end of the bar is pulsed to produce an intense reversed local field localized at that end. A reverse domain is nucleated and propagates down the bar at the velocity determined by the magnitude of the uniform reversed field. The transit time of the wall is measured using two pick-up coils spaced 2.5 cm apart along the bar axis.

Plots of velocity versus drive field at room temperature for the Néel, Bloch and Head-to-Head walls are given in Figures 1a, b, and c. The linear region at low fields for the Bloch and Néel walls can be characterized with mobilities of

$$\mu_B = 6.16 \times 10^3 \text{ cm/sec-Oe} \qquad \text{(Bloch)}$$
$$\mu_N = 5.8 \times 10^3 \text{ cm/sec-Oe} \qquad \text{(Néel)}$$

These two mobilities are nearly equal and are only about 1/7 the Head-to-Head wall mobility reported last year. These values are somewhat higher than those reported by Shumate[3] ($\mu_B = 4.8 \times 10^3$ cm/sec-Oe, $\mu_N = 4.3 \times 10^3$ cm/sec-Oe) but the ratio $\mu_B/\mu_L = 1.06$ is the same and agrees with Rosencwaig's computed value based on orthorhombicity effects on the wall stiffness[4].

At higher velocities there are notches and plateaus in the velocity-drive field plots. We have concluded from acoustic velocity measurements that these irregularities indicate interactions of the moving wall with various acoustic waves when their velocities coincide. Apart from small differences due to orthorhombicity, the shear mode acoustic velocities along the three axes are all near $4 \times 10^5$ cm/sec, and the longitudinal modes near $7 \times 10^5$ cm/sec. We note from Figure 1a that the Néel wall couples mainly to the shear modes and from Figure 1b that the Bloch wall interacts significantly with the longitudinal modes also.

We have conducted measurements at temperatures between 250°K and 600°K. As temperature goes up, mobilities go down monotonically, coupling with the shear modes is diminished, while coupling to the longitudinal mode increases. The temperature dependence of the mobility for Bloch and Néel walls is shown in Figure 2; the temperature dependence of mobility is seen to be virtually identical in the two cases. The linear decrease of mobility with increasing temperature in the 400 to 600°K region is to be contrasted with the exponential $e^{-\alpha T}$ dependence found for Head-to-Head walls.

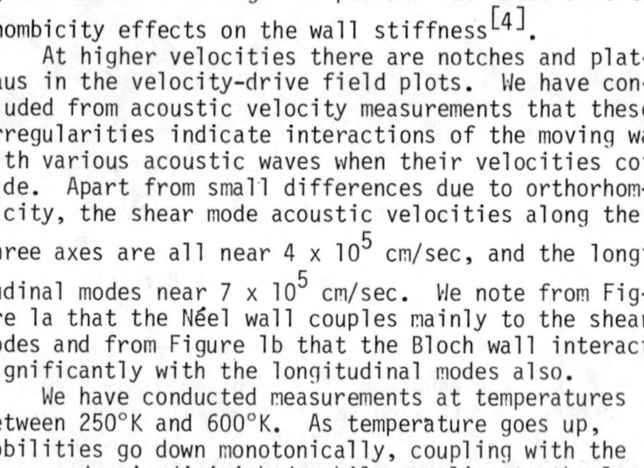

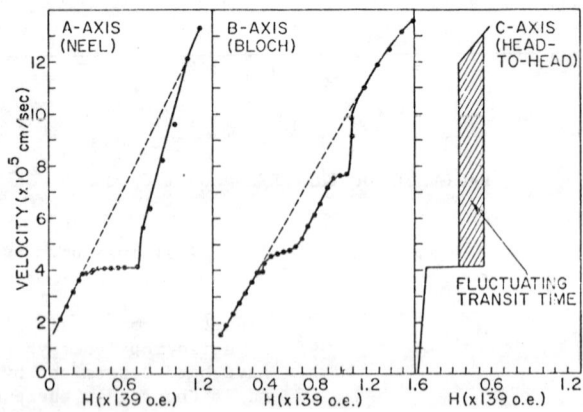

Figure 1a, b, c. Velocity vs. applied field at room temperature.

Out of the whole spectrum of information obtained from our experiments, the values and temperature dependences of the mobilities of the three types of domain walls has constituted the focus of our attention. If we assume a damping process characterized by a Gilbert parameter $\alpha$, and that the wall conserves its shape in motion we can obtain Walker-like solutions for the moving walls. This was done by Gyorgy and Hagedorn[5] for the Bloch and Néel walls, with the results

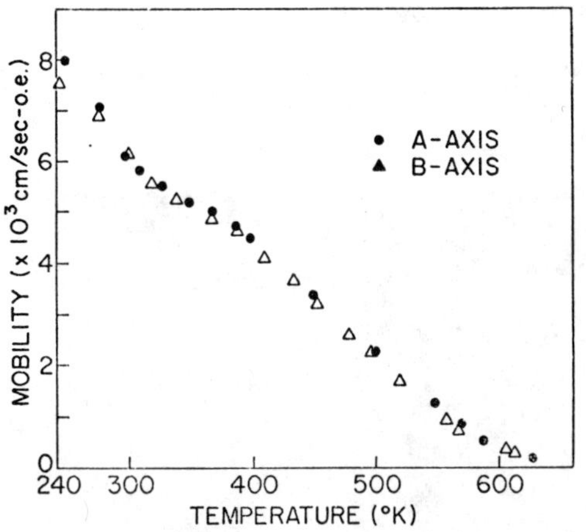


Figure 2. Mobility vs. temperature for Néel and Bloch walls in $YFeO_3$.

$$\mu_B = \frac{\gamma}{\alpha}\sqrt{\frac{A}{K^*}} \qquad \mu_N = \frac{\gamma}{\alpha}\sqrt{\frac{A}{K^* + 2\pi M^{*2}}}$$

We have extended this treatment to the case of Head-to-Head walls, obtaining

$$\mu_H = \frac{\gamma}{\alpha}\sqrt{\frac{A}{K^* - 2\pi M^{*2}}}$$

where A is the isotropic exchange constant, $K^*$ is the effective anisotropy in the ac plane and $M^*$ is the net magnetization. From these calculations we expect $\mu_H > \mu_B > \mu_N$. For $YFeO_3$, however, $2\pi M^{*2}/K^* \approx 10^{-3}$, so all three velocities should be virtually identical if we assume equal $\alpha$.

The question therefore arises whether or not $\alpha$ should be the same for all three cases. We anticipate that, for the surface to volume ratio of the samples used, $\alpha$ comes mainly from bulk damping mechanisms which we believe to be coupling from a uniformly translating wall state to various spinwave states. We have therefore studied the spin wave spectrum of a uniformly magnetized two-sublattice antiferromagnet with canting produced by antisymmetric exchange. The results show two modes of excitation, a rocking mode and a twisting mode, similar to the $\gamma = 0$ and $\sigma = 0$ modes of Paul's[6] analyses of spin-waves in the presence of a Bloch wall in an antiferromagnetic material. The $k = 0$ bottom of these modes correspond to the $\omega_{XY}$ and $\omega_Z$ modes, respectively of Herrmann[7], and for $YFeO_3$ are expected to be quite similar in energy. We have also solved for the modes of a canted antiferromagnet containing a stationary domain wall. For all three kinds of walls each mode splits into a low energy branch of the spin waves bound to the wall and a free spin-wave branch similar to the modes of the uniformly magnetized material. The differences among the spectra of the three kinds of walls lie primarily in the conbtribution of the magnetostatic energies; since the magnetization is only a few e.m.u. in $YFeO_3$ these effects are dwarfed by the energies due to crystalline anisotropy and exchange, both symmetric and antisymmetric. We therefor find no significant difference in the spin wave spectra associated with the three kinds of walls, and from Thiele's[8] work on the spectra of moving walls anticipate no significant differences in the spin wave spectra of the three kinds of walls when moving. Our present theoretical effort centers on examining the mechanisms which couple the moving domain wall to the spin wave spectrum for the three cases.

We have also been led by our theory to re-examine the data which shows Head-to-Head wall mobilities some seven times larger (at room temperature) than the Néel and Bloch mobilities. We have taken photographs of the signal induced in the pick-up coils by the Head-to-Head wall in passing and find evidence that the wall structure may be several millimeters long. Since the static wall thickness is only a few hundred Angstroms, it may be that the domain wall is conical or multi-conical in shape when propagating. If so, the conical faces may be nearly Bloch or Néel walls propagating obliquely with their characteristic mobilities, giving large domain tip velocities. On the other hand, this model of the motion is hard to reconcile with the very different temperature dependence of the mobilities observed, roughly 1/T for Block and Néel and $e^{-\alpha T}$ for Head-to-Head. Experiments are being set up to establish and verify the shape of the moving domain wall, to see if conical domains exist, and to see if systematic variations of the cone angle with temperature can reconcile the two apparently different temperature dependencies of mobility.

REFERENCES

*This work supported by the Office of Naval Research through contract number N0014-75-C-0290.

1. C. H. Tsang, R. L. White, AIP Conference Proceedings, 20th Annual Conference on Magnetism and Magnetic Materials (1974).

2. K. J. Sixtus and L. Tonks, Phys. Rev. 37, 930 (1931); 42, 419 (1932); 43, 70, 931 (1933).

3. P. W. Shumate, Jr., J. Appl. Phys., Vol. 42, No. 13, pp. 5770-5772, 1971.

4. A. Rosencwaig, J. Appl. Phys., Vol. 42, No. 13, pp. 5773-5775, 1971.

5. E. M. Gyorgy, and F. B. Hagedorn, J. Appl. Phys., Vol. 39, No. 1, pp. 88-90, 1968.

6. David I. Paul, Phys. Rev., Vol. 126, No. 1, pp. 78-82, 1962.

7. G. F. Herrmann, J. Phys. Chem. Solids, Vol. 24, pp. 597-606, 1963.

8. A. A. Thiele, Phys. Rev. B., Vol. 7, No. 1, pp. 391-397, 1973.

# MAGNETIC PROPERTIES OF (110)[001] ORIENTED LOW ALLOY IRON

D. R. Thornburg, K. Foster, and G. C. Rauch
Westinghouse Research Laboratories
Pittsburgh, Pennsylvania 15235

## ABSTRACT

Magnetic characteristics have been determined for several 6 mil thick samples of low alloy iron with strong (110)[001] orientations. These textures have been obtained by primary recrystallization and normal grain growth in iron base alloys with small additions of Si, Cr, and Mn. This presentation will discuss how highly oriented specimens were obtained by the control of chemical composition, processing, and final annealing. Both the texture and magnetic properties of these alloys will be described. The strong (110)[001] textures were confirmed by torque measurements, magnetic properties, reflective X-ray pole figures, and quantitative domain analyses.

## INTRODUCTION

Highly textured commercial oriented silicon steels are obtained by a secondary recrystallization process. In these alloys, the primary recrystallized texture is usually complex with a minor (110)[001] component that grows at the onset of secondary recrystallization.[1] In a previous investigation,[2] we have found that a variety of iron-cobalt alloys, containing from 10 to 27% Co, can be obtained by primary recrystallization and normal grain growth to obtain both the (110)[001] and (100)[001] textures. In the present study, we have determined that in iron with small alloying additions good (110)[001] texture can also be obtained by primary recrystallization and normal grain growth. Strongly oriented materials were obtained by controlling the chemical composition and subsequent processing to final gage. Although a large number of alloys have been successfully textured, this paper will describe the processing of heats of only two compositions and the resulting magnetic characteristics.

## EXPERIMENTAL PROCEDURE

Both experimental heats (5,000 lbs) were air induction melted using A101 electrolytic iron, followed by vacuum-arc remelting, forging into slabs, and finally hot rolling at between 1025 and 1050°C to 4.2 mm. The above processing was all done using commercial facilities. The hot bands for both heats were processed in the laboratory to final gage (0.15 mm). The chemical composition of the hot bands is given in Table I. After cleaning, the hot band of heat A was annealed 5 h-850°C in $H_2$; heat B was given no hot band anneal. Both samples were then rolled to 2.0 mm, and annealed at 850°C in $H_2$, heat A for 5 h; heat B for 1 h. The samples were then rolled to 0.5 mm and again annealed in $H_2$; heat A 1 h-850°C, heat B 2-3 min at 900°C. Finally both heats were rolled to final gage (0.15 mm). Samples were given a final anneal in $H_2$ of 48 h; heat A at 850°C, heat B at 900°C, to develop final orientation. The samples were evaluated using torque and magnetic measurements, as well as metallographic examination, X-ray pole figures, quantitative domain analysis, and electrical resistivity measurements.

TABLE I

Composition of Alloys
(Weight Percent)

| Alloy | % Si | % Cr | % Mn | % S | % C | % O |
|---|---|---|---|---|---|---|
| A | 0.04 | 0.60 | 0.12 | .0024 | .020 | .019 |
| B | 0.50 | 0.03 | 0.15 | .0033 | .010 | .002 |

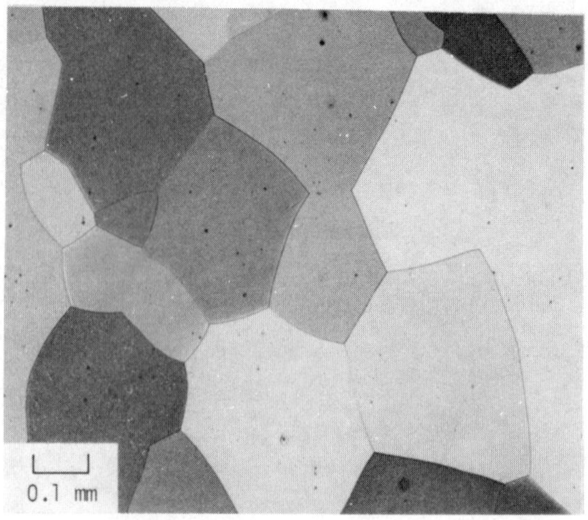

Fig. 1 Photograph of the surface of textured Alloy B (0.5% Si) annealed 48 h-900°C. Etchant 2% Nital.

## RESULTS AND DISCUSSION

The chemical analyses in Table I indicate that heat A is a 0.6% Cr alloy, while heat B contains 0.5% Si. Both alloys also have minor amounts of other elements. The sulfur levels of both alloys (Table I) were much lower than normally found in commercial alloys. In fact, there is some indication that strong (110)[001] cannot be obtained by primary recrystallization and normal grain growth in similar alloys containing sulfur contents above 0.012%. The carbon contents (Table I) are also low for commercial alloys, but are probably obtainable through the use of vacuum degassing techniques. A metallographic examination (see Fig. 1 for photomicrograph of alloy B) of the alloys indicated that both had average grain sizes after final annealing of 0.15 mm (the sheet thickness). This, combined with kinetic studies,[3] indicates that the (110)[001] orientation developed by primary recrystallization and normal grain growth. The mechanism of (110)[001] texture formation by primary recrystallization and normal grain growth observed in this investigation is different from that reported by Littmann[4] on commercial thin-gage oriented silicon-iron. The alloys studied in our work never contained a strong (110)[001] orientation, developed previously by secondary recrystallization, at any stage prior to final annealing.[3]

The torque and dc magnetic characteristics (Table II) definitely show the high degree of (110)[001] orientations obtained in both heats A and B. In both heats, the peak torque values exceeded 200,000 erg/cm$^3$, while $B_{10}$ was above 19 kG. Both the torque and $B_{10}$ values indicate the high degree of texture formation in these alloys. Table III shows both the electrical resistivity and 60 Hz ac characteristics of both

TABLE II

Torque and DC Magnetic Properties

| Alloy | Peak Torque (erg/cm$^3$) | Ratio | $B_{10}$ (kG) | $B_{100}$ (kG) | $H_c$ (Oe) |
|---|---|---|---|---|---|
| A | 209,000 | .41 | 19.3 | 21.3 | .29 |
| B | 220,000 | .42 | 19.6 | 21.4 | .15 |

TABLE III

Electrical Resistivity and 60 Hz Magnetic Properties at 6 Mils

| Alloy | ρ (μΩ-cm) | $P_c$ (W/lb) 15 kG | 17 kG | 19 kG | $P_z$ (VA/lb) 15 kG | 17 kG | 19 kG |
|---|---|---|---|---|---|---|---|
| A | 14.5 | .81 | 1.06 | 1.45 | .91 | 1.44 | 12.9 |
| B | 17.9 | .61 | .76 | 1.03 | .66 | .91 | 7.3 |

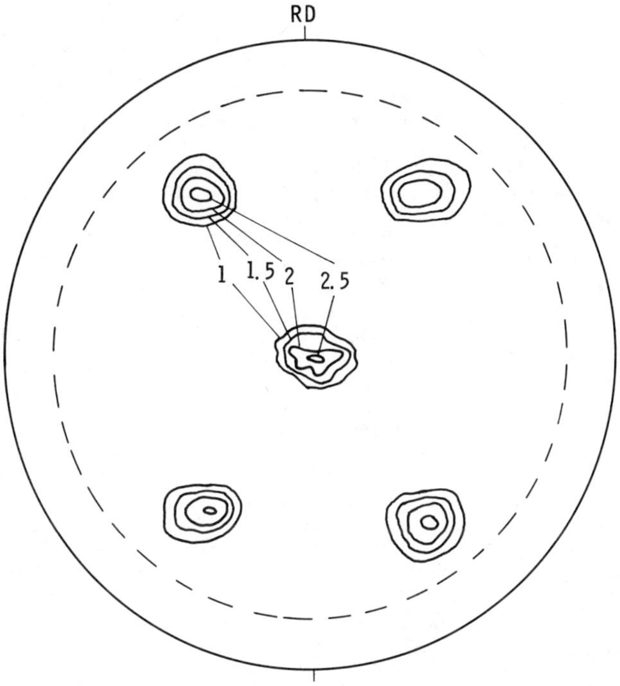


Fig. 2—(110) Pole figure of a 0.5% Si alloy annealed 48h-900°C, showing a strong (110) [001] or Goss texture. Contours are 1 to 2.5 times random intensity.

materials. Alloy A (0.6% Cr) had somewhat higher losses than alloy B. However, considering that its resistivity was low (14.5 μΩ-cm) and its $H_c$ value high (see Table II), the higher losses were not unexpected. On the other hand, heat B, with a resistivity of only 17.9 μΩ-cm had excellent losses and exciting characteristics, especially at the higher induction levels (Table III). The torque and magnetic characteristics shown in Tables II and III indicate the high degree of preferred orientation developed in these alloys.

Figure 2, a (110) X-ray pole figure of alloy B, indicates that a strong (110)[001] orientation has been developed in that alloy after the 48 h anneal at 900°C. Figure 3, a (200) X-ray pole figure of the same alloy indicates that in addition to the (110)[001] orientation, there is a minor (100)[001] texture component in this material. A quantitative analysis of the pole figure data indicated that alloy B had 67% of (110) and 20% of (100) planes within 10° of the sheet surface. Quantitative domain analyses on alloy B indicated that 61 volume % of (110) and 25 volume % of (100) planes were within 12° of the sheet surface. In addition, the domain analyses showed that, for the oriented grains, the [001] direction had an average deviation of only 6° from the rolling direction. Alloy A showed similar results using X-ray and domain analyses.

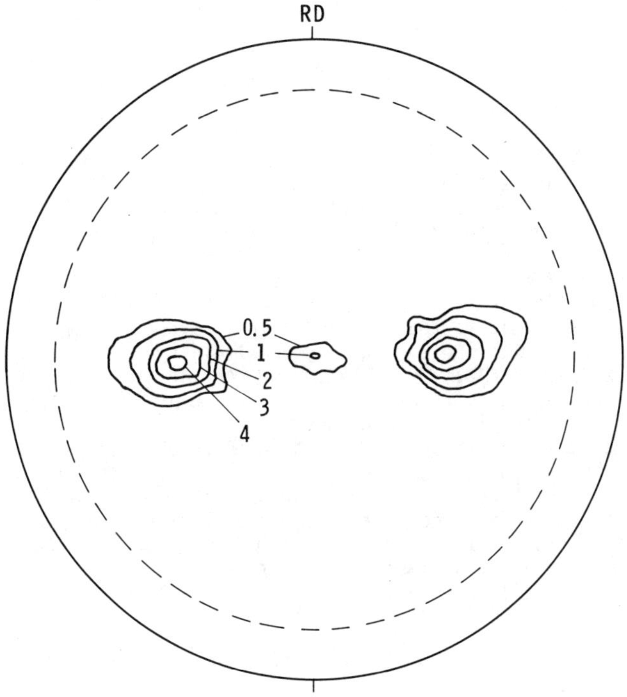


Fig. 3—(200) Pole figure of a 0.5% Si alloy annealed 48h-900°C, showing strong (110) [001] and minor (100) [001] textures. Contours are 0.5 to 4 times random intensity.

In summary, we have shown that excellent (110)[001] orientations can be obtained in low alloy iron samples by primary recrystallization and normal grain growth. Furthermore, the samples had high $B_{10}$ values and, in the case of the 0.5% Si alloy, low 60 Hz ac core losses at the higher induction levels.

REFERENCES

1. B. D. Cullity, "Introduction to Magnetic Materials," Addison-Wesley Publishing Co., Reading, MA., (1972).
2. K. Foster and D. R. Thornburg, AIP Conf. Proc. 24, 709 (1975).
3. G. C. Rauch, D. R. Thornburg, and K. Foster, to be published.
4. M. F. Littmann, JAP, Vol. 36, pp. 1225-1227, (1965).

# EFFECT OF CHROMIUM ON THE MAGNETIC PROPERTIES AND TEXTURE OF NON-ORIENTED STEEL

P.K. Rastogi, Inland Steel Research Laboratories, 3001 East Columbus Drive, East Chicago, Indiana 46312


## ABSTRACT

Little information is available in the literature regarding the A.C. permeability and core loss at intermediate and high inductions for Cr-bearing non-oriented steels. The present work was designed to investigate the influence of Cr up to 1.75 Wt% on these magnetic properties and the texture of low carbon steels. The results indicate that the 15 kG core loss decreases linearly with increasing Cr concentration at a rate of 0.54 watts/lb/Wt% Cr. The improvement in the property is primarily attributed to the decrease in the classical eddy current loss through the increase in resistivity. The permeability at 15, 17 and 18 kG is adversely affected by the addition of Cr, showing a minimum at ∿1.0 Wt%. Since the saturation magnetization decreases linearly as a function of Cr concentration, the observed changes in the permeability values are quantitatively explained in terms of a texture model. This model was developed from the pole densities of various orientations studied in these steels.


## INTRODUCTION

It has been suggested[1] that the presence of Cr in dilute concentrations is conducive to the development of {110} and {200} textural components at the expense of {211} and {222} components in cold rolled and annealed low carbon, low manganese steels. This type of annealing texture is considered favorable in attaining superior magnetic properties. On the basis of these results, an investigation was designed to examine the influence of Cr at levels up to 1.75 Wt% on the texture and magnetic properties in non-oriented low carbon steels.

## MATERIAL PREPARATION AND EXPERIMENTAL PROCEDURE

Six 40 lb vacuum melted alloys of low carbon steel containing 0.05/0.06 Wt% C, 0.54/0.57 Wt% Mn, 0.016/0.018 Wt% S and Cr levels in the range of 0.02 to 1.75 Wt% were used in this investigation. The alloys were hot rolled to about 0.09" thick plates and cold rolled to 0.025" thick sheets. Standard longitudinal Epstein samples, 28.0 cm x 3.0 cm, were cut from the cold rolled sheet and decarburized to a carbon level < 0.006 Wt%. These Epstein samples were subsequently employed for magnetic measurements. The A.C. permeability was measured at 15, 17 and 18 kG, while the core loss was measured only at 15 kG. In addition, measurements were made to determine the saturation magnetization and resistivity.

The grain size was characterized in terms of the average number of intercepts/mm which is inversely proportional to the average grain size. Texture was monitored by the inverse pole figure technique in terms of the pole density, $(I/I_R)_{hkl}$, of eight planes measured at the mid-plane of the decarburized samples.

## RESULTS AND DISCUSSION

It is apparent from Fig. 1 that the saturation magnetization decreases linearly with increasing Cr concentration at a rate of 0.25 kG/Wt% Cr, while resistivity increases linearly with Cr addition at a rate of 4.40 μ-ohmcm/Wt% Cr. These rates are found to be consistent with estimated rates for binary Fe-Cr alloys,[2] which suggest that Cr behaves as a simple diluent in Cr-bearing steels.

The data in Table I indicate that the grain size in the decarburized steels is essentially independent of composition. However, the cold rolled decarburized texture is significantly affected by the presence of Cr, as can be seen from Table I and Fig. 2. It can be seen that the {211} and {222} components attain maximum values at 1.06 Wt% Cr, while the opposite is true with respect to {110} and {310} planes. The remaining orientations exhibit a smaller dependence on composition. Since there is an insignificant change in grain size with Cr content, it is believed that the different annealing textures are generated by the influence of Cr on recovery and thus on nucleation orientation rather than by an effect on texture generation through grain growth. In addition, it is important to note that the discrepancies between these texture results and those reported by Goodman[1] are primarily attributed to the differences in steel composition and heat treatment.

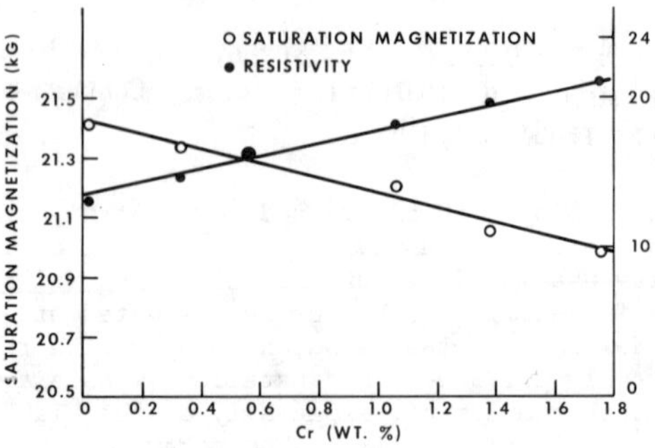


**FIGURE 1** EFFECT OF Cr CONTENT ON SATURATION MAGNETIZATION AND RESISTIVITY

TABLE I

Mid-Plane Pole Densities, $(I/I_R)_{hkl}$, and Grain Size as A Function of Cr Content in Cold Rolled and Decarburized Cr-Bearing Steels

| Cr (Wt%) | Average Grain Boundary (Int/mm) | ± 2σ | 110 | 200 | 211 | 310 | 222 | 321 | 420 | 332 |
|---|---|---|---|---|---|---|---|---|---|---|
| 0.02 | 19.1 | 2.7 | 1.80 | 0.52 | 1.59 | 0.53 | 6.04 | 0.49 | 0.73 | 1.88 |
| 0.30 | 21.6 | 2.7 | 1.01 | 0.65 | 2.00 | 0.36 | 7.16 | 0.47 | 0.88 | 1.46 |
| 0.56 | 24.2 | 2.4 | 0.74 | 0.42 | 2.17 | 0.31 | 8.17 | 0.43 | 0.33 | 2.12 |
| 1.06 | 22.8 | 2.3 | 0.59 | 0.40 | 2.42 | 0.21 | 10.10 | 0.43 | 0.35 | 2.01 |
| 1.38 | 21.9 | 2.3 | 0.86 | 0.45 | 2.02 | 0.55 | 6.50 | 0.79 | 0.43 | 1.75 |
| 1.75 | 22.4 | 2.5 | 1.14 | 0.36 | 1.87 | 0.47 | 6.08 | 0.63 | 0.57 | 1.69 |

(Columns 110 through 332 are grouped under the heading PLANE.)

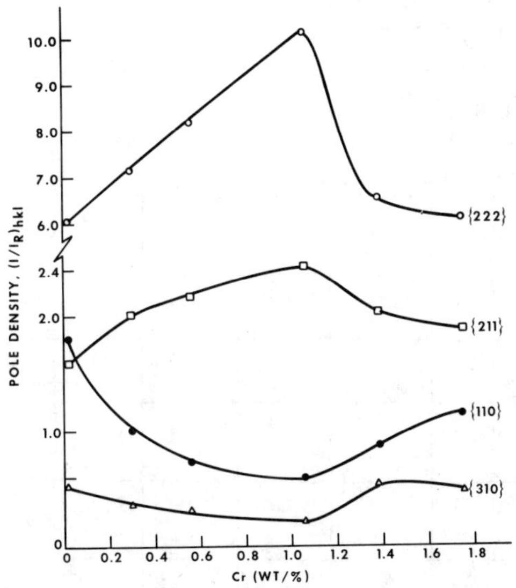


**FIGURE 2** EFFECT OF Cr CONTENT ON POLE DENSITIES OF VARIOUS ORIENTATIONS

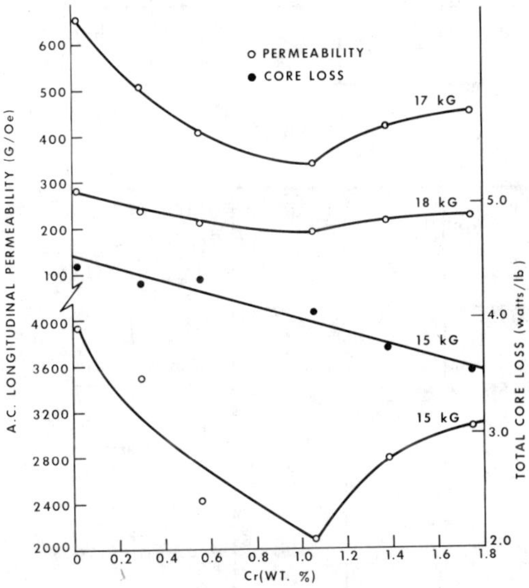


**FIGURE 3** EFFECT OF Cr CONTENT ON PERMEABILITY AND CORE LOSS AT DIFFERENT INDUCTIONS

It can be seen from Fig. 3 that the Cr has a significant detrimental effect on the 15, 17 and 18 kG permeabilities, exhibiting a minimum value at a Cr level of 1.06 Wt%. It is known that the permeability of non-oriented steel at 17 and 18 kG is primarily dependent upon texture and, to some extent, on saturation magnetization, while at 15 kG it is a function of both texture and grain size. A plot of permeability against a texture parameter, $T_p$, is shown in Fig. 4, which indicates that in spite of the reduction in saturation magnetization with increasing Cr content, the permeability at 17 and 18 kG is a linear function of $T_p$. A similar trend is also observed at 15 kG, since the grain size is unaffected by Cr. The above results, thus, clearly suggest that permeability at these inductions is primarily controlled by texture, which is qualitatively consistent with the observations of an earlier investigation.[3]

The influence of Cr on core loss is shown in Fig. 3, where a linear decrease is observed at a rate of 0.54 watts/lb per Wt% of Cr. The observed improvement in core loss is primarily attributed to the reduction in classical eddy current loss[4] through the increase in resistivity.

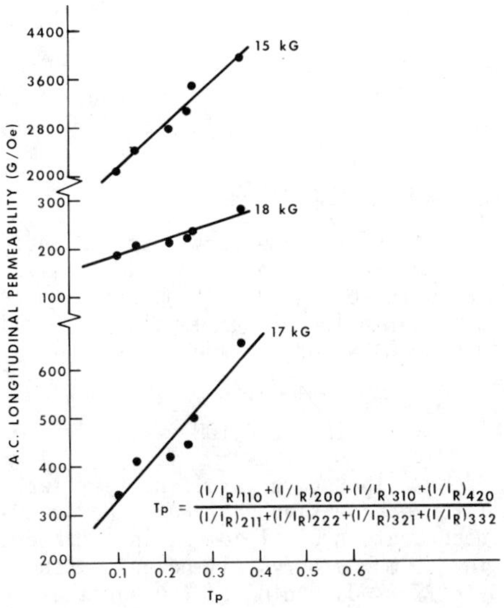


**FIGURE 4** A CORRELATION BETWEEN PERMEABILITY AND TEXTURE PARAMETER (Tp)

## CONCLUSION

The following conclusions are drawn from this investigation:

1. The saturation magnetization and resistivity decreases and increases linearly with increasing Cr concentration, respectively.
2. Cr has a detrimental effect on 15, 17 and 18 kG A.C. permeability values through the formation of unfavorable textures.
3. A linear correlation is observed between permeability and texture parameter, ($T_p$).
4. The 15 kG core loss decreases linearly with increasing Cr content and the improvement in core loss is primarily due to the decrease in eddy current loss through the increase in resistivity.

## REFERENCES

1. S. R. Goodman, Met. Trans; 2, 2051 (1971).
2. R. M. Bozorth, Ferromagnetism, D. Van No Strand Co., Inc., N.Y., 228 (1961).
3. P. K. Rastogi, AIP Conference Proceeding No. 24, 724 (1975).
4. ASTM Special Technical Publication, No. 371, 31 (1964).

# ELEMENTARY COUPLING ENERGY OF $CO^{2+}$ ION TO THE LATTICE IN MIXED NICKEL-ZINC FERRITES DOPED WITH CO.

A. MARAIS, T. MERCERON, M. PORTE

Laboratoire de Magnétisme, C.N.R.S.
1, Place Aristide Briand - 92190 MEUDON (France)

## ABSTRACT

The study of elementary coupling energy w of $Co^{2+}$ ion to the lattice has been made in mixed Ni-Zn ferrites doped with Co. The variation of Zn concentration results in variation of Curie point $T_C$. A linear relationship between w and $T_C$ is observed.

## INTRODUCTION

Amongst the polycrystalline ferrites we have studied so far, we have seen that $Fe^{2+}$, $Mn^{3+}$, $Cu^{2+}$, $Co^{2+}$, with degenerate orbital moment in octahedral sites contribute to the magnetocrystalline anisotropy [1]. Thus their energy differs depending on the particular site they occupy in the lattice with respect to the direction of spontaneous magnetization. Each of the site occupied by the above ions is coupled to the magnetic lattice by an energy of the form $E = w \cos^2 \phi$ [2] where $\phi$ is the angle between the anisotropy direction and the magnetization, w is the elementary coupling energy depending on the nature of the ion. Using the induced anisotropy Ku and the relation :

$$Ku = \frac{2}{15} c\, w^2/kT,$$

due to Neel and Taniguchi [3,4] we have evaluated w for different ions in different ferrites, where Ku is the uniaxial anisotropy induced with a field of 10 kOe at a temperature T where the time constants of the ionic or electron migrations are such that a superstructure could be formed, c the concentration of the anisotropic ion and k the Boltzmann constant.

The purpose of this work is to find out a relation, if it exists, between w and certain fundamental properties of the ferrite in which the ion is introduced. We started with the w of $Co^{2+}$ by the study of the exchange $Co^{2+} \leftrightarrow Co^{3+}$. We chose NiZnCo ferrite with a small deficit of iron which is to induce the formation of $Co^{3+}$, indispensable for the electron exchange $Co^{2+} \leftrightarrow Co^{3+}$. This type of ferrites has been studied by several including us since 1958. This type of ferrites being interesting for practical applications in filters for telecommunication due to their low losses. The band of after effect due to electron exchange has been found by us to have a maximum of DA near 220K. This range of T enable one to measure $K_u$ statically with a liquid $N_2$ cryostat. Further the quantity of cobalt added is very small and remains constant such that the electron exchange is maintained sensibly the same as also the $Co^{2+}$ content ought not to vary much. The samples we have chosen are of compositions $Fe_{1,985}\, Ni_{0,975-x}\, Zn_x\, Co_{0,04}\, O_{4+\gamma}$ with

x = 0., 0.075, 0.175, 0.275, 0.375, 0.475, 0.575

resulting in seven samples I, II, III, IV, V, VI, VII. Thus while maintaining the electron exchange constant, the surroundings of $Co^{2+}$ (which contribute to anisotropy) is modified considerably. Besides zinc being non magnetic its addition changes very much magnetization and the Curie temperature.

## EXPERIMENTAL

The samples were prepared by the usual ceramic technology. They were sintered at 1200 C for 4 hours in pure oxygen atmosphere.

The final products were analysed to determine the $Co^{3+}$ concentration and the respective compositions are given in table I. The degree of oxydation is very nearly zero for all the samples.

| Samples | $Ni^{2+}$ | $Zn^{2+} = x$ | $Co^{2+}$ | $Co^{3+}$ | $Fe^{3+}$ | $\gamma$ |
|---|---|---|---|---|---|---|
| I | 0,40 | 0,575 | 0,030 | 0,010 | 1,985 | - 0,0025 |
| II | 0,50 | 0,475 | 0,025 | 0,015 | 1,985 | 0,0000 |
| III | 0,60 | 0,375 | 0,022 | 0,018 | 1,985 | 0,0015 |
| IV | 0,70 | 0,275 | 0,024 | 0,016 | 1,985 | 0,0005 |
| V | 0,80 | 0,175 | 0,025 | 0,015 | 1,985 | 0,0000 |
| VI | 0,90 | 0,075 | 0,030 | 0,010 | 1,985 | - 0,0025 |
| VII | 0,975 | 0,000 | 0,035 | 0,005 | 1,985 | - 0,0050 |

Table I

| Samples | $K_{umax}$ erg/cm$^3$ | $T_{1K_{umax}}$ | C.$10^{-20}$ $Co^{3+}$ ions/cm$^3$ | w.$10^{15}$ erg/ion | $\sigma$ uem/g | $T_c$ °K |
|---|---|---|---|---|---|---|
| I | 3.100 | 240 | 1,27 | 2,42 | 83 | 465 |
| II | 5.500 | 235 | 1,90 | 2,66 | 87 | 555 |
| III | 7.700 | 232 | 2,28 | 2,82 | 88 | 635 |
| IV | 10.300 | 215 | 1,98 | 3,36 | 82 | 695 |
| V | 11.000 | 200 | 1,84 | 3,50 | 78,3 | 765 |
| VI | 10.300 | 210 | 1,22 | 4,24 | 62 | 825 |
| VII | 5.000 | 230 | 0,66 | 4,26 | 50 | 895 |

Table II

As we are interested in only the electron exchange at low temperatures the variation in $\gamma$ in such small degree does not interfere. The magnetic relaxation was evaluated by the measurement of the permeability disacommodation as a function of temperature and that of loss angle (arc tan $\delta$) as a function of temperature and frequency. The induced anisotropy Ku was measured by a torque balance. The measurements were carried out a 77K after treatments under a field at temperatures successively increasing $T_1$ followed by cooling under a field at 77K. One gets curves of $K_u = f(T_1)$. The Curie temperature and magnetization were also measured, the latter being by vibration sample magnetometer.

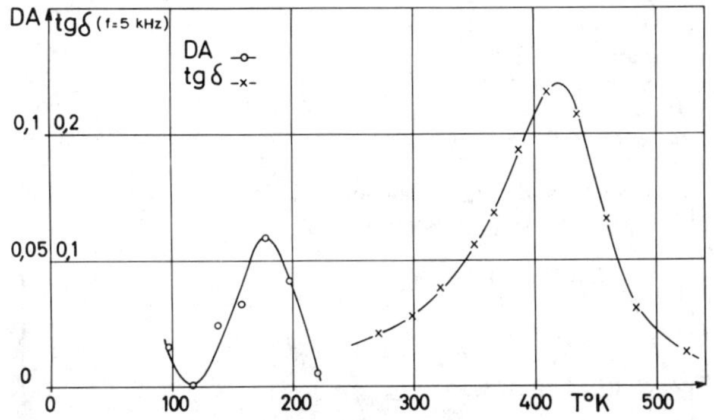


Fig. 1 : Variation of DA and tgδ of sample IV as function of temperature.

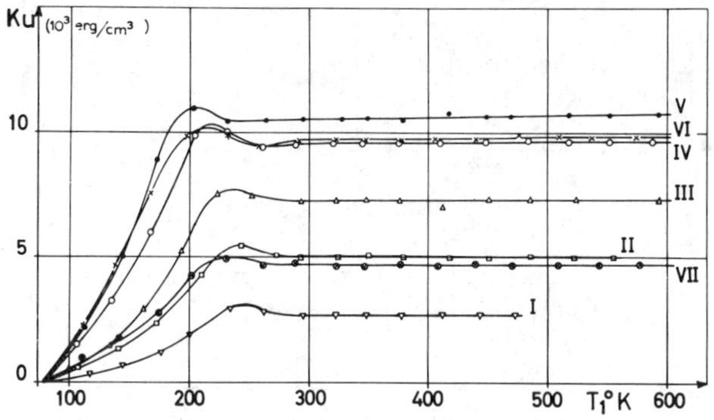


Fig. 2 : Variation of Ku of seven samples as function of treatment temperature.

## RESULTS

The thermal spectrum of DA of sample IV $Fe_{1,985}\ Ni_{0.70}\ Zn_{0.275}\ Co^{2+}_{0.024}\ Co^{3+}_{0.016}\ O_{4+0.0005}$ shows a maximum at 183K and the tan $\delta$ for f = 5 000 Hz as well at 435K (fig. 1). The two above parameters are manifestations of the electron exchange $Co^{2+} \leftrightarrow Co^{3+}$. The activation energy deduced from tan $\delta$ measurements for different frequencies is 0.8 eV in good agreement with the value found earlier by workers [5]. $K_u = f(T_1)$ is shown in fig. 2, where $T_1$ is the temperature of the treatment. Table II shows the values of different parameters for the seven samples. $K_u$ max is the maximum value of $K_u$ obtained in the range 200 to 240K, $T_{K_{umax}}$ is the temperature at which this maximum occurs, c the concentration per cm$^3$ of migrating elements, these being electrons $e^-$ between $Co^{2+}$ and $Co^{3+}$. It is hence the lower concentration of these two ions which control the number of electrons that migrate. In our case it is the $Co^{3+}$ concentration that controls the exchange. w is calculated from the relation :

$$K_u = 2/15\ c\ w^2/KT_1 \quad (1).$$

The formula must be rigourously written as :

$$K_u = 2/15\ c\ w(T)\ w(T_1)/KT_1 \quad (2).$$

where T is the measuring temperature and $T_1$ the treatment temperature. For evaluating w, we have supposed $w(T) \not\neq w(T_1)$, the error ought to be small as T and $T_1$ are not very much different, as compared to the change in w that comes about from the composition variation. $\sigma$ is the magnetization at $T_1$ and $T_c$ the Curie temperature. The linear variation of w with $T_c$ is shown in fig. 3.

## DISCUSSION

The coupling energy w was introduced by Neel in order to explain the directional order (or orientated superstructures) in alloys with two components A and B : $E = w \cos^2 \phi$, w being the elementary coupling energy of a pair A-B and $\phi$ the angle between the directions AB and spontaneous magnetization.

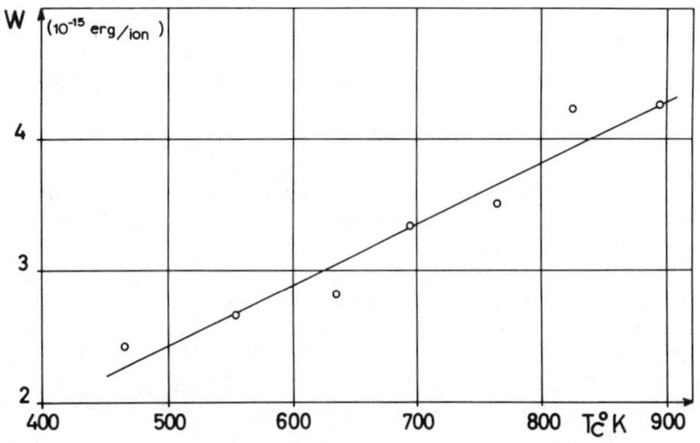


Fig. 3 : Variation of w as function of Curie point $T_c$.

In the case of ferrites, it is not possible to speak about a pair of metallic ions, the structure being different, with oxygen ions forming a cubic close packed structure, there are no direct cationic bonds. Nevertheless an individual $Co^{2+}$ has a minimum anisotropy energy while the magnetization is directed along the trigonal axis of the site it occupies. It will thus have a tendency to migrate in a manner so as to make coincide the trigonal axis with the direction of magnetization In place of pair model we have a model of preferential sites but the elementary energy w is of the same type. It is seen that w varies very much with the composition. We could not find a simple relation between the variation of w with $\sigma$, on the contrary w varies linearly with $T_c$ (fig. 3). Thus w is closely related to the molecular field coefficient or the exchange energy.

The above result confirms that the directional order model of Neel and Taniguchi applies well also to ferrites. In the model the ion position is coupled to the direction of spontaneous magnetization and we can add that the absolute value of w is related linearly to the exchange energy. Thus we see that w varies not only with nature of the ion ($Fe^{2+}$, $Cu^{2+}$, $Mn^{3+}$, $Co^2$) but also with the ferrite composition in which it is introduced. For the same ion w will be different in different ferrites, because the exchange energy is different also.

## REFERENCES

|1| A. MARAIS and T. MERCERON : Phys. stat. sol. 24, 635 (1967).

|2| L. NEEL, J. Phys. Radium, 13, 249 (1952).

|3| L. NEEL, J. Phys. Rad., 15, 225 (1954).

|4| S. TANIGUCHI and M. YAMAMOTO : Sci. Repts. Research Insts. Tohoku Univ. Ser. A6, 330 (1954)

|5| T. MERCERON : Ann. Phys. (France) 10, 121 (1965)

# THE INDUCED MAGNETIC ANISOTROPY IN DOUBLE HCP Co-Fe ALLOYS

T. Wakiyama*, G. Y. Chin, M. Robbins, R. C. Sherwood, and J. E. Bernardini
Bell Laboratories, Murray Hill, New Jersey 07974
*Permanent Address: Tohoku University, Sendai, Japan

## ABSTRACT

The magnetic anisotropy induced by magnetic annealing through the fcc → hcp or the fcc → double hcp phase transformation has been investigated for the Co-Fe alloys containing less than 4 at.% Fe. The sign of the induced uniaxial anisotropy was found to change from negative for the hcp phase below 1 at.% Fe to positive for the double hcp phase above 1 at.% Fe. The very large induced anisotropy (∿ $10^6$ erg/cc) was obtained for the double hcp polycrystalline alloys, while the sintered alloys showed a smaller anisotropy (∿ $10^4$ erg/cc). The result is discussed in terms of a crystallographic texture produced by cooling through the phase transformation in a magnetic field.

## INTRODUCTION

The double hexagonal close-packed (dhcp) structure in cobalt-rich Co-Fe alloys was discovered by Chikazumi, Wakiyama and Yosida[1] during a study of the magnetocrystalline anisotropy in 3d transition metals and alloys. This new dhcp phase, described by the stacking sequence ABAC, has unique ferromagnetic properties, namely, the direction of easy magnetization of the magnetocrystalline anisotropy at room temperature is perpendicular to the c-axis[1,2] in contrast with the hcp pure Co. Detailed structural and magnetic properties of this alloy system have been reported by one of the present authors.[2] Recently, an extensive study on the crystal structures and phase transformations of cobalt alloys containing less than 8 at.% Fe have been made by Onozuka et al.[3,4] by means of X-ray diffraction, electron microscopy and heat capacity measurements, and the stability range of the dhcp phase has been determined. A recent NMR study in the Co-Fe system by Kawakami et al.[5] has also confirmed the existence of this new phase; a magnetic phase diagram of alloys at low Fe concentration was proposed.

The fact that the dhcp phase of Co-Fe alloys possesses an easy basal plane of magnetization suggests some interesting possibilities for magnetic annealing. An induced anisotropy associated with the fcc → hcp transformation in a magnetic field was first found by Takahashi and Kono[6] for pure Co and Co-Ni alloys, and was subsequently studied by Graham[7] and by Sambongi and Mitui.[8] In the case of the dhcp Co-Fe alloy, we can expect the large negative uniaxial magnetocrystalline anisotropy to play an important role in the development of the anisotropy induced by magnetic annealing.

## EXPERIMENTAL

Polycrystalline disk samples were cut from ingots melted in argon atmoshpere by means of an induction furnace. The raw materials used were Co of 99.93% purity, containing 0.038% Ni 0.009% C and 0.005% Zn as main impurities, and Fe of 99.99% purity. The content of Fe in the samples was determined by chemical analysis. Specimen dimensions ranged from 9.31 to 15.83 mm in diameter and from 0.373 to 0.797 mm in thickness. Disk samples were also prepared by sintering powder, pressed under approximately 10,000 psi pressure, at 950°C for 90 min. in $H_2$. The starting powders were prepared by hydroxide coprecipitation and reduction in $H_2$ for four hours at 400 and 600°C for $Co_{0.985}Fe_{0.015}$ and $Co_{0.980}Fe_{0.020}$, respectively. The dimension and density of the sintered disks were 12.8 mm dia. × 0.47 mm and 8.16 g/cc for $Co_{0.985}Fe_{0.015}$, and 11.1 mm dia. × 0.34 mm and 7.38 g/cc for $Co_{0.980}Fe_{0.020}$. All samples were first annealed at 950°C for 90 min in $H_2$, and then cooled to room temperature in zero field or in a field of 1600 Oe.

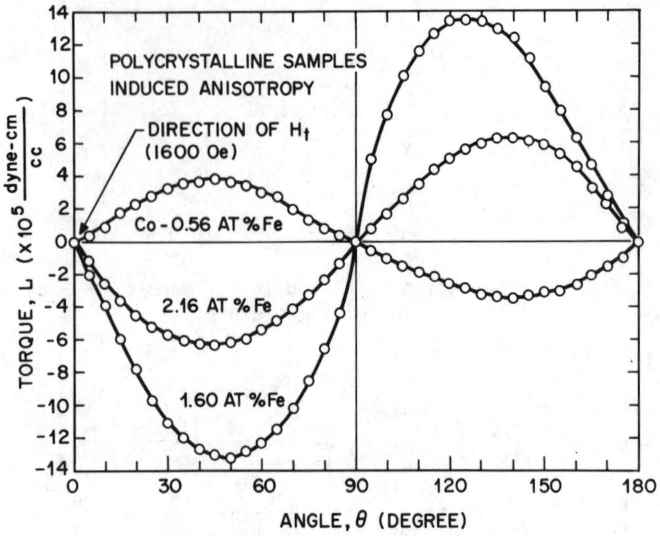


Fig. 1 Torque curves for the induced anisotropy of 0.56, 1.60 and 2.16 at.% Fe-Co polycrystals.

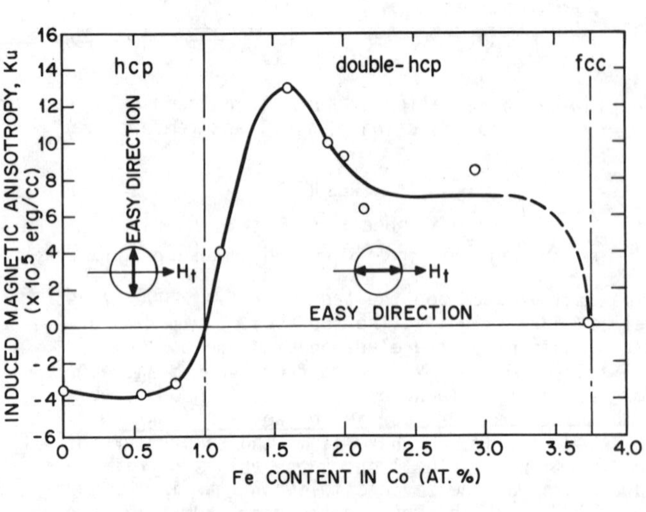


Fig. 2 Induced uniaxial constant as a function of the composition of Co-Fe Alloy.

The cooling rate was approximately 14°C/min. in the temperature range 450 to 200°C. The magnetic anisotropy was determined at room temperature by measuring the torque in a field of 10,000 Oe, which was sufficient for saturation. Torque curves attributed to the induced anisotropy were obtained by subtracting those measured on samples cooled in zero field from those on the same samples cooled in a field.

## RESULTS AND DISCUSSION

Figure 1 shows torque curves of the induced anisotropy for three different Fe concentrations. It is interesting to note that they have a uniaxial character and that the sign changes depending on the Fe content. Values of the induced anisotropy constant $K_u$ determined from curves such as these are plotted as a function of Fe concentration in Fig. 2. The salient results are as follows: (1) The sign of $K_u$ changes from negative to positive at about 1 at.% Fe, corresponding to the phase boundary between the hcp and dhcp structures.[4] That is, in the hcp region the easy direction of the induced anisotropy is perpendicular to the annealing field $H_t$, whereas in the dhcp region it is parallel to $H_t$. (2) The values of $K_u$ for the dhcp phase are large as contrasted with those for the hcp phase. Indeed, the value of $1.3 \times 10^6$ erg/cc for the 1.6 at.% Fe alloy is believed to be the largest value reported for magnetic annealing of polycrystalline material.

Previously, a mechanism has been proposed for the magnetic annealing effect associated with the fcc → hcp transformation in pure Co and Co-Ni alloys.[7,8] The same mechanism appears applicable to the present results also if we take into account the difference in the behavior of magnetocrystalline anisotropy between the hcp and dhcp phases of Co-Fe alloys.[2,5] According to the magnetic phase diagram obtained by the NMR study,[5] in the hcp region below 1 at.% Fe, the c-axis at the transformation temperature is the hard axis as dictated by magnetocrystalline anisotropy. Therefore during the fcc → hcp phase transformation it tends to become aligned in the plane perpendicular to the annealing field $H_t$. In this way, a texture is produced in the polycrystalline sample. At room temperature, however, the c-axis becomes the easy axis, so that the field direction becomes a hard axis as shown in Fig. 2. In the dhcp region above 1 at.% Fe, the magnetic annealing results in a similar texture as in the case of the hcp region, since the c-axis of the dhcp grains is also the hard axis at the transformation temperature.[5] In contrast to the hcp phase, however, the easy axis of the dhcp phase remains perpendicular to the c-axis down to below room temperature.[2,5] Therefore, the produced texture gives rise to the induced anisotropy with the easy direction parallel to $H_t$ as shown in Fig. 2. Initial attempts at detecting the texture by X-ray pole figure technique have been unsuccessful. There are several experimental complications to be resolved. First, the samples generally contain mixtures of fcc, hcp and dhcp phases and hence there is overlap from some of the lines. Secondly, the magnitude of $K_u$ suggests less than 10 percent alignment of the grains.

Concerning the large magnitude of the induced anisotropy $K_u$ of the dhcp phase, it is mainly attributed to the large magnetocrystalline anisotropy of this phase[2] as compared with that of the hcp phase.[2] The absence of the induced anisotropy for the 3.74 at.% Fe alloy is due to the fact that the phase transformation did not occur down to room temperature. This was confirmed by measurements of the temperature dependence of resistivity,[9] which showed a distinct change at the transformation temperature for the other samples.

The sintered samples also showed induced anisotropy. In this case, the value of $K_u$ was small ($\sim 10^4$ erg/cc) as compared with the cast samples. The porosity and the small grain size of the sintered materials may have inhibited the fcc → dhcp transformation.[10]

## REFERENCES


1. S. Chikazumi, T. Wakiyama and K. Yosida, Proc. Int. Conf. Magnetism, Nottingham, 756 (1964).
2. T. Wakiyama, A.I.P. Conf. Proc. 10, 921 (1973).
3. T. Onozuka, S. Yamaguchi, M. Hirabayashi and T. Wakiyama, J. Phys. Soc. Japan 33, 857 (1972).
4. T. Onozuka, S. Yamaguchi, M. Hirabayashi and T. Wakiyama, J. Phys. Soc. Japan 37, 687 (1974).
5. M. Kawakami, T. Hihara, Y. Kōi and T. Wakiyama, Proc. Int. Conf. Magnetism, Moscow 2, 158 (1973).
6. M. Takahashi and T. Kōno, J. Phys. Soc. Japan 15, 936 (1960).
7. C. D. Graham, Jr., J. Phys. Soc. Japan 17 B-I, 321 (1962).
8. T. Sambongi and T. Mitui, J. Phys. Soc. Japan 18, 1253 (1963).
9. T. Wakiyama, F. J. Di Salvo, J. V. Waszczak and G. Y. Chin, unpublished.
10. G. A. Fritzlen, W. H. Faulkner, B. R. Barrett and R. W. Fountain, Precipitation from Solid Solution, American Society for Metals, Metals Park, Ohio, 449 (1959).


# HOT PRESSED CERAMIC FERRITES: MAGNETIC-MECHANICAL-MICROSTRUCTURAL INTERACTIONS


L.S. Brissette, E.A. Grossi, J.M. Titlar, K. Cherven
National Micronetics, Inc., W. Hurley, New York 12491

R.M. Spriggs
Lehigh University, Bethlehem, Pa. 18015


## ABSTRACT


Utilizing unique ceramic hot pressing processing technologies, custom hot-pressed NiZn and MnZn ferrites have been prepared having near-theoretical densities and fine-grain-size microstructures coupled with superior magnetic and mechanical properties. The interactions among magnetic and mechanical properties and microstructural features to be presented include: Mechanical-Microstructural (bend strengths increase with decreasing grain size - up to 45,000 p.s.i. at 7 microns for NiZn and 40,000 p.s.i. at 10 microns for MnZn); Magnetic-Microstructural (initial permeabilities and figures of merit show sharp increases with decreasing grain sizes and as function of increasing drive levels, e.g., μQ of 130,000 at 10 gauss (100 KHZ) and 10 microns grain size for MnZn); Thermal-Magnetic (initial permeability and figure of merit as a function of temperature at three frequencies [0.5 MHZ, 1.75 MHZ and 3.5 MHZ] and different grain sizes, measured for "as machined" and thermally annealed toroids show several unusual responses); and Thermal-Microstructural and Magnetic-Mechanical interactions. Greater permeabilities and higher strengths have been observed for the MnZn ferrite of this study than previously reported, even at larger grain sizes.

# THE INVAR CHARACTERISTICS ON Co-Fe ALLOYS

M. Takahashi, F. Ono and K. Takakura
Department of Applied Physics, Tohoku University, Sendai, 980, Japan

## ABSTRACT

The Invar characteristics has been examined for Co-Fe alloys. The measurements of the magnetization and the thermal expansion over the temperature 20°C∿1050°C carried out for the alloys containing 59 to 100 at.%Co. The concentration dependence of the saturation magnetization around 850°C shows a sharp minimum at about 68 at.%Co. The thermal expansion coefficient of 77 at.%Co alloy is invariant over a temperature range from about 900°C to 750°C and the value of the coefficient decreases and the temperature range becomes narrower with decreasing Co concentration. The small thermal expansion coefficient and the extremely large volume magnetostriction estimated from the expansion curves are found around 74 at.%Co alloy at about 820°C. These characteristics are quite similar to that of Fe-Ni Invar alloys at room temperature, and may be explained by the local fluctuation model in martensitic phase transition.

## INTRODUCTION

The Invar alloy has been accepted practically as a material with a very small thermal expansion coefficient at room temperature.[1] In addition, the concentration dependence of the saturation magnetization Ms around the Invar composition shows an anomalous behavior and large spontaneous volume magnetostriction, high field susceptibility, and pressure effect to Ms and Curie temperature. Furthermore, the Invar characteristics commonly appear (1) close to the composition of $\gamma\rightarrow(\alpha+\gamma)$ phase boundary and (2) at the outer electron numbers of about 8.6 atom. These interesting characteristics are the so-called Invar problem and much attention has been paid to it.

The phase diagrams of Co-Fe alloys containing more than 60 at.%Co and Fe-Ni alloys containing less than about 30 at.%Ni are very similar to each other. The Fe-Ni Invar alloy around 30 at.%Ni and Co-Fe alloys with about 75∿90 at.%Co are located in $(\alpha+\gamma)$ mixture phase at room temperature. Also, the curves of Ms vs composition at 0°K for both Fe-Ni and Fe-Co alloys exhibit a kink at a certain composition in $(\alpha+\gamma)$ mixture phase. Based on this similarity, it is interesting to study whether Co-Fe alloys show the Invar characteristics or not.

## EXPERIMENTAL PROCEDURE

The temperature dependence of Ms was measured for the alloys with various Co concentration between 59 to 100 at.% by using a vibrating sample magnetometer in an external field 14 kOe.

The measurements of thermal expansion have been made with a differential method by using a strainmeter from room temperature up to 1050°C. As a reference material, a quartz rod (5ϕx53 mm) was used and the thermal expansion coefficient, $0.58\times10^{-6}$ $deg^{-1}$ is assumed to be constant[2] over the temperature range up to 1050°C. The rod specimens (3.0ϕ x53.0 mm) formed by forging and grinding were used after being annealed for 1 hour at 1100°C.

The crystal structure and the lattice parameter were determined from room temperature to 1050°C by using a Co-target X-ray diffractmeter.

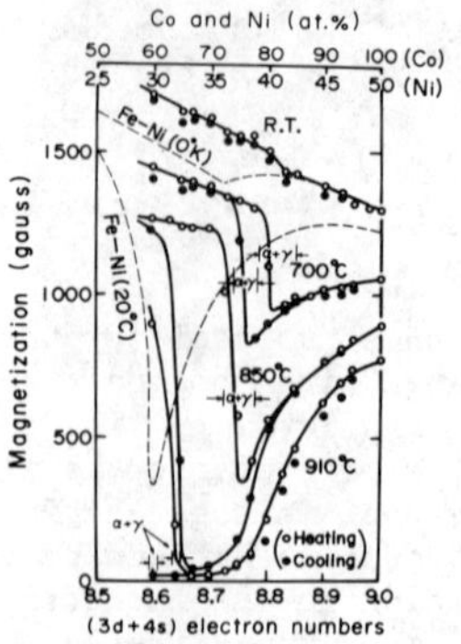


Fig.1 Concentration dependence of Ms of Co-Fe alloys. The dependence for Fe-Ni alloys at 0°K and at 20°C are also plotted for comparison.

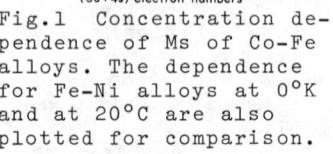

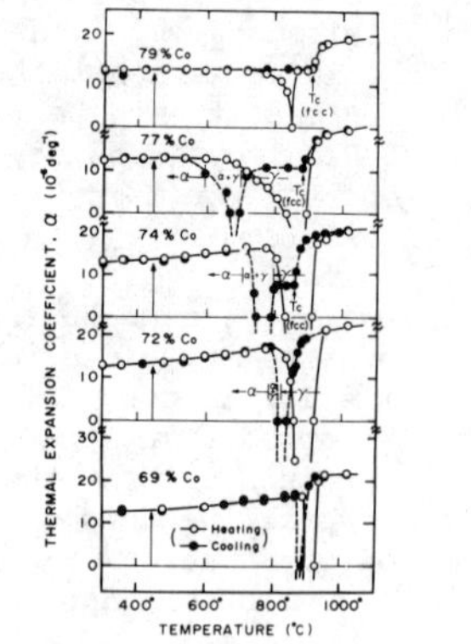


Fig.2 Temperature dependences of α obtained from the thermal expansion measurements.

## EXPERIMENTAL RESULTS

(1) Compositional dependence of the saturation magnetization: The temperature dependence of the saturation magnetization was measured from room temperature up to 1050°C at a heating and cooling rate 100°C/hr. On heating, Ms for 59∿74 at.%Co-Fe alloys disappeared abruptly and became zero at about 900°C ($\alpha\rightarrow\gamma$ transition temperature), while for the alloys more than 77 at.%Co, Ms decreased gradually with increasing temperature and disappeared at the Curie temperature for γ-phase at about 1000°C. On cooling, Ms appeared suddenly for 59∿69 at.% Co alloys having a temperature hysteresis about 10∿100°C, while for 72∿79 at.%Co alloys Ms appeared at about 900°C and increased gradually and then abruptly at about 800∿700°C. For 82∿90 at.%Co alloys, Ms changed reversibly on heating and cooling. Fig.1 shows Ms vs concentration near the temperatures at which the irreversible change in Ms with temperature occurred. As seen in Fig. 1, the dependence of Ms on composition exhibits a kink at room temperature around the composition of 82 at.%Co. The value of Ms for alloys in the γ-phase, however, decreased remarkably from the Co side at a temperature higher than 700°C with increasing Fe concentration. For example, the sharp minimum appeared on cooling at about 77 and 70 at .%Co at 700°C and 850°C, respectively, and was no longer observed at 910°C. It is interesting that these compositions where the minimum takes place nearly correspond to those for $\gamma\rightleftarrows(\alpha+\gamma)$ phase boundary at those temperatures, and shift toward lower concentration of Co with increasing temperature

This behavior of Ms vs composition as a function of temperature is quite similar to that for Fe-Ni alloys near the Invar composition. Namely, in the case of Fe-Ni alloys, there is a slight kink around 36 at.%Ni at 0°K and a sharp minimum at about 30 at.%Ni at 20°C as shown in Fig. 1. The compositional dependence of Ms(o) at T=0 for f.c.c. Co-Fe alloys is obtained by extrapolating the Ms vs T curve in γ-phase. As the result, it is found that Ms(o) slightly increases with decreasing Co content up to about 80 at.%Co (outer electron number 8.8/atom) and then takes a tendency to decrease. This behavior is similar to that of

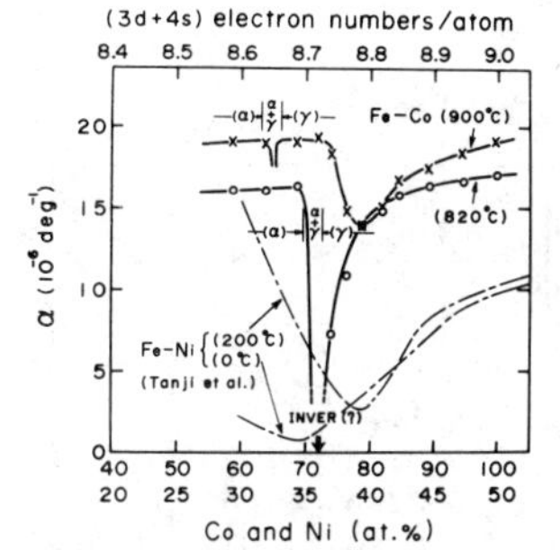


Fig.3 Concentration dependences of α for Co-Fe alloys at 820 and 900°C. The values of Fe-Ni alloys are also shown against the same outer electron numbers per atom.

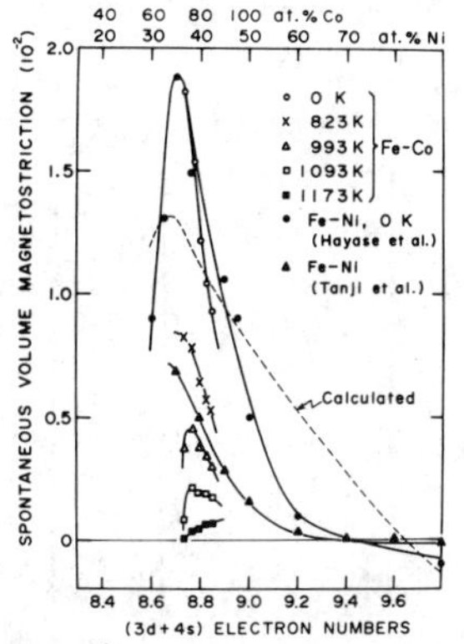


Fig.4 Concentration dependences of $\omega_s(T)$ in Co-Fe alloys. The estimated values for Fe-Ni alloys by Hayase et al.[3] and Tanji et al.[4] and the calculated values by using the rigid band model are also shown.

Fe-Ni alloys around 40 at.%Ni (8.8/atom). On the other hand, Tc decreases with decreasing Co concentration from 1070°C to 730°C for 74 at.%Co-Fe alloy.

(2) Compositional dependence of thermal expansion coefficient: The measurements of the thermal expansion were carried out at the heating and cooling rate of 50°C/hr. As the results, the abrupt and gradual contraction and expansion of about 0.3% were observed for 59∿69 at.%Co-Fe alloys and 72∿79 at.%Co-Fe alloys, respectively at the temperatures corresponding to the ($\alpha \rightleftarrows \gamma$) phase transformation on heating and cooling. There was no observed discontinuity for the alloys more than 80 at.%Co.

The thermal expansion coefficient, α was obtained from the measurement of the inclination of the tangential line at each point on the thermal expansion curve and shown in Fig.2. It is seen in this figure that α for 74∿79 at.% Co alloys is invariant on cooling and constant over the temperature range from the Curie point for γ-phase to temperatures beyond the $\gamma \rightarrow (\alpha+\gamma)$ transition temperature. The values of the constant decreases from 11 to $7 \times 10^{-6}$/deg and the temperature range becomes narrower from 150 to 50°C with decreasing concentration from 77 to 74 at.%Co.

The dependence of α on composition is given in Fig.3. As seen in this figure, the sharp minimum exists near 77 at.%Co at 900°C. This composition is about 8% higher than the composition where Ms is a minimum at the same temperature. At 820°C, the constant value sharply decreases with decreasing Co content and it seems to become zero (Invar) at about 72 at.%Co. This composition is higher by about 4% than that for the minimum of Ms. There is a good correspondence for these characteristics between Fe-Co and Fe-Ni Invar as shown in the same figure.

## DISCUSSION

As the Invar alloys have a extremely large spontaneous volume magnetostriction, $\omega_s(T)$, it is interesting to examine $\omega_s(T)$ for Co-Fe alloys. The value of $\omega_s(T)$ can be estimated from the observed thermal expansion curve using the Grüneisen relation[3] in the usual way. The concentration dependence of $\omega_s(T)$ is shown in Fig.4. As seen in this figure, $\omega_s(T)$ shows a peak at about 77 at.%Co at 720°C and it becomes large anomalously, as the temperature decreased. The values of $\omega_s(o)$ obtained by using the $\omega_s(T)$ vs $T^2$ relation[4] are also shown in Fig.4. The relation of $\omega_s(o)$ vs composition corresponds to those of Fe-Ni Invar alloys obtained by Hayase et al.[3] and Tanji et al.[5]

To explain the anomalous behavior of $\omega_s(o)$ in Co-Fe alloys, the latent antiferromagnetism and the low spin state models proposed by Kondorsky et al.[6] and Weiss[7] are considered. In these models, the numbers of the Fe-Fe atom pair play a significant role to the anomalies of $\omega_s(o)$ and Ms(o) and these models explain the anomalies in Fe-Ni Invar alloys reasonably. However, it is difficult to explain nearly the same order of anomalies found in 20 at.%Fe-Co alloys, because the number of the Fe-Fe atom pair in 20 at.%Fe-Co alloy is much less than that for 70 at.%Fe-Ni Invar alloy.

The collective electron model, which is successfully applied to the Invar alloys[8] is taken into account. By this model, Ms(o) and Tc should be zero at the electron numbers 8.6/atom. However, both Ms and Tc for Co-Fe alloys are large at the expected numbers of electrons as in the case of Fe-Pd alloys.[9] This fact can be hardly explained by this theory.

Therefore, the model for local fluctuation[10] in the (α+γ) mixture phase may explain these anomalies. When a specimen is cooled down from a high temperature, the α-phase which has a large atomic volume (11.982 $Å^3$ at 900°C) precipitates dispersibly in the γ-phase (11.860 $Å^3$ at 900°C). Therefore, the usual contraction by the lattice could be supressed by the appearance of the α-phase and this causes α to be smaller and $\omega_s(T)$ to be larger. But a more detailed study is necessary for quantitative discussions.

## ACKNOWLEDGMENT

The authors express their thanks to Prof. T. Ikeda, Dr. Y. Tanji and Assistant Prof. T. Suzuki for their kind discussions. The thanks are extended to Mr. S. Kadowaki for the preparation of the specimen, and to Mr. M. Aihara for the X-ray diffraction measurements.

## REFERENCES

1) C. H. Guillaume: Compt. Rend. 125, 235 (1897).
2) J. Strong: Modern Physical Laboratory Practice, Blackie and Son Ld., (1950).
3) M. Hayase et al.: J. Phys. Soc. Japan, 34, 925 (1973).
4) E. P. Wohlfarth: J. Phys. C. (Solid St. Phys.) 2, 68 (1969).
5) Y. Tanji and Y. Shirakawa: J. Japan Inst. Met., 34, 228 (1970). (in Japanese)
6) E. I. Kondorsky and V. L. Sedov: J. Appl. Phys., 31, 331S (1960).
7) R. J. Weiss: Proc. Phys. Soc., 82, 281 (1963).
8) T. Mizoguchi: J. Phys. Soc. Japan, 25, 904 (1968); M. Shimizu and S. Hirooka: Phys. Letters, 27A, 530 (1968).
9) H. Fujimori and H. Saito: J. Phys. Soc. Japan, 20, 293 (1964).
10) S. Kachi and H. Asano: J. Phys. Soc. Japan, 27, 536 (1969).

# THE STABILITY OF AUSTENITIC STAINLESS STEELS UNDER DEFORMATION

A. Riley, J.G. Booth and R.S. Tebble
Department of Pure and Applied Physics, University of Salford, Salford M5 4WT, England

## ABSTRACT

The effect of severe deformation (up to 90% reduction in area) on the magnetic properties of a wide range of austenitic stainless steels is reported. The range includes the commercially available 300 and 200 series together with some special steels. The amount of strain induced martensite formed under deformation has been studied by magnetisation measurements. The general response of a steel to deformation appears to be related to its effective nickel and chromium contents through a phase-field diagram not unlike Schaeffler's diagram for welded steels. A dependence is also found between strain-induced transformations and the severity with which the deformation is carried out as well as the composition. Considerable differences are found between mill-drawn wires and steels of identical compositions deformed to the same cross-section in the laboratory but under isothermal conditions.

## INTRODUCTION

Some applications of stainless steel wires require a stable, low magnetic permeability in addition to the high tensile strengths and corrosion resistance. Low permeabilities result from the retention during fabrication and subsequent service of the f.c.c. austenitic $\gamma$-phase. However, variations in the composition of austenite may profoundly affect the readiness with which a martensitic transformation to the b.c.c. ferromagnetic $\alpha$-phase can occur. Llewellyn and Murray[1] in a detailed description of the mechanical properties of austenitic stainless steels only briefly considered magnetic properties in presenting the martensite content in cold-rolled samples. Eberly[2] has discussed the variation in permeability after cold-reduction (of thickness) of some 300 grade steels and has pointed out that narrow compositional limits are required for accurate predictions of the magnetic properties of cold-worked commercial stainless steels.

This paper reports the effects of up to 90% reduction in area on the magnetic properties of a wide range of austenitic stainless steels. Although the investigation has been primarily concerned with the commercial grades of the 300 and 200 series, some special steels (Nitronic 32 and 50) and some experimental steels have also been studied.

## EXPERIMENTAL PROCEDURE

Samples of 305, 316, 304, 302, Armco 18/2 (Nitronic 32) and Armco 22/13/5 (Nitronic 50) were available in the form of mill-drawn wires of various diameters (1 to 6 mm). Wires from most of these grades were also produced in the laboratory from thick commercially supplied rods along with samples of grades 202 and 310. For the laboratory reductions deformation was achieved with square-section laboratory rollers using the lightest possible passes at a rolling speed of 10 cm $s^{-1}$ and at the same time preventing appreciable temperature rises by water cooling. Small pieces for magnetic analysis were cut away by spark erosion as the reduction proceeded.

A number of experimental alloys were formed by arc melting mixtures of commercial grades in such proportions as to form compositions occupying appropriate regions of the modified Schaeffler diagram to be described below.

Magnetisations ($\sigma$) in fields up to 17 kG were obtained using a vibrating sample magnetometer. Specimens were formed into the shape of approximate cylinders 1 mm x 3 mm by spark erosion and

| Type | Mn | Ni | Cr | Mo | C | N | $\alpha$ | $\mu$ | r.i.a |
|---|---|---|---|---|---|---|---|---|---|
| 304 | 1.79 | 8.01 | 18.5 | 0.32 | 0.05 | 307 | 71 | 25 | 70% |
| 302 | 1.6 | 8.09 | 17.55 | 0.24 | 0.11 | 407 | 32.5 | 2.2 | 70% |
| 316 | 1.69 | 10.12 | 16.62 | 2.43 | 0.06 | 160 | 28 | 2.1 | 70% |
| 305 | 1.30 | 10.92 | 17.8 | 0.17 | 0.06 | 476 | 1.75 | 1.2 | 70% |
| N32 | 12 | 1.47 | 18.53 | <0.05 | 0.03 | 1200 | 0.55* | 1.02 | 49% |
| 202 | 9.44 | 5.9 | 17.8 | 0.11 | 0.07 | 2115 | 0.45 | 1.04 | 81% |
| B | 7.47 | 9.25 | 19.45 | 1.12 | 0.10 | 1200 | 0.102 | 1.012 | 91% |
| E | 1.68 | 12.83 | 18.67 | 1.82 | 0.06 | 150* | 0.014 | 1.008 | 90% |
| N50 | 5.37 | 12.53 | 20.86 | 2.14 | 0.04 | 2900 | <0.01 | <1.004 | 90% |
| 310 | 1.68 | 21.0 | 24.9 | 0.09 | 0.06 | | <0.01 | <1.004 | 90% |

N in ppm
* estimated value

Table I. Compositions (wt%) and magnetic properties of the laboratory reduced steels; $\alpha$ is the percent of martensite formed by $\gtrsim$70% r.i.a.; $\mu$ is the initial permeability at 250 Oe as a result of the r.i.a. indicated.

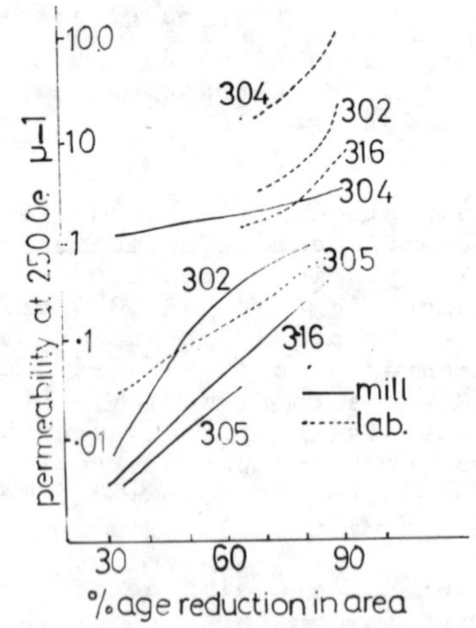


Figure 1 Permeability at 250 Oe for mill-drawn and laboratory reduced steels as a function of reduction in area (% r.i.a.)

electropolishing to remove any surface contamination. Demagnetising fields were taken into account where necessary. Chemical analyses of all the steels examined are shown in Table I.

## RESULTS

The mill-drawn samples show initial permeabilities $\mu = 1+4\pi\rho(\Delta\sigma/\Delta H)$; after high reduction ranging from 1.02 to 2.0 (Figure 1). However, as also shown in Fig 1, when steels of the same compositions were carefully reduced in the laboratory much higher permeabilities resulted, only the 305 sample showing values below 2.0 whilst the other grades acquired permeabilities up to 200. Despite the dramatic differences between the two methods of reduction the same hierarchy of permeability values is found so that, under both conditions of reduction the order of increasing permeability and decreasing austenite stability is 305, 316, 302, 304.

Figure 2 shows the variation in martensite content i.e. the amount of transformed material for the mill-drawn and laboratory-reduced specimens. The quantity of martensite present is found from the value $\sigma_0$ obtained by extrapolation of the linear high-field part of the magnetisation curve to zero applied field. The martensite content c is then given by $\sigma = cM_0$ where $M_0$ is the intrinsic magnetisation of the martensite calculated by the

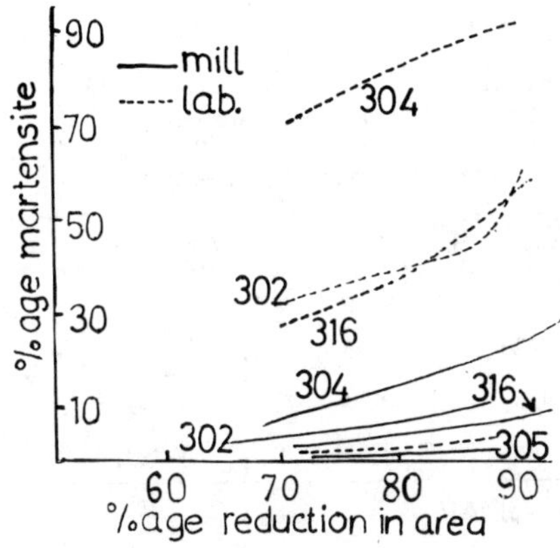


Figure 2 Martensite content of mill drawn and laboratory reduced steels as a function of reduction in area.

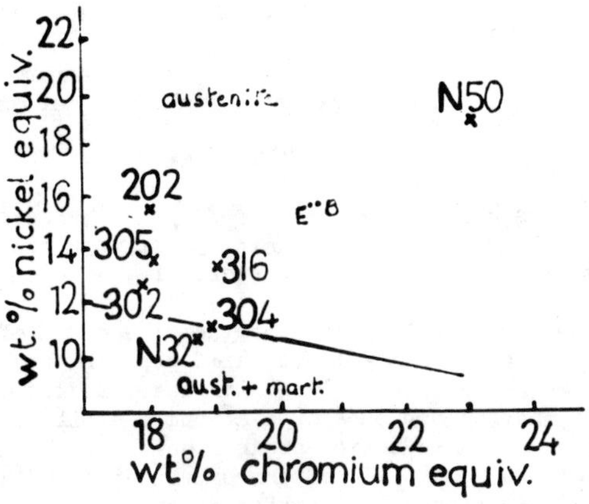


Figure 3. Adapted Schaeffler diagram showing effective Ni and Cr compositions for the steels examined, assuming nickel equivalents of 30x for carbon, 10x for nitrogen and 0.6x for Mn.

usual method of considering the reduction in the intrinsic magnetisation of pure iron produced by alloying additions[3]. The same order of stability is again exhibited for the mill-drawn and laboratory-reduced specimens if the martensite content is taken as the criterion of stability rather than the permeability (Fig 2). The alloy 304, reduced isothermally, transformed to up to 85% martensite at higher reductions whilst its mill-drawn counterpart transformed to only 25% . On the other hand the 305 grade showed less than 10% martensite to be produced under both conditions of deformation.

202 and Nitronic 32 are examples of steels in which the main austenite stabiliser, Ni, has been partially replaced by manganese and nitrogen. Samples were available only in the form of relatively thin sheet or wire so that high reduction was not possible. The magnetic measurements (see Table 1) show that these steels are of comparable stability, both being more stable than 305. No increase in martensite content could be detected in mill-drawn or laboratory reduced samples of the N50 and 310 steels. Taking into account experimental errors this implies that the permeability is less than 1.004 under all conditions of deformation at room temperature.

## DISCUSSION

Bloom and White[4] have reported some magnetic measurements on mill-drawn wires of the 300 series and found that a given wire had a lower permeability if the r.i.a. was achieved using either a high speed of pass or a small number of passes i.e. there is an inverse relationship between and the "violence" with which the reduction was achieved. For example a 302 wire after 80% r.i.a. produced in 10 passes had $\sim$ 7. This value was approximately three when the reduction was achieved in 6 passes. In our work major differences between mill-drawn and laboratory reduced material also indicate that for a given steel the permeability reflects the violence with which the reduction is carried out. The most obvious reason why this should be so is that the mill-drawing operations are usually carried out at successively increasing temperatures as a result of the heat of working at each die. Further experiments are being performed to isolate the mechanisms which could be responsible for this.

In attempting to rationalise the behaviour under deformation of the steels in terms of their varied compositions we found it useful to construct a diagram not unlike the well-known Schaeffler diagram[5] (in which the phases to be expected in stainless steel welds are predicted in terms of effective Ni and Cr contents). In constructing this diagram (in our case indicating the stability of the fcc form with respect to deformation) some knowledge of the relative effectiveness of the austenizing elements is required. A comparison of pairs of steels of similar Cr content and approximately equal stability[6] suggests that Mn is 0.6 times as effective as Ni. Schaeffler's original consideration required C to be counted as 30 times as effective as Ni in stabilising austenite in welds. As all our alloys contained comparable C contents we were unable to isolate the effect of this element and have provisionally used Schaeffler's value. Tisinai and Samans[7] find N to be 40 times as effective as Ni in stabilizing austenite in solution annealed steels. We find from a consideration of some experimental alloys (indicated in table I as E and B) that a N effectiveness of between 5 and 10 times that of Ni would probably be more appropriate. Using these values the diagram for all the steels we have examined is reproduced as Fig 3. The sloping full line in this figure indicates the austenite/(austenite + martensite) phase boundary of the Schaeffler weld diagram. We assume that a similar boundary will exist in approximately the same position for deformation induced phase fields. In general, for the small number of steels we have examined the nearer an austenitic steel is to this boundary the more likely martensite is to be produced by deformation . Only the steel N32 occupies an anomalous position using the potency values given, and it may be that the extraordinary concentration of Mn in this alloy renders it non-comparable with the rest. We propose to investigate this apparent relationship further in future work.

## ACKNOWLEDGEMENT

This work has been supported by the Ministry of Defence (AUWE) and material was supplied by the British Steel Corporation and Bridon Wires Ltd. The authors are also grateful to NSWC (White Oak Laboratory), the Armco Corporation and to the USAF (AFOSR) for helpful information and advice.

## REFERENCES

1. D.T. Llewellyn and J.D. Murray Bisra ISI Conference 197-212 (1964)
2. W.S. Eberley. Electrical Manufacturing Sept (1958) vol. 62 pt 3. 90-94
3. K. Hoselitz: Ferromagnetic Properties of Metals and Alloys (Oxford, 1952)
4. F.K. Bloom and J.S. White Wire & Wire Products October 1952 1036
5. A.L. Schaeffler Metal Progress 56 680-688 (1949)
6. A. Riley, J.G. Booth and R.S. Tebble (to be published)
7. G.F. Tisinai and C.H. Samans. Trans ASM 51 27 (1959)

# A STUDY ON THE OCCURRENCE OF HIGH PERMEABILITY IN GRAIN ORIENTED SILICON STEEL - HI-B

K. Takashima, T. Sato, and F. Matsumoto
Process Technology R. & D. Laboratories, Nippon Steel Corp., Kitakyushu, Japan

## ABSTRACT

The recrystallization behavior of grain oriented silicon steel was investigated using 3% silicon steel containing AlN, Sb-Se and MnS respectively, as grain growth inhibitors.

The most important factors in the production of grain oriented silicon steel with high permeability ($\mu_{10}$>1900) were the cold rolling with heavy reductions of over 80% and the presence of a strong inhibitor.

The heavy cold reduction brought on the small deviation of the [001] axis of the primaries with the (110) planes almost parallel to the rolling direction. This texture provides a large concentration of secondary nuclei.

The heavy cold reduction induced an increase in (200) and (222) intensities in the primary texture and made the primary grains fine and favored grain growth.

The presence of a strong inhibitor restrained effectively an easy grain growth.

## INTRODUCTION

Recently, demands for grain oriented silicon steel with high permeability have become quite strong from the standpoint of decrease in energy loss.

Some studies have been reported so far concerning the processing of the production of grain oriented silicon steel with high permeability.[1,2] However, there is a lack of report regarding the mechanism of the occurrence of high permeability, namely, the sharp (110)[001] secondary texture.

The purposes of this study are to extract the most important factors in the production of grain oriented silicon steel with high permeability and to reveal the mechanism of the occurrence of high permeability.

## EXPERIMENTAL PROCEDURE

Three kinds of 2.3mm thick hot rolled 3% silicon steel sheets with inhibitors and 0.9mm thick 3% Si-Fe single crystals with (110) planes obtained by secondary recrystallization method[1] were used. Compositions of these specimens are shown in Table I.

For the purpose of making the constant final thickness; 0.30mm, hot rolled sheets were cold rolled to the intermediate thicknesses corresponding to the final cold reductions. Before final cold rolling, specimens A1 and C were annealed for 2min at 1100°C and quenched. Specimens A2 and B were annealed for 2min at 900°C and slowly cooled. Then, they were cold rolled to reductions from 30 to 87.5% and annealed for 2min at 850°C in wet $H_2$+$N_2$. After this anneal they were given a final heat treatment for 20hrs at 1200°C with a heating rate of 50°C/hr in dry $H_2$+$N_2$.

Crystal D1 whose [001] axis is parallel to the rolling direction and crystal D2 whose [001] axis is 13° away from rolling direction were cold rolled to 60 and 85% reductions and then annealed for 2min at 850°C in dry $H_2$+$N_2$.

## RESULT AND DISCUSSION

The relations of final cold reduction to permeability for heat treated specimens are shown in Fig. 1. The permeability of samples A1 and C increased with increasing cold reduction and high permeability values ($\mu_{10}$>1900) were obtained over 80% in reduction. In specimens A2 and B the permeability showed maximum values in the reduction range of 60 to 70% and went down over 70% in reduction, but high permeability values were not obtained.

This suggests that the heavy final cold reduction over 80% and the presence of an inhibitor in an adequate condition are necessary for obtaining high

Table I. Chemical composition of specimens. (%)

| SPECIMEN | C | Si | Mn | S | Al | N | Sb | Se |
|---|---|---|---|---|---|---|---|---|
| A1, A2 | 0.050 | 2.95 | 0.08 | 0.028 | 0.030 | 0.007 | — | — |
| B | 0.031 | 3.15 | 0.08 | 0.025 | 0.002 | 0.004 | — | — |
| C | 0.042 | 3.00 | 0.04 | 0.008 | 0.001 | 0.005 | 0.03 | 0.03 |
| D1, D2 | 0.002 | 2.95 | 0.11 | 0.005 | 0.041 | 0.019 | — | — |

* Acid sol. Al

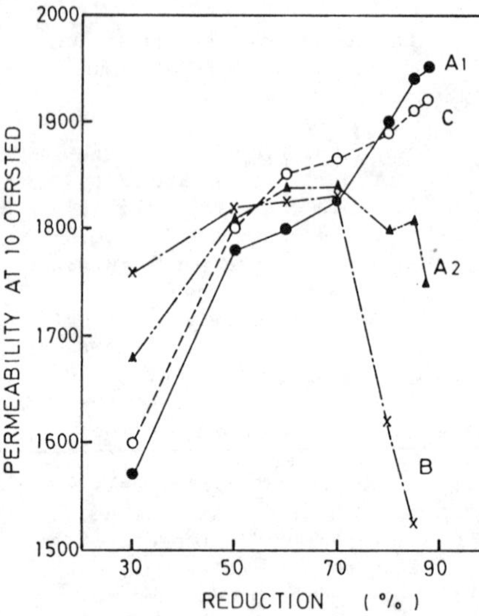


Fig. 1. Effect of final cold reduction on permeability for final annealed products of specimens A1, A2, B and C.

permeability products.

To clarify the effect of this inhibitor in an adequate way, the growth behavior of primary grains in some specimens during final heat treatment were observed. As shown in Fig. 2, in case of the 85% final cold reduction grain growth scarcely took place in specimen A1 until the beginning of secondary recrystallization. On the other hand, grain growth took place in specimen A2 and markedly in specimen B in which secondary recrystallization did not occur. In case of the 60% reduction, grain growth did not take place in all specimens. Since the grain growth becomes

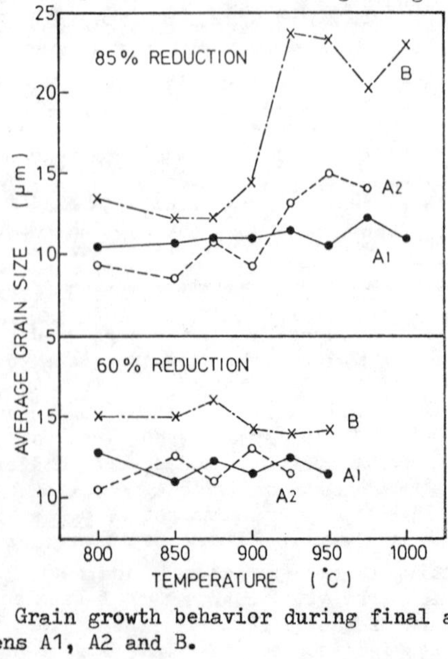


Fig. 2. Grain growth behavior during final anneal in specimens A1, A2 and B.

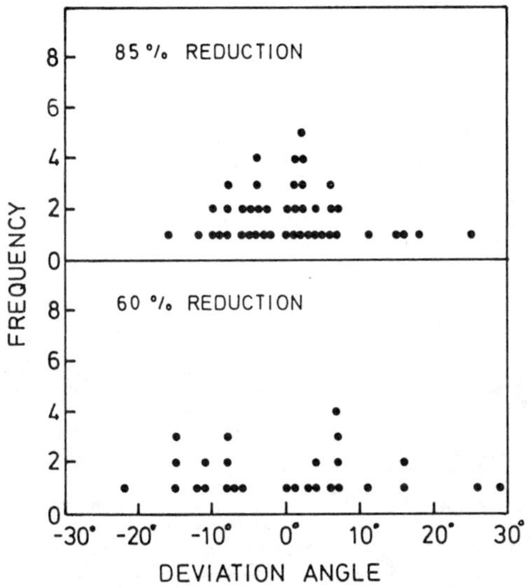


Fig. 3. Deviation of [001] axis of primaries with near (110) plane from rolling direction in specimen A1 annealed at 850°C for 2 min.

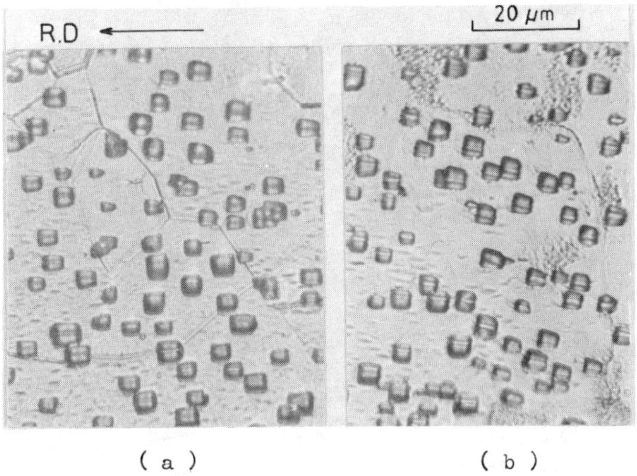


Fig. 4. Micro-etched surfaces of crystal D2 after cold rolling and anneal. (a) 85% reduction (b) 60% reduction

easier with increasing final cold reduction, the stronger inhibitors are needed at the higher reduction levels.

Next, the necessity of the heavy final cold reduction over 80% is discussed.

In specimen A1 with a strong inhibitor, the deviation of [001] axis of primaries with near (110) planar orientation to the rolling direction was smaller in case of the 85% reduction than that in case of the 60% reduction, as shown in Fig. 3. This fact was supported by the single crystal results.

In crystal D2, the [001] axis of primaries with near (110) planes came close to the rolling direction after 85% reduction, whereas for 60% reduction a rotation of the [001] axis toward rolling direction was smaller as shown in Fig. 4. In crystal D1 the [001] axis in primaries with near (110) planar orientation were parallel to the rolling direction in either reduction.

During the final heat treatment, changes in the primary structure and texture could not be observed until the beginning of secondary growth in specimen A1.

These facts suggest that the sharp (110)[001] primaries will bring about the secondary texture sharpness.

The (110)[001] primaries were often observed to exist in the form of a colony or a group in the 85% reduced A1 specimen as shown in Fig. 5. It is probable that the colony of the (110)[001] primaries develops into a secondary nucleus.

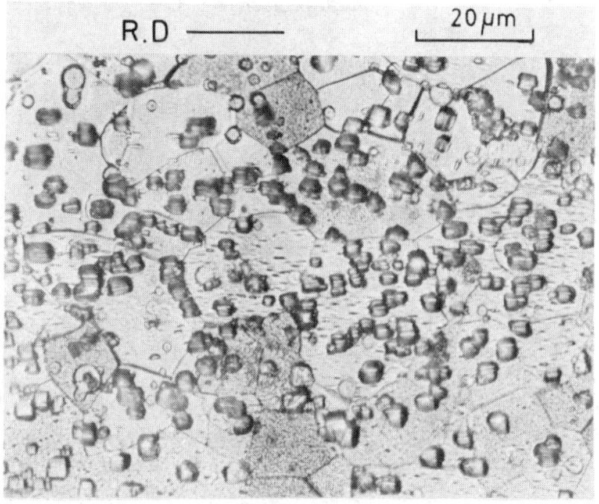


Fig. 5. A group of the primaries with near (110) [001] orientation in specimen A1 final cold rolled 85% and annealed at 850°C for 2 min.

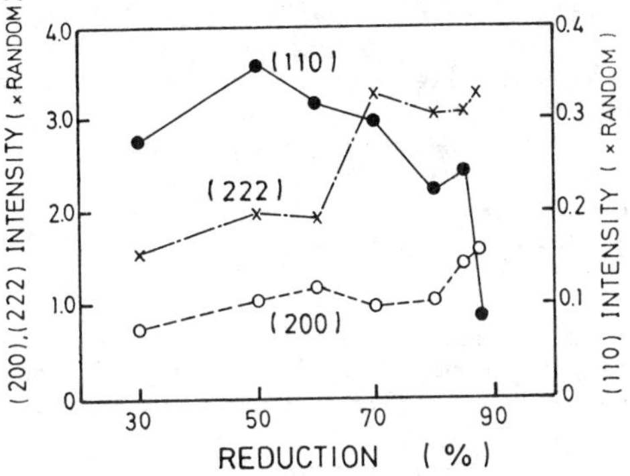


Fig. 6. Effect of final cold reduction on the pole densities of primaries in specimen A1 annealed at 850°C for 2 min.

Furthermore, the heavy reduction induced two phenomena in the primary matrix of specimen A1.

First, primary grains became finer with increasing reduction. For example, the average diameters of 60 and 85% reduction were 12μm and 10μm, respectively. It seems that the secondary nuclei which started first to grow grow easier into large secondaries at the expense of the surrounding grains in case of the heavy reduction.

Secondly, (200) and (222) intensities increased and the (110) intensity decreased in the primary texture with increasing reduction, as shown in Fig. 6. If there are particular angular relations between the orientation of secondaries and that of consumed major components, the secondary texture can be stronger with increasing intensity of major components. Therefore, it is also considered that the heavy reduction promotes the sharp secondary texture. A decrease in the (110) intensity with increasing reduction implies a decrease in the secondary nuclei and consequently large secondary grains in final products.

ACKNOWLEDGEMENT

The authors are grateful to Dr. T. Ichiyama, Mr. T. Wada and Mr. K. Kuroki for their valuable suggestions regarding preparation of the present study.

REFERENCES

1. S. Taguchi, A. Sakakura, and H. Takashima, U.S. Patent 3287183 (1966)
2. T. Ichiyama, M. Koizumi, T. Sato, and Y. Kato, Japanese Patent 747939 (1974)

# DEPENDENCE OF LOSSES ON STRESS AND ORIENTATION OF ORIENTED SILICON-IRON

W. M. Swift
Westinghouse Research Laboratories
Pittsburgh, Pennsylvania 15235

## ABSTRACT

The dependence of 60 Hz-1.0 T losses on uniaxial tensile and compressive stress states was characterized for (110)[001] 3% Si-Fe sheet as a function of θ, the angle between the [001] direction and the applied field and stress direction. Tensile stress produced lower losses compared to the zero stress state for 0°≤θ≤10° and 55°≤θ≤90°. Compressive stress generally increased losses compared to zero stress and to tensile stress of the same magnitude, with two noteworthy exceptions. For θ=55° losses were approximately independent of stress, while losses under compressive stress were less than losses under tensile stress of the same magnitude for 30°≤θ<55°.

## INTRODUCTION

Several investigators[1,2] have studied the relationship between losses during 60 Hz magnetization and stress applied parallel to the [001] direction in polycrystalline (110)[001] 3% Si-Fe sheet. Tensile stress tends to decrease losses while compressive stress produces large increases in losses. These observations have been accounted for by magnetoelastic theory coupled with observations of stress-dependent domain structures.[3-5] The work reported herein is concerned with the dependence of losses on stress as a function of θ, the angle between the direction of stress and the [001] direction in polycrystalline (110)[001] sheet, for magnetization parallel to the direction of applied stress.

## EXPERIMENTAL

The material used in this study was initially in the form of cold rolled-decarburized, 0.27 mm thick 3% Si-Fe sheet. Eight Epstein sets were sheared at various angles (θ) to the rolling direction and annealed to bring about selective (110)[001] grain growth. The annealed strips were tested for dc-magnetic properties and 60 Hz-1.0 T losses by standard Epstein frame methods. A single-strip loss tester together with apparatus for applying a uniform tensile or compressive stress parallel to the long dimension of a 30 mm x 150 mm (half-length Epstein strip) was used to measure stress-dependent losses. The salient features of the electronic apparatus for measuring losses have been described elsewhere.[6]

## RESULTS AND DISCUSSION

Unstressed magnetic properties based on Epstein frame testing are summarized in Table I for the eight

TABLE I

Epstein Frame Measured Magnetic Properties of (110)[001] 3% Si-Fe Sheets

| θ (Degrees From Rolling Direction) | $B_{800}$ (T) | $B_{8000}$ (T) | $B_r^*$ (T) | $H_c^*$ (A/m) | $P_c$ 1.0 T (W/kg) |
|---|---|---|---|---|---|
| 0° | 1.88 | 2.00 | 1.59 | 7.2 | 0.57 |
| 10° | 1.57 | 1.99 | 1.40 | 8.3 | 0.59 |
| 20° | 1.51 | 1.90 | 1.03 | 10.2 | 0.96 |
| 30° | 1.36 | 1.71 | 0.48 | 12.8 | 1.03 |
| 45° | 1.24 | 1.53 | 0.20 | 17.5 | 1.36 |
| 55° | 1.22 | 1.50 | 0.12 | 20.9 | 1.53 |
| 75° | 1.30 | 1.59 | 0.08 | 27.7 | 1.57 |
| 90° | 1.39 | 1.64 | 0.28 | 15.9 | 0.98 |

*Peak magnetizing field of 800 A/m.

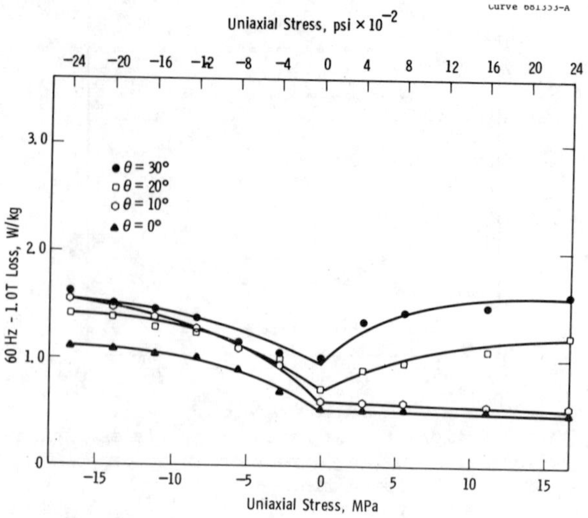


Fig. 1-Dependence of 60 Hz-1.0 T loss on stress for several angles between the [001] direction of (110)[001] Si-Fe and the direction of uniaxial stress

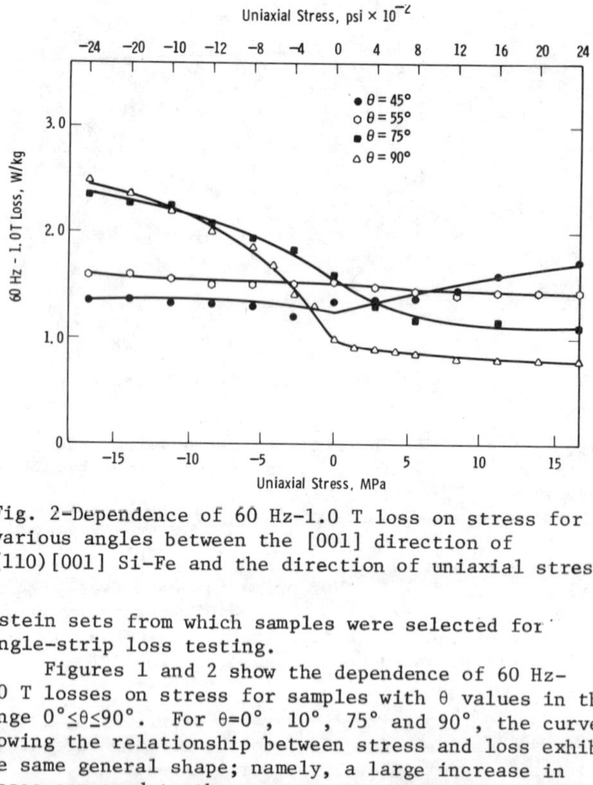


Fig. 2-Dependence of 60 Hz-1.0 T loss on stress for various angles between the [001] direction of (110)[001] Si-Fe and the direction of uniaxial stress

Epstein sets from which samples were selected for single-strip loss testing.

Figures 1 and 2 show the dependence of 60 Hz-1.0 T losses on stress for samples with θ values in the range 0°≤θ≤90°. For θ=0°, 10°, 75° and 90°, the curves showing the relationship between stress and loss exhibit the same general shape; namely, a large increase in losses compared to the zero stress state occurs with the application of uniaxial compressive stress, while tensile stress decreases losses compared to zero stress. The shape of the loss versus stress curves changes significantly in the range 20°≤θ≤55°. For θ=20° the shape of the curve is almost symmetric about the origin (zero stress). For θ=30° and θ=45°, losses are larger for tensile stress compared to a compressive stress of the same magnitude, and, for θ=55°, losses are nearly independent of stress. These relationships are shown in the polar plot of Fig. 3 where zero stress losses are compared to losses for compressive and tensile stress states of 16.6 MPa.

It is useful to qualitatively explain the aforementioned loss versus stress curves in terms of the stress-dependence of Type I (180° bar-like domains arrayed parallel to the [001] direction) and Type II

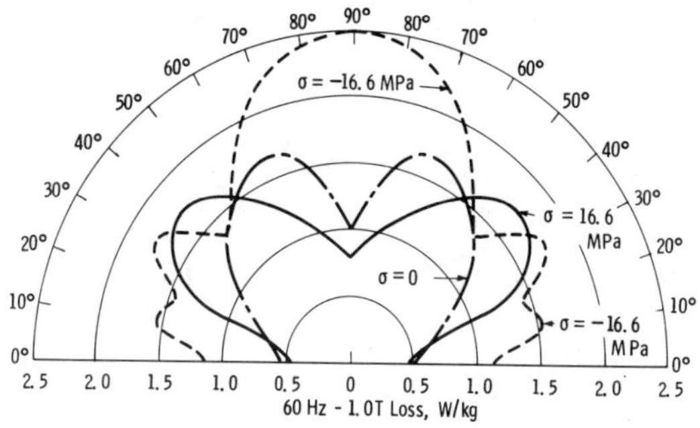


Fig. 3-Dependence of 60 Hz-1.0 T loss on the angle θ for σ = -16.6, 0 and +16.6 MPa

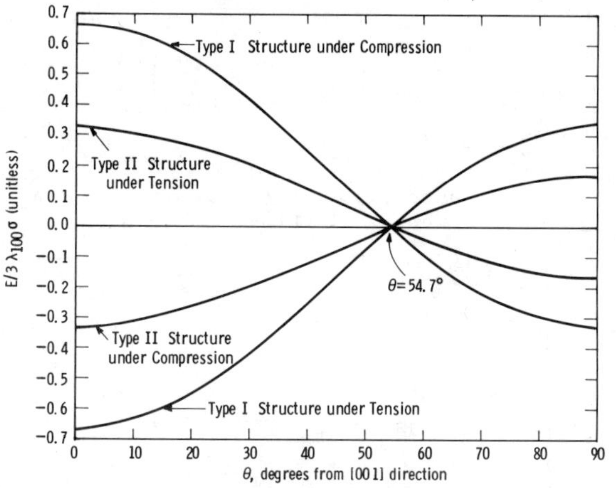


Fig. 4-Dependence of reduced magnetoelastic energy on the angle θ for domain structures occurring in thin-sheet (110)[001] silicon-iron

(domains arrayed parallel to the transverse ±[001], ±[100] directions with surface flux closure domains) domain structures, and magnetoelastic theory. The reader is referred elsewhere for detailed discussions of these domain structures.[3-6]

Application of magnetoelastic theory[2] to a Type I domain structure predicts the following relation between magnetoelastic energy ($E_I$) and θ:

$$E_I = -\frac{3}{2}\lambda_{100}\sigma\ (\cos^2\theta - \frac{1}{3})\ \text{ergs/cm}^3 \qquad (1)$$

where $\lambda_{100}$ is the magnetoelastic strain for magnetization parallel to a <100> direction, a positive quantity for Si-Fe, and σ is the applied stress (positive for tension, negative for compression). For a Type II structure, the corresponding energy[2] is

$$E_{II} = +\frac{3}{4}\lambda_{100}\sigma\ (\cos^2\theta - \frac{1}{3})\ \text{ergs/cm}^3 \qquad (2)$$

In Fig. 4, equations (1) and (2) are shown in reduced terms, i.e., $E/3\lambda_{100}\sigma$, for the case of tensile and compressive stresses. Note that these curves intersect at approximately θ=55°. At this angle, the magnetoelastic energy is zero for both Type I and Type II domain structures. This result explains why 60 Hz-1.0 T losses were independent of stress for the θ=55° curve of Fig. 2. For θ=0°, the Type I domain structure is stabilized by tension, which will result in either lower losses or little change in losses. Compression stabilizes the Type II structure for θ=0° and raises losses for reasons noted elsewhere.[2,7] The same arguments essentially apply for the case of θ=10°.

The application of compressive stress to (110)[001] sheet with θ=90° stabilizes the Type I structure relative to the Type II structure, which in essence stabilizes 180° bar-like domains transverse to the applied field direction. Conversely, the transverse <100> directions, which are needed in part for magnetization parallel to H when θ=90 are destablized by compression. Consequently, higher magnetizing fields compared to zero stress or tensile stress are required for magnetization using the transverse <100> directions. Tensile stress, on the other hand, favors the stabilization of the Type II structure, but in this case, as opposed to the θ=0° case, the transverse <100> directions have a net component parallel to the field direction. One would therefore expect a higher permeability and a lower loss magnetization process. The same reasoning applies to the θ=75° case, where the decrease in losses under tensions as well as the increase in losses under compression are more pronounced than the θ=90°, presumably due to a greater degree of misalignment of the transverse <100> directions with respect to the field direction.

The range 10°<θ<55° represents a region where losses under tension gradually become larger than losses under compression. Moreover the stress-induced losses always exceed the unstressed losses. From Fig. 4 it is seen that the Type I structure under tension is of lower energy than Type II under compression in the range 0°≤θ≤54.7°, but that the difference in energy between these two states decreases with increasing θ. The change from a lower energy magnetization process under compression compared to losses under tension in this range of θ values is determined by the changes in domain structures during the dynamic cycle. Thus, observations of dynamic domain processes would therefore be needed to clarify the mechanism(s) giving rise to lower losses under compression in range 30°≤θ≤55°.

## CONCLUSIONS

The following conclusions apply to stress-dependent 60 Hz-1.0 T losses of (110)[001] 3% Si-Fe as a function of θ, the angle between the [001] direction and the applied field direction:

(1) Tensile stress produced lower losses than zero stress or compressive stress states for 0°≤θ≤10° and 55°<θ≤90°.

(2) Losses were stress-independent for θ=55° as predicted by magnetoelastic theory.

(3) Compressive stress produced lower losses than tensile stress of the same magnitude for 30°≤θ<55°; however, unstressed losses were the lowest in this range of θ values.

## REFERENCES

1. S. M. Pegler, "The Magnetic Properties of Japanese Hi-B (110)[001] Grain Oriented Silicon Iron," Proc. 20th Annual Conference on Magnetism and Magnetic Materials, No. 24, pp. 718-720, 1975.
2. A. J. Moses, "Effects of Stress On Magnetic Properties of Silicon-Iron Laminations," J. Material Sci., Vol. 9, pp. 217-222, 1974.
3. L. J. Dijkstra and W. M. Martius, "Domain Pattern In Silicon-Iron Under Stress," Rev. Mod. Phys., Vol. 25, No. 1, pp. 146-149, 1953.
4. J. W. Shilling and G. L. Houze, Jr., "Magnetic Properties and Domain Structure In Grain-Oriented 3% Si-Fe," IEEE Trans. On Magnetics, Vol. MAG-10, No. 2, pp. 195-223, 1974.
5. W. D. Corner and J. J. Mason, "The Effect of Stress On The Domain Structure of Goss Textured Silicon-Iron," Brit. J. Appl. Phys., Vol. 15, pp. 709-718, 1964.
6. J. W. Shilling, "Domain Structures In 3% Si-Fe Single Crystals With Orientations Near (110)[001]," IEEE Trans. on Magnetics, Vol. MAG-9, No. 3, pp. 351-356, 1973.
7. W. M. Swift and G. F. Wolfe, "Influence of Compressive Stress on Magnetic Properties of A (110)[001] Single Crystal of 3% Si-Fe," (to be submitted for publication in IEEE Trans. on Magnetics).

# STRESS FIELDS AND STRAINS ENERGIES ASSOCIATED WITH CLOSURE DOMAINS

J. N. Pryor and J. J. Kramer
University of Delaware
Newark, Delaware 19711

## ABSTRACT

The stress fields caused by the presence of wedge-like closure domains in a single crystal of iron with {100} surface are identical to the stress fields of a wedge disclination at a free surface (disclination line parallel to surface). Using this observation, the exact solutions of these stress fields are derived. From these stress fields expressions for the strain energy associated with each closure domain as well as the elastic interaction between closure domains are calculated. The self energies of the closure domain are calculated to be considerably different from the values found by previous investigators who have used the model of a wedge disclination in a cylinder.[1] Finally, domain refinement due to applied external stresses is explained on the basis of elastic interaction with the closure domains.

## INTRODUCTION

Several authors have come forth with models to determine the strain energy associated with closure domains in {100} single crystals of iron with edges parallel to <100> type directions. The domain structure of such a crystal is shown in Fig. 1.

Recognizing that the strain energy associated with the closure domain is equal to the energy required to introduce the unstrained wedge ABC into the smaller wedge-shaped hole A'B'C' (shown in Fig. 2), Kittel[2] has assumed that all of the resulting strain exists only in the closure domain. This assumption coupled with the strain relation $\varepsilon = 3/2\ \lambda_{100}$ ($\lambda_{100}$ is the magnetostriction constant) leads to the following relation for the strain energy per

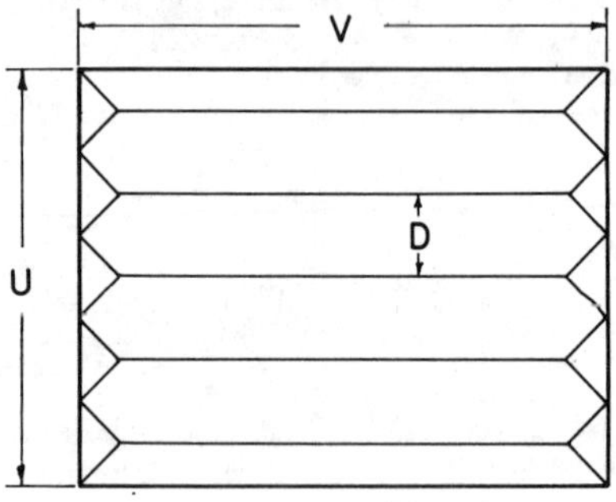


Fig. 1: Domain geometry for {100} iron crystal

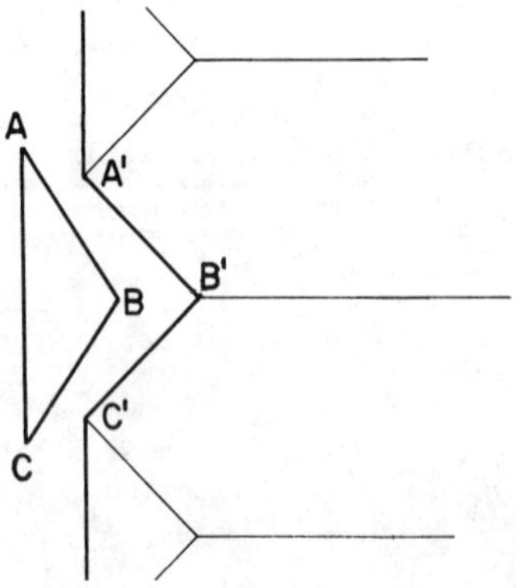


Fig. 2: Unstrained closure region ABC to be fitted into wedge A'B'C'

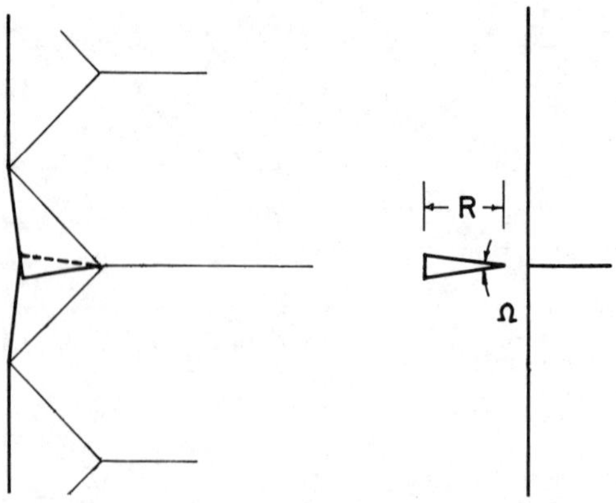


Fig. 3: Closure domain stresses modeled by a wedge disclination

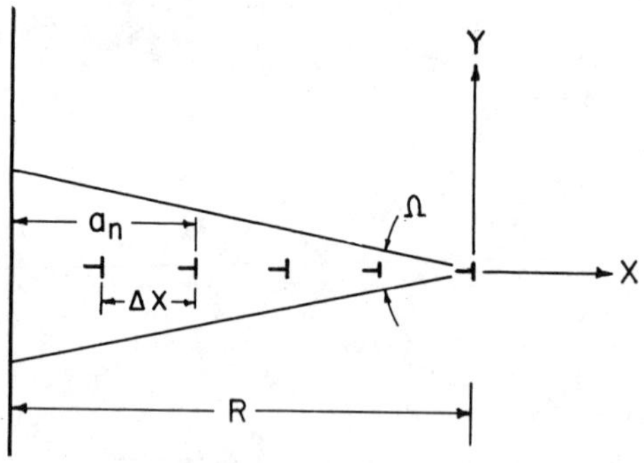


Fig. 4: Wedge disclination at free surface modeled by array of edge dislocations

closure domain per unit thickness of material:

$$E = \frac{9}{32} E_y \lambda_{100}^2 D^2 \tag{1}$$

where $E_y$ is Young's modulus and D is the domain width.

Miltat and Kléman[1] have used the stress fields of a wedge disclination in a cylinder to determine the strain energy associated with the general 90° domain wall - 90° wall - 180° wall junction. If applied to a closure domain, their equation yields the following relationship:

$$E = \frac{9}{128} \frac{E_y \lambda_{100}^2 D^2}{\pi(1-\nu^2)} \tag{2}$$

where $\nu$ is Poisson's ratio.

## NEW MODEL

The process of introducing wedge ABC into the smaller wedge-shaped hole A'B'C' shown in Fig. 2 can be equivalently thought of as introducing a smaller wedge of angle $\Omega$ into a slit at the edge of the crystal as long as $\Omega$ (which equals $\measuredangle$ABC minus $\measuredangle$A'B'C') is very small. Fig. 3 shows the validity of this statement although the magnitude of $\Omega$ is greatly exaggerated. Wedge ABC is first cut in half and placed in the wedge-shaped hole (Fig. 3a). It is then clear that eliminating the overlap of these two segments is equivalent to introducing the smaller wedge of angle $\Omega$ (Fig. 3b). At this point it can be concluded that the stress fields associated with a closure domain are equivalent to those of a wedge disclination at a free surface.

The stress fields were derived by observing that a disclination can be approximated by a linear array of edge dislocations as shown in Fig. 4. Using the stress fields given by Head[3] for an edge dislocation at a free surface ($\vec{b}$ parallel to surface) in an isotropic material, the stress fields $\sigma$, for a wedge disclination at a free surface were calculated.

$$\sigma(x,y,R) \text{ for wedge disclination at surface} \approx \lim_{\Delta x \to 0} \sum_{n=0}^{R/\Delta x} \sigma_\perp(x+n\Delta x, y, R-n\Delta x) = K\Omega \int_0^R f(x+Q,y,R-Q)\,dQ \quad (3)$$

where K is a collection of elastic constants, Q is an integrating variable, and $R-n\Delta x$ equals $a_n$, the distance between the $n^{th}$ dislocation and the free surface. The following solutions were derived:

$$\sigma_{xx} = \frac{G\Omega}{2\pi(1-\nu)}\left\{\frac{1}{2}\ln\left[\frac{(x+2R)^2+y^2}{x^2+y^2}\right] - \frac{y^2}{x^2+y^2} - \frac{2R(x+2R)^2(x+R) - (x^2+6xR+6R^2)y^2 - y^4}{\left[(x+2R)^2+y^2\right]^2}\right\} \quad (4)$$

$$\sigma_{yy} = \frac{G\Omega}{2\pi(1-\nu)}\left\{\frac{1}{2}\ln\left[\frac{(x+2R)^2+y^2}{x^2+y^2}\right] + \frac{y^2}{x^2+y^2} - \frac{2R(x+2R)^2(x+3R) + (x^2+10xR+14R^2)y^2 + y^4}{\left[(x+2R)^2+y^2\right]^2}\right\} \quad (5)$$

Here G is the shear modulus.

## APPLICATION OF STRESS EQUATIONS

The energy associated with each closure domain in a crystal such as the ones shown in Fig. 1 is equal to the sum of the self energy plus the elastic interaction with one-half the stress fields of the remaining closure domains along that side of the crystal. The self energy is given by the equation:

$$E_s = \int_0^{-R}\int_0^{\Omega} -x\sigma_{yy}\,d\Omega\,dx = \frac{G\Omega^2R^2}{4\pi(1-\nu)} \quad (6)$$

and the interaction energy by:

$$E_I = \int_0^{-R} \Omega(-x)\sigma_{yy}\big|_{y=L}\,dx = \frac{G\Omega^2R^2}{2\pi(1-\nu)}\left\{1 - \frac{L^2}{4R^2}\ln\left[\frac{4R^2}{L^2}+1\right]\right\} \quad (7)$$

where L is the separation distance between closure domains. Therefore, the total strain energy associated with a closure domain is given by:

$$\frac{G\Omega^2R^2}{4\pi(1-\nu)}\left\{1 + \sum_{n=1}^{\infty}\left[1-n^2\ln\left[\frac{1}{n^2}+1\right]\right]\right\} = \frac{.402G\Omega^2R^2}{\pi(1-\nu)} \quad (8)$$

An infinite number of terms were included in the interaction summation because D (or 2R) is usually very small compared to the crystal dimensions and the series converges rapidly. The equilibrium domain spacing for a crystal such as the one shown in Fig. 1 can now be calculated by minimizing the sum of strain and domain wall energies and recognizing that $\Omega = 3\lambda_{100}$

$$D_{eq} = \frac{2}{3\lambda_{100}}\left[\frac{\pi(1-\nu)V\gamma_{180}}{.402G}\right]^{\frac{1}{2}} \quad (9)$$

where $\gamma_{180}$ is the surface energy associated with a 180° domain wall.

If a uniform external tension, $T_x$, is applied horizontally to the crystal shown in Fig. 1, domain refinement will occur. This results because domain refinement reduces the total closure domain area, thus more lattice cells have their long axis aligned in the horizontal direction and the crystal is effectively lengthened in the direction of the applied tension. As a result, domain refinement lowers the potential energy of the crystal under the applied tension. The average total displacement of the vertical edges of this crystal as a function of D, the domain width, is given by:

$$\text{displ.} = \frac{\Omega D}{2} - \frac{2U}{D}\int_{\infty}^{\infty}\int_{-R}^{\infty} \frac{1}{U}\varepsilon_{xx}\,dx\,dy \quad (10)$$

The first term arises from the rotation of the long axes of unit cells in the formation of closure domains and the second from the elastic response of the material to this rotation. This expression can be simplified to

$$\text{displ.} = \frac{3\lambda_{100}}{2}D\left[1 - \frac{(\pi-2)}{4\pi(1-\nu)}\right] \quad (11)$$

Therefore the change in potential energy of the crystal in presence of $T_x$ due to changing domain width, D, is given by

$$\Delta E_p = \frac{3\lambda_{100}}{2}D\,T_xU\left[1 - \frac{(\pi-2)}{4\pi(1-\nu)}\right] \quad (12)$$

The equilibrium domain spacing of the crystal shown in Fig. 1 under applied tension, $T_x$, is given by the minimization of strain, domain wall, and potential energies.

$$D_{eq} = \left[\frac{4\pi(1-\nu)\gamma_{180}V}{9\lambda_{100}^2(.402G) + \frac{3\lambda_{100}}{2}(3\pi-4\pi\nu+2)T_x}\right]^{\frac{1}{2}} \quad (13)$$

## SUMMARY AND CONCLUSIONS

Within the limitation of assuming an isotropic medium and ignoring stress relaxation due to finite sheet thickness, the preceding analysis provides an exact solution of the stress fields, strain energies, and elastic interactions associated with closure domains. The resultant strain energy is six times larger than that predicted by Miltat and Kléman and approximately one-half of that predicted by Kittel. But this is expected because the Miltat and Kléman model unrealistically restricts the crystal dimensions in the case of multiple domains. Similarly, the Kittel model yields strain energies that are too large because it assumes that strains are confined to the closure domain and, hence are much too large.

With applied tension in the direction of the long domains the equilibrium domain spacing varies as

$$D_{eq} = \left[\frac{a}{b + cT_x}\right]^{\frac{1}{2}}$$

## ACKNOWLEDGEMENT

This work was supported by NSF under Grant #DMR74-20580.

## REFERENCES

1.) J. E. A. Miltat and M. Kléman, Phil. Mag. 28 1015 (1973)
2.) Charles Kittel, Rev. Mod. Phys. 21 566 (1949)
3.) A. K. Head, Proc. Phys. Soc. 66 797 (1953)

# OBSERVATION OF DOMAIN STRUCTURE IN SOFT-MAGNETIC MATERIALS BY MEANS OF HIGH VOLTAGE SCANNING ELECTRON MICROSCOPE

T. YAMAMOTO AND K. TSUNO, JEOL LTD., Nakagami Akishima, Tokyo 196, Japan.

## ABSTRACT

The applicability of a 200 kV SEM to observation of the magnetic domains in bulk specimens is presented. The minimum and maximum material depths contributing to magnetic contrast were experimentally determined at 50-200 kV. The range of depths most effective for the contrast was determined as well. The resolution of domain images was also measured at these voltages. A method of observing dynamically changing domain structures is further described.

## INTRODUCTION

Ever since the SEM method for observing magnetic domains was established by Fathers et al,[1] there have been increasing reports on SEM applications to studies of magnetic materials,[2,3] using a SEM operable at 30 - 50 kV. We have so far engaged in observing domains with a 200 kV SEM and found that increased accelerating voltage not only enhanced the magnetic contrast,[1] but also reduced surface projection and grain contrasts[4]. Besides we have noted a strong dependence of the magnetic contrast on the take-off angle at which backscattered electrons are received, thus establishing an optimal observing condition.[5] The increased voltage reduces the image resolution as it increases electron diffusion in the material. However, increased electron penetration allows discerning the domains in materials having unusual surfaces, e.g., with a deformation layer, plated with a metallic material, etc. The aim of this paper is first to quantitatively clarify the domain visibility in such materials and second to show the applicability of SEM to analysis of the dynamical behaviours of the domain structures in specimens with such surfaces.

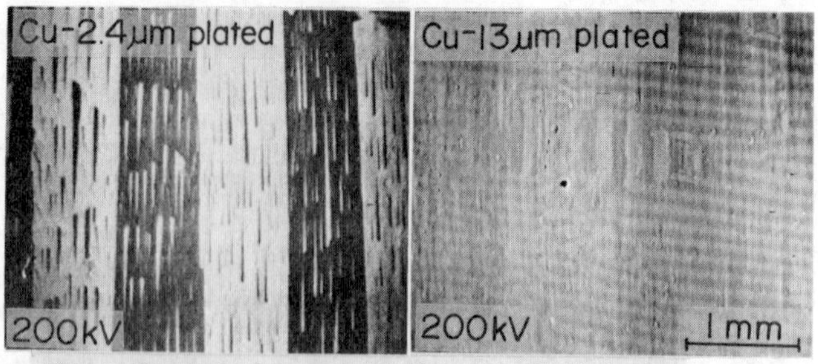


Fig.1. Domain images in Cu-plated sheets

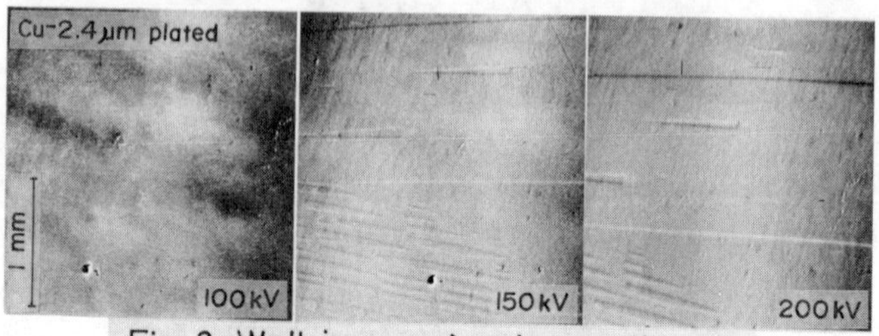


Fig 2 Wall image at various voltages

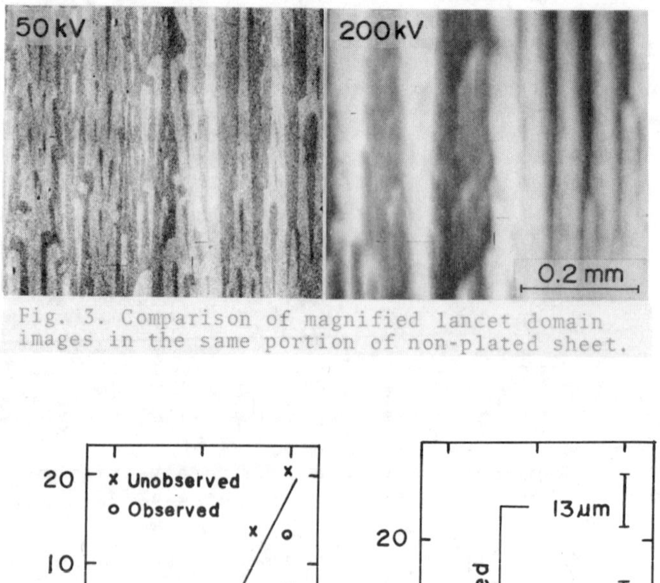


Fig. 3. Comparison of magnified lancet domain images in the same portion of non-plated sheet.

## EXPERIMENTAL METHODS, RESULTS AND DISCUSSION

As specimens, we used grain-oriented 3% Si-Fe sheets which were plated or vacuum-evaporated with copper (thickness: 0.3 - 20 µm), and determined the domain visibility by changing the accelerating voltage of the SEM (the JSEM-200) from 50 to 200 kV in the backscattered electron mode[4](see Fig. 1). In this case, each specimen was tilted 45° and the probe current was about $10^{-7}$A. The resolution of the domain images was also measured from the width of a line image (see Fig. 2) which appeared on a wall when the specimen was mounted so as to ~~exhibit no magnetic contrast of antiparallel domains~~ ( the width provided an exact measure of the resolution[6]) The minimum contrast depth and the most effective depth range were determined by the varied image width of a small lancet domain in a non-plated specimen[5](see Fig. 3). The range was further checked by the Cu thickness predictable from the measured width of the above mentioned wall image in the plated specimens. The results are summarized in Figs. 4a and b. The minimum and maximum depths were approx. 4 and 15 µm at 200 kV and the most effective range was 5 - 8 µm. The resolution at 200 kV was better than 25 µm.

The domain structures were successfully observed at more than 150 kV in a mechanically polished 3% Si-Fe sheet with #1500 emery paper and in a commercial Si-Fe sheet, used mainly as a transformer core, which was covered with a 5 µm thick glass-like coating (composition being MgO, $SiO_2$ and $P_2O_5$; and the averaged atomic number about 10). However, no domain was observed at 100 kV in both specimens. The electron penetration of the coating is 6 times that of copper. Thus, the domains must be discerned even at 50 kV. The invisibility at 100 kV suggests the existence of a non- or hard magnetic material interface whose thickness corresponds to about 4 µm Cu thickness.

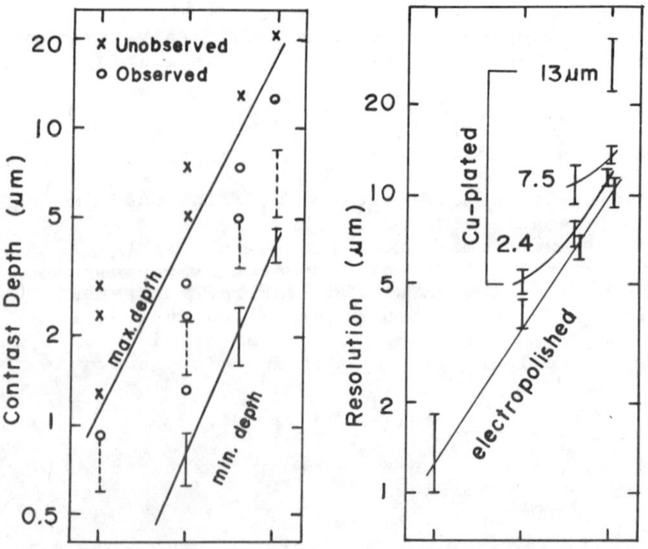


Fig. 4 a(left). Voltage dependence of contrast depths
Dotted lines: the most effective depth range.
Fig. 4 b(right). Voltage dependence of domain resolution.

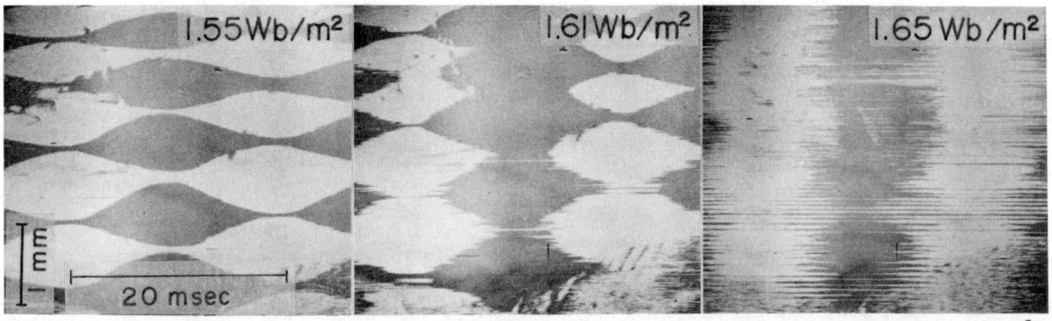


Fig. 5. Wall motion in electropolished sheet at 50 Hz magnetic field. Scanning direction is parallal to the 180° wall.

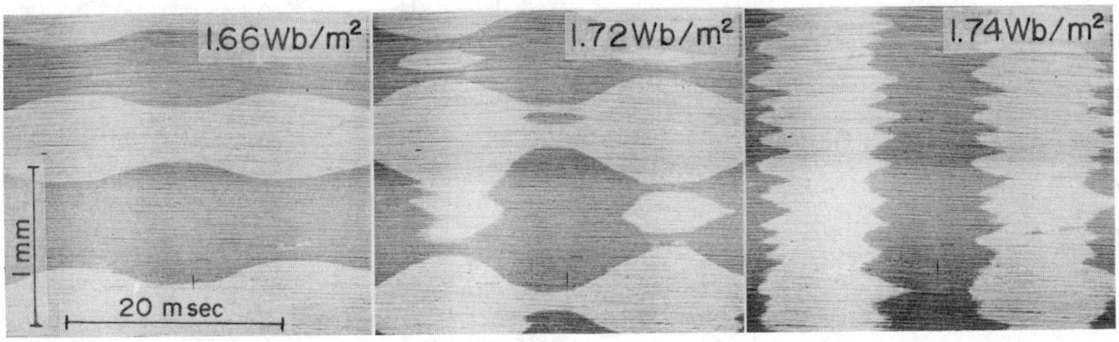


Fig. 6. Wall motion in commercial silicon steel with an insulating coating at 50 Hz magnetic field.

When a periodically changing external excitation is applied to a specimen, the domain structure at each phase are displayed continuously on a line in a single micrograph by synchronizing the excitation with the probe scanning. The sweep time of a raster can be varied from 5 to 40 msec. and the observation technique was applicable up to 1 kHz. Figs. 5 and 6 show wall movements induced by a 50 Hz sinusoidal magnetic field in electropolished and glass-coated 3% Si-Fe sheets. In Fig. 5, the wall movement is relatively smooth when the domains with the same magnetization direction separate each other, whereas the movement becomes discontinuous when the domains contact. However, such movement is considerably reduced for the coated sheet. The periodic modulation of a magnetized area reveals that domain nucleation ocurrs at the same spacing. The discontinuous movement is also moderated for the mechanically polished sheet mentioned above. The interface beneath the coating and surface imperfections caused by the polishing may play a certain role in domain nucleation as well as in wall movement. Furthermore, we preliminarily tried to directly observe wall bowing by changing the accelerating voltage. The domain structure at a 50 Hz magnetic field exhibits almost the same image at 100 and 200 kV within an observable limit of 50 μm. The most effective ranges in Fig. 4a thus imply that the bowing is not significant at depths between 2 and 6 μm.

REFERENCES

1. D.J.Fathers, J.P.Jakubovics, D.C.Joy, D.E.Newbury and H.Yakowitz, Phys. Stat. Sol., a20, 535(1973); ibid, a22, 609(1974).
2. D.E.Newbury and H.E.Yakowitz, Conf. on Magnetism and Magnetic Materials, A.I.P. Conf., Procceedings, Vol. 18, Part 2, 1974, p 1372.
3. H.J.Leamy, S.D.Ferris, G.Norman, D.C.Joy, R.C.Sherwood, E.M.Gyorgy and H.S.Chen, Appl. Phys. Lett., 26, 259(1975).
4. T.Yamamoto, H.Nishizawa and K.Tsuno, J. Phys. D ( Appl. Phys.), 8, L113(1975).
5. T.Yamamoto, H.Nishizawa and K.Tsuno, to be published (Phil. Mag.).
6. T.Yamamoto and K.Tsuno, to be published(Phil. Mag.).

# EFFECT OF INSULATING COATING ON DOMAIN STRUCTURE IN GRAIN ORIENTED 3% Si-Fe SHEET AS OBSERVED WITH A HIGH VOLTAGE SCANNING ELECTRON MICROSCOPE

T. Irie and B. Fukuda
Research Laboratories, Kawasaki Steel Corp., Chiba, Japan

## ABSTRACT

Direct observation of the domain structure in (110) [001] grain oriented 3% Si-Fe sheets with insulating coating of magnesium silicate and phosphate up to 6μm thickness has been successfully made by using a high voltage (200KV) scanning electron microscope. Comparison of two adjacent portions, the one with coating and the other without, in a single grain shows that the coating removes surface closure domains and refines 180° wall spacings. On the concave surface of a bent specimen without coating and that with coating of magnesium silicate and conventional phosphate, transversal stress pattern of ±[100] and ±[010] domains appears which has been observed eleswhere on uncoated sheets under simple longitudinal compressive stress. Application of a new phosphate of lower thermal expansivity in place of conventional one effectively suppressed the generation of stress pattern. Also, specimens with the new coating are found to give lower magnetostriction than uncoated and the conventionally coated specimens. This indicates that the new coating gives stronger tensile stress to the sheet and counterbalances the compressive stress arising from bending, suppressing the generation of ±[100] and ±[010] domains.

## INTRODUCTION

Some of the insulating coatings for grain oriented silicon-iron sheet have been known to give tensile stress owing to the difference in contraction between the sheet and coatings during cooling after heat treatment and to improve magnetostriction and iron loss [1] [2]. These improvements will be referred to the variation of magnetic domain structure. However, direct observation of domains through such coatings has been unsuccessful with conventional methods, e.g, Kerr effect, powder pattern technique, and etc.

A method for observing magnetic domains with a scanning electron microscope, SEM, was established by Fathers et al. [3]. Recently, increased accelerating voltage SEM has enabled one to observe the domains in roughly polished 3% silicon-iron sheet [4]. Deeper penetration resulting from the increased voltage is expected to overcome even the masking effect of the coatings.

This paper shows the domains in oriented silicon-iron sheet observed through the coatings with a high voltage SEM, and describes the influence of coating on the domain structure.

## EXPERIMENTAL PROCEDURE

Grain oriented 3% silicon-iron sheets with high degree of (110)[001] texture (RG-H 0.30 mm thick) were used. The following specimens, two of which possess a double-layer coating, were investigated:

specimen I : without coating (removed by pickling)
specimen II : magnesium silicate covered with conventional phosphate top coating.
specimen III : new phosphate top coating in place of the conventional one.

Table I Constitution of insulating coatings

| magnesium silicate coating | phosphate coating | |
|---|---|---|
| | conventional | new |
| $2MgO \cdot SiO_2$ | $P_2O_5$, $MgO$ $Cr_2O_3$, $Al_2O_3$ | $SiO_2$, $P_2O_5$ $MgO$, $Cr_2O_3$ |

The magnesium silicate coating was formed during high temperature annealing for secondary recrystallization. The phosphate coatings were then applied and baked. The thickness of the double coated layer was 4μm, about 1.5 μm being that of the silicate coating. Constitution of the coatings is listed in Table I.

Domain observation was made with a high voltage SEM (JSEM-200, backscattered electron made) at accelerating voltages of 100, 150 and 200 KV. A scintillation detector was placed at a take off angle of 55°, facing perpendicular to the incident beam. The current and aperture angle of incident beam were $3 \times 10^{-8}$A and $2.5 \times 10^{-3}$ rad., respectively. No surface treatment was made on the specimens before the observation. Bending of the specimens was done by using templates of various curvatures within the limit of elastic deformation. Magnetostriction was also measured with strain gauges pasted on the specimens which undergo bending.

## RESULTS AND DISCUSSION

Magnetic contrast from the coated surface of specimen III is shown in Fig.1. No magnetic contrast appeared at 100 KV, then fuzzy stripe domain pattern showed up at 150 KV. A distinct stripe domain pattern was observed at 200 KV. This voltage dependence of the contrast remained the same for thinner (2μm) and thicker (6μm) coatings. When magnetized, the dark stripes increased in width as shown in Fig.2. This verifies that each of the bright and dark stripes corresponds to a magnetic domain in the coated specimens.

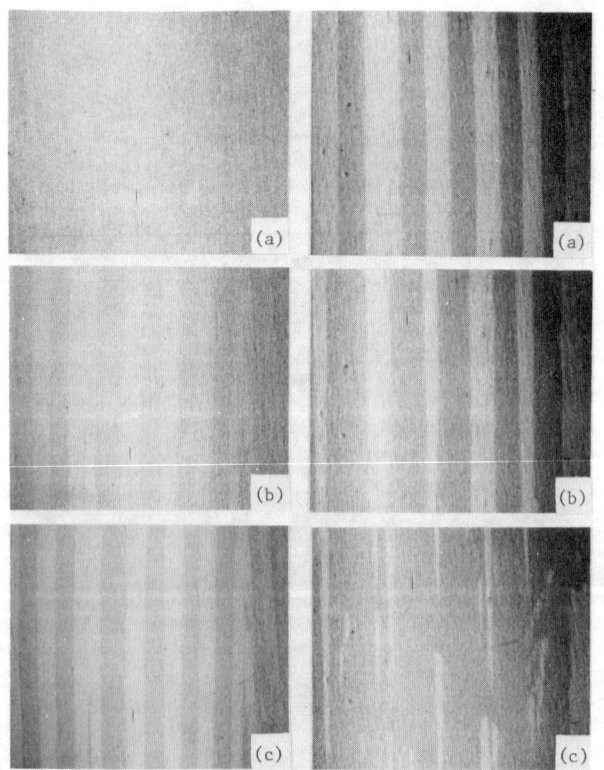

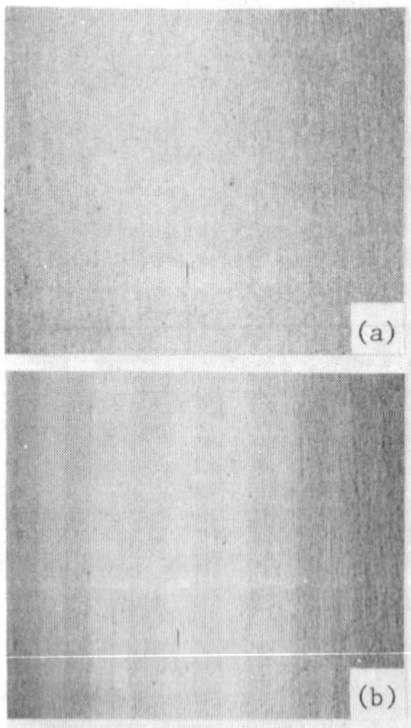


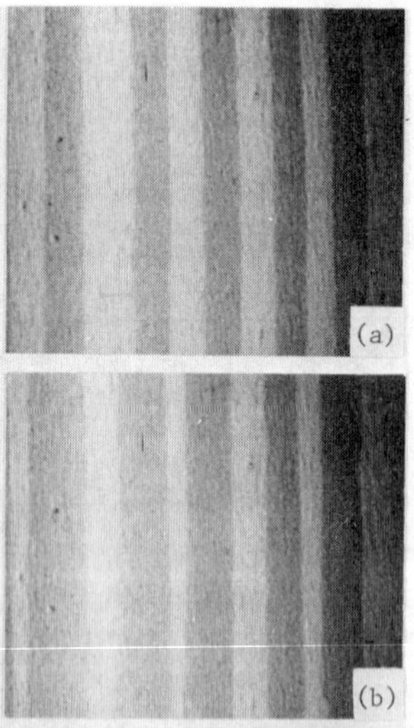


1mm

Fig.1 Effect of accelerating voltage on domain image in specimen III with coating of 4μm thickness:
(a)100KV (b)150KV (c)200KV

Fig.2 Change in domain image with DC magnetization: (a)demagnetized (b)magnetized at 1.3T (c)magnetized at 1.5T

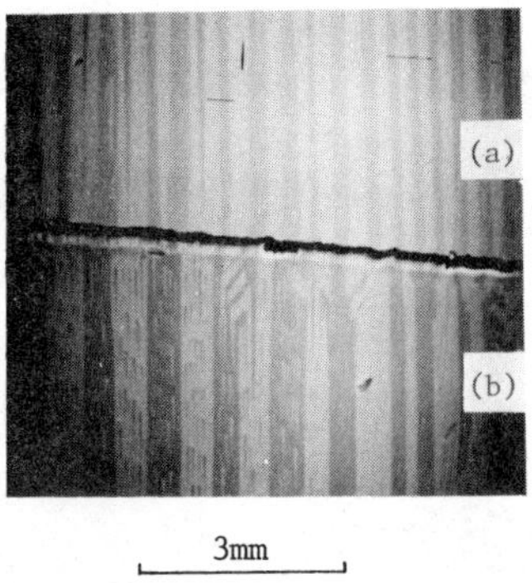


Fig.3 Change in domain pattern by removing coating: (a) with coating, (b) without coating.

Partial removal of the coating on specimen III by dipping in fused caustic soda and subsequent electrolytic polishing resulted in the domain pattern shown in Fig.3. A line which runs across the figure is the boundary dividing the coated and uncoated area in a single grain. Patterns (a) and (b) in Fig.3 indicate that surface closure domains are removed and 180° wall spacings are made finer by the existence of the coating. Externally applied londitudinal tensile stress has been reported to exert the same effects on uncoated silicon-iron sheet [5,6,7]. The coating is concluded to have applied tensile stress to the specimen and changed domain structure, accordingly.

The domain pattern in bent specimens was investigated as shown in Fig.4 to observe the effects of coating on the bending stress sensitivity of magnetic properties.

Before bending, all specimens exhibited only 180° domain structure. The wall spacing of specimen I is considerably broader than that of the other two, that of specimen III being the narrowest. After bending, concave surface of specimens I and II showed remarkable changes, i.e., the stress pattern which is similar to that observed under compressive stress applied to rolling direction [5,6,8]. On the other hand, somewhat blurred 180° domains are still retained in specimen III with some indication of the stress pattern. Compressive stress has been shown to generate domains with magnetization parallel to $\pm[100]$ ($=[100]$ and $[\bar{1}00]$) and to $\pm[010]$ ($=[010]$ and $[0\bar{1}0]$) directions to reducing magnetoelastic energy. These domains induce surface closure domains with magnetization parallel to $\pm[001]$ ($=[001]$ and $[00\bar{1}]$) directions to reducing magnetostatic energy at the surface of the sheet [8]. If a silicon-iron sheet with such domain structure is magnetized, movement of 90° wall will occur in the sheet to induce significant magnetostriction. As shown in Table II, the magnetostriction at 1.7T in specimen I and II is larger than that in specimen III especially after bending. This fact is consistent with the characteristics of the domain structure mentioned above. This also indicates that the new phosphate gives a stronger tensile stress and hence cancels the compressive stress caused by bending. The difference in tensile stress will be attributed mainly to the difference in thermal expansivity [9] between the new phosphate ($4\times10^{-6}$)-magnesium silicate ($11\times10^{-6}$) combination and the conventional phosphate ($8\times10^{-6}$)-magnesium silicate combination relative to 3% silicon-iron base ($13\times10^{-6}$).

Table II DC magnetostriction in bent specimens

| | specimen | | |
|---|---|---|---|
| | I | II | III |
| DC magnetostriction B=1.7T R=2m | $10.5\times10^{-6}$ | $9.0\times10^{-6}$ | $1.0\times10^{-6}$ |

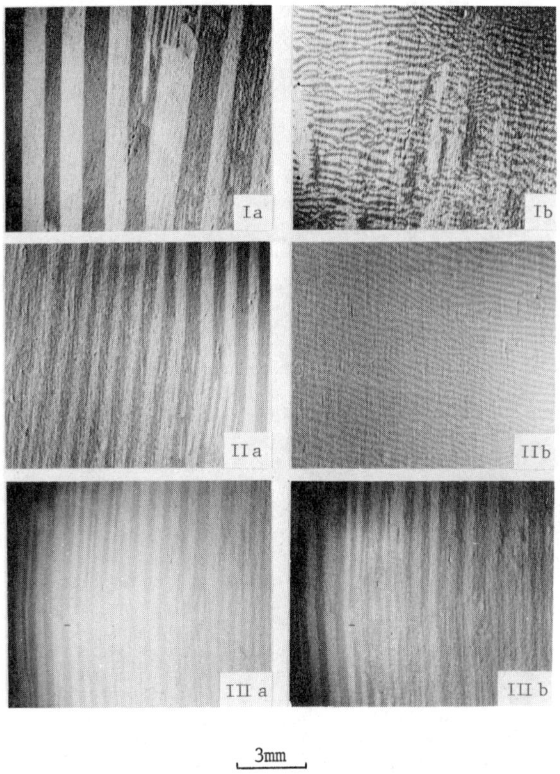


Fig.4 Change in domain pattern by bending.
I: specimen without coating.
II: specimen with magnesium silicate and conventional phosphate coating.
III: specimen with magnesium silicate and new phosphate coating.
suffix a and b denote before bending and after bending(concave surface, R=2m), respectively.

AKNOWLEDGMENT

The authors are grateful to Messrs. T. Yamamoto and K. Tsuno of JEOL Ltd. for their assistance in SEM operation.

REFERENCES

1. T.Yamamoto, S.Taguchi, A.Sakakura and T.Nozawa, IEEE Trans. Mag. M8 (1972)
2. A.J.Moses, S.M.Pegler and J.E.Thompson, Proc. IEE 119, 1222 (1972)
3. D.J.Fathers, J.P.Jakubovics, D.C.Joy, D.E.Newbury and H.Yakowitz, Phys.Stat.Sol. (a) 24, 609 (1974)
4. T.Yamamoto, H.Nishizawa, K.Tsuno, J.Phys.D 8 L113 (1975)
5. J.J.Gniewek, J.Appl.Phys. 34, 3618 (1963)
6. G.L.Houze,Jr, J,Appl.Phys. 40, 1090 (1969)
7. J.W.Shilling, J.Appl.Phys. 42, 1787 (1971)
8. L.J.Dijikstra and U.M.,Martius, Rev.Mod.Phys. 25, 146 (1953)
9. H.Shimanaka, I.Matoba, T.Ichida, S.Kobayashi and T.Funahashi, EPS Conference, Soft Magnetic Material 2 Cardiff, U.K., April 1975

# MAGNETIC PROPERTIES OF $SnCo_2O_4$ SPINEL*

Erika Hermon, D.J. Simkin and R.J. Haddad
Chemistry Dept., McGill University

W.B. Muir
Physics Dept., McGill University, Montréal,
Québec, Canada H3C 3G1

ABSTRACT

The compound $SnCo_2O_4$ is an inverted IV-II spinel with the Co ion equally distributed among the octahedral and tetrahedral sites and the Sn ion occupying the remaining octahedral sites[1]. We would therefore expect this compound to order antiferromagnetically. Magnetization measurements however show ferrimagnetic ordering with $T_{FN} = 44°K$. These results are consistent with the Néel molecular-field model for a two sublattice ferrimagnet. Below $T_{FN}$ the magnetization goes through a maximum and then falls to a value of about $2x10^{-2}\mu_B$ as the temperature decreases towards zero. This magnetization is consistent with an equal distribution of the Co ions among the octahedral and tetrahedral sites and the difference of the magnetic moment of $Co^{2+}$ on each site. The high coercive force, $H_c$>8KOe, at low temperatures is attributable to the high magnetocrystalline anisotropy of the $Co^{2+}$ ion. Mössbauer spectra on the $^{119}Sn$ nucleus taken at 4.2°K are typical of a magnetically ordered compound. A computer analysis of the spectra showed that the internal field produced by the $Co^{2+}$ ions at the $^{119}Sn$ sites is about 80KOe.

## INTRODUCTION

This compound has been measured as one of the end members in a proposed study of the solid solution series $ZnFe_2O_4$ - $SnCo_2O_4$. Previous X-ray and magnetic susceptibility measurements[1] have shown that $SnCo_2O_4$ crystalizes in the inverted spinel structure with the $Co^{2+}$ ions equally distributed among the octahedral and tetrahedral sites. The magnetic measurements were made above 77°K and gave a slight indication of ferrimagnetic ordering with an estimated ferrimagnetic Néel temperature, $T_{FN}$, of 54°K. The present measurements were undertaken in an attempt to clarify the magnetic properties of the material.

## METHOD

$SnCo_2O_4$ was prepared using standard ceramic techniques. Stoichiometric amounts of $SnO_2$ and $CoCO_3$ were fired for 24 hours at 800°C in air and then allowed to cool to room temperature by turning off the furnace. The charge was thoroughly ground and fired at 1200°C for 24 hours. The sample was then cooled at a rate of 1°C/min. The final color of the polycrystalline sample was dark green. The X-ray powder pattern is typical of a single-phase spinel structure and gave a = 8.64 Å in agreement with Poix[1].

Magnetic measurements were made in fields up to 20 KOe using a previously described vibrating-sample magnetometer and gas-flow cryostat[2]. The results of the measurements are shown in figure 1. The Mössbauer measurements were made using a 10mC Ba $^{119m}SnO_3$ source. The constant-acceleration spectrometer and cryostat have been previously described[2]. The Mössbauer data were fit using the least squares program described in reference 2. Data were taken at 300°K and 4.2°K. At 300°K a typical $Sn^{4+}$ one-line spectrum having an isomer shift

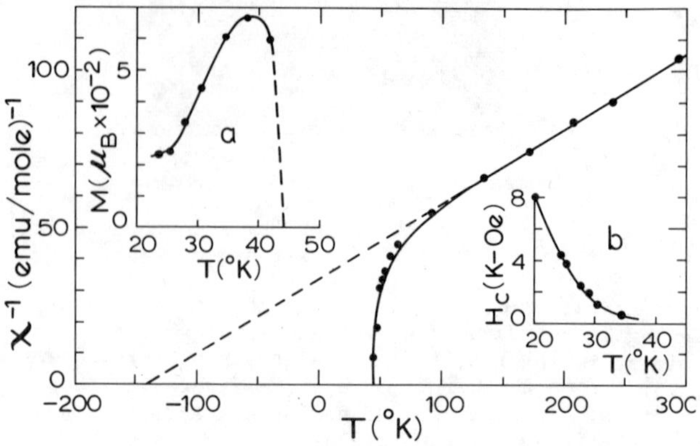


Fig. 1: Reciprocal magnetic susceptibility vs. temperature for $SnCo_2O_4$. Inset (a) : Saturation magnetic moment vs. T. Inset (b) : Coercive force vs. T.

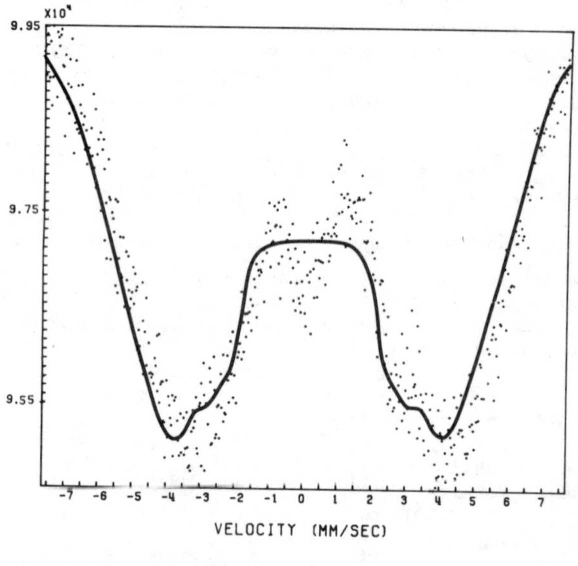


Fig. 2: Mössbauer spectrum of $^{119}SnCo_2O_4$ at 4.2°K. Source: $Ba^{119m}SnO_3$. (ref. 4)

of 0.215mm/sec with respect to the $BaSnO_3$ source was obtained[3]. The low temperature spectrum shown in figure 2 consists of two broad lines[4]. We assume this spectrum is due to magnetic splitting from the molecular-field at the Sn nucleus which results in two groups of three unresolved lines. On this basis the computer fit of the data gave the following parameters: isomer shift 0.216mm/sec, line width (unresolved) 1.2mm/sec, and molecular field 80 KOe.

## DISCUSSION

Above $T_{FN}$ the magnetic susceptibility, Figure 1, is characteristic of ferrimagnetic materials. For a two-sublattice ferrimagnet Néel[5] has shown that

$$1/\chi = T/C + 1/\chi_o - \sigma/(T-\theta') \quad (1)$$

The solid line in figure 1 is the result of fitting this expression to the data and gives $C = 4.35$ emu/mole, $1/\chi_0 = 35.1$ (emu/mole)$^{-1}$, $\theta' = 41^oK$, $\sigma = 158$ (emu/mole)$^{-1}$ and $T_{FN} = 44^oK$. Extrapolation of the linear portion of the curve gives a Weiss temperature $\theta = -145^oK$. The effective magneton number of the $Co^{2+}$ ions derived from the Curie constant is $5.0 \pm 0.2$, which agrees with the usually measured value of 4.8[6]. The parameters C, $1/\chi_0$, $\theta'$ and $\sigma$ are related to the molecular-field coefficients and a straightforward calculation yields[5] $N_{AA} =$ 38 mole/emu, $N_{BB} = 15$ mole/emu and $N_{AB} = -45$ mole/emu. The positive signs of $N_{AA}$ and $N_{BB}$ and the negative sign of $N_{AB}$ suggest ferromagnetic ordering within the sublattice and antiferromagnetic ordering between the sublattices. It has been pointed out by Smit and Wijn[7] that the molecular-field model tends to overestimate the interaction in the A sublattice. For spinels in general structural considerations[8] suggest $N_{BB}>N_{AA}$. For the present material this is in disagreement with the high-temperature results, but in agreement with the Néel P-type magnetization data below $T_{FN}$.

Inset (a) of figure 1 shows the magnetization below $T_{FN}$. The very small value of M, about $2x10^{-2}\mu_B$, below $20^oK$ suggests that we have two nearly equally magnetized sublattices which are antiferromagnetically coupled. If as we expect the $Co^{2+}$ ions are equally distributed among the tetrahedral A sites and the octahedral B sites[1], then we must look for different moments on the two sites. The $Co^{2+}$ ion in tetrahedral coordination has a $^4A_2$ ground state and no orbital contribution to the angular momentum is expected[9]. The magnetic moment should be given by the spin only value of $3\mu_B$. In octahedral coordination the ground state is $^4T_{1g}$ and an orbital contribution to the angular momentum is expected. It does not seem unreasonable that an extra $.04\mu_B$ of magnetic moment per B site $Co^{2+}$ ion should be available from this source and thus account for the observed low-temperature magnetization.

In an attempt to clarify the structure of the ordered state, Mössbauer measurements on the $^{119}Sn$ nucleus were made. The observed 80 KOe internal field is due to the resultant of the dipolar fields from two $Co^{2+}$ sublattices and a calculation of this field is underway. Also, we are preparing a Mössbauer source experiment involving a $Sn\ ^{57}Co_2O_4$ source which will be cooled to $4.2^oK$ and a room-temperature stainless steel absorber. This should allow measurements of the internal field at the $Co^{2+}$ ions.

The coercive force is shown in inset (b) of Figure 1. Its value, which is in excess of 8 KOe below $20^oK$, is consistent with the known large magnetocrystalline anisotropy of materials containing $Co^{2+}$ ions. This large coercive force is the reason that magnetization measurements were not carried out below $20^oK$ as we found it was impossible to convincingly close the hysterisis loop with the available 20 KOe field at $15^oK$.

## CONCLUSIONS

$SnCo_2O_4$ is a ferrimagnetic material with $T_{FN} = 44^oK$. Below $T_{FN}$ the magnetization appears to be due to two equally populated sublattices colinearly aligned and antiferromagnetically coupled. The low-temperature magnetization of $2x10^{-2}\mu_B$ can be accounted for by a small $.04\mu_B$ increase over the spin-only value of the magnetic moment of the B site $Co^{2+}$ ions. This is possible because the B-site ions are octahedrally coordinated and have a $^4T_{1g}$ ground state where a contribution from the orbital as well as the spin angular momentum is expected.

The resultant of the dipolar fields from the two $Co^{2+}$ sublattices results in an internal field of 80 KOe on the $^{119}Sn$ nuclei as measured by the Mössbauer effect. The high coercive force which is in excess of 8 KOe below $20^oK$ is consistent with the known large magnetocrystalline anisotropy of materials containing $Co^{2+}$ ions.

## REFERENCES

* This work was done in partial fulfilment of the Ph.D. requirements of one of the authors (R.J. Haddad) and it was supported by the National Research Council of Canada and the Direction Générale de l'Enseignement Supérieur de Québec, subvention F.C.A.C.

1. P. Poix, Ann. Chim. 9, 274 (1964).
2. Erika Hermon, R. Haddad, D. Simkin, D.E. Brandão, W.B. Muir, submitted to Can. J. Phys.
3. V.I. Goldanskii, The Mössbauer Effect and its Applications to Chemistry, Cons. Bureau, N.Y., 1964.
4. The small zero velocity line is due to unreacted Sn impurities and has been omitted from the computer fit.
5. L. Néel, Ann. Phys. Paris 3, 137 (1948).
6. A.H. Morrish, The Physical Principles of Magnetism, John Wiley and Sons, N.Y. 1965, pg. 67.
7. J. Smit and H.P. J. Wijn, Ferrites, Philips Technical Library, Eindhoven, 30, 1959.
8. E.W. Gorter, Phillips Res. Rept. 9, 295 (1954).
9. B.N. Figgis, Introduction to Ligand Fields, Interscience Publishers, 1966.

# EPITAXIAL $NiFe_2O_4$ FILMS DEPOSITED ON $Nd_3Ga_5O_{12}$ SUBSTRATE

P. GIBART and G. SURAN
Laboratoire de Magnétisme, C.N.R.S.
92190 MEUDON-BELLEVUE (France)

## ABSTRACT

Epitaxial $NiFe_2O_4$ deposited by CVD on MgO exhibits large strain as shown by FMR whereas $NiFe_2O_4$ deposited on $Nd_3Ga_5O_{12}$ has magnetic properties closed to the bulk material and exhibit SWR spectra.

## INTRODUCTION

In epitaxial growth, strains associated with lattice and thermal expansion mismatch between the film and the substrate limit the quality of an epitaxial magentic oxide films(1). In spite of many experimental works(2) the epitaxy of spinel films - with magnetic properties close to the bulk ferrite - is yet to be solved. The main limitation in getting good films is to have good single crystal substrates available. In fact the only non magnetic spinel usually available in single crystal form is $MgAl_2O_4$ but its lattice parameter $a_0$=8.08 Å is too different from the magnetic spinel of interest for which $8.2 < a_0 < 8.6$. MgO (a = 4.213 Å) was also widely used as a substrate, but there is a limited number of ferrite with $a_0$ close to 2x4.213 Å. In order to find out new substrates for epitaxial growth of ferrite, we tried non magnetic garnets : $NiFe_2O_4$ was deposited in the same run on <111> MgO and <111> $Nd_3Ga_5O_{12}$ (NdGG) and even on <111> GGG, so that cristallographic and magnetic properties of deposits could be easily compared.

## EXPERIMENTAL

$NiFe_2O_4$(3) films were grown by chemical vapor deposition (CVD) in open system. The reactor was formed of concentric tubes for the purpose of gas supply : chlorine diluted in argon over the source, then $O_2$ and argon in the outer tubes. Ferrite oxides were used as source and this starting ferrite was prepared from coprecipitated hydroxydes. The source temperature $T_2$ was choosen between 900 and 1100°C and the deposition temperature $T_1$ between 800 and 1000°C. Deposition at temperatures higher than 1000°C cannot be achieved, this results from the thermodynamic feasibility of the reaction as discussed elsewhere (3). The CVD equipment is quite simple and the deposition rate is high($\lesssim$10µm/hour), but this method has several limitations : the deposition temperature being rather low, the as grown ferrite is not completely formed, the films contain some $Fe^{2+}$ and a slight departure from stoichiometry (Fe/Ni≃2.04) is observed. In order to improve the magnetic properties of the film a post annealing under pure oxygen during several hours is necessary.

The structural features were checked by X-rays Laue back reflection, X-rays and electron diffraction, SEM and surfaces examination by electron microscopy.

The magnetic properties were measured by ferromagnetic resonance (FMR), the measurements being performed at 17.3 GHz. Assuming $H_\perp$ and $H_\parallel$ are the resonance field perpendicular and parallel to the <111> plane of the film we have (2):

$$\frac{\omega}{\gamma} = H_\perp - 4\pi M_s - \frac{4K_1}{3M_s} - H_{st} \qquad (1)$$

$$\frac{\omega}{\gamma} = \sqrt{H_\parallel \left(H_\parallel + 4\pi M_s + \frac{K_1}{M_s} + H_{st}\right)} \qquad (2)$$

$$H_{st} = \frac{6C_{44}\left(1 + \frac{2C_{12}}{C_{11}}\right)}{1 + \frac{2}{3}\,\frac{C_{12}+2C_{44}-C_{11}}{C_{11}}}\;\frac{\lambda_{111}}{M_s}\left|(1-\eta)(\alpha_f-\alpha_s)\Delta T + \eta\,\frac{a_s-a_f}{a_f}\right| \qquad (3)$$

where $H_{st}$ is the strain induced anisotropy field, $C_{ij}$, the elastic constants, $\eta$ the fractional stress relieve, $\alpha$ the thermal expansion coefficient, $a_s$ and $a_f$ are the lattice parameter corresponding respectively to the film and the substrate. The other parameters have their usual meaning. $K_1$, $M_s$ and g are well known in the bulk $NiFe_2O_4$ ($K_1/M_s$ = - 250 Oe, g = 2.19 respectively) $4\pi M_s$ and $H_{st}$ can be easily deduced from equations (1) and (2).

## EPITAXY

It is generally admitted that epitaxial growth of one crystals over the other is possible if they join along a plane boundary which differs only slightly in their structural feature or/and cell dimensions.(5) MgO and spinel have the same oxygen close packing structure. Eight unit cells of MgO ($a_0$ = 4.213 Å) fit exactly an unit cell of spinel with $a_0$ = 8.426 Å.

The lattice parameters of garnet and spinel are quite different, but large similarities between atomic spacing exist. This features should be favorable for epitaxial growth of ferrite on garnet. In practice, two cases can be taken into consideration : epitaxial growth of ferrite on (100) or (111) garnet plane. It can be (4) easily shown that in a (100) plane if the $(1\bar{1}0)$ axis of the spinel lattice is parallel to the (010) axis of the garnet, and if the lattice parameter are in the ratio $a_{garnet} = \sqrt{2}\, a_{spinel}$, then the position of metal atoms in both structure can be fitted exactly. Octahedral sites in ferrite correspond to octahedral sites in garnet, and tetrahedral sites alternatively to tetrahedral and dodecahedral sites in garnet (4). In a (111) plane a good fit of both structure is obtained for the ratio $a_{garnet} = 3/2\, a_{spinel}$, two unit cells of garnet correspond to three unit of spinels. A detailed study shows(6) that different types of (111) planes in the garnet structure have some similarities with (111) planes in $NiFe_2O_4$ with the ratio 3:2 several metal atoms correspond exactly, distances between metal ions are almost identical. Deposition of a ferrite film on a (111) plane should lead to a smooth surface as (111) planes are natural growing surface of single ferrite crystals. Some experimental features confirm this assumption. (2)

## RESULTS AND DISCUSSION

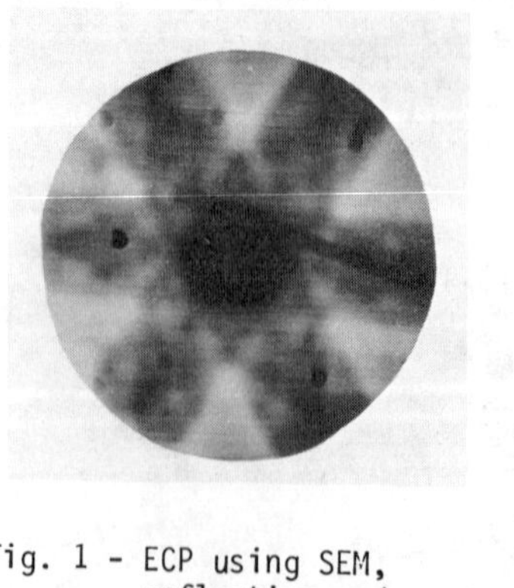

Fig. 1 - ECP using SEM, reflective mode.

$NiFe_2O_4$ films deposited on (111) plane of MgO were generally epitaxial as it was shown by X-ray Laue back reflection and investigation by S.E.M. Fig. 1 presents an Electron Channeling Pattern (ECP) using Scanning Electron Microscope (SEM). On Fig. 2 is reported X-ray reflections spectra around the (222) reflection of MgO. The peak corresponds to the (222) reflection of MgO($a_0$ = 4.213) the two other to (444) reflections of ferrites with $a_0$ = 8.402 and $a_0$ = 8.410 corresponding respectively to $NiFe_2O_4$ and an unknown phase. This third peak ($a_0$ = 8.41) could be attributed to an Mg rich ferrite film formed by interdiffusion between the film and the substrate. This result shows that in $NiFe_2O_4$ deposited on MgO the large lattice and thermal expansion mismatch (Table I) is mainly accomodated by elastic deformation, the stress relieved by cracks or misfit dislocations being about $\eta = 0.2$. This large strain perturbes strongly the magnetic properties as it is observed by FMR both on as deposited and annealed films. We observed one or two broad resonance lines with a resonance linewidth of 800 - 1000 Oe in the as deposited state and 600-800 Oe in annealed state.

Table I
Lattice constant and thermal expansion of bulk material

| | $NiFe_2O_4$ | MgO | $Nd_5Ga_5O_{12}$ | $Gd_3Ga_5O_{12}$ |
|---|---|---|---|---|
| $a_o$ (Å) | 8.34 | 4.213 | 12.506 | 12.386 |
| $10^6/K°$ | 10 | 13.8 | 9.2 | 9.2 |

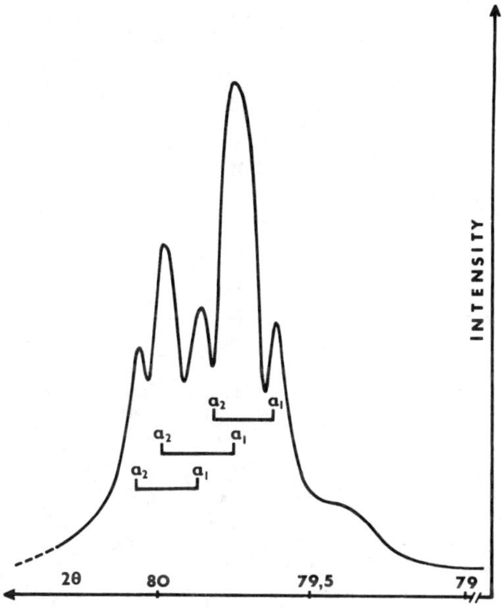


Fig. 2 - X-rays diffraction around the 222 reflection of MgO.

This large value of $\Delta H$ can be attributed to some local dislocations induced by strain and to the interdiffusion between films and substrate. The total strain induced anisotropy field was found to vary between -2000 Oe and -3000 Oe for various samples, and the as deduced strain release $\eta$ varies between 0.52 and 0.39. Discrepancies with X-ray measurements could be attributed to some plastic deformation.

In order to investigate the mechanism of epitaxy proposed, $Nd_3Ga_5O_{12}$ was choosen as non magnetic garnet substrate because the ratio of lattice parameter of NdGG and $NiFe_2O_4$ is exactly $^3/_2$. Furthermore the thermal expansion of $NiFe_2O_4$ and NdGG are very close (Table I). Some experiments were also performed on GGG substrate and $NiFe_2O_4$ deposited on this sample were always polycrystalline with a resonance linewidth of 20% to 50% higher than films simultaneously grown on NdGG. Films deposited on NdGG were epitaxial but the size of crystallites depended largely upon deposition parameters and film thickness. In several cases crystallites with various orientation in the (111) plane were found. The crystallites had threefold symmetry with edges parallel to the <110> direction. After annealing the in plane size of crystallites increased up to 10 - 1000 μm.

Fig. 3 - Electron diffraction pattern of a $NiFe_2O_4$ film deposited on NdGG.

In some favorable cases large surface single crystal films were obtained (~1 cm²) as it is shown on fig. 3. The lattice parameter determined by X-ray measurements was $a_o$ = 8.33Å value which is practically the same as in bulk samples.

Heat treatments permitted to improve highly the magnetic properties. On films annealed at 1100°C during 3 hours and slowly cooled to room temperature (10°C/hour) the as measured $4\pi M_s$ and $\Delta H$ is respectively 3350 G and 90 Oe values which can be compared with bulk single crystal (3400 G and 10-20 Oe). (Fig. 4) The $\Delta H$ of main line can be attributed to surface imperfection of substrate and some residual $Fe^{2+}$. The strain induced anisotropy field was found to be negligible.

On best samples annealed at 1100°C well defined spin wave resonance modes were observed (fig. 5). The higher order modes obey the well known spin wave disper-

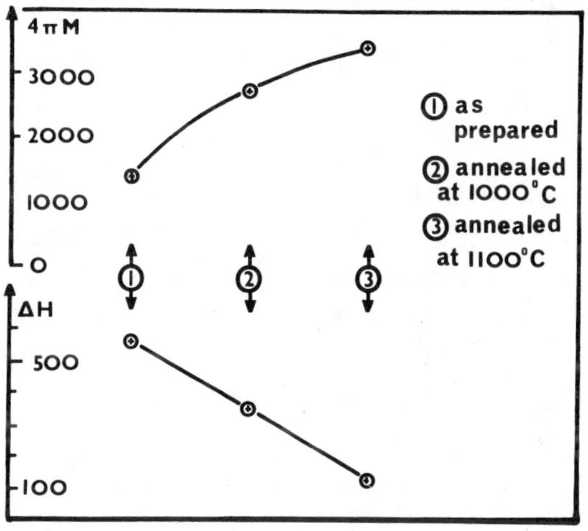


Fig. 4 - Influence of annealing temperature on $4\pi M_s$ and $\Delta H$ in a $NiFe_2O_4$ film deposited on a NdGG substrate.

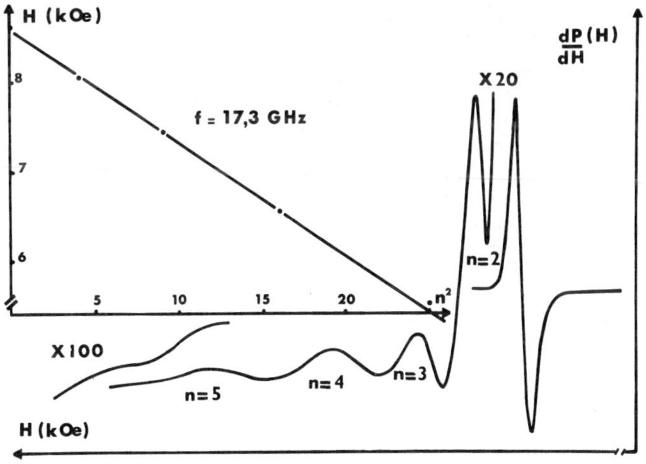


Fig. 5 - SWR spectrum of $NiFe_2O_4$ film for perpendicular orientation and the corresponding resonance field vs square of the SWR mode number.

sion relation

$$\frac{\omega}{\gamma} = H_1 - 4\pi M_s - \frac{4K_1}{3M_s} + Dk^2 \quad k = \frac{n\Pi}{d} \qquad (4)$$

As the spin wave modes are weakly excited the effective surface anisotropy can be considered to be small and hence n is an integer. Both even and odd modes are excited corresponding to asymmetric surface conditions. The n = 1 mode was not well resolved because of the overlapping with the main resonance. Fig.5 shows that with the present numbering the spin wave modes obey fairly well a quadratic dispersion law. The increasing linewidth with increasing mode order, however not completely understood, is partly due to geometrical inhomogeneities in the thickness of the film and some compositional inhomogeneities along the film plane.

As a conclusion, epitaxial spinel films were obtained on non magnetic garnet substrates when the ratio of lattice constant of garnet and spinel is equal to 3/2. Further improvments on magnetic properties can be achieved by improving deposition processes.

REFERENCES


(1) P.J. Besser, J.E. Mee, P.E. Elkins, D.M. Heinz, Mat. Res. Bull. 6 111 (1971)

(2) G. Suran, Soft Magnetic Materials Conf. 2, Cardiff (1975) to be published.

(3) P. Gibart, M. Robbins, A.B. Kane, J. Crystal Growth 24/25,166 (1974), and references therein.

(4) G.R. Pulliam, J. Appl. Phys. 38 1120 (1967).

(5) K.L. Chopra, Thin Film Phenomena (McGraw-Hill, New York 1969) p.224.

(6) X. Oudet, Acta Phys. Pol. A47, 789 (1975).


# LEAD FREE BISMUTH SUBSTITUTED GARNET FILMS BY L.P.E.*


T. R. Johansen, F. G. Hewitt, E. J. Torok, and D. L. Fleming
Sperry Univac
St. Paul, Minnesota 55165



## ABSTRACT

The growth of lead-free bismuth substituted rare earth garnet films by liquid phase epitaxy was studied for stripe domain light deflector use. Lead-free fluxes ($Bi_2O_3$,$Bi_2O_3$-BaO-$B_2O_3$) were used. Growing such films is extremely difficult because of (1) flux adhesion and (2) a large thermal expansion mismatch relative to GGG both of which tend to cause severe cracking. Curves of normal and in-plane lattice match versus growth temperature together with curves of lattice match versus temperature and details of melt composition are presented.


## INTRODUCTION

The motivation for this work on bismuth substituted garnets is to provide a suitable material for the deflector crystal for a stripe domain two dimensional light deflector. The calculated deflection efficiency for various materials is shown in figure 1.[1-4] High deflection efficiency requires a high Faraday coefficient and low absorption. The best such materials are $RbFeF_3$ in the visible, and bismuth substituted garnet in the infrared.

Bismuth substituted garnets grown in lead fluxes tend to include a considerable amount of lead which increases the optical absorption.[5-8] For this reason, our work has been principally with lead-free fluxes.

The simplest lead free flux is $Bi_2O_3$. Other lead free fluxes have in addition BaO, $B_2O_3$, $Na_2O$, or $CeO_2$, $GeO_2$, or $SiO_2$. These lead free fluxes are more difficult to work with than lead oxide fluxes. Flux adhesion is larger. Film cracking occurs due to adhesion and to differential expansion. Also, $Bi_2O_3$ has such a tendency to dissolve platinum that for some melts platinum crystals grow at nearly the same rate as garnet crystals. As noted by other investigators, the lower the melt temperature during deposition (the greater the supersaturation), the greater the percentage of bismuth in the resulting crystal.[7,8] The larger the percentage of bismuth, the larger the lattice constant. As shown below, the lattice constant of crystals grown in $Bi_2O_3$ fluxes is such a strong function of growth temperature that very accurate temperature control is necessary to obtain growth in region I (defined as the region in which the in-plane lattice of film and substrate are the same)[9-11]. In order to determine whether a film has been grown in region I or II (defined as the region in which the in-plane lattice constants of film and substrate are different[8,9]), it is necessary to measure the in-plane lattice mismatch between film and substrate, $\Delta a\|$, with x-rays. The measurement of the perpendicular lattice mismatch, $\Delta a\perp$, is not sufficient.

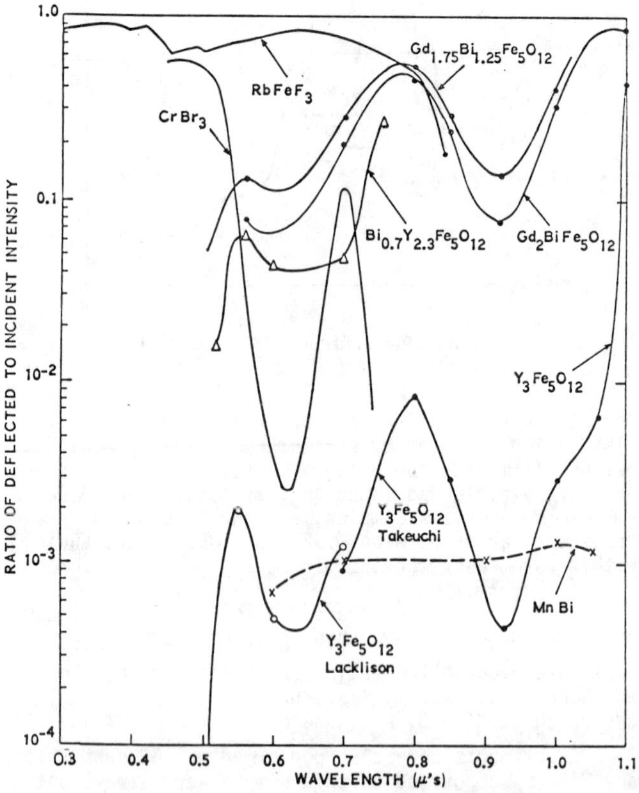


Fig. 1. Ratio of deflected to incident intensity vs wavelength for BiGdIG, from Takeuchi[2], for BiYIG from Lacklison et al[3], for $RbFeF_3$ from Chen et al[4] for $CrBr^3$ from Dillon et al[13] and YIG[2,3].

Lattice parameter difference measurements were made by diffraction with a $K\alpha_1$ copper x-ray beam. Rocking curves were taken for the (8,8,8), (0,8,8), (4,6,6), (12,8,0), and (12,6,0) diffraction planes. The difference in diffraction angles for each diffraction plane gives a linear relation between the in-plane and normal lattice mismatches[11]. Thermal expansion measurements were taken by mounting a modified Ungar #4035 Heating Unit on the Nonius eucentric goniometer.

## THE MELTS

Crystals were grown from two types of $Bi_2O_3$ melts, from $Bi_2O_3$-$B_2O_3$ melts from $Bi_2O_3$-BaO melts and from $Bi_2O_3$-PbO melts. The composition of each melt is given in table 1:

TABLE 1. Bi GARNET MELTS
(Concentrations in Milimoles)

| Melt | $Bi_2O_3$ | PbO | BaO | $B_2O_3$ | $Ga_2O_3$ | $Fe_2O_3$ | $Y_2O_3$ | $Yb_2O_3$ | $Tm_2O_3$ |
|---|---|---|---|---|---|---|---|---|---|
| A | 100 | - | - | - | - | 60 | - | 4.8 | - |
| B | 186 | - | - | - | - | 27 | - | 4.3 | - |
| C | 190 | 954 | - | - | 15 | 102 | - | - | 3 |
| D | 96 | 477 | - | - | - | 58 | - | 1.5 | - |
| E | 20 | - | 350 | 261 | - | 100 | 40 | - | - |
| F | 135 | - | - | 29 | - | 52 | - | 12 | - |

Melt A is a high temperature $Bi_2O_3$ melt in which the principal precipitate is garnet. Figure 2 shows the perpendicular and in-plane lattice mismatches, $\Delta a\perp$ and $\Delta a\|$, between film and GGG substrate as a function of growth temperature for (111) films from this melt. Note that the in-plane mismatch is zero in the region between 1050°C and 1070°C. This constitutes region I. Above that region, films grow with such a deficiency of bismuth that the crystals grow with an in-plane lattice smaller than that of the substrate, rather than supporting the high tensile strain of an in-plane match. The resulting lattice imperfections make the magnetic properties of such region II films unsuitable. Likewise, films grown at temperatures lower than 1050°C have such a surplus of bismuth substitution that films grow with in-plane lattice larger than that of the substrate, rather than supporting the high compressive stress of region I growth.

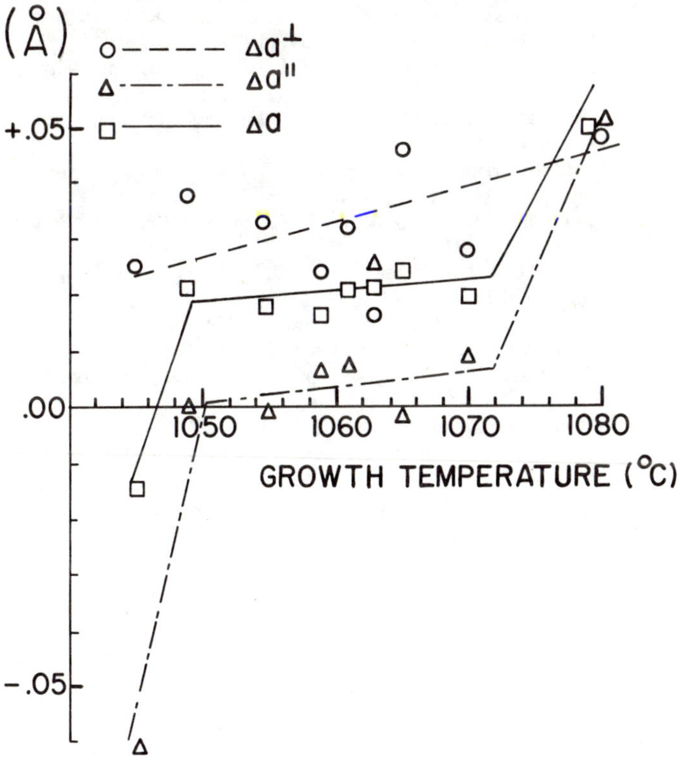


Fig. 2. $\Delta a\perp$, $\Delta a\|$, $\Delta a$ vs growth temperature for BiYbIG crystals grown on GGG in a $Bi_2O_3$ garnet phase melt of type A, Table 1. Region I is 1050°C to 1070°C.

Note that, with the exception of one borderline data point, all the films in figure 3 have a smaller lattice in the direction normal to the substrate than parallel to the plane. This means that at room temperature, where the data was taken, all the films are under tension. This occurs because the coefficient of expansion of the film is larger than that of the substrate. This is further shown in figure 3 in which the normal lattice difference between films and substrate, $\Delta a\perp$, is measured as a function of temperature. This large differential expansion coefficient causes all melt A films to be under tension at room temperature, even films grown at the compressive limit of region I. The films in figure 3 were grown in region I and show a differential coefficient of expansion, $\overline{\Delta\alpha}$, of 3.3 x $10^{-6}$/°K in the range from 23° to 500°C. The corresponding value for YIG on GGG is 2.0 x $10^{-6}$/°K in the same range, based on the work of Geller et al.[10]. The $\Delta\alpha$ for the melt A films over the temperature range 23°C $\leqslant$ T $\leqslant$ 1060°C can be estimated by adding the strain limit of Makino et al[11] at the lower end of region I to the observed $\Delta a$ at room temperature of 0.018 Å; the result is $\overline{\Delta\alpha} \approx$ 3 x $10^{-6}$/°K over the region 23°C $\leqslant$ T $\leqslant$ 1060°C. The corresponding difference between bulk YIG crystals and GGG crystals over the same temperature range is 1.25 x $10^{-6}$/°K.[12] At the time of this writing, it is not established whether the higher differential dilatation of the L.P.E. films as opposed to that of the bulk crystals is due to the added bismuth. However, the difference in coefficient of expansion between film and substrate is large enough to crack the films, provided the film thickness exceeds one micron, and indeed all the films in figure 2 are cracked. Uncracked melt A films ranging in thickness from 10 to 30 microns were grown on elastically deformable GGG substrates thinned to 10 microns by Syton polishing and phosphoric etching.

In order to overcome the thermal cracking problem, a low temperature melt was sought to decrease the excursion from growth to room temperature. This melt, melt B of table 1, has a low

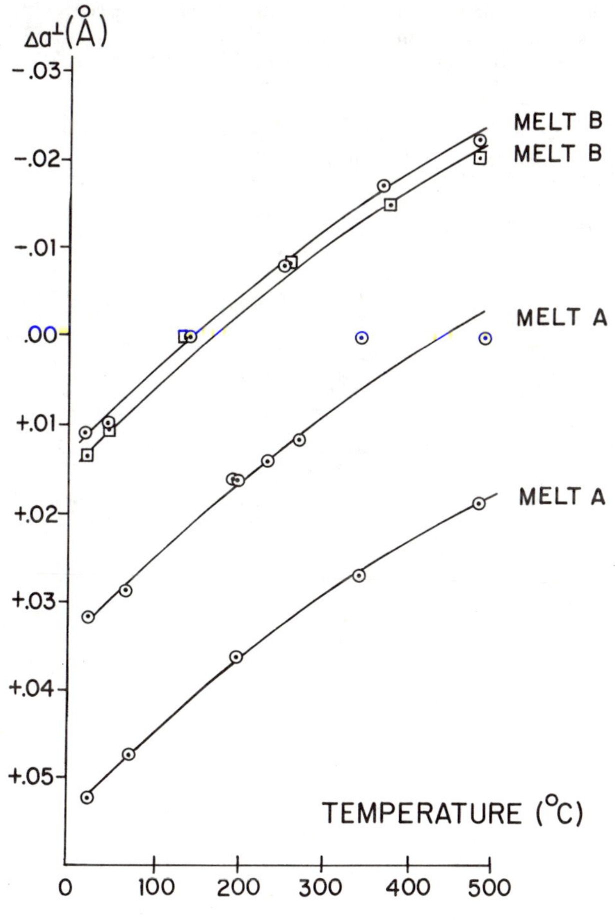


Fig. 3. $\Delta a\perp$ vs crystal temperature for BiYbIG crystals grown on GGG in melts of type A and type B.

enough saturation temperature so that crack-free region I films were obtained. Curves of $\Delta a\perp$, $\Delta a\|$, and $\Delta a$ versus growth temperature are shown in figure 4. Region I spans 765°C to 785°C and is limited on the high end by the saturation temperature. Films grown at the low temperature end of region I are nearly strain free at room temperature. Thermal dilation curves for two melt B films are plotted in figure 3; they are similar to those of melt A films. For the two melt B films $\overline{\Delta\alpha}$ = 3.50 x $10^{-6}$/°K and 3.58 x $10^{-6}$/°K over the range 23°C $\leqslant T \leqslant$ 500°C.

The principal precipitate from melt B is an unidentified non-garnet. The thickness of L.P.E. garnet films grown in melt B is limited by the formation of this precipitate in the melt.

Two $Bi_2O_3$-PbO[6-8] melts were investigated: type C and D in Table 1. Strain free BiTmFeGa films were obtained in melt C. The thermal mismatch measured for two melt C films is 1.7 x $10^{-6}$/°K and 2.26 x $10^{-6}$/°K in the 23°C to 500°C range. Similar results were obtained for films of BiYbFe garnet grown in a $Bi_2O_3$-PbO melt of type D: $\Delta\alpha$ = 2.82 and 3.14 x $10^{-6}$/°K. The films grown in the PbO melts have excellent mechanical quality; the flux adhesion is not a problem. X-ray measurements show that some region I films with compressive stress at room temperature were obtained.

An attempt was made to increase the coefficient of expansion of the substrate by substituting Bi in GGG. However, flux grown BiGGG did not accept Bi in significant amounts (a = 12.376 Å); attempts to substitute Bi in (111) YGG and GGG by liquid phase epitaxy were also unsuccessful.

The $Bi_2O_3$-$B_2O_3$ system (melt F) and the BaO-$B_2O_3$-$Bi_2O_3$ system (melt E) were investigated. L.P.E. films grown in these systems could not withstand the increased flux adhesion damage due to the $B_2O_3$. Although films grown in melt F had 0.8 Bi per formula, crystals grown in melt E had only 0.1 Bi atoms per formula. Attempts to obtain higher bismuth substitution in melt E by diluting the melt and lowering the growth temperature were unsuccessful.

Currently under investigation are the $Bi_2O_3$-$Na_2O$ and the $Bi_2O_3$-$K_2O$ systems which do not suffer from the excessive flux adhesion damage of melts E and F. Crack free region I L.P.E. films have been grown from a garnet phase melt in the former system.

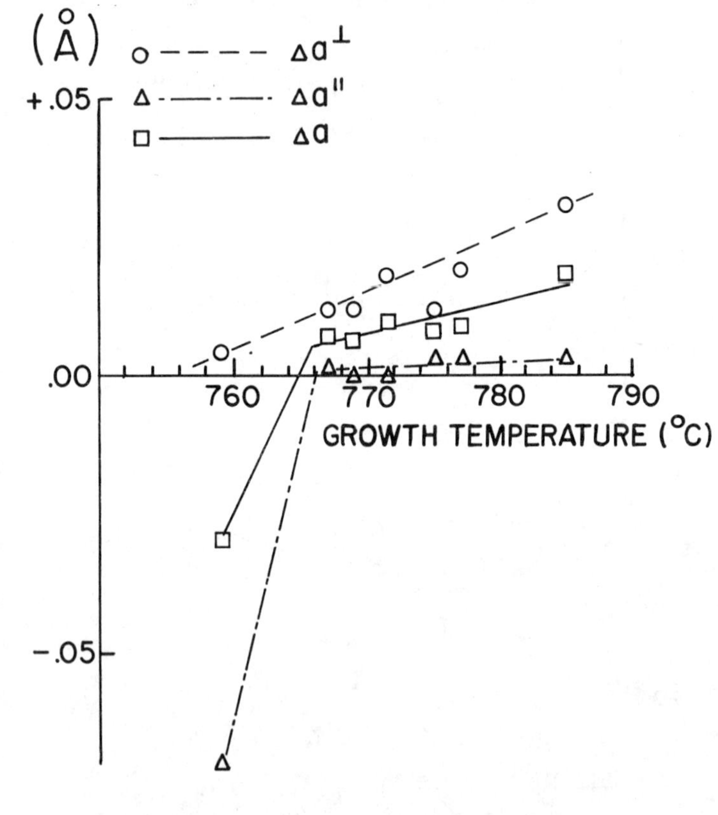


Fig. 4. $\Delta a\perp$, $\Delta a\|$ and $\Delta a$ vs growth temperature for BiYbIG crystals on GGG grown in a non-garnet $Bi_2O_3$ melt of type B, Table 1. Region I is 765°C to 785°C.

## DISCUSSION

Figures 2 and 4 show that the lower the growth temperature in a given melt, the higher the lattice constant and the higher the bismuth content, in agreement with other work in other melts. The fact that the lattice mismatch versus temperature curves have a plateau over region I indicates that the substrate lattice size strongly influences the bismuth content in region I. Evidently the energy of a growing film with a lattice that nearly matches the substrate lattice is enough lower than that of a strained lattice to favor selection of atoms of the proper size for nearly strain free growth.

The lead free L.P.E. films demonstrate the importance of differential thermal expansion in the two region epitaxial growth model, and the importance of measuring in-plane lattice mismatch as well as perpendicular. The growth of lead-free bismuth substituted garnet films by L.P.E. is more difficult than growth in lead fluxes; however, the feat can be performed, and the increase in optical efficiency over lead containing films make the task eminently worthwhile for our goal of a 0.80 efficient $10^6$ spot solid state light deflector as well as other optical devices.

## ACKNOWLEDGEMENTS

The authors wish to acknowledge the help of M. Kestigian of Sperry Corporate Research Center for supplying the substrates and for many helpful discussions, and the help of J. A. Krawczak, W. J. Simon, J. E. Kovacic, and G. L. Nelson for depositing films and making measurements.

## REFERENCES

*This effort was sponsored in part by Wright Aeronautical Laboratories Contract F33615-75-C-1099, and Office of Naval Research Contract N00014-73-C-0308.

1. T. R. Johansen, D. I. Norman, E. J. Torok, J. Appl. Phys. 41, 1342-3, (1970).
2. H. Takeuchi, S. Ito, I. Mikami, S. Taniguchi, J. Appl. Phys. 44, 4789, (1973).
3. D. E. Lacklison, G. B. Scott, H. I. Ralph and J. L. Page, I.E.E.E. Trans. Magn., MAG-9, 457, (1973).
4. F. S. Chen, H. J. Guggenheim, H. J. Levinstein, S. Singh, Phys. Rev. Letters 19, 948, (1967).
5. S. Wittekoek, T. J. A. Popma, P. E. Bongers, A.I.P. Conf. Proc. 18, 944, (1973).
6. J. M. Robertson, P. K. Larson, P. F. Bongers, I.E.E.E. Trans. Magn. MAG-11 5, 1112, (1975).
7. A. Akselrad, R. E. Novak, D. L. Patterson, A.I.P. Conf. Proc. 18, 949-53, (1973).
8. S. Wittekoek, J. M. Robertson, T. J. A. Popma, P. F. Bongers, A.I.P. Conf. Proc. 10, 1418, (1972).
9. P. J. Besser et al, A.I.P. Conf. Proc. 5, 125-9, (1971).
10. P. J. Besser et al, Mat. Res. Bull. 6, 1111, (1971).
11. H. Makino, T. Hibiya, K. Matsumi, A.I.P. Conf. Proc. 18, 80, (1973).
12. S. Geller et al, Mat. Res. Bull. 7, 1219, (1972).
13. J. F. Dillon Jr et al, J. Phys. Chem. Solids 27, 1531, (1966).

# OXYGEN STABILIZED RARE-EARTH IRON INTERMETALLIC COMPOUNDS*

M. P. Dariel†, M. Malekzadeh and M. R. Pickus
Inorganic Materials Research Division, Lawrence Berkeley Laboratory
Berkeley, California 94720

## ABSTRACT

A new, oxygen-stabilized intermetallic compound was identified in sintered, pre-alloyed rare-earth iron powder samples. Its composition corresponds to formula $R_{12}Fe_{32}O_2$ and its crystal structure belongs to space group $Im\underline{3}m$. The presence of these compounds has been observed, so far, in several R-Fe-O systems, with R = Gd, Tb, Dy, Ho, Er and Y.

## INTRODUCTION

Rare-earth iron intermetallics and in particular the $RFe_2$ (R = rare-earth metal) Laves phase compounds possess interesting properties such as huge magnetostriction, a strong dependence of the Young modulus on an applied field ($\Delta E$ effect) and high magnetocrystalline anisotropy. Similar to the $RCo_5$-type permanent magnet materials, the R-Fe intermetallic compounds are extremely brittle. Powder metallurgical techniques seemed therefore appropriate for preparing suitably sized and shaped specimens. In the course of the characterization of the sintered powder compacts, the presence of a previously unreported compound was encountered. This compound was subsequently identified, synthesized and its crystal structure determined. It represents the first reported case of a rare-earth transition metal, oxygen stabilized compound with apparently metallic properties.

## COMPOUND PREPARATION

A series of binary $RFe_2$ and ternary $R^1_{1-x}R^2_xFe_2$ alloys, was prepared by arc-melting on a water cooled copper hearth under a Zr gettered argon atmosphere. After a 48h long homogeneization anneal at 1000°C in evacuated quartz capsules, the alloys were pulverized by ball milling under toluene in a planetary ball mill. The 40-50μ particule size powder was rinsed and dried in an argon-containing glove box. Rubber tubing, 1" long, was manually filled with powder and isostatically compressed at 75,000 psi. The cold pressed samples were wrapped in Ta foils and sintered in a dynamic vacuum furnace. The temperature was increased stepwise, each 100 C to 1100 C. Each temperature increase resulted in an abrupt pressure increase in the system, due to the sample degassing. The samples were maintained for approximately 6 h at the highest temperature in a vacuum of $2 \times 10^{-6}$ torr. The sintered products were examined by optical microscopy and electron probe microanalysis. Powder diffraction patterns were taken using a Picker diffractometer with an X-ray monochromator.

In several instances the X-ray diffraction patterns included a series of lines which could not be attributed to any of the well known R-Fe intermetallics or to other compounds (e.g. oxides, nitrides), which could have been formed presumably, as a result of atmospheric contamination of the high specific surface powder.

In an effort to identify the unknown compound, which will hereafter be referred to as the β-compound, a series of Dy-Fe alloys was prepared in the 60-80 at. % Fe range, at 2 at. % increments. These alloys were submitted to the various powder preparation and sintering stages as described above. The alloy corresponding to approximately 72 at. % Fe yielded a nearly single β-phase diffraction pattern, which was subsequently used for the structure determination, with a few additional low intensity lines due to some residual $DyFe_2$ phase.

## STRUCTURE DETERMINATION

The diffraction pattern of the single phase alloy was successfully indexed on the basis of a body-centered-cubic unit cell with a 8.91Å lattice parameter (Table I). A literature search revealed that the only known crystal structures commensurate with that symmetry, lattice parameter and composition ratio of the $2_1$ components were those of the binary compound $Ca_3Ag_8$ and the similar ternaries $R_3Ni_6Si_2$.

Table I. OBSERVED* d-VALUES OF THE Dy-BASED β-compound ($Dy_{12}Fe_{32}O_2$)

| h k l | d | I | h k l | d | I | h k l | d | I |
|---|---|---|---|---|---|---|---|---|
| 1 1 2 | 3.622 | 11 | 5 1 0 | 1.739 | 6 | 4 4 4 | 1.2797 | 12 |
| 2 2 0 | 3.162 | 4 | 5 2 1 | 1.6187 | 16 | 7 1 0 | 1.2539 | 4 |
| 3 1 0 | 2.805 | 30 | 4 4 0 | 1.5666 | 20 | 7 2 1 | 1.2064 | 21 |
| 2 2 2 | 2.561 | 51 | 5 3 0 | 1.5203 | 18 | 7 3 0 | 1.1639 | 10 |
| 3 2 1 | 2.371 | 43 | 6 0 0 | 1.4775 | 15 | 7 3 2 | 1.1259 | 15 |
| 4 0 0 | 2.217 | 31 | 6 1 1 | 1.4381 | 28 | 8 0 0 | 1.1080 | 6 |
| 4 1 1 | 2.090 | 100 | 6 2 2 | 1.3364 | 38 | 8 1 1 | 1.0962 | 17 |

*Up to 2Θ = 90° with Cu-K radiation.

XBL 7511-9407

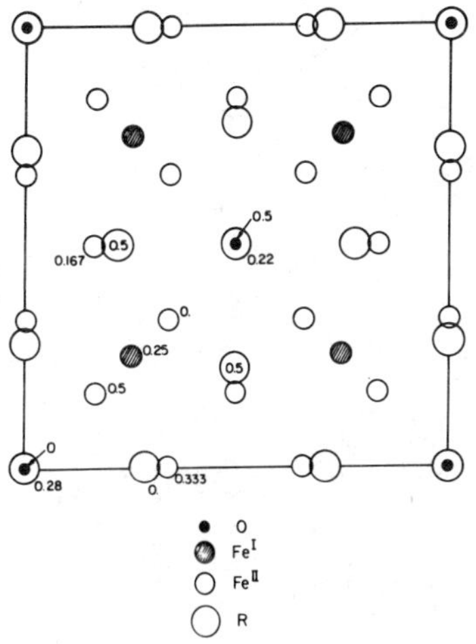


Fig. 1. Projection on the (001) plane.

In order to ascertain the crystal structure of the β-compound, its experimental diffraction pattern was compared to computer generated ones. The computer program used as input the assumed space group, the atom position coordinates and the lattice constants. It computed the diffraction line positions are intensities. The intensities of the individual reflections were calculated according to $I = p|F|^2LP$, with p the multiplicity, $|F|^2$ the structure factor and LP the combined Lorentz-Polarization factor, corrected for the presence of the X-ray monochromator.

Good agreement between the experimental and the computer results was obtained by assuming for the β-compounds a component ratio corresponding to the formula $Dy_3Fe_8$ and the space group $Im\underline{3}m$ (Int. Tab. No. 229), with 4 formula units per unit cell. No such compounds however, have ever been reported in binary R-Fe systems and it seemed unlikely, that they should have been overlooked. It was assumed therefore, that the β-compounds had been stabilized by the presence of a ternary component. Chemical analysis of the single phase Dy-Fe β-compound revealed the presence of 0.94 w.%O, 0.05 w% C and 0.09 w% N. Crystal structure considerations, as well as the results of the chemical analysis suggest that the actual formula of the β-compound should be $Dy_{12}Fe_{32}O_2$, space group $Im\underline{3}m$, with the atoms distributed on the following sites: (Fig. 1).

Table II. LATTICE PARAMETERS OF $R_{12}Fe_{32}O_2$ COMPOUNDS

| R | a(Å) |
|---|---|
| Gd | 8.919 ± 0.002 |
| Tb | 8.885 ± 0.002 |
| Dy | 8.8692 ± 0.0005 |
| Ho | 8.8435 ± 0.001 |
| Er | 8.815 ± 0.002 |
| Y | 8.8832 ± 0.0005 |

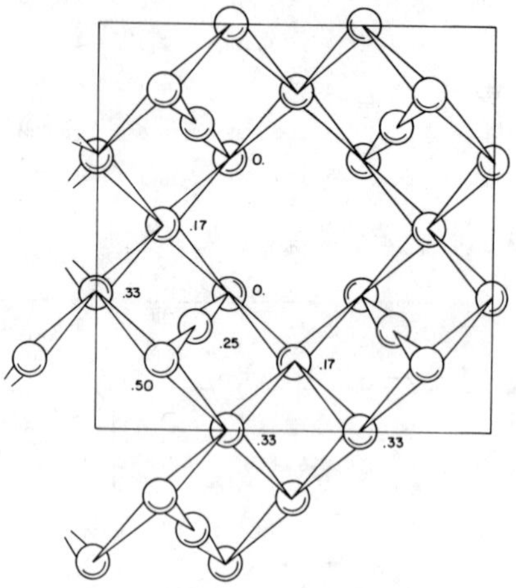


Fig. 2. Network of iron atoms projected on the (001) plane.

2 0 in a
12 Dy in e (x = 0.28)
8 $Fe_I$ in c
24 $Fe_{II}$ in h (x = 0.667)

The atomic position parameters for sites e and h were determined by trial and error, observing the best fit between the experimental and computed diffraction patterns (reliability factor 0.15). The positions assigned to the oxygen atoms are tentative only, since the X-ray spectra are insufficiently sensitive to the presence of the light atoms.

Efforts were made to determine the occurrence of the β-compounds in other than Dy, R-Fe systems. Single phase samples were successfully prepared for R = Gd and Ho. The presence of the β-compound was observed, though not as a single phase, for R = Tb, Er and Y, the additional phases being $RFe_3$, $R_6Fe_{23}$ intermetallics and some cubic $R_2O_3$ sesquioxides and RN nitrides. No trials have been made, so far, for preparing light rare-earth containing β-compounds. The lattice parameters of the β-compounds that have been observed, are shown in Table II.

The spatial arrangement of the atoms in the β-compound structure can be viewed as a network of $Fe_{II}$ atoms located on pairs of parallel {110} planes. The two planes of each pair are interconnected at regular intervals by the $Fe_I$ atoms (Fig. 2). This network leaves empty spaces around the (0,0,0) and (1/2,1/2,1/2) positions, which are filled by regular R octahedra with an oxygen atom at their center. The molar volume of the β-compound is reduced by 11% with respect to that of its constituents, this figure being very similar to that in the $RFe_2$ Laves compounds.

DISCUSSION

Metalloid stabilized intermetallic compounds have been observed in intra-transition metal systems. Notable examples are the $Ti_2Ni$ type phases, a large number of which are oxygen stabilized derivatives of nonoccurring binary phases. The effect of oxygen is to multiply by a factor of three the number of binary combinations for which the $Ti_2Ni$ structure is stable.[3]

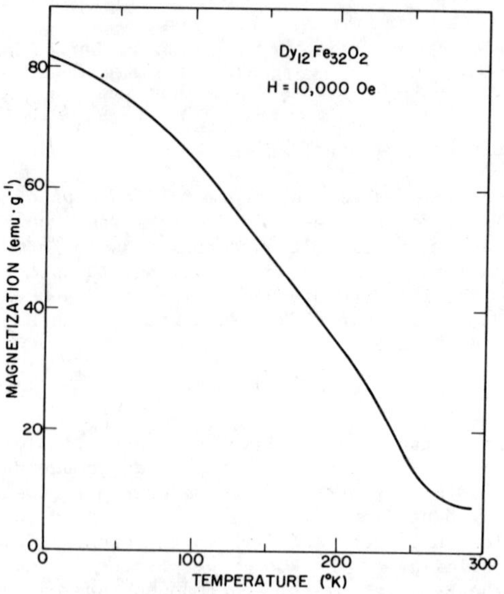


Fig. 3. Temperature dependence of the magnetization in nearly single phase $Dy_{12}Fe_{32}O_2$.

In the $Ti_2Ni$ type phases, oxygen has a variable solubility, in the β-compounds this point has not been verified. The $T^1_3T^2_3O$ and $T^1_3T^2_3C$ phases[4,5] ($T^1$, $T^2$ = transition metals) are further examples of metalloid stabilized compounds which have no binary counterparts. Similar nitrogen or carbon stabilized compounds involving rare-earth metals with Group III-A and IV-A metals (Al, Ga, In, Tl, Sn and Pb) have been reported[6]. No report has yet been made, to the best of the authors' knowledge, of oxygen stabilized rare-earth transition metal intermetallic compounds.

Ternary rare-earth, transition metal, oxygen systems have been subject to many investigations. The best known ternary compounds are the rare-earth iron garnets ($R_3Fe_5O_{12}$) and orthoferrites ($RFeO_3$). Both series have a relatively high oxygen concentration and are non-metallic. The β-compounds have a low oxygen concentration and a metallic luster. It is not clear at present whether they are thermodynamically stable, or only metastable, with respect to their decomposition into a rare-earth oxide (presumably $R_2O_3$) and a binary R-Fe intermetallic. If the latter were the case, it would account for the β-compounds not having been observed in previous investigations. In the course of ternary R-Fe-O system studies, oxygen is usually introduced in the form of rare-earth sesquioxides. It is not implausible that the high stability of these oxides would prevent their partial reduction to form a β-compound. In the present case, the method of preparation was different, in that high specific area prealloyed R-Fe compound powder was exposed to a low oxygen pressure at elevated temperatures.

Preliminary measurements below room temperature, using a VSM indicate that the Dy-based β-compound orders magnetically at 265±5 K (Fig. 3). The residual magnetization after the abrupt decrease at 260 K is probably due to the presence of the minority $DyFe_2$ phase. Assuming a De Gennes like dependence of the magnetic ordering temperature on J of the rare-earth component, similar to that followed by the $RFe_2$ and $RCo_2$ compounds, the Tb and Gd-based β-compounds might, in principle, be still ordered at room temperature. The value of the magnetization at 4.2 K suggests that the magnetic moments of Dy and Fe are aligned antiferromagnetically. The crystal structure of the β-compounds being different from either that of R-Fe intermetallic compounds ($RFe_2$, $RFe_3$, $R_6Fe_{23}$, $R_2Fe_{17}$) or other high oxygen content compounds (orthoferrites, garnets), new magnetic structures can be expected. Both the intermetallics and the oxygen containing compounds possess highly interesting properties, many of which have found practical application. Magnetometric and Mossbauer effect studies of the β-compounds are being initiated with the purpose of determining some of their interesting and, perhaps useful, properties.

REFERENCES

* This work was supported by the U.S. Energy Research and Development Administration.

† On leave from the Nuclear Research Center-Negev and the Department of Materials Engineering, Ben-Gurion University, Beer-Sheva, Israel.


1. L. D. Calvert and C. Rand, Acta Crystallogr. 17, 1175 (1969).
2. E. L. Hladyschewsky, P. L. Krypiakewytsch and O. L. Bodak, Z. Anorg. Allg. Chemie 344, 95 (1966).
3. M. V. Nevitt, Electronic Structure and Alloy Chemistry of the Transition Elements, Ed. by P. A. Beck, Interscience, New York, (1963), p. 127.
4. N. Schönberg, Acta Chem. Scand. 8, 932 (1954).
5. K. Kuo, Acta Met. 1, 301 (1953).
6. H. Nowotny, Seventh Rare-Earth Research Conference, Coronado, California, 1968, p. 309.


# MAGNETIC BEHAVIOR OF SOME RARE EARTH GERMANIDES OF THE TYPE $RFe_2Ge_2$*


S. K. Malik,† S. G. Sankar, and V. U. S. Rao
Department of Chemistry
University of Pittsburgh
Pittsburgh, Pennsylvania 15260

and

R. Obermyer
Department of Physics
Pennsylvania State University
McKeesport, Pennsylvania 15132


## ABSTRACT


Magnetic properties of tetragonal rare earth germanides of the type $RFe_2Ge_2$, where R = Pr, Gd, Tb, Er and Th, have been investigated in the temperature interval 4.2 to 273 K and in magnetic fields up to 20 kOe. The magnetization of $ThFe_2Ge_2$ is small and weakly temperature dependent. The compounds with Pr, Gd and Tb show antiferromagnetic ordering with Néel temperature of 14 K, 11 K and 7.5 K, respectively. The Pr sublattice in $PrFe_2Ge_2$ and the Tb sublattice in $TbFe_2Ge_2$ both undergo a transition, which at 4.2 K occurs at a field strength of about 15 kOe. The Gd compound does not show a transition up to a field of 20 kOe. The compound $ErFe_2Ge_2$ is paramagnetic in the temperature range investigated. However, some magnetization from Fe sublattice seems to be present in all the compounds.


## INTRODUCTION

The present work on $RFe_2Ge_2$ (R = rare earth and Th) compounds is a continuation of the systematic investigation of magnetic, electronic and transport properties of $RM_2X_2$ compounds, where M = Cr, Mn, Fe, Co, Ni, Cu, Ag and Au and X = Si or Ge. These compounds crystallize in $ThCr_2Si_2$ type structure isomorphous with tetragonal $BaAl_4$ type structure with Ba sites occupied by rare earth atoms, Al(1) sites occupied by M atoms and Al(2) sites occupied by X atoms. From earlier investigations of $RCo_2Ge_2$ compounds, McCall *et al.*[1] concluded that the Co atoms do not carry a localized moment. The $RMn_2Ge_2$ and $RMn_2Si_2$ compounds are ordered ferromagnetically and Mn atoms carry a localized moment.[2,3] In this paper, we report magnetic properties of some $RFe_2Ge_2$ compound with R = Pr, Gd, Tb, Er and Th.

## EXPERIMENTAL

The $RFe_2Ge_2$ compounds were prepared by induction melting of the stoichiometric amounts of constituent elements in a water-cooled copper boat under a purified atmosphere of argon gas. The purity of the starting materials was as follows: rare earth 99.9%, Fe 99.99% and Ge 99.9999%. The ingots were turned and melted several times and the weight loss during melting was negligible in every case. The powder X-ray diffraction patterns were taken on a Picker X-ray unit using $CuK_\alpha$ radiation. The magnetic measurements were carried out using Faraday method in fields up to 20.6 kOe and in the temperature interval 4.2 to 273 K.

## RESULTS AND DISCUSSION

All the $RFe_2Ge_2$ compounds investigated in this paper were found to be of single phase and the powder X-ray diffraction patterns could be indexed on the basis of a body-centered tetragonal structure of $ThCr_2Si_2$ type. The lattice constants obtained by an iterative least square fit of the observed $\sin^2\theta$ values are listed in Table I and are in good agreement with those reported[4,5] for some of these.

The variation of magnetization as a function of temperature at two different applied fields is shown in Figures 1-3. It is to be noted that the compounds with Pr, Gd and Tb show distinct antiferromagnetic ordering at low fields. The Néel temperature obtained are respectively 7.5 K for $TbFe_2Ge_2$, 11 K for $GdFe_2Ge_2$ and 14 K for $PrFe_2Ge_2$. The compounds with Er and Th do not show ordering in the temperature range investigated. The magnetization of $ThFe_2Ge_2$ is small and only weakly temperature dependent; the moment values for applied field of 20.6 kOe being 0.3 $\mu_B$ at 4.2 K and 0.19 $\mu_B$ at 273 K. Figure 4 shows the variation of magnetization as a function of applied field at 4.2 K. $PrFe_2Ge_2$ and $TbFe_2Ge_2$ undergo either a spin-flop or metamagnetic transition which occurs at a field strength of about 15 kOe.

Above the respective Néel temperatures, the susceptibility of $RFe_2Ge_2$ compounds is found to be larger than expected on the basis of trivalent rare earth ions. A similar situation was found earlier[5] in the case of $RFe_2Si_2$ compounds (R = La, Pr, Nd, Sm,

Table I. Lattice constants and Neel temperature ($T_N$) of $RF_2Ge_2$ compounds

| Compound | Lattice constant a(Å) | c(Å) | $T_N$(K) |
|---|---|---|---|
| $PrFe_2Ge_2$ | 4.055 | 10.524 | 14 |
| $GdFe_2Ge_2$ | 3.986 | 10.478 | 11 |
| | 3.989 | 10.485 (Ref. 4) | |
| | 3.899 | 10.48 (Ref. 5) | 11 |
| $TbFe_2Ge_2$ | 3.976 | 10.479 | 7.5 |
| $ErFe_2Ge_2$ | 3.941 | 10.414 | |
| $ThFe_2Ge_2$ | 4.102 | 10.166 | |

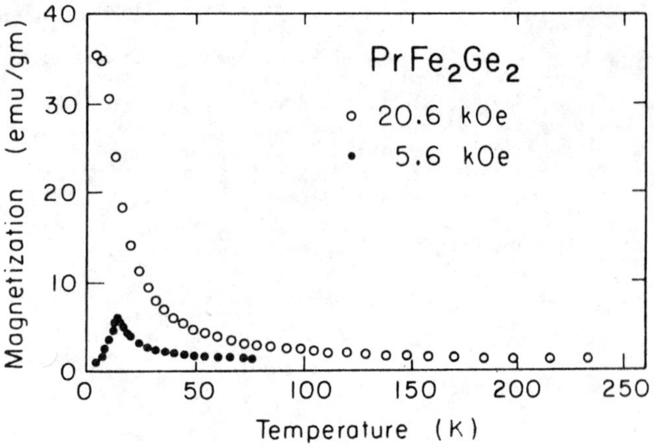


Fig. 1 Magnetization versus temperature for $PrFe_2Ge_2$ at two different applied fields.

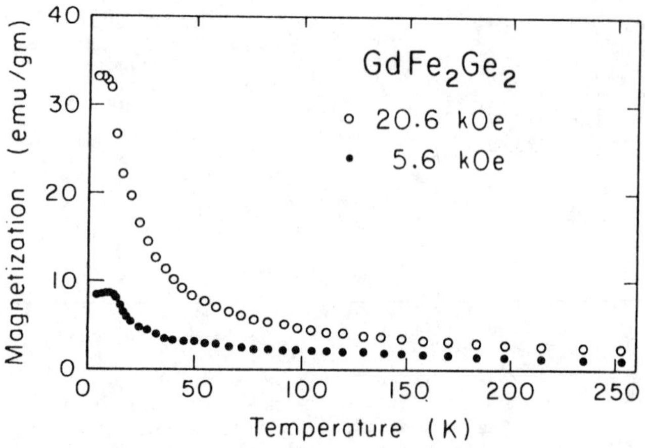


Fig. 2 Magnetization versus temperature for $GdFe_2Ge_2$ at two different applied fields.

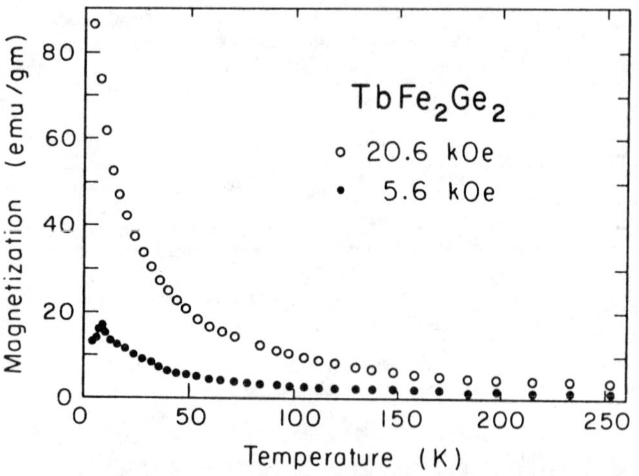


Fig. 3 Magnetization versus temperature for $TbFe_2Ge_2$ at two different applied fields.

Gd and Dy) and $RFe_2Ge_2$ compounds (R = Ce, Nd and Gd) and has been interpreted in terms of a spontaneous magnetization from Fe sublattice superimposed on the rare earth sublattice. This seems to be consistent with $Fe^{57}$ Mössbauer results in $GdFe_2Si_2$ in which the presence of both magnetic and nonmagnetic iron was observed. However, neutron diffraction studies[6] on $NdFe_2Si_2$ failed to detect an appreciable ferromagnetic moment. Thus, the origin of weak ferromagnetism in these compounds is not clear as yet. Further experiments are underway to understand the nature of magnetic moment on iron atoms.

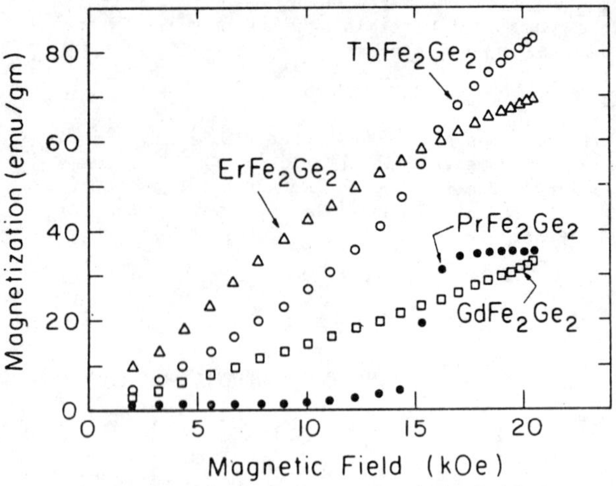


Fig. 4 Magnetization versus applied field for $RFe_2Ge_2$ compounds at 4.2 K.

## References

* Supported by the U. S. Energy Research and Development Administration.

† On leave of absence from Tata Institute of Fundamental Research, Homi Bhabha Road, Bombay 400 005, INDIA.

1. W. M. McCall, K. S. V. L. Narasimhan, and R. A. Butera, J. Appl. Phys. 44, 4724-4726 (1973).
2. K. S. V. L. Narasimhan, V. U. S. Rao, R. L. Bergner, and W. E. Wallace, to appear in J. Appl. Phys.
3. K. S. V. L. Narasimhan, V. U. S. Rao, W. E. Wallace, and I. Pop (this conference).
4. W. Rieger and E. Parthe, Mh. Chem. 100, 444-454 (1969).
5. I. Felner, I. Mayer, A. Grill, and M. Schieber, Solid State Commun. 16, 1005-1009 (1975).
6. H. Pinto and H. Shaked, Phys. Rev. B7, 3261-3266 (1973).

## $Co^{2+}$-$Co^{2+}$ INTERACTION IN $CoNb_2O_6$

I. Yaeger and A.H. Morrish
Department of Physics, University of Manitoba, Winnipeg, Canada
and
B.M. Wanklyn
Clarendon Laboratory, Oxford, England

ABSTRACT

Neutron diffraction studies of $CoNb_2O_6$ powder at 2°K indicate an ordered antiferromagnetic structure of the $Co^{2+}$ spins(1). The transition temperature has not been observed in the susceptibility and field induced magnetization measurements we took on flux grown single crystals of $CoNb_2O_6$ from 3.3°K to 300°K. The crystallographic space group of $CoNb_2O_6$ is Pbcn. Group theoretical analysis of $Co^{2+}$-$Co^{2+}$ bilinear interactions within the Pbcn space group indicates that both symmetric and antisymmetric $Co^{2+}$-$Co^{2+}$ interactions are allowed. We show that the salient features of our experimental results at temperatures below 50°K can be accounted for in terms of the interaction between $Co^{2+}$ ions occupying the lowest lying Kramers' doublet. The local principal axes of the g tensor for the $Co^{2+}$ ground doublet at all sites are found practically to coincide with the orthorhombic crystallographic axes giving $g_a = 4.1\pm0.2$, $g_b = 4.4\pm0.2$ and $g_c = 5.7\pm0.3$. The $Co^{2+}$-$Co^{2+}$ antisymmetric interaction happens to be negligible as compared to the symmetric one. From the temperature dependence of the paramagnetic susceptibility it is found that the dominant $Co^{2+}$-$Co^{2+}$ interaction along the a and c directions is the intrasublattice ferromagnetic coupling rather than the intersublattice antiferromagnetic interaction.
A complete account of this work will be published elsewhere.

REFERENCES

1. H. Weitzel and S. Klein, Solid State Commun. 12, 113 (1973).

## INFLUENCE OF ANTIFERROMAGNETIC T-DOMAINS ON TRANSPORT IN PURE NiO†

J.E. Keem, L.L. VanZandt, Dept. of Physics

J.M. Honig, Dept. of Chemistry, Purdue Univ.

ABSTRACT

We report a series of observations of Seebeck effect, resistivity, and heat capacity on high quality single crystals of antiferromagnetic insulating NiO as a function of temperature and stoichiometry. The magnitude of the resistivity is shown to be proportional to the stoichiometry and over the entire temperature range studied a quantitative association is obtained between the temperature dependence of the resistivity and the magnetic ordering. The heat capacity shows evidence of a lambda transition at the Néel point and is only very slightly affected by stoichiometry. The Seebeck effect data are insensitive to both the magnetic transition and stoichiometry and, but for an anomaly to be discussed, may be associated with the ionization energy of a hole from an $Ni^{+2}$ vacancy. We also report the observation of a 1/2(volt/K) anomalous thermoelectric effect which appears below 400K in the most stoichiometric annealed samples. The anomalous time dependence, relative to the establishment of a temperature gradient, magnitude, saturation with increased temperature gradient, and temperature dependence are tentatively explained in terms of elastomagnetic interactions of carriers with mobile antiferromagnetic T-domain walls.

†Work supported by NSF Grant #GH34314.

## NEW MAGNETIC COMPOUNDS WITH HEUSLER AND HEUSLER-RELATED STRUCTURES

James C. Suits
IBM Research Laboratory, San Jose, CA 95193

ABSTRACT

A number of new compounds have been found which have the Heusler crystal structure or a crystal structure derived from the Heusler structure. Many of these compounds are ferromagnetic at room temperature. $Rh_2MnSn$ exhibits the fully ordered Heusler structure, has a Curie temperature of 410°K and a moment at 5°K of 3.1 $\mu_B$ per formula unit. Substitution of Pb for Sn increases the moment and lowers the Curie temperature. Substitution of group IIIB elements for Sn causes disorder on the Mn-IIIB sub-lattice and the magnetic properties indicate that nearest neighbor Mn-Mn interactions are antiferromagnetic. Substitutions from group VB and VIB have also been made. Replacement of Mn in $Rh_2MnSn$ can be accomplished with any of the 3d transition elements. Some of these compounds, e.g., $Rh_2FeSn$ show a new structure which is a large tetragonal distortion of the Heusler structure. The c/a ratio is approximately 1.2. The appearance of this tetragonal modification is believed due to a band type of Jahn-Teller effect. A consequence of this is that a tetragonal to cubic structure transformation should occur at elevated temperature and this has been observed for $Rh_2CoSn$.

# HIGH FIELD MAGNETIZATION STUDIES ON $Y_{1-x}Th_xFe_3$ COMPOUNDS

K. S. V. L. Narasimhan, R. A. Butera and C. J. Kunesh,
Department of Chemistry, University of Pittsburgh,
Pittsburgh, PA 15260

## ABSTRACT

Saturation magnetization measurements were made on $Y_{1-x}Th_xFe_3$ compounds up to a field of 120 kOe. The moment decreases linearly with x. The results on fine particles of $Y_{.1}Th_{.9}Fe_3$ indicate that the anomaly observed in this material does not arise due to the magnetic anisotropy.

## INTRODUCTION

The effect of varying electron concentration on the magnetic ordering of rare earth compounds with transition metals has been undertaken in our laboratory. During this study it was found that an anomalous magnetic transition exists in the $R_{1-x}Th_xFe_3$ compounds over a certain composition if R is a trivalent metal similar in size to that of thorium.[1] $R_{1-x}Th_xFe_3$ show an antiferromagnetic to ferromagnetic transition as temperature is increased from 4.2 K to room temperature. In the case of $Y_{1-x}Th_xFe_3$ compounds the anomalous region extends from $x \cong 0.35$ to 0.90.[2] The purpose of the present study is to collect additional information as to the nature of this transition and to obtain saturation magnetic moments under an applied field of 120 kOe.

## EXPERIMENTAL

The preparation of the compounds and annealing procedures have been discussed in detail in our previous publication.[2] The ingots were ground to a fine powder in a mortar and pestle under toluene. The powders of $Y_{.1}Th_{.9}Fe_3$ were sifted through 75 μm, 45 μm and 37 μm sieves in a glove bag under helium atmosphere. The various fractions, namely > 75 μm, < 75 μm but greater than 45 μm, < 45 μm but greater than 37 μm and ≤ 37 μm particles were sealed in pyrex tubes for magnetic measurements using the Faraday method. The measurement of magnetization in high fields up to 120 kOe was carried out on coarse powders (>> 75 μm) packed in a plastic sample holder. An Intermagnetic General 120 kOe $Nb_3Sn$ superconducting magnet was used in conjunction with an elevator method. In this method the sample was pulled through a pair of opposing precision wound coils and the time integral of induced voltage was measured using an integrating digital voltmeter. The method used is similar to that of Giauque et al.[3] but adapted for use with small samples and a superconducting magnet.

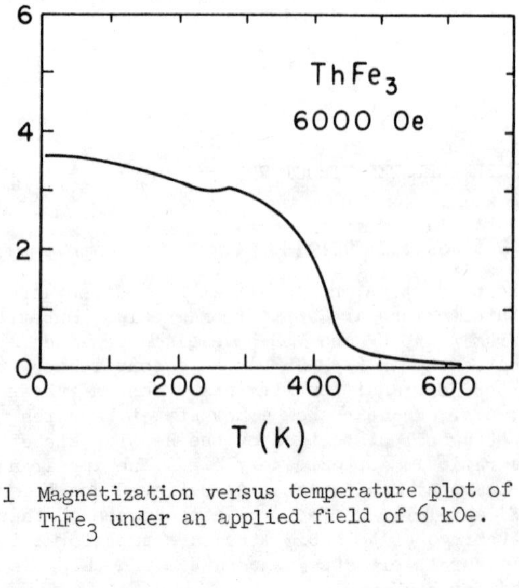


Fig. 1 Magnetization versus temperature plot of $ThFe_3$ under an applied field of 6 kOe.

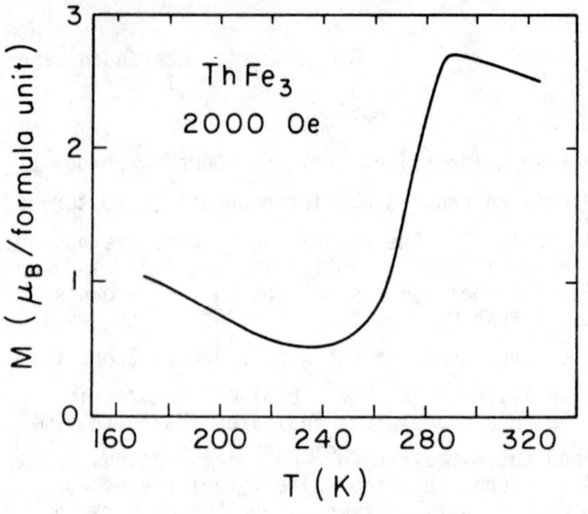


Fig. 2 Magnetization versus temperature plot of $ThFe_3$ under an applied field of 2 kOe.

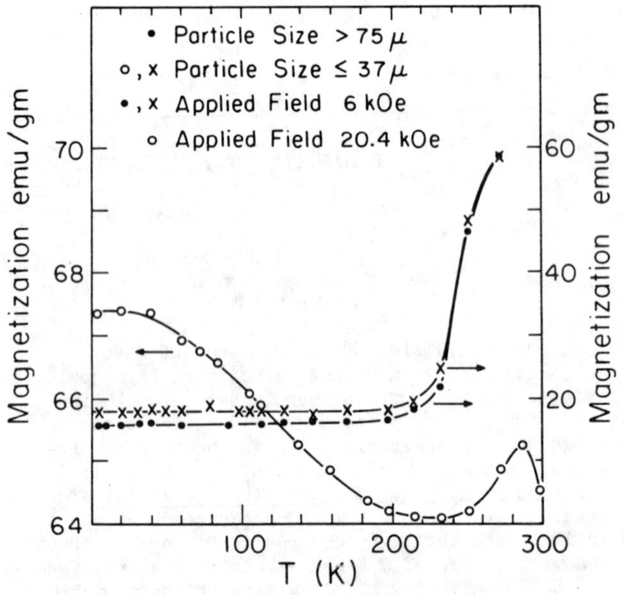


Fig. 3 Magnetization versus temperature plots for fine particles of $Y_{.1}Th_{.9}Fe_3$. Note the ordinate scale for 20 kOe measurements.

## RESULTS AND DISCUSSION

Figs. 1 and 2 show the magnetization versus temperature plots of $ThFe_3$ under an applied field of 6000 Oe and 2000 Oe. A sharp magnetic transition is observed in applied fields of 2000 Oe and the transition is nearly absent under an applied field of 6000 Oe. This transition is intensified as yttrium is substituted for thorium.[2] Fig. 3 shows the plots of magnetization versus temperature on ≤ 37 μm and ≥ 75 μm particles of $Y_{.1}Th_{.9}Fe_3$. Similar results were obtained with particles of other sizes. Magnetization versus field measurements on these particles show a magnetic transition similar to that reported earlier.[2] If the transition is due to a change in the magnetic hardness of the material then one would expect the magnetic moment of the smaller particles (≤ 37 μm) to orient more easily along the field direction than the larger particles (≥ 75 μm). This would result in a large increase in the magnetization and lower intensity of the transition for

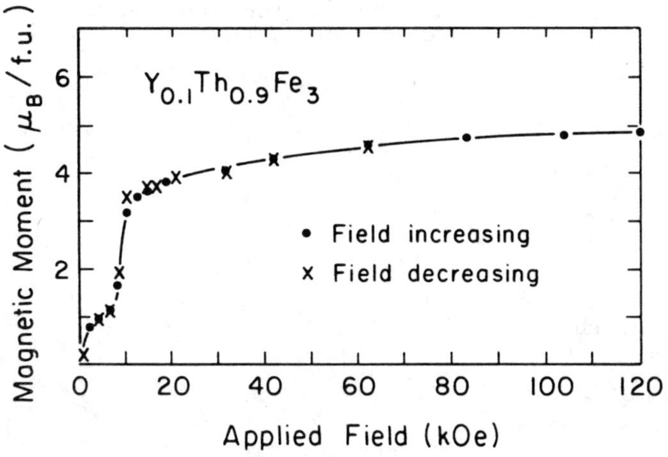


Fig. 4 Magnetization versus applied field for $Y_{.1}Th_{.9}Fe_3$ at 150 K.

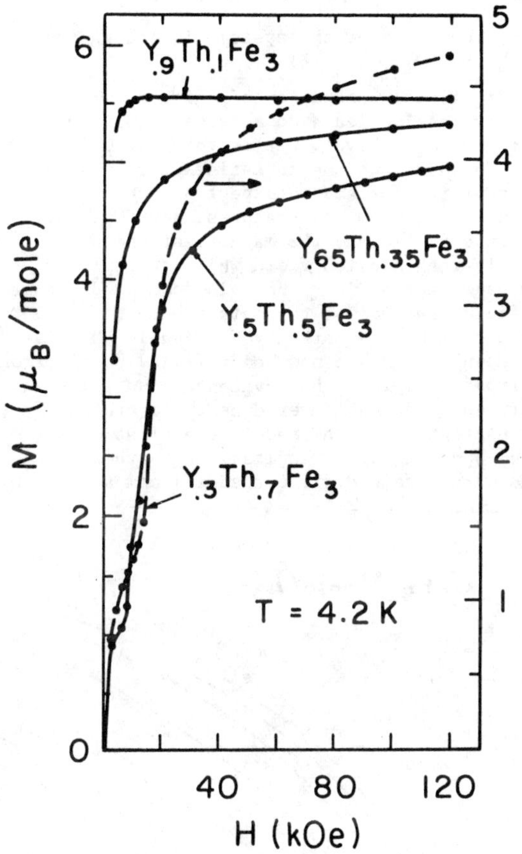


Fig. 5 Magnetization versus applied field for some $Y_{1-x}Th_xFe_3$ compounds at 4.2 K.

the small particles. No such effect was observed in the present study. Further, loosely packed powders of $Y_{.1}Th_{.9}Fe_3$ do not show any hysteresis in the magnetization versus field data at 150 K when the field is reduced from 120 kOe to near zero field (see Fig. 4). The above observations suggest that the anomalous transition does not arise from the changes in the magnetic anisotropy. We postulated that the anomalous transition arises from one of the possible two iron sublattices, namely, 3(b)-6(c) and 18(h), changing its magnetic order from ferromagnetic to antiferromagnetic as the temperature is lowered. The rise in the magnetization as the temperature is decreased (Fig. 3) under an applied field of 20 kOe below the transition can be thought of as arising from the saturation of the ferromagnetic sublattice [3(b)-6(c)] and partial destruction of antiferromag-

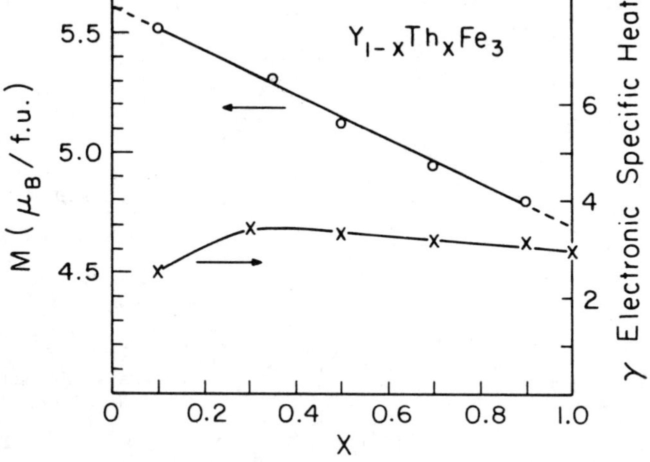


Fig. 6 Variation of magnetic moment and electronic specific heat (from ref. 4) as a function of x.

netic order of the 18(h) sublattice due to the applied field.

The results of high field magnetization for some of the compounds are shown in Fig. 5. All the compounds studied were difficult to saturate unless the applied field exceeded 60 kOe. Even at 120 kOe perfect saturation was not reached. Saturation moments were collected by extrapolating magnetization versus inverse field plots from above 60 kOe to infinite field. A plot of magnetic moment versus composition is shown in Fig. 6 and the plot also shows the electronic specific heat versus composition which have been reported earlier.[4] The moment decreases linearly with x and if we assume the linearity to extend to x = 0 and 1, then a difference of one Bohr magneton exists between $YFe_3$ and $ThFe_3$ (work on $ThFe_3$ and $YFe_3$ is currently in progress). Since thorium has one valence electron more than yttrium we can conclude that the difference of nearly one Bohr magneton arises due to the filling of the 3d band by this excess electron. This would also suggest that the spin up 3d-subband is nearly full at $YFe_3$ and the spin down 3d-subband begins to populate as thorium is substituted for yttrium. Electronic specific heat[5] which can be correlated with the density of states at the Fermi surface also shows an increase as thorium is substituted for yttrium, indicating that the spin down band is being populated.

## CONCLUSION

The anomalous transition observed in $Y_{1-x}Th_xFe_3$ series for x > .35 has been found to be intrinsic to the material and may be accounted for by considering that in the magnetically ordered region the Fe moments on the 18(h) site atoms are aligned antiferromagnetically below the transition temperature and ferromagnetically above. The moments on the 3(b)-6(c) sites are ferromagnetic at all temperatures. The results of this work have shown that the occurrence of transition is not the result of a change in magnetic anisotropy. The variation of the saturation moment has been accounted for by a band filling mechanism wherein the iron 3d band absorbs the extra electrons provided by the substitution of a +4 valence atom (Th) for a +3 valence atom (Y).

## REFERENCES

1. C. J. Kunesh, Ph.D. thesis, University of Pittsburgh, Pittsburgh, PA 15260, 1973.

2. C. J. Kunesh, K. S. V. L. Narasimhan and R. A. Butera, AIP Conf. Proceedings, No. 10, 1065, 1973.

3. W. F. Giauque, E. W. Hornung, R. A. Fischer, G. E. Brodale and R. A. Butera, Rev. Sci. Instr. 35, 213, 1964.

4. C. J. Kunesh, K. S. V. L. Narasimhan and R. A. Butera, J. App. Phys. 46, 1349, 1975.

5. N. F. Mott, Adv. Phys. 13, 325, 1964.

# MAGNETIZATION OF SINGLE CRYSTAL $(Ti,V)_2O_3$

C. F. Eagen
Research Staff, Ford Motor Company
Dearborn, Michigan 48121

N. C. Koon
Naval Research Laboratory
Washington, D. C. 20375

and

L. L. VanZandt
Purdue University
West Lafayette, Indiana 47907

## ABSTRACT

The field dependent magnetization of single crystal $Ti_2O_3$ doped with 2% and 5% vanadium was measured down to 1.6K. The results disagree with the predicted magnetic behavior based on the VanZandt impurity level model proposed for this system.

A model for the electronic structure of semimetallic $(Ti_{1-x}V_x)_2O_3$ has been proposed by VanZandt[1]. This model has been successfully employed to explain the anomalously large linear term in the low temperature specific heat[2] and the observed conductance in inelastic electron tunneling through Schottky barriers[3]. The model is also consistent with observed ultrasonic properties[4,5,6]. A critical test of this model, as VanZandt has suggested[7], is the behavior of the low temperature magnetization with varying magnetic fields. In this note we present preliminary single crystal magnetization data on two samples with nominal vanadium concentrations of 2 and 5 at.% which casts serious doubt on the validity of the VanZandt model.

Single crystal samples of $Ti_2O_3$, doped with 2 and 5% $V_2O_3$, were drawn from the melt in molybdenum crucibles at a rate of roughly 5 mm/hour. Small cubes, roughly 5 mm on a side, were cut from the boulis and oriented by back scattered x-rays. The magnetization of these cubes was measured using a conventional vibrating sample magnetometer mounted in a superconducting solenoid capable of producing fields up to 65 KGauss at temperatures down to 1.6K. No corrections were made for the small demagnetization effects.

The basic features of the VanZandt model are that the $V^{3+}$ ions substitute for the $Ti^{3+}$ ions in the host lattice. For each host band derived from the $Ti^{3+}$ levels there is a corresponding V impurity state lying somewhat below the appropriate host band. Hence, the two d electrons characteristic of $V^{3+}$ occupy $a_{1g}$ like V states which lie below the filled $a_{1g}$ valence band of semiconducting $Ti_2O_3$. The crucial point of the argument is that some of the otherwise empty V $e_g$ levels, which lie below the host conduction band, also lie slightly below the upper edge of the valence band. These impurity levels form a band which, being fractionally occupied, introduces holes into the host valence band. These light holes are presumably responsible for the normal transport properties of the system, including the positive Hall coefficient[8].

The fractional occupation of these spin degenerate impurity levels leads to a large density of states at the Fermi level. It is this large density of states which allows one to satisfactorily account for the specific heat and to expect pronounced magnetic effects. At low temperatures, and for small applied magnetic fields, the magnetization should rise sharply with applied field while the spin up impurity band empties into the spin down band. For applied fields sufficient to empty the spin up impurity band one expects a marked reduction in $d\vec{M}/dH$ as the spin down states are now being filled by electrons derived from the spin up host valence band which is characterized by a much lower density of states.

The parameters of the model were first adjusted to fit the specific heat results; the further successes were obtained without additional adjustment.

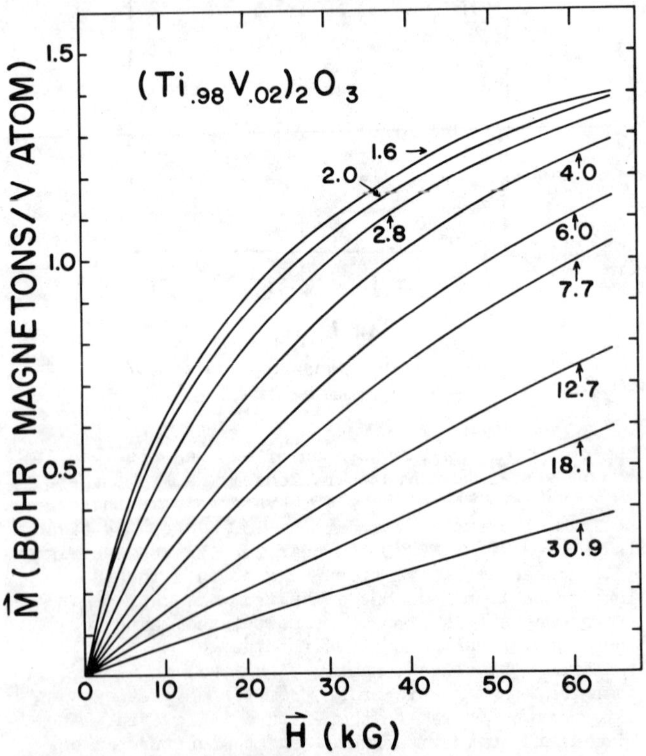


FIG. 1. Field and temperature dependent magnetization of a nominal 2% vanadium doped $Ti_2O_3$ sample in units of Bohr magnetons per vanadium atom.

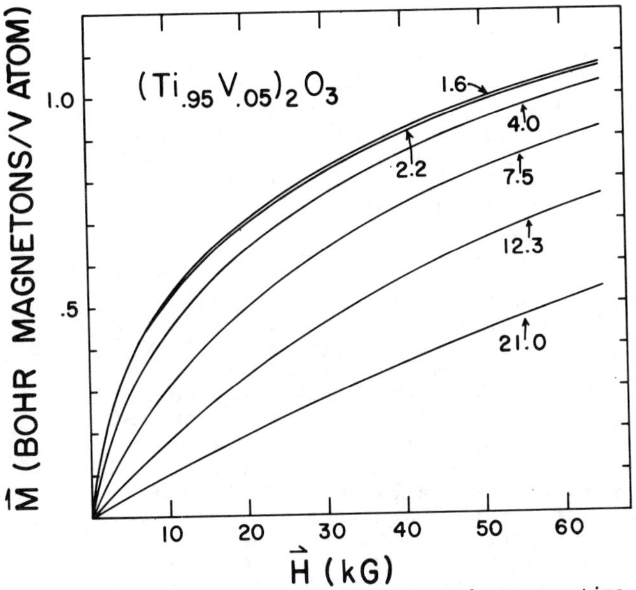


FIG. 2. Field and temperature dependent magnetization of a nominal 5% vanadium doped $Ti_2O_3$ sample.

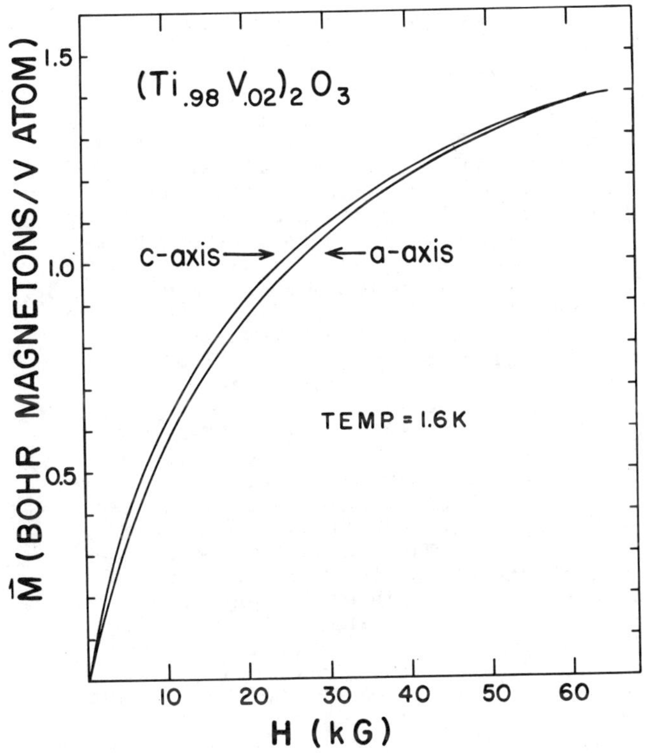


FIG. 3. Magnetization of a 2% vanadium doped $Ti_2O_3$ sample parallel and perpendicular to the c axis.

These parameters then give a total saturated moment of 0.24 spins or, with the assumption that g = 2, 0.24 Bohr magnetons per vanadium ion.

Figure 1 shows the magnetization isotherms for our nominal 2% V sample. 2% is within the range where the model has met with success, but the lower concentration limit where a maximum amount of structure in χ (H) is expected[7]. Other than a general trend toward saturation at the lowest temperatures, the results do not agree with predictions. The observed magnetizations are larger by approximately a factor of 4 or 5 than the predictions of the model. Also, the model predicts a more definite trend towards saturation than occurs experimentally.

Figure 2 shows the same data for 5% V doping. The same factor of 4 discrepancy between observation and prediction occurs. No significant trend toward saturation is observed or predicted. Figure 3 compares the magnetization parallel and perpendicular to the conventional c axis. The measurements were made on separate specimens cut from the same boule. The model predicts no anisotropy and very little is seen.

Our measured magnetization curves are consistent with the higher temperature work of Dumas et al[9] who have observed concentration dependent effective paramagnetic moments near or exceeding the value appropriate for S =1/2 per V atom. Recently, Dumas et al[10] have applied a "spin glass" type of theory to this system. They have measured the magnetization of single crystal specimens and observe remnant magnetizations and hysteresis at low fields and temperatures. The model of VanZandt contains no local moments on Vanadium ions. The reason for this was the insistence of the specific heat experimenters that there was no sign of them. The model describes quite well the specific heat in moderate magnetic fields as well as in their absence. Even if there were no other discrepancies, however, the discovery of remnance and hysteresis is instantly and unarguably fatal to a model which--like VanZandt's--contains no permanent, localized moments.

However, when Dumas et al. adjust the parameters of their spin glass model to fit the observed magnetization, they obtain a specific heat which is too large by the same factor of 4 as above. (in this concentration range) They attribute this discrepancy to the inapplicability of their model in this relatively high concentration regime.

There is thus a serious discrepancy between what the experimental specific heat and the magnetic susceptibility seem to be saying which no currently available model explains.


1. L.L. VanZandt, Phys. Rev. Lett. 31, 598 (1973).
2. M.E. Sjöstrand and P.H. Keesom, Phys. Rev. B7, 3558 (1973).
3. P.C. Eklund and L.L. VanZandt, Phys. Rev. B11, 784 (1975).
4. T.C. Chi and R.J. Sladek, Phys. Rev. B7, 5080 (1973).
5. T.C. Chi and R.J. Sladek, Proc. 5th Int. Conf. Int. Friction and Ultrason. Attn., Aachen (1973).
6. J.G. Bennett and R.J. Sladek, J. Sol. St. Chem. 12, 370 (1975).
7. L.L. VanZandt, Mat. Res. Bull. 9, 593 (1974).
8. J.M. Honig, L.L. VanZandt, T.B. Reed, and J. Sohn, Phys. Rev. 182, 863 (1969).
9. J. Dumas, C. Schlenker, and R.C. Natoli, Sol. St. Comm. 16, 493 (1975).
10. J. Dumas, C. Schlenker, J.L. Tholence, and R. Tournier (to be published).

# SINGLE CRYSTAL GROWTH OF $HoFe_2$ AND $ErFe_2$

J. B. Milstein
Naval Research Laboratory
Washington, DC 20375

## ABSTRACT

We report the first single crystal growth of centimeter-dimension specimens of $HoFe_2$ and $ErFe_2$. These materials are cubic Laves phase compounds, which, as a class, exhibit such properties as magnetic anisotropy, large magnetostriction, and spin reorientations. The present crystals, of as much as 15 gm mass, make possible single crystal neutron diffraction and magnetostriction measurements.

The starting compounds prepared from 99.9% rare earth ingot and 99.95% iron rod, were melted together in electric arc or radio frequency induction furnaces. Single crystals were grown by the Czochralski technique, using an electric arc-powered furnace, under gettered argon atmosphere. $HoFe_2$ and $ErFe_2$ behave as congruently melting compounds, and grow under similar conditions.

The crystals produced have been examined by x-ray and neutron diffraction, which show 1) the absence of structure on the neutron diffraction rocking curve peaks and 2) mosaic spreads of not more than 0.3°. The very high quality of the specimens obtained may be deduced from these results.

## INTRODUCTION

Since the discovery of cubic Laves phases having the compositions $HoFe_2$ [1,2], and $ErFe_2$ [2,3], there has been some interest in elucidating the crystallographic [4-6] and magnetic [6-8] properties of these compounds. More recently, the large magnetostrictive strains exhibited by these compounds at room temperature have been investigated [9-11]. In addition, equilibrium phase diagram studies of the holmium-iron system [12,13] and the erbium-iron system [14-16] have been performed. A reference to the growth of single crystals of $ErFe_2$ (and $TbFe_2$) appears in the paper of Clark, Cullen and Sato [10], but the specimens used in that study were single crystal grains located in and cut from polycrystalline boules [17]. The preparation of centimeter dimension single crystals of $HoFe_2$ or $ErFe_2$ prior to this work has not been reported.

## EXPERIMENTAL DETAILS

The starting materials used in this work were 99.9% Ho ingot and 99.9% distilled Er metal purchased from Research Chemicals and 99.95% iron rod purchased from Materials Research Corporation. The materials were handled in a manner similar to that reported earlier [18], with certain modifications. The rare earth (RE) metal and iron rod were cut to prepare starting compositions corresponding to the formula $REFe_{1.98}$, to suppress formation of second phase. Typically, a total mass of tens of grams was weighed out, to an accuracy of $\pm$ 0.001 gm. Buttons of the starting compositions were prepared by repeated arc melting alternated with inversion of the solidified mass and by radio frequency induction melting with magnetic levitation (r-f levitation melting). The second procedure was carried out in a Hukin water cooled crucible [19] fitted to an Arthur D. Little model MP crystal growing furnace, powered with a 20 KW, 450 kc Taylor-Winfield generator. The material prepared by r-f levitation melting was held molten for at least 15 minutes to ensure complete dissolution and reaction. The furnace was operated at power levels 20% to 30% higher than was necessary for the material to melt completely, with the molten material probably not appreciably hotter than 1500°C. Nassau and coworkers stated that for melts which they estimated to have been heated to 2000°C, only 10 seconds were required to achieve apparent homogeneity [1]. The loss of mass observed after the button preparation typically amounted to about one part per thousand when arc melting was employed, and about two parts per thousand using r-f levitation melting. Such losses, even if totally ascribed to one element, do not significantly change the button composition.

## RESULTS AND DISCUSSION

Polycrystalline seeds were cut from the buttons for the initial crystal growth attempts in the arc furnace described previously [18]. The crystals were grown by the Czochralski method, using pull rates of 5-7 mm/hr, with seed rotation of approximately 20 rpm and hearth rotation of approximately 60 rpm in the opposite sense. The erbium compound could be grown at a slightly higher rate than that used for $HoFe_2$. Titanium-gettered argon at several psig overpressure was used as a protective atmosphere. Single crystal regions measuring several millimeters across by some centimeters length were found in the first boules grown in each instance. Single crystal seeds were cut from these first boules and have been used for further crystal growth experiments.

The equilibrium phase diagram work on the Ho-Fe and Er-Fe systems is somewhat contradictory. The paper of O'Keefe, Roe and James [12] states that only the Ho and $HoFe_2$ phases occur in melts containing more than 33 at. % Ho and that $HoFe_2$ melts at 1335° $\pm$ 15°C. In a later paper, Roe and O'Keefe [13] report that $HoFe_2$ melts incongruently at 1288°C. From the phase diagram presented in the second paper, it would appear that a liquid containing approximately 36.5 at. % Ho would be required to give $HoFe_2$ as the stable solid. The papers of Buschow and van der Goot [14] and Meyer [15] state that $ErFe_2$ melts congruently, while Kolesnikov, Trekhova and Savitskii [16] state that it melts peritectically. Melting points from 1300°C to 1360°C were reported.

Some observations on the crystals produced suggest that $HoFe_2$ and $ErFe_2$ are in fact congruently melting. When pulling a noncongruently melting material, one expects the composition of the first portion of material to solidify to differ from that of the starting material, and the composition of the solid to vary as a function of fraction of melt pulled, ultimately reaching that of the starting material. For a pure, congruently melting material, no change in composition as a function of fraction of melt pulled is expected. The specimens produced have not exhibited any change in composition when examined by x-ray or neutron diffraction techniques. Such changes would be expected to manifest themselves as variations in lattice parameter or the appearance of second phase material. No tests have been made for impurities, as previous experience has shown that the crystal growth procedure used does not result in contamination of the crystals grown above that present in the starting materials, nor does a significant impurity gradient develop in the crystals. Metallic impurities would not be expected to modify diffraction results if present in low concentration, because they would not cause lattice distortion, and do not contribute much diffractive power.

From the observations made on materials melted initially by arc furnace and by r-f levitation methods, it is clear that the latter method has several significant advantages:

1. The fact that r-f levitation allows complete homogenization in one operation eliminates the danger of preferential loss of one of the components upon shattering during cooling or remelting cycles, as may occur when using an arc melting method.

2. By means of the r-f levitation technique the entire mass is maintained at an elevated temperature, which eliminates the possibility of providing a "cold-trap" for volatile contaminants. In fact, the material prepared by r-f levitation exhibits less surface dross than that prepared by arc melting, using starting materials from the same lots. No analytical data have yet been obtained to support or refute this observation, but the fact that r-f levitation results in a somewhat greater loss of mass seems to support this assertion. The use of purer starting materials might be instructive.

3. Without tedious analytical work, one can never be certain that an arc melted specimen has been melted sufficiently to have been homogeneized, especially when the mass of material becomes large.

Routine x-ray and neutron diffraction measurements have been performed to assess the crystallographic quality of some of the specimens. Typically, neutron diffraction rocking curve peaks taken with the full crystal illuminated have exhibited no structure and have been sufficiently sharp that a mosaic spread of not greater than $0.3^{o}$ has been determined. Rocking curve peaks observed for several diffraction lines on the same specimen have been measured and yield consistent results. On some crystals, low angle grain boundaries have been present in the region where the seed joins the grown boule. In such a case, the rocking curve may appear slightly skew. Upon sectioning and x-raying such a specimen, these low angle grain boundaries are invariably found to be confined to the surface of the boule, and generally "grow out" over a few millimeters distance. Furthermore, the orientation of the seed and the boule are found to coincide, which strongly suggests that these low angle grains are spuriously nucleated, perhaps by the presence of dross on the melt surface. In this regard, one might conjecture that nonmetallic impurities, principally oxygen and nitrogen, should be rejected from the growing crystal on the basis of relative ionic size and the close packed structure which exists in the Laves phase. These ions and their compounds should have a great tendency to decorate grain boundaries or to congregate as inclusions of second phase material. This would imply that the crystals grown in this and the previous work are purer with regard to nonmetallic elements than the starting material. Unfortunately, the requisite analytical techniques are not currently available to us. Although interesting, the study of impurity behavior and phase relations per se have not been warranted as part of the present program.

Specimens of single crystal $HoFe_2$ and $ErFe_2$ are presently being used in magnetostrictive strain [20], spin wave [21] and magnetization measurements [22].

ACKNOWLEDGEMENTS

The author would like to thank Dr. J. J. Rhyne of the National Bureau of Standards for taking the neutron diffraction rocking curve data and for useful discussion, and Mr. L. R. Johnson for assistance in setting up the r-f levitation system.

REFERENCES

1. K. Nassau, L. V. Cherry and W. E. Wallace, J. Phys. Chem. Solids 16, 123 (1960).
2. J. H. Wernick and S. Geller, Trans. Met. Soc. AIME 218, 866 (1960).
3. A. E. Dwight, Trans. Am. Soc. Metals 53, 479 (1961).
4. R. C. Mansey, G. V. Raynor and I. R. Harris, J. Less-Common Metals 14, 329 (1968).
5. A. E. Ray, Proc. 7th Rare Earth Research Conf., Vol. II, 473 (1969).
6. E. Burzo, F. Angew, Physik 32, 127 (1971).
7. K. H. J. Buschow and R. P. van Stapele, J. Appl. Phys. 41, 4066 (1970).
8. M. P. Dariel, U. Atzmony and R. Guiser, J. Less-Common Metals 34, 315 (1974).
9. (a) N. C. Koon, A. I. Schindler and F. L. Carter, Phys. Lett. 37A, 413 (1971).
   (b) A. E. Clark and H. S. Belson, A.I.P. Conf. Proc. 5, 1498 (1972).
10. A. E. Clark, J. R. Cullen and K. Sato, A.I.P. Conf. Proc. 24, 670 (1975).
11. M. P. Dariel and U. Atzmony, Int. J. Magnetism 4, 213 (1973).
12. T. J. O'Keefe, G. J. Roe and W. J. James, J. Less-Common Metals 15, 357 (1968).
13. G. J. Roe and T. J. O'Keefe, Met. Trans. 1, 2565 (1970).
14. K. H. J. Buschow and A. S. van der Goot, Phys. Stat. Sol. 35, 515 (1969).
15. A. Meyer, J. Less-Common Metals 18, 41 (1969).
16. V. E. Kolesnikov, V. F. Trekhova and E. M. Savitskii, Izv. Akad, Nauk SSR, Neorg, Mat., 7, 495 (1971).
17. A. E. Clark, private communication.
18. J. B. Milstein, N. C. Koon, L. R. Johnson and C. M. Williams, Mat. Res. Bull. 9, 1617 (1974).
19. D. A. Hukin, "A New Design of Cold Crucible and Its Application to the Growth of Rare Earth Metal Single Crystals", Ref. 24/71, Clarendon Laboratory, Univ. of Oxford, Oxford, Great Britain.
20. N. C. Koon, C. M. Williams and J. B. Milstein, to be published.
21. (a) J. J. Rhyne, N. C. Koon and J. B. Milstein and H. A. Alperin, to be published.
   (b) R. M. Nicklow, N. C. Koon, C. M.

Williams and J. B. Milstein, A.I. P. Conf. Proc. 24, 165 (1975).

22. J. H. Schelleng, N. C. Koon and J. B. Milstein, to be published.

# MAGNETIC PROPERTIES OF $RMn_2X_2$ COMPOUNDS* (R=RARE EARTH, Y OR Th AND X=Ge, Si)

K. S. V. L. Narasimhan, V. U. S. Rao and W. E. Wallace
Department of Chemistry, University of Pittsburgh
Pittsburgh, Pennsylvania 15260

and

I. Pop
Department of Physics
Cluj University, Cluj, ROMANIA

## ABSTRACT

Magnetic properties of $RMn_2Ge_2$ and $RMn_2Si_2$ compounds crystallizing in the $BaAl_4$ type structure have been investigated over a wide temperature range (4.2 K to 1200 K). $LaMn_2Ge_2$, $CeMn_2Ge_2$, $PrMn_2Ge_2$, $NdMn_2Ge_2$ and $LaMn_2Si_2$ all possess Curie temperatures above 300 K and the easy direction of magnetization have been found to be along the c-axis. When R is a heavy rare earth antiferromagnetic coupling was obtained at 4.2 K between the rare earth and manganese moments whereas when R is a light rare earth a ferromagnatic coupling of moments was obtained. $YMn_2Si_2$, $ThMn_2Si_2$ and $ThMn_2Ge_2$ possess a small moment $<\cdot 2\mu_B$ at 4.2 K. High temperature susceptibilities on $RMn_2Si_2$ compounds indicate a Curie-Weiss behavior.

## INTRODUCTION

Magnetic Studies on compounds containing Mn was undertaken in order to understand the effect of varying the Mn-Mn distance on the magnetic ordering, Curie temperature and anisotropy. A wide variety of compounds crystallizing in the $ThCr_2Si_2$ type structure which is isomorphous with $BaAl_4$ type structure has been reported in the literature.[1,2] Omjec and Ban[3] reported the magnetic properties of some of the thorium containing silicides and germanides crystallizing in this structure. The unusual magnetic behavior of the Mn containing compounds prompted us to investigate the rare earth containing silicides and germanides. $RCo_2Ge_2$ (R=rare earth) compounds were investigated in detail[4] and it was found that cobalt in these compounds does not carry a moment. Investigations on the $RFe_2Ge_2$ compounds have shown[5,6] that iron does carry a small magnetic moment. In a previous investigation[7] it was reported that in $RMn_2Ge_2$ compounds Mn carries a moment of nearly 1.5 Bohr magnetons per atom. It was also found[7] that when R=La, Ce, Pr, Nd, the compounds have curie temperatures above room temperature but for the heavier rare earths the ordering temperatures are below 100 K. We report in this paper our results on the $RMn_2Si_2$ compounds and some further investigations on the $RMn_2Ge_2$ compounds.

## EXPERIMENTAL

The compounds were prepared by mixing the elements in stoichiometric proportions and induction melting them in a water-cooled copper boat under an atmosphere of purified argon. The loss of manganese during melting due to its volatility was compensated by adding the required excess and remelting. Tm and Yb compounds were contained in a MgO crucible, which was placed in a tantalum tube and then sealed under a partial pressure of argon. The tantalum tube formed the susceptor for subsequent induction melting. All the compounds were wrapped in a tantalum foil and annealed for 15 days at 750°C. X-ray diffraction patterns were taken with a $CuK_\alpha$ radiation using a Picker X-ray diffractometer. All the compounds except $YbMn_2Si_2$ were found to be single phase. $YbMn_2Si_2$ contained extra lines that did not belong to the $BaAl_4$ type. Magnetic measurements on loose powders sealed in pyrex and quartz tubes were carried out using the faraday method over a temperature range of 4.2 K to 1200 K in applied fields up to 20 kOe.

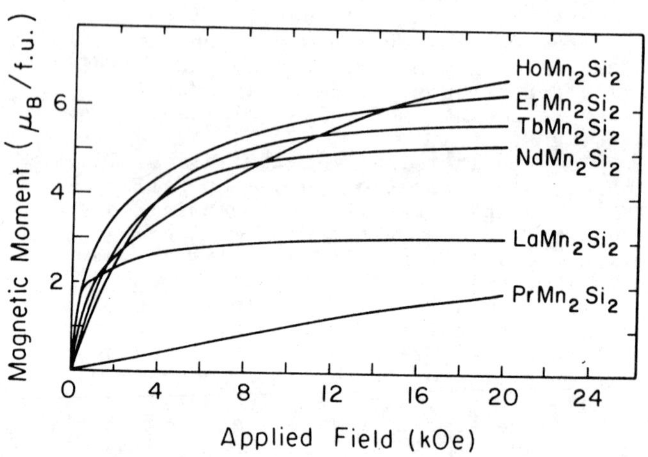


Figure 1. Magnetization versus field plots for some $RMn_2Si_2$ compounds at 4.2 K.

Table I

| | Magnetic moment $\mu_B$/FU 4.2 K, 20 kOe | Lattice Parameters a(Å)±.005 | c(Å)±.005 |
|---|---|---|---|
| $LaMn_2Si_2$ | 3.08 | 4.114 | 10.611 |
| $CeMn_2Si_2$ | 0.15 | 4.010 | 10.523 |
| $PrMn_2Si_2$ | 1.79 | 4.031 | 10.559 |
| $NdMn_2Si_2$ | 5.14 | 4.015 | 10.542 |
| $GdMn_2Si_2$ | 4.37 | 3.951 | 10.479 |
| $TbMn_2Si_2$ | 5.60 | 3.930 | 10.468 |
| $DyMn_2Si_2$ | 6.55 | 3.923 | 10.453 |
| $HoMn_2Si_2$ | 6.65 | 3.909 | 10.424 |
| $ThMn_2Si_2$ | † | 4.019 | 10.483 |
| $YMn_2Si_2$ | 0.09 | 3.923 | 10.446 |
| $LaMn_2Ge_2$ | 3.0 | 4.198 | 10.985 |
| $CeMn_2Ge_2$ | 3.10 | 4.135 | 10.928 |
| $PrMn_2Ge_2$ | 3.90 | 4.117 | 10.902 |
| $NdMn_2Ge_2$ | 6.00 | 4.105 | 10.912 |
| $GdMn_2Ge_2$ | 3.60,* 3.0 | 4.031 | 10.900 |
| $TbMn_2Ge_2$ | 6.02 | 3.999 | 10.850 |
| $DyMn_2Ge_2$ | 3.50 | 3.980 | 10.837 |
| $HoMn_2Ge_2$ | 5.50 | 3.977 | 10.845 |
| $ErMn_2Ge_2$ | 6.00 | 3.973 | 10.798 |
| $ThMn_2Ge_2$ | 0.10 | 4.090 | 10.907 |

*Measured at 120 kOe, 4.2 K., †Antiferromagnetic (Ref. 8).
FU = formula unit

## RESULTS AND DISCUSSION

$LaMn_2Si_2$ is a simple ferromagnet with a Curie temperature of 303 K and saturation magnetic moment of 3.08 $\mu_B$/molecule at 4.2 K. Magnetization versus field plots for some of the compositions are shown in Fig. 1. Ce, Pr, Y, Tm show a linear increase in the magnetiation as the field is increased. The rest of the compounds show a tendency to saturate at the highest field (≈20 kOe). Measured moments at 4.2 K are shown in Table 1. The moment observed for Nd, Gd, Tb, Dy, Ho and Er can be explained by assuming antiparallel spin coupling of rare earth and Mn atoms which gives rise to antiferromagnetic moment coupling between these atoms if R is a heavy rare earth and ferromagnetic coupling if it is a light rare earth. Ce, Pr and Tm compounds do not show ferromagnetic ordering down to 4.2 K and hence the moments measured are not the saturation moments. $YMn_2Si_2$ and $ThMn_2Si_2$ also do not show any magnetic ordering between room temperature and 4.2 K. $ThMn_2Si_2$ shows a small maximum in the magnetization at 457 K in an applied field of 6 kOe. Neutron diffraction measurements by Ban et al.[8] indicate that $ThMn_2Si_2$ is antiferromagnetic with a Neel temperature of 483 K. The results on $RMn_2Ge_2$ compounds[7] have shown that the Mn sublattice in $NdMn_2Ge_2$ is ferromagnetic (Curie temperature $T_C$=334 K) whereas in $GdMn_2Ge_2$ ($T_C$=97 K) the Mn sublattice is antiferromagnetic above 97 K with a Neel temperature of 365 K. Examination of the lattice parameters of the $RMn_2Ge_2$ and $RMn_2Si_2$ compounds (Table 1) shows that when the a lattice parameter is less than about 4.10 Å the Mn sublattice is antiferromagnetically ordered or the Curie temperature of the compounds are much below room temperature. Also the Mn sublattice possesses only a weak moment (∿0.1 $\mu_B$) at 4.2 K when a <4·10 Å

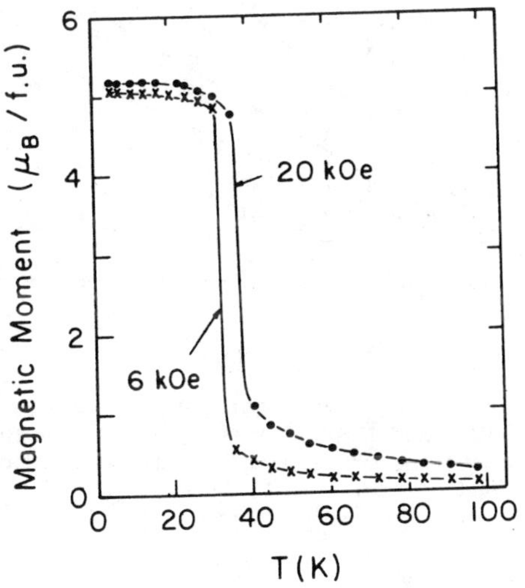


Figure 2. Magnetization versus temperature plot for $NdMn_2Si_2$ at 6 and 20 kOe.

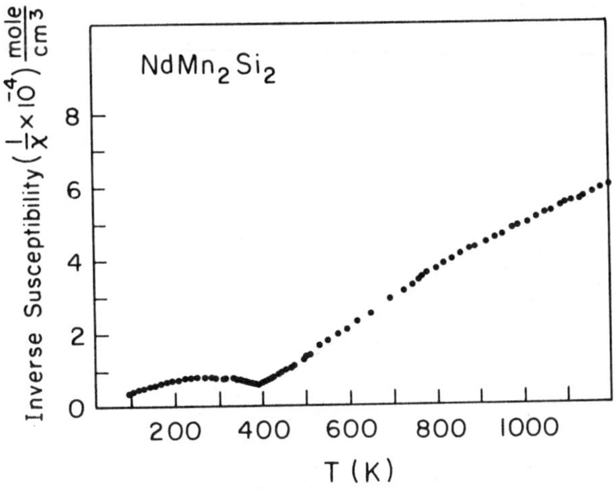


Figure 3. Inverse susceptibility versus temperature plot for $NdMn_2Si_2$.

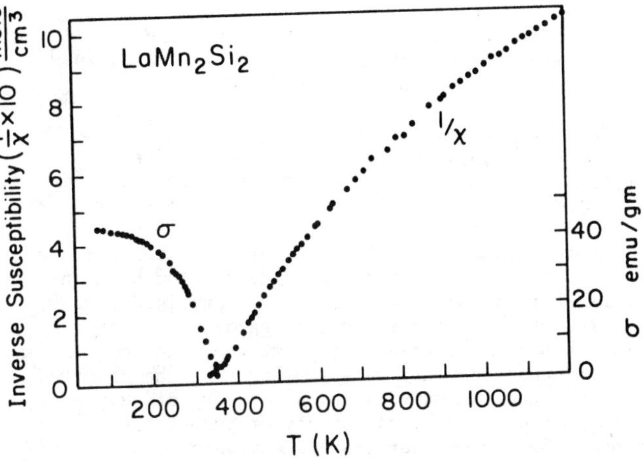


Figure 4. Magnetization (σ) and inverse susceptibility for $LaMn_2Si_2$.

unless the R atom is a magnetic rare earth. This can be seen by comparing the results obtained on $YMn_2Si_2$, $ThMn_2Si_2$ and $ThMn_2Ge_2$ at 4.2 K and the other $RMn_2X_2$ compounds in Table 1.

Figures 2 and 3 show the results obtained on $NdMn_2Si_2$ compounds. The magnetization versus temperature curve shows a sudden drop at 32 K in a field of 6 kOe and at 36 K in a field of 20 kOe. The inverse susceptibility versus temperature shows a broad hump from 100 K to 380 K and above this temperature Curie-Weiss behavior was obtained. All the remaining $RMn_2Si_2$ compounds with the exception of $LaMn_2Si_2$ (Figure 4) show a similar 1/χ vs. T behavior except that the hump in 1/χ vs. T curve extends to a different range of temperature for each compound. We postulate that Mn moments in $NdMn_2Si_2$ are ordered antiferromagnetically at high temperatures (∿100 K to ∿380 K) and the Nd sublattice is magnetically disordered. Below 36 K ordering occurs within the Nd sublattice which then couples ferromagnetically with Mn sublattice. At this temperature the Mn sublattice itself changes from antiferro to ferromagnetic ordering. Such a model is consistent with the saturation moments measured at 4.2 k.

A significant finding in the study of these materials is that $LaMn_2Ge_2$, $NdMn_2Ge_2$, $CeMn_2Ge_2$, $PrMn_2Ge_2$ and $LaMn_2Si_2$ all possess an easy direction of magnetization along the c-axis. This was determined from an examination of the X-ray diffraction of oriented fine powder at room temperature. Measurement of anisotropy field, coercive force are currently in progress.

REFERENCES

* This work was carried out through a grant from Army Research Office-Durham.

1. Z. Ban and M. Sikirica, Acta Cryst. 18, 594-598 (1965).
2. W. Rieger and E. Parthe, Mh. Chem. 100, 444-454 (1969).
3. L. Omjec and Z. Ban, Z. Anorg. Allg. Chem. 380, 111-117 (1971).
4. W. M. McCall, K. S. V. L. Narasimhan and R. A. Butera, J. App. Phys. 44, 4724-4726 (1973).
5. I. Felner, I. Mayer, A. Grill and M. Schieber, Solid State Commun. 16, 1005-1009 (1975).
6. S. K. Malik, S. G. Sankar, V. U. S. Rao, R. Obermeyer (this conference).
7. K. S. V. L. Narasimhan, V. U. S. Rao, R. L. Bergner and W. E. Wallace, J. App. Phys. 46, 4957-60 (1975).
8. Z. Ban, L. Omjec, A. Szytula and Z. Tomkowicz, Phys. Stat. Solids A27, 333-338 (1975).

# MAGNETOELASTIC CONTRIBUTIONS TO THE ELASTIC CONSTANTS OF HOLMIUM, DYSPROSIUM, AND TERBIUM

S. Gama, B.M. Kale, M.S. Torikachvili, O. Ferreira, M. Arellano, D.G. Pinatti, and P.L. Donoho*
Instituto de Física "Gleb Wataghin", Universidade Estadual de Campinas, Campinas, S. P., Brasil

## ABSTRACT

The magnetoelastic contributions to the elastic constants of Ho, Dy, and Tb have been determined at temperatures in the paramagnetic range in applied fields up to 75 kG. Magnetization measurements were also made in order to facilitate the comparison of the experimental results with theory. Initially the results were analyzed in terms of a theory due to Southern and Goodings, but the agreement was poor. The field dependence predicted by this theory, using magnetoelastic coupling constants estimated from static magnetostriction data, was generally much smaller than the observed dependence, and, in many cases, the theory did not predict the correct qualitative field or temperature dependence. The reasons for these discrepancies are, apparently, the use by Southern and Goodings of first-order perturbation theory, together with an inappropriate application of finite-strain theory. We have carried out a calculation of the magnetoelastic contribution to the elastic constants, employing the molecular-field approximation, but without using perturbation theory and using only the conventional small-strain tensor. This approach, similar to that of Freyne, is in much better agreement with the experimental results.

## INTRODUCTION

Strong magnetoelastic anomalies have been observed in the temperature dependence of the elastic constants of several of the rare-earth elements[1-5], and the field dependence of some of the elastic constants has also been reported for Er[6] and Dy[7]. Southern and Goodings[8] and Freyne[9] have made theoretical calculations of the magnetoelastic contributions to the elastic constants of rare-earth materials, and the work of Southern and Goodings led these authors to a number of interesting predictions concerning the field and temperature dependences of the elastic constants. This paper reports a number of measurements on the field and temperature dependences of certain of the elastic constants of Ho, Dy, and Tb, primarily in the paramagnetic temperature range of each element. The experimental results are not in agreement with the theory of Southern and Goodings in most respects, leading to the conclusion that certain of the assumptions and approximations of this theory are not appropriate to the present problem. A thermodynamic approach, based on the molecular-field approximation, but avoiding the use of perturbation theory, similar in some respects to that of Freyne[9], has been employed in the work reported here to calculate the magnetoelastic contributions to the elastic constants. This approach leads to an agreement with the experimental results which, considering the well-known limitations of the molecular-field approximation, is surprisingly good.

## EXPERIMENTAL RESULTS

The elastic constants were determined using a conventional ultrasonic pulse-echo "overlap" method, in which the velocities of 20-MHz elastic waves were measured for various directions of the propagation and polarization vectors as functions of temperature and applied magnetic field. For the results which can be presented here, the temperature was in the paramagnetic range for each material, and fields up to 75 kG were applied. The single-crystal samples were obtained from a commercial supplier. In order to permit a better correlation between the elastic-constant data and theoretical predictions, the sample magnetization was measured at the same temperatures and fields employed for the elastic-constant data.

An example of the results obtained for Tb is given in Fig. 1, where the dependence of $c_{33}$ on magnetization is plotted for several temperatures, with the applied field in the basal plane. These results, all of which

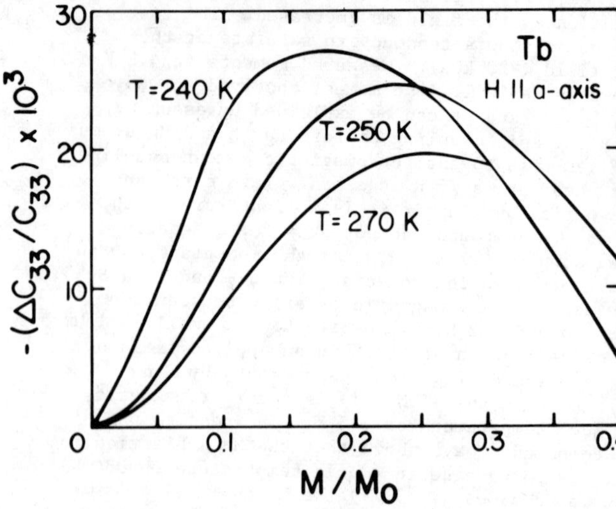


Fig. 1 $\Delta c_{33}/c_{33}$ vs a-axis magnetization for terbium at temperatures from 240 K to 270 K.

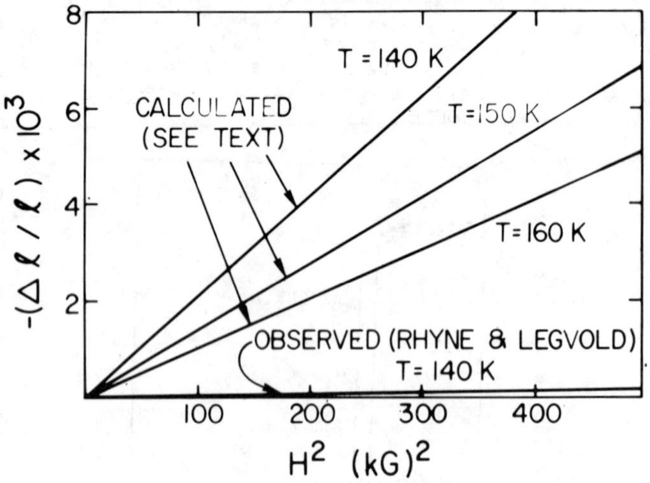


Fig. 2 Comparison of magnetostriction of holmium observed by Rhyne and Legvold with calculated value using present results and theories of Southern and Goodings and Callen and Callen.

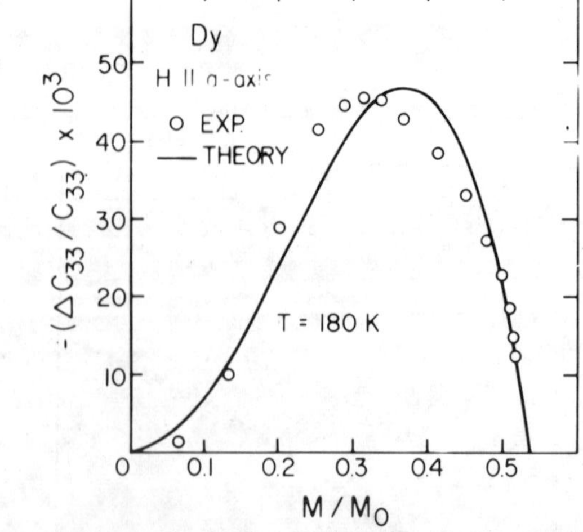


Fig. 3 Comparison of experimental and theoretical values for $\Delta c_{33}/c_{33}$ vs magnetization for Dy at T = 180 K.

exhibit a peak followed by a change in slope, also exhibit a change in sign at temperatures just above $T_N$ for sufficiently high magnetization. This behavior is quite different from that predicted by the theory of Southern and Goodings[8], for which no temperature dependence is expected when the elastic-constant is plotted as a function of magnetization. It should also be pointed out that the theory of Southern and Goodings, to the same degree of approximation, predicts no dependence on magnetization for $c_{44}$ and $c_{66}$, yet the field dependence for these elastic constants is comparable to that observed for $c_{11}$ and $c_{33}$.

It can be shown that the expression for the field-dependent part of $c_{33}$, namely $\Delta c_{33}/c_{33}$, obtained in the paramagnetic range using the theory of Southern and goodings is very close to that obtained for the static magnetostriction, $\Delta\ell/\ell$, along the c-axis in the well-known theory of Callen and Callen[10]. The data of Rhyne and Legvold show a much smaller value for the field-dependent magnetostriction than the results of Fig. 1, with no indication of the qualitative shape of the data shown here. The comparison between the experimental results and the theory of Southern and Goodings is, perhaps, better made in Fig. 2. Here, a value for $\Delta\ell/\ell$ along the c-axis of Ho, estimated from the data of Rhyne et al[12] is compared with the value calculated by using experimental data on $\Delta c_{33}/c_{33}$ and $\Delta c_{11}/c_{11}$, together with the theories of Southern and Goodings and Callen and Callen. In this figure it can be seen more explicitly that the theory of Southern and Goodings leads to an estimated magnetoelastic coupling which is much too large to explain the magnetostriction data.

A final example of experimental results is given in Fig. 3, in which the dependence of $\Delta c_{33}/c_{33}$ on magnetization is compared with the prediction of a theory described below. Here it can be seen that both the experimental results and the theoretical curve exhibit the same qualitative features, although the position of the calculated peak is too high. This small discrepancy is perhaps due to the use in the theory of the molecular-field approximation, since the theoretical curve involves a combination of single-ion and two-ion magnetoelastic coupling terms.

## THEORY

It is evident from the preceding discussion of the experimental results that, as suggested earlier[6], the theory of Southern and Goodings[8], as developed for the paramagnetic temperature range, does not adequately describe the field and temperature dependences of the elastic constants of the materials considered here. In essence, Southern and Goodings used a first-order perturbation calculation to obtain the internal energy, U. Since the elastic constants can be regarded as the second partial derivatives of U with respect to the strain components, the internal energy must exhibit at least a quadratic dependence on the strain components. In the present case, only a linear dependence is found. Southern and Goodings introduced a quadratic dependence on strain by a simple replacement of the usual small-strain components by the components of the well-known finite-strain tensor, which include terms quadratic in the small-strain tensor components. They then expressed the elastic constants as second derivatives of the internal energy with respect to the small-strain tensor. There appears to be no justification for this procedure, and the experimental results seem to bear out this conclusion. It should be emphasized, however, that the paper of Southern and Goodings, except for this point, provides an excellent contribution to the general area of magnetoelastic effects in the rare earths.

In an effort to obtain better agreement with experiment, we have developed an approach whose starting point, like that of Southern and Goodings, is the theory of magnetoelastic interactions of Callen and Callen. In our approach, however, we attempt to calculate such thermodynamic quantities as the internal energy or the free energy without treating the anisotropy and magnetoelastic terms of the model Hamiltonian as small perturbations to the exchange and Zeeman terms. Initially, in order to simplify the calculations, we have employed the molecular-field approximation to describe the exchange energy, including anisotropic exchange. With this approximation, the Hamiltonian can be diagonalized exactly. The internal energy and its derivatives may then be computed without recourse to perturbation theory. Before computing the internal energy and its derivatives, we adjusted the crystal-field and exchange parameters of the model Hamiltonian to give the best possible fit to the magnetization, anisotropy energy, and static magnetostriction. Because perturbation theory is not used in this approach, the final elastic-constant results depend strongly on the choice of these parameters. Detailed computations have been made only for the case of Dy at this time, and, since no attempt has yet been made to take explicit account of the helical ordering phase, calculations below $T_N$ are expected to be meaningful only for applied basal-plane fields sufficient to suppress the helical ordering.

The theoretical curve of Fig. 3 involves three second derivatives of the internal energy. The first results from only single-ion interactions, the second only from two-ion interactions, and the third term involves both single-ion and two-ion interactions. The fit to the experimental data was made by means of a two-parameter least-squares method. The value obtained for the single ion coupling constants from this fit is the following:

$$|B_1^{\alpha,2} + 2\,B_2^{\alpha,2}| = 45 \pm 5 \text{ K / ion.}$$

These coupling constants differ by a small numerical factor from those defined by Callen and Callen, but otherwise they have the same meaning. A rough estimate of the same combination of coupling constants obtained from the magnetostriction data of Clark et al[13] is 25 K/ion, but the data do not permit a really accurate estimate. The corresponding two-ion constant, which includes both $\ell = 0$ and $\ell = 2$ terms is 50% larger in absolute value. The above value is within a factor of two of the rough estimate from the magnetostriction data. Considering the quality of the overall fit of Fig. 3, it is suggested that the present theoretical approach, despite its use of the molecular-field approximation, provides a reasonably good description of the magnetoelastic behavior of the elastic constants in the cases to which it has been applied. Further calculations on other rare-earth materials are under way.

## REFERENCES

*On sabbatical leave from Rice University, Houston, TX.

1. M. Long, A. R. Wazzan, and V. R. Stern, Phys. Rev. 178, 775 (1969).
2. M. Rosen and H. Klimker, Phys. Rev. B 1, 3748(1970).
3. K. Salama, F. R. Brotzen, and P. L. Donoho, J. Appl. Phys. 43, 3254 (1972).
4. K. Salama, F. R. Brotzen, and P. L. Donoho, J. Appl. Phys. 44, 180 (1973).
5. M. Rosen, D. Kalir, and H. Klimker, Phys. Rev. B 8, 4399 (1973).
6. W. C. Hubbell, P. L. Donoho, C. L. Melcher, and K. Salama, AIP Conf. Proc. 18, 1263 (1974).
7. B. M. Kale, P. L. Donoho, D. G. Pinatti, and O. Ferreira, AIP Conf. Proc. 24, 651 (1965).
8. B. W. Southern and D. A. Goodings, Phys. Rev. B 7, 534 (1973).
9. F. Freyne, Phys. Rev. B 5, 1327 (1972).
10. E. Callen and H. B. Callen, Phys. Rev. 139, A445 (1965).
11. J. J. Rhyne and S. Legvold, Phys. Rev. 138, A507 (1965).
12. J. J. Rhyne, S. Legvold, and E. T. Rodine, Phys. Rev. 154, 266 (1967).
13. A. E. Clark, B. F. DeSavage, and R. Bozorth, Phys. Rev. 138, A216 (1965).

# MAGNETOPLASTIC DEFORMATION OF Dy CRYSTALS

H. H. Liebermann and C. D. Graham, Jr.
University of Pennsylvania, Philadelphia, Pa. 19174

## ABSTRACT

Several authors have reported that if single crystals of Dy or Tb are magnetized in fields of order 100 kOe parallel to a magnetic hard direction at 4.2K, a permanent plastic deformation occurs. Neither the cause nor the mechanism of this deformation has been found. We have investigated this phenomenon in Dy crystals, using optical and electron microscopy, x-ray diffraction, and magnetic measurements. Plastic deformation is found to occur generally at 70 to 90 kOe at 4.2K in Dy; the mechanism is mechanical twinning on the $\{10\bar{1}2\}$ $\langle 10\bar{1}1\rangle$ system, which was determined by independent mechanical tests to be the primary twinning system. The magnetic properties (coercive field, apparent anisotropy) are permanently changed by the deformation, as would be expected. The most likely driving force for the deformation appears to be the decrease in magnetic energy of the twins, which are oriented with an easy $\langle 10\bar{1}0\rangle$ direction almost parallel to the field. A calculation shows that this energy difference leads to an equivalent stress comparable to the yield stress of Dy. Another possibility is that the magnetostrictive strain simply exceeds the yield strain. The experiments and their interpretation are made difficult by the large magnetic torque acting on a sample that is not perfectly aligned in the field.

## INTRODUCTION

Magnetocrystalline anisotropy is notably high in the hcp rare earth metals Dy and Tb. In Dy, the anisotropy at 4.2K is about $-5\times10^8$ erg/cm$^3$, the negative sign indicating that the $\langle 0001\rangle$ hexagonal axis is the hard axis magnetization.[1,2] Anisotropy in the basal plane is relatively weak, with $\langle 11\bar{2}0\rangle$ easy directions. Below the Curie temperature of 85K, a magnetostrictive strain distorts the basal plane and lowers the symmetry[3], but this has no direct bearing on the present work. Dy also has high magnetostriction, with strains approaching $10^{-3}$ at low temperatures[4]. Attempts to directly measure the anisotropy and magnetostriction by applying high magnetic fields along the hard axis have led to the remarkable discovery that fields of the order of 100 kOe at 4.2K cause a permanent plastic deformation in Dy and Tb[5-7]. No detailed study of this phenomenon has been reported, but Rhyne et al[6] observed twinning in deformed samples. Chikazumi et al.[7] suggest that slip occurs in Tb subjected to pulsed fields up to 620 kOe at 77K.

This paper reports a further investigation of this field-induced deformation, using x-ray diffraction, optical and scanning microscopy, and high-field and low-field magnetic measurements at 4.2K and 77K, respectively.

## EXPERIMENTAL PROCEDURE AND RESULTS

Course-grained polycrystalline Dy samples were grown from nominal 99.9% distillation-purified Dy by a modified Bridgman method. A polycrystalline sample supported on a water-cooled stainless steel tube was passed slowly several times through a planar rf heating coil, in an Ar atmosphere. Single crystal samples were spark-cut from the resulting button.

A major experimental problem was to devise a satisfactory method for holding the sample in the magnetic field. The crystal had to be free to deform, but also had to be held against large mechanical torques resulting from even slight misalignment. For example, if the applied field deviates only $1^\circ$ from the $\langle 0001\rangle$ hard axis, the torque on the sample is of order $10^7$ dyne-cm or 10 $kg_f$-cm per cm$^3$. After a number of preliminary experiments, the method adopted was to make the sample cylindrical with the $\langle 0001\rangle$ direction parallel to the axis, and hold it in a closely-fitting cylindrical hole. This prevented rotation about any axis normal to $\langle 0001\rangle$, while permitting elongation parallel to $\langle 0001\rangle$. Samples of about 3 mm$^3$ volume were spark-trepanned from the buttons, and two orthogonal faces were spark-planed parallel to $\{11\bar{2}0\}$ and $\{10\bar{1}0\}$ for two-surface analysis of deformation markings. These surfaces, and the ends of the cylinder, were first mechanically polished and then electropolished and lightly etched in a perchloric acid-methanol mixture at 196K. This gave a finish that was satisfactory for optical and scanning electron microscopy. The surfaces were also examined by back-reflection x-ray diffraction before and after exposure to high fields, to check on orientation and crystal perfection.

Magnetic measurements at 77K in fields to about 7 kOe were made along easy and hard axes, before and after deformation, using a vibrating-sample magnetometer.

High magnetic fields, up to 130 kOe, were applied at 4.2K using the Bitter magnets of the Laboratory for Research on the Structure of Matter, University of Pennsylvania. The saturation magnetization of Dy at 4.2K was found to be 348 emu/gm. The slope of the initial portion of the hard-axis magnetization curve gave anisotropy values of $-4.4\pm1.4\times10^8$ erg/cm$^3$ at 4.2K and $-3.7\pm0.7\times10^8$ erg/cm at 77K, in reasonable agreement with values found in the literature.[6]

Hard-axis magnetization was measured using a low-frequency vibrating-sample magnetometer as high fields were applied. An upward departure from the linear curve was observed at fields in the range 65-100 kOe. Fig. 1, curve 1 is an example. The change in slope is irreversible, and is taken to indicate the beginning of plastic deformation. Application of higher fields leads to further changes in the magnetic behavior and to further plastic deformation, as shown in curves 2 and 3 of Fig. 1. The changes in magnetic behavior can be simply summarized; the easy axis becomes harder, and the hard axis easier. The coercive field in the hard direction increases from approximately zero to 500 Oe; the field required to reach saturation in the easy direction increases from about 1 kOe to more than 50 kOe. Analogous changes in magnetic properties are observed at 77K, and all changes persist unchanged if the sample is heated to room temperature and re-cooled.

Examination of the samples at room temperature showed permanent plastic elongation of 1 to 4% along $\langle 0001\rangle$ after exposure to hard-axis fields of up to 130 kOe. X-ray diffraction, optical microscopy, and scanning-electron microscopy show that the principal deformation mode is mechanical twinning on the $\{10\bar{1}2\}\langle 10\bar{1}1\rangle$ system. This is the twinning mode ex-

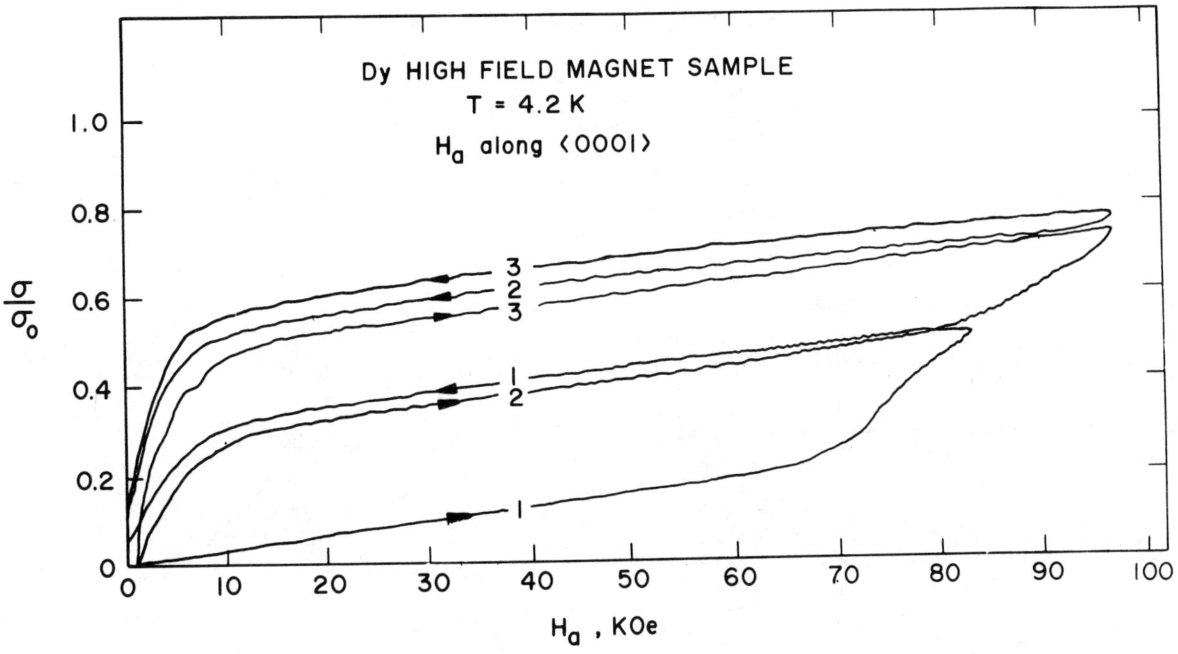


Fig. 1 Magnetization of Dy at 4.2K with applied field along ⟨0001⟩

pected for extension along the ⟨0001⟩ axis. In an early experiment, evidence for {11$\bar{2}$2} ⟨11$\bar{2}$3⟩ slip was also observed, but this does not seem to be common.

Several Dy crystals were plastically deformed by mechanical compression along the principal axes, at room temperature and at 77K, by amounts ranging from 1 to 25%. Deformation was mainly by {10$\bar{1}$2} ⟨10$\bar{1}$1⟩ twinning, and the magnetic properties at 77K changed in a similar way to those of samples deformed by magnetic fields. In samples deformed both magnetically and mechanically, the deformation markings were more or less non-uniform, being more concentrated and less regular near the surfaces of mechanical contact, i.e., the compression faces of the mechanically-deformed samples and the points of maximum mechanical torque force near the ends of the magnetically-deformed samples.

Generally we can say that high fields applied along the ⟨0001⟩ direction in Dy cause plastic elongation in this direction, and that the mechanism of the deformation is mechanical twinning. The deformation and its effects on the magnetic properties are the same as those obtained by similar strains produced by ordinary mechanical deformation. No evidence for any field-induced phase transformations or other exotic phenomena was observed.

We suggest that the driving force for the deformation is the lowering of the magnetostatic and anisotropy energies resulting from formation of twins. The twinned regions have a ⟨10$\bar{1}$0⟩ easy direction within about $5^{o}$ of the applied field, and at an applied field of 60 kOe this results in a magnetic energy difference between the untwinned and twinned regions of about $2\times 10^{8}$ erg per $cm^{3}$ of twinned material. Since twinned regions are elongated in the field direction by about 9%, this energy difference is equivalent to a tensile stress of $23\times10^{8}$ dyne/$cm^2$, which is comparable to the yield stress of polycrystalline Dy at room temperature and could reasonably cause plastic deformation in a single crystal at 4.2K. This simple energy argument predicts that the crystals should convert completely to the twinned state, but the normal occurence of work hardening raises the stress required for new twins to form, and limits the fraction of twinned material.

The observed changes in magnetic properties are easily attributed to the appearance of a substantial volume of twinned material in new crystallographic orientation, plus the occurrence of elastic and plastic strains at twin intersections, non-metallic inclusions, etc.

Two other possible origins of deformation are a) that the magentostrictive strain in high fields exceeds the elastic limit, so that plastic deformation occurs; and b) that the deformation is simply the result of mechanical torque forces acting on the sample due to imperfect alignment of the hard axis with the field. Although the evidence cannot eliminate either of these possibilities, we think they are less likely than the model proposed above.

ACKNOWLEDGEMENTS

We thank P. J. Flanders for help with the experiments, V. V. Vorobyov for preliminary work in the growth of Dy crystals, and J. J. Rhyne for helpful suggestions during the course of the work and for comments on the manuscript.

This work was supported by the National Science Foundation under NSF G-33633 through the Laboratory for Research on the Structure of Matter, University of Pennsylvania.

REFERENCES

1. D. R. Behrendt, S. Legvold, F. H. Spedding, Phys. Rev. 109, 1544 (1958).
2. J. J. Rhyne, A. E. Clark, J. Appl. Phys. 38, 1379 (1967).
3. F. J. Darnell, Phys. Rev. 130, 1825 (1963), 132, 1098 (1963).
4. S. Legvold, J. Alstad, J. J. Rhyne, Phys. Rev. Letts. 10, 509 (1963).
5. J. J. Rhyne, S. Foner, E. J. McNiff, R. Doclo, J. Appl. Phys. 39, 892 (1968).
6. R. J. Elliott, Magnetic Properties of Rare Earth Metals, Plenum Press (1972), p. 140.
7. S. Chikazumi, S. Tanuma, I. Oguro, F. Ono, K. Tajima, IEEE Trans. Mag-5, 265 (1969).

# PRECIPITATION HARDENED RE-Co-MAGNETS

A. Menth
Brown Boveri Research Centre
CH-5401 Baden / Switzerland

ABSTRACT

Magnetic and metallographic investigations are reported on $Sm(Co,Cu)_z$ with $5 \le z < 8.5$ and Cu-contents up to 35 % of the transition metal fraction. The as-cast as well as the heat treated materials were evaluated. Two typical compositions $Sm(Co_{0.65}Cu_{0.35})_{5.6}$ as an example of a single phase 1:5 type magnet material and $Sm(Co_{0.84}Cu_{0.16})_{6.9}$ as an example of a two phase magnet material, namely a primary 2:17 phase in a 1:5 type matrix, are considered in detail. Both materials produced, after an appropriate high temperature annealing, rectangular ideal demagnetization curves, implying that the macroscopic 2:17 primary phase is magnetically hard. Low temperature heat treatments resulted in two maxima in the coercivity $_IH_C$ as a function of T. The data are discussed based on metallurgical investigations of the phase diagrams and electron microscopy results.

## INTRODUCTION

The hardmagnetic potential of the intermetallic compounds $RECo_5$, mainly $SmCo_5$, has been exploited based on two different technological processes. They obey two different physical mechanisms for the coercivity. So far the powdermetallurgical approach, using sufficiently small magnetic particles (≈ 5 μm diameter) which resist nucleation of reversed domains, is dominant in the commercial production of RE-Co permanent magnets. In contrast, precipitation hardened material relies on the pinning mechanism of Bloch walls and shows hardmagnetic properties on bulk samples.

Nesbitt and coworkers[1,2] have shown that permanent magnetic properties can be obtained on bulk pieces of $RE(Co,Cu)_5$-compounds. Leamy and Green[3] demonstrated that small precipitates with the rhombohedral $Th_2Zn_{17}$ structure are present in $CeCo_{3.8}Fe_{0.5}Cu_{0.9}$ and evidence[4] was obtained for the pinning of domain walls by these precipitates. Whereas increasing Cu-content reduces the magnetization M, it increases the coercivity $_IH_C$. Coercivity values up to 30 kOe have been achieved. In fact, such high coercivity values are not obtained in strictly stochiometric 1:5 compounds. Perry and coworkers[5] showed that the coercivity of a stochiometric 1:5 compound could not be raised in bulk materials above 7 kOe, irrespective of heat treatment and Cu-content. The high coercivity values reported earlier may be due to deviations in the compositions before and after preparation of the materials, i.e. due to the loss of RE through evaporation and oxidation.

To obtain optimum size and concentration of precipitates for magnetic pinning of the Bloch walls the alloys have to be heat treated. Two coercivity maxima occur with annealing temperature in both the Ce-Co-Cu-[6] and the Sm-Co-Cu-system[5].

Tawara and Senno[7] and more recently Oiwa et al[8] have shown that for low copper concentrations the maximum coercivity value is obtained for a RE/TM ratio of about 1:7. The lower RE/TM ratios resulted in an increase of the remanence. These authors reported on a powder metallurgically prepared $Sm_{0.8}Ce_{0.2}(Co_{0.79}Fe_{0.05}Cu_{0.16})_{7.2}$ magnet an energy product of 20 MG Oe with a coercivity value of about 7 kOe. Such magnets are competitive with $SmCo_5$-magnets, especially in applications where high energy products but not the high coercive fields of $SmCo_5$ are used.

We investigated the hard magnetic properties of Sm-Co-Fe-Cu alloys as a function of composition in the phase space between the 1:5 and 2:17 compounds. The aim was to clarify systematically the hard magnetic potential offered and to increase the understanding of the relation between magnetic, structural and microstructural properties.

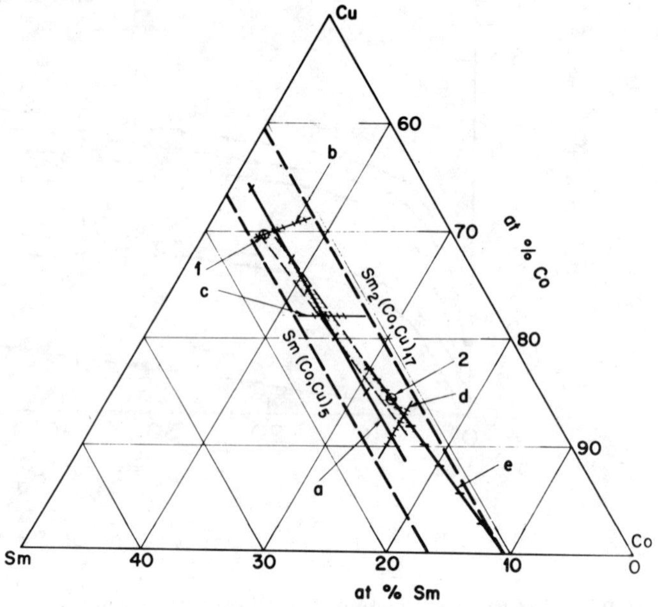


Fig. 1: Phase diagram of the Sm-Co-Cu-system reproducing the investigated alloys. The shaded area marks region of useful magnetic properties.

The preparation of the alloys and the experimental details of the heat treatment are reported in another paper[9]. A number of compositional cuts in the ternary system Sm-Cu-Cu (Fig. 1) were investigated. Besides a cut along the 1:6 tie-line (a in Fig. 1), three cuts were made between the $Sm(Co,Cu)_5$ and the $Sm_2(Co,Cu)_{17}$ lines: at a constant Co/Cu ratio of 13/7, (b in Fig. 1), at a constant Cu content of 22 at% (c in Fig. 1) and at a constant Co content of 75 at% (d in Fig. 1). Finally a cut was made between the composition of $Sm(Co_{0.65}Cu_{0.35})_{5.6}$ and $Sm_2Co_{17}$ (e in Fig.1).

## MAGNETIC PROPERTIES

For the measurements of the bulk material properties either single crystalline or homogeneously aligned samples were taken from the alloys. These were ground to spheres with diameters between 1 and 2 mm. After magnetizing in fields up to 45 kOe their properties were measured in a vibrating sample magnetometer in fields up to 21 kOe.

Alloys of cut a are not discussed here, since they are of limited interest as permanent magnetic materials due to the fact that either the remanence or the coercivity are low (their properties are reported by Nagel et al[9]). In Fig. 2 the magnetic properties of the as cast materials along the cuts b, c and d are reproduced. A coercivity maximum was observed along each of the three cuts. These maxima lie on a compositional line (cut e in Fig. 1). Fig. 3 shows the magnetic properties of the corresponding samples, measured in the as-cast state. The coercivity goes through a maximum around 60 at% Co with the remanence increasing over the entire range. These alloys seem to exhibit, due to their high remanence values a potential for permanent magnet use. The low realized energy products in relation to $B_r$ and $_IH_C$ stem from the fact, that the slopes of the demagnetization curves do not correspond to the demagnetization factor of a sphere (M = 3 H). The slope

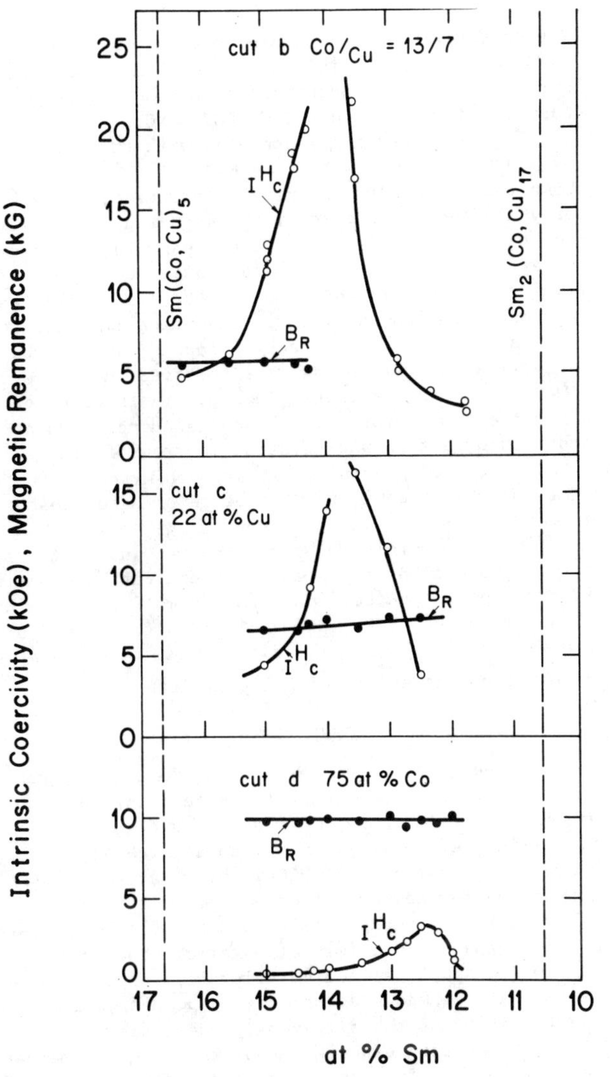


Fig. 2: Coercivity $_IH_C$ and remanence $B_r$ along the cuts b, c, and d of samples in the as-cast condition.

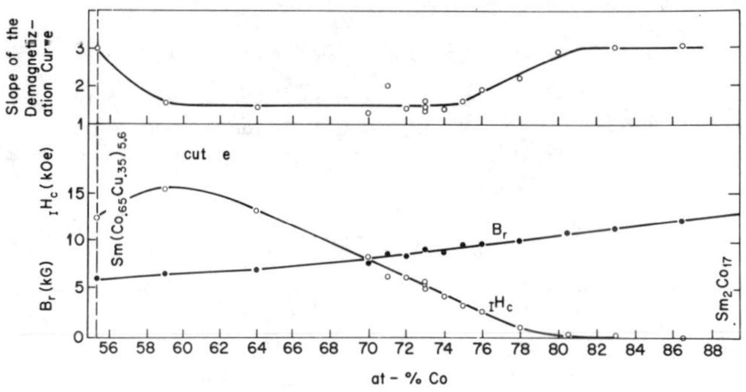


Fig. 3: $_IH_C$ and $B_r$ along cut e between alloy 1 $Sm(Co_{0.65}Cu_{0.35})_{5.6}$ and $Sm_2Co_{17}$, together with the slope of the demagnetization curve for as-cast smaples.

is reduced and varies between 1.5 and 3 (Fig. 3). By an appropriate heat treatment at elevated temperature the slope becomes 3 and the demagnetization curves exhibit well defined knee-values $H_k$ ($M_{(H_k)} = 0.9\ B_r$) comparable to $_IH_C$. This effect is, as we shall see later, tied to the metallurgical configuration of these materials. A second heat treatment at lower temperatures gave rise to two peaks in coercivity as a function of temperature. The effect of temperature on $_IH_C$ is demonstrated for

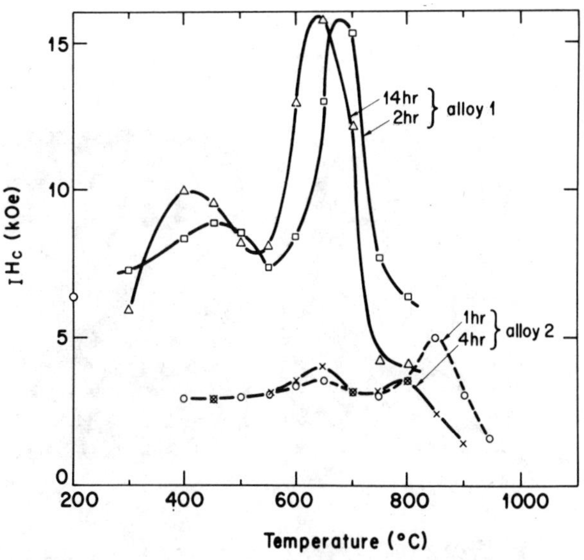


Fig. 4: Effect of low temperature heat treatment for alloy 1 $Sm(Co_{0.65}Cu_{0.35})_{5.6}$ and alloy 2 $Sm(Co_{0.84}Cu_{0.16})_{6.9}$.

alloy 1 of Fig. 1 $Sm(Co_{0.65}Cu_{0.35})_{5.6}$ in Fig. 4. Decreasing the Cu content resulted in a shift of the two peaks to higher temperatures, illustrated in Fig. 4 by alloy 2 $Sm(Co_{0.84}Cu_{0.16})_{6.9}$ of Fig. 1, namely 650°C and 850°C respectively. Increasing the holding time decreased the temperature of the second peak, whereas the first peak was not changed within experimental error.

## MICROSTRUCTURE

The main features of the microstructure will be described by means of the same two alloys 1 and 2, chosen already for the description of the magnetic properties. The metallographic investigation showed that single phase material was only obtained along the tieline between $SmCo_5$ and $SmCu_6$ and within a narrow compositional range around $Sm_2Co_{17}$. The as-cast microstructure of alloy 1 is of a modulated single phase 1:5 type. The modulation being long range segregations with constant Sm content but fluctuating Co/Cu ratio[5,9]. Therefore we can conclude that the change of slope in the demagnetization curve is related to this microstructural change and is due to the modulation of composition and with this in coercivity. Annealing at 1180°C for 3 hours causes the modulation to disappear resulting in single phase 1:5 grains with small Cu-rich intergranular layers. Extended treatment at 650°C created optically visible 2:17 precipitates, annealing at 800°C caused massive incoherent 2:17 precipitation to occure.

Detailed investigation by Perry[10] have shown, that Cu appreciably expands the phase space of the 1:5 phase towards the transition metal rich side at temperatures below the peritectic temperature. Therefore one can conclude that the high temperature heat treatment tend to augment and homogenize the 1:5-structure fracture of the alloys.

In the phase space between 1:5 and 2:17, the alloys form at least two phases. Their structures[5,9] turn out to be mainly of the 1:5 and 2:17 type. Alloy 2 $Sm(Co_{0.84}Cu_{0.16})_{6.9}$ is more transition metal rich than alloy 1 and represents a typical two phase compound. Fig. 5a shows the microstructure of the as-cast alloy, the light etching phase being the 2:17 primary phase in a 1:5 matrix. Fig. 5d illustrates the domain structure of the as-cast alloy, reflecting the different magnetic properties of the 1:5 and 2:17 phases. The matrix possesses the Co/Cu fluctuations analogous to alloy 1 with comparable heat treatment responses (Fig. 5b) and a

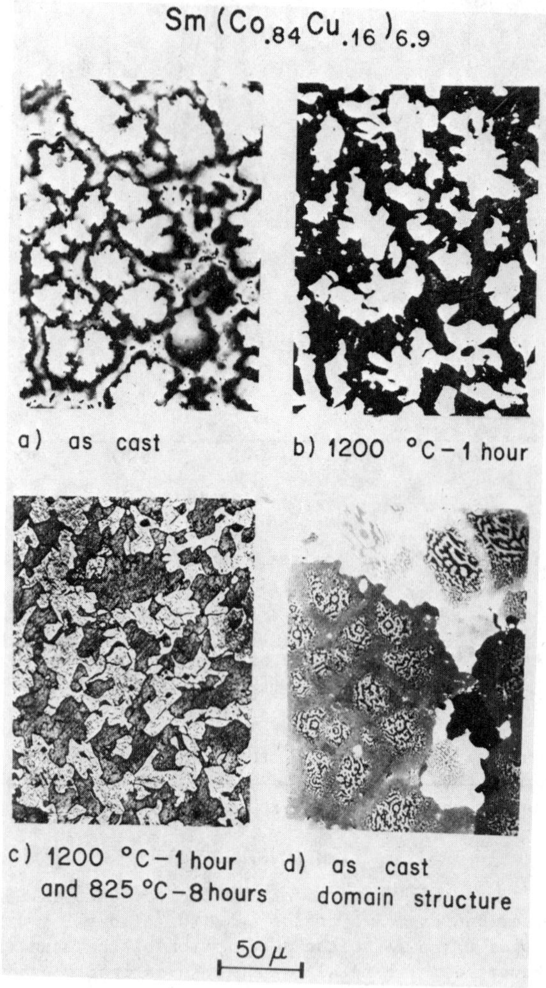


Fig. 5: Optical microstructures of alloy 2 $Sm(Co_{0.84}Cu_{0.16})_{6.9}$ for the indicated stages of heat treatment. Light etching phase corresponds to the 2:17 phase.

composition of $Sm(Co_{0.81}Cu_{0.19})_{5.9}$. The primary 2:17 phase, $Sm(Co_{0.87}Cu_{0.13})_{7.5}$ shows after extended annealing at 850°C visible precipitates (Fig. 5c). The two compositions give clear evidence of the extension of the 1:5 and 2:17 phase space at elevated temperatures, as has been confirmed by Perry[10].

Despite the two phases an ideal rectangular demagnetization curve was observed. Therefore the question arises as to whether the primary 2:17 phase possesses a coercivity or not? Perkins et al[13] have investigated an idealized model of a second magnetic phase in the form of spheres embedded in a permanent magnetic material of different coercivity. The results demonstratete clearly that both phases have to exhibit coercivity in order to realize a rectangular demagnetization loop, and that their coercivity values have to be similar. This means that the 2:17 phase is magnetically hardened and therefore hardened by a pinning mechanism. This opens up new prospects for new permanent magnetic materials on the 2:17 base which we shall discuss separately[14].

Substituting Cu by Fe results in the expected increase of $B_r$ and for concentrations smaller than 10 at.% of the transition metal fraction, the properties are not fundamentally changed[9].

Electron microscopy studies have shown[3,4,11,12] that small precipitates with the rhombohedral 2:17 structure are present within the 1:5 phase in the as cast state. These precipitates are very stable below 650°C but coarsen rapidly at 800°C. Melton and Perkins[13] ascribe this to the occurence of a coherent precipitate phase boundary between 650°C and 800°C.

The matrix strain associated with the precipitates influences the coercivity. Perkins et al[13] have shown theoretically that precipitate size, shape and array and the difference of the magnetic properties of the bulk and the precipitates can explain the hard-magnetic properties of the 1:5 phase. On the other hand alloy 2 contains in addition to the 1:5 a 2:17 fraction on a macroscopic scale. It is magnetically hard, for reasons not yet known. Stacking faults, anti phase boundaries and twin bands were observed and are severely affected by heat treatment. An analysis is complicated, since the two macroscopic phases influence each another through the inhomogeneous internal field. No microscopic precipitates were observed and therefore the coercivity must be related to these defects.

The two coercivity peaks observed in the temperature dependence of the heat treatment can be explained in terms of the eutectoidal transformation of the $RECo_5$ phases into the $RE_2Co_7$ and $RE_2Co_{17}$ as found by Den Broeder and Buschow[15]. The differences in temperatures result from the effect of alloying with Cu.

## CONCLUSION

In the phase space between the $Sm(Co,Cu)_5$ and $Sm_2(Co,Cu)_{17}$ permanent magnetic materials can be achieved with either single or two phases with a wide variety of properties, e.g. coercivity values as high as 30 kOe or anergy products up to 24 MGOe. This feature is very attractive for commercial magnet production, since essentially one technological process allows the engineering of a wide variety of material properties.

On the physical and metallurgical aspects one can claim that the knowledge of the phase diagramm and of the formation of the different phases worked out allow a rather detailed description of the relevant mechanism. In additions the fact that the 2:17 phase is magnetically hardened opens up new prospects for materials with higher magnetic energy products, without having to produce these very fine particles otherwise required by the lower magnetic anisotropy fields of the uniaxial 2:17 materials.

## ACKNOWLEDGEMENTS

I wish to thank Dr. H. Nagel whose work provided the background of our permanent magnet work and Drs. R.S. Perkins, K.N. Melton and A.J. Perry for their valuable contributions.

## REFERENCES

1. E.A. Nesbitt, R.H. Willens, R.C. Sherwood, E. Bühler, J.H. Wernick, Appl. Phys. Letters 12, 361 (1968).
2. E.A. Nesbitt, J. Appl. Phys. 40, 1259 (1969).
3. H.J. Leamy and M.L. Green, IEEE Trans. Magn. 9, 205 (1973).
4. H.J. Leamy and M.L. Green, AIP Conf. Proc. 10, 593 (1973).
5. A.J. Perry, H. Nagel, A. Menth, Proc. 3rd European Conf. on Hard Magn. Mats., Amsterdam 1974, p. 149.
6. G.Y. Chin, M.L. Green, H.J. Leamy, AIP Conf. Proc. 24, 691 (1975).
7. Y. Tawara, H. Senno, Japan. J. Appl. Phys. 7, 966 (1968); IEEE Trans. Mag., 1974, 10, 313.
8. I.T. Oiwa, M. Honshiwa, E. Kikachi, Proc. 11th Rare Earth Res. Conf., Traverse City Michigan 1974, p. 353.
9. H. Nagel, A.J. Perry, A. Menth, Submitted to J.Appl. Phys.
10. A.J. Perry, Internal Lab. report, Brown Boveri Res. Centre, will be published.

11. K.N. Melton, R.S. Perking, Submitted to J. Appl. Phys.
12. J.E. Smeggil, P. Rao, J.A. Livingston, E.F. Koch, AIP Conf. Proc. 18, 1144 (1974).
13. R.S. Perkins, J. Bernasconi, H.J. Wiesmann, Submitted to J. Appl. Phys.
14. A. Menth, H. Nagel. In preparation.
15. F.J.A. den Broeder, K.H.J. Buschow, J. Less. Common Metals 29, 65 (1972).

# MAGNETIC PROPERTIES OF SINTERED $Sm_2TM_{17}$ MAGNETS

H. Nagel, Brown Boveri Research Center
CH-5401 Baden, Switzerland

## ABSTRACT

$Sm_2TM_{17}$ magnets have been produced by powder metallurgical techniques, with TM standing for combinations of 3d transition metals. The compositional dependence as well as the influence of the different process parameters, as for instance particle size and sintering temperature, have been investigated. The maximum hard magnetic values obtained so far are a magnetic remanence of 11.3 kG and a coercivity field of 11 kOe. However, maximum $B_R$ and $_IH_c$ could not be obtained simultaneously so far. All the magnets present demagnetization curves typical for a chemically and structurally homogeneous composition of the material. The hard magnetic properties observed are discussed with respect to the primary magnetic properties.

## INTRODUCTION

Investigations of the alloy systems $RE_2(Co_{1-x}Fe_x)_{17}$ have shown[1-3] that $Sm_2(Co_{1-x}Fe_x)_{17}$ is the most interesting for permanent magnet use. It shows uniaxial magnetic behaviour in the range $0 \leq x < 0.5$ with optimum primary magnet properties between $0.1 < x < 0.2$. The saturation magnetization lies between 13.5 and 14 kG and the crystal anisotropy field $H_A$ is about 65 kOe as based on single crystal measurements[3]. Only a few attempts have been undertaken to prepare powder magnets from these materials[3-5]. However so far the primary magnetic potential could be used only to a limited and rather unsatisfactory degree.

Very recently it was shown in this laboratory[6] that the magnetic anisotropy of $Sm_2(Co_{1-x}Fe_x)_{17}$ compounds can be increased by substituting some of the Co or Fe with Mn or Cr. For instance, $H_A$ of $Sm_2Co_{17}$ is increased from 65 kOe to 96 kOe if 15 % of the Co is substituted by Mn. Even more pronounced is the influence of Cr. A substitution of only 5 % of the Co yields an $H_A$ of 105 kOe. However this increase of $H_A$ is accompanied by a decrease in $M_s$. $M_s$ decreases strongly with the Cr-content, whereas the reduction of $M_s$ is small for Mn-additions up to 20 % of the TM fraction. One aim of the present work has been to find whether the primary magnetic potential of $Sm_2TM_{17}$ can be used to a similar degree to that of the $RECo_5$ compounds. The second aim has been to optimize the TM composition to achieve $B_R$ values $> 11$ kG and $_IH_c$ values $> 3$ kOe together with a linear demagnetization curve. Those magnets would be an alternative to the high remanent Alnico.

## EXPERIMENTAL RESULTS AND DISCUSSION

The production process developed for $RECo_5$ compounds was the basis for the present investigations. The 2:17 compounds were melted together in boron nitride crucibles under an argon atmosphere by inductive heating. The purity of Sm was 99.9 %, that of the transition metals was better than 99.99 %. Only single phase 2:17 alloys or those containing a very small amount of 1:5 phase were used.

The alloys were jet milled, with nitrogen as working gas, together with some percent of a Sm-rich alloy to compensate for the Sm loss during the milling and sintering steps. The powders were aligned magnetically in a 50 kOe field, isostatically pressed at 6000 atm. and then sintered for 1/2 hour and subjected to a post sintering heat-treatment. The M-H loops of the resulting cylindrical samples were measured in fields of up to 50 kOe.

The grain size of the powder was optimized to achieve as high a magnetization as possible combined with a reasonable coercivity $_IH_c$ of $> 3$ kOe. The investigations showed that averaged grain sizes lying between 2.3 and 3.0 $\mu$m, determined with Fisher sub sieve sizer, yielded good magnets. Smaller particle sizes result in too low $B_R$ values whilst larger ones yield an insufficient coercive field. The second critical process parameter is the sintering of the samples. The sintering temperature ($T_S$) necessary to obtain a density of about 95 % of the theoretical value was found to lie between 1370 K and 1470 K depending strongly on the composition. This reflects the different melting temperatures of the compositions studied. 2:17 magnets containing Mn or Cr in addition to Co and Fe were prepared with this process. In Fig. 1a the demagnetization curves of Mn containing magnets are shown for a constant cobalt content. $B_R$ decreases weakly with increasing Mn content. As can be seen, the maximum coercive field of 13.4 kOe was obtained in a magnet with a Mn/TM ratio of 0.125. This magnet also possesses the highest energy product of 28 MGOe. The magnet with Mn/TM=0.1, possessing a $B_R$ of 11.2 kG and a $_IH_c$ of 4.4 kOe, is also an interesting one for practical use. The effect of chromium is also very pronounced (Fig. 1b). Unfortunately the increase of $_IH_c$ with increasing Cr/TM-ratio is accompanied by a strong decrease of $B_R$.

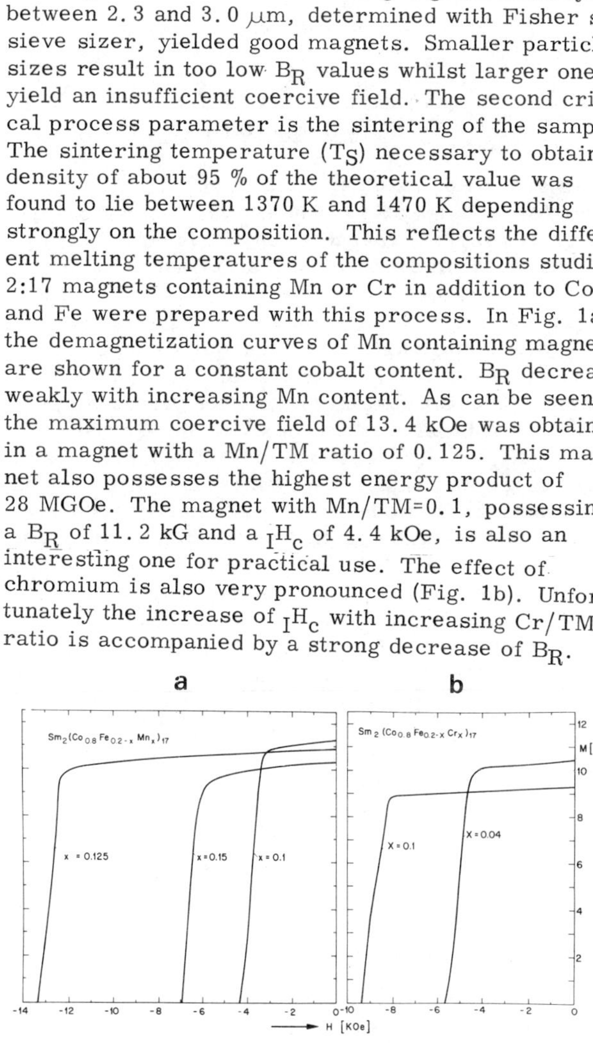


Fig. 1: Demagnetization curves of sintered $Sm_2TM_{17}$ magnets.

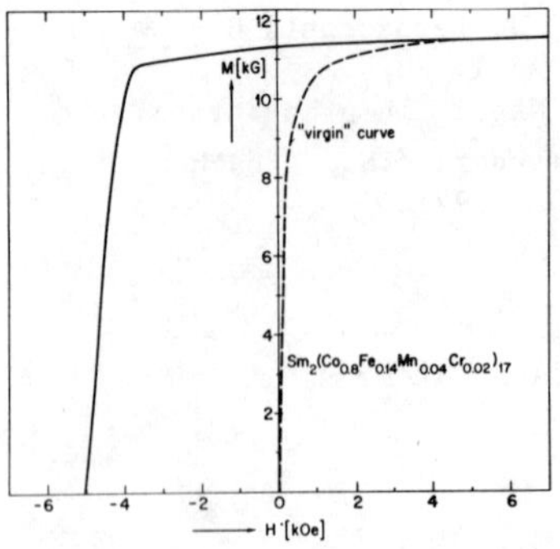


Fig. 2: "Virgin" magnetization curve and demagnetization curve of a $Sm_2(Co_{0.8}Fe_{0.14}Mn_{0.04}Cr_{0.02})_{17}$ magnet.

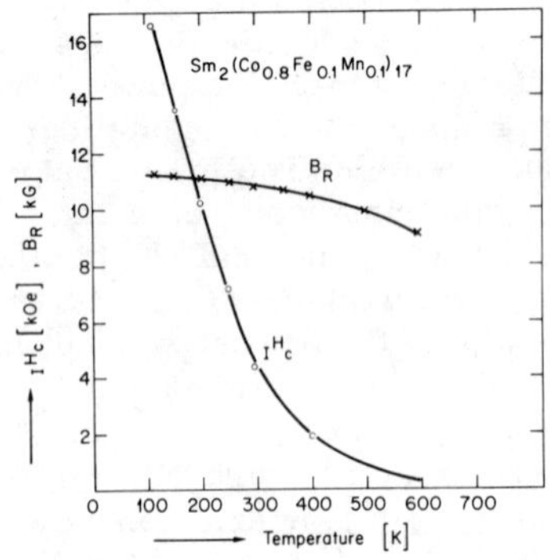


Fig. 3: Temperature dependence of $B_R$ and $_IH_c$ of a $Sm_2(Co_{0.8}Fe_{0.1}Mn_{0.1})_{17}$ magnet.

Values of 9.5 kG, as obtained for Cr/TM=0.1, are far too low with respect to the aims of this work. Finally we prepared magnets containing both Mn and Cr. The hysteresis loop of such a magnet is shown in Fig.2. It combines a relatively high $B_R$ of 11.3 kG with $_IH_c$ of 5.05 kOe. None of the demagnetization curves of the magnets showed signs of the steplike behaviour which would be typical for a two phase material. With their well defined knee values the curves are indicative rather of a chemically and structurally homogeneous material. The virgin curves of magnetization observed (Fig. 2) are typical for magnets in which the coercivity is determined by nucleation of reversed magnetized domains not by a pinning mechanism of Blochwalls. The hard magnetic properties obtained may be compared with the single crystal measurements on the same compounds[6]. This yields for example (Fig. 1a) $B_R/M_s$ ratios of 0.84 and 0.83 for Mn concentrations of Mn/TM=0.1 and 0.15 respectively. The corresponding figures for $_IH_c/H_A$ are 0.05 and 0.08. These values mean that the primary magnetic potential is used to nearly the same degree as in $RECo_5$ compounds ($B_R/M_s$ between 0.85 and 0.9) with respect to the saturation magnetization[7]. However, this is not the case for $_IH_c/H_A$ taking into account the values of 0.08-0.1 obtained with $RECo_5$ compounds. From Fig. 1 and the $H_A$ measurements it may be concluded that with increasing Mn or Cr content the magnetic anisotropy is used to a larger extent.

The temperature dependence of both magnetic remanence and coercivity of a $Sm_2(Co_{0.8}Fe_{0.1}Mn_{0.1})_{17}$ magnet, the same as in Fig.1a, are shown in Fig. 3. The curves are characteristic for all magnets reported here, for the preparation of which the process parameter had been constant. $_IH_c$ decreases monotonically with temperature in the investigated range of temperature. The dependence is stronger than in sintered $RECo_5$ magnets. For instance, the ratio of $_IH_c$ (400 k)/$_IH_c$(300 K) is about 0.4 compared with 0.6 and 0.5 for $SmCo_5$ and $MMCo_5$ respectively. However, the investigations so far have indicated that the temperature dependence is strongly related to preparational details which are not yet completely optimized. The value of $dB_R/dT$ at room temperature as derived from Fig. 3 is 0.004 kG $K^{-1}$ or 0.04 % $K^{-1}$.

This value decreases with decreasing Co-content because the Curie temperature of $Sm_2Co_{17}$ is lowered by addition of Fe as well as Mn and Cr[6]. However, it is comparable with that of $SmCo_5$ magnets[9] and better than that observed in $MMCo_5$ magnets.

## CONCLUSIONS

The present investigations have shown that $Sm_2TM_{17}$ magnets can be produced with useful hard magnetic properties. For that purpose the liquid phase sintering process, developed for the $RECo_5$ compounds, has been optimized. The resulting values depend strongly on the composition of the TM fraction especially on the Mn and Cr-content reflecting the change of the primary magnetic properties. So far the most promising magnet obtained was $Sm_2(Co_{.8}Fe_{.14}Mn_{.04}Cr_{.02})_{17}$ with a $B_R$=11.3 kG and a $_IH_c$= 5.05 kOe if one has in mind a substitution of Alnico magnets by 2:17 magnets. A further improvement of the hard magnetic properties, i.e. the preparation of magnets with energy products of up to 30 MGOe should be possible by investigating the influence of both the process parameters and the compositions more completely than it has been done to date.

## ACKNOWLEDGEMENTS

The author is indebted to Drs. A. Menth, R. S. Perkins and A. J. Perry for their experimental and theoretical help. He also gratefully acknowledges the technical assistance of B. Rennhard.

## REFERENCES

1) A. E. Ray, K. J. Strnat; USAF Materials Laboratory, Wright-Patterson Air Force Base, Ohio, AFML-TR-71-53(1971); 71-210(1971); 72-99(1972); 72-202(1972); 73-112(1973)
2) H. F. Mildrum, M. S. Hartings, K. J. Strnat, J. G. Tront; AIP Conf. Proc. 10, 618(1973)
3) R. S. Perkins, S. Gaiffi, A. Menth; IEEE Trans. Mag. MAG-11, 1431(1975)
4) R. C. Carriker, A. S. Rashidi; AIP Conf. Proc. 10, 608(1973)
5) J. P. Heinrich, H. Garrett, R. P. Allen; J. Appl. Phys. 54, 1973(1974)
6) R. S. Perkins, S. Strässler, A. Menth; Paper these proceedings
7) H. Nagel; Proc. 3rd Europ. Conf. Hard Mag. Mat. Amsterdam 1974, p. 153
8) M. G. Benz et. al; Report AFML TR-142, Air Force Materials Laboratory, Wright Patterson Air Force Base, Ohio, (1971)
9) K. Bachmann; AIP Conf. Proc. 18, 1168(1974)

# BASAL PLANE ANISOTROPY IN THE PSEUDOBINARY COMPOUND $Y_{0.25}Nd_{0.75}Co_5$

A.E. Miller and T. D'Silva
University of Notre Dame, Indiana 46556

and J.P. Heinrich
FTD/PDSX, Wright-Patterson AFB, Ohio 45433

ABSTRACT

The pseudobinary compound $Y_{0.25}Nd_{0.75}Co_5$ has an easy c-axis at 250°K and above. Below 150°K the easy direction is in the basal plane. The transition from c-axis to basal plane occurs continuously over a temperature span of 100°K.

Anisotropy in the basal plane was first detected as a peak at 210°K in the measured saturation magnetization of a single crystal sphere of $Y_{0.25}Nd_{0.75}Co_5$ aligned at room temperature in a magnetic field. The single crystal was subsequently aligned by the Laue method and an angular scan of magnetization in the basal plane, at an applied field of 6700 Oe, showed six-fold anisotropy from 77.4°K to about 250°K.

The anisotropy field along the semi-hard b-axis is so large that the field dependence of magnetization appears to come to saturation at a value much lower than the saturation magnetization. In the transition temperature range, where the semi-hard and easy axes lie along the surface of a cone, the anisotropy constant $K_4$ could be obtained by extrapolating the field required to saturate along the semi-hard axis. The anisotropy constant $K_4$ is $1.5x10^6$ ergs/cc at 178°K and decreases exponentially as the semi-cone angle decreases.

The basal plane anisotropy of the compounds $Y_{0.5}Nd_{0.5}Co_5$ and $Y_{0.75}Nd_{0.25}Co_5$ is under investigation.

# BEHAVIOR OF A DOMAIN WALL IN $Dy_2Co_{17}$

C. W. Allen, A. E. Miller and B. D. Cullity
Dept. of Metallurgical Engineering and Materials Science
University of Notre Dame, Notre Dame, Indiana 46556

ABSTRACT

A variety of faulted polytypic forms which depend on thermal history and probably on chemical composition have been observed for $Dy_2Co_{17}$. For a sample having predominantly a mixture of 4H and 6H intergrowths, a ferrimagnetic domain wall has been observed in TEM which moves easily with changes of field and whose orientation depends on local foil thickness. For a Bloch wall in this easy-basal-plane intermetallic one expects the wall to be parallel to the basal plane. In the situation observed, however, the wall is curved in regions of specimen thickness variation, reaching a maximum deviation from the basal plane of about 15 degrees in the thickest area observable. Attemps by magnetometry to observe a cone structure suggested by this behavior have consistently led to the easy-basal-plane model for magnetization. However, while the anisotropy energy is large, it varies slowly with orientation within 20 deg of the basal plane. It is concluded that the substantial field arising from the free pole structure of the wedge-shaped foil together with this peculiar anisotropy is responsible for large deviations of the domain magnetizations from the basal plane, which the domain wall orientation reflects. It is important, therefore, in deducing magnetic details of such a material from TEM observations, to take due account of possible effects of the thin wedge geometry of the area observed.

# MAGNETOCRYSTALLINE ANISOTROPY OF COMPOUNDS WITH COMPOSITIONS NEAR $Gd_2Co_{17}$

T.Katayama, Y.Koizumi*, K.Kawanishi*, T.Shibata, and T.Tsushima
Electrotechnical Laboratory, Tanashi, Tokyo 188, Japan
*College of Sci. Engineer., Nihon Univ., Chiyoda-ku, Tokyo 101

ABSTRACT

The magnetocrystalline anisotropy of Gd-Co phases in the stable region of the $Th_2Zn_{17}$-type structure changes very sensitively from negative to positive as Gd/Co increases above the stoichiometric ratio. Magnetic properties of single crystals with the compositions of $Gd_2Co_{17}$ and $Gd_{2.1}Co_{16.8}$ were measured. $K_1$ and $K_2$ of $Gd_2Co_{17}$ at room temperature are $-2.7\times10^6$ and $+2.5\times10^5$ erg/cm³, respectively. While, $K_1$ and $K_2$ of $Gd_{2.1}Co_{16.8}$ at the same temperature are $3.0\times10^6$ and $4.7\times10^5$erg/cm³, respectively. This change of anisotropy is discussed in terms of the crystalline field around Gd and Co sites.

## INTRODUCTION

There are two crystal structures of $Gd_2Co_{17}$: $Th_2Zn_{17}$-type (hexagonal, $P6_3/mmc$) and $Th_2Zn_{17}$-type (rhombohedral $R\bar{3}m$).[1] The former is stable at high temperatures and the latter is stable below 1200 - 1300°C. $Gd_2Co_{17}$ crystallizes usually in the low temperature phase. This type exists within a narrow homogeneous region near the composition of 2:17. It is known that the stoichiometric compound $Gd_2Co_{17}$ with the $Th_2Zn_{17}$-type is a ferrimagnet with easy plane in (00·1).[2]

It is shown, however, that isostructural phases in the Gd-rich region such as $Gd_{2.1}Co_{16.8}$ have the easy magnetization along the <00·1> axis, and that the magnetocrystalline anisotropy of the phases is very sensitive to the Gd content similar to hexagonal phases in $Y_{2+x}Co_{17-2x}$.[3] This sensitive change of anisotropy is discussed in terms of the crystalline field around Gd and Co sites.

## EXPERIMENTAL

Single crystals of $Gd_2Co_{17}$ and $Gd_{2.1}Co_{16.8}$ were prepared by the Bridgman method using BN-coated crucibles.[4] These compositions were identified by means of the electron probe micro-analysis. Powder specimens with the nominal compositions of 23.9, 24.9, and 25.9 wt% Gd-Co were prepared from arc melted alloys. If one Gd substitutes for a pair of Co in $Gd_2Co_{17}$ according to the formula of $Gd_{2+x}Co_{17-2x}$, these nominal compositions of the arc melted alloys correspond with $Gd_2Co_{17}$, $Gd_{2.09}Co_{16.82}$, and $Gd_{2.18}Co_{16.64}$, respectively. The deviations from the nominals are estimated to be less than 0.5 wt% in Gd during the meltings and the heat-treatments. The purities of the raw metals were 99.9%. In order to obtain specimens of single phase $Th_2Zn_{17}$-type, the grown single crystals and the arc-melted alloys were homogenized for several hours at about 1100°C in a pure argon atmosphere, and cooled to room temperature at the rate of 1°C/min. The powder specimens were oriented in a field of 20 kOe and fixed with resin. X-ray powder diffraction patterns were taken from two surfaces of each oriented specimen : one surface parallel to the magnetizing field (H) and the other perpendicular to H.

A Foner-type vibrating sample magnetometer and a magnetic balance were employed for measurements of the magnetization. Magnetocrystalline anisotropy constants ($K_1$ & $K_2$) of spherical single crystals were determined from the magnetization curves along the directions of easy and hard magnetizations according to Sucksmith's method.[5]

## EXPERIMENTAL RESULTS

In Fig.1 are shown X-ray diffraction patterns of the specimens with compositions of 23.9, 24.9, and 25.9 wt% Gd-Co, whose surfaces are perpendicular to the magnetizing field (H). The diffraction intensities from the (30·0) and the (22·0) planes decrease with increasing Gd content, but the diffraction intensities from the

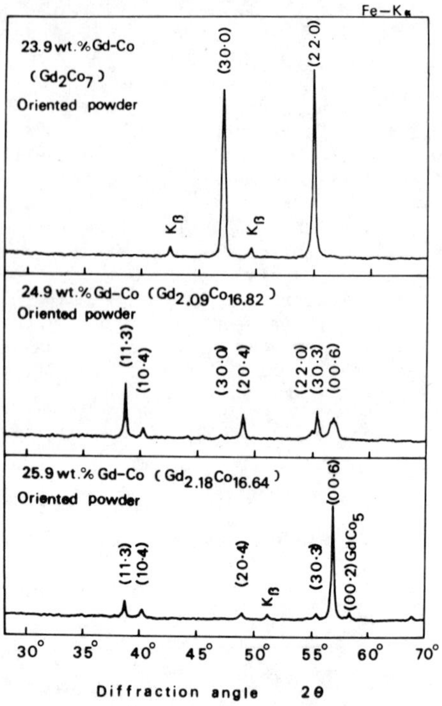


Fig.1 X-ray powder pattern traces for oriented specimens with the nominal compositions of 23.9, 24.9, and 25.9 wt% Gd-Co. The middle trace suggests that the magnetic anisotropy is almost zero.

(00·6) planes show the contrary behavior.

Figure 2 shows the changes of the X-ray intensity ratios of (30·0)/(30·3) and (00·6)/(30·3) against the deviation from the stoichiometric composition $Gd_2Co_{17}$. For example, in the case of diffraction from the surface perpendicular to H {Fig.2(A)}, high intensity ratio of (30·0)/(30·3) was obtained for the stoichiometric composition (this means that the compound has the easy magnetization plane), and the intensity ratio (30.0)/(30·3) decreases according to the increase of the Gd content (X). On the other hand, the intensity ratio of (00·6)/(30·3) from the same surfaces incraces with increasing Gd content {Fig.2(B)}.

These results suggest that the easy direction of

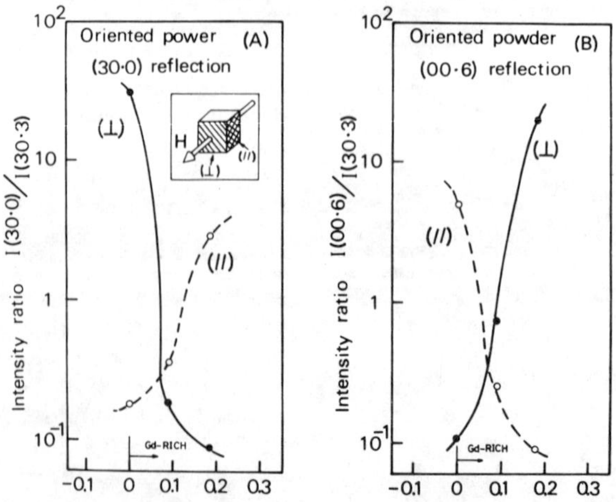


Fig.2 Change of the X-ray intensity ratios of (30·0)/(30·3) {(A)} and (00·6)/(30·3) {(B)} against the compositional deviation X from stoichiometric $Gd_2Co_{17}$. (⊥) is for specimen surfaces perpendicular to H, and (//) is for surfaces parallel to H.

Tab.1 CRYSTALLOGRAPHIC DATA AND SOME MAGNETIC PROPERTIES OF $Gd_2Co_{17}$ AND $Gd_{2.1}Co_{16.8}$ SINGLE CRYSTALS

| COMPOUNDS | HEAT-TREATMENT | EASY DIRECTION | $a(A)$ | $c(A)$ | $c/a$ | $\rho(g/cm^3)$ | $M_s(emu/cm^3)$ | $T_c(K)$ |
|---|---|---|---|---|---|---|---|---|
| $Gd_2Co_{17}$ | 1060°C,3hrs. 1°C/min. | Basal plane | 8.349 | 12.24 | 1.47 | 8.87 | 578 (R.T) | 1199 |
| $Gd_{2.1}Co_{16.8}$ | 1100°C,5hrs. 1°C/min. | c-axis | 8.431 | 12.09 | 1.43 | 8.81 | 459 | 1075 |

magnetization changes continuously from the basal plane to the c-axis with increasing Gd content within the stable region of the $Th_2Zn_{17}$-type structure near the composition of 2:17. And it is found that the magnetic anisotropy ($K_1+2K_2$) becomes zero for $Gd_{2+x}Co_{17-2x}$ at the crossing point of X=0.07.

Two single crystals with the compositions $Gd_2Co_{17}$ and $Gd_{2.1}Co_{16.8}$, whose structures are the rhombohedral $Th_2Zn_{17}$-type, were prepared, and their magnetic properties measured. Lattice constants ($a$ & $c$), $c/a$, density ($\rho$), saturation magnetization ($M_s$), and Curie temperature ($T_c$) of each single crystal are summarized in Table 1. Thus, $a$ of $Gd_{2.1}Co_{16.8}$ is larger slightly than that of $Gd_2Co_{17}$, but $c$ of $Gd_{2.1}Co_{16.8}$ is smaller than that of $Gd_2Co_{17}$. And, $M_s$ and $T_c$ of $Gd_{2.1}Co_{16.8}$ are lower than those of $Gd_2Co_{17}$.

In Fig.3 is shown the temperature dependences of magnetocrystalline anisotropy constants $K_1$ and $K_2$, anisotropy field $H_A$, and saturation magnetization $4\pi M_s$ of the $Gd_2Co_{17}$ single crystal. $K_1$ and $K_2$ are $-2.7\times10^6$ and $2.5\times10^5 erg/cm^3$ at room temperature, and $-2.8\times10^6$ and $-3.1\times10^5 erg/cm^3$ at 80 K, respectively. That is, the easy magnetization lies in the basal plane. $K_1$ does not change so much with temperature, and $K_2$ is one order smaller than $K_1$ at these temperatures.

Figure 4 shows the temperature dependences of $K_1$, $K_2$, $H_A$, and $4\pi M_s$ of the $Gd_{2.1}Co_{16.8}$ single crystal. $K_1$ and $K_2$ are $3.0\times10^6$ and $4.7\times10^5 erg/cm^3$ at room temperature, and $3.0\times10^6$ and $3.4\times10^5 erg/cm^3$ at 80 K, respectively. $K_1$ and $K_2$ are positive and almost constant with temperature. That is, the direction of easy magnetization of the crystal lies along the c-axis.

Thus, magnetocrystalline anisotropy of the single crystals in the stable region of the $Th_2Zn_{17}$-type changes also from negative to positive according to the increase of the Gd content from stoichiometric $Gd_2Co_{17}$.

## DISCUSSION

The rhombohedral $Gd_2Co_{17}$ structure is derived from the $CdCo_5$ structure by the substitution of one third of the Gd atoms by pairs of Co atoms having smaller radii than Gd in each of the alternate planes.[6] In the case of compounds such as $Gd_{2.1}Co_{16.8}$ containing small excess amounts of Gd compared with the stoichiometric ratio of 2:17, the lattice constant $a$ expands slightly and the lattice constant $c$ shrinks slightly in comparison with those of $Gd_2Co_{17}$ (Tab.1). It can be considered that the excess Gd atoms occupy some of the sites of Co pairs aligned parallel to the c-axis in the rhombohedral $Gd_2Co_{17}$ structure similar to the case of hexagonal phase in $Y_{2+x}Co_{17-2x}$.[3] And, the crystal environment of the Gd sites in the rhombohedral $Gd_2Co_{17}$ ($Th_2Zn_{17}$-type), considering only the nearset and next nearest Gd neighbors, is changed locally to one of the two kinds of crystal environments of the Gd sites in hexagonal $Gd_2Co_{17}$ ($Th_2Ni_{17}$-type) by the substitution of a Co pair by a Gd atom.[7] No change in the X-ray diffraction patterns resulting from this local "phase transition" was observed. Only a slight increase of the diffraction intensities was detected from (11·3) and (30·0) planes. This increase is caused mainly by the difference of the atomic scattering factors between Gd and Co. Thus, all of the compounds with the compositions between $Gd_2Co_{17}$ and $Gd_{2.1}Co_{16.8}$ exhibit only the rhombohedral $Th_2Zn_{17}$-type structure. For this reason, it may be considered that the significant changes observed in the magnetocrystalline anisotropy originate from the changes of the crystal environments of the Gd sites caused by the substitution of the Co pairs by the excess Gd atoms.

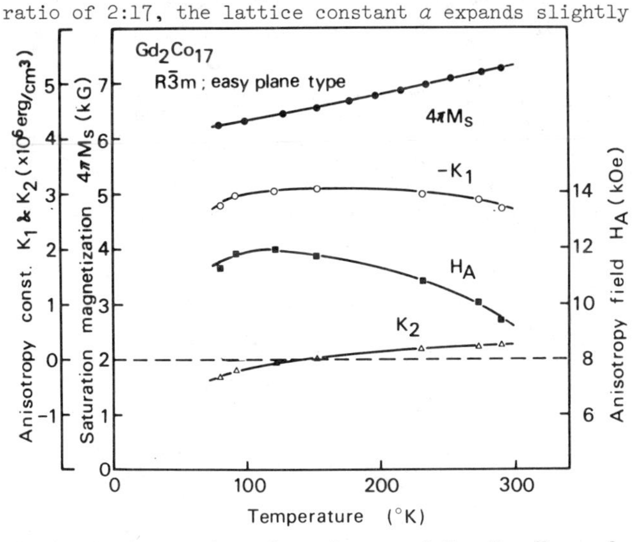


Fig.3 The temperature dependences of $K_1$, $K_2$, $H_A$, and $4\pi M_s$ for the $Gd_2Co_{17}$ single crystal.

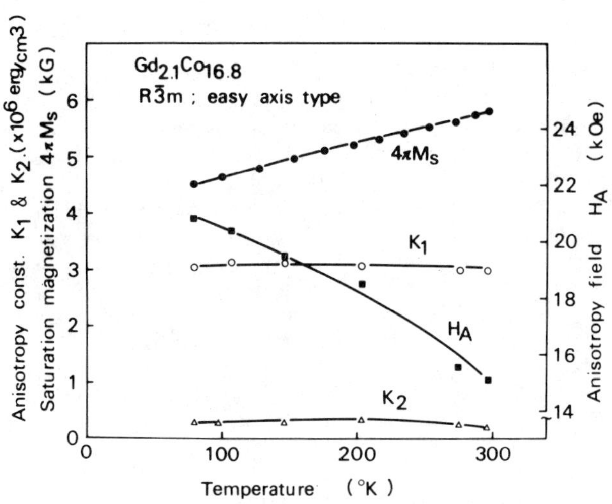


Fig.4 The temperature dependences of $K_1$, $K_2$, $H_A$, and $4\pi M_s$ for the $Gd_{2.1}Co_{16.8}$ single crystal.

We would like to thank Dr.M.Hirano of our laboratory for useful discussions.

## REFERENCES


1) K.H.J.Buschow and A.S.Van Der Goot, J.Less-Common Metals 17, 249 (1969)
2) K.J.Strnat, AIP Conf. Proc. NO.5, 1047 (1971)
3) S.Yajima, M.Hamano, and H.Umebayashi, J.Phys.Soc. Japan 32, 861 (1972)
4) T.Katayama and T.Shibata, J.Crystal Growth 24/25, 396 (1974)
5) W.Sucksmith and J.E.Thompson, Proc.Roy.Soc. (London) A225, 362 (1954)
6) A.E.Ray and R.S.Harmer, Proc.9th Rare-Earth Research Conf. 368 (1971)
7) J.E.Greedan and V.U.S.Rao, J.Solid State Chem. 6, 387 (1973)

# HIGH FIELD MAGNETIC MEASUREMENTS ON SINTERED $SmCo_5$ PERMANENT MAGNETS

Stanley R. Trout and C. D. Graham, Jr.
Department of Metallurgy and Materials Science and
Laboratory for Research on the Structure of Matter
University of Pennsylvania, Philadelphia, Pa 19174

ABSTRACT

Hysteresis loops in fields to 100 kOe have been measured at 300, 77, and 4.2K parallel and perpendicular to the alignment axis in a series of sintered $SmCo_5$ magnets with compositions varying from 16.24 to 17.08 at% Sm. Analysis of the data taken perpendicular to the alignment axis permits evaluation of the effective anisotropy constant $K_1$, and also the degree of misorientation of the individual particles. Intrinsic coercive fields $H_{ci}$ varied from 0.34 to 16.4 kOe at room temperature. In all samples, $H_{ci}$ and $K_1$ increased rapidly with decreasing temperature, roughly doubling between room temperature and 77K. This confirms more generally the result reported for two samples by Benz and Martin.[1] The values of $H_{ci}$ and $(BH)_{max}$ depend strongly on composition, but the anisotropy does not. Variations in permanent magnet properties are therefore not directly related to variations in the bulk anisotropy.

## INTRODUCTION

Benz and Martin[1] measured the magnetic properties of several sintered $SmCo_5$ magnets as a function of temperature and made the surprising observation that $H_{ci}$ increased linearly with decreasing temperature, following quite closely the linear increase in $K_1$. The coercive field in these magnets is usually attributed to domain wall nucleation or pinning effects that are structure sensitive[2] and not necessarily linearly dependent on the bulk crystal anisotropy.

The present investigation was undertaken to see whether this temperature dependence of $H_{ci}$ is general for $SmCo_5$ magnets, and more broadly to add to the understanding of the coercive field and the permanent magnet properties of $SmCo_5$ and related materials. The experiments also gave information about the degree of alignment of the individual particles in sintered magnets.

## SAMPLES AND EXPERIMENTAL PROCEDURE

A series of samples covering the composition range from 16.24 to 17.08 at% Sm ($SmCo_{5.16}$ to $SmCo_{4.85}$) was obtained from D. L. Martin of the GE Research and Development Center. These are the same samples whose room-temperature properties were reported by Martin, Benz, and Rockwood[3]; they show a wide range of quality as permanent magnets. Magnetization curves and hysteresis loops were measured parallel and perpendicular to the alignment axis at 300, 77, and 4.2K on cube samples of 3.2 mm size in fields to 100 kOe using a mechanically-driven vibrating sample magnetometer. The demagnetizing factor was taken as $4\pi/3$, a value confirmed by measurements on an iron cube in the same apparatus. Magnetic saturation could not be attained in all samples at low temperatures even at 100 kOe, so the low temperature saturations were obtained from the room temperature values and the single-crystal temperature dependence.[4]

Conventional permanent magnet properties were obtained from the easy axis magnetization data, and the effective bulk anisotropy was determined from the hard-axis data. A perfectly-oriented single crystal with anisotropy described by a single constant ($E_K = K_1 \sin^2\theta$) has a linear hard-axis magnetization curve with slope

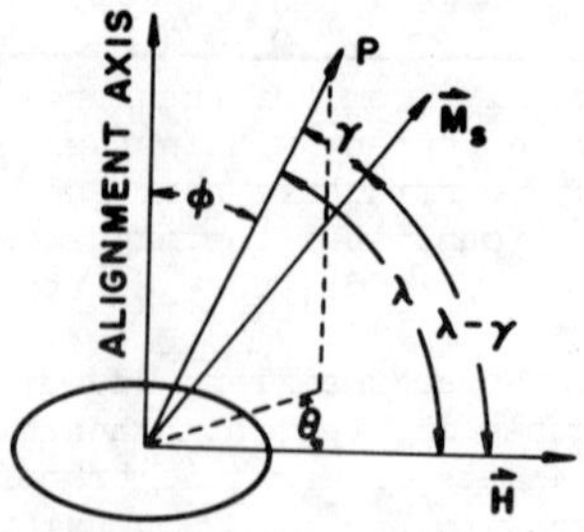


Fig. 1. Angles used in calculation of hard-axis magnetization curves. P is the axis of an individual particle. Angles $\emptyset$ and $\lambda$ are not coplanar; angles $\gamma$ and $\lambda$ are coplanar.

$M_s^2/2K_1$. In sintered magnets, the alignment of the individual particles is not perfect, so more complex magnetization curves result. The linear hard-axis curve reported by Benz and Martin[1] were obtained by first magnetizing the samples in the easy direction to minimize domain wall motion. We chose instead to fit the hard axis curves measured in decreasing fields to a calculated curve in which there are two adjustable parameters: the uniaxial anisotropy constant $K_1$ and an angle $\beta$ which measures the distribution of the polar angles $\emptyset$ (see Fig. 1) between the particle axes and the alignment axis. The angle $\emptyset$ is assumed to be described by a spherical normal distribution, $f(\emptyset) = k\exp(-\emptyset^2/\beta^2)$, where k is a normalization constant. Direct observation by Martin[5] show this assumption to be reasonable. The distribution of particle axes is assumed independent of the azimuthal angle $\theta$. To a first approximation, the standard deviation of $\emptyset$ is related to $\beta$ by $\sigma = \beta(1-\frac{\pi}{4})$.

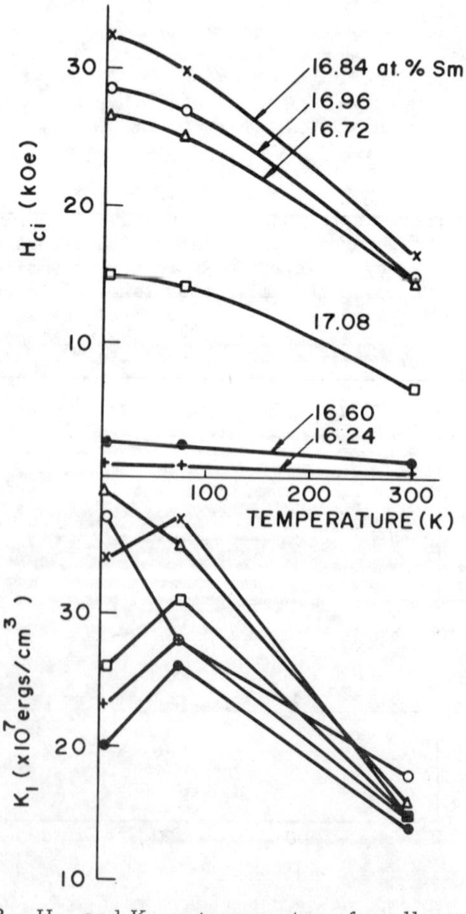


Fig. 2. $H_{ci}$ and $K_1$ vs temperature for all samples.

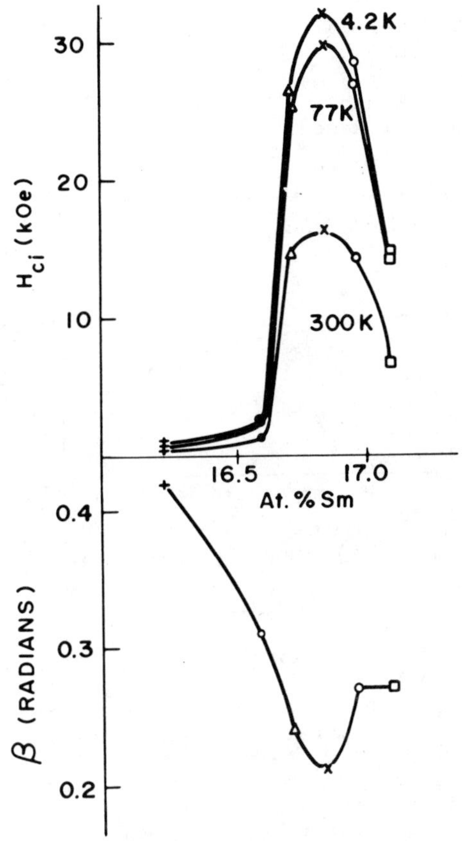


Fig. 3. $H_{ci}$ (three temperatures) and average β vs composition.

A field applied perpendicular to the alignment axis causes the magnetization of the particle to rotate by an angle γ towards H. The angle γ is determined by a balance between the torque from the field $L=M_sH\sin(\lambda-\gamma)$ and the torque from the anisotropy $L=-dE_K/d\phi=-K_1\sin 2\gamma$. The resulting magnetization is given by

$$M=M_s\int_0^{2\pi}\int_0^{\pi/2}f(\phi)\cos(\lambda-\gamma)\sin\phi\,d\phi\,d\theta=f(H,K_1,\beta). \quad (1)$$

Tables of $M/M_s$ were calculated for a series of values of $h=HM_s/2K_1$ and β, and the experimental hard-axis curves were fitted to the calculated values to obtain the quantities $K_1$ and β. In the fitting, greatest weight was given to the high field data, since in this region the magnetization should change by rotation rather than by wall motion, and the calculated curves assume only rotation.

## RESULTS AND CONCLUSIONS

Numerical data are given in Table I and plotted in Figs. 2 and 3. The following points are worthy of note.

The coercive field increases rapidly with decreasing temperature in all samples, approximately doubling between 300 and 77K. The absolute value of $H_{ci}$, however, is strongly dependent on composition, varying by almost a factor of 50 from the best to the worst sample. Maximum $H_{ci}$ is observed at about 16.8 at% Sm, in agreement with Martin, Benz, and Rockwood,[3] and the composition for maximum $H_{ci}$ does not change with temperature of measurement.

The room-temperature anisotropy is almost independent of composition, and the numerical value of about $1.5\times10^8$ erg/cm$^3$ is in reasonable agreement with single-crystal values.[4,6] The measured $K_1$ increases with decreasing temperature, approximately doubling between 300 and 77K. There is serious scatter in the values of $K_1$ at 4.2K. This may be experimental error caused by the increasing difficulty of saturating

TABLE I

| at%Sm | $H_{ci}$ kOe | $(BH)_{max}$ MGOe | $K_1$ $10^8$ erg/cm$^3$ | β rad |
|---|---|---|---|---|
| | 0.34 | 3.77 | 1.5 | 0.42 |
| 16.24 | 0.52 | 4.88 | 2.8 | 0.39 |
| | 0.84 | 5.23 | 2.3 | 0.40 |
| | 1.08 | 8.24 | 1.4 | 0.31 |
| 16.60 | 2.24 | 15.6 | 2.6 | 0.29 |
| | 2.44 | 16.6 | 2.0 | 0.32 |
| | 14.8 | 21.0 | 1.6 | 0.24 |
| 16.72 | 25.2 | 23.1 | 3.5 | 0.22 |
| | 26.6 | 21.5 | 3.9 | 0.21 |
| | 16.4 | 22.7 | 1.5 | 0.21 |
| 16.84 | 30.0 | 24.5 | 3.7 | 0.22 |
| | 32.4 | 23.8 | 3.4 | 0.19 |
| | 14.4 | 19.7 | 1.8 | 0.27 |
| 16.96 | 27.0 | 18.6 | 2.8 | 0.24 |
| | 28.6 | 19.2 | 3.7 | 0.25 |
| | 6.6 | 19.9 | 1.5 | 0.27 |
| 17.08 | 14.1 | 22.0 | 3.1 | 0.28 |
| | 14.7 | 21.8 | 2.6 | 0.25 |

The three sets of values for each sample are at 300, 77, and 4.2K.

the magnetization as the anisotropy increases; however, other investigators have found the anisotropy to drop at low temperatures.[1,4,6]

The degree of particle misalignment β is minimum in the samples of highest $H_{ci}$, but it is hard to say if there is any causal relation between the two quantities.

The results show clearly that $H_{ci}$ does not depend directly on the bulk anisotropy, contrary to the suggestion of Benz and Martin.[1] If $H_{ci}$ is controlled by a domain wall nucleation or pinning event, it is reasonable that $H_{ci}$ would have the temperature dependence of the domain wall energy, which will be approximately the same as the temperature dependence of $K_1^{\frac{1}{2}}$. But the magnitude of $H_{ci}$ will depend on some highly local structure and need not correlate with the magnitude of $K_1$.

## ACKNOWLEDGEMENTS

This work was supported by the National Science Foundation through the Laboratory for Research on the Structure of Matter under contract DMR 72-03025. We are grateful to D. L. Martin for providing the samples and to P. J. Flanders for assistance with the experiments.

## REFERENCES

1. M. G. Benz and D. L. Martin, J. Appl. Phys. 43 4733 (1972).
2. J. D. Livingston, AIP Conf. Proc. 10 643 (1973).
3. D. L. Martin, M. G. Benz, and A. C. Rockwood, AIP Conf. Proc. 10 583 (1973).
4. E. Tatsumoto, J. Okamoto, H. Fuji, and C. Inoue, Suppl. J. de Physique 32 C1-550 (1971).
5. D. L. Martin, private communication.
6. S. G. Sankar, V. U. S. Rao, E. Segal, W. E. Wallace, W. G. Frederick, and H. J. Garrett, Phys. Rev. 11 435 (1975).

- 0 -

# UPON INFLUENCING THE MAGNETOCRYSTALLINE ANISOTROPY OF $RE_2TM_{17}$ COMPOUNDS

R.S.Perkins, S.Strässler and A.Menth
Brown Boveri Research Centre, CH-5401 Baden, Switzerland

## ABSTRACT

A simplified form of the theoretical single-ion, crystal field contribution to the magnetocrystalline anisotropy of $RE_2TM_{17}$ compounds is given. Through an evaluation of this contribution in $Sm_2(Co_{1-x}Fe_x)_{17}$ compounds, two methods of favourably altering their anisotropy were suggested. The first involves substitution of Sm by Tb in order to compensate the adverse influence of Fe beyond $x \approx 0.2$. The expected effect is observed. Substitution of Co by Mn or Cr was used in an attempt to avoid the consequences of structural features of the Fe-containing alloys upon the anisotropy. These new compounds yield up to 40% improvement in the anisotropy, and at higher magnetizations, compared to $Sm_2Co_{17}$.

## INTRODUCTION

$RECO_5$ alloys possess a very large uniaxial magnetocrystalline anisotropy[1], $K_1$, whilst the corresponding $RE_2Co_{17}$ compounds show a much smaller, if not planar, anisotropy[1]. Consequently much effort has gone into producing permanent magnets of the former, rather than the latter, material. The discovery[2,3] that substitution of Co by Fe in 2:17 compounds substantially increases the magnetization, $M_s$, and causes a transition of otherwise planar, $-K_1$, to uniaxial, $+K_1$, anisotropy, rejuvinated interest. A continuing aim of studies of 2:17 compounds has been to improve upon either $M_s$[4] or $K_1$. Recent measurements of the $Sm_2(Co,Fe)_{17}$ system[5] suggest that higher $M_s$ reduce the probability of obtaining useful coercivity. In contrast, increasing $K_1$ increases this probability.

For the latter reason, our investigations have concentrated on understanding and improving the anisotropy of 2:17 compounds.

## EXPERIMENTAL

The results to be described below are measurements of $M_s$ and $K_1$, on single crystals. Material was produced by melting in BN crucibles in an RF furnace, followed by heat-treatments at approximately 1150°C. In most cases a small quantity of a second phase, probably a Co-based eutectic, remained. This did not substantially affect the results. Magnetic measurements were carried out in a vibration magnetometer, and analysed as described earlier[6].

## RESULTS AND DISCUSSION

Fig.1 shows the crystal field contribution to the anisotropy of $Sm_2(Co_{1-x}Fe_x)_{17}$ compounds at room temperature. The data are produced from the subtraction[7,8] of the transition-metal sublattice anisotropy[6], of $Y_2(Co_{1-x}Fe_x)_{17}$, from the total anisotropy of the samarium alloys[5]. The data show that this contribution falls to zero at $x \approx 0.5$. The

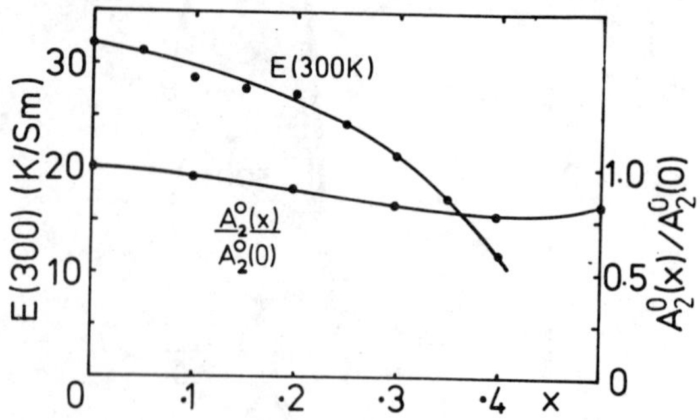


FIG. 1: VARIATION OF THE STABILISATION ENERGY PER SM-ATOM AT 300 K, AND THE FRACTIONAL CHANGE IN $A_2^0$, FROM LATTICE PARAMETER CHANGES ALONE, IN THE SERIES $Sm_2(Co_{1-x}Fe_x)_{17}$.

crystal field anisotropy in 1:5 and 2:17 compounds has been succesfully predicted theoretically. These exact treatments are, however, not amenable to straightforward interpretation of experimental results of varied compounds. We have considered a simplified version of the same theory which yields agreement to within about 20% of the exact treatment. Assuming that the Sm-Co exchange interaction is large, the crystal field may be treated as a perturbation. The stabilisation energy at 0°K, E, defined as the difference in free energy of a Sm-ion with moment lying along the z-axis and in the basal plane, becomes

$$E \simeq \alpha_J(1-\sigma)\langle r^2\rangle A_2^0 \; 3J(J-\tfrac{1}{2}) \qquad (1)$$

where $\alpha_J$ is the Stevens' multiplicative factor, $\langle r^2\rangle$, the mean square 4f shell radius, J, the Sm quantum state, $\sigma$, a 4f electron shielding constant, and

$$A_2^0 = e^2 \sum_k \frac{Z_k}{|\vec{R}_k|^3} (3\cos^2\theta_k - 1) \qquad (2)$$

where $Z_k$, $\vec{R}_k$ are the charge and vector to the kth neighbour of the source atom. $\theta_k$ is the angle between $\vec{R}_k$ and the c-axis.

In this simplified case, the crystal field anisotropy is proportional to the term, $A_2^0$. This depends upon nearest neighbour charge and distance, which vary in a series of compounds such as $Sm_2(Co,Fe)_{17}$. In order to reduce $A_2^0$ to zero from lattice parameter changes alone, a c/a-ratio of 1·51 would be needed. From the lattice constants[10] however, the c/a ratio at $x \approx .5$ is 1.466.

The crystal field anisotropy falls much quicker to zero than structure changes alone would predict. Since this behaviour is also reproduced at lower temperatures, the phenomenon is not related to temperature dependence effects. A possible explanation is that the neighbouring transition-metal atoms possess a charge, in addition to that assumed for the Sm-neighbours. Alternatively, the screening of the Sm charge may be dependent upon the Co:Fe ratio. However, the fact remains that E changes sign at higher Fe concentrations. The sign of E is determined by that of $\alpha_J$ and of J, in addition to $A_2^0$. Since

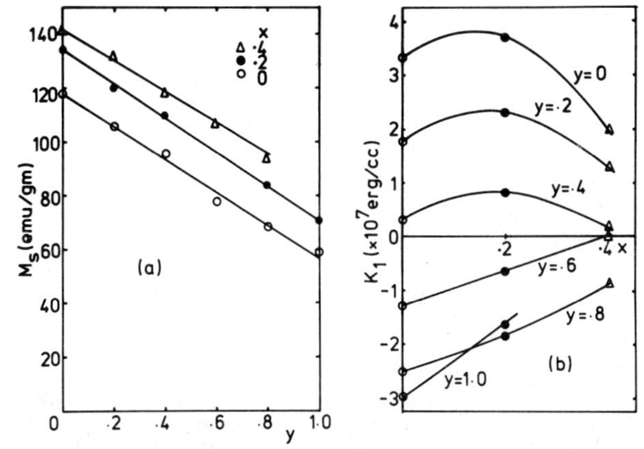


FIG. 2: COMPOSITION DEPENDENCE OF (A) THE SATURATION MAGNETISATION, AND (B) THE ANISOTROPY CONSTANT IN THE COMPOUNDS $(Sm_{1-y}Tb_y)_2(Co_{1-x}Fe_x)_{17}$.

the former quantities change sign according to the rare-earth concerned, the possibility exists in principal to correct the changing sign of E in $Sm_2(Co,Fe)_{17}$ by substitution of Sm by, for example, Tb or Pr which have the highest negative $\alpha_J$ values ($\alpha_J$(Sm)=+0.0412, $\alpha_J$(Tb)=-0.0101, $\alpha_J$(Pr)=-0.0201). For this reason the system $(Sm,Tb)_2(Co,Fe)_{17}$ was studied. Pr was not chosen due to the possibility of unfavourable moment orientations occuring, and the Sm basis was used since they possess the highest $+K_1$ values of the light rare-earths.

Figs.2a,b show the dependence upon Tb and Fe concentration of $M_s$ and $K_1$. It is seen that the $K_1(x)$ curves all approach zero at high x values. Additionally the slopes are greater beyond 50% Tb. $K_1(x,y)$ will be positive for $0.5 < y < 1.0$ and $0.4 < x < 1.0$. However, although $K_1(1.0,1.0)$ could exceed $4\times10^7$erg/cc, $M_s$ will probably not exceed 8kGs. The form of some Tb-rich magnetization curves suggest other than simple colinear moment alignment.

From (1) and (2) E is dependent upon the charge of neighbouring atoms, electronic screening, and lattice parameters. The increasing c/a ratio in $Sm_2(Co,Fe)_{17}$ is due to the faster rise of c compared to a[10]. This has been interpreted[12] as being related to the Fe-Fe interactions in the "dumb-bell" transition-metal sites of the 2:17 structures. It had, however, not been established whether this phenomenon also occurs in the corresponding Mn and Cr containing compounds. Additionally the possibility exists that the atomic charges and their screening would be altered by the presence of other transition metals.

Figs.3a,b show the results obtained for $Sm_2(Co,Mn)_{17}$. These compounds have not been previously studied, although aligned-powder measurements exist for Ce, Th and Nd[3]. The data indicate a 40% rise in $K_1$ below 15% Mn. The magnetization drops below that of $Sm_2Co_{17}$ beyond 15% Mn. A compromise can therefore be sought between the rise in $M_s$ with Fe present, and the rise in $K_1$ with Mn present. Figs.4a,b show these data for $Sm_2(Co,Fe)_{17}$ compounds. These materials clearly present better properties than $Sm_2(Co,Fe,Mn)_{17}$ and afford an improved potential for permanent magnet development. That this potential is realised is clear from the work also published here[13].

The cause of the $K_1$ rise in the Mn compounds has been intensively studied. It is in fact found to be related to changes in the Sm-transition metal exchange field. This

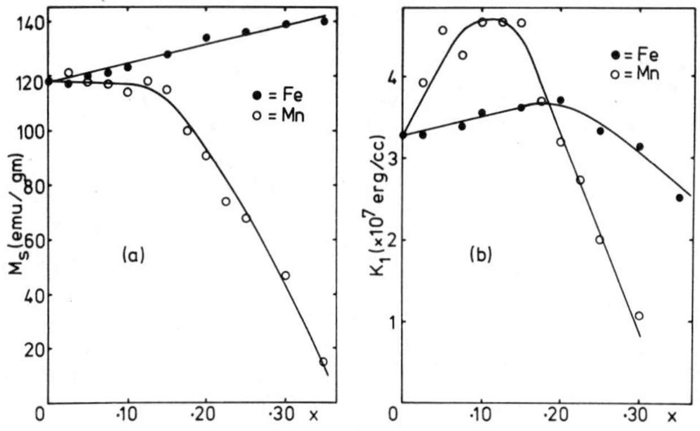


FIG. 3: COMPARISON OF (A) THE MAGNETISATION, AND (B) THE ANISOTROPY CONSTANT VARIATIONS IN $Sm_2(Co_{1-x}TM_x)_{17}$ FOR TM = FE, MN.

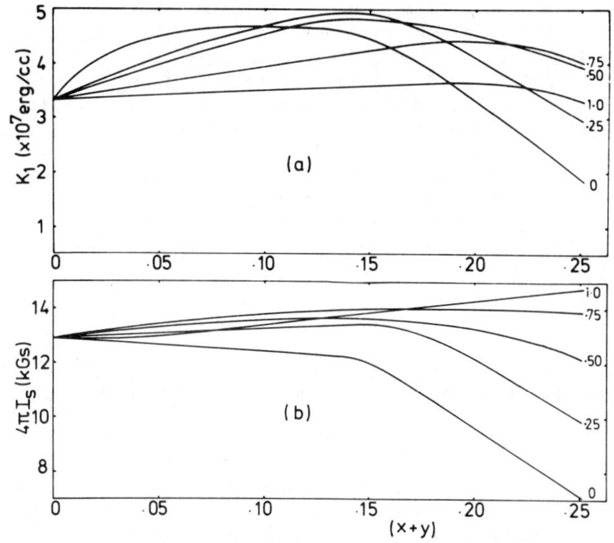


FIG. 4: VARIATIONS OF (A) ANISOTROPY AND (B) MAGNETISATION IN $Sm_2(Co_{1-(x+y)}Fe_xMn_y)_{17}$ ALLOYS. THE ABSCISSA, (X+Y) IS TOTAL PROPORTION OF CO SUBSTITUTED; THE VARIOUS CURVES ARE THE FRACTION OF THIS WHICH IS FE, X/(X+Y).

alters the temperature dependence of $K_1$ substantially so that at room temperature the $K_1$ values rise with Mn concentration. The Cr compounds exhibit a similar behaviour, but have the disadvantage of severely reducing the magnetization. $K_1$ rises by 22% at 5% Cr, whilst $M_s$, drops by 12%.

REFERENCES


1. G.Hoffer and K.Strnat, J.Appl.Phys.38, 1377-1378 (1967).
2. H.J.Schaller, R.S.Craig and W.E.Wallace, J.Appl.Phys.43, 3161-3164 (1972).
3. A.E.Ray and K.J.Strnat, IEEE Trans.Mag.8, 516-518 (1972).
4. K.S.V.L.Narasimhan and W.E.Wallace, J.Sol.State Chem.13, 315-318 (1975).
5. R.S.Perkins, S.Gaiffi, A.Menth, IEEE Trans. Mag.11, 1431-3433 (1975).
6. R.S.Perkins and H.Nagel, Physica 80B, 143-152 (1975).
7. K.H.J.Buschow, A.M.van Diepen and H.W.de Wijn, Sol.State Comm.15, 903-906 (1974).
8. S.G.Sankar, V.U.S.Rao, E.Segal, W.E.Wallace, W.G.D.Frederick and H.J.Garrett, Phys.Rev. B11, 435-439 (1975).
9. J.E.Creedan and V.U.S.Rao, J.Sol.State Chem.6, 387-395 (1973).
10. A.E.Ray and R.S.Harmer, Proc.9th Rare Earth Research Conf., Blacksburg, Virginia, Vol.I, 368, Oct.1971.

11. A.V.Deryagin and N.V.Kudrevatykh, Phys. Stat.Sol.(a)30, K129 (1975).
12. D.Givord and R.Lemaire, IEEE Trans.Mag.10, 109-113 (1974).
13. H.Nagel, Paper these proceedings.

# ORIENTATION AND REMANENT MAGNETIZATION OF $SmCo_5$ MAGNETS

W. M. Swift, W. T. Reynolds, R. M. Schrecengost and D. V. Ratnam*
Westinghouse Research Laboratories
Pittsburgh, Pennsylvania 15235

## ABSTRACT

The Schulz x-ray method of planar orientations was used to characterize the statistical distribution of (0001) plane orientations for five $SmCo_5$ magnets having energy products in the range 4.5-19 MGOe. X-ray intensity data were used to calculate the volume fraction of grains having their (0001) planes located in eighteen equiarea polar intervals defined by $\alpha$, the tilt angle of the (0001) plane with respect to the sample plane, and $\phi$, the azimuth angle defined from an arbitrary direction in the sample plane. From these data and assumptions about magnetization reversal, calculations of remanent magnetizations and energy products were made.

## INTRODUCTION

It is well established that magnetic field alignment of $SmCo_5$ particles is one of several essential procedures employed to achieve high energy product permanent magnets.[1-3] The magnetic field produces a torque which, owing to a large c-axis magnetocrystalline anisotropy, tends to align $SmCo_5$ particles with their c-axis parallel to the applied field direction. Complete alignment by a magnetic field is never achieved, however, since other factors, such as particle size distributions and particle morphologies, will also be important determinants of the overall alignment obtained when a powered sample is pressed in a magnetic field. Since magnetic properties will be related to the degree of particle alignment,[4] it is important to establish methods for characterizing the crystallographic texture of the finished product. The purpose of this paper is to describe the results of x-ray determinations of the statistical distribution of grain orientation for several $SmCo_5$ magnets, and how these data are related to magnetic properties.

## EXPERIMENTAL

Several $SmCo_5$ magnets with energy products levels in the range 4.5-19.0 MGOe were prepared using the powder metallurgical sintering technique. The alignment field was varied between 0-20,000 Oe during the pressing operation in order to achieve magnets with a range of orientations. All magnets were prepared from the same blended powder mixture and had a composition of 36.8 wt. pct. Sm and 63.2 wt. pct. Co after sintering. Samples were magnetized in a pulsed field of 60,000 Oe and B-H loops measured in a tip field of 28,000 Oe.

Plane orientations were determined by the Schulz x-ray method employing Cr-$k_\alpha$ radiation.[5] Samples measuring about 20 mm square by 2 mm thick were polished and etched and subsequently seated in a Siemens texture goniometer. Intensities of reflections from (0002) planes were continuously recorded as $\alpha$, the angle of reflecting plane with respect to the sample surface, and $\phi$, the azimuth angle from an arbitrary direction in the sample plane, were changed at constant rates. The intensity data were used to obtain computer plotted pole figures as well as intensity histograms.

TABLE I

Experimental and Calculated Magnetic Properties of $SmCo_5$ Magnets

| Sample | $\rho$* | $B_r$** (G) | $B_r$(I) (G) | EP (MGOe) | EP(I) (MGOe) |
|---|---|---|---|---|---|
| A | 0.915 | 4500 | 6880 | 4.5 | 11.8 |
| B | 0.924 | 6450 | 7850 | 10.0 | 15.4 |
| C | 0.973 | 7850 | 8970 | 15.0 | 20.1 |
| D | 0.963 | 8050 | 9030 | 16.0 | 20.3 |
| E | 0.954 | 8680 | 9620 | 19.0 | 23.0 |

*Assumed density of $SmCo_5$ = 8.60 gms/cm$^3$ [Ref. 1].

**Assumed $B_s$ of $SmCo_5$ = 11,000 G [Ref. 7].

## RESULTS AND DISCUSSION

Table I summarizes the experimetally measured remanent magnetizations, $B_r$, energy products, EP, and packing fractions, $\rho$, for five $SmCo_5$ magnets. Also shown in Table I are calculated remanent magnetizations and energy products based on x-ray diffraction data and assumptions about the mode of magnetization reversal. The means by which x-ray data were coupled to magnetization reversal modes in order to calculate $B_r$ and EP will be brought out shortly.

Figure 1 is a (0002) pole figure for sample E in Table I and shows contours of constant intensity about the (0002) pole located at the origin of the pole figure. The contours are associated with grains having their (0001) plane off-oriented with respect to the sample plane, or equivalently, having their c-axis misaligned with respect to a normal to the sample surface. The contours are symmetric about the origin, but somewhat elliptically shaped. For the other samples listed in Table I, however, contours of constant inten-

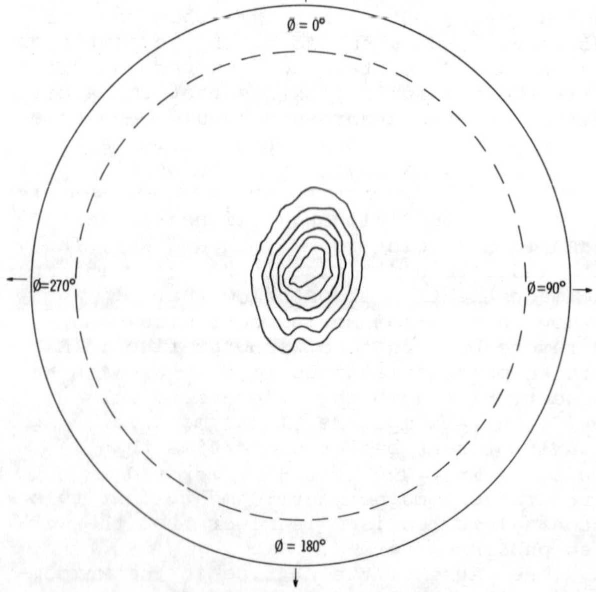


Fig. 1-(0002) pole figure for sample E (19MGOe $SmCo_5$ magnet). Contours times random intensity: X, 1.5X, 2X, 2.5X, 3X, 3.5X

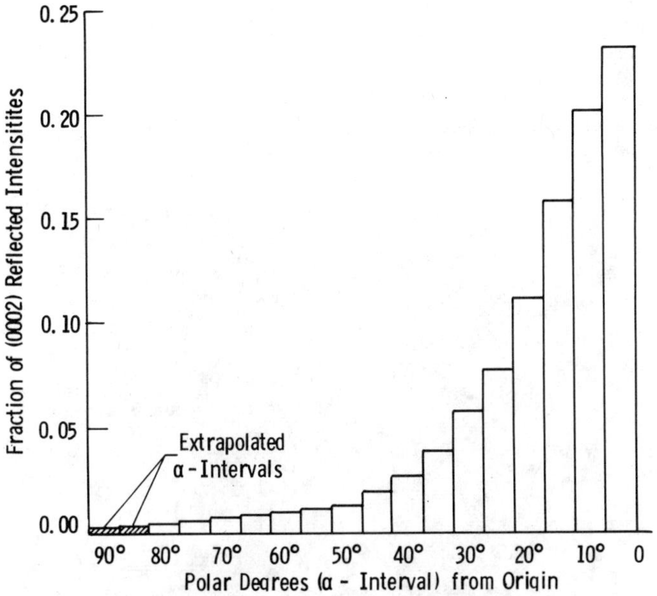


Fig. 2-Intensity histogram showing fraction of total repleted (0002) intensities as a function of the polar angle α for sample E (19MGOe $SmCo_5$ magnet).

sity were circular about the origin, indicating that the deviations from perfect (0001) plane orientation were independent of the azimuth angle, ϕ, for constant pole angle, α. This latter statement, although not rigorously correct for the plane deviations of sample E, is taken to be approximately valid for sample E since the small amount eccentricity of this samples' intensity contours does not measurably influence subsequent discussions and calculations of $B_r$ and EP.

Figure 2 is the intensity histogram associated with the pole figure for sample E. The intensity histogram shows the fraction of total reflected intensities (all intensity counts occurring in the range $0° \leq \alpha \leq 80°$, $0° \leq \phi \leq 360°$) occurring in each of sixteen equal polar areas. Each area, or α interval, comprises a 5° range in α and the full 360° range in ϕ. If there were no grains having their basal planes oriented in the range $80° < \alpha < 90°$, the intensity histogram of Fig. 2 would also give the volume fraction, f, of grains located in each interval. For most samples in Table I, the exception being the random standard (sample A),

TABLE II

Volume Fractions of (0001) Plane Orientations For $SmCo_5$ Magnets*

| α Interval (Polar Degrees) | f (A) | f (B) | f (C) | f (D) | f (E) |
|---|---|---|---|---|---|
| 0- 5 | 0.058 | 0.108 | 0.123 | 0.128 | 0.232 |
| 5-10 | 0.059 | 0.106 | 0.119 | 0.124 | 0.202 |
| 10-15 | 0.059 | 0.098 | 0.116 | 0.119 | 0.159 |
| 15-20 | 0.058 | 0.097 | 0.104 | 0.014 | 0.113 |
| 20-25 | 0.059 | 0.091 | 0.096 | 0.096 | 0.078 |
| 25-30 | 0.059 | 0.081 | 0.085 | 0.086 | 0.058 |
| 30-35 | 0.059 | 0.075 | 0.072 | 0.072 | 0.039 |
| 35-40 | 0.059 | 0.065 | 0.060 | 0.059 | 0.027 |
| 40-45 | 0.059 | 0.059 | 0.050 | 0.049 | 0.020 |
| 45-50 | 0.057 | 0.051 | 0.041 | 0.038 | 0.013 |
| 50-55 | 0.056 | 0.042 | 0.033 | 0.030 | 0.012 |
| 55-60 | 0.056 | 0.033 | 0.026 | 0.025 | 0.011 |
| 60-65 | 0.055 | 0.027 | 0.021 | 0.022 | 0.010 |
| 65-70 | 0.055 | 0.022 | 0.018 | 0.017 | 0.009 |
| 70-75 | 0.052 | 0.018 | 0.015 | 0.015 | 0.008 |
| 75-80 | 0.048 | 0.013 | 0.012 | 0.010 | 0.006 |
| 80-85* | 0.046 | 0.009 | 0.007 | 0.005 | 0.003 |
| 85-90* | 0.046 | 0.005 | 0.002 | 0.003 | 0.002 |

*Each column shows the volume fraction, f, occurring in each of twenty α-intervals. Sample identifications are given in parentheses.

only a small fraction of grains are located in the interval $80° < \alpha < 90°$. It was possible to extrapolate the raw intensity data into the range beyond α=80°, and thereby obtain an estimate of the volume fraction of grains in each of the eighteen equal polar areas. These data in terms of volume fraction are summarized in Table II, and in essence describe the statistical distribtuions of (0001) plane orientations for these samples.

Factors other than the statistical distribution of grain orientations in a $SmCo_5$ magnet are required to calculate $B_r$ and EP. The mode of magnetization reversal as a function of misorientation angle α will control the magnitude of $B_r$ in a given grain. Assume that all grains are magnetized to saturation ($B_s$) and that magnetization reversal is accomplished by an irreversible process in which the magnetization vector flips from the $+B_s$ state to the $-B_s$ state. For this model, or Model I, the sample $B_r$ is given by

$$B_r(I) = \rho \, B_s \sum_{i=1}^{i=18} f_i \cos\bar{\alpha}_i \qquad \text{(Model I)} \tag{1}$$

where ρ is the packing fraction and $f_i$ is the volume fraction of grains in the i-th α interval, where $\bar{\alpha}_i$ is the average misorientation angle in the i-th α interval. The calculated $B_r$ values from equation (1) are given in Table I as are calculated energy products, EP(I), given by

$$EP(I) = (B_r(I))^2/4 \tag{2}$$

The agreement between experimental $B_r$ and EP values and those calculated from equations (1) and (2), respectively, is poor although the agreement is generally better for the high energy product permanent magnets (magnets C,D,E). There are obvious reasons for a lack of agreement with the proposed model. First, all grains will not be saturated, and as the angle α increases the departure from saturation will undoubtedly increase. Second, the model ignores rotational magnetization reversal processes which are expected to be particularly important as the departure from perfect alignment increases.[6] If, in fact, these factors were known as a function of α, it should be possible to more accurately calculate $B_r$ and EP from orientation data.

## CONCLUSIONS

1. The Schulz x-ray method of planar reflections was used to characterize the statistical distribution of (0001) planar orientation for several $SmCo_5$ magnets.

2. Contours of constant intensity in $SmCo_5$ magnets were symmetric about the (0001) pole and for most samples essentially circular in shape.

## REFERENCES


1. K. Strnat, G. Hoffer, T. Olson, W. Ostertag, and J. J. Becker, J. Appl. Phys., 38, 1001 (1967).
2. D. K. Das, IEEE Trans. Magnetics, 5, 214 (1969).
3. M. G. Benz and D. L. Martin, Appl. Phys. Letters, 17, 176 (1970).
4. L. M. Magat and M. M. Korotkova, Fiz. Metal. Metalloved., 35, No. 5, 1110 (1973).
5. L. G. Schulz, J. Appl. Phys. 20, 1030 (1949).
6. E. C. Stoner and E. P. Wohlfarth, Phil. Trans. Roy. Soc. (London), A-240, 599 (1948).
7. C. W. Searle, W. D. Frederick, and H. J. Garret, Trans. IEEE, MAG-9, 164 (1973).


*Crucible Materials Research Center, Pittsburgh, PA. 15230.

# A METALLOGRAPHIC METHOD FOR THE DETERMINATION OF CRYSTAL ALIGNMENT IN Co-R PERMANENT MAGNETS

D. L. Martin
General Electric Corporate Research & Development
Schenectady, New York 12301

ABSTRACT

A metallographic method has been developed for determining the alignment of grains in sintered cobalt-rare-earth permanent magnets. The basis for the method is to use the eutectoid transformation of the $Co_5R$ phase to decorate the basal planes, and thus enable one to determine some information on the orientation of the grain with respect to the alignment direction.

The method has been used to study the degree of alignment in aligned and unaligned Co-Sm magnet samples. It is shown that it is possible to obtain quantitative grain alignment results which compare favorably with alignment data obtained from magnetic measurements and x-ray diffraction.

## INTRODUCTION

Alignment of the crystal grains so that their c-axes are parallel prior to densification is an important step in the processing of cobalt-rare-earth permanent magnets. A high degree of crystal alignment is needed in order to optimize the magnetization of the magnet specimen. The saturation and remanent magnetization are directly influenced by the degree of crystal alignment whereas the coercive force ($_bH_c$) and $(BH)_{max}$energy product are indirectly related to the degree of alignment through their relations to $B_r$.

Previously, information on the degree of crystal alignment in Co-Sm magnets has been obtained from magnetic measurements and from x-ray orientation studies. The ratio of remanent magnetization to the specimen saturation value; that is, $A = B_r/B_{is}$, has been found to be a useful parameter for quality control evaluation of alignment.

In regard to the use of x-ray diffraction techniques, Swift et al. have shown that quantitative orientation results can be obtained from the x-ray diffraction method which enabled them to calculate energy product values of Co-Sm samples in good agreement with measured values.[1]

A new method for determining the degree of crystal alignment is described in this report. It is based on the fact that the lamellae, formed within the $Co_5R$ grains by a eutectoid transformation, are parallel to the basal planes of the $Co_5R$ phase.[2,3] It is this orientation relationship which provides information on the alignment of the grains in a sintered magnet. In Fig. 1a, the lamellar structure of a well-aligned Co-Sm magnet is shown. Note that most of the grains have basal plane traces nearly perpendicular to the alignment direction. This structure with the lamellae perpendicular to the alignment direction is indicative of a high degree of crystal alignment. In contrast, the lamellae of the different grains in an unaligned sintered magnet sample do not have a common direction (Fig. 1b).

Since first reported by den Broeder and Buschow,[2] the eutectoid decomposition of $Co_5R$ compounds has been studied extensively.[3-6] Characteristic of the eutectoid structure in all the Co-R systems is the lamellar structure similar to that shown in Fig. 1. Thus, the metallographic method should be applicable to $Co_5R$ magnet alloys containing Ce, La, Pr, Gd, as well as Sm.

## EXPERIMENTAL RESULTS AND DISCUSSION

In preparation for metallographic examination, the samples were transformed by holding ten days at 750°C. Protection against oxidation during the decomposition treatment was obtained by sealing the sample in a welded container. A section cut parallel to the alignment direction was polished and etched, and reference lines

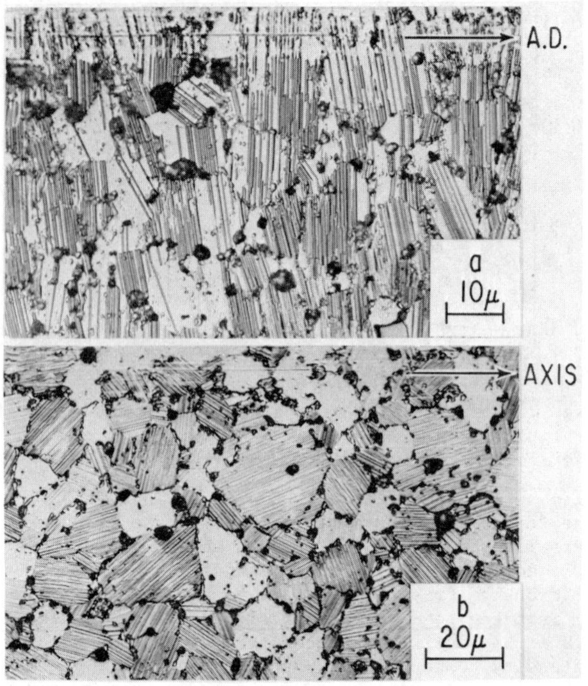


Fig. 1 -Lamellar structure in Co-Sm alloy samples held 10 days at 750°C. (a) aligned, (b) unaligned. The sample axis or alignment direction is indicated. Etchant: 3 glycerine, 1 $CH_3COOH$ and 1 $HNO_3$.

parallel to the alignment direction were scribed on the surface before metallographic examination and photographing.

The basis for the analysis was to measure the angle $\theta$ between the grain lamellae and the alignment direction. The procedure followed was to place a transparent grid with a 0.4 inch (1 cm) spacing over the micrograph, and measure the basal plane trace angle at each grid intersection. If the grid point was located on a void, inclusion, or grain boundary, no angle measurement was made. In some samples, angle measurements were not made at as many as 15% of the grid points. In a few samples, the lamellar eutectoid structure did not develop. Severe loss of samarium during the 750°C treatment due to oxidation seemed to explain the absence of the lamellar structure in several samples that were not sealed in welded tubes, but not in other samples. Perhaps in those samples there was an excessive amount of $Co_7Sm_2$ since Zijlstra and den Broeder have reported an absence of the platelets in samples containing high samarium.[7]

On a typical 8" x 10" print, more than 150 angle measurements were made. For a magnet sample, where eight or more locations were examined, more than 1200 measurements were required. To reduce the labor involved in reading and recording the large number of angle measurements needed, a data-logging system was constructed which utilized an optical shaft encoder and counter for the angle measurements. The counter data was transferred to a paper tape through a data-logging system so that it could be fed into a computer.

A statistical analysis of the data was made by using a time-sharing computer program called STATSYSTEM.[8] A histogram of the angle $\theta$ data for a well-aligned, hydrostatically pressed, sintered magnet sample is given in Fig. 2a. While it shows a high concentration of $\theta$ values near 90°, there is a small, but significant frac-

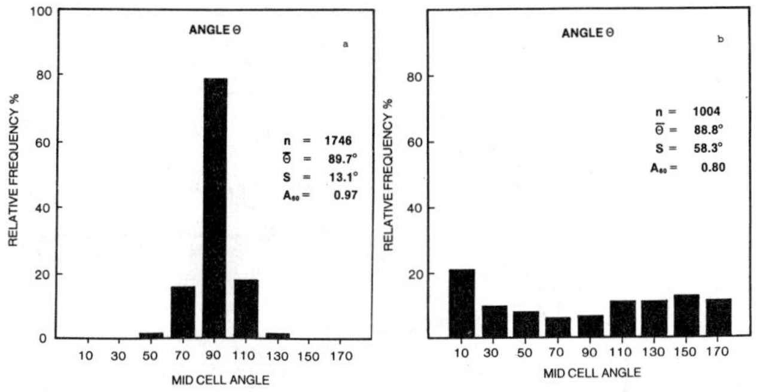


Fig. 2 - Histogram of angle θ for (a) aligned and (b) unaligned Co-Sm samples.

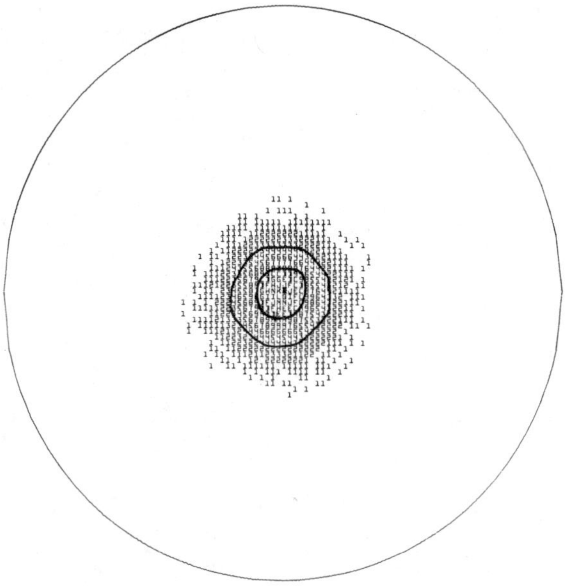

Fig. 3 - Basal plane pole figure for an aligned Co-Sm magnet. The projection plane is normal to the alignment direction. The intensity levels are in times random units.

TABLE I
SUMMARY OF TEST DATA

| Sample | $B_r$ kG | $H_c$ kOe | $(BH)_{max}$ MGOe | $A_{60}$ | n Obs. | θ mean | s |
|---|---|---|---|---|---|---|---|
| A* | 4.4 | 4.3 | 4.7 | 0.80 | 1004 | 88.8 | 58.3 |
| B | 8.7 | 7.9 | 17.4 | 0.970 | 1746 | 89.7 | 13.1 |
| C | 9.7 | 9.5 | 23.3 | 0.988 | 2049 | 89.3 | 10.0 |

*Unaligned

tion of poorly aligned grains. A histogram for the unaligned sample, Fig. 2b, shows a more uniform distribution of θ values, although there is a slight peak near $0^o$.

The angle θ is not sufficient to define the orientation of the grains. Basal plane trace angle data for other sections are needed to define the texture. Examination of other planes parallel to the alignment axis on one of the bars showed a similar distribution of the angle θ. Thus, these results indicate that the basal plane pole density is uniformly distributed around the alignment axis. In a well-aligned sample, most of the basal poles are within $20^o$ of the alignment axis. In other words, there is a [0001] fiber texture. This has been confirmed by x-ray pole figure data on an aligned, sintered bar (Fig. 3), and is not an unexpected result for a sample in which the initial powder particles were aligned in a magnetic field followed by isostatic pressing. The standard deviation of the angle θ can be used as a measure of the degree of alignment. Thus, in Fig. 2, note that the well-aligned sample has a standard deviation of $13.1^o$ compared to $58.3^o$ for the unaligned bar.

A summary of test data for three sintered, isostatically pressed bars is given in Table 1. Sample A was not aligned. The results show that the standard deviation is a good measure of the degree of alignment, and in general agreement with the $B_r/B_s$ ratio value.

## SUMMARY

A metallographic method has been described which can yield quantitative information on the degree of alignment of Co-R permanent magnet samples. The basis for the method is the lamellar structure developed during the eutectoid decomposition of the $Co_5Sm$ phase. The method was used to study the alignment in aligned and unaligned bars. It was shown that a [0001] fiber texture exists in aligned, isostatically pressed bars with over 90% of the basal plane poles located within $20^o$ of the alignment axis.

The metallographic method has the advantage of presenting an alignment picture of the grains, and does offer the possibility of studying specific regions of a sample, such as near the surface. It should prove to be a useful technique supplementing magnetic methods.

## ACKNOWLEDGMENTS

The author expresses his thanks to John Geertsen for making most of the angle measurements, to A. Ritzer for the optical metallography and to R. Goehner for x-ray texture determinations, and to Drs. James Livingston and Lyman Johnson for their valuable technical discussions. The author is indebted to R. Laforce and D. Sorensen for the design and assembly of the angle measuring and data logging equipment which saved innumerable hours of measurement time.

## REFERENCES

1. W.M. Swift, W.T. Reynolds, R.M. Schrecengost, and D.V. Ratnam, AIP Proc., Magnetism and Magnetic Materials - 1975; to be published in 1976.
2. F.J.A. den Broeder and K.H.J. Buschow, J. Less Common Metals 29, 65 (1972) and 33, 191 (1973).
3. J.G. Smeggil, P. Rao, J. Livingston, and E.F. Koch, AIP Conf. Proc. 18, 1144 (1974).
4. K.H.J. Buschow, J. Less Common Metals 37, 91 (1974).
5. D.L. Martin and J.G. Smeggil, IEEE Trans. on Magnetics MAG-10, 704 (1974).
6. D.L. Martin, J.G. Smeggil, W. Hatfield, and R. Bolon, IEEE Trans. on Magnetics MAG-11, 1420 (1975).
7. H. Zijlstra and F.J.A. den Broeder, Proc. of 1973 International Conf. on Magnetism V, 304 (1974).
8. G.J. Hahn and W.B. Nelson, "Introduction to STATSYSTEM," GE Information Services, 401 Washington St., Rockville, Md. 20840. (Document 4203,01-4).

# SPIN REORIENTATION IN $NdCo_5$ SINGLE CRYSTALS

M.Ohkoshi*, H.Kobayashi*, T.Katayama, M.Hirano, and T.Tsushima
Electrotechnical Laboratory, Tanashi, Tokyo 188, Japan
*Dept. of Appl. Phys., Waseda Univ., Shinjuku-ku, Tokyo 160

## ABSTRACT

A magnetic study was done precisely on single crystals of $NdCo_5$ from a standpoint of the spin reorientation (SR). The easy axis rotates continuously from the b-axis <10·0> at lower SR temperature ($T_{SR1}$ at 245 K) to the c-axis <00·1> at upper SR temperature ($T_{SR2}$ at 285 K) in the bc-plane. The temperature variations of the magnetocrystalline anisotropy constants $K_1$, $K_2$, and $K_4$ determined precisely from the magnetization and torque curves show that $K_1$ changes its sign from negative to positive just at $T_{SR2}$ and $K_2$ from positive to negative at a little higher temperature than $T_{SR2}$, but that $(K_1+2K_2)$ reverses its sign exactly at $T_{SR1}$. Further, it is found that the anisotropy constant in the basal plane $K_4$ vanishes near $T_{SR2}$, though it is fairly large ($\sim 10^7 erg/cm^3$) at low temperatures. On account of this anisotropy change, the magnetization curve along the a-axis <11·0> has a peak at a temperature slightly higher than $T_{SR1}$ when effective field of more than 10 kOe is applied.

## INTRODUCTION

Intermetallic compound $NdCo_5$ having the hexagonal $CaCu_5$-type (space group P6/mmm) is known as a ferromagnet[1]. Some of the magnetic properties such as saturation magnetization and magnetic anisotropy have been already reported[2]. One of the most characteristic properties is the sharp change of easy direction within a narrow temperature region[3] which accompanies an anomaly in specific heat[4].

These phenomena in $NdCo_5$ may be understood in terms of spin reorientation (SR) phenomena. Physical properties of the SR phenomenon have been recently studied very well in several kinds of weak ferromagnets such as $RMO_3$[5] (R=rare earth, M=$Fe^{3+}$ or $Cr^{3+}$), $Fe_3BO_6$[6], and $\alpha$-$Fe_2O_3$[7]. The SR phenomenon in a ferromagnet such as $NdCo_5$ is interesting from both points of view of phase transition and its application, because it accompanies a very large flux change. It might be useful for a thermomagnetic transducer, for example. We have studied very precisely some of the magnetic properties of $NdCo_5$ single crystals from a standpoint of SR.

## EXPERIMENTAL

Single crystals of $NdCo_5$ were prepared by a Bridgman method using a BN coated recrystallized $Al_2O_3$ crucible[8]. The purity of each raw metal was 99.9%, and the starting compositions were slightly Nd rich compared to the stoichiometric. The lattice constants *a* and *c* were determined as 5.012 Å and 3.965 Å, and the X-ray density ρ as 8.45 g/cm³, though they are somewhat different from the previous values[1]. The grown crystals were shaped into spheres with diameters between 1 and 3 mm. Magnetization and magnetic torque were measured from 77 K to $T_c$ (925 K by the present determination).

## RESULTS AND DISCUSSION

Saturation magnetization $M_s$ of $NdCo_5$ single crystals measured by keeping the sample free to rotate decreases monotonically without any anomaly with increasing temperature. However, there has been reported on aligned powders an anomalous peak in the magnetization near room temperature depending on an applied field[9]. It is explained by considering a remarkable decrease of the anisotropy energy at around room temperature.

Temperature dependences of the magnetization components, $M_a$, $M_b$, and $M_c$ along the principal hexagonal axes a<11·0>, b<10·0>, and c<00·1> are shown in Fig.1 for effective fields of 0, 1, 5, and 10 kOe. It is clear from the figure that the magnetic moment is parallel exactly to the b-axis below 245 K, and that it begins to deviate from the b-axis at 245 K and reaches the c-axis at 285 K. From 285 K to the $T_c$ the c-axis is the easy. These characteristic temperatures are the same for the specimens from several ingots.

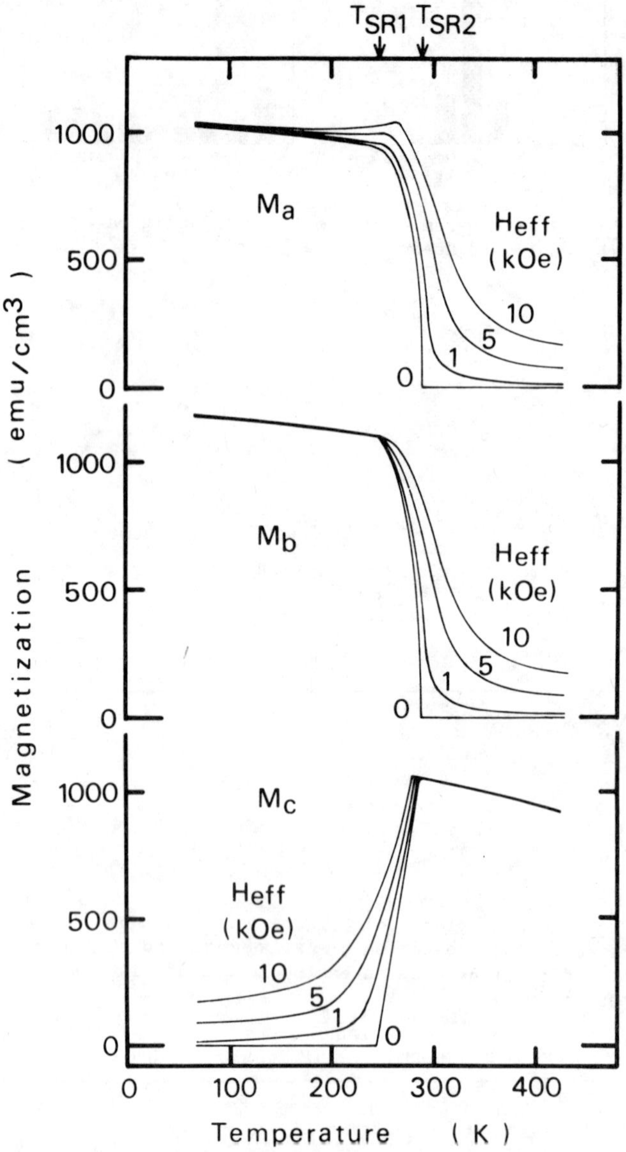


Fig.1 Temperature dependences of the $M_a$, $M_b$, and $M_c$ for various effective fields.

It is also observed that the $M_a$ is just equal to $M_b \cdot \cos(\pi/6)$ in zero effective field at all temperatures. The anisotropy constant $K_4$ in the c-plane, which is as large as $\sim 10^7 erg/cm^3$ at low temperatures, decreases rapidly with increasing temperature. It becomes almost zero in the spin rotating region without reversing its sign. So that, the $M_a$-T curve has a peak at a temperature slightly higher than 245 K, when effective field of more than 10 kOe is applied.

It is thus concluded that the ferromagnetic moment of $NdCo_5$ rotates in the bc-plane from 245 K ($T_{SR1}$) to 285 K ($T_{SR2}$) keeping the spin configuration ferromagnetic. Such phenomenon that the coherent rotation of magnetization takes place continuously in a certain temperature region below $T_c$ (or $T_N$) is similar to the so-called rotational-type SR observed in $SmFeO_3$[10].

Fig.2 shows temperature dependences of the rotation angle of the magnetic moment from the c-axis with or

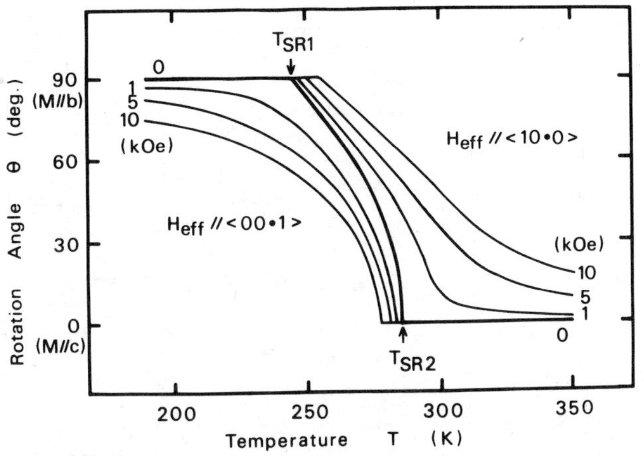


Fig.2 Temperature dependence of the rotation angle θ for various effective fields.

without effective field applied along the c- or the b-axis. The rotation angle in zero effective field is roughly agreeable with the neutron diffraction data[9]. A field induced SR is observed, and $T_{SR1}$ is shifted by $+1.2\times10^{-3}$deg/Oe in the b-axis, and $T_{SR2}$ by $-0.8\times10^{-3}$ deg/Oe in the c-axis.

Torque curves in the c-plane near the SR temperature are composed of a complete $\sin(6\phi)$ term when the applied field exceeds over 12 kOe, but their amplitudes increase with applied field (Fig.3). These results indicate that the magnetization component in the c-plane rotates in the same phase with applied field, though the magnetization is deviated still from the c-plane. The deviating angle decreases with a further increase of applied field and the saturated torque amplitude corresponds to the value of $K_4$.

The magnetocrystalline anisotropy contants K's for a hexagonal symmetry are defined by the following expression for the free energy E :

$$E = K_0 + K_1\sin^2\Theta + K_2\sin^4\theta + (K_3 + K_4\cos 6\phi)\sin^6\theta, \quad (1)$$

where θ is the polar angle from the c-axis and φ the azimuthal angle from the a-axis. Then, the following two equations hold for the $M_c$ and $M_b$ provided that the magnetization lies always in the bc-plane :

$$\frac{H}{M_c} = -\frac{2}{M_s^2}[K_1+2K_2+3(K_3-K_4)] + \frac{4}{M_s^4}[K_2+3(K_3-K_4)]M_c^2 - \frac{6}{M_s^6}(K_3-K_4)M_c^4 . \quad (2)$$

$$\frac{H}{M_b} = \frac{2K_1}{M_s^2} + \frac{4K_2}{M_s^4}M_b^2 + \frac{6}{M_s^6}(K_3-K_4)M_b^4 . \quad (3)$$

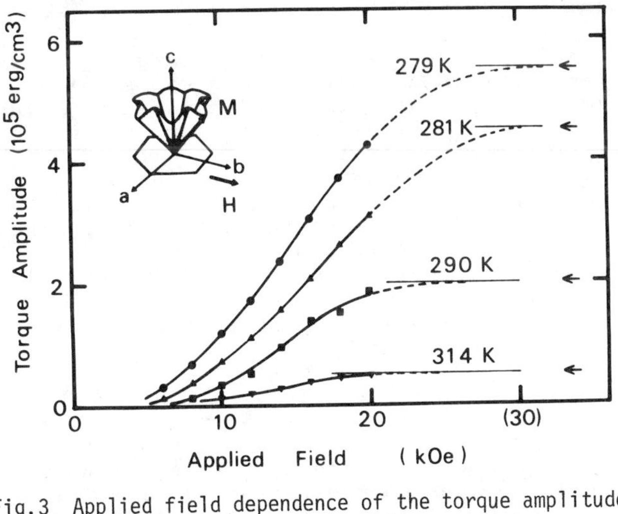


Fig.3 Applied field dependence of the torque amplitude in the c-plane for various temperatures. The arrows are the estimated values from the $M_a$ and $M_b$ curves.

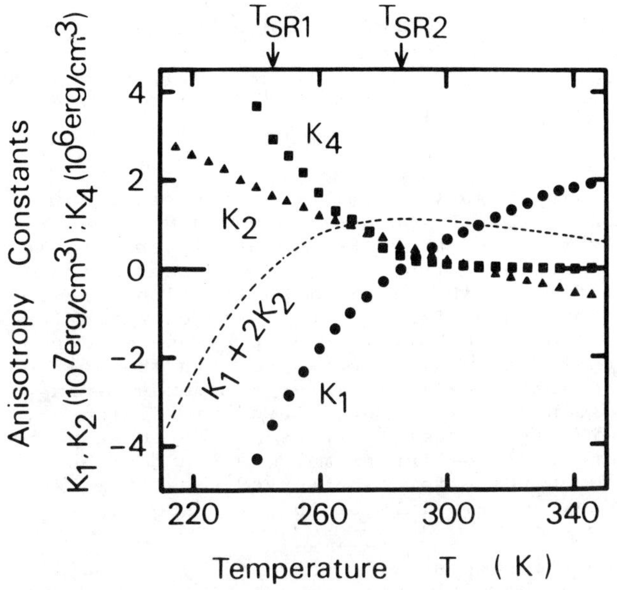


Fig.4 Temperature dependences of $K_1$, $K_2$, and $K_4$.

Each plot of $H/M_c$ *vs* $M_c^2$ and $H/M_b$ *vs* $M_b^2$ is found to be almost a straight line. There is observed a slight concave tendency for the $M_c$ and a convex one for the $M_b$ near the SR temperature. It indicates that the contribution of $(K_3-K_4)$ is negligibly small compared with that of $K_1$ or $K_2$, though $K_4$ itself obtained from the torque measurement is fairly large. Then, $K_1$ and $K_2$ are obtained accurately from the slope and the intercept of these straight lines. Fig.4 shows the temperature dependences of $K_1$, $K_2$, and $K_4$ near SR region. Both the value of $K_3$ and its temperature dependence are thought to be almost the same as those of $K_4$. It is confirmed in Fig.4 that $K_1$ changes its sign from negative to positive just at $T_{SR2}$, while $(K_1+2K_2)$ reverses its sign exactly at $T_{SR1}$.

## CONCLUSION

We have studied the magnetic anisotropy on single crystals of $NdCo_5$ from a standpoint of the SR, and have shown that $NdCo_5$ shows a rotational-type SR near room temperature. Below 245 K ($T_{SR1}$), the spins are parallel to the b-axis, and they rotate in the bc-plane between $T_{SR1}$ and $T_{SR2}$ (285 K). Above $T_{SR2}$, they are parallel to the c-axis up to $T_c$.

The hexagonal anisotropy constants have been precisely obtained around the SR temperature, and these constants are correlated well with the observed SR in $NdCo_5$.

## REFERENCES

1. R.Lemaire: Cobalt 32, 132 (1966)
2. E.Tatsumoto, T.Okamoto, H.Fujii, and C.Inoue: J. de Phys. (Paris) 32, C1-550 (1971)
3. W.G.D.Frederick, C.W.Searle, and M.Hoch: AIP Conf. Proc.18, 1197 (1973)
4. D.A.Keller, S.G.Sanker, R.S.Craig, and W.E.Wallace: AIP Conf. Proc.18, 1207 (1973)
5. T.Yamaguchi: J. Phys. Chem. Solids 35, 479 (1974)
6. M.Hirano, T.Okuda, T.Tsushima, S.Umemura, K.Kohn, and S.Nakamura: Solid State Commun. 15, 1129 (1974)
7. T.Moriya: Magnetism I, ed. Rado and Suhl (Academic Press, New York, 1963) p.85
8. T.Katayama and T.Shibata: J. Cryst. Growth 24/25, 396 (1974)
9. H.Bartholin, B.Van Laar, R.Lemaire, and J.Schweizer: J. Phys. Chem. Solids 27, 1287 (1966)
10. R.L.White: J. Appl. Phys. 40, 1061 (1969)

# NARROW BLOCH WALLS IN $RCo_5$-TYPE RARE EARTH COBALT COMPOUNDS

K.H.J. Buschow and M. Brouha
Philips Research Laboratories, Eindhoven, The Netherlands

ABSTRACT

The magnetic isotherms of various compounds of the type $RCo_5$ in which Co is partially replaced by either Cu or Ni were studied at 4.2 k as a function of Co concentration. The form of the virgin magnetization curve, the presence of large intrinsic coercive forces and the observation of pronounced thermomagnetic history effects indicate the presence of Bloch walls in these compounds having a width of only a few interatomic distances. It is argued that the presence of these narrow Bloch walls is the primary reason for the high coercive forces observed usually in solid pieces of materials of the type $RCo_{5x}Cu_{5-5x}$.

------

The samples were prepared by arc melting from 99.9% pure starting materials. The arc-cast buttons were homogenized for 10 days at 1000°C in vacuum. X-ray diffraction and microscopic examination showed that after this treatment they were of a single phase. All compounds investigated were found to be of the hexagonal $CaCu_5$-type of structure. The magnetic measurements were performed on powder particles ( > 100 μm) which were aligned at room temperature and fixed in this position by means of glue.

Magnetic isotherms at 4.2 K for several $YCo_{5x}Ni_{5-5x}$ compounds investigated are shown in Fig. 1. The coercive force is seen to increase initially with Co concentration. Near x = 0.4 it reaches a maximum of about 14 kOe. For the compounds with the highest Co concentrations ($x \geqslant 0.6$) the magnetic isotherms do not show any hysteresis. The present results were reached with samples consisting of relatively large powder particles in which no coercive force had been introduced by prolonged milling. It is clear that a prolonged milling treatment would lead to an appreciable coercive force also for the Co-rich compounds, including $YCo_5$. The values of $_IH_c$ rather reflect the strengths of the intrinsic coercive force.

Similar results as those presented above were also attained with $LaCo_{5x}Ni_{5-5x}$[1], $ThCo_{5x}Ni_{5-5x}$[2], $SmCo_{5x}Ni_{5-5x}$ and $SmCo_{5x}Cu_{5-5x}$. For all these compounds the temperature dependence of the magnetization, measured in a directions parallel to the easy axis, ($\sigma$) is rather typical: when the samples are cooled in zero field to 4.2K and $\sigma$ is measured upon heating the samples, $\sigma$ is rather low at low temperatures but suddenly rises to a much higher value. The temperature at which this rise in $\sigma$ occurs is higher according as the measuring field is lower. These features are absent if the sample is cooled to 4.2K in a magnetic field. Examples of this behaviour are shown for $SmCo_4Cu$ and $SmCo_3Ni_2$ in Figs. 2 and 3.

In the same manner as in previous investigations [1,2] it can be argued that the results shown above are strong indications for the presence of Bloch walls having a width of only a few interatomic distances. For such narrow walls there exists a difference in energy between the positions in which the center of the wall coincides with a lattice plane and the positions in which the center of the wall is situated just in between such planes.

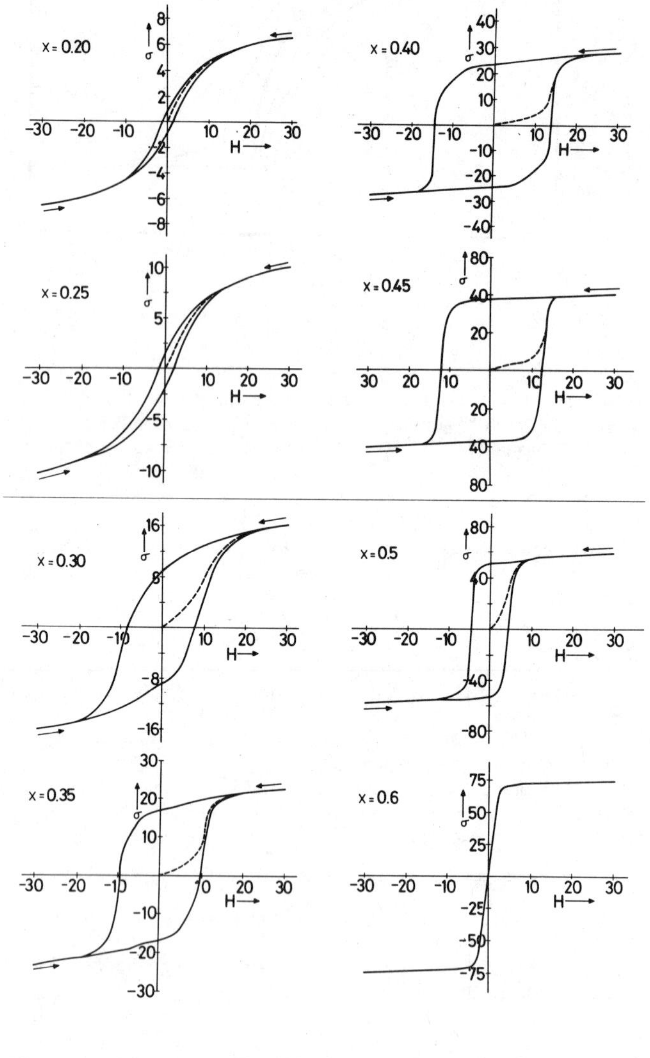


Fig. 1. Magnetic isotherms at 4.2 K for various $YCo_{5x}Ni_{5-5x}$ compounds. The broken line represents the virgin magnetization curve. The magnetization is in emu/g; the field strength is in kOe.

This leads to the following features for systems containing narrow Bloch walls: [3,5]

i homogenous wall pinning throughout the whole crystal gives rise to an intrinsic coercive force.

ii the curves of primary magnetization leave the origin in an almost horizontal manner. At a field strength sufficiently high to be able to propagate the walls ($H_p$) the magnetization strongly rises, reaching a value close to the saturation magnetization. In general $H_p$ is close to $_IH_c$.

iii The occurrence of a potential barrier between the stable positions of the Bloch walls implies that the magnetization process is thermally activated.

All the above mentioned features were observed in the proper concentration range of various pseudobinary rare earth compounds based on $RCo_5$, in which Co was partially replaced by Cu or Ni.

In general the Bloch wall thickness de-

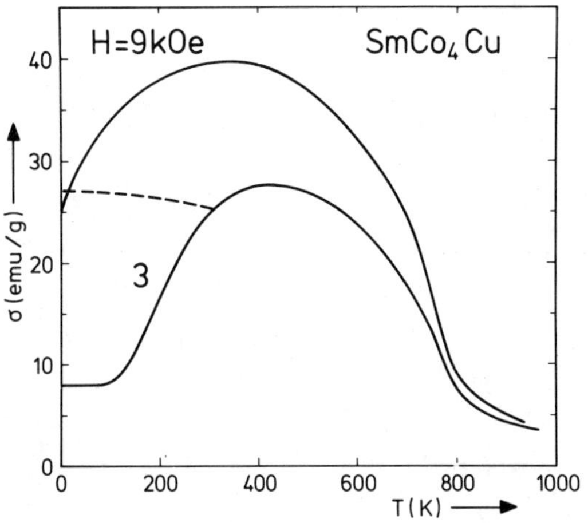


Fig. 2. Temperature dependence of the magnetization of $SmCo_4Cu$ after zero field cooling. The broken curve was obtained after cooling in a field of 3 kOe.

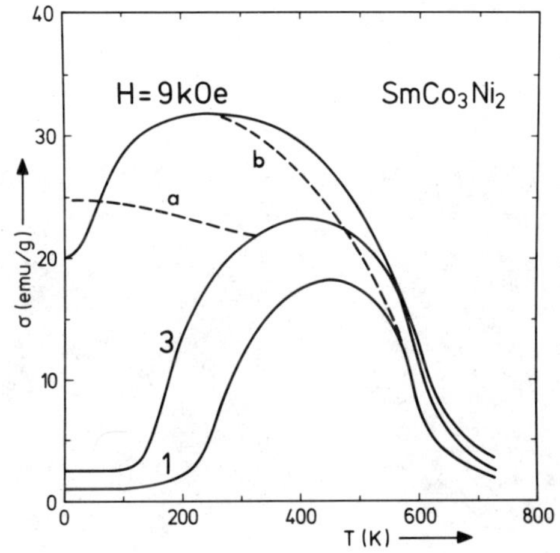


Fig. 3. Temperature dependence of the magnetization of $SmCo_3Ni_2$ after zero field cooling. The broken lines a and b were obtained after cooling in fields of 3 and 1 kOe, respectively.

pends on the relative strengths of the ferromagnetic coupling energy C between neighbouring spins and the magnetocrystalline energy K. For a uniaxial ferromagnet this latter quantity can conveniently be expressed as $K = \frac{1}{2} H_A I_s$, were $H_A$ and $I_s$ represent the anisotropy field and the saturation magnetization.The coupling energy C is usually taken as being proportional to the Curie temperature $T_c$. Since earlier investigations have revealed that the relation $C \propto T_c$ is less well obeyed in this class of materials [6] we will take C in this study to be proportional to $\mu_s^2$ ($\mu_s$ is the saturation moment per formula unit) the ratio K/C is then proportional to $H_A/\mu_s$. In general it can be shown that this ratio increases with decreasing Co concentration, leading to an increase in the intrinsic coercive force. For low Co concentration this property is lost again since here one comes into the regime where the exchange splitting of the 3d band collapses and the compounds become paramagnetic [7,8]. From the results attained in the present study it could be inferred that narrow Bloch walls are likely to be present when the ratio $H_A/\mu_s$ becomes higher than about 30.

Inspection of the literature data on $\mu_s$ and $H_A$ available for the various $RCo_5$ compounds [9,10] shows that $H_A/\mu_s$ is mostly well below 30; the only possible way to increase $H_A/\mu_s$ seems to reduce $\mu_s$ by magnetic dilution. Only the compounds $SmCo_5$ (and possibly $CeCo_5$) has a $H_A/\mu_s$ ratio of the required magnitude. The values of $H_A$ reported for $SmCo_5$ range from 210-290 kOe [10]. Even with the lower limit $H_A$ = 210 kOe one derives, with $\mu_s = 7.2\ \mu_B$ per formula unit, a value equal to 29 for the ratio $H_A/\mu_s$. It is clear that if in $SmCo_5$ ($CeCo_5$) the magnetization is reduced by magnetic dilution with for instance Cu or Ni, the conditions for the presence of narrow Bloch walls will be met rather easily, since the values for $H_A$ remain approximately constant [11]. This also follows from the results shown in Figs. 2 and 3. Compounds of the type $SmCo_{5x}Cu_{5-5x}$ and $CeCo_{5x}Cu_{5-5x}$ are unique in so far as they were reported to have a relatively large $_IH_c$ already in as-cast conditions [9]. The coercive force can further be increased by annealing at low temperatures. As a reason for the large $_IH_c$ values the presence of a complex microstructure has been suggested, involving phase separations resembling spinoidal decomposition [9,11]. However, it would appear from the present investigation that the primary reason for the large coercive forces is the presence of narrow Bloch walls, since we have found now high $_IH_c$ values even in single phase samples (as can be deduced for example from the results shown in Fig. 2) and furthermore also in systems as $SmCo_{5x}Ni_{5-5x}$ or $YCo_{5x}Ni_{5-5x}$ which do not show a complex microstructure nor give rise to precipitates. In systems containing narrow Bloch walls the effect of precipitates is expected to be quite large due to the strong pinning of the walls associated with their high wall energy.

In $SmCo_{5x}Cu_{5-5x}$ and $CeCo_{5x}Cu_{5-5x}$, when these have been annealed at low temperatures, it is the simultaneous presence of narrow Bloch walls and precipitates which makes these materials so favourable for the attainment of high coercive forces.

REFERENCES

1. Brouha M., and Buschow K.H.J., IEEE Trans. MAG-11, 1358 (1975).
2. Buschow K.H.J., Brouha M., and Elemans J.B.A.A., Phys. Stat.Sol. (a) 30, 177 (1975)
3. van den Broek J.J., and Zijlstra H., IEEE Trans. MAG-7, 226 (1971); Zijlstra H., Rare Earth and Actinide Conference, Durham 1971; Conference Digest No. 3, The Institute of Physiscs, London (1971).
4. Barbara B., Bécle C., and Lemaire R., J. Physique 32, C-1, 299 (1971).
5. Egami T., and Graham C.D., J. Appl. Phys. 42, 1299 (1971).
6. Livingston J.D., and McConnell M.D., J.Appl. Phys. 43, 4756 (1972)
7. Brouha M., and Buschow K.H.J., J. Phys.(F) Metal Phys. 5, 543 (1975).
8. Buschow K.H.J., and Brouha M., J.Appl. Phys. (in the press).
9. Nesbitt E.A., and Wernick J.H., "Rare Earth Permanent Magnets", Academic Press, New York (1973).
10. Buschow K.H.J., and Velge W.A.J.J. Z. Angew. Phys. 26, 157 (1969).
11. Katayama T., and Shibata T., Japan J. Appl. Phys. 12, 762 (1973).
12. Hofer F., IEEE Trans. MAG-6, 221 (1970).

# Fe-Cr-Co DUCTILE MAGNET WITH (BH)max ≈ 8 MGOe

H. Kaneko , M. Homma and M. Okada
Department of Materials Science, Faculty of Engineering, Tohoku University, Sendai, JAPAN 980 ;
S. Nakamura and N. Ikuta
Miyagi Technical College, Natori, JAPAN

## ABSTRACT

A single crystal of an Fe-30wt%Cr-23wt%Co-1wt%Si alloy was prepared by a recrystallization process.

The magnetic properties of <100> , <110> and <111> single crystals were measured after the following heat treatment. The crystals were solution-treated at 1300 °C for 30 min and then quenched into water. They were tempered in a magnetic field of 2 KOe at 640 °C, and were swaged. They were then step- tempered. It was found that the best magnetic properties were obtained with the <100> single crystal and were given as Br = 13 KG, bHc = 980 Oe and (BH)max = 8 MGOe.

## INTRODUCTION

Fe- Cr-Co permanent magnet alloys are known to be potential magnets with energy products which are comparable to those of Alnico 5 and with a good ductility. (1)~(5) The magnetic hardening of these alloys is performed by tempering after solution treatment. Upon tempering, an iron-rich phase ($\alpha_1$) and chromium-rich phase ($\alpha_2$) precipitate out of the super-cooled α phase by spinodal decomposition. It is well known that magnetic properties of permanent magnets depend on the crystal orientation. However, in this system the magnetic properties of single crystals have not yet been measured. The present authors prepared the single crystals by a recrystallization process. In this paper the magnetic properties and the microstuctures of the single crystals of Fe-Cr-Co permanent magnet alloys are presented.

## EXPERIMENTAL PROCEDURE

Fe-30wt%Cr-23wt%Co-1wt%Si alloys were melted from 99.9 % electrolytic iron, 99.9 % electrolytic chromium, 99.5 % cobalt and 99.5 % silicon in an induction furnace and cast into a sand mould with an inside diameter of 8.5 cm. In order to prepare the single crystals, a recrystallization process was used. The ingots were pre-heated at 900°C for 1 hr ( phase in this state : α+γ+σ) and were moved at the rate of 4 cm/hr into a furnace with temperature gradient of about 30°C/cm from 1200°C to 1300°C, and then were quenched into water ( phase in this state : α ). The preparation procedure of the single crystal is schematically shown in Fig. 1. The single crystals of the ingot-size could be prepared by this process, although a few fine grains remained at the surface of the ingots. In order to produce the ingots with a certain crystal orientation, seed crystals were welded at the top of the ingots.

The single crystals were tempered at 640°C for 40 min in magnetic field of 2 KOe (thermomagnetic treatment ). Following the thermomagnetic treatment, they were tempered at 600°C, 580°C, 560°C for 1 hour each, respectively, and subsequently tempered at 540°C for 5 hours ( step-tempering ). Some of them were swaged after thermomagnetic treatment, and then were step-tempered.

The magnetic properties were measured with an automatic flux meter. The microstructures were observed by electron microscopy.

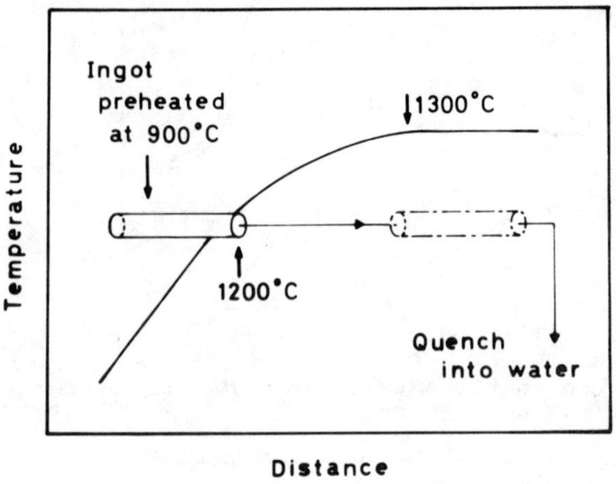


Fig. 1. Schematic diagram of preparation of single crystal.

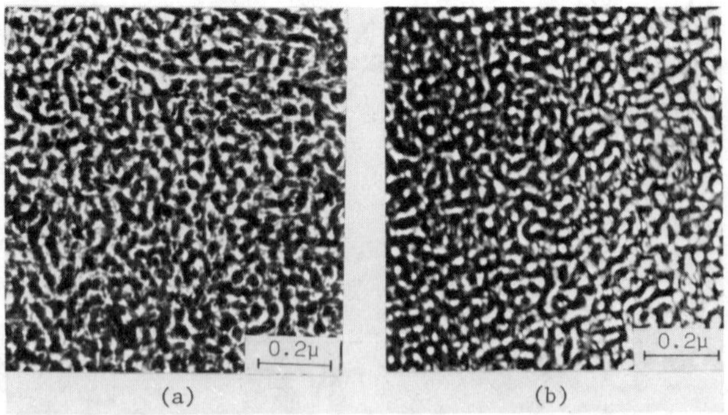


Photo. 1. Electron micrographs of an Fe-30wt%Cr-23wt% Co-1wt%Si alloy tempered at 640°C for 40 min without magnetic field and step-tempered.
(a) : (100) (b) : (110)

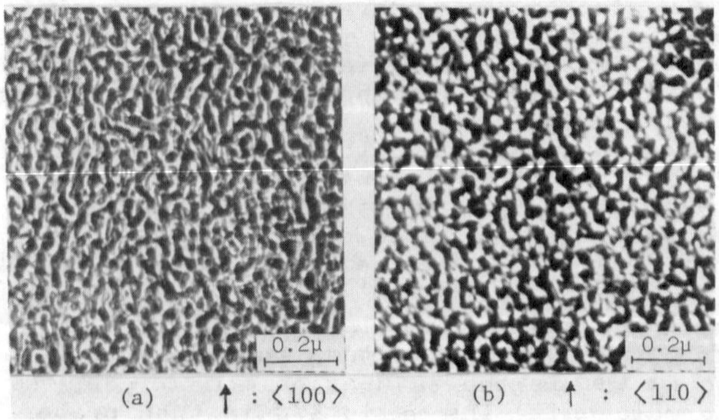


Photo. 2. Electron micrographs of Fe-30wt%Cr-23wt%Co-1wt%Si single crystals tempered at 640°C for 40 min with magnetic field and step-tempered. The plane of foils : (100), direction of magnetics field,
(a) : <100> (b) : <110>

## RESULTS AND DISCUSSION

Photo. 1 (a) and (b) are microstructures of the (100) and (110) plane of the alloys tempered without magnetic field. The modulated structures can be seen in the micrographs. A bright precipitate is an iron-rich phase ($\alpha_1$) and a dark precipitate is a chromium-rich phase ($\alpha_2$). The rod-like precipitate is about 150 A in diameter. However, any noticeable difference between (a) and (b) is not recognized.

Table 1. Magnetic properties of single crystals of an Fe-30wt%Cr-23wt%Co-1wt%Si alloy .

| Specimens | Br (KG) | bHc (Oe) | (BH)max (MGOe) |
|---|---|---|---|
| Isotropic (tempered at 640°C without a magnetic field) | | | |
| <100> | 9.8 | 660 | 2.6 |
| <110> | 9.6 | 630 | 2.4 |
| <111> | 9.4 | 610 | 2.1 |
| Anisotropic | | | |
| <100> | 11.4 | 750 | 4.9 |
| <110> | 11.1 | 770 | 4.5 |
| <111> | 10.9 | 740 | 4.0 |

Thermomagnetic treatment : 640°C, 40 min, H = 2 KOe.
Step-tempering : 600°C, 580°C and 560°C,1 hour each, 540°C, 5 hours.

Photo.2 (a) and (b) are the microstructures of the single crystals tempered with magnetic field along the ⟨100⟩ and ⟨110⟩ axis respectively. The precipitates in both micrographs seem to be elongated zigzag in the direction of the magnetic field shown by the arrow. A similar structure was observed in the micrograph of the ⟨111⟩ single crystals. The alignment of the precipitates is not affected very much by the crystal orientation. Therefore, it can be assumed that the magnetic properties do not depend on the crystal orientation. In fact, there is no practical difference in the magnetic properties as shown in Table 1.

Since the Fe-Cr-Co permanent magnet alloys which are thermomagnetic tempered at 640°C can be still forged, it may be expected that the precipitates are elongated to the direction of forging, and that the magnetic properties are improved with the elongation. Following the thermomagnetic treatment, the single crystals were swaged at room temperature in the same direction of magnetic field, and then were step-tempered. The magnetic properties of them are shown in Fig. 2. They are remarkably increased. The <100> single crystal swaged at 70 % reduction has a good magnetic property ; that is Br = 13 KG, bHc = 980 Oe and (BH)max = 8 MGOe. The microstructures of the swaged <100> single crystal are shown in Photo. 3 (a) and (b). The precipitates are remarkably elongated in the direction of swaging. The increase of the magnetic properties of the swaged alloys may be associated with the elongation of precipitates. Such an effect of swaging was recognized in the polycrystal alloys. As an example, the magnetic properties of the Fe-28wt%Cr-23wt%Co-1wt%Si polycrystal alloys which were swaged at 40 % reduction are given as Br=13.5 KOe, bHc =700 Oe and (BH)max =6 MGOe.

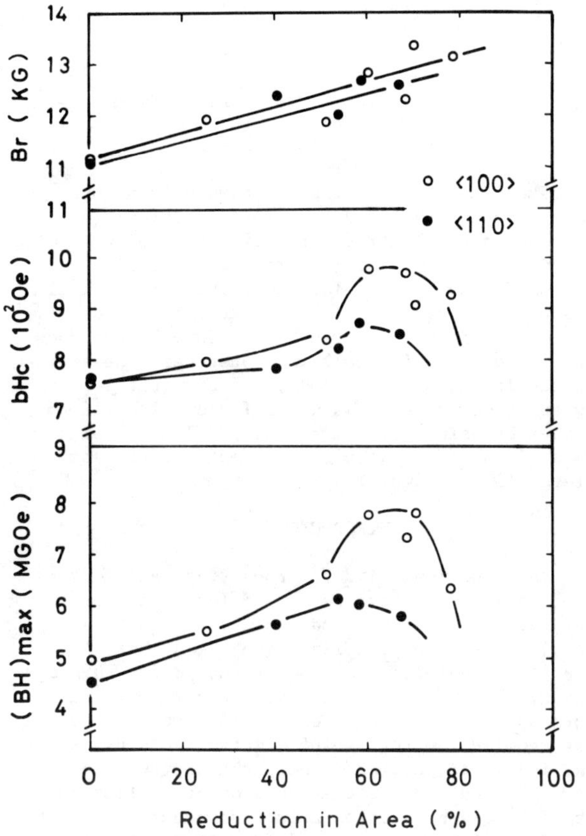


Fig. 2. Effect of swaging on the magnetic properties of Fe-30wt%Cr-23wt%Co-1wt%Si single crystals.

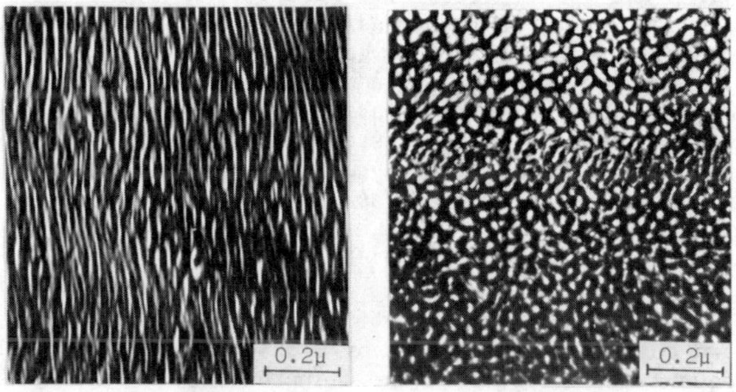


(a) (b)

Photo. 3. Electron micrographs of an Fe-30wt%Cr-23wt%Co-1wt%Si ⟨100⟩ single crystal which was tempered at 640°C for 40 min with a magnetic field, and were swaged to 70% reduction, and were then step-tempered.
(a): parallel , (b): normal to the direction of swaging.

## CONCLUSION

The magnetic properties of Fe-30wt%Cr-23wt%Co-1wt% Si alloys which are tempered in a magnetic field followed by step-tempering do not depend practically on the crystal orientation. However, it is found that the magnetic properties of the <100> single crystal are much improved by swaging after the thermomagnetic treatment.

## REFERENCES

1 H. Kaneko, M. Homma and K. Nakamura, AIP. Conf. Proc., NO.5, 1088(1971).
2 H. Kaneko, M. Homma, K. Nakamura and M. Miura, IEEE Trans. Magnetics, MAG-8, 347(1972).
3 A.Higuchi, M. Kamiya and K. Suzuki, 3rd European Conf. Proc. on Hard Mag. Mat. Amsterdam, 201(1974).
4 H. Kaneko, M. Homma, T. Fukunaga and M. Okada, IEEE Trans. Magnetics, MAG-11, 1440(1975).
5 M. McCaig, IEEE Trans. Magnetics, MAG-11, 1443(1975)

# MAGNETIC INK JET

G. J. Fan
IBM Research Center, Yorktown Heights, N.Y. 10598

## ABSTRACT

The application of a fluid jet used in printing has been proposed in the past. Most of the previous work was based on electrostatic deflection of the jet, i.e., the droplets in the jet are charged and deflected in an electric field. In this talk we shall discuss a technology which uses a magnetic fluid jet and magnetic forces for deflection. This system uses fine particle magnetite colloidal suspension as an ink and deflection is accomplished by passing the jet through an inhomogeneous magnetic field. It will be shown that by using some very basic deflection principles a class of ink jets can be designed and some examples will be shown.

## INTRODUCTION

As the functions of data processing equipment become more powerful, there is a need for an output printer that is of high speed, quiet, and that can handle graphical and pictorial data. A number of non-impact technologies have been studied for this application. Ink jet has been a well known contender for this application. However, until recently the ink jet was confined to the use of electrostatic principles, i.e., the jet, or rather, the droplets in a jet were charged and the charged droplets deflected by the field of an electrostatic transducer to achieve printing[1,2]. In fact several different electrostatic ink jets were developed in the last few years. The field has been surveyed by Kamphoefner[3].

Some of the technical problems of the electrostatic technology are the use of high voltage electronics and the electrostatic interaction of the charged droplets during transit. For the above reasons it seemed of interest to investigate the possibility of an ink jet system using a magnetic ink and a magnetic force for its deflection. This study led to several magnetic ink jet systems and it seems that in certain applications the magnetic ink jet has definite advantages over the electrostatic systems.

## FLUID JET

For a given fluid jet of velocity V, it has been shown that if it is excited by a disturbance at a given frequency f, for some f, the disturbance would grow as it propagates down the jet. When the amplitude of the disturbance becomes comparable to that of the jet diameter it is clear that the jet would break up into droplets. It is well known that for a single frequency of disturbance the jet would break up into a uniform stream of droplets. The flow of the droplets is given by the relationship

$$V = fd, \tag{1}$$

where d is the spacing between the droplets.

The fluid mechanical aspect of the magnetic ink jet is similar to the electrostatic deflected jet. In this work it was found that to achieve a stable stream one can only use very fine colloidal suspension for the ink. The fluid used in this work and its properties are given in a subsequent paper[5]. One major difference between magnetic ink jet and electrostatic ink jet is that the viscosity of the magnetic fluid at low shear rate is much higher than conventional ink. The high viscosity generally results in a slightly higher pressure for a given velocity of the jet. It also has a secondary effect on the dynamics of the wave propagation in the fluid jet. In general the rate of growth of the wave in the jet is slower in a high viscosity medium.

## MAGNETIC INK JET

In a magnetic jet the force on a given droplet is given by:

$$F = M\nabla H, \tag{2}$$

where $\nabla H$ is the applied field gradient and H is the magnetic moment of droplet. Since the magnetic moment has no long term steady state remanence the magnetic moment of the droplet is a function of an applied field, the force on the droplet in the region that M is proportional to H is actually proportional to $H\nabla H$. The magnetization curve of the fluid is given in Fig. 1. To achieve a linear deflection one can either operate the jet in a saturated field region and vary the field gradient or have the jet in a constant gradient and vary the field. The experiments performed use a transducer that gives a field above $H_{sat}$ and deflection is achieved by changing the field gradient. One way to achieve a region of constant field gradient is given by a Wedged Gap as shown in Fig. 2a. The field at the center of the gap is shown in Fig. 2b, in that the field for a limited region can be approximated by

$$H = H_o - a(i)x \tag{3}$$

where a(i) is the field gradient as a function of applied current.

As a droplet passes the transducer, the jet would be deflected by an angle,

$$\theta = \frac{\sigma\, a(i)D}{V^2} \tag{4}$$

where $\sigma$ is the magnetic moment of the ink, D is the path length and V is the stream velocity.

As an example, the typical values for the system are

$$\sigma = 15 \text{ emu/gm}$$
$$a = 10^5 \text{ gauss/cm}$$
$$V = 10^3 \text{ cm/sec}$$

The deflection angle for a given transducer length L in cm is $\theta = 1.5$ D radian and the current used to achieve this gradient in the wedge shape deflector is about 50 ma.

For a typical printing application, the minimum amount of deflection needed would be about 0.1 radian.

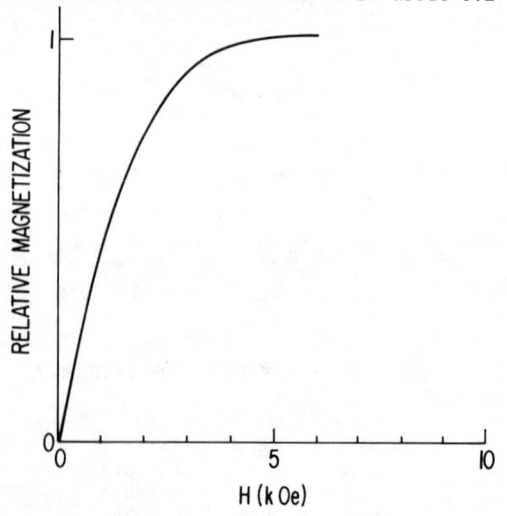

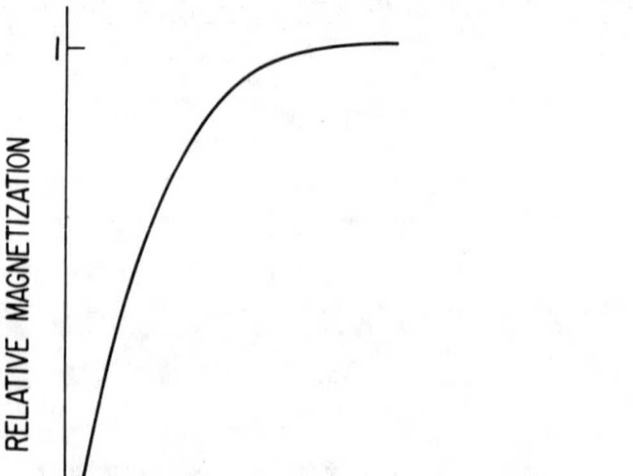


Fig. 1. Magnetization of $Fe_3O_4$ Ferrofluids

For a synchronous jet, to eliminate the interaction problem, the path length in a transducer has to be shorter than d, i.e., D < d. Given this condition and the practical limits of the highest magnetization achievable in ink and the largest magnetic field gradient, the operation of the jet hasto be limited to very low velocity. In fact in such low velocity the jet itself tends to be unstable. An alternative way of operation is given by Johnson[4] where an asynchronous jet is used, i.e., only one droplet is emitted into the transducer for deflection at a given time. The asynchronous jet is also a relatively slow process and it is generally difficult to achieve high printing rate with this method.

Another way to achieve such deflection is given in Fig. 3 where an additional deflector is used. The first deflector is used for the selection of the droplet and the second one generates a vertical raster. The deflection in the x direction $\perp$ to plane of drawing is achieved by paper motion. In this case the selector only has to split the stream in an orthogonal direction so that the unwanted ink can be deflected into a gutter to be recirculated, and more than one droplet can be present in the deflector since it only generates a constant raster. One can use a high velocity synchronous jet and achieve high speed printing with this raster deflection scheme. At an excitation of 25 KHz using a 9 x 13 matrix, printing speed in excess of 100 cps can be achieved. A higher printing rate can be achieved by increasing the drop rate of the jet.

Since deflection of the magnetic ink jet is accomplished by changing of a field gradient, it is possible to put an orthogonal deflector to achieve two dimensional deflection as shown in Fig. 4. In that case three deflectors are employed. The vector mode of printing is of interest because it provides a better utilization of the droplets. For the same jet speed with the vector mode one can obtain in excess of four times the printing speed. It also has some advantages

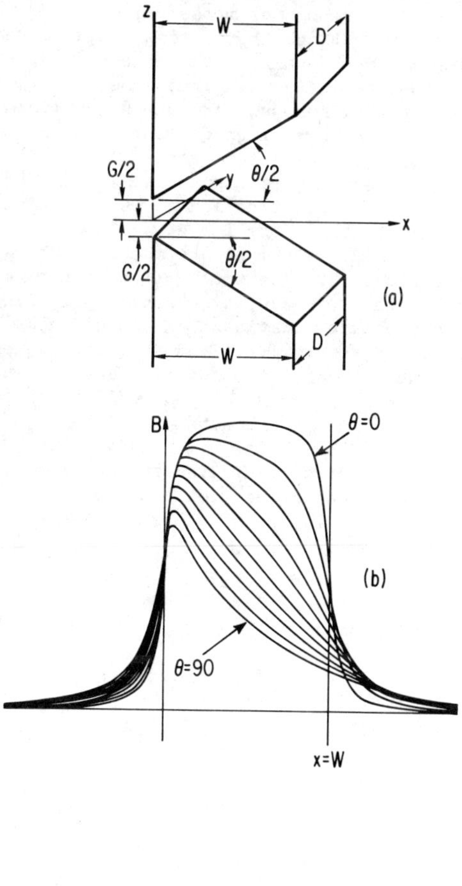


Fig. 2 a-b. Field in the x Direction from a Wedged Pole Piece as a Function of θ.

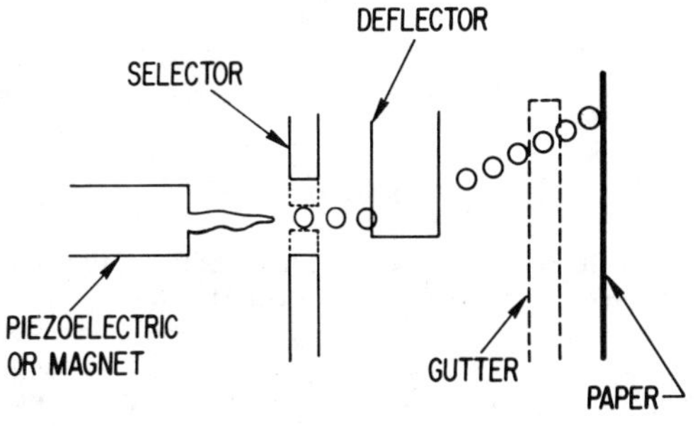


Fig. 3. Raster Magnetic Ink Jet

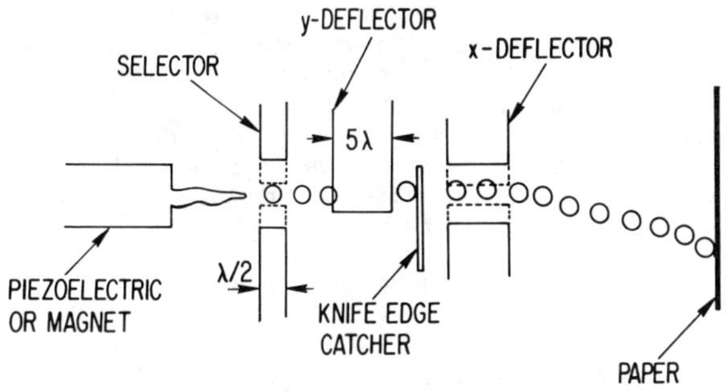


Fig. 4. Vector Magnetic Ink Jet

such that it can trace a smoother curve as compared with the raster mode where quantization is inevitable. The disadvantage of this method is that it needs more mechanical as well as electronic components.

## CONCLUSIONS

Two magnetic ink jet printing schemes are discussed in this paper. It appears that these methods can achieve high printing speeds that are consistent with some of the requirements of high speed input/output terminals.

## ACKNOWLEDGMENT

Most of the work reported here was done in collaboration with Dr. R. Toupin and without his effort this work would have been impossible. The author is grateful to Dr. Z. Kovac and Dr. C. Sambucetti who made the fluid. Thanks are also due to Dr. D. Lo, Mr. H. Seitz and Mr. J. Mitchell for their work on the vector jet and to Dr. D. Quales for field calculation.

## REFERENCES

1. R. G. Sweet, "High Frequency Recording with Electrostatically Deflected Ink Jets," Rev. Sci. Inst., Vol. 36, pp. 131-136, Jan. 1965.
2. A. M. Lewis and A. D. Brown, "Electrically Operated Character Printer," U.S. Patent 3,298,630, 1/10/67.
3. F. J. Kanphoefner, "Ink Jet Printing," IEEE Trans. on Electronic Devices, Vol. 19, No. 4, Apr. 1972.
4. C. E. Johnson, "Magnetic Ink Jet," U.S. Patent 3,510,878,5/19/70.
5. Z. Kovac and C. Sambucetti, "Preparation of Water Based Magnetic Ink," same Conference.

# CROSSTIE MEMORY SIMPLIFIED BY THE USE OF SERRATED STRIPS*

L. J. Schwee, H. R. Irons, and W. E. Anderson
Naval Surface Weapons Center, White Oak Laboratory
Silver Spring, Maryland 20910

## ABSTRACT

Serrated strips of Permalloy about 350 Å thick and 10μm wide are used to position domain walls, crossties, and Bloch lines, and to simplify propagation, detection, and fabrication. The serrated strips are etched so that they align along the easy axis of the film. The saw-teeth on each edge of the strip give the strip a cuneated or wedge-shaped appearance. After a magnetic field is applied normal to the strip length, the magnetization along each edge rotates back to the nearest direction which is parallel to the edge. Two domains are thus formed with a domain wall along the center of the strip. The magnetization along the edges lies parallel to the edges to minimize stray fields. The edge effects are large compared to the induced anisotropy of the film. The wall formed at the center of the strip can be positive or negative Neel or a crosstie wall. Crossties form at the necks of the strip. Bloch lines locate themselves in potential wells between the necks. These properties simplify propagation and detection. The crosstie memory can be fabricated on silicon wafers or chips with integrated drive electronics, detector amplification, and decoding.

## INTRODUCTION

In previous crosstie-Bloch line studies, walls were placed on Permalloy films about 350 Å thick by using special coils[1], or special deposition fields[2], or special annealing procedures[3,4]. Using an annealing technique[4], walls could be placed to within ±1μm over a length of 2.5 cm. Subsequent alignment with such walls could be accomplished using the Kerr magneto-optic effect. To simplify procedures so that conventional mask aligners could be used, the effects of etched boundaries were investigated and the results are here reported.

## SERRATED STRIPS

As a first experiment, a mask was made using

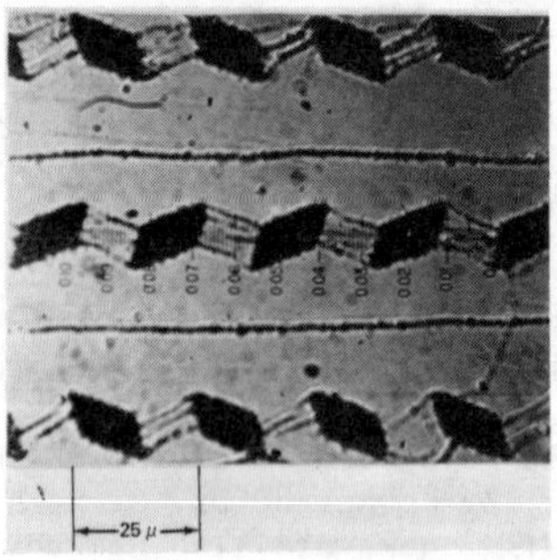


Figure 1 - Wall Placement Using Parallelograms

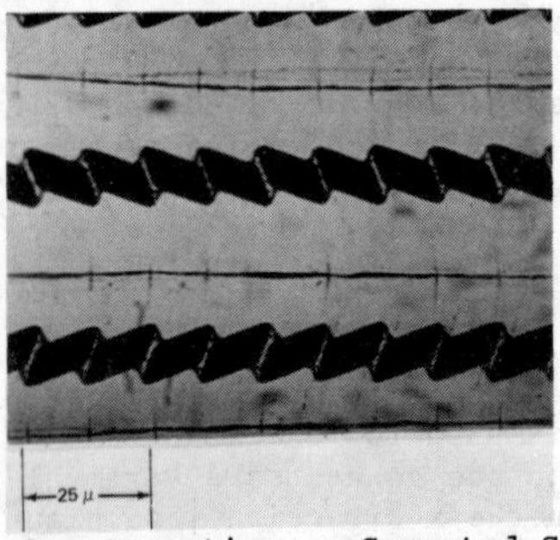


Figure 2 - Crossties on Serrated Strips

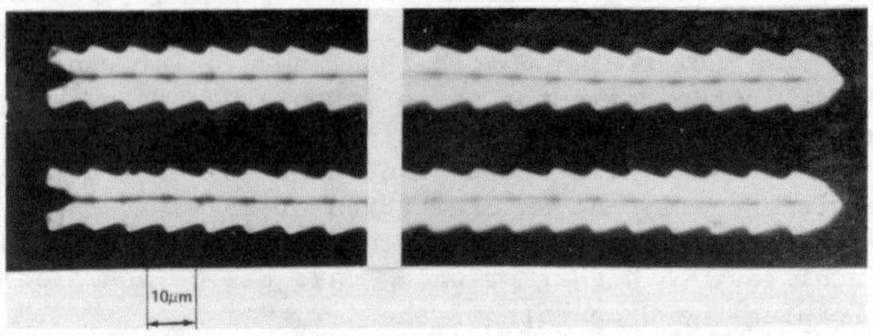


Figure 3 - Neel Walls on Serrated Strips

parallelograms and a film was etched as shown in Figure 1. As expected, after a hard axis field was applied (up in the figure), walls would form as shown. The tiny reversal domains between parallelograms were also expected[5]. This type of mask was generated because it was expected that easy propagation would result. However, when the field was reversed to form crossties the walls became impossibly crooked. It appeared that symmetrical boundaries would be desirable to avoid crooked walls. Consequently, a double exposure was made using the same mask. The serrated strips which result are shown in Figure 2. The serrated strips are aligned along the easy axis of the film and the walls form after a hard axis field is applied. A fast magnetic pulse (1-10nsec) along the opposite hard axis direction forms crossties provided the film thickness is between about 300 and 450 Å. The crossties form at the necks of the strip as shown in Figure 2. The width of the strip neck is 25μm. The Bloch lines are in potential wells in between the crosstie locations. A second mask was generated with several widths and sawtooth angles. An example of one of these is shown in Figure 3. Here the width is 10μm at the neck. The wall terminates at the beginning and end of the strip as desired. The width of the strip has a large influence on the properties of the crossties. The static properties are compared with the properties observed in a large unetched film in Table I. All these films were 350 Å thick.

Table I

STATIC MAGNETIC PROPERTY COMPARISON

| Type of Film | Wall Placement Field | Crosstie Nucleate Field | Bloch Line Movement Field | Crosstie Annihilate Field |
|---|---|---|---|---|
| Unetched film | 4 Oe | 0.8 Oe | 0.1 Oe | 1.6 Oe |
| Serrated, 25μm Width | 80 Oe | 2.3 Oe | 1.0 Oe | 6.5 Oe |
| Serrated, 10μm Width | ~500 Oe | 7.5 Oe | 6.2 Oe | 19.0 Oe |

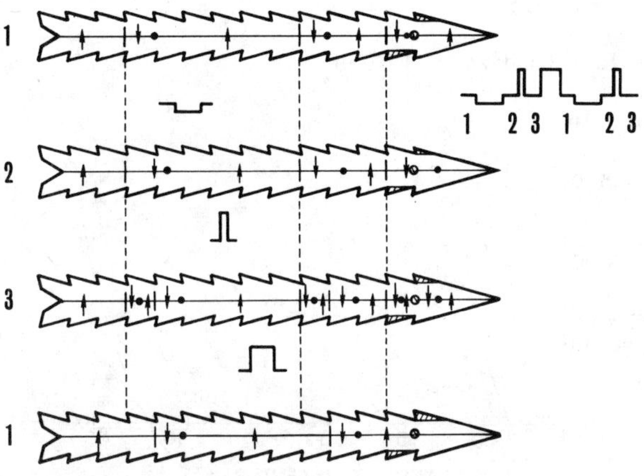


Figure 4 - Single Conductor Propagation

In the unetched film which can be considered an infinite plane sheet, the Bloch line has a coercive force of about 0.1 Oe. In the serrated strips the coercive force remains the same but the potential wells and the demagnetizing effects of the strip increase the field required to force a Bloch line into a neighboring well or along the wall. The magnitude of the field required to move the Bloch line in different directions differs slightly as does the magnitude required to nucleate and annihilate crosstie-Bloch line pairs when positive or negative fields are used. These differences are small and average values are shown in Table I. For easy wall placement the serrations should be shaped as in Figure 2 rather than as in Figure 3. An easy axis field up to 2 Oe can be applied before the wall will move. This corresponds to the normal coercive force of the wall.

## SINGLE CONDUCTOR PROPAGATION

Suppose a single conductor (not shown) is available above and along the strip shown in Figure 4 so that the illustrated magnetic fields can be applied (up and down in the figure). The Bloch lines in Step 1 of Figure 4 are displaced from their normal equilibrium positions in their potential wells by the repulsive force of the neighboring crossties. The repulsive force is about 0.75 Oe if the Bloch line is 3μm away from the crosstie, about 0.5 Oe if it is 4μm away, and 0.25 Oe if it is 5μm away[6]. A negative magnetic pulse large enough to move the Bloch line over the potential hump is applied. The Bloch line then stops in the next potential well as shown in Step 2 of Figure 4. Next a positive short pulse of amplitude large enough to generate a crosstie-Bloch line pair is applied[7]. The new crosstie now displaces the lead Bloch line because of its repulsive force. The resulting situation is shown in Step 3 of Figure 4. Next an annihilation pulse is applied. This pulse is large enough in amplitude and long enough in duration to annihilate the trailing crosstie-Bloch line pair but not the lead crosstie-Bloch line pair.

Several experiments to confirm this method of propagation were done on the 25μm wide strips but the 12.5μm distance between serrations was too long for consistent selective annihilation. Future masks will be designed to improve these operating margins. By using two conductors a scheme which is similar to a previously demonstrated propagation method[8] can be designed.

## MAGNETORESISTANCE DETECTION

Referring to Figure 5, a method of detection is illustrated which takes advantage of the high sensitivity angle of 45°. Here the cross-hatched areas are

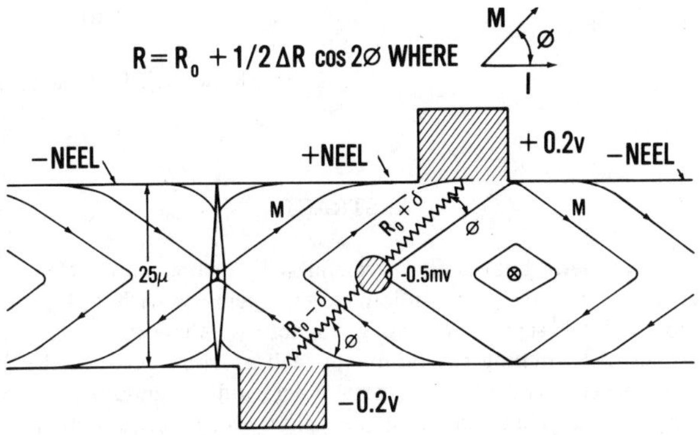


Figure 5 - Magnetoresistance Detection

locations where three separate conductors make contact with the Permalloy. If, as illustrated, the large connections are held at plus and minus 0.2 V, a current will flow in the Permalloy along the resistance path. The center connection is the signal contact. The magnetization directions are exaggerated to make the concept clear. Notice that between the signal contact and the +0.2 V contact that the magnetization is almost parallel to the current and between the signal contact and the -0.2 V contact the magnetization is almost perpendicular to the current. Thus the upper resistance arm has higher resistance than the lower resistance arm. If a bit were not present at the detector a negative Neel wall would be present and instead of -0.5 mV at the center contact, +0.5 mV would be seen. Thus a 1 mV signal difference can be expected[9]. The resistance between the outer contacts is normally about 22 ohms so that the Johnson noise level is ~8μ V. About a 40 dB signal to noise ratio can be expected.

## PERMALLOY ON SILICON

Experiments indicate that Permalloy can be satisfactorily deposited on semiconductor grade silicon. To produce films with zero magnetostriction the iron content of the Ni-Fe source must be increased by 0.6% compared with the 82.8% Ni - 17.2% Fe source normally used for glass substrates. By fabricating the crosstie memory on silicon chips or wafers the amplifiers for the detectors can be fabricated on the spot so that most likely only a fraction of a millivolt will be required at the detector output. Also the drive electronics and decoding can be integrated so that the memory will appear to the user as a convenient semiconductor memory except that it will also be nonvolatile.

## REFERENCES

*This work was supported by the Naval Air Systems Command, Washington, D. C. 20362.

1. L. J. Schwee, AIP Conf. Proc. 10, 996-1000 (1973).
2. L. J. Schwee, H. R. Irons, and W. E. Anderson, AIP Conf. Proc. 18, 1383-1387 (1974).
3. D. S. Lo and M. C. Paul, IEEE Trans. Magn., Mag-10, 1079-1081 (1974).
4. L. J. Schwee, H. R. Irons, W. E. Anderson, R. S. Sery, and K. P. Scharnhorst, Naval Ordnance Laboratory Technical Report, NOLTR 74-176 (1974).
5. L. J. Schwee, AIP Conf. Proc. 10, 378-382 (1973).
6. D. S. Lo and M. C. Paul, Univac, St. Paul, MN (private communication).
7. R. S. Sery, IEEE Trans. Magn., Mag-11, 29-30 (1975).
8. L. J. Schwee, H. R. Irons, and W. E. Anderson, IEEE Trans. Magn., Mag-10, 564-566 (1974).
9. L. J. Schwee, H. R. Irons, W. E. Anderson, R. S. Sery, and O. J. Van Sant, Naval Surface Weapons Center Technical Report, NSWC/WOL/TR 75-167 (1975).

# BUBBLES AS LATRIX ELEMENTS

M. M. Hanson, F. G. Hewitt, A. D. Kaske, R. E. Lund, and E. J. Torok
Sperry Univac
St. Paul, Minnesota 55165

## ABSTRACT

A magnetic bubble film overlaid with orthogonal slotted drive lines can function as a random access memory with optical read and/or optical signal processing capability. Memory operation is obtained by switching the bubble in each aperture between the four stable corner positions. The write operation is random access bit organized. The read operation is facilitated by the use of a mask and is word organized. A one dimensional readout array can be implemented by using a cylindrical lens for focusing all word lines on a linear detector array. Such a system may be useful for realizing a latrix.

## INTRODUCTION

An optical page composer is a matrix of storing cells, each cell acting as a light valve which passes light or shuts off light according to the state of the cell. In an optical memory as envisioned by Rajchman[1], writing is done by storing in a medium (e.g. film thermoplastic, MnBi) a holograph image of the light pattern emerging from the composer. When the composer is equipped with an array of photosensors for reading out an image of an area of the storage medium, and a random access memory, Rajchman called the resulting combination a latrix[1]. The following describes how a bubble film can be used as a random access memory with capability for optical signal processing and as such may be useful for implementing a latrix.

## DESCRIPTION

When polarized light is passed through a bubble supporting crystal, the polarization of the light is rotated by the Faraday effect. An analyzer can be placed to attenuate all the transmitted light except that which has passed through the bubbles. Thus, when laser light is passed through the crystal and analyzer, the bubbles appear as bright spots of light. Such a crystal may be used as a latrix.

Consider two orthogonal sets of slotted strip lines as shown in figure 1. These lines are deposited on a crystal capable of supporting bubbles. At each square aperture formed by the intersection of two slots is a bubble. As will be explained below, the bubble can be moved to any one of the four corners of the aperture by appropriate currents in the strip lines. The information in each aperture is stored by the position of the bubble.

Figure 2 is a plot of the normal field from a slotted line. The current, of course, goes in the same direction on each side of the slot. The gradient of the normal field is the quantity that causes a bubble to move. When the current is positive, as in figure 2, the bubble goes from left to right. When the current is negative, the bubble goes from right to left. The bubble stops at a point where the slope (or gradient) becomes small enough so that the net force on the bubble is insufficient to overcome the coercivity. The bubble does not stop at the bottom of the hill, but near the bottom.

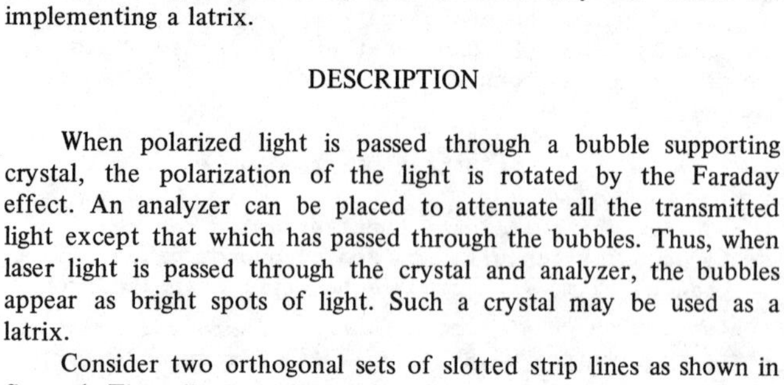

Fig. 1: Two orthogonal sets of slotted lines deposited on the surface of a bubble crystal (not shown).

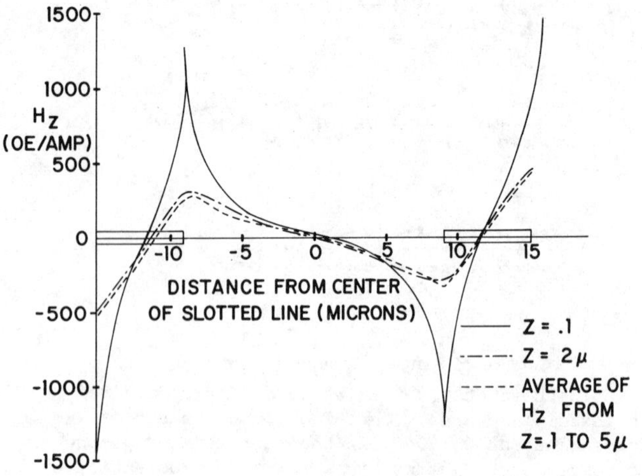


Fig. 2: Normal field, $H_Z$, from current in a slotted strip line. The line is 30 u wide and the slot is 18 u wide. The three curves correspond to the field in a plane z = 0.1 u from the line, z = 2 u from the line, and the average through a 4.9 u thick film separated by 0.1 u from the strip line.

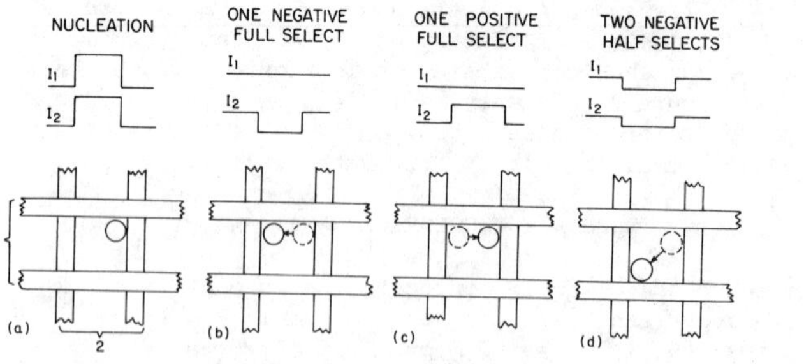


Fig. 3: Nucleation and manipulation of a bubble by two intersecting slotted lines.

Next, consider the case where there are two slotted strip lines intersecting at right angles as in figure 3. A bubble can be nucleated in the interior of the intersection by a coincidence of currents in both lines. Once nucleated, a bubble can be moved back and forth from side to side by a full select current in one of the lines. Or, on the other hand, a bubble can be moved back and forth along a diagonal by two half select currents in both lines. The amplitude of these half select currents is 0.707 of the full select current.

Let three of the four possible bubble positions be covered with a deposited opaque shield, while the upper right hand corner position in each case remains uncovered. The bubbles can be operated as bit organized coincident current random access memory by using the half select current mode only. The bubbles that are in the upper right hand uncovered corners are digital "ones" and the bubbles in covered corners are digital "zeros". A single bubble in an array can be switched along a diagonal by two half select currents without disturbing any of its neighbors. If an array of photodiodes is placed behind the bubble array, the presence of light on a photodiode indicates a digital "one", and absence, a "zero". However the cost of the sensing array can be greatly decreased if word readout is used. If the device has $10^6$ bubbles, one requires not a square array of $10^6$ photodiodes, but a linear array of $10^3$. This can be done as follows:

Let us note that so far we have talked about using only two of the four possible bubble locations. We have not considered using the upper left hand corner or the lower right. Suppose, after each write,

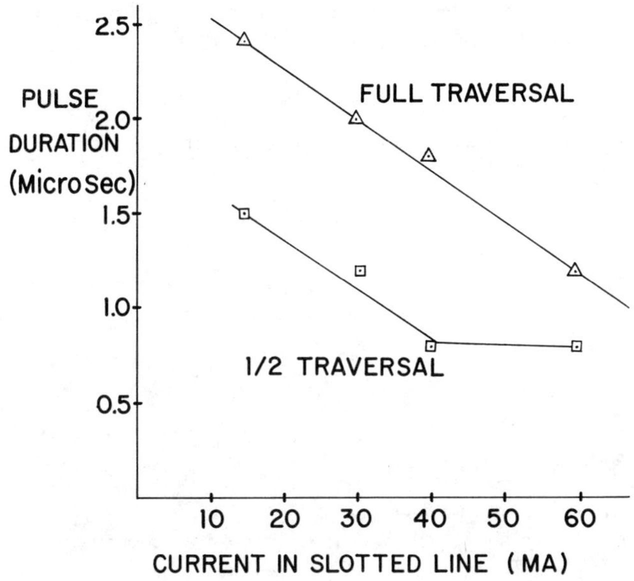


Fig. 4: Pulse duration required to obtain the desired bubble traversal as a function of the drive current.

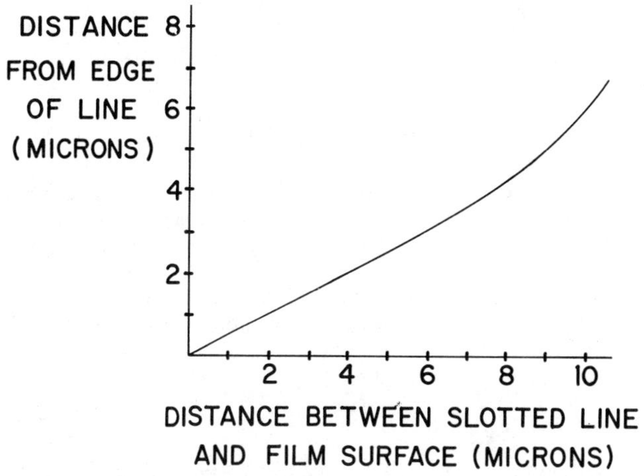


Fig. 5: Distance between the edge of the slot and the point where the gradient of the z component of the field is zero as a function of the separation between conductor and film.

the word line is given a full select current, not a half select current, but a current big enough to move a bubble already in the "one" state (in the upper right hand corner) to the upper left hand corner. Any bubble in the lower left hand corner is not moved. This shields the bubble. If this procedure is always followed, no bubbles are exposed. To read a word, the word line is given a full select current. All the "one" bubbles in that word are moved from the upper left across to the upper right and become exposed. All "zero" bubbles in that word are moved from the lower left to the lower right and remain under the shield. Thus only the "one" bubbles of that single word line are exposed. If the whole bubble array is illuminated with parallel light and a cylindrical lens is placed a focal length away from a string of photodiode sensors, the information in the word is read out. The function of the cylindrical lens is to focus all word lines on the same string of photodiodes. After interrogation, the word line is given another full-select current of opposite polarity to move the bubbles back behind their shield. Another read using a different word line can then begin.

## OPTICAL EFFICIENCY

The optical efficiency of the page composer can be greatly increased by combining it with a holo-lens which focuses incoming light on the bubbles. The efficiency is then the product of the efficiency of the holo-lens and the efficiency of the bubble film.

The efficiency of bleached holograms can be 60%, while that of holograms in dichromated gelatin approaches 100%. The efficiency of ordinary non-bismuth substituted garnets in the visible is only a tenth of one percent; but with the substitution of one bismuth atom per formula, the efficiency can be increased to about 10%. There is a decrease in efficiency when the light is incident at large angles from the normal; however, the large index of refraction of the garnets (n=2.2) causes the path of the ray in the garnet to be nearer the normal. For example a light ray incident at 30° from the normal passes at 13° to the normal inside the film.

The above considerations show that the total efficiency of the page composer for visible light will be slightly less than 10%. This is sufficient for certain memory materials such as thermoplastics (which have sensitivities of 10 to 100 microjoules per square centimeter) but not for others, e.g. MnBi.

The page composer can also be operated in a reflective mode with a mirror on one side of the bubble film and with strip lines on the other side of the mirror.

## EXPERIMENTAL APPROACH

Experimental magneto-optic page composer circuitry was fabricated on YEuTmIG material, which provides nominal bubble diameters of six microns. This material was chosen because it was easily available, in order to investigate switching operation. A bismuth substituted garnet is required for high optical efficiency. Prior to fabrication of the circuitry a stress relief layer of 4000 Å of silicon dioxide was rf sputtered onto the YEuTmIG surface. The circuitry consists of two layers of orthogonal gold conductor patterns separated by a $SiO_2$ insulator layer. Typical conductor patterns are as shown in figure 1 with 18 microns between lines. The conductor patterns were formed by the vapor deposition of 400 Å Cr and 5000 Å Au, spin coating of 8000 Å Shipley AZ1350 photoresist, exposure of the conductor pattern via a Kasper mask aligner, and ion milling of the gold lines with an Argon ion beam current of 100 microamps. Silicon dioxide, rf sputtered to a thickness of 7000 Å, is used as the insulator between the orthogonal drive circuitry.

## EXPERIMENTAL RESULTS

A Faraday effect microscope was used to test the operation of an experimental device. We found that all the operations shown in figure 3 could be performed with pulse amplitudes of 10 mA or higher.

Another experiment was performed to determine the required duration of the bubble propagation pulse to obtain desired bubble movement at various drive currents. The rise time, fall time, and pulse amplitude of the propagation pulse were held constant. A slotted drive line 30 microns wide with a 18 micron slot was used. Full traversal and half traversal designates bubble movement of two and one bubble diameters respectively. Figure 4 shows that for a specific required traversal a near linear dependence of pulse duration on drive current exists for currents from 0 to 60 mA.

## PROBLEMS AND SOLUTIONS

There are certain problems in the device as outlined above: (1) the bubble may be partially hidden by the strip line; (2) the bubble may escape under a strip line; and (3) the bubble may become stuck and not move unless the pulse amplitude is increased.

Problem 1 can be ameliorated by increasing the distance between the line and the film surface; this causes the maximum and minimum of the field to move away from the strip line edge as shown in figure 5. Note also that the bubble does not stop at the minimum field, but at the field where the gradient becomes too small to overcome the coercivity. This fact also helps to position the bubble farther from the strip line where it will not be hidden; unfortunately, it also contributes to the problem of bubble sticking. If the bubble stops at the point where the field gradient is too small to overcome the coercivity, then when the polarity of the pulse is reversed in order to shift the bubble back, the gradient is of opposite sign, but still too small to overcome the coercivity. There are two means of overcoming this difficulty: shaping the pulse into a quasi-sawtooth so that its amplitude decreases in time, and coating the strip lines with a permanent magnet layer (e.g. cobalt) that repels the bubble. In either case the field that starts the bubble is greater than the field when the bubble stops, so the difficulty is overcome.

## SYSTEM ARCHITECTURE DISCUSSION

The proper place in a memory hierarchy for a random access bubble latrix with $10^6$ bits and access time of approximately two microseconds is not as a high speech cache (which requires one tenth the number of bits and one twentieth the access time) but as a second level memory in a three level hierarchy. This consists of a semiconductor cache, the latrix, and the optical mass store.

The work on the bubble latrix is in an early phase; however, the present work indicates the feasibility of the device.

## REFERENCES

1. Rajchman, J. A. "Promise of Optical Memories" J. Appl. Phys. 41, 1376 (1970).

# MAGNETIC GAS SENSOR

Martin Rayl, Peter J. Wojtowicz and Harold D. Hanson
RCA Laboratories, Princeton, N. J. 08540

ABSTRACT

The sensor consists of a porous core of some magnetic material with an induction coil wrapped around it. Exposure of the core to oxidizing or reducing gases causes chemical changes to occur. These result in changes in the core's magnetic properties which are observed as changes in the inductance of the coil. Measurements of the time rate of change of the inductance are used to determine the concentration of the active species in the gas stream. Among the materials which have been used to demonstrate these principles are magnetite to detect $O_2$ or other oxidizing gases and $\alpha$-$Fe_2O_3$ for CO, $H_2$ and other reducing gases.

One advantage of the magnetic gas sensor over other solid-state gas sensors is the fact that the coil can be separated physically from the core and gas stream by an insulating coil form, making it feasible to use this device at elevated temperatures and in corrosive environments.

Considerable work has been done in the past several years to develop pollution detectors, especially gas sensors. One of the areas of great interest is detection of oxygen, carbon monoxide, nitrogen oxides, or hydrocarbons at levels characteristic of automotive exhaust. At present, gas sensors fall into three classes. The electro-optic (E-O) types are typically large, delicate instruments, many of which are beyond the development stage and are commercially available, and which can be made to respond to specific gases in the presence of others. Liquid gas detectors of many types are also commercially available, can be smaller than E-O types, and can also be quite specific. Usually, liquid types operate only very near room temperature and are subject to fluid loss and poisoning problems. Solid state gas detectors promise exciting improvements over the E-O and liquid types in the areas of size and ruggedness, as well as relative insensitivity to temperature and other details of the ambient. However, solid state sensors tend to be non-specific in their response to gases, being primarily oxidation-reduction state response devices.

One kind of solid state device is an electrical resistor made of one or several metallic oxides, usually heated about 200 degrees above room temperature, which changes its resistance in the presence of a reducing gas, either by oxidizing the gas or by adsorption changing the conduction mechanism in the sensor. Another is a high temperature electrolytic cell based on a solid electrolyte in which the mobile ion or the corresponding vacancy is the ion of the gas which is to be detected. Changing the concentration of this gas changes the EMF of the cell. In addition to the specificity problems, these detectors are used with difficulty at elevated temperatures because of problems associated with thermally cycling the electrical contacts to the sensor to temperatures above the melting point of solder.

One approach to solving the contact problem is to make the sensor a magnetic material which changes its magnetization by contact and reaction with the gas to be detected. The magnetization change is caused by varying the oxidation state of some of the magnetic ions by contact and reaction with an oxidizing ($O_2$) or reducing ($H_2$ or CO) gas. The sensor can be the core of an inductor and the change of inductance or quality factor can be the useful signal from the detector. Since it is straightforward to make a coil and coil form which can by cycled easily between 20 C and 600 C, such a device would not have contact problems.

Within the context of the magnetic gas sensor idea, one materials system stands out as the leading candidate for use as the core material - the oxides of the transition metal ions and especially the oxides of iron. It is well-known that the various oxides of iron can be prepared by controlling the $CO/CO_2$ ratio in the gas in which iron is oxidized or, say, $\alpha$-$Fe_2O_3$ is heated at temperatures above 250 C. It should be possible, for example, to switch reversibly from hematite to magnetite with a large accompanying change in magnetization by adding CO or $O_2$ to an inert ambient gas at about 300 C.

We have made devices in which the core material was either magnetite ($Fe_3O_4$) or hematite ($\alpha$-$Fe_2O_3$). A single layer $8\mu H$ coil was made from .010" diameter platinum wire wound on a ceramic form. The inner diameter of the form was 1.0 cm; the length of the coil was 2.2 cm. Cores were 0.95 cm diameter by 1.2 cm long cylinders cold pressed from powder material at about 1000 psi. Inductance measurements were made at 10 MHz with a Boonton Radio Corp. Type 250-A RX Meter in a container in which the ambient could be controlled. A flowing gas apparatus was used in which calibrated amounts of $O_2$, CO, or $H_2$ could be added to the carrier gas, dry nitrogen. Data were obtained at various controlled temperatures between 350 C and 600 C. Figure 1 shows the data for a series of experiments done at 550 C in which a core of $\alpha$-$Fe_2O_3$ was used to detect CO gas at the concentrations shown. After each experiment the core was flushed with $O_2$ gas until the inductance returned to its initial value, characteristic of the starting $\alpha$-$Fe_2O_3$. For this core, the total change in inductance varied between 34 and 43%. The time over which the inductance change occurs is of the order of tens of minutes and varies with the concentration of the active gas. Additional experiments performed on cores of different geometry demonstrated that the characteristic time is inversely proportional to the surface area of the core.

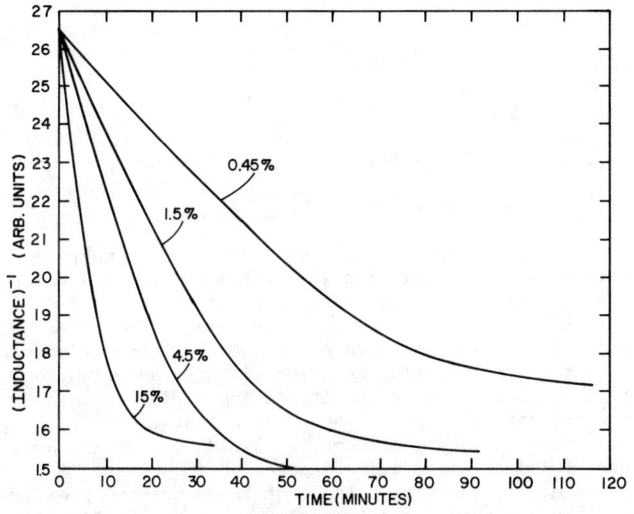


Figure 1. Curves showing the time dependence of the inductance of a coil containing an $\alpha$-$Fe_2O_3$ core for various concentrations of CO in dry nitrogen. The temperature is 550 C in all cases.

As shown in Fig. 1, the active gas concentration is not simply obtained from the total inductance change. Instead, it is determined from the initial slope of the curve of the time dependence of the inductance. Figure 2 is a log-log plot showing that the initial slope is proportional to some power of the active gas concentration. This observed relationship between the dynamics of the inductance change and the concentration of the active gas species is what makes this type of device appear to have promise as a gas sensor.

The relationship determined in Fig. 2 is

$$-\frac{dL^{-1}}{dt} \propto c^n, \qquad (1)$$

where L is the inductance, c is the concentration of the active species, and n is the slope of the line. The CO data are those of Fig. 1; the $H_2$ data were obtained from similar experiments done using another core of $\alpha$-$Fe_2O_3$. For CO, $n \approx 1/2$, while for $H_2$, $n \approx 2/3$. Assuming that the observed $-dL^{-1}/dt$ is proportional to the rate at which the oxidation-reduction reactions occur in the core material, equation (1) shows that the reaction rate is proportional to $c^n$.

According to the empirical Freundlich isotherm[1], the variation of the coverage of a solid surface by an active species with concentration is sublinear. Identifying our observed (sublinear) power laws with surface coverage, we conclude that the reaction rate of this system is governed by the surface coverage of the core. This result is consistent with those found in other chemical systems undergoing heterogeneous reactions. In each case[1], the observation of fractional order kinetics indicates that the surface coverage governs the reaction rate.

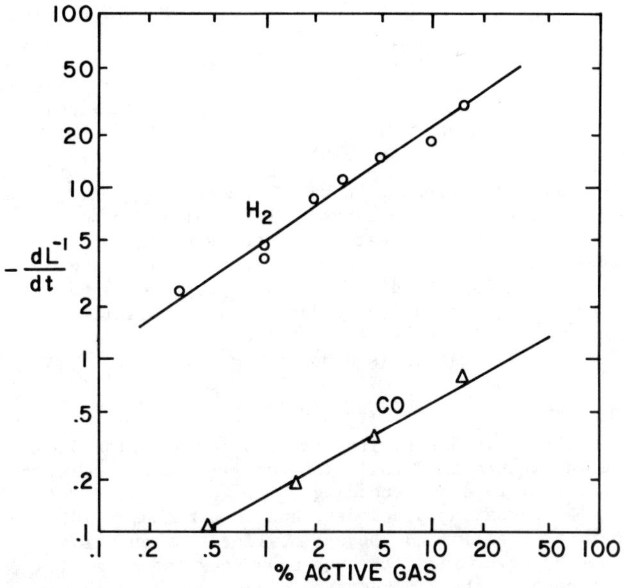


Figure 2. Initial sensor response as a function of gas concentration. The lower curve was determined from the data of Fig. 1; the upper curve represents a similar experiment using $H_2$ gas and another $\alpha$-$Fe_2O_3$ core.

The prototype device which we have described above demonstrates that a gas sensor can be based on changes in the oxidation-reduction state of a magnetic material. The advantages of this type of sensor are that the coil and contacts can be kept separate from high temperatures and corrosive gas environments. Problems with the device include the need for periodic rejuvenation of the initial inductance, variations of inductance and resistance with temperature, and the slow response. For the particular sensor which we made, the characteristic times would have to be improved at least by two orders of magnitude for the device to be practical. Unfortunately, the demonstrated inverse linear relationship between response time and the surface area of these cores indicates that changes in geometry are unlikely to provide the required degree of improvement. What is required is a different solid system, in which the oxidation-reduction reaction is a bulk rather than surface process.

REFERENCES


1. S. W. Benson, The Foundations of Chemical Kinetics, (McGraw-Hill Book Co., Inc., New York, 1960), pp. 25, 621-626.

# MAGNETIC ORIENTATION OF DISC MEDIA

Ronald D. Weiss
Nashua Corporation
Nashua, New Hampshire 03060

ABSTRACT

Magnetic orientation ratios in excess of two have been achieved in discs. These values are comparable to, if not greater than those reported for tapes. Orientation is shown to be related to magnetizing fields and dispersion quality. The effects of disc orientation on recording characteristics such as signal amplitude and noise are reported. Off-axis orientation is discussed also.

## INTRODUCTION

The derived benefits from tape orientation have been described in detail elsewhere[1,2] and are equally beneficial to disc recording.

Disc coating processes, and in particular spin-coating, has limited the use of linear orientation. It has been determined, through squareness measurements, that non-oriented spin-coated discs possess a preferential radial orientation where the orientation ratio is approximately one. The role of disc orientation is to provide the alignment of acicular magnetic particles such that their longitudinal axis coincides with the circumferential recording direction.

## EXPERIMENTAL PROCEDURE AND MEASUREMENTS

Aluminum substrates were spin-coated with magnetic dispersion consisting of finely dispersed particles of gamma ferric oxide in an epoxy binder. Subsequent to coating, orientation, and curing, discs were polished and evaluated. The discs were then die-punched for corresponding orientation data in either a BH tester[3] or vibrating sample magnetometer.

For disc media the orientation ratio is defined by the squareness in the circumferential direction ($S_c$) divided by the squareness in the radial direction ($S_r$).

## FIELD DEPENDENCE OF THE ORIENTATION

Linear tape orientation can be achieved from mechanical and magnetic processes[4,5] associated with longitudinal coating. In disc coating, where centrifugal forces result, the natural state of the magnetic particles corresponds to a radial direction or minimal orientation (orientation ratio $\approx 1$).

The magnitude of disc orientation, in the circumferential direction, was found to be greater than that reported for tapes in the longitudinal direction.[5,6] Disc orientation as a function of the orientation field is shown in Figure 1. Orientation saturation occurs at ~1000 oersteds--which is consistent with similar iron oxides.[6,7]

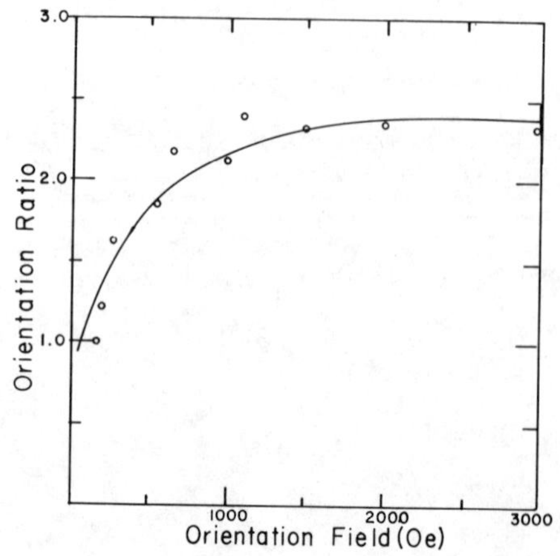


Figure 1 - Disc Orientation Ratio as a Function of Orientation Field. The dispersion viscosity was 1400 centipoise.

## TIME DEPENDENCE OF THE ORIENTATION

The time dependent behavior of acicular magnetic particles in a dispersion was determined to be a function of viscosity[8] and particle-to-particle interaction.[9] It has been determined that magnetic dispersions in the viscosity range of 500 to 2000 centipoise, have a 1-2 millisecond reaction time to the orientation field. Experiments with alternating fields up to 500 Hz have confirmed this responsive orientation.

High frequency alternating fields ($>500$ Hz) have been found to cause oxide separation in magnetic dispersions. A suitable balance of frequency and rheology must be achieved for alternating or pulsed field orientation.

The use of alternating polarity fields as well as pulsed single polarity fields for disc orientation suggested certain intrinsic limitations. In a previous work,[10] a deterioration of the dispersion orientation was suggested for a secondary (opposing) field, after the primary orientation occurred. This effect has not been duplicated for low frequency field pulsing (~1 Hz) but the dispersion sensitivity to a 'critical field strength' did confirm the original 'Dorf Effect'. Two significant differences were noted from the original work. First, the presence of the Dorf Effect was strongly influenced by field application times. Second, the pulsed field application times used in this study did not destroy the orientation nor reduce the maximum orientation with the secondary field (see Figure 2).

## DISPERSION DEPENDENCE OF THE ORIENTATION

The influence of dispersion rheology on the orientation has been described.[7,9] The orientation ratio was found to be approximately linear in the viscosity range of 500-2000 centipoise. The data suggests that high viscosities are partially responsible for the orientation ratio achievements in disc media.

It has been shown that orientation ratios can be maintained even with poor quality dispersions--specifically, when particle agglomeration takes place. The effect of agglomeration on orientation is shown in Figure 3. Agglomeration is represented by changes in amplitude, viscosity, and coated disc surface finish. Experiments have demonstrated that as magnetic particles form aggre-

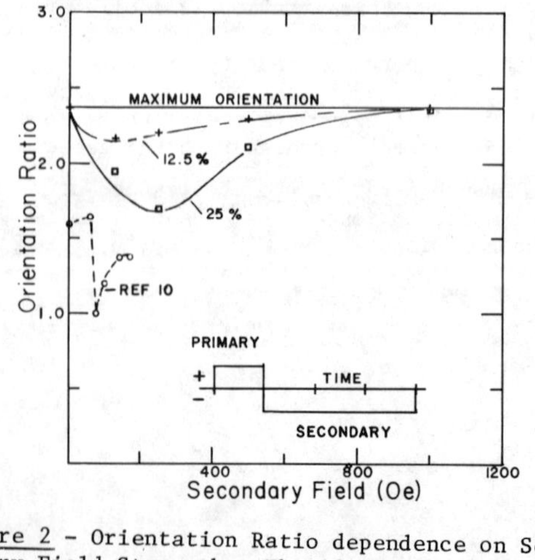


Figure 2 - Orientation Ratio dependence on Secondary Field Strength. The 12.5% and 25% primary Duty Cycle conditions are shown and compared to the maximum orientation ratio achieved with a 100% Duty Cycle. The magnitude of the primary field was 1000 Oe.

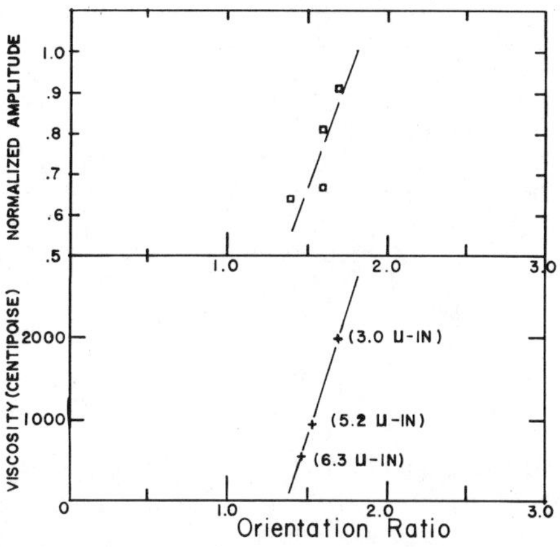


Figure 3 - Disc Amplitude (F = 3.5 MHz) and Dispersion Viscosity dependence on Orientation Ratio for an agglomerated dispersion.

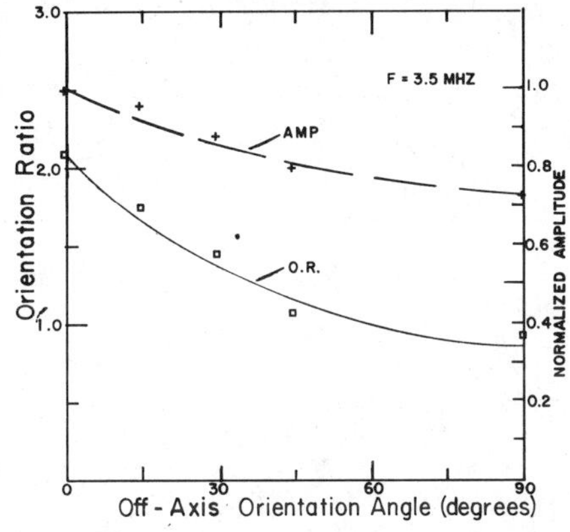


Figure 4 - The influence of the Off-Axis Orientation Angle on Disc Amplitude and Orientation Ratio. The Orientation Field was 1000 Oe.

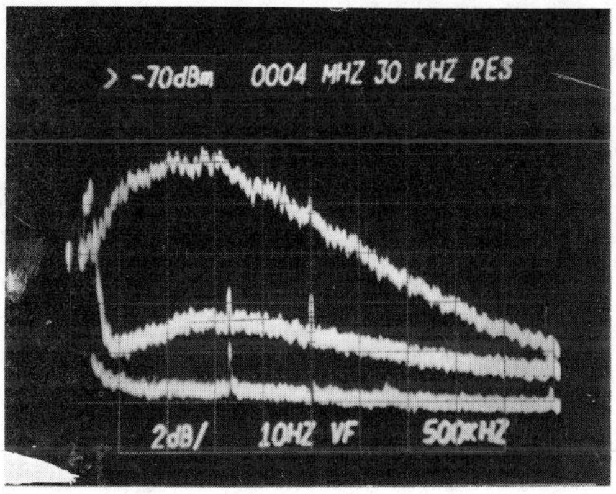


Figure 5 - The Saturated Noise Spectra of the measurement system (lower), a typical oriented disc (middle), and an oriented disc with particle agglomeration (upper).

gates, the mean particle size increases, raising surface finish and reducing viscosity.

## OFF-AXIS ORIENTATION

It has been reported,[6,11] that off-axis orientation can improve magnetic performance. In disc recording, off-axis orientation is defined by the angle that the orientation assumes with respect to the tangent to a circumferential line. Zero degree ($0^o$) orientation is defined as circumferential orientation and $90^o$ off-axis orientation is radial orientation--also present as the non-oriented state. Figure 4 shows the resulting orientation ratio and amplitude as a function of the off-axis orientation angle. The isolated pulse width was found to decrease linearly as the off-axis angle increased.

## SATURATED NOISE STUDIES ON ORIENTED DISCS

Studies of the dc-erased noise have been used to obtain information about the quality of magnetic dispersions.[12,13,14]

Oriented coatings having an agglomerated particle condition will result in satisfactory orientation ratios, notably lower signal outputs and S/N ratios, and higher noise levels. Figure 5 illustrates the effects of an oriented agglomerated dispersion on the saturated noise spectrum. The change in the noise spectrum at low frequencies has been attributed to inhomogenities in the medium.[13]

Noise levels from oriented and non-oriented discs were compared. The absolute value of the saturated noise was higher in oriented discs by $\sim$1.5-2 db.

## DISCUSSION

Orientation ratios in excess of 2 and even 3 have been achieved regularly and appear to be representative of disc orientation.

This work has shown that optimum orientation is not always associated with high signal outputs. Particle agglomeration resulting during orientation does not significantly affect orientation ratios--although other recording parameters such as Amplitude and Noise do suffer.

The influence of viscosity on the orientation of spin-coated discs has been described. The lowest energy state for spin-coated acicular particles has been identified with radial orientation (orientation ratio $\approx$1). This condition is believed to be a result of the localized viscosity influence on acicular particles undergoing radial translation during the presence of centrifugal forces.

Disc orientation has been achieved with single polarity, pulsed single polarity, and alternating polarity fields. The choice of field type and viscosity are system integrated. Various adverse effects of alternating fields ( >500 Hz) have been described as have the influence of secondary fields and times.

Off-axis disc orientation has confirmed the pulse-width/amplitude compromise established for tape media.

The increase in the saturated noise of oriented coatings is believed related to the induced particle interaction. Orientation provides an increase in the noise level at lower frequencies and enhances the signal-dependence of the noise.

In summary, oriented disc media exhibit similar performance characteristics as tape media. The critical balance between orientation and viscosity appears to be more sensitive in discs than tapes. This criticality is believed to be related to the forces placed on the particles during spin-coating. These forces however, are believed to be responsible for the higher orientation ratios.

## ACKNOWLEDGEMENT

The author would like to thank D. Marro and G. Starkeson for their expert technical assistance in the experimental and theoretical aspects of the project and E. Dionne for his help in sample preparation and measurements.

## REFERENCES

1. G. Bate and J. K. Alstad, IEEE Trans. Magn. MAG-5, 821 (1969).
2. G. Bate and L. P. Dunn, IBM J. Res. Develop. 18, 563 (1974).

3. R. D. Weiss, Tech. Rep. USASI Committee X3.2.7B/1 (1970).
4. C. H. Arrington, Jr, U.S. Pat. 3,080,319 (1963).
5. H. Roller et.al., U.S. Pat. 3,836,395 (1974).
6. G. Bate, H. Templeton, J. Wenner, IBM J. Res. Develop. 6, 348 (1962).
7. P. H. Lissberger and R. L. Comstock, IBM Tech Rep. TR 02, 479 (1970).
8. J. J. Newman and R. B. Yarbrough, J. Appl. Phys. 39, 5566 (1968).
9. P. Smaller and J. Newman, IEEE Trans. Magn. MAG-6, 804 (1970).
10. R. R. Pearce, Nature, 212, 1566 (1966).
11. G. Bate, J. Appl. Phys. 32, 239S (1961).
12. I. Stein, J. Appl. Phys. 34, 1976 (1963).
13. I. Mikami, J. Appl. Phys. 33, 1591 (1962).
14. J. L. Su and M. L. Williams, IBM J. Res. Develop. 18, 570 (1974).

## A BUBBLE DOMAIN MEMORY CELL

P. J. Hayes, NASA Langley Research Center, Hampton, Virginia 23665; and I. J. Walker, LTV Aerospace Corporation, Hampton, Virginia 23666

### ABSTRACT

The present conception of a 100 M bit bubble domain digital data recorder for spacecraft applications requires the storage to be compartmentalized in cells containing a group of bubble domain chips. Arranging the memory chips in cells minimizes coil size and power by requiring only a small group of chips to be driven at any given time, and also permits "graceful failure" in that failure of a cell represents the loss of only a small portion of the memory. A cell has been designed and constructed in order to explore deviations from simple coil and magnetic bias plate theory resulting from the close proximity of coils and bias plates. The cell size was designed to accommodate eight(8), 100 K bit chips having the dimensions 0.24" X 0.24" X 0.020" thick; however, the cell operation was actually tested with 10 K bit chips of an earlier design. Before and after the coil assembly was placed in the bias structure, measurements were made of coil ac resistance and oe/amp at the coil center. From these measurements the coil power at nominal operating field was calculated. Both the ac resistance and oe/amp increase upon installation of the coil into the bias structure. Since R increases more than the square of oe/amp, the result is an increase in power. However, the relative percent change in coil power due to the magnetic bias structure is of the order of only 10% for the cell described here. The 10 K bit bubble chips used in the cell were purchased as a matched set. They were margin tested prior to being installed in the cell. The cell has been operated at 150 kHz and margins are within 10% of initial measurements. At 45 oe peak field and 12.5% operational duty cycle (maximum value for a storage module of 8 cells), the inner and outer coils dissipate 5.9 and 8.1 watts, respectively, and the temperature of the chip area of the cell stabilizes after 30 minutes at approximately 11°C above ambient.

## PREPARATION OF WATER BASED MAGNETIC INK

Z. Kovac and C. Sambucetti
IBM Research Laboratory, Yorktown Heights, New York 10598

### ABSTRACT

The motivation of this work was to provide an ink for magnetic ink jet research. Calculations based on energies of interaction among magnetic particles in aqueous system and their role in preparation of stable colloidal dispersion of $Fe_3O_4$ and Fe in water will be presented and the conditions of the stability for the fluid given.

Some chemical precipitation methods for the synthesis of $Fe_3O_4$ with particle size ranging between 100-200 Å have been developed and will be presented. Experimental data on the physical chemistry of the fluid will also be discussed.

# HIGH FIELD-HIGH GRADIENT MAGNETIC SEPARATION: A REVIEW

F. E. Luborsky
General Electric Corporate Research and Development, Schenectady, N. Y. 12301

ABSTRACT

Magnetic separation technology was given new impetus with the introduction of a separator using a matrix of steel wool in a high field combined with an efficient magnetic circuit for the electromagnet. This technology is economically attractive for a wide variety of new separations. The development and status of these applications will be briefly reviewed. This commercial interest stimulated theoretical interest to develop mathematical descriptions of the separation process. The status of the theoretical calculations will be reviewed and their predictions will be compared to the available experimental work. The limitations of the present calculations, and the future directions of the modeling work, will be discussed.

## I. INTRODUCTION

In 1971 Kolm et al[1,2] reported on their development of a high field, high gradient, magnetic separator (HGMS). The HGMS utilizes a filamentry matrix of ferromagnetic material in a field sufficient to saturate the filaments. The resultant high gradient and high field along the edges of the filaments provided very effective regions for trapping magnetic particles. It was demonstrated that even weakly paramagnetic particles, such as CuO, could be efficiently trapped in such a system.

In this first disclosure, a number of potential applications, being evaluated in the MIT Francis Bitter National Magnet Laboratory, were described. These included the beneficiation of low grade iron ores, removal of inorganic sulfur from coal, and the purification of water. In addition, the first industrial scale application of this technology was mentioned; the brightness beneficiation of kaolin clays.

I would like to outline the ingredients involved in this breakthrough in the old and well developed technology of magnetic separation. What appear to be trivial innovations when looked at individually, when taken together, become significant. These innovations have been described[1-6] by a number of the participants in this development, from their own viewpoints. The first ingredient was the recognition of the need for sharp edges to create high field gradients. Frantz[7] as far back as 1937, had patented a filter composed of screens of magnetic ribbons. However the equipment was designed to operate at low fields, below saturation, because of economic considerations. This and similar types of filters are used now in a number of separations. The second ingredient was the recognition of the need for high fields. Around 1955 a high field separator was designed and built by Jones[8]. It consisted of a box filled with grooved steel plates. Fields of up to 20 kOe were developed along the peaks of the grooves where the high gradients exist. However the collecting area and the high field volume were very limited, the fabrication cost high, and the magnet to generate the high fields was costly.

The introduction[3] of large diameter iron-bound solenoids was the next ingredient although again Frantz[7] had earlier described the advantage of iron-bound structures. Improvement in the economics of operation was achieved by increasing the solenoid diameter, d, since the capacity increases as $d^2$ but the power required increases only as d. The iron cladding further reduced the power required by a factor of ∿2. It now was possible to use the filamentary packings, such as the wire screens and ribbons, which could not be economically magnetized before. Of particular interest, partly because of their low cost, was the use of steel wool. This type of matrix has a number of advantages: It is typically 95% void so that the resistance to flow is minimal. The active surface area available for particle capture is maximum. The edges are sharp and thus interact optimally with the very small particles most minerals are ground to in order to liberate the magnetic phase. The sharp edges are non-uniform so that the extent of the high field gradient is variable resulting in interaction with a wide range of particles. Gradients in the range of 1 kG/micron are developed on these edges; orders of magnitude higher than in conventional separators.

The final ingredient was the recognition of the role of retention time [5,20] and thus control of fluid velocity, to reduce drag forces and improve the separation. This clearly relates material throughput and magnet volume.

These then were the ingredients; large area, inexpensive matrices with sharp edges, economically generated fields high enough to saturate these matrice, and control of the fluid velocity.

The first review of HGMS was published in 1973[9] covering the preliminary applications work and physics. In 1974 a detailed history of the development of HGMS was published[10], emphasizing wet separation of weakly magnetic minerals. In the next sections of this review we will briefly describe the new applications resulting from this advance and then summarize the status of the mathematical modeling of the separation process in this system. We will not be considering any of the developments in flow-through magnetic separations such as the quadrupole[11] or straight wire[12-13] configurations.

## II. APPLICATIONS

### A. Application in Commercial Production: Brightness Beneficiation of Kaolin Clays[5,15-19]

White clays are used as fillers and coating for paper. Extensive deposits of suitable clays are available but they are disclored by a few percent of iron-stained anatase ($TiO_2$). This impurity exists as micron and submicron very weakly magnetic particles; their susceptibility is about $10^{-5}$ emu/g at 13 kG.

Flotation and selective flocculation techniques for purification were commercialized but because of their high cost their application was limited to producing premium, high brightness clays. Early work[20] on magnetic separation of this impurity demonstrated its technical feasibility but the long retention times required in the available Frantz separator made the economics unfavorable. Continued work in the J. M. Huber Corp. laboratories led to a critical fusion of technology in 1967[5]. A separator was constructed using a canister filled with high gradient producing Frantz screens installed in the high field gap of a Jones separator generating nearly 10 kG in an 8 in. diameter, 20 in. high gap. This resulted in an increase in throughput of a factor of ten for the same separation efficiency. At this point the process had reached the stage of commercial practicality but further improvements resulted from consultations and work with the National Magnet Laboratory and Magnetic Engineering Associates. The first production prototype was placed on stream in 1969 at Huber. It consisted of a 20 in. diameter canister, 12 in. high, with a field of 20 kG requiring 200 kwatts of power. Production units, placed in service in a number of clay companies in 1973 and 1974, have working diameters of 84 in. and lengths of 20 in. Working at 20 kG they require about 500 kW of power. Throughputs are in the range of 15 to 66 tons of clay/hr for each unit depending on input quality used and output quality required. The U.S. clay industry is now processing more than 4 million tons/year; a significant portion of this is processed by HGMS.

The use of superconducting magnets, instead of electromagnets is discussed for application to clay beneficiation[21] and for other applications[22,22a,23]. It would appear that in large installations such as discussed here, the superconducting magnet system with its own liquid helium cryostat could result in cost savings, only if the initial investment is not increased since depreciation charges account for ~50%, and power accounts for ~20% of the operating costs of present system.

No other major application of the new HGMS technology has yet reached full production. This operating experience is useful in evaluating the problems to be expected, and the economic potential, of other applications. These applications will be discussed briefly in the next section as well as some applications for which the older, low field, high gradient magnetic separators proved to be technically and economically viable.

B. Prototype and Laboratory Demonstrated Applications

(1) Removal of sulfur from coal

The useful constituents of coal are diamagnetic while a good proportion of the undesirable sulfur is in paramagnetic iron compounds. These paramagnetic components can be held in the HGMS and thus separated. Other components in the coal, normally ending up as ash, are also paramagnetic and are removed together with the Fe-S. The Bureau of Mines[24] analyses of coal beds throughout the U.S. indicates that a large proportion of the coals contain 1% to 4% total sulfur of which about 75% is in paramagnetic compounds and is therefore removeable by magnetic separation. The organically bound sulfur will not be removed by any magnetic separation technique.

The first attempt to remove sulfides from finely ground coal, by magnetic means, was described in a 1957 German patent[25] issued as an addition to a 1956 U.S. patent.[26] Russian work,[27], also in 1958, support the results. Pretreatments of the coal at 200° to 300°C in air or steam were used to form films of ferrimagnetic oxides on the surface of the paramagnetic sulfides to increase their magnetic susceptibility and thus enhance their removal efficiency. More recent studies of the same process were reported in 1965 by Kester.[28]

Beneficiation of untreated powdered coals appear to have been observed[29] in 1961, confirmed[30] in 1965 and studied in more detail[31] in 1967. From 54% to 86% of the original pyritic sulfur was removed from four different coals. HGMS was applied to the removal of inorganic sulfur and ash from one type of coal, in a water slurry[32-35]. Typically, one pass through the separator decreased the total sulfur in the coal from 1.3 down to 0.8 wt % with a recovery of 80% of the coal. An economic analysis suggested total processing costs would be in the range of 36-63 cents/ton for a large production installation. These costs are very attractive compared to the costs of other technologies for obtaining clean coal or for removal of sulfur from stack gases (Table I). More thorough studies are needed to determine the optimum processing procedures and the applicability of this process to the wide variety of coals mined.

(2) Beneficiation of weakly magnetic iron ores

The mining industry will soon exhaust their iron ore deposits containing sufficient magnetite to make magnetic drum beneficiation effective. There are, however, vast deposits of oxidized taconites which could be used if an economic beneficiation method were available. These ores consist mainly of small grains of hematite, goethite and iron silicates mixed with chert. To liberate the iron containing minerals from the chert the ore must be ground to a particle size less than 30-40μ. This can then be beneficiated, to an iron level acceptable for steel making, by a flotation process. Alternatively the same wet slurry can be beneficiated in the HGMS process. Previous conventional magnetic separators were ineffective because of the small particle size and the low susceptibilities of the iron containing minerals.

Both batch tests on several different ores[36] and continuous tests[37] using a Carousel Separator[38, 39] have demonstrated significant beneficiation with HGMS. In the batch tests ores containing 33% iron were upgraded to 66% iron with total recovery of ~85%. In the work with the continuous separator[37, 39], the objective was to provide a partially beneficiated oxidized taconite ore with minimum loss in iron, which would then be beneficiated by a final flotation process. Recoveries as high as 95% were achieved and the original ore containing 33% iron was beneficiated to about 50% iron. This improvement is predicted to provide substantial savings in the subsequent flotation process. New magnetic separators designed specifically for this process are the next step in this evaluation[37].

TABLE I. TYPICAL COSTS OF SULFUR REMOVAL (1974)

| | Investment $/KW | Processing* $/ton |
|---|---|---|
| MINE MOUTH BENEFICIATION | | 0.25-1.5 |
| TRANSPORTATION | | |
| 100 miles by train | | 1.0 -2.6 |
| 200 miles by train | | 1.5 -3.3 |
| 300 miles by train | | 2.4 -3.7 |
| MAGNETIC DESULFURIZATION OF FEED COAL | 2.5 | 0.3 -0.7 |
| LIMESTONE BOILER INJECTION | 4-8 | 1.0 -1.5 |
| STACK GAS DESULFURIZATION | 30-80 | 1.4 -4.2 |
| TALL STACK DISPERSAL | 1-4 | 0.07-0.3 |

*each $/ton ≈ 0.15 mils/KW-hr power generating cost.

(3) Cleaning liquid streams

There are two types of applications of HGMS in cleaning fluids. The simplest is the direct removal of suspended magnetic solids, as for example, iron oxides from steel mill waste water. The other applications involve precipitating dissolved contaminants, or coagulating ultrafine particle contaminants, onto magnetic "seed" particles. The basic advantage of HGMS over conventional water treatment is the high speed and large capacity of the HGMS even for the removal of very small particles.

(a) Removal of Magnetic Particles from Water

Experiments have been carried out on various stages of the steel mill process[40-42]. Waste waters from steel mills are produced at rates of 600-3000 gal/min, while processing of this water, to meet present standards, employs sedimentation in large clarifiers operating at less than 2 gal/min/ft$^2$. The laboratory tests using HGMS, shown in Table II, show the feasibility of using 10 to 40 times higher rates. The American Institute of Steel Industries has planned pilot tests of this new technology[40].

(b) Removal of Particles from Process Streams

There are a variety of applications in which magnetic particulates must be removed from process streams. Many of the easier applications have already been realized industrially using commercially available equipment operating at low fields. For example, in (1) liquid ammonia manufacturing to prevent loss of catalyst activity during oxidation to nitric acid[43]; (2) synthesis of ammonia, where these contaminants cause foaming and consequent loss of reactants; (3) oil lubrication and hydraulic systems to reduce wear, to reduce oxidation of the oil, and to reduce valve sticking; (4) processing porcelain powders used for fine china and high voltage insulators; (5) processing foods such as milk and liquid chocolate and in many other applications. In all of these applications conventional filters either cannot be used at all if there are other particulates present, or they are very inefficient.

TABLE II. HGMS OF VARIOUS STEEL MILL EFFLUENTS[41]

| | Magnetic field kG | Number of passes | Flow velocity gal/min/ft$^3$ | Suspended solids FEED mg/l | Suspended solids TREATED mg/l |
|---|---|---|---|---|---|
| BLAST FURNACE SCRUB WATER | 10 | 2 | 88 | 1340 | 13 |
| VACUUM AND ELECTRIC FURNACE | 10 | 2 | 53 | 300 | 2.5 |
| BASIC OXYGEN SCRUB WATER | 9.6 | 1 | 64 | 4500 | <1 |
| MILL SCALE PIT OVERFLOW | 19 | 2 | 69 | 150 | 13 |
| ROLLING MILL WATER | 11.5 | 1 | 19 | 47.6 | 14 |

(c) Clean-up of Fuel Lines in Power Plants

A recent installation of a combined cycle gas turbine-steam turbine plant by the General Electric Co. uses Frantz Co. filters with permanent magnet fields. They are installed in both the fuel oil and steam lines feeding a grid type mechanical atomizer to scavange iron oxide particulates and other magnetic debris, to prevent clogging of the burner nozzles. Conventional filters were found to be ineffective in this application.

(d) Municipal and Industrial Waste Treatment

Perhaps the most important potential application of HGMS is in the treatment of municipal and industrial waste-waters. As in the conventional treatment of waste water the contaminants are precipitated, or coagulated, by the addition of various chemical reagents but in HGMS a finely divided iron oxide seed material is used. The contaminants precipitate, or agglomerate, onto this seed material and both are removed in the filter. The basic advantages of this new approach are (1) the greatly reduced reaction time necessary because of the seeding material and because of the high efficiency of the filter even for very small particles and (2) the extremely high filtration rates. Thus municipal plants using sand filter beds run at rates of 2 to 20 $gal/min/ft^2$ while HGMS experiments[40, 44-48] have been run at greater than 200 $gal/min/ft^2$. Efficient reduction in concentrations of dissolved organics, bacteria, orthophosphates, and suspended solids, and reductions in color and turbidity were demonstrated. (Table III). After filtration the seed material can be freed of adsorbed impurities[46] by back flushing while still held in the filter. The seed can then be released for reuse. The life of this seed material will strongly effect the economics of the process and has yet to be evaluated. Preliminary cost evaluations[40] using a conceptual design for a large plant yield costs in the order of 15-20 cents/1000 gal of water treated. This compares favorably with conventional tertiary treatment plants, uses less energy and considerably less space. A low field Frantz Ferro-filter has been in operation in Levittown, N. J. for several years [5,48a].

(4) Beneficiation of ores (other than iron ores)

A number of ores have been tested in a preliminary way. These include tailings of Mo and W minerals, rare earth ores and uranium ores. Conceptually there are a large number of ores which might benefit from HGMS[49].

(5) Biological applications

When red blood cells are in their deoxygenated state their susceptibility, relative to water, is expected to be $0.3 \times 10^{-6}$ $emu/cm^3$; about two orders of magnitude lower than CuO. Even so, these red blood cells 'ave been filtered out using HGMS, and without any observable biological damage[50]. The filter performs as expected requiring much lower velocities than the CuO. This technique could be applied to either producing plasma with low red blood cell levels or to preparing very pure red blood cell populations.

Another application involves the use of magnetic particles as a support material in immobilized enzyme reactors[51]. The HGMS then removes the immobilized enzyme with the particle, from the process stream so that it can be returned to the reactor. Immobilized enzyme reactors on conventional supports are now used industrially in a variety of processes. The use of a removable immobilized enzyme will greatly extend the applicability of these processes as well as the yield of some now in use.

TABLE III. TYPICAL RESULTS OF HGMS TREATMENT OF WATER[44]

| | CHARLES RIVER SAMPLE | | DEER RIVER SEWAGE SAMPLE | |
|---|---|---|---|---|
| | control | treated | control | treated |
| COLIFORM BACTERIA, (/100 ml) | 16,000 | 300 | $2.8 \times 10^6$ | 18,000 |
| TURBIDITY, (JTU units) | 1,700 | 1 | 50 | 3 |
| COLOR, (color units) | 3,700 | 1 | 100 | 20 |
| SUSPENDED SOLIDS, (mg/liter) | 690 | 5 | 45 | 9 |
| ORTHO PHOSPHATE, (P parts/billion) | | | | |
| no added clay | 330 | 180 | 3,000 | 300 |
| with bentonite clay | 330 | 60 | 3,000 | <100 |

C. Speculative Applications

There are a number of waste products in the form of airborne dusts which could be removed by HGMS. For example fly ash, from coal fired power plants, contains a large percentage of magnetic particles[52]. A preliminary analysis done in our laboratory indicates that these could be economically removed by HGMS thereby greatly reducing the load on the more expensive electrostatic precipitators and therefore reducing the size and investment required in them. Other process dusts e.g. from steel making and uranium oxide production could similarly be recovered without using wash water or fabric filters.

## III. MATHEMATICAL MODELING OF MATRIX SEPARATION

The basic incentive of the modeling is the need to develop the conceptual understanding, as well as the quantitative models, necessary to predict the technical performance of a particular system so that the economics can be predicted without extensive trial-and-error testing. In the next sections we will outline the model as developed to date, and compare its predictions to pertinent experimental results.

The problem is to calculate the motion of a paramagnetic particle traveling in a fluid stream in the presence of a magnetized fiber. First calculations[53, 54] assumed a fiber with circular cross-section, with its long axis perpendicular to the field and velocity vectors, and magnetized uniformly to saturation. These analyses followed the analogous analysis[55] carried out for electrostatic filters. In order to arrive at an exact solution to the equations of motion only the magnetic and drag forces were considered. Simplifying assumptions were made by neglecting fluid boundary layer changes around the fiber. For the fluid flow, two cases were considered; (1) for high Reynolds numbers, where the inertial forces are much greater than the frictional forces the potential flow equation is valid, and (2) for Reynolds number <<1 where the inertial forces are included in an approximate way[54].

Considering only the case of potential flow here, the flow potential is given by

$$\Psi(\text{flow}) = Vr\cos\Theta + (Va^2/r)\cos\Theta \qquad (1)$$

and the field potential

$$\Psi(\text{field})) = -rH\cos\Theta + (2\pi M_s a^2/r)\cos\Theta \qquad (2)$$

using the coordinate system and definitions shown in Fig. 1. The fluid velocity components are then

$$\begin{aligned} V_r &= \delta\Psi/\delta r = V\cos\Theta - (Va^2/r^2)\cos\Theta \\ V_\Theta &= \delta\Psi/\Theta\delta r = -V\sin\Theta - (Va^2/r^2)\sin\Theta \end{aligned} \qquad (3)$$

and the field components

$$\begin{aligned} H_r &= H\cos\Theta + (2\pi M_s a^2/r^2)\cos\Theta \\ H_\Theta &= -H\sin\Theta + (2\pi M_s a^2/r^2)\sin\Theta \end{aligned} \qquad (4)$$

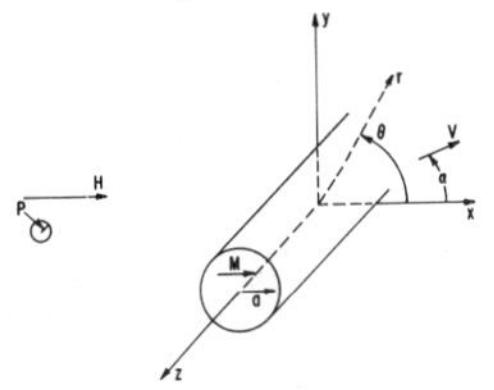


Fig. 1. Coordinate system used to calculate interaction of particle with radius P with fiber of radius a.

The force on the particle is then

$$F = \frac{\delta}{\delta r}\int M\cdot H dV = \frac{\delta}{\delta r}\int(\chi_p - \chi_f)H^2 dV \qquad (5)$$

$$F = (\chi_p - \chi_f)V\delta H^2/\delta r \qquad (6)$$

for a particle small compared to the fiber, and

$$\begin{aligned} F_r &= (16\pi^2\chi M_s P^3 a^2/3r^3)[(2\pi Ma^2/r^2) + H\cos 2\Theta] \\ F_\Theta &= -(16\pi^2\chi M_s P^3 a^2/3r^3)\sin 2\Theta, \end{aligned} \qquad (7)$$

where $\chi_p$ and $\chi_f$ are the particle and fluid susceptibilities. The simple equations of motion for this system

$$dr/dt = F_r/6\pi\eta P + V_r \qquad (8)$$

$$rd\Theta/dt = F_\Theta/6\pi\eta P + V_\Theta \qquad (9)$$

have a closed form solution when the term in $r^{-2} = 0$ i.e. when $2\pi M_s/H = 0$. Then the trajectory of a particle entering at a reduced y coordinate, $y_a = y/a$, at $x = \infty$ is given by

$$y_a^\infty = r_a[1 - (1/r_a^2)]^2\sin\Theta - (V_m/2Vr_a^2)\sin 2\Theta. \qquad (10)$$

The coordinate, for which the particle is just captured, the critical entering coordinate, is

$$\begin{aligned} \zeta = y_a^{c,\infty} &= (V_m/2V) && \text{if} \quad V_m/V < \sqrt{2} \\ &= \text{constant of singularity} && \text{if} \quad V_m/V > \sqrt{2} \end{aligned} \qquad (11)$$

We thus have the effective cross-section per unit length of fiber for capture of particles; namely $2\zeta$.

Typical trajectories are shown in Fig. 2 and the dependence of $\zeta$ on $V_m/V$ are shown in Fig. 3. The equations of motion have also been solved[56,57] for a few values of the short range term, $2\pi M_s/H$, by numerical integration. The effect on some of the calculated trajectories is shown by the dashed line in Fig. 2. The resultant critical coordinates are modified as shown in Fig. 3. Note that there is some discrepancy between the values reported. The capture of particles throughout the filter is now obtained by assuming a suitable lattice of fibers and summing the capture of particles in each filter element. The fractional recovery of particles is then given by

$$R = 1 - \exp(-4FL\zeta/3\pi a) \qquad (12)$$

where a square lattice of fibers has been assumed so that only 2/3 of the fibers are oriented correctly for capture. The capture on incorrectly oriented fibers has been ignored although their contribution may be significant. This effect has been considered analytically[57] but it was concluded that the differences in $\zeta$ for the two orientations studied could not

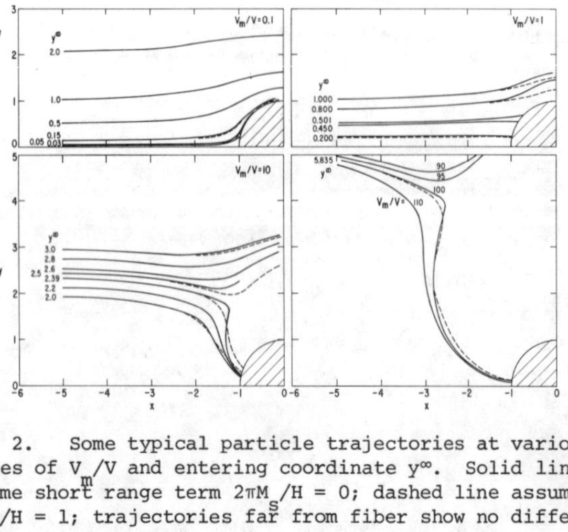


Fig. 2. Some typical particle trajectories at various values of $V_m/V$ and entering coordinate $y^\infty$. Solid lines assume short range term $2\pi M_s/H = 0$; dashed line assumes $2\pi M_s/H = 1$; trajectories far from fiber show no difference.

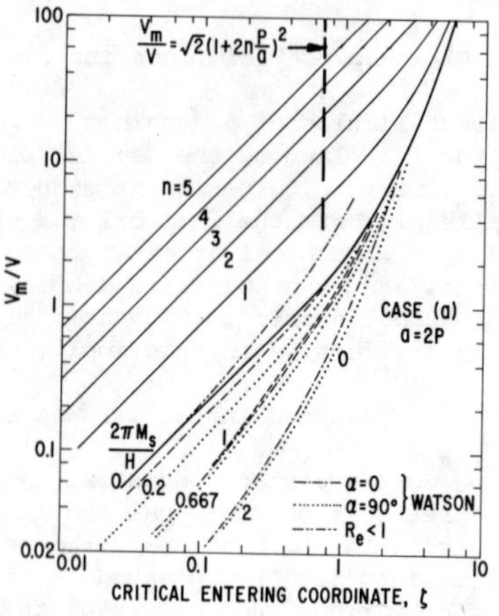


Fig. 3. Relations between critical entering coordinates and $V_m/V$. Calculated by Luborsky[56] ———, - - -, and — · ·; calculated by Watson[59] · —— · —— and · · ·. Heavy curve for $2\pi M_s/H = 0$ is the same from both sources.

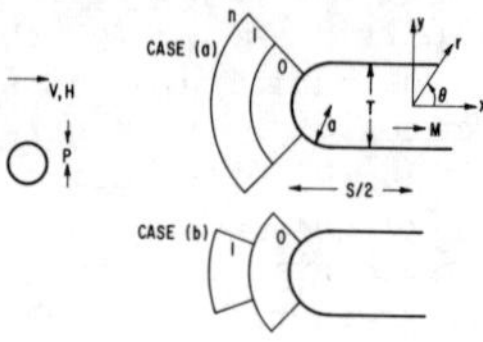


Fig. 4. Two geometries assumed for the build-up of particles on the fiber.

account for the difference observed experimentally[58] using a system of monodispersed $Fe_2O_3$.

It was fortunate that at the beginning of our interest in this problem there had just been published a body of experimental results[9, 59], with sufficient details to allow a meaningful comparison with the analytical predictions[56]. It was immediately apparent, from the mass of particles captured in the filter, that the first mathematical model would have to be extended to take into account the build-up of many layers of particles on the fibers. In addition, from examination of the results extrapolated to zero field, it was strongly suggested that some particles were also being captured non-magnetically. Further, since the experimental work was done with steel wool fibers, the ribbon-like shape of the fiber would have to be considered.

Two simple geometries were considered for the build-up of particles, Fig. 4, based on preliminary experimental observations just published [60]. In this calculation, the capture cross-section of the fiber changes as each layer of particles is completed. These rapidly changing values of $\zeta$ with $\eta$ are shown in Fig. 3. More recent direct observation of the build-up[61, 62] serve to illustrate the difficulty of obtaining experimental information and the complexity of accurately describing the build-up. Watson[63] has now also considered the calculation of particle build-up in a similar way but using a continuum model for the build-up instead of the discrete layer model. The extension of the original analysis[56] to take into account the particle build-up, mechanical trapping and ribbon-like shape of the fibers was used to compare quantitatively to the experimental results on the model systems. The fractional recovery of paramagnetic particles was derived as

$$R = 1 - \exp[-fFL(\zeta+\xi)/3S] \qquad (13)$$

where f is adjusted to give a best fit to the experimental results but may be thought of as a capture efficiency; and $2\xi$ is the reduced cross-section for mechanical capture of particles. The mechanical capture was written as

$$R_m = 1 - \exp[-fFL\xi/3S] \qquad (14)$$

where $\xi$ was assumed to have the form

$$\xi = Q'(S + 2P)/2Va. \qquad (15)$$

$Q'$ is a factor adjusted to give a best fit for both recovery of the paramagnetic particles, R, and for the purity of the recovered particles, G. From the mixture of paramagnetic CuO and diamagnetic $Al_2O_3$ introduced into the separator, G is easily shown to be given exactly by

$$G = Rw_p/(Rw_p+R_m w_p) \qquad (16)$$

The comparison between the entire body of experimental results and the results of the calculations are shown in Figs. 5, 6, and 7. The number of layers of particles required for this fit is given in Fig. 8. Although the correlation between the calculations and the experimental results is good, it does involve two adjustable parameters. The value of f might approach 1.0 if the model included the factors responsible for loss of particles from the fiber; e.g. the loss due to fluid drag or the loss due to the kinetic energy of the incoming particles. The poor agreement with the purity factor G suggests the need for introducing other mechanisms for capture of the daimagnetic $Al_2O_3$ e.g. agglomeration of $Al_2O_3$ with CuO particles. Even without these additions we must explore whether a better fit might be obtained by considering (1) the magnetic potential expressions for the ribbon; (2) the particle size distribution and the change in the distribution as the particles travel down the filter bed; (3) the distribution of fiber dimensions; (4) the changing flow conditions in the fluid boundary layer near the fiber; (5) the changing flow potential for the different flow velocities and inclusion of the inertial term; (6) the exact solution of the equations of motion over the entire $V_m/V$ regime even using only the two forces considered; (7) the equations of motion with inclusion of other forces present such as gravitational forces and London forces[60, 64]; and (8) the orientation distribution between the fibers in the matrix and the field and flow directions.

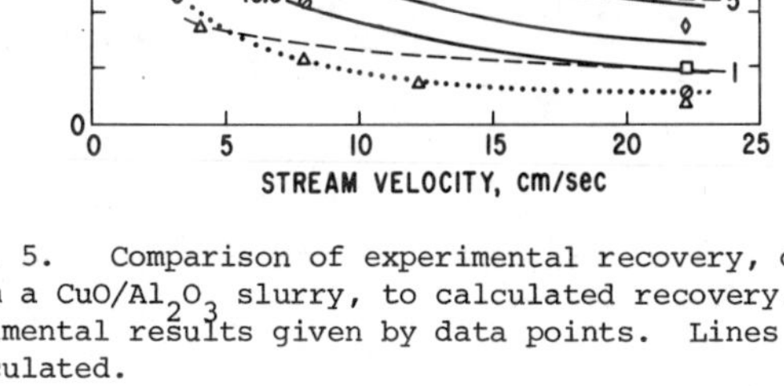

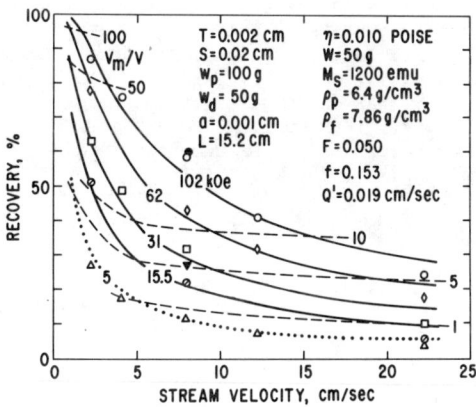

Fig. 5. Comparison of experimental recovery, of CuO from a $CuO/Al_2O_3$ slurry, to calculated recovery. Experimental results given by data points. Lines all calculated.

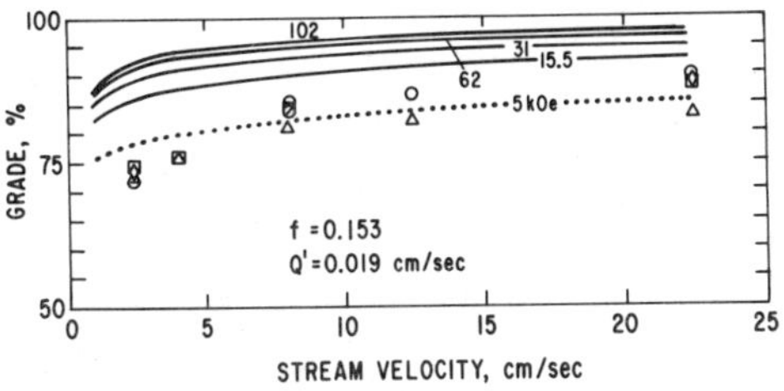


Fig. 6. Comparison of experimental and calculated purity of the CuO recovered in Fig, 5.

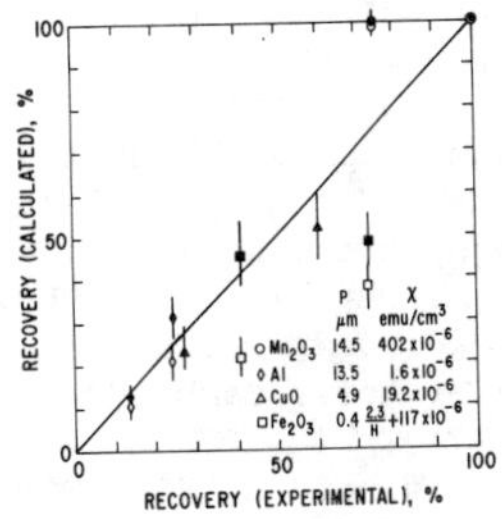


Fig. 7. Comparison of experimental and calculated recoveries of a variety of different particles.

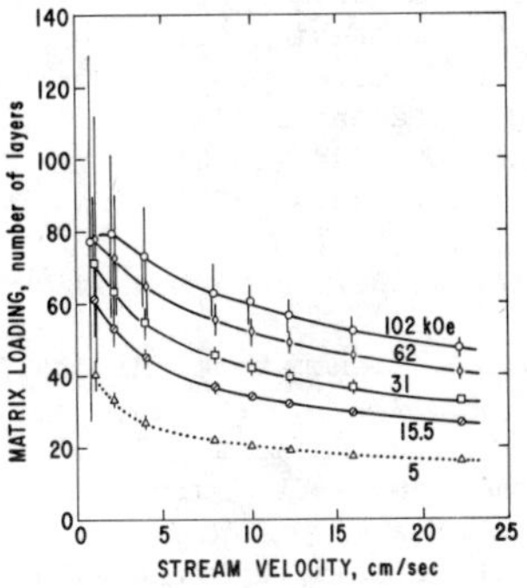


Fig. 8. Calculated build-up of particles on the matrix from the $CuO/Al_2O_3$ slurry in Fig. 5. Bars indicate top-to-bottom change in the filter.

The particle trajactory model has recently been applied to obtaining a more detailed understanding of clay purification[19,21]. Rather than trying to model these complex but practical systems, it is of considerable help to model and simulate experimentally a simple system. In the practical system we have steel wool which is very non-uniform, we use an ill-defined size distribution of particles, and these are mixed with diamagnetic particles which may coagulate with the paramagmetic particles. Watson[58] has recently reported such work using wires for the matrix and monodispersed, uniform size fractions of $Fe_2O_3$, with no diamagnetic particles. The filter invariably performed better when H and V directions were parallel rather than perpendicular. But their calculations[57] of capture cross-sections, given in Fig. 3, show that the capture cross-section is larger for V and H perpendicular rather than paralle. They conclude that the hydrodynamic drag on the particles after capture sweeps the particles off in the perpendicular case. Although this effect should be calculable it has not been done yet. The build-up of particles was reported to agree with theory.

## IV. SUMMARY AND CONCLUSIONS

HGMS technology has developed from its initial conception to large scale application for the processing of clay, in an extremely short time. Other large scale applications are still in the evaluation stage. These are developing more slowly for several reasons, depending on the particular application: lack of management commitment; difficulty in obtaining clear-cut technical evaluations because of the complexity of the materials to be processed, the variety of the processing alternatives, and the absence of the use of the mathematical models to help in evaluating the performance; and lack of venture capital for new process work.

The first-order mathematical model, wit the adjustable parameters, is satisfactory in predicting the performance of the model system of $CuO$-$Al_2O_3$ and the real clay system. A more quantitative model will require more theoretical work as well as experimental work to evaluate the applicability of the extended model. The pay-off will come in the ability to obtain the technical and economic answers, for a particular application, faster and with more confidence in the results.

REFERENCES


1. H. Kolm, E. Maxwell, J. Oberteuffer, D. Kelland and C. deLatour, Magnetism and Magnetic Materials, 1971. AIP Conf. Proc. No. 5, p. 949 (1972).
2. H. Kolm, Proceedings of the High Gradient Magnetic Separation Symposium, May 22, 1973. Ed. by J. Oberteuffer and D. Kelland, MIT Francis Bitter National Magnet Laboratory, Cambridge, Mass., June 22, 1973, p. 17.
3. P. Marston, Proceedings of the HGMS Symposium, ibid., p. 25.
4. P. Marston, U.S. Patent 3,527,678, Dec. 14, 1971.
5. J. Iannicelli, Nat'l. Inform. Tech. Service Document PB 240 880 15WN (1974); presented at 1975 Intern'l. Clay Conf., Mexico City, July 1975.
6. H. Kolm, Trans. Magnetics, MAG-11, 1567 (1975).
7. S. G. Frantz, U.S. Patent 2,074,085, 1937; Chem. & Met. Eng. 45, 274 (1938).
8. G. H. Jones, Brit. Patents 767,124 and 768,451 (1957).
9. J. A. Oberteuffer, IEEE Trans. Magnetics MAG-9, 303 (1973).
10. J. E. Lawver and D. M. Hopstock, Minerals Sci. Eng. 6 (3), 154 (1974).
11. W. M. Aubrey and R. M. Funk, the 32nd Annual Mining Symposium, Jan. 1971, Duluth, Minn. Publ. by Univ. of Minn., p. 55.
12. Y. M. Eyssa, Ph.D. Thesis, Univ. Wisconsin, Sept. 25, 1975, Nuclear Engr. Dept.
13. Y. M. Eyssa and R. W. Boom, IEEE Trans. Magnetics MAG-11, 1585 (1975).
14. J. A. Good, Digests of Intermag Conf. - 1975, No. 24-8. IEEE Publ. 75CHO 932-4-MAG.
15. R. R. Oder, Proceedings of the HGMS Symposium, ibid., p. 55.
16. R. R. Oder and C. Price, Tappi 56 No. 10, 75 (1973).
17. R. R. Oder, Pulp and Paper Magazine of Canada 75, No. 10, T 366 (Oct. 1974).
18. J. Iannicelli, Canad. Patent 935,126, Oct. 9, 1973.
19. J. H. P. Watson and D. Hocking, IEEE Trans. Magnetics, MAG-11, 1588 (1975).
20. J. Iannicelli, N. Millman and W. J. D. Stone, U.S. Patent 3,471,011, Oct. 7, 1969.
21. J. H. P. Watson, N. O. Clark and W. Windle, Proc. Eleventh Intern. Mineral Processing Congr. 1975. Publ. by Istituo di Arte Mineraria - Univ. di Cagliari, Cagliari, Italy, 1976. Paper 29.
22. Z. J. J. Stekly, IEEE Trans. Magnetics MAG-11, 1594 (1975).
22a. P. G. Marston, Proc. Fifth Intern. Conf. on Magnet Technology, Rome, Italy, April 1975. Publ. by Laboratori Nazionali CNEN, Frascati, Italy, p. 424.
23. Z. J. J. Stekly, G. Y. Robinson, Jr., G. J. Powers, Cryogenics and Industrial Gases, July/August, 1973.
24. Bureau of Mintes, Information Circular 8301, 1966. See also L. Hoffman et-al. National Technical Information Service PB211505. May 1972.
25. S. Siddiqui, German Patent 1,005,012, March 28, 1957.
26. S. Siddiqui, U.S. Patent 2,272,265, Nov. 27, 1956.
27. A. Yurovsky and I. Remesnikov, Coke and Chemistry 12, 9 (1958).
28. W. Kester, Min. Engin. 17, No. 5, 72 (1965).
29. W. Kester, Thesis, West Virginia University School of Mines, 1966.
30. R. D. Harris, FallMeeting Society of Mining Engin. of AIME. Oct. 7-9, 1965, Phoenix, Arizona.
31. W. Kester, J. Leonard, and E. Wilson, Min. Cong. J. 53, No. 7, 70 (1967).
32. S. C. Trindade, MIT Dissertation 1329, April 1973.
33. S. C. Trindade, Proceedings of the HGMS Symposium ibid., p. 102.
34. S. C. Trindade and H. H. Kolm, IEEE Trans. Magnetics, MAG-9, 310 (1973).
35. S. C. Trindade, J. B. Howard, H. H. Kolm and G. J. Powers, Fuel 53, No. 3, 178 (1974).
36. D. R. Kelland, IEEE Trans. Magnetics MAG-9, 307, (1973).
37. D. R. Kelland and E. Maxwell, IEEE Trans. Magnetics, MAG-11, 1582 (1975).
38. D. Kelland, Proceedings of the HGMS Symposium, ibid., p. 71.
39. J. Oberteuffer, IEEE Trans. Magnetics MAG-10, 223 (1974).
40. R. R. Oder and B. I. Horst, Second National Conf. on Complete Water Reuse, Chicago, May 8, 1975.
41. J. A. Oberteuffer, I. Wechsler, P. G. Marston, and M. J. McNallan, IEEE Trans. Magnetics, MAG-11, 1591 (1975).
42. M. J. McNallan, Masters Thesis, MIT, Dept. of Metallurgy and Materials Science, May 1974.
43. G. R. Gillespie and D. Goodfellow, Chem. Engin. Progress 70, 81 (1974).
44. C. deLatour, Trans. Magnetics MAG-9, 314 (1971).
45. C. deLatour, Second National Conf. Complete Water Reuse, Paper 30a, Chicago, May 8, 1975.
46. C. deLatour and H. Kolm, IEEE Trans. Magnetics MAG-11, 1570 (1975).
47. P. G. Marston, 47th Annual Conf., Water Pollution Control Federation, Denver, Oct. 10, 1974.
48. G. Bitton, R. Michel, C. deLatour and E. Maxwell, Water Research 8, 107 (1974).
48a. Chemical Week, p. 39, July 28, 1971.
49. I. Wechsler, Proceedings of the HGMS Symposium, ibid., p. 38.
50. D. Melville, E. Paul and S. Roath, Digests of Intermag. Conf. - 1975, No. 29-4.
51. P. A. Monro, P. Donnell and M. D. Lilly, IEEE Trans. Magnetics MAG-11, 1573 (1975).
52. J. D. Watt and D. J. Thorne, J. Appl. Chem., 15, 585 (1965).
53. C. P. Bean, Bull. Am. Phys. Soc. 16 350 (1971).
54. J. H. P. Watson, J. Appl. Phys. 44 4209 (1973).
55. G. Zebel, J. Colloid Science 20 522 (1965).
56. F. E. Luborsky and B. J. Drummond, Trans. Magnetics MAG-11, XXX (1975).
57. J. H. P. Watson, Trans. Magnetics, MAG-11, 1597 (1975).
58. J. H. P. Watson and P. W. Riley, Digests of Intermag Conf. - 1975, No. 24-4.
59. J. A. Oberteuffer, Proceedings of the HGMS Symposium, ibid., p. 86.
60. D. Himmelblau, Masters Thesis, M.I.T., 1973.
61. C. Cowen, F. J. Friedlaender and R. Jaluria, Trans. Magnetics, MAG-11, 1600 (1975).
62. E. Maxwell and D. Kelland, Digests of Intermag Conf. -1975, No. 24-5, IEEE Publ. 75CHO 932-4-MAG.
63. J. H. P. Watson, Digests of Intermag Conf. - 1975, No. 24-3. IEEE Publ. 75CHO 932-4-MAG.
64. G. J. Powers and D. L. Cummings, Digests of Intermag Conf. - 1975. No. 24-1.

# A VERSATILE MAGNETOSTRICTIVE DISPLACEMENT TRANSDUCER

Ivan J. Garshelis
Research Associates, Inc. Linden, New Jersey 07036

## ABSTRACT

A magnetostrictive displacement transducer of exceptional versatility is obtained by forming a magnetostrictive wire into a helical spring. Displacements producing extension or compression of the spring are transformed by this configuration into torsional strains within the wire element. The biaxial principal stresses establish preferred domain orientations which are helically disposed around the wire axis. The sense of the magnetization within these domains is switched by alternating current pulses through the wire. The degree of anisotropic orientation is sensed as the emf induced in a solenoid wound along the wire. Simplified theory predicts a range of approximately linear response which is supported by measurements on prototypes. The versatility of the device derives from the wide range of attainable compliance. This flexibility allows the construction of simple devices with useful displacements in the range $10^{-3}$ to $10^{2}$ mm indicating applications similar to those of LVDTs.

## INTRODUCTION

Many displacement transducers, i.e., devices which present an electrical analog to some mechanical displacement, operate on electromagnetic principles. Generally their output signal reflects the time rate of change of the magnetic flux linking conductors within appropriately placed coils. The time period during which the flux change is completed is usually established by the (fixed) frequency of some driving source. The magnitude of the changing flux may respond to the displacement by the alteration either of the relative position of ferromagnetic elements within the magnetic circuit or of the susceptibility of "fixed" members as by their elastic deformation. This use of the phenomenon of inverse magnetostriction has been confined to applications wherein the displacement is only an incidental accompaniment to a force due e.g., to a pressure, weight or acceleration.[1] Such restricted utility reflects both the limited compliance of magnetostrictive materials under direct tension or compression and the unwieldy constructions required for the direct application of bending or torsion. By utilizing a configuration which transforms the deflection mode of an elastic member, a magnetostrictive displacement transducer with almost any desired compliance can be constructed within a package size and shape which is similar to such moving element devices as LVDTs.

## DESCRIPTION

Such a configuration is obtained by forming a magnetostrictive wire into a helical spring. In this configuration, shown in Fig.1, axial displacements producing extensions or compressions of the spring are transformed into proportional torsional strains within the wire element. The merit of this configuration may be realized by comparing its compliance with that of a directly loaded rod-like element of the same size and type of material. By applying (5) from Fig.1 to a spring of typical proportions ($D/d \simeq 10$, $K \simeq 2$), the compliance gain is found to be nearly four orders of magnitude. Although the limiting elastic deformations of typical materials under direct loading is under 1%, the helical configuration allows deflections even exceeding the initial length of the device. The deflection in compression is often limited by the contact of adjacent coils rather than elasticity. Modifications of the helix to a conical envelope or ultimately to a flat spiral avoid this limitation.

## OPERATION - THEORY

Operation as a displacement transducer stems from the crystalline distortions which collectively make up the torsional strain due to deflection, since they introduce an additional anisotropy to the direction of the local magnetization. Depending on the type and magnitude of other anisotropic influences, the principal strains associated with the torsion establish preferred magnetic domain orientations which are more or less helically disposed around the element axis. This displacement oriented magnetization can be sensed by a variety of means, but one which is especially appropriate for torsionally strained slender elements is the inverse Wiedemann effect.[2] (This is the appearance under torsion of an axial component of magnetization accompanying an applied circular field.) The arrangement by which this effect is utilized is shown in Fig.2(a). Here is seen a small section of the wire element through which is passed an alternating current and over which is placed a solenoidal winding.

In the absence of helical anisotropy (and fields parallel to the element axis) there is no net axial component to the changing magnetization within the element, Fig.2(b). Although the helical anisotropy induced by strain has by its symmetry no net initial axial component, the unique directionality of the instantaneous circular field abolishes the equivalence of the antiparallel directions. Thus, the oscillations of the magnetization by the alternating circular field present a varying axial flux which links the turns of the solenoidal winding. The induced emf is in general a nonlinear function of the rate of change of current, since irreversible magnetization changes take place at threshold values of field and proceed at internally limited dynamic rates. The approximate analysis for this emf made by Granath[3] for a straight torqued rod may not be used here since his idealizing assumptions ignore these realities of ferromagnetism. Although the instantaneous emf for arbitrary excitation currents is

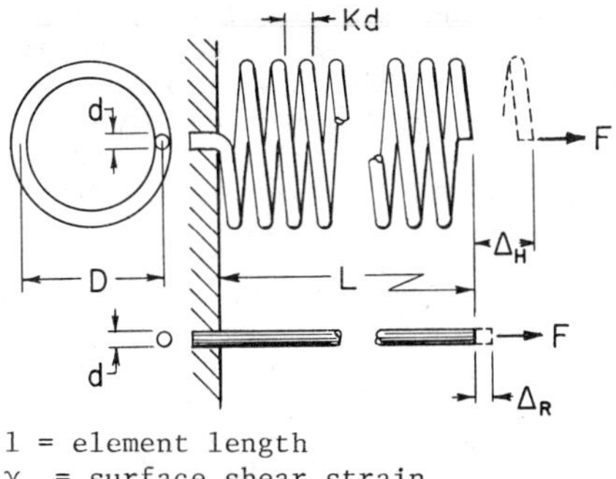


$$\Delta_H = \frac{D l \gamma_s}{d} \quad (1)$$

$$l = \pi D N \quad (2)$$

$$\frac{\Delta_H}{FL} \simeq \frac{8D^3}{Kd^5G} \quad \text{for } NK \gg 1 \quad (3)$$

$$\frac{\Delta_R}{FL} = \frac{4}{\pi d^2 E} \quad (4)$$

$$\frac{\Delta_H}{\Delta_R} \simeq \frac{16}{K}\left(\frac{D}{d}\right)^3 \quad (5)$$

$l$ = element length
$\gamma_s$ = surface shear strain
$N$ = number of (active) turns
$G$ = Modulus of Rigidity   $E$ = Modulus of Elasticity

Fig. 1. Compliance gain of helical configuration

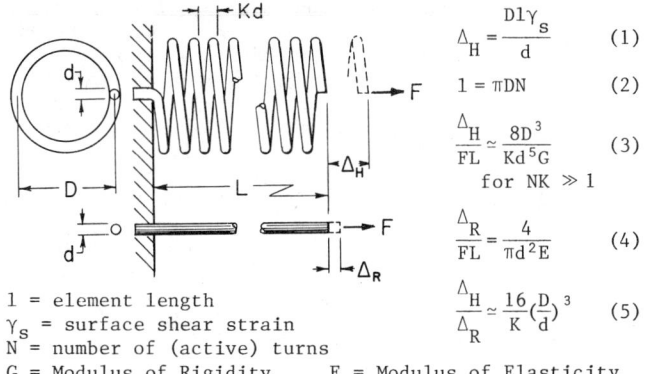

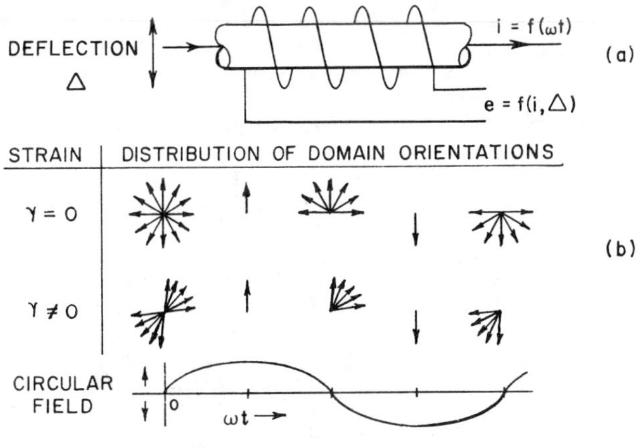


Fig. 2. Method for obtaining emf from deflection (a) Effect of torsional strain on domain orientations (b)

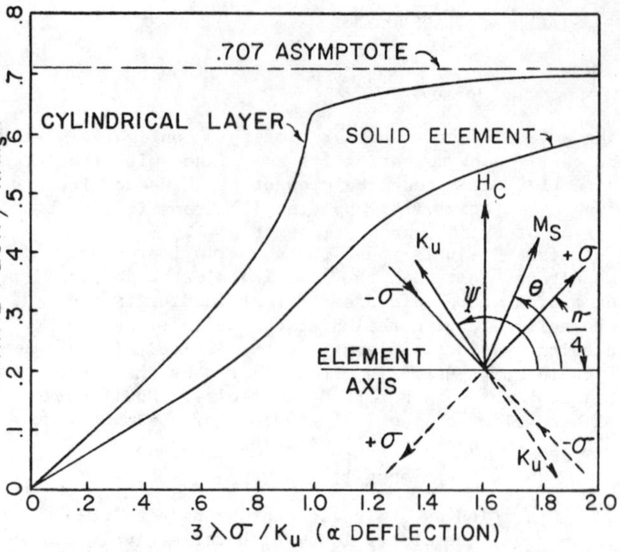


Fig. 3. Theoretical transfer function

not readily predictable, its time integral, i.e., the total axial flux change, may be determined for quasi-static saturating currents. The salient features of the transfer function for displacement transducers operating in this mode can thus be established.

Refering now to the vector diagram in Fig. 3, this analysis proceeds by first finding the stable positions for the local magnetization vectors, $M_s$, as they "fall back" from alignment with the saturating circular field, $H_c$, as the current goes to zero. These stable positions minimize the total free energy density relative to the magnetoelastic and other effective anisotropic influences. The analysis is simplified but not vitiated by assuming an isotropic positive magnetostriction, $\lambda$, and that the magnetization in any unstrained cylindrical layer is randomly oriented by a uniaxial anisotropic influence, $K_u$, having negligible radial components. Under these conditions and using the conventional representation of magnetoelastic energy based on principal stresses, $\sigma$ (rather than strains), the energy density of the magnetization vector in the position shown is

$$E = \frac{3}{2}\lambda\sigma\sin^2(\theta - \frac{\pi}{4}) - \frac{3}{2}\lambda\sigma\sin^2(\frac{3\pi}{4} - \theta) + K_u\sin^2(\psi - \theta) \quad (6)$$

The two magnetoelastic energy terms are due to the biaxial principal stresses characterizing the torsion. From $\partial E/\partial\theta = 0$,

$$\theta = \frac{1}{2}\tan^{-1}(\frac{3\lambda\sigma}{K_u}\sec 2\psi + \tan 2\psi) \quad (7)$$

The axial flux per unit area A, from a thin cylindrical layer is proportional to $\overline{\cos\theta}$ taken over the appropriate range of $\psi$. For the assumed isotropic distribution of $K_u$, the flux density is found to vary with the stress (in rationalized units) as shown in Fig. 3. In a solid element, the stress at any radius is proportional to the radius which requires that the flux density be integrated over the cross section to determine the appropriate transfer function. This result is also shown in Fig. 3, the abscissa now representing the stress at the surface.

OPERATION - PRACTICE

In practice, the peak circular field need not reach the saturating magnitude assumed in the foregoing analysis since much of the magnetization reversal occurs by far less energetic processes than direct vector rotation, e.g., domain wall motion. This mode of reversal will take place mostly during the current rise, as the coercive field is reached at progressively increasing depths within the wire element, with a relatively small contribution from the "fall back" as the current goes to zero.

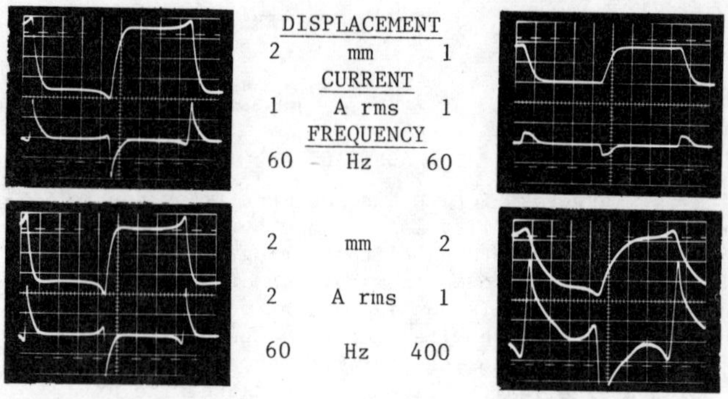


Fig. 4. Axial flux (upper trace) and emfs from prototype transducer under varying conditions

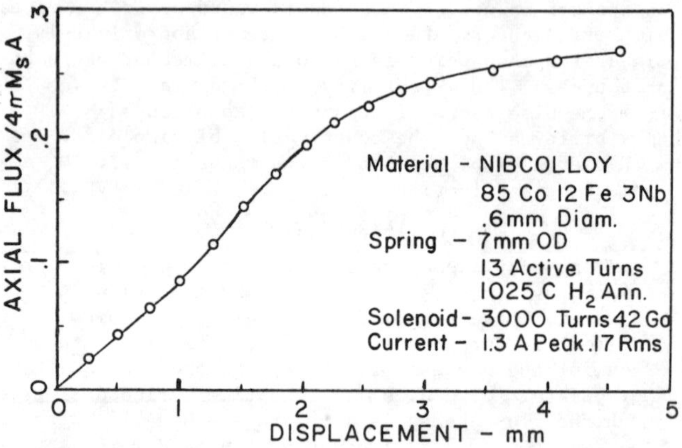


Fig. 5. Transfer characteristic of prototype transducer

These features as well as the actual dynamics of the reversal process are clearly seen in the oscillograms of Fig.4. These were made with sinusoidal excitation currents crossing zero near the center of the "fall back" peaks.

The transfer function for the prototype transducer used for these oscillograms is shown in Fig.5. It is seen to possess the same salient features as the theoretical curve of Fig.3. Although both curves are symmetrical about the origin, operation with large scale stress reversals magnifies hysteresis. This consequence of imperfect elasticity can be minimized by a differential construction in which symmetrical portions of the active element(s) are caused to undergo opposite changes from an initially established strain. Such a construction accommodates displacements in both directions without the tension/compression crossover.

Spring dimensions for use in a displacement transducer having any desired range may be calculated from data of the type shown in Fig.5, for various materials, by using the appropriate equations from Fig.1.

The linear range for any material will expectedly depend on those properties which affect the dimensionless ratio $3\lambda\sigma/K_u$. Experimentally this range is found from its implicit equivalent $\gamma_s$ to be $.0001<\gamma_s<.002$ in typical materials. Using casual estimates of reasonable combinations of $3<D<50$mm, $.2<d<4$mm, $1<N<50$ turns and $\gamma_s$, it can be seen that conveniently sized transducers with this simple construction can be applied to linear displacements in the range $10^{-3}$ to $10^2$mm.

REFERENCES

1. K. Ara and M.J. Brakas IEEE Trans. Mag. 11 1352-1354 1975
2. I.J. Garshelis IEEE Trans. Mag. 10 344-358 1974
3. J.A. Granath J. Appl. Phys. 31 178s-180s 1960

# HGMS: MATHEMATICAL MODELING OF COMMERCIAL PRACTICE

R. R. Oder
The Bechtel Corporation, San Francisco, California 94119
and
C. R. Price
J. M. Huber Corporation, Huber, Georgia 31040

## ABSTRACT

The HGMS method is understood with the aid of a simple mathematical model which emphasizes commercial requirements. With this model, a single magnetic separation parameter $M\chi(d/D)^2 H\tau X(1-X)/\eta$ is sufficient to specify ideal separator performance and to suggest optimum HGMS process capacity. The model, which is applicable where direct impingement on the collection surface dominates particle capture, has been tested extensively in the cleaning of kaolin clay where commercially important color improvements can be achieved by magnetic removal of up to 2% by weight of ¼ to 8 micron sized discolored mineral impurities rich in titanium dioxide. Process flows up to 8.5 cm/sec, magnetic fields up to 1.4 T and filter beds composed of 50 to 200 micron mean diameter stainless steel wools packed to 70% void volumes have been investigated. Clays of mean particle diameters between 0.3 and 3 microns have been processed as stable suspensions in water containing up to 60% by weight solids.

## INTRODUCTION

The capture of magnetic particles by high gradient magnetic fields has been studied theoretically by others using the equations of motion of magnetic particles flowing around a single magnetized strand.[1] The performance of an ideal filter composed of many such fibers is then given by a filter performance equation similar in form that that first given by Bean.[2] Neither Bean's result nor the more recent analyses were available in the early stages of commercialization so a simple physical model was developed which, although less rigorous, has sufficient resolution to describe important characteristics of magnetic filtration in terms only of conditions at the surfaces of the wires and of the porosity of the packed bed. With this model the gross characteristics of unloaded filter performance can be assessed easily without requiring sophisticated calculations. The purpose of this paper, then, is to present this physical model and its experimental verification in kaolin clay processing with the belief that its simplicity and ease of interpretation will aid the assessment of possible new applications of HGMS methods.

To date the only commercial application of High Gradient Magnetic Separation (HGMS) technology is in the cleaning of kaolin clay[3] which is a white aluminosilicate mineral occurring as a very finely divided particulate dispersion. The whitest kaolins are utilized in the manufacture of quality paper products.

The color of naturally occurring kaolin is improved by removing micron-sized mineral impurities (nominally anatase, mica, rutile, iron pyrite) which are present in dilute concentrations. Most of these color-bodies are liberated from the clay particles and can be magnetically separated from wet slurries of kaolin on the basis of their paramagnetism.

The HGMS units employed in the kaolin application operate like automatic backflushing filters with magnetic entraining forces which can be turned on and off at will. The filter bed is composed of very fine strands of magnetic material (stainless steel wool) which are loosely packed within the working volume of a surrounding electromagnet.[4] The magnetized strands distort the background magnetic field and create intense magnetic field gradients localized over their surfaces. These regions of high field gradients provide capture sites uniformly distributed throughout the magnetized volume.

## SCATTERING MODEL

The physical principles of capture on the magnetized strand surfaces in the filter bed can be understood with the aid of a simple scattering model which ignores details of particle trajectories and emphasizes fundamental coustraints placed upon the process by the economic and technical requirements of commercial practice. With this model the filter transmittance, T, relating the concentration, $C_0$, of magetic particles at the output of a magnetic filter of length L to that of the particles at the input, C, is given in terms of a probability of direct collision with the collection surface expressed per particle per length of travel, $p_c$, and a conditional probability of sticking once having collided, $P_{s/c}$,

$$T \equiv \frac{C_0}{C} = e^{-p_c p_{s/c} L} \tag{1}$$

The luxury of migration of magnetic particles across the process flow to a collection surface, while possible in principle, need not be required in practice to develop significant commercial separations. In the commercial case, process economics will demand the maximum throughput consistent with performance requirements and the quasi random filter packings employed will result in a tortuous flow path which ensures ultimate particle capture, even at high throughput, if the filter is long enough. This is equivalent to a statement that collection surface impact scattering will dominate the capture of magnetic particles. For this case, the scattering probability, $p_c$, will be given in terms of the fractional amount of magnetic collection surface available, X, in a segment of length equal to a strand diameter, D, with fractional void volume (1-X),

$$p_c \approx \frac{X(1-X)}{D} \tag{2}$$

Further, the requirement that micron-sized and weakly magnetic particles be captured at maximum process flow necessitates the use of an open, porous filter structure which, while providing a high density of capture sites, will not adversely affect process flow. More precisely, this second requirement means that the chance of the captured magnetic particles' being swept away, after having collided with the capture surface, will be dominated by flow through the packed bed. This second requirement can be stated as a conditional probability of sticking, $P_{s/c}$, assumed given by the ratio of the magnetic retaining force on the stand surface to the viscous force tending to drag the particles away.

If one assumes a Stokes Law drag (viscosity $\eta$) on the particle of volume magnetic susceptibility $\chi$, and diameter, d, flowing through the bed whose volume is packed with magnetic strands of magnetization M oriented transverse to the magnetizing field, H, then the transmittance takes the form,

$$T \equiv \frac{C_0}{C} = e^{-4\pi\chi M(d/D)^2 H\tau X(1-X)/9\eta} \tag{3}$$

which is similar to the expression given earlier by Bean.[2] In Eq. (3) the sticking probability has been maximized by orienting the undisturbed flow parallel to the magnetizing field which, in turn, is oriented transverse to the strand length. This choice correlates stagnation points in the flow pattern (minimum viscous force) with maximum field gradient (magnetic force) points on the stand surface. The maximum magnetic field gradient developed at the surface of a smooth strand is $8\pi M/D$. The retention time in the filter bed, $\tau$, is given by the ratio of filter length to flow velocity. The appearance of a retention time effect in this simplification is a result of the assumption of a Stokes'

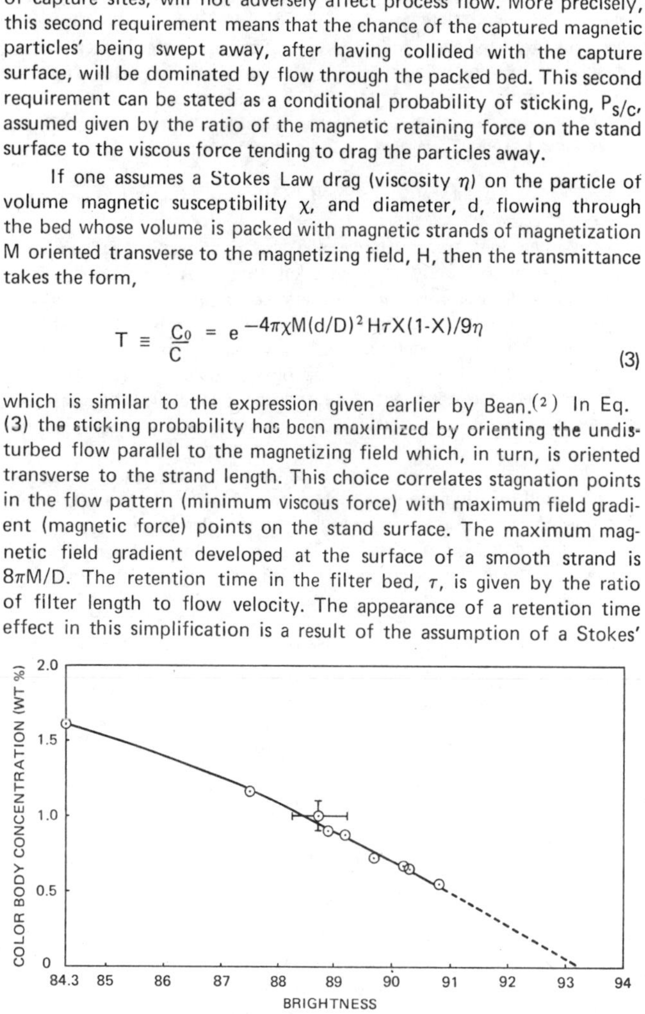


FIG. 1 RELATIONSHIP BETWEEN COLORBODY CONCENTRATION AND CLAY BRIGHTNESS FOR A CLAY OF 0.5μ MEAN PARTICLE SIZE. BRIGHTNESS IS MEASURED BY REFLECTION OF 457 Mμ LIGHT.

solution for the motion of the micron sized particles. This assumption is justified for commercial clay processing, where the appropriate Reynolds number is small compared to 1.

## EXPERIMENTAL CONFIRMATION

The simplified model has received extensive testing in measurements of the removal of colorbodies from a variety of kaolin clays processed with HGMS separators in commercial operation employing magnetic field strengths up to 1.4 T with retention times between 6 and 480 seconds, (flow velocities between .1 and 8.5 cm/sec) with filter beds composed of strands of stainless steel wool ranging from 50 to 200 microns mean diameter packed from 70 to 98% void volumes. All measurements reported here were made on clay samples taken from a single source. In this manner, magnetic susceptibility, particle size and slurry viscosity were constant for all the measurements. The clay particles were ½ micron mean diameter (90% of the particles $< 2$ micron diameter) and were prepared as stable suspensions in water containing about 30% by weight solids. The slurry viscosity was $9.7 \times 10^{-2}$ poise and the magnetic susceptibility[5] of the magnetic colorbody isolates[6] taken from the clay was $X_g = 5 \times 10^{-7}$ cgs/gm. The experimental samples consisted of the first 250 cc's of material to issue from the magnetic separator of void volume up to 54,000 cc's. By sampling in this manner, collection surface loading was not an important factor and ideal separator performance could be directly studied. Colorbody concentrations were determined by X-ray fluorescence spectroscopy and clay color was determined by reflectance of 457 mμ wavelength light.

The transmittance of liberated colorbodies is shown in Fig. 2 for varying degrees of magnetic separation as given by the condensed magnetic separation parameter, $H\tau X(1\text{-}X)$. It is important to note that any particular value of the magnetic separation parameter could have been achieved in a variety of ways; it is only required that the overall combination of magnetic field strength, retention time, and strand packing density as given by the condensed separation parameter shall have the particular value in question. In Fig.2 the term liberated pertains only to those colorbodies which are free to be removed by magnetic means without the use of other chemical or mechanical forces. The straight line fit shown in Fig. 2 yields the numerical determination,

$$4\pi M\chi(d/D)^2/9\eta = 9.1\text{X}10^{-6} \qquad \text{(overstead-sec)}.^{-1} \quad (4)$$

The measured transmittance and the relationship between clay brightness and colorbody concentration (Figure 1.) can be utilized to predict brightness improvements effected by magnetic separation. Such a comparison between measurement and theory is shown in Fig.3 where the degree of magnetic separation is again given by the magnetic separation parameter $H\tau X(1\text{-}X)$. The observations of Figs. 2 and 3 verify the predicted exponential dependence of the filter transmission on the condensed magnetic separation parameter for the range of values studied.

The fit between theory and experiment for kaokin clay can be further tested utilizing the numerical expression given in Eq. (4). Using the value of $M_s$= 1300 gauss measured for stainless steel wool, theory predicts that colorbodies of mean diameters ranging between .8 and 3 microns will be extracted by an HGMS unit packed with strands of

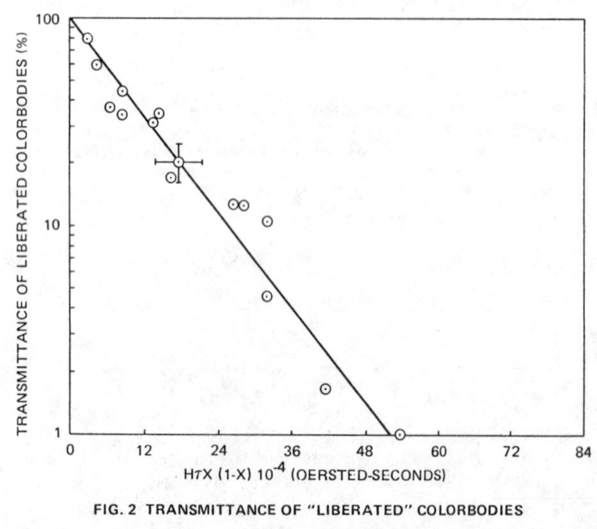


FIG. 2 TRANSMITTANCE OF "LIBERATED" COLORBODIES

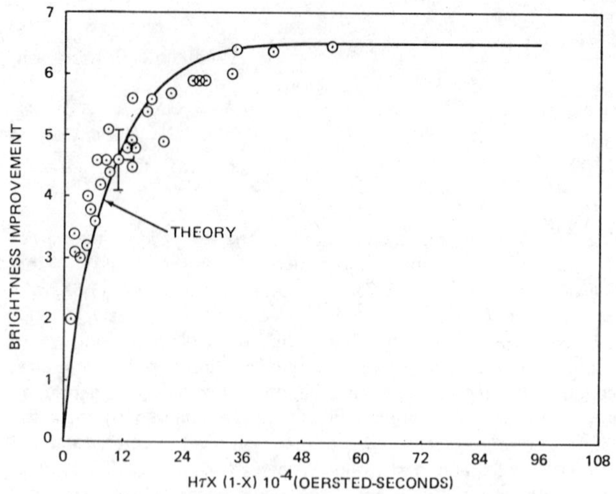


FIG. 3 BRIGHTNESS IMPROVEMENT ACHIEVED WITH MAGNETIC SEPARATION

the range of diameters employed here. This observation is in excellent agreement with determinations of colorbody size made on these materials by Sayin and Jackson[6] who report anatase down to ¼ micron and rutile up to 8 micron mean diameter. The density of the mineral anatase has been used in converting the measured gram susceptibility to the volume susceptibility appearing in Eq. (4).

The model suggests that brightness improvement achievable by HGMS processing is determined by specification of a single magnetic separation parameter $\chi M(d/D)^2 H\tau X(1\text{-}X)/\eta$. Once the performance level, as given by the transmittance, has been determined, the accessible parameters, H, $\tau$ M, X, D and $\eta$ can be traded off according to the model so as to achieve optimum magnetic separator throughput.

TABLE I. BRIGHTNESS VERSUS RETENTION TIME FOR THREE LEVELS OF PACKING DENSITY

| Retention Time (min.) | Brightness 2 | Brightness 4 | Brightness 8 |
|---|---|---|---|
| Packing Volume (%) | | | |
| 4.4 | 87.5 | 88.7 | 89.9 |
| 8.7 | 88.9 | 90.2 | 90.3 |
| 15.5 | 89.4 | 89.9 | 89.9 |

This point is illustrated in Table 1 where the brightness values connected by diagonals correspond to processing with nearly identifical values of $X(1\text{-}X)\tau$. By adjusting packing volume, it is possible within limits to increase flow rate for a predetermined brightness level without sacrifice of performance. Tradeoffs on this type can be made for all variables which make up the separation parameter.

## CONCLUSIONS

The scattering model is a simple and intuitive method which has guided the early commercialization of HGMS methods. In the HGMS cleaning of kaolin clay, the model suggests and experiment verifies that a single magnetic separation parameter, similar to that given by Bean and later confirmed by particle trajectory models, is sufficient to describe ideal separator performance. The model can be extended by including such effects as the inertia of the flowing particles, the spatial variation of the magnetic force and flow fields, and the multiple-layer collection of magnetic particles so as to account for non-ideal magnetic separator performance as well. The simple approximation given here, however, is sufficient for most purposes of predicting separator performance in the processing of kaolin clay. The model is expected to be useful in assessing the technical and economic feasibility of HGMS applications outside the kaolin area.

## REFERENCES


1. Proc. Intermag 75, London, England (April 1975). See F. E. Luborsky, Paper 3D-6, 21st Annual Conference on Magnetism and Magnetic Materials, Philadelphia, Pa. December 10, 1975.
2. Bean, C.P., Bull. Am. Phys. Soc. 16, 350 (1971).

3. Oder, R.R., and Price, C.R., TAPPI 56, 75-78(1973).
4. Proc. HGMS Symposium, Francis Bitter National Magnet Laboratory, MIT, Cambridge, MA (May 1973).
5. Oder, R.R., and Livesay, W., unpublished.
6. Sayin, M., and Jackson, M. L., To be published in Clays and Clay Minerals 24 (1976).


# MICROWAVE LOSSES IN GGG


J D Adam, J H Collins
Department of Electrical Engineering, University of Edinburgh, Edinburgh EH9 3JL, Scotland

and D B Cruikshank
M.E.S.L., Newbridge EH28 8LP, Scotland



## ABSTRACT

Realisation of low-loss microwave devices using YIG grown by LPE on gadolinium gallium garnet (GGG) imposes severe constraints not only on the dielectric and magnetic losses in the YIG but also in the single crystal GGG substrate. Measurements of dielectric loss in GGG show that $\tan \delta_d < 2 \times 10^{-4}$ at X-band. In contrast, the magnetic loss in GGG is found to be both frequency and bias field dependent. At 30 GHz a full resonant absorption is observed in a cylindrical GGG resonator with a linewidth of approximately 7k Oe and a 'g' splitting factor of approximately 2. These results are consistent with broad paramagnetic resonance of the $Gd^{3+}$ ion. For devices operating in X-band the effective $\tan \delta$ is thereby restricted to $10^{-3}$, whereas for devices operating at 2 GHz the effective $\tan \delta$ is $2 \times 10^{-4}$.


## INTRODUCTION

High quality magnetic garnet films of specialised compositions are grown on single crystal gadolinium gallium garnet (GGG) substrates by liquid phase epitaxy (LPE) for bubble domain memory applications. Here GGG has proved to be an almost ideal substrate. Yttrium iron garnet (YIG) films grown by LPE on GGG have received attention for microwave delay line and filter applications. However, to date, no information is available on the loss characteristics of GGG at microwave frequencies. Such microwave losses lead to increased insertion loss for devices utilising epitaxial YIG on GGG. They can be of particular significance if the necessary microstrip circuitry is integrated onto the GGG substrate, as well. For example, during an investigation of composite epitaxial YIG on GGG-single crystal rutile dielectric resonators for ultra-low loss X-band bandpass filters, significant losses attributable to GGG were observed. The object of this paper is to give specific data on GGG losses obtained from this investigation.

## DIELECTRIC LOSS MEASUREMENTS

Initial dielectric loss, $\tan \delta_d$, measurements were performed using the ASTM standard method; designation: F131-70(1). Here the permittivity and $\tan \delta_d$ data are derived from the perturbation in resonant frequency and Q when a thin rod of the dielectric is inserted into an E field maximum of a $TE_{103}$ waveguide cavity. GGG samples of dimensions 15 mm x 1 mm x 1 mm were measured at 9.3 GHz. The $\tan \delta_d$ of GGG was $3 \times 10^{-4}$. In comparison the $\tan \delta_d$ of single crystal yttrium aluminium garnet (YAG) rod of the same dimensions was better than the sensitivity of the apparatus, $2 \times 10^{-4}$. The difference between $\tan \delta_d$ for GGG and $\tan \delta_d$ for YAG may be due to a small r.f. magnetic field in the sample vicinity. Nevertheless, these measurements establish that GGG is an acceptable microwave dielectric material.

## MAGNETIC LOSS MEASUREMENTS

Magnetic bias field dependent losses of GGG were measured at X-band in the structure shown in Figure 1. Input coupling to the 12 mm diameter cut-off circular

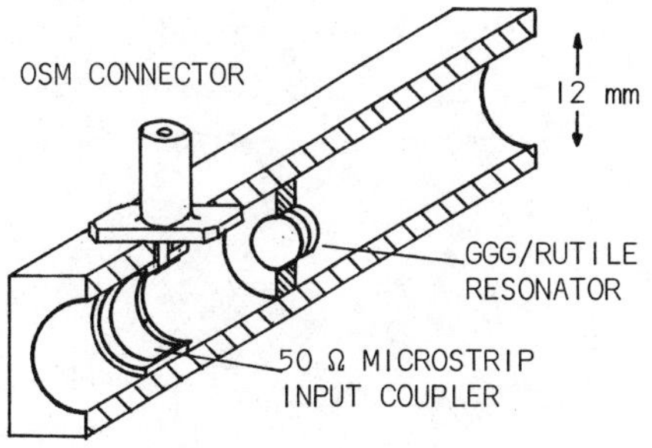


FIGURE 1 X-BAND RESONATOR MEASUREMENT STRUCTURE

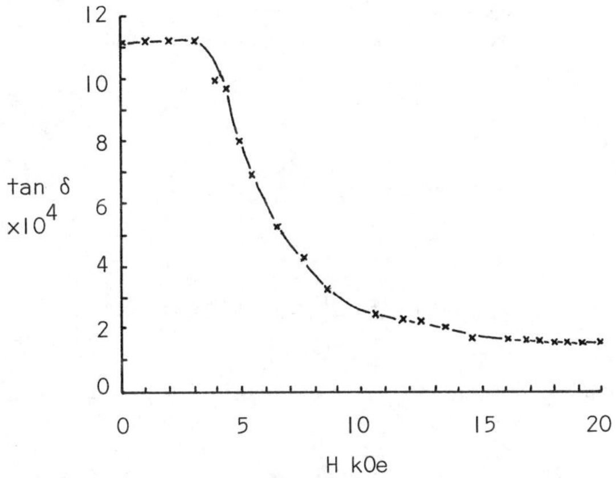


FIGURE 2 PLOT OF TAN δ VERSUS BIAS FIELD, H, AT 8.74 GHz FOR COMPOSITE GGG-RUTILE RESONATOR OF DIAMETER 6.25 mm - GGG THICKNESS, 0.5 mm, RUTILE THICKNESS, 0.7 mm

waveguide was by 50 Ω microstrip of electrical length λ/3(2). Coupling to the composite GGG-rutile resonator under test for the $TE_{10\delta}$ mode was adjusted by the microstrip to resonator spacing. To reduce conduction losses resonator diameters were chosen to be 6.25 mm. Unloaded $Q(Q_u)$ measurements were obtained from an HP 8410A network analyser with polar display, where $\tan \delta$ was derived as $1/Q_u$. Figure 2 shows the variation of $\tan \delta$ as a function of magnetic bias field for a composite GGG-rutile resonator formed from a 0.5 mm thick disc, cut from the <110> plane, and bonded to an 0.7 mm thick single crystal rutile disc. The in-plane magnetic bias field was applied along the GGG <100> direction.

The rutile disc alone resonated with a $Q_u > 6000$ ($\tan \delta < 1.6 \times 10^{-4}$) at 9.04 GHz. This was the limit for the sensitivity of the apparatus. The GGG disc, at zero bias field, reduced the $TE_{10\delta}$ resonant frequency to 8.74 GHz and the $\tan \delta$ to $1.1 \times 10^{-3}$. Note from Figure 2 that $\tan \delta$ monotonically decreases to $1.6 \times 10^{-4}$ at bias fields above 18k Oe which approximates the $\tan \delta$ of the rutile disc. The GGG magnetic loss, $\tan \delta_m$, versus magnetic bias field can be calculated by subtracting the high field loss from the measured $\tan \delta$.

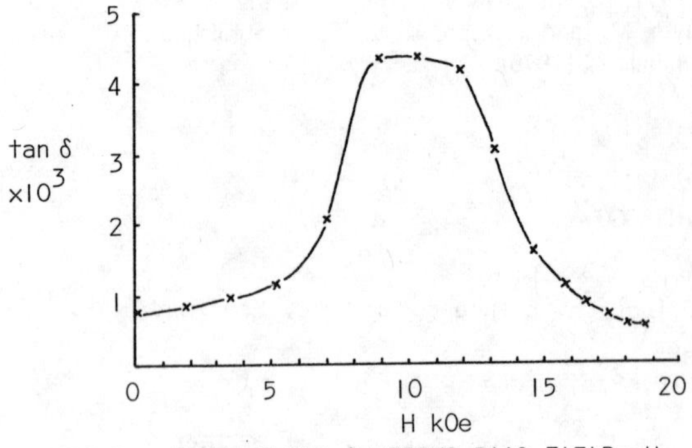


FIGURE 3 PLOT OF TAN $\delta$ VERSUS BIAS FIELD, H, AT 30.07 GHz FOR GGG RESONATOR OF DIAMETER 2mm AND LENGTH 3 mm

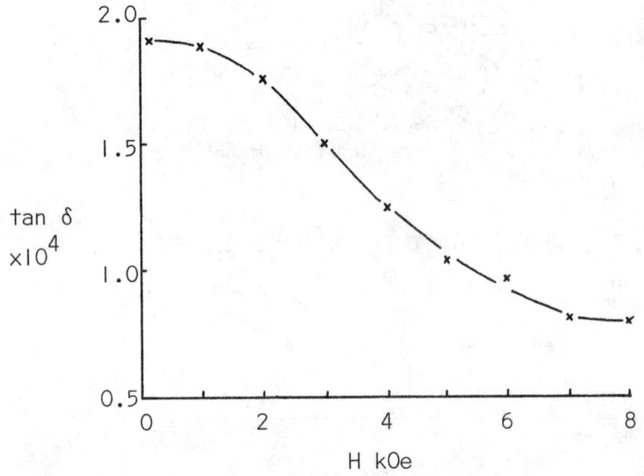


FIGURE 4 PLOT OF TAN $\delta$ VERSUS BIAS FIELD, H, AT 2.038 GHz FOR COMPOSITE GGG-RUTILE RESONATOR - GGG THICKNESS, 2 mm, AND DIAMETER 18 mm, RUTILE THICKNESS, 6 mm, AND DIAMETER 20 mm

No dependence of tan $\delta$ on orientation of the bias field in the <110> plane was found.

Similar results were obtained from single crystal GGG cylinders having a diameter 6.25 mm and 7.65 mm length, which resonated alone in the $TE_{101}$ mode at 11.5 GHz. Since Czochralski grown GGG (3) has typically an excess of 0.7% Gd, polycrystalline GGG resonators of stoichiometric and gadolinium excess composition were prepared by conventional sintering and pressing techniques. Loss measurements on these samples showed identical behaviour to the single crystal GGG and no detectable dependence upon excess Gd. Measurements were repeated with YAG resonators of the same dimensions which gave a tan $\delta$ of $1.6 \times 10^{-4}$ independent of magnetic bias field.

On the basis of the X-band measurements the bias field dependent loss was attributed tentatively to paramagnetic resonance of the $Gd^{3+}$ ion. To confirm this, measurements were performed at 30 GHz using a similar test structure to that shown in Figure 1 but with iris coupling to the WR28 feed waveguide and a cut-off waveguide of diameter 4.4 mm. Single crystal GGG resonators of 2 mm diameter and 3 mm length were used. These resonated at 30.07 GHz in the $TE_{101}$ mode. Unloaded Q values were deduced from measurements of the maximum return loss and the 3 dB bandwidth of the resonance. The variation of tan $\delta$ with magnetic bias field is shown in Figure 3 and demonstrates the full resonant absorption behaviour. The field for maximum absorption agrees with the 2.8 MHz/Oe (equivalent to a splitting factor g of 2) field dependence expected from the $Gd^{3+}$ ion(3). The measured linewidth of the absorption is 7k Oe, explaining the absence of a peak at X-band.

A YAG resonator of the same dimension as the GGG showed a tan $\delta$ of $3.25 \times 10^{-4}$ independent of bias field. Since the tan $\delta$ of GGG was $5 \times 10^{-4}$ at 18k Oe, magnetic losses may not have been completely removed from the GGG at this field.

Microwave devices using epitaxial YIG on GGG substrates are at present studied in the range 1 to 12 GHz. Measurements on GGG at 2 GHz have shown that losses here are significantly lower than at X-band. Measurements of $Q_u$ using an HP network analyser with polar display were performed on a composite resonator formed from a 2 mm thick by 18 mm diameter <111> oriented GGG disc bonded to a 6 mm thick by 20 mm diameter single crystal rutile disc. The test jig was similar to that shown in Figure 1 but scaled up for operation at 2 GHz. The rutile disc alone resonated at 2055 MHz with a $Q_u$ of 13000. The GGG disc reduced the resonant frequency to 2038 MHz. Figure 4 shows the variation of tan $\delta$ with bias field Note that tan $\delta$ decreases from $2 \times 10^{-4}$ at zero bias field to $8 \times 10^{-5}$ at bias fields above 8k Oe. Tan $\delta_m$ for GGG at 2 GHz is calculated to be $1.3 \times 10^{-4}$ at zero bias field.

## DISCUSSION

Electron paramagnetic resonance of $Gd^{3+}$ in diamagnetic garnets(4), specifically YGG and YAG, has received attention in relation to the anisotropy of GdIG. Results have shown that $Gd^{3+}$ in the garnet structure exhibits seven or more well-defined spectral lines with a splitting factor g of ~2, depending on the orientation. In the measurements reported in this paper, the broad linewidth is attributed to close coupling between $Gd^{3+}$ ions in the GGG leading to a merging of the single ion spectral lines into a broad absorption of linewidth ~7k Oe and an effective g factor of ~2, without orientation dependence.

Losses measured in GGG at low microwave frequencies, 2 GHz, (tan $\delta_m \doteqdot$ tan $\delta_d \doteqdot 10^{-4}$) will not significantly effect the performance of microwave delay lines or filters. At X-band frequencies and above the magnetic loss (tan $\delta_m = 10^{-3}$) may significantly effect device performance particularly in view of the ultra-low f.m.r. linewidth (<0.2 Oe)(5) reported recently for epitaxial YIG films grown by LPE.

## ACKNOWLEDGEMENTS

The authors wish to thank the Department of Chemistry of Edinburgh University for use of their 'Varian' electromagnet.

This work has been carried out with support of the Procurement Executive, Ministry of Defence, sponsored by DCVD.

## REFERENCES

1. ASTM Designation: F131-70.
2. W H Harrison, IEEE Trans, MTT-16, 4, 210-218, 1968.
3. C D Brandle and R L Barns, J Crystal Growth, 26, 169-170, 1974.
4. J Overmeyer et al, 'Paramagnetic Resonance in Solids', Academic Press, 224-233, 1960.
5. H L Glass and M T Elliott, to be published in J Crystal Growth.

# SYNTHESIZED LINEARLY DISPERSIVE MICROWAVE YIG DELAY LINE WITH WIDE INSTANTANEOUS BANDWIDTH

A. Platzker and F.R. Morgenthaler
Chu Associates, Littleton, Mass. 01460 and Department of Electrical Engineering and Center for Materials Science and Engineering
Massachusetts Institute of Technology, Cambridge, Mass. 02139

ABSTRACT

Our recently reported magnetic field synthesis procedure was utilized to construct a series of broad band, low loss linearly dispersive magnetoelastic microwave delay lines. High performance device operation was obtained over the entire tested frequency range of .5 to 4 GHz. The delay lines consisted of a fractional centimeter long (100) YIG rod, immersed in a nonuniform d.c. magnetic field, generated by soft iron pole pieces whose shape was uniquely synthesized to provide the desired delay characteristics. Typical operating parameters at a fixed d.c. magnetic field bias are: second echo-free instantaneous bandwidth 1000 MHz; frequency dispersion .5 - 1.5 μsec/GHz; two way untuned insertion loss 32 ± 2.5 db and input/output isolation 15-20 db.

The most striking features of the newly developed device, are the very broad instantaneous bandwidth, coupled with a very high degree of linearity of the delay time/frequency relation. These highly desirable features were achieved without sacrifice in insertion loss which remains relatively low and uniform across the operating frequency band.

## INTRODUCTION

Several attempts have been made in the recent past to build wide band, linearly dispersive magnetoelastic delay lines operating at microwave frequencies.[1,2] Such devices if realizable, would find extensive use in many applications dealing with signal forming and processing.

The approximate total delay time (elastic and spin wave) of the first echo is

$$T = \frac{2}{v}\left(z + \frac{\omega}{\gamma H'(z)}\right) \qquad (1)$$

where v, z, ω, γ and H'(z) are respectively the transverse elastic velocity, the distance between the turning point and the front face of the sample, the radian frequency, the gyromagnetic ratio and the spatial derivative of the z-directed axial magnetic field $H_z(r=0)$ at the turning point. As seen from (1), the simplest axial field profile needed to achieve the desired delay characteristics is a linearly varying one where H'(z) is the constant H'. Since the internal magnetic field profile of a ferromagnetic rod immersed in a uniform d.c. magnetic field bias is highly nonlinear, a naked rod cannot be utilized. Previous workers resorted to the use of other geometries such as whole or truncated pyramids or cones in order to provide an internal magnetic field that varied approximately linearly with position. In one class of approaches, single crystals of these geometries were used; in another, single crystal rods were surrounded by ceramic magnetic sleeves and/or truncated cones to achieve the hoped for conditions over a portion of the single crystal rod. The operating characteristics of all such devices revealed several shortcomings, namely: high insertion loss with large variations of such losses over the operating band; narrower than expected useable frequency bandwidth and for certain applications, a still intolerably high degree of nonlinearity of the delay time vs. frequency relation. Due to the tight tolerances and relatively large volumes required, the fabrication costs of the single crystal pyramids and cones were very high. The methods employing ceramic sleeves and cones, on the other hand, turned out to require trial and error. Small variations in otherwise identical configurations gave rise to internal fields which caused holes in the operating frequency band. Such gaps were due to either inefficient electromagnetic/spin wave conversion or to spin wave defocusing effects.

In contrast, our recently reported field synthesis procedure[3] enabled us to construct broad band, low loss, linearly dispersive delay lines by utilizing YIG rods immersed in a nonuniform d.c. magnetic field generated by soft iron pole pieces. The shape of these pole pieces was uniquely determined by the synthesis procedure according to the desired operating parameters.

## DESIGN CONSIDERATIONS AND RESULTS

For linearly varying internal field profiles

$$\frac{\omega}{\gamma} \simeq H_z(r=0) = H_o + zH' \qquad (2)$$

and by combining (1) and (2) we obtain

$$T = \frac{2}{vH'}\left(\frac{2\omega}{\gamma} - H_o\right) \qquad (3)$$

From (3) we readily obtain the relation between the dispersion factor D and the field gradient H' as $D = dT/d\omega = 4/(\gamma v H')$. For YIG, $D \simeq 370/H'$ μsec/GHz. The spin wave focusing condition[4] is met for all values of H' but other considerations impose upper and lower bounds on its magnitude. To ensure efficient spin wave/elastic conversion, we require $H' < H'_{crit}$ where $H'_{crit}$[5] increases linearly with ω. For YIG at 1 GHz, it equals 1.7 KOe/mm. If H' assumes too low a value on the other hand, the e.m./spin wave conversion efficiency decreases. This occurs because the turning point, which with decreasing H' moves inward from the front face eventually becomes too distant from the antennas.

The width of the second echo-free frequency band $\Delta\omega = \omega_2 - \omega_1$ is also obtained from (3) as $\Delta\omega = \omega_1 - \frac{1}{2}\gamma H_o$. $\omega_2$ is the frequency at which the delay time of the first echo equals the delay time of the second echo (or twice the delay of the first echo) at frequency $\omega_1$. To maximize the bandwidth, $H_o$ should be chosen to be as low as possible.

Based on our earlier experience with ceramic cones, we reasoned that the value of $H_o$ should be chosen to be compatible with the condition that the entire bandwidth lies within the magnetostatic manifold boundaries. This leads to the inequalities $\gamma H_o \leq \omega_1$ and $\omega_2 \leq \gamma\sqrt{H_o(H_o + M)}$ where M is the magnetization, 1780 Oe for YIG. Since the e.m. wave passes through the magnetostatic region before ultimately converting to a spin wave at the turning point, we believed that a cutoff condition would increase the insertion loss to an unacceptable level. However, since as we report below, acceptable losses have been achieved even upon a violation of these conditions, the restrictions they impose are obviously not as critical as we had thought.

Our goal was to build a two port device operating at L band with a high degree of linearity of the delay time/frequency relation over the widest possible second echo-free bandwidth. A two-way untuned insertion loss of 35 db with overall variations of no more than 5 db across the band was acceptable. The dispersion factor D was to lie in the range of .5-1 μsec/GHz.

To meet the above requirements, we used our synthesis procedure to design soft iron pole pieces that would generate an internal field of 300 +300z(Oe) where z is in mm., on the z-axis of a 3 mm. diameter (100) YIG rod. Briefly the synthesis procedure calculates the field and the magnetization on the rod surface from its given value on the axis. This is done by properly solving the equations governing the d.c. portions of $\vec{H}$ and $\vec{M}$ and not by any approximations similar to the famous Sommerfeld approximation. The procedure then matches the air/material boundary conditions and generates equipotential surfaces. A high permeability material such as soft iron, cut along two of these surfaces, will generate the desired field on the axis.

A cross-section of the device is shown in Fig.1. The YIG rod is placed inside a Teflon piece which separates the two soft iron pole pieces. The input and output antennas are loops of thin wire soldered to the end of semirigid cables at quadrature and pressed by a Teflon disc against the front face of the rod (only one cable is shown for clarity). The whole structure was placed between the pole pieces of an electromagnet.

Several sets of pole pieces with different diameters were built and tested for a dependence of the device performance on the diameter. No degradation in performance was observed when the pole pieces diameter was reduced from 2 1/2" to 1" or when the length of the YIG rod was halved from 1 cm. to .5 cm. These dimensions do not represent the limit; work is currently under way to further miniaturize the device.

The delay time vs. frequency relation of a 0.5 cm long rod immersed in 1" diameter pole pieces is shown in Fig. 2A. A comparable figure for a naked rod immersed in a d.c. field is shown in Fig. 2B. The figures show a series of echoes (from single frequency narrow pulses) vertically displaced by an amount linearly proportional to the frequency. The horizontal axes depict the delay times.

As seen in Fig. 2A, the delay time/frequency relation is extremely linear over the bandwidth of 800 MHz. No deviation from linearity could be detected with a scope resolution of 5 nsec. The insertion loss across the band was 32 $\pm$ 2.5 db and input/output isolation 15-20db. By fitting the curve in the figure to (3) we obtain the value of

$$H_z(r=0)=73 + 740z \quad (\text{Oe}) \qquad (4)$$

for the field along the z-axis. The dispersion factor D = 0.5 μsec/GHz. Equation (4) for $H_z(r=0)$ differs appreciably from the value of 300 + 300z(Oe) assumed in our synthesis procedure. This discrepancy however did not prevent us from meeting our design goals, especially the high degree of linearity over a wide bandwidth.

The reason for the discrepancy lies not in our synthesis procedure but in its practical implementation. In particular, the design did not take into account the hole in the front pole piece which was drilled to accomodate the antennas. The presence of this hole sufficiently distorts the field in the active area of the device which extends only 2/3 of a mm. from the rod face into the crystal. An amended synthesis procedure that takes this into account is currently being implemented. Notice also that the value of $H_o$ = 73 Oe. violates the inequality $\omega_2 \leq \gamma\sqrt{H_o(H_o + M)}$ but causes no degradation in the insertion loss.

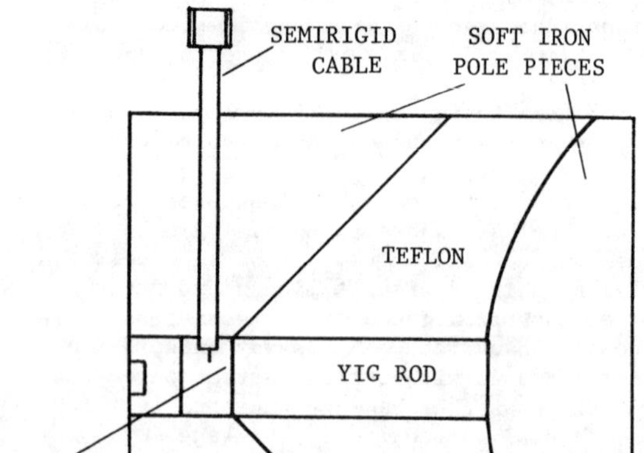


Fig.1 Cross Section of the Delay Line

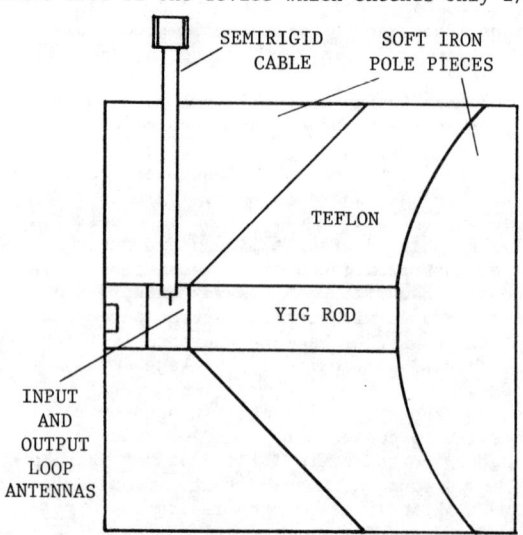

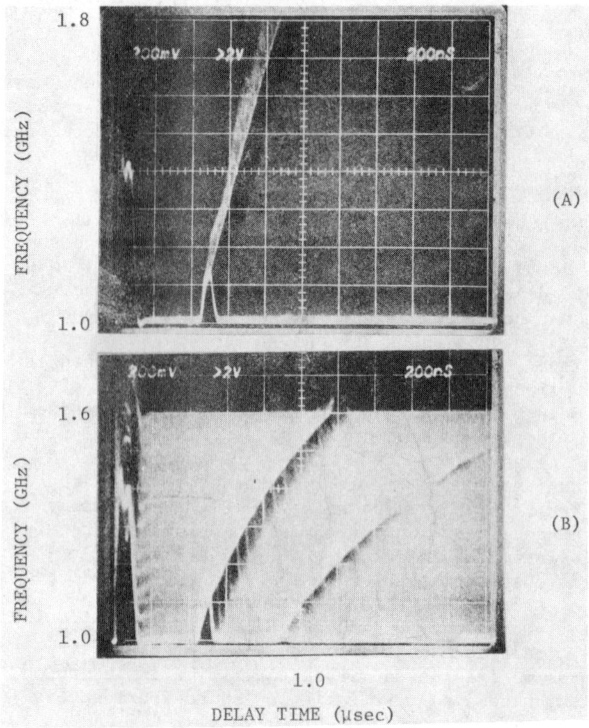


Fig. 2 Frequency vs. Delay Time

(A) YIG Rod with synthesized pole pieces. Excellent linearity, second echo-free bandwidth 800 MHz.

(B) YIG rod alone. Poor linearity, second echo-free bandwidth 400 MHz.

The same device was tested at high frequencies, through 4 GHz, and its performance in this range was comparable to its performance at L band. Noticeable was the extremely high degree of linearity and broad instantaneous bandwidth which increased to 1.3 GHz at the high frequency end. At the same time the dispersion factor D decreased to ∿ 0.22 μsec/GHz. The insertion loss did not increase however and was actually marginally lower. Plans are now underway to test similar devices at higher frequencies. In an earlier part of our program we tested a large number of ceramic cone configurations to achieve the same design goals outlined above. Our best effort resulted in a device of poorer linearity over the narrower instantaneous bandwidth of 550 MHz. Our synthesis procedure has already enabled us to build far better devices than this.

REFERENCES

1. For a general review and reference to pre-1970 work see B.A. Auld in Appl. Solid State Science, 2, Academic Press (1971).
2. H. Dötsch, JAP, 43, 1923 (1972).
3. F.R. Morgenthaler, AIP Conference Proc., No. 24, Magnetism and Magnetic Materials, 503 (1974).
4. F.R. Morgenthaler, IEEE, MAG-8, 550 (1972).
5. E. Schlömann, R.I. Joseph, JAP, 35, 2382 (1964).

# MAGNETOOPTICAL STUDIES ON SPIN-REORIENTATION IN RARE EARTH ORTHOFERRITES

N.Koshizuka, K.Hayashi, M.Suzuki*, and T.Tsushima
Electrotechnical Laboratory, Tanashi, Tokyo 188, Japan


## ABSTRACT

Several types of spin-reorientation (SR) in some of the $RFeO_3$ are studied by Faraday rotation measurements ; rotational SR of $\Gamma_4 \rightarrow \Gamma_2$ type in $(ErSm)FeO_3$, ($Co^{2+}$, $Ti^{4+}$) doped $YFeO_3$, and abrupt SR of $\Gamma_4 \rightarrow \Gamma_1$ type in $DyFeO_3$. Observations of SR by Faraday rotation were made in these crystals with incident light parallel to the optical axis. In relation with the decrease of $Fe^{3+}$ ion's anisotropy at $T_{SR}$, an abrupt decrease of the coercive force are found in these systems.

In general, Faraday rotation in $RFeO_3$ originates from $Fe^{3+}$ ions in the visible and near IR regions, while $R^{3+}$ ion's contribution to the Faraday rotation was observed for the wavelengths corresponding to the electronic transitions of $R^{3+}$ ions in $ErFeO_3$ and $DyFeO_3$ at low temperatures. In $DyFeO_3$, a large contribution of $Dy^{3+}$ ions was observed at ∿ 1.2 μm in the Faraday spectrum, and it is confirmed that the $Dy^{3+}$ moments are polarized along the c-axis in zero applied field above $T_{SR}$. Magnetic field induced SR was also observed in $DyFeO_3$, and the temperature dependence of the critical field was obtained as $H_{SR} \propto |T-T_{SR}|^{3/4}$.


## INTRODUCTION

Spin-reorientation (SR) phenomena of the rare earth orthoferrites ($RFeO_3$) have been studied by many experimental techniques such as neutron diffraction, Mössbauer effect, optical absorption, specific heat, and conventional magnetic measurements[1]. Recently we have reported on the SR in $(ErSm)FeO_3$ studied by Faraday rotation (FR) measurements with the incident light parallel to the optical axis in order to avoid the birefringence in such biaxial crystals[2]. In this work, further studies on the SR in $(ErSm)FeO_3$, $Co^{2+}$ and $Ti^{4+}$ doped $YFeO_3$, and $DyFeO_3$ are presented.

From the experimental and theoretical works on SR in $RMO_3$(M:Fe,Cr), it is convinced that the anisotropic exchange interactions between the $R^{3+}$ and $M^{3+}$ ions play an important role in the SR of these crystals[3]. Following this mechanism, the increase of the $R^{3+}$ spin polarization causes the SR of $Fe^{3+}$ or $Cr^{3+}$ ions in $RMO_3$.

It seems interesting to know the temperature and magnetic field dependences of the $R^{3+}$ moments in the ordered spin system of $Fe^{3+}$ ions in $RFeO_3$ by FR measurements.

The easy axes of magnetization of $(ErSm)FeO_3$[4] and $YFeO_3$:(Co,Ti)[5,6] are parallel to the c-axis at $T > T_2$, and rotate from the c-axis to the a-axis with decreasing temperature between $T_2$ and $T_1$, and coincide with the a=axis at $T < T_1$ (rotational type SR of GxFz→FxGz). While in $DyFeO_3$[7], the spontaneous magnetization is also parallel to the c-axis at $T > T_{SR}$, but it disappears abruptly at $T_{SR}$ and it takes an antiferromagnetic spin configuration at $T < T_{SR}$ (abrupt SR of GxFz→AxGy).

## EXPERIMENTAL

Single crystals were made by a floating zone technique using an IR heating type furnace. Single crystals with various compositions of $Er_xSm_{1-x}FeO_3$ (x=0, 0.3, 0.5, 0.7, and 1.0) and $YFe_{1-x}Co_{x/2}Ti_{x/2}O_3$ (x=0, 0.0042, 0.0068, 0.0102, and 0.0125) were grown in order to get the specimens with various SR temperatures[4,5,6]. FR measurements were made using an automatic recording polarimeter (Jasco MOE-7). Low temperature measurements down to 10 K were made with use of a cryochip refrigerator (Air Products and Chemicals). The coercive force $H_c$ was determined from the magnetic field dependences of FR (H=0 ∿ ±8 kOe).

The optical axes of these crystals lie in the a=plane and are inclined about 50° from the c-axis in the red[8,9] and near IR regions[9]. The directions of the propagation vector of the light and magnetic field were set parallel to one of the optical axes of the crystals

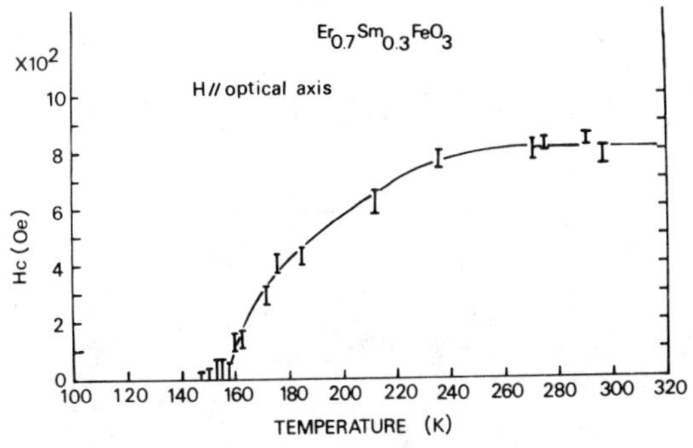


Fig.1 Temperature dependence of coercive force Hc obtained from the applied field dependence of Faraday rotation in $Er_{0.7}Sm_{0.3}FeO_3$ ($T_2$=160K)

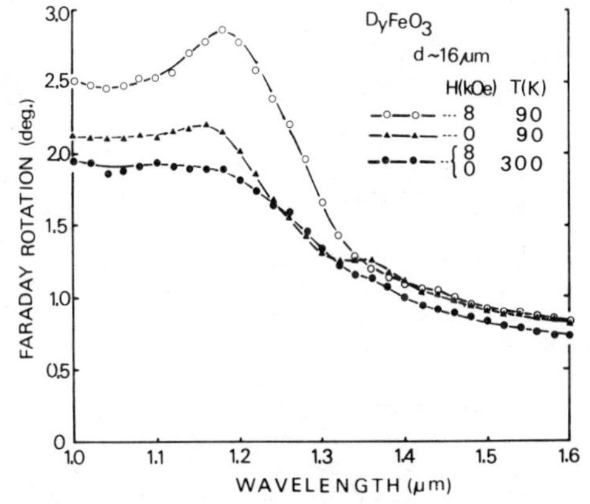


Fig.2 Faraday rotation spectra of $DyFeO_3$. H and T are applied magnetic field and temperature, respectively. d is the sample thickness.

## RESULTS AND DISCUSSIONS

1) $(ErSm)FeO_3$ system : The magnitude of FR for λ = 0.63 μm is 4000 ∿ 4500°/cm above $T_2$, while it becomes zero within the experimental error below $T_1$ in zero applied field[2]. When a magnetic field is applied, FR becomes observable even at low temperatures below $T_1$ because of the spin polarization of $Fe^{3+}$ ions along the c-axis. However, in $ErFeO_3$, for instance, when cooled down to 10 K, the FR originated from the $Fe^{3+}$ ions becomes very small and the FR for ∿ 1.5 μm from the $Er^{3+}$ ions is enhanced with the increase of spin polarization of the $Er^{3+}$ ions along the magnetic field. Any hysteresis phenomena of the FR during a cycle of temperature change were not found as far as the measurements were done after the magnetization of $Fe^{3+}$ ions was saturated at $T > T_2$. It is interesting to get temperature dependences of $H_c$ related with the variation of $Fe^{3+}$ ion's anisotropy energy near $T_{SR}$. In $Er_{0.7}Sm_{0.3}FeO_3$, $H_c$ is about 800 Oe at T ∿ 260 K, but it decreases with temperature and becomes 50 Oe or less at $T_2$= 160 K as shown in Fig.1. This phenomenon corresponds to the decrease of anisotropy energy of $Fe^{3+}$ ions near $T_{SR}$.

2) $YFeO_3$:(Co,Ti) system : Rotational type SR was also observed by the FR measurements. Wide temperature regions between $T_1$ and $T_2$ ($T_1$ = 281 K, $T_2$ = 333 K for x=0.0125, for example) were observed consistently with the magnetic measurements[5,6]. In contrast to the cases of $(ErSm)FeO_3$, hysteresis loops were found during a cycle of temperature change.

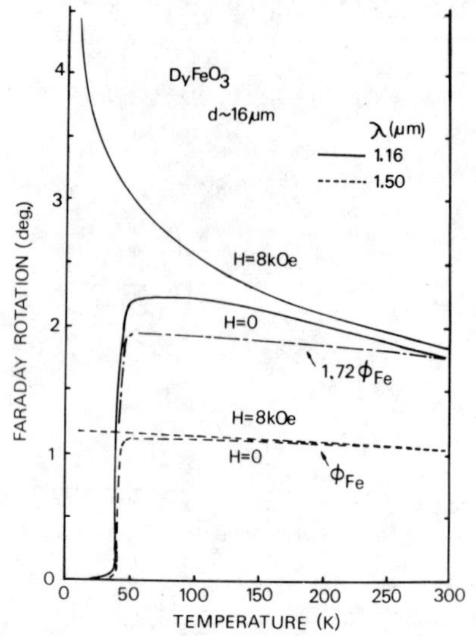


Fig.3 Temperature dependence of Faraday rotation for $DyFeO_3$. $\lambda$ is the wavelength of light.

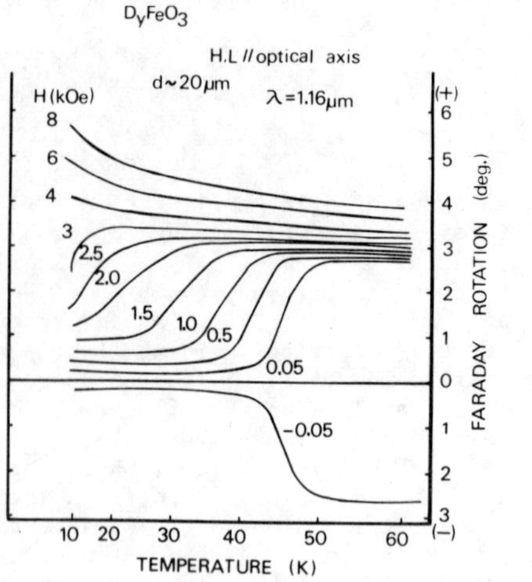


Fig.4 Temperature dependence of Faraday rotation for $DyFeO_3$ near $T_{SR}$ in various applied fields. L is the propagation vector of light.

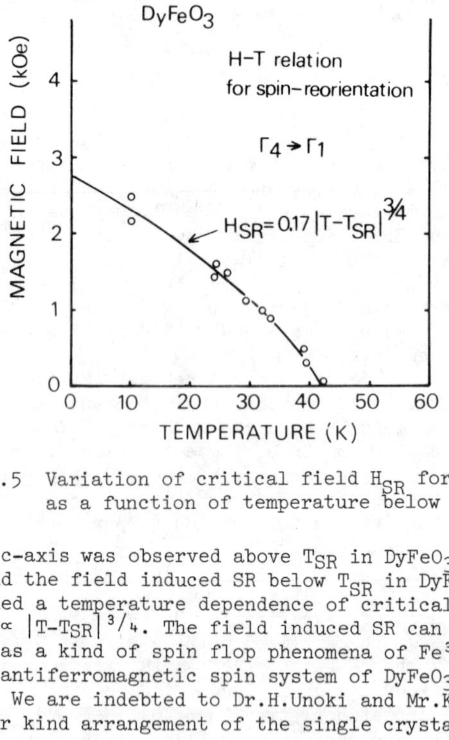


Fig.5 Variation of critical field $H_{SR}$ for $DyFeO_3$ as a function of temperature below $T_{SR}$.

3) $DyFeO_3$ : FR spectra of $DyFeO_3$ are shown in Fig. 2. A dispersion type anomaly exists at ∿ 1.2 μm which corresponds to the $^6H_{15/2} \rightarrow {}^6H_{9/2}$ transition of $Dy^{3+}$ ions[9]. From the magnetic field and temperature dependences of the FR spectra, it becomes evident that $Dy^{3+}$ ions contribute to the FR at ∿ 1.2 μm. In Fig.3, temperature dependences of the FR for $\lambda$ = 1.16 and 1.50 μm are shown taking the applied field as the parameter. From this figure the SR is found to occur abruptly at ∿ 40 K, and the FR for $\lambda$ = 1.16 μm is believed to originate from both contributions of $Fe^{3+}$ and $Dy^{3+}$ ions in contrast to the sole contribution of $Fe^{3+}$ ions for $\lambda$= 1.50 μm. A dot-dashed line (1.72 $\phi_{Fe}$) means the estimated temperature dependence of FR due to $Fe^{3+}$ ions for 1.16 μm. The difference between the solid line for H=0 and the dot-dashed line shows the contribution of $Dy^{3+}$ spin polarization along the c-axis in the exchange field generated by $Fe^{3+}$ ions. Comparing with the magnetization data on $DyFeO_3$[7], we find that net FR at 1.16 μm $\phi \propto M_{Fe} + 0.1M_{Dy}$, where $M_{Fe}$ and $M_{Dy}$ are the magnetization of $Fe^{3+}$ and $Dy^{3+}$ ions, respectively. The FR for $\lambda$ = 1.16 μm and H = 8 kOe abruptly increases at low temperatures, which is accounted for by the increase of magnetization of $Dy^{3+}$ ions along the b-axis. Temperature dependences of FR for $\lambda$ = 1.16 μm in various applied fields are shown in Fig.4. It appears that there exist temperature widths in the SR of $DyFeO_3$, however, these widths may be due to a demagnetization effect. With the increase of applied field, the SR occurs at lower temperatures than $T_{SR}$ for H=0. Magnetic field dependences of FR were also measured, and we observed the field induced SR of $DyFeO_3$ below $T_{SR}$. Critical fields $H_{SR}$ where SR is induced at a fixed temperature are given as a function of temperature in Fig.5, and it turns out that $H_{SR} \propto |T-T_{SR}|^{3/4}$. Since the antiferromagnetic spin axis of $Fe^{3+}$ ions is parallel to the b-axis below $T_{SR}$ (AxGy), the field induced SR can be accounted for as a kind of spin flop phenomena of $Fe^{3+}$ ions due to the increase of field component along the b-axis : the spin flop may be explained by a free energy consideration on the $Fe^{3+}$ spin system in the applied field, the single ion anisotropy field of $Fe^{3+}$ ions, and the effective field due to the anisotropic exchange interaction between $Dy^{3+}$ and $Fe^{3+}$ ions in $DyFeO_3$[3].

## CONCLUSION

The SR phenomena were studied by the FR measurements in $(ErSm)FeO_3$, $YFeO_3$:($Co^{2+}$,$Ti^{4+}$), and $DyFeO_3$. We found appreciable contributions of $R^{3+}$ ions to the FR in the wavelength regions associated with the electronic transitions of $R^{3+}$ ions in $ErFeO_3$ and $DyFeO_3$ at low temperatures. The spin polarization of $Dy^{3+}$ ions along the c-axis was observed above $T_{SR}$ in $DyFeO_3$. We also found the field induced SR below $T_{SR}$ in $DyFeO_3$ and obtained a temperature dependence of critical field as $H_{SR} \propto |T-T_{SR}|^{3/4}$. The field induced SR can be accounted for as a kind of spin flop phenomena of $Fe^{3+}$ ions in the antiferromagnetic spin system of $DyFeO_3$.

We are indebted to Dr.H.Unoki and Mr.K.Oka for their kind arrangement of the single crystal preparation.

*Research fellow from Tokyo Electrical Engineering College.

## REFERENCES

1) R.L.White, J.Appl.Phys. 40, 1061 (1969)
2) N.Koshizuka, M.Hirano, T.Okuda, S.Nakamura, H.Hiruma and T.Tsushima, AIP Conf.Proc. No,24, 61 (1975)
3) T.Yamaguchi, J.Phys.Chem.Solids 35, 479 (1974)
T.Yamaguchi and K.Tsushima, Bussei 14, 483 (1973)
4) R.C.Sherwood, L.G.Van Uitert, R.Wolfe, and R.C.LeCraw, Phys.Letters 25A, 297 (1967)
5) E.M.Gyorgy and L.G.Van Uitert, Proc.International Conference on Ferrites, Kyoto, 376 (1970)
6) M.Makino and Y.Hidaka, IEEE Trans.Mag. MAG-8, 449 (1972)
7) G.Gorodetsky, B.Sharon, and S.Shtrikman, J.Appl.Phys. 39, 1371 (1968)
8) W.J.Tabor, A.I.Akhutkina, and L.G.Van Uitert, J.Appl. Phys. 41, 3018 (1970)
9) M.V.Chetkin, J.U.S.Didosjan, and A.I.Akhutdina, IEEE Trans.Mag. MAG-7, 401 (1971)

# ANISOTROPIC LINEAR MAGNETIC BIREFRINGENCE AND MODULATION OF LIGHT IN SOME MAGNETIC COMPOUNDS.

J.P. JAMET and TRAN KHANH VIEN
Equipe de Magnétisme et d'Optique des Solides
92190 - BELLEVUE (FRANCE)

ABSTRACT

A linear modulation of light intensity by second-order magnetooptical interaction (linear magnetic birefringence (L.M.B.) is reported. Theoretical predictions and experimental curves are given for two cubic crystals: YIG and TbIG and for a rhomboedral crystal : $\alpha$ - $Fe_2O_3$ (Hematite).

## 1 - INTRODUCTION

Up to now, a linear modulation of the intensity of light is obtained by using the Faraday effect in magnetic crystals |1|,|2|. We have shown |3,4| that the same linear modulation can be performed by using the anisotropy of the linear magnetic birefringence (L.M.B.) (or Cotton-Mouton |5,6|) in magnetic crystals.

## 2 - LINEAR MODULATION OF LIGHT INTENSITY BY COTTON-MOUTON ANISOTROPY IN CUBIC CRYSTALS.

In transparent magnetic crystals, it has been shown |5,6| that the L.M.B. is described by a fourth rank tensor $f_{ijkl}$. The tensor form is dependent on the crystallographic symetry of the crystal. In cubic crystals it is dependent on three independent coefficients $f_{11}$, $f_{22}$ and $f_{44}$ and the anisotropy of the L.M.B. is characterized by the magnetooptical anisotropy coefficient: $\Delta f = f_{11} - f_{12} - 2f_{44}$.

As an example, figure 1 shows the evolution of the static L.M.B. in the case of a light propagation normal to a (110) plane of a YIG sample. The Cotton-Mouton phase difference is :

$$\phi_{CM} = \frac{q_0 f_{44}}{\sqrt{\varepsilon_r}} \left\{ \left[1 - \frac{\Delta f}{4 f_{44}}(3\sin^2\varphi - 1)\right]^2 + \frac{\Delta f}{f_{44}} \sin^2\varphi (3\sin^2\varphi - 1)\right\}^{\frac{1}{2}} M_0^2 \qquad (1)$$

where $\varphi$ is the angle between static magnetization $M_0$ and the [010] direction in the (110) plane.

Under this experimental condition, by adding a dynamic magnetic field $h_{rf}$ perpendicular to $\vec{M}_0$ and $\vec{k}_{ph}$ (light propagation vector), a linear modulation of light intensity is obtained as indicated by the relation :

$$\frac{\Delta I(t)}{I_0} \# \frac{\frac{\Delta f}{2f_{44}} \sin 2\varphi \left[2(12\sin^2\varphi - 5) + 3\frac{\Delta f}{2f_{44}}(3\sin^2\varphi - 1)\right]}{1 + \left(\frac{\Delta f}{4f_{44}}\right)^2 (3\sin^2\varphi - 1)^2 - \frac{\Delta f}{2f_{44}}(3\sin^2\varphi - 1)\cos 2\varphi} \frac{\Delta M(t)}{M_0} \qquad (2)$$

Fig. 1 shows both the orientation dependence of static L.M.B. and the light modulation in the (110) plane for YIG at 300K.

In the case of light propagation normal to a (111) plane, it can be shown that light modulation is zero if $h_{r.f.}$ is perpendicular to $\vec{k}_{ph}$ ; but $h_{r.f.}$ is parallel to $k_{ph}$ :

$$\frac{\Delta I(t)}{I_0} \# \frac{\sqrt{2} \cdot \frac{\Delta f}{2f_{44}} \cdot \sin 3\varphi}{6\left[1 + \frac{1}{3}\frac{\Delta f}{2f_{44}}\right]} \cdot \frac{\Delta M(t)}{M_0} \qquad (3)$$

Curve (———) of figure 2 shows the results obtained with a (111) disc of TbIG.

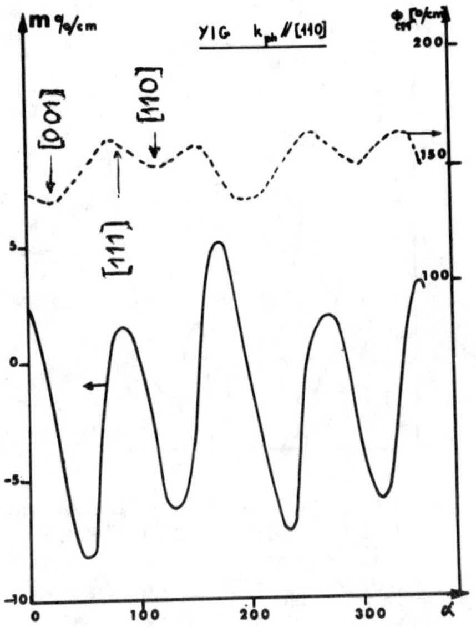


Fig. 1. - Orientation dependence of static L.M.B. (- - - -) and of the light modulation (———) in the (110) plane for YIG at 300°K and $\lambda$ = 1,15 $\mu$. $\alpha$ is the angle between $M_0$ and a reference direction in the plane .

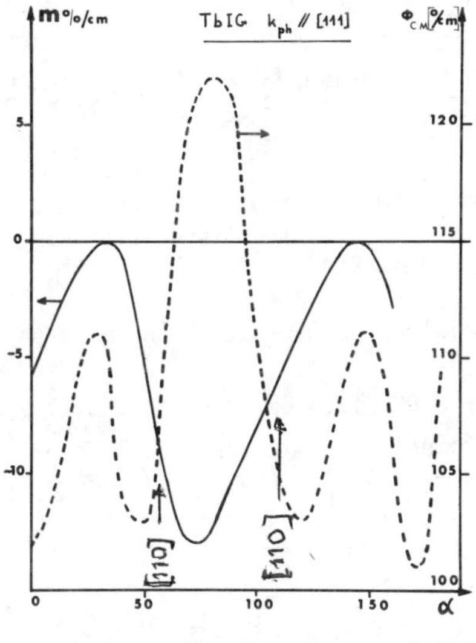


Fig. 2. - Orientation dependence of light modulation (———) in the (111) plane of TbIG at 300°K and $\lambda$ = 1,15 $\mu$. $\alpha$ is an angle between $M_0$ and a reference direction in the plane .

## 3 - STATIC L.M.B. AND LIGHT MODULATION IN RHOMBOEDRAL MAGNETIC CRYSTALS ($\alpha$ - $Fe_2O_3$ case).

Hematite $\alpha$ - $Fe_2O_3$ is a rhomboedral crystal |9| with a magnetic phase transition at the Morin temperature ($T_m \neq 260$°K). For $T < T_M$, $\alpha$ - $Fe_2O_3$ is antiferromagnetic with spins lying along the c-axis and for $T > T_M$, $\alpha$ - $Fe_2O_3$ is a weak-ferromagnet with canted spins in the c-plane |7|. We are concerned here with this high temperature phase.

### 3-1. L.M.B. of $\alpha$ - $Fe_2O_3$ at 300K.

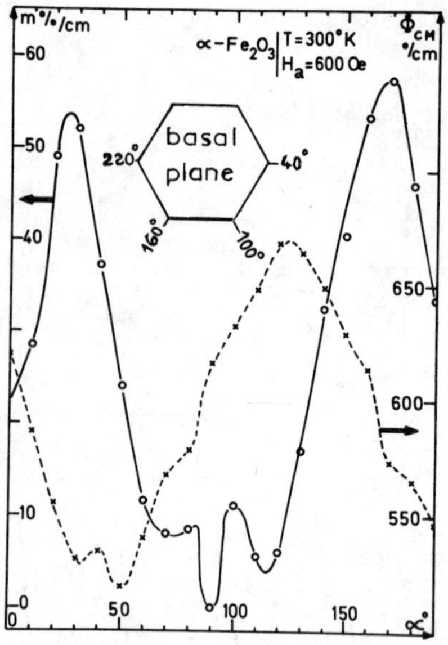


Fig. 3. - 180° periodicity of the orientation dependence of static L.M.B. (- - - -) and modulation (———) in the easy plane of $\alpha Fe_2O_3$ with $\vec{k}_{ph}$ making and angle of 1° with c axis. T = 300°K $\lambda$ = 1,15 $\mu$. $\alpha$ is an angle between $M_0$ and a reference direction of the basal plane .

By using the same calculation procedure as in |6| one can show that the L.M.B. is, for $\bar{3}m$ crystallographic symetry and for light propagation nearly perpendicular to the c-plane :

$$\Phi_{CM} \# \left[\frac{A^2}{16} - 2AE\theta(\cos 2\varphi - 2\cos 4\varphi)\right]^{1/2} \quad (4)$$

where A and E are dependent upon the tensor elements and $\theta$ is the small angle between c-axis and light propagation vector $\vec{k}_{ph}$ and $\varphi$ caracterising the magnetization direction in the basal plane.

This relation shows that L.M.B. must be isotropic if $\theta = 0$.

3-2. LINEAR MODULATION OF LIGHT INTENSITY.

By introducing, as in the case of cubic crystals, a time modulated magnetic field $h_{r.f.}$ perpendicular to $M_0$ in the c-plane, it is possible to obtain a linear intensity modulation :

$$\frac{\Delta I(t)}{I_0} \# \frac{2AE\theta(\sin 2\varphi - 4\sin 4\varphi)}{\frac{A^2}{16} - 2AE\theta(\cos 2\varphi - 2\cos 4\varphi)} \frac{\Delta M(t)}{M_0} \quad (5)$$

This relation shows that modulation exists only if $\theta \neq 0$, that is, if L.M.B. is anisotropic.

3.3. EXPERIMENTAL RESULTS.

We have used a disc-shaped sample of synthetic $\alpha$ - $Fe_2O_3$ (1.24 mm thickness and 3 mm - diameter). The c-axis makes a 1° angle with the normal to plane-faces. Figures (3) and (4) show two different experimental configurations.

As shown in figure (3), there is a strong anisotropy of the L.M.B. with a 180° - period dependence upon $\vec{M}$ orientation in the basal-plane. In this configuration, light propagation vector is perpendicular to the plane faces, that is $\theta \neq 1°$. In this case light modulation has the same 180° - period.

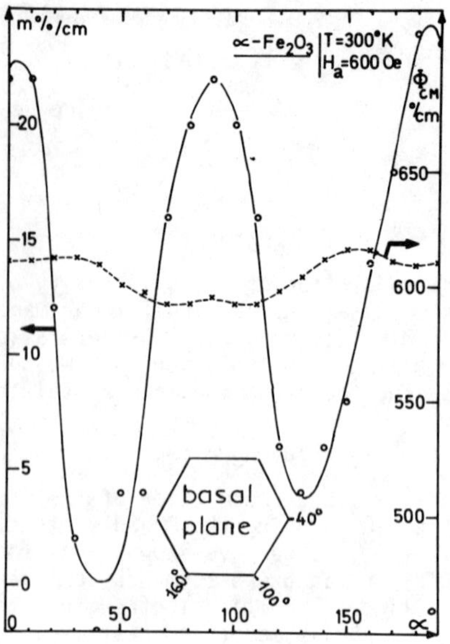


Fig. 4. - Light modulation rate (———) and $\Phi_{CM}$ (- - -) $\vec{k}_{ph}$ very close to c axis - $\lambda$ = 1,15 $\mu$ T = 300°K. $\alpha$ is an angle between $M_o$ and a reference direction of the basal plane

As shown in figure (4), there is also an anisotropy of the L.M.B. with a 180° - period component much smaller than in the figure (3) case and also 60° - period and 90° - period components. In this case, light propagation vector was not normal to the plane faces, but inside the sample, $\theta < 1°$. But in this case light modulation has a 90° - period.

In both cases, no modulation has been observed when $h_{r.f.}$ is parallel to $k_{ph}$. This is due to the very strong anisotropy in the c-direction.

We have used $\bar{3}m$ model in such a way that in relations (4) and (5) magnetization is always parallel to applied magnetic field : because of the strong anisotropy in the c-direction, it is not the case for magnetic field intensities we have used in the experiments. So the disagreement between the experiments and theoretical calculations performed by using $\bar{3}m$ model can be due to :

- angular misorientations of the sample.
- the fact that $\bar{3}m$ model does not take into account of a particular behaviour of magnetization in hematite. So that $\bar{3}m$ model should not be a good model to describe magnetooptical properties of hematite |8|.

CONCLUSION

Light modulation technique by introducing magnetization spatial variations is an important tool to study magnetic behaviour of magnetic crystals.

The authors wish to acknowledge R. Krishnan, N. Boccara and X. Oudet for helpful discussions and M. Malmanche for X-ray orientations.

REFERENCES

|1| R.C. Le Craw, I.E.E.E. Trans. Magn. 2 (1966), 304.

|2| R.W. Cooper, J.L. Page, Radio Electron. Eng. 39 (1970), 302.

|3| Tran Khanh Vien, J.P. Jamet, H. Le Gall and R. Krishnan, Opt. Comm. 13 (1975), 70.

|4| Tran Khanh Vien, J.P. Jamet, H. Le Gall and R. Krishnan, Revue de Physique Appliquée, tome 10, Nov. 75, 57.

|5| R.W. Pisarev, I.G. Sini, N.W. Kolpakova and Y.M. Yakolev, J.E.T.P. 33 (1971) 1175.

|6| Tran Khanh Vien, Thèse 3ème cycle, Paris 1975.

|7| C.G. Shull, W.A. Strauser and E.D. Wollaw, Phys. Rev. 83 (1951), 333.

|8| Tran Khanh Vien (to be published).

|9| L.Pauling, S.B. Hendricks, J. Amer. Chem. Soc. 47 (1925) 781.

# LIGHT SCATTERING IN $FeCl_2$ AT THE METAMAGNETIC TRANSITION

E. Yi Chen, J. F. Dillon, Jr. and H. J. Guggenheim
Bell Laboratories, Murray Hill, New Jersey 07974
Richard Alben
Yale University, New Haven, Connecticut 06520

ABSTRACT

Light scattering by coexisting magnetic phases, that is domain structure, has been observed previously in various ferro- and ferrimagnetic crystals, $CrBr_3$, EuO, MnBi and $Y_3Fe_5O_{12}$ and in the antiferromagnets $FeCl_2$ and $Dy_3Al_5O_{12}$. Recently Griffin[1] et al. reported that in $FeCl_2$ at the metamagnetic transition the scattering differed for the two senses of circular polarization. They used this apparent "magnetic circular dichroism" to determine features of the magnetic phase diagram. Though an explanation was offered, the origin of this polarization asymmetry has remained a vexing puzzle. We have studied light scattering in (0001) sublimation sheets (20-100μm thick) of $FeCl_2$ with [0001] parallel to both the applied field and the direction of propagation of light. Asymmetry in the scattering of two circular polarizations was encountered throughout the spectral range 0.4 - 1.1 μm and the temperature range 2.1 - 20.0K. The diffraction and its dependence on circular polarization are explained in terms of the refractive index contrast between the two states, antiferromagnetic and paramagnetic. We have found that the scattering can provide a sensitive method of determining the mixed phase boundaries; but that the apparent dichroism is not an intrinsic property of antiferromagnets,[2] and at some wavelengths can give exceedingly deceptive information about the phase diagram. A full report will be published elsewhere.

1. J. A. Griffin et al., Phys. Rev. B <u>10</u>, 1960-6 (1974).

2. In the course of studies on another Ising-like antiferromagnet, $Dy_3Al_5O_{12}$, we have encountered the same effects reported here for $FeCl_2$. The scattering by the mixed phase is prominent. On the other hand the apparent dichroism was not seen.

# FARADAY ROTATION AND OPTICAL ABSORPTION IN $FeBr_2$

J. A. Griffin* and J. D. Litster
Department of Physics
Massachusetts Institute of Technology
Cambridge, Massachusetts 02139

## ABSTRACT

$FeBr_2$ is a layered, anisotropic antiferromagnet which has been shown to exhibit behavior in a magnetic field consistent with multicritical behavior. We present measurements of Faraday rotation and optical absorption in $FeBr_2$ and illustrate their sensitivity to the magnetic ordering.

## INTRODUCTION

Recently there has been considerable interest in multicritical phenomena which occurs in the field-dependent behavior of anisotropic antiferromagnets. One such antiferromagnet is $FeBr_2$.[1] $FeBr_2$ is composed of antiferromagnetically coupled ferromagnetic layers of $Fe^{++}$ ions. The easy axis of antiferromagnetic order is the c axis, normal to the layers, and the Neel temperature is $T_N \approx 14$ K. Such magnetic structure is particularly interesting due to its behavior in the presence of an external magnetic field applied in a direction perpendicular to the layers. At low temperature $T << T_N$ and at a critical magnetic field, there is a first order spin-flip transition from the antiferromagnetic alignment directly to a saturated paramagnetic alignment, in which all ferromagnetic layers align parallel to the magnetic field. With increasing temperature the discontinuity in total magnetization that results from this first order transition is found to decrease to zero, at which point the spin-flip transition becomes second order up to the Neel temperature. This type of field-dependent behavior has been labelled metamagnetic, and in another ferrous dihalide, $FeCl_2$, the intersection of the first order line with the second order line occurs at a magnetic tricritical point.[2]

$FeBr_2$, however, is known to differ from $FeCl_2$ in several interesting respects. At the tricritical point in $FeCl_2$ the line of first order transition intersects with a continuous slope the line of second order transitions in the internal magnetic field versus temperature phase diagram. Previous measurements in $FeBr_2$ indicate that these two lines intersect with a discontinuous slope, and, in addition, the antiferromagnetic phase can be entered from the paramagnetic phase by increasing the temperature at constant internal magnetic field.[3] Neither of these features are expected at a tricritical point. Another feature of the tricritical behavior in $FeCl_2$ is that the observed ratio of the tricritical temperature $T_t$ to the Neel temperature $T_N$ is within 10% of the mean field theory prediction calculated from the antiferromagnetic and ferromagnetic coupling strengths determined by neutron scattering. In $FeBr_2$ the ratio of the reported[3] "tricritical" temperature $T_t \sim 5$ K to $T_N \sim 14$ K is too small to be consistent with the mean field theory of ferromagnetic in planar coupling and antiferromagnetic between plane coupling. Kincaid has proposed, still within the framework of mean field theory, a phase diagram (no longer a tricritical point) for $T_t/T_N < 4/9$ which appears consistent with the previous experimental data on $FeBr_2$ but requires that the ratio of the intra-planar ferromagnetic coupling to the inter-planar antiferromagnetic coupling be $Z_F J_F/|Z_{AF}J_{AF}| < 0.6$.[4] Recent neutron scattering measurements[4] show, however, $Z_{AF}J_{AF} = -0.49$ meV and $Z_F J_F = 2.48$ meV.

Thus it appears that metamagnetic behavior in $FeBr_2$ may not be associated with a tricritical point, or else, unlike previously studied systems, is not consistent with the mean field model predictions. For this reason we have begun magneto-optical measurements of magnetic ordering in $FeBr_2$. In this paper we present measurements of the temperature and field dependence of the Faraday rotation at 632.8 nm and the zero field optical absorption between 500 nm and 560 nm.

## EXPERIMENTAL

Different experimental arrangements were used for the optical absorption and the Faraday rotation. In both arrangements the light propagated parallel to the c axis of the crystal which was perpendicular to the plane of the sample, a 1 cm × 0.5 cm × 0.02 cm plate. The absorption measurements were made using a Spex 1 meter monochromator and a Tungsten lamp with the sample mounted on a copper cold finger in the vacuum space of an optical cryostat; sample temperature was controlled to ~ 50 mK. The Faraday rotation measurements were carried out in a Bitter solenoid with the magnetic field along the crystal c axis. The $FeBr_2$ was contained in a sample holder filled with $He^4$ exchange gas whose temperature was controlled by a servo using a $SrTiO_3$ capacitance thermometer as a sensor. This thermometer is not affected by the magnetic field but slowly drifts in time. The sample temperature was measured with a calibrated Ge resistance thermometer before and after each field scan, and the sample temperature was found to have changed less than 5 mK.

The linearly polarized beam from a 632.8 nm He-Ne laser was attenuated to ~ 10 μW and passed through about the central 25% of the sample area. The plane of of polarization of the incident light was modulated at a frequency $\omega/2\pi = 50$ kHz, and light passing through the sample was analyzed with a linear polarizer before detection by a photomultiplier tube. The signal at frequency $2\omega$ was proportional to $\cos 2(\beta - \phi)$, where $\phi(T,H)$ was the rotation produced by the sample and $\beta$ the angle of the analyzer with respect to the incident polarization. This was detected by a lock-in amplifier and maintained at zero by a feedback loop with a stepping motor to rotate the analyzer. Counting of the pulses to the stepping motor provided a digital readout of the Faraday rotation accurate to .05 degrees. The magnetic field was also determined by digital monitoring of the magnet current.

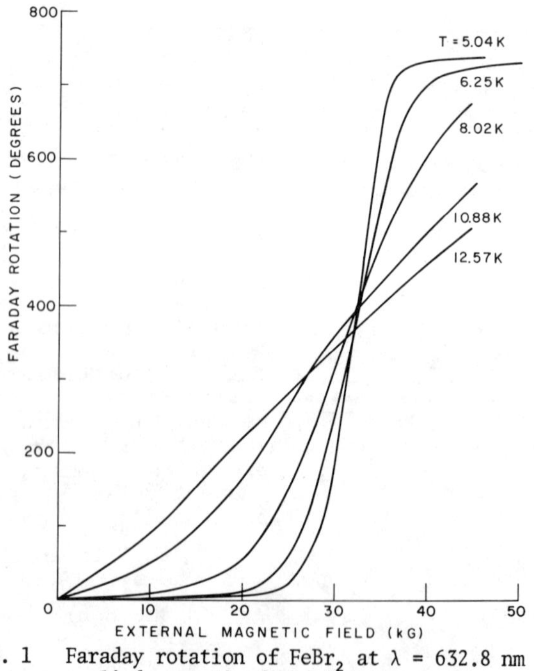


Fig. 1 Faraday rotation of $FeBr_2$ at $\lambda = 632.8$ nm versus applied magnetic field for temperatures $5K \leq T \leq 12.6K$.

## FARADAY ROTATION IN $FeBr_2$

Fig. 1 illustrates the Faraday rotation of $FeBr_2$ at 632.8 nm for temperatures between ~ 5 K and ~ 13 K as a function of applied field. At the lowest temperature, T ~ 5 K, three regions of field dependence are apparent. For fields up to ~ 23 kG the Faraday rotation is nearly independent of magnetic field. This behavior is characteristic of the antiferromagnetic phase in which one sublattice produces a Faraday rotation which is largely cancelled by the rotation of the opposite sign produced by the other sublattice. The net rotation which is produced results because the sublattice magnetization in the layers parallel to the magnetic field is enhanced and that in the opposite sublattice is decreased. For magnetic fields larger than ~ 37 kG the Faraday rotation saturates. This saturation occurs because all ferromagnetic layers have become aligned parallel to the magnetic field, and the Faraday rotation from the two sublattices add. For intermediate values of applied magnetic field the Faraday rotation appears to be linear in the external magnetic field. In a first order transition from the antiferromagnetic to saturated paramagnetic state saturated paramagnetic domains are spatially intermixed with the antiferromagnetic regions; the internal field is constant, and the applied field is $H_i + 4\pi NM$, where M is the net magnetization and N a demagnetizing factor. This type of behavior is reflected in the Faraday rotation. At the next temperature, T ~ 6 K, the antiferromagnetic and saturated paramagnetic Faraday rotations are similar to that at T ~ 5 K, but the slope of the intermediate region has decreased and is no longer constant. This is indicative of the onset of the second order transition. The next isotherm, T ~ 8 K, illustrates more clearly the behavior of the Faraday rotation in the region of second order transitions. There is a rapid initial increase in slope which continues up to a field of ~ 34 kG, followed by a decrease in slope as saturation is approached. The maximum in slope of this isotherm indicates the occurrence of the second order transition. In the next isotherm, T ~ 11 K, the second order transition occurs at ~ 27 kG, and in the final isotherm, T ~ 13 K, this transition occurs at ~ 15 kG. These last three isotherms illustrate the decrease in magnetic field required for the occurrence of the second order transition as temperature increases, and thus allow the determination of the lambda line. The five isotherms in Fig. 1 illustrate qualitatively the change in Faraday rotation and thus total magnetization at various regions in the magnetic phase diagram. Analysis of measurements of this type will enable us to determine the magnetic phase diagram of $FeBr_2$ with a sensitivity in M of ~ .01% of the saturation magnetization. Unfortunately quantitative analysis of the data here is not worthwhile as the sample developed a structural defect that caused the notable rounding of the first order transition regions. Quantative analysis of data on a better sample will be reported as soon as measurements are complete.

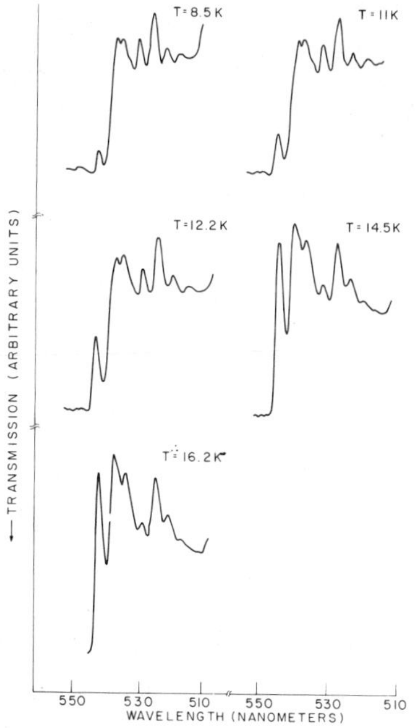


Fig. 2 Transmission of $FeBr_2$ from 550 nm to 510 nm for temperatures between $8K \leq T \leq 16K$.

## OPTICAL ABSORPTION

We have also made zero field optical absorption measurements in $FeBr_2$; as can be expected the general feature of the absorption spectrum in $FeBr_2$ and $FeCl_2$ are similar. In the near IR there is a broad absorption band consisting of two peaks, one centered at ~ 1.47 microns and the other centered at ~ 1.65 microns. As was the case in $FeCl_2$ the low energy peak becomes more pronounced with respect to the other peak as temperature is decreased. This transition is a spin allowed, parity forbidden transition from the $^5T_{2g}$ ground level to the two-fold degenerate $^5E_g$ level. In the visible region of the spectrum the absorption in $FeBr_2$ is dominated by a group of narrow absorption bands which are strongly temperature dependent. This type of behavior occurs in one group of bands centered at ~ 485 nm and in another weaker group centered at ~ 540 nm. Fig. 2 illustrates the zero field transmission of $FeBr_2$ from 560 nm to 500 nm in the temperature range from 8.5 K to 16 K. Particularly interesting in this group of bands is that which occurs at ~ 542.5 nm. At the lowest temperature in the antiferromagnetic phase, this band is relatively weak. With increasing temperature the relative absorption of this band increases quite rapidly, resulting in an eight-fold increase in oscillator strength between 8.5 K and 14 K. Above 14 K this temperature dependence appears to saturate, as is seen from data at 30 K and 40 K. As was the case in $FeCl_2$ at 427 nm, it appears as though the oscillator strength of this absorption in $FeBr_2$ is proportional to the ferromagnetic nearest neighbor disorder. At low temperatures the weakness of the absorption is due to the relatively ordered ferromagnetic layers, whereas at higher temperature more thermal disorder within the layers produces a stronger absorption. If, in fact, this oscillator strength is proportional to the nearest neighbor correlation, its temperature dependence may be used to optically measure the magnetic specific heat.

## ACKNOWLEDGMENTS

We are grateful to Professor S. E. Schnatterly for the use of the temperature control apparatus, to Professor Y. Farge for the sample of $FeBr_2$ and to D. R. Nelson for assistance in computer analysis.

## REFERENCES

*This work was performed while the author was a Guest Scientist at the Francis Bitter National Magnet Laboratory, which is supported at M.I.T. by the National Science Foundation.

1. I. S. Jacobs and P. E. Lawrence, Phys. Rev. 164, 866-878 (1967); I. S. Jacobs and P. E. Lawrence, J. Appl. Phys. 35, 996-997 (1964).

2. R. J. Birgeneau, G. Shirane, M. Blume and W. Koehler, Phys. Rev. Lett. 33, 2078-2080 (1974); J. A. Griffin and S. E. Schnatterly, Phys. Rev. Lett. 33, 1576-1578 (1974).

3. A. R. Fert, P. Carrara, M. C. Lianusse, G. Mischler and J. P. Redoules, J. Phys. Chem. Solids 34, 223-230 (1973); C. Vettier, H. L. Alberts and D. Bloch, Phys. Rev. Lett. 31, 1414-1417 (1973).

4. J. M. Kincaid and E. G. D. Cohen, Phys. Lett. 50A, 317-318 (1974); W. B. Yelon and C. Vettier, J. Phys. C8, 2760-2768 (1975).

# OPTICAL STUDIES OF THE MAGNETIC PHASE DIAGRAM OF $MnCl_2$ AND $MnBr_2$

M. REGIS, Y. FARGE
Laboratoire de Physique des Solides
associé au C.N.R.S.
Université Paris-Sud
91405 ORSAY (France)

B.S.H. ROYCE
Materials Lab.
Princeton University
PRINCETON, N.J. 08540 U.S.A.

ABSTRACT

Manganous chloride and bromide are transparent ionic insulators which exhibit magnetic ordering in zero field at temperatures below 1.96 K and 2.30 K. Optical absorption measurements of internal transitions of the $Mn^{2+}$ ion are combined with dichroism studies to determine the magnetic phase diagrams. In each compound, three regions of magnetic order are found and studied as a function of magnetic field and temperature. The correlation between magnetic circular dichroism and the magnetization is calculated. The linear dichroism, induced by the magnetostriction, provides a measurement of the magnetic energy, and consequently of the specific heat. The present data are compared to previous neutron diffraction studies and a model of the magnetic ordering is discussed.

## INTRODUCTION

$MnBr_2$ and $MnCl_2$ are two transparent ionic insulators, which exhibit antiferromagnetic ordering in zero field at temperatures below 2.3 and 1.96°K (1). They cristallize in the layer type hexagonal structure of, respectively, $CdI_2$ and $CdCl_2$. The neutron data (2) suggest that the $Mn^{2+}$ spins lie in the basal planes of the orthorhombic unit cell, in the antiferromagnetic phases.

In this paper, optical absorption spectra at liquid helium temperature are reported and the magnetic circular dichroism is measured at temperatures between 1,6 K and 4,18°K and applied magnetic fields up to 25 kOe, parallel to the $\vec{c}$ axis of the sample. This MCD signal is proportional to the transverse magnetization. A linear dichroic signal is detected when the $\vec{c}$ axis of the sample is inclined at 45° with respect to the light beam and magnetic field. This linear dichroic signal is related to the magnetostriction and thus to the magnetic energy. With optical informations, the magnetic phase diagrams of the two salts are plotted, and $MnBr_2$ as well as $MnCl_2$ are found to exhibit two magnetically ordered regions.

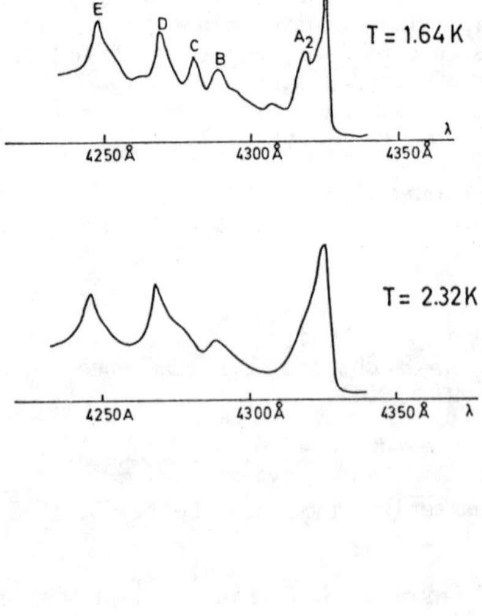


Fig 1 - Optical absorption of $MnBr_2$ corresponding to the ${}^6A_{1g} \to [{}^4E_g, {}^4A_{1g}]$ transitions.

## I - ABSORPTION SPECTRA

Five series of absorption lines have been studied, at liquid helium temperature in the visible and ultraviolet, for the two salts. Here, we report only the band between 4 200 and 4 400 Å corresponding to the ${}^6A_{1g}$ (${}^6S$) $\to [{}^4Eg, {}^4A_{1g}]$ (${}^4G$) transition of the $Mn^{2+}$ ion in the cubic field description, which will be used to study the magnetic properties of the two salts. Fig 1 shows the optical absorption of $MnBr_2$ above and below $T_N$ in the 4 300 Å region. The $A_1$ and $A_2$ peak presumably correspond to exciton magnon transitions. The ${}^4E_g$ level is splitted by second-order spin-orbit interaction, and the two peaks are magnon sidebands corresponding to the two purely excitonic transitions. The same behaviour is observed in the case of $MnCl_2$, although without as good resolution. The $A_2$ peak is a cold band which disappears at the transition temperature (at the high transition temperature in $MnCl_2$ and at the low transition temperature in $MnBr_2$).

## II - MAGNETIC CIRCULAR DICHROISM

The MCD signal for the transition ${}^6A_{1g} \to {}^4E_g, {}^4A_{1g}$ is calculated in the case where a magnetic field $H_x$ is applied in the $\vec{x}$ direction. All the $Mn^{2+}$ spins lying in the $\vec{z}$ direction. So the MCD signal has proved to be proportional to the quantity $\frac{H_x}{H_z}$ ($H_z$ being the exchange field), and thus proportional to the transverse magnetization. The calculation predicts also that there is

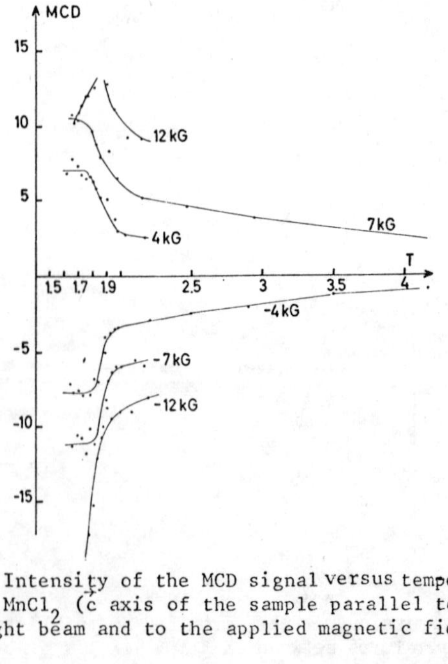


Fig 2 - Intensity of the MCD signal versus temperature in $MnCl_2$ ($\vec{c}$ axis of the sample parallel to the light beam and to the applied magnetic field).

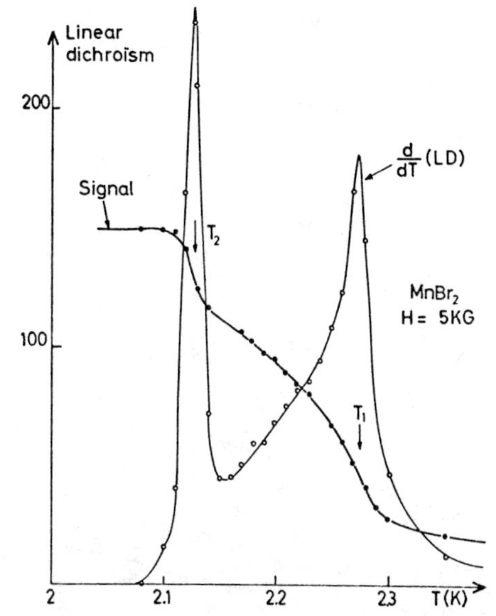


Fig 3 - Linear dichroism of the $A_1$ peak in $MnBr_2$, and its first derivative versus temperature (applied field and light propagation at 45° of the $\vec{c}$ axis).

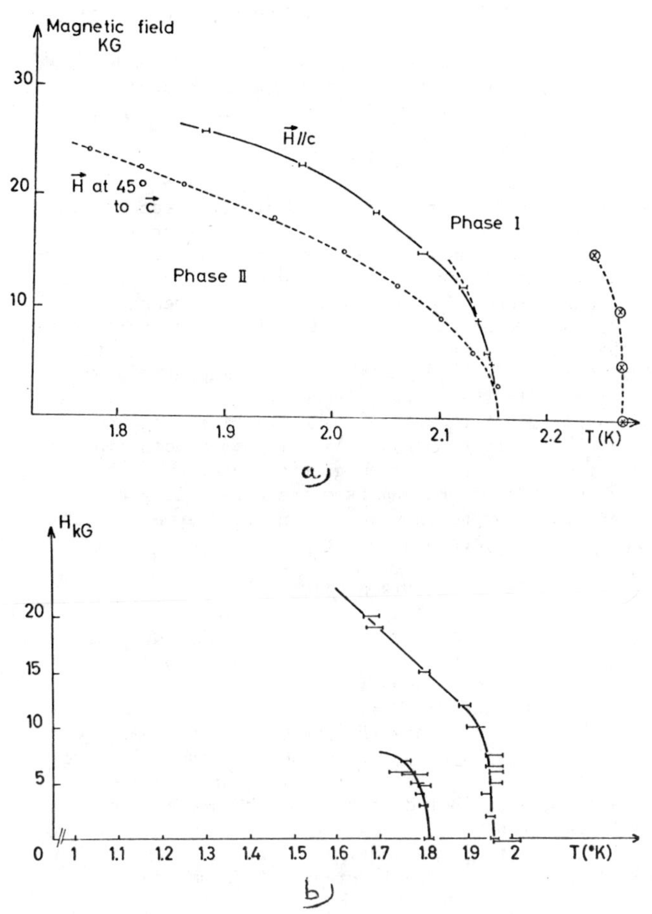


Fig 4 - Optically determined magnetic phase diagrams of :

a) $MnBr_2$ (++ indicates neutron measurements of Wollan et al with the applied field parallel to the $\vec{c}$ axis).

b) $MnCl_2$

negligible shift induced by the Hx field on the pseudo Zeeman sublevels of the ground state in the ordered phase, and that the contribution to the MCD of the two sublattices has same sign. Thus the MCD signal has mainly the shape of a single peak (zeroth-moment change).

Fig 2 shows the intensity of the MCD signal as a function of temperature, for various $H_x$ fields in $MnCl_2$. The same behaviour is found for $MnBr_2$. At the first transition temperature ($T_1$ = 1,96°K in zero field in $MnCl_2$), the magnetization, or MCD signal, changes abruptly and increases when the temperature decreases. In the phase II, reached for the smallest values of the $H_x$ field (T < 1,81°K in $MnCl_2$), the magnetization becomes constant with decreasing temperature, at constant field. Thus the perpendicular susceptibility is constant, as is expected in an antiferromagnetic material.

## III - LINEAR DICHROISM

When the $\vec{c}$ axis of the crystal is inclined at 45° with respect to the light beam and the applied magnetic field , a linear dichroism appears , which is very probably coming from the magnetostriction which reduces the local symmetry and acts as a distortion. Because this distortion is very small, we can assume that the linear dischroism is proportional to the magnetostriction. This effect on the imaginary part of the refractive index is exactly the same as the linear magnetic birefringence, coming from the real part of the refractive index. Following Jahn (3), and Kleemann(4), this magnetostriction is proportional to the magnetic energy Um. Thus, the first derivative of the linear dichroism versus temperature is proportional to the magnetic heat capacity $C_m$ of the crystal. On fig 3 are plotted the linear dichroism and its first derivative for $MnBr_2$. The derivative clearly shows two peaks, corresponding to two transition temperatures.

## IV - MAGNETIC PHASE DIAGRAM

All these optical properties :

1°) maxima of the derivative of the linear dichroism,
2°) changes in the behaviour of the MCD signal,
3°) disappearance of the $A_2$ peak in the aborption spectrum,

are used to plot the magnetic phase diagram of the two salts (fig 4).

$MnCl_2$ is found to exhibit two magnetic phase transitions as previously reported (1). $MnBr_2$ also exhibits two magnetic phase transitions. Wollan et al (2) are previously detected only one of these transitions in their neutron experiments. The strong similarities between the two salts behaviour could indicate that the magnetic structures are the same. Phase II is certainly an ordered antiferromagnetic phase, with the spins lying perpendicular to the $\vec{c}$ axis, and phase I a partly disordered phase.

The sharpness of the specific heat peak at 2.15°K in $MnBr_2$ is an indication of a first order phase transition. Further experiments, in neutron scattering, and also in X ray scattering, would be of interest to resolve the structure of the phase I in $MnBr_2$, and also to look for the lattice distortions.

## REFERENCES

(1) R.B. MURRAY - Phys. Rev., 100, 1071, (1955).
- Phys. Rev., 128, 1570, (1962).
R.A. BUTERA and W.F. GIAUQUE - J. of Chem. Phys., 40, 2379, (1964).
R.B. MURRAY and L.D. ROBERTS - Phys. Rev., 100, 1067, (1955).
W.F. GIAUQUE - J. of Chem. Phys., 42, 9, (1965).

(2) M.K. WILKINSON, J.W. CABLE, E.O. WOLLAN, W.C. KOEHLER - Oak Ridge National Lab. Reports, 2430, 8th April 1958, p. 65. 2501, 10th June 1958,p. 37.
E.O. WOLLAN, W.C. KOEHLER, M.K. WILKINSON - Phys. Rev., 110, 638, (1958).

(3) I.R. JAHN and K. BITTERMANN - Sol. St. Comm., 13; 1897, (1973).

(4) W. KLEEMANN and Y. FARGE - J. Phys. Lettres, 35, L 135, (1974).

# ANALYSIS OF THE MORIN PHASE TRANSITION IN HEMATITE FROM THE LINEAR MAGNETIC BIREFRINGENCE.

H. Le Gall[+], E.G. Rudashewsky[++], C. Leycuras[+] and D. Minella[+].
+ Laboratoire de Magnétisme et d'Optique des Solides, C.N.R.S., 92190 Meudon-Bellevue, France.
++ P.N. Lebedev Physical Institute, Moscow 117 924, U.R.S.S.

ABSTRACT

The specific linear magnetic birefringence of a single crystal of hematite ($\alpha$-$Fe_2O_3$) is reported at 1.15 micron wavelength as a function of the d.c. magnetic field and temperature near the Morin phase transition temperature. When the temperature increases from the region corresponding to the antiferromagnetic state to the weak ferromagnetic state, the linear magnetic birefringence is zero for low applied magnetic field, then grows slowly and finally reaches a maximal value. The variation of the linear magnetic birefringence near and at the Morin temperature is discussed both from the evolution of the antiferromagnetic vector with the d.c. field and temperature and from the possibility to detect the growth with the temperature of surface weak ferromagnetic states accross the crystal.

## INTRODUCTION

Hematite or $\alpha$-$Fe_2O_3$ has a first-order phase transition at the Morin temperature $T_M$. Below $T_M$ the crystal is antiferromagnetic (AF) and above $T_M$ it shows a weak ferromagnetism (WF). In the former case the spins are directed along the three-fold $C_3$ rhombohedral axis (or C axis) and in the second case the spins are canted in the basal plane (111). The Morin transition is generally attributed to the sign change of the C-axis anisotropy.

A theoretical explanation of the WF structure has been given first by Dzyaloshinsky |1| from symmetry considerations and later by Moriya |2| from a quantum mechanical treatment which imply an antisymmetrical superexchange interaction of the form $\vec{D}.\vec{M}_1 x \vec{M}_2$ where $\vec{D}$ is a constant vector in the |111| direction and $\vec{M}_1$ and $\vec{M}_2$ are the two equivalent magnetic sublattices. Above the Morin temperature, this additional exchange interaction induced a small canting of the spins producing a small net moment.

Most of experiments have been performed on hematite from pure magnetic measurements (perpendicular and parallel susceptibilities, magnetostriction, A.F. resonance and neutron scattering). Being proportional to the square of the spin operator $S^2$, the linear magnetic birefringence (LMB) may provide interesting information on the evolution of the magnetic structure and magnetic phase transitions in AF crystals as first applied in hematite by Pisarev |3|. The magnetooptical investigation of this crystal has been used recently by the authors |10| in order to determine the magnetic (pure Cotton-Mouton effect) and elastic (via the magnetostriction) contributions to LMB in the WF phase. For that the LMB was measured at room temperature as a function of d.c. field $H_a$ and under stress applied perpendicularly to the $H_a$. We have showed that the LMB has negative values under stress due to the associated rotation of the antiferromagnetic vector $\vec{L}=\vec{M}_1-\vec{M}_2$ and that about 98 % of the LMB is due to a pure Cotton-Mouton effect. In what follows the magnetooptical investigation of hematite is extended to the low temperature phase transition by using d.c. fields up to 20 kOe.

## RESULTS AND DISCUSSION

In all the experiments a laser beam at 1.15 micron wavelength was applied perpendicularly to a single-crystal platelet of hematite of 0.32 mm thickness. The light was propagating along the optical axis corresponding to the |111| direction. The figure 1 shows typical evolution of LMB in hematite near the Morin temperature as a function of the d.c. magnetic field applied perpendicularly to the C-axis. Above T=254°K the LMB has a saturation amplitude for increasing magnetic field corresponding to the WF state. At zero d.c. field the LMB vanishes when T<241°K due to the AF state of the spins with the antiferromagnetic vector oriented along the |111| direction. At T=241°K the LMB has a small increase with the d.c. field up to 6 kOe and then the slope of the curve has a stronger value until the saturation value of the birefringence is reached. For T<241°K the slope of the LMB is first negative for low d.c. field values and then becomes positive for higher fields. The higher positive slope parts of the curves are shifted toward increasing values of the d.c. field when the temperature is decreasing. Such an evolution of the LMB is associated with the Morin transition and the field induced spin-flop as discussed later. On the figure 2 the LMB has been recorded as a function of temperature with magnetic fields up to 20 kOe applied perpendicularly to the |111| axis. The rotation of the AF vector $\vec{L}$ away from the |111| axis and toward the (111) basal plane, corresponding to the Morin transition, produces a strong increase of the LMB from zero in the AF state to about 700°/cm in the WF state as described by the lower field curve ($H_a$=0.3 kOe) in the Fig. 2. The Morin transition in this case is

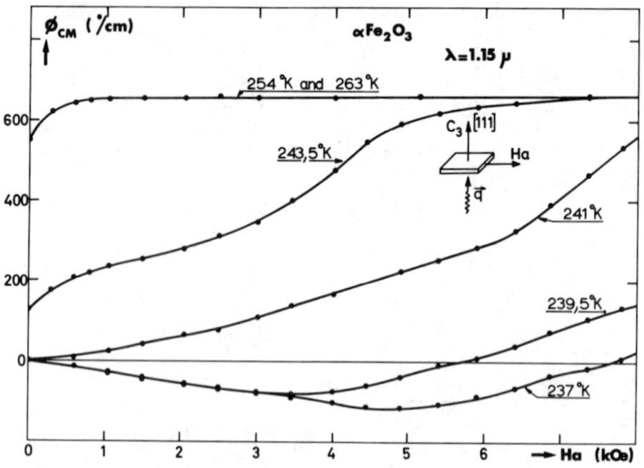


Fig. 1 : Magnetic field evolution of the LMB of hematite near the Morin transition.

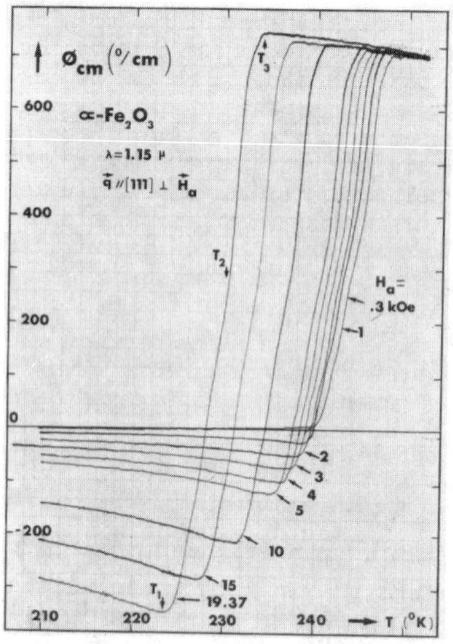


Fig. 2 : Temperature dependence of the LMB of hematite at the Morin transition and under field induced transitions.

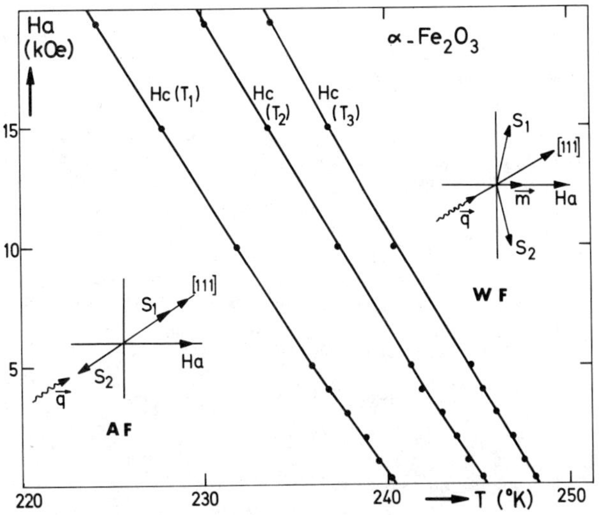


Fig. 3 : Magnetic phase diagram of hematite near the Morin transition.

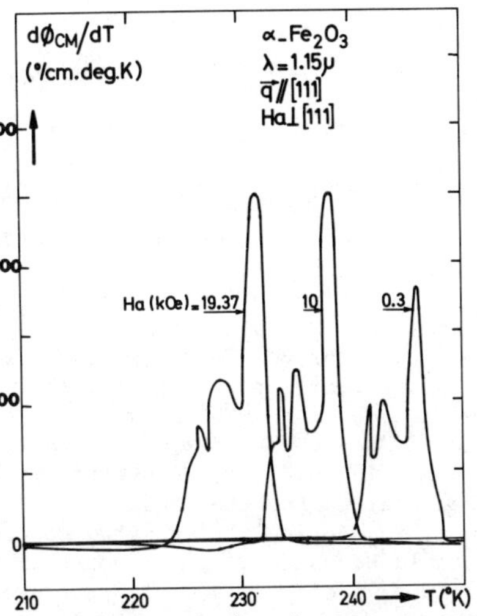


Fig. 4 : Temperature derivative of the LMB in hematite at and near the Morin temperature.

spread over a small range of ±4°K from $T_1$=241°K to $T_3$=248°K. The value of the Morin temperature is smaller than those measured from very pure crystal ($T_M$≃260°K) by others authors |4| and this difference can be attribute to some impurities added in our crystal.

When increasing the magnetic field perpendicular to the |111| axis we observe a shift of the transition, now field induced, toward lower temperatures. In addition a negative value of the LMB increases in the low temperatures region with the d.c. field. In this case the evolution of the birefringence is induced first by a sublattice rotation toward $H_a$ which produces the negative values of the Cotton-Mouton effect and then by a gradual rotation of the AF vector from the C axis and toward the basal plane (111) as described by the sharp rise of the LMB between $T_1$ and $T_3$. The amplitude of the Cotton-Mouton effect in the WF state is independent of the d.c. field up to 20 kOe but decreases when the temperature increases. In addition the curves of the Fig.2 show a strong change of the slope for a well defined temperature $T_2$ which decreases when the magnetic field increases. The Fig. 3 gives the phase diagram H-T where linear evolutions of the critical temperatures $T_1$, $T_2$ and $T_3$ as a function of the magnetic field $H_a$ are observed. The region under the line $H_c(T_1)$ corresponds to a canting of the magnetic sublattices toward $H_a$ in the low temperatures region. This canting vanishes at zero field (pure AF state). The region above the line $H_c(T_3)$ is the usual WF state observed in hematite. For a given d.c. field the AF vector $\vec{L}$ seems to rotate from the |111| axis toward the basal plane following two main steps corresponding to the regions $T_2$-$T_1$ and $T_3$-$T_2$. Such an evolution during either the Morin transition, or a field induced transition in the low temperatures region, has not been previously observed in our knowledge. Recently Galperin et al |5| indicated the possibility to detect the simultaneous existence of two kinds of spin-flop at the Morin transition from the use of Mössbauer experiment. It is interesting to note that recent theoretical studies have shown the possibility to obtain two types of spin configurations in the Morin transition. According to Nagai et al |6|, the renormalized acoustic magnons energy, obtained from the use of anisotropy energy restricted to the second-order terms, does not vanish as suggested previously by Herbert |7| at the Morin temperature. On the other hand, they showed that two kinds of spin configurations are stable in the range ΔT between $T_M$ and $T_M(1+H_A/2H_E)$ where $H_E$ and $H_A$ are the exchange and dipolar anisotropy fields respectively. Such temperatures could correspond to $T_1$ and $T_2$ in the fig. 2 ($T_1$-$T_2$ ≃ 4°K) but that temperature interval is one order higher than the calculated value |6|. This difference between the theoretical and experimental values can be attributed partly to the possibility of added impurities which spread the low temperature transition in our crystal. A different description of the spin behavior during the Morin transition has been given by Chow and Keffer |8|. By using both the second-order and the four-order anisotropy fields they described the Morin transition from spin-flop nuclei corresponding to surface WF state generated near the surface of the sample in the low temperatures region by the softening of surface magnons. The region of the surface WF increases slowly with temperature and then has a fast spread inward accross the crystal for a well defined temperature. Such a critical temperature could correspond again to $T_2$ as reported on the Fig. 2.

Finally a fine structure has been detected in the AF→WF transition as shown on the Fig. 4 which corresponds to the temperature derivative of the Cotton-Mouton effect deduced from the Fig. 2. This derivative which can be correlated to the magnetic heat capacity associated with the spin-fluctuations |9|, shows a strong maximum in the high temperature region of the Morin transition corresponding to $T_3$ and two smaller peaks in the low temperature region. Further discussions on this more complex structure will be reported in the near future.

Acknowledgement : The authors wish to thank Academician A.M. Prokhorov of the P.N. Lebedev Physical Institute of Academy of Sciences of the USSR and Dr. J. Winter Director at the C.N.R.S. (France) for support of this joined work and for fruitfull discussions.

REFERENCES

|1| I.E. Dzyaloshinsky, Soviet Phys. JETP, 5, 1259, (1959) and J. Phys. Chem. Solids, 4, 241, (1958).

|2| T. Moriya, Phys. Rev., 117, 635, (1960); 120, 91, (1960).

|3| R.V. Pisarev, Proc. Intern. Conf. Magnetism, Grenoble (1970), J. Phys. C1, 32, 1051, (1971).

|4| S. Foner and Y. Shapira, Phys. Letters, 29 A, 276, (1969).

|5| F.M. Galperin, A.N. Salugin, A.A. Saigin and N.V. Elistratov, Phys. Stat. Sol. (a), 22, K7, (1974).

|6| O. Nagai, N.L. Bonavito and T. Tanaka, J. Phys. C : Solid State Phys., 6, 470, (1973).

|7| D.C. Herbert, J. Phys. C : Solid State Phys., 3, 891, (1970).

|8| F. Keffer and H. Chow, Phys. Rev. Letters, 31, 1061, (1973).

|9| H. Dachs, Proc. 18th Conf. Magnetism and Magnetic Materials, Denver 1972, AIP Conf. Proc., nomber 10, 854, (1973).

|10| U.S. Merkoulov, E.G. Rudeshewsky and H. Le Gall, J.E.T.P. Letters, 22, 140, (1975).

# STUDIES OF MAGNETOOPTICAL EFFECTS IN GARNETS THIN FILM WAVEGUIDES+

G. Hepner, J.P. Castéra and B. Désormière
Laboratoire Central de Recherches THOMSON-CSF
Domaine de Corbeville - 91401 ORSAY (France)

## ABSTRACT

Mode conversion in Gd-Ga substituted YIG films grown on GGG are studied for guided light at 1.15 μm. The films are magnetized either parallel or perpendicular to propagation, so that conversion is obtained through the Faraday or the Cotton-Mouton effect. Depending on the substitution ratio in YIG, easy in-plane or bubble films are obtained. For easy in-plane films, the birefringence induced by epitaxy is found to favour the mode conversion. In this case, conversion ratios of 75% and 50% were obtained for the Faraday and Cotton-Mouton configuration, respectively. These two kinds of mode converters could be used to build a non-reciprocal integrated optical device, as discussed in a previous paper.

## INTRODUCTION

Epitaxial garnet thin films have been extensively studied for "magnetic bubble" applications. As first reported by Tien et al.[1], they are also suitable as optical waveguides in the near infrared region. Light modulation through the Faraday effect was demonstrated. But this has to compete with other modulation techniques in Integrated Optics, for instance that using electro-optical effects. On the other hand, non reciprocal devices can be designed by using the gyromagnetic effects, like in the Microwave technology. Our interest has been focussed on this second type of application. The analysis we performed led us to define a new structure for designing a non-reciprocal mode converter, which is thought to constitute the basis of isolators and circulators in Integrated Optics. This structure is composed of two mode converters, one gyromagnetic and one anisotropic, as proposed by Yamamoto[2]. The gyromagnetic converter is obtained through the Faraday effect, with the magnetization parallel to the propagation. As previously proposed[3], the anisotropic converter is obtained through the Cotton-Mouton effect with the magnetization transverse to the propagation. In contrast with the Faraday effect, the Cotton-Mouton effect is reciprocal since it is quadratic in magnetization. Furthermore it is known that, in garnets, the permittivity change due to this Cotton-Mouton effect is of the same order of magnitude as that of the Faraday effect[4,5]. Maximum conversion is observed here for a magnetization nearly 45 degrees out of the film plane.

To reach a complete mode conversion, the two coupled modes must be degenerated or phase matched by a periodic magnetic structure. The degeneracy, if achieved is a preferred solution since it needs a comparatively smaller interaction length. We have recently reported[6] the possibility of obtaining this degeneracy by making use of the birefringence induced by the liquid phase epitaxy process.

## EXPERIMENTAL

Experiments were performed using a 1.152 μm laser light propagated in gallium-gadolinium substituted YIG waveguides. The in and outcoupling of the light was achieved through rutile prisms. We measured the coupling angles for the different TE and TM modes. Using these values and the mode equations, we determined the refractive index and the thickness of the film.

To study the mode conversion through the Faraday and the Cotton-Mouton effects, we allowed magnetic fields to be applied either along the propagation direction (Hz), or perpendicular to this direction and either in the film plane (Hy) or perpendicular to it (Hx). Applying a large field Hy suppresses conversion and the losses due to optical absorption can be determined for the different modes. We find losses of 5 dB/cm, typically. The Faraday conversion was measured with an Hz field sufficient to saturate the film in this direction. The Cotton-Mouton conversion was obtained while orienting the magnetization perpendicular to the propagation, by applying fields in the xy plane corresponding to the (110) plane. With the $TE_0$ as the incident mode, the power of the converted $TM_0$ mode was measured versus the distance z separating the two prisms.

The power conversion $R_{(z)}$, given by :

$$R(z) = I_{TM}(z) \,/\, I_{TE}(z)$$

was plotted together with a fitted theory using the following expression [7].

$$R(z) = \frac{4K^2}{4K^2 + \Delta\beta^2} \sin^2 \left\{ (4K^2 + \Delta\beta^2)^{1/2} \, z/2 \right\} \quad (1)$$

where $I_{TM}(z)$ and $I_{TE}(z)$ are the measured intensities of the converted mode and the incident mode, suppressing conversion.

K and $\Delta\beta$ are respectively the coupling constant modulus and the difference of the propagation constants for the two modes[7]. Assuming an isotropic film, the propagation constants $\beta_1$, $\beta_2$ associated to the $TE_0$ and the $TM_0$ modes respectively, can be determined. The epitaxial linear birefringence n is then inferred by means of the relation :

$$\Delta n = (\beta_2 - \beta_1 - \Delta\beta)/k \quad (2)$$

where k is the wavenumber in the vacuum.

Relation (2) is a good approximation for well confined modes in optically anisotropic films[8]. As shown below, the values so obtained for $\Delta n$ can be compared with those computed from the lattice mismatch which gives the stress, together with the photoelastic tensor elements. The lattice mismatch was measured by X rays with double diffraction.

## RESULTS

For an (111) easy in-plane film, the experimental results are shown in fig.1. The maximum conversion rates are 75% and 50% for the Faraday and the Cotton-Mouton effect, respectively.

An Hy field of 5 Oe was applied in the case of

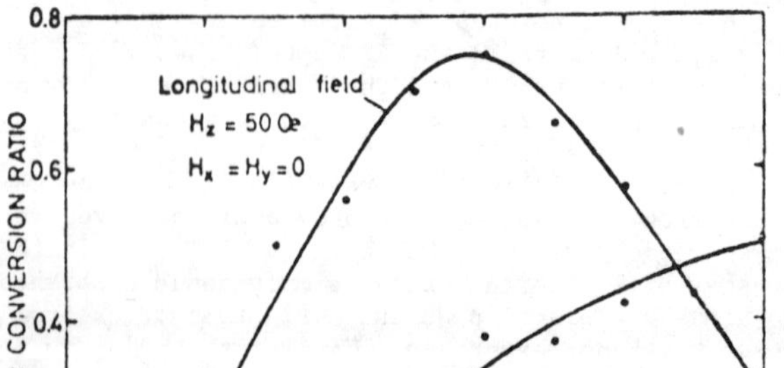

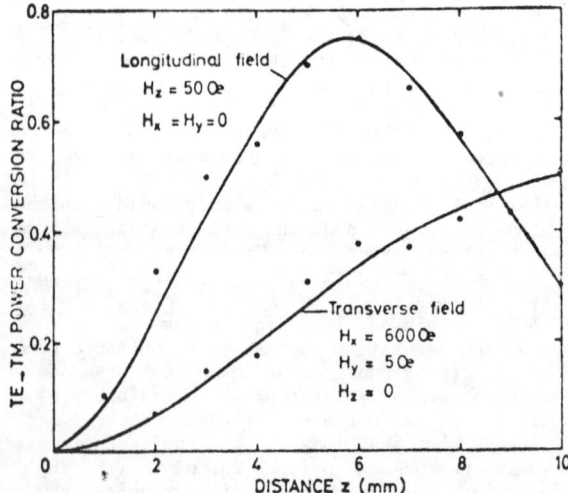

Fig.1 Measured conversion ratio for an easy in-plane film for Faraday (upper curve) and Cotton-Mouton effect (lower curve)

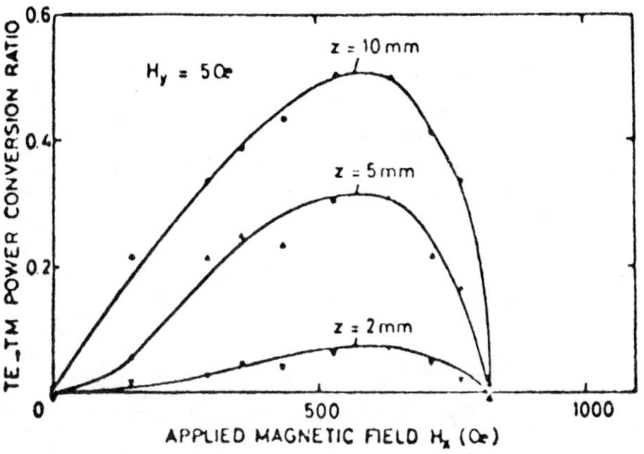


**Fig. 2** Conversion ratio for an easy in-plane (No 1 in Table I) versus the Hx field for different values of the distance. Maximum conversion is obtained for 600 Oe, when the magnetization is 45° out of plane.

Cotton-Mouton conversion to ensure that the magnetization was transverse. Figure 2 shows that the Cotton-Mouton conversion is maximum when Hx ≃ 600 Oe. The magnetization is then at 45° out of the plane. The thickness of the film was 5.5 µm. Using relation (1) we obtain the fitted constants :

$K_F$ = 130 deg/cm for the Faraday rotation.

$K_{CM}$ = 75 deg/cm for the Cotton-Mouton phase shift

(giving a half-wavelength plate when KL = π) and, $\Delta\beta/k = 4.5 \times 10^{-5}$.

The epitaxial induced linear birefringence is $5.5 \times 10^{-5}$ and favours the conversion.

We studied films of YIG with different substitutions of gallium and gadolinium, grown by liquid phase epitaxy. For films of nominal composition $Gd_{0.5}Y_{2.5}Fe_{4.1}Ga_{0.9}O_{12}$ and of thickness 5.5 µm, we measured conversion ratio exceeding 90% for the Faraday conversion. For bubble films the lattice mismatch and the birefringence change sign thus leading to a small conversion ratio. This is shown on fig.3 where the experimental points for Faraday conversion are compared to a theoretical curve, with different losses for the two modes. The epitaxial birefringence is $-3 \times 10^{-4}$. Table I shows for 4 films, the lattice mismatch $\Delta a \equiv a_f - a_s$, the stress σ calculated assuming elastic deformation of the film so that it fits the substrate[9], the birefringence Δn' due to σ calculated from the piezooptic constants given by Pisarev and al[10] and the birefringence Δn calculated from Integrated Optics. Owing to the approximations made, the agreement between Δn and Δn', is satisfactory.

CONCLUSION

We have shown that high conversion rates can be obtained in uniformly magnetized films for Faraday and Cotton-Mouton conversions when the induced epitaxial birefringence is made favourable by proper design of the films. This opens a possibility for building practical non-reciprocal devices for Integrated Optics. The magnetic field necessary for setting the magnetization at 45° out of the plane is high with the film presently used. Thus it seems necessary to look for films with an epitaxial birefringence still favouring conversion ("degenerated" films) and needing low fields for Cotton-Mouton conversion (oblique easy axis). Studies are now in progress in this direction.

REFERENCES

+ Work supported by the Direction des Recherches et Moyens d'Essais - FRANCE

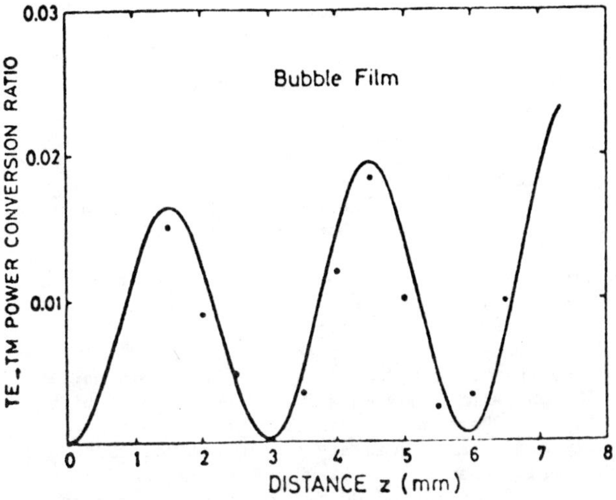


**Fig. 3** Faraday conversion ratio for a bubble film (No 4 in Table I). The increasing of the maximums is explained by the difference between the TE and TM losses.

TABLE I

Comparison between the epitaxial linear birefringence determined from mode conversion (Δn) and that deduced from lattice mismatch (Δn') for different $Gd_yY_{3-y}Fe_{5-x}Ga_xO_{12}$ **films.** Δa is the film-substrate lattice mismatch, σ is the stress.

| Film | x | y | Δa ($10^{-4}$ Å) | σ·$10^{-8}$ (dyn/cm²) | Δn' ·$10^5$ | Δn ·$10^5$ |
|---|---|---|---|---|---|---|
| 1 | 0.9 | 0.45 | 18 | -2.3 | 5.0 | 5.4 |
| 2 | 1.0 | 0.50 | 8 | -1.1 | 2.4 | 3.0 |
| 3 | 1.0 | 0.40 | 0. | 0. | 0.0 | 0.0 |
| 4 | 1.2 | 0.40 | -10.6 | 13.3 | -29.2 | -30.2 |

1. P.K. Tien, R.J. Martin, R. Wolfe, R.C. Le Craw and S.L. Blank, Appl. Phys. Lett. 21,394 (1972)

2. S. Yamamoto and T. Makimoto, J. Appl. Phys. 45,882 (1974)

3. G. Hepner and B. Désormière, Appl. Opt. 13,2007 (1974) and Revue Technique THOMSON-CSF, 6,1115 (1974)

4. R.V. Pisarev, I.G. Sinii, N.N. Kolpakova and Yu. M. Yakovlev, Sov. Phys. JETP 33,1175 (1971)

5. J.F. Dillon Jr, J.P. Remeika and C.R. Staton, J. Appl. Phys. 40,1510 (1969), J. Appl. Phys. 41,4613 (1970)

6 G. Hepner, B. Désormière and J.P. Castéra, Appl. Opt. 14,1481 (1975)

7. A. Yariv, IEEE, J. Quantum Electron. QE-9,919 (1973)

8. S. Yamamoto, Y. Koyamada and T. Makimoto, J. Appl. Phys. 43,5090 (1972)

9. P.J. Besser, J.F. Mee, P.E. Elkins and D.M. Heinz Mat. Res. Bull. 6,1111 (1971)

10. R.V. Pisarev, N.N. Kolpakova, A.G. Titova and L.M. Dashevskaya, Sov. Phys. Solid State 17,31 (1975)

# PHOTOINDUCED MAGNETIC SURFACE ANISOTROPY FIELD IN YIG THIN FILMS*

T. S. Stakelon, P. Yen, H. Puszkarski†, P. E. Wigen
Physics Department, Ohio State University, Columbus, Ohio 43210

## ABSTRACT

Photoinduced changes in the spin wave spectrum of annealed pure YIG thin films have been measured at 23.3GHz and positively identified as arising from a photoinduced magnetic surface anisotropy field. The effect depends upon the thermal history of the YIG thin film sample. For example, after cooling to 20 K, illumination with white light (.2 joules/cm$^2$ or greater) reduces the resonance field separation of a surface spin wave mode and nearest body spin wave mode by 50% while the body mode incurs negligible shift. At 20 K, a photoinduced magnetic surface energy of .0027 erg/cm$^2$ is necessary to account for these and other measurements sensitive to the SWR boundary conditions. The measured photoinduced surface energy is an order of magnitude smaller at 175 K.

## INTRODUCTION

Substitution at low levels of tetravalent and divalent ions into yttrium iron garnet (YIG) has been shown to result in a large class of photoinduced magnetic effects[1,2,3]. This includes the modification by infrared and optical radiation of the uniaxial anisotropy, linear dichroism, coercive force and initial permiability of doped YIG. Observations of these photoinduced changes have usually been limited to techniques of a type which weight the magnitude of the change over the entire sample and would be insensitive to a different response at or near the surface. In contrast, the details of spin wave resonance spectra of thin epitaxial YIG films are selectively sensitive to magnetic properties of the film surfaces.[4] This paper reports the observation using spin wave resonance of a photoinduced anisotropy field at the surface of epitaxial thin films of YIG.

## EXPERIMENTAL METHOD

Six YIG films were used in this study[5], all grown epitaxially on gadolinium gallium garnet substrates and ranging in thickness from 0.56 to 1.2 microns. The films were annealed from 2 to 6 hours at 1100 C to 1300 C in oxygen.

Spin wave resonance spectra were studied at 23.3GHz, at temperatures between 292K and 4.2K. The spectrometer was a cavity reflection type; field modulation at 400 Hz and phase-sensitive detection were employed. A typical measurement period consisted in first cooling the sample in darkness from room temperature within five minutes. The spin wave resonance spectrum was recorded several times, the sample was illuminated with a tungsten lamp, and then the spectrum was again recorded in darkness several times. Irradiation was repeated to determine that the response was saturated.

## MEASUREMENT OF PHOTOINDUCED SURFACE ENERGY

YIG films annealed as described above are known to exhibit two surface modes in their spin wave spectra.[6] These surface modes are described by imaginary wavenumbers (k) and hence their amplitudes decay exponentially into the film. The necessary and sufficient condition for their existence is that the effective field acting on spins at the surface is less than for the bulk spins. The surface mode exchange energy (Dk$^2$, where D is the exchange stiffness constant) is nearly proportional to this difference in the effective field. In contrast, the spin wave with the smallest real wavenumber (first body mode) has a finite amplitude extending over the entire film and has an exchange energy which varies more slowly with the surface field between the limits 0 and $D(\pi/L)^2$ (L is the film thickness). Changes in the exchange energy, magnetization or magnetocrystalline anisotropy field can be measured using the resonance field shift of the spin wave and the dispersion relation at constant frequency. For example, with the applied field $H_o$ in the {100} direction in the film plane the dispersion relation is

$$(\omega/\gamma)^2 = \left[H_o + 4\pi M_s + H_K + Dk^2\right]\left[H_o + H_K + Dk^2\right]$$

where $M_s$ is the saturation magnetization (139 Gauss), $H_K$ is the first order cubic magnetocrystalline field and $\gamma$ is the gyromagnetic ratio (1.759 x 10$^7$/Gauss-sec).

Figure 1(a) indicates part of the resonance spectrum of a YIG film before and after irradiation at 104 K. Irradiation results in no measureable shift in the first body mode and therefore no change in the bulk properties has occurred. The shift in the surface mode positions are therefore due to changes in their exchange energy resulting from a photoinduced surface anisotropy field which changes the total surface energy. Surface energies giving rise to a spin wave resonance spectrum can be calculated from the resonance field splittings and relative intensities of the spin waves.[4] The calculated photoinduced anisotropy energies at the free and substrate surfaces at 104 K are $1.1 \times 10^{-3}$ and $0.43 \times 10^{-3}$ erg/cm$^2$ respectively.

Figure 1(b) shows the results of the same measurement at 20 K where only one surface mode is visi-

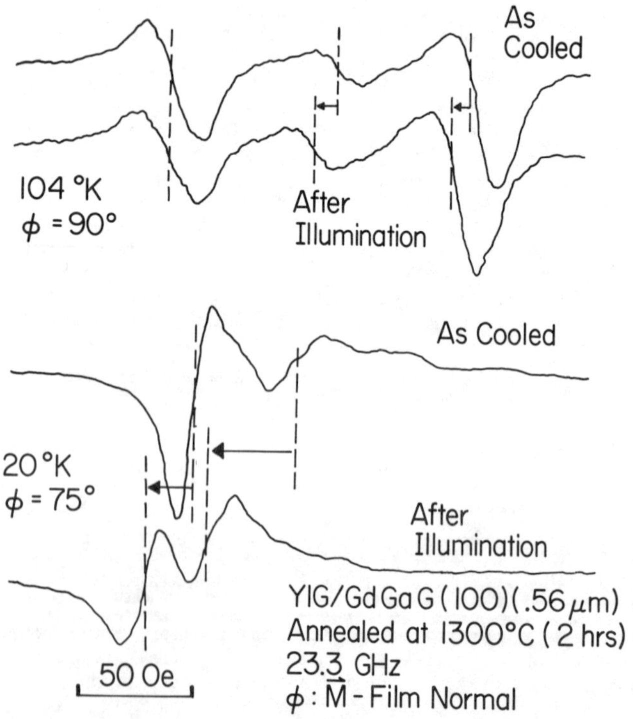


Figure 1(a). From left to right, the first body mode, free surface and substrate surface mode absorption derivatives and the resonance field shifts resulting from irradiation at 104 K. (b) The first body mode and remaining surface mode absorption derivatives incur changes in both resonance field positions and relative intensities at 20 K during irradiation. The applied field is approximately 7400 Oe.

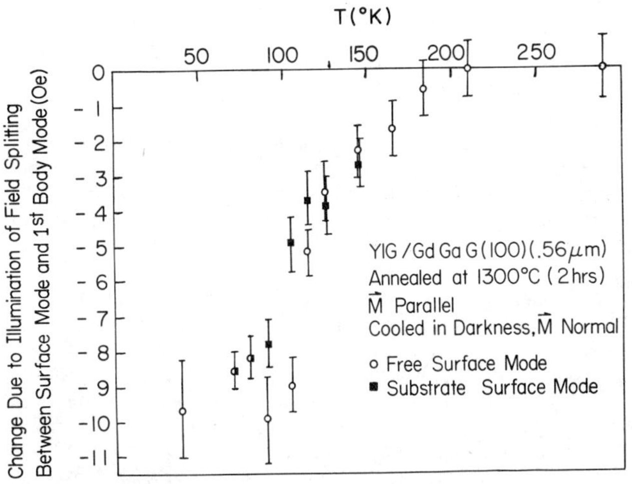


Figure 2. Temperature dependence of the resonance field splitting between the first body and surface modes. As the temperature is lowered the substrate surface mode continuously transforms into the first body mode.[4]

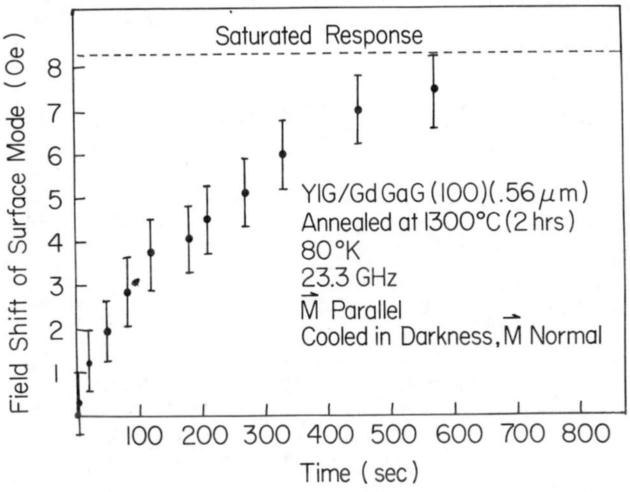


Figure 3. Resonance field splitting of the surface mode from the first body mode during irradiation of approximate intensity .2mW/cm$^2$.

ble. Irradiation results in a shift in the resonance field of both body and surface modes as well as a change in their relative intensities by a factor of three. Such changes can be understood in terms of a photoinduced surface anisotropy energy of $2.7 \times 10^{-3}$ erg/cm$^2$ at the free surface. Bulk property changes cannot account for the observed behavior of the spin wave intensities.

The temperature dependence of the photoinduced shift of the surface modes relative to the first body mode is shown in Figure 2. At about 175 K the photoinduced shift is negligible. Figure 3 indicates the approach to equilibrium of the mode splitting during irradiation of approximate intensity .2mW/cm$^2$. However, the spin wave resonance is stable for at least several hours before and after irradiation at all temperatures. No measureable dependence of the photoinduced surface energy on the orientation of the magnetization during cooling was observed.

Higher order body modes are also sensitive to the photoinduced surface energy but to a lesser extent than surface modes. The surface energy in annealed YIG films has been shown to be strongly dependent on the orientation of the magnetization about the film normal.[4] By suitably orienting the applied magnetic field, the sharp condition is obtained in which all the body modes have wavenumbers given by the solution to tan kL = 0. Under this condition all odd spin wave modes have a minimum (ideal-

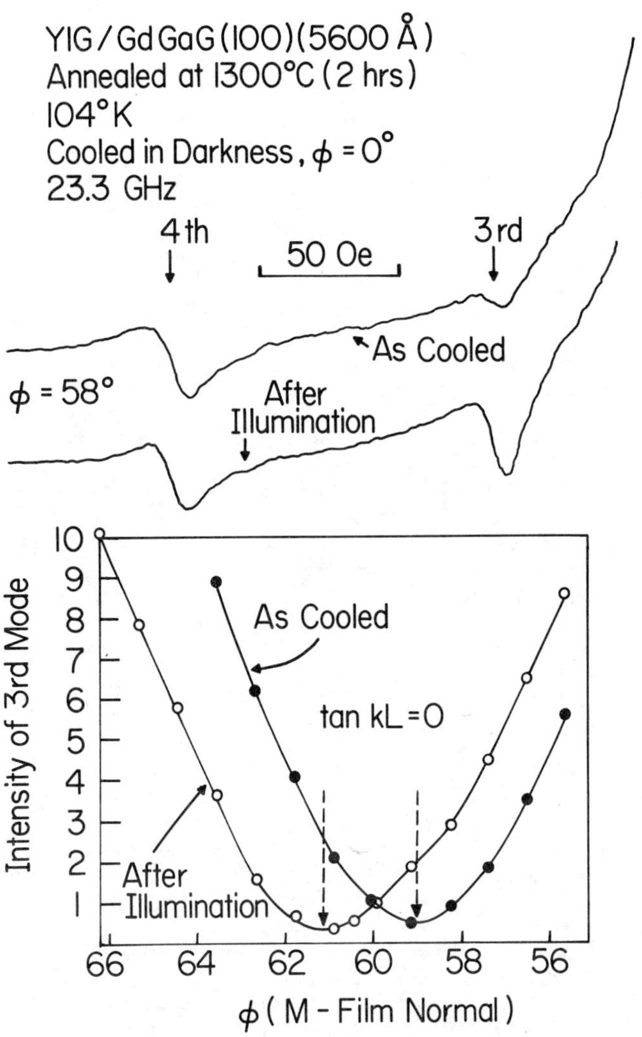


Figure 4. Shift due to irradiation of the position of the odd mode resonance intensity minimum as the film magnetization is rotated with respect to the film normal (angle $\phi$).

ly, zero) intensity. Figure 4 indicates how the photoinduced surface energy changes the necessary magnetization orientation for this condition to occur. Near the minimum the odd mode intensities are altered considerably by irradiation but the even modes are undisturbed.

The photoinduced surface energy is most likely induced by a concentration of defect centers at or near the surface of the YIG film. Low levels of some divalent or tetravalent impurity present in the bulk of the film may concentrate near or at the surface by diffusion during the anneal. It is also possible that the migration of oxygen vacancies during the anneal which then concentrate at the surfaces may result in photo-active $Fe^{2+}$ replacing either the tetrahedral or octahedral $Fe^{3+}$.

REFERENCES

* supported by NSF

† Permanent address: A. Mickiewicz Univ., Poznań, Poland.

1. Gyorgy, E. M. et. al., IBM J. of Research, 14, 321 (1970).
2. Hisatake, K., Japan. J. Appl. Phys., 13, 2067 (1974).
3. Antonini, B. et.al., AIP Conf. Proc. 24, 182 (1975).
4. Wigen, P., Stakelon, T. S., Puszkarski, H., Yen, P., this volume.
5. These were grown by chemical vapor deposition and supplied by Rockwell International, Anaheim, Cal.
6. Yu, J. T., et.al., Phys Rev. B, 11, 420 (1975).

# MULTIPHONON INELASTIC LIGHT SCATTERING: A RESULT OF HOT RECOMBINATION

J. Vitins and P. Wachter,
Laboratorium für Festkörperphysik, ETH Hönggerberg, 8049 Zürich,
Switzerland.

ABSTRACT

We have investigated the multiphonon inelastic light scattering from the magnetically ordering Eu and the diamagnetic Yb chalcogenides. Due to the similarities between the infrared luminescence and this multiphonon light scattering we suggest, that this light scattering is best described by a "hot recombination" process. We consider this scattering to be a bulk property which is largely governed by the polaronic behavior of the excited "hot" electron.

Whereas Raman scattering in semiconductors with incident excitation energies E, which do not coincide with optical transitions, is rather well understood within the framework presented by Loudon[1)], there have been more difficulties in interpreting and predicting the scattering which arises for E near or in resonance with optical transitions[2,3)]. Characteristic in many polar crystals for such excitation conditions, $E \gtrsim E_g$, is the existence of long overtone sequences in the emission spectra[4,5)]. This multiphonon light scattering has been most thoroughly investigated experimentally and theoretically for CdS, however, a conclusive result appears to be not as yet available.[6)] Recently, multiphonon light scattering has also been observed in the Eu and Yb chalcogenides[7-11)]. In this paper we propose a model which explains the behavior of the scattering in these compounds and which also may be applied to other crystals.

The Eu chalcogenides are magnetic semiconductors with the magnetic moment arising from the sharply localized $4f^7$ states. The Yb chalcogenides, on the other hand, are diamagnetic ($4f^{14}$), however, they exhibit a very similar band structure to the Eu compounds. In both the Eu and Yb chalcogenides the ground state of the localized 4f levels lies between the p valence bands of the anion and the 5d conduction bands of the cation[12,13)]. Optical transitions in the visible are predominantly between the 4f and 5d levels.

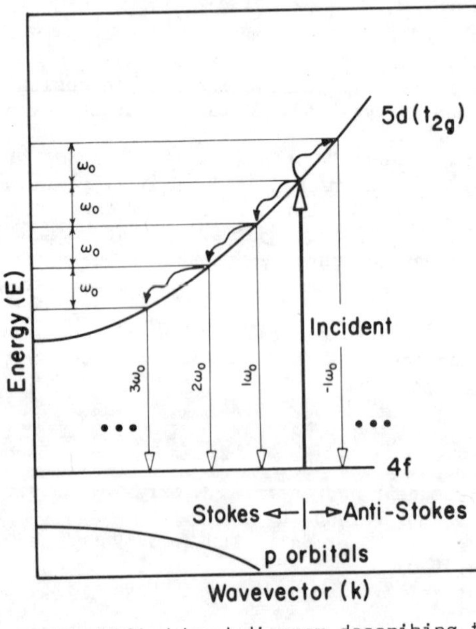


Fig. 1 Simplified band diagram describing the multiphonon inelastic light scattering in the Eu and Yb chalcogenides. The excited "hot" electron relaxes successively to real band states separated by $\omega_0$. From each state there is a small probability of recombination.

Although these compounds crystallize in the fcc structure, the light scattering has been found to be of first order[7)]. For an interpretation in terms of a Raman scattering process it is necessary to incorporate a mechanism which lifts the inversion symmetry. Earlier we have proposed that defects inherent in the samples were the cause for the breakdown of the selection rules[9)]. However, as we have found no corresponding correlation of the scattering intensity to the defect concentration (varied by doping), we conclude, that this is not the essential symmetry breaking mechanism. Furthermore, the very low scattering intensity in EuO cannot be explained by a corresponding low defect concentration. In EuS and EuSe the scattering intensity is quenched for temperatures near and below the magnetic ordering temperature[7,9,10,11)]. Following this experimental fact Tsang et al.[10)] have proposed that the inversion symmetry is lifted by a spin disorder mechanism. Our investigations on EuTe, however, indicate that this mechanism cannot be dominant, as the scattering intensity does not decrease for $T < T_N$ but rather the contrary[11)]. Moreover, as the diamagnetic Yb chalcogenides exhibit very similar multiphonon scattering, we conclude, that spin disorder is not essential for the light scattering in these compounds. In doped EuTe the light scattering has been proposed to be a surface effect[6)]. However, as the penetration depth of visible light in these compounds is in the order of 1000 Å we believe that the scattering is a bulk property.

We propose that the multiphonon light scattering in the Eu and Yb chalcogenides is best described by a mechanism of "hot recombination". Martin and Varma[14)] have proposed a "cascade model", also based on recombination. In this model the photo-electron is excited into real band states from which it decays to the bottom of the conduction band through the successive emission of LO phonons. At each step of the relaxation process there is a small probability of radiative recombination of the excited not yet thermalized electron with its own hole. This leads to a series of emission lines on both Stokes and anti-Stokes side for $T > 0$. As the relaxation of "hot" electrons occurs predominantly through the emission of LO phonons the emission lines are shifted by multiples of LO phonon frequencies. A more elaborate description of this model has been suggested by Zeyher[15)]. For the Eu and Yb chalcogenides the cascade model needs modification, as the recombination does not occur between a free electron-hole pair, but rather between a more or less localized "hot" 5d electron and its strictly localized hole in the 4f-shell. The relaxation process describing the multiphonon line sequence in these compounds is shown in fig. 1. For sake of simplicity the bands in this diagram are only schematically drawn. Due to the localization of the 4f levels indirect absorption processes can be neglected for excitations between the 4f and 5d states. In the cascade model as, proposed by Martin and Varma, the scattering process can be factored into identical single step transitions which occur in succession. Hence, one can expect the scattering intensities of the harmonics to decrease (due to dissociation or trapping) approximately exponentially, and the linewidths to increase roughly linearly. This is indeed observed in these compounds as is shown in fig. 2. Here the experimental data are plotted for YbS using E = 2.71 eV (4579 Å) laser excitation. The relative

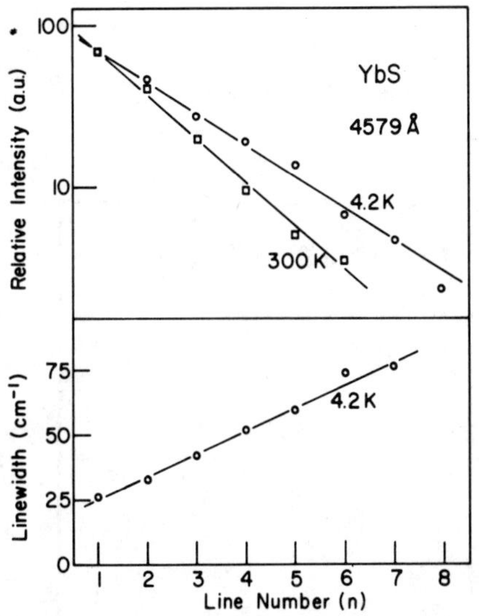


Fig. 2 Relative scattering intensities (upper half) and corresponding linewidths (lower half) of the overtone line sequence in YbS for 4.2 K and 300 K.

scattering intensities of the first line at 4.2 and 300 K have been set equal. One observes, that at 4.2 and 300 K the scattering intensities decrease at a similar rate. In EuS we have found that the intensities decrease with the same slope at 4.2 and 300 K. These observations are in agreement with the cascade model[14,15]. Further, it is interesting to note, that the slope of the scattering intensities of the harmonics is also dependent on excitation energy. This was found to be most drastic in YbS and YbSe, where also the intensity of the first Stokes line exhibits a strong dependence on excitation energy ($2.3 < E < 2.8$ eV). The change in slope with variing E may possibly be explained by the dispersion of the 5d band[14].

The breakdown of selection rules in resonant Raman scattering has variously been observed and has mostly been explained for excitations into exciton states[16,17]. Thus, for large radius exciton intermediate states and for large electron-phonon coupling constants one expects usual selection rules to be no longer applicable. In the model of "hot recombination" the 4f electron is excited into real band states and the lifetime of the "hot" electron is expected to be of the order of $10^{-9}$ sec. The infrared luminescence in the Eu chalcogenides is governed by the polaronic behavior of the excited electron in the 5d band (combined electronic and magnetic polaron)[18]. Due to its finite lifetime we believe that also the "hot" photo-electron is best described by a polaronic behavior with large orbital radius. Hence, this scattering must not follow usual selection rules and can also occur in crystals with $O_h$ symmetry[16]. Furthermore, as is known from the infrared luminescence, there is a strong relaxation of the lattice around the localized hole in the 4f shell [18]. Due to this lattice relaxation we expect that the q-selection rule for the phonons is strongly relaxed. Thus phonons from throughout the Brillouin-zone can take part in this scattering. This explains the large linewidths (20-50 cm$^{-1}$) generally observed in the Eu and Yb chalcogenides.

The polarization of the scattered light in the Eu and Yb chalcogenides also exhibits a behavior not usual for Raman scattering. Hence, in the Eu chalcogenides and in YbTe we observe a polarization of the scattered light which is perpendicular to the incident polarization for $2.3 < E < 2.8$ eV. However, with the same excitation energies the scattering in YbS and YbSe is found to be parallel. In both cases the scattering geometry was identical and the E vector of polarization of the incident light was in the plane of incidence. Leaving the scattering geometry unchanged no dependence of the polarization on crystal orientation was found. Whereas in the Eu chalcogenides and YbTe the excitation occurs into the 5d ($t_{2g}$) band, in YbS and YbSe the excitation is into either a $4f^{13}5d(t_{2g})$ multiplet or a $5d(e_g)$ band[13]. From these observations we conclude, that the polarization of the emission lines is not primarily dependent on the crystal symmetry but more so on the nature of the electronic transition.

In YbS and YbSe the scattering intensities of the emission lines have been found to decrease strongly with increasing temperature[19]. This has been explained by a thermally activated quenching process as has also been employed for the infrared luminescence[18]. Similar quenching of multiphonon light scattering has been observed in CdSe[20]. This similarity between the infrared luminescence and the multiphonon light scattering further justifies the description of the latter by a mechanism of hot recombination.

ACKNOWLEDGEMENT

The authors are grateful to Dr.E. Kaldis for supplying the crystals.

REFERENCES:

1) R. Loudon, Advan. Phys. 13, 423 (1964)
2) B. Bendow, J.L. Birman, A.K. Ganguly, T.C. Damen, R.C.C. Leite, and J.F. Scott, Optics Commun. 1, 267 (1970)
3) C.A. Ferrari, J.B. Salzberg, and R. Luzzi, Solid State Commun. 15, 1081 (1974)
4) R.C.C. Leite, J.F. Scott, and T.C. Damen, Phys. Rev. Lett. 22, 780 (1969)
5) J.F. Scott, R.C.C. Leite, and T.C. Damen, Phys.Rev. 188, 1285 (1969)
6) J.L. Birman, Proced. of the 12th Int. Conf. on the Phys. of Semiconductors, Stuttgart 1974
7) R.K. Ray, J.C. Tsang, M.S. Dresselhaus, R.L. Aggarwal, and T.B. Reed, Phys. Lett.37A, 129 (1971)
8) G.D. Holah, J.S. Webb, R.B. Dennis, and C.R. Pidgeon, Solid State Commun. 13, 209 (1973)
9) A. Schlegel and P. Wachter, Solid State Commun. 13, 1865 (1973)
10) J.C. Tsang, M.S. Dresselhaus, R.L. Aggarwal, and T.B. Reed, Phys. Rev. B9, 984 (1974)
11) J. Vitins and P. Wachter, Solid State Commun. 17, 911 (1975)
12) G. Güntherodt, Phys. cond. Matter 18, 37 (1974)
13) R. Suryanarayanan, Ph.D. Thesis, Orsay Paris, 1973
14) R.M. Martin and C.M. Varma, Phys. Rev. Lett. 26, 1241 (1971)
15) R. Zeyher, Solid State Commun. 16, 49 (1975)
16) E. Mulazzi, Phys. Rev. Lett. 25, 228 (1970)
17) R.M. Martin and T.C. Damen, Phys. Rev. Lett.26, 86 (1971)
18) P. Streit, Phys. kondens. Materie 15, 284 (1973)
19) J. Vitins and P. Wachter, to be published
20) E. Gross, S. Permogorov, YA Morozenko, and B. Kharlamov, phys. stat. sol.(b)59, 551 (1973)

# ELECTRONIC STRUCTURE OF THE IONIC FERROMAGNET AND CATALYST $La_{1-x}MnO_3$ BY SPIN POLARIZED, UV AND X-RAY PHOTOEMISSION SPECTROSCOPY.

S.F. Alvarado, W. Eib, P. Munz and H.C. Siegmann
Laboratorium für Festkörperphysik, ETH, 8049 Zürich, Switzerland
and
M. Campagna and J.P. Remeika
Bell Laboratories, Murray Hill, New Jersey 07974.

ABSTRACT

We report ultraviolet photoemission spectra and spectra of the spin polarization (SSP) of photoemitted electrons from $La_{1-x}Pb_xMnO_3$ for photon energies $\hbar\omega < 11$ eV and X-ray photoemission density of states using monochromatized AℓKα radiation. Contributions from high spin $Mn^{3+}$ and $Mn^{4+}$ ions to the photocurrent are unambiguously identified in the binding energy range $0 = E_F < E_B < 3.5$ eV, while the bonding levels generated by the covalent admixture of O(2p) and Mn(3d) states are located between 3 and 7 eV below $E_F$, overlapping with Pb(6s) levels. We compare these results with the predictions of the two limiting models for the description of outer d-electrons in perovskite, the ligand field approach and the generalized LCAO interpolation scheme of Mattheiss. From the structure in the SSP we deduce that the correlation energy $U_{eff}$ of 3d-electrons in these ionic ferromagnets cannot exceed 2-2.5 eV.

The $La_{1-x}Pb_xMn^{3+}_{1-x}Mn^{4+}_xO_3$ compounds exhibit an interesting correlation between ferromagnetism and electrical conductivity. The most striking features being a semiconductor to metal transition observed at $T \sim T_c$ ($T_c$= 340 K for x = 0.3) and the almost linear relationship between the magnetization and the electrical resistivity. Responsible for this peculiar behavior is the presence of the $Mn^{3+}$ and $Mn^{4+}$ ions which are in a high spin configuration in octahedral oxygen coordination. It has been proposed that below $T_c$ the electron states near the Fermi energy are made up of 2p(O) and $e_g$(Mn) wave functions, and that the spins of the more localized $t_{2g}$(Mn)-levels are coupled via the Zener mechanism[2]. For the description of these electron states both a generalized LCAO scheme[3], i.e. a collective behavior and a localized picture[4] have been put forward, but experimental evidence in support of either of these approaches has been scarce so far. The difficulty arises from the fact that there are many types of electron states in a narrow energy range. The aim of the present study is to show how these difficulties may be overcome by combining X-ray (XPS) and ultraviolet (UPS) photoemission spectroscopy, and the measurement of the photoelectron spin polarization (photo-ESP). The experiments were done on surfaces of single crystals obtained by cleaving in UHV. Techniques and sample preparation have been already discussed[5,6,7].

We experimentally determine the energy of the d-states as well as their position relative to the unpolarized O(2p) levels. The intensity of the 3d(Mn), 2p(O) and 6s(Pb)emission are analysed and compared with results on insulating perovskites, specially $SrTiO_3$. It is concluded that the $(3d)t^3_{2g}$-levels are centered at 2.9 ± 0.3 eV below the Fermi level $E_F$, overlapping the O(2p) band; the $e^2_g$ states populated with 0.7 electron/Mn atom form a band of ~ 1 eV. The results can be understood only by assuming that the Pb(6s) levels form a band, extending up to near $E_F$.

In fig. 1 we show the XPS photoemission spectrum of the valence band region of conducting $La_{0.7}Pb_{0.3}MnO_3$ and of insulating $SrTiO_3$[8]. The difference between both curves gives an idea of the relative emission intensities of the s, p and d contributions. The XPS of $SrTiO_3$ should give the contribution of the 2p(O) states to the total density of states (DOS) since the covalent p-d mixing is expected to be quite small[5]. The mixing is

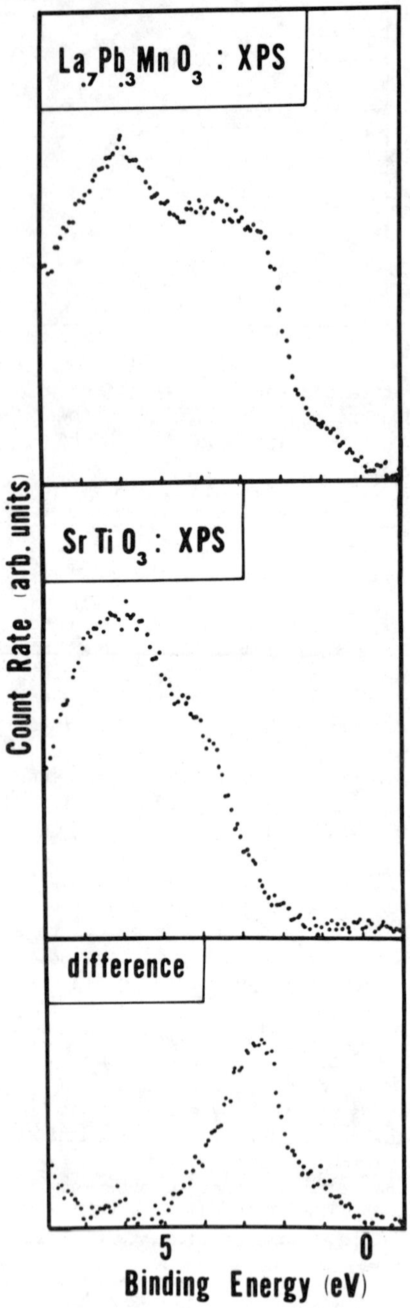


Fig.1 X-ray photoemission spectra of the valence band region of $La_{.7}Pb_{.3}MnO_3$, $SrTiO_3$, and the difference, see text and Ref.10.

larger in the LaPb-Manganite ($LaMnO_3$ is an antiferromagnet with $T_{Néel}$ = 100 K). Although XPS may overemphasize the d-contribution in the experimental DOS, this comparison allows to locate the region of predominant 3d(Mn) and 6s(Pb) emission in the range $0=E_F$ down to ~ 5 eV. For XPS the theoretical cross sections for $3d(Mn^{4+}_{0.3}+Mn^{3+}_{0.7})$, $2p(O_3)$ and $6s(Pb_{0.3})$ are proportional to ~ 1050, ~ 750 and ~ 300 respectively[9].

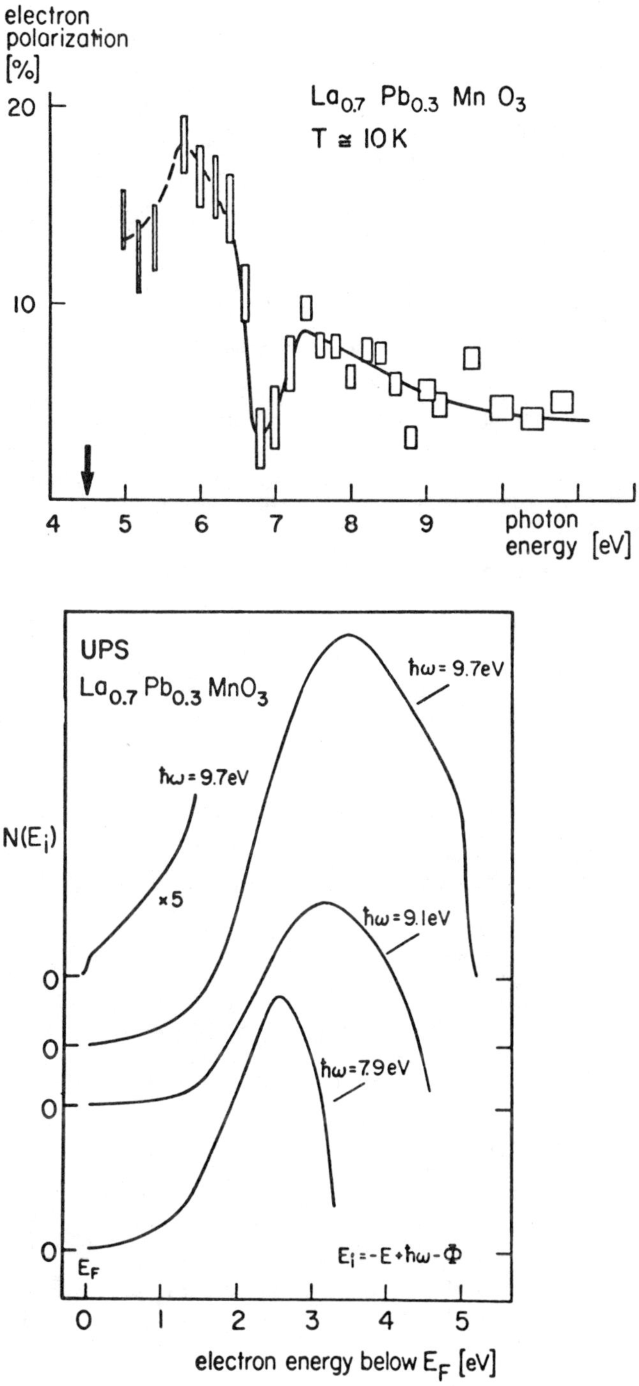


Fig.2 Dependence of Photo-ESP in % on Photon energy for $La_{0.7}Pb_{0.3}MnO_3$ at the temperature 10 K and at a magnetic field 16.4 kOe. Photoelectric threshold is indicated by arrow. Lower Part: UPS-energy distribution at various photon energies. The number of emitted electrons $N(E_i)$ is in arbitrary units. Discontinuity at $E_F$ appears on amplification only (x5).

The low intensity of the EDC at $E_F$ indicates that $t^3_{2g}$-levels cannot be located at $E_F$. The best fit is obtained by assigning a strength of ~ 200 to the $e_g$ band and locate it within ~ 1.5 eV from $E_F$, and ~ 850 to the much more localized $t^3_{2g}$-levels near 2.5 - 3.5 eV below $E_F$. In order to fully account for the relative emission intensities we still have to assume a non negligible 15% mixing of both $t_{2g}$ and $e_g$ with O(2pπ) and (2pσ)-states respectively

This explanation becomes unique because it is the only plausible one that fits all three measurements.

The UV and ESP-photoemission spectra are shown in fig. 2. The shape of the UPS spectrum does not depend on photon energy and also is quite similar to the XPS spectrum. This shows that the density of final states of the photoelectron is not important to the interpretation. The non-zero electron spin polarization observed near threshold and the weakening of the UV-DOS relative to the X-ray DOS in the same region of binding energy confirms the presence of d-electrons near $E_F$. Emission from d-states is known to decrease with ħω because of the centrifugal barrier. A pronounced shoulder rises from 1.6 eV below $E_F$ (UPS); the decrease of the ESP at this very same energy shows that the onset of this shoulder must be due to an increase in the emission of unpolarized p(6s)- or O(2p)-electrons. The ESP starts to increase again at 2.2 eV below $E_F$, thus indicating the threshold for the $t^3_{2g}$ emission. The ESP is highest near $E_F$ in accordance with a 100% polarized $e_g$-band. Since only ~ 20% polarization is observed one concludes that the emission of unpolarized p(6s)-electrons starts already near threshold. A more localized electron approach, assuming strongly correlated d-levels in analogy to the 4f-levels of metallic Rare Earth compounds in the interconfigurational fluctuation (ICF) state and small crystal field splitting, cannot account for the results because it is in contrast with the observed line intensity. In this approach one assumes that photoemission is faster than the time corresponding to the bandwidth of the $e_g$-levels. On such a fast time scale the $Mn^{4+}$ and $Mn^{3+}$ are found to coexist, in analogy to the 2 ionic configuration of the $Sm^{2+}$- and $Sm^{3+}$-ions in the metallic phase of SmS[10]. Then the two peaks in the ESP-spectrum, whose thresholds are separated by ~ 2.2 eV, are identified as the final states of the transition $Mn^{3+} \rightarrow Mn^{4+}$ + photoelectron and $Mn^{4+} \rightarrow Mn^{5+}$ + e respectively. The first transition has two final levels namely $^5E \rightarrow {}^4T_2$ + e and $^5E \rightarrow {}^5A_2$ + e. The second transition is $^4A_2 \rightarrow {}^3T_1$ + e. The difference between both transition thresholds is an upper limit to the effective intraionic correlation energy $U_{eff}$ = 2.2 eV. The model assuming localized $t_{2g}$-levels and delocalized $e_g$-electrons fits therefore best the results.

In conclusion we have shown how information on d-levels can be obtained by combining various photoemission techniques even in the case of the very complex transition metals catalysts. This demonstrates for the first time the promising aspects of these techniques for investigating the detailed behavior of electrons in the critical but interesting range, where neither the uncorrelated band model nor the strong correlated Hubbard model are the correct approach. Temperature dependent studies and investigations involving materials with analogous crystal structure, but containing 3-d-electrons with different degree of correlation are in progress.

Helpful discussion with Dr.'s P.W. Anderson, L.F. Mattheis, J.M. Rice, G.K. Wertheim and J.A. Wilson are gratefully acknowledged. Part of this work was supported by the Schweizerischer Nationalfonds.

REFERENCES

1. L.K. Leung, A.H. Morrish, and C.W. Searle, Can. J. of Physics 47, 2697 (1969)
2. Clarence Zener, Phys. Rev. 82, 3 (1951)403
3. L.F. Mattheis, Phys. Rev. B2, 3918 (1970) and Phys.Rev. B6, 4718 (1972)
4. J.B. Goodenough, Progr. Sol.State Chem. 5, 145 (1971)
5. G.K. Wertheim, L.F. Mattheis, M. Campagna and T.P. Pearsall, Phys.Rev.Lett. 32, 997 (1974)
6. P. Cotti and P. Munz, Phys.Cond.Matter 17, 307 (1974)

7. S.F. Alvarado, W. Eib, F. Meier, D.T. Pierce, K. Sattler, H.C. Siegmann and J.P. Remeika, Phys. Rev. Lett. 34, 319 (1975) and G. Busch, M. Campagna and H.C. Siegmann, J. Appl. Phys. 41, 1044 (1970)
8. The XPS spectrum of insulating $SrTiO_3$ has been measured with the help of a flood gun (HP model 18623 A). The spectrum has been shifted in energy so as to obtain alignment of the main O(2s) levels around 24 eV $E_F$ for purpose of comparison.
9. J.S. Scofield, UCLR Report No 51326, 1973.
10. M. Campagna, E. Bucher, G.K. Wertheim, and L.D. Longinotti, Phys.Rev.Lett. 33, 165 (1974)
11. L.A. Pedersen and W.F. Lubby, Science 176, 1355 (1972); R.J.H. Voorhoeve, J.P. Remeika and D.W. Johnson, Jr. Science 180, 62 (1973)
12. R.J.H. Voorhoeve, J.P. Remeika and L.E. Trimble, to be published in Ann. New York Acad. Sci. (1975).

# MEASUREMENT OF THE DOMAIN WIDTHS OF A MAGNETIC THIN FILM IN AN IN-PLANE FIELD

**S. Kern*, P. V. Cooper, and D. J. Craik**
**Department of Physical Chemistry, University of Nottingham, Nottingham, England**

## ABSTRACT

When an in-plane field is imposed upon a magnetic thin film, the domains of one magnetic direction generally grow at the expense of those magnetized in the opposite direction. We describe a simple and sensitive method for measuring the change in domain widths that uses the Faraday rotation produced by the domains on a polarized laser beam.

After the in-plane field is applied, an ac field of frequency, f, is applied normal to the film and the light intensity of the beam, after passing through an analyzer, is detected and displayed on an oscilloscope. The analyzer is adjusted until the display consists only of the second harmonic, 2f. The zero-order diffraction intensity is then measured as the sample is saturated in one direction by a dc field applied perpendicular to the film plane, and then in the other direction. The difference in the light intensity for the two directions provides a direct determination of the difference in widths of the oppositely magnetized domains in the in-plane field.

For a magnetic thin film with perpendicular uniaxial anisotropy the imposition of a field normal to the plane will cause the previously equally wide, alternately-magnetized, domains to change their relative widths. As several authors have shown[1], an in-plane field applied to a thin film having an anisotropy with other than uniaxial symmetry will also produce this effect and measurements of the change have been used to determine the magnitude of the anisotropy. We wish to describe a new, simple, and sensitive method for measuring the domain width changes produced by such a field.

Our apparatus is as shown in Figure 1. The polarized output from an argon-ion laser passes through the thin film sample at normal incidence. Provision is made for applying an in-plane field using an electromagnet (whose poles are labelled N and S in the Figure) or a normal dc or ac field using the coils (indicated by C). The light is Faraday - rotated by the domains of the film and then analyzed prior to detection. If $\alpha$ is the angle between the analyzer and the "crossed polarizer-analyzer" position, then adding the light amplitudes from the oppositely magnetized domains gives a zero-order diffraction intensity of[2,3,4,5]

$$I_p(\alpha) = I_o[(2p - 1)\cos\alpha\sin\beta + \sin\alpha\cos\beta]^2 \quad (1)$$

where $\beta$ is the Faraday rotation angle and p is the fraction of a domain period that is magnetized in one direction; that is, $p = \frac{1}{2}$ in the absence of an externally applied field. The crossed polarizer-analyzer position ($\alpha = 0$) in the absence of an external field can be found by applying an ac field, $H = H_o \sin\omega t$ to the coils. This causes p to become $p + a\sin\omega t$ and then.

$$I_{\frac{1}{2}}(\alpha) = I_o\cos^2\alpha\sin^2\beta\left[\frac{a^2}{2}(1 - \cos 2\omega t) + 2a\tan\alpha\cot\beta\sin\omega t + \tan^2\alpha\cot^2\beta\right] \quad (2)$$

The dc and $\omega$ terms quickly become large with changes in $\alpha$ and it is, therefore, easy to obtain the $\alpha = 0$ position, for which only the $2\omega$ term survives, by displaying the detector output on an oscilloscope. For $\alpha = 0$

$$I_p(0) = I_o(2p - 1)^2\sin^2\beta \quad (3)$$

and the detected intensity is proportional to the square of the sample magnetization: $I_{\frac{1}{2}}(0) = 0$, (no applied field), $I_1(0) = I_0(0) = I_o\sin^2\beta$ (corresponding to saturation in either normal direction) - see Figure 2a.

With the addition of an in-plane field p becomes $p + \Delta$ where $\Delta$ is some fractional amount of, say, increase.[1] We can again use the $2\omega$ condition to obtain an equality in response (representing a minimum = 0 in the $\omega$ response) by altering $\alpha$. Minimizing

$$I_p^H(\alpha) = I_o\left\{[2(p + \Delta) - 1]\cos\alpha\sin\beta + \sin\alpha\cos\beta\right\} \quad (4)$$

gives (for small $\alpha$ and $\beta$)

$$\Delta = -\alpha/2\beta_\Delta \quad (5)$$

where $\beta_\Delta$ is the Faraday rotation produced by the domains in the presence of the in-plane field.

Of the several experimental techniques that can be employed we here describe two. If the sample is first saturated in one direction and then the other, the detected intensities are,

$$I_1(\alpha) = [\cos^2\alpha\sin^2\beta + \sin^2\alpha\cos^2\beta + 2\sin\alpha\cos\alpha\sin\beta\cos\beta]\, I_o \quad (6a)$$

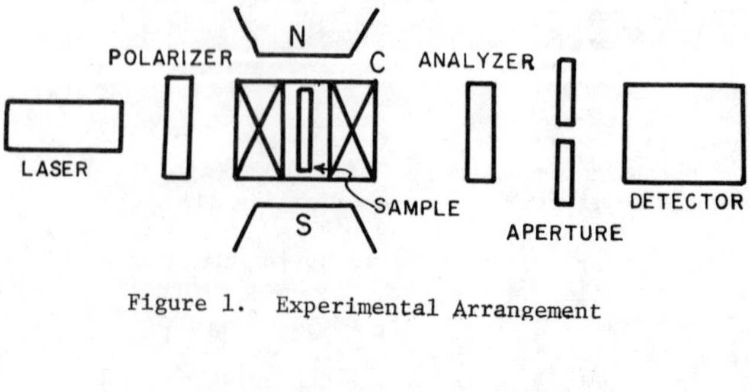


Figure 1. Experimental Arrangement

$$I_o(\alpha) = [\cos^2\alpha \sin^2\beta + \sin^2\alpha \cos^2\beta - 2 \sin\alpha \cos\alpha \sin\beta \cos\beta]\, I_o \qquad (6b)$$

with a difference,

$$D = I_1(\alpha) - I_o(\alpha) = 4\, I_o \sin\alpha \cos\alpha \sin\beta \cos\beta \qquad (7)$$

which, for small angles and an in-plane field gives

$$D \cong 4\, \alpha\, \beta_\Delta\, I_o = 8\, \Delta\, \beta_\Delta^{\,2}\, I_o \qquad (8)$$

The dependence on $I_o$ may be eliminated by taking the ratio

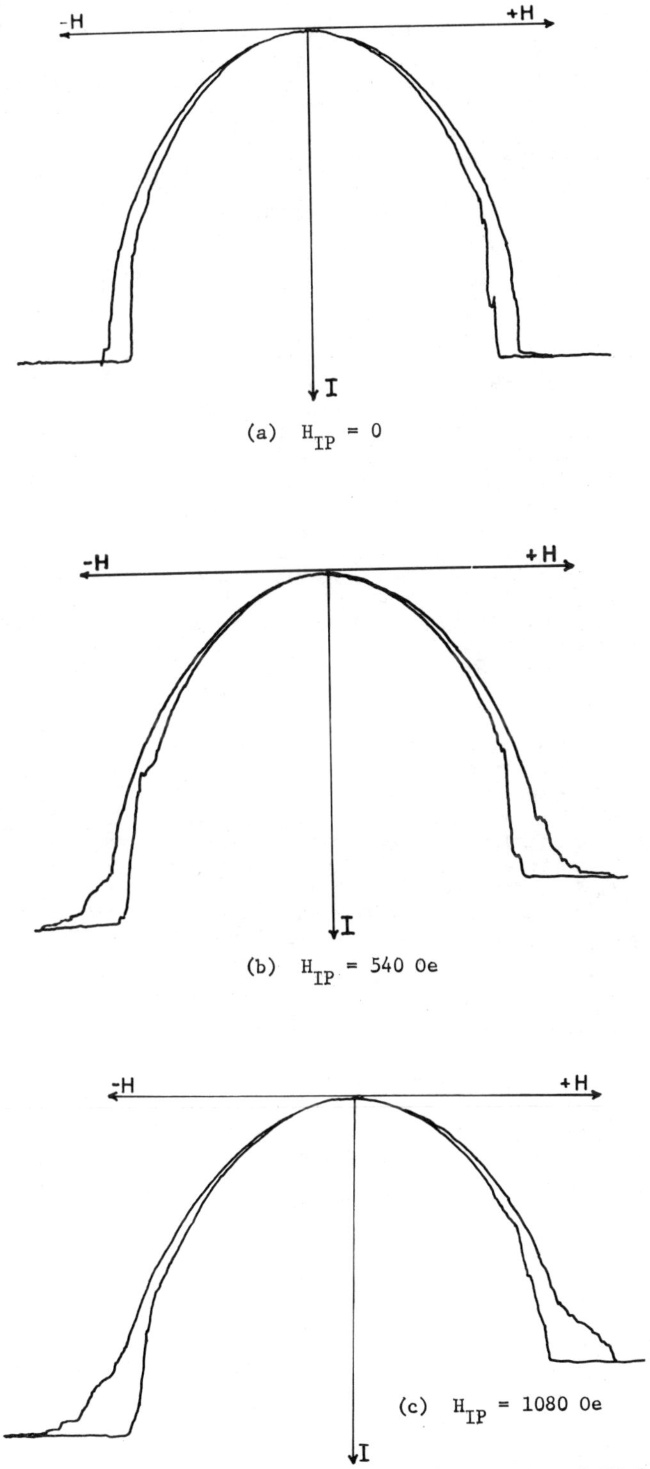


(a) $H_{IP} = 0$

(b) $H_{IP} = 540$ Oe

(c) $H_{IP} = 1080$ Oe

Figure 2. Zero Order Diffraction Intensity vs. Normal Field

$$\frac{D}{I_1(0)} = \frac{8\Delta\beta_\Delta^{\,2} I_o}{\beta_o^{\,2} I_z(0)} = \frac{8\Delta\beta_\Delta^{\,2}}{\beta_o^{\,2}} \qquad (9)$$

so measurements must be made of the Faraday rotation in the presence and absence of the in-plane field.

A second, and simpler, method is to use the average of the saturation intensities

$$\langle I \rangle = \frac{I_1(\alpha) + I_o(\alpha)}{2} = [\cos^2\alpha \sin^2\beta + \sin^2\alpha \cos^2\beta]\, I_o \qquad (10)$$

which for small angles and an in-plane field gives,

$$\langle I \rangle \cong \left[\beta_\Delta^{\,2} + \alpha^2\right] I_o = \beta_\Delta^{\,2} \left[1 + 4\Delta^2\right] I_o \qquad (11)$$

Then,

$$\frac{D}{\langle I \rangle} = \frac{8\Delta}{1 + 4\Delta^2} \qquad (12)$$

which expression is independent of $\beta_\Delta$ as well as $I_o$. If we let $\xi \equiv D/\langle I \rangle$, then

$$\Delta = \frac{1}{\xi}\left[1 \pm \sqrt{1 - \frac{\xi^2}{4}}\right] \qquad (13)$$

with the - sign being physically significant here. Using the data from Figure 2 we obtain $\Delta = .016$ for $H_{IP} = 540$ Oe and $\Delta = .033$ for $H_{IP} = 1080$ Oe giving a change in $\Delta$ of $6.2 \times 10^{-5}$/Oe for this sample in this orientation.

It is clear that this method provides great sensitivity, since changes in relative domain widths of 1% or less are easily measured, something not easily done by photographic techniques. We are planning to use this method to investigate the appropriateness of a previously derived formulation of the magnetostatic properties of these thin films in the presence of an in-plane field.[6]

We wish to thank the Plessey Company for the sample used in these experiments.

REFERENCES

*Permanent address: Department of Physics, Colorado State University, Fort Collins, Colorado, 80523.

1. See, for example,
J. W. F. Dorleijn, W. F. Druyvesteyn, G. Bartels and W. Tolksdorf Philips Res. Repts. 28 152 (1973)
Y. Shimada, H. Kojima and K. Sakai J. Appl. Phys. 45 4598 (1974)
Y. Shimada J. Appl. Phys. 45 3154 (1974)
M. H. Yang and M. W. Muller J. Appl. Phys. 45 4130 (1974)
M. W. Muller, S. K. Chung and M. H. Yang IEEE Trans Mag 11 1121 (1975)
2. H. Boersch & M. Lambeck z. Phys. 177 157 (1964)
3. H. M. Haskal IEEE Trans. Mag. 6 542 (1970)
4. K. R. Papworth Phys. Stat. Sol. (a) 22 373 (1974)
5. S. Kern and D. J. Craik (Submitted to J. Phys. D)
6. P. V. Cooper and D. J. Craik J. Phys. D: 6 1393 (1973)

# UV PHOTOEMISSION STUDIES OF YTTRIUM IRON GARNET

P.K. Larsen and R. Metselaar, Philips Research Laboratories, Netherlands, and B. Feuerbacher, Surface Physics Division, European Space Agency, Noordwijk, Netherlands

ABSTRACT

Ultraviolet photoemission measurements on thin epitaxially grown layers of n- and p-type conducting yttrium iron garnet have been performed in the photon energy range 7.7 - 21.2 eV. The samples were cleaned in situ by ion bombardment followed by an annealing treatment. Photoemission spectra were recorded in a 4π geometry using an ac modulation technique. The photoelectron energy distribution curves show the absence of narrow iron 3d states.

## INTRODUCTION

In recent years the electronic structure of iron garnets has been a matter of considerable interest and has motivated numerous, mostly optical, investigations. This has led to a basic understanding of the optical crystal-field transitions of the trivalent iron ions in garnets.[1,2] The proper assignment of charge-transfer transitions observed in optical and magneto-optical spectra is, however, still a subject of discussion and different energy level schemes have been proposed for iron garnets.[3,4] An alternative energy level scheme has also been proposed based on electrical conductivity studies of YIG ($Y_3Fe_5O_{12}$).[5]

Photoemission studies of solids have in the past been very useful in obtaining information on the band structure and energy levels of solids. The energy distribution curve (EDC) of the photoemitted electron has appeared to relate to the density of states, e.g. a region with a high density of states is often seen as a peak in the EDC.[6] In order to obtain new information on the energy level schemes of garnets we have performed photoemission experiments on YIG in the photon energy range, hν = 7.7-21.2 eV.

## EXPERIMENTAL PROCEDURE

The photoemission experiments were performed in ultra high vacuum ($p \leq 10^{-9}$ torr) using a retarding field analyser.[7] Single crystalline thin films of YIG made by liquid phase epitaxy (LPE) on substrates of $Gd_3Ga_5O_{12}$ were used. In order to avoid charging effects $Si^{4+}$ or $Pb^{2+}$ substituted films were studied.[8] Chemical analysis by microprobe indicated the concentrations for a n-type conducting film (sample I) with a room temperature resistivity of $10^4$ Ωcm to be [Si] = 0.046 and [Pb] = 0.025 per formula unit. For a p-type film (sample II) of resistivity $7.10^6$ Ωcm at room temperature [Pb] = 0.32 and [Si] = 0.04 per formula unit. The film thickness was 4.6 μm for sample I and 3.1 μm for sample II. Both film surfaces were (111) planes. A 0.5 μm thick platinum layer was sputtered over parts of the sample surface leaving 8x8 $mm^2$ uncovered. This Pt layer served as an electrical contact to the YIG layer and further in the photoemission experiments as a reference giving the position of the Fermi-energy. The sample was attached to a heater at the end of a support in the vacuum chamber. After cleaning, the sample could be positioned in situ to the center of a 10 cm diameter gold-coated sperical collector (≃ 4π) for the photoemission experiments.

The cleaning of the sample surface was found necessary in order to get reproducible photoemission results. Auger measurements of uncleaned films showed a coverage of mainly carbon, which could be removed by ion bombardment. We found that ion bombardment with oxygen ions accelerated to 450 volts ($i_{sample} \simeq 10^{-7}$ A/$cm^2$ in 30 minutes) followed by an annealing treatment at the same oxygen partial pressure as used for the ion bombardment ($p_{O_2} \simeq 10^{-5}$ torr) gave reproducible photoemission. The sample annealing temperature was approximately 700°C and the annealing time was 15-30 minutes.

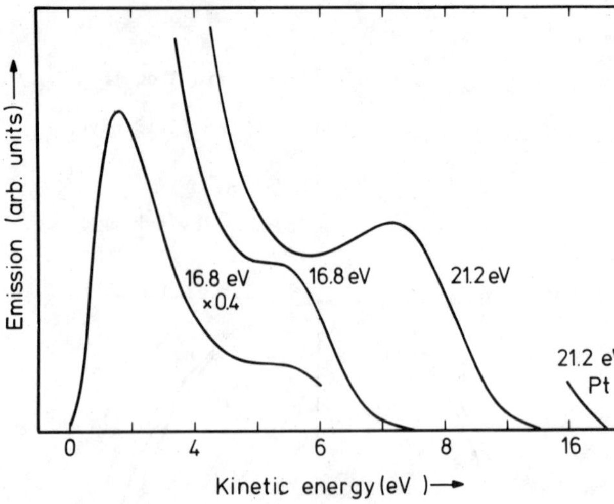


Figure 1 - Number of photoemitted electrons per eV as a function of their kinetic energy. n-type YIG, sample I at two different photon energies.

Measurements of the EDC's were performed by superposing a small ac modulating voltage onto the retarding emitter voltage, $V_r$. The in-phase component of the collected, ac-modulated photocurrent was than measured as $V_r$ was swept.[7] The energy resolution was better than 0.30 eV.

The resonant lines of $H_2$ at 10.2 eV, and 7.7 eV we were used in combination with windows of LiF, $MgF_2$ or quartz to provide the incident radiation. A windowless, differentially pumped lamp was used for the resonance lines of Ne at 16.8 eV and of He at 21.2 eV. The illuminated area on the sample surface was approximately 10 $mm^2$.

## EXPERIMENTAL RESULTS

The photoemission spectra were found to be rather similar for samples I and II and, with a single exception, spectra from sample II will be omitted. Fig. 1 show EDC's measured at the photon energies 16.8 and 21.2 eV. For small retarding voltages a strong peak (not shown for hν = 21.2 eV) is observed. For both photon energies this peak has its maximum for photoelectrons emitted with kinetic energies of about 1.5 eV ($V_R \simeq 1.5$ V) and is due to secondary electrons formed by inelastic scattering.[9] At higher kinetic energies a peak (or a strong shoulder) is observed about 4.5 eV below the high energy edge of each EDC. At energies above this peak the spectra are quite similar not showing much structure. However, it should be noted that for increasing energies the EDC's show a rather rapid decrease in a range of about 2 eV followed by a much slower drop-off. This drop-off may be compared with the sharp high energy edge of one Pt EDC also shown in Fig. 1. The Fermi-level $E_F$ corresponds to the energy at which the Pt emission drops to zero. The higher lying peak in the two spectra is seen to be displaced in energy by about 4.4 eV, which corresponds to the difference in photon energies. The spectra, excluding the region with the secondary electron contribution, therefore represent the initial energy distribution of the photoexited electrons and can be plotted on an initial energy scale with $E_F$ chosen as the zero of energy. That is, we relate the initial energy $E_i$ to the kinetic energy in vacuum, E, by $E_i = E + \Phi - h\nu$, where Φ is the work function.

Fig. 2 shows EDC's of sample I as a function of

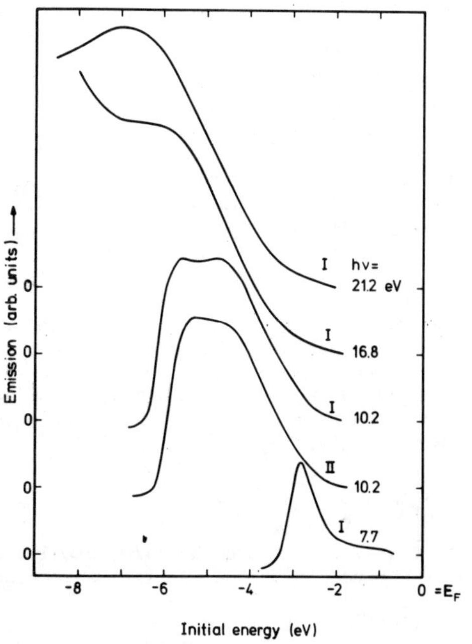


Figure 2 - Photoemission energy distribution curve for n-type YIG (sample I) and p-type YIG (sample II at four different photon energies.

$E_i$ for the photon energies 21.2 eV, 16.8 eV, 10.2 eV and 7.7 eV. Also an EDC of sample II at hν = 10.2 eV is shown. For the photon energies 10.2 eV and 7.7 eV the decrease of the EDC's for $E_i$ less than -5.5 eV and -3 eV resp. is due to the onset of the work function escape threshold. The peak in the hν = 7.7 eV spectrum therefore is not significant to our study. The 10.2 eV EDC's also are modified by the threshold function. However, some structure not observed at the higher photon energy spectra is present at initial energies between -4 eV and -6 eV. If we identify the new structure with the peak seen at high hν for sample I we see that the peak position seems to shift with the photon energy; the total shift from hν = 21.2 eV to 10.2 eV is estimated to be about 2 eV. It should be noted that the spectra of samples I and II at hν = 10.2 eV are only slightly shifted in energy and that the hν = 7.7 eV spectrum shows a photo-emission onset at $E_i$ = -0.7 eV.

Spectra at hν = 10.2 eV were also recorded at temperatures above the Curie temperature(289 °C) and were found quite similar to spectra at room temperature. The magnetization therefore seems not to influence the EDC's. This can partly be due to the 4π geometry used and partly to the formation of domains eliminating the magnetic fields outside the samples.

## DISCUSSION

The onset of the photoemission for sample I found at $E_i \simeq -2$ eV corresponds fairly well with the top of the valence band as determined by conductivity studies. The band gap energy at room temperature is approximately 2.6 eV [10]. The Fermi-level in sample I, which is a partly compensated n-type material, is pinned by the Si donors and is situated approximately 0.3 eV below the bottom of the conduction band.[5] If no occupied states in the bandgap were present that onset would be expected at $E_i$ = -2.3 eV. The presence of acceptor levels of Pb acceptors of a fairly high concentration ($10^{20}$ $cm^{-3}$) and further presumable of intrinsic defects of oxygen vacancies, which act as donors, but are deeper than the Si donors[5], give a reasonable explanation of the onset at $E_i \simeq -2$ eV. The tail observed in fig. 2 for hν = 7.7 eV up to $E_i \simeq -0.7$ eV is also readily explained by these states in the bandgap. This tail could not be investigated on the same scale for the higher photon energies due to weaker signals. The onset for sample II is expected at $E \simeq -0.5$ eV but is found near -1.75 eV. We ascribe this shift to downwards band-bending near the surface due to oxygen vacancies, which normally are present in p-type YIG[8]. The apparent gradient in defect concentration may have been caused by the cleaning procedure.

The onset at $E_i \simeq -2$ eV gives the uppermost energy of filled band states. These states can be iron 3d states and oxygen 2p states. The iron ions in YIG occupy octahedral and tetrahedral sites, and four subbands are assumed to be formed.[4] These d-bands are generally assumed to be narrow bands.

Assuming the oxygen 2p band is at least several eV's broad we first consider the possibility, that the d-bands are narrow bands (with bandwidths less than 0.5 eV). In photoemission d-bands with a small energy dispersion will be seen in EDC's as sharp structures. The absence of such a sharp structure from the onset at $E_i \simeq -2$ eV to below -3 eV gives a strong indication that no narrow d-bands are present as the uppermost filled band states. This agrees with results on impurity induced optical absorption[8] and with high temperature electrical conductivity results[10], both indicating the presence of a broad valence band as the highest lying filled band in YIG.

We assume that the highest band is the oxygen valence band upon which, for energies $E_i \leq -3$ eV broadened d bands are superposed. The structure in the 10.2 eV spectra could indicate the presence of the subbands mentioned above. The absence of this structure at higher photon energies may be due to different final states.

## ACKNOWLEDGEMENT

The authors want to thank J.M. Robertson for the supply of the garnet films, P.A. Paans for the microprobe analysis, and J. Verhoeven for the auger measurements. The technical assistance of M. Adriaens of European Space Agency is gratefully acknowledged. One of us (PKL) is grateful to M.M. Traum and N.V. Smith of Bell Laboratories for enlightening and stimulating discussions.

## REFERENCES

1. WOOD, D.L. and REMEIKA, J.P., J. Appl. Phys. 38. 1038 (1967).
2. SCOTT, G.B., LACKLISSON, D.E. and PAGE, Y.L., Phys. Rev. B10, 971 (1974).
3. WEMPLE, S.H., BLANK, S.L., SEEMAN, Y.A. and BIOLSI, W.A., Phys. Rev. B9, 2134 (1974).
4. WITTEKOEK, S., POPMA, T.Y.A., ROBERTSON, J.M. and BONGERS, P.F., Phys. Rev. B, to be published.
5. LARSEN, P.K. and METSELAAR, R., J. Solid State Chem. 12, 253 (1975).
6. SMITH, N.V., Crit. Rev. Solid State Sciences, 2, 45 (1971).
7. DERBENWICK, G.F., PIERCE, D.T. and SPICER, W.E. in "Methods of Experimental Physics", vol. 11, Academic Press, New York and London, 1974.
8. LARSEN, P.K. and ROBERTSON, J., J. Appl. Phys. 45, 2867 (1974).
9. BERGLUND, C.N. and SPICER, W.E., Phys. Rev. 136, A1030 (1964).
10. LARSEN, P.K. and METSELAAR, R., unpublished.

# DETERMINATION OF COMPLEX MAGNETIC SURFACE ENERGIES FROM SWR SPECTRA*

P. E. Wigen, T. S. Stakelon, H. Puszkarski†, and P. Yen
Physics Department, Ohio State University, Columbus Ohio 43201

## ABSTRACT

The correct analysis of spinwave resonance spectra in ferromagnetic films must properly account for the conditions which determine the rf magnetization at the film surfaces. Over the past twenty years, the treatment of the surface condition has evolved from the Kittel condition of complete surface pinning, through a partial pinning condition to the inclusion of anisotropy energies acting at the film surfaces. Recently, it has been shown that to properly account for the angle dependence of the spinwave spectra in homogenious garnet films, the surface energies must be expanded in the Neél expression having the form:

$$K_s(\phi) = \Sigma_\ell b_\ell P_\ell(\cos\phi)$$

where $\phi$ is the angle between the surface normal and the direction of the magnetization and $P_\ell(\cos\phi)$ are the Legendre polynomials. Using a new method of analysis based on the pinning parameter model, the coefficients $b_\ell$ have been determined for each surface by using detailed measurements of the angle dependence and intensity ratios of the surface modes and the body modes of the spinwave spectra of YIG films on GdGaG substrates. A spin Hamiltonian is presented which includes microscopic interactions that predict surface energies consistent with the angular, frequency and temperature dependence of the data.

## INTRODUCTION

Since the prediction[1] and first observation[2] of standing spinwave modes in thin ferromagnetic films, the understanding of experimental spinwave spectra has often been hindered by the problem of separating the roles of the surface magnetic properties from that of inhomogeneities in the bulk magnetic properties. Results obtained from vacuum deposited metal films have often varied from film to film due to the irreproducible nature of these internal field inhomogeneities. Nevertheless, the ideal subject in thin film spinwave resonance experiments has been regarded as a film which will submit to an analysis based on perfectly uniform bulk properties. Since 1958, therefore, these analyses have depended primarily on the surface magnetic boundary conditions which have grown in complexity due to the demands of the experimental results.

Initially, Kittel[1] assumed that the surface spins were completely pinned by some unknown mechanism giving a boundary condition of the form

$$m_s = 0 \qquad 1)$$

where $m_s$ is the transverse component of the radio-frequency (rf) magnetization at the film surfaces.

Pincus[3] extended the notion of the surface condition by introducing a mixed boundary condition

$$am_s + b\nabla_z m_s = 0 \qquad 2)$$

where the z-direction is taken as normal to the film plane. Soohoo[4] introduced the concept of a uniaxial surface anisotropy which produces an angle dependent boundary condition at the film surface given by:

$$\frac{1}{m_\psi}\nabla_z m_\psi = -\frac{K_s}{A_{ex}}(\cos^2\phi)$$

and $\qquad 3)$

$$\frac{1}{m_\phi}\nabla_z m_\phi = -\frac{K_s}{A_{ex}}(\cos 2\phi)$$

where $\psi$ is the angle within the film plane and $\phi$ is the angle between the magnetization film and the z-direction, $K_s$ is the surface anisotropy energy and $A_{ex}$ is the exchange constant.

About this same time, a volume inhomogeneity[5,6,7] in the internal field was introduced to account for the non-quadratic behavior of lower order modes, the position of the critical angle $\phi_c$, where only one spinwave mode is excited and the lack of a significant number of high order spinwave modes at the parallel orientation.

The surface inhomogeneity model[8] proposed the concept of a unidirectional surface anisotropy and predicted the presence of a non-propagating surface mode in the spinwave spectra. In the limit of long wavelength spinwaves, the surface boundary condition for this model is given by

$$\frac{1}{m_s}\nabla_z m_s = \frac{K_p}{A_{ex}}\{\cos(\alpha_o - \phi)\} \qquad 4)$$

where $\alpha_o$ is the angle between the film normal and the direction of the unidirectional surface anisotropy field, $\bar{K}_p$.

Nonpropagating surface modes have been observed in yttrium iron garnet (YIG) films[9-11] and their analyses indicate that the surface anisotropy must be expressed in terms of a tensorial model[10] expressed by

$$\frac{1}{m_s}\nabla_z m_s = \frac{K_\perp\cos^2\phi + K_{||}\sin^2\phi}{A_{ex}} \qquad 5)$$

where $K_\perp$ is the magnitude of the surface anisotropy energy when the magnetization is normal to the film $\phi = 0$, and $K_{||}$ is the in plane contribution, $\phi = 90^o$. At this time, the quality of YIG films and the data obtained from them are such that it is possible to write the surface anisotropy energy in the general form:

$$\frac{1}{m_s}\nabla_z m_s = \frac{K_s(\phi)}{A_{ex}} \qquad 6)$$

$$K_s(\phi) = \sum_\ell b_\ell P_\ell(\cos\phi)$$

where $b_o$ is the coefficient of an isotropic term, $b_1$, for the unidirectional term, $b_2$ for the uniaxial term, etc.

This paper discusses recent efforts at applying this general form of the surface anisotropy to spinwave resonance experiments with thin YIG films and testing its consequences. From the angle dependence of the spinwave spectra, the $b_\ell$ coefficients in Equation 6 associated with each film surface have been evaluated. In addition, a microscopic explanation of the magnetic surface interactions is proposed which accounts for the origin of contributions to the surface energy having different symmetries.

## SURFACE PINNING PARAMETER AND SURFACE MODES

In analyzing the spinwave resonance data, a surface pinning parameter[8] has been introduced to relate the propagation constant of the spinwave mode and therefore its position and intensity to the surface anisotropy energies. As originally introduced in the surface inhomogeneity model, which assumed a unidirectional surface anisotropy field, $\bar{K}_p$, the surface pinning parameter has the form:

$$A = 1 - \frac{g\mu_B}{2SzJ}\hat{\gamma}\cdot\bar{K}_p \qquad 7)$$

where $2SzJ/g\mu_B$ is the exchange field acting on a spin due to the interactions of that spin with its z neighbors in the adjacent plane and $\hat{\gamma}$ is a unit vector parallel to the magnetization. For the tensorial surface field,[10]

$$A = 1 - \frac{g\mu_B}{2SzJ} \hat{\gamma} \cdot \underline{\underline{K}} \cdot \hat{\gamma} \qquad 8)$$

where $\underline{K}$ has diagonal elements $K_{||}$, $K_{||}$, and $K_{\perp}$ given in Equation 5.

In terms of the general expression for the surface anisotropy energy as given in Equation 6, the surface pinning parameter will have the form

$$A = 1 - \frac{K_s}{A_{ex}} \qquad 9)$$

For a film having symmetric boundary conditions, the following general properites can be noted about the surface parameter. If A is less than unity, the surface anisotropy field assists the applied field and the surface spins are "pinned". For this case a family of spinwave body modes may be excited by a uniform rf field. When A is equal to unity, the surface field has no contribution and the boundary condition reduces to $\nabla_z m_s = 0$. For this case, only the uniform mode will be excited by a uniform rf field. For A greater than unity, the surface field is opposed to the applied field and in addition to the spinwave body modes of the film, a symmetric and an antisymmetric nonpropagating surface mode are also normal modes of the system. A surface spinwave mode is characterized by imaginary wavenumbers and therefore has an amplitude which decays exponentially away from the surface at which it is localized. The charateristic decay length decreases with increasing surface anisotropy energy. Body modes, of course, have real wavenumbers and their amplitudes vary sinusoidally across the entire film.

In general, the surface interactions present at each surface are not symmetrical, especially for annealed single crystal epitaxial films grown on a single crystal substrate. To completely evaluate the influence of the surface anisotropy energies[12] on the spinwave spectra it is thus necessary to analyze the data in terms of two independent surface parameters:

$$A_s = 1 - \sum_\ell a_\ell^s P_\ell(\cos\phi)$$

and $\qquad\qquad\qquad\qquad 10)$

$$A_f = 1 - \sum_\ell a_\ell^f P_\ell(\cos\phi),$$

where $A_s$ is the pinning parameter associated with the substrate surface of the film and $A_f$ the free surface of the film. If $A_s$ or $A_f$ is greater than unity and the film thickness is greater than the characteristic decay length of the surface mode, the properties of the surface mode will be determined primarily by the surface anisotropy energy at that surface.

## EXPERIMENTAL RESULTS

This section describes, after a brief review, the results of recent experiments which use spinwave resonance to measure magnetic surface anisotropy fields of thin YIG films. The experiments also offer experimental confirmation of a number of resonance effects predicted[12] but previously unobserved.

In order to probe the surfaces of a thin magnetic film by spinwave resonance it is desirable that the interior of the film be homogeneous since volume inhomogeneities in the bulk magnetic properties result in spinwave spectra that are determined only partly or not at all by the surface conditions.[7] Extremely homogeneous films of YIG[13] on the order of one micron thick can be routinely produced by epitaxial growth on gadolinium gallium garnet substrates. The low intensity of higher-order spinwave modes[14] relative to the uniform mode intensity in etched and unannealed liquid phase epitaxial (LPE) films is evidence that the pinning parameters $A_s$ and $A_f$ of Equation 9 are nearly unity indicating that negligibly small surface energies are present in as grown films. In chemical vapor deposited films (CVD) or annealed LPE films, significant surface anisotropies are present.

A non-propagating surface spinwave mode in iron garnet films was first observed using spinwave resonance in YIG films grown by CVD.[9] This mode was found to be localized at the YIG-substrate interface. Subsequent annealing[10] in oxygen of the as-grown film at temperatures from 600°C to 1300°C increased the localization of the surface mode. At the highest annealing temperature (1100°C to 1300°C for six hours) a second surface mode primarily localized at the free surface of the film was also observed. It has been found that the surface conditions necessary for the existence of surface modes can be induced in a similar fashion by the annealing of LPE films.

Both the free surface and substrate interface surface modes display a strong dependence on the orientation of the film saturation magnetization, $\overline{M}$, with respect to the film normal. When $\overline{M}$ is oriented parallel to the film plane the surface mode resonances have a maximum field splitting from the first body mode resonance. The localization of the surface modes and therefore the surface anisotropy energy is then at its maximum value. As $\overline{M}$ is rotated out of the film plane, the field splitting between the surface and body modes decreases and is accompanied by large changes in their relative intensities. The surface modes are observed to transform into body modes as $\overline{M}$ is rotated towards the film normal. The spinwave energy and intensity of the surface mode is extremely sensitive to changes in the surface anisotropy. In addition the body modes undergo rapid changes in intensity when the surface anisotropies are passing through zero. These two factors have made it possible to accurately measure the variation of the surface anisotropy energy with the orientation of $\overline{M}$ about the film normal.

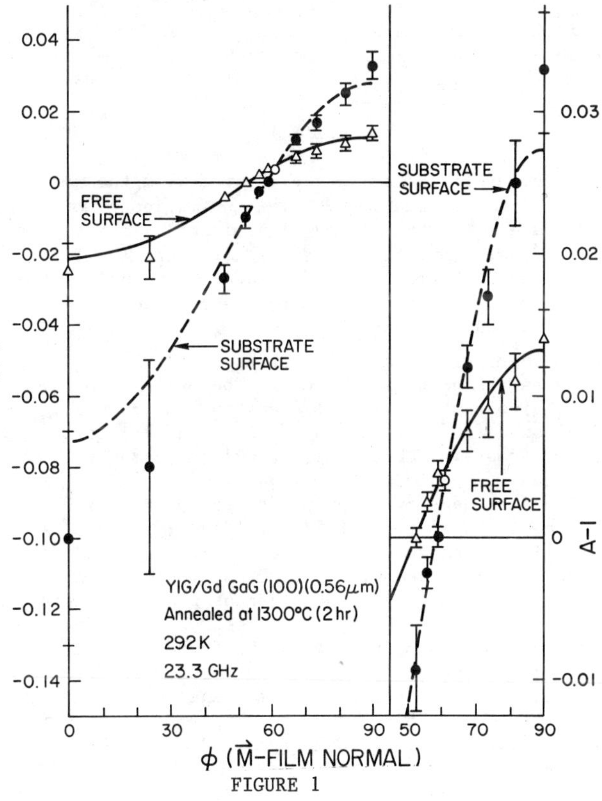


FIGURE 1

The surface pinning parameter, A, at the substrate and free surfaces of an annealed YIG film measured using the spin wave spectra. The modes associated with each surface are determined unambiguously by annealing and etching experiments as described in Ref. 10. $\phi$ is the angle between the film magnetization and the film normal. The curves shown are of the form given by Equation 10 using coefficients, $a_\ell$, from Table 1. The curves on the right hand side are the data from 50° to 90° shown on an expanded vertical scale.

Figure 1 indicates the results of a series of experiments in which the pinning parameter at each surface and therefore the surface anisotropies were measured as a function of the orientation of $\overline{M}$ in a (100) plane. The YIG film used in the measurements had a (100) orientation and a thickness of approximately 0.56 micrometers. It had been annealed at $1300^{\circ}$C for two hours and surface modes localized at both the substrate and free surfaces were observed. The spectra were measured at 23.3 GHz and field modulation and phase sensitive detection were employed. Relative intensities of the modes were measured by twice integrating the absorption derivatives using a Nicolet 1072 signal averager. The theory[12] of asymmetric boundary conditions was used to find the values of $A_s$ and $A_f$ consistent with these quantities at each angle.

The curves shown in Figure 1 are the least squares fit to the data of the form given by Equation 10. Terms with unidirectional symmetry (odd $\ell$) were eliminated from the expansion since the spinwave spectrum remained unchanged when the direction of $\overline{M}$ was reversed. It was also found that only two terms in Equation 8, namely $P_0$ and $P_2$ were necessary to obtain a good fit to the data. Thus the surface parameter can be written

$$A_s = 1 - a_0^s - a_2^s \left(\frac{3\cos^2\phi - 1}{2}\right)$$
$$A_f = 1 - a_0^f - a_2^f \left(\frac{3\cos^2\phi - 1}{2}\right) \qquad 11)$$

| $\ell$ | $a_\ell\,(\times 10^{-3})$ | $\sigma b_\ell\,(10^{-3}\,\mathrm{ergs/cm^2})$ |
|---|---|---|
| SUBSTRATE INTERFACE | | |
| 0 | 5.94 ± 0.70 | 3.23 ± 0.38 |
| 2 | 66.8 ± 4.2 | 36.03 ± 0.22 |
| 4 | (7.2 ± 7.5) | |
| FREE SURFACE | | |
| 0 | -1.37 ± 0.21 | - 0.75 ± 0.11 |
| 2 | 22.9 ± 1.00 | 12.4 ± 0.54 |
| 4 | (1.4 ± 3.2) | |

TABLE 1

The coefficients, $a_\ell$, evaluated from the data in Figure 1 using Equation 9. The coefficients for $\ell = 4$ are given but do not significantly improve the fit and have low statistical significance. Only the terms $\ell = 0, 2$ are retained. The values $\sigma b_\ell$ are the corresponding surface anisotropy energy densities as explained in the text.

These coefficients and the corresponding surface energies are given in Table 1. The coefficient of $P_4$ is given but is neglected due to its low statistical significance. The surface energy density values shown in Table 1 are calculated using the relation $b_\ell = 2SzJa_\ell$ from Equations 6 and 10. The surface density of spins, $\sigma$, for a (100) YIG surface is taken to be $6.53 \times 10^{13}/\mathrm{cm}^2$. The effective exchange energy of one unit cell, 2SJ, is $8.33\times 10^{-15}$ ergs and, for a (100) surface, z, the number of neighbors in the adjacent plane, is unity. Surface anisotropy energies necessary to account for spinwave spectra of thin metallic ferromagnetic films[15] are typically on the order of one erg/cm$^2$. The much smaller surface energies encountered in YIG result from the lower density of uncompensated spins in the ferromagnetic YIG lattice and the smaller bulk exchange constant.

The existence of the large angle dependent terms in the surface anisotropies at each surface allows the experimental realization of a variety of asymmetric boundary conditions simply by rotating the applied magnetic field and therefore the film magnetization with respect to the film normal. This has allowed the systematic study of several spinwave resonance effects whose observation has not previously been reported. These include the repulsion between two nearly degenerate surface spinwave modes, the alternate disappearance of even or odd numbered modes and the existence of an increasing intensity of even numbered modes with increasing mode number.

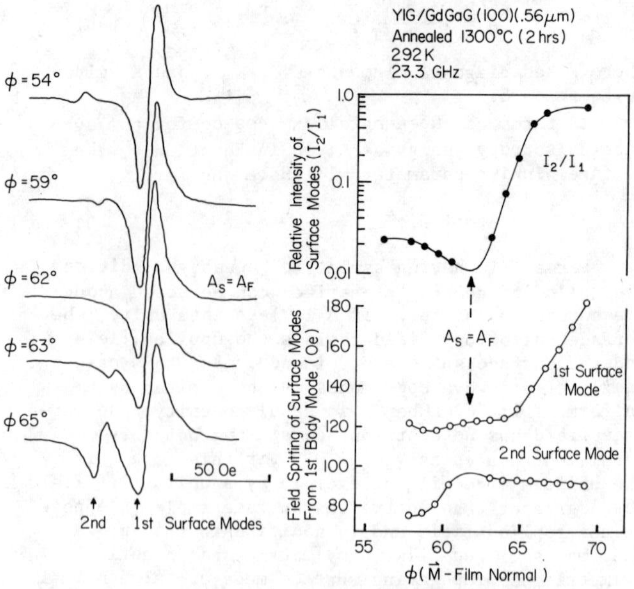


FIGURE 2

Spinwave spectra of an annealed YIG film near the orientation of $\overline{M}$ ($\phi = 62^{\circ}$) where the boundary conditions at the two film surfaces become equal ($A_s = A_f$). The intensity ratio of the two surface modes passes through a minimum and the spinwave exchange energies repel one another at the symmetric boundary condition.

As $\overline{M}$ is rotated from an orientation within the film plane towards the normal, the surface anisotropy energy decreases more rapidly at the substrate interface than at the free surface as indicated in Figure 2. At $\phi = 62^{\circ}$ the surface anisotropy energies become equal ($A_s = A_f$). Since the spinwave exchange energies of each of the surface modes is determined predominantly by the anisotropy energy at each surface, the modes approach degeneracy as the boundary conditions become equal. The surface excitations must then be described as a symmetric and an antisymmetric combination of spinwave amplitudes which decay exponentially from the surfaces. As Figure 2 indicates, the spinwave exchange energies of these two surface mode branches are repelled near the symmetric condition. This repulsion is due to the overlap of the surface mode amplitudes in the bulk of the film.

Spinwave mode intensities are determined by the transverse rf magnetic moment of the spinwave mode integrated across the entire film thickness. Hence, the antisymmetric mode has a zero amplitude when the symmetric boundary conditions, $A_f = A_s$, are achieved. These results are shown in Figure 2. Also at this orientation of $\overline{M}$, all even numbered body modes have antisymmetric symmetry and therefore their intensities pass through zero.

When the surface anisotropy energy of at least one surface is small, intensities of body spinwave modes as a function of mode number, n, can be approximately described by some negative power of the mode number ($n^{-p}$). In general the power law describing the intensity envelopes for even and odd numbered modes is different. As shown in Figure 3, when $\overline{M}$ is rotated through the angle $\phi = 53^{\circ}$, the surface anisotropy energy at the free surface vanishes and the even and odd mode intensity envelopes become equal. At $\phi = 61^{\circ}$ this condition of equal intensity envelopes is again observed, corresponding to the vanishing of the surface anisotropy energy at the substrate surface. The surface mode was observed to transform into the first body mode when the associated surface anisotropy energy vanishes and becomes negative ($A \leq 1$).

Figure 4 indicates that when $\phi = 56^{\circ}$ the third mode intensity is observed to pass through a minimum. In fact, all odd body mode intensities were observed

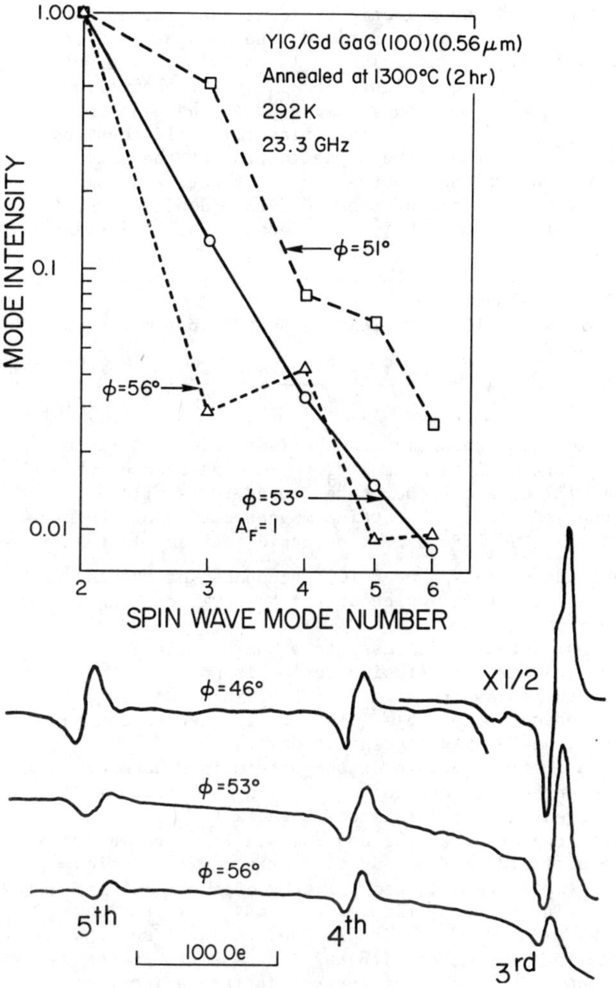


FIGURE 3

Spinwave spectra of an annealed YIG thin film in the orientation of $\bar{M}$ at which the anisotropy at the free surface vanishes ($A_f = 1$). At this orientation ($\phi = 53^o$) the intensities of the even and the odd modes exhibit the same dependence on mode number.

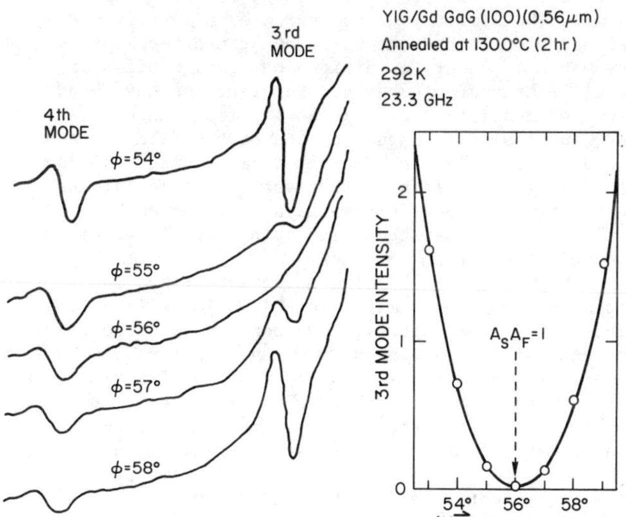


FIGURE 4

The disappearance of odd numbered body modes from the spinwave spectrum occurs at the origin of the film magnetization ($\phi = 56^o$) for which the surface anisotropies are such that $A_sA_f = 1$. The behavior of the third mode intensity is shown.

to pass through a minimum at this orientation. This orientation corresponds to the condition $A_sA_f = 1$ (corresponding approximately to $K_s = -K_f$). Under this condition the wavenumbers (k) of the body modes can be shown to be given by tankL = 0, where L is the film thickness. Odd numbered body modes then have an integral number of wavelengths in the film thickness though not antisymmetric about the central plane of the film. Nevertheless they will have zero intensity.

When the film magnetization is oriented along or near the film normal ($\phi \simeq 0^o$) the spinwaves are strongly pinned at both surfaces ($A_s$, $A_f < 1$ but unequal). For this case the intensities of the even numbered modes are observed to increase with increasing mode number until an intensity maximum is reached as shown in Figure 5. For example, at $\phi = 24^o$ the

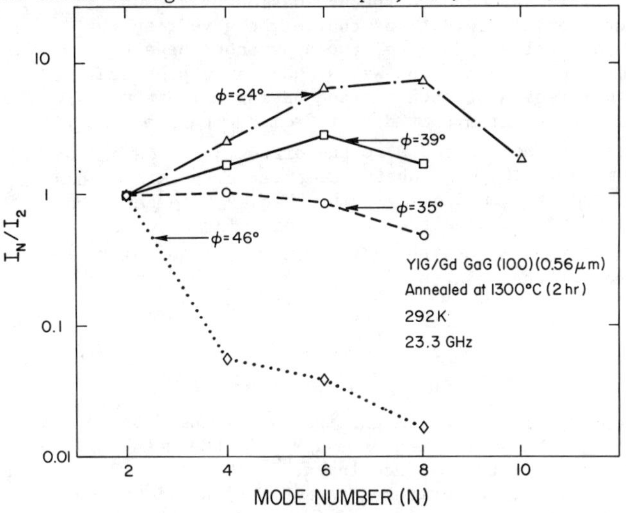


FIGURE 5

The intensities of even numbered spinwave modes in an annealed YIG film is shown normalized to the second mode intensity. As the angle between the film magnetization and the film normal is decreased, the surface anisotropy energy increases, resulting in a reversal in the even mode intensity ratios.

eighth mode has the greatest intensity of all the even numbered modes. This behavior can be understood as follows. In the limit of equal and large surface anisotropy energies, it can be shown that the intensity of even numbered modes increases linearly with the mode number. Of course, all the even modes become weaker in absolute intensity as the anisotropies become equal. On the other hand, if one of the surface anisotropy energies is zero, the intensities of even modes decreases with mode number. What is being observed in Figure 5 is the even mode behavior in an intermediate range of surface anisotropies in which the intensity envelope exhibits both trends.

Such a striking reversal in intensity versus mode number will occur over a large range of surface conditions and should be widely observed in spinwave spectra if sufficient sensitivity is available.

In addition to allowing the experimental manipulation of boundary conditions, the separation of the surface anisotropy energy into terms of isotropic and uniaxial symmetry supplies information about the microscopic origin of the magnetic surface interactions. A strong indication of the origins of the substrate coefficients $a^s_o$ and $a^s_2$ in Equation 10 is the temperature dependence of the spinwave spectrum. As the sample temperature is lowered, the critical angle $\phi^s_c$, defined as the angle at which $A_s = 1$, increases from 61$^o$ at room temperature and reaches 90$^o$ at about 90K (the critical temperature of the sample). Below this temperature no surface mode is observed. According to Equation 11 the critical angle is given by

$$\phi_c^s = \arccos \frac{1}{3} (1-2a_o^s/a_s^s)^{\frac{1}{2}} \qquad 12)$$

This behavior of the critical angle can be explained by a continuous increase in $a_o^s/a_2^s$ with decreasing temperature. At the critical temperature, $a_o^s = a_2^s/2$.

On the other hand, the free surface surface mode has an exchange energy and critical angle which varies little as the temperature is lowered to 4.2°K. Therefore the coefficients $a_o^f$ and $a_2^f$ are nearly temperature independent.

## MICROSCOPIC MODELS

The detailed analysis of the data in terms of the magnitude, angle dependence and temperature dependence of the surface anisotropy energy allows a realistic approach to the evaluation of various microscopic interactions that might give rise to these results. To date, two approaches have been undertaken. Ramer and Wilts (RW)[16] have proposed a thin region at each surface having uniform magnetic properties that differ from the bulk of the film. It is assumed that due to the diffusion of Gd and Ga from the $Gd_3Ga_5O_{12}$ substrate, this surface layer has a magnetization and a g-value that have temperature dependences that differ from those of the bulk.

The effective g-value in the doped surface layer is given by the relation

$$g_{layer} = \frac{M_{Fe} - M_{Gd}}{\frac{M_{Fe}}{g_{Fe}} - \frac{M_{Gd}}{g_{Gd}}} \qquad 13)$$

where $M_{Fe}$ is the net magnetization of the iron sublattice in the surface layer and $M_{Gd}$ is the magnetization of the Gd in the surface layer.

At room temperature, magnetization values and thicknesses can be obtained for each surface that allow a good fit to the spinwave spectrum. At reduced temperature where the surface layer approaches magnetic compensation, $g_{eff}$ approaches infinity and the critical angle is predicted to approach the parallel orientation of $\overline{M}$ ($\phi = 90°$) as observed in the data. However, the differences in $g_{Fe}$ and $g_{Gd}$ used in the analysis are much too small to fit the data.

Below the compensation temperature, the RW model also predicts a critical angle emerging from the perpendicular orientation of $\overline{M}$ ($\phi = 0°$). The surface mode then being a normal mode at the perpendicular orientation. However, such surface modes have not been observed below the critical temperature of the samples discussed above.

In any real system, diffusion will not produce a uniform layer but rather an inhomogeneous layer with the compensated spins being confined to a very thin region within it. This compensation wall will move within the surface region as a function of temperature but will be present at all temperatures up to 290K[17]. Assuming a uniform surface layer having a well defined compensation temperature and $g_{eff}$ may be a serious oversimplification of a realistic diffusion region when considering the influence of a property so sensitive to Gd concentration as $g_{eff}$.

A second approach[18] treats the surface as a single atomic layer. The Hamiltonian consists of two terms. The first is a single ion anisotropy term of the form $D(S^z)^2$ which has an easy axis normal to the film plane. The second term is a temperature and frequency dependent term arising from the molecular field coupling between the iron spins at the substrate surface of the YIG film and the gadolinium spins in the substrate. This term is essentially isotropic.

With these terms the Hamiltonian has the form

$$\hat{H} = -2\Sigma_{\ell\ell'jj'} J_{\ell\ell'} \hat{\underline{S}}_{\ell j} \cdot \hat{\underline{S}}_{\ell' j'} - g\mu_\beta \overline{H}_{int} \cdot \Sigma_{\ell j} \hat{\underline{S}}_{\ell j} - \Sigma_{\ell j} \{D(\hat{S}^z_{\ell j})^2 + B_{eff} \langle S^\gamma \rangle_{Gd} S^\gamma_{\ell j}\} \delta_{\ell 0} \qquad 14)$$

where $\overline{H}_{int}$ is the magnetic field at the site of the spins due to the applied field, the demagnetization and bulk anisotropy fields, D is the single ion anisotropy field coefficient, $\langle S^\gamma \rangle_{Gd}$ is the thermal expectation value of the $\hat{\gamma}$ component of the gadolinium spin, $B_{eff}$ is the effective molecular field coupling coefficient between the magnetization of the gadolinium ions at the substrate and the net iron moment ions at the surface of the YIG film and $\delta_{\ell 0}$ is the Kronecker delta. The index $\ell$ denotes the atomic layer in the plane of the film, with $\ell = 0$ the substrate surface layer, while the index j denotes the position within the layer.

This Hamiltonian yields a surface parameter that has the form

$$A_s = 1 - \frac{B_{eff}\langle S^\gamma \rangle_{Gd}}{2SzJ} - \frac{D(S-\frac{1}{2})}{2SzJ}(3\cos^2\phi - 1) \qquad 15)$$

In this approximation the isotropic or $a_o$ term is proportional to $B_{eff}\langle S^\gamma \rangle_{Gd}$. The magnitude of this term will have a temperature dependence similar to that of the gadolinium sublattice magnetization of the mixed garnet $Y_{1.5}Gd_{1.5}Fe_5O_{12}$. The rapid increase in the magnitude of $\langle S^\gamma \rangle_{Gd}$ near 100K accounts for the change in the ratio $a_o/a_2$ which shifts the critical angle to $\phi = 90°$.

While this model assumes a sharp surface layer effect, a narrow diffusion region is not expected to alter these results significantly since the temperature dependence of $\langle S^\gamma \rangle_{Gd}$ is not expected to change significantly with concentration.

A possible origin of the D term in Equation 14 is proposed[18] to be due to an $Fe^{2+}$ ion associated with an oxygen vacancy localized at the interface. The model suggests that the non-S state $Fe^{2+}$ ion in a crystal field gradient at the surface will contibute to a single ion anisotropy energy of the required form. The concentration of such oxygen vacancy defects at the surfaces is proposed to increase with annealing consistent with the increase of $a_2$ observed on annealing. The photoinduced surface anisotropy that has been observed[19] does indicate that a change in the charge state occurs for some ions located at or near the surfaces of annealed YIG films.

## CONCLUSIONS

Spinwave resonance experiments with epitaxial garnet films of high quality have allowed magnetic surface properties to be investigated in annealed YIG films. The symmetry of the surface anisotropy has been completely characterized and the effects of asymmetric boundary conditions has been studied in detail. Microscopic surface interactions have been proposed consisting of a molecular field interaction between dissimilar magnetic ions at the film surface and a crystal field anisotropy term arising from non-S state ions interacting with crystal field gradients at the film surface. The analysis presented here would also be consistent with thin diffusion regions located at the film surfaces.

The notion of inducing magnetic surface interactions in homogeneous magnetic thin films with negligible surface anisotropy, and then studying the interactions using the spinwave resonance spectra may be generally applicable. Investigations of LPE YIG films coated with metals or subjected to low energy ion bombardment are currently in progress. In metal single crystal platelets of ferromagnetic metals grown from the vapor phase approximate ideal film properties to a greater degree than vapor deposited metal films, and may allow similar studies of metal surfaces.

REFERENCES

* Supported by NSF
† Permanent address, A. Mickiewicz University, Poznań, Poland.

1. C Kittel, Phys. Rev. 110 1295 (1958).
2. M. H. Seavey, Jr. and P. E. Tannenwald, Phys. Rev. Letters 1, 168 (1958).
3. P. Pincus, Phys. Rev., 118 658 (1959).
4. R. F. Soohoo, Phys. Rev., 131 594 (1963).
5. P. E. Wigen, C. F. Kooi, M. R. Shanabarger and T. D. Rossing, Phys. Rev. Letters. 9 206 (1962).
6. A. M. Portis, App. Phys. Letters 2, 69 (1963).
7. C. F. Kooi, P. E. Wigen, M. R. Shanabarger and J. V. Kerrigan J. App. Phys. 35 791 (1964).
8. H. Puszkarski, Acta Phys. Polonica 38 217 and 899 (1970).
9. S. D. Brown, R. D. Henry, P. E. Wigen and P. J. Besser, Sol. St. Comm. 11, 1179 (1972).
10. J. T. Yu, R. A. Turk and P. E. Wigen Phys. Rev. B. 11 420 (1975).
11. R. D. Henry, P. J. Besser, D. M. Heinz and J. E. Mee. IEEE Trans. Magn. 9, 535 (1973).
12. H. Puszkarski, Phys. Stat. Sol. (b) 50 87 (1972).
13. R. D. Henry, S. D. Brown, P. E. Wigen and P. J. Besser, Phys. Rev. Letters 28, 1272 (1972).
14. C. Vittoria and J. J. Krebs; AIP Conf. Proc. 24 486 (1974).
15. C. F. Kooi, W. R. Holmquist, P. E. Wigen and J. T. Doherty; J. Phys. Soc. Japan, 17 B-J 599 (1962).
16. O. G. Ramer and C. H. Wilts, Phys. Stat. Sol. (b) (to be published).
17. A. Vassiliev, J. Nicolas and M. Hildebrandt; C. R. Acad. Sci. 252 2529 (1961).
18. P. E. Wigen and H. Puszkarski, Sol. St. Comm. (to be published).
19. T. S. Stakelon, P. Yen, H. Puszkarski and P. E. Wigen (in this proceedings).

# EFFECTS OF LEAD INCORPORATION ON THE FERROMAGNETIC RESONANCE LINEWIDTHS OF LIQUID PHASE EPITAXIAL GROWN YTTRIUM IRON GARNET*

M. T. Elliott
Electronics Research Division, Rockwell International
Anaheim, California 92803

## ABSTRACT

Single crystal thin films of yttrium iron garnet (YIG) were grown on gadolinium gallium garnet (GGG) substrates by the isothermal dipping method of liquid phase epitaxy. The concentration of Pb in the films was adjustable through selection of the growth temperature. FMR linewidths were measured at 9.1 GHz. The linewidths exhibited a correlation with Pb content. The minimum linewidth was found to occur when the aggregate impurity distribution was charge compensated. The narrowest linewidth observed, 0.14 Oe at 25°C, appears to correspond to the Kasuya-LeCraw intrinsic limit. For samples not charge compensated, the linewidths exhibited distinct temperature peaks characteristic of valence-exchange loss mechanisms. Detailed analysis of the FMR spectra including surface modes of the Damon-Eshbach type, revealed the transverse wavevector to be quantized according to the dimensions of the finite samples.

## INTRODUCTION

Yttrium iron garnet (YIG) has been an important microwave material for a number of years. One of its most significant properties is an extremely narrow ferromagnetic resonance (FMR) linewidth. Devices which make use of this property generally employ YIG in the form of a polished single-crystal sphere. An alternative configuration, which is both easier to fabricate and compatible with planar electronics technology, is a single crystal thin film. Two areas needing improvement before YIG thin films become accepted alternatives to bulk crystals are the linewidth and the temperature stability. This paper is concerned with the reduction in the FMR linewidth. The temperature stability will be treated in another publication.

Early work on YIG films which were epitaxially grown on GGG substrates by chemical vapor deposition showed the linewidths to be narrow[1]; but not as narrow as those of the best spheres. In addition, imperfections were present which resulted in broadening of the resonance. More recently, advances in the liquid phase epitaxial (LPE) growth process have resulted in the growth of single crystal YIG films of exceptionally high quality. However, it has been reported that unavoidable strains or contamination by substrate constituents may prevent attainment of intrinsic linewidths in LPE YIG[2] grown by the "tipping" method. In a recent publication[3] the results of a study on Pb incorporation into YIG films were presented. Here the linewidths of films with varying quantities of Pb were measured. A minimum in the linewidth was observed to occur at approximately 0.5 wt % Pb or about 0.02 atoms of Pb per formula unit ($Y_3Fe_5O_{12}$). A mechanism was hypothesized whereby the minimum in linewidth corresponded to the point at which the aggregate impurity distribution is charge compensated. Experiments were conducted to test this idea. These results are reported.

## EXPERIMENTAL TECHNIQUES AND RESULTS

The films were grown by the isothermal dipping technique on <111> GGG substrates. The film thicknesses ranged from 0.55μm to 16μm. Pb from the flux was incorporated, along with Pt, as an impurity in these YIG films. The Pb content was adjustable by proper selection of the growth temperatures. The Pb content was determined by x-ray diffraction measurements of the associated increase in the film lattice parameter. The Pt content did not vary appreciably with growth temperature and was essentially constant for all samples. The concentration of Pt was in the vicinity of 0.5 - 1.0 wt %, as determined by x-ray fluorescence spectroscopy.

The samples were prepared by photolithographic etching of 0.125, 0.25 and 0.50 mm diameter discs from the YIG thin films. Additional samples were made into 1 x 2 mm rectangles. FMR measurements were made at 9.1 GHz using a $TE_{102}$ cavity. The FMR apparatus allowed for measurements over the temperature range of -200°C to 240°C. The linewidths of all the samples were determined in both the perpendicular ($H_a$ along film normal) and parallel ($H_a$ along film surface) orientation. In parallel resonance the broadening due to imperfections becomes important as a result of the 2 magnon scattering into the large density of available states. Rotation of the sample in the parallel configuration allowed the applied field to be directed along either <112> or <110> directions. A NMR probe was used to determine the field values at the half-power points.

The results of the linewidth measurements exhibit a very pronounced minimum when the Pb content approximately corresponds to 0.5wt % Pb or about 0.02 atoms of Pb per formula unit ($Y_3Fe_5O_{12}$). The narrowest line measured was 0.14±0.02 Oe at 25°C. The lineshape of this resonance was observed to be a slightly modified Lorentzian. The linewidth measurements have been corrected for the finite modulation amplitude used in this derivative detection system. For samples with little or no Pb the linewidth was observed to increase to approximately 0.8 Oe. Samples with greater than 0.5 wt % Pb again exhibited linewidth increasing with increasing Pb content. The linewidths were also measured as a function of temperature. Only the high temperature region was investigated for the samples containing either an excess or a deficit in Pb. In the case where the samples were in large compression (high Pb content), a distinct peak in the linewidth measurements was observed at 140°C along with a smaller peak at approximately 80°C. For samples in tension (very little Pb), the linewidths were maximum at approximately 60°C with another reduced peak at 150°C.

Analysis of the FMR spectra on discs made from the thinner films allowed the exchange constant to be calculated as a function of the Pb content. The exchange constant was observed to decrease with Pb content. The largest value measured was A = 3.5 x $10^{-7}$ ergs/cm. The modes for the low Pb films were pinned on one surface whereas the modes for the high Pb films were quasi-pinned (even modes observed). Significant deviations from the quadratic dependence occurred in films of thickness ~1μm. The Neel temperatures for these thin films did not correlate as well with Pb content as was seen for thicker films[3]. FMR spectra for films approximately 10μm thick allowed the magnetostatic modes to be observed. Measurement of the field at the various magnetostatic resonances together with the calculation of the in-plane component of the wavevector (roots of the zeroth order Bessel function) allowed the slope of the dispersion curve to be determined. A value of 0.246 was obtained for a thick film (11.06μm) at a $\Delta a^{\perp} \approx 0$. This value was observed to decrease for the thinner film with the same lattice mismatch. Typical values for films of 0.6-0.8μm thickness was 0.17. These values are corrected for the observed undercutting of the discs during etching.

The FMR spectra obtained from the 1 x 2 mm rectangles were also studied. Here the slope of the field dispersion curve for the 11.06μm film in perpendicular resonance was 0.253. The magnetostatic modes were found to correspond to in-plane wavevector components quantized according to the sample geometry. All the modes could be identified in terms of the $(n_L, 1)$ series where $n_L \equiv n\pi/L$ and n is any integer greater than 1. There were, however, modes missing from this series. The spectra in the parallel configuration exhibited magnetostatic resonances of the bulk modes and, in addition, numerous well defined surface modes of the Damon and Eshbach variety[4]. The field dispersion curve for the bulk case had a slope of 0.193 and for the surface case the initial slope was 0.12. Again the resonances were quantized according to the sample geometry and some expected modes were absent.

## DISCUSSION

For pure, perfect YIG crystals, the principal contribution to the linewidth arises from the Kasuya-LeCraw processes[5]. These mechanisms, which are intrinsic to the material, impose a lower limit on the linewidth. The theoretical value at the temperature and frequency of applicability is 0.19 Oe. The measured value of 0.14 ± 0.02 Oe thus appears to be in excellent agreement with the assumption that the intrinsic limit has been reached in these films. The fit of the lineshape to a slightly modified Lorentzian corroborates this point of view. The measurement in the parallel configuration demonstrates that LPE YIG films can be grown with essentially defect-free surfaces and no compositional variation.

The observed correlation between linewidth and Pb content is not unexpected. Indeed, incorporation of Pb is known to cause several effects in YIG[3]. Certain magnetic parameters, such as the magnetization, will be altered. The lattice parameter of the film and, therefore, the film/substrate misfit stress will be altered. These effects are expected to be homogeneous over the film volume and should not contribute to line broadening although they will produce a shift in the position of the resonance peaks. Another effect of Pb incorporation, which is expected to contribute to an increase in the linewidth, arises from the requirement to maintain charge neutrality. Since Pb enters YIG as a divalent substitutional impurity and since Y and Fe are both trivalent in pure YIG, some charge compensating mechanism must come into play. Such a mechanism, for example a change in the oxidation state of Fe, will produce a broadening of the resonance through an electron hopping (valence-exchange) process[5]. However, as has been pointed out, Pt is also present in appreciable concentrations. Since Pt is tetravalent, there will be a mutual charge compensation when the Pb and the Pt atomic concentrations are equal. This will occur somewhere around 0.5 wt % Pb. This corresponds to the Pb content of the sample exhibiting the 0.14 Oe linewidth. The observation of the peaks in the linewidths as a function of temperature supports the charge compensation model as previously discussed[6].

The decrease in the exchange constant with increasing Pb can be partially explained. The observed change over the total variation of Pb contents used (0 to 2.5 wt %) was approximately 10%. Such a large decrease is not expected from a pure dilution model where only 2% of the iron is being replaced by Y. Valence changes in the iron to compensate for the Pb may be responsible for part of the change observed. Another possibility is that the slopes of the dispersion curves are anomolous. The departure from a quadratic behavior for the lower order modes can be explained[7] by the resonance involving spins from only part of the sample. In other words, the magnetic "thickness" changes with mode number, approaching the true thickness only for large wave numbers. Measurements of the exchange constant involving only the first few resonances will be erroneously large. For films containing little Pb the exchange constant was observed to decrease from ~5 x $10^{-7}$ergs/cm to a constant value of 3.5 x $10^{-7}$ergs/cm as the measurements were made at progressively larger wave numbers.

A simple interpretation of the field dispersion curve slopes is not possible. Wolfram and DeWames[8] calculated the values to be 0.250 and 0.203 for the perpendicular pinned and unpinned cases respectively. In parallel resonance the values that apply here can be shown to be 0.184 and 0.203 for the unpinned and pinned bulk modes and 0.13 and 0.11 for the unpinned and pinned surface modes. Consequently the results presented here are seen to lie somewhere between the pinned and unpinned case for parallel resonance and agree well with the unpinned model for perpendicular resonance on thick YIG films. The quantization in terms of the sample geometry has been previously reported[9]. Here, however, the quantization has been extended to include the surface Damon and Eshbach modes. The absence of certain modes of the $(n_L, 1)$ series cannot as yet be explained and would seem to be suggestive of an interference effect. This area continues to be investigated.

## ACKNOWLEDGEMENTS

The author is indebted to F. S. Stearns for his assistance in film preparation and to R. D. Henry and H. L. Glass for many helpful discussions.

## REFERENCES

*This research was supported by the Air Force Office of Scientific Research (AFSC) under contract F44620-75-C-0045.

1. J. E. Mee, G. R. Pulliam, J. L. Archer, and P. J. Besser, IEEE Trans. Mag., MAG-5, (1969), 717.
2. R. C. Linares and E. L. Sloan III, J. Crystal Growth, 27, (1974), 249.
3. H. L. Glass and M. T. Elliott, J. Crystal Growth, 27, (1974), 253.
4. R. W. Damon and J. R. Eshbach, J. Phys. Chem. Solids, 19, (1961), 308.
5. M. Sparks, "Ferromagnetic-Relaxation Theory," McGraw-Hill, New York (1964).
6. H. L. Glass and M. T. Elliott, to be published in J. Crystal Growth.
7. B. Hoekstra and J. Robertson, paper 6B-1 (this conference).
8. R. G. DeWames and T. Wolfram, J. Appl. Phys., 41, (1970), 987.
9. M. Sparks, B. R. Tittman, J. E. Mee and C. Newkirk, J. Appl. Phys., 40, (1969), 1518.

# MAGNETIC ANISOTROPY OF $Ir^{4+}$ IN $NiFe_2O_4$ CRYSTALS.

R. Krishnan
Laboratoire de Magnétisme, C.N.R.S.
92190 Meudon-Bellevue, France.

ABSTRACT

Single crystals of $NiFe_2O_4$:Ir were prepared by the flux method and their magnetic moment and anisotropy were studied. The decrease of the magnetic moment with Ir concentration suggests $Ir^{4+}$ with a predominante octahedral site occupation. The single ion contribution at 4K to $\Delta K_1$, $\Delta K_2$ turns out to be + 2.5 and - 10 $cm^{-1}$ respectively. This is explained in terms of $Ir^{4+}$ in octahedral site with an axial distortion along |111|.

## INTRODUCTION

The metals ions of the second ($4d^n$) and the third ($5d^n$) transition series by virtue of high spin orbit coupling constant (500 - 2500 $cm^{-1}$) are expected to contribute strongly to magnetic anisotropy and magnetostriction. The magnetic studies of these ions present in a ferrimagnetic host crystals started since only recently. It was shown for example that ruthenium ions ($4d^n$) present in $Y_3Fe_5O_{12}$ (YIG) contribute strongly to the magnetic anisotropy with $K_1 > 0$ and $K_2 < 0$ and it is generally accepted that $Ru^{3+}$ ions are involved |1 2|. However the possibility of the influence on $K_1K_2$ of $Ru^{4+}$ whose presence cannot be neglected is not very clear. Thus if one looks at the effect of Ru on the magnetic anisotropy of $NiFe_2O_4$, it is quite the opposite of YIG with $K_1 < 0$ and $K_2 < 0$ |3|. The theoretical model developed for $Ru^{3+}$ is not adequate to explain this result. The author has evoked in this case the possibility of $Ru^{4+}$ playing an active role |3|. Hansen has shown that as regards $Ir^{4+}$ in YIG the contribution to $K_1$ and $K_2$ is positive |4|. It is interesting to note that in the case of iridium which is in the low spin state in an octahedral site, $Ir^{3+}$ would be diamagnetic thus leaving $Ir^{4+}$ as the only candidate magnetic to be considered. The purpose of this work is to find out how Ir would behave in a spinel host crystal considering the anomaly mentioned above for Ru. We expected hopefully to be able to throw some light on these problems, by studying the magnetic anisotropy in iridium doped $NiFe_2O_4$ crystals and which we report here.

## EXPERIMENTAL

Single crystals of $NiFe_2O_4$:Ir were prepared by the flux method using PbO-$B_2O_3$ as flux. The raw materials used were of 5N purity except $IrO_2$ which was 4N (Johnson Mathey company). The crystallisation was carried out in the temperature range 1180 - 900C. The crystals after separation were subjected to a heat treatreatment consisting of heating to 900°C and cooling to 600C at about 3°C/h in order to eliminate any $Ni^{2+}$ from tetrahedral (A) site. This is quite important since A site $Ni^{2+}$ are known to be strongly anisotropic |5|. Ferromagnetic resonance techniques were used to determine the anisotropy constants $K_1$ and $K_2$ in the X-band and in the temperature range 4 - 300K. Magnetisation measurements were also carried out on these samples using vibrating sample magnetometer. Several samples from the same batch were used for the above two investigations.

## RESULTS

The crystals were analysed chemically to get the actual composition. Ir and Pt were determined by emission spectrograph method. The Ir content in the samples were 0, 0.50, 2.50, 3.0 and 4.0 wt % whereas Pt was in the range 0 - 0.5 wt %. It is seen that only a very small fraction of Ir added to the melt enteres the crystal. However Ir content variation in different crystals from the same batch was less than ± 3 %. An analysis of the oxygen content of these crystals showed them to be strongly oxidising which indicates the presence of $Ir^{4+}$ ions. This is a very important result and will be recalled for the interpretation of the properties.

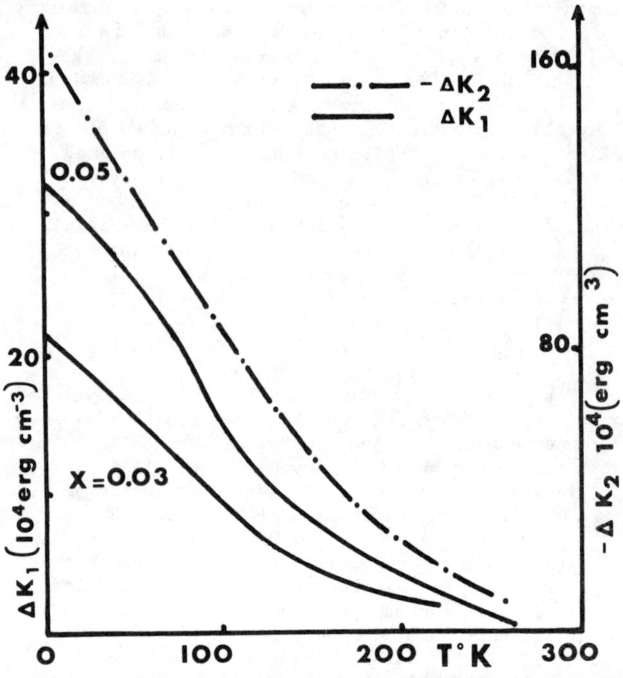


Fig. 1 Temperature dependence of Ir contributions to magnetic anisotropy ($\Delta K_1 \Delta K_2$)

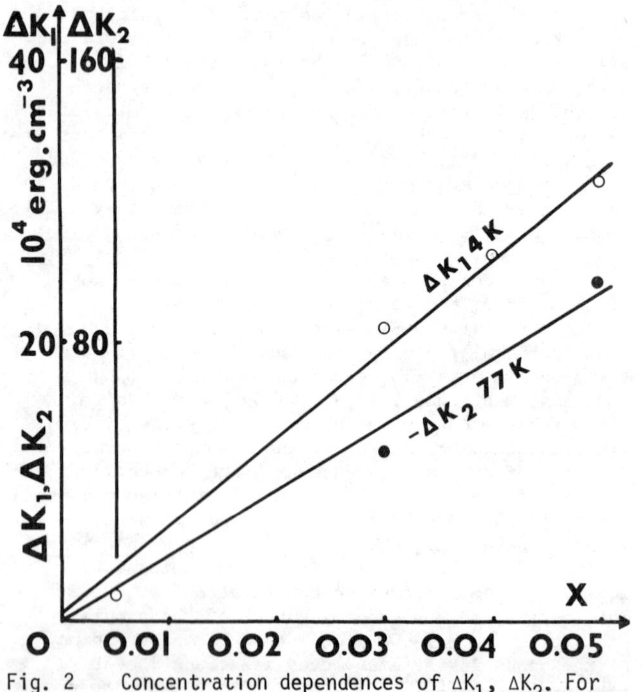


Fig. 2 Concentration dependences of $\Delta K_1$, $\Delta K_2$. For convenience of scale $\Delta K_2$ at 77K is considered.

Magnetic anisotropy $K_1$ is negative at 293K for all the samples studied with |001| and |111| as the hard and easy axes respectively whereas at lower temperatures $K_1$ becomes positive with |110| and |001| as the hard and easy axes respectively. Thus $|K_1|$ decreases with increasing concentration near 293K, but at lower T getting strongly positive, with $|K_1|$ increasing with the Ir content. $K_2$ which is negligible for $NiFe_2O_4$ becomes strongly negative with increasing Ir content and at low temperatures. Fig. 1 shows the Ir contribution

($\Delta K_1$ and $\Delta K_2$) to anisotropy for two compositions as a function of temperature. The maximum spread in the $K_1$ and $K_2$ measured on different samples of the same batch was less than 10 %. Here the x values are calculated based on the formule $Ni_{1+x}Fe_{2-2x}Ir_xO_4$ which is justified later in this work. Fig. 2 shows the concentration dependences of $\Delta K_1$ and $\Delta K_2$ ($\Delta K_1$ is the difference $K_1(NiFe_2O_4) - K_1(NiFe_2O_4{:}Ir)$, $\Delta K_2$ is just the $K_2$ of $NiFe_2O_4$:Ir itself) which is found to be quite linear, showing the validity of one ion contribution. Fig. 3 shows the variation of magnetic moment $n\mu_B$ at 4K with the concentration, $n_{\mu B}$ was calculated from the relation $n_{\mu B} = \frac{\sigma \times \text{mol.wt}}{5585}$. The molecular weight was calculated for each concentration and for the two possible valence states of Ir. $n_{\mu B}$ shows a clear decrease as x increases. The spread in $n_{\mu B}$ for different samples of the same batch amounted to less than 2 %.

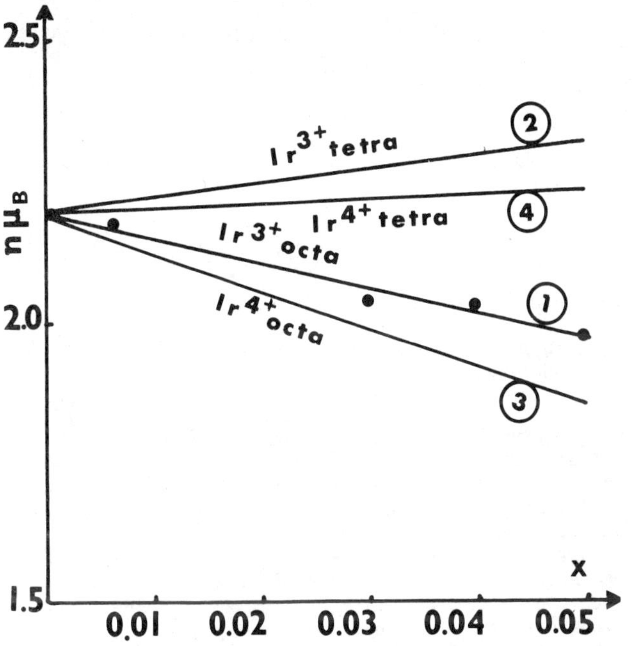


Fig. 3 Magnetic moment variation at 4K with Ir concentration. The numbers correspond to the different cases discussed is the text. The closed circles are experimental points.

DISCUSSION

It would be advantageous to start the discussion with the magnetization data which will help settling the question of the valence state of Ir. As both $Ir^{3+}$ and $Ir^{4+}$ might be present, one can expect two types of substitutions like $NiFe_{2-x}Ir^{3+}_xO_4$ and $Ni_{1+x}Fe_{2-2x}Ir^{4+}_xO_4$. It is also possible to conceive a mixture of the above two. Magnetic moments were calculated for the above two compositions, in the following way using always the strong field scheme and spin only magnetic moment, the latter assumption cannot alter the trend of the results.

Case 1. $Ir^{3+}$ octahedral sites S = 0.

Case 2. $Ir^{3+}$ tetrahedral sites S = 1n = $2\mu_B$.

Case 3. $Ir^{4+}$ octahedral sites S = 1/2n = $1\mu_B$.

Case 4. $Ir^{4+}$ tetrahedral sites S = 1/2 n = $1\mu_B$.

These four cases are shown in fig. 3 with the appropriate numbers. It is seen that cases 2 and 4 (Ir in tetrahedral sites) lead to an increase in $n\mu_B$ in disagreement with the experiment and the experimental points fall in line close with the case 1 that is $Ir^{3+}$ in B sites. However this case has to be ruled out since this ion being diamagnetic cannot contribute to the large $K_1$ and $K_2$ encountered here. So the next close agreement is obtained with $Ir^{4+}$ in B sites (case 3). However this underestimates the magnetic moment by about 5 % in the extreme case. This discrepancy can be remedied if a small fraction of $Ir^{4+}$ is supposed to occupy the A site. What is relevant here for our case is that iridium is present as $Ir^{4+}$ (as indicated also by the oxidising nature of these crystals) and most of them say 90 % are present in the octahedral site. In this connection it might be mentioned that magnetisation measurements have been particularly valuable for these crystals, because a similar study of $NiFe_2O_4$:Ru crystals by Jonker |6| could not help much in deciding a similar problem there. One of the reasons for the above is that both $Ru^{3+}$ and $Ru^{4+}$ are paramagnetic unlike the Ir case.

With the above arguments it can be thus concluded that the composition of the crystals be better described as $Ni_{1+x}Fe_{2-2x}Ir_xO_4$ and the x values were calculated accordingly. It was also supposed that $Ir^{4+}$ is mostly present in the octahedral site. Hansen |4| has considered the contribution of octahedral low spin $5d^5$ ($Ir^{4+}$) to anisotropy within the scope of the single ion model. Due to high crystalline field splitting (10 Dq ∽ 35 000 $cm^{-1}$) in garnet and equally valid in spinel, the mixing with higher cubic energy terms are neglected. Also the ground terms are split by the trigonal field, spin orbit coupling and the exchange field leading to the Hamiltonian of the form :

$$= V_t(\vec{r}) - \zeta\vec{L}.\vec{S} + g\mu_B.\vec{H}_e.\vec{S},$$

where $V_t(\vec{r})$ is the local crystalline field energy, the second term is the spin-orbit coupling energy and the third is the exchange energy which is considered isotropic. As the low spin $d^5$ configuration $t^5_{2g}$ is equivalent to a hole in the shell, spin orbit coupling constant $\lambda$ is replaced by the one electron parameter $-\zeta$. All the other symbols have the usual meaning. The spin orbit coupling is relatively high here and of the order of 3500 $cm^{-1}$. The |111| direction is chosen as the quantization axis, $K_1$ and $K_2$ from $Ir^{4+}$ ions are found to be positive in YIG in contrast to $Ru^{3+}(4d^5)$ where $K_1 > 0$ and $K_2 < 0$. This anomaly in the case of $Ir^{4+}$ needs the trigonal field to be positive which is in contradiction to ESR in $Y_3Ga_5O_{12}$ |7| |8|. However in our present case such a problem does not arise as $K_2 < 0$ indicating the negative trigonal field. It is seen that the ratio $\Delta K_2/\Delta K_1$ is independent of Ir concentration and for T = 0K, it is also independent of the exchange field and is only governed by $v/\zeta$ where v is the single ion trigonal parameter. Hansen |4| has given a plot of $\Delta K_2/\Delta K_1$ as a function of $v/\zeta$ and one can thus obtain the atomic parameters from the experimental values of $\Delta K_1/\Delta K_2$ of $Ir^{4+}$. Now looking at fig. 1 one finds that $\Delta K_2 > \Delta K_1$ and that the former shows no sign of saturation near 4K. Also from fig. 3 it is seen that the concentration dependence of $\Delta K_1$ and $\Delta K_2$ is linear. The one ion contribution from $Ir^{4+}$ (pure octahedral site occupation) to magnetic anisotropy constants at 4K can be evaluated as $\Delta K_1/10N = 2.5 \pm 0.2$ $cm^{-1}$ and $\Delta K_2 = -10 \pm 2 cm^{-1}$. Besides the change in the sign for $\Delta K_2$ the one ion contributions are an order of magnitude smaller than in YIG |4|. The ratio $\Delta K_2/\Delta K_1$ at 4K is found to be roughly - 5 as against + 16 in YIG. If one now refers to the variation of $\Delta K_2/\Delta K_1$ with $v/\zeta$ for the octahedral site from ref. 4, then for $\Delta K_2/\Delta K_1 = -5$, one gets $v/\zeta \simeq -2$, which again is the value found for $Ru^{3+}$ ($4d^5$) in YIG |1|. Thus there is surprisingly a great similarity in the behaviours of $Ir^{4+}$ and $Ru^{3+}$ both $d^5$ metal ions in the octahedral site though one is in the spinel and the other in the garnet. This is rather interesting and would indicate that these two octahedral sites are very similar though qualitatively. At this point we can recall that the $K_1$ and $K_2$ from other ions like $Co^{2+}$ and $Cr^{3+}$ have also been shown to be the same sign both in YIG and $NiFe_2O_4$ |8 9 10 11|. Pushing these arguments further would then mean to say that the radical change in the observed behaviour of Ru with $K_1 > 0$ in YIG and with $K_1 < 0$ in $NiFe_2O_4$ would then indicate that octahedral $Ru^{3+}$ are not the only ones to be involved and one has to eventually consider the possible role of $Ru^{4+}$

also along with the tetrahedral site occupancy for $Ru^{3+}$. $Ru^{4+}$ will be diamagnetic in the strong field scheme in the tetrahedral site. Now we might consider the possible influence on $K_1$, $K_2$ of a fraction of $Ir^{4+}$ which could eventually be present in the tetrahedral sites, a case which we had evoked to fit better the magnetic moment data. The transformation properties of the ground states of $d^5$ and $d^9$ are identical in the strong field limit and in the tetrahedral site they have $^2T_2$ as the ground state. Sturge et al |11| have shown that such a case would give rise to both $K_1$ and $K_2$ negative. So this would mean that in our case where a pure octahedral occupation for $Ir^{4+}$ has been assumed, any presence of $Ir^{4+}$ in the tetrahedral sites will lead to an underestimate of $\Delta K_1$/ion and an overestimate of $\Delta K_2$/ion.

In conclusion, in $NiFe_2O_4$ iridium enters predominantly as $Ir^{4+}$ and in the octahedral site as seen by the magnetization measurements. This cannot at this stage of the work eliminate the possibility of finding $Ir^{4+}$ in tetrahedral sites and eventually the presence of $Ir^{3+}$. The $Ir^{4+}$ contribution to $\Delta K_1$ and $\Delta K_2$ in $NiFe_2O_4$ is quite similar to that from $Ru^{3+}$ also a $d^5$ ion in YIG. More work is needed on say, directional order, or photomagnetic effects to obtain further information.

X-ray orientation by A. Malmanche and magnetization measurements by Dr. V. Cagan are greatfully acknowledged.

REFERENCES

1. R. Krishnan, Phys. Stat. Sol. 1, K17 (1970).
2. P. Hansen, Phys. Rev.B 3, 862 (1971).
3. R. Krishnan, Phys. Stat. Sol.(a) 4, K177 (1971).
4. P. Hansen, J. Schuldt and W. Tolksdorf, Phys. Rev. B 8, 4274 (1973).
5. R. Krishnan, J. Appl. Phys. 39, 1340 (1968).
6. H.D. Jonker, J. Sol. State Chem. 10, 116 (1974).
7. E.L. Offenbacher and H. Waldman, Bull. Am. Phys.Soc. 13, 435 (1968).
8. M.D. Sturge, E.M. Gyorgy, R.C. Le Craw and J.P. Remeika, Phys. Rev. 180, 413 (1969).
9. S. Miyamoto, N. Tanaka and S. Iida, J. Phys. Soc. Jap. 20, 753 (1965).
   R. Krishnan, Thèse Doctorat d'Etat, Paris (1964).
10. R. Krishnan and V. Cagan, J. Appl. Phys. 42, 1639 (1971).
11. R. Krishnan, AIP Conf. Proc. No. 10, 112 (1972).

## FERROMAGNETIC RESONANCE IN PERMALLOY PLATELETS

J. H. Liaw and R. C. Barker, Department of Engineering and Applied Science, Yale University, New Haven, Connecticut 06520

ABSTRACT

Ferromagnetic resonance experiments at x-band have been performed on single-crystal permalloy platelets etched into discs and rectangles. We have found that the experimental mode positions for a 6.5μ thick disc in perpendicular resonance fall between the spin-pinned and spin-unpinned field-mode number curves corresponding to the magnetoexchange theory.[1] The initial slope[1] is closer to the unpinned theory. Similar excellent agreement for magnetostatic modes in rectangular samples has been obtained. In both configurations the experimental curve differs widely from the simpler insulator magnetostatic mode theory in which the exchange energy is ignored. From the intersection of the experimental curve with the field axis, assuming no knowledge of the surface pinning, Mo can be estimated to within 0.5% for a given γ and thickness d. To obtain information about the surface pinning from initial slope, however, the model must be extended to include a specific model surface boundary condition. In any case, d must be known to high accuracy. In rectangular samples, the mode spectra for both perpendicular and parallel resonance can be compared with the magnetoexchange theory. Furthermore, for thinner samples, e.g., 2000 Å-6000 Å, we have found spin wave spectra on the low field side of the parallel resonance spectrum. In such cases the magnetostatic modes are extremely weak. We shall discuss the self-consistency of these spectra.

1. T. Wolfram and R. E. De Wames, Physical Review B4, 3125(1971).

# FERROMAGNETIC RESONANCE IN NI-CO ALLOY PLATELETS AT ROOM TEMPERATURE*

C.Y. Wu, H.T. Quach, and A. Yelon
Department of Engineering Physics, Ecole Polytechnique,
Montreal, Canada H3C 3A7

## ABSTRACT

We have measured the resonance field and linewidths of single-crystal Ni-Co alloy platelets at 9.45, 24.82 and 35.49 GHz, at room temperature as a function of field direction. The linewidth anomaly observed in similar experiments was quite small in one such sample. Indirect experimental evidence that the anomaly is due to sample handling is presented. Experimental results are compared with theory based on Landau-Lifshitz equation. The agreement in the case of the sample without the anomaly is better than in previous studies, but some discrepancies remain.

## INTRODUCTION

As part of a continuing effort to investigate the ferromagnetic resonance (FMR) behavior of metallic platelets, we have been studying Ni-Co alloys. Our samples are produced by vapor decomposition of bromides [1], and grow as highly perfect single crystals with typical dimensions 0.5 mm × 0.5 mm × 1 — 20 μm. They are thus considerably thicker than the Ni platelets grown by similar techniques [2]. As in the case of Ni platelets [1], the NiCo has a [100] orientation with ⟨100⟩ and ⟨110⟩ edges. Electron microprobe and Auger electron spectroscopy measurements indicated that the composition is uniform to better then 1% [3] except in the first few atomic layers. Typical compositions examined thus far are on the order of 10 - 30% Co.

Ferromagnetic resonance experiments were at X-, K-, and Q- band frequencies at room temperature. We have measured the resonance fields and linewidths in the (001) sample plane, in the (100), (110) as well as some arbitrary planes perpendicular to the samples. The direction of propagation of the microwave field in our experiments was always normal to the films. Most of the Ni-Co platelets gave results similar to those obtained for nickel except that magnetostatic modes were very readily observed near (±10°) perpendicular resonance at X- and K- bands for samples with > 20% cobalt. For other angles, a single line was observed near the uniform precession mode. Samples with lower cobalt content either did not exhibit multimode standing waves, or the spectra were not resolved.

Owing to the small sample size and the absence of substrate, the methods which have been developed for handling and mounting these samples for FMR experiments have seldom been completely satisfactory. We have therefore attempted several different sample handling methods during the past year, and have investigated their effect on line-position and line-shape. While up to the present we still do not have a completely satisfactory way of mounting the sample, we have nevertheless obtained sufficient data to strongly suggest that the linewidth anomaly found in most of our out-of-plane FMR experiments on NiCo platelets, analogous to that observed in earlier experiments on Ni, could be a consequence of sample handling. Further, we now have measured one NiCo sample whose oblique angle line-width anomaly can be considered to be absent or at least far smaller than previously reported.

In the next section, details of the experiments are given and the results compared with calculations employing the Landau-Lifshitz equation of motion coupled with electromagnetic boundary conditions [2,4,5]

## EXPERIMENTS, RESULTS, AND DISCUSSION

The microwave spectrometer used was the JEOL model JES-ME operated at X-, K-, and Q- band frequencies; the cavity modes were the cylindrical TE011, TE011, and TE012 respectively. The sample rod is inserted from the top of the cavities and can be rotated. Thus its axis is always perpendicular to the horizontal magnetic field. For the out-of-plane experiments, the sample plane was always kept vertical; different orientation of the sample surface with respect to the sample rod axes would then allow one to choose any general angle between the applied field and the crystal.

While, as indicated above, a number of different ways have been attempted to mount the samples on the sample rod, we will focus here on the two methods which led to fruitful results. Method One (M1) was to position the sample between two small, thin pieces of mica held together by minute trace of grease. Method Two (M2) was to have a mica sheet on one side of the sample and nail polish on the other side; if the sample were not perturbed by the solidification of the nail polish, one would have an excellent way of handling the sample. In either case, the free side of the mica sheet was then glued to a flat vertical surface of the quartz sample rod.

M1 was the method used in most earlier measurements [2,6-8]. However, owing to the small dimension of our Q- band sample rod, it was difficult to avoid

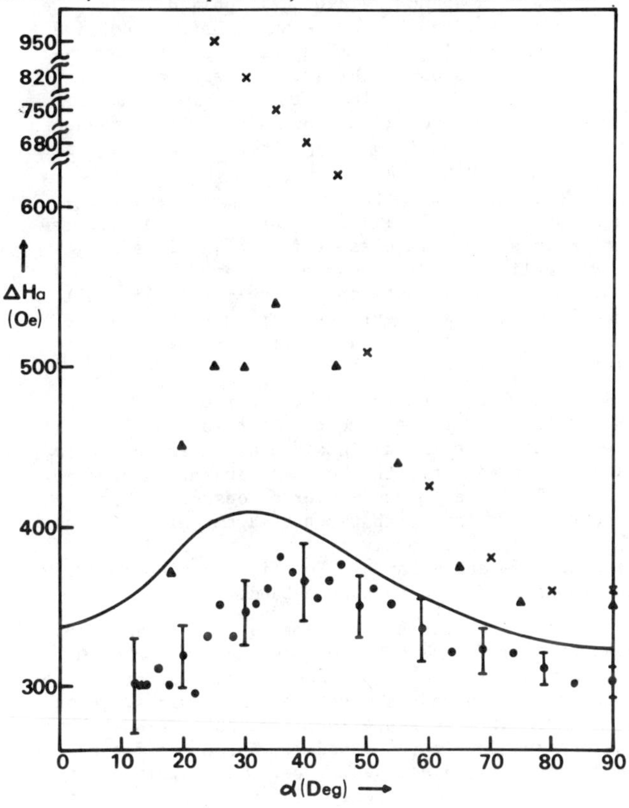


Figure 1 : X- band (9.451 GHz) applied field linewidth ΔHa versus angle α, between applied field and sample normal. •: data for Sample X, error bars are selectively inserted to indicate uncertainties in linewidth measurements; ▲: data for Sample T1; ×: data for Sample T1 after considerable handling; ■ : data for Sample T2; solid line is the calculated curve using the parameters given in Table 1.

having grease eventually touching the sample, causing a mechanical aftereffect manifested by a gradual shift in resonance field and line shape [8]

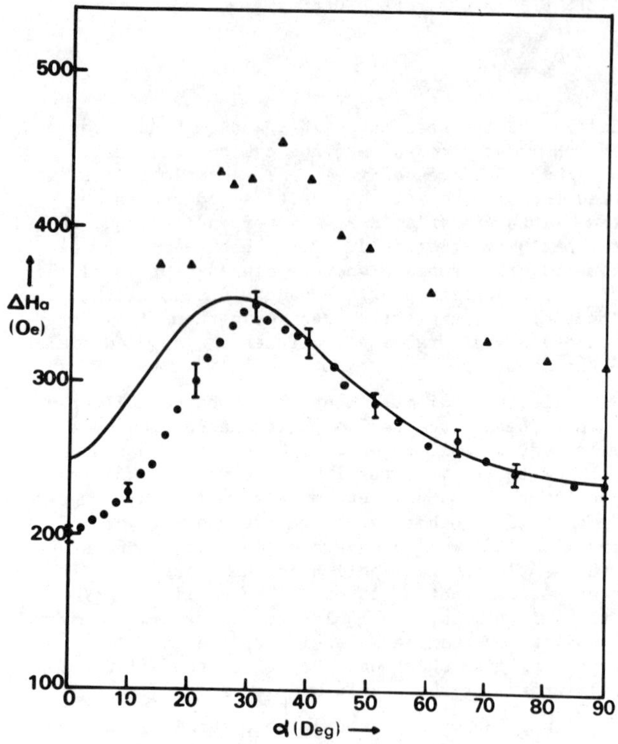


Figure 2 : K- band (24.820 GHz) applied field linewidth ΔHa versus angle between applied field and sample normal α. For explanation of the symbols, refer to the caption for Figure 1. Data for Sample T1 are not given for $0 \leq \alpha \leq 15^{o}$, as a magnetostatic mode spectrum is observed in this range.

We then tried method M2. Typical results for a fresh sample (Sample T1) mounted this way are represented by solid triangles in figures 1, 2, and 3, where applied field linewidths ΔHa are plotted as a function of angles between the applied field and the normal to the sample surface. These points may be compared with the calculated curves, shown as solid lines in the figure.

Even though the theoretical curve has not been chosen for best fit to the data for sample T1, it is clear that the peak in linewidth at intermediate angles cannot be fit with the same parameters used to fit parallel and perpendicular resonance. This result is similar to that reported for Ni [2,6], although here, the anomaly is more pronounced at X-band then K- and Q-bands. For Ni, a larger anomally was found at K-band than at X-band.

Repetition of the experiments on the same sample gave the same results as long as the sample had not been moved relative to the sample rod. After sample handling, performed as carefully as possible, we may have a dramatic increase in linewidths as shown by the crosses in Figures 1 and 3. These are for the same sample T1, discussed above. The magnitude of line broadening is markedly different at different angles. The linewidth is not appreciably changed at parallel or perpendicular, but the anomalous peak is greatly enhanced. The resonance field is not appreciably modified at all angles, and is always in good agreement with theory.

Given the fact that even careful sample handling can result in a substantial change in the linewidth, it is appropriate to inquire whether even the triangles represent an undamaged sample. As a result, we performed measurements on a large number of samples at X- band, using both M1 and M2. Since the cavity and sample holder are large for X- band, it was easy to avoid aftereffect with M1, by using a large piece

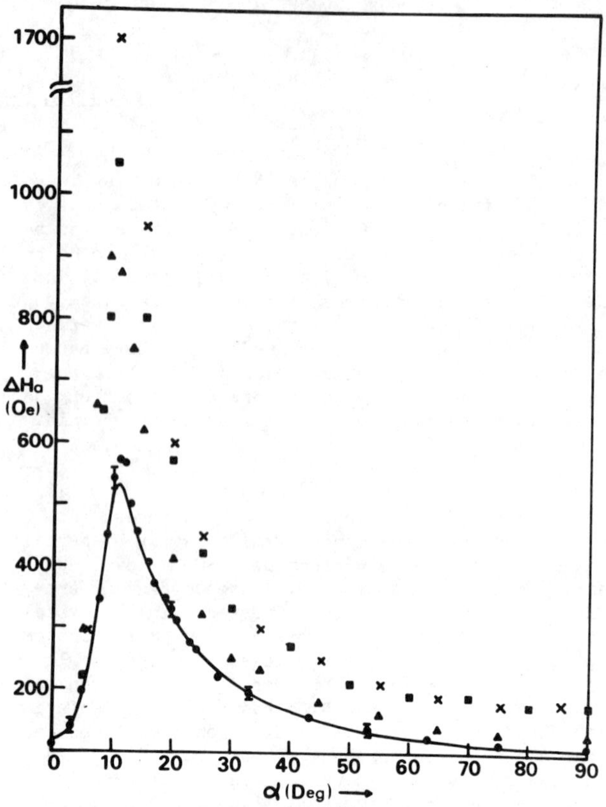


Figure 3 : Q- band (35.490 GHz) applied field linewidth ΔHa versus angle between applied field and sample normal α. For explanation of the symbols, refer to the caption for Figure 1. The absence of data points below certain values of α were either due to poor signal/noise ratio or insufficient magnetic field intensity.

of mica. It was found that method M2 was generally not as satisfactory as M1 in that it often resulted in larger linewidths. It was also found that samples which exhibited relatively large linewidths at parallel and perpendicular resonance did not necessarily give large linewidth maxima at $\alpha \approx 11^{o}$. Data for such a sample, T2, mounted by method M1 are represented by squares in Figure 1.

From among about 20 samples, all produced in the same batch, we were able to mount one, Sample X, whose maximum oblique angle linewidths at all three frequencies are considerably lower than those of the other samples. Data for this sample are represented by points in the three figures. The sample was mounted by method M1. In order to perform Q band experiments, we were obliged to cut the mica down, but we have verified that this did not introduce aftereffect or modify the X- band results.

We have compared the experimental results for Sample X with theory, and have attempted to achieve a best fit assuming spin-unpinned boundary conditions, and frequency independent paramaters. The optimum values obtained to date (the theoretical curves in the figures), are given in Table 1. The values for M and K in the Table correspond to roughly 15% Co. It is clear that the match for X- band is excellent.

While considerable discrepancies at the K- and Q-band frequencies still exist between theory and experiment, it is clear that the out-of-plane linewidth anomaly is absent or at most quite small for this sample. From our preceeding discussion, it seems highly likely that what makes Sample X different from the other samples we have looked at so far is that by good fortune, it has not been malhandled. Such malhandling is difficult to avoid owing to extreme small sample

TABLE 1

| Saturation Magnetization M (Oe) | Landau-Lifshitz Damping Constant λ(rad/sec) | Spectroscopic Splitting ratio γ(rad/sec - Oe) | Exchange Constant A(erg/cm) | Conductivity σ(m/sec) | Cubic Anisotropy Constants (erg/cm$^3$) $K_1$ | $K_2$ |
|---|---|---|---|---|---|---|
| 637 | $2.18 \times 10^8$ | $-1.94 \times 10^7$ | $0.8 \times 10^{-6}$ | $1.28 \times 10^{17}$ | $5 \times 10^3$ | 0 |

size. This would be even more so for the Ni samples, whose thicknesses are in the range of 1000Å to 10000Å [2]. Such a viewpoint is further encouraged by the fact that calculated and measured oblique angle FMR linewidth are in good agreement for bulk single crystal Ni [9].

## CONCLUSIONS

We believe that the oblique angle linewidth anomaly observed with most Ni and NiCo samples is a result of sample handling. Further experiments are called for to support this conclusion, but two points are clear. First, one sample has been studied which exhibits essentially no anomaly. Second, even for this sample, there are discrepancies between theory and experiment. We have been unable to fit both parallel and perpendicular linewidth for K band (and extrapolated perpendicular linewidth for Q band) with the same parameters. These discrepancies call for further theoretical investigation. At present, we are studying the effect of surface anisotropy on our results.

## ACKNOWLEDGMENTS

We would like to thank Y.J. Liu and R.C. Barker for helpful discussions. The generosity of Y.J. Liu in making his computer program for calculating FMR field and linewidth available to us is gratefully acknowledged. We also wish to thank Mr. R. Bornais for his work on Auger electron spectroscopy of Ni and NiCo platelets.

## REFERENCES


* Partly supported by the National Research Council of Canada.

(1) R. W. DeBlois, J. Appl. Phys., 36, 1948 (1965); J. Vac. Sci. 3, 146 (1966).

(2) Y. J. Liu, Ph.D. Thesis, Yale University (1974), unpublished.

(3) R. Bornais, unpublished report.

(4) C. Vittoria, R. C. Barker, and A. Yelon, J. Appl. Phys. 40, 1561 (1969).

(5) C. Vittoria, G. C. Bailey, R. C. Barker, and A. Yelon, Phys. Rev. B, 7, 2112 (1973)

(6) C. Vittoria, R. C. Barker, and A. Yelon, Phys. Rev. Letters, 19, 792 (1967).

(7) C. Vittoria, Ph.D. Thesis, Yale University (1970), unpublished.

(8) A. Yelon, H. T. Quach, G. Spronken, and C. Y. Wu, The Seventh International Colloquium of Magnetic Thin Films, Regensburg, 1975.

(9) J. R. Anderson, S. M. Bhagat, and F. L. Cheng, Phys. Stat. Solidi (b) 45, 357 (1971).

# CRYSTALLINE ANISTROPY OF COBALT FERRITE FROM OBSERVATIONS OF FERROMAGNETIC RESONANCE*

L. Assadourian+ and L. Silber
Polytechnic Institute of New York, Brooklyn, New York 11201

## ABSTRACT

The magnetocrystalline anistropy coefficients $K_1$, $K_2$ and $K_3$ and the magnetogyric ratio of cobalt ferrite at temperatures between 22°C and 37°C have been determined from observations of FMR at millimeter wavelengths. The only previously reported measurements on cobalt ferrite by the FMR technique are those of Tannewald at 90°C, who reported g=2.7 ± .3 and $K_1 \sim 1 \times 10^6$ ergs/cm³. Our measured value of the g factor was 2.6 ± .2. The values of the anistropy coefficients are given in the table below:

| TEMP. (°C) | $K_1$ ($10^6$ ergs/cm³) | $K_2$ ($10^6$ ergs/cm³) | $K_3$ ($10^6$ ergs/cm³) |
|---|---|---|---|
| 22 | 2.6 | 0.1 | -0.8 |
| 32 | 2.2 | 0.8 | -0.5 |
| 37 | 2.0 | 0.8 | -0.4 |

The value of $K_1$ at 22°C agrees well with that of Perthel, etal (2.7 x $10^6$ ergs/cm³), who derived their value from analysis of static magnetization curves. He was not able to determine $K_2$ and $K_3$. It was not possible from our results to separate the effect of anistropy of the magnetization from the magnetocrystalline anistropy, but assuming the values obtained by Perthel, the error introduced in the determination of $K_1$ is less than 10%.

## INTRODUCTION

Cobalt ferrite is of interest because of its very large magnetocrystalline anisotropy, due to the unquenched spin-orbit coupling of the cobalt ion. There have been several measurements of this anistropy reported in the literature, as well as theoretical studies to calculate this anistropy from first principles. Previous measurements have been made by analysis of static torque curves(1) and static magnetization curves(2). There has been only one measurement by means of ferromagnetic resonance, at an elevated temperature(3). In ferrites containing small amounts of cobalt, discrepancies were observed between static and microwave measurements of the anisotropy; these were attributed to the "rapid relaxation" of the cobalt ion contribution. It was felt of interest to try to determine the crystalline anisotropy of cobalt ferrite by ferromagnetic resonance at lower temperatures. The large spin-orbit coupling of cobalt also leads to an anisotropy in the saturation magnetization, predicted by Slonczewski(4), and observed experimentally by Perthel et al(2). It should be possible in principle to observe this also by FMR. Thus, our object was to measure the phenomenological anisotropy parameters of cobalt ferrite by means of FMR over as large a range of temperature as possible.

## THEORY

The model we use is a phenomenological one, in which we assume that the applied static magnetic field is large enough to magnetize the sample (taken to be a sphere) as a single domain. We further assume that at resonance the spins precess in phase, and remain parallel to one another (the "ferromagnetic" mode of resonance). We assume that the magnetization is anisotropic, but must have the symmetry of the lattice, which is cubic. Under these assumptions, the equilibrium orientation of this single domain is determined by the condition that the free energy density is a minimum. The frequency of small oscillations of the magnetization about this equilibrium direction is then given by the Smit-Suhl expression involving the second derivatives of the free energy.

The terms in the free energy density which depend on the orientation of the magnetization are given by

$$F = F_Z + F_A + F_{ME} + F_E + F_D$$

Where $F_Z$, $F_A$, $F_{ME}$, $F_E$, and $F_D$ are, respectively, the Zeeman, magnetocrystalline, anisotropy, magnetoelastic, elastic, and demagnetization energies.

$$M = M_0 + M_1(\alpha_1^2\alpha_2^2 + \alpha_2^2\alpha_3^2 + \alpha_3^2\alpha_1^2) + M_2(\alpha_1^2\alpha_2^2\alpha_3^2) \quad (1)$$

$$F_Z = -\vec{M} \cdot \vec{H} \quad (2)$$

$$F_A = K_1(\alpha_1^2\alpha_2^2 + \alpha_2^2\alpha_3^2 + \alpha_3^2\alpha_1^2) + K_2(\alpha_1^2\alpha_2^2\alpha_3^2) + K_3(\alpha_1^4\alpha_2^4 + \alpha_2^4\alpha_3^4 + \alpha_3^4\alpha_1^4) \quad (3)$$

$$F_{ME} = b_1(\alpha_1^2 e_{11} + \alpha_2^2 e_{22} + \alpha_3^2 e_{33}) + b_2(\alpha_1\alpha_2 e_{12} + \alpha_1\alpha_3 e_{31} + \alpha_2\alpha_3 e_{23}) \quad (4)$$

$$F_E = \frac{1}{2} C_{11}(e_{11}^2 + e_{22}^2 + e_{33}^2) + \frac{1}{2} C_{44}(e_{12}^2 + e_{23}^2 + e_{31}^2) + C_{12}(e_{11}e_{22} + e_{22}e_{33} + e_{33}e_{11}) \quad (5)$$

$$F_D = +\frac{1}{2}\vec{M} \cdot \overleftrightarrow{N} \cdot \vec{M} \quad (6)$$

In the above expressions, the α's are the direction cosines of the magnetization with respect to the crystallographic axes, the e's are elastic strains, b's are magnetoelastic coefficients, c's are elastic constants. It may be shown(5), when the appropriate derivatives are taken, that the magnetoelastic term has the same angular form as the first order term in the phenomological magnetocrystalline anisotropy energy. Thus these two terms may be confined, resulting in an "effective" first-order anisotropy coefficient $K_1' = K_1 + b_1^2(S_{11} - S_{12}) - \frac{1}{2} b_2^2 S_{44}$ where $S_{ij}$ are elements of the elastic compliance tensor. The experiments were performed with the static magnetic field lying in a (110) or (100) plane. Solution of the equilibrium equation shows that for H sufficiently large $\vec{M}$ will also lie in this plane, though it will not in general be parallel to $\vec{H}$. The resonance equation has been derived for arbitrary orientation of H in one of these crystallographic planes(6). For arbitrary orientation of H, this equation must be solved simultaneously with the equilibrium equation, to find values of magnetization orientation and field which satisfy both equations. This procedure must be done by successive approximations, since the equilibrium equation is transcendental. This procedure was carried out using a digital computer to compare theoretical and experimental results. In the principal crystallographic directions, [100], [111], and [110], the resonance equation can be expressed in closed form. The resulting equations involve the material parameters γ (the magnetogyric ratio), $K_1'$, $K_2$, $K_3$ (the magnetocrystaline anisotropy parameters), and $M_0$, $M_1$, $M_2$, (the anisotropic magnetization parameters of Eq. 1).

## EXPERIMENTAL TECHNIQUES

Samples of cobalt finite were grown from a flux using the procedure of Banks et al(7). They were ground into spheres, annealed to relieve any stress induced in the grinding, and the spheres were oriented by the Laue back-reflection technique. Each sample was mounted on a glass rod such that the sample could be rotated around a [110] axis. The samples were mounted in a shorted waveguide several diameters from from the short, so that the rf and static magnetic fields, perpendicular to one another, lay in the (110) plane, and by rotating the crystal, the direction of the static field in the (110) plane could be varied. At fixed frequency, the field for resonance was observed as a function of θ, the angle between $\vec{H}$ and a [100] direction in the (110) plane. The frequency was chosen high enough that the fields were always large enough to assure that M lay in this same plane. In practice, frequencies from 50 to 70 GHz were employed for the rotation curves. In addition, the field for resonance in a [100] direction was measured as a function of frequency over the frequency range of 47 to 81 GHz, in order to determine the magnetogyric ratio $\gamma$. Measurements were made at temperatures of 22°C, 32°C, and 37°C. The temperature range was limited at the low end by the very large magnetic fields required to obtain resonance in the hard [111] direction, and at the high end to avoid the possibility of magnetic annealing. Despite the large linewidth (~10KOe), the field for resonance in a particular direction could be determined within about 0.2KOe.

## EXPERIMENTAL RESULTS

We report here detailed results for a sample 0.44mm in diameter. Figure 1 shows the rotation curve at 22°C, while figure 2 shows these data at 37°C. Data were taken over the entire range of $\theta$ of 360°. The results were periodic, with period of 180°, and symmetric about $\theta$ = 90°, as the theory indicates, so we plot only the range $\theta$ = 0 to 90°.

In order to fit the experimental points to the theory, values for the material parameters $K_1'$, $K_2$, $K_3$, $M_0$, $M_1$, $M_2$, and $\gamma$ must be chosen. From the resonant fields in the principal directions, three of these parameters may be determined. We took the value of $M_0$ from the static measurements of Perthel(2). Our own measurement of $M_0$ using a vibrating sample magnetometer was consistent with this value. We then chose values of the other parameters to obtain as good a fit as possible between the measured and calculated rotation curves. While the calculation of the rotation curve neglected damping, this can be included approximately by correcting the gyromagnetic ratio $\gamma$ for the relaxation time. This correction is strictly valid only if the line-width is isotropic, which is probably untrue for our material, but the error introduced is small. The values of $M_1$ and $M_2$ we found seemed ridiculously large - five to ten times larger than those obtained by Perthel. Although there is no reason to assume that the microwave values of $M_1$ and $M_2$ should be the same as their static values, for reasons discussed below we do not believe our experimental results are sufficiently precise to determine $M_1$ and $M_2$. We therefore chose the best values of $K_1$, $K_2$, and $K_3$ to fit the rotation curves.

The theory described above indicates that at fixed frequency, the largest field should be required in the [111] direction, $\theta$ = 54°44°. Figure 1 shows

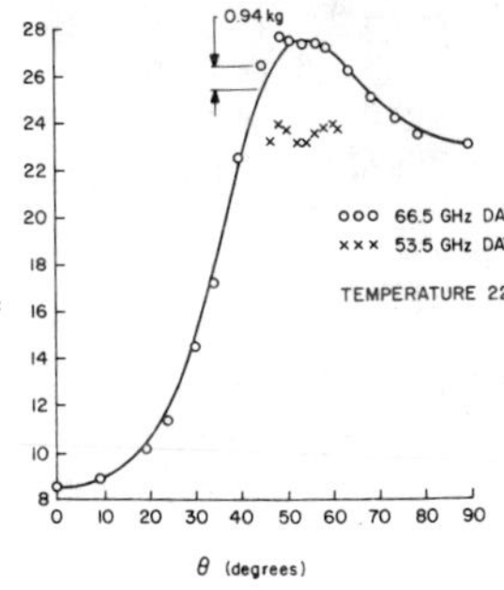


FIG. 1 FIELD FOR RESONANCE ($H_R$) VS FIELD ORIENTATION ($\theta$)

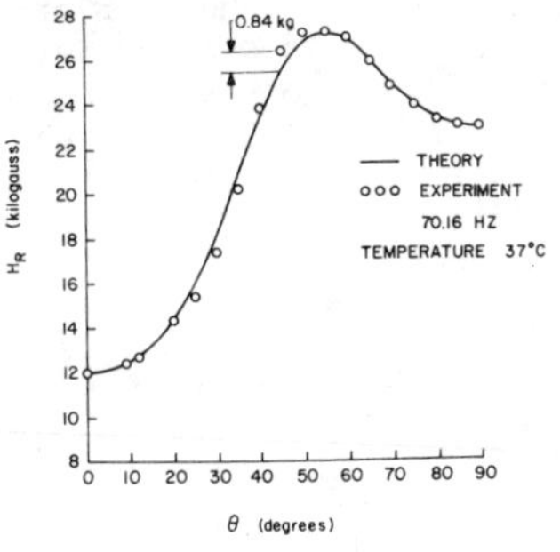


FIG. 2 FIELD FOR RESONANCE ($H_R$) VS FIELD ORIENTATION ($\theta$)

that this was not observed experimentally. Indeed the data taken at a frequency of 53.56Hz show a local minimum in this direction. These results suggest that the sample is not truly cubic. In spite of this, a fair fit can be obtained between theory and experiment. At 22°C, the rms deviation between theory and experiment is 0.33 KOe, and the largest deviation is 0.94KOe. At 37°C, the rms deviation is 0.27KOe, and the largest deviation 0.75KOe. In both instances, this maximum deviation is observed at $\theta$ = 45°. Table I summarizes the results obtained at the three temperatures

Table I

| | | | |
|---|---|---|---|
| Temperature (°C) | 22 | 32 | 37 |
| $K_1'$($10^6$ ergs/cm$^3$) | 2.6 | 2.18 | 2.03 |
| $K_2$($10^6$ ergs/cm$^3$) | 0.1 | 0.8 | 0.8 |
| $K_3$($10^6$ ergs/cm$^3$) | -0.8 | -0.5 | -0.4 |
| rms deviation (KOe) | 0.33 | 0.31 | 0.27 |
| largest deviation (KOe) | 0.94 | 0.84 | 0.75 |

As an estimate of the reliability of the values of the parameters, we varied each one until the rms deviation was doubled. On this basis, we believe the precision of $K_1$ is of the order $\pm$.2 x $10^6$ ergs/cm$^3$, while that of $K_2$ and $K_3$ is $\pm$ 1.0 x $10^6$ ergs/cm$^3$. Thus, only the value of $K_1'$ may be considered to be reliable.

## DISCUSSION

Careful examination of the Laue photographs of this sample showed that the Laue spots in a 110 zone are split, rather than symmetrical. This suggests that there is a microscopic crack or strain in the sample. Thus, it is not possible to apply a static magnetic field so that it is in a 110 or 111 direction in all parts of the sample. This was confirmed by rotation curves taken in two (100) planes. In this plane, the field should be a maximum in the 110 direction, 45° from the 100 direction, and symmetric about the 110 direction. In one 100 plane this was observed to be true; in another, say 010 plane, a dip was found in the 110 direction. We thus conclude that our sample was not perfect, but we believe that the data taken in the easy direction, [100], is valid. Thus we can determine the quantity $K_1'$, if we neglect $M_1$.

Attempts were made to prepare better crystals by a chemical transport technique. X-ray fluorescence analysis of these crystals showed them to contain an excess of cobalt ferrite. The flux-grown crystal, within the precision of the fluorescence analysis, was stoichiometric. One conclusion to which we come is that it is extremely difficult to prepare "good" cobalt ferrite.

## REFERENCES

(1) H. Shenker, Phys. Rev. 107, 1246-1249 (1957).
(2) R. Perthel, G. Elbinger, and W. Keilig, Phys. Stat. Sol. 17, 151-154 (1966).
(3) P. Tannenwald, Phys. Rev. 99, 463 (1955).
(4) J. C. Slonczewski, Phys. Rev. 110, 1341-1348 (1958).
(5) C. Vittoria, J. N. Craig, and G. C. Barley, Phys. Rev. 1310, 3945-3956 (1974).
(6) To be published.
(7) W. Kunnmann, A. Wold, and E. Banks, J. Appl. Phys. 33, 1364-13655 (1962).

*This work was supported in part by the Joint Services Electronics Program, and in part by the National Science Foundation. It is based on a dissertation by L. Assadourian submitted to the Department of Electrical Engineering and Electrophysics, Polytechnic Institute of New York, in partial fulfillment of the requirements of the Ph.D. degree.

+Present address: Picatinny Arsenal, Dover, N.J.

# MAGNETIC RESONANCE OF Gd IN $LaNi_5$ and $LaNi_5$ HYDRIDE

W. M. Walsh, Jr., L. W. Rupp, Jr., P. H. Schmidt and
L. D. Longinotti, Bell Labs, Murray Hill, New Jersey 07974

ABSTRACT

The influence of hydrogen uptake on the magnetic properties of $LaNi_5$ has been investigated by observing the electron spin resonance of small amounts of Gd substituted for La. Hydrogenation causes a large exchange induced g-shift to decrease very strongly in magnitude and to reverse sign. Only ESR signals corresponding to the unhydrided or fully hydrided states are observed, indicating very little range of solid solution. These results are similar to the cases of ESR of Gd and Mn in the Pd, H system but some distinctions are drawn. In particular appreciable microwave loss in fully hydrided $LaNi_5$ shows it to be only weakly conducting rather than metallic. Repeated hydrogenation cycles produce an increasingly intense ferromagnetic resonance signal suggesting precipitation of metallic Ni.

## INTRODUCTION

Among the many elements and intermetallic compounds which form hydrides one of the most remarkable is $LaNi_5$. After initial "activation" at moderately high hydrogen pressure (~100 atm) which causes the bulk material to break into a fine powder $LaNi_5$ can take up six or slightly more hydrogen atoms per formula unit at ~ 2 atm. pressure at room temperature and releases the hydrogen at $\gtrsim$ 1.5 atm.[1,2] We have examined the effect of hydrogen incorporation on the electron spin resonance (ESR) of 0.1% Gd substituted for La.[3] In addition to very striking changes in the Gd resonance upon hydrogenation, large microwave losses were observed suggestive of a much reduced mobile carrier concentration. Repeated hydrogenation cycles produced a ferromagnetic resonance of increasing intensity which we presume to indicate formation of metallic nickel.

## EXPERIMENTAL PROCEDURES AND RESULTS

The compound $LaNi_5$ was prepared in bulk form by arc melting stoichiometric proportions of La and Ni metal in an argon atmosphere on a cooled copper hearth. Weighing indicated no loss and X-ray analysis revealed only the hexagonal $CaCu_5$ structure. Gd was introduced by making an initial $La_{0.99}Gd_{0.01}$ alloy. The intermetallic compound reacts slowly in air so was stored in dry argon. The surface regions of the arc-melted buttons proved weakly magnetic and were etched away. Slices cut from the buttons were examined at 12 GHz and were studied further only if no ferromagnetic resonance was observed. The paramagnetic resonance line of Gd at g = 1.86 was readily observed below 20 K in the bulk compound.[3] The slice was then crushed into coarse powder in argon and exposed to hydrogen at ~ 130 atm. for 10 hours or more at room temperature. The resultant very fine powder was mixed with a small amount of silicone grease and spread in a thin layer between sheets of inert hydrocarbon film. This "package" was then quickly mounted in the microwave cavity (12, 17 or 35 GHz) and cooled to 77 K where no evidence of hydrogen release was observed even in a vacuum for several days. This low temperature stability is consistent with the absence of appreciable proton diffusion below 140 K.[4]

The microwave cavity Q was severely degraded at all temperatures by the $LaNi_5$-hydride powder relative to unhydrided powder or comparably-sized powders of aluminum or graphite. The implication is that the hydrided material has quite a low carrier concentration, i.e., has become semiconducting or barely metallic.

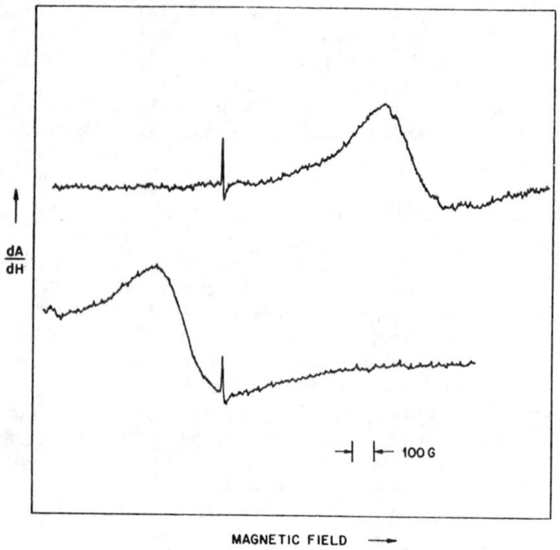


Fig. 1 Absorption derivative ESR spectra of $La_{0.99}Gd_{0.01}Ni_5$ taken at 34.8 GHz and 4.2 K. The sharp signal is due to conduction electron spin resonance in Li metal marking the free-electron g-value. The upper trace shows the Gd signal at g = 1.86 in the unhydrided phase. The lower trace shows the Gd signal at g = 2.05 in the hydride.

That a significant change in electronic properties results from the incorporation of hydrogen is also clearly shown by the ESR of Gd which now lies at g = 2.05 (see Fig. 1). Since the centroid of a $Gd^{3+}$ spectrum occurs at g = 1.99 in nonmagnetic insulating hosts the g-shift in unhydrided $LaNi_5$ is large and negative (-0.13) and changes sign upon hydrogen uptake (+0.06).

Hydrogen is released rather slowly if the hydride is brought to room temperature but appears to be rapidly and completely removed by warming to ~ 50°C. This was monitored by repeated return of the sample to the liquid helium bath where the g = 2.05 signal decreased in intensity as the g = 1.86 signal reappeared. No signals at intermediate g-values were ever observed which we interpret to imply that very little solid solution of hydrogen in $LaNi_5$ is stable on time scales of 2-3 minutes under our sample handling conditions. The absence of intermediate g-shifts which would be expected for solid solutions of small concentrations of H in $LaNi_5$ is very similar to Shaltiel's results for Gd in the Pd, H system.[5]

Repeated uptake-release cycles yielded very reproducible results in the Gd spin resonance behavior. The uptake of hydrogen at room temperature did not appear to be impeded by the silicone grease coating the $LaNi_5$ particles.

It was noted, however, that repeated cycling led to a monotonic increase in intensity of a broad signal at lower fields ("g" ~3.4). This signal was almost independent of temperature up to room temperature and was quite similar to the signals occasionally seen in material where a trace of ferromagnetic nickel was suspected. This phenomenon was reproduced in $LaNi_5$ without any Gd doping and uncontaminated by contact with grease. We carried out only six cycles so do not know whether the effect can eventually result in significant degradation in the material's hydrogen uptake and release properties.

Attempts to observe resonance due to Mn doped in via a $Ni_{0.99}Mn_{0.01}$ alloy were unsuccessful in $LaNi_5$ in both the unhydrided and hyrided state. This is in contrast to the results of Alquie et al. for Mn in the Pd-H system.[6]

## CONCLUSIONS

The large negative g-shift of Gd in $LaNi_5$ indicates appreciable exchange coupling of the $Gd^{3+}$ local moment to the band electronic susceptibility of the host.[3] The effect is essentially identical to the case of Gd in Pd metal.[5] The magnitude of the g-shift supports the inference of appreciable exchange enhancement of the host band susceptibility.[3,7] The uptake of hydrogen produces a very significant change in both the conductivity and the magnetic behavior of the compound. The conductivity must decrease by several orders of magnitude to produce the strong microwave loss observed in the hydride phase. Similarly the decreased magnitude of the Gd g-shift implies a considerable reduction in band susceptibility in qualitative agreement with the ESR studies of the Pd-H system.[5,6] The sign reversal is particularly interesting in this regard since it indicates there is still an appreciable host susceptibility though its character may well have changed from d-like to more s,p-like. The variety of electronic effects upon hydrogen uptake as well as the great ease with which uptake and release occur suggest $LaNi_5$ and its hydride deserve a careful band-structure analysis. The analogy with palladium may be an oversimplification based on our emphasis on local magnetic moment behavior.

We wish to acknowledge useful discussions with Professor D. Davidov.

## REFERENCES

1. J. H. N. van Vucht, F. A. Kuijpers and H. C. A. M. Bruning, Philips Res. Repts. 25, 133 (1970).

2. Considerable information is summarized in F. A. Kuijpers, Philips Res. Repts. Suppl. 2 (1973).

3. D. Davidov and D. Shaltiel, Phys. Rev. Letters 21, 1752 (1968).

4. T. K. Halstead, J. Sol. State Chem. 11, 114 (1974). (1974).

5. D. Shaltiel, J. Appl. Phys. 34, 1190 (1963).

6. G. Alquie, A. Kreisler, G. Sadoc and J. P. Burger, J. de Phys. Letters 35, L-69 (1974).

7. I. D. Weissman, L. H. Bennett, A. J. McAlister and R. E. Watson, Phys. Rev. B11, 82 (1975).

AUTHOR INDEX

AUTHOR INDEX

AIP Conference Proceedings

| | | L.C. Number | ISBN |
|---|---|---|---|
| No. 1 | Feedback and Dynamic Control of Plasmas (Princeton) 1970 | 70-141596 | 0-88318-100-2 |
| No. 2 | Particles and Fields - 1971 (Rochester) | 71-184662 | 0-88318-101-0 |
| No. 3 | Thermal Expansion - 1971 (Corning) | 72-76970 | 0-88318-102-9 |
| No. 4 | Superconductivity in d- and f-Band Metals (Rochester 1971) | 74-18879 | 0-88318-103-7 |
| No. 5 | Magnetism and Magnetic Materials - 1971 (2 parts) (Chicago) | 59-2468 | 0-88318-104-5 |
| No. 6 | Particle Physics (Irvine 1971) | 72-81239 | 0-88318-105-3 |
| No. 7 | Exploring the History of Nuclear Physics (Brookline, 1967, 1969) | 72-81883 | 0-88318-106-1 |
| No. 8 | Experimental Meson Spectroscopy - 1972 (Philadelphia) | 72-88226 | 0-88318-107-X |
| No. 9 | Cyclotrons - 1972 (Vancouver) | 72-92798 | 0-88318-108-8 |
| No.10 | Magnetism and Magnetic Materials - 1972 (2 parts) (Denver) | 72-623469 | 0-88318-109-6 |
| No.11 | Transport Phenomena - 1973 (Brown University Conference) | 73-80682 | 0-88318-110-X |
| No.12 | Experiments on High Energy Particle Collisions - 1973 (Vanderbilt Conference) | 73-81705 | 0-88318-111-8 |
| No.13 | π-π Scattering - 1973 (Tallahassee Conference) | 73-81704 | 0-88318-112-6 |
| No.14 | Particles and Fields - 1973 (APS/DPF Berkeley) | 73-91923 | 0-88318-113-4 |
| No.15 | High Energy Collisions - 1973 (Stony Brook) | 73-92324 | 0-88318-114-2 |
| No.16 | Causality and Physical Theories (Wayne State University, 1973) | 73-93420 | 0-88318-115-0 |
| No.17 | Thermal Expansion - 1973 (Lake of the Ozarks) | 73-94415 | 0-88318-116-9 |
| No.18 | Magnetism and Magnetic Materials - 1973 (2 parts) (Boston) | 59-2468 | 0-88318-117-7 |
| No.19 | Physics and the Energy Problem - 1974 (APS Chicago) | 73-94416 | 0-88318-118-5 |
| No.20 | Tetrahedrally Bonded Amorphous Semiconductors (Yorktown Heights, 1974) | 74-80145 | 0-88318-119-3 |
| No.21 | Experimental Meson Spectroscopy - 1974 (Boston) | 74-82628 | 0-88318-120-7 |
| No.22 | Neutrinos - 1974 (Philadelphia) | 74-82413 | 0-88318-121-5 |
| No.23 | Particles and Fields - 1974 (APS/DPF Williamsburg) | 74-27575 | 0-88318-122-3 |
| No.24 | Magnetism and Magnetic Materials - 1974 (20th Annual Conference San Francisco) | 75-2647 | 0-88318-123-1 |
| No.25 | Efficient Use of Energy (The APS Studies on the Technical Aspects of the More Efficient Use of Energy) | 75-18227 | 0-88318-124-X |
| No.26 | High-Energy Physics and Nuclear Structure - 1975 (Santa Fe and Los Alamos) | 75-26411 | 0-88318-125-8 |
| No. 27 | Topics in Statistical Mechanics and Biophysics: A Memorial to Julius L. Jackson (Wayne State University-1975). | 75-36309 | 0-88318-126-6 |
| No. 28 | Physics and Our World: A Symposium in Honor of Victor F. Weisskopf (M.I.T. 1974) | 75-7207 | 0-88318-127-4 |
| No. 29 | Magnetism and Magnetic Materials - 1975 (21st Annual Conference, Philadelphia) | 76-10931 | 0-88318-128-2 |